RADIATION RESEARCH

Biomedical, Chemical, and Physical Perspectives

The Fifth
International Congress of Radiation Research

Under the Patronage of
The President of the United States of America

The Fifth International Congress of Radiation Research was organized by a corporation, the Fifth International Congress of Radiation Research, Inc., with Officers appointed by the Radiation Research Society (of the United States) and approved by the International Association for Radiation Research.

V. P. Bond, *President*
J. L. Magee, *Vice President*
W. K. Sinclair, *Secretary-General*

ORGANIZING COMMITTEE: *The Officers and*

H. I. Adler	C. W. Edington	H. S. Kaplan
E. L. Alpen	M. M. Elkind	H. M. Patt
C. L. Dunham	E. J. Hart	H. H. Rossi

Nancy S. Weers, R. J. Burk, Jr., *Assistants to the Secretary-General*

INTERNATIONAL ASSOCIATION FOR RADIATION RESEARCH

A. Hollaender (U.S.A.)–*Honorary President*
J. W. Boag (U.K.)–*President*

RADIATION RESEARCH SOCIETY

V. P. Bond–*President*

It is a pleasure to acknowledge the financial support provided to the Congress by the organizations listed on page xxi. The contributions of some of these organizations, and especially the USAEC (now USERDA), helped to defray the cost of these Proceedings.

Radiation Research

Biomedical, Chemical, and Physical Perspectives

Proceedings of the
Fifth International Congress of Radiation Research
Held at Seattle, Washington, U.S.A.
July 14-20, 1974

Edited By

Oddvar F. Nygaard
Case Western Reserve University, Cleveland Ohio

Howard I. Adler
Oak Ridge National Laboratory, Oak Ridge, Tennessee

Warren K. Sinclair
Argonne National Laboratory, Argonne, Illinois

With the assistance of Mary Jane Loop
Oak Ridge National Laboratory

ACADEMIC PRESS, INC. *New York San Francisco London 1975*
A Subsidiary of Harcourt Brace Jovanovich, Publishers

ACADEMIC PRESS RAPID MANUSCRIPT REPRODUCTION

ACADEMIC PRESS, INC.
111 Fifth Avenue, New York, New York 10003

United Kingdom Edition published by
ACADEMIC PRESS, INC. (LONDON) LTD.
24/28 Oval Road, London NW1

Library of Congress Cataloging in Publication Data
International Congress of Radiation Research, 5th, Seattle, 1974.
Radiation research.

Sponsored by the International Association for Radiation Research and the Radiation Research Society.
Bibliography: p.
Includes index.
1. Radiobiology--Congresses. 2. Radiology, Medical --Congresses. 3. Radiation--Congresses. I. Nygaard, Oddvar F. II. Adler, Howard I. III. Sinclair, Warren K. IV. International Association for Radiation Research. V. Radiation Research Society. VI. Title.

[DNLM: 1. Radiation--Congresses. W3 IN558 1974r / WN100 I514 1974r]
QH652.A1Is6 1974 539.2 75-40324
ISBN 0-12-523350-7

PRINTED IN THE UNITED STATES OF AMERICA

CONTENTS

PREFACE

The basic aspects of radiation effects on living systems include not only examination of these effects at all levels in such systems—molecular, cellular, organ, and whole animal—but also consideration of the processes by which radiation interacts with matter and generates chemical species capable of subsequent interaction with biologically important molecules. Recognition of the need to consider these fundamental aspects lies at the root of radiation research as a field. Consequently, meetings and discussions among physicists, chemists, and biomedical scientists have played a special role in the development of this area of research. Initially such meetings were small, and took place at Oberlin, Ohio (1948), and subsequently in London, Aarhus, Liege, Stockholm, Cambridge, and Venice. The interdisciplinary character of the subject was emphasized, as well as the necessity for physicists, chemists, and biologists to meet and to discuss their different approaches to the common problems confronting them. At the same time concern was felt about the uncertainties associated with long-term exposure to radiation delivered even at very low dose rates, and the need to understand fundamental processes in order to evaluate the potential hazards.

During this same period a number of societies devoted to radiation research were founded in different countries, including the United States, Great Britain and other European countries, and Japan. Additional societies are still being formed as interest in this challenging field continues to grow.

This national and international activity resulted in the First International Congress of Radiation Research, which was held at Burlington, Vermont, in 1958. This congress was an important development, as not only did it open the eyes of many to the broad implications of research in radiation chemistry, physics, and biology, as well as in medicine, but also it provided a base for the development of international cooperation. This base was firmly established at the Second International Congress of Radiation Research, held at Harrogate, England, in 1962, where it was decided to form the International Association of Radiation Research (IARR). A principal responsibility of the IARR is to act as sponsor for these international congresses. In addition, the Association provides for their continuity, decides upon appropriate hosts and locations, fosters the development of international cooperation in the field, and encourages smaller meetings. Under IARR sponsorship, the Third Congress was held in Cortina, Italy, in 1966; the Fourth in Evian, France, in 1970; and the present Fifth Congress in Seattle, Washington, USA, July 14-20, 1974.

The development of modern biology and the explosive increase of information on basic structures in the nucleus and cytoplasm, the discovery of the structure of nucleic acids and of the mechanisms of transcription and translation of information within cells have opened many new avenues to the biologists. This accumulation of basic information has been paralleled by many significant new developments in mammalian radiobiology, basic biochemistry, radiation chemistry, and physics.

Radiation genetics and radiation cytology have provided some of the keys for the development of modern molecular biology. At the same time theories of radiation action, models of possible low-level effects in populations, and comparisons of radiation with other hazards have been generated.

A number of Symposia and many contributed papers at the Seattle Congress emphasized the significance of the above mentioned areas of radiation research. Public awareness of the importance of the uses and effects of radiation had increased appreciably. While the focus of the present Congress was on fundamental research, there were several well-attended sessions on the practical aspects of radiation research as it relates to radiotherapy, central station power generation by both nuclear fission and fusion, and the environment.

The scientific sessions and other activities of the congress were held at the Seattle Center, a large parklike area in the heart of Seattle. Leading scientists, many of whom were from fields not directly involved with radiation research, accepted invitations as speakers in the Plenary Sessions and Symposia. The fresh points of view and the range of experiences which these speakers provided were important factors in the overall success of the Congress.

The recent transformation in the United States of the former Atomic Energy Commission into the broader framework of a new agency, the Energy Research and Development Administration, provides the investigator with new challenges. Those disciplines which have contributed so much to radiation research are now facing the additional challenge of considering other environmental insults and especially those resulting from the various aspects of energy generation. These new hazards must be evaluated, modified, and reduced, or perhaps even eliminated. Some of the Plenary Sessions, Symposia and sessions for contributed papers which were convened at this Congress addressed themselves to such questions, and we can confidently expect more attention to be devoted to these subjects in the future.

This was a full and busy congress, attendance was excellent, and no less than 36 countries were represented. Especially important to this congress was the presence of so many young investigators, both from within the USA and abroad, many of whom were assisted by travel grants from the Congress and its sponsors.

The enormous number of developments that have taken place in radiation research in recent years were reflected best at Seattle, not so much in the fundamental discoveries that have accumulated in recent times, but rather by the greatly broadened scope of the subject, and the evidence of the role it must play in the general area of science as well as in the public arena. An air of excitement and expectancy prevailed throughout and a common remark was "There is so much going on that is interesting, I cannot possibly take it all in." But this was said with a feeling of challenge rather than chagrin, as if the truly broad dimensions of the field were just being fully realized. To the Congress organizers and their sponsors, the IARR and the Radiation Research Society, this feeling was especially encouraging as it demonstrated the healthy state of radiation research and the central role it has played, and is playing, in advancing knowledge in certain areas of basic biology, chemistry, physics, and medicine.

We hope that these proceedings will capture the optimistic feeling of challenge for the future which we think pervaded the Seattle Congress.

V.P. Bond, *President*
W.K. Sinclair, *Secretary-General*

5th International Congress of
Radiation Research

EDITORS' NOTE AND ACKNOWLEDGMENTS

These Proceedings have been produced by typing edited manuscripts on a photocomposition system operated by Maryville-Alcoa Engravers, Inc. The completed copy and the diagrams, prepared by Academic Press, were printed from the camera-ready copy by the offset method.

The editorial tasks involved in the preparation of this volume were greatly facilitated by the dedicated assistance of Mary Jane Loop who handled much of the logistics and did the major share of proofreading; the Editors are deeply grateful to her for her invaluable contribution to the publication of these Proceedings.

We further wish to express our appreciation to Mrs. Paulette Snowden for editorial assistance, and to the staff of the Health Center Library, Case Western Reserve University, for its tireless help in verifying and completing a great many of the references.

This volume will serve as a record of the Plenary Sessions and Symposia of the Seattle Congress. The Editors regret that it was not possible to publish the Contributed Papers as well. However, abstracts of all papers, including those reproduced in this volume, are available as a permenant record in the July, 1974 issue of the periodical *Radiation Research* (Vol. *59*, pp. 1–315).

A separate list of contributors to the volume has not been provided since full addresses of the authors appears with their contributions.

O. F. Nygaard
H. I. Adler
W. K. Sinclair

EDITORS' NOTE AND ACKNOWLEDGMENTS

[illegible]

The Officers and Organizing Committee of the Fifth International Congress of Radiation Research

Wish to acknowledge the generous assistance provided
By the following Agencies of the United States government:

Atomic Energy Commission (Now, ERDA)
National Institutes of Health
Environmental Protection Agency
Defense Nuclear Agency
National Aeronautics and Space Administration

By the Commission of the European Communities (Euratom)

By the following Scientific Societies:

Radiation Research Society
International Association for Radiation Research
Radiological Society of North America
American Roentgen Ray Society
American Society of Therapeutic Radiologists
(Anna Hamann Memorial Fund)

By the following Foundations and Laboratory organizations:

Mallinckrodt Foundation
Battelle (Pacific Northwest Laboratory)
Argonne Universities Association

By the following commercial firms:

Varian Associates
Commonwealth Edison
Picker Corporation
G.D. Searle & Co.
Schwarz/Mann
New England Nuclear Corporation
DeSoto Incorporated
Packard Instrument Co.

We also wish to acknowledge, with pleasure, the cooperation of co-sponsors in certain of the symposia who have joined with us in arranging speakers and travel support. These include:

The International Atomic Energy Agency, Vienna, Austria (IAEA)
The International Commission on Radiation Units and Measurements (ICRU)
The Comite Internationale de Photobiologie (CIP)

Travel support for invited participants on the program was mainly provided from funds made available by the U.S. government agencies listed above, and for specific symposia in part by the co-sponsoring bodies.

Travel support for a limited number of other participants was provided from private funds contributed by many of the private organizations listed above, upon recommendations of the Travel Support Committee.

SPECIAL ADDRESSES

DIXY LEE RAY, CHAIRMAN

U.S. Atomic Energy Commission

1973–1974

Speech at Opening Ceremonies of the Fifth International Congress on Radiation Research July 14, 1974 Seattle, Washington

Dr. Dixy Lee Ray
Chairman, U.S. Atomic Energy Commission
Washington, D. C. 20545, USA

Dr. Bond, distinguished people here on the platform with me, participants, guests of this Congress, it is doubly a pleasure for me to be able to speak with you today and spend a little time here at home. As Dr. Bond and Councilman Benson have indicated, this is my home territory, and a wonderful opportunity to come back to the "real Washington," even for a short time. I appreciate those kind words that my friend Vic Bond has said. Sometimes in these very interesting months in which I have been associated with the Atomic Energy Commission, I have wondered if I would ever be able to accommodate enough homework to feel reasonably comfortable in the many areas of interest that this great Agency has. And I do have one boast. I can tell myself when things are tough: "But anyhow, I know more about nuclear science and radiation research than any other marine biologist that I know."

There is another reason why it is a pleasure to be here today and that is to see the institution in which I spent nearly ten years; it's up at the other end of the walk from here, that beautiful facility called the Pacific Science Center. And if you will forgive me for a little propaganda, it's a pretty nice place and I commend it to you. I hope you will take some time to visit it as well as other places nearby while you are here. And if you haven't yet signed up for your salmon barbecue—well, you won't ever be the same once you've experienced a salmon barbecue at the Pacific Science Center; so I hope you will indeed take advantage of that and enjoy it.

This Congress, I am told, is quite remarkable. It is the fifth International Congress for Radiation Research. The participation is among the largest for any of the five congresses; as we heard, over 1700 people are coming, mainly of course, from the host country, but 36 nations will be represented. This is a testimony once again to the international universality of science. This Congress is also remarkable for its support. It is supported in fact by five Federal agencies, by EURATOM and the International Atomic Energy Agency, by five scientific societies, by three foundations and organizations, and by eight commercial concerns. What better evidence could there be of broad and diverse interest than that kind of support for your Congress!

The program, too, is remarkable. You have scheduled for yourselves six plenary sessions, twenty-six symposia, one hundred twenty-two sessions of scientific papers, and one lecture for the public. It's a tremendous program. These six plenary sessions will cover such things as Energy Needs, Nuclear Power and the Environment; Prospects for Fusion Power; and Technological Applications of Radiation. The subjects of the symposia range from radiation effects on both physical and biological systems to new ways of estimating genetic risks in man; and the use of heavy particles in radiation therapy of cancer to electron effects in liquids; and so on. As to the one hundred and twenty-two sessions of scientific papers, the range and the diversity of subject is almost beyond comprehension.

These Congresses that began less than 20 years ago have come to a point where the sheer volume of scientific fact, observation, and experimental results to be reported and discussed during this coming week would have seemed unimaginable two to three decades ago. Although radiation research has its roots deep in the end of the Nineteenth Century, the great expansion of information, the great growth of understanding and the beginnings of some wisdom with respect to utilization of information in this field is a product of the present day. It is an edifice of understanding of radiation which is built by you who are here and participating in this Congress. You and your colleagues and fellow workers in laboratories throughout the world, you are the builders of this structure of knowledge. You are the contributors to an understanding of one of the most awesome of all the phenomena in science: Radiation!

Although the primary purpose of this gathering of talent is to exchange information and opinions and to learn from each other, I would like to express the hope that the wider public too will mark the significance of these meetings and the contribution of this Congress to the greater understanding of this remarkable force.

The ways this energy from radiation have promptly been put to work subserving our many fundamental needs are truly a modern phenomenon. It is being applied so many ways in medicine; its uses in industry expand continuously by virtue of the basic research going on; and most notable now are the researches in physics and engineering developing the knowledge to build the reactors that are capable of converting great concentrations of this energy into a more easily controllable form of energy. Is this not truly a modern phenomenon?

When we think of the range of information that has come from this field and the ways in which it is used, and the rapid expansion and application of that knowledge, it is surely an experience unlike any that has touched any single generation of people before. We hear often enough of the rapid growth of knowledge in the present time. We hear often enough of the problems, as well as the benefits, of our present scientific technological age. Indeed, sometimes we are led to believe that perhaps the rate of increase of knowledge is too great, as if somehow we could ever know enough. And there are those who despair, because we have not always, as a society, used our knowledge and applied our information with full foreknowledge of all possible consequences. But I would like to say we have come a long, long way. And when we stop to think about the significance of man's coming to know the awesome power of the atom—nothing like that has ever happened in human society since the dawn of recorded history.

Now it was a very long time before we human beings first learned how to control a source of energy: That source was the energy of a combustion process which we call fire. We don't know exactly how our remote ancestors came to understand fire and to control it. Certainly it is a fearsome and a terribly difficult energy to control, but somehow it did happen.

Myths and legends tell us that there was a young fellow, a god named Prometheus, who stole fire from the other gods and gave it to human beings, because he kind of thought we were rather nice people or something. He wasn't very well thought of among his contemporaries for having done that and, indeed, the other gods put a punishment upon him, which to read about it is something rather frightful. But since the days of Prometheus and human understanding of fire, there probably is no single event to compare with the development of knowledge of atomic energy and the advent of understanding how to apply nuclear energy to create the stock of useful, as well as destructive, things that human beings now command.

Now we have before us the possibility of using an energy source that does not burn up fossil organic fuels, one that is totally different and totally new. Is it any wonder then, with this new awesome force abruptly coming into the hands of this generation of mankind, that our contemporaries sometimes wonder whether it has been a wise thing. Sometimes they set forth—those of us in the Atomic Energy Commission came to feel—a punishment not unlike that that was meted out to Prometheus. We have been much in the public eye of late, and there is much that we have to learn from the questions, the concerns and the worries of a public affected by developments in science but not always well informed either of what those discoveries are or of what they may mean.

And so I would like to comment for a moment about scientific knowledge, which like plants and animals, grows best when nurtured and when it has the benefits of a congenial environment. But first I would like to mention the role of the United States Atomic Energy Commission in nurturing knowledge of the nuclear science. In this field, as in other areas of science and engineering, the tools, the instruments, the facilities with which people work, are essential for developing further understanding and more information. And also there is the great need for the support of the creative minds of the scientists and engineers that use these facilities. Filling these needs has been a duty of the Commission and despite a widespread attitude of skepticism for government at institutions, in the United States at least, the result is a large collection of outstanding laboratories of radiation research in this country. Oak Ridge, Los Alamos, Lawrence Livermore, Lawrence Berkeley, Brookhaven, Argonne, Ames, and a score of others, these are all parts of the vast complex of institutions that belong to the Atomic Energy Commission. The accelerators, these great machines that have told us so much about the fundamental nature of matter, from the 200 BEV proton accelerator, now known as the Fermi National Accelerator Laboratory, to SLAC—horrible name, isn't it?—the Stanford Linear Accelerator Center, to controlled thermonuclear reaction machines called SCYLLAC, TOKAMAK, and dozens of other interesting names, these are the machines which probe the nature of matter, and they were built by the Atomic Energy Commission. Each one of these buildings, institutions, laboratories, really ought to have a plaque on it saying something like this: "This installation, or instrument or laboratory or whatever it is, was constructed from public funds made available by the United

States Atomic Energy Commission, acting on behalf of all of the people of the United States." Neither high energy physics nor radiation biology would exist today in the form that they do, without the strong support and encouragement from the program divisions of the United States Atomic Energy Commission.

As for the people who work in these places, and use these instruments, over the period of time from 1948 to 1973, a quarter of a century, the United States AEC has developed training programs for undergraduate students, for graduates, for postdoctorals, for members of the profession, for faculty, that average out to more than 1600 people each year, a total of over 40,700 individuals from the United States alone who have benefited from these training programs.

Beginning in 1954 when President Eisenhower conceived the Atoms for Peace plan, teams of scientists have gone from the United States with lecture programs, short course training materials, exhibits, and instruments of great variety to visit in nearly every nation in the world.

And citizens from other countries, 85 Western non-Soviet nations in all, have traveled to the United States to participate in AEC research programs in the AEC's national laboratories, a total of 10,513 citizens of the world.

More recently, begun in 1968, and now growing, we have developed a program with the nations of the Soviet Bloc. Five nations already have been involved, and this coming year one more will be added, for a total of 108 people from that part of the world taking part in international study and cooperative programs. And so I think we can say that AEC has had a vital role in the development of a worldwide rapport within the scientific community.

But outside of the scientific community, and especially at home, the United States Atomic Energy Commission has not fared so well with the public at large. And those of you who came in through the front entrance of this lovely building may have seen some evidence of this.[1] —I'm told they didn't even spell my name right.—We have to be prepared for this type of response to science in general and to the technological applications of science especially. And for radiation research in particular, the problems of sympathetic communication with an uninformed public are very great indeed. It is very easy to raise worry to almost hysterical and panic proportions when one is dealing with honest, well-meaning, but ignorant citizens. And when one at the same time has to be concerned with a force that does have hazards that must be handled properly, we can see full well how simple it is for people to become increasingly worried and concerned as they hear accusations that are not always very well, or sometimes too defensively, responded to.

Historically, the task of informing the public in scientific matters has fallen to those who are outside the mainstream of scientific research. But this situation is no longer tolerable. Recall for a moment for the coming week: Six plenary sessions, twenty-six symposia, one hundred twenty-two sessions of scientific papers and—I counted them—a total of about one thousand scientific papers, to be presented during this week.

And one for the general public.

Yet it is the public that is affected by what happens in science. Let's take for example the topics of five of your plenary sessions. I'll give you the titles: "Energy

[1]Peaceful demonstrations against nuclear power were in progress outside the building at the time.

Needs, Nuclear Power and the Environment," "Prospects for Fusion Power," "Human Radiobiology," "Technological Applications of Radiation," and "Evaluation of Hazards of Radiation Exposure Relative to Other Environmental Agents." Scientists in this field are interested in those questions that deeply concern the public. It is human beings and society at large which draw up just such a list of topics when asked: What would you like to know, what do you feel you need to know, about radiation research?

And so I would like to express a hope. I would like to urge that the essence of the information presented in these important sessions—and I named the plenary sessions only as an example—could somehow be put into plain language; and that someone, some one of you who is not afraid to use simple, common words and be reasonably accurate and give the right spirit to it, might bring this information out of the totally technical caste and into another kind of proceeding that would serve to enlarge the intellectual horizons of the citizen, not only of this country but of the world as well.

The United States Atomic Energy Commission is deeply committed to better communication with the public and will assist in any effort to improve this dialogue. Just as this government agency will listen carefully to the new information and the data coming from these scientific sessions, and will heed the words of wisdom, and warning, that come from this Congress made up of talented participants, so we would like to leave with you this thought, a message to keep in your hearts during the coming days, and perhaps give it some further thought. The words were beautifully put some years ago, 1959 to be exact, by the French scientist Jean Rostand on the occasion of his accepting the Kalinga prize in India. He wrote at that time: "Any distinction between the man of science and the ordinary man is no longer admissible, any more than a form of segregation based upon an inequality of knowledge. Whether we like it or not, the laboratory henceforth opens right into the street. Science not only affects us at any given moment of our day-to-day existence, it dogs us, it pursues us. Have we not, all of us, been transformed into involuntary guinea pigs ever since atomic fission, without asking our opinion, began to plant particles in our bones?

"The obligation to endure gives us the right to know.

"The time is clearly coming"—that was 1959 and I think we can say today the time is clearly here—"when the man in the street will have his say with regard to the great social, national, international and moral issues latterly raised by certain applications of science; and it may be that the specialist himself, weary of bearing on his own the weight of his too-heavy responsibilities, will rejoice in finding understanding and support in public awareness."

Thank you all very much. I know you're going to have a great week and learn a lot. And I hope you will enjoy our great city of Seattle.

Thanks.

Twelfth Failla Memorial Lecture
The Time Scale in Radiobiology

J. W. Boag

Department of Physics Institute for Cancer
Research, Sutton, Surrey, Great Britain

I knew Dr. Failla only during the last 10 years of his life and there are many here today who knew him intimately. There are many others, however, who know of him only by his reputation and his published work, and to them I should like to convey, by a few reminiscences, some impression of the man himself. Fig. 1 shows Dr. Gino Failla about 1960. When he started his career in radiation

FIG. 1. Dr. G. Failla, about 1960.

research it was possible for an individual to make original contributions to many different aspects of the field, and Failla did so. He was, of course, a founding member of the Radiation Research Society of America and consequently of this International Association. He saw the Radiation Research Society, which he served as President in 1953, as a means of preventing the fragmentation of the subject into water-tight compartments of physics, chemistry, biology and medicine, and as a forum where the basic effects, the useful applications and the hazards of ionizing radiation could be discussed by workers trained in many different scientific disciplines.

Although he had been a leader in the field for more than a generation when I first met him, and was then over 60 years of age, his biological age was much less than that. I well remember a meeting in Washington in 1953 which was chaired by Philip Abelson and was about neutron-induced cataract. There had been some controversy about dosimetry in the questions following one lecture, and towards the end of the discussion the Chairman, seeing Failla at the back of the hall, said "I am now going to call on that wise old man, Professor Failla, to come forward and tell us exactly what the truth is about these measurements of neutron dose." Failla tripped rapidly up to the front, mounted the rostrum, and turning to the Chairman said, in his high-pitched voice, "I should like to thank you very much, Mr. Chairman, for the first of these two epithets but not for the second!"

Fig. 2 gives a brief glimpse of the time-scale of Failla's career (*1*,*2*). He took his first degree in Electrical Engineering in 1915 at Columbia and completed graduate school two years later with an M.A. degree. He joined the staff of Memorial Hospital in 1915 and remained there for 28 years, until he moved to the Chair of Radiology (Physics) at the College of Physicians and Surgeons of Columbia University in 1943. His first scientific paper on "Radium technique at The Memorial Hospital" appeared in 1920 and was followed by a constant stream of publications on many topics until his last paper appeared posthumously in 1964.

Edward de Bono (*3*), in his popular writings on psychology, has drawn attention to what he calls "lateral thinking"—the daydreaming of a child, or of an inventor, or of a theoretical physicist, which so often follows inconsequential pathways to reach an idea or a hypothesis that could hardly have been attained by any more conventional route. Shakespeare described the seven ages of man by their physical characteristics. De Bono recognizes only three phases in our mental approach to life. The first is characterized by the question "why?" and is that of the infant to whom everything is new. This phase may last up to the age of about five years. In the second phase the question becomes "why *not?*" and the adventurous child tries out everything, often with highly interesting when not actually disastrous results. This phase may extend for a further five to seven years only to make way then, as a rule, for the 'third age of man' which is characterized by the universal answer "because. . . .". He '*knows*' everything! Only a few preserve the ability to continue asking the questions "why?" and even "why not?" throughout their lives, but it is to these few that we owe new literature, new art, new science, new political systems. Failla was one of these.

I remember him showing me around his laboratory one day in the summer of 1953. There were many interesting and important things to show, all yielding the

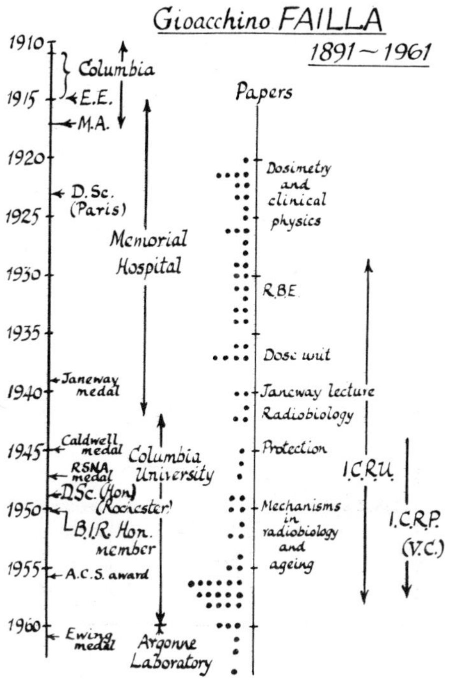

FIG. 2. The time scale of Failla's career.

expected results, but the thing which was puzzling and intriguing him that day, and which he most wanted to discuss with me, was a very simple experiment. He had designed a flat circular ionization chamber to measure a square array of spots of radioactive material (Fig. 3). The circular chamber covered a number of spots simultaneously and as it was moved over the square grid of points this number fluctuated and with it the current from the chamber. To reduce this fluctuation which can be measured as the ratio (max-min)/mean he had made a considerably larger diameter ion chamber, averaging over many more spots, and found, to his surprise, that the relative fluctuations were somewhat greater. Eventually we worked out the associated geometrical problem and found that if one takes the ratio of the circle radius to the spacing of rows in the array there are, indeed, as shown in Fig. 3, certain narrow ranges of this ratio (r/a) which yield abnormally low fluctuations in the number of spots covered, and other ranges where the fluctuations are much larger. Following this up later I found that we had stumbled on a simple two-dimensional analogue of a well known problem in topology. Failla was delighted with this small but non-trivial solution which could add to the accuracy and elegance of his experiments. He could not let an unexpected, even if apparently unimportant, observation lie around unexplained.

On my way to this meeting I visited London, Ontario, and found inscribed on the wall of their new medical school a quotation from Francis Bacon, in which he had set down without any false modesty and yet surely without exaggeration, a list of the reasons why he thought himself fit to inquire into natural phenomena.

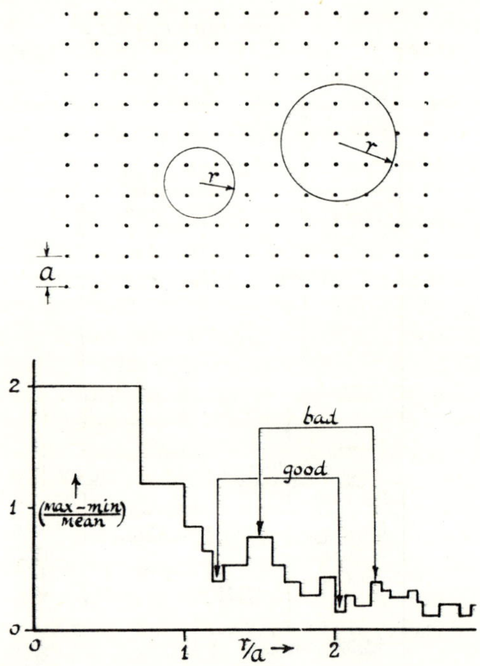

FIG. 3. Fluctuations in the ionization current from a circular chamber exposed to a square array of radioactive sources.

His claims are such as any scientist may aspire to, but Gino Failla whose memory we honor tonight, measured up to them better than most. "A mind nimble and versatile enough to catch the resemblances of things—which is the *chief* point—and at the same time steady enough to fix and discern their subtle differences: endowed by nature with the desire to seek, patience to doubt, fondness to meditate, slowness to assert, readiness to reconsider, carefulness to set in order—and neither affecting what is new nor admiring what is old, and hating every kind of imposture".

In my view—and I believe it would be Failla's too—the most useful purpose that a large Congress of this kind can serve is to enable scientists working in different parts of the whole field to see the relationships between their own special areas of research and those of other workers with different interests and different disciplinary training. If through this lecture I can persuade some of the biologists present to attend selected radiation chemical sessions and vice versa, at least one of my aims will have been achieved.

The last twenty years have seen enormous progress in studying the dynamic mechanisms of radiation action on both simple and complex systems. Although the time resolution attained has been highest in the radiation chemical field, ingenious ways have also been devised of carrying fast analytical techniques right into the more complex biological field. I shall not attempt, in this lecture, to pursue a straight and narrow path but rather meander from one part of the field to another, seeing where some important advances in kinetic studies have been made—in physical techniques, in chemical mechanisms and again in cellular studies. I have nothing new to say to the experts in the field of fast reactions, but shall try to convey to those not working in this field not only the excitement of seeing things happening in a few millionths of a millionth of a second (i.e. in picoseconds) but also the relevance of some of the results won in this way to the whole subject of radiobiology.

The time scale of the phenomena we are dealing with in radiation research is a long one which can only be presented compactly on a logarithmic scale ranging from the 10^{-16} second electron orbiting time in an excited molecule, through the subsequent chemical kinetics to the development of the final biological effects which may take years. There is a succession from the fast physical phenomena in Fig. 4 through almost as fast chemical processes to much slower biological effects, but with a good deal of overlap, for many of these processes are open-ended, as shown by the arrows—they can be slowed down by control of the environment or, in the case of chemistry by changing the concentration of the reactants. In this logarithmic presentation, microseconds lie near the centre of the range and, as far as radiobiological effects in wet systems are concerned, much of the action is still concentrated in this region though advance parties have penetrated right up to the frontier. We note that for radical reactions in which no activation energy is required the half-life of a transient species may be only a few picoseconds (10^{-12} sec) if a capturing solute is present in molar concentration.

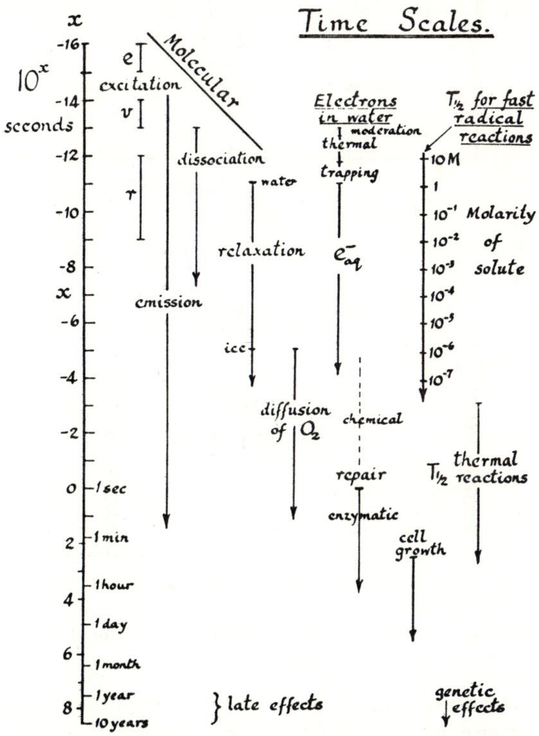

FIG. 4. Time scales for a variety of molecular, chemical and biological processes.

In the biological time scale we have to take account of both thermal reactions and enzymatic processes, and of the limiting effects of diffusion in cellular system, as well as the longer time scales for cell growth, reproduction and genetic effects.

The earliest experimental attack on this problem of the time required for the development of radiation damage was by way of varying the dose rate of the radiation and giving equal doses in shorter or longer times. Any effect which depends upon dose rate is, ipso facto, one in which latent radiation damage may be repaired or, alternatively, rendered irreversible by a later part of the radiation energy delivered. There must be some cooperation between different portions of the total dose in spite of the time interval between them (Fig. 5). Any wholly irreversible portion of the radiation damage should be dependent only upon the total doses delivered and not upon its distribution in time. It is, however, evident that, under continuous irradiation, the damage inflicted by any one small portion of the total dose will interact simultaneously with that due to all the earlier and all the later portions delivered during a lengthy irradiation. Consequently, from studies made at different dose-rates one can draw only indefinite conclusions about the time scale for the development of damage.

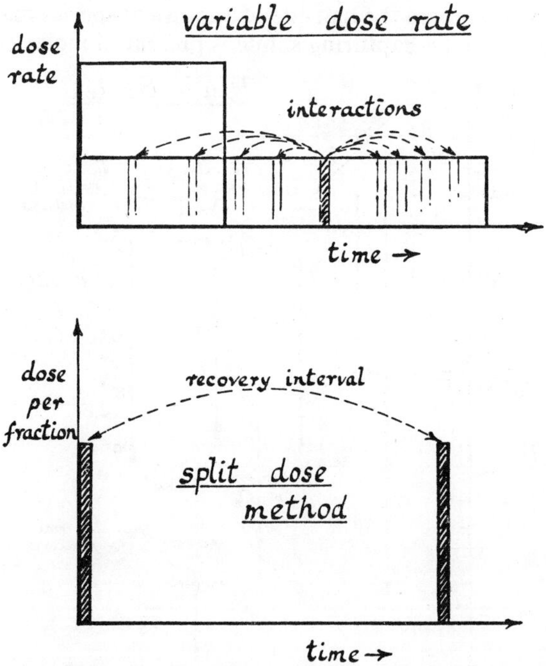

FIG. 5. The advantage of the split-dose method of assessing the time scale of recovery processes.

I do not know who was the first to recognize that the 'split-dose' technique offered a more precise way of studying the time scale for the development of radiobiological damage (Fig. 5). Certainly McComb and Quimby were aware of this when they adopted the technique in their studies of skin erythema in 1936 (*4*), as was Karl Sax in 1939 when he applied a split-dose method to investigate the time during which broken chromosomes were capable of restitution (*5*). His work, indeed, stimulated a careful mathematical study by Fano and Marinelli (*6*) of the way in which one could logically deduce from such experiments information not only about the time dependence of recovery processes, but also about their dose dependence. This type of experiment has become popular again following its use by Elkind and Sutton (*7*) in their classical demonstration of the existence of repair processes in mammalian cells. In most of these split-dose biological experiments each fraction of the dose was delivered in a few minutes and the interval between them ranged from minutes to hours. We shall be discussing presently split-dose experiments in which each fraction of the dose was delivered in a few nanoseconds and the interval between them was measured in microseconds.

Let us look then at some of the experimental methods now in use to investigate this wide time scale (Fig. 6). Whenever the interactions we wish to study are *hidden processes*, i.e. are not directly observable at the time they occur, but only through their influence on events which occur much later, we must introduce two disturbances into the system and vary the interval between them. One of these will be the radiation pulse, and the second may be another radiation pulse (as in the split-dose method) or some rapid change in the system such as admission of oxygen, rapid freezing, rapid mixing or some other perturbation.

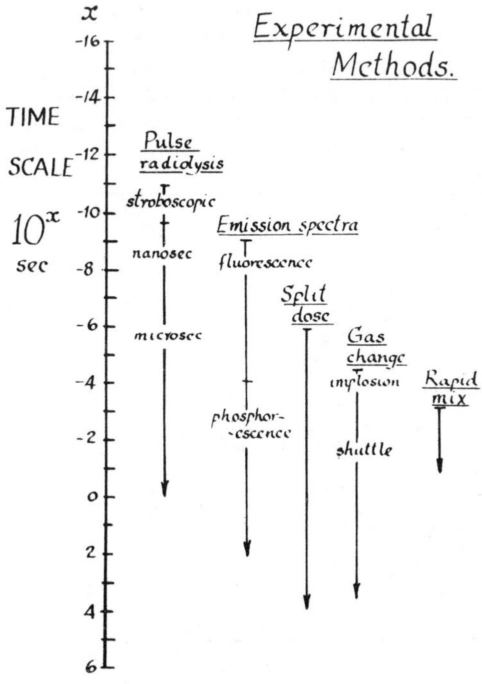

FIG. 6. The range of times over which various experimental methods are useful.

By varying the interval between the two perturbations we are then able to study the lifetime of the transient states responsible for the hidden interactions. We shall see how this has been done by split-dose, gas change, and rapid mix techniques.

There are, however, some transient reactions which are not hidden but patent because the participating molecules identify themselves by clear code signals. Various codes are available for carrying the signals but the simplest and still the most widely used are the optical absorption or emission spectra of the molecules involved. Either of these codes allows rapid signals to be received, and is usually sufficiently sensitive. They run into difficulties only when several interesting species with overlapping spectra are present. Even in that case the signals may be separable if, for instance, the overlapping spectra have widely different lifetimes.

In pulse radiolysis we use the absorption spectrum of one or other of the reactants or products. The technique, then, is to give a short pulse of radiation to the solution containing the solute whose reaction with the primary radicals is to be studied, and to observe the formation and decay of transient species by the changing optical absorption of the solution.

Fig. 7 shows a few of the words in this code—a small selection of optical absorption spectra belonging to some of the transient species which may be produced in an irradiated aqueous solution if the appropriate solutes are present. These spectra are, in general, broad absorption bands with considerable overlap and in the early days of pulse radiolysis it was not always easy to disentangle the spectra. With the much more sophisticated equipment in use today the individual spectra can be recognized, not only by more precise location of the peak, but also from the shape of their shoulders and tails. Once the code signal of a transient species is plainly established its chemical properties and reaction rates can be studied in detail by forming the species in a simple chemical system in which it is exposed to a single solute at a time and then measuring its reaction rate constant

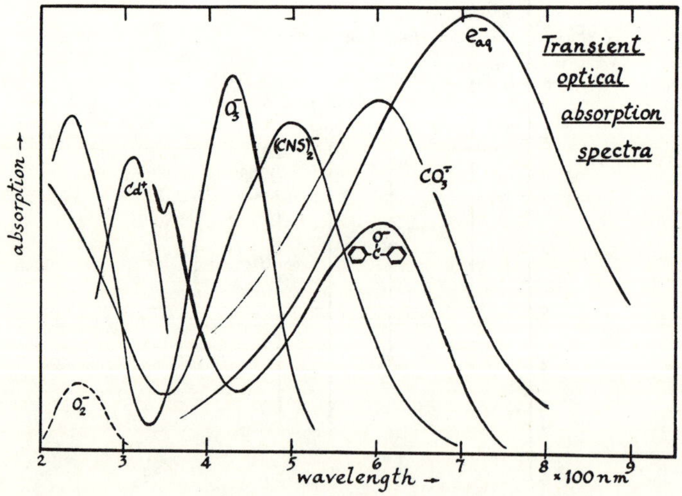

FIG. 7. A selection of absorption spectra of transient ionic species in solution.

by the speed with which its absorption spectrum disappears. In cases where the primary radical itself has only a very weak absorption spectrum—as, for instance, in the important cases of the OH radical and the H atom—the reaction can be followed by the rate of *appearance* of the absorption due to one or other of the reaction products. When oxygenated water at high pH is irradiated, neither the solute—O_2—nor the primary radical formed—O^-—shows any useful absorption but the radical product—O_3^-—absorbs strongly in the blue and the reaction is readily followed by tuning in to this absorption peak at 430 nm.

$$O_2 + O^- \rightarrow O_3^-$$

The radiation chemistry of aqueous solutions begins with the formation of the primary radicals, H, OH and the unattached but solvated electrons around the tracks of secondary particles. The initial spatial grouping of these species has been discussed vigorously by radiation chemists for decades and much theoretical work has gone into simplified models of the 'spurs' within which a high proportion of the radicals formed by low-LET radiation are presumed to be formed. Only those escaping from radical-radical reactions within the spurs are available to react with solutes present in low concentration.

Let us look first at the most exhaustively studied of all these radiation-induced transient species—the hydrated electron. It enjoys this special position firstly (and most importantly) because it has such a strong absorption (*8*) and is therefore very easy to study! Secondly, because its formation and behavior is of great interest to theoretical chemists; and thirdly, because it is one of the three principal species formed in water and is therefore intimately involved in many radiobiological reactions. The radiation chemists present will, I hope, excuse me if I illustrate its nature by a simple analogy and model. The molecules of any polar liquid, such as water, possess an unsymmetrical distribution of electric charge and this unbalanced 'electric moment' causes them to line up in any external electric field.

We can represent such molecules by floating discs each carrying an embedded *magnet* which will respond to an external *magnetic* field. If now we swing another magnet above this assembly to represent a wandering electron, we find that as long as it is moving rapidly (Fig. 8a) the inertia of the floating discs prevent them from following its motion. As the moving magnet slows down, however, they have time to respond and ultimately the moving magnet is 'captured' by the ordered assembly which it has itself created (Fig. 8b). It remains mobile, but much less so than before, since its movements are impeded by the inertia of its captors which must turn to preserve the same radial configuration. This is a two-dimensional model of an electron trapped by its own hydration shell of oriented water molecules—the transient species christened "hydrated electron" in 1953 by the late Dr. Platzman (*9*) whose death a year ago was such a sad loss to his friends and to theoretical radiation chemistry. In very pure water the hydrated electron can live for about a millisecond but it reacts at the maximum possible rate, i.e. at every encounter, with many solutes and its half-life is therefore only a few microseconds in a 100 micromolar concentration of electron-scavenging material. When a stable solute reacts with an electron the adduct acquires radical or ionic nature and these secondary radicals can react, usually more slowly, and often much more selectively, with other molecular

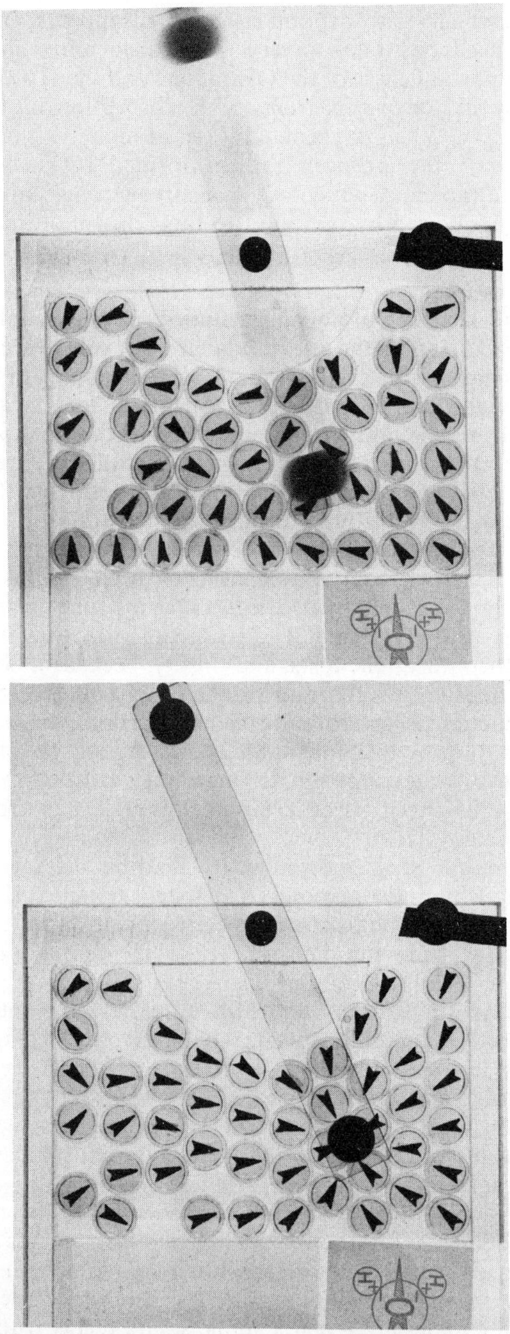

FIG. 8. (a) The moving electron (blurred) is too fast to organize the molecules in its vicinity. (b) As it slows down the electron is trapped by an ordered assembly of molecules.

species which may be present. They can, in fact, be used as if they were standard reagents, not stored in bottles but manufactured pure in the solution when required.

These electron adducts can transfer their excess electron to other molecules which they meet in the solution and an elegant demonstration of a cascade transfer of this kind has been given by Adams, Michael and Willson (*10*). They irradiated an aqueous solution containing a high concentration of acetone, a low concentration of acetophenone and just a trace of benzophenone in the proportions indicated by areas in the upper part of Fig. 9. All three solutes react strongly with hydrated electrons, the acetone rather more rapidly than the other two solutes so that virtually all the electrons rapidly attach themselves to acetone molecules and the resulting electron adduct has no color in the visible region. However, only two microseconds after the radiation pulse was given the solution shows the color characteristic of the acetophenone electron adduct and 50 μsec later this has changed chameleon-like to that of the ketyl-radical ion arising from benzophenone. The electron has been transferred in succession to the scarcer solutes as soon as it met them in a close enough encounter. If a trace of oxygen were present this would ultimately rob even the benzophenone of its electron.

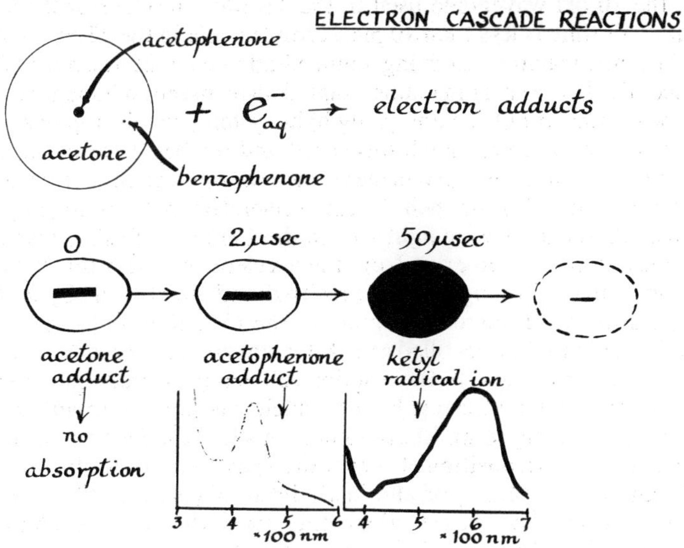

FIG. 9. The electron initially captured by the abundant scavenger can be handed on successively to substances with higher electron affinity, even though present in small concentration.

To prove that this kind of cascade transfer can occur equally well in a cellular system Adams et al. irradiated solutions containing various bases. In one experiment they took a solution containing cytosine, uracil and thymine, each at a concentration of 10 millimolar, with 400 micromolar orotic acid and only 12.5 micromolar paranitroacetophenone (PNAP) and they irradiated the mixture by a pulse. Virtually all the electrons must have been trapped first on the three abundant bases but 4 μsec later the spectrum detected was that of the adduct of the relatively scarce orotic acid and 100 μsec later this had given place to the spectrum of PNAP, present only in micromolar concentration. Evidently electron-hopping can proceed very effectively inside cells.

It is characteristic, perhaps, of the present epoch in scientific work that one should have to consider the question 'of what use is this beautiful demonstration of cascade electron transport?' Well, in this case the answer can be given quite briefly. It was on a basis of electron affinities that Adams and Dewey (*11*) first suggested a new group of electron-affinic sensitizers of anoxic cells and it is largely through this type of radiation-chemical 'screening' that these sensitizers have been progressively improved, and new families found. A great deal will be heard about them at this Congress.

I mentioned earlier that one reason for the extensive studies of e^-_{aq} was the interest of theoreticians in its formation, its initial yield, and its immediate reactions with other radicals in the spurs. It is in the pursuit of these rather fundamental questions that the time resolution of the pulse radiolysis technique has been pressed furthest—in fact, into the picosecond range. The electron bunches which ride on the wave fronts in the waveguide of a linear accelerator are very compact indeed. Each normal pulse emerging from the machine contains tens, hundreds or even thousands of such bunches spaced one guide wavelength apart. In the 10 cm waveguide used in the Toronto accelerator the length of a single bunch in time is less than 10 picoseconds while in the 25 cm waveguide of the new Argonne machine, carrying a much higher current, the bunch lasts some 40 picoseconds. In the latter machine isolated 40 picosecond bunches of electrons can now be extracted but the analyzing light system, which employs a laser and photodiodes, cannot quite match this speed and the best time resolution is still obtained by adopting the very elegant stroboscopic method first devised by Bronskill and Hunt (*12*) for their 10 cm accelerator in Toronto (Fig. 10). This method uses as the analyzing light source the Cerenkov flash generated by the electron pulse itself as it emerges from the accelerator, strips this light pulse off the electron bunch by a mirror and steers it around a rectangular path by means of lenses and movable mirrors. In practice the Cerenkov pulse is enhanced by having it formed in a Xenon-filled cell. After traversing the rectangular path the light flash is then injected into the reaction cell in any desired phase relationship with respect to a later electron bunch which has come straight through the mirrors. A time-sharing rotary shutter interrupts first the electron beam, then the light path and in a third position allows both to pass. This procedure creates three clases of light signals falling on the final photocell (see Fig. 10). The first class arises from the electron bunch alone traversing the cell and emitting some Cerenkov radiation. The second class consists of the strong Cerenkov pulse from the Xenon cell traversing unirradiated solution. The third class consists of the direct Cerenkov flash from the cell plus the delayed analyzing flash attenuated by absorption in the iradiated solution. With this so-called 'stroboscopic' system the absorption suffered in the cell by the analyzing light flash can be derived by addition or subtraction of the three classes of light signals recorded, without the need for impossibly fast electronics. By these means the time scale from 50 picoseconds up to 1 nanosecond after the pulse can be explored in the Argonne machine, and the interval from 18 picoseconds to 350 picoseconds in the Toronto accelerator. It is encouraging that, with such highly sophisticated experimental systems, the results obtained in the two laboratories are in such good agreement

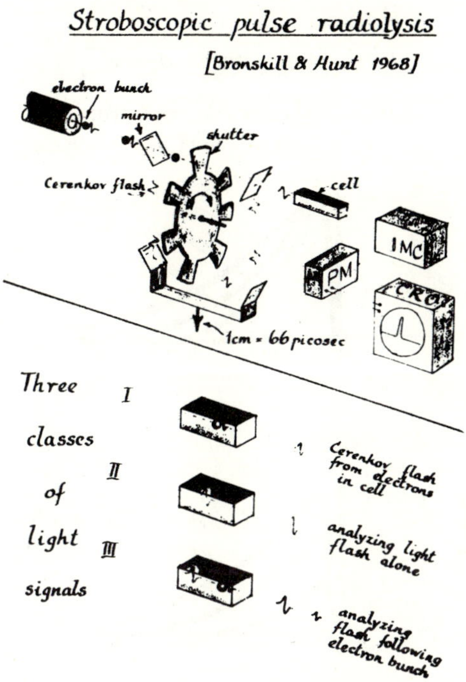

FIG. 10. Diagramatic sketch of the stroboscopic arrangement for using the Cerenkov light from electron bunches to detect very shortlived absorbing species.

that they have been able to report some of them (*13*) in joint publications! They both find that the optical absorption due to the hydrated electron is already fully developed at the end of the shortest pulse available, that is to say in less than 10 picoseconds. Recently an even smaller limit of 4 picoseconds has been given from another laboratory for the solvation time of photogenerated electrons (*14*). Both teams agree also on the initial yield of hydrated electrons, which must include those formed within "spurs", and they agree that the initial decay rate of the electron concentration is significantly slower than is predicted by the current "spur diffusion model" of an electron track in water. This evidence will doubtless lead to a reassessment of the parameters commonly accepted for the "spur" model.

An experimental observation of great interest which has come out of the picosecond work is illustrated in Fig. 11. Hunt and his colleagues observed that whereas strong acids at high concentration did not reduce the initial yield of hydrated electrons —i.e. the concentration formed by a given dose—but only increased their rate of disappearance as shown in the upper part of Fig. 11, some other solutes—acetone, NO_2^-, etc.—cause a progressive reduction in the initial yield as shown in the lower part of Fig. 11 (*15*). They interpreted this as being due to the reaction of the solute with a "precursor" of the hydrated electron—in fact, with the "subexcitation electrons" whose chemical role had also been predicted by Platzman long ago (*16*). Colloquially, these have been dubbed "dry electrons". If subexcitation electrons can react before they have acquired a hydration shell

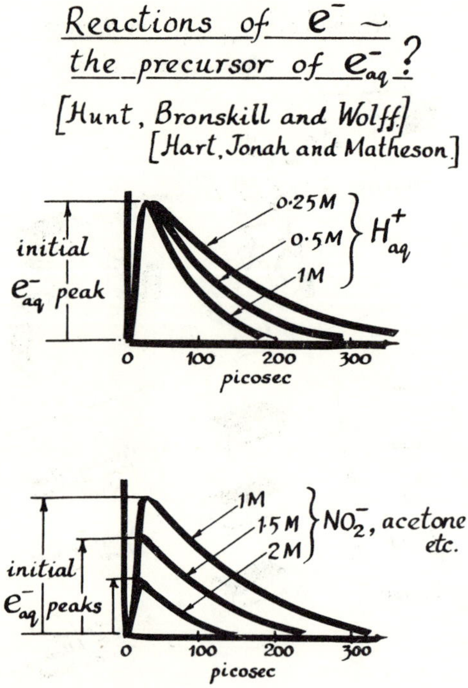

FIG. 11. Distinction between solutes which react only with the hydrated electron (above) and others (below) which can capture its precursor.

then of course the initial yield of e^-_{aq} will be lower at high scavenger concentrations.

An alternative explanation of the reduced yield has been put forward by Czapski and Peled (*17*), following and extending an explanation due to Schwarz (*18*). They point out that if a scavenging solute is present at 2 molar concentration in water this implies that about 1 molecule in 27 is a solute molecule. When a hydrated electron is formed at a random point in such a solution, the chance that one of the 10 to 20 molecules which are in its immediate trapping shell shall be a solute molecule is quite high and the two are then already involved in an "encounter" without having to find one another by diffusion. Provided the solute is one which reacts at virtually every encounter the electron will vanish without trace in less than a picosecond after it is formed and this will appear experimentally as a reduction in the initial yield. Solutes like the H^+_{aq} ion in the strong acids, on the other hand, which do not react at each encounter, may allow many of these "encounter pairs" to separate again. This explanation would account for the reduced initial yields without assigning a chemical role to the subexcitation electron and even the kinetic equation to which it leads turns out to be identical with that derived on the assumption that the dry electron is the reacting species. There are, however, some solutes—of which benzene is one (Hart, 1974, private communication)—which react only slowly with hydrated electrons and yet reduce their initial yield. Evidently, in such cases reaction with the subexcitation electrons is the simplest explanation. Here, as so often, we have to keep our minds open to alternative explanations—practicing T. C. Chamberlin's "Method of Multiple Hypotheses" (*19*).

I have mentioned emission spectra as another code in which signals about the development of radiation damage can be transmitted. A fast electron produces not only ions but also excited molecules along its path. Some of these excited molecules may at once break up, but others will emit the excess energy as luminescence—either as immediate 'fluorescence' or as delayed 'phosphorescence'—the latter arising from triplet states. The spectrum of the luminescence is here the code signals which identifies the emitting molecule. This technique has recently been applied in an imaginative way by Fielden and Lillicrap (*20*) to study the transfer of energy between different DNA bases and the preferential localization of this energy at abnormal bases incorporated into the DNA chain. Let us consider only one of their experiments. They incorporated BUdR into bacterial DNA at various concentrations from a mere 0.2% up to almost 100% of the thymidine positions. The extracted and dried DNA was irradiated at low temperature by a pulse of fast electrons and the intensity of the emitted luminescence was recorded electronically over many decades of the time scale.

The curve of intensity of emission against time has one form for natural DNA and quite a different form for pure BUdR (Fig. 12). As the percentage of BUdR incorporated into the DNA is reduced the intensity falls progressively but the intensity vs. time curve retains the shape characteristic of the BUdR component down to very low values of percentage incorporation. This must mean that the absorbed energy is being channeled to the BUdR bases and emitted from them. Remembering that the energy of the fast electrons is initially distributed at random throughout the molecule one can calculate that a significant proportion of the excitation energy has had to travel across some 100 bases in order to reach a BUdR site from which emission can occur.

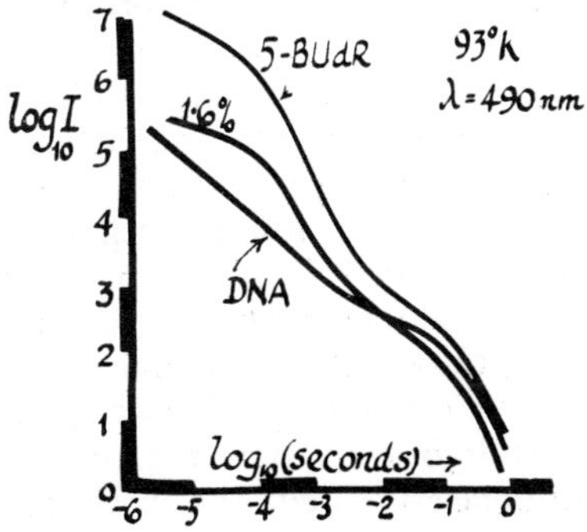

FIG. 12. Intensity versus time curve for the emission of visible light from irradiated DNA, 5-BUdR, and from DNA in which only 1.6 per cent of the thymidine has been replaced by 5-BUdR.

In the remaining minutes I should like to move to biological work and see how fast kinetic methods are now being applied in this more difficult and more complex field.

We do not as yet know with certainty the precise chemical mechanisms by which oxygen sensitizes cells to ionizing radiation nor yet the organelles or structures within cells on which it acts. We do, however, know as the result of many studies that the effect is by no means a simple one. In the fundamental studies by Powers and coworkers (*21*), using as experimental material the spores of *Bacillus megaterium*, several different components of oxygen-sensitive damage have long been recognized. These are revealed by their selective response to changing temperature, to moisture content, and to various free radical scavengers. In the dry spore system some of these oxygen-sensitive species have lifetimes of many minutes even at room temperature or above, as indicated by the post-effect of oxygen after the end of the irradiation. What can we say about the lifetime of the same, or similar, species in wet-cell systems?

This question was one of the first to be attacked by a technique which was fast by the standards of twenty years ago. Howard-Flanders and Moore (*22*) put a wet bacterial culture onto a small shuttle and fired the latter through a membrane from an oxygen-free to an oxygen-rich atmosphere and vice versa immediately after irradiation. With the time resolution they were able to achieve in this way—approximately 20 milliseconds—they could detect no effect of exposing to oxygen after irradiation. The species reacting with O_2 in wet vegetative bacteria evidently had a lifetime shorter than 20 milliseconds. The same problem has recently been taken up again by two different groups using techniques with much higher time-resolution—1 microsecond in the one case and about 100 microseconds in the other. There is some agreement and some disagreement between the findings of these two groups, who used different bacteria and different pulse techniques.

Epp and his coworkers (*23*) employed the split-dose technique in its most elegant and sophisticated form. With two separate pulsers, each delivering 3 nanosecond pulses of 600 kV electrons, they were able to irradiate their cultures of *E. coli* B/r attached to mylar foils and exposed to various low oxygen concentrations (Fig. 13). With this arrangement they set out to measure the time necessary for oxygen to penetrate far enough into bacterial cells to give any selected oxygen enhancement factor. The first pulse was made large enough to bind all the dissolved oxygen in the cells and thus to carry the survival curve past the 'break point' representing the change to anoxic conditions which they had investigated very thoroughly in previous experiments. The interval between pulses allowed some oxygen to diffuse into the cell and the second pulse then tested the radiation sensitivity which this oxygen had conferred on the cells. By comparison with their own previous calibration experiments this sensitivity assay could be used to assess the effective oxygen concentration at the sensitive sites within the cell at the time the second pulse was given and thus, on certain assumptions about the location of these sites, the amount of oxygen which had penetrated into the cell could be estimated. The curves they obtained, of which a sample is shown in the lower part of Fig. 13, showed several interesting features. All their curves, for different initial oxygen concentrations, showed that a significant effect of oxygen could be detected—i.e. that some oxygen had diffused to the sensitive sites in the bacterial cell—at about 100 microseconds. Now the well-established fact that breaking survival curves are obtained on this bacterium when the radiation dose is increased above a certain level is proof that the transient species, whose reaction with oxygen produces lethal damage, must

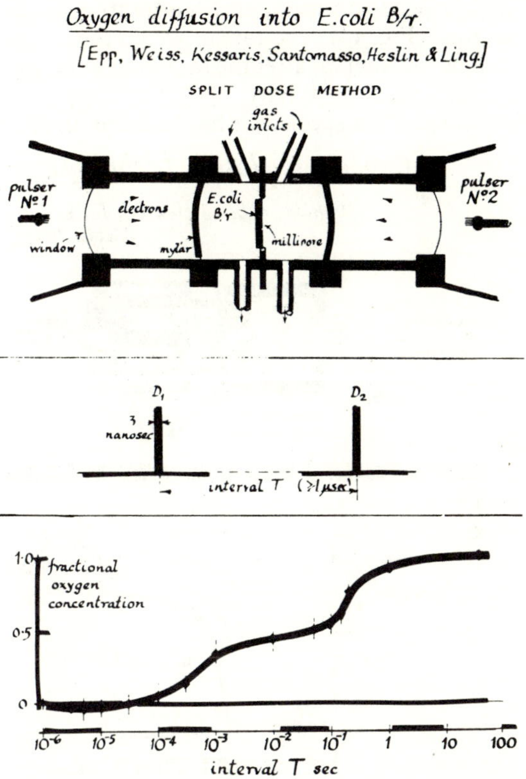

FIG. 13. A study of oxygen diffusion times for the bacterium *E. coli* B/r.

have decayed before any oxygen could diffuse to the sensitive sites. The interval of 100 microseconds can therefore be taken as an upper limit for the lifetime of the oxygen-sensitive damage in *E. coli* B/r under the conditions of these experiments. The other interesting feature of this diffusion curve is the very much slower penetration of oxygen into the cell than would be expected from simple diffusion theory. In an accompanying paper, Kessaris, Weiss and Epp (*24*) consider various possible explanations of this phenomenon and put forward the interesting hypotheses that the radiation dose creates a high concentration of oxygen-scavenging species within the cell, which then impede the inward diffusion of oxygen. This explanation is analogous, although on a smaller scale, to the 'oxygen debt' due to accumulated lactic acid which A. V. Hill showed many years ago could explain the slow diffusion of oxygen into exhausted muscle fibres (*25*). Their model predicts concentration-time curves so similar in shape to those actually observed, that they constitute strong evidence although not, perhaps, proof of the correctness of their hypothesis.

In a comparable experiment, Michael and his colleagues in the Gray Laboratory used an implosion technique to bring pure oxygen at several atmospheres pressure into contact with the bacteria—in this case *Serratia marcescens*—before or after irradiation by a 2 microsecond pulse of fast electrons (*26*). Their trigger arrangements for admitting the gas gave them a time resolution of 100 μsec. (Fig. 14). The rate at which the oxygen-sensitive species decayed is reflected by the interval after the radiation pulse which was required to reestablish the anoxic level of sensitivity. They found approximately exponential decay of the oxygen-sensitive species with a half-life of 500 microseconds.

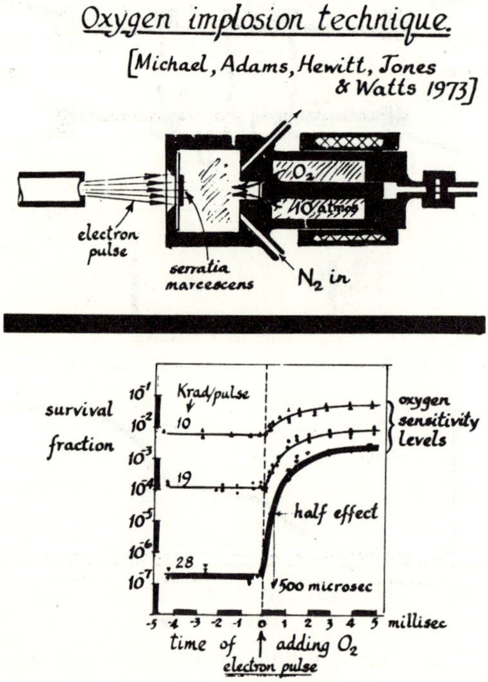

FIG. 14. A study of the duration of oxygen-sensitive species in the bacterium *Serratia marcescens* following a 2 microsecond pulse of electrons.

The apparent quantitative disagreements between these two experiments which are models of care, both in execution and interpretation, are interesting. Epp et al. (*23*) found an upper limit of 100 μsec for the lifetime of the active species and a lengthy time scale for diffusion. Michael et al. (*26*) found a half-life of 500 μsec for their active species and they obtained virtually the full sensitizing effect when oxygen contact with the cells occurred at the same instant as the radiation pulse. Diffusion of oxygen into the cell was therefore considerably faster than the 100 μsec time-resolution of their system. There is no *direct* conflict between the results since they were using different bacteria. *E. coli* B/r is a radiation resistant organism and might therefore enjoy not only a better enzymatic repair system but also a better chemical scavenging system for the oxygen-sensitive species, which would then have a shorter lifetime. Moreover, as regards diffusion time the bacteria in the experiments of Michael et al. were suddenly exposed to several atmospheres of pure O_2 and so the fractional concentration to be attained at the sensitive sites need only be about one-fortieth of that required in Epp et al.'s experiment: this could reduce the effective diffusion time by a large factor. However, the differences are intriguing and are likely to stimulate further experiments which may well bring to light some unexpected explanation.

We have discussed thus far techniques applicable to chemical solutions, to crystalline bases and dry DNA, or to bacterial cells. What experimental methods can be used to study the hidden fast processes taking place in the much more fragile mammalian cell? The method of rapidly mixing two solutions, developed first for radiation chemical work and applied later to bacteria, has proved to be gentle enough for mammalian cells. The sudden mixing of the cell suspension either before or after irradiation with another solution containing some sensitizer or protector is here the necessary second perturbation. The apparatus designed by Adams et al. (*27*) for doing this is shown in simplified form in Fig. 15. In

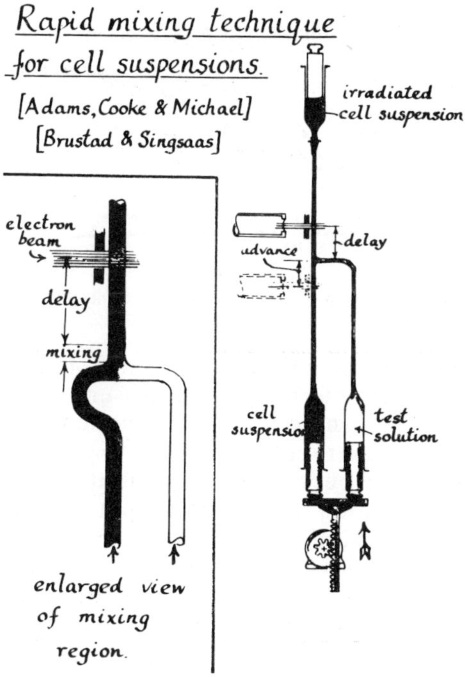

FIG. 15. Apparatus for studying the time scale of oxygen and other sensitizers or protectors by mixing the cell suspension with the test solution before or after irradiation.

practice an additional capillary fed with dosimetric solution from a third syringe catered for the necessary accuracy in dosimetry. Irradiation can be either pulsed or continuous since the time scale is set not by the pulse length or spacing but by the flow time between the zone of mixing and the zone of irradiation. Indeed, continuous irradiation by a uniform electron beam is more convenient since the slugs of solution irradiated by separate pulses must overlap adequately to avoid any danger of cells passing through unirradiated and this demands a high pulse repetition frequency. The time resolution which can be achieved depends, therefore, upon the maximum flow rate of the solution and this depends, in practice, upon the bursting pressure of the syringes. Flow rates of up to 2.7 m/sec have been achieved, giving a time-resolution of a few milliseconds. Perhaps the most interesting result obtained with this apparatus to date is the time scale for the oxygen effect in mammalian cells. I am indebted to Drs. Adams and Michael for permission to refer to their results before they have appeared in print (*28*). Briefly, at 2 milliseconds after mixing an enhancement ratio of 1.7 was obtained (Fig. 16) whatever the oxygen concentration in the test solution over the range used in the experiments (1% to 50%). However, the development of the full oxygen effect up to an enhancement ratio of 2.8 took many milliseconds and was much slower with the more dilute solutions. One is tempted to speculate that the immediate component of amount 1.7 must relate to damage in or close to the outer cell membrane and the remaining diffusion-dependent component to damage in some internal structure in the cell.

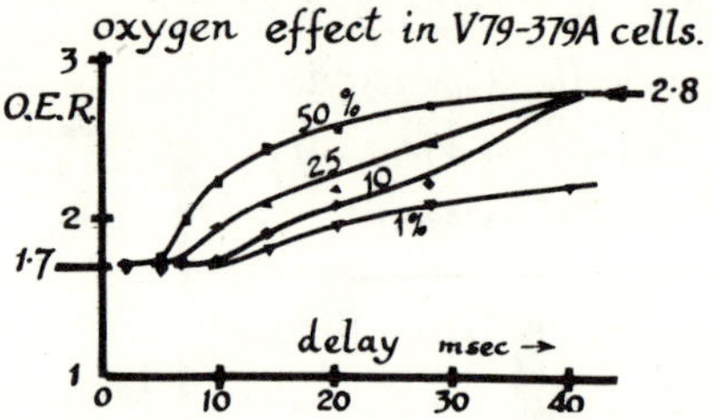

FIG. 16. The lifetime of oxygen-sensitive species in a mammalian cell as measured in the rapid mixing apparatus of Fig. 15.

The oxygen effect has been the target for most of the biological experiments I have referred to. These fast techniques for measuring the time scale for oxygen sensitization within the living cell are revealing new complexities in this old phenomenon. Young workers entering the field need have no fear that all the interest has been squeezed out of this subject. Rather, the scene is now set for a full and definitive explanation which must integrate the radiation chemistry with the growing knowledge of cell structure and function.

I warned you that in this talk I would cover the material in breadth rather than in depth. I have, of course, been unable to deal *adequately* with any of the many topics which I have used as illustrations but if I have at least aroused your interest I shall be content and you can hear more about all these topics at this Congress. I hope that next time you may contemplate a fast kinetic experiment in your own biological specialty you will not rule it out "because" it seems too difficult but rather say "why not?" I am sure that would have been Failla's response in any such situation.

REFERENCES

1. L. H. Gray *Brit. J. Radiol.* **35**, 216 (1962).
2. H. H. Rossi, *Am. J. Roentgenol. Radium Ther. Nucl. Med.* **87**, 608 (1962).
3. E. De Bono, *The Use of Lateral Thinking.* Pelican Books, Harmondsworth, England, 1971.
4. W. S. MacComb and E. H. Quimby, *Radiology* **27**, 196 (1936).
5. K. Sax, *Proc. Nat. Acad. Sci. U.S.A.* **25**, 225 (1939).
6. U. Fano, and L. D. Marinelli, *Proc. Nat. Acad. Sci.* **29**, 59 (1943).
7. M. M. Elkind and H. Sutton, *Radiat. Res.* **13**, 556 (1960).
8. E . J. Hart and J. W. Boag, *J. Am. Chem. Soc.* **84**, 4090 (1962).
9. R. L. Platzman, in *Basic Mechanisms in Radiobiology* (J. L. Magee, M. D. Kamen, and R. L. Platzman, Eds.) U. S. National Research Council Publication No. 305, Washington D. C., 1953.
10. G. E. Adams, B. D. Michael, and R. L. Willson, in *Radiation Chemistry, Int. Conference on Radiation Chemistry*, 1968 **Vol. 1,** Advances in Chemistry Series No. 81 (E. J. Hart, Ed.) p. 289, American Chemical Society, Washington, D. C. 1968.
11. G. E. Adams, and D. L. Dewey, *Biochem. Biophys. Res. Common.* **12**, 6 (1963).
12. M. J. Bronskill and J. W. Hunt, *J. Phys. Chem.* **72**, 3762 (1968).
13. J. W. Hunt, R. K. Wolff, M. J. Bronskill, C. D. Jonah, E. J. Hart, and M. S. Matheson, *J. Phys. Chem.* **77**, 425 (1973).
14. P. M. Rentzepis, R. P. Jones, and J. Jortner, J. *Chem. Phys.* **59**, 766 (1973).
15. K. Y. Lam and J. W. Hunt, *Int. J. Radiat. Phys. Chem.* **7**, in press (1975).
16. R. L. Platzman, *Radiat. Res.* **2**, 1 (1955).
17. G. Czapski and E. Peled, *J. Phys.* Chem. **77**, 893 (1973).
18. H. A. Schwarz, *J. Chem. Phys.* **55**, 3647 (1971).
19. T. C. Chamberlin, *J. Geol.* **5**, 837 (1897).
20. E. M. Fielden and S. C. Lillicrap, *Curr. Top. Radiat. Res. Q.* **7**, 133 (1972).
21. E. L. Powers and A. Tallentire, in *Actions Chimiques et Biologiques des Radiations* **Vol. 12** (H. Haissinsky, Ed.) pp. 3–67. Masson et Cie., Paris, 1968.
22. P. Howard-Flanders and D. Moore, *Radiat. Res.* **9**, 422 (1958).
23. E. R. Epp. H. Weiss, N. D. Kessaris, A. Santomasso, J. Heslin, C. C. Ling. *Radiat. Res.* **54**, 171 (1973).
24. N. D. Kessaris, H. Weiss, and E. R. Epp, *Radiat. Res.* **54**, 181 (1973).
25. A. V. Hill, *Proc. R. Soc. Ser. B* **104**, 39 (1929).
26. B. D. Michael, G. E. Adams, H. B. Hewitt, W.B.G. Jones, and M. E. Watts, *Radiat. Res.* **54**, 239 (1973).
27. G. E. Adams, M. S. Cooke, and B. D. Michael, *Nature (London)* **219**, 1368 (1968).
28. G. E. Adams, and B. D. Michael, these Proceedings.

ENERGY NEEDS, NUCLEAR POWER AND THE ENVIRONMENT[1]

Chairman: B. Lindell

[1]Plenary Session A–1

Energy Needs, Nuclear Power, and the Environment [1]

Alvin M. Weinberg
Office of Energy Research and Development, Federal Energy Office, Washington, D. C. 20500, USA

During 1973 the world used more than 200 mQ [1 milli-Q (mQ) $\equiv 10^{15}$ Btu $= 1.06 \times 10^{18}$ joules] of energy; the United States used more than 70 mQ. Each person in 1973 used, on the average, about 60×10^6 Btu (or 64×10^9 joules). The total energy output of the world during the past decade has increased at a rate of more than five percent per year; the per capita increase has been around three percent per year. Though predictions of future energy growth are risky (there have been more than 133 predictions of future energy demand and supply since 1952), almost all who have studied the matter estimate that the world's energy demand will continue to rise, for example, to a total of 800 mO per year, and a per capita expenditure of 130×10^6 Btu (140×10^9 joules) by 2000 A.D. (*1*).

As the recent oil embargo brought so forcefully to our attention, such growth in demand represents a real, not an imaginary, threat to our economic and social well-being. The discrepancy between demand and supply is likely to grow; and most countries, including the United States, now try to plan rather than accept their energy future. We in the U.S. have committed ourselves to "Project Independence" (P.I.) to make the United States adequately self-sufficient in energy by the early 1980's. The Federal Energy Administration (FEA) is now preparing detailed plans for achieving this goal (as well as defining the goal itself), and expects to have the plan ready by November of this year (*2*).

In the United States during the past year we imported about 12 mQ of energy—mainly some 6×10^6 barrels of oil per day. By 1985 some estimates suggest that the difference between petroleum we shall need and the petroleum we shall produce domestically will amount to as much as 36 mQ per year or the equivalent of 18×10^6 barrels per day. To bring our supply into better balance with demand, we are depending on three strategies.

1. *Conserve energy*—We estimate that as much as 7×10^6 barrels of oil per day can be saved by 1985 if we use smaller cars, eliminate waste in energy-using industries, and generally avoid unnecessary use of energy.

2. *Increase production of domestic oil and gas*—This would involve tapping reserves in Alaska, the Outer Continental Shelf, and possibly the oil shales in the West.

3. *Substitute coal and uranium for oil and gas*—Such substitution would have to occur in industrial processes, and in utility boilers, and by conversion

[1]Congress presentation A-1-1.

of coal into oil and gas through hydrogenation. In general, this strategy probably would imply a gradual shift toward electrification of the energy system.

These measures are, on the whole, relatively short-term, say for the next 10 to 30 years. But the total world oil reserve (recoverable with today's technology and including shale and tar sands) of perhaps 25 Q will probably be almost gone within 80 years; even coal resources, of which the world possesses perhaps six times as much, may be used up within a few hundred years (*3*). Thus, as we contemplate the vast expanse of man's future, we are forced to consider the inexhaustible energy sources: solar, geothermal, fusion, and fission via the breeder reactor.

It would be foolhardy at this stage to dismiss any of these sources of energy; hopefully, our ultimate energy system will be a mixture of all, with fossil fuel used as it is available. But if we are fully realistic, we must recognize that the scientific—let alone technological—feasibility of fusion is still to be demonstrated; and that the total geothermal flux, representing as it does only 1/20,000 of the sun's insolation, can hardly be counted on to supply the bulk of man's long-term energy, though the heat stored in the earth represents a very large reserve if we only could tap it extensively, say by mining heat from hot dry rocks. According to these views of the future (and I warn again about how difficult is predicting the future), this leaves solar energy and its children—winds, ocean thermal gradients, biosynthesis—and the fission breeder—based on burning the enormous amounts of uranium and thorium in the granitic rocks—as the two most likely candidates for ultimate major prime suppliers of man's energy system.

Solar energy is in most ways the most benign energy source. Unfortunately, it is very dilute; and it remains unclear how cheaply we can convert such a dilute source of energy into an economical one. The average solar flux on the earth's surface is 150 watts per m^2; thus, in principle, the entire projected energy demand of the United States in 2000 A.D. (say 180 mQ) might be supplied with 150,000 square miles of solar collectors converting heat into electricity at 10 percent efficiency. But these estimates, in my view, are hard to be taken very seriously, simply because the cost of solar energy with any of the schemes now contemplated is so high. At the moment, I would estimate that central electricity generated in a large solar farm based on flat-plate collectors might cost three times more than nuclear energy generated with fission breeder reactors. The cost of solar electricity as we now understand the technology probably represents an upper limit to the cost of energy generated on a large scale. Thus, insofar as one can project the cost of energy generated by newish technology (the fission breeder) or yet-to-be demonstrated technology (large-scale solar), *on strictly economic grounds* the fission breeder is most likely to become the world's ultimate source of prime energy.

I realize that in making such a prediction I belie a certain nuclear partiality. New developments that would reduce the cost of photovoltaic collector plates, or greatly increase photosynthetic yield, or clearly demonstrate both scientific and technical feasibility of fusion, may very well invalidate my prediction as to the central role of the fission breeder. But we cannot yet *count* on these eventualities, and it seems at least highly plausible that man will have no choice but to learn to live comfortably with fission breeders.

CAN MAN LIVE WITH FISSION? (*4*)

Let us, then, ask what technological or scientific tasks remain to be accomplished if indeed man is to live comfortably, and for a very long time, with fission. The potential problems posed by fission have been belabored so often that it is by now trite to list them: low-level radioactive emissions either in the reactor or in the rest of the fuel cycle (including transport); possibility of catastrophic accident; possibility of diversion of fissile material; and ultimate disposal of radioactive waste.

At the outset, one must recognize that a *properly* operating nuclear reactor, with its sub-systems, imposes a very small burden on the environment. Thus, the regulations of the U.S. Atomic Energy Commission limit the emission of radioactivity from light water reactors, and from breeders, to an equivalent dose of five mrem per year by any ecological pathway. The hazard from such low dose is hardly enough to be concerned with; if one assumes the doubling dose for genetic effects in man is about 30 rads, then five mrem would increase the mutation frequency by 1.6×10^{-4}, assuming the dose response is linear at low dose and low dose rate. Is it possible that we can eventually determine whether or not the dose-response at very low doses is linear? Some progress in establishing this relation at very low dose may be expected from the findings of Seymour Abrahamson, et al. (*5*) on the proportionality of mutation per locus per rad to the total DNA content of cells, ranging from bacteria to man. Thus it is at least conceivable that experiments done on huge bacterial populations at low dose may some day throw light on the effect of very low dose on man: that a scientific question which I at one time had deemed impossible to answer in principle might ultimately turn out to be susceptible to resolution by science.

I do not believe that the very small routine emissions from reactors and their sub-systems could reasonably explain the public concerns about nuclear energy that are now being expressed and that conceivably could jeopardize the nuclear option. It is, rather, the other three issues—safety, diversion, waste disposal—that are the primary concerns. What can we say about each of these, and do we see ways of reducing both the real concerns and the publicly perceived concerns?

Reactor safety—A1000-MWe reactor in steady state contains almost 10^{10} curies of radioactive fission products; a liquid metal fast breeder of this capacity contains about 2500 kilograms of ^{239}Pu; a light water reactor with Pu recycle contains perhaps 600 kilograms of plutonium. Despite these enormous inventories of radioactivity, practically no activity escapes from the containment shell. This is an extraordinary engineering tour de force.

It would be wrong to deny that under no circumstances can a reactor accident occur which would breach the containment, but the probability of this occurring and causing extensive loss of life is certainly extremely small. The whole issue of the safety of pressurized water reactors has recently been examined by a task force headed by Professor Norman C. Rasmussen of the Massachusetts Institute of Technology. Rasmussen subjected light water reactors to fault-tree analysis—that is, he constructed various conceivable sequences of events that might lead to an accident, and he then assigned a probability to each event in the sequence. These probabilities are based on the best knowledge that is available as to the reliability of, say, a particular valve or instrument. The probability of the entire accident sequence is the product of all the independent probabilities of the events leading to the particular accident.

Rasmussen's study (*6*) concludes that the probability of a loss-of-coolant accident in a light water reactor that would breach the containment and would cause more than 100 deaths is 10^{-6} per reactor per year. A probability of 10^{-6} per reactor per year suggests that if, say, 20,000 reactors are eventually deployed throughout the world, one could expect, on the average, an accident that would breach the containment and cause more than 100 deaths once every 50 years. Obviously, the seriousness of a radioactive release depends upon the wind and weather conditions at the time of the release, the population density and its distribution near the reactor, the form in which the radioactivity is released, and so on. Thus, Rasmussen computes that by far the larger share of uncontained accidents would cause relatively few casualties; he calculates the probability of a catastrophe with several thousands dead to be 1000 times less probable than the 10^{-6} per reactor-year for the lesser accidents. Thus, if the world eventually depends on 20,000 reactors for its primary energy, an event of this seriousness could be expected not oftener than once in some 5×10^4 years.

Rasmussen's study, I believe, is an important contribution to our understanding of the hazards of nuclear reactors; but it is, by its nature, incomplete. For example, though common-mode failures were considered, there is always the chance that totally unexpected common-mode failures could occur. Yet, despite such easy criticisms, the Rasmussen study lends support for the belief that, first, an uncontained reactor accident is a very unlikely event and, perhaps more important, that the probability of immense damage from such an event is very much lower.

Diversion of fissile material—This is a concern that has received much attention in recent months, largely as the result of the writings of Theodore Taylor (*7*). One way to make diversion all but impossible is to never decontaminate the fissile material during its circuit from reactor to chemical plant to fuel fabrication plant and back to the reactor. This would require remote handling throughout the fuel cycle and would add to the cost of nuclear power.

Of course, in assessing dangers of diversion, one must recognize that the danger of stealing an already built nuclear weapon may well be greater than the danger of diverting fissile material and fashioning a crude weapon from it.

Waste disposal — Again, so much has been written about waste disposal that I shall be brief. James Conant in the early 1950's before the American Chemical Society claimed that the disposal of radioactive wastes would prove to be so intractible that, as he put it, nuclear power would not be worth the candle (*8*). This is a position which the community of nuclear experts rejects, and for what I believe are substantial reasons. At present, the U.S. Atomic Energy Commission has opted for waste management on site—that is, the radioactive wastes will be stored in concrete structures until such time as an acceptable permanent method of disposal has been devised. My personal conviction is that the original idea of storing solidified wastes in salt can be made acceptable. Bedded salt laid down some 200 million years ago has shown no sign of intrusion by natural waters. If one can assure that man will not intervene over extremely long periods, such as 200,000 years, during which time the ^{239}Pu will have decayed, this disposal method should be considered acceptable.

But 200,000 years is a very long time, and who can say that our social institutions will survive that long? Can we therefore find a technological fix (*9*) that would reduce the span of time over which the radioactive wastes require surveillance? In principle, this can be done if we can separate the actinides very sharply from the lanthanides and other fission products—say reduce the residual long-lived plutonium in the wastes to 10^{-2} percent; the separated Pu would be returned to the reactor and burned out. The hazard of the wastes themselves would then be dominated by the ^{137}Cs (half-life = 33 years) and the ^{90}Sr (half-life = 20 years). Thus, a potential hazard of 200,000 years is reduced to one of 1000 years which is still a long time; but it is not so long as to defy conception. Social institutions—for example, the Catholic Church—have lasted this long.

TECHNOLOGICAL FIXES FOR FISSION

The burden of my argument is: first, that in all likelihood the fission breeder will be the primary source of mankind's energy for a very long time; and, second, that we shall have to devise additional technological fixes to make the breeder fully acceptable. Two such fixes I have already mentioned—sharp separation of actinides from lanthanides, and close coupling between all parts of the fuel cycle so that fissile material is never decontaminated. Let me now describe two other fixes that would further increase the acceptability of fission.

Perhaps the most important single development is the energy center (*10*). In retrospect, it seems that we nuclear experts misunderstood the true character of nuclear energy when we visualized the nuclear reactor simply as a replacement for a conventionally-fired steam boiler. But this is a rather shortsighted view of the matter. Actually, a nuclear reactor, and particularly a breeder reactor, is inoperable unless it is supported by a fuel cycle system that involves transport of radioactivity, fuel fabrication, chemical reprocessing, and waste disposal or at least waste management. Each step in the fuel cycle poses a potential, though small, risk. Would it not seem reasonable to accept the nuclear system indeed as a system, and reduce the damage that could conceivably accompany a malfunction by confining all parts of the operation to a relatively few nuclear energy centers or parks? Instead of contemplating in the United States some 500 separate locations where high-level radioactivity is handled in 2000 A.D., would it not be much better to reduce the number of such locations to perhaps 100 nuclear parks, with each center producing perhaps 10,000 MWe at first, and ultimately possibly three times this amount? And would not the world, with its ultimate complement of perhaps 20,000 reactors, do well to confine these—together with chemical plants and fuel fabrication plants—to some 1000 sites, each site with 20 reactors located on it?

If such centers are feasible, their social advantages are compelling. First, the centers would be carefully guarded, and there would be no external lines of communication: diversion of fissile material would be made very difficult. Second, the centers would be built over a long time: the construction force would be stable, and the reactors would be better built; they would be safer and probably cheaper. Third, should an accident occur at a center, one would be able to draw on a much larger cadre of experts to control the situation than would be the case in an isolated installation. Finally, the confinement of radioactive operations to relatively few locations on our planet would surely reduce the amount of land that fission could ever conceivably affect. Because reactors are clustered, the area of land at risk per kilowatt of power produced is less than if the reactors were isolated.

Just how feasible are such large reactor centers? Would the local climate be changed intolerably by such concentrations of energy production? Would the energy system be unacceptably vulnerable to the loss of a single such center? Could blocks of electrical energy as large as 30,000 megawatts be transported; or would the center have to generate hydrogen and transport H_2, not electricity? Could the utility industry, fragmented as it is in some parts of the world, particularly in the U.S., accommodate to such centers?

These are questions that are now being studied; and, indeed, Congressman Mike McCormack has submitted legislation to the U.S. Congress formally proposing such studies (*11*).

Assuming that the technical problems of energy centers are soluble, how could we go from here, with our nuclear systems widely scattered, to the clustering of facilities that seems to me so sensible? Here I think we can take advantage of the shift from light water reactors (LWRs) to breeders. The light water reactors will presumably be displaced by breeders within the next 50 years. Is it not reasonable, at the least, to decide now to site breeders only in energy centers so that when the last LWR is decommissioned, all of our fission energy system will be deployed rationally in relatively few complete sites? In the United States the fission breeders built or planned are on sites—in Idaho, Hanford, Oak Ridge— that are already nuclear parks. Thus, we have not yet pre-empted the siting policy of breeders. I can hardly think of a more important decision with respect to nuclear energy than to adopt centralized siting.

If the world's energy future follows along the path that I have outlined, with fission breeders in central sites, then I should think the environmental impact of the fission energy system would be reduced to a minimum. Yet, even with nuclear parks, and improvements in construction and operation of reactors, it would be too much to claim that a world based on fission will be less radioactive than a world based on solar energy or even on coal. It seems inevitable that despite our improvements in technology and in siting, the biosphere will be exposed to a small additional radiation insult above background. This is the ultimate risk that we must accept in opting for fission.

Is it conceivable that even this minute risk can eventually be reduced? I believe there is such a possibility, and this brings us in a way to what this International Congress of Radiation Research is all about. Much of the biomedical work reported here is aimed at assessing radiation hazards; but much is also aimed at combatting detrimental effects of radiation. Let us suppose that over the span of time I envisage we shall learn how to cure radiation-induced disease—solid cancers, leukemias, for example—and to identify and control genetic defects induced by radiation. Our situation with respect to residual effluents from reactors would, as has been suggested by Howard I. Adler of the Oak Ridge National Laboratory (personal communication), then resemble the situation we now have with respect to control of disease induced by pathogens. The most important step in the control of pathogen-induced disease was the introduction of public health measures; this reduced our exposure to pathogens. But public health measures could never reduce such exposure to zero. The remaining disease induced by pathogens was brought fully under control only after antibiotics were discovered.

In corresponding fashion, the residual radiation insult may be dealt with, never perfectly to be sure, through breakthroughs in the treatment and cure of radiation-induced disease. Nor is this really an idle dream. The discovery of enzymes that repair damaged DNA (*12*) suggests that organisms possess repair mechanisms at the cellular level to protect against environmental insult, in analogy with the cellular mechanisms (immune system) that protect against pathogenic disease. Such mechanisms were necessary to preserve life against chemical and physical insults in the environment over evolutionary history, just as they were necessary to preserve life against pathogens in the environment. The hope for control and cure of environmentally-induced disease is therefore based not merely on blind faith, but also on the observations that life itself possesses stability against such environmental attack.

Let me close on a broader note. I concede that my energy scenario based on fission breeders may prove wrong; in any event, in the reasonably near future our expanding energy system will be based—to a large extent—on hydrocarbons, many of them derived from coal, and these will put carcinogens and mutagens into the environment.

This International Congress of Radiation Research was conceived largely as a means of mobilizing the world's scientific enterprise to measure, control, and diminish the environmental effects of a particular energy system: nuclear fission. I would suggest that, with the broadening of the world's approach to its energy crisis, Congresses such as this, and the scientific community they represent, may also expand their concerns: to measure, control, and diminish the effects of other environmental insults that result from our expanded attempts to provide society with adequate energy. Such an extension of interest by the Radiation Research community appears to me to be a natural development. As a member of the technological energy community that seems always to be inventing new and more serious environmental insults even as we devise new ways of coping with man's increasing demand for energy, I can only say we would be very grateful for any help you can offer in coping with these hazards.

REFERENCES

1. S. Schurr (Ed.), *Energy, Economic Growth and the Environment,* Johns Hopkins University Press, Baltimore, 1972 .
2. P. L. Auer, *Science* **184**, 295 (1974).
3. M. K. Hubbert, *Sci. Amer.* **225**, 60-70 (September 1971).
4. Alvin M. Weinberg, in *Proceedings of IIASA Planning Conference on Energy Systems,* IIASA-PC-3, pp. 262–274, Laxenburg (July 17-20, 1973).
5. S. Abrahamson, M. Bender, A. D. Conger, and S. Wolff, *Nature (London)* **245**, 460 (1973).

6. N. C. Rasmussen, Chairman, *AEC Task Force on Light Water Reactor Safety,* U.S. Atomic Energy Commission, Washington. Task Force report in press as of June 22, 1974.
7. John McPhee, *Profiles: The Curve of Binding Energy–Theodore Taylor, The New Yorker,* **44**, Dec. 3, pp. 54–145 (1973); *ibid.*, Dec. 10, pp. 50–108; *ibid.*, Dec. 17, pp. 60–97.
8. J. B. Conant, *Chemical and Engineering News* **29**, 3847-3849 (1951).
9. A. M. Weinberg, *The University of Chicago Magazine* **59**, 6-10 (October 1966); also *Bull. Atom. Sci.* **22**, 4-8 (December 1966).
10. A. M. Weinberg, *Nuclear News* **14**, 33-37 (December 1971).
11. Mike McCormack, *"State Environmental Review of Proposed Reactor Sites,"* H. R. 12823, U.S. House of Representatives, Washington, February, 14, 1974.
12. R. B. Setlow, in *Regulation of Nucleic Acid and Protein Biosynthesis*, (V. Koningsberger and L. Bosch, Eds.), pp. 51–62. Elsevier Publ. Co., Amsterdam, 1967.

PROSPECTS FOR FUSION POWER[1]

Chairman: H. K. Forsen

[1]Plenary Session B–1

Status and Future Prospects for Fusion Power[1]

Robert L. Hirsch and William L. R. Rice
Division of Controlled Thermonuclear Research
United States Atomic Energy Commission
Washington, D. C. 20545, USA

INTRODUCTION

For over 20 years fusion researchers have sought to establish that the energy released by thermonuclear fusion may be used for practical purposes. During the past few years their work has been very successful and so it now appears likely that fusion energy can be developed as a new prime energy source. While a great deal of research and development remains to be done, we believe that fusion power could become a commercial reality by the end of this century, assuming the reasonable attainment of scientific and engineering goals and the timely commitment of necessary funds.

In the following, we review some of the basic requirements for fusion power, and identify the central features of the near- and long-term development programs. Some of the safety and environmental characteristics of first-generation fusion reactors are described. These suggest that fusion power will have some extremely attractive features, as well as some limitations.

THE PRINCIPAL GOAL

The principal goal of the fusion program is to develop the technology necessary for the safe and economical production of electricity from fusion energy. First-generation fusion power plants will embody many of the features of conventional fossil-fired or nuclear power stations. In present day electrical power plants, the chemical or nuclear energy released ultimately manifests itself as heat energy, which is transferred to a coolant that directly or indirectly produces steam to power a turbine generator. A fusion reactor will provide another such thermal energy source.

The energy source for a fusion power plant will be an extremely hot (100,000,000°C) gas, called a fusion plasma. In a plasma comprised of equal quantities of the hydrogen isotopes deuterium and tritium, fusion reactions would liberate about 20% of their energy in charged particles that would stay with the plasma and help maintain its temperature. About 80% of their energy would be carried by neutrons that would immediately leave the reaction volume and enter a moderator region surrounding the plasma where they would give up their energy to the moderator, which would accordingly be heated. A coolant pumped through the heated moderator region would be used to produce steam, which would produce electrical power by well known techniques.

[1] Congress presentation B–1–1.

THE FUSION PLASMA PROBLEM

Many thermonuclear fusion reactions are possible (*1*). However, there is general agreement that the reaction which will most probably be used in the first generation of fusion reactors is that of deuterium with tritium

$$^{2}D_{1} + {}^{3}T_{1} + 5 - 150\ \text{keV} \rightarrow {}^{4}He_{2}\ (3.5\ \text{MeV}) + {}^{1}n_{o}\ (14.1\ \text{MeV}).$$

Deuterium is abundant in the oceans. Tritium is not found in significant quantities in nature but it can be produced by absorption of the DT reaction neutron in lithium, which can be placed in the moderator region surrounding the plasma. Lithium occurs abundantly in both land-based brines and in sea water, and appears abundant enough at low cost for millions of years of DT fusion power production.

The energy gain in the DT reaction is substantial and may be several hundred times greater than the energy invested in initiating and sustaining the reaction process. For practical electrical power production this provides a comfortable margin even considering various expected losses.

Since the DT reaction involves the fusing of two nuclei, it is necessary that the mutual electrostatic repulsion of the nuclei be overcome. This can best be accomplished by giving the nuclei a good deal of energy, which means heating a DT gas to a very high temperature. Well before reaching the temperatures required for fusion, the deuterium and tritium lose their electrons and the fuel gas becomes fully ionized. This assemblage of positively charged nuclei and negatively charged electrons is called a plasma. Thus, the central problem in fusion research is one of producing a hot plasma and holding it long enough for more energy to be released than was used to create and sustain it (see Fig. 1).

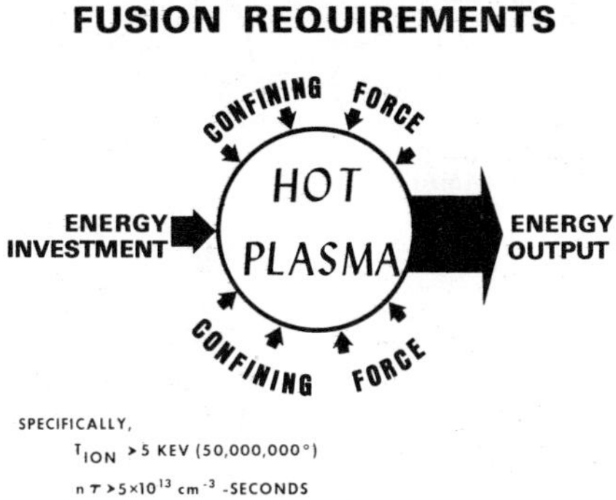

Figure 1

The British scientist J. D. Lawson formulated crude criteria for a power-producing fusion reactor in the late-1950's (*2*). Minimum levels of plasma temperature, density and confinement time are specified for which the fusion energy yield would be sufficient to maintain expected plasma energy losses. The usual way of expressing these conditions is to give the product of the plasma density and plasma confinement time for a given ion temperature. For example, at a plasma temperature (energy) of 5 keV (about 50,000,000° C), the product of plasma density times confinement time is $n\tau \sim 5 \times 10^{13}$ cm^{-3}-second. This criterion, shown on the bottom of Fig. 1, has stood as a major milestone for the fusion program since the criteria were first formulated.

The Lawson criteria are based on a Maxwellian plasma particle distribution, which is the distribution of particle velocities (or energies) that occurs in any plasma when it is in thermal equilibrium at a given temperature. When non-Maxwellian distribution functions are utilized, the minimum level plasma conditions for $n\tau$ can be relaxed by about a factor of ten. Recent analyses of non-Maxwellian systems indicate that fusion research now is closer to break-even conditions than had been appreciated, and it is likely that this milestone achievement will be reached in hydrogen plasmas later this decade (*3–5*).

APPROACHES TO FUSION POWER

In the United States, four major approaches are being pursued to produce practical fusion power (Fig. 2). Three (tokamak, mirror, and theta pinch) utilize magnetic fields to confine a plasma for either pulsed or steady state operation. Magnetic confinement takes advantage of the fact that in a strong magnetic field, individual charged particles (D^+, T^+, e^-) are constrained to move along magnetic field lines in tight helical trajectories. This isolates the plasma from the chamber walls, thereby avoiding excessive cooling of the plasma below the thermonuclear reaction temperature by the wall. The tokamak and theta pinch are toroidal in shape, while the mirror can be thought of as a magnetic field configuration of roughly tubular shape that confines the plasma by means of fields that are much stronger at the ends of the tube than at its center. The tokamak approach accounts for about 60% of the magnetic confinement research, while theta pinches and mirrors represent about 20% each.

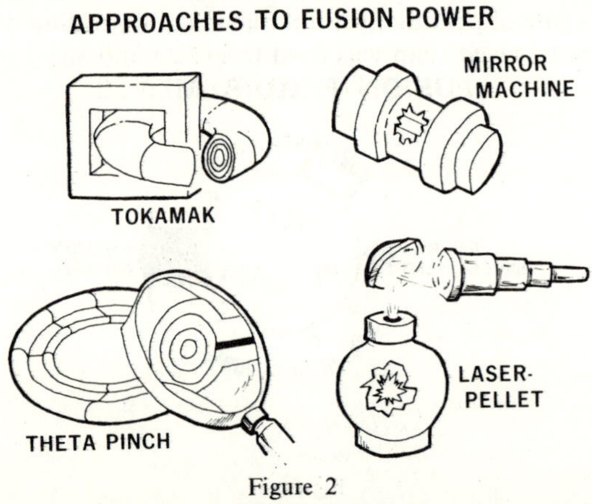

Figure 2

The laser-fusion concept is based on the use of high energy, short pulse laser beams focused on suitable fuel pellets to heat and compress the fuel. This heating and compression can in principle initiate thermonuclear reactions on a very short time scale. It is envisioned that the process could be repeated with a succession of pellets in order to realize a significant cumulative release of energy. Laser-pellet fusion reactions are envisioned to occur so rapidly that inertial forces provide adequate confinement during the thermonuclear reaction. Laser fusion is in the relatively early stages of research, wherein the work is devoted predominantly to theoretical analyses, fabrication of the requisite laser irradiation facilities, and first-generation pellet irradiation studies (*6*).

The major problem in magnetic confinement has been plasma instabilities that cause premature and excessive escape of plasma from the various magnetic confinement systems. This problem has required a major theoretical and experimental physics effort to develop an appropriate understanding. Research on the problem has been under way since 1952. In the 1950's, plasma production and heating techniques were developed, and reactor level plasma densities were produced relatively easily. In the early 1960's, thermonuclear reaction temperatures were achieved for the first time. Plasma confinement began to improve dramatically beginning about 1965, and by 1971 a variety of confinement concepts had closely approached the maximum confinement theoretically possible.

Further progress in fusion research depends upon larger plasma confinements, some of which are now under construction or being planned. These new devices will have the larger capacities needed to achieve reactor-like conditions and will answer critical questions of scaling to larger size.

For the past two years, experimental results have been extremely encouraging. Experiments have been designed depending heavily on theory, and most of the results have agreed with predictions. In tokamaks in particular, most of the critical problems that can be addressed in relatively small experiments have been investigated and this area of research and development is now shifting from small to medium size systems.

An example of a "small" fusion device is the Oak Ridge Tokamak (ORMAK) at the Oak Ridge National Laboratory (Fig. 3). ORMAK recently demonstrated size scaling for tokamaks and is now testing high power neutral beams as a means of raising plasmas to higher temperatures. The Princeton Large Torus (PLT) is a medium size tokamak now under construction at the Princeton Plasma Physics Laboratory (Fig. 4). PLT is expected to provide a reactor-grade hydrogen plasma when it operates in late 1975.

Another device to be constructed is the Doublet III (for "double tokamak") at General Atomic (Fig. 5). It is expected to provide an even more energetic plasma than PLT and is aimed at scaling the successful results of Doublet II, which showed that non-circular cross-section tokamaks require a much lower and therefore less expensive magnetic field than circular cross-sections. Doublet III is expected to operate in late 1977.

THE RESEARCH AND DEVELOPMENT PROGRAM

In view of the recent progress in magnetic confinement and excellent prospects for the future, a plan was recently developed to outline the essential elements of a program leading to operation of a demonstration fusion power plant (*7*). The major elements of the program are summarized in Fig. 6.

The program recognizes the need for parallel and interdependent efforts in confinement systems (experimentation with increasingly larger confinement experiments), research (including confinement theory and computer simulation) and engineering for both component development and fusion reactor technology. Systems are projected to operate about 1980 that will for the first time produce significant fusion energy by the controlled fusion of deuterium-tritium gas mixtures. Thereafter, two subsequent Experimental Power Reactors are envisioned to produce electrical power and to provide vehicles to study the

FIGURE 3

engineering and control features necessary for a Demonstration Fusion Power Plant. The latter is expected to begin operation in the mid to late 1990's and serve as the basis for fusion power commercialization. These reactor systems are believed necessary to provide the knowledge and experience required to demonstrate safety and economics and hence lay the base for a smooth transition of fusion power into the utility environment.

THE ENERGY SYSTEM

Figure 7 is an illustration of the general features of a fusion power plant based on the DT fusion reaction. In such a plant, helium would be employed as the coolant, graphite as the moderator, and lithium as the fertile material for tritium production. The plasma would be confined by means of magnetic fields generated by coils surrounding the blanket-shield region. Other features are as illustrated. The energy-conversion portion would be relatively conventional, employing steam generated in a heat exchanger to drive a turbine-generator system.

FIGURE 4

It is expected that on the order of 5% of the fuel would be burned per pass through the reactor, so provision must be made to separate the unburnt fuel from the fusion-produced helium and subsequently recycle this fuel with makeup fuel back into the reactor. Tritium bred in the blanket region would be continuously removed and either fed into the fuel injection system or stored for subsequent use. The neutrons that result from the DT reaction provide energy for electrical power generation and are most often absorbed in lithium, thereby producing the necessary tritium fuel. However, these neutrons will damage the reactor first wall as they pass through, and many of them will activate the blanket-shield structure. This absorption in the structure can induce intense radioactivity, and the level of this radioactivity is dependent upon the materials of construction. This question will be discussed further below.

The bulk of world fusion research to date has been devoted to study of plasma problems. Hence, the first wall, blanket, and power conversion systems have thus far only been the subject of analytical study. Such studies have focused on individual problems in fusion power plants (*8,9*) as well as the potential net worth of fusion power in the world economy (*10*). In recent months, more detailed studies of complete conceptual fusion power plants have been completed (*11,12*). These have served to focus attention on scientific and engineering problems that appear important to the development of the respective concepts as practical power-producing systems. The studies have been relatively sophisticated because codes and data from fission reactor research and development provide a strong starting point.

DOUBLET III

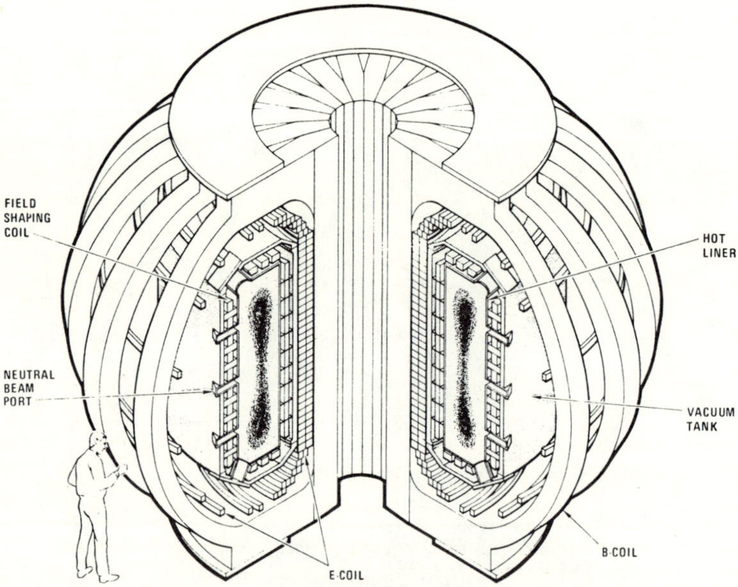

FIGURE 5
A drawing of the Doublet III tokamak to be built at General Atomic Corporation.

ENVIRONMENTAL IMPACT OF FUSION POWER

Although the first electrical power-producing fusion system probably will not operate before 1985, the Atomic Energy Commission has embarked on a detailed study of the environmental impact of such plants. The wealth of engineering data from fission reactor development and the experience from the production of tritium in the weapons program permits a fairly comprehensive assessment of the environmental characteristics of fusion power plants at this early phase in their development. One important benefit from such early studies is the identification of significant potential problems, thus providing ample opportunity to conduct necessary research to reduce the severity of such problems. A few of these will be discussed.

It is well recognized that tritium must be contained safely during routine plant operation and in the event of accidents. Further, plant components contaminated by exposure to tritium must be designed to permit necessary maintenance. Fortunately, tritium has been and continues to be studied in great detail. The state of the art was summarized in 1968 (*13*) and analyses have been conducted to scope tritium handling requirements in fusion power plants (*1,14*). A recent conference considered sources, measurement, movement and effects of tritium (*15*), but in spite of the significant base of knowledge, the need for information on long-term effects of low concentrations of tritium was indicated.

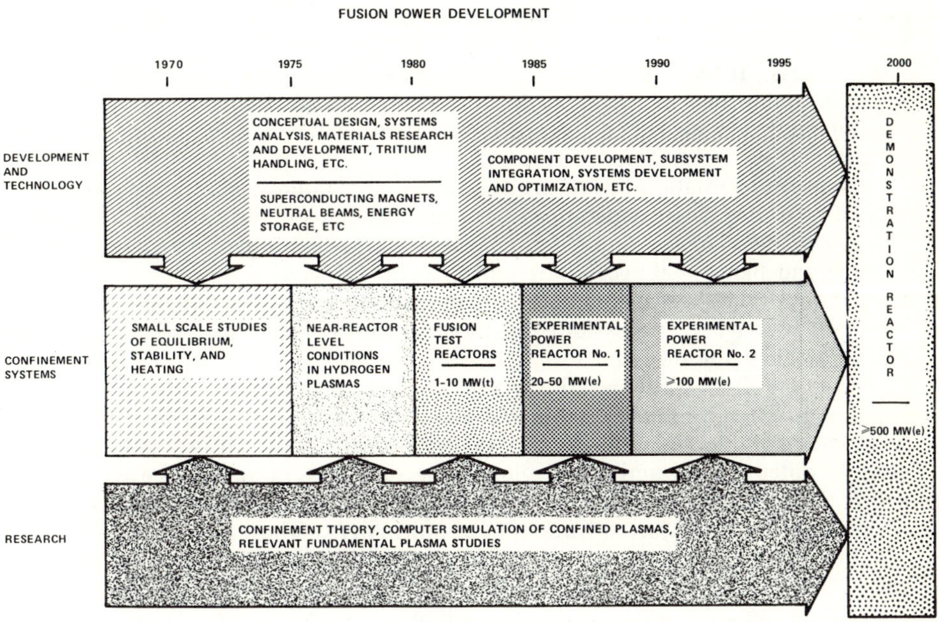

FIGURE 6

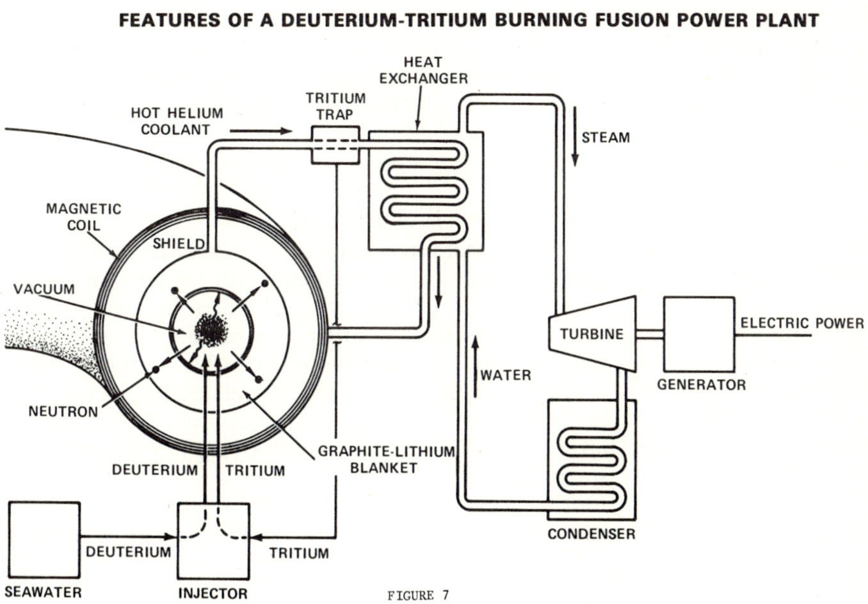

FIGURE 7

Fortunately, tritium is one of the less hazardous radioactive nuclides. It does not constitute a significant radiation hazard external to the human body, because it decays by emission of a low energy beta particle, which has low penetrating power. Internally, it can only be ingested in quantity in the form of HTO, which passes through the body with an average biological half-life of about 10 days. Nevertheless, tritium will constitute a significant handling problem since kilogram quantities are projected to exist in such plants. Planned research and development will be aimed toward demonstrating the safe containment and handling of such quantities of tritium.

Neutron irradiation effects are a second area of interest. Radiation damage to the first wall will vary depending on the neutron flux and the wall composition. The first wall will have a finite lifetime, which will probably be less than the plant lifetime and hence will necessitate replacement. Depending on alloy composition, the wall will become more or less radioactive due to neutron absorption. This radioactivity will necessitate the use of remote maintenance during planned and unplanned shutdowns.

For the first wall and the blanket structure, an alloy that meets all engineering strength requirements and also shows minimum neutron activation would be desirable. Recognizing this, the AEC began studies in 1973 intended to identify potential low activation alloys. One such class of materials are the so-called SAP (sintered aluminum product) alloys, which have been assessed by scientists at the Brookhaven National Laboratory (*16*). These consist of micron-size aluminum oxide particles in an aluminum matrix. Calculations indicate that use of SAP would reduce long-lived residual activity many orders of magnitude lower than the levels associated with niobium or stainless steel alloys.

Other options are also under study, since SAP has limited high-temperature strength and poses fabrication problems. Refractory metal alloys are of interest, for they possess excellent high-temperature strength and are resistant to liquid metal corrosion (although they are susceptible to corrosion by oxygen). The most interesting of these are vanadium-based alloys. It has been calculated that vanadium would display substantially less induced radioactivity than niobium, which was used in early fusion reactor conceptual designs. Also, vanadium alloys would permit superior tritium breeding characteristics compared with use of niobium structure (*17*).

Figure 8 illustrates how Relative Biological Hazard (the activity per unit thermal power divided by the maximum permissible concentration (MPC) for the radioisotope) might vary as a function of time-after-discharge for the niobium from a fusion reactor incorporating a niobium structure and the vanadium from the same reactor incorporating a structure of vanadium containing 100 parts per million niobium impurity[2]. To calibrate these characteristics a curve showing the decay of the high level waste from an LMFBR is shown. The Relative Biological Hazard is not a direct measure of environmental impact, but it is a useful index for quantifying the magnitude of the radioactive waste disposal problem.

Figure 9 illustrates another approach to analyzing the magnitude of induced radiation, in this instance by means of comparison of afterheat decay energies. It

[2] D. Steiner, Oak Ridge National Laboratory, personal communication, June 1974.

illustrates again how conceptual fusion power plants would have considerably less problem with afterheat removal through selection of vanadium alloys (*10*).

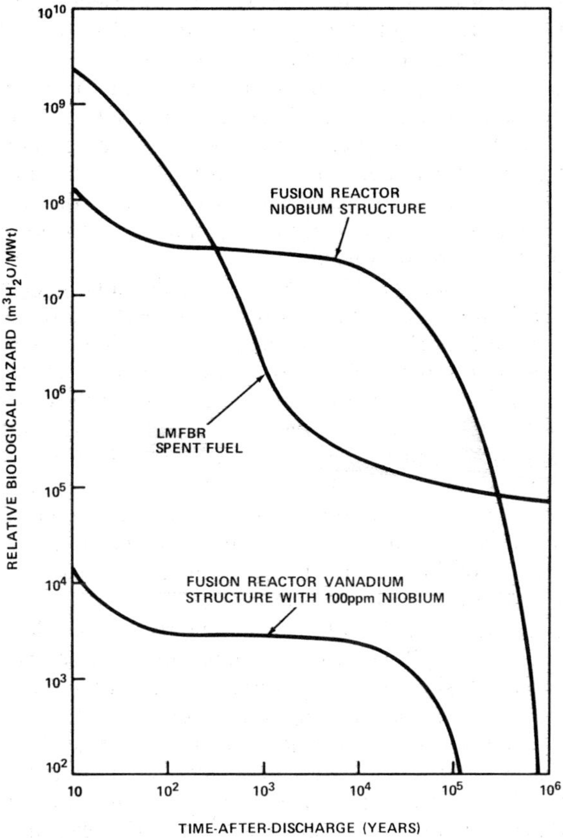

FIGURE 8 - RELATIVE BIOLOGICAL HAZARD AS A FUNCTION OF TIME-AFTER-DISCHARGE

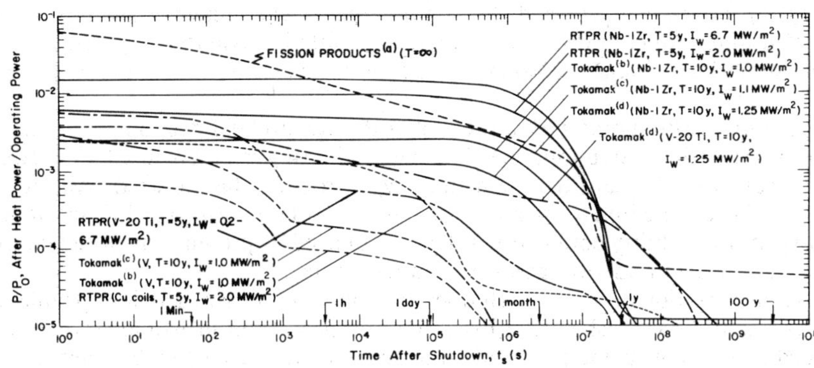

FIGURE 9

RELATIVE AFTERHEAT POWERS FOR FUSION AND FISSION POWER PLANTS AS FUNCTIONS OF TIME AFTER SHUTDOWN. T = OPERATING TIME I_w = 14-Mev NEUTRON WALL LOADING

Fusion reactors will be inherently safe against nuclear runaway because the amount of deuterium and tritium present in the reactor chamber at any time is only on the order of a gram and since the nature of plasmas is to always tend toward disassembly. Since fusion reactors will not involve the use of weapons grade materials (e.g., uranium or plutonium), there will be no possibility of their diversion. Because the decay power density in a fusion reactor can be made quite low, there is no chance of a melt-down associated with a loss of coolant. Because the radioactive hazards and safety characteristics of fusion reactors appear so attractive, fusion plants may be acceptable for urban siting, which would reduce transmission costs and allow the reject heat to be used for heating and air conditioning. Of course, because fusion is a nuclear process, no chemical combustion products will be produced.

The most significant sources of energy that could cause accidental failures are the magnetic coils, the high pressure coolant (if used), and the lithium fertile material, which is combustible unless it is appropriately alloyed or compounded. By careful plant design, it is believed that all of these potential energy sources can be suitably protected. Further, the total energy content of these sources is such that their release could damage the reactor, resulting in economic loss, but would probably not cause significant danger to the general public.

The off-site requirements of fusion power plants relate primarily to procurement of fuels and construction materials. The fuel deuterium is obtained from water and is essentially an infinite fuel resource. Tritium must be bred by neutron absorption in lithium, hence the lithium reserves dictate the ultimate availability of the tritium fuel. Lithium is found abundantly in heavy brines and in pegmatite ores in the United States. Assessments of the adequacy of U.S. lithium reserves have concluded that the presently known and inferred reserves are sufficient to meet fusion power requirements for times on the order of hundreds of years. When extracted from sea water, lithium reserves would be sufficient for millions of years. The adequacy of terrestrial reserves seems assured on any time scale of present concern. In fact, such concerns may be of secondary importance, since later fusion power plants operating on the deuterium-deuterium fusion cycle are likely to be developed after DT reactors. With tritium no longer important as a fuel, there would be no continued need for lithium.

Ore requirements for other fusion power plant materials will depend on the concept(s) ultimately selected for commercial development. Early studies indicated that copper and lead requirements would necessitate the greatest land usage, should a reference tokamak design be the basis for use of fusion power (*18*).

Helium is also a material of interest, although no land despoilment results from its extraction. Assuming use of superconducting magnets in fusion power plants, substantial quantities of helium will probably be required. Although large-scale use of helium for fusion would not take place until after the turn of the century, its availability could very well be influenced by present-day decisions on helium extraction and storage in the United States.

Transportation of fusion fuels does not appear to present a serious problem. Deuterium is easy to ship as a high pressure gas. Tritium would have to be shipped to any new fusion power plant to provide fuel for startup, but thereafter, tritium would be generated on-site and only makeup lithium would have to be shipped to the plant.

Activated structural materials will have to be shipped to reprocessing and/or disposal sites. Replacement intervals for the first wall will depend on useful life, which will be determined by many factors, such as alloy composition, neutron flux on the wall, wall stresses, corrosion by coolants, etc. It is believed that with proper alloy selection, induced radioactivity should be minimal and it will be nonvolatile and should not present a long-term storage problem. Further, because of materials shortages projected for the 21st century, fusion structural materials will probably be refabricated for reuse, minimizing the volume of radioactive material to be stored.

FUSION AS A SOLUTION TO OTHER PROBLEMS

There have been a number of applications of fusion energy identified that do not involve generation of electricity. For example, Powell and coworkers are assessing the technology, economics and environmental effects of directly producing synthetic fuels (hydrogen and methanol) in fusion reactor systems (*19*). They conclude that with base load operation and higher efficiencies, a synthetic fuels energy system based on fusion reactors appears substantially cheaper than nuclear electric-fossil fuel systems. Further, environmental emissions are cut almost to zero and the strip mining of large portions of western states is eliminated.

Perhaps the most dramatic non-electric application of fusion reactor systems could be their support of the present expanding fission reactor economy. Two issues of significance in these power plants are the availability of fuel (uranium-235) and the disposal of the radioactive fission products resulting from their operation. Fusion reactors have the potential of providing answers to both of these concerns.

Breeding of fissionable elements in fusion reactors has been the subject of a number of studies (*20*). In the DT reaction, the neutrons emitted are not necessary for continuation of fusion reactions, as they are in fission chain reactors. They could be slowed down as desired in a blanket region and absorbed in a fertile material. This could be ^{238}U, from which ^{239}Pu could be bred, or it could be ^{232}Th, from which fissionable ^{233}U could be bred. The plutonium fuel could be cycled in water reactors or used as fuel makeup for liquid metal-cooled fast breeder reactors. The uranium-233 could be used as fuel in existing light water reactors. In either event, a fusion reactor operating solely as a producer of fuel could be used to supply the existing and planned fission reactors.

The actinide elements and fission products formed during operation of fission reactors present a difficult long-term solid waste disposal problem. The primary approach employed thus far is to store the wastes in containers, while research continues on effective long-term storage techniques. Kubo and Rose (*21*) and earlier workers have analyzed the recycle and burnup of actinide elements in fission reactors. Wolkenhauer and coworkers (*22*) have conducted a similar study on burnup of actinides by means of the neutrons emanating from fusion reactor systems. They find that actinide elements could be transmuted in such systems.

In their study, Wolkenhauer, et al. have studied the feasibility of transmutation of high-level radioactive waste in fusion systems. They conclude that troublesome fission products such as ^{90}Sr, ^{137}Cs and ^{129}I can theoretically be transmuted. While further analyses must be undertaken to balance factors such as wall transmutation during waste exposure, possible power generation schemes in such systems, and operating hazards, these encouraging initial results indicate considerable promise for this solution to radioactive waste disposal. Indeed, fusion reactors appeared most attractive of various means recently studied in assessing waste elimination by transmutation (*23*).

In total, fusion reactor systems offer the very real possibility of a true synergism with fission power plants. They could generate necessary fission reactor fuel and could transmute the solid radioactive waste products to less hazardous forms.

SUMMARY

Fusion research in the United States has seen excellent progress in the past several years. As a consequence, plans have been formulated leading to progressively larger plasma confinement experiments, ultimately intended to establish plasma conditions equivalent to those in a power-producing system. Such experiments will include study of thermonuclear reactions in deuterium-tritium gas mixtures, as a precursor to the design and construction of electrical power-producing reactors.

The environmental characteristics of fusion reactor concepts have been estimated by analyses of conceptual designs. It appears that tritium, one constituent of the fuel and the only volatile radioactive material in a fusion power plant, can be safely contained. Careful selection of materials of construction will minimize radioactivity induced by neutron activation. This activity can be held to levels that would eliminate problems due to shut-down temperature rise due to decay afterheat buildup. Because the induced structural radioactivity is non-volatile, ultimate disposal of structure should not prove to be a major problem.

The fusion fuel deuterium is effectively infinite in supply. Lithium, required to breed tritium, is believed to be in adequate supply for millions of years of DT fusion power. Structural materials for fusion reactors may in some cases be in short supply in the 21st century and resource limitations may force certain appropriate design choices.

The primary purpose of fusion development is to produce a new prime source of electrical power. In addition, significant applications of fusion energy may include the production of synthetic fuels and the complementary support of fission reactor systems by breeding of fissile fuel and burning of radioactive waste products.

REFERENCES

1. L. D. Hansborough and E. L. Draper, Jr., *Overall Tritium Considerations for Controlled Thermonuclear Reactors,* Technical Report ESL-10, Energy Systems Laboratories, The University of Texas at Austin, Texas, 1973.
2. J. D. Lawson, *Proc. Phys. Soc.*, Sec. B, Vol. **70**, p. 6 (1957).
3. J. M. Dawson, H. P. Furth and F. H. Tenney, *Phys. Rev. Lett.*, **26**, 1156 (1971).

4. D. L. Jassby and H. P. Furth, *Fusion-Reaction-Injected and Neutral-Beam Injected Suprathermal Ions in the Tokamak*, TM-271, Plasma Physics Laboratory, Princeton University, Princeton, New Jersey, 1974.
5. D. L. Jassby and H. P. Furth, *Abstract, Meeting of the Am. Phys. Soc.* (1974).
6. J. L. Emmett, J. Nuckolls and L. Wood, *Sci. Am.* **230, No. 6**, 24 (June 1974).
7. U.S. Atomic Energy Commission, *Fusion Power by Magnetic Confinement*, WASH-1290, National Technical Information Service, U. S. Department of Commerce, Springfield, Virginia 22151.
8. Institute of Electrical and Electronics Engineers, *Fifth Symposium on Engineering Problems of Fusion Research*, Princeton University, 1973. IEEE Pub. No. 73 CH0843-3-NPS, IEEE, New York, N. Y., 1974.
9. U. S. Atomic Energy Commission, *First Topical Meeting on the Technology of Controlled Nuclear Fusion*. CONF-740402, 1974, USAEC (to be published).
10. R. F. Post and F. L. Ribe, *Fusion Reactors as Future Energy Sources*, LA-UR-73-1684, Los Alamos Scientific Laboratory, Los Alamos, New Mexico, 1973.
11. B. Badger, et al, *Wisconsin Tokamak Reactor Design*, UWFDM-68, **Vol. 1**. University of Wisconsin (November 1973).
12. W. C. Gough, *Fusion Reactors Systems Studies Review Meeting*, WASH-1278, Division of Controlled Thermonuclear Research, U. S. Atomic Energy Commission, Washington, D. C., 1973.
13. D. G. Jacobs, *Sources of Tritium and Its Behavior Upon Release to the Environment*, TID-24635. Clearinghouse for Federal Scientific and Technical Information, National Bureau of Standards, U. S. Department of Commerce, Springfield, Virginia, 22151, 1968.
14. J. D. Lee, in *The Chemistry of Fusion Technology* (D. M. Gruen, Ed.), pp. 51–90. Plenum Publishing Corp., New York, 1973.
15. A. A. Moghissi and M. W. Carter (editors), *Tritium*. Messenger Graphics, Phoenix, Arizona, 1973.
16. J. R. Powell, F. T. Miles, A. Aronson and W. E. Winsche, *Studies of Fusion Reactor Blankets with Minimum Radioactive Inventory and with Tritium Breeding in Solid Lithium Compounds: A Preliminary Report*. BNL 18236, Brookhaven National Laboratory, Upton, New York, 1973.
17. D. Steiner, *Nuc. Fusion*, **14**, p. 33 (1974).
18. U. S. Atomic Energy Commission, *Fusion Power: An Assessment of Ultimate Potential*. WASH-1239, Division of Controlled Thermonuclear Research, USAEC, Superintendent of Documents, U. S. Government Printing Office, Washington, D. C., 1973.
19. J. R. Powell, F. Salzano, W. Sevian, K. Horrman and P. Bezler, *An Evaluation of the Technical, Economic and Environmental Features of a Synthetic Fuels Economy based on Fusion Reactors*. BNL 18430, Brookhaven National Laboratory, Upton, New York, 1973.
20. Battelle Pacific Northwest Laboratory, *Annual Report for 1973 on Controlled Thermonuclear Reactor Technology*. BNWL-1823, National Technical Information Service, U. S. Department of Commerce, Springfield, Virginia, 22151, 1974.
21. A. S. Kubo and D. J. Rose, *Science*, **182**, 1205 (1973).
22. W. C. Wolkenhauer, B. R. Leonard, Jr., and B. F. Gore, *Transmutation of High-Level Radioactive Waste with a Controlled Thermonuclear Reactor*. BNWL-1771, Battelle Pacific Northwest Laboratories, Richland, Washington, 1973.
23. U. S. Atomic Energy Commission, *High-Level Radioactive Waste Management Alternatives*, pp. 58–60. WASH-1297, National Technical Information Service, U. S. Department of Commerce, Springfield, Virginia, 1974.

Fusion Research in Japan.[1]

Shoichi Yoshikawa
Department of Physics
University of Tokyo
Tokyo, Japan 113

The energy need of Japan in the next several decades is expected to grow rapidly, even if the present trend of moderate population growth is to continue. The reasons are several. One is that the standard of living is expected to go up with accompanying increase in power need. The second is that the worldwide shortage of mineral resources would force Japan to depend more on low grade ores which usually means a larger energy need. Thirdly, the increase in the population in other countries would inevitably reduce the amount of food available to Japan through import. This results in the need of some kind of synthesis of food material and/or cattle feed, which cannot be accomplished without the use of energy.

Presently the major part of the energy need of Japan is satisfied by oil imported chiefly from the Middle East. Practically no oil is produced in Japan. Other resources such as hydro-electric power and coal are not adequate. Because the population density of Japan is the highest in the world, if due consideration is given to the fact that the habitable land is a small fraction, the power generated per unit area is higher than in the U.S. This fact means atmospheric pollution in Japan is perhaps the worst in the world. These reasons are strong incentives to seek alternate solutions to the energy need of Japan. In a sense, the difficulties that Japan faces now would be prevalent difficulties that other countries will face in years to come.

In spite of the efforts by many individuals and groups, the nuclear power generating capacity is not expected to be a large part of Japanese energy producing capacity until about 1985 to 1990. Even then conventional fission reactors would not be adequate, as Japan has no significant amount of uranium deposit and at present lacks the facilities to enrich uranium. Therefore, it appears that Japan has to depend mainly on imported fossil fuel in this century.

The future of Japan depends on how Japan can solve the energy problem. If Japan wishes to be independent in terms of the energy, the choices are rather few. The hydro-electric power and geothermal power are not sufficient, unless the geothermal heat is mined deep in the crust which might cause unpleasant side effects such as earthquakes. The solar energy is certainly a possibility, especially if the surface of the ocean can somehow be used. A rather ambitious program called "Project Sunshine" has been started in Japan to consider various alternatives related to the utilization of the solar energy. At

[1]Congress presentation B-1-2.

present it is not clear whether this solar energy will satisfy all the energy needs of Japan, especially from the standpoint of cost.

The other alternative is of course nuclear power. Here again the large population density is a detrimental factor for developing nuclear power stations. In particular, reprocessing of spent fuels and disposal of radioactive waste may eventually determine how acceptable the nuclear power is to Japan. Japan does not have geologically stable areas for waste disposal.

Nonetheless, intensive efforts are being made to make nuclear power acceptable. The breeder program was started about ten years ago and a prototype reactor is being constructed. Other than the public acceptance of the nuclear power, the question of how to shorten the doubling time of plutonium is a difficult problem.

The fusion power, though nuclear, is considered to be the ultimate solution to the energy need of Japan. It is clear that there is no problem of securing the energy source, and the radioactive waste is expected to be no more than 1/10th of the fission reactor for the same power generation. Furthermore since the fusion reactor tends to have a surplus of neutrons, they could be used to convert dangerous radioactive material into something less harmful. The doubling time of tritium is expected to be less than one year. Of course, the fusion power has not been controlled and there are some people who maintain that the use of a miniature hydrogen bomb is the only practical way to release the energy. However, there are people who try to achieve a controlled thermonuclear reaction either by means of magnetic confinement of high temperature plasmas or by inertial confinement. In Japan, as in other countries, a program to achieve controlled thermonuclear reaction (CTR) has been in existence for some time.

The CTR effort of Japan has been rather small compared with that of the U.S. and the U.S.S.R. The main work was carried out in two governmental laboratories and various universities. The research effort is mainly in the line of so called low β toroidal confinement devices. Related to it, the study of fundamental plasma physics is being carried out. These activities result in laying the foundation for starting out a major effort on fusion. In the following, a brief account of what was achieved and then the future plan will be described.

A toroidal device of the type of tokamak, called JFT-2, has been in operation for two years. This device was designed in 1968–9, about the time that the Soviet tokamak was showing good results, demonstrating for the first time that the plasma confinement exceeds the Bohm scaling law. This device is located at Japan Atomic Energy Research Institute (JAERI) at Tokai, about 100 kilometers north of Tokyo. The size of the device is 90 cm major radius and 25 cm minor radius with a magnetic field of 10 kilogauss which is to be increased to 18 kilogauss sometime around the end of 1974. The size is similar to that of Ormak at Oak Ridge National Laboratory (U.S.A.). The device was constructed by the Hitachi Company.

The experimental result shows that the plasma confinement follows the scaling law found in both the Soviet tokamak, T-3 and the U.S. toroidal device, ST, even though JFT-2 is considerably larger in size and lower in magnetic field.

Two toroidal devices of the type of the stellarator first invented in the

U.S.A. are currently in operation. One is at the Institute of Plasma Physics at the Nagoya University and another one is at Kyotò University. They are smaller in size than JFT-2. Yet the experimental results show that the plasma confinement is better than that predicted by the Bohm scaling law and perhaps explained by the neo-classical law, if the effect of asymmetrical magnetic field is taken into account.

Coming next to JFT-2, there is a device, JFT-2A, also at JAERI, which has a poloidal divertor to divert plasma particles to a different chamber, so that the particles will not return to the main chamber to cool the high-temperature plasmas. This machine should be in operation by August of 1974. Another bigger stellarator is being built at Nagoya to be completed towards the end of 1975.

Experiments on plasma heating are also being done. Plasma heating by rf, turbulence etc., is widely known. The neutral injection is also being prepared for the further heating of tokamak-type plasmas.

In high β toroidal plasma research, there is the Electro-Technical Laboratory. Toroidal theta pinch experiments are carried out there. Some laser works are also started at Osaka University. The aim is to produce a break-even condition before this decade is over. With all these activities, a considerable number of people (about 100 physicists and engineers) are now available engaged, together with approximately an equal number of graduate and post-graduate students. The construction of the research devices is usually done by industries.

Recently, the government has been taking a more favorable view towards the fusion research program. The present fusion 5-year program is going to expire in March, 1975. In view of recent progress of fusion research, both in Japan and in other countries, the next 5-year plan is expected to be substantially more extensive. While the current 5-year plan has a total budget of approximately \$20 million, the next 5-year plan probably will be funded in the range of several hundred million dollars.

With the indicated increase in the budget, it is hoped that a tokamak-type low β plasma confinement device will be constructed either in 1978 or in 1979. The device now under consideration would have a minor radius of 100 cm and a major radius of 300 cm with a magnetic field strength of 50 kilogauss. With this, under the most pessimistic plasma confinement scaling law, the product of the density and the energy confinement time would reach approximately $5 \times 10^{13} \text{sec/cm}^3$. This number should be sufficient to satisfy the reactor condition with the use of the two-component reactor concept developed by the Princeton group.

The experiment will be conducted in a new laboratory under the auspices of the Japan Atomic Energy Commission. The laboratory will eventually have a staff of approximately 500. It is also necessary to have a wide range of various other, smaller experiments to support this major effort.

There are several factors which may delay this plan. One is the fact that Japan is not accustomed to undertake big research and development programs. Naturally this leads to confusion as to how this fusion program is to be organized. The second is that the public awareness of the future energy needs seems to be less than that in the U.S.; thus it is somewhat more difficult to convince the public about the need for a fusion research program of this magnitude.

In spite of these difficulties, the fusion research program of Japan will hopefully flourish in the future, and as the time passes the energy need is felt more strongly. It would seem that the future of Japan will very much depend on whether or not fusion power is available in year 2000.

TECHNOLOGICAL APPLICATIONS OF RADIATION[1]

Chairman: L. B. Sztanyik

[1]Plenary Session D–1

Technological Applications of Radiation: An Introduction[1]

Laszlo B. Sztanyik
National Research Institute for Radiobiology and Radiohygiene
Budapest, Hungary

This Plenary Session is going to discuss the present status and foreseeable trends in technological applications of ionizing radiations.

Utilization for practical purposes of the physical, chemical, and biological effects of radiation has always been among the primary goals of scientists interested in radiation research. However, it was only after the Second World War, when large radiation sources became readily available, that a more intense work on technological applications of radiation could be initiated. As a result of this work, ionizing radiation in kilorad and megarad doses has found steadily increasing applications in chemical syntheses, in polymer and petroleum industry. Radiation induced graft-polymerization, co-polymerization, and cross-linking seem to have the best prospects among the industrial applications of radiation, and these applications will be presented by our first speaker today.

The microbicidal activity of ionizing radiation is one of the radiobiological effects that is of considerable interest for the health and food industries.

Radiation sterilization of medical products has developed very fast during the past 15 to 20 years to the point where there are now approximately 50 commercial and semi-commercial gamma sterilization plants and numerous accelerator-based plants all over the world. Over the past two decades, radiation-sterilization of medical products has become one of the most successful large-scale applications of atomic energy for peaceful purposes, being second only to power generation.

Radiation offers several advantages as a sterilization method that makes it attractive in a growing number of situations. The assortment of the major articles sterilized by radiation to date includes a large variety of pre-packed disposable medical products, sutures and other implants, medicaments and pharmaceutical starting materials, cosmetics and washing powder, biological tissues and preparations of biological origin.

Several applications of radiation for *preservation of food* are also based on full or partial sterilization of the products. Although the elimination of all microbial activity in food appears more difficult and complicated than it was originally supposed, and the verification of the wholesomeness of irradiated food has necessitated an almost endless series of experiments, the large-scale radiation-preservation of some food products looks promising provided certain technical, economical and marketing problems are overcome.

[1]Presented at Plenary Session D-1.

In addition to the growth of microorganisms, other mechanisms by which food can be lost include sprouting, the effects of food enzymes, insect infestation, and the presence of parasitic organisms, for instance, *Trichinella spiralis*. It is assumed that prospects are generally favorable for the application of radiation-processing to decrease food loss by these mechanisms, too.

Sterilization of medical products and preservation of food by radiation technology will be reviewed by two distinguished experts in these fields.

Another way of achieving better food supply is the *creation of improved strains of agricultural products* through the application of mutagenic agents, like radiation. The dose of radiation needed to produce genetic mutations in seeds is much smaller than the doses needed for chemical synthesis or sterilization. A relatively small radiation source of a few hundred curies can produce enough mutants to keep several biologists busy for years analysing, selecting the few good strains from the many poor ones, and developing the new good strains. Nevertheless, the value of this radiation application to agriculture might be very substantial. It is a great pleasure to have here one of the pioneers of this research field, who will give us a comprehensive picture of the recent developments.

Of the miscellaneous applications of radiation technology which will not be reviewed today, because of the rather limited time that is available, I would like to call your attention to two more.

First, the application of ionizing radiation for *insect eradication* or *control*. This is based on the principle that radiation can cause genetic damage, notably lethal mutations in the sperm of the target species, thus rendering inviable zygotes resulting from the union of a genetically impotent sperm and a normal ovum (or vice versa). Several other effects are, of course, possible and indeed often accompany the production of dominant lethality in the germ cells.

Developing forms of insect pests are irradiated with a dose of several kilorads that does not kill them, but is sufficient to render them incapable of reproduction. Under certain conditions, the entire population of insects can be destroyed by releasing a large number of radiation-sterilized, infertile males so that they outnumbered the potent wild males during the mating season. The success of the sterile insect release method for insect control depends, among other things, upon the ability of sterilized males to compete with wild flies and to live long enough to inseminate females.

This technique is potentially applicable to the single-mating pests such as screwworm fly, tse-tse fly, and a few others.

The preparation of *radiation-attenuated vaccines* against parasitic diseases of domestic animals is another application of radiation technology to be mentioned here. The principle involved in this technology is now well established and appreciated. Like other living organisms, helminths are also damaged by ionizing radiation. If infective stages are treated with a suitable dose of radiation and administered to the host, this damage manifests itself by a reduction in the number of parasites which develop to maturity, and an impairment of their reproductive capacity. The pathogenic effect is usually correspondingly reduced. An immunogenic effect is, however, still exerted if a sufficient number of parasites persist in the host for long enough at the appropriate stages of development. The magnitude of changes is related to the dose of radiation used and, therefore, the preparation of a vaccine depends on one being able to choose a dose which will reduce the pathogenic effect of the parasite to an acceptable level without a significant reduction of immunogenic effect.

Radiation-attenuated vaccines have been developed against a number of animal parasitoses and some of these have already been in practical use for several years, for instance, "Dictol" for *Dictyocaulus viviparus* in cattle, "Difil" for *D. filaria* in sheep, and the recently developed vaccine against *Ancylostoma caninum*, the causative agent of hookworm disease in dogs.

Accordingly, ionizing radiation is becoming a tool of considerable importance in agriculture, certain branches of food and processing industry, and medicine. In many cases, radiation technology proved superior to the conventional methods of technology, for instance, in sterilization of heat sensitive plastic articles, or products of biological origin. On the other hand, insufficient specificity of action, the generally low yields, and the high cost involvements are still among the chief difficulties found. However, radiation technology is relatively young and much more room remains for skillfully oriented research that will certainly accelerate developments and improve the prospects for further applications.

REFERENCES

1. Radiation: *A tool for industry*. Arthur D. Little, Inc., Cambridge, Mass. 1959.
2. W. Mulligan, The preparation of radiation-attenuated vaccines against helminthic infections. In *Nuclear Techniques in Helminthology Research* (to be published by I.A.E.A. in 1974)
3. E. D. Offori, Neutron irradiation and its possible application for insect control. In *Radiobiological Applications of Neutron Irradiation*, I.A.E.A., Vienna, 1972.
4. L. B. Sztanyik, Application of ionizing radiation to sterilization. (to be published in 1974).

Applied Industrial Radiation Chemistry of Monomers and Polymers[1]

Allan S. Hoffman
University of Washington,
Seattle, Wash. 98195, USA

INTRODUCTION

The most significant effects of ionizing radiation on monomers and polymers are due to the reactions of the free radicals formed during irradiation. Thus it is not surprising that the molecular radiation reactions of most importance to the polymer industry are based on free radical-initiated crosslinking or degradation of polymer molecules, or on polymerization or copolymerization of monomers ± polymers. Tables 1–3 summarize the various

TABLE 1
Processes Involving Irradiation of Polymers

Predominant Effect	Polymer Examples	Preferred Radiation	Examples of Products
Crosslinking	Polyethylene Rubber/Poly-Ethylene	β	Upgraded Wire and Cable Insulation; Shrinkable Film and Tubing Foam
	Rubbers (Bulk, Latex)	β (γ)	Adhesives, Upgraded Resins & Fabricated Items
	Polyethylene Oxide (Dilute Aqueous Solution)	β (γ)	Highly Hydrated Gels
Crosslinking, End Group Stabilization	Cellulose, Hemi-Cellulose, Lignin (Wood Chips)	β	Retardation of Microbial Attack and Increased Pulping Yields
Degradation	Polyethylene Oxide (Bulk)	γ (β)	Controlled Molecular Weight Distribution Resin
	Cellulose	γ (β)	Controlled Molecular Weight Distribution Resin

[1]Congress presentation D–1–1

TABLE 2
Processes Involving Irradiation of Polymers + Monomers

Predominant Effect	Polymer Examples	Monomer Examples	Preferred Radiation	Examples of Products
Crosslinking	Polyvinylchloride Polypropylene Rubber Polyvinyllidene Fluoride	Allylmethacrylate Triallylcyanurate Ethylene Glycol Dimethacrylate	β	Upgraded Wire Insulation, Adhesives
Bulk Grafting	Polyethylene	Acrylic Acid	γ	Improved Adhesion, Membranes
	Cellulose	N-Methylol Acrylamide	β	Durable Press Fabrics
	Silicone Rubber	N-Vinyl Pyrrolidone	γ (β)	Biomedical Materials (e.g. Contact Lens)
	Starch	Acrylamide	γ	Flocculant
Surface Grafting	Nylon Polyesters Tetrafluoroethylene Polyethylene Silicone Rubber Polyurethanes	Acrylic Acid N-Vinyl Pyrrolidone Hydroxyethyl Methacrylate Sodium Styrene Sulfonate Acrylamide, etc.	β, γ	Modified Surface Wettability, Adhesion and Chemical or Biochemical Reactivity
Curing via Copolymerization	Polyunsaturated Polymers	Acrylics Styrene	β, U.V.	Metal, Plastic and Wood Coatings; Adhesives; Printing Inks
Imbedding of Porous Matrices via Copolymerization	Polyunsaturated Polymers	Acrylics, Styrene	γ	Composites of Wood-Polymer, or Concrete-Polymer

TABLE 3
Processes Involving Irradiation of Monomers

Predominant Effect	Monomer Examples	Preferred Radiation	Examples of Products
Polymerization Within Porous Matrices	Acrylics, Styrene, Ethylene Glycol Dimethacrylate	γ	Composites of Wood-Polymer, Concrete-Polymer or Paper-Polymer
Emulsion Polymerization	Vinyl Acetate Butyl Acrylate Acrylics	γ	Polymer Latex
Solid State Polymerization	Trioxane	β	Polymer Resin
Polymerization at Elevated Pressure	Ethylene	γ	Polymer Resin

processes with these species which could become or have already become industrially significant. In 1972 the USAEC estimated the total value of irradiated products in the U.S. at 200 million dollars. Over 90% of the installed radiation capacity is electron beam accelerators, the rest being cobalt-60 gamma sources. This is mainly due to the high dose rates available from accelerators, resulting in a wide variety of high speed processes which have been adapted to use electron accelerators. (Fig. 1 emphasizes the differences between accelerators and gamma sources.)

In the following sections of this paper specific processes will be described according to each particular product area. A selected reference list will be presented at the end of the article; Table 4 presents information on the various companies involved in commercial radiation processes.

CROSSLINKED WIRE AND CABLE INSULATION

This is probably the most economically significant radiation process in existence; the Raychem Corp. pioneered in this field over 15 years ago. Today over 20 electron accelerators are involved in this one application, by more than four companies. The primary product here is electron irradiation-crosslinked polyethylene, filled with pigments, aging and oxidation stabilizers, reinforcing agents, flame-retardation agents, etc. The product line tends to emphasize the smaller diameter wire and cables, where radiation has a distinct advantage over chemical techniques. The major upgrading attained due to irradiation is the improved thermal stability of the insulation, thus permitting greater current loads and higher operating temperatures. Solder iron resistance is also improved. The product is competitive with higher priced insulation (e.g. Teflon) having greater initial thermal stability.

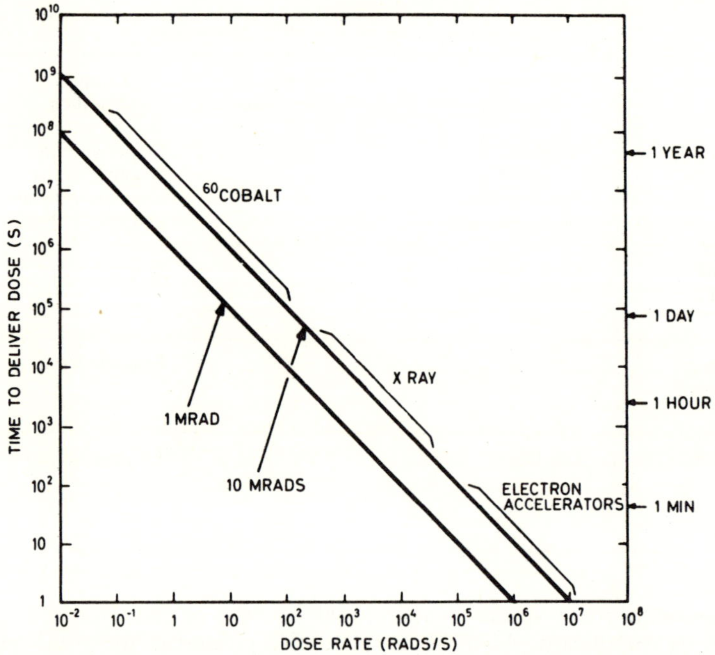

FIG. 1. Effect of dose rate on time required to deliver 1 or 10 Mrad. doses, for radiation of different penetration abilities.

An important modification of this process is based on the combination of a multifunctional monomer (e.g. see Table 5) with a polymer resin before extruding the wire or cable insulation. The monomer enhances the radiation reactivity, and can reduce significantly the required crosslink dose for pure polymers as polyethylene (~10 Mrads). The mechanism of this action is presumably due to formation of three-dimensional crosslinked matrix throughout the polymer, although covalent bonds to the matrix polymer are probably also formed; the overall result is an improvement in the thermal stability of the product. This process modification permits the use of polymers which might otherwise have degraded or become embrittled upon irradiation; examples include the inexpensive polyvinyl chloride resins, polypropylene and polyvinylidene fluoride.

HEAT SHRINKABLE FILM, TUBING AND MOLDED OBJECTS

A unique property of radiation-crosslinked semi-crystalline resins is their potential for viscoelastic memory. If the product is warmed and stretched just after moderate crosslinking and then quickly cooled in the stretched configuration, it will return to its original shape or dimensions upon subsequent heating as certain of the quenched crystallites melt out and the crosslinks cause a rubber-like retraction of the molecules. In this manner shrinkable films, tubings and molded objects of a variety of shapes have been developed commercially for uses

TABLE 4
Monomer/Polymer Radiation Processes in the United States

A. COMMERCIAL	
PROCESS OR PRODUCT	MAJOR COMPANIES INVOLVED
Crosslinked Wire and Cable Insulation	Raychem Corp. I.T.T. Judd Wire/Electronized Chemicals Corp. Western Electric (Bell Telephone)
Heat Shrinkable Film, Tubing and Specialty Items	Raychem Corp. W. R. Grace (Cryovac Div.) Electronized Chemicals Corp. Markell Corp. RAI Research Corp.
Crosslinked Polyethylene Foam	Voltek Corp.
Durable Press Fabrics (± stain release)	Deering-Milliken Corp.
Curing of Coatings	Ford Motor Co. PPG Industries Beta-Cure (O'Brien Paint Corp.)
Wood-Plastic Composites	ARCO Chem. Corp. Radiation Technology Corp. Radiation Machinery Corp.
Controlled Molecular Weight Degradation of Polyethylene Oxide	Union Carbide Corp.
Grafted Acrylamide/Starch Polymeric Flocculant	Atlas Chem. (ICI)
Grafted Polyethylene Battery Separator, Membranes	RAI Research Corp.
Polymeric Product "Z"	Confidential
B. CLOSE TO OR SMALL SCALE COMMERCIALIZATION	
PROCESS OR PRODUCT	MAJOR COMPANIES INVOLVED
Wood Chip Irradiation	Radiation Development Co., Ltd.
Acrylic Acid Grafted Polyethylene Adhesion Promotor	Dow Chem. Co.

TABLE 5

FORMULATION OF POLYFUNCTIONAL MONOMERS

REACTIVE DOUBLE BONDS

$CH_2 = CH-$	Vinyl
$CH_2 = C<$	Methylene (or Vinylidene)
$CH_2 = CH-CH_2-$	Allyl
$-CH = CH-$	Vinylene

SOME EXAMPLES

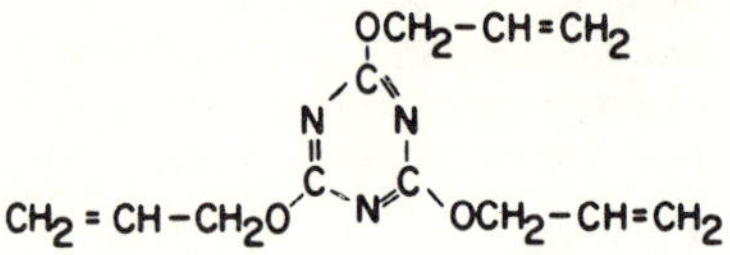	Triallylcyanurate
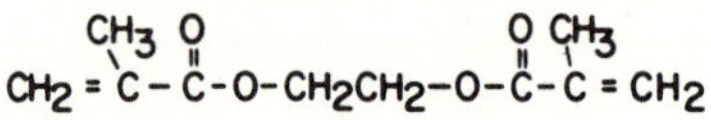	Ethylene Glycol Dimethacrylate
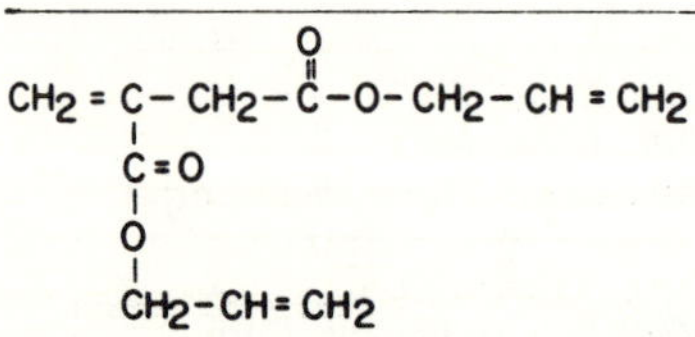	Diallyl Itaconate
$CH_2=CH-CH_2-O-\overset{O}{\overset{\|}{C}}-CH=CH-\overset{O}{\overset{\|}{C}}-O-CH_2-CH=CH_2$	Diallyl Maleate
$CH_2=CH-CH_2-O-\overset{O}{\overset{\|}{C}}-CH=CH_2$	Allyl Acrylate
$(CH_2=CH-CH_2-O)_3P=O$	Triallyl Phosphate
$CH_2=CH-CH_2-O-CH_2-\overset{O}{\overset{\|}{C}}-O-CH_2-CH=CH_2$	Diallyl Glycolate
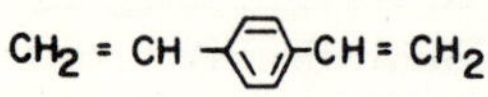	Divinyl Benzene

in food wrapping, electronics, construction, toys, etc. Raychem was a pioneer in the shrinkable tubing area while W. R. Grace's Cryovac Division led the way in shrinkable films. Many accelerators, operating continuously, are involved in this application.

In order to make heat-shrinkable items from non-crystalline polymers such as rubbers, one possible alternative is to pre-blend a crystalline polymer, e.g. polyethylene, with the rubber and then electron irradiate, warm, expand, and quench-cool. Up to 50% of the crystalline polymer may be needed. A summary of all of the polymers processed under B and C above is presented in Table 6.

CROSSLINKED POLYETHLENE FOAM

A relatively new process is also based on the radiation cross-linking of polyethylene, in this case prior to expansion of the polymer into a foamed product. Sekisui Chem. Co. (Japan) developed the process and the Voltek Corp. is applying it here in the U.S. In this process the polyethylene resin is mixed with a blowing agent, pigments, stabilizers, flame retardants, etc., then continuously extruded into a long sheet at a temperature below the decomposition point of the blowing agent. Following this, and in a continuous line, the sheet is irradiated by an electron accelerator to crosslink the polyethylene, then passed through ovens to expand to a foamed sheet product. Because of the crosslinking, it is possible to limit the foam cell size to less than 250 μm at volumetric expansions as great as thirty times the initial volume. Relatively uniform cell size is also accomplished, resulting in a product with a smooth surface appearance and feel, and uniform thickness. Modified polyethylene resins, as ethylene-vinyl acetate (EVA) copolymers and chlorinated polyethylene may be used to achieve softer, more tear resistant foams (from EVA) or flame retardant foams.

Major applications include products in the automotive, sports, construction, boating and appliance industries. The products are competitive with chemically crosslinked polyethylene foams and also have some property advantages.

TEXTILES

Only one industrial scale radiation process is currently running in the textile industry, introduced by the Deering Milliken Corp. It is a combined radiation and chemical process, the first step of which is the radiation grafting of a monomer, N-methylol acrylamide (NMA) to a polyester/cotton blended fabric.

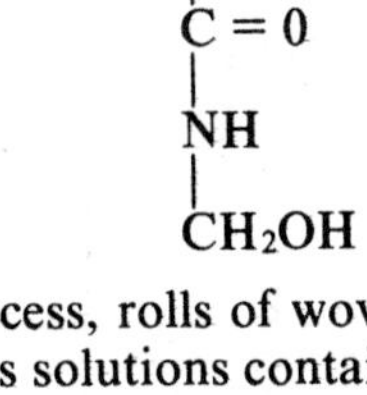

(NMA)

In this step of the process, rolls of woven fabric are continuously padded with (immersed in) aqueous solutions containing the monomer, partially dried, and irradiated by an electron accelerator to bond NMA to the cotton cellulose molecules. Sometimes stain-release resins are also padded onto the fabric.

TABLE 6
Radiation Crosslinked Polymer Materials & Applications (*14*)

BASE POLYMER MATERIAL	SHRINKABLE TUBING	SHRINKABLE PARTS	SHRINKABLE TAPE	DIELECTRIC ROD. SHEET	WIRE INSULATION JACKETING
Polyethylene	0	0	0	0	0
Polyvinyl Chloride	0	–	–	–	0
Polyproplyene	–	–	–	×	–
Polyamide	–	–	–	–	×
Polyvinylidene Fluoride	0	–	–	–	0
Ethylene-Fluorocarbon Copolymer	×	–	–	–	0
Fluoropolymer	0	–	–	–	×
Ethylene-Vinyl Acetate	0	–	×	–	–
Thermoplastic Polyimide	–	–	–	–	0
Chlorinated Polyethylene	–	–	–	–	0
Neoprene	0	0	×	–	–
Butyl	0	0	–	–	–
Silicone Elastomer	0	0	–	–	–
Fluoro Elastomer	0	0	–	–	–
Ethylene Propylene Rubber	0	0	–	–	0
Polyurethane	×	–	–	–	–

0 = Current Application
× = Experimental

Subsequent to this, garments are made from the fabrics, gently pressed, and then oven cured to crosslink the cellulose chains of cotton via the chemical reaction of pendant N-methylol groups of the previously-grafted NMA monomer with cellulose OH groups. This produces a "durable press" article, holding its press over many washing cycles. If stain release resins have been added to the fabric, they are also cured onto the polyester fibers and within the cotton during the oven treatment, yielding a durable press, stain releasing product. Several electron accelerators are in use at Deering Milliken Corp. continuously producing the radiation-grafted fabric. The product is competitive with (100%) chemically crosslinked cotton blend fabrics. A similar process, with reversed chemical and radiation steps, has not proven to be industrially feasible to date.

Additional processes involving radiation modification of textiles are in the research and development stages. They involve radiation-grafting of flame retardant or dye accepting monomers, and curing of non-woven fabric binders, crimp stabilizers, flocking adhesives, and pigment print binders. Another process involving corona discharge grafting of acrylic acid onto polyester fabrics for improved dyability and soil release was promoted for a while, but chemical grafting techniques may be more economical in this case.

CURING OF COATINGS AND INKS

Radiation-curable coatings for factory pre-finished wood panels and furniture, printing inks, metal coils for canned goods, steel plates for large construction (as ships) and plastic parts for automobiles have achieved various levels of commercial production during the past ten years since the Ford Motor Company introduced the process. They are all based on the use of polyfunctional monomers and polymers, sometimes thinned with multi-functional vinyl or allyl monomers (Table 5). Most of the processes have involved special synthesis of radiation sensitive compositions, and the patent literature abounds with examples of these. The major attractions for radiation-cured coatings (vs. chemical or heat curing) are low pollution (solvent-less) coatings, high speed process, low energy requirements, room temperature curing and lower plant area.

Electron curing facilities for prefinished wood panels and coil coatings have operated on a commercial scale within the past few years in this country; however, the current activity in these areas is greatly diminished. The major electron curing facilities in the U.S. today are at the Ford Motor Co. for curing coatings on plastic automotive parts. A number of accelerators are in use for this application; the irradiation is accomplished in inert atmosphere. The trend towards use of large banks of U.V. lamps has increased significantly. This trend is based mainly on lower capital investment but there are a number of other factors which enter into consideration, and these are summarized in Table 7.

The future hopes for electron curing of coatings are based on the need to compete with U.V. for curing rapidly the thin coatings used, for example, in printing inks. Towards this end lower energy (e.g. 100 kV) higher current (e.g. 500–1000 mA) machines are under development.

TABLE 7
Comparison of Curing Methods (*29*)

	ELECTROCURE	ULTRAVIOLET	OVEN
Capital Cost	High	Low	Med.
Length (ft.)	10–30	10–30	90–300
On-Off Operation	Yes	Yes	No
Inert Atmosphere	Yes	No	No
Cure Temp. (°F)	Ambient	100–120	130+
Cure Time	1 Sec or Less	Seconds	Minutes
Cure Coatings on Plastics	Yes	Yes	No
Cure Pigmented Coatings	Yes	No	Yes
Require Catalyst	No	Yes	Sometimes
Coating Solids (%)	90–100	90–100	35–65
Coating Coverage, Sq. Ft./Mil	1000–1400	1000–1400	400–500
Coating Cost	High	High	Low
Cost Per Mil -Sq. Ft.	Low–Med.	Low	Med.
Hazard	X-rays (can be avoided)	UV	Heat, Solvent

WOOD-PLASTIC COMPOSITES

Upgraded wood products may be produced by imbibing a radiation-polymerizable liquid into the pores of shaped wood objects followed by curing with gamma radiation. This is one of the most successful commercial polymer radiation processes using gamma radiation. ARCO Chemical Corp. has a major share of the market (see Table 4). The process involves a batch, multi-step procedure described as follows for the example of floor tile: (1) cut and stack wood into bales of tiles, (2) load bales into processing cans, (3) load processing cans into resin-impregnating vessel, (4) evacuate vessel and then impregnate with monomer under pressure, (5) using an inert gas, pressure pump excess monomer out of vessel, (6) transfer vessel to water-immersed cobalt-60 irradiation facility and irradiate to cure monomer, (7) remove vessel from radiation facility, remove processing cans from vessel and then remove wood-plastic tiles from cans, (8) sand and finish wood-plastic tiles.

The major use of wood-plastic composites has been for parquet flooring where improved wearing and low maintenance properties are important. The choice of wood will influence not only the physical properties of the product but also its appearance and cost. A very porous wood such as pine or birch can imbibe over two times its own weight in monomer while a harder wood such as oak only picks up half its own weight in monomer. The monomer of preference is methyl methacrylate. This choice is based on considerations of low heat of polymerization, low dose requirement for (close to) complete curing, low dose rate sensitivity, low volatility and toxicity, and good physical properties. Sometimes multifunctional monomers such as trimethylol propane trimethacrylate are added to accelerate the curing. Flame retardants, dyes and other additives may also be employed depending on the end use.

This process has also been applied to upgrade inexpensive cellulose fiber materials such as waste wood products, sawdust, bagasse and the like. The product would presumably find some use in low cost construction. The economics are not generally very favorable due to the large amount of expensive monomer needed to upgrade these materials.

WOOD CHIP IRRADIATION

A new process for the irradiation of wood chips is in the advanced stage of development by Radiation Development Co., Ltd. of Canada. In this process wood chips are continuously conveyed by air at high speed through a pipe which passes under the beams of several electron accelerators; a dose of only 0.1–0.2 Mrad is delivered. The chips continue on through the pipe to be dumped into large chip storage piles adjacent to the pulp mill from where they are eventually conveyed to the mill for pulping (Fig. 2). There are two major economic benefits of such a process: first, there is a significant retardation in the extent of microbial decay of the cellulose in the wood, which normally occurs in outside chip storage piles upon long-term storage (6–12 months); second, when the wood chips are pulped a significant pulp yield increase is observed. These results have been seen in both lab pilot scale and mill scale trials with Douglas fir wood chips and Kraft pulping (Fig. 3).

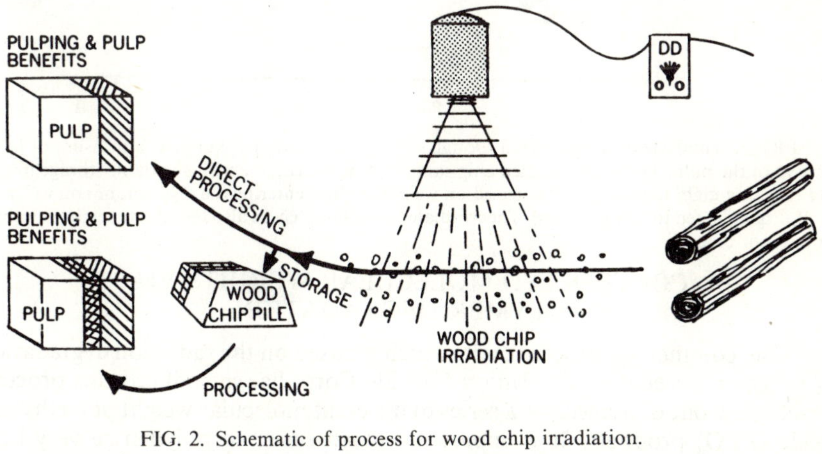

FIG. 2. Schematic of process for wood chip irradiation.

The retardation in chip pile degradation is probably related to two factors: (1) radiation destruction of microorganisms and (2) oxidative destruction of microorganisms due to the ozone produced in the chip during irradiation. The mechanism is not clearly established. The mechanism of the pulping yield increase is even more complex. There are three major components in wood: cellulose, hemicellulose and lignin. A comparison of pulping yields at constant lignin contents shows 3–6% higher yields for the irradiated chips. Therefore, the combined cellulose and hemicellulose content of the pulp from irradiated chips must be higher than that in pulp from control chips, when these pulps are compared at constant lignin contents. Since little change in cellulose content occurs upon pulping, it has been concluded that the yield increase is due to an increased retention of hemicelluloses. This has been confirmed by sugar analyses of the different pulps. The mechanism is not well understood at this time, and is being intensely pursued.

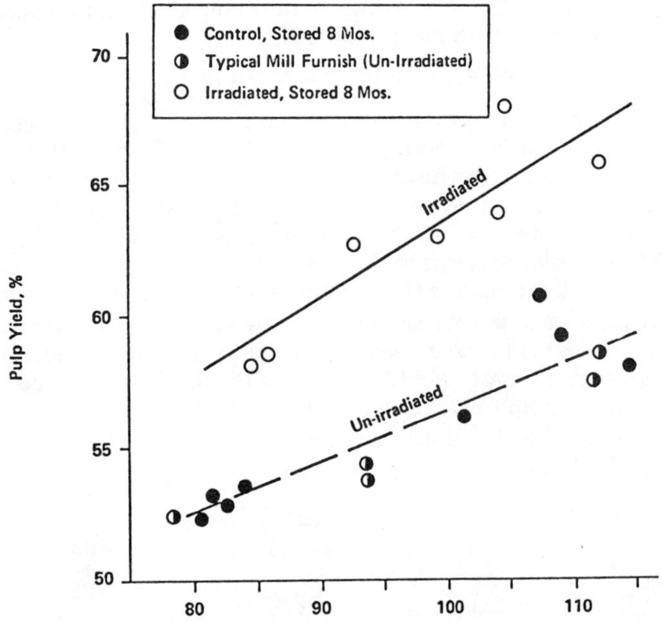

FIG. 3. Total yield of pulp from Douglas fir chips vs. Kappa Number, a measure of lignin content of the pulp. Two un-irradiated chip samples were tested, one with a variable storage history and one after eight months (yield is based on weight of chips entering the digestor, not on weight of chips originally fed to storage piles). The irradiated chip sample was also tested after eight months in an outdoor chip pile. (*32*)

CONTROLLED MOLECULAR WEIGHTS VIA POLYMER DEGRADATION

One commercial process exists which is based on the radiation degradation of polymer molecules. The Union Carbide Corp. has introduced this process, which grew out of a need for a series of different molecular weight polyethylene oxide (PEO) products. It is apparently much easier to polymerize very high molecular weight PEO than it is to stop the polymerization at fixed, lower molecular weights. Thus, cobalt-60 gamma radiation is used to degrade the high molecular weight PEO to specific, lower molecular weights.

MISCELLANEOUS PRODUCTS BASED ON CROSSLINKED AND GRAFTED HYDROPHILIC POLYMERS

A growing number of botanical, chemical, biological and biomedical speciality product applications are based on the radiation gelation of dilute aqueous polymer solutions or the radiation grafting of hydrophilic polymers onto polymer substrates. One example is the radiation grafting of acrylamide onto starch molecules to produce a pure, very high molecular weight polymer suitable as a flocculating agent in aluminum ore processing (Atlas

Chemicals/ICI). In another example, soft contact lenses may be made by grafting N-vinylpyrrolidone within a silicone rubber matrix (Essilor, France). Surface grafts of hydroxyethyl methacrylate and N-vinyl-pyrrolidone onto silicone rubber materials have shown improved blood compatibility over the untreated silicone base material. Presumably, the more aqueous a material becomes, the more tolerant will be the biological environment towards it. These products are in various stages of development, ranging from laboratory curiosities to close to actual commercialization.

Another process for grafting a hydrophilic monomer (acrylic acid) onto a hydrophobic support (polyethylene powder) has been licensed for commercialization (Dow Chem. Co.). The product enhances adhesiveness when added to untreated polyethylene before extrusion. A similar system—acrylic acid grafted to polyethylene sheets—has been marketed as a battery separator (RAI Research Corp.). Miscellaneous acidic and/or basic monomers have also been grafted to hydrophobic sheets to yield ion exchange or ion perm-selective membranes (RAI Research Corp.).

MISCELLANEOUS POLYMERIZATION PROCESSES

Gamma radiation-induced emulsion polymerization of vinyl acetate and other monomers has been brought to a high degree of technical sophistication through the pilot plant stage. Although it has not yet been commercialized, it remains an attractive commercial possibility due to the improved control of polymerization kinetics and resultant higher conversion yields. Purity of the product is also claimed to be an important plus.

High pressure gamma radiation-initiated polymerization of ethylene has been carried through the pilot plant stage. Again, as in the case of emulsion polymerization, advantages include better reaction control and a purer product. This process has not yet achieved commercialization to the author's knowledge.

A third polymerization process of note is the gamma radiation polymerization of monomer-impregnated concrete. The product has improved strength and chemical resistance over the untreated concrete. Other possible mineral matrices include red mud and clay. None of these products has yet reached commercialization.

A process has been developed for the electron beam solid state polymerization of trioxane crystals to yield polyformaldehyde. The product is pure and of higher molecular weight. Although a large-scale pilot facility has been operated, the process has not been fully commercialized (JAERI, Takasaki, Japan).

RUBBER VULCANIZATION

Several processes have been studied for the radiation vulcanization of rubber lattices or bulk rubber $\pm$ solvent to produce various rubber products, such as fabricated goods (e.g. gloves, tapes), adhesives, tires, etc. None of these as yet has reached a significant commercial level, to the author's knowledge.

REFERENCES

A. GENERAL OR REVIEW PAPERS

1. A. Charlesby, *Atomic Radiation and Polymers*, Pergamon Press, N.Y., 1960.

2. A. Chapiro, *Radiation Chemistry of Polymeric Systems*, Interscience Publ., N.Y., 1962.
3. *Large Radiation Sources for Industrial Processes*, proceedings of symposium held in Munich, Aug., 1969. International Atomic Energy Agency, Vienna, 1969.
4. K. Motojima, et al., "Industrial Application of Radiation and Radioisotopes in Japan," as abstracted in *Nucl. Sci. Abstr.*, **25**, 5033 (1971)
5. D. E. Harmer, and D. S. Ballantine, Radiation Processing, *Chem. Eng. (N.Y).*, **78** (9), 98 and (10), 91 (1971).
6. 4th International Conference of the Peaceful Uses of Atomic Energy, U.N., Geneva. **Vol. 14**: *Applications of Nuclear Techniques in Industry and in the Study of Natural Resources.* United Nations, Geneva, 1971.
7. M. Dole (Ed.), *The Radiation Chemistry of Macromolecules*, **Vols. I, II**, Academic Press, N.Y., 1972.
8. J. Deitch and S. Padwo, *Commercial Uses of Ionizing Radiation—First Interim Report*, U.S. Dept. of Commerce, Bureau of Domestic Commerce, Office of Business Research and Analysis; November, 1973.
9. Y. Tabata, *Industrial Application of Radiation Chemistry in Japan*, presented at American Nuclear Society Meeting in San Francisco, Nov. 15, 1973.
10. J. E. Wilson, *Radiation Chemistry of Monomers, Polymers, and Plastics*, M. Dekker, N.Y., 1974.

B. CROSSLINKED WIRE AND CABLE

11. G. Odian and B. S. Bernstein, Monomers Improve Radiation Crosslinking in Polymers, *Nucleonics*, **21**, 80 (1963).
12. B. S. Bernstein and J. Lee, Upgrading of By-Product Polypropylene, *Ind. Eng. Chem. Prod. Res. Dev.*, **6**, 213 (1967).
13. Z. S. Yegorova et al., The Effect of Polyfunctional Monomers on Radiation Crosslinking of Polyethylene, *Vysokomol. soyed.*, **A9**, 1451 (1968).
14. L. F. Rossetti, Irradiated Insulation Materials: An Overview, *Wire and Wire Products*, Feb. 1973.
15. J. Silverman, M. Joshi, and K. Singer, Radiation Crosslinking of Polyethylene in the Presence of Acrylic Acid, presented at American Nuclear Society Meeting, San Francisco, Nov., 1973.
16. W. A. Salmon, Radiation Crosslinking of Poly(vinyl chloride) with Acrylic Monomers, *ACS Polymer Preprints*, **14**, 111 (1974).

C. HEAT-SHRINKABLE FILM, TUBING AND MOLDED OBJECTS

17. G. Odian and B. S. Bernstein, Memory Effects in Irradiated Polyethylene, *J. Appl. Polymer Sci.*, **8**, 1853 (1964).
(See also A. 3, 5, 6, 8, 9.)

D. CROSSLINKED POLYETHYLENE FOAM

18. N. Sagane, Radiation Industry of Foamed Polyethylene, presented at Regional Training Course on Industrial Radiation Processing (IAEA, JAERI), Tokyo, Nov., 1970.
19. D. A. Trageser, Crosslinked Polyethylene Foam, *Modern Plastics encyclopedia*, McGraw-Hill, N.Y., 1973.

E. TEXTILES

20. V. Stannett and A. S. Hoffman, Radiation Techniques in the Textile Industry, *Amer. Dyestuff Reporter*, **57**, 25, 91 (1968).
21. A. S. Hoffman, Radiation Modification of Textiles, in *Large Radiation Sources for Industrial Processes*, p. 301, IAEA, Vienna, 1969.

22. F. J. Miller and P.S. Baker, *Radioisotopes and Radiation in the Textile Industry*, Report ORNL-11C-29, Oak Ridge National Laboratory, Feb., 1972.
23. W. K. Walsh, E. Bittencourt, and H. A. Rutherford, Radiation-Induced Addition of Flame Retardants to the Double Bond in Acrylamidomethyl Cellulose, *Text. Res. J.* **43**, 590 (1973).

F. CURING OF COATINGS AND INKS

24. J. E. Hinsch, Electron Beam Curing, *Met. Prog.* **95**, 72 (1969).
25. E. Rotkirch, Fibre-Board Coating Economics, in *Large Radiation Sources for Industrial Process*, p. 529. IAEA, Vienna, 1969.
26. W. B. Pedersen, and K. A. Singer, Electron Beam Curing of Surface Coatings—Recent Experimental Results, *Eurisotop Report No. 39*, presented at Conference on Radiation and Isotope Techniques in Civil Engineering, Brussels, Oct. 1970.
27. A. S. Hoffman, Electron Curing of Coatings, *At. Energy Rev.* **9**, 347 (1971).
28. R. H. Chandler, Ltd., *Patents on Electron Beam Curing*, R. H. Chandler, Ltd.; London, England, April, 1971.
29. S. B. Levison, *Electrocoat, Powder Coat, Radiate—Which and Why*? Report prepared for Federation of Societies for Paint Technology by David Litten Laboratories, N.Y., 1972.
30. S. H. Schroeter, The Ultraviolet Curing of Coatings, *Am. Chem. Soc. Polymer Prep.* **13**, 401 (1972).

G. WOOD PLASTIC COMPOSITES

31. A. E. Witt, and J. E. Morrissey, Economics of Making Irradiated Wood-Plastic Products, *Mod. Plast.* **49**, (1), 78 (1972).
(See also A. 3, 5, 6, 8, 9.)

H. WOOD CHIP IRRADIATION

32. K. V. Sarkanen, G. W. Mitchell, L. Paszner, and A. S. Hoffman, Increased Kraft Pulp Yield by Wood Chip Irradiation (presented at T.A.P.P.I. meeting, Miami, Fla., January, 1974).

I. CONTROLLED MOLECULAR WEIGHTS VIA POLYMER DEGRADATION

(See A. 3, 5, 6, 8, 9.)

J. MISCELLANEOUS PRODUCTS BASED ON CROSSLINKED AND GRAFTED HYDROPHILIC POLYMERS

33. A. Charlesby and P. M. Kopp, Radiation Protection in Aqueous Polymer Solutions, *Intern. J. Radiat. Biol.*, **5**, 521 (1962).
34. J. Laizier, and G. Wajs, Resines Silicones Hydrophiles Obtenues par Greffage Radiochimique, in *Large Radiation Sources for Industrial Processes*, IAEA, p. 205. Vienna, 1969.
35. A. S. Hoffman et al., Preparation and Application of Radiation-Grafted Hydrogels as Biomaterials, *Am. Chem. Soc. Coatings and Plastics Prepr.*, **34**, 568 (1974).
36. B. D. Ratner, A. S. Hoffman, and J. D. Whiffen, Blood Compatibility of Radiation-Grafted Hydrogels, *Biomater. Med. DEVICES Artif. Organs*, in press.

K. MISCELLANEOUS POLYMERIZATION PROCESSES

37. M. Steinberg et al., "Concrete-Polymer Materials" and "Composite Materials from Solid Waste," in *Isotopes Development Programs—1972*, p. 102, 105. Div. of Appl. Tech., USAEC, Wash., D.C., 1973.
38. V. T. Stannett, and E. P. Stahel, Large-Scale Radiation-Induced Chemical Processing, *Annu. Rev. Nucl. Sci.*, **21**, 397 (1971).

L. RUBBER VULCANIZATION

39. J. Laizier et al., Radioreticulation des Elastomeres Naturels et Synthetiques en Emulsions ou Dispersions, in *Large Radiation Sources for Industrial Processes*, p. 165. IAEA, Vienna, 1969.
40. J. R. Puig, Continuous Radiation Vulcanization, *Rev. Gen. Caoutch. Plast.*, **49**, 289 (1972).

Mutations in Plant Breeding—A Glance Back And a Look Forward[1]

Åke Gustafsson
Institute of Genetics
Lund University, Lund, Sweden

PLANT BREEDING—SCIENCE OR ART?

Plant breeding is often declared to be a kind of art rather than a branch of science. Such a statement is incorrect, of course. In his daily work a plant breeder must combine methods from numerous branches of science—among them genetics, ecology, biochemistry—with agricultural procedures relating to fertilizers, irrigation and mechanization. He has to consider problems of food digestibility and taste, as well as many capricious demands from the general public and from politicians. Plant breeders have to know their crop plants in detail, applying not only methods of electrophoresis and gas chromatography, but must also use their eyes, nose, and gustatory talents in selection and evaluation. In this sense plant breeding combines science with artistry. But in its requirements of expert knowledge, keen intelligence and versatility, plant breeding is not less exacting than other branches of science.

Here artificial mutagenesis also enters the field of plant improvement.

INCREASING COMPLEXITY OF PLANT BREEDING

The double cross method of hybrid corn, founded on long-time empiricism, was proposed by Jones in 1918 (*1*) with enormous economic consequences. An even more elaborate technique, equally ingenious and profitable, concerns the sugar beet, *Beta vulgaris* (*2*). Its improvement partially rests on inbreeding and outcrossing, as in hybrid corn, leading to conspicuous heterosis. Cytoplasmic male sterility is introduced into the female inbred lines. The male parent is a tetraploid bulk cross. The hybrid consists of a triploid rather uniform population, generally well adapted and vigorous in its root properties. Formerly the cultivated sugar beet had a serious weakness in its *multigerm* seed ball, causing extensive labor in the field and consequent high costs. Mutations to the *monogerm* characteristic have been successfully included in the diploid lines.

These examples from maize and sugar beet indicate how progressively more and more sophisticated methods are being adopted in plant improvement. In sugar beet they are based on chromosomal and cytoplasmic genes, in addition to the application of polyploidy and spontaneous mutation. Advanced husbandry techniques have been employed, both in field cultivation and in laboratory evaluations.

[1]Congress presentation D–1–2.

DOMESTICATION OF WILD SPECIES

The way in which wild species are turned into crop plants is brilliantly illustrated by von Sengbusch's work with lupine species (*3*). The wild or semiwild species of *Lupinus luteus, angustifolius* and *albus* were used early in Europe for fodder and green manures. The protein of these species is of a high nutritive quality with crude-protein contents of 40 per cent or more of the dry matter of seed. Oil content is also high. However, the use of wild forms of lupines in cattle feeding was hazardous owing to alkaloid "lupinosis" often appearing, with chronic liver damage, severe cramps, and even death as a consequence.

In a lecture in 1927 the famous geneticist Erwin Baur pointed out that there would be no success in lupine breeding until quick mass methods of alkaloid analysis were elaborated. This was indeed the starting point for Sengbusch's successive domestication of the wild species, involving the following improved characters: 1) low alkaloid content, 2) soft seed coat, 3) non-shattering pods, 4) rapid juvenile development, 5) white seeds, 6) high yields—characters of essential importance to lupine agriculture.

A varying degree of self-incompatibility and outcrossing keeps the gene pools of many *Lupinus* species in a fair state of heterozygosity. After a rapid enlargement of population size, rare recessive genes, as well as new mutations, became accessible as homozygotes for selection, although the frequency of the useful genes was low. This is the Sengbusch method of plant breeding (*4*), which is also applied in other crops for mass detection of mutations with changed and improved protein qualities, for the isolation of rare cases of disease resistance, or for the detection and elimination of harmful metabolic products.

In the original selection work using *Lupinus luteus*, 6–7 million plants resulted in 40-50 individuals low in alkaloids, a frequency of approximately 1:100,000. Soft seed coat was detected in two out of 20,000 plants, whereas 8-10 million yellow lupine plants gave rise to only one that was nonshattering.

The chief domestication of lupine species was accomplished in ten years. In 1927, the first sweet lupine was isolated. In 1931, its cultivation area comprised two hectares. In 1934, the first sale to farmers took place (3800 hectares), and in 1937 the first sale abroad. In 1938, the German cultivation area had risen to more than 110,000 hectares.

It is salient that several mutant types of practical interest have been induced in lupine species also by radiation (*4*).

Similar methods of mass reproduction and testing have been applied in *Melilotus* for the isolation of mutants low in glucoside content (*5*), employing radiation to increase mutation rate. Hybrids of strains with low-glucoside content were still low in glucosides but they often displayed pronounced heterosis. So, according to the author, hybrid breeding seemed practicable.

Similar "blockade mutants" of wild and semiwild characters have been produced in many species, both through artificial mutagenesis and spontaneously. They are also used in well-established crop plants as *Zea mays* and *Pisum sativum* (*6*). A case of such blockade improvement was recently described in poppy, *Papaver somniferum* (*7*). Poppy oil is outstanding in its fatty acid quality. International agreements prohibit the common farming of morphine-rich cultivars in Sweden. However, morphine synthesis can be blocked at many steps.

Systematic analysis by means of gas chromatography led to the isolation of genotypes that were low in morphine.

Consequently, the rather simple but ingenious isolation of blockade mutations, begun so successfully by Sengbusch in the late 1920's, has turned out to be profitable in higher plants, as well as in micro-organisms (*Aspergillus, Penicillium, Claviceps*, cf. *8*, referring *inter alia* to the radiation work by Hollaender in the middle 1940's). The practical approach was not based on guesswork. It was derived by definite considerations emanating from Vavilov's law of parallel variation and homologous series (*9*). One especially interesting result of the early work in *Lupinus* was the finding that a non-shattering mutant isolated in mass cultures differed from non-shattering lines of other leguminous species and genera. Superficially, the phenotypic result was identical, but anatomically the lupine mutant fell outside the homologous series of variation (*4*). It is now a well-established fact that similar phenotypes can originate from alterations in numerous gene loci. This is particularly obvious in the eceriferum and erectoides phenotypes of barley (*10,11*), where a series of independent gene changes lead to parallel, gross morphological expression, but with the fine structure of the wax crystals or spike details clearly distinguishable.

DIRECT MUTATION IN BARLEY IMPROVEMENT

The cultivated barley species has been in the center of radiobiological interest since the 1930's. It has a low chromosome number (x=7), with the individual chromosomes distinguishable; it is self-fertile, and it possesses numerous characteristic traits and genetic markers. In addition, it is an important crop plant in the northern hemisphere, both for human consumption and for animal feeding purposes. The plant was used by Stadler (*12*) in his pioneering research of X-ray-induced mutations. The numerous scientists working with barley cannot be listed here. The reader is referred to Symposia arranged by the International Atomic Energy Agency, Vienna, or to the Proceedings of the First and Second International Barley Genetics Symposia.

The successful application of induced mutation in barley improvement is based on the following facts:

1) Induced mutants can, after the necessary official testing, be directly released as new cultivars. 2) Highly productive, as well as inferior, mutants can with great advantage be utilized in a continuous recombination breeding programme. That this has not been done to a much greater extent than up to now probably is due to the conservative attitude of many plant breeders and a lack of close scientific co-operation. 3) All sorts of characters, quantitative and qualitative, biochemical, physiological and ecological, can be changed in a positive or negative direction. 4) Radiations, as well as chemical mutagens, are useful agents in this transformation. 5) However, contrary to common opinion, the mutation process is definitely not random. Different mutagenic agents often cause mutation at specific gene loci, as indicated by the above mentioned eceriferum and erectoides mutations in barley (*10,11*).

The best-studied cases of directly useful mutations consist of erectoides 32 = $ert\text{-}k^{32}/ert\text{-}k^{32}$ and early 8 = $mat\text{-}a^{8}/mat\text{-}a^{8}$ in barley, released under the names of Pallas (*13*) and Mari (*14*). They were officially approved as new varieties in 1958 and 1960, respectively.

Their extraction was based on results from previous experiments. *ert-k*^{32} arose as a parallel mutation to *ert-c*^{1} (erectoides 1 from Gull barley, analogous but not homologous), isolated in 1937, and numerous other erectoid mutants in various barley strains. The variety Bonus, from which *ert-k*^{32} arose, is a well-tillering, productive variety but unsatisfactory in lodging resistance. This inferior characteristic was amended in several *ert*-mutants—the best one being *ert-k*^{32} (Pallas), having arisen after X-irradiation. After its release, it became widely cultivated in Sweden, Denmark and Great Britain. It is still widely spread in Spain.[2]

Mari, on the other hand, was isolated as a parallel mutant to erectoides 16 and is remarkably early in heading and ripening. The latter mutant was induced in 1943 from Maja barley and evoked a great deal of practical interest, but it did not stand the severe testing preceding a release. Also in this case, Bonus was a suitable parent, being rather late in maturating. A vital mutation increasing earliness would be of immediate practical value. In fact, before *mat-a*^{8} appeared, another early mutant ("early 2") was isolated, which was stiff-strawed and highly productive. A true homologue of ert-16 is *mat-a*^{8}, since both are mutations of the same locus, *mat-a*. Soon after this extraction, several homologues were isolated from Bonus: *mat-a*^{8}, or Mari, is an X-ray mutant; *mat-a*^{11} and *mat-a*^{12} arose after γ and proton irradiation, respectively. Later on, a series of other *a*-mutants have been isolated and have now been tested in different parts of the world.

So far, *mat-a*^{8}, *mat-a*^{11} and *mat-a*^{12}, as well as the mutant "mat-13" (possibly of locus *mat-a*), all four having arisen from Bonus, seem to be performing well both as homozygotes and in recombination. The Maja mutant erectoides 16, on the other hand, is fairly good in homozygous performance, but has failed in recombination work.

Around *ert-k*^{32} and *mat-a*^{8}, an intensive crossing and backcrossing programme was started, utilizing these alleles in numerous ways (Figs. 1 and 2). Remarkable results of mutant gene recombination have been obtained through the work of Drs. A. Hagberg, G. Persson and K. Wiklund of the Swedish Seed Association at Svalöv and Undrom (Fig. 3). In fact, their results stress the necessity of a continuous integration of theoretical and practical aspects in mutation theory and plant breeding, not only in barley but in other crop plants as well.

The reason why Mari, especially, has become important in Swedish breeding programmes relates to its day-length tolerance (*15,16*). In phytotron and greenhouse cultivation it produces abundant seed also under short photoperiods of 12 and 8 hours. Its parent Bonus, as well as most of the investigated Scandinavian varieties, become highly sterile under these photoperiods or do not head at all. The consequence is that Mari can be cultivated in extensive areas outside its place of origin, in fact from northernmost Scandinavia down to the Mediterranean, even in Morocco, Lebanon, Ethiopia, the Canary Islands, and Mexico and Argentina. Numerous mutants of the *a*-locus, independent of ancestry, display a similar photoperiod reaction. Of 20 putative *a*-mutants at least 14 manifest the photoperiod response and reaction to low temperature characteristic of Mari and the Japanese variety Kinai 5 (*17*). The confirmed *a*-mutants have arisen after X, γ, proton and neutron irradiation, as well as after EMS and EI treatments.

Several mutants of barley have been released into farming. They are listed in a report by Sigurbjörnsson and Micke (*18*). Of special interest is the English variety Golden Promise (γ radiation) and the American variety Luther (diethyl sulphate).

In the quoted article, a series of direct mutations from other species are included. Apart from mutant ornamentals, which often are of a bizarre nature, mutants have been released in such important crop plants as rice, bread wheat, durum wheat, oats, soybeans, beans, peas, groundnuts, castor bean, etc., in diploids as well as polyploids, especially of cereal and leguminous species. The paucity of progress in *Triticum aestivum*—with some interesting exceptions—is challenging. To the results in *Triticum durum* a special heading is devoted below.

RECOMBINATION OF INDUCED MUTANT ALLELES

The barley mutation work at Svalöv and Undrom, Sweden (Fig. 3) was mentioned above. It has involved mutations of positive and negative character. The first released recombinant using the Pallas allele (Fig. 1) was Hellas, a cross product of Pallas and the widespread cultivar Herta. The yield of Hellas is very high, the resistance to lodging and straw breakage is outstanding, and the quality properties are good. However, neither of the parental participants (Pallas with its Bonus genes and Herta) is resistant to the spectrum of Swedish mildew races. Crossing and backcrossing Pallas with the primitive barley type "Long glumes" resulted in the partially mildew-resistant Visir barley, sold for export (*19*). It is now replaced by the varieties Rupal and Senat. The former is a Pallas backcross with the "Rupee" gene for mildew resistance. The cultivar Senat is a more complex release combining Pallas and Hellas with the mildew-resistant line "Triple-awn lemma".

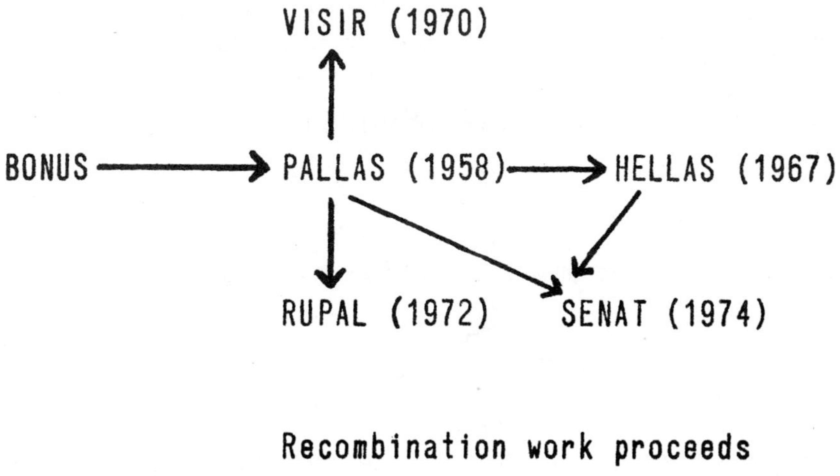

FIG. 1. The PALLAS allele (ert-k^{32}) in recombination breeding.

Rupal is a medium-late, short-strawed, two-row barley with complete mildew resistance to date. It shows no field sprouting in the case of overmaturity and is little attacked by smut or yellow rust. It has a high-quality straw and gives a maximum grain yield in southern Sweden and Denmark.

According to official tests, Senat is the stiffest two-row variety that has been produced in southern Sweden and has a maximum yield in the southeastern parts of the country. It has quality properties similar to those of Hellas, including high resistance to field sprouting. After Pallas was approved in 1958, at least four recombinant varieties possessing the *ert-k*^{32} allele have been released as "originals". Two varieties have been widely cultivated (Pallas and Hellas), and two more (Rupal and Senat) are highly promising.

The successful use of the Mari allele (*mat-a*^{8}) is a fascinating event. The original mutant is still in cultivation. It certainly has some of the negative sides of its progenitor Bonus; for instance, a great susceptibility to mildew and considerable straw breakage at overmaturity. In addition, the straw of Mari is often too short under dry conditions. On the other hand, it has the high-yield potential and the tillering ability of Bonus combined with striking earliness. The inferior characteristics were successively remedied.

The recombination and backcrossing of the Mari allele (Fig. 2) has resulted in a diversity of new varieties, involving gradations from a very early to medium-late maturity. Owing to its new climatic reactions Mari was included in breeding programs at two different stations: Svalöv—56° N and Undrom—63° N (Fig. 3). Mari and Mona were isolated at Svalöv, together with a series of new recombinants that await further testing. The three varieties Kristina, Eva and Salve, on the other hand, were produced at Undrom, though they were also tested farther south.

Kristina (*19*) is a malting barley from the cross Domenx Mari, with high-enzyme activity in the germinating seed. The onset of germination is rapid, a property that limits its wide cultivation in regions with high precipitation at harvest. Recently, it has been considered for malt production in suitable climates.[2] Kristina is high in yield and resistant to lodging (Fig. 4). Its malting quality and straw properties are outstanding.

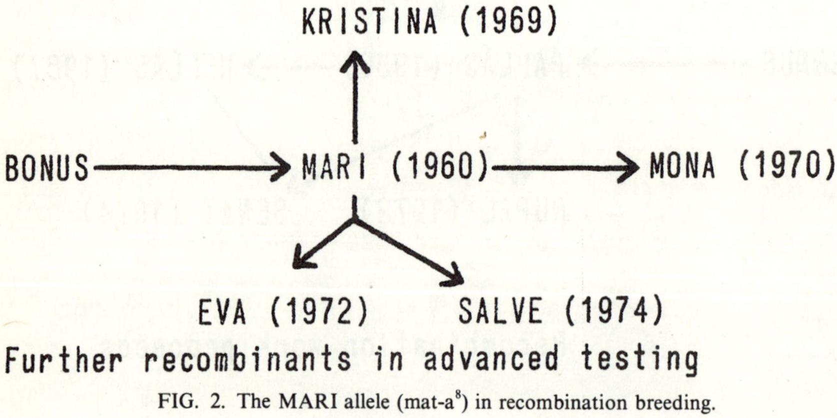

FIG. 2. The MARI allele (mat-a^{8}) in recombination breeding.

[2] K. Wiklund, Madrid, personal communication.

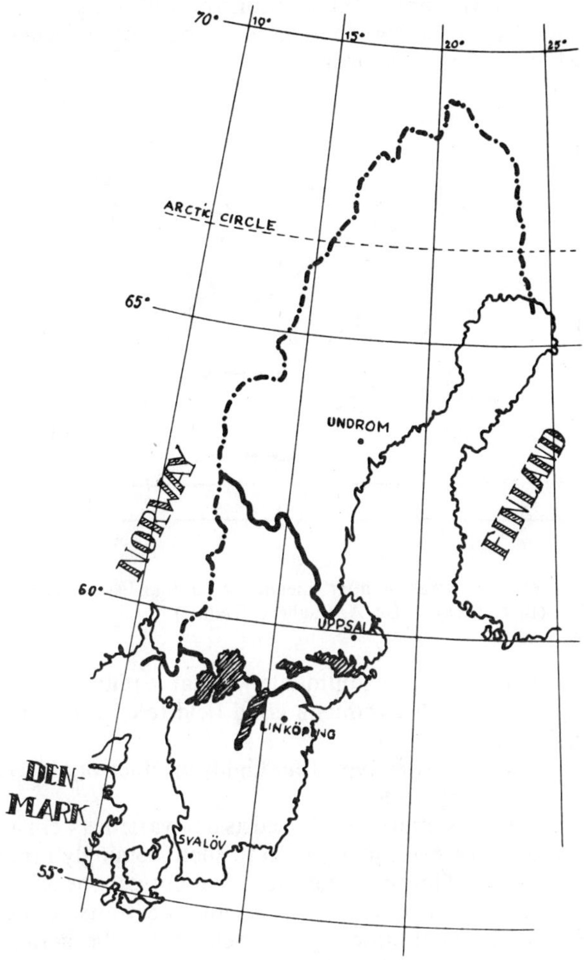

FIG. 3. Sweden divided into three regions of cereal cultivation. The Swedish Seed Association has its chief station at Svalöv and three out of several more branch stations at Linköping, Uppsala (Ultuna) and Undrom.

Mona is a backcross product of Mari and the primitive variety "Monte Cristo" (*19*). It is high-yielding, early, has good straw properties and is resistant to mildew. It is especially suitable in the rotation as a precrop for oil-crop sowings and has, in a few years time, become widespread in Sweden, as well as abroad. It has a high recombination potential and is used in the cross-breeding work including protein- and lysine-rich strains and mutants. Like Mari, it is photoperiod insensitive.

Eva and Salve are varieties obtained from a single cross: Mari x Birgitta. The last mentioned variety is large-kernelled, rather early and particularly adapted to central and northern Swedish conditions. Eva is early like Birgitta, but stiffer-strawed and with a higher seed production. It is recommended to replace the

Birgitta parent over a wide area of cultivation and will compete with the mutant recombinant Gunilla (*v.* below) in regions of northern Sweden, where it is sufficiently early for reliable maturity.

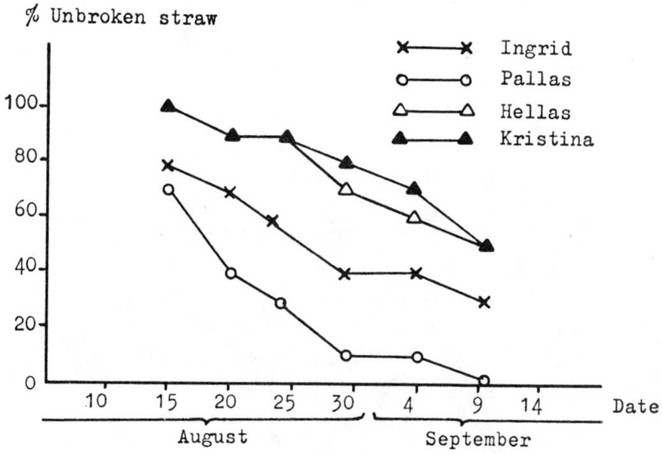

FIG. 4. Process of straw breakage after ripening of the four barley varieties Ingrid, Pallas, Hellas and Kristina. (By courtesy of Dr. A. Hagberg, Svalöv.)

The sister variety Salve is medium-late with large plump kernels, stiff straw and high seed production. It exhibits a good field tolerance to mildew but no specific race resistance.

Thus in 14 years' time, after 1960, four highly productive Mari recombinants have become officially approved.

In the discussed cases, mutants released as new varieties were included in the gene recombination. However, in addition to such seemingly rare events, there are numerous examples of induced mutations that possess one or more improved qualities but which are also loaded with inferior traits, either caused by pleiotropic effects of the mutated gene itself or by the action of parallel, deleterious mutations. This explanation was indeed suggested already by Stadler (*20*) but is also valid with regard to primitive strains of the world crop collections. No conflict can exist concerning the value of natural *and* induced variation. Both sources must be considered in plant improvement. In the coming years, the number of induced variants will increase owing to better methods of induction, mass analysis and selection. This conclusion is not only valid with regard to "blockade mutations", but also with regard to such fundamental properties as disease resistance, protein and amino acid content and composition.

The use of inferior mutants in recombination breeding was well illustrated in an article by Schiemann (*21*), also in barley. Another interesting case of repeated improvement was described in the navy pea bean (*22*, also *18*). Here, a defective, bush-like X-ray mutant was used as the crossing partner in combination with a variety resistant to races of *Colletotrichum lindemuthianum*. In this way the new bush bean Sanilac was produced, later followed by the varieties Seaway, Gratiot and Seafarer—all released by the Michigan State University (*18*). For this accomplishment, Allard (*50*, p. 447) has the following comment: "It should be noted that early-maturing bush types are abundant in the common-bean species; hence the fact that the gene for bush habit of growth carried by Sanilac was from an X-ray-induced mutation is only incidental to the breeding of this variety." This is not necessarily correct. The success may depend on the choice of the original variety (Michelite), from which the bush-like mutant arose, and, of course, on the respective mutant allele. Compare, for instance, the choice of Bonus barley for irradiation work and the induced alleles *ert-k*^{32} and *mat-a*^{8}, so successful in recombination.

A case in barley where an induced, new character was transferred from a defective mutant, giving rise to a highly valuable variety, was recently reported (*19*). The original X-ray mutant, described as "very bushy, extremely waxy and with an outstanding straw-strength," was defective in grain characters. It was isolated at Svalöv (56° N) in 1939 from a southern Swedish variety and incorporated into northern Swedish materials at Undrom (63° N) through complex out- and backcrossings. The entire time, the characteristic straw properties were retained. After a final crossing to the large-kernelled variety Birgitta, with subsequent intense selection, the barley strain named Gunilla was obtained. The new variety, approved in 1970, is the earliest two-row variety in the collection of the Swedish Seed Association. It can be cultivated up to the far north of Sweden and yields about 10 per cent more than the best six-row variety of the northern districts (*23*).

Gunilla itself has been used as a crossing partner with the aim of a further increase of earliness and productivity, retaining the mutant characteristics.

The original mutant was quite extraordinary (*24*, p.33): "It differs much from all barley lines growing at present at Svalöf." It also showed "adaptability to the North-Swedish climate" (*19*). In this case the mutant (44/3) was definitely expedient in the complex transformation of the northern barley materials.

A fresh turn of events took place in plant breeding after the discovery of high-lysine mutants in maize (*25,26,27*). Subsequently, a mass screening for high-protein and lysine strains were carried out in many crop plants employing the so-called DBC method (*28*). The spontaneous maize mutants opaque-2 and floury-2, isolated in the 1920's, have in recent years been widely used in corn breeding, apparently with success. The same holds true for the Hiproly strain of barley (*29*), high in protein and lysine, but useless in its primitive Ethiopian form. The protein-lysine gene(s) have now by repeated backcrossing been transferred to modern northern European genotypes. Yield has increased, though not yet sufficiently for varietal release.

In the meantime, mutagenic experimentation (*30-32*) has led to the isolation of a mutant (Risö No. 1508) having a lysine content of seed protein increased by some 45 per cent compared with its parent Bomi barley. The new amino acid composition is in this instance, as in opaque-2 and Hiproly, caused by a single allelic change, apparently a different gene from that of Hiproly. The mutant 1508 was induced by ethylenimine treatment. Lysine mutants, though not so drastic, were also isolated after γ-irradiation. In fact, six mutations in the offspring of 6000 mutagenically treated barleys suggest that mutations causing an increase of lysine are not particularly rare.

In leguminous species induced mutations have been studied for many years. An X-ray-induced mutant was actually released by Gelin (*33*). The protein problems of such species were recently taken up for further analysis (cf., for instance, *34*).

Numerous induced mutants will certainly possess some inferior traits in combination with the rare positive ones. Referring to mutations for disease resistance (mildew), but equally valid also for changed protein and amino acid composition, the following quotation can aptly conclude this chapter. "So far, there has been a very limited use of the *ml-o* mutants in practice and commercial varieties with this type of resistance are not yet released. Apparently, the difficulties mentioned in connection with the very first mutant, M 66, have had a discouraging effect . . . By a suitable backcrossing program, however, it should be possible to separate the mutant gene from the other undesirable characters" (*35*).

INDUCED MUTATIONS IN TRITICUM DURUM

Durum wheat, which is a balanced tetraploid species (2n=28), has been thoroughly studied by Italian geneticists around Scarascia-Mugnozza, under the auspices of FAO, IAEA, and the Italian Nuclear Energy Committee (*36-39*). The conclusion of his introductory work was that it is possible in *durum* wheat to improve the studied varieties through mutation, both with regard to lodging resistanc

, mutations do not break down the framework of positive characters. The mutagenic agents used in *durum* wheat consisted of X-rays, γ and neutron irradiation, as well as alkylating agents.

In the years 1965-1969, mutant yield trials were carried out in a number of localities in 16 countries of the Near East (*37*). The data obtained showed that many mutant lines not only yielded more than the local varieties and the common varieties in the trials, but often had greatly improved characters in other respects; for instance shorter straw, better lodging resistance, increased earliness, higher volume weight and increased quality of grain. For the total experimentation, a uniform recording system was designed covering both genetic and environmental data. These were edited for computer analysis. The significant improvements were most obvious in reduced plant height and in lodging resistance. Several mutant lines combined improved characters. At the time of publication, three mutants had been released in Italy as new varieties. In the latest survey available (*18*), four mutant lines are listed, three of them having arisen after neutron irradiation and one after X-rays. All four lines have improved yielding ability and lodging resistance. One of them, Castelnuovo, is characterized by increased adaptability.

In the computer records, varietal and environmental causes of variation were analysed. Great differences appeared with regard to characters such as yield, height, volume weight, heading and ripening time. Effects of genotype-environment interaction were fairly conspicuous, especially for characters related to the developmental physiology of the wheat plant.

In recent studies (*38,39*) induced mutations of a dominant type were included in cross-breeding programmes. Mutations present, according to the authors, an advantage over other sources of variability. The mutation method makes use of carefully selected parental strains as the basic foundation for new variants, displaying simple gene differences easily followed in recombination procedures. Single genes are able to shift, in the desired direction, fundamental characters also connected with the expression of high yield. This is valid, of course, for both natural and induced variants.

Triticum durum and *T. aestivum* are now used on an extensive scale for the production of high-yielding *Triticale* strains. Especially the hexaploid *Triticale* strains are planned to fill an ecological gap in the world-wide cultivation of cereal crops. The integration of wheat and rye chromosomes into a harmonious whole seems to offer some breeding problems of great phylogenetic interest. To what extent chromosome rearrangements and gene mutations will have a part in this successive crop transformation remains to be studied.

The international *durum* programme has offered a good deal of valid information on positive mutant induction. Unfortunately, *aestivum* programmes are at present not oriented to a similar, uniform goal. Except for some Indian and Siberian cases (*18*), few positive changes resulting in new varieties have been reported.

CHROMOSOME MANIPULATION IN PLANT BREEDING

In the first chapter of this survey the increasing complexity of modern plant breeding was pointed out. The examples mentioned dealt with cross-fertilizers as maize and sugar beet. The economic impact of the new hybrid methods has given rise to serious expectations in also altering common self-fertile cereals (wheat and barley) into hybrid populations with changed reproductive systems. In wheat, cytoplasmic factors for male sterility are planned to work together with an open-flowering structure in wind pollination. Since no definite screening technique is as yet sufficiently advanced to *induce* genes for cytoplasmic male sterility (with restorer genes), plant breeders have to rely on the transfer of such properties from related species and genera by means of distant hybridization and systematic backcrossing.

In barley another technique has been proposed that uses nuclear genes effecting male sterility, with a breeding system involving tertiary chromosome trisomy and diploid chlorophyll lethality. This method, proposed by Ramage and Wiebe (*40,41,19*), seems to work in certain dry climates (Arizona) with suitable barley genotypes. The method can be adjusted to local demands in many ways. Also in this system, new pollination characteristics may be carried over from one species to another by planned crossing as in the cytoplasmic case discussed above.

In fact, the idea of manipulating the germ plasm of common cultivars by transfer of genes and gene segments through distant hybridization was outlined by Sears in an ingenious paper (*42,43*). A gene for leaf-rust resistance was transferred from the primitive *Aegilops umbellulata* to bread wheat. This involved first the production of a hexaploid amphidiploid (2n=42) from the cross of *A. umbellulata* (2n=14) with *T. dicoccoides* (2n=28). The new hexaploid was then backcrossed twice to *aestivum* wheat (2n=42). The segment carrying the gene for rust resistance in an extra chromosome was, after X-irradiation, chromosome breakage and translocation, transferred to a wheat chromosome. Of 6091 offspring plants studied, 132 were resistant to leaf rust and 40 of these carried a translocated chromosome with at least 17 different cases of translocation. In one such instance the selfed offspring plants were distinguishable from normal control plants only by their resistance and slightly later maturity. The transferred segment was evidently rather small, with few extra genes besides the gene for resistance. These results have then been followed by a series of similar experiments. No doubt, the entire *Triticinae*, inclusive *Triticale*, afford challenging opportunities for future genome repatterning.

Actually, the use of chromosomal translocation in plant breeding has scarcely begun. As far as inversions and reciprocal translocations within a definite barley genotype are concerned, these are mostly fully viable in the homozygous condition (*44,45,19*). Random translocations generally lack pronounced position effects, although they show a series of minor changes in morphology and physiology, acting as genetic modifiers. They may lead to reconstructional events, especially in polyploids (*46*), giving rise to duplications and deficiencies, which in their turn disturb gene balance. Small deficiencies and duplications are probably partially instrumental in the transfer of wild into domesticated species and in their further improvement.

The structural contrast to genome repatterning is afforded by the gene mutations, in the extreme consisting of individual base substitutions. The effects of gene mutations range from large to very small ones. Gaul and his co-workers (*47*) have proposed to use "micromutations" in an integrated method of "maintenance breeding," a sort of advanced elite selection by means of induced mutation. The presented data seem to indicate that this may be an auxiliary method in the improvement of individual varieties in yielding ability (or in other characters as well). The unfortunate limitation of the method is, of course, the sizable statistical variance in a character such as yield, where fluctuating climatic and ecological conditions conceal small genetic effects.

There is at present, I consider, no severe disagreement among mutationists in their views on the mutual importance of major and minor mutations in plant breeding and plant evolution. The barley work, which implies a sort of "green evolution" (*19*), indicates this in full.

Finally, DNA manipulation of the germ plasm has also started in crop plant mutationism. The data here are meagre, however. There are indications of "transformation processes" in *Arabidopsis* and *Petunia* (*48*). Further suggestions of such phenomena have recently been obtained in barley with the "waxy gene" (*wx* in chromosome 1).[3] Especially with regard to problems of resistance, protein content and amino acid composition such directing influences would have far-reaching importance.

CONCLUDING REMARKS

Mutation and gene recombination are the basic mechanisms in evolution. The scant use of induced mutations in plant improvement up to now is surprising and disappointing. Even if 99 per cent of all mutations—spontaneous or induced—would be less vigorous than their parent in homozygous condition, their heterozygous effects, which are not necessarily negative, still remain. We may even propose the occurrence of beneficial effects of some deleterious mutations. Furthermore, a series of homozygous mutations—consisting of reciprocal translocations and inversions, possibly also gene changes of enzyme

[3]N. V. Turbin, Moscow, personal communication.

and protein characteristics—constitute neutral events. A certain small number of mutants are directly superior when homozygous; even more so are numerous mutant recombinants.

In addition, we have to realize more fully that genes and mutations of an organism do not act in a vacuum; their ultimate viability reactions, their increase or disappearance relate to past and present conditions of climates and soils, not to forget the innumerable interactions with other organisms, animals and plants. Vigor *is* a relative concept.

It is indeed astonishing, when glancing back, what a multitude of experiments have been devoted in universities and colleges to the blind induction of mutations in higher plants, creating "worthless junk" instead of meaningful selections. This may in its turn have influenced the attitude of many plant breeders, who do not consider mutation a worthwhile principle in plant improvement. But nobody can avoid the use of heterosis in his corn-breeding programme or dismiss the complex methods used in sugar beet breeding. Likewise, he can no longer ignore the methods developed for mutant mass screening, for instance in disease resistance and amino acid patterns.

The "green evolution" of crop plants has successfully utilized natural variation from all corners of the earth. The hereditary characters incorporated into modern crops can nowadays be induced as easily by the available mutagens. This is true of dwarf and semidwarf characters, causing adaptation to fertilizers and irrigation, of long- and short-day or photo-insensitivity reactions, of properties in disease resistance, of changes in protein and amino acid characteristics. Blocking the metabolic pathways of alkaloid formation in lupines was a radical procedure at the time, relying on residual heterozygosity and recent mutation. This can nowadays be done by induced mutation. Natural and induced variation do not compete, they complement each other. Now one source had better be used, now another. And the value of induced mutations is increasing, in the same degree as natural sources of variation unfortunately disappear.

This brief retrospect of previous shortcomings and procedures really indicates the need for a better co-ordination and co-operation in plant breeding. We now know which mutagens are essential in practical mutation work, and we include radiations as well as chemicals. We now know that there is no principal difference between natural and induced variation. We can easily induce mutational events covering the gap between gross chromosomal rearrangements and DNA-base substitutions. We also know that induced variation may in the future fill and replace the loss or lack of natural variability, or may even extend the limits of variation. We know that mutation and recombination—as evidenced, for instance, by the successful work in barley–will be united into routine procedures, also leading to new breakthroughs in plant improvement (*49*). Moreover, ingenious techniques of mass testing are under way, combining traditional and prospective ideas. The mutation method has come to stay, not alone but with gene recombination, with heterosis and with polyploidy.

We say that the present is the past's future. To glance back is then to look forward; failures give rise to progress. A synthesis of breeding methods is around the corner.

REFERENCES

1. D. F. Jones, *Conn. Agric. Exp. Stn., New Haven, Bull.* **207,** 1 (1918).
2. Swedish Sugar Co./Hilleshög Seed Co., *Sugar Beet in Sweden.* Hilleshögs Frö AB, Box 302, 261 23 Landskrona, Sweden (1968). (Also in the pamphlet "Hilleshög" from the same firm).
3. R. von Sengbusch, *Landwirtsch. Jahrb.* **91,** 723 (1942).
4. Å. Gustafsson and I. Gadd, *Hereditas* **53,** 15 (1965).
5. A. Micke, *Fifth Congress of the European Association for Research on Plant Breeding,* Milano, pp. 262–268. Pavia, 1968.
6. Å. Gustafsson, *Hereditas* **50,** 211 (1963).
7. U. Nyman and O. Hall, *Hereditas* **76,** 49 (1974).
8. Å. Gustafsson, in *Mechanisms in Radiobiology* (M. Errera and A. Forssberg, Eds.), **Vol. I,** pp. 477–497. Academic Press, New York, 1961.
9. N. I. Vavilov, *The Origin, Variation, Immunity and Breeding of Cultivated Plants.* Chronica Botanica Co., Waltham, Mass. 1951.
10. U. Lundqvist, P. von Wettstein-Knowles and D. von Wettstein, *Hereditas* **59,** 473 (1968). (Also unpublished data.)
11. Å. Gustafsson, in *Induced Mutations in Plants,* pp. 9–31. International Atomic Energy Agency, Vienna, 1969.
12. L. J. Stadler, *Science* **68,** 186 (1928).
13. G. Borg, K. Fröier and Å. Gustafsson, in *Second United Nations International Conference on the Peaceful Uses of Atomic Energy,* **Vol. 27:** *Isotopes in Agriculture,* pp. 341–349. United Nations, Geneva, 1958.
14. Å. Gustafsson, in *Recent Plant Breeding Research* (E. Åkerberg, A. Hagberg, G. Olsson, O. Tedin, Eds.), pp. 89–104. Almqvist and Wiksell, Uppsala, 1963.
15. I. Dormling, Å. Gustafsson, H. R. Jung and D. von Wettstein, *Hereditas* 56, 221 (1966).
16. I. Dormling and Å. Gustafsson, *Theor. Appl. Genet.* **39,** 51 (1969).
17. R. Takahashi and S. Yasuda, in *Barley Genetics II* (R. A. Nilan, Ed.), pp. 388–408. Washington State University Press, 1971.
18. B. Sigurbjörnsson and A. Micke, in *Polypoidy and Induced Mutations in Plant Breeding,* pp. 303–343. International Atomic Energy Agency, Vienna, 1974.
19. Å. Gustafsson, A. Hagberg, G. Persson and K. Wiklund, *Theor. Appl. Genet.* **41,** 239 (1971).
20. L. J. Stadler, *Hered.* **21,** 2 (1930).
21. E. Schiemann, *Ber. Dtsch. Bot. Ges.* **48,** 477 (1930).
22. E. E. Down and A. L. Andersen, *Science* **124,** 223 (1956).
23. G. Persson, *Sver. Utsädesfören. Tidskr.* (in press) (1974).
24. Å. Gustafsson, *Hereditas* **33,** 1 (1947).
25. E. T. Mertz, L. S. Bates, O. E. Nelson, *Science* **145,** 279 (1964).
26. O. E. Nelson, E. T. Mertz and L. S. Bates, *Science* **150,** 1469 (1965).
27. O. E. Nelson, in *Genes, Enzymes and Populations* (A. M. Srb, Ed.), pp. 303–311. Plenum Press, New York, 1973.
28. R. Mossberg, in *New Approaches to Breeding for Improved Plant Portein,* pp. 151–161. International Atomic Energy Agency, Vienna, 1969.
29. L. Munck, *Hereditas* **72,** 1 (1972).
30. H. Doll, *Hereditas* **74,** 293 (1973).
31. A. Tallberg, *Hereditas* **75,** 195 (1973).
32. J. Ingversen, A. J. Andersen, H. Doll and B. Köie, in *Nuclear Techniques for Seed Protein Improvement,* pp. 193–198. International Atomic Energy Agency, Vienna, 1973.
33. O. Gelin, *Agri. Hort. Genet.* **13,** 183 (1955).
34. H. Müller and W. Gottschalk, in *Nuclear Techniques for Seed Protein Improvement,* pp. 235–253. International Atomic Energy Agency, Vienna, 1973.
35. A. Wiberg, *Hereditas* **75,** 83 (1973).
36. G. T. Scarascia-Mugnozza, in *Induzierte Mutationen und ihre Nutzung* (H. Stubbe, Ed.), pp. 205–210. Abhandl. Deutsch. Akad. Wiss. Berlin, Akademieverlag, Berlin, 1967.

37. T. B. Bogyo, G. T. Scarascia-Mugnozza, B. Sigurbjörnsson and D. Bagnara, in *Induced Mutations in Plants*, pp. 699–716. International Atomic Energy Agency, Vienna, 1969.
38. G. T. Scarascia-Mugnozza, D. Bagnara and A. Bozzini, in *Induced Mutations in Plants*, pp. 183–198. International Atomic Energy Agency, Vienna, 1972.
39. A. Bozzini, *Theor. Appl. Genet.* **44,** 304 (1974).
40. R. T. Ramage, *Crop Sci.* **5,** 177 (1965).
41. G. A. Wiebe and R. T. Ramage, in *Barley Genetics II* (R. A. Nilan, Ed.), pp. 287–291. Washington State University Press, 1971.
42. E. R. Sears, in *Genetics in Plant Breeding, Brookhaven Symposia in Biology*. No. 9, pp. 1–22 (1956).
43. A. Hagberg, in *Manual on Mutation Breeding*. pp. 177–180. Technical Reports Series No. 119, International Atomic Energy Agency, Vienna, 1970.
44. Å. Gustafsson, U. Lundqvist and I. Ekberg, *Hereditas* **56,** 200 (1966).
45. N. Nybom, *Acta Agric. Scand.* **4,** 633 (1954).
46. J. Mac Key, *Hereditas* **40,** 65 (1954).
47. H.P.K. Gaul, E. Ulonska, C. Zum Winkel and G. Braker, in *Induced Mutations in Plants*, pp. 375–398. International Atomic Energy Agency, Vienna, 1969.
48. R. A. Nilan, in *Genes, Enzymes and Populations* (A. M. Srb, Ed.), pp. 205–222. Plenum Press, New York, 1973.
49. T. Berg, *Hereditas* **69,** 288 (1971). (Also unpublished data.)
50. R. W. Allard, in *Principles of Plant Breeding*, pp. 447. Wiley, New York, 1960.

The Use of Ionizing Radiation for Preservation of Food and Feed Products [1]

Edward S. Josephson, Ari Brynjolfsson, and Eugen Wierbicki
US Army Natick Laboratories, Natick, Mass. 01760, U.S.A.

INTRODUCTION

Exposing food to ionizing radiation can contribute to closing the worldwide food deficit by reducing food spoilage losses, by making available more food of higher nutritional quality (animal protein food) to more people, and by keeping prices down by reducing losses. Because ionizing radiation kills disease-causing organisms, it can reduce the incidence of food-borne diseases. It also reduces our dependence upon some of the chemical additives, such as nitrites and nitrates, now being questioned by health authorities to control food spoilage and food-borne diseases.

The three basic types of ionizing radiation used for processing of food are electrons (10 MeV maximum energy), X-rays (5 MeV maximum energy) produced by electrons in an X-ray target, and gamma rays from ^{60}Co and ^{137}Cs. Electrons, X-rays, and gamma rays cause ionization in the food by either the primary electrons or by the secondary electrons resulting from gamma or X-ray interactions in the food with little rise in temperature and little total chemical change. The ionized and activated molecules form unstable secondary products that kill the organisms. Another effect is to slow down post-harvest growth and maturation in some fruits and vegetables.

Objectives

Sreenivasan (*1*) has reported five distinct objectives which can be achieved by exposing food to ionizing radiation: (a) radappertization—the total elimination of food-spoilage or disease-causing organisms, thus making the prepackaged food stable without refrigeration; (b) radurization—the reduction of spoilage organisms sufficiently to retard spoilage; (c) radicidation—inactivation of all organisms posing public-health hazards; (d) destruction of insects infesting stored cereals, cereal products, grains, and fruits; and (e) control of physiological processes to delay post-harvest ripening in fruits and inhibit sprouting in tubers and bulbs, thus extending shelf life. Some of these applications have been reported by Ley (*2*) and are shown in Table 1.

The use of ionizing radiation to preserve foods is considered to be a "cold" process because there is only a slight temperature rise in the food when exposed to ionizing energy. It is precisely this feature which makes the process attractive

[1] Congress presentation D–1–3.

because, with the absence of a marked temperature rise, adverse changes in the food associated with other food processes such as off-flavor, off-color, loss in texture and impairment of nutritional quality are minimized. The irradiated food, therefore, has the characteristics of fresh food which it approaches in appearance, taste, and quality. The process can be used to preserve a wide variety of foods and animal feeds in a range of sizes and shapes from large crates to thin slices.

TABLE 1
Some Possible Applications of Ionizing Radiation to the Treatment of Food

Application	Dose range (Mrad)
Sterilization to allow long-term unrefrigerated storage, e.g., for meats and meat products	4–6
Sterilization of food ingredients, e.g., spices	1–2
Extension of refrigerated storage (0–4°C), e.g., meat and fish	0.05–1.0
Elimination of specific pathogens, e.g., salmonellae from meat, poultry, and eggs, and animal feeds	0.5–1.0
Reduction of mold-yeasts, e.g., on fruits and vegetables	0.1–0.5
Destruction of parasites, e.g., in meat	0.01–0.2
Disinfestation, e.g., cereals	0.01–0.1
Inhibition of sprouting, e.g., potatoes and onions	0.005–0.015
Sterilization of laboratory-animal diets	2–5

Source: F. J. Ley (*2*).

RADAPPERTIZATION

The scientific feasibility of using ionizing radiations to preserve highly perishable animal protein foods, such as meat and poultry, for long periods of time under non-refrigerated conditions has been proven under the US Army Radiation Preservation of Food Program. Technology is well advanced for radappertized ham, bacon, pork, pork sausage, beef, corned beef, chicken, codfish cakes and shrimp and, except for the determination of the irradiation dose requirements, for lamb, turkey, and ground beef with binders (*3,4*).

The radappertization process basically involves preirradiation treatment with heat (65–75°) to inactivate autolytic enzymes, packaging under partial vacuum in a sealed container impermeable to moisture, air, light, and microorganisms, bringing the food package to the temperature at which it will be irradiated, and exposing the food package to ionizing radiation until the absorbed dose is obtained.

Table 2 summarizes some of the major technological parameters of radappertized foods which have been successfully produced in the laboratory. Bacon irradiated at temperatures below 25°C is of excellent quality. Other products develop off-flavors that are greatly reduced by irradiation at low temperatures. Low temperature irradiation is a basic requirement for producing acceptable beef (*4*). Other foods can be improved by irradiating them while frozen at −30° ± 10°C (*5*). However, as temperature is lowered below 0°C, higher irradiation doses are required to achieve the same degree of biocidal effect. Also, the cost of chilling increases as temperature is lowered below the limit of mechanical refrigeration which is about −30°C. Therefore, the most favorable balance of quality, cost, and required irradiation dose appears to be at about −30 ± 10°C.

TABLE 2
Minimum Required Doses (MRD) for Radappertization[a]
(Preliminary Data)

Food	Irradiation temperature (°C)	MRD (Mrad)
Bacon	5 to 25	2.3
Beef	−30 ± 10	4.7
Beef	−80 ± 10	5.7
Beef[b]	−30 ± 10	3.7
Chicken	−30 ± 10	4.5
Ham	5 to 25	2.9
Ham	−30 ± 10	3.7
Ham[c]	−30 ± 10	3.3
Pork	5 to 25	4.6
Pork	−30 ± 10	5.1
Shrimp	−30 ± 10	3.7
Codfish cakes	−30 ± 10	3.2
Corned beef	−30 ± 10	2.4
Pork sausage	−30 ± 10	2.7

[a]Based on 10^{12} reduction in numbers of spores of *Cl. botulinum* (12-D)
[b]With the additives: 0.75% NaCl and 0.375% sodium tripolyphosphate
[c]Low $NaNO_2/NaNO_3$ (25/100 mg/kg)

Source: Mr. Abe Anellis, US Army Natick Laboratories.

The minimum radiation doses (MRD) given in Table 2 were obtained in accordance with the 12D concept of microbiological safety. The MRD data indicate the radiation dose in megarads needed to reduce the numbers of viable spores by a factor of 1×10^{12}, based upon the recovery data of the most radiation resistant strains of *Clostridium botulinum* used in inoculated pack studies with the individual foods in sealed cans as the substrate (*6*). As the data indicate, the MRD values depend on the food and its temperature during irradiation. The foods containing curing agents or spices have lower MRD's (ham, pork sausage, bacon) than foods without these ingredients.

Mixtures of about 0.75 percent sodium chloride and 0.25 to 0.5 percent food grade phosphates, such as sodium tripolyphosphate (TPP), are excellent binding agents both for radappertized hamburgers (ground beef) and for formed rolls of beef, chicken, pork, and lamb. Weight loss during enzyme inactivation was reduced from the normal 30−35 percent loss with no additives to 10−15 percent with these additives (*7*). This improved their juiciness. All products retained their shape through extended room temperature storage and during kitchen preparation. The meat rolls may be readily sliced after reheating. In addition, as shown in Table 2, the MRD for beef was reduced from 4.7 to 3.7 Mrad when 0.75% NaCl and 0.375% TPP were added to the beef prior to enzyme inactivation and irradiation.

To protect the radappertized foods from bacterial recontamination after irradiation and long term non-refrigerated storage, durable packaging of the food prior to irradiation is required. Two program goals have guided progress in the field: (a) determining reliability of commercially available metal containers for low temperature radiation sterilization of prepackaged foods, and (b) developing flexible lightweight containers capable of withstanding rough handling and storage, retaining protective qualities during storage without any adverse effects on the food contained therein. There is no problem in irradiation of tinplate containers at doses up to 7.5 megarads at temperatures as low as −90°C, provided the can enamels used are of the epoxyphenolic or phenolic types and the end-sealing compounds are the blend of cured and uncured butyl elastomers, the blend of polychloroprene and butadienestyrene elastomers, or blend of polychloroprene and uncured butyl elastomers (*8*).

The US Food and Drug Administration approved four plastic films as food contactants for radappertized foods: polyethylene, polyethylene terephthalate, vinyl chloride-vinyl acetate copolymer, and nylon 6. Other films being investigated for FDA approval are the ethylene-butene copolymer, vinylidene chloride-vinyl chloride copolymer, polystyrene, plasticized polyvinyl chloride, nylon 11, and a blend of ethylene-butene copolymer and polyisobutylene. These films are used as the food contactant layer in a laminated structure with aluminum foil (middle layer) as a moisture and oxygen barrier, and either polyethylene terephthalate or nylon 6 as the outside layer to give strength to the laminate in the form of pouches. The laminated flexible package consisting of chemically bonded polyethylene terephthalate and medium density polyethylene as the food contactant, aluminum foil (middle layer), and nylon 6 (outside layer) was found to be very reliable for packaging irradiated foods (*8*). Over 350,000 such flexible packages were used during the past two years for vacuum packaging of more than 40,000 kg of beef for experiments to prove safety for consumption of radappertized beef with less than 0.01% failures after vacuum packaging and irradiation. Both metal containers and flexible packages have to be sealed under vacuum to prevent rancidity of the lipids in the foods packaged for radappertization.

Table 3 shows quality of radappertized hams using the 9-point hedonic scale for preference (*9*). In the case of meat and poultry products the rating of 5 ("neither like nor dislike") is considered to be the threshold of acceptability. A rating of 7 or above indicates a highly acceptable product.

It is of more than passing interest that the ham from the Experiment No. 72/80 in Table 3 was used at the request of the National Aeronautics and Space Administration (NASA) by the astronauts of the Apollo 17 flight to the moon in December 1972. The ham slices, 12 mm in thickness and weighing approximately 105 ± 5g, were eaten at three meals in sandwiches made with radurized bread (50,000 rads) using radiation insect-disinfested rye flour (50,000 rads). The verdict of the astronauts was most encouraging. They reported: "the juicy, chewy (irradiated) ham and cheese on (irradiated) rye was one of the space culinary delights enjoyed by the Apollo 17 astronauts" (*10*). Radappertized ham slices were also orbited in Skylab III as emergency backup food.

In response to a NASA request for foods for the Apollo-Soyuz Test Program (ASTP), the Army Natick Laboratories recently provided

TABLE 3
Acceptance of Radappertized Ham[a]
(Hams stored at room temperature for 1 to 12 months prior to serving; testing period: October 1971–May 1974)

Expt. No.	Recipe	No. Raters	Average Acceptance Rating[b]
71/90	Grilled	16	8.1
71/123	Baked	16	7.5
72/80[c]	Baked	17	7.7
72/80[c]	Baked	45	7.1
72/80[c]	Baked	17	7.9
72/101	Baked	25	6.6
73/17	Baked	22	7.4
72/180	Baked	14	7.9
73/46	Baked	45	7.4

[a]3.7–4.3 Mrad at −30 10°C, based on 10^{12} reduction in numbers of spores of *Cl. botulinum* (12-D).
[b]9-point hedonic scale: 9 is "like extremely"; 5 is "neither like nor dislike"; 1 is "dislike extremely."
[c]APOLLO 17 ham.

(Source: Mrs. Lucy J. Rice, U. S. Army Natick Laboratories)

radappertized ham, corned beef, shrimp, turkey slices, cooked salami and beef steaks for evaluation by prospective pilots, both Russian and American, of the scheduled ASTP flight.

Data in Table 4 show good acceptance by test panels of radappertized shrimp, pork sausage and beef. Similar data were obtained on other foods radappertized in the frozen state (*4,11*).

In the United Kingdom, considerable experience is available on animal feeds irradiated at 2.5 Mrad for pathogen-free feeds and at 5.0 Mrad for germ-free feeds (*1,2*). Irradiated feeds are more nutritious than thermally sterilized feeds. Thermal sterilization (autoclaving) lowers considerably the concentration of individual amino acids, particularly the essential amino acids, lysine, methionine and tryptophan. This was shown by animal feeding studies with rats and mice using thermally sterilized versus irradiation-sterilized animal feeding mixtures (*12*).

Massa (*13*) reported that irradiation of hoof and mouth disease virus in the dry state with a 4 Mrad dose reduced the number of virus particles by 10^7; the same degree of reduction in the liquid state was achieved by 3 Mrad. Radappertization at cryogenic temperatures in sealed containers in the absence of oxygen offers a means of eliminating this virus in many infected animal products.

In 1969 the Health Authorities of the United Kingdom and The Netherlands approved radappertized foods for hospital patients in reversed barrier isolation. No nutritional or toxicological problems have been reported as a consequence, even though patients subsist entirely on these foods for several months. These patients have either received organ transplants or are being treated for leukemia. Their immune responses to materials foreign to their bodies have been

TABLE 4
Acceptance of Radappertized Shrimp, Pork Sausage, and Beef[a]

Expt. No.	Product	Recipes	No. Raters	Average Acceptance Rating[b]
71/156	Shrimp[c]	Cocktail	21	7.5
71/156	Shrimp[c]	Cocktail	17	7.6
73/19	Shrimp[c]	Cocktail	31	7.4
73/53	Shrimp[c]	Cocktail	46	6.9
72/116	Pork Sausage[d]	Fried	28	7.5
72/116	Pork Sausage[d]	Fried	23	7.4
72/116	Pork Sausage[d]	Fried	17	7.2
73/65	Pork Sausage[d]	Fried	23	7.4
71/135	Beef[e]	Brown Gravy	35	5.8
71/104	Beef[e]	Brown Gravy	17	6.9
72/14	Beef[e]	Brown Gravy	16	6.8
72/109	Beef[e]	Mushroom Gravy	17	6.0

[a]Non-refrigerated storage: 1 to 7 months.
[b]9-point hedonic scale where 9 is "like extremely"; 5 is "neither like nor dislike"; 1 is dislike extremely.
[c]3.8–4.9 Mrad at −30 ± 10°C.
[d]2.7–3.3 Mrad at −30 ± 10°C.
[e]4.7–6.1 Mrad at −30 ± 10°C.

(Source: Mrs. Lucy J. Rice, U.S. Army Natick Laboratories.)

suppressed to minimize rejection. Because the suppression of the immune response makes the patients hypersusceptible to bacterial infections, they are kept in a sterile environment and are fed sterile diets. Although heat can sterilize diet, it is not a suitable method for all foods and limits the variety of foods the patients can eat. Radappertization permits a much wider selection of foods and helps stimulate the patient's appetite with improvement in morale and nutritional condition.

Radurization and Radicidation

It is in the range of 0.1 and 1.0 Mrad doses that radurization and radicidation of most foods are achieved. The radiation process has been used with success to reduce in food the populations of yeasts, molds and non-spore forming bacteria which cause spoilage or disease. The foods most extensively studied in this dose range are meat, poultry, fin fish, shell fish, eggs, fruits and berries.

The process involves exposure of the raw product to ionizing radiation, preferably as soon as possible after slaughter or harvest since spoilage once begun can only be arrested but not reversed. Extension of shelf life in chilled products for periods from a few days to several weeks is achieved. One application, being considered by fishing interests in the Federal Republic of Germany, is to install radiation facilities on board fishing vessels to process the catch daily. This will permit the boats to remain at sea for longer periods and still land a catch of excellent quality. Technology and proof of safety to the consumer have been sufficiently well established for the Health Officials of Canada in 1973 to give the "green light" to permit test marketing of cod and haddock radurized

(150 Krad maximum) for shelf life extension (*14*).

Another important application is the use of ionizing radiation to eliminate salmonellae in poultry, eggs, red meats, and animal feeds. The following quotation from the newspaper, *The Boston Globe,* dated November 8, 1971, indicates the extent of the salmonella problem in the United States. "The American Public Health Association claims that fresh meat bearing the seal 'Inspected for Wholesomeness by US' is no assurance that it is wholesome. The Association is threatening to sue the US Agriculture Department if it doesn't start requiring labels warning consumers that inspected meat can cause disease. At the crux of the dispute are microbiological contaminants, primarily salmonella, which are transmitted through meat and which cause poisoning. The National Academy of Sciences estimates that salmonella alone sickens 2 million Americans a year. Most recover quickly, but infants and elderly victims sometimes die."

Irradiation with a dose of 0.5 to 1.0 Mrad eliminates salmonellae in animal feed which is frequently infected during bagging, storage and distribution.

In 1965 the US Army Natick Laboratories, the US Atomic Energy Commission, and Atomic Energy of Canada, Ltd., began a joint research program on radurization and radicidation of chilled and frozen chicken. Technology and proof of safety to the consumer having been well established, in 1973 the Food and Drug Directorate of the Canadian Department of National Health and Welfare gave permission for test marketing of chicken radicidized (700 Krad maximum) to control salmonellae (*14*).

During recent years scientists at the University of Nebraska and Michigan State University have done research on radurization and radicidation of fresh red meat. Because chilled fresh red meat in retail cuts is very perishable, final cutting and packaging is done in the retail store which is quite inefficient. Urbain (*15*) has estimated that centralized cutting of red meat could save as much as eleven cents per kilogram. With radiation (50–150 Krad) and use of proper packaging and addition of phosphates, the retail meat cuts could be centrally processed and kept for at least 3 weeks at 4.4° C, a sufficient time for distribution to the retail outlets and final sale. The irradiation dose range used would also destroy any parasites infesting the meat. The potential annual amount of meat in the USA which could be radurized is 7.5 million metric tons.

Scientists in the United Kingdom showed that frozen blocks of beef, horsemeat, and animal feed can be radicidized without the need for thawing as required by other disinfecting methods. This minimizes further spread of the infected product without measurable loss in food quality.

Radurization doses have been demonstrated to be very effective in delaying mold spoilage in strawberries, oranges, grapes, and other fruits and berries.

DISINFESTATION OF INSECTS AND PARASITES; RETARDATION OF MATURATION

Doses below 100 Krads may be used to insect-disinfest grain, fruits, dried fish; to eliminate parasites, such as *Trichinella spiralis* from pork and other meat items; to inhibit sprouting of potato tubers and onion bulbs; to delay ripening of fruits such as mangoes and bananas; to delay mold spoilage in partially-baked bakery products and in bread used in the Apollo flights to the moon. Another important application, related to food preservation but not to food processing, is the sterile male technique to control or eliminate several insect species which destroy food — e.g. the Mediterranean fruit fly and the screw worm.

The egg phase of the life cycle of insects is usually the most sensitive to irradiation, followed by the larval, pupal, and adult stage in that order (*16–18*). *Oryazephilus surnamensis L.* may serve as an example. No adult will develop from eggs receiving 6 Krad, larvae receiving 8 Krad, or pupae receiving 15 Krad. All adults receiving 20 Krads will have died after 2 weeks. At slightly lower doses, the eggs, the larvae and the pupae may develop to the adult stage, but these adults will be sterile.

According to Tilton and Brower (*17*) most insects are sterilized at doses between 5 and 20 Krad. Some adult moth species are, however, very resistant and some will survive 100 Krad, but with low fertility and fecundity index and the few progeny they produce, even at 50 Krad, are sterile.

The amount of grain stored in silos in several harbors in the United States ranges from 10^5 to 2×10^6 metric tons. Similar grain storage facilities are located or are being built in other countries. Cost of disinfestation by irradiation at these large storage facilities would be equivalent to or less than by chemicals. The chemicals used for disinfestation are usually very toxic to humans. Phosphine, a chemical commonly used for disinfestation, cannot be used at concentrations above 0.05 ppm because of its toxicity to humans. It kills only the adult stage and must, therefore, be used repeatedly. It reacts with coenzymes, metals and cysteine (*19*). In the period 1963–68 the annual use of insecticides in the United States was approximately 100,000 metric tons or 500 grams per capita. While these pesticides have been great blessings, irradiation, which does not leave any residues and causes less than one chemical bond change per 10^7 chemical bonds of product, should be considered where practical.

Insects infesting certain fruits such as the papaya have caused quarantine barriers to be raised against their entering the 48 contiguous states of the United States. Application of ionizing radiation to disinfest papayas grown in Hawaii has been sponsored by the US Atomic Energy Commission and the US Department of Agriculture in order to seek permission for export of this fruit to the United States mainland.

Sprout inhibition of potatoes by use of a 5–15 Krad dose gives a similar overall effect on storage stability as chemical treatments using CIPC (chloro-N-isopropyl-phenyl carbamate) and maleic hydrazide. Lately the interest in using ionizing radiation has been renewed because of the discovery of a beneficial synergistic effect of irradiation and a spray of dilute solutions of salicylic acid (*20*). Irradiation decreases very significantly the wound-healing mechanisms of potatoes and makes them more susceptable to rot. The irradiation treatment should, therefore, be delayed until after thorough post-harvest suberization or wound healing. The potatoes should be treated gently after irradiation. A small amount of salicylic acid significantly reduces the rotting (*20*) which in irradiated potatoes was the limiting factor in their storage stability.

Soon after irradiation of potatoes, the sucrose content increases and may reach the level of 1–1.5% 6–12 days after storage. Subsequently the sucrose content decreases, and in 2–3 months it will have returned to the level of fresh or untreated potatoes (*21*). After 3–9 months' storage the sucrose content of irradiated potatoes is lower than in the untreated potatoes. The high sucrose content just after irradiation causes reduced color quality of potato chips made from irradiated potatoes. In the practical application this is, however, unimportant, because only those potatoes that will be stored for a long period will be irradiated.

Storage stability of onions can also be extended very considerably by low doses (5–15 Krad). After 10 months of storage 88% of the bulbs receiving a 9 Krad dose were rated as first-grade bulbs. When they were pre-immersed in 1500 ppm salicylic acid solution, close to 100% received first-grade rating after 10–12 months' storage (*20*).

Trichinae (*Trichinella spiralis*) and tapeworms (*Cystecercus bovis* and *Echinococcus granulosus*) can be eliminated or controlled by low doses of irradiation. Alicata and Burr (*22*) showed that the trichinellae can be controlled by low doses in the order of 15 Krad. Their studies were confirmed and extended in scope by Gomberg et al (*23*). In spite of the thorough and extensive meat inspection and meat handling control in the United States and other highly developed countries, parasites in meats are still a problem (*24,25*).

Delay by 2–3 days in the ripening of several varieties of mangoes may be obtained with doses of 25–35 Krads (*26*) without affecting the quality of the fruit. Ripening of bananas may similarly be delayed 7–8 days by a dose of 35 Krad.

Nutrition Aspects

The effect of ionizing radiation on the nutritional value of foods is not markedly different in degree from that of other methods of preservation. Protection of nutrients is improved by holding the food at low temperature during irradiation and reducing or excluding free oxygen from the radiation milieu. Data in Table 5 indicate that retention of thiamine, riboflavin, niacin, and

TABLE 5
Effect of Processing on the Vitamin Content of Shelf-Stable Canned Ham

Vitamin	Treatment	mg/100 g[a]	% retention
Thiamine	Control	3.82 ± 0.38[b]	—
	4.5 Mrad at −80°± 5°C	3.25 ± 0.79	85
	Thermally processed	1.27 ± 0.36	32
Riboflavin	Control	1.01 ± 0.18[b]	—
	4.5 Mrad at −80°± 5°C	1.25 ± 0.09	123
	Thermally processed	1.10 ± 0.24	109
Niacin	Control	31.5 ± 0.81[b]	—
	4.5 Mrad at −80°± 5°C	23.8 ± 2.92	76
	Thermally processed	14.6 ± 4.49	46
Pyridoxine	Control	1.11 ± 0.15[b]	—
	4.5 Mrad at −80°± 5°C	1.02 ± 0.12	92
	Thermally processed	0.64 ± 0.03	57

[a]Moisture, fat, salt-free basis.
[b]Average ± S.D. Three samples per treatment.

(Data furnished by Mrs. Miriam H. Thomas, Nutrition Div., Natick Labs.)

pyridoxine in ham radappertized at cryogenic temperature is as good as, or better than, the retention of these vitamins in the same ham thermally sterilized under time and temperature conditions used commercially.

The great body of evidence accumulated during the past two decades, with few exceptions, indicates that irradiated foods are as nutritious as thermally processed foods (*27*). Raica and Howie (*28*) reported no changes in digestibility of proteins, fats, and carbohydrates exposed to doses of 5.6 Mrad. Read et al. (*29*) found no differences in metabolizable energy and available protein, fat, and carbohydrate between radappertized (5.6 Mrad) and nonirradiated control composite diets fed to rats for 4 generations. Kennedy (*30*) observed little change in nutritive value of animal feeds (protein concentrate) using doses of 0.5 and 1.0 Mrad and no nutritional changes with frozen eggs irradiated at 0.5 and 5.0 Mrad. Ley (*31*) found no adverse effect on the nutritive value of animal feeds irradiated at 0.1 and 0.5 Mrad. He concluded that radiation at 2.0 Mrad is superior to heat processing with respect to retention of protein quality. He also described excellent results with radappertized feed for germ-free rat and mouse colonies maintained for 5 years. Radappertization and radicidation have been used in preference to thermal sterilization to sustain germ-free and specific pathogenfree rats, mice, pigs, and chickens (*32–36*).

There are very few reports dealing with the effects of radiation on the nutritive aspects of amino acids and proteins in fruit. Loaharanu (*37*) stated that no significant differences were detected in terms of nutritive values between mangoes, papayas, rumbatans, and longans gamma irradiated for insect disinfestation and their nonirradiated controls. Vakil et al. (*38*), using rats, found no significant changes in protein, fat, and mineral content of wheat, gamma irradiated at 20 and 200 Krad for insect disinfestation. Approximately 80 to 90% retention of thiamine, riboflavin, and niacin was obtained in irradiated wheat at either 20 or 200 Krad (*38*). Jaarma and Henricson (*39*) reported that the nutritional adequacy of irradiated (14–15 Krad for pigs and 200 Krad for rats) and control potatoes was equal. Similar conclusions were drawn by Varela and Urbano (*40*), Lang and Bassler (*41*) and Fujimaki et al. (*42*). No data pertaining to the nutritional quality of proteins irradiated (5-15 Krad) for sprout inhibition during storage were noted. Gamma irradiation of bleached, enriched, hard wheat

TABLE 6

Effect of Radiopasteurization on Ascorbic Acid Retention (%) in Fruit

Product	Dose Krad	Retention Percent
Oranges, Temple	100	97
	200	72
Tangerines	40	104
	80	94
	160	94
Tomatoes	100	86
	200	86
	300	91
Papayas	125	110

Source: Calculated from Dennison and Ahmed (*44*) and Wenkam and Moy (*45*) by Miriam H. Thomas, Nutrition Div., Natick Laboratories.

flour in the range 30 to 50 Krad had no detrimental effect on the thiamine, riboflavin, niacin, or pyridoxine content (*43*).

Calculations from data on oranges, tangerines, tomatoes and papayas reported by Dennison and Ahmed (*44*) and Wenham and Moy (*45*) are shown on Table 6.

A comprehensive review of the effects of ionizing radiation on nutrients has been written by E. S. Josephson, M. H. Thomas, and W. K. Calhoun as a chapter of the book entitled, *"Nutritional Evaluation of Food Processing"* to be published by the Avi Publishing Company and will be summarized in Madrid in September 1974 (*46*).

REDUCED DEPENDENCE ON CHEMICAL FOOD ADDITIVES

Recently, the Army Natick Laboratories researchers investigated the possibility of reduction in the additions of nitrite and nitrate in cured meats, such as ham and bacon. Nitrite and nitrate benefit the organoleptic qualities such as characteristic flavor and pink color of cured meats. Nitrite, in combination with other curing agents, also inhibits toxin production by *Cl. botulinum* in thermally processed meats. The use of these curing agents, however, has been under reappraisal by the meat industry and the health regulatory agencies because under certain conditions nitrite may react with free amines in food forming nitrosamines, which are carcinogenic (*47–50*). In addition, the residual nitrite left in cured meats after processing may react in the gastrointestinal tract with free amines, forming carcinogenic N-nitroso compounds. Model experiments with laboratory animals have shown that high concentrations of nitrite and certain amino compounds induced tumors characteristic to the corresponding N-nitroso compounds (*49,51*). Because of the formation of nitroso compounds from nitrite and amines in the stomach, there is a definite need to reduce the intake of nitrite as much as possible and in particular reduce the amount added to our foods. This was strongly recommended by the toxicology experts of the International Symposium of Nitrite in Meat Products that took place in Zeist, The Netherlands, September 11–14, 1973.

The experiments conducted by the Natick Laboratories researchers show that the additions of nitrite to cured, smoked ham and bacon can be reduced from 156 mg/kg, the amount commonly used by the meat industry, to 25 mg/kg in radappertized meats without affecting the characteristic color, odor, flavor and overall acceptance of the products, and with the guarantee that no *Cl. botulinum* toxin will be formed (*52,53*). Table 7 shows the preference data for the low and high nitrite radappertized ham. The data indicate the high quality products containing only 25 mg/kg sodium nitrite and 100 mg/kg sodium nitrate added to the products during curing, instead of the commonly used 150-156 mg/kg nitrite and 500-600 mg/kg nitrate. The important factor in achieving this notable reduction by 83% is the fact that radappertization destroys *Cl. botulinum,* thus eliminating the need for that amount of nitrite required for controlling *Cl. botulinum* in nonirradiated cured meats.

Spices generally have high bacterial counts. When other foods are seasoned with spices, the spices are the foci for rapid bacterial growth. Although ethylene oxide has been used, it leaves an undesirable residue (*54*). Heat is unsatisfactory because it drives off or destroys the desirable volatiles which are characteristic of

TABLE 7
Acceptance of High and Low Nitrite Radappertized Ham[a]
(Consumer-Type Panel: n = 32)

Sample Series No.	ppm added		Irradn. Source	Acceptance Ratings	
	$NaNO_2$	$NaNO_3$		10[c] (Days)	30[c]
1	150[b]	600	Cobalt-60	6.8 ± 1.5	6.2 ± 1.7
2	25[b]	100	Cobalt-60	6.4 ± 1.5	6.0 ± 2.2
3	150	600	Electrons	6.5 ± 2.0	6.8 ± 1.4
4	25	100	Electrons	7.0 ± 1.2	6.8 ± 1.4
LSD (Least Significant Difference):				NS	0.7

[a]3.7–4.4 Mrad at −30° ± 10°C
[b]Paired sets of samples, Cobalt-60 vs. Electrons for both the low and high Nitrite - Nitrate hams.
[c]Storage time without refrigeration after radappertization.

(Source: Mrs. Lucy J. Rice, U. S. Army Natick Laboratories)

spices. Work by Polish (*55*) and Hungarian (*56*) investigators has shown that radappertization is highly effective and can be substituted for ethylene oxide.

The onset of mold spoilage on orange peels is delayed by biphenyl. This chemical has been suspected of being a carcinogen. Radurization (75–300 Krad) of citrus fruit has been demonstrated to achieve an equivalent delay in onset of mold spoilage and can be used as a substitute for this chemical.

SAFETY FOR CONSUMPTION (WHOLESOMENESS)

Exposure of food and animal feed to ionizing radiation is the first major food processing method to come on the scene since food regulatory agencies were established at the national level in many countries. Some countries including the United States have created a legal fiction by defining ionizing radiation as a food additive in order to control the food radiation process. In common with all new food additives, ionizing radiation is banned unless a petition is submitted to the appropriate national regulatory agency (the US Food and Drug Administration in the United States) requesting approval for an exception to the ban on a case-by-case basis. The petition must contain experimental evidence to indicate safety to the consumer (wholesomeness) of the specific irradiation-processed food. The major wholesomeness aspects which must be addressed in the evidence in support of the petition are nutritional quality, inability to measure induced radioactivity, freedom from pathogens and their toxins, and absence of toxic, mutation-inducing or cancer-forming compounds.

The challenge has been successfully met for the first three of the four aspects involved in proof of wholesomeness through control of the processing procedures, including the food itself, and limiting the use of radiation sources to the gamma rays of ^{60}Co and ^{137}Cs, X-rays (below 5 MeV) and electrons (below 10 MeV). Additional evidence, if required in support of any particular petition, can readily be obtained experimentally, quickly and at low cost.

It is the 4th aspect, demonstration of the absence of toxic, carcinogenic, and mutagenic compounds that is difficult, time-consuming, and costly. In the

United States the Army Medical Department spent over 6 million dollars during a 10-year period (1955–1965) studying 21 irradiated foods, representing all classes of foods in the diet of North Americans, and concluded that these foods are wholesome. The Food and Drug Administration (FDA), however, ruled that experiments in even greater depth are needed.

In 1970 agreement was reached by the Army Medical Department and the FDA on a new research protocol to demonstrate the absence of unwholesome radiolysis compounds in radappertized beef. This study involves chemical analyses and feeding of radappertized beef and appropriate nonirradiated controls to dogs (3 years), rats (4 generations) and mice (4 generations). The study, which began in 1971 and will end in 1976, involves 1500 dogs, 27,000 rats, and 20,000 mice at a cost of approximately $5 million. At this stage of the study, the animals fed radappertized beef have been doing as well as or better than the animal groups fed frozen or heat-sterilized beef.

Many nations do not have the resources, facilities, and trained personnel to conduct studies of the dimension undertaken by the Army Medical Department. The European Nuclear Energy Agency (ENEA) of the OECD, in collaboration with the International Atomic Energy Agency (IAEA) and the Food and Agriculture Organization (FAO), brought into being on 1 January, 1971, a collaboraive effort, supported by funds from 19 countries (now increased to 21 countries), known as the "International Wholesomeness Project in the Field of Food Irradiation." The research sponsored by the International Project to prove safety for consumption is done by contract and coordinated by a project leader located at the Federal Research Institute for Food Preservation at Karlsruhe. When wholesomeness for a particular food is proved, it is hoped that a recommendation from the World Health Organization (WHO) will be accepted by the health authorities of the member nations. The first foods selected for study in the International Program are white potatoes, wheat and wheat flour, a mixture of three species of fin fish, and rice.

Notwithstanding all the difficulties in obtaining proof of wholesomeness of irradiated foods, fourteen countries have approved at least one of seventeen irradiated foods and the FDA has approved a number of packaging materials as food contactants for radiation processing. These approvals are shown in Tables 8 and 9. The major research effort in food irradiation in the United States and abroad is now being directed to obtain the scientific evidence that irradiated foods are safe to eat.

The law pertaining to Food Additives in the United States, as implemented ever since it was enacted in 1958 and in several other countries, requires a petition for approval for each and every food to be irradiation-processed. This is an extremely costly procedure and has delayed receipt of approvals for a broad enough spectrum of irradiated foods to make the process commercially attractive.

There was reason for going the "food-by-food" petition route for food additives early in the history of food irradiation. Wholesomeness data were scarce and the public was concerned about the possibility of adding radioactivity to foods exposed to ionizing radiation.

The situation now has changed. We know how to irradiate foods without adding any measurable radioactivity to the food. The data accumulated during

TABLE 8

GENERAL SURVEY OF IRRADIATED FOOD PRODUCTS CLEARED FOR HUMAN CONSUMPTION IN DIFFERENT COUNTRIES

(Grouped according to product) December 1973

Compiled by Dr. K. Vas, International Atomic Energy Agency, Vienna

PRODUCT	COUNTRY	PURPOSE OF IRRADIATION	SOURCE OF RADIATION	DOSE (Krad)	DATE OF APPROVAL
Fruits & Vegetables					
POTATOES	USSR	sprout inhibition	^{60}Co	10	14 March 1958
	Canada	sprout inhibition	^{60}Co	10 max.	9 November 1960
				15 max.	14 June 1963
	USA (white potatoes)	sprout inhibition	^{60}Co	5 - 10	30 June 1964
			^{137}Cs	5 - 10	2 October 1964
			^{60}Co + ^{137}Cs	5 - 10	1 November 1965
	Israel	sprout inhibition	^{60}Co	15 max.	5 July 1967
	Japan	sprout inhibition	^{60}Co	15 max.	30 August 1972
	WHO **)	sprout inhibition	^{60}Co or ^{137}Cs	15 max.	12 April 1969
	Spain	sprout inhibition	^{60}Co	5 - 15	4 November 1969
	Hungary *)	sprout inhibition	^{60}Co	10	23 December 1969
			^{60}Co	15 max.	10 January 1972
	Denmark	sprout inhibition	10 MeV electrons	15 max.	27 January 1970
	Netherlands	sprout inhibition	^{60}Co	15 max.	23 March 1970
			4 MeV electrons	15 max.	23 March 1970

*) experimental batches
**) temporary acceptance for 5 years

PRODUCT	COUNTRY	PURPOSE OF IRRADIATION	SOURCE OF RADIATION	DOSE (Krad)	DATE OF APPROVAL
<u>Fruits & Vegetables</u> (Cont'd)					
POTATOES (Cont'd)	Bulgaria *)	sprout inhibition	^{60}Co		1971
	Uruguay *)	sprout inhibition	^{60}Co		1971
	Philippines **)	sprout inhibition	^{60}Co	15 max	13 September 1972
	France **)	sprout inhibition	^{60}Co	7,5 - 15	8 November 1972
	Italy	sprout inhibition	^{60}Co or ^{137}Cs	7,5 - 15	30 August 1973
ONIONS	Canada	sprout inhibition	^{60}Co	15 max.	25 March 1965
	USSR *)	sprout inhibition	^{60}Co	6	25 February 1967
	Israel	sprout inhibition	^{60}Co	10 max.	25 July 1968
	Netherlands *)	sprout inhibition	^{60}Co	15 max.	5 February 1971
			4 MeV electrons	15 max.	5 February 1971
	Thailand	sprout inhibition	^{60}Co	10 max.	20 March 1973
	Italy	sprout inhibition	^{60}Co or ^{137}Cs	7,5 - 15	30 August 1973
GARLIC	Italy	sprout inhibition	^{60}Co or ^{137}Cs	7,5 - 15	30 August 1973
DRIED FRUITS	USSR	insect disinfestation	^{60}Co	100	15 February 1966
FRESH FRUITS & VEGETABLES	USSR *)	radurization	^{60}Co	200 - 400	11 July 1964

*) experimental batches
**) temporary acceptance for 5 years

PRODUCT	COUNTRY	PURPOSE OF IRRADIATION	SOURCE OF RADIATION	DOSE (Krad)	DATE OF APPROVAL
<u>Fruits & Vegetables</u> (Cont'd)					
MUSHROOMS	Netherlands	growth inhibition	^{60}Co	250 max.	23 October 1969
			4 MeV electrons	250 max.	23 October 1969
ASPARAGUS	Netherlands *)	radurization	^{60}Co	200 max.	7 May 1969
STRAWBERRIES	Netherlands *)	radurization	^{60}Co	250 max.	7 May 1969
			4 MeV electrons	250 max.	7 May 1969
COCOABEANS	Netherlands *)	insect disinfestation	^{60}Co	70 max.	7 May 1969
			4 MeV electrons	70 max.	7 May 1969
SPICES & CONDIMENTS	Netherlands *)	radicidation	^{60}Co	800 - 1000	13 September 1971
			4 MeV electrons	800 - 1000	13 September 1971
<u>Grain & Grain Products</u>					
GRAIN	USSR	insect disinfestation	^{60}Co	30	1959
WHEAT & WHEAT FLOUR (changed on 4 March 1966 from WHEAT PRODUCTS)	USA	insect disinfestation	^{60}Co	20 - 50	21 August 1963
			^{137}Cs	20 - 50	2 October 1964
			5 MeV electrons	20 - 50	26 February 1966
WHEAT, FLOUR, WHOLE WHEAT FLOUR	Canada	insect disinfestation	^{60}Co	75 max.	25 February 1969
WHEAT & GROUND WHEAT PRODUCTS	WHO **)	insect disinfestation		75 max.	12 April 1969

*) experimental batches
**) temporary acceptance for 5 years

PRODUCT	COUNTRY	PURPOSE OF IRRADIATION	SOURCE OF RADIATION	DOSE (Krad)	DATE OF APPROVAL
Meat & Fish					
SEMI-PREPARED RAW BEEF, PORK & RABBIT PRODUCTS (in plastic bags)	USSR *)	radurization	^{60}Co	600 - 800	11 July 1964
POULTRY, EVISCERATED (in plastic bags)	USSR *)	radurization	^{60}Co	600	4 July 1966
	Netherlands *)	radurization	^{60}Co	300 max.	31 December 1971
CULINARY PREPARED MEAT PRODUCTS (FRIED MEAT, ENTRECOTE) (in plastic bags)	USSR *)	radurization	^{60}Co	800	1 February 1967
SHRIMPS	Netherlands *)	radurization	^{60}Co	50 - 100	13 November 1970
			4 MeV electrons	50 - 100	13 November 1970
Other Products					
DRY FOOD CONCENTRATES	USSR	insect disinfestation	^{60}Co	70	6 June 1966
ANY FOOD for consumption by patients who require a sterile diet as an essential factor in their treatment	UK	radappertization			1 December 1969
DEEP-FROZEN MEALS	Netherlands ***)	radappertization	^{60}Co	2500 min.	27 November 1969
FRESH, TINNED & LIQUID FOODSTUFFS	Netherlands ***)	radappertization	^{60}Co	2500 min.	8 March 1972

*) experimental batches
***) for hospital patients in reversed barrier isolation

TABLE 9

APPROVALS BY THE US FOOD & DRUG ADMINISTRATION FOR RADIATION PRESERVATION OF FOOD PACKAGING MATERIALS *
July 1971

Packaging Materials

PACKAGING MATERIAL	PETITIONER	SOURCE	DOSE	FOOD ADDITIVE PETITION NO.	FILING DATE	FILING FED. REG. VOL.	FILING FED. REG. PAGE	REGULATION DATE	REGULATION FED. REG. VOL.	REGULATION FED. REG. PAGE
Nitrocellulose Coated Cellophane	AEC	gamma	1 megarad	1297	2-8-64	29	2318	8-14-64	29	11651
Glassine Paper	AEC	gamma	1 megarad	1297	2-8-64	29	2318	8-14-64	29	11651
Wax Coated Paperboard	AEC	gamma	1 megarad	1297	2-8-64	29	2318	8-14-64	29	11651
Polypropylene Film with or without Adjuvants	AEC	gamma	1 megarad	1297	2-8-64	29	2318	8-14-64	29	11651
Ethylene-Alkene-1 co-polymer	AEC	gamma	1 megarad	1297	2-8-64	29	2318	8-14-64	29	11651
Polyethylene Film	AEC	gamma	1 megarad	1297	2-8-64	29	2318	8-14-64	29	11651
Polystyrene Film with or without Adjuvant Substances	AEC	gamma	1 megarad	1297	2-8-64	29	2318	8-14-64	29	11651
Rubber Hydrochloride with or without Adjuvant Substances	AEC	gamma	1 megarad	1297	2-8-64	29	2318	8-14-64	29	11651
Vinylidene Chloride - Vinyl-Chloride Copolymer Film (Saran Wrap)	AEC	gamma	1 megarad	1297	2-8-64	29	2318	8-14-64	29	11651
Polyolefin Film with or without Adjuvant Substances or Vinylidene Chloride coatings (Saran)	AEC	gamma	1 megarad	5M1674	7-30-65	30	9551	3-19-68	33	4659
Polyethylene Terephthalate with or without Adjuvant Substances or Vinylidene Chloride coatings (Saran) or Polyethylene coatings	AEC	gamma	1 megarad	5M1674	7-30-65	30	9551	3-19-68	33	4659
Nylon 11	AEC	gamma	1 megarad	6M1820	9-8-65	30	11400	3-19-68	33	4659
Vinylidene Chloride Copolymer (Saran) Coated Cellophane	AEC	gamma	1 megarad	5B1670	2-18-65	30	9116	6-11-65	30	7599
Vegetable Parchment	US Army	gamma	6 megarad	5M1622	1-15-65	30	547	3-12-65	30	3354
Polyethylene Film with or without Adjuvants	US Army	gamma	6 megarad	5M1645	7-21-65	30	9116	6-10-67	32	8360
Polyethylene Terephthalate with or without Adjuvants	US Army	gamma	6 megarad	5M1645	7-21-65	30	9116	6-10-67	32	8360
Nylon 6 Films with or without Adjuvants	US Army	gamma	6 megarad	5M1645	7-21-65	30	9116	6-10-67	32	8360
Vinyl Chloride - Vinyl Acetate Copolymer Film with or without Adjuvant	US Army	gamma	6 megarad	5M1645	7-21-65	30	9116	6-10-67	32	8360
Kraft Paper	US Army	gamma	50,000 rads	7M2172	5-23-67	32	7877	7-19-67	32	10567

* Compiled by Mr. Frank Leone, US Atomic Energy Commission.

the past 20 years from long and careful experiments have not demonstrated the existence of a danger to the consumer (*57*).

What is now needed is a systematic compilation of all the scientific data pertaining to the toxicological aspects of irradiated food, with microbiological aspects considered separately. The purpose of this compilation is to determine if food irradiation can be considered by the health authorities as a process for which clearance can be given overall. This view, which has been expressed by Ley (*2*), would obviate the need for the expensive, long-term toxicological testing for each irradiated food.

In 1972 a panel of experts met in Bombay and "recommended that, in giving clearances, regulatory agencies consider food irradiation as a process, and not as a food additive, and that FAO and the IAEA convene a panel, with WHO's participation, to determine what needs to be done to implement the recommendation" (*57*). I hope this recommendation will be implemented promptly.

Food irradiation, when approved as a process, would bring a number of immediate benefits to the people of the world. It would reduce significantly the incidence of food-borne salmonella. Even in the United States 2 million cases of salmonella infections are estimated to occur annually. Bloom (*58*) cited L. W. Brownell's belief that if *all* pork were irradiated with ^{60}Co, trichinosis would be wiped out at a cost of but a penny a pound. Beef tapeworms, which are a problem in many countries where beef is undercooked or eaten raw, would be destroyed. Liver fluke and fish tapeworm infections, which are common in Japan and some Scandinavian countries where raw fish is eaten, would disappear. Since radappertization destroys *Cl. botulinum* spores, it can reduce by 83% the amounts of the suspected precarcinogens, nitrate and nitrite, needed to prevent formation of botulinum toxin in canned cured meats. Irradiation-processing of food can replace or reduce the need for some of the chemical fumigants, sprout inhibitors, and bacterial and mold inhibitors. Several of these have been suspected as carcinogens, mutagens, or toxicants. Food irradiation can help upgrade the diets of populations from substandard or marginal to good, quantitatively by making available more food by reducing spoilage losses, and qualitatively by making it possible to provide nutritionally better food such as animal protein food to those who need it most.

SUMMARY AND CONCLUSIONS

1. Radiation-processing provides a new dimension to food preservation because these foods retain almost completely the desirable characteristics of fresh food in appearance and taste.

2. The irradiation process can destroy the pathogens in food to bring under control many of the diseases which afflict the human race and the animals and plants which comprise its food.

3. The nutritional quality of foods preserved by ionizing radiation is as good as or better than foods preserved by the older, well-established and approved food preservation processes.

4. Radiation-processing of foods can reduce the dependence upon chemical food additives currently needed for preservation or control of disease. Continued use of some of these chemicals is under review by health authorities because they

may be dangerous to the health of the consumer.

5. Processing food with ionizing radiation can upgrade the nutritional status of populations by providing more food and nutritionally better food (animal protein) which otherwise would not be available.

6. Laws in some countries define irradiation as an additive when applied to food processing. The ways in which the laws have been interpreted and implemented require approval by the health authorities of irradiated foods on an individual "food-by-food" basis. Obtaining the scientific evidence for proof of wholesomeness of irradiated foods which is convincing to the health authorities at national levels on a "food-by-food" basis remains the number 1 problem. This has delayed commercial application.

7. The experimental data on wholesomeness of irradiated foods obtained during the past 20 years has not demonstrated danger to the consumer.

8. It is recommended that all existing wholesomeness data be compiled and reviewed by experts under the aegis of FAO, IAEA, and WHO to determine if food irradiation can be considered as a process for which clearance by the health authorities can be given overall.

REFERENCES

1. A. Sreenivasan, in *Proceedings SOS/70 Third International Congress Food Science and Technology*, Washington, D.C., August 1970 (G. L. Stewart and C. L. Wiley, Eds.), pp. 595–603. Institute of Food Technologists, Chicago, 1971.
2. F. J. Ley, in *Proceedings SOS/70 Third International Congress Food Science and Technology*, Washington, D.C., August 1970 (G. L. Stewart and C. L. Willey, Eds.), pp. 806–815. Institute of Food Technologists, Chicago, 1971.
3. E. Wierbicki, M. Simon and E. S. Josephson, in *Radiation Preservation of Foods* (Publication 1273, National Academy of Sciences–National Research Council), pp. 383–409. Washington, D.C., 1963.
4. E. S. Josephson, A. Brynjolfsson, E. Wierbicki, D. B. Rowley, C. Merritt, Jr., R. W. Baker, J. J. Killoran and M. H. Thomas, in *Radiation Preservation of Food*, pp. 471–490. International Atomic Energy Agency, Vienna, Austria, 1973, (STI/PUB/317).
5. E. S. Josephson, A. Brynjolfsson and E. Wierbicki, *Trans. N. Y. Acad. Sci.* **30**, 600 (1968).
6. A. Anellis, D. Berkowitz, W. Swantak and C. Strojan, *Appl. Microbiol.* **24**, 453 (1972).
7. G. W. Shults, D. R. Russell and E. Wierbicki, *J. Food Sci.* **37**, 860 (1972).
8. J. J. Killoran, in *Radiat. Res. Reviews*, pp. 369–388. Elsevier Pub. Co., Amsterdam, The Netherlands, 1972.
9. D. R. Peryam and F. J. Pilgrim, *Food Technol.* **11**, (Supplement) 9 (1957).
10. Anonymous, in *AEC Authorizing Legislation, Fiscal Year 1974*. Hearings, Joint Committee on Atomic Energy, Congress of the United States, 93rd Congress, part 4, March 20 and 22, 1973, page 2275. US Government Printing Office, Washington, D.C., 1973.
11. F. Heiligman and L. Rice, *J. Food Sci.* **37**, 420 (1972).
12. B. O. Eggum, *Z. Tierphysiol., Tierernaehr. u. Futtermittelkunde* **25**, 204 (1969).
13. D. Massa, in *Food Irradiation* (IAEA, STI/PUB/127), pp. 329–347. International Atomic Energy Agency, Vienna, Austria, 1966.
14. K. F. MacQueen, Atomic Energy of Canada, Ltd., Ottawa. Personal Communication, 1973.
15. W. M. Urbain, in *AEC Authorizing Legislation, Fiscal Year 1974*. Hearings, Joint Committee on Atomic Energy, Congress of the United States, 93rd Congress, part 4, March 20 and 22, 1973, page 2467. US Government Printing Office, Washington, D.C., 1973.
16. P. B. Cornwel (Ed.), *The Entomology of Radiation Disinfestation of Grain*, p. 236. Pergamon Press, New York, 1966.
17. E. W. Tilton, J. H. Brower, in *Radiation Preservation of Food*, Proceedings of a Symposium in

Bombay 13–17 nov. 1972, pp. 295–309. International Atomic Energy Agency, Vienna 1973. (STI/PUB/317.)

18. M. S. Hoedaya, D. Hutabarat, S. I. Sastradihardja and S. Soetrisno, in *Radiation Preservation of Food,* Proceedings of A Symposium in Bombay 13–17 Nov. 1972, pp. 281–294. Publication International Atomic Energy Agency. Vienna, 1973 (STI/PUB/317.)
19. S. K. Majournder, in *Proceedings SOS/70 Third International Congress Food Science and Technology,* Washington, D.C. (G. L. Stewart and C. L. Willey, Eds.), pp. 518–531. Institute of Food Technologists, Chicago, 1971.
20. H. M. Roushdy, K. Shurty, and A. A. Mahmoud, in *Radiation Preservation of Food,* Proceedings of a Symposium in Bombay, 13–17 Nov., 1972, pp. 105–125, International Atomic Energy Agency, Veinna, 1973. (STI/PUB/317).
21. T. A. Trulsen, in *Radiation Preservation of Foodstuffs.* Proceedings of the Second Scandinavian Meeting on Food Preservation by Ionizing Radiation, Stockholm, 9–11 Sep. 1963. IVA Meddelanda Nr. **138** (1964).
22. J. E. Alicata and G. O. Burr, *Science* **109**, 595 (1949).
23. H. J. Gomberg and S. E. Gould, *Science* **118**, 75 (1953).
24. W. H. Wright, K. B. Kerr, and L. Jacobs, *Public Health Report* **58**, 1293 (1943).
25. W. J. Zimmerman, J. H. Steele, and I. G. Kagan, *Public Health Report* **83,** 957 (1968).
26. P. M. Nair, P. Thomas, K. K. Ussuf, K. K. Surendranathan, S. P. Limaye, A, N. Srirangarajan, and S. R. Padwal Desai, in *Radiation Preservation of Food,* Proceedings of a symposium in Bombay 13–17 Nov. 1972, pp. 347–365. International Atomic Energy Agency, Vienna, 1973. (STI/PUB/317.)
27. N. Raica, Jr., J. Scott, and W. Nielsen, *Radiat. Res. Rev.* **3,** 447 (1972).
28. N. Raica, Jr. and D. L. Howie, in *Food Irradiation,* Proc. Symp. Karlsruhe, 6–10 June 1966, pp. 119–135. International Atomic Energy Agency, Vienna 1966.
29. M. S. Read, H. F. Kraybill, W. S. Worth, S. W. Thompson, Jr., G. J. Issac, and N. F. Witt, *Toxicol. Appl. Pharmacol.* **3,** 153 (1961).
30. T. S. Kennedy, *J. Sci. Food Agr.* **16,** 433 (1965).
31. F. J. Ley, *Food Irradiation Information No. 1,* pp. 8–22 (1972).
32. T. Sato, *Shokuhin Kogyo* **13,** 89 (1970).
33. A. Schoen and H. H. Hiller, *Z. Tierphysiol. Tierernachr. Futtermittelk.* **27,** 338 (1971).
34. H. Udes, H. H. Hiller, and N. C. Juhr, *Z. Versuchstierk.* **13,** 160 (1971).
35. F. J. Ley, J. Bleby, M. E. Coates, and J. S. Patterson, *Lab. Anim.* **3,** 221 (1969).
36. M. E. Coates, R. Fuller, G. F. Harrison, M. Lev, and S. F. Suffolk, *Brit. J. Nutr.* **17,** 141 (1963).
37. P. Loaharanu, in *Disinfestation of Fruit by Irradiation* (Proc. Panel on Use of Irradiation to Solve Problems in the Intern. Fruit Trade, Honolulu, 1970). International Atomic Energy Agency, Vienna, 1971.
38. U. K. Vakil, M. Aravindakshan, H. Srinivas, P. S. Chauhan, and A. Sreenivasan, in *Radiation Preservation of Food* (Proc. Symp. Bombay, 1972), pp. 673–702. International Atomic Energy Agency, Vienna, 1973.
39. M. Jaarma and B. H. Henricson, *Acta Vet. Scand.* **5,** 238 (1964).
40. G. Varela and G. Urbano, *Atti simp. Int. Agrochim.* **8,** 563 (1971).
41. K. Lang and K. H. Bässler, in *Food Irradiation,* Proc. Symp. Karlsruhe, 6–10, June 1966, pp. 167–169. International Atomic Energy Agency, Vienna, 1966.
42. M. Fujimaki, T. Makoto, and T. Matsumoto, *Agr. Biol. Chem.* **32,** 1228 (1968).
43. F. Heiligman, L. J. Rice, L. W. Smith, Jr., M. H. Thomas, N. J. Kelley, and E. Wierbicki, 33rd Annual Meeting, Institute of Food Technologists, 1973.
44. R. A. Dennison and E. M. Ahmed, *Isotopes and Radiation Technol.* **9,** 194 (1971–1972).
45. N. S. Wenkam and A. P. Moy, Animal Report, 1 June 1967–31 May 1968, Hawaii Univ., Col. of Tropical Agri. AEC Report UH-235-P-5-4, pp. 126–135 (1968).
46. E. S. Josephson, M. H. Thomas, and W. K. Calhoun, Fourth International Congress of Food Science and Technology, Madrid, Sept. 22–27, 1974.
47. I. A. Wolff and A. E. Wasserman, *Science* **177,** 15 (1972).
48. J. G. Sebranek and R. G. Cassens, *J. Milk Food Technol.* **36,** 76 (1973).
49. R. Preussmann, in *Proc. Int. Symp. Nitrite in Meat Prod.* (PUDOC), pp. 217–226. Wageningen, The Netherlands, 1974.

50. L. Friedman, in *Proc., Meat Ind. Res. Conf.*, pp. 11–20. Chicago, Illinois, 1973.
51. J. Sander, in *Proc. Int. Symp. Nitrite in Meat Prod.* (PUDOC), pp. 243–250. Wageningen, The Netherlands, 1974.
52. E. Wierbicki and F. Heiligman, in *Proc. Int. Symp. Nitrite Meat Prod.* (PUDOC), pp. 189–212. Wageningen, The Netherlands, 1974.
53. E. Wierbicki, F. Heiligman and A. Wasserman, in *Proc. 20th European Meeting Meat Res. Workers*, Dublin, Ireland, September 15–20, 1974 (in press).
54. W. M. Urbain, in *AEC Authorized Legislation, Fiscal Year 1974*. Hearings, Joint Committee on Atomic Energy, Congress of the United States, 93rd Congress, Part 4, March 20 and 22, 1973, page 2468. US Government Printing Office, Washington, D.C., 1973.
55. S. Bachman and J. Gieszczynska, in *Aspects of the Introduction of Food Irradiation in Developing Countries*, Panel Proc. series, pp. 33–41. International Atomic Energy Agency, Vienna, Austria, 1973. (STI/PUB/362.)
56. J. Farkas, J. Beczner and K. Inche, in *Radiat. Preservation of Food*, pp. 389–402. International Atomic Energy Agency, Vienna, Austria, 1973. (STI/PUB/317.)
57. *Aspects of the Introduction of Food Irradiation in Developing Countries.* Proceedings of a Panel, Bombay, 18–22 November 1972, organized by the Joint FAO/IAEA Division of Atomic Energy in Food and Agriculture. International Atomic Energy Agency, Vienna, 1973.
58. M. T. Bloom, in *Think*, **30** (5), 3–7 September–October 1964. International Business Machine Corporation (IBM), Armonk, New York 10504.

Radiation Sterilization—An Industrial Process[1]

F. J. Ley
Irradiated Products Limited
Elgin Estate, Elgin Drive, Swindon
Wiltshire, England

INTRODUCTION

A new dimension has been added to the use of radiation in the medical field with the introduction of radiation as a sterilizing agent. Its use in diagnosis through radioactive tracers or X-rays and in therapy administered with the most sophisticated of electrical machines or radioisotope units, is familiar in the hospital world, being well established therein. In contrast, the application of radiation sterilization is in industry where the installation of large radiation sources is already commonplace in many countries.

The beginnings in the early 1950's centred on the Van de Graaff machine and linear accelerators and the pioneering efforts of Ethicon Inc. here in the United States must be recognized. However, although sterilization with electron beams is still current practice in a number of plants, the use of gamma rays from cobalt-60 is preferred. The first steps in this direction were taken by the U.K.A.E.A. which, in common with similar organizations elsewhere, was attempting to exploit the tremendous potential for cobalt-60 production arising through the rapid construction of nuclear reactors. The first full-scale commercial gamma plant was commissioned in the U.K. in 1960. It reached a loading of 500,000 curies before its demolition after twelve years of operation.

The process gained rapid acceptance (*1*) within industry and approval by health authorities because it provided a 'cold' sterilization method combining the property of lethal effect with penetration. Its immediate impact occurred in the introduction of disposable products making it possible, for example, to use heat-labile plastics and new packaging materials and package designs. Certainly, the technique has proved complementary to sterilization methods based on heat and to the use of chemical agents, in particular ethylene oxide gas.

ISOTOPE SOURCES AND PLANT

Type of Source

The development of commercial plants using gamma radiation has been centred around cobalt-60 because it is the cheapest source combined with reasonable half-life of 5-1/4 years. A low cost of production can sometimes be gained by using cobalt as a flux flattener, e.g. for controlling the reactivity of the

[1]Congress presentation D–1–4.

core of a nuclear reactor. Cesium-137 appears to be a serious rival to cobalt-60 in the long term but the single gamma ray of 0.66 MeV in 86% of disintegrations puts it at a serious disadvantage compared with cobalt, which has two gamma rays per disintegration, 1.17 and 1.33 totalling 2.5 MeV/disintegration. Cesium-137 is present in the large quantities of fission product liquor but the cost of separation plant is very high. However, this isotope does have the advantage of long half-life, about 30 years, and shielding can be accomplished with comparative ease.

Plant Design

The main objectives in designing an irradiation plant are a high utilization of the radiation energy, i.e. high efficiency, and uniform dose throughout the product. Other considerations include a minimum hold-up time in the process and ease of replenishing the radioactive source. To achieve a high efficiency of utilization of the gamma radiation energy, it is necessary for the product being treated to occupy a large fraction of the solid angle surrounding the source and to be present in sufficient depth so that the gamma rays are considerably attenuated in passing through it. This latter presents some difficulty when the bulk density of the material being processed is low.

Figure 1 shows the elements of a design in which packages of standard size are carried on a roller conveyor system into the concrete cell. Once inside they are pushed by hydraulic rams to and fro across one face of the radioactive source,

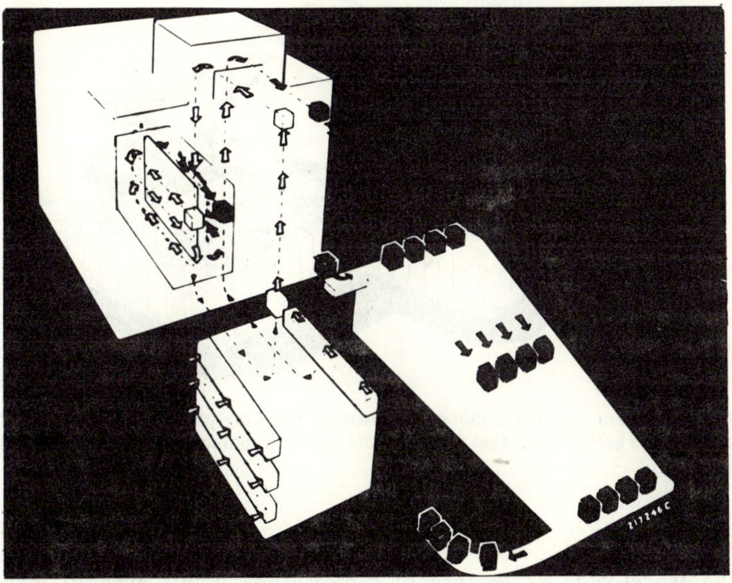

FIG. 1. Diagram showing the package flow through the original U.K.A.E.A. radiation plant. Cartons enter the concrete shield from the storage rack in the foreground and accumulate after irradiation on the unloading ramp.

and they then move to the other face for similar treatment before emerging on to the post-treatment storage rack. The system is made more efficient by using a second parallel conveyor system so that two thicknesses of box are presented to the source at the same time in an attempt to improve efficiency. Figure 2 shows boxes being moved on a monorail system, the dark area representing the concrete shield. Eight lines of monorail are traversed by each carrier of product. The source is in the center so that the radiation is presented with four layers of product on each side.

The safe store position for the source can be either 'wet' or 'dry'. The former is a pond of water about 6m deep. The frame holding a number of stainless steel tubes containing the cobalt-60 is lowered to the bottom. Transfer of cobalt-60 in and out of the frame may be done using long handling tongs. The 'dry' store is merely a narrow shielded trench. A shielding plug is integral with the source frame and seals the top of the trench when the source is lowered. Active rods are transferred to the source frame through a hollow loading tube set in the concrete shield, the transport flask being brought up flush on the outside.

The majority of existing plants are designed on the general ideas illustrated but some are quite different; for example, one involves the packages moving in a spiral around a centrally located cobalt-60 source. Many 'batch' plants have been built; here the product is loaded into the cell and rotated around the source for a fixed time and then unloaded.

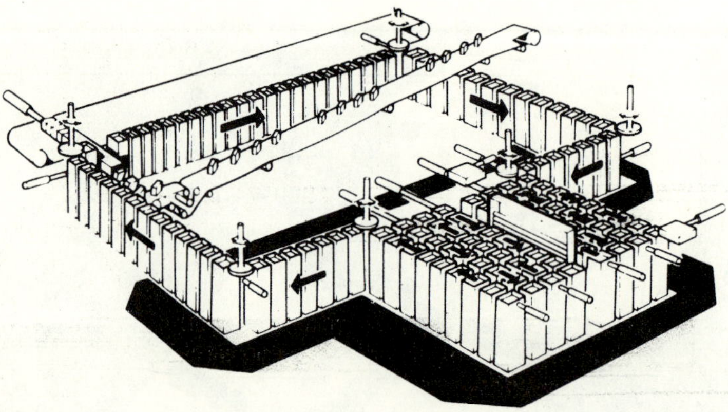

FIG. 2. Diagram showing a monorail system of handling packages. Each carrier contains several cartons and in this case makes 8 passes of the radiation source. The shaded area represents the concrete shield.

Industrial Growth

The world-wide acceptance of radioisotope plants even of megacurie capacity can be seen from Table 1, indicating not only approval of this sterilization method but also of the health and safety aspects of plant construction and operation. It is quite acceptable in most countries for radiation plants to operate within general industrial estates. Such a scene is shown in Fig. 3.

TABLE 1
Distribution of Large Scale ^{60}Co Sterilization Plants

Country	Number	Total Capacity (Ci) $\times 10^{-6}$
Argentina	1	1.0
Australia	3	4.0
Belgium	1	0.25
Canada	4	2.0
Czechoslovakia	1	1.0
Denmark	2	2.0
France	1	1.0
Greece	1	0.25
Hungary	1	0.5
India	1	1.0
Israel	1	1.0
Italy	5	3.5
Japan	1	0.5
Netherlands	2	1.5
New Zealand	1	1.0
Poland	1	0.25
Republic of Ireland	1	1.0
South Africa	1	1.0
Spain	1	0.75
Sweden	2	2.0
Switzerland	1	0.5
United Kingdom	7	5.0
United States	8	6.0
USSR	1	1.0
West Germany	3	3.0
Total	52	41.0
Installed Curies (estimated)		20.0×10^6

It seems likely that many more plants will yet be built following the increasing use of medical disposable products throughout the world. Certainly the curve shown in Fig. 4 would bear out this prediction. As far as electrical machines are concerned, there are probably about 8 involved in medical sterilization work with 2 in Denmark and others in U.K., Poland, France, Japan and the U.S.S.R.

Costs

Considerable throughput is needed before investment in radiation plant is warranted. The fact that the radioisotope source is subject to unavoidable decay adds to the normal factors requiring high capital cost machinery to be operated continuously. Up to a source loading of about 350,000 curies a 'batch' type system is attractive, particularly with high density products. Without the source this type of plant would involve investment of about \$125,000, and for the 'continuous' type designed to take a source capacity of a megacurie the investment increases to about \$350,000. Cobalt-60 is normally purchased as required, keeping pace with increasing throughput; an average figure for cost is \$500 per kilocurie.

FIG. 3. Twin cobalt-60 radiation plants, one 'batch' and one 'continuous' type, attached to a large warehouse handling mainly medical supplies. The operation is located on a normal industrial estate.

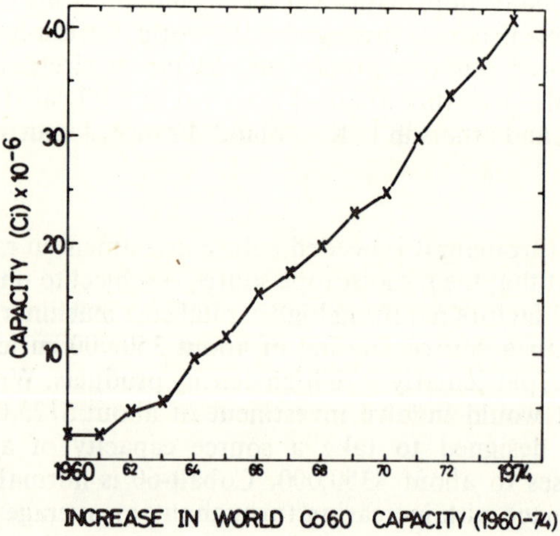

FIG. 4. Increase in world cobalt-60 capacity during the years 1960–1974.

Many large radiation facilities offer a sterilization service with customers delivering and collecting cartoned products regularly, very often on a weekly basis. For a minimum dose of 2.5 Mrad prices vary enormously between different countries, reaching as low as about \$25 per m^3 in the U.K. where there is considerable competition between facilities, to \$100 per m^3 elsewhere. Factors such as density of product, varying between 0.1 and 1.0, convenience of carton size, and size and frequency of consignments are all taken into account in pricing.

RADIATION STERILIZATION DOSE

Background Studies

The choice of sterilization dose for disposable medical products is founded on the results of basic and applied research in microbiology and supported by now with considerable practical experience. It has been most fortunate that a wealth of quantitative data on the radiation sensitivity of organisms has emerged from academic studies in the field of radiation biology. A program of applied research has been superimposed with recognition of the importance of parameters such as initial population levels, environmental conditions during and after irradiation and inherent resistance, on the radiation inactivation of organisms.

A series of investigations in the U.S. in 1956 were aimed to give supporting evidence for the choice of sterilization dose. These gave an insight into the comparative resistance of a wide range of bacterial spores (*2*), vegetative organisms (*3*) and yeasts and molds (*4*). Similar experiments in the U.K. (*5*) confirmed the highest resistance in spores, of the aerobic variety, *Bacillus pumilus* E601 was again outstanding. Attempts were now made to simulate the practical conditions of product contaminants. *B. pumilus* spores were air dried from various suspending media, representing inorganic, organic and biological material, onto the surface of various materials such as plastics used in product manufacture (*6*). In this latter work dose-survival curve data are presented, allowing estimation of the extent of inactivation achieved. Considerable ingenuity was used to artificially contaminate actual products prior to irradiation at different doses followed by sterility testing. Such studies involved catheters, rubber gloves, needles and plastic tubing (*7*) and sutures (*8,9*). In Denmark, organisms isolated from dust and random samples of medical products were examined for resistance (*10–12*). Dose-survival curves illustrated the high resistance of *Streptococcus faecium* A21 when dried into serum and of *B. sphaericus* which is inherent (*13*). Figure 5 illustrates the wide range of inherent resistance to be considered and data in Table 2 confirms the marked influence of environmental conditions as might occur in practice.

Choice of Dose

It is now generally recommended that a minimum dose of 2.5 Mrad is required for medical products. For example, this dose is quoted in the national specifications of the U.S., U.K., Australia, Belgium and Czechoslovakia. In Scandinavia a minimum dose of 3.5 Mrad is considered more appropriate. The International Atomic Energy Agency has been active in promoting a code of practice (*14*, *15*) which emphasizes the necessity to (a) manufacture products under hygienic conditions, (b) ensure bacteria-proof packaging, and (c) select a radiation dose to give a margin of safety which will meet requirements defined by individual national health authorities. The code covers other aspects of radiation plant operation as will be referred to later.

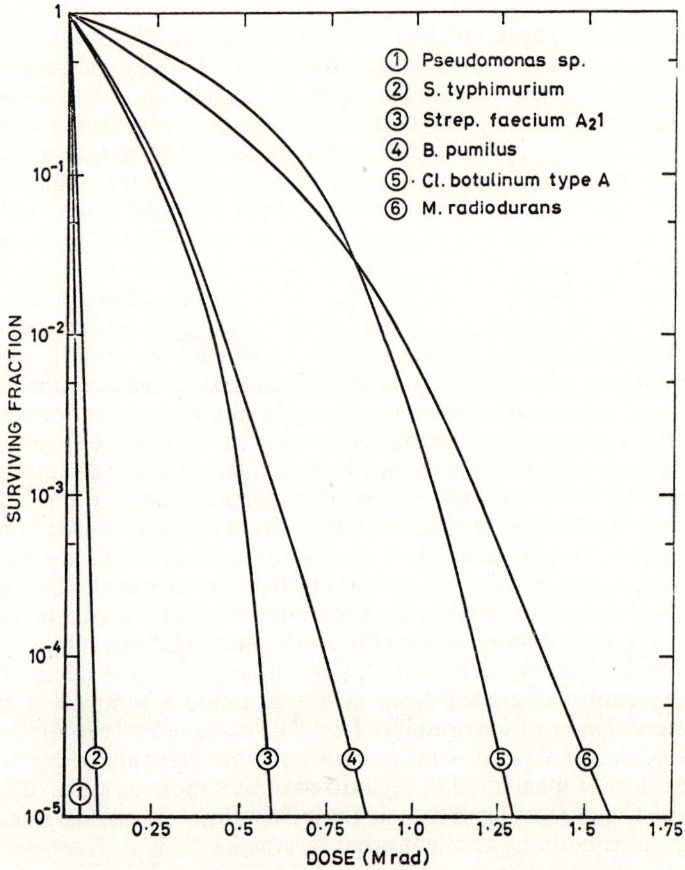

FIG. 5. Radiation sensitivity of different bacteria illustrated by dose-survival curves prepared after irradiation of the organisms in buffered suspension in the presence of air.

TABLE 2
The Influence of Environment During Irradiation on the Dose Required for Reduction by a Factor of 10^5 of Populations of Pertinent Organisms

Organism	Dose (Mrad)	Conditions during irradiation
S. typhimurium	0.1	Buffer suspension in air
	0.3	Buffer suspension anoxic
	0.1	Dried onto kaolin powder in air
	0.5	Dried in bone meal in air
Strept. faecium $A_2 1$	0.6	Buffer suspension in air
	1.9	Buffer suspension anoxic
	0.7	Dried onto kaolin powder in air
	3.0	Dried in serum in air
B. pumilus spores	0.9	Buffer suspension in air
	1.5	Buffer suspension anoxic
	0.7	Dried onto kaolin powder in air
	1.7	In grease in air

PROCESS AND PRODUCT CONTROL

Dosimetry

Film systems, usually clear or red polymethyl methacrylate, are widely used to monitor the radiation dose in routine plant operation and at commissioning. At the latter time the pattern of dose distribution within standard packages and the relationship between dose and target density are established by experimental runs. In routine operation dosimeters are regularly inserted in carriers being placed in areas where the dose is expected to be at a minimum. Red 'perspex' dosimeters (*16*), complete with calibration curve, are purchased in the U.K. from Harwell or clear 'perspex' from the U.K. Gamma and Electron Panel, and from similar organizations in other countries. Records of readings are carefully maintained for inspection by health authorities. Records of source strength are also maintained and commissioning dosimetry repeated after 'topping up'. Finally, a pen-recording system continuously monitors conveyor speed and source position in an automatic plant and source position in the batch type. For the convenience of the plant operator and the recipient of the products, small adhesive radiation indicator labels are affixed to the outside of each carton. The label changes from yellow to red during irradiation and helps to ensure against confusing irradiated with unirradiated product. The label is based on PVC impregnated with an acid-sensitive dye (*17*); a similar label which changes from green to brown has been developed in The Netherlands.

Microbiological Measurements

The only direct microbiological check on plant efficiency as recommended in the International code (*16*) is at commissioning. It is considered appropriate to test that the inactivation of a standard bacterial preparation at several doses is identical to that obtained using an approved radiation source. This is intended to cover any possibility that plant design changes may influence lethal effect in a quantitative manner. Test preparations based on *Strep. faecium* A_2l and *B. spaericus* have been developed in Denmark. However, it is doubtful whether differences in lethal efficiency between gamma radiation plants exist at all, and the difference demonstrated between gamma and electron irradiation (*18*) might be accounted for in the physical makeup of the bacterial monitor.

The question of the use of biological indicators to check the efficiency of radiation sterilization in a routine way often arises. Obviously, absorbed dose is more readily measured by physical or chemical dosimeters; an indicator in the form of a bacteriological test piece is quite superfluous. Such test pieces are used with ethylene oxide sterilization for example, because no other monitor is available which will measure simultaneously at the site of the product the variety of parameters controlling the process.

The sterility test on products after irradiation appears an ideal way of monitoring the process but unfortunately it is impracticable to test a sufficient number of items to be able to detect other than gross contamination. However, it has been suggested that the sensitivity of 'sterility testing' could be improved by deliberately using low radiation doses on samples of production items, so increasing markedly the probability of obtaining positives in the subsequent sterility test. Both theoretical (*19*) and practical (*20*) studies have given encouraging results.

In current practice physical and chemical dosimetry as a routine plant control procedure is approved by most health authorities, with microbiological investigation confined to monitoring the hygiene of production and packaging of items prior to sterilization.

RANGE OF APPLICATION

Disposable Products

Some examples of the many different items of radiation-sterilized equipment now marketed are shown in Fig. 6. The range has increased as the move to the use of disposables in hospital practice gains momentum. The list includes sutures, syringes, needles, scalpels, catheters, gloves, petri dishes and cellulosics including dressings, towels, swabs and drapes. Recently a sterile procedure pack has been designed based on radiation as the sterilizing agent. It consists of items such as hemostatic sponges, cotton gauze dressings, towels, forceps and a sachet of antiseptic solution, contained in a molded compartmentalized tray, hermetically sealed with a compressed 'tear off' closure. The whole is enclosed, together with its disposable bag, in a heat-sealed polyethylene envelope. This illustrates the tremendous flexibility in manner of packaging.

Packaging materials themselves and containers destined for use in aseptic handling areas are irradiated after bulk packaging in bags and cartons. Materials successfully treated include polyethylene, nylon, cellophane, assorted laminates and coated papers. The penetrating properties of gamma rays are used to advantage in the routine sterilization of reels of foil some 35 cm in diameter and 18 cm wide weighing 16 kg each. The overdose ratio with such products is necessarily high being about 1.6 in this latter example.

In designing a disposable gauntlet isolator (*21*) for use in human surgery, it was anticipated that gamma radiation would be used for sterilization. The plastic tent-like structure complete with gloves and tubes is folded and packed into a cardboard transport container suitable for the conveyor system of the radiation facility. At the hospital the tent is erected and requires no permanent installation. The tent and accessories and even the environment for the operation has been sterilized by radiation. Similarly, radiation has been used to sterilize the plastic incubator used in the Caesarean delivery of a germfree child.

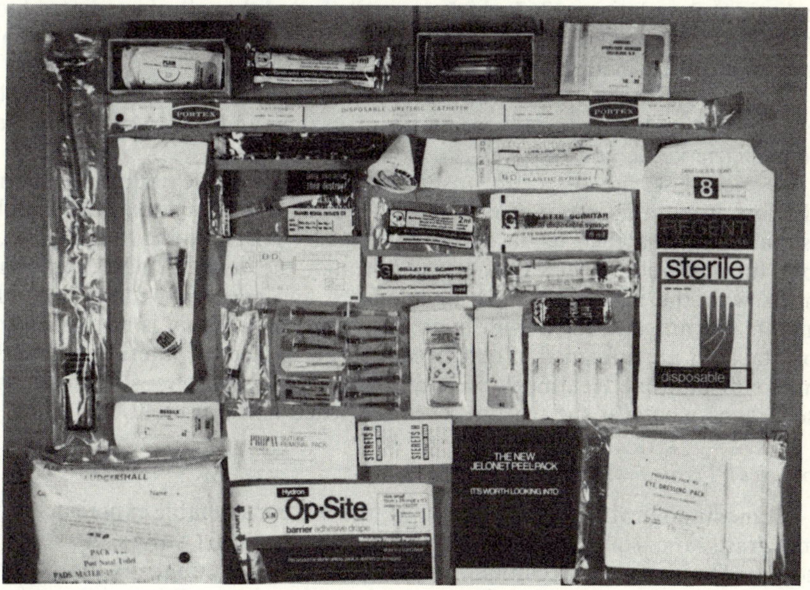

FIG. 6. Some examples of the wide variety of medical products currently being radiation-sterilized.

Diets

Several hundreds of tons of diets for specific pathogen-free animals are radiation sterilized in the U.K. each year suing a 2.5 Mrad dose. These are animals of many species including rats, mice and poultry and they usually are being bred for use in the testing of pharmaceuticals, food additives, etc. Radiation has proved most effective for microbiological control with little or no effect on nutritional quality or palatability (*22*). Diet is double-wrapped in polyethylene bags in about 12 kg quantities and transported and irradiated either in cardboard cartons or in stout paper sacks.

For several years gamma radiation has been used for the sterilization of a variety of food items intended for patients whose natural resistance to infection has been impaired. This state may be seen during treatment with high doses of cytotoxic or immuno-suppressive drugs, and may also occur spontaneously, for example in aplastic anemia and agranulocytosis. Many of the patients benefiting have undergone renal transplantation. The patient is maintained in a pre-sterilized room or plastic tent for a prolonged period in circumstances of the greatest attainable microbiological sterility. Irradiation supplements autoclaving of foods to ensure a palatable and varied diet. Treated foodstuffs include meats, fish, vegetables, potato, rice crispies, cornflakes, bread, jam, marmalade, sugar, tea, instant coffee, biscuits and cake (*23*). Whole frozen meals are often used, being irradiated at 2.5 Mrad at −15° C and maintained frozen until use.

Special regulations to make formal the use of irradiated food in this way came into operation in the U.K. in 1969. These amended the Food (Control of Irradiation) Regulations 1967 which generally forbid the sale of irradiated food for human consumption, requiring the submission of evidence of safety before exemption can be made.

Apart from food, a variety of other items are gamma-irradiated for use by these patients including certain cosmetics, magazines and cutlery and extending to an electric kettle and toaster.

A special service direct to hospitals illustrates the flexibility of gamma radiation. Apparatus impossible to sterilize in other ways can be packed up at the hospital and dispatched to the radiation plant by post or vehicle. If by post, a rapid turn-around service is achieved through the sender double-wrapping parcels with the inside wrap already stamped and addressed for return. The contents are not interfered with in any way since unpacking is not necessary. Hospital staff are informed of care needed in the choice of materials and of the dangers of multiple treatment. Re-sterilization of disposables is very strongly discouraged whatever the reason—even if opened and not used.

Biological Tissues and Pharmaceuticals

Some biological tissues are sterilized; there is considerable experience with bond grafts and some success with blood vessels, heart valves, nerves and tendons. The radiation technique is widely used, for example at the Central Tissue Bank in Warsaw. Certain prepared vaccines have all been successfully sterilized.

As far as pharmaceutical products are concerned, few are marketed and these are mainly antibiotic preparations, particularly for ophthalmic use. So far, there has been no real need in this area since industry has solved many sterilization problems by investment in aseptic handling areas, many equipped with laminar flow systems. Furthermore, the application of radiation to individual pharmaceuticals normally requires approval from health authorities, and it is necessary to produce detailed experimental evidence of non-toxicity; this requirement applies to any change in the method of manufacture. Such evidence is very expensive to accumulate although sufficient data may be forthcoming by direct quantitative study of chemical change. For example, detailed work has been done on sulphacetimide (*24*).

FUTURE PROSPECTS

There will be wider recognition of gamma radiation in particular, as a sterilizing agent and it will be seen as complementing the use of existing traditional chemical agents and heat. The influence of factors such as product and package design in control of infection practice may be seen as more important than the actual microbial inactivation step in manufacture, and this will lead to the fuller exploitation of the unique properties of gamma radiation. More discussion of dose requirements is necessary, with health authorities being prepared to adopt a more flexible attitude to meet individual situations. For example, gamma radiation at comparatively low doses could be applied more widely to certain products, particularly pharmaceuticals, to add to the margin of safety already achieved by aseptic handling techniques and filtration.

More information is needed on the effect of irradiation on materials such as plastics, giving a clearer indication of the formulations necessary to prevent radiation degradation. For example, suitable grades of PVC should be readily available which do not show brown discoloration, and the means of inhibiting embrittlement of polypropylene needs clarification.

Many more radiation facilities will be built worldwide and the continuing progress in the nuclear energy field assures the supply of radiation source. Harmony of legislation between countries is a necessary step to free trade in sterilized medical products which has been stimulated by the introduction of disposable 'ready to use' equipment.

It is hoped that this subject will continue to arouse the interest of academic and applied scientists and to attract the help of the international organizations such as IAEA. A recent IAEA publication (*25*) provides detailed accounts of developments in all those areas touched on in this paper and includes other topics relevant to radiation sterilization. In addition, the proceedings of a U.S.P. conference held in Washington in 1972 provide a valuable appraisal (*26*).

REFERENCES

1. R. M. Longstaff and J. Grant, *J. Br. Nucl. Energy Soc.* **10,** 229, (1971).
2. R. E. Pepper, N. T. Buffa, and V. L. Chandler, *Appl. Microbiol.* **4,** 149 (1956).
3. W. Y. Koh, C. T. Morehouse, and V. L. Chandler, *Appl. Microbiol.* **4,** 143 (1956).
4. A. E. Bridges, J. P. Olivo, and V. L. Chandler, *Appl. Microbiol.* **4,** 147 (1956).
5. E. M. Darmady, K.E.A. Hughes, M. M. Burt, B. M. Freeman, and D. B. Powell, *J. Clin. Pathol.* **14,** 55 (1961).
6. M. M. Burt and F. J. Ley, *J. Appl. Bacteriol.* **26,** 484 (1963).
7. R. Oliver and A. H. Tomlinson, *J. Hyg.* **58,** 465 (1960).
8. W. Van Winkle, P. M. Borick, and M. Fogarty, in *Radiosterilization of Medical Products,* pp. 169–180. IAEA, Vienna, 1967.
9. T. Bartha, E. Mérő, and J. Szita, *Acta Microbiol. Acad. Sci. Hung.* **16,** 31 (1969).
10. E. A. Christensen, *Acta Path. Microbiol. Scand.* **61,** 483 (1964).
11. E. A. Christensen and N. W. Holm, *Acta Path. Microbiol. Scand.* **60,** 253 (1964).
12. E. A. Christensen, N. W. Holm, and F. A. Juul, in *Radiosterilization of Medical Products,* pp. 265–283. IAEA, Vienna, 1967.
13. E. A. Christensen, in *Sterilization and Preservation of Biological Tissues by Ionizing Radiation,* pp. 1–13. IAEA, Vienna, 1969.
14. *Radiosterilization of Medical Products, Pharmaceuticals and Bioproducts.* IAEA Tech. Rep. Series No. 72. IAEA, Vienna, 1967.
15. *Radiosterilization of Medical Products and Recommended Code of Practice.* Pub. No. STI/PUB/157, IAEA, Vienna, 1967.
16. B. Whittaker, in *Manual of Radiation Dosimetry.* (N. W. Holm and R. J. Berry, Ed.) p. 363. Marcel Dekker, New York, 1970.
17. J. J. Farrell and R. L. Vale, *Nucleonics* **21,** 78 (1963).
18. C. Emborg, *Acta Path. Microbiol. Scand.* **80,** 367 (1972).

19. A. Tallentire, J. Dwyer, and F. J. Ley, *J. Appl. Bacteriol.* **34**, 521 (1971).
20. F. J. Ley, B. Winsley, P. Harbord, A. Keall, and T. Summers, *J. Appl. Bacteriol.* **35**, 53 (1972).
21. P. C. Trexler and M. F. Pilcher, in *Germfree Research* (J. B. Heneghan, Ed.). Academic Press, New York and London, 1973.
22. F. J. Ley, J. Bleby, M. E. Coates, and J. S. Paterson, *Lab. Anim.* **3**, 221 (1969).
23. R. Harrison, *Nutrition* **16**, 105 (1962).
24. G. O. Phillips, D. M. Power, and M. Sewart, *Radiat. Res.* **46**, 236 (1971).
25. *Manual on Radiation Sterilization of Medical and Biological Materials.* IAEA Pub. No. STI/DOC/10/149, IAEA, Vienna, 1973.
26. *Proceedings of the U.S.P. Conference on Radiation Sterilization,* Washington, D.C. 1972, *Int. J. Radiat. Steril.* 1 (1973).

HUMAN RADIOBIOLOGY[1]

Chairman: D. W. van Bekkum

[1]Plenary Session C–1

The Late Effects of Acute External Exposure to Ionizing Radiation in Man[1]

S. Jablon
National Research Council
Washington, D. C., USA

When I began to prepare this talk, I found that I was embarrassed by two distinct problems: First, that data on late effects on humans accumulate very slowly indeed, so that I don't have anything really new to report to you, and, secondly, that the subject is so large that it is impossible to summarize it within the time available here. As it happens, however, as I think you all know, late in 1972 the United Nations Scientific Committee on Effects of Ionizing Radiation, or UNSCEAR, published a two-volume work entitled "Ionizing Radiation: Levels and Effects" (*1*), and shortly thereafter the National Academy of Sciences–National Research Council published a two-hundred-page report by the Advisory Committee on the Biological Effects of Ionizing Radiation (BEIR) entitled "The Effects on Populations of Exposure to Low Levels of Ionizing Radiation" (*2*). These two reports, taken together, are virtually encyclopedic in their coverage of the subject; in fact they address a rather broader set of problems than will be discussed here. My object, then is to distill some of the essentials from the UNSCEAR and BEIR reports, add what information has become available in the years since they appeared, and try to interpret what we now know that is probably true, what we "know" that may not be true, and where we know that there are gaps in our knowledge.

I will not attempt to discuss whether it is proper to assess risks at low levels of radiation by linear interpolation of risks measured at high doses. There is some reason to hope that new data which bear on this point may soon become available, but at present there are none that I know of.

The state of knowledge of the subject at this time is such that our approach must be empiric. There has been some very interesting work by Rossi and Kellerer, and others, which may, in time, enable us to interpret what we see in the whole man in terms of the radiobiological events that occur within the cell, or within a part of a cell. Unfortunately, this approach, though exciting in terms of prospects for the future, is not yet at the point where we can rely on it with confidence for such tasks as estimating the risks to man from particular levels of exposure. For this, we are compelled to make the best of the relatively few existing long-term studies of humans who have been irradiated, and even so, the calculations must depend upon one or another of various possible assumptions about the reasonableness of which there has been, to say the least, vigorous debate.

[1]Congress presentation C–1–1.

It will be convenient to begin with a consideration of effects on adults and children and discuss separately the consequences of pre-natal irradiation.

One of the earliest of the so-called "late effects" of radiation to be noted among the Japanese atomic bomb survivors was radiation cataract. However, disabling cataracts are apparently quite rare. The most recent ophthalmologic examinations were done about ten years ago by Miller and his colleagues (*3*). About 2,400 persons were examined with a slit-lamp biomicroscope. Axial opacities were found in about 6 percent of persons with doses exceeding 90 rads but only 1.3 percent of controls. Most of the opacities, however, were quite small, and did not interfere with vision. There is no evidence that these opacities have been progressing with time.

An interesting finding was an association between radiation and what is described as iridescent or polychromatic changes in the axial posterior subcapsular area of the lens, shown in Fig. 1.

POLYCHROMATIC SHEEN IN THE AXIAL POSTERIOR SUBCAPSULAR AREA OF THE LENS 1963–64

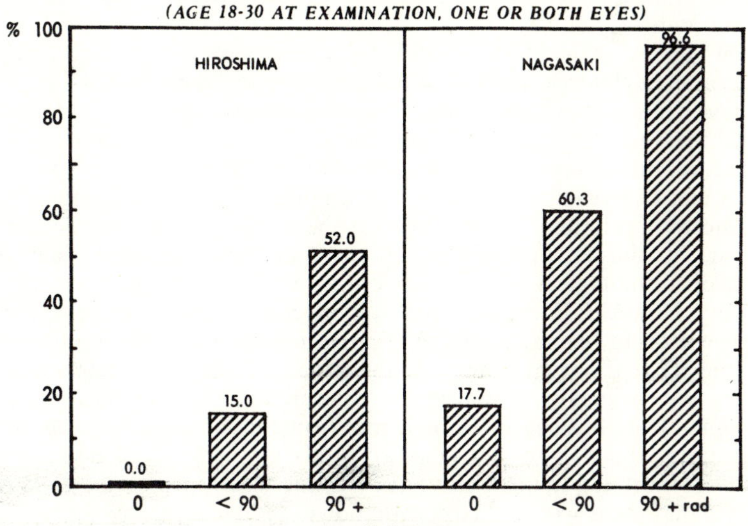

FIG. 1. Prevalence of Polychromatic Sheen in the Axial Posterial Subcapsular Area of the Lens. (Nefzger, Miller, and Fujino, Ref. *3*)

In each city the prevalence of polychromatic changes increased markedly with increasing radiation dose; further, at every dose level, the changes were found more frequently in Nagasaki than in Hiroshima. Interestingly, there are very few conditions which are more prevalent in Nagasaki than in Hiroshima, for the fixed dose. The Nagasaki weapon was a source of gamma rays almost exclusively, while in Hiroshima perhaps 20 percent of the dose was received from fast neutrons, and this is thought to account for the greater prevalence of radiation effects in Hiroshima. However, the findings with respect to polychromatic sheen are contrary. It cannot be proved, but it seems at least quite possible that the comparison between cities has been compromised by a failure of standardization of observers.

An effect of radiation which has been sought among the Japanese survivors, *but not found*, is impairment of fertility (*4,5*). Evidently, whole-body radiation doses compatible with human survival can produce, at most, transient alterations in fertility. Closely related to the subject of impairment of fertility is that of genetic effects. This is a subject on which I am not expert and so shall dispose of by saying that although no genetic effects of radiation have thus far been found among the offspring of the Japanese survivors, the techniques for seeking such effects have, hitherto, been very crude. However, there is some hope that new techniques, enabling the identification of variant proteins, may yet enable estimates to be made of the genetic effects of radiation on man.

Closely related to the problem of genetic effects is the observation of chromosomal aberrations which have been found in examinations of peripheral leukocytes of survivors. As Awa and his colleagues reported four years ago at the IVth International Congress of Radiation Research, the proportions of cells with such stable aberrations as reciprocal translocations and pericentric inversions reached some 30 times control values among survivors who had received more than 500 rads (*6*).

On Sunday, many of you heard Dr. Conard report on a finding of impaired growth among children exposed to fallout in the Marshall Islands.[2] Belsky and Blot have examined this question among the Japanese survivors, and also found that, 20 years after exposure, children who had received doses exceeding 100 rads were a few centimeters shorter than controls (*7*). The effect was greatest among those under five years at exposure, and diminished with increasing age.

Another radiation effect that has been sought among the Japanese survivors, but thus far unsuccessfully, is non-specific "acceleration of aging". The notion of a "biological age," to be contrasted with the chronological age, is beguiling. However, if one attempts really to come to grips with the question of exactly what is to be understood, in an operational sense, by the term "biological age," the concept proves to be very elusive (*8*). In any case, it appears that the authors who originally advanced the concept really meant non-specific life-shortening. Does this occur? The evidence from the Japanese A-bomb survivors is that it apparently does not (Fig. 2). Radiation certainly is associated with premature mortality in man; but if one excludes deaths from cancer, there is no evidence that those who had large radiation doses had increased mortality rates from other diseases, at least up to 25 years after exposure.

On the other hand, it is abundantly clear that at least large doses of radiation are associated with subsequent increased occurrence of cancer and, especially, leukemia (Fig. 3).

Among the Japanese survivors, the leukemia mortality ratio increased regularly with the radiation dose (*9*). Among those with doses exceeding 200 rads, averaging about 350 rads, the leukemia death rate was about 30 times normal. Mortality from neoplasms other than leukemia was also increased among survivors who had received large doses, but for the solid tumors, the increase was only about 60 percent among the survivors with the largest doses.

[2]Late Effects Group, paper L-1-3. *Radiat. Res.* **53**, 313 (1974).

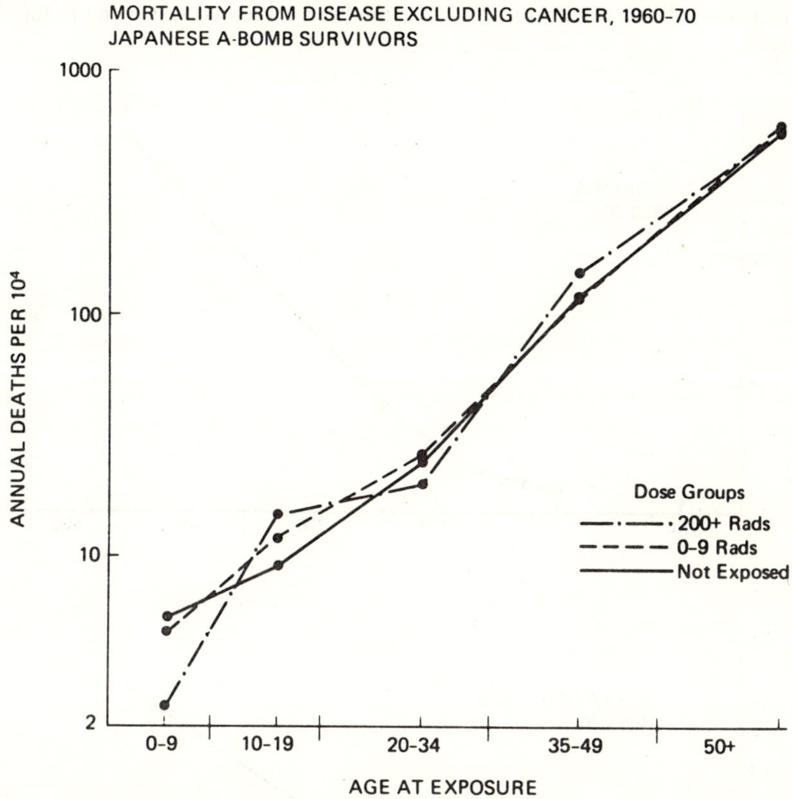

FIG. 2. Mortality from Disease, Excluding Cancer 1960–1970, Among Japanese A-Bomb Survivors. (Jablon and Kato, Ref. *9*)

The characteristic latent periods for leukemia and for other cancers are different (Table 1). Doll and Smith's study of women irradiated for metropathia haemorrhagica (*10*) showed that the leukemia rate was remarkably elevated by five years after irradiation, then slowly declined. On the other hand, the rate for the solid tumors, whether of heavily irradiated or other sites, was most greatly increased more than 15 years following irradiation.

A similar pattern occurred among the Japanese A-bomb survivors (Fig. 4). The leukemia mortality ratio was no less than 50 during the fifth through ninth years after exposure, for those who had doses exceeding 100 rads. Leukemia deaths then declined constantly but were still several times normal 25 years after exposure. The mortality ratio for other neoplasms, however, moved somewhat erratically up to 1960, that is, 15 years after exposure, then began to climb.

In view of the very large differences between leukemia and other cancers, both as to relative risk and as to latent period, the question naturally arises whether, among the solid tumors, there are differences as to their susceptibility to induction by radiation. Are only some forms of cancer induced? In this connection, it is worth noting that in no reported human experience has the occurrence of chronic lymphocytic leukemia been increased following radiation exposure.

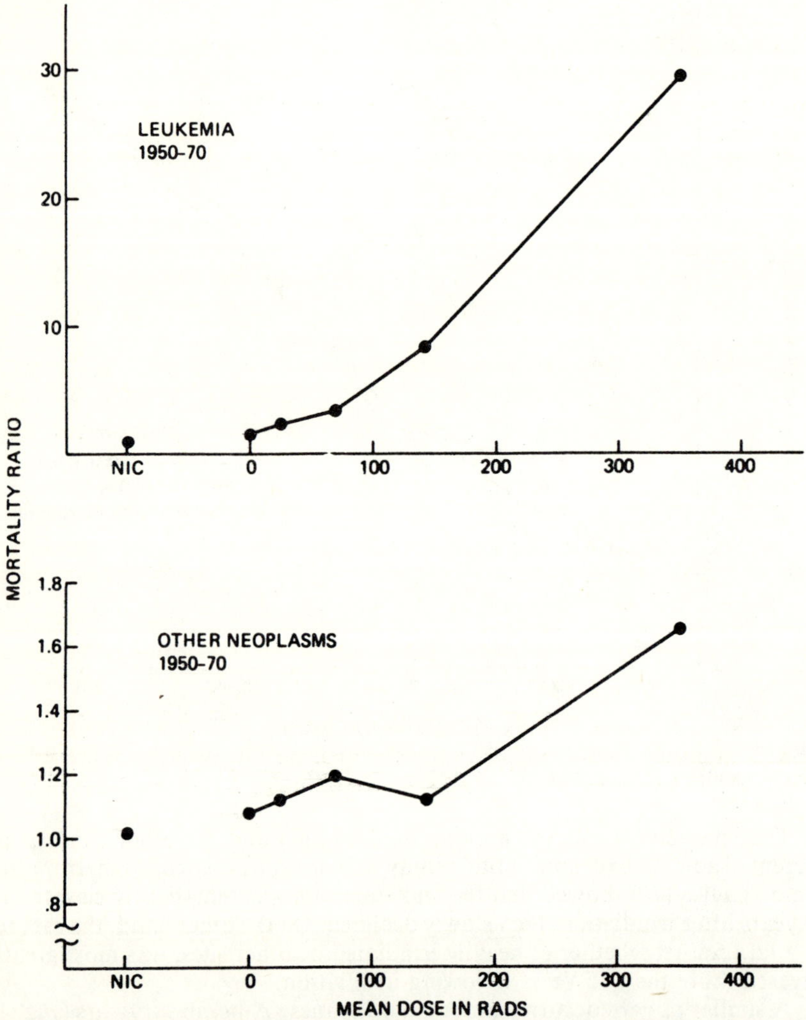

FIG. 3. Mortality Ratios for Leukemia and for Other Neoplasms by Radiation Dose Among Japanese A-bomb Survivors. (Jablon and Kato, Ref. *9*)

There is some evidence in the Japanese experience on the question of specific tumor induction (Fig. 5). The data here refer to the relative risk of death from cancer. Hence, cancers which do not often cause death are not represented. A notable case in point is thyroid carcinoma which is certainly increased following external radiation, but which is usually a relatively indolent tumor which seldom causes death. There is some variation in relative risk for the different types of cancer shown, from about two for lung, breast, and the two miscellaneous classes, ranging down to no increase in risk whatever for cancer of the cervix and uterus. Cervical cancer rates are high among Japanese women. Similarly, stomach cancer is the most frequently occurring form in Japan, and it shows only a small and non-significant increase among heavily irradiated survivors. There are at least two possible explanations of the differences among the sites: One, of course, is that there *are* real differences between tumor types as to their susceptibility to induction by radiation. The second possibility is that induction of any specific cancer by radiation is independent of the background rate of occurrence of that cancer. Thus, cancer of the cervix is about three times as frequent as breast cancer among Japanese women, and stomach cancer four times as common as lung cancer among men and women. An increase of a few cases might be noticeable among the breast or lung cancers, but undetectable among the much larger number of cervical or stomach cancers.

TABLE 1
Relative Risk of Death From Cancer By Time After X-Irradiation

	Time since first treatment		
Cause of death	5–9 years	10–14	15 or more
Leukemia	8.57	6.06	4.00
Cancer of heavily irradiated sites	1.57	1.55	2.07
Other cancers (excluding breast cancer)	0.50	1.19	1.64

Data from Doll and Smith, Ref. *10.*

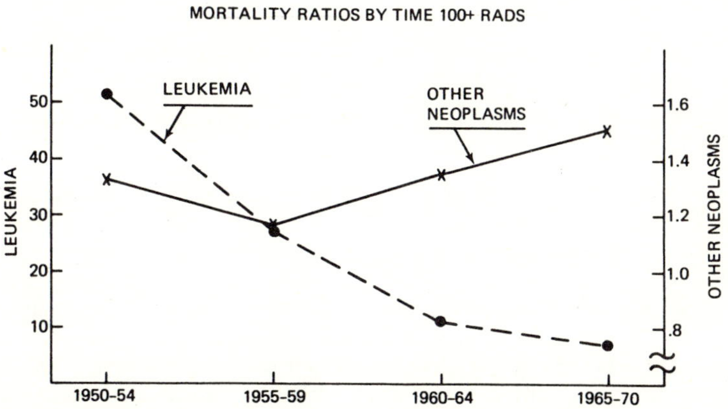

FIG. 4. Mortality Ratios for Leukemia and for Other Neoplasms by Time After Irradiation Among Japanese A-bomb Survivors. (Jablon and Kato, Ref. *9*)

Thyroid cancer was surely increased among the survivors as a result of radiation, even though the increase is not ordinarily noticeable in mortality data (Fig. 6).

The data from Wood et al. (*11*) show the results of a survey of the Japanese survivors, and the progressive increase in the prevalence of thyroid cancer with increasing radiation dose is clearly apparent, as is the fact that, at every dose level, the prevalence among females is something more than twice that among males. An increase in thyroid carcinoma has also been reported by Hempelmann and colleagues (*12*) among children irradiated for thymic enlargement and, recently, by Modan et al. (*13*) among children irradiated for tinea capitis. Modan also reported finding radiation induction of brain tumors, and of tumors of the parotid gland. Salivary gland tumors were also reported to occur among the Japanese survivors by Belsky et al. (*14*) and, at this Congress, by Dr. Takeichi and his colleagues.[3]

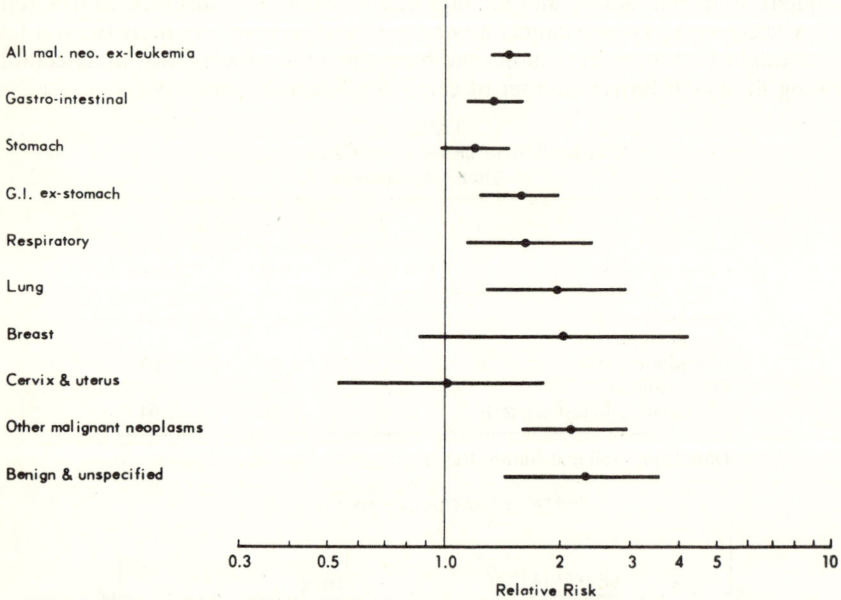

FIG. 5. Relative Risk of Death From Cancer by Site Among Japanese A-bomb Survivors. (Jablon and Kato, Ref. *9*)

When measured in terms of relative risk, the variation in age sensitivity is rather different between leukemia and the other cancers (Fig. 7). For leukemia, children and older adults appear to be at greatest relative risk, and young adults at the smallest. For other cancers, however, although children are again at high relative risk, age variation among adults is not very impressive. It should be emphasized that, if we consider *absolute* risk rather than relative risk—that is, consider the number of excess cases induced per million person-year-rads—then older persons were at higher risk than children. However, because cancer incidence is so very much higher among older people, the excess induced cases constitute only a small relative increase.

It would be natural to assume that the fetus is even more sensitive to the effects of ionizing radiation than is the young child—and this does appear to be true. Ten years after the Hiroshima and Nagasaki atomic bombs, Miller (*15*) reported that fetuses at gestational ages of seven to fifteen weeks at the time of exposure showed a teratogenic effect consisting of a small head circumference,

[3]Congress presentation B–16–6. *Radiat. Res.* **59**, 103 (1974).

accompanied by mental retardation. Wood and his colleagues (*16*) followed up at 20 years and prepared a figure which shows the effect clearly (Fig. 8).

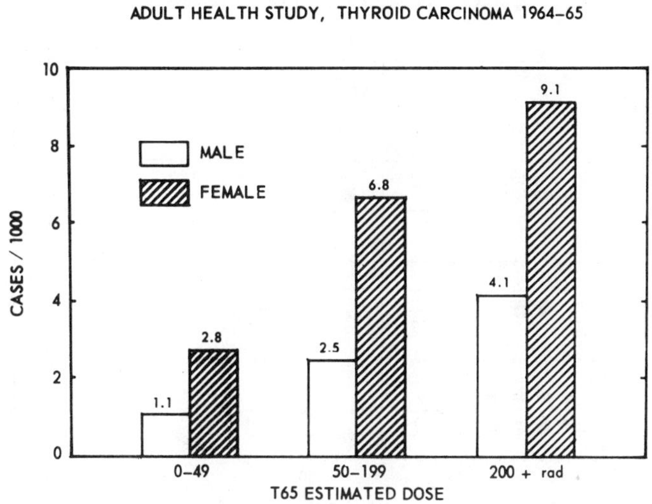

FIG. 6. Prevalence of Thyroid Carcinoma by Radiation Dose and Sex Among Japanese A-bomb Survivors. (Wood, Tamagaki, et al., Ref. *11*)

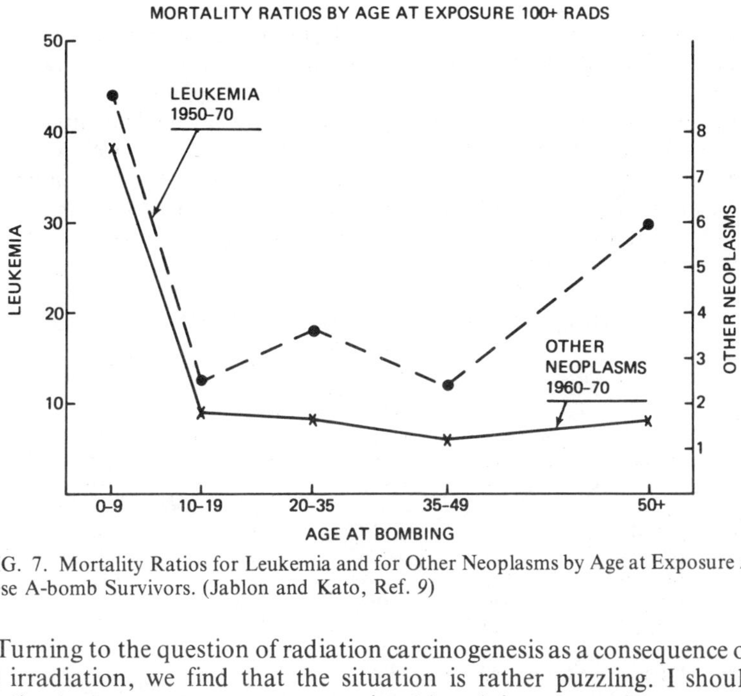

FIG. 7. Mortality Ratios for Leukemia and for Other Neoplasms by Age at Exposure Among Japanese A-bomb Survivors. (Jablon and Kato, Ref. *9*)

Turning to the question of radiation carcinogenesis as a consequence of prenatal irradiation, we find that the situation is rather puzzling. I should say immediately that not everyone agrees with this opinion. In particular, Dr. Mole, as he told us yesterday morning[4], is satisfied that the reported effects are real, but I am not quite convinced.

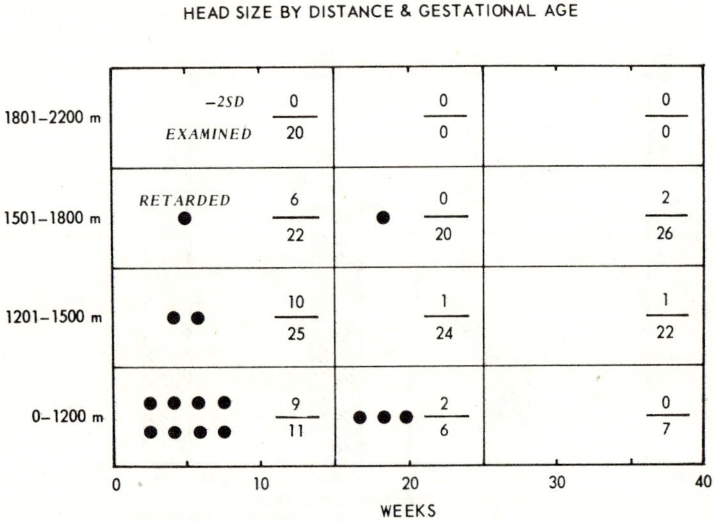

FIG. 8. Diminished Head Circumference and Mental Retardation at Age Twenty Among Japanese Survivors Irradiated Prenatally. (Reproduced from Ref. *16*, J. W. Wood, K. G. Johnson and Y. Omori, *Pediatrics* **39**, 385 (1967) with the authors' and publisher's permission.)

Publications from the Oxford Survey of Childhood Cancers are well known to you, and I will refresh your memories with one table (Table 2). This shows data of Stewart and Kneale (*17*). The relative risks refer to the chance that a child who was X-rayed prenatally will die within ten years of birth from leukemia or other cancer. For the whole period of about 20 years the excess risk among X-rayed children was 56 percent, falling from 105-percent excess in the earliest years of birth to a mere 12 percent in 1960–1965. This reduction in risk is presumed to result from a progressive lowering of doses per film as well as a possible reduction in the number of films taken.

TABLE 2
Relative Risk of Cancer Among British Children Irradiated Prenatally

Year of birth	Number of pairs	Number X-rayed: Cancer deaths	Number X-rayed: Controls	Relative risk
1943-49	1,822	212	110	2.05
1950-54	2,735	437	316	1.46
1955-59	2,153	381	248	1.65
1960-65	939	111	100	1.12
Total	7,649	1,141	774	1.56

Data from Stewart and Kneale, Ref. *17*.

[4]These proceedings.

Stewart's original reports, which dated from the 1950's, stimulated a number of other investigators to examine the question. A very careful study by MacMahon (*18*) obtained results very similar (Table 3) to those previously reported by Stewart.

TABLE 3
Relative Risk of Cancer Among Prenatally Irradiated Children Born in New England

	X-rayed		Not X-rayed		
	Number (thousands)	Deaths	Number (thousands)	Deaths	Relative risk
All cancer	77	85	647	471	1.52
Leukemia	77	47	647	257	1.54
1st births	51	35	212	100	1.46
2+ births	26	12	435	157	1.28

Data from MacMahon, Ref. *18*.

Stewart's total series gave a relative risk of 1.56 for leukemia and other cancer together; MacMahon obtained 1.52 for all cancers, and 1.54 specifically for leukemia. Of some interest here is the fact that children who were first births had a much greater relative risk for leukemia than those who were later births, and remarkably, the relative risks among first births and among later births are *both* smaller than the overall relative risk! This result comes about because first-born children have a higher risk of leukemia than later births even in the absence of radiation; and first-born children are also about four times as likely to be X-rayed. This is an illustration of the kinds of problems that can arise in surveys on human subjects, and which demand very careful analysis. I'm sorry to say that they don't always get the care they deserve.

So far the picture looks reasonably clear. Let us examine some of the problems (Table 4). In one of their earliest papers, Stewart et al. had looked at a number of factors besides prenatal X-ray. The data here are taken from their 1958 paper (*19*) and show some very queer results. Non-abdominal X-ray to the mother, at all periods before birth, was associated with a 28 percent excess risk of childhood cancer. Most of the X-ray exposures would have occurred prior to conception. The numbers are fairly large, and the excess is statistically significant. Although Stewart and her colleagues have not returned to this issue, others have examined it (Table 5).

Graham and his colleagues (*20*), in the Tri-State Study of childhood leukemia found, remarkably, that after standardization for year of birth, age of mother and birth order, the relative risk for *pre-conception* irradiation of the mother was almost identical with the risk assignable to X-ray *during* pregnancy. This, in turn, was almost identical with the risk found by MacMahon and by Stewart et al. Graham et al. found that *pre-conception* irradiation of the fathers was also associated with an increase in leukemia risk to the child! The excess risk associated with paternal irradiation was of only borderline statistical significance but can hardly be ignored.

On the other hand, with respect to the Hiroshima–Nagasaki survivors, there was only one cancer death in the first ten years among the children who had been irradiated prenatally. Expectation, at Japanese mortality rates, was about 0.75 deaths. The difference was not, of course, statistically significant, but more importantly, the observed number of deaths was much smaller than would be expected if the carcinogenic effects of X-ray reported by Stewart and by MacMahon were extrapolated upwards to the levels of dose received by the survivors. With respect to pre-conception irradiation, a report by Okada et al. (*21*) on more than 35,000 children of A-bomb survivors, conceived and born *after* the exposure, showed no increase in leukemia mortality among these children up to 1969, at which time all had been followed for at least ten years.

TABLE 4
Relative Risk of Cancer Among British Children in Relation to Maternal Radiation Exposure History

Kind and time of X-ray	Cancer deaths	Controls	Relative risk
Total	1,299	1,299	
Abdominal X-ray, during relevant pregnancy	178	93	2.06
Abdominal X-ray, before marriage	44	26	1.72
Abdominal X-ray, after marriage but before conception	109	121	0.89
Other than abdominal, all periods before birth	531	456	1.28

Data from Stewart, Webb, and Hewitt, Ref. *19*.

TABLE 5
Relative Risk of Leukemia Among American Children in Relation to Parental Radiation Exposure History

	Percent X-rayed		
	Leukemia deaths	Controls	Relative risk
Mother			
X-ray during pregnancy	30.1	22.5	1.56
X-ray prior to relevant conception	46.0	36.4	1.55
Father			
X-ray prior to relevant conception	26.3	22.4	1.31
Either parent (prior)	54.3	45.3	1.47
Both parents (prior)	18.2	14.4	1.49

Data from Graham, Levin, et al., Ref. *20*.

An interesting study was done in Baltimore by Diamond, Schmerler, and Lilienfield (*22*). The Baltimore study, unlike most studies of prenatal irradiation, was *prospective*, so the authors could obtain data about deaths from all causes among the children, not merely about deaths from leukemia and other cancers (Table 6).

TABLE 6
Relative Risk of Death Prior to Age Ten Among Baltimore Children in Relation to Prenatal Irradiation

	X-rayed	Not X-rayed	Relative risk
White children			
Person—years	79,800	156,000	
Leukemia	6	4	2.94
Other cancer	4	9	0.87
All other deaths	136	140	1.90
Nervous system	13	13	1.90
Respiratory system	42	32	2.57
Digestive system	11	11	1.96
Congenital malformations	25	36	1.36
Accidents	27	26	2.03
Black children			
Person—years	72,600	123,300	
Leukemia	0	3	0.00
Other cancer	3	7	.73
All other deaths	151	272	0.94

Data from Diamond, Schmerler and Lilienfeld, Ref. *22*.

Among the exposed white children, there was excessive mortality from leukemia, but not from other cancers. More surprising was the fact that among the exposed white children, mortality was excessive from all other causes of death, including even accidents! This latter difference is significant at the one-percent level. Among comparable numbers of black children, on the other hand, there was no excessive mortality from any cause of death among those X-rayed prenatally

To summarize—some authors find prenatal irradiation associated with excess risks on the order of 50 percent for death before age ten from leukemia or other cancer. Another author finds about the same risk for leukemia associated with pre-conception irradiation. If both of these associations are causal, then there is the remarkable coincidence that the risk of childhood leukemia as a *somatic* effect of radiation is the same as the risk as a *genetic* effect! However, neither effect is confirmed by the experience of the Japanese atomic bomb survivors. Another study shows the leukemia effect for whites, but not for blacks, and shows an "effect" for accidental death among whites! My personal view is that there is no plausible way to reconcile these findings if they are to be considered as being causally related to X-ray. However, recalling that none of the studies cited were experimental—all were, necessarily, surveys in which there was no assurance whatever that the irradiated and non-irradiated children came from comparable families—we are led to the thought that perhaps the results might be explained by the idea that families have varying social and medical characteristics; that some of these characteristics may be associated with varying probabilities of leukemia or other cancer or death from accidents in their offspring; and that these characteristics may be associated with the probability that the mother will be X-rayed, either during pregnancy or at other times. It may be significant that the black mothers in the Baltimore study were practically all clinic patients: That is, they were of a fairly uniform social class, and all entered the medical care stream in the same way. Among these mothers, the chance that X-rays would be made did not depend upon their economic circumstances, or upon other factors than the purely medical indication for X-ray.

Since, as has been seen, there appears to be no question that ionizing radiation is carcinogenic with respect to both the child and the adult, why should there be any question about carcinogenicity for the fetus, which is, presumably, much more sensitive than the adult? The problem is that the effect reported by Stewart et al. for prenatal irradiation differs by an order of magnitude from effects found even in young children. The BEIR report gives an estimate of risk to children under age ten as something less than four deaths per million person-year-rads from all forms of cancer, including leukemia, while the risk for fetuses, reported by the Oxford Survey of Childhood Cancers is 57—more than ten times as large. Besides this large difference in magnitude of effect, the latent periods are different, since the effects reported for fetuses are for the first ten years after exposure, while for children, the deaths from solid tumors do not begin to occur until after ten years following exposure. Since most X-ray exposure of fetuses is in the third trimester—in fact, pelvimetry, when done, is usually done within days of birth—it is a little difficult to see why the effects of radiation should be so different as to both magnitude and latency.

These thoughts are, of course, speculative, and they are presented to balance what appear at this time to be equally speculative arguments that X-ray delivered to the fetus in the course of pelvimetry actually *causes* all of the subsequent excess cancer in the child. More studies, of different kinds, in varying populations are needed before this issue can be satisfactorily resolved. It will perhaps not surprise you to learn that we, at the National Research Council, have actually begun such a study with the support of the Bureau of Radiological Health and the National Cancer Institute. We expect to survey more than 1,000 children who died of cancer, and who were born in military hospitals to dependent wives of servicemen. There will, therefore, be some control of both social class and reasons for X-ray, and we hope to shed additional light on this vexing question.

REFERENCES

1. United Nations Scientific Committee on the Effects of Atomic Radiation, *Ionizing Radiation: Levels and Effects.* **Vol. 1:** *Levels,* **Vol. 2:** *Effects.* Publication No. E.72IX.17, E.72IX.18. United Nations, New York (1972).
2. Advisory Committee on the Biological Effects of Ionizing Radiation, *The Effects on Populations of Exposure to Low Levels of Ionizing Radiation.* National Academy of Sciences–National Research Council, Washington, D. C. (1972).
3. M. D. Nefzger, R. J. Miller, and T. Fujino, *Am. J. Epidemiol.* **89,** 129 (1969).
4. D. G. Siegel, *Radiat. Res.* **28,** 278 (1966).
5. W. J. Blot and H. Sawada, *Am. J. Hum.* Genet. **24,** 613 (1972).
6. A. A. Awa, S. Neriishi, et al., *Lancet* **2,** 903 (1971).

7. J. L. Belsky and W. J. Blot, *Am. J. Public Health* (in press).
8. J. W. Hollingsworth, A. Hashizume, and S. Jablon, *Yale J. Biol. Med.* **38,** 11 (1965).
9. S. Jablon and H. Kato, *Radiat. Res.* **50,** 649 (1972).
10. R. Doll and P. G. Smith, *Br. J. Radiol.* **41,** 362 (1968).
11. J. W. Wood, H. Tamagaki, et al., *Am. J. Epidemiol.* **89,** 4 (1969).
12. L. H. Hempelmann, *Science* **160,** 159 (1968).
13. B. Modan, D. Baidatz, et al., *Lancet* **1,** 277 (1974).
14. J. L. Belsky, K. Tachikawa, et al., *J. Am. Med. Assoc.* **219,** 864 (1972).
15. R. W. Miller, *Pediatrics* **18,** 1 (1956).
16. J. W. Wood, K. G. Johnson, and Y. Omori, *Pediatrics* **39,** 385 (1967).
17. A. Stewart and G. W. Kneale, *Lancet* **1,** 1185 (1970).
18. B. MacMahon, *J. Nat. Cancer Inst.* **28,** 1173 (1962).
19. A. Stewart, J. Webb, and D. Hewitt, *Br. Med. J.* **1,** 1495 (1958).
20. S. Graham, M. L. Levin, et al., *Nat. Cancer Inst. Mongr.* **19,** 347 (1966).
21. H. Okada, T. Tomiyasu, et al., *Risk of Leukemia in Offspring of Atomic Bomb Survivors, Hiroshima and Nagasaki, May 1946–June 1969*. Atomic Bomb Casualty Commission Technical Report No. 30-72 (1972).
22. E. L. Diamond, H. Schmerler, and A. M. Lilienfeld, *Am. J. Epidemiol.* **97,** 283 (1973).

The Risk of Malignancy from Internally-Deposited Radioisotopes[1,2]

R. E. Rowland
Center for Human Radiobiology
Argonne National Laboratory, Argonne, Illinois 60439, USA

The effects of internally-deposited radioisotopes are of vital importance to us today, for, as we enter a nuclear era, the human body will have the possibility of acquiring a host of new radioisotopes. Now, of course, internally-deposited radioisotopes are not new to mankind; ^{14}C, ^{40}K, ^{226}Ra, and ^{228}Ra exist in all of us as a consequence of the natural levels of these isotopes in our environment. We know, however, essentially nothing about the effects of these particular isotopes on man at environmental levels, let alone the effects of the new radioisotopes that are entering our environment.

There are two major problems that face those who investigate the internal emitters in man; first, the dosimetry is usually extraordinarily complex, and second, for ethical and practical reasons one cannot test the effects of such noxious agents by experiment. The second of these forces the investigator to take advantage of human experience which took place unexpectedly, perhaps decades earlier, so that often not only does he not know the levels of radiation involved, but even the identities of the exposed individuals. Nevertheless, the need to know is so overwhelming that our efforts must be so directed that all possible information can be derived from the scattered cases of internal radioisotope contamination.

It is my intention to consider the somatic effects of the internal emitters, and to accept as proven the hypothesis that the major long-term somatic effect of radiation on man is cancer. The question before us then relates to the risk of malignancy; how is the incidence related to the dose delivered to the sensitive cells at risk? I will also assume that while a linear hypothesis is and perhaps must be used for radiation protection purposes, it is our desire here today to deduce the shape of the dose-response function from the data itself.

The problems associated with the internal emitters are well illustrated by the studies of Conard and his associates of the Marshallese who were exposed on Rongelap Island in 1954 (*1*). Eighty-two people were exposed; 64 on Rongelap received an estimated whole-body gamma dose of about 175 rad and 18 others on a nearby island about 69 rad. Over and above this whole-body dose the inhalation of iodine isotopes and consumption of iodine-contaminated food resulted in total thyroid doses on the order of 135 to 335 rads in adults to as high as 1200 to 1400 rads in small children. Twenty years later, 28 people

[1]Congress presentation C-1-2
[2]Work performed under the auspices of the U. S. Atomic Energy Commission.

have developed thyroid abnormalities, three of which are malignant. In addition, one 19-year-old boy died of acute myelogenous leukemia in 1972 which may well have been the result of his radiation exposure (Conard, private communication).

In this study we find the classic problems of the internally-deposited radionuclides. First, in addition to the internal iodine dose, there was also an external total body gamma dose. Second, neither the external gamma nor the internal iodine doses are accurately known. Third, the population is small. Fourth, the range of dose values is narrow—this is an important point to which I will return later. Fifth, there were malignant effects, but they did not occur for quite a few years after the event, and, since only 20 years have elapsed since the exposure, it is not unlikely that more malignancies may appear in the future.

The problems associated with this study of the Marshallese are common to most studies of the effects of internal emitters in man. These include difficult dosimetry, the fact that the population under study is always too small (and, incidentally, being human, the members may elect not to be part of any follow-up, thereby playing havoc with our statistics!), the range of dose values over which there is significant experience is limited, and the follow-up time by necessity is the lifetime of the exposed individual.

From Conard's studies of the Marshallese we can obtain quantitative estimates of the risk of thyroid malignancy from exposure to radioiodine, but only if we assume a linear relationship between dose and effect.

Are there any exposed populations which provide the required number of cases, well-documented dosimetry, and wide range of dose values so that risk estimates can be made without recourse to a preconceived dose-response function? Consider the studies of Thorotrast in man. Several excellent follow-up studies are underway, particularly in Portugal, Denmark, Japan, and more recently, Germany (*2*). The population under study is quite large, on the order of several thousand individuals, making it perhaps the largest internal-emitter study. There is a high incidence of liver tumors in this group, particularly hemangioendotheliomas, yet the only risk estimates that can be made are those based on a linear hypothesis. The quantities of Thorotrast administered ranged only from about 10 ml to 100 ml, with most cases receiving doses in the range of 15 ml to 75 ml. This implies too narrow a dose range for us to deduce the shape of a dose response function, and, therefore, we must base the calculation of risk on an assumed shape for this function.

The studies with which I am most intimately acquainted are the studies of radium in man, and it should come as no surprise that I consider them to have the best probability of fulfilling the above requirements. The study of ^{224}Ra in Germany by Spiess and Mays is a good starting point (*3*). These investigations have identified some 1811 individuals who received ^{224}Ra in the period 1944–51. Of these, 898 have been followed-up, or almost 50% of the series. The doses calculated from the injection records range from 6 to 5750 rads, calculated as the average dose to the total skeleton. The highest dose in the series is thus almost a factor of 1000 greater than the lowest, so considerable leverage exists for determination of the shape of a dose-response curve.

Unfortunately (and there is always an "unfortunately" in internal-emitter studies), the exposed population included children of all ages as well as adults, and the highest doses were found in the lowest age groups. Spiess and Mays have broken the study into two segments, based on age, and reported their results for all juveniles and separately for adults. I have chosen to combine their data into one single group, and have used 815 cases, for which complete injection data and follow-up history have been published, for my presentation today.

The breakdown of the cases into dose ranges is shown in Table 1.

TABLE 1
^{224}Ra Cases [Spiess and Mays, 1971 (*4*)]

Dose range (rads)	Average dose (rads)	Cases	Sarcomas	Sarcoma fraction
2000–5750	3332	21	8	0.381 ± 0.135
1000–1999	1352	72	19	0.269 ± 0.061
500–999	692	121	9	0.074 ± 0.025
200–499	313	221	8	0.036 ± 0.013
90–199	139	205	3	0.015 ± 0.008
6–89	54	175	0	0.0 ± 0.006
		815	47	

The end point used for analysis in Table 1 is the bone tumor, 47 of which have been identified in the 815 cases used in this analysis. One wonders about the other 50% of the injected population. Do they represent the same distribution of doses? Do they contain the same number of bone tumors? Or is it possible that most of the tumor cases have been included in the study group? This is a rather likely possibility, which would have the result of biasing upward our risk estimates. Since we don't know the answers to these questions, they will serve to stand as warnings, and we will proceed under the assumption that the data are valid as given.

These data are shown in graphical form in Fig. 1 with a linear dose-response function which has been fitted by the method of least squares. I have taken one liberty with the data here, for my own convenience. Since I am going to use a number of figures of the same type, I have chosen to use the same scale on each of these illustrations. Therefore, to spread the ^{224}Ra dose values out on a scale actually designed for ^{226}Ra dose values, each dose has been arbitrarily multiplied by the factor 3.5. You should recognize that this procedure in no way changes the shape of the distribution.

These data points can also be fitted with a dose-squared-exponential function, which has worked well with the ^{226}Ra population, and such a fit is illustrated in Fig. 2 using semilogarithmic coordinates.

The fitted dose-response function provides us with the desired measure of risk. This set of data, even though it covers a range of doses varying by a factor of about 1000, does not effectively discriminate between the two functions I have chosen to test. It is obvious that more cases are needed, particularly in the dose region between 10 and 100 rads.

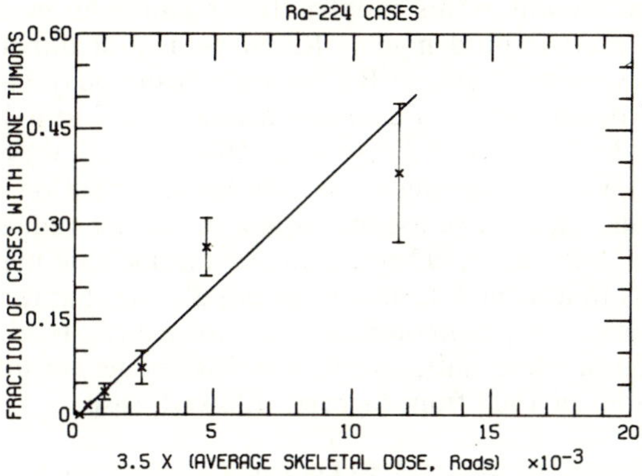

Fig. 1. Fraction of cases in each dose range with bone tumors plotted as a function of average skeletal dose from ^{224}Ra and daughters. The actual dose values have been multiplied by the arbitrary factor 3.5 to spread the dose levels. The error bars represent one standard deviation in tumor fraction. A linear dose-response function has been fitted to the data points.

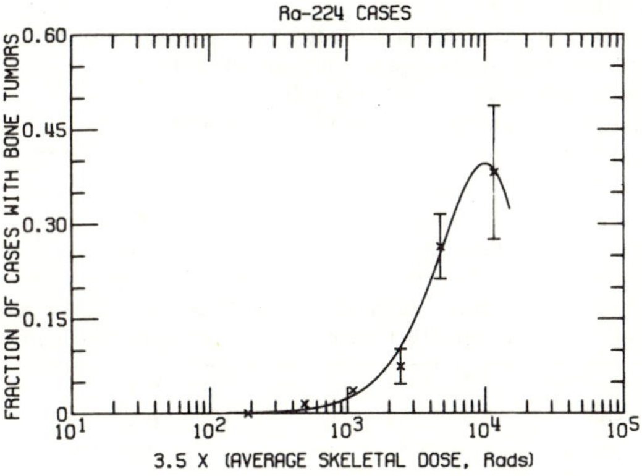

Fig. 2. Fraction of cases in each dose range with bone tumors plotted as a function of average skeletal dose from ^{224}Ra and daughters. The actual dose values have been multiplied by the arbitrary factor 3.5 to spread the dose levels. The error bars represent one standard deviation in tumor fraction. A dose-squared exponential function has been fitted to the data points.

Table 2 shows the fitted functions, and Table 3 shows the predicted number of tumors, under the listed conditions for several low doses. It is readily apparent that, given a significant number of cases between 50 and 100 rads, it should be possible to clearly distinguish between these two functions. Spiess and Mays have pointed out that there exists a yet-unstudied series of 1000 to 2000 ^{224}Ra cases in Germany in which the average skeletal dose is calculated to be 56 rads (*3*). Therefore, given the time and effort, it should be possible to obtain from these cases a definitive resolution to this question of the shape of the dose-response function for ^{224}Ra under these conditions of administration.

Let us now examine the study of the effects of radium on man in this country, a study with which I am very familiar. Unfortunately, both ^{226}Ra and ^{226}Ra mixed with varying quantities of ^{228}Ra were acquired internally by those exposed to radium in this country. It is possible for us to document very accurately the internal burden of ^{226}Ra in man both during life and, from suitable samples, after death. ^{228}Ra, however, can today only be accurately measured after death, for, as a consequence of its 5.7 year half-life, and the fact that those who acquired it did so some 50 years ago, very little remains in the body today. We, in assigning doses, add the contribution from ^{228}Ra to the ^{226}Ra dose. Thus, tacitly, we assume equality, rad for rad, for the emissions from the two different decay schemes, an assumption that may not be tenable. Doing so results in assigned doses to our population that range from 1 rad to 44,000 rads. For my discussion today I have arbitrarily used only those cases with doses of 30 or more rads, so that the dose range considered varies by a factor on the order of 1500 from highest to lowest dose.

TABLE 2
Risk Estimates for ^{224}Ra in Man

Conditions:
- Multiple injections over periods from weeks to a few years
- Calculated for period 19–25 years after first injection
- Risk calculated as:
  - Fraction of cases with a bone tumor as a function of dose
- Dose calculated as average to mineralized skeleton
- Linear hypothesis: $F = 1.5 \times 10^{-4} \cdot D$
- Dose-square-exponential: $F = 3.7 \times 10^{-7} \cdot D^2 e^{-D/1400}$

TABLE 3
Bone Tumors Predicted From ^{224}Ra

Dose (rads)	Linear function tumors/1000 cases	Dose-squared-exponential tumors/1000 cases
100	15	3.5
50	7.5	0.9
10	1.5	3.6×10^{-2}

The current status of the radium population under study today by the Center for Human Radiobiology at Argonne National laboratory is shown in Table 4.

TABLE 4
Status of the Radium Studies, April 1974

Population	Known	Studied	Number with doses>30 rads	Malignancies Bone	Malignancies Head
Total	3803	1572	621	54	26
Listed employees of three dial-painting plants before 1938	720	377	235	25	7
Above less cases found as a consequence of malignancies			221	11	4

Of the 3803 names of potentially-exposed individuals known to us, we have studied only 1572, or 41%. These cases were exposed by all routes—dial painting, medical injections, and laboratory work—over the period from the discovery of radium until today. Yes, even today radium dial painting goes on, but, let me hasten to add, apparently under well-controlled conditions. Thus, while we can and do examine the results from this entire population, it is appropriate to divide the group into various segments. One segment constitutes a well-defined population of employees of three dial painting plants, who worked prior to 1938, and all of whose names were obtained from employment lists. You might be interested to know that some of these fifty-year-old lists just came into our hands within the last few years. As shown in Table 4, we have located and studied 377 of the 720 listed individuals, or 52% of them, and this group had 25 of the known 54 bone tumors in our population. Since many of these cases were known to us as a consequence of their malignancies, one might suggest that such tumor cases should be eliminated from this group until we achieve close-to-complete coverage of the entire population. If we do this, however, the number of bone tumors in this series is markedly reduced.

The effect is shown even more clearly in Table 5. Here the cases are divided into dose groups, first for the total population, then the listed population, and finally, the number of acceptable tumors remaining if those found on the basis of symptoms are eliminated. You will note that it is the high-dose tumors that are eliminated; these were often added to the study by selective exhumation of known tumor cases.

TABLE 5
U.S. ^{226}Ra and ^{228}Ra Cases

	Total Population		Listed Population: 3 Dial Plants		
Dose range (rads)	Cases	Bone tumors	Cases	Bone tumors	Less cases found by symptom
10000–44400	42	15	20	11	1
5000–9999	44	15	16	6	3
2500–4999	57	13	25	6	5
1000–2499	59	10	20	2	2
450–999	55	1	22	0	0
30–449	364	0	132	0	0

In Fig. 3 the bone tumors in the total population under study are fitted with a dose-response function of the dose squared exponential form. In Fig. 4 this same curve is shown, but the data points are now from the listed population of the three dial painting plants. Three sets of points are shown. The X's are the observed tumor frequencies in the population. The *'s are the frequencies left when we remove the cases found as a consequence of their malignancy. This procedure, unfortunately, almost completely removes the bone tumors at high doses. The O's indicate lower limit to the bone tumor fraction, based on two assumptions. The first assumption is that the dose distribution in the unmeasured cases is identical to that in the measured cases. The second is that, for each dose interval, there are no bone tumors in the unmeasured population. However, we know of 16 more bone tumors in the unstudied portion of this population. Therefore, at this time all we can say is that the true distribution most likely lies somewhere between the points indicated by the X's and the O's. It is undoubtedly pure coincidence, but the dose-response function derived from the total population under study (Fig. 3) appears to fall within these two distributions.

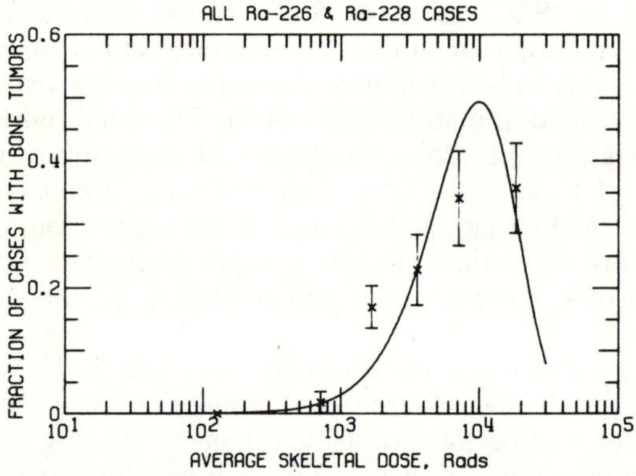

Fig. 3. Fraction of the cases in each dose range with bone tumors plotted as a function of average skeletal dose from ^{226}Ra and ^{228}Ra. The error bars represent one standard deviation. The fitted dose response function is of the form $F = \text{const} \times D^2 \times eD^{-D/D_0}$

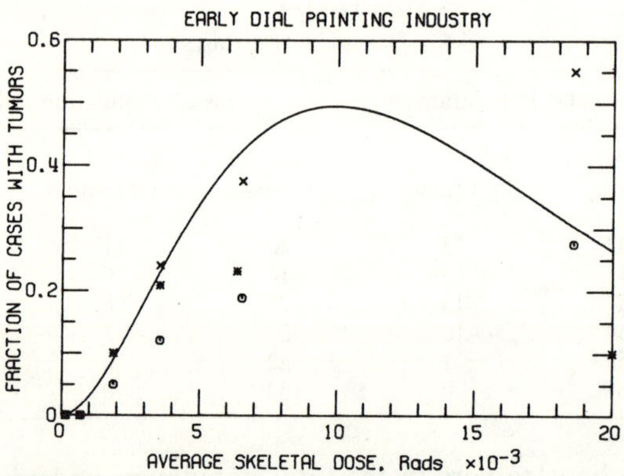

Fig. 4. Fraction of the listed cases in three U.S. dial painting plants with bone tumors in each dose range plotted as a function of average skeletal dose. X represents the fraction when all bone tumors are included. * represents the fraction when cases found as a consequence of a malignancy are removed. O represents the fraction when the number of cases in each dose range are doubled, to account for the unmeasured cases, on the assumption that there will be no more bone tumors in each range. The true fraction probably lies somewhere between the X's and the O's. The solid curve is the dose response shown in Fig. 3.

It is not possible to get a meaningful linear fit for a dose-response function to these radium data, either the bone tumor fraction from the entire study or from the selected employees of the three dial painting plants. The dose-squared exponential function derived from the entire population, since it fits the selected population adequately, is our best estimation for the bone tumor risk from internally-deposited ^{226}Ra or mixed ^{226}Ra and ^{228}Ra acquired from the dial painting industry. The function fitted to the bone tumor data is shown in Table 6.

TABLE 6
Risk Estimates for ^{226}Ra in Man

Conditions:
Exposure in dial painting industry
Calculated for period 35–59 years after first exposure
Risk calculated as:
Fraction of cases with a bone tumor as a function of dose
Dose calculated as average to mineralized skeleton

$$F = 3.7 \times 10^{-8} D^2 e^{-D/4970}$$

This function predicts 30 bone tumors in 1000 cases, each with a skeletal dose of 1000 rads, exposed under the described conditions. If we use this risk at this level, we can compare the expected number of such tumors at lower doses, depending upon whether, below this level, the dose-response function is linear or continues to follow a dose-squared exponential. These predictions are shown in Table 7.

TABLE 7
Bone Tumors Predicted from ^{226}Ra and ^{228}Ra

Dose (rads)	Linear function tumors/1000 cases	Dose-squared-function tumors/1000 cases
1000	30	30
500	15	8.4
100	3	0.4

I've now described, briefly, due to time limitations, the status of the two radium studies currently underway, and my estimates of the bone tumor risks derived from each. Would it be possible to extend our knowledge by combining the two studies? I consider it appropriate at this session on Human Radiobiology to indulge in some speculation, and, if we wish to extract information on the risk of malignancy, some speculation is required. Let us consider how the dose to the skeleton from ^{224}Ra and its daughters might be related to that from ^{226}Ra and ^{228}Ra and their daughters. Spiess and Mays have done so, pointing out that we probably should compare the dose delivered to the first 10 microns of soft tissue adjacent to bone (*3*). From their calculations of the dose to this 10-micron layer we learn that, when the ^{226}Ra and ^{228}Ra dose is expressed as the average skeletal dose, then the average skeletal dose for ^{224}Ra must be multiplied by a factor of 14.2 to obtain a set of comparable dose levels. That is, the dose to the 10-micron layer from the skeletal content of ^{226}Ra that yields an average skeletal dose of 1000 rads is the same as the dose to this layer from the quantity of ^{224}Ra that delivered 1000/14.2 or 70 rads to the skeleton.

However, if we multiply all the ^{224}Ra dose levels by this factor of 14.2, we find that a plot of the two sets of data on the same set of coordinates is meaningless. This may indicate that the dose from the relatively short exposures from ^{224}Ra in this series (from a few weeks to a few years) cannot be compared to the originally high, then subsequently lower lifetime dose from ^{226}Ra in a dial painter. On the other hand, it might also hint that our calculations of the doses are not correct. My colleague, John H. Marshall, is now actively investigating this latter possibility.

We can, however, compare the two sets of data empirically, and this I have done in Fig. 5. Here are the two sets of data plotted together after the ^{224}Ra data have been multiplied by the factor 3.5. I think you will agree with me that the two sets of data fit remarkably well, and you now see why I chose to use this particular factor when presenting the ^{224}Ra data.

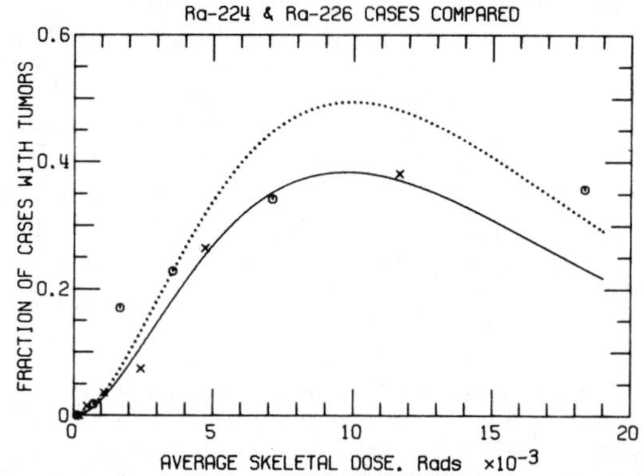

Fig. 5. Superimposed tumor fractions for ^{224}Ra and ^{226,228}Ra as functions of average skeletal dose. O shows the tumor fractions for ^{226,228}Ra series, fitted (---) by $F = \text{const} \times D^2 \times e^{-D/D_0}$. X shows the ^{224}Ra series, similarly fitted (—).

If we now can find a theoretical basis for a combination of this nature, it might be possible to consider combining the two studies into one enlarged population. It would be premature to do more than suggest this possibility now. However, it is tempting, for the essentially doubled population would provide greater predictive power for the shape of a dose-response function than either study by itself.

My purpose has been to point out where we are today in our attempts to quantitate the risk of malignancy from internal emitters. It appears that only from the two large radium studies will we ever have a large enough population with a significant spread in doses to attempt to formulate risk estimates from the population itself, that is, by letting the data itself define the shape of the dose response relationship.

I have taken care to point out the weaknesses of thse current studies; their far-from-completeness of follow-up, their built-in tumor bias that will only disappear as we approach 100% follow-up; and, as a consequence of the long latent period in man, the time it takes to complete these studies. I do not wish to leave you, however, with only a negative impression of these studies. It may well be that only these studies have the power to show the true shape of a dose-response relationship in man. For this reason alone these most difficult and complex studies must continue, with their greatest emphasis directed toward achieving 100% follow-up of well-defined groups. Only by achieving this primary goal will we ever be able to relate dose to risk. However, if we can determine this relationship for radium in man, then we should be able to extrapolate, from properly formulated models, the risk to man as a function of dose from those isotopes of the future—plutonium, americium and the other transuranic elements, isotopes we hope will never contaminate man at high enough levels to show effects.

REFERENCES

1. R. A. Conard, B. M. Dobyns, and W. W. Sutlow, *J. Amer. Med. Assoc.* **214,** 316 (1972).
2. M. Faber (Ed.), *Proc. Third Int. Meet Toxicity of Thorotrast,* Danish Atomic Energy Commission, Risø Report no. 294, Denmark, 1973.
3. H. Spiess and C. W. Mays, *Health Phys.* **19,** 713 (1970).
4. H. Spiess and C. W. Mays, in *Radionuclide Carcinogenesis* (C. L. Saunders, R. H. Busch, J. E. Ballou, and D. D. Mahlun, Eds.) U.S. Atomic Energy Commission, Conf–720505 (1973). (Available from National Technical Information Service, U.S. Depart. of Commerce, Springfield, Va. 22151).

RE-EVALUATION OF HAZARDS OF RADIATION EXPOSURE RELATIVE TO OTHER ENVIRONMENTAL AGENTS

Chairman: U. H. Ehling

[1]Plenary Session E–1

Chemical and Radiation Carcinogenesis in Man and Experimental Animals[1,2]

James A. Miller and Elizabeth C. Miller
McArdle Laboratory for Cancer Research
University of Wisconsin Medical Center
Madison, Wisconsin 53706, USA

INTRODUCTION

It is now well established that some cancer in man results from exposures to certain chemicals and radiations, both ultraviolet and ionizing radiations. These chemical and physical agents are also carcinogenic in experimental animals and, where adequately tested, in mammalian cell cultures. However, only very limited data are available on the relative roles of and the interrelationships, if any, between these various environmental agents in the causation of the majority of the cancers in man. Nothing is known of the relationship between these agents and possible carcinogenic viral information in the etiology of cancer in man. Furthermore, little is known about the molecular mechanisms by which chemicals and radiations induce cancers in either man or experimental animals. The objective of this brief review is to present certain aspects of chemical and radiation carcinogenesis in man and experimental animals and some of the problems in the elucidation of their roles in carcinogenesis in the human.

ENVIRONMENTAL FACTORS IN CANCER INDUCTION IN THE HUMAN

Studies of the incidences of cancers at various sites in man have revealed that these incidences may differ greatly as a function of either the geographical location of the individual populations or of time. Thus, the incidence of stomach cancer among the U.S. white population has now decreased to one-third of its 1930 level, while during the same period the incidence of bronchogenic carcinomas has increased approximately 15-fold among men and 6-fold among women (Fig. 1) (*3*). During this same period cancer incidences for the colon or large intestine and the mammary gland have remained relatively constant. The populations of various countries also show strikingly different incidences of particular cancers. For instance, cancer of the stomach is approximately 8 times as frequent among the Japanese as among the U.S. white population, while cancer of the colon is only one-fifth as frequent among the Japanese (Fig. 2) (*4*).

[1] Congress presentation E–1–1.

[2] The subject of chemical carcinogenesis in man and experimental animals has been reviewed in more detail by the authors (*1,2*).

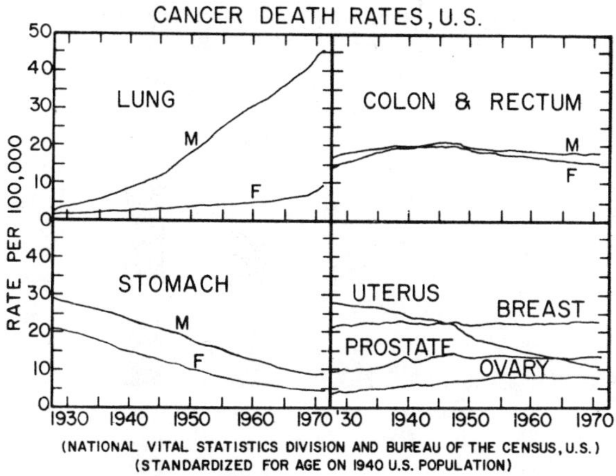

FIG. 1. Changes in the mortalities from cancer at various sites in white U. S. males and females in recent decades (*3*).

Furthermore, as exemplified by the data on liver cancer in some parts of Asia and Africa and by data on esophageal cancer in the Caspian littoral zone, the incidence of a particular cancer can differ markedly from place to place within a relatively limited area (*5,6*). These geographical differences appear to have little, if any, genetic component, as exemplified by the data on the descendants of Japanese migrants to the U.S. (Fig. 3) (*7*).

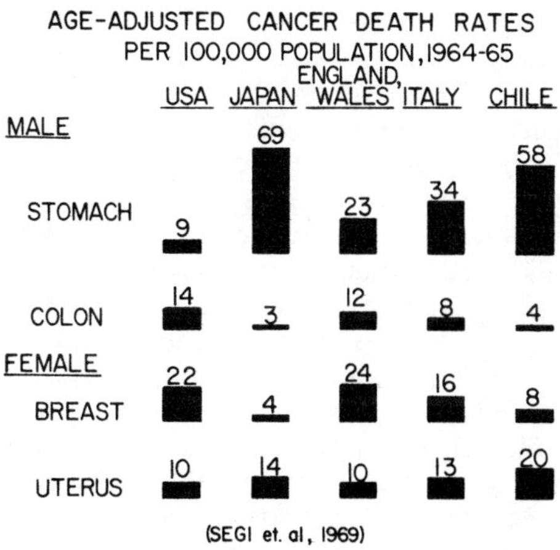

FIG. 2. Differences in the mortalities from cancer at various sites in selected countries (*4*).

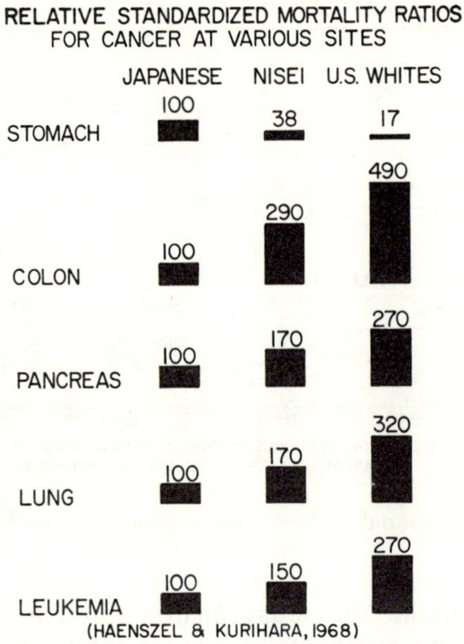

FIG. 3. Relative mortalities for cancer at various sites in second generation Japanese migrants to the United States as compared to Japanese in Japan (*7*).

The conclusion that has been drawn from the above and other data is that environmental factors play an important role in determining the incidences of an important share of human cancers. Indeed, Higginson concludes that as much as 90% of the human cancers have strong environmental factors in their etiology (*8*). This deduction receives support from the fact that those tissues (skin, lungs, gastrointestinal tract) which are in immediate contact with the environment are generally more likely to develop cancer than more protected tissues (*4*). Ultraviolet light has been identified as the major etiological agent in the induction of cancer of the skin of the exposed parts of the body in light-skinned individuals (*9*). In industrial situations cancers of the skin have also been related to exposures to certain soots, oils, and tars. Bronchogenic carcinoma is now related primarily to exposure to cigarette smoke, and the incidence increases markedly if certain industrial exposures (radon, asbestos) are superimposed on cigarette smoking (*10–12*).

A number of other chemicals have been shown to be etiological agents for human cancer (Table 1). In most of these cases, which involve either industrial agents or drugs, the exposed populations have been relatively small and, although the risks among the exposed populations have been high, these agents are not known to be important factors in the over-all incidence of cancer in the human population. The industrial chemical carcinogens include various aromatic amines (benzidine, 2-naphthylamine, and 4-aminobiphenyl) which are potent bladder carcinogens. They also include a number of chemicals which cause increased incidences of carcinoma of the respiratory tissues—bis(2-chloroethyl)sulfide, nickel compounds, chromium compounds, chloromethyl

TABLE 1

Chemicals Recognized as Carcinogenic in the Human and Experimental Animals

Chemical	Site
2-Naphthylamine, benzidine 4-aminobiphenyl, 4-nitro-biphenyl	Urinary bladder
N,N-bis(2-chloroethyl)-2-naphthylamine (chlornaphazine)	Urinary bladder
bis(2-Chloroethyl)sulfide	Respiratory tract
Diethylstilbestrol	Vagina
Chloromethyl methyl ether	Lungs
Vinyl chloride	Liver
Soots, tars, oils	Skin, lungs
Cigarette smoke	Lungs, urinary bladder
Betel nuts	Buccal mucosa
Chromium compounds	Lungs
Nickel compounds	Lungs, nasal sinuses
Asbestos	Lungs, pleura

methyl ether, asbestos, and radon. Vinyl chloride has recently been implicated in the induction of angiosarcomas in the liver. The drugs which are carcinogenic for the human are chlornaphazine, which induced carcinomas of the urinary bladder, and diethylstilbestrol, which has been implicated in the occurrence of carcinoma of the vagina.

A number of studies in which ionizing radiations were used either therapeutically or diagnostically, as well as analysis of the neoplasms which developed in the survivors of the atom bomb explosions in Japan, have shown ionizing radiations to be capable of inducing a variety of cancers in man (*13*). However, there appear to be no data at the present time which provide convincing support for the idea that background ionizing radiations are a major factor in the over-all incidence of any type of human cancer. The epidemiological and other problems involved in determining the carcinogenic hazards of low levels of ionizing radiations in human populations have been discussed recently in some detail (*14*).

CHEMICAL AND RADIATION CARCINOGENESIS IN EXPERIMENTAL ANIMALS

As in man, both ultraviolet and ionizing radiations can induce tumors in experimental animals (*9,13*). Ultraviolet light induces tumors essentially only in the skin, the target tissue that it can reach, while ionizing radiations induce neoplasms in a wide variety of tissues. In the latter case the sites of tumor formation appear to depend on the sites of localization and physical properties of the radiations. The tumors frequently develop at the site of irradiation, but they may develop at secondary sites as a consequence of a radiation damage-induced alteration in hormonal balance and its effect on the cellular growth and differentiation of other tissues (*15*).

Likewise, chemical carcinogens can induce tumors in a wide variety of tissues in experimental animals. These carcinogens comprise a chemically very diverse group of compounds, some examples of which are shown in Fig. 4. Some of these carcinogens characteristically induce tumors entirely or largely at sites of application, while others induce tumors primarily in tissues distant from the site of application. While most of the known chemical carcinogens are synthetic products, some, including the very potent hepatocarcinogen aflatoxin B_1, are metabolic products of living organisms (*16*).

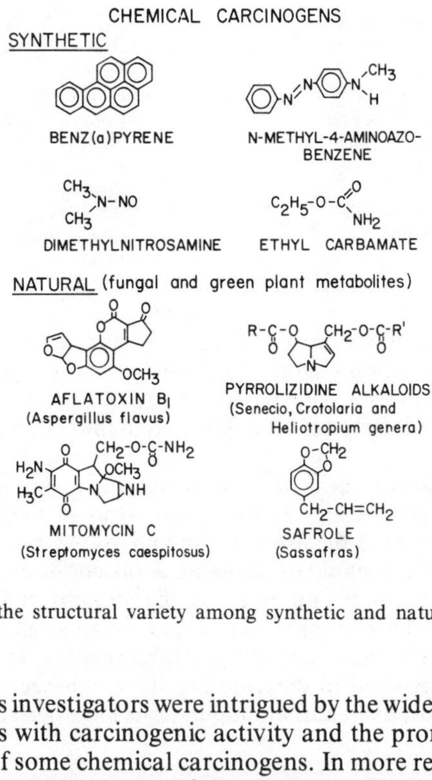

FIG. 4. Some of the structural variety among synthetic and naturally occurring chemical carcinogens (*1,16*).

For many years investigators were intrigued by the wide variety of seemingly unrelated structures with carcinogenic activity and the pronounced species and tissue specificities of some chemical carcinogens. In more recent years, however, it has become apparent that many of these compounds are not carcinogenic as such. Rather they are precarcinogens, which are metabolized by the host to reactive and carcinogenic forms known as ultimate carcinogens (Fig. 5). Thus, the relative extents of metabolism via carcinogenic and noncarcinogenic pathways are of importance in determining the carcinogenic activity of a chemical under a given set of conditions.

In the same manner as other pharmacologic agents, chemical carcinogens must induce their effects through interaction(s) with critical cellular components. Since the conversion of normal cells to malignant cells appears to be, at least largely, irreversible and is certainly heritable, the informational macromolecules of the cell (DNA's, RNA's, and proteins) have been implicated as the most likely sites of these interactions that lead to reduced or altered control of cellular growth. In fact, a wide variety of chemical carcinogens do bind covalently to all three types of macromolecules in tissues which are susceptible to tumor

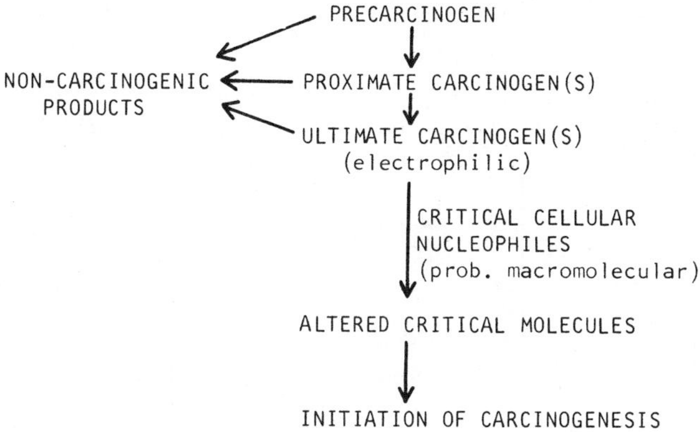

FIG. 5. The general metabolism of chemical carcinogens.

formation (*1,17,18*). In general, the levels of bound carcinogens are in the range of one carcinogen residue per 10^4-10^6 nucleotide or amino acid residues, and in some cases there are good correlations between the amounts of these bound carcinogen residues and the likelihood of tumor formation. In other cases, non-susceptible tissues may have as much total bound residues as susceptible tissues. Much more molecular detail is needed to pin-point the critical reactions for carcinogenesis.

It is now becoming apparent that some of the chemical carcinogens yield more than one reactive species and that each can react with the informational macromolecules (*2*). However, there is a unifying feature in that most, if not all, of these ultimate reactive forms are electrophilic derivatives that can react with nucleophilic sites in the cell. In Fig. 6 the type reaction is written as a S_N2 reaction in which charge separation and the attack of the nucleophilic reactant occur simultaneously. S_N1 reactions in which ionization of the electrophilic reactant occurs first and mixed S_N1/S_N2 reactions are also possible. In other cases, a bond in the carcinogen may be cleaved in homolytic or in redox reactions to yield free radicals, which are also electrophilic reactants. The formation of free radicals from chemical carcinogens may be particularly relevant to a discussion of carcinogenesis by ionizing radiation, since free radical formation is a primary consequence of irradiation of biological material.

Electrophilic reactants find a variety of nucleophilic targets in the nucleic acids and proteins, and those nucleophilic sites which have been demonstrated to react with chemical carcinogens in vivo are listed in Fig. 6. While a variety of studies suggest that the electrophilic reactive forms of carcinogenic chemicals are also ultimate carcinogenic forms, the critical interactions have not been defined. A number of studies have indicated that reaction at the N-7 position of guanine residues, usually the most frequent site of reaction by alkylating agents, may be of less consequence in carcinogenesis than alkylation at other sites (*19*). Correlations between the total amounts of alkylation at other sites on the nucleic acids and carcinogenesis have also been difficult, but recent studies by Goth and Rajewsky (*20*) have pointed to the rate of repair or removal of portions of DNA

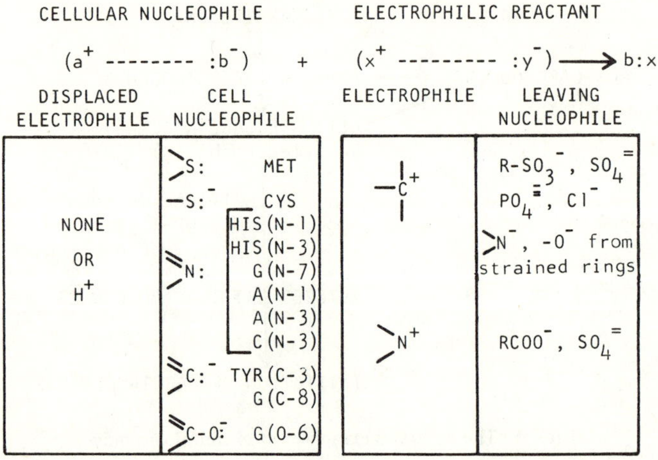

FIG. 6. Some of the nucleophilic targets in vivo of chemical carcinogens.

containing alkylated O-6 guanine residues as a possible key factor. The additional probability that reactions at particular portions of the biopolymers may be much more important determinants of carcinogenesis than reactions at other sites makes the analysis of the critical carcinogen–macromolecule interactions much more difficult.

The natures of the electrophilic reactive forms of chemical carcinogens in vivo are now becoming apparent, and some examples are shown in Fig. 7. The carcinogenic alkylating and acylating agents are electrophilic derivatives as administered and do not require metabolic activation. The alkylating agents encompass a relatively wide variety of structures—epoxides, ethylene imines,

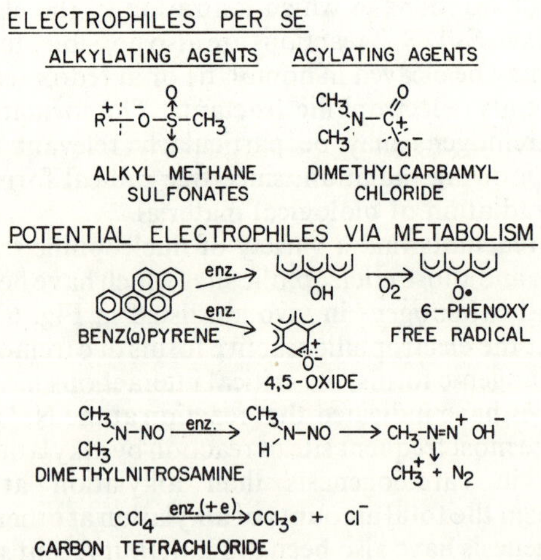

FIG. 7. The electrophilic reactive forms of some chemical carcinogens.

strained lactones, alkylsulfonates, N- and S-mustards, etc., while the known acylating carcinogens are rare. In contrast, the major share of the chemical carcinogens appear to require metabolic activation in vivo. The examples shown in Fig. 7 include the metabolism of benz(a)pyrene, an ubiquitous compound in man's environment from the incomplete combustion of organic material, to either the reactive 4,5-(K-region) epoxide or to 6-hydroxybenz(a)pyrene which is readily oxidized to a reactive free radical (*21,22*). Dimethylnitrosamine is one of a number of carcinogenic dialkylnitrosamines which can be oxidatively dealkylated to an unstable monoalkylnitrosamine and thus give rise to an alkylating species (*23*). Carbon tetrachloride, an hepatocarcinogen, is reduced in vivo to the very reactive trichloromethyl free radical (*24,25*). A variety of other examples have been presented elsewhere (*1,2*). Many of these reactions are catalyzed by P_{450}-NADPH enzyme systems in the endoplasmic reticulum of cells.

CHEMICAL CONSEQUENCES OF THE CELLULAR ABSORPTION OF IONIZING AND ULTRAVIOLET RADIATIONS

The initial events in the absorption of ionizing radiation by cells have been reviewed in detail elsewhere (*26,27*), but it is important to emphasize that the initial events in the cellular absorption of ionizing radiations also lead to electrophilic reactants. Thus, a profusion of free radicals are generated from macromolecules both directly and indirectly via the radicals generated in the radiolysis of water (Fig. 8). Molecular oxygen, which has diradical character, can also generate and react with free radicals to form a variety of oxidized products (*26–28*). As with other electrophilic reactants the most likely targets of these radiation-derived free radicals would appear to be base residues in the nucleic acids and amino acid residues in proteins.

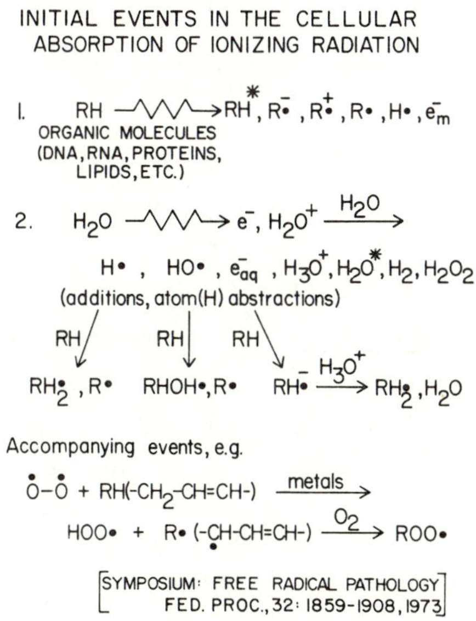

FIG. 8. Some of the initial events in the absorption of ionizing radiation by cells (*26,27*).

Carcinogenic ultraviolet light appears to be rather more subtle in its effects on the cell. It is absorbed by unsaturated bonds and, for example, induces pyrimidine dimer formation in DNA (*29*). Limited ability to excise these dimers has been correlated with susceptibility to skin tumor formation through study of Xeroderma pigmentosum patients (*30,31*). Setlow and Hart (*32*) at these meetings have provided an incisive experiment which strongly suggests that carcinogenesis by UV light in fish cells involves a mutagenic process.

MECHANISMS OF CARCINOGENESIS BY CHEMICALS AND RADIATIONS

Even with some knowledge of the ultimate reactive carcinogenic derivatives of chemical carcinogens or of radiations, we are still far from a detailed molecular knowledge of these carcinogenic processes (Fig. 9). In at least some situations, especially with ultraviolet light, carcinogenesis appears to result directly from mutational events which result in cellular information that is different from that of the normal parent cell. However, permanent lines of cells with reduced responsiveness to normal host controls can also be envisioned to develop as a consequence of reaction of an electrophilic reagent with either RNA's or proteins. These interactions may ultimately yield a genetic change (such as would occur if an abnormal DNA polymerase had an increased error frequency) or epigenetic events similar to those which are apparently of importance in the stable differentiation of an organism. In some or many cases activation of the expression of carcinogenic viral information through electrophilic attack on the cell may also be a critical component of a carcinogenic process.

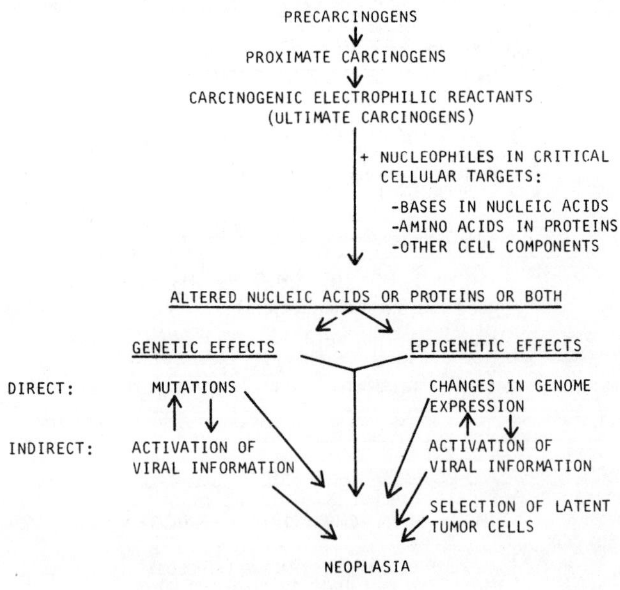

FIG. 9. Possible mechanisms of action of chemical carcinogens.

CONCLUSIONS

These brief considerations of cancer in man and experimental animals and the discussion of the initial steps in carcinogenic processes have emphasized the apparent similarity in the basic mechanisms of carcinogenesis by radiations and chemicals. Further understanding of these processes rests to an important extent on the elucidation of the relative roles of genetic and epigenetic events in the initiation of neoplastic cells and on more detailed information on the nature of promoting or cocarcinogenic events (e.g., the two-step induction of skin tumors in mice (*33*)). Of prime importance is an understanding of the extent to which viral information is necessary for, or enhances the development of, particular tumors in the human (*34*).

Equally important for the control of cancer in man are better means of detecting and estimating the carcinogenic hazards to man of environmental agents. In the case of viral information definitive approaches are probably not possible prior to the development of clear evidence for this component in the etiology of human cancers. Control of exposure to ultraviolet light and to ionizing radiations above background levels in our environments appears reasonably feasible at the present time. For radiations the major problems appear to be the question of the doses which can be considered to have acceptable benefit-to-risk ratios for the general population and the education of the public with respect to these hazards. Control of exposure to all of the naturally occurring and man-made chemical carcinogens appears to be much more difficult in view of the relatively large number of such chemicals, our lack of knowledge with respect to the identity or presence of some of them, and the greater difficulties of extrapolating from data on experimental animals or from in vitro systems to the human organism. However, an important recent advance with considerable promise has been the development of host and host-tissue mediated mutagenicity assays as sensitive indicators of the metabolism of chemicals to electrophilic reactants and, hence, of potential carcinogenic activity (*35–38*). Furthermore, in an important contribution Ehrenberg et al. (*39*) have recently demonstrated how the genetic effects of chemicals can be expressed as radiation-equivalent doses from a determination of their mutagenic effects. They have further suggested that the exposure doses of human tissues may be estimated from the levels of macromolecule-bound derivatives and the levels of macromolecular binding as a function of dose in experimental animals. These developments again suggest fundamental similarities between chemical and radiation carcinogenesis and provide further reason for optimism for real advances in the understanding of mechanisms of action and of control of these agents in our environments in the near future.

REFERENCES

1. J. A. Miller, *Cancer Res.* **30,** 559 (1970).
2. E. C. Miller and J. A. Miller, in *The Molecular Biology of Cancer* (H. Busch, Ed.), pp. 377–402. Academic Press, New York, 1974.
3. E. Silverberg and A. I. Holleb, Cancer statistics, 1974—Worldwide epidemiology. *CA—A Cancer Journal for Clinicians* **24,** 2 (1974).

4. M. Segi, M. Kurihara, and T. Matsuyama, *Cancer mortality for selected sites in 24 countries*, No. 5 (1964–1965). Tohoku University School of Medicine, Sendai, Japan, 1969.
5. M.S.R. Hutt, in *Liver Cancer* (IARC Scientific Publications No. 1), pp. 21–29. International Agency for Research on Cancer, Lyon, 1971.
6. E. Mahboubi, J. Kmet, P. J. Cook, N. E. Day, P. Ghadirian, and S. Salmasizadeh, *Brit. J. Cancer* **28,** 197 (1973).
7. W. Haenszel and M. Kurihara, *J. Natl. Cancer Inst.* **40,** 43 (1968).
8. J. Higginson, *Can. Cancer Conf.* **8,** 40 (1969).
9. F. Urbach, in *The Biologic Effects of Ultraviolet Radiation (with Emphasis on the Skin)* (F. Urbach, Ed.), pp. 635–650. Pergamon Press, Oxford and New York, 1969.
10. E. L. Wynder and K. Mabuchi, *Prev. Med.* **1,** 300 (1972).
11. E. C. Hammond and I. J. Selikoff, in *Biological Effects of Asbestos* (P. Bogovski, J. C. Gilson, V. Timbrell, J. C. Wagner, and W. Davis, Eds.), pp. 312–317. International Agency for Research on Cancer, Lyon, 1973.
12. D. A. Holaday, *Health Phys.* **16,** 547 (1969).
13. A. C. Upton, in *Cancer Medicine* (J. F. Holland and E. Frei, III, Eds.), pp. 90–101. Lea and Febiger, Philadelphia, 1973.
14. A. C. Upton, J. M. Brown, R. F. Kallman, M. A. Kastenbaum, and J. A. Miller, in *Research Needs for Estimating the Biological Hazards of Low Doses of Ionizing Radiations* (Report of an Ad Hoc Panel of the Committee on Nuclear Science, National Research Council), pp. 21–45. National Academy of Sciences, Washington, D.C., 1974.
15. J. Furth, *Cancer Res.*, **23,** 21 (1963).
16. J. A. Miller, in *Toxicants Occurring Naturally in Foods* (Committee on Food Protection, Food and Nutrition Board, National Research Council), pp. 508–549. National Academy of Sciences, Washington, D.C., 1973.
17. E. C. Miller and J. A. Miller, *Pharmacol. Rev.* **18,** 805 (1966).
18. P. Brookes, *Cancer Res.* **26,** 1994 (1966).
19. P. D. Lawley, in *Chemical Carcinogens* (C. E. Searle, Ed.), American Chemical Society Monograph, *in press*.
20. R. Goth and M. F. Rajewsky, *Proc. Nat. Acad. Sci., U.S.A.* **71,** 639 (1974).
21. P. L. Grover, A. Hewer, and P. Sims, *Biochem. Pharmacol.* **21,** 2713 (1972).
22. W. Caspary, S. Lesko, R. Lorentzen, and P. Ts'o. *Fed. Proc. Fed. Am. Soc. Exper. Biol.* **33,** 1500 (1974).
23. P. N. Magee and J. M. Barnes, *Adv. Cancer Res.* **10,** 163 (1967).
24. J. A. Castro and M. I. Diaz Gomez, *Toxicol. Appl. Pharmacol.* **23,** 541 (1972).
25. G. Krishma, I. G. Sipes, and J. R. Gillette, *The Pharmacologist* **15,** 260 (1973).
26. W. A. Pryor, *Fed. Proc. Fed. Am. Soc. Exper. Biol.* **32,** 1862 (1973).
27. L. S. Myers, Jr., *Fed. Proc. Fed. Am. Soc. Exper. Biol.* **32,** 1882 (1973).
28. H. B. Demopoulos, *Fed. Proc. Fed. Am. Soc. Exper. Biol.* **32,** 1859 (1973).
29. R. B. Setlow, *Progr. Nucleic Acid Res. Mol. Biol.* **8,** 257 (1968).
30. J. E. Cleaver, *Nature* (London) **218,** 652 (1968).
31. D. Bootsma, M. P. Mulder, F. Pot, and J. A. Cohen, *Mutat. Res.* **9,** 507 (1970).
32. R. B. Setlow and R. R. Hart, these proceedings.
33. R. K. Boutwell, *Progr. Exp. Tumor Res.* **4,** 207 (1964).
34. M. Benyesh-Melnick, and J. S. Butel, in *The Molecular Biology of Cancer* (H. Busch, Ed.), pp. 403–485. Academic Press, New York, 1974.
35. M. S. Legator and H. V. Malling, in *Chemical Mutagens. Principles and Methods for Their Detection* (A. Hollaender, Ed.), **Vol. 2,** pp. 569–589. Plenum Press, New York, 1971.
36. B. N. Ames, in *Chemical Mutagens. Principles and Methods for Their Detection* (A. Hollaender, Ed.), **Vol. 1,** pp. 267–282. Plenum Press, New York, 1971.
37. W. E. Durston and B. N. Ames, *Proc. Nat. Acad. Sci., U.S.A.* **71,** 737 (1974).
38. E. C. Miller and J. A. Miller, in *Chemical Mutagens. Principles and Methods for Their Detection* (A. Hollaender, Ed.), **Vol 1,** pp. 83–119. Plenum Press, New York, 1971.
39. L. Ehrenberg, K. D. Hiesche, S. Osterman-Golkar and I. Wennberg. *Mutat. Res.* **24,** 83 (1974).

Problems in the Reevaluation of Genetic Risks from Radiation and Other Environmental Hazards[1]

Per Oftedal
Institute of General Genetics
University of Oslo, Norway

Till recently—say 3 or 5 years ago—there was a wide gap between the accepted norms for radiation protection and the levels of population exposure due to industrial uses of ionizing radiation and to releases of radioactive substances from reactors.

Today, this wide gap between what the ICRP calls Maximum Permissible Doses—or Dose Limits—and the prospective doses is not so indisputably clear. On the one hand, the growth of the energy demand has led to the realization that an increasing fraction of the MPD will be called upon in the foreseeable future. On the other hand, questions have been asked about what *is* an acceptable risk and an acceptable dose level. Through increased public awareness of the levels and possible effects of radiation, the pressure towards reduced releases and decreased permissible dose levels has been felt quite strongly, and has led to political decisions.

So, we are in a relatively tight position between the needs of the energy producers on the one hand, and the protectionists on the other. Both have in common, however, the need for clear answers and exact figures upon which to base technical and practical decisions.

One of the most striking simplifications of the presentation of experimental data on genetic effects of radiation is the Abrahamson-Bender-Conger-Wolff (*1*) paper in *Nature* in the fall of 1973. These authors discovered that there is a linear relationship between the efficiency of mutation induction by radiation—i.e. forward mutations per locus per rad—and the DNA content in the haploid genome of the organism in question.

This relationship is interesting in many different ways, and gives a very convincing ring to the main conclusion of the authors, namely that it is possible and justified to extrapolate, on the basis of animal and plant data, to man. This is obviously important, and serves to increase the confidence in a method which has been generally applied, though at times with some doubt. The relationship also indicates that one can utilize, for instance, Drosophila data with the same confidence–or scepticism–as mouse data.

But when it comes to understanding the relationship itself, it is very difficult to find an interpretation of general appeal. It might be taken to mean that the cell nucleus *in toto* determines the radiation effectiveness, and not the individual loci. In fact, the general approach has much in common with Sparrow's (*2*) concepts of radiotaxons, based on interphase chromosome volume. Another interpretation —likewise pointed out by the authors—is that the greater complexity of higher organisms should come about by essentially the same number of radiogenetic targets being present, but being larger and more complex in higher organisms. This sounds vaguely plausible, though I can think of no other field in which humans can be quantified as being 30 percent more complex than mice, or—for that matter—four times less complex than barley.

I will also mention Grant Brewen's interpretation (*3*) of chromosome breakage frequencies in irradiated cells from six different mammalian species, based on chromosome arm numbers, which again leads to increased confidence in the correlation between animals and man.

Translocation data have been generated at a great rate, and though the variability is considerable among experimenters and with experimental conditions, the general trend is rather uniform (Fig. 1).

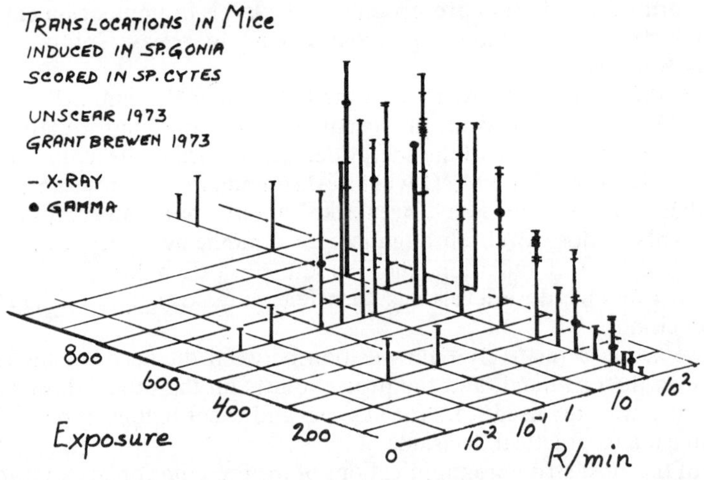

FIG. 1. Incidence of translocations in the mouse, induced in spermatogonia and observed in primary spermatocytes. [Data from Refs. *3* and *21*.]

The general shape of the dose-effect curve indicates a saturation plateau at 500–600 R acute exposure, and also a dependence on dose rate. The diagram demonstrates some aspects of research tactics: the number of experiments being done in a few years when a cheap, efficient and dramatic technique becomes available; the tendency to do experiments in the same dose range—there are 16 observations at 500–600 R with about 100 R/min exposure rate; and the questionable value of doing dose-rate experiments in a dose range where the acute response is already complex—as a plateau would indicate. The dose rate effect of 100 vs 0.1 R/min is twice as high at 600 R as at 300 R. Due to a saturation at the higher dose it may in reality be higher. At low doses, one wonders if there will be a dose rate effect at all. The data are simply lacking.

Permit me now to go back to some of my own work. I shall be referring to the humped dose-effect curve as it appears in experiments with Drosophila spermatogonia. A humped dose-effect curve can be constructed very easily on the assumption of heterogeneous sensitivities to mutation and killing in the exposed cell populaton, and a coincidence of the high phase of both sensitivities (*4*). This present model was designed on the basis of an interpretation of experimental data from acutely-irradiated Drosophila spermatogonia (*5*).

The model was tested by protracted irradiation, on the logic that protraction should lead to greater survival of sensitive cells, and therefore higher mutation yield (*6,7*). This would be a good test, because it is just the opposite of the commonly accepted view that protracted doses are less efficient, based on the impressive data from Russell's (*8*) group working with mice. In our experiments protracted irradiation did indeed give much higher yield of mutations, about five times higher than was expected on the basis of Abrahamson and Friedman's (*9*) data for exposures to several thousand R. But again, at 550 R, there is no yield above controls (Fig. 2).

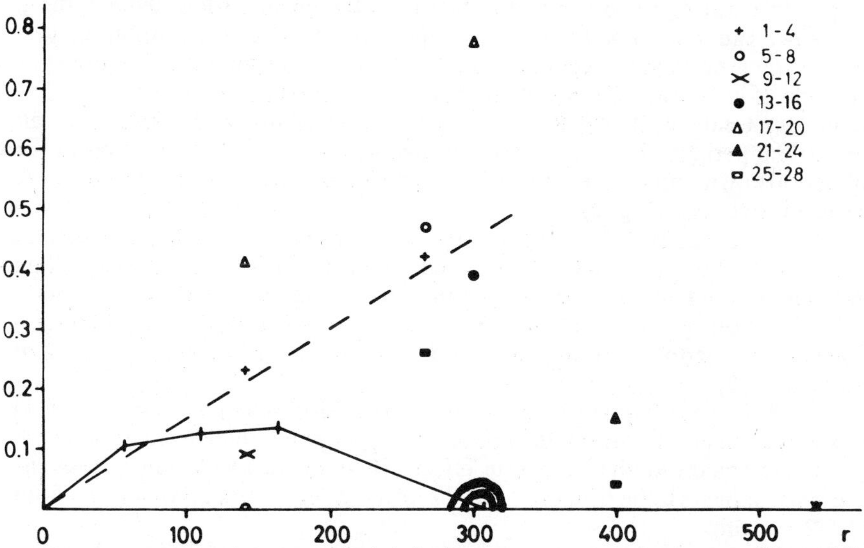

FIG. 2. Yield of sex-linked recessive lethals after acute or protracted exposures of Drosophila spermatogonia. [From Refs. *5* and *6*, with the addition of Gonzales' results (*9*) for 300 R acute and protracted exposures.]

It was therefore a great pleasure to hear that Dr. F. W. Gonzales, a competent research worker at Brookhaven, had undertaken to repeat the experiments as closely as possible. He completed the experiments, but unfortunately became ill and died, so he did not write up and publish the data. Gonzales measured the effect of 300 R acute and 300 R protracted irradiation over 8 hours, as I had done. But since he was testing just this most significant dose, he could do better than I, using the same egg collection for the two exposure groups etc. Gonzales' data[2] for 300 R acute exposure confirmed my results. He had no effect above controls (Fig. 2).

But for the protracted doses, however, our results were not readily reconciled. Gonzales found no effect at all of 300 R protracted exposure, and there is apparently a clear contradiction between his data and mine (Fig. 2).

[2]F. W. Gonzales (being prepared for publication by Professor S. Abrahamson).

In searching for differences between the experimental set-ups, I found only one aspect which differed, namely that of temperature control. Drosophila is cultured at 25° C. At the relevant developmental stage, the spermatogonia divide very rapidly. The mitotic period is about two hours, as is known from experiments on mutation yield in synchronized eggs (*10*). The periods between sensitive stages are about two hours, and this fits with the estimates of cell proliferation based on histological observation (*11*).

Gonzales did his irradiation in a gamma room: Acute exposures close to a cesium source, protracted ones at a distance of several meters. The lab journal indicates that the Victoreen dose measurements were made at 21 degrees. There is no mention of further temperature control. I therefore wrote Professor Rossi, who discussed the matter with Dr. Gonzales' technician who confirmed that no special temperature control had been used in the gamma room, and that the temperature during exposure had in fact been the normal room temperature.

Now, can this explain the diverging results? On one assumption, yes. Namely, that the important parameter is not the total dose but rather the dose per cell cycle. If in Gonzales' experiment the cell cycle time has been doubled, so that he in fact treats with 300 R over two cell generations, and 150 R per cell generation, perhaps the best basis for comparison would be my earlier result for 550 R—also with no mutation yield above controls—and also with about 150 R per cell generation (Fig. 2).

Thus the humped dose-effect curve after protracted irradiation indicates that the height of the hump is determined not only by the sensitivities to killing and to mutation induction, but also by the dose per cell cycle. This is a parameter which has been priorly used in discussions of somatic effects on proliferating tissues by Lamerton (*12*), but only briefly in discussions of genetic effects of radiation (*6,7*).

In 1972 Lyon, Papworth and Philips (*13*) published an analysis of the yield of specific locus mutations in mouse after spermatogonial irradiation, with specific reference to the dose rate effect. Their main conclusion is that the efficiency appears to be rising again at the lowest rates, from a minimum at about 0.01 R per min.

Dose rate is of course a complex parameter made up of the two elements, total dose and duration of exposure. Fig. 3 shows a tridimensional chart of the three elements pertaining to the mouse experiments. A quantitative analysis has not been attempted. The trough has now become a valley, with a minimum just above ten days' exposure. The rise is somewhat variable but fairly convincing at exposure times of 40 days or more, and the impression is one of a minimum efficiency of radiation when the dose is spread over a single cell cycle—which is on the order of two weeks in the mouse spermatogonia—and an increase in efficiency when the dose is spread over several cell cycles. This impression is strengthened by data presented at this Congress on the mutagenic efficiency of protracted irradiation from tritium in the mouse indicating an RBE approaching 1, in comparison with acute X-rays or gamma rays (*14*).

The minimum efficiency might either be the result of a maximally stimulated and yet not damaged repair system, or of elimination of sensitive cells and a corresponding pile-up of resistant cells in the most resistant stage of the cell cycle. The long sterility period after protracted exposures with about 100 R per cell cycle (*13*) would appear to indicate that killing and pile-up may be the more plausible interpretation. With exposures at about 10 R per cell cycle there is no sterility.

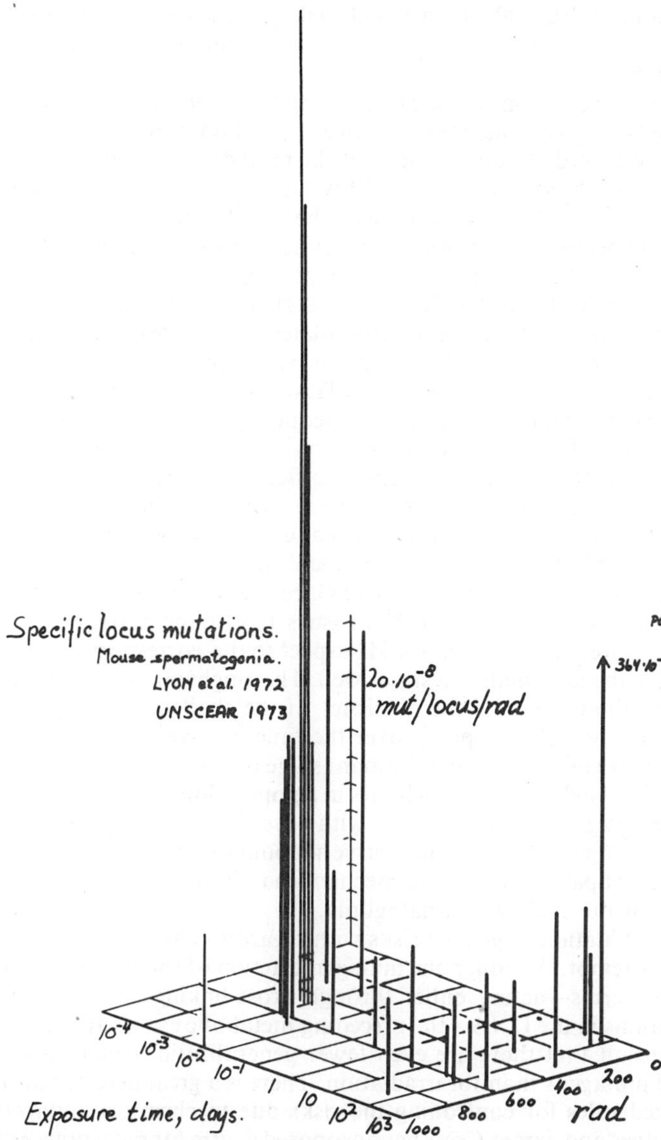

FIG. 3. Specific locus mutations induced in mouse spermatogonia. Efficiency of acute protracted and fractionated exposures. The group of four tall columns at 500 rad represent the efficiency of challenging doses 24 hr after a conditioning dose of 100 or 500 rad. [Data from Refs. *13*and *21*.]

One additional aspect is brought out in the dose range covered by the acute exposures, from 300 to 1000 R: The efficiency is steadily decreasing with increasing dose. Ordinarily, the situation is described as a linear yield increase up to 600 R and a fall at 1000 R. One might ask if it be justified to extrapolate back to zero dose, which leads to an efficiency estimate of about $35 \cdot 10^{-8}$ mutations per locus per rad.

We should also consider the fascinating fractionation effect found by Dr. W. L. Russell (*14*). Two doses may—when given 24 hours apart—have the efficiency just about doubled, on the average, for the total dose. The picture becomes even more interesting if one analyses this fractionation effect in the terms of conditioning and challenging doses. The conditioning dose would have the efficiency expected on the basis of other acute exposures, the challenging dose would have the efficiency necessary to explain the increased effect. As a result of conditioning with 100 or 500 R 24 hours earlier, the efficiency of the challenging dose is up to twice as high as the extrapolated zero acute dose efficiency (Fig. 3).

This phenomenon may be interpreted in various ways. In any case it must be taken to indicate that some cell populations under some conditions have this genetic response and transmit it to subsequent generations, but the explanation of this important phenomenon is really quite open still. It is perhaps in order to point out that this array of mutagenic efficiency values is condensed into two values in the ABCW procedure, with a factor of three—for protraction—between them. Interlocus and intersex variabilities are very large, and come as additional factors of non-uniformity.

A proliferative pattern containing G_0 cells in addition to those cells normally proliferating is known from other tissues (e.g. bone marrow, and skin). On ground of pure analogy, one would suspect that a corresponding pattern might be found in the mammalian testis as well. This might also resolve the difficulty of explaining the erratic genetic efficiency of the challenging dose. It is easier to understand a variable response over the time involved—the first 24 hours—in terms of cells coming out of a dormant stage due to the conditioning dose, than by invoking synchrony after 24 hours in a population with a two weeks' turnover time, or damage to a repair system with doses of one hundred or a few hundred rads. The trough in the efficiency curve at about two weeks seems to indicate that the important parameter for this mechanism to become operative is the dose per cell cycle of the cycling spermatogonia.

A reevaluation of genetic risks with regard to chemical mutagenesis I shall not even attempt. Although the mutagenic action of chemicals may be simple in molecular terms—acting either through cross-linkings, base substitutions or intercalations in the DNA—the preceding metabolic events are so important for the end result that there are even fewer generalizations that can be made for chemical mutagens than for irradiation. There is a great need for the creation of generalized rules for controlling the risks due to chemical mutagenesis. Both Bryn Bridges and James Crow have proposed a term for rad equivalents, in order to allow a formulation of Maximum Permissible Doses, derived permissible concentrations, etc., and I think we all hope that a formula of the type 'mole seconds per liter' might be used for summing and evaluating the total chemical mutagen load, in analogy with 'Ci seconds per cubic meter' used for exposure to a radioactive gas. The most promising approach, at least for alkylating agents, appears to come from Ehrenberg's group in Stockholm (*15*), using alkylations of the histidine moiety of long-lived proteins in the blood (e.g. hemoglobin) as an exposure measurement, and RBE factors for calculation of the mutagen load.

I want to discuss some examples of the complexities involved when the effects of exposure to chemical mutagens are to be quantified.

I refer first to a dose-effect curve published some ten years ago by Zimmermann and Schwaier (*16*), showing survival and $ad_{6\text{-}45}$ reversions in *Saccharomyces cerevisiae* after treatment with 0.003% N-nitroso-N-methyl-acetamide for various times. The survival curve is heterogeneous, and the reversion curve sharply humped: 30 seconds' treatment gives 95% survival and the same frequency of reversions as does 30 minutes' treatment, with 60% survival. The data may be interpreted as a result of coincident heterogeneities in killing sensitivity and mutation induction sensitivity, either as a true variation in the two sensitivities or as a variation in one metabolic parameter—say permeability—which infuences the efficiency of the chemical both as a killing agent and as a mutagen. However, with a chemical, it is also possible to view the reaction pattern as a result of two processes: On the one hand the killing and mutagenic action of the chemical on an unadapted organism, and on the other hand the time-dependent development of a more resistant metabolism which reduces both the killing efficiency and the mutagenic efficiency by a large factor.

Even with chemicals, a given dose can be administered at a high or a low rate, with quite different results. In a recent study by Igali, von Borstel and one of their students (*17*) on the killing efficiency of the anti-schistosomiasis agent hycanthone on yeast cells, it has been found that when a constant concentration is applied for variable times, the killing efficiency is much higher than when the dose is varied by changing the concentration with a constant exposure time of 8 hours. But in the same experiments, the compound IA-5 (which represents a fairly simple change in the hycanthone molecule) is much more efficient in killing with a constant exposure time of 8 hours and a variable concentration, than with a constant concentration and variable exposure time, for the same milligram-hour exposure.

Now let me say a few words about interaction between chemicals and radiation. I choose as an example the action of caffeine, which is a purine derivative, and which appears to modify the effect of radiation by the blocking of repair systems ordinarily active in the cell. This has been demonstrated very neatly by one of our students, Anne Marstokk (*18*), using a test system proposed by Dr. Angus Bateman in Manchester, that of the empty Drosophila testis: If *Drosophila melanogaster* males are exposed to doses between 5 and 10 kilorads, the proliferating spermatogonia may survive or be killed off, and the testis may be dissected out and scored two weeks later as being full, or as being empty. Dose-effect curves may be obtained in the usual way.

If the males are fed caffeine, killing of the spermatogonia is enhanced. In the dose-effect curves the slope is unchanged, only the D_Q being affected (Fig. 4). This corroborates the findings of several other investigators, indicating that it is mainly repair capacity which is affected by caffeine.

Dr. Trygve Eklund at our Institute has used the humped dose effect curve in yeast for a similar type of analysis. In starved yeast cell cultures, the cells are homogeneously sensitive. With heterogeneous survival, i.e. growing cultures, a humped dose-effect curve is obtained for adenine reversions, but not for threonine reversions (*19*) (Fig. 5). The mutations are of different types. The adenine mutation is a base substitution, while the threonine is a frameshift mutation. According to Eklund's results, the variation in sensitivity to killing and to mutagenesis thus appears to be coincident with regard to reversion of base subtitution mutations, but not with regard to reversion of frameshift mutations.

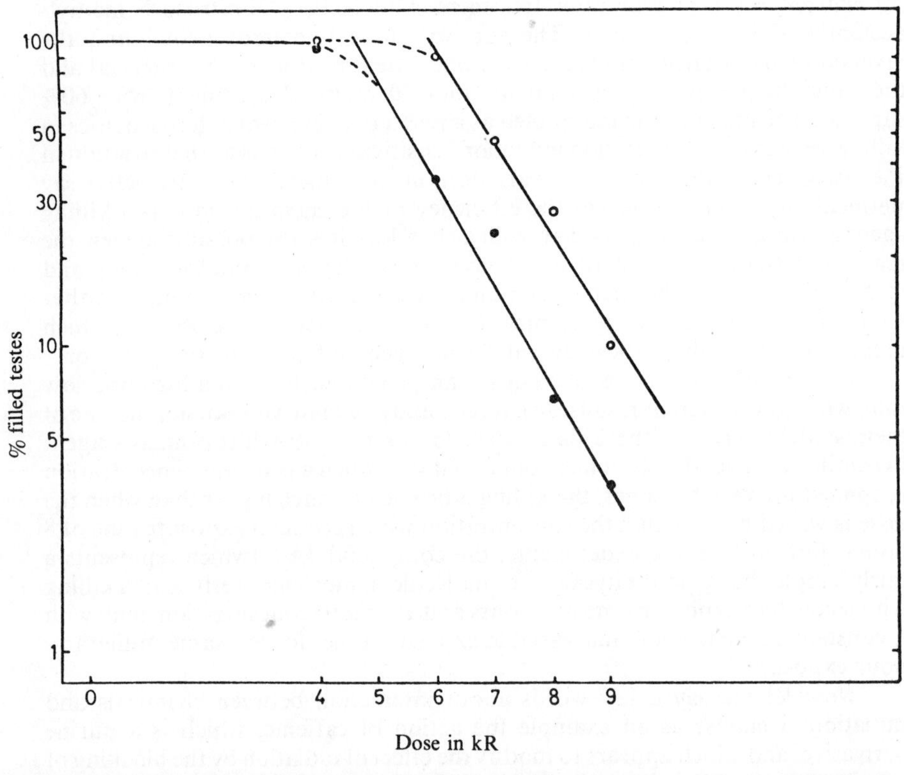

FIG. 4. Sterilization of *Drosophila* testis with and without caffeine feeding; ●: with caffeine (1%), o: without caffeine [Anne Marstokk, unpublished thesis.]

With caffeine there is in these cells no effect on the killing pattern (Fig. 6). The same heterogeneity prevails; however, there is again a differential effect on the two reversion types, but the picture is now more complex. With threonine reversions caffeine has no effect, but in the adenine reversion curve, the reversion frequency is reduced by a factor of about 2 and the hump may also be reduced (Fig. 7). Thus, the difference between the two types of reversion is again apparent, but the relationship between killing effect and adenine reversion effect is not clear. Based on the conventional interpretation of caffeine as a repair inhibitor, the reduction in adenine reversions may be interpreted as a result of a blocking of those repair processes which revert base substitutions, and are themselves mainly base substitutions. Apparently these processes do not determine the survival of yeast cells after radiation damage, in either the sensitive or the resistant fraction of the cell population.

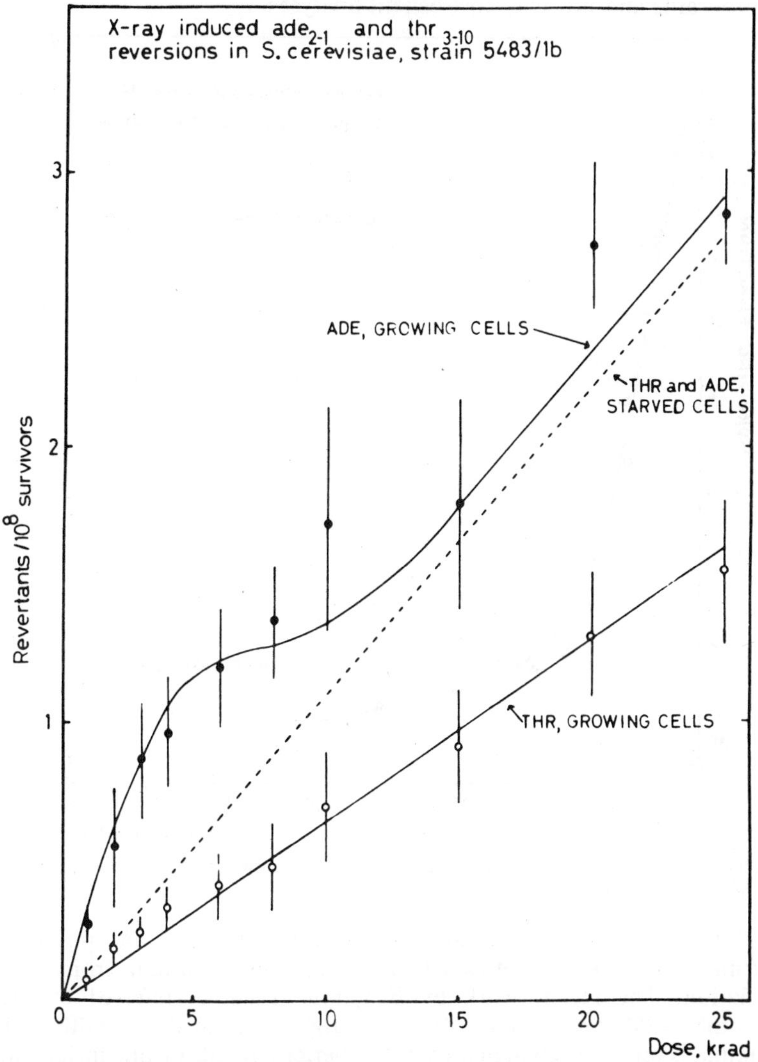

FIG. 5. Yield of ad_{1-2} and thr_{4-1} reversions after X-irradiation of *Saccharomyces cerevisiae* in growing cultures. [T. Eklund, unpublished.]

Obviously, with a dose-effect relationship of this kind, it becomes very difficult to develop a permissible exposure concept for chemical noxae. In addition to the exposure concept (like mg h/1), and an efficiency factor (like RBE), an exposure pattern factor becomes necessary, as well as a factor for interaction. The solution appears to be either to set the allowable intake so low that every variable is unimportant due to the large safety factors used, or of adopting an ad hoc and individual evaluation for each chemical. It is not necessary to point out the implications of the latter approach when it comes to test programs and intricate cost-benefit analyses.

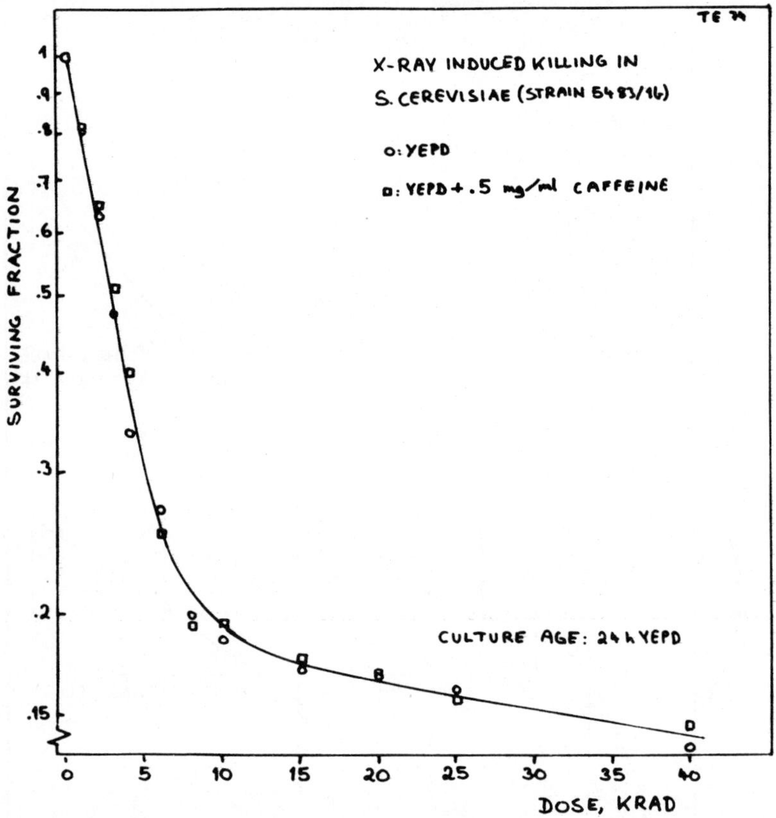

FIG. 6. Survival of growing *S. cerevisiae* cells irradiated with caffeine present or absent in the medium. [T. Eklund, unpublished.]

But the proof of the pudding comes with the eating, and no amount of experimental work can outweigh a corresponding datum from a human population. The recent work by Benjamin Trimble, referred by Howard Newcombe (*20*), on the level of genetically determined ill health in British Columbia is very important in providing a better estimate of, and insight into, the dynamics of our genetic load. Following this line of thought, more effort should be put into the study of human populations of more or less well-defined risks. The OECD/NEA in Paris is now attempting to organize a multinational study of the possible relationship between preconception exposure to diagnostic X rays of the human female and the frequency of Down's syndrome in her offspring. With increasing probability of survival from certain cancer forms (such as Hodgkin's disease, and childhood leukemia), small and heavily exposed patient groups become available for study, and also their progeny. Workers in chemical industries may provide useful groups. In some situtations, such as with hycanthone treatment in the tropics, the AF-2 food additive in Japan, and with vinyl chloride industrial releases, large population groups have been exposed, and should be studied. It is somewhat disappointing that the WHO has not been more active in these matters at an early stage.

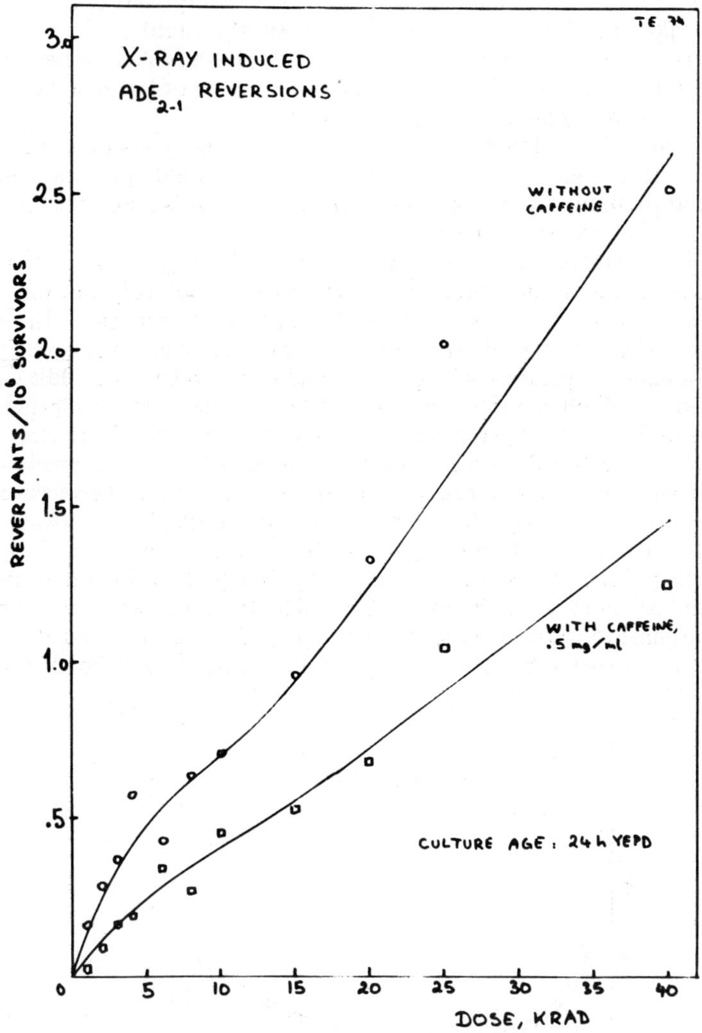

FIG. 7. Yield of ad $_{1\text{-}2}$ reversions after X-irradiation of *S. cerevisiae* in growing cultures with caffeine present or absent in the medium. [T. Elklund, unpublished.]

Genetic risk estimates have been arrived at by two very eminent committees during the last couple of years (*21, 22*). The risk estimates have wide ranges, but overlap to a high degree. Basically, of course, the underlying observational data are the same, and there is some overlap in the committee memberships, so it is perhaps not altogether unexpected that rather similar estimates should result.

One surprising aspect is that there is such repeated emphasis of the reduction by two thirds in radiation efficiency due to a presumed low dose rate or multiple small fraction exposure of human populations. My surprise stems partly from the uncertainty with which I regard the relevance of the experimental findings, and the lack of coherent explanation for it. But it is also surprising that this factor of three is taken so explicitly into account, against a doubling dose estimate which covers an order of magnitude, or a fifty-fold (UNSCEAR) or twenty-five or eighteen fold (BEIR) range for risk estimates for more or less specific genetic conditions. Is there a political wind blowing here?

Apart from all this, I think I should end by reminding you of the shape of a cost-benefit curve. Again, a sort of Mac-Namarian rational optimism tends to dominate our thinking, and we expect that when all the cards are on the table, the decision will be clear-cut and easy to make.

In reality, the trough of the cost-benefit curve is flat. Fig. 8 shows one drawn with reference to radiation protection and cancer induction (*23*), but the shape will not be different for mutagenesis, or for effects of chemicals: There are obvious areas where the cost of restriction or permissiveness are excessive, and all reasonably intelligent persons will agree to avoid these, and in the middle is a flat area with no well-defined minimum. An expert with responsibility in protection will move into this trough from the area of overcostly restriction, and stop when the costs for protection become acceptable. An expert with responsibility in production will move into the trough from the area of overcostly permissiveness, and will stop when the costs for damage become acceptable. The twain shall never meet. The resolution of the conflict is a political problem and will be solved according to political rules: In some cases by dictatorship, in some through intrigue, and at times through reasonable deliberation and weighing of many kinds of arguments. One may complain about the lack of insight among responsible non-experts, but the blame for this condition should be placed with great care.

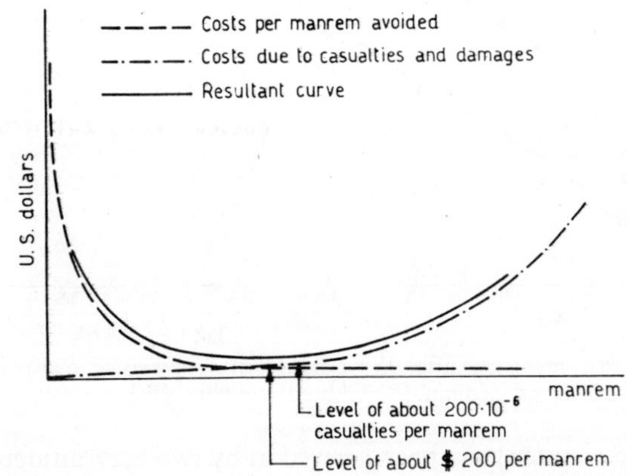

FIG. 8. Schematic presentation of a cost-benefit curve for radiation damage (*23*).

REFERENCES

1. S. Abrahamson, M. A Bender, A. D. Conger and S. Wolff, *Nature (London)* **245**, 460 (1973).
2. A. H. Sparrow, A. G. Underbrink and R. C. Sparrow, *Radiat. Res.* **32**, 915 (1967).
3. J. G. Brewen, R. J. Preston, K. P. Jones and D. G. Gosslee, *Mutat. Res.* **17**, 245, (1973).
4. P. Oftedal, *Hereditas* **60,** 177 (1968).
5. P. Oftedal, *Genetics* **49**, 181 (1964).
6. P. Oftedal, *Hereditas* **51,** 13 (1964).
7. P. Oftedal, *Mutat. Res.* **1**, 63 (1964).
8. W. L. Russell, in *Repair from Genetic Radiation Damage* (F. H. Sobels, Ed.), pp. 205–217. Pergamon, London, 1963.
9. S. Abrahamson and L. D. Friedmann, *Genetics* **49**, 357 (1964).
10. P. Oftedal, *Scand. J. Clin. Lab. Ingest.* **22**, **Suppl. 106**, 97 (1968).
11. B. P. Sonnenblick, in *Biology of Drosophila* (M. Demerec, Ed.), pp. 62–167. John Wiley and Son, New York, 1950.
12. L. F. Lamerton, *Phys. Med. Biol.* **13**, 1 (1968).
13. Mary F. Lyon, D. G. Papworth and Rita J.S. Phillips, *Nature (London)* **238,** 101 (1972).
14. R. B. Cumming, W. L. Russell and G. A. Sega, *Radiat. Res.* **59**, 63 (1974) (Abstract).
15. L. Ehrenberg, K. D. Hiesche, S. Osterman-Golkar and I. Wennberg, *Mutat. Res.* **24**, 83 (1974).
16. F. G. Zimmermann and R. Schwaier, *Z. Vererbungsl.* **94**, 261 (1963).
17. Sandor Igali and R. C. von Borstel, *Mutat. Res.* **26**, 452 (1974).
18. Anne Marstokk and Per Oftedal, *Mutat. Res.* **21**, 194 (1973).
19. T. Eklund, *Genetics* **74**, (June Suppl., No. 2, Part 2) 70 pp. (1973).
20. Howard Newcombe, These Proceedings and *Radiat. Res.* **59**, 74 (1974).
21. *Ionizing Radiation: Levels and Effects*. United Nations, New York, 1972.
22. *The Effects on Populations of Exposure to Low Levels of Ionizing Radiation*. NAS–NRC, Washington D.C., 1972.
23. P. Oftedal, *Curr. Top. Radiat. Res. Q.* **9,** 66 (1973).

THE BASIC PHYSICS OF THE INTERACTIONS OF RADIATION WITH MATTER

Chairman: M. Inokuti

[1]Symposiums B–10 and B–29

Introduction to the Symposium on the Basic Physics of the Interactions of Radiation with Matter[1,2]

Mitio Inokuti
Argonne National Laboratory
Argonne, Illinois 60439, USA

This Symposium was originally planned to take place under the chairmanship of Professor Robert L. Platzman. Shortly after his untimely death on 2 July, 1973, I was asked to take over his task. At that time Professor U. Fano gave me advice and encouragement, and handed me some notes by Platzman on the organization of the Symposium. Therefore I have followed Platzman's ideas as closely as possible in the choice of subjects and speakers.

Let me present my interpretation of his ideas and simultaneously depict a road map for the audience to follow along. Our general topic is the nature of ionization in gases and the behavior of gaseous ions. In the elucidation of elementary mechanisms of radiation actions—the problem which absorbed Platzman's devotion virtually throughout his scientific life—the ionization of gases plays a unique role. Ions are always one of the major *initial* species produced under any mode of irradiation. Among many possible indices of radiation action, the (total) ionization in gases is often the easiest to measure, and has long been used as a basis of dosimetry. Quite apart from its practical merit, the ionization in gases has a fundamental significance. A full analysis of the ionization offers a prototype of the more general problem Platzman kept attacking. That problem is to predict absolute yields for all initial species produced in a given substance, and thus to provide a firm logical starting point of kinetic considerations of radiation chemistry and biology. This idea is apparent in many of Platzman's papers, most notably in his lecture (*1*) at Cortina d'Ampezzo in 1966. Since then, much progress has been achieved in the understanding of ionization phenomena, as we shall hear today.

Dr. Berkowitz will tell us about photoionization, i.e., the simplest and therefore best understood class of ionization phenomena. The simplicity arises for two reasons. First, a photon most often gets completely absorbed by a molecule, and thus we know how much energy went into it. Second, for photon energies at which absorption is strong, the interaction is predominantly of the dipole type; in other words, every spatial part of the molecule feels the same strength of the photon's electric field. Despite the apparent simplicity, photoionization processes already show many intricacies of ionization in general. For example, the role of superexcitation and preionization,

[1]Congress presentation B-10-1.
[2]Work performed under the auspices of the U.S. Atomic Energy Commission.

which Platzman first pointed out, has been best demonstrated in studies on photoionization. Also, analyses of photoelectrons, a subject of intense current studies, tell us a great deal about alternative ionic states that are left behind.

Dr. van der Wiel will discuss electron-impact ionization, i.e., a second category of ionization. It is especially important for radiation research, because much of the radiation action in general originates from secondary electrons generated in the medium. Although the total ionization cross section for electron impact is a classic object of study, recent work brings out great details of the process by analyzing scattered electrons or ejected electrons, and sometimes even both in coincidence. The wealth of information obtained in this way awaits theoretical analysis, as well as applications to radiation research.

Dr. Kim will describe a systematic analysis of secondary-electron spectra arising from a single collision. His method is based on the classification of ionizing collisions into two classes, i.e., hard collisions and glancing collisions. For hard collisions the struck electron may be treated as weakly bound initially. For glancing collisions at high incident velocities there is a close relation to photoionization, which enables one to test the compatibility of data from independent sources. Dr. Kim's analysis closely follows Platzman's tradition of scrupulous and comprehensive evaluation of data as a prerequisite for application.

Dr. Niehaus' lecture will concern one of less usual classes of ionization phenomena, namely, ionization by energy transfer from excited species, including metastable states and high Rydberg states. This family of processes exemplify the ways in which the energy initially stored gets used later to cause a new product —a question of cardinal importance to radiation chemistry. As Platzman discussed on various occasions, a dramatic manifestation of ionization by excited species is seen in the total ionization of slightly impure rare gases (the Jesse effect).

Dr. Smith will treat ionization by collisions of ions having velocities comparable to, or even smaller than, those of atomic-electron velocities. In this context I wish to quote a remark (*2*) by Platzman at the Oberlin Symposium in 1950: "To say that a heavy particle of velocity v_0 ($v_0 = c/137$ = velocity of the electron in the lowest orbit of the hydrogen atom), which has a most appreciable energy (25 keV for a proton, 99 keV for an alpha particle), *does not* ionize with good efficiency is not justified by any established knowledge and is probably also incorrect." Interestingly, this prophetic remark was addressed to Fano, who later pointed out (*3*) the mechanism of efficient ionization, i.e., interpenetration of electron shells.

The last two lectures will deal with the behavior of gaseous ions, a subject whose importance to radiation chemistry is now well recognized. Dr. Ottinger will treat spontaneous dissociation of isolated ions. Because this process hinges upon the internal energy deposited in an ionization event, the subject is closely related to the lectures by Dr. Berkowitz and by Dr. van der Wiel. The same comment applies to Dr. Kebarle's subject, i.e., ion-molecule reactions. In addition to studies at low gas pressures, he will discuss the ion-molecule reactions at high pressures, a subject of strong interest to condensed-phase radiation chemists and biologists.

As we shall hear from Dr. Ottinger and Dr. Kebarle, much is now known about the behavior of ions. The major reason for this is that ions can be studied by means of mass spectroscopy. In contrast, neutral species that are initially produced have so far defied detailed scrutiny, owing to the lack of appropriate instrumentation. Of course, exceptions are long-lived excited species, which Dr. Niehaus will tell about, and some electronically excited species that emit light. Nevertheless, such excited species are only a fraction of all neutral initial products. Let us hope that at a future Congress we shall be able to discuss the behavior of neutral species in as nearly a complete way as we now discuss the behavior of ions.

Before concluding this introduction, I wish to say a few words about Dr. Platzman, with whom I had the privilege of working for several years. I first met him in the summer of 1962 during a visit to Argonne National Laboratory. After an extremely courteous reception in his office, I was seated in front of him across a big table. He pulled out of a cabinet a file, which contained several reprints of my papers. As soon as he laid one of them flat on the table, I was utterly astonished at the amount of remarks he had written virtually on every page, in ink of various colors (which according to his habit showed that he had studied the papers many times). As he articulated his remarks, softly but firmly, I quickly realized that he had insight far more penetrating than mine into each topic of my papers. To most of his criticisms I had no choice but to surrender. That was the most awesome experience I have ever had in my scientific career. Since that day I learned numerous lessons from him.

The most impressive lesson was his rigorous intellectual attitude, which reminded me of Socrates'. He often started out a scientific discussion by sharply distinguishing what is known from what is unknown. His thought process may be exemplified by his persistent question of the first day I met him. It was roughly as follows: "Is that statement of yours on molecule M compatible with everything else we know about M?" In other words, his intellectual code was that any new statement one makes on an object be tested against every other piece of solid knowledge one already has access to about that object. In my opinion, the adherence to this code was a key to his monumental achievements in radiation physics and chemistry.

The memory of Platzman should remain with radiation scientists for many years to come. His various contributions are excellently summarized in the obituaries published in the June issue of *Radiation Research* (*4*). Also a forthcoming issue of *International Journal of Radiation Physics and Chemistry* will be another tribute to Platzman. In addition to many papers by his friends throughout the world, that issue will include a complete list of his publications, many of which will remain a source of inspiration and insight for all of us.

Finally I should add that Platzman's work in his last years at the University of Chicago was just as successful as that in his earlier years. Together with several graduate students he pursued the most detailed analyses of initial processes of radiation actions in the strategic examples he had chosen carefully — helium, molecular hydrogen and argon. Results of these analyses will be published shortly (*5–7*).

REFERENCES

1. R. L. Platzman, in *Radiation Research Proceedings of the Third International Congress of Radiation Research, Cortina d'Ampezzo, Italy 1966,* (G. Silini Ed.), pp. 20–42. North-Holland Publishing Company, Amsterdam, 1967.
2. R. L. Platzman, in *Symposium on Radiobiology. The Basic Aspect of Radiation Effects on Living Systems, Oberlin College, June 14–18, 1950,* (J. J. Nickson Ed.), pp. 20–21. John Wiley & Sons, New York.
3. U. Fano and W. Lichten, *Phys. Rev. Lett.* **14,** 627 (1965).
4. U. Fano and E. J. Hart, *Radiat. Res.* **58,** 559 (1974).
5. D. A. Douthat, *Radiat. Res.* (in press).
6. D. E. Gerhart, *J.. Chem. Phys.* (in press).
7. E. Eggarter, *J. Chem. Phys.* (in press).

Photoionization[1,2]

J. Berkowitz
Physics Division, Argonne National Laboratory
Argonne, Illinois 60439, U.S.A.

INTRODUCTION

Although physicists have known how to describe the photoionization process mathematically since the early days of wave mechanics in the late 1920's and early 1930's, it is only in the past 15 years that a great mass of experimental data has provided us with a detailed knowledge of the variation of this interaction from one molecule to another, and across the electromagnetic spectrum. The experimental approaches that have been employed may be cataloged as

1. Measurement of absolute photoabsorption cross sections.
2. Determination of the fraction of this absorption which leads to ionization, or photoionization yield.
3. Use of photoelectron spectroscopy to determine the states of the ion formed, and their relative abundance.
4. Application of mass analysis to determine whether the ionized products survive as parent ions or decompose into fragments, and the dependence of these processes on the wavelength, or photon energy.
5. Measurement of mass-analyzed photoions and energy-analyzed photoelectrons in coincidence, enabling one to deduce the fragmentation pattern that results when a specific internal energy is imparted to the molecular ion by the initial photoionization act.

By examining the simplest systems first, and gradually building upon this foundation, we shall have occasion to note how these various techniques have enabled us to understand the behavior of larger and more complex systems.

Photoionization is the limiting case of photoabsorption. In the simplest case of atomic hydrogen, only a single electron is available to absorb the incident electromagnetic radiation. There exists a threshold energy (the ionization potential, I.P.) below which only discrete, well-defined photon energies $h\nu$ can be absorbed by this system according to the well-known Bohr frequency condition

$$h\nu = E_n - E_0,$$

where E_n is one of a set of excited states of the hydrogen atom and E_0 is the ground state.

The spectrum of these states can be described by the Rydberg formula,

[1] Congress presentation B-10-2.
[2] Work performed under the auspices of the U.S. Atomic Energy Commission.

$$E_n = \text{I.P.} - R/n^2,$$

where R is called the Rydberg constant and I.P. is the ionization potential, which is equal to R for atomic hydrogen. If one were to plot the probability of light absorption as a function of the photon energy, one would observe no absorption except in the regions where one of these E_n states could be produced. In those energy regions one would note sharp peaks, the integrated area of which is proportional to a quantity called oscillator strength. For atomic hydrogen, the succesive oscillator strengths, or f_n, vary as n^{-3}. Although the values of f_n for very large n approach 0, the average value of oscillator strength per unit energy, $f_n/\Delta E$ becomes a smooth function df/dE at the ionization threshold.

For all other systems besides atomic hydrogen (many-electron systems) one can isolate absorption processes which can be envisioned as single-electron excitations, although quantum mechanics has taught us that the excited states formed are states of the entire system, not those of one electron in the system. When these states can be described approximately by the Rydberg formula, we often refer to them as Rydberg states even for many electron systems. In general, for systems containing N electrons, there can be many states of the singly positive ion, or N − 1 electron system. There exist a set of Rydberg states converging to each one of these ion states and the average value of the oscillator strength for each of these sets of Rydberg states smoothly merges into the partial cross section (expressed in units of df/dE) for formation of that particular ionic state.

Examples of this behavior, in the cases of atomic hydrogen and atomic lithium, are shown in Fig. 1.

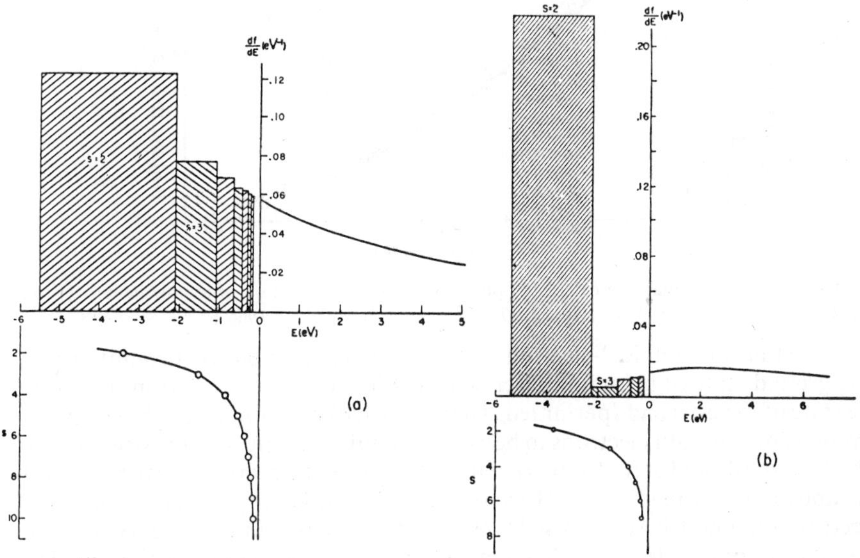

FIG. 1. Oscillator strength distribution in the discrete spectrum, shown merging smoothly with the continuum at the ionization potential for a) H atom and b) Li atom. [Reproduced from U. Fano and J. W. Cooper, *Rev. Mod. Phys.* **40**, 441 (1968), with the permission of The American Institute of Physics. Note that an error in the ordinate scale in the original figure has been corrected.]

What can we say about the behavior of the photoabsorption cross section in the post-threshold region? For atomic hydrogen (where most of our information is based on theory), the cross section just beyond threshold (*1*) declines as $E^{-8/3}$, and asymptotically as $E^{-7/2}$ (where E is the photon energy). If one were to observe the ion produced, rather than the light absorbed, there would be an abrupt rise in the photoionization cross section at threshold, and then a decline with excess photon energy in the manner described above. Since the electron that leaves the scene can assume a continuous range of kinetic energies, the set of states of the ionization region is a continuum and the Bohr frequency condition can be met with no apparent discontinuity in the photoabsorption cross section.

AUTOIONIZATION IN ATOMIC GASES

Atomic helium, which can be studied experimentally far more readily than atomic hydrogen, behaves approximately in the way we have just described. The experimental and calculated photoabsorption cross section of atomic helium are shown in Fig. 2.

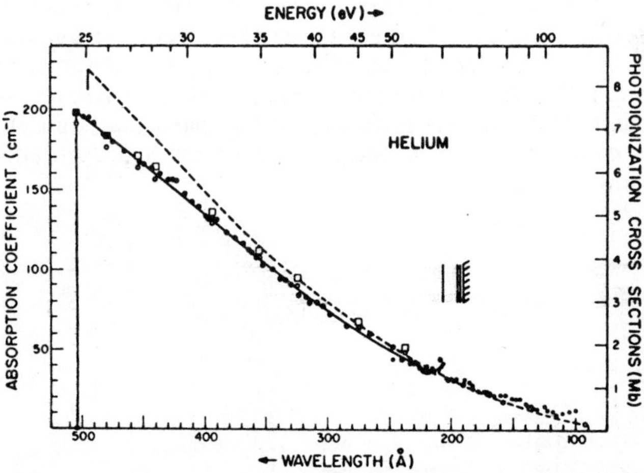

FIG. 2. Experimental and calculated photoionization cross section of atomic helium. [From J.A.R. Samson, *Adv. At. Mol. Phys.* **2**, 177. Academic Press, N.Y., 1966.]

Even in this simple 2-electron system, we encounter our first complication. The ground state of helium is described as $(1s)^2$, i.e., the two electrons populate the state of lowest n and *l* permitted. There is a small probability, on the order of a few percent, for both electrons to be excited simultaneously by photoabsorption. The lowest allowed state for this excitation is 2s2p, the energy of which is ~60.1 eV above the ground state. However, the threshold for ejection of a single electron (the first I.P.) occurs at 24.59 eV ≡ 504.3 Å as seen in Fig. 2. Hence, this first 2-electron excited state lies somewhere in the ionization continuum. The process of photoabsorption in the 60.1 eV region can be viewed as instantaneous population of this 2s2p state, followed by a configuration interaction between this state and the underlying continuum, or equivalently, as an interference between two channels of photoabsorption. In the former view, the temporary (or quasi-discrete) 2s2p state is formed and survives for $\sim 1.6 \times 10^{-14}$ sec before decaying into He^+ and an electron. The process is referred to as autoionization, and usually manifests itself as pronounced structure in the photoabsorption or photoionization spectrum. In the helium system, there exists a series of such states, which we can simplistically describe as 2s3p, 2s4p, etc., each of which will manifest this autoionization structure. One such study of this structure, due to Madden and Codling (*2*), is shown in Fig. 3.

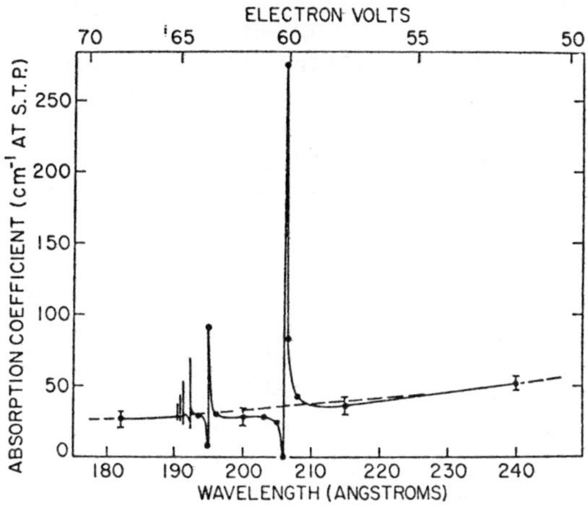

FIG. 3. The absorption coefficient of helium in the 60.1 eV region. [Reproduced from R. P. Madden and K. Codling, *Astrophys. J.* **141**, 364 (1965), with the permission of The University of Chicago Press. Copyright 1965 by The University of Chicago.]

In general, the shape of this structure can take many forms, depending upon the relative probability of photoabsorption and of the interaction between quasi-discrete state and underlying continuum. In one extreme, when the latter interaction is weak and the photoabsorption process to the quasi-discrete state is strong, one simply observes a peak superposed upon a more or less flat continuum. In the other extreme, when photoabsorption to the quasi-discrete state is extremely weak but the configuration interaction is strong, a huge dip, or inverse peak, is seen in the photoabsorption curve. (See below for the 3s excitation in argon.) This latter behavior is referred to as a "window resonance." In intermediate cases, one observes an asymmetric structure (*3*), as typified by the situation in helium. In some simple cases, it is possible to make *a priori* calculations about the relative strengths of the two effects which govern this shape.

Now that we have had our first encounter with autoionization, we are in a position to examine more complex systems in which such structure is prominent. Another noble gas, argon, provides us with two additional examples of such behavior.

In the usual atomic orbital nomenclature, the ground state of argon is written

$$(1s)^2(2s)^2(2p)^6(3s)^2(3p)^6.$$

The first ionization potential (the minimum energy necessary to remove an electron from the outermost 3p shell) is 15.759 eV. Since the resulting Ar^+ has an incompletely filled 3p orbital, there is a net orbital angular momentum and net spin angular momentum, and these can couple in two ways, giving rise to the two states $^2P_{3/2}$ and $^2P_{1/2}$. These states differ in energy by 0.177 eV, the more stable one (the ionic ground state) being $^2P_{3/2}$. As we have noted above, we can anticipate a Rydberg series converging not only to the ground state, but also to the $^2P_{1/2}$ excited state. Those members of the series with energy above 15.759 eV will be energetically capable of autoionizing, the mechanism in this case involving a readjustment of the coupling between orbital and spin angular momentum. Figure 4 is a spectrum of the Ar^+ ions produced in this energy region by photoabsorption. It shows a large number of these Rydberg members, each with an asymmetric structure, and smoothly converging upon the $^2P_{1/2}$ excited state. This type of autoionization has also been observed in krypton and xenon, and should also occur in neon.

Returning once again to argon, we can see that in the single-electron approximation it should be possible to excite electrons from each of the inner shells 3s, 2p, etc. to their respective Rydberg levels. All of these levels will lie above the first I.P. Let us consider those excitations in which an electron from the 3s orbital is excited to a set of states which the selection rules require to be P states. The reorganization of the electronic distribution of the Ar* (excited argon atom) results in an electron refilling the 3s orbital, and another electron being ejected. Within the framework of quantum mechanics, it is not possible to specify whether it is the outermost (excited) electron or a 3p electron which refills the 3s orbital, or which electron is ejected. The dynamical behavior of this process can be seen in the photoabsorption spectrum of Fig. 5. From the distinct "window resonances" we can readily infer that photoexcitation of 3s electrons is very weak, and the configuration interaction is strong.

Recently, Nakamura et al. (*4*) have photographically observed resonances in argon at much higher energy (49.9 Å $< \lambda <$ 50.8 Å) corresponding to the excitation of electrons from the 2p orbital to Rydberg levels, with resulting autoionization.

Heretofore, we have been implicitly assuming that excitation of atomic systems to states above the first I.P. will lead to ionization. This is usually the case, since re-emission of the absorbed light is characterized by lifetimes of the order of $10^{-6} - 10^{-9}$ sec, whereas autoionization in the helium case noted above occurs in $\sim 10^{-14}$ sec, and in many other cases has been observed in the $10^{-11} - 10^{-14}$ sec range.

However, there are some situations in which the autoionization can be slowed down because the selection rules are unfavorable, or the configuration interaction is very weak. One such example which we have recently observed (*5*) occurs in atomic oxygen. Two of the most prominent autoionization peaks in this spectrum, at 878–879 Å and 791–793 Å (see Fig. 6) have also been observed to re-radiate, and hence the two processes must be competitive. An analysis (*5*) has shown that the autoionization process in these transitions is retarded by unfavorable selection rules, bringing its characteristic lifetime into the 10^{-8} sec range. As a consequence, the oscillator strength for photoabsorption is not identical to that for photoionization for these peaks, but this behavior is the

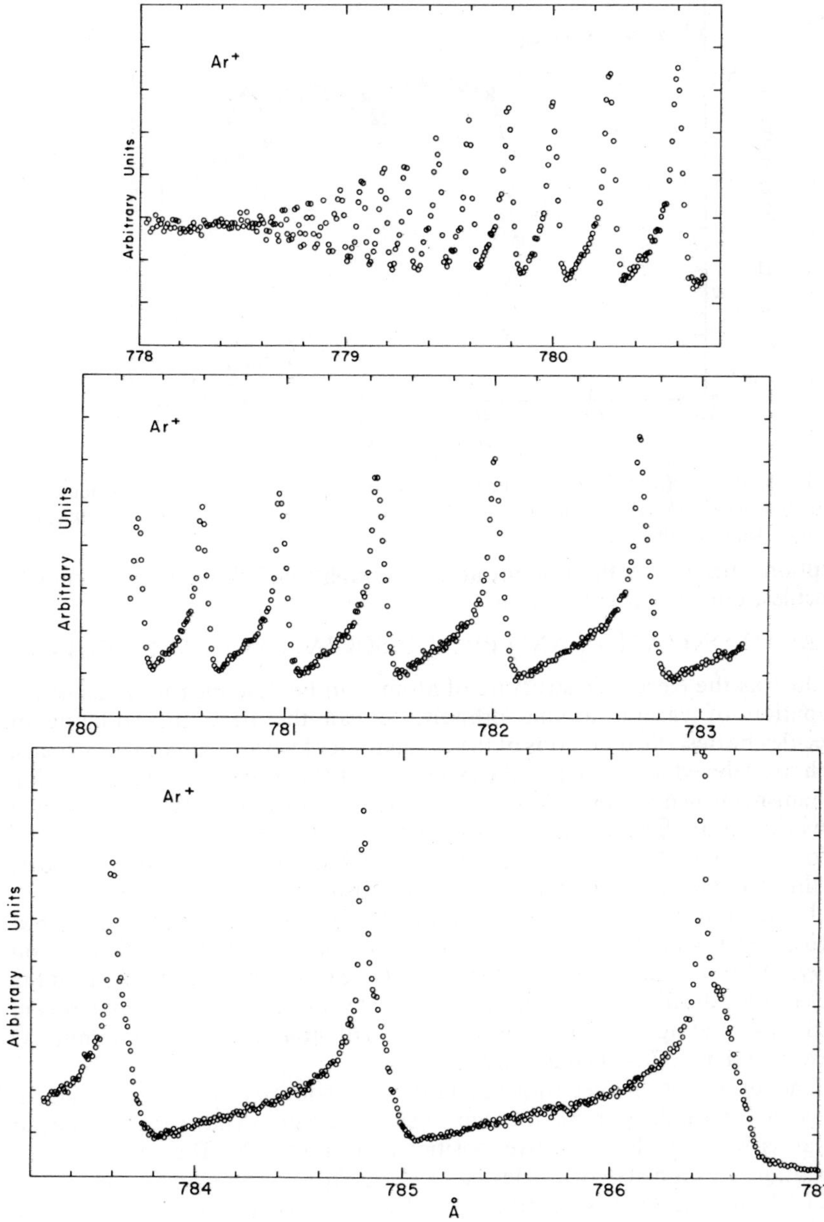

FIG. 4. Prominent autoionization structure in argon between the $^2P_{3/2}$ and $^2P_{1/2}$ ionic states, as observed in photoionization mass spectrometric studies. [From J. Berkowitz, *Adv. High Temp. Chem.* **3**, 123. Academic Press, N.Y., 1971.]

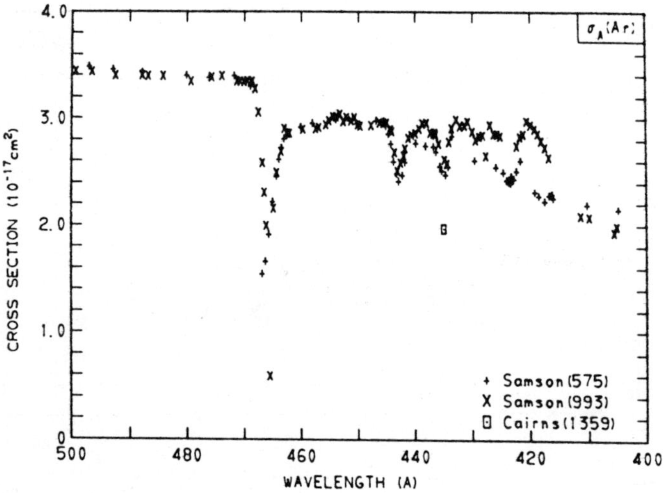

FIG. 5. Photoabsorption cross section of argon in the "window resonance" region due to excitation of 3s electrons. [Reproduced from R. D. Hudson and L. J. Kieffer, *At. Data* **2**, 205 (1971). Copyright Academic Press, Inc.]

exception rather than the rule for atoms, though (see below) it is common in molecules, but for another reason.

AUTOIONIZATION AND PRE-DISSOCIATION IN MOLECULES

Just as the electronic structure of atoms can be described in terms of the occupation of various atomic orbitals, so can the electronic structure of molecules be described in terms of the occupation of various molecular orbitals, which are labeled according to the symmetry of the molecule. The interaction mechanisms which we have outlined that give rise to autoionization structure in atoms can be applied equally well to molecules. However, molecules have the additional capability of absorbing energy into vibrational and rotational modes, and this provides them with alternate mechanisms for autoionization.

Let us consider the simplest molecular gas, H_2. Its ground state is described as $(1\sigma_g)^2$. The bonding of the two atoms in this molecule (i.e., the dissociation energy, D_0) is accurately known *(6)* to be 4.478 eV. Upon removal of one of the $1\sigma_g$ electrons, the dissociation energy is reduced *(7)* to 2.651 eV. The usual way in which this binding is depicted is a plot of potential energy vs. internuclear distance, and Fig. 7 is such a diagram.

The processes of electronic excitation or photoionization occurring in molecules are usually so rapid that the much heavier nuclei have not had time to substantially alter their relative positions or momenta. The mathematical description of this behavior is referred to as the Born-Oppenheimer approximation. In the diagram of Fig. 7, this is equivalent to drawing a vertical line from the equilibrium internuclear distance of H_2 to its intersection with the potential energy curve of H_2^+. Actually there is some range of allowed internuclear distance for H_2 in its lowest vibrational state, and this gives rise to a range of intersections of vertical lines with the H_2^+ curve. The higher the point of intersection on the H_2^+

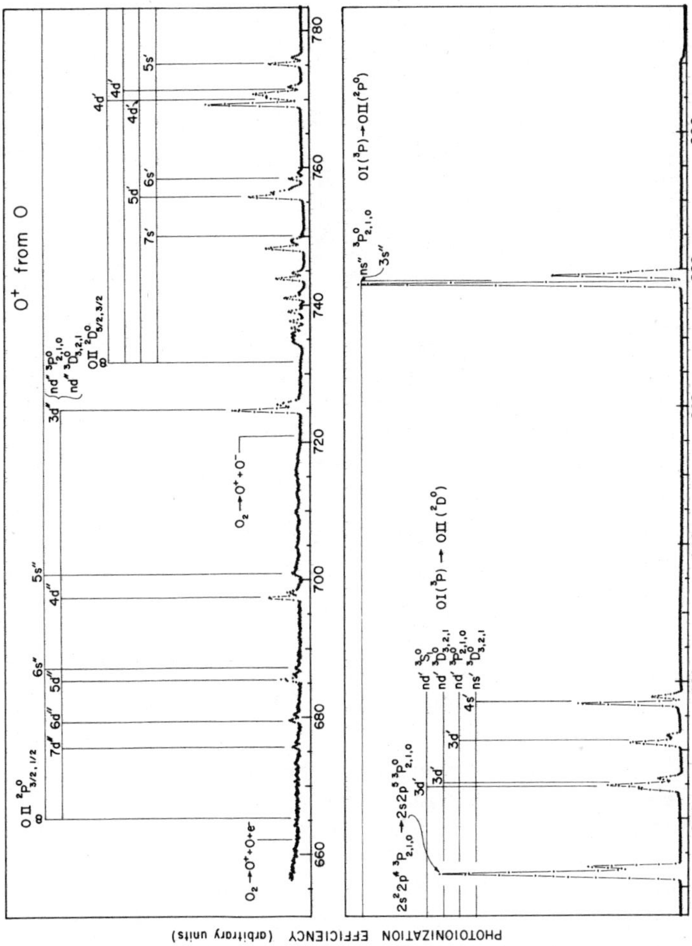

FIG. 6. Relative photoionization cross section of atomic oxygen. The peaks at 878–879 Å and 791–793 Å have ionization yields less than unity. [Reproduced from P. M. Dehmer, J. Berkowitz and W. A. Chupka, *J. Chem. Phys.* **59**, 5777 (1973), with the permission of The American Institute of Physics.]

curve, the more rapidly will the resulting H_2^+ ion vibrate in its potential well.

The technique called photoelectron spectroscopy enables us to measure the relative probability of forming H_2^+ in various vibrational states. The experiment involves sending monochromatic radiation of sufficiently high energy $h\nu$ through the target gas (H_2) and measuring the kinetic energy of the electrons released in the photoionization process with some type of energy analyzer. The equation describing this process is simply a statement of conservation of energy, and is similar to the Einstein photoelectric equation

$$h\nu = (\mathrm{I.P})_n + (\mathrm{K.E.})_n.$$

Since the left-hand side of the equation is a constant, there is a one-to-one relationship between the flux of electrons of a particular kinetic energy and the

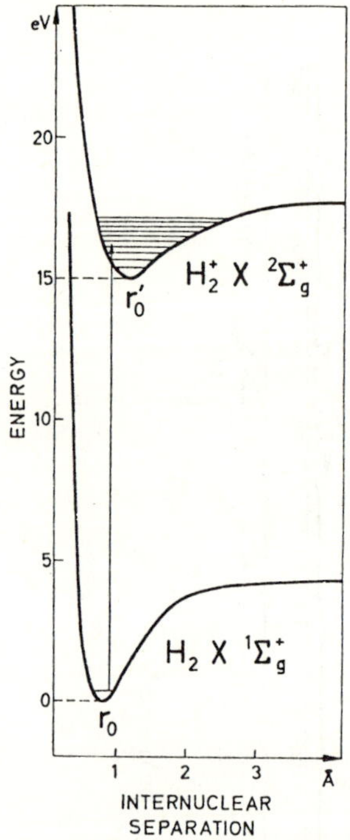

FIG. 7. Potential curves of H_2 and H_2^+ in their ground states, indicating a typical photoionization transition. [From K. Siegbahn, et al., "ESCA", Almquist & Wiksells, Uppsala, 1967.]

flux of H_2^+ ions in a particular vibrational state. The result of such an experiment, with $h\nu = 21.2$ eV, is shown in Fig. 8 for H_2, HD and D_2. We can see that a large number of vibrational levels are appreciably populated, the most probable ones being labeled $v' = 2$ for H_2^+ and HD^+, and $v' = 3$ for D_2^+.

Now let us backtrack a bit. We know that there are Rydberg states converging to each one of the ionization potentials $(I.P)_n$ for H_2 as well as for atoms. These Rydberg states involve excitation of one of the $1\sigma_g$ electrons in H_2 to a distant orbit. This distant orbit will have little effect on the binding between the two hydrogen nuclei, and hence most of the bonding will be dependent upon the one electron still remaining in the $1\sigma_g$ orbital. Therefore, we can anticipate that the potential energy curve of such a Rydberg state will look very much like the H_2^+ curve in Fig. 7. This, in turn, implies that there is a large probability of exciting vibrational motion when one excites to such a Rydberg state, for the same reason that vibrational excitation is imparted to H_2^+. It should now be obvious that it is possible to excite a Rydberg state, and concomitantly excite vibrations in that Rydberg state, such that the total absorbed energy (electronic plus vibrational) is greater than the first I.P. of H_2 at 15.425 eV. Once again we have a situation in which a quasi-discrete state is formed, but the mechanism of autoionization now involves the transfer of vibrational energy to the "Rydberg electron." In effect, the occurrence of this process is a manifestation of the

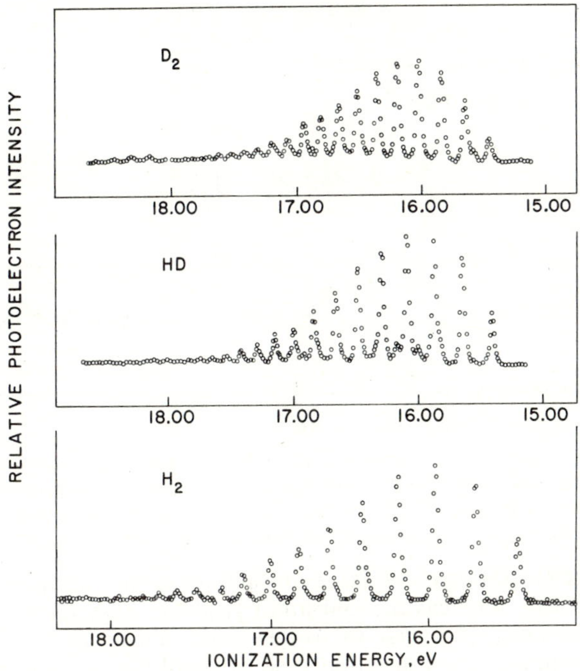

FIG. 8. Photoelectron spectra for the transition X $^1\Sigma_g^+$ ($v''=0$) + $h\nu \rightarrow$ X $^2\Sigma_g^+(v')'\,e^-$ in H_2, HD and D_2. [From J. Berkowitz and R. Spohr, *J. Electron. Spectrosc.* **2**, 143 (1973). North-Holland Publ. Co., Amsterdam.]

breakdown of the Born-Oppenheimer approximation, which attempts to separate nuclear and electronic motions on the basis of one being much faster than the other. One may think of the Rydberg electron in its large orbit as moving more slowly, enabling it to more readily exchange energy with the slow moving nuclei. The photoionization spectrum of H_2 (Fig. 9) in the first eV above threshold is dominated by this vibrational autoionization process.

There is still another mode of autoionization appropriate to molecules, which is referred to as rotational autoionization. Thus far, it has been explored only in a narrow energy range in H_2, and hence is too specialized for our present purposes.

Intimately connected with the vibrational degrees of freedom available to molecules is the capability (in the limit of large vibrational excitation) of dissipating this energy by dissociation. As we have seen, D_0 for H_2 is 4.478 eV, while I.P. (H_2) = 15.425 eV. Hence, when H_2 has absorbed enough energy to cause ionization, it has more than ample energy to dissociate. This is generally true for almost all molecules. In particular, when a quasi-discrete state H_2^* of the Rydberg type described above is formed, it can decay by autoionization or by the analogous neutral channel called pre-dissociation. (Prof. Platzman stressed the importance of these states about a decade ago (*8*), and gave them the name "super excited states.") It turns out that these processes have characteristic rates which are competitive, and as a consequence only a fraction of the absorbed

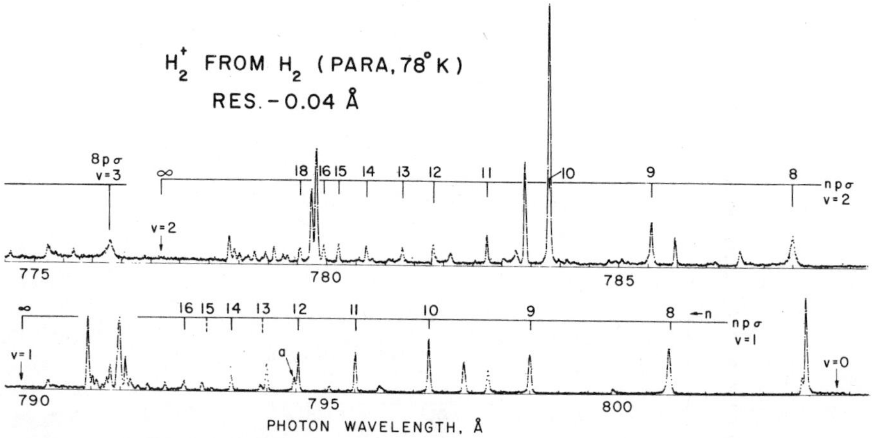

FIG. 9. Relative photoionization cross section of H_2 in the first ~0.5 eV above threshold. The peaks are attributable to the vibrational autoionization process. (As noted, this particular experiment was performed with para-H_2 at 78° K.) [Reproduced from W. A. Chupka and J. Berkowitz, *J. Chem. Phys.* **48**, 5726 (1968), with the permission of The American Institute of Physics.]

energy above the ionization potential results in ionization. This fraction is called the ionization yield. Whereas the ionization yield for most atoms under most circumstances is unity (the case of oxygen atom mentioned above being a noteworthy exception), the ionization yield of small molecules is significantly different from unity just above the threshold but asymptotically approaches unity at ~20–25 eV. For larger molecules, it is not even clear whether this asymptotic behavior is attained. Examples are given in Figs. 10–13 of four rather

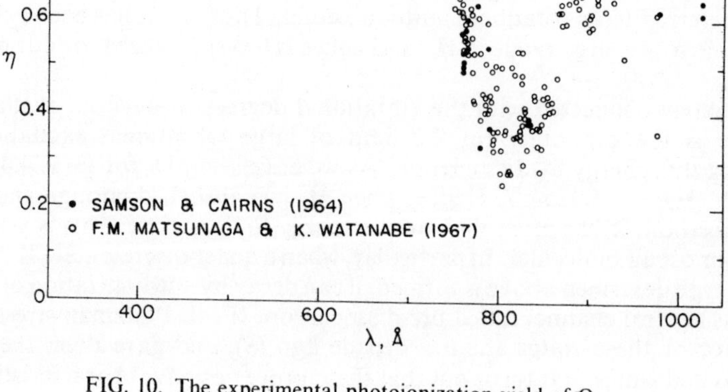

FIG. 10. The experimental photoionization yield of O_2.

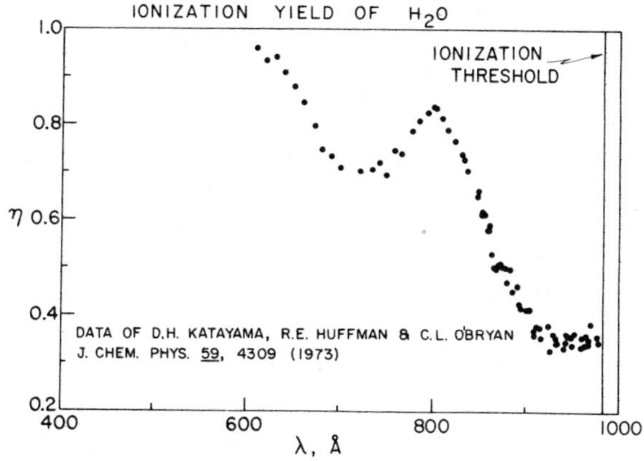

FIG. 11. The experimental photoionization yield of H_2O.

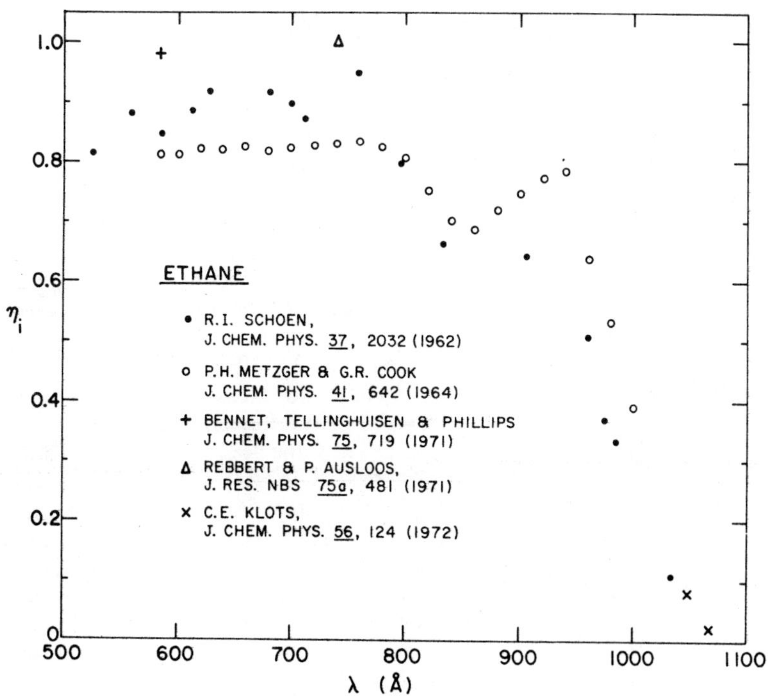

FIG. 12. The experimental photoionization yield of C_2H_6.

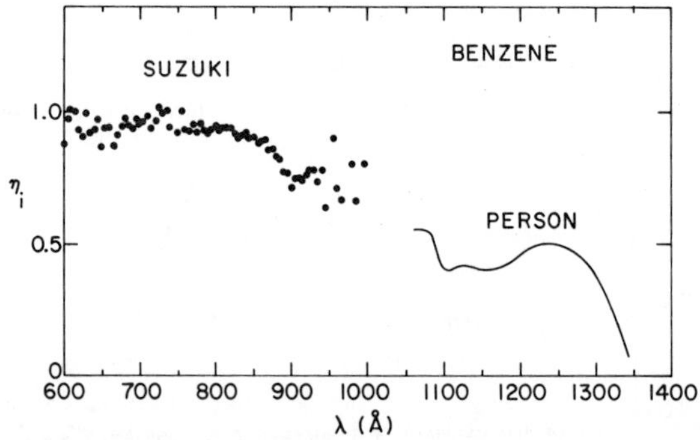

FIG. 13. The experimental photoionization yield of C_6H_6. [From M. Yoshino, J. Takeuchi and H. Suzuki, *J. Phys. Soc.*, Japan, **34**, 1039 (1973).]

different molecules—O_2, H_2O, C_2H_6 and C_6H_6—where the asymptotic approach to unity appears to occur, but the presently available evidence for CH_2O (*9*) and SF_6 (*10*) suggests a yield substantially lower than unity at ~600 Å.

THE VARIATION OF TOTAL AND PARTIAL CROSS SECTION WITH PHOTON ENERGY

The molecule CO_2 will serve as our example in this section. Its sequence of occupied molecular orbitals can be written as follows:

$$(1\sigma_g)^2(1\sigma_u)^2(2\sigma_g)^2(3\sigma_g)^2(2\sigma_u)^2(4\sigma_g)^2(3\sigma_u)^2(1\pi_u)^4(1\pi_g)^4.$$

The $1\sigma_g$ and $1\sigma_u$ are combinations of oxygen 1s orbitals, essentially the K-shell of atomic oxygen. Similarly, $2\sigma_g$ represents the slightly perturbed K-shell of atomic carbon. The $3\sigma_g$ orbital involves 2s orbitals on oxygen and carbon, while $2\sigma_u$ involves oxygen 2s and carbon $2p_z$. Both $3\sigma_g$ and $2\sigma_u$ have strong constructive overlap and can properly be called bonding orbitals.

The uppermost four orbitals involve various combinations of 2p orbitals on oxygen and carbon. The $1\pi_u$ orbital is bonding, the other three relatively non-bonding. These inferences can be gleaned from Ref. *11*.

The photoelectron spectra obtained at 21.2 eV (Fig. 14) and 1253.6 eV (Fig. 15) verify the bonding and non-bonding nature of these orbitals, but also provide valuable insight about the relative probability of ejecting electrons from these orbitals with different incident energies. Thus, the 21.2 eV spectrum displays the four energetically accessible states (labeled X, A, B, C) of which only the $A^2\pi_u$ state is broad, indicating removal of an electron from a bonding orbital. The 1253.6 eV spectrum of Fig. 15 (note the changes in ordinate scale) is dominated by the high energy peaks corresponding to ejection from $1\sigma_g$, $1\sigma_u$ and $2\sigma_g$, the K-shells of oxygen and carbon, respectively. The probability of ejecting electrons from the four uppermost orbitals has become very small in the X-ray energy range. The intermediate $3\sigma_g$ and $2\sigma_u$ orbitals, composed of 2s contributions, have not declined as much in relative probability.

From this example and other studies (*12*) we have learned that the partial cross section for ionization from orbitals having large "p" character declines more rapidly than from orbitals having large "s" character. Also, ionization in

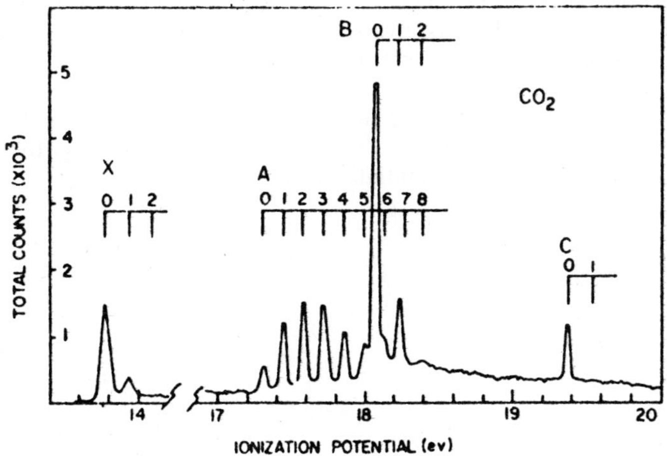

FIG. 14. The 21.2 eV photoelectron spectrum of CO_2. [Reproduced from J. A. R. Samson, J. L. Gardner and J. E. Mentall, *J. Geophys. Res.* **77**, 5560 (1972). Copyright by American Geophysical Union.]

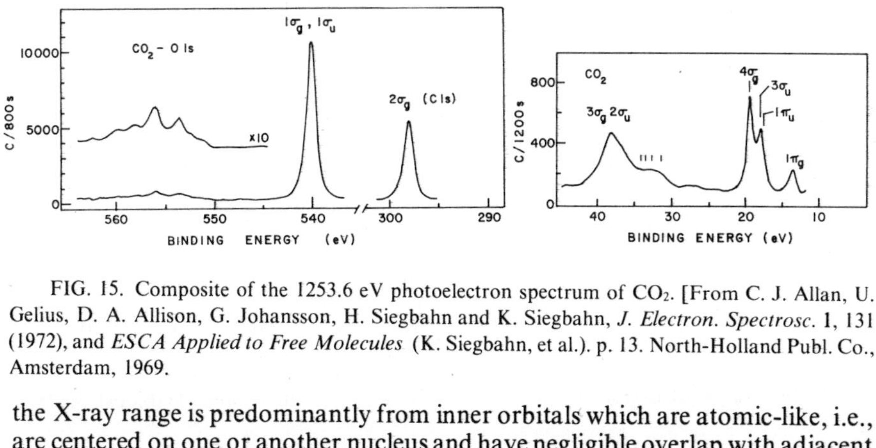

FIG. 15. Composite of the 1253.6 eV photoelectron spectrum of CO_2. [From C. J. Allan, U. Gelius, D. A. Allison, G. Johansson, H. Siegbahn and K. Siegbahn, *J. Electron. Spectrosc.* 1, 131 (1972), and *ESCA Applied to Free Molecules* (K. Siegbahn, et al.). p. 13. North-Holland Publ. Co., Amsterdam, 1969.

the X-ray range is predominantly from inner orbitals which are atomic-like, i.e., are centered on one or another nucleus and have negligible overlap with adjacent atomic orbitals. For this reason, it seems plausible that the photoionization cross sections of molecules in the X-ray range can be estimated as the sum of the cross sections of the constituent atoms.

With this additional piece of information, it is possible to construct a more-or-less complete picture of the photoabsorption cross section across the entire electromagnetic spectrum for a number of molecules (*13*). Such a graph is shown for CO_2 in Fig. 16, where the logarithm of photoabsorption cross section (σ_a) is plotted against wavelength (λ). (The region below the ionization threshold is not shown). The steep decline below ~300 Å is a noteworthy feature of this graph.

There are a number of theorems, or sum rules, which can be used to test this oscillator strength distribution. One of these, the Thomas-Reiche-Kuhn sum rule, or S(0), states that the integral of (df/dE)—which is proportional to σ_a—plotted against E—which is inversely proportional to λ—is equal to the number of electrons in the molecule, if the integral is carried out over all E. Another sum rule, S(-2), relates the integral of σ_a vs. λ to the electric dipole polarizability α of the molecule, a quantity which is usually accurately known from independent measurements. In the case of CO_2, we note in Fig. 16 that two sets of measurements differ substantially in the 550–650 Å region.

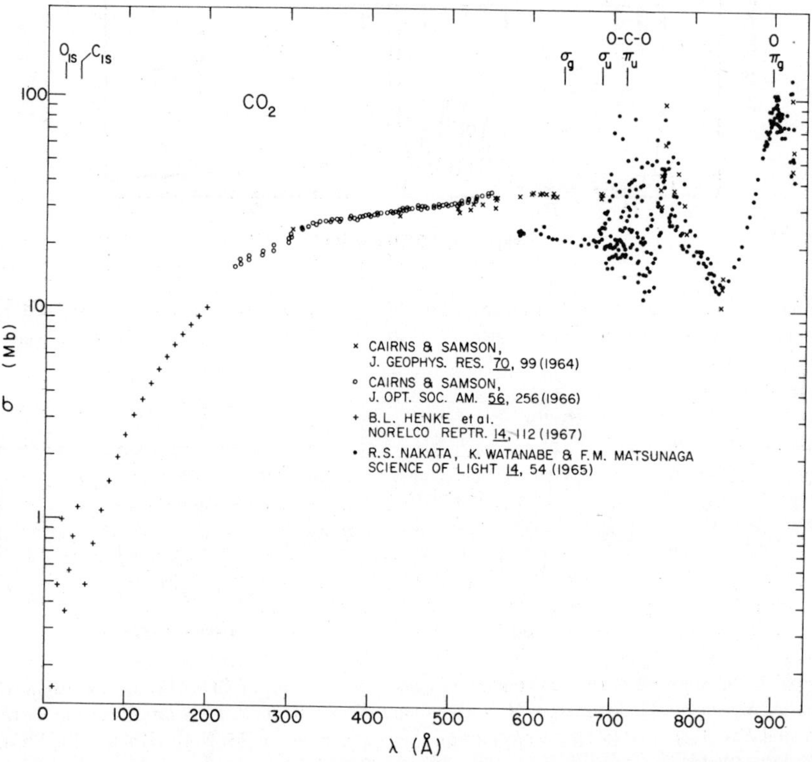

FIG. 16. The photoabsorption cross section of CO_2 from ionization threshold through the "K-edge."

Using the set of data with lower cross sections, we obtain S(0) = 21.1, and $\alpha = 2.47 \times 10^{-24}$ cm^3. From the set of data with higher cross sections, we deduce S(0) = 21.82 and $\alpha = 2.82 \times 10^{-24}$ cm^3. The total number of electrons in CO_2 is 22, and the best experimental values (*14*) of α lie between 2.59–2.91 × 10^{-24} cm^3. Hence we conclude that the higher cross sections are more nearly correct. The experimental plot of Fig. 16 tends to suggest this choice, as do more recent measurements. Many other examples of such oscillator strength distributions and analysis of the data in terms of sum rules will be available in a forthcoming monograph (*13*).

One further useful sum rule (S_i(-1)) relates the integral of (1/E)(df/de) vs. E to the slope of the absolute cross section for ionization by fast electrons plotted against $E_{elec.}$, where $E_{elec.}$ is the energy of incident electrons. Here we have one example (others will be presented at this symposium) of the intimate relation between photoionization and ionization by fast charged particles. The value of $S_i(-1)$ (obtained from the ionization component of photoabsorption only) deduced from the photoionization data is 5.60. The directly measured (*15*) electron impact ionization value is 5.75 ± 0.073. In this instance, the three sum rules provide us with a satisfactory check on the oscillator strength distribution.

PARTIAL CROSS SECTIONS AND MASS SPECTRA

Finally, we turn our attention to a problem the solution of which is still in its infancy. From the oscillator strength distribution (curves such as Fig. 16) and measured photoelectron spectra at a number of wavelengths, it is possible to deduce the partial cross sections for formation of various states of the molecular ion as a function of photon energy. If one knew the subsequent life history of each molecular ion state (whether it survives as parent ion or decomposes during the time of the experiment) then, with simple bookkeeping, one could synthesize a mass spectrum. The use of coincidence technique is beginning to provide solutions to this problem.

In a recent experiment (*16*), CO_2 was irradiated with 21.2 eV photons. The photoelectrons produced were energy-analyzed and the ions were mass- and energy-analyzed. The photoelectrons and photoions were then investigated with a coincidence circuit, in order to deduce which ions resulted from particular states of CO_2^+. It was found that the three lowest states led only to parent ions, while the state formed by ejection of an electron from the $4\sigma_g$ orbital led to O^+ + CO when the lowest vibrational levels of this ionic state was formed, and to CO^+ + O in higher vibrational levels.

At very much higher photon energy (930 eV), mass spectroscopic measurements have been performed (*17*) showing the following product distribution.

Ion	%
C^+	25 ± 0.2
O^+	52 ± 6
CO^+	8.9 ± 0.5
CO_2^+	5.4 ± 0.6

The observations can be semi-quantitatively explained by the following analysis. At 930 eV, the predominant ionization is from $1\sigma_g$, $1\sigma_u$ and $2\sigma_g$ (the K-shells of O and C) as already shown. The "hole-states" formed by K-shell ionization are highly excited and can in principle de-excite by radiation, dissociation, or re-population of the K-hole from an electron in a higher orbital, the excess energy then being available to eject other more weakly bound electrons. This latter event is called an Auger process. For molecules containing atoms of low atomic number (<30) this is the dominant process. On a time scale of $\sim 10^{-15}$ sec, the molecular ion finds itself multiply charged as a consequence of the Auger process. This in turn will cause the atomic constituents of the multiply-charged ion to be strongly repelled from one another by the Coulombic repulsion. If the molecular ion is triply charged, two O^+ ions and one C^+ ion will rapidly leave the interaction zone. This analysis is seen to explain the gross features of the 930 eV mass spectrum. The other ions observed can be explained by taking into account the relatively small probability of forming other states of the molecular ion.

The above description may have profound significance for the behavior of biological systems subjected to X-ray irradiation. Prof. Platzman had long ago (*18*) focused attention on K-shell ejection (induced by charged particle ionization, in that study) as an important energy deposition process in radiobiology. If K-shell electrons are ejected and a Coulombic explosion ensues, a large molecule such as DNA may be locally disrupted, and upon subsequent charge neutralization and chemical re-adjustment, a modified molecular species may be formed. An obvious area for experimental activity is the examination of such systems with photon energies just above and just below that necessary to eject K-shell electrons.

REFERENCES

1. H. Bethe and R. Jackiw, *Intermediate Quantum Mechanics*, p. 236. Benjamin, New York, 1968.
2. R. P. Madden and K. Codling, *Astrophys. J.* **141**, 364 (1965).
3. U. Fano, *Phys. Rev.* **124**, 1866 (1961).
4. M. Nakamura, M. Sasanuma, S. Sato, M. Watanabe, H. Yamashita, Y. Iguchi, A. Ejiri, S. Nakai, S. Yamaguchi, T. Sagawa, Y. Nakai and T. Oshio, *Phys. Rev. Lett.* **21**, 1303 (1968).
5. P. M. Dehmer, J. Berkowitz and W. A. Chupka, *J. Chem. Phys.* **59**, 5777 (1973).
6. G. Herzberg, *J. Mol. Spectrosc.* **33**, 147 (1970).
7. G. Herzberg, in *Commentarii Pontificia Academia Scientiarum*, **Vol. II**, N. 49, pp. 1–36, Vatican City State, 1972.
8. R. L. Platzman, *The Vortex* 23, 372 (1962).
9. J. M. Mentall, E. P. Gentieu, M. Krauss and D. Neumann, *J. Chem. Phys.* **55**, 5471 (1971).
10. M. Sasanuma, E. Ishiguro, Y. Morioka and M. Nakamura, *3rd Int. Conf. on Vacuum Ultraviolet Radiation Physics*, Physical Society of Japan, Tokyo, 1971.
11. A. D. McLean and M. Yoshimine, *Tables of Linear Molecule Wave Functions*, IBM, 1967.
12. W. C. Price, A. W. Potts and D. G. Streets, in *Electron Spectroscopy* (D. A. Shirley, Ed.), North-Holland Publ. Co., Amsterdam, 1972.
13. J. Berkowitz, *Photoabsorption, Photoionization and Photoelectron Spectroscopy*, (to be published, Academic Press.)
14. A. A. Maryott and F. Buckley, *Nat. Bur. Stand.* (*U.S.*) *Circ. 537* (1953).
15. F. F. Rieke and W. Prepejchal, *Phys. Rev.* **A6**, 1507 (1972).
16. J.H.D. Eland, *Int. J. Mass. Spectrom. Ion Phys.* **9**, 397 (1972).
17. T. A. Carlson and M. O. Krause, *J. Chem. Phys.* **56**, 3206 (1972).
18. R. L. Platzman, in *Symposium on Radiobiology. The Basic Aspects of Radiation Effects on Living Systems*, (J. J. Nickson, Ed.) p. 106. J. Wiley & Sons, New York, 1952.

Electron Energy Deposition in Matter at the Molecular Level[1]

M. J. van der Wiel
FOM Institute for Atomic & Molecular Physics
Amsterdam, The Netherlands

INTRODUCTION

The special nature of the material that is dealt with in radiation research enables us to restrict our discussion of its interaction with electrons in a number of respects:

—biological matter, even in the condensed phase, does not to any large extent form collective band structures. So there is no immediate necessity to consider collective effects like excitation of plasmons or phonons.
—the frequency of molecular collisions, at most 10^{-12} sec^{-1}, is too low to affect the excitation and the predominant, rapid decay via (auto-) ionization and dissociation.
—most molecules of interest consist of low-Z atoms, which means we can disregard energy transfer into bremsstrahlung.

These restrictions then leave us with the interaction of projectile electrons with the electron shells of isolated molecules.

In a symposium to honor the late Dr. Platzman it seems appropriate to take as the central theme the relevance of electron impact experiments to the study of superexcited states of molecules. These states, discrete states above the first ionization potential, were first recognized by Platzman (*1–3*) more than a decade ago as being of great importance for radiation chemistry. The reason is that, in competition with auto-ionization, a superexcited state may undergo dissociative decay into often excited fragments with considerable kinetic energy; in other words, highly reactive species that can initiate a great variety of unusual chemical reactions. A recent, well-documented review on the subject of superexcited states and their role in radiation chemistry is that of Makarov and Polak (*4*).

Starting with a high-energy projectile electron, one knows that its energy degradation down to values somewhere between 1 keV and 100 eV is governed completely by Bethe's expressions for the relevant cross sections. In particular, it appears that the main fraction of the energy is expended in dipole (or optically allowed) transitions. Indeed, a major part of this paper is devoted to a demonstration of this analogy between photon and electron impact, and its use in areas where progress has been hampered by the lack of photon sources, variable over a wide wavelength range (up into far U.V. and soft X-ray).

[1]Congress Presentation B–10–3.

In the low energy range, below one or a few hundred eV, typically the domain of electrons liberated in ionizing events, also optically forbidden and electron exchange transitions make a significant contribution to the stopping of electrons. In a later section is reported recent progress in the study of those aspects of fragmentation that are essential to the evolution of radiation action: the state of excitation and kinetic energy of the fragments. The final section deals with inner-shell excitation, to both discrete and continuum states and the specific type of decay products.

The scope of this paper does not permit any attempt at completeness. Wherever possible, recent review papers are referred to, new developments are indicated and some representative results are discussed.

HIGH INCIDENT ENERGY

Let us define a high energy electron as having a velocity much larger than the orbital velocities of the molecular electrons involved in the inelastic collision, i.e. energies of a few hundred eV and higher for outer shell excitations. For such an electron the energy deposition is determined completely by f(K,E), the generalized oscillator strength (GOS), a function of momentum and energy transfer, K and E, respectively. Being the basic quantity in the Bethe approximation, the GOS is related to the differential scattering cross section by:

$$(d^2\sigma/d\Omega dE)(E,K) = (2/E)(k_n/k_o)(1/K^2)\ f(K,E), \tag{1}$$

where k_0 and k_n are the magnitudes of the incident and scattered momenta. For a survey of the theory the reader is referred to the extensive review paper by Inokuti (*5*). Very briefly, one might say that in the Bethe domain the only, but very formidable, problem that remains to be solved is the construction of dependable molecular wave functions for the ground and excited states. A possible guide towards solution of this problem for large molecules would be the direct measurement of ground state wave functions. This is feasible in electron impact; under conditions of very large K and E (binary encounter), the GOS for ionization is related to the electron momentum distribution in a very direct fashion. Contributions of the various shells may be resolved on the basis of differences in binding energy. This requires measurement of the energies of both outgoing electrons in coincidence. Experiments of this type have been reported for solid (*6*) as well as gaseous targets (*7*) (see Fig. 1).

As regards the experimental determination of f (K,E), the situation has also been summarized recently (*8*). The method one applies is that of electron spectrometry, i.e. measurement of energy loss spectra at different angles, a widespread technique which developed over the last decade from the pioneering work of Lassettre et al. (*9*) and of Boersch et al. (*10*).

As an example Fig. 2 shows a recent result (*11*) for the GOS of H_2, measured over a wide range of E and K. Of importance is that the use of this technique is not restricted to molecules having sufficient vapor pressure for gas phase scattering. Experiments have also been reported on solid or condensed phase samples, using electron microscopy in conjunction with energy loss analysis (*12–14*).

While specifying the probability for energy and momentum transfer to a target molecule, the GOS for transitions to states above the first ionization

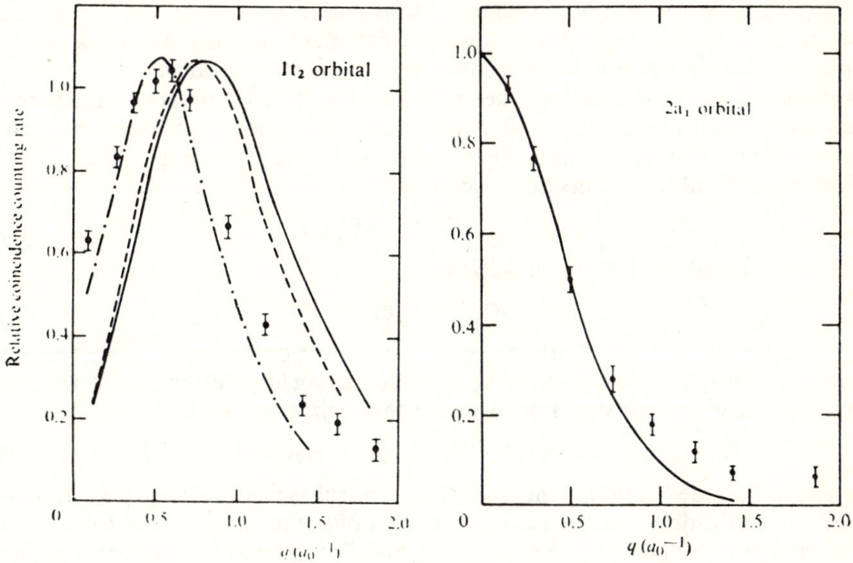

FIG. 1. Square of electron momentum distribution for the two valence orbitals ($1t_2$ and $2a_1$) of CH_4, measured in an electron-electron coincidence experiment and compared with various calculated distributions [Hood et al. (*7*)].

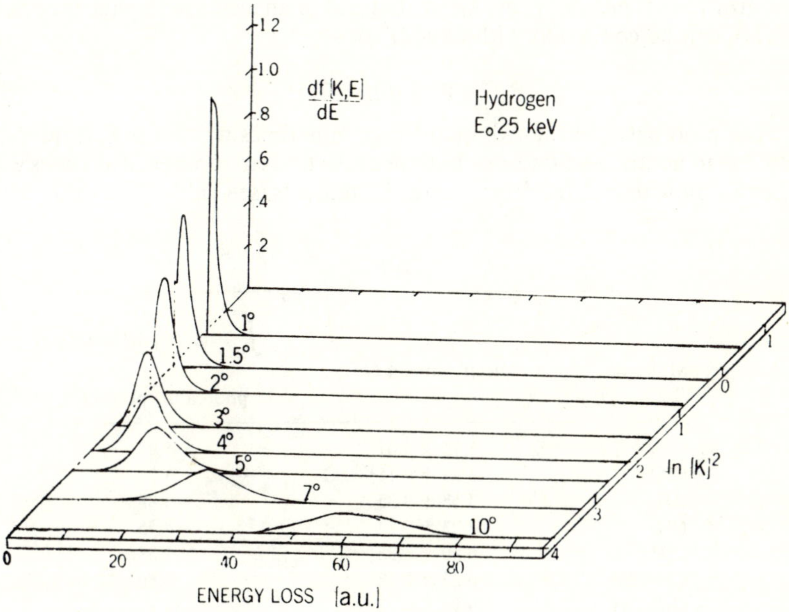

FIG. 2. Generalized oscillator strength (Bethe surface) of H_2, measured with 25 keV incident electron energy [Ulsh et al. (*11*)].

potential does not in general contain information on details of the primary products of the energy absorption. Measurement of such detail has been confined so far—except for the coincidence work described in the following section—to total cross sections, i.e. integrals of eq. (1) for excitation, ionization or dissociation. Again very briefly, in the Bethe theory total cross sections work out to have a strikingly simple dependence on the impact energy (E_o). For optically allowed transitions we have[2]

$$\sigma(E_o) = [A \ln E_o + B]\, E_o^{-1}, \tag{2}$$

and for optically forbidden transitions

$$\sigma(E_o) = C\, E_o^{-1}. \tag{3}$$

The constants A, B and C are molecular quantities, depending in some way on the GOS. Thus, A is a weighted integral of the leading (dipole) term of the GOS at small momentum transfer. For instance, for ionization one has:

$$A_i = \int_{IP}^{\infty} \eta_i(E)\,(R/E)(df/dE)dE, \quad (R = \text{Rydberg energy}) \tag{4}$$

where $\eta_i(E)$ is the fraction of the continuous optical oscillator strength df/dE that leads to ionization and the integral extends over the entire photo-absorption continuum from the IP up. From eqs. (2) and (3) it is immediately apparent that a simple plot of σE_o vs. $\ln E_o$ (Bethe plot) tells us the nature of the process under study (see Fig. 3a).

What can one learn about superexcited states from high energy cross sections? Firstly, suppose the spectrum of df/dE of a molecule is known, but not the ionization efficiency. Then A_i, as obtained in an electron impact experiment (eq. (2)), can be compared with (e.g. (*15*))

$$\int_{IP}^{\infty} (R/E)(df/dE)dE$$

The ratio between the two quantities, sometimes written as $\bar{\eta}$, differs from unity due to neutral excitations, that are excitations to superexcited states which do not autoionize but decay into neutral fragments (see Table 1). In other words,

TABLE 1

Molecule	A_i electr. impact (*16*)	$\int_{IP}^{\infty}$ (R/E)(df/dE)dE photoabsorption (*15*)
CO_2	5.75 ± 0.07	5.97
H_2O	3.24 ± 0.15	3.21
NH_3	3.58 ± 0.35	3.98
CH_4	4.23 ± 0.13	4.55
C_2H_6	6.80 ± 0.36	7.38
CH_3OH	6.22 ± 0.18	7.56
C_2H_5OH	9.94 ± 0.27	11.8

[2] Or the appropriate relativistic formulae (*5*) in case E_0 is above roughly 10 keV.

one obtains a spectral average of the contribution of non-ionizing superexcited states (averaged in a way that is relevant for electron impact); or rather, a lower limit to that contribution because of the competition between autoionization and neutral dissociation.

Secondly, it is evident that of these two alternate decay routes mentioned, the second (dissociation) will be affected much more strongly than the first by changes in mass of the fragment. Indeed, strong isotope effects have been observed in ionization and dissociation cross sections (*16–18*).

Thirdly, if the light emission from a molecular fragment is found to have a threshold above the first IP but too low for the other fragment to be ionic, one knows a superexcited state is involved. Its nature can be learned from the high energy behavior of the cross section (eqs. (2) and (3)). For instance, production of excited H fragments from alkanes appears (Fig. 3) to proceed exclusively via optically forbidden transitions (*19*). It should be emphasized that, although non-dipole processes can perfectly well be studied in the high-energy range, on the whole they do not contribute significantly to the energy degradation of electrons in that range.

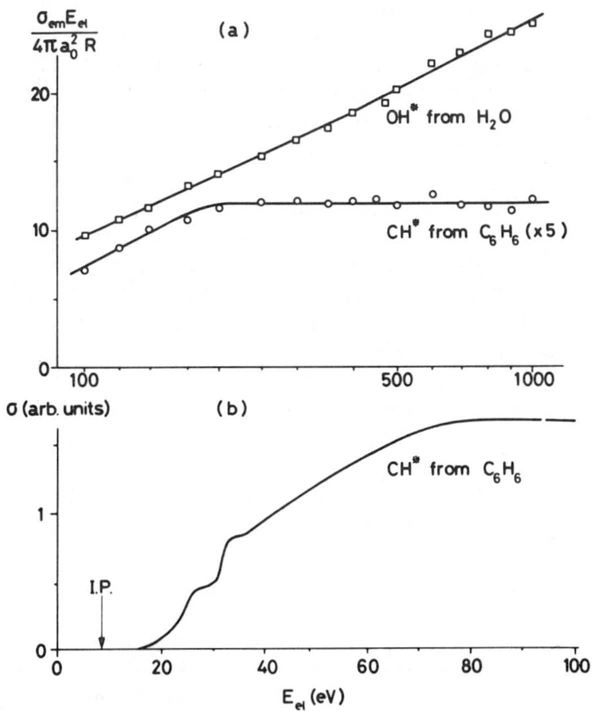

FIG. 3. Dissociation and emission cross sections of molecules [Beenakker (*19*)], a) Bethe plots of cross sections for $OH(A^2\Sigma \rightarrow X^{2P})$ emission from H_2O (optically allowed; eq. 2) and for CH $(A^2\Delta \rightarrow X^2p)$ emission from C_6H_6 (optically forbidden; eq. 3). b) Threshold behavior of CH-radiation emission from C_6H_6.

DIPOLE APPROXIMATION

From eq. (2) it is apparent that at high energy the cross section is dominated by the dipole term $A \cdot \ln E_o$. In many applications, therefore, approximation of $\sigma(E_o)$ by just this term provides sufficient accuracy. The information one needs in order to make this dipole approximation is knowledge of the spectral distribution of df/dE. This section discusses the possibilities of obtaining such distributions from differential electron scattering experiments.

The first method, used mostly at impact energies below 1 keV, is to measure f(K,E) as a function of K, i.e. at a variety of angles, and then extrapolate to K=0 to find the optical oscillator strength f(E) (or df/dE for a continuum). In this regard the work of Lassettre et al. (*9*) should be mentioned. The second, more convenient method, is to employ considerably higher impact energies (several to several tens of keV's) and zero or near-zero scattering angle. Under these conditions K_{min} (equal to $\sqrt{E^2/2E_0}$) is sufficiently small to neglect all but the first term in the expansion of f(K,E) around K=0:

$$f(K,E) = f(E) + aK^2 + bK^4 + \ldots \quad (5)$$

In practice the error due to this dipole approximation, i.e. the neglect of the higher terms in eq. (5), appears to be on the order of 5–10% for values of E up to a few hundred eV (*8*). This means the electron impact experiment provides optical constants of rather good accuracy. One obvious advantage over true optical measurements is that the continuum of energy transfers occurring in an electron impact experiment represents a continuum of photon energies extending from the infrared to at least the soft X-ray region, the only restriction being that $E \ll E_0$.

This method has been used extensively over the past years in the determination of the gross features—without much spectroscopic detail—of optical oscillator strength distributions in atoms and molecules. References to the work of various groups are given in (*8*). Two groups have developed this method further by measuring the energy loss (= photon energy) in coincidence with one of the products of the energy absorption. This could be either a mass-separated ion, an ejected electron or a photon. Thus, one simulates each one of the experiments possible with a photon source of variable energy (see Table 2 and references therein). The coincidence experiments are particularly relevant for radiation research in the amount of detail they provide on the nature of the primary product as a function of the energy transfer. From a complete set of such measurements on one molecule one obtains (see Table 2 for numbers):

TABLE 2

	low-K electron impact	ref.	equivalent to:
1	energy loss	*20–22*	photoabsorption
	coinc. with eject. electr.:		
2	– without energy analysis	*22*	photoionization
3	– with energy analysis	*20,21*	photoelectron spectrosc.
4	coinc. with ions	*22–25*	photoioniz. mass spectr.
5	coinc. with photons	*26*	photofluorescence

1) the total absorption;
2) the fraction that leads to ionization;
3) the relative contribution of each molecular orbital to the ionization;
4) the distribution over the fragmentation channels;
5) the state of excitation of ions and neutral fragments, insofar as they undergo radiative decay.

Evidently, this type of data would have been obtained optically if photon sources had been available of the same characteristics as in the coincidence experiments. In fact, in a number of laboratories using electron synchrotron radiation, studies of the type mentioned are in progress presently.

A survey of the results obtained with the electron coincidence technique, including a critical comparison with the optical alternative, has been given in ref. (*8*) (see also refs. in Table 2). For a more detailed discussion I have selected measurements (*22*) in CH_4 of the types 1, 2 and 4 (Table 2). Fig. 4a puts into evidence the production of neutral fragments above the first IP, i.e. proof of the occurrence of superexcited states, in two energy regions. The first, in accordance with earlier measurements, lies between the IP and 16 eV. An assignment of the superexcited states involved as vibrationally excited high Rydberg states of the outermost orbital is due to Nishikawa and Watanabe (*28*). The second region, above 23 eV, must arise from multiple electron transitions, since the highest one-electron binding energy of the valence shell amounts to 22 eV. The multiply excited states, moreover, must be so strongly repulsive that dissociation can compete effectively with autoionization. The probability for neutral excitation to these states is surprisingly large when compared to ionization. The fragmentation into the various ionic channels is shown in Fig. 4b. Note that the contribution of H^+ fragments is larger than one usually finds, due to the use of an ion extraction system that transmits ions with kinetic energy up to 20 eV without discrimination.

Finally, Fig. 5 shows a result obtained with the electron-electron coincidence method 3 in Table 2, namely the contribution of the four valence orbitals of H_2O to the ionization, at two energy transfers (*21*). The relative peak heights at 21 eV agree with those obtained in photoelectron spectrometry. This constitutes a further proof of the equivalence of the two methods.

LOW INCIDENT ENERGY

In contrast to the situtation in the Bethe domain, the scattering theory for low or intermediate energy electrons (except very near threshold) is still a matter which has not been solved satisfactorily. Even an enumeration of the various approximations (e.g. see review by Bransden (*29*)) is beyond the scope of this paper. Experiments in this range, however, are favored by large cross sections and have been described in great numbers in the literature. Out of this wealth I have selected a few studies, which pertain directly to the existence of superexcited states. Subject of the discussion will be measurements of the ionization efficiency, neutral formation, kinetic energy, angular distribution and detection of excitation state of dissociation fragments.

The existence of discrete states above the first IP can be demonstrated from fine-structure in the low-energy ionization cross section (*30,31*). However, more

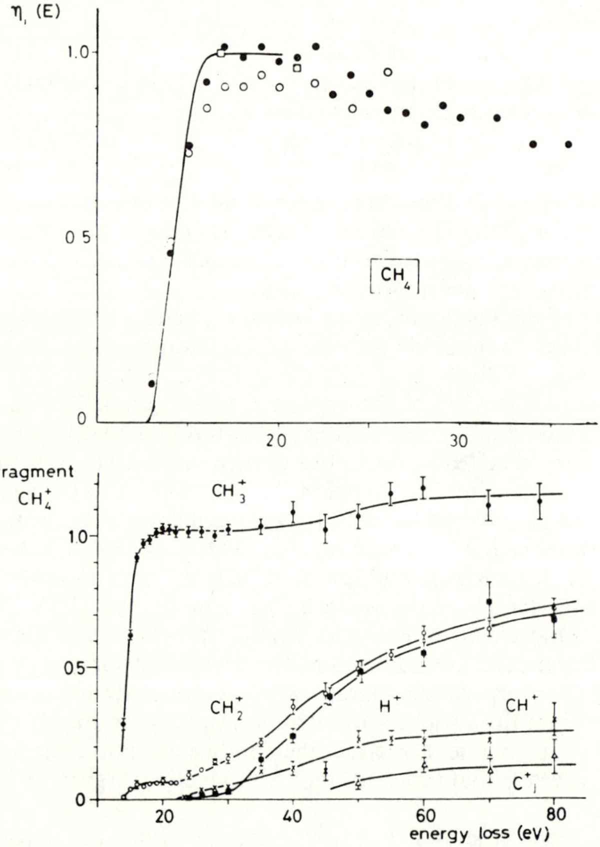

FIG. 4. a) Ionization efficiency of CH_4. ●—high energy electron-electron coincidence measurement [Backx and Van der Wiel (*22*)]; O—low energy el-el coincidence experiment [Ehrhardt and Linder (*32*)]; □—photo-ionization [Rebbert and Ausloos (*27*)]; solid curve—theory [Nishikawa and Watanàbe (*28*)].

b) Fragmentation of CH_4 as a function of energy transfer (= photon energy, see Table 2), measured in an electron-ion coincidence experiment (*22*). Ion separation was performed in a time-of-flight spectrometer free of discrimination for fragments with kinetic energies up to 20 eV.

quantitative information on their relative importance is gained from an experiment as the one by Ehrhardt and Linder (*32*) on electron-electron coincidences in alkanes. They establish (see Fig. 6) the ionization efficiency by comparing energy absorption (energy loss intensity) with ionization (intensity of coincidences between scattered and ejected electrons). Note that this low-energy (70 eV) experiment is complementary to the one described in the previous section in that it provides the contribution of neutral fragmentation due to both optically allowed and forbidden transitions. The main conclusions are that the mean excitation energy for large alkanes is generally a few volts above the IP, while of all neutral excitations up to 80% takes place via superexcited states (Fig. 6). Though being a very promising approach, the assumption of isotropy for the

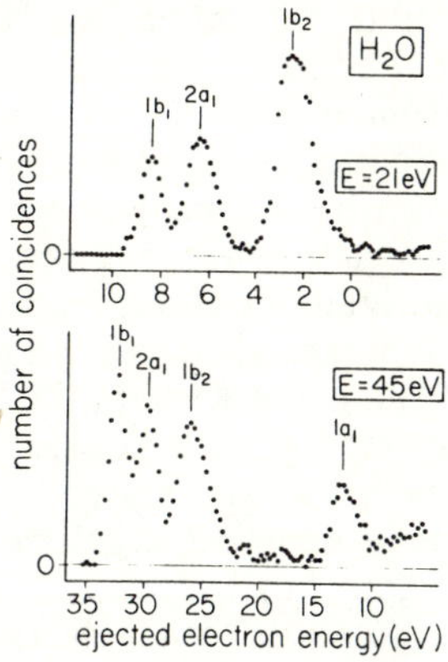

FIG. 5. Apparent photoelectron spectra of H_2O at energy transfers (= photon energies, see Table 2) of 21 and 45 eV, measured in an electron-electron coincidence experiment [Branton and Brion (*21*)].

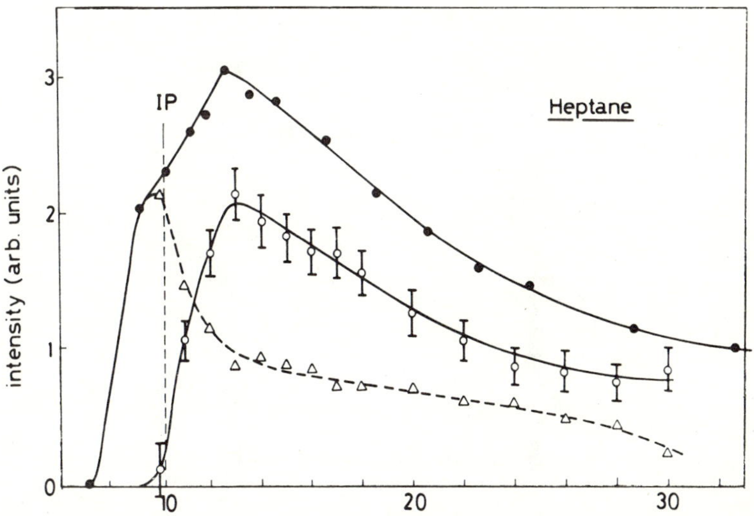

FIG. 6. Energy spectra of 70 eV electrons scattered by n-heptane [Ehrhardt and Linder (*32*)]. ●—energy loss; O—electron-electron coincidence; Δ—neutral excitations (= energy loss minus coincidence spectrum).

ejected electrons needed further investigation. This led to a comprehensive study of the angular correlation between the two outgoing electrons in the ionization of He and Ar (*33–35*). The experiment provides a severe test for the fine details (without integration over angle and energy of one or both outgoing electrons) of any theoretical description of the ionization event.

Neutral formation via superexcited states also emerges as a probable mode of excitation in large hydrocarbons from the work of Niehaus (*36*). The technique employed is the following: A pulsed electron beam produces ions and neutral fragments. Neutrals are allowed to drift into a second ionization region, where they are ionized by a second pulsed beam and analyzed in a mass spectrometer. The method allows determination of mass and —in some cases—internal energy of a neutral fragment. It is not quantitative, however, since obviously the fraction that is ionized by the second beam depends on the kinetic energy of the fragment.

This kinetic energy of dissociation has long been the subject of studies in mass spectrometers, often cleverly adapted for the purpose. However, it might be more fruitful to focus on recent progress, the "second generation" of this type of experiment, as introduced by Dunn and Kieffer (*37*). Basically, the dissociation fragments are formed in a field-free region, while energy and ion mass selection is performed as a function of the angle with respect to the primary beam. The combined knowledge of kinetic energy and angular distribution of a fragment generally suffices for identification of the excited state involved, at least for two-atomic molecules. (For larger molecules the kinetic energy distribution tends to smear out and the angular behavior becomes rather isotropic.) Recently, the method was refined by Crowe and McConkey (*38*) to give an energy resolution of 0.1 eV. Fig. 7 gives an example of the detailed character of the data one obtains. The results led to the identification of a whole series of superexcited states in H_2.

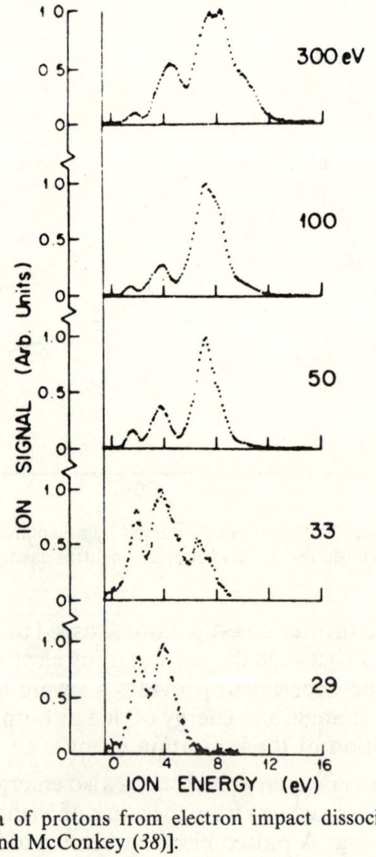

FIG. 7. Energy spectra of protons from electron impact dissociation of H_2, at the impact energies indicated [Crowe and McConkey (*38*)].

The state of excitation of a fragment can be observed in a variety of ways. If a metastable state is involved, the fragment will have a reduced ionization threshold (*36*) or may be detected through Auger emission from metal surfaces (*39*). If the state radiates, the simplest quantity to measure is the emission cross section as a function of impact energy. The position of the threshold and possible structure in the excitation function indicate (Fig. 3b) whether the excitation occurs via superexcited states. The distinction between optically allowed or forbidden transitions takes place via energy variation in the Bethe domain (see earlier section), while the typically low-energy process of electron exchange (spin-forbidden) is characterized by a very steep threshold behavior and an E_o^{-3} decrease at higher energies (*40*). Extensive work of this type on molecules has been reported recently (*19*), of which we mention a few conclusions for H_2O and aliphatic hydrocarbons:

—the majority of excited fragments are formed via superexcitation.
—transitions leading to excited H and OH from H_2O and CH from hydrocarbons are optically allowed.
—transitions leading to excited H from hydrocarbons are optically forbidden and have a constant cross section for different lengths of the carbon chain.

INNER SHELL PROCESSES

A molecule containing a vacancy in an inner shell of one of its constituent atoms is in a superexcited state: a state of the neutral molecule in case the inner electron is promoted to a molecular orbital; a state of the ion in case the inner electron is removed from the molecule. The predominant mode of decay is that of Auger emission of an outer electron, fluorescence (*41*) being improbable for $Z < 40$. To the author's knowledge no fragment has been detected which still carried an inner hole, meaning that the Auger transition in all cases is more rapid than dissociation. Since inner shells of low-Z atoms are usually well separated from the outer ones, the total "stopping" of a large molecule by inner shell processes is obtained by simple addition of a limited number of atomic inner shell cross sections. Such cross sections for the K-shells of C, N and O have been determined by Glupe and Mehlhorn (*42*), using an integral measurement of the Auger electron spectrum. This spectrum is also of interest for the investigation of doubly ionized states (*43*), the major decay product of an inner shell ionized molecule.

While inner shell ionization depends mainly on atomic properties, molecular aspects come into the picture if inner electrons are excited to normally unoccupied outer orbitals of the molecule. Energy loss spectroscopy has demonstrated (*13*, *44–46*) that a considerable fraction of the oscillator strength is connected with such transitions (see Fig. 8). It is also known from experiments of the type described in a previous section (*25*), and from ESCA studies (*47*), that the decay of these states frequently leads to population of highly excited ionic states, which are hardly populated from the ground state. An example is the production of N_2^+ ions in the $C^2\Sigma^+$ state via K-shell excitation.

A further point of interest is the following observation: insofar as each inner electron is well localized on one nucleus, the valence shell is equally affected by removal of such an electron as it is by raising the charge on the corresponding nucleus. For instance, promotion of a carbon K-electron in CH_4 to an outer orbital creates a transient species having a valence shell isoelectronic with that of NH_4 and the same effective nuclear charge. Indeed, Wight and Brion (*48*) found that a strong correlation exists between the K-shell spectrum of CH_4 and the valence shell spectrum of NH_4 (see Fig. 8b). This "core analogy" model seems to apply generally in those cases in which the molecular geometry of the two species compared is similar.

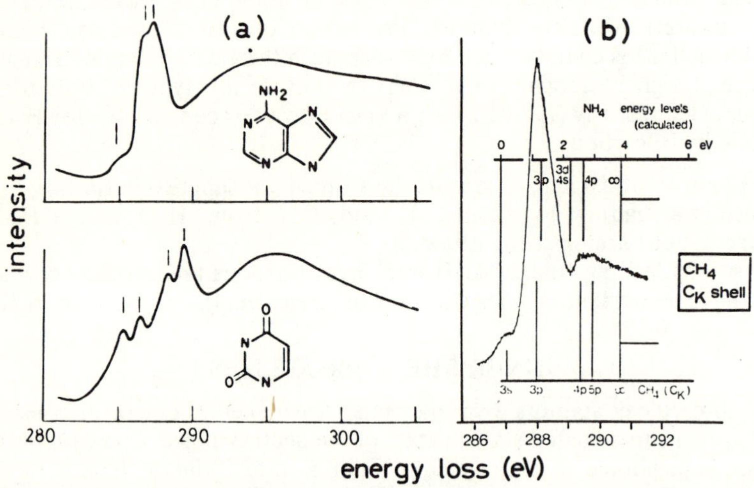

FIG. 8. a) Energy loss spectra of 25 keV electrons in the region of carbon K-shell excitation in thin films of adenine and uracil [Isaacson (*13*)]. b) Energy loss spectrum of 2.5 keV electrons in CH_4 in the carbon K excitation region [Wight and Brion (*48*)]. Application of the "core analogy" model (see section on Inner Shell Processes) leads to information on the valence shell excitation spectrum of the radical NH_4.

Quite spectacular effects occur if deep inner shells of a high-Z atom in a molecule are ionized. To the extent that the decay takes the form of a cascade of Auger transitions, a very high degree of ionization is attained by the time the cascade reaches the molecular bonding electrons. The evident result is that the molecule explodes into a number of highly charged fragments, each of which carries a large amount of Coulomb repulsion energy. Such explosions of CH_3I molecules have been observed in X-ray absorption by Carlson and White (*49*).

CONCLUSIONS

Some aspects of the study of superexcited states by means of electron impact have been reviewed. All recent data support Platzman's early conclusion that superexcitation accounts for a major fraction of the energy lost by electrons. We have encountered examples of superexcited states of every possible nature:

—Rydberg states, which are part of series converging on the various IP's of a molecule (single valence electron excitation). A special case are high vibrationally excited Rydberg states near the first IP.

—multiply excited states (multiple valence electron excitation);

—inner shell vacancy states (inner electron transitions to bound or continuum states). c

The extension of the optical and U.V. range to higher energy transfer by means of the electron coincidence experiments (Table 2) proved to be of particular importance for the study of multiple electron transitions. Clearly much work remains to be done in the investigation of neutral fragments.

ACKNOWLEDGEMENT

This work was performed as part of the research program of the Stichting voor Fundamenteel Onderzoek der Materie (Foundation for Fundamental Research on Matter) and was made possible by financial support from the Nederlandse Organisatie voor Zuiver-Wetenschappelijk Onderzoek (Netherlands Organization for the Advancement of Pure Research).

REFERENCES

1. R. L. Platzman, *J. Phys. Radium* **21**, 853 (1960).
2. R. L. Platzman, *The Vortex* **23**, 372 (1962).
3. R. L. Platzman, *J. Chem. Phys.* **38**, 2775 (1963).
4. V. I. Makarov and L. S. Polak, High Energy Chem. (USSR) **4**, 1 (1970) (transl. from Khimiya Vysolikh Energii **4**, 1 (1970)).
5. M. Inokuti, *Rev. Mod. Phys.* **43**, 297 (1971).
6. R. Camilloni, A. G. Guidoni, R. Tiribelli and G. Stefani, *Phys. Rev. Lett.* **29**, 618 (1972).
7. S. T. Hood, E. Weigold, I. E. McCarthy and P.J.O. Teubner, *Nature (London)*, (Phys. Sci.) **245**, 65 (1973).
8. M. J. van der Wiel, in *The Physics of Electronic and Atomic Collisions, Invited Papers and Progress Reports*, VIII ICPEAC (B. C. Cobic and M. V. Kurepa Eds.), p. 417. Beograd, 1974.
9. E. N. Lassettre and co-workers (10 papers) *J. Chem. Phys.* **40**, 1208–1275 (1964).
10. H. Boersch, J. Geiger and H. J. Reich, *Z. Physik* **161**, 296 (1961).
11. R. C. Ulsh, H. F. Wellenstein and R. A. Bonham, J. *Chem. Phys.* **60**, 103 (1974).
12. M. Isaacson, *J. Chem. Phys.* **56**, 1803 (1972).
13. M. Isaacson, *J. Chem. Phys.* **56**, 1813 (1972).
14. C. Colliex, University of Orsay, Thesis no. 716, 1970.
15. J. Berkowitz, M. Inokuti and J. C. Person, in *Electronic and Atomic Collisions, Abstracts of Papers*, VIII p. 561. ICPEAC, Beograd, 1973.
16. F. F. Rieke and W. Prepejchal, *Phys. Rev. A* **6**, 1507 (1972).
17. D. A. Vroom and F. J. de Heer, *J. Chem. Phys.* **50**, 573 (1969).
18. D. A. Vroom and F. J. de Heer, *J. Chem. Phys.* **50**, 580 (1969).
19. C.I.M. Beenakker, Thesis, University of Leiden, 1974.
20. M. J. van der Wiel and C. E. Brion, *J. Electr. Spectr.* **1**, 443 (1973).
21. G. R. Branton and C. E. Brion, *J. Electr. Spectr.* **3**, 129 (1974).
22. C. Backx and M. J. van der Wiel, *J. Phys. B*, to be published (1974).
23. M. J. van der Wiel, *Physica* **49**, 411 (1970).
24. M. J. van der Wiel and G. Wiebes, *Physica* **53**, 225 (1971).
25. Th.M. El-Sherbini and M. J. van der Wiel, *Physica* **59**, 433 (1972); idem **59**, 453 (1972).
26. R. E. Rebbert and P. Ausloos, *J. Res. Nat. Bur. Stand.* Sect. A: Ph. Ch. **75**, 481 (1971).
27. C. Backx, M. Klewer and M. J. van der Wiel, *Chem. Phys. Lett.* **20**, 100 (1973).
28. S. Nishikawa and T. Watanabe, *Chem. Phys. Lett.* **22**, 590 (1973).
29. B. H. Bransden, in *The Physics of Electronic and Atomic Collisions, Invited Papers and Progress Reports*, VIII ICPEAC (B. C. Cobic and M. V. Kurepa, Eds.) p. 399. Beograd, 1974.
30. P. Marmet and L. Kerwin, *Can. J. Phys.* **38**, 972 (1960).
31. J. W. McGowan, M. A. Fineman, E. M. Clarke and H. P. Hanson, *Phys. Rev.* **167**, 52 (1968).
32. H. Ehrhardt and F. Linder, *Z. Naturforsch.* **22a**, 444 (1967).
33. H. Ehrhardt, K. H. Hesselbacher, K. Jung and K. Willmann, *J. Phys. B* **5**, 1559 (1972).
34. H. Ehrhardt, K. H. Hesselbacher, K. Jung, Z. Schulz and K. Willmann, *J. Phys. B* **5**, 2107 (1972).
35. H. Ehrhardt, K. H. Hesselbacher, K. Jung, E. Schubert and K. Willmann, *J. Phys. B* **7**, 69 (1974).
36. A. Niehaus, *Z. Naturforsch.* **22a**, 690 (1967).

37. G. H. Dunn and L. J. Kieffer, *Phys. Rev.* **132**, 2109 (1963).
38. A. Crowe and J. W. McConkey, *Phys. Rev. Lett.* **31**, 192 (1973).
39. M. J. Mumma, M. Misakian, W. M. Jackson and F. L. Faris, *Electronic and Atomic Collisions, Abstracts of Papers*, VIII ICPEAC, p. 370. Beograd, 1973.
40. V. I. Ochkur, *Sov. Phys. JETP* **18**, 503 (1964).
41. R. W. Fink, R. C. Jopson, H. Mark and C. D. Swift, *Rev. Mod. Phys.* **38**, 513 (1966).
42. G. Glupe and W. Mehlhorn, *Phys. Lett.* **25A**, 274 (1967).
43. D. Stahlherm, B. Cleff, H. Hillig and W. Mehlhorn, *Z. Naturforsch.* **24A**, 1728 (1969).
44. M. J. van der Wiel, Th.M. El-Sherbini and C. E. Brion, *Chem. Phys. Lett.* **7**, 161 (1970).
45. G. R. Wight, C. E. Brion and M. J. van der Wiel, *J. Electr. Spectr.* **1**, 457 (1973).
46. G. R. Wight and C. E. Brion, *J. Electr. Spectr.* **3**, 191 (1974).
47. K. Siegbahn and co-workers, *ESCA Applied to Free Molecules*, p. 66. North-Holland Publ. Co., Amsterdam, 1969.
48. G. R. Wight and C. E. Brion, *Chem. Phys. Lett.*, to be published (1974).
49. T. A. Carlson and R. M. White, Oak Ridge National Lab. Report ORNL-P-591.

Secondary Electron Spectra[1,2]

Yong-Ki Kim
Radiological and Environmental Research Division,
Argonne National Laboratory,
Argonne, Illinois 60439, U.S.A.

INTRODUCTION

The unique "style" of science advocated by Robert L. Platzman has been well known among his colleagues as is pointed out in the obituary by Fano (*1*). To study the energy deposition by fast charged particles, Platzman emphasized the need for a *consistent* set of cross sections for the primary and subsequent events that take place in the irradiated medium (*2*). Major components in such a set of cross sections are (a) optical oscillator strengths both for ionization and for excitation (i.e., transition to all states below the first ionization threshold), (b) ionization and excitation cross sections by charged-particle impact, and (c) cross sections for secondary electrons. Various sum rules both for oscillator strengths (*3*) and for inelastic-scattering cross sections by fast charged particles (*4*) play indispensable roles in establishing the consistency of the cross sections.

Reliable data on the oscillator-strength distribution and on the ionization by photons and electrons became available in the last decade such that they no longer deter the application of the "Platzman approach" to radiation physics. The current understanding of the oscillator-strength distribution has been summarized well by the two preceding speakers of this symposium (*5,6*) and also by Fano and Cooper (*3*). An excellent review on the ionization cross sections by electron-impact is given by Kieffer and Dunn (*7*).

Secondary electrons are produced whenever ionizing radiation (e.g., α, β, and γ rays) passes through a medium. Obviously, secondary-electron spectra form essential parts in the development of electron degradation spectra. The experimental study on the emission of secondary electrons by electrons and protons has experienced rapid progress in the past decade (*8*). The purpose of the present talk is to report on a theoretical method (a) that checks the consistency of the secondary-electron spectra and (b) that enables one to extrapolate and interpolate the secondary-electron spectra to various regions of incident-particle and secondary-electron energies in gaseous media. The method utilizes the relationship between the oscillator-strength distribution and the Born cross sections for charged-particle impact as well as the fact that the Rutherford and Mott cross sections become valid for secondary electrons with kinetic energies far in excess of their binding energies (*8*). Empirical data on the oscillator-strength distribution are used in the present method, and hence no large scale computation is involved.

In contrast, the poor knowledge of the excitation cross sections by charged-

[1]Congress presentation B–10–4.
[2]Work performed under the auspices of the U. S. Atomic Energy Commission.

particle impact still remains a major obstacle to the understanding of the physical stage of radiation interaction defined by Platzman (*9*). In the present work, we focus our attention only on the energy distribution of the secondary electrons as this is most relevant to problems in radiation research.

THEORY

Theoretical studies of secondary-electron spectra have been mostly for fast incident particles based on the Born approximation with atomic wavefunctions of moderate accuracy. The main difficulty here comes from the lack of reliable continuum wavefunctions for atoms and molecules with nontrivial shell structures. Furthermore, no *successful and tractable* theory has been developed for slow incident particles where the Born approximation fails.

According to the Born approximation (*10*), the cross section for the production of secondary electrons with kinetic energy W by an incident particle of charge ze and velocity $v = \beta c$ with the speed of light c is given by

$$d\sigma/dW = 4\pi a_0^2 z^2 \alpha^2/\beta^2 R \times \Sigma_j \left\{ (df/dE_j)(R^2/E_j)\,[\ln(\beta^2/1-\beta^2) - \beta^2] + B(\beta^2, E_j) \right\}, \quad (1)$$

where R is the Rydberg energy (13.6 eV), a_0 is the Bohr radius (0.529 Å), α is the fine structure constant, and df/dE_j is the dipole oscillator-strength density for ionization with energy transfer E_j. The summation in Eq. (1) is over different binding energies, i.e., $E_j = W + B_j$, where B_j is the binding energy of the jth orbital. The first term in the curly brackets of Eq. (1) represents the contribution from the glancing collisions to $d\sigma/dW$. The glancing-collision term is dominated by the contributions from the outer orbitals resulting in slow secondary electrons. For large energy transfers $(dF/dE_j)/E_j$ diminishes rapidly as $E_j^{-4.5}$ or faster (*11*). The last term in Eq. (1), $B(\beta^2,E_j)$, contains contributions both from the glancing and knock-on collisions, but it becomes independent of β^2 for fast incident particles. For fast secondary electrons, $B(\beta^2,E_j)$ approaches the Rutherford cross section, i.e.,

$$B(\beta^2,E_j) \xrightarrow[E_j \gg B_j]{} N_j(E_j/R)^{-2}, \quad (2)$$

where N_j is the electron occupation number.

Platzman suggested (private communication) to plot the ratio of the experimental $d\sigma/dW$ to the Rutherford cross section for the valence orbital

$$Y(E) = (d\sigma/dW)(\beta^2/4\pi a_0^2 z^2 \alpha^2)\,E^2/R \quad (3)$$

as a function of R/E, where

$$E = W + B_1, \quad (4)$$

with the lowest binding energy B_1. Figure 1(a) is an example of such a plot for the water molecule. The ratio Y(E) has been extensively used also by Bichsel (*12*) in his study of stopping power.

In the Platzman plot, the area under the curve is proportional to the ionization cross section

$$\sigma_{ion} = \int_0^{W_{max}} (d\sigma/dW)\,dW, \quad (5)$$

where the maximum energy for the secondary electron, W_{max}, is determined from the conservation of energy as well as the usual definition for the secondary electron being the slower one after the collision if the incident particle is an electron.

From Eq. (1)–(3), we realize that (a) for small E, say $E < 10R$, Y(E) should resemble (df/dE)E in shape, and (b) for large E, Y(E) should approach an effective number of electrons participating in the ionizing collision. These features of the Platzman plot have been utilized already in the thesis work of one of Platzman's students (*13*) to supplement the lack of reliable data on $d\sigma/dW$ of He at that time.

The oscillator-strength distribution to be compared with Y(E) must be that for ionization. Furthermore, a partitioning of the df/dE for ionization is necessary when alternative ionic states are available for a given energy transfer, as is frequently the case for molecules with several molecular orbitals of similar binding energies. The Auger electrons resulting from inner shell excitations appear as sharp peaks in the Platzman plot, and they can be separated easily from other parts of the spectra.

The secondary-electron spectra for H_2O measured by Opal, Beaty, and Peterson (*14*) with 500-eV incident electron is compared with the corresponding df/dE for ionization compiled by Berkowitz (private communication) in Figs. 1(a) and 1(b). The spectra for secondary electrons of $W \gtrsim 8$ eV clearly resemble the optical spectra, and the spectra for fast secondary electrons approach the Mott cross section (*15*). The shaded area in Fig. 1(a) represents the number of secondary electrons with kinetic energies too low to ionize further. Parenthetically, we note that the optical and electron-impact data for the ionization of H_2O in Fig. 1(a) do not support the undulating structures for $E \gtrsim 25$ eV shown earlier by Platzman (*9*). Similar comparisons are made on a large number of atoms and molecules elsewhere (*8,16*).

NORMALIZATION AND EXTRAPOLATION OF EXPERIMENTAL DATA

As is discussed in the preceding section, the area in the Platzman plot gives $\sigma_{ion}\beta^2/4\pi a_0^2 z^2 \alpha^2$. If accurate values of σ_{ion} are known, therefore, the secondary-electron spectra can be normalized against σ_{ion}. Since most secondary electrons ejected are slow (about 2/3 of them have kinetic energies less than the lowest binding energy), the shape of the spectra for slow secondary electrons must be known well. Unfortunately, most experimental data on $d\sigma/dW$ are not reliable for slow secondary electrons ($W \lesssim 30$ eV). Comparison and extrapolation of $d\sigma/dW$ for slow secondaries to make it consistent with the optical data are essential for the normalization of $d\sigma/dW$, as is illustrated in Figs. 2(a) and 2(b) for N_2.

A further cross examination of $d\sigma/dW$ thus normalized can be made by checking the dependence of $d\sigma/dW$ on the incident particle velocity, at a given W. As β^2 increases the graph of Y(E) [Eq. (3)] as a function of

$$X = \ln(\beta^2/1-\beta^2) - \beta^2 \tag{6}$$

becomes a straight line with the slope [see Eq. (1)–(4)]

$$A = [\Sigma_j(df/dE_j)(1/E_j)]E^2 \tag{7}$$

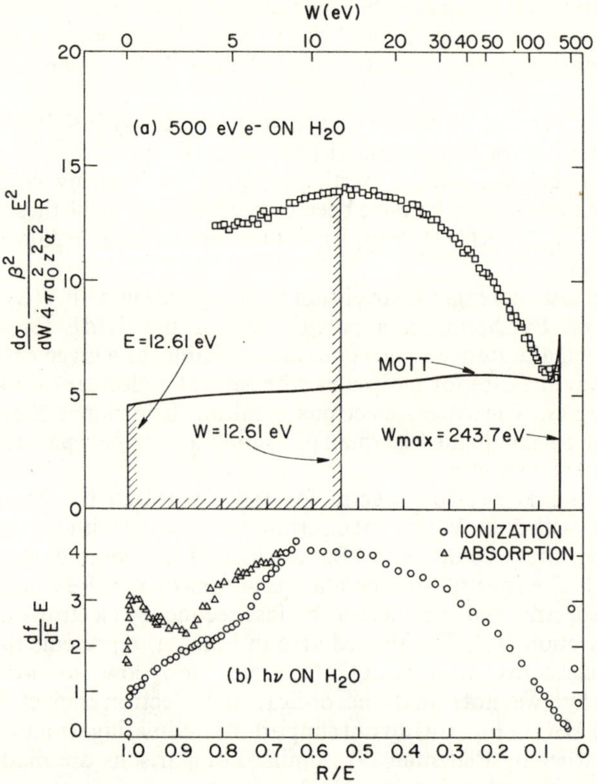

FIG. 1. The Platzman plot for the electron-impact (a) and photoionization (b) cross sections of H_2O. The squares in Fig. 1(a) represent the electron-impact experimental data by Optal et al. (*14*), and the curve marked MOTT represents the Mott cross section (*8,15*) for 500-eV incident electrons. The area bounded by the shaded lines and the measured spectra in Fig. 1(a) is proportional to the number of secondary electrons emitted with kinetic energies lower than the lowest ionization potential, viz., 12.61 eV. In Fig. 1(b), the ionization yield is assumed to be unity for $h\nu \gtrsim 20$ eV. The photon data are those compiled by Berkowitz (*5*, and private communication).

and the intercept given by B in Eq. (1). This graph is known as the Fano plot (*17*); it is a powerful method not only for showing the convergence of $d\sigma/dW$ toward correct asymptotic behavior, but also useful for checking the magnitudes of cross sections at lower incident energies.

Two Fano plots are presented for N_2 in Fig. 3; one for σ_{ion} to be used for normalization of $d\sigma/dW$ and another for $d\sigma/dW$ at $W = 13.1$ eV ($R/E = 0.475$). Experience shows that irregularities both in $d\sigma/dW$ and in σ_{ion} at different X are easily detected in the Fano plot because most cross sections have monotonic behavior in such a plot. It is simple to extrapolate and interpolate $d\sigma/dW$ for arbitrary incident-electron energies at a fixed W with the aid of the Fano plot. A set of the Fano plots for N_2 corresponding to various R/E values has been constructed from the electron-impact data by Opal et al. (*14*). The experimental data have been renormalized to be consistent with the adopted values of σ_{ion} in

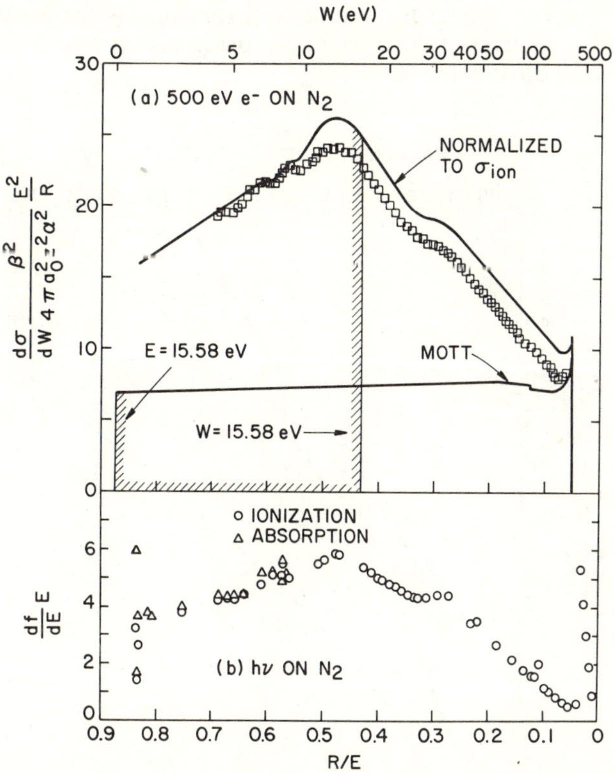

FIG. 2. Electron-impact (a) and photoionization (b) cross sections of N_2. In Fig. 2(a), the secondary-electron cross section normalized to σ_{ion} (see text and Fig. 4) is included. The small step at $R/E = 0.13$ in the Mott cross section is due to the fact that the K electrons can be ejected by 500-eV incident electron with a maximum kinetic energy of ~90 eV (K-shell binding energy ≈ 410 eV). Sharp structures in the optical data near the first ionization threshold have been omitted from Fig. 2(b) for clarity. For other legends, see the caption for Fig. 1. As in Fig. 1(a), the shaded area is proportional to the number of secondary electrons with kinetic energies less than the first ionization potential, viz., 15.58 eV.

Fig. 3. The adopted σ_{ion} is based on the electron-impact measurement by Tate and Smith (*18*), with a slight reduction of magnitude to account for likely errors in the pressure measurement (*7*). The adopted σ_{ion} is also in excellent agreement (4–7%) with the measurement by Rieke and Prepejchal (*19*) using fast incident electrons and positrons (0.1–2.7 MeV).

Figure 4 is the Platzman plot of the secondary-electron spectra from N_2 by incident electrons of various energies. The data for the incident energies T = 50–2000 eV are based on the experiment by Opal et al. (*14*), but the curve for T = 10 keV is an extrapolation based on the set of the Fano plots at various R/E. The asymptotic slope A given by Eq. (7) for high values of X [Eq. (6)] can be obtained from optical data. In practice, the inner-shell contributions to A are negligible compared to that from the valence orbital so that $A \approx (df/dE)E$ where E is the photon energy in the optical data. Figure 2(b) shows $(df/dE)E$ as a function of E.

From Figs. 2–4, one should be able to construct $d\sigma/dW$ for N_2 for arbitrary primary- and secondary-electron energies, using σ_{ion} in Fig. 3 to check the normalization of $d\sigma/dW$ for consistency.

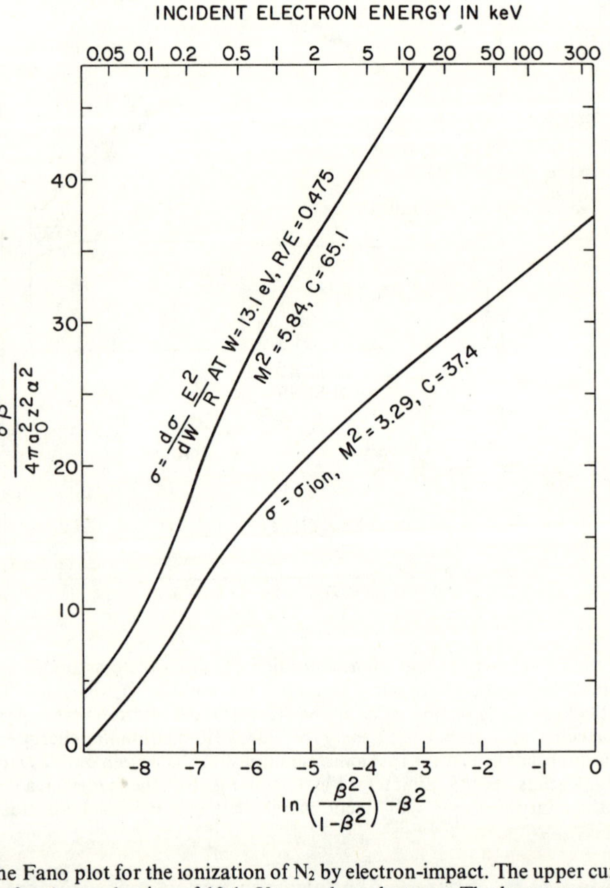

FIG. 3. The Fano plot for the ionization of N_2 by electron-impact. The upper curve represents the cross section for the production of 13.1 eV secondary electrons. The lower curve represents the total ionization cross section [Eq. (5)]. Both curves become straight lines for large X [Eq. (6)] of the form $Y = M^2X + C$, where Y is the ordinate, and the values of M^2 and C are given in the figure.

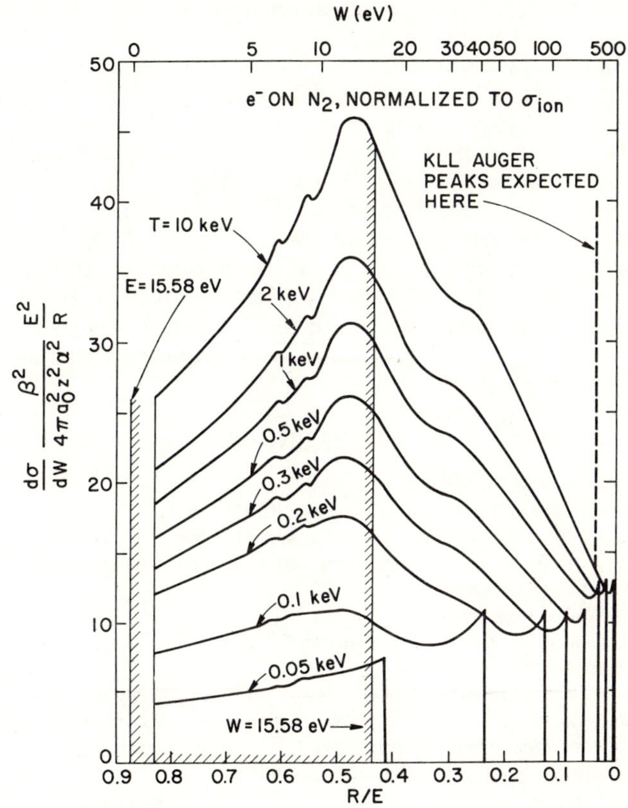

FIG. 4. Secondary-electron spectra for N_2 for electron-impact. The area under each curve has been normalized to σ_{ion} in Fig. 3 for corresponding incident energies T. The expected position and width (but not the amplitude) of the KLL Auger peaks are indicated. The area under the Auger peaks should be proportional to the sum of the total cross sections for ionization and excitation of K electrons. The spectra between the ionization threshold and $R/E = 0.832$ ($W = 0.77$ eV) would depend on the energy resolution owing to the sharp autoionization peaks in that region. See Ref. *8* for discussions on the features seen in the spectra. For the meaning of the shaded area, see the caption for Fig. 2.

CONCLUDING REMARKS

To apply the method prescribed above more effectively, it is desirable to have the following experimental data on various molecules of interest in radiation research:

(a) Photoionization cross sections on an absolute scale;
(b) Ionization cross sections for charged-particle impact on an absolute scale;
(c) Cross sections for the production of very slow secondary electrons, i.e., $d\sigma/dW$ at $W \approx 0$, on an absolute scale as was done on He, Ne, and Ar by Grissom et al. (*20*);
(d) Secondary-electron spectra on a relative scale, with emphasis on slow secondary electrons; and

(e) Partial cross sections leading to alternative ionic states for a given energy transfer.

For the application of the Platzman approach to radiation physics, the cross sections for the excitations to states below the first ionization threshold both by photon and charged-particle impact appear as the most uncertain component. Major effort, both theoretical and experimental, therefore, should be directed toward the determination of excitation cross sections. Also, recent success in the application of the Platzman approach to He (*21*), H_2 (*22*), and Ar (*23*) indicates that we may soon be able to substantially broaden our understanding of some problems in radiation physics such as the nearly constant ratio of the W value (energy to produce an ion pair) to the first ionization potential (*2*) and the sensitivity of electron degradation spectra to the accuracy of cross sections used.

I am grateful to Dr. J. Berkowitz for providing me with his compilation of the oscillator strengths.

REFERENCES

1. U. Fano, *Radiat. Res.* **58,** 559 (1974).
2. R. L. Platzman, *Int. J. Appl. Radiat. Isotopes* **10,** 116 (1961).
3. U. Fano and J. W. Cooper, *Rev. Mod. Phys.* **40,** 441 (1968).
4. M. Inokuti, Y.-K. Kim, and R. L. Platzman, *Phys. Rev.* **164,** 55 (1967).
5. J. Berkowitz, this symposium.
6. M. van der Wiel, this symposium.
7. L. J. Kieffer and G. H. Dunn, *Rev. Mod. Phys.* **38,** 1 (1966).
8. Y.-K. Kim, *Radiat. Res.* **61,** 21 (1975).
9. R. L. Platzman, in *Radiation Research.* Proc. 3rd Intern. Congress Radiat. Res., Cortina d'Ampezzo, 1966; (G. Silini, Ed.) pp. 20–42. North-Holland, Amsterdam, 1967.
10. M. Inokuti, *Rev. Mod. Phys.* **43,** 297 (1971).
11. A.R.P. Rau and U. Fano, *Phys. Rev.* **162,** 68 (1967).
12. H. Bichsel, *Phys. Rev. B* **1,** 2854 (1970).
13. W. F. Miller, Ph.D. Thesis, Purdue University, 1956.
14. C. B. Opal, E. C. Beaty, and W. K. Peterson, *At. Data* **4,** 209 (1972).
15. N. F. Mott, *Proc. Roy. Soc.* London *Ser. A* **126,** 259 (1930).
16. Y.-K. Kim, Argonne National Laboratory Report No. ANL-8060-I (1973), pp. 55–73; Y.-K. Kim and T. Noguchi, submitted to *Int. J. Radiat. Phys. Chem.*
17. U. Fano, *Phys. Rev.* **95,** 1198 (1954).
18. J. T. Tate and P. T. Smith, *Phys. Rev.* **39,** 270 (1932).
19. F. F. Rieke and W. Prepejchal, *Phys. Rev. A* **6,** 1507 (1972).
20. J. T. Grissom, R. N. Compton, and W. R. Garrett, *Phys. Rev. A* **6,** 977 (1972).
21. D. A. Douthat, *Radiat. Res.* **61,** 1 (1975).
22. D. E. Gerhart, submitted to *J. Chem. Phys.*
23. E. Eggarter, submitted to *J. Chem. Phys.*

Ionization by Energy Transfer from Excited Species[1]

A. Niehaus
Fakultät für Physik, Universität Freiburg/Br.
W. Germany

INTRODUCTION

If matter is subjected to high energy radiation, the largest part of the primary energy is dissipated by a large number of processes each involving only energy quantities on the order of a few eV. At this level the ionization by collisional transfer of energy stored in the form of electronic excitation plays an important role. The final result of high energy radiation on matter may therefore to a large extent be determined by the characteristics of such processes, and it is necessary to study these processes in detail.

It is the aim of this paper to discuss the different types of ionization processes occurring in thermal collision, by transfer of electronic excitation, and to give an outline of our present understanding of their mechanisms. No attempt is made to give a complete review of work done in the field, but rather to emphasize those aspects which are most relevant with respect to radiation physics.

We may distinguish the following types of collisional ionization processes:

(1) Ionization by electronically excited species in metastable states.
(2) Ionization by electronically excited species in states which are coupled to their ground states by allowed dipole transitions.
(3) Ionization occurring in collisions with species in high Rydberg states.
(4) Ionization occurring in slow collisions with ions of high recombination energy.

These four types of processes will be discussed below. By far the most reliable and detailed experimental and theoretical information is available for process (1)—also called Penning (*1*) ionization—and therefore most of the available space will be devoted to this process.

PENNING IONIZATION

A. Atoms

The process of ionization of atoms B in thermal collisions with excited metastable atoms A* may be written as

$$A^* + B \rightarrow A + B^+ + e^- \qquad (1a)$$

$$\rightarrow AB^+ + e^- \qquad (1b)$$

[1]Congress presentation B–29–1

It may be considered as the autoionization of the molecular collision system, occurring if the excitation energy E* of A* is larger than the ionization energy E_i of B. Most experimental information on process (1) is available for the cases where A* is a metastable rare gas atom [$He^*(1s2s)^{1,3}S$, $Ne^*(2p^5 3s)^3P_{2,0}$, $Ar^*(3p^5 4s)^3P_{2,0}$] (*2,3,4*).

A good description of the mechanism of (1) is given by the potential curve model (*5–10*) (Fig. 1) which is based on the Born-Oppenheimer approximation. The electronic states of the molecular system before and after ionization are described by the potential curves $V_*(R)$ and $V_+(R)$, and the probability $W(t) = \frac{\Gamma(R(t))}{\hbar}$, for the $V_* \rightarrow V_+$ transition is given by the square modulus of the electronic coupling matrix element between the discrete state and the ionization continuum. In the vertical transition electronic energy and energy of heavy particle motion are separately conserved, which leads to the basic relations for Penning Ionization

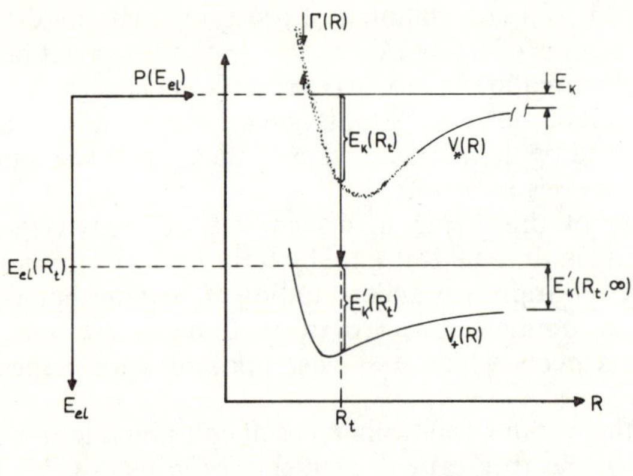

FIG. 1. Potential curve model for Penning ionization.

$$E_{el}(R_t) = V_*(R_t) - V_+(R_t) \tag{2}$$

$$E'_K(R_t,\infty) = E_K(\infty) + V_*(\infty) - V_*(R_t) - [V_+(\infty) - V_+(R_t)] \tag{3}$$

connecting, for a process with collision energy $E_K(\infty)$, occurring at distance R_t, the energy of the ejected electron $E_{el}(R_t)$, as well as the final relative kinetic energy of the heavy particle system $E'_K(R_t,\infty)$, in a very simple way to the energy difference between the potential curves, $V_*(R_t) - V_+(R_t)$.

The quantitative formulation of the problem, connecting the basic physical quantites (V_*, V_+, Γ) to the quantities accessible in an experiment, like cross section for ionization $\sigma(E_K)$, electron energy distribution $P(E_{el}, E_K)$, kinetic energy distribution $P(E'_K, E_K)$, and cross section ratios for the associative path (1b) to total ionization, $\sigma_A(E_K)/\sigma_{tot}(E_K)$, has been given by Miller (*10*) in classical, semiclassical, and quantum mechanical treatment of the heavy particle motion. The predictions of the theory have been tested (*3,4*) by experiment in several cases—mainly by comparison with results of measurements of $P(E_{rl})$—and no significant inconsistencies were observed. Therefore, the general mechanism of reactions (1) may be regarded as established, at least in their simplest form, when no excitation transfer channels are open.

From relation (3) it is clear that the associative path (1b)—characterized by $E'_K(\infty) < 0$ — can only be chosen by the system, if

$$E_K + V_*(\infty) - V_*(R_t) < V_+(\infty) - V_+(R_t) \quad (4)$$

within the accessible range of R_t. This will in general be the case if E_K is small and V_+ strongly attractive. A strongly decreasing ratio $\sigma_A(E_K)/\sigma_{tot}(E_K)$ with E_K is expected. Recent experiments[2], (*3*) show this behavior quite clearly. Examples are given in Fig. 2.

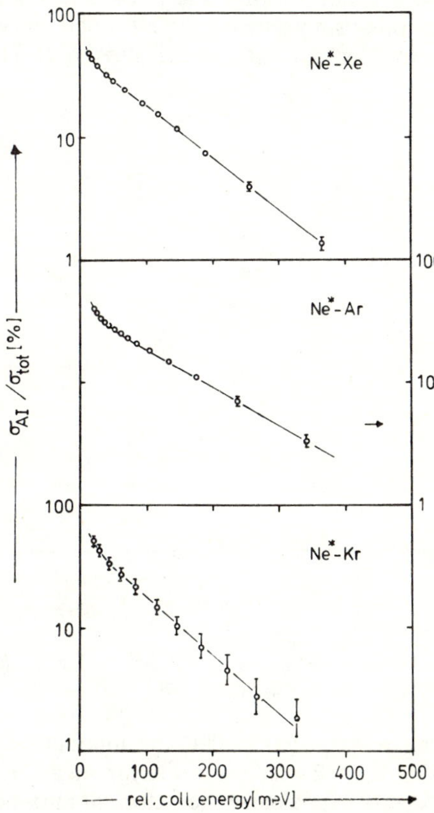

FIG. 2. Variation of the ratio of associative to total ionization with collision energy for the systems Ne*/Ar, Kr, Xe.

[2] L. Kaufhold, Dissertation, Freiburg.

Most important with regard to radiation physics is of course the size of the cross sections of process (1). Within the frame of the classical version of the theory sketched above, the cross sections can be calculated from the quantities V_*, V_+, Γ in the following way. First the transition probability for a collision with energy E_K and impact parameter b is calculated by

$$P(b,E_K) = 1 - \exp\{-\int_{R_c(b)}^{\infty} [2\Gamma(R)dR/\hbar v_{rad}(R,b)]\} \quad (5)$$

with $v_{rad}(R,b) = [(2/\mu)(E_K - V_*(R) - E_K b^2/R^2)]^{1/2}$ the radical velocity, and $R_c(b)$ the classical turning point given by the condition $v_{rad}(R_c,b) = 0$. The total cross section is then obtained as

$$\sigma_{tot}(E_K) = 2\pi \int_0^{\infty} bP(b,E_K)db \quad (6)$$

Instead of (5) and (6), quantum expressions may also be used. Only for the system $He^*(2^3S)/H$ *ab initio* calculations of σ_{tot} have been carried out (*11,12*). For the same system calculations using semi-empirical functions V_* and Γ^4 have been performed, and the cross section was also measured (*13*). The agreement is good as shown in Fig. 3.

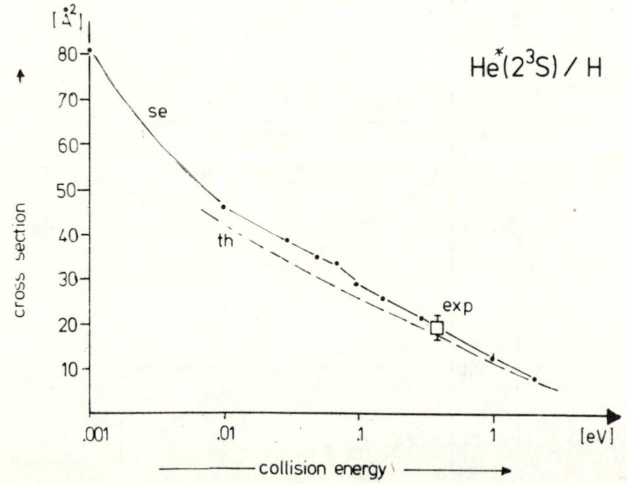

FIG. 3. Comparison of "*ab initio*" calculated and semiempirically determined ionization cross section with the experimental value for the system $He(2^3S)/H(^2\Sigma)$.

The He*/H system is special in that the interaction potential V_* is very strong with a well depth $\epsilon = 2.4$ eV (*14*). Measurements of $P(E_{el})^4$ show that in cases where He* interacts with a close shell atom or molecule, ϵ is typically only on the order of thermal energies. In these more normal cases the cross section dependence on E_K is usually similar to the special case shown in Fig. 4, where experimental data (*15,16*) are compared with a calculation (*17*), in which relations (5), (6), and model functions V_* and Γ are used.

The behavior of $\sigma(E_K)$ seen in Fig. 4 is quite general in that $\sigma(E_K)$ decreases in a region of $E_K \ll \epsilon$, and increases in a region $E_K \gg \epsilon$ with E_K. In the region $E_K \ll \epsilon$,

$\sigma(E_K)$ may be approximated by the close collision cross section, as

$$\sigma_{tot}(E_K) \approx \eta \pi b^2_{max} \tag{7}$$

with η the efficiency per close collision, which in the energy region considered should be nearly independent of E_K, and b_{max} the largest impact parameter for which a close collision surmounting the rotational barrier of $V_*(R) + E_K b^2/R^2$ is possible. b_{max} can be calculated from the long range forces between A* and B incorporated in V_*. In case of the van der Waals attraction, $b_{max} \propto v^{-2/3}$ so that $\sigma(E_K) \propto (E_K)^{-1/3}$ and the rate constant $k(T) \propto (T)^{1/6}$. In the region $E_K \gg \epsilon$, where $\sigma(E_K)$ increases with E_K, the total cross section may be approximated by

$$\sigma_{tot}(E_K) \approx \int_{R_c(0)}^{\infty} [4\pi R^2 \Gamma(R)/\hbar][\mu/2E_K]^{1/2}[1 - V_*(R)/E_K]^{1/2} dR \tag{8}$$

This approximation involves the neglection of attraction in $V_*(R)$ and the assumption that $P(b) \ll 1$ (see (5)). Relation (8) can be evaluated if the model functions $\Gamma(R) = A \cdot \exp(-\alpha R)$ and $V^*(R) = B \cdot \exp(-\beta R)$ are assumed, and if $V_*(R)$ is expanded around the turning point for central collisions $R_c(0)$ and only the first two terms used. Taking advantage of the fact that $\alpha \cdot R_c(0) \gg 1$ one obtains the simple result

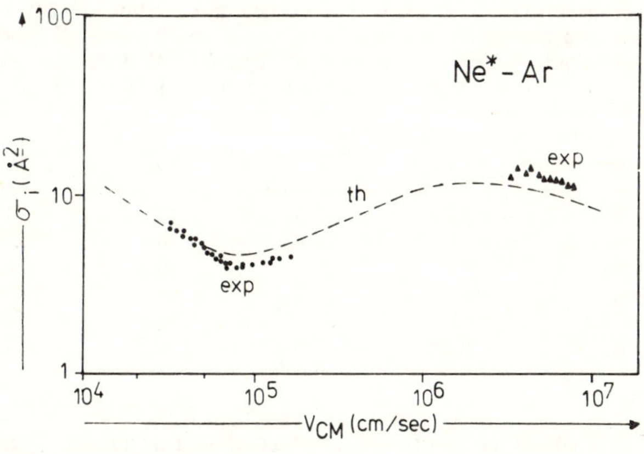

FIG. 4. Dependence of total ionization cross section on collision energy for the system Ne*/Ar. Comparison of calculation (-, ref. *17*) with experiment (• ref. *15*; △ ref. *16*).

$$\sigma_{tot}(E_K) \approx (2A/\hbar)(\mu/2B)^{1/2}[\pi/(\alpha \cdot \beta)]^{3/2}(\ln E_K/B)^2(E/B)^{\alpha/\beta - 1/2} \tag{9}$$

Relation (9) is not thought to be a very good approximation to the cross section, but it shows how the "hardness" of the potential V_*, given by β, and the "steepness" of the decrease of $\Gamma(R)$, given by α, determine the energy dependence of the cross sections. Just to get a rough idea of the value of α/β one may approximate $\alpha/\beta \sim \sqrt{E_i(T)/E_i(P)}$, where $E_i(T)$ and $E_i(P)$ are the ionization energies of target and metastable projectile. The decrease of the slope of $\sigma(E_K)$ at high E_K observed in Fig. 4 is due to the fact that $P(b) \sim 1$ at stronp penetration

Recent flowing afterglow measurements of the quenching cross section of $He(2^3S)$ by several gases in the temperature range 300–900° K(*18*) show indeed a strong increase of $\sigma_q(E_K)$ with E_K, in rough agreement with (9), and a value of α/β estimated as described above. Also preliminary results[3] of measurements of $\sigma(E_K)$ in velocity selected beam experiments for the systems He(2^3S)/ Ar, Kr, Xe confirm this rather strong increase, in contradiction to earlier beam results (*19*). On the other hand, our beam experiments show that $\sigma(E_K)$ for ionization by He(2^1S) stays nearly constant in the energy range considered (300 − 900° K). The strong rise of the He(2^3S) cross sections as compared to He(2^1S) cross sections resolves a discrepancy between absolute cross section results obtained in flowing afterglow experiments, and beam measurements (*20*) which had been performed at different temperatures, and therefore agreed in the He(2^1S) values, but disagreed in the He(2^3S) values by a factor of 2–3. If the flowing afterglow He(2^3S) values are taken at the temperature (~ 800° K) of the beam experiments, also here agreement is found (*18*). This is seen from Table 1. The agreement between the two sets of data adds much to our confidence in the reliability of the values and also proves that the ionization process (1) is the main quenching process for the systems considered.

TABLE 1
Comparison of quenching cross sections obtained in flowing afterglow experiments (F.A. ref. *18*) with ionization cross sections obtained in beam experiments (B. ref. *20*) for collisions of He(2^3S) and He(2^1S) with some targets. FA-data are adapted to temperature of B-measurements (~ 800° K) in case of the He(2^3S)-data

Target	Q(2^3S) [Å^2]		Q(2^1S) [Å^2]	
	FA	B	FA	B
N_2	13.3	15.5	12.5	21.2
Ar	13.6	16.9	16.4	23.02
Xe	19.0	20.0	36.4	34.21
O_2	28	29.4	44.6	43.65
CO_2	48	63	84.6	59.37
H_2	3.25	3.34	2.25	2.90

The fact that the He(2^1S) cross sections do not rise as steeply as the He(2^3S) cross sections may be explained either by a small value of α in the singlet case, or by the condition that the transition probability P(b) is already at low collision energies for most of the hard sphere collisions close to unity. This latter explanation is favored by recent results of elastic scattering experiments (*21*).

The transition probability $\Gamma/\hbar$ is proportional to the square of the matrix element which couples the discrete state V· to the continuum. Formally it may be divided into two parts which correspond to two different types of ionization mechanisms (*8*), namely[4]

[3] E. Illenberger, A. Niehaus (Freiburg) to be published.

[4] Arabics (1), (2) denote the participating electrons.

$$A^*(1) + B(2) \rightarrow A(2) + B^+ + e^-(1) \quad (10a)$$

$$A^*(1) + B(2) \rightarrow A(1) + B^+ + e^-(2) \quad (10b)$$

Process (10b) involves the electronic transitions $A^* \rightarrow A$, and $B \rightarrow B^+ + e^-$, of which the former is forbidden in the case of metastables. In fact, experimental data, as well as numerical calculations performed for the system $He(2^3S)/H$ (*12*), confirm that the process (10a) is the dominant mechanism. From this follows the important practical consequence that the dipole selection rules are not valid for the population of the different electronic states of the ion formed in Penning ionization. It follows also that $\Gamma(R)$ should decrease with R at large R as an exponential, as was actually found experimentally (*4*) and in numerical calculations (*12*).

So far only the simplest case of Penning ionization has been discussed, in which the molecular state V_* can be regarded as isolated over the whole range of accessible distances. If other molecular states cross or "pseudocross" the incoming V_*, the important question of excitation transfer as a competing process arises. At thermal energy, and in the case of atoms as targets, such processes do not seem to be important. However, at elevated energies the situation changes, as has been shown for the He*/Xe system (*22*). At energies of only a few eV the processes

$$He^* + Xe \rightarrow Xe^+ + He + e^-(P(E_{el})) \quad (11a)$$

$$\rightarrow Xe^{**} + He$$

$$\llcorner\!\!\rightarrow Xe^+ + e^-(E^{**} - E_i) \quad (11b)$$

compete. The two processes can be distinguished by their electron spectra, thus allowing to measure the branching ratio as a function of collision energy. The result is shown in Fig. 5 for the system He*/Xe, Kr, Ar.

B. Molecules

Regarding Penning Ionization of molecules by He-metastables, most of the characteristics follow from the theory of "simple" Penning ionization of atoms outlined above. Electronic excitation of the molecular Penning ions is in most cases found to be substantially different from that in photoionization, due to the different type of transition matrix element. Vibrational and rotational excitations, on the other hand, are very similar to those in photoionization (*23–30*), due to the fact that both processes are vertical. Deviations from the populations observed in photo-ionization can be induced by the presence of the collision partner during the ionization, which may perturb the molecular potential curves. This perturbation obviously is in most cases small, as can also directly be seen from measured $P(E_{el})$ (*27*). Strong deviation from Franck-Condon behavior was observed in optical measurements of the fluorescence light emitted from molecular Penning ions for the systems He*/HCl, HBr (*26*). On the other hand, recent electron spectroscopic measurements again show that the vibrational population is Franck-Condon like[5]. A comparison of the populations observed by optical and by electron spectroscopic method with the population

[5] H. Hotop (private communication).

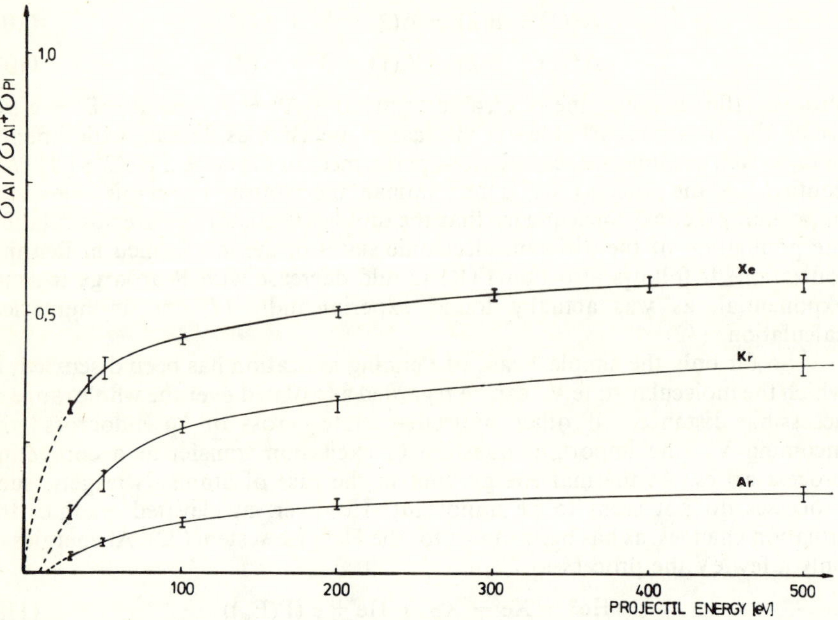

FIG. 5. Branching ratio for excitation transfer (followed by autoionization) and Penning ionization, as a function of collision energy, for the systems He*/Ar, Kr, Xe.

observed for photoionization is shown in Fig. 6. The discrepancy is resolved by the assumption that the molecular ion formed by Penning ionization may convert vibrational energy into kinetic energy of relative motion of the collision partners. Direct evidence of such a conversion has been found for the system He*/H_2 (*31,6*).

From the potential curve model for Penning ionization described above it is also clear that there will be the possibility of a reaction of the Penning ionized system. Depending on the kinetic preparation of the system, these reactions will be very similar to the reactions occurring in the corresponding ion-molecule collision system at low kinetic energies. Evidence to this is given by the fact that for the Penning reactions

$$\begin{aligned} He^* + HD &\rightarrow HD^+ \quad (87.2\%) \\ &\rightarrow HeD^+ + H \quad (7.1\ \%) \\ &\rightarrow HeH^+ + D \quad (4.7\%) \\ &\rightarrow HeHD^+ \quad (1\%) \end{aligned} \tag{12}$$

the intensity ratio $[HeH^+]/[HeD^+]$ is identical with that obtained by extrapolating the results of beam type experiments for the system (HD^+ + He) to zero kinetic energy (*6*). Other examples of ion-molecule reactions in the exit channel of Penning ionization are given in refs. *6* and *32*.

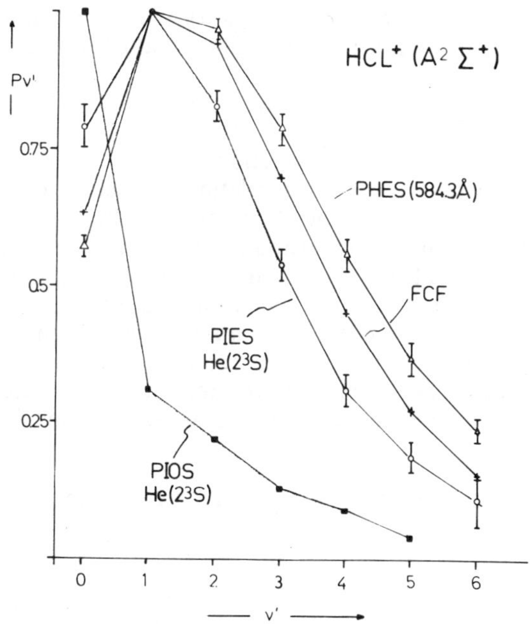

FIG. 6. Comparison of vibrational populations of $HCl^+(A^2\Sigma^+)$ observed by Photoelectron Spectroscopy (PhES), by Penning electron spectroscopy (Pies), and by optical spectroscopy of the Penning ion formed in collision with $He(2^3S)$ (Pios).

Concerning the fragmentation of a large molecular ion formed by Penning ionization, it may be noted that it should be very similar to the fragmentation following photoionization with quanta of energy similar to the energy of the metastable, because vibrational excitation is similar. Due to the fact that there are no selection rules—except for spin conservation—for Penning transitions, there may in some cases be the tendency of stronger fragmentation.

If the metastable excitation energy is not much greater than the first ionization energy of the molecule—as is the case for $Ar^*(^3P_{2,0})$ of $E^* \approx 11.7$ eV—the situation becomes much more complicated. Due to the high density of molecular states in this region, there usually will exist the possibility of excitation transfer, which then may be followed by autoionization or dissociation. Measurements of the quenching cross sections of Ar* by molecules with ionization potentials below and above E* do not show a systematic increase close to E*, indicating that here the Penning ionization is not the dominant process (*33*). The very mechanism of the reactions competing with Penning ionization in the case of Ar* is not clear. The fact, however, that there are some "weak quenchers" of Ar* which at the same time are species with relatively low density of states, in the region of interest (for instance Kr, H_2, N_2, CO) (*33*), indicates the importance of the density of available states in the region of E*. Currently the long range dipole-dipole quenching mechanism (*34–36*) is invoked to explain the quenching of $Ar^*(^3P_{2,0})$ (*33*). This mechanism was initially proposed (*34*) to explain the quenching of atoms in excited states which are coupled to the ground state by an allowed dipole transition. It predicts that the quenching cross section should vary as

$$\sigma_q \propto (\mu^{1/2}\sigma_a)^{2/5} \quad (13)$$

where σ_a is the photo absorption cross section of the quencher at the energy E*. In fact, in those cases where σ_a is known, relation (13) seems to give a fairly good description of the behavior of the different quenchers as shown in Fig. 7. However, observing the facts (1) that application of the theory implies that A* be coupled to the ground state by an allowed dipole transition, which is not the case for Ar*($^3P_{2,0}$), and (2) that the deviations from the behavior predicted by (13) are considerable, the proposed mechanism cannot be regarded as certain. It should also be noted that the long-range dipole-dipole mechanism completely neglects the exchange mechanism which leads to cross sections of the same order of magnitude (see Table 1) as the quenching cross sections for Ar($^3P_{2,0}$).

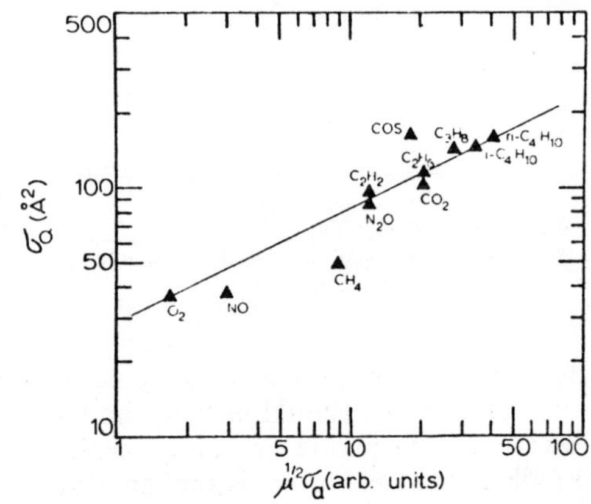

FIG. 7. Ar*($^3P_2$) metastable quenching cross sections vs $\mu^{1/2}\sigma$. The line drawn through the data has a slope of 2/5.

IONIZATION BY ATOMS IN RESONANCE STATES

This process cannot be studied in beam type experiments so that experimental information until now has been mainly restricted to total cross sections for ionization or quenching.

An approximate theoretical description of the process was given by Katsuura (*34*), and by Watanabe and Katsuura (*35*). The procedure is similar to the one outlined above for Penning ionization. It differs in the following aspects.

(1) In the expansion of the electronic coupling operator, only the dipole-dipole term is considered whereas the monopole term, which was the dominant one in Penning transitions, is neglected.

(2) In the integration of the contributions from different separations of the collision partners, a straight line trajectory is assumed and thus the effect of

the interaction potential V_* on the cross section neglected.
In this way the total ionization cross section is determined solely by the dipole transition matrix elements μ_A and μ_B of the two transitions involved, namely $A^* \rightarrow A$ and $B \rightarrow B^+ + e^-$, and of course by the collision velocity v:

$$\sigma_{tot} = 13.88 \, (\mu_A^2 \, \mu_B^2 / \hbar v)^{2/5} \tag{14}$$

The advantage of this formula is that the quantities μ_A and μ_B are directly measurable in separate experiments, thus allowing the prediction of cross sections without ab initio information. In ref. *35* a number of cross sections obtained in this semiempirical way, using formula (14), are reported. Generally these cross sections for ionization of atoms or small molecules are $\sim 50 - 100$ Å^2, roughly independent of the type of resonance state involved (numbers for He, Ne, Ar, Hg, Cd resonance states are given).

If the ionization cross sections are measured directly, the long-range dipole-dipole interaction model leading to (14) can be tested. Such direct measurements of the sensitized ionization of a large number of molecular species by $Ar^*(^1P_1)$ and $Ar^*(^3P_1)$ have been performed (*37*). The results show, with a few exceptions, good agreement with the predictions, within the precision of the measurements of $\sim\pm 10\%$.

Within the frame of the model one may also obtain the total quenching cross section σ_Q, by substituting for μ_B the dipole transition matrix element for total absorption, μ_B^Q, in (14), which in addition to $B \rightarrow B^+ + e^-$ also accounts for transitions $B \rightarrow B^*$, not leading to ionization. μ_B^Q can be obtained from absorption measurements and thus σ_Q calculated. It may be worth noting that the validity of the model implies that the chemical consequences are the same when the light quantum is absorbed directly, or when it is transferred in collision with A*.

IONIZATION IN COLLISIONS WITH HIGH RYDBERG ATOMS

As an electron bound to an atom or molecule A is excited to higher and higher Rydberg states, the coupling matrix elements involved in the ionizing collisions between A** and B (discussed in the previous two sections) become weaker and weaker, leading to a decrease of the corresponding cross sections with principal quantum number of the excited electron. On the other hand, the very loosely bound electron may gain enough energy by superelastic scattering from the collision partner B to become ionized, thus leading to the process

$$A^{**} + B^{\ddagger} \rightarrow A^+ + B^{\ddagger\prime} + e^- \tag{15}$$

In (15), it is the small change in internal thermal energy of B, caused by the scattering $e^- + B^{\ddagger} \rightarrow e^- + B^{\ddagger\prime}$, which is responsible for ionization, rather than the large electronic excitation energy of A**. Processes of the type (15) were only very recently observed (*38*) in beam experiments. For rare gas atoms A** in Rydberg states with n approximately in the range $15 \lesssim n \lesssim 50$, cross section for ionization of some molecules are listed (*38*) in Table 2. The cross sections are rather large due to the "size" of the Rydberg atoms. It is interesting to note that only the dipolar molecules appearing in Table 2 have such large cross sections. Cross sections for ionization by H_2, N_2, O_2, NO, CH_4 are found to be much smaller (*38,6*).

TABLE 2
Absolute cross sections for collisional ionization of Rydberg atoms by dipolar molecules (ref. *38*)

Rydberg atom	cross sections in 10^4 Å^2; error ~ 30%			
	H_2O	NH_3	SO_2	C_2H_5OH
He**	0.29	0.15	0.11	0.27
Ne**	0.78	–	0.27	0.50
Ar**	1.2	0.74	0.40	0.80

Matsuzawa (*39–41*) has formulated a theoretical description of the process. In essence this description is based on the fact that the main process occurring in (15) is the superelastic collision of a free electron with a molecule, in which the electron gains energy from the rotation of the molecule by coupling to its dipole moment. In this way the difference found between dipolar molecules and others is explained. Also the absolute values calculated by a refined version of the theory (*41*) are in rather good agreement with the experimental values. Unpublished results[6] directly confirm that the ionization in collisions with dipolar molecules is caused by transfer of rotational energy to the Rydberg electron. Other unpublished results[7] show that the cross sections in such collisions increase with quantum number n approximately as shown in Fig. 8 for the systems (rare gas)**/H_2O.

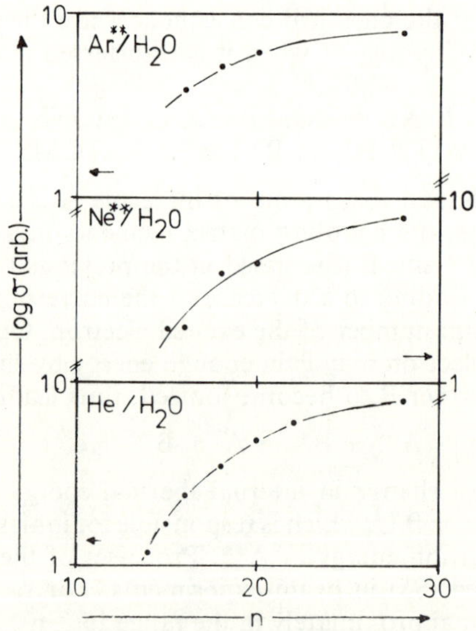

FIG. 8. Approximate dependence of ionization cross section of Rydberg atoms He**, Ne**, Ar** by H_2O, on principal quantum number n.

[6] W. A. Chupka, privately communicated by M. Matsuzawa.

[7] H. G. Schäfer (Diplomarbeit, Freiburg).

Fowler and Preist (*42*) and Fowler (*43*) calculated the cross sections for ionization (15) of highly excited alkali atoms by quadrupolar molecules.

If the high Rydberg atom is not ionized at large separation from the collision partner by rotational energy transfer, it may be ionized at smaller separations by diabatic coupling effects by which the electron can gain energy from the relative motion. Such ionization processes have been observed (*6*) in collisions of Ar** and Kr** with H_2. As in the case of the other ionization processes discussed, these ionizations at smaller distances may of course be followed by ion-molecule reactions as evidenced by experiment (*6*).

IONIZATION IN THERMAL COLLISIONS WITH AN ION OF HIGH RECOMBINATION ENERGY

If an atomic ion of charge state larger than 1 forms a collision system with another neutral atom or molecule, in most cases the system formed is "superexcited" and can autoionize by electronic coupling to one or more of the lower lying electronic continua. Such processes have been proposed to occur (*44*), but until now have not been investigated at low collision energies, where ionizations following diabatic excitations are not possible, and where they constitute the only possible mechanism. Since multiply-charged ions are produced in irradiated matter in considerable numbers, by fast electrons via Auger processes following inner shell ionization, and since the cross sections at low collision energies should be of the same order as Penning cross sections, ionization by the proposed mechanism may be of practical importance.

First results have been obtained for collisions of He^{++} with Hg and Xe (*45*) in beam type experiments. The main ionization mechanism observed at low collision energies may be described as

$$He^{++} + Hg \rightarrow He(ls^2) + Hg^{+++} + e^- \quad (16a)$$

$$He^{++} + Xe \rightarrow He(ls^2) + Xe^{+++} + e^- \quad (16b)$$

The cross section for (16) rises steeply, as shown in Fig. 9, with decreasing collision energy, as expected. From the observed electron spectra arising in processes (16) it is clear that the ejection of the electron occurs *during* collision.

One interesting aspect of reactions (16) with molecules as targets is the fact that here multiply-charged molecular ions may be formed which then should fragmentate in a different way than ions formed by the other important ionization processes. Also, fairly high kinetic energy fragments may be produced by strong Coulomb repulsion.

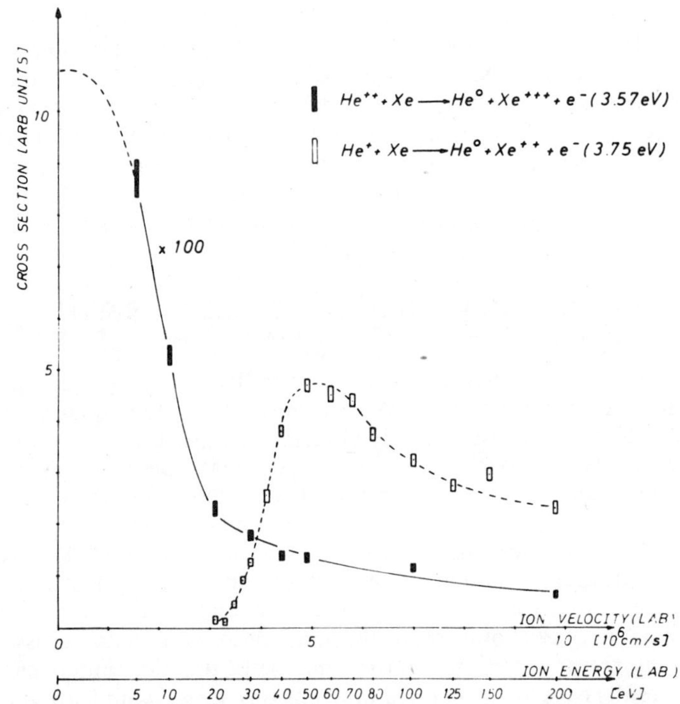

FIG. 9. Velocity dependence of cross section for ionization of Xe by He^{++}. At low velocities the cross section is of the order Å^2.

REFERENCES

1. F. M. Penning, *Naturwissenschaften XV* 818 (1927).
2. R. D. Rundel and R. F. Stebbings, in *Case Studies In Atomic Collision Physics II* (McDaniel, McDowell, Ed.), p. 549. North-Holland Pub. Co., Amsterdam, 1972.
3. A. Niehaus, *Ber. Bunsenges. Physik. Chemie* **77** 632 (1973).
4. A. Niehaus, in *Proc. VIIIth. ICPEAC* (Invited talks), p. 649. Belgrade, 1973.
5. Z. Herman and V. Čermak, *Coll. Czech. Chem. Comm.* **31**, 649 (1966).
6. H. Hotop and A. Niehaus, *Z. Phys.* **215**, 395 (1968).
7. E. E. Muschlitz Jr., *Science* **159**, 599 (1968).
8. H. Hotop and A. Niehaus, *Z. Phys.* **228**, 68 (1969).
9. H. Hotop and A. Niehaus, *Z. Phys.* **238**, 452 (1970).
10. W. H. Miller, *J. Chem. Phys.* **52**, 3563 (1970).
11. J. S. Cohen and N. F. Lane, *J. Phys. B* **6**, L 113 (1973).
12. W. H. Miller, C. H. Slocomb, and H. F. Schafer, III, *J. Chem. Phys.* **56**, 1347 (1972).
13. J. S. Howard, J. P. Riola, R. D. Rundel, and R. F. Stebbings, *J. Phys. B* **6**, L 109 (1973).
14. H. Hotop, E. Illenberger, H. Morgner, and A. Niehaus, *Chem. Phys. Lett.* **10**, 493 (1971).
15. S. Y. Tang, A. B. Marcus, and E. E. Muschlitz, Jr., *J. Chem. Phys.* **56**, 1347 (1972).
16. J. T. Moseley, J. R. Peterson, D. C. Lorents, and M. Hollstein, *Phys. Rev. A* **6**, 1025 (1972).
17. R. E. Olson, *Phys. Rev. A* **6**, 1031 (1972).
18. W. Lindinger, A. L. Schmeltekopf, and F. C. Fehsenfeld, *J. Chem. Phys.* (1974) to be publ.

19. A. Pesnel, H. Hourdin, G. Watel and C. Manus, *J. Phys. B* **6**, L 326 (1973).
20. J. P. Riola, J. S. Howard, R. D. Rundel, and R. F. Stebbings, *J. Phys. B* **7**, 376 (1974).
21. C. H. Chen, H. Haberland, and Y. T. Lee, (to be publ.).
22. G. Lantschner and A. Niehaus, *Chem. Phys. Lett.* **23**, 223 (1973).
23. V. Čermak, *Coll. Czech. Chem. Comm.* **33**, 2739 (1968).
24. W. W. Robertson, *J. Chem. Phys.* **44**, 2456 (1966).
25. A. L. Schmeltekopf, F. C. Fehsenfeld, and E. E. Ferguson, *J. Chem. Phys.* **48**, 2966 (1968).
26. W. C. Richardson, D. W. Setser, *J. Chem. Phys.* **58**, 1809 (1973).
27. H. Hotop, A. Niehaus, *Int. J. Mass Spectrom. Ion Phys.* **5**, 415 (1970).
28. V. Čermak, J. B. Ozenne, *Int. J. Mass Spectrom. Ion Phys.* **7**, 399 (1971).
29. H. Hotop, in *Advances in Mass Spectroscopy,* (A. Quayle, Ed.) Vol. **5**, 116, Pergamon Press, New York, 1971.
30. C. E. Brion, C. A. McDowell, and W. B. Stewart, *Chem. Phys. Lett.* **13**, 79 (1972).
31. H. Hotop, *Radiat. Res.* **59**, 379 (1974).
32. H. Hotop, F. W. Lampe, and A. Niehaus, *J. Chem. Phys.* **51**, 593 (1969).
33. L. G. Piper, J. E. Velazeo, and D. W. Setser, *J. Chem. Phys.* **59**, 3323 (1973).
34. K. Katsuura, *J. Chem. Phys.* **42**, 3771 (1965).
35. T. Watanabe and K. Katsuura, *J. Chem. Phys.* **47**, 800 (1967).
36. B. M. Smirnov and O. B. Firsov, *Soviet Phys. – JETP Lett.* **2**, 297 (1965).
37. C. E. Klots, *J. Chem. Phys.* **56**, 124 (1972).
38. H. Hotop and A. Niehaus, *J. Chem. Phys.* **47**, 2506 (1967).
39. M. Matsuzawa, *J. Phys. Soc. Japan* **32**, 1088 (1972).
40. M. Matsuzawa, *J. Phys. Soc. Japan* **33**, 1108 (1972).
41. M. Matsuzawa (to be published).
42. G. N. Fowler and T. W. Preist, *J. Chem. Phys.* **56**, 1601 (1972).
43. T. W. Preist, *J. Chem. Soc. Faraday Trans.* I **68**, 661 (1972).
44. L. M. Kishinewsky and E. S. Parilis, *Sov. Phys. JETP* **28**, 1020 (1969).
45. A. Niehaus and M. W. Ruf (paper to be presented at IVth. Intern. Conf. At. Phys. Heidelberg, W. Germany, 1974).

Ionization by Interpenetration of Electron Shells[1,2]

Felix T. Smith and A. Salop
Molecular Physics Center
Stanford Research Institute
Menlo Park, California 94025, USA

INTRODUCTION

At the symposium (*1*) that preceded this one, eight years ago, Professors Fano and Kistemaker both pointed out the potential importance of the inner-shell ionization and excitation processes arising from the interpenetration of more or less filled electron shells in two colliding atoms or ions. This subject was then a rather new one to which attention had been drawn by experimental measurements at Leningrad (*2*) and at the University of Connecticut (*3,4*), and the explanation for these processes in terms of the electron promotion mechanism of molecular orbital theory had just been given by Fano and Lichten (*5*). In the eight years that have intervened since that time, a great deal of work has been done that has shed light on the details of this process, but it is impressive to look back to the earlier symposium and see how sound were the statements on this topic and the predictions of its importance.

Actually, the interpenetration of electron shells as a source of excitation and ionization is a significant process even for the outer shells of atoms and molecules—as can be seen in collisions of closed-shell atoms or molecules at energies as low as a few hundred eV. The same process applies at somewhat higher energies (10 or 20 keV and above) for the ionization of inner shells—such as the K shells of O, N, and C in biological materials. As we shall see later, these processes may be quantitatively rather significant; perhaps 10 or 20% of the energy available may be deposited in inner-shell ionization events.

The phenomena under consideration here are confined to a situation where two particles collide, both of which are carrying one or more electrons forming the shell structure characteristic of stable atomic or ionic states. It is not surprising that such collisions were long neglected by research workers concerned with the effects of the radiation produced by charged particle interactions. Radioactive sources and, for a long time, accelerators produced only very simple particles, such as photons, electrons, protons, deuterons, and alpha particles. Of course, heavy ion recoil, whether from radioactive decay, neutron bombardment, or fission, has long been recognized. Nevertheless, until very recently, when heavy ion bombardment has become an available technique for research in radiobiology, the collisions of complicated species such as

[1]Congress presentation B–29–2.

[2]Work supported by the Atomic Energy Commission, Division of Biology and Medicine.

partially ionized heavy ions were generally left to be studied by atomic physicists and given comparatively little attention with regard to their radiation effects in biological systems.

For such collision systems, one can refer to three main classes of excitation and ionization phenomena: coulomb ionization, which is most important at relatively high incident velocities; charge transfer processes, which can, when exothermic, be quite important even at very low collision energies; and shell interpenetration effects, which are especially important at intermediate energies.

The principal purpose of this paper is to review the third of these mechanisms, the very important ionization and excitation processes associated with those collisions in which the partners get so close that various atomic shells interpenetrate. Fortunately, the brief discussion that can be given here can be supplemented by the two excellent review articles of Kessel and Fastrup (*6*) and Garcia, Fortner, and Kavanagh (*7*), and also by much detailed work reported in the 1972 *Proceedings of the International Conference on Inner Shell Ionization Phenomena* (*8*).

COULOMB IONIZATION

When a charged particle passes through matter, it creates a rapidly changing electromagnetic field. The electrons in the atoms of the medium respond to this time-dependent field and are thereby excited and frequently removed from their immediate environment in an ionization event. This type of ionization is often called coulomb ionization, because it is a simple response to the time-dependent coulomb field. Since a rapidly varying electromagnetic field is involved, coulomb ionization by fast particles bears a close relationship to photoionization, a relationship that has been well exploited in studies of excitation and ionization by fast heavy particles such as protons and alpha particles and even heavier nuclei, especially when they have been substantially stripped of their accompanying electrons. The theory of the coulomb excitation mechanism is very well developed (*7,9,10*) and will not be discussed in detail here, but it is an important point of departure for comparisons with other theories. According to this model, the ionization cross section peaks at a particle velocity that matches approximately the electron velocity in the atomic shell involved and decreases rapidly with decrease in projectile velocity.

Figure 1 is a plot of experimentally obtained K-shell ionization cross sections vs impact energy for proton and alpha particle impact on various heavy target atoms, taken from a recent paper by W. Brandt (*9*), which illustrates very clearly the general behavior of coulomb ionization.

The experimental results, when plotted in the appropriate reduced units, lie very close to a theoretical universal curve valid in the region where the projectile is slower than the K-shell electron. The universal curve is derived from the predictions of the plane wave Born approximation (PWBA), with corrections made for the effects of coulomb deflection of the projectile and the variation of K-shell binding with internuclear distance. The scaling reflects the velocity-matching requirements referred to previously and takes into consideration the dependences of the cross sections on projectile and target nuclear charges.

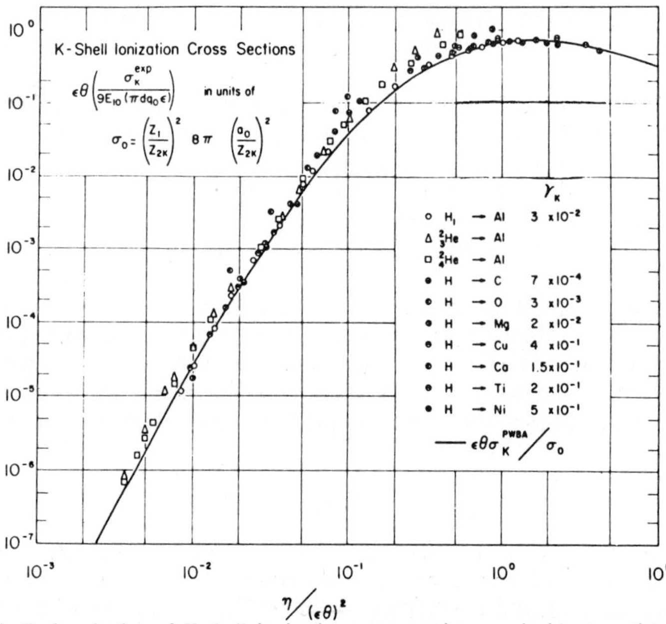

FIG. 1. Reduced plot of K-shell ionization cross sections vs. incident particle energy for structureless charged particles on various atoms. The theoretical curve (solid line) is derived from plane wave born approximation theory with corrections made for effects due to coulomb deflection of the projectile and variation of K-shell binding with internuclear distance. [From W. Brandt, Ref. *9*.]

CHARGE TRANSFER IONIZATION

Charge transfer processes resulting in the production of excited ions have already been discussed in this symposium by Professor Niehaus, and we need not go into details about them here. We need only remember that these processes, when they are sufficiently exothermic, may have large cross sections, even down to the lowest velocities of collision. Furthermore, a multiply-charged bombarding ion may rob an atom or molecule of an electron from an inner shell. This event can be followed by an Auger emission of another electron or by an X-ray event. The cross sections and energy dependence of these processes are, of course, totally different from the behavior of the coulomb ionization process.

SHELL INTERPENETRATION PHENOMENA

Figure 2 shows how interpenetrating electron shells can cause ionization with a cross section far above that for coulomb ionization (*9,11*). It shows the contrast between the bare charged particles (protons, alphas) and singly ionized heavy atoms (C,N,O,Ne). The results for the heavy ions exceed the universal coulomb curve by factors of 10^4 and greater. Clearly, these heavy particles are ionizing predominately by a mechanism entirely different from the coulomb process. The closed inner shells of the collision partners interpenetrate, and forces associated with the Pauli exclusion principle are set up during the collision. Some of the electrons are forced into transient, highly excited, quasimolecular configurations during the collision. When the atoms separate, these electrons have a high probability of being left in excited states, often with ample energy to cause ionization in subsequent Auger transitions. This is basically the mechanism for which Fano and Lichten (*5*) developed their molecular orbital model.

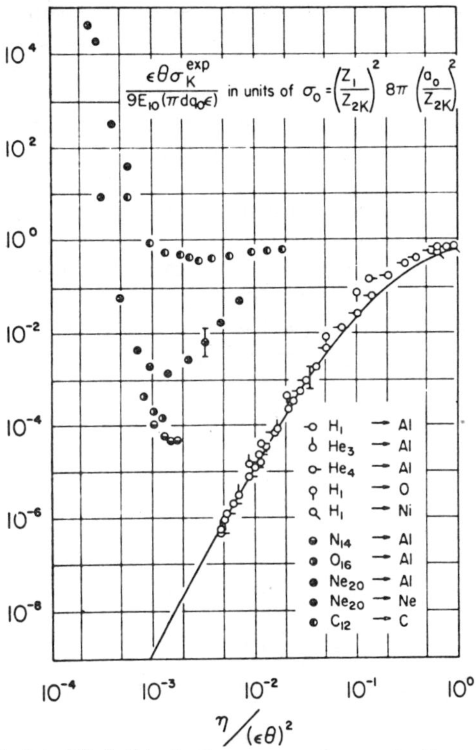

FIG. 2. Reduced plot of K-shell ionization cross sections vs. incident particle energy for both structureless charged particles and heavy ions (with associated electron shell structure) on various atoms. The theoretical curve is the same as in Fig. 1. [From W. Brandt, Ref. *9*.]

Before we discuss the interpenetration phenomena in more detail, we will consider some additional examples of the general disparity between heavy ion data and the predictions of theories applicable to incident structureless particles. Figures 3 and 4 show similar cross sections taken from Garcia et al. (*7*) for excitation of K-, L-, and M-shell electrons during collisions of several heavy ions with various targets. The data are compared with calculations of point charge coulomb ionization obtained by using binary encounter theory (BEA) (roughly comparable with PWBA calculations) and plotted with slightly different reduced units. Again, most of the experimental cross sections are greatly in excess of theoretical values for an incident point charge and reflect the dominance of shell interpenetration ionization.

Figure 5, taken from Ogurtsov's work (*12*), is a very elegant picture that demonstrates the fundamental difference between the K-shell excitation produced by a heavy ion collision and that due to photon, electron, and proton interactions. Auger electron spectra, corresponding to transitions involving an initial K-shell vacancy in neon, are shown for excitation produced by 1.5-keV photon, 3.2-keV electron, 300-keV proton, and 200-keV Ne^+ ion bombardment. The photon and electron spectra are identical, reflecting the fact that the electron excites the K shell via an electromagnetic pulse. In the case of proton excitation, in which the mechanism is similar, the same dominant transitions are noted, but a few other weak lines are observed, which can be attributed to additional excitation in the neon atom produced as a result of the proton interaction. The fourth spectrum for excitation by incident neon ions, however, is dramatically different from the other three. The Auger electron spectrum is considerably smeared out, with a very large increase in the numbers of lower energy electrons emitted, indicating the dominance of L-shell interpenetration excitation processes. These examples are certainly sufficient to indicate how important shell interpenetration effects are with regard to the amount and pattern of energy deposition. They produce large amounts of excitation involving considerable energy transfer which often leads to multiple ionization through Auger cascades, electron shake-off, and other reorganizational processes.

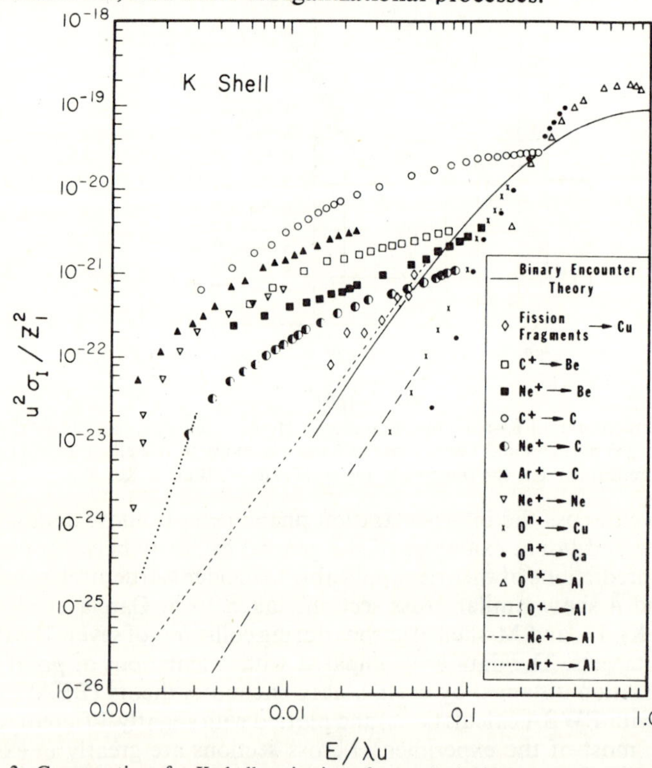

FIG. 3. Cross sections for K-shell excitation of atoms by incident heavy ions plotted in terms of the reduced parameters used in the binary encounter theory. E and u are respectively the ion energy and the K electron binding energy, λ is the ratio of the projectile to electron mass, and Z_I is the projectile atomic number. The solid curve is derived from binary encounter theory. [From J. D. Garcia, R. J. Fortner and T. M. Kavanagh, *Rev. Mod. Phys.* **45**, 111 (1974), with the permission of the American Institute of Physics.]

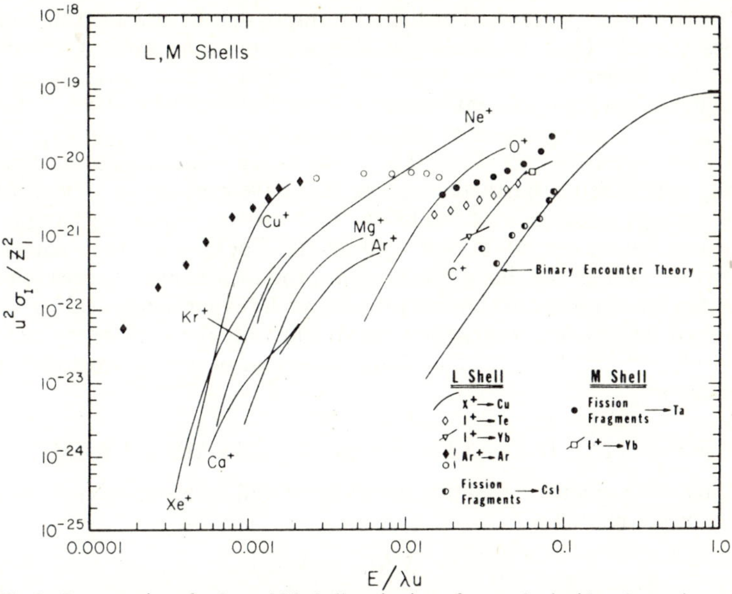

FIG. 4. Cross sections for L- and M-shell excitation of atoms by incident heavy ions plotted in terms of the reduced parameters used in the binary encounter theory. The description is the same as that for Fig. 3. [From J. D. Garcia, R. J. Fortner and T. M. Kavanagh, *Rev. Mod. Phys.* **45,** 111 (1974), with the permission of the American Institute of Physics.]

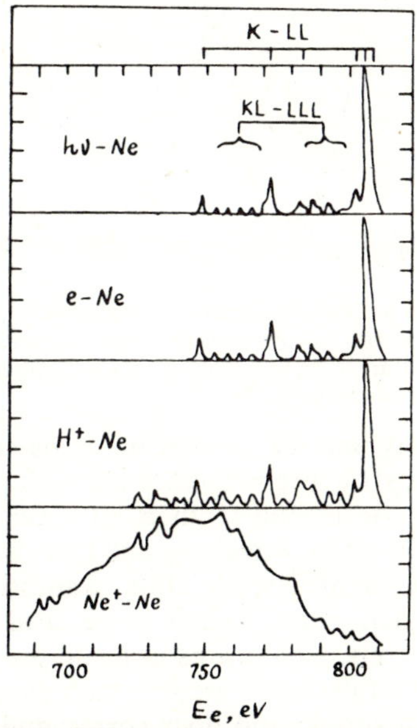

FIG. 5. Comparison of Ne K-shell Auger electron spectra produced by (a) 1.5 keV photons, (b) 3.2 keV electrons, (c) 300 keV protons, and (d) 200 keV Ne^+ ions. [From G. N. Ogurtsov, *Rev. Mod. Phys.* **44,** 1 (1972), with the permission of the American Institute of Physics.]

How can these processes be studied in more detail? We have seen several examples already. One can study the X-rays or photons emitted, or one can examine the Auger spectra in experiments that are in a sense "integrated" measurements, giving useful information about the excitation produced over a wide range of scattering angles or impact parameters. However, the experiments that have provided the most insight into inner-shell excitations are differential scattering experiments, particularly those that have been set up to measure the inelastic energy loss of the collision for various scattering angles (or, equivalently, distances of closest approach). These are often coincidence experiments in which the charge states of the product ions are also measured. We shall concentrate for the next few paragraphs on the experiments dealing with the collisions of Ar^+ on Ar, as these very nicely illustrate the shell interpenetration phenomena and are the sort of measurements that originally led to the Fano-Lichten theory.

DIFFERENTIAL INELASTIC ENERGY LOSS MEASUREMENTS

Figure 6 is a spectrum (*13*) that shows three well-resolved energy-loss peaks observed in the reaction

$$Ar^+ + Ar \rightarrow Ar^{2+} + Ar^{2+} + 3e$$

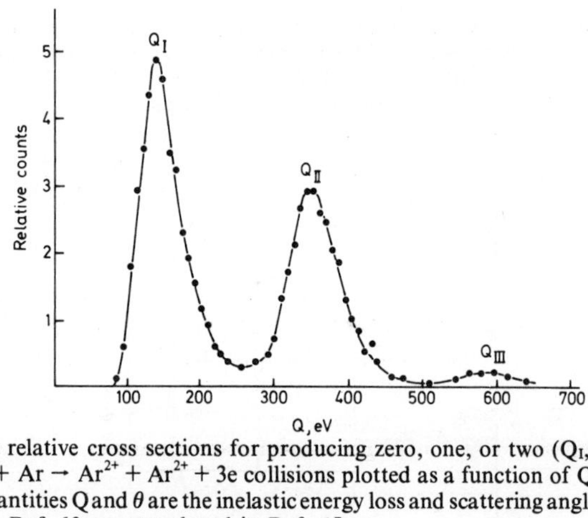

FIG. 6. The relative cross sections for producing zero, one, or two (Q_I, Q_{II}, or Q_{III}) L-shell vacancies in $Ar^+ + Ar \rightarrow Ar^{2+} + Ar^{2+} + 3e$ collisions plotted as a function of Q for 25 keV, $\theta = 16°$ collisions. The quantities Q and θ are the inelastic energy loss and scattering angle respectively. [From Afrosimov, et al., Ref. *13*, as reproduced in Ref. *6*.]

at 25-keV collision energy and 16° scattering angle. Fano and Lichten first explained this three-peak structure in connection with earlier measurements in their classic 1965 paper. According to their interpretation, Q_I, Q_{II}, and Q_{III} correspond to collisions in which zero, one, and two electrons, respectively, are raised out of the inner L shells of either Ar atom, in addition to a variable number of M-shell excitations. Each vacancy tends to be filled by an Auger transition involving two outer electrons. One moves in to fill the vacancy, and the other leaves the atom with an energy slightly smaller than the relevant L-shell binding energy (~200 eV), which should and does correspond to the energy interval between Q_I and Q_{II} and between Q_{II} and Q_{III}. Q_I peaks at about 180 eV, which is insufficient for L-shell ionization but which corresponds to appreciable M-shell excitation. The reason for the appearance of only three peaks will become clear when we discuss the Fano-Lichten model.

By examining these processes as a function of scattering angle at a given collision energy, one can determine the distances of closest approach at which these L-shell excitation processes occur. Figure 7 shows relative probabilities for the inelastic energy-loss events corresponding to Q_I, Q_{II}, and Q_{III} of Fig. 6 as a function of distance of closest approach for 50-keV Ar^+-Ar collisions (*14*). We see that below 0.27 Å essentially only outer-shell excitation processes are occurring. The onset for L-shell ionization at about 0.27 Å is quite sharp, and by 0.20 Å the relative probability of double L-shell excitation is almost 100%.

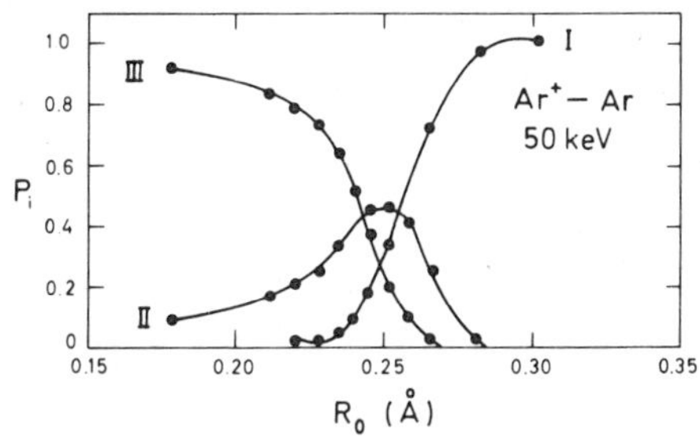

FIG. 7. The relative probabilities P_i for producing zero, one, or two (I, II, III) L-shell vacancies plotted as a function of distance of closest approach for 50 keV Ar^+ - Ar collisions. [From B. Fastrup, G. Hermann and K. J. Smith, *Phys. Rev. A.* **3**, 1591 (1971), with the permission of the American Physical Society.]

The average inelastic energy loss $\overline{Q}$ vs. the distance of closest approach for Ar^+-Ar collisions in the range of collision energies from 25 keV to 1.5 MeV is shown in a comprehensive plot in Fig. 8 (*15*). This figure illustrates clearly the striking increases in excitation that occur when various shells start to interpenetrate, as characterized by various critical values of R_o, the distance of closest approach. Thus, the step increases in $\overline{Q}$ observed for the 25-keV data at $R_o \approx 0.25$ Å occur when the L shells start to interpenetrate, as discussed previously. Another marked increase occurs at about 0.10 Å, presumably because of the production of additional L vacancies. Finally, an approximately 3-keV increase in $\overline{Q}$ is observed for R_o values between 0.02 and 0.01 Å. This can be attributed to the interpenetration of the argon K shells and the consequent production of a K-shell vacancy.

FANO-LICHTEN MODEL

Figure 9 is a reproduction of the diabatic molecular orbital diagram first presented by Fano and Lichten to illustrate the electron-promotion excitation process for the Ar^+-Ar system. We obviously cannot go into all the details here; a good deal of literature is now available on this subject, and the reader is referred, in particular, to Refs. *6* and *7* and to the original papers of Fano and Lichten (*5,16*). With reference to the experimental Ar^+-Ar data, however, consider collisions in which the energy is high enough so that L shells can interpenetrate, corresponding to internuclear distances somewhat smaller than 0.25 Å (or ~ 0.5 atomic unit on the Fano-Lichten diagram). It can be seen that the diabatic 4fσ molecular orbital, which rises sharply with decrease in internuclear separation in this region, carries two 2p electrons through a series of closely-spaced crossings. It is thus likely that transitions will occur and that after the collisions the electrons will find themselves left in highly excited orbitals with one or two L-shell vacancies appearing in the separated argon ions. Ionization through Auger transitions can then follow. Thus, the presence of three excitation peaks and the sharp onset of L-shell excitation at distances of closest approach of about 0.25 Å in the experimental data discussed previously are explained by this model. Similarly, the onset of K-shell excitation at about 0.02 Å is explained as due to a transition between the $2p\sigma$ and $2p\pi$ molecular orbitals, this time through the mechanism of rotational coupling (*6*), which will be described shortly.

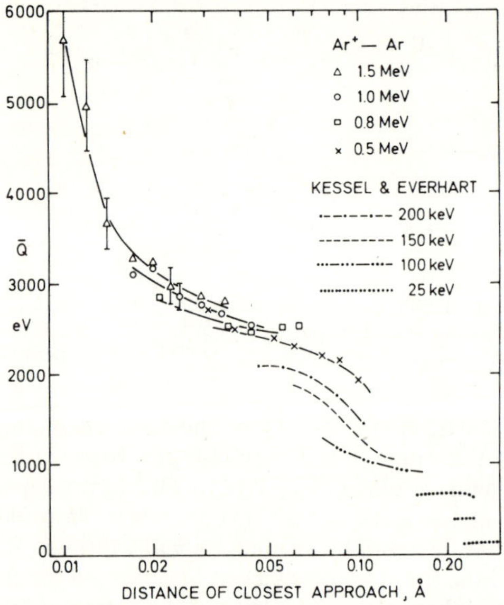

FIG. 8. The average inelastic energy loss $\overline{Q}$ for Ar^+ - Ar collisions plotted as a function of distance of closest approach with the incident energy as a parameter. From Q. C. Kessel, P. H. Rose and L. Grodzins, *Phys. Rev. Lett.* **22,** 1031 (1969), with the permission of the American Physical Society.]

In general, this electron-promotion molecular orbital description of the interaction should have its maximum validity when the orbital energies in the two atoms are not very different and when the projectile velocity is considerably smaller than the orbital electron velocity in question. Symmetric or almost symmetric collisions form one of the most important cases of the first condition, but it may apply also in very asymmetric cases, for instance, where the K-shell energies in one atom are close to the L-shell energies in another. The velocity criterion means also that the molecular orbital description may be valid for inner-shell interactions at velocities at which it may be totally inapplicable for the outer shells in the same collision. At high velocities in a condensed medium, the bombarding ion soon becomes very highly stripped, and coulomb excitation will probably become the dominant process, although if the ion is heavy, electron capture and loss may also be an important mechanism.

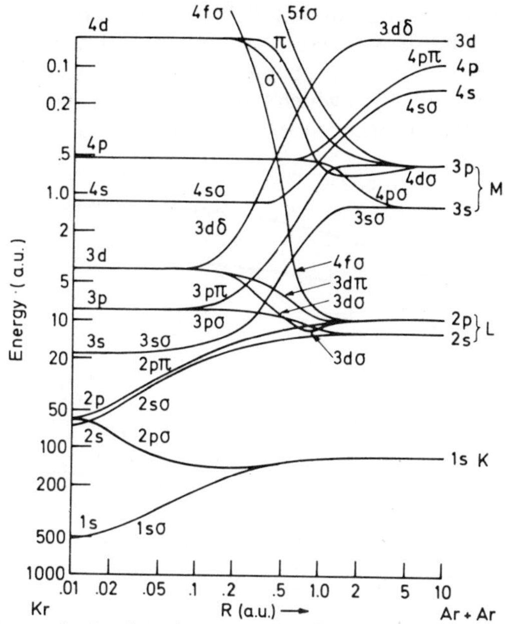

FIG. 9. Semi-quantitative diabatic molecular orbital energy level diagram for the Ar_2 molecule as proposed by Fano and Lichten. [From U. Fano and W. Lichten, *Phys. Rev. Lett.* **14**, 484 (1965), with the permission of the American Physical Society.]

COUPLING MECHANISMS

We will briefly discuss the various types of energy-level crossings and interactions occurring in these collisions by referring to Fig. 10, a schematic representation of the lowest one-electron energy levels of two atoms A and B differing slightly in nuclear charge $Z_A > Z_B$. (These curves are actually appropriate for analyzing K-shell excitations.) In region I, two σ states run approximately parallel for some distance. In such a region a transition can take place even though there is no curve crossing. The coupling matrix element can be assumed to have an exponential dependence on separation, and the coupling between the two levels is often referred to as Demkov (*17,18*) coupling. In region II, a crossing occurs between the $2p\sigma$ and $2p\pi$ levels. Similar crossings between levels of different angular momentum symmetry can occur at various internuclear separations, and the coupling between the relevant levels is collectively referred to as rotational coupling. (In region II, where the crossing takes place in the united atom limit, it is sometimes termed 'united atom coupling'.) For some time it was thought that because of the different angular momentum symmetries involved, coupling of the two states would be unimportant at these crossings; however, because the internuclear line is rotating during the collision process, there is actually an angular momentum coupling between the levels. In fact, the $2p\sigma$-$2p\pi$ transition represents an extremely important process for K-shell excitation in heavy ion collisions. Finally, in region III we have a case where two σ states cross. These are called radial couplings; the coupling is due to the radial motion, and the cross sections can be estimated by the well-known Landau-Zener formula (*19,20*). It should be noted that electron promotion to higher orbitals will take place only if an actual vacancy exists in these orbitals; otherwise, the Pauli principle excludes such transitions. Thus, for example, for atoms with $Z > 10$ in which the L shells are complete, the $2p\sigma$-$2p\pi$ rotational interaction cannot occur unless a previous collision has removed one or more L-shell electrons.

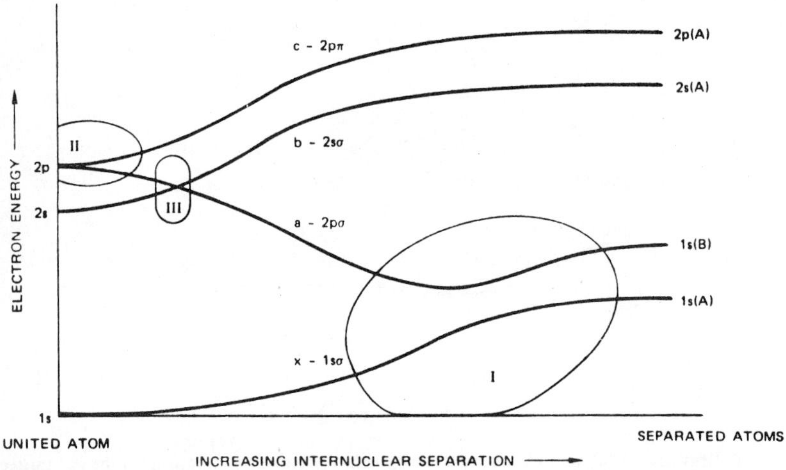

FIG. 10. Schematic representation of the lowest one-electron energy levels of two atoms A and B differing slightly in nuclear charge with $Z_A > Z_B$.

Consider once more Fig. 9, the Fano-Lichten diagram for Ar^+-Ar levels and, in particular, the $4f\sigma$ molecular level as the internuclear separation is decreased slightly from the value $R_s \approx 0.5$, where it begins its sharp rise. During a comparatively small change in internuclear separation, the level makes crossings with a number of states to which transitions can occur through either radial or rotational coupling. Such a situation exists for many collision systems and, in fact, a particular level may pass through a manifold of states within a small region of internuclear separations. Under these conditions the probability is close to 100% that some transition will occur and that at least one of the atoms will be in an excited or ionized state following the collision.

ENERGY THRESHOLDS FOR INNER-SHELL EXCITATION

We consider now, in a very approximate manner, the energy thresholds required for inner-shell excitations. In general, whether we think in terms of the situation involving a large number of crossings in a restricted range of internuclear separations characteristic of L-shell interpenetration effects or of the characteristic K-shell mechanism involving the $2p\sigma$-$2p\pi$ transition, one can say roughly that the cross section approaches a value that is on the order of πR_s^2, where R_s is a critical internuclear separation (basically the sum of the relevant interpenetrating shell radii). To reach this distance of approach, the collision energy must be high enough to overcome the repulsive force keeping the two nuclei apart and also to supply the energy I_s required for excitation (or ionization). One can roughly characterize the magnitude and energy dependence of the cross section by a crude formula:

$$\sigma \approx \pi R_s^2 [1-(I_s+Z_1Z_2e^2/R_s)/E] = \pi R_s^2 [1-(E_s/E)] \quad , \tag{1}$$

where E is the collision energy, and Z_1 and Z_2 are the effective nuclear charges of the target and projectile atoms, respectively (nuclear charge reduced by the effective screening of the electron shells inside R_s).

For inner-shell interactions, assuming $Z_1 \approx Z_2 = Z$, the repulsive potential $Z_1Z_2e^2/R_s$ (which is proportional to Z^3, since R_s is proportional to $1/Z$) dominates I_s (which is proportional to Z^2). The threshold for the inner-shell excitation process occurs, therefore, at a collision energy many times larger than the characteristic inner-shell energy I_s and the cross section gradually approaches πR_s^2 as the energy increases further.

For outer shell transitions, because of screening effects, the nuclear repulsion loses its importance, and the threshold is determined essentially by I_s, the outer shell excitation (or ionization) potential.

ENERGY DEPOSITION IN INNER-SHELL PROCESSES

We proceed now to obtain a rough estimate of the relative amount of energy expended in inner-shell excitation and ionization as compared to outer-shell processes. For a given shell at energies well above threshold, the cross section for excitation of a given electron approaches the quantity $\sigma_s = \pi R_s^2$. The amount of energy that can be deposited in a given shell labeled by the principal quantum number *s* is related to N_s, the number of electrons in the shell, and also to the relevant excitation (or ionization) potential. This we can approximate in the hydrogenic form as

$$I_s \cong (e^2Z_s^2/2a_0s^2) \quad . \tag{2}$$

The simplest approximation available for R_s is

$$R_s \cong 2a_0s/Z_s \quad . \tag{3}$$

The energy per atom deposited in a given shell by ionization will then be given roughly by

$$E_s = \sigma_sN_sI_s \cong \pi R_s^2I_sN_s \cong 2\pi a_0e^2N_s \, . \tag{4}$$

This expression should be used only for order-of-magnitude estimates and only at energies several times the effective threshold given by E_s in Eq. (1). It is useful, however, in providing a measure of the amount of energy deposited in a given shell by relatively fast bombarding particles.

Thus, when the collision energies get sufficiently large, the fraction of the energy absorbed in exciting the shell is determined by the number of inner-shell electrons compared to the number of outer-shell electrons. On the basis of this very rough model, one could say that, in a biological medium containing nitrogen, oxygen, or carbon, something on the order of 10 to 20% of the energy associated with primary ionization is likely to go into K-shell excitation processes in these atoms and the rest into outer-shell excitations. This is quite a significant energy deposition. The vacancies produced will result in Auger transitions, with consequent ejection of large numbers of energetic electrons with a high potential for producing localized biological damage.

CONCLUDING REMARKS

We have briefly outlined current understanding of shell interpenetration excitation and ionization phenomena and other excitation processes associated with heavy ion-atom collisions at relatively low energies. In a sense, one purpose of this paper has been to counter the common impression that, once the velocity of a charged particle falls below the velocity of atomic valence electrons in the medium, collisions are largely elastic and rarely lead to electronic excitation. We have examined experimental evidence to the effect that low energy excitation cross sections in such systems are many orders of magnitude larger than the cross sections predicted on the basis of the conventional coulomb-ionization model. The deeper understanding of inner-shell excitation processes and of electron capture reactions that is being achieved will certainly find application in several areas of radiation research. For example, on the basis of the results discussed here, it seems clear that the energy loss of a charged particle in matter should have much greater statistical fluctuations (straggling) than would be predicted from earlier conceptions. More generally, findings such as these will aid immeasurably in elucidating the energy-loss mechanisms associated with the Bragg peak. Finally, the demonstrated importance of inner-shell excitation processes in these collisions focuses the spotlight on the attendant Auger processes with their high potential for localized radiation damage via coulombic explosions and the production of high-LET electrons.

REFERENCES

1. U. Fano, in *Radiation Research,* Proc. 3rd International Congress Radiation Res., Cortina d'Ampezzo, 1966; (G. Silini, Ed.) pp. 20–42, 68–95. North-Holland, Amsterdam, 1967.
2. V. V. Afrosimov, Yu. S. Gordeev, M. N. Panov, and N. V. Fedorenko, *Zh. Tekh. Fiz.* **34,** 1613, 1624, 1637 (1964) [*Sov. Phys.-Tech. Phys.* **9,** 1248, 1256, 1265 (1965)].
3. E. Everhart and Q. C. Kessel, *Phys. Rev. Lett.* **14,** 247 (1965).
4. Q. C. Kessel, A. Russek, and E. Everhart, *Phys. Rev. Lett.* **14,** 484 (1965).
5. U. Fano and W. Lichten, *Phys. Rev. Lett.* **14,** 627 (1965).
6. Q. C. Kessel and B. Fastrup, *Case Stud. At. Phys.* **3,** 139 (1973).
7. J. D. Garcia, R. J. Fortner, and T. M. Kavanagh, *Rev. Mod. Phys.* **45,** 111 (1974).
8. *Proceedings of the International Conference on Inner Shell Ionization Phenomena and Future Applications,* U.S. Atomic Eenergy Commission, Technical Information Center, Oak Ridge, Tenn., 1972 (CONF-720404. Available from National Technical Information Service, Springfield, Va. 22151.)
9. W. Brandt, see Ref. *8,* p. 948.
10. E. Merzbacher, see Ref. *8,* p. 915.
11. W. Brandt and R. Laubert, *Phys. Rev. Lett.* **24,** 1037 (1970).
12. G. N. Ogurtsov, *Rev. Mod. Phys.* **44,** 1 (1972).
13. V. V. Afrosimov, Yu. S. Gordeev, A. M. Polyanskii, and A. P. Shergin, *Zh. Eksp. Teor. Fiz.* **57,** 806 (1969) [*Sov. Phys. JETP* **30,** 441 (1970)].
14. B. Fastrup, G. Hermann, and K. J. Smith, *Phys. Rev. A* **3,** 1591 (1971).
15. Q. C. Kessel, P. H. Rose, and L. Grodzins, *Phys. Rev. Lett.* **22,** 1031 (1969).
16. W. Lichten, *Phys. Rev.* **164,** 131 (1967).
17. Yu. N. Demkov, *Zh. Eksp. Teor. Fiz.* **45,** 195 (1963) [*Sov. Phys. JETP* **18,** 138 (1964)].
18. R. E. Olson, *Phys. Rev. A* **6,** 1822 (1972).
19. L. D. Landau, *Phys. Z. Sowjetunion* **2,** 46 (1932).
20. C. Zener, *Proc. R. Soc. London Ser. A* **137,** 696 (1932).

Spontaneous Break-up of Gaseous Ions[1]

Ch. Ottinger[2]
Max-Planck-Institut für Strömungsforschung
Göttingen, Germany

INTRODUCTION

Ionization of a molecule normally leaves the product ion in an excited state. Often the excess energy is used to subsequently dissociate the ion. The resulting characteristic pattern of fragment masses is the basis of mass spectrometric analysis. Usually the fragmentation follows the ionization almost instantly, too fast to be observable as a separate process. Occasionally, however, and in particular with organic molecules, the decomposition can be delayed by as much as several microseconds. This is long enough to study these processes in a mass spectrometer. Ions with observably long decomposition lifetimes are called metastable ions. The majority of modern metastable ion studies continue to operate within the classical lifetime range around a few microseconds, although the last 8 to 10 years have seen an extension down to 10^{-12} and up to 10^{-3} sec. The study of delayed spontaneous dissociations has lately experienced a real boom and has developed from an esoteric branch of mass spectrometry into a subject in its own right. Its bearing ranges from quantum mechanical and statistical problems in theoretical unimolecular kinetics to analytical organic mass spectrometry and computerized apparatus.

In the following, let us first see how the "ordinary" metastables, with lifetimes in the microsecond region, are observed experimentally. In the next three sections we will discuss what metastables can teach us about the chemistry, the energetics, and the kinetics of ionic decompositions.

INSTRUMENTAL

Consider a metastable ion of mass m_1 which is accelerated out of a mass spectrometer ion source and holds together until it has reached the field-free region (see Fig. 1a), but then decomposes in flight:

$$m_1^+ \rightarrow m_2^+ + \Delta m \quad .$$

Then the fragment m_2^+ will continue to travel with the velocity of m_1^+. Its kinetic energy is therefore now $m_2/m_1 \times E_{kin}(m_1^+)$, reduced by the mass ratio m_2/m_1 from the normal ion kinetic energy for which the instrument is designed.

a. The classical detection mode is based on the fact that the mass scale

[1]Congress presentation B–29–3.

[2]Visiting Fellow, 1973–74, Joint Institute for Laboratory Astrophysics, University of Colorado and National Bureau of Standards, Boulder, Colorado 80302.

calibration is proportional to the ion energy. The m_2^+ fragments considered are therefore recorded at a mass

$$m_2 \cdot m_2/m_1 = m_2^2/m_1 = m^* \quad .$$

m* is in general a non-integral apparent mass value. This explanation of the unorthodox peaks at m* is due to Hipple and Condon's work done 30 years ago (*1*). In the jargon of the trade, a peak at m*, rather than the ion m_1^+, is also often referred to as the "metastable."

From a measurement of m*, together with the condition that m_1 and m_2 have integral values, it is easy to find m_1, m_2 individually. Thus, metastables afford a unique identification of the precursor ion m_1^+, in cases where m_2^+ might be formed in different ways. This is an important aid in the elucidation of ionic breakdown pathways and is one of the major motivations of most metastable studies.

b. Instead of measuring the energy reduction factor m_2/m_1 on the mass scale through m*, a direct energy measurement by means of an electric sector field has often been used in recent years. Löhle and Ottinger (*2,3*) installed such a sector between the exit slit of a conventional magnetic spectrometer and the detector (see Fig. 1b). A similar instrument is now available commercially. Wacks, Bente, and McLafferty arrived at an equivalent geometry by simply reversing a commercial double-focusing spectrometer (*4*). In these arrangements the magnetic field determines $m^* = m_2^2/m_1$ and the electric field determines m_2/m_1. From these two variables m_1 and m_2 are then unambiguously known. Particularly elegant is another mode of operation, where one observes spontaneous decomposition in the *second* field-free region. The magnet singles out a certain mass m_1, and a scan of the electric sector voltage then gives all the fragments m_2, into which m_1^+ decomposes ["DADI" = Direct Analysis of Daughter Ions, (*5*)].

c. Very popular is also the use of an unmodified double-focusing spectrometer for the detection of metastables (see Fig. 1c). In order to observe the reaction $m_1^+ \rightarrow m_2^+ + \Delta m$ in the first field-free region, either of two methods can be used. In the Major technique (*6*) the fields are set as above: the electric field to transmit ions of energy in m_2/m_1 eU, and the magnetic field to mass m_2^2/m_1. Alternatively, in the Barber-Elliott technique (*7*), the acceleration voltage U is raised to m_1/m_2 times its normal value, with the electric and magnetic sectors set to transmit normal ions m_2^+. The latter technique has the advantage of operating always at the same ion energy throughout the spectrometer, with the exception of the acceleration section. The former technique, conversely, maintains fixed focusing conditions at the ion source, but is more susceptible to energy discrimination effects along the ion path.

In all instruments with an electric sector a two-dimensional array of operating parameters has to be covered if one wishes to obtain a complete metastable decomposition pattern of a given substance. Coutant and McLafferty (*8*) have described a fully-automated metastable data acquisition system which displays the metastable pattern graphically as a three-dimensional plot (Fig. 2).

Metastable ions are usually of low abundance compared to ordinary fragment intensities in a mass spectrum (typically 10^{-3} to 10^{-4}). It is an especially advantageous feature of instruments employing an electric sector that they automatically discriminate against all ordinary fragment ions (with full kinetic energy). A metastable detection sensitivity as low as 2×10^{-8} of the total ionization (i.e. the summed intensity of all peaks of the ordinary mass spectrum) has been achieved in this way (*3*). As a result, many previously unknown spontaneous dissociations were found (*3,4*).

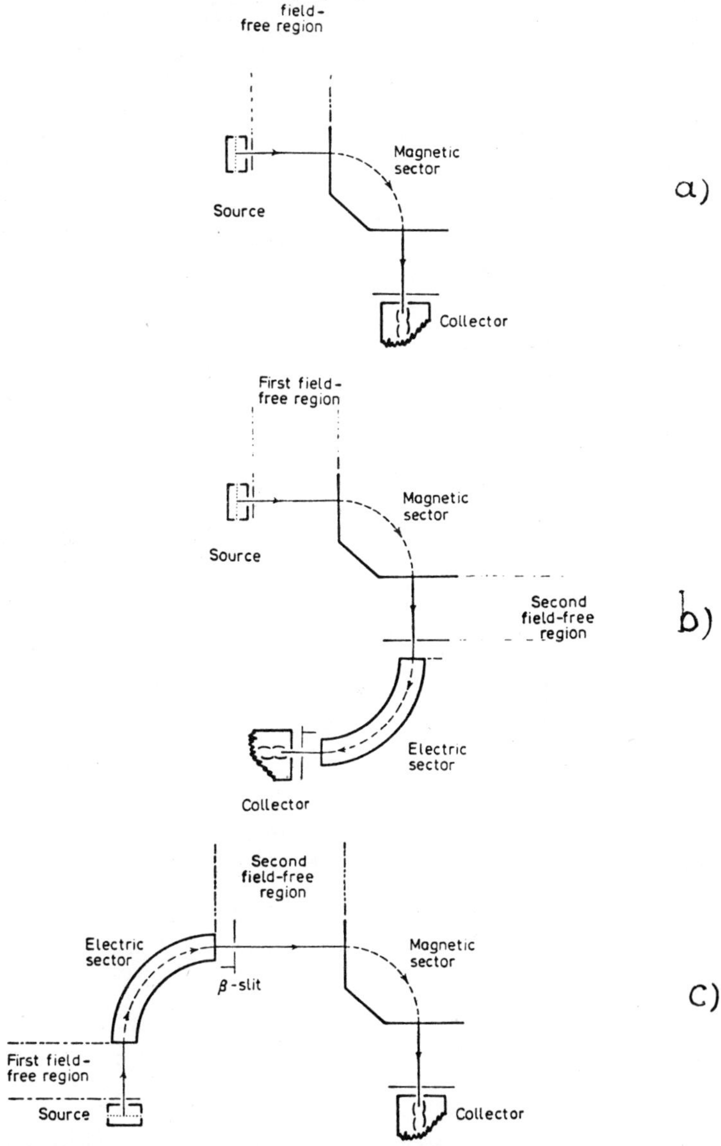

FIG. 1. Techniques for detecting spontaneous ion dissociations [adapted from J. L. Holmes and F. M. Benoit (*13*)] a) Single focusing sector field mass spectrometer. b) Same as a), but with electric sector added. c) Double focusing spectrometer.

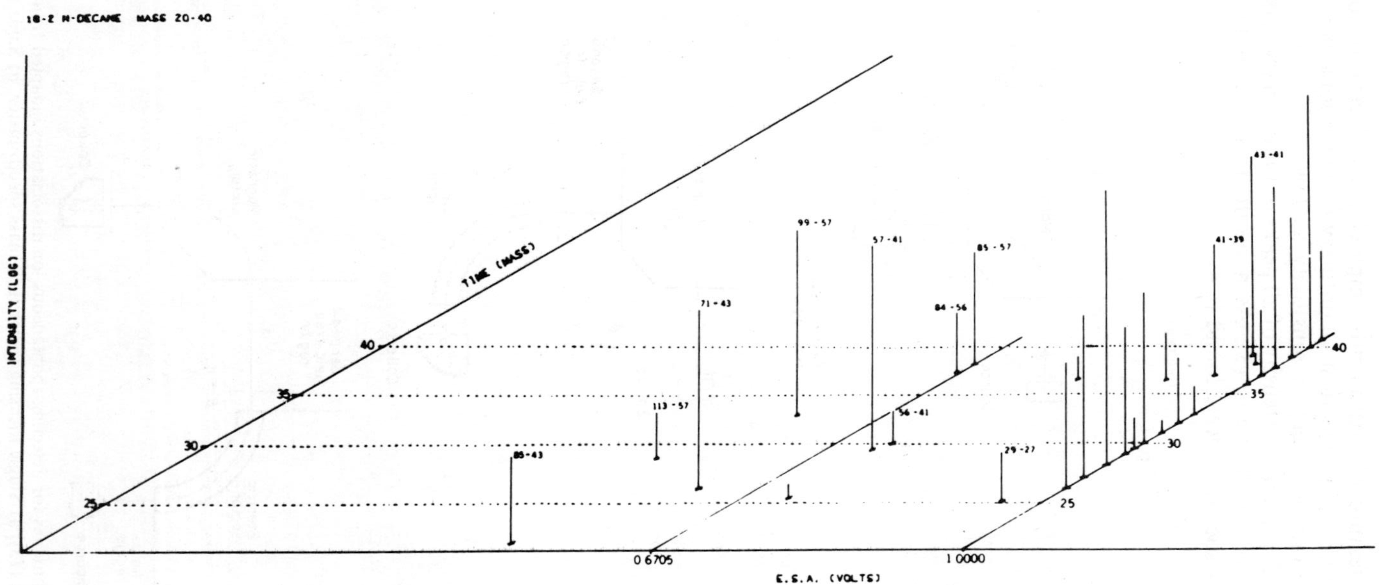

FIG. 2. Three-dimensional plotter output of an automatic data acquisition system for metastable ions [from J. E. Coutant and F. W. McLafferty, *Int. J. Mass Spectrom. Ion Phys.* **8**, 323 (1972)]. A double focusing spectrometer was operated in the Major mode of simultaneous scanning of electrostatic analyzer voltage (X axis) and magnet current (Y axis). Ion currents are displayed logarithmically on the Z axis. X = 1.000 gives the normal mass spectrum (in this case, of *n*-decane). All other peaks are metastables, labeled by the mass of precursor and product ion.

For use with simple instruments of type a), special devices have recently been invented which enhance the metastable peaks relative to the ordinary mass spectrum: In Daly's metastable detector (*9*) special ion optics allow only ions with less than the full energy eU, i.e. the metastables, to reach a secondary electron emitter. The released secondary electrons are detected in a scintillator-photomultiplier combination (Fig. 3).

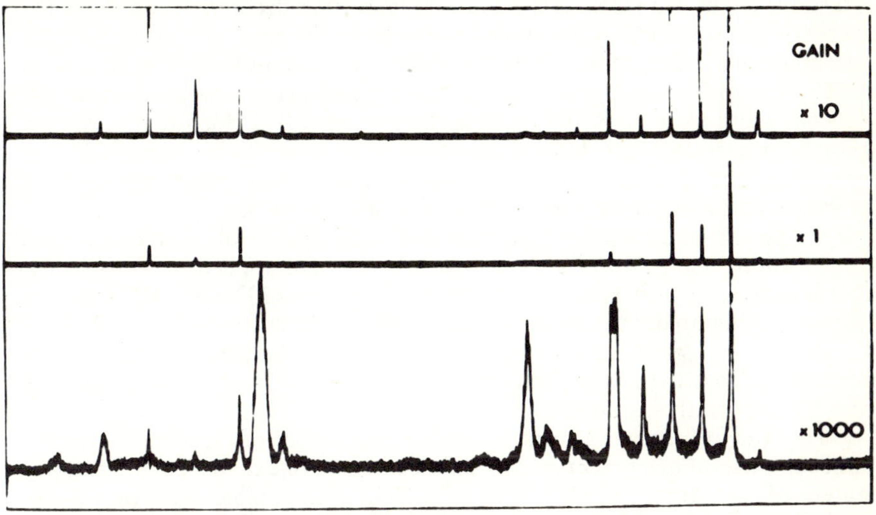

FIG. 3. Enhancement of metastable peaks by Daly's detector [from N. R. Daly, A. McCormick, R. E. Powell and R. Hayes, *Int. J. Mass Spectrom. Ion Phys.* 11, 255 (1973)]. Bottom trace: Predominantly metastable peaks, with the ordinary mass spectrum largely suppressed; middle and top trace: ordinary mass spectrum, with suppressor turned off.

In instruments with an electric sector field, either preceding or following the magnetic sector, an especially interesting type of study is possible: one can select an ordinary metastable decomposition in the first field-free region, follow the fragment ion formed into the second field-free region and see if it, too, is metastable and undergoes a second delayed decomposition. Such *consecutive* metastables are all of very low intensities. They were first discovered by Jennings (*10*). A high sensitivity search by Löhle and Ottinger (*3*) revealed as many as 48 such two-step metastables in simple hydrocarbons, 17 of them in heptane alone. Consecutive metastables are now an established fact, although from a theoretical standpoint their occurrence is quite surprising (*11*).

A possible source of error in all metastable studies is dissociation caused by collisions of the m_1^+ ions with the residual gas. In order to evaluate the contribution to a metastable peak from truly spontaneous processes, it is therefore imperative to measure the peak intensity as a function of pressure in the mass spectrometer tube (*3,12*).

CHEMICAL APPLICATIONS

Metastables, although only a minor constituent in the mass spectrum, are indicators of how the bulk of the fragment ions in the spectrum are formed.

Without metastables our knowledge of decomposition paths would be quite uncertain. For example, it would not be clear if in the breakdown of propane the propylene ions are formed from the molecular ions by loss of H_2 or only by consecutive losses of two H atoms. The metastable transitions tell us that both modes occur. Caution has to be exercised, though, in relating the metastable peak intensity to the importance of a particular fragmentation. Although intense metastable peaks probably always indicate important decomposition steps, the inverse of this statement is not true. For example, the ethyl ion, mass 29, is the most abundant in the propane mass spectrum. It is certainly formed by the loss of a methyl group from the parent ion. The corresponding metastable, however, $44^+ \rightarrow 29^+$, is quite weak (2×10^{-4} of total ionization) (*3*). Evidently this process generally occurs very fast. Even with extremely high sensitivity, loss of CH_2 has never been observed in hydrocarbons larger than ethane (*13,2*), so that this, somewhat surprisingly, appears to be an unlikely process.

Studies of metastables become an even more powerful technique for the identification of break-down mechanisms if they are applied to isotopically-labeled molecules. For example, in $CD_3CH_3^+$ the process $33^+ \rightarrow 30^+$ (HD loss) is at least several hundred times more intense than the alternative processes $33^+ \rightarrow 31^+$ (H_2 loss) or $33^+ \rightarrow 29^+$ (D_2 loss) (*14,2*). This shows that molecular hydrogen elimination from the ethane ion proceeds by abstraction of one H atom each from the two methyl groups.

The study of isotope effects is another area where metastables are useful, thanks to their unique precursor ion identification. For example, in the partially deuterated methanes, CH_3D, CH_2D_2, and CHD_3, metastables were only found for the loss of H from the parent ion, never of D (*15,16*). The small difference of about 50 meV between the C-H and the C-D zero point energies plays the decisive role here (*17*). Metastable ions, possessing specially low excitation energy, respond particularly sensitively to a small energy bias.

Metastables can also help to solve the problem of intramolecular atom migration, the so-called scrambling. For example, it was found that $CD_3CH_3^+$ ions decompose to give CD_3^+ (+CH_3) and CH_3^+ (+CD_3) in about equal amounts, but about 10 times less CDH_2^+ or CDH_2^+ (*2*). This indicates that some, but not much, scrambling of D and H occurs prior to the C-C rupture. From the known time scale of this rupture, given by the ion flight time in the apparatus, one finds a rate constant of about $5 \cdot 10^4$ atom interchanges per second. On the other hand, in the HD/H_2/D_2 losses compared above, the overwhelming intensity of HD loss indicates that there is no scrambling in that case. This is not a contradiction, since HD loss occurs from weakly excited CD_3CH_3 ions, C-C rupture requires more excitation energy. The more highly excited parent ions thus attain a greater degree of internal scrambling in the given reaction time, as is to be expected.

It has even been possible to observe directly how scrambling in an ion progresses as time goes on. Yeo et al. (*18*) compared one and the same reaction, a particular McLafferty rearrangement, occurring as an ordinary fast decomposition within the ion source, as a delayed process occurring in the first field-free region, and finally in the second field-free region. The degree of scrambling was found to increase in this order, i.e. with the available time (Fig. 4).[3]

In fragment ions scrambling can be much faster than in parent ions. For example, the losses of CH_2, CDH, and CD_2 (in metastable reactions) from

[3]See the next page for footnote 3.

$C_2H_2D_3^+$ occur in about the statistical ratio 1 to 6 to 3, indicating that the H-D scrambling in this fragment ion is complete on the experimental time scale (*2*). For other observations on scrambling, including C atom scrambling, see (*13*).

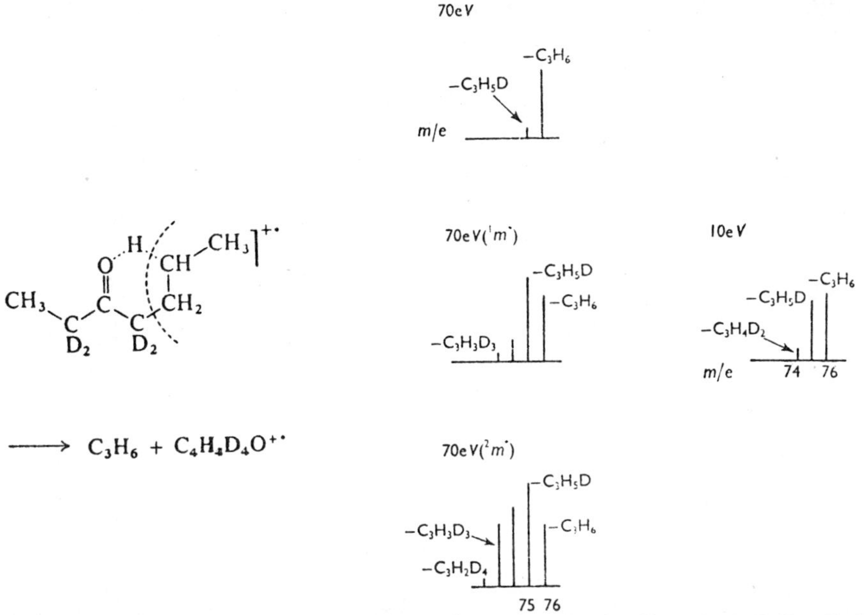

FIG. 4. Demonstration of H/D scrambling progressing with time [from A. N. H. Yeo, R. G. Cooks and D. H. Williams, *Chem. Commun.* 1269, (1968)]. Mass spectra for the process shown at left: 70 eV: ordinary mass spectrum; 70 eV (^{1}m*): metastable peaks from the first field free region; 70 eV (^{2}m*): metastable peaks from the second field free region. Scrambling increases in this order. Lower electron energy (10 eV, see the *ordinary* spectrum at right) slows down break-up and results in increased scrambling, resembling 70 eV(^{1}m*).

Metastable peaks can in certain cases give interesting information on the *structure* of the decomposing ion. In many cases a fragment ion with the same net formula can be prepared from a variety of substances. If such an ion undergoes two (or more) metastable decompositions, the branching ratios into these various channels can indicate whether the decomposing ion remembered its origin structurally. A classic example is a study by Rosenstock et al. (*19*) who measured the branching ratios of the two processes $C_6H_{13}^+ \rightarrow C_4H_9^+$ and $C_6H_{13}^+ \rightarrow C_3H_7^+$ where the $C_6H_{13}^+$ ions were prepared from ten different substances. Since the energy content as well as the structure of the $C_6H_{13}^+$ ions affects the branching ratio, one cannot expect perfectly constant yield ratios even for identical structures. Nevertheless, in this study near constancy was found, which proves that the $C_6H_{13}^+$ ions had identical structures in the ten cases. Many similar examples are given in the review by Holmes and Benoit (*13*). They include cases

[3]The last paragraph seemingly contradicts the preceding one: Since higher internal energy E accelerates decomposition, the short-lived ions might be expected to be *more* scrambled. The problem is resolved by considering that the rate constants $k_s(E)$ for scrambling and $k_d(E)$ for decomposition may increase with E quite differently. An energy increment ΔE causes a much larger increase in k_d than in k_s. Decomposition terminates the scrambling, so that short-lived ions are less scrambled.

in which the branching ratios did divide into several categories, indicating different structures of the precursor ion. In particular this is true for the branching ratios of molecular ion (rather than fragment ion) dissociations from parent molecules possessing different (isomeric) structures (*20*). An especially striking example of the effect of initial preparation on the metastable decomposition branching ratio was reported by Löhle and Ottinger (*3,21*). The branching studied was between the metastable processes $C_3H_5^+ \rightarrow C_3H_3^+$ and $C_3H_5^+ \rightarrow CH_3^+$ in cycloheptane. Normally, i.e. for ordinary fragment ions $C_3H_5^+$ as they leave the ion source, the second process is 2500 times weaker than the first. If, however, the fragment $C_3H_5^+$ is itself formed by a metastable decomposition, (either from $C_5H_9^+$ or from $C_4H_8^+$), then the *consecutive* metastable decay to CH_3^+ becomes half as likely as the one to $C_3H_3^+$. Apparently some multistep decomposition chains have to go along certain lanes, prescribed by something like "selection rules for decomposition." In the above case, it has not yet been analyzed just which structural differences in the $C_3H_5^+$ ion are responsible for the change of 2500:1 to a 2:1 branching ratio.

ENERGETICS

Where does the initial excitation energy of a molecular ion go? Some of it is used up in breaking bonds. Other parts are retained as vibrational energy in the fragments, possibly enabling them to undergo spontaneous dissociation at a later time. But some energy also appears as translational energy of the fragment ions. This has been known from measurements employing the so-called deflection method, as developed in particular by Taubert (*22*). The total energy released into translation by such an "explosion" of an ion can amount to several electron volts, but is commonly only a few tenths of a volt. In the case of a metastable decomposition, a particularly elegant and precise method can be used to measure these small amounts of energy.

Let T be the energy released in the decomposition. Then the fragments m_2^+ and Δm fly apart with a relative velocity

$$u = \sqrt{2T/\mu}$$

in the reference frame of the ion m_1^+ decomposing in flight (μ is the reduced mass of m_2^+ and Δm). The fragment m_2^+ receives the velocity increment $\Delta u = \Delta m/m_1 \cdot u$.

The velocity of m_2^+ in the laboratory system is obtained by adding vectorially Δu to the velocity v_0 of the c.m. itself, which is the velocity of m_1^+:

$$v = v_0 + \Delta u \text{ , with } v_0 = \sqrt{2eU/m_1} \text{ .}$$

The mass spectrometer measures not the velocity, but the energy of m_2^+, which is

$$(m_2/2)v^2 = (m_2/2)v_0^2 + m_2 v_0 \Delta u + (m_2/2)\Delta u^2 \text{ .}$$

Neglecting the small last term, we obtain an energy increment, $m_2 v_0 \Delta u$. Although Δ is small, it is multiplied with the large velocity, v_0, so that the energy increment can be a sizeable fraction of the normal fragment energy $(m_2/2)v_0^2$. The energy increment turns out to be

$$\Delta E = 2 \cdot \sqrt{(\mu/m_1)eU \cdot T} = (m_2/m_2)eU \cdot \sqrt{(\Delta m/m_2)(T/eU)} = (m_2/m_1)eU \cdot \beta \text{ .}$$

In the latter expression β is the fractional change $2 \cdot \sqrt{(\Delta m/m_2) \cdot T/eU}$ of the energy $m_2/m_1 \cdot eU$ of m_2^+. For example, with a total energy release of T=0.3 eV, an acceleration voltage U = 3kV and a typical mass ratio of $\Delta m/m_2 = 1/2$, the increment of translational energy of m_2^+ as measured in the lab system is 28.3 eV. The total energy spread between ions m_2^+ split off in the forward ($\Delta u > 0$) or backward ($\Delta^u < 0$) direction with respect to the direction of flight of m_1^+ would then be 56.6 eV. So the small energy of 0.3 eV appears greatly amplified.

The energy measurement can again be done, as described above for the mass identification, either on the mass scale or by using a special energy selector. Both these methods were proposed and applied in 1965, the former by Beynon et al. (*23*), the latter by Ottinger (*24*). A comparison of the two techniques is given in Fig. 5. It shows energy and mass spectra of the CO^+ ion from the metastable decomposition $CO_2^{++} \rightarrow CO^+ + 0^+$. The peaks are split into two components, corresponding to forward and backward velocity components Δu. Fragment ions with sideways oriented Δu are discriminated against by slits in the ion optics. In addition, with increasing pressure a collision-induced peak appears in the center. The pressure-dependent process evidently releases no measurable kinetic energy. From the separation of the outer two peaks, fragmentation energies of T = 5.1 eV (from the mass spectrum) and 5.3 eV (from the energy spectrum) were calculated. This good agreement demonstrates the equivalence of the two methods. Incidentally, Fuchs and Taubert (*25*) also measured T for ordinary CO^+ fragments directly from the ion source, using the deflection technique, and found 5.5 eV. This points to the same mechanism being responsible for the energy release both in the fast and in the slow decomposition.

The peak shapes, of which Fig. 5 gives an example with extreme discrimination against sideways velocity components, were studied analytically in great detail by Beynon et al. in 1967 on the mass scale (*26*) and in 1972 on the energy scale (*27*). In particular, the latter study investigates carefully the peak shape, as measured with a double focusing spectrometer in the Major scanning mode, with different degrees of discrimination. In the calculation the release of only one sharply defined amount of energy T was assumed. Strong discrimination by a short, narrow slit then produces a peak doublet; longer slits give a dish-shaped peak, and an infinitely long slit finally gives a broad, flat-topped peak. The peak shapes observed in practice range from strongly dish-shaped over flat-topped to round-topped. Evidently, the actual energy releases are distributed over a range. The distribution of T can then be deduced from the measurements, using the theoretical results for any given T. The accuracy of this method is potentially very high; energy releases as low as 0.2 meV have already been detected.

Useful chemical information can be gained from the measurement of energy release. For example, metastable H_2 elimination from the heavier heptane and butane fragment ions (i.e. $C_3H_7^+ \rightarrow C_3H_5^+ + H_2$) usually gives a broad metastable peak, corresponding to an energy release of T = 0.3 to 0.4 eV (*28*). By contrast, H_2 elimination from the light fragments $C_2H_5^+$ and $C_2H_4^+$ gives a narrow peak with T = 0.03 eV (*28,2*). This indicates differences in the detailed reaction mechanism. H elimination always gives a narrow peak. In isotopic studies, this fact helps to distinguish an H_2 elimination (usually broad) from D elimination (narrow) (*29*).

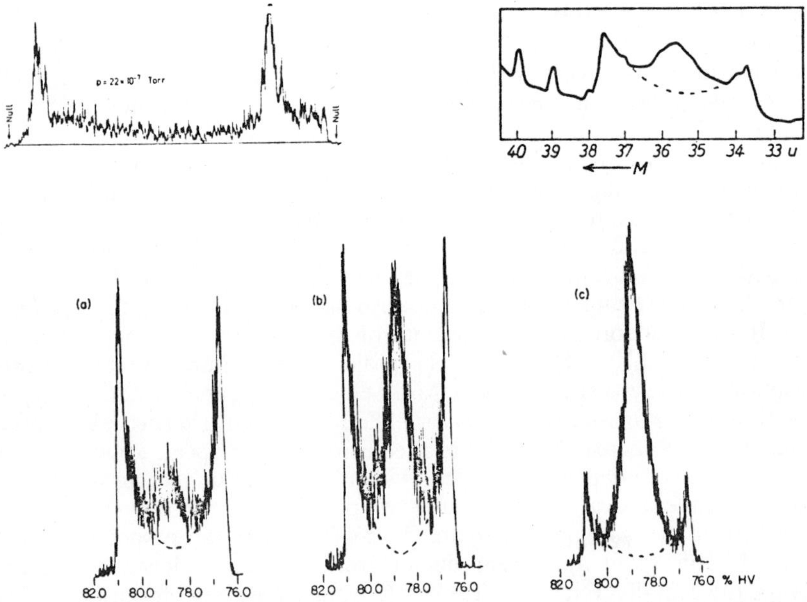

FIG. 5. Release of translational energy in the process $CO_2^{++} \rightarrow CO^+ + O^+$ occurring in flight. Top left: kinetic energy spectrum of CO^+ [from Ch. Ottinger, *Z. Naturforsch.* **22a,** 20 (1967)]. Top right: mass spectrum of CO^+ [from R. Fuchs and R. Taubert, *Z. Naturforsch.* **20a,** 823 (1965)]. In addition to the two components at 33.8 and 37.6 m.u., a central peak appears due to the collision-induced process $CO_2^{++} \rightarrow CO_2^+ \rightarrow CO^+ + O$. Bottom: energy spectra of CO^+ (from R. G. Cooks, J. H. Beynon, R. M. Caprioli and G. R. Lester, *Metastable Ions*, Elsevier, Amsterdam, 1973) demonstrating the appearance of the central peak with increasing pressure (from left to right).

Different isomers can give quite dissimilar energy releases for analogous reactions. For example, NO loss from the parent ion of nitrophenol gives broad peaks (T = 0.75 eV) with the ortho- and para-, but not with the meta-isomer (*23*). Very broad peaks are usually associated with charge-separating decompositions of doubly charged ions. The case of CO_2^{++} has been discussed above, another well-known example is $C_6H_6^{++} \rightarrow CH_3^+ + C_2H_5^+$. Here T = 2.67 eV (*30*). If this release is entirely due to electrostatic repulsion of the two charges starting off at a distance d from each other, then d is about 5.5 Å. This is consistent with the assumption that $C_6H_6^{++}$ has a chain structure, with the two charges sitting at the opposite ends of the chain (*31*) (Fig. 6).

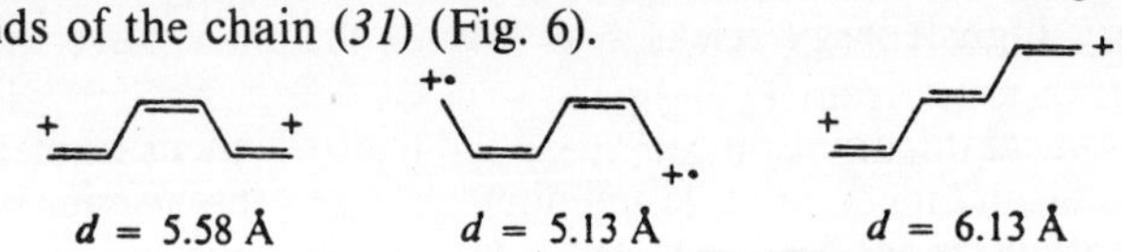

$$C_6H_6^{2+} \rightarrow CH_3^+ + C_5H_3^+$$

FIG. 6. Possible structures of the $C_6H_6^{++}$ ion consistent with the measured energy release in its decomposition [from J. H. Beynon and A. E. Fontaine, *Chem. Commun.* 717 (1966)].

KINETICS

Let us now look at the time dependence of unimolecular decompositions. This problem is of fundamental as well as practical interest: Fundamentally, because one would like to know how a molecule can store energy above its dissociation limit, and what determines its eventual break-up; and from a practical point of view, it is important to know how long energized primary ions can survive, and act as potential reactants in some environment, or what radicals they release in their break-up, which in turn may react further.

Ever since Hipple's discovery of metastables, attempts have been made to measure their lifetime. In early work, the time lag between ionization and entry into the field-free region, the so-called ion source residence time, was varied. Of course, the measured metastable yield dropped as this time lag was made longer (*32*). Analyses in terms of simple exponential decay laws, or possibly superpositions of such laws, were contradictory (*33*), probably due to uncontrollable ion optical effects. The lifetime range accessible to these early experiments was naturally prescribed by the time the ions take in a mass spectrometer to fly from the ion source to the first field-free region, i.e. a few microseconds. Progress toward shorter lifetimes was made by observing spontaneous decompositions of ions during their acceleration. In special ion sources the regions of ionization can be arranged to lie within the strong ion draw-out field. A freshly formed ion is then immediately accelerated, and the long ion residence time lag is cut out. The principle of detecting delayed decomposition is the same as before, they betray themselves by kinetic energy loss. Previously (see *Instrumental* section), the energy loss associated with decomposition in the field free region was

$$eU - (m_2/m_1)eU = (\Delta_m/m_1)eU \quad ,$$

where U is the full acceleration voltage. In the case of decomposition during acceleration the formula still applies, except that now U is only that portion of the accelerating voltage which the precursor ion traversed up to the instant of its decomposition. Accordingly, the fragment ions will exhibit a *range* of kinetic energies, depending on where in the accelerating field, or how long after the ionization, they were formed.

This kinetic energy tail can again be detected either on the mass scale or on the energy scale. Spectacularly short lifetimes can be measured using field ionization rather than electron impact ionization. The high accelerating fields (10^6 or 10^7 V/cm) in field ionization sources lead to measurable voltage drops over lifetimes as short as 10^{-11} sec, see Fig. 7 (*34–36*). At still shorter times, corresponding to the extremely high field near the ionizing tip, the ion would no longer dissociate in a truly spontaneous fashion, but rather would be torn apart by the electric field (*35,37*). Most of the field ionization kinetic studies have been done by evaluating the peak tailing on the mass scale, although measurements on the energy scale, using double-focusing mass spectrometers, have been reported [(*37*) and references therein]. In this case, both the Barber-Elliott and the Major technique can again be used.

With electron-impact and charge-exchange ionization, quantitative measurements on short-lived metastables have been done on the *energy scale* (*38,28,39–41*). A special molecular beam ion source permitted measurements down to 3 to 5 nsec. Observations of short-lived ion species allow one to trace the formation of a mass spectrum back towards its beginning. In the usual mass spectrometer the development of the fragmentation pattern is essentially complete by the time mass analysis is made, i.e. several microseconds after the initial ionization. However, at times as short as several nanoseconds after the ionization, considerable fragmentation is still going on. As an example, Fig. 8 compares a portion of two benzonitrile mass spectra, one as commonly measured (5 μsec after ionization), the other as measured in the special mass spectrometer of Ref. (*39*) only 10 nsec after ionization. Mass 76 is the fragment most slowly formed. It could be shown that 2/3 of all mass 76 fragment ions in the usual mass spectrum result from break-ups later than 10 nsec after the electron impact! This figure can be contrasted with a field ionization measurement by Tenschert and Beckey (*42*) of neopentane where, as a result of the unusual instability of the parent ions, 95% of them decompose within the first 10^{-12} sec.

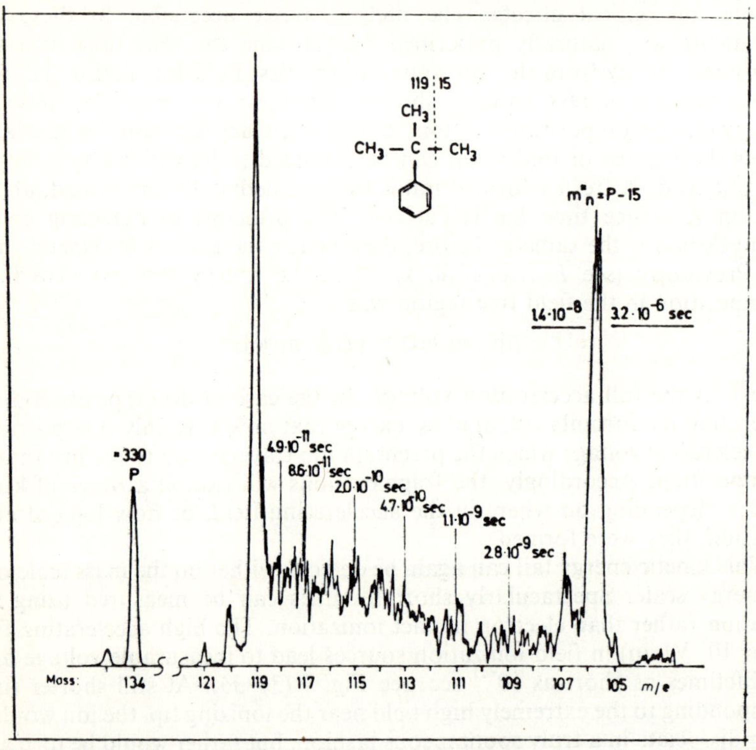

FIG. 7. Peak tailing on mass scale of field ionization spectrum, due to "fast metastables" [from H. D. Beckey, H. Hey, K. Levsen and G. Tenschert, *Int. J. Mass Spectrom. Ion Phys.* **2**, 101 (1969)]; note indicated lifetimes. The "slow metastable" peak at right is an integral over the life-time range give.

Both from the field ionization and electron impact measurements the result emerged clearly that the ionic decomposition is non-exponential. Figure 9 shows the rate of a heptane decomposition, as a function of time, from field ionization (FI) and electron impact (ES) measurements. The agreement is remarkable, the small difference in slope can be attributed to different excitation energies of the parent ions.

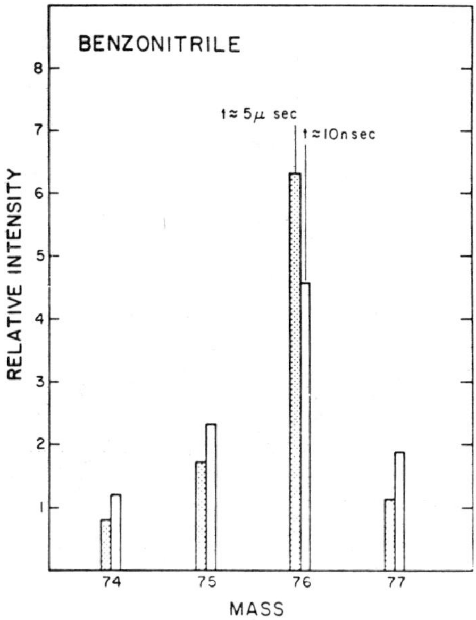

FIG. 8. Portion of mass spectrum of benzonitrile, as measured 10 nsecs after ionization, compared with ordinary mass spectrum [after I. Hertel and Ch. Ottinger, *Z. Naturforsch.* **22a,** 1141 (1967)].

The decay curve can be approximated by an inverse-power law $dN/dt \sim 1/t^r$. An especially simple case is the benzonitrile decomposition $C_6H_5CN^+ \rightarrow C_6H_4^+ + HCN$, the rate of which decreases, between $3 \cdot 10^{-9}$ and $6 \cdot 10^{-6}$ sec, very nearly like $1/t$ (*39,43,44*). Such empirical decay laws can of course be expanded into a superposition of exponentials. The decay constants of these exponentials then have a continuous distribution function, $\Pi(k)$, in the case of benzonitrile, for example, $\Pi(k) = 1/k$. In the benzonitrile case it was possible to observe directly how the lifetime changes as the energy of the ionizing electrons is varied (*39*), see Fig. 10. At 19 eV electron energy, the fall-off of the decomposition rate is much lower and the proportion of long-lived ions much higher than at 30 eV. A decomposition with a very unusual behavior was found by Levsen and Beckey (*45*). Here the decay rate goes through a *maximum* at around 10^{-11} sec as was verified using two different field geometries (Fig. 11). Apparently the break-up has to be preceded by a rate-determining induction step, the nature of which is, however, difficult to infer.

The existence of a continuum of rate constants is predicted by the generally-accepted theory for the dissociation of organic ions. This is the quasi-equilibrium theory (QET), which states that dissociation occurs through an activated complex of the ion. The dissociation rate constant is then given statistically by (*46*)

$$k(E) = 1/h\,[N^{\ddagger}(E-E_0)/\rho(E)]$$

where N‡ is the number of vibrational states of the activated complex accessible at the energy E, E_0 is the threshold energy, and ρ is the density of the states of the ion. N, ρ and k all depend strongly on the excitation energy (*47,48*).

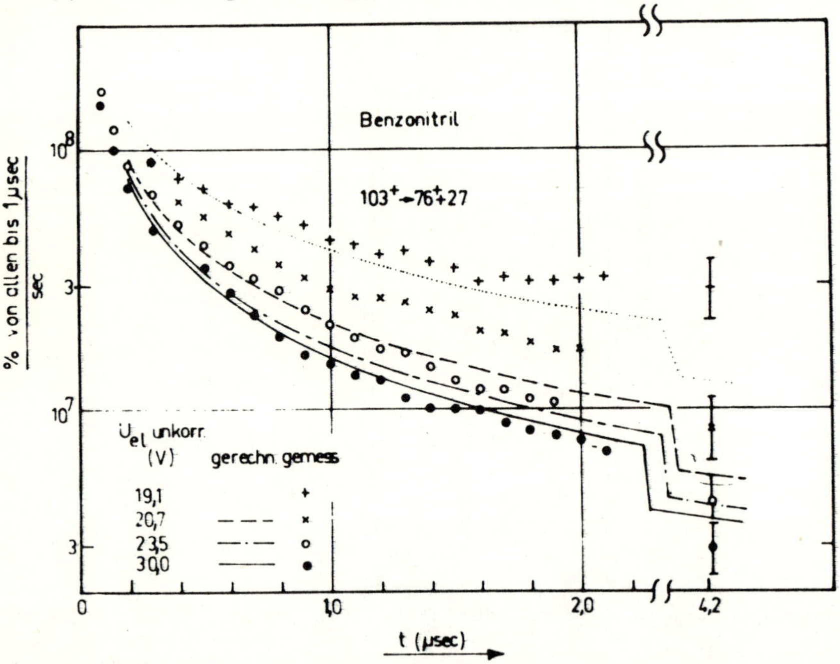

FIG. 9. Decomposition rate of $C_7H_{16}^+ \rightarrow C_5H_{11}^+ + C_2H_5$, as a function of time, both for field (FI) and electron impact (ES) ionization [from G. Tenschert and H. D. Beckey, *Int. J. Mass Spectrom. Ion Phys.* **7,** 97 (1971)].

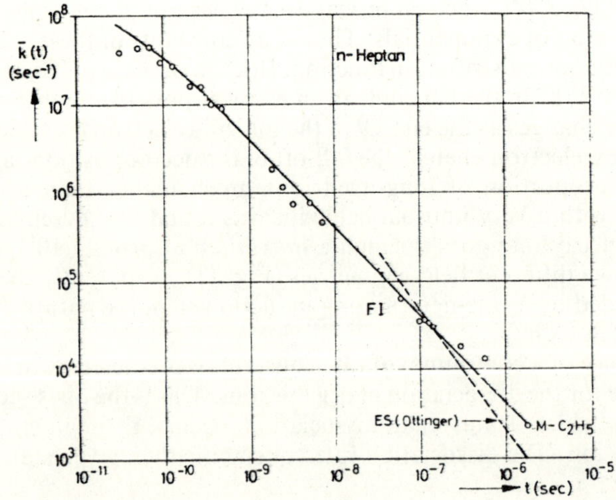

FIG. 10. Decomposition rate of $C_7H_5N \rightarrow C_6H_4^+ + HCN$, as a function of time, for different ionizing electron energies [from I. Hertel and Ch. Ottinger, *Z. Naturforsch.* **22a,** 1141 (1967)].

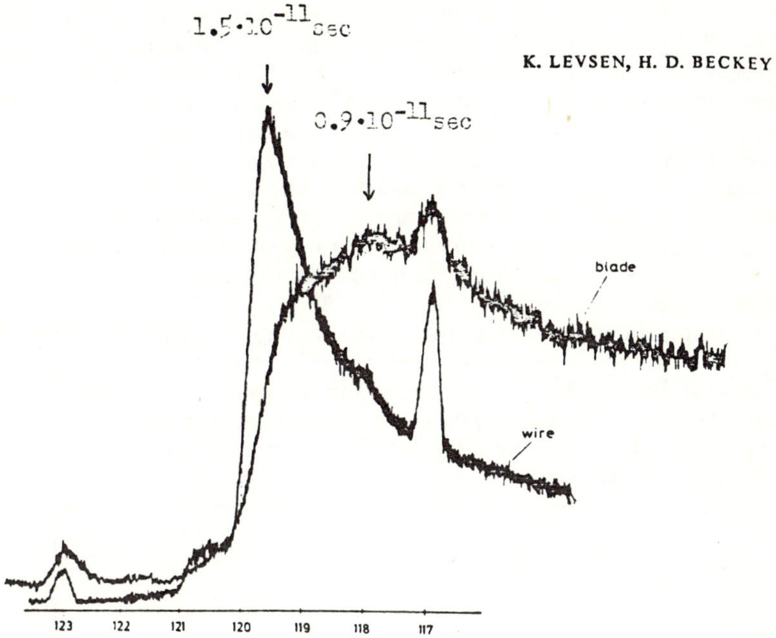

Valerophenone. Part of FI spectrum. The rearrangement peak at $m/e = 120$ displays a long peak tailing.

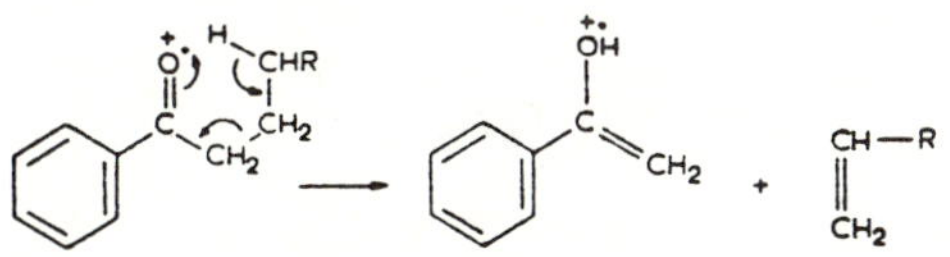

($m/e = 120$)

FIG. 11. Field ionization spectrum, showing fast metastable peak with induction period [from K. Levsen and H. D. Beckey, *Int. J. Mass Spectrom. Ion Phys.* **7**, 341 (1971)].

Both electron impact and field ionization impart to the molecular ions excitation energies E which vary over a certain range. Consequently, the rate constants k(E) are also distributed continuously. This explains the non-exponential decay curves. The function k(E) is the key feature of this theory. Once it is known, the development of a mass spectrum as a function of time can be predicted.

In the most widely used form of the QET, $N^{\ddagger}$ is calculated on the basis of reasonable guesses for the parameters of the activated complex. [For a recent example, see (*49*).] An alternative formulation of the QET by Klots circumvents this guesswork (*50*). Klots relates by microscopic reversibility the rate constant for unimolecular dissociation to the cross section for the reverse, the association of the two fragments (e.g. to the Langevin cross section). Instead of the known properties of the activated complex, this theory only uses the measurable properties of the separated fragments.

Experimentally, only a few determinations of k(E) curves have been made (*44,39;40,41*). In the case of electron impact experiments, assumptions were necessary about the distribution P(E) of excited states in the ion, although these assumptions could be supported by measurements of the ion yield near threshold electron energy (*44*), or by photoionization data (*39*). A much more precise control over P(E) can be obtained with ionization by charge exchange from atomic ions, where the energy transferred to the organic ion is, within very narrow limits, equal to the difference of the two ionization potentials. Thus, P(E) approximates a δ-function in this case. Correspondingly, the decay curve is no longer the result of a continuum of contributing rate constants, but is now an almost pure exponential.

Using different atomic ions, one can step through the energy scale and measure the resulting decay constants k(E) separately at discrete excitation energies. This method has only been applied so far to three molecules, among them benzene (*40,41*). Here, the two parent ion decompositions $C_6H_6^+ \rightarrow C_6H_5^+ +$ H and $\rightarrow C_4H_4^+ + C_2H_2$ were studied. The resulting k(E) curves are shown in Fig. 12. Two facts deserve comment:

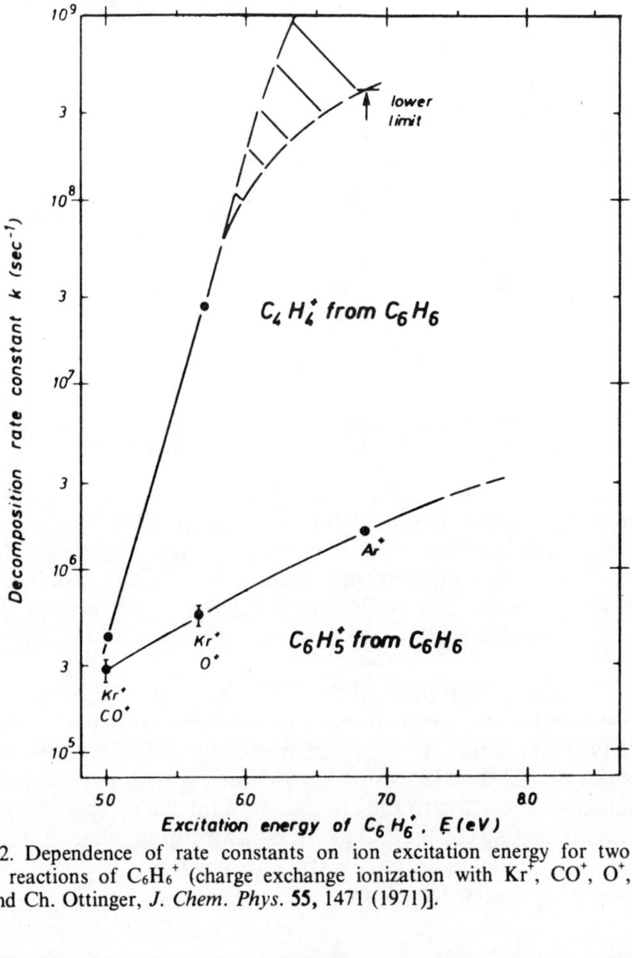

FIG. 12. Dependence of rate constants on ion excitation energy for two unimolecular dissociation reactions of $C_6H_6^+$ (charge exchange ionization with Kr^+, CO^+, O^+, Ar^+ [from B. Andlaner and Ch. Ottinger, *J. Chem. Phys.* **55,** 1471 (1971)].

$C_4H_4^+$ formation, after charge exchange of C_6H_6 with Ar^+, proceeds very rapidly, too fast to be time-resolved in the apparatus used. On the other hand, $C_6H_5^+$ (also from $C_6H_6 + Ar^+$) is formed very slowly, with $k = 1.5 \cdot 10^6\ sec^{-1}$. These two decompositions of $C_6H_6^+$ can therefore not be in competition with each other. This is a definitive answer to a long-standing problem in the QET and shows that organic ions can have isolated electronic states each with their own, separate, decomposition channels. Rosenstock et al. have analyzed the same problem theoretically (*49*). They compared the measured yield curves of benzene fragment ions in the threshold region of photoionization experiments with the yield calculated from QET. Agreement could only be obtained if it was assumed that only the $C_6H_5^+$ ion is formed from the $C_6H_6^+$ ground state, while the $C_4H_4^+$ fragment originates from the first excited state of $C_6H_6^+$, 2.25 eV above the ground state. There is no competition between these two channels.

The second interesting result from the charge-exchange lifetime measurements is the exceedingly slow rise of k(E) for the $C_6H_5^+$ formation. Over an energy interval of 2 eV, k only increases by a factor of 5 to 6 and never exceeds $3 \cdot 10^6\ sec^{-1}$. Compared to most other ionic fragmentations this is certainly a very sluggish one. The slow rise of k(E) has been confirmed by photoionization measurements (*51*). Lifshitz et al. have recently again verified the anomalous nature of k(E) for this process (*52*). Using a novel pulsed ionization plus ion storage (up to 1 msec) technique, they were able to study selectively only those $C_6H_5^+$ ions which originated with a given delay. This delay time was then correlated with the ionizing electron energy.

Theoretically the unusually slow benzene decomposition remains unexplained, see Fig. 13. Rosenstock et al. (*49*) were unable to reconcile simultaneously the measured k(E) and the measured photo-ion threshold curve with the QET calculations. Similarly, Klots' theory fails to match the observation, if one assumes reasonable association cross sections.

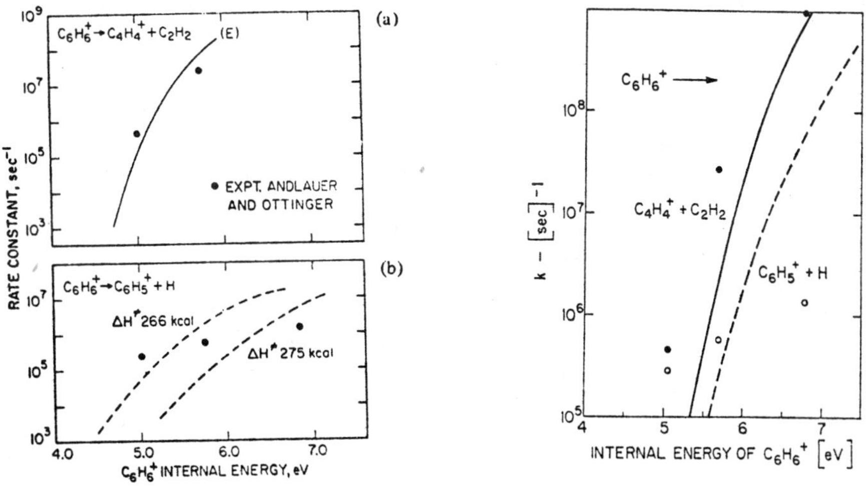

FIG. 13. Two different attempts at theoretical fits of the data shown in Fig. 12. Left: from H. M. Rosenstock, J. T. Larkins and J. A. Walker, *Int. J. Mass Spectrom. Ion Phys.* **11**, 309 (1973); right: from C. E. Klots, *Z. Naturforsch.* **27a**, 553 (1972).

Incidentally, the slow increase of k(E) with E would lead one to predict an unusually intense metastable peak in the ordinary electron impact mass spectrum, because ions formed with excitation energies anywhere within a broad range (more than 1 eV wide) will all decompose slowly enough to contribute to the metastable peak (i.e. $3 \cdot 10^4 \leqslant k \leqslant 10^6$ sec). Indeed, this peak was found to have an intensity of more than 1% of the base peak (M=78) (*53*). This is an illustration of a general rule of thumb: intense metastable peaks probably indicate a slow rise of k(E).

The QET predicts steeply rising functions k(E) for small molecules (*16*). According to the rule of thumb, these should therefore have very weak metastable peaks, if any at all. The observation of metastable ions of methane (*15,16*) was explained by Klots on the basis of a special effect, namely tunneling through the centrifugal potential barrier (*17*). Klots was able to explain nicely (*54*) the observed relative intensities for delayed loss of an H atom within the series of deuterated methanes.

Metastable ions have been observed for even smaller, and indeed diatomic, molecules.

In CO_2: $CO_2^{++} \rightarrow CO^+ + O^+$ (*25,28*),
in CO: $CO^{++} \rightarrow C^+ + O^+$ (*25*).

Singly charged ions:

In N_2: $N_2^+ \rightarrow N^+ + N$ (*55*)
in NO: $NO^+ \rightarrow O^+ + N$ (*55*)
in H_2S: $H_2S^+ \rightarrow S^+ + H_2$ (*56,57*)
$\rightarrow HS^+ + H$ (*56*)
$HS^+ \rightarrow S^+ + H$ (*56*)
in NO_2: $NO_2^+ \rightarrow NO^+ + O$ (*58*)
in N_2O: $N_2O^+ \rightarrow NO^+ + N$ (*59,58*)

For molecules as small as these, the QET is most likely not applicable. Instead, their dissociations are amenable to discussion in terms of potential curves. Let us consider the example $N_2O^+ \rightarrow NO^+ + O$ (Fig. 14). The ionization energy of N_2O is 12.9 eV. The lowest energy process for NO^+ formation is a spin-forbidden dissociation from $N_2O^+(^2\Pi)$ via the $N_2O^+(^4\Sigma^-)$ curve to $NO^+(^1\Sigma)$ + $N(^4S)$. Thermochemically this asymptote lies 1.3 eV above the N_2O^+ ground state. The measured appearance potential for ordinary NO^+ fragments is 15.01 eV (*60*), which places the $^2\Pi/^4\Sigma$ intersection at 0.8 eV above the $^4\Sigma$ asymptote. The delayed fragmentation can then be interpreted as a predissociation, slowed down by the inter-system transition requirements, of ground state N_2O^+ ions containing vibrational energy up to the crossing point.[4] Consistent with this picture is the measured fragmentation energy $T = 1.05 \pm 0.05$ eV (*58,59*). The alternative mechanism of initial excitation of some higher N_2O^+ state, followed by radiative decay—as the rate-determining step—to the dissociating state, can *in this case* (!) be discarded on account of the appearance potential of the metastables. The measured values of 15.1 ± 0.2 eV (*59*) and 15.7 ± 0.5 (*58*) point again to the region of the $^2\Pi/^4\Sigma$ crossing.

[4]Ryan (*61*) has, by charge exchange, prepared N_2O^+ ions with vibrational excitation above the crossing point, which nevertheless did not dissociate within about 10^{-5} sec.

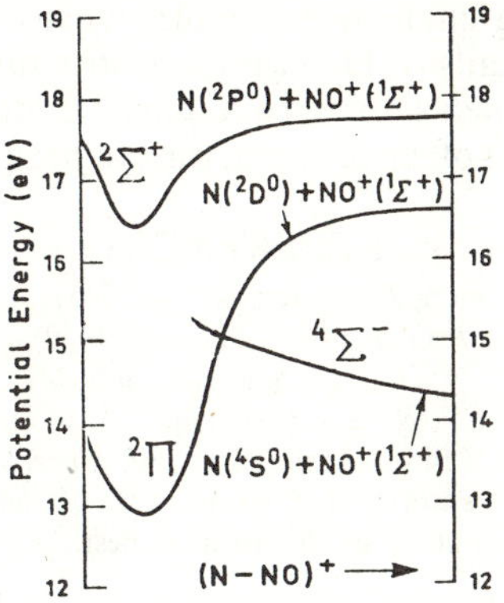

FIG. 14. Potential curves pertinent to the predissociation $N_2O^+ \rightarrow NO^+ + O$ [Adapted from K. R. Ryan, *J. Chem. Phys.* **57**, 271 (1972)].

CONCLUSION

The importance of the study of spontaneous ion decompositions is not easily summarized, because this field relates to so many other areas.

Modern instrumentation has made available metastables as a useful aid in organic analysis. We have discussed in some detail the practical relevance to radiation chemistry: Measurements in the mass spectrometer have revealed the important internal rearrangements as well as decomposition paths of ions (see section on *Chemical Applications*). They have demonstrated that ions can carry energies of up to several electron volts for periods of microseconds, and that in the eventual break-up excited radicals can be formed capable of further fragmentation (see *Energetics*). They have established the time scale of unimolecular ion decompositions, in particular in the regime of short times of interest in fairly condensed systems (see *Kinetics*). From a fundamental viewpoint, of course, these three areas of chemistry, energetics and kinetics are all related. Theory is challenged to provide a complete, unifying understanding.

Metastable ions have stimulated the development of statistical formalisms, like enumeration of quantum states (*47,48*). Another theory inspired by metastables is that of the lifetime of a quantum state imbedded in a continuum (*62*). A problem of current interest is the threshold behavior. One would like to know whether at the threshold decomposition sets in with a finite rate constant, or whether the continuum of rate constants extends all the way down to zero.

Experiments with extremely long-lived metastables should help to clarify this (*43,63*). Newly developed techniques of suspending ions in a vacuum for periods of hours may open up completely new possibilities (*64*). Other experimental refinements in the future will involve better control over the excitation energy. Photoionization, in conjunction with coincidence measurements of energy-analyzed electrons, may allow more precise measurements of k(E) on a large scale (*65–67*). From this information and lifetime measurements of the type described above, one could then infer the energy deposition function for various ionizing agents, another important quantity (*68*). Mass spectrometric, photoelectron, electron impact, and optical studies of predissociating small molecular ions will continue to complement each other and will reveal details of the electronic structure.

It is evident that metastables have come a long way since Hipple and Condon's explanation (*1*) of some small, diffuse peaks in certain mass spectra.

REFERENCES

1. J. A. Hipple and E. U. Condon, *Phys. Rev.* **68,** 54 (1945).
2. U. Löhle and Ch. Ottinger, *J. Chem. Phys.* **51,** 3097 (1969).
3. U. Löhle and Ch. Ottinger, *Int. J. Mass Spectrom. Ion Phys.* **5,** 265 (1970).
4. T. Wachs, P. F. Bente III, and F. W. McLafferty, *Int. J. Mass Spectrom. Ion Phys.* **9,** 333 (1972).
5. See, e.g. J. F. Ryan and F. J. Biros, *Abstr. 21st Conf. Amer. Soc. Mass Spectr.*, San Francisco 1973, p. 340. [The title of this work illustrates the range of usefulness of metastable ion studies: "Determination of fragmentation mechanisms of pesticides by metastable defocusing and DADI."]
6. F. W. McLafferty, J. Okamoto, H. Tsuyama, Y. Nakajima, T. Noda and H. W. Major, *Org. Mass Spectrom.* **2,** 751 (1969).
7. M. Barber and R. M. Elliott, *ASTM E-14 12th Annual Conference on Mass Spectrometry*, Montreal, 1964.
8. J. E. Coutant and F. W. McLafferty, *Int. J. Mass Spectrom. Ion Phys.* **8,** 323 (1972).
9. N. R. Daly, A. McCormick, R. E. Powell and R. Hayes, *Int. J. Mass Spectrom. Ion Phys.* **11,** 255 (1973).
10. K. R. Jennings, *Chem. Commun.*, 283 (1966).
11. L. P. Hills, J. H. Futrell, and A. L. Wahrhaftig, *J. Chem. Phys.* **51,** 5255 (1969).
12. C. E. Melton, M. M. Bretscher and R. Baldock, *J. Chem. Phys.* **26,** 1302 (1957).
13. J. L. Holmes and F. M. Benoit, in *MTP Int. Rev. of Science, Phys. Chemistry Series One.* **Vol. 5,** *Mass Spectrometry*; (A. D. Buckingham and A. Maccoll, Eds.) p. 259. Butterworths, London, 1972.
14. C. A. Lifshitz and R. Sternberg, *Int. J. Mass Spectrom. Ion Phys.* **2,** 303 (1969).
15. Ch. Ottinger, *Z. Naturforsch.* **20a,** 1232 (1965).
16. L. P. Hills, M. L. Vestal, and J. H. Futrell, *J. Chem. Phys.* **54,** 3834 (1971).
17. C. E. Klots, *Chem. Phys. Lett.* **10,** 422 (1971).
18. A.N.H. Yeo, R. G. Cooks, and D. H. Williams, *Chem. Commun.* 1269 (1968).
19. H. M. Rosenstock, V. H. Dibeler, and F. N. Harllee, *J. Chem. Phys.* **40,** 591 (1964).
20. F. W. McLafferty and T. A. Bryce, *Chem. Commun.* 1215 (1967).
21. U. Löhle and Ch. Ottinger, *Abstr. 17th Conf. Amer. Soc. Mass Spectr.* Dallas, 1969, p. 80.
22. R. Taubert, *Z. Naturforsch.* **19a,** 484 and 911 (1964).
23. J. H. Beynon, R. A. Saunders, and A. E. Williams, *Z. Naturforsch.* **20a,** 180 (1965).
24. Ch. Ottinger, Physics Lett. **17,** 269 (1965); see an Erratum in: Ch. Ottinger, *Z. Naturforsch.* **22a,** 20 (1967).
25. R. Fuchs and R. Taubert, *Z. Naturforsch.* **20a,** 823 (1965).
26. J. H. Beynon, and A. E. Fontaine, *Z. Naturforsch.* **22a,** 334 (1967).
27. J. H. Beynon, A. E. Fontaine and G. R. Lester, *Int. J. Mass Spectrom. Ion Phys.* **8,** 341 (1972).
28. Ch. Ottinger, *Z. Naturforsch.* **22a,** 20 (1967).
29. Ch. Ottinger, *J. Chem. Phys.* **47,** 1452 (1967).
30. W. Higgins and K. R. Jennings, *Chem. Commun.* 99 (1965).
31. J. H. Beynon and A. E. Fontaine, *Chem. Commun.* 717 (1966).
32. J. A. Hipple, *Phys. Rev.* **71,** 594 (1947).
33. N. D. Coggeshall, *J. Chem. Phys.* **37,** 2167 (1962); J. C. Schug, *ibid.* **40,** 1283 (1964).
34. H. D. Beckey and H. Knöppel, *Z. Naturforsch.* **21a,** 1920 (1966).
35. H. D. Beckey, H. Hey, K. Levsen and G. Tenschert, *Int. J. Mass Spectrom. Ion Phys.* **2,** 101 (1969).
36. B. W. Viney, *Int. J. Mass Spectrom. Ion Phys.* **8,** 417 (1972).

37. A. M. Falick, P. J. Derrick and A. L. Burlingame, *Int. J. Mass Spectrom. Ion Phys.* **12,** 101 (1973).
38. O. Osberghaus and Ch. Ottinger, *Phys. Lett.* **16,** 121 (1965).
39. I. Hertel and Ch. Ottinger, *Z. Naturforsch.* **22a,** 1141 (1967).
40. B. Andlauer and Ch. Ottinger, *J. Chem. Phys.* **55,** 1471 (1971).
41. B. Andlauer and Ch. Ottinger, *Z. Naturforsch.* **27a,** 293 (1972).
42. G. Tenschert and H. D. Beckey, *Int. J. Mass Spectrom. Ion Phys.* **7,** 97 (1971).
43. Ch. Ottinger, in *Proc. IVth Internat. Congr. Radiation Research*, Evian 1970, *"Physics and Chemistry."* (to be published).
44. J. Hertel and Ch. Ottinger, *Z. Naturforsch.* **22a,** 40 (1967).
45. K. Levsen and H. D. Beckey, *Int. J. Mass Spectrom. Ion Phys.* **7,** 341 (1971).
46. H. M. Rosenstock, *Advan. Mass Spectrom.* **4,** 523 (1968).
47. W. Forst, *Chem. Rev.* **71,** 339 (1971).
48. M. R. Hoare and Th. W. Ruijgrok, *J. Chem. Phys.* **52,** 113 (1970); M. R. Hoare, *ibid.* **52,** 5695 (1970); *ibid.* **55,** 3058 (1971).
49. H. M. Rosenstock, J. T. Larkins, and J. A. Walker, *Int. J. Mass Spectrom. Ion Phys.* **11,** 309 (1973).
50. C. E. Klots, *J. Phys. Chem.* **75,** 1526 (1971); *Z. Naturforsch.* **27a,** 553 (1972).
51. W. A. Chupka, J. Berkowitz, and S. I. Miller, in *Abstr. 20th Conf. Amer. Soc. Mass Spectr.*, Dallas, 1972.
52. C. Lifshitz, A. M. Peers, M. Weiss and M. J. Weiss, in *Abstr. Mass Spectrometry Conference*, Edinburgh, 1973.
53. Ch. Ottinger, *Z. Naturforsch.* **20a,** 1229 (1965).
54. C. E. Klots, *Abstr. 19th Conf. Amer. Soc. Mass Spectr.*, Atlanta, 1970.
55. A. S. Newton and A. F. Sciamanna, *J. Chem. Phys.* **50,** 4868 (1969).
56. V. H. Dibeler and H. M. Rosenstock, *J. Chem. Phys.* **39,** 3106 (1963).
57. E. G. Jones, J. H. Beynon and R. G. Cooks, *J. Chem. Phys.* **57,** 3207 (1972).
58. A. S. Newton and A. F. Sciamanna, *J. Chem. Phys.* **52,** 327 (1970).
59. R. J. Coleman, J. S. Delderfield, and B. G. Reuben, *Int. J. Mass Spectrom. Ion Phys.* **2,** 25 (1969).
60. V. H. Dibeler, J. A. Walker, and S. K. Liston, *J. Res. Nat. Bur. Std.* **A71,** 371 (1967).
61. K. R. Ryan, *J. Chem. Phys.* **57,** 271 (1972); compare also: J. H. D. Eland, *Int. J. Mass Spectrom. Ion Phys.* **12,** 389 (1973).
62. F. H. Mies and M. Krauss, *J. Chem. Phys.* **45,** 4455 (1966); G. Alexandru, *Int. J. Mass Spectrom. Ion Phys.* **11,** 17 (1973).
63. H. Tatarczyk and U. von Zahn, *Z. Naturforsch.* **27a,** 1646 (1972).
64. F. L. Walls and G. H. Dunn, *J. Geophys. Res.*, 1974, (in press).
65. B. Brehm and A. V. Puttkamer, *Z. Naturforsch.* **22a,** 8 (1967).
66. J.H.D. Eland, *Int. J. Mass Spectrom. Ion Phys.* **8,** 143 (1972).
67. J.H.D. Eland and C. J. Danby, *Int. J. Mass Spectrom. Ion Phys.* **8,** 153 (1972).
68. G. G. Meisels and R. H. Emmel, *Abstr. 20th Conf. Amer. Soc. Mass Spectr.*, Dallas, 1972.

Ion Molecule Reactions[1]

P. Kebarle
Chemistry Department
University of Alberta
Edmonton, Alberta, Canada

The chemical consequences of an ion molecule collision depend on the relative velocity of the interacting pair. Most common and thus important, in radiation chemistry are interactions at thermal velocities. The fate of the persistent reaction complex formed at thermal energies also depends on the occurrence of collisions with third molecules which may remove excess energy. The nature of the products is therefore also dependent on pressure. Apart from collisional deactivation and thermal activation, high pressure leads to long reaction sequences, occurrence of reactions with small rate constants and the achievement of ionic equilibria. The mass spectrometric study of ionic reactions under these conditions has proven extremely rewarding. The field has developed to such an extent that now it produces important information on energetic and stability of ions. This information is of great significance also to outside fields like physical organic and inorganic chemistry and to the theory of ionic solvation.

Thus, the observations on ion solvent molecule complexes answer questions pertaining to the importance of hydrogen bonding by protic solvents and modification of acidity, basicity and reactivity of ions in the presence of solvent molecules.

[1]Congress presentation B–29–4.

PARTICLE PENETRATION PHENOMENA

Chairman: W. Brandt

[1]Symposium E–2

Energy Loss and Ranges of Charged Particles in Matter[1]

P. Sigmund
H. C. Ørsted Institute, DK-2100 Copenhagen, Denmark

INTRODUCTION

Niels Bohr (*1*), in 1948, summed up about 40 years of intense research in particle penetration phenomena—by himself and many others—in a comprehensive article that included an illuminating discussion of the (then well established) classical and quantal theory of the stopping of point charges, and a qualitative outline of the more complex mechanism of energy loss of heavy atomic particles in matter. The report of a N.R.C. subcommittee on penetration of charged particles (*2*) demonstrates that, in 1963, the knowledge of the stopping of light charged particles (not heavier than alpha particles) at not too low velocities (corresponding to a proton energy of at least several MeV) had arrived at a fairly quantitative stage. The report, as well as Fano's critical review (*3*), specified a number of goals for further experimental and theoretical research, aiming at a precision of, say, 1% or better in range and energy loss data. Such accuracy was found necessary because of the significance of stopping data in the analysis of nuclear and elementary-particle reactions. More basically, possible shortcomings of the fundamental quantal stopping theory could be uncovered only at such a level of accuracy. The same report (*2*) indicates an increasing attention to stopping of high-mass, low-velocity particles. There had been a long-standing need for such attention, in particular from the fields of radiation effects in inorganic and organic materials, as well as low-energy nuclear physics. In 1963, a general accuracy of 20–30% was to be considered a major break-through in low-velocity stopping.

Since 1963, research in penetration phenomena has been expanding, and a number of major developments have taken place. Perhaps the most outstanding one is the discovery and thorough study of charged-particle channeling, which is being covered by another talk in this symposium (*4*). In the field of light-ion stopping, high-precision measurements have been performed, a number of significant extensions to Bethe's basic stopping theory have been made, and considerable effort has been invested in calculating ionization potentials, the basic parameter entering the Bethe formula. In the low-velocity range, ion implantation has become an important tool in solid-state research and technology, and much progress has been achieved in the understanding of the basic physics involved, in particular the penetration phenomena, (*6–8*). Ion beams at intermediate velocites (e.g., 1 MeV alpha particles) have become a useful tool in surface analyses (*9*). The stopping mechanism in this velocity

[1]Congress presentation E–2–1.

range is complex, and a considerable experimental effort is being made at present to supply reliable stopping data. The theoretical description is mostly semi-empirical. The same statement applies to the stopping of heavier ions at comparable or higher velocities (*10*).

In this report, I should like to comment rather briefly on some of the points mentioned above. I shall concentrate on developments after 1963 and on ions (rather than electrons) in the nonrelativisitc velocity range. A more extensive report with similar scope—but containing a broad introductory part—has been prepared recently (*5*).

GENERAL

The key quantity is the *stopping power*, ($-dE/dx = \overline{\Delta E}/\Delta x|_{\Delta x=0}$), where $\overline{\Delta E}$ is the mean energy loss of a monoenergetic particle beam (energy E, velocity v) after passing a homogeneous layer of target material with thickness Δx. The width of the transmitted energy spectrum can be characterized by the *straggling* $\Omega^2 = \overline{(\Delta E - \overline{\Delta E})^2}$.

dE/dx and Ω can be measured in *transmission* experiments by use of thin foils or gas targets. The analysis is particulary simple in the case when the transmitted spectra are gaussian. Stopping powers can also be measured in *reflection*, i.e., by means of Rutherford scattering from thick targets. In combination with thin evaporated or implanted layers, thick-target measurements can provide quite accurate stopping data, in particular relative stopping powers (compared to some standard target). The up to now most precise stopping-power data have been found by low-temperature calorimetry of the deposited energy in solid targets (*11*).

We shall consider "random" (i.e., amorphous or polycrystalline) targets in the following. Energy-loss spectra depend on the orientation of the target with respect to the beam when the target is a single crystal, because of channeling (*4*). Channeling effects tend to *compensate* in polycrystals (*12*), but compensation may be incomplete at low energies. Recent experimental evidence points to the direction of energy loss being less sensitive to lack of compensation than other effects such as multiple scattering (*13*).

Electronic excitation is the prominent mechanism of energy loss. Qualitatively, the *electronic stopping power* $(dE/dx)_e$ behaves friction-like ($\propto v$) for $v \lesssim Z_1^{2/3}c/137$,(*14*), where Z_1 is the atomic number of the ion and c the velocity of light. After passing through a broad maximum, $(dE/dx)_e$ decreases roughly like v^{-2}. At relativistic velocities, the stopping power approaches a minimum and eventually increases again, and additional stopping mechanisms such as bremsstrahlung become significant. Energy transfer to target nuclei is important at low velocities, in particular so for *heavy* ions. The *nuclear stopping power*, $(dE/dx)_n$, has its maximum around $v \approx (1/M_1+1/M_2)^{1/2} (Z_1Z_2e^2/a)^{1/2}$, where M_1 and M_2 are the nuclear masses of the projectile and target, respectively, and a is the screening radius characterizing the elastic interaction potential between the nuclei. Roughly, for very heavy ions, nuclear stopping dominates up to around 10^6 eV, and electronic stopping takes over slowly, going through a maximum around 10^8 eV. For very light ions, electronic stopping has its maximum around 10^5 eV, and nuclear stopping is significant (on a relative scale) only below 10^3 eV. For ions of intermediate mass,

$(dE/dx)_n$ dominates in the eV range, and $(dE/dx)_e$ in the MeV range, while the two compete in the keV range. Even in case of negligible nuclear stopping, nuclear collisions may be quite important in determining the straggling and higher moments of the energy-loss spectrum. Moreover, large-angle *scattering* of an ion is determined by nuclear rather than electronic collisions.

LIGHT-ION STOPPING

In this section we consider the energy range above and down to the maximum stopping power. The extended Bethe formula for the stopping power of a light ion—that is approximated as a point charge Z_1e and assumed to have an infinite mass—reads

$$- dE/dx = 4\,\pi N\,(Z_1^2Z_2e^4/mv^2)\,[\log(2mv^2/I) - \log(1 - v^2/c^2) - v^2/c^2 - C/Z_2 - \delta/2] + Z_1^3J, \qquad (1)$$

where N is the number of target atoms per unit volume, m the electron mass, I the mean ionization potential, C/Z_2 the innershell correction term, $-\delta/2$ the density correction, and J a correction arising from the second term in the Born perturbation series. For a critical discussion of eq.(1) see ref. (*3*).

The key parameter is I, which is defined as a sum involving oscillator strengths. It is only very recently that an evaluation of I by means of realistic wave functions for a great number of excited states has become possible for other than the lightest target atoms (*15*). Even now, the most reliable I-values come from experiments. However, experimental values, found by fitting stopping-power data to eq. (1), hinge upon proper expressions for the various correction terms in eq. (1). Some caution is appropriate right now, in particular with regard to the influence of the Z_1^3-term, (*11,16*). Sternheimer's (*17*) modified Bloch relation, $I/Z_2 = (9.76+58.8Z_2^{-1.19})eV$ describes the gross behavior of I rather well for $Z_2 \gtrsim 10$, but substantial structure (oscillatory behavior) in the I versus Z_2 relation is now well documented, (*18–20,11,15*). A number of calculations of I, (18,19), have been based upon the dielectric theory of stopping, that hinges on the excitation spectrum of a free electron gas (*21*). Such calculations appear numerically successful, in particular when carried out by means of Hartree or Hartree-Fock charge distributions rather than Thomas-Fermi ones, although the lattter would be more appropriate from a consistency point of view. Note that an even more rigorous Thomas-Fermi type of approach, which is based on Bloch's dynamical Thomas-Fermi model, gave the rather poor result $I/Z_2 = 4.95eV$ (*22*).

One striking advantage of the dielectric theory (*21,23*) is the simultaneous outcome of inner-shell corrections C/Z_2. These arise from those electrons that have orbital velocities comparable to or greater than v. This correction is quite substantial even far above the stopping-power maximum (*2*). More rigorous calculations of C/Z_2, (*24,2*) are restricted mainly to K and L electrons. A recent attempt (*25*) to utilize the (classical) binary-encounter approximation to estimate shell corrections has so far had limited success.

The Z_1^3 correction, found experimentally and identified as a deviation from the first Born approximation by Barkas et al. (*26*) is an important and now well documented correction at low velocities. It amounts to 3–4% in dE/dx for 10 MeV alpha particles in Ta (*27*), and correspondingly more at lower velocities. The relative contribution goes approximately as v^{-2}, (*27*). Calculations have been based on classical, (*28,29*) or quantal (*30*) oscillator models.

The density correction $-\delta/2$ is rather small at nonrelativistic velocities. However, Brynjolffson (*31*), on the basis of a semiclassical model of a dense gas of oscillators, advocates a substantial density correction, $-\Delta_3-\Delta_4$, even at nonrelativistic velocites. Although the calculated apparent ionization potentials agree strikingly well with measured ones, a compelling demonstration of the very existence of this correction on the basis of quantal stopping theory would appear desirable.

With the increasing interest in surface analysis by ion beams, the additivity of stopping powers in compound targets (Bragg's rule) has received renewed interest (*9*). Examples of deviations from additivity have been known for a long time, (*3,32*). They were typically 1% effects, and assumed to arise from a change in the contribution of valence electrons to the ionization potential due to bonding. No definite deviations from Bragg's rule appear to be known for metals. Drastic effects have been found for He^+ beams in the 1 MeV range incident on solid and gaseous targets (*9*), but most of the data need to be confirmed. It seems documented, however, that measurable chemical effects occur in hydrocarbons (*3,33*).

For energy-loss *straggling*, Bohr's (*1*) classical expression is valid to a much better level of accuracy than the corresponding expression for the stopping power. This is well corroborated by experiment. For details the reader may consult refs. (*3*) and (*32*). Some recent data may be found in refs. (*9*) and *34*).

HEAVY-ION AND LOW-VELOCITY STOPPING

For not fully stripped ions, the stopping power can be assumed, as a first approximation, to be proportional to the square of the ionic rather than the nuclear charge (*1*), i.e.,

$$dE/dx \cong Z^{*2}(dE/dx)_{p,v} \tag{2}$$

where $(dE/dx)_{p,v}$ is the stopping power of a (bare) proton at velocity v. By assuming the effective charge state Z^* to be independent of the target material, one can generate stopping power tables for a wide range of ions, targets, and velocities (*10*), in particular when interpolation is made between reliable experimental data. However, eq. (2) is only a rough extension of eq. (1) and not well justified theoretically. It is hardly well understood why the use of eq. (2) to *define* an effective charge state from measured values of dE/dx produces a universal Z^*-function of some scaled velocity, (*35,36*), that indeed turns out to resemble the actually measured charge state, at least for *gas* targets. Nevertheless, eq. (2) yields a semi-empirical description of dE/dx down to energies somewhat below the stopping-power maximum.

For velocities far below the maximum, the projectile charge is heavily screened since all or most target electrons have orbital velocities greater than v. Because of the latter, stopping-power expressions for the low-velocity range usually yield velocity-proportional stopping power, (*14,37*). The Lindhard &

Scharff (*14*) expression—which has had great impact on ion implantation—is based on elastic scattering of "free" target electrons in the static field of a screened point charge. Firsov's (*37*) expression is based on a simple geometric model of charge exchange between two interpenetrating, but noninteracting atomic systems. Neither model takes into account the quasi-molecular nature of heavy-particle collisions at low velocities (e.g. *38*). The general behavior of the stopping power with regard to magnitude and energy dependence is described reasonably well by either model. The same can be said, with some caution, about the observed structure in $(dE/dx)_e$ as a function of Z_1 and Z_2 (actual evaluations have normally been based on the Firsov model, [cf. (*5,39–41*), for the key references]. However, it seems important in this context to refer to stopping measurements in single crystals—done essentially at fixed impact parameter—where Z_1-oscillations with a maximum-to-maximum ratio of up to 9 have been observed (*40*), and Z_2-structure of similar magnitude (*41*). It is this author's impression that such large effects have not been explained by the modified Firsov theories employing realistic atomic charge distributions instead of Thomas-Fermi ones (*42*). Similar evidence can be drawn from ion-gas collisions (*43*). An extension of the free-electron gas model of low velocity stopping (*14*) by Lindhard & Finnemann (unpublished) seems capable of producing large oscillations, but difficult to adapt to real atomic systems.

Finally we note that our comments on low-velocity stopping of heavy ions also hold, at least qualitatively, for protons, including the occurrence of Z_2-structure. [Cf. *(5)* for details.]

NUCLEAR STOPPING

According to Lindhard et al. (*14*), electronic and nuclear stopping are separable approximately even in the low-velocity region where nuclear stopping may be sizable and even dominating. The reason is that the dominating contributions to electronic stopping stem from comparatively distant collisions where nuclear energy losses are small, and (to a lesser extent) vice versa. This has the consequence that electronic stopping can be measured in transmission experiments even at low velocities. Nuclear stopping, however, is connected with large-angle deflections. Therefore, nuclear-stopping measurements require either a wide-angle detection system, or an extensive data analysis. The most fundamental procedure is to go back to study atomic interaction forces via *differential* scattering measurements.

In the energy range between one order of magnitude below and above the maximum in the nuclear stopping power, where colliding particles approach each other to distances of typically $10^{-1}–10^{1}$ screening radii (if contributing essentially to nuclear stopping), Lindhard's (*14*) or Firsov's (*44*) Thomas-Fermi type screened-Coulomb interaction potentials describe the interaction with reasonable accuracy, as experienced by differential scattering measurements (*45–47*). At lower velocities, a more heavily-screened interaction appears appropriate (*47,48*). Rather direct measurements of the nuclear stopping power have been reported recently by Sidenius (*49*) who used wide-angle detection. Högberg's (*50*) measurements refer to ions scattered at zero angle, and thus produce a partial stopping power mainly.

Occasionally, experimental evidence has been presented for pronounced oscillations in nuclear stopping, similar to Z_1 and/or Z_2 oscillations in electronic stopping (*51*). However, there is also strong evidence against oscillations with an amplitude greater than, say, 10%. In multiple-scattering measurements, the width of a distribution is correlated with $(dE/dx)_n$ (*1*). And in sputtering measurements, the sputtering yield is proportional to the energy deposited in recoiling atoms (*53*). In both types of measurement, the relation of the experimental quantity to $(dE/dx)_n$ is not sufficiently close to allow *absolute* stopping-power measurements. However, the *relative* behavior appears to present compelling evidence against pronounced oscillations of the nuclear stopping power (*13,52*).

For a number of reasons, investigations of nuclear stopping and scattering have concentrated on obtaining universal relations such as stopping power as a continuous function of atomic numbers. A number of systems, in particular pairs of colliding rare-gas atoms (ions), have been studied in much more detail, both with regard to nuclear and electronic energy loss (*47,54*).

RANGES

An approximate measure of the total range of a particle is found by integration of the inverse stopping power, i.e.,

$$R(E) = \int_0^E dE'/(dE'/dx) \tag{3}$$

This expression is often called the c.s.d.a. (continuous slowing down app.) range. Whether or not it is sufficiently accurate, depends on the type of application.

In the high-velocity region, ranges are used to determine the energies of nuclear reaction products, and the range is defined experimentally as the length of a visible particle track. Eq. (3) needs, then, a (negative) correction that accounts for the drop-off in ionization efficiency at low velocity. Moreover, there is a (positive) correction arising from the discrete nature of the energy loss (*14*). In the evaluation of eq. (3) it is necessary to take into account both electronic and nuclear stopping, but possibly only a partial nuclear stopping power may be used in case where tracks containing large-angle deflections have been eliminated in the evaluation of range statistics. Existing tabulations of light-ion ranges (*32,55*), have been based on eq. (1), except that the Z_1^3 correction was not known until quite recently (*28,30,56*). Heavy-ion ranges for high velocities have been tabulated on the basis of semi-empirical stopping powers (*10*).

Heavy-ion ranges in solids at low energies ($1-10^3$keV) have been studied thoroughly since about 1960, both experimentally and theoretically, (*6–8*). A major challenge has been the development of new solid-state devices and materials by ion implantation. Because of the small path lengths of low-energy ions (on the order of 10^1-10^4 Å), and the low ionization efficiency, the length of a particle track can usually not be measured accurately. Instead, the end position of an ion can be determined by means of radioactive-tracer bombardment and subsequent stripping of thin ($\gtrsim$1Å) successive layers (*6*). Alternatively, several recently developed tools of surface analysis (destructive or nondestructive) can be used to trace depth profiles of implanted ions (*9*). Thus, the distribution in *projected range* (the projection is on the surface normal) is measured most often.

Because of the importance of nuclear stopping and scattering, low-energy range profiles may be quite broad. Occasionally, the width exceeds the depth (*7,8,57*). Consequently, a sizable fraction (up to around 30%) of the bombarding ion beam may actually be reflected via one or more large-angle nuclear collisions, provided the projectile ions are much lighter than the target atoms (*57*).

In the low-energy region, measurement of the range is the most important tool for determining stopping powers. Therefore, and because of many applications, the various factors influencing range profiles have been studied in considerable detail. The most intriguing one was the occurrence of highly penetrating tails that eventually lead to the verification of channeling (*58,4*). Other factors are diffusion after slowing-down—which may lead to "super tails" (*59*)—and various artefacts of the subsequent analysis, such as inhomogeneous layer-stripping, etc. The theory of low-energy ion ranges was outlined mainly by Lindhard et al. (*14*), who specified the electronic and nuclear stopping cross sections and the relevant range concepts, and derived a fairly universal range-energy relationship for the c.s.d.a. range. The latter has been tabulated in a convenient form by Winterbon (*60*). Accurate penetration profiles have been constructed later on the basis of the same physical model (*61,62*). The agreement with measured range profiles is quite satisfactory (~10%) in case of many amorphous targets (in particular metal oxides), but considerable discrepancies remain in case of polycrystalline targets (*63*). These discrepancies would most likely be eliminated if there were available a comprehensive theory of ion ranges in single crystals. Such a theory is still missing although specific situations can be handled reasonably well by computer simulation (*64*).

REFERENCES

(Key references with important quotations of earlier work are marked by an asterisk.)

* 1. N. Bohr, The Penetration of Charged Particles through Matter. Mat. Fys. Medd. Dan. Vid. Selsk. **18**, no. 8, pp.1–144 (1948).

* 2. Studies in penetration of charged particles in matter. NAS-NRC publication no. **1133**, pp.1–388 (1964).

* 3. U. Fano, Penetration of protons, alpha particles, and mesons. *Ann. Rev. Nucl. Sci.* **13,** 1 (1963).

* 4. S. Datz, Channeling of charged particles in crystals. These proceedings.

* 5. P. Sigmund, Energy loss of charged particles in solids. In *Radiation damage processes in materials* (C. H. S. Dupuy, Ed.) in press. Available from the author as Monograph no. 73–20, 1–186 (1973).

* 6. J. W. Mayer, L. Eriksson, and J. A. Davies, *Ion Implantation in Semiconductors.* Academic Press, N. Y., 1970.

* 7. P. Sigmund, *Rev. Roum. Phys.* **17,** 823, 969, and 1079 (1972).

* 8. H. E. Schiøtt, in *Interaction of charged particles with solids* (A. N. Goland, Ed.) pp. 6–31 (1973). Natl. Techn. Inform. Serv., 5285 Port Royal Rd., Springfield, Va., 22151.

* 9. J. W. Mayer and J. F. Ziegler, (Eds.), *Ion Beam Surface Layer Analysis.* Elsevier, Lausanne (1974); *Thin Sol. Films* **19,** 1–463 (1973).

*10. L. C. Northcliffe and R. F. Schilling, *Nucl. Data Tables* **7,** 233 (1970).

*11. H. Sorensen and H. H. Andersen, *Phys. Rev.* **B8,** 1854 (1973).

12. J. Lindhard, *Mat. Fys. Medd. Dan. Vid. Selsk.* **34, no. 14,** (1965).

*13. H. H. Andersen, J. Bøttiger, H. Knudsen, P. M. Petersen, and T. Wohlenberg, Phys. Rev. (in press).
14. J. Lindhard, M. Scharff, and H. E. Schiøtt, *Mat. Fys. Medd. Dan. Vid. Selsk.* **33, no. 14,** 1 (1963). Also see pp. 1–16 in ref. (2).
15. J. L. Dehmer, R. P. Saxon, and M. Inokuti. To be published.
*16. J. C. Ashley, *Phys. Rev.* **B9,** 334 (1974).
*17. R. M. Sternheimer, *Phys. Rev.* **145,** 247 (1966).
*18. C. C. Rousseau, W. K. Chu, and D. Powers, *Phys. Rev.* **A4,** 1066 (1971).
19. W. K. Chu and D. Powers, *Phys. Lett.* **40A,** 23 (1972).
20. H. H. Andersen, H. Sørensen, and P. Vajda, *Phys. Rev.* **180,** 373 (1969).
21. J. Lindhard and A. Winther, *Mat. Fys. Medd. Dan. Vid. Selsk.* **34, no. 4,** pp. 1–22 (1964); J. Lindhard and M. Scharff, ibid. **27, no. 15,** pp. 1–31 (1953).
*22. J. A. Ball, J. A. Wheeler, and E. L. Firemen, *Rev. Mod. Phys.* **45,** 333 (1973).
23. E. Bonderup, *Mat. Fys. Medd. Dan. Vid. Selsk.* **35, no. 17,** 1 (1967).
24. M. C. Walske, *Phys. Rev.* **88,** 1283 (1952); ibid. **101,** 940 (1956).
25. J. H. Harberger, R. E. Johnson, and J.W. Boring, *Phys. Rev.* **A9,** 1161 (1974).
26. W. H. Barkas, J. N. Dyer, and H. H. Heckman, *Phys. Rev. Lett.* **11,** 26 (1963).
27. H. H. Andersen, H. Simonsen, and H. Sørensen, *Nucl. Phys.* **A 125,** 171 (1969).
28. J. D. Jackson and R. L. McCarthy, *Phys. Rev.* **B6,** 4131 (1972).
29. J. C. Ashley, R. H. Ritchie, and W. Brandt, *Phys. Rev.* **B5,** 2393 (1972).
30. K. W. Hill and E. Merzbacher, *Phys. Rev.* **A9,** 156 (1974).
*31. A. Brynjolfsson, *Some Aspects of the Interaction of Charged Particles with Matter.* Harvard Coop. Soc., Cambr., Mass. pp. 1–180 (1973).
*32. H. Bichsel, in *American Institute of Physics Handbook*, pp. 8–142, McGraw-Hill, N.Y. (1972).
*33. D. Powers, A. S. Lodhi, W. K. Lin, and H. L. Cox, *Thin Sol. Films* **19,** 205 (1973).
34. E. Bonderup and P. Hvelplund, *Phys. Rev.* **A4,** 562 (1971).
*35. M. D. Brown and C. D. Moak, *Phys. Rev.* **B6,** 90 (1972).
36. T. E. Pierce and M. Blann, *Phys. Rev.* **173,** 390 (1968).
37. O. B. Firsov, Zh. Eksp. Teor. Fiz. **36,** 1517 (1957).
*38. M. Barat and W. Lichten, *Phys. Rev.* **A6,** 211 (1972).
*39. B. Fastrup, P. Hvelplund, and C. A. Sautter, *Mat. Fys. Medd. Dan. Vid. Selsk.* **35, no. 10,** pp. 1–28 (1966); P. Hvelplund and B. Fastrup, *Phys. Rev.* **165,** 408 (1968).
40. F. H. Eisen, *Can. J. Phys.* **46,** 561 (1968); J. Bøttiger and F. Bason, *Rad. Eff.* **2,** 105 (1970).
*41. J. L. Whitton, *Can. J. Phys.* **52,** 12 (1974).
*42. Ref. in sect. 4.5 of ref. (*5*); and F. F. Komarov and M. A. Kumakhov, Zh. Tekh. Fiz. **43,** 1329 (1973); V. S. Kessel'man, ibid. **42,** 1161 (1971); B. E. Baklitsky, E. S. Parilis, and V. K. Ferleger, *Rad. Eff.* **19,** 155–160 (1973).
*43. D. J. Bierman, W. C. Turkenburg, and C. P. Bhalla, *Physica* **60,** 357 (1972).
*44. O. B. Firsov, *Zh. Eksp. Teor. Fiz.* **33,** 696 (1957).
*45. H. H. Andersen, J. Bøttiger, and H. Knudsen, *Phys. Rev.* **A7,** 154 (1973).
46. A. v. Wijngaarden and W. E. Baylis, *Phys. Rev.* **A7,** 937 (1973).
*47. P. Loftager and G. Claussen, MIT Conf. ICPEAC, Book of Abstracts, 518 (1969); V. V. Afrosimov, Yu. S. Gordeev, V. K. Nikulin, A. M. Polyanskii, and A. P. Shergin, *Zh. Eksp. Teor. Fiz.* **62,** 848 (1972).
*48. P. Sigmund & U. Lillemark, *Physica* **71,** 258 (1974).
49. G. Sidenius, *Mat. Fys. Medd. Dan. Vid. Selsk.* **39,** in press.
50. G. Högberg, *Phys. Stat. Sol.* **(b)48,** 829 (1971).
51. R. S. Nelson, *Phys. Lett.* **28A,** 676 (1969) C. Foster, M. W. Thompson, and I. M. Wilson, *J. Phys.* **B5,** 1332 (1972); G. W. Neilson, B. W. Farmery, & M. W. Thompson, *Phys. Lett.* **46A,** 45 (1974).
52. H. H. Andersen and H. L. Bay, *Rad. Eff.* **13,** 67 (1972).
53. P. Sigmund, *Phys. Rev.* **184,** 383 (1969).
*54. Q. C. Kessel and B. Fastrup, *Case Stud. Atom. Phys.* **3,** 137 (1973).
55. W. H. Barkas and M. J. Berger, pp. 103–172 in ref. (*2*).
56. J. C. Ashley, R. H. Ritchie, and W. Brandt, *Phys. Rev.* **8,** 2402 (1973).
*57. P. Sigmund, in *Phys. Ionized Gases* (M. Kurepa, Ed.) Inst. Phys. Belgrade, pp. 137–195 (1972).
*58. G. R. Piercy, J. A. Davies, F. Brown, and M. McCargo, *Phys. Rev. Lett.* **10,** 399 (1963).

*59. E. V. Kornelsen, F. Brown, J. A. Davies, B. Domeij, and G. R. Piercy, *Phys. Rev.* **136A,** 849 (1964).
60. K. B. Winterbon, Report AECL-3194, pp. 1–37 (1968). Chalk River Nucl. Lab.
61. K. B. Winterbon, P. Sigmund, and J. B. Sanders, *Mat. Fys. Medd. Dan. Vid. Selsk.* **37, no. 14,** (1970).
62. K. B. Winterbon, *Rad. Eff.* **13,** 215 (1972).
*63. J. L. Whitton, in *Phys. Ionized Gases* (M. Kurepa, Ed.) Inst. Phys. Belgrade, pp. 283–304 (1972).
*64. D. P. Jackson and D. V. Morgan, *Contemp. Phys.* **14,** 25 (1974).

Channeling: A Tool for the Study of the Interactions of Energetic Particles Penetrating Solids[1,2]

Sheldon Datz
Oak Ridge National Laboratory
Oak Ridge, Tennessee 37830, U.S.A.

INTRODUCTION

The effect of the crystal lattice on the motion of charged particles has been intensively studied by many groups since the modern discovery of channeling in about 1962. The phenomenon is sufficiently well understood to permit its application in many branches of science and technology ranging from the measurement of nuclear lifetimes to the location of foreign atoms and radiation-induced effects in solids. A recent review by Gemmel (*1*) contains over 700 references and a mere listing of the subjects studied would take more time than is here allocated to the subject. Instead I will just describe the concept of channeling and show its application to the determination of interatomic potentials and stopping powers (*2*) and to the definition of atomic states of fast ions penetrating solids (*3*,*4*).

The general concept of channeling is illustrated in Fig. 1. Consider a projectile atom aimed for a hard collision with an atom contained in a closely packed atomic row in a crystal. The projectile will be slightly deflected by the repulsive potential of the first atom, again by the second, *et cetera*. The final collision will thereby be considerably softened, and, depending upon the angle with respect to the atomic *row*, it may not collide at all with the original target atom. This is the basis of channeling; i.e., within an angle with respect to an atomic row or plane, determined by the interatomic potential and spacing of atoms, the particle will be deflected by a continuum potential made by properly summing the atomic potentials. These angles can be on the order of several degrees and, as can be seen in Fig. 2, a considerable part of all directions of incidence in the crystal can in fact be effective in channeling.

The primary result of channeling is to prevent small impact parameter atomic collisions from occurring. This leads to the essential elimination of nuclear stopping and restricts electronic interactions to only valence and conduction electrons.

The group of experiments which we will discuss today will center on the measurement of energy loss of channeled ions. The setup is pictured in Fig. 3 and consists simply of an ion beam, an orientable thin single crystal and an energy-sensitive particle detector which subtends a small solid angle in the forward direction.

[1]Congress presentation E–2–2.

[2]Research sponsored by the U.S. Atomic Energy Commission under contract with the Union Carbide Corporation.

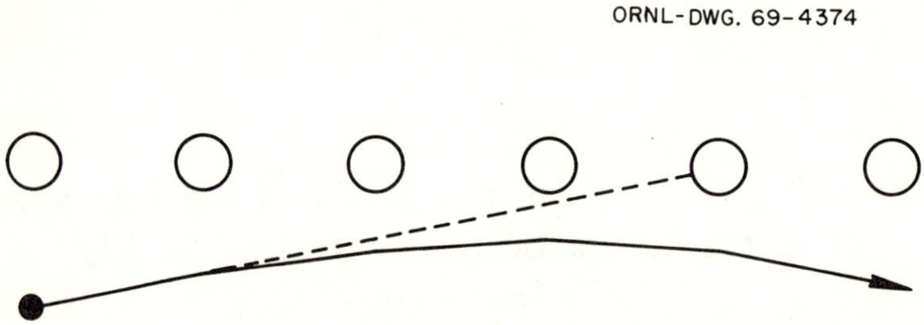

FIG. 1. Schematic showing the undeflected path of an ion in collision with a row atom compared with the deflected path due to a row repulsive potential.

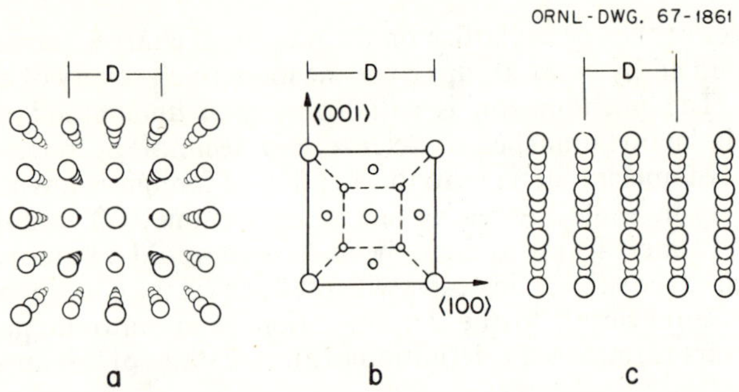

FIG. 2. Several views through a face-centered cubic crystal illustrating axial and planar channels.

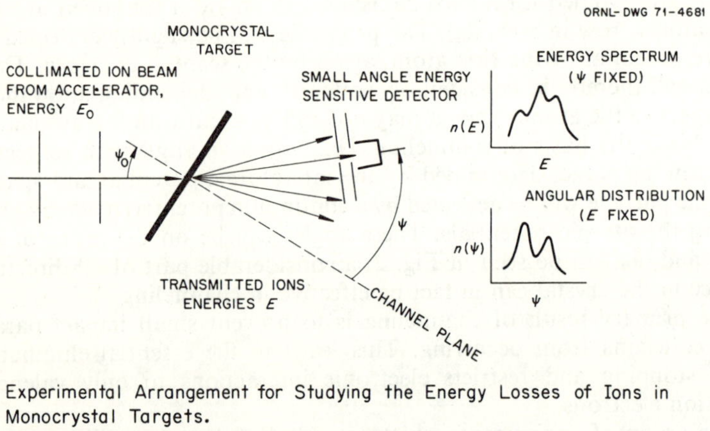

FIG. 3. Outline of the experimental arrangements for studying the energy losses of ions in thin single crystal targets.

ATOMIC INTERACTIONS FROM PLANAR CHANNELING MEASUREMENTS

If we carry out this experiment in a planar channeling direction we observe spectra of the type shown in Fig. 4. Here we see discrete peaks in energy loss. The discreteness of the energy loss can be understood from the following model.

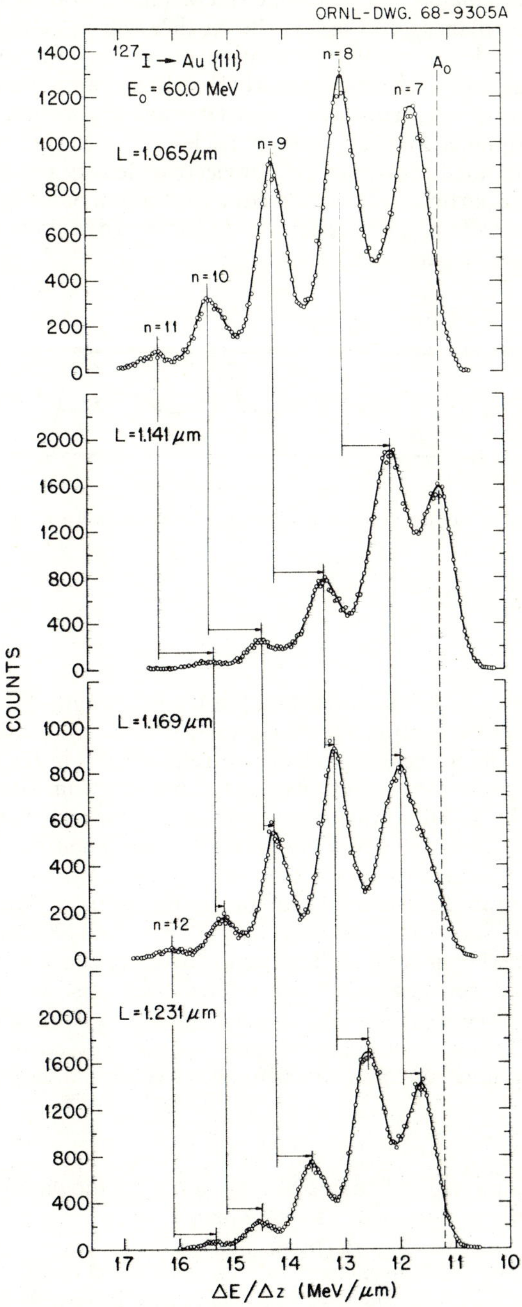

FIG. 4. Energy spectra observed in ^{127}I ions transmitted through (111) channels of an Au crystal as a function of the pathlength.

An ion enters a planar potential which has a minimum at the center and rises toward the atomic planes (see Fig. 5). The particle will be deflected toward the center and will undergo oscillatory motion on its way through the crystal. Particles entering far from the midplane will have higher amplitude and, since the potential is an harmonic, will have a shorter wavelength than those entering closer to the midplane. Further, these particles will suffer greater energy loss because they penetrated regions of greater electron density and they do so more often. If all exiting particles were observed, a continuum of energy loss would be observed. But if we accept only particles in the forward direction we observe *only* particles with wavelengths such that

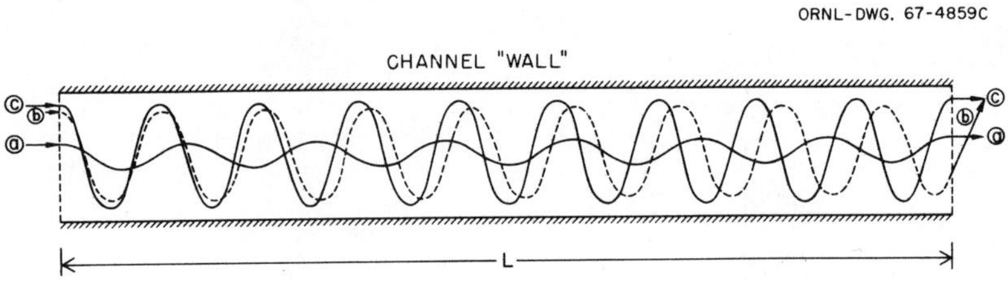

FIG. 5. Trajectories for three entrance points. Particles a and c will reach a detector in line with the beam.

$$n\lambda = L \tag{1}$$

or

$$(n+1/2)\lambda = L$$

where n is the number of oscillations and L is the pathlength. For reasons given elsewhere (*2*) the half-integral solutions do not lead to experimentally observable groups because of imperfections (mosaic spread) present in the crystal.

The wavelength of the oscillations can be obtained from information of the type shown in Fig. 4. The position marked A_0 corresponds to the minimum-energy-loss group. The particles in this group have been confined to trajectories near the channel midplane where the potential is almost harmonic. The peaks of the intermediate groups move with changing path-length as predicted. Plots of the peak positions versus $\Delta E/\Delta z$ for 60 MeV ^{127}I in the (100) and (111) planar channels of Au are shown in Fig. 6. The vertical distance between lines represents the energy spacing between groups at a fixed thickness. To first order, the horizontal interval between the n and (n+1) groups is the wavelength of the particles in the n group at the corresponding value of $\Delta E/\Delta z$ on the abscissa.

A better and more convenient measure has been found to be the oscillation "frequency"

$$\omega \equiv n(2m)^{1/2}v/L \tag{2}$$

When properly corrected, all values of n should yield the same relationship of initial stopping power E_0 to oscillation frequency. An example of this is shown for 60 MeV I ions in Au in Fig. 7. The linearity of the curves obtained was at first surprising (and quite useful in easing data treatment). Although we have used the case of 60 meV I in Au, the linearity of the ω vs (dE/dz) curves is by no means unique and has been found for all systems studied (see Table 1). Hence, we have the empirical relationship $(-dE/dz)_{E=E_0} = \alpha + \beta\omega$.

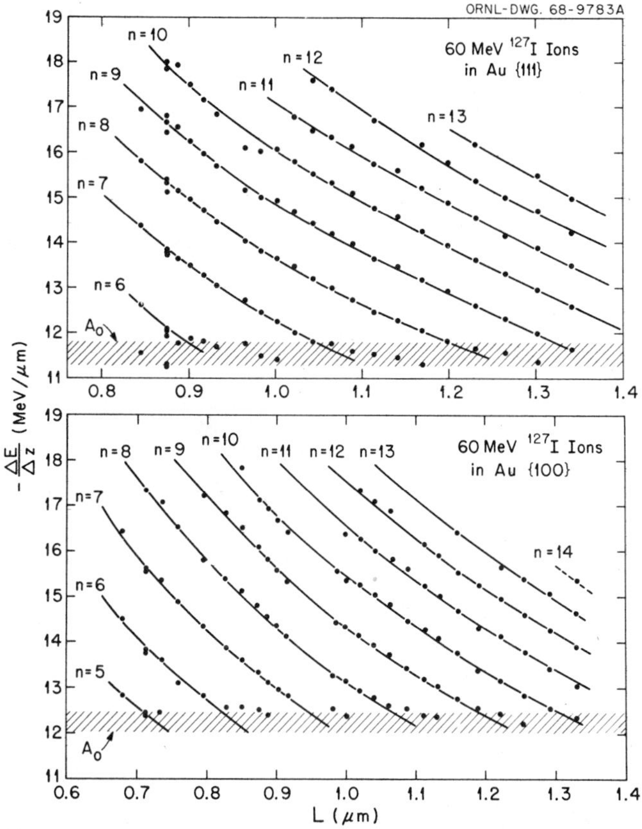

FIG. 6. Energy loss at the group peak positions for 60 MeV I ions channeled in (111) and (100) planes of Au as a function of pathlength. The symbol n designates the number of particle oscillations along the path.

Robinson (*5*) developed a method for the extraction of atomic and planar potentials from the data obtained in these experiments.

$$V_o(\bar{x}) = 4\pi\kappa\rho l \int_x^\infty rV(r)dr \tag{3}$$

where V(r) is the interaction potential between an ion and a lattice atom separated by a distance r, ρ is the atomic density of the target, l is the halfwidth of the planar channel, x is the length of the normal from the ion to the plane of lattice atoms, and κ is a factor allowing for the possibility that there may be parallel channels of different widths in some crystal structures (e.g., Si). The channeled ions move between a pair of planes so that the planar channel potential (see Fig. 2)

$$V_2(x) = V_1(l + x) + V_1(l - x), -l \leqslant x \leqslant l \tag{4}$$

The ion oscillates with a frequency

$$\omega^{-1} = 2 \int_0^{x_m} [V_2(x_m) - V_2(x)]^{-1/2}dx, \tag{5}$$

where x_m is the oscillation amplitude. The model stopping power is

$$S(x,E) = s_0 + s_1[\sigma(x) - 1] \tag{6}$$

where s_0 and s_1 depend on the ion energy. The spatial dependence of the stopping

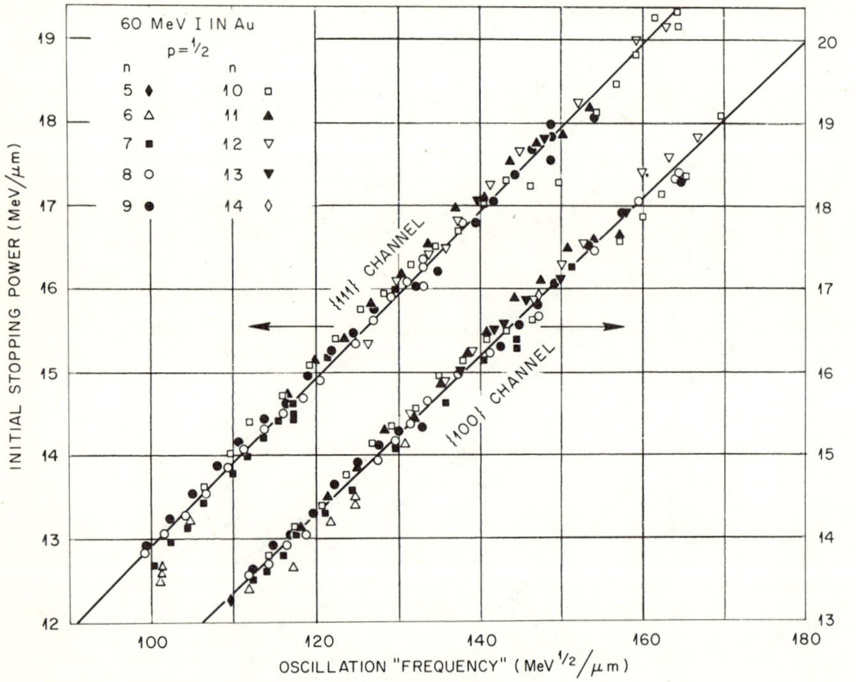

FIG. 7. The dependence of the stopping power of 60 MeV I ions on their transverse oscillation frequencies in two planar channels of Au.

TABLE 1.
Curvature Parameters, Screening Lengths and Ionic Charges Deduced for Various Ions in Au, Ag, and Si

Target	Ion	Energy (MeV)	Curvature Parameter (eV/Å^3) y_{111}	y_{100}	y_{111}/y_{100}	Screening Length (l/b) Å	Derived Ion Charge Z_2
Au	^{4}He	3	64	108	0.59	0.31	2.0
	^{16}O	10	154	253	0.61		4.8
	^{127}I	15	837				26.
		21.6	915	1600	0.57		30.
		60	1288	2077	0.62		40.
Ag	^{127}I	21.6	833	1420	0.59	0.31	33.
		60	1042	1696	0.62		41.
			y_{111}	y_{110}			
Si	^{1}H	0.4	14.5	30.1		0.45	0.9
			10.9				

power is described by the function $\sigma(x)$, normalized so that $\sigma(0)=1$. The value of s_o, the stopping power at the channel center, may be taken directly from the experiments by measuring the minimum energy loss (leading edge) of the spectrum. The stopping power of the channeled ions is found by averaging Eq. 6 over the oscillatory motion, weighting each portion of the channel cross section by the time which the ion spends there.

Empirically, it has been found that the data are well represented by the straight line (see listing in Table 1)

$$(-dE/dz)_{E=E_o} = \alpha + \beta\omega \tag{7}$$

as illustrated by Fig. 7. Because of the apparent generality of this observation, it has been adopted as an empirical first principle, the consequences of which are to be explored. When Eq. 6 is averaged over the motion of the ion and the result is compared with Eq. 7, it is found that as long as the number of half-oscillations executed by the ion is integral, the intercept and slope of the straight line are given by

$$\alpha = s_0 - s_1, \tag{8}$$

$$\beta = 2s_0 \int_0^{x_m} \sigma(x)\,[V_2(x_m) - V_2(x)]^{1/2}\,dx \tag{9}$$

From Eq. 9 it is found that, as long as β is constant,

$$\sigma(x) - d/dx\,\{[2/V_2''(0)][V_2(x) - V_2(0)]\}^{1/2}, \tag{10}$$

where the primes represent differentiation with respect to x.

The experimental data may be used to evaluate the curvature parameter,

$$\begin{aligned} y &\equiv 2\pi^2(s_0 - \alpha)^2/\beta^2 \\ &= V_2''(0)/l = 2V_1''(l)/l \\ &= -8\pi\rho\kappa[V(l) + lV'(l)] \end{aligned} \tag{11}$$

where the empirical form is given on the first line and three equivalent theoretical forms are given on the following lines. Given data for a sufficient number of channels with distinct values of l, the potential itself can be evaluated by integration of the last of Eqs. 11. For more limited data, the parameters of an assumed potential function can be evaluated.

Experimental curvature parameters are listed in Table 1 for each combination of ion, energy, and channel. The values are in no way dependent on assumptions about the potential function. The values of y are strongly dependent upon the ion and its energy, showing that the planar-channel potential must also depend on these variables. However, the ratio of the values for the two channels appears to be essentially independent of the ion and its energy within experimental error ± 10%. If the interaction potential between the channeled ion and the individual lattice atoms is of the screened Coulomb type, the curvature parameter is

$$y = -8\pi\rho Z_1 Z_2 e^2 b\ \phi'(bl), \tag{12}$$

where Z_1e and Z_2e are the (effective) nuclear charges of the two particles, $\phi(bl)$ is the screening function, and b is a constant. The ratio of the values for two channels determines the value of the screening constant, b. The constancy of the ratio y_{111}/y_{100} for Au in Table 1 implies the similar constancy of b, which is thus shown to be a property of the gold crystal. The values of b shown in Table 1 are therefore derived from the weighted mean values of the curvature parameter ratios of all probing ions. Physically, the implication is that the nuclear charge of the ion is always screened by the same number of electrons, whatever its position in the channel. As it approaches a target atom, it never gets so close that this screening is reduced significantly. Thus, it acts always as a simple charged particle which does not much perturb the medium. That is, all three ions may be considered as test charges. The strong variation of the experimental curvature parameters with the ion and its energy results merely from changes in the magnitude of the ionic charge with these variables.

After consideration of numerous forms of the potential energy it was shown by Robinson (Fig. 8) that the Hartree potential for the lattice atom describes the interaction extremely well in the region of impact parameters sampled by planar channeling measurements.

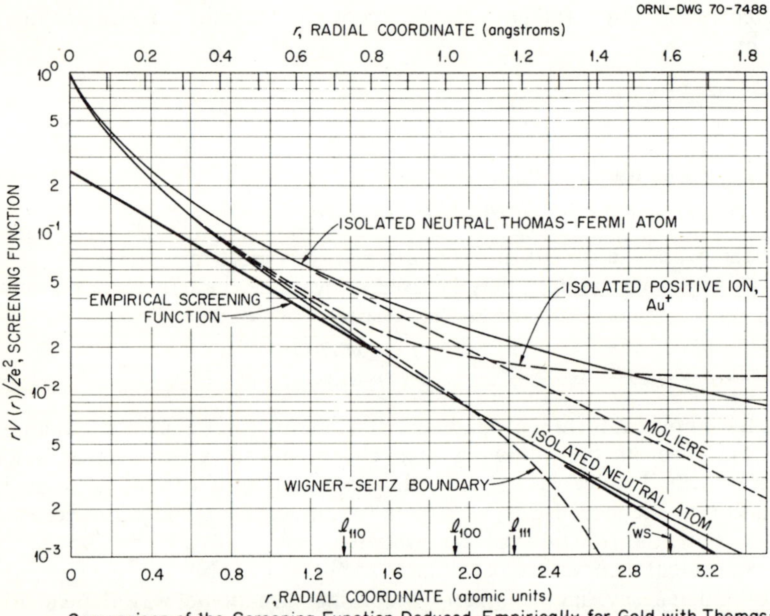

Comparison of the Screening Function Deduced Empirically for Gold with Thomas-Fermi and Relativistic Hartree Calculations.

FIG. 8. Comparison of a screening function deduced from planar channeling experiments in Au with Thomas-Fermi and relativistic Hartree calculations.

HYPERCHANNELING

From the interatomic potentials determined by this method, the effective potentials of the atomic rows ("strings") in the lattice may be developed. Such a potential surface is shown in Fig. 9. Since solid state effects on the ion-atom interaction potentials occur at large distances from the atoms, it is desirable to investigate the largest impact parameter events possible. In the planar channeling experiments discussed so far, the maximum impact parameters are determined by the planar spacings. It is possible by doing experiments in an axial channeling geometry to probe even larger impact parameter interactions provided one can study ions confined to only one particular axial channel. In planar channeling, a channeled ion travels in a single channel between adjacent atomic planes, whereas in ordinary axial channeling the majority of channeled ions wander from one axial channel to another. Axially channeled ions of a limited class, however, are expected to have transverse energies less than the potential barrier between adjacent rows, and accordingly should be constrained to travel through the crystal in a single axial channel. This phenomenon, called hyperchanneling or proper axial channeling, was recognized in the early computer calculations of Robinson and Oen (*6*) and was observed by Eisen (*7*) as a low-energy loss "tail" in axial energy loss distributions. In recent experiments (*8*) hyperchanneling was observed as a distant well-separated energy loss group, and systematic investigations of the effect indicate that solid state effects may in fact be present.

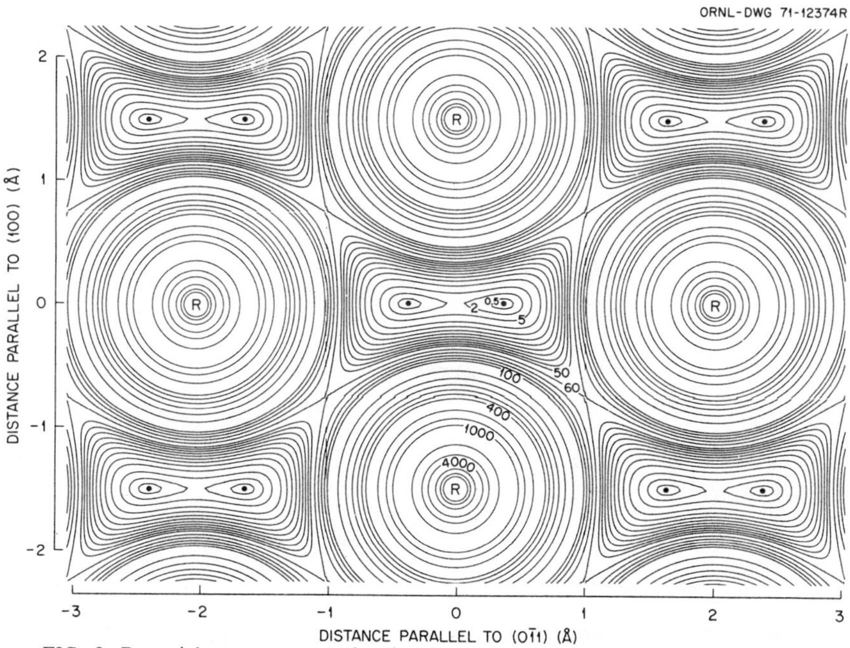

FIG. 9. Potential energy contours for I ions in the continuum potentials of (011) rows (R) of Ag atoms. All values are in eV and are relative to the minimum (●), which has an absolute value of 72 eV.

Hyperchanneled ions are expected to have at least two distinctive characteristics, a smaller acceptance angle than ordinary axial channeling and lower energy losses because they sample larger average impact parameters than ions which wander from channel to channel. This effect was studied by Appleton et al. (*8*) using a highly collimated beam of 21.6 MeV I ions onto a 0.60 μm thick (001) Ag crystal. Energy spectra measured with the beam incident at various angles to (011) in (111) are shown in Fig. 10. The prominent group on the right, lying within a narrow angular range and centered on a dE/dx value of 3.15 MeV/μm, is due to hyperchanneling. The major peaks in the energy loss distributions in Fig. 10 taken at ±0.45 relative to (011) are nearly identical to those obtained at angles 5 to 10° from (011) in (111), i.e., normal planar channeling. As the incidence angle with respect to (011) is decreased, the energy loss distributions in Fig. 10 lose the characteristic (111) shape at about 0.30°, and a new group of ions with significantly lower energy losses is dominant within ~0.12° of (011). This low loss group has the two characteristics expected of hyperchanneled ions, i.e., those ions have been contained in the closed equipotentials of Fig. 9.

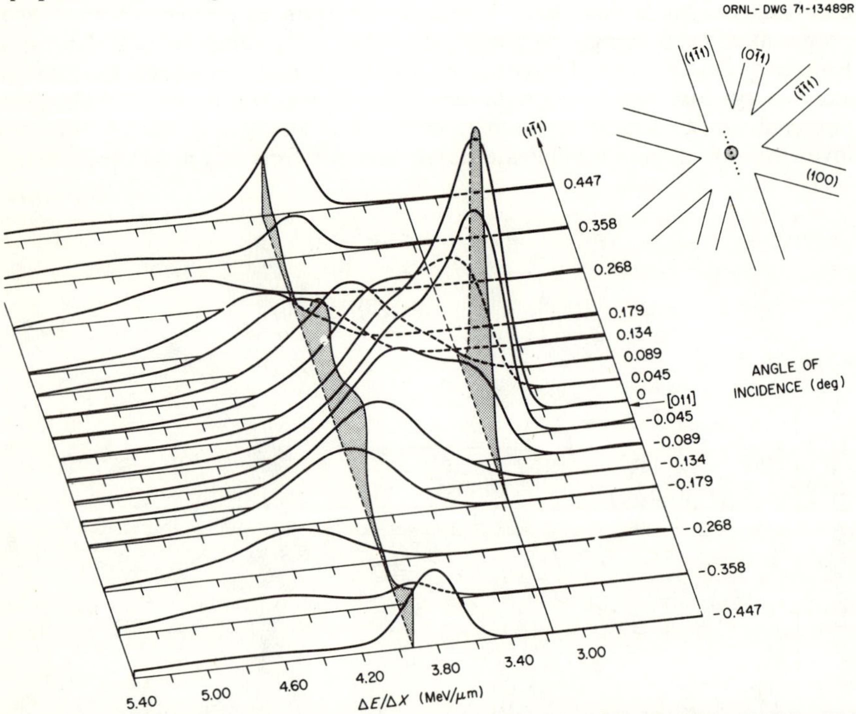

FIG. 10. Normalized energy loss distributions at various incidence angles to (011) for 21.6-MeV I ions transmitted through 0.85 μm of Ag. The insert shows the angular extent of the measurements (dotted line) and the hyperchanneling region (shaded central spot).

The additional group seen at intermediate angles arises from particles which have spent their time in the region of the potential maximum between rows. From an analysis of the transverse energies required to remove particles from the hyperchanneled group it has been shown that the free atom Hartree potential predicts too low a saddle point between rows and that, in fact, a correction such as that given in the Wigner-Seitz model is necessary to describe the data.

CHARGE EXCHANGE OF CHANNELED IONS

When fast ions enter a medium, they undergo a series of charge capture and loss collisions. As the charge on the ion increases, the probability for electron capture increases relative to electron loss and, ultimately, an "equilibrium" charge-state distribution is attained. The most probable charge state is generally characterized by equal cross sections for electron capture and loss. Typically, this dynamic equilibrium stage is easily achieved within a hundred atom layers in a polycrystalline solid medium. However, when the ion is channeled in a thin single crystal, it is constrained to large impact-parameter collisions with the atoms in the solid, and the cross sections for capture and loss of electrons by the ions may be affected by the density and velocity of the electrons they encounter.

The exit charge-state distributions (CSDs) obtained from injection of 40-MeV oxygen ions in initial charge states ranging from 5+ to 8+ into a <110> axial-channeling direction in a 3300 Å thick Au crystal are shown in Fig. 11, together with the CSD obtained from injection in a "random" direction (*3*). The random distribution (dashed curve) is representative of charge-state equilibrium since it is independent of a pathlength (.143 to .663 μm) in the crystal and independent of input charge state (6+, 7+, or 8+). The channeled-ion CSDs are clearly non-equilibrium cases. Since the 8+ fraction is enhanced for all initial charge states, it is immediately obvious that the equilibrium CSD for channeled ions will contain a higher fraction of higher charge states than the random case. Moreover, it is clear that some of the charge-changing cross sections for the channeled ions in attaining equilibrium must be quite small, i.e., if we assume that equilibrium is attained in *ca* 10 mean-free paths for charge exchange, then charge-changing cross sections of $<<\sim 5\times 10^{-18}$ must be involved if equilibrium is not achieved in .5 μm with target densities of $\sim 5\times 10^{22}$ atoms/cm^3.

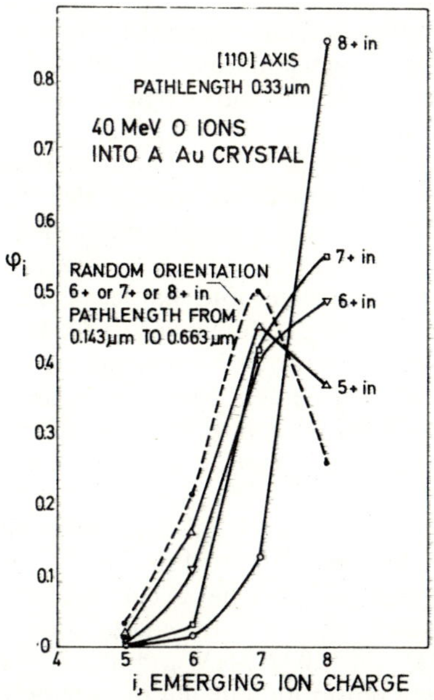

FIG. 11. Emergent charge state distributions obtained for random and (110) channeled oxygen ions with an input energy of 40 MeV. Input charges and pathlengths are indicated.

Since the emergent CSDs for channeled ions are clearly far removed from equilibrium, it should be possible to adduce the relevant charge-changing cross sections from variations in CSD with change in pathlength and, in some cases, from the additional information obtained from different input charge states. For planar channels, a wide range of pathlengths is available within a single crystal by simply tilting the plane of the crystal with respect to the beam direction. The points shown in Fig. 12 were obtained for channeling of 40-MeV 0^{8+} ions varying in pathlengths along a (100) planar channel.

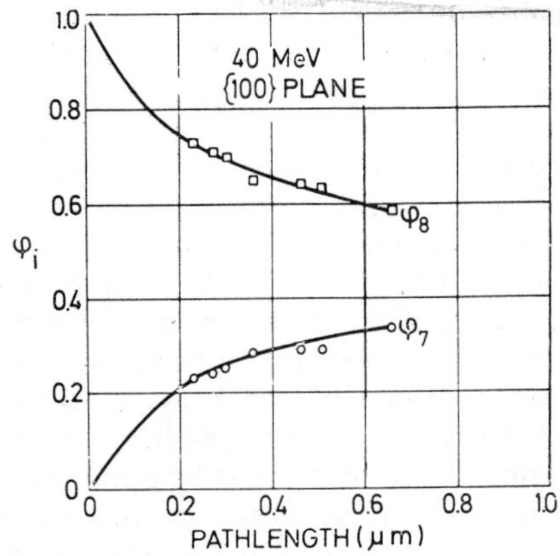

FIG. 12. Emergent fractions of 8+(ϕ_8) and 7+(ϕ_7) obtained from injection of 40 MeV 0^{8+} ions into (100) planar channel as a function of pathlength through the channel.

Such studies have shown that the relevant charge-changing cross sections are on the order of 10^{-19} cm^2, which is at least two orders of magnitude lower than that expected from atomic collisions with particles of medium or large atomic number, but comparable to those observed for helium. This again reflects that channeled ions cannot interact with strongly bound electrons.

With such small cross sections operative it is clear that a large fraction of ions with initial charges of 7+ and 8+ pass through the crystal with *no* charge exchange. Hence, for the first time we are able to measure the stopping powers for ions in solids as a function of their ionic charge (*4*). Such measurements have been carried out and some results are shown in Fig. 13. Here beams of 27.8 MeV 0^{6+}, 0^{7+}, and 0^{8+} have been injected into a silver crystal channel, and the energy loss spectrum of that portion of the emergent beam which is in the same charge state as the initial beam is recorded. A major portion of these particles have not undergone charge exchange and it can be seen that the peaks lie at different energy losses.

The most probable energy loss for all permutations of input and output charge are plotted in Fig. 14 vs the square of the input charge state. Considering the cases for which no charge exchange has occurred ($\Delta q = 0$), the energy loss is accurately proportional to the square of the charge state (i.e., $s = kq^2$ where $k =$ .024 MeV/μm at 40 MeV and .032 MeV/μm at 27.8 MeV). These k values correspond to the stopping power anticipated for a hypothetical singly charged channeled oxygen ion.

Other experiments utilizing these properties of channeled ions are presently under way and promise further detailed information on the penetration of energetic particles into solids.

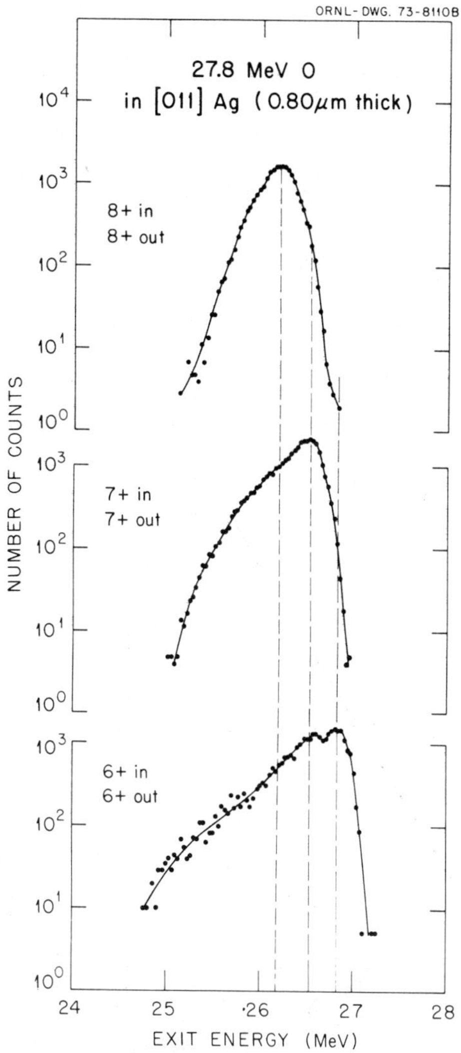

FIG. 13. Emergent energy spectra for 27.8 MeV 0 ions channeled in Ag (011), 0.8 μm pathlength, $\Delta q = 0$.

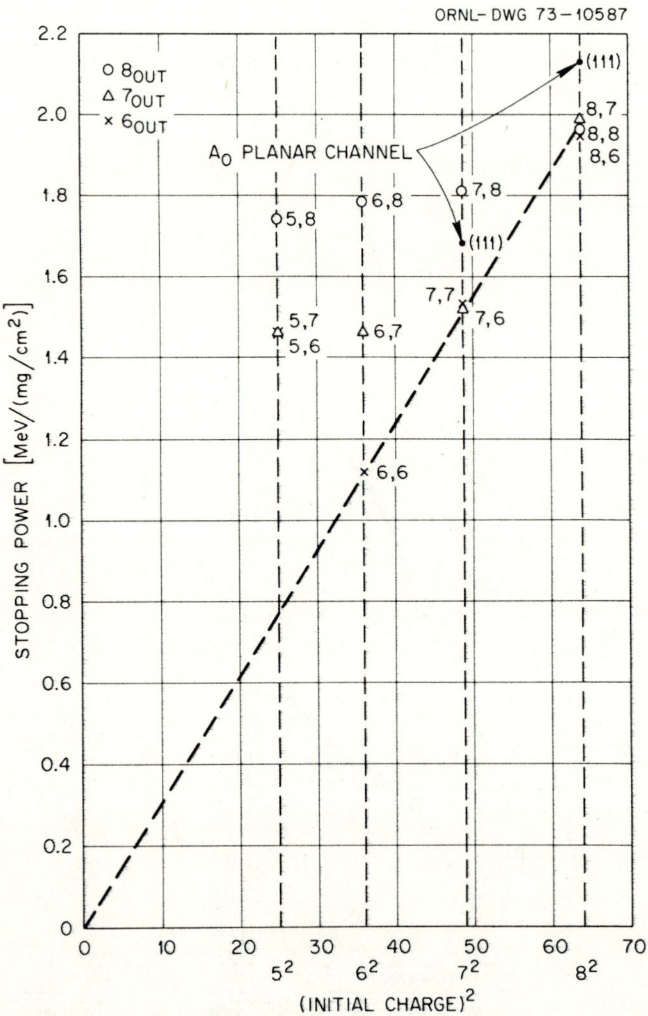

FIG. 14. Stopping powers of 27.8 MeV 0 ions in the (011) axis of Ag for various charge combinations q_{in} and q_{out}. (The points labeled (111) are the A_o group losses in planar (111) channels).

REFERENCES

1. D. S. Gemmel, *Rev. Mod. Phys.* **46**, 129 (1974).
2. S. Datz, B. R. Appleton and C. D. Moak, in *Channeling* (D. V. Morgan, Ed.) pp. 153–180. Wiley, New York, 1974.
3. S. Datz, F. W. Martin, C. D. Moak, B. R. Appleton and L. Bridwell, *Radiat. Eff.* **12**, 163 (1972).
4. S. Datz, B. R. Appleton, J. A. Biggerstaff, M. D. Brown, H. F. Krause, C. D. Moak, and T. S. Noggle, in *Atomic Collisions in Solids V* (S. Datz, B. R. Appleton and C. D. Moak, Eds.) pp. 63–79. Plenum Press, New York, 1975.
5. M. T. Robinson, *Phys. Rev.* **B4**, 1461 (1971).
6. M. T. Robinson and O. S. Oen, *Phys. Rev.* **132**, 2385 (1963).
7. F. H. Eisen, *Phys. Lett.* **23**, 401 (1966).
8. F. H. Eisen, *Can. J. Phys.* **46**, 561 (1968).
9. B. R. Appleton, C. D. Moak, T. S. Noggle and J. H. Barrett, *Phys. Rev. Lett.* **28**, 1307 (1972).

Transmission of Fast Molecules Through Solids[1]

Joseph Remillieux
Institut de Physique Nucleaire
Universite Claude Bernard, Lyon, France

INTRODUCTION

Let us consider a beam of protons, in the MeV energy range, bombarding a thin foil, such that only a very small fraction of the incident energy is lost in penetration. The proton-solid interaction is now well understood in amorphous solids and in crystals under channeling conditions. This paper describes observed changes in these interactions when molecular beams composed of H_2^+ or H_3^+ molecular ions are used. We compare interactions corresponding to incident beams of H^+, H_2^+ and H_3^+ ions of the same velocity.

The first part is a review of the anomalies observed with molecular beams in stopping power, angle and energy distributions inside and beyond the foil, channeling properties, charge-exchange processes and beam-foil excitation. The second part describes the phenomena of molecular transmission by which H_2^+ ions can emerge in a bound state from a foil. Such transmission has been studied extensively with very thin gold crystals in channeling configurations. An interpretation of this phenomenon requires a molecular recombination between two protons resulting from the breakup of the incident ions as they emerge from the foil.

The study of the interaction of fast molecular systems with solids promises to be an interesting tool in atomic-collision physics in that the phenomena require close space and time correlations between ion-atom collisions.

ANOMALIES OBSERVED WITH MOLECULAR BEAMS

Molecular States in the Incident Beam

In the following sections we will assume that the incident beam is composed of H_2^+ ions. The case of H_3^+ incident beams will be discussed in a separate section.

In a beam of energy E_1 an incident H_2^+ molecular ion is described by its vibrational level v, its internuclear distance R_0 and the angle θ_0 between the internuclear axis and the beam axis. With usual electrostatic accelerators the time of flight between the accelerator source and the target is sufficiently short ($\sim 10^{-7}$ sec) to assume that the (v, R_0, θ_0) distribution is the same at the entrance surface of the target as inside the source; the angles θ_0 are randomly distributed and the population P(v) of the 19 vibrational levels can be calculated from the Franck-Condon principle. The validity of this calculated population has been

[1] Congress presentation E-2-3.

demonstrated experimentally by Walters et al. (*1*). From the population P(v), we can calculate the distribution in internuclear distances $N(R_0)$ shown in Fig. 1. This distribution, with a mean value $\bar{R}_0 \sim 1.29$ Å and a width 0.9 Å FWHM, is clearly different from the distribution describing the ground state centered at $R_0(v = 0) = 1.08$ Å.

The beam currents considered are always so low that only one H_2^+ molecule is in the foil at any one time.

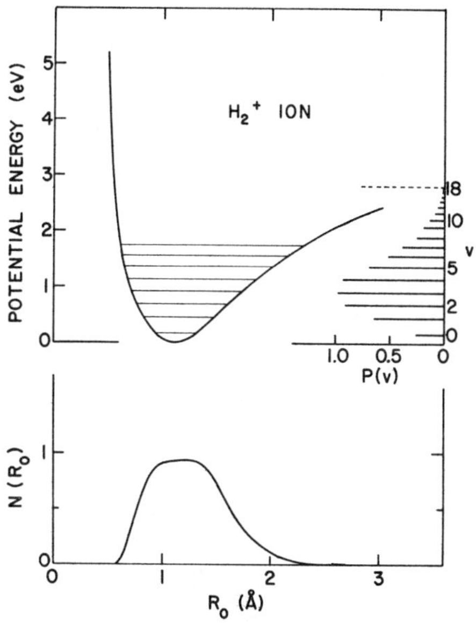

FIG. 1. Distributions of vibrational excitations and internuclear distances of the incident H_2^+ ions.

Molecular Effects Inside the Solid

In a large-angle collision with a target electron the binding electron of the incident H_2^+ ion is stripped away. This breakup occurs in the first atomic layers of the target in so short a time that the instantaneous kinematic state of the two protons, we will refer to as the 'breakup products', is still that of the vibrational level v. A Coulomb explosion begins and a momentum p_m of molecular origin is transferred along the internuclear axis to each proton,

$$\vec{p}_m = \vec{p}_v + \vec{p}_c(R_0,t).$$

p_v is the vibrational momentum of the incident ion at the breakup time; p_c is the momentum transferred after a time t to each proton by the mutual Coulomb repulsion.

Inside the solid the momentum of each proton of mass M_1 becomes $\vec{p} = \vec{p}_m + \vec{p}_1$, where $p_1 = \sqrt{E_1 M_1}$ is the incident momentum of each proton.

From the distributions P(v) and $N(R_0)$ shown in Fig. 1, we can calculate the relative energies associated with the vibration and the repulsion processes. In the center of mass of the two protons the mean energy due to the vibration of the incident molecule is < 1 eV/proton, and the total Coulomb explosion liberates on an average 6.2 eV/proton; although the conversion of the Coulomb energy to kinetic energy is rather slow (at the time scale of the time of flight of the proton in the solid), p_c quickly becomes greater than p_v after the breakup event. We note that the energy associated with the molecular rotation of the incident ions is negligible in this study.

Molecular effects inside the solid can be separated into two groups: first, kinematic effects determined by the longitudinal component $p_m\cos\theta$ and the transverse component $p_m\sin\theta$ of the momentum p_m; second, effects of mutual perturbation between the two protons when they are interacting with electrons of the solid.

Kinematic Effects

Longitudinal momentum (energy distribution)—In the range of beam energies considered, the total momentum of each proto after the breakup event becomes

$$p \sim p_1 + p_m\cos\theta.$$

The molecular component shifts the energy of each proton by the value

$$\Delta E \sim (p_1 p_m / M_1)\cos\theta, \text{ when } p_m \ll p_1.$$

Molecular energies, although in the eV range in the center of mass of the moving proton pair, can induce in the laboratory system energy shifts in the keV range, as long as $E_1 \sim 1$ meV. This energy shift has been observed by the line shape of nuclear resonances, mostly (p,γ), either in the fundamental study of the nuclear reactions, or by using these as energy-calibration standards for particle accelerators (*1,2*). In any case, observed resonances with incident molecular beams are broadened.

Transverse momentum (angular distribution—channeling)—The transverse component $p_m\sin\theta$ of the molecular momentum determines the angular distribution of the breakup products relative to the beam direction. This component takes a particular importance in channeling configurations, in which the 'transverse energy' of the channeled particle relative to the axial or planar channel determines the trajectory, and the stability of the trajectory of the particle inside a crystal. The additional transverse energy of molecular origin is of the same order of magnitude as the critical transverse energies which determine the angular acceptance of the channels. The modification of the transverse energy distribution of channeled protons due to molecular effects has been observed in different ways.

In backscattering experiments, Caywood et al. (*3*) measured shallower and narrower axial and planar dips. Eisen and Uggerhøj (*4*), measuring the scattering yield from interstitial dopants, observed a reduction of the 'flux peaking' effect. Golovchenko (*5*), measuring the angular distribution of channeled particles emerging from a thin crystal, observed a modification of the 'donut pattern.' The dechanneling rate of molecular breakup products is observed to be faster than that of mononuclear projectiles (*6,7*). This result is explained in terms of 'alignment' of the two proton internuclear axis with the channeling direction: the

Coulomb explosion being prevented along the transverse direction by the planar or axial potential results in an increase of the longitudinal distance between the two protons.

Electronic Interactions

The close distance between the two protons after the breakup perturbs the individual interactions of each proton with the electrons of the target, particularly the energy-loss process and the charge-exchange process.

Stopping power—Assuming a Z_1^2 dependence of the stopping power with the effective charge Z_1 of the incident ion, one should expect that the stopping power per proton increases by a factor of two when the distance R between the two protons inside the solid decreases from very large to very small values. In fact, for internuclear distances in the Å range, the energy loss suffered by each of the two protons, as a function of R and θ, in close collisions and in distant collisions has to be considered separately.

This excess of energy loss of the breakup products has been observed recently by Dettmann et al. (*8*) They found, for example, that for a 270-keV H_2^+ ion, bombarding a 500-Å thick carbon foil, the energy loss per proton is 30% higher than for the corresponding H^+ beam.

Charge exchange—The close distance between the two protons when they emerge from the foil can perturb the processes of electron capture near the exit surface.

i) Emerging neutral fraction—Since the electron capture cross section varies as some power of Z_1, one can predict a significant increase in the production of neutral H atoms when the breakup products emerge at close distances. With 200-keV H_2^+ ions, Meggitt et al. (*9*) measured a 28% increase in the neutral beam fraction transmitted through a 5 $\mu g/cm^2$ carbon foil. With thicker foils, the mean distances between the emerging protons increase and indeed weaker effects are observed.

ii) Charge exchange to the continuum—By this process electrons can emerge from the solid with the proton, and at the same velocity, but in an unbound state. Dettmann et al. (*8*) have shown that the cross section for this process is higher with incident molecular ions than for monoatomic ions. For example, with 215-keV H_2^+ ions they measured twice the number of these electrons per emerging proton, through a carbon foil of thickness 500 Å.

iii) Beam-foil excitation—A proton emerging from the foil with a bound electron, as H, has a certain probability to be in an excited state. The population distribution of these states is currently measured by the beam-foil spectroscopy techniques. Bukow et al. (*10*) show that when the incident beam is molecular, this distribution is shifted towards higher excited states and becomes equal to that observed with protons when the molecular beam is excited by several successive foils.

Molecular Effects Beyond the Foil

The effects due to the proximity of the two breakup protons during their flight between the foil and the detectors can be again separated into kinematic and atomic effects.

Kinematic Effects

In most cases the breakup products emerge from the foil as H^+ ions and then the Coulomb explosion continues in the vacuum. This residual Coulomb energy can modify the angular and energy distributions of the emerging particles before they reach the detectors. Golovchenko and Laegsgaard (*11*) have studied this downbeam Coulomb repulsion by using very thin dissociative foils, so that most of the repulsion occurred after the foil. With 2-MeV H_2^+, and a 100-Å thick carbon foil, they measured the angle-energy distribution shown in Fig. 2. Inside the solid the repulsion is screened by the target electrons. The two protons become independent when separated by a distance $R \gg V_1/\omega_p$, where V_1 is the proton velocity and ω_p is the plasmon frequency of the electron gas (*12*). As there is no screening effect involved in this downbeam repulsion, the final distribution of the detected protons can be rigorously related to the corresponding distribution of the protons as they emerge from the foil. In fact, the authors got a good description of this experimental distribution by assuming that all the incident ions have an internuclear distance $R_0 = 1.06$ Å, taking into account the multiple scattering effects inside the foil and the energy resolution of the detectors.

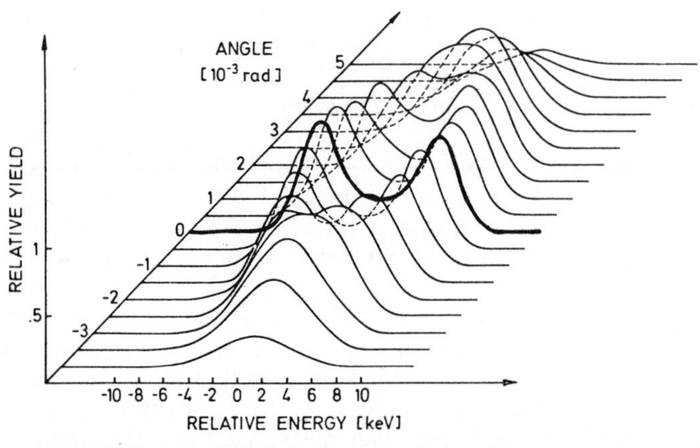

FIG. 2. Angle-energy distribution of protons transmitted through a 2 $\mu g/cm^2$ carbon foil from a 2-MeV H_2^+ incident beam. [Reproduced from J. Golovchenko and E. Laegsgaard, *Phys. Rev.* **9A**, 1215 (1974) with the permission of American Physical Society.]

Atomic Effects

When one of the two protons emerges as a foil-excited neutral atom, its in-flight decay can also be perturbed by the close proximity of the other proton. For example, Meggitt et al. (*9*) report that Andrä noticed the destruction of the 'quantum beats' usually observed in a beam-foil arrangement when an incident molecular beam is used.

MOLECULAR TRANSMISSION

In 1970 bound H_2^+ molecular ions were discovered among the particles transmitted through a carbon foil bombarded by a H_2^+ beam (*13*). The probability for this 'molecular transmission' was very low, typically decreasing from 10^{-4} to 10^{-7} through carbon foils of thickness ranging from 3 to 50 $\mu g/cm^2$ for 2-MeV H_2^+ incident ions. Following this experiment a systematic study of this molecular transmission phenomenon was undertaken at the University of Lyon with amorphous foils and thin crystals aligned in channeling configurations. All the results are consistent with a molecular recombination occurring between the two protons with a concomitant electron capture at the back surface of the foil.

Experimental Procedure

Details of the experimental problems related to such measurements, including thin crystal preparation, are described in previous publications (*14–16*). A typical setup is shown in Fig. 3. The various components transmitted through the target (H^+,H^o and H_2^+-D^+) are separated by magnetic analysis. Then the H_2^+ component is separated from the D^+ ions polluting the incident beam by backscattering analysis. We wish to find the total probability P per incident H_2^+ ion, for an incident molecular ion to emerge from the solid in a H_2^+ state. However, experimentally we measure only the fraction T (called transmission yield) of the H_2^+ ions emerging inside the acceptance angle ψ_a of the detection system. In most of our experiments T is only slightly less than P. To be sure that the geometrical effects of the detection are small, we measure also an upper bound, f. The experimental values of f and T are always so that the probability P is bounded as $f > P \gtrsim T$.

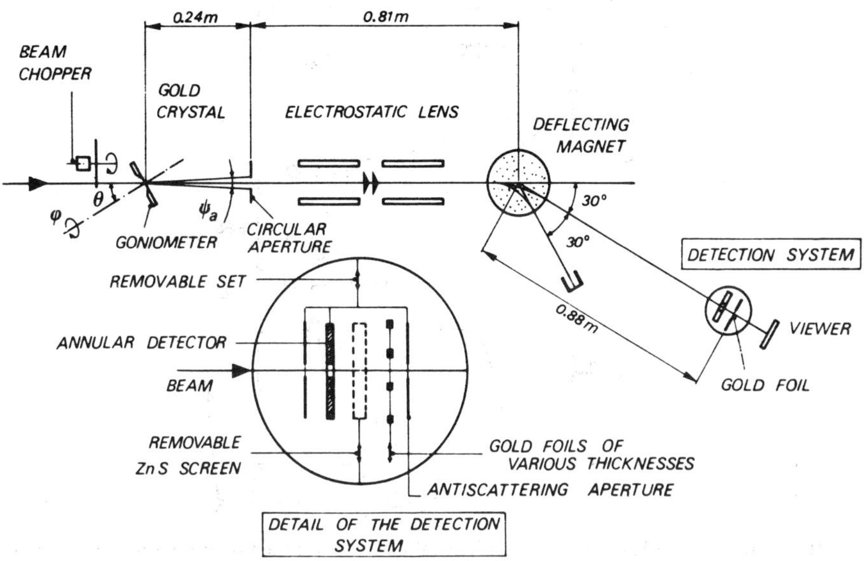

FIG. 3. Experimental setup used to measure molecular transmission yield.

Transmission in Planar Channeling

Since the transmission phenomenon requires close correlations, in space and in relative momentum, between the products of the molecular breakup as they emerge from the foil, the strong governing properties of planar channels must clearly favor these correlations and allow a detailed study of the molecular transmission, as in fact the experiments show.

Three stages have to be considered in the history of the transmitted molecule:

i) the probability $P_{o(CH)}$ for both breakup products to be channeled at the entrance of the crystal. For Au planar channels covering the size range of the incident molecules, the relative probabilities for the two protons to be channeled between the same atomic planes (1 plane) and on either side of an atomic plane (2 planes) are plotted on Fig. 4. The latter probability reaches a maximum at an interplanar distance $d_p \sim 1.1$ Å and decreases for the most open planar channels.

ii) the probability for both protons to remain channeled until they emerge from the crystal and, finally,

iii) the probability for a recombination event at the back surface of the foil. This event firstly imposes kinematic conditions between the two emerging protons, and secondly requires an electron capture by the two-proton system from the exit surface.

These three stages can be studied more or less independently in planar-channeling configurations by varying successively the planar spacing d_p, the path length x of the ions inside the crystal, the tilt angle ψ between the incident beam and the planar direction, the ion energy E_1 and the incident molecular system H_2^+ or H_3^+.

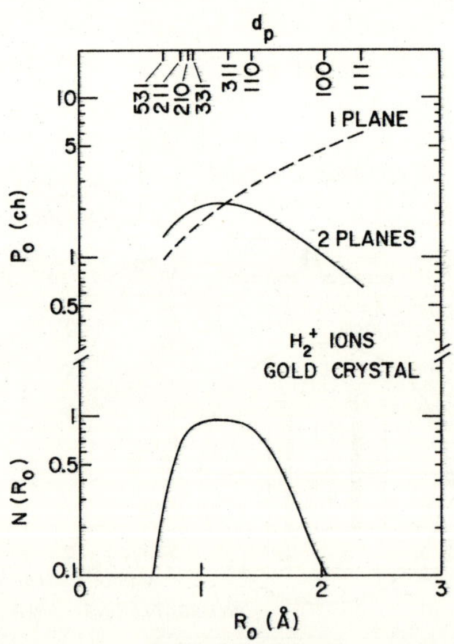

FIG. 4. Relative probabilities, after a H_2^+ breakup event, for the two protons to be channeled between the same atomic planes (1 plane) or on either side of an atomic plane (2 planes), in a gold crystal (in arbitrary units).

Transmission Parameters

For a given energy and path length of the ions, Fig. 5 shows that the molecular transmission is strongly enhanced in a planar-channeling alignment. From the 'random' value, the transmission tends to increase with the planar spacing, reaches a maximum and then decreases along the most open planes. The most open planes, where $d_p > \bar{\bar{R}}_0$ (see Fig. 4), are not the best 'molecular transmitter' because here the two protons are not restricted enough in their relative motion to remain closely correlated. On the other hand, along less open planes, where $d_p \sim \bar{\bar{R}}_0$, the relative kinematics of the two channeled protons is restricted throughout the penetration and kept more favorable to a recombination event when they leave the crystal. In planar channels, where $d_p < \bar{\bar{R}}_0$, dechanneling becomes important and the H_2^+ yield decreases. This

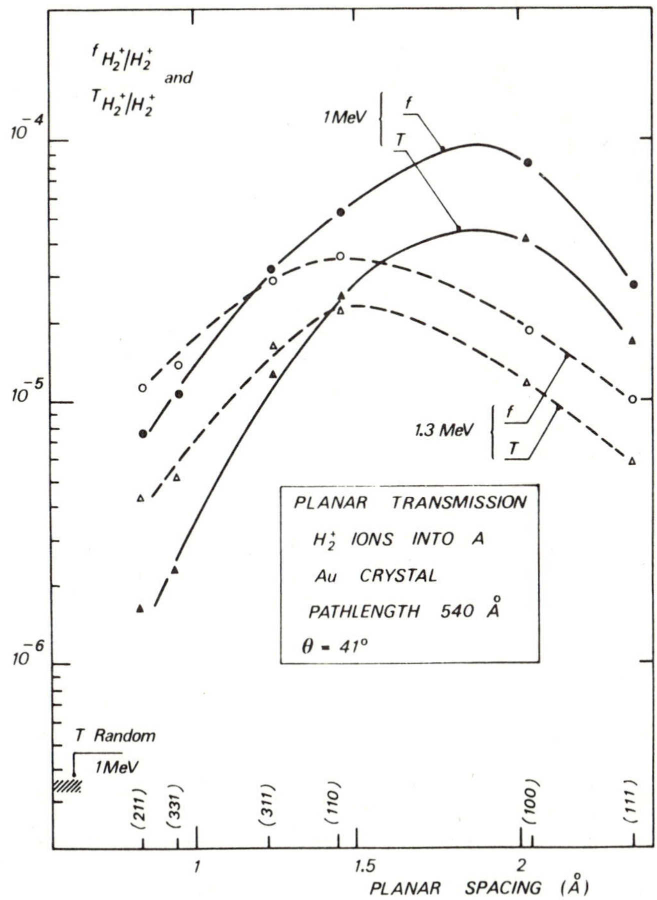

FIG. 5. Planar-spacing dependence of the molecular transmission. The similarity between the curves f and T shows that the fraction of H_2^+ ions which emerge from the crystal with an angle higher than the acceptance angle of the detection system is very low.

competition between space available to the channeled particles and dechanneling probability determines the planes with maximum transmission. Figure 5 shows that the transmission maximum is shifted towards the narrower planes when the ion energy increases, mainly because at higher energy small planes are less affected by the dechanneling process.

The importance of the longitudinal component of the Coulomb repulsion can be studied by increasing the time of flight of the ions inside the crystal, e.g. by increasing the penetration thickness. The transmission yield diminishes with the path length, as shown in Figs. 6 and 7. The reduction is smaller for the most open planar channels than for the narrower channels, which is connected with differences in the path-length dependence of dechanneling in the various channels. The molecular recombination process for the open planes (111) and (110) varies with the energy as $\sim E_1^{-4}$, whereas the atomic electron capture cross section varies as $\sim E_1^{-3}$.

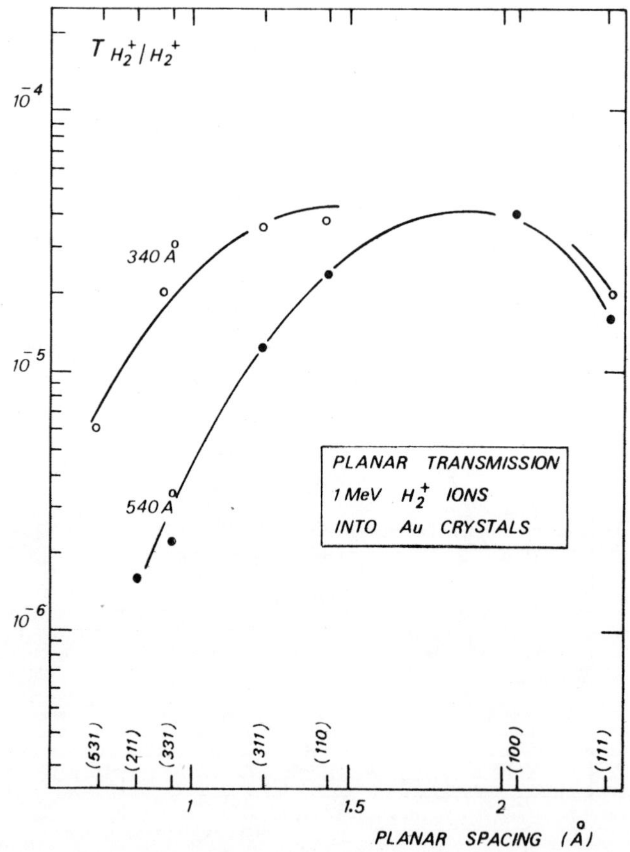

FIG. 6. Planar-spacing dependence of the molecular transmission for two different path lengths.

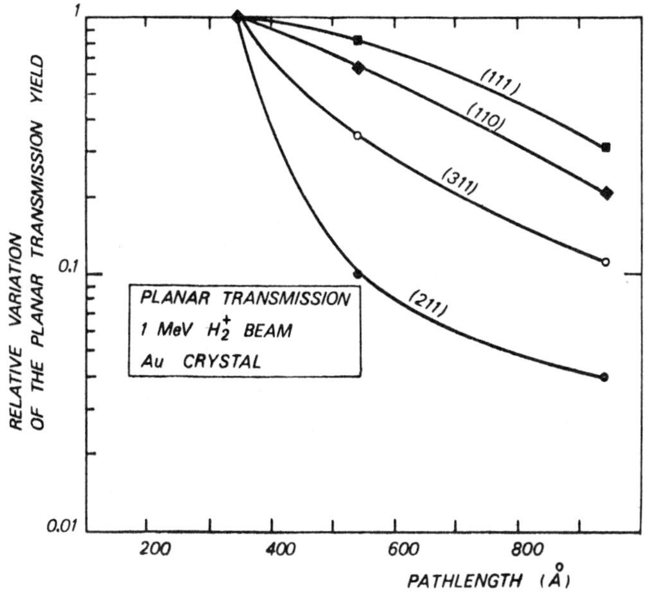

FIG. 7. Path-length dependence of the transmission yield T along planar channels. The T values are normalized to a path length of 340 Å.

Transmission in Axial Channeling

Most of the particles channeled along axes wander from string to string occupying a large volume inside the crystal. Thus the spatial correlation between the protons after the breakup event is less efficiently preserved by the string potential than by the planar potential. Moreover, the Coulomb repulsion between the protons is opposed more by planar than axial potentials. In fact, the molecular transmission along the most open axial channels, where the initial channeling probability is very large and where the dechanneling probability is very low, has never been found higher than the corresponding maximum planar transmission. The axial transmission yield exhibits no clear dependence on the interatomic spacing d along the rows, but is sensitive to the geometrical arrangement of the rows surrounding the channel. Another striking difference between axial and planar transmission lies in their angular widths. Planar critical angles for the transmission phenomena are some 20%–30% smaller than the classical channeling critical angles; but in axial transmission they are very much smaller, suggesting that only well-channeled protons contribute to the recombination process. A correlation between axial transmission and 'hyperchanneling' appears (*17*).

Experiments With an H_3^+ Incident Beam

Our understanding of the H_2^+ transmission phenomenon has been verified and improved by studying the probability for an incident H_3^+ ion to produce an H_2^+ ion downstream from the target. [This probability is called $T(H_2^+/H_3^+)$ in analogy with the corresponding value $T(H_2^+/H_2^+)$ measured with an incident H_2^+ beam.] The 'transmission' of an H_3^+ ion involving a two-electron capture, even if it exists, is evidently nonmeasurable here. As far as the H_2^+ production process is concerned the major differences between H_2^+ and H_3^+ incident beam are

i) a $N(R_0)$ distribution slightly shifted towards smaller internuclear distances,

ii) a higher repulsion effect inside the crystal, and

iii) a recombination process involving three 'candidates' but perturbed by the close proximity of the 'third proton.' In fact the 'transmission yields' measured with H_3^+ ions and H_2^+ ions are surprisingly alike. Figure 8, for example, shows the similarity of the planar transmission yield vs d_p, which could suggest that the 'bulk' effects, channeling and dechanneling, are not very different. However, the more pronounced energy dependence of the transmission with H_3^+ ions, along weak and strong planes, indicates that it is the recombination process that is the most perturbed by the third proton. Molecular transmission yields

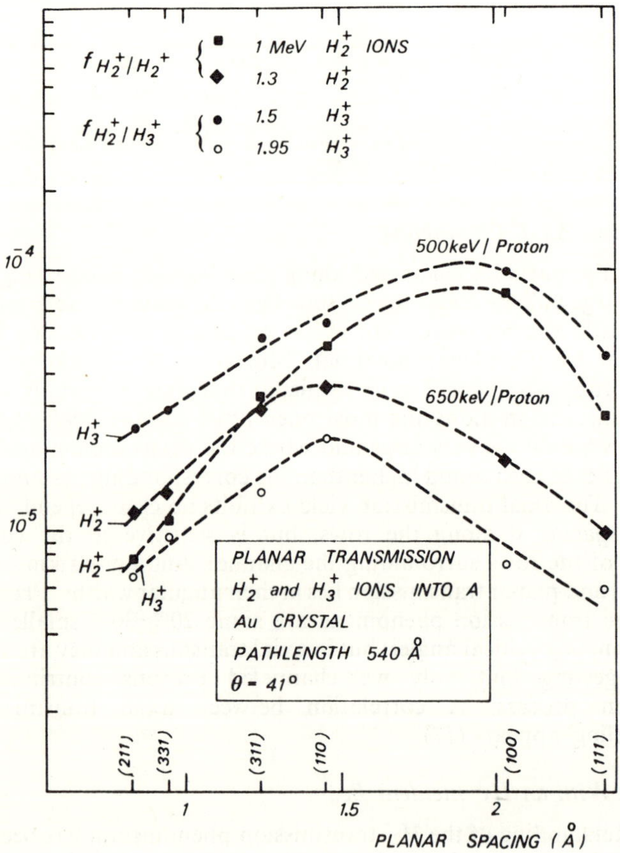

FIG. 8. Planar transmission probabilities for incident H_2^+ and H_3^+ ions of equal velocity.

measured along axial and random directions exhibit also a strong similarity between H_2^+ and H_3^+ data. Results shown in Fig. 9 suggest that the molecular transmission phenomena through very thin amorphous foils involve also a recombination process between two breakup products.

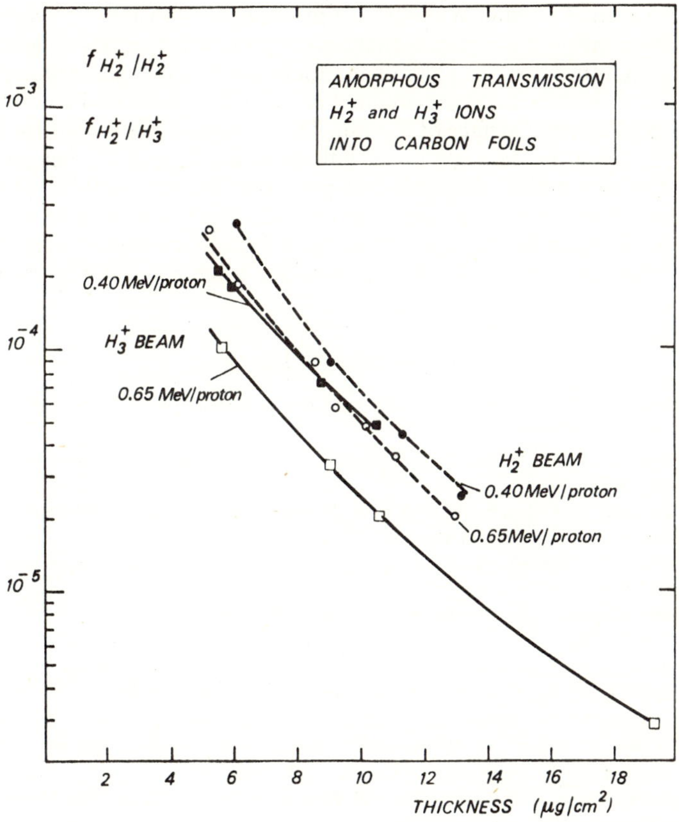

FIG. 9. Path-length dependence of the transmission yield through carbon foils for incident H_2^+ and H_3^+ ions of equal velocity.

CONCLUSION

Proton-solid interactions have been shown to be notably different when studied with a proton beam and with a molecular H_2^+ or H_3^+ beam. Angle and energy distributions of the particles inside the solid are modified, particularly the channeling properties inside a crystalline solid. The charge-state distribution and the beam-foil excitation of particles transmitted through thin targets are different depending on whether the incident beam is composed of H^+, H_2^+ or H_3^+ ions.

Among the transmitted particles, H_2^+ ions can be detected when the incident beam is an H_2^+ or H_3^+ beam. When the beam is aligned with a thin crystal target in a channeling configuration, this transmission process is strongly enhanced. The fact that neither the most open planar channels nor the axial channels are the best molecular 'transmitters' can be explained in terms of a molecular recombination occurring near the exit surface of the foil. This latter process is confirmed by the similarity between the data obtained with incident H_2^+ and H_3^+ beams of same velocity. The channeling configuration proves to be very useful to study such phenomena. More theoretical and experimental work is needed to understand the molecular recombination event using different molecular ions, different emerging surfaces and looking at a possible alignment of the recombined molecules.

REFERENCES

1. W. L. Walters, D. G. Costello, J. G. Skofronick, D. W. Palmer, W. E. Kane, and R. G. Herb, *Phys. Rev.* **125**, 2012 (1962).
2. A. D. Bacher, E. A. McClatchie, M. S. Zisman, T. A. Weaver, and T. A. Tombrello, *Nucl. Phys. A***181**, 453 (1972).
3. J. M. Caywood, T. A. Tombrello, and T. A. Weaver, *Phys. Lett.* **37A**, 350 (1971).
4. F. H. Eisen and E. Uggerhøj, *Radiat. Eff.* **12**, 233 (1972).
5. J. Golovchenko, *Channeling in Very Thin Crystals.* Thesis, Rensselaer Polytechnic Institute, New York, 1972, p. 50.
6. T. A. Tombrello and J. M. Caywood, *Phys. Rev. B***8**, 3065 (1973).
7. L. Little, Thesis, New York University, 1974 (unpublished).
8. K. Dettmann, K. G. Harrison, and M. W. Lucas, *J. Phys. B***7**, 269 (1974).
9. B. T. Meggit, K. G. Harrison, and M. W. Lucas, *J. Phys. B***6**, 362 (1973).
10. H. H. Bukow, H. V. Buttlar, D. Haas, P. H. Heckmann, M. Holl, W. Schalgheck, D. Schürmann, R. Tielert, and R. Woodruff, *Nucl. Instrum. Methods* **110**, 89 (1973).
11. J. Golovchenko and E. Laegsgaard, *Phys. Rev. A***9**, 1215 (1974).
12. W. Brandt, in *Atomic Collisions in Solids* (S. Datz, B. R. Appleton, and C. D. Moak, Eds.) pp. 261–288. Plenum Press, New York, 1975.
13. J.-C. Poizat and J. Remillieux, *Phys. Lett.* **34A**, 53 (1971).
14. J.-C. Poizat and J. Remillieux, *J. Phys. B***5**, 94 (1972).
15. J.-C. Poizat and J. Remillieux, *J. Phys. (Paris)* **33**, 1013 (1972).
16. M. J. Gaillard, J.-C. Poizat and J. Remillieux, in *Atomic Collisions in Solids* (S. Datz, B. R. Appleton, and C. D. Moak, Eds.) pp. 681–692. Plenum Press, New York, 1975.
17. S. Datz, this Symposium.

Primary Processes and Track Effects in Irradiated Media[1,2]

R. H. Ritchie and Werner Brandt

Health Physics Division, Oak Ridge National Laboratory,
P.O. Box X, Oak Ridge, Tennessee 37830;
Department of Physics, University of Tennessee
Knoxville, Tennessee 37916
and
Department of Physics, New York University
New York, N.Y. 10003, USA

INTRODUCTION

When heavy charged particles penetrate condensed matter at velocities v_1 large compared to the Bohr velocity $v_0 = e^2/\hbar$, energy is deposited in different modes throughout two spatially distinct domains which we refer to as the infratrack and ultratrack (*1*). The upper part of Fig. 1 gives a schematic

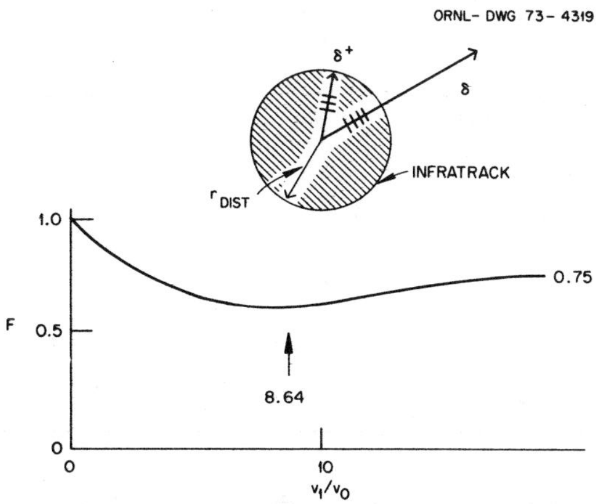

FIG. 1. Upper part: Schematic sketch of the extent of the infratrack and ultratrack surrounding the trajectory of a charged particle propagating into the plane of the figure. The scored region marks the infratrack of radius $r_{dist} \approx v_1/\Omega_{res}$, where coherent collective excitations occur in dense targets. The delta rays issue from collisions close to the trajectory, at $r_{close} \approx \hbar/2m\, v_1$. Some of them have ranges less than r_{dist}, others have ranges $\gg r_{dist}$, which define the extension of the ultratrack. Lower part: The fraction, F, of energy deposited in the infratrack as a function of v_1/v_0. It is composed, approximately, of 0.5 in the form of collective excitations and 0.2 in the form of incoherent excitations from delta rays passing through or stopping in the infratrack.

[1]Congress presentation E–2–4.

[2]Research sponsored by the U.S. Atomic Energy Commission under contract with the Union Carbide Corporation.

representation of these domains for a projectile with trajectory propagating into the plane of the figure. The lower part shows the dependence of the fraction, F, of the total energy deposited in the infratrack on the projectile velocity as a function of v_1/v_o where $v_o = e^2/\hbar = 2.18 \times 10^8$ cm/sec. Note that in terms of the projectile energy, E_1, this quantity follows simply from the relation $v_1^2/v_o^2 = 40\ E_1$ (in MeV)/M_1, where M_1 is the mass of the projectile in units of the proton mass (amu) which, in turn, is equal to the number of nucleons in the projectile.

In studying the interaction of a charged particle with a target, one distinguishes loosely between close collisions and distant collisions. Close collisions are direct Rutherford scattering encounters between the primary particle and a target electron. Distant collisions refer to coherent excitations of target electrons by the long-range electric field of the passing particle. From the viewpoint of the target electrons, a passing point charge constitutes an electrical impulse. The duration of the impulse increases with the distance b from the particle trajectory as b/v_1. When the pulse duration becomes appreciably larger than the response time $\sim\Omega_{res}^{-1}$ of the target electrons, which resonate with frequency Ω_{res}, they follow adiabatically the electric field set up by the moving particle. Then, the medium merely polarizes under the influence of the particle, with negligible energy absorption. The dividing surface between the resonating domain and the rest of the medium is determined by the radius $r_{dist} \simeq v_1/\Omega_{res}$ from the trajectory. This inner domain is referred to hereinafter as the *infratrack* (*1*).

Close collisions, over distances $r_{close} = \hbar/2mv_1$ from the trajectory, impart large momenta to target electrons. The struck electrons emerge from such encounters as energetic delta rays in directions primarily perpendicular to the trajectory. The range of many delta rays exceeds r_{dist}. Energetic delta rays can produce secondary delta rays. We refer to the domaf prima delta rays of range > r_{dist}, with the associated showers of secondary, tertiary, etc. delta rays, as the *ultratrack*. Much work has been done recently on the properties of the ultratrack and its relation to the irradiation conditions that prevail when electrons are set in motion by X or gamma rays (*2,3*).

The energy loss per unit path length traversed by the projectile is called the stopping power of the target, or the LET of the projectile-target system. It is given by

$$- dE/dx = (4\pi Z_1^2 e^4/mv_1^2) nZ_2 \int_{r_{close}}^{r_{dist}} (1/b^2) b\,db \qquad (1)$$

where Z_1e is the charge of the moving particle, nZ_2 the electron density in the target, and b the impact parameter of the trajectory with regard to a given electron. Inserting the infratrack parameters $r_{close} = \hbar/2mv_1$ and $r_{dist} = v_1/\Omega_{res}$, one obtains for the integral in Equation (1)

$$L = \ln(2mv_1^2/\hbar\Omega_{res}) , \qquad (2)$$

which is referred to as the stopping number per target electron.

We have verified that the simple cutoff criteria r_{close} and r_{dist} for the impact parameter b are, in fact, the properly averaged cutoff values for the fully developed impact-parameter-dependent excitation function, $G(b\Omega_{res},v_1)$ extending from $b = 0$ to $b = \infty$. Its properties are displayed in Fig. 2, as calculated for a

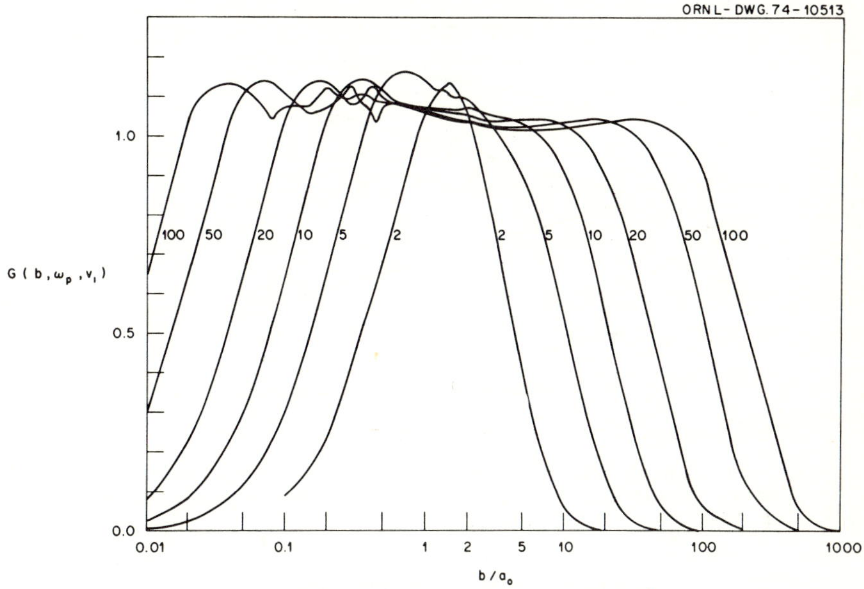

FIG. 2. The impact-parameter-dependent excitation function, $G(b,\omega_p,v_1)$, of an electron gas of plasma resonance frequency $\omega_p = (4\pi e^2 n_o/m)^{1/2}$. Integration of G/b^2 over all impact parameters, b, from 0 to ∞ yields Eq. (2) with $\Omega_{res} = \omega_p$; ω_p was taken equal to 0.5 a.u. in this calculation.

proton penetrating an electron plasma, of resonance frequency $\Omega_{res} = \omega_p = (4\pi n_o e^2/m)^{1/2}$, where n_o is the electron density in the plasma.[3] Here $\omega_p = 1/2$ a.u. The integration

$\int_0^\infty b^{-2}\, G(b,\Omega_{res},v_1)db$ yields precisely L as given in Equation (2) when $v_1 >> v_o$.

It is evident from Fig. 2, and the values of $r_{close} = \hbar/2mv_1$ and $r_{dist} = v_1/\Omega_{res}$, that the track structure changes with v_1 as sketched in Fig. 3 as a function of E_1/M_1. The locus of the projectile trajectory is uncertain only to order of the de Broglie wavelength of the projectile, shown for a proton $\lambda_p = 2\pi\hbar/1837\, mv_1$. The infratrack between r_{close} and r_{dist} shrinks with decreasing projectile velocity, and diminishes faster than the dimension of the ultra-track. Concurrently, the stopping power, Eq. (1), rises. But the fraction of energy deposited in the infratrack stays nearly constant (Fig. 1), and the energy density in the infratrack increases. The frequency distribution of the collective excitation is determined by the medium. As a consequence, the mode of energy transfer from the collective excitations in the infratrack to create primary localized radiation effects is insensitive to the properties Z_1 and v_1 of the projectile. Only the density of excitation, and its spatial extension depend on Z_1 and v_1.

[3]Courtesy of G. Basbas.

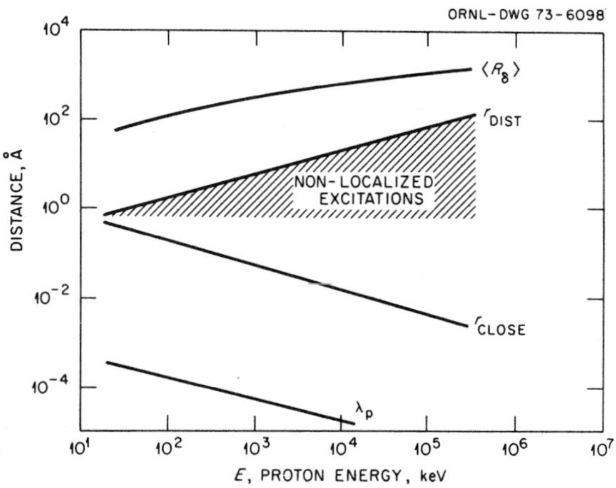

FIG. 3. Dependence of track structure on the velocity of the projectile expressed as $E_1/M_1 = 1/2\ v_1^2$. The infratrack extends to $r_{dist} = v_1/\Omega_{res}$, the ultratrack to the range of the delta rays that originate near $r_{close} = \hbar/2mv_1$.

COLLECTIVE EXCITATIONS

The excitation of *plasmons* or collective electron motion in the domain of the infratrack, then, is of prime importance. This term refers to the elementary excitations of a many-electron system induced by the longitudinal electric field set up by a moving charged particle. One may discuss plasmons in terms of the familiar single-electron excitation, leaving a "hole" in the ground state. A moving particle excites electron-hole pairs coherently in the infratrack. The electron-hole pairs recombine and create other particle-hole pairs. These "virtual" electron-hole pairs interact via the Coulomb field between them, creating a quantum mechanical system analogous to a set of coupled oscillators. The electron-hole pairs, because of the many-body coupling in condensed media, never resonate at the frequency for single electron-hole excitation, ω, but at a higher value, Ω_{res}, given approximately by

$$\Omega_{res} = (\omega^2 + f\omega_p^2)^{1/2}, \tag{3}$$

where $f \lesssim 1$ is the oscillator strength of single-particle hole pairs. After some time, the plasmons are damped by incoherent scattering processes and decay into localized excitations. If the coupling is weak, $f\omega_p << \omega$, the resonance frequency Ω_{res} is shifted only slightly relative to ω, approaching the situation in gases. If coupling is strong, $f\omega_p >> \omega$, collective behavior dominates the infratrack. The best known examples of plasmon modes are found in metals, of course, where the conduction electrons are nearly free, $\omega \sim 0$, while $\hbar\Omega_{res} \sim \hbar\omega_p \sim 5 - 15$ eV.

We can make contact between the energy loss rate of a charged particle and the energy absorption into collective excitation by noting a well-known result of linear response theory. The medium can be characterized by a dielectric function, $\epsilon = \epsilon_1 + i\epsilon_2$, where Re $\epsilon \equiv \epsilon_1$, the real part of the ϵ, is related to the optical refractive index and Im$\epsilon \equiv \epsilon_2$, the imaginary part, to the absorption of light in the medium. From a knowledge of ϵ_1 and ϵ_2 through optical measurements, one can derive the *energy loss function* for charged particles, $\mathrm{Im}(-1/\epsilon) = \epsilon_2/(\epsilon_1^2 + \epsilon_2^2)$. This function has a peak at $\Omega_{res} \simeq (\omega^2 + f\omega_p^2)^{1/2}$ in spectral ranges of ω where ϵ_2 is small, and $\epsilon_1 \ll \epsilon_2$ and increasing with ω. One can also measure $\mathrm{Im}(-1/\epsilon)$ directly, e.g., through characteristic energy losses of energetic electrons in thin foils under single collision conditions. Wherever this has been possible, the excitation functions so measured and derived from optical data were found to agree within experimental error (*4*).

Collective excitations are most pronounced in metals. But recent work has shown that they are a pervading phenomenon also in insulators. Although of shorter lifetime, i.e., of larger width in the resonance curve, than in metals, these collective effects have been observed in a number of different substances. For example, Fig. 4 displays

$$\epsilon_1,\ \epsilon_2,\ \text{and}\ \mathrm{Im}(-1/\epsilon)$$

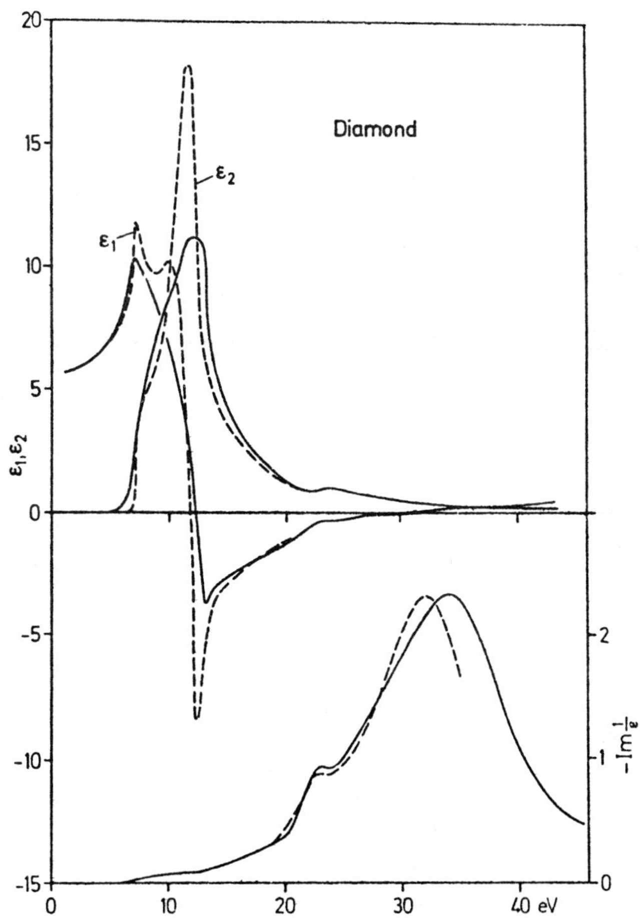

FIG. 4. Optical data of diamond, and the energy loss function $\mathrm{Im}(-1/\epsilon)$, with a plasmon peak at $\hbar\Omega_{res}$ = 34eV. [From ref. *4*.]

for the valence crystal diamond, with clear evidence for a collective peak in $\mathrm{Im}(-1/\epsilon)$ (*4*). Fig. 5 show the same quantities for the van der Waals solid Xe, which exhibits similar collective properties (*5*).

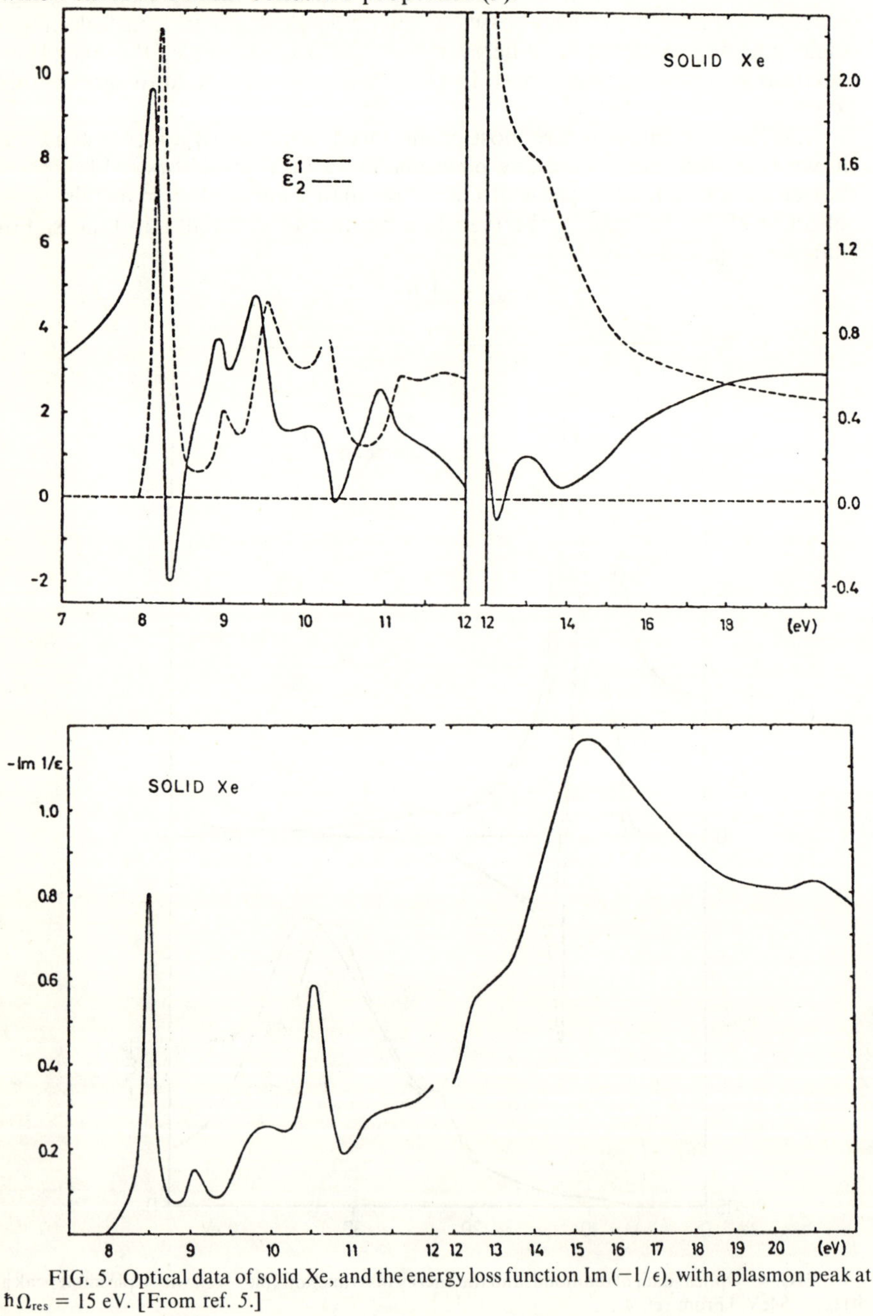

FIG. 5. Optical data of solid Xe, and the energy loss function $\mathrm{Im}\,(-1/\epsilon)$, with a plasmon peak at $\hbar\Omega_{res} = 15$ eV. [From ref. *5*.]

Water is of prime interest in biological contexts. One can estimate readily from Equation (3) that if plasmons can be sustained in water, i.e., if their damping rate is less than their frequency, the resonance energy for water should be 21 eV (*1*). Only very recently have optical constants for water been measured to sufficiently high energies to see whether a collective resonance occurs at that energy. Fig. 6 shows the results measured by Heller, Hamm, Birkhoff and Painter (*6*). A pronounced peak appears in the excitation function just at 21 eV. Following this accomplishment, it clearly becomes important that such data be obtained over a wide range of energies for biological media. But one should expect that, in first approximation, their response will be similar to that of water, suggesting that damped collective oscillations may be sustained in such substances.

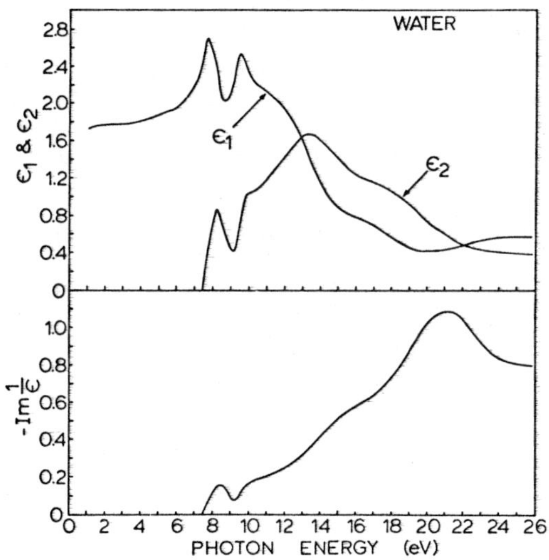

FIG. 6. Optical data of liquid H_2O, and the energy loss function $\mathrm{Im}(-1/\epsilon)$, with a plasmon peak at $\hbar\Omega_{res} = 21$ eV. [From ref. *6*.]

ENERGY LOCALIZATION

The collective excitations in the infratrack induced by a charged particle present to the absorption modes in the medium an excitation spectrum with a frequency distribution peaked at Ω_{res}, which is determined by the dielectric properties of the medium, not by the particle. One can show (*1*) that the spatial density distribution $\rho(\vec{r})$, of the excitation in the radial direction from the trajectory is approximately given by

$$\rho(\vec{r}) = (Z_1^2/r^2)\exp(-2\Omega_{res}r/v_1) \qquad (4)$$

when $\Omega_{res}\, r/v_1 >> 1$. It follows that $\rho(\vec{r})$ depends on the relevant properties of the projectile, Z_1 and v_1, only in the sense that

$$Z_1^{-2}\, v_1^2\, \rho(\vec{r}) = f(\vec{r}/v_1;\, \Omega_{res}) \qquad (5)$$

is a function which is the same for all particles of a given velocity v_1 in a medium characterized by Ω_{res}.

We consider the molecules of a condensed substance as localized "resonators." By absorbing the energy proffered to them by the collective excitations in the infratrack, they dampen these excitations and localize the energy in a manner which is determined by the coupling between the absorbers and the collective excitation field.

A program to elucidate energy localization processes in biological media requires extensive experimental data and rather elaborate interpretations (*7*). The problem is complicated for inhomogeneous media, because structural entities can cause local energy accumulations from the collective excitation field, leading to structural specificity of radiation effects. To see what is required, we can illustrate the localization processes in terms of a simple model (*8*). Suppose a coherent excitation field has been created by a particle in the infratrack, of a distribution in frequency, ω, centered at $\omega = \Omega_{res}$, and of intensity $I(\vec{r},\omega)$ at some point $\vec{r}$. As Fig. 7 indicates schematically, this collective state is degenerate, i.e., overlaps in frequency, with the absorption spectra $g_1(\omega)$ and $g_2(\omega)$ in the continuum above the ionization limit of molecules, say, with bond 1 and bond 2 in the vicinity of $\vec{r}$. Then, the relative probability, that is, the *specificity*, $S_{local}^{1,2}$, of exciting bond 1 and bond 2 in its vicinity is given by

$$S_{local}^{1,2} = [\int d\omega I(\vec{r},\omega) g_1(\omega) / \int d\omega I(\vec{r},\omega) g_2(\omega)] \approx [g_1(\Omega_{res}) / g_2(\Omega_{res.})] \ , \qquad (6)$$

independent of the moving particle and of $\vec{r}$. The energy transfer dampens the plasmons and excites the electrons forming the bonds. The efficiency of the transfer depends on the magnitude of $g(\omega)$ at Ω_{res}. The phenomenon is akin to the

ORNL-DWG. 74-10514

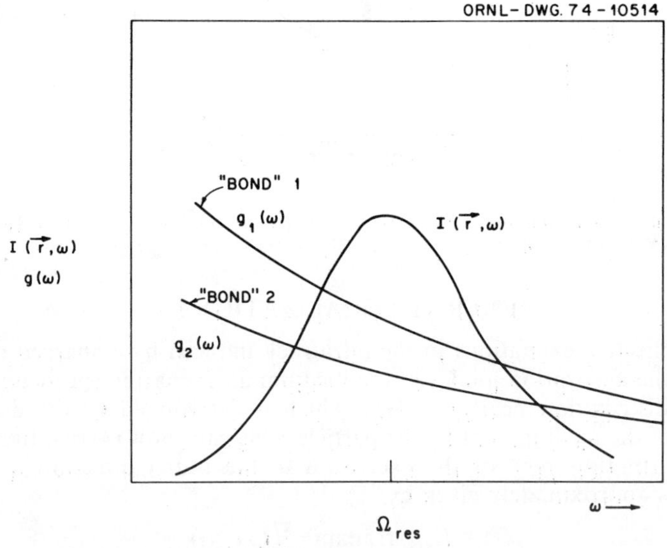

FIG. 7. Overlap of the intensity distribution in frequency for the collective excitation in the infratrack, $I(\vec{r},\omega,)$, peaked at Ω_{res}, with the continuum absorption spectra of localized bonds 1 and 2 (schematic). This leads to specificity of energy localization, as indicated in Eqs. (6) and (7) (schematic).

autoionization in atoms (*8,9*). Since Ω_{res} is a material constant given by the dielectric properties of the medium, the local specificity is a property of the medium itself.

If the distribution of absorbers is inhomogeneous, one may approximate the absorption spectrum of jth type of bond by $\Sigma_\nu g_j(\omega)\,\delta(\vec{r}-\vec{r}_{\nu j})$ where $r_{\nu j}$ denotes the position of the bond j in a molecule ν. Then

$$S_{local}^{1,2}(v_1) = [\sum_\nu \int d\omega I(\vec{r}_{\nu 1},v_1,\omega)g_1,(\omega)/\sum_\nu \int d\omega I(\vec{r}_{\nu 2},v_1,\omega)g_2(\omega)] \qquad (7)$$

depends through $f(\vec{r}/v_1,\ \Omega_{tres})$, Eq. (5), on the particle velocity. It implies that bonds in structural inhomogeneities (e.g., in the nucleus of a cell) are affected by this process of energy localization only if they fall within the range $r_{dist} = v_1/\Omega_{res}$ of the infratrack.

In molecular media, the absorption spectra $g_j(\omega)$ appearing in Eqs. (6) and (7) may to some extent (and with caution) be taken from optical data of molecular gases. In addition, there are a number of localization modes which depend, themselves, on the many-body properties of condensed media (*1*). Mention of some examples may suffice here. Low-energy delta rays in propagating through the infratrack may be captured into the wake trailing a moving charged particle, shown schematically in Fig. 8 (*10*). Electrons captured in wake-riding states are swept along with the projectile, which reduces the probability of electron-geminate ion recombination, particularly for the low energy delta rays of importance for this recombination process. A primary local ionization may cause a sudden electric pulse in its immediate proximity, leading to intermolecular electron shake-off or, in effect, to intermolecular electron transfer. Auger electrons in dense media have been observed to be coupled to plasmons. This may imply that primary ionizations excite secondary plasmons, which are set in motion locally by the medium to screen the newly created ion charge. They subsequently decay into secondary local excitation processes. A familiar analogue to this phenomenon is the solvation of initially "dry" electrons, just after ejection in a primary ionization, by the electrostrictive response of the medium to the charge of the electron after thermalization.

ORNL DWG. 73-11634

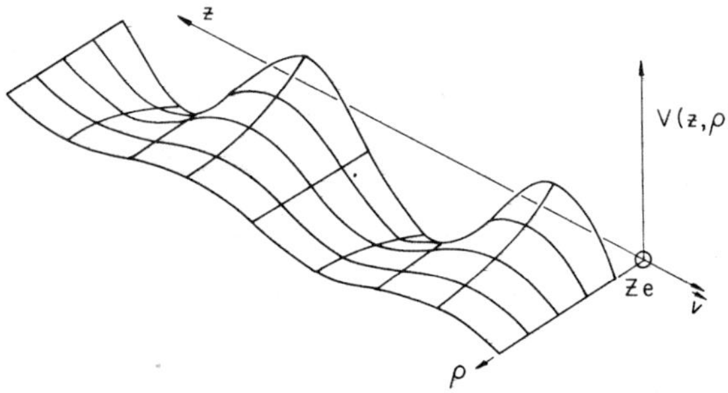

FIG. 8. Schematic representation of the wake trailing a moving charged particle in a dense medium. Shown is the potential energy of an electron in this wake. The period of the wake in the z direction (opposite to the direction of $\vec{v}$) is $2\pi v/\Omega_{res}$. The lateral extension in the ρ direction is $\sim r_{dist} = v/\Omega_{res}$. [From ref. *10*.]

CONCLUSION

The tentative conclusions presented here concerning the role of the infratrack in primary radiation processes in condensed media open the intriguing possibility of obtaining specificity functions for energy localization in infratracks from the study of the dielectric functions of the media. This has become possible with the advent of synchrotron light sources (*11*). The dielectric response determines the probability distribution of the kinds of chemical bonds into which the energy is channelled. Such excited bonds can cause, e.g., biological effects directly or create distributions in kind and concentration of reactive species leading to chemical and, eventually, to chemically induced biological effects. The specificity of energy localization in the infratrack is expected to be nearly the same for all particles of equal velocity in a given medium. This specificity is given by the medium in a self-consistent way through its dielectric properties, and influenced by microscopic structural inhomogeneities.

REFERENCES

1. W. Brandt and R. H. Ritchie, in *Physical Mechanisms in Radiation Biology* (R. D. Cooper and R. W. Wood, Eds.) pp. 20–46. Technical Information Center, Office of Information S United States Energy Commission, 1974. CONF-721001 (Available from National Technical Information Service, Springfield, Virginia 22151.)
2. R. Katz, B. Ackerson, M. Homayoonfair, and S. C. Scharma, *Radiat. Res.* **47**, 402 (1971).
3. J. Fain, M. Monnin, and M. Montret, *Radiat. Res.* **57**, 379 (1974).
4. J. Daniels, C. V. Festenberg, H. Raether and K. Zeppenfeld, *Springer Tracts Mod. Phys.* **54**, 77 (1970).
5. P. Kiel, *Z. Physik* **214**, 251 (1968).
6. J. M. Heller, Jr., R. N. Hamm, R. D. Birkhoff, and L. R. Painter, *J. Chem. Phys.* **60**, 3483 (1974).
7. For solids, F. C. Brown, *Solid State Phys.* **29**, 1 (1974).
8. W. Brandt and S. Lundqvist, *Arkiv Fysik* **28**, 399 (1965).
9. W. Brandt, L. Eder, and S. Lundqvist, *J. Quant. Spectrosc. Radiat. Transfer* **7**, 185 (1967).
10. V. N. Neelavathi, R. H. Ritchie, and W. Brandt, *Phys. Rev. Lett.* **33**, 302 (1974).
11. M. L. Perlman, E. M. Rowe, and R. E. Watson, *Phys. Today* **27**, (Nov. 7) 38 (1974).

ELECTRONS IN LIQUIDS[1]

Chairmen: G. R. Freeman and R. Schiller

[1]Symposiums E–10 and E–28

The Optical Absorption Spectrum of the Solvated Electron in Polar Liquids and in Binary Solutions[1,2]

Leon M. Dorfman and James F. Gavlas
Department of Chemistry, The Ohio State University
Columbus, Ohio 43210, USA

INTRODUCTION

In our investigations of the optical absorption spectrum of the solvated electron (*1–6*) in polar liquids we have, over the past four years, focused much of our attention on the important question of whether, and to what extent, short-range interactions may be a dominant factor, along with the bulk solvent properties, in determining the optical properties of the solvated electron. We have approached this question with three different types of experiments: (a) a broad survey of the spectrum of e^-_{sol} in different classes of compounds (*3–6*) with a view to determining whether the spectral grouping (*3,7*) of these absorptions showed any marked dependence upon class of compound; (b) determination of the spectrum of e^-_{sol} in binary solutions (*3,4*), and notably of the dependence of λ_{max} upon composition (principally for solutions of polar liquids) in a search for evidence of selective short-range interaction; (c) determination of the effect upon the spectrum of coupling (*8,9*) of e^-_{sol} with alkali metals in certain liquids.

In the brief time available to us in this symposium I would like to discuss the results of the first two of these different types of experiments. Some new observations of the spectrum of e^-_{sol} in three liquid amides (*6*) and two amines have been added to this survey; the results for the amides, as will be seen, resolve a question we were faced with in attempting some classification of the spectral grouping. I would then like to indicate how these results for the amides fail to fit an empirical correlation which had been recently proposed (*10*) for λ_{max} of the spectra of e^-_{sol}. Finally I should like to touch briefly on the question of whether structure is observable in the optical absorption spectrum of e^-_{sol} at room temperature, a question of considerable theoretical interest which has received some attention recently, and which I hope will be resolved by some recent measurements we have made.

SPECTRA OF e^-_{sol} IN AMIDES

The optical absorption spectrum of the solvated electron has been determined, by the pulse radiolysis method, in each of three, pure liquid amides at room temperature: N,N-dimethylformamide (DMF), N,N-

[1]Congress presentation E-10-1.
[2]This work was supported by the U.S. Atomic Energy Commission.

dimethylacetamide (DMAd) and N,N-diethylformamide (DEF). These absorption spectra, over the range 800 to 2100 nm, are shown in Fig. 1. The absorption maximum, in each case, occurs at the following wavelengths: DMF, 1680±30 nm; DMAd, 1800±50 nm; DEF, 1775±150 nm. Reactivity of the transient species responsible for these absorptions toward N_2O and toward anthracene (with formation of anthracenide ion), together with the nature of the spectrum, is sufficient evidence to identify the species as the solvated electron.

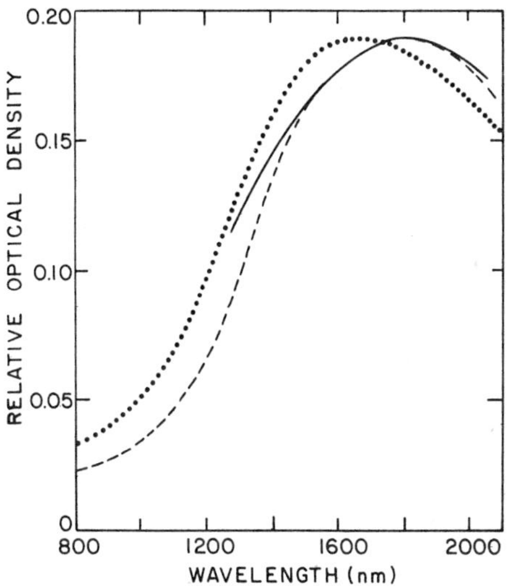

FIG. 1. Optical absorption spectra at room temperature of the solvated electron in liquid amides. DMF....., DMAd – – –, DEF ——.

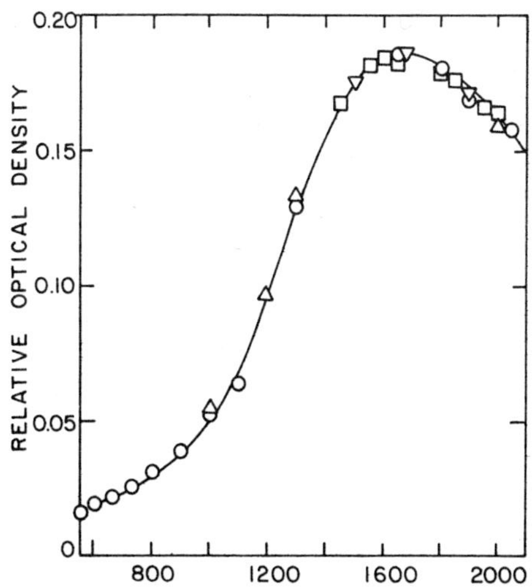

FIG. 2. Optical absorption spectrum at room temperature of the solvated electron in DMF showing individual data points. Separate points represent individual experiments with separate samples; ∇ represents a superposition of such separate experiments.

These amides were of particular interest to us in examining the spectral grouping of the absorption of e^-_{sol} with class of compound, since Hayashi et al. (*11*) had recently reported absorption maxima of 650 nm for e^-_{sol} in DMF, and 625 nm in DMAd. Such values would have placed λ_{max} for these amide electrons directly in the spectral grouping of the hydroxy compounds, a result of significance for any concept of spectral grouping with type of compound. As is clear, however, our results for these amides are sharply at variance with the findings of Hayashi et al. as we find no peaks at 650 nm. As is shown in Fig. 2, in which the data for DMF from 580 nm to 2100 nm are presented, we fail to find any structure in the visible range. We are at a loss to explain such a drastic discrepancy in the results, although the fall-off, toward 900 nm, in the sensitivity of the detector used in the earlier work may have been a contributing factor. The peaks we do find for the three amides are in a region of the infrared which seems consistent with the spectra of amines which have been reported.

EMPIRICAL CORRELATION OF FREEMAN

Our data for the absorption maxima in DMF, DMAd and DEF were also of interest to us as a further test of an empirical correlation recently proposed by Freeman (*10*) which directs attention to the role of short-range forces (particularly with functional groups of the molecule) in determining the nature of e^-_{sol}. Freeman has correlated the transition energy, E_{max}, for the optical absorption with a purely empirical term which involves the molecular polarizability and the Kirkwood correlation parameter (*12,13*), g_K, so that the importance of short-range order in the liquid is taken into account. g_K is obtained from the equation: $[(D_s-1)(2D_s+1)/9D_s](M/d) = (4\pi N/3)[\alpha + (g_K\mu^2/3kT)]$ where M is the molecular weight, d the density, N Avogadro's number, α the optical polarizability, μ the molecular dipole moment, k Boltzmann's constant, and T the absolute temperature. A plot of E_{max} vs the purely empirical term $D_s d g^3_K \alpha_p$, where the power of g_K, namely 3, was obtained by fitting the data for a variety of compounds, and α_p is the polarizability of the polar group rather than the molecular polarizability, is shown in Fig. 3. As Freeman contends, this

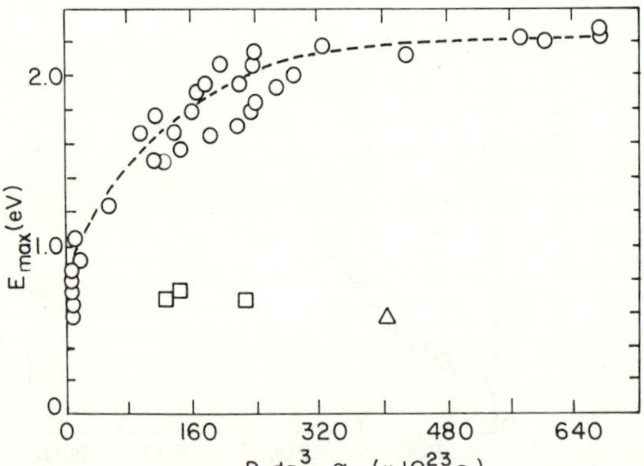

FIG. 3. Correlation of Freeman for transition energy for spectrum of e^-_s with liquid properties. E_{max} is plotted against $D_s dg^3_K \alpha_p$. O, various liquids at different temperatures; □, amides; △ HMPA.

correlation fits reasonably well for water, ammonia, some alcohols and ethers. It fails completely for the three amides for which we have determined λ_{max}, and for hexamethyl-phosphoric triamide (*5*) ($\lambda_{max} = 2200$ nm), as may be seen from Fig. 3. The values of D_s for the amides, it will be noted, are considerably higher than for other liquids having similar values for E_{max}.

SPECTRAL GROUPING OF ABSORPTION MAXIMA

The variety of types of compound in which the spectrum of the solvated electron is now known has been greatly expanded with the development of fast infrared detection (*3,14*) systems. In addition to the spectra in ammonia (*15*), water (*16*), and a number of alcohols (*1*), which have been known for quite some time, the optical absorption bands have been determined in a number of ethers (*3,4*) and amides and amines (*3–6*), systems in which λ_{max} may lie at wavelengths as long as 2300 nm. A number of these spectra are shown in Fig. 4.

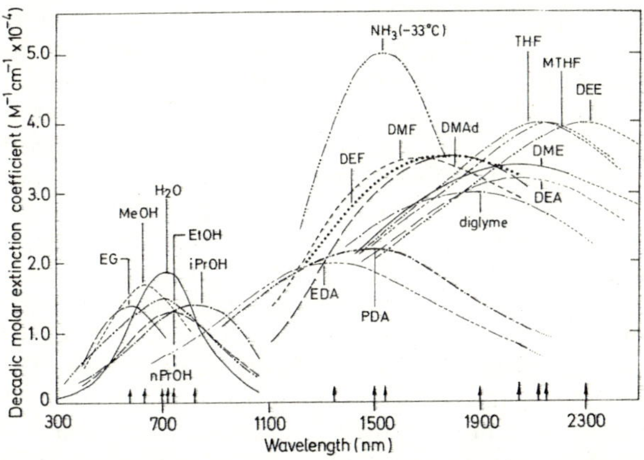

FIG. 4. Optical absorption spectra of the solvated electron in a variety of compounds. EG ethylene glycol, MeOH methanol, PDA propanediamine, DME dimethoxyethane, DEA diethylamine, DEE diethyl ether. The extinction coefficients are not known for the amides, DEF, DMF and DMAd, the spectra for which are arbitrarily positioned in this figure.

Examination of such data indicates that there is indeed some spectral grouping (*7*) of λ_{max} with class of compound, even though there is, to be sure, substantial overlap among some of these different compounds. This may be seen somewhat more clearly in Fig. 5 in which the wavelength of λ_{max} is shown for a large number of compounds. The hydroxy compounds include water (*16*), several aliphatic alcohols from our earlier work (*1*), the recently determined spectrum in 2-methoxyethanol (*17,6*) and the interesting cases of 2-methyl-2-propanol and 2-methyl-2-butanol recently reported by Hentz and Kenney-Wallace (*17*). The data shown for the ethers, all far out in the infrared, are from our reports of the past few years (*3,4*). The amides shown are the three described in the first section of this talk.

The liquids for which λ_{max} is shown in Fig. 5 have static dielectric constants ranging from 78.3 to 3.6 at room temperature. Although there is substantial overlap in the values of λ_{max} at room temperature, the overriding factor in the spectral grouping, as we have pointed out (*7*), seems to be the class of compound.

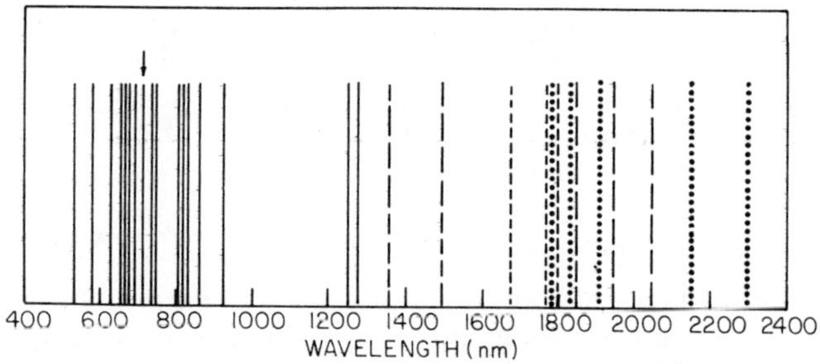

FIG. 5. Optical absorption maxima for the solvated electron in various liquids at room temperature: —— hydroxy compounds (—— ← is for water and 2-methoxyethanol; – – ammonia and amines; - - - amides; ..... ethers.

The hydroxy compounds are mostly in the visible to the near-infrared range, with 2-methyl-2-propanol and 2-methyl-2-butanol (*17*) slightly beyond 1200 nm (3-methyl-3-pentanol and 3-ethyl-3-pentanol (*17*), with λ_{max} as yet undetermined, appear to be beyond 1300 nm). It is noteworthy that no other compound yet reported has λ_{max} in the visible to the near infrared overlapping with the OH-compounds shown, and it seems reasonable to conclude, therefore, that the optical properties of e^-_{sol} in this class of compound are determined by properties of the hydroxy group in the solvent structure. Hentz and Kenney-Wallace (*17*) have drawn attention to this and have suggested that the OH dipole configuration in a single solvation shell may be the dominating factor with the effect of branching structure of the alcohol superimposed. Such short-range effects are reflected in different ways in molecular models (*18*), in which short-range forces must be a dominant parameter, as well as in the cavity-continuum model (*19*) in which cavity radius is an important parameter. The amides and amines shown are in an intermediate region from about 1400 to 2050 nm. The ethers are farther into the infrared with λ_{max} ranging from 1800 to 2300 nm. Our experimental data for THF were compared with calculations based on the cavity-continuum model of Copeland et al. (*19,20*). We find agreement between the experimental value for E_{max} and the calculated value, for an effective cavity radius of 4Å (which is different from the radius of the stable configuration and from the void radius) and a coordination number of 6 or 8.

e^-_{sol} IN BINARY SOLUTIONS

The spectrum of the solvated electron in binary solutions (*2,3*) and the concentration dependence of λ_{max} in particular should provide evidence of selective solvation as well as furnishing further tests of the validity of theoretical models. Spectra of e^-_{sol} in a variety of binary systems have now been reported. In all such systems which we have investigated, only a single absorption band is seen, no matter how dissimilar the absorption maxima for the individual components (as, for example, water and THF (3), at 720 nm and 2120 nm respectively). A complete range of behavior is seen in the dependence of λ_{max} upon composition. In some systems the dependence is very nearly linear, e.g. ethylenediamine-tetrahydrofuran (*3*), or slightly sigmoidal in shape as for ammonia-water or ethylenediamine-water (*2*). The optical properties are not dominated by one component in these systems. In other systems the composition dependence of λ_{max} is strongly dependent upon one component, as for water-tetrahydrofuran (*3*) in which water is the dominating component. In yet other systems, the wavelength of λ_{max} is so strongly dominated by one component as to be almost independent of the presence of the second. This is true for certain alcohol-alkane (*21–22*) solutions and for ethylenediamine-ethanol (*21*) in which the dominating component, ethylenediamine, has the *lower* static dielectric constant of the pair. A summary of some of the data illustrating these different types of dependence is presented in Table 1. In those systems in which λ_{max} is dominated by one component, that component is underlined.

TABLE 1
Composition Dependence of λ_{max} of e^-_{sol} in Binary Solutions

Linear or Near Linear Dependence	λ_{max} Strongly Affected or Dominated by One Component[a]
EDA-THF	<u>H_2O</u>-THF[c]
H_2O-NH_3[b]	<u>EDA</u>[d]-C_2H_5OH
H_2O-EDA[b]	<u>C_2H_5OH</u>-Hexane
EDA-HMPA	<u>C_2H_5OH</u>-DiMeButane

[a]The compound underlined dominates the observed λ_{max}.
[b]Sigmoidal dependence departing slightly from linearity.
[c]Dielectric constant dominated by THF.
[d]EDA has lower dielectric constant of the pair.

Calculated data for the binary liquid system water-ammonia using the molecular model of Howat and Webster (*18*) have been compared (*18*) with experiment. For this tetramer model the calculated wavelength dependence is somewhat closer to linear than the experimental data, but the agreement is not unreasonable.

STRUCTURE IN THE ABSORPTION SPECTRUM OF e^-_{sol}

The question of whether structure exists in the absorption spectrum of the solvated electron at room temperature, which has generally been reported as a broad, structureless band, is of theoretical interest. For a continuum model, one may speculate as to whether bands of the nature of a Rydberg series may exist. For a molecular model there may conceivably be structure related to molecular properties of the immediate molecular configuration around the electron.

Attention has recently been focused on this interesting question in a series of papers (*23,24*) on ethanol and other alcohols in which it is reported that such structure is *observable*. This observation is in conflict with earlier data. We have, accordingly, repeated (*25*) the determination of the spectrum of e^-_{sol} in ethanol at room temperature. Our data, representing several determinations at each wavelength, are shown in Fig. 6 along with the data of Vannikov and Marevtsev (*23*). Our data fail to sustain the claim (*23,24*) of observable structure greater than a detection sensitivity of 1.3% of the total absorption. Moreover, comparison of the more recent report (*24*) on structure with the earlier paper (*23*) indicates also that random signal fluctuation may have been misinterpreted as structure.

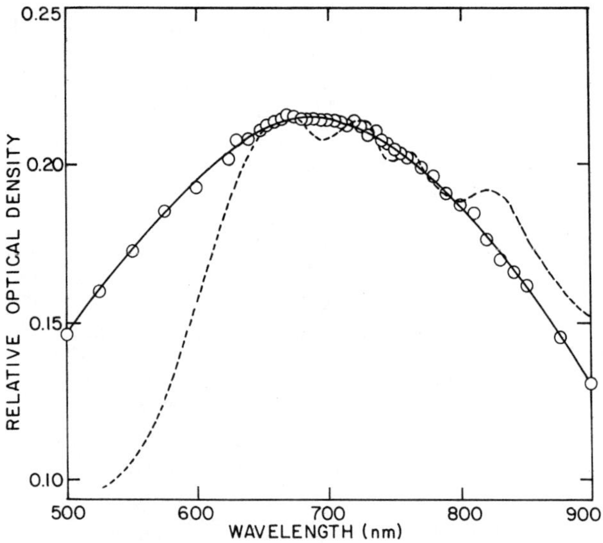

FIG. 6. Optical absorption spectrum of the solvated electron in ethanol at 296° K. The open circles represent our experimental data, each point being the average of two to seven individual data points. The dashed line represents the data of Vannikov and Marevtsev (*23*) normalized to our data at 680 nm.

REFERENCES

1. M. C. Sauer, Jr., S. Arai, and L. M. Dorfman, *J. Chem. Phys.* **42**, 708 (1965).
2. J. L. Dye, M. G. DeBacker, and L. M. Dorfman, *J. Chem. Phys.* **52**, 6251 (1970).
3. L. M. Dorfman, F. Y. Jou, and R. Wageman, *Ber. Bunsenges. Phys. Chem.* **75**, 681 (1971).
4. F. Y. Jou and L. M. Dorfman, *J. Chem. Phys.* **58**, 4715 (1973).
5. E. A. Shaede, L. M. Dorfman, G. J. Flynn, and D. C. Walker, *Can. J. Chem.* **51**, 3905 (1973).
6. J. F. Gavlas, F. Y. Jou, and L. M. Dorfman, *J. Phys. Chem.* **78**, 2631 (1974).
7. L. M. Dorfman and F. Y. Jou, in *Electrons in Fluids* (J. Jortner and N. R. Kestner, Eds.), pp. 447. Springer-Verlag, Berlin, 1973.
8. J. L. Dye, M. G. DeBacker, J. A. Eyre, and L. M. Dorfman, *J. Phys. Chem.* **76**, 839 (1972).
9. B. Bockrath and L. M. Dorfman, *J. Phys. Chem.* **77**, 1002 (1973).
10. G. R. Freeman, *J. Phys. Chem.* **77**, 7 (1973).
11. N. Hayashi, E. Hayon, T. Ibata, N. N. Lichtin, and A. Matsumoto, *J. Phys. Chem.* **75**, 2267 (1971).
12. J. G. Kirkwood, *J. Chem. Phys.* **7**, 911 (1939).
13. G. Oster and J. G. Kirkwood, *J. Chem. Phys.* **11**, 175 (1943).
14. L. M. Dorfman, in *Techniques of Chemistry*, **Volume 6, Part 2** (A. Weissberger and G. Hammes, Eds.), pp. 463–519. John Wiley and Sons, Inc., New York, 1973.
15. M. Gold and W. L. Jolly, *Inorg. Chem.* **1**, 818 (1962).
16. E. J. Hart and J. W. Boag, *J. Am. Chem. Soc.* **84**, 4090 (1962).
17. R. R. Hentz and G. A. Kenney-Wallace, *J. Phys. Chem.* **78**, 514 (1974).
18. G. Howat and B. C. Webster, *J. Phys. Chem.* **76**, 3714 (1972).
19. D. A. Copeland, N. R. Kestner, and J. Jortner, *J. Chem. Phys.* **53**, 1189 (1970).
20. N. R. Kestner, in *Electrons in Fluids* (J. Jortner and N. R. Kestner, Eds.) pp. 1–28. Springer-Verlag, Berlin, 1973.
21. J. R. Brandon and R. F. Firestone, *J. Phys. Chem.* **78**, 792 (1974).
22. B. J. Brown, N. T. Barker, and D. F. Sangster, *J. Phys. Chem.* **75**, 3639 (1971).
23. A. V. Vannikov and V. S. Marevtsev, *Int. J. Radiat. Phys. Chem.* **5**, 453 (1973).
24. V. S. Marevtsev, A. V. Vannikov, and N. A. Bakh, *Khim. Vys. Energ.* **8**, 52–55 (1974).
25. J. F. Gavlas and L. M. Dorfman, *Int. J. Radiat. Phys. Chem.* **7**, in press (1975).

Theory of Optical Spectra of Solvated Electrons[1]

Neil R. Kestner
Chemistry Department
Louisiana State University
Baton Rouge, Louisiana 70803, USA

The theory of optical spectra is, and must be, but one aspect of the study of solvated electrons. By solvated electrons we shall mean, in this paper, that the electron is localized about some point in space and has an electronic wave function which decreases exponentially with distance from the center of this trap which we will refer to as a cavity. Our cavity will have some finite, although possibly small size. While we will primarily deal in this paper with the optical spectra of this "species," one must be very careful in discussing theories and models of this property to remember that these same models must explain a variety of other properties in order to be really satisfactory. In addition we must state the obvious, namely that this area is very complicated and our models can not be expected to provide exact quantitative agreement with experimental data. They must, however, predict qualitative trends, the major factors or parameters involved, and lead to at least semiquantitative predictions of new phenomena, not known when the model was originally designed. There probably are some experimental quantities, however, which are too subtle for a given model to predict.

While one could discuss in this paper localized electrons in a variety of media (inert gases, hydrocarbons, solids, etc.), we shall concentrate on fluids made up of small polar molecules since much experimental work has been done on these.

The first model for solvated electron, in particular the metal-ammonia solutions, assumed a 'particle in a box' model with only unspecified strong short-range interactions (*1*). Jortner (*2*) was able to adopt polaron theory (*3*) and show that at large distances on the cavity the electron-medium potential is dominated by a potential

$$-\beta e^2/r \text{ where } \beta = (1/D_{op} - 1/D_s)$$

where D_{op} is the optical and D_s the static dielectric constant. Jortner had to introduce an ad hoc parameter to represent the cavity size, R_o, which had to be made zero or at least very small in order to explain the hydrated electron. This theory in both the simplest "adiabatic" form or the more elaborate "self consistent" version (*4*) was quite successful despite its parameters and despite its

[1]Congress presentation E–10–2.

inability to explain any other features of the optical spectrum. It did emphasize the need to include the correct long-range potential.

More elaborate models considered only the short-range interactions via quantum mechanical calculations. They were usually too specific so, using ideas developed in treating electrons in rare gases, Jortner and Kestner (*5*) and especially Copeland, Kestner, and Jortner (CKJ) (*6*) developed a model which was hoped to contain all of the essential features necessary to explain the stability as well as all optical and other properties of solvated electrons in a polar liquid. It was purposely kept simple so the effects due to the various molecular parameters could be easily studied. The adiabatic version for electron-polar medium interaction was used. Kevin, Fueki, Feng and co-workers (*7,8*) have improved this model by using the more complicated self-consistent interaction. In all cases studied thus far the two methods lead to very similar results and therefore, because of the widely differing interests of our two groups, these models have been used to study a wide variety of situations. Gaathon and Jortner (*9*) have also improved upon the CKJ model.

We shall ignore the details of the theory in this paper and concentrate on the results, predictions, and limitations. The equations used in each model have been well documented in the references (*10*). However, in order to understand the model a few general comments are needed. Basically we assume that the surroundings of the electron cavity consists of two regimes: immediately around the cavity is a first coordination layer of some fixed number of molecules (4,6,8, and 12 being studied for convenience); beyond this the medium is treated as a continuum. In the outer continuum polaron theory is assumed to be applicable, but in addition a term called V_0 is added to include the effects of electron medium interaction not involving the polar nature of the medium. This energy V_0 is also the energy of a quasi-free electron (*11*). The molecules in the first layer are represented by point dipoles and an effective intermolecular repulsion (or size). The orientation of the dipoles is dictated by the effective field generated by the trapped electron and thermal fluctuations (roughly according to the Langevin formula). In solving this problem one minimizes the total energy of the system, i.e. the electronic energy and the medium reorganization energy. In cases of mixed solvents and dense gases extra terms must be added so that one can properly minimize the free energy of the system.

Very recently several calculations have been done to remove the inherent simplicity of our model as regards the short-range electron molecule interactions. These range from various types of semi-empirical molecular orbital calculations to the very extensive ab initio calculations of Newton (*12*). In our review we shall refer to these, as appropriate, in representing the restrictions and limitations of our more elementary models.

The most gross parameter of optical spectrum is the position of the band maximum. Assuming that the model correctly explains the properties of the unexcited species, then one can simply calculate the energy to excite the system to the first excited state usually called a $2p$ state by convention. However, the ground and excited state differ greatly from a Coulombic Hamiltonian. The ground state is closer to a 'particle in a box' but with a long-range Coulombic tail. The fact that the excited s state lies above the first p state confirms this (*13*). Most of the results calculated have been based on the one-term exponential

(Coulombic-like) wave function and must be viewed with some caution. Slightly better results are obtained by numerical integration (*14*), using spherical Bessel functions, two-term exponentials, or even gaussian functions (*15*).

In Table 1 we have collected some of the calculations made with these semi-continuum theories. The results are extremely good considering the simplicity of the model.

TABLE 1
Band Maximum

Ammonia	
Experimental (203° K)	0.80 eV
Theoretical	
Copeland, Kestner, Jortner	0.94[a]
Kestner, Logan	1.09[b]
Logan, Kestner	0.92[c]
Kevin, Fueki, Feng	1.18[d]
Gaathon, Jortner (300° K)	0.89[e]
Water	
Experimental (298° K)	1.72 eV
Theoretical	
Copeland, Kestner, Jortner	2.70[a]
Gaathon, Jortner	2.0 [e]
Logan, Kestner	2.02[c]
Fueki, Feng, Kevin	2.15[d]
Ice[d]	
Experimental (77° K)	1.9
Theoretical	
Fueki, Feng, Kevin	1.84
Methanol[d]	
Experimental	2.3(77° K) 1.87(298° K)
Theoretical	
Fueki, Feng, Kevin	2.09 1.85(298° K)
Ethanol[d]	
Experimental	2.3(77° K) 1.80(298° K)
Theoretical	
Fueki, Feng, Kevin	2.15 1.79
Tetrahydrofuran[f]	
Experimental	.585 eV (Room Temperature)
Theoretical	
Jou, Dorfmann	.583 eV (Vo = −0.5 eV)
2-methyl tetrahydrofuran[d]	
Experimental	1.0 (77° K)
Theoretical	
Fueki, Feng, Kevin	1.04
Triethylamine[d]	

Experimental	0.75 (77° K)
Theoretical	
Fueki, Feng, Kevin	0.83

[a]See reference *6*
[b]See reference *13*—small cavity, two-term wave function
[c]See reference *20*—large cavity for water
[d]See reference *8*—small water cavity
[e]See reference *9*—large cavity for water
[f]See reference *21*.

There is one basic difference between the Fueki, Feng, and Kevin results (*8*) and the others. In their calculations the intermolecular repulsions have been ignored except for ammonia. This leads to small cavities in most liquids. No data is available on this point for most liquids. In metal ammonia solutions, volume expansion data confirms a cavity of a few Ångströms' size, but in water the data is not so clear. The recent ab initio molecular plus continuum calculations of Newton (*12*, and private communication) suggest that the cavity in ammonia is only slightly larger than in water. Also, the pressure-shift experiments done by Schindewolf and Olinger (*17*) and by Farhataziz, Perkey, and Hentz (*18,19*) when analyzed in terms of electron volts show larger shifts in the hydrated electron spectra with pressure than does the ammoniated electron. When better representations of the water repulsion are introduced as in the calculations of Gaathon and Jortner (*9*) and Logan and Kestner (*20*), good agreement with experimental data is obtained even with the increased cavity size. This in itself points out the limitations of models, the results in terms of any one observable are not necessarily unique.

The models seem to be consistent with the wide range of experimental results obtained in widely differing fluids. Within any one series of compounds, one can often predict trends based purely on molecular sizes *if* the bulk dielectric properties are similar. For example, one expects the more bulky butanol to have a lower transition energy than its straight chain homolog. This is indeed observed. Likewise, diethyl ether is at a lower energy than dimethyl ether, and methyl tetrahydrofuran is at a slightly lower energy than tetrahydrofuran (*21*), because the molecules are more bulky and thus must have slightly larger cavities, slightly smaller excitation energies.

An improved continuum model which properly allows variations (in a linear fashion only) in the orientational polarization independent of the electronic energy has been developed by Tachiya, Tabata, and Oshima (*22*). They have applied their model to the hydrated electron and obtained a band maximum of 2.04 eV. This model is nice and simple but because it neglects the short range molecular interactions, it is limited in its usefulness.

Several molecular orbital calculations have been carried out on solvated electrons: some recent work includes the important work of Raff and Pohl (*23*); Natori and Watanabe (*24*); Ray (*24*); Aldrich, Gary, and Cusachs (*25*); Ishimaru, Yamabe, Fukui, and Kato (*26*); and Weissmann and Cohan (*27*). The most elaborate semi-empirical calculations have been made by Howat and Webster (*28*). They used the INDO-MO formalism on both the dimer and tetramer models. In their stable tetramer the cavity size was smaller for water than ammonia, but not zero in either case. The excitation energies were 2.08 eV for water and 0.72 eV for ammonia. These calculations represent the most successful application thus far of molecular orbital theory. However, since only short-range interactions are included these results must be carefully considered.

In the case of ammonia and water a reference point is becoming available. Newton (*12*) has carried out very elaborate ab initio calculations on these systems. He finds that $(H_2O)_4^-$ is only barely stable (i.e. less than kT at room temperature). Moreover, he has added to his calculation the long-range components required in the models of Kestner et al. (*10*) and Kevan et al. (*7,8*). In particular, he has used the self-consistent long-range interaction. These factors change both the qualitative and quantitative nature of the results. The model of $(H_2O)_4^-$+ continuum is stable as is the $(NH_3)_4^-$+ continuum. Furthermore, the cavity sizes in the two cases are not very different (ammonia is slightly larger). This is encouraging and suggests that possibly an INDO calculation including the long-range continuum contributions might lead to good agreement with experiment and could provide the intermediate step between the CKJ type models and ab initio calculations (*29*). It would also provide a proper way to introduce temperature, pressure, and mixed solvent effects.

A most interesting series of experiments have been carried out by Schindewolf and Olinger on ND_3 (*30*) and Gaathon, Czapski, and Jortner (*31*) on D_2O. They have studied the optical spectrum of solvated electrons produced by pulse radiolysis in dense gases as a function of gas density and temperature. The observed band maxima change but slowly as a function of density. In supercritical ammonia it changes rapidly below 0.15 g/ml (about 1/4 of the liquid density). Along the vapor-liquid isotherm the D_2O solvated electron shows only a very weak dependence on density down to 0.02 g/ml. It is possible that the low density D_2O results will be modified when supercritical isothermal experiments are completed, but in any case it is clear that over a wide range of densities the spectrum is only moderately affected by density changes. This indicates that the short-range interactions are extremely important.

Gaathon and Jortner (*9*) were able to show that their model (an improved version of CJK) predicted this dependence very well. Since we know that H_2O^- and NH_3^- are unstable, theory should predict that the solvated electron should be unstable below some density. This requires a proper calculation of the entropy necessary to form this complex in the gas phase. Work on this is in progress (*32*) and the initial predictions are that the hydrated electron could be stable down to densities as low as 0.02 g/ml with the ammoniated electron stable to 0.03 g/ml. Better entropy calculations may raise these critical densities somewhat.

Another set of experiments which theoretical work must explain are those involving mixed solvents. Dorfman and co-workers (*21,33,34,35*) have investigated large numbers of binary systems. The optical spectra exhibit very diverse dependences on the mole fraction of the species. In this case there are many theoretical complications since not only must one consider a wide variety of species which can exist but one must properly evaluate their *relative* stabilities. Howat and Webster (*28*) in their INDO calculations considered *only* the species corresponding to the bulk concentration, although this is not necessarily correct. In that case the transition energy varied almost linearly with the mole fraction, not quite in agreement with experimental data.

We have also done calculations on this system but for all possible species of the type $(NH_3)_{4-n}(H_2O)_n$ at five different mole fractions including all factors common to CKJ theory as modified by Gaathon and Jortner. These results are shown in Fig. 1 (*20*). In addition, we have attempted to evaluate the relative stabilities by including entropy of demixing terms, entropy of solvation, and the energy to break solvent bonds in ammonia relative to water (estimated from the heats of vaporization). These simple calculations lead to the dotted curve on Fig. 1. However, the various species at each mole fraction have such similar free energies that better estimates of solvation effects could easily reproduce the experimental curve. It is important to point out that our reasonable agreement is only obtained when the water cavity is made larger by the introduction of intermolecular repulsions in the first coordination layer which are similar for both water and ammonia. Without this feature (now substantiated by Newton's calculations) the $(H_2O)_4$ species would dominate the spectrum at almost all mole fractions contrary to the experimental data. This demonstrates the complexity and the subtle features of the mixed solvent results.

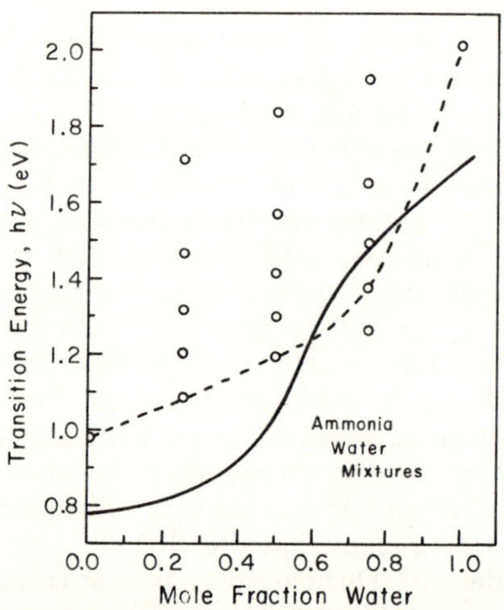

FIG. 1. Ammonia-Water Mixed Solvents: Band Maximum versus Water Mole Fraction. The circled points refer to calculations on various four-coordinated cavities: the highest energy point refers to an $(H_2O)_4$ cavity, the next lowest to $(H_2O)_3NH_3$, and so on to the $(NH_3)_4$ cavity in these solvents. The dashed line connects points based on a preliminary minimum free energy criterion (see text). The experimental results of Dye, DeBacker, and Dorfman (reference *34*) is the solid line.

More subtle features of the optical spectra involve the pressure and temperature dependences of the band maximum. In the early work by Jortner (*2*), the only way to explain the temperature dependence of the spectrum was to arbitrarily assume a change in cavity radius with temperature. However, since temperature is an integral part of the new models, one can calculate temperature dependences directly. In Table 2 we list some results by the CKJ model and its

TABLE 2
Calculations of Solvated Electrons

	NH_3		H_2O	
	exp.	theory	exp.	theory
$h\nu$(eV)[a]($1s \rightarrow 2p$)	0.8	0.89	1.72	2.0
$\Delta_{1/2}$(eV)[a]	0.46	0.18+.07[b]	0.92	0.42+.09[b]
Temp. coeff. of $h\nu$[a,c]	-0.2×10^{-3}	-0.3×10^{-3}	-1.0×10^{-3}	-0.3×10^{-3}
radius(Å)[a]		2.87		2.29
ΔH(eV)[a] cavity heat of solution	1.7±.7	1.02	1.7	1.5
Photo-electric threshold (eV)[a]	1.6	1.94		3.0
V_0(eV)[a]		−0.22		+0.19
Oscillator strength[c] ($1s \rightarrow 2p$)	0.77		0.65	
Momentum formula		0.89		0.80
Dipole length formula		0.64		0.73
Transition energy[d] ($1s \rightarrow 2s$) (eV)		1.34		

[a] See reference *9*
[b] Polaron contribution via Levich formula. See reference *37*.
[c] See reference *15*
[d] See reference *13*

improvement by Gaathon (*9*). When properly calculated, the temperature dependence observed experimentally is accurately reflected in the theoretical results. Of practical importance, most of the change with temperature arises from the change of dipole orientation (Langevin-like factor) in addition to small variations of the bulk parameters.

Schindewolf and co-workers (*17*) and Hentz and co-workers (*18,19*) have also determined pressure shifts of the optical spectrums of the hydrated and ammoniated electron. In Table 3 we show some of the predictions of the CKJ model (as slightly modified by Logan, *20*) for the ammoniated electron. Again the pressure effects are almost completely accounted for by changes in V_0 values (extrapolated from calculations by Gaathon and Jortner) and cavity size changes with pressure. The cavity size does change somewhat as a natural result of the calculation. Gaathon, Kestner, and Jortner (*32*) have found similar results for the hydrated electron and Feng, Fueki, and Kevan (*36*) were able to explain pressure shifts in methanol and ethanol in similar routine calculations.

Another potentially useful piece of experimental data is the total integrated oscillator strength of the absorption band. However, while experiment and theory are in reasonable agreement here, there are theoretical (and probably experimental) problems. To accurately calculate the oscillator strength, accurate

TABLE 3
Pressure Effects
(Ammonia)

At 1.06 kbar (296° K)	
Experimental	0.75[a] eV
Theoretical[b]	0.857 + 0.025[c] = 0.882 eV
($V_o = -0.17$ eV)	
At 6.7 kbar (296° K)	
Experimental	0.91[a] eV
Theoretical[b]	0.923 + 0.090[c] = 1.013 eV
(V_o=0.0 eV)	

[a]See Reference *19*
[b]See Reference *20*
[c]Corrections for cavity size reduction based on 3.2 Å effective radius

wave functions must be used. Kestner et al. (*15*) found that the one-term representation of the wave function usually used is not sufficient to yield reliable data. This is confirmed for continuum models in recent work by Carmichael and Webster (*14*). The best estimates of theoretical predictions for ammonia and water are listed in Table 2.

Up until this point the simple theoretical models are in quite good agreement with the experimental results and predict trends very well. Let us now consider even more subtle properties for which theory is less successful, namely the line width and line shape.

It is only with the recent models that we have been able to even attempt to calculate line shapes for solvated electrons. This is possible because one obtains potential curves of the energy of the various states as a function of some nuclear parameters. In most of the work one has used the variation with cavity size about the equilibrium position (i.e. a totally symmetric vibration), but Kestner and Jortner (*37*) have also considered some dipole orientation fluctuations.

From potential curves or the more elaborate surfaces of reference *37*, we can calculate the line shape via semiclassical theory. This can be done because all of the width originates from the spherically symmetric modes. In addition to this, there are fluctuations in the long-range polaron interactions. For ammonia they contribute about 0.07 eV to the half-width (0.09 eV for water) at normal conditions. These contributions are similar to the polarization fluctuations studied by Tachiya et al. (*22*) except we only include the effects of the medium beyond the first coordination layer.

The first observation we can make is that line widths predicted are much smaller than experimental results (see Table 2 for example). At least as disturbing, if not more so, is the shape. Experimentally almost all solvated electrons have optical spectra which are skewed to the high energy side and are not symmetrical as the theory suggests (Fig. 2). In addition, if we calculate the half-width by any of the new models, it will increase roughly as the square root of the temperature. Elaborate experimental studies on electrons in ammonia show that the line width is almost independent of the temperature. Furthermore, the line width does not appear to be dependent on solvent density ruling out any strong competition among cavities of various coordination numbers.

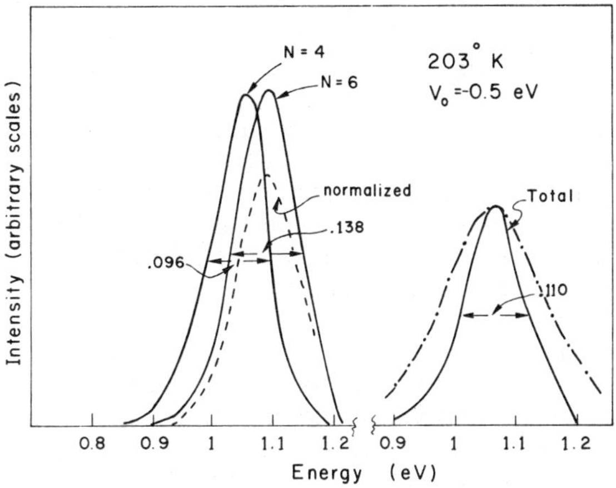

FIG. 2. Calculated Line Shapes: for specific coordinations (left) and for the total sum (right). The dot-dash line shows the effect of including long-range polaron fluctuations.

There are several possible explanations for this qualitative discrepancy but almost all can be eliminated. These include a larger variety of coordination numbers than we have calculated here, with an almost continuous distribution of energies; contributions of higher excited states, even the continuum. The latter can be estimated and should have almost no oscillator strength (see Fig. 3). These excited levels should also be greatly affected by the density of the fluid since they depend directly on β, and yet lines in very low density water vapor are as broad as those in the liquid.

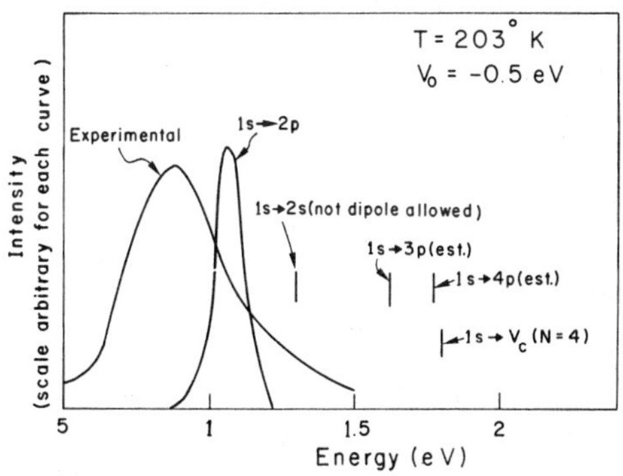

FIG. 3. Comparison of calculated and experimental line shapes as well as the position of higher excited states. Calculated curves do not include long-range polaron fluctuations. [Reprinted with permission from N. R. Kestner and J. Jortner, *J. Phys. Chem.* **77,** 1040 (1973). Copyright by the American Chemical Society, Washington, D.C.]

Thus the question of the extremely broad lines remain. Two other explanations have been proposed which have not been fully evaluated as yet: asymmetric fluctuations (estimates indicate that asymmetric modes have a very small force constant; one calculates that *at most* they could contribute $kT \ln 2$ or 0.012 eV to the half-width) and permanent distortion of the cavity from its spherical shape (this could arise from the rapid exchange of solvent molecules between the first coordination layer and the bulk medium. At any one instant one would have a variety of cavity shapes). Although we cannot treat this latter problem easily without major changes in our model, one can argue that this could lead to asymmetric lines with little temperature dependence if non-spherical cavities had larger transition energies as well as different transition energies for the $1s \rightarrow 2p_0$ and $2p_{\pm 1}$ states (a fact which we have not been able to verify thus far; similar results can be found in reference *22*). The problem of line shape in mixed solvents is even less well explained by our theoretical models.

There has been one proposal for the observed line shape due to Kajiwara, Funabashi, and Naleway (*38*) which yields the correct line shape but involves two drastic and extremely radical assumptions: that the bound electron should be treated as a particle in a finite box with no long-range interactions, and that there is only one bound state in the cavity. Both assumptions can not be justified in any complete model.

In addition to the normal solvated electron, one can also discuss the two-electron cavity species especially in ammonia. The work of Fueki, Feng, and Kevin (*40*) and the corrections of the Copeland and Kestner (*41*) paper by Logan and Kestner (*20*) predict a stable species, i.e. more stable than two one-electron cavities. Using only a one-term wave function the spectra of the one- and two-electron species are similar (Table 4), but when more accurate wave functions are used the two electron cavity becomes larger. At the moment there is little experimental data to unambiguously confirm any findings of the models.

TABLE 4
Electrons in Ammonia[a]
(One-term wave function, 203° K)

One-Electron Cavity (N=4)	
ΔH_1, heat of solution =	−0.9413 eV
transition energy =	0.926 eV
Two-Electron Cavity	
ΔH_2, heat of solution =	−2.601 eV
transition energy =	1.21 eV (1.29 eV)
$\Delta H_{2_1} = (1/2)\Delta H_2 - H_1$ =	−0.259 eV (stability of two electron species)

[a]See reference *20*

Models containing both a discrete solvation layer and approximating the remainder of the liquid by a continuum are remarkably successful in treating the main features of electrons in liquids, but in their present form do not do so well on details such as line shape and width. Encouraging for the future are semi-empirical calculations modified to include long-range effects. Detailed calculations are still needed to check out a few cases accurately.

Experimentally we need more work on mixed solvent systems including studies in the vapor state (via optical and mobility measurements) as well as studies to unravel the contributions to the optical line shape and locate the positions of the higher excited states and emission. Major new theoretical work is needed on time-dependent phenomena: mobility (*42*), trapping rates (*43*), time development of spectra, and reactivity since these could provide some critical tests of theoretical models.

REFERENCES

1. R. A. Ogg, *J. Am. Chem. Soc.* **68,** 155 (1940).
2. J. Jortner, *J. Chem. Phys.* **30,** 839 (1959).
3. See, for example, G. R. Allcock, *Advan. Phys.* **5,** 412 (1956).
4. J. Jortner, *Mol. Phys.* **5,** 257 (1962). Calculations are contained in J. Jortner, S. A. Rice, and E. G. Wilson, in *Solutions Metal-Ammonia: Colloque Weyl* (G. Lepoutre and M. J. Sienko, Eds.) pp. 222–276. Benjamin, New York, 1964.
5. J. Jortner and N. R. Kestner in *Metal-Ammonia Solutions: Colloque Weyl II* (J. J. Lagowski and M. J. Sienko, Eds.) pp. 49–103. Butterworths, London, 1970.
6. D. A. Copeland, N. R. Kestner, and J. Jortner, *J. Chem. Phys.* **53,** 1189 (1970). Some of this theory is contained in the earlier work of D. E. O'Reilly, *J. Chem. Phys.* **41,** 3736 (1964); and R. H. Land and D. E. O'Reilly, *ibid.* **46,** 4496 (1967).
7. See, for example, K. Fueki, D.-F. Feng, L. Kevan, and R. Christofferson, *J. Phys. Chem.* **75,** 2291 (1971); and other references in this paper and reference *8*.
8. K. Fueki, D.-F. Feng, and L. Kevan, *J. Am. Chem. Soc.* **95,** 1398 (1973).
9. A. Gaathon and J. Jortner in *Electrons in Fluids. The Nature of Metal-Ammonia Solutions* (J. Jortner and N. R. Kestner, Eds.) pp. 429–446. Springer-Verlag, Berlin, 1973.
10. See, especially N. R. Kestner in *The Nature of Metal Ammonia Solutions* (J. Jortner and N. R. Kestner, Eds.) pp. 1–25. Springer-Verlag, Berlin, 1970, or references *6* and *9*.
11. B. E. Springett, M. H. Cohen, and J. Jortner, *J. Chem. Phys.* **48,** 2720 (1968).
12. M. Newton, *J. Chem. Phys.* **58,** 5833 (1973); and private communication.
13. J. Logan and N. R. Kestner, *J. Phys. Chem.* **76,** 2738 (1972).
14. I. Carmichael and B. Webster, submitted to *J. Chem. Soc. Faraday Trans. II.* **70,** 1570 (1974).
15. N. R. Kestner, J. Jortner, and A. Gaathon, *Chem. Phys. Lett.* **19,** 328 (1973).
16. D.-F. Feng, D. Ebbing, and L. Kevan, *J. Chem. Phys.* **61,** 249 (1974).
17. U. Schindewolf and R. Olinger, reported by U. Schindewolf in *Metal Ammonia Solutions: Colloque Weyl II* (J. J. Lagowski and M. J. Sienko, Eds.) pp. 199–215. Butterworths, London, 1970.
18. R. R. Hentz, Farhataziz, and E. M. Hansen, *J. Chem. Phys.* **55,** 4974 (1971). (H_2O data)
19. Farhataziz, L. M. Perkey, and R. R. Hentz, *J. Chem. Phys.* **60,** 4383 (1974). (NH_3 data)
20. J. Logan and N. R. Kestner (work in progress).
21. F. Y. Jou and L. M. Dorfman, *J. Chem. Phys.* **58,** 4715 (1973).
22. M. Tachiya, Y. Tabata, and K. Oshima, *J. Phys. Chem.* **77,** 263 (1973); **77,** 2286 (1973). This is based on an improved model due to K. Iguchi, *J. Chem. Phys.* **48,** 1735 (1968). In addition their polarization term contains an adjustable parameter.
23. L. Raff and H. A. Pohl, *Advan. Chem.* Series No. 50, 173 (1965).
24. M. Natori and T. Watanabe, *J. Phys. Soc. Japan* **21,** 1573 (1966).
25. M. S. Aldrich, L. P. Gary, and L. C. Cusachs (unpublished work, 1970).
26. S. Ishimaru, H. Kato, T. Yamabe, and K. Fukui, *J. Phys. Chem.* **77,** 1450 (1973); *Chem. Phys. Lett.* **17,** 264 (1972); and S. Ishimaru, H. Tomita, T. Yamabe, K. Fukui, and H. Kato, *Chem. Phys. Lett.* **23,** 106 (1973).

27. M. Weissmann and N. V. Cohen, *Chem. Phys. Lett.* **7**, 445 (1970); *J. Chem. Phys.* **59**, 1385 (1973).
28. G. Howat and B. C. Webster, *J. Phys. Chem.* **76**, 3714 (1972).
29. In reference *27* this was attempted via a perturbation approach.
30. R. Olinger, S. Hahne, and U. Schindewolf, *Ber. Bunsenges. Phys. Chem.* **76**, 349 (1972).
31. A. Gaathon, G. Czapski and J. Jortner, *J. Chem. Phys.* **58**, 2648 (1973).
32. A. Gaathon, N. R. Kestner, and J. Jortner (work in progress).
33. See, for example, L. M. Dorfman and F. Y. Jou in *Electrons in Fluids. The Nature of Metal-Ammonia Solutions* (J. Jortner and N. R. Kestner, Eds.) pp. 447–457. Springer-Verlag, Berlin, 1973.
34. J. L. Dye, M. G. DeBacker, and L. M. Dorfman, *J. Chem. Phys.* **52**, 6251 (1970).
35. L. M. Dorfman, F. Y. Jou, and R. Wageman, *Ber. Bunsenges. Phys. Chem.* **75**, 681 (1971).
36. D.-F. Feng, K. Fueki, and L. Kevan, *J. Chem. Phys.* **57**, 1253 (1972).
37. N. R. Kestner and J. Jortner, *J. Phys. Chem.* **77**, 1040 (1973).
38. T. Kajiwara, K. Funabashi, and C. Naleway, *Phys. Rev.* **6**, 808 (1972).
39. The continuum level is much higher in energy. See, for example, the excellent photoelectron emission (PEE) experimental work of H. Avlich, B. Baron, P. Delahay, and R. Lugo, *J. Chem. Phys.* **58**, 4439 (1973) and their earlier papers.
40. D.-F. Feng, K. Fueki, and L. Kevan, *J. Chem. Phys.* **58**, 3281 (1973).
41. D. A. Copeland and N. R. Kestner, *J. Chem. Phys.* **58**, 3500 (1973).
42. Initial attempts in this have been made by M. Tachiya and A. Mozumder, *J. Chem. Phys.* **60**, 3037 (1974).
43. Interesting ideas in this area are contained in the work of G. Nillson, *J. Chem. Phys.* **56**, 3427 (1972); and K. Fueki, D.-F. Feng, and L. Kevan, *J. Phys. Chem.* **78**, 393 (1974).

Electron Yields and Reaction Kinetics in Polar Liquids[1]

John W. Hunt, K. Y. Lam and W. J. Chase
Department of Medical Biophysics
University of Toronto
and the Ontario Cancer Institute
Toronto, Ontario
Canada
M4X 1K9

INTRODUCTION

In polar solvents the electrons produced by high energy radiation are localized or trapped by the surrounding solvent molecules in a delicate balance of short and long-range interactions. The pulse radiolysis studies of Baxendale and Wardman in cold alcohols (*1,2*) have shown that in the infrared there is an earlier intermediate species called the damp electron, e^-_{damp}, about which the polar solvent molecules relax to form the solvated electron, e^-_{sol}.

In μsec times a homogeneous distribution of e^-_{sol} has been observed from which the homogeneous or free ion yield, $g(e^-_{sol})_{fi}$, has been determined. In water $g(e^-_{aq})_{fi}$ has been measured to be 2.7/100 eV by several techniques. Initially the ions are clustered around the tracks of the ionizing particles to form spurs. Using nsec and psec techniques, Jonah et al. (*3*) and Wolff et al. (*4*) have studied the yields and spur decay of e^-_{aq}. In their studies, the yield (at 30 psec) of e^-_{aq} is 4.0 ± .2/100 eV (*4*). More recently Wolff et al. (*5*) have shown that the total yield of electron products, $G(S^-)_{total}$, is 4.8/100 eV, somewhat larger than $g(e^-_{aq})$ measured at 30 psec and indicating that fast recombinations occur in the spur.

In concentrated aqueous solutions of scavengers, fast electron reactions have been studied at psec times (*6,7*) using a stroboscopic pulse radiolysis (SPR) technique (*8*). The rate constant, $k(e^-_{aq} + S)$ for the reactions,

$$e^-_{aq} + S \longrightarrow S^- \tag{I}$$

are different in dilute and concentrated solutions. The main differences are due to the ionic strength of the solutions (*9*), while minor differences can be explained by the time-dependent rate constants described in detail by Schwarz (*10*). As well, the observed initial yield of e^-_{aq} is reduced markedly in concentrated (0.1 to 5 *M*) solutions of scavenger (*6*). Hamill (*11*), and Hunt and his co-workers (*4,7*), suggested that there is a fast electron process in concentrated solutions which can be explained by assuming a "dry" electron, e^-_{dry}, which diffuses quickly through the medium and reacts with solute molecules

[1] Congress presentation E–10–3.

along its path before e^-_{aq} is formed. Similarly, Freeman (*12*) has explained the neutralization processes using a mobile electron, $e_{quasi\text{-}free}$, for his analysis. Alternatively, a time-dependent rate constant described by Schwarz (*10*), or the encounter pair model of Czapski et al. (*13*) might explain these reactions.

This paper is an extension of the psec SPR studies in water to studies in alcohols. The reduced yield of e^-_{sol} in concentrated solutions of solutes has been followed in alcohols in wavelengths of 300 to 1050 nm, with strong evidence being found for fast electron processes for precursors of the e^-_{sol}.

This paper also describes the yield of e^-_{sol} and the total yield of electron products in alcohols. The yields of e^-_{sol} are not as well known as that in water—the early studies of Sauer et al. (*14*) indicate that $g(e^-_{sol})_{fi}$ is 1.1/100eV for methanol and 1.0/100 eV for ethanol. However, recent evidence suggests that the yields of e^-_{sol} are larger, with values of 2.0/100eV for methanol and 1.7/100eV for ethanol (*15*). Recently, Kenney-Wallace et al. (*16*) have estimated decays of e^-_{sol} from the spur. In our paper, yields of e^-_{sol} at 150 psec were determined relative to the yields of e^-_{aq}. The total yield of electron products were explored in alcohols and other organic solvents using the SPR technique.

EXPERIMENTAL

The SPR system has been used to detect the transient species produced by the fine structure pulses (spaced 350 psec apart) of the linear accelerator at the University of Toronto. Small absorption signals ($\geqslant 0.2\%$) can be observed with a time resolution as short as 20 psec. Wavelengths of 300 to 1050 nm have been studied using a solid state detector, the PIN-10 manufactured by United Detector Technology. The detection and signal processing system have been described (*8*). It should be mentioned that the observed absorption signals are the summation of the signals produced by the train of fine structure pulses by the linear accelerator. This simplified analysis is correct only for first order kinetics. Therefore, some inaccuracies may occur in complex reactions in the analysis of the spur decays. All of the observed yields are compared to the absorption signals in e^-_{aq}, and assuming the value of $g(e^-_{aq})_{30\ psec} = 4.0$ (*4*).

All chemicals were used as received from the supplier. The solutions were bubbled with argon gas before irradiation.

RESULTS AND DISCUSSION

Correlation of the efficiency of initial e^-_{sol} *yield reduction with* $k(e^-_{sol} + S)$

The initial yield of e^-_{sol} was obtained by extrapolating the decay observed on a picosecond kinetic tract back to the midpoint of the time response of the fine structure pulse. A reduced yield of e^-_{aq} has been obtained in aqueous solutions containing high concentrations of scavenger. A more marked reduction of e^-_{sol} occurred in alcohols. In both cases a plot of $\ln([e^-_{sol}]_S/[e^-_{sol}]_o)$ vs $[S]$ produces a straight line; the subscripts are the concentrations of the scavengers. The $[S_{37}]$ value is defined as the scavenger concentration which reduces the yield of e^-_{sol} to the 37% level. Figs. 1 and 2 show the relationship between the $[S_{37}]$ values and the rate constants of the reaction (I), Fig. 1 is the scavenging process in water. There is little correlation between the $k(e^-_{aq} + S)_{dil}$ in

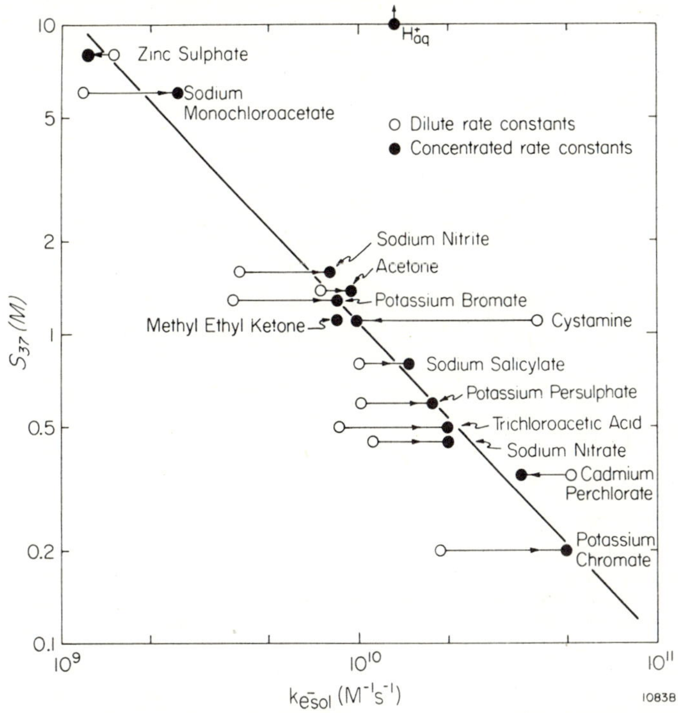

FIG. 1. Plot of $[S_{37}]$ values of scavengers in water and their reaction rate constants $k_{e^-_{aq}}$ with hydrated electron on logarithm-logarithm scale. The open circles, o, are points using $k_{e^-_{aq}}$ measured in dilute solutions taken from literature, and black circles, •, are points using $k_{e^-_{aq}}$ measured in concentrated solutions as obtained in our laboratory. The straight line is drawn with a slope of −1. Note that the arrows joining the o and • points are only for illustrating the directions of changes of k's from dilute to concentrated solutions. Error bars of the points are not shown in the plot. Estimates of errors in parameters measured are shown in Table 1. (Reprinted with permission from K. Y. Lam and J. W. Hunt, *J. Phys. Chem.*, in press. Copyright by the American Chemical Society.)

dilute solutions (open circles) and the $[S_{37}]$ values, while there is a good correlation when the $k(e^-_{aq} + S)_{conc}$ in concentrated solutions (solid circles) are considered. The relationship between $[S_{37}]$ and $k(e^-_{sol} + S)_{conc}$ can be expressed by the equation

$$[S_{37}] \cdot k(e^-_{sol} + S)_{conc} = Q \tag{1}$$

in which $Q = 1.0 \pm 0.2 \times 10^{10}$ sec^{-1} H_2O. The only scavenger which does not fit with this relationship is the solvated proton, H^+_{sol}.

Fig. 2 compares the scavenging efficiency of three solvents—H_2O, ethylene glycol and ethanol. Scavengers in ethanol have a smaller $[S_{37}]$ than they do in water. Again there is a fair relationship between the $[S_{37}]$ and $k(e^-_{sol} + S)_{conc}$, with the constant Q having the smaller value of $1.3 \pm 0.3 \times 10^9$ sec^{-1} in ethanol. The scavenging constant Q in ethylene glycol is intermediate between those in water and ethanol.

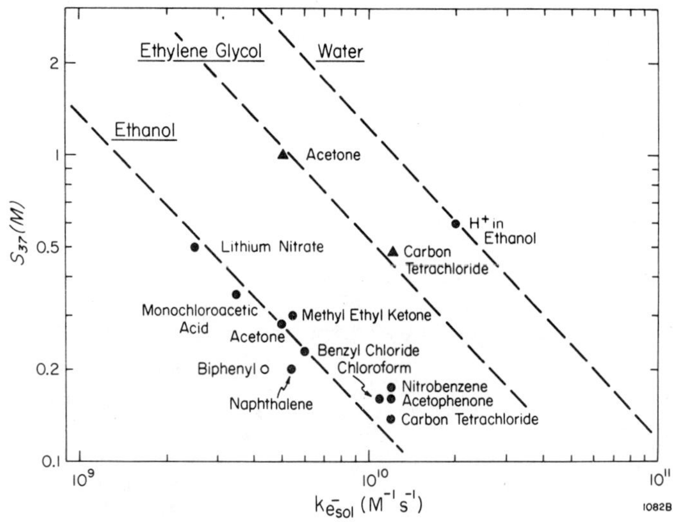

FIG. 2. Plot of [S_{37}] values of scavengers in ethanol and their reaction rate constant k with e^-_{sol} on logarithm-logarithm scale. The black circles, ●, are k measured in concentrated solutions as obtained in our laboratory, and the open circle is the k measured in dilute solutions (taken from the literature). Error bars of the points are not shown in the plot. Estimates of errors in the parameters measured are shown in Table 3. The line labelled "water" is the curve in Fig. 1. The lines labelled "ethylene glycol" and "ethanol" are best fitting lines drawn with slope −1. (Reprinted with permission from K. Y. Lam and J. W. Hunt, *J. Phys. Chem.*, in press. Copyright by the American Chemical Society.)

Table 1 is a selected list of [S_{37}] values in different polar solvents. Several points should be noted. First, for a given scavenger, the scavenging efficiency increases towards the right (the scavenging efficiency is proportional to $1/[S_{37}]$). Secondly, the scavenging process is not as efficient for H^+_{sol} as compared with other scavengers. The [S_{37}] values for H^+_{sol} fall well outside the family of curves shown in Figs. 1 and 2. Thirdly, the [S_{37}] values change in a regular way with different solvents. There is no simple relationship, however, between the [S_{37}] values and the molecular weight of the solvent. As well, [S_{37}]

TABLE 1

Observed [S_{37}] Values in Different Polar Solvents at 700 to 800 nm.

Solute	H_2O	Ethylene Glycol	Methanol	Ethanol	1-Propanol
Acetone	1.4±0.1	1.0 ±0.1	0.5 ±0.05	0.3 ± 0.03	0.25 ± 0.05
Methyl Ethyl Ketone	1.1±0.1	0.8 ±0.2	0.35±0.05	0.3 ± 0.05	0.21 ± 0.03
Carbon Tetrachloride		0.36±0.1	0.17±0.03	0.14± 0.05	0.11 ± 0.02
Biphenyl			0.3 ±0.03	0.19± 0.02	0.18 ± 0.04
H^+_{sol}	≥10	1.3 ±0.2	0.7 ±0.1	0.6 ± 0.1	0.5 ± 0.1

values in comparable scavengers are smaller in alcohols than in water. There is a fair relationship between the formation time, τ_{sol}, of e^-_{sol} in different solvents (*17*) and $1/[S_{37}]$.

Formation and Reactions of e^-_{sol} in Cold Solutions

Some recent SPR experiments have been carried out to examine the relationship between τ_{sol} and the scavenging process. These experiments consisted of observing absorption traces at various temperatures and wavelengths. The absorption spectra in water and ethanol are shown in Fig. 3. The spectrum at 30 psec in water is identical to the well-known spectrum in μsec times. In

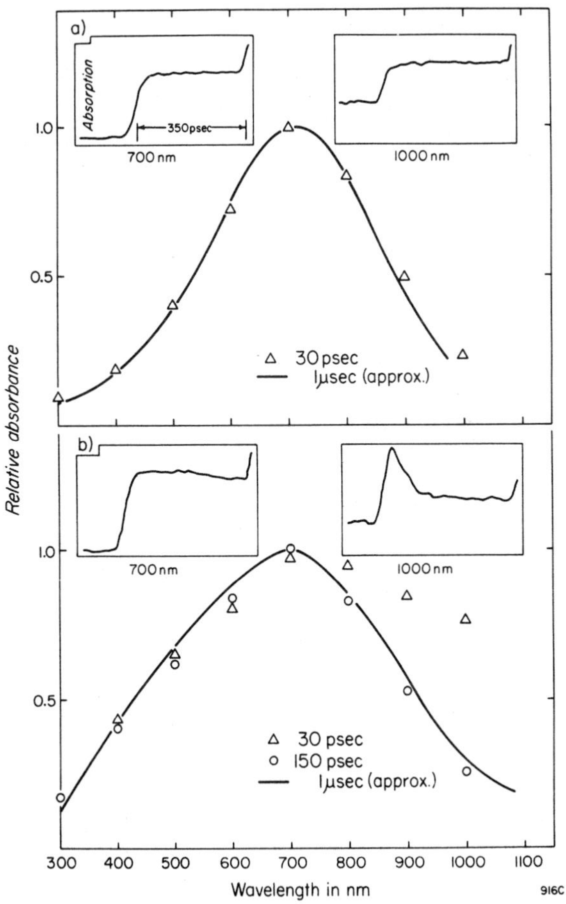

FIG. 3. Picosecond spectral changes for e^-_{sol} in water and ethanol at room temperature. The microsecond spectra are redrawn from
a. M. S. Matheson, in *Solvated Electron*: A Symposium Sponsored by the Division of Physical Chemistry (E. J. Hart, Ed.) pp. 45-54. Advan. in Chem. Ser., No. 50, American Chemical Society, Washington, D.C. 1965.
b. M. Dorfman, *ibid.* pp. 36-44. The insets show the formation and decay of the absorption signals at 700 nm and 1000 nm.

ethanol, the absorption signals change with time as shown in the insets. At 30 psec the absorption spectrum contains signals in the infrared which decay quickly, leaving a spectrum at 200 psec which is identical to that formed in μsec times. The 30 psec infrared spectrum is similar to that of Baxendale and Wardman (*1*,*2*), which they attribute to a precursor of e^-_{sol}, e^-_{damp}.

Fig. 4 is a family of psec traces at 1050 nm, with temperatures from −50 to +45°C, showing the slower relaxation time from e^-_{damp} to e^-_{sol} at lower temperatures. Fig. 5 shows the [S_{37}] values in ethanol containing the scav-

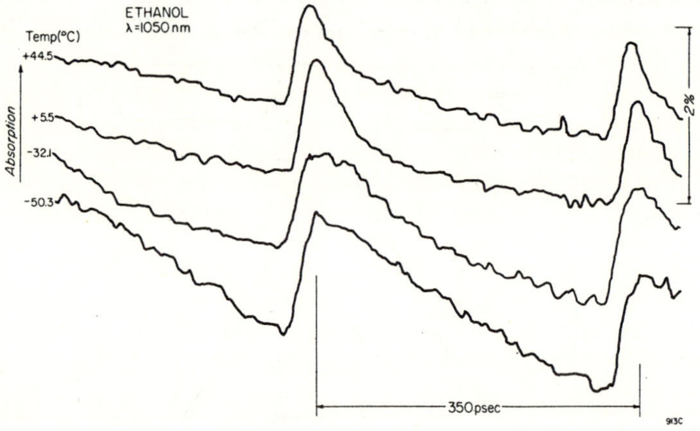

FIG. 4. Picosecond kinetic absorption traces in ethanol at a wavelength of 1050 nm showing the increase in decay rates of e^-_{damp} with increasing temperature.

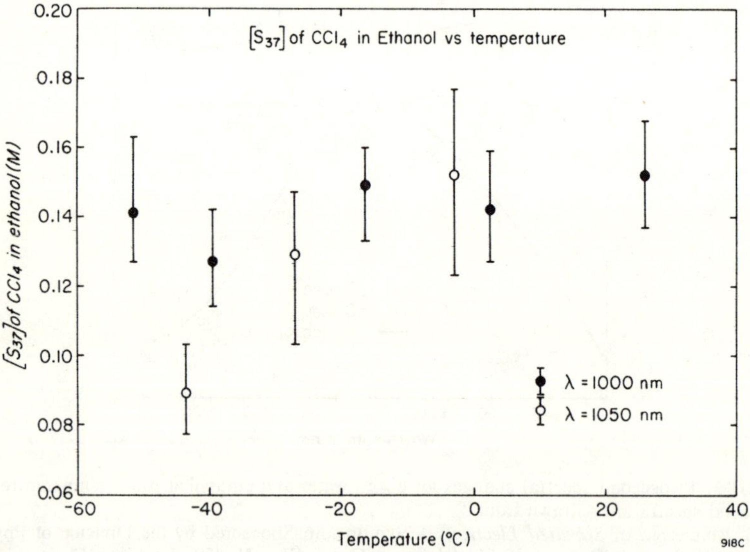

FIG. 5. Plot of [S_{37}] of CCl_4 in ethanol vs temperature. The preliminary date at wavelengths of 100 and 1050 nm suggests that there are no marked changes in the [S_{37}] with temperature change. At room temperature, a [S_{37}] value of 0.14 was also obtained at 750 nm.

enger. These were observed at various temperatures. The observed results suggest that the $[S_{37}]$ value is relatively independent of temperature. Along this temperature range, the τ_{sol} changes by a factor of >5. This indicates that the fast electron process is not related directly to the formation time of e^-_{sol}, but instead from a precursor of e^-_{sol}, such as e^-_{damp} or an electron which occurs in earlier steps of the solvation process.

The Initial Yield of Electrons

Yields of e^-_{sol}

The yield of e^-_{sol} has been redetermined relative to the $g(e^-_{aq}) = 4.0$ at 30 psec. The low yield of e^-_{sol} in alcohol of ~ 1.5/100eV obtained by Bronskill et al. (*18*) and Wolff et al. (*4*) was in error for several reasons. As shown in Fig. 3, the spectrum of e^-_{sol} at 30 psec should not be used, because the e^-_{damp} spectrum interferes with the true e^-_{sol} signal. Secondly, Jha et al. (*15*) have determined considerably smaller molar optical extinctions for the e^-_{sol} in alcohols compared to the previously used values of Sauer et al. (*14*). The former extinctions were used in this paper. Thirdly, more accurate dosimetry corrections have been used. The new values of $g(e^-_{sol})_t$ are summarized in Table 2.

TABLE 2

Yield of Electrons

	$\epsilon(M^{-1}cm^{-1})$	$g(e^-_{sol})_{free\ ion}$	$g(e^-_{sol})_t$	t	$G(S^-)$
H_2O	$\epsilon_{575}(e^-_{aq}) = 1.04 \times 10^4$ [a]	2.7 ± 0.1	4.0 ± 0.3	30 psec	4.8 ± 0.3
Methanol	$\epsilon_{\lambda_{max}}(e^-_{sol}) = 10.2 \times 10^3$ [b]	2.0	3.3 ± 0.3	150 psec	4.0 ± 0.4
Ethanol	$\epsilon_{\lambda_{max}}(e^-_{sol}) = 9.4 \times 10^3$ [b]	1.7	3.2 ± 0.3	150 psec	4.0 ± 0.4
Cyclohexane					4.2 ± 0.4

[a] E. M. Fielden, E. J. Hart, *Rad. Res.*, **32**, 564 (1967).
[b] See Reference *15*.

Total Yield of Electron Products

The intense absorption spectra of the electron adducts of several scavengers have been studied. In concentrated solutions (.01 to 1 *M*), the absorption signal consists of two components—a fast formation of S^-, followed by reaction (1). The fast component increased with scavenger concentration. At the highest concentrations at least 90% of S^- are scavenged.

The scavengers used were sodium salicylate, Cd^{++} and biphenyl in alcohols, and biphenyl in cyclohexane and cylohexane-ethanol mixtures. In all cases the psec spectra were identical to μsec spectra and were the same from solvent to solvent (*19*). The total yield $G(S^-)_{total}$ is similar to that of water as shown in column 4 of Table 2. There appears to be a close agreement between the yield of ions in vapors and in liquids. The yield of the products also suggests that most of the electrons occur by direct ionization rather than excitation.

In agreement with the studies of Wolff et al. in aqueous solutions, there is a larger yield of electron adducts, $G(S^-)_{total}$ than $g(e^-_{sol})_{150psec}$. This can be explained by assuming a fast neutralization and reaction of the electrons before the stabilized e^-_{sol} have been formed.

Models of the Fast Electron Processes

The reduced yields of e^-_{sol} and the fast electron attachment to the scavengers do not fit with any normal kinetic mechanisms. The two most likely candidates to explain the abnormal fast processes are the "dry" electron models (*6,7,11*) and the large effective radius of the electron proposed by Czapski (*13*) and by Miller (*20*).

In the first model, the secondary electrons ejected by radiation diffuse some distance through the medium and react with the solute molecules on the way. Finally, the polar molecules align around the electron to form the relaxed e^-_{sol}. The scavenging efficiency from medium to medium would depend upon the drift distance of the electrons before they form the e^-_{damp}. The weakness of this model is the close relationship between $1/[S_{37}]$ and the $k(e^-_{sol} + S)_{conc}$. One might expect large differences in the relative scavenging efficiencies between e^-_{dry} and e^-_{sol} from solute to solute molecule. Only one, H^+_{sol}, does not agree with this general relationship.

Czapski explains the fast process in water by defining an encounter pair volume around the electron in which a reaction between the $(e^-_{aq} + S)$ occurs in times shorter than the resolution of the SPR system ($<10^{-11}$ sec). An effective reaction radius, r_{eff}, defines an encounter pair volume around the e^-_{sol}, and explains the reduced yields of e^-_{sol}. A similar analysis can be used for a quantum mechanical tunnelling model from the electron to the scavengers proposed by Miller (*20*). The reduced yield of e^-_{sol} and the r_{eff} in alcohols can be given by the equations,

$$[e^-_{sol}]/[e^-_{sol}]_o = \exp(-[S]/[S_{37}]) \tag{2}$$

$$\text{and } r_{eff} = 7.35/[S_{37}]^{1/3} \text{ in Å} \tag{3}$$

These equations are simplifications of the analysis of Czapski et al. (*13*), making use of the $[S_{37}]$ values and practical units. For example, the $[S_{37}]$ is 0.14 *M* for CCl_4 in ethanol, from which $r_{eff} = 14$Å, was calculated from equations (2) and (3). Similar r_{eff} values have been found in efficient scavengers in water. The encounter pair volume for water is quite plausible, but Czapski's model does not explain certain crucial details of the fast electron processes in other media, such as the large encounter pair volume, and the relationship between the $[S_{37}]$ values and $k(e^-_{sol} + S)$. For example, the rate constant $k(e^-_{sol} + S)$ of CCl_4 in ethanol is $1.25 \times 10^{10}\ M^{-1}\text{sec}^{-1}$. One can calculate an r_{eff} from the Smoluchowski equation for neutral solutes:

$$k = 4\pi N r_{eff}(D_{e^-_{sol}} + D_s)/1000 \tag{4}$$

The diffusion coefficient De^-_{sol} in ethanol has been calculated by the conductivity studies of Fowles (*21*), at $0.7 \times 10^{-5}\text{cm}^2\text{sec}^{-1}$, and the D_s for CCl_4 in ethanol has been calculated by the work of Hammond and Stokes (*22*), at $1.5 \times 10^{-5} cm^2 sec^{-1}$. The calculated r_{eff}, 7.5Å, is smaller than the previously calculated value of 14Å. This is not surprising since the experimental evidence shown in this paper indicates that the fast reactions occur from an electron precursor (i.e. a e^-_{damp} or earlier). This is in general agreement in the pulse radiolysis studies of Baxendale and Wardman in cold alcohols (~−100° C), in which a similar reduction of the yield of e^-_{damp} has been found (*2*).

The new facts described in this paper suggest that none of these models gives an accurate picture of the observed data. Some of the crucial points are as follows:

1) Close relationship between the fast electron processes and the slower reaction, $e^-_{sol} + S$.
2) In alcohols, a reduction of the initial yields of e^-_{damp}, ($[S_{37}]$values), occur in the presence of scavengers. These values are similar to the values obtained in the studies of e^-_{sol} at room temperature.
3) The $[S_{37}]$ values appear to be fairly independent of the e^-_{sol} formation time τ_{sol} for the different temperatures in the same solvents.
4) The order of the scavenging constant $[S_{37}] \cdot k(e^-_{sol} + S)_{conc} = Q$, is similar to the order of $1/\tau_{sol}$.

The observed results, summarized above, might be explained by this model:

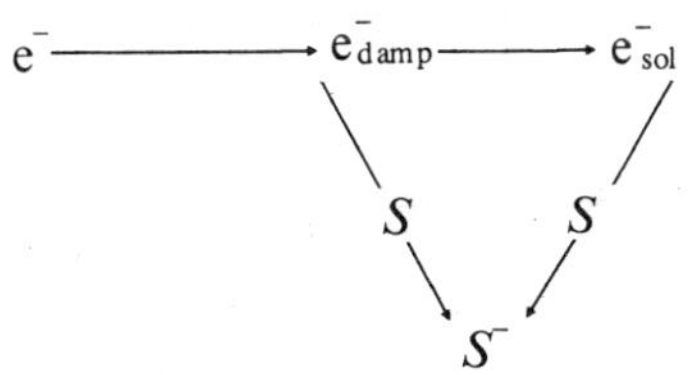

(II)

The fast processes could be explained by a long-range interaction between the electron and the solute molecule S, not from the e^-_{sol}, but instead, from the e^-_{damp}. The long-range interaction could be a result of the large effective radius of the reaction, either from a quantum mechanical electron tunnelling process (*20*), or a large encounter pair volume of the reaction (*13*). The fast electron reactions might be expressed by a competition between the reaction, $e^-_{damp} + S$, and the solvation process, τ_{sol}. One weakness of this model is that point 3) above, cannot easily be explained satisfactorily. However, a more detailed model of τ_{sol} and $[S_{37}]$, might explain the fast electron processes.

A quite different model is proposed here which seems to explain most fast electron data.

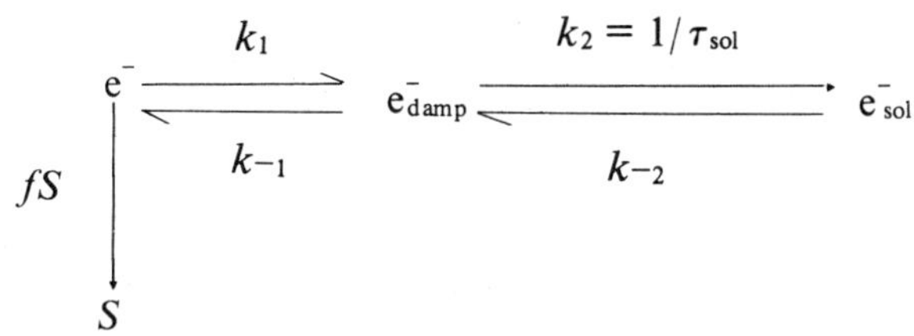

(III)

Two main assumptions could explain both the normal diffusion reactions and the fast electron processes.

1) The e^-_{sol} is unstable, with back reactions k_{-1} and k_{-2} to e^- (e^- could equally be the $e^-_{quasi\text{-}free}$ or e^-_{dry}).

2) All of the electron reactions occur only from the e^-, and not from the e^-_{sol} or e^-_{damp}.

This proposal is essentially the same as the model of Schmidt and Allen (*23*) for a weakly polar solvent, *n*-hexane. The intermediate mobilities of the electrons can be explained by making the assumption that the electron spends part of its lifetime as a localized state of the e^-_{sol}, and part of its lifetime in a mobile state. A similar model might be envisioned in water and alcohols, in which there is a small but finite probability that the electron can escape from the electron trap. This is due to molecular fluctuations around the electron.

The normal diffusion reactions would be limited mainly by the back reactions, k_{-2} and k_{-1}. The rate constants would include an efficiency factor, *f*, for the reaction, $e^- + S$. The mobile electrons will move some distance until the electron finds a trap in the medium and becomes the e^-_{damp} or, alternatively, the electron, on the way, reacts with a solute molecule. The reduced yield of e^-_{sol} can also be explained by the stochastic models of Freeman (*12*) and Wolff (*6*), which agree with logarithmic scavenging processes expressed in equation (2). This model could also explain the general relationship between the fast electron reaction and $k(e^-_{sol} + S)_{conc}$ in different scavengers, since all of the reactions would occur from the same species, the e^-.

There are several important weaknesses of this model as well. In the present theories for the structure of the solvated electron, the energetics for an electron escaping from a trap is very small. The backward reactions, k_{-2} and k_{-1}, might be extremely small. As discussed earlier, the observed $[S_{37}]$ values seems to be fairly independent of temperatures, while a large change of τ_{sol} occurs. This could be explained by assuming a fairly constant ratio between the forward and backward reactions $k_1/k_{-1} \cong k_2/k_{-2}$ over this temperature range in each medium. To fit the different scavenging constant *Q*, from the various media in this model, the ratio of the forward and backward reactions might be different (i.e., $k_2/k_{-2} \propto Q$ in each medium). Alternatively, there might be different paths for the electrons in the mobile states.

CONCLUSION

It is obvious that large inroads in the problem of fast electron reactions in water and alcohols have now been achieved. However, many vexed questions are still open to explanation and to understanding the details of the initial steps of the electrons produced by high energy radiations in polar solutions.

ACKNOWLEDGMENTS

The authors thank S. G. Chenery, E. Rasburn and H. B. Michaels for their helpful suggestions, and A. Worthington for his technical assistance. We also thank Mr. T. Horrigan and the staff of the Linac Laboratory of the University of Toronto for their co-operation and the National Research Council for its financial support. The experimental work was supported by the Medical Research Council of Canada and the National Cancer Institute of Canada.

REFERENCES

1. J. H. Baxendale and P. Wardman, *Nature (London)*, **230**, 449 (1971).
2. J. H. Baxendale and P. Wardman, *J.C.S. Faraday Trans.*, I, 69, 584 (1973).
3. C. D. Jonah, E. F. Hart and M. S. Matheson, *J. Phys. Chem.*, **77**, 1838 (1973).
4. R. K. Wolff, M. J. Bronskill, J. E. Aldrich and J. W. Hunt, *J. Phys. Chem.*, **77**, 1350 (1973).
5. R. K. Wolff, J. E. Aldrich, T. Penner and J. W. Hunt, *J. Phys. Chem.*, in press.
6. R. K. Wolff, M. J. Bronskill and J. W. Hunt, *J. Chem. Phys.*, **53**, 4211 (1970).
7. J. E. Aldrich, M. J. Bronskill, R. K. Wolff and J. W. Hunt, *J. Chem. Phys.*, **55**, 530 (1971).
8. M. J. Bronskill and J. W. Hunt, *J. Phys. Chem.*, **72**, 3762 (1968); M. J. Bronskill, W. B. Taylor, R. K. Wolff and J. W. Hunt, *Rev. Sci. Instrum.*, **41**, 333 (1970); J. E. Aldrich, R. Foldvary, J. W. Hunt, W. B. Taylor and R. K. Wolff, *Rev. Sci. Instrum.*, **43**, 991 (1972).
9. M. Anbar, *Advan. Chem. Ser.*, **50**, 55 (1965).
10. H. A. Schwarz, *J. Chem. Phys.*, **55**, 3657 (1971).
11. W. H. Hamill, *J. Phys. Chem.*, **73**, 1341 (1969).
12. G. R. Freeman, *J. Chem. Phys.*, **46**, 2822 (1967).
13. G. Czapski and E. Peled, *J. Phys. Chem.*, **77**, 893 (1973).
14. M. C. Sauer, S. Arai and L. M. Dorfman, *J. Chem. Phys.*, **42**, 708 (1965).
15. K. N. Jha, G. L. Bolton and G. R. Freeman, *J. Phys. Chem.*, **76**, 3876 (1972).
16. R. R. Hentz and G. A. Kenney-Wallace, *J. Phys. Chem.*, **78**, 514 (1974).
17. L. Gilles, J. E. Aldrich and J. W. Hunt, *Nature Phys. Sci. (London)*, **243**, 70 (1973).
18. M. J. Bronskill, R. K. Wolff and J. W. Hunt, *J. Chem. Phys.*, **53**, 4201 (1970).
19. K. Y. Lam, Ph.D. dissertation, University of Toronto, 1974.
20. J. R. Miller, *J. Chem. Phys.*, **56**, 5173 (1972).
21. P. Fowles, *Trans. Faraday Soc.*, **67**, 428 (1971).
22. B. R. Hammond and R. H. Stokes, *Trans. Faraday Soc.*, **41**, 1641 (1955).
23. W. F. Schmidt and A. O. Allen, *J. Chem. Phys.*, **52**, 4788 (1970).

Reactions Rates of Electrons at Short Times[1]

Gidon Czapski
Department of Physical Chemistry
The Hebrew University, Jerusalem, Israel

and

E. Peled
Department of Chemistry
Tel-Aviv University, Israel

In recent years Hunt and his co-workers developed the picosecond stroboscopic pulse radiolysis technique (*1–4*). The development of this very elegant and powerful tool enabled the observation of very early events in the radiation chemistry of aqueous solutions down to the picosecond range.

Using this technique (*5*) and other ones (*6*), the solvation time of electrons, or their upper limit, were determined.

Other studies in the psec. range were mainly concerned with following the decay rate of e_{aq} at this early time range, in the absence, but mainly in the presence of various solutes (*1–4*). Studies, where the lifetime of e_{aq} is very short, were done by direct observations (*1–4*) and by indirect methods (*7–10*). Most of the work was performed in very concentrated solutions. These results raised several questions, challenging existing theories (*11–13*), and gave rise to new theories, such as the role of dry electrons (*14–15*).

The direct determination of $G(e_{aq}) \sim 4.0$ at 10^{-11} sec after the pulse, and the lack of any appreciable decay of e_{aq} between 10^{-11}–10^{-9} sec in pure water (*4*), show that both of these observations are in conflict with previous predictions of the spur diffusion model (*11–13*). Nevertheless, even these observations indicate clearly that the decay of e_{aq} up to times of $\sim 10^{-7}$ sec is in regions where e_{aq} is initially distributed in a non-homogeneous manner.

Part of the conflict between the experimental observations and the spur diffusion model can be accounted for when several corrections are made, such as a final hydration time of a few psec for the electron (*16*) and the different time dependence of rate constants of the different reactions of e_{aq} (*17–18*). Additional modifications in the spur diffusion model may be added, such as the introduction of *initially* formed species other than e_{aq}, H^+ and OH, like H, H_2 (*13*), or even H_2O_2 and O. Moreover, one may assume that the typical spur may have a different average number of radicals and initial radius (*19*). This last modification could account for the "delayed" decay of e_{aq} as found experimentally, but at the same time does not account correctly for the influence of scavengers on the molecular yields.

Since at early stages after the deposition of the radiation energy dry

[1]Congress presentation E–10–4.

electrons and holes are most probably formed, one should consider the fate and role of these species. The dry species may of course solvate and this will take no longer than $\sim 10^{-12}$ sec in aqueous media (*5,6*). During this time period, the dry ions may recombine or react with solutes prior to solvation, and this in competition with the solvation process. From the lack of any appreciable effect of high concentrations of scavengers for the dry species, on $G(e_{aq})$ and G(OH) (*20,21*), it seems that even if these scavengers react with the dry species, this process is not in competition with the annihilation of the dry species, but at most may be in competition with the solvation process.

$$e^-_{dry} + S \rightarrow S^- \tag{1}$$

$$e^-_{dry} = H_2O \rightarrow e^-_{aq} \tag{2}$$

The principal deviations in the behavior of concentrated solutions when compared to dilute ones, can be summarized as follows:

a) When two solvated electron scavengers are in competition, the relative rate constants depend, sometimes strongly, on the concentrations (*2,7*).

b) The extrapolated initial yield of e^-_{aq} in concentrated solutions in the presence of various scavengers gave values which were much lower than expected (*3*) from the known values of $k_{e^-_{aq}+S}$. The low extrapolated yields at ~10 psec after the pulse, for concentrated solutions of several scavengers, suggested rate constants as high as $2.5 \times 10^{11}\ M^{-1}\ sec^{-1}$. These high values were attributed by Hunt to reactions of e^-_{dry} (*2,3*) by assuming that *S* and H_2O compete for e^-_{dry} through reactions (1) and (2).

We will try to show that the above-described phenomena may be interpreted without necessarily taking into consideration the reaction of e^-_{dry}. We believe that one cannot extrapolate the initial yields of e^-_{aq} in concentrated solutions. This will be the case even after correcting for the time-dependence and the concentration-dependence of rate constants and their ratios, of the reactions of e^-_{aq} with electron scavengers. The time-dependence has been discussed by Schwartz (*17*): his theoretical treatment accounted for only part of the anomalies found in rate constants in the psec range, both for diffusion-controlled reactions of e^-_{aq} and for somewhat slower reactions, such as $H^+ + e^-_{aq}$.

The reactions of e^-_{aq} with most scavengers *S* are diffusion-controlled. The process can be visualized to occur in two steps, the first one being the transport of *S* towards e^-_{aq} yielding a reactive configuration (an encounter pair $(e^-_{aq}:S)$); the second step involves the transformation of this configuration (encounter pair) into the product S^-. In addition, the encounter pair may separate through diffusion without reacting.

For fast reactions the overall rate is limited by the transport process by which e^-_{aq} and *S* meet: i.e., is controlled by diffusion. The typical diffusion-controlled reactions are those where practically every encounter yields a product.

$$e^-_{aq} + S \underset{k_{-1}}{\overset{k_1}{\rightleftharpoons}} (e^-_{aq}:S) \xrightarrow{k_2} S^- \tag{3}$$

For these reactions $k_{-1} < k_2$ (otherwise not every encounter would yield S^-), while for $e^-_{aq} + H^+$, $k_{-1}/k_2 = 5$ (*17*) (since for H^+, k_{obs} is about five times smaller than the diffusion-controlled value would be). For reactions which are diffusion-controlled, τ_{en} (the encounter pair lifetime) is at most 10 psec. The theories of diffusion-controlled reactions were developed by Smoluchowski (*22*). The model assumes that *S* (uncharged) has to diffuse toward e^-_{aq} and, when the distance between their centers reaches a value *R*, they react.

The solution of the diffusion equation with the boundary conditions:

$$S(r,O) = S_O \text{ for } r > R$$

$$S(r,t) = O \text{ for } r \leqslant R$$

yields

$$k = (4\pi DRN/1000)[1 + R/(\pi Dt)^{1/2}] \tag{4}$$

Equation (4), the time dependent form of k_{obs} which reduces at longer times to the stationary form

$$k_{obs} = 4\pi NDR/1000 \tag{5}$$

For charged particles Debye (*23*) extended the theory taking the interaction into account. The solution of the diffusion equations in this case yields an equation similar to (5) but *R* is replaced by $r_{effective}$ given by

$$r_{effective} = [\int_R^\infty \exp(U/KT)(dr/r^2)]^{-1} \tag{6}$$

(*U* is the coulombic interaction energy)

For dilute solutions this integration yields

$$r_{effective} = R[Q/(e^Q - 1)] \tag{7}$$

where $Q = Z_s e^2/\epsilon kTR$ (Z_s is the charge number of *S*, e is the electron charge, ϵ is the dielectric constant and *k* is Boltzman's constant).

R in equations 4–7 is the distance between e^-_{aq} and *S* at which they react. *R* is often taken as the sum of the crystallographic radii, ($r_s + r_{e^-_{aq}} = R$) although there is no proper foundation for this. For example, *R* may exceed $r_s + r_{e^-_{aq}}$ if the reaction occurs via tunneling as suggested by Anbar and Hart (*24*) for these reactions. Furthermore, for several reactions it was found experimentally that *R* exceeds $r_s = r_{e^-_{aq}}$ by a factor of up to three, both for anions and neutral molecules (*24*).

Let us examine whether the assumptions for the diffusion-controlled reactions are valid in very concentrated solutions:

The boundary condition in Smoluchowski's equation—*S*(r,O) = 0 for r ⩽ R—assumes that no e^-_{aq} is present at $t = 0$ in the spheres of radius *R* around *S*. In concentrated solutions of *S*, the volume occupied by *S* molecules may be very appreciable.

We can calculate the probability *P* that *n* molecules of *S*, each of an "encounter volume" $v = (4/3)\pi R^3$, do not cover a random spot in the solution of volume *V*.

If we assume that $R^3 > r_s^3$, *P* will be given by

$$P = (1 - v/V)^n \simeq e^{-[S]\bar{v}} \tag{8}$$

where [*S*] is the concentration in Mole/liter and $\bar{v}$ in the partial molar volume of *S* in liters/mole.

Table 1 gives the concentrations of *S* for various values of *R* with the respective values of *P*.

TABLE 1
The Concentration Dependence of the Probabilities of Initial Formation of Encounter Pairs on the Reaction Cross Section (R)**

	P* =								
R Å	0.9	0.8	0.7	0.6	0.5	0.4	0.3	0.2	0.1
3	1.45	3.2	5.07	7.25	10.	13.	17.4	28.3	33.
4	0.41	0.9	1.44	2.07	2.85	3.72	4.95	6.64	9.5
5	0.32	0.7	1.12	1.59	2.2	2.87	3.83	5.12	7.36
6	0.18	0.4	0.64	0.92	1.27	1.66	2.21	2.96	4.24
7	0.11	0.25	0.41	0.58	0.8	1.04	1.39	1.86	2.66
8	0.08	0.17	0.27	0.39	0.54	0.71	0.94	1.26	1.79
9	0.055	0.12	0.19	0.28	0.38	0.44	0.66	0.88	1.27
10	0.04	0.09	0.146	0.20	0.276	0.36	0.48	0.64	0.92

* P defined in equation (8).
** All concentrations are in moles/liter.

Since for diffusion-controlled reactions of e^-_{aq} the average value of R can be taken as ~ 6 Å, it follows that in ~ 1 M solution of S, P ~ 0,5, or half of e^-_{aq} which would be introduced randomly into this solution would be formed initially as encounter pairs (e^-_{aq} :S). Thus, if $S(r,O) = 0$ for $r \leqslant R$, this means that half of e^-_{aq} will disappear instantaneously as soon as it is incorporated into encounter pairs. Therefore, any yield of e^-_{aq} as extrapolated to $t = 0$ will account only for the fraction of e^-_{aq} which is not initially formed as encounter pairs. This analysis assumes a zero lifetime of the encounter pairs (e^-_{aq} : S). The implications of the encounter pair lifetime exceeding 10^{-11} sec have been discussed elsewhere (*24*).

The relative yield of e^-_{aq} , as given by P in equation (8), has the identical functional form as derived by Wolff et al. (*2*) for a model where dry electrons compete in reactions (1) and (2).

If we attribute the linear decrease of log[$G(e^-_{aq})$] with [S], as found experimentally (*2*) due to equation (8), the following values of R are found: $R = 6$ Å, 6.5 Å, 9 Å and 9.5 Å for H_2O_2, acetone, NO_3^- and Cd^{2+}, respectively. Values of R ~ 6 Å are within the average value of R for diffusion-controlled reactions of e^-_{aq} (*25*). A value of R of 9 Å or 9.5 Å is high, although such values are suggested for several solutes in their reactions with e^-_{aq} (*25*). If, with these values of R we calculate the rate constants of the reactions of e^-_{aq} with these solutes from equations (5)–(7), the computed values are only about twice the value of the measured ones.

Thus, the results of Wolff et al. (*2*) seem to be interpreted reasonably well without having to assume any reaction of e^-_{dry} with these solutes.

High values of R are not unknown and may be due to tunneling or long-range electron transfer (*25*).

In Hunt's experiments (*1–4,21*) e^-_{aq} is observed at t ~ 30 psec, and the yield of e^-_{aq} is integrated from $t = 0$. In our opinion this integration is invalid, even if the time dependence of k_{obs} is taken into account, as was done by Schwarz (*17*), since this integration considers only the decay which is due to encounter pairs formed through diffusion and neglects those which were formed initially. Such a mathematical treatment necessarily gives too high values of $G(e^-_{aq})$ depending on the fraction of encounter pairs formed initially. Hence, in concentrated solutions the error may be very large (Table 1). As $G(e^-_{aq})$ found experimentally was lower than expected it was suggested that dry electrons react efficiently with S at times shorter than 10^{-11} sec.

The high values of the decay rate constants as derived by Hunt (*3*) from the fraction of e^-_{aq} which decayed during his pulse ($t < 30$ psec), and attributed by him to reactions of dry electrons with S, could be, in fact, solely due to the decay of encounter pairs initially formed.

Fig. 1 shows the relative yield of e^-_{aq} as a function of τ—the lifetime of e^-_{aq} with S—as computed by Aldrich et al. (*3*). The observed yields of e^-_{aq} in concentrated solutions of various solutes were much lower than predicted from Fig. 1 and the known values of τ for the reactions of e^-_{aq} with these solutes (Table 2, reference *3*).

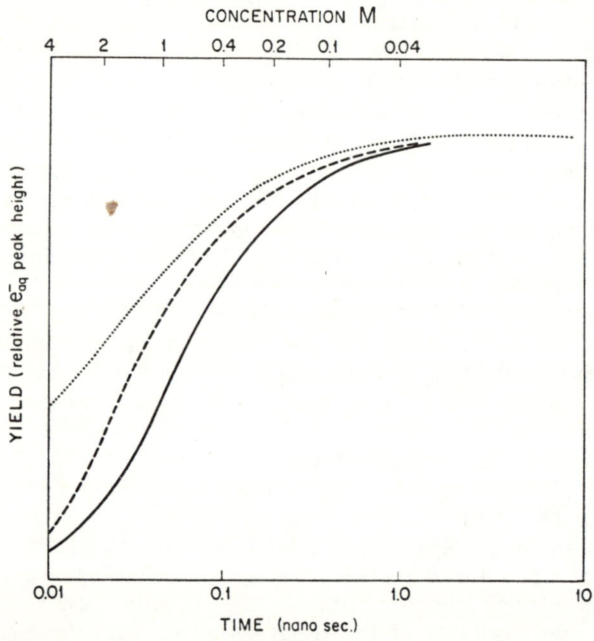

FIG. 1. Relative yields of e^-_{aq} as a function of τ or $[S]$. τ is the lifetime of $e^-_{aq} + S$. $[S]$ is for a reaction with $D = 6 \times 10^{-5}$ cm^2 sec^{-1}, $R = 6$ Å and $k_{e^-_{aq}+S} = 2.5 \times 10^{10}$ M^{-1} sec^{-1}.
..... predicted by Wolff et al. (*2*)
----- predicted by equation (8)
—— correction of Wolff's (*2*) line with equation (8).

Values of τ which could have been consistent with the observed yields were up to twenty times shorter than the known values of τ for the reaction of e^-_{aq} with these same solutes. The line computed by Aldrich et al. (*3*) did not incorporate a time dependence of $k_{e^-_{aq}+S}$; a correction to this effect would have shortened $\tau(e^-_{aq}+S)$ in concentrated solutions of S by a factor of up to about two. We added to Fig. 1 the yields of e^-_{aq} as predicted by equation (8) and then corrected Aldrich's (*3*) curve accordingly. The corrected curve predicts much lower yields of e^-_{aq} for a given value of $\tau_{e^-_{aq}+S}$ than did the original (*3*) correction. The time dependency would even lower further the yield of e^-_{aq}. For example, if $k_{e^-_{aq}+S} \sim 2.5 \times 10^{10}$ M^{-1} sec^{-1}, $D = 6 \times 10^{-5}$ cm^2 sec^{-1}, $R = 6$ Å and $[S] = 4$ M, the relative yield of e^-_{aq} predicted by Aldrich et al. (*3*) is 0.45—while with our correction, it is less than 0.05.

When Aldrich's correction is included as well as ours, the yield is expected to drop, and this drop is accounted for without necessarily assuming that dry electrons react with S and compete through reactions (1) and (2).

This analysis does not rule out at all the possibility that dry electrons react with S and compete through reactions (1) and (2). What is shown is that the low yield of e^-_{aq} in concentrated solutions of efficient e^-_{aq} scavengers at 10^{-11} sec after the pulse could, in many cases, be explained without assuming the reactivity of dry electrons.

In any case, even if dry electrons react with an efficient scavenger S, in order to extrapolate the initial yield of e^-_{aq} from the measured one at 10^{-11} sec after the pulse, the following treatment has to be made:

In addition to the mathematical treatment Hunt has made, which took into account the decay within the pulse, one has to correct for the time dependence of this decay (*17*) as well as calculate the fraction of encounter pairs formed initially. If the yield of e^-_{aq} at 10^{-11} sec after the pulse is still too low, which is not the case with most scavengers in aqueous solutions, then one may look for the possibility that the missing yield may have reacted as dry electrons. Of course, one has also to correct for the final solvation time of the electron as demonstrated by Rzad and Schuler (*16*). The low efficiency of H^+ on lowering $G(e^-_{aq})$ at 10^{-11} sec after the pulse, even at molar concentration, was used as additional support for dry electrons which were assumed not to react with H^+. It was shown that since H^+ is reacting somewhat slower than by a diffusion-controlled rate, this effect may also be accounted for if we assume that the encounter pair of H^+ and e^-_{aq} has a lifetime of several psec (*24*). The same explanation may be true for some other scavengers studied whose rate constants with e^-_{aq} are lower than for diffusion-controlled rates, and which seem to decrease $G(e^-_{aq})$. In this case, again one has to assume a lifetime of the encounter pair of a few psec, but the question arises: what are the optical properties of these encounter pairs?

e^-_{aq} is measured by absorption and the encounter pair (e^-_{aq}:S) may have a similar absorption as S^-. In this case the absorption measurements will give an apparent decrease of e^-_{aq} which will exceed the one expected from the rate constant of the reaction $e^-_{aq} + S$. Furthermore, our approach explains also that for two solutes, S_1 and S_2, which both react with e^-_{aq} with equal diffusion-controlled rate constants in dilute solutions, and shows in concentrated solutions of these solutes a different relative efficiency (*24*). If S_1 and S_2, in their encounter pair with e^-_{aq}, have somewhat different lifetimes than in a simultaneous encounter pair of S_1 and S_2 with e^-_{aq}, the probability of the reaction of e^-_{aq} with S_1 or S_2, respectively, will depend on the ratio of the lifetimes of their encounter pairs (e^-_{aq} : S_1) and (e^-_{aq} : S_2) (*24*).

These effects may explain most of the anomalies in competition studies of concentrated solutions of e^-_{aq} scavengers and of the kinetic studies in the picosecond range. They may explain the low $G_{e^-_{aq}}$ observed in some of the concentrated solutions, and seem to be able to explain the results without having to have recourse to novel chemical species such as dry electrons. This, however, does not rule out the possibility of reactions of dry electrons prior to solvation, but indicates this assumption is not yet a necessity, especially in aqueous media.

The existence of dry electrons in alcohols was demonstrated when the solvation process was observed (*26*). Therefore, it is almost sure that they will also occur in aqueous solutions.

The lower values of S_{37} in alcohols as compared to these values in water (*27*), seemed to indicate that in these media at least, the dry electron model has to be accepted rather than the encounter pair we have suggested, and which is outlined schematically as follows:

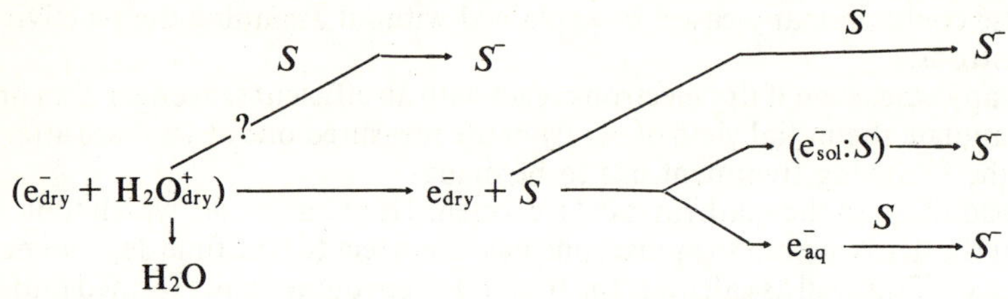

In alcohols it was demonstrated (*28*) that solvation times are much longer, therefore e^-_{dry} has a longer time and a higher probability to react with S which is not the case in water. Our model as outlined above, questions, at least in water, whether scavenger S can compete with annihilation processes of e^-_{dry} as well as whether S can react with those dry electrons which escaped annihilation, before they get solvated.

Additional observations in alcohols seemed to favor the dry electron model. It was found that scavengers which react in alcohols with rates lower than diffusion-controlled rates are very efficient quenchers of $G(e^-_{sol})$ $_{(30\ \text{psec})}$ (*29*). Furthermore, it was found that some of these solutes decrease the yield of the damp electrons, thus suggesting that these solutes have to scavenge the precursor of the damp electron which ought to be the dry electron.

An additional feature found was that in various solvents, for electron scavenger S, the product $k_{e^-+S} \approx S_{37}$ was a constant in any given solvent, although its values varied with solvent (*27*). This last correlation, between the efficiency of scavenging dry electrons with the rate of reaction of the solvated electron for various solutes is somewhat surprising. It does not seem to be expected that such a correlation between the reactivity of dry and solvated electrons for different scavengers will occur, nor that the reactivity of dry electron with a given solute will depend on the solvent. Nevertheless, this correlation is not predicted by the encounter pair model either; on the contrary, the product should be proportional to the square of the encounter radius.

Hunt estimated that a dry electron will encounter more 300 molecules in water prior to solvation (private communication). This provided the efficiency in the reaction of dry electrons with scavenger molecules approaches unity, otherwise the collision number will even increase. As the efficiency of reaction of e^-_{dry} with S should not vary by much in various solvents, and as solvation times in alcohols exceed by more than an order of magnitude those in water, also the collision frequency between dry electron and solute molecules will increase by a similar extent in alcohols as compared to water.

This would predict S_{37} values in alcohol to be even much lower than observed (*27*).

In order to explain the behavior in alcohols as well as in water we will modify our model in view of the observations of damp electrons (*27,29,30*).

The scheme would be as follows:

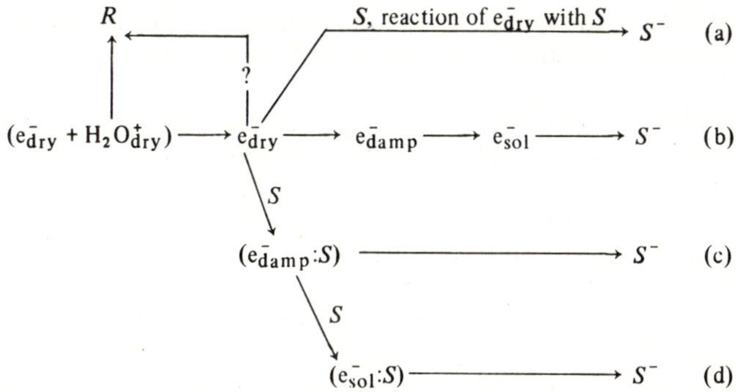

This scheme assumes in all the solvents that dry electrons which do not annihilate and get into the media may solvate. This solvation is a continuous process as observed in alcohols (*30*), and we designated as "damp electrons" the partially solvated electrons, the main feature of which should be that they are trapped in the medium and thus have mobilities comparable to e^-_{sol}.

According to this scheme also damp electrons in presence of scavengers will be formed in concentrated solutions as encounter pairs, which would disappear even faster than encounter pairs of *S* with a fully solvated electron.

This scheme predicts that in solvents where solvation is slower we would be able to observe damp electrons as found in alcohols. Furthermore, as encounter pairs of *S* with e^-_{damp} are formed initially, one expects that scavenger, at 30 psec will reduce e^-_{damp} without having to assume that these scavengers react with dry electrons. The last feature of this model is that it may both explain the lower S_{37} values in alcohols as compared to water as well as the fact that low S_{37} values in alcohols are found even for scavenger reacting somewhat slower than in diffusion controlled reactions. Miller (*31–33*) has suggested in organic glasses a very similar model; he accounted for efficient scavenging of the electrons over long distances by tunneling. The tunneling distance depends both on the well depth of the damp electron as well as on its lifetime.

As in alcohols the lifetime of damp electrons exceeds that in water by much, we would expect larger tunneling distances to be effective. Therefore S_{37} would be lower in alcohols than in water and this effect may also be observed for reactions of $e^-_{sol} + S$ which are somewhat slower than diffusion-controlled.

We believe that the existence of e^-_{dry} in water has no effect on the validity of the diffusion spur model, although this was previously questioned (*14,15*) for the following reasons:

a) The solvation process is completed before any appreciable chemical reaction and diffusion takes place.
b) It has been shown that the decay of e^-_{aq} in the time range $10^{-9} - 10^{-7}$ sec (*27*) is nonhomogeneous, thus suggesting the existence of spurs.
c) It has been shown that even if dry electrons are scavenged in concentrated solutions, this does not prevent any annihilation.

Therefore, at the utmost, the effect would be the formation of S^- in the spurs at an early stage.

The two main difficulties with the diffusion model are the low initial yield of e^-_{aq} ($G(e^-_{aq}) = 4.0$) and the spur half-lifetime, which is too long by about an order of magnitude. These difficulties are not solved by attributing reactivity to dry electrons.

The long half-life of the spur could be explained if one could change the spur parameters. If the typical spur has a larger radius and contains more radicals but has a lower initial concentration of radicals, then in such a case one could fit these parameters and account for the *G* values of the radicals and molecular products as well as for the half-life of the spur or e^-_{aq} (*19*).

The disadvantage of this assumption is that this model with the revised parameters will predict a different dependence of the molecular yield on solute concentrations. By prolonging the spur lifetime by an order of magnitude one could expect that the effect of scavengers on the yield of molecular products would occur at concentrations of scavengers lower by an order of magnitude as compared to the observed behavior.

The low yield of e^-_{aq}, $G(e^-_{aq})_{30\ \mathrm{psec}} \sim 4.0$ is surprising, especially as recently Wolff et al. (*21*) measured the yield of electron attachment products, using Cd^{2+} and RSSR, and found at 30 psec an electron adduct yield of 4.8, which equals the expected initial e^-_{aq} yield calculated by Schwarz (*13*). Wolff et al. assume that P—the electron adduct—is formed from two processes from the reaction of e_{dry} with the scavenger and from the reaction of e^-_{aq} with the same scavenger. The difference of 0.8 units between $G(P)_{30\ \mathrm{psec}}$ and $G(e^-_{aq})_{30\ \mathrm{psec}}$ suggests that ~ 20% of the initial dry electrons recombine with H_2O^+ (*14,15,21*) or, alternatively, OH + e^-_{aq} occurs in very dense regions or spurs. This process has to be completed in the absence of scavengers within less than 30 psec, and high concentrations of scavengers are assumed to interfere with it (*21*).

The fast electron attachment within less than 30 psec observed by Wolff et al. (*21*) in concentrated solutions of Cd^{2+} and RSSR was explained as being due to the reaction of these solutes with dry electrons (*21*). This explanation is mathematically indistinguishable from our encounter pair model, and therefore cannot be taken as evidence for their hypothesis. Both the dry electron scavenging model or the encounter pair model may explain the efficient scavenging within less than 30 psec. However, both models are unable to account for the difference between $G(P)_{30\ \mathrm{psec}}$ and $G(e^-_{aq})_{30\ \mathrm{psec}}$ when one considers that molar concentrations of halide ions, which are assumed to be efficient H_2O^+ or OH scavengers, do not increase $G(e^-_{aq})$ either in the picosecond (*21*) or the nanosecond (*20*) range. Neither model explains the source of molecular products (H_2 and H_2O_2) or the effects of e^-_{aq} or OH scavengers on these products.

If H_2 is found in the time region of 10^{-9} - 10^{-7} sec where e^-_{aq} decays from the *G* value of 4.0 to 2.7, then the scavengers should affect G_{H_2} much more than observed. Alternatively, one has to assume that H_2 is formed in less than 30 psec after the pulse; in this case, the effect of scavengers on G_{H_2} would occur at higher concentrations than observed, even if the scavenger would react with dry electrons assuming they are the precursors of H_2.

In our opinion, neither the diffusion spur theory, nor the dry electron or encounter pair model are consistent, at the same time, with the effects of scavengers on G_{H_2} and on $G(e^-_{aq})$ at psec or nsec, or with $G(e^-_{aq\ 30\ psec}) = 4.0$, and $G(P)_{30\ psec} = 4.8$, or with the observed half life of spurs being about 10^{-8} sec. Each hypothesis may explain only part of the observations.

In the abstract of this conference Jonah et al. (*34*) reported new recent data in which e^-_{aq} decay up to ~ 100 psi was observed. This decay was reported to bring the initial value of $G(e^-_{aq})$ up to 4.5, which is not any more far from the value of 4.8 assumed in the spur diffusion calculations (*11–13*).

If these observations will be confirmed it seems that there are no real disagreements between the diffusion model and the experimental results. The fact that the spur decay of e^-_{aq} is over the time range of 10^{-11}–10^{-7} sec is too long as compared to theoretical predictions (*12–13*). This behavior may be explained if one assumes, as was suggested earlier (*35*) to explain some scavenging properties of the molecular y $ield of H_2, that there is a distribution of spurs with various initial concentrations. If such a distribution exists, these spurs will show different lifetimes and may spread the lifetime of e^-_{aq} as found recently (*34*).

REFERENCES

1. M. J. Bronskill, R. K. Wolf and J. W. Hunt, *J. Chem. Phys.* **53**, 4201 (1970).
2. R. W. Wolf, M. J. Bronskill and J. W. Hunt, *J. Chem. Phys.* **53**, 4211 (1970).
3. J. E. Aldrich, M. J. Bronskill, R. K. Wolf and J. W. Hunt, *J. Chem. Phys.* **55**, 530 (1971).
4. R. K. Wolf, M. J. Bronskill, J. E. Aldrich and J. W. Hunt, *J. Phys. Chem.* **77**, 1350 (1973).
5. M. J. Bronskill, R. K. Wolf and J. W. Hunt, *J. Phys. Chem.* **73**, 1175 (1969).
6. P. M. Rentzepis, R. P. Jonas and J. Jortner, *Chem. Phys. Lett.* **15**, 480 (1972).
7. E. Peled and G. Czapski, *J. Phys. Chem.* **75**, 3626 (1971).
8. T. Sawai and W. H. Hamill, *J. Phys. Chem.* **74**, 3914 (1970).
9. P.L.T. Bevan and W. H. Hamill, *Trans. Faraday Soc.* **66**, 2533 (1970).
10. S. Khorana and W. H. Hamill, *J. Phys. Chem.* **75**, 3081 (1971).
11. A. H. Samuel and J. L. Magee, *J. Chem. Phys.* **21**, 1080 (1953).
12. A. Kuppermann, in *Radiation Research.* Proc. 3rd Int. Congress of Radiat. Res. (G. Silini, Ed.) pp. 212–234. North-Holland Publishing Co., Amsterdam, 1967.
13. H. A. Schwarz, *J. Phys. Chem.* **73**, 1928 (1969).
14. W. H. Hamill, *J. Phys. Chem.* **73**, 1341 (1969).
15. T. Sawai and W. H. Hamill, *J. Chem. Phys.* **52**, 3843 (1970).
16. S. J. Rzad and R. H. Schuler, *J. Phys. Chem.* **77**, 1926 (1973).
17. H. A. Schwarz, *J. Chem. Phys.* **55**, 3647 (1971).
18. R. M. Noyes, in *Progress in Reaction Kinetics,* (G. Porter, Ed.) Vol. 1, p. 129. Pergamon Press, Oxford, 1961.
19. A. Kuppermann, Report CALT-767-P.4-127 (California Institute of Technology, Pasadena.)
20. E. Peled, D. Meisel and G. Czapski, *J. Phys. Chem.* **76**, 3677 (1972).
21. R. K. Wolf, J. E. Aldrich, T. L. Penner and J. W. Hunt, *J. Phys. Chem.* **79**, 210 (1975).
22. M. V. Smoluchowski, *Z. Phys. Chem.* **92**, 129 (1917).
23. P. Debye, *Trans. Electrochem. Soc.* **82**, 265 (1942).
24. G. Czapski and E. Peled, *J. Phys. Chem.* **77**, 893 (1973).
25. E. J. Hart and M. Anbar, *The Hydrated Electron*, p. 187. Wiley Interscience, New York, 1970.

26. C. D. Jonah, E. J. Hart and M. S. Matheson, *J. Phys. Chem.* **77**, 1838 (1973).
27. K. Y. Lam and J. W. Hunt, *Int. J. Radiat. Phys. Chem.* **7** (in press) (1975).
28. L. Gilles, J. E. Aldrich and J. W. Hunt, *Nature (London)* **243**, 70 (1973).
29. J. W. Hunt and K. Y. Lam, These Proceedings. [Abstract *Radiat. Res.* **59**, 255 (1974)].
30. J. H. Baxendale and P. Wardman, *Nature (London)* **230**, 449 (1971).
31. J. R. Miller, *J. Chem. Phys.* **56**, 5173 (1972).
32. J. R. Miller and J. E. Willard, *J. Phys. Chem.* **76**, 2641 (1972).
33. J. R. Miller, *Chem. Phys. Lett.* **22**, 180 (1973).
34. C. D. Jonah, E. J. Hart, J. R. Miller and M. S. Matheson, *Radiat. Res.* **59**, 110 (1974) (Abstract).
35. E. Peled and G. Czapski, *J. Phys. Chem.* **74**, 2903 (1970).

Measured Mobilities of Electrons[1]

Gordon R. Freeman
Chemistry Department
University of Alberta
Edmonton, Canada T6G 2G2

INTRODUCTION

The second part of this brief Symposium on Electrons in Liquids deals with electron mobility and the energy of the conduction state. Although there is some unavoidable overlap between the four papers, the first two present experimental values of electron mobilities and conduction state energies in liquids. The third paper is about the theory of mobility, and it theoretically correlates the mobility with the conduction state energy. The fourth paper contrasts electron mobilities in liquids with those in the corresponding gases. Differences between behavior in the liquid and gas phases are enlightening. Several workers are studying mobilities as a function of fluid density (*1–14*), for example, in liquids at high temperatures or in gases at high pressures.

Values of electron mobilities are necessary to the understanding of electron reaction kinetics. They are also a measure of the interactions between the electron and the fluid. The mobility is related to the freedom of motion of electrons in a medium, which is in turn related to the strength of the interactions between the electrons and the medium. By measuring mobilities one obtains information about the interactions.

If one has an electron in a vacuum and one puts an electric field across the vacuum, the electron will be continuously accelerated as long as it is in the field. The drift velocity in a constant field E increases linearly with time (Fig. 1a).

If the electron is travelling through a gas, it will be decelerated by collisions with the molecules (Fig. 1b). The electron is accelerated between collisions and decelerated during collisions. It eventually acquires a constant average velocity. A second electron would behave similarly (Fig. 1c). A large number of electrons have a smooth average behavior (Fig. 1d). The mobility u is defined as the steady state drift velocity w divided by the applied field strength:

$$u = w/E \tag{1}$$

The force F exerted on an electron in an electric field is equal to the product of the charge ξ on the electron and the applied field strength E: $F = \xi E$. In a vacuum the drift velocity of the electron in the direction of the field, averaged over the time t that it has been in the field, is

$$\overline{V_{d,t}} = at/2 = \xi Et/2m \tag{2}$$

[1]Congress presentation E–28–1.

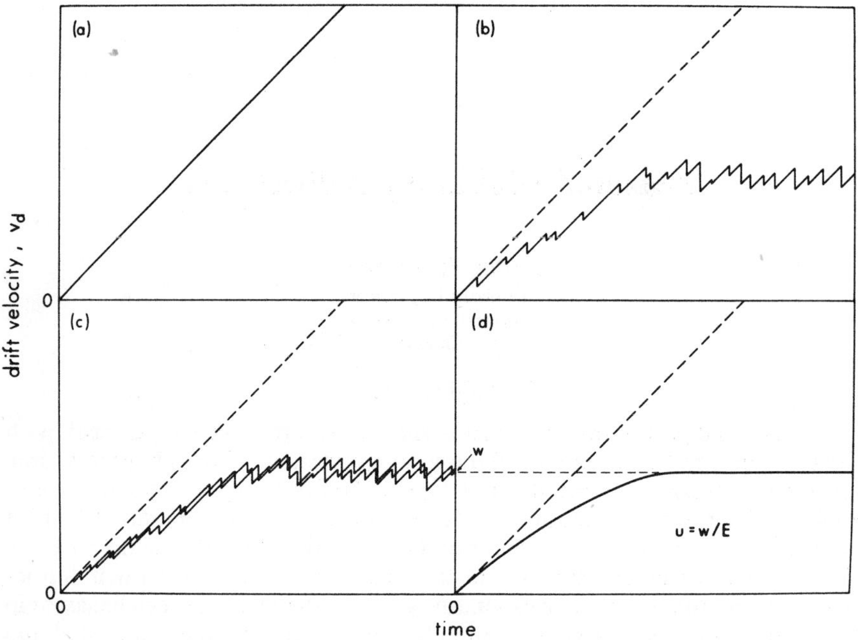

FIG. 1. Drift velocity of an electron as a function of time in a constant electric field, E. (a) In a vacuum. (b) In a gas. (c) Plots for two electrons. (d) Average behavior of many electrons.

where a is the acceleration.

In the presence of a gas the average distance between collisions, the mean free path (mfp), is equal to the inverse of the product of the number n of molecules/cm^3 and the collision cross section σ per molecule:

$$\text{mfp} = 1/n\sigma \tag{3}$$

The average time τ between collisions is the mean free path divided by the relative velocity $\bar{v}_r$ of the electrons with respect to the molecules:

$$\tau = \text{mpf}/\bar{v}_r = 1/n\sigma\bar{v}_r \tag{4}$$

The drift velocity v_d is usually much smaller than the relative velocity $\bar{v}_r$, and the discussion in this article will be limited to conditions, such as low field strengths, where $v_d \ll \bar{v}_r$. At steady state the average gain of drift velocity in time τ equals the average loss of drift velocity during a collision. The steady state velocity w is therefore:

$$w = a\tau/2 = \xi E/2mn\sigma\bar{v}_r \tag{5}$$

The mobility is:

$$u = w/E = \xi/2mn\sigma\bar{v}_r \tag{6}$$

For a Maxwellian distribution of velocities (*15*):

$$u = 4\xi / 3(2\pi mkT)^{1/2} n\sigma \tag{7}$$

Thus for low density gases one has $u \propto 1/T^{1/2}n\sigma$, which at 298K and 1 torr pressure reduces to $u(\text{cm}^2/\text{Vs}) = 4.3 \times 10^{-9}/\sigma(\text{cm}^2)$.

Representative values of σ determined in this way are listed in Table 1. The collision cross section increases as one proceeds from methane to hexane, as might be expected on the basis of molecular size. The electron-molecule collision cross section is usually called the electron scattering cross section of the molecule. For these simple molecules the scattering cross sections are proportional to the molecular polarizabilities α (Table 1), although this correlation does not hold for more complex molecules.

TABLE 1
Electron Mobilities in Alkane Gases

Gas	U_{298} (10^5 cm^2 torr/Vs)[a]	$\sigma(\text{Å}^2)$[b]	$\alpha(\text{Å}^3)$[c]	$(\sigma/\alpha)(\text{Å}^{-1})$
CH_4	93	4.6	2.6	1.8
n-C_4H_{10}	33.1	13.0	8.1	1.6
n-C_6H_{14}	20.6	20.8	11.8	1.8

[a]Reference *11*.
[b]Electron-molecule collision cross section, or electron scattering cross section of the molecule.
[c]Molecular polarizability.

Now let us compare the electron mobilities in the liquids of these compounds (*16,17*). The measurements were not all made at the same temperature, but temperatures as close as possible to each other were chosen (Table 2). If one attempts to use the simple gas phase equation to estimate the scattering cross sections, one obtains unrealistic values (Table 2). They are violently different from the gas phase values in Table 1. The gas phase equation cannot be used for the liquid phase. The unsurprising conclusion is that a liquid is not a gas. Anyone who has had water in his lungs or gas in his stomach could verify that.

TABLE 2
Electron Mobilities in Alkane Liquids

Liquid	T(K)	U (cm^2/Vs)	$\sigma^*(\text{Å}^2)$[c]
CH_4	140	430[a]	0.23
n-C_4H_{10}	219	0.013[a]	16,000
n-C_6H_{14}	220	0.006[b]	43,000

[a]Reference *16*.
[b]Reference *17*.
[c]$\sigma^* = 1.39 \times 10^8/nu$, cm^2, hypothetical value based upon gas phase theory. These values are unrealistic.

The model for electron mobilities in gases at low densities (*15*) is basically sound. We need a model for electron transport in liquids. In fact, we need several models for the liquid phase. For example, the apparent cross section $\sigma^* = 0.23$ Å^2 in liquid methane is an order of magnitude *smaller* than the gas phase cross section $\sigma = 4.6$ Å^2, while the values of σ^* in liquid butane and hexane are three orders of magnitude *greater* than those of σ in the gas phase (Table 1 and 2).

Mobilities have been measured in more than fifty liquids. Values in the diverse compounds range continuously from 10^{-3} to 10^3 cm^2/Vs. The two extremes correspond to the high mobilities of quasifree electrons and the low mobilities of localized electrons. Models have been developed to partially describe the behavior of quasifree electrons (*13* and references therein) and of those with intermediate mobilities (*17–19*). However, the mobilities of localized electrons have been adequately interpreted only for one special case, namely, for electrons localized in bubbles in helium (*20–22*). Several different effects appear to contribute to electron localization in fluids and much more experimental work is needed to sort them out.

For an empirical description of electron mobilities in liquids, the liquids may be conveniently divided into polar and nonpolar groups. The nonpolar liquids may be subdivided into saturated and unsaturated compounds.

POLAR LIQUIDS

In polar liquids, such as ammonia and water, electrons form localized states (*23*). Mobilities in all these liquids are low (*24,25*).[2] In Table 3 the mobilities are compared at the normal boiling points of the liquids to partially compensate for differences in liquid properties. For example, the viscosities of most liquids, whether polar or nonpolar, are ~2 *mP* at the normal boiling points (*26,27*).

The mobilities in ammonia, diethyl ether and water are similar to each other. The value in the ethers increases with increasing length of the hydrocarbon chain. Thus, the mobility in ethers with long alkyl chains would approach the value in a long-chain hydrocarbon, at their respective boiling points.

The Arrhenius temperature coefficients of mobility, E_u, in polar liquids are 3–5 kcal/mol (Table 3).

TABLE 3
Electron Mobilities in Polar Liquids

Liquid	T_b(K)[a]	U (cm^2/Vs)[b]	E_u (kcal/mol)	η(mP)
NH_3	240	0.0091[c]	—	2.5
$(C_2H_5)_2O$	308	0.007[d]	3.4	2.0
$(n\text{-}C_3H_7)_2O$	364	0.023[d]	4.3	2.2
H_2O	373	0.008[e]	4.1	2.8
$(n\text{-}C_4H_9)_2O$	415	0.055[d]	5.3	2.0

[a]Normal boiling point.
[b]Mobility at T_b.
[c]Reference *24*.
[d]Reference *26*.
[e]Extrapolation of data in reference *25*.

[2]J.-P. Dodelet and G. R. Freeman, Electron mobilities and ranges in liquid ethers. (To be published).

NONPOLAR LIQUIDS

In nonpolar liquids one can obtain the whole range of behavior, from the low mobility in ethylene to the high values in methane, xenon and krypton (Fig. 2). The mobility in hydrocarbons depends strongly on molecular structure.

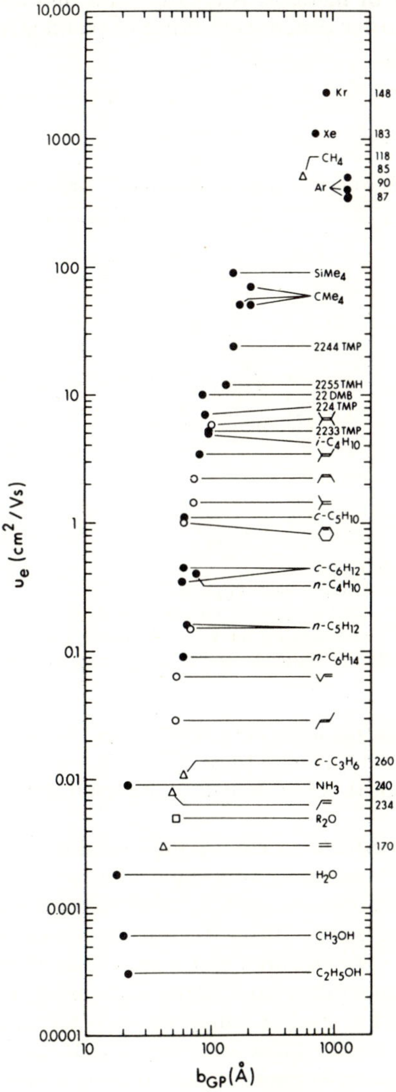

FIG. 2. Thermal electron mobilities in various liquids, plotted against secondary electron ranges b_{GP} in the same liquids. The temperature was 296 ± 3K unless otherwise mentioned on the right margin. [The data references are given in reference *29*, except for ethylene, propylene and cyclopropane (*16*).]

SATURATED COMPOUNDS

The mobility in methane at its boiling point is similar to that in argon at its boiling point (Table 4). However, the mobility in argon becomes dependent on the field strength at fields greater than 200 V/cm (*29*), whereas that in methane remains field-independent up to 1800 V/cm (*16*). This indicates that the energy transfer cross section of methane is greater than that of argon, which in turn correlates with the shorter penetration range of secondary electrons in methane than in argon (Fig. 2).

TABLE 4
Electron Mobilities
in Nonpolar Liquids: Saturated Compounds

Liquid	T_b(K)[a]	U (cm^2/Vs)[b]	E_u (kcal/mol)[c]
Ar	87	500. (*5*)	−0.3
CH_4	109	510. (*16*)	−0.1
CH_3CH_3	185	0.56 (*16*)	2.8
$CH_3CH_2CH_3$	231	0.40 (*16*)	3.0
i-C_4H_{10}	263	5.0[d]	1.9
n-C_4H_{10}	273	0.16 (*16*)	4.2
neo-C_5H_{12}	283	68. (*17,31*)	0.5 (*17*), 0.0(*31*)
n-C_5H_{12}	309	0.21[d]	4.2
n-C_6H_{14}	342	0.2 (*17*)	4.3
c-C_6H_{12}	354	1.1 (*32*)	2.9

[a]Normal boiling point.
[b]Mobility at T_b (references in parentheses).
[c]Arrhenius temperature coefficient of mobility near T_b; E_u varies with temperature. References are the same as those for u.
[d]J.-P. Dodelet, K. Shinsaka and G. R. Freeman (to be published).

Electrons are quasifree in liquid methane, as they are in argon. The lowest energy state of an extra electron in these liquids is a conduction band. For extra electrons there is no localized state below the conduction band in methane or argon.

On going from methane to ethane the mobility drops by three orders of magnitude (Table 4). It decreases only slightly further on going from ethane to propane to butane, then remains approximately constant in the larger *n*-alkanes at their normal boiling points (Table 4). For extra electrons in ethane and the larger *n*-alkanes, there are localized states below the lowest conduction band.

The Arrhenius temperature coefficient of mobility, E_u, tends to decrease as the mobility increases (Fig. 3). The temperature coefficients in the straight chain hydrocarbons are similar to those in polar liquids (cf. Tables 3 and 4). The coefficients in methane and argon near their boiling points are actually negative; the limiting negative value would presumably be that equivalent to the $T^{-3/2}$ temperature dependence in crystalline semiconductors, due to electron scattering by phonons (*29*).

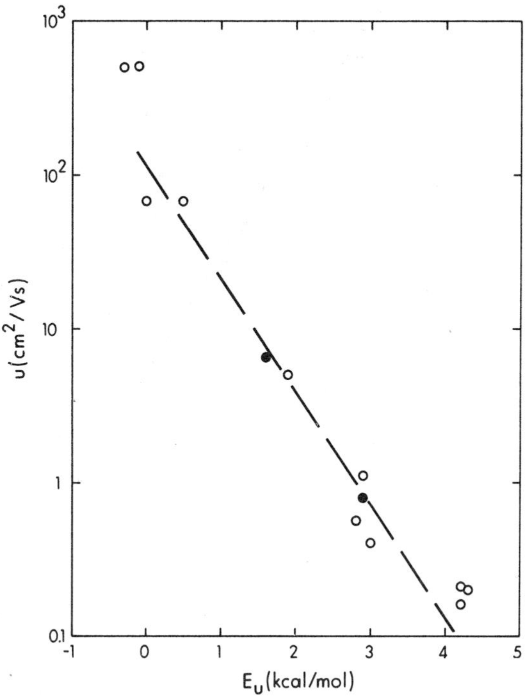

FIG. 3. Electron mobilities in nonpolar, saturated compounds plotted against the Arrhenius temperature coefficient E_u. O, data from Table 4; ●, data from Figs. 4 and 5 for *n*-hexane-neopentane solutions. The dashed line corresponds to hypothetical systems for which $u = 110\ e^{-E/RT}$ at $T = 293K$, and is drawn only as a line of reference.

For liquids in which the mobility has intermediate values, such as isobutane and neopentane, E_u is intermediate as well (Table 4). This also applies to mixed solvents (Fig. 3). It has been shown *(17)* that in binary solutions of neopentane and *n*-hexane the mobility u_{mix} at 300K follows equation (8):

$$u_{mix} \approx u_{np} \exp(-x_h E_h / RT) \quad (8)$$

where $u_{np} = 70\ cm^2/Vs$ is the mobility in pure neopentane, x_h is the mole fraction of hexane in the solution, $E_h = 4.3$ kcal/mol is the Arrhenius temperature coefficient in *n*-hexane and R is the gas constant. The exponential dependence of u_{mix} on x_h is demonstrated in Fig. 4. Furthermore, the Arrhenius temperature coefficient E_{mix} in the solutions is given by $E_{mix} = x_h E_h$, in agreement with equation (8) (see Fig. 5).

The mobility behavior in mixed solvents is consistent with the model that electron traps exist in *n*-alkanes and that each trap consists of several molecules.

By contrast, electrons in neopentane are quasifree. The main difference between the mobilities in neopentane and argon at their respective boiling points is attributable to a $T^{-3/2}$ phonon-mode factor: $70(283/87)^{3/2} = 411\ cm^2/Vs$, compared to the reported value of 500 in argon (Table 4).

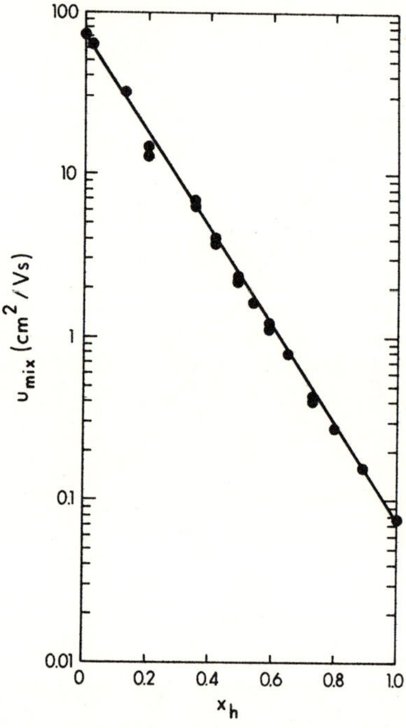

FIG. 4. Mobility in solutions of *n*-hexane in neopentane at 300K. x_h is the mol fraction of hexane. [Reference *17*.]

A further similarity between electron behavior in neopentane and argon is that the mobility *increases* by a factor of two upon freezing the substances (*29,30*). The mobility varies as $T^{-3/2}$ in both solids. The behavior in neopentane is displayed in Fig. 6.

In seeding an interpretation of the dependence of the mobility on molecular structure it was found that there was no correlation between mobility and molecular polarizability or dipole moment in weakly polar molecules. For simple alkanes there was a correlation between mobility and the degree of sphericity of the molecules. The degree of sphericity increases in the series of molecules $(CH_3)_2CH_2 < (CH_3)_3CH < (CH_3)_4C$, and the mobilities increase in the same order (Table 4). Other factors also contribute for larger molecules. For example, the range of the electron interaction in a given molecule appears to be limited to about two C-C bonds in series, which includes groups up to neopentyl in size (*31*). Molecular packing in the liquid might also be important.

The correlation between mobility and the degree of sphericity of the molecules seems to indicate that the orientation of anisotropic molecules is important in the formation of electron traps. The correlation may be linked to the anisotropy of polarizability of the molecules and to the ease of rotation of the molecules in the liquid.

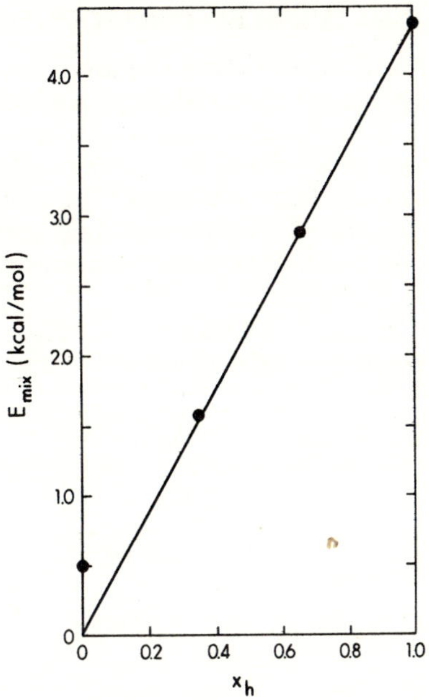

FIG. 5. Variation of mobility Arrhenius temperature coefficient E_{mix} with mole fraction of *n*-hexane, x_h, in hexane-neopentane solutions. [Reference *17*.]

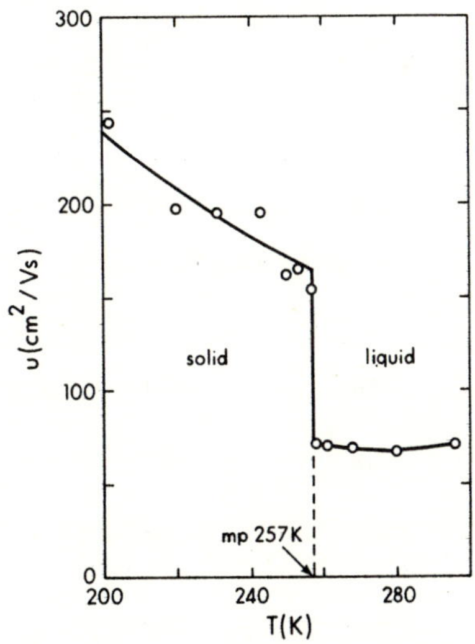

FIG. 6. Mobility of electrons in solid and liquid neopentane. The curve through the points for the solid has the equation $u = 240(200/T)^{3/2}$ cm^2/Vs. [Reference *31*.]

UNSATURATED COMPOUNDS

The mobilities of electrons in ethylene and propylene (*16*) are two orders of magnitude smaller than those in ethane and propane (Tables 4 and 5). In general, mobilities in terminal olefins are smaller than those in the corresponding alkanes.

The same applies to internal olefins in the *trans* configuration (Tables 4 and 5). However, mobilities in *cis* olefins are larger. The *cis-trans* effect is largest in (*28*). The mobility in *cis*-butene-2 is two orders of magnitude larger than that in the *trans* isomer (Table 4), and is an order of magnitude greater than that in *n*-butane.

TABLE 5
Electron Mobilities in Unsaturated Compounds

Liquid	T_b(K)	U (cm^2/Vs)	E_u (kcal/mol)
		OLEFINS	
	169	0.003[a]	—
	226	0.005[a]	~5
	266	~0.8[b]	—
	267	~0.03[b]	—
	274	0.015[b]	5.4
	277	1.5[b]	3.7
	356	2.9[b]	3.6
	359	8.6[c]	1.3
		AROMATICS[b]	
	353	~0.9	~6
	384	~0.4	~4
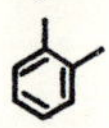	418	~0.2	~4
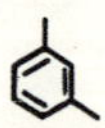	412	~0.5	~4
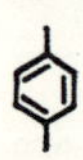	411	~0.5	~4

[a]Reference *16*.
[b]Reference *29*.
[c]K. Shinsaka and G. R. Freeman (to be published).
By extrapolation of data in reference *33*.

The low mobilities in terminal olefins might be associated with the formation of transient negative ions. It is difficult to understand how the same phenomenon could explain the larger mobility in *cis*-butene-2 than in *n*-butane. The difference between the mobilities in the C_4 hydrocarbons is one of the most striking clues that has yet been obtained about electron behavior in liquid hydrocarbons. It has yet to be deciphered.

Thermal electrons are captured by conjugated olefins and by acetylenes. Electron mobilities cannot, therefore, be measured in these liquids.

Example mobilities in aromatic liquids (*32*) are listed in Table 5. It does not seem worthwhile to attempt a detailed interpretation of these values until the *cis-trans* effect in olefins is understood.

REFERENCES

1. J. L. Levine and T. M. Sanders, Jr., *Phys. Rev.* **154,** 138 (1967).
2. R. Grünberg, *Z. Naturforsch.* **23a,** 1994 (1968).
3. B. Huber, *Z. Naturforsch.* **24a,** 578 (1969).
4. W. Legler, *Phys. Lett.* **31A,** 129 (1970).
5. J. A. Jahnke, L. Meyer and S. A. Rice, *Phys. Rev.* **3A,** 734 (1971).
6. H. Renkert, F. Hensel and E. U. Franck, *Ber. Bunsenges. Phys. Chem.* **75,** 507 (1971).
7. J. P. Hernandez, *Phys. Rev.* **5A,** 635 (1972).
8. R. M. Ostermeier and K. W. Schwarz, *Phys. Rev.* **5A,** 2510 (1972).
9. U. Even and J. Jortner, in *Electrons in Fluids* (J. Jortner and N. R. Kestner, Eds.) pp. 361–381. Springer-Verlag, Berlin, 1973.
10. A. Bartels, *Phys. Lett.* **44A,** 403 (1973).
11. L. G. Christophorou, R. P. Blaunstein and D. Pittman, *Chem. Phys. Lett.* **18,** 509 (1973).
12. J. A. Jahnke and M. Silver, *Chem. Phys. Lett.* **19,** 231 (1973).
13. J. Lekner and A. R. Bishop, *Phil. Mag.* **27,** 297 (1973).
14. T. Kimura and G. R. Freeman, *J. Chem. Phys.* **60,** 4081 (1974).
15. W. P. Allis, in *Handbuch der Physik* (S. Flügge, Ed.), pp. 383–444. Springer-Verlag, Berlin, 1956.
16. M. G. Robinson and G. R. Freeman, *Can. J. Chem.* **52,** 440 (1974).
17. R. M. Minday, L. D. Schmidt and H. T. Davis, *J. Phys. Chem.* **76,** 442 (1972).
18. R. Schiller, *J. Chem. Phys.* **57,** 2222 (1972).
19. N. R. Kestner and J. Jortner, *J. Chem. Phys.* **59,** 26 (1973).
20. B. E. Springett, *Phys. Rev.* **155,** 139 (1967).
21. M. H. Cohen and J. Jortner, *Phys. Rev.* **180,** 238 (1969).
22. R. A. Young, *Phys. Rev.* **2A,** 1983 (1970).
23. J. Jortner, *Actions Chim. Biol. Radiat.* **14,** 7 (1970).
24. D. S. Berns, in *Metal-Ammonia Solutions* (G. Lepoutre and M. J. Sienko, Eds.) pp. 146–157. Benjamin, New York, 1964.
25. G. C. Barker, P. Fowles, D. C. Sammon and B. Stringer, *Trans. Faraday Soc.* **66,** 1498 (1970).
26. R. W. Gallant, *Physical Properties of Hydrocarbons.* **Vols. 1** and **2.** Gulf Publishing Co., Houston, 1968.
27. *Handbook of Chemistry and Physics.* (R. C. Weast, Ed.) 50th Ed. Chemical Rubber Co., Cleveland, 1969.
28. J.-P. Dodelet, K. Shinsaka and G. R. Freeman, *J. Chem. Phys.* **59,** 1293 (1973).
29. L. S. Miller, S. Howe and W. E. Spear, *Phys. Rev.* **166,** 871 (1968).
30. K. Shinsaka and G. R. Freeman, *Can. J. Chem.* **52,** 3556 (1974).
31. J.-P. Dodelet and G. R. Freeman, *Can. J. Chem.* **50,** 2667 (1972).
32. K. Shinsaka and G. R. Freeman, *Can. J. Chem.* **52,** 3495 (1974).

Energies of Conduction Bands in Dielectric Liquids[1,2]

Richard Holroyd
Department of Chemistry
Brookhaven National Laboratory
Upton, New York 11973 USA

INTRODUCTION

Dielectric liquids such as hydrocarbons are normally insulators. Excess electrons when introduced into these liquids can occupy a band of states in which the electron can be either localized (trapped) or extended. The mobility of the electron is determined by the populations of these two types of states. The trapped electron is relatively immobile compared to the electron in the extended state. The energy corresponding to the lowest extended state is a mobility edge and it is this energy which is the subject of this paper. The energy of this state relative to the energy of an electron in vacuum is defined as V_o.

The influence of the energy of the conducting state on the properties of the electron is dramatically shown by the data on excess electrons in liquid helium and argon. In helium the conduction band is at +1.0 eV (*1*), as a consequence of dominant repulsive interactions. As a result, the electron creates a void of 30 Å diameter around itself and therefore becomes relatively immobile. In argon the conduction band is at −0.33 eV (*2*) because of the dominance of long-range polarization interactions. There are no electron traps and the electron moves relatively freely ($\mu = 440\ cm^2/V\text{-sec}$).

In dielectrics such as hydrocarbons, the energy of the conducting band also influences the properties of excess electrons. Values of V_o for hydrocarbons range from +0.2 to −0.4 eV (*3–5*), that is, closer to the value for argon, and bubble formation is energetically unfavorable in these liquids. It has been shown that electron mobility correlates with V_o (*3,6*) in that the mobility is low for high values of V_o and increases as the conduction band becomes lower. This correlation will be discussed in detail later. The energy of the conducting band (V_o) may also influence chemical reactions involving electrons: as will be discussed, the threshold of photoionization of solutes depends on V_o (*5*). It is also anticipated that the rates of other processes involving electrons, such as electron attachment reactions, will depend on V_o.

PHOTOELECTRIC MEASUREMENTS OF V_o

The main method used to measure the energies of conduction bands (V_o)

[1]Congress presentation E-28-2.

[2]Research performed under the auspices of the U.S. Atomic Energy Commission.

is the photoelectric effect. It was applied originally to liquid argon (*2*) and helium (*1*) and more recently to hydrocarbons (*3,4*). In this method the work function of a metal is measured while the metal is immersed in a liquid and compared to the work function of the metal in vacuum; the difference is by definition V_o. The experiment requires measurement of the current from a photocell as a function of wavelength (see Fig. 1). The data are then fit by least squares to Fowler's function to determine the work function, ϕ. Fowler's function gives the correct wavelength dependence (near threshold) for the photocurrent from a metal in vacuum and has been assumed to apply in dielectric liquids as well. This assumption is justified by the results; as seen in Fig. 1 the data fit Fowler's function quite well. A theoretical study indicates that this function should apply to dielectric liquids, but a different wavelength-dependence is anticipated for polar liquids (*7*).

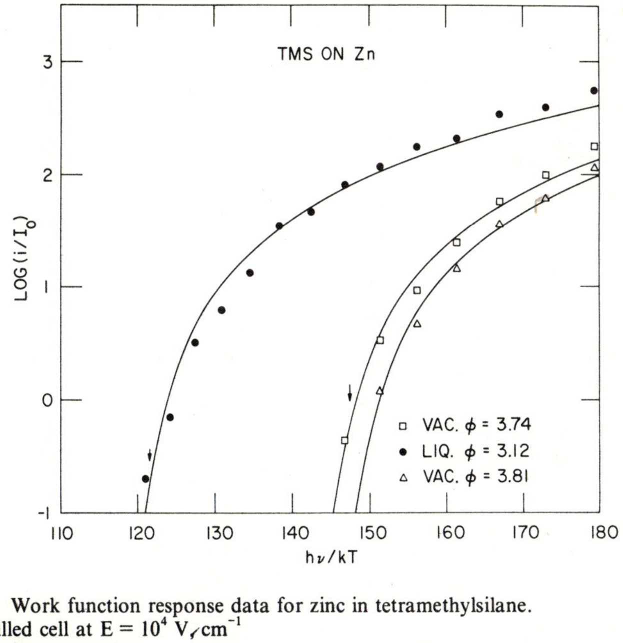

FIG. 1. Work function response data for zinc in tetramethylsilane.
● liquid filled cell at $E = 10^4$ V/cm^{-1}
▣ vacuum response before
△ vacuum response after
[Reproduced from R. A. Holroyd and M. Allen; *J. Chem. Phys.* **54**, 5014 (1971) with the permission of the American Institute of Physics.]

The results obtained by this method are included in Table 1. V_o is close to zero for *n*-pentane and hexane, and lower for the highly branched hydrocarbons. The lowest value is that obtained for tetramethylsilane ($V_o = -0.6$ eV). The reproducibility of the results is about 0.05 eV and an accuracy of this order of magnitude is claimed for this method.

TABLE 1
Conduction Band Energies (V_0)

Solvent	Temp. °C	V_0	Method[a]	Ref.
n-Alkanes				
n-Tetradecane	22°	+0.21	ΔI	*(5)*
n-Decane	22°	0.18	ΔI	*(5)*
n-Octane	22°	0.13	ΔI	*(5)*
n-Hexane	22°	0.06	Δφ	*(3,4)*
	22°	0.1	ΔI	*(5)*
	22°	0.16	Δφ	*(10)*
n-Pentane	22°	0.01	ΔI	*(5)*
	22°	−0.02	Δφ	*(3,4)*
	22°	~−0.2	ΔI	*(11)*
n-Butane	−2°	0.12	ΔI	*(5)*
Propane	−51°	−0.07	ΔI	*(5)*
Ethane	−91°	0.02	Δφ	*(12)*
Cycloalkanes				
Methylcyclohexane	22°	0.08	ΔI	*(5)*
Cyclohexane	22°	0.01	ΔI	*(5)*
Cyclopentane	22°	−0.21	Δφ	*(3,4)*
	22°	−0.21	ΔI	*(5)*
1,3-Dimethyladamantane	22°	−0.30	ΔI	*(5)*
Branched alkanes				
2-Methylpentane	22°	+0.02	Δφ	*(4)*
	22°	+0.01	ΔI	*(5)*
	22°	−0.20	Δφ	*(10)*
3-Methylpentane	22°	−0.03	Δφ	*(4)*
	22°	0.01	ΔI	*(5)*
	22°	−0.14	Δφ	*(10)*
2,3-Dimethylbutane	22°	−0.1	Δφ	*(10)*
2,2-Dimethylbutane	22°	−0.26	ΔI	*(5)*
	22°	−0.15	Δφ	*(10)*
2,3,4-Trimethylpentane	22°	−0.05	ΔI	*(5)*
2,2,5-Trimethylhexane	22°	−0.10	ΔI	*(5)*
2,2,4-Trimethylpentane	22°	−0.18	Δφ	*(3)*
	22°	−0.17	ΔI	*(5)*
2,2,4,4-Tetramethylpentane	22°	−0.33	ΔI	*(5)*
Neopentane	22°	−0.43	Δφ	*(3)*
	22°	−0.35	ΔI	*(5)*
Misc.				
2,3-Dimethylbutene-2	22°	−0.25	ΔI	*(5)*
Benzene	22°	−0.14	Δφ	*(10)*
	22°	−0.5	ΔI	*(11)*
Toluene	22°	−0.22	Δφ	*(10)*
Tetramethylsilane	22°	−0.62,−0.55	Δφ	*(3,4)*
	22°	−0.60	ΔI	*(5)*

[a] Δφ is photoelectric, ΔI is photoionization of TMPD.

EFFECT OF V_o ON PHOTOIONIZATION OF SOLUTES

The observed differences in conduction band energies are expected to influence reactions involving electrons in dielectric liquids. For example, the threshold for photoionization of a solute such as tetramethyl-*p*-phenylenediamine (TMPD) shifts as V_o changes (*5*,*8*).

$$TMPD + h\nu \rightarrow TMPD^{+} + e^{-} \quad (1)$$

In the vacuum the energy required for reaction 1 is defined as the ionization potential (I_g). In the liquid the requirement is changed by the energy necessary to transfer each species into the liquid (equation 2):

$$I(P) = I_g + P(TMPD^{+}) + V_o - \Delta H_{soln}(TMPD) \quad (2)$$

The polarization energy of the positive ion $P(TMPD^{+})$ is estimated to be approximately 1.3 eV in hydrocarbons; since it is a function of the dielectric constant and since ϵ is close to 2.0 for most hydrocarbons, $P(TMPD^{+})$ will not change much from liquid to liquid. The heat of solution of TMPD is small and can be neglected. Thus shifts in ionization thresholds in the liquid reflect primarily shifts in V_o and, furthermore, can be used to measure this quantity.

Some typical results obtained for the photoionization of TMPD are shown in Fig. 2. For each of these liquids the photocurrent onset occurs at a different energy. Above onset the current increases a factor of ~30 for each 0.1 eV increase in photon energy. Since the curves near threshold are roughly parallel, V_o is determined by the shifts in the curves on the energy axis (equation 3)

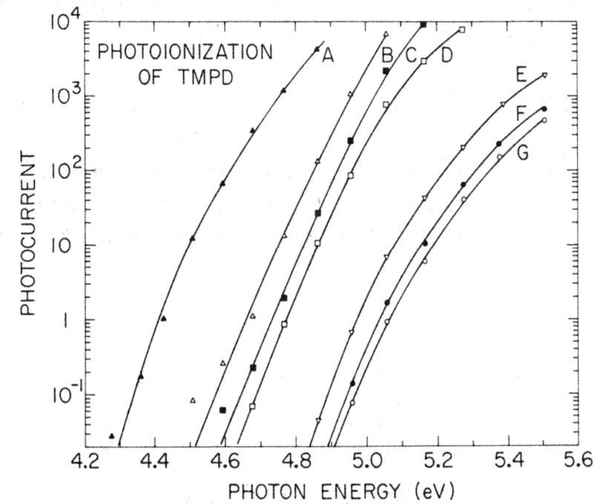

FIG. 2. Photocurrent (in picoamperes) vs photon energy (in electron volts), normalized to equal absorbed light intensity, for photoionization of TMPD in various solvents: A: TMS ▲; B: neopentane △; C: cyclopentane ■; D: 2,2,4-trimethylpentane ⊡; E: 3-methylpentane △; F: *n*-pentane ● G: *n*-hexane ʘ. Applied field 3 kV-cm. All solutions were air saturated. Concentration of TMPD: 25–50 μM. [Reproduced from R. A. Holroyd, *J. Chem. Phys.* **57**, 3007 (1972) with the permission of the American Institute of Physics.]

$$\Delta V_o = \Delta I(P) - \Delta P(TMPD^+) \quad (3)$$

This method is relative in that if V_o is known for one liquid it can be determined for any other.

Results obtained for 20 different liquids by this method are included in Table 1. The results are very reproducible in that onsets always occur at the same energy. There is very good agreement of photoionization shifts with V_o values obtained from the photoelectric method.

MODELS FOR CONDUCTION BAND ENERGIES

For liquids as complex as hydrocarbons a theory has not been developed which permits accurate predictions of conduction band energies. However, for monatomic liquids (e.g., liquid argon) there have been theories advanced to account for this and other excess electron properties (*13,14*). Predicted values of V_o are in reasonable agreement with experiment (see Table 2). Extension of these calculations to methane has been less successful however (Table 2). To derive V_o from first principles requires using gas phase scattering data to predict liquid behavior. Although this approach was successful for the inert gases, application to non-spherical molecules like hydrocarbons may be justifiably questioned. A more useful approach has been to use the model and invert the procedure to calculate liquid phase scattering cross sections from V_o values (*19,20*). In the model, $V_o = T + P$, where P is the long-range polarization energy of the electron in the liquid and T is the kinetic energy term ($\hbar^2k^2/2m$). Thus, T can be calculated for liquids for which V_0 has been determined experimentally. Since P is about the same (approx. −2.4 eV) for many of the liquids, the differences in V_0 imply differences in kinetic energy. For example, T for *n*-pentane (2.43 eV) is larger than for neopentane (T = 2.07). From T the radius (*a*) of the hard-core sphere from which the electron is excluded by the model, and the scattering cross section $\sigma_s = 4\pi a^2$, can be derived.

Values of *a* calculated from V_o for the liquids studied range from a low of 1.87 Å for cyclopentane to a high of 3.03 Å for tetradecane (*5*). The value of *a* increases with chain length for the *n*-alkanes and decreases with branching (*19*). In general, *a* is found to be given by the relation (*21*) $a = 0.303$ (critical volume)$^{1/3}$, which emphasizes the dependence of V_o on the free volume available to the electron.

TABLE 2

Liquid	T (°K)	V_0 Exptl.	V_0 Theory	μ
He	4.2°	+1.05 (*1*)	+1.32 (*15*)	0.22 (*16*)
Ar	82°	−0.33 (*2*)	−0.46 (*13*) −0.63 (*15*)	440.
Kr	117°	−0.78 (*15*)	–	1800. (*15*)
Methane	109°	0.0 (*12*)	−0.75 (120°) (*17*)	400. (*18*)

[Reference numbers are given in parentheses.]

RELATION OF V_0 TO THE ELECTRON MOBILITY (μ)

When conduction band energies for hydrocarbons were first measured (*3*), a correlation of V_0 with the electron mobility was indicated. Recent data have supported the interdependence of these two properties. Since the electron mobility is discussed in other papers in this symposium (*22,23*) it will not be dealt with in detail here. We wish to note, however, the nature and significance of the $\mu - V_0$ correlation and to point out that two of the proposed models of mobility predict quantitative relationships of μ to V_0 and to examine how well the available data fit these models. The experimental relationship of μ to V_0 is shown in Fig. 3. Overall there is a roughly linear dependence of log μ on V_0.

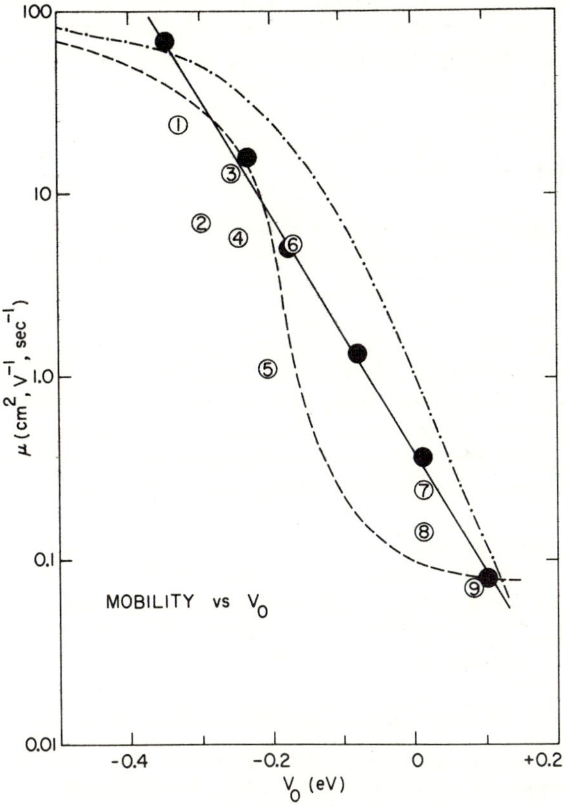

FIG. 3. Mobility vs V_0 for hydrocarbons; solid points are for the neopentane-*n*-hexane binary mixture (results from refs. *23* and *24*); open points 1) 2,2,4,4-tetramethylpentane, 2) 1,3-dimethyladamantane, 3) 2,2-dimethylbutane, 4) 2,3-dimethylbutene-2, 5) cyclopentane, 6) 2,2,4-trimethylpentane, 7) cyclohexane, 8) *n*-pentane, 9) methylcyclohexane
Model calculations: dashed curve—ref. *7*; dotted curve—ref. *10*.

In the percolation model of Kestner and Jortner (*6*) regions of low ($\mu \sim 0.06$ cm^2, V^{-1}, sec^{-1}) and high ($\mu \sim 100$ cm^2, V^{-1}, sec^{-1}) mobility are assumed to exist, and the overall mobility is determined by the relative importance of each region as given by fluctuation theory. This model predicts an S-shaped dependence of μ on V_o; μ is large when V_o is quite negative, and near a certain critical value of V_o there is a large decrease in the mobility (see dashed line in Fig. 3) to a lower limit of $\mu = 0.06$.

In the bubble model of Schiller (*9,10*) the mobility is a function of the energy of the localized electron, as estimated from data on positronium in these liquids. In this model the predicted value of μ is also high at low values of V_o; as the conduction band energy increases there is a rapid dropoff in μ above the trap energy (dotted line in Fig. 3). The trap energy is assumed to be constant in all liquids.

Measurements of V_o for mixtures of neopentane and *n*-hexane (*24*) were obtained to test the predictions of these models. The experimental findings showed V_o to be linearly dependent on the mole fraction of *n*-hexane, X_h, (equation 4). A similar result was obtained for mixtures of TMS and *n*-hexane.

$$V_o(\text{mixt}) = -0.35 + 0.45\, X_h \tag{4}$$

The electron mobility has also been measured for the neopentane mixtures (*25,26*), and these data have been included in Fig. 3. Over the entire concentration range the dependence of log μ on V_o is nearly linear. A linear dependence is expected if equation 4 is considered, in view of the empirical fact that log μ is proportional to X_h for these mixtures (*27*). These results may be expressed in another way if we note that since both the mobility activation energy (E_A) and V_o are proportional to X_h for these mixtures, then E_A must be linearly related to V_o. To test this relationship the available experimental data have been plotted in Fig. 4. Data are included for eight pure liquids as well as two binary mixtures of *n*-hexane-neopentane. Although there is a limited amount of data, there appears to be a correlation in that E_A is small when V_o is low and increases to larger values as V_o is raised. It is informative to now locate the trap energy in various liquids on an absolute scale. It follows from the above and the data in Fig. 4 that the lowest trapped state in *n*-hexane is at −0.1 eV, in 2,2,4-trimethylpentane at −0.22 eV, in 2,2-dimethylbutane at −0.32 eV, and in neopentane at −0.37 eV.

TEMPERATURE EFFECTS ON V_o

Some information is available both experimentally and from theory on the effect of temperature on the conduction band energy. In terms of the model of Springett et al. (*14*) the effect of a decrease in temperature would be mitigated through an increase in density, moving the molecules closer together on the average which would tend to increase the kinetic energy term (T). However, the polarization energy also increases with increasing density and thus V_o, which depends on the difference in these two large energy terms, is very sensitive to the model. The calculated increases in V_o from the room temperature values for a 100° drop in temperature are 0.06 for tetramethyl-

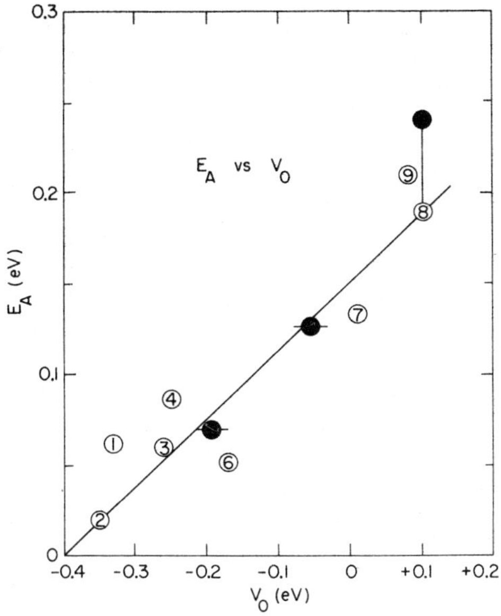

FIG. 4. Activation energy of mobility vs V_o; open points pure liquids:
1) 2,2,4,4-tetramethylpentane
2) neopentane
3) 2,2-dimethylbutane
4) 2,3-dimethylbutene-2
6) 2,2,4-trimethylpentane
7) cyclohexane
8) *n*-hexane
9) methylcyclohexane
solid points – are *n*-hexane-neopentane mixtures (refs. *23* and *25*)

silane, 0.19 for 2,2,4-trimethylpentane, and 0.31 for *n*-hexane. The experimental data available for comparison at present is that obtained a) from the photoionization of TMPD (*5*) and b) from work function measurements (*28*). In the photoionization case, changes in V_o are complicated by variation in ion-separation probabilities and threshold behavior with temperature. If we judge V_o on the change in ionization onset alone, the results show that when the temperature is decreased to −80°, V_o increases by 0.11 for tetramethylsilane, 0.14 for 2,2,4-trimethylpentane, and 0.16 for *n*-hexane. Preliminary work function measurements indicate an increase of 0.05 for tetramethylsilane and 0.18 for 2,2,4-trimethylpentane over this temperature range. The agreement with theory is qualitative in that the evidence supports increases in V_o with decreasing temperature and differences between liquids are suggested.

EFFECT OF V_o ON ELECTRON REACTIONS

As noted above, thresholds for photoionization of solutes can change with changes in V_o. For similar reasons one would expect that reactions such as 5 would also be similarly affected.

$$e^- + \text{solute} \underset{k_{-5}}{\overset{k_5}{\leftrightarrow}} \text{solute}^- \qquad (5)$$

The forward reaction is favored if the conduction band is high (as, for example, in hexane) while the reverse reaction is favored if V_o is low. Some experimental information is available on this point. It has been noted *(29)* for the reaction of the electron with ethyl bromide that the rate constant depends on V_o in much the same way that the rate of this reaction varies with kinetic energy in the gas phase. A similar behavior has recently been noted for other solutes *(30)*. Another example is the reaction of the electron with biphenyl. In this case the rate constant of the forward reaction increases (less than proportionately) with the mobility *(31)*. However, in tetramethylsilane in which μ is very high the decomposition of the biphenyl anion (the reverse of reaction 5) is observed,[3] which is interpreted here to be a direct consequence of the low value of V_o in this solvent.

In conclusion we note that the change in work function of a metal when immersed in a liquid is at present the best measure of the conduction band energy. We have shown how the position of this band affects various properties of the excess electron in liquids. The separation between the conduction band and trapped state is a major factor determining the electron mobility. The value of V_o can influence the photoionization threshold of solutes and the dynamics of electron attachment reactions.

REFERENCES

1. M. A. Woolf and G. W. Rayfield, *Phys. Rev. Lett.* **15,** 235 (1965).
2. B. Halpern, J. Lekner, S. A. Rice and R. Gomer, *Phys. Rev.* **156,** 351 (1967).
3. R. A. Holroyd and M. Allen, *J. Chem. Phys.* **54,** 5014 (1971).
4. R. A. Holroyd, B. K. Dietrich and H. A. Schwarz, *J. Phys. Chem.* **76,** 3794 (1972).
5. R. A. Holroyd and R. L. Russel, *J. Phys. Chem.* **78,** 2128 (1974).
6. N. R. Kestner and J. Jortner, *J. Chem. Phys.* **59,** 26 (1973).
7. A. M. Brodskii and Y. Y. Gurevich, *Sov. Phys. JETP* **27,** 114 (1968).
8. R. A. Holroyd, *J. Chem. Phys.* **57,** 3007 (1972).
9. R. Schiller, *J. Chem. Phys.* **57,** 2222 (1972).
10. R. Schiller, *Int. J. Radiat. Phys. Chem.* **5,** 491 (1973).
11. Y. Nakato, M. Ozaki and H. Tsubomura, *J. Phys. Chem.* **76,** 2105 (1972).
12. S. Noda and L. Kevan, *J. Chem. Phys.* **61,** 2467 (1974).
13. J. Lekner, *Phys. Rev.* **158,** 130 (1967).
14. B. E. Springett, J. Jortner and M. H. Cohen, *J. Chem. Phys.* **48,** 2720 (1968).
15. J. Jortner, *Ber. Bunsenges. Phys. Chem.* **75,** 696 (1971).
16. L. Meyer, H. T. Davis, S. A. Rice and R. J. Donnelly, *Phys. Rev.* **126,** 1927 (1962).
17. K. Fueki, *Can. J. Chem.* **50,** 3379 (1972).
18. G. Bakale and W. F. Schmidt, *Z. Naturforsch.* **289,** 164 (1973).
19. K. Fueki, D-F. Feng and L. Kevan, *Chem. Phys. Lett.* **13,** 616 (1972).
20. H. T. Davis, L. D. Schmidt and R. M. Minday, *Chem. Phys. Lett.* **13,** 413 (1972).
21. H. T. Davis and L. D. Schmidt, *Can. J. Chem.* **51,** 3443 (1973).
22. G. R. Freeman, this Symposium.
23. R. Schiller, this Symposium.
24. R. A. Holroyd and W. Tauchert, *J. Chem. Phys.* **60,** 3715 (1974).
25. R. M. Minday, L. D. Schmidt and H. T. Davis, *Phys. Rev. Lett.* **26,** 360 (1971).
26. H. T. Davis, L. D. Schmidt and R. M. Minday, *Chem. Phys. Lett.* **13,** 413 (1972).
27. R. M. Minday, L. D. Schmidt and H. T. Davis, *J. Phys. Chem.* **76,** 442 (1972).

[3]J. M. Warman, E. Zador, M. P. De Haas and A. Hummel, Concerning the equilibrium e + biphenyl ↔ biphenyl$^-$ in liquid tetramethylsilane. Private communication.

28. R. A. Holroyd, S. Tames and A. Kennedy, to be published in *J. Phys. Chem.*
29. A. O. Allen and R. A. Holroyd, *J. Phys. Chem.* **78,** 796 (1974).
30. A. O. Allen, R. A. Holroyd and T. E. Gangwer, *J. Phys. Chem.* **79,** 25 (1975).
31. G. Beck and J. K. Thomas, *J. Chem. Phys.* **57,** 3649 (1972).

Excess Electrons and Energy Fluctuations in Non-Polar Liquids[1]

Robert Schiller
Central Research Institute for Physics
1525 Budapest, P.O.B. 49, Hungary

INTRODUCTION

Electrons moving around in a liquid medium without belonging to any individual atom, ion or molecule were produced long before electrons were discovered (*1*). The properties, formation and reactions of these excess electrons if in dipolar liquids such as water, ammonia or alcohols are described in great detail and are theoretically well understood (cf. papers of this Symposium). This class is called solvated electrons; they behave very similarly to massive negative ions so far as thermodynamic properties, mobilities and reaction rates are concerned. Chemists are often tempted to say that they represent the simplest sort of negative ions.

Theoreticians (*2,3*) offer for the explanation of these observations the notion of electron localization: the charge density of the solvated electron has a maximum at a certain site in the liquid and increasing distance decreases the density asymptotically. Localization is brought about by self-trapping. The electron polarizes its environment, orienting the dipolar molecules of the liquid. The mutual attraction between electron and medium is sufficiently strong to trap the electron in the electric field of the oriented dipoles, so it moves together with the molecules which form the potential well as does a massive ion which has its own solvation shell. Kestner's (*4*) paper in this Symposium gives a detailed account of the theories based on this model.

Localization, however, is not the only possibility for an excess electron to stabilize in a liquid. An electron in a field-free space can be described as a travelling plane-wave. In certain liquids excess electrons preferentially exist in such a form. This happens if long-range attraction due to polarization and short-range repulsion owing to the Pauli principle balance each other, thereby making the resultant field acting on the electron rather weak. In such cases the electron propagates as a plane-wave scattered by the atoms or molecules of the medium (*2,3*). Quasi-free electrons, as the above described species is denoted, have a much higher mobility than do localized ones and the temperature-dependence of the mobility, rather than being exponential, exhibits a $T^{-3/2}$ proportionality.

Mobilities of this kind are comparatively rare which is evident if one realizes that only the delicate balance of attractive and repulsive forces maintains the quasi-free electron state. Electrons localize in dipolar liquids where the main interaction is static attraction. However, they do the same in non-polar and,

[1]Congress presentation E–28–3.

moreover, in weakly-polarizable liquids where the faint attraction cannot overcome the strong short-range repulsion. As a result the excess electron repels the neighboring atoms or molecules, rearranges the environment, forms a low-density region—called a bubble, and localizes around its center.

One would, on the basis of this picture, expect the following situation. There are dipolar liquids, such as water, and very weakly polarizable liquids, for example liquid helium, in which excess electrons are ready to localize, hence one observes a mobility of the order $\mu_L \approx 10^{-3}\ cm^2/Vs$, predictable from knowing the radius of the solvation-shell or bubble and from Stokes law. There are, however, other liquids, such as liquid argon, in which polarization and repulsion are mutually balanced. Here the electrons move as plane-waves and the mobility, being defined by the scattering properties of the liquid (*5–7*), is on the order of $\mu_F \approx 100\ cm^2/Vs$.

Freeman (*8*) has summarized in this Symposium our knowledge regarding electron mobilities. Whereas water, helium and argon are seen to meet the expectations, a host of substances, virtually non-polar liquids, exhibit mobilities falling between μ_F and μ_L. This is difficult to understand within the framework of the models outlined above.

It has also been observed (*9–18*) that high electron mobility is always accompanied by high radiation chemical yield of free ions. By this latter is understood the yield of those ions which can be collected by vanishingly low electric fields ("zero field") or extremely dilute solutes (scavengers).

Holroyd (*19*) reports a third parameter which decreases with increasing mobility and yield, and this is the energy of the conductivity state or quasi-free electron state. This is the energy change which is coupled with an electron being transferred from vacuum into the liquid so rapidly that no structural rearrangement can take place, in which case the electron remains quasi-free and its energy is the same as that of the bottom of the conductivity band. The conductivity-state energy is observed to decrease whenever mobility and free-ion yield increase.

The object of the present paper is to suggest a physical model which, involving fairly simple mathematics, is able to account for the observations summarized above: the continuous variation of electron mobilities from liquid to liquid, and the interdependence of mobility, conductivity state (quasi-free state) energy and radiation chemical yield of electrons.

ELECTRON MOBILITY

Our model (*20,21*) is based on the two-state picture of the electrons: they are either in a quasi-free and high mobility state or in a localized and low mobility state. We assume, however, that both states occur in each and every liquid, and it is only their proportion that varies from substance to substance. Let the high mobility be again denoted by μ_F and the low mobility by μ_L. The observed mobility, μ can be given by the weighted average

$$\mu = \mu_F(1 - P) + \mu_L P \approx \mu_F(1 - P) \tag{1}$$

where P is the probability of finding an electron in the localized state (the relationship $\mu_F \gg \mu_L$ has been made use of). If localization probability changes continuously, so also does mobility in terms of the present picture. The real task is to construct a theory for P.

To this end let us compare a portion of a liquid containing a quasi-free electron with another portion of the same liquid containing a localized electron. The two portions differ in structure, energy, and volume of interaction.

Localized electrons were said to rearrange the structure of their environment, whereas quasi-free ones allow it to remain virtually unchanged. The energy of introducing a localized electron from the vacuum, E_t, is consequently a sum of the electron energy, E_e, measured from the vacuum level, and of the medium rearrangement energy, E_M, measured from the energy of the unperturbed liquid. The energy of introducing a quasi-free electron from the vacuum, V_o, does not involve any term of medium rearrangement energy.

As to the volume of interaction between liquid molecules and an electron in the localized state, one has to have recourse to positron annihilation experiments (*22*). Positrons are known to form positronium atoms, i.e. positron-electron pairs, before becoming annihilated. The lifetime of positrons in the form of positronium depends on the electron density of the environment. The lifetimes measured for liquid He and also for hydrocarbons indicate positronium atoms to be localized in a vacuum bubble formed as the result of repulsion between atomic and positronium electrons. The bubble radius in He was found to be about 15 Å, in hydrocarbons some 5 Å. The bubbles containing positroniums may differ in size from those containing electrons. It seems to be safe, however, to state that localized electrons interact with only few electrons and the bubble-size in hydrocarbons is on the order of the molecular volume (i.e. molar volume per Avogadro number).

Quasi-free electrons on the other hand are spread over a large number of molecules interacting, strictly speaking, with the entire liquid system at the same time.

This difference in the volumes of interaction makes statistical fluctuations important. Let us consider a liquid system of macroscopic volume and of given temperature, density and energy content. Let us imagine the volume to be divided into a large number of equally-sized subsystems. With the total energy of the system, U, the average energy of each subsystems is $\bar{E} = U/n$, where n denotes the number of subsystems. If the subsystems are small enough, their actual energy content, E, usually differs from $\bar{E}$; it might be higher or lower than that. This spread in energy is called energy fluctuation (*23,24*). The probability, p(E)dE, of finding a cell with an energy between E and E + dE is equal to

$$p(E)dE = (2\pi\sigma^2)^{-1/2} \exp[-(E - \bar{E})^2/2\sigma^2]dE \,. \tag{2}$$

Density (and in multi-component systems, concentration) also exhibits fluctuations. The extent of fluctuation is characterized by the modulus σ. For the case of energy fluctuations, neglecting energy changes due to density fluctuations, one has $\sigma^2 = kT^2C_v$, where C_v is the heat capacity of a subsystem, i.e. the product of the heat capacity per molecule and the number of molecules in the system.

Let us compare the energy of a liquid containing a localized electron with that containing a quasi-free electron. The total energy of the liquid without an electron is $n\bar{E}$; introducing a quasi-free electron it changes to $U_F = n\bar{E} + V_o$. If, however, a localized electron is introduced one has to consider the actual energy, E, of the subsystem which will contain the electron; hence the total energy can be written as $U_L = (n-1)\bar{E} + E + E_t$.

Adopting the law that it is always the lower energy state that is the more stable, the condition of localization is given by the inequality

$$U_L < U_F \qquad \text{i.e.} \quad E + E_t < \bar{E} + V_o \qquad \text{or} \quad E - \bar{E} < V_o - E_t \tag{3}$$

The probability of finding a subsystem which satisfies Eq. (3) is equal to the probability of an excess electron to localize, P. This can be expressed by the integral

$$P = \int_{-\infty}^{(V_o - E_t)} p(E)dE = (1/\sqrt{2\pi}) \int_{-\infty}^{(V_o - E_t)/\sigma} \exp(-t^2/2)dt \,. \tag{4}$$

In order to evaluate the above integral one has to know the values of V_o, σ and E_t. No measurement problems are presented by V_o, it has been determined for a series of liquids, mainly hydrocarbons (*19*). The parameter σ is given byhe size of the subsystem and the specific heat of the liquid. E_t has not, as yet, been measured and no simple theoretical method seems to be available for its calculation.

Experimental mobilities should be compared with theoretical values computed according to Eqs. (1) and (4). This check of the theory requires the knowledge of a further quantity, μ_F, the value of which is also unknown at present. Hence, the theory of electron mobility leaves two parameters, E_t and μ_F, free to fit the experimental results.

FREE-ION YIELD

An attempt is being made to rationalize the observed parallelism between electron mobilities and free-ion yields in terms of the two-electron state model outlined above. Ions which are free in the sense that an extremely weak electric field can collect them are certainly at a considerable distance from any other ion of opposite sign; otherwise they would recombine. Radiation always produces positive ions and electrons in pairs, one ion and one electron stemming from the same molecule. The recombination of these geminate charge-carriers is a very effective process, in fact that is the route of ion recombination at low dose-rates. Accordingly, the yield of free ions must be equal to the yield of electrons which escape geminate recombinations.

Onsager (*25*) developed a theory for the geminate recombination of massive ions which obey classical laws of diffusion. He expressed the escape probability, w_p, by the form

$$w_p = 4\pi \int_0^\infty r^2 F(r) \exp(-e^2/\epsilon rkT)dr \,, \tag{5}$$

where e is the elementary charge, ϵ the dielectric constant and F(r) is the distribution of distances between the ions of a pair. This latter expression means that at the instant of attaining thermal energy the fraction of ion-pairs with an interionic distance between r and r + dr equals $4\pi r^2 F(r)dr$.

We attribute the following life-history to an electron ejected from a molecule. Usually the electron leaves its parent molecule with excess kinetic energy. Interacting with the medium, it slows down and attains the thermal energy of its surroundings. At this moment it is thought to be quasi-free, the very short period of moderation supports this assumption. Its geminate ion attracts the quasi-free thermal electron, trying to drag it back. It can be shown by simple quantum mechanics (*26*) that the excape-probability of quasi-free electrons is 1/2, irrespective of the properties of the actual substance.

Thus, every second ion-pair disappears during the early quasi-free period of electron life. The other half of the electrons, however, escape and localize, the rate of localization being controlled by P, the probability of finding a trapping site. The higher the value of P, the smaller the distance between ion and localized electron.

Localized electrons move as do massive, classical particles, hence Eq. (5) applies to them, the sole difference depending on the meaning of F(r) which denotes the distribution of distance between ions and localized electrons.

It is apparent from the foregoing consideration that F(r), and hence w_p, according to this model will be a function of P. A simple calculation leads to the following expression of overall escape probability:

$$G_{fi}/G_o = [\alpha^3/4] \int_0^\infty x^2 \exp[-(1/x + \alpha x)]dx \; . \qquad (6)$$

Here $x = e^2/\epsilon rkT$, $\alpha = (\sigma_t/v)(e^2/\epsilon kT)P$, v is the volume and σ_t the absorption cross section of the trap. The escape probability has been identified with the ratio of the free-ion yield to the yield of total primary ionization, G_o.

A new adjustable parameter has appeared in Eq. (6) in addition to E_t, and this is the ratio σ_t/v. For the value of G_o one may take the primary ionization yields measured in the vapor phase of the substances.

COMPARISON OF THEORY WITH EXPERIMENT

The quantities μ, V_o and G_{fi} can be directly measured; their interdependence as defined by our theory is expressed by Eqs. (1), (4) and (6). Two of the equations—Eqs. (1) and (4)—give the dependence of μ on V_o; another two—Eqs. (4) and (6)—describe the G_{fi} *vs* V_o function; whereas the third combination—Eqs. (1) and (6)—determines the interrelation of G_{fi} and μ. Three adjustable parameters are involved in the three expressions, being μ_F, E_t, and σ_t/v. Since these parameters generally vary from one substance to another, one has to choose a class of substances, the members of which are closely related to each other. Here one may assume that the adjustable parameters are the same for the whole class, although μ, V_o and G_{fi} differ markedly.

This approximation was applied in the numerical evaluation of the theory, saturated hydrocarbons being the test substances. Calculated curves and measured data are presented in Figs. 1–3. Conventional curve-fitting procedure resulted in the values $\mu_F = 65\ cm^2/Vs$, $E_t = -0.28$ eV and $(\sigma_t/v)(e^2/\epsilon kT) = 10$. The curves, each of which contains two adjustable parameters, seem to be in reasonable agreement with the observations.

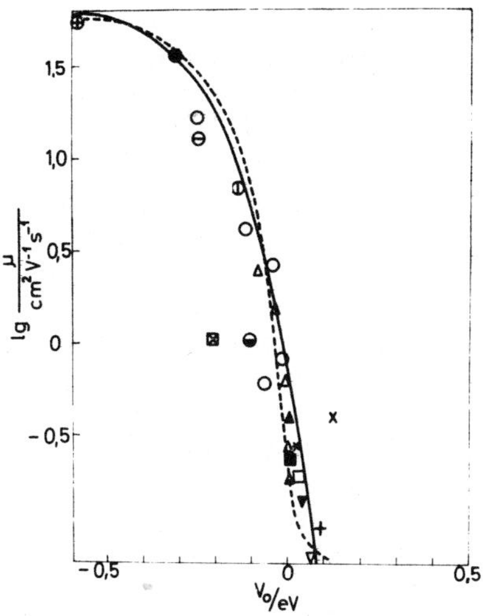

FIG. 1. Mobility as a function of conductivity state energy. Continuous line: theoretical curve according to Eqs. (1) and (4). Dotted line: theoretical curve according to Ref. (*29*). Both curves are fitted to the experimental data.
⊕ neopentane (18)[a]; ● 2,2,4,4-tetramethylpentane (14)[b]; O normal-hexane–neopentane mixtures (30)[a]; △ normal-hexane–iso-octane mixtures (31)[a]; Θ 2,2-dimethylbutane (17)[b,d]; ⏀ iso-octane (18)[a]; ☒cyclo-pentane (18)[b]; ▲ propane (32)[b]; × normal-butane (18)[b]; ■ cyclo-hexane (14)[b]; ▼ normal-pentane (18)[b]; + normal-hexane (18,21,33); ▽ methyl-cyclo-hexane[b]; ◒ 2,3-dimethylbutane (21)[c]; ⋇ 2-methylpentane[b,c,d]; □ 3-methylpentane[b,c,d].

[a]L. Nyikos and R. Schiller, Conductivity state energy and electron localization in hydrocarbon mixtures. To be published.

[b]R. A. Holroyd, personal communication.

[c]J. P. Dodelet and G. R. Freeman, Electron mobilities and ranges in liquid ethers. To be published.

[d]Following the advice of an unknown referee the V_o data in footnote *b* were used instead of those in Ref. (21).

DISCUSSION

The ideas propounded here are not claimed to be either original or unique. Many of the physical principles and processes indicated in the foregoing considerations have already been suggested in one form or another: Freeman (*9*) wrote about electron-trapping in pre-formed traps; Schmidt and Allen (*18*) mentioned the possible role of rotational fluctuations in electron transport; Funabashi and Kajiwara (*27*) have shown the trapping potentials to be related to energy fluctuations and the trap depth and density to depend on the shape of the molecules; Abell and Funabashi (*28*) assumed the Onsager relation, Eq. (5), to be valid also for quasi-free electrons, and, finding an exponential form for F(r), they stressed the fact that different liquids differ in density and scattering cross section of the traps. Their work implies a relationship between free-ion yield and electron mobility.

Kestner and Jortner (*29*) did not regard energy fluctuations as a factor that changes the electron states in a homogeneous liquid but rather as one that influences the transparency of the medium towards electrons. This effective medium theory defines the proportion of low-transparency zones in the same way as P is given by Eq. (4); the mobility depends, however, on this quantity in a manner more intricate than the proportionality in Eq. (1). The $\mu(V_o)$ function can be computed in this theory by four adjustable parameters whereas Eq. (1) and (4) require only two. A curve based on this theory and fitted to all the available experimental points is given in Fig. 1. It is difficult to tell which of the two theories results in a better description of the observations. Moreover, the activation energies of the mobilities can also be calculated from both theories (*20,21,29*) and the agreement with experiment is equally good in both cases.

Competing and eventually conflicting models and theories have been developed in the last few years in an attempt to understand the properties of excess electrons in non-polar liquids. The question is far from being settled. New experimental data regarding rate constants and mobilities await interpretation. Adjustable parameters, electron energies and mobilities require theoretical explanations in terms of the structural properties of the liquids.

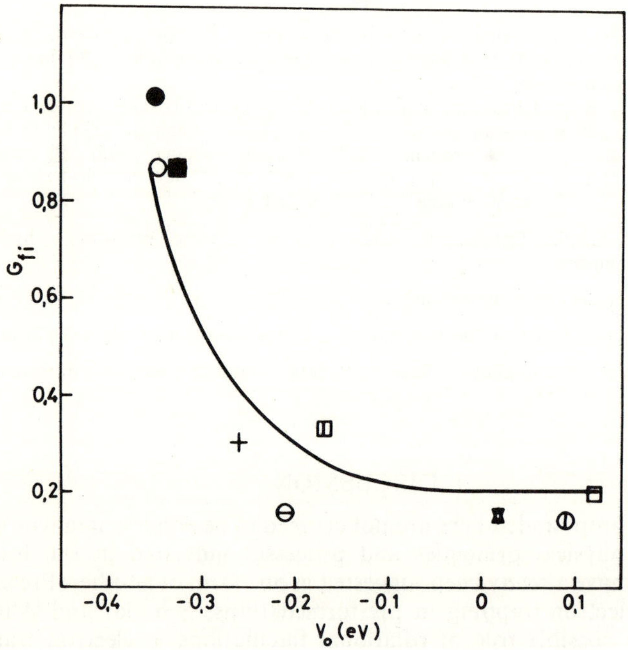

FIG. 2. Free-ion yield as a function of conductivity state energy. Continuous line: theoretical curve according to Eqs. (4) and (6)
● neopentane (14,33); O neopentane (16,33); ■ 2,2,4,4-tetramethylpentane (14)[b]; + 2,-dimethylbutane (16)[b]; ◫ 2,2,4-trimethylpentane (16,33); ϴ cyclo-pentane (16,33); ∇ cyclo-hexane (16)[b]; △ normal-pentane (16,33); Ⓘ normal-hexane (16,21)[b]; □ normal-butane (16)[b]
[b]See legend to Fig. 1.

[Reproduced from R. Schiller and Sz. Vass, *Int. J. Radiat. Phys. Chem.* 7, 193 (1975) with the permission of Pergamon Press, Ltd.]

Incomplete as these theories might be, two basic features of the phenomena have been realized by them almost without exception: a) thermal electrons in non-polar liquids cannot be regarded as classical particles even long after their being moderated; b) fluctuations in energy and density play a central role in the life-history of the electron. We believe these aspects to be valid and important, and it is this that makes us think that the problem of excess electrons represents a conceptually new chapter in physical chemistry.

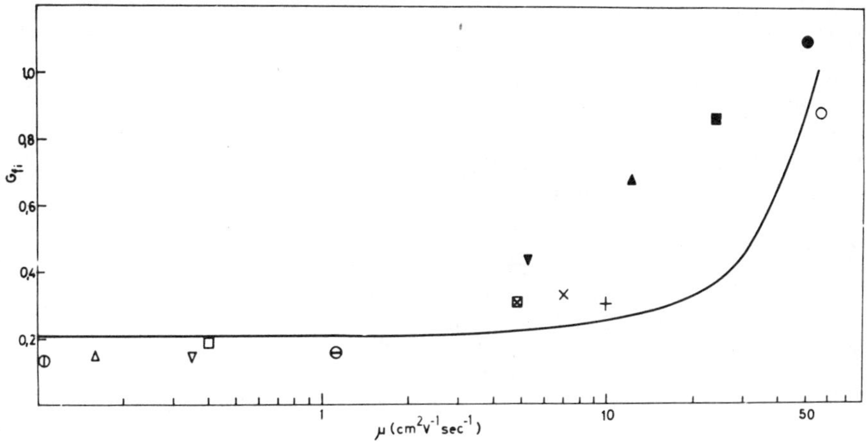

FIG. 3. The interdependence of free-ion yield and mobility. Continuous line: theoretical curve according to Eqs. (1) and (6).
● neopentane (14); O neopentane (16,18); ■ 2,2,4,4-tetramethylpentane (14); ▲ 2,2,5,5-tetramethylhexane (14); ▼ 2,2,3,3-tetramethylpentane (14); × 2,2,4-trimethylpentane (16,18); + 2,2-dimethylbutane (16,18); ⊠ 2-methylpropane (13); Θ cyclopentane (16,18); □ normal-butane (16,18); ∇ cyclo-hexane (16,18); △ normal-pentane (16,18); ϕ normal-hexane (16,18)

[Reproduced from R. Schiller and Sz. Vass, *Int. J. Radiat. Phys. Chem.* **7,** 193 (1975) with the permission of Pergamon Press, Ltd.]

ACKNOWLEDGEMENT

The author is indebted to M. L. Nyikos for his cooperation. He is also grateful to Dr. G. R. Freeman and Dr. R. A. Holroyd for making available their unpublished data.

REFERENCES

1. W. Weyl, *Ann. Phys. Chem.* **121,** 601 (1964).
2. J. Jortner, in *Actions Chimiques et Bioligiques des Radiation* (M. Haissinsky, Ed.) **Vol. 14** pp. 7–72. Masson, Paris, 1970.
3. J. Jortner, *Ber. Bunsenges. Phys. Chem.* **75,** 696 (1971).
4. N. Kestner, this Symposium.
5. M. H. Cohen and J. Lekner, *Phys. Rev.* **158,** 305 (1967).
6. J. Lekner, *Phys. Rev.* **158,** 130 (1967).
7. H. T. Davis, L. D. Schmidt and R. M. Minday, *Phys. Rev.* **A3,** 1027 (1971).
8. G. R. Freeman, this Symposium.

9. P. H. Tewari and G. R. Freeman, *J. Chem. Phys.* **49,** 954 (1968).
10. P. H. Tewari and G. R. Freeman, *J. Chem. Phys.* **49,** 4394 (1968).
11. K. Horacek and G. R. Freeman, *J. Chem. Phys.* **53,** 4486 (1970).
12. M. G. Robinson, P. G. Fuochi and G. R. Freeman, *Can. J. Chem.* **49,** 3657 (1971).
13. P. G. Fuochi and G. R. Freeman, *J. Chem. Phys.* **56,** 2333 (1972).
14. J. P. Dodelet and G. R. Freeman, *Can. J. Chem.* **50,** 2667 (1972).
15. W. F. Schmidt and A. O. Allen, *Science* **160,** 301 (1968).
16. W. F. Schmidt and A. O. Allen, *J. Chem. Phys.* **52,** 2345 (1970).
17. W. F. Schmidt and A. O. Allen, *J. Chem. Phys.* **50,** 5037 (1969).
18. W. F. Schmidt and A. O. Allen, *J. Chem. Phys.* **52,** 4788 (1970).
19. R. A. Holroyd, this Symposium.
20. R. Schiller, *J. Chem. Phys.* **57,** 2222 (1972).
21. R. Schiller, Sz. Vass and J. Mándics, *Int. J. Radiat. Phys. Chem.* **5,** 491 (1973).
22. A. Butchikin, V. I. Goldanskii, A. O. Tatoor and V. P. Shantarovich, *Zh. Eksp. Teor. Fiz.* **60,** 1136 (1971). [In Russian.]
23. N. Davidson, *Statistical Mechanics.* McGraw-Hill, New York, 1962.
24. L. D. Landau and E. M. Lifshitz, *Statistical Physics.* Pergamon, London, 1958.
25. L. Onsager, *Phys. Rev.* **54,** 554 (1938).
26. R. Schiller and Sz. Vass, *Int. J. Radiat. Phys. Chem.* **7,** 193 (1975).
27. K. Funabashi and T. Kajiwara, *J. Phys. Chem.* **76,** 2726 (1972).
28. G. C. Abell and K. Funabashi, *J. Chem. Phys.* **58,** 1079 (1973).
29. N. R. Kestner and J. Jortner, *J. Chem. Phys.* **59,** 26 (1973).
30. R. M. Minday, L. D. Schmidt and H. T. Davis, *Phys. Rev. Lett.* **26,** 360 (1971).
31. G. Beck and J. K. Thomas, *J. Chem. Phys.* **57,** 3649 (1972).
32. H. T. Davis, L. D. Schmidt, and R. G. Brown, in *Electrons in Fluids* (J. Jortner and N. R. Kestner, Eds.). Springer, Berlin, 1973.
33. R. A. Holroyd and M. Allen, *J. Chem. Phys.* **54,** 5014 (1971).

Mobilities of Slow Electrons in Low- and High-Pressure Gases and Liquids[1,2]

L. G. Christophorou[3]
Health Physics Division, Oak Ridge National Laboratory
Oak Ridge, TN 37830

Mobilities of slow (thermal and epithermal) electrons in low- ($\lesssim$ 500 Torr) and high- (~500 to ~34,111 Torr) pressure gases are discussed and are related to the molecular structure and to the mobilities of thermal electrons in liquid media.

ELECTRON MOBILITIES IN LOW-PRESSURE GASES

The mobility μ of thermal electrons in low-pressure gases of polar molecules decreases with increasing magnitude of the permanent electric dipole moment D (*1*). This is shown in Fig. 1 where μ_{Torr}[4] is plotted as a function of D. The solid line in Fig. 1 is a linear least squares fit to the experimental data and is represented by the relationship

$$\mu_{\mathrm{Torr}} = 2.74 \times 10^5 \ \mathrm{D}^{-1.48} \ . \tag{1}$$

The broken line in Fig. 1 is a theoretical estimate of the dependence of μ_{Torr} on D and is based on a calculation by Altshuler (*2*) who considered the problem of electron scattering by a point dipole in the Born approximation. Although the theoretical prediction

$$\mu_{\mathrm{Torr}} = 4.51 \times 10^5 \ \mathrm{D}^{-2} \tag{2}$$

differs from the experiment, the differences are not drastic and the magnitude of μ_{Torr} is seen to be determined primarily[5] (*3*) by the electron-electric dipole scattering potential. For D $\gtrsim$ 0.5 Debye, Eq. (1) can be used to predict μ_{Torr} from a knowledge of D.

For small values of D ($\rightarrow$ 0), μ_{Torr} decreases with increasing static polarizability α (Fig. 2).

In addition to the molecular parameters D and α, distinct submolecular

[1] Congress presentation E–28–4.

[2] Research sponsored by the U.S. Atomic Energy Commission under contract with Union Carbide Corporation.

[3] Also, Department of Physics, University of Tennessee, Knoxville, Tennessee 37916.

[4] μ_{Torr} is defined as S/P_T (= 1 Torr) where S is the slope of the w vs E/P_T curves in the region where the drift velocity w varies linearly with the pressure-reduced electric field E/P_T.

[5] For a discussion of the detailed behavior of the scattering cross section and hence μ_{Torr} especially around D = 1.625 debye, see ref. (*3*). See also Chapter 4 in ref. (*5*).

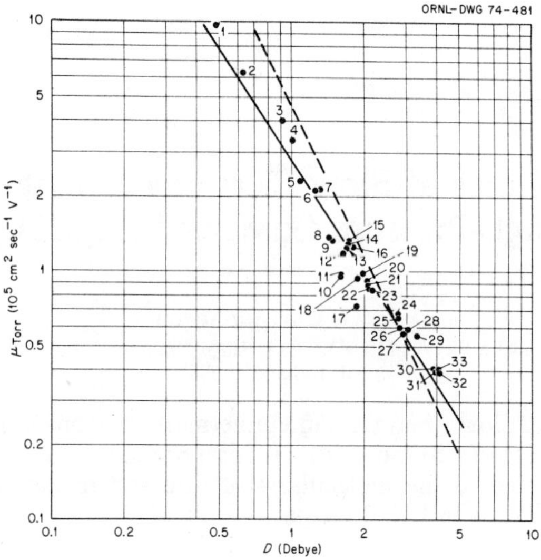

FIG. 1. μ_{Torr} vs D (see Ref *1* for identification of molecules 1–33).

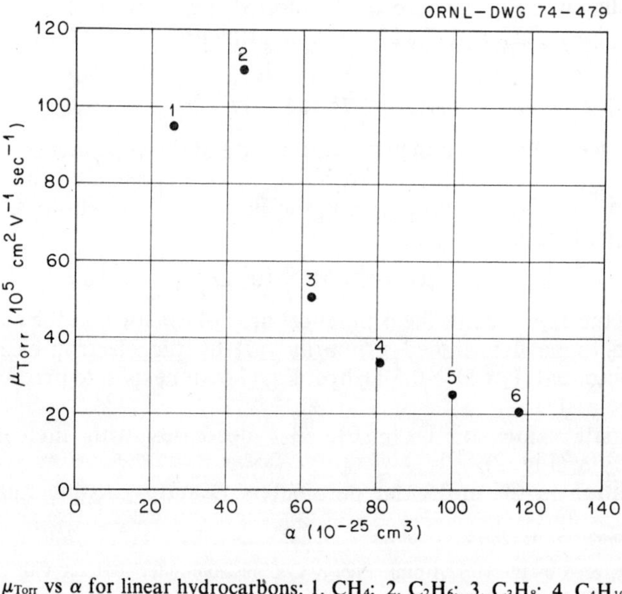

FIG. 2. μ_{Torr} vs α for linear hydrocarbons: 1. CH_4; 2. C_2H_6; 3. C_3H_8; 4. C_4H_{10}; 5. C_5H_{12}; 6. C_6H_{14} (see ref. *1*).

characteristics of the molecular structure affect μ. Thus μ was found to decrease with increasing number of doubly-occupied π-electron orbitals for a number of linear, cyclic and aromatic hydrocarbons (Fig. 3). For linear hydrocarbons with two doubly-occupied π-electron orbitals, μ increases with increasing separation of the two π-orbitals (*1,4*). Additionally, μ was found[6] to depend on the cis- and trans-forms of a given molecule. Thus the mobility of thermal electrons in trans-butene-2 is 2.8 times as large as that for the cis-butene-2.[7]

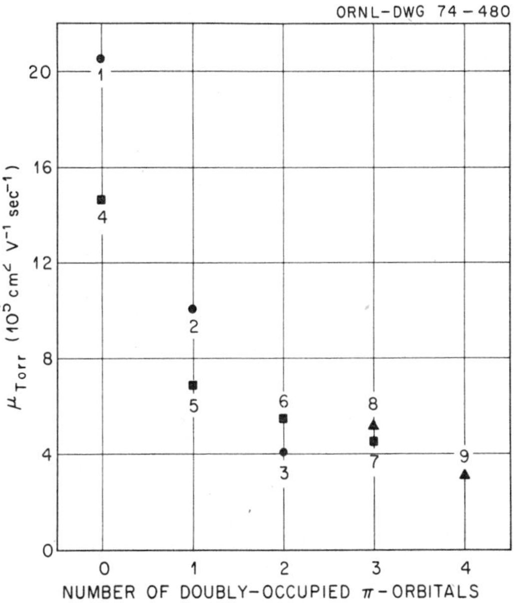

FIG. 3. μ_{Torr} vs number of doubly-occupied π-orbitals. ● Linear hydrocarbons: 1. $CH_3(CH_2)_4CH_3$; 2. $CH_3(CH_2)_3CH = CH_2$; 3. $CH_3CH_2CH_2CH = C = CH_2$. ■ Cyclic nonaromatic hydrocarbons: 4. cyclohexane; 5. cyclohexene; 6. 1,4-cyclohexadiene; 7. 1,3,5-cycloheptatriene. ▲ Aromatic hydrocarbons: 8. Benzene; 9. naphthalene. (See Refs. *1* and *4*).

ELECTRON MOBILITIES IN HIGH-PRESSURE GASES

Until recently the mobility of slow electrons in gases was considered to be independent of the gas pressure. This, apparently, was a consequence of the fact that mobility measurements were made at relatively low pressure, quite generally below ~ 1 atm. However, recent studies—especially by the Hamburg workers (see Ref. *1*)—at much higher pressures have revealed large and varying effects of pressure on the electron mobility. These are dramatized in Figs. 4 and 5 where $\mu_{P'}/\mu_{\leqslant 500}$ is plotted as a function of $[\epsilon_e]_M$ for a number of pressures P′. The quality $\mu_{P'}$ denotes the electron mobility at a pressure P′ (corrected for compressibility, so that the number density is proportional to P′), $\mu_{\leqslant 500}$ denotes the electron

[6] L. G. Christophorou, K. Gant and D. Pittman (unpublished results).

[7] For information on the dependence of μ on $[\epsilon_e]_M$ for some organic molecules see ref. (*1*).

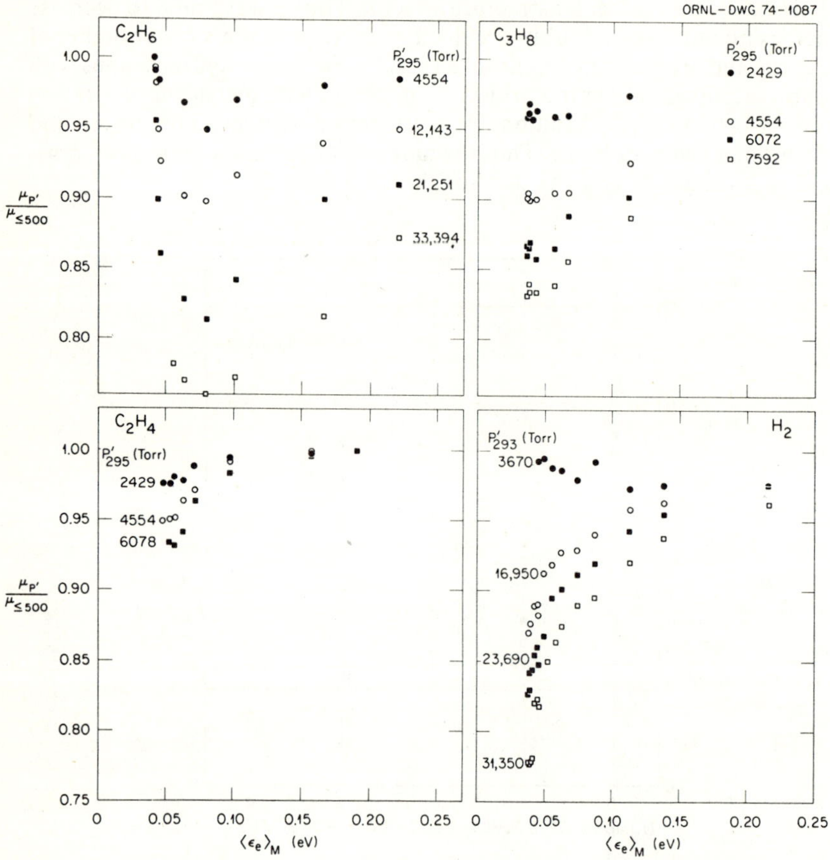

FIG. 4. $\mu_{P'}/\mu_{455}$ vs $[\epsilon]_M$ for a number of P'_{295} (see Ref. *1* as to the sources of original data).

mobility at $P \leqslant 500$ Torr and $[\epsilon_e]_M$ is the mean electron energy when the electron energy distribution is represented by a Maxwellian distribution function.

For all four molecules in Fig. 4 the electron mobility at pressure P is smaller than that at 500 Torr, the difference increasing with increasing P′. The dependence of $\mu_{P'}/\mu_{500}$ on $[\epsilon_e]_M$ indicates a resonance-like behavior with a minimum at ~0.075 eV for C_2H_6 and at ~0.045 eV for C_3H_8. For C_2H_4 and H_2 this minimum apparently occurs at or below thermal energies. These observations have, quite generally, been attributed to the delay in electron drift due to temporary capture of the electrons by the molecules (*1*).

An electron capture and loss mechanism, however, cannot be invoked to explain the experimental finding that the electron mobility in CH_4 increases (rather than decreases) with increasing CH_4 density. (A similar behavior has been reported for $C(CH_3)_4$ and Ar at low energies; see Ref. *1*). The striking dependence of $\mu_{P'}/\mu_{500}$ on $[\epsilon_e]_M$ and P′ for CH_4 is shown in Fig. 5. The data in Fig.

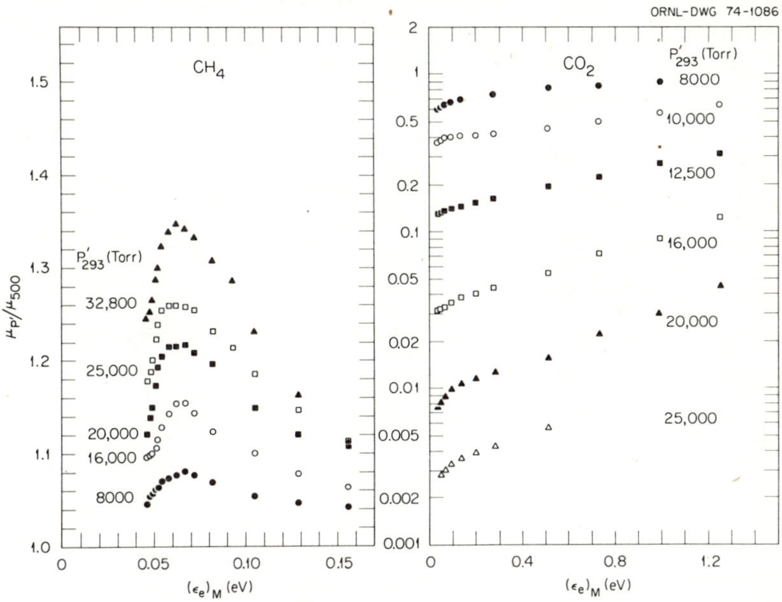

FIG. 5. $\mu_{P'}/\mu_{500}$ vs$[\epsilon]_M$ for a number of P'_{293} for CH_4 and CO_2 (see Ref. *1* as to the sources of original data).

5 indicate (see also following section) that in this case it is necessary to look for factors which in the dense gas reduce rather than increase the electron scattering cross section (the electron mobility is approximately inversely proportional to the scattering cross section) (*1*,*5*).

The simple capture and loss mechanism is also inadequate to explain the pressure dependence of μ_{Torr} for CO_2. The electron mobility in CO_2, contrary to the findings for C_2H_6, C_3H_8, C_2H_4, and H_2 at room temperature, decreases much faster than $(P')^{-1}$ and indicates that the effects other CO_2 molecules have both in stabilizing CO_2^{-*} (to form stable CO_2^- ions) and in decreasing its rate of destruction (autoionization) have to be considered in detail. The probability of autodetachment and capture is very often a complex function of the number density N. Recent studies on electron attachment to O_2 (*6*) and C_6H_6 (*7*) in high densities of various gaseous media clearly showed the strong and complex dependences of the probabilities of attachment and autodetachment on the density and the nature of the gaseous medium.

The three distinct types of pressure dependences of the electron mobility—examplified by the behavior of CH_4, C_2H_6 and CO_2—are shown in Fig. 6. In Fig. 6 the quantity μ_L/μ_G is the ratio of the mobility of thermal electrons in the liquid and μ_G is the corresponding gaseous electron mobility adjusted for the change in density between the gas and the liquid (see next section).

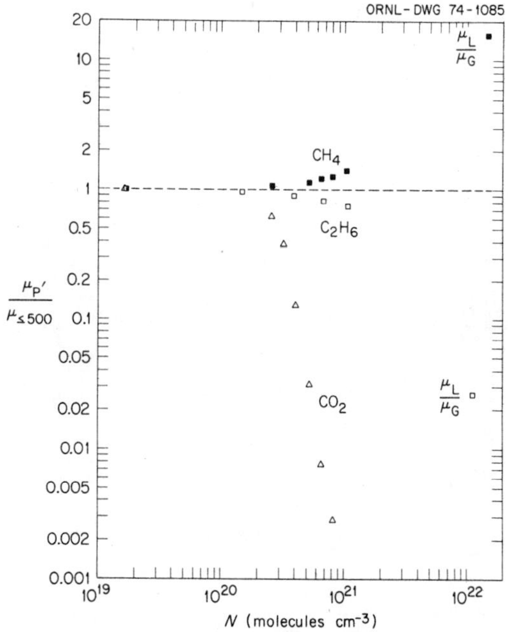

FIG. 6. $\mu_{P'}/\mu_{\leqslant 500}$ vs N for CH_4, C_2H_6, and CO_2.

ELECTRON MOBILITIES IN GASES AND LIQUIDS

The mobilities of thermal electrons in low- and high-pressure gases have been compared with those in the corresponding liquids in an effort to link together knowledge on electron motion in gases with that on electron motion in liquids. Firstly, the long-range electron-electric dipole interaction, seen in the first section to describe reasonably well the variation of the electron mobility with the electric dipole moment for gases, seems to be responsible for the low values of the electron mobility measured in polar liquids (*8*). Secondly, from our studies on low-pressure gases the quantity μ_G defined as

$$\mu_G \equiv S\,(N_{Torr})/N_L \tag{3}$$

was determined and compared with the electron mobility in the liquid μ_L. In Eq. (3) N_{Torr} is the number of molecules per cm^3 per Torr at the particular gas temperature and N_L is the number of molecules per cm^3 of the corresponding liquid at the temperature at which the measurement of μ_L was made. The quantity μ_G is regarded as the gaseous electron mobility adjusted for the change in density between the gas and the liquid. From this study it was found that for a number of linear saturated hydrocarbons, except methane, $\mu_L/\mu_G \ll 1$ (0.006–0.026), and for a number of cyclic hydrocarbons $\mu_L/\mu_G < 1$ (0.047–0.26). These low (< 1) values of the ratio μ_L/μ_G were attributed to the delay in electron drift in these molecules at high densities and in the liquid phase caused by an electron capture and loss process. In contradistinction with these results it was found that for a number of relatively spherical and branched hydrocarbons and for the heavier rare-gas atoms $\mu_L > \mu_G$. Experimental evidence suggests that for these systems, thermal electrons become more mobile the denser these gases are and attain their largest mobility in the liquid. On the basis of the μ_L/μ_G data (*1,4*) thermal electrons are about 6, 12, 16, 4, 119, and 730 times more mobile in liquid 2-methyl butene-2, 2,2-dimethyl propane, methane, argon, krypton and xenon, respectively, than in the corresponding low-pressure gases. In this connection, it is important to note that neither an electron capture and loss process, nor a (collective) electron trapping mechanism involving more than one molecule, nor an inelastic electron scattering process characteristic of the high pressure and/or the condensed phase can explain values of $\mu_L/\mu_G > 1$. Additional inelastic processes would increase the scattering cross section and thus decrease rather than increase the electron mobility. It is thus critically important to look for factors which in the dense gas and the liquid reduce rather than increase the scattering cross section, such as suitable effective potentials leading to coherent and forward electron scattering.

In our opinion the findings in the last two sections (see also Refs. *1* and *4*) suggest that for the relatively spherical organic molecules and the heavier rare-gas atoms, the electron scattering potential in the dense gas and the liquid is such that at thermal and near-thermal electron energies it leads to small scattering cross sections and thus to large values of the electron mobility, in a similar fashion that spherically symmetric and strong potentials give rise to the Ramsauer-Townsend effect in the heavier rare-gas atoms and methane at low pressures. It is emphasized that the experimental evidence which leads to this conclusion is strong, in spite of the fact that it can be argued that the observed increases in the electron mobility with increasing gas density and the high ($<$ 1) values of the ratio μ_L/μ_G may result from low ($>$ 1) values of the so-called "structure factor".

Finally, it is stressed that an integrated approach whereby low-pressure information (usually better known and understood) is related to that in the liquid phase (usually less known and understood) through the intermediate density region ("quasi-liquid" regime) which allows a study of the gradual transition from isolated to condensed-phase behavior, is a new and a promising approach for understanding this and similar problems.

REFERENCES

1. L. G. Christophorou, *Int. J. Radiat. Phys. Chem.* (in press) 1975.
2. S. Altshuler, *Phys. Rev.* **107,** 114 (1957).
3. L. G. Christophorou and A. A. Christodoulides, *J. Phys.* **B2,** 71 (1969).
4. L. G. Christophorou, R. P. Blaunstein and D. Pittman, *Chem. Phys. Lett.* **18,** 509 (1973).
5. L. G. Christophorou, *Atomic and Molecular Radiation Physics*. Wiley-Interscience, New York. 1971.
6. R. E. Goans and L. G. Christophorou, *J. Chem. Phys.* **60,** 1036 (1974).
7. L. G. Christophorou and R. E. Goans, *J. Chem. Phys.* **60,** 4244 (1974).
8. A. K. Pikaev and A. M. Brodskii, *High Energy Chem. (USSR) (Eng. Transl.)* **6,** 201 (1972).

RADIATION EFFECTS IN FROZEN MEDIA[1]

Chairman: J. Kroh

[1]Symposium A–6

Ground State Structure of Trapped Electrons in Glassy Matrices[1]

Larry Kevan
Department of Chemistry, Wayne State University
Detroit, Michigan 48202, USA

INTRODUCTION

Excess electrons produced by radiolysis or photoionization can be stabilized or trapped in a variety of liquids and solids with which the electrons react only slowly if at all (*1*). In solids, glassy matrices have proved most efficient for trapping electrons. Efficient trapping matrices include 10 M NaOH glass, 5 M K_2CO_3 glass, methanol, ethanol, isopropanol, methyltetrahydrofuran (MTHF), triethylamine, and 3-methylhexane, as a representative series of glasses spanning a range of polarity in order of decreasing polarity. The basic features of the energy level structure (*2,3*) of trapped electrons in these matrices and of the electron mobility (*4,5*) of photoexcited electrons in these matrices seem at least partially understood. Here we will focus on the ground state geometrical structure of trapped electrons in glassy matrices. We must first consider the nature of the electron stabilization process.

Essentially, the electrons are stabilized by a solvation process in which molecular dipoles are oriented toward the electron. In liquids this solvation process can be seen directly by pulse radiolysis (*6*) and by laser photolysis (*7*) in terms of a spectral blue shift of the trapped electron optical absorption band. The time scale of this process can be semiquantitatively understood on the basis of a simple model for the rate of dipole rotation towards a suddenly introduced charge (*8*). In glassy matrices this solvation process can also be seen directly on a slower time scale by pulse radiolysis (*9*). In addition, irradiation at 4 K to produce trapped electrons followed by optical absorption or electron paramagnetic resonance (EPR) measurements support this solvation process (*10*).

If the electron orders the solvent around it, our problem is to deduce the ground state geometrical structure experimentally and then to try to understand the degree of structural reorganization around the electron theoretically. Here we have attacked this problem experimentally by using EPR to study the proton arrangement around trapped electrons in several glassy matrices.

NUMBER OF FIRST SHELL NUCLEI FROM SATELLITE LINE ANALYSIS

We first show that careful examination and analysis of "forbidden" EPR

[1]Congress presentation A–6–1.

transitions gives new information about the number of and distance to first shell nuclei around trapped electrons; this work was done in collaboration with Michael K. Bowman and Robert M. Schwartz. The origin of these forbidden EPR transitions is shown in Fig. 1. The ordinary allowed transitions are shown by the solid lines which correspond to the selection rule that the electron spin flips and the nuclear spin does not. The forbidden lines are shown by the dashed lines which correspond to a flip of the nuclear spin in addition to the electron spin. Whenever the dipolar interaction between the unpaired electrons and surrounding nuclear spins is significant, the forbidden lines become allowed to a greater or lesser degree. As shown at the bottom of Fig. 1, the forbidden lines are separated from the main allowed line by approximately the nuclear frequency.

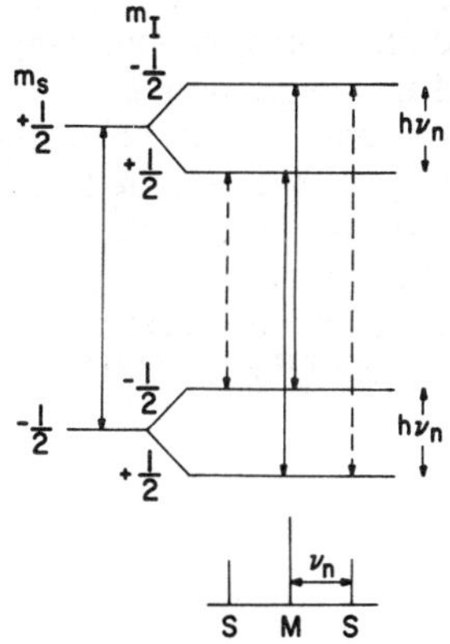

FIG. 1. Energy level diagram for an electron spin ($S = 1/2$) and nuclear spin ($I = 1/2$) system showing allowed transitions (solid lines) and forbidden transitions (dashed lines). The symbols S and M stand for satellite and main line observed transitions respectively.

Satellite lines are well known for trapped hydrogen atoms in sulfuric acid glass at 77 K (*11*), but have never been analyzed to obtain the maximum information available (*12*). The ratio of the satellite line intensity, T_1, to the observed main line intensity, $2\ T_2$, is given by

$$\frac{T_1}{2T_2} = \frac{3}{20}\frac{g^2\beta^2}{H^2}\left\langle \sum_{i=1}^{n} r_i^{-6} \right\rangle \tag{1}$$

where g is the electron g-factor, β is the Bohr magneton, H is the applied magnetic field and r_i is the distance from the trapped electron to the ith matrix proton (*13*). This formula has been averaged over all angles so it applies to trapped radicals in glassy matrices. Note that from the ratio of measured intensities a unique value of r is not obtained unless a particular number of matrix nuclei is assumed. However, there is an additional relation involving the energy separation between the main line and satellite line transitions which can be used in conjunction with equation 1 to determine both the number of nearest interacting protons and their distance (*12*). With the energy separation given by ΔE this expression is

$$(\Delta E)^2 = (g_N\beta_N H)^2 + \frac{3gg_N\beta\beta_N}{4r^3} \qquad (2)$$

where the subscript N refers to the interacting nucleus. Note that no summation over the number of interacting matrix nuclei appears in equation 2. The distance r is here an average distance to all of the nearest matrix nuclei interacting with the unpaired electron spin. In contrast, equation 1 involving the intensities includes a distance r which depends on a specific number of matrix nuclei at distance r interacting with the unpaired electron spin. By using equation 1 one can obtain different distances $r_{eff}(n)$ which depend on the assumed number of closest interacting protons. Thus by combining equations 1 and 2 it is possible to determine a unique distance to a specific number of closest matrix nuclei interacting with the unpaired electron spin. One simply finds the value of n that makes $r_{eff}(n) = r_{\Delta E}$ obtained from equation 2.

The satellite line information is not normally observable for trapped electrons in 10 M NaOH glassy ice because the line width is too broad and the satellite lines are buried within the linewidth. However, if one increases the microwave frequency at which one normally observes EPR from 9 GHz to 35 GHz, then the forbidden satellite lines associated with trapped electrons in 10 M NaOH glassy ice can be discerned (Fig. 2) (*14*).

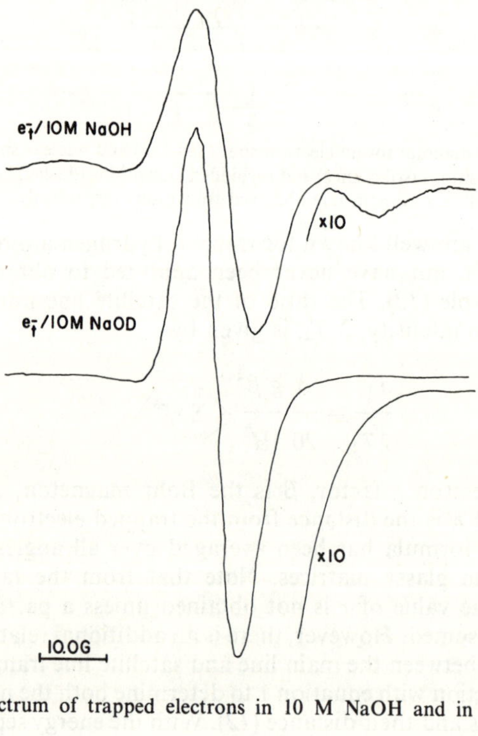

FIG. 2. EPR spectrum of trapped electrons in 10 M NaOH and in the deutrated matrix observed at 35 GHz.

TABLE 1
Distances to Nearest Matrix Protons Around Trapped Hydrogen Atoms and Trapped Electrons in the Matrices Indicated

System	$r_{\Delta E}$	$r_{eff}(n)$
H/8 M H_2SO_4	2.16 Å	1.71 Å (1)
		2.15 (4)
H/12 M H_3PO_4	2.41 Å	1.88 Å (1)
		2.37 (4)
e_t^-/10 M NaOH	~2.1 Å	1.5 Å (1)
		1.8 (4)
		2.1 (8)

Table 1 shows an analysis of the distance to nearest protons for several irradiated systems. For H atoms trapped in acid glasses, the separation between the main line and satellite line can be measured quite accurately and an accurate value of $r_{\Delta E}$ can be obtained. It can be seen that $n = 4$ for H atoms trapped in sulfuric acid and phosphoric acid glasses. Furthermore, the average distance to these four protons is somewhat larger in the phosphoric acid glass which probably reflects the longer PO bonds, since the H atoms are trapped close to the oxyanions (*15*).

For electrons in the sodium hydroxide ice we can only obtain an approximate value of $r_{\Delta E}$ since the satellite and main lines overlap somewhat and the separation in energy is not determined as accurately. However, the analysis does indicate that n is very likely equal to 8 protons at a distance of ~2.1 Å in the first solvation shell around the electron in this matrix. This would of course correspond to a tetrahedral model with four oxygens of four water molecules located at the corners of the tetrahedron and the four water dipoles oriented toward the electron located in the center of the tetrahedron. Unfortunately, this kind of analysis cannot be carried out for trapped electrons in other systems, even at 35 GHz, because the position of the satellite lines always overlaps with a background line of the matrix. It is pertinent to note that the structure obtained here is in agreement with calculations based on a semicontinuum model for the trapped electron in ice matrices (~2 Å) (*16*).

ELECTRON SPIN ECHO BACKGROUND

Electron spin echo experiments have been only little exploited in chemistry and we would like to show here how they may be of use with reference to the particular trapped electron problem at hand. The electron spin echo work was done in collaboration with M. K. Bowman, Ian M. Brown, Yu. D. Tsvetkov and V. Y. Yudanov. Spin echo spectrometry is well known in nuclear magnetic resonance but is less well known in electron magnetic resonance. A spin echo is a consequence of a game one can play with classically precessing dipoles. First, one applies a microwave pulse of sufficient length and magnitude to flip the electron spins 90° from the Z-axis magnetic field direction into the XY plane where the spins precess. In this XY plane the spin system is subject to various time-dependent magnetic interactions with the matrix which cause different spins to precess apart. The spins precess apart for a time τ in this XY plane and are then subjected to a 180° microwave pulse which flips the spins in the other direction in the XY plane. The spins then precess together and at time 2τ after the first pulse, the spin vectors all line up and a small echo or burst of microwave energy is observed on an oscilloscope. This is a two-pulse spin echo experiment which corresponds to most of the experiments reported here. One can also do 3-pulse spin echo experiments in which two 90° pulses separated by time τ are applied followed by a third 90° pulse at time T later; the echo appears at time τ after the third pulse. Essentially the second 90° pulse of this sequence puts the precessing spins into a holding pattern in the $-Z$ direction for time T before flipping them up to the XY plane with third pulse to finally produce an echo. One application of a three-pulse spin echo experiment will be given.

Normally, one studies the spin echo amplitude as a function of the time between the pulses, and the decay of the echo amplitude is related to the various magnetic relaxation processes present in the matrix (*17,18*). However, in this work we are interested in the modulation of this echo decay envelope (*19*). The modulation contains information about the dipolar interactions between the trapped electrons and magnetic nuclei in the matrix and, hence, about the distances to the nearby magnetic nuclei in the matrix (*20*). The origin of electron spin echo modulation can be seen by referring to Fig. 3. The same four-level spin system is shown as in Fig. 1 and the allowed and forbidden transitions are denoted similarly. When one pulse of microwaves of sufficient amplitude, denoted by H_1, is applied to excite spins from both lower levels corresponding to different orientations of nuclear spins, one can obtain modulation effects. Essentially, the pulse excites one set of spins from level C into level B and another set of spins from level D into the same level B. These two sets of spins do not precess at precisely the same frequencies since they come from different energy levels with different Larmor frequencies. Consequently, these two sets of spins beat against one another to produce a modulation effect. The value of H_1 needed is less if one is looking at deuterons instead of at protons, since the deuteron magnetic moment is smaller, so most of the experiments were done with deuterated matrices.

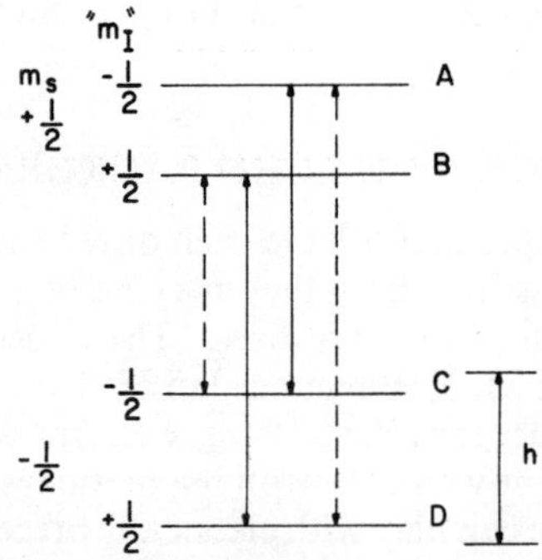

FIG. 3. Energy level diagram for electron spin ($S = 1/2$) and nuclear spin ($I = 1/2$) system with allowed transitions (solid lines) and forbidden transitions (dashed lines) shown. For explanation of how this figure illustrates the origin of electron spin echo modulation refer to the text.

ELECTRON SPIN ECHO OF TRAPPED ELECTRONS IN 10 M NaOD ICE

Figure 4 shows the modulation of the spin echo envelope for electrons in glassy deuterated sodium hydroxide ice. The decay of the spin echo envelope has been subtracted out and only the modulation effect is shown. The dashed line is the calculated modulation (*20*) averaged over all angles for a glassy matrix (*19*) based on the model described below. The experimental modulation can be roughly characterized by the frequency and depth of the modulation. The frequency of the modulation is rather sensitive to the dipolar interaction distance between the electron and a matrix deuteron. It appears that the distance to the nearest deuteron must be greater than 3 Å to obtain approximately the right frequency. This distance is greater than the 2 Å distance to the nearest protons in the first solvation shell around the electron in this matrix. Thus we believe that the second shell deuterons in this matrix are contributing in a major way to the electron spin echo modulation observed. The second shell deuterons begin at 3.5 Å as described below. We also find that we can only fit the observed modulation depth if we assume that about 20% of the water dipoles in the second solvation shell are oriented toward the electron (*19*). So 20% of the second shell deuterons are taken at 3.5 Å. The other 80% of the water dipoles and their deuterons are taken to be uniformly distributed throughout the second solvation shell and the distances to their deuterons are taken correspondingly.

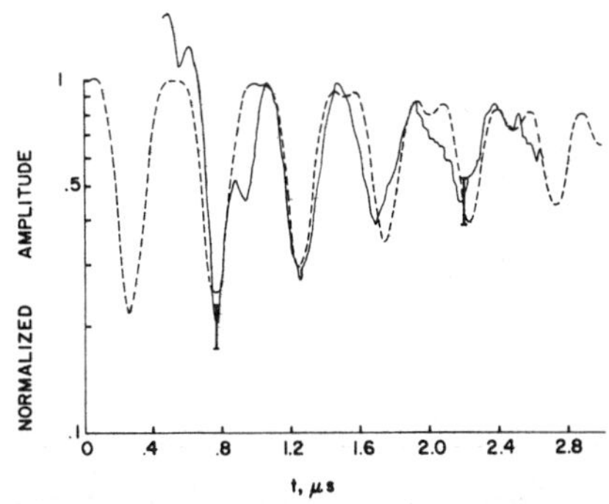

FIG. 4. Experimental (solid line) and simulated (dashed line) spin echo modulation for trapped electrons in glassy 10 M Na OD/D_2O. The simulation corresponds to interaction of the electron with deuterons beginning at 3.5 Å in the second solvation shell and extending throughout the second solvation shell. In addition it is assumed that 20% of the water dipoles in the second solvation shell are oriented toward the electron. For further details see text.

In addition to the main features of the experimental curve shown in Fig. 4, there are two prominent glitches which are not reproduced by the theoretical calculations. These glitches can be seen prominently at about 0.6 and 0.9 μs. More detailed simulations seem to show that these glitches are due to the first solvation shell deuterons at a distance of about 2 Å. Much of the effect of the first solvation shell dipoles seems to be averaged out due to the different magnitude of these dipole interactions and due to quadrupole interactions.

The distance shown in Fig. 4 of 3.5 Å is the distance to the beginning of the second solvation shell. This was taken somewhat arbitrarily from semicontinuum model calculations as a good fit distance in the original simulations (*19*), but we have now found a way to determine this distance independently. This can be done from a three-pulse echo experiment. The three-pulse echo is strongly modulated and the modulation changes phase at about 3.6 μs. This change in phase can be related to the distance of the nuclei that contribute to the three-pulse echo. It is argued that these are only deuterium nuclei in the second solvation shell because the time scale of the three-pulse echo is much longer than that of the two-pulse echo. Calculation shows that this change in phase at 3.6 μs corresponds to 3.5 Å. We summarize our results for trapped electrons in glassy sodium hydroxide ice by Fig. 5.

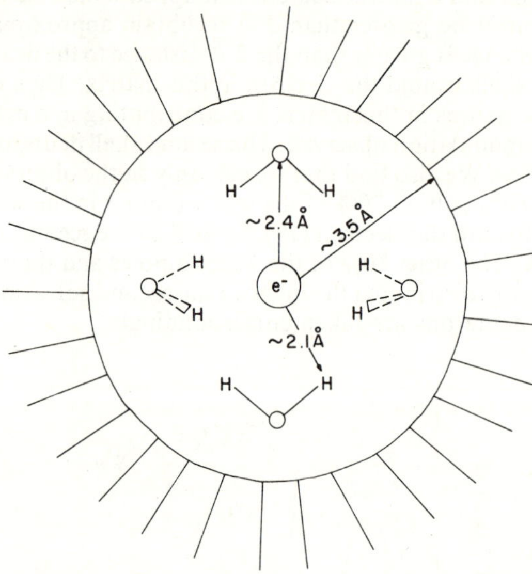

FIG. 5. Model for the first solvation shell structure of water dipoles around a trapped electron in glassy ice and distance (~3.5 Å) to the beginning of the second solvation shell. In addition it has been deduced that ~20% of the water dipoles in the second solvation shell are oriented toward the electron.

ELECTRON SPIN ECHO OF TRAPPED ELECTRONS IN MTHF

We have also made some preliminary deductions about the geometrical structure of matrix molecules around trapped electrons in MTHF. These experiments were made possible because of the synthetic expertise of Dr. Robert Boeckman of Wayne State University who was able to prepare the specifically deuterated MTHF molecules shown in Table 2. Analysis of the EPR linewidths of the trapped electron in MTHF shown in Table 2 indicates that the electron interacts about equally with all of the protons directly connected to the ring carbons. However, the electron interacts distinctly more weakly with the methyl protons. Electron spin echo modulation results on these same deuterated compounds indicate the same conclusions as from the linewidth analysis. Preliminary studies of simulation of the electron spin echo results suggest that the average distance from the electron to the deuterons or protons attached to the ring carbons is about 3.8 Å. The depth of modulation suggests that the electron interacts with three or four MTHF molecules in its first solvation shell. Because the distances are much greater in MTHF than in the sodium hydroxide ice, only interactions to the first solvation shell contribute to the electron spin echo modulation in MTHF. Quantitative analysis of the EPR linewidths in terms of their second moments (*21*) is consistent with the simulations of the electron spin echo modulation. If only 2 MTHF molecules exist in the first solvation shell, the distance to their ring protons is indicated to be a few tenths of an Å less than the 3.8 Å value. Although our analysis is as yet incomplete, we offer Fig. 6 as a preliminary schematic of the electron trap in MTHF.

TABLE 2
EPR Linewidths for Trapped Electrons in Variously Deuterated MTHF Molecules

Matrix		Linewidth, ΔH_{pp}
MTHF—d_0	$CH_2CH_2CH_2CH(CH_3)O$	4.0 ± 0.1 *G*
MTHF—d_3	$CD_2CH_2CH_2CD(CH_3)O$	3.3
MTHF—d_4	$CH_2CD_2CD_2CH(CH_3)O$	3.1
MTHF—d_{10}	$CD_2CD_2CD_2CD(CD_3)O$	1.6

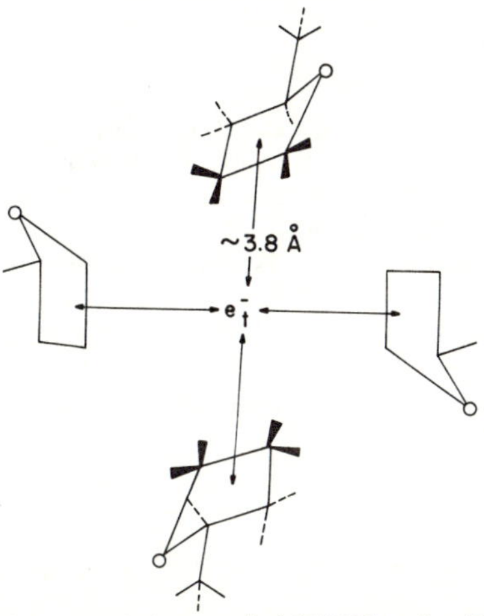

FIG. 6. Schematic model for trapped electrons in MTHF based on EPR linewidth and electron spin echo modulation data. The four MTHF molecules are arranged tetrahedrally around the electron and the indicated distance is from the electron to the plane of the ring carbons. According to the present preliminary analysis less than four MTHF molecules may be in the first solvation shell; see text for details.

SUMMARY

In summary it appears that we have the beginning of a detailed picture of the proton arrangement and hence the molecular arrangement around trapped electrons in several different glassy matrices. This has been made possible by careful analysis of the effects of forbidden EPR transitions both by direct observation of these transitions (satellite lines) and by their effect on electron spin echo envelopes. Although direct observation of the forbidden EPR transitions seems limited to only a few matrices, the electron spin echo method seems broadly applicable. It appears that rather detailed molecular structure information out to approximately 4 Å from trapped electrons can be obtained, even in disordered matrices, by use of the techniques described herein.

ACKNOWLEDGEMENT

This research was supported by the U. S. Atomic Energy Commission under contract AT(11-1)-2086 and in part by the US-USSR Scientific Exchange Program. I also want to specifically acknowledge the excellent work of my collaborators as named in the text.

REFERENCES

1. L. Kevan, *Adv. Radiat. Chem.*, **4**, 181 (1974).
2. L. Kevan, *J. Phys. Chem.*, **76**, 3830 (1972).
3. S. Noda and L. Kevan, *J. Phys. Chem.*, **78**, 2454 (1974).
4. T. Huang, I. Eisele and L. Kevan, *J. Chem. Phys.*, **59**, 6344 (1973).
5. T. Huang and L. Kevan, *J. Chem. Phys.*, **61**, 4660 (1974).
6. J. H. Baxendale and P. Wardman, *J. Chem. Soc. Faraday Trans., I,* **69**, 584 (1973).
7. P. M. Rentzepis, R. P. Jones and J. Jortner, *J. Chem. Phys.*, **59**, 766 (1973).
8. K. Fueki, D. F. Feng and L. Kevan, *J. Phys. Chem.*, **78**, 393 (1974).
9. L. Kevan, *J. Chem. Phys.*, **56**, 838 (1972).
10. H. Hase, M. Noda and T. Higashimura, *J. Chem. Phys.*, **55**, 5411 (1971).
11. H. Zeldes and R. Livingston, *Phys. Rev.*, **96**, 1702 (1954).
12. M. Bowman, L. Kevan and R. N. Schwartz, *Chem. Phys. Lett.*, **30**, 208 (1975).
13. G. T. Trammell, H. Zeldes and R. Livingston, *Phys. Rev.*, **110**, 630 (1958).
14. M. Bowman, L. Kevan, R. N. Schwartz and B. L. Bales, *Chem. Phys. Lett.*, **22**, 19 (1973).
15. J. Zimbrick and L. Kevan, *J. Chem. Phys.*, **47**, 5000 (1967).
16. K. Fueki, D. F. Feng and L. Kevan, *J. Am. Chem. Soc.*, **95**, 1398 (1973).
17. I. Brown, *J. Chem. Phys.*, **58**, 4242 (1973).
18. A. D. Milov, K. M. Salikhov and Yu. D. Tsvetkov, *Sov. Phys.–Solid State,* **14**, 1956 (1973).
19. M. K. Bowman, I. M. Brown and L. Kevan, *Chem. Phys. Lett.*, **22**, 16 (1973).
20. L. G. Rowan, E. L. Hahn, W. B. Mims, *Phys. Rev.*, **137**, 61 (1965); W. B. Mims, *Phys. Rev. B,* **5**, 2409 (1972).
21. B. L. Bales, J. Helbert and L. Kevan, *J. Phys. Chem.*, **78**, 221 (1974).

Some Aspects of Electron Dynamics in Solid Alkanes[1]

Ing-yih Cheng and Koichi Funabashi
Department of Chemistry and the Radiation Laboratory[2]
University of Notre Dame
Notre Dame, Indiana 46556, USA

INTRODUCTION

Understanding of the early processes ($10^{-9} \sim 10^{-12}$ sec) in radiation chemistry of liquids and solids requires some knowledge of the dynamic properties of thermal and epithermal electrons. The purpose of this paper is to discuss the relationship between the molecular properties of the aggregate and the excess electron dynamics. The term thermal (or thermalized) electrons has been used in the literature loosely to mean those electrons with kinetic energy on the order of kT. Such a definition is of little use and significance when the excess electron is in a highly fluctuating potential so that the zero-point kinetic energy alone may be much greater than kT, and becomes completely meaningless when the Born-Oppenheimer separation of electron and nuclei motions is not permissible. The latter situation arises when the "electron-phonon" interaction is important such as in the polaron formation. The usual definition of thermalization is, of course, that the statistical distribution of the total energy of the system, which includes the electronic energy, "phonon" energy and their interaction, follows the Boltzmann distribution. The most important state, for the description of thermalized electrons at low temperatures, is the ground state. In the next section, we present some properties of the ground state of a model electron-phonon system, which consists of a linear chain of diatomic molecules and an excess electron, over the entire range of electron-phonon coupling strength. Generalizing this model, we shall make a conjecture on what the most important "phonon" mode may be in determining the magnitude of electron mobility in alkanes.

In the last section, we shall discuss some experimental results related to epithermal electrons in solid alkanes and their theoretical interpretation in terms of molecular properties.

THERMALIZED ELECTRONS

— Ground State of a Model Electron-Phonon System —

The most remarkable manifestations of the effect of molecular properties on the thermal electrons are in the magnitude of the electron drift mobili-

[1]Congress presentation A–6–2.
[2]Operated under contract with U.S. Atomic Energy Commission. Report No. C00-38-959.

ties (*1–3*) in various liquids. Neopentane (*1*), for example, has the electron drift mobility of 50 ~ 70 $cm^2/v.s.$ at 23° C, which is greater than that in n-pentane (~ 0.14 $cm^2/v.s.$) by more than two orders of magnitude. Similarly, the electron mobility (*4*) in *cis*-butene-2 is 2.2 $cm^2/v.s.$ and that in *trans*-butene-2 is 0.029 $cm^2/v.s.$, both at 20° C. It is not the purpose of this section to *explain* these mobility data or to *propose* a compatible transport mechanism. Some attempts (*5–9*) along these lines have been made in the literature, but there are still too many unknown factors (or parameters) in these alkane liquids to make a theory unique. The purpose of this section is to understand the basic physics behind some mechanism by which a relatively small change in the molecular property may lead to a large difference in the electron dynamics such as in the magnitude of mobilities. There is apparent correlation between the geometry of the molecule and the mobility; the more spherical the molecule, the greater the mobility. One might then postulate that electron traps are deeper and more numerous in non-spherical molecules. But, why should the geometry of the molecule be related to the trap potential? Under some conditions, one may argue that the non-sphericity of the molecule leads to a more fluctuating electron-potential, because of the anisotropy of molecular polarizability and fluctuation in transfer energies. It is easy to show that such an argument is not useful and probably not valid in general. Consider, for example, an excess electron in liquid- or solid-CCl_4, which has the same symmetry and geometry as CH_4. Although there has been no published report on the mobility in pure CCl_4, the most likely fate of the excess electron would be a dissociative capture by CCl_4 molecule and the resulting negative-charge mobility would be much smaller than that in CH_4 (*10,11*) (~ 400 $cm^2/v.s.$ at 111° K). This example suggests strongly that the concept of a static electron potential, even after allowing for the non-sphericity, is not necessarily a useful one for the purpose of understanding the thermal electrons.

Instead of a static potential which is completely determined, in principle, by the molecular arrangement in the system, the excess electron interacts dynamically with the nuclear motions of the surrounding molecules ("phonon") and deforms the local molecular configuration. The configurational deformation reacts back on the excess electron in such a way as to change the electron wave-function. The electron and nuclear motions are coupled in a self-consistent way, and neither one of them is independent of the other. Electron dynamics in this case is best described by that of polaron. Polaron dynamics in non-polar molecular crystals has been studied by Holstein (*12*), Munn and Siebrand (*13*) and Toyozawa (*14*).

When the electron-phonon interaction is small (i.e., large polaron), the mobility of the polaron is controlled essentially by the scattering processes (*15*) and its effective mass is almost equal to that of an undressed electron in the conduction band. When the electron-phonon interaction is increased, the polaron dynamics changes from that of successive scatterings in the conduction band to that of hopping (*12,16*) from site to site. The effective mass of the polaron then becomes quite large in comparison to the band-electron mass. We demonstrate that this transition from the large-to-small polaron dynamics takes place rather sharply as the electron-phonon interaction is increased.

Holstein's model (*12*) consists of a linear chain of diatomic molecules and an excess electron. The wavefunction for this system is expressed as

$$\psi (r,x_1 \ldots x_N) = \Sigma a_n (x_1 \cdots x_N) \varphi_n (r,x_n), \quad (1)$$

where ϕ_n is the negative-ion wavefunction at site n and the coefficient a_n, which is a function of all the internuclear coordinates, is to be determined by the following equation.

$$\left[\sum_m \left(-\frac{\hbar^2}{2m}\frac{\partial^2}{\partial x_m^2} + \frac{1}{2} M w_0^2 x_m^2\right) - A x_n\right] a_n - J (a_{n-1} + a_{n+1}) = E a_n , \quad (2)$$

where M is the reduced mass of the diatomic molecule, ω_0 the frequency of the harmonic vibration, A the coupling constant for electron and nuclei motion, and J the electronic transfer energy which is assumed to be independent of the nuclei coordinates x_n. It may be noted that the wavefunction (1) does not have the product form of electronic and vibrational wavefunctions, except for the case where a_n is a common factor for all n. The latter situation is realized only in the limit of $A \to 0$ or $J \to \infty$. This means that the wavefunction (1) allows for a possible breakdown of the Born-Oppenheimer approximation. This point was also emphasized by Merrifield (*17*) in connection with the exciton-phonon coupling problem of molecular crystals.

We determine the coefficient a_n using the variational form of Lee-Low-Pines (*18*);

$$a_n = \frac{1}{\sqrt{N}} X (x_1 - x_1^n)\ X (x_2 - x_2^n) \ldots X (x_n - x_n^n) \ldots , \quad (3)$$

where $X(x_i - x_i^n)$ is the ground-state eigenfunction for a harmonic oscillator with frequency ω_0 and with its center of oscillation at x_i^n. Our variational parameters are, therefore, a set of numbers $[x_i^n]$.

Substituting (3) into (2) and minimizing the energy E with respect to variation of x_i^n, one can obtain the ground-state energy E in the following closed form (*19*):

$$E/\hbar\omega_0 \equiv \epsilon = N/2 - 3g/2(\sigma/g)^{1/3} + \sigma/2 - \exp(-\sigma)/\gamma, \quad (4)$$

and

$$g = \sigma (1 + 2 \exp(-\sigma)/\gamma)^{3/2}, \quad (5)$$

where

$$\gamma = \hbar\omega_0/2J, \quad (6)$$

$$g = A^2/2\hbar M \omega_0^3. \quad (7)$$

Since the parameter σ can be eliminated from equations (4) and (5), the ground-state energy E can be expressed uniquely as a function of characteristic parameters γ and g. It is important to note that the ground-state energy E and its wavefunction depend upon the *relative* magnitudes of J and A in units of ω_0. One does not get the adiabatic limit of Holstein (*12*), for example, simply by making $\omega_0 \to 0$.

Instead of eliminating σ from (4) and (5), it is convenient to look into the behavior of ϵ and g as a function of σ. Fig. 1 shows such relationships for $\gamma = 0.2$. Using Fig. 1, the relationship between g and ϵ can be plotted as in Fig. 2, where the arrow indicates the direction of increasing σ. The loop in the g–ϵ curve disappears for γ which is greater than a critical value γ_c. Using (5), it can be shown that $\gamma_c = 3\exp(-5/3)$.

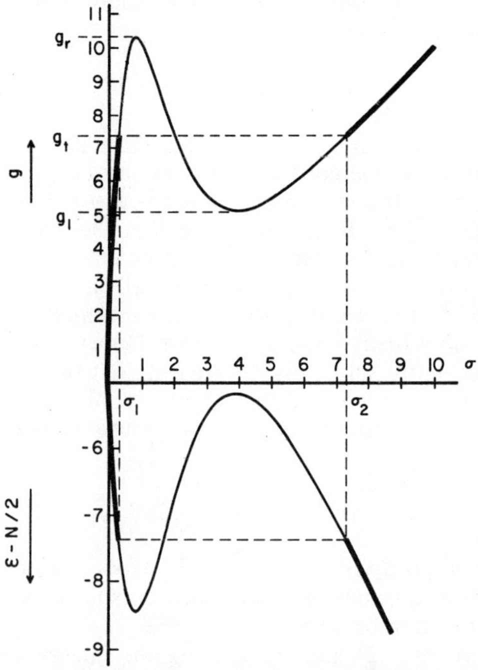

FIG. 1. Plots of g and ϵ as a function of σ for $\gamma = 0.2$, where g is the "electron-lattice" coupling parameter, ϵ the energy in units of the vibrational quantum, and γ is the ratio of the vibrational quantum to the half band-width.

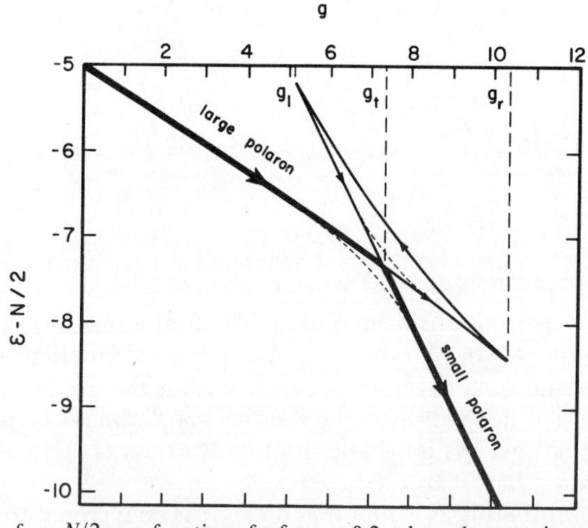

FIG. 2. Plot of $\epsilon - N/2$ as a function of g for $\gamma = 0.2$, where the meanings of the symbols ϵ, g, and γ are the same as Fig. 4, and $N/2$ represents the sum of zero-point vibrational energies in units of $\hbar\omega_0$.

Equations (4) and (5) refer to the ground state of the polaron. For small wave-number states, these equations retain the same form, provided that the wave-number dependence of σ and γ are taken into account. The effective mass m^* can be calculated from (4) and (5) to be

$$m^* = \hbar^2 \exp(\sigma)/2Ja^2 = m_c \exp(\sigma), \tag{8}$$

where a is the lattice constant. The ratio m^*/m_c is plotted against g in Fig. 3, where it can be seen that the effective mass ratio jumps three orders of magnitude at $g \simeq 7.3$ for $\gamma = 0.2$. According to this result, the transition from the large- to small-polaron state is sudden. As Shore and Sanders (*20*) pointed out, however, this sudden change of the effective mass is an artifact resulting from the particular form of the variational function used here. In a more complicated variational function, the discrete change is smoothed out. However, the effective mass ratio remains to be a highly non-linear function of the electron-phonon coupling parameter g.

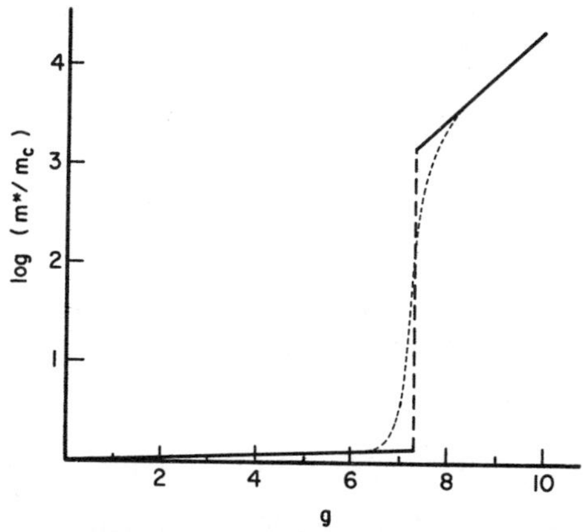

FIG. 3. Effective mass ratio, m^*/m_c, as a function of g for $\gamma = 0.2$, where $m_c = h^2/2Ja^2$.

In order to apply the polaron model for alkanes, one must first choose some mode of nuclear motion which is coupled to the electron motion in a significant way. It is clear that in general all the normal modes for the aggregate must be considered for electron-phonon coupling. But, if one is interested in understanding the difference of the effective mass in spherical (e.g. neopentane) and non-spherical (e.g. *n*-pentane) molecules, the most important nuclear motion is probably the molecular rotation or libration. The coupling parameter g, in this case, is the ratio of the relaxation energy, which arises from molecular reorientation in the presence of an excess-electron, to the rotational quantum.

One way to test the applicability of the polaron dynamics to a particular system is to study the temperature dependence of the mobility. According to Holstein (*12*), the transport mechanism of a small polaron changes from that of activated hopping with a positive temperature coefficient to that of tunneling with little or negative temperature coefficient at about one half of the

Debye temperature of the solid. The temperature dependence of electron drift mobility in glassy 3-methylpentane (*21*) seems to fit this description, as shown in Fig. 4. Unfortunately, mobility data for other alkanes at low temperatures are not known too well. More experiments in this area are needed before one can make a more definite statement about the possible transport mechanism for excess electrons in solid alkanes.

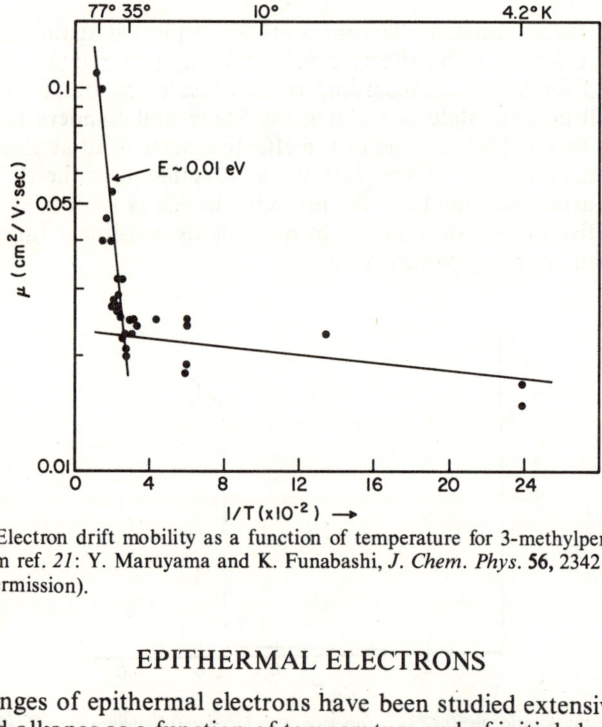

FIG. 4. Electron drift mobility as a function of temperature for 3-methylpentane glass (reproduced from ref. *21*: Y. Maruyama and K. Funabashi, *J. Chem. Phys.* **56,** 2342 (1972), with the publisher's permission).

EPITHERMAL ELECTRONS

The ranges of epithermal electrons have been studied extensively (*1,3,22–24*) in liquid alkanes as a function of temperature and of initial electron energy. Similar studies (Y. C. Chang and W. B. Berry, unpublished) have been made using alkane thin films at 77° K. The purpose of this section is to report these experimental data in the frozen systems and also to discuss the physical meaning of the "range" or the attenuation constant b, which is defined by the *empirical* equation

$$I = I_0 \exp(-r/b), \tag{9}$$

where I is the intensity of electron beam at the distance r from the film surface and I_0 is the intensity of electron beam at the surface.

As was noted by Holroyd, Dietrich and Schwarz (*23*), the experimental points could have been represented with comparable accuracy by a gaussian function of distance, $\exp(-r^2/b^2)$. Freeman (*4*) further noted that a better description of the range requires an additional term of the form, $0.5\,(b^2/r^3)$, to the gaussian or exponential functions. There is little doubt that the "true" distribution of the electron range can not be a simple analytical function such as exponential or gaussian. Figure 5 shows a typical experimental relationship between the electron beam intensity as a function of the film thickness. The basic reason for our choice of the form (9) is that the approximate physical meaning for the constant b can be given within the framework of the multiple scattering theory (*25*), while physical meaning of any additional term, even though the curve-fitting is improved, may not always be unambiguous.

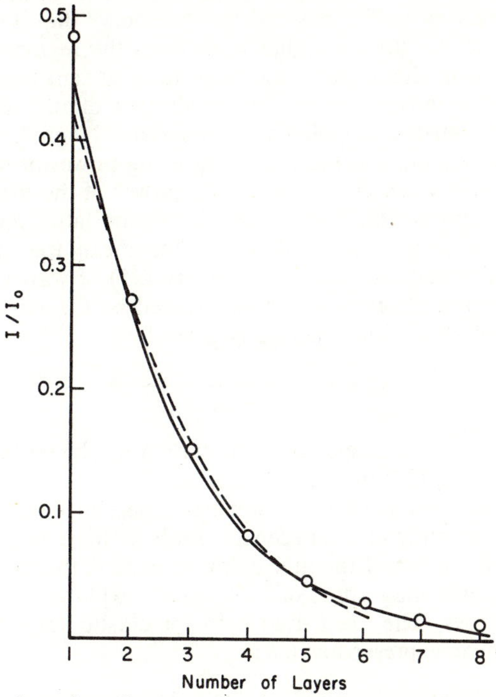

FIG. 5. Electron beam intensity I as a function of film thickness (one layer $\simeq$ 20 Å). Circles represent measured values, the solid curve is the exponential fit and the dotted curve is the gaussian fit for tetramethylsilane at 77° K (Y. C. Chang and W. B. Berry, unpublished).

If one uses the interpretation that the parameter b in either an exponential or gaussian distribution is essentially the "thermalization" distance, one must define the term "thermalization" properly. As was pointed out in the preceding section, it is not simply a matter of slowing down the translational kinetic energy of the epithermal electron to kT. "Thermalization" implies a distribution of approximate eigenstates according to the Boltzmann distribution and in each of these eigenstates the electron-kinetic energy is not necessarily a constant even in an approximate sense—the motions of electrons and nuclei may be coupled so strongly that only the sum of electronic and nuclear-motion energies is approximately conserved.

It is difficult to understand the significance of a procedure in which one calculates b within the framework of a diffusion model for the electron motion with the knowledge of inelastic scattering cross-section and the average energy loss per collision. It is clear that such a procedure is a meaningful approximation if one is interested in the range of hot electrons whose initial energy is on the order of MeV.

An alternative interpretation of the constant b in the context of the thin-film experiment is the following. Consider a beam of electrons with kinetic energy E (in vacuum) impinging perpendicularly upon the film surface. The electrons are scattered elastically and inelastically as soon as they enter the medium. At a point b inside the film sufficiently distant from the surface, the average electron behaves as if it has lost the "memory" that it originally came from the direction of the surface rather than from the region further deep in the film. At such point the forward and backward current become equal and there is no net flow of current inward in the absence of an external field. The distance b from the surface is called the "coherence length", which is determined by the total (elastic and inelastic) scattering power of the medium but can be approximated by the elastic scattering power of the medium when the inelastic scattering cross-section is small. When the latter approximation is valid, the coherence length is just equal to the mean free path for elastic scattering. Using the multiple scattering theory (*25*) of waves, one can show that the coherent propagation of a wave is damped by the exponential expression (9). If the low-field mobility (*26*) μ is given by

$$\mu = \frac{4}{3} \frac{eb}{(2\pi mkT)^{1/2}} \tag{10}$$

the measurement of b gives the direct information on the mobility of electron at some quasi-free state (*27*) (not necessarily at V_o).

Table 1 lists the constants b for some alkanes, together with measurements made by other methods. As seen in Table 1, there is a correlation between the magnitude of b and the molecular geometry; the more spherical the molecule, the larger the magnitude of b. A simple explanation of this correlation can be as follows. The mean-free path for elastic scattering b for low-energy electron can be expressed as (*28*)

$$b^{-1} = 4\pi n\,[\langle a^2\rangle - \langle a\rangle^2 + \langle a\rangle^2 S(0)] \tag{11}$$

TABLE 1
Attenuation Parameters* of Alkanes

	b_{I}	b_{II}	b_{III}	b_{IV}	b_{V}
Neopentane	175(140)	147	178	250	212
Cyclopentane	70(56)	77	69	56	61
3 MP	38(30)	50	70		61
n-Hexane	38(30)	42	67	56	58
n-Pentane	38(30)	36	72		62
Nonane	9(7)				

*All numbers refer to those at 300° K, except those in parentheses.
b_{I}: Exponential distribution, numbers in parentheses refer to those at 77° K (Y. C. Chang and W. B. Berry, unpublished).
b_{II}: Gaussian distribution, photo-injected (0.1 ~ 0.5 eV) electrons (ref. *23*).
b_{III}: Gaussian distribution, X-ray electrons (ref. *35*).
b_{IV}: Exponential distribution, calculated (ref. *36*).
b_{V}: Exponential distribution, calculated (ref. *37*).

where n is the density of the scatterers, a the scattering length and $S(0)$ is the structure factor. In (11), $\langle a \rangle$ is the average scattering length over different orientations of the molecule with respect to the direction of electron incidence. It is clear that the difference, $\langle a^2 \rangle - \langle a \rangle^2$, is a measure of the non-sphericity of the molecule and vanishes for a completely spherical molecule. If $\langle a \rangle$ and $S(0)$ are about the same, as in the case of n-pentane and neopentane, the difference in b comes from the fluctuation, $\langle a^2 \rangle - \langle a \rangle^2$. Because $S(0)$ is on the order of 10^{-2} in a typical alkane liquid[3], $[\langle a^2 \rangle - \langle a \rangle^2]/\langle a \rangle^2$ may be as small as 4×10^{-2} to account for the difference of factor 5 in b.

As Table 1 shows, the differences in b for photo-injected and X-ray generated electrons are surprisingly small. The main contribution for the coherent spreading of the electron wave seems to come from the low energy region. This observation has already been made by Mozumder and Magee (*29*) in a different theoretical framework.

Although the magnitude of b is relatively insensitive to the energy of epithermal electrons in the thin-film experiment, we are not certain that the measured b's indeed correspond to those at the bottom of the quasi-free state V_0 at this time. The mean-free path of scattering may change rather steeply near the bottom of the conduction band. In fact, for a highly disordered system, Mott (*30*) conjectures there is an energy region of ~ 0.2 eV above V_0 (E_c in Mott's notation) where the mean-free path is approximately equal to the "lattice" constant, as shown in Fig. 6. in such a system, an electron near V_0 will not be able to propagate significantly, but will be trapped (localized) quickly. It is likely that in the case of low-mobility entities, such as n-pentane, the magnitude of b in Table 1 corresponds to that of electrons with energy above E_L of Fig. 6.

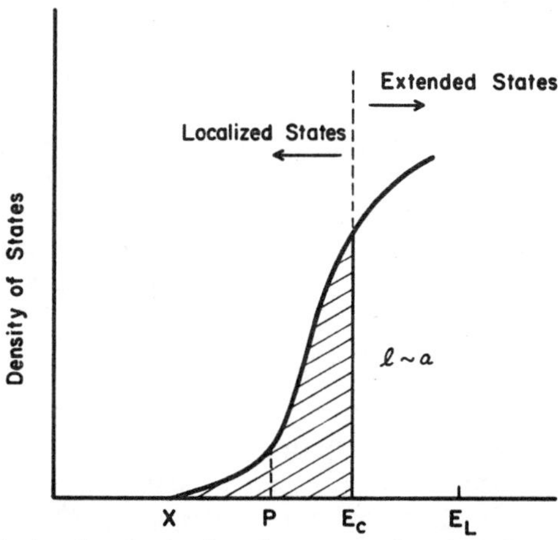

FIG. 6. Schematic drawing for density of states as a function of energy, according to Mott (ref. *30*). The mean-free path l is of the order of the lattice constant a above the bottom of the quasi-free states E_c, for a highly disordered material, where the electron wavefunction at each site is not the S-type.

[3]K. D. Luks and J. Kohn, Department of Chemical Engineering, University of Notre Dame, private communication.

The dynamics of quasi-free electrons must reflect the density of states below the bottom of conduction band V_o in general. The electron mobility at the quasi-free state V_o is likely to be correlated with the thermal mobility itself. The correlation of b with thermal mobilities has been investigated extensively by Freeman (*4*) and his coworkers. In any event, the so-called "quasi-free mobilities" are not likely to be independent of the nature of the solvent. The same observation was made by Allen and Holroyd (*31*).

Finally, let us consider the correlation of the mobility (*32*) and V_o; for a certain group of alkanes, the higher the mobility, the lower the energy level V_o. The understanding of this correlation would be a little simpler if one can talk about the electron mobility at V_o ("quasi-free mobility") rather than the thermal mobility itself, because the quasi-free mobility is basically related to the "width" of the energy level V_o and the correlation between the real and imaginary part of a kinetic coefficient is not accidental [Kramers-Kronig relation (*33*)]. The Springett-Jortner-Cohen (SJC) (*27*) description of V_o, however, is not intended for this correlation; the energy level V_o in SJC model has no width and, consequently, contains no information concerning the magnitude of mobility associated with V_o.

The difficulty in relating V_o and the quasi-free mobility in a quantitative way is then a common problem in the multiple scattering theory for an excess electron in a liquid or in a disordered material: the problem of calculating the real and imaginary part of the self-energy (*34*). A qualitative relationship, however, can be understood in terms of the following model of Mott (see reference *30*, p. 12).

Consider a spherical volume v whose diameter is the mean-free path l for the excess electron such that

$$v = 4\pi l^3/24. \tag{12}$$

The phase of the electron wavefunction inside such a sphere is *statistically* uncorrelated with that of the outside. This means that the coherent part of the electron wavefunction is effectively confined to this sphere on the average, although the total wavefunction is extended incoherently throughout the system at V_o. An electron confined in the volume (12) has a zero-point kinetic energy of

$$\Delta V_o = h^2/2ml^2, \tag{13}$$

which can be thought of as the scattering correction to the zero-order V_0 of the SJC model. It can be seen from (13) that the smaller the mean-free path l, the larger the correction. If all the parameters in the SJC model are equal, the observed difference in the quasi-free energy level can be ascribed to the difference in the mean-free path as in (13). Such assumption is difficult to justify at this time. However, it is also difficult to justify a procedure (*5,7*) which is designed to calculate the radius of the hard-core scattering potential of the SJC model so as to give the correct magnitude for V_o, without the scattering correction such as (13), which can be seen to be non-negligible for small mean-free paths.

REFERENCES

1. W. F. Schmidt and A. O. Allen, *J. Chem. Phys.* **52,** 4788 (1970).
2. R. M. Minday, L. D. Schmidt and H. T. Davis, *J. Chem. Phys.* **54,** 3112 (1971).
3. J. P. Dodelet and G. R. Freeman, *Can. J. Chem.* **50,** 2667 (1972).
4. J. P. Dodelet, K. Shinsaka and G. R. Freeman, *J. Chem. Phys.* **59,** 1293 (1973).
5. H. T. Davis, L. D. Schmidt and R. M. Minday, *Chem. Phys. Lett.* **13,** 413 (1972).
6. R. R. Hentz, in *Proceedings of the International Meeting on Primary Radiation Effects in Chemistry and Biology,* Buenos Aires, Argentina, 1970 (M. B. A. Crespi and M. A. Molinari, Eds.), pp. 147–165. Comision Nacional de Energia Atomica, Buenos Aires, Argentina, 1971.
7. K. Fueki, D. F. Feng and L. Kevan, *Chem. Phys. Lett.* **13,** 616 (1972).
8. R. Schiller, *J. Chem. Phys.* **57,** 2222 (1972).
9. N. R. Kestner and J. Jortner, *J. Chem. Phys.* **59,** 26 (1973).
10. P. G. Fuochi and G. R. Freeman, *J. Chem. Phys.* **56,** 2333 (1972).
11. G. Bakale and W. F. Schmidt, *Z. Naturforsch.* **28,** 511 (1973).
12. T. Holstein, *Ann. Phys.* **8,** 325 (1959).
13. R. W. Munn and W. Siebrand, *J. Chem. Phys.* **52,** 6391 (1970).
14. Y. Toyozawa, in *Polarons in Ionic Crystals and Polar Semiconductors* (J. T. Devreese, Ed.), pp. 1–27. North-Holland Publishing Co., Amsterdam-London, 1972.
15. L. Friedman, *Phys. Rev.* **140,** 1649 (1965).
16. J. Yamashita and T. Kurosawa, *J. Phys. Soc. Japan* **15,** 802 (1960).
17. R. E. Merrifield, *J. Chem. Phys.* **36,** 2519 (1962).
18. T. D. Lee, F. Low and D. Pines, *Phys. Rev.* **90,** 297 (1953).
19. K. Funabashi and I. Y. Cheng, *Int. J. Radiat. Phys. Chem.*, in press.
20. H. B. Shore and L. M. Sander, *Phys. Rev.* **B7,** 4537 (1973).
21. Y. Maruyama and K. Funabashi, *J. Chem. Phys.* **56,** 2342 (1972).
22. M. G. Robinson, B. G. Fuochi and G. R. Freeman, *Can. J. Chem.* **49,** 984 (1971).
23. R. A. Holroyd, B. K. Dietrich and H. A. Schwarz, *J. Phys. Chem.* **76,** 3794 (1972).
24. N. E. Houser and R. C. Jarnagin, *J. Chem. Phys.* **2,** 1069 (1970).
25. M. Lax, *Phys. Rev.* **85,** 621 (1952).
26. H. Margenau, *Phys. Rev.* **69,** 508 (1946).
27. B. E. Springett, J. Jortner and M. Cohen, *J. Chem. Phys.* **48,** 2720 (1968).
28. See, for example, C. Kittel, *Quantum Theory of Solids,* p. 373. John Wiley and Sons, Inc., 1964.
29. A. Mozumder and J. L. Magee, *J. Chem. Phys.* **47,** 939 (1970).
30. N. F. Mott and E. A. Davis, in *Electronic Processes in Non-crystalline Materials,* p. 32. Clarendon Press, Oxford, 1971.
31. A. O. Allen and R. A. Holroyd, *J. Phys. Chem.* **78,** 796 (1974).
32. R. A. Holroyd and M. Allen, *J. Chem.* Phys. **54,** 5014 (1971).
33. See, for example, L. D. Landau and E. M. Lifshitz, in *Statistical Physics,* p. 391. Pergamon Press Ltd., 1958.
34. See, for example, L. M. Roth, *Phys. Rev.* **B7.** 4321 (1973).
35. W. F. Schmidt and A. O. Allen, *J. Phys. Chem.* **72,** 3730 (1968).
36. G. C. Abell and K. Funabashi, *J. Chem. Phys.* **58,** 1079 (1973).
37. A. Mozumder, *J. Chem. Phys.*, in press.

Deferred Luminescence in Organic Matrices at Low and Very Low Temperatures[1]

P. Cordier, F. Kieffer, C. Lapersonne-Meyer and J. Rigaut

Laboratoire de Physico-Chimie des Rayonnements
Associé au CNRS
Université de Paris-Sud, 91405 Orsay, France

INTRODUCTION

In earlier studies (*1–4*) involving irradiation at 77 K it was found that methylcyclohexane (MCH) forms a particularly interesting glassy matrix for luminescence studies, especially in the presence of solutes such as biphenyl. Upon warming it emits a thermoluminescence with two well-separated glow peaks, at 90 K and 95 K. It was shown that the former is due to the recombination of cations with trapped electrons and the latter to their recombination with anions (*1,2*).

Moreover, at the temperature of irradiation an isothermal luminescence (ITL) was observed which must also be attributed to a cation-electron recombination, since it was suppressed by optical bleaching of the trapped electrons like the first glow peak. Was this ITL the tail of a glow peak whose maximum could lie at 77 K or below, or did the 90 K peak merely arise from a thermal acceleration of the ITL observed at 77 K?

Experiments on the effect of solute concentration invalidated the second assumption: the light output of ITL, integrated over a certain time, increased linearly with biphenyl concentration between 10^{-4} and 10^{-2}mole/dm^3, while the integrated emission of the 90 K peak went through a maximum around 10^{-3} mole/dm^3. Thus, the same population of trapped electrons could not be involved (*3*). Furthermore, experiments carried out at 4.2 K revealed the existence of an ITL at this temperature too, with a decay law identical with that of ITL at 77 K (*3,4*). It was concluded that in MCH, at both these temperatures, an identical mechanism must be chiefly responsible for ITL and that the only plausible mechanism was electron tunnelling.

In the present paper we discuss the significance of this decay law and of that observed in a softer glass, and we bring forth direct experimental evidence for the exclusive presence of tunnelling at very low temperatures and for its predominance up to 77 K in MCH. This evidence is obtained by irradiating at 77 K and cooling down to 4.2 K, while recording luminescence intensity during the whole temperature change.

[1]Congress presentation A–6–3.

EXPERIMENTAL TECHNIQUE

Chemicals were analytical reagent grade and some were further purified: MCH and 3-MP (3-methylpentane) by chromatography over freshly activated silica gel and biphenyl by multiple zone refining.

Samples were degassed by the freeze-pump-thaw technique, except in the pulse radiolysis experiments where the solution was flushed with nitrogen before congealing in the atmosphere of boiling liquid nitrogen. For γ-irradiations at 77 K, either the copper cell described before (*5*) could be used or, to make conditions identical with those prevailing in the work at 4.2 K, a cryostat system consisting of two glass Dewar vessels; the inner one was closed by a lid carrying various inlets and in the center a stainless steel tube, one meter long, plunging into the coolant. In this case, the sample (generally 0.5 ml) was contained in a small brass cup screwed to the lower end of the tube with an indium washer as an air-tight seal. A thermocouple was fitted below the brass cup and a thin copper cylinder, on which a resistance heater was wound, surrounded the cup and the lower part of the tube. The tube was polished inside and served as a light guide; its upper end was closed by a quartz window. This set-up was also employed for work with a van de Graaff accelerator.

In experiments involving irradiation at 77 K and subsequent cooling, this cryostat was set in position on the γ-ray source; the inner Dewar vessel was empty, but a small amount of liquid nitrogen was left in the copper cylinder surrounding the sample holder. The helium tank and transfer tube were connected to the cryostat before irradiation. After irradiation the photomultiplier tube was put in position and helium transfer was started as soon as the ITL decay at 77 K had been recorded for a length of time sufficient to obtain a good extrapolation.

For pulse radiolysis the samples were frozen on a stainless steel foil (thickness 12 μm) which constituted the bottom of an open metal cell projecting through the wall of a styrofoam box filled with liquid nitrogen. Thus the samples were bathed in liquid nitrogen and the electron beam only crossed the thin foil.

Radiation sources employed were: a cobalt-60 source (maximum dose rate $\approx$ 0.35 Mrad/h), a van de Graaff electron accelerator (energy: 3 MeV, current: 200 μA) with conversion of the electron beam to X rays by a refrigerated copper cylinder surrounding the cryostat, and a Febetron 706 accelerator supplying about 10^{14} electrons of 600 keV in a pulse of 3 ns.

The total luminescence from the sample, i.e. fluorescence and phosphorescence of the solute, was monitored indifferently by a variety of photomultiplier tubes; in the pulse radiolysis experiments, biphenyl fluorescence only was measured, phosphorescence being entirely eliminated by means of a Corning 7.54 filter and a small Bausch and Lomb monochromator.

RESULTS AND DISCUSSION

Decay laws

Methylcyclohexane—As was previously reported (*4*), in methylcyclohexane glass containing biphenyl, ITL was found to decay according to a law [1] which was first observed by Bagdasar'yan (*6*):

$$I_o/I(t) = 1 + k(t - t_o) \quad [1]$$

where I_o is the luminescence intensity at time t_o of the beginning of observation, $I(t)$ the intensity at time t, and k a coefficient which has the dimension (time^{-1}) of a rate constant.

Fig. 1 shows that the value of this coefficient k, i.e. the slope of the linear plot of $I_o/I(t)$ vs time, decreases as irradiation time increases. In these plots, after irradiations lasting from 3 ns to 38 min, t_o was always taken at 2.5 min. It seems that the variation of slope with irradiation time is simply due to the hyperbolic form of the decay law, as an inherent property of the hyperbola. When the slopes of these lines are plotted vs irradiation time, a hyperbola is obtained again (Fig. 2).

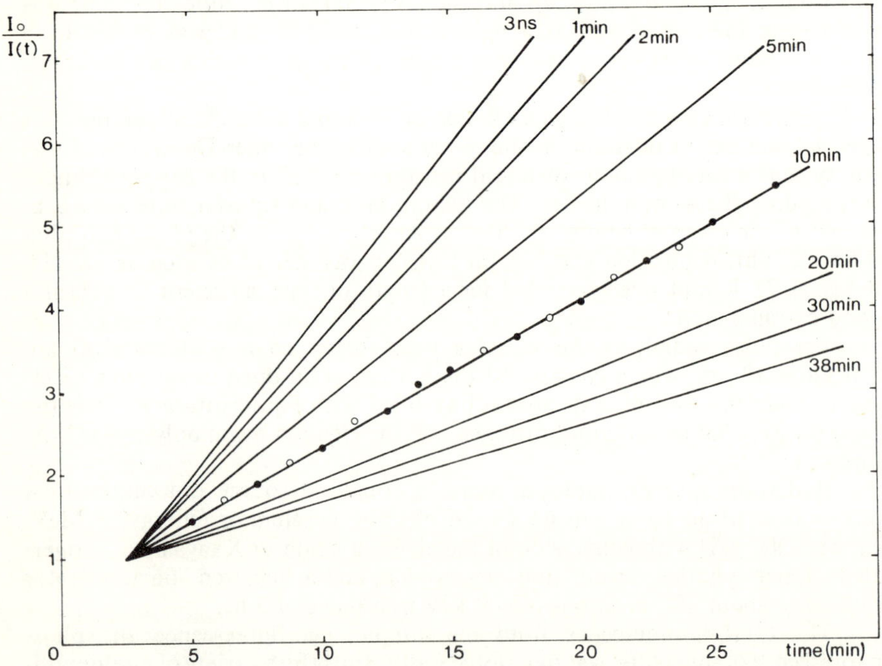

FIG. 1. Plots of $I_o/I(t) = f(t)$ for ITL decay of MCH containing biphenyl (10^{-3}mol/dm^3) after different irradiation times. All curves are normalized at time t_o=2.5 min after the end of irradiation. The 10 min curve shows 2 sets of experimental data:

● dose rate ≈ 5 krad/min

O dose rate ≈ 0.16 krad/min

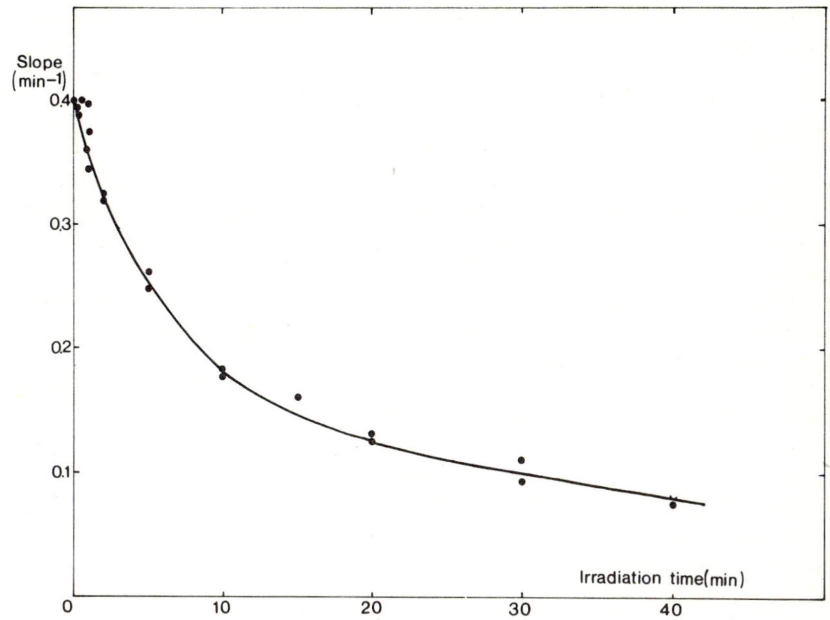

FIG. 2. Plot of the slopes of $I_o/I(t) = f(t)$ curves (i.e. of the values of coefficient k of equation [1]) vs irradiation time. Several points for one irradiation time represent different experiments.

A similar variation of slope is observed when several plots are made from the same experimental decay curve, taking different times as t_o, but in this case they all extrapolate back to the same point on the time scale for $I_o/I(t) = 0$. For irradiations lasting less than one minute, such an extrapolation practically intersects the abscissa at the time corresponding to the end of irradiation, while for longer irradiations the intersect lies further and further back in the time span of irradiation. A qualitative explanation is that detrapping occurs already during irradiation.

In Fig. 1, the plot for an irradiation time of 10 min shows two series of experimental points, obtained with dose rates of ≈ 5 and 0.16 krad/min, and thus with doses of ≈ 50 and 1.6 krad. It is obvious from this plot that the value of k is independent of dose rate and of dose, at least over this range.

Similar results have been obtained for sodium hydroxide glass containing phenol (*7*). Thus it seems to be generally true that the coefficient k varies only with time of irradiation and time elapsed between irradiation and observation, but is independent of dose rate and dose. Earlier results (*3,4,7*) showed that ITL decays observed after irradiation at 77, 66 and 4.2 K can be superposed by simply normalizing initial intensities at identical t_o time. Hence, k also has the same value after irradiation at these three temperatures, although a change of temperature after irradiation causes a change of luminescence intensity.

In the pulse radiolysis work the decay of ITL could be observed over many decades. For this reason the results were plotted on a log-log scale which offers a convenient representation of the decay law [2] of Debye and Edwards (*8*):

$$I(t) = f(t^{-\alpha}) \qquad [2]$$

The Bagdasar'yan law [1] can be considered as a special case of [2] when the exponent α equals 1.

This extension of our work towards shorter times of irradiation and smaller delays t_o between iradiation and observation was aimed at determining the limits of validity of the predominant recombination mechanism; the shortest time t_o reached before was 1 min after irradiation with the van de Graaff accelerator. After a 3 ns pulse from the Febetron accelerator it was possible to measure the ITL decay from 2 μs to 20 s by taking pictures of the oscilloscope screen over different time sweeps. Some overlap between these made possible the normalization of successive pictures. The inset in Fig. 3 shows such an oscillogram with a sweep rate of 1 μs per division. The three linear plots of Fig. 3 show the log-log relationship between fluorescence intensity and time between 10^{-6} and 10^{3} s and over an intensity range of more than 9 orders of magnitude; curve 1 covers times from 2.10^{-6} to 10^{-2} s, curve 2 from 10^{-2} to 20 s; curve 3 refers to results obtained with a γ-irradiation of 1 min and covers times from 60 to 3000 s. The slope α of all these curves equals 1 ± 0.1. Hence the decay law [1] is valid over the whole range studied.

Other glasses—To sum up the results obtained so far in methylcyclohexane glass, the same kinetics of ITL decay have been observed at 4.2 K over a period from 1 min to an hour, and at 77 K from 2μs to an hour. Identical kinetics were observed in aqueous NaOH glass at 4.2 K and at 77 K from 2.5 min to over 3 h (*7*). In 3-MP glass the same kinetics were observed only at 4.2 K (*4*), while at 77 K—the glass transition temperature T_g of this matrix—a linear plot was not obtained with equations [1], but with equation [3]

$$[\, I_o/I(t)\,]^{1/2} = 1 + k\,(t - t_o) \qquad [3]$$

This decay law was observed by Myers (*9*) after photoionization of tetramethyl *p*-phenylene diamine in 3-MP at 77 K and was interpreted in terms of thermal detrapping. Such a mechanism seems of course quite likely at T_g and is generally accepted for 3-MP.

Relationship between ITL decay and trapped electron decay—It seems worthwhile to compare the above data with the kinetic equations for the decay of trapped electrons as measured by absorption spectroscopy. As luminescence intensity at any instant is proportional to the rate of disappearance of trapped electrons de_t^-/dt, it should be possible to find equations [1] and [3] by differentiating the corresponding equations for trapped electron decay.

In 3-MP a linear relationship [4] can be deduced from literature data (*10,11*):

$$[e_t^-]_t/[e_t^-]_o = f(t^{-1}) \qquad [4]$$

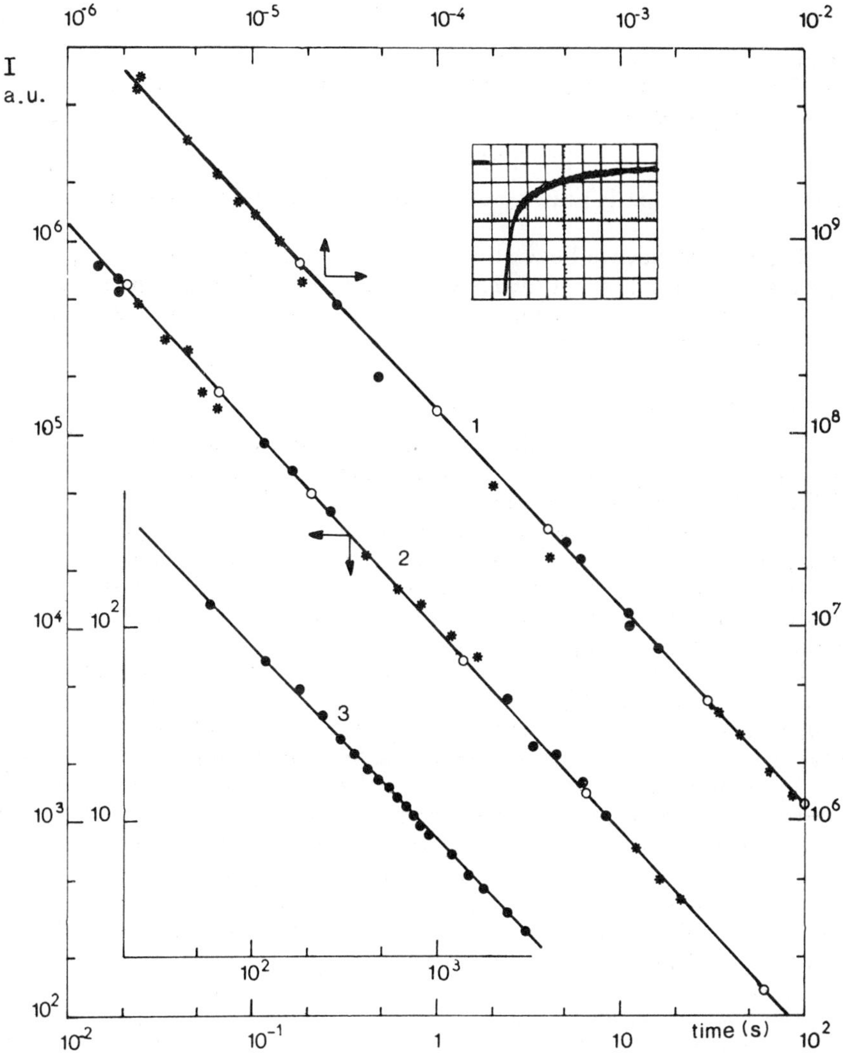

FIG. 3. Decay of ITL at 77 K in MCH containing biphenyl (10^{-3}mole/dm^3) as a log-log plot of intensity in arbitrary units vs time. Curve 1 (fluorescence decay between 2.10^{-6} and 10^{-2}s) and curve 2 (fluorescence decay between 10^{-2} and 60 s): after a 3 ns pulse of 600 keV electrons. Curve 3 (decay between 60 and 3000 s) : after a 1 min γ-irradiation.

o points of normalization between oscillograms • • • *** • • • points from successive oscillograms.

Inset: an oscilloscope trace of a decay (sweep rate: 1 μs per division).

where $[e_t^-]_t$ and $[e_t^-]_o$ are trapped electron concentrations at time t and at time t_o. Differentiation yields a linear relationship of the form

$$I_o/I = f(t^2) \qquad [5]$$

which is equivalent to equation [3].

For MCH no data are available, but in alkaline glass the relationship [6] has been proposed *(12)*:

$$[e_t^-]_t/[e_t^-]_o = 1 - A \ln(t/t_0) \qquad [6]$$

from which equation [1] can be obtained by differentiating and putting $k = 1/t_0$.

The correspondence between equations [1] and [6], and between [3] and [5] shows that both types of studies are concerned with the same population of trapped electrons whose disappearance by recombination with cations is responsible for light emission.

Activation energies

It has been shown (*12*) that the decay of trapped electrons according to relation [6] is compatible with a tunnelling mechanism, just as we have shown (*4*) the compatibility of tunnelling with the decay of ITL according to equation [1]. However, no direct proof of a tunnelling mechanism was as yet available. Such proof could be given if it were possible to show that the process has zero activation energy. Experiments involving irradiation at 77 K and subsequent cooling could demonstrate this if the decay of luminescence intensity proved to be unaffected by the temperature change.

Fig. 4 shows how luminescence intensity changes with temperature for an MCH glass containing biphenyl (10^{-3} mole/dm^3), irradiated 10 min at 77 K (dose rate ≈ 5 krad/min). The initial hyperbolic portion of the luminescence curve is the usual ITL; then, as soon as the temperature (upper curve) begins to drop, luminescence intensity decreases sharply down to a temperature of about 62 K where a small shoulder appears on the temperature curve, due to the freezing of the small amount of liquid nitrogen surrounding the sample. At lower temperatures no further reduction of light intensity is observed. During warm-up, similarly, practically no change of luminescence intensity is recorded until all the nitrogen is melted, i.e. above 62 K; then the intensity increases with temperature and a new decay is observed after a constant temperature of 77 K has been reached.

The three linear plots on Fig. 4 show that the decay law [1] applies whenever a constant temperature is achieved (successively at 77, 4.2 and 77 K). Very similar results have been obtained by Ershov and Kieffer (submitted to *Nature*) with aqueous alkaline glass.

Luminescence intensity at any instant being proportional to the reaction rate of the cation-electron recombination, it can be considered proportional to the rate constant when temperature is the only parameter which changes. Hence, it appears from Fig. 4 that the rate constant practically does not change with temperature between 60 K and 4.2 K, and we can conclude that the activation energy of the recombination reaction is practically zero over this temperature range. From the Arrhenius equation an average activation energy value of ≈ 0.04 eV can be calculated for the temperature range between 60 and 77 K. In fact, there is probably some contribution from a thermal process above 60 K, and its relative importance increases with increasing temperature; in this way the apparent activation energy should increase up to the glow peak at 90 K where the tunnelling process should be negligible in comparison to the thermal process.

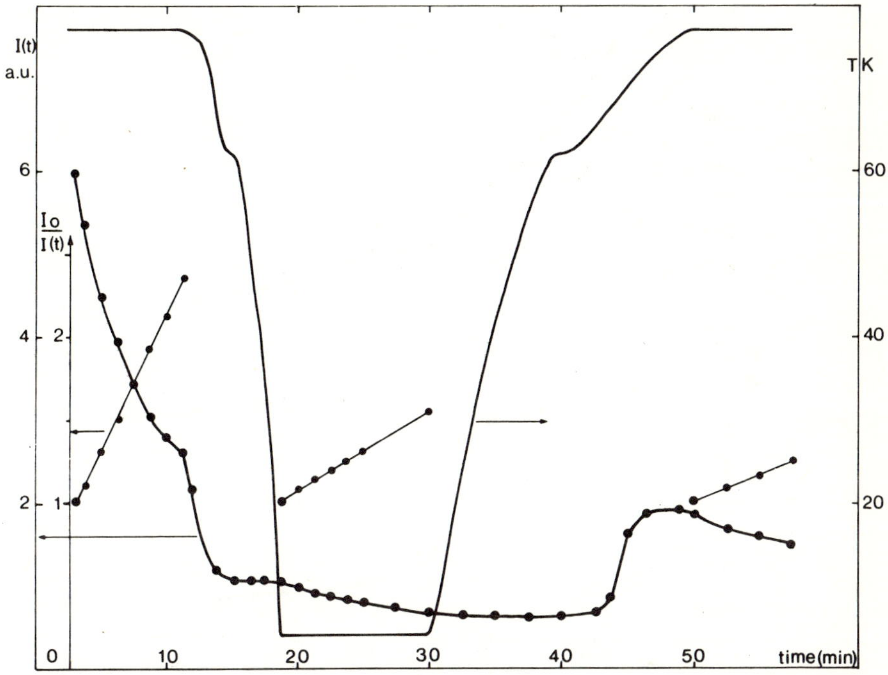

FIG. 4. Effect of temperature on the decay of luminescence intensity of MCH containing biphenyl (10^{-3}mole/dm^3) after γ-irradiation (10 min) at 77 K. Upper curve: temperature (right hand ordinate scale). Lower curve: luminescence intensity (left hand ordinate scale). The three straight lines are plots of $I_o/I(t) = f(t)$ for the isothermal portions of the luminescence curve.

A quantitative interpretation of these curves seems hazardous, as the effect of cooling an irradiated sample from 77 K to 4 K depends moreover on solute concentration. This is shown in Fig. 5 where curves A to C represent experiments with increasing biphenyl concentrations. Curves B and C are normalized to the same arbitrary units as curve A at t_o and are shifted upwards by 2, respectively 3 arbitrary units, for greater clarity of the drawing. The downward-pointing arrows indicate the beginning of the temperature drop and the upward-pointing arrows the point where liquid helium temperature is reached. Table 1 gives the intensity ratios $I_{4\ K}/I_{77\ K}$ calculated from corresponding points on the experimental curve after cooling to 4.2 K and on the very precise extrapolation of the 77 K decay curve. Again it is difficult to attribute a quantitative meaning to these figures, for various factors such as the cooling rate which is not too well controlled might have some influence. However, the existence of a concentration effect is quite certain. At the present

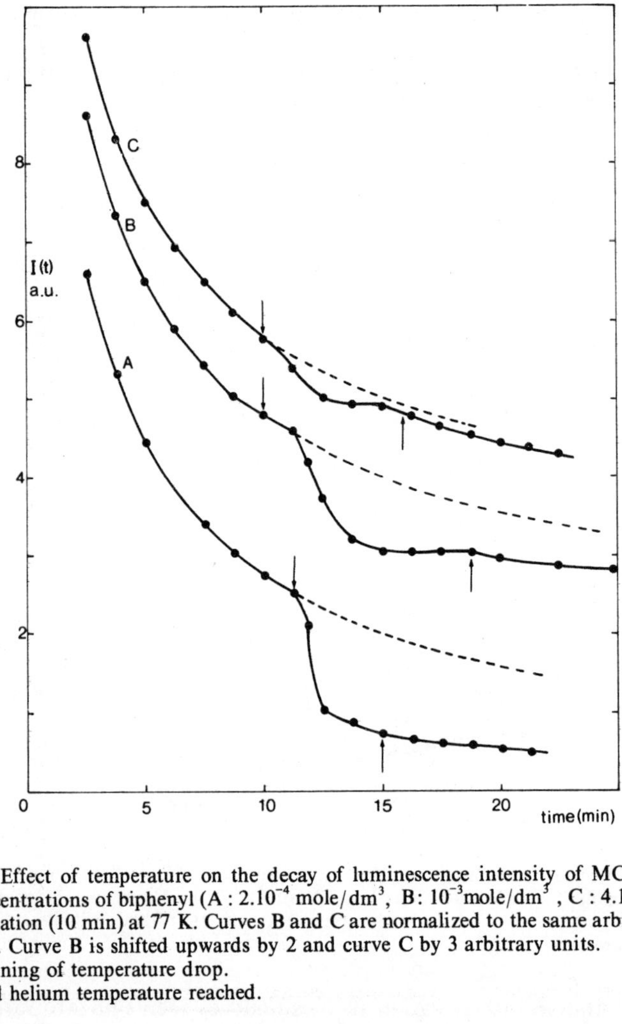

FIG. 5. Effect of temperature on the decay of luminescence intensity of MCH containing different concentrations of biphenyl (A : 2.10^{-4} mole/dm^3, B: 10^{-3}mole/dm^3 , C : 4.10^{-3}mole/dm^3) after γ-irradiation (10 min) at 77 K. Curves B and C are normalized to the same arbitrary units as curve A at t_o. Curve B is shifted upwards by 2 and curve C by 3 arbitrary units.

↓ : beginning of temperature drop.

↑ : liquid helium temperature reached.

TABLE 1

Curve	Biphenyl concentration (mole/dm^3)	I_{4K}/I_{77K}
A	2.10^{-4}	0.37
B	10^{-3}	0.64
C	4.10^{-3}	0.95

time we have no explanation to offer, but we think that it may be connected with a competition between tunnelling of the electron towards a cation or towards biphenyl molecule. A complementary study of the effect of temperature on the rate of formation of biphenyl radical-anions, correlated with the rate of trapped electron decay, at different biphenyl concentrations, is planned in the near future.

CONCLUSION AND OUTLOOK

Deferred luminescence studies have proved to be a very convenient technique for studying reactions involving electron transfer in glassy solids at low temperatures. They have contributed considerable evidence for the role of electron tunnelling. However, all calculations so far had to be based on simplifying assumptions, some of which are almost certainly wrong, e.g. the assumption of a single trap depth. A more refined treatment by Brocklehurst (*13*) seemingly leads to practically the same results as the simpler calculations.

In the future it should prove useful to undertake parallel studies of tunnelling towards trapped cations and neutral scavenger molecules, to compare results obtained with ionizing radiation and two-photon photoionization, and finally to work with organic monocrystalline solids for which calculations should be much easier. Unfortunately, deferred luminescence is very weak in the latter and a study of ITL in single crystals may prove to be impossible.

REFERENCES

1. C. Deniau, A. Deroulede, F. Kieffer and J. Rigaut, *J. Lumin.* **3**, 325 (1971).
2. A. Deroulede, *J. Lumin.* **3**, 302 (1971).
3. F. Kieffer, C. Meyer and J. Rigaut, *Chem. Phys. Lett.* **11**, 359 (1971).
4. F. Kieffer, C. Lapersonne-Meyer and J. Rigaut, *Int. J. Radiat. Phys. Chem.* **6**, 79 (1974).
5. J. Bullot, A. Deroulede and F. Kieffer, *J. Chem. Phys.* **63**, 150 (1966).
6. Kh. S. Bagdasar'yan, R. I. Milyutinskaya and Yu. V. Kovalev, *Khim. Vysok. Energii.* **1**, 127 (1967).
7. B. G. Ershov and F. Kieffer, *Chem. Phys. Lett.* **25**, 576 (1974).
8. P. Debye and J. O. Edwards, *J. Chem. Phys.* **20**, 236 (1952).
9. D. L. Myers, *The Decay Kinetics of the Infrared Stimulated Recombination Luminescence in a Rigid Organic Solution*. Ph.D. Thesis. Cornell University, Ithaca, N.Y. 1971.
10. D. W. Skelly and W. H. Hamill, *J. Chem. Phys.* **44**, 2891 (1966).
11. K. Funabashi, C. Hebert and J. L. Magee, *J. Phys. Chem.* **75**, 3221 (1971).
12. K. I. Zamaraev, R. F. Khairutdinov, A. I. Mikhailov and V. I. Gol'Danskii, *Dokl. Akad. Nauk SSSR.* **199**, 640 (1971).
13. B. Brocklehurst, *Chem. Phys.* **2**, 6 (1973).

Trapped Electrons and Anions in Rigid Organic Glasses at 4° K[1]

Akira Namiki, Masato Noda and Takenobu Higashimura
Research Reactor Institute, Kyoto University
Kumatori-cho, Sennan-gun, Osaka-fu, Japan

INTRODUCTION

Radiolysis of organic glasses at 77° K was first employed for studying the intermediate species by means of ESR (*1–3*). The solvated electron was found by this method although it was confirmed later by the pulse radiolysis method (*4–6*). In 1967, the radiolysis at 4.2° K was carried out by D. R. Smith and J. J. Pieroni and the trapped electron, the state before the electron becomes to be solvated, was verified by ESR (*7*). Optical spectroscopy of the trapped electron was started in 1969 by both pulse radiolysis at 77° K (*8*) and gamma-ray radiolysis at 4.2° K (*9*). The trapped electron is a precursor of the solvated electron[2]. The lifetime of the trapped electron is too short to be measured at room temperature. At 77° K it can be observed by the pulse radiolysis technique, but at 4.2° K it is stable enough to be measured by steady state radiolysis.

For the study of other intermediates appearing in radiation chemical processes, radiolysis of organic glasses at low temperature has been used widely. Many cations and anions which disappear very quickly at room temperature were found (*10*), giving useful informations on electronic structures of various aromatic molecules. Radiolysis at lower temerature than 77° K was carried out and a new species which does not appear at 77° K was discovered (*11*). It is natural to suppose that the intermediates with much shorter lifetime will be found by the radiolysis at very low temperature.

There are two alternative methods for studying the very fast processes in radiation chemistry: the low temperature pulse radiolysis and the very low temperature radiolysis. Almost the same results have been obtained by both methods on the trapped electron and the ionic intermediates. The low temperature pulse radiolysis has its advantage in studying reaction kinetics, while the very low temperature radiolysis has advantages in studying the molecular structure of unstable intermediates because ESR and optical bleaching measurements can be carried by this method.

TRAPPED ELECTRONS

Trapped electrons prior to solvation are observed in polar, slightly polar and also non-polar matrices by the very low temperature radiolysis, the low

[1] Congress presentation A–6–4.
[2] In this article, this pre-solvated electron is referred to as "trapped electron" and the electron observed by the steady-state radioloysis at 77° K is called "solvated electron," though the latter appears in matrices of the solid phase.

temperature pulse radiolysis and the room temperature picosecond pulse radiolysis. They show optical absorption spectra which differ dramatically from those of solvated electrons in respective matrices observed at 77° K. Absorption spectra of solvated electrons are structureless and abnormally wide. Those of trapped electrons are also structureless and very wide, extending from the infrared to the near ultraviolet. Roughly speaking, the absorption spectra of trapped electrons in polar matrices rather resemble those of solvated electrons in non-polar matrices. In Fig. 1, trapped electron spectra in three different kinds of matrices are shown. In the case of ethanol, the trapped electron spectrum shows a remarkable difference from that of the solvated electron. Ethylene glycol-water (50/50 by vol.) glass belongs to this class of matrices. In the case of 2-methyltetrahydrofuran, where the solvated electron spectrum lies in the infrared region, the trapped electron spectrum shows a red shift from that of the solvated electron. Although the solvated electron spectrum is composed of two peaks, the one attributed to a discrete transition and the other to a continuum transition (*12*), the trapped electron spectrum shows only one peak.

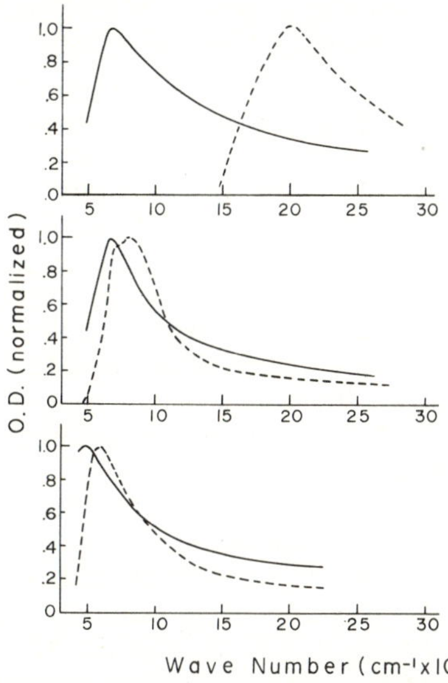

FIG. 1. Absorption spectra of trapped electrons for the irradiation and measurement at 4° K (full lines) and of solvated electrons for the irradiation at 77° K and the measurement at 4° K (broken lines). Upper figure: ethanol; Middle figure: 2-methyltetrahydrofuran; Bottom figure: methylcyclohexane. Spectra are normalized with respect to optical density.

The trapped electron is photobleached easily and the photodetached electron is captured by the electron scavenger molecules when the matrix contain the scavenger. Therefore, the absorption spectra of the trapped electron is attributed to a transition to the continuum state.

Illumination with light of different wavelengths bleaches out different components of the trapped electron spectrum in ethanol glass. Therefore, there should be electrons stabilized in traps of different depths. Abnormally broad absorption spectra of the trapped electrons are caused by these two characteristics, the bound-free transitions and electrons in various traps.

The width, ΔH_{ml}, of the ESR spectrum of the trapped electron shows a remarkable difference from that of the solvated electron. ΔH_{ml} of the trapped electron in the ethanol glass is 6.7 G whereas that of the solvated electron is 16.0 G (*13*). In the methylcyclohexane glass, the former is 5.9 G and the latter is 4.3 G (*14*). These changes can be interpreted to mean that, upon solvation, the distance between the electron and the protons of the surrounding molecules becomes small in the case of ethanol and it becomes large in the case of methylcyclohexane. The solvated electron in methylcyclohexane shows paramagnetic saturation at lower microwave power than that of the trapped electron (*14*). This can also be interpreted by a change of the above distance occurring upon solvation.

We can make the following image of the structure of the trapping site (*13*). In the liquid and also in the glass, dipolar molecules or bond dipoles in the molecules are arranged in equilibrium according to the laws of thermodynamics. The potential produced by surrounding dipoles obeys a statistical distribution determined by both the temperature of the system and the sum of the interaction energies between each set of surrounding dipoles. When the liquid is cooled to make a glass, the instantaneous arrangement of the dipoles is frozen and leaves a number of stable potential peaks and potential wells of different depths in the glass. The depth of potential wells thus formed distribute continuously. There should exist a number of wells deep enough to trap the electron. However, most of them are shallower than the solvated state. When the thermalized electron is captured by these traps, the bond dipoles move to relax the Coulombic interaction between the trapped electron and the dipoles if the glass is not frozen so rigidly. The distance between the trapped electron and the protons in the molecules becomes small or large depending on the sign of the bond dipole moment. Thus the trapped electron transforms to the solvated electron.

In Table 1, some optical characters of the trapped electrons are summarized. In the matrices shown below the dotted line in the table, the trapped electron spectra are almost similar to those of the solvated electron. Solvation may be complete even at 4° K in these matrices.

The ratio of the yield of the trapped electron for irradiation at 4° K to the yield of the solvated electron for irradiation at 77° K is larger than unity in some matrices and smaller than unity in some other matrices. However, they do not deviate greatly from unity. It is difficult to estimate with enough accuracy the yield of the trapped electrons at 4° K because the ESR signal of the trapped electron shows paramagnetic saturation with very small microwave power. Therefore, we can say only that the yields of the trapped electron at 4° K and the solvated electron at 77° K are "about" the same.

It must be noted that, in some glasses, a low energy threshold occurs in the trapped electron spectra. If the molecular dipoles forming the trap are arranged at random in thermodynamic equilibrium, there should be many shallow traps, causing no threshold in the absorption spectra. Slight orientation at 4° K may be allowed or the electrons in shallow traps tunnel to the neighboring traps with depths larger than the threshold (*18*).

ANIONIC INTERMEDIATES

Some aromatic anions are observed in the stable form in glasses of aliphatic hydrocarbon, but they cannot be stable in alcohol glasses because they abstract a proton from the matrix. However, they become stable at 4° K as follows.

TABLE 1
Trapped electrons in organic glasses for the irradiation and the measurement at 4.2° K

	$\lambda_{max}(4°)$ (nm.)	$\lambda_{max}(77°)$ (nm.)	λ_{thresh} (nm.)	η	Reference
C_2H_5OH	1500	545	2500	1.0	(13)
Ethylene glycol - Water	1900, 714	610			
2-Methyltetrahydrofuran	1540	1180, 1330	2500	0.8	(17)
2-Methyl-n-amylamine	1380	1180	2400		
sec-Butylamine	1220	1130	2000	1.0	(16)
3-Methylpentane	2300	1650		1.2	(14)
Methylcyclohexane	2100	1700		0.9	(14)
CH_3OH	605	510			
10 M NaOH Water	650	590			
N-Methyl-n-butylamine	1210	1175		1.3	(16)
C_2H_5OD	545	540			
Triethylamine	1680	1680		1.4	(16)
Tripropylamine	1680	1680		1.3	(16)
1,2-Propanediamine	980	980			
3-Methyhexane	1800	1650			

$\lambda_{max}(4°)$: Peaks of absorption spectra of trapped electrons for the irradiation and the measurement at 4.2° K. Electrons in matrices below the broken line may be solvated ones.

$\lambda_{max}(77°)$: Peaks of absorption spectra of solvated electrons for the irradiation at 77° K and the measurement at 4.2 ° K.

λ_{thresh} : Threshold wavelengths at the long wavelength side of absorption spectra of trapped electrons for the irradiation and the measurement at 4.2° K.

η : Ratios of the yield of trapped electrons at 4° K to that of solvated electrons at 77° K.

Reagent grade ethanol containing 0.1 mole % of nitrobenzene or its halogen derivatives is cooled to form a clear glass and is irradiated with ^{60}Co gamma rays at 4° K. Fig. 2 shows the spectra of these glasses irradiated for 15 min. with the dose rate of 1×10^{18} eV/g.min. Spectral peaks of nitrobenzene, 19.0, 21.2, 30.2 and 31.0 kK., are in good agreement with the peaks of nitrobenzene anion, 17.0, 21.2, 30.6 kK, observed in the 2-methyltetrahydrofuran glass irradiated at 77° K (*19*). Therefore, they can be attributed to the nitrobenzene anion. Spectra of *p*-chloro, *o*-chloro and *p*-bromo nitrobenzene resemble that of nitrobenzene in the overall wavelength region. Because the interaction of π electrons on the aromatic ring with the chlorine atom and bromine atom may be weak (*20*), these spectra should be attributed to anions of the respective halogen derivatives. If the glasses containing these compounds are irradiated at 77° K, there appears no absorption peak except a very weak, broad absorption in the range between 20 and 25 kK. At 77° K, the anions are unstable, abstracting a proton from the matrix.

The absorption spectrum of the ethanol glass with 0.05 mol % of 2-chloronaphthalene irradiated at 4° K is shown in Fig. 3. The wavelengths of the peaks, 326, 372, 750 and 838 nm, are in good agreement with those of the naphthalene anion appearing in the 2-methyltetrahydrofuran glass with naphthalene irradiated at 77° K (*21*). In the pulse radiolysis experiment of ethanol containing 2-chloronaphthalene at 100° K, peaks of 380, 760 and 840 nm were observed and identified to be the 2-chloronaphthalene anion (*22*). Therefore, we can say that the 2-chloronaphthalene anion remains stable for the irradiation at 4° K. When this glass is warmed rapidly to 77° K and is cooled again to 4° K, a new peak with a wavelength of 339 nm appears, and the peak at 326 nm decreases in intensity. This result is consistent with that for pulse radiolysis of 2-methyltetrahydrofuran at 100° K (*22*). The 339 nm absorption is attributed to a hydronaphthyl radical which is formed by the proton abstraction of the anion from the matrix. Protonation should not occur at 4° K but it occurs at 77° K.

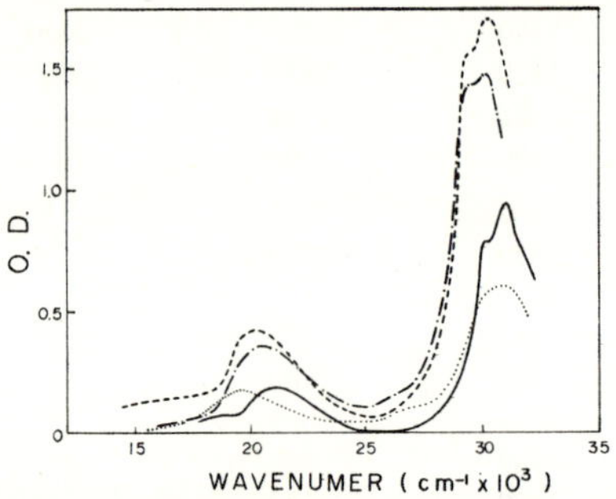

FIG. 2. Absorption spectra of ethanol glass containing 0.1 mol % of nitrobenzene and its halogen derivatives irradiated at 4° K with the dose of 1.5×10^{19} eV/g. min. ——: $C_6H_5NO_2$; ----: *p*-$C_6H_4ClNO_2$; -·-·-: *p*-$C_6H_4BrNO_2$; ·····: *o*-$C_6H_4ClNO_2$.

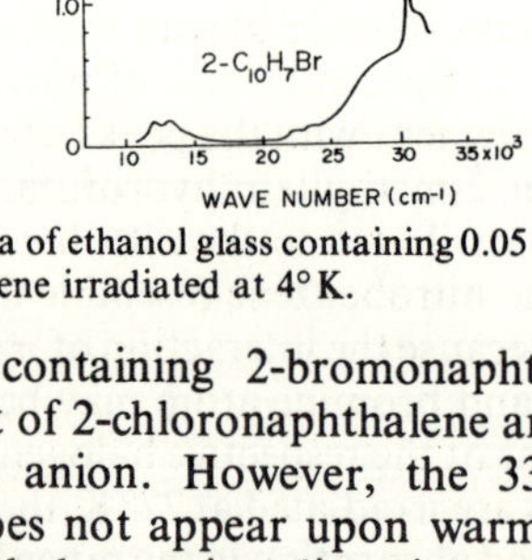

FIG. 3. Absorption spectra of ethanol glass containing 0.05 mol % of 2-chloronaphthalene and 0.1 mol % of 2-bromonaphthalene irradiated at 4° K.

The ethanol glass containing 2-bromonaphthalene shows an almost identical spectrum to that of 2-chloronaphthalene and proves the formation of the 2-bromonaphthalene anion. However, the 339 nm absorption of the hydronaphthyl radical does not appear upon warming the irradiated glass to 77° K. The 2-bromonaphthalene anion dissociates to naphthalene radical and bromine anion upon warming, and the former then reacts with ethanol to produce ethanol radical.

When the ethanol glass containing benzyl chloride is irradiated at 4° K, absorption bands arise in the wavelength region between 300 and 400 nm. Sharp peaks at 306, 308 and 318 nm are attributed to the benzyl radical; these are also observed by the pulse radiolysis at 100° K (*22*), and the widths of the absorption peaks are larger than those observed at 77° K. A broad peak at 340 nm is attributed to an anionic precursor of the benzyl radical because this peak disappears upon warming the glass to 77° K with an accompanying growth of the benzyl radical bands (*23*). This precursor is not detected by the pulse radiolysis at 100° K within 1 microsecond after the pulse. It may be the benzyl chloride anion (*22*) or a charge transfer complex composed of the benzyl radical and the chloride anion (*24*). The precursor band appears in the case of C_2H_5OD glasses as is shown in Fig. 4. It is interesting that this band is bleached by the illumination with a tungsten lamp while the same band in the C_2H_5OH glass cannot be bleached with the same light.

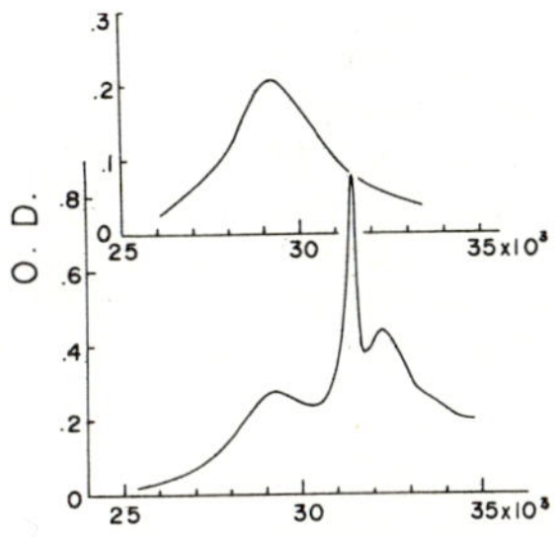

Wave Number (cm^{-1})

FIG. 4. Absorption spectrum of deuterized ethanol containing benzyl choloride irradiated at 4° K. Upper figure is the bleached fraction attributed to the precursor of the benzyl radical.

In the case of many halogenated aromatic compound, dissociative electron attachment requires a positive activation energy on the order of 10^{-1} eV in the gas phase experiment. The activation energy in this case is determined from the difference between the zero-point energy for the molecule and the intersection of the molecular negative-ion and dissociative negative-ion potential energy curves or from the difference between the zero-point energy for the molecule and the intersection of the molecular and molecular negative-ion potential-energy curves (*25*). In the case of our experiment, however, the activation energy is the difference between the zero-point energy for the molecular negative-ion and the intersection of the molecular negative-ion and the dissociative negative-ion potential-energy curves. There was no experimental estimation of this activation energy. Stability of the molecular negative-ion at 4° K increases very much compared with that at 77° K, if the activation energy is slightly positive.

There remain various problems which will be studied successfully by means of the very low temperature irradiation: the temperature dependence of the radical yields and the electron scavenging (*23*), the mechanism of the electron capture in the condensed phase, migration of hole (*11*), etc.

REFERENCES

1. R. S. Alger, T. H. Anderson and L. A. Webb, *J. Chem. Phys.* **30**, 695 (1959).
2. C. Chachaty and E. Hayon, *J. Chem. Phys.* **61**, 1115 (1964).
3. D. R. Smith and J. J. Pieroni, *Can. J. Chem.* **43**, 876 (1964).
4. E. J. Hart and J. W. Boag, *J. Am. Chem. Soc.* **84**, 4090 (1962).
5. J. W. Boag and E. J. Hart, *Nature* (*London*) **197**, 45 (1963).

6. J. P. Keen, *Nature* (*London*) **197**, 47 (1963).
7. D. R. Smith and J. J. Pieroni, *Can. J. Chem.* **45**, 2723 (1967).
8. J. T. Richards and J. K. Thomas, *J. Chem. Phys.* **53**, 218 (1970).
9. T. Higashimura, M. Noda, T. Warashina and H. Yoshida, *J. Chem. Phys.* **53**, 1152 (1970).
10. W. H. Hamill, in *Radical Ions* (E. T. Kaiser and L. Kevan, Eds.) pp. 321–416. Interscience Publishers, New York, 1968.
11. A. Ekstrom and J. E. Willard, *J. Phys. Chem.* **74**, 1708 (1970).
12. W. H. Hamill, *J. Chem. Phys.* **53**, 473 (1970).
13. H. Hase, T. Warashina, M. Noda, A. Namiki and T. Higashimura, *J. Chem. Phys.* **57**, 1039 (1972).
14. H. Hase, T. Higashimura and M. Ogasawara, *Chem. Phys. Lett.* **16**, 214 (1972).
15. H. Hase, T. Warashina, M. Ogasawara and T. Higashimura, *J. Chem. Phys.* (in press).
16. T. Ito, K. Fueki, A. Namiki and H. Hase, *J. Phys. Chem.* **77**, 1803 (1973).
17. H. Yoshida, M. Ogasawara, T. Warashina and T. Higashimura, *J. Chem. Phys.* **56**, 4238 (1972).
18. T. Higashimura, *Annu. Rep. Res. Reactor Inst. Kyoto Univ.* **6**, 38 (1973)
19. T. Shida and S. Iwata, *J. Phys. Chem.* **75**, 2591 (1971).
20. K. Kimura and S. Nagakura, *Mol. Phys.* **9**, 117 (1965).
21. S. Iwata and T. Shida, *J. Am. Chem. Soc.* **95**, 3473 (1973).
22. A. Arai, S. Tagawa and M. Imamura, *J. Phys. Chem.* **78**, 519 (1974).
23. T. Higashimura, A. Namiki, M. Noda and H. Hase, *J. Phys. Chem.* **76**, 3744 (1972).
24. M. Irie, M. Shimizu and H. Yoshida, *Chem. Phys. Lett.* **25**, 102 (1974).
25. J. C. Steelhammer and W. E. Wentworth, *J. Chem. Phys.* **51**, 1802 (1969).

Reactions of Electrons in 3-Methylhexane Glass[1]

Hugh A. Gillis, Norman V. Klassen, and George G. Teather
Division of Physics, National Research Council of Canada
Ottawa, Canada K1A 0S1

INTRODUCTION

An important aspect of the radiation chemistry of solid hydrocarbons is their propensity to trap electrons and cations in large yields. The literature on the nature and reactions of trapped electrons and positive ions has been summarized in a few reviews *(1–3)*. Obviously an important reaction of these species is neutralization. In hydrocarbons, especially at low temperature, most electrons are thermalized at sufficiently small distances from a positive ion in the same spur so that the coulombic energy between the pair is greater than the kinetic energy. The ion-pair is described as geminate, and undergoes geminate recombination unless one or both members of the pair is scavenged by solute. In this paper we report measurements on the kinetics of geminate recombination in 3-methylhexane (3MH) following pulse irradiation. Our measurements were made with liquid 3MH between 112 K and 140 K, rather than with a hard glass. At 112 K the viscosity of 3MH is calculated from an extrapolation of data given by Ling and Willard *(4)* to be about 10^2 poise. We also report measurements on growth of pyrene anion and cation following pulse-irradiation of solutions of pyrene in 3MH glass at 76 K.

GEMINATE RECOMBINATION IN COLD LIQUID 3MH

We have shown that the spectrum of the trapped electron in 3MH glass at 76 K undergoes a blue-shift with time, with λ_{max} being approximately 2000 and 1700 nm at 3 μsec and 380 sec, respectively *(5)*. The ratio O.D.(2100 nm)/O.D.(1525 nm) is 1.14 and 0.81 at 3 μsec and 380 sec, respectively. In order to reliably follow electron decay by absorption measurements at a particular wavelength, it must be shown that spectral changes are not occurring simultaneously. Therefore, it was considered important to show that the kinetics were the same within experimental error at two or more wavelengths. Good agreement was found between the kinetics followed at 900 and 1525 nm, or at 1525 and 2100 nm in a split-analyzing beam system for temperatures of 112 K and higher. A typical result is shown in Fig. 1. The ratio O.D.(2100 nm)/O.D.(1525 nm) at any time for the experiment of Fig. 1 was 1.18, which shows that in cold liquid 3MH the electron has a spectrum very similar to the initial (3 μsec) spectrum in the glass at 76 K. At temperatures lower than 112

[1]Congress presentation A–6–5.

K we find that the initial decay at 900 nm and 2100 nm is faster than at 1525 nm, which indicates that spectral changes are significant.

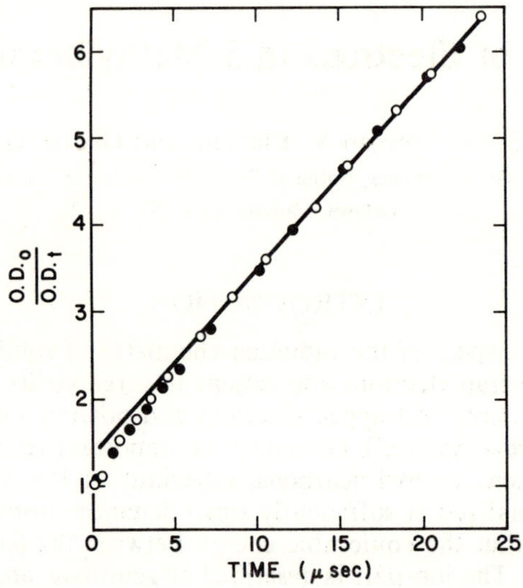

FIG. 1. Decay of electrons in liquid 3MH at 121 K after an 80 nsec electron pulse, followed at 1525 nm ●, and 2100 nm O. $O.D._t$ and $O.D._o$ are the optical densities at time *t*, and the extrapolated value at mid-pulse, respectively.

The average $G\epsilon$ at 1525 nm for 7 series of experiments (7 different samples of pure 3MH) was 4.75×10^4, independent of temperature over our range. If it is assumed that ϵ_{max} is the same as has been found for ϵ_{max} for the long-time spectrum in 3MH glass at 77 K, *i.e.* $3 \times 10^4\ M^{-1}\ cm^{-1}$ *(1)*, then ϵ_{1525} here is $2.6 \times 10^4\ M^{-1}\ cm^{-1}$ and $G(e_s^-)$ is 1.8. This is considerably higher than the yield of 0.9 reported for the glass at 77 K following a Co^{60} γ-irradiation *(6)*.

As indicated in the Introduction, most of the electrons in cold liquid 3MH are expected to be geminate. The *G* of free ions (those that escape the coulombic field of the geminate positive ion) in 3MH is probably comparable to that in *n*-pentane, which decreases with temperature and is only 0.02 at 160 K *(7)*. Confirmation of this low free ion yield was established by following electron decay in two experiments at the same apparent temperature (126 K) but at doses differing by a factor of 7.2. The decay rate divided by dose was $\leqslant$10% slower at the lower dose even after 85% of the electrons had reacted. The difference might have been due to a small temperature difference, but even if it was a concentration effect due to second-order neutralization of free ions, it indicates that the initial yield of free electrons is at most a few percent of the total yield. It is therefore merely a coincidence that the electron decay of Fig. 1 obeys a second-order plot after about 50% decay.

At all temperatures in our study, provided that there were no spectral shift effects, plots of $O.D._o/O.D._t$ *vs.* ($C \cdot$ time) gave identical plots for appropriate choice of the scaling factor *C*. This is illustrated in Fig. 2 where the time scales differ by a factor of 200. We also found that a very similar plot describes electron decay in liquid propane at 88 K[2], where the initial half-life is 54 nsec. This suggests that electrons in many hydrocarbons at low temperature might show the same type of decay kinetics.

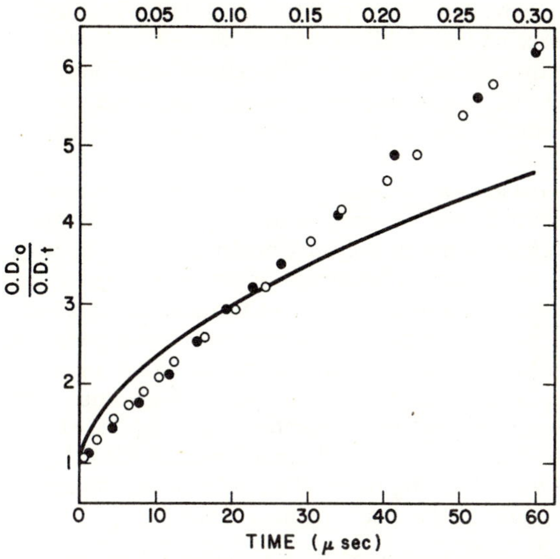

FIG. 2. Electron decay in pulse-irradiated liquid 3MH at two temperatures. For points ● T = 140 K, pulse length = 15 nsec, λ= 900 nm and the upper time scale was used; for points O T = 119 K, pulse length = 100 nsec, λ = 1525 nm and the lower time scale was used. The solid curve is a plot of expression (1) (see text).

Rzad, et al., *(9)* have derived a functional description of the life-times of geminate electrons produced in the radiolysis of pure hydrocarbons from the concentration dependence observed for ion scavenging. Their description is

$$G(t) = G_{gi} \exp(\lambda t) \operatorname{erfc}(\lambda t)^{1/2} \quad (1)$$

where G(t) is the yield of electrons still present at time t, G_{gi} is the total yield of electrons, λ is a parameter with the dimensions of frequency and erfc $(\lambda t)^{1/2}$ is the error function complement of $(\lambda t)^{1/2}$. $G(t)/G_{gi}$ should be given by $O.D._t/O.D._o$ in our system, and expression (1) is plotted as the solid line in Fig. 2 with λ chosen to give agreement with experiment around $O.D._o/O.D._t = 3$. It is seen that agreement between expression (1) and our results is not good. If expression (1) is fitted to our results at long times, it predicts a much faster initial decay than is found experimentally.

Expression (1) is derived *(9)* from the empirical scavenging expression

$$G(P) = G_{fi} + G_{gi} \frac{(\alpha c)^{1/2}}{1 + (\alpha c)^{1/2}} \quad (2)$$

where $G(P)$ is the yield of product resulting from the reaction of ions with a

[2]For details of earlier work on liquid propane with a slower detection system, see reference *(8)*.

scavenger at concentration *c*, G_{fi} and G_{gi} are the yields of free and geminate ions, respectively, and α is a parameter proportional to the rate constant of the scavenging reaction. Expression (2) has been found to describe a large amount of data at room temperature in several solvents and up to concentrations of solute high enough to scavenge over 50% of the geminate electrons *(10)*. Of course it might not hold in 3MH at low temperature, in which case expression (1) would not be expected to apply. However, Infelta *(11)* has found that expression (2) does describe the yields of methyl radicals produced by electron scavenging with methyl bromide in 3-methyloctane solutions between room temperature and 113 K. He found the geminate ion yield to be proportional to absolute temperature, and to decrease from 5.0 at room temperature to 1.8 at 113 K.

A possible reason for the failure of expression (1) to describe our results is that it should describe the lifetimes of all electrons that react with scavengers, which might include some presolvated or dry electrons as well as the solvated electrons which we observe. Evidence for dry electrons in water and alcohols has been presented by Hamill *(12,13)*. In the absence of solute, part of the dry electrons would undergo geminate recombination and part would become solvated. This competition would explain the fact that we observe a *G* of solvated electrons of about 1.8, whereas treatment of scavenging experiments with expression (2) leads to total *G*'s of electrons of 4–5 *(10,14)*.

Rzad, et al. *(9)*, Mozumder *(15)*, and Freeman *(16)* all predict *G*(t) should be proportional to $t^{-1/2}$ at long times. We find that this describes electron decay only after about 60% of the electrons have decayed.

PULSE RADIOLYSIS OF PYRENE IN 3MH AT 76 K

It is well known, especially from the work of Hamill and co-workers *(1)* that large yields of anions result from the irradiation of hydrocarbon glasses containing dissolved electron scavenger. The mechanism by which these anions are produced is still not clear. Thus Oka, et al. *(17)*, have assumed that scavenging of electrons in 3MH glass at 77 K takes place before trapping, and have obtained reasonable competition plots on this basis. On the other hand, Miller *(18–20)* has presented evidence in favor of a model in which electrons become physically trapped in the matrix before encountering a scavenger molecule, following which they may tunnel to such a molecule. Also to be considered is a hopping mechanism, according to which the electron is solvated or trapped, then can be thermally detrapped with the quasi-free electron moving to a new trapping site or to a scavenger molecule. With the hope of learning something about the mechanism involved, solutions of pyrene in 3MH glass at 76 K were pulse irradiated. The spectra found in the presence and absence of triethylamine added as a positive ion scavenger are shown in Fig. 3. The main peak at 492 nm must be due to pyrene$^-$, in agreement with the literature *(21)*. Pyrene$^+$ absorbs around 450 nm, but some absorption due to pyrene$^-$ has also been reported there *(22)*. Since the presence of triethylamine considerably reduces the relative height of the 450 nm peak, it must be mainly due to pyrene$^+$.

The changes with time of yields of pyrene$^-$, pyrene$^+$ and the trapped electron from a 3MH glass at 76 K containing 1.9×10^{-2} M pyrene are shown in Fig. 4. For times longer than 5 μsec a monochromator was placed between the

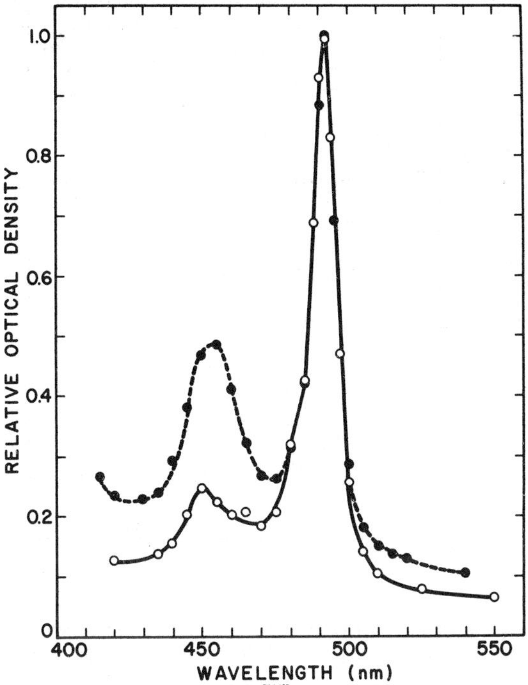

FIG. 3. Spectra of solutions of pyrene in 3MH glass at 76 K, 1 μsec after the start of a 100 nsec pulse. For points ●, [pyrene] = 14 mM; for points O [pyrene] = 20 mM and the glass also contained 2.4 mole % triethylamine.

analyzing light source and the sample to avoid bleaching. The absence of bleaching effects was confirmed by performing experiments at different light intensities. G(pyrene$^-$) was calculated by taking $\epsilon_{492} = 4.95 \times 10^4$ M^{-1} cm^{-1} *(17)*, G(pyrene$^+$) was calculated by assuming all absorption at 450 nm is due to pyrene$^+$ and by taking $\epsilon_{450} = 4.4 \times 10^4$ M^{-1} cm^{-1} *(18)*,[3] and G(e$^-$) was found from $\epsilon_{1525} = 2.6 \times 10^4$ M^{-1} cm^{-1} as assumed above.

Our direct observation of a growth in pyrene$^+$ in 3MH glass where diffusion must be extremely slow, is confirmation of hole migration involved by Hamill *(1)*. Similarly Zador et al. *(23)* found the formation kinetics of pyrene$^+$ in cyclohexane solution at room temperature indicates the formation of positive charge carriers with mobilities far in excess of those expected for molecular positive ions.

Unfortunately the results of Fig. 4 for pyrene$^-$ do not allow us to decide among the mechanisms discussed above. Pyrene$^-$ grows in after the pulse, probably by either a tunneling or a hopping mechanism, and the electron decays at about the same rate as pyrene$^-$ grows. But the pyrene$^-$ measured at 10^{-8} sec could result mainly from reactions of presolvated electrons, or by rapid tunneling or hopping of electrons that happen to be trapped close to solute molecules. The treatment of tunneling developed by Miller *(18–20)* shows that tunneling can be efficient even at times less than 10^{-8} sec.

[3]This value is somewhat uncertain.

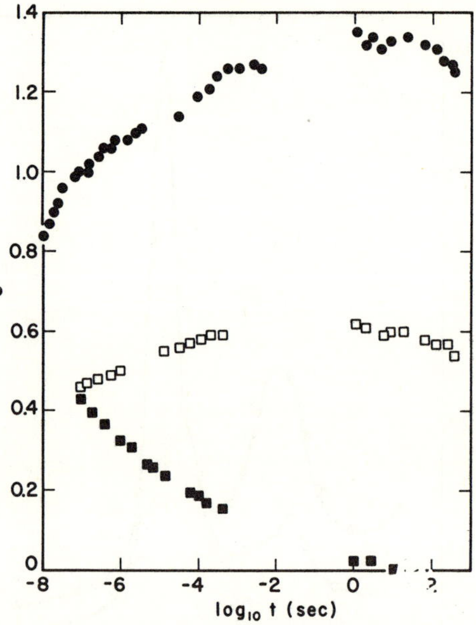

FIG. 4. Yields at various times after a pulse for a 19 mM pyrene solution in 3MH glass at 76 K: ●, pyrene$^-$ followed at 492 nm; □, pyrene$^+$ followed at 450 nm; ■ e_s^- followed at 1525 nm.

REFERENCES

1. W. H. Hamill, in *Radical Ions* (E. T. Kaiser and L. Kevan, Eds.), pp. 321–416. Wiley-Interscience, New York, 1968.
2. J. E. Willard, in *Fundamental Processes in Radiation Chemistry* (P. Ausloos, Ed.), pp. 599–650. Wiley-Interscience, New York, 1968.
3. L. Kevan, in *Actions Chimiques et Biologiques des Radiations,* Vol. 15 (M. Haissinsky, Ed.), pp. 81–143. Masson et Cie, Paris, 1971.
4. A. C. Ling and J. E. Willard, *J. Phys. Chem.*, **72,** 3349 (1968).
5. N. V. Klassen, H. A. Gillis, and G. G. Teather, *J. Phys. Chem.*, **76,** 3847 (1972).
6. J. Lin, K. Tsuji, and F. Williams, *J. Am. Chem. Soc.*, **90,** 2766 (1968).
7. W. F. Schmidt and A. O. Allen, *J. Chem. Phys.*, **52,** 2345 (1970).
8. H. A. Gillis, N. V. Klassen, G. G. Teather, and K. H. Lokan, *Chem. Phys. Lett.*, **10,** 481 (1971).
9. S. J. Rzad, P. P. Infelta, J. M. Warman, and R. H. Schuler, *J. Chem. Phys.*, **52,** 3971 (1970).
10. R. H. Schuler and P. P. Infelta, *J. Phys. Chem.*, **76,** 3812 (1972).
11. P. P. Infelta, *Some considerations on ion recombination processes in the radiolysis of hydrocarbons: application to studies at low temperature*. Ph.D. Dissertation, Carnegie-Mellon University, Pittsburgh, 1971.
12. W. H. Hamill, *J. Phys. Chem.*, **73,** 1341 (1969).
13. H. Ogura and W. H. Hamill, *J. Phys. Chem.*, **78,** 504 (1974).
14. S. J. Rzad and K. M. Bansal, *J. Phys. Chem.*, **76,** 2374 (1972).
15. A. Mozumder, *J. Chem. Phys.*, **55,** 3020 (1971).
16. G. R. Freeman, *Int. J. Radiat. Phys. Chem.*, **4,** 237 (1972).
17. K. Oka, H. Yamazaki, and S. Shida, *Bull. Chem. Soc.* Japan, **46,** 2347 (1973).
18. J. R. Miller, *J. Chem. Phys.*, **56,** 5173 (1972).

19. J. R. Miller and J. E. Willard, *J. Phys. Chem.*, **76,** 2641 (1972).
20. J. R. Miller, *Chem. Phys. Lett.*, **22,** 180 (1973).
21. J. Jagur-Grodzinski, M. Feld, S. L. Yang, and M. Szwarc, *J. Phys. Chem.*, **69,** 628 (1965).
22. W. I. Aalbersberg, G. J. Hoijtink, E. L. Mackor, and W. P. Weijland, *J. Chem. Soc.*, 3049 (1959).
23. E. Zador, J. M. Warman, and A. Hummel, *Chem. Phys. Lett.* , **23,** 363 (1973).

Formation and Decay of Trapped Electrons in Frozen Media[1]

Johan Moan[2]
Department of Biophysics
Norsk Hydro's Institute for Cancer Research
Montebello, Oslo 3, Norway

INTRODUCTION

Formation and decay of trapped electrons (e_t^-) is, indeed, a comprehensive subject. Numerous investigations have been made to elucidate the formation and decay mechanisms of e_t^- in various media, and several review articles have been written *(1–4)*. The present article is not intended to be a complete survey of the field. Only a few aspects of the subject will be discussed, and mainly polar matrices will be considered.

FORMATION OF e_t^-

The electron ejected in an ionization process is probably free or quasi-free for a certain time. During this time it loses energy by collisions, inducing ionizations and excitations in its surroundings and interacting with vibrational modes in the matrix. It may react with solute molecules present (scavengers), with solvent molecules, or become stabilized or trapped in the matrix. The electron may also recombine with its mother ion both prior and subsequent to trapping. Such a recombination process frequently leads to the formation of an excited molecule which may emit luminescence under certain conditions. Charge recombination seems to be one of the main mechanisms of production of excited states by ionizing radiation *(5)*, and occurs with a rather large probability. Thus the yield of X-ray-induced recombination luminescence of tryptophan in ethyleneglycol/water (EG/H_2O) glass is around 4.2 as measured by Steen *(5)*.

Several investigators have observed a larger phosphorescence to fluorescence ratio during X- or γ-irradiation than during UV-irradiation *(6–8)*. This indicates that triplet states are formed directly by charge recombination without intersystem crossing from the singlet manifold. Therefore, the time between ionization and recombination must be long enough to allow spin flipping of the electron. The appropriate spin relaxation time has not been measured in the absence of a magnetic field. However, it is almost certainly orders of magnitude larger than the recombination time of free ions which is shorter than 10^{-11}sec *(9)*. This consideration leads us to conclude that a substantial

[1]Congress presentation A–6–6.

[2]Fellow of the Norwegian Research Council for Science and the Humanities.

fraction of the "spontaneously" recombining electrons are temporarily trapped for a time long enough to allow spin flipping to occur. There will probably be a competition for these unstably trapped electrons between recombination, scavenging and stable trapping. When the concentration of electron scavenger increases, the yield of recombination luminescence decreases, as shown on Fig. 1. Correspondingly, the total yield of radicals increases (Fig. 1). From our scavenger studies *(10)* we conclude that roughly 50% of the electrons recombine with their mother ions in EG/H_2O at 77K.

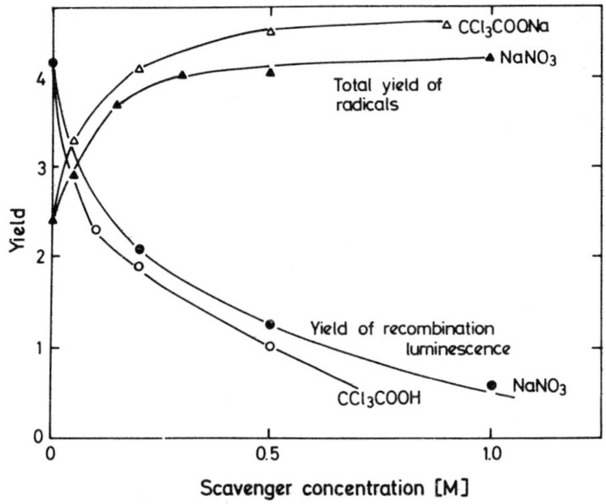

FIG. 1. The total yield of radicals (rel. units) in γ-irradiated EG/H_2O and the yield of X-ray-induced luminescence of indole in EG/H_2O as a function of scavenger concentration. The data for the total yields are taken from ref. *10* and those for the luminescence are taken from ref. *5* and ref. *57*.

Electrons in shallow traps may become stably trapped by forcing the surrounding polar solvent molecules to rotate to make the traps deeper. The mobility of molecular dipoles in a glass increases when the temperature increases. Thus, one would expect that the probability of stable trapping increases with increasing temperature while the probability of recombination and scavenging decreases. This is observed experimentally (Figs. 2, 3 and 4). Fig. 2 shows that the scavenger efficiency in EG/H_2O is markedly lower at 123K than at 77K. Fig. 3 shows the results of Higashimura et al. for ethanol *(11)*. The scavenger efficiency is around 4 times greater at 4K than at 77K, and decreases with decreasing wavelength of observation. The electrons absorbing in the infrared part of the spectrum are probably those which are in the shallowest traps and therefore most exposed to scavenging. Fig. 4 shows that the yield of e_t^- in EG/H_2O increases while the yield of recombination luminescence decreases with increasing temperature. From this figure the probability P_R of recombination may be estimated:

$$P_R^{77K} = 1.2\ P_R^{133K}$$
$$(1 - P_R^{77K}) = 0.75\ (1 - P_R^{133K})$$

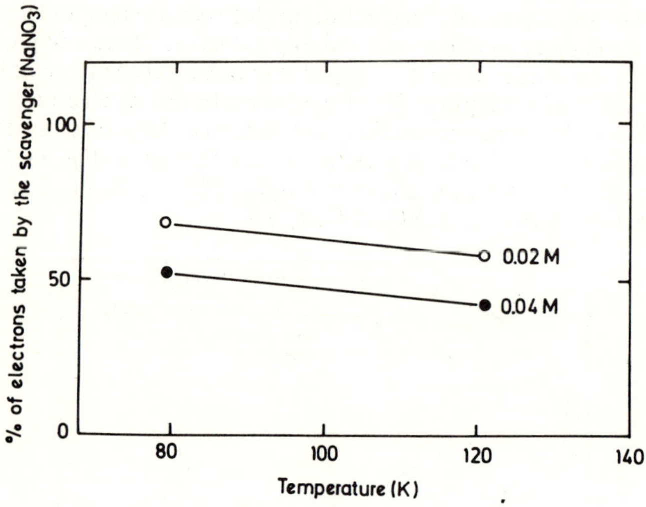

FIG. 2. Reduction in the yield of photoinduced e_t^- in EG/H_2O containing $5 \cdot 10^{-4}$M tryptophan by $NaNO_3$. Data taken from ref. *12*.

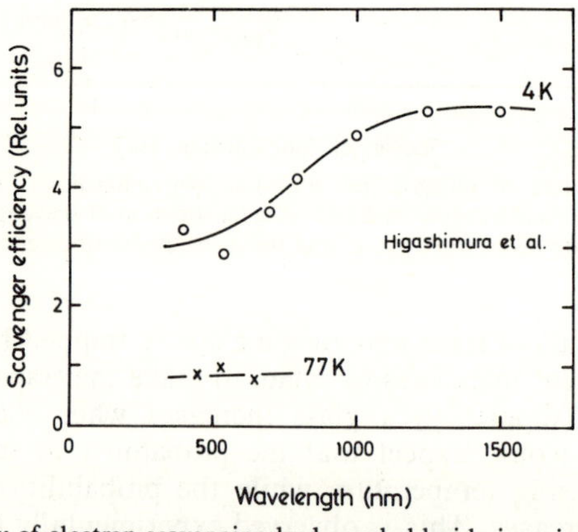

FIG. 3. Efficiency of electron scavenging by benzyl chloride in γ-irradiated ethanol at 4 and at 77K as a function of the wavelength at which the absorption is measured. Data taken from ref. *11*.

This gives $P_R^{77K} = 0.65$ which is in satisfactory agreement with the value found from scavenger studies *(10)*. The yield of thermoluminescence of tryptophan in EG/H_2O increases even more than the yield of e_t^- in the temperature region 77–120K *(12)*. Thus, the fraction of the electrons which are trapped close to their mother ions increases with increasing temperature, i.e. the mean distance between the trapped electrons and the positive ions decreases.

When exposed to visible light, e_t^- frequently disappears (bleaching). The efficiency of bleaching usually decreases with increasing temperature *(4,13,14)* indicating that the retrapping process becomes more efficient.

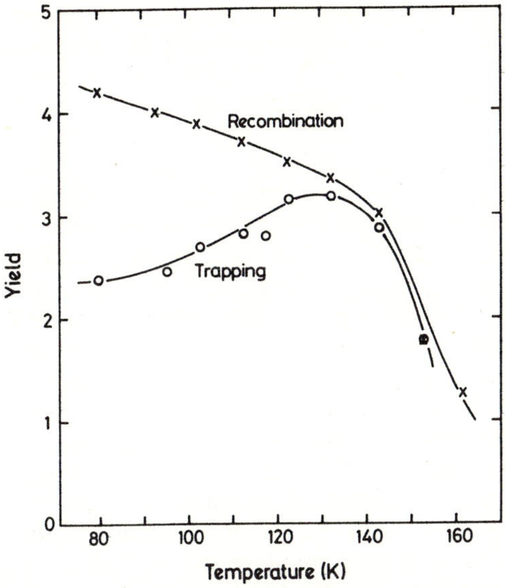

FIG. 4. The yield of radioluminescence from charge recombination in EG/H_2O containing tryptophan and the yield of trapped electrons in the same matrix as a function of the irradiation temperatures. The electron yields, induced by pulse radiolysis, were monitored by ESR-spectroscopy. The luminescence data are taken from ref. *5*.

Higashimura, Hase and coworkers have studied electron trapping at 4K and at 77K *(15–17)*. Their work definitely supports the electron trapping model with pre-existing shallow traps. In alcohols the electrons in such traps are stable at 4K and show a narrow ESR-line of width 3-6 gauss and an optical absorption spectrum extending into the infrared. On rapid warming to 77K these electrons become stably trapped, by forcing the surrounding dipoles to rotate to conform to the lowest energy configuration. In such a relaxed trap the electron is bound by around 2 eV in the alcohol matrices and have relatively strong interaction with their surroundings, as indicated by the width of the ESR-spectrum (12-15 gauss). On slow warming from 4K to 77K a substantial fraction of the electrons disappear, probably by thermally stimulated recombination with their mother ions.

Stable trapping by molecular reorientation can probably not compete with recombination if the temperature is too far below the glass transition point, i.e. if the viscosity is too high. Hase et al. *(17)* found that the yield of e_t^- in ethanol by γ-radiolysis is the same at 4K and at 77K. [The glass transition temperature of ethanol is around 95K *(18)*.] In accordance with this Dainton *(4)* reports that in 10M NaOH the yield of e_t^- is constant between 77 and 150K, while it increases by a factor of 1.6 from 150 to 210K.

The model with shallow traps and molecular reorientation is also in accordance with the pulse radiolysis work of Richards and Thomas *(19)*, of Kevan *(20)*, and of Baxendale and Wardman *(21)*. These authors found that the optical absorption spectra due to electrons in alcohols shift toward the blue on a microsecond timescale. The spectral shifts at 77K seem to occur more rapidly in the more polar matrices *(20)*. This is somewhat surprising since it is in the opposite direction of what would be expected based on the relative matrix viscosities. However, the viscosity is a macroscopic property of the matrix while the trapping process is obviously concerned with a relaxation on the microscopic scale. This was also the conclusion of Baxendale and Wardman *(21)* who studied the solvation process in alcohols near their melting points. These authors found that there is some correlation between the decay time of the infrared part of the spectrum and the expected relaxation time of the solvent molecules in the field of a unit charge. However, deviations are expected at the lower temperatures *(21)*. Thus, another explanation of the spectral shifts of e_t^- is possible, namely migration from shallow to deeper traps. In correspondence with this explanation is the observation that the decay rate of the infrared absorption is more influenced by a scavenger than the decay rate of the blue absorption *(21)*. Furthermore, there are probably pre-existing deep traps in the alcohols since electrons in such traps are observed at 4K *(17)*. However, these observations do not conclusively distinguish between the reorientation and the migration model, since the shallowly trapped electrons may tunnel to the scavenger, and molecular relaxation may occur at 4K. A possible way to attack the problem would be to create a great number of pre-existing deep traps in a matrix at 4K and compare the yield of e_t^- in this matrix with the yield in an untreated matrix. The traps might be generated by irradiating the matrix at 77K, cooling it to 4K and removing the electrons from the traps by bleaching with visible light. We performed such an experiment in EG/H^2O at 77K, but found the same yields of e_t^- in both samples. This shows either that the molecules surrounding an electron in a deep trap rotate back to random position to destroy the trap when the electron is removed or that there is no migration from shallow to deep traps. Further experiments are needed to clarify this point.

In conclusion, the msodel of electron trapping by capture of electrons in shallow pre-existing traps seems to be in correspondence with most of the experimental data. Recombination of shallowly trapped electrons with their mother ions in polar matrices at 77K seems to be competing with relaxation of the shallow traps to deeper and more stable ones, the probability of the two processes being of the same magnitude. However, the electron migration model is not excluded by the experimental data.

In future treatment of electron trapping the following possibilities should be considered: 1) The existence of voids in the matrices, and the expansion of such voids with the temperature *(4)*. The temperature dependence of the shape of the absorption spectra *(4)* and possibly also the temperature dependence of the yields of e_t^- may be related to the thermal expansion of the matrix. 2) The probability of recombination and scavenging of thermal electrons may depend on the kinetic energy. Suppose that the electron moves an appreciable part of its distance prior to trapping with a velocity corresponding to the energy kT; the crucial radius r_c where the thermal energy of the electron equals the Coulomb interaction with the positive ion will increase with the temperature. Thus, if r_c is comparable with the mean distance r_t travelled by the electrons, the probability of recombination will decrease with increasing r_c, i.e. with increasing temperature. In the polar matrices irradiated at temperatures around 77K r_t is probably much smaller than r_c *(12)*. Thus, it seems that the temperature dependence of the yields of e_t^- cannot be explained in this way. This conclusion is supported by the fact that the yield of trapped electrons in ethanol is the same at 4 and 77K *(17)*. If the interaction of mobile electrons and scavenger molecules is of the induced dipole type, the cross section of scavenging will be inversely proportional to the velocity of the electrons, i.e. to the square root of the temperature if the mobile electrons are thermalized *(11)*. Thus, the ratio of the scavenger efficiency at 4K to that at 77K should be $\sqrt{77}/\sqrt{4} = 4.4$ which is close to the experimentally observed ratio *(11)* (Fig. 3). However, the same argument may apply to the interaction of electrons and trapping centers, in which case this mechanism cannot explain the temperature dependence on the scavenger efficiencies. 3) The traps are surrounded by polar molecules oriented favorably to form potential wells, and around the potential wells there will probably be potential barriers *(11)* that increase in height with increasing depth of the corresponding wells. The probability of crossing the barriers will be dependent on the kinetic energy of the electrons. Thus, trapping in shallow traps will probably be favored notably at low temperatures. 4) The interaction of mobile electrons with lattice phonons. Since the phonon field increases with increasing temperature, this interaction will tend to diminish the mobility of electrons as the temperature increases. It seems that the data obtained in recent electron drift mobility and Hall mobility experiments on γ-irradiated 10M NaOH-glass are well characterized by a band model, where the electrons are scattered by acoustical and optical phonons.*(22)*.

DECAY OF e_t^-

The decay of trapped electrons involves a displacement of the electrons from the traps to the reaction centers which may be solute molecules, solvent molecules or cations. Three types of mechanisms for electron transfer will be discussed, namely thermal detrapping, transfer of e_t^- as an entity through a viscous medium, and tunnelling.

Thermal detrapping

The electron traps at 77K are about 2eV deep in polar matrices and 0.5 to 0.8 eV in nonpolar matrices. This may be seen from the photodetachment thresholds *(22–25)*. For comparison the magnitude of kT is $6.6 \cdot 10^{-3}$eV at 77K and $3.4 \cdot 10^{-4}$eV at 4K. The thermal activation energies determined from thermoluminescence glow curves in EG/H_2O and from the temperature dependence of the decay of e_t^- in EG/H_2O above 123K are only a few tenths of an electron volt *(26)*. In other cases the activation energy is even zero, as will be discussed below. Our conclusion is, therefore, that the thermal detrapping of electrons is improbable, notably in polar matrices.

Viscosity-dependent decay

It was previously believed that polar and nonpolar matrices at 77K were completely rigid, i.e. no molecular motion could occur. However, it seems obvious that molecular relaxation can take place in most of the trapping matrices investigated. This is clearly shown by the ESR- and optical spectra at 77K and at 4K *(15)* as well as by pulse radiolysis at 77K *(19,20)*. It is also clear that in many matrices, peaks of thermoluminescence glow curves are correlated with increased molecular motion in certain temperature regions. In Table 1 the glass transition temperatures of some matrices are compared with the temperatures of the peaks of the glow curves in the same matrices. There is obviously a correlation. A similar conclusion was reached by Mele et al. *(35)* who studied phase transitions of irradiated fluorinated polymers by thermoluminescence measurements and differential thermal analysis.

TABLE 1

Matrix	T_{TL}(K) Glow peaks		T_g(K) Glass transition
1 : 1 Ethyleneglycol/H_2O	90[a]	138[a]	143[b]
3 : 1 Methanol/H_2O	92[c]	115[c]	113[b]
Methanol	98[c] 104[d]	145[d]	103[b]
Ethanol (96%)	90[c]	(100[c])	95[e]
Water	108[f]	155[fd]	135[b] (162[g])
9M LiCl/H_2O	91[h]	142[h]	143[i]

Glass transition temperatures and temperatures at peaks of thermoluminescence glow curve in various matrices.

a) ref. *12*, *b)* ref. *27*, *c)* ref. *28*, *d)* ref. *29* and ref. *30*, *e)* ref. *18*, *f)* ref. *31*, *g)* ref. *32*, *h)* ref. *33*, *i)* ref. *34*.

Several investigators stress the effect of viscosity on electron decay. Perhaps the most clearcut demonstration of this is the work of Lin et al. *(36)*. Fig. 5 shows some of their data. The time for half decay of trapped electrons in 3-methylpentane-isopentane mixtures at 77K is dependent on the half-power of viscosity. A similar dependence was found for the luminescence decay of triphenylamine in matrices of much lower viscosities (1–1000 P) *(37)*. Ogasawara et al. *(38)* and Ichikawa et al. *(39)* found that the electron decay rate in 2-methyltetrahydrofuran was temperature dependent around 93K. The activation energy for the decay in the temperature range 92-95K was around 0.8 eV *(39)*. This is close to the activation energy 0.98 eV which is associated with viscosity changes in this matrix *(40)*. The decay of electrons in EG/H_2O at temperatures above 123K is temperature dependent with an activation energy of 0.45 eV *(26)*. The same activation energy was found for the thermal quenching of tryptophan phosphorescence in EG/H_2O. It is tempting to associate this activation energy with rotational mobility of the solvent molecules. Typical activation energies of viscosity changes of alcohol matrices are just of this magnitude, i.e. 0.5–0.9 eV as determined by Ling and Willard *(41)*.

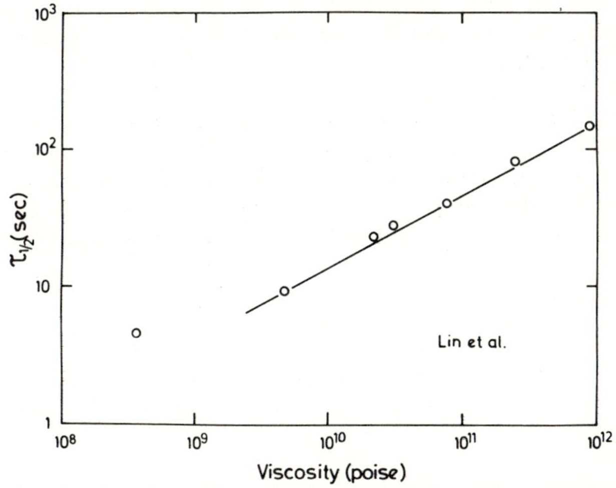

FIG. 5. Viscosity dependence of $t_{1/2}$, the time for half decay of trapped electrons in 3-MP-isopentane mixtures at 77K. Measurements by ESR spectroscopy by Lin et al. *(36)*.

The recombination of trapped electrons with positive tryptophan ions in EG/H_2O is temperature dependent even down to 77K *(26)*, which is 66 degrees below the glass transition point. We proposed a recombination model which considers e_t^- moving as an entity through the medium in the field of its mother ion *(26)*. This model was supported by the observation that H^+, which is an inefficient scavenger of mobile electrons, turned out to be very efficient in scavenging the electrons that recombined in the temperature region 110K–140K *(42)*.

The influence of an electric field on the motion of trapped electrons in organic matrices at 77K was demonstrated by the electropholuminescence experiments of Bullot and Albrecht *(43)*. Their work shows that trapped electrons in 3-methylpentane exhibit no diffusion in the θ-space but only inward diffusion in the radial direction controlled by the central field of the cation. Steen and Kongshaug *(44)* compared the scavenging efficiencies of a number of solutes towards mobile electrons at 77K with the scavenging efficiencies towards trapped electrons at 123 K in EG/H_2O. Table 2 shows data from their work. It may be noted that the normalized scavenging efficiencies of the two positive scavengers H^+ and Cu^{2+} is larger towards trapped than towards mobile electrons, while the opposite is the case of the negative scavengers NO_3^- and CCl_3COO^-. This observation also indicates that there is some effect of an electric field on the transfer of trapped electrons.

It seems that glass transition points and peaks of thermoluminescence glow curves usually are in a temperature region where the viscosity is 10^9–10^{13}P. Bernas and Truong *(45)* studied the phosphorescence lifetime τ_p of tetramethylparaphenylenediamine and benzene in various glasses as a function of the viscosities of the glasses. While τ_p was nearly constant at viscosities above 10^9P, it decreases abruptly below this viscosity. It is possible that at about 10^9P the time a molecular dipole needs to adjust itself in the field near a unit charge is of the same order of magnitude as the typical observation times in such experiments. Ling and Willard *(46)* studied the electrical conductivity of irradiated and unirradiated 2-methyltetrahydrofuran (MTHF). They found that in the temperature range of 87–100K the MTHF dipoles can be significantly oriented by an electric field of 10^2–10^4V cm^{-1}. This is approximately the field 100–1000 Å from a unit charge in MTHF. The viscosity of MTHF in the temperature region 87–100K varies from 10^{13} to 10^7P *(40)*. These considerations indicate that the mobility of solvent dipoles is high enough to allow a solvated charge to "slide" continuously through the medium, even at high viscosities.

TABLE 2

Solute	Relative scavenging efficiency Mobile electrons (77K)	Trapped electrons (123K)
HCl	0.11	3.7
$CuCl_2$	3.5	4.0
H_2O_2	1	1
$NaNO_3$	1.3	0.37
CCl_3COONa	2.2	0.6

Relative scavenging efficiencies for mobile and trapped electrons in EG/H_2O–glass from ref. *44*. The values are normalized so that the value for H_2O_2 is unity.

Electron tunnelling

There seems to exist only one type of clearcut evidence for electron tunnelling in frozen matrices, namely temperature independence of the decay rate. However, a number of other kinds of observations have been reported which indicate that tunnelling occurs.

Temperature independence is to be expected for a tunnelling process since kT is orders of magnitude smaller than the trap depth and since the depth of a relaxed trap is only slightly dependent of the temperature over a wide temperature range. It must be recalled, however, that a major fraction of the trapped electrons generated at temperatures around 4K are in shallow unrelaxed traps while complete relaxation occurs at 77K resulting in significantly deeper traps. With one exception (3-methylhexane, which exhibits the same absorption spectrum after irradiation at 73K and at 4K) this is true for all the polar and nonpolar matrices investigated *(47)*.

Kieffer et al. *(48)* studied the isothermal γ-induced recombination luminescence of biphenyl in methylcyclohexane (MCH) glass at 4.2, 66 and 77K and found that the decay rate was identical at the three temperatures. Later they included 3-methylpentane (3MP) glass in the investigation *(49)* and found the same decay rate as in MCH at 4K but a faster decay rate at 77K. This indicates that the recombination is a tunnelling process at 4K in both matrices but only in MCH at 77K. The polarity of the two matrices, as well as the optical absorption spectrum of e_t^- in each of them, is identical *(47)*. Therefore, it is reasonable that the same decay rate is found at 4K. The viscosity at 77K, however, is widely different, being $2.2 \cdot 10^{12}$P in 3MP and $1.2 \cdot 10^{18}$P in MCH *(41)*. The faster decay rate in 3MP at 77K could be due to thermal detrapping. However, if thermal detrapping occurs in 3MP it would supposedly also occur in MCH, since the trap depth seems to be identical in the two matrices. Thus, the recombination at 77K in 3MP is probably a viscosity dependent process. The absorption spectrum of e_t^- in MCH is different at 4 and 77K, indicating that the mean trap depth is different *(47)*. This taken into consideration, it is surprising that the same decay rate is found at 4 and at 77K. A complicating factor is also that the electrons seem to travel further prior to trapping at 4K than at 77K, as seen from scavenging studies *(11,50)*. Furthermore, it is confusing to note that while the yield of e_t^- at 4K usually is smaller or equal to that at 77K *(47,50)*, the yield of recombination luminescence is a factor 3-4 larger at 4K than at 77K *(49)*.

Temperature independence of the decay of e_t^- in alkaline glasses is reported by Tamaraev et al. *(51)* and by Kroh and Plonka *(14)*. Tamaraev et al. found that $[e_t^-]$ decreased linearly with the logarithm of the decay time and at the same rate at 4.2 and 77K. Corresponding results were found by Kaalhus (*52*) for the decay of e_t^- in EG/H_2O in the temperature range 77-123K. In this matrix e_t^- decays by reaction with the ethyleneglycol molecules, the total number of radicals being constant below 123K. Kroh and Plonka *(14)* found the same decay rate of γ-induced e_t^- in 8M NaOH at 113 and 173K. This is somewhat in disagreement with the findings of Dainton and Gopinathan *(54)*, which show that the decay is temperature dependent at least above 163K. However, the discrepancies may be due to errors in the temperature measurements. Ershov and Kieffer *(54)* found that the decay rate of isothermal recombination luminescence of phenol in 10M NaOH was identical at 4 and 77K.

The time dependence of the decay may indicate the mechanism of decay. Thus, in a tunnelling process the decay rate decreases strongly with increasing time, unless there is a wide distribution of trap depths or unless all the electrons tunnel nearly the same distance. This may be illustrated by the following example. According to Miller *(55)* the tunnelling rate constant in 3 MP should be approximately given by

$$\log_{10}k = 15 - 0.443a\sqrt{V_o - E}\ ,$$

where a is the barrier width in Å and $V_o - E$ is the photobleaching threshold in eV. During a decay of 1 hour, 55% of the trapped electrons decay in this matrix at 77K and the rate constant decreases from $6 \cdot 10^{-4}s^{-1}$ to $1.2 \cdot 10^{-4}s^{-1}$ *(36)*. If we assume that there is a distribution of the photobleaching thresholds between 0.5 and 0.6 eV, we see that 55% of the electrons must be trapped 45 to 59Å from the centers to which they tunnel. This is unlikely and it seems that tunnelling is not the main mechanism of decay in this matrix, a conclusion which is in agreement with the observation that the decay is temperature dependent around 77K *(36,49)*. In other matrices, however, a decay is observed that is compatible with electron tunnelling. Thus the decay of e_t^- in 10M Na

OH at 4.2K and at 77K *(51)*, and the decay in EG/H_2O at 77K and at 123K *(52)* approximately fit the expression $[e_t^-] = [e_t^-]_o (K_1 - K_2 \ln t)$. Furthermore, the decay rate of e_t^- in EG/H_2O is faster at temperatures below ca. 130K than would be expected if the decay were thermally stimulated or purely a viscosity dependent process *(26)*.

In matrices where almost no decay of electrons occurs in the absence of scavengers, transfer of electrons from traps to scavenger moslecules has been observed. This was found by Miller *(55)* in MTHF at 77 and 87K, by Steen in EG/H_2O in the temperature region 77–113K (manuscript in preparation), and by Kroh and Plonka *(14)* in 8M NaOH at 113 and at 173K. This effect is observed for positively as well as for negatively charged scavengers, and can therefore not be explained only by the motion of the electrons in the electric field of the scavengers.

Miller *(55)* proposed a model for electron scavenging in organic glasses which involves tunnelling of the trapped electrons to the scavengers. This model is consistent with the trapping/recombination model described in the present work, and seems to be in accordance with several experimental observations, such as a) the ability of a scavenger of low electron affinity to increase the yield of the anion of a scavenger of high electron affinity *(55)*, b) the linear dependence of the logarithm of the yields of e_t^- on the scavenger concentration *(42,55)*, c) the observed time dependence of electron capture *(55)* and, finally, d) the observations that scavenger efficiencies usually decrease with increasing temperature *(11,12,* and H. B. Steen, in preparation*)* (Figs. 2 and 3) and with increasing polarity of the matrix *(56)*. The depths of the pre-existing traps will probably increase with increasing polarity of the matrix. Correspondingly, the rate of tunnelling of the shallowly trapped electrons to the scavenger will decrease and the probability of permanent trapping by dipole relaxation will increase with increasing polarity.

In conclusion, several reported experimental observations indicate that electron tunnelling occurs in low temperature glasses, but few of them are to be regarded as conclusive. Tunnelling is probably important in matrices below their glass transition temperatures, notably in matrices with high viscosities.

ACKNOWLEDGEMENTS

The author appreciates helpful discussions with Lic. techn. Olav Kaalhus.

REFERENCES

1. B. G. Ershov and A. K. Pikaev, *Radiat. Res. Rev.* **2**, 1 (1969).
2. A. Ekström, *Radiat. Res. Rev.* **2**, 381 (1970).
3. J. E. Willard, *Science* **180**, 553 (1973).
4. F. S. Dainton, *Chemie* **75**, 608 (1971).
5. H. B. Steen, *Israel J. Chem.* **8**, 227 (1970).
6. B. Brocklehurst, G. Porter and J. M. Yates, *J. Phys. Chem.* **68**, 203 (1964).
7. H. B. Steen, *Photochem. Photobiol.* **8**, 47 (1968).
8. P. I. Høynnås and H. B. Steen, *Photochem. Photobiol.* **11**, 67 (1970).
9. A. H. Samuel and J. L. Magee, *J. Chem. Phys.* **21**, 1080 (1953).
10. O. Kaalhus and J. Moan, *Int. J. Radiat. Phys. Chem.* (1975, in press).

11. T. Higashimura, A. Namiki, M. Noda and H. Hase, *J. Phys. Chem.* **76**, 3744 (1972).
12. J. Moan, *Acta Chem. Scand.* **26**, 897 (1972).
13. J. Moan, *Israel J. Chem.* **9**, 637 (1971).
14. J. Kroh and A. Plonka, *Int. J. Radiat. Phys. Chem.* **6**, 211 (1974).
15. T. Higashimura, M. Noda, T. Warashina and H. Yoshida, *J. Chem. Phys.* **53**, 1152 (1970).
16. H. Hase, M. Noda and T. Higashimura, *J. Chem. Phys.* **54**, 2975 (1971).
17. H. Hase, T. Warashina, M. Noda, A. Namiki and T. Higashimura, *J. Chem. Phys.* **57**, 1039 (1972).
18. C. A. Angell, *J. Chem. Ed.* **47**, 583 (1970).
19. J– T. Richards and J. K. Thomas. *J. Chem. Phys.* **53**, 218 (1970).
20. L. Kevan, *J. Chem. Phys.* **56**, 838 (1972).
21. J. H. Baxendale and P. Wardman, *J. Chem. Soc.* **69**, 584 (1973).
22. T. Huang, I. Eisele and L. Kevan, *J. Chem. Phys.* **59**, 6334 (1973).
23. J. Moan, *Chem. Phys. Lett.* **17**, 565 (1972).
24. A. Bernas, D. Grand and T. B. Truong, *J. Chem. Soc. D: Chem. Commun.* **13**, 759 (1972).
25. J. R. Miller and J. E. Willard, *J. Phys. Chem.* **76**, 2341 (1972).
26. J. Moan, *J. Lumin.* **6**, 256 (1973).
27. D. H. Rasmussen and A. MacKenzie, *J. Phys. Chem.* **75**, 967 (1971).
28. J. Moan, Cand. Real. Thesis, Trondheim University (1970).
29. B. G. Ershov, S. A. Puntezhis, A. K. Pikaev and V. I. Spitsyn, *Dokl. Phys. Chem. SSSR* **209**, 889 (1973).
30. W. Roszak and J. Kroh, *Radiochem. Radioanal. Lett.* **7**, 253 (1971).
31. L. I. Grossweiner and M. S. Matheson, *J. Chem. Phys.* **22**, 1514 (1954).
32. A. A. Miller, *Science* **163**, 1325 (1969).
33. J. Moan, *J. Chem. Phys.* (1975, in press).
34. C. A. Angell and E. J. Sare, *J. Chem. Phys.* **52**, 1058 (1970).
35. A. Mele, A. Delle Site, C. Bettinali and A. Di Domenico, *J. Chem. Phys.* **48**, 3297 (1968).
36. J. Lin, K. Tsuji and F. Williams, *J. Am. Chem. Soc.* **90**, 2766 (1968).
37. J. A. Leone and W. H. Hamill, *J. Chem. Phys.* **49**, 5294 (1968).
38. M. Ogasawara, K. Ohno, K. Hayashi and J. Sohma, *J. Phys. Chem.* **74**, 3221 (1970).
39. T. Ichikawa, H. Yoshida and K. Hayashi, *Bull. Chem. Soc. Japan* **46**, 812 (1973).
40. A. C. Ling and J. E. Willard, *J. Phys. Chem.* **72**, 1918 (1968).
41. A. C. Ling and J. E. Willard, *J. Phys. Chem.* **72**, 3349 (1968).
42. J. Moan, *Acta Chem. Scand.* **26**, 399 (1972).
43. J. Bullot and A. C. Albrecht, *J. Chem. Phys.* **51**, 2220 (1969).
44. H. B. Steen and M. Kongshaug, *J. Phys. Chem.* **76**, 2217 (1972).
45. A. Bernas and T. B. Truong, *Photochem. Photobiol.* **15**, 311 (1972).
46. A. C. Ling and J. E. Willard, *J. Phys. Chem.* **73**, 2408 (1969).
47. H. Hase, T. Higashimura and M. Ogasawara, *Chem. Phys. Lett.* **16**, 214 (1972).
48. F. Kieffer, C. Meyer and J. Rigaut, *Chem. Phys. Lett.* **11**, 359 (1971).
49. F. Kieffer, C. Meyer and J. Rigaut, *Int. J. Radiat. Phys. Chem.* **6**, 79 (1974).
50. H. Yoshida, M. Ogasawara, T. Warashina and T. Higashimura, *J. Chem. Phys.* **56**, 4238 (1972).
51. K. I. Zamaraev, R. F. Khairutdinov, A. I. Mihailov and V. I. Gol'danskii, *Dokl. Akad. Nauk. USSR* **199**, 640 (1971).
52. O. Kaalhus, Kjeller Report KR146, 17 (1972).
53. F. S. Dainton and C. Gopinathan, *Trans. Farad. Soc.* **65**, 151 (1969).
54. B. G. Ershov and F. Kieffer, *Chem. Phys. Lett.* **25**, 576 (1974).
55. J. R. Miller, *J. Chem. Phys.* **56**, 5173 (1972).
56. T. Sawai, Y. Shinozaki and G. Meshitsuka, *Bull. Chem. Soc. Japan* **45**, 984 (1972).
57. H. B. Steen, *J. Phys. Chem.* **74**, 4059 (1970).

Electron Tunneling in Rigid Media[1]

J. Kroh and Cz. Stradowski
Institute of Radiation Chemistry
Technical University
Łódź, Wróblewskiego 15, Poland

INTRODUCTION

Frozen aqueous solutions of electrolytes are well known as the matrices efficiently stabilizing electrons (*1,2*). It has been commonly assumed that at sufficiently low temperatures they can trap electrons for practically infinitely long periods of time. The "mobilization" of trapped electron (e_t^-) must be assisted either by the absorption of photon with adequate energy or by heating the matrix and consequently destructing the trap (*3,4*). However, in the presence of certain electron acceptors the slow decay of e_t^- is observed at 77°K (*5–7*), without any contribution of light or thermal destruction of the trap. Reaction of trapped electrons, e_t^-, proceeds under conditions which do not allow reagent collisions in a classical sense, since neither the diffusion of scavenger molecule nor of electron together with its trap seem probable. Several authors (*8,9*) have paid attention to the fact that in solid state the life-time of closely distributed reagent pairs is relatively long and their separation remains constant for long periods of time. This situation favors certain transfer mechanisms which hardly can be taken into account for gases or liquids. One of the often postulated mechanisms responsible for the transfer of electrons to a scavenger molecule in rigid media is quantum mechanical tunneling. Numerous unusual kinetic features of electron decay have been recently explained in terms of subbarrier transfer (*6,7,10–12*). The most important are: time and scavenger concentration dependence of e_t^- decay (*6,7,10,11*), temperature dependence (*12*) and wavelength dependence (*13*). The aim of the present work is to elaborate the kinetic model of electron tunneling and to test its applicability for reactions of e_t^- in frozen aqueous solutions.

EXPERIMENTAL

Aqueous solutions of NaOH, ethylene glycol, K_2CO_3, LiCl, $NaClO_4$ and HCOONa were prepared from analar reagents and doubly distilled water. All other reagents were of analar grade and used without further purification. The decay of e_t^- was examined in the presence of various electron scavengers such as NO_3^-, CrO_4^{-2} and $C_6H_5O^-$. Disc ice samples, about 2 mm thick and 12 mm in diameter, were prepared at 77°K. The samples containing scavengers were not deaerated, since deaeration does not affect the results. The samples were supported in washerlike rings without optical windows. The irradiations were carried out either with ^{60}Co γ-rays at a dose 7.9×10^{-2} Mrad/hour or with

[1]Congress presentation A–6–7.

X-rays at a dose rate 2.4 Mrad/hour. The total dose was 0.2 Mrad. The optical absorption spectra were recorded by using an optical glass Dewar vessel and a Beckman DK 2A spectrophotometer.

All the operations with irradiated samples were performed at 77° K in the dark.

For the pulse radiolysis experiments the samples were frozen in suprasil cells with thickness 0.5 cm or 1 cm. The sample was immersed in liquid nitrogen in a quartz Dewar vessel with suprasil windows. A linear accelerator was used as the source of 13 MeV electron pulses. The conventional analyzing system consisted of 400 W Xenon lamp, filter, system of mirrors, quartz monochromator and P-28 phototube. Between the Xenon lamp and the sample the glass filters with transmission maxima in the range 360 to 720 mm were placed. In order to check the influence of the intensity of analyzing light, metal nets reducing the light intensity by a factor 2.5 were placed in front of the filter.

The duration of the pulse was 0.5 to 5.5 μs. The dose per 2.5 μs pulse was ca. 37 krad, as measured by KCNS dosimeter.

THEORY

The Rate Constant

The probability, *T*, of tunnel transfer through potential barrier, as shown in Fig. 1, may be expressed (*14*) as (Eq. 1):

$$T = \frac{16(E-V_1)^{1/2}(E-V_3)^{1/2}\exp\left\{-2a\left(\frac{8\pi^2 m}{h^2}(V_2-E)\right)^{1/2}\right\}}{\left((E-V_1)^{1/2}+(E-V_3)^{1/2}\right)^2+\left((V_2-E)^{1/2}-\frac{(E-V_1)^{1/2}(E-V_3)^{1/2}}{(V_2-E)^{1/2}}\right)^2}$$

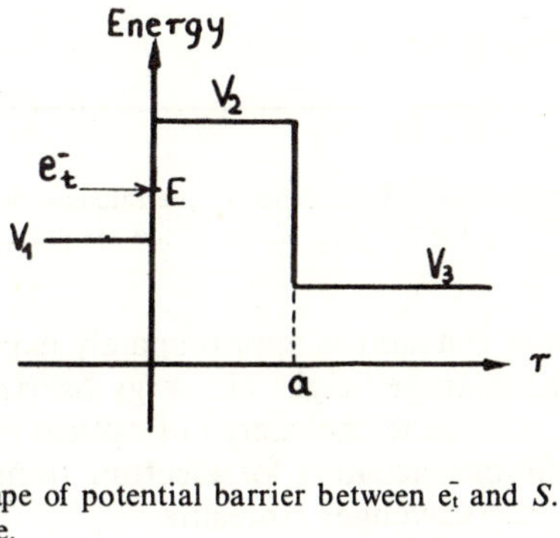

FIG. 1. Assumed shape of potential barrier between e_t^- and *S*. *V* — potential energy. *E* — energy of e_t^- in ground state.

The preexponential term changes only slightly with the parameters of the barrier. The probability of electron transfer, *T*, is determined by the exponential term, especially by the barrier width, *a*.

The rate constant of tunneling, k, may be represented as a product of frequency of electron collisions with the barrier, ν, and the transfer probability, T.

$$k = \nu \cdot T \tag{2}$$

The factor ν corresponds to the kinetic energy of electron in its ground state.

In fact, the situation of electron in the trap is more complicated since the trap is three-dimensional and the barrier is not rectangular. More realistic three-dimensional treatment[2] based on the semicontinuum model of e_t^- (*15*) allowed us to calculate with better approximation the rate constant of tunneling as a function of the width of the barrier, a. The calculations have been performed for frozen 8 *M* NaOH solution at 77° K containing nitrate ion as a scavenger. The scavenger and trapped electron radii have been assumed to be 1.5 Å and 2.2 Å, respectively. The illustration of the rate of tunneling, k, as plotted vs. pair separation, r, in 8 *M* NaOH is given in Fig. 2.

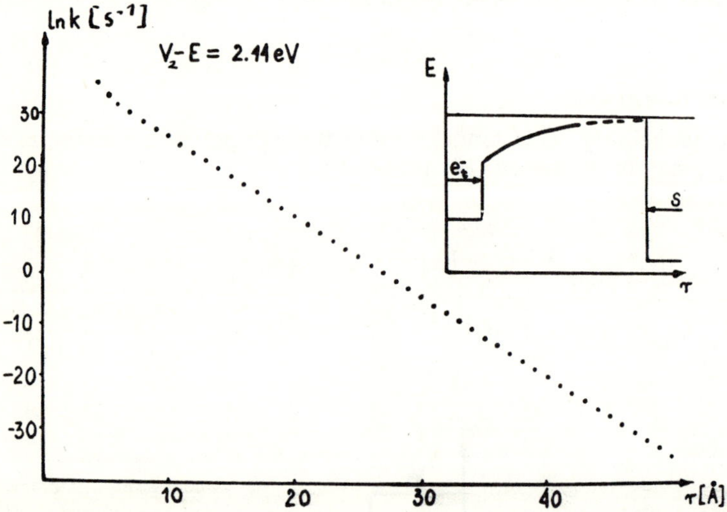

FIG. 2. The theoretically predicted dependence of the rate constant of electron tunneling to NO_3^- ions in 8 *M* NaOH at 77°K. Inset: The shape of the assumed potential barrier between electron and scavenger molecule.

As it is seen, the rate constant is approximately exponentially dependent on r for $r > 10$ Å, and an average height of energy barrier may be determined as 2.45 eV. This energy is close to the energy of optical excitation, 2.2 eV. The above mentioned calculations account for electron transfer to only one free resonant energy level in the scavenger molecule.

In fact, the scavenger molecules have often many electronic states in close vicinity of the resonant level because of electron oscillations, vibrations and electron phonon interactions. Thus, the rate constant k should be multiplied by

[2]W. Bartczak, J. Kroh, and Cz. Stradowski; unpublished calculations.

a coefficient representing the number of energy states in a scavenger molecule corresponding to energy of e_t^- in its ground state.

The Kinetics

In the irradiated system we have no possibility to investigate the reactions between reagents, separated by the same and only one distance. In contrast, the electron-scavenger separation varies within certain limits and its distribution (i.e. distribution of barrier widths) determines to a large extent the rate of tunneling. The first dependence representing the distribution of pair e_t^--S separations was based on reference (*10*). However, this type of distribution is adequate rather to the case of nearest neighbor distance between molecules of the same kind. The illustration of such distribution is given in Fig. 3. The exponential dependence derived in reference (*10*) is true if all the probabilities of electron scavenging by successive scavenger molecules are treated as corresponding to quite independent and equivalent events. Having this in mind we have suggested another distribution, shifted towards shorter pair separation.[3]

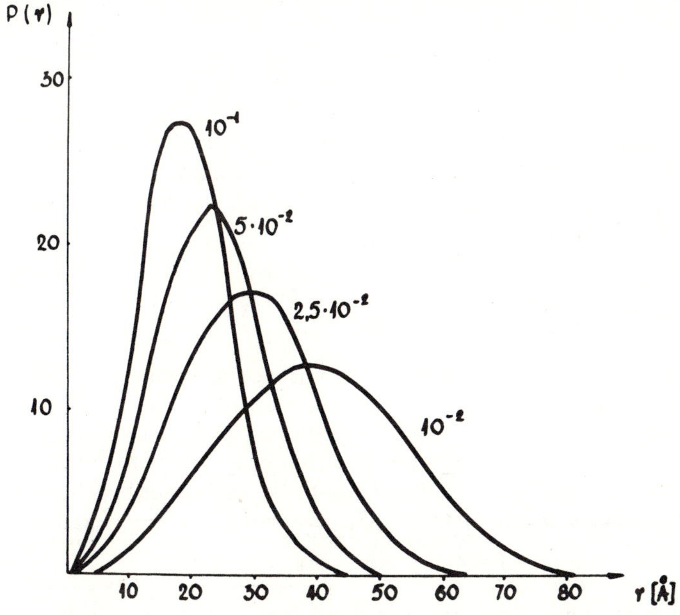

FIG. 3. Probability of electron stabilization at distance r from scavenger as a function of r.

Let us assume that each group of reagent pairs, with separation r, react according to first order kinetics, and that the corresponding rate constant is given by Eq. 2. This rate constant, $k(r, V_2-E)$, exponentially decreases while the distance r and square root of the height of the barrier $V_2 - E$, increases. The measurable electron concentration $[e_t^-]$ represents the total sum of electrons belonging to all the different groups.

[3]W. Bartczak, J. Kroh, and Cz. Stradowski, to be published.

Let us denote the total electron concentration at time $\tau = 0$ as $[e_t^-]^o$. The concentration of electrons with separation r after time τ, $[e_t^-]_r$, may be given as:

$$[e_t^-]_r = [e_t^-]^0 \, P(r) \, e^{-k(r, V_2 - E)} \tag{3}$$

where $P(r)$ is the probability of electron being stabilized at a distance r from the scavenger molecule. The measurable concentration of e_t^- after time τ may be expressed as:[4]

$$\frac{[e_t^-]}{[e_t^-]^0} = \int_0^\infty P(r) \, e^{-k(r, V_2 - E)\tau} \, dr \tag{4}$$

Let us take simultaneously into account the possibility of scavenging "dry" electrons before they become trapped. This process proceeds according to equation (*5*):

$$\frac{G^0}{G} = 1 + \frac{k_s[S]}{k_t[T]} \tag{5}$$

where S and T denote scavenger and trap respectively, and k_s and k_t rate constants of scavenging and trapping of "dry" electrons.

Scavenging of both dry and trapped electrons may be described by the following equation:

$$\frac{[e_t^-]}{[e_t^-]_0^0} = \frac{1}{1 + (k_s[S]/k_t[T])} \int_0^\infty P(r) \, e^{-k(r, V_2 - E)\tau} \, dr \tag{6}$$

where $[e_t^-]_0^0$ is e_t^- concentration in the absence of scavenger. There are many indications that the trap depth and, consequently, the height of the barrier varies also according to a certain distribution G(E). Such distribution could be estimated from the absorption spectrum of e_t^-. For 8 M NaOH it has been found that the "red" part of the spectrum is well described by a gaussian of type:

$$A = A_{max} \exp[-b(\epsilon_{max} - \epsilon)^2] \tag{7}$$

where A is the optical density, ϵ is the energy of the light absorbed, A_{max} and ϵ_{max} are the optical density and energy of light corresponding to the maximum, respectively. Eq. 7 is illustrated in Fig. 4. Knowing the distribution of the trap depths one can introduce it to the kinetic model of scavenging.

The final expression, taking into account all the above mentioned effects, has the following form:

$$\frac{[e_t^-]}{[e_t^-]_0^0} = \frac{1}{1 + (k_s[S]/k_t[T])} \int_0^\infty \int_0^\infty P(r) \, G(E) \, e^{-k(r, V_2 - E)\tau} \, dr \, dE \tag{8}$$

[4]J. Kroh and Cz. Stradowski, to be published in *Int. J. Radiat. Phys. Chem.*

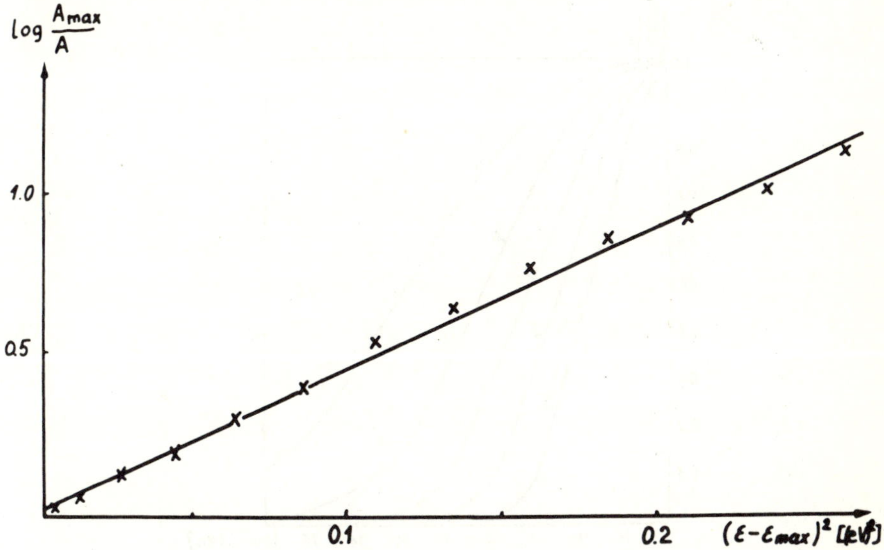

FIG. 4. The graphical test of Eq. 7. for red part of e_t^- spectrum of e_t^- in 8 *M* NaOH at 77°K.

For any given matrix and scavenger the right side integral, I(S, τ), depends only on scavenger concentration and time. Thus, one can write Eq. 8 in a simplified form as:

$$\frac{[e_t^-]}{[e_t^-]_o^o} = \frac{I(S,\tau)}{1 + (k_s[S]/k_t[T])} \tag{9}$$

The function I (S, τ) has been calculated numerically and for 8 *M* NaOH containing NO_3^- ions, its graphical representation is shown in Fig. 5.

According to Eq. 1, if $E < V_3$, the tunneling is impossible and I (S, τ) = 1. Such situation is characteristic for selective "dry" electron scavengers. If $k(r, V_2 - E)$ = const., i.e. if the decay kinetics is of classic collision type, one obtains the equation:

$$\frac{[e_t^-]}{[e_t^-]_o^o} = \frac{e^{-k\tau}}{1 + (k_s[S]/k_t[T])} \tag{10}$$

where the numerator describes the decay of electrons in time and the denominator represents the diminution of initial concentration of electrons (in respect to $[e_t^-]_o^o$) in consequence of competition between scavenging and trapping.

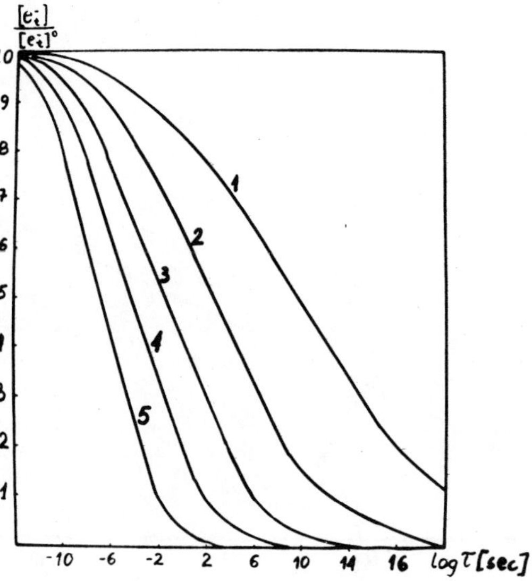

FIG. 5. Graphical representation of I (S, τ): theoretically predicted dependence of $[e_t^-]$ on logarithm of time. 1 — $10^{-2}M$; 2 — 2.5×10^{-2} M; 3 — 5×10^{-2} M; 4 — 10^{-1} M; 5 — 0.2 M.

RESULTS

The yield of trapped electrons in frozen 8 M NaOH, $G_{e_t^-} = 2.2 / G_{e_t^-} \times \epsilon_{max} = 4.2 \times 10^4 \pm 0.2 \times 10^4$, $\epsilon_{max} = 19{,}000$ (*16*) diminishes in the presence of various scavengers. In order to establish the influence of scavengers we have examined the dose dependence of optical density. The typical electron accumulation curves for two groups of scavengers are shown in Figs. 6 and 7. Large

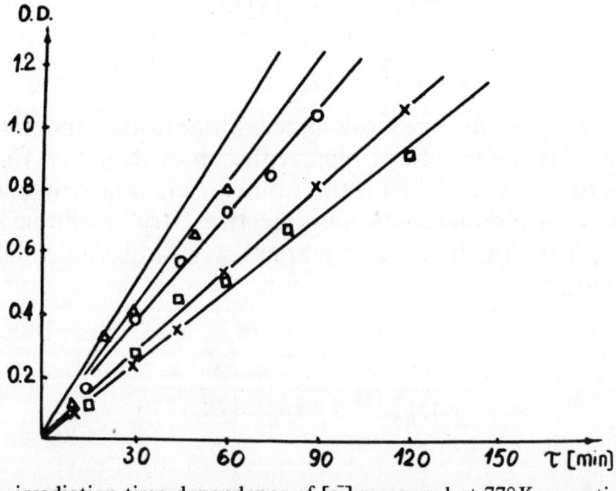

FIG. 6. γ — irradiation time dependence of $[e_t^-]$ measured at 77°K as optical density, O.D., for 585 nm. Solid line: 8M NaOH ice. Remaining lines correspond to the following concentrations of sodium phenolate added; × — 0.1 M; o — 0.25 M; + — 0.5 M; □ — 1.0 M. The dose rate was $7.9 \cdot 10^{-2}$ Mrad/h.

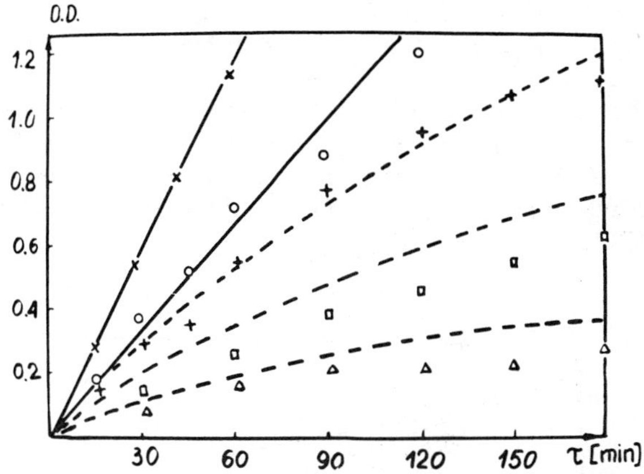

FIG. 7. γ — irradiation time dependence of $[e_t^-]$ measured for alkaline ice as optical density, O.D., at 585 nm. Concentration of added $NaNO_2$: × — 0, o — $5 \cdot 10^{-3}$ *M* $NaNO_2$; + — 10^{-2} *M* $NaNO_2$; □ — $2.5 \cdot 10^{-2}$ *M* $NaNO_2$; Δ — $5 \cdot 10^{-2}$ *M* $NaNO_2$. The dose rate was $7.9 \cdot 10^{-2}$ Mrad/h.

additions of inefficient scavengers, such as phenolate and chloroacetate, give the linear dependence (Fig. 6). In contrast to this, efficient scavengers, such as NO_3^-, NO_2^-, CrO_4^{-2} $[Fe(CN)_6]^{-3}$ exhibit different behavior (Fig. 7).

During storage of the irradiated sample in the dark at liquid nitrogen temperature, the optical density corresponding to e_t^- decreases. Such effect has not been observed in the absence of scavenger in the sample.

Thus, the decay must be due to the scavenging of trapped electrons. The typical kinetics of this process is shown in Fig. 8.

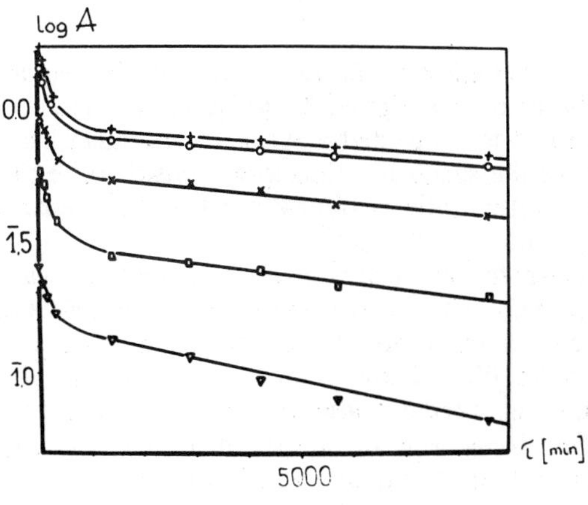

FIG. 8. The decrease of optical density O.D. for e_t^- in 8 *M* NaOH containing $2.5 \cdot 10^{-2}$ *M* $NaNO_2$ as measured at different wavelengths: + — 600 nm; o — 650 nm; × — 700 nm; □ — 750 nm; Δ — 800 nm.

The samples with NO_3^-, NO_2^-, and $[Fe(CN)_6]^{-3}$ exhibit similar behavior, while CH_2ClCOO^- and $C_6H_5O^-$, even added in high concentrations approaching 1 *M*, were much less effective towards e_t^-.

The decrease of total $[e_t^-]$ calculated as integrated optical density with the logarithm of time, is demonstrated in Fig. 9, for six aqueous matrices containing 2.5×10^{-2} *M* NO_2^-. The rate of e_t^- decay increases in the following sequence: 50% ethyleneglycol, 5 *M* K_2CO_3, 8 *M* HCOONa, 8 *M* NaOH, 15 *M* LiCl, 8 *M* $NaClO_4$.

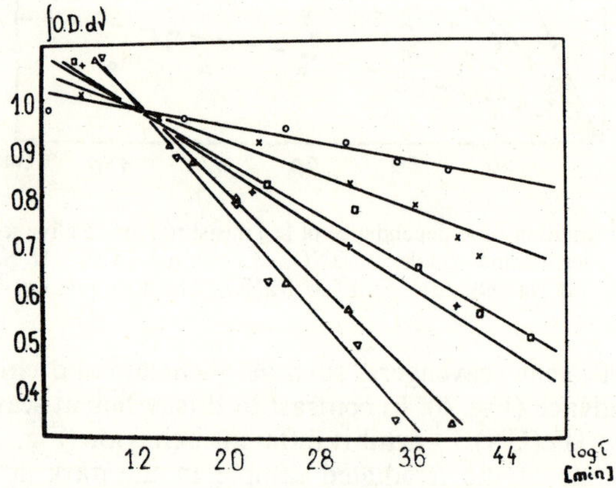

FIG. 9. The decrease of total electron concentration measured as the area under the optical density curve, ∫ O.D. dν, during the storage of irradiated, frozen aqueous matrices containing 2.5×10^{-2} *M* NO_2^-: o — 50% ethylene glycol; x — 5 *M* K_2CO_3; □ — 8 *M* HCOONa; + — 8 *M* NaOH; Δ — 15 *M* LiCl; ∇ — 8 *M* $NaClO_4$.

The decay is normalized to the same level at time equal to 20 min. The decay of e_t^- may be also investigated by pulse radiolysis. The spectrum of the trapped electron has been recorded after a 0.5 μs pulse. As it is seen in Fig. 10, the "red" part of the spectrum undergoes considerable changes in the first seconds after the pulse until it finally reaches the "normal" shape known from γ-radiolysis work.

Fig. 11 demonstrates the decrease of $[e_t^-]$ in μs time-scale for 8 *M* NaOH containing 0.10 *M* NO_3^-. The decay curves were recorded with (a) and without (b) the filter placed in the front of the sample in order to reduce intensity analyzing light. With filtered analyzing light the decay has been markedly slowed down. However, further reduction of light intensity did not affect the decay kinetics. The decay of e_t^- in the presence of CrO_4^{-2} appears to be more sensitive towards light, as it may be seen in Fig. 12.

On a μs time-scale and in the case of filtered light the decay is completely hampered; however, it is still observed in ms times. In contrast to this, e_t^- is completely stable in the system containing $C_6H_5O^-$ ions independently of the time of storage and scavenger concentration up to 1 *M*.

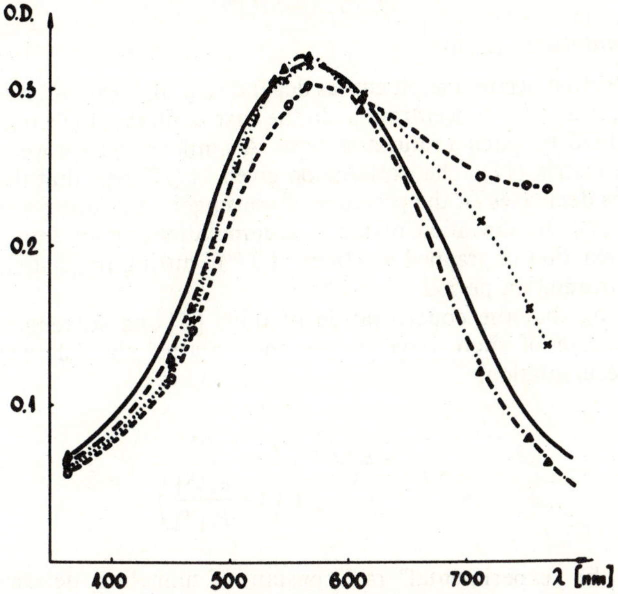

FIG. 10. The changes of e_t^- spectra after 0.5 μs pulse. Solid line — the spectrum obtained from γ — radiolysis; o — after 5 μs; x — after 0.1 s; Δ — after 3 s.

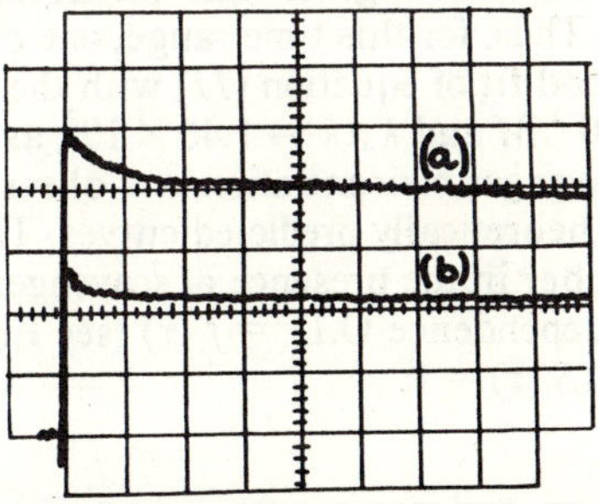

FIG. 11. Oscilloscope trace of decay of e_t^- in 8 M NaOH with 0.1 M NO_3^- at 565 nm
(a) without filtering, optical path 1 cm, 100 μs/div, 100 mV/div.
(b) with glass filter, optical path 1 cm, 100 μs/div, 50 mV/div.
The light intensity is four times reduced as compared to (a).

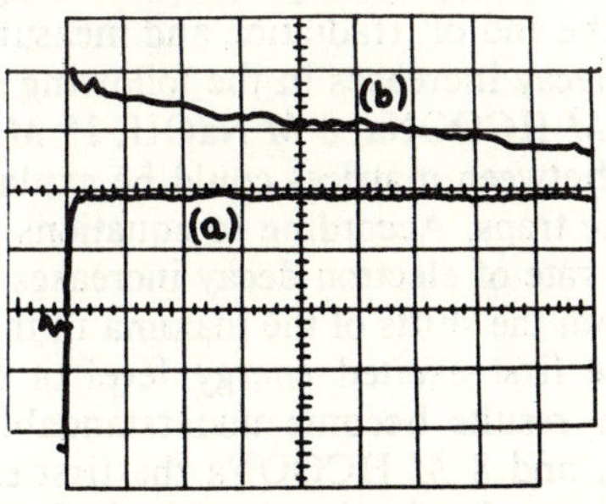

FIG. 12. Oscilloscope trace of decay of e_t^- in 8 M NaOH with 10^{-2} M CrO_4^{-2} at 565 nm, with filtered analyzing light. Optical path 1 cm, 20 mV/div; (a) 10 μs/div, (b) 0.5 s/div.

DISCUSSION

The Accumulation Curves

The electron accumulation curves depend to a high extent on the nature of the scavenger and its concentration. In the case of efficient electron scavengers the curves tend to reach a saturation level. A similar situation was observed in perchlorate matrix (*17*). The explanation given in (*17*) was that the concentration of traps decreases in the presence of scavenger. According to our original suggestion (*6*), the deviation of the e_t^- accumulation curves from linearity is caused by reaction of trapped electrons at 77° K during irradiation and in the early post-irradiation period.

Assuming that the concentration of traps $[T]$ and scavenger $[S]$ do not change because of their large excess one obtains[4] the following equation for the e_t^- accumulation:

$$[e_t^-] = \frac{[e_t^-]_0^0}{\tau} \frac{1 - e^{-k\tau}}{k\left(1 + \dfrac{k_s[S]}{k_t[T]}\right)} \tag{11}$$

where k is the "experimental" rate constant of tunneling, determined as the slope of the decay curve in a semilogarithmic plot against time (τ). As it is seen from Fig. 8, for a certain range of time the decay is linear when represented in semilog scale. Thus, for this time range, one can calculate the experimental rate constant. Good fit of equation (*11*) with the experimental data may be obtained for $[T] = 10^{-2}$ *M* and $k_s/k_t = 1.46 \times 10^2$, as may be seen from Fig. 7. However, for high scavenger concentration one observes deviation of experimental points from the theoretically predicted curves. This may be caused by a decrease of the trap number in the presence of scavenger. In the case of phenolate one obtains linear dependence O.D. $= f(\tau)$ (see Fig. 6) since tunneling of e_t^- does not occur and I $(S, \tau) = 1$.

The Decay Curves

One may conclude on the basis of Fig. 5 that a major part of electrons should decay linearly with the logarithm of time. The decrease of total $[e_t^-]$, calculated as integrated optical density and plotted against the logarithm of time elapsed between the end of irradiation and measurement, is demonstrated in Fig. 9. The rate of decay increases in the following sequence: 50% ethylene glycol, 5 *M* K_2CO_3, 8 *M* HCOONa, 8 *M* NaOH, 15 *M* LiCl and 8 *M* $NaClO_4$. The above differences between matrices could be explained by differences between the depths of the traps. According to equations 2 and 4, the trap depth should decrease as the rate of electron decay increases. However, such a trend can not be deduced from the shifts of the maxima in the corresponding optical spectra of e_t^-. Still, the first excited energy level of e_t^- may be bound with different energies. Our results become understandable by assuming that in 50% EG, 5 *M* K_2CO_3, and 8 *M* HCOONa the first excited energy level lies below the bottom of the conduction band, while in 8 *M* NaOH 15 *M* LiCl and 8 *M* $NaClO_4$ it is located in the conduction band or very close to it.

Spectral Changes

In contradiction to the results demonstrated in Fig. 10, the previous pulse radiolysis observations of the spectra of e_t^- in 10 *M* NaOH at 77°K did not indicate any spectral changes (*18*) or only slight shifts in the "red" part of the spectrum (*19,20*). Absorption in the red part of the spectrum is probably due to the electrons located in relatively shallow traps. Diminution of optical density in this part of the spectrum is not compensated by equivalent enhancement of optical density at maximum and shorter wavelengths. Similar situation has been observed in alcohols (*21*). One possible explanation is that the weakly trapped electron easily reacts with the closely situated positive hole.

Such a reaction is highly probable, because of the lowering of the energy barrier by the coulomb field of the positive charge. In 8 *M* NaOH the role of positive holes is played by the radical anion $O^{\cdot -}$ with the neighbor excess cation Na^+. A reaction of this type could be particularly susceptible to photobleaching because electrons in the conduction band are easily attracted by positive centers and their retrapping is less probable. This fact may be responsible for the difficulties in detecting such electrons by nanosecond pulse radiolysis: the intensity of analyzing light emitted by boosted lamps is much greater than in our experiments and consequently the photobleaching may play a more significant role. The different behavior of various parts of the spectrum, i.e. of electrons trapped more or less deeply, may be also demonstrated in γ-radiolysis experiments, as seen in Fig. 8. It has been previously shown by us that, based on electron tunneling, one can expect the following correlation:

$$\log k\lambda \propto \lambda^{-1/2} \quad (12)$$

The superposition of the wavelength dependence obtained from Fig. 8 for alkaline ice and for 3-methylpentane is presented in Fig. 13. The experimental points lie approximately on the same curve. Thus one concludes that when k is determined for the same range of time, the λ dependence has fairly universal character and is independent of the matrix.

The Kinetics of Electron Decay

The decay of e_t^-, studied at 650 nm in 8 *M* NaOH containing phenolate, practically does not occur in μsec and msec time scale. Thus, one must assume that $k(r, V_2-E) = 0$ and $I(S,\tau) = 1$. Indeed, reciprocal optical density depends linearly on phenolate concentration for any time-range. The efficiency of scavenging of dry electrons $F_{e^-} = k_s/k_t\,[T]$, estimated by us on the basis of the above dependence, is equal to 0.42 l mol^{-1}.

The decay of e_t^- in the presence of NO_3^- ions, as studied in the wide timescale, is shown in Fig. 14. Similar measurements have been carried out by Miller (*11*) with a nanosecond technique; however, nothing is known about the intensity of the analyzing light in his detection system. A comparison between the dependence predicted by Eq. 8 and experimental results is given in Fig. 14.

The deviation of the experimental points for short times ($\leqslant 10^{-3}$ sec) is most probably due to our neglecting, in the calculations, the distribution of trap depths. Indeed electron-acceptor pairs with very small separation prob-

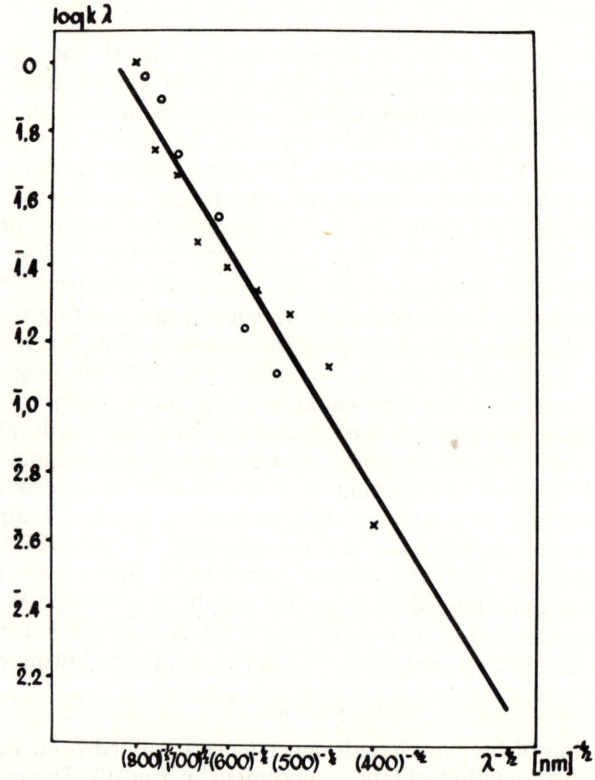

FIG. 13. Correlation between the rate constant of e_t^- decay, k, and the wavelength λ. o — 3 — methylpentane, x — 8 M NaOH with 2.5×10^{-2} M $NaNO_3$.

ably correspond to relatively deep traps (otherwise they would not survive very early stage of tunneling). This should slow down the decay of electrons for small separations and short times.

The value of I (S,τ) from Eq. 8 has been calculated numerically, taking $r_{NO_3^-} = 1.5$ Å and the number of vibronic states in the scavenger molecule equal to unity. The estimated efficiency of scavenging of dry electrons F_{e^-} in the case of NO_3^- equals 2.3 l mol^{-1} and is of the same order as that calculated from Eq. 11.

In contrast to this, the decay of e_t^- in the presence of CrO_4^{-2} is much faster than expected from calculations of I(S,τ). The estimated F_{e^-} is equal to 100 l mol^{-1}. The possible explanation of the fast e_t^- decay may be that the number of vibronic states in the CrO_4^{-2} ion is much greater than in the case of NO_3^-. Very rough estimations indicate that the rate constant, $k(r, V_2-E)$, should be multiplied by a factor of 50 in order to reach agreement between theoretical and experimental data. It is worth stressing that F_{e^-} is also ca. 50 times greater for CrO_4^{-2} than for NO_3^-. The influence of scavenger on the photobleaching rate exhibits a similar trend. Thus it seems that the number of free vibronic states in a scavenger molecule may characterize the scavenging effi-

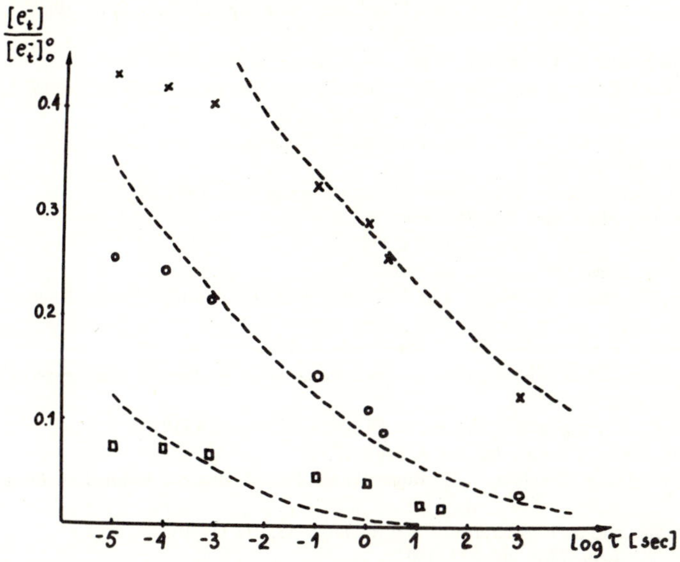

FIG. 14. Decay of e_t^- in 8 *M* NaOH at 565 nm, with filtered analyzing light. Broken curves are calculated from Eq. 8. x — 5×10^{-2} *M* NO_3^-; o — 0.1 *M* NO_3^-; □ — 0.2 *M* NO_3^-.

ciency in respect to dry, trapped and photoexcited electrons. However, further investigations are needed in order to support the above suggestion.

CONCLUSIONS

The data presented above indicate that the kinetic model of scavenging both dry and trapped electrons fairly well explains the dependence of e_t^- decay on time, scavenger concentration and wavelength in the case of O.D. measurements. The tunnel term I (S, τ), is strongly influenced by the nature of the matrix (characterized by different $V_2 - E$ values) and less by the nature of efficient scavenger, i.e. by its radius and number of vibronic states on resonant level. In contrast to this the competitive term, given in Eq. 5, depends to a large extent on the type of scavenger and is certainly very sensitive towards disturbances caused by photobleaching.

The suggested treatment allows one to determine the value F_{e^-} with better approximation than by previously applied methods. In order to check further the validity of the described kinetic model, one should determine F_{e^-} for a larger variety of scavengers and stabilizing matrices and compare the obtained values with the data for corresponding liquid systems.

REFERENCES

1. B. G. Ershov, A. K. Pikaev, P. Ja. Glazunov, and V. I. Spitsyn, *Dokl. Akad. Nauk SSSR* **149**, 363 (1963).
2. J. D. Schulte-Frohlinde and K. Eiben, *Z. Naturforsch.* **17a**, 445 (1962)
3. F. S. Dainton and C. Gopinathan, *Trans. Faraday Soc.* **65**, 151 (1969).

4. C. Elston and L. Kevan, 154th Am. Chem. Soc. Meeting, Chicago, Sept. 1967, Abstract V 77.
5. B. G. Ershov and E. L. Tseytlin, *Khim. Vys. Energij*, **4**, 186 (1970).
6. J. Kroh and Cz. Stradowski, *Proc. 3rd Tihany Symposium on Radiation Chemistry*, (J. Dobo and P. Hedvig, Eds.) Akademiai Kiado, Budapest, Hungary 1972.
7. R. F. Khairoutdinov, A. Y. Mikhailov, and K. J. Zamaraev, *Dokl. Akad. Nauk SSSR, fiz. khim.* **200**, 905 (1971).
8. A. Y. Mikhailov, *Dokl. Akad. Nauk SSSR, ser. khim.* **197**, 136 (1971).
9. D. L. Dexter, *Phys. Rev.* **93**, 985 (1954).
10. J. R. Miller, *J. Chem. Phys.* **56**, 5173 (1972).
11. J. R. Miller, *Chem. Phys. Lett.*, **22**, 180 (1973).
12. F. Kiefer, M. Magat, C. Meyer, and J. Rigaut, *Nature (London) Physical Science*, **232**, 130 (1971).
13. J. Kroh and Cz. Stradowski, *Int. J. Radiat. Phys. Chem.* **5**, 243 (1973).
14. H. Margenau and G. M. Murphy, in *The Mathematics of Physics and Chemistry*, p. 373. P.W.N. Poland, 1962.
15. K. Fueki, D. F. Feng, and L. Kevan, *J. Phys. Chem.* **74**, 1976 (1970).
16. K. Eiben, *Angew. Chemie, Int. Ed.* **8**, 619 (1970).
17. J. A. Batyuk, E. N. Borisova, L. T. Bugaenko, and V. M. Byakov, *Khim, Vys. Energij*, **4**, 143 (1970).
18. J. K. Thomas and J. T. Richards, *J. Chem. Phys.* **53**, 218 (1970).
19. L. Kevan, *J. Chem. Phys.* **56**, 838 (1972).
20. R. F. Dainton, *Ber. Bunsen Ges.* **75**, 608 (1971).
21. J. H. Baxendale, *J. Chem. Soc. Faraday Trans. I*, **69**, 584 (1973).

FAST RESPONSE TECHNIQUES IN CHEMISTRY AND BIOLOGY[1]

Chairman: J. W. Hunt

[1]Symposium C–10

Rapid-Mixing Studies on the Time-Scale of Radiation Damage in Cells[1]

G. E. Adams, B. D. Michael, J. C. Asquith
M. A. Shenoy*, M. E. Watts, and D. W. Whillans
Gray Laboratory, Cancer Research Campaign,
Mount Vernon Hospital
Northwood, Middlesex HA6 2RN, England

*Bhabha Atomic Research Centre,
Biology and Agriculture Division
Trombay-Bombay 85, India

INTRODUCTION

There is evidence that the sensitizing effects of oxygen and other chemical dose-modifying agents on the response of cells to ionizing radiation involve fast free-radical processes. Early electron-spin resonance studies on bacterial spores irradiated in oxygen (*1*) demonstrated the probable role of free radicals in this system.

Subsequent application of fast response techniques using bacterial systems have shown that extremely fast processes occur when the bacteria are exposed to oxygen or other dose-modifying agents during irradiation. The time scales observed were consistent with the involvement of fast free-radical reactions in the expression of these effects. One technique (*2*) involved irradiation of samples of the bacterium *Shigella flexneri* under conditions where the gaseous environment of the cells could be changed very rapidly, either before, or after, irradiation with 1.8 MeV electrons from a Van de Graaff accelerator. The results showed that the maximum life-time, in anoxia, of the oxygen-dependent damage was <10 milliseconds, which was the time resolution of the apparatus.

The experimentally attainable time range between irradiation of the cells and exposure to oxygen, either before or after irradiation, has since been shortened. Epp and co-workers (*3*) currently use a double-pulse technique to irradiate a thin layer of bacteria with two separate pulses of electrons. The pulse-separation time can be pre-set over a time range from microseconds to seconds or longer. The principle of the experiment is that the first pulse is designed to consume, by radiolytic action, the intra-cellular oxygen. More oxygen then diffuses from the surrounding gas phase back to the critical radiosensitive sites within the cells and, providing this occurs before the second pulse is given, the cells will show an oxic response to the second pulse. By varying the time interval it was shown that, under these conditions, the life-time of oxygen-dependent damage in *E. coli* B/r is less than 100 microseconds.

[1] Congress presentation C–10–1.

Two fast-response techniques are in use in the Gray laboratory for measuring the time scale for interaction of cellular damage with oxygen, hypoxic cell sensitizers and other substances which modify the overall cellular response. One technique involves exposing cells, mounted on a surface, to oxygen exploded from a high pressure chamber. The cells can be irradiated with a 2 microsecond electron pulse at various times before or after exposure to the oxygen, with a precision of about ± 100 microseconds. In a study with the bacterium *Serratia marcescens* (*4*) it was found that, under the conditions of the experiment, the oxygen-dependent damage appeared to decay with a half-life of about 0.5 milliseconds. Although this technique has the advantage of fast time resolution, it is limited in that it can only be used to study dose-modification by substances which are volatile.

The other technique, a liquid fast-flow rapid-mix method, permits the study of the time scale of interaction of cellular radiation damage with a wide range of water-soluble dose-modifying agents including oxygen. In this technique, anoxic suspensions of cells are mixed in a capillary junction with a solution containing oxygen, or any other dose-modifying substance, either just before or just after irradiation. Experiments with *Serratia marcescens* (*5*) showed that when the bacteria were mixed with oxygen 40 milliseconds before irradiation, the full oxygen enhancement ratio (OER) was obtained. Similar experiments have been carried out with other sensitizers including N-ethylmaleimide (*5*), triacetoneamine-N-oxyl (TAN) (*6*), and the radioprotective agent glycerol (*7*).

The method has been modified to permit the study of fast dose-modifying processes in suspensions of mammalian cells cultured in vitro. This paper outlines some of the applications of the Rapid-Mix technique illustrated by examples from some current research projects carried out in the Gray Laboratory.

THE APPARATUS

The apparatus is shown in Fig. 1. Syringe A contains a de-aerated suspension of either bacteria or mammalian cells, and syringe B contains a solution of the dose-modifying agent dissolved in the same medium. The syringes are connected to two glass capillary tubes C and D of internal cross-section 0.75 mm^2 which intersect at E with a third tube F of cross-section 1.5 mm^2. Tube F terminates in a collector syringe G. The plungers A and B are driven upwards by a linear actuator H which displaces the contents of the syringes at a constant rate of 2 cm^3/second. The liquids mix rapidly at the junction E and are collected in syringe G for subsequent analysis of cell survival. The mixture is irradiated with an electron beam from a 1.8 MeV linear accelerator delivering 2 microsecond pulses at a repetition frequency of 560 $seconds^{-1}$. The beam passes through an aperture I of dimensions 6.5 mm × 12 mm placed close to the tubes. The vertical position of the irradiation zone is variable with respect to the mixing junction E so that the tubes can be irradiated at any distance from 6 to 130 mm above or below the junction.

The liquid flow-rate of about 3 meters/second is such that 90% of the cells receive one pulse of electrons. The remaining 10% receive an 'overlap' dose of two pulses. Although the overlap reduces slightly the surviving fraction at any dose, it ensures that no cell passes through the apparatus without being irradiated.

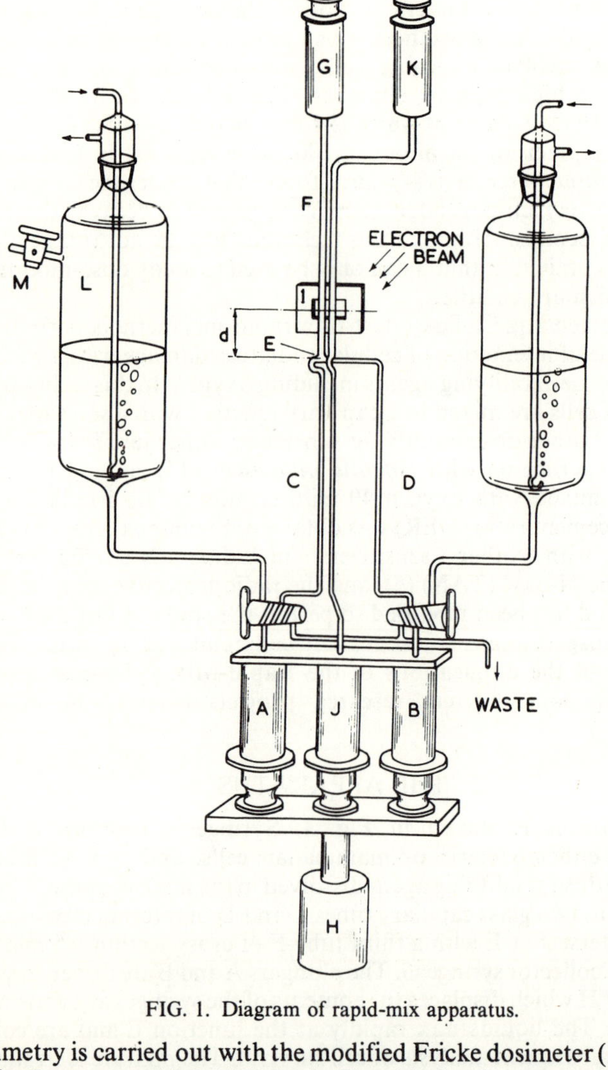

FIG. 1. Diagram of rapid-mix apparatus.

Dosimetry is carried out with the modified Fricke dosimeter (10^{-2} *M* ferrous sulphate in O_2-saturated 0.8 *N* H_2SO_4). The solution is contained in the central syringe J connected by a capillary tube which passes through the irradiation zone to a collector syringe. The actuator drives all three syringes simultaneously, thus a dose measurement is obtained for each irradiation.

The methods of filling the apparatus with the de-oxygenated bacteria or mammalian cells have been described in detail elsewhere (*8*).

TIME SCALE OF THE OXYGEN EFFECT

Figures 2, 3, and 4 illustrate some of the data obtained in a recent study of the time scale of the oxygen effect in irradiated suspensions of the bacterium,

Serratia marcescens, and in irradiated suspensions of Chinese Hamster cells, line V79–379A (*8*).

Serratia marcescens

Control experiments showed that suspensions of the bacteria can be passed through the apparatus, in the absence of irradiation, and collected in the top syringe, G, without loss of viability. The reproducibility of the radiation-induced cell-killing was tested by irradiating bacterial suspensions equilibrated with oxygen and mixed with oxygenated medium several msec before irradiation. Similar experiments were carried out mixing anoxic bacteria with anoxic medium followed by irradiation. In all cases, exponential survival curves were obtained for bacteria irradiated either in phosphate buffer or in nutrient broth. Further, the observed oxygen enhancement ratios (OER) were identical to those found for the bacteria irradiated under similar conditions in conventional γ-ray experiments.

Figure 2 shows survival data for the bacteria when *anoxic* suspensions in buffer were irradiated 5 and 40 msec after mixing with buffer equilibrated with either one atmosphere oxygen (100%) or with 20% oxygen (50% and 10% after mixing). The slopes of the survival curves for 50% oxygen are identical to that for the bacteria when irradiated after prolonged contact with oxygen. This shows that pre-irradiation contact times as short as 5 msec are sufficient to give rise to the full OER. For 10% oxygen, the data show that whereas 40 msec contact time is sufficient to produce the full OER, the value for 5 msec is reduced slightly from 3.1 to 2.5.

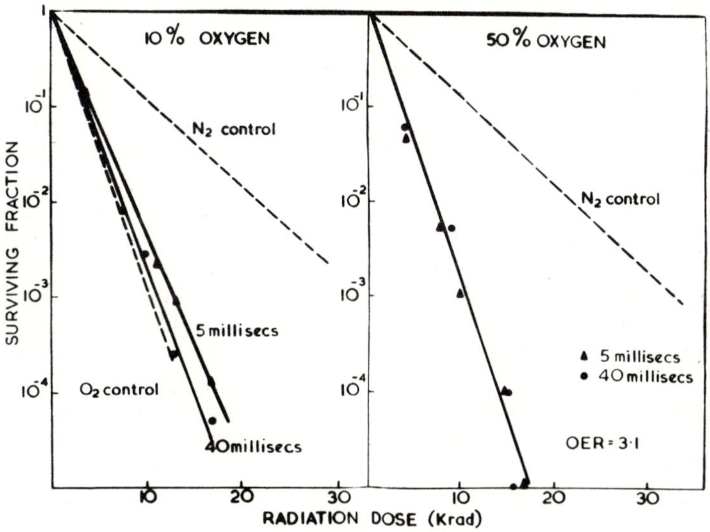

FIG. 2. Survival data for *Serratia marcescens* irradiated in buffer 5 and 40 msec after mixing with buffer containing either 20% or 100% oxygen.

Some experiments (data not shown) were carried out in which the anoxic bacteria were irradiated *before* they were mixed with buffer equilibrated with one atmosphere of oxygen. A small, but significant oxygen effect was observed provided the oxygen was added not more than 10 milliseconds after irradiation of the anoxic bacteria. The OER were 1.24 for 5 msec and 1.1 for 10 msec.

From these and other experiments carried out in suspensions both in phosphate buffer and in nutrient broth, it was found that the very short contact times required for the development of the full oxygen effect were in broad agreement with results obtained with other fast response techniques. The results demonstrated that, in bacteria, oxygen is able to diffuse extremely rapidly to the damaged site—or sites—within the cell where the oxygen effect is manifest. Further, the fact that oxygen must be present within 10 msec after irradiation in order to give rise to even a slight post-effect, confirmed that oxygen interacts with damaged sites which have only a very short life-time, thus providing further evidence that the molecular damage is of a free radical nature.

One of the objects of these experiments was to see if the overall oxygen effect had more than one component and to see whether they could be resolved in time. However, the short time required for diffusion and the short life-time of the chemical lesion on which oxygen acts are such that the time resolution available with this apparatus was insufficient to observe any resolvable components in the OER for this bacterium even if they should exist. This was not so for the experiments with cultured mammalian cells.

Chinese Hamster Cells

Several control experiments showed that, like bacteria, the more fragile mammalian cells are able to survive the strong shearing forces caused by passage at high velocity through the capillary system. The plating efficiency for the cells was generally at least 70%.

One series of radiation experiments showed that when the cells were irradiated in the apparatus either in anoxia or after prolonged contact with oxygen, both the absolute radiosensitivities (D_0) and the oxygen enhancement ratio, OER, were the same as the corresponding values obtained for similar cultures irradiated with X rays at conventional low dose-rates.

In the short contact-time experiments, anoxic cells were mixed with buffer containing oxygen at various concentrations and irradiated at times from 2 to 40 msec after mixing. Some of the survival data from these experiments are shown in Fig. 3. In all cases, oxygen appears to be dose-modifying irrespective of both the oxygen concentration and the pre-irradiation contact times. The observed oxygen enhancement ratios were dependent, however, upon both parameters. Also shown in Fig. 3 is the effect of adding oxygenated buffer 5 msec *after* irradiation of the anoxic cells. Although there is a significant degree of oxygen sensitization under these conditions, the OER is very small (1.1). This indicates that, as in bacteria, the life-time of the chemical lesion, or lesions, upon which oxygen acts, is very short.

Figure 4 shows the values of the OERs from survival curves obtained when oxygen at various concentrations is added to anoxic cells 2, 5, 7, 10, 14, 20, 28, and 40 msec *before* irradiation of the mixture. It is clear that the rise of OER with increasing pre-irradiation contact time is 'biphasic'. The OER for 2 msec, which is the shortest contact time available, is 1.7 for all the oxygen concentrations tested. For 1% O_2, the OER shows little change up to about 10 msec contact time and then increases gradually to 2.2 at 40 msec. For 10% O_2, the OER rises more rapidly from the 2 msec value of 1.7 and reaches 2.8 at 40 msec. For the two

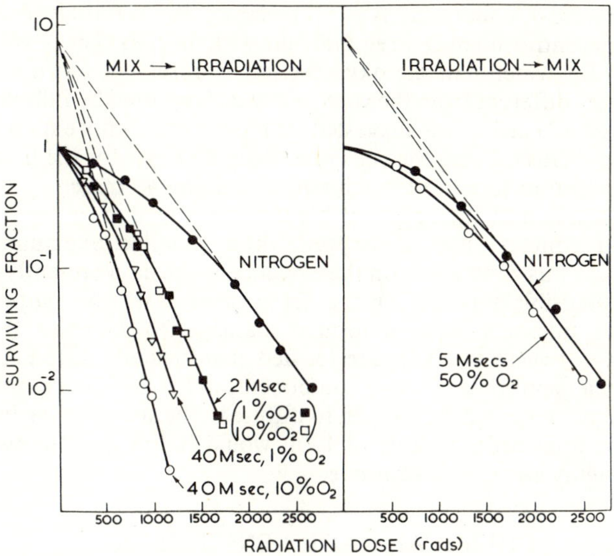

FIG. 3. Survival data for hypoxic V79 cells mixed with hypoxic or oxygenated Hanks' salts solution.

a) before and b) after irradiation

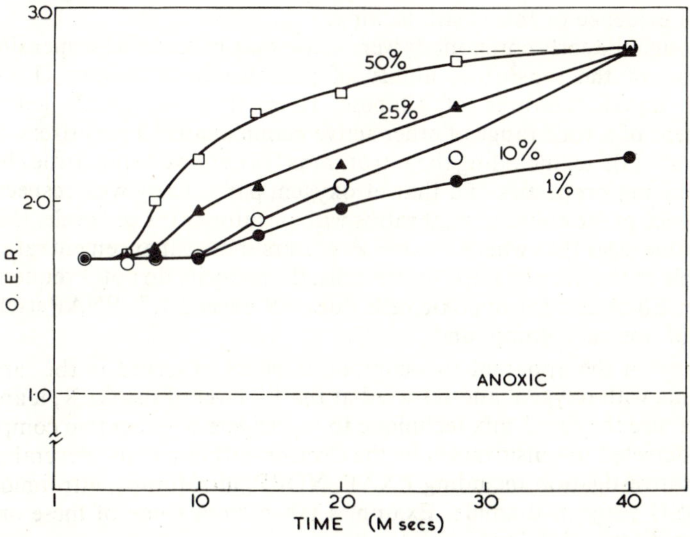

FIG. 4. Change in oxygen enhancement with time between mixing and irradiation in different concentrations of oxygen.

higher concentrations of 25% and 50% the rate of rise of the OER is correspondingly greater.

Values of the OER were also obtained for nine different concentrations of oxygen ranging from 6.8 to about 280 μM in experiments where the oxygen was added 40 msec before irradiation. The data were constant with the equation derived by Howard-Flanders and Alper (*9*) relating the magnitude of the oxygen effect to the concentration of O_2 in equilibrium with the cells. The K-value, the O_2 concentration for which half the oxygen-effect is observed, was about 13 μM which is not very different from the value of 8-9 μM reported by Elkind et al. (*10*) for X-irradiated V79 cells. This suggested that in the rapid-mix experiments, the rate of oxygen diffusion into the cells is sufficiently rapid for the intra-cellular oxygen concentration to almost reach equilibrium by 40 msec.

In the discussion of the apparent biphasic relationship of the OER to the pre-irradiation contact time of oxygen, three possible explanations were considered (*8*). It was concluded, on the evidence of the data presently available, that the most plausible hypothesis is that there are at least two regions in the cell where oxygen interacts with molecular damage to increase the overall radiosensitivity. If these two sites are located at sufficiently different distances from the plasma membrane, the two components in the OER curve could reflect the different times required for oxygen to diffuse to the sites. On the basis of this model and the intermediate value of 1.7 it would follow that the two oxygen effects are roughly equal in magnitude.

CHEMICAL RADIOSENSITIZERS

There are now many chemical compounds which have the ability to differentially sensitize hypoxic cells to radiation-induced lethality. Most of these compounds, the electronic-affinic group, appear to act like oxygen although conclusive evidence of this is still lacking.

Although the early compounds were active only in bacterial suspensions, the observation of the sensitizing ability of paranitroacetophenone (PNAP) in irradiated anoxic cultures of Chinese Hamster cells (*11,12*) led to the development of a wide range of other active mammalian cell sensitizers. Studies with several of these, including the nitrofurans (*13*), showed similarities between their sensitizing properties and that of oxygen, particularly with respect to the independence of the enhancement ratios with position in the cell cycle. However, it has become clear that whereas some sensitizers show enhancement ratios (ER) comparable to the normal OER for the cells, the majority do not. Frequently, the maximum ER found for hypoxic cells does not exceed 1.7: PNAP itself, is an example of one such compound.

In view of the apparent two-compound effect observed in the rapid-mix experiments with oxygen, and resolved at the 1.7 level of the OER, it appeared possible to use the rapid-mix technique to investigate whether two components could be detected in sensitization by the electron-affinic agents. Several systems are under investigation including PNAP, NDPP, nitrofurans, nitroimidazoles, and the -SH reagent, diamide. Examples taken from some of these on-going studies are discussed below.

Paranitroacetophenone (PNAP)

Figure 5 shows the sensitization of anoxic V79 cells irradiated in Hanks' solution and mixed with equal volumes of a deaerated solution of 3 m*M* PNAP 2 and 40 msec before irradiation. Sensitization is clearly evident and the respective enhancement ratios are 1.2 and 1.43. However, for the same anoxic cells irradiated with conventional low dose-rate X rays, an enhancement ratio of 1.7 is obtained at 0.4 m*M* and the ratio of 1.43 is attained at concentrations less than about 0.2 m*M*. It appeared therefore, that 40 msec contact time is sufficient to allow only about 10 - 15% of the PNAP to penetrate the cell to the site where sensitization occurs. When the PNAP was added to the anoxic cells 5 msec *after* irradiation, no sensitization was observed.

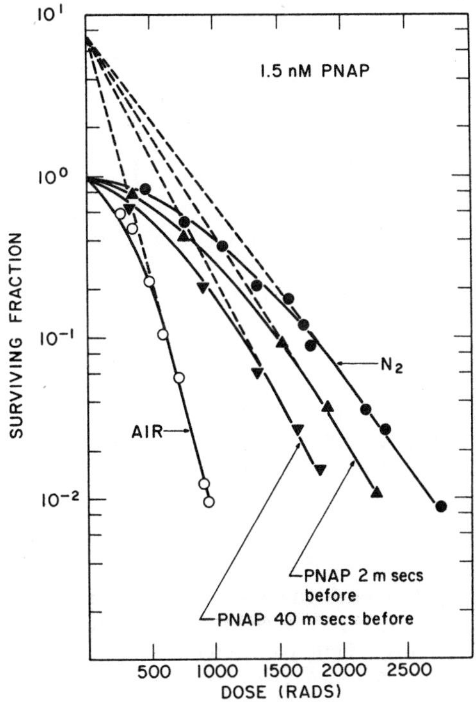

FIG. 5. Sensitization of hypoxic V79 cells by PNAP added 2 and 40 msec before irradiation. Final concentration: 1.5 m*M*.

These experiments were inconclusive with regard to the question of whether PNAP was acting in a manner similar to that of oxygen. They showed however, that mammalian cell sensitization by electron-affinic agents is an extremely rapid process as had been found previously in bacterial systems.

The following experiment (Asquith, unpublished) produced a result which was more definitive. Hypoxic cells which had been in contact with 1.5 m*M* PNAP for at least 30 minutes, were irradiated in the rapid-mix apparatus 2 msec after mixing with oxygen-saturated buffer. The data are shown in Fig. 6. The left-hand

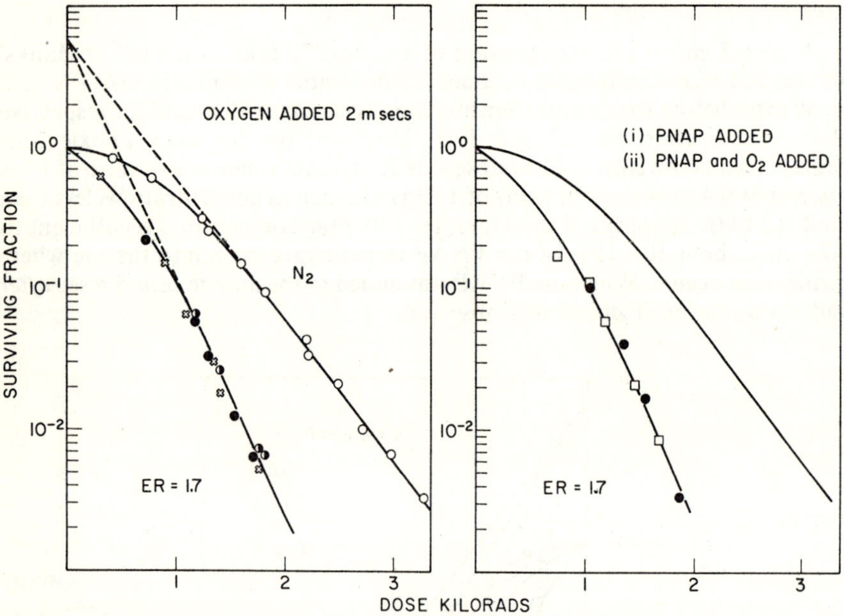

FIG. 6. a) The 2 msec OER of 1.7 for V79 cells.
O N_2 control
● 1% O_2 at 2 msec
X 10% O_2 at 2 msec
◐ 50% O_2 at 2 msec
b) Solid line: N_2 control curve
● PNAP sensitization of V79 cells after prolonged contact.
□ O_2-saturated buffer added to V79 cells containing PNAP, 2 msec before irradiation.

figure reproduces the control data for oxygen sensitization when various concentrations of oxygen are added to the anoxic cells 2 msec before irradiation. The OER is 1.7 in all cases. In the right-hand figure, the solid circles are the survival data for the anoxic cells irradiated in the apparatus after prolonged contact with PNAP. The ER of 1.7 is in agreement with the earlier value obtained with X rays. The open squares represent the data when the cells in contact with PNAP were mixed with the oxygen 2 msec before irradiation. There is clearly no sensitization over and above that found in the PNAP control experiment, i.e., the ER remains at 1.7. These results indicate therefore that the intracellular damage which interacts with oxygen to produce the fast oxygen effect is the same as that which interacts with those electron-affinic sensitizers which give a maximum ER of 1.7. The nature of the site remains to be identified.

The 2-Nitroimidazole, Ro-07-0582

The substituted 2-nitroimidazole, (Ro-07-0582, Roche Products, Ltd.) of the structure:

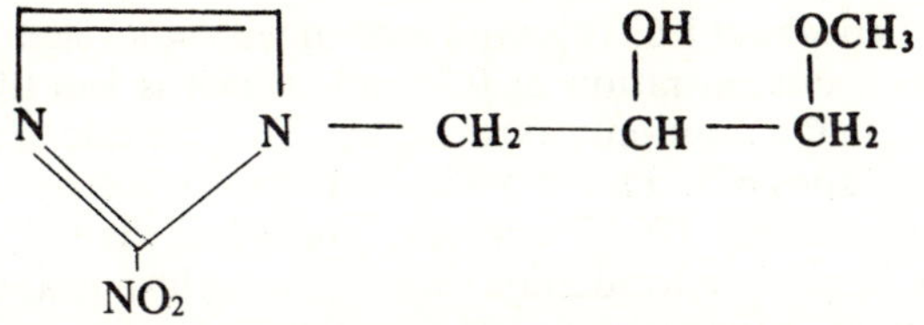

is an example of an electron-affinic sensitizer which gives an enhancement ratio of almost the full OER in Chinese Hamster cells irradiated in anoxia (*14*). It is also a sensitizer which has been shown recently to be effective in various systems in vivo including solid animal tumors containing hypoxic cells (*15*), (Fowler, Begg, Foster, Sheldon, McNally, Peters, Rauth, private communications).

In vitro, the maximum enhancement ratio of ~2.5 is attained at concentrations of about 3 m*M* and above. Further, the sensitizer is very water-soluble and has a low toxicity. The cells can be exposed, therefore, to concentrations very much greater than the minimum required to give the full sensitizing effect. This prompted a rapid-mix investigation to see whether the relationship between enhancement ratio and time would show for this sensitizer, a resolution into two compounds similar to that seen for oxygen.

Anoxic V79 cells were mixed with anoxic buffer containing 100 m*M* 0582 at times from 2–40 msec before irradiation. Fig. 7 shows some of the preliminary data. There is evident sensitization and the enhancement ratio does increase with increasing contact time. However, even at 30 msec, this very high concentration

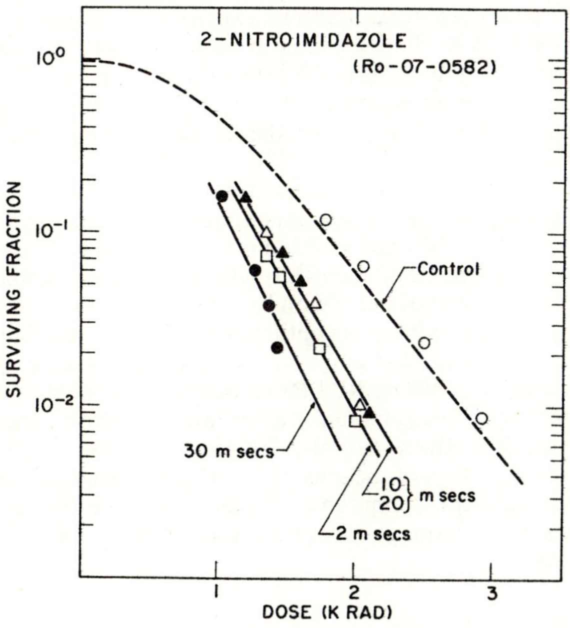

FIG. 7. Sensitization of hypoxic V79 cells by the 2-nitroimidazole 0582 added 2, 10, 20, and 30 msec before irradiation [0582] = 50 m*M*.

leads to an ER of only about 1.6. This ratio is obtained, under standard low dose-rate conditions at a concentration of 0.35 m*M* which is less than 1% of the concentration used in the rapid-mix experiments. Thus the rate of penetration of this sensitizer to the sites of interaction within the cells is less by about an order of magnitude than that for PNAP and very much less than that for oxygen. Clearly, the use of longer pre-irradiation times in the rapid-mix apparatus will be necessary to detect any time resolution in the overall sensitizing effect of 0582.

NDPP

A Mannich base of paranitroacetophenone (NDPP) of structure:

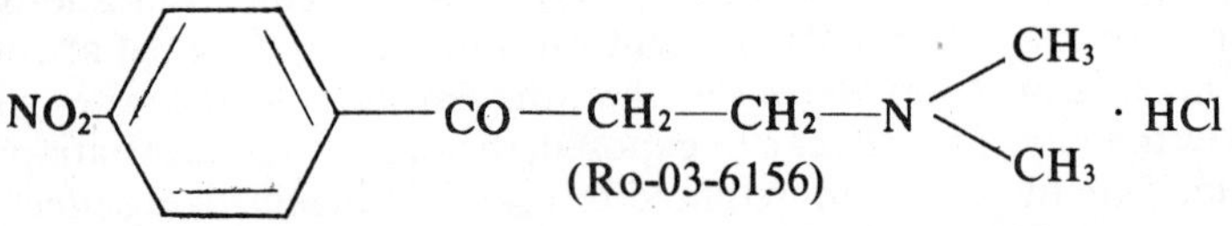

(Ro-03-6156)

is a potent sensitizer of hypoxic cells in vitro but has some properties which are not usually found in the electron-affinic group. The compound was originally synthesized as a water-soluble derivative of PNAP for testing in vivo. It sensitizes hypoxic cells at very low concentrations in vitro(*16,17*) and shows some effect in vivo (*18*). NDPP is an unusual compound in that its sensitizing efficiency is increased considerably on the protein content of the medium (*17,* and Asquith, unpublished) and, in low serum concentration, is dependent on the position of the cells in the mitotic cycle (Asquith, unpublished).

Fig. 8 shows some data from rapid-mix experiments carried out with a low concentration of NDPP in Hanks' balanced salts solution containing a small amount of serum (<1%). When 100 μM NDPP in anoxic solution is added to the anoxic cell suspension 2 msec before irradiation, the enhancement ratio is 1.36. After 40 msec pre-irradiation contact time, the enhancement ratio is considerably increased (ER = 2.0).

This result contrasts with the data from the mammalian cell experiments with the other sensitizers where the rate of penetration appears to be sufficiently slow so that only a relatively small fraction of the concentration can reach the required site within the time-range available with this rapid-mix apparatus. As is found with rapid-mixing experiments with bacteria for other sensitizers, NDPP appears to be able to show almost its optimum effect in mammalian cells even though the contact time before irradiation is as short as 40 msec.

If the sensitive site with which NDPP interacts is inside the cell nucleus, then the rate of penetration of the compound is of the same order of magnitude as that of oxygen. Alternatively, the apparently fast time response may be due to a radiation-induced interaction of the sensitizer with the plasma membrane of the cell. If this is so, then the compound affords a means of investigating the importance of membrane damage in contributing to radiation-induced lethality in mammalian cells.

Diamide

The reagent diamide, $(CH_3)_2N{-}CO{-}N{=}N{-}CO{-}N(CH_3)_2$, has been shown to be a sensitizer for mammalian cells (*19*). The compound is a specific -SH reagent for reduced glutathione and on this basis was investigated as a sensitizer which could act by supressing SH levels within the cell.

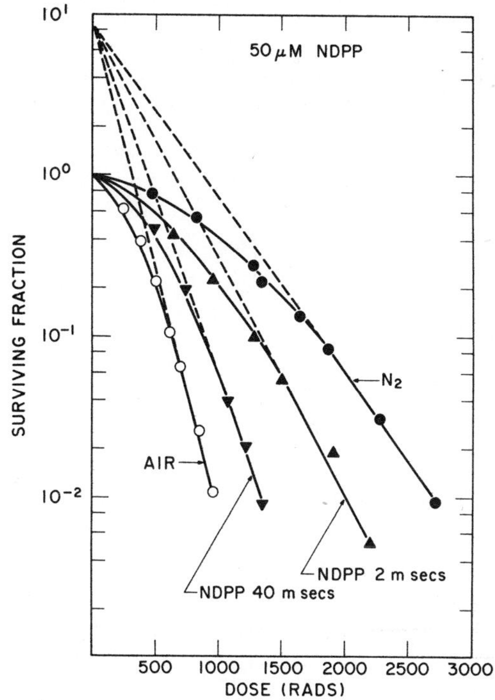

FIG. 8. Sensitization of hypoxic V79 cells by NDPP added 2 and 40 msec before irradiation [NDPP] = 50 μM.

In hypoxic cells, diamide produced some sensitization by a small change in D_0 but, in addition, it was shown to suppress the extrapolation number in both oxic and hypoxic cells. The structure of the compound and its evident oxidizing properties suggests that it might exhibit some of the sensitizing properties associated with the electron-affinic sensitizers. Accordingly, the compound is being studied by the rapid-mix technique using both the bacterium *Serratia marcescens* and Chinese Hamster cells (Watts and Whillans, work in progress).

Fig. 9 shows survival curves from rapid-mix experiments with suspensions of bacteria in phosphate buffer. Two concentrations were used, 3 and 30 mM, which are non-toxic to the unirradiated bacteria under the conditions of the experiment. For 3 mM diamide in contact with the bacteria for 4, 20, and 40 msec before irradiation, a constant enhancement ratio of 1.6 is obtained. This is to be compared with the value of 2.5 found in comparable experiments using ^{60}Co γ-irradiated suspensions. For 30 mM, the value at 4 msec is the same as for 3 msec but then increases to 2.1 at 40 msec. Diamide added 4 msec *after* irradiation of the bacteria produced no sensitization.

These results are qualitatively similar to those obtained from a rapid-mix study of bacterial suspensions containing N-ethylmaleimide (*5*) which is also a sulfhydryl reagent. In this study, it was concluded that, in anoxia, the pre-irradiation contact time was too short for NEM to appreciably affect the SH levels in the bacteria and that the large degree of sensitization observed was due to the involvement of fast free-radical processes. Similarly with diamide, it is unlikely that suppression of intra-cellular -SH levels can occur during such a very short contact time and that, under these conditions, diamide is sensitizing by a mechanism involving fast free-radical processes.

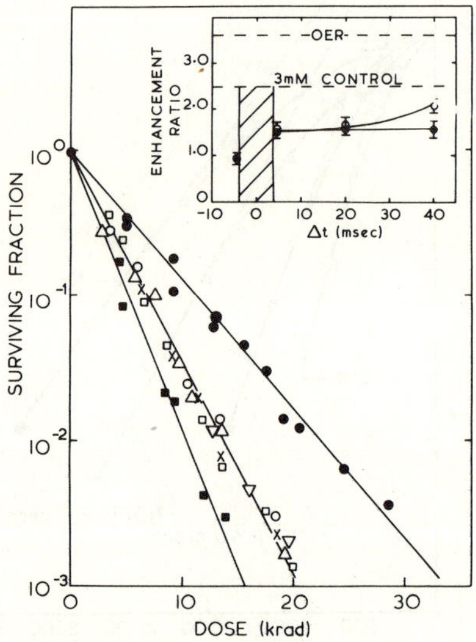

FIG. 9. Sensitization of hypoxic bacteria (*Serratia marcescens*) by diamide at different concentrations.

● Control
X 4 msec, ▽ 20 msec, O 40 msec } 3 m*M*
△ 4 msec, □ 20 msec, ■ 40 msec } 30 m*M*

Inset.
● 3 m*M*.
○ 30 m*M*.

Fig. 10 shows some preliminary data from experiments with suspensions of V79 cells. An anoxic solution of diamide (200 μM) was added to the anoxic cells 40 msec before irradiation. The experimental points fall on the anoxic control curve showing that there is no effect either on the slope or on the extrapolation number. The solid line shows the corresponding survival curves for the cells when irradiated with X rays in the presence of 100 μM diamide. A final concentration of 100 μM is clearly insufficient to permit penetration of this reagent at a concentration necessary to produce even a slight change in radiosensitivity. Although diamide is cytotoxic at higher concentrations, it is possible to use considerably higher concentrations in the rapid-mix experiments where the *overall* contact time is much less. Experiments are in progress to see if such concentrations can produce the drop in extrapolation number when the *pre-irradiation* contact time is very short.

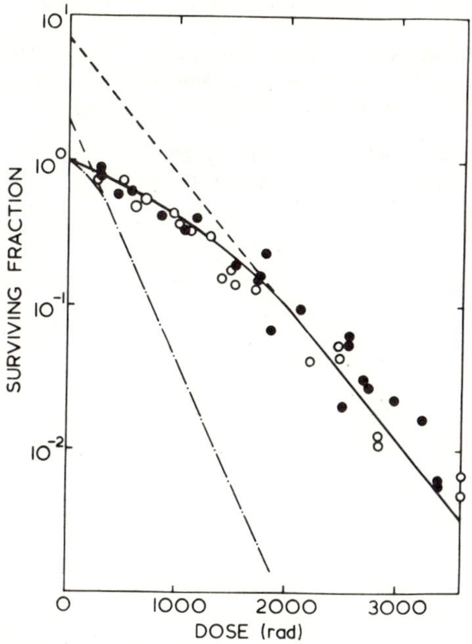

FIG. 10. Failure of 100 μM diamide to affect radiation survival of V79 cells when added 40 msec before irradiation.

O Control
O 100 μM diamide.
—·—·— Sensitizing effect of 100 μM diamide after prolonged contact.

REFERENCES

1. C. F. Ehret, B. Smaller, E. L. Powers, and R. B. Webb, *Science* **132**, 1968 (1960).
2. P. Howard-Flanders and D. Moore, *Radiat. Res.* **9**, 422 (1958).
3. E. R. Epp, H. Weiss, N. D. Kessaris, A. Santomasso, J. Heslin, and C. C. Ling, *Radiat. Res.* **54**, 171 (1973).
4. B. D. Michael, G. E. Adams, H. B. Hewitt, W. B. G. Jones, and M. E. Watts, *Radiat. Res.* **54**, 239 (1973).
5. G. E. Adams, M. S. Cooke, and B. D. Michael, *Nature* (*London*) **219**, 1368 (1968).
6. T. Brustad, *Scand. J. Clin. Lab. Invest.* **22**, 31 (1968).
7. T. Brustad and B. Singsaas, *Radiat. Res.* **45**, 94 (1971).
8. M. A. Shenoy, J. C. Asquith, G. E. Adams, B. D. Michael and M. E. Watts, *Radiat. Res.*, **62**, in press, (1975).
9. P. Howard-Flanders and T. Alper, *Radiat. Res.* **7**, 518 (1957).
10. M. M. Elkind, R. W. Swain, T. Alescio, H. Sutton, and W. B. Moses, in *Cellular Radiation Biology*, pp. 442–466. Williams and Wilkins, Baltimore, 1965.

11. G. E. Adams, J. C. Asquith, D. L. Dewey, J. L. Foster, and R. L. Willson, *Int. J. Radiat. Biol.* **19**, 575 (1971).
12. J. D. Chapman, R. G. Webb, and J. Borsa, *Int. J. Radiat. Biol.* **19**, 561 (1971).
13. J. D. Chapman, A. P. Reuvers, J. Borsa, A. Petkau, and D. R. McCalla, *Cancer Res.* **32**, 2616 (1972).
14. J. C. Asquith, M. E. Watts, K. Patel, C. E. Smithen, and G. E. Adams, *Radiat. Res.* **60**, 108 (1974).
15. J. Denekamp, B. D. Michael, and S. Harris, *Radiat.* Res. **60**, 119 (1974).
16. G. E. Adams, J. C. Asquith, M. E. Watts, and C. E. Smithen, *Nature* (*London*) **239**, 23 (1972).
17. G. F. Whitmore, A. M. Rauth, S. Gulyas and R. Bush in *Proc.XIII Int. Cong. Radiol. Madrid*, 1973, in press.
18. J. Denekamp and B. D. Michael, *Nature* (London), **239**, 21 (1972).
19. J. W. Harris and J. A. Power, *Radiat. Res.* **56**, 97 (1973).

Application of Fast Polarography in Pulse Radiolysis[1]

M. Grätzel, K. M. Bansal and A. Henglein
Hahn-Meitner Institute, Berlin, West Germany

INTRODUCTION

The identification of short-lived intermediates formed during photochemical and radiation chemical reactions has been greatly facilitated by the advent of the flash photolysis and pulse radiolysis techniques. Predominantly three methods, fast kinetic spectroscopy, conductivity and ESR resonance have been applied to follow their decay kinetics. More recently, another analytical technique, polarography, has been successfully combined with the pulse radiolysis and flash photolysis methods (*1,2*). The principle of this method consists in the detection of short-lived intermediates, formed by pulse irradiating or flash photolyzing a sample solution, on the basis of their redox behavior at a mercury drop electrode. The advantages of the technique of short-time polarography are:

(1) It provides a further independent technique to determine rate constants for fast radical reactions in solution.

(2) Short-time polarograms of radicals may be used for the identification of transitory species in pulse radiolysis and flash photolysis.

(3) The redox behavior of short-lived transients which have biological importance may be investigated.

(4) The electro-kinetic analysis of the polarographic current-time curves yields values for the rate of the electron transfer from the electrode to the electroactive species in solution and the transfer coefficient. The latter parameter reflects on the distribution of occupied and unoccupied electronic levels of the participants in the redox reaction.

The present paper first describes the experimental setup of the technique followed by a brief introduction in the theoretical background of the method and the evaluation of the data. Finally, polarograms and current-time curves of radicals from the domain of inorganic, organic and biological chemistry are discussed, with emphasis on biologically important intermediates.

EXPERIMENTAL

A schematic representation of the Notre Dame pulse-polarography layout (*3*) is given in Fig. 1. This is an improved version of the apparatus described earlier (*1,2,4*). A three-electrode arrangement is used consisting of a hanging mercury drop electrode, a mercury pool which serves as a counter electrode, and a saturated calomel element (SCE) as reference electrode. The liquid under

[1] Congress presentation C–10–2.

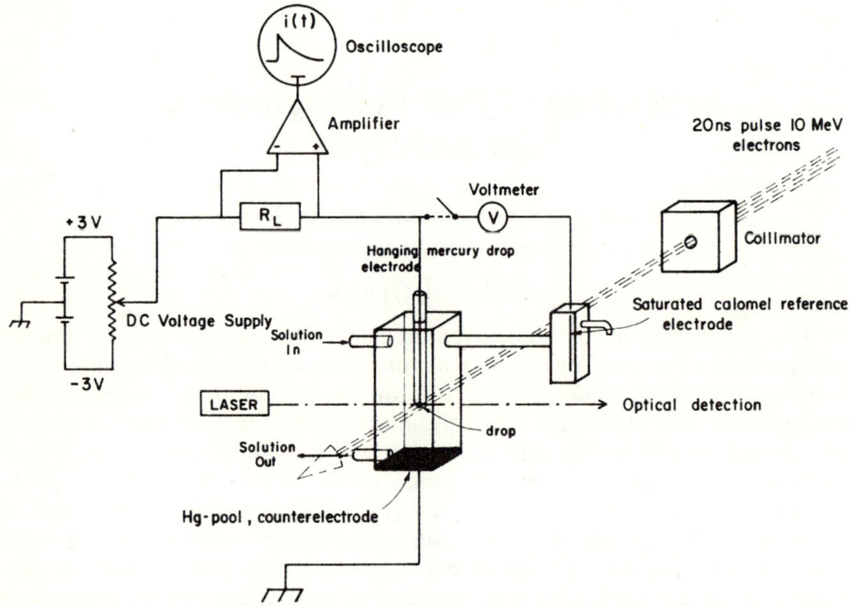

FIG. 1. Measuring device for short-time polarography.

investigation flows through a 1 cm square suprasil sample cell, thereby passing the hanging mercury drop (working electrode) and establishing electrical contact to the SCE reference electrode. The solution contains an inert electrolyte (Na_2SO_4) and the solute which is to react with the primary radicals from the solvent radiolysis. The latter radicals are formed in homogeneous distribution around the mercury drop by a 20 ns electron pulse from a 10 MeV linac. If the intermediate produced by reaction of the primary radicals with solute molecules can be reduced or oxidized at the potential which is applied to the drop, then a faradaic current will flow across the double layer separating mercury drop and electrolyte. The time function of this current is recorded oscillographically as a voltage drop across a variable load resister R_l. The measurement is repeated at different potentials which are provided by a D.C. voltage supply and measured with a voltmeter against the SCE reference electrode.

THEORY

The polarographic registration of short-lived species formed by the radiation pulse in the vicinity of the Hg-drop comprises the recording of current-time curves at given electrode potentials. A prerequisite for an adequate interpretation of these measurements is the knowledge of the basic factors determining the shape of a polarographic current-time curve. Figure 2 shows schematically the three important kinetic processes which have to be considered:

a) the diffusion of the transient towards the electrode;

b) the chemical reaction of the transient in the bulk solution and in the diffusion layer; and

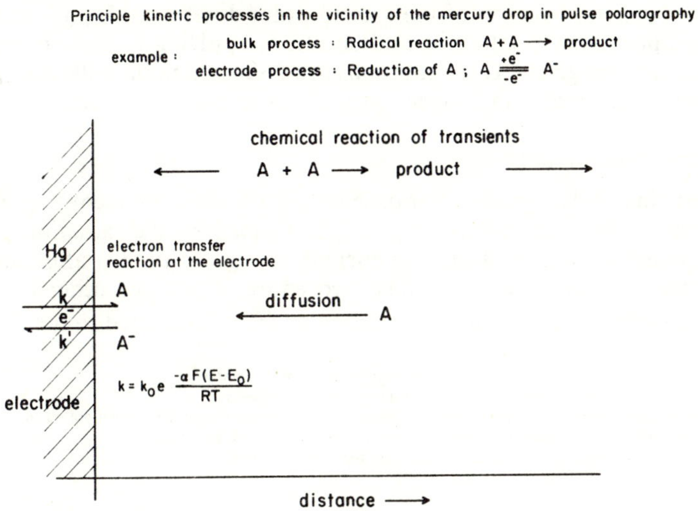

FIG. 2. Kinetic processes in the vicinity of the Hg electrode after pulse irradiation.

c) the rate of transfer of an electron across the electrode/electrolyte interface. This is a heterogeneous process with a rate constant depending on the electrode potential

$$k = k_0 \exp[-\alpha F(E - E_0)/RT] \tag{1}$$

where E is the applied potential, E_0 the reversible redox potential of the system, and k_0 the rate constant at the latter potential. The transfer coefficient α can assume values between 0 and 1. For many electrode reactions α is close to 0.5.

A theoretical description of polarographic current-time curves is possible if the function $c_A(t,x)$ can be determined. $c_A(t,x)$ describes the dependency of the transient concentration on the time (t = 0 is the time of the radiation pulse) and on the distance x from the electrode surface. The current time function i(t) is readily derived from $c_A(t,x)$ by partial differentiation with respect to x and multiplication with a constant:

$$i(t) = nFAD(\partial c_A/\partial x)_{x=0} \tag{2}$$

where n is the number of electrons participating in the electrode reaction, F is the Faraday constant, and A is the surface area of the electrode. In order to obtain $c_A(t,x)$, solutions for the differential equation

$$\partial c/\partial t = D\partial^2 c/\partial x^2 - 2k_r c_A^2 \tag{3}$$

have to be found. It is assumed in Eq. (3) that the transient, A, disappears in the bulk solution via a second-order reaction with the rate constant 2 k_r. This is the case for the majority of radicals in a pulse radiolysis experiment. Since the radicals are initially produced in homogeneous distribution by the radiation pulse, all solutions to Eq. (3) have to obey the boundary condition

$$c(x,t = 0) = c^0 \tag{4}$$

Although a general solution of the second-order differential Eq. (3) cannot be obtained, approximate solutions and solutions applying to certain restricted conditions may be given. These solutions are discussed in the following section and are referred to later in connection with the experimental data.

Short-Time Polarograms

Short-time polarograms of transients are obtained by recording current-time curves from a short-lived depolarizer formed by the radiation pulse at various potentials and plotting the current at a given time t_1 after the pulse against the respective potential. The procedure is outlined schematically in Fig. 3. The time interval t_1 may be chosen as short compared to the first half-

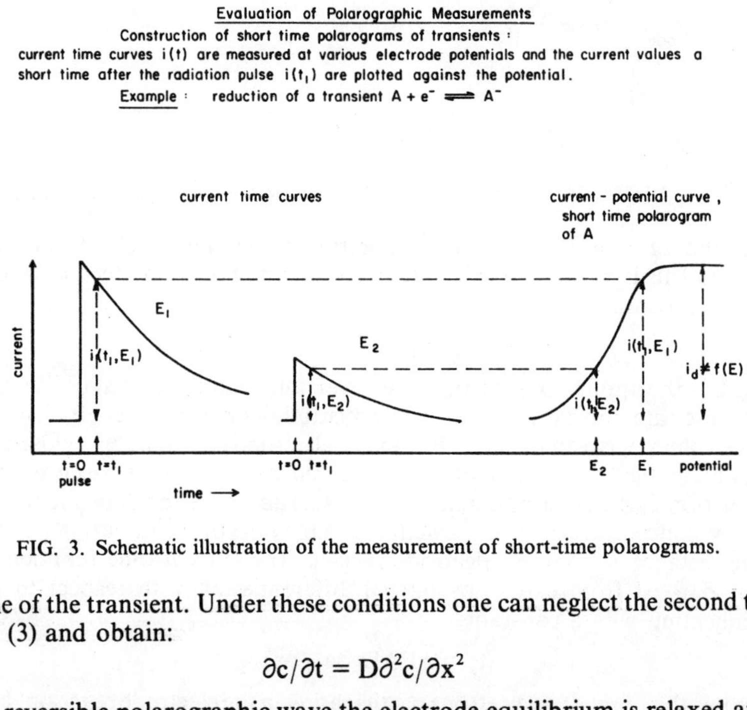

FIG. 3. Schematic illustration of the measurement of short-time polarograms.

lifetime of the transient. Under these conditions one can neglect the second term in Eq. (3) and obtain:

$$\partial c/\partial t = D\partial^2 c/\partial x^2 \tag{5}$$

For a reversible polarographic wave the electrode equilibrium is relaxed at the time of observation. The only additional restriction on the analytical solution of Eq. (5) is then the second boundary condition

$$c(t,x = \infty) = c_b \tag{6}$$

and the current is controlled only by the diffusion of the transient. One obtains from Eq. (5) via Laplace transformation and integration

$$i(t_1) = nFA(c_b - c_0)(D\pi t_1)^{1/2} \tag{7}$$

and for the current in the limiting current region where the concentration at the electrode $c_a = 0$

$$i(t_1) = nFA(c_b)(D\pi t_1)^{1/2} \tag{7a}$$

Since the concentration at the electrode surface c_b obeys the Nernst-equation

$$E = E_0 + 0.058\ n\ \log(c_0/c_b - c_0)$$

one finally obtains for the current potential function describing the polarographic wave

$$E = E_0 + 0.058/n\ \log[i_d(t_1) - i(t_1)/i(t_1)] \tag{8}$$

In the case of an irreversible polarographic wave the electrode equilibrium at $t = t_1$ is not yet relaxed. Hence, the electrode kinetics contribute to the overall kinetics. This is expressed by imposing another restriction on the solution of Eq. (5)

$$k \cdot c_o = D(\partial c/\partial x)_{x=0} \tag{9}$$

where k is the rate constant of the transfer reaction which depends on the electrode potential according to Eq. (1). With these boundary conditions one obtains for the current time function

$$i(t)/i_d(t_1) = \pi^{1/2}\lambda \exp(\lambda^2)\operatorname{erfc}\lambda \tag{10}$$

where the dimensionless parameter λ is defined as

$$\lambda = k t_1^{1/2}/D^{1/2} \tag{11}$$

Analogous to conventional polarography one observes in pulse-polarography three different kinds of waves. These waves are shown schematically in Fig. 4.

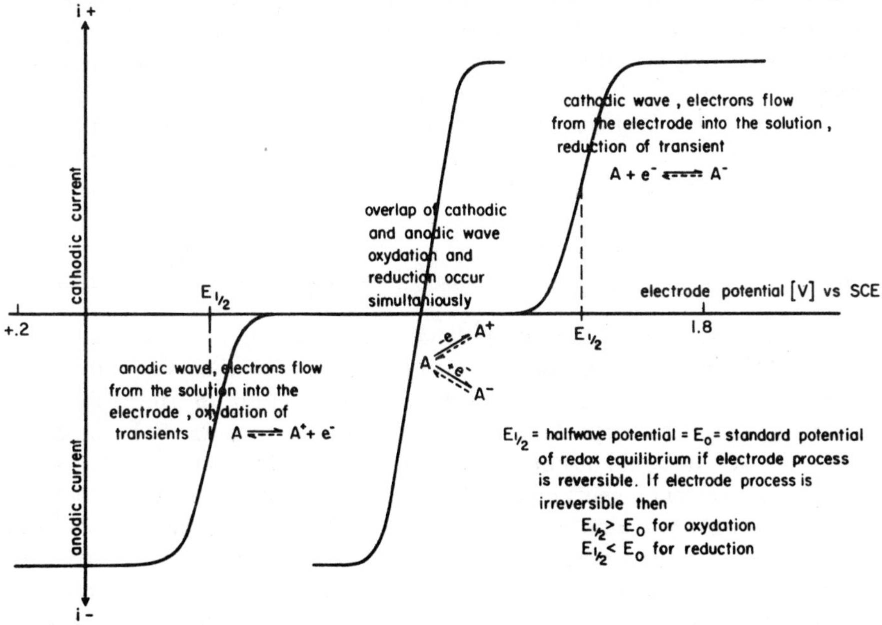

FIG. 4. Schematic representation of polarographic waves feasible in pulse polarography.

A cathodic wave represents an electrode process where the electrons are flowing from the electrode into the solution. Hence, the transient is reduced. Conversely, an anodic wave indicates that the electron flows from the solution to the electrode. Hence, the transient is oxidized. If both processes occur in the same potential range, a composite wave with an anodic and cathodic branch is observed. The halfwave potential is defined in the usual way as the potential at which the current is at one-half of its maximum value in the limiting current region. For reversible waves the halfwave potential is equal to the thermodynamic standard potential of the redox equilibrium in which the transient participates. For an irreversible wave the halfwave potential has a more positive value than the standard potential if the wave is anodic; it is at more negative potentials if the wave is cathodic. The total potential region accessible to investigations with a mercury electrode is +.2 to −1.8 V vs. SCE.

The majority of the short-time polarograms of transient species are irreversible waves. This is readily understood from the nature of Eqs. (10) and (11). The expression on the right hand side of Eq. (10) can assume values between 0 and 1 and increases with increasing λ. Since the time t_1 at which the current is measured is only several microseconds, the parameter λ is small. In particular this applies for the foot of the polarographic wave where the rate constant k has its smallest values. Therefore, the measured current $i(t_1,E)$ not only is controlled by the diffusion process as is required for obtaining a reversible wave, but is also affected by the transfer reaction. This leads to a shift of the irreversible wave to more negative (positive) potentials for a cathodic (anodic) process. However, at the top of the irreversible wave the k value is large enough to ensure control of the current exclusively by diffusion of the transient. The ratio of the current in the wave and of the limiting current $i(t_1)/i_d(t_1)$ may be used to determine the parameter λ from Eq. (10) at different potentials. Using Eq. (11) one obtains k as a function of potential, which finally allows one to evaluate the transfer coefficient α by means of Eq. (1).

Theoretical Description of the Complete Current-Time Curves

Long-Lived Transients

If the transients formed by the radiation pulse are long-lived with respect to the time domain of measurement, then the current-time curves within the polarographic wave are described by Eq. (10) and (7) for an irreversible and reversible process, respectively. In the limiting current range Eq. (7a) is valid for both processes. It should be noted that under these conditions the equations not only give the current a short time t_1 after the radiation pulse but throughout the entire observed time region. Eq. (11) also predicts that the λ value increases with time. Hence, at longer times, the ratio $i(t)/i_d(t)$ in Eq. (10) approaches unity. This implies that the current measured within an irreversible wave is lower than the diffusion-controlled current at shorter times but eventually merges with i_d. This offers another possibility to determine the dependency of k on the electrode potential. The deviation of $i(t,E)$ from $i_d(t,E)$ at shorter times is described by Eq. (10) and may be measured as a function of the electrode potential. Eq. (11) is then used to determine the rate constants at different potentials. From a semilogarithmic plot of k vs. the potential according to Eq. (1) the transfer coefficient is obtained.

Short-Lived Transients, Determination of Rate Constants for the Chemical Reaction of the Transients

Polarographic current-time curves give information about the kinetics of chemical reactions in the bulk solution. A suitable potential region for such investigations is the limiting current region of the polarographic wave of the transients where diffusion and chemical reaction are the only kinetic processes which have to be considered. For a second-order reaction of the transients the solutions of Eq. (3) have to obey the new boundary condition

$$c_b(t) = c_b(t = 0)/(c_b(t = 0)2k_r t + 1) \tag{12}$$

As was already noted above, this differential equation has no direct analytical solution. In the analogous case involving first order chemical disappearance of the radicals, an explicit solution to the boundary problem exists (*5*)

$$i_{(t)} = i_d \exp(-k_1 t) = i_d c_b(t)/c_b(t = 0) \tag{13}$$

where k_1 is the first-order rate constant. Birk and Perone (*6*) postulated a limiting form of the solution to the second-order case. Using Eq. (13) and substituting $c_b(t)/c_b(t = 0)$ by Eq. (12) they obtained the following relationship:

$$i_d(t)/i(t) = 1 + 2k_r c^0 \cdot t. \tag{14}$$

From Eq. (7a) and (14) one obtains:

$$1/i_{(t)}t^{1/2} = (\pi^{1/2}/nFAD^{1/2}c^0) + (2k_2\pi^{1/2}/nFAD^{1/2})t\ . \tag{15}$$

A plot of $1/i_d t^{1/2}$ vs. t should therefore yield a straight line. Since c^0 is known in a radiation chemical experiment, D as well as $2k_2$ can be obtained from the intercept with the ordinate axis and the slope of the straight line. The validity of Eq. (15) can be checked by comparing the value of $2k_2$ obtained with the value derived from the optical measurements.

RESULTS AND DISCUSSION

During the last few years an extensive study of the polarographic behavior of a large number of radicals produced by pulse irradiation of aqueous solutions has been carried out. The limited scope of this presentation does not permit a discussion of all the available data, and the reader is referred to earlier publications (*1,2,7–10*) and to a forthcoming review article (*11*). In the following sections we shall confine ourselves to the presentation of few illustrative examples underlining the validity of the equations derived above. In addition, some more recent investigations (*3*) will be reported which were performed to shed light on the electrochemical and kinetic behavior of rare valency states of metal ions, large detergent molecules, and intermediates of biological redox processes.

Polarographic Investigations of Inorganic Intermediates

One of the primary radicals formed by interaction of radiation with water is the OH radical. Figure 5 shows a short-time polarogram of the OH radical,

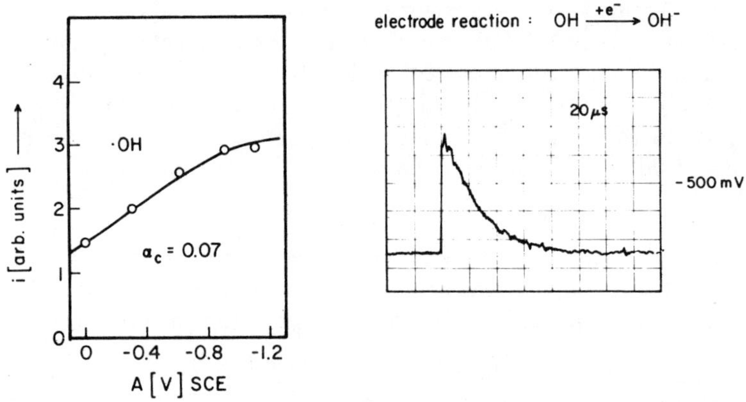

FIG. 5. Cathodic wave from the OH radical reduction; inserted oscillogram: current-time curve obtained at −.5 V, solution: 0.2 M Na_2SO_4, N_2O saturated, initial radical concentration 10^{-5} M.

obtained from currents at $t_1 = 20$ μs. As is expected from its strongly oxydizing properties, the OH radical is readily reduced at the mercury drop over the whole potential region

$$OH + e^- \longrightarrow OH^- \tag{16}$$

The decrease in the height of the current signal at voltages more positive than −300 mV denotes the onset of a flat irreversible wave. Evaluation according to the procedure described in the Theory section yields a transfer coefficient $\alpha = .07$. Included in Fig. 5 is an oscillogram showing a current-time curve obtained from the plateau region of the wave. From the rapidly rising signal after the pulse a time constant < 5 μs for the detection system is derived. The subsequent decrease of the current is due to the rapid recombination of OH radical and the growth of the diffusion layer.

Figure 6 illustrates the polarographic behavior of Tl^{2+} which is formed in aqueous N_2O-saturated solutions of thallous sulfate by reaction with OH radicals (*12*)

$$Tl^+ + OH \longrightarrow Tl^{2+} + OH^- \tag{17}$$

A cathodic wave is observed at relatively positive potentials which is indicative of the strongly oxidizing properties of Tl^{2+}. Apparently Tl^{2+} prefers the pathway of

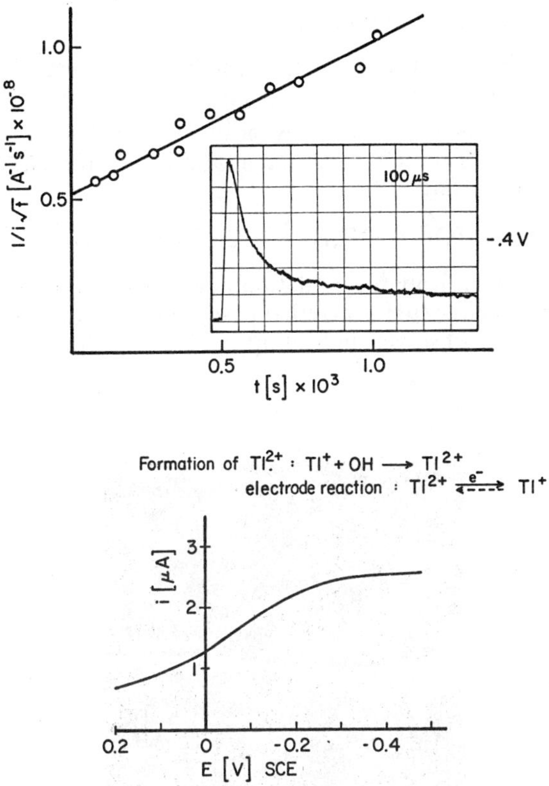

FIG. 6. Upper part: cathodic wave from the Tl^{2+} reduction; inserted oscillogram: current-time curve obtained at $-.3$ V; lower part: kinetic plot according to Eq. (11) of the current-time curve; initial radical concentration 5×10^{-6} M; solution: 02. M Na_2SO_4, 10^{-4} M Tl_2SO_4, N_2O saturated.

returning to the +1-valency state instead of giving up an electron to the electrode to form the stable thallic ion Tl^{3+}. This finding underlines the greater stability of the +1-valency state of thallium with respect to the +3 state. The inserted oscillogram in Fig. 6 shows again a transient current-time curve obtained from the limiting current region of the Tl^{2+} reduction. Compared to the OH curve the current decays towards zero at a much slower rate indicating a much longer lifetime of the Tl^{2+} ion in the solution than of the OH radical. Indeed, the value for the biomolecular rate constant for OH recombination $2k_r$ is 10^{10} M^{-1} s^{-1} while that for the reaction

$$Tl^{2+} + Tl^{2+} \longrightarrow Tl^{+} + Tl^{3+} \tag{18}$$

is only 1.9×10^{8} $M^{-1}s^{-1}$ (*12*). The kinetic evaluation of the current-time curve is also presented in Fig. 6. From a plot of $1/i\sqrt{t}$ against t according to Eq. (15) and from the initial concentration $[Tl^{2+}]_{(t=0)} = 5 \times 10^{-6}$ we obtain $k = 2 \times 10^{8}$ M^{-1} s^{-1} which is in excellent agreement with the literature value.

Investigations of Organic Intermediates

A class of radicals which have strongly reducing properties and hence show anodic waves at fairly negative electrode potentials are α-alcohol radicals of the type –C·–OH where the lone electron is located in α-position to the OH-group. These radicals are formed in N_2O-saturated solutions of primary and secondary alcohols according to Eq. (19)

$$> CHOH + OH,H \rightarrow > C - OH + H_2O, H_2 \qquad (19)$$

Oscilloscope traces showing the oxidation of ethanol anion radicals formed by pulse irradiating alkaline (pH = 14) ethanol solutions (*13*) are presented in Fig. 7. Negative signals are observed indicating oxidation of the anion radicals according to

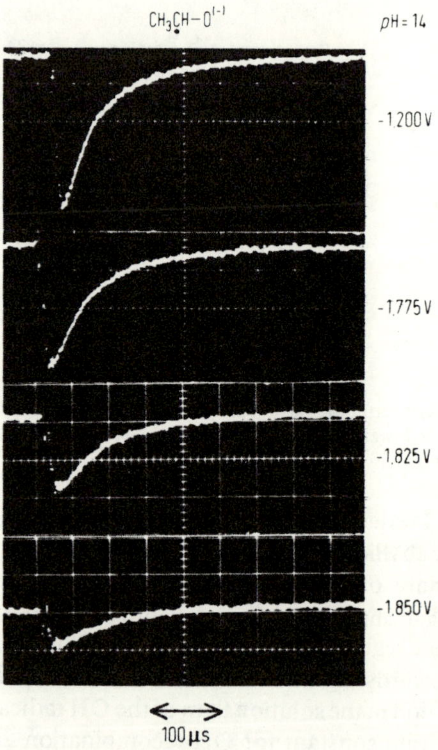

Anodic current (negative signal) vs. time at *p*H = 14 and different potentials ($2 \cdot 10^{-3}$ M ethanol; $2.5 \cdot 10^{-2}$ M N_2O; 1 M NaOH; electrode surface: 0.015 cm^2; initial radical concentration: $5 \cdot 10^{-6}$ M)

FIG. 7. Anodic current (negative signal) vs. time at pH = 14 and different potentials (2×10^{-3} M ethanol; 2.5×10^{-2} M N_2O; 1 M NaOH; electrode surface: 0.015 cm^2; initial radical concentration: 5×10^{-6} M).

$$CH_3 - CH - O^- \leftrightarrow CH_3 - CH = 0 + e^- \tag{20}$$

The anodic currents decrease at more negative potentials. In Fig. 8 the current

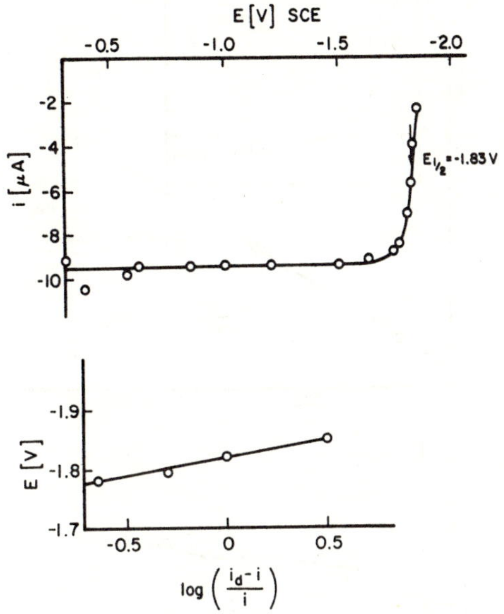

FIG. 8. Upper part: plot of current vs. potential at pH 14; lower part: plot of E vs. log [*4d*–i)/i] at pH 14; solution same as for Fig. 7.

at $t_1 = 20\ \mu s$ after the pulse is plotted as a function of the electrode potential. A halfwave potential of −1.83 V is obtained for the ethanol anion radical oxidation which is close to $E_{1/2} = -1.83$ V for the reduction of acetaldehyde. The coincidence of the halfwave potentials for oxidation and reduction in a redox equilibrium is a criterion for the reversibility of an electrode process. It may be concluded that the oxidation of the ethanol anion radical is a very fast process and hence is reversible even at as short a time as 20 μs. This implies that the halfwave potential 1.83 V is the true thermodynamic standard potential of the ethanol anion radical/acetaldehyde redox equilibrium. The current potential curve in Fig. 8 then should obey Eq. (8). Indeed, a plot of E vs. $\log[(i_d - i)/i]$ which is included in Fig. 8 gives a straight line. From the slope of this line one evaluates $n = 0.9$ which is in good agreement with the expected value $n = 1$.

The technique of pulse polarography is particularly suited to application in situations where radicals are produced by the radiation which cannot be detected by the customary techniques such as kinetic spectroscopy and conductivity. Such transients are, for example, methylene radical formed by reaction of OH or H with molecules containing a long aliphatic chain. Aqueous solution of detergent molecules provide useful model systems for studies of this type of radicals since soap molecules, due to their amphipatic nature, are very soluble in water. Figure

9 shows polarographic data from the pulse radiolysis of N_2O-degassed solutions

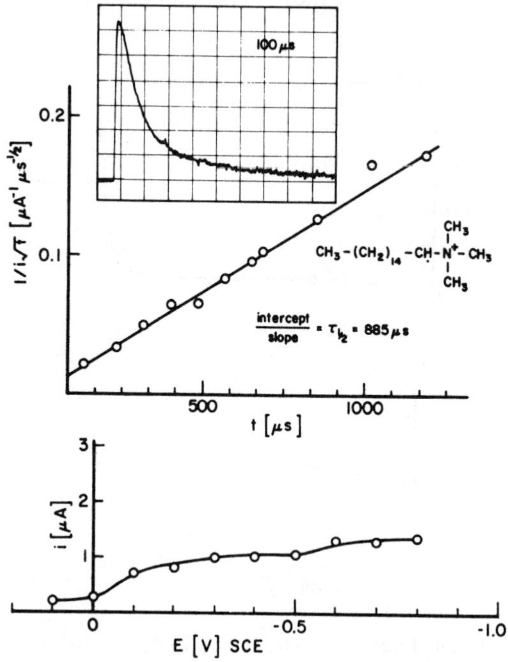

FIG. 9. Short-time polarogram of CTAB radicals; current-time curve at E = −.7 V; kinetic plot of this curve; solution: 5×10^{-4} M CATAB, 0.1 M Na_2SO_4; initial radical concentration: 10^{-5} M.

of cetyltrimethylammonium bromide (CTAB). It is reasonable to suppose that the principal reaction of OH radicals with CTAB is the abstraction of an OH radical from the α–CH_2 group

$$CH_3 - (CH_2)_{14} - CH_2 - \overset{+}{N}(CH_3)_3 + OH \rightarrow CH_3 - (CH_2)_{14} - \dot{C}H - \overset{+}{N}(CH_3)_3 + H_2O \quad (21)$$

In addition a small percentage of other methylene radicals may be formed in the reaction. The ease of reduction of CTAB radicals at the Hg electrode is indicated by the fact that a cathodic current is observed over the whole potential region

$$-\dot{C}H - \overset{+}{N}(CH_3)_3 + e^- \xrightarrow{H^+} - CH_2 - \overset{+}{N}(CH_3)_3 \quad (22)$$

the inserted oscillogram in Figure 9 shows a current-time curve of the CTAB radical reduction at E = −.7 V. The current decays over several hundred microseconds, including a relatively long lifetime of the CTAB radical. The rate of disappearance of CTAB radicals can be determined from a plot of $1/i\sqrt{t}$ vs. t which is also included in Figure 9. This plot gives a straight line. The ratio of the ordinate intercept of this line and the slope is equal to the first half lifetime $\tau_{1/2}$ of the CTAB radicals as is apparent from inspection of Eq. (15). One obtains a value

$\tau_{1/2} = 885\ \mu s$. From this value and an initial concentration of radicals of 10^{-5} M one calculates the second-order rate constant for the CTAB radical decay $k_r \simeq 10^8\ M^{-1}\ s^{-1}$. The excellent linearity of the kinetic plot in Fig. 9 suggests that the transport of radical sites towards the electrode occurs by diffusion of the entire detergent molecule. Conversely, in solutions of sodium lauryl sulfate (NaLS) the electrode is covered by NaLS molecules over a wide potential range and the observed currents reflect only segmented diffusion of radical sites formed by OH attack of NaLS in the immediate surface region of the electrode.

We now focus our attention on three molecules which participate in biologically important reactions: ascorbic acid, nicotinamide-adenine dinucleotide (NAD), and riboflavin. All three molecules act as coenzymes for important physiological redox processes and in all three cases the overall redox reaction is a 2-electron process. The intermediates of these reactions can be produced by the pulse radiolysis method and the technique of short-time polarography allows for the investigation of their electrochemical and kinetic properties. The pertinent redox reaction of ascorbic acid:

$$\text{ascorbic acid} \leftrightarrow \text{dehydro-ascorbic acid} + 2e^- + 2H^+ \quad (23)$$

is quasireversible at the mercury electrode with a halfwave potential of 0.22 V at pH = 1.8 and 0.19 V at pH 10.7. The biologically relevant intermediate radical anion (As·) is produced by radiation in solutions containing ascorbic acid and Br^- ions

$$Br + OH \xrightarrow{Br^-} Br_2^-,\ Br_2^- + As \rightarrow 2Br^- + As\cdot \quad (24)$$

Figure 10 shows the recorded current vs. time curves for a pulsed solution

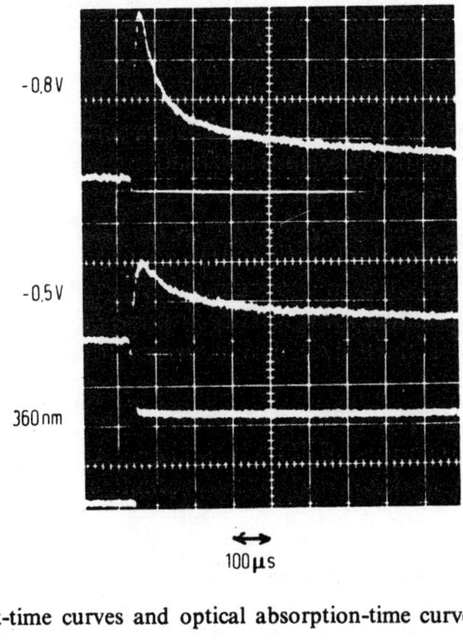

FIG. 10. Current-time curves and optical absorption-time curve from the reduction of the ascorbic acid radical.

containing 0.35 M Na_2SO_4 as base electrolyte, 5×10^{-2} M KBr, 2.5×10^{-2} M N_2O and 10^{-3} M ascorbic acid. A cathodic current is observed indicating reduction of the ascorbic acid radical in the accessible potential region. The current signal increases with increasing negative potential at the electrode, and beyond −.8 V becomes independent of the potential. While the current decreases with time due to the growth of the diffusion layer, the optical absorption of the radical at 360 nm remains constant during the time of observation, since the radical in its anionic form is rather long-lived in alkaline solution. Figure 11 shows the short time polarogram of the ascorbic acid radical at $t_1 = 8$ μs and a plot of the current vs. $t^{-1/2}$ for different potentials. At −0.8 V, a straight line over the whole time

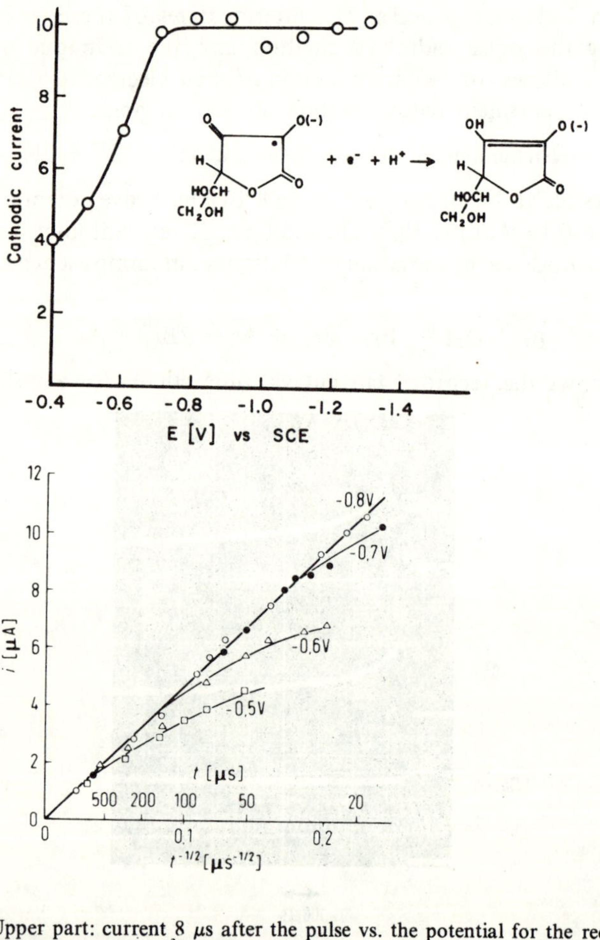

FIG. 11. Upper part: current 8 μs after the pulse vs. the potential for the reduction of the ascorbic acid radical at pH = 10 (10^{-3} M ascorbic acid; 2.5×10^{-2} M N_2O; A = 0.023 cm^2; $c^0 = 1.5 \times 10^{-5}$ M). Lower part: plot of i vs. $t^{-1/2}$ at different potentials for the reduction of the ascorbic acid radical (10^{-3} M ascorbic acid; 2.5×10^{-2} M N_2O; A = 0.023 cm^2; $c^0 = 1.5 \times 10^{-5}$ M).

range is obtained which indicates that essentially diffusion alone determines the current here. This straight line therefore gives i_d as a function of time. The curves for the other potentials are straight and have the same slope as the curve for −0.8 volts at longer times. Deviations from the straight line occur at shorter times, this effect becoming more pronounced with decreasing negative potential. For each curve, λ was obtained according to Eq. (10) by forming the ratio i/i_d at a given time. The value of $k/D^{1/2}$ was then calculated using Eq. (11). The values of $k/D^{1/2}$ obtained from λ values determined at different times for a given potential were in good agreement. D could be calculated as 6.2×10^{-6} cm^2 s^{-1} from the slope of the straight line in Fig. 11 at longer times using Eq. (7a). The absolute rate constant k is plotted in Fig. 12 vs. the potential in a semilogarithmic plot. The transfer coefficient α of the cathodic process as defined by Eq. (11) is obtained as 0.23 ± 0.06 from the slope of the straight line.

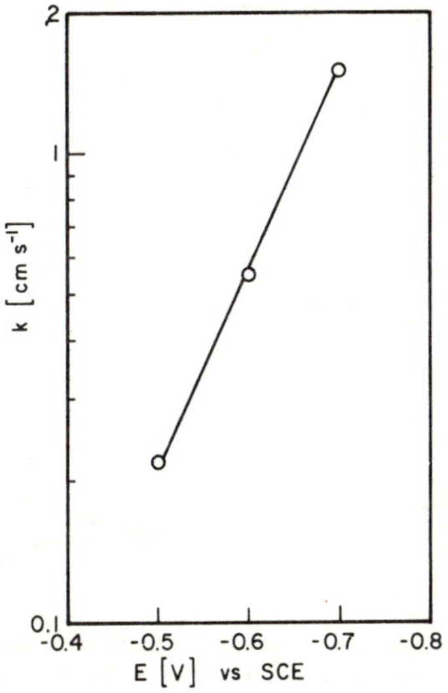

FIG. 12. Semilogarithmic plot of the transfer rate constant k vs. the potential for the reduction of the ascorbic acid radical.

Flavin-adenine dinucleotide is a coenzyme which operates in numerous physiological redox processes, e.g. the respiratory chain. These reactions involve 2-electron reduction of the riboflavin ring to dihydro-riboflavin and reoxidation. This process is shown in Fig. 13. Valuable information about the electrochemical behavior of the transitory intermediates formed by one-electron reduction of riboflavin may be obtained from short-time polarography. The reduction of riboflavin at the mercury electrode is a two-electron reaction with a halfwave potential of ca. 0.6 V at pH 10. Figure 13 shows a conventional polarogram of 2×10^4 M riboflavin, 2×10^{-2} M sodium formate and .2 M Na_2SO_4 in aqueous solution. The halfwave potential is at −.63 V SCE under these conditions. These solutions were of exactly the same composition as used in the pulse-polarography experiments. The samples were degassed prior to the measurements with CO_2 gas. In an aqueous solution containing CO_2 and formate, all the primary radicals are converted into formate radical ions. The latter radical transfers an electron to riboflavin, thereby forming the biologically relevant intermediate. This reaction is practically completed a few microseconds after the pulse. Thus, only the behavior of the riboflavin radical (II) is registered polarographically. The current $i(t_1 = 20 \cdot s)$ observed from this radical is plotted in Fig. 13 vs. the electrode

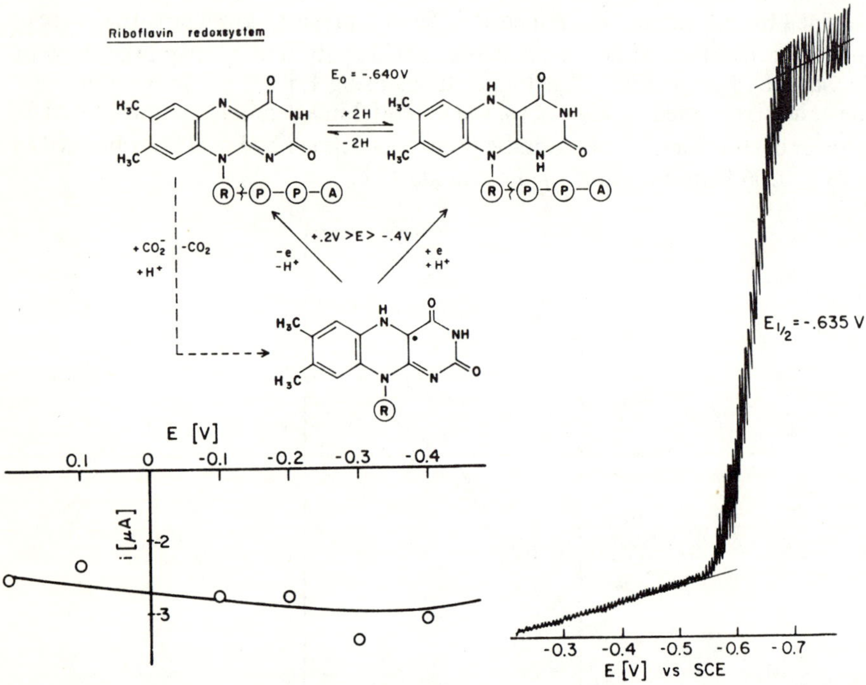

FIG. 13. Redox reactions in the riboflavin/dihydroriboflavin systems; conventional polarogram of riboflavin, short-time polarogram of the riboflavin radical. Solutions: 2×10^{-4} M riboflavin, 0.2 M Na_2SO_4, COI_2 saturated.

potential. This curve shows that the radical II is reoxidized to riboflavin in the accessible potential range of $+2\,V < E < -.4$ V. This indicates that radical II has relatively strong reducing properties. Figure 14 shows an oscilloscope trace from the oxidation of the riboflavin radical. From the kinetic evaluation of this trace a first half-lifetime of the radicals $\tau_{1/2} = 213$ μs is obtained. This is in excellent agreement with the value $\tau = 200$ μs calculated from the literature rate constant for the radical reaction and their initial concentration.

Finally, the intermediate of the NAD^+/NADH redox coenzyme was examined polarographically. The redox reaction and the structure of the intermediate radical are shown in Fig. 15 together with the short-time polarogram of the intermediate. The radical was formed by irradiating N_2O degassed solutions of NAD^+ containing sulfate electrolyte and isopropanol. In such a system all the primary radicals from the radiolysis of water are converted into isopropyl radicals which transfer an electron to NAD^+. The product NAD· radicals have strongly reducing properties. This is indicated by an anodic wave extending towards potentials more negative than −1 V.

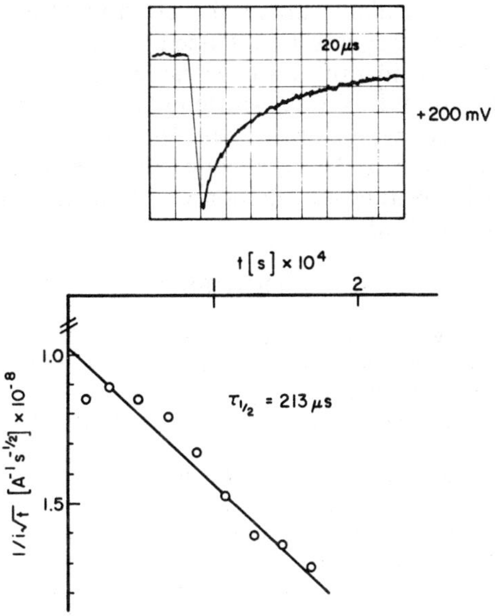

FIG. 14. Current-time curve of the riboflavin radical at +.2 V; kinetic evaluation; initial radical concentration: 5×10^{-6} M.

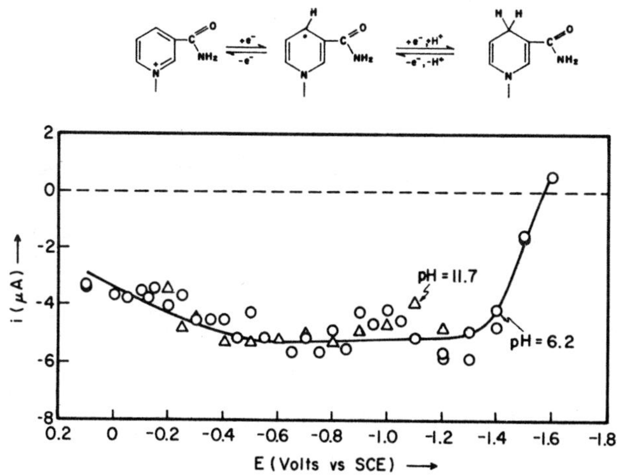

FIG. 15. Redox reactions in the NAD^+/NADH system; short-time polarogram of NAD·; solution: 2×10^{-4} M NAD^+, .05 M isopropanol, N_2O saturated.

CONCLUSIONS

The technique of short-time polarography has been established as a valuable tool in pulse radiolytic investigations. It is now possible to identify short-lived intermediates by their electrochemical properties. Considerable information has become available about the electron transfer properties of organic radicals. This work may now be extended to the study of more complex systems. For example, the identification of radical sites formed during the radiolysis of biological molecules should be attempted. The electrokinetic feature of the technique is connected with the interpretation of irreversible polarographic waves of radicals. The application of Gurney's theory of electron tunneling may reveal interesting information on the distribution of electronic energy levels within the radicals.

REFERENCES

1. J. Lilie, G. Beck and A. Henglein, *Ber. Bunsenges. Phys. Chem.* **75**, 458 (1971).
2. M. Grätzel, A. Henglein, J. Lilie and M. Scheffler, *Ber. Bunsenges. Phys. Chem.* **76**, 67 (1972).
3. M. Grätzel and T. Chen, to be published.
4. M. Grätzel and A. Henglein, *Ber. Bunsenges. Phys. Chem.* **77**, 2 (1973).
5. K. E. Reinert and H. Berg, *Monatsber. Dtsch. Akad. Wiss. Berlin* **4**, 26 (1962).
6. J. R. Birk and S. P. Peroue, *Anal. Chem.* **40**, 436 (1968).
7. M Grätzel, K. M. Bansal and A. Henglein, *Ber. Bunsenges. Phys. Chem.* **77**, 11 (1973).
8. K. M. Bansal, M. Grätzel, A. Henglein and E. Janata, *J. Phys. Chem.* **77**, 16 (1973).
9. M. Grätzel, A. Henglein and K. M. Bansal, *Ber. Bunsenges. Phys. Chem.* **77**, 6 (1973).
10. K. M. Bansal, A. Henglein, E. Janata and R. M. Sellers, *Ber. Bunsenges. Phys. Chem.* **77**, 1139 (1973).
11. A. Henglein in *Electroanal. Chem.*, in press.
12. H. A. Idwarsk and R. W. Dodson, *J. Phys. Chem.* **78**, 892 (1974).
13. K.-D. Asmus, A. Henglein, A. Wigger and G. Beck, *Ber. Bunsenges. Phys. Chem.* **70**, 756 (1966).
14. E. J. Land and A. J. Swallow, *Biochemistry* **8**, 2117 (1963).

Catalysis of Electron and Electron Transfer Reactions in Micellar and Protein Systems[1]

M. Grätzel, M. Cooper and J. K. Thomas
Radiation Laboratory[2] and Department of Chemistry
University of Notre Dame, Notre Dame, Indiana 46556

INTRODUCTION

The rate constants for reaction of hydrated electrons e^-_{aq}, with many solutes have been investigated in aqueous solution (*1*). In many systems rate constants $> 10^{10}\ M^{-1}\ sec^{-1}$ are observed, and it is assumed that the reaction of e^-_{aq} with charged solutes is also well understood, and is satisfactorily interpreted
with charged solutes is also well understood, and is satisfactorily interpreted
by the Debye-Hückel relationship (*2*). However, unusual reaction rates have been observed for reactions of e^-_{aq} with solutes in micellar systems. For example, the rate of reaction of e^-_{aq} with biphenyl, ϕ_2, dissolved in sodium lauryl sulphate, NaLS, micelles, is ~ 1/50 that observed in a simple homogeneous solution (*3*). It is suggested that the negative surface charge of the sulfate groups on the micelle repel e^-_{aq} thereby severely decreasing the rate of reaction of e^-_{aq} with ϕ_2 in the micelle. Biphenyl dissolved in the aqueous phase does react with e^-_{aq}. Only about 2% of the total ϕ_2 concentration is in the aqueous phase compared to ~ 98% in the micellar phase, which leads to a lower rate of reaction. The rate of reaction of e^-_{aq} with pyrene is decreased even further (to less than 1/200) which reflects on the relative solubility of pyrene in the two phases. Similar effects have been observed with the reaction of e^-_{aq} with benzene (*4*).

In systems containing cationic micelles such as cetyltrimethylammonium bromide, CTAB, a net positive charge is built up on the surface of the micelle. These micelles exert an attractive force on e^-_{aq}. It is noted that in the presence of CTAB micelles (*3*) the rate constant for reaction of e^-_{aq} with cupric ions decreases from $3 \times 10^{10}\ M^{-1}\ sec^{-1}$ in water to 1.6×10^{10}. This decrease is larger than that calculated for a salt effect, and may be due to a competition of CTAB and Cu^{++} for e^-_{aq}. As e^-_{aq} does not react with CTAB but is just associated with the positively charged micelle, then the eventual rate of reaction with Cu^{++} is decreased. The micelle may be reactive with e^-_{aq} as in the case of cetyl pyridinium chloride (CPC) micelles (*5*). The net positive charge of the micelle then catalyzes the rate of reaction of e^-_{aq} with the pyridinium group, and rate constants of ~ $10^{12}\ M^{-1}$ are observed. It is concluded that cationic micelles attract e^-_{aq}. In the case of CTAB micelles alone, the lifetime

[1]Congress presentation C-10-3.
[2]The Radiation Laboratory of the University of Notre Dame is operated under contract with the U.S. Atomic Energy Commission. This is AEC Document No. COO-38-964.

of e^-_{aq} is not greatly affected as the micelle itself is not reactive. However, the micelle may compete with other solutes for e^-_{aq} by trapping the hydrated electron in its positive potential field. If the micelle is reactive as in the case of CPC, then an abnormally large rate of reaction of e^-_{aq} is observed.

In the present work the net surface charge on micelles and proteins is changed by the addition of salt or surfactants. Simple surfactants such as NaLS bind with a protein such as serum albumin to give a net negative surface charge to the protein. A similar process with CTAB produces a net positive charge on the protein. Thc effect of charge on the rates of reaction of e^-_{aq} and CO_2^- with micelles and proteins is then investigated.

EXPERIMENTAL

The technique of pulse radiolysis is used to investigate the reactivity of charged species with solutes in or on CTAB micelles. This technique has been described previously (*6,7*), and it suffices here to outline the method. The samples are deaerated by prolonged bubbling with nitrogen gas or carbon dioxide gas. Concentrations of several μM of e^-_{aq} were produced in the samples by irradiation with 10 nsec pulses of 7 MeV electrons from the Notre Dame Arco Linac. The variation of the e^-_{aq} concentration with time was observed by fast (< 2 nsec response) spectrophotometric methods (*8*). The formate radical ion CO_2^-, was produced by capturing e^-_{aq} with CO_2 which was present at saturation point in the systems. In some solutions the hydroxyl radical OH was removed from the sphere of reaction by 0.1 M t-butanol since it was found that the OH reaction with CTAB did not change the kinetics of e^-_{aq} reactions. The reaction of OH with CTAB produces a long-lived transient absorbing at wavelengths below 400 nm.

Cetyltrimethyl bromide was purchased from Fluka Co. and purified by repeated recrystallization from ethanol-ether solution. Pyrene was purified by chromatography. Pyrene 1-sulfonic acid was obtained from Eastman Chemical Co. and recrystallized in ethanol. Pyrene butyric acid was obtained from Eastman Chemical Co. and used without further purification. Crystalline bovine serum albumin (essentially fatty acid free) was purchased from Sigma Chemical Co.

RESULTS

Reactions of e^-_{aq}

Fig. 1 shows the rate of growth of pyrene sulfonic acid (PSA) anion, observed at 493 nm, in the pulse radiolysis of this solute in deaerated water and in 0.05 M CTAB. The pulse radiolysis of both systems gives rise to e^-_{aq} which react with the PSA to form the anion:

$$e^-_{aq} + \text{PSA} \rightarrow \text{PSA}^- \tag{1}$$

In both cases the concentration of PSA is 5×10^{-5} M but the rate of growth of the anion is much faster in the micellar solution.

Kinetic traces such as those shown in Fig. 1 may be used to calculate the rate constant for reaction, by plotting the logarithm of the absorption versus time.

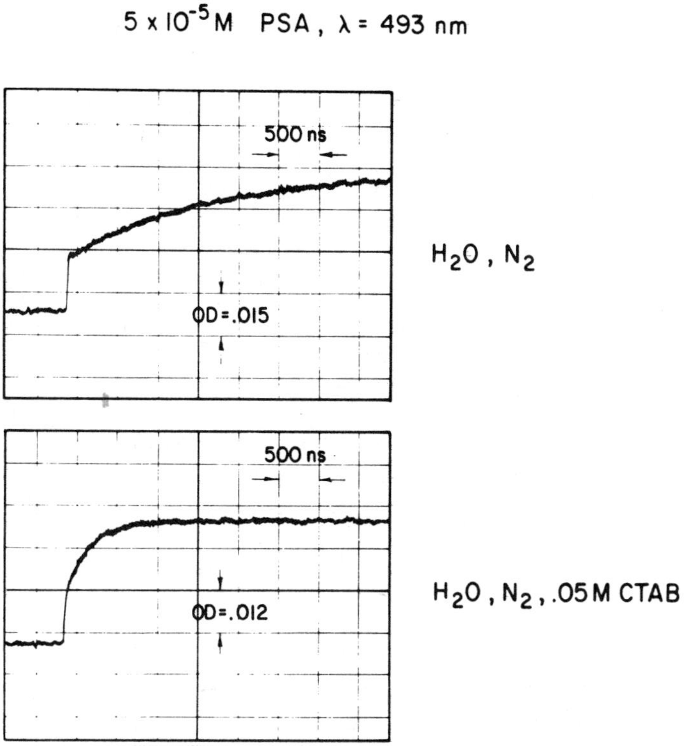

FIG. 1. Oscilloscope traces of PSA anion formation monitored at 493 nm; upper trace. 5×10^{-5} *M* PSA in aqueous solution, lower trace 5×10^{-5} *M* PSA, [CTAB] = 0.05 *M*.

The slope of the line gives a half-life of reaction $t_{1/2}$. The second order rate constant at any pyrene concentration $[P]$ is then given by

$$k = (0.69/t_{1/2}[P])\ M^{-1}\ \text{sec}^{-1} \tag{2}$$

as $[e^-_{aq}] \ll [P]$ pseudo first order conditions prevail in the experimental systems.

In simple aqueous solution the rate constant for reaction (1) is 1.6×10^{10} M^{-1} sec^{-1}, and it may be as high as 2×10^{11} M^{-1} sec^{-1} in the presence of CTAB. Cationic micelles drastically increase the rate of reaction (1). However the rate constant k, for reaction (1) is dependent on the PSA concentration in the micellar system. Fig. 2 illustrates the variation in k with PSA concentration for reaction (1). A continuous decrease in k is observed with increasing PSA concentration. Similar effects are observed for pyrene butyric acid (PBA) in simple aqueous solution, and also when solubilized in CTAB micelles.

The rate of reaction of e^-_{aq} with pyrene solubilized in CTAB micelles is also much larger than that encountered in simple homogeneous solutions. Fig. 3 shows the variation of the rate constant with pyrene concentration at constant CTAB concentration of 0.02 *M*. Up to about 10^{-4} *M* pyrene the rate

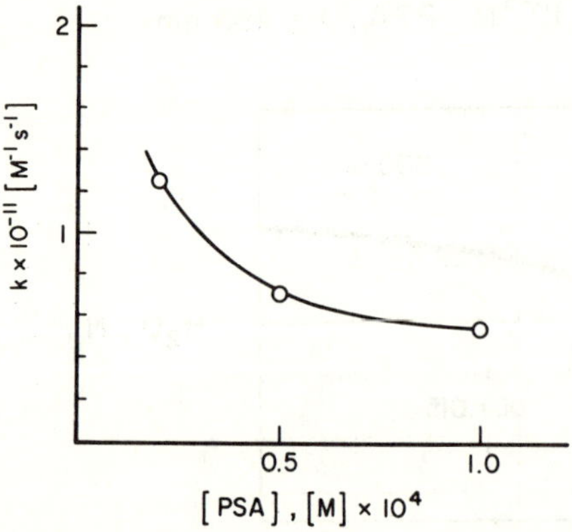

FIG. 2. Rate constants for hydrated electron reaction with PSA as a function of PSA concentration; [CTAB] = 0.02 *M*.

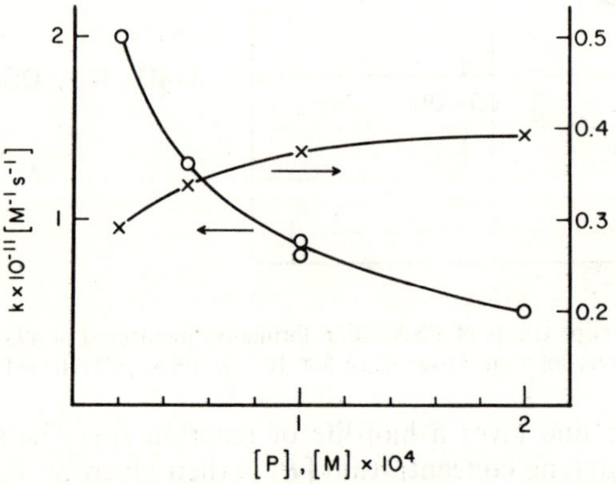

FIG. 3. Effect of pyrene concentration on the pyr$^-$ yield and the rate of pyr$^-$ formation; [CTAB] = 0.02 *M*.

constant decreases. The yield of the pyrene anion P^-, formed by the reaction of e^-_{aq} with pyrene may be expressed as the ratio $[P^-]/[e^-_{aq}]$. Both the concentration of P^- formed and the concentration of e^-_{aq} initially formed by the 10 nsec electron pulse are easily determined from the extinction coefficients of both species at 493 nm ($\epsilon(e^-_{aq}) = 5000\ M^{-1}\ cm^{-1}$ (*1*); $\epsilon(P^-) = 5 \times 10^4\ M^{-1}\ cm^{-1}$ (*9*). A continuous increase in this ratio is observed on increasing the pyrene concentration. The rate constant for reaction of e^-_{aq} with pyrene also depends on the micelle concentration. Fig. 4 shows that the rate constant at a pyrene concentration of 5×10^{-5} *M* continually decreases as the CTAB concentration increases. The yield of pyrene anions, again expressed as the ratio $[P^-]/[e^-_{aq}]$ shows a slight decrease over this range of CTAB concentration.

It appears that the rate constant for reaction of e^-_{aq} with pyrene varies with both pyrene and CTAB or micelle concentration. The ratio of pyrene to

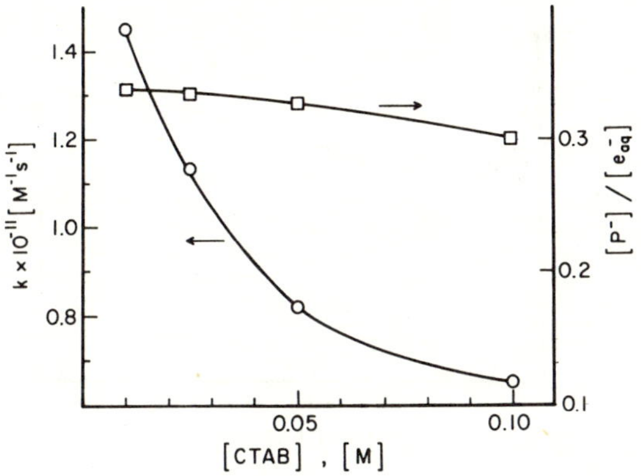

FIG. 4. Effect of CTAB concentration on the pyr$^-$ yield and the rate constant of pyr$^-$ formation; [pyrene] = 5×10^{-5} *M*.

CTAB may be kept constant and the total concentration of solutes varied. The rate constant of e^-_{aq} with pyrene decreases on increasing the total solute concentration, Fig. 5. The pyrene anion yield expressed as $[P^-]/[e^-_{aq}]$ shows a small increase over the same concentration range.

Although the reaction of e^-_{aq} with pyrene is a reaction of a charged species with a neutral one, a strong effect of ionic strength is noted in micellar systems. In Fig. 6 the rate constant of e^-_{aq} with pyrene is plotted versus sodium chloride concentration, at a constant pyrene concentration of 5×10^{-5} *M* and [CTAB. = 0.025 *M*. The rate constant decreases with increasing [NaCl], while at the same time the yield of pyrene anions, $[P^-]/[e^-_{aq}]$ increases.

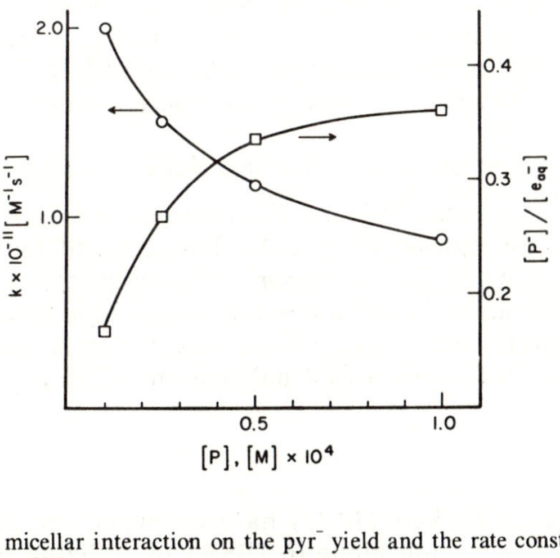

FIG. 5. Effect of micellar interaction on the pyr$^-$ yield and the rate constant of pyr$^-$ formation; [CTAB]/[pyrene] = 500.

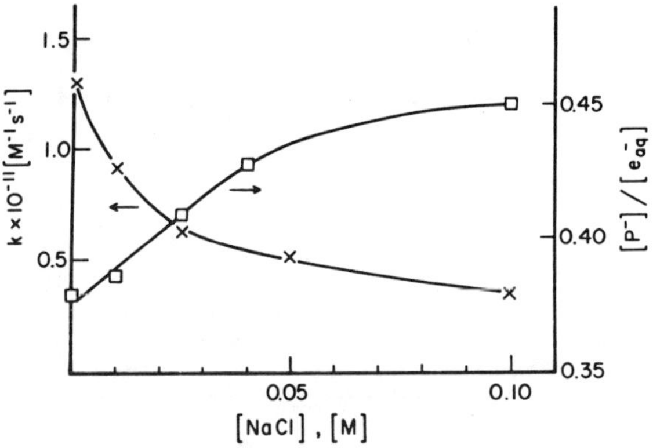

FIG. 6. Salt effect on the reaction rate and yield of e_{aq}^- with solubilized pyrene; [CTAB] = 0.025 *M*, [pyrene] = 5×10^{-5} *M*.

Reactions of CO_2^-

The following section of data is concerned with the reactions of CO_2^- with pyrene, PSA and PBA. The anion CO_2^- is generated in a pulse radiolysis experiment in the presence of CO_2 so that the e_{aq}^- reactions of CO_2^- are monitored by observing the anions of pyrene, PSA or PBA, at a wavelength of 493 nm.

In simple aqueous solutions CO_2^- does not react with PSA. However in the presence of CTAB micelles an efficient electron transfer reaction occurs to give the anion of PSA. A typical oscilloscope trace is shown in Fig. 7. The radical ion CO_2^- also transfers an electron to pyrene solubilized in a CTAB micelle. However there is practically no reaction with PBA solubilized in CTAB micelles.

The variation of the rate constant of the reaction of CO_2^- with pyrene in CTAB micelles with pyrene concentration is shown in Fig. 8. The rate constant decreases with pyrene concentration up to 10^{-4} *M*. The yield of pyrene anions $[P^-]/[e_{aq}^-]$ also increases with pyrene concentration.

Lifetime of the anions of pyrene, PSA and PBA in CTAB micelles

The anions of pyrene, PSA and PBA formed in CTAB micelles either by reaction with e_{aq}^- or transfer from CO_2^-, have long lifetimes usually greater than 5 msec as shown in Fig. 7*c* for $pyr^{\bar{\cdot}}$. This contrasts sharply with the rates of formation of the anions which are in the nsec or μsec time range, depending on the actual experimental conditions. In water the PSA^- anion disappears via a second order reaction with a first half lifetime of 50 μsec at $[PSA^-] = 6 \times 10^{-6}$ *M*.

Serum albumin

Phillips and co-workers (*10,11*) have examined the binding of various penicillins to bovine serum albumin (BSA) by utilizing the reaction of e_{aq}^-

with this protein. It is suggested (*12, 13*) that e_{aq}^- eventually ends up on the sulfur linkages on the protein. The species formed has an absorption spectrum with a maximum at 410 nm. We have studied the effect of the surface charge on this reaction.

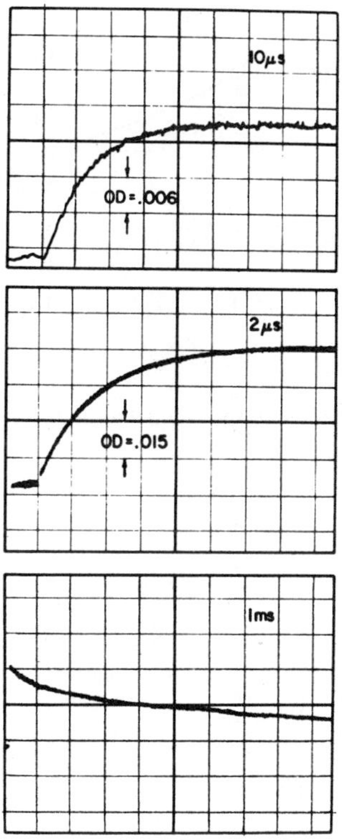

a

b

c

FIG. 7. Oscilloscope traces of electron transfer from CO_2^- to pyrene and PSA in 0.05 *M* CTAB solution λ = 493 nm. *a*) [PSA] = 5 × 10^{-5} *M*; *b*) [pyrene] = 5 × 10^{-5} *M*; *c*) decay of pyrene negative ion formed by electron transfer of e_{aq}^- reaction, same conditions as (*b*).

Negative charge

Fig. 9*a* shows the effect of addition of NaLS on the rate constant for reaction of e_{aq}^- with BSA. At low concentrations of NaLS a slight increase in rate is noted which levels off and then decreases at higher NaLS concentration. The yield of the sulfur electron adduct at λ> 410 nm roughly follows the same behavior Fig. 9*b*.

Positive charge

Fig. 10 shows the effect of addition of CTAB in the rate constant for reaction of e_{aq}^- with BSA. An increase is noted at low CTAB concentrations. This shows a maximum at ~ 4 × 10^{-5} *M*, and a minimum at 7 × 10^{-5} *M*. Above 7 × 10^{-5} *M* the rate constant continually increases to 2 × 10^{-3} *M*, the critical micelle concentration of CTAB. The maximum rate constant for e^-_{aq} + BSA is 6×10^{11} M^{-1} sec^{-1}. The yield of the sulfur/e^-_{aq} adduct is more or less constant in this series of experiments.

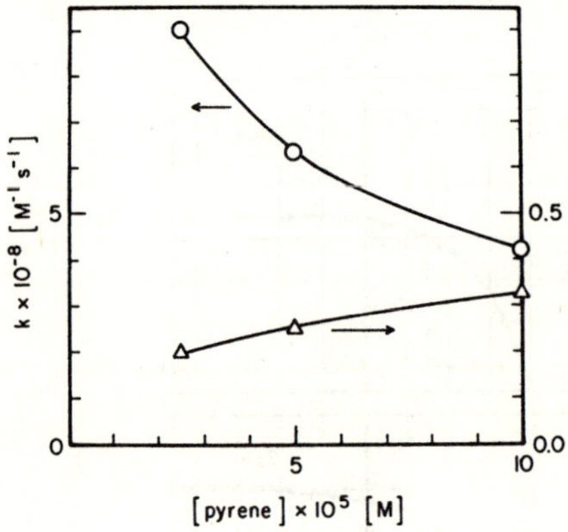

FIG. 8. Effect of pyrene concentration on the rate of the electron transfer from CO_2^- and the yield of pyr$^-$. [CTAB] = 0.05 M.

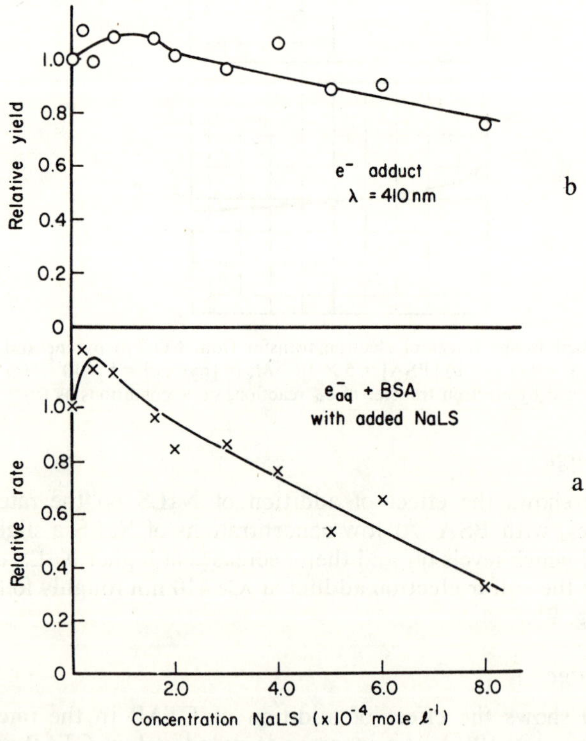

FIG. 9. *a*) **X** Effect of NaLS on the rate of reaction of e^-_{aq} with BSA. *b*) **O** Also the effect of NaLS on the yield of the 410 nm absorption.

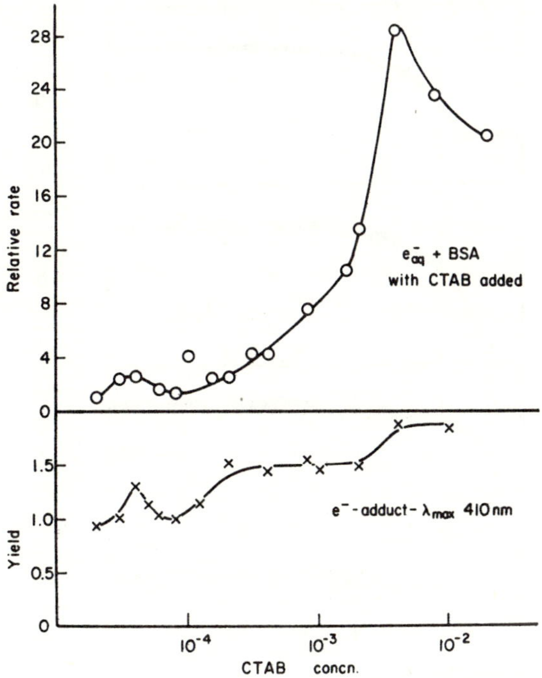

FIG. 10. **O** Effect of CTAB on the rate of reaction of e_{aq}^- with BSA. **X** Yield of 410 nm species.

Electron transfer from CO_2^-

The anion CO_2^- reacts with BSA to produce the sulfur/e_{aq}^- adduct. This is shown in Fig. 11 when observed at 410 nm. The addition of NaLS decreases the rate of formation of this species, while CTAB increases the rate of formation.

DISCUSSION

The data clearly show the catalytic effect of cationic micelles in the reactions of two anions, e_{aq}^- and CO_2^-. The CTAB micelle is nearly spherical and contains CTAB molecules. The micelle surface consists of a large number of positively charged quaternary ammonium groups, of which 80% are neutralized by the Br^- counter ion. The interior of the micelle is non-polar and consists of cetyl hydrocarbon chains. At the concentrations used pyrene is situated in the interior of the micelle in the hydrocarbon region (*14*). However, we have shown earlier that this solute has a significant mobility in the micelle and does move to the micelle surface (*15*). The pyrene is in dynamic motion in the interior of the micelle. The anionic side groups of PSA and PBA are associated with the quaternary ammonium head groups of the micellized surfactant molecules and the pyrene ring protrudes into the micellar palisade

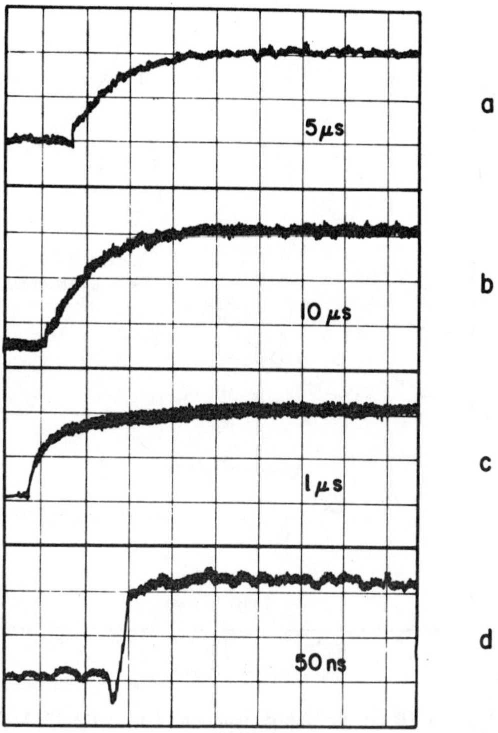

FIG. 11. Reaction of CO_2^- with BSA. Growth of the 410 nm species. *a*) BSA (sweep 5 μsec/cm); *b*) 5×10^{-4} *M* NaLS (sweep 10 μsec/cm); *c*) 6×10^{-4} *M* CTAB (sweep 1 μsec/cm); *d*) 10^{-2} *M* CTAB (sweep 50 ns/cm).

layer. Since in the case of PBA the pyrene ring is separated from the carboxylate group by three methylene groups, its penetration is deeper than for PSA where solubilization occurs in the immediate surface region.

The pulse radiolysis of aqueous solutions initially produces hydrated electrons in homogeneous distribution. In the presence of cationic micelles it is to be expected that e_{aq}^- will undergo rapid trapping in the micellar positive potential field which extends well into the surrounding aqueous phase.

$$e_{aq}^- + M \underset{k_3'}{\overset{k_3}{\rightleftarrows}} (M \cdot e_{aq}^-) \tag{3}$$

Since the CTAB surfactant molecule itself is rather unreactive toward e_{aq}^-, the hydrated electron eventually disappears by reaction with itself, other radicals formed by the radiation, or impurities

$$e_{aq}^- + e_{aq}^- \rightarrow H_2 + 2\,OH^- \tag{4}$$

The half-lifetime of e_{aq}^- is relatively long under these conditions. A probe such as pyrene, PSA or PBA in or on the micelle reacts rapidly with e_{aq}^- leading to a greatly diminished half-lifetime. The initial driving force of the reaction is again the attraction of e^-_{aq} towards a micelle occupied with a scavenger molecule

$$e^-_{aq} + M_0 \underset{k_3'}{\overset{k_3}{\rightleftarrows}} (M_0 \cdot e^-_{aq}) \tag{5}$$

Subsequent intramicellar reaction with the solubilized probe occurs rapidly:

$$(M_0 \cdot e^-_{aq}) \xrightarrow{k_6} (P^-)_M \tag{6}$$

Such a mechanism is suitable to describe the rather unusual kinetic observations displayed in Figs. 1-6. In particular, the dependency of the rate constant and P^- yield on the probe concentration shown in Fig. 2 and 3 can be explained. The decrease of $[P^-]/e^-_{aq}$ with progressive dilution of the probe is expected, since the fraction of micelles occupied by pyrene becomes smaller and these compete less efficiently for e^-_{aq} with other empty micelles. Furthermore, the above mechanism predicts that the half-lifetime of P^- formation should depend only slightly on the probe concentration as indeed is observed. Since Eq. (2) is used to calculate the rate constant this leads to an increase in k with decreasing probe concentration. If the electron adduct was formed in a simple pseudo-first order reaction via

$$e^-_{aq} + (P)_M \rightarrow (P^-)_M \tag{7}$$

then the half lifetime of P^- formation should be always inversely proportional to the probe concentration. Consequently, the product $t_{1/2}[P]$ in Eq. (2) and the rate constant k would not depend on $[P]$. Also, the yield of electron adduct expressed as $[e^-_{aq}]/[P]$ should be close to 1 since e^-_{aq} would react almost quantitatively with the probe. Our experimental findings, however, support competitive trapping of e^-_{aq} by occupied and empty micelles followed by intramicellar electron transfer. Although it is difficult to discriminate the individual contributions of k_3 and k_6 to the overall rate constant, k values like 2×10^{11} at [pyrene] = 2×10^{-5} M indicate that both processes are very rapid. Reaction 3 has already been investigated for solutions of cetylpyridinium chloride micelles where the micellar head groups act as electron scavenger and an unusually large rate constant ($k_3 \geqslant 10^{12}$ M^{-1} s^{-1}) has been obtained. More striking is the experimental evidence that process (6), the penetration of a solvated anion such as e^-_{aq} into the hydrocarbon-like micellar core, followed by reaction with solubilized pyrene can occur efficiently and very rapidly. Questions regarding the detailed nature of this process, for example whether the hydrated electron loses its hydration shell upon entry into the micelle, cannot be answered yet. However, it is reasonable to suppose that the high electrical field in the vicinity of the micelle-water interface facilitates the transfer of an electron from the aqueous phase into the micelle interior as will be elucidated further below. It should be noted that the yield of pyrene negative ion in Fig. 3, although much smaller than 1, is still bigger than the fraction of occupied micelles. For example, at 5×10^{-5} M pyrene in 0.02 M CTAB about one-third of the e^-_{aq} initially produced form P^- while only one-sixth of the micelles have pyrene incorporated. This observation suggests that electron trapping by CTAB micelles is not a completely irreversible process and that exchange of e_{aq}^- between different micelles is possible as is expressed by writing Eq. (3) and (5) as equilibrium reactions. Hydrated electrons then are capable of "hopping" from one micelle to another until a micelle with a solubilized scavenger is found where reaction 6 can occur. This explains why the half-lifetime of P^- formation is not completely independent of probe concentration, but decreases with progressive dilution of probe.

The decisive role of the host micelles in the scavenging reaction of e_{aq}^- by pyrene can also be inferred from Fig. 4, which shows the effect of micelle concentration on the rate constant of P^- formation. Apparently, with a constant pyrene concentration the k values and the yield of P^- both increase with decreasing CTAB concentration indicating a higher probability of reaction in solutions with a larger fraction of occupied micelles. However, by decreasing the number of micelles one not only enlarges the fraction of occupied micelles but one also reduces the possible interaction between individual micelles. Alternatively, at higher concentration there is a more extensive overlap between the micellar diffuse layers (these phenomena are dealt with in more detail below). It will be shown presently that the electrostatic environment in the vicinity of the micelles is changed, and e_{aq}^- is not as readily attracted toward the micellar surface. This would account for the decrease in k and at the same time explain why the yield decreases only slightly at higher CTAB concentration.

The effect of micellar interaction is more clearly demonstrated in Fig. 5 where measurements of $[P^-]/[e_{aq}^-]$ and k at constant pyrene/CTAB ratio but varying total solute concentration are reported. The rate constant increases at higher concentrations reflecting the smaller attraction of e_{aq}^- towards the micellar surface. At the same time the yield of pyrene negative ions increases. This is to be expected if e_{aq}^- has a greater freedom to move from one micelle to another at higher micellar concentrations, where the extension of the electrical potential field into the aqueous phase is diminished. A similar effect and a reduction of the micellar surface potential may be achieved by adding electrolyte to the micellar solutions. Fig. 6 exhibits similar features to Fig. 5. Increasing the ionic strength strongly decreases the rate constant, while $[P^-]/[e_{aq}^-]$ increases. Apparently, k_3 and the equilibrium constant of reaction 3 are subject to a pronounced salt effect. This observation clearly demonstrates the importance of the micellar surface potential and the range of the diffuse double layer for reactions of e_{aq}^- with the solubilized molecules.

The transfer of an electron from CO_2^- in the aqueous phase to pyrene and PSA, solubilized in or on a CTAB micelle, further illustrates the role played by the charged interface of the micelle. In a micelle-free solution CO_2^- does not transfer an electron to PSA. The large positive potential promotes the reaction which occurs readily on a CTAB micelle. The rate of transfer is about five times faster with pyrene than with PSA, which reflects on the nature of the reaction. Presumably, CO_2^- may not enter the micelle as it does not react with PBA. Hence pyrene has to diffuse to the surface in order for electron transfer to occur. This explains why the formation of the pyrene negative ions with CO_2^- is ca. 20 times slower than in the case of the e_{aq}^- reaction. PSA is associated with the surface of CTAB and surface diffusion towards CO_2^- precedes the electron transfer. This process seems to be slower than the diffusion of pyrene within the micelle. The pyrene group of PBA, which is restricted in its movement toward the surface, reacts only very inefficiently with CO_2^-. Such behavior underlines the fact that CO_2^- cannot enter the interior of the micelle and that therefore the electron transfer reaction has to occur at the surface of the micelle. This is an important difference to the behavior of e_{aq}^- which reacts with all three solubilizates, PSA, PBA and pyrene, at approximately the same rate. It shows that e_{aq}^- can readily penetrate the hydrocarbon phase, which is bounded by a positively charged interface. Hence, in this respect the hydrated electron is different to other "normal" anions such as CO_2^-.

The foregoing discussion applies equally well to the case in which a negative species such as e_{aq}^- or CO_2^- reacts with a large protein such as BSA. The negative potential created by absorption of NaLS inhibits reaction. Absorption of CTAB creates a positive potential which promotes the reaction.

The above results suggest that the surface potential of micelles or proteins and the electrostatic environment in the neighborhood of these entities are of considerable importance in influencing the kinetics of electron-transfer reactions. As noted earlier, negative surface potentials greatly enhance the reaction rate, even favoring electron-transfer processes which do not occur in homogeneous aqueous solution. These measurements in model systems suggest that membranes may also exhibit similar kinetic behavior. Thus it might be possible to influence the radiation damage to living cells by suitably controlling the potential of the cell's membranes.

REFERENCES

1. M. Anbar and E. J. Hart, *The Hydrated Electron.* Wiley, New York, 1070
2. M. Matheson, in *Solvated Electron*, (E. J. Hart, Ed.) Advances in Chemistry Series A.C.S. Publication No. 50, p. 45. American Chemical Society, Washington, D.C., 1965.
3. S. C. Wallace and J. K. Thomas, *Radiat. Res.* **54,** 49 (1973).
4. E. J. Fendler and T. H. Fendler, *Advan. Phys. Org. Chem.* **66,** 1472 (1970).
5. M. Grätzel, J. K. Thomas and L. K. Patterson, *Chem. Phys. Lett.* (1974), in press.
6. T. Kajiwara and J. K. Thomas, *J. Phys. Chem.* **76,** 1700 (1972).
7. R. McNeil, J. T. Richards and J. K. Thomas, *J. Phys. Chem.* **74,** 2290 (1970).
8. J. Hunt and J. K. Thomas, *Radiat. Res.* **32,** 149 (1967).
9. D. Gill, J. Japur-Grodzinski and M. Swarcz, *Trans. Faraday Soc.* **60,** 1424 (1964).
10. G. O. Phillips, D. M. Power, C. Robinson and J. V. Davies, *Biochim. Biophys. Acta* **215,** 491 (1970).
11. G. O. Phillips, D. M. Power, C. Robinson and J. V. Davies, *Biochim. Biophys. Acta* **295,** 8 (1973).
12. G. E. Adams, R. L. Willson, J. E. Aldrich and R. B. Cundall, *Int. J. Radiat. Biol.* **16,** 343 (1969).
13. G. E. Adams, in *Advances in Radiation Research*, (M. Burton and J. L. Magee, Eds.) **Vol 3,** p. 125. J. Wiley, New York, 1972.
14. M. Grätzel, K. Kalyanasundaram and J. K. Thomas, *J. Am. Chem. Soc.* (1975), in press.
15. M. Grätzel and J. K. Thomas, *J. Am. Chem. Soc.* **95,** 6885 (1973).

The Picosecond Reactions of Electrons with Biologically Important Molecules[1]

K. Y. Lam and J. W. Hunt

Department of Medical Biophysics,
University of Toronto
and the Physics Division
The Ontario Cancer Institute
500 Sherbourne St., Toronto, Ontario
Canada, M4X 1K9

INTRODUCTION

The technique of pulse radiolysis has been used to investigate the chemistry of molecular damage by ionizing radiation for over a decade. Mainly due to the response time limitation, most of the studies were made in dilute solutions. In the late 1960's Bronskill et al. (*1*) developed a stroboscopic pulse radiolysis (SPR) system which can detect transient signals in the time interval from 30 to 350 psec between the fine structure pulses from the Toronto Linear Accelerator. This has enabled pulse radiolysis studies at a much higher solute concentration (~1 *M*).

Preliminary studies with the SPR system (*2–4*) have shown that reactions occurring in concentrated solutions differ from those in dilute solutions in mainly two respects: (i) the rate constant, $k_{e^-_{aq}}$, for the reaction of e^-_{aq} with electron scavenger, S, as

$$e^-_{aq} + S \xrightarrow{k_{e^-_{aq}}} S^-$$

differs from those measured in dilute solutions for many scavengers depending on the charge of the scavenger (*4*). (ii) the initial yield of e^-_{aq} at 30 psec is drastically reduced by the presence of most scavengers (*4*) and there is a similarly fast process producing the electron adduct S^- (*i.e.*, a fast reduction of the yield of e^-_{aq} and a corresponding fast formation of S^- (*5*)).

As the biological cell is analogous to a concentrated solution (containing up to 20% by weight of solute), the above phenomena may be very important in the study of radiation damage. This paper presents a summary of all reactions of electrons with biologically important molecules observed in the picosecond time scale.

EXPERIMENTAL

The SPR system used for this work has been described in detail elsewhere (*6*). The electron pulse obtained from the University of Toronto linac had a half-width of about 6 ns. Doses in the range from 3 to 8 krad per pulse were obtained. Repetitive pulses with a rate of 60 pulses per second were used and the pulse to pulse amplitude fluctuation was normally about 2%.

[1]Congress presentation C-10-4.

The signal-to-noise ratio of the SPR system has recently been improved considerably so that in a typical experiment signals as small as 0.1% of the analyzing light can be detected.

The amino acid (General Biochemical Corp.) and nucleoside 5′-monophosphate (Sigma Chemical Corp.) solutions were made in triply-distilled water and any required pH adjustments were done by the addition of perchloric acid or sodium hydroxide.

The analysis of the reduced yields of e^-_{aq} and the measurement of the bimolecular rate constants for the reactions of e^-_{aq} with scavengers have been discussed in detail by Wolff et al. (*3*) and by Aldrich et al. (*4*). An outline of the method of measurements of initial yields and reaction rate constant is illustrated in Fig. 1. The concentration of scavengers in moles/liter required to reduce the initial e^-_{aq} yield to 37% compared to that in pure water is defined as the C_{37} value of the scavenger. Since concentrated solutions were studied, a small correction (<5%) was made to account for the differences in the electron densities of the solutions.

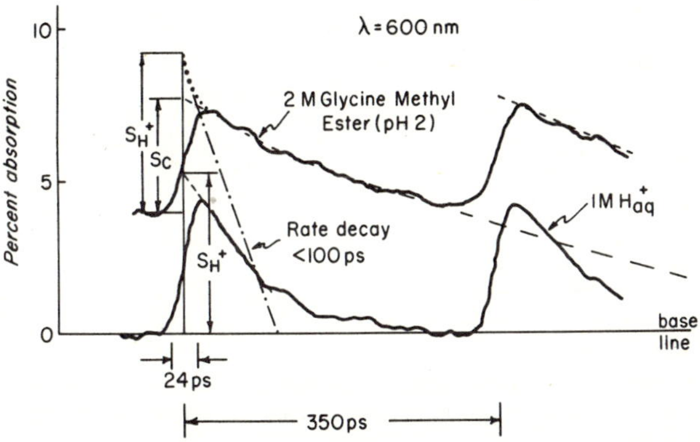

FIG. 1. The traces show the e^-_{aq} absorption signals in the presence of 2 *M* glycine methyl ester (GME) and 1 *M* perchloric acid. The relative initial yields of e^-_{aq} are obtained by extrapolating the absorption traces back to the center of the radiation pulse (dashed lines, and the amplitude of the signals $S_H^+ - S_c$). The difference, $S_H^+ - S_c$, is the reduced initial yield of e^-_{aq} in the 2 M GME relative to that in 1 *M* perchloric acid. The dotted curve..... is the minimum rate of reaction during this 24 psec period. As discussed in other papers, the rate of reaction is probably much faster than this (*2–4*).

RESULTS AND DISCUSSION

a) e^-_{aq} *initial yield reduction:*

It has been shown (*3*) that the initial yield of e^-_{aq} in the presence of scavengers can be described mathematically as follows:

$$[e^-_{aq}] / [e^-_{aq}]_o = \exp(-[S] / C_{37}(S)) \quad (1)$$

where $[e^-_{aq}]$ and $[e^-_{aq}]_o$ are the observed initial concentrations of e^-_{aq} with and without the scavenger S, respectively; [S] is the concentration of scavenger in moles/liter and $C_{37}(S)$ is the C_{37} value of the scavenger. This relation has been found to be true for all the electron scavengers studied in our laboratory. Figure 2 shows the reduction of e^-_{aq} initial yield and the semilog plot for acetone. The mechanism of these fast e^-_{aq} initial yield reductions is not yet clear, even though many models (*7–10*) are capable of explaining part of the observation. Similar results are obtained in other solvents such as the alcohols (*11*). The model most compatible with these observations involves the existence of a chemically reactive precursor to the solvated electron (*11*). This precursor has not yet been directly observed in water. For simplicity, it will be referred to as the dry electron in the rest of this paper.

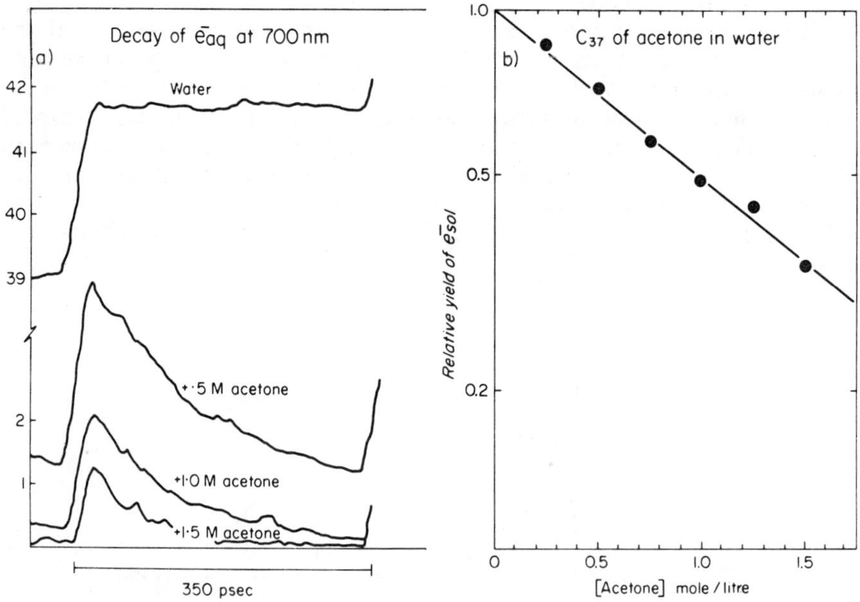

FIG. 2. a. Kinetic traces of the absorption signals of e^-_{aq} in the presence of acetone. Note that for pure water and for low concentrations of acetone, the kinetic traces ride on top of a background absorption.

b. Semilogarithmic plot of initial yield of e^-_{aq} vs acetone concentration. The C_{37} value of acetone is the concentration of acetone required to reduce the relative yield of e^-_{aq} to 37%.

b) C_{37} values and $k_{e^-_{aq}}$:

In early studies in our laboratory (*4*), many amino acids have been found to reduce the initial yields of e^-_{aq} significantly. For instance, the protonated form of cystine $(\text{-S-CH}_2\text{-CH(NH}_3)^+\text{–COOH})_2$ was one of the most effective compounds for reducing the initial yield of e^-_{aq}. A comprehensive study has been undertaken with most of the amino acids that are soluble enough to be studied at high concentration. In addition, nucleic acid components, the mononucleotides, have also been investigated to see whether these fast reactions are involved in cellular radiation damage. Some of the results of initial e^-_{aq} yield reductions are shown in Fig. 3. A list of the C_{37} values and the reaction rate constants with e^-_{aq} for these biologically important compounds at various pH is shown in Table 1.

In the study of many inorganic scavengers in water, Lam et al. (*11*) have demonstrated a correlation between the C_{37} value of scavengers and their reaction rate $k_{e^-_{aq}}$ with e^-_{aq} measured at high concentrations (with the exception of H^+_{aq}). A plot of $\ln C_{37}$ vs $\ln k_{e^-_{aq}}$ for the biological compound studied demonstrates a close relationship between the C_{37} values and $k_{e^-_{aq}}$ values (Fig. 4.). However, the most striking observation as displayed in Fig. 4 is the change of the reaction rate constants from those measured in dilute solutions to those measured in concentrated ones. Large increases in rate constants at high scavenger concentrations are observed for negatively charged scavengers such as guanosine 5′-monophosphate (5′-GMP) and uridine 5′-monophosphate (5′-UMP) at pH 8. In contrast to this, there are large decreases in the rate constants for positively charged scavengers such as cystamine at pH 3.7. These rate constant changes are expected due to the ionic shielding effect of charged scavengers at high concentration.

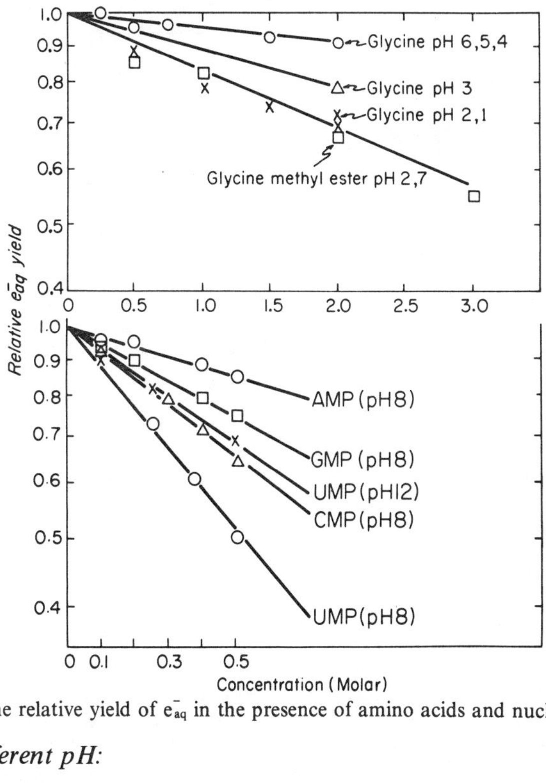

FIG. 3. The relative yield of e^-_{aq} in the presence of amino acids and nucleotides.

c) C_{37} at different pH:

The reactivity of many organic compounds with e^-_{aq} has been shown to depend strongly on the pH (*12*). This variation has been attributed to the configurational changes of the scavenger at different pH. The results in Table 1 show that the C_{37} values of scavengers are also pH dependent. Some of the compounds ionize to different configurations at different pH's (e.g. glycine and oxalic acid) while some do not (e.g. glycine methyl ester and dimethyl cystine hydrochloride). In general, it is observed that when the $k_{e^-_{aq}}$ decreases at lower pH's the C_{37} value increases, and vice versa. On the other hand, compounds whose configurations are independent of pH have constant C_{37} values. This pH dependence suggests that the dry electron and the hydrated electrons are attacking the same functional group or reactive site of the solute molecule. This is in agreement with the general observation that the product of the reaction of scavenger with the dry electron is the same as that with the hydrated electron (K. Y. Lam and J. W. Hunt, manuscript in preparation).

The e_{aq}^- reactivity of large polymer molecules like t-RNA from yeast (molecular weight about 25,000) and the enzyme lysozyme (molecular weight 14,300) is very low. Moreover, at the highest concentrations possible (0.4 *M* per nucleotide or per amino acid component) there was no appreciable decrease on the hydrated electron yield. This is expected because polymers will have a lower reaction rate merely because of a lower collision frequency with the electron, as discussed by Braams et al. (*13*) and Shragge et al. (*14*).

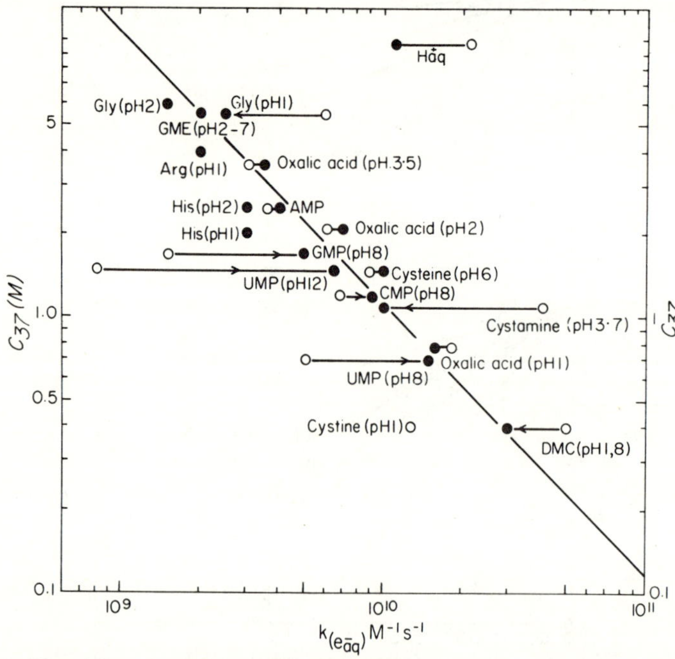

FIG. 4. Plot of C_{37} values of scavengers and their reaction rate constants k with e_{aq}^- on log-log scale. The open points, o, are k measured in dilute solutions taken from literature; •, points with k measured in concentrated solutions as obtained in our picosecond pulse radiolysis system. Note that the arrows joining the o and • points only serve to illustrate the direction of change of k's measured in concentrated solution from dilute solution. Error bars of the points are not shown in the plot. Estimates of error in the parameters measured are shown in Table 1.

TABLE 1

Solute	pH	C_{37} (*M*)	k_{pico} ($M^{-1}s^{-1}$)	References (k_{pico})	k_{micro} ($M^{-1}s^{-1}$)	References $k_{(micro)}$
Glycine (Gly)	1	5.6 ± 1	2.5 ± 1 × 10^9	(*4*)	3.9 × 10^9	(*15*)
Glycine (Gly)	2	6 ± 1	1.5 ± 1 × 10^9	(*16*)	2.5 × 10^9	(*17*)
Glycine (Gly)	3	~10	~1 × 10^9	(*16*)	5.4 × 10^8	(*17*)
Glycine (Gly)	4	>10	⩽10^8	(*16*)	4.6 × 10^7	(*17*)
Glycine (Gly)	6	>10	⩽10^8	(*16*)	8.3 × 10^6 (pH 6.4)	(*18*)
Glycine Methyl Ester (GME)	2	5.5 ± 1	2 ± 1 × 10^9	(*16*)	6.8 × 10^9 (pH 5.3)	(*19*)
	7	5.5 ± 1		(*16*)		
Cystine	1	.39		(*4*)	1.3 × 10^{10}	(*13*)
Cystamine	3.7	1.1	1.0 × 10^{10}	(*4*)	4 × 10^{10}	(*18*)
Dimethyl Cystine	1	.4 ± .1	3 ± 1 × 10^{10}	(*16*)	—	
(DMC)	8	.4 ± .1	3 ± 1 × 10^{10}	(*16*)	5 × 10^{10}	(*20*)
Cysteine	6	1.4 ± .2	1 ± .3 × 10^{10}	(*4*)	8.7 × 10^9	(*18*)
Arginine (Arg)	1	4 ± 1	2 ± 1 × 10^9	(*16*)	—	
	5.3	>10	⩽10^8	(16)	1.5 × 10^8 (pH 6.1)	(*21*)
	14.5	>10	⩽10^8	(*16*)	6.3 × 10^7 (pH 11.5)	(*21*)
Histidine (His)	1	2 ± 1	3 ± 1 × 10^9	(*16*)	—	
	3	2.5 ± 1	3 ± 1 × 10^9	(*16*)	7 × 10^9 (pH <5)	(*21*)
	11	>10	⩽10^8	(*16*)	~10^7	(*21*)
Adenosine 5′-monophosphate (5′-AMP)	8	2.5 ± .5	4 ± 1 × 10^9		3.8-3.6 × 10^9	(*14,22*)
Cytosine 5′–monophosphate (5′-CMP)	8	1.2 ± .2	9 ± 2 × 10^{10}	(*16*)	6.8 × 10^9	(*14*)
Guanosine 5′–monophosphate (5′-GMP)	8	1.7 ± .5	5 ± 1 × 10^9	(*16*)	1.5 × 10^9	(*14*)
Uridine 5′–monophosphate	8	.7 ± .1	1.5 ± .3 × 10^{10}	(*16*)	5 × 10^9	(*14*)
(5′-UMP)	12	1.5 ± .3	6.5 ± 1 × 10^9	(*16*)	8 × 10^8	(*14*)
RNA		>10	~10^9/nucleotide unit	(*16*)	—	
Lysozyme		>10	~10^9/amino acid unit	(*16*)	1.2 × 10^9	*a*
Oxalic acid	1	7. ± .1	1.5 ± .5 × 10^{10}	(*11*)	1.7 × 10^{10}	*b*
	2	2 ± .2	7.5 ± 1 × 10^9	(*11*)	.67 × 10^{10}	*b*
	3.5	3.5 ± .5	3.5 ± .5 × 10^9	(*11*)	.3 × 10^{10}	*b*

[a] P. C. Shragge, unpublished data.
[b] Calculated from the equation given in ref. *12*.

CONCLUSION

Our results indicate that reactions of the hydrated electron with biologically important molecules, such as the amino acids and nucleotides, at high concentration, are quite different from those occurring in dilute solution. The reaction with the dry electron accounts for most of the reaction in concentrated solutions. Fig. 4 demonstrates that there is very little correlation between the reaction rate constants measured in dilute solution and the efficiency of the dry electron reaction ($1/C_{37}$). Consequently, we suggest that there will be no direct correlation between the reactions in dilute solutions and those in the more concentrated solutions found in a living cell. A complete understanding of radiation damage to living cells can only be obtained by direct observation of reactions at high solute concentrations.

REFERENCES

1. M. J. Bronskill and J. W. Hunt, *J. Phys. Chem.*, **72,** 3762 (1968).
2. M. J. Bronskill, R. K. Wolff and J. W. Hunt, *J. Chem. Phys.* **53,** 4201 (1970).
3. R. K. Wolff, M. J. Bronskill and J. W. Hunt, *J. Chem. Phys.*, **53,** 4211 (1970).
4. J. E. Aldrich, M. J. Bronskill, R. K. Wolff and J. W. Hunt, *J. Chem. Phys.*, **55,** 530 (1971).
5. R. K. Wolff, J. E. Aldrich, T. Penner, and J. W. Hunt, *J. Phys. Chem.*, **79,** 210 (1975).
6. M. J. Bronskill, W. B. Taylor, R. K. Wolff and J. W. Hunt, *Rev. Sci. Instrum.*, **41,** 333 (1970).
7. H. A. Schwarz, *J. Chem. Phys.*, **55,** 3657 (1971).
8. W. H. Hamill, *J. Phys. Chem.* **73,** 1341 (1969).
9. G. Czapski and E. Peled, *J. Phys. Chem.*, **77,** 893 (1973).
10. J. R. Miller, *J. Chem. Phys.*, **56,** 5173 (1972).
11. K. Y. Lam and J. W. Hunt, *Int. J. Radiat. Phys. Chem.*, **7,** in press (1975).
12. F. A. Peter and P. Neta, *J. Phys. Chem.*, **76,** 360 (1972).
13. R. Braams, *Radiat. Res.*, **27,** 319 (1966).
14. P. C. Shragge, H. B. Michaels and J. W. Hunt, *Radiat. Res.*, **47,** 598 (1971).
15. P. Neta, M. Simic and E. Hayon, *J. Phys. Chem.*, **76,** 3507 (1972).
16. J. E. Aldrich, K. Y. Lam, P. C. Shragge and J. W. Hunt, *Radiat. Res.*, **63,** in press (1975).
17. J. V. Davies, M. Ebert and A. J. Swallow, in *Pulse Radiolysis* (M. Ebert, J. P. Keene, A. J. Swallow and A. J. Baxendale, Eds.) pp. 165–170. Academic Press, New York, 1965.
18. G. E. Adams, G. S. McNaughton and B. D. Michael, in *The Chemistry of Ionization and Excitation* (G. R. A. Johnson and G. Scholes, Eds.), pp. 281–293. Taylor and Francis Ltd., London, 1967.
19. M. Simic and E. Hayon, *Radiat. Res.*, **48,** 244 (1971).
20. M. Z. Hoffman and E. Hayon, *J. Am. Chem. Soc.*, **94,** 1950 (1972).
21. R. Braams, *Radiat. Res.*, **31,** 8 (1967).
22. E. J. Land and A. J. Swallow, *Biochim. Biophs. Acta*, **162,** 427 (1968).

CURRENT TOPICS IN DOSIMETRY[1]

Chairman: H. O. Wyckoff

[1]Symposium D–7

International Intercomparison of Neutron Dosimetry[1,2]

Randall S. Caswell
National Bureau of Standards, Washington, D. C. 20234, and
Leon J. Goodman and Richard D. Colvett
Radiological Research Laboratories, College of Physicians and Surgeons
Columbia University, New York, New York 10032, U.S.A.

INTRODUCTION

The International Commission on Radiation Units and Measurements (ICRU) has been concerned for many years with the accuracy and comparability of radiation measurements. With the increased interest in the use of neutrons for radiation therapy and radiation biology, it is most important that hospitals and laboratories use compatible measurements. The ICRU, therefore, decided to sponsor an intercomparison of neutron kerma and absorbed dose measurements. This International Neutron Dosimetry Intercomparison (INDI) was carried out using the Radiological Research Accelerator Facility (RARAF) which was made available through the cooperation of Brookhaven National Laboratory and Columbia University. The stated purpose of the intercomparison was: "This International Neutron Dosimetry Intercomparison will be conducted to compare the results obtained by various individuals and / or groups in performing absolute fast neutron dosimetry using similar or different techniques in situations approximating those generally encountered in radiotherapy and radiobiology. Dosimetry for contaminant gamma rays will also be performed. A long-range goal will be to identify the most accurate method(s) of performing absolute fast neutron dosimetry. This Intercomparison is not intended to include instruments designed primarily for radiation protection." It is hoped that out of this intercomparison information will be gained about the state-of-the-art in neutron dose measurement both in air and in phantom, and about the advantages, disadvantages, corrections, and systematic errors involved in various methods of neutron radiation measurement. The intercomparison is intended not to be simply a comparison of instruments (dosimeters), but to actually intercompare neutron dosimetry as practiced in the participating hospitals and laboratories. In a sense, this intercomparison is 11 different comparisons in one (11 irradiation conditions in air and in phantom), or 22 if one counts the gamma ray measurements separately. The experimental measurements were carried out at Brookhaven during the period January to November 1973. This present report will discuss the experimental arrangements

[1]Congress presentation D–7–1.

[2]Supported in part by the United States Atomic Energy Commission and by the Commission of the European Communities (Euratom)

of the intercomparison and give some preliminary results of analysis of the intercomparisons. Results will be given by method only since the participants (listed in Appendix A) have not had an opportunity to review and approve the results. In any case, the results reported should be considered preliminary. Nevertheless, much information about neutron dosimetry methods and practices is already available and will be given here. A much more complete report will be published later.

EXPERIMENTAL ARRANGEMENTS

The RARAF facility utilizes a 4 MV Van de Graaff positive ion accelerator to generate neutrons for a variety of radiobiology and radiological physics studies. A low-scatter facility for neutron measurements in air is available which was used for the intercomparison measurements. Groups came with instruments for two-week periods, typically three groups at a time. Quantities intercompared were: (1) tissue kerma rate in free air at 15.5, 5.5, 2.1, and 0.67 MeV neutron energies and for the ^{252}Cf fission neutron spectrum; and (2) tissue absorbed dose rate at depths of 5, 10, and 20 cm in a distilled water phantom for the 15.5 and 5.5 MeV neutron energies only (as representative of neutron radiation therapy energies). All participants did not do measurements at every neutron energy and position in the phantom. The quantity to be reported was kerma or absorbed dose in ICRU muscle tissue as specified in ICRU Report 10b (*1*), i.e., 10.2% H, 12.3% C, 3.5% N, 72.9% O, and 1.1% other elements, by weight. The nominal radiation quantities and kerma rates are given in Table 1. The experimental arrangement used for the intercomparison is shown in Fig. 1. All dosimetry measurements are made at 0° with respect to the incident accelerator beam. Neutrons are produced by the reactions listed in Table 1 when the accelerator beam strikes the target located at a reference distance of 0 cm. For the ^{252}Cf neutrons, the target is removed and a ^{252}Cf source in a tube is pneumatically driven to the same location. An air-filled, tissue-equivalent-wall transmission ionization chamber is the primary monitor of neutron production. Kerma measurements are made at 30 cm from the accelerator target. For the phantom measurements, an open-top acrylic plastic cube of outside dimensions 30 cm and

TABLE 1
Nominal Radiation Quantities and Kerma Rates

Reaction	Neutron energy, MeV	Energy spread, ± per cent	Max. total tissue kerma rate in air 30 cm from source, rad h^{-1}	Approx. gamma ray tissue kerma rate, per cent of total tissue kerma rate
$^3H(d,n)^4He$	15.4	4	40	5
$^2H(d,n)^3He$	5.5	7	80	5
$^3H(p,n)^3He$	2.1	5	20	5
$^3H(p,n)^3He$	0.67	20	5	5
fission	^{252}Cf spectrum	–	6[a]	40[a]

[a]For a ~ 2mg ^{252}Cf source.

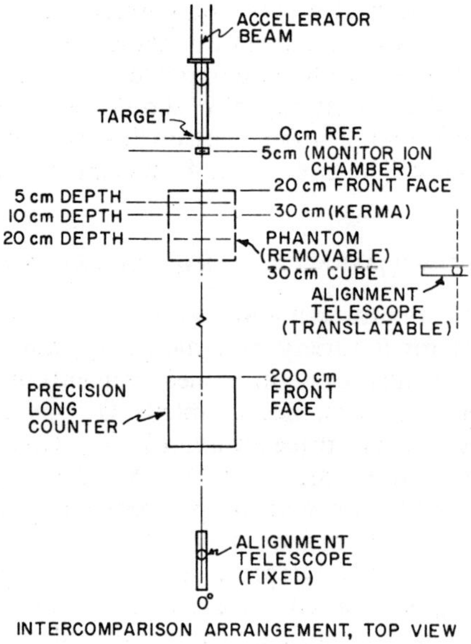

FIG. 1. Experimental arrangement for the intercomparison, top view.

walls 0.64 cm thick held distilled water, chosen over tissue-equivalent liquids because of the high degree of purity and reproducibility possible. A precision long counter (*2*) which served as a neutron fluence monitor was located at 200 cm. Two alignment telescopes were used, a fixed telescope above the beam line at 0° for alignment along a beam axis, and a translatable telescope for distance and height measurements. The optical measuring system was extremely accurate which meant that distance measurement and reproducibility problems could be neglected in the intercomparison. Other quantities monitored were: beam charge, gamma-ray dose with a neutron-insensitive gamma-ray GM counter (*3*), and an acrylic plastic block containing boron counters at three depths which were monitored separately. Neutron spectrometry on the three highest monoenergetic neutron energies was performed with a liquid scintillator with rise-time discrimination.

A 10 Ci ^{137}Cs gamma-ray source enclosed in a housing with collimator, and calibrated by the National Bureau of Standards, was provided for the use of participants desiring such calibrations. A frequent method for calibration of TE ionization chambers used in neutron dosimetry is to calibrate the chambers with gamma rays and obtain the neutron response from the ratio of W_n/W_e, i.e. the ratio of the energies per ion pair in the counter gas for secondary particles produced by neutrons and secondary electrons produced by gamma rays.

PRELIMINARY RESULTS AND DISCUSSION

Results reported below include the measuring system felt to be most accurate by each participant where that information was available. Results are also given for other systems, even though not the preferred system, in order to study the results of different measurement methods. The neutron measurement results are identified by the neutron measurement method used, even though different gamma-ray dosimeters were used by various workers with the same neutron measurement method (e.g., tissue-equivalent ionization chamber). Sometimes results were given both for the assumption of constant W (mean energy per ion pair) and W variable with neutron energy. In the future, both results will be studied, but only the constant W results are reported here.

For ionization chamber instruments, W is probably the major uncertainty in obtaining values for kerma and absorbed dose. Other corrections include neutron sensitivity of the gamma-ray dosimeter, k_u; kerma values for tissue and chamber materials; relative stopping powers; subtended angle of the detector at the measurement position; dead-time corrections and counting statistics; inverse square correction; attenuation and scatter corrections (including void and buildup corrections); saturation, polarity, pressure, and temperature corrections. In addition, uncertainties exist in the absolute calibration methods for dosimeter systems such as the ^{137}Cs source for ionization chamber calibration, the Pu-Be(α,n) neutron source for the polyethylene-ethylene counter calibration, neutron sources for calibration of the precision long counter, and so forth.

15.5 MeV Neutron Energy

Results for kerma in air at 15.5 MeV neutron energy are shown in Fig. 2. The check for stability of results vs time throughout the year for a single system (indicated by the black triangles) indicates reproducibility within a spread of 2%. The 11 tissue-equivalent (TE) ionization chamber results are within a spread of 11%. The polyethylene-ethylene ionization chamber results appear somewhat lower than the average of the TE ionization chamber, the acetylene-polystyrene equivalent ionization chambers are slightly higher than the TE chamber average but not inconsistent, while the silicon diode and precision long counter are definitely higher, although not inconsistent in view of the large error bars given.

The absorbed dose measurements in the phantom at 5 cm show two new characteristics: a much larger spread in values than for the kerma values in air, and the polyethylene-ethylene ionization chambers and more particularly the polyethylene-ethylene counter read lower than the TE chambers (see Fig. 3). The reason for this difference is not known at this time and should be investigated. The reader is cautioned against any bias in favor of the TE chambers simply because they are the most popular system. We, of course, need to understand the physical cause of the difference. A summary of the air and phantom measurements at 15.5 MeV neutron energy is shown in Fig. 4. The measurements shown do not represent depth dose measurements since the measurement indicated by "air" and shown at 0 cm is the kerma measurement in free air (at 30 cm actual distance) and is not a measurement on the front surface of the phantom. Since the ordinate in this figure is logarithmic, one can immediately see the greater percentage spread in the phantom measurements as compared to the air measurements, and the tendency of the C_2H_4 systems to give lower dose than the TE systems.

Gamma-ray dose measurement results at 15.5 MeV neutron energy are presented in Fig. 5. Several observations may be made concerning this figure: (1) the gamma-ray dose is relatively low compared to the total dose both in air and in

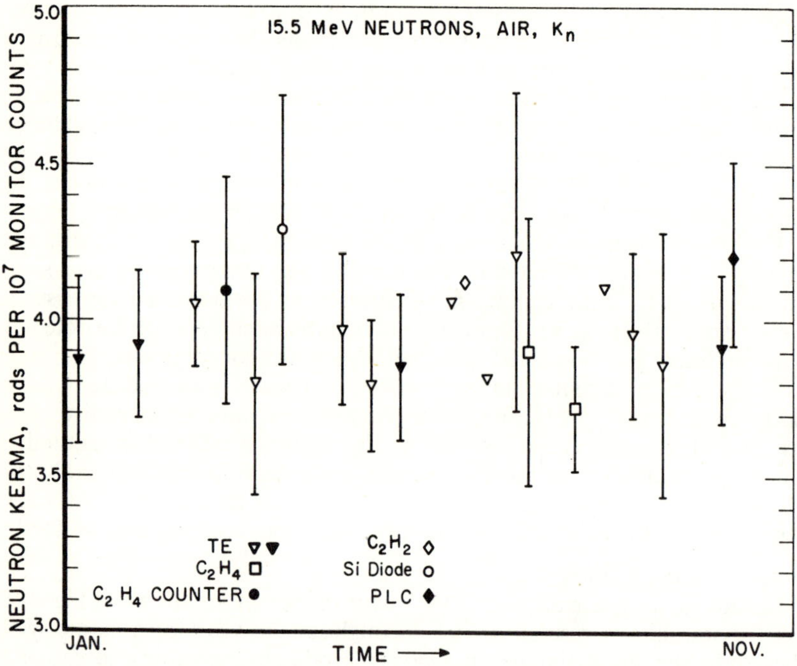

FIG. 2. Results for measurements of neutron kerma in air at 15.5 MeV neutron energy. Results are given for tissue-equivalent ionization chambers (TE), polyethylene-ethylene ionization chambers (C_2H_4), polyethylene-ethylene proportional counter (C_2H_4 counter), acetylene-polystyrene equivalent ionization chamber (C_2H_4), silicon diode (Si diode), and the precision long counter used as a fluence monitor with kerma obtained from fluence-to-kerma factors (PLC). The solid triangles indicate measurements made throughout the year as a test of the overall reproducibility of the measurements. Error bars shown are the estimates of overall error by the participants. Where error bars are not shown, the estimate was not available at the time of writing (the errors may *not* be assumed negligible). The time scale on this and other similar figures extends from January to November, 1973, with the main measurements by groups being made in March, May, July, and September. The time scale is therefore only approximate, but would show a trend if results increased or decreased during the year.

the phantom, probably less than 5% in air and probably less than 15% at all depths—uncertainties in this measurement would usually not be significant radiobiologically due to the higher RBE of neutrons; (2) the gamma-ray dose is very hard to measure when it is such a small fraction of the total dose—some negative results were obtained, here reported as zero; (3) there is a wide variation of results for given methods, indicating a need to see whether there are problems with some of the methods—geometry, varying use of corrections, etc.

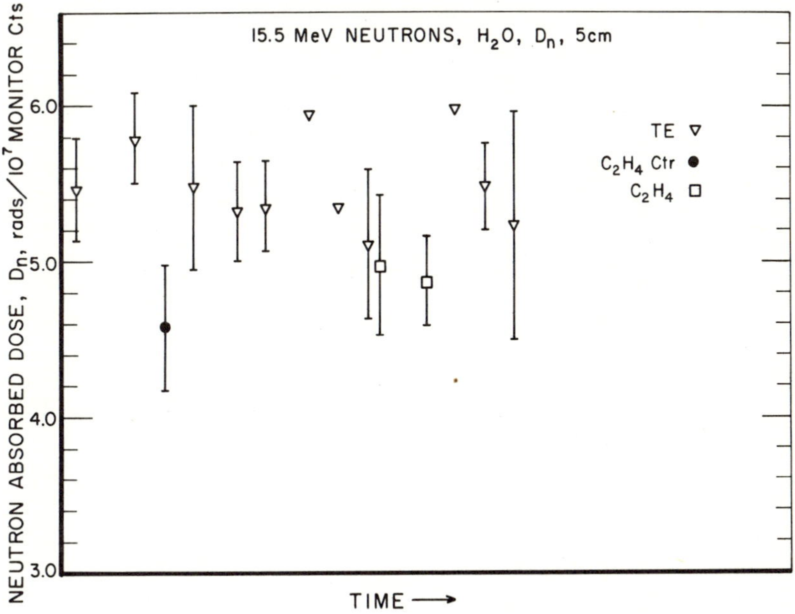

FIG. 3. Results for measurements of neutron absorbed dose at 5 cm depth in the water phantom at 15.5 MeV incident neutron energy. Symbols are as in Fig. 2.

5.5 MeV Neutron Energy

The kerma measurements in air are shown in Fig. 6. Note the stability check throughout the year within a spread of 4%. The Si diode and the precision long counter tend to read high, while the C_2H_4 systems, particularly the counter, read lower than the average. In Fig. 7 are shown the results at 10 cm depth in the phantom. As in previous results shown, there is a very large overlapping of error bars indicating either conservative estimates of error on the part of the participants or for most of the systems large, common errors, such as the uncertainty in the mean energy per ion pair, W, for TE gas. A summary of the 5.5 MeV neutron energy measurements is shown in Fig. 8. Note the increasing spread of the measurements with increasing depth in the phantom. At 20 cm the 5.5 MeV neutrons are greatly attenuated, leading to low, relatively hard-to-measure dose levels.

The gamma-ray dose measurements, shown in Fig. 9, indicate that the gamma-ray dose at 5.5 MeV neutron energy is a somewhat larger fraction of the total dose than at the higher incident neutron energy (15.5 MeV). There usually seems to be a consistency in that instruments which read high where the gamma-ray dose is a small percentage of the total dose, also tend to read high at large depth in the phantom where the gamma-ray dose is on the order of 20 percent of the total dose.

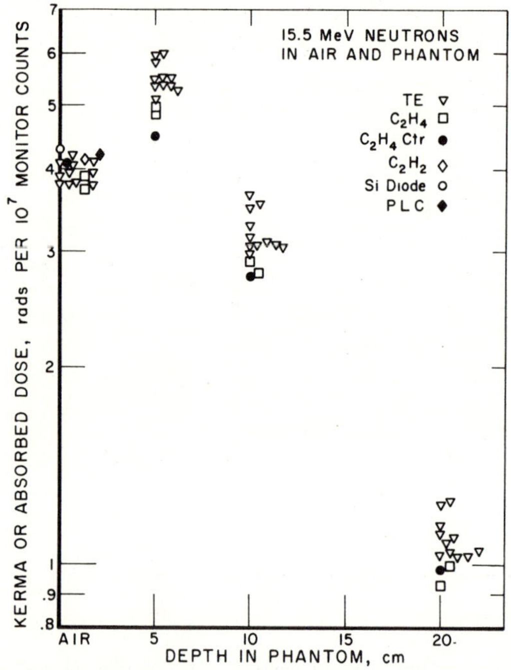

FIG. 4. Summary of results for 15.5 MeV incident neutrons in air and in the water phantom. Since the ordinate is logarithmic, vertical spread of results in a group is a measure of the fractional spread of the actual results. Note the increased spread of the phantom measurements compared to the measurements in air. There is no significance to the lateral spread of points within a group.

2.1 MeV, 0.67 MeV, and ^{252}Cf Neutron Energies

The measurements in air at 2.1 MeV neutron energy, shown in Fig. 10, are similar to those reported in air at the higher energies. There is a spread of 20% among tissue-equivalent chamber results, the Si diode is higher than average while the C_2H_4 counter is lower than average. The precision long counter result is near the mean. Roughly similar behavior is shown for 0.67 MeV neutrons (Fig. 11) and for ^{252}Cf fission spectrum neutrons (Fig. 12) which have a mean energy of about 2.15 MeV and a dose mean energy of about 3 MeV. There are 20 MeV gamma rays due to the ^{3}H(p,γ) reaction present during the 2.1 MeV and 0.67 MeV neutron irradiations which might account for the higher response of the Si diode.

A summary of the neutron measurements in air vs energy is presented in Fig. 13. Note increasing spread of results at the lower neutron energies (probably partly because the dose levels are lower) and the tendency of the TE ionization chambers to read higher than the C_2H_4 systems, but not than the C_2H_2 chamber. Finally, a summary of the gamma-ray measurements in the presence of a predominantly neutron radiation field is given in Fig. 14. Note that all monoenergetic radiation fields are relatively pure neutron fields, containing less than about 4% gamma rays.

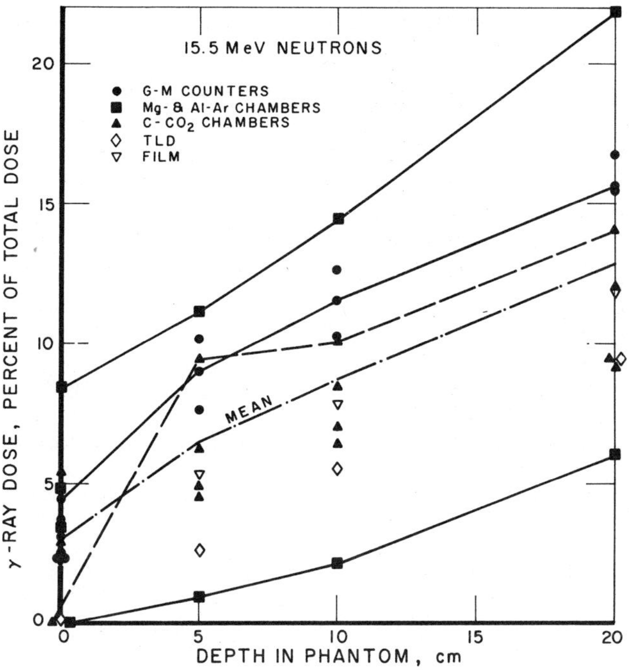

FIG. 5. Gamma-ray dose measurements in air and the phantom for 15.5 MeV incident neutron energy.

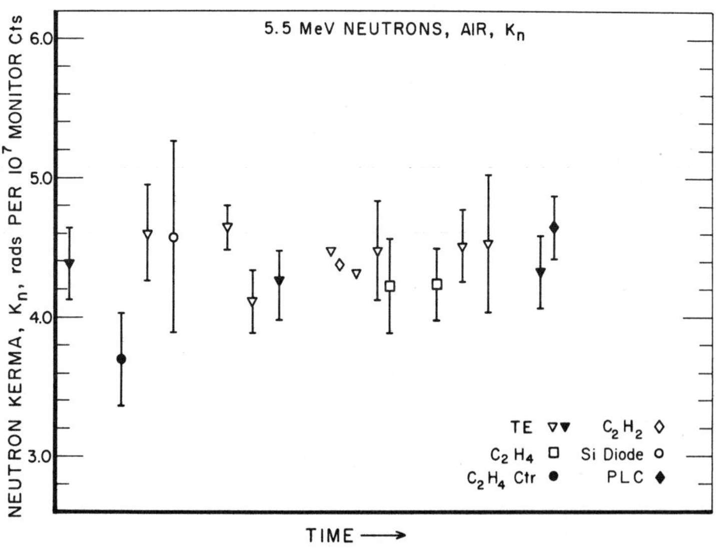

FIG. 6. Results for neutron kerma measurements in air at 5.5 MeV neutron energy.

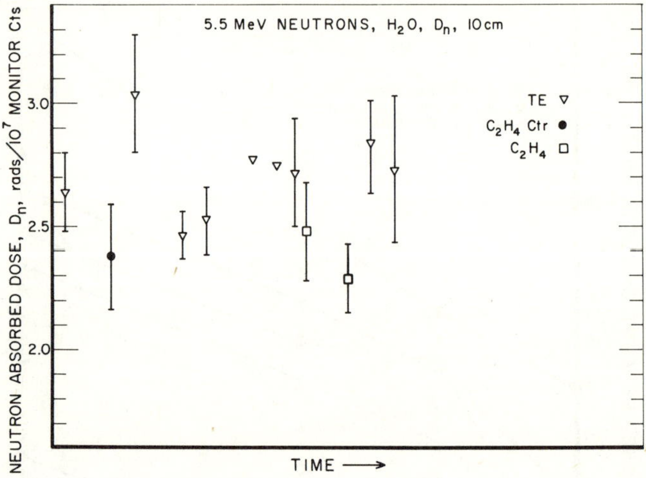

FIG. 7. Results for measurements of neutron absorbed dose at 10 cm depth in the water phantom at 5.5 MeV incident neutron energy.

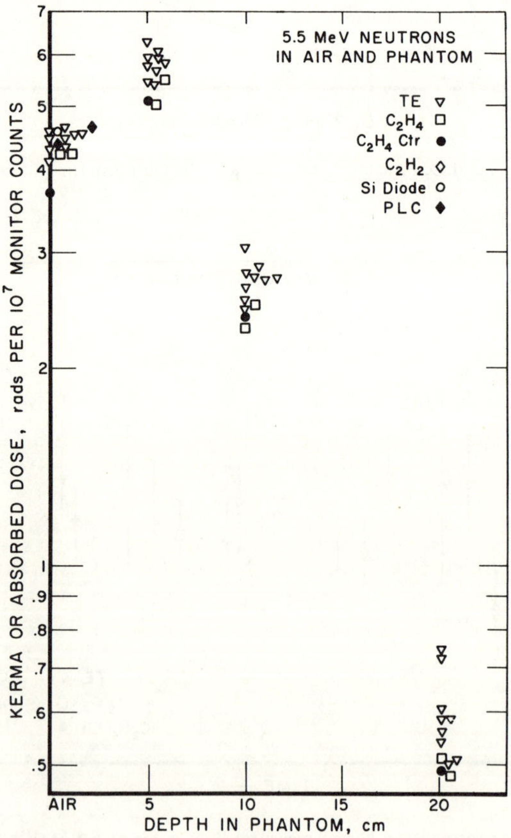

FIG. 8. Summary of results for 5.5 MeV incident neutron in air and in the water phantom.

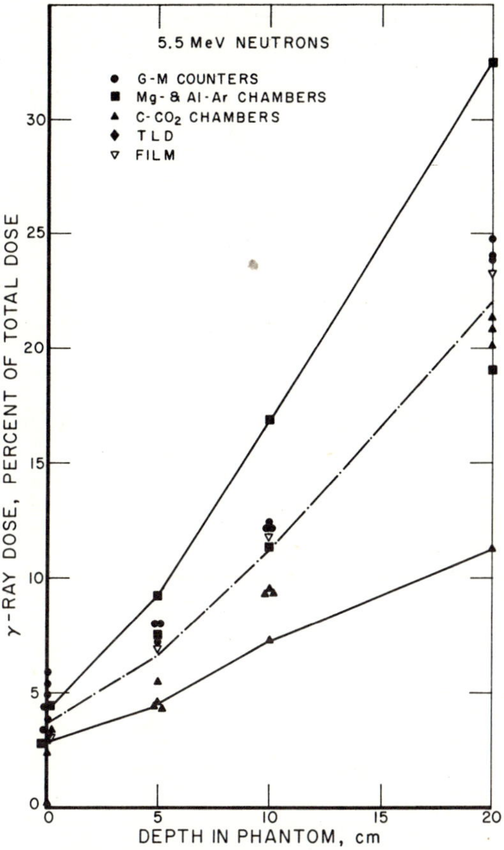

FIG. 9. Gamma-ray dose measurements in air and the water phantom for 5.5 MeV incident neutron energy. The dashed line is the mean of all measurements.

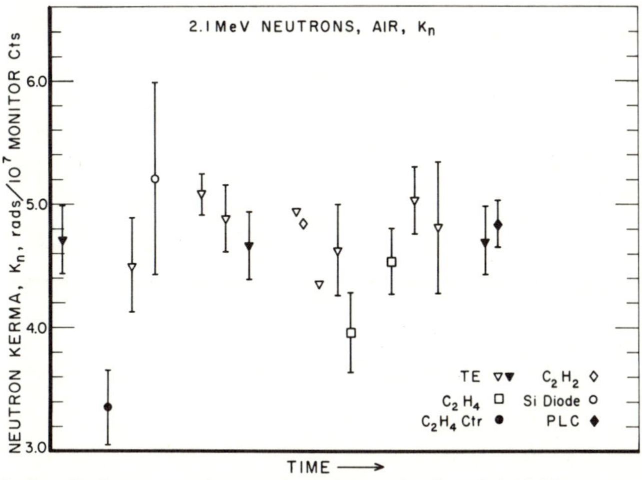

FIG. 10. Results for neutron kerma measurements in air at 2.1 MeV neutron energy.

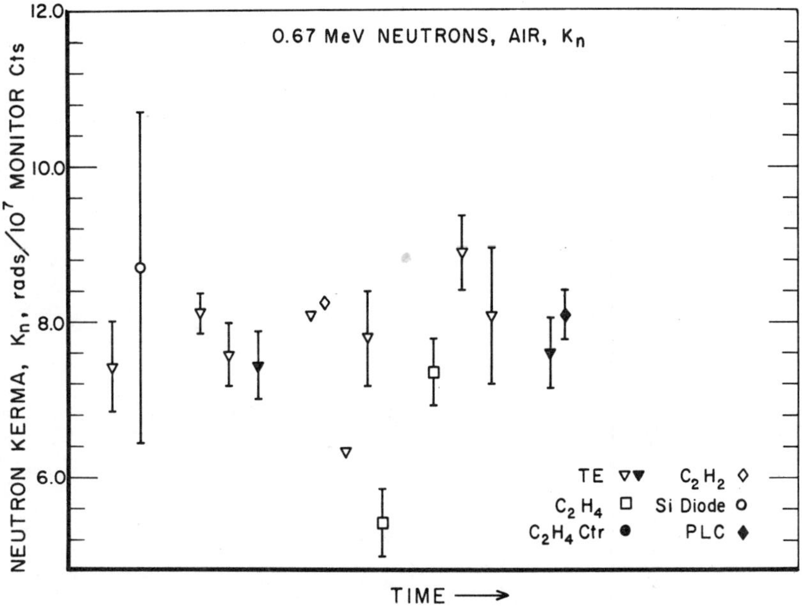

FIG. 11. Results for neutron kerma measurements in air at 0.67 MeV neutron energy.

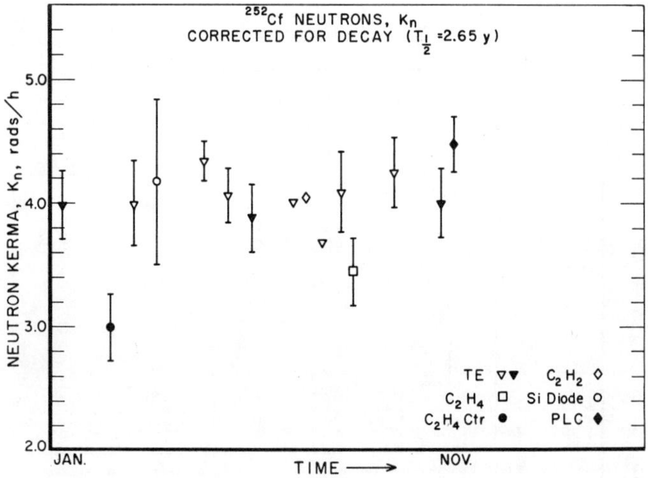

FIG. 12. Results for neutron kerma measurements in air for ^{252}Cf fission spectrum neutrons. The results are corrected for decay assuming a 2.65 year half-life. The maximum correction is about 20%.

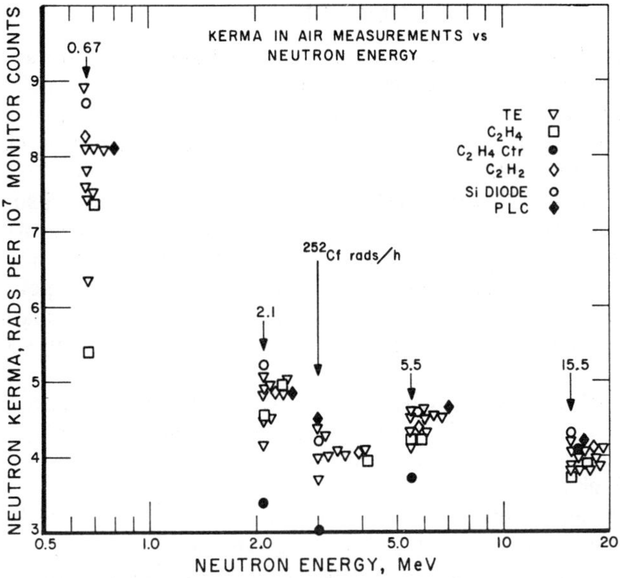

FIG. 13. Summary of the results of neutron kerma measurements in air versus neutron energy. For presentation ^{252}Cf fission neutrons are shown at an assumed dose average energy of 3 MeV.

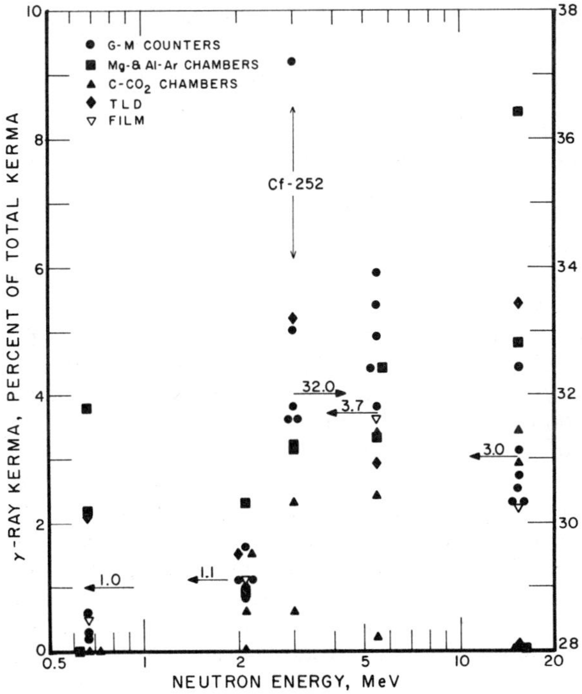

FIG. 14. Gamma-ray kerma measurements in air versus incident neutron energy. All measurements refer to the scale on the left except the ^{252}Cf fission neutron source measurements which refer to the scale on the right.

CONCLUSIONS

Although the results presented here are preliminary and the analysis of the data is not complete, some tentative conclusions may be drawn:

(1) There are no gross discrepancies (e.g., factor-of-2 from the mean) in the measurement of neutron kerma in air or neutron absorbed dose in a phantom among the participants.

(2) Agreement to better than 5% in neutron kerma and absorbed dose, which is desirable in order to have tumor doses known and consistent to 5% (widely recognized as desirable), is achieved in the majority of cases, but not in all cases.

(3) Although the gamma-ray dose uncertainties are larger percentage-wise than the neutron dose uncertainties, measurement of gamma-ray dose is probably satisfactory for those effects where the RBE of neutrons is known to be large compared to unity.

(4) Use of methods which are very different in principle, for example TE ionization chambers vs the silicon diode and precision long counter, give similar, though not precisely identical results (although the sampling of devices other than the TE chambers was small).

(5) The uncertainties of all methods used are sufficiently large that agreement is usually consistent within the given uncertainties. This is not always true, however, for example, at depths of 10 and 20 cm in water for 5.5 MeV neutrons.

(6) The predominantly used method, the tissue-equivalent ionization chamber, shows significant variation as used by the various participants. A study needs to be made of the corrections used by the participants to see if uniform corrections, particularly for W, the mean energy per ion pair, would improve this situation.

(7) The TE ionization chambers and C_2H_4 ionization chambers read differently. The TE chambers tend to read higher, the C_2H_4 chambers lower. This effect is even more pronounced for the C_2H_4 counter at low neutron energies. The reason for these differences needs to be investigated.

(8) Measurements in the phantom show a wider spread in general than measurements in air. Studies need to be made of the corrections adopted to see if this can be improved through uniform or better corrections. This may be related to inadequate corrections for changes in the neutron spectrum in the phantom.

(9) Monitoring measurements indicated that reproducibility of the neutron irradiations given to the instruments being compared was better than ±2%. It is difficult to establish how much better it may be than this figure.

(10) The question of constant W (the mean energy per ion pair) vs various formulations of W variable with neutron energy needs to be investigated experimentally and theoretically for the gases in neutron dosimetry.

(11) The participants are to be commended for the very great amount of effort which participation in an intercomparison such as this one entails. It is hoped that the analysis in progress will answer many of the now unanswered questions, and will point to directions for improvement of neutron dose measurements and guidance for research toward better neutron dosimetry.

APPENDIX A

List of Participants in the International Neutron Dosimetry Intercomparison (INDI)

P. R. Almond, J. B. Smathers, A. R. Smith, V. A. Otte	Department of Nuclear Engineering Texas A&M University College Station, Texas 77834 U.S.A. –and– University of Texas at Houston M. D. Anderson Hospital 6723 Bertner Avenue Houston, Texas 77025 U.S.A.
F. H. Attix	Nuclear Sciences Division U.S. Naval Research Laboratory Washington, D. C. 20375 U.S.A.
D. K. Bewley	MRC Cyclotron Unit Hammersmith Hospital Ducane Road London W12 OHS United Kingdom
J. J. Broerse	Radiobiological Institute of the Organization for Health Research TNO 151 Lange Kleiweg Rijswijk (Z.H.), The Netherlands
G. Burger, E. Maier	Gesellschaft für Strahlen-und Umweltforschung MBH D-8042 Neuherberg Ingolstadter Landstrasse München, West Germany
M. Chemtob, R. Soulie	Department de Protection Service de Protection Sanitaire Commissariat a l'Energie Atomique Fontenay-aux-Roses, France
H. W. Dickson, D. J. Christian	Health Physics Division Oak Ridge National Laboratory P. O. Box X Oak Ridge, Tennessee 37830 U.S.A.
L. J. Goodman, R. D. Colvett	Radiological Research Laboratories Columbia University 17 Cornell Avenue Brookhaven National Laboratory Upton, New York 11973 U.S.A.
D. Greene	Christie Hospital and Holt Radium Institute Withington, Manchester M20 9BX United Kingdom
K. Kawashima, K. Hoshino, T. Hiraoka, H. Matsuzawa	National Institute of Radiological Science 9-1 4-Chome Anagawa Chiba-shi, Japan
H. Kühn	Gesellschaft für Strahlen-und Umweltforschung MBH 6000 Frankfurt/Main Paul-Ehrlich Str 15.u.20 West Germany

D. Shosa, W. E. Kiker	Radiological Physics Division Defense Nuclear Agency Armed Forces Radiobiology Research Institute Bethesda, Maryland 20014 U.S.A.
G. E. Tumbraegel	Erprobungsstelle 53 der Bundeswehr 3042 Munster Postfach 342 West Germany
F. S. Williamson, G. L. Holmblad, J. Spokas	Division of Biological and Medical Research Argonne National Laboratory 9700 South Cass Avenue Argonne, Illinois 60439 U.S.A. –and– Illinois Benedictine College Lisle, Illinois 60532 U.S.A.

APPENDIX B

ICRU International Neutron Dosimetry Intercomparison Committee

Members:
D. K. Bewley, Hammersmith Hospital, London W12, England
J. J. Broerse, Radiobiological Institute, TNO, Rijswijk, Netherlands
R. S. Caswell, National Bureau of Standards, Washington, D.C., U.S.A. (Chairman)
K. W. Geiger, National Research Council, Ottawa, Canada
L. J. Goodman, Columbia University–Brookhaven National Laboratory, New York, U.S.A.
J. W. Müller, Bureau International des Poids et Mesures, Sèvres, France

Consultant: R. D. Colvett, Columbia University–Brookhaven National Laboratory, New York, U.S.A.

Sponsors from ICRU:
A. Allisy, Bureau International des Poids et Mesures, Sèvres, France
H. H. Rossi, Columbia University, New York, New York, U.S.A.
W. K. Sinclair, Argonne National Laboratory, Illinois, U.S.A.
H. O. Wyckoff, International Commission on Radiation Units and Measurements, Washington, D.C., U.S.A.

REFERENCES

1. International Commission on Radiation Units and Measurements, Report 10b. *Physical Aspects of Irradiation. Natl. Bur. Stand. (U.S.) Handb.* **85**, 4 (1964).
2. J. De Pangher and L. L. Nichols, *A Precision Long Counter for Measuring Fast Neutron Flux Density*. Battelle Northwest Laboratory Report BNWL-260 (1966). [Available from Clearinghouse for Federal Scientific & Technical Information, Springfield, Va. 22151.]
3. E. B. Wagner and G. S. Hurst, *Health Phys.* **5**, 20 (1961).

Dosimetric Parameters and Terminology Relevant to Radiobiology[1]

R. Oliver
Royal Postgraduate Medical School
University of London
DuCane Road
London W12 OHS, U.K.

BACKGROUND

In 1963 the ICRU produced its Publication 10e on "Radiobiological Dosimetry" (NBS Handbook 88). A small committee has now been set up under Dr. V. Bond[2] to draft a revised edition, extending the recommendations to cover the present requirements of radiobiology, and this paper is intended as a review of some of the problems considered and the proposals made, for early consideration and comment by both the physicists and the biologists concerned.

Two main sections are envisaged. The first, as the title would suggest, will deal with the specification and measurement of radiation dose, some consideration being given to the special problems of micro-dosimetry as these are of concern in radiobiology. The second will make some proposals on terminology related to the specification of dose response and its modification, as, in discussion, it is clear that recommendations on dosimetry must be closely linked with the applications for which they are required, and unclear terminology can lead to confusion.

THE SPECIFICATION OF DOSE

In the first instance, the biological effect is related to the absorbed dose, that is the ratio of the average energy deposited by the ionizing particles per unit mass measured in $J.kg^{-1}$ or in rad (with one rad equal to 0.01 $J.kg^{-1}$). Dosimetry is the technique of measuring or predicting this energy deposition.

Providing the irradiation conditions are repeatable and the *relative* doses are monitored with sufficient accuracy, the *form* of the dose response can be investigated, and this has sufficed for many earlier studies. But comparison of similar results obtained in different centers, or even with different biological systems in the same laboratory, requires an attention to *absolute* accuracy in the dosimetry which has not always been available in biological departments. The experimenter should be able to make his own dose measurements or calculations, and sections dealing with both these are proposed. But he should

[1] Congress presentation D–7–2.

[2] S. Curtis, M. M. Elkind, Patricia Lindop, R. Oliver and W. Pohlit.

also provide adequate *specification* of the irradiation conditions, so that these can be reproduced by other workers and independent estimates made of the doses involved.

The *quality* of the radiation should be stated, at least in terms of the photon or particle energies, or for orthovoltage X-ray equipment the peak kilovoltage together with the filtration or half value thickness for the radiation beam. For more detailed considerations it is necessary to provide information on radiation quality in terms of the track or dose averages for energy transfer by the ionizing particles (LET), and reference to relevant specialist publications will be given (e.g. ICRU Publication 16; ref. *1*).

For radiobiological studies, the exposure (in roentgens) has often been measured in air at the point to be occupied by the surface or the mid-line of the irradiated object. Provided that adequate information on field size, distance, thickness of object, etc. are given, such a statement of the exposure will usually allow the irradiation to be repeated and the doses estimated. However, a more detailed statement of dose should, in general, be attempted. If the radiation and field arrangements correspond to the conditions of full scatter used in radiotherapy, depth dose distribution and surface dose can be deduced from standard data and isodose charts. However, for many radiobiological situations such conditions cannot be provided, and it is then necessary to attempt a measurement of the dose in the particular tissue or tissues of interest. Considerable care is necessary to achieve sufficient accuracy in these determinations, and it must be re-emphasized that a statement of the absorbed dose in rad (or in the future in $J.kg^{-1}$) requires with it a statement of the tissue concerned.

The complete irradiation may be delivered with one or a combination of radiation fields, but will not, in general, be completely uniform within the target tissue or volume. If, therefore, the mean target dose (or alternatively the modal dose) is stated, some further information on the dose variation should be included, at least indicating the maximum and minimum dose within this region. In radiotherapy $\pm$ 10% of the average dose has been regarded as acceptable, although it is often possible to reduce this to $\pm$ 5% when using high energy photon radiation. Such uniformity may not be practicable in some radiobiological experiments, but the irradiation technique should be considered in detail at the planning stage and designed with this problem in mind. Maximum uniformity should be aimed at, but this should not depend too critically on the alignment of the fields involved. In this connection Sinclair (*2*) has defined irradiation as *uniform* when there is a maximum dose ratio of not more than 1.10 within the target volume, *moderately uniform* for ratios between 1.10 and 1.30 and *non-uniform* for conditions with ratios above 1.30.

Although with high energy photon radiation the energy absorbed (dose) does not vary rapidly with the atomic number of the material, with lower X-ray energies this dependence is often important. Absorbed dose may then vary from tissue to tissue, and change suddenly at, for instance, the interface between bone and soft tissue, or between glass and a layer of attached cells. Experimenters should also be aware of the corresponding problem of dose "build up" near an interface with high energy (e.g. ^{60}Co) radiation. Specimens may need to be covered so as to provide secondary particle equilibrium and accurately predictable doses. Irradiation procedures should, so far as possible, avoid such discontinuities within the target region, and, in any case, the details given should allow any problems of this type to be appreciated.

Taking account of these points the specification of dose should include:

The *nature and energy* of the radiation used (e.g. 250 kVp X-radiation with 2 mm Cu h.v.l, ^{60}Co gamma radiation, fission neutron beam, etc.). For some purposes a statement of the LET of the radiation will be required.

The *dimensions and arrangement* (source-skin distance, relative angulation, etc.) of the treated fields relative to the object irradiated.

The *average dose* (rad or $J.kg^{-1}$) delivered to the target volume, preferably with an estimate of the accuracy of this estimate. A less desirable alternative is a statement of the exposure at the position of the mid-line of the object to be irradiated, together with dimensions enabling the required target dose to be estimated.

The *maximum and minimum* dose in this target region, or other information about uniformity (e.g. ± 10% of average).

Details of any *inhomogeneities* likely to produce local dose variations within the target volume.

The corresponding dose received at any *other relevant sites* as these may, in some circumstances, affect the overall biological effect observed.

SPECIFICATION OF THE IRRADIATION PROCEDURE

As recovery of potentially lethal or sub-lethal radiation damage is possible, in general, the effect of a given dose on a cell system or organism is reduced for irradiation at a low dose rate or in a series of fractional exposures spaced in time, compared with that for a single short exposure. Specification of the procedure should therefore include relevant information on fractionation and dose rate, and the following terms are suggested:

The *total dose* given in N fractions = D_N (rad)

(D_1 and D_2, the equivalent doses given in one or two fractions are particular values often referred to in radiobiology and must be distinguished from the common use of D_1 in description of the initial slope of the survival curve for which $D_{0,1}$ is now recommended.)

The dose in the ith of the N fractions = ${}_iD_N$ (rad).

The overall *duration* (the *protraction*) of the procedure = T (days or hours).

The spacing between the ith and the $(i + 1)$th fractions = ${}_it_N$ (days or hours).

The average *dose-rate* during exposure = $\dot{D}$. (Further details may be necessary if there is significant variation of the dose rate within or between fractions, for instance for a pulsed radiation or when a short-lived radionuclide source is used.)

Where the stated uniformity of the physical dose is achieved by the use of several radiation fields, the biological effect in different regions will depend on whether or not this dose uniformity is maintained for each fractional exposure, or only obtained by arithmetical summation of doses delivered on separate occasions. Thus, details of the fractionation *pattern* will also be relevant to a proper description of the complete procedure.

Biological "recovery" will include the effects of "repair" associated with intracellular processes, and of "repopulation", by division or migration of cells into the region. It must be appreciated, therefore, that even a detailed specification of the physical parameters of the irradiation will not completely specify the biological end effect. For this purpose information would also be required on parameters which might affect the ability of the cells or organism to cope with radiation damage—factors such as the culture medium used, the temperature, the extent of physiological stress, etc.

THE MEASUREMENT OF DOSE

Two general techniques of dose measurement are available and will be discussed in the Publication—the use of a small *probe* to measure the variation of dose from point to point, and the use of a *sample exchange* system where the dosimetric material replaces the whole volume of interest and yields a reading proportional to the *average dose*. In the former case the probe will usually be a small ionization chamber. For the latter procedure, ferrous sulfate provides a convenient chemical dosimeter. Thermoluminescent dosimetry (TLD) may be used in suitable circumstances for either technique, and this system is now proving very useful in radiobiological studies. An outline of practical dosimetric procedures is proposed, but for more detail reference will be made to other ICRU publications, particularly numbers 17(*3*) and 21(*4*).

In the probe technique the small radiation detector is placed as near to the point of interest as possible. For instance, this may be on the surface or in a body cavity, or alternatively at corresponding points within a phantom which is made as life-like as possible from the point of view of dimensions, effective atomic number and electron density. The detector must not disturb significantly the radiation field, and the dose must be relatively uniform over the volume of the detector. The suggested rule is that the linear dimension of the ionization chamber should not exceed 0.1 of that of the irradiated object.

If then the signal obtained from the dosimeter is proportional to the energy absorbed in the probe, the dose in the surrounding medium may be calculated, provided the detector has a wall of similar material with the following dimensional requirements:

Either (1) The wall thickness is greater than the range of the secondary particles, so that local secondary particle equilibrium is established, but not so thick as to introduce significant attenuation of the primary photon fluence, e.g. represents less than 5 percent of the corresponding mass absorption coefficient.

A wall thickness of 0.2 mm of unit density material falls within these limits for photons from about 20 keV to some 300 keV. For 1 MeV radiation a wall thickness of about 2 mm is required.

or (2) For higher photon energies where these conditions cannot be met, the wall thickness is greater than the range of the tertiary particles or delta rays (so that equilibrium is established with these), but does not appreciably attenuate the flux (e.g. is less than 5 percent of the range) of the secondary electrons which are here treated as the primary radiation component. On this basis a wall thickness 0.2 mm is adequate for 3 MeV photons or, of course, for an electron beam of corresponding equivalent energy.

For an air ionization chamber with air equivalent walls in the first category, the dose in the surrounding tissue or other medium is given by the exposure in roentgens multiplied by a factor f_m, which contains the ratio of the mass energy absorption coefficients for the medium and for air respectively. A table (*4*) will be provided for f_m listed against photon energy or half value thickness of the X-ray beam. In the second category, the dose in the surrounding medium is given by the ionization per unit mass of gas and a factor, g_m, which contains the ratio of the mass collision stopping powers for the secondary electrons in the medium and in air, respectively. Average values of this factor, calculated by Pohlit for this publication, will be listed against the maximum photon energy of the X-ray beam or the mean energy for a beam of fast electrons.

Ferrous sulfate solution with the lower concentration of sulfuric acid recommended here by Pohlit[3], and lithium flouride or borate as used for TLD are all effectively water or soft tissue equivalent. Absorbed dose in tissue then corresponds to the energy absorbed per g in the dosimeter. However, the higher *density* of LiF (2.16 $g.cm^{-3}$) must be borne in mind if many such dosimeters are to be used at a time to explore dose distribution. The rule is suggested that the increase in mass due to the dosimeters should not exceed 1% of the total mass of the irradiated object.

MICRODOSIMETRY

The simple concept of absorbed dose related to the biological effect is relevant, for instance, in radiotherapy where the dose is high and essentially uniform throughout the tumor. However, in relation to microscopic targets, it is important for the biologist to appreciate that the energy is transferred by discrete secondary particles, which lose energy not uniformly but stochastically along their tracks. Thus, the energy actually deposited in very small volumes within the target region of nominally uniform dose varies because of:

(1) non-uniform deposition of energy along each particle track,
(2) variation in the number of such particles crossing the particular volume, and
(3) variation in the track length of the particles within the volume.

The lower the average dose for the region or the smaller the volume element considered, the greater is the variation in the locally deposited energy per unit mass which is referred to as the *specific energy*, (*2*), and expressed in rad or $J.kg^{-1}$. If the spread in these values is sufficiently small, say not more than 20%, the concept of a single value of absorbed dose is adequate. On this definition, Table 1 lists for various target sizes the dose, below which the non-uniformity of specific energy must be considered, these data being compiled by Pohlit for this Publication.

At low average dose it will be unusual for a small volume element to be crossed by two or more ionizing particles; in fact, most such volumes will

[3]Stock solution for storage 0.784 g $FeSO_4 \cdot (NH_4)_2 SO_4$ + 0.117 g NaCl + 253 g $CsSO_4$ + 100 ml 2N H_2SO_4, diluted 1/10 for dosimetry.

suffer no energy loss at all. But, however low the average dose, the volume elements which *are* traversed by a single particle will show the *same* distribution of specific energy (local dose). Calculations from Kellerer (*5*) quoted by Pohlit for single secondary electrons from absorption of ^{60}Co gamma radiation crossing a sphere of water or soft tissue 1 μm in diameter give specific energy values mainly between 1 and 10 rad, although more than 100 rad will occur occasionally. Table 2 summarizes data for ^{60}Co radiation and a 1 μm target size, quoted by Pohlit from calculations by Kellerer (*5*). At an average dose of 100 rad, 30% of targets will have specific energies above this average, and 1% above 500 rad, but, in contrast, 10% less than 10 rad. Even for a dose of only 10 rad, 1% of targets will receive more than 100 rad, although in this case 48% will receive less than 1 rad and 40% no energy at all. Table 3 lists similar data (*5*) for volumes of 12 μm diameter, corresponding perhaps to a whole cell. For average doses of 1 and 0.1 rad, 0.5% of targets receive more than 2.5 and 0.7 rad respectively, and in the latter case some 25% of targets are not crossed by any ionizing particle.

TABLE 1
The Specific Energy Concept

Type of Organism	Typical Volume	Dose below which variation in specific energy more than ± 20%
Mammalian Cells	10^{-9} cm^{-3}	3 rad
Cell Nuclei	10^{-11} cm^{-3}	200 rad
Bacteria	10^{-12} cm^{-3}	1000 rad
Viruses	10^{-15} cm^{-3}	3×10^{5} rad

TABLE 2
Range of Specific Energy with ^{60}Co Gamma Radiation
Tissue Volumes 1 μm Diameter

Average Dose	Specific Energy not more than	Percentage of 1 μm diameter volumes
100 rad	10 rad	10%
	50 rad	30%
	100 rad	70%
	500 rad	99%
10 rad	0 rad	40%
	1 rad	48%
	10 rad	80%
	100 rad	99%

In general, the spread in z values is wider for, say, a 5 MeV neutron beam than for ^{60}Co gamma radiation. Thus, if a dose close to the average dose is required to produce the effect of interest, the neutron beam will be relatively less efficient, as in this case a greater proportion of the total energy will be wasted in volumes with too low or unnecessarily high local doses. On the other hand, if the effect requires a specific energy above the average dose, the neutron beam will provide a greater number of volumes with this high local dose and will therefore appear the more efficient irradiation procedure.

It is clear, therefore, that these variations in specific energy may well be significant in many biological situations and cannot be ignored in the interpretation of the biological response at low average dose and the relative efficiency (RBE) of different radiations.

TABLE 3
Range of Specific Energy with ^{60}Co Gamma Radiation
Tissue Volumes 12 μm Diameter

Average Dose	Specific Energy not more than	Percentage of 12 μm diameter volumes
1 rad	0.4 rad	0.5%
	2.5 rad	99.5%
0.1 rad	0 rad	25 %
	0.15 rad	75 %
	0.7 rad	99.5%

RESPONSE MODIFICATION

Radiobiology is concerned with the relationship between physical dose and biological response, and with the study of the many factors, physical, chemical and physiological, which affect this relationship. If these results are to be clearly understood by all workers in the field, it would seem to be useful for some conventions in terminology be agreed to as appropriate by both the biologists and those concerned with dosimetry. It is for this reason that it is proposed to consider the specification of dose response and modification as part of radiobiological dosimetry.

Obviously, when studying the influence of a particular parameter on the response, all other parameters should, ideally, be kept constant. But this is not always possible. For instance, different total doses, producing the same biological end effect under different conditions, *cannot* simultaneously be delivered at the same *dose rate* and in the same *overall time,* and these are both factors which may affect response. Also, it should be appreciated that the same end result (e.g. cell or animal lethality) following two different procedures may, in fact, involve differences in the chemical and biological processes contributing to the expression of this radiation lesion.

The response modification when the irradiation procedure is changed may be quite complicated and is best appreciated by comparison of the full dose response curves under the two conditions of interest. However, in many cases it is not possible to quantitate the biological response, but only to match the level of effect and compare the corresponding absorbed doses used. This ratio has generally been referred to as the *dose modifying factor*. But these procedures do not modify the *dose*, which is a physical matter—the average energy dissipated by ionizing particles per unit mass of medium; they only modify the resultant *biological response* for a given dose. It is a *response modification factor* which is determined, expressed as a dose ratio, in fact as *isoeffect dose ratio,* and it is recommended that this term be used. Or possibly the expression *modification dose ratio,* symbolized as MDR, or simply M, may be more acceptable.

In general this dose ratio, M, will vary with the level of response, S, and for condition y relative to condition x will be expressed as:

$$ {}_{y}M_{x}(S) = \frac{D_x}{D_y} $$

where D_x and D_y are the doses producing the same response level, S, and the numerator always refers to the reference, condition. M will be greater or less than unity according to whether irradiation under the test or reference conditions is the more effective. In some special cases, the *shapes* of the response curves are the same over the complete dose range, so that this response modification dose ratio is a constant, conditions under which the difference in irradiation procedure would have been said to be "simply dose modifying".

Two classes of response modification have been given particular attention. The first, relating to the use of different types of radiation, has been considered as expressing the *Relative Biological Effectiveness* (RBE) of the test radiation compared with some reference conditions. It has been general to refer to orthovoltage (200–300 kVp) X-radiation, but it would now be more rational to adopt ^{60}Co gamma radiation as the reference photon radiation. Such radionuclide sources are widely available and the narrow, high energy spectrum allows accurate dosimetry and a good dose distribution with a minimum of variation in local dose with atomic number. For the reference conditions the dose rate should be sufficiently high for radiobiological "repair" to be negligible during irradiation, but not so high that the environment is altered, for instance, by producing significant radiochemical depletion of oxygen in the medium.

Secondly, there is interest in the *oxygen effect,* the generally observed enhancement of the biological effect in the presence of oxygen. The response modification dose ratio with hypoxic conditions as the reference, ${}_{\mathrm{o}}M_{\mathrm{H}}$, is conventionally referred to as the *Oxygen Enhancement Ratio* (OER). Thus:

$$ {}_{\mathrm{o}}M_{\mathrm{H}}(S) = \frac{D_{\mathrm{H}}}{D_{\mathrm{o}}} = OER_{\mathrm{S}} $$

where D_{H} and D_{o} are the doses for response level S under hypoxic and oxygenated conditions respectively.

As aerated conditions are normal, it would be reasonable to use this as the reference condition and express the inverse ratio:

$$ {}_{\mathrm{H}}M_{\mathrm{o}} = \frac{D_{\mathrm{o}}}{D_{\mathrm{H}}} $$

But providing that the suggested convention of symbols is maintained, no confusion need occur with the use of either of these ratios.

During our discussions Professor Lindop referred to the rather different dose ratios which are relevant to radiotherapy. Here the concern is with the relative tumor response under alternative irradiation regimens, each limited by its corresponding *tolerance* normal tissue reaction, which incidentally does not necessarily arise in both cases from the same sequence of events or even in the same tissue. Two ratios of doses, matched for biological response, may be defined in this connection, although they will not, in general, be both known in relation to an individual patient's treatment. With condition A as reference:

$$_{B}M_{A}\ (\text{tolerance}) = \frac{\text{tolerance dose for A}}{\text{tolerance dose for B}} = \frac{D_A}{D_B}$$

and correspondingly:

$$_{B}M_{A}\ (\text{tumor}) = \frac{\text{dose for given effect with A}}{\text{dose for same effect with B}}$$

Using this second relationship, from the point of view of tumor response, a dose D_B with conditions B will correspond to dose D_A' with conditions A where:

$$D'_A = D_B \cdot {}_{B}M_{A}\ (\text{tumor})$$

The ratio of equivalent tumor doses at normal tissue tolerance is therefore:

$$\frac{D'_A}{D_A} = \frac{D_B \cdot {}_{B}M_{A}\ (\text{tumor})}{D_A} = \frac{_{B}M_{A}\ (\text{tumor})}{_{B}M_{A}\ (\text{tolerance})}$$

This final ratio is a sort of "therapeutic gain factor", being greater than unity if normal tissue tolerance corresponds to a bigger tumor effect under conditions B than for the reference conditions A.

A similar requirement exists in relation to hypoxia. The interest is usually in the relative response of hypoxic tumor cells for matched survival of oxygenated cells in normal tissue. Then the dose ratio of interest is given by:

$$\frac{_{B}M_{A}\ (\text{hypoxic})}{_{B}M_{A}\ (\text{oxygenated})} = \frac{D_{AH}}{D_{BH}} \cdot \frac{D_{Bo}}{D_{Ao}} = \frac{OER_A}{OER_B}$$

This combined ratio is greater than unity if the same oxygenated cell response corresponds to a *lower* survival of hypoxic cells in the test conditions B than under the reference conditions A, and, as indicated, is given by the quotient of the corresponding oxygen enhancement raios.

By analogy with the discussion above, comparing equivalent tumor doses for matched normal tissue tolerance, it may be possible to express the relative hazard of two procedures for radiation protection purposes. Thus, if under the test (B) and reference (A) conditions at a matched level of cell survival, the number of mutations correspond to those resulting from reference condition doses of D'_A and D_A respectively:

$$\frac{D'_A}{D_A} = \frac{_{B}M_{A}\ (\text{mutations})}{_{B}M_{A}\ (\text{survival})}$$

This ratio will be greater than unity if, at a given level of cell survival, a greater number of mutations is produced under conditions B than for the reference conditions A, the former then being considered to be more "hazardous".

Finally, there could usefully be discussion of the following terms proposed by Professor Lindop in connection with response modification. *Enhancement* (inverse *protection*) refers to an increase in the observed biological effect by the use of a drug (or other treatment) which could not itself have produced such an effect, whereas *potentiation* (inverse *attenuation*) describes a synergistic effect, where both procedures contribute to the effect but the combined effect is greater than when either is used alone.

REFERENCES

1. I.C.R.U. Publication 16. *Linear Energy Transfer*. ICRU, Washington, 1970.
2. W. K. Sinclair, in *Radiation Dosimetry*, p. 643. Academic Press. New York, 1969.
3. I.C.R.U. Publication 17. *Radiation Dosimetry—X-rays Generated at Potentials of 5–150 kV*. ICRU, Washington. 1970.
4. I.C.R.U. Publication 21. *Radiation Dosimetry—Electrons with Initial Energies 1–50 MeV*. ICRU, Washington, 1972.
5. A. M. Kellerer, in *Mikrodosimetrie Grundlagen einer Theorie der Strahlenqualität*. Report GSF-B1 Gessellschaft für Strahlenforschung, 1968.

Microdosimetery of Auger Electrons[1]

Y. Feige and A. Gavron
Soreq Nuclear Research Centre
Yavne, Israel

INTRODUCTION

The last decade has witnessed a large increase in the use of radioisotopes which decay by electron capture for diagnostic purposes. Some have also been applied for therapeutic uses and in radiobiologic research. It was believed, for quite a while, that the radioisotopes decaying by electron capture were relatively safe, when used instead of other, beta emitting isotopes. However, radiobiological experiments on the one hand, and improved dosimetric calculations on the other, indicate that the Auger cascade resulting from electron capture decay may impart excessive damage to cellular components due to the highly localized energy deposition involved and because of molecular disruption.

This presentation is basically a review of work done in this field with some examples out of our previous publications. This review will briefly cover the various physical parameters involved in an adequate dosimetry of Auger electrons. The biological significance of these findings will be evaluated and compared with the results obtained in various experiments where ^{125}I was incorporated in the thyroid gland of mice and rats, as well as in DNA molecules. It seems impossible to decide at this stage which specific effect (or combination of effects) of the Auger cascade is responsible for the enhanced biological effectiveness observed.

THE AUGER CASCADE

Auger electrons and characteristic X-rays are emitted when vacancies in an inner electronic shell are filled. The primary vacancy in the electronic shell of the atom is usually produced by orbital electron capture, by internal conversion, or by the photoelectric absorption of gamma or X-ray photons. Auger transition and X-ray emission are competing and/or complementary de-excitation processes.

The probability of X-ray emission due to an initial vacancy, the fluorescence yield, is a function of the atomic charge Z, and the electronic shell involved. The K-fluorescence yield ω_K, which is well known, increases with the atomic number according to:

$$\omega_K(Z) = Z^4/(b_K + Z^4)$$

where $b_K = 1.12.10^6 \simeq 32^4$. The K-Auger yield a_K decreases as $a_K = 1 - \omega_K$.

As can be seen from Fig. 1, the ejection of an Auger electron is the major relaxation process for excited states produced by a K shell vacancy in low-Z

[1]Congress presentation D–7–3.

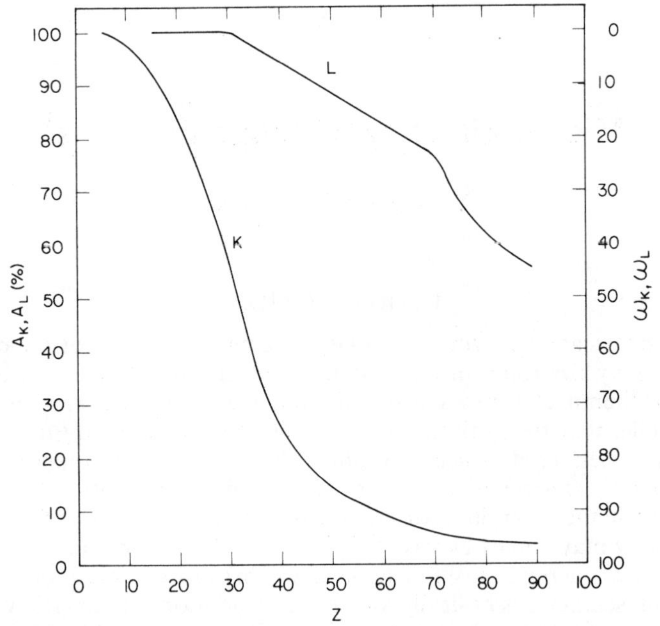

FIG. 1. Auger and fluorescent yields for K and L vacancies as functions of the atomic number Z.

elements. At about Z=32 (Ge), the emission of Auger electrons and X-rays is equally probable. As Z increases, the predominant process is the emission of X-rays, and the Auger effect becomes relatively unimportant for K shell vacancies.

The de-excitation of an initial or secondary vacancy in the L shell is more complicated and consequently less well known in detail. L-fluorescence yields are insignificant below Z=40 (Zr), and reach about 20% at Z=64 (Gd), as is also shown in Fig. 1. The average L fluorescence yield $\bar{\omega}_L$ may be approximated by a similar expression:

$$\bar{\omega}_L (Z) = Z^4 / (b_L + Z^4)$$

where $b_L \sim 10^8 = 100^4$.

Vacancy transfers, the so-called Coster Kroning transitions, may occur within the same electronic shell. An L subshell vacancy may be filled by an outer L subshell electron and the difference in subshell binding energy is transferred to a low energy electron ejected from a higher shell. No distinction will be made in this paper between these Coster-Kroning and Auger electrons.

Very little information is available regarding the fluorescence yields in the M and higher shells. For U (Z=92), the M-fluorescence yield is only about 5%. Thus the filling of M and higher shell vacancies involves Auger electrons, almost exclusively.

A detailed description of the theoretical and empirical aspects of the Auger effect may be found in the relevant literature (*1–8*). A special conference was devoted to the recent developments and applications of inner shell ionization phenomena (*9*).

We may conclude this brief survey by stating that as the atomic number increases, the fraction of the energy carried away as X-ray photons increases, at the expense of the Auger electrons. For higher Z atoms, the L shell fluorescence yield should also be taken into account. Furthermore, as the de-excitation process spreads toward the outer shells of the atom, the fluorescence yield decreases significantly, by about a factor of 5 per shell, and therefore shell vacancies and Auger electrons become more numerous. The energies of the M and N Auger electrons are obviously very low but as their number is relatively large, the total energy involved in these "radiationless" transitions may be of importance in specific cases.

Some of the many vacancies created by the Auger cascade in the outermost shells of the now multiple-charged atom may be filled by electrons from other atoms of the same molecule. This intramolecular charge transfer, which occurs during the final stages of the electronic readjustments, may easily disrupt molecular bonds. The positive charge distribution throughout the molecule may cause the dissociation of the molecule due to Coulomb repulsion forces. In gaseous compounds, multiple-charged molecular fragments have been observed after K-shell vacancies were produced by various means (*10–15*). There is no reason to doubt that molecular decomposition is a frequent and general consequence of the Auger cascade, under other physical conditions, as well.

DOSIMETRY

Our brief review of the Auger cascade should convince us that it is necessary to deal separately with the two distinct types of low energy radiations involved, namely the fluorescent X-ray photons and the Auger electrons, which have quite different ranges in tissue for the same energies.

Until a few years ago, the contribution of Auger electrons to the locally absorbed radiation dose was usually either completely ignored, or accounted for only very approximately. It was generally assumed that the entire energy released per decay was carried by the X-rays alone. This was a conservative attitude as long as only external radiation sources were involved. The dosimetry of internally deposited radionuclides decaying by electron capture, or internal conversion, for which this assumption would lead to underestimating the absorbed dose, was significantly improved by Smith et al. (*16–17*). Their methods of calculating the mean energy per disintegration enables one to determine the mean radiation dose delivered to organs or tissue by internally deposited radionuclides.

The requirements for the dosimetry of Auger electrons at the cellular or subcellular level were first discussed in some detail by Wrenn et al. and applied to the determination of the dose within erythrocytes from incorporated ^{55}Fe (*18–20*). They found a ten-fold higher dose within the erythrocytes containing the ^{55}Fe-labelled hemoglobin than the dose delivered to blood plasma.

Feige et al. discussed the implications of ^{125}I decay in the thyroid gland and calculated the microscopic dose distribution within the thyroid cell (*21–23*). Grieg et al. in Glasgow (*24*) and Feinendegen's group in Germany (*25–26*) independently developed similar approaches to the microdosimetry of ^{125}I in order to clarify biological and clinical findings.

By now, the following significant advances have been made in the field of dosimetry which together permit an accurate detailed assessment of the local energy deposition for non-homogeneous distributions of internally deposited radionuclides, even at the cellular and subcellular levels.

First, a basic and consistent framework was developed, under the auspices of the MIRD Committee, for calculating absorbed doses. The main concepts involved are the "absorbed fraction," ϕ, of the emitted energy actually deposited within the target of interest and the "equilibrium dose constant," Δ, namely the total energy released by the radionuclide per disintegration.

$$\Delta = \sum_i \Delta_i = 2.13 \sum n_i E_i \quad (\text{g-rad}/\mu\text{Ci-hr})$$

where n_i, E_i and Δ_i are the intensity, energy and equilibrium absorbed dose constants, respectively, of the i-th type of radiation per disintegration (*17,27–29*).

Dillman (*30–32*) developed computer methods to calculate these parameters separately for each of the K and L X-rays, conversion electrons and Auger electron groups down to the M shell, associated with electron capture decay and internal conversion of gamma rays. The information characterizing the decay scheme is used directly as input data. As an example, the output data for the relatively simple case of ^{55}Fe, which decays by electron capture directly to the ground state of ^{55}Mn, is reproduced (*32*) in Table 1. One obtains the 5.9 and 6.5 keV K X-rays in only 23.4% of the disintegrations; they carry about 23% of the energy released. The 5.8 and 5.2 keV electrons occur in 66.3% of the decays and carry about 57% of the energy. About 20% of the energy emitted is carried by the many very low energy electrons which accompany the ^{55}Fe decay.

TABLE 1
Output data for Iron-55

Radiation (i)		Mean number/ disintegration (n_i)	Mean energy (MeV) ($\bar{E}_i$)	Δ_i $\left(\frac{\text{g-rad}}{\mu\text{Ci-h}}\right)$
Electron capture				
X-rays:	Kα-1	0.140	0.0059	0.0018
	Kα-2	0.0708	0.0059	0.0009
	Kβ-1	0.231	0.0065	0.0003
	Kβ-2	0.0006	0.0065	0.0000
Auger electrons:	KLL	0.536	0.0052	0.0059
	KLX	0.127	0.0058	0.0016
	LMM	1.50	0.00006	0.0019
	MXY	3.17	0.0001	0.0007

[From Dillman, *J. Nucl. Med.* **11, Suppl. 4**, 15 (1970).]

In order to evaluate the absorbed fraction for Auger electrons, their range and energy deposition must be known. Berger's recent publications extending the data on the range and the absorbed fraction for low energy electrons below 10 keV down to 0.5 keV (*33–35*), made this possible. Fig. 2 compares the 90% absorbed dose ranges (R_{90}) of low energy electrons to those of photon radiations of equal energy (*36*). One sees that for 2 keV electrons this range is about 0.1 μm in water, compared with 40 μm for a 2 keV X-ray, i.e. they differ by a factor of 500. For 10 keV radiations the ratio of the ranges is about 2200, (2 μm for electrons in water compared with 4.4 mm for 10 keV X-rays). Thus, the volume in which practically all the energy is absorbed is many orders of magnitude smaller for electrons than for X-ray photons of comparable energy.

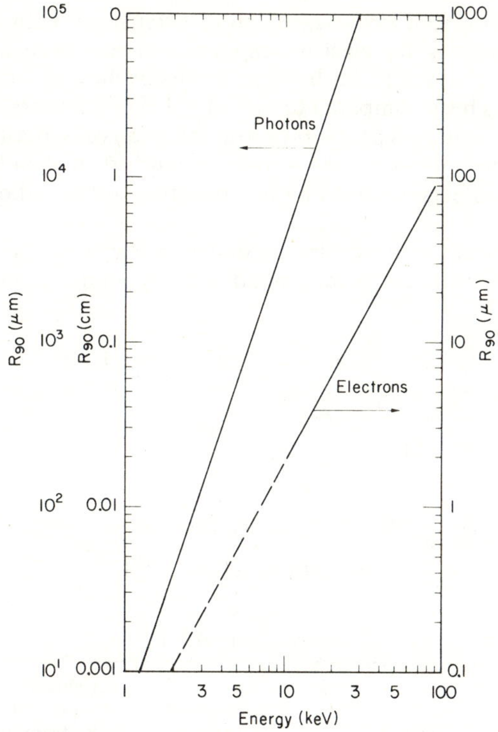

FIG. 2. 90% absorbed dose range for low-energy electrons and photons in water. [From Martin and Blickert-Toft, in *Nuclear Data Tables,* **A8**, Appendix IV, pp. 163–168 (1970).]

It is evident, therefore, that for radionuclides decaying by electron capture, which localize in specific cells or cellular components, it is essential to distinguish between the energy released in the form of X-rays and that fraction of the energy per disintegration which is distributed among several low-energy and short-range Auger electrons. While the X-rays escape from the cell and may thus be utilized for external counting, the Auger electrons are totally absorbed locally.

Physically we have a clear model, and we have also the mathematical means for deriving the microscopic energy deposition, which may be extremely inhomogeneous for these kinds of radionuclides. We have retained the microscopic concepts, such as mean absorbed dose, to apply to the dosimetry of microscopic regions of cellular and subcellular dimensions.

Berger (*35*) recently showed the relationship between this macroscopic, or classical approach, and parameters such as the "specific energy" Z and the "event-size" distribution, introduced by Rossi and Kellerer into the field of microdosimetry (*37,38*).

Let us turn now to some biological examples where this more specific approach has been found useful.

THE BIOLOGICAL EFFECTIVENESS OF ^{125}I

Iodine-125 has been investigated more extensively than any of the other radionuclides decaying by electron capture and has been applied to many different biological systems. Its biological effectiveness in animal and human thyroid glands has been compared to that of ^{131}I. Furthermore, incorporation of ^{125}I into DNA molecules of proliferating mammalian cells in culture, as well as of bacteria and bacteriophages, has also been studied in detail and the results compared to the effects observed by incorporating other radionuclides, mostly ^{3}H, ^{14}C, or ^{32}P.

^{125}I decays by electron capture, as shown in Fig. 3, to the 35.5 keV excited level of ^{125}Te which decays to its ground state by internal conversion or direct

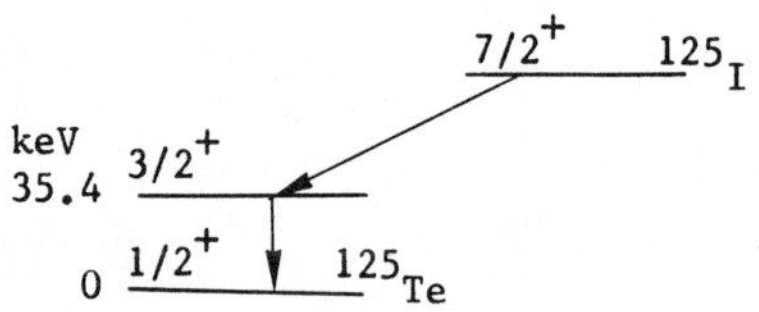

35.4 keV: 7%(e_k/γ=11; K/L + M + ..= 4)

FIG. 3. Decay scheme of ^{125}I.

TABLE 2
Decay scheme of ^{125}I

Radiation	Origin	Energy E (keV)	Abundance A per 100 disintegrations	(A×E)/100 (keV)
Photons	Unconverted γ-rays	35.5	6.8	2.41
	K_β X-rays	31.1	24.0	7.46
	K_α X-rays	27.4	111.6	30.58
	L X-rays	3.8	21.5	0.82
		Total	163.9	41.27
Electrons	M internal conversion	34.7	8.0	2.77
	L internal conversion*	30.9	11.6	3.59
	K L X Auger electrons	26.4	5.8	1.53
	K L L Auger electrons	22.7	13.7	3.11
	K internal conversion	3.7	74.6	2.76
	L M M Auger electrons	2.9	149.0	4.32
	M X Y Auger electrons	0.8	359.0	2.87
		Total	621.7	20.95

Total energy per disintegration = 62.22 keV

*Includes K X Y Auger electrons.

[From Feige et al., in *Proc. IAEA Symposium on Biophysical Aspects of Radiation Quality*, Sydney 1971, pp. 383–404. STI/PUB/286, IAEA Vienna, 1971.]

gamma emission. The Auger and conversion electrons resulting from this decay are shown in Table 2 (*22, 24, 26, 31, 39, 40*). While 41 keV of the energy per decay is associated with the unconverted γ ray and the K X-rays, about one third of the energy, 21 keV per disintegration, is carried by the conversion and Auger electrons. Fig. 4 shows the electron radiation dose per decay of ^{125}I as a function

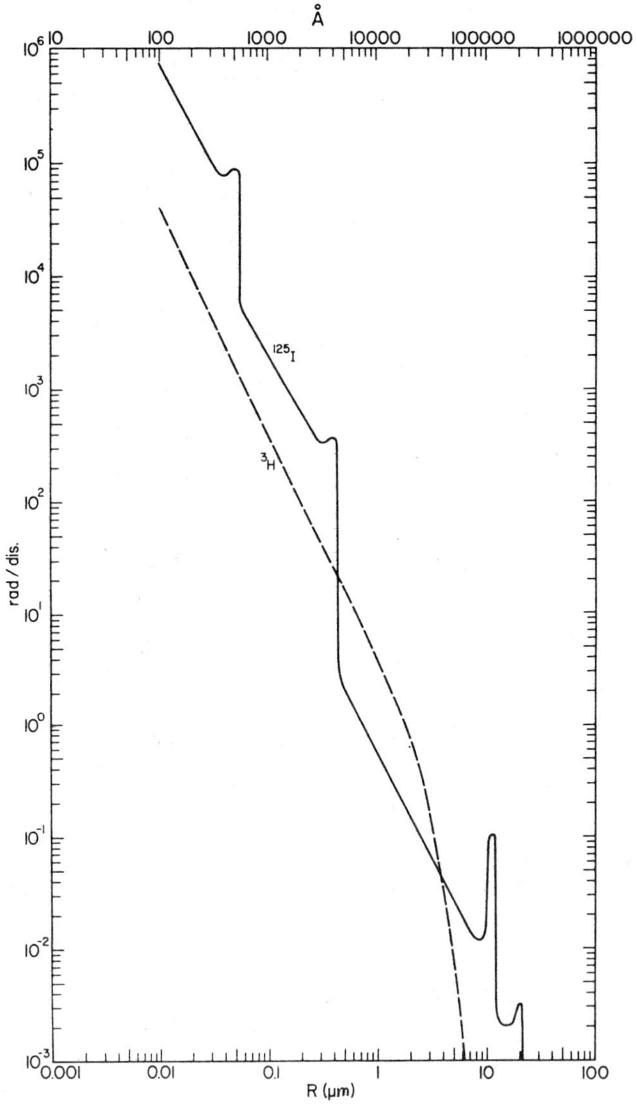

FIG. 4. The dose per decay as function of the distance from a point source for ^{125}I (and ^{3}H). [From Feige and Gavron, in *Proc. Third Symposium on Microdosimetry,* Stresa 1971 (H. G. Ebert, Ed.) pp. 389–398. EUR-4810, Brussels, 1972.]

of the distance from a point source. The small peaks at the end of each group appear as the result of the simplified energy loss function assumed in our calculations (*23*). We grouped the electrons into four groups as shown in Table 3.

TABLE 3
Electron groups of ^{125}I decay

Energy E(keV)	Abundance A (%)	Range R(μm)	LET = E/R (keV/μm)	E·A/100 (keV)
32.4	19.6	20.9	1.55	6.36
23.8	19.5	12.1	1.96	4.64
3.2	220	0.44	7.2	7.08
0.8	360	0.056	14.2	2.87

[From Feige et al., in *Proc. IAEA Symposium on Biophysical Aspects of Radiation Quality*, Sydney 1971, pp. 383–404. STI/PUB/286, IAEA Vienna, 1971.]

The thyroid gland consists of numerous follicles, more or less rounded in shape of about 100 μm in diameter. The shell of each follicle consists of a layer of closely packed living cells, 5 to 10 μm thick. The nucleus occupies a significant portion of the cell volume. The space inside the follicle is filled with the colloid gel in which all the iodine is contained almost entirely in the colloid, rather than distributed homogeneously throughout the whole thyroid gland; the colloid fraction is of particular importance for the evaluation of the dose due to ^{125}I. In the normal thyroid the colloid occupies between 50 and 75% of the volume of the gland. In the thyrotoxic gland, most follicles are relatively small and even the larger ones contain only small amounts of colloid (*21–23*).

Fig. 5 shows the contribution of the various electron components to the dose-rate as a function of the distance from the colloid-cell interface for 1 μCi ^{125}I per gram of normal thyroid tissue. The range of the low-energy electrons is small compared with the dimensions of a thyroid cell, thus giving rise to a non-uniform dose distribution across the individual thyroid cell. The cell nucleus is subject to a relatively weak radiation field of 8 mrad/h, while the colloid-cell interface, where the hormone synthesis is completed, is subject to much higher radiation doses (32 mrad/h). The actual dose depends on the exact dimensions of the cell and of the colloid in which the iodine is stored (*22–23*).

Recently, calculations of the dose rate at points external to spherical homogeneous electron sources performed by J. F. Valley et al., (*41*) using Berger's tabulated scaled point kernel data, showed fair agreement with our results.

One can now explain the seemingly controversial effects of ^{125}I decay on various biological functions in the thyroid glands of rats and mice, such as hormone production, response to goitrogens and TSH, and the proliferating capacity of thyroid cells when compared with the effects produced by ^{131}I (*42–47*). Experimental data show, for instance, that ^{125}I damages the system for the production of thyroid hormone without affecting the system responsive to TSH (*42,44*). All these findings may be accounted for by the inhomogeneous distribution of the ^{125}I dose within the thyroid follicle, namely the relatively high energy deposition by the Auger electrons of ^{125}I at the colloid-cell interface as compared with the much lower dose to the cell nucleus only a few μm away from that interface. Although the *distribution* of ^{131}I is similarly non-uniform within the follicle, the *dose is* uniformly distributed because of the relatively long range of the ^{131}I beta rays.

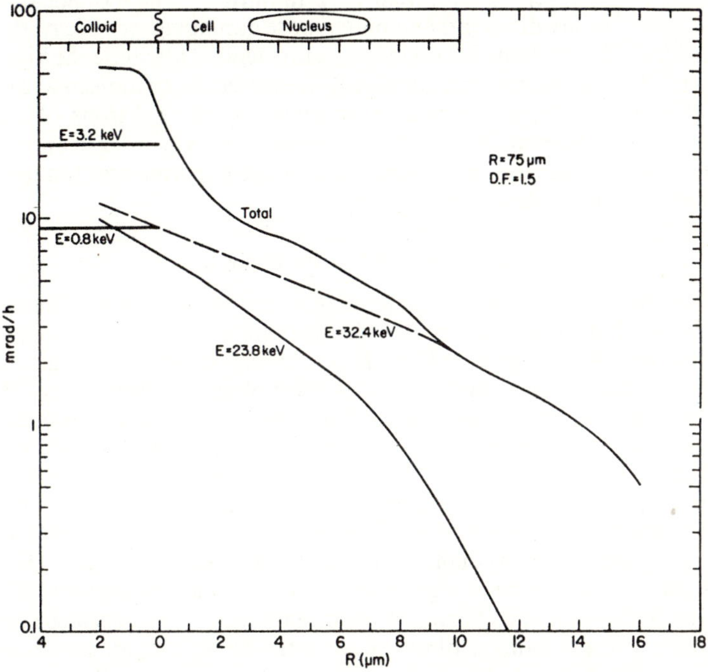

FIG. 5. Dose rate distribution of the electron groups in a normal thyroid cell, for 1 μCi ^{125}I/g [From Feige et al. in *Proc. IAEA Symposium on Biophysical Aspects of Radiation Quality*, Sydney 1971, pp. 383–404. STI/PUB/286, IAEA, Vienna, 1971.].

Because of these differences in dose distribution, a few medical centers preferred the use of ^{125}I for the treatment of thyrotoxic patients in the hope that, while a sufficient dose will be delivered to the hormone synthesis site, the relatively low radiation to the cell nucleus would reduce the incidence of hypothyroidism following treatment (*48–52*). It is still too early to assess the therapeutic advantages of this approach, as many difficulties have been encountered in practice.

In the highly differentiated system of thyroid structure and hormonal function, the ^{125}I is concentrated *outside* the cells in the colloid fluid. Only in those cases where the ^{125}I is incorporated within structures of specific biological importance, such as in the cell nucleus itself as part of the DNA molecule, will the biological consequences of molecular disruption manifest themselves.

High specific activity-labelled iododeoxyuridine (IUdR is generally used in order to introduce ^{125}I into the DNA of proliferating cells, bacteria and phages.

IUdR is an analogue of thymidine and both are specific precursors of DNA. Many authors report that the toxicity of ^{125}I is much higher than could be anticipated from the findings with other radionuclides in similar molecular positions, especially with regard to lethality (survival), and that practically every ^{125}I decay causes a double-strand break. Hoffer and Hughes (*57*) labelled leukemia cells with various doses of ^{125}IUdR and found survival curves resembling those observed with high LET radiations. The biological effects of intranuclear ^{125}I were different, and much more damaging, per decay, than those produced by ^{131}I or tritium. Many investigators tend to attribute the high specific effects of ^{125}I decaying within a labelled DNA molecule, to molecular disruption rather than to the dense ionizations produced by the electrons of the Auger cascade (*53–62*). One should consult the original papers in order to appreciate the extreme care necessary to overcome some of the experimental limitations and obstacles inherent in this type of study.

CONCLUDING REMARKS

One would like to be able to determine which of the various physical effects (i.e., energy deposition, molecular disruption, or chemical transmutation) associated with Auger electron cascade is responsible for the biological damage in DNA. It is known, for instance, that all the biological effects due to ^{3}H labeled DNA can be explained as resulting from the beta radiation and that the absorbed dose fraction is the relevant parameter down to subcellular dimensions (*63–67*). The killing effects of ^{14}C and ^{32}P when incorporated into DNA are mainly attributed to chemical transmutations, rather than to ionizing radiations (*67–69*). Since the local energy deposition due to the electrons emitted from ^{125}I is greater (per disintegration) than that of a beta particle emitted from ^{3}H (see Fig. 4), this may certainly be a significant factor in the damage to DNA (*54*). Molecular disruption and chemical transmutation to $^{125-}$Te may also contribute, either directly to the damage observed, or interfere with repair processes. Local energy deposition alone cannot account for the high biologic effectiveness of ^{125}I observed in biologic systems.

I cannot conclude this review without mentioning at least some of the important topics which could not be treated because of space limitations but which definitely merit discussion.

Should a LET-dependent quality factor (Q.F.) be considered for the low energy Auger electrons whose LET exceeds 3.5 keV/μm in tissue (*22,23*)? Some findings, such as the absence of a shoulder in ^{125}I-labelled survival curves and that the biological effects of ^{125}I resemble the effects produced by high LET radiations, are not inconsistent with the assumption of a quality factor. However, since the enhanced effects produced by ^{125}I can in part be attributed to other peculiarities of the Auger cascade, further experimental data should be furnished before an LET-dependent Q.F. for low energy electrons be re-established. At the present stage of our understanding there are no undisputed experimental data to prove that low energy electrons can cause specific biological damage in excess of that attributed to ionizations (*64–69*). The ICRP amended its previous recommendations on this subject and suggested a Q.F. of 1 for all electrons down to the lowest energies (*70*). An appropriate dose modifying factor could be useful in the future to characterize specific biological effects of this class of radionuclides (*69*).

The possibility of initiating Auger cascades in specific biological structures by means of the photoelectric absorption of characteristic external X-rays, opens up a vast new field of research possibilities and practical applications. This new approach will be discussed by Feinendegen et al. at this Congress (*71*).

Last but not least, a word of caution to practitioners in nuclear medicine. The use of radionuclides which decay by electron capture such as ^{51}Cr, ^{67}Ga, ^{75}Se, ^{85}Sr, ^{123}I and ^{197}Hg, to name a few, is being extended rapidly in the field of nuclear medicine. These radionuclides have been considered especially suited for in vivo diagnostic studies, as their decay scheme does not involve the emission of charged (beta or alpha) particulate radiations, which would deliver radiation doses to the investigated organ without contributing towards the diagnostic information obtained by external photon counting or scanning. Taking the radiation dose delivered by Auger electrons adequately into account, as outlined in this paper, may lead to some revision in this attitude in specific circumstances, especially when alternative diagnostic methods are available.

REFERENCES

1. E.H.S. Burhop, *The Auger Effect and Other Radiationless Transitions*. Cambridge University Press, 1952.
2. I. Bergström and C. Nordling, in *Alpha-, Beta- and Gamma-Ray Spectroscopy*. (K. Siegbahn, Ed.), **Vol. 2**, pp. 1523–1543. North-Holland Publishing Co., Amsterdam, 1965.
3. R. W. Fink, R. C. Jopson, H. Mark and C. D. Swift, *Rev. Mod. Phys.* **38**, 513 (1966).
4. W. Mehlhorn, *The Auger Effect*, University of Nebraska, Lincoln, 1969.
5. K. D. Sevier, *Low Energy Electron Spectrometry*, J. Wiley & Sons, Inc., 1971.
6. E.H.S. Burhop and W. N. Asaad, in *Advances in Atomic and Molecular Physics* (D. R. Bates and I. Esterman, Ed.), **Vol. 8**, pp. 163–284, Academic Press, N.Y., 1972.
7. W. Bambynek et al., *Rev. Mod. Phys.* **44**, 716 (1972).
8. N. A. Dyson, *X-rays in Atomic and Nuclear Physics*, Longman Group Ltd., London, 1973.
9. *Proc. Int. Conf. on Inner Shell Ionization Phenomena and Future Applications* (R. W. Fink et al., Ed.). Conf. 720404, USHEC, T.I.C., Oak Ridge, 1973.
10. A. H. Snell, F. Pleasonton and T. A. Carlson, in *Chemical Effects of Nuclear Transformations* (1960), **Vol. I**, pp. 147–153. IAEA, Vienna, 1961. (STI/PUB/34.)
11. T. A. Carlson and R. M. White, in *Chemical Effects of Nuclear Transformations* (1964), **Vol. I**, pp. 23–33. IAEA, Vienna, 1965. (STI/PUB/91.)
12. A. H. Snell. See Ref. *2*, pp. 1545–1555.
13. T. A. Carlson and R. M. White, *J. Chem. Phys.* **44**, 4510 (1966).
14. A. Ore, in *Radiation Research* (G. Silini, Ed.), pp. 54–67. North Holland Publ. Co., Amsterdam, 1967.
15. S. Wexler, *Science* **156**, 901 (1967).
16. E. M. Smith, C. C. Harris and R. H. Rohrer, *J. Nucl. Med.* **7**, 23 (1965).
17. E. M. Smith, in *Handbook of Radioactive Nuclides* (Yen Wang, Ed.), pp. 167–200. Chem. Rubber Co., Cleveland, 1969.
18. M. E. Wrenn et al., *Health Phys.* **24**, 645 (1973).
19. M. E. Wrenn and N. Cohen, *Health Phys.* **13**, 1075 (1967).
20. M. E. Wrenn, The dosimetry of ^{55}Fe in: *Proc. First I.R.P.A. Congress* (W. S. Snyder et al., Eds.) **Vol. 2**, pp. 843–850. Rome, 1968.
21. Y. Feige, A. Gavron and J. Gross, in *Proc. European Congress on Radiation Protection*, Menton, 1968 (P. Pellerin, Ed.), pp. 351–356. Le-Vesinet, 1971.

22. Y. Feige et al., Local energy deposition in thyroid cells due to the incorporation of ^{125}I. in *Proc.* IAEA *Symposium on Biophysical Aspects of Radiation Quality*, Sydney 1971, pp. 383–404. STI/PUB/286, IAEA Vienna, 1971.
23. A. Gavron and Y. Feige, *Health Phys.* **23**, 491 (1972).
24. F. C. Gillespie, J. S. Orr and W. R. Greig, *Brit. J. Radiol* **43**, 40 (1970).
25. H. H. Ertl and L. E. Feinendegen, in *Proc. Second Symp. on Microdosimetry* Stresa (1969) (H. G. Ebert, Ed.), pp. 787–800. EUR-4452, Brussels 1970.
26. H. H. Ertl, L. E. Feinendegen and H. J. Heiniger, *Phys. Med. Biol.* **15**, 447 (1970).
27. R. Loevinger and M. Berman, *J. Nucl. Med.* **9, Suppl. 1,** 7 (1968).
28. E. M. Smith, in *Medical Radionuclides: Radiation Dose and Effects,* (R. J. Cloutier et al., Ed.), pp. 17–31. AEC Symposium Series 20. Conf.-691212. USAEC T.I.C. Oak Ridge, 1970.
29. R. Loevinger. See Ref. *28*, pp. 487–489.
30. L. T. Dillman. See Ref. *28*, pp. 55–62.
31. L. T. Dillman. *J. Nucl. Med.* **10, Suppl. 2,** 7 (1969).
32. L. T. Dillman. *J. Nucl. Med.* **11, Suppl. 4,** 5 (1970).
33. M. J. Berger, *J. Nucl. Med.* **12, Suppl. 5,** 5 (1971).
34. M. J. Berger, in *NBS IR,* pp. 73–107. National Bureau of Standards, Washington, 1973.
35. M. J. Berger, in *Proc. Fourth Symp. on Microdosimetry,* Verbania Pallanza (J. Booz et al., Ed.), **Vol. 2,** pp. 695–715. EUR-5122 Luxemburg, 1974.
36. M. J. Martin and P. H. Blickert-Toft, in *Nuclear Data Tables,* A8, Appendix IV, pp. 163–168 (1970).
37. H. H. Rossi, Microscopic energy distribution in irradiated matter, in *Radiation Dosimetry* (F. H. Attix and W. C. Roesch, Eds.), **Vol. I.,** pp. 43–92. Academic Press, N.Y., 1968.
38. A. M. Kellerer and H. H. Rossi, in *2nd Microdosimetry Symp.* (H. G. Ebert, Ed.), pp. 843–853. EUR-4453, Brussels, 1970.
39. H. H. Ertl, *Dosimetrie von inkorporiertem Tritium und Jod-125.* Kernforschunsanlage Jülich, Report Jü-688-ME, 1970.
40. D. E. Charlton, *Radiat. Res.* **50**, 455 (1972).
41. J. F. Valley, A. P. Kushelevsky and P. Lerch, *Health Phys.* **26**, 295 (1974).
42. J. Gross, M. Ben-Porath, A. Rosin and M. Bloch, in *Thyroid Neoplasia* (S. Young and D. R. Inman, Eds.), pp. 291–306, Academic Press, London, 1968.
43. W. R. Greig. et al, *Brit. J. Radiol.* **43**, 542 (1970).
44. A. L. Vickery and E. D. Williams, *Acta Endocrinol. (Copenhagen),* **66,** 201 (1971).
45. W. J. Jongejan and L. M. Van Putten, *Int. J. Radiat. Biol.* **22**, 489 (1972).
46. Z. Lewitus, in *Proc. of Regional Conf. on Rad. Prot.*, Jerusalem (Y. Feige and T. Schlesinger, Eds.), **Vol. 1,** pp. 433–439. Israel AEC, 1973.
47. A. R. Reddy and K. K. Nagpal, Paper 179, in *Proc. Third International Congress IRPA,* Washington, 1973.
49. Z. Lewitus et al., *Semin. Nucl. Med.* **1**, 411 (1971).
50. I. R. McDougall, W. R. Greig, and F. C. Gillespie, *New Engl. J. Med.* **285**, 1099 (1970).
51. W. R. Greig, I. R. McDougall and K. E. Halnan, *Brit. Med. Bull.* **29,** 63 (1973)..
52. Z. Lewitus, et al, in *Peaceful Uses of Atomic Energy, Proc. of the Fourth International Conference, Geneva,* 1971, **Vol. 13,** pp. 337–348. IAEA, Vienna and UN, New York, 1972. (STI/PUB/300.)
53. L. E. Feinendegen. See Ref. *22*, pp. 419–430.
54. Y. Feige and A. Gavron, in *Proc. Third Symposium on Microdosimetry,* Stresa 1971. (H. G. Ebert, Ed.), pp. 389–398. EUR-4810, Brussels, 1972.
55. R. Roots, L. E. Feinendegen and V. P. Bond. See Ref. *54*, pp. 371–388.
56. G. Ahnström et al. *Mutat. Res.* **10**, 247 (1970).
57. K. G. Hoffer and W. L. Hughes, *Radiat. Res.* **47**, 94-109 (1971).
58. R. E. Krisch, *Int. J. Radiat. Biol.* **21**, 167 (1972).
59. R. E. Krisch and R. D. Ley, *Int. J. Radiat. Biol.* **25**, 21 (1974).
60. A. Schmidt and G. Holtz, *Int. J. Radiat. Biol.* **24**, 307 (1973).

61. J. J. Burki, Paper E–13–4, this Congress. *Radiat. Res.* **59** (1974). Abstract.
62. B. R. Young, H. J. Burki and R. B. Painter, Paper E–13–5, this Congress. *Radiat. Res.* **59** (1974). Abstract.
63. V. P. Bond and L. E. Feinendegen, *Health Phys.* **12**, 1007 (1966).
64. L. E. Feinendegen, in *Biological Effects of Transmutation and Decay of Incorporated Radioisotopes*, pp. 1–5. IAEA, PUB/183, Vienna, 1968.
65. S. Person, see Ref. *64*, pp. 29–64.
66. R. Oliver, see Ref. *64*, pp. 165–185.
67. See Ref. *64*, General Conclusions, pp. 233–247.
68. R. E. Krisch, *Int. J. Radiat. Biol.* **18**, 259 (1970).
69. V. P. Bond, see Ref. *28*, pp. 399–411.
70. Progress report from I.C.R.P., *Health Phys.* **17**, 389 (1969).
71. L. E. Feinendegen, G. T. Lentulis and V. P. Bond, *Radiat. Res.* **59** (1974). Abstract.

Particle Dosimetry by Track Etching, with Applications to Apollo Astronauts and to Heavy Radionuclide Mapping in Biological Systems[1]

Robert L. Fleischer[2]
General Electric Research and Development Center
Schenectady, New York 12305, USA

INTRODUCTION

Particle track etching is a tool that has been applied in dosimetry and many other areas during the last decade (*1,2*). Dosimetry uses extend from neutron and heavy particle counting to determinations of micro-distributions of radionuclides in bone and tissue. Personnel uses have been in such widely separated sites as the depths of uranium mines and in distant space, where Apollo capsules have orbited the moon. Nuclear reactors and locations where significant exposures to alpha emitting nuclides are possible are somewhat more conventional locations.

Particle detection in solids depends on the fact that paths of intense damage only a few tens of atoms across are produced in insulating materials by the passage of heavily ionizing nuclear particles, such as fission fragments or alpha particles. The damage tracks may be revealed for optical examination by treatment with a properly chosen chemical reagent that rapidly and preferentially attacks the damaged material in such a way as to create a hole along each etched track. Figure 1 shows both the diversity of materials that are suitable—crystals, glasses, and plastics—and the wide range of sizes to which a track can be etched—from 50 Å diameter tracks where only the damaged material has been removed to tracks that are 200 μm across and therefore visible even to the naked eye. In the latter case the undamaged material surrounding the narrow damage trail has been removed to enlarge the hole for easy observation.

The dosimetric uses of track etching may be separated into two major categories, depending on whether the radiation itself is detected or whether the presence and distribution of radionuclides in biological systems are measured. These two cases are considered under the headings Particle Dosimetry and Radionuclide Distributions.

[1]Congress presentation D-7-4.
[2]Also visiting Scientist, National Center for Atmospheric Research and Atmospheric Physics and Chemistry Laboratory, National Oceanic and Atmospheric Administration, Boulder, Colorado.

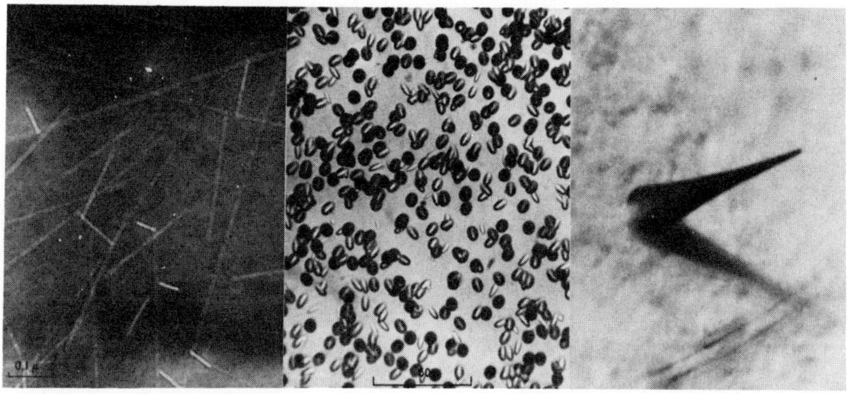

FIG. 1. Etched particle tracks in diverse materials, etched to different sizes. Left: A mineral, muscovite mica, with etched fission tracks 0.0050 μm in diameter; center: ordinary glass with fission tracks etched to 8 μm diameters; right: cosmic ray iron nucleus etched to 200 μm diameter in Lexan polycarbonate plastic helmet. The tracks on the left are visible only in an electron microscope, those in the center are visible in an ordinary light microscope, and that on the right can be seen by the naked eye.

PARTICLE DOSIMETRY

The possibility of doing dosimetry with particle track detectors was noted in 1963 (*3*), and progress in this area has been reviewed at various stages of development (*1*,*2*,*4*,*5*). Slow heavy particles such as alpha particles or fission fragments are detected directly by their tracks, while others such as neutrons or energetic electrons and photons are counted by means of secondary heavy particles that they produce by inducing fission, or alpha particle reactions or by recoiling nuclei in a solid track detector or adjacent to it.

Special qualities of track detectors are that they are generally simple, compact, rugged, passive, integrating devices with low background, relatively low sensitivity to environmental extremes, and considerable ease of readout. They can be compositionally similar to biological material. They require no power source while acquiring data. As noted in Table 1, a wide variety of particles can be detected. Reference (*2*) gives details and an extensive overview of the special qualities of particle track dosimetry and describes the extensive variety of automatic readout and scanning procedures that have been devised.

Secondary Reactions: Neutron Detection

By far the most extensive dosimetric use of track etching has been in neutron detection carried out by recording the heavy fragments that are released in neutron-induced fission. For slow (thermal) neutrons ^{235}U is the most usual nuclide to be employed; the fission fragments are either detected by a low-uranium solid such as Lexan polycarbonate placed next to the uranium so as to record the ejected fission fragments (*3*) or by fission of uranium within the detector itself, which may have a natural uranium content that is at a useful level (*6*) or may be doped to a desired uranium content (*7*).

TABLE 1.
Particle Track Dosimetry

Type of Radiation	Mode of Observation	Detector Material	Sensitivity [tracks/cm^2 per particle/cm^2]	Notes
Fission Fragments	Direct	Any Dielectric Solid	1.0	
Heavy Cosmic Rays; Heavy Ions from Accelerator	Direct	Most Plastic Detectors	1.0	For Ions at sufficiently low energies
Alpha Particles	Direct	Cellulose plastics or UV-irradiated Lexan	0.1–1.0	Most efficient at energies below 4 MeV
Protons	Direct	Cellulose Nitrate or Acetate	0.1–1.0	Only at very low energy (<.55 MeV)
Protons	Induced fission	Any Dielectric Solid	$\sim 10^{-5}$	High energy; depends on fission cross section of fissioned nuclide
Fast Neutrons	Induced fission	Any Dielectric Solid	$\sim 10^{-5}$	"
Thermal Neutrons	Induced fission	Any Dielectric Solid	$\sim 5 \times 10^{-3}$	for ^{235}U fission plates
Thermal Neutrons	(n. alpha) reaction	Cellulose Nitrate or Acetate	$\sim 2 \times 10^{-2}$	^{10}B reaction plates
Electrons	Induced fission	Any Dielectric Solid	10^{-8} to 10^{-7}	energy 200–500 MeV
Photons	Induced fission	Any Dielectric Solid	$\sim 2\text{–}3 \times 10^{-6}$	energy >20 MeV

Where spectral information is desired, separate detectors with additional fissioning nuclides such as ^{237}Np and ^{232}Th may be included. Since these elements are only fissionable with fast neutrons (> 0.5 MeV and >1.0 MeV respectively), the three counts for the number of tracks adjacent to the U, Np, and Th give a three-parameter energy spectrum for the neutrons (*8*). A similar but simplified system is regularly used at the reactor of the Nuclear Research Center, Würenlingen, Switzerland to record both routine and accidental doses to personnel (*9*). Similarly, the nuclear safeguards device sketched in Fig. 2 makes use of the same principle to record the fluxes of neutrons at different energies as a

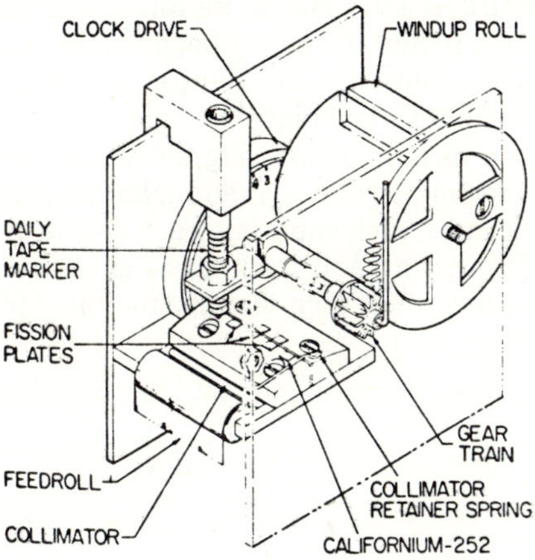

FIG. 2. Nuclear safeguards device to monitor use of nuclear reactors. The neutron energy spectrum is recorded as a function of time by fission tracks induced in ^{235}U, ^{238}U, ^{237}Np, and ^{232}Th and recorded on a moving polycarbonate tape. A ^{232}Cf fission source provides a known flux of particles to permit calibration of the tape speed (*10*). Similar procedures make possible a compact neutron dosimeter for personnel use at nuclear reactors (*8,9*).

function of time on a detector tape that is translated past a set of fission plates (*10*). The continuous recording is designed to give a tamperproof check on the power level at which reactors are operated. The device is designed to be left unattended over long periods of time for later retrieval.

Indirect Detection of Other Particles

As Table 1 makes evident, a variety of other particles and radiations can readily be counted by their property of inducing fission; high-energy charged particles including protons, electrons of a few hundred MeV, and photons of more than ~20 MeV can be detected. Such dosimetry is particularly useful around high-energy accelerators and is also applicable to recording doses of the lightly ionizing nucleonic portion of the cosmic rays that are encountered at high altitudes or in space. Often detector pairs with different fissioning nuclides will be needed if neutron effects are to be separated from those of the charged particles of interest.

Direct Particle Detection

Most charged nuclei may be detected directly. The lighter nuclei such as protons and alpha particles are observable only at low energies, $\leqslant$5 MeV for alphas and $\leqslant$ 0.55 MeV for protons. This threshold property allows the natural alpha activity of such nuclides as ^{238}U, ^{235}U, ^{232}Th, and ^{226}Ra to be seen directly, while nuclides such as ^{222}Rn, which emits alpha particles with higher energies, are best recorded by allowing alpha particles to be slowed down through an air gap or a thin degrader foil before entering the track detector.

Radon-222, an alpha-active gas, has been the subject of extensive track dosimetry. A simple piece of cellulose nitrate is a convenient recorder of the alpha activity in air from ^{222}Rn and its alpha active daughters, which commonly become attached to small aerosol particles. Radon-contaminated air, such as can be found in homes constructed on the tailings from uranium mining and in uranium mines, can be monitored with cellulose nitrate (*11,12*).

A recent application of this technique is to measure the rate at which radon emanates from soil, as a means for locating the subsurface uranium deposits, which are the usual source regions for the radon. In this procedure (see Fig. 3) an array of inverted plastic cups, each containing a cellulose nitrate sheet, is placed in the soil for a specified period and then removed, etched, and scanned. The measured contours of radon flux then identify the most likely sites to drill for uranium ore (*5*).

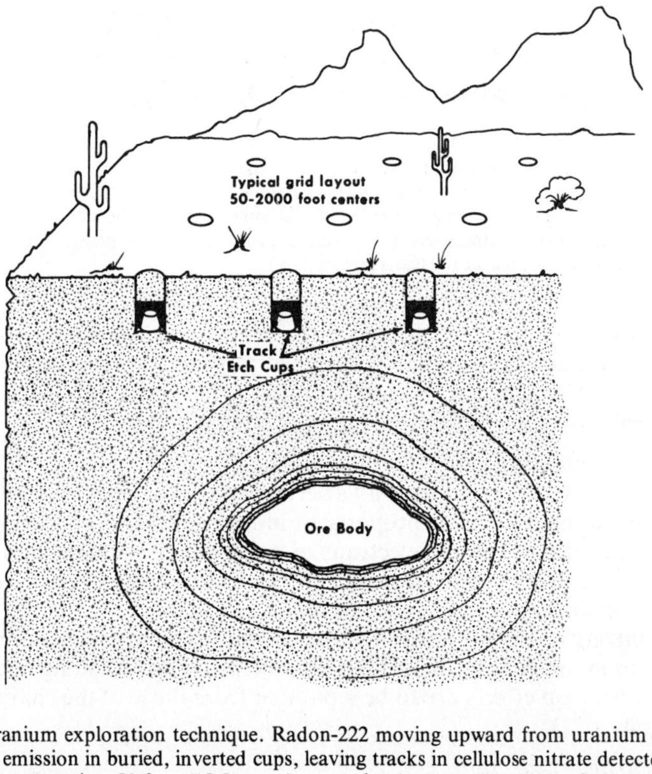

FIG. 3. Uranium exploration technique. Radon-222 moving upward from uranium deposits decays by alpha emission in buried, inverted cups, leaving tracks in cellulose nitrate detectors. In a typical grid layout there is a 50 foot (15.5 meter) separation between the center of one cup and its nearest neighbor. The contours of the track density reveal likely sites for more detailed mining exploration (*5*). Cellulose nitrate detectors are also being used to monitor radon in homes built on tailings from uranium mining.

Heavy Ion Dosimetry

Although only about 1% by number of cosmic ray nuclei consist of ions that are at least as heavy as carbon, these particles are of considerable potential importance in space because of their destructiveness and the fact that their biological effects have yet to be evaluated adequately. It is known that for certain cells (for example T-1 cultured kidney cells) a cosmic ray iron nucleus will leave lethal damage in each cell nucleus it crosses in the last ~3 mm of its range. Iron and its near neighbors in the chart of the elements are in fact the cytologically most important of the heavy cosmic rays.

A particular plastic, Lexan polycarbonate, is a uniquely suitable detector for assessing biological damage from heavy ions since its detection threshold lies at the ionization level where inactivation of T-1 cells acquires unit probability for an ion that intersects a cell nucleus (*13*).

For this reason Lexan detectors have been utilized as dosimeters on each Apollo mission (*13–16*) regardless of whether they were deliberately installed for that purpose. Some of these detectors have been sent as small film packs, attached to the astronauts' space suits (*14*). Other experiments have utilized existing space hardware. For example, Fig. 4 shows a detector worn by astronaut

FIG. 4. Cosmic rays have caused flashes in the eyes of astronauts and also left their marks in their helmets. As they slow down and come to rest, many of these heavy, charged particles (nuclei of atoms as heavy as iron) leave damage tracks in the plastic of which space helmets are made. Chemical etching reveals these tracks and allows them to be counted and measured, so that the number of tracks can be observed and used as a record of the heavy particle exposure of the crews throughout Apollo missions. The doses measured on the Apollo 8 and 12 missions are probably of negligible biological importance, but on extended deep space missions the doses could become a matter of concern. This dosimeter is the helmet worn by astronaut Pete Conrad during the Apollo 12 mission.

Pete Conrad on the Apollo 12 mission—a space helmet. This and other helmets were inverted, partially filled with sodium hydroxide, and etched to reveal the tracks of heavy particles at the inner surfaces of the helmets (*13*). A similar experiment was done using the Lexan frame (Fig. 5) of an Apollo 14 experiment, in which tracks were etched to easily visible size (*15*).

The experiments were directed at assessing the damage to non-generative cells. Estimates of the fraction of such cells that were deactivated in typical Apollo missions were low, typically less than one cell in 10^4 even for giant cells (*13*). However, when extrapolated to the two-year time-span that would be required for a Mars mission, losses would rise to a few percent, which is presumably no longer a negligible number. Observations of the marked effects of spacecraft shielding do give promise that by judicious design and rearrangement of the interior equipment the heavy ion flux could be significantly reduced (*15*).

Because of the short interaction mean free path of heavy ions, the flux of heavy cosmic rays that reaches ground level is negligible at all points on the earth. The attenuation is such, however, that at the altitudes used by supersonic aircraft the long-term fluence of heavy nuclei should be monitored (*16*).

The character of the etched tracks studied in the Apollo experiments is illustrated in Fig. 6. Changes in etching rate along the tracks allow the local ionization to be measured and the individual nuclei to be identified (*17*) as noted

FIG. 5. Lexan parts from the Apollo 14 electrophoresis experiment serve as dosimeters for avy cosmic rays that penetrated the command module (*15*). The tracks have been etched to sizes re they are readily visible to the naked eye. See also Fig. 6.

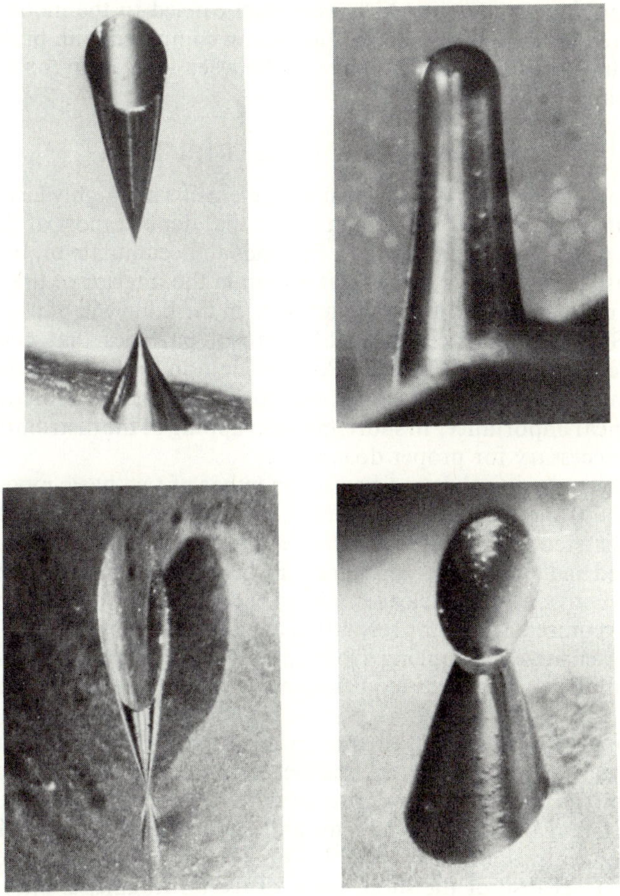

FIG. 6. Etched tracks of cosmic rays that penetrated the device pictured in Fig. 5. The taper and curvature of the tracks of the particles allow them to be individually identified. For example the particle at the upper left was an argon ion, that at the upper right iron.

in the caption. The direction of particle motion can be inferred from the track shape; for example, nuclei such as the blunt tracks shown in Fig. 1 (right) and Fig. 6 (upper right) stopped while moving into the etched surface. Since the numerous such tracks seen in the helmets were therefore coming from their interiors, they must in general have entered from the opposite sides. It has been correctly noted that although we do not know all that the astronauts were thinking during the mission, we do have some direct, quantitative information on what was going through their minds.

A more spacially specific examination of cellular brain damage from heavy cosmic ray nuclei is the mouse experiment (*18*) on Apollo 17. Lexan polycarbonate brain covers were surgically implanted in California desert mice

(*Perognathus longimembris*) which were then carried to the moon and back. Tracks etched in the Lexan detectors are being compared with brain lesions to establish what quantitative relation exists between ionization (or linear energy loss) levels and cellular damage.

RADIONUCLIDE DISTRIBUTION

The biological effects of alpha-emitting elements are highly localized due to the short range in tissue of the helium nuclei that deposit most of the energy of radioactive decay. If, for example, ^{239}Pu tends to accumulate on bone surfaces (*19*), the dose there will be vastly greater than in the interior of the bone, and a wholebone dose calculated from the total Pu in the bone will of itself have little meaning. Similarly, inhaled, insoluble aerosol particles that contain alpha emitters such as ^{239}Pu, ^{232}Th, or ^{210}Po provide non-uniform deposits on lung tissue, so that total lung burden of these nuclides will not adequately describe their biological importance. In short, micro-mapping of the distribution of alpha emitters is necessary for proper dosimetry.

Particle tracks in solids can be used in either of two basic modes: as alpha detectors (*2,20*) and as fission detectors (*2,21*). In the alpha mode natural alpha-decay is registered in an alpha-sensitive detector placed adjacent to the material to be mapped and left for sufficient time for tracks to accumulate. This procedure is similar to conventional nuclear emulsion procedures for alpha-mapping, both in the time required and in the resolution available. It does have the advantages of being insensitive to fogging from intense backgrounds of less intensely ionizing radiation and of not requiring the detector to be shielded from visible light.

Fission mapping (*21*) is a still more powerful tool. If the alpha emitter has a high cross section for fission by thermal neutrons (examples are ^{235}U or ^{239}Pu), its distribution can be sensitively displayed in a section of tissue by firmly sandwiching it against a track detector and exposing the assembly in a nuclear reactor. Concentrations below 10^{-14} atoms/atom can be measured in this manner, and because of the short range of fission fragments the resolution is superior to that of alpha-mapping. Mapping of heavy nuclides that are not fissioned by slow neutrons are in many cases possible with the use of fast neutrons or energetic charged particles. An example is Th; details are given in reference (*2*).

An unexpected bonus in fission mapping with Lexan polycarbonate is a photographic-like image of the section being mapped (*22*). This image normally results from roughening of the surface from minute alpha particle and/or ^{7}Li tracks from (n, alpha) reactions from boron or lithium in the material being mapped. A waiting period of at least a week after irradiation or ultra-violet irradiation of the detector is needed to reveal this structure adequately (*23*).

Figure 7 is a graphic example of the clarity of the plutonium distribution and the imaging of the underlying bone structure (*24*). The high density of alpha reactions in the bone itself suggests that the relative biological effect of thermal neutrons should be expected to be higher for the bone than for the adjacent tissue. The plutonium is highly specific to bone surfaces.

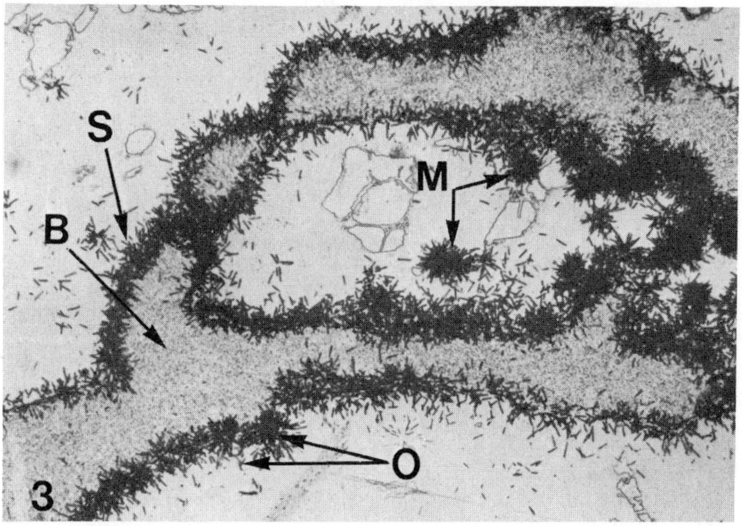

FIG. 7. A fission radiograph showing the location of ^{239}Pu in a biological section. A Lexan polycarbonate detector was pressed against the section, neutron-irradiated, and then etched to produce this image, which shows three trabeculae with surface deposits of ^{239}Pu (S), osteoclasts with ^{239}Pu aggregates on bone surfaces (O), macrophages with star-producing aggregates in bone marrow (M) and bone image (B) in the lumbar vertebral body from a dog given 2.7 μCi of ^{239}Pu/kg and killed 28 days later. Typical track lengths are ~10 μm. [Courtesy W.S.S. Jee.]

SUMMARY

Particle track detectors make possible a range of particle dosimetry applications—in uranium mines, in residential homes, in Apollo spacecraft, at nuclear reactors and particle accelerators, and at nuclear weapons plants. The detailed micromapping property of particle track detectors allows localized distributions of alpha emitters in biological substances to be determined and therefore makes possible the calculation of doses that are specific to the biological structures where they are actually received.

ACKNOWLEDGMENT

I am indebted to W. S. S. Jee for Figure 7 and permission to reproduce it.

REFERENCES

1. R. L. Fleischer, P. B. Price and R. M. Walker, *Science* **149,** 383 (1965).
2. R. L. Fleischer, P. B. Price and R. M. Walker, *Nuclear Tracks in Solids*. University of Calif. Press, Berkeley, 1975.
3. R. M. Walker, P. B. Price and R. L. Fleischer, *Appl. Phys. Lett.* **3,** 28 (1963).
4. K. Becker, in *Radiation Dosimetry*, F. H. Attix (Ed.), Vol 1, pp. 79–142. Academic Press, New York, 1972.

5. R. L. Fleischer, H. W. Alter, S. C. Furman, P. B. Price and R. M. Walker, *Science* **178,** 255 (1972).
6. P. R. Prevo, R. E. Dahl and H. H. Yoshikawa, *J. Appl. Phys.* **35,** 2636 (1964).
7. R. L. Fleischer, P. B. Price and R. M. Walker, *J. Nucl. Sci. Eng.* **22,** 153 (1965).
8. S. Pretre, E. Tochilin and N. Goldstein, *Proc. First Int. Cong. Radiation Protection*, pp. 491–505. Pergamon Press, N. Y., 1968.
9. S. Pretre, *Radiat. Eff.* **5,** 103 (1970).
10. B. Weidenbaum, D. B. Lovett and H. D. Kosanke, Flux monitor utilizing track-etch film for unattended safeguards application, *Trans. Am. Nucl. Soc.* **13,** 524 (1970).
11. K. Becker, *Health Phys.* **16,** 113 (1969).
12. D. B. Lovett, *Health Phys.* **16,** 623 (1969).
13. G. M. Comstock, R. L. Fleischer, W. R. Giard, H. R. Hart, Jr., G. E. Nichols and P. B. Price, *Science* **172,** 154 (1971).
14. E. V. Benton and R. P. Henke, *High-Z Cosmic ray Exposure of Apollo 8-14 Astronauts*. U.S. Air Force Weapons Lab. Rept. AFWL-TR-72-5, June (1972).
15. R. L. Fleischer, H. R. Hart, Jr., G. M. Comstock, M. Carter and A. Renshaw, *Science* **181,** 436 (1973).
16. O. C. Allkofer, W. Enge, W. Heinrich and H. Rohrs, in *Int. Conf. on Protection against Accelerator and Space Radiation*; Geneva, April 1971; CERN Rept. 71-16, Vol. 1, pp. 512–523. European Council for Nuclear Research, Geneva, 1971.
17. P. B. Price and R. L. Fleischer, *Annu. Rev. Nucl. Sci.* **21,** 295 (1971).
18. O. T. Bailey, et al., *Biocore experiment*, Apollo 17 Preliminary special publication SP-330, Chapter 26, pp. 1–14. (1973).
19. B. Bleaney, *Brit. J. Radiol.* **42,** 51 (1969).
20. R. L. Fleischer, P. B. Price, R. M. Walker and E. L. Hubbard, *Phys. Rev.* **133A,** 1443 (1964).
21. P. B. Price and R. M. Walker, *Appl. Phys. Lett.* **2,** 23 (1963).
22. J. D. Kleeman and J. F. Lovering, *Science* **156,** 512 (1967).
23. P. B. Price and R. L. Fleischer, *Radiat. Eff.* **2,** 291 (1970).
24. W.S.S. Jee, R. B. Dell and L. G. Miller, *Health Phys.* **22,** 761 (1972).

PHOTODYNAMIC INACTIVATION OF MACROMOLECULES AND CELLS

Chairman: L. R. Caldas

[1]Symposium E–3

Psoralen Cross-Links in DNA: Biological Consequences and Cellular Repair[1]

Ronald S. Cole and Richard R. Sinden
Department of Biochemistry
Department of Microbiology
University of Georgia
Athens, Georgia 30602, USA

Psoralens, or furocoumarins, are a class of compounds particularly noted for their sensitizing action on organisms and nucleic acids in the presence of long wavelength ultraviolet light (*1–5*). Some biological consequences of psoralen-plus-light exposures include killing of cells and viruses (*3–5*), mutagenesis (*2*), induction of malignancies in animals (*6*), and stimulation of pigmentation in skin cells (*7*). Based on this latter property, psoralens have been used clinically to treat patients suffering from skin disorders such as vitiligo and to increase the skin's tolerance to sunlight by stimulation of pigment production (*7,8*).

The molecular basis for the photosensitizing action of psoralens appears to be a consequence of their binding rather specifically to native DNA in the presence of light, forming adducts to single strands (*9–12*) and interstrand cross-links (*13–15*). The molecular structure and numbering system of the parent compound, psoralen, is shown in Fig. 1. Psoralens have two known photochemically reactive sites, the 3,4-pyrone and 4',5'-furan double bonds, which react specifically with the 5,6-double bond of pyrimidine bases, forming cyclobutane-type adducts between one psoralen and one pyrimidine base (*10,11*). These 1:1, or monoadducts correspond to the psoralen adducts to single strands of DNA. In addition to monoadducts, psoralens form covalent cross-links between the complementary strands of native DNA by a two-step photochemical reaction (*12*). A molecular mechanism of how psoralens bind to DNA and form monoadducts and cross-links is shown in Fig. 2, and proceeds as follows. Psoralen molecules first intercalate between the stacked base pairs of native DNA. Upon exposure to 360 nm light, intercalated psoralens absorb light and react with adjacent pyrimidine bases, forming adducts to single strands. Upon further exposure to light, some of the monoadducts rapidly react with bases in the opposite DNA strand, as shown, forming double adducts which act as interstrand cross-links. These reactions are evidently facilitated by the fortuitous arrangement of reactive sites in both DNA and the psoralen molecule. A psoralen molecule intercalated between two DNA base pairs having pyrimidines in opposite strands, can be aligned so that its two photochemically reactive sites are simultaneously in close proximity to the 5,6-double bonds of both pyrimidine bases (*13*). Using scale molecular models made up of 10 base pairs, we have assembled the intercalated, monoadduct, and cross-linking structures. All of

[1] Congress presentation E-3-2

these structures are easily assembled, and the distortions in the native DNA structure required to form the cross-linking diadduct appear to be comparable to that required to form a pyrimidine dimer in one strand.

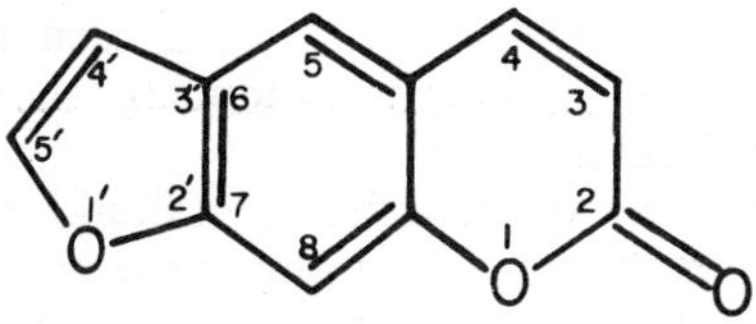

FIG. 1. Molecular structure and numbering system of psoralen.

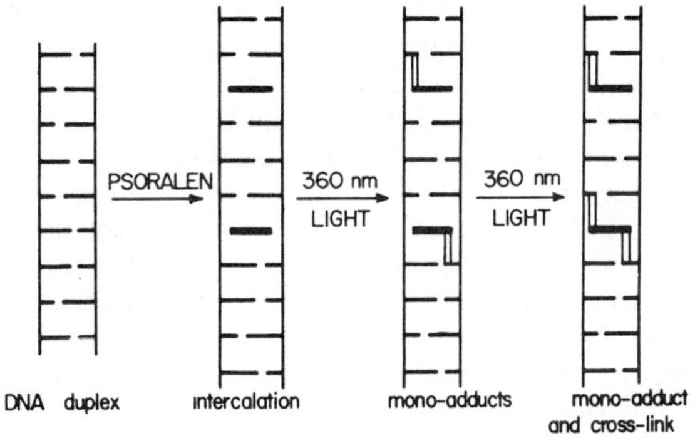

FIG. 2. Molecular mechanism of how psoralens bind to DNA to form adducts to single strands and interstrand cross-links. Psoralen molecules first intercalate between the stacked base pairs of duplex DNA. This weak binding occurs in the dark, is reversible, and appears to produce no biological consequences. Upon exposure to 360 nm light, psoralen molecules photochemically react with adjacent pyrimidine bases, and become covalently linked as adducts to single strands. Upon further exposure to light, some of these monoadducts further react to form diadducts which act as cross-links between the complementary DNA strands.

The above mechanism appears to be well supported by the following experimental evidence. Psoralens, in the dark, bind reversibly to DNA. This has been demonstrated by a variety of techniques which include the solubilization of psoralens by solutions of DNA, equilibrium dialysis, and viscosity increases when psoralens are added to solutions of DNA. This reversible binding probably occurs in the DNA of cells, but appears to have little or no biological consequence. Following exposure to light, psoralens are found to be covalently bound to DNA (*1,12*), and the chemical structures of several of the monoadducts have been determined (*10,11*).Under conditions ordinarily used for biological inactivations and for clinical treatments, the yield of cross-links is about one-third that of adducts to single strands (*12*). The photochemical binding of psoralens to synthetic polynucleotides of known sequence has been reported, and consistent with the suggested mechanism, the relative binding rates to poly d(A:T), calf-thymus DNA, and poly dA:dT are about 10, 5, and 1, respectively (*16*).

Biological effects of psoralen-plus-light treatments appear to be mediated primarily by the formation of cross-linking products in DNA. Comparing a number of psoralen derivatives with various substituent groups, it was found that compounds which react most efficiently with native DNA to form cross-links, also show the most pronounced sensitizing action on skin cells (*9*) and bacteria (*3,5*). Under conditions where DNA repair is minimized, *E. coli* and phage λ are killed by psoralen-plus-light exposures producing about one cross-link and several monoadducts per genome (*5*).

A priori, it might be expected that cross-linking defects, rather than monoadducts in DNA, might be considerably more potent damages leading to loss of biological activities. A cross-link remaining in cellular DNA would prevent complete separation of the complementary strands, and thus block normal semiconservative replication. Other DNA-related metabolic events which may require temporary strand separation, such as transcription, might also be impaired. Consistent with this interpretation, psoralen-plus-light treated DNA has a greatly reduced template activity in an in vitro transcription system (*16*), and RNA synthesis is blocked in *E. coli* cells containing unexcised psoralen cross-links (R. Sinden and R. S. Cole, unpublished results). In addition to blocking strand separation, formation of a cross-link necessarily damages the intact base sequence in both DNA strands at nearly the same point along the helical axis. Because of these properties, it has been difficult to imagine how cells containing DNA cross-links were able to recover and form normal progeny. However, it is known that cells survive from treatments which produce many cross-links in the DNA of each cell, suggesting that cross-linked DNA is successfully repaired. For example, animal cells in tissue culture recover from treatment with nitrogen or sulfur mustards producing thousands of cross-links per cell (*17*), and bacteria maintain their colony-forming ability after exposure to cross-linking agents producing 50–70 cross-links per genome (*5,18*).

The repair of cross-linked DNA is evidently under genetic control, and is mediated by enzymes which are involved in the repair of other DNA damages, such as pyrimidine dimers induced by 254 nm light and ionizing-radiation-induced DNA defects. *E. coli* strains carrying mutations in genes affecting repair or recombination are more sensitive to DNA cross-linking agents than are wild-type cells. For example, wild-type, excision-defective (*uvr*$^-$), and recombination-deficient (*recA*$^-$) bacteria lose their colony-forming ability after psoralen-plus-light or nitrogen mustard treatments which produce approximately 60, 20, and 4 cross-links per genome, respectively (*5,18*). A double mutant which is both *uvr*$^-$ and *recA*$^-$ is extremely sensitive to cross-linking agents, and is killed by exposures producing not more than one cross-link per genome (*5*). Cells from human patients suffering from xeroderma pigmentosum are deficient in the excision of pyrimidine dimers and are also more sensitive to psoralen-plus-light (*19*). Evidently, the repair of cross-linked DNA depends upon a combination of excision and recombination.

Cross-links induced in bacterial DNA are specifically removed during an incubation in growth medium after cross-linking treatments, and this reaction is controlled by the *uvr A*, *B*, and *C* genes (*29,33*) which code for enzymes catalyzing the excision of UV-light-induced pyrimidine dimers (*20*). The complete removal of cross-links from the DNA of mammalian cells has also been demonstrated, and appears to occur in a two-step process (*21*).

It is also known that cross-linking agents induce extensive genetic exchanges in eukayrotic cells (*22–27*) and in phage λ infecting *E. coli* carrying a λ prophage (*28*). Thus, it appears that cross-links in cellular DNA cause discrete structural changes which result in extensive genetic exchanges (*32,34*), and these induced genetic exchanges play an important, if not essential, role in the repair of cross-linked DNA.

Several mechanisms have been proposed for the repair of cross-linked DNA. For example, a segment containing the cross-link could be deleted (*27*). This would allow the strands to separate during the next cycle of replication, but would cause the loss of at least one base pair for each cross-link removed. Kohn, Steigbigel and Spears (*29*) have suggested another mechanism in which one arm of the cross-link is first excised and the resultant gap is filled by repair synthesis. Excision-repair of the second arm would complete the process. A difficulty with this mechanism is the proposed use of a single-stranded region, still containing the partially excised cross-link, as a template for repair polymerization. However, Rupp and Howard-Flanders (*30*) have shown that damaged bases remaining in template strands block polymerization of a continuous complementary strand, so that during replication a gap is left in the daughter strand at the position of each damaged base in the template strand. This type of DNA defect, which should be structurally analogous to a partially excised cross-linking residue, is quite stable in cells, and is thought to be corrected by a recombinational event with an intact region of a sister duplex (*31*).

Recently, another mechanism has been proposed, in which excision and recombination function sequentially to restore the intact base sequence to a cross-link damaged region (*32*). This mechanism is shown in Fig. 3, and proceeds as follows. Two strand cuts are first placed near each cross-link. A nuclease widens the excision gap, exposing a single-stranded region. The partially excised cross-linking residue still attached to the single-stranded region blocks repair polymerization, thus stabilizing the excision gap. Strand exchanges with a homologous duplex then insert a segment into the gap, restoring the intact base sequence and bihelical structure to the damaged region. Excision-repair of the second arm of the cross-linking residue completes the repair sequence.

A mechanism of this type is supported by a wide variety of biophysical, chemical, and genetical evidence. After treatment of *E. coli* with psoralen-plus-light, interstrand cross-links in the cellular DNA are removed during a 30-minute incubation. During cross-link excision the DNA is cut by incision nucleases controlled by the *uvrA* and *B* genes, so that upon denaturation, the DNA strands separate into segments having a molecular weight about equal to twice the single-strand cross-link spacing (*32,33*). Both cuts at each cross-link are placed in the same strand, as shown at the top of Fig. 3 (*33*). Strand cuts at cross-links are quite stable in cells, and remain open for more than 20 min. One interpretation of these results is that normal repair polymerization and strand rejoining are blocked by the partially excised cross-linking residue still attached in the single-stranded region.

We have further characterized some structural properties of cross-linked DNA isolated at various stages of repair and recombination, by testing its susceptibility to DNAses known to degrade specific structures. The sequence of

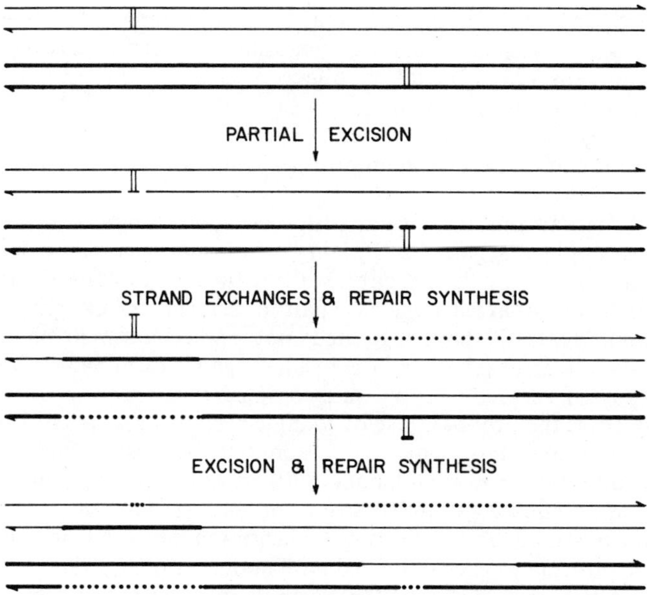

FIG. 3. One possible mechanism for the repair of cross-linked DNA is shown here to illustrate the types of intermediates detected and the sequence of events involving partial excision followed by recombination. Strands of the two separate homologous duplexes shown here are represented as thin and thick lines, and at the top of the diagram, each duplex contains one cross-link. In the first step of the repair sequence, two cuts are placed in one DNA strand near each cross-link. A nuclease then widens the excision gap, exposing a single stranded region still containing the partially excised cross-linking residue. Strand exchanges with an intact region of the homologous duplex restore the bihelical structure and intact base sequence to the cross-link damaged regions. When this is completed, the second arm of the cross-linking residue is excised. Dashed lines indicate regions of repair synthesis.

strand cutting at cross-links, and the products of this reaction are shown in Fig. 4. Strand cutting during cross-link removal appears to take place in two discrete steps, with the first cut by the UV endonuclease (product of the *uvrA* and *B* genes) being made much faster than the second, which appears to be made by the 5' and 3' exonuclease activity of DNA polymerase I. For sufficiently short incubations, only the first cut is made at most of the cross-links. Although the strands have been partially cut, the DNA is reversibly bihelical after denaturation. These denatured molecules had covalent links between their complementary strands were primarily double-stranded, but in addition had single-stranded "tails" which amounted to about 24% of their composition. These "tails" terminated in a 3' hydroxyl group, suggesting that the UV endonuclease made the first cut on the 5' side of one arm of the cross-link, and that a hydroxyl group was formed at the 3' terminus at the strand cut. This result is in accord with the products of the UV endonuclease making strand incisions in DNA containing pyrimidine dimers (*20*). After the second cut is made in cross-linked DNA, the stabilized excision gap is enlarged to a single-stranded region of 500–700 bases. This was determined using the S_1 nuclease of *Aspergillus oryzae*, a DNAse highly specific for single-stranded regions of DNA (*38,39*). After cross-link removal, S_1 nuclease cut the DNA into double-stranded segments having a molecular weight about equal to the average cross-link spacing, and about 600 bases were released for each fragment produced.

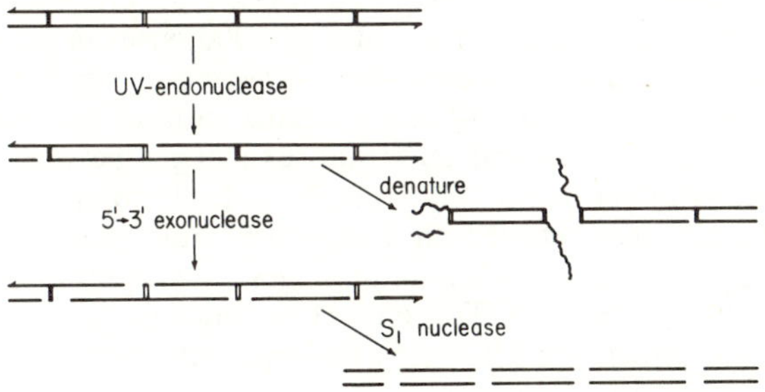

FIG. 4. Suggested mechanism for the removal of cross-links, and characterization of intermediates and products of this process. The UV-endonuclease (*uvrA* and *B* gene products) first makes an incision on the 5' side of one arm of the cross-linking residue. This partially cut structure, upon denaturation, dissociates into double-stranded segments having long single-stranded tails with 3' hydroxyl termini. A second strand cut by a 5' to 3' exonuclease removes the covalent link between the strands, permitting complete denaturation. The excision-gap is then widened by an exonuclease to a length of about 600 bases. This product is susceptible to digestion by S_1 nuclease, and the products are double-stranded segments having a mw about equal to the cross-link spacing and acid-soluble oligonucleotides from the single-stranded regions.

In wild type cells the cut strands eventually become joined into high molecular weight DNA comparable to that from intact cells. This strand rejoining apparently depends upon genetic recombination, as recombination-deficient strains carrying mutations at *recA* or *recB*, *C recF sbcB* are unable to undergo this process (*33*). In Rec$^-$ strains the DNA strands remain as low molecular weight molecules. Recombinational events between different DNA duplexes labeled with radioactive and density isotopes were also detected during the period of strand rejoining, and are controlled by the *recA* gene (*32*). These exchanged regions are on the order of several thousand nucleotides in length (*33*), and are thus similar to those detected as a result of recombinational or post-replication repair of pyrimidine dimers in excision-defective *E. coli* cells (*31*).

Extensive repair-polymerization also occurs during the period of strand rejoining and strand exchanges, and the majority of this newly synthesized DNA is in the form of long segments (ca. greater than 3000 nucleotides) covalently attached to molecules synthesized before the cross-linking treatment with psoralen-plus-light (*33*). This repair-polymerization appears to be of the long segment type controlled by *recA*, as first described by Cooper and Hanawalt (*35*).

Extensive DNA degradation is characteristic of cells exposed to cross-linking agents. This breakdown in psoralen-plus-light treated cells occurs during the period of strand rejoining and is controlled by the *recB*, *C* genes (Cole and Sinden, unpublished results) which code for the subunits of exonuclease V(*36*).

There is also substantial genetic evidence for a sequential repair mechanism involving excision and recombination. As described earlier, DNA cross-linking agents induce extensive mitotic recombination in eukaryotic cells. More recently, Howard-Flanders, Lin, and Bardwell (*37*) have shown that phage λ containing interstrand cross-links undergoes extensive genetic exchanges with a prophage in lysogenic strains of *E.* coli. When phage replication was blocked, these exchanges occurred in wild-type, but not excision-defective host strains, demonstrating that excision, but not replication, was required for the genetic exchanges. These exchanges also appeared to be on the order of 1000 bases in length.

Evidently, *E. coli* cells correct cross-linking defects induced in their DNA by psoralen-plus-light or other cross-linking agents. Succcssful repair evidently involves partial excision followed by strand exchanges between homologous duplexes. As bacterial cells normally contain multiple copies of their haploid genome, strand exchanges between these genomes could restore the intact base sequence to a duplex region containing damages to both strands at nearly the same point along the helical axis. This type of mechanism may also be operative in eukaryotic cells, which are known to undergo extensive mitotic exchanges after treatments with cross-linking agents.

ACKNOWLEDGMENTS

The authors thank Dr. Sidney Kushner for his valuable assistance and advice on some of the work described here. This work was supported by United States Public Health Service Grant CA 14396.

REFERENCES

1. L. Musajo and G. Rodighiero, in *Photophysiology* (A. C. Giese, Ed.) pp. 115–147. Academic Press, New York, 1972.
2. J. W. Drake and J. McGuire, *J. Virol.* **1**, 269 (1967).
3. E. L. Oginsky, G. S. Green, D. G. Griffiths, and W. L. Fowlks, *J. Bacteriol.* **78**, 821 (1959).
4. L. Musajo, G. Rodighiero, G. Colombo, V. Torlone, and F. Dall'acqua, *Experientia* **21**, 22 (1965).
5. R. S. Cole, *J. Bacteriol.* **107**, 846 (1971).
6. A. C. Griffin, R. E. Hakim, and J. Knox, *J. Invest. Dermatol.* **31**, 289 (1957).
7. R. M. Lerner and A. B. Lerner, in *Dermatologic Medications,* **2nd Ed.**, pp. 98–99. Year Book Publishers, Chicago, 1960.
8. J. Africk and J. Fulton, *Br. J. Dermatol.* **84**, 151 (1971).
9. G. Rodighiero, L. Musajo, F. Dall'acqua, S. Marciani, G. Caporale, and L. Ciavatta, *Biochim. Biophys. Acta* **217**, 40 (1970).
10. L. Musajo, F. Bordin and R. Bevilaqua, *Photochem. Photobiol.* **6**, 927 (1967).
11. L. Musajo, F. Bordin, G. Caporale, S. Marciani and G. Rigatti, *Photochem. Photobiol.* **6**, 711 (1967).
12. R. S. Cole, *Biochim. Biophys. Acta* **254**, 30 (1971).
13. R. S. Cole, *Biochim. Biophys. Acta* **217**, 30 (1970).
14. R. S. Cole and D. Zusman, *Biochim. Biophys. Acta* **224**, 660 (1970).
15. F. Dall'acqua, S. Marciani, and G. Rodighiero, *FEBS Lett.* **9**, 121 (1970).
16. P. Chandra, S. Marciani, F. Dall'acqua, D. Vedaldi, G. Rodighiero, and R. K. Biswa, *FEBS Lett.* **35**, 243 (1973).
17. A. R. Crathorn and J. J. Roberts, *Nature (London)* **211**, 150 (1966).
18. P. D. Lawley and P. Brookes, *Nature (London)* **206**, 480 (1965).
19. H. P. Baden, J. M. Parrington, J. D. Delhanty, and M. A. Pathak, *Biochim. Biophys. Acta* **262**, 247 (1972).
20. A. Braun, in *Molecular Mechanisms for the Repair of DNA* (P. C. Hanawalt and R. B. Setlow, Eds.). Plenum Press, New York, 1975.
21. B. R. Reid and I. G. Walker, *Biochim. Biophys. Acta* **179**, 179 (1969).
22. G. Morpurgo, *Genetics* **48**, 1259 (1963).
23. M. Shaw and M. Cohen, *Genetics* **51**, 181 (1965).

24. J. German and J. LaRock, *Tex. Rep. Biol. Med.* **27**, 409 (1969).
25. H. J. Evans and D. Scott, *Proc. Roy. Soc. Ser. B.* **173**, 491 (1969).
26. R. Holliday, *Genetics* **50**, 323 (1964).
27. M. Schewe, D. Suzuki, and U. Erasmus, *Mutat. Res.* **12**, 255 (1971); *ibid.* **12**, 269 (1971).
28. P. Howard-Flanders and P-F. Lin, *Genetics* **73 Suppl.**, 85 (1973).
29. K. Kohn, N. Steigbigel, and C. Spears, *Proc. Nat. Acad. Sci. USA* **53**, 1154 (1965).
30. W. D. Rupp and P. Howard-Flanders, *J. Mol. Biol.* **31**, 291 (1968).
31. W. D. Rupp, C. E. Wilde, D. Reno, and P. Howard-Flanders, *J. Mol. Biol.* **61**, 25 (1971).
32. R. S. Cole, *Proc. Nat. Acad. Sci. USA* **70**, 1064 (1973).
33. R. S. Cole and R. R. Sinden, in *Molecular Mechanisms for the Repair of DNA* (P. C. Hanawalt and R. B. Setlow, Eds.). Plenum Press, New York, 1975.
34. P. Howard-Flanders, R. S. Cole, and P-F. Lin, *Stud. Biophys.* **36–37**, 211 (1973).
35. P. K. Cooper and P. C. Hanawalt, *Proc. Nat. Acad. Sci. USA* **69**, 1156 (1973).
36. J. Tomizawa and H. Ogawa, *Nature (London) New Bio.* **239,** 14 (1972).
37. P. Howard-Flanders, P-F. Lin and E. Bardwell, in: *Molecular Mechanisms for the Repair of DNA* (P. C. Hanawalt and R. B. Setlow, Eds.). Plenum Press, New York, 1975.
38. T. Ando, *Biochim. Biophys. Acta* **114**, 158 (1966).
39. R. C. Wiegand, G. N. Goodson, and C. M. Radding, *Fed. Proc. Fed. Am. Soc. Exper. Biol.* **33,** 1351 (1974).

Chemical Aspects of Photodynamic Action in the Presence of Molecular Oxygen[1]

Klaus Gollnick

Institut für Organische Chemie der Universität München,
D-8 München 2, Germany

Irradiation with UV light of wavelengths shorter than about 3300 Å is generally destructive to living systems because of photolabile substances which absorb in this spectral region. Whereas these destructive effects of the UV irradiation are independent of the presence of molecular oxygen, irradiation of living matter with light of longer wavelengths than 3300 Å requires the presence of molecular oxygen in order to cause what is also fundamentally a destructive process. In addition to the presence of molecular oxygen, irradiation with near UV and visible light requires the presence of a photosensitizer, either present as a natural constituent of the cell or externally applied, which absorbs the radiation and thus makes available the absorbed energy to the biological system by mechanisms discussed below. The photosensitized process which needs the presence of oxygen was termed "Photodynamic Action" by H. F. Blum (*1*), and it is this process with which we are concerned in the following discussion.

Photosensitizers such as eosin, erythrosin, rose bengal, methylene blue, hematoporphyrin and other porphyrins, methyl cholanthrene, chlorophyll or hypericin, to mention only a few of the more important sensitizers, cause the following photodynamic effects in the presence of molecular oxygen and light: killing of bacteria and protozoa, hemolysis of erythrocytes, stimulation of skeletal muscles, sensitization of the human skin, inactivation of bacteriophages, destruction of enzymes and nucleic acids, and denaturation of proteins (*1–5*).

Relatively early during the studies of the photodynamic action it was recognized that the general characteristics of "photosensitized oxygenation reactions" correspond to those of photodynamic actions: approximate temperature independence, irreversibility, independence of the intensity of radiation and of the exciting wave-length, and quantum yields approaching unity (*1*).

It seems, therefore, justified to take a closer look at the chemistry of photosensitized oxygenation reactions that take place in well-defined systems in solution, to examine the mechanisms involved and to discuss product formation with various classes of organic compounds that may be model compounds for photodynamic actions.

Absorption of a light-quantum raises the sensitizer from its singlet ground-state ($^1S_0$) into one of the electronically excited singlet states ($^1S_1$*, $^1S_2$*, ...) depending upon the energy of the absorbed photon. Generally, molecules in

[1] Congress presentation E-3-3.

higher electronically excited states have lifetimes shorter than about 10^{-11}sec during which they return to the lowest of the electronically excited singlet states, i.e. to the "first excited singlet state" ($^1S_1$*) that possesses a lifetime of some 10^{-9} sec or longer. ^1S*, thus formed with a quantum yield of unity, disappears mainly by monomolecular (or pseudo-monomolecular) reactions such as fluorescence (F), internal conversion (IC) and intersystem crossing (ISC) to the lowest triplet state $^3S_1$*). Since triplet states are generally many orders of magnitude longer-lived than their corresponding first excited singlet states, bimolecular reactions are much more favored with triplet state sensitizers. Fig. 1 shows a simplified energy level diagram in which the processes pertinent to the discussion of photooxygenation reactions in solution are included.

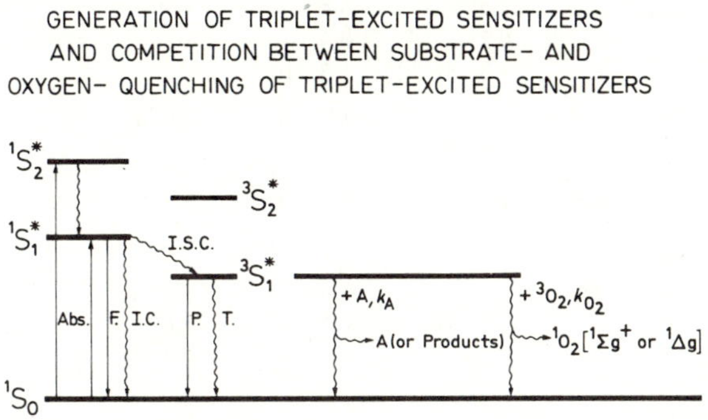

FIG. 1. Generation of Triplet Excited Sensitizers and Competition between Substrate- and Oxygen-Quenching of Triplet Excited Sensitizers

The efficiency of a bimolecular reaction of $^3S_1$* with either a substrate A or molecular oxygen will thus depend on how effectively these processes will compete with each other and with the monomolecular (radiationless) deactivation to the sensitizer's singlet ground state (process T). (Phosphorescence (P) normally does not occur at room temperature in fluid solutions.)

According to Fig. 1, the quantum yields of triplet formation (Φ_T) and of triplet sensitizer quenching by substrate A (Φ_A) or by molecular oxygen (Φ_{O_2}) are given as

$$\Phi_T = k_{ISC}/(k_F + k_{IC} + k_{ISC}) \quad (1)$$

$$\Phi_A = \Phi_T \times k_A[A]/(k_T + k_A[A] + k_{O_2}[O_2]) \quad (2)$$

$$\Phi_{O_2} = \Phi_T \times k_{O_2}[O_2]/(k_T + k_A[A] + k_{O_2}[O_2]) \quad (3)$$

As is seen from the competition ratio,

$$\Phi_A/\Phi_{O_2} = k_A[A]/k_{O_2}[O_2], \quad (4)$$

the appearance of a substrate activation or of an oxygen activation (or both) will decisively depend on the reaction conditions, i.e. on the substrate and oxygen concentrations applied, provided that the rate constants k_A and k_{O_2} do not differ by several orders of magnitude. Triplet energies (E triplet), quantum yields of triplet formation (Φ_T) and triplet lifetimes (τ_T) as well as the triplet quenching constants k_{O_2} and k_A are given Table 1 for alcoholic solutions of the dyes eosin, rose bengal and methylene blue and for the standard ketonic sensitizer, benzophenone.

An approximately diffusion-controlled rate constant is found for oxygen-quenching of triplets in solution (*6–8*) and the factor of about 1/9th by which k_{O_2} is actually reduced (i.e. $k_{O_2} \approx 1/9 \cdot k_{diff}$ for various solvents) is the spin statistical factor anticipated from a more detailed quenching sequence (*9*).

TABLE 1.
Generation of Triplet Excited Sensitizers and Competition Between Substrate- and Oxygen-Quenching of Triplet Excited Sensitizers

	Eosin[a)]	Rose Bengal[a)]	Methylene Blue[b)]	Benzo-phenone[c)]
$E_{TRIPLET}$(kcal/mol)	43.2[d)]	39.5[d)]	34[e)]	68.5[f)]
Φ_T	0.25	0.8	~1	1.0[g)]
τ_T (sec)	$4.5 \cdot 10^{-4}$	$1.5 \cdot 10^{-4}$		
k_{O_2} ($M^{-1}sec^{-1}$)	$1.2 \cdot 10^9$	$1.2 \cdot 10^9$	$2 \cdot 10^9$	
k_{A_2} ($M^{-1}sec^{-1}$):				
β-Carotene		$1.3 \cdot 10^9$ [h)]		
Cyclooctatetraene	$1.4 \cdot 10^9$	$5.0 \cdot 10^8$		
Alcohol	$\sim 10^2$ [i)]	~1[j)]		$1.8 \cdot 10^6$ [k)]
Benzidine		$3.6 \cdot 10^9$ [l)]		
Dimethyl Aniline		$1.0 \cdot 10^9$ [l)]	$2.0 \cdot 10^9$	$2.7 \cdot 10^9$
Aniline	$1.4 \cdot 10^9$ [m)]	$2.5 \cdot 10^9$ [n)]	$\sim 10^9$	$5.0 \cdot 10^9$ [o)]
Tryptophan	$9 \cdot 10^8$ [m)]		$2.0 \cdot 10^9$	
Tyrosine	$7 \cdot 10^7$ [m)]			
Histidine			10^6	
Guariosirve-monophosphate			10^9	
Methionine			$3 \cdot 10^8$	
Di-n-butyl Sulfide				$6.6 \cdot 10^8$ [p)]

a) in methanol, Ref *6*). *b*) in methanol/water (1:1), Ref. (*8*). *c*) in alcohols. *d*) Ref. (*10*). *e*) Ref. (*11*). *f*) Ref. (*12*). *g*) Ref. (*13*). *h*) Sensitizer: Chlorophyll a, Ref. (*14*). *i*) Calc., Ref. (*15a*). *j*) calc. from Ref. (*15b*). *k*) in isopropanol, Ref. (*16*). *l*) Ref. (*17*). *m*) Ref. (*18*). *n*) Ref. (*19*). *o*) Ref. (*20*). *p*) Ref. (*21*).

Using $2 \cdot 10^9$ $M^{-1}sec^{-1}$ as an average value for k_{O_2} in various solvents, a log[A] vs. log[O_2] plot for various k_A-values is shown in Fig. 2 which allows us to determine the substrate concentration [A] at which, for a given oxygen concentration, the rates of triplet quenching by A and O_2 are the same. In oxygen-saturated methanol, the substrate concentration has to be about forty times larger than in air-saturated water in order to achieve equal rates of substrate and oxygen quenching of triplet sensitizers. Thus, even if all the other experimental conditions are kept constant, a change from an oxygen-saturated organic solvent to an air-saturated aqueous system may be accompanied by a change of the prevailing mechanism (i.e. from oxygen activation to substrate activation).

Interaction of triplet excited sensitizers with substrates may result in energy transfer if the triplet energy level of the substrate is equal to or lower than that of the sensitizer. Since β-carotene has an E_T below 22 kcal/mole (see below), quenching of chlorophyll triplets (Table 1) and other dye-triplets is almost certainly due to energy transfer exclusively (*14,22*). Cyclooctatetraene probably deactivates triplet eosin and triplet rose bengal by the same energy transfer mechanism (*7*). If, however, the triplet energy level of the substrate exceeds that of the sensitizer, the energy transfer becomes very inefficient and other processes, such as hydrogen abstraction or electron transfer resulting in the formation of free radicals and radical ions, respectively, come into play. Ketonic sensitizers such as benzophenone having $^3(n,\pi^*)$-states are especially apt to dehydrogenate substrates and dyes seem to acquire such an aptitude with increasing triplet energies above about 40 kcal/mole.[2] Electron transfer seems to be the result of triplet quenching by anilines, and this process appears to be nearly independent of the sensitizer's triplet energy.

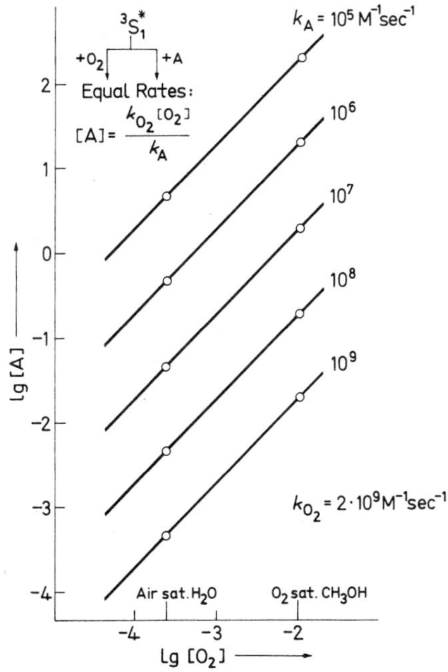

FIG. 2. Equal Rates of Triplet Sensitizer Interactions with Substrates having rate constants $k_A = 10^5$–10^9 M^{-1} sec^{-1} and $^3O_2$ ($k_{O_2} = 2\cdot10^9$ $M^{-1}sec^{-1}$)

Of the some twenty amino acids that occur in typical proteins, only tryptophan, tyrosine, hystidine, methionine and cysteine react at appreciable rates on irradiation in the presence of photodynamic sensitizers and oxygen (*2,3*),

[2] When 3β-hydroxy-Δ^4-cholestene is photooxidized in the presence of various sensitizers, the ratio of the two main products, Δ^4-cholestenone-(3) and 4,5-α-epoxi-cholestanone-(3), increases with increasing triplet energy of the sensitizers used (*23,24*). Whereas the latter product is assumed to be produced by an interaction of the lower excited singlet molecular oxygen ($^1\Delta g$) with the substrate, the former product was thought to be formed via the higher excited singlet molecular oxygen ($^1\Sigma g^+$) (*11,24*). However, this species is so efficiently quenched to $^1\Delta_gO_2$ in solution that it can hardly react in bimolecular reactions (*25*). Increasing efficiency in dehydrogenating substrates with increasing triplet energies of the dye-sensitizers, however, do explain the observed change of product ratio (*7*) and such a view has been confirmed with other dyes and substrates (*10*). Among the dyes used as photodynamic sensitizers, riboflavin seems to be especially able to abstract hydrogen, although it may also transfer energy to oxygen.

and of the bases of the nucleic acids, only guanine is particularly sensitive (*4,26,27*). All of these compounds were found to quench triplet sensitizers (*8,18*), probably by electron transfer from the substrate to the triplet excited dye. If such an electron transfer is followed by transfer of a proton, the overall effect again is hydrogen abstraction.

Interaction of triplet excited sensitizers with molecular oxygen ($^3\Sigma g^- O_2$) gives rise to the formation of singlet excited oxygen (*7,28,29*) as was originally suggested by Kautsky (*30,31*). If the energy of the triplet sensitizer is high enough (i.e. above about 40 kcal/mole) the primarily formed singlet state of O_2 is suggested to be $^1\Sigma g^+$ which possesses an excitation energy of 37.7 kcal/mole (*32*). Sensitizers which have triplet energies below this value must produce singlet oxygen in the $^1\Delta g$-state (22.6 kcal/mole above the triplet ground state, $^3\Sigma g^-$) directly. Since singlet oxygen in the $^1\Sigma g^+$-state is very short-lived in solution (less than 10^{-11}sec, (*25*) as a consequence of the highly efficient transition to the lower $^1\Delta g$ state, sensitizers of triplet energies above 23 kcal/mole produce $^1\Delta g$ O_2 independent of their actual triplet energies although with different yields (*10*).

Instead of energy transfer from triplet sensitizers to triplet oxygen molecules, an electron transfer to $^3O_2$ generating the superoxide ion radical $\cdot O_2^-$ may occur (*33,34*). However, with eosin as a sensitizer, less than one out of a thousand encounters between the triplet sensitizer and oxygen leads to $\cdot O_2^-$. In proton donating solvents, $\cdot O_2^-$ may react to give H_2O_2 and O_2, but there is evidence (*35*) that the oxygen molecule is not produced in the excited singlet state $^1\Sigma g^+$ as was earlier suggested (*36*).[3] The superoxide ion radical may react with H_2O_2 to hydroxyl radicals (*37*); with amines, nitric oxides are reported to be formed (*35*) in contrast to the production of demethylated and N-formylated amines from reactions of N-methyl amines with singlet oxygen (*38–40*) (see below).

The triplet quenching processes so far discussed are shown in Fig. 3.

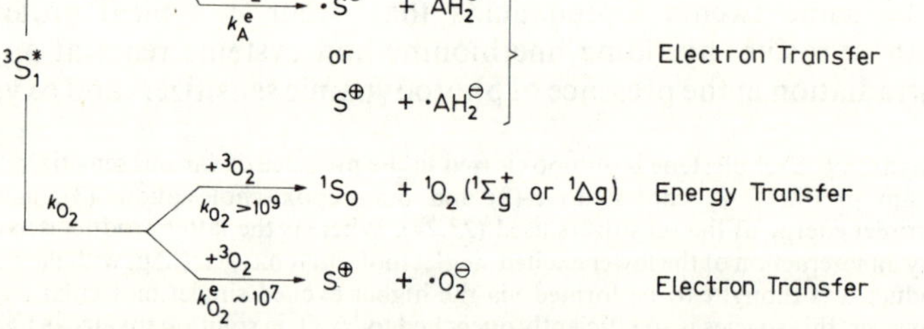

FIG. 3. Substrate- and Oxygen-Quenching of Triplet Excited Sensitizers

[3] Note added in proof: However, production of singlet oxygen in its $^1\Delta_g$ state is indicated by some very recent experiments in which $\cdot O_2^-$, electrolytically produced in dry acetonitrile, is reacted with 1,3-diphenylisobenzofuran as a $^1O_2$-trapping agent after water was introduced in the system (*120*).

To complicate matters, the triplet excited sensitizer may react with the sensitizer in the singlet ground state by electron transfer thus producing one molecule of reduced dye and one molecule of an oxidized dye (*33,34*):

$$^3S_1^* + {}^1S_0 \rightarrow \cdot S^- + \cdot S^+ \tag{5}$$

Back transfer of an electron may regenerate the two sensitizer molecules in their singlet ground states; $\cdot S^-$ may be oxidized to $^1S_0$ by oxygen, whereas $\cdot S^+$ may oxidize the substrate to give $^1S_0$ and AH_2^+ or $\cdot AH$.

Summarizing, we may state that interaction between $^3S_1^*$ and substrate A (or AH_2 of Fig. 3) mainly results in the formation of radical species whereas interaction of $^3S_1^*$ and $^3O_2$ mainly give rise to singlet excited molecular oxygen in its reactive $^1\Delta g$ state (referred to as $^1O_2$ in the following). The former process has been classified as a "Type I," the latter as a "Type II" process (*7,41*).

So far, we have seen that the occurence of substrate radicals and of singlet oxygen will depend on the competition ratio Φ_A/Φ_{O_2} (equation 4) which in turn is governed by our choice of sensitizer and solvent as well as by the substrate and oxygen concentrations applied. However, whether we shall observe product formation according to a Type I or according to a Type II process will furthermore depend on how efficient the subsequent reactions, namely

$$\cdot AH + {}^3O_2 \rightarrow \rightarrow \text{oxygenation product} \tag{6}$$

on the one hand, and

$$^1O_2 + AH_2 = \text{oxygenation product} \tag{7}$$

on the other, will be under the chosen experimental conditions.

Thus, if $\cdot AH$ (or $\cdot AH_2^+$, $^3AH_2^*$ etc.) undergoes mainly a back reaction to AH_2, product formation may be only due to a Type II process although $^1O_2$ is formed with a rather low quantum yield; interaction of the substrate with the triplet sensitizers leads only to quenching of the Type II process in this case. On the other hand, the production of singlet oxygen with a rather high quantum yield may only lead to quenching of the Type I process if the substrate is not suited to undergo any reaction with $^1O_2$.

Aliphatic alcohols such as isopropanol and ethers do not react with singlet oxygen[4]; however, hydrogen abstraction with ketonic or quinoid sensitizers occur with appreciable quantum yields (*45,46*).

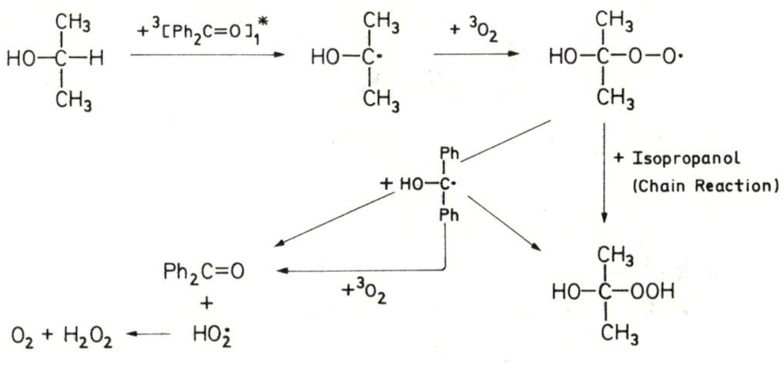

[4] Alcoholates, however, do react under conditions of Type II photosensitized oxygenation reactions (*42,43*) and with singlet oxygen produced by microwave discharge (*44*).

With isopropanol, a quantum yield of about unity is reported for the 2-hydroxy-2-hydroperoxypropane production, whereas with ethers such as tetrahydrofuran or dioxane the quantum yields of the hydroperoxide formations approach a value of twenty. The latter findings indicate the occurrence of a radical chain reaction as may be anticipated from the well-known autoxidation reaction sequence.

Dehydrogenation of olefins by sensitizers such as benzophenone produce allyl radicals that react with $^3O_2$ in a radical chain reaction to rearranged and unrearranged allyl hydroperoxides (*47*).

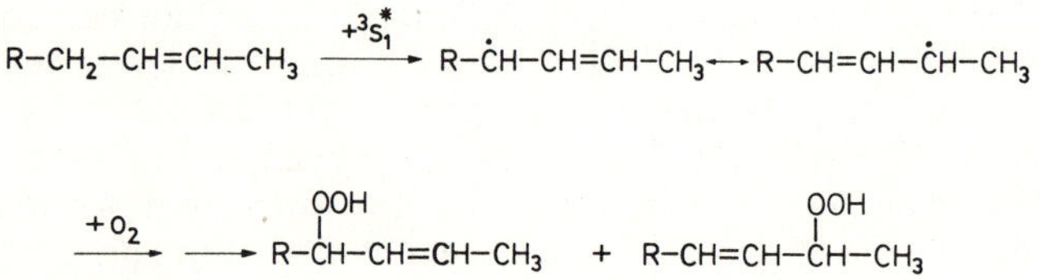

However, singlet oxygen may also be formed with quantum yields as high as 0.5 in benzophenone-sensitized reactions and thus may give rise to appreciable competition between Type I and Type II processes (*10*) (for product formation: see below).

Hydrogen abstraction or electron transfer followed by a loss of a proton may also occur with phenols, amines, and sulfides. Tertiary amines were reported to give dealkylated amines as well as amides (*48*); however, there is evidence that these products are formed by interaction of singlet oxygen with the amines rather than by the radical pathway (*38–40*). Sulfides give rise to the production of α-hydroxysulfides and sulfoxides (*49*) whereas only sulfoxides are reported as products of $^1O_2$ reactions with sulfides (*7,50,51*) (see below).

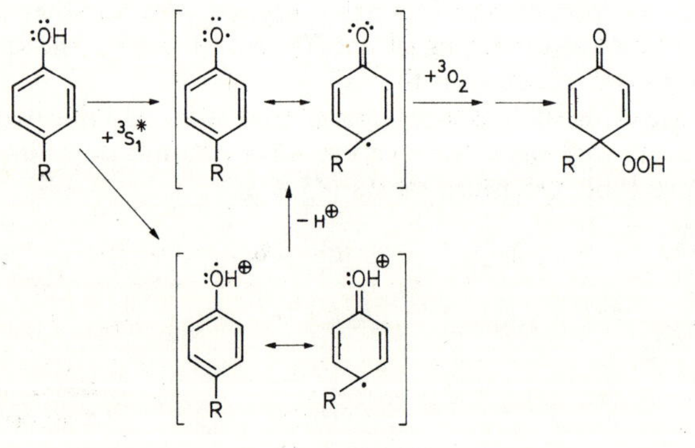

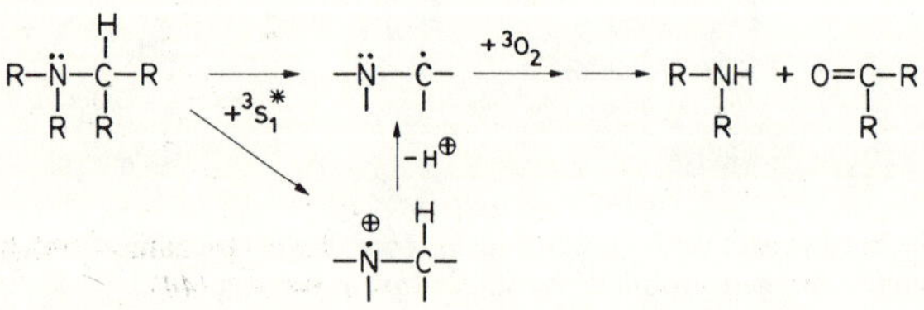

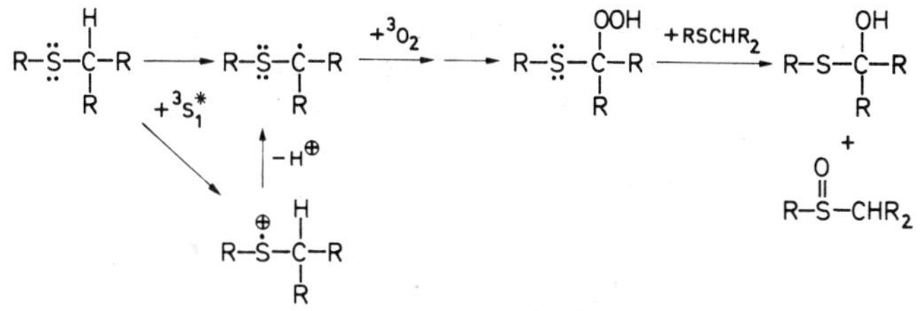

With regard to singlet oxygen reactions, we have to consider three competing processes that can occur with $^1O_2$: spontaneous deactivation of $^1O_2$ in a (pseudo)-monomolecular reaction to the triplet ground state, and the acceptor interaction with $^1O_2$ that may lead to either product formation and/or to quenching of singlet oxygen.

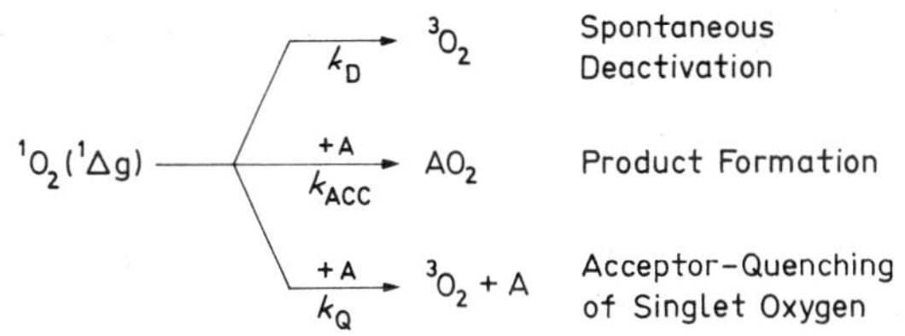

$$\frac{k_D}{k_{ACC} + k_Q} = \beta'(M) \qquad \frac{k_D}{k_{ACC}} = \beta\ (M)$$

The lifetime of singlet oxygen, $\tau_{^1O_2}$, equal to the reciprocal spontaneous deactivation rate constant k_D, was recently measured in various solvents and found to be solvent-dependent (*8,52–54*):

$\tau_{^1O_2}$ in H_2O	:	$\sim\ 10^{-6}$ sec
$CH_3OH/H_2O(1{:}1)$	:	$3.5\cdot10^{-6}$ sec
$CD_3OD/D_2O(1{:}1)$	:	$3.5\cdot10^{-5}$ sec
CH_3oh	:	$7.14\cdot10^{-6}$ sec
Benzene	:	$2.4\cdot10^{-5}$ sec
Pyrdine	:	$3.2\cdot10^{-5}$ sec

With the aid of the laser pulse technique, rate constants k_Q and k_{ACC} may be determined directly (*8,52,54*). Using the recently measured $^1O_2$ lifetimes, one may also calculate the bimolecular rate constants from the β-values defined as

$$\beta = k_D/k_{ACC}\ (\text{mole/l}) \quad \text{or} \quad \beta' = k_D/(k_{ACC} + k_Q)\ (\text{mole/l}).$$

The β- and β'- values are easily determined by a kinetic analysis of steady state photooxygenation experiments (*7,28*). The quantum yield of product formation according to a Type II process is given by

$$\Phi_{ox.P} = \Phi_T \times k_{O_2}[O_2]/(k_T + k_A[A] + k_{O_2}[O_2]) \times \times k_{ACC}[A]/(k_D + (k_{ACC} + k_Q)[A]) \qquad (9)$$

Cisoid 1,3-diene systems (cyclopentadienes, cyclohexadienes, polycyclic anthracenes, heterocycles) (*63*), olefins (*7,28*), enamines and vinyl ethers (*64–66*), sulfides and sulfoxides (*7,50,51,67*), disulfides (*68–70*), amines (*7,38–40,71*), phenols (*72,73*), and alcoholates (*42–44*) have been shown to interact with singlet oxygen, and some representatives of these classes are included in Table 2.

TABLE 2.
Interaction of $^1O_2$ ($^1\Delta_g$) with Substrates.
$\beta(\beta')$ - Values and Rate Constants.

Substrate (in Methanol)	$\beta(\beta')$ (M)	$k_Q + k_{ACC}$ ($M^{-1}sec^{-1}$)	k_{ACC} ($M^{-1}sec^{-1}$)	Ref.
2,5–Dimethylfuran	$1 \cdot 10^{-3}$		$1.4 \cdot 10^{8}$	28
Ergosterol[a)]	$4.5 \cdot 10^{-3}$		$9.3 \cdot 10^{6}$	55
Anthracene[b)]	$4.2 \cdot 10^{-1}$		$1.0 \cdot 10^{5}$	56
2,3–Dimethyl–2–butene	$3 \cdot 10^{-3}$		$4 \cdot 10^{7}$	57,58
2–Methyl–2–butene	$5.5 \cdot 10^{-2}$		$2.5 \cdot 10^{6}$	57,58
2–Methyl–2–pentene	$1.8 \cdot 10^{-1}$		$7.8 \cdot 10^{5}$	57,58
1–Methylcyclohexene	1.2		$1.2 \cdot 10^{5}$	57,58
Cholesterol[a)]	0.90		$4.7 \cdot 10^{4}$	55
Cyclohexene	26		$5.4 \cdot 10^{3}$	57,58
Diethyl Sulfide	$2 \cdot 10^{-2}$		$7 \cdot 10^{6}$	59
Methionine	$5 \cdot 10^{-3}$	$3 \cdot 10^{7}$	$5 \cdot 10^{6}$	8
Tryptophan[c)]	$4 \cdot 10^{-3}$	$4 \cdot 10^{7}$	$4 \cdot 10^{6}$	8
Histidine[c)]	$3 \cdot 10^{-3}$	$5 \cdot 10^{7}$	$7 \cdot 10^{6}$	8
Benzyl Amine[d)]	1.62	$1 \cdot 10^{6}$	$6 \cdot 10^{5}$	60
sec.-Butyl Amine[d)]	2.0	$2.1 \cdot 10^{7}$	$7 \cdot 10^{6}$	60
Triethyl Amine[a)]	$3 \cdot 10^{-3}$	$9.3 \cdot 10^{7}$	$1.3 \cdot 10^{7}$	61
Nicotine	0.10	$6.4 \cdot 10^{6}$	$1.4 \cdot 10^{6}$	17
β-Carotene[e)]		$1.3 \cdot 10^{10}$		62

a) in pyrydine. b) in benzene. c) in methanol/water (1:1). d) in water.
e) quenching of singlet oxygen only.

1,3-Diene systems and olefins react with $^1O_2$ with rate constants that clearly indicate the electrophilicity of singlet oxygen (*7,28,58*); Hammett plots, obtained from singlet oxygen reactions with *m*- and *p*-substituted α,β,β-trimethylstyrenes (*74*) and p,p′-disubstituted α,α'-dimethylstilbenes (*75*, and K. Gollnick and F. N. Kotsonis, unpublished), provided ρ-values of −0.9 and −1.3, respectively, which prove the electrophilic nature of singlet oxygen. There is no indication so far that 1,3-dienes and olefins quench $^1O_2$ by another path than by reaction to photooxygenation products. Amines, however, were found to quench $^1O_2$ to its triplet ground state in addition to their ability to react with them (*17,60,61*), and there exist other amines which are only powerful quenchers of singlet oxygen, such as diazabicyclooctane (DABCO) (*76*), hexamethylenetetramine (*71*), t-butylamine (*71*), and N-methyl-2,2,6,6-tetramethylpiperidine (*77*).

All three amino acids for which singlet oxygen oxidation has been shown to occur (*8*) are also catalysts for the transition of singlet oxygen to its triplet ground state. Whereas the reaction of singlet oxygen takes place at the sulfur atom or at the heterocyclic ring in methionine, histidine and tryptophan, respectively, it is tempting to suggest (from comparison of k_Q-data of Table 2) that singlet to triplet conversion in O_2 is due to interaction of $^1O_2$ with the α-amino group of the amino acids. However, such a deactivation can only occur with the unprotonated amino group (i.e. at higher pH-values) since ammonium compounds are neither quenchers nor acceptors of singlet oxygen (*7,71*).

The most powerful quencher of singlet oxygen so far detected is β carotene (*9,78,79*). The high efficiency is due to energy transfer from singlet oxygen to β-carotene which therefore should possess a triplet energy well below 23 kcal/mole. In contrast the amines which are at least three orders of magnitude less effective $^1O_2$-quenchers execute their interaction with $^1O_2$ probably by transferring an electron to $^1O_2$ (*76,80,81*).

The outstanding capability of β-carotene as one of the most efficient quenchers of singlet oxygen and of triplet excited sensitizers is thus due to energy transfer to its low-lying triplet state. We may therefore anticipate a similar behavior for all molecules which have triplet energies below 23 kcal/mole. Furthermore, we may expect that raising the energy gradually from below 23 kcal/mole to higher values will extinguish firstly the molecule's singlet-oxygen quenching capabilities and thereafter its triplet-sensitizer quenching abilities. Decreasing the number of conjugated double bonds from eleven as in β-carotene to below nine conjugated double bonds (and thus raising the energy of the triplet state) drastically diminishes the quenching of singlet oxygen whereas the number of conjugated double bonds has to drop to seven in order to afford a sharp decrease in the quenching of chlorophyll a (*78,79*). Singlet oxygen quenching has been suggested to be the main cause for the protective role of β-carotene in photosynthetic organisms (*82*), and the recent finding that a minimum of nine double bonds in carotenoids are necessary for efficient protection from light damage (*83*) supports this suggestion. Further support of this hypothesis is lent by the fact that mice can be protected against photosensitization by hematoporphyrin when treated with β-carotene (*84*). There seems to be a good chance that carotenoids (with nine conjugated double bonds or more) as well as other compounds with triplet energies below 23 kcal/mole will play an important role in many diseases caused by photosensitization in the presence of oxygen (*1,85,86*).

With regard to the products of singlet oxygen reactions with the various classes of compounds mentioned above, the 1,3-diene systems and olefins are the most extensively studied and best understood acceptors. 1.4-Cycloaddition occurs with 1,3-diene-systems such as anthracene and ergosterol delivering the isolable and well-defined trans-annular (endo-) peroxides (*63*).

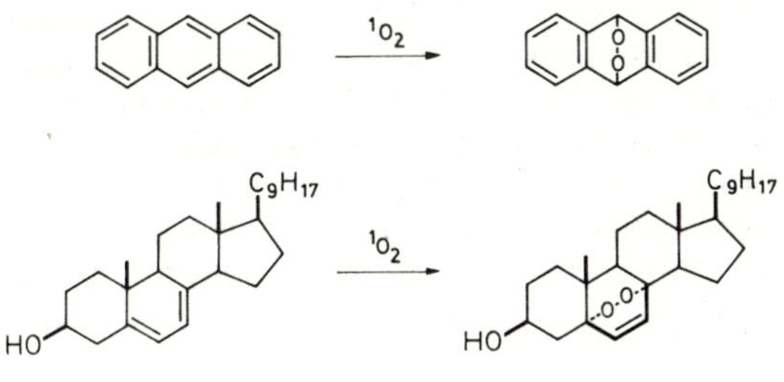

Furan and its derivatives such as 2,5-dimethylfuran afford ozonides which are stable only at low temperatures; at room temperature they react with the solvent (*7,28,63*). With methanol, a methoxy hydroperoxide is formed. Imidazol is reported to yield an oxidation product in which two solvent molecules (methanol) are incorporated; a 1.4-addition of $^1O_2$ is assumed to proceed (*87*). Methylimidazol reported to yield acetylurea (*2*) may also be produced via a 1.4-$^1O_2$-addition followed by a split of the O-O bond and hydrolysis.

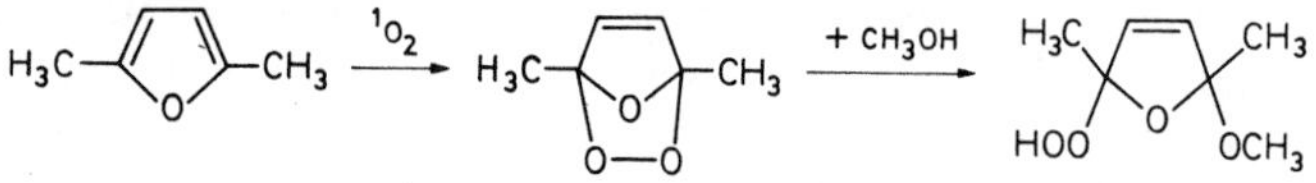

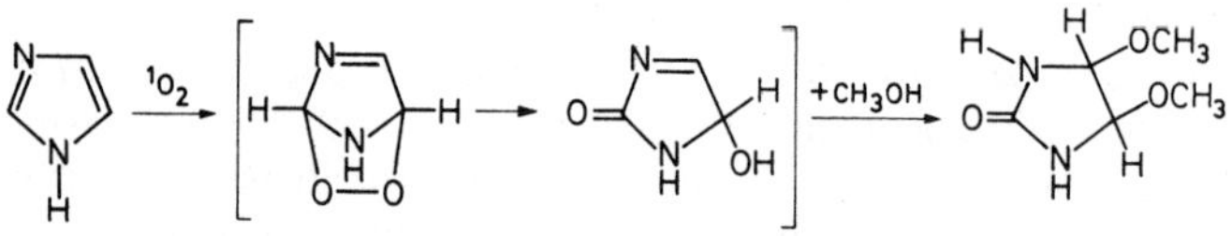

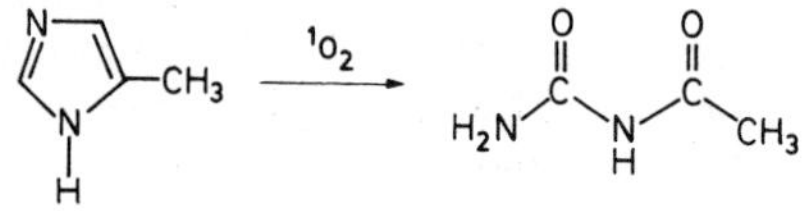

Olefins which possess allylic hydrogens generally interact with $^1O_2$ according to an ene-reaction to give exclusively the rearranged allyl hydroperoxides (*7,28*).

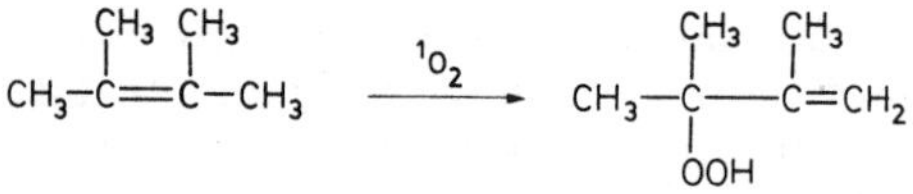

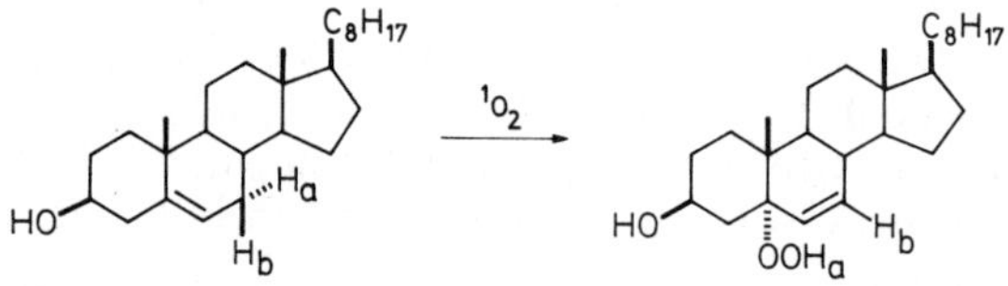

Since the rate of $^1O_2$-attack on the olefinic double bond varies considerably with its alkylation (Table 2), this reaction appears to be highly selective; steric hindrance and conformational effects contribute significantly to the selectivity (*6,7,57,58*).

Cholesterol reacts with $^1O_2$ to give 3β-hydroxy-5α-hydroperoxy-Δ^6-cholestene (*55*) by using only the 7α-hydrogen as is required by the ene mechanism (*88*), whereas antoxidation via free radicals produce several other hydroperoxides but not the $^1O_2$-product (*89*). Incorporation of the singlet oxygen product into the membranes of normal red blood cells leads to hemolysis which has characteristics very similar to the photohemolysis of blood cells in the presence of porphyrins and oxygen (*90,91*). β-carotene orally applied to patients who suffer from erythropoetic protoporphyria reduces greatly the photosensitivity (*84*) probably by acting as a dark anti-oxidant in the red blood cell membrane (*91*).

Unsaturated fatty acids and esters such as methyl linoleate can also undergo singlet oxygen oxygenation reactions to the corresponding rearranged allylic hydroperoxides (*92*).

Enamines and vinyl ethers may add singlet oxygen to give 1,2-dioxetanes which easily decompose to the corresponding carbonyl compounds, sometimes accompanied by the emission of light (*64–66*). 1.2-Cycloaddition of $^1O_2$ is also likely to occur with tetraphenylimidazol and N-benzoylhistidine methyl ester (*87,93*).

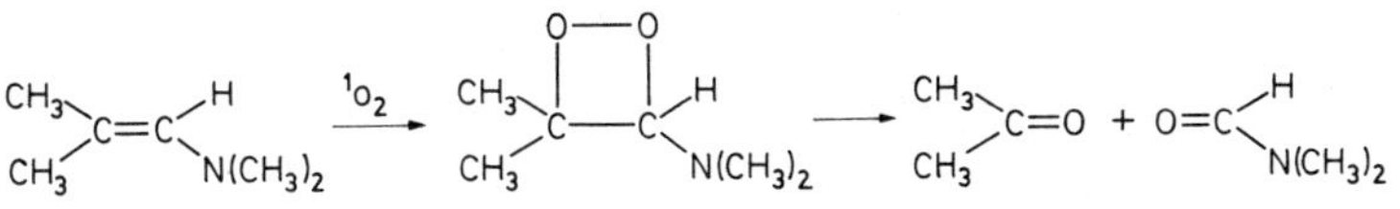

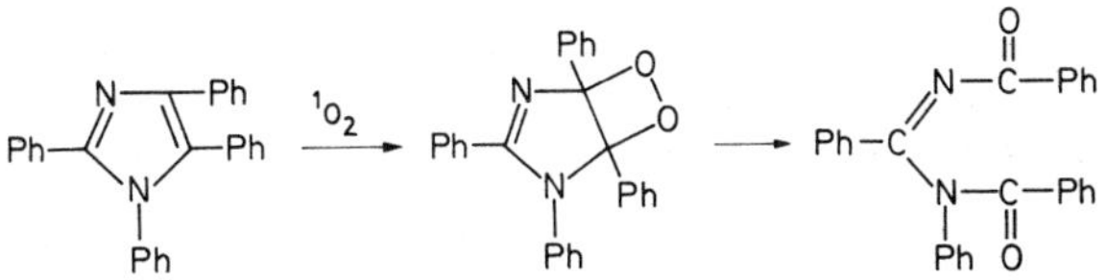

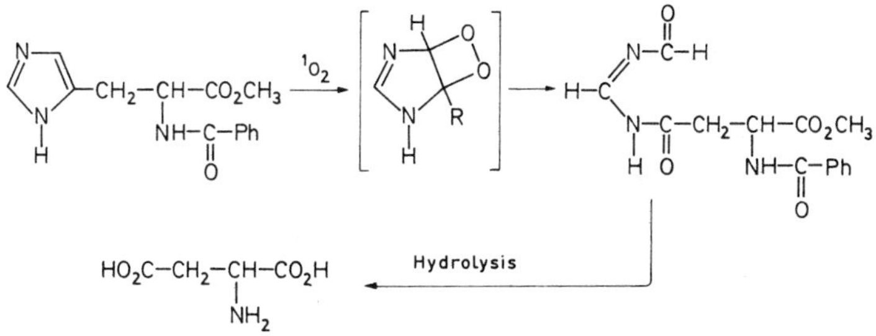

1,4-Addition as well as 1,2-addition of singlet oxygen thus seem to be possible pathways in reactions with imidazoles. Histidine is one of the few amino acids that is efficiently attacked in dye-sensitized photooxygenation reactions of either the free amino acids or when contained in proteins and enzymes; methylene blue sensitization of histidine oxygenation yields four unidentified products (*2*). Inhibition of histidine photooxygenation by azide ions and by 2,3-dimethyl-2-butene, as well as a newly developed test of singlet oxygen reactions involving a comparison of photooxygenation rates in normal and perdeuterated solvents, established the singlet oxygen pathway (*8*).

In the same manner, 3-methylindole reacts (probably via 1,2-cycloaddition to an unstable dioxetane) to *o*-formamidoacetophenone (*94*), and tryptophan oxygenation photosensitized by methylene blue in methanol-water (1:1) seems to occur at least partly by a singlet oxygen reaction (*8*). In benzene, singlet oxygen oxidation of tryptophan is reported to yield a 3a-hydroxy-hexahydropyrroloindole via a presumed hydroperoxide whereas N-formylkynurenine is formed among a series of other products in aqueous solutions (*95*), presumably via a dioxetane intermediate.

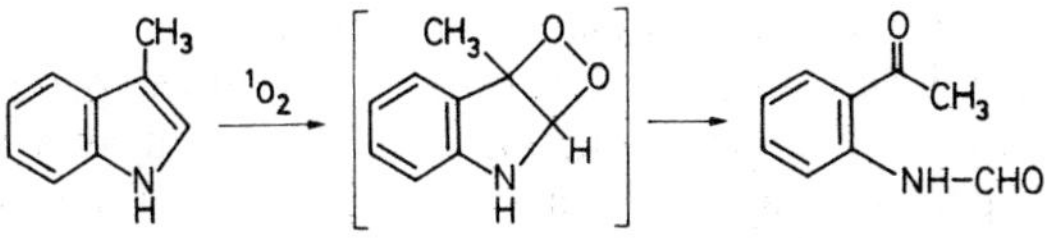

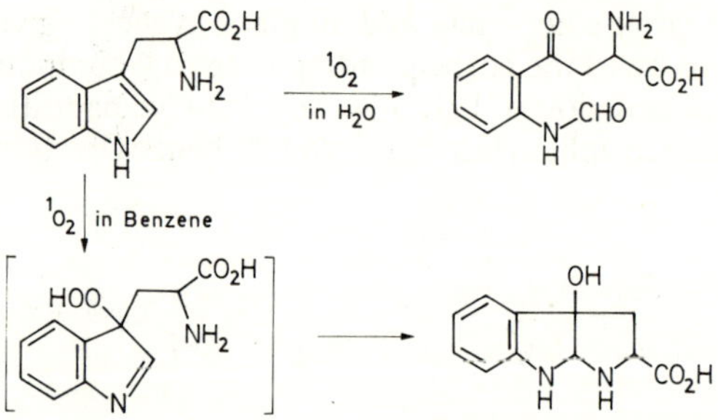

However, N-formylkynurenine may also be formed by a free radical mechanism since azobisisobutyronitrile-induced Type I oxidation of 2-isopropyl-3-methylindole in hexane yielded a hydroxyacylaminoacetophenone derivative (*96*).

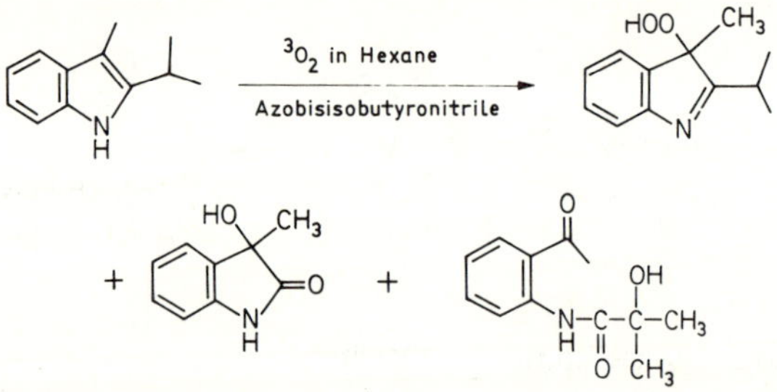

Type I mechanisms have also been suggested for tryptophan photosensitized oxygenation by some authors (*34,97*) and others have concluded that a change from a Type I to a Type II process occurs when the oxygen concentration is increased to values above $2 \cdot 10^{-4}$ *M* O_2 (*98*). Using the rate constants for tryptophan of Tables 1 and 2, Fig. 4 shows how the quantum yields of tryptophan activation (Φ_A), oxygen activation ($\Phi_{^1O_2}$), and product formation by a singlet oxygen mechanism ($\Phi_{ox.P.}$) vary with the oxygen and tryptophan concentrations. Interestingly enough, photosensitized oxygenation in a Type II process seems to prevail in air-saturated methanol-water (1:1)-mixtures at $2 \cdot 10^{-3}$ *M* tryptophan (*8*) although under these conditions triplet sensitizer quenching by tryptophan is two times larger than by oxygen and in spite of the fact that only one out of about 75 singlet oxygen molecules gives rise to product formation. It should be mentioned that the quantum yield of product formation, calculated to be 0.005 under these conditions, falls in the range of quantum yields that are typical for the photodynamic inactivation of enzymes (*2*).

To complicate matters, one of the tryptophan products, N-formylkynurenine, appears to be an additional sensitizer of tryptophan photooxygenation (*99*).

As was mentioned above, among the bases of nucleic acids only guanine appears to be rather sensitive toward photosensitized oxygenation (*4,26*). Although relatively little is known about the nature of the primary products of photodynamic action on nucleic acids, a Type I as well as a Type II process affording a hydroperoxide as a primary product in both cases (with perhaps

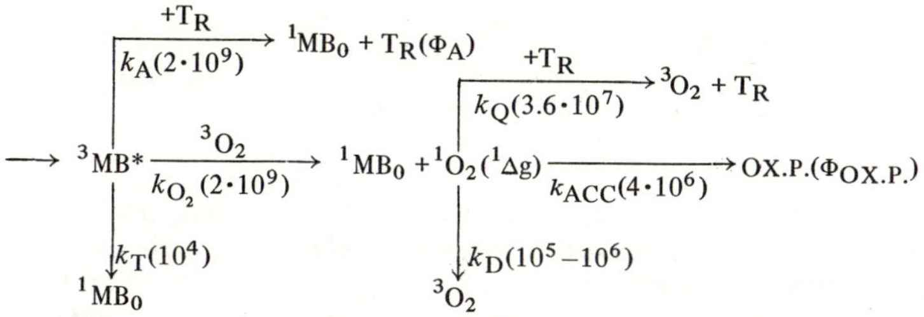

		Φ_A	$\Phi_{^1O_2}$	$\Phi_{OX.P.}$
CH_3OH, O_2-sat. $[O_2] = 10^{-2}$ M				
	$[A] = 10^{-2}$ M	0.5	0.5	0.03
	$= 10^{-3}$ M	0.09	0.91	0.01
	$= 10^{-4}$ M	0.01	0.99	0.001
CH_3OH/H_2O (1:1) $[O_2] = 6 \cdot 10^{-4}$M				
(Air-sat.)	$[A] = 10^{-2}$ M	0.94	0.06	0.003
	$= 2 \cdot 10^{-3}$ M	0.63	0.37	0.005
	$= 10^{-4}$ M	0.15	0.85	0.001

FIG. 4. Methylene Blue Sensitized Oxygenation of Tryptophan in methanol/water (1:1). Rate constants from Table I and Table II used for calculation.
Experimental conditions: $[A] = 2{\cdot}10^{-3}$ *M* and $[O_2] = 6{\cdot}10^{-4}$ *M:* Ref.(*8*).

endoperoxide formation preceding the hydroperoxide formation in the Type II process) may be envisaged.

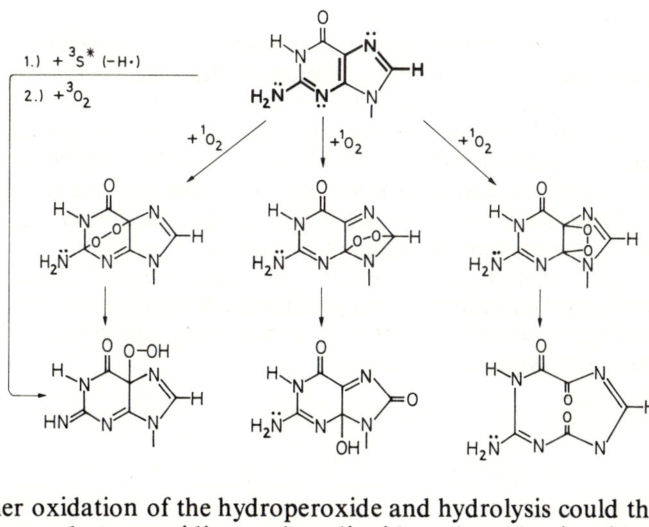

Further oxidation of the hydroperoxide and hydrolysis could then lead to the reaction products guanidine, carbon dioxide and parabanic acid which were observed when (^{14}C-labelled, indicated by *) guanine was submitted to lumichrome-sensitized oxygenation (*26,100*).

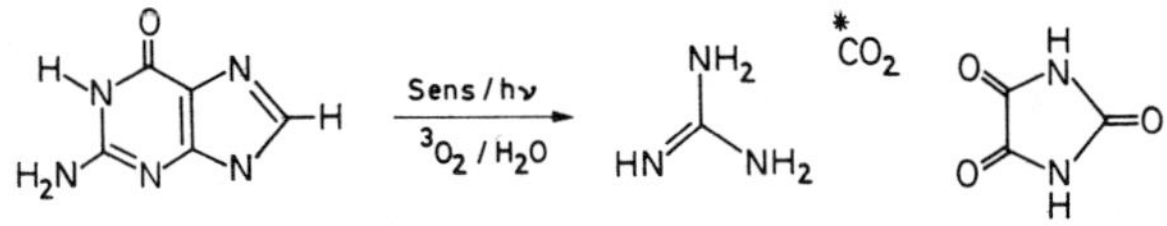

In view of the fact that chemically generated singlet oxygen as well as singlet oxygen generated by microwave discharge parallel photosensitized oxygenations of various purines and pyrimidines with respect to efficiency and (so far as determinable) to products (*101,102*), and, furthermore, since uric acid derivatives such as tetramethyluric acid are very likely to be oxidized in a Type II process (*103–105*), singlet oxygen oxygenation of guanine, guanosine and guanosine moieties of nucleic acids is a real possibility.

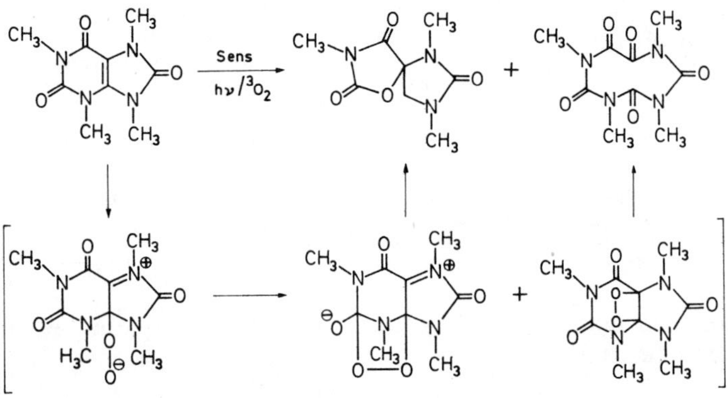

However, if the photosensitized oxygenation is executed on living organisms, the situation is drastically changed in that the dye must be bound to active sites on the nucleic acid in order to become effective as a photodynamic sensitizer (*4,106,107*). Binding of the sensitizer, however, seems to be accompanied by an enhanced photoreduction of the dye (*106,107*) which is expected to favor a Type I process. Care has therefore to be observed when a mechanism established for a certain system under special conditions shall be applied to the same or similar systems under changed conditions.

Sulfides are oxidized to sulfoxides by singlet oxygen (*7,50,51*); further oxidation by $^1O_2$ yields sulfones although at a much reduced rate (*67*). Similarly, disulfides afford thiolsulfinates (*68–70*), probably by the same mechanism by which sulfides are oxidized.

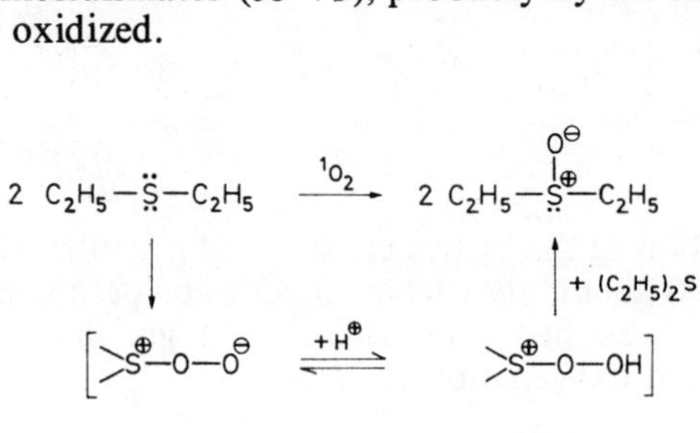

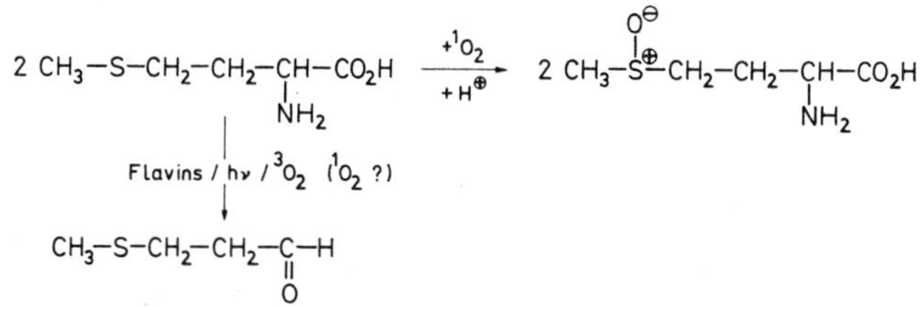

The intermediacy of a zwitterion $>S^{+}\text{-}OO^{-}$ was proposed that may be preceded by a charge transfer complex, $>S^{+}\cdots\cdots O_2^{-}$; the zwitterion should then react with another sulfide molecule to produce two molecules of the sulfoxide (*82*). However, attack of a nucleophilic oxygen site of the zwitterion on the nucleophilic sulfur of a sulfide molecule appears to be a rather unlikely process. We suggest, therefore, that protonation of the zwitterion occurs and that the $>S^{+}OOH$ thus formed is attacked by the nucleophilic sulfide to give one molecule of sulfoxide plus one molecule of a protonated sulfoxide which immediately looses its proton. The solvent-dependance of sulfoxide formation (*51,82*) favors this view. We would suggest that this mechanism also applies to the photosensitized oxygenation of methionine at lower pH values. For a methanol-water (1 : 1) solvent, methionine oxygenation sensitized by methylene blue has been shown to occur via the Type II process (*8*). However, at a higher pH value (about 8) and with methylene blue as a sensitizer, the stoichiometry changes from 2 molecules of methionine per one molecule of oxygen consumed to a one to one ratio; furthermore, using flavins such as lumiflavin, riboflavine or flavine mononucleotide as sensitizers, the major product appeared to be methional (*2*). Riboflavin, as was suggested above, seems to possess an enhanced capability to dehydrogenate organic compounds, and thus a Type I mechanism is rather likely to occur.

Alkylamines such as sec-butylamine (*60*), triethylamine (*61*) and N,2,2,5,5-pentamethylpiperidine (*77*) exhibit the lone-pair electrons at nitrogen as the only nucleophilic site, at which interaction with electrophilic $^1O_2$ can occur. The same seems to apply to a series of other amines such as benzylamines (*60*), nicotine (*17*), pseudopelletierine and N-methyl-granatanine (*38,39*) as well as for tropinone (*38,39*) and codeine (*40*) whose photooxygenation (and singlet oxygen oxygenation in a O_2-microwave discharge reaction) are shown below.

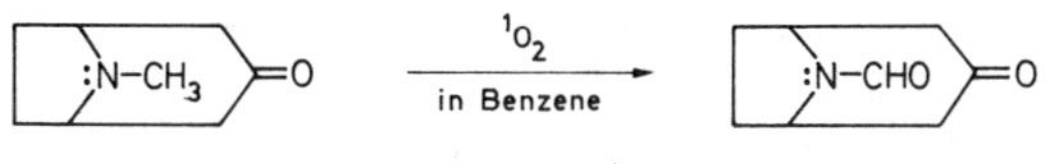

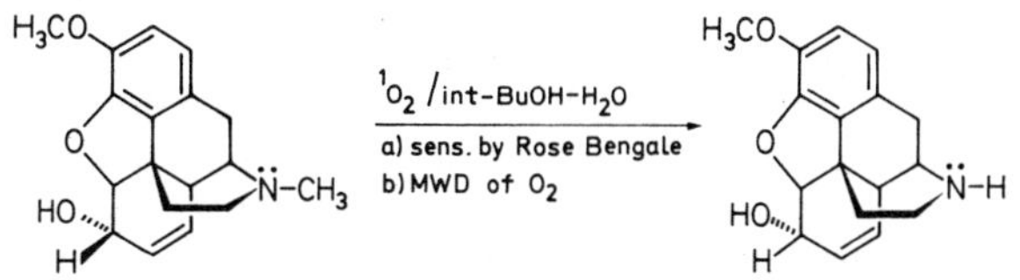

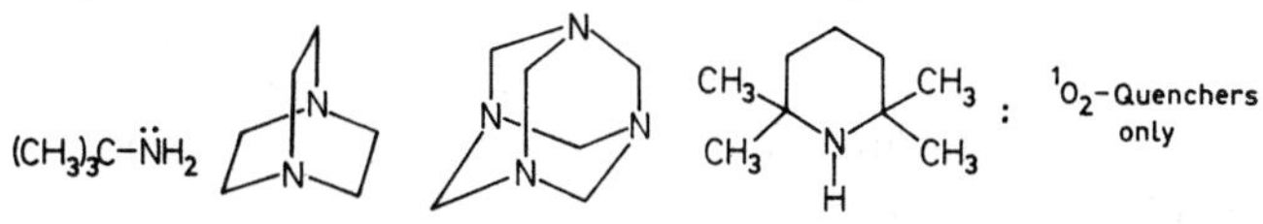

Amines which do not have an α-C-H group such as *t*-butyl amine and 2,2,5,5-tetramethylpiperidine as well as amines which cannot form a C = N-double bond (Bredt-rule) such as DABCO and hexamethylenetetramine do not give products with singlet oxygen but execute rather efficient quenching of $^1O_2$ to its triplet ground state. Ammonium compounds neither react with $^1O_2$ nor quench singlet oxygen. All these facts lead us to propose the following mechanism (see also *71*):

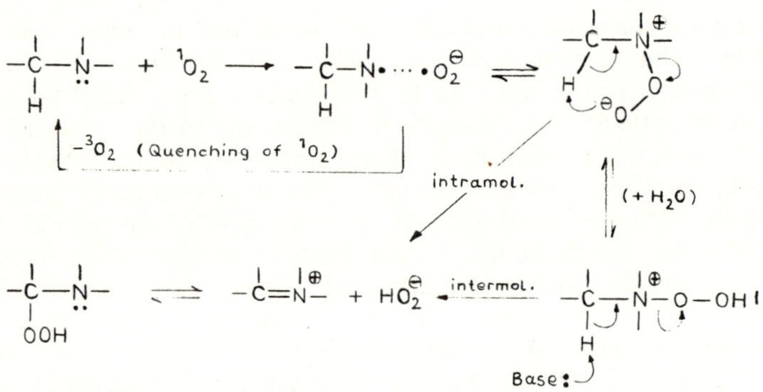

in aqueous solution:

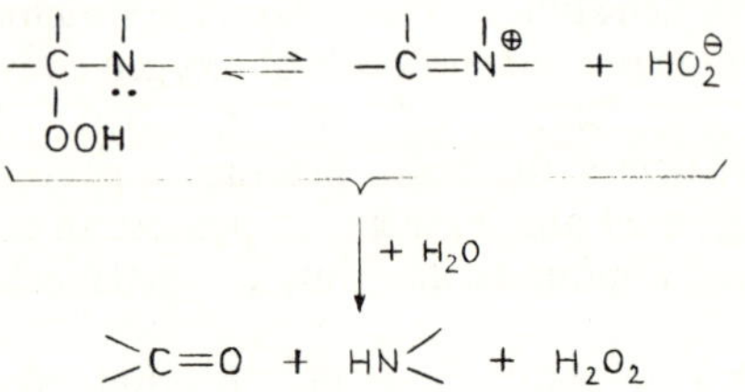

in non-protic solution (e.g. benzene) :

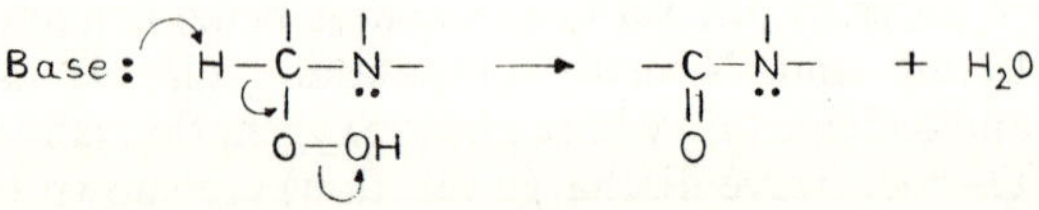

Demethylation observed with codein (*40*), N,2,2,5,5-pentamethylpiperidine (*77*) and 20 α-dimethylaminopregnan (*108*) in alcoholic solutions as well as amide production when tropinone, pseudopelletierine and N-methylgranatanine are submitted to napthalene-photosensitized oxygenation in benzene (*38,39*) are well accounted for by the proposed mechanism.

Azide-quenching of singlet oxygen is proposed among others as a tool for deciding whether a Type I or a Type II mechanism prevails since quenching of $^1O_2$ by N_3^- is rather efficient whereas quenching of triplet sensitizers by N_3^- is not (*8*). However, quenching of $^1O_2$ by N_3^- is accompanied by a reaction in the presence of appropriate compounds such as olefins (*109–111*) and 1,3-dienes (*110*) yielding hydroperoxy azides as primary products that are easily reduced to the corresponding hydroxy azides under the experimental conditions. A careful analysis of the products formed a) in photosensitized oxygenation reactions in the presence of N_3^- and α-terpinene, b) during microwave discharge of $^3O_2$ in the presence of N_3^- and α-terpinene, and c) during electrolysis of N_3^- in the presence of $^3O_2$ and α-terpinene leads us to propose the intermediacy of a N_3OOH-species which attacks the olefinic double bond to give a hydroperoxy cation that subsequently reacts with N_3^- (or other nucleophiles such as water etc.) to the final products.

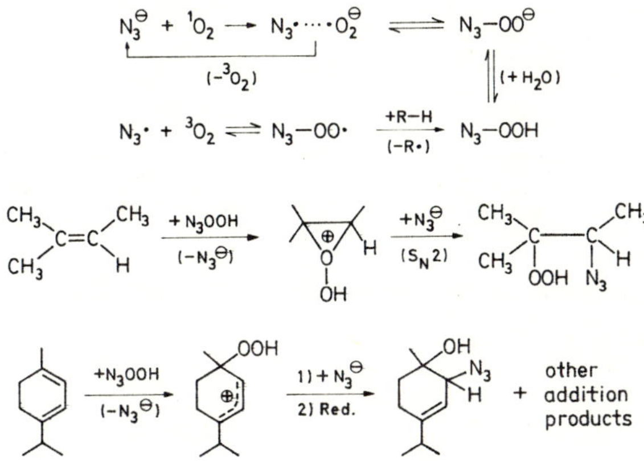

Similarly, alcoholates (RO^-) which interact with $^1O_2$ to give carbonyl compounds (*42–44*) are assumed to be attacked by $^1O_2$ at the nucleophilic site to give RO_3^- (*43,71,112*) that may undergo intra- and intermolecular reactions similar to those proposed for the amine-$^1O_2$-reactions:

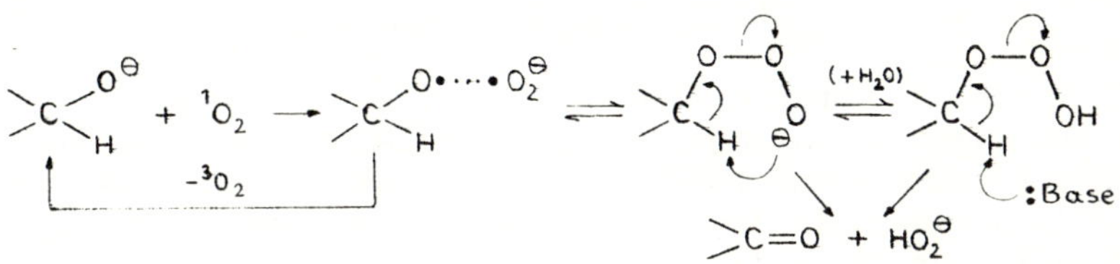

It is thus proposed that nucleophilic species such as sulfides, sulfoxides, disulfides, amines, azid ions, and alcoholates (and, probably, phenols and other electron-donating species) all interact with electrophilic singlet oxygen with virtually the same mechanism: formation of a charge-transfer complex (which may decompose to substrate $+^3O_2$: quenching) followed by bond formation and protonation; the protonated form undergoes subsequent reactions to the final products.

Phenols are well-known radical scavengers and thus may be expected to react mainly via a Type I process in photooxygenation reactions. However, evidence for a singlet oxygen reaction was recently obtained with α-tocopherol (*72,73*). It is assumed that α-tocopherol executes its protective role in photodynamic actions by this reaction (*73*).

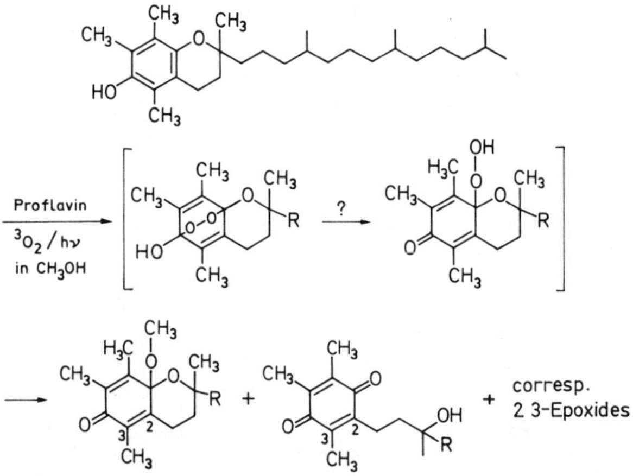

Finally, a list of sources of singlet oxygen is given in Fig. 5.

SOURCES OF $^1O_2$ ($^1\Delta_g$)

1. $^3S_1^* + {}^3O_2 \rightarrow {}^1S_o + {}^1O_2$
2. $H_2O_2 + Na\ OCl^- \rightarrow {}^1O_2 + Na\ Cl + H_2O$ (*113*)
3. $(C_2H_5O)_3PO_3 \rightarrow (C_2H_5O)_3PO + {}^1O_2$ (*114*)
4. $4\ Cr\ O_8^{3-} + 2H_2O \rightarrow 7\ {}^1O_2 + 4\ CrO_4^{2-} + 4\ OH^-$ (*115*)
5. Anthracene Endoperoxide $\xrightarrow{\Delta}$ Anthracene + $^1O_2$ (*116*)
6. O_3—Photolysis $\rightarrow {}^1O_2$ (*117*)
7. $^3O_2$—Microwave Discharge $\rightarrow {}^1O_2$ (*118*)
8. CO_2—Microwave Discharge $\rightarrow {}^1O_2$ (*119*)
9. N_2O—Microwave Discharge $\rightarrow {}^1O_2$ (*44*)

FIG. 5. Sources of Singlet Oxygen ($^1\Delta_gO_2$).

In addition to kinetic methods such as competitive inhibition with an established singlet oxygen acceptor or quencher (*8,28*), or comparison of rates in normal and deuterated solvents (*8*), product formation and product distribution should be investigated with other singlet oxygen sources than that of photosensitization whenever possible. Independence of the products formed and of the product distribution may then be taken as a strong indication that a singlet oxygen mechanism prevails in the particular reaction investigated. Likewise, whenever possible other methods for producing free radicals (Fenton's reagent, electrolysis, high energy radiation etc.) should be applied and the products and product distributions be compared in order to establish a Type I mechanism.

ACKNOWLEDGEMENT

The author wishes to express his graditude to the Fonds der Chemischen Industrie for the continuous support of his work.

REFERENCES

1. H. F. Blum, *Photodynamic Action and Diseases caused by Light*. Hafner Publ. Co., New York, 1964.

2. J. D. Spikes and M. L. MacKnight, in *Research Progress in Organic, Biological and Medicinal Chemistry* (U. Gallo and L. Santamaria, Eds.), **Vol. III, part 1,** pp.124–136. North-Holland Publ. Co., Amsterdam, 1972.
3. J. D. Spikes and M. L. MacKnight, *Ann. N. Y. Acad. Sci.* **171,** 149 (1970).
4. A. Knowles, in *Research Progress in Organic, Biological and Medicinal Chemistry* (U. Gallo and L. Santamaria, Eds.), **Vol. III, part 1,** pp. 183–213. North-Holland Publ. Co., Amsterdam, 1972.
5. L. Santamaria and G. Prino, in *Research Progress in Organic, Biological and Medicinal Chemistry* (U. Gallo and L. Santamaria, Eds.), **Vol. III, part 1,** pp. XI–XXXV. North-Holland Publ. Co., Amsterdam, 1972.
6. K. Gollnick and G. O. Schenck, *Pure Appl. Chem.* **9,** 507 (1964).
7. K. Gollnick, *Advan. Photochem.* **6,** 1 (1968).
8. R. Nilsson, P. B. Merkel, and D. R. Kearns, *Photochem. Photobiol.* **16,** 117 (1972).
9. B. Stevens, *Acc. Chem. Res.* **6,** 90 (1973).
10. K. Gollnick, T. Franken, G. Schade, and G. Dörhöfer, *Ann. N.Y. Acad. Sci.* **171,** 89 (1970).
11. D. R. Kearns, R. A. Hollins, A. U. Khan, R. W. Chambers, and P. Radlick, *J. Am. Chem. Soc.* **89,** 5455 (1967).
12. J. G. Calvert and J. N. Pitts, Jr., *Photochemistry*. John Wiley and Sons, New York, 1966.
13. W. M. G. Moore, G. S. Hammond, and R. P. Foss, *J. Am. Chem. Soc.* **83,** 2789 (1961)
14. R. Livingston and K. E. Owens, *J. Am. Chem. Soc.* **78,** 3301 (1956).
15a. C. S. Foote, *Science* **162,** 963 (1968).
15b. G. O. Schenck and R. Wolgast, *Naturwissenschaften* **48,** 737 (1961).
16. S. G. Cohen and A. D. Litt, *Tetrahedron Lett.* 837 (1970).
17. G. O. Schenck and K. Gollnick, *J. Chim. Phys.* **55,** 892 (1958).
18. J. Chrysochoos and L. I. Grossweiner, *Photochem. Photobiol.* **8,** 193 (1968).
19. R. Chapelon, G. Perichet, and B. Poyet, *Mol. Photochem.* **5,** 77 (1973).
20. M. Santhanam and V. Ramakrishnan, *Chem. Comm.*, 344 (1970).
21. J. Guttenplan and S. G. Cohen, *Chem. Comm.*, 247 (1969).
22. E. Fujimora and R. Livingston, *Nature* (London) **180,** 1036 (1957).
23. A. Nickon and W. L. Mendelson, *J. Am. Chem. Soc.* **85,** 1894 (1963).
24. D. R. Kearns, R. A. Hollins, A. U. Khan, and P. Radlick, *J. Am. Chem. Soc.* **89,** 5456 (1967).
25. S. J. Arnold, M. Kubo, and E. A. Ogryzlo, *Advan. Chem. Ser.* **77,** 133 (1968).
26. A. Wacker, in *Research Progress in Organic, Biological and Medicinal Chemistry* (U. Gallo and L. Santamaria, Eds.), **Vol. III, part 1,** pp. 107–123. North-Holland Publ. Co., Amsterdam, 1972.
27. M. I. Simon and H. Van Vunakis, *Arch. Biochem. Biophys.* **105,** 197 (1964).
28. C. S. Foote, *Acc. Chem. Res.* **1,** 104 (1968).
29. D. R. Kearns, *Chem. Rev.* **71,** 395 (1971).
30. H. Kautsky, *Biochem. Z.* **291,** 271 (1937).
31. H. Kautsky, A. Hirsch, and W. Baumeister, *Ber. Dtsch. Chem. Ges.* **64,** 2053 (1931).
32. K. Kawaoka, A. U. Khan, and D. R. Kearns, *J. Chem. Phys.* **46,** 1842 (1967).
33. M. Nemoto, Y. Usui, and M. Koizumi, *Bull. Soc. Chem. Japan*, **40,** 1035 (1967).
34. A. Kepka and L. I. Grossweiner, *Photochem. Photobiol.* **14,** 621 (1971).
35. R. Poupko and I. Rosenthal, *J. Phys. Chem.* **77,** 1722 (1973).
36. A. U. Khan, *Science* **168,** 476 (1970).
37. I. Fridovich, *Acc. Chem. Res.* **5,** 321 (1972).
38. M. H. Fisch, J. C. Gramain, and J. A. Olesen, *Chem. Commun.*, 13 (1970).
39. M. H. Fisch, J. C. Gramain, and J. A. Olesen, *Chem. Commun.*, 663 (1971).
40. J.H.E. Lindner, H. J. Kuhn, and K. Gollnick, *Tetrahedron Lett.*, 1705 (1972).
41. G. O. Schenck, *Ind. Eng. Chem.* **55,** 40 (1963).
42. G. O. Schenck and C. H. Krauch, unpublished results. C. H. Krauch, *Neue Akzeptorsysteme der photosensibilisierten O_2-Übertragung*. Diss. Univ. Göttingen, 1960.
43. H. H. Wasserman and J. E. VanVerth, *J. Am. Chem. Soc.* **96,** 585 (1974).

44. K. Gollnick and D. Haisch, unpublished results.
45. G. O. Schenck and H. D. Becker, *Angew. Chem.* **70,** 504 (1958).
46. G. O. Schenck, H. D. Becker, K. H. Schulte-Elte, and C. H. Krauch, *Chem. Ber.* **96,** 509 (1963).
47. F. R. Mayo, *Acc. Chem. Res.* **1,** 193 (1968).
48. R. F. Bartholomew and D. S. Davidson, *Chem. Commun.*, 1174 (1970).
49. D. Barnard, L. Bateman, and J. J. Cunneen in *Organic Sulfur Compounds* (N. Kharasch, Ed.), **Vol. 1,** pp. 229–247. Pergamon Press, London, 1961.
50. G. O. Schenck and C. H. Krauch, *Angew. Chem.* **74,** 510 (1962).
51. C. S. Foote and J. W. Peters, *J. Am. Chem. Soc.* **93,** 3795 (1971).
52. P. B. Merkel and D. R. Kearns, *J. Am. Chem. Soc.* **94,** 1029 (1972).
53. D. R. Adams and F. Wilkinson, *J. Chem. Soc., Faraday Trans. 2,* **68**, 586 (1972).
54. R. H. Young, D. Brewer, and R. A. Keller, *J. Am. Chem. Soc.* **95,** 375 (1973).
55. G. O. Schenck, K. Gollnick, and O. A. Neumüller, *Liebigs Ann. Chem.* **603,** 46 (1957).
56. E. J. Bowen, *Trans. Faraday Soc.* **50,** 97 (1954).
57. G. O. Schenck and K. Gollnick, *Chemismus und Kinetik der durch Xanthenfarbstoffe photosensibilisierten O_2-Übertragungen. Forsch. Ber., Land Nordrhein-Westfalen,* **No. 1256,** Westdeutscher Verlag, Köln und Opladen, 1963.
58. K. Gollnick, *Advan. Chem. Ser.* **77,** 78 (1968).
59. C. S. Foote and J. O. Peters, in *23rd Int. Congress of Pure and Applied Chemistry, Special Lectures,* **Vol. 4,** pp. 129–154. Butterworth's, London, 1971.
60. K. Gollnick and J.H.E. Lindner, unpublished results.
61. W. F. Smith, *J. Am. Chem. Soc.* **94,** 186 (1972).
62. A. Farmilo and F. Wilkinson, *Photochem. Photobiol.* **18,** 447 (1973).
63. K. Gollnick and G. O. Schenck, in *1.4-Cycloaddition Reactions in Heterocyclic Syntheses* (J. Hamer, Ed.), pp. 255–344. Academic Press, New York, 1967.
64. C. S. Foote and J. W.-P. Lin, *Tetrahedron Lett.* 3267 (1968).
65. S. Mazur and C. S. Foote, *J. Amer. Chem. Soc.* **92,** 3225 (1970).
66. P. D. Bartlett and A. P. Schaap, *J. Am. Chem. Soc.* **92,** 3223 (1970).
67. G. O. Schenck and C. H. Krauch, *Chem. Ber.* **96,** 517 (1963).
68. R. W. Murray and S. L. Jindal, *Photochem. Photobiol.* **16,** 147 (1972).
69. R. W. Murray and S. L. Jindal, *J. Org. Chem.* **37,** 3516 (1972).
70. B. Stevens, S. R. Perez, and R. D. Small, *Photochem. Photobiol.* **19,** 315 (1974).
71. K. Gollnick and J.H.E. Lindner, *Tetrahedron Lett.*, 1903 (1973).
72. T. Matsuura, N. Yoshimura, A. Nishinaga, and I. Saito, *Tetrahedron* **28,** 4933 (1972).
73. G. W. Grams, K. Eskins, and G. E. Inglett, *J. Am. Chem. Soc.* **94,** 866 (1972).
74. C. S. Foote and R. W. Denny, *J. Am. Chem. Soc.* **93,** 5162 (1971).
75. F. N. Kotsonis, *A Hammett Study on the Termination Step of the Indirect Photo-oxygenation of trans-4,4'-Disubstituted α,α'-Dimethylstilbenes.* M. S. Thesis, The University of Arizona, Tuscon, Arizona, 1969.
76. C. Ouannes and T. Wilson, *J. Am. Chem. Soc.* **90,** 6527 (1968).
77. D. Bellus, H. Lind, and J. F. Wyatt, *J. Chem. Soc., Chem. Commun.*, 1199 (1972).
78. C. S. Foote, Y. C. Chang, and R. W. Denny, *J. Am. Chem. Soc.* **92,** 5216 (1970).
79. C. S. Foote, Y. C. Chang, and R. W. Denny, *J. Am. Chem. Soc.* **92,** 5218 (1970).
80. E. A. Ogryzlo and C. W. Tang, *J. Am. Chem. Soc.* **92,** 5034 (1970).
81. R. H. Young, R. L. Martin, D. Feriozi, D. Brewer, and R. Kayser, *Photochem. Photobiol.* **17,** 233 (1973).
82. C. S. Foote, R. W. Denny, L. Weaver, Y. Chang, and J. Peters, *Ann. N. Y. Acad. Sci.* **171,** 139 (1970).
83. M. M. Mathews-Roth, T. Wilson, E. Fujimori, and N. I. Krinsky, *Photochem. Photobiol.* **19,** 217 (1974).
84. M. M. Mathews, *Nature (London)* **203,** 1092 (1964).
85. I. A. Magnus, in *Research Progress in Organic, Biological and Medicinal Chemistry* (U. Gallo and L. Santamaria, Eds.), **Vol. III, part 2,** pp. 571–600. North-Holland Publ. Co., Amsterdam, 1972.

86. H. Ippen, in *Research Progress in Organic, Biological and Medicinal Chemistry* (U. Gallo and L. Santamaria, Eds.), **Vol. III, part 2,** pp. 601–622. North-Holland Publ. Co., Amsterdam, 1972.
87. H. H. Wasserman, K. Stiller, and M. B. Floyd, *Tetrahedron Lett.* 3277 (1968).
88. A. Nickon and F. Bagli, *J. Amer. Chem. Soc.* **83,** 1498 (1961).
89. L. L. Smith, J. I. Teng, M. J. Kulig, and F. L. Hill, *J. Org. Chem.* **38,** 1763 (1973).
90. A. A. Lamola, T. Yamane, and A. M. Trozzolo, *Science* **179,** 1131 (1973).
91. A. A. Lamola, T. Yamane, A. M. Trozzolo, S. R. Fahrenholtz, and F. H. Doleiden, Molecular Mechanisms in the Photohemolysis of Porphyric Red Blood Cells. 167th ACS Nat. Meeting, Preprints ORGN 1, Los Angeles, April 1974.
92. H. R. Rawls and P. J. vanSanten, *Ann. N. Y. Acad. Sci.* **171,** 135 (1970).
93. H. H. Wasserman, *Ann. N. Y. Acad. Sci.* **171,** 108 (1970).
94. N. A. Evans, *Aust. J. Chem.* **24,** 1971 (1971).
95. M. Nakagawa, T. Kaneko, K. Yoshikawa, and T. Hino, *J. Am. Chem. Soc.* **96,** 624 (1974).
96. H. H. Wasserman and M. B. Floyd, *Tetrahedron Lett.* 2009 (1963).
97. A. Knowles and S. Gurnani, *Photochem. Photobiol.* **16,** 95 (1972).
98. Y. Usui, C. Iwanaga, and M. Koizumi, *Bull. Chem. Soc. Japan* **42,** 1231 (1969).
99. P. Walrant and R. Santus, *Photochem. Photobiol.* **19,** 411 (1974).
100. J. S. Sussenbach and W. Berends, *Biochim. Biophys. Acta* **95,** 184 (1965).
101. F. R. Hallett, B. P. Hallett, and W. Snipes, *Biophys. J.* **10,** 305 (1970).
102. I. Rosenthal and J. N. Pitts, Jr., *Biophys. J.* **11,** 963 (1971).
103. T. Matsuura and I. Saito, *Tetrahedron* **25**, 541 (1969).
104. T. Matsuura and I. Saito, *Tetrahedron* **25,** 549 (1969).
105. T. Matsuura and I. Saito, *Tetrahedron* **25,** 557 (1969).
106. J. S. Bellin, *Photochem. Photobiol.* **4,** 33 (1965).
107. J. S. Bellin, *Photochem. Photobiol.* **8,** 383 (1968).
108. F. Khuong-Huu-Laine and D. Herlem-Gaullier, *Tetrahedron Lett.* 3649 (1970).
109. W. Fenical, D. R. Kearns, and P. Radlick, *J. Am. Chem. Soc.* **91,** 3396 (1969).
110. K. Gollnick, D. Haisch, and G. Schade, *J. Am. Chem. Soc.* **94,** 1747 (1972).
111. C. S. Foote, T. T. Fujimoto, and Y. C. Chang, *Tetrahedron Lett.* 45 (1972).
112. K. Gollnick, D. Haisch, and G. Schade, unpublished results.
113. C. S. Foote and S. Wexler, *J. Am. Chem. Soc.* **86,** 3879 (1964).
114. R. W. Murray and M. L. Kaplan, *J. Am. Chem. Soc.* **90,** 537 (1968).
115. J. W. Peters, J. N. Pitts, Jr., I. Rosenthal, and H. Fuhr, *J. Am. Chem. Soc.* **94,** 4348 (1972).
116. H. H. Wasserman and J. R. Scheffer, *J. Am. Chem. Soc.* **89,** 3073 (1967).
117. R. P. Wayne, *Advan. Photochem.* **7,** 311 (1969).
118. E. J. Corey and W. C. Taylor, *J. Am. Chem. Soc.* **86,** 3881 (1964).
119. K. Gollnick and G. Schade, *Tetrahedron Lett.* 857 (1973).
120. E. A. Mayeda and A. J. Bard, *J. Am. Chem. Soc.* **96,** 4023 (1974).

The Effects of Photodynamic Action Involving Oxygen upon Biological Systems[1]

Aubrey Knowles
MRC Vision Unit
University of Sussex
Brighton, BN1 9QG
Sussex, England

INTRODUCTION

Photodynamic action—the effect of visible light and a sensitizer—upon biological systems has been thought of as a scientific curiosity for many years, but now we have some real understanding of the possible mechanisms of sensitized photooxidations, and this, coupled with new techniques for the analysis of biological materials, has meant that photodynamic action is coming of age. This insight into mechanism has led to the use of sensitized oxidation as a novel and very specific tool for the biologist and biochemist, and I would like to describe two important new fields that are developing—therapeutic applications of the modification of macromolecules, and the selective oxidation of amino acids as a probe of protein structure and function.

Singlet oxygen appears to be the reactive intermediate in many sensitized oxidations, but recent findings suggest that it may also mediate other types of reaction, for example, the *cis-trans* isomerization of polyenes.

A complementary aspect of photodynamic action is the fact that most animal and plant cells contain colored substances known to be sensitizers from in vitro studies, such as hematoporphyrin, chlorophyll, flavins, and so on. In the normal cell, these do not appear to have any damaging effect. The last section of this paper will discuss some current ideas about protective mechanisms against such endogenous sensitizers.

THERAPEUTIC USES OF PHOTODYNAMIC ACTION

Many viruses are susceptible to the lethal effect of illumination in the presence of dyes, and Wallis and Melnick (*1*) have studied the inactivation of a wide variety of types. When dyes are added to suspensions of some viruses, the dye molecules pass through the viral coat and are bound internally so tightly that excess dye can be removed by passing the suspension down a cation-exchange column. The resulting virus particles appear colorless, but show extreme photosensitivity; the result of complexing herpesvirus with toluidine blue in this way is shown in Fig. 1.

[1]Congress presentation E–3–4.

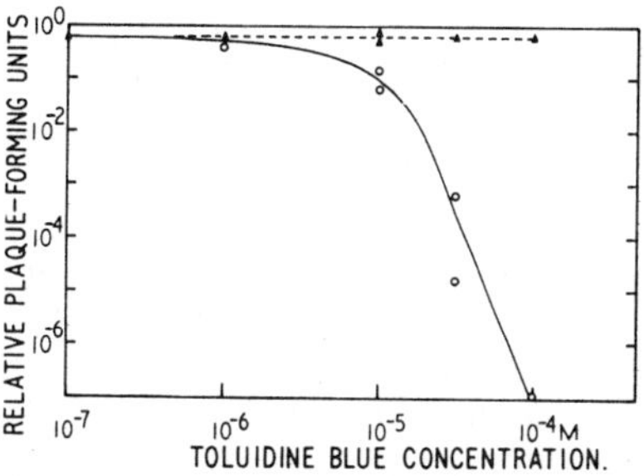

FIG. 1. The photosensitivity of herpesvirus treated with different concentrations of toluidine blue, unbound dye removed and then exposed to white light for 5 mins at pH 9. Redrawn from Wallis and Melnick (*1*).

At first, photodynamic inactivation was thought to be an ideal method for the production of vaccines, for the dye was only bound inside the virus and acted exclusively upon the nucleic acid, leaving the antigenic properties of the outer coat unchanged. However, it proved too difficult to remove the dye after the inactivation, and the idea was put aside until recently it has been shown that simian virus (SV 40) and herpes simplex virus inactivated by photodynamic treatment with neutral red are still able to transform hamster cells, and specific antigens for the two viruses were identified in the two cells (*2,3*). This is consistent with the idea that sensitized oxidation brings about degradation of single purine bases, leading to single-strand breaks in the DNA (see Knowles (4)). This will disrupt the replication cycle of the virus, but leave it sufficiently intact to bring about cell transformation. This again raises exciting possibilities for vaccine production, especially since herpesvirus is thought to be associated with cervical cancer.

Direct attack on a replicating virus in situ has been successfully applied to herpes keratitis in the rabbit cornea (*5*) and in the therapy of human herpes simplex (*6*). In the latter treatment, the vesicular skin lesions were ruptured and treated with neutral red. The lesion was then exposed to a fluorescent lamp for two 15-minute periods. This simple technique was very effective, with rapid healing of the lesions and a low rate of recurrence.

Hiatt (*7*) found that proflavine added to replicating poliovirus is bound to the RNA before the encapsidation process and so cannot be removed from the intact virus. This means that it is a very effective way of 'tagging' the virus, for in the dark, the dye has no effect upon the infectivity of the virus, but when illuminated for 5 mins, there is less than 0.001% survivors. This labelling technique was used to demonstrate the degree of replication of the virus in different media, for if replication occurs, the original virus retains its photosensitivity while the progeny is not affected by light (Fig. 2).
In this way, it could be demonstrated that poliovirus can multiply in non-primate cells, while a theory that it can replicate in the house-fly was disproved.

SENSITIZED OXIDATION IN THE SELECTIVE MODIFICATION OF PROTEINS

The use of dye-sensitized oxidation as a probe of enzyme structure was first employed by Ray and Koshland (*8*) who followed the rate of inactivation of phosphoglucomutase by methylene blue in light at pH 7, and related this to the destruction of the amino acid residues. The initial rapid phase of the inactivation corresponded to the loss of methionine and histidine residues, known to be the most susceptible to photodynamic attack from studies of the free amino acids (*9*). When about half of these residues had been destroyed, the rate fell to 10% of the initial value, and it was concluded that residues on the outside of the folded molecule were rapidly destroyed while the slow phase represented the destruction of residues less accessible to the dye. This suggests that the dye molecules must approach close to the site of action and that the oxidizing species cannot migrate very far. The only other groups found to be degraded in this reaction were four sulfhydryl groups, which were all lost at the same rate and are thus sited in similar locations. These findings are summarized in Table 1.

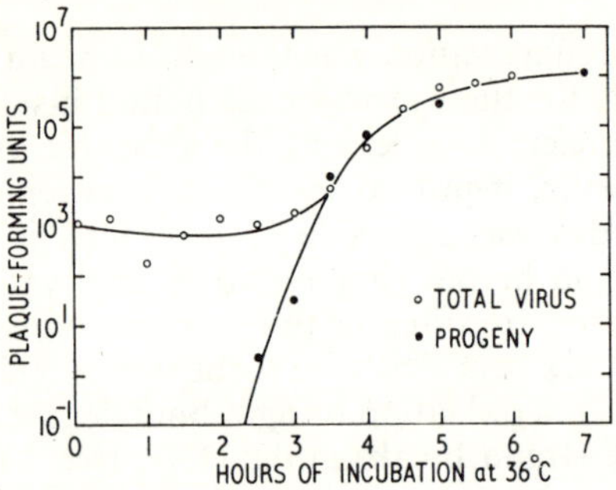

FIG. 2. Growth curve for poliovirus tagged with proflavine in monkey kidney cell cultures in the dark. After illumination, only the untagged progeny virus was found in the cell harvest. Redrawn from Hiatt (*7*).

TABLE 1

Rate and extent of sensitized oxidation of amino acids in phosphoglucomutase From Ray and Koshland[18]

Residue	First-order rate constant for destruction Min^{-1}	Number of groups oxidized	Number of groups essential for activity
Accessible histidine	0.65	5	1
Accessible methionine	0.37	6	1
Sulphydryl groups	0.07	4	-
Buried histidine	0.05	8	-
Buried methionine	0.05	6	-
Tyrosine	0.01	-	-
Tryptophan	0.01	-	-

The initial portion of the inactivation curve for the enzyme could be fitted on the basis of the assumption that, of the oxidizable groups, only one histidine and one methionine were essential to the active center of the enzyme.

This approach has been developed in several laboratories, and applied to an increasing number of enzymes. The properties of a number of sensitizing dyes have been investigated, so that by a choice of dye and pH, it is possible to degrade specific groups in an enzyme at will. Analysis of the data is generally based on the concept of the reactivity of a particular residue being related to its ease of access by the dye. In this way, the conformation of the molecule in solution can be compared with that from X ray analysis of the crystal. The oxidation is quite mild, and an 87% inactivation of ribonuclease by methylene blue and light can be completely reversed by chemical reduction of the enzyme with thioglycolic acid (*10*).

In most cases, the dye seems to act preferentially on amino acids near the active center, but this does not necessarily mean that the dye molecule is bound in the substrate binding-site, or that the inactivation arises *directly* from the oxidation of one or two key residues. The degradation of a single residue can lead to the unfolding of the protein structure, and it is this secondary effect that causes the loss of activity. In the case of ribonuclease U_1, oxidation of just two histidine residues will inactivate the enzyme, while the circular dichroism (CD) spectrum—and hence the conformation—is unchanged (*11*). More drastic reaction conditions will also oxidize tyrosine and cystine, and then the structure does change. If a substrate analogue, guanosine-2′-phospate is bound to the enzyme, the two histidines are protected from oxidation and the activity is unchanged.

Another very specific inactivation is the effect of rose bengal and light upon RNA polymerase (*12*). The site of oxidation is again thought to be histidine, and its loss inhibits the principal function of the enzyme in the elongation of DNA chains, although it can still bind DNA and the bases, and can still catalyze the exchange of pyrophosphtate. Apparently large sections of the molecule are unaltered, but perhaps due to a structural change, it is incapable of bringing the bases together.

On the other hand, changed conformation does not always mean a loss of enzyme activity, for changes in the conformation of bovine trypsin and chymotrypsin resulting from photodynamic action are not necessarily related to the inactivation of the enzymes (*13*). Under mild conditions, sensitized oxidation will give a molecule that is less thermally stable, is vulnerable to urea and has a changed fluorescence, although the activity is unchanged.

The choice of dye and reaction conditions determine the site of attack, the effect of pH being the most important factor (*14*). This is because of the difference in pk_a between the residues; for example, histidine is very susceptible to oxidation by most dyes, but only if it is in the unionized form, that is, in a medium of pH greater than 6. Fig. 3 shows the effect of variations of pH and dye upon the three most vulnerable amino acids in ribonuclease A.

Under alkaline conditions, all three dyes destroy histidine, and there is a proportional loss of activity. In acid, there is no loss of histidine, and so the fall in activity is due to the degradation of some other residue. Examination of the methionine plot shows that it is loss of this residue at low pH that causes the inactivation. However, flavin mononucleotide is unique in destroying tyrosine at all pH values.

Some specificity can be attained by using a sensitizer that binds at a specific point on the molecule. Rippa and Pontremoli (*16*) used pyridoxal phosphate to sensitize a dehydrogenase: this binds at a lysine, but did not oxidize this residue, and the inactivation of the enzyme was due to the degradation of two neighboring histidines in the active center of the enzyme.

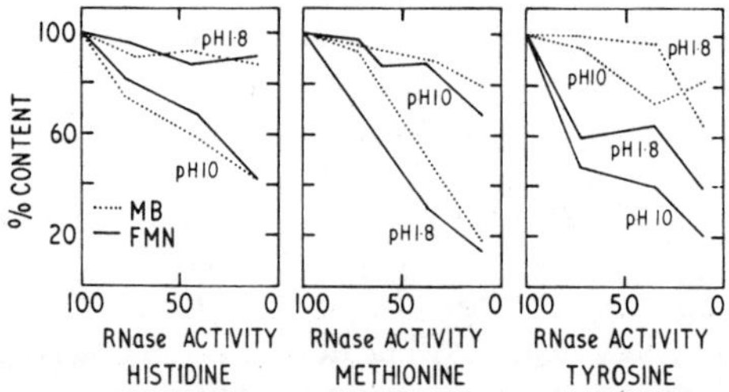

FIG. 3. The decrease in ribonuclease activity with histidine, methionine and tyrosine content upon sensitized photooxidation with methylene blue (MB) and flavin mononucleotide (FMN) at pH 1.8 and 10. Eosin Y was also tested and gave results similar to those from methylene blue. Adapted from MacKnight and Spikes (*15*).

Another specific effect can be obtained by self-sensitization in proteins that are colored. Horse heart cytochrome *c* contains iron and when illuminated under alkaline conditions, two residues—histidine-18 and methionine-80—were oxdized (*17*). When the molecule is bound to cardiolipin—a model of the binding site in the mitochrondrion—tyrosine-48 was also oxidized under the same conditions. This suggests a conformational change in the protein, which is borne out by changes in the CD spectrum. An interesting corollary of this effect is that hematoporphyrin is resistant to self-sensitization of this sort. This is thought to be a consequence of the different spin-state of the iron atom in the two proteins—low in the ferrocytochrome and high in the hematoporphyrin, the latter having a quenching effect on the excited-state intermediate in the oxidation. The quenching effects of the paramagnetic ions Co^{++}, Cu^{++}, and the lanthanides have also been used to modify the pattern of amino acid modification in enzymes (*18,19*). When Cu^{++} is co-ordinated onto ribonuclease A, methionine-79 is selectively oxidized; this group could not be picked out by any other technique. The inhibiting effect of the paramagnetic ion can extend over the whole molecule, but is greatest nearest the ion. Hence, residues nearest to the site of coordination are the least likely to be oxidized.

PHOTODYNAMIC ISOMERIZATION

Sensitization of the *cis-trans* isomerization of polyenes by low-energy sensitizers is a well-known phenomenon that has been thoroughly investigated during the past ten years (see Turro (*20*)). Two types of mechanism have been proposed, one involving triplet-triplet energy transfer from sensitizer to polyene, and the second based on radical formation by the sensitizer. A third mechanism has not been put forward for the isomerization of 15,15′-*cis*-*β*-carotene sensitized by methylene blue and light (*21*). The same reaction takes place when chemically produced singlet oxygen is passed through a solution of *cis*-*β*-carotene, which suggests that it is singlet oxygen generated by the triplet-excited methylene blue that causes the isomerization. The reaction has two curious features: (i) the reaction only goes from *cis*-*β*-carotene to the all-*trans* isomer; no other isomers are formed and the reaction is not reversible, and (ii) the methylene blue-sensitized reaction still took place even when oxygen was removed from the solution by flushing with nitrogen.

I have studied the isomerization of another long-chain polyene, 11-*cis*-retinal, sensitized by rose bengal, a dye known to be an efficient generator of singlet oxygen (*22*). The rate of the isomerization is increased by lowering the concentration of oxygen in the solution (Fig. 4) and the reaction is not inhibited by the singlet oxygen quencher, diazabicyclooctane. It must be concluded that singlet oxygen is *not* intermediate in the isomerization of retinal.

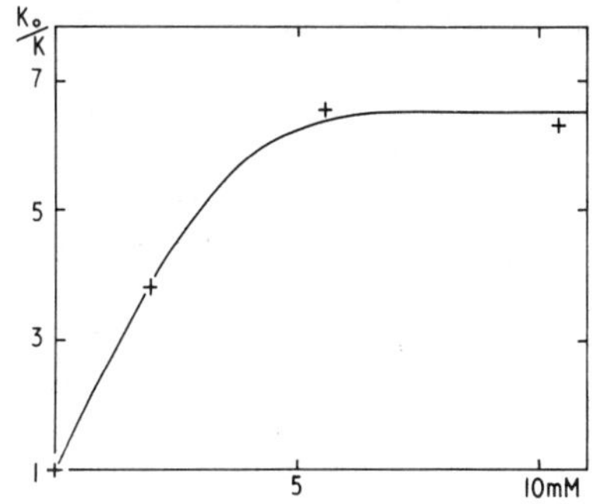

FIG. 4. Oxygen inhibition of the rose bengal-sensitized isomerization of 11-*cis*-retinal in ethanol.

PHOTOSENSITIZATION OF BIOLOGICAL SYSTEMS BY ENDOGENOUS SENSITIZERS

a. Hematoporhyrin

Porphyrins are known to be photosensitizers from in vitro experiments, for example, hematoporphyrin efficiently sensitizes the oxidation of methionine in aqueous solution (*23*). Interperitoneal injection of hematoporphyrin into mice at a dose of only 100mg/kg body weight will cause 50% of deaths after 24 hr illumination (*24*). The related human disease, erythropoietic porphyria, gives rise to an extreme photosensitivity of the skin, and apparently arises from faulty metabolism of protoporphyrin. Erythrocytes taken from patients suffering from the disease have extremely high protoporphyrin levels, and readily hemolyse when illuminated with 400 nm light. Thus, an excess of a naturally occurring pigment can cause extreme photosensitization.

Addition of a radical scavenger, dithiothreitol, protected the erythrocytes, apparently by trapping porphyrin, peroxy or hydroxyl radicals. Similarly, this compound protected the sensitized mice against the lethal effects of light, though this result was more equivocal as the scavenger is toxic. Another type of protective agent has been used in the treatment of humans suffering this disease: β-carotene given orally will give a desensitization (*25*). It will be shown in the next section that this polyene is a very efficient quencher of singlet oxygen.

The actual mechanism of the lysis of the erythrocytes is still uncertain. Goldstein and Harber (*26*) found that oxygen is consumed when protoporphyrin-containing cells are illuminated and concluded that the rupturing of the cell membrane is due to the oxidation of both lipids and the amino acids of the structural protein. On the other hand, Schothorst, Steveninck, Went and Suurmond (*27*) confirmed that the membrane is damaged and the cation-permeability increased, but concluded that the membrane lipid is not affected, and only the membrane proteins suffer degradation.

Xeroderma pigmentosum is a disorder also causing skin photosensitivity, but this is not due to excessive sensitivity, but a lack of normal repair mechanisms. Illumination of fibroblasts from patients with xeroderma pigmentosum causes modifications to the bases in the DNA that cannot be repaired (*28*). Treatment of the DNA with bromouracil deoxyriboside causes a different sort of lesion: strand breaks which the cells are capable of repairing, and so the diseased cells appear to lack one part of the normal repair mechanism, namely the enzyme that cuts the DNA chain around a defective base as a first step in its replacement. This means that lesions that give altered bases without strand breaks cannot be repaired. Mild photodynamic lesions are of this kind, and so natural protection against the effects of endogenous sensitizers may be simply one of repair.

It had been thought that UV-repair mechanisms were not effective against the effects of photodynamic action, but Harm (*29*) has found a relationship between the efficiency of UV-repair in a series of strains of *Escherichia coli* and their resistance to dye-sensitized inactivation (Fig. 5).

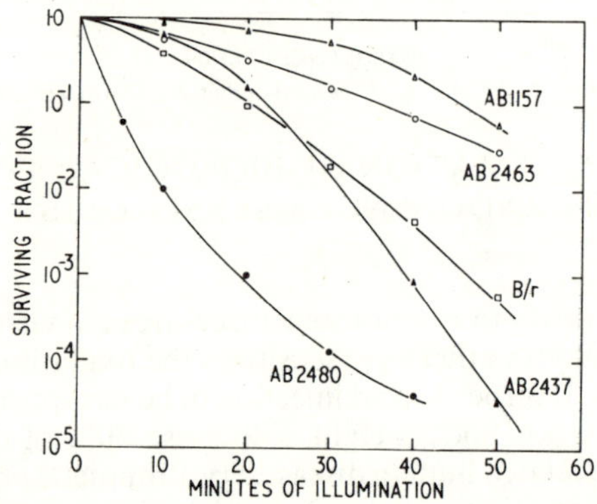

FIG. 5. The resistance of 5 strains of *Escherichia coli* to photosensitized oxidation by acridine orange. Adapted from Harm (*29*).

Strain AB1157 is of 'wild type' and has both *uvr* and *rec* repair systems. AB2463 and AB2437 are single mutants, lacking respectively *rec* and *uvr* mechanisms, and AB2480 is a double mutant lacking both systems. Thus both UV repair systems are to some extent effective in the repair of photodynamic lesions. Repair in the dark after acridine orange-sensitized attack has been demonstrated for *E. coli* B/r while the radiation-sensitive strain, B_{s-1}, shows no recovery (*30*). More recent studies have shown the relationship between the systems for the repair of UV and photodynamic damage to be quite complex; for example, Imray and MacPhee (*31*) have studied several repair-deficient strains of *Salmonella* and *E. coli* and conclude that the *rec* system is of minor importance in recovery from photodynamic attack by acridine orange. The most important factor for repair is the presence of the enzyme DNA-polymerase I.

There is also the possibility that the cell is protected by agents which react with the oxidizing species generated by light. Tocopherols have a protective action against the hemolysis of erythrocytes that can be related to their ease of dye-sensitized oxidation (*32*). The enzymes catalase and superoxide dismutase are thought to be able to protect cells against damage by singlet oxygen (*33*).

b. Chlorophyll

Chlorophyll is also an efficient photosensitizer in solution and it is remarkable that there is no apparent degradation of chloroplasts on exposure to light. One interesting case of natural sensitivity to red light has been reported (*34*). Fig. 6 shows that in oxygen, long-wavelength light will degrade the chlorophyll of *Euglena gracilis* and produces a stable chlorophyll-free mutant, apparently by modification of DNA. This seems to be a loss of chlorophyll sensitized by itself, and possibly has an evolutionary importance.

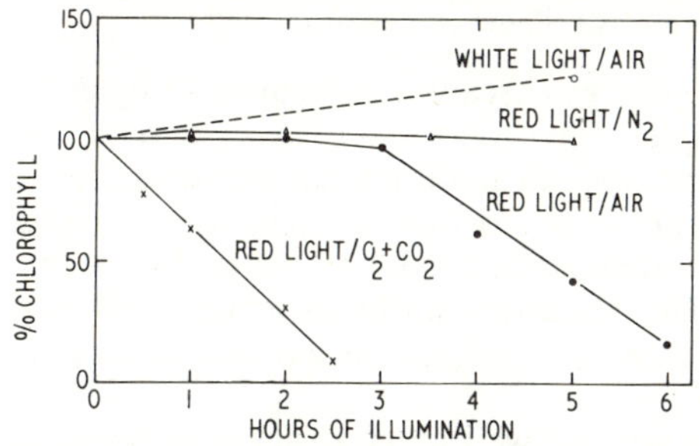

FIG. 6. The effect of light on the chlorophyll content of *Euglena gracilis* cells in the logarithmic phase of growth. The cells were exposed to (i) white light of optimum intensity for chlorophyll synthesis, and (ii) red light of λ> 610 nm in atmospheres of nitrogen, oxygen and a 95:5 mixture of oxygen and carbon dioxide. Redrawn from Leff and Krinsky (*34*).

In general, it is now accepted that one function of the carotenoids in the chloroplast is to protect the cell from photodynamic damage, and carotenoid-free mutants of several types of photosynthetic organism are killed by light (*35*). Carotenoids act primarily as auxiliary pigments, extending the absorption range of chlorophyll and passing absorbed light energy *to* chlorophyll molecules by singlet-singlet energy transfer. A second class of carotenoid, also sited in the chloroplast, acts as a trap for unwanted excitation energy, either directly, by triplet-triplet transfer *from* excited chlorophyll molecules or, indirectly, by reaction with the singlet oxygen formed by interaction of molecular oxygen with chlorophyll triplets.

The carotenes have just the right properties for this dual role, for their lowest singlet levels lie at relatively high energies, making them good donors, while the first triplet levels are low, making them good triplet acceptors. In addition, the two functions are separated by a very low probability of singlet-triplet intersystem crossing (*36*). For β-carotene, the singlet level lies at 64 kcal mole^{-1}, while that of chlorophyll-b is 44 kcal mole^{-1}; the β-carotene triplet is at about 22 kcal mole^{-1} while that of chlorophyll-b is at about 33 kcal mole^{-1} (*37,38*). Witt (*39*) has shown that as the intensity of illumination is increased, the photosynthetic system saturates, while the yield of carotenoid triplet—which must be generated *via* excited chlorophyll—steadily increases (Fig. 7).

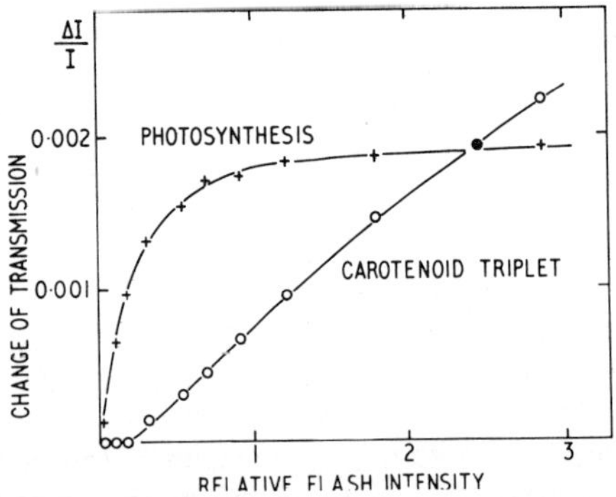

FIG. 7. The relative yields of photosynthesis and triplet carotenoid formation at different light intensities in spinach chloroplasts. Based on data from Witt (*39*).

Each step of the energy cascade—chlorophyll singlet to triplet, chlorophyll triplet to carotene triplet—involves a loss of energy, so that by the carotene triplet stage, the energy of the original quantum has been degraded to a safe level, for the energy of triplet carotene is too low to excite singlet oxygen. The carotenoid triplet may undergo radiationless decay to the ground stage, or isomerize or be oxidized to an epoxide. The latter mechanism apparently takes place in *Euglena gracilis* in the light, and a dark-mechanism exists to reduce the carotenoid back to its original state (*35*).

The carotenoids are also well suited to act as traps for singlet oxygen, either by triplet energy transfer, by the isomerization of the carotenoid (see earlier section) or by epoxide formation. Here again, the remarkably low triplet energies of the carotenes enable them to quench singlet oxygen very efficiently (*37,40*).

Another mechanism for the desensitizing effect of carotenoids seems to operate in *Halobacterium cutirubrum,* a non-photosynthetic bacterium that can survive hostile conditions of salinity and intense UV radiation by virtue of a very efficient photoreversal system (*41*). Inactivation of the carotenoids in the bacterium increases the lethal effect of UV, but their protective function acts by increasing the rate of photoreactivation rather than by lowering the UV sensitivity. Thus carotenoids may also be important in cellular photoreactivation.

c. *Photodynamic Action in the Retina*

The visual photoreceptor cell has a stack of membranes containing the visual pigment molecules arranged to give the maximum degree of absorption. The commonest vertebrate visual pigment, rhodopsin, has been the most widely studied, and it is thought that 70% of the light absorbed by rhodopsin is utilized in the generation of the visual neural signal. There are many examples of retinal lesions caused by high light intensities, and animal experiments show that continuous illumination of the retina at normal intensities for quite short periods can cause complete breakdown of the receptor cells. Noell and Albrecht (*42*) have reviewed the possible causes of such degeneration, including photodynamic damage sensitized by rhodopsin. The potential of rhodopsin as a sensitizer is probably low, as the inter-system crossing yield to the triplet stage or the isolated chromophoric group of the molecule is only 0.13 (*43*).

An alternative explanation of the photodegradation could be the excessive loss of rhodopsin found under conditions of continuous illumination, for it is also the main structural protein of the receptor membranes. However, inherited retinal dystrophy also involves a loss of membrane structure that is accelerated by light, even though a symptom of the disease is an abnormally *high* concentration of rhodopsin in the retina (*44*).

The products of free radical oxidation have been detected in illuminated receptor membranes and it has been suggested that these are generated by light-excited rhodopsin (*45*). Receptor cells have a remarkably high tocopherol content and this may function as an antioxidant to prevent oxidation of the membrane lipids (see Daemen (*46*)). A high degree of protection may not be necessary, for there is a very rapid renewal of photoreceptor membranes; for example, the entire receptor membrane assembly in the rat is replaced every 9 days (*47*).

REFERENCES

1. C. Wallis and J. L. Melnick, *Photochem. Photobiol.* **4,** 159 (1965).
2. F. Rapp, J. H. Li, and M. Jerkofsky, *Virology* **55,** 339 (1973).
3. N. H. Seemayer, K. Hirai and V. Defendi, *Int. J. Cancer* **12,** 524 (1973).
4. A. Knowles, in *Research Progress in Organic, Biological and Medicinal Chemistry,* **Vol. 3, Pt. 1** (U. Gallo and L. Santamaria, Eds.), 183–213, North-Holland, Amsterdam, 1972.
5. C. Moore, C. Wallis, J. L. Melnick and M. D. Kuns, *Infect. Immunol.* **5,** 169 (1972).
6. T. D. Felber, E. B. Smith, J. M. Knox, C. Wallis and J. L. Melnick, *J. Am. Med. Assoc.* **223,** 289 (1973).
7. C. W. Hiatt, Photodynamic tagging of poliovirus with proflavine, in *Research Progress in Organic, Biological and Medicinal Chemistry,* **Vol. 3, Pt. 1** (U. Gallo and L. Santamaria, Eds.), 320–331, North-Holland, Amsterdam, 1972.
8. W. J. Ray and D. E. Koshland, *J. Biol. Chem.* **237,** 2493 (1962).
9. J. D. Spikes and R. Livingston, in *Advances in Radiation Biology,* **Vol. 3** (L. Augenstein, R. Mason and M. Zelle, Eds.), pp. 29–121. Academic Press, New York, 1969.
10. G. Jori, G. Galiazzo, A. Marzotto and E. Scoffone, *Biochim. Biophys. Acta* **154,** 1 (1968).
11. J. Hashimoto, K. Takahashi and T. Uchida, *J. Biochem. (Tokyo)* **73,** 13 (1973).
12. A. Ishihama and J. Hurwitz, *J. Biol. Chem.* **244,** 6680 (1969).
13. T. R. Hopkins and J. D. Spikes, *Radiat. Res.* **53,** 315 (1973).
14. J. D. Spikes and M. L. MacKnight, *Ann. N.Y. Acad. Sci.* **171,** 149 (1970).
15. M. L. MacKnight and J. D. Spikes, *Boll. Chim. Farm.* **109,** 659 (1971).
16. M. Rippa and S. Pontremoli, *Arch. Biochem. Biophys.* **133,** 112 (1969).
17. G. Jori, A. M. Tamburro and A. Azzi, *Photochem. Photobiol.* **19,** 337 (1974).
18. G. Jori, G. Gennari, G. Galiazzo and E. Scoffone, *Biochim. Biophys. Acta* **236,** 749 (1971).
19. G. Jori, M. Folin, G. Gennari, G. Galiazzo and O. Buso, *Photochem. Photobiol.* **19,** 419 (1974).
20. N. J. Turro, *Photochem. Photobiol.* **9,** 555 (1969).
21. C. S. Foote, Y. C. Chang and R. W. Denny, *J. Am. Chem. Soc.* **92,** 5218 (1970).
22. A. Knowles, Photo-sensitized isomerization of retinal by dyes. *Nature (London), New Biol.* **236,** 202 (1972).
23. G. Jori, G. Galiazzo and E. Scoffone, *Biochemistry* **8,** 2868 (1969).
24. L. C. Harber, J. Hsu, H. Hsu and B. D. Goldstein, *J. Invest. Dermatol.* **58,** 373 (1972).
25. M. M. Mathews-Roth, M. A. Pathak, T. B. Fitzpatrick, L. C. Harber and E. H. Kass, *New Eng. J. Med.* **282,** 1231 (1970).
26. B. D. Goldstein and L. C. Harber, *J. Clin. Invest.* **51,** 892 (1972).
27. A. A. Schothorst, J. Steveninck, L. N. Went and D. Suurmond, *Clin. Chim. Acta* **39,** 161 (1972).
28. J. E. Cleaver, *Proc. Nat. Acad. Sci. U.S.A.* **63,** 428 (1969).

29. W. Harm, *Biochem. Biophys. Res. Commun.* **32,** 350 (1968).
30. J. Das, B. Bagchi and U. Chaudhuri, *Photochem. Photobiol.* **19,** 317 (1974).
31. F. P. Imray and D. G. MacPhee, *Mol. Gen. Genet.* **123,** 289 (1973).
32. G. W. Grams and K. Eskins, *Biochemistry* **11,** 606 (1972).
33. A. F. Agro, P. De-Sole, G. Rotilio and B. Mondovi, *Ital. J. Biochem.* **22,** 217 (1973).
34. J. Leff and N. I. Krinsky, *Science* **158,** 1332 (1967).
35. N. I. Krinsky, in: *Photophysiology,* **Vol. 3** (A. C. Giese, Ed.), pp. 123–195. Academic Press, New York, 1968.
36. T. G. Truscott, E. J. Land and A. Sykes, *Photochem. Photobiol.* **17,** 43 (1973).
37. C. S. Foote, Y. C. Chang and R. W. Denny, *J. Am. Chem. Soc.* **92,** 5216 (1970).
38. C. A. Parker and T. A. Joyce, *Photochem. Photobiol.* **6,** 395 (1967).
39. H. T. Witt, *Q. Rev. Biophys.* **4,** 365 (1971).
40. A. Farmilo and F. Wilkinson, *Photochem. Photobiol.* **18,** 447 (1973).
41. M. A. Hescox and D. M. Carlberg, *Can. J. Microbiol.* **18,** 981 (1972).
42. W. K. Noell and R. Albrecht, *Science* **172,** 76 (1971).
43. R. Azerad, R. Bensasson, M. B. Cooper, E. A. Dawe and E. J. Land, Abstract E.1., *International Conference on the Excited States of Biological Molecules,* Lisbon, 1974. To be published in full by Wiley-Interscience, New York.
44. J. E. Dowling and R. L. Sidman, *J. Cell. Biol.* **14,** 73 (1962).
45. V. E. Kagan, A. A. Shvedova, K. N. Novikov and V. P. Kozlov, *Biochim. Biophys. Acta* **330,** 76 (1973).
46. F. J. M. Daemen, *Biochim. Biophys. Acta* **300,** 255 (1973).
47. R. W. Young, *J. Cell. Biol.* **33,** 61 (1967).

THE EFFECTS OF UV LIGHT ON GENETIC FUNCTIONS[1]

Chairman: R. Latarjet

[1]Symposium D–8

Dissipation of UV Energy in Nucleic Acids and Nucleoproteins[1,2]

J. W. Longworth
Biology Division
Oak Ridge National Laboratory
Oak Ridge, Tennessee 37830, USA

A fluorescence and phosphorescence is found from all of the heterocyclic bases of nucleic acids, and this permits an investigation of the channels of energy dissipation in nucleic acids. The luminescence is subject to thermal activated quenching and is only readily detected at low temperatures. The separate heterocyclic bases have individual features allowing characterization in complex heteropolymers, particularly by the phosphorescence, spectra and triplet decay times. The series nucleotide, dinucleotide, homopolymer and heteropolymer (random and sequential) has been studied to follow the perturbation imposed when residues become incorporated into nucleic acids. Two striking features are present in dinucleotides which also apply to polymers. The fluorescence of many dinucleotides is subject to a significant red shift and is featureless. This is termed an exciplex fluorescence and results from contiguity of heteroaromatic rings. The phosphorescence originates in a singlet base, and is the base with the lowest energy triplet state. Excitation spectra show both bases contribute to this phosphorescence; there has been a transfer of electronic energy. DNA in both the duplex and the single stranded stacked conformation exclusively phosphoresces from thymidylyl residues (T). A phage DNA, PBS2, which replaces T with uridylyl (U) exclusively phosphoresces from U. The absence of emission from the other bases is attributed to transfer and this is supported by excitation spectral studies. The fluorescence lifetime of duplex DNA is 10 ns at 77° K, reflecting the exciplex nature. This is an unusually long lifetime and suggests that considerable electronic transfer would occur within a polymer. Anisotropy of fluorescence of polyA is depolarized, but only to the same extent as ApA, indicating that there is only a small degree of energy migration. Exogenous ligands which quench the singlet and triplet show there is only a small amount of electronic energy transfer in DNA at singlet and triplet level. The predominant process of energy dissipation in nucleic acids is radiationless and thermally activated. To derive information about DNA at 300° K it is necessary to study DNA at much lower temperatures and extrapolate. The internal quenching of the fluorescence of heterocyclic bases occurs through a temperature independent and a temperature dependent channel. The temperature dependent internal conversion is subject to

[1]Congress presentation D-8-1.
[2]Research sponsored by the U.S. Atomic Energy Commission under contract with the Union Carbide Corporation.

a compensation law, the pre-exponential rate linearly correlates with the activation energy. Intersystem crossing to the triplet is a relatively small process at 77° K and for several of the bases is significantly increased at 300° K. Photochemical reactions proceed from both the singlet and triplet levels and occur in C, T and A in homopolymers. In DNA T and C are photochemically reactive. The limited studies on nucleoprotein complexes have not disclosed any significant interaction of the excited states of nucleic acids and protein.

Genetic Effects of UV on *Escherichia coli*—a Model for Prokaryotes[1]

B. A. Bridges
MRC Cell Mutation Unit, University of Sussex
Falmer, Brighton BN1 9QG, England

INTRODUCTION

The notion that base-pair substitution mutations induced by radiations arise as mistakes in repair was established by the work of Witkin (*1*) who showed that *Escherichia coli* B strains believed to be deficient at the locus *exr* (now called *exrA*, or *lexA* in the K-12 family) were both more sensitive to UV and at the same time non-mutable by UV. Such bacteria were also non-mutable by γ-radiation and thymine starvation (*2*). It subsequently transpired that these strains (made by "one-step" mutagenesis) were not as genetically "clean" as had been believed, but near-isogenic strains made by transduction clearly confirmed Witkin's original conclusion (*3*).

Since UV, ionizing radiation and thymine starvation cause mutations via the $exrA^+$ pathway, it was proposed that this is an error-prone pathway capable of repairing single-strand gaps in DNA whether produced directly or indirectly (*2,4*).

ExrA$^+$ REPAIR AS A MINOR OPTIONAL PATHWAY FOR REPAIR OF ANY SINGLESTRAND GAP

Three facts published at about the same time combined to take attention away from this idea of a common error-prone pathway. They were the facts (i) that in excision-deficient bacteria mutations arise as a result of replication of DNA containing pyrimidine dimers (*5*), (ii) that DNA synthesized after UV irradiation in such strains contains gaps opposite dimers which are subsequently filled in $recA^+$ bacteria by recombinational exchanges between daughter strands (*6,7*), and (iii) that *recA* bacteria are non-mutable by UV(*8,9,10*). It was natural, therefore, that mutations should come to be thought of as mistakes in genetic recombination, particularly during daughter strand exchanges (*4,11*). As will be discussed below, I do not believe this proposition has yet been established and I would like briefly to reaffirm the status of the $exrA^+$ pathway as an optional pathway for the repair of single-strand gaps, however formed.

With aerobic gamma irradiation there are two major types of single-strand break. One type persists for up to 40 min. in growth medium and normally

[1] Congress presentation D–8–2.

requires the $recA^+$ and $exrA^+$ genes for repair. It is a reasonable assumption that mutations arise during repair of these gaps, in which case it is clear that only a small fraction of these repair events leads to mutations (*12*). The error-prone pathway can be completely abolished and repair of these gaps occur in the absence of mutagenesis, for example in certain suppressed Exr^- strains or if DNA replication is brought to the end of a cycle in an *exrA* strain before irradiation (*13,14*). Another class of break produced by ionizing radiation is repaired exceedingly rapidly by DNA polymerase I, often within the period of irradiation (*15*). As breaks of this type are repaired even in *exrA* and *recA* strains (which are almost non-mutable by ionizing radiation), they are repaired essentially in an error-free manner. Presumably they do not persist long enough for a significant fraction to enter the $exrA^+$ pathway even when it is functional. That they can enter this pathway is, however, indicated by the fact that in *polA* bacteria, in which there is no DNA polymerase I activity, they persist longer (*15*) and a greater proportion of them enter the $exrA^+$ pathway and lead to mutations (*16*).

After UV irradiation, mutations may arise after unexcised pyrimidine dimers pass through the replication fork (*15*). Miss Jorunn Eyfjord in this laboratory has shown that when gaps in newly-synthesized DNA are allowed to be filled with the bacteria held in phosphate buffer, no mutations arise (as determined by loss of photoreversibility); repair under these conditions is thus error-free. Error-free gap-filling also occurs during the second replication cycle after UV irradiation of an excision-deficient strain in medium enriched with Casamino acids and during the first 15 min immediately after UV (Eyjford, Doubleday, Green and Bridges, unpublished data). In *exrA* strains, error-free filling of daughter-strand gaps occurs almost normally at doses below about 40 ergs/mm^2 (*17*). Clearly, daughter-strand gaps can be filled in both error-free and error-prone ways.

Gaps resulting from excision of pyrimidine dimers are almost always filled in an error-free way as evidenced by the much greater resistance to UV mutagenesis of excision-proficient compared with deficient strains. This is entirely consistent with the present hypothesis since excision-gaps are known to be very short-lived. It is now clear, however, that at some loci mutations do arise as a result of excision-gaps being repaired by the $exrA^+$ pathway at a very low frequency (*18–20*). It may well turn out that a substantial proportion of UV-induced mutants in wild-type bacteria are the result of excision-initiated $exrA^+$ repair events, albeit occurring with a very low probability.

All these results tend to establish the proposition that most if not all single-strand DNA gaps can be repaired in an error-free manner but that there exists an optional error-prone pathway of repair in $exrA^+$ cells that has a greater chance of operating, the more persistent the gap. As will be seen later, however, there may be additional factors about which we know little, the effect of which is to render certain gaps immune to error-prone repair.

THE INVOLVEMENT OF RECOMBINATION

There is at present little evidence for, and none against the hypothesis that recombinational exchanges may be involved in $exrA^+$—dependent mutation. There are, however, two pieces of evidence against the hypothesis that they are *necessarily* involved. The first is the demonstration that mutations may arise at normal rates prereplicatively after γ-irradiation and (as a result of excision) after UV in chemostat-grown bacteria, a large proportion of which contain only a single copy of the genome (*20*). In such bacteria there would not have been any homologous duplex with which the duplex containing a single-strand gap could interact to initiate a recombinational exchange.

The other line of evidence comes from work with bacteriophages. Certain phages, e.g., λ, also require the *exrA*$^+$ and *recA*$^+$ genes of their host in order to mutate after UV. They also require the host bacterium itself to be irradiated (*21-23*). Devoret *et al.* (*24*) have argued cogently that the "Weigle-reactivation" of λ that occurs in an irradiated bacterium is not a recombinational process. The single-stranded DNA phage S13 also shows UV-mutagenesis in an irradiated host at low multiplicities of infection (*25*) where it is very difficult to imagine any sort of recombination process taking place. Bleischrodt (personal communication) has confirmed this with reversion of an amber mutant of the closely related ϕX174, and has shown that it is dependent upon the *recA*$^+$ gene.

Why should non-recombinational mis-repair events be blocked by the *recA* gene? I believe the reason is due to the *recA*$^+$ gene being necessary at a very early stage for the stabilization of the type of single-strand gap that cannot be sealed quickly by DNA polymerase I. It is known that, in *recA* bacteria, DNA degradation after gamma irradiation begins immediately and daughter strand gaps are not sealed even if the opposing dimer is removed by photoreversal (*17*). The very early requirement suggested by these results is consistent with such a gap being unavailable for any type of repair in the absence of the *recA*$^+$ gene product.

THE "DOSE-SQUARED" RESPONSE CURVE—THREE HYPOTHESES

It is characteristic of UV mutagenesis of bacteria under many conditions and at many loci that the dose response kinetics are non-linear, often corresponding closely to a dose-squared function. Thus a given increment of dose has a greater effect the greater the dose that has already been given. Three hypotheses are current to explain this effect. One states that two lesions may cooperate to produce a mutation with a higher probability than either alone (*26-28*). The second supposes that there is competition for the lesions between an error-prone and error-free pathway and that the latter is progressively less efficient with increasing dose (*29*). The third states that in addition to producing premutational lesions, UV produces damage which causes progressive "induction" of the *exrA*$^+$-dependent repair system (*30*). The possibility that this system might be inducible was suggested (*23*, and Radman (1972), personal communication) on the basis of work with phage λ where *exrA*$^+$-dependent mutagenesis is greatly enhanced if the bacteria are pre-irradiated (*21-23*). A recent experiment in our laboratory (J. Eyfjord, unpublished data) has, however, failed to uncover any evidence for a persisting effect of a hypothetical induced state in excision-proficient *E.coli* WP2. The experiment consisted of a split-dose treatment in which the two doses, each of 25 ergs, were separated by periods of growth of between 0 and 60 min. It should be emphasized that these doses did not cause a significant amount of lethality or slowing down of growth or DNA synthesis. It was found that the interaction of the two doses to give a greater than additive mutation frequency was substantially diminished if the doses were separated by

10 min and had completely disappeared by about 30 min. This is completely contrary to the prediction of the induction hypothesis which would be that the effect of the second dose if given 10 or 20 min later should be as great as or greater than that with no delay.

A further difference between Weigle-reactivation and mutagenesis of λ and the hypothetical induction of a derepressed error-prone state responsible for increasing the mutagenic effectiveness of UV with regard to the bacterial chromosome is that the former can be induced indirectly by conjugational introduction of an UV-irradiated plasmic (*31*,*29*), whereas the latter cannot (*29*).

Doubleday, Bridges and Green (to be published) have recently completed some observations with an excision-deficient strain *E.coli* WP2 *uvrA* which reveal some significant new features of UV mutagenesis and also fail to show any evidence that physiological induction is responsible for the dose-squared response curve. They followed loss of photoreversibility of mutations after UV during growth in different media. In minimal medium, photoreversibility was lost over about two cycles of replication and in medium supplemented with Casamino acids it was lost over one cycle. In Casamino acids medium containing pantoyl lactone, photoreversibility was lost even more slowly than in minimal medium (Fig. 1). In all three media, the total number of induced mutants

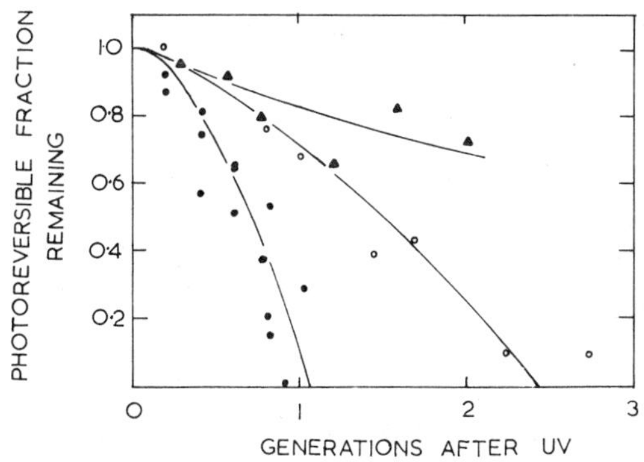

FIG. 1. Change with time of incubation in different growth media after UV irradiation in the fraction of mutations subject to photoreversal. The experimental treatment actually gave about 0·85 photoreversibility at zero time, all observed values have been normalized to 1·0 at zero time on the assumption that all mutants are in principle photoreversible (*31*). Glucose salts-tryptophan medium (o); supplemented with 0·4% Casamino acids (●); or with 0·4% Casamino acids and 0·1*M* pantoyl lactone (▲).

recoverable without photoreversal was the same during the first two cycles of replication. These and other results indicate that the probability that a premutational lesion will enter the error-prone pathway in a given replication cycle and give rise to a mutation was close to 1·0 in Casamino acids medium, about 0·7 in minimal medium and about 0·1 in Casamino acids medium with pantoyl lactone. Physiological conditions obviously have a profound influence on the probability that repair will go through the error-prone pathway.

The results also strongly suggest that, although potentially mutagenic lesions may be replicated and repaired in an error-free manner (e.g., in the presence of pantoyl lactone), they persist and give rise to mutations if subsequently transferred to optimal conditions (e.g., plates containing Casamino acids without pantoyl lactone). Simple arithmetic indicates that the probability that a pyrimidine dimer at any given site will give rise to a mutation is low, perhaps around 1% even at the top of the dose-squared curve (*32*). The fact that the composition of the growth medium during the first few replication cycles does not alter the ultimate mutation yield even though the probability of mutation-induction in any given replication cycle may vary considerably, suggests that there may be a very limited number of pyrimidine dimers that are capable of initiating error-prone repair. If so, then in plates containing Casamino acids they must do so in the first replication cycle with a probability of unity. That there are indeed a limited number of potentially mutagenic pyrimidine dimers is clearly seen if one considers the second replication cycle after UV in Casamino acids medium. No mutations are being formed (photoreversibility is already completely lost) but examination of the molecular weight of the DNA being synthesized at that time shows that daughter-strand gaps are still being formed by the replication of pyrimidine dimers (J. Eyfjord, personal communication). These dimers (constituting the majority of those formed in the DNA) must be those which are not potentially mutagenic, all the latter having given rise to mutations during the first replication cycle.

The fact that there are some dimers that never give rise to mutations suggests that my original contention, that all single strand gaps are in principle substrate for error-prone repair, may be an over-simplification. It is possible to construct a plausible model in which most daughter strand gaps are immune to error-prone repair (M.H.L. Green, personal communication), and the high mutation rates observed with some genetic backgrounds (e.g. *tif*, *polA*, and *dnaB* (*30,35,36*) may reflect a lifting of this immunity.

On the basis of the induction hypothesis for the dose-squared response one might argue that the premutational pyrimidine dimer might be distinguished from other dimers by being in an "induced" fraction of the bacterial population. Since Casamino acids would be expected to enhance an "induction" process and since pantoyl lactone is known specifically to inhibit other "induction" phenomena dependent on the *exrA*$^+$ gene (e.g. filamentation (*33*), prophage induction in a *tif* strain (*34*), one would expect that these treatments would affect the number of premutational dimers if these are in fact determined by their situation in an induced cell. In fact, these treatments affect neither the total potential yield of mutants nor do they alter the dose-squared nature of the response curve. On the contrary, their effect is more consistent with the notion that some sort of "induction" process might be involved in determining the probability that a mutation will arise in a given replication cycle from a premutational lesion, although what this "induction" might mean at the molecular level is not clear.

Thus, although the level of "induction" may drastically alter the probability of an error-prone event in any given cycle, it has no effect on the dose-response kinetics. Moreover, since there seems to be a limited number of premutational lesions, the second hypothesis would also seem to be unlikely. Inhibiting error-free repair at higher doses should not increase the number of mutations since all premutational lesions appear to be able to initiate mutations under our experimental conditions even at low doses.

If one were to exclude both these hypotheses (and this would probably be premature), one would be forced to conclude that there are relatively more premutational lesions per erg/mm^2 after high than after low doses and that the first hypothesis (collaborative interaction of two photoproducts to yield a premutagenic lesion) is the most attractive explanation for this.

ACKNOWLEDGEMENTS

I thank Dr. M. H. L. Green, Miss Jorunn Eyfjord and Mr. O. P. Doubleday for permission to refer to unpublished results and for much stimulating discussion.

REFERENCES

1. E. M. Witkin, *Brookhaven Symp. Biol.* **20**, 17 (1967). (Available from Ofc. of Tech. Serv., Dept. of Commerce, Wash., D.C.).
2. B. A. Bridges, J. Law and R. J. Munson, *Mol. Gen. Genet.* **103**, 266 (1968).
3. S. G. Sedgwick and B. A. Bridges, *Mol. Gen. Genet.* **119**, 93 (1972).
4. E. M. Witkin, *Annu. Rev. Genet.* **3**, 525 (1969).
5. B. A. Bridges and R. J. Munson, *Proc. Roy. Soc. B*, **171**, 213 (1968).
6. W. D. Rupp and P. Howard-Flanders, *J. Mol. Biol.* **31**, 291 (1968).
7. W. D. Rupp, C. E. Wilde, D. L. Reno and P. Howard-Flanders, *J. Mol. Biol.* **61**, 25 (1971).
8. S. Kondo, H. Ichikawa, K. Iwo and T. Kato, *Genetics* **66**, 187 (1970).
9. A. Miura and J. Tomizawa, *Mol. Gen. Genet.* **103**, 1 (1968).
10. E. M. Witkin, *Mutat. Res.* **8**, 9 (1969).
11. B. A. Bridges, *Annu. Rev. Nucl. Sci.* **19**, 139 (1969).
12. R. J. Munson and B. A. Bridges, *Biophysik* **6**, 1 (1969).
13. B. A. Bridges, W. J. H. Gray, M. H. L. Green, M. A. Rothwell and S. G. Sedgwick, *Genetics, Suppl.* **73**, 123 (1973).
14. M. H. L. Green, W. J. H. Gray, S. G. Sedgwick and B. A. Bridges, *J. Gen. Microbiol.* **77**, 99 (1973).
15. C. D. Town, K. C. Smith and H. S. Kaplan, *Science* **172**, 851 (1971).
16. B. A. Bridges and R. P. Mottershead, *Heredity* **29**, 203 (1972).
17. B. A. Bridges and S. G. Sedgwick, *J. Bacteriol.* **117**, 1077 (1974).
18. H. Nishioka and C. O. Doudney, *Mutat. Res.* **8**, 215 (1969).
19. H. Nishioka and C. O. Doudney, *Mutat. Res.* **9**, 349 (1970).
20. B. A. Bridges and R. P. Mottershead, *Mutat. Res.* **13**, 1 (1971).
21. J. J. Weigle, *Proc. Natl. Acad. Sci. U.S.A.* **39**, 628 (1953).
22. R. Devoret, *C.R. Acad. Sci.* **260**, 1510 (1965).
23. M. Defais, P. Fauquet, M. Radman and M. Errera, *Virology* **43**, 495 (1971).
24. R. Devoret, M. Blanco, J. George and M. Radman, in *Molecular Mechanisms for the Repair of DNA* (P. Hanawalt and R. B. Setlow, Eds.). Plenum Press, N.Y., in press (1975).
25. E. S. Tessman and T. Ozaki, *Virology* **12**, 431 (1960).
26. B. A. Bridges, *Mutat. Res.* **3**, 273 (1966).
27. C. O. Doudney and C. S. Young, *Genetics* **47**, 1125 (1962).
28. H. D. Mennigmann, *Mol. Gen. Genet.* **117**, 167 (1972).
29. B. A. Bridges, M. A. Rothwell and M. H. L. Green, *An. Acad. Bras. Cienc.*, in press (1974).
30. E. M. Witkin and D. L. George, *Genetics, Suppl.* **73**, 91 (1973).
31. J. George, R. Devoret and M. Radman, *Proc. Nat. Acad. Sci. U.S.A.* **71**, 144 (1974).
32. B. A. Bridges, R. E. Dennis and R. J. Munson, *Genetics* **57**, 897 (1967).
33. P. Van de Putte, C. Westenbrock and A. Rörsch, *Biochim. Biophys. Acta* **76**, 247 (1963).
34. E. P. Kirby, W. L. Ruff and D. A. Goldthwait, *J. Bacteriol.* **111**, 447 (1972).
35. E. M. Witkin, *Proc. Nat. Acad. Sci. USA*, **71**, 1930 (1974).
36. E. M. Witkin, in *Molecular mechanisms for the repair of DNA* (P. Hanawalt and R. B. Setlow, Eds.), Plenum Press, N.Y., in press (1975).

The Present Status of DNA Repair Mechanisms in UV Irradiated Yeast Taken as a Model Eukaryotic System[1]

E. Moustacchi, R. Waters, M. Heude and R. Chanet
Fondation Curie–Institut du Radium, Biologie
Orsay, France

INTRODUCTION

The repair mechanisms of altered DNA are generally less well understood for eukaryotes than they are for prokaryotes and bacteriophages. For mammalian cell lines cultured in vitro the specific labelling of DNA has allowed the biochemical analysis of some of the steps of the repair processes whereas the determination of their genetic controls is, with a few exceptions (*1,2*), obviously difficult. On the other hand, with fungi and more specifically with yeast taken as a model unicellular eukaryotic system, the genetic approach has been extensively explored: radiosensitive mutants are readily detected and genetically analyzed, double and multiple mutants can be constructed and from their responses to irradiation the number of repair pathways involved can be suggested. The lack of thymidine kinase in these organisms has hampered for a certain time the biochemical analysis of repair. However, the recent isolation of yeast strains capable of taking up and incorporating thymidine 5'-monophosphate into their DNA (*3–5*) opens new possibilities for the future. In spite of this difficulty, attempts to measure the induction and removal of UV-induced pyrimidine dimers were performed by several groups during the last three years (*6–9*).

Most of the work done with the radiosensitive mutants of *Ustilago* (*10*), *Aspergillus* (*11*), *Neurospora* (*12*) and the yeasts *Saccharomyces cerevisiae* (*13–20*) and *Schizosaccharomyces pombe* (*21,22*) concerned UV-induced lethality, mutagenesis and genetic recombination in comparison with the responses of the original wild types.

The genetic analysis had demonstrated that several recessive genes are involved in the control of these responses. For instance, in the yeast *S. cerevisiae* the number of independent genes affecting the radiosensitivity is higher than the 30 *rad* loci identified. This indicates that the genetic control of radiosensitivity is, even in such a simple eucaryote, much more complex than in the case of bacteria. It can be calculated that around 50 loci are probably involved in yeast (*15*). In comparison to this multiplicity of controls it can be noticed that, for instance, the *rev* genes (*23,24*) govern the UV-induced mutagenic response with no interference with the recombinogenic UV-induced pattern, whereas the *rec* genes

[1]Congress presentation D–8–3.

(*25*) modify the UV-induced intragenic recombination with no effect on the mutagenic response.

A classification of all the genes conferring radiosensitivity has been attempted on the basis of an analogous rationale: the behavior of the yeast mutants was compared to that of the well documented *Escherichia coli* radiosensitive mutants. This includes studies of

a) the liquid holding recovery under non-growing conditions after irradiation in association with the effect of visible light taken as a specific process for removing pyrimidine dimers (*26–28*);

b) the effects of drugs, such as caffeine, known for their inhibitory effects on the repair processes (*29,30*);

c) the cross-sensitivity of the mutants to ionizing radiation, chemicals and heat (*31–36*);

d) the response to lethal damages of double (*22,37,38*) or multiple mutants (*38,39*) as compared to single mutants;

e) the influence of a block due to a nuclear mutation (*rad*, *rev* or *rec* genes) on the response to mitochondrial genetic damage (*16,40,41*). In this respect, the yeast *S. cerevisiae* offers the experimental advantage that the mitochondrion can be dispensed with, the cells still survive as "petite colonie" cytoplasmic mutants which are respiratory deficient (*42*). In this system the interrelations between the control of repair of damaged nuclear and mitochondrial DNA's can be studied.

From these approaches, the following main ideas arose:

1. The two main repair pathways described for *E. coli*, i.e. the excision-resynthesis and post-replicative recombinational repair pathways, do exist in yeast. The existence of the former pathway is supported not only by indirect evidence but also by biochemical analysis. The rad_1 and rad_2 mutants for instance have been shown to be blocked in the excision of UV-induced pyrimidine dimers. Other loci are epistatic to rad_1 and rad_2 (rad_3, rad_4) and are likely to act on this excision pathway (*38*).

2. It is necessary to postulate the existence of at least two other repair pathways in order to explain some of the radiobiological data obtained on multiple mutants (*39,43,44*) and on synchronized cells (*43,45*). The 4 distinct repair pathways appear to be interrelated for some of the steps (*35,43,45,46*).

3. The genetic control of the mitochondrial response to a UV treatment involves nuclear genes and mitochondrial determinants.

Since the second point has been thoroughly discussed in a recent symposium this year by R. Haynes (*46*), we will focus our attention on certain aspects of the first and third points raised here.

THE FATE OF UV-INDUCED PYRIMIDINE DIMERS IN THE NUCLEAR AND MITOCHONDRIAL DNA'S OF *SACCHAROMYCES CEREVISIAE*

The repair mechanisms of the major lesions induced by UV light in DNA, namely pyrimidine dimers, have been extensively investigated in recent years. One such repair system, that of excision-resynthesis has been investigated in phages, bacteria, fungi and mammalian cell lines. However, the existence of a comparable repair system has been questioned in many members of the plant kingdom, as excision-resynthesis cannot be demonstrated in some of its members following the post-irradiation incubation which allowed it to function in both bacterial and human cell lines.

To date two systems have been utilized in attempts to measure such events in *S. cerevisiae*.

Resnick and Setlow (*7*), by measuring the competition of UV-induced dimers in yeast total DNA with UV irradiated transforming DNA for photoreactivating enzyme, suggested that postirradiation incubation in nutrient medium of logarithmic phase cells resulted in the removal of such induced pyrimidine dimers. Stationary phase cells did not exhibit any such competition following similar treatments and it was thus suggested they could not remove these photoproducts.

Unrau et al. (*6*), by labelling the DNA with a general nucleic acid precursor and developing a chromatographic technique to separate the pyrimidine dimers from the pyrimidines, indicated that wild type stationary phase cells do excise after an incident dose of 6600 ergs/mm^2 up to 60-70% of the UV-induced dimers from the total cell DNA following the postirradiation incubation of such cells in nutrient medium for 4 hours or more. No indications were given about survival levels in this work. Both the forementioned groups stated that UV-sensitive mutants of the *rad*$_1$ and *rad*$_2$ loci could not remove pyrimidine dimers (py$\hat{}$py) from their DNA upon postirradiation incubation in nutrient medium.[2]

As these results where somewhat ambiguous and since these estimates of the removal of py$\hat{}$py from the total DNA of UV-irradiated *S. cerevisiae* cells were for total DNA, it was decided to investigate the removal of UV-induced py$\hat{}$py from both the nuclear and mitochondrial DNA's at well defined growth stages of haploid and diploid cultures of this organism. It was established whether such postirradiation treatments as the application of visible light and dark liquid holding in nutrient medium or buffer influenced these events and whether such results were related to survival and cytoplasmic "petite" induction.

DNA from cells labelled with ^{3}H-6-uracil, a general precursor of nucleic acids, was extracted and purified in a CsCl gradient according to Williamson et al. (*47*). An example of the type of separation of nuclear and mitochondrial DNA's obtained is shown in Fig. 1. A further separation was afforded by separately pooling the nuclear and mitochondrial portions of each gradient and rerunning each portion in a second CsCl gradient similar to the first.

The purity of each type of DNA is verified by estimating the cytosine/thymine ratios of the hydrolsed DNA (0.66 for nuclear DNA and 0.25 for mitochondrial DNA). The hydrolysis of the DNA and the chromatography of the bases and pyrimidine dimers (py$\hat{}$py) were done according to Unrau et al. (*6*).

An example of the amounts of py$\hat{}$py found on the immediate extraction of nuclear and mitochondrial DNA's following UV irradiation with 5000 ergs/mm^2 incident dose is given in Fig. 2.

1) Nuclear DNA

Fig. 3 gives the dose-related yields of two types of py$\hat{}$py induced in DNA (T$\hat{}$T

[2]In the text, pyrimine dimers will be referred to as pypy, with T$\hat{}$T indicating thymine-thymine dimer and C$\hat{}$T cytosine-thymine dimer, etc. In some of the figures, the latter have also been indicated by T[]T and C[]T, respectively.

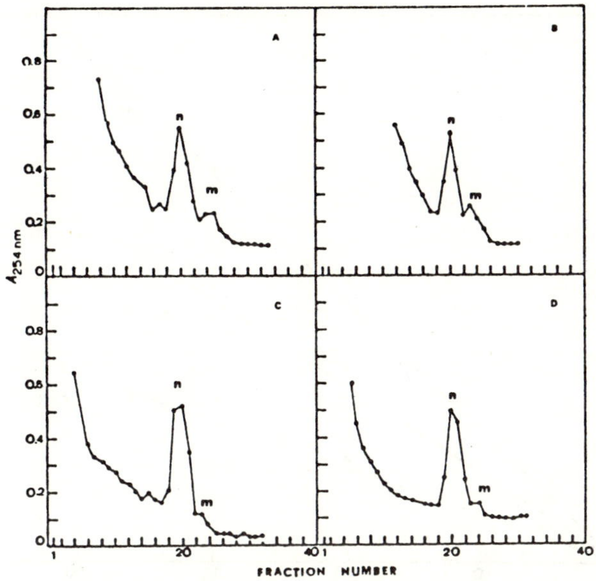

FIG. 1. The optical density of fractions from cell extracts run in cesium chloride gradients where "n" denotes the nuclear DNA peak (density 1.700 g/cm^3) and "m" the mitochondrial peak (density 1.685 g/cm^3). A: Unirradiated cells; B: irradiated with 5000 ergs/mm^2, immediate extract; C: irradiated with 5000 ergs/mm^2 and incubated for 72 hr in saline prior to extraction; D: 5000 ergs/mm^2 followed by 24 hr in nutrient medium.

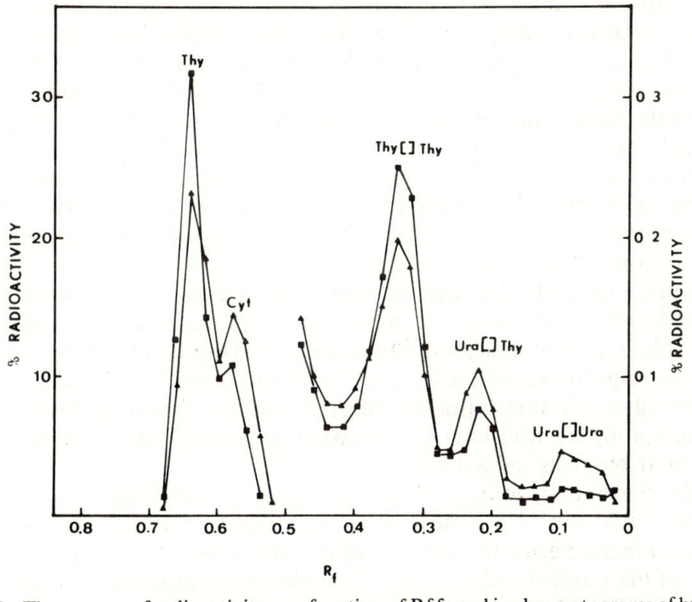

FIG. 2. The percent of radioactivity as a function of Rf found in chromatograms of hydrolyzed DNA extracted from stationary yeast cells UV-irradiated with 5000 ergs/mm^2. Left hand side: thymine, cytosine region; right hand side: dimer region. (■) Mitochondrial DNA; (▲) Nuclear DNA.

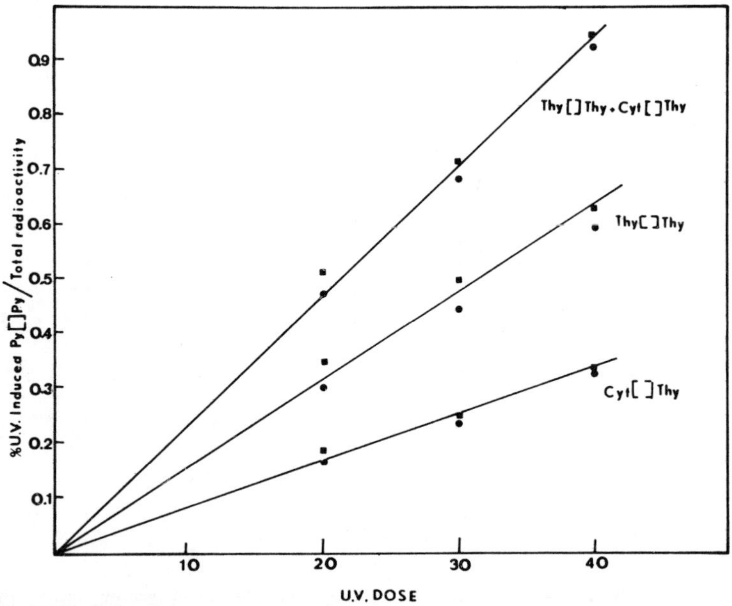

FIG. 3. The induction by ultraviolet light of pyrimidine dimers in the nuclear DNA of haploid (■) and diploid (●) Saccharomyces cerevisiae.

and ĈT) by 254 nm UV light in haploid and diploid cells. The total yield increases linearly with dose and no striking difference between haploid and diploid cells could be detected. Consequently, it appears that the amount of shielding of the DNA by the cytoplasm (and especially by RNA which is abundant in these cells) is much the same for both degrees of ploidy. The yield of UV-induced pŷpy is related to the base composition of the organism (*48*). From the data presented here, which are in good agreement with those reported (*48*), it can be estimated that about 27 dimers/genome/ergs mm^2 are formed in *S. cerevisiae*. In *E. coli* about five times fewer dimers are induced per erg, per genome (*49*), whereas in human fibroblasts about 1000 times more dimers are induced per erg, per genome as compared to yeast (*50,51*).

The percentage of pŷpy removed from the nuclear DNA of both haploid and diploid cells upon postirradiation incubation in nutrient medium is equal to that found in saline. In both cases it is dose-dependent (Fig. 4): the higher the dose the lower percentage of excision is found. The stimulation of ^{14}C-2-uracil incorporation observed on the dark liquid holding in saline of UV-treated stationary phase cells (undergoing no division) as compared to unirradiated controls (Table 1) suggests that repair is indeed involved.

Under these conditions, the excision system appears to be saturated at doses around 900 ergs/mm^2. UV doses low enough to determine the dose at which the zero order kinetic began for the plot of dimers excised for 10^7 nucleotides as a function of the number induced for 10^7 nucleotides could not be used owing to limitations of the chromatographic technique used. However, as the number excised following both 2 and 3×10^3 ergs/mm^2 is the same (9×10^3) (data not shown), it is extremely likely to represent the number of pyp̂y maximally excisable. Further support for this figure is afforded by the 56% excision of pyp̂y following 1.5×10^3 ergs/mm^2 found by Nasim (personal communication) who used a more sensitive electrophoretic technique (*52*). This represents an excision of approximately 9.4×10^3 pyp̂y/10^7 nucleotides.

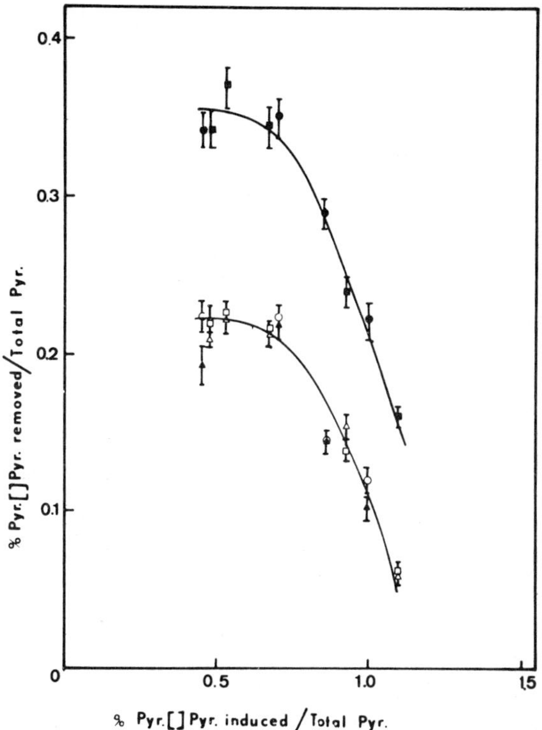

FIG. 4. The excision of pyrimidine dimers from the nuclear DNA's of haploid and diploid *S. cerevisiae*. following a) 4 hr post irradiation incubation in nutrient medium (□) haploid, (O) diploid; b) 72 hr post irradiation incubation in saline (△) haploid, (▲) diploid; c) 72 hr post irradiation incubation in saline followed by 4 hr incubation in nutrient medium (■) haploid, (●) diploid.

TABLE 1

The proportion of counts/min in stationary phase yeast DNA prelabelled with ^{3}H-6-uracil on ultraviolet irradiation and incubation in saline containing ^{14}C-2-uracil for 72 hours.

Treatment	Counts/min ^{3}H in C+T	Counts/min ^{14}C in C+T	Counts/min ^{3}H in TT̂	Counts/min ^{14}C in TT̂
No ultraviolet irradiation	60,164	1668	99	5.0
5000 ergs/mm^2	44,679	2670	410	4.2

Table 2 compares this figure with previous estimates of the saturation of excision on other organisms. The estimate for *S. cerevisiae* is seen to be higher than that found for prokaryotes and identical to that for human fibroblasts.

TABLE 2

Species	Number of pŷpy required to saturate excision/ 10^7 nucelotides	D_{37} ergs/ mm^2	Genome size
E. coli (1)	$2,3.10^3$	810	3.10^7
P. mirobilis (10)	$3,5.10^3$	400	1.10^7
S. cerevisias	9.10^3	800	1.10^{10}
human fibroblasts (3)	$9.10^{3(x)}$	30	$3,6.10^{12}$(*4*)

$^{(X)}$calculated by assuming 300 ergs/ mm^2 saturates the excision system (*3*), and a D_{37} of 30 ergs/ mm^2 (2) gives 7.5×10^5 pŷpy per genome (*3*).

The incubation in nutrient medium of UV-treated cells, previously subjected to liquid holding in saline, increases the amount of pŷpy excized as compared to that seen on incubation in saline or nutrient medium alone (Fig. 4) (These conditions mimic those used for observing a liquid holding recovery of survival). This feature was apparent at all doses and may be well one phenomenon responsible for the increase in survival seen following the liquid holding of UV-irradiated cells. Parry and Parry (*53*) have suggested, via physiological and genetical experiments in yeast, that the number of UV-induced lesions being repaired by an excision repair mechanism is less on direct plating than the number repaired following liquid holding in saline. Concomitantly, the number of UV-induced lesions repaired via a recombinational pathway is greater on direct plating on nutrient medium as compared to that following liquid holding.

The results presented in Fig. 4 partially substantiate this hypothesis. As expected, survival was also increased by these treatments (Fig. 5). The postirradiation treatments applied resulted in approximately the same percentage removal of pŷpy following a specific dose regardless of ploidy. Hence, on considering the survival curves in Fig. 5, the differences seen in survival according to ploidy are not attributable to either

i) less pŷpy being induced per UV dose in the diploid, or

ii) a more efficient excision of pŷpy occurring in diploid, as compared to haploid cells.

As suggested by various authors (*29,54*), the increased resistance of diploid cultures may thus be due to a DNA repair mechanism other than that of excision resynthesis. In order to determine whether this increase in resistance is i) a result of an actual removal of pŷpy; or ii) perhaps a recombinational event requiring DNA synthesis, it should be feasible to measure the dimer content of UV-irradiated cells incubated in nutrient medium but inhibited in nuclear DNA synthesis, as compared to irradiated cells allowed to undergo several replications in nutrient medium.

The influence of photoreactivating light on the survival of wild-type and radiation-sensitive mutants after UV-irradiation, and its relationship to other irradiation treatments, has been reported (*26,28*).

A visible light treatment when applied to UV-treated cells prior to dark liquid holding, has no effect on the percentage survival obtained following doses of 500 and 1000 ergs/ mm^2. However, when cells received doses ranging from 1500 and 2500 ergs/ mm^2, survival is considerably enhanced in cultures illuminated prior to liquid holding as compared to liquid holding alone (Fig. 6). This observation can be partly explained by assuming that photoreactivation operates by reducing the number of lesions repairable by the hitherto saturated excision system. It also suggests that non-photoreactivatable lesions occur which are repairable during liquid holding, these possibly being pŷpy which are inaccessible to the photoreactivating enzyme, or UV-induced lesions other than pŷpy (*8,28*).

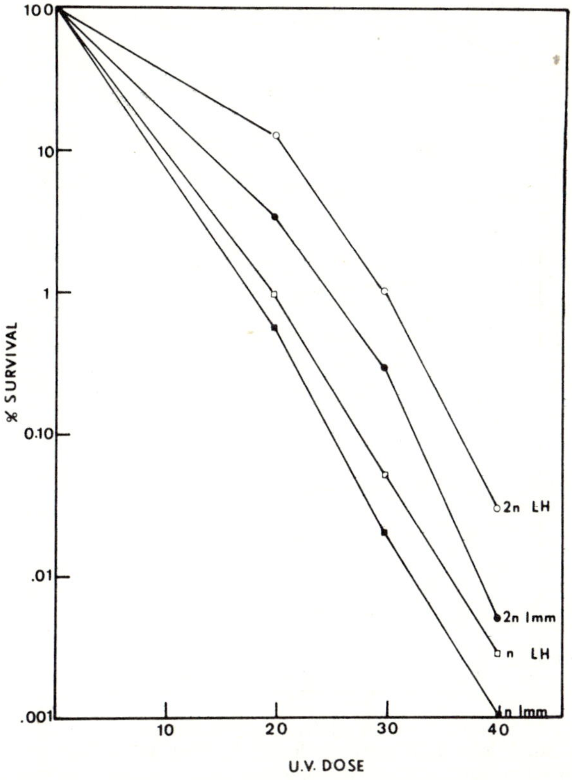

FIG. 5. The survival of haploid and diploid saccharomyces cerevisiae following immediate plating after UV irradiation (■) haploid; (●) diploid; or after 72 hr liquid holding in saline prior to plating (□) haploid, (O) diploid.

Survival of a strain carrying a *rad*$_1$ mutation, seen to be excision-deficient (*6–8*), is still photoreactivatable after liquid holding (*26,27*). On the other hand, in wild type cells, photoreactivation following liquid holding for 48 to 72 hr does not increase the survival as compared to liquid holding alone (*8,26,28*). However, after a dose of 5000 ergs/mm^2 and liquid holding in saline for 72 hr, which removes approximately 10% of the pŷpy, illumination was seen to still remove about 50% of the remaining pŷpy. In other words, despite this splitting of dimers by visible light, cell survival was not increased in the wild type cells, whereas, it is enhanced in *rad*$_1$. This mutant has been shown to be deficient in the process of excision of pŷpy; and as it has been stated that incised pŷpy preclude photoreactivation (*55*), *rad*$_{1-3}$ may well be incision deficient. If so, it is this process which has determined survival during liquid holding in the wild type and precluded the increase of survival by photoreactivation. One possibility is that incision occurs during liquid holding but is not the rate limiting step in the process.

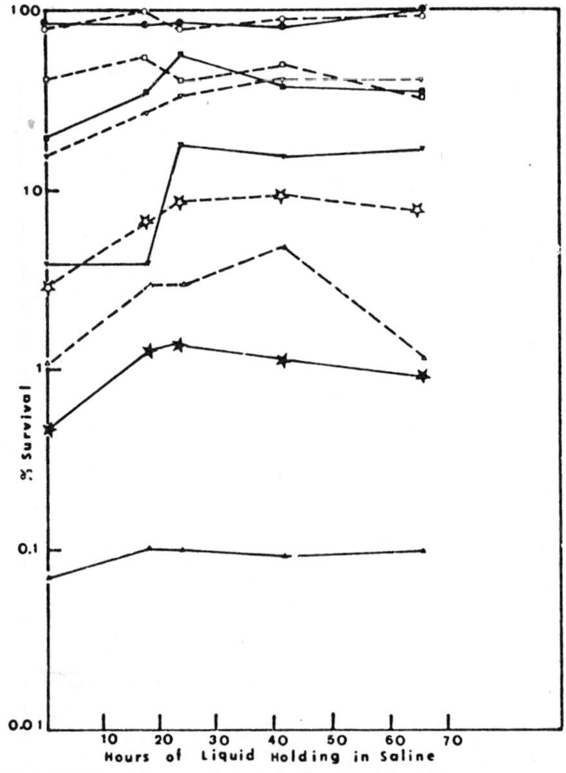

FIG. 6. Effects of photoreactivation prior to liquid holding alone on survival as a function of time liquid held. Solid lines (—) refer to liquid holding alone and dotted lines (---) to photoreactivation prior to liquid holding following ultraviolet doses of (O,●) 500 ergs/mm^2; (□,■) 1000 ergs/mm^2; (▽,▼) 1500 ergs/mm^2; (☆ ★) 2000 ergs/mm^2; (△,▲) 2500 ergs/mm^2.

2) *Mitochondrial DNA*

A primary factor encountered on attempting to study the removal of UV-induced py$\hat{}$py from the mitochondrial DNA following the above treatments was the loss of a considerable portion of the mitochondrial DNA capable of banding at a density of 1.685 g/cm^3 in a cesium chloride density gradient (Figs. 1 and 7). This initially appears to resemble the ethidium bromide-induced degradation of mitochondrial DNA (*56*).

Despite the number of counts/min being considerably diminished, it was possible to analyze any remaining mitochondrial DNA banding at a density of 1.685 g/cm^3 for py$\hat{}$py content. The fact that this DNA always contained the same percentage of py$\hat{}$py as originally induced indicates the absence of a controlled excision resynthesis process operating on this DNA (*57*) (Fig. 8). Hence, one can discount this mechanism when discussing any possible repair processes involved in the changes of ρ^- induction seen in many biological experiments discussed below.

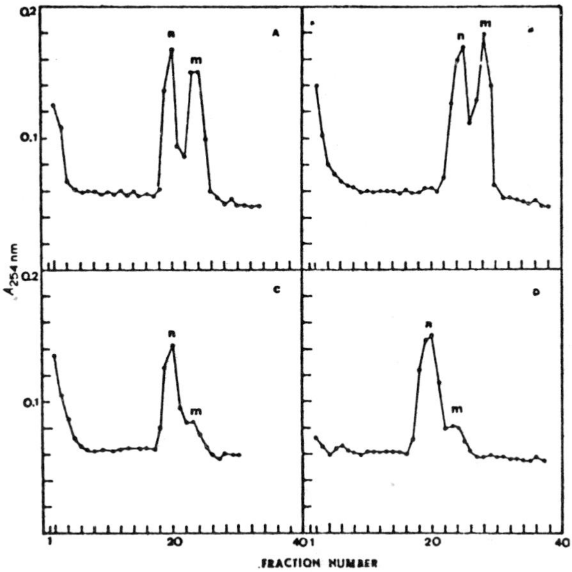

FIG. 7. The optical density of 254 nm of fractions collected from a cesium chloride gradient containing the "mitochondrial DNA" sections pooled from a preliminary CsCl gradient of total cell DNA. "n" refers to residual nuclear DNA occurring in the sample (density 1.70 g/cm^3); "m" refers to mitochondrial DNA (density 1.685 g/cm^3).

A. Unirradiated sample.

B. UV-irradiated with 5000 ergs/mm^2, immediate DNA extract,

C. Cells irradiated with 5000 ergs/mm^2 of UV + 72 hours in saline prior to DNA extraction,

D. Cells irradiated with 5000 ergs/mm^2 of UV + 24 hours in nutrient medium prior to DNA extraction.

One of the nucleus-determined repair processes (*58*) operating on UV-induced $\hat{py}$py, namely that of photoreactivation, operates also on mitochondrial DNA's of *S. cerevisiae* (Fig. 9).

To summarize, the removal of $\hat{py}$py via a controlled excision repair mechanism appears to be limited to nuclear DNA. Mitochondrial DNA in UV-irradiated cells is seen to undergo an extensive degradation.

PHYSIOLOGICAL AND GENETIC FACTORS WHICH MODIFY UV-INDUCED CYTOPLASMIC GENETIC DAMAGES

Particular interest in our group is given to the interrelations between the genetic controls of the repair of nuclear damage and mitochondrial damage as expressed by the induction of the cytoplasmic "petite" (or ρ^-) mutation. The maintenance of the respiratory ability requires, besides the functioning of specific nuclear genes (*42,59,60*), the presence of a cytoplasmic genetic determinant referred as the ρ factor (*59*). In recent years, ρ has been equated to mitochondrial

DNA. Exposure to various agents (intercalating dyes, UV light, thermal shocks, etc...) induces alterations in the base composition of mitochondrial DNA (*61*) or its loss (*56,62*). This results in a pleiotropic mutation, the cytoplasmic respiratory-deficient "petite" mutation.

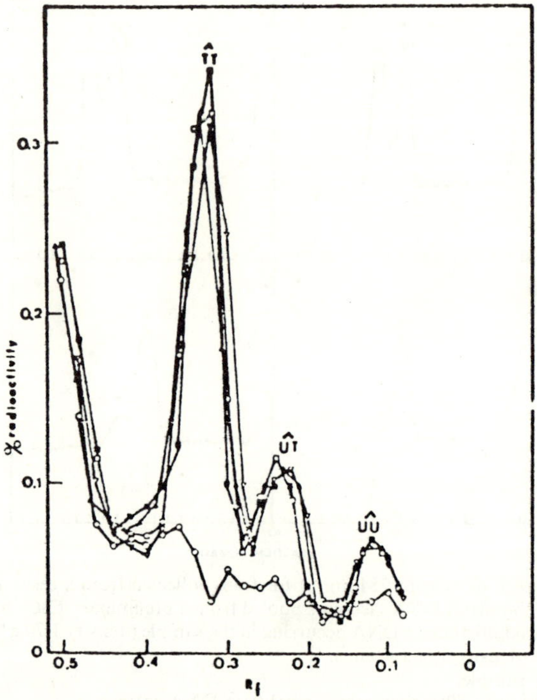

FIG. 8. The percent of radioactivity as a function of Rf found in the dimers region of chromatograms of mitochondrial DNA (O) non irradiated cells (■) immediate extract after a 4000 ergs/mm^2 UV irradiation; (▲) 4000 ergs/mm^2 of UV followed by 72 hr incubation in saline prior to extraction; (□) the same but 24 hr incubation in nutrient medium; (▽) the same but 72 hr in saline and subsequent 12 hr incubation in nutrient medium prior to extraction.

1) Physiological Studies:

UV light is an efficient agent for the induction of the "petite" mutation in *S. cerevisiae*. Pyrimidine dimer formation appears to be the essential lesion since this effect is reversible by visible light (*63*). On the other hand, depending upon the physiological conditions, this mutation is reversible by dark holding (*64*). The use of the split-dose technique allows one to demonstrate a recovery of the ρ^+ phenotype as compared to a single dose (*65*). When exponentially growing cells are UV-treated and then kept in the dark in non-growth medium, a substantial decrease in the frequency of mutants over survivor is found in the delayed plating as compared to immediate plating (Fig. 10). It can be seen that when visible light is applied after dark holding of the cells, no further change in the frequency is observed. This may be interpreted as if the photoreactivable lesions were already removed by the postirradiation incubation. A similar reversibility is also observed when a heat treatment follows a contact with ethidium bromide (*66*), one of the most powerful inducers of ρ^- mutants (*67*).

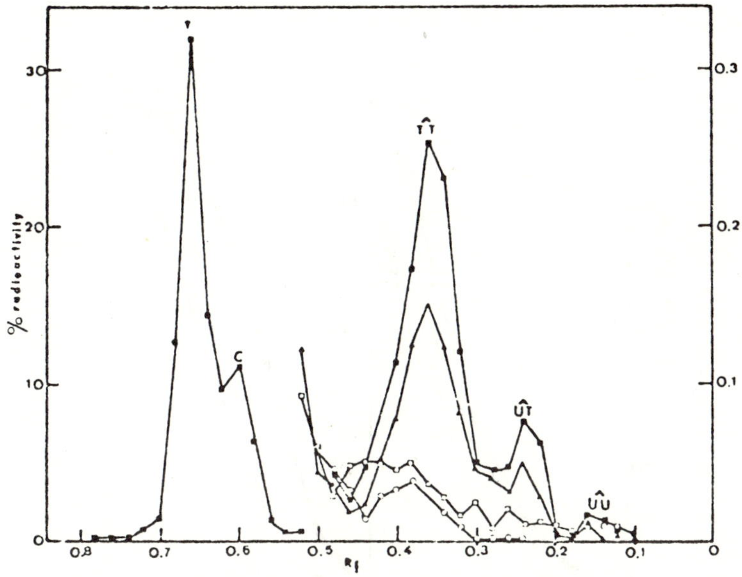

FIG. 9. The effect of visible light treatment after UV irradiation on the percent of radioactivity as a function of Rf found on chromatograms of hydrolysed mitochondrial DNA extracted from cells irradiated with 4000 ergs/mm^2. (O) non irradiated cells; (□) non irradiated cells given visible light treatment; (■) UV irradiated cells; (▲) UV irradiated cells followed by visible light treatment.

On the other hand, when stationary-phase cells are UV-irradiated and then held in the dark, the frequency of ρ^- over survivors increases when delayed plating is compared to immediate plating (*29,64*) (Fig. 10). In that case too it appears that illumination after liquid holding does not change the frequency of ρ^-'s. Hence, an event not reversible via photoreactivation has occurred during the liquid holding of UV-irradiated cells.

The contribution of nucleus-directed protein synthesis in the repair of lethal and mitochondrial genetic damage after UV-irradiation of exponential and stationary phase haploid cells was examined. This was carried out by using cycloheximide, a specific inhibitor of nuclear protein synthesis. The inhibitor was added prior to or/and after the UV treatment. It appears that nuclear protein synthesis is required for both the increase in survival seen after the liquid holding of cells at all stages and for the "petite" recovery seen after the liquid holding of exponential phase cells. The characteristic negative liquid holding observed for the UV-induction of "petite" in stationary phase remained following all the treatments which inhibited nuclear protein synthesis (Fig. 11).

By use of drugs such as erythromycin and chloramphenicol, the contribution of mitochondrial protein synthesis in these phenomena was examined. It was shown that mitochondrial proteins are necessary for the recovery of survival of UV-treated exponential-phase cells and are implicated in the mechanism leading to the restoration of the ρ^+ phenotype in UV-irradiated dark liquid-held

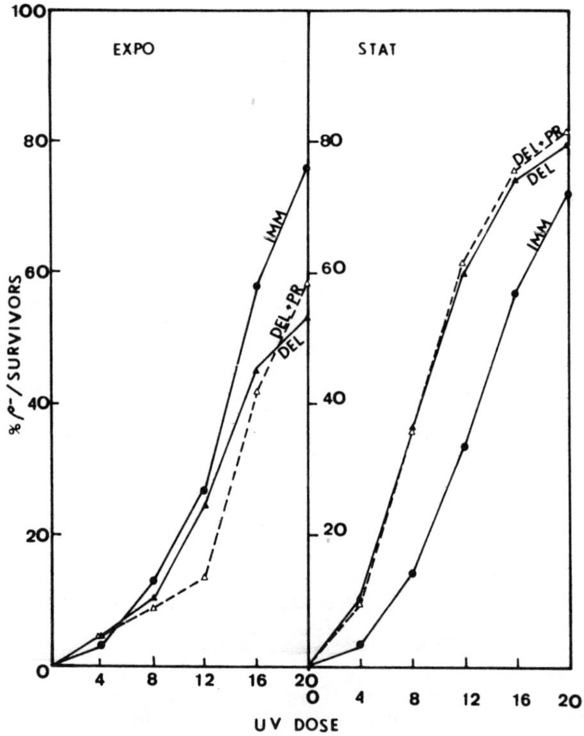

FIG. 10. Effect of dark liquid holding alone or followed by a visible light treatment on the frequency of "petite" induction as a function of the UV dose in exponential and stationary phase cells. (●) immediate plating; (▲) delayed plating, 48 hours of liquid holding; (△) delayed plating, 48 hours of liquid holding followed by an illumination.

exponential-phase cells. Here again, in stationary-phase cells, mitochondrial enzymes do not seem to participate in the negative liquid holding process for the ρ^- induction. The effect of these drugs on survival is not surprising since it is known that recovery can be suppressed by withholding oxygen or by inhibiting oxidative metabolism during the dark liquid holding (*68*). Moreover, a respiratory-deficient ρ^- mutant does not show liquid holding recovery unless an energy-supplying substrate is added to the incubation medium (*69*).

The nature of the negative liquid holding of ρ^- induction may be questioned and two possibilities not exclusive of each other may be considered:

a) the increase in ρ^- frequency during dark holding may result from the extensive degradation of mitochondrial DNA (*57*). This degradation may occur in stationary-phase cells after the completion of initial steps of the repair processes leading to ρ^+ in exponential-phase cells. The nuclease(s) likely to be involved in the degradative process, may escape inhibition by the drugs if they have long lifetimes. If so, this approach does not allow us to assign a nuclear or a mitochondrial origin to these enzymes.

b) alterations in the binding of mitochondrial DNA to the mitochondrial membrane have been frequently mentioned as events which may lead to the conversion of the ρ^+ to the ρ^- genotype (*66,70*). It is known that the biogenesis of mitochondria depends on complex interactions between the mitochondrial-directed and the nucleus-directed protein-synthesis of the mitochondrial membranes. Such modifications may alter the binding of mitochondrial DNA to the membrane, leading to anomalies in its replication or in its susceptibility to further enzymatic attack. This last event should have to take place after the transfer of the cells to the plating medium without the inhibitory drugs.

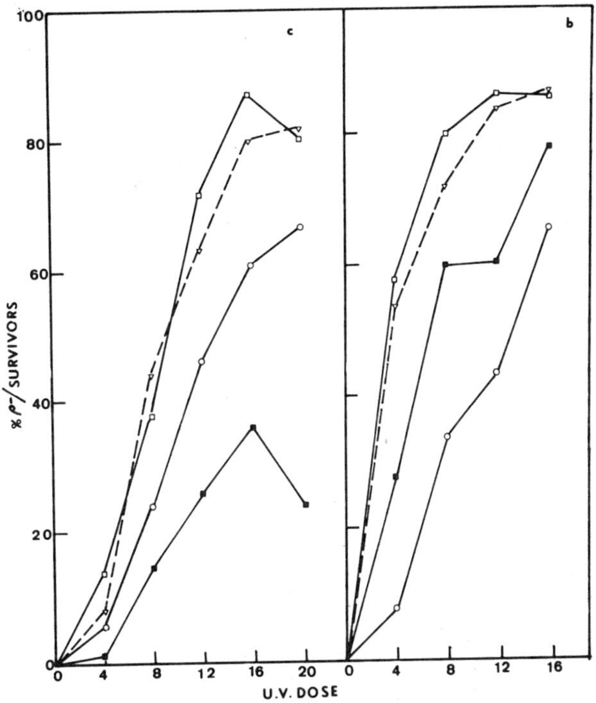

FIG. 11. Influence of pre- and post-treatment with cycloheximide (CH) (1 μg/ml) on the UV induction of "petites" in dark liquid held. a) exponential phase cells; b) stationary phase cells. (O) immediate plating of cells pre-treated with CH; (■) delayed plating after storage in water of cells pre-treated with CH; (□) delayed plating after storage in CH of cells pre-treated in CH; (△) delayed plating after storage in CH followed by a visible light treatment.

These two possibilities are currently being examined by the biochemical analysis of mitochondrial DNA degradation in suitable UV sensitive mutants. This study is being carried out with the use of protein synthesis inhibitors.

Special attention was given by our group to the variations in sensitivity to radiations during a synchronized cell cycle (*71–73*). A variation in the lethality and in the cytoplasmic "petite" induction was encountered following UV-irradiation at different stages of the cycle. Lethality was at a minimum during the G_2 phase and this was not concomitant with the minimum degree of "petite" induction, the latter being at a minimum during the end of G_1 (*71*).

These variations of sensitivity observed at different cell cycle stages have generally been interpreted as resulting from variations in efficiency of the repair mechanisms. A good support for this idea comes from the work of Fabre (*30,43*) who demonstrated that such variations of lethal damage are abolished in some UV-sensitive mutants of *Schizosaccharomyces pombe* blocked in certain types of repair processes. Mutants specifically sensitive to the cytoplasmic "petite" induction (see below) also demonstrate a constant response to the UV induction as compared to the original wild type (*71*). This might indicate that in the case of this cytoplasmic genetic damage, wild type cells are able to repair a fraction of the damage at certain stages of the cell cycle. Moreover, it has been shown that dark liquid holding of irradiated cells increases the magnitude of cyclic variations of "petite" induction.

2) Genetic Studies:

In relation to the problem of the degree of autonomy of the mitochondrion *vis-a-vis* the nucleus, we have tried to determine to what extent some of the mechanisms governing the repair of UV-induced nuclear lesions act on mitochondrial genetic lesions. The rationale behind this study was the following: if the putative repair system of mitochondrial damage is not controlled by a given *rad* nuclear gene, the same dose-effect relationship for the "petite" induction is expected in both wild type and *rad* mutant. If, on the other hand, the loss of a given repair step controlled by one *rad* locus operates on both nuclear and mitochondrial lesions, a modified sensitivity to "petite" induction is expected in the *rad* mutant.

The systematic analysis of 21 independent *rad* mutants have shown us that both situations are encountered (*40*). In other words, it appears that some of the steps of the repair of mitochondrial damage are controlled by the nucleus. In this respect it is interesting to notice that two mutants blocked in UV-induced intragenic recombination (*25*) also demonstrate a diminution in the frequency of UV-induced "petites" (*41*) as compared to the original wild type (for instance, a *rec_5* strain is about 4 times less sensitive at an incident dose of a 1000 ergs/mm^2).

On the other hand, it is also clear that some of the steps of the repair pathways controlled by the nucleus, and for repair of nuclear damage, do not affect the reversibility of mitochondrial genetic damages. These steps are possibly governed by some other genes acting on mitochondrial lesions, these genes being located on the chromosomes or on the mitochondrial genome.

If this interpretation were correct, mutants with an almost normal sensitivity to the lethal effect of UV but particularly sensitive to "petite" induction would be expected to be found. A selective method for this type of mutant was devised and 5 mutants corresponding to these criteria were found after a mutagenic treatment. They are referred as "uvsρ".

As compared to the original wild type, the uvsρ mutants are characterized by a high frequency of spontaneous "petites". Survival and "petite" induction curves are shown on Fig. 12. The mutants still present curves of complex shapes, but the extrapolation numbers are generally smaller than in the wild type and doses 2 to 10 times smaller are sufficient to induce the same proportion of "petite" mutants.

The genetic analysis demonstrates that three of the uvsρ mutations (uvsρ5, uvsρ6 and uvsρ34) segregate like mendelian genes whereas two of the mutants demonstrate a non-mendelian pattern of segregation. For these latter mutants it has been shown that the loss of the ρ factor is accompanied by the loss of the uvsρ marker as seen in subsequent crosses with a wild type strain.

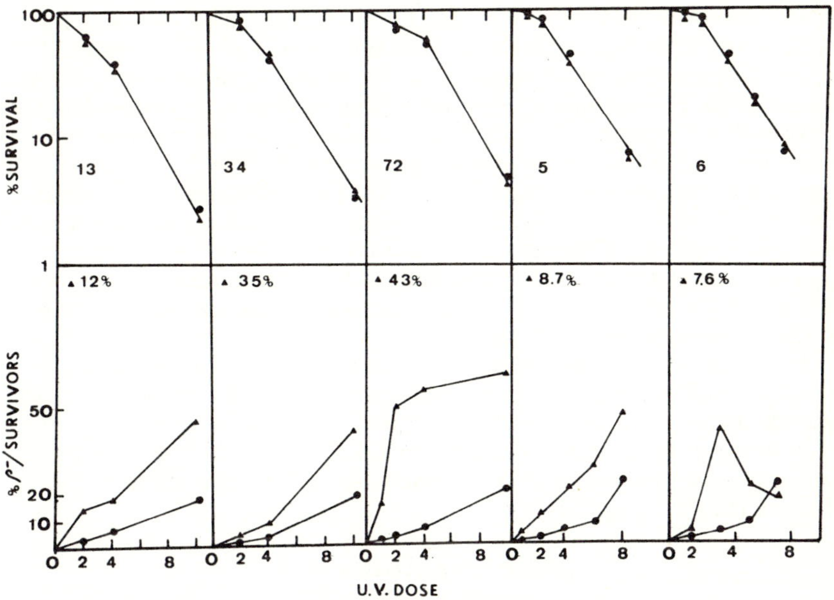

FIG. 12. Survival and "petite" induction in uvsρ mutants as a function of UV dose (●) wild type strain N123; (▲) derived uvsρ mutants. In the lower part of the graph the spontaneous frequency of ρ^- mutants for the uvsρ strains is given.

It is then likely that the mitochondrial genome codes for a product involved in the ultimate response of wild type cells to UV-induced cytoplasmic damage (repair enzyme, cofactor of a nuclear repair enzyme or modifier of the permeability of the organelle to a nuclear enzyme).

The ρ^- mutation is by no means comparable to a single nuclear mutation since the spontaneous and UV-induced ρ^- frequencies are much higher than the rates of induction of recessive nuclear mutations for a given dose. Moreover, the cytoplasmic respiratory deficiency has been shown to be accompanied by a total loss of, or altered base composition of, mitochondrial DNA (*56,61,62*).

Although the biochemical experiments give no evidence for a repair mechanism, other than photoreactivation, acting on mitochondrial DNA, various biological data indicate that the fate of the respiratory character is not irrevocably determined immediately after mutagenic treatments (*40,64-66,72*). Some mechanism responsible for curing mitochondrial DNA damage is suggested by such observations. In principle, the recovery of the ρ^+ phenotype may occur by at least two main mechanisms, one being the maintenance and subsequent selective multiplication of undamaged mitochondrial DNA molecules, the other being an enzymatic repair of damaged mitochondrial DNA molecules.

The first suggestion would not appear to be applicable to the UV doses used in these experiments, as an average of 60-200 dimers per mitochondrial DNA molecule would be induced by a dose of 1000-3000 ergs/mm^2 assuming 50 mitochondrial DNA molecules of a size of 21.10^6 daltons occur per cell (*73*). Hence, the second possibility for the repair mechanism remains.

The mitochondrial genome has been shown to exhibit a recombination of genetic markers (*74,75*), and more recently both ρ^+ and ρ^- mitochondrial DNA appear to be capable of recombination (*76–78*). Hence, in an eukaryotic organism in which the mitichondrion is dispensable, a recombinational process may well be involved in determining whether an intermediary state of the mitochondrial genome is expressed as ρ^+ or ρ^- following UV irradiation. A prerequisite for this suggestion is that mitochondrial DNA is the primary target for ρ^- induction. Although the existence of a nuclear master copy of mitochondrial DNA cannot be completely refuted (*79*), many experiments have failed to demonstrate such a copy (*73*). The fact that ethidium bromide at low concentrations is a potent ρ^- inducer and is unable to induce nuclear DNA alterations, also supports the idea that the mitochondrion is the main target.

In conclusion, the absence of an excision process acting on the mitochondrial DNA of UV-treated cells indicates that any possible repair mechanisms suggested by the previously cited biological data should involve both degradative and recombinational events. A similar conclusion was reached by the genetic analysis of the patterns of loss of mitochondrial genetic markers for drug resistance (*80*).

ACKNOWLEDGMENTS

I am grateful to Dr. R. Latarjet for constant support and advice. The experimental work was partly supported by the Délégation Générale a'la Recherche Scientifique et Technique and by an Euratom grant.

REFERENCES

1. J. E. Cleaver, *Nature (London)*, **218**, 652 (1968).
2. J. Epstein, J. R. Williams, and J. B. Little, *Proc. Nat. Acad. Sci. U.S.A.* **70**, 977 (1973).
3. S. Jansen, E. R. Lochman, and E. R. Megnet, *FEBS Lett.*, **8**, 113 (1970).
4. M. Brendel and R. H. Haynes, *Molec. Gen. Genet.* **117**, 39 (1972).
5. R. B. Wickner, *J. Bacteriol.*, **117**, 252 (1974).
6. P. Unrau, R. Wheatcroft, and B. S. Cox, *Molec. Gen. Genet.*, **113**, 359 (1971).
7. M. A. Resnick and J. K. Setlow, *J. Bacteriol.*, **109**, 979 (1972).
8. R. Waters and E. Moustacchi, *Biochim. Biophys. Acta.* **366**, 241 (1974).
9. F. Fabre and E. Moustacchi, *Biochim. Biophys. Acta*, **312**, 617 (1973).
10. R. Holliday, *Mutat. Res.*, **4**, 275 (1967).
11. J. Fortuin, *Mutat. Res.*, **11**, 149 (1971).
12. A. L. Schroeder, *Molec. Gen. Genet.*, **107**, 291 (1970).
13. S. Nakai and S. Matsumoto, *Mutat. Res.*, **4**, 129 (1967).
14. R. Snow, *J. Bacteriol.*. **94**, 571 (1967).
15. B. S. Cox and J. M. Parry, *Mutat. Res.*, **6**, 37 (1968).
16. E. Moustacchi, *Mutat. Res.*, **7**, 171 (1969).
17. M. A. Resnick, *Genetics*, **62**, 519 (1969).
18. I. A. Zakharov, T. N. Kozina, and I. V. Federova, *Mutat. Res.*, **9**, 31 (1970).
19. D. Averbeck, W. Laskowski, F. Eckardt, and E. Lehman-Brams, *Molec. Gen. Genet.*, **107**, 117 (1970).

20. J. C. Game and B. S. Cox, *Mutat. Res.*, **12,** 328 (1971).
21. M. Schupbach, *Mutat. Res.*, **11,** 361 (1971).
22. F. Fabre, *Molec. Gen. Genet.*, **110,** 134 (1971).
23. J. J. Lemontt, *Mutat. Res.*, **13,** 319 (1971).
24. J. J. Lemontt, *Genetics*, **68,** 21 (1971).
25. U. S. Rodarte-Ramon and R. K. Mortimer, *Radiat. Res.*, **49,** 133 (1972).
26. J. M. Parry and E. M. Parry, *Mutat. Res.*, **8,** 545 (1969).
27. B. S. Kilbey and S. M. Smith, *Molec. Gen. Genet.*, **104,** 253 (1967).
28. J. M. Parry and B. S. Cox, *Gen. Res.*, **12,** 187 (1968).
29. E. Moustacchi and S. Enteric, *Molec. Gen. Genet.*, **109,** 69 (1970).
30. F. Fabre, *Molec. Gen. Genet.*, **117,** 153 (1972).
31. F. K. Zimmerman, *Molec. Gen. Genet.*, **102,** 247 (1968).
32. M. Brendel, N. A. Khan and R. H. Haynes, *Molec. Gen. Genet.*, **106,** 289 (1970).
33. W. Laskowski and E. Lehman-Brams, *Biophysik*, **10,** 51 (1973).
34. R. Waters and J. M. Parry, *Molec. Gen. Genet.*, **124,** 135 (1973).
35. M. Brendel and R. H. Haynes, *Molec. Gen. Genet.*, **125,** 197 (1973).
36. A. Schenberg-Frascino and E. Moustacchi, *Molec. Gen. Genet.*, 115, 243 (1972).
37. N. A. Khan, M. Brendel, and R. H. Haynes, *Molec. Gen. Genet.*, **107,** 376 (1970).
38. J. C. Game and B. S. Cox, *Mutat. Res.*, **16,** 353 (1972).
39. J. C. Game and B. S. Cox, *Mutat. Res.*, **20,** 35 (1973).
40. E. Moustacchi, *Molec. Gen. Genet.* **114,** 50 (1972).
41. E. Moustacchi, *J. Bacteriol.*, **115,** 805 (1973).
42. B. Ephrussi, *Nucleo-Cytoplasmic Relations in Micro-Organisms.* Oxford: At the Clarendon press (1953).
43. F. Fabre, *Radiat. Res.*, **56,** 528 (1973).
44. E. G. Hunnable and B. S. Cox, *Mutat. Res.*, **13,** 297 (1971).
45. R. Chanet, M. Heude, and E. Moustacchi, *Molec. Gen. Genet.*, **132,** 23 (1974).
46. R. H. Haynes, Proceedings of the Squaw Valley Conference on "Molecular mechanisms of repair of DNA" (P. Hanawalt and R. Setlow Ed.) (in the press) (1974).
47. D. H. Williamson, E. Moustacchi, and D. Fennell, *Biochim. Biophys. Acta*, **238,** 369 (1971).
48. P. Unrau, R. Wheatcroft, B. S. Cox, and T. Olive, *Biochim. Biophys. Acta*, **312,** 626 (1973).
49. J. M. Boyle and R. B. Setlow, *J. Mol. Biol.*, **51,** 131 (1970).
50. J. E. Cleaver, *Int. J. Radiat. Biol.*, **18,** 557 (1970).
51. J. E. Cleaver and J. E. Trosko, *Photochem. Photobiol.*, **11,** 547 (1970).
52. M. C. Birnboim and A. Nasim, *Molec. Gen. Genet.*, **130,** 291 (1974).
53. J. M. Parry and E. M. Parry, *Genet. Res.*, **19,** 1 (1972).
54. R. Waters and J. M. Parry, *Molec. Gen. Genet.*, **124,** 145 (1973).
55. M. H. Patrick and H. Harm, *Photochem. Photobiol.*, **18,** 371 (1973).
56. E. S. Goldring, L. I. Grossman, D. Krupnick, D. R. Cryer, and J. Marmur, *J. Mol. Biol.*, **52,** 323 (1970).
57. R. Waters and E. Moustacchi, *Biochim. Biophys. Acta* (in the press) (1974).
58. M. A. Resnick, *Photochem. Photobiol.*, **9,** 307 (1969).
59. F. Sherman, *Genetics*, **48,** 375 (1963).
60. F. Sherman and P. Slonimski, *Biochim. Biophys. Acta*, **90,** 1 (1964).
61. J. C. Mounolou, H. Jakob and P. Slonimski, *Biochem. Biophys. Res. Commun.*, **24,** 218 (1966).
62. P. Nagley and A. W. Linnane, *Biochem. Biophys. Res. Commun.*, **39,** 989 (1970).
63. D. Pittman and P. R. Pedigo, *Exp. Cell Res.*, **17,** 359 (1959).
64. M. Heude and E. Moustacchi, *C. R. Acad. Sci. Paris*, **277,** 1561 (1973).
65. N. G. Maroudas and D. Wilkie, *Biochim. Biophys. Acta*, **168,** 681 (1968).
66. H. R. Mahler and P. S. Perlman, *J. Supramol. Struct.*, **1,** 105 (1972).
67. P. Slonimski, G. Perrodin, and J. Croft, *Biochem. Biophys. Res. Commun.*, **30,** 232 (1968).
68. M. H. Patrick, R. H. Haynes, and R. B. Uretz, *Radiat. Res.*, 28, 144 (1964).
69. E. N. De Languth and U. Gelos, *Experientia*, **28,** 1413 (1972).

70. P. O. Weislogel and R. A. Butow, *J. Biol. Chem.*, **246,** 5113 (1971).
71. R. Chanet, D. H. Williamson and E. Moustacchi, *Biochim. Biophys. Acta*, **324,** 290 (1973).
72. R. Chanet and M. Heude, *Molec. Gen. Genet.*, **131,** 21 (1974).
73. P. Borst and R. A. Flavell, in *Mitochondria and Biomembranes*, pp.1–19. North-Holland, Amsterdam, 1972.
74. D. Y. Thomas and D. Wilkie, *Biochem. Biophys. Res. Commun.*, **30,** 368 (1969).
75. M. Bolotin, D. Coen, J. Deutsch, B. Dujon, P. Netter, E. Petrochilo, and P. Slonimski, *Bull. Inst. Pasteur, Paris*, **68,** 215 (1971).
76. E. P. Sean, PhD Thesis, University of Wisconsin (1972).
77. D. H. Williamson and D. Fennell, *Molec. Gen. Genet.*, **131,** 193 (1974).
78. G. Michaelis, E. Petrochilo, and P. Slonimski, *Molec. Gen. Genet.*, **123,** 51 (1973).
79. D. Wilkie, *J. Mol. Biol.*, **7,** 527 (1963).
80. J. Deutsch, B. Dujon, P. Netter, E. Petrochilo, P. P. Slonimski, M. Bolotin-Fukuhara, and D. Coen, *Genetics*, **76,** 195 (1974).

Radiology Applied: Mapping Transcriptional Organization in Pro- and Eukaryotes[1]

Walter Sauerbier
Department of Biophysics and Genetics
University of Colorado Medical Center
4200 East Ninth Avenue
Denver, Colorado 80220, USA

INTRODUCTION

In this contribution I shall pursue two objectives. First, I want to describe how UV irradiation of the template DNA disturbs the normal process of transcription, and secondly, I want to show how these observations provide us with a most powerful and widely applicable method of investigating the transcriptional organization of chromosomes.

The biologist, who seeks to understand the mechanisms of control of gene expression in terms of time and amount, is faced with a confusing multitude of conceivable and possibly realized molecular control devices. These devices can act on the transcriptional, post-transcriptional, and translational levels. Knowledge of the transcriptional linkage of genes, i.e., which genes are co-transcribed into the same RNA molecule, becomes therefore requisite to all further investigations of the control of gene expression. As we shall see later on, the transcriptional organization of chromosomes is best studied by observing the perturbed states of transcription. But, before discussing the perturbed states, I shall present a simplified scheme for the normal transcriptional process. I shall include only those steps which are common to transcription of DNA by RNA polymerases from bacteria, algae and mammalian cells, and by bacteriophage-induced RNA polymerases.

The first goal of an RNA polymerase is to find the DNA template through three-dimensional diffusion which, after collision of the polymerase with the DNA, is presumably reduced to a "one-dimensional" diffusion along the DNA molecule until the RNA polymerase arrives at a polymerase binding, or promotor, site. Here it is held in place more tightly. The next step appears to be a uni-directional drift along a few dozen base pairs with or without denaturing the double-stranded DNA, until the RNA chain initiation site is reached. At this point, RNA chains are initiated (always with a purine triphosphate) and, thereafter, chain elongation through polymerization of purine and pyrimidine triphosphates proceeds along the codogenic DNA strand up to a terminating base sequence, the terminator. Here, the completed RNA chain and the RNA polymerase are released, and for the RNA polymerase a new cycle of

[1] Congress presentation D-8-4.

transcription can begin. Many RNA polymerase molecules, separated only by a dozen or so nucleotide pairs, may transcribe the same DNA segment. That segment of DNA, which is bounded on the codogenic strand by a promotor and in the 5' direction from the promotor by a terminator, defines a scripton. A scripton beginning with a given promotor may vary in length through activation or inactivation of differently located terminators. In prokaryotes and in ribosomal RNA region of eukaryotes, scriptons contain several cistrons, or genes, as a rule.

The preceding scheme shows how the RNA polymerase, while transcribing the codogenic strand of double-stranded DNA, must proceed through a sequence of predetermined steps. Most of the determinants of this program are encoded in the DNA in terms of unique base sequences. Proper spatial arrangement of these determinant-base-sequences on the DNA template, in conjunction with the state of the RNA polymerase and the cellular milieu (or the conditions of the in vitro reaction), trigger the RNA polymerase into executing its transcribing functions in a sequentially ordered fashion. Since absorption of UV light photochemically alters the DNA bases, we may expect that the transcriptional template functions of DNA can be modified by UV light to give faulty information to the transcribing RNA polymerase molecule. A discussion of the transcriptional consequences of such faulty information will be the first objective of this contribution.

The second objective is to present a universally applicable method for determining transcriptional linkage of genes. We can base such a method on the following, simple rationale: if the probability of transcription of a given gene could be made inversely proportional to the distance of its 5'-end from its promotor (on the codogenic DNA strand), then distances of genes from their respective promotos could be directly measured by quantifying the rates of gene expression. In the next section I shall discuss how UV irradiation of the template DNA interferes with the various steps of the transcriptional process, and I shall conclude that UV irradiation of the DNA reduces the expression of every gene with an efficiency directly proportional to the distance between the 5'-end of a gene and its respective promotor.

UV EFFECTS ON THE TRANSCRIPTIONAL TEMPLATE FUNCTIONS OF DNA

In this part I shall discuss the UV effects on individual transcriptional functions of DNA in the sequence of their utilization by the RNA polymerase.

Binding of RNA Polymerase to UV Irradiated DNA

Binding of RNA polymerase to UV irradiated DNA has been measured by RNA polymerase-mediated retention of radioactive DNA on nitrocellulose filters (*1,2*), or by competitive inhibition of in vitro RNA synthesis by addition of UV irradiated DNA (*3–5*). The conclusions drawn from such experiments are: (a) Binding of various RNA polymerases to various DNA templates is neither reduced, nor increased by prior UV irradiation of the DNA. This statement appears to be generally valid for all DNAs to which RNA polymerase binds specifically (*3,6,7*).

(b) In the case of calf thymus DNA, to which bacterial RNA polymerase binds nonspecifically (*8*), UV irradiation leads to increased polymerase binding (*4*, *5*). Experiments with UV irradiated calf thymus DNA templates yield results which are not always in agreement with those from DNA templates that provide for specific polymerase binding.

The Effect of Template Irradiation on the Rate of RNA Chain Initiation

The loss of in vitro RNA chain initiation following UV irradiation of template DNA has been quantitated by either γ-^{32}P-ATP labeling of the 5'-ends of RNA chains (*3,2*) or by determinations of the total number of RNA chains on sucrose gradients or polyacrylamide gels (*3,2*) which were initiated during a given reaction period. It was observed that the rate of in vitro RNA chain initiation by a bacterial RNA polymerase is reduced with pseudo-first order kinetics by UV irradiation of the template DNA (bacteriophage T4 DNA, ref. *3*; and for a dose up to 2000 erg/mm^2 with calf thymus DNA, ref. *2*). However, reduction of the rate of RNA chain initiation by UV irradiation is very inefficient compared to the reduction in RNA chain length and the rate of overall RNA synthesis.

In vivo initiation of RNA chains following UV irradiation of bacteria (*9*) and of cultured mammalian cells (*10*) (nucleolar RNA synthesis) is also little affected by whole cell UV irradiation.

The most precise in vivo observations on UV effects on RNA chain initiation come from studies on the transcription of UV irradiated bacteriophage T7 (*11*). They show little or no reduction in the rate of RNA chain initiation with UV doses up to about 400 erg/mm^2.

Thus, for a low UV dose range (less than 400 erg/mm^2) it is safe to completely neglect the UV-induced loss of initiated RNA chains when UV damage on the transcriptional level is considered. (It should be kept in mind, however, that above 10^4 erg/mm^2 initiation of RNA chains can be greatly reduced, and with UV doses approaching 10^5 erg/mm^2 initiation can be practically abolished; ref. 7.)

Length of RNA Chains Synthesized from UV Irradiated DNA Templates

The average length of RNA chains synthesized in vitro from UV irradiated T4 DNA (*3*), T7 DNA (*12*), calf thymus DNA (*2*), and Lambda DNA (*13*) is progressively reduced as the UV dose to the DNA template is increased. The experimentally observed reduction in RNA chain length as a function of UV dose to the DNA template can be described by

$$L_{RNA}^{UV} = \sum_{k=0}^{\infty} [1/(1 + k)] \cdot (D/k!)e^{-D} \quad ; \text{ref. } 3 \tag{1}$$

or in closed form by

$$L_{RNA}^{UV} = (1 - e^{-D})/D; \text{ ref. } 14 \tag{2}$$

In eqs. 1 and 2, L_{RNA}^{UV} refers to the number average RNA chain length, D to the mean number of transcription terminating UV lesions per scripton, and k to the actual number of transcription-terminating UV lesions for a given scripton. Equations 1 and 2 also give the average distance between a promotor and the transcription-terminating UV lesion that is next to the promotor. In these considerations we, of course, assume scriptons to be of equal size; however, eq. 2 can be readily modified to include 2 or more scriptons of different length. These kinetics can be utilized to obtain limited information on the spread of a paucidisperse distribution of scripton sizes for a given DNA template.

I find it convenient to refer to the dose dependency, as expressed by eqs. 1 and 2, as $1/(1 + D)$ kinetics (*3,12*). This dose-effect relationship is of major importance for describing UV inactivation of transcriptional or replicational template functions of linear molecules.

The Rates of Overall RNA Synthesis From UV Irradiated DNA Templates

The long-term rates (in excess of 20 min) of in vitro and in vivo RNA syntheses from UV irradiated DNA templates show UV dose dependencies that are identical to the observed reductions in RNA chain length both in terms of kinetics ($1/(1 + D)$) and in terms of UV cross sections (*2,3,9,10,15*). We may, therefore, assume that the rate of overall RNA synthesis from UV irradiated DNA templates is determined by the average distance between the promotor and the transcription-terminating UV lesions closest to the promotor. Such a correlation would be expected only if the steps of finding the RNA polymerase binding site plus RNA chain initiation were slow compared with the time required for RNA chain polymerization. This is indeed the case, as independent studies show.

RNA chain lengths and the overall RNA synthesis are reduced by UV irradiation of the DNA template with 10 to 50 times higher efficiencies than RNA chain initiation. (Comparison of the two differing quantities is meaningful only in a low UV dose range, since chain initiation is lost with pseudo-first order kinetics, and RNA chain length and the rate of RNA synthesis are reduced with $1/(1 + D)$ kinetics.)

The Preferential Inhibition of Transcription of the Promotor-Distal Portions of Scriptons Following UV-Irradiation of the Template DNA

The loss of homology between RNA synthesized from UV irradiated T4 DNA templates and from nonirradiated templates has been extensively invesitgated by RNA-DNA hybridization-competition experiments (*3*). Quantitative analysis of the dose response of RNA homology loss reveals that the kinetics and the UV cross sections are identical to the ones observed for RNA chain shortening and reduction in the rates of RNA synthesis (*14*). We (*3*), therefore, concluded that only the portions of scriptons between promotors and the first "downstream" transcription-terminating UV lesion are transcribed. Those portions of scriptons, which lie beyond the first transcription-terminating lesion and up to the natural terminator, fail to be transcribed.

RELEASE OF RNA POLYMERASE AT THE TRANSCRIPTION-TERMINATING UV LESION

RNA polymerase is not trapped at transcription-terminating UV lesions but released with a short delay. This has been demonstrated in vitro for bacterial (*3*) and bacteriophage-induced RNA polymerases (Sauerbier and Millette, unpublished) in reactions utilizing excess template DNA and allowing for 8 to 10

RNA chain initiations per RNA polymerase molecule. The same conclusion has been arrived at for in vivo transcription of UV irradiated bacterial DNA (*9,10*) and UV irradiated mammalian DNA (nucleolar transcription, ref. *10*).

From the in vitro observations with T4 DNA (*3*) and T7 DNA (Sauerbier and Millette, unpublished), and in vivo observations with mammalian cells (*10*) we know that every polymerase molecule is released directly by the transcription-terminating UV lesion with a probability of 95 percent, or greater. We can not rule out that some 5 per cent, or less, of the polymerase molecules may be arrested at certain UV lesions and eventually proceed to elongate the RNA up to a transcription-terminating UV lesion and are released thereafter.

Miscoding in the Process of Transcribing UV Irradiated DNA Templates

The preceding observations do not exclude the existence of photochemically altered DNA bases that lie ahead of the transcription-terminating UV lesion and which might lead to a missense, or nonsense message in the messenger RNA (mRNA). This event, if it occurred at all, would have to be infrequent since gene activity (as measured by gene complementation) and gene-specific protein synthesis (as measured by ^{14}C-amino acid labeling) are lost following UV irradiation of template DNA at the same rates (*16*). Also, gene-specific RNA synthesis and gene-specific protein synthesis are UV inactivated in T7 at the same rates, which rules out the possibility that nonsense base sequences appear at a measurable frequency in the mRNA transcribed from UV-irradiated DNA (*17*).

The Nature of the Transcription-Terminating UV Lesion(s)

We do not know much about these lesions except that, in the ribosomal RNA cistrons of *E. coli*, about 70 per cent of the transcription-terminating UV lesions can be photoreactivated (*10*). Therefore, most of them seem to be pyrimidine dimers.

Transcription-terminating UV lesions are not repaired for several hours after UV irradiation in the nucleoli of mouse L cells and of HeLa cells (*10*).

A Model for the Transcription of UV-Irradiated DNA Templates

A very simple picture of the UV effects on the transcriptional functions of template DNA has emerged from the preceding studies. Fig. 1 presents its predominant features. Inspection of the diagram (Fig. 1) shows immediately that the probability for any gene to be transcribed in its entirety from UV-irradiated DNA is inversely proportional to the distance between the 5'-terminus of the gene (codogenic strand) and its promotor. It is, also, obvious that specific gene expression in a large population of homologous chromosomes will be UV inactivated with pseudo-first order kinetics provided that each undamaged gene has a constant probability of being expressed; in other words, that the rate of gene-specific RNA and protein synthesis correctly reflects the number of undamaged gene copies. We, therefore, have found a technique to induce the desired correlation between the rate of gene expression and the transcriptional distance. The only further requirement for mapping transcriptional linkages through UV irradiation of the template DNA is a quantitative assay for the gene products (RNA or proteins).

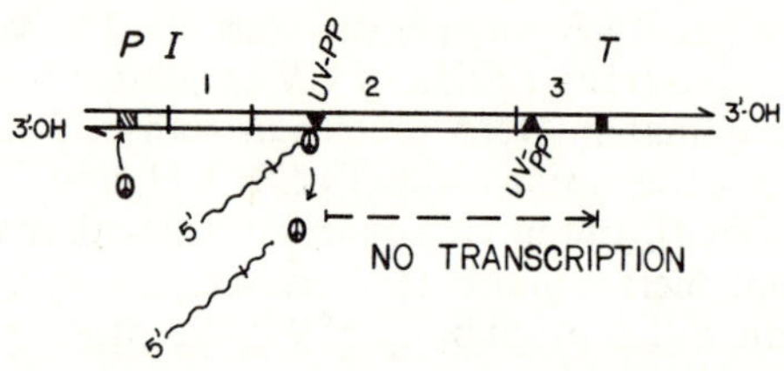

FIG. 1. Transcription from a UV-damaged scripton. Solid lines represent the strands of template DNA. P denotes the promotor, or RNA polymerase binding site, I the site for the initiation of RNA chains, T the natural terminator site, and UV-PP the transcription-terminating UV photoproducts. At the site of a transcription-terminating photoproduct, indicated by a filled wedge, transcription terminates and both RNA chain and RNA polymerase are released.

UV STUDIES ON THE TRANSCRIPTIONAL ORGANIZATION OF PRO- AND EUKARYOTIC CELLS

The first correlation of the UV sensitivity of gene expression with gene location on a chromosome was made by Starlinger and Kolsch (*18*) in the galactose operon of *E. coli.* After the transcription-terminating nature of UV lesions became known (*3*, *9*), the ribosomal RNA cistrons in yeast were mapped through the UV sensitivities of their expression by van den Bos, Retel and Planta (*19*). Later on, Hausmann and Härle (*20*) and Herrlich, Rahmsdorf, Pai and Schweiger (*21*) used the same concept in mapping the transcriptional sequence of some of the early genes of T3 and T7. In spite of these encouraging successes of "UV mapping" of transcriptional linkage, the usefulness of the method has not been generally realized. It was obvious from our in vitro studies (*3*) that mapping the transcriptional organization of chromosomes through UV dose-effect observation had the potential of becoming an accurate and versatile method. We, therefore, decided to test its accuracy and the assumptions inherent in its in vivo application (*10,11,16,17*).

UV Mapping of Scriptons on Bacteriophage T7

Our problem was to apply an "uncalibrated yardstick" to measuring an undetermined distance. This task could certainly be facilitated by using many reference points and by making overlapping measurements back to a common point. For calibrating our "yardstick" we selected T7 DNA since molecular biologists had provided us with the total length of the T7 chromosome, with the length of the early region of this chromosome, and with the approximate positions of about 28 of the T7 genes (see abscissa of Fig. 4). Furthermore, the sizes of most T7-induced proteins and the sizes of the mature T7 early RNAs were known (*22–24*). The transcriptional linkage was undetermined at that time. We measured the UV-induced loss of T7 early gene expression in multiply-infected, dark-repair-deficient, *Escherichia coli* B_{s-1} (*25*) by quantifying the T7 early proteins through ^{14}C-amino acid labeling and autoradiography. We observed pseudo-first order loss of expression for each of the 4 gene products that can be readily resolved on polyacrylamide gels (Fig. 2). When the UV inactivation cross sections for the T7 early genes are plotted vs their map position it is seen that all points fall onto a straight line which intercepts the ordinate at, or very close to,

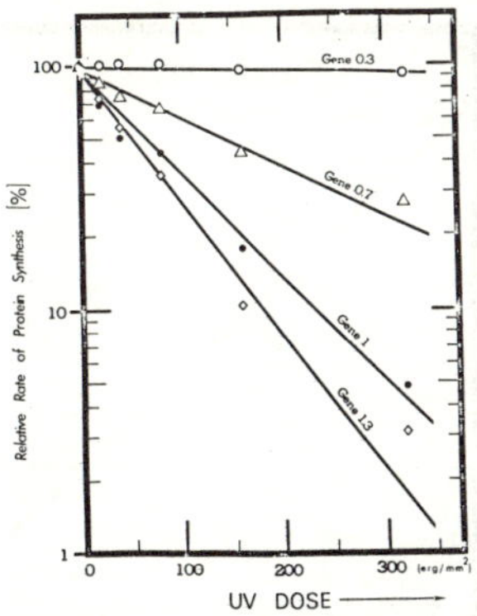

FIG. 2. Loss of synthesis of T7 early proteins in multiply-infected *E. coli* B_{s-1} as a function of UV dose to the infecting phage. T7 proteins were [^{14}C]-labeled, separated on SDS-polyacrylamide gels, autoradiographed and quantified by densitometry of autoradiograms. Ordinate: normalized rate of sythesis of individual T7 early proteins. Abscissa: UV dose to the infecting phage. [From Bräutigam and Sauerbier *J. Virol.* **12**, 882 (1973) with permission of ASM Publications Office.]

the origin (Fig. 3). It is, therefore, shown that the UV sensitivity of each gene is almost exclusively determined by its distance from the left hand end of the T7 chromosome. This was shown to be the promotor site for in vivo and in vitro transcription of the T7 early genes (*26,27*). By using a T7 with a deletion of the early region terminator (T7am342, LG37, ref. *22*) we have expanded the scripton length from the 7,200 nucleotide pairs of the early region to 36,000 base pairs, the entire T7 chromosome. Now the predictions of the transcription model could be tested in this immensely long scripton by assaying for the expression of some 15 genes. Fig. 4 shows the UV inactivation cross sections for all T7 genes that are readily identifiable on SDS-polyacrylamide gels. Again, the values for the UV cross sections fall onto a straight line with the same slope as was displayed by the early region genes (Fig. 4).

The same type of experiment was performed for the T7 early region genes, only this time the quantifying of gene expression was done at the level of RNA synthesis (*17*). The UV inactivation cross sections for the early T7 genes observed in these experiments are the same as those determined from the protein assays (*11,17*).

Two major conclusions can be drawn from the preceding observations with T7. (1) UV mapping of transcriptional linkage can be successfully applied to the genes of T7 (and as we know, to the genes of any other organism tested so far). (2) In spite of multiple bacteriophage infection (many homologous chromosomes per cell) the rate of gene expression correctly reflects the number of undamaged gene copies in the same cell (tested for multiplicities up to 10).

Applying UV mapping to transcription of the T7 late region by T7-induced RNA polymerase (gene 1.0 product) indicates that there are some independent scriptons and there are some overlapping scriptons which have different promotors but a common terminator (Bräutigam and Sauerbier, unpublished; the latter observations agree with in vitro experiments of Golomb and Chamberlin, ref. *28*).

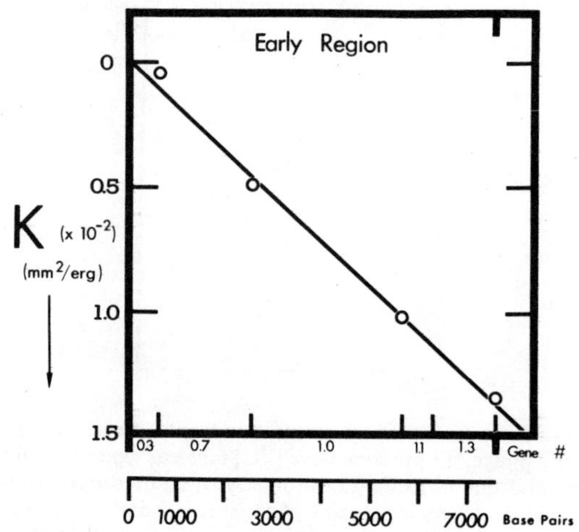

FIG. 3. UV cross sections for the inactivation of expression of T7 early genes vs map position of genes. Data from Fig. 2. [From Bräutigam and Sauerbier *J. Virol.* **12,** 882 (1973) with permission of ASM Publications Office.]

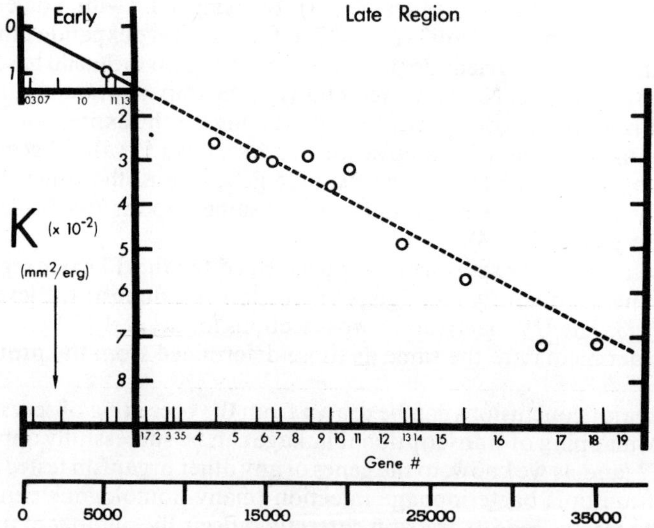

FIG. 4. UV cross sections for the inactivation of expression of T7 early and late genes vs position on the genetic map. In these experiments both early and late genes were transcribed by host RNA polymerase since the infecting phage was deficient in synthesizing T7 RNA polymerase and carried a deletion of the early region terminator (mutant T7 am 342, LG37).[From Bräutigam and Sauerbier *J. Virol.* **12,** 882 (1973) with permission of ASM Publications Office.]

UV Mapping of Scriptons in Bacteriophage T4

The early regions of the T4 chromosome are much more complex than the T7 early region. They are comprised of over 50 genes, whose expression is controlled by a number of control devices which, thus far, have defied description in molecular terms. To this complicated system, UV mapping introduces the potential of studying the transcriptional linkage of any given T4 gene at any time during the intracellular development of the bacteriophage. Through such experiments a switch from distant, "immediate early" promotors to proximal "delayed early" promotors was discovered for the genes 43 and 45 by Kathleen Hercules (*29*) (see Fig. 5). It was, further, demonstrated through her experiments that some mutants of T4 (tsGl, Tom Mattson, unpublished; far P14, Dwight Hall, ref. *30*) indeed fail to activate the delayed early promotors under nonpermissive conditions, as was suspected by Tom Mattson (personal communication) from timing of gene expression. Thus, UV mapping in T4, in conjunction with the use of genetically altered phage, will eventually provide the information on transcriptional regulation which is necessary for elucidating the mechanics of T4 gene regulation on the transcriptional level.

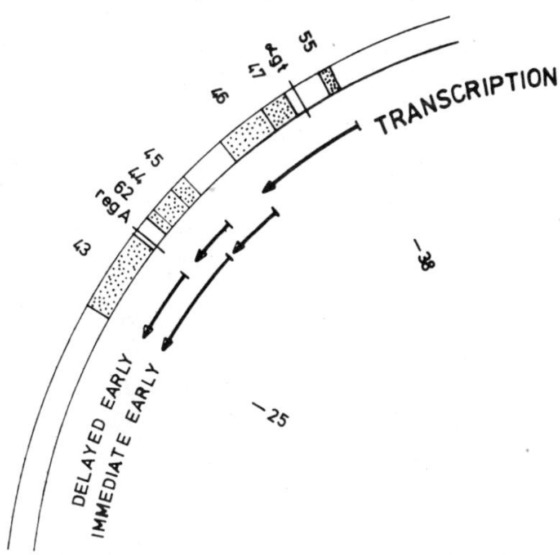

FIG. 5. Immediate early and delayed early modes of transcription in the early region of the bacteriophage T4 chromosome as determined by ultraviolet light mapping. Arrows indicate co-transcribed cistrons in the direction of transcription. [From Hercules and Sauerbier *J. Virol.* **12**, 872 (1974) with permission of ASM Publications Office.]

Mapping the tRNA Scriptons in T4

These mapping studies are performed independently by Judith Olah at Denver and by Sarabhai and Abelson at San Diego. I want to present their results to indicate the lower limit of resolution of the UV mapping technique. The objective is to resolve the transcriptional sequence of adjacent cistrons which are each about 80 nucleotide pairs long. They are, thus, about 400 times shorter than the longest scripton we have tested (the complete T7 chromosome). It is found that the transcriptional organization of the T4 tRNA region can be mapped by the UV technique. The tentative transcriptional order in vivo is indicated in Fig. 6. From this experience we believe that the transcriptional sequence of two adjacent cistrons of about 100 nucleotide pairs each can be unambiguously determined by UV mapping, if these cistrons are rather close to a promotor. The transcriptional sequence of two such small units is no longer resolved by UV mapping when both units are 500 or more nucleotide pairs away from the common promotor. (It should be kept in mind that we usually apply UV mapping to genes 600 to 3,000 nucleotide pairs long.)

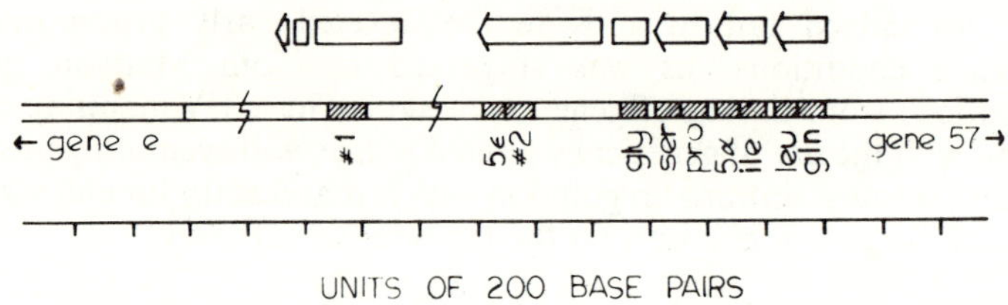

FIG. 6. Tentative transcriptional organization of the transfer RNA cistrons of bacteriophage T4. Scripton assignment through ultraviolet light mapping was made by Judith Olah (unpublished) and by Sarabhai and Abelson (personal communcation). Arrows indicate direction of transcription and length and positions of scriptons. Uncertainties in scripton assignments are indicated by interruptions of arrows.

The Transcriptional Organization of the Ribosomal RNA Genes in Mouse L Cells and in HeLa Cells

Very little is known about the transcriptional organization of the mammalian cell chromosomes, and introduction of UV mapping to cultured mammalian cells could lead to a major breakthrough in this area. Hackett (*10*) has found that the effects of untraviolet radiation on the transcription in mammalian cells resemble those described for prokaryotic systems, i.e., the predominant effect appears to be premature termination of RNA chains and a deletion of transcription of promotor-distal portions of UV-damaged scriptons. He has, therefore, utilized his findings to investigate the transcriptional organization of the ribosomal RNA genes in mouse L cells and in HeLa cells. He arrives at the following conclusions:

(1) Each 45S precursor RNA (from which the mature 18S and 28S rRNAs are produced by cleavage and processing) is transcribed from an independent scripton with a promotor and terminator of its own. This contrasts with the prevailing view that several hundred cistrons for the 45S RNA are arranged in tandem behind a common promotor (*31*).

(2) The mature 18S and 28S rRNAs of mouse L cells and of HeLa cells are arranged on the 45S precursor RNA in the order of 18S, then 28S in the direction of transcription (Fig. 7) This again is in conflict with the prevailing view that 28S is first and 18S is next (*32*).

These corrections of our knowledge of the transcriptional arrangements in mammalian cells, as achieved through the UV mapping technique, are perfect illustrations of the need for and the potential of this new technique for the study of transcriptional linkage in eukaryotic systems. Naturally, the technique should be used to investigate the transcriptional organization of some of the heterogeneous nuclear RNA molecules.

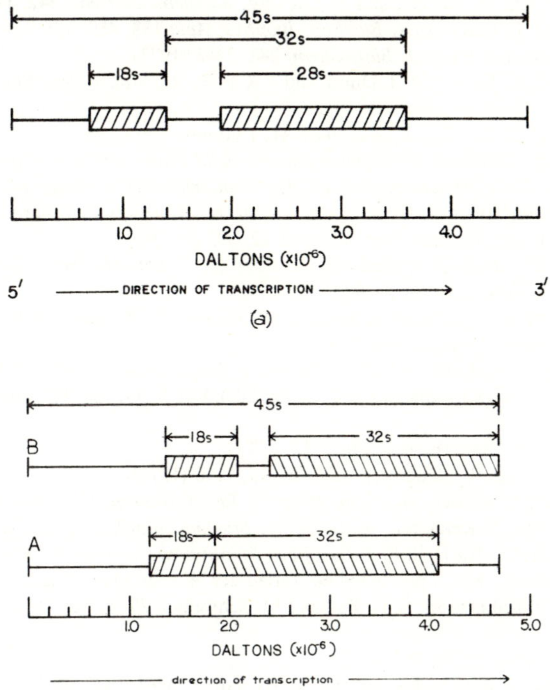

FIG. 7. (a) Location of 18S and 28S rRNA on the mouse L cell 45S ribosomal precursor as determined by UV mapping. The locations of 18S and 28S rRNAs are shown by the hatched areas, the nonconserved RNA by the solid lines. The direction of transcription has been assumed to be 5' to 3'. [Data from P. B. Hackett, ref. *10*; See also P. B. Hackett and W. Sauerbier, *J. Mol. Biol.* **91**, 235 (1975).]

(b) Location of the 18S and 28-32S rRNA on the HeLa cell ribosomal RNA transcription unit. Two possibilities for the arrangment of the rRNAs (hatched areas) and nonconserved spacer RNAs (solid lines), based on UV inactivation data, are shown in maps A and B. In constructing both maps two constraints were utilized (1) the 45S precursor RNA is 4.7×10^6 daltons and (2) the ratio of target sizes of the cistrons coding for 18S and 28-32S is $K_{28-32}/K_{18}=2.2$. [From P. B. Hackett, ref. *10*.]

REFERENCES

1. O. W. Jones and P. Berg, *J. Mol. Biol.* **22**, 199 (1966).
2. U. Hagen, M. Ullrich, E. E. Peterson, E. Werner, and H. Kröger, *Biochim. Biophys. Acta* **199**, 115 (1970).
3. W. Sauerbier, R. L. Millette, and P. B. Hackett, Jr., *Biochim. Biophys. Acta* **209**, 368 (1970).
4. U. Hagen, K. Keck, H. Kröger, F. Zimmermann, and T. Lücking, *Biochim. Biophys. Acta* **95**, 418 (1964).
5. F. K. Zimmermann, H. Kröger, and T. Lücking, *Biochem. Zeitschr.* **342**, 115 (1965).
6. A. Ishihama and T. Kameyama, *Biochim. Biophys. Acta* **138**, 480 (1967).
7. M. Chamberlin and J. Ring, *J. Biol. Chem.* **248**, 2245 (1973).
8. R. A. Burgess, A. A. Travers, J. J. Dunn, and E.K.F. Bautz, *Nature New. Biol.* (London) **221**, 43 (1969).
9. H. Michalke and H. Bremer, *J. Mol. Biol.* **41**, 1 (1969).
10. P. B. Hackett, Jr., *Inhibition of Deoxyribonucleic Acid Transcription by Ultraviolet Irradiation in Mammalian Cells: Determination of the Transcriptional Linkage of the 18S and 28S Ribosomal Ribonucleic Acid Genes.* Doctoral thesis, University of Colorado, 1974.
11. A. R. Bräutigam and W. Sauerbier, *J. Virol.* **12**, 882 (1973).
12. W. Sauerbier, in *Advances in Radiation Biology* (J. T. Lett and H. I. Adler, Eds.), **Vol. 6**. Academic Press, New York (in press).
13. K. Nozu, in *Aspects of Cellular and Molecular Physiology* (K. Hamaguchi, Ed.), **Vol. 19**, pp. 147–165. University of Tokyo Press, 1972.
14. S. E. Bresler, V. L. Kalinin, and D. A. Perumov, *Biopolymers* **2**, 135 (1964).
15. R. B. Painter, in *Photophysiology* (A. C. Giese, Ed.), **Vol. 5**, pp. 169–189. Academic Press, New York, 1970.
16. K. Hercules and W. Sauerbier, *J. Virol.* **12**, 872 (1973).
17. A. R. Bräutigam and W. Sauerbier, *J. Virol.* **13**, 1110 (1974).
18. P. Starlinger and E. Kölsch, *Biochem. Biophys. Res. Commun.* **17**, 508 (1964).
19. R. C. van den Bos, J. Retel, and R. J. Planta, *Biochim. Biophys. Acta* **232**, 494 (1971).
20. R. Hausmann, and E. Härle, in *First European Biophysical Congress* (E. Broda, A. Locker, and H. Springer-Lederer, Eds.), pp. 467–488. Verlag der Wiener Med. Akad., 1971.
21. P. Herrlich, H. J. Rahmsdorf, S. H. Pai, and M. Schweiger, *Proc. Nat. Acad. Sci. USA* **71**, 1088 (1974).
22. F. W. Studier, *Science* **176**, 367 (1972).
23. W. C. Summers, T. Brunovskis, and R. W. Hyman, *J. Mol. Biol.* **74**, 291 (1973).
24. J. J. Dunn and W. F. Studier, *Proc. Nat. Acad. Sci. USA* **70**, 3296 (1973).
25. R. F. Hill, *Biochim. Biophys. Acta* **30**, 636 (1960).
26. R. W. Hyman, *J. Mol. Biol.* **61**, 369 (1971).
27. R. L. Millette, C. D. Trotter, P. Herrlich, and M. Schweiger, *Cold Spring Harbor Symp. Quant. Biol.* **35**, 135 (1970).
28. M. Golomb and M. Chamberlin, *Proc. Nat. Acad. Sci. USA* **71**, 760 (1974).
29. K. Hercules, and W. Sauerbier, *J. Virol.* **14**, 341 (1974).
30. J. R. Johnson and D. H. Hall, *J. Virol.* **13**: 666 (1974).
31. R. P. Perry and D. E. Kelley, *J. Cell. Physiol.* **76**, 127 (1970).
32. P. K. Wellauer and J. B. Dawid, *Proc. Nat. Acad. Sci. USA* **70**, 2827 (1973).

THE EUKARYOTIC CHROMOSOME[1]

Chairman: B. J. McCarthy

[1]Symposium E–9

The Replication of *Drosophila* DNA: The Periodic Distribution of Replication Origins[1]

Alan B. Blumenthal
Laboratory of Radiobiology
University of California
San Francisco, California 94143, USA

INTRODUCTION

Eukaryotic chromosomes may contain only one very large DNA molecule, up to many centimeters in length (*1–3*). Completion of DNA replication in a 6–10 hr S phase involves replication from many origins along the molecule (*4–6*). In HeLa and Chinese hamster cells, an estimated $10^4 - 10^5$ origins for replication are active at a given instant in S phase (*5*) and some origins used during one S phase of a cell may not be used in subsequent S phases of its progeny (*7*). Similarly, Callan (*6*) has shown that in cells of the same organism, different numbers of replication origins may be active in S phase and this number seems to be larger with shorter S phases. The objectives of the present study were to demonstrate the factors affecting the length of S phase and the principles of the linear organization of replication origins in chromosomal DNA of the fruit fly, *Drosophila melanogaster*.

EXPERIMENTAL MATERIALS, METHODS, AND DEFINITIONS

The two cell types from *Drosophila melanogaster* used in this work were early embryos (0–2 hr after fertilization) and somatic cells in culture. The cultured cells are a continuous line (No. 2) derived by Schneider (*8*) from well-differentiated *Drosophila* embryos just before they hatched to first instar larvae. Early embryonic nuclei divide in a syncytium with a doubling time of 9–10 min at 24° C. S phase is estimated from the length of interphase to be ≤ 3.4 min (*9*). The cultured cells, however, have a doubling time of about 24 hr at 25° C and an S phase of about 10 hr (*12*). In both cases, the normal diploid DNA complement is replicated, though the length of the S phases differs by a factor of about 200.

DNA isolated from early embryos was observed by electron microscopy to have closely spaced replicating structures (see ref. *10* for details). However, replication sites occurred too infrequently in cell culture DNA to be observed by electron microscopy, so replicating DNA was visualized by DNA fiber autoradiography. In this method, replicating DNA is first labeled with ^{3}H-thymidine and then released from the cell by detergent lysis and drawn into linear fibers on a glass slide. An overlay of photographic emulsion reveals the sites of

[1]Congress presentation E–9–1.

DNA replication after long exposure to the ^{3}H-containing DNA (see ref. *11* for details).

The unit of replication, called a replicon by other authors, will be defined here as the segment of DNA duplicated from a single origin by two fork-like structures moving in opposite directions from the origin. Lengths of DNA will be specified in kilobases (kb), referring to 1000 bases or base pairs of single- or double-stranded DNA, respectively. Three kb is approximately equal to 1 μm of DNA.

SUMMARY OF PREVIOUS RESULTS

Embryo DNA (10, 11)

Drosophila DNA molecules were frequently observed to contain several bubbles, or "eyes" (Fig. 1). These eyes were composed of two branch points, or

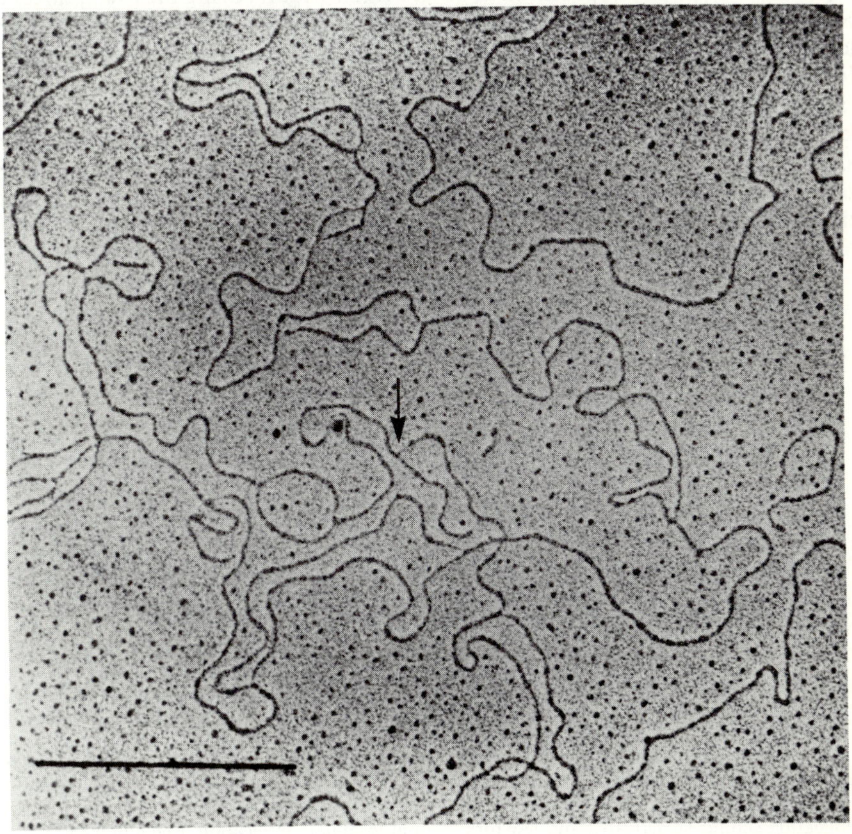

FIG. 1. Electron micrograph of replicating DNA purified from early *Drosophila* embryos. The DNA sample was spread in a hyperphase of 40% formamide revealing single-stranded regions in the DNA next to some of the replication forks (see arrow). The marker is 3 kb. [Photograph courtesy of H. J. Kriegstein, Stanford University.]

"forks," and two equal-length, double-stranded daughter DNA molecules which contained the newly replicated DNA. When these molecules were treated with 40% formamide (Fig. 1), single-stranded regions were revealed next to the forks of some of the eyes. These regions indicate a process of discontinuous DNA strand elongation. Analysis of these single-stranded regions (*10*) demonstrated that the eyes were produced by DNA replication which had proceeded bidirectionally from an origin within the eye. In other words, DNA synthesis occurs at both forks of an eye.

Eye lengths and the distances between centers of neighboring eyes (eye-to-eye distances) determined from many such molecules gave means of 4.1 kb and 9.7 kb, respectively. These values represent molecules found all through the 3.4-min S phase, some just beginning replication and others nearly completed. Analysis of molecules with respect to their extent of replication (i.e., the percentage of a DNA fragment occurring in eye structures), or progress through S phase, showed that replication at most origins began during the first quarter of S phase. Eyes expanded linearly with time through the middle half of S and then lengthened rapidly by merging with each other in the last quarter of S. Merger of forks from adjacent eyes appeared to result in termination of their activity. The rate of eye expansion in the middle half of S gave a mean replication rate of 2.6 kb/min/fork and a mean origin-to-origin distance of 7.9 kb, in contrast to the slightly larger 9.7-kb eye-to-eye distance determined empirically for the whole S-phase population. Most of the embryo DNA replicated in about 1.25 min, though a few origins were far enough apart (about 20 kb) to require about 3.7 min to complete replication.

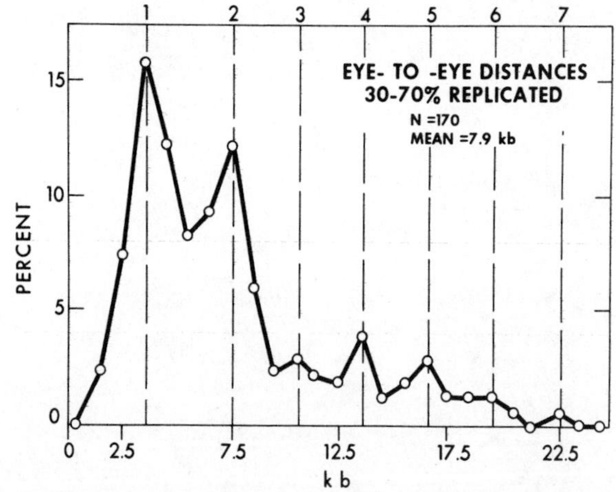

FIG. 2. Distribution of eye-to-eye distances of replicating DNA from early *Drosophila* embryos. These measurements were derived from DNA molecules which were 30–70% replicated; that is, 30–70% of their length was occupied by eyes. In this part of the S phase, few origins of replication should be activated and few forks should terminate replication. The numbered dashed lines represent multiples of 3.4 kb. The interval for this histogram is 1 kb. [Reprinted from A. B. Blumenthal, H. J. Kriegstein and D. S. Hogness, *Cold Spring Harbor Symp. Quant. Biol.* **38**, 205 (1974) with the permission of the Cold Spring Harbor Laboratory.]

When eye-to-eye distances were considered from molecules selected from a limited part of S phase during which eye growth was not affected by mergers (e.g., 30–70% replicated) (Fig. 2), the histogram showed a striking periodicity based on multiples of about 3.4 kb. This indicates that replication origins were not randomly distributed about a mean of 7.9 kb but were spaced at intervals of 3.4 n, where n was observed to be as large as 7.

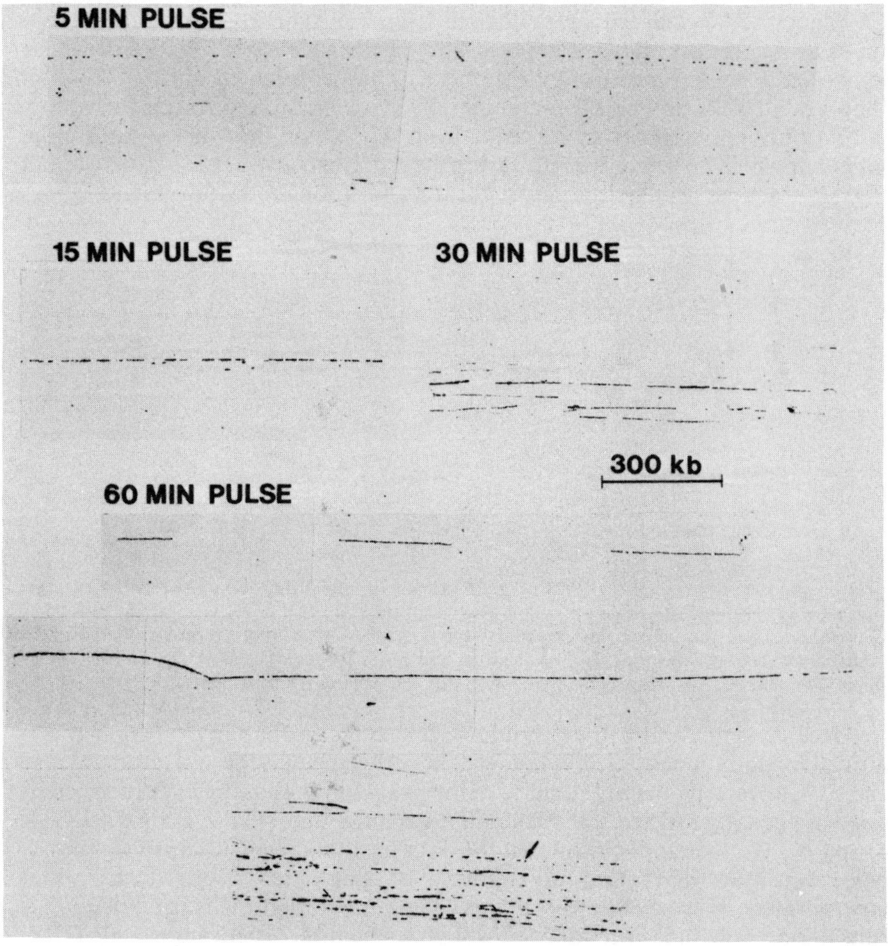

FIG. 3. Autoradiographs of DNA derived from *Drosophila* cells in culture showing grain track patterns from 5- , 15- , 30- , and 60-min samples. Cells in exponential growth in a suspension culture (2.9 × 10^6 cells/ml) were pulse labeled with ^{3}H-thymidine (New England Nuclear, 200μCi/ml, 55 Ci/mole) for 5, 10, 15, 30, and 60 min. The 60-min sample shows the distinction between the grain densities resulting from one labeled daughter strand of an eye and those from two which have collapsed together during spreading. The arrow in the bottom photo indicates small grain tracks and c-c distances after a 60-min pulse. In the same field are the longer grain tracks and c-c distances expected from complete replication of a cluster of origins (see text). The AR 10 film was exposed for 7.5 months. The grain density for one labeled daughter molecule was about 3.3 grains/μm. The lowest grain density observed in the various experiments (Table 1) was 1.7 grains/μm.

Cell Culture DNA (11)

Autoradiographs of cell culture DNA show "grain tracks" occurring in linear arrays, reflecting the tandem, linear distribution of replication units in the molecules (Fig. 3). The grain tracks, produced by ^{3}H incorporated into the replicating DNA, are separated by gaps containing only background grains.

Since the labeled cells were derived from an asynchronous culture, grain tracks arose from replication units which had been initiated before the pulse began, had been initiated during the pulse, or had terminated during the pulse. Thus, grain tracks can arise from the activity of one or two replicating forks (Fig. 4), and only grain tracks which arise from replication units active before the pulse, but which do not terminate during the pulse, can give a true estimate of μm of grain track/min of pulse time/replication fork, or replication rate.

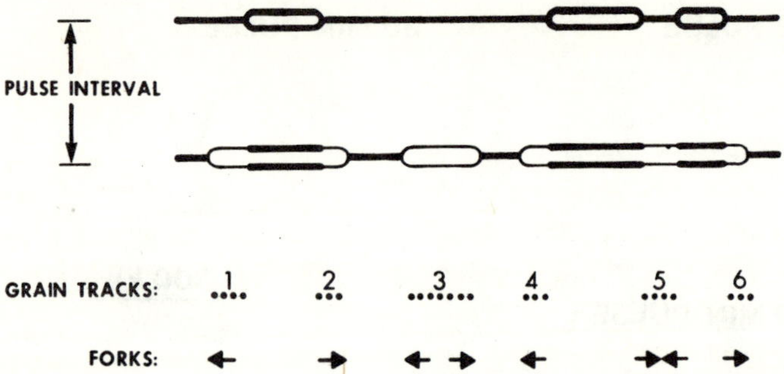

FIG. 4. Illustration of how a single grain track can arise from one or two forks during a pulse label of ^{3}H-thymidine. The heavy lines represent the state of the DNA before the pulse with ^{3}H-thymidine, and the lighter lines represent DNA synthesized during the pulse interval. [Reprinted from A. B. Blumenthal, H. J. Kriegstein and D. S. Hogness, *Cold Spring Harbor Symp. Quant. Biol.* **38**, 205 (1974) with the permission of Cold Spring Harbor Laboratory.]

Reducing the pulse time should decrease the probability of grain tracks formed by initiation or termination events during the pulse (Fig. 4, grain tracks 3 and 5). Conversely, it should increase the probability of grain tracks formed by single replicating forks active throughout the pulse time (Fig. 4, grain tracks 1, 2, 4, and 6). We therefore considered the data resulting from successively shorter pulse times to determine the activity of single replicating forks which incorporated ^{3}H-thymidine throughout the pulse period. This is analogous to considering only those eyes active through the middle of S in embryonic DNA, when few initiation events or mergers occur.

Measurements of many autoradiographs for each pulse time gave mean values for grain track lengths and distances between the centers of adjacent grain tracks (c-c distances). Extrapolation of these data to zero pulse time (see Fig. 11 of Ref. *11*) gave an average replication rate estimate of 2.6–3.2 kb/min/fork and a mean eye-to-eye distance of about 40 kb.[2]

If all the DNA of *Drosophila* chromosomes were replicated concurrently

[2]We can directly measure eye-to-eye distances from electron microscope pictures of replicating embryonic DNA but for DNA autoradiographs we must derive eye-to-eye distances from c-c distances.

from origins 40 kb apart at 2.6 kb/min/fork (or 5.2 kb/min/eye), the chromosomes would be replicated in about 8 min. Since this is not the case (S ≅ 10 hr), it was necessary to hypothesize clusters of origins, containing an average origin spacing of 40 kb, in which the individual origins of a cluster were simultaneously active.

If clusters of origins about 40 kb apart completed replication in less than 10 min and the individual eyes of a cluster merged to form a giant eye, then a ^{3}H-thymidine pulse longer than 10 min would produce mostly long grain tracks, representing replication of the whole cluster. The c-c distances would be large because of both the long grain tracks and the spacing between the clusters. In fact, with much longer pulse times (30, 60, or 75 min), most of the small c-c distances observed would have arisen from clusters which had incorporated ^{3}H-thymidine during only part of their active period (less than 10 min). These clusters would have either terminated replication early (e.g., Fig. 4, grain track 5) or initiated replication late (e.g., Fig. 4, grain track 3) in the pulse interval. The resulting grain tracks would indicate the fork merging site between neighboring origins or the initiation sites of replication. C-c distances, then, would give actual eye-to-eye and origin-to-origin distances within a cluster.

The histogram of c-c distances after a 75-min pulse (Fig. 10 of Ref. *11*) shows two peaks at the low end of the distribution with means of 28 and 56 kb. As in embryonic DNA, origins of replication may be spaced at periodic intervals of 28 n kb, where only n = 1 and n = 2 are observed in the data. (The weighted mean of these two peaks is about 38 kb, which is close to the earlier estimate of 40 kb for mean eye-to-eye distance.)

ADDITIONAL RESULTS

Autoradiographic experiments were repeated with asynchronous cells growing in suspension culture (Fig. 3) rather than surface cultures as in the previous experiments. This was done to see if changed growth conditions, selection of autoradiographs, or longer exposure times would affect the data. When grain track lengths and c-c distances were measured (Fig. 5), mean grain track lengths were similar to those obtained in the earlier experiment (Table 1), even when the exposure time was increased to 7.5 months. However, the mean c-c distances for the two experiments cannot be compared because the data in Experiment 2 were selected to contain long gaps between autoradiographs. This selection gave rise to high mean c-c distances. Whereas the averages of c-c distances were heavily weighted by a number of larger values, the histograms for longer pulse times (Fig. 5, 30 and 60 min) showed small c-c distances, which appeared in the autoradiographs (Fig. 3, bottom panel) presumably from labeling only at the beginning or end of the pulse time. These small c-c distances formed several peaks with average values of 34, 63, and 94 kb for the 30-min data and 32 and 72 kb for the 60-min data. This indicates an average repeat period for origins of replication of about 33 kb, slightly higher than the 28 kb derived from Experiment 1.

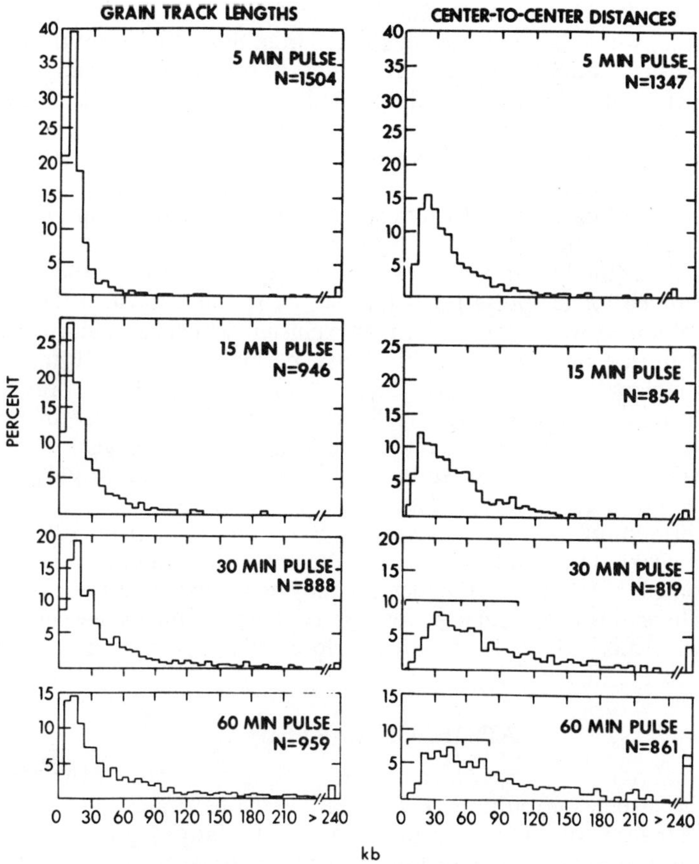

FIG 5. Histograms of grain track lengths and center-to-center (c-c) distances of adjacent grain tracks after the indicated pulse labeling times. The data were pooled from Experiments 2 and 2a (see Table 1). The data in these histograms are a random sample of the data used to calculate the averages shown in Table 1. The brackets shown in histograms of 30- and 60-min c-c distance data indicate the areas selected for computation of average peak values (see text).

DISCUSSION

Determination of S-Phase Length

The 100- to 200-fold difference in the time needed to replicate chromosomal DNA (Table 2) is not accounted for by the difference in fork replication rate (1- to 2-fold) or the difference in spacing of active origins (about 4-fold). The long S phase of cell culture DNA could be explained (a) by the delayed activation of clusters of origins during S; (b) by large distances between clusters, the replication time for these long stretches of DNA determining the length of S; or (c) by a small number of origins, perhaps in heterochromatin or satellite DNA, which are very far apart (600 min $\times$ 2 $\times$ 2.6 kb/min $\cong$ 3100 kb) and require 600 min for replication. These alternatives cannot be clearly resolved with the available data.

TABLE 1
Average Grain Track Length as a Function of Pulse Labeling Time with ^{3}H-Thymidine

Pulse labeling time (min)	Mean grain track lengths[a,b]					
	Experiment 1		Experiment 2		Experiment 2a	
	kb	n	kb	n	kb	n
5	13.8	837	12.4	1205	13.2	534
10	—	—	15.9	1431	15.9	537
15	25.0	1134	20.3	635	20.3	401
30	32.6	427	35.0	786	35.9	229
60	—	—	52.0	450	40.3	235
75	48.8	283	—	—	—	—

[a]Experiment 1–surface culture, Experiments 2 and 2a–suspension cultures.
[b]Exposure times: Experiments 1 and 2 – 3.3 months, Experiment 2a – 7.5 months.

The replication of embryonic DNA in 3.4 min can be adequately explained by the simultaneous activation of these closely spaced origins at the onset of S phase and their replication at 2.6 kb/min/fork.[3] Though most replication units can be completed in less than 1.5 min, a few very long spacings between active origins could explain the 3.4-min S phase. By analogy, the S phase of cell culture DNA might be explained by the second or third alternatives above.

Origin Spacing and Chromosome Structure

The 4-fold difference in mean eye-to-eye distances (Table 2) for embryo and cell culture DNA implies that not all the origins used in embryonic DNA are used in cell culture DNA but that replication proceeds through presumptive origins. Callan (*6*) has observed similar effects in various amphibian cells. These observations imply that termination sites for replication [postulated by Huberman and Riggs (*4*)] cannot be obligatory signals in the DNA.

The periodic spacing of replication origins indicates an underlying periodicity in the chromosome structure. This idea is strengthened by the occurrence, in *Drosophila* salivary glands, of polytene chromosomes with condensed regions (bands) separated by smaller, uncondensed regions. Rudkin (*13*) demonstrated that the DNA content of these bands is 26–32 kb per haploid chromosome copy, and Keyl and Pelling (*14*) concluded that each band acts as a unit of replication. The correspondence of our numbers and these observations supports the idea of bands as replication units and led to a model (*11*) with the number and spacing of replication origins available in a given cell type related to the periodic packing of DNA (chromatin) into bands. Thus, we predicted that embryonic chromosomes are composed of condensed regions containing an average of 3.4 kb of DNA corresponding to the periodic spacing of origins observed in this DNA (Table 2). Not every 3.4- or 30-kb unit, however, would contain a functional replication origin, as shown by the spacings of origins at integral multiples of 3.4 or 30 kb.

This model further suggests that origins which can be used for replication

[3]The rate of 2.6 kb/min (or 0.9 μm/min) at 25° C, derived by independent methods for the two cell types, is similar to the replication rate for most mammalian cells grown at 37° C (*5*).

are in the interband or non-condensed regions of the chromosome and that condensation of chromatin occludes origins for replication. Heterochromatin, which is highly condensed, may then have few available origins and the distance between these origins might regulate the length of S phase.

Differentiated *Drosophila* cells (including cultured cells) contain large heterochromatic regions, but short-S-phase embryonic nuclei seem to have no heterochromatin (*15*). This supports our hypothesis and also raises the possibility that the organization of active replication units is related to the transcriptional activity of the DNA and the differentiation of this activity in specialized cell types.

TABLE 2
Summary of Data on Drosophila DNA Replication

DNA source	Nuclear doubling time	S phase (min)	Mean eye-to-eye distance (kb)	Mean origin-to-origin distance (kb)	Periodic origin spacing (kb)	Mean replication rate at 25° C (kb/min/fork)
Embryos (0–2 hr)	9–10 min	⩽3.4	9.7	7.9	3.4	2.6
Cell cultures	~24 hr	~600	40	–	28–33	2.6–3.2
Difference	~150	100–200	~4	–	8–10	1.0–1.2

ACKNOWLEDGMENT

Work supported by the Damon Runyan Memorial Fund and the U.S. Atomic Energy commission.

REFERENCES

1. C. D. Laird, *Chromosoma* **32,** 378 (1971).
2. R. Kavenoff and B. H. Zimm, *Drosophila. Chromosoma* **41,** 1 (1973).
3. T. D. Petes, C. S. Newlon, B. Byers, and W. L. Fangman, *Cold Spring Harbor Symp. Quant. Biol.* **38,** 9 (1974).
4. J. A. Huberman and A. D. Riggs, *J. Mol. Biol.* **32,** 327 (1968).
5. R. B. Painter and A. W. Schaefer, *J. Mol. Biol.* **45,** 467 (1969).
6. H. G. Callan, *Proc. Roy. Soc. London Ser. B* **181,** 19 (1972).
7. F. Amaldi, M. Buongiorno-Nardelli, F. Carnevali, L. Leoni, D. Mariotti, and M. Pomponi, *Exp. Cell Res.* **80,** 78 (1973).
8. I. Schneider, *J. Embryol. Exp. Morphol.* **27,** 353 (1972).
9. M. Rabinowitz, *J. Morphol.* **69,** 1 (1941).
10. H. J. Kriegstein and D. S. Hogness, *Proc. Natl. Acad. Sci. USA* **71,** 135 (1974).
11. A. B. Blumenthal, H. J. Kriegstein, and D. S. Hogness, *Cold Spring Harbor Symp. Quant. Biol.* **38,** 205 (1974).
12. S. Dolfini, A. M. Courgeon, and L. Tiepolo, *Experientia* **26,** 1020 (1970).
13. G. T. Rudkin, *Results Probl. Cell Differ.* **4,** 59 (1972).
14. H. G. Keyl and C. Pelling, *Chromosoma* **14,** 347 (1963).
15. A. P. Mahowald, *J. Exp. Zool.* **167,** 237 (1968).

Mammalian Chromosome Structure: Ultrastructural Aspects of Specialized Regions and Chromosome Aberrations[1]

B. R. Brinkley, Manley McGill, and Myles L. Mace
Division of Cell Biology, Department of Human Biological Chemistry and Genetics
The University of Texas Medical Branch at Galveston
Galveston, Texas 77550
and the Department of Biology,
The University of Texas M. D. Anderson Hospital and Tumor Institute at Houston
Houston, Texas 77025, USA

INTRODUCTION

The deleterious effects of various natural and synthetic agents such as X rays, UV, drugs, and viruses on the genetic apparatus of mammalian cells cannot be properly evaluated until a thorough knowledge of eucaryotic chromosome structure is attained. In spite of two decades of significant progress in cellular and molecular genetics, relatively little has been learned about the organization of genes in mitotic and meiotic chromosomes. Within recent years, however, new and significant progress has been made by both light and electron microscopic investigations. The purpose of this report will be to review some of these findings and to correlate them with existing concepts of chromosome architecture. Specific emphasis will be placed upon the structure of specialized chromosome regions such as the kinetochore, nucleolar organizer, and telomere. The structure and distribution of heterochromatin on mitotic and meiotic chromosomes will be considered along with some aspects of chromosome aberrations as evaluated by electron microscopy.

STRUCTURAL SIGNIFICANCE OF CHROMOSOME BANDING

Until recently, light microscopic studies contributed sparingly to knowledge of metaphase chromosome structure. Aside from providing information on arm length, location of kinetochores (centromeres), secondary constrictions, and satellites, very little could be derived from light microscopic investigations concerning the arrangement of DNA and associated nucleoproteins within the chromosome. The discovery of new staining procedures which produce specific banding patterns on metaphase chromosomes has essentially revolutionized the field of cytogenetics and provided important new insight into chromosome organization (see reference *1* for review). Although the chemical mechanism of most banding procedures is not understood, the techniques have served to identify two important features of metaphase chromosome structure: DNA base

[1]Congress presentation E-9-2.

sequences are arranged in a nonrandom order along the chromatid, and blocks of heterochromatin occupy specific sites on the chromosomes.

Among the first indications that specific DNA base sequences were arranged in bands along the metaphase chromosomes was provided by Caspersson and co-workers (*2–4*). By staining chromosomes with fluorescein-tagged alkylating agents (quinicrine mustard), they were able to detect light and dark bands by UV microscopy. Bands were present on all chromosomes examined, and each pair displayed a specific banding pattern. Their method, now known as the Q-band procedure, is thought to stain regions which are rich in adenine and thymine bases. Using indirect immunofluorescence, Dev et al. (*5*) demonstrated that antibodies to adenosine (Anti-A) produce a specific pattern of light and dark bands along the chromosome which was identical to the pattern produced by the Q-band procedure. More recently, AT rich regions of metaphase chromosomes have also been detected by indirect immunofluorescence using Anti-C follosing selective destruction of guanine residues with methylene blue (*6*). The Anti-C fluorescence pattern is the reverse of Q-banding or Anti-A banding. Thus, the interband regions of Q-banded chromosomes are stained after Anti-C immunofluorescence. The banding pattern provided by Anti-C is also identical to the so-called R-bands following special staining with Giemsa (*7*) for bright-field microscopy.

Other banding procedures have now been developed which give additional reference for the identification of metaphase chromosomes. Several cytological procedures produce banding patterns which are essentially identical to Q-banding but use Giemsa stain instead (*8–16*). The pattern which is known as G-bands can be quickly and simply observed by bright-field microscopy.

In addition to providing information on the arrangement of DNA and nucleoproteins within metaphase chromosomes, G-band patterns are excellent for identifying chromosome homologies between different species. As shown in Fig. 1a, three macrochromosomes shared by the domestic pigeon, chicken, and ring-necked dove can be identified by their G-bands (photograph courtesy of Dr. A. Dean Stock). Also, the method allows rapid identification of structural rearrangements such as deletions, duplications, inversions, and translocations.

Perhaps the most compelling evidence for the clustering of highly repetitious DNA base sequences and the localization of heterochromatin came from the classic experiment of Pardue and Gall (*17*). Using *in situ* hybridization, they were able to demonstrate that all of the satellite DNA of the mouse was localized in the centromere region. Since mouse satellite DNA is composed of highly repetitious base sequences and is associated with constitutive heterochromatin, their study was the first to provide evidence for the localization of heterochromatin at specialized regions of mitotic chromosomes. These investigators also noted that the treatment used in producing *in situ* hybrids led to specific centromeric staining when Giemsa was applied to the preparations. Subsequently, Arrighi and Hsu (*18*) perfected the centromeric staining method which is now known as the C-band procedure (Fig. 1b). A comparison of Q-, C-, and G-banding patterns of marker chromosomes of a human malignant melanoma cell line is shown in Fig. 2 (courtesy of Dr. T. R. Chen).

Although the molecular mechanism of Giemsa C-banding is not completely understood, it must involve some feature of the highly reiterated DNA in the

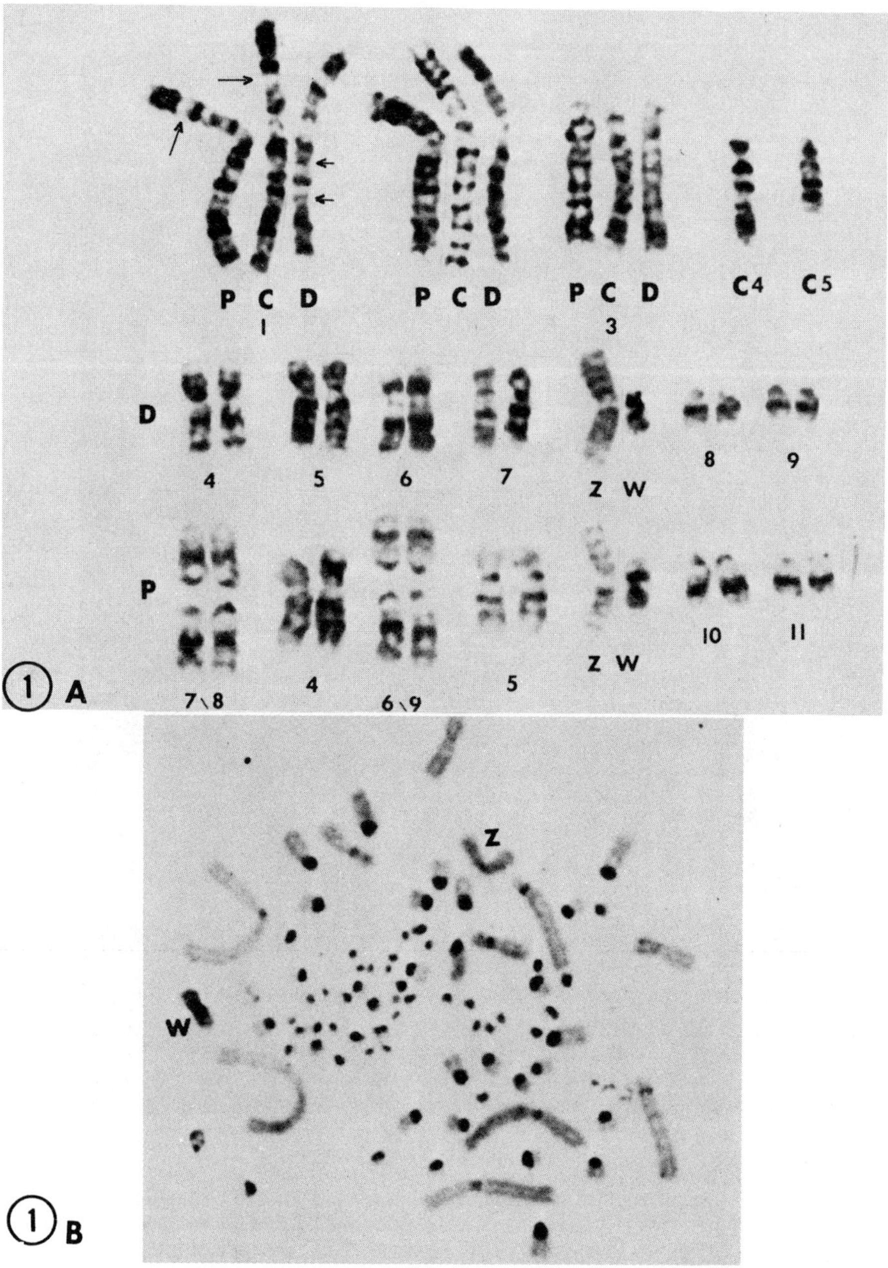

FIG. 1. *a*, G-bands in the major chromosomes of the chicken (C), dove (D), and pigeon (P). Top row compares the three largest chromosomes of the three species. Middle and bottom rows compare the macrochromosomes and largest microchromosomes of dove and pigeon, respectively. *b*, C-banded metaphase chromosomes of the pigeon. Note dark staining heterochromatin at the centromere of most chromosomes. [Courtesy of A. D. Stock, S. E. Arrighi and K. Stefos, *Cytogenet. Cell Genet.* **13**, 410 (1974).]

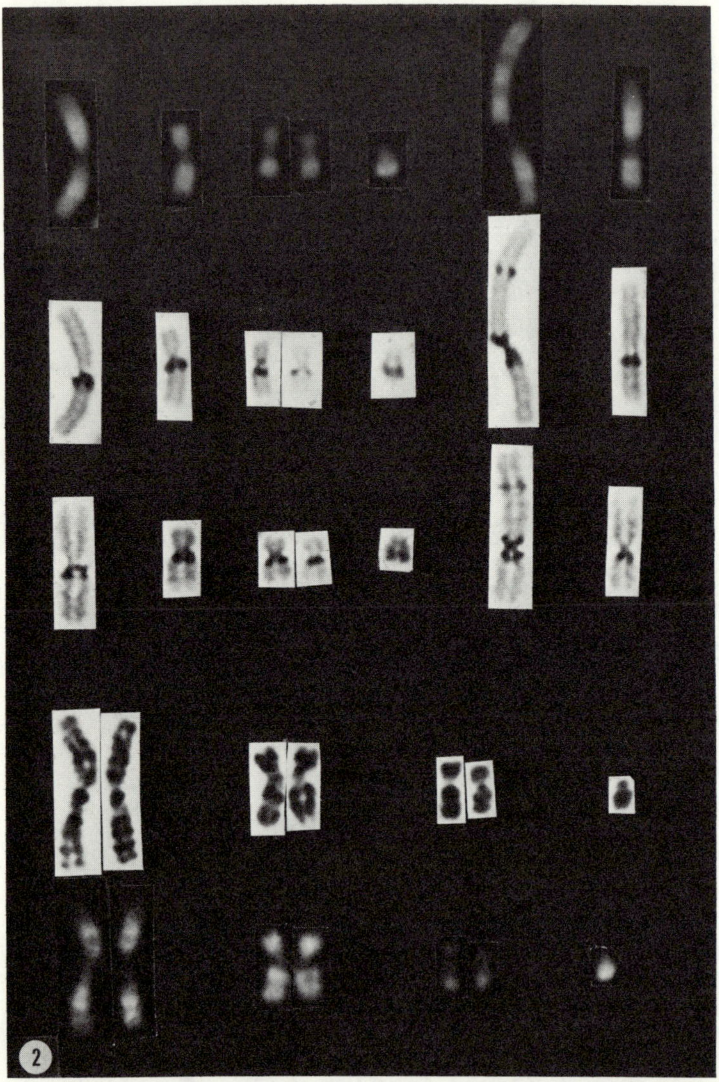

FIG. 2. Comparison of Q-bands (rows 1 and 5), C-bands (rows 2 and 3), and G-bands (row 4) in selected chromosomes of a human malignant melanoma cell line (C-32). Chromosomes 1, 9, 16, and the Y, along with two marker chromosomes, are shown in rows 1-3. The 1, 9, 16, and Y chromosomes are shown in rows 4 and 5. (Courtesy of T. R. Chen.)

centromeric region. Experiments in our own laboratory (*19,20*) have been aimed at exploring further the nature of DNA and heterochromatin in the centromere region with the use of acridine orange fluorescence, immunofluorescence, and electron microscopy.

Acridine orange binds specifically to the negatively charged phosphate groups of DNA. The mode of binding to nucleic acids with highly ordered double-stranded form is monomeric, with an emission maximum of 530 nm, while single-stranded nucleic acids bind acridine orange in an aggregate emitting at 590 nm. Since it has been shown that a similar shift in emission spectra occurs with nucleoproteins (*21*) and is a direct index of the degree of thermal denaturation, an assessment of strandedness of regions of denatured and renatured chromosomes seemed feasible. When air-dried chromosome preparations of mouse L-cells were stained with acridine orange, the entire chromosome was uniformly stained and the emission was clearly green, indicating that the DNA in such preparations was not denatured. If the chromosomes were denatured with 95% formamide and prevented from renaturing by adjusting the denaturing solution to 2% formaldehyde prior to staining, the chromosomes appeared red as expected if denaturation was complete. If the denatured chromosomes (without formaldehyde) are transferred to SCC (Standard Saline Citrate) at 65°C for varying times to allow reassociation, a pattern of differential reassociation is seen. The centromere regions were green indicating reassociation of DNA strands, while the chromosome arms remained red. The centromere region fluoresced green within 60–120 sec, and longer incubation at 65°C did not change the fluorescence pattern until 18–20 hrs, at which time the entire chromatid fluoresced green. These results are consistent with the notion that centromeric regions contain highly reiterated base sequences which reassociate at a rapid rate. The DNA in the chromatids, which contain moderately repetitious DNA and unique sequences, reassociate at a slower rate.

These results are not unexpected in view of the work of Pardue and Gall (*17*). Further confirmation for the localization of constitutive heterochromatin at the centromere was obtained from direct immunofluorescence using antibodies to single-stranded DNA (*19*). For these studies, sera from various patients with the disease systemic lupus erythematosis (SLE) were pooled and tested for antibodies with specificity to single-stranded DNA by the Farr ammonium sulfate precipitation technique under conditions described by Pincus (*22*). When the sera containing single-strand, DNA-specific antibody were incubated with air-dried, undenatured chromosomes followed by treatment with fluorescent-tagged anti-human lgG, the chromosomes failed to fluoresce. However, when chromosomes were denatured with either 0.1 *N* NaOH or 95% formamide in the presence of 0.5% formaldehyde, they stained brightly throughout. When the preparations were allowed to renature in 0.1 × SSC at 66°C for a few minutes, differential fluorescence such as shown in Fig. 3a and b was observed. The centromeric DNA, having reassociated, remained unstained while the chromosome arms fluoresced brightly due to the binding of antibody to single-stranded DNA. The immunofluorescent staining pattern was identical to the C-banding staining pattern (Fig. 3b).

The various banding patterns have many important implications. In regard to chromosome structure, they clearly show that the DNA of metaphase chromosomes is essentially double-stranded with specific nucleotide sequences organized in patterns which are characteristic for each chromosome in the complement. This information imposes certain restrictions on chromosome

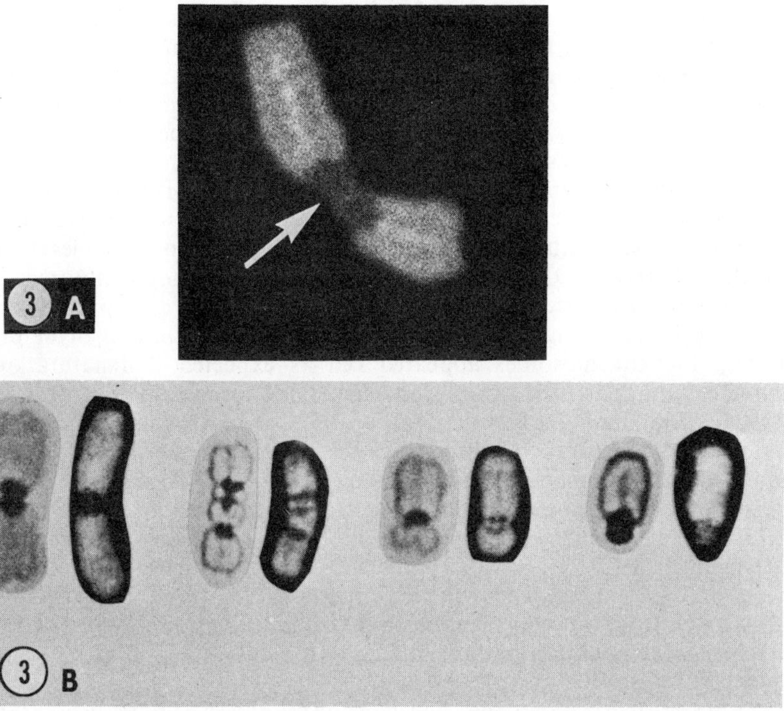

FIG. 3. *a*, Metaphase chromosome of mouse L-cell treated for C-banding and stained with SLE antisera. Absence of fluorescence at centromere (arrow) indicated that DNA is double stranded in this region. Fluorescence in the arms suggests that most of the DNA in this region is still single-stranded. *b*, Comparison of C-banding pattern (left chromosome) and SLE staining (right chromosome) in selected chromosomes of mouse L-cells. [From M. L. Mace, Jr., S. S. Tevethia and B. R. Brinkley, *Exp. Cell Res.* **75**, 521 (1972).]

model-building. Thus, earlier models derived from EM observations in which chromosomes were made up of a single randomly folded 250 Å fiber (*23*) are clearly in error. In order for bands to be produced across the chromatid, the fiber or fibers must have some lateral association at various sites which bring specific sequences into register. Moreover, the C-banding pattern can be equated with the distribution of constitutive heterochromatin on metaphase chromosomes. These new findings prompted us to reexamine metaphase chromosomes with the electron microscope.

ULTRASTRUCTURE OF METAPHASE CHROMOSOMES

In many respects, the electron microscope has not yet lived up to its potential as a tool for the analysis of chromosome structure. Although most modern instruments are capable of resolving small molecules and, in some instruments, even atomic structure, technical limitations have greatly restricted the usefulness of EM in chromosome research. Perhaps the major contribution of electron microscopic studies has been to identify the fibrous nature of mitotic and meiotic chromosomes and the structure of specialized regions along the chromosome. In most studies, chromosomes are either observed in thin sections or whole-mount preparation following surface-spreading or after isolation.

Ultrathin Sections

When mitotic cells are fixed in buffered glutaraldehyde, post-fixed in osmium tetroxide, and embedded in Epon or related resins, they display the appearance shown in Fig. 4. In these cells, mitosis was arrested with 0.06μg/ml Colcemid (*24*) prior to fixation and embedment. The individual chromatids are rarely observed, and the boundary between them is not clearly indicated. The chromosome arms appear fibrous but, because of their compactness, it is impossible to make accurate measurements of fiber dimensions.

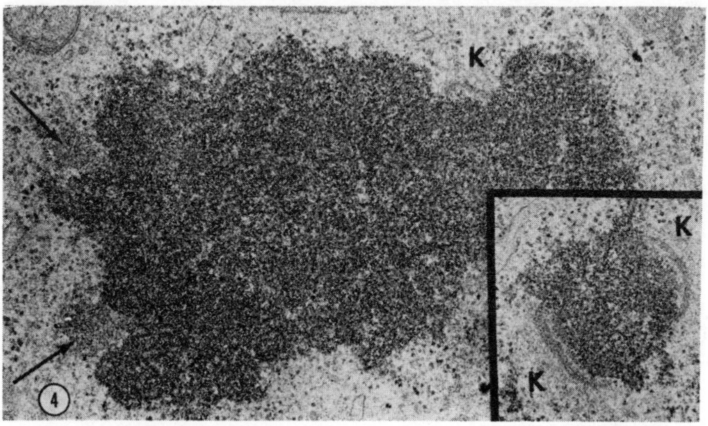

FIG. 4. Electron micrograph of sectioned metaphase chromosome of Chinese hamster cell. Although relatively little information is available on fiber dimension, specialized regions, kinetochore (K) and telomeres (arrows), are apparent. Inset shows cross-section of chromosome at the kinetochore region. × 40,000.

In general, relatively little information can be derived about the fibrous organization of mitotic chromosomes from ultrathin sections. When fibers can be resolved, they measure 80–350 Å in diameter. Thinner fibers can be observed in regions which display gaps or attenuations as indicated in a subsequent section of this report. Also, we have been able to partially "unfold" the metaphase chromosome by treating mitotic cells with culture medium made hypotonic with distilled H_2O (3 parts water: 1 part medium) prior to fixation in glutaraldehyde, adjusted to the same toxicity (*25*). As shown in Fig. 5, the fibers of the chromosome can be more clearly resolved after hypotonic pretreatment and measured 80–130 Å in diameter. When cells were returned to isotonic culture media for 10–15 min after hypotonic swelling, the chromosomes resumed the appearance shown in Fig. 4. The viability of cultured cells was not affected by brief hypotonic treatment as indicated by normal cloning efficiency following treatment (Hsu, unpublished observations). Ultrathin sections have provided important new information on the structure of specialized regions, especially the kinetochore, as will be discussed in a subsequent section.

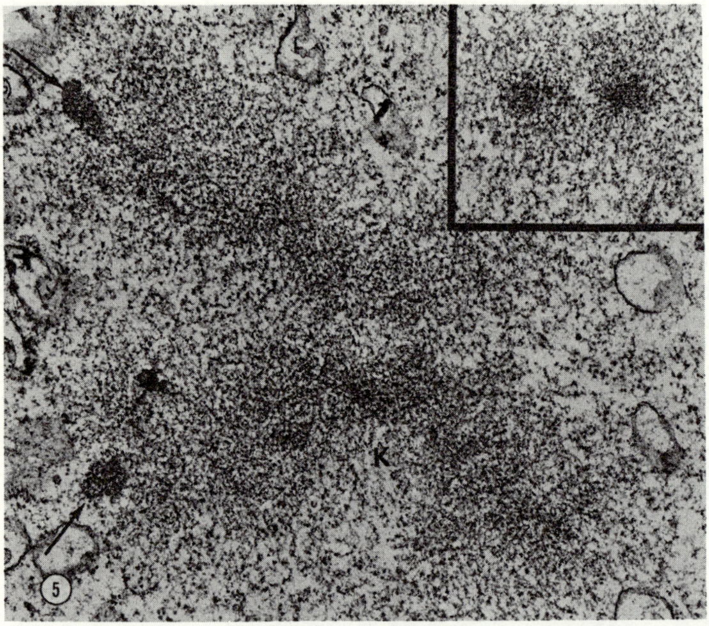

FIG. 5. Same chromosome as in Fig. 4 except cell has been treated for 15 min in hypotonic medium (3 parts water: 1 part medium) prior to fixation in hypotonic glutaraldehyde. Bodies at telomeres (arrows) are still condensed, but chromosome fibers are more easily observed. Inset shows cross-section of chromosome. × 20,000.

Whole-Mount Preparations

Perhaps one of the most frequently used procedures for observing the fibrous texture of chromosomes involves spreading of mitotic or meiotic cells onto an aqueous hypophase. The method, which was initially developed by Gall (*26*) for observing interphase chromatin, relies upon surface tension forces to disperse the chromosome. As shown in Fig. 6, the chromosome morphology is retained and the packing of fairly uniform-sized filaments can be seen in the chromatids. These measure approximately 250 Å in diameter and can be seen to loop out and return to the chromosome axis throughout the length of the chromosome. In most well-spread preparations, the fibers within the chromatid appear to display relatively little order, and the classical "chromonema" are not observed. Also, there is no indication of discrete "half-chromatids" or other levels of organization frequently referred to by those who interpret the chromosome as a polyneme structure. In spite of the large probability of producing artifacts by such harsh dispersion methods, several models for metaphase chromosome structure have been derived from such preparations (*23,27,28*)

Perhaps a more reliable method for observing whole-mount chromosomes in the EM comes from more gentle methods of isolation such as the procedure developed by Wray and Stubblefield (*29*) and Stubblefield and Wray (*30*). By this

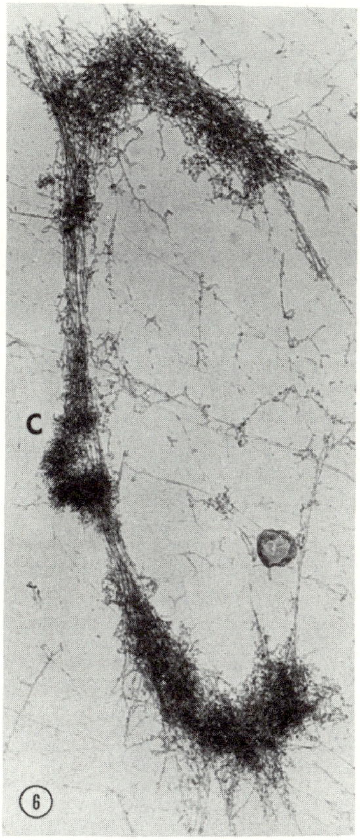

FIG. 6. Electron micrograph of water-spread mouse chromosome after Anderson critical-point drying. The two chromatids can be clearly seen and are still held together at the centromere (C). Note numerous 250 Å fibrils in arms. × 20,000.

method, Colcemid-arrested metaphase cells are lysed in a solution containing 1.0 *M* hexylene glycol, 0.5*mM* $CaCl_2$, and 0.1*mM* "Pipes" buffer. By adjusting the isolation buffer to pH 10.5, Wray et al. (*31*) were able to demonstrate DNA on alkaline sucrose gradients with molecular weights for approximately 2×10^7 up to 2×10^8 daltons. Except for the method of Corry and Cole (*32*), most other chromosome isolation procedures cause serious degradation of DNA. As shown in Fig. 7, chromosomes isolated by this procedure display a morphology which closely resembles the best light microscopic images of chromosomes. Both chromatids are visible and are composed of numerous 250 Å fibrils which loop out along the arms. The centromere region (primary constriction) can be clearly recognized, although no microtubules or kinetochores can be observed in such preparations. As discussed later, chromosomes with intact kinetochores with microtubule attachments have now been isolated by a modification of the Wray and Stubblefield procedure (McGill and Brinkley, in preparation).

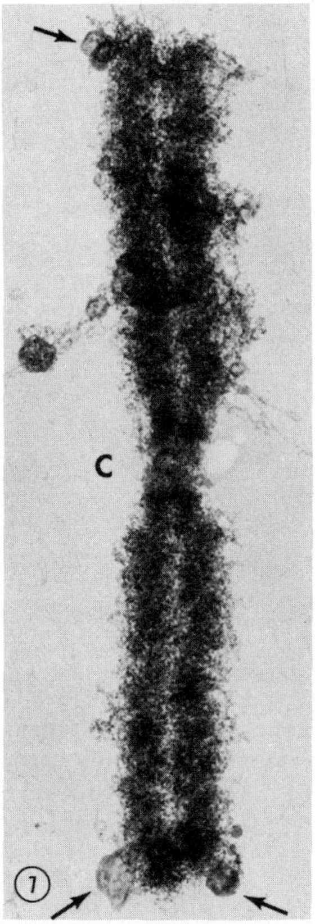

FIG. 7. Electron micrograph of chromosome from mouse L-cell isolated in hexylene glycol-Pipes buffer at pH 10.5. The chromatids are separated and composed of 250 Å fibrils. Chromosome morphology is not distorted and the centromere (C) is clearly visible. Note membranous fragments attached to the telomeres (arrows). × 20,000.

Composition and Dimension of Chromosome Fibers

Although the most frequently reported diameter for fibers of surface-spread chromosomes is 250 Å, fibers in the range of 30–300 Å have been reported (*26,34–36*). Several obvious questions arise concerning the fibers. Are they real or preparative artifacts? What is their composition in regard to DNA, histone, and nonhistone protein? How many fibers are present in the metaphase chromatid and how are they arranged? Most of these questions have been dealt with in previous reviews (*37-39*). Needless to say, there is general disagreement regarding fiber diameters. Wolfe and Grim (*40*) and Solari (*41*) have evidence that thin 180 Å chromosome fibers *in situ* are converted to thick 250 Å fibers upon isolation due to changes in ion concentration at the time of cell lysing. Ris (*42,43*) has interpreted the 250 Å fiber as a composite of two 100 Å fibers. His evidence is based upon the action of chelating buffers which appeared to convert the 250 Å fibril to a pair of 100 Å structures. Griffith and Bonner (*44*) also concluded that the 250 Å fiber is composed of a pair of 100 Å fibrils. The controversy is an important one which must be resolved before the basic structure of the chromosome can be agreed upon.

In regard to DNA within the fibers, DuPraw (*23,45*) demonstrated that digestion with trypsin reduced the diameter of the 250 Å fibril at various intervals giving a "beaded" structure held together by thin, trypsin-resistant threads of approximately the diameter of one Watson-Crick DNA molecule. DuPraw and Bahr (*46*) determined dry mass per unit length on water-spread chromosomes and concluded that the ratio of fiber-length to DNA-length in metaphase chromosomes was 100:1. However, the reliability of this value depends upon the amount of alteration of fibers during chromosome preparation. Recently, Lang (*47*) has produced thick 120–130 Å fibers from purified DNA subjected to 95% ethanol solutions. Since most chromosome preparations are subjected to high alcohol concentrations during dehydration, it is not hard to visualize serious artifacts in the structure of DNA due to loss of water from the preparations.

Arrangement of Fibers Within Chromosomes

How are the fibers observed by electron microscopy arranged into the metaphase chromatid? This question is basic to the issue of chromosome strandedness (see reference *48* for review). One school of investigators proposes that the chromosome is composed of a single folded 250 Å fibril which extends the length of the chromatid. Originally, DuPraw (*23*) proposed that the 250 Å fiber was randomly folded along the chromatid. As a result of newer chromosome banding techniques, the random folded fiber model is almost certainly in error (see discussion earlier in this paper).

A second group of investigators view the chromosome as a polynemic structure consisting of two or more 250 Å fibrils along its length. By observing numbers of slightly "unfolded" preparations of isolated Chinese hamster chromosomes, Stubblefield and Wray (*30*) proposed that each chromatid was multistranded, consisting of a central "core" with associated epichromatin (see reference *39* for review). Thus, the old argument of uninemy versus polynemy still lingers with chromosome researchers; and unless new innovations in methodology are developed, the issue of chromosome strandedness is not likely to be resolved by electron microscopy. Indeed, improved analysis of chromosomal DNA by molecular techniques may resolve the issue of strandedness long before it can be determined by EM. Studies of DNA size in gentle lysates of *Drosophila* chromosomes by viscoelastic techniques have led Kavenoff and Zimm (*49*) to conclude that "in its simplest form" each chromosome is composed of one long DNA molecule (see also reference *50*). A similar conclusion has been reached by Petes et al., (*51*) from sedimentation analysis of DNA from *Saccharomyces cerevisiae*.

ULTRASTRUCTURE OF SPECIALIZED CHROMOSOME REGIONS

By conventional light microscopic staining procedures, metaphase chromosomes display several differentiated loci along their length, i.e., the kinetochore (centromere), nucleolar organizer, and telomere. In addition, the use of banding techniques have served to identify numerous other differentiated sites on metaphase chromosomes (*1*). Electron microscopic studies have provided additional information on the structure of specialized regions.

Heterochromatin

Although heterochromatin was initially described as a differentially stained region within interphase nuclei, it was also known to exist at specific sites on metaphase chromosomes. Prior to the development of C-banding, however, it was essentially impossible to detect heterochromatin at metaphase because of the degree of condensation of both hetero- and euchromatin. It is now apparent that large blocks of heterochromatin are localized at the kinetochore region. In addition, smaller masses may be distributed throughout the chromosome arms. By carefully comparing C-band patterns of mouse chromosomes with ultrathin sectioned and surface-spread chromosomes from the same material, we have been able to observe heterochromatin regions consisting of tightly packed 250 Å fibrils with greater electron density than adjacent euchromatin regions (Fig. 8). Analysis of ultrathin sections of the synaptinemal complex from mouse primary spermatocytes has also allowed us to observe the centromeric heterochromatin in meiotic chromosomes by electron microscopy (*52*).

Kinetochore

Ultrastructure studies have also been useful in defining the structure of the mammalian chromosome kinetochore (*53–56*). As shown in Figs. 9 and 10, the kinetochore has a characteristic structure consisting of three distinct dense layers at the surface of the chromosome along the primary constriction. Spindle microtubules (chromosomal fibers) attach to the outermost layer of the kinetochore. Interpretation of the three-dimensional structure of the kinetochore has been somewhat controversial. Brinkley and Stubblefield (*53*) originally described the kinetochore as a filament. The filament is composed of a dense axial element which measures 200–300 Å in diameter and is approximately 0.1 to 0.2μm long. Numerous thin 70 Å fibrous threads were observed to extend out from the axial element in a lampbrush-like arrangement. Analysis of serial sections suggested that two kinetochore filaments were contained on each chromatid.

Other investigators (*54–56*) have described the kinetochore as a multilayered disc or plaque. Recently, we reinvestigated mammalian kinetochore structure by serial-section analysis of cross sections of preselected Indian muntjak chromosomes (McGill and Brinkley, in preparation). Sections cut perpendicularly to the long axis of the chromosome in the kinetochore region clearly showed that the kinetochore was, in fact, a disc and not a filament as we originally proposed (*53*). In an elegant study of kinetochore structure and development in primitive Protista, Ris and Kubai (*57*) and Kubai (*58*) have obtained good evidence that the kinetochore of primitive organisms arises as a specialization of the nuclear envelope surface. Their study also suggests that the kinetochore of these organisms is organized into a disc or plaque.

In preliminary studies, McGill and Brinkley (unpublished observations) have attempted to isolate mammalian chromosomes with intact, functional kinetochores. By lysing cells in an isolation buffer consisting of 15% hexylene

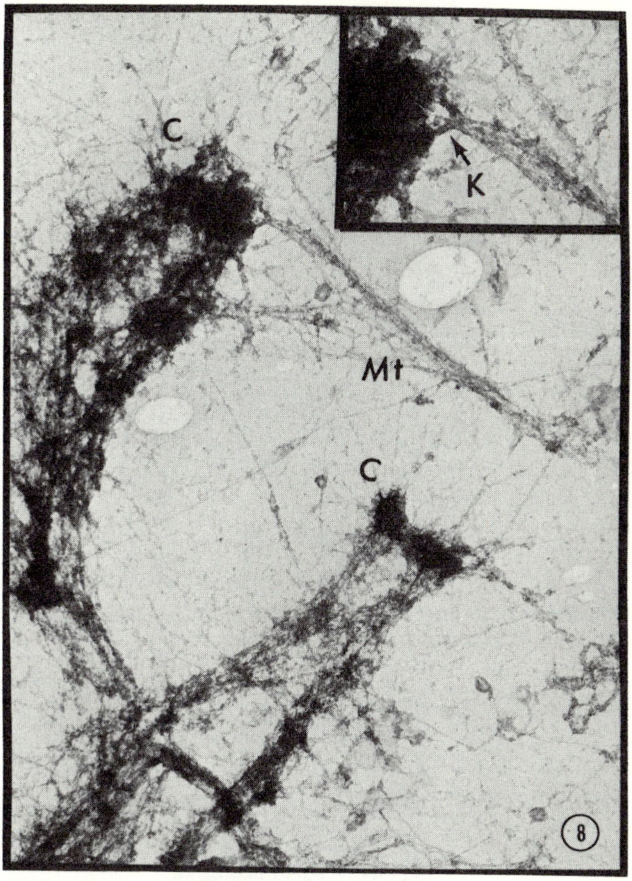

FIG. 8. Electron micrograph of mouse chromosomes spread on 0.1 *M* ammonium acetate. Fibers in arm are more relaxed, and dense heterochromatin at the centromere (C) is clearly visible, Microtubules (Mt) are preserved by this procedure and, in some chromosomes, their attachment to the kinetochore (K in inset) is visible. × 20,000.

glycol, 2×10^{-3} *M* $MgCl_2$ and 1×10^{-3} *M* MES buffer adjusted to pH 10.7, chromosomes with intact kinetochores can be observed in whole-mount preparations or thin sections. Microtubules are retained and the kinetochores have a disc-like appearance similar to those seen in intact cells. Further studies with isolated kinetochores should permit a more systematic analysis of their structure and composition as well as provide knowledge of their function in chromosome movement.

Nucleolar Organizer

One of the first indications that some repetitious gene sequences were clustered at specific loci on metaphase chromosomes comes from studies of the nucleolar organizer (NO) in a variety of eucaryotic cells. In mammalian cells, the

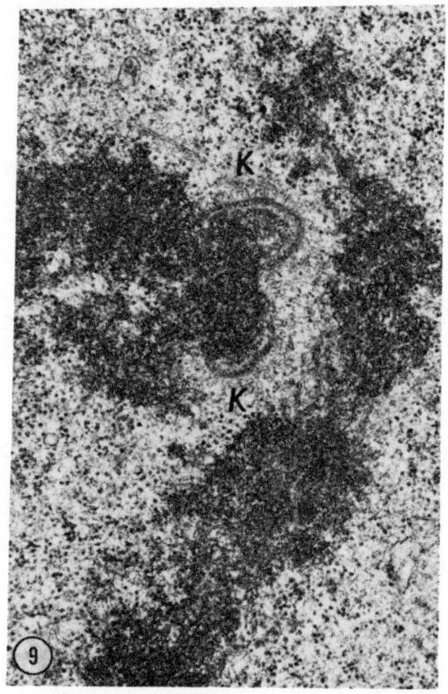

FIG. 9. Kinetochores (K) are clearly visible on these Chinese hamster chromosomes arrested with rotenone. × 40,000. (Micrograph courtesy of S. S. Barham.)

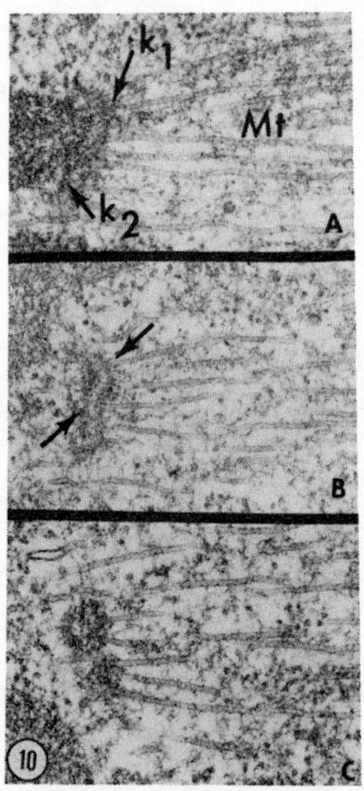

FIG. 10. *a*, Normal kinetochore from rat kangaroo cell showing outer (k_1) and inner (k_2) dense element. *b*, Damaged kinetochore from cell treated with mitomycin C. Kinetochore elements are clearly separated from chromosome. *c*, Adjacent serial section to a. × 50,000. [From B. R. Brinkley and W. N. Hittelman, *Int. Rev. Cytol.* **42,** 49 (1975).]

nucleolar organizer consists of hundreds of copies of r-DNA which code for 18 and 28 S ribosomal RNA (*59*). Recently, elegant electron micrographs of amplified r-DNA in amphibian oocytes by Miller and colleagues (*60,61*) have provided convincing morphological proof for the molecular scheme of r-RNA transcription. Although the NO is clearly visible at metaphase by conventional light microscopic techniques, relatively few studies have been carried out on this component by electron microscopy. Hsu et al., (*62*) used a light microscopic preselection technique to identify the NO region on the X-chromosome of rat kangaroo cells (strain PtK_1) prior to ultrathin sectioning for electron microscopy. As shown in Figs. 11a-b, the NO of mammalian chromosomes consists of an electron-dense chromatin core surrounded by less-dense fibrils 50–70 Å in diameter. In sectioned chromosomes, the width of the NO is the same as that of the chromatid arm, indicating that this region is not really constricted as suggested by light microscopy but is filled with fibrils of different sizes and staining characteristics. The structural similarity of the NO to the kinetochore is discussed by Brinkley and Stubblefield (*63*). The simplest interpretation of its structure on metaphase chromosomes is that of a lampbrush-like filament with thin fibrous loops extending out from a dense axial core.

Telomere

Except for telocentric chromosomes which have a kinetochore at the end of one arm, the ends of chromosomes (telomeres) have been given special significance in cytological and genetic studies. Ultrastructural analysis of chromosomes may or may not detect unusual features at the ends of chromosomes. In whole-mount, isolated chromosomes, the telomeres of most chromosomes contain fibers 250 Å which loop out and back into the chromosome arm. Telomeres often associate with the nuclear envelope in mitotic and meiotic prophase. Stubblefield (*39*) has described membranous attachments at the telomeres of isolated metaphase chromosomes and assumed that they were derived from the nuclear envelope (see Fig. 7). Hsu et al. (*62*) were able to detect nucleolar organizer-like material at the telomere of some metaphase chromosomes of Chinese hamster, but most chromosomes have no such differentiated structures at their tips.

In general, relatively little can be said about the ultrastructure of telomeres of metaphase chromosomes. There may be special features of telomeres which permit them to attach to nuclear envelope membranes and thereby insure order within the interphase nucleus, facilitate pairing and synapsis at meiosis, and perhaps regulate DNA synthesis. Presently, ultrastructural studies provide only a limited amount of information regarding these options.

ANALYSIS OF CHROMOSOME ABERRATIONS BY ELECTRON MICROSCOPY

In recent years, several laboratories have resorted to transmission and scanning electron microscopy to evaluate damage to mammalian chromosomes by ionizing radiation, drugs, and viruses (*64–72*). Although the advantages of electron optics for the study of chromosome damage are yet to be fully exploited, the superior resolving power of EM should permit a more critical analysis of

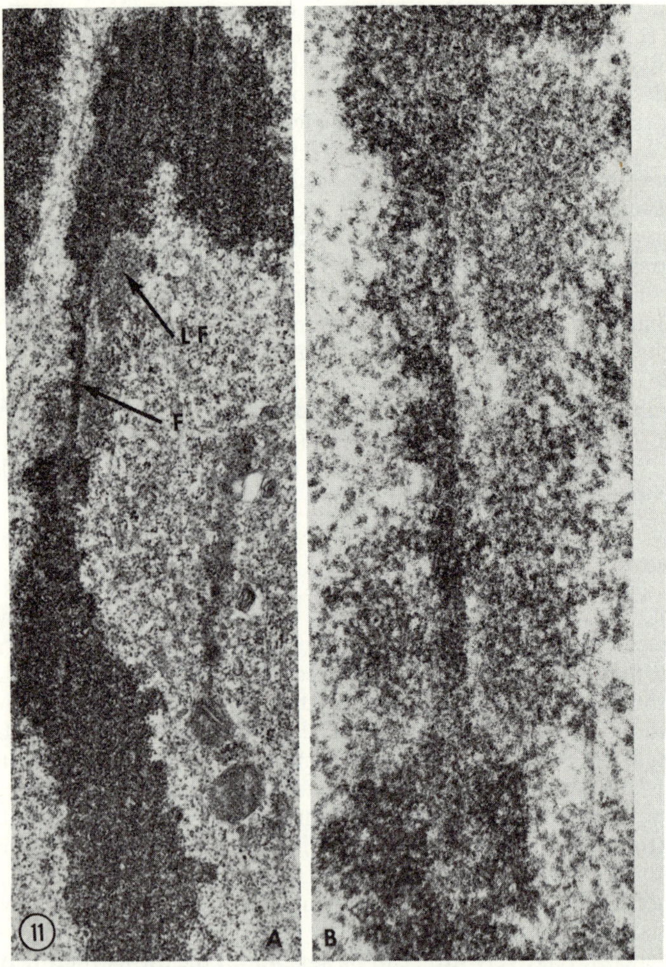

FIG. 11. *a*, Electron micrograph of X-chromosome of rat kangaroo cell showing stretched nucleolar organizer (NO) in long arm. A dense filament (F) is surrounded by less-dense (LF) fibrous elements. *b*, Higher magnification of the NO region showing details of dense filament and less-dense fibrous elements. a, × 25,000; b, ×90,000. [From R. M. Humphrey and B. R. Brinkley, *J. Cell Biol.* **42**, 745 (1969).]

aberrations, help to identify the target for damage, and elucidate the problem of how a lesion in DNA or protein is expressed as an aberration at metaphase. A lengthy discussion of the ultrastructure of chromosome aberrations is beyond the scope of this report. For more details, the reader should consult the above references or the recent review of Brinkley and Hittelman (*73*). Their findings are summarized in Table 1. In general, ultrastructural analysis of chromosome breaks, gaps, and exchanges confirms analyses made by light microscopy; however, there are notable exceptions. Some aberrations which would have been scored as breaks by light microscopy were found to have fibers extending across the lesion. Conversely, and perhaps of lesser significance, aberrations which would have been scored as gaps by light microscopy were, in fact, true breaks when viewed by EM. The ultrastructure of a gap indicates that this lesion may be composed of fibrils ranging in diameter from 50 to 800 Å The larger fibers are probably aggregates of two or more 250 Å fibrils.

Analysis of some lesions suggested a bineme organization of the chromatid. Greatly attenuated gaps in the chromatid often appeared to be composed of two parallel fiber bundles. Side-arm bridges, which were produced by x-irradiation of late G_2 and prophase cells, consisted of thin "sub-chromatid" fibrils connecting two daughter chromatids in anaphase. Such bridges have been cited as evidence for half-chromatids (see reference *73* for review). When unstretched, the bridges were approximately one-half the diameter of the chromatid. Ultrastructural analysis was, however, insufficient to clearly establish the "half-chromatid" nature of side-arm bridges.

TABLE 1
Comparative Aspects of Chromosome Damage Observed with Light and Electron Microscopy

Type of damage	Light microscope	Electron microscope
I. Breaks		
A. Aligned	Apparent structural discontinuity, broken ends aligned	1. True structural discontinuity, broken ends aligned, sister chromatid connections 2. True structural continuity
B. Unaligned	Apparent structural discontinuity, broken ends unaligned	1. True structural discontinuity, broken ends unaligned, sister chromatid connections 2. True structural continuity
II. Gaps		
A. Chromatid	Achromatic lesions with apparent structural continuity	1. True structural continuity a. 50–80 Å fibrils b. 200–250 Å fibrils c. 500–800 Å fibrils 2. True structural discontinuity
B. Isochromatid	Achromatic with apparent structural continuity; may appear like exaggerated secondary constrictions or nucleolar organizer regions	True structural continuity, but not like nucleolar organizers
III. Bridges		
A. Chromosome	Dicentric with chromosome stretched across metaphase plate; acentric fragments often present	Same as light microscope except stretched regions often double, revealing "half-chromatids"
B. Side-arm	Stickiness at sharply bent regions of anaphase arms; apparent structural continuity between associated arms	Same as light microscope except true structural continuity of sub-chromatid fibers
IV. Stickiness	Adherence of chromosome arms at anaphase; no acentric fragments	Thin chromatin exchanges connecting various chromosome arms at metaphase and anaphase

Chromosome "stickiness" can be produced by drugs such as actinomycin D and ethidium bromide if applied to cells in the S or G_2 phase of the cell cycle (*74*). Stickiness appears to be produced by entanglement of fibers at the sub-chromatid level of organization. Such lesions are expressed as chromatid-type aberrations in subsequent mitosis. These data suggest that "stickiness" and side-arm bridges may be one and the same.

Many features of chromosome damage can be analyzed in greater detail when viewed by electron optics. In general, however, interpretation of chromosome aberrations by electron microscopy is greatly limited by a lack of understanding of chromosome organization. Hopefully, with progress being made in several areas of molecular and cellular biology, the complete architecture of mammalian chromosomes will be worked out in the present decade.

ACKNOWLEDGMENT

We gratefully acknowledge micrographs contributed by Dr. A. Dean Stock and Dr. T. R. Chen of The University of Texas M. D. Anderson Hospital and Tumor Institute at Houston. Appreciation is also extended to Drs. T. C. Hsu and Frances Arrighi for their advice and criticisms and to Mrs. Betti Ledlie for secretarial assistance. These investigators were supported in part by a research grant DHEW 1R01 CA 14675 to BRB.

REFERENCES

1. T. C. Hsu, *Annu. Rev. Genet.* 7, 153 (1973).
2. T. Caspersson, S. Farber, G. Foley, J. Kudynowski, E. J. Modest, E. Simonsson, U. Wagh and L. Zech, *Exp. Cell Res.* **49**, 219 (1968).
3. T. Caspersson, L. Zech, E. J. Modest, G. E. Foley, U. Wagh and E. Simonsson, *Exp. Cell Res.* **58**, 128 (1969).
4. T. Caspersson, L. Zech, E. J. Modest, G. E. Foley, U. Wagh and E. Simonsson, *Exp. Cell Res.* **58**, 141 (1969).
5. V. G. Dev, D. Warburton, O. J. Miller, D. A. Miller, B. F. Erlanger and S. M. Beiser, *Exp. Cell Res.* **74**, 288 (1972).
6. R. R. Schreck, D. Warburton, O. J. Miller, S. M. Beiser and B. F. Erlanger, *Proc. Natl. Acad. Sci. U.S.A.* **70**, 804 (1973).
7. B. Dutrillaux and J. Lejeune, *C. R. Acad. Sci.* **272**, 2638 (1971).
8. M. E. Drets and M. W. Shaw, *Proc. Natl. Acad. Sci. U.S.A.* **68**, 2073 (1971).
9. S. R. Patil, S. Merrick and H. A. Lubs, *Science* **173**, 821 (1971).
10. W. Schnedl, *Chromosoma* **34**, 448 (1971).
11. M. Seabright, *Lancet* 971 (1971).
12. H. C. Wang and S. Fedoroff, *Nature (London) New Biol.* **235**, 52 (1972).
13. M. Seabright, *Chromosoma* **36**, 204 (1972).
14. H. Kato and K. Moriwaka, *Chromosoma* **38**, 105 (1972).
15. T. Utakoji, *Chromosomes Today* **4**, (in press).
16. V. G. Dev, D. Warburton and O. J. Miller, *Lancet* 1285 (1972).
17. M. L. Pardue and J. G. Gall, *Science* **168**, 1356 (1970).
18. F. E. Arrighi and T. C. Hsu, *Cytogenetics* **10**, 81 (1971).
19. M. L. Mace, Jr., S. S. Tevethia and B. R. Brinkley, *Exp. Cell Res.* **75**, 521 (1972).
20. M. L. Mace, Jr., *Heterochromatin and Repeated Sequences in Mammalian Chromosomes.* Ph.D. Thesis, The University of Texas Graduate School of Biomedical Sciences at Houston (1974).
21. M. R. Alvarez, *Exp. Cell Res.* **69**, 72 (1971).
22. T. Pincus, *Arthritis Rheum.* **14**, 623 (1971).

23. E. J. DuPraw, *Proc. Natl. Acad. Sci. U.S.A.* **53,** 161 (1965).
24. B. R. Brinkley, E. Stubblefield and T. C. Hsu, *J. Ultrastruct. Res.* **19,** 1 (1967).
25. B. R. Brinkley, *J. Cell Biol.* **35,** 17 (1967). (Abstract)
26. J. G. Gall, *Science* **139,** 120 (1966).
27. D. E. Comings and T. A. Okada, *Cytogenetics* **9,** 436 (1970).
28. J. G. Abuelo and D. E. Moore, *J. Cell Biol.* **41,** 73 (1969).
29. W. Wray and E. Stubblefield, *Exp. Cell Res.* **59,** 469 (1970).
30. E. Stubblefield and W. Wray, *Chromosoma* **32,** 262 (1971).
31. W. Wray, E. Stubblefield and R. Humphrey, *Nature (London) New Biol.* **238,** 237 (1972).
32. P. M. Corry and A. Cole, *Nature (London) New Biol.* **245,** 100 (1973).
33. S. L. Wolfe, in *Biological Basis of Medicine* (E. E. Brittar, Ed.), pp. 3-42. Academic Press, New York, 1969.
34. Ernest J. DuPraw, *DNA and Chromosomes.* Holt, Rinehart and Winston, New York, 1970.
35. A. Cole, in *Theoretical and Experimental Biophysics* (A. Cole, Ed.), pp. 305–368. Dekker, New York, 1967.
36. B. R. Zirkin and S. L. Wolfe, *J. Ultrastruct. Res.* **39,** 496 (1972).
37. J. A. Huberman, *Annu. Rev. Biochem.* **42,** 355 (1973).
38. D. E. Comings and T. A. Okada, in *Perspectives in Cytogenetics* (S. W. Wright, B. F. Crandall and L. Boyer, Eds.), pp. 223–250. Charles Thomas Company, Springfield, Illinois, 1972.
39. E. Stubblefield, *Int. Rev. Cytol.* **35,** 1 (1973).
40. S. L. Wolfe and J. N. Grim, *J. Ultrastruct. Res.* **19,** 382 (1967).
41. A. J. Solari, *Exp. Cell Res.* **67,** 161 (1971).
42. H. Ris, in *Handbook of Molecular Cytology* (A. Lima-de-Faria, Ed.), pp. 221–250. North-Holland, Amsterdam, 1969.
43. H. Ris and D. F. Kubai, *Annu. Rev. Genet.* **4,** 263 (1970).
44. J. Griffith and J. Bonner, in *Adv. Cell Biol.*, (in press).
45. E. J. DuPraw, *Develop. Biol.* **13,** 188 (1969).
46. E. J. DuPraw and G. F. Bahr, *Acta Cytol.* **13,** 188 (1969).
47. D. Lang, *J. Mol. Biol.* **78,** 247 (1973).
48. S. Wolff, in *International Review of Cytology* (G. H. Bourne and J. F. Danielli, Eds.), 279–296. Academic Press, New York, 1969.
49. R. Kavenoff and B. H. Zimm, *Chromosoma* **41,** 1 (1973).
50. R. Kavenoff, L. C. Klotz and B. Zimm, *Cold Spring Harbor Symp. Quant. Biol.* **38,** 1 (1973).
51. T. D. Petes, B. Byers and W. L. Fengman, *Proc. Natl. Acad. Sci. U.S.A.* **70,** 3072 (1973).
52. T. C. Hsu, J. E. Cooper, M. L. Mace, Jr. and B. R. Brinkley, *Chromosoma* **34,** 73 (1971).
53. B. R. Brinkley and E. T. Stubblefield, *Chromosoma* **19,** 28 (1966).
54. P. T. Jokelainen, *J. Ultrastruct. Res.* **19,** 19 (1967).
55. D. E. Comings and T. A. Okada, *Exp. Cell Res.* **67,** 97 (1971).
56. U-P. Roos, *Chromosoma* **41,** 195 (1973).
57. H. Ris and D. F. Kubai, *J. Cell Biol.* **60,** 702 (1974).
58. D. F. Kubai, *J. Cell Sci.* **13,** 511 (1973).
59. E. H. McConkey and J. W. Hopkins, *Proc. Natl. Acad. Sci. U.S.A.* **51,** 1197 (1964).
60. B. A. Hamkalo and O. L. Miller, Jr., *Annu. Rev. Biochem.* **42,** 379 (1973).
61. O. L. Miller and B. R. Beatty, *Science* **164,** 955 (1969).
62. T. C. Hsu, B. R. Brinkley and F. E. Arrighi, *Chromosoma* **23,** 137 (1967).
63. B. R. Brinkley and E. Stubblefield in *Advances in Cell Biology* (D. M. Prescott, L. Goldstein and E. McConkey, Eds.), pp. 119–185. Appleton-Century-Crofts, New York, 1970.
64. R. M. Humphrey and B. R. Brinkley, *J. Cell Biol.* **42,** 745 (1969).
65. B. R. Brinkley and R. M. Humphrey, *J. Cell Biol.* **42,** 827 (1969).
66. B. R. Brinkley and M. W. Shaw in *Genetic Concepts and Neoplasia*, pp. 313–345. The Williams and Wilkins Company, Baltimore, 1970.
67. W. Scheid and H. Traut, *Mutat. Res.* **10,** 159 (1970).
68. W. Scheid and H. Traut, *Mutat. Res.* **11,** 253 (1971).
69. A. Brogger, *Humangenetik* **13,** 1 (1971).
70. H. M. Golomb, G. F. Bahr and D. S. Borgaonkar, *Chromosoma* **34,** 94 (1971).
71. C. K. Yu, *Can. J. Genet. Cytol.* **13,** 690 (1971).
72. B. Wahren, F. Lampert and O. Goetz, *Exp. Cell Res.* **75,** 271 (1972).
73. B. R. Brinkley and W. N. Hittelman, *Int. Rev. Cytol.* **42,** 49 (1975).
74. M. McGill, S. Pathak and T. C. Hsu, *Chromosoma* **47,** 157 (1974).

Structure and Replication of the Yeast Chromosome[1]

Walton L. Fangman
University of Washington
Seattle, Washington, USA

The yeast, *Saccharomyces cerevisiae*, is a eukaryotic microorganism with very small chromosomes. The haploid genome consists of about 9×10^9 daltons of DNA distributed among seventeen chromosomes observed by genetic analysis. The average chromosome, therefore, contains only about 5×10^8 daltons of DNA, some 100-fold less than an average mammalian chromosome and only 4-fold larger than the T4 bacteriophage chromosome. Studies of DNA molecules released from the yeast nucleus by sedimentation and by electron microscopy have shown that the average DNA molecule has a mass of 4 to 6×10^8 daltons, the size expected if a single DNA duplex corresponds to a chromosome. Individual molecules range in size from about 10^8 to 10^9 daltons. In spite of this small size, yeast chromosome replication is similar to that in animal cells: it is initiated at multiple sites along a DNA molecule, and initiation sites are spaced along the DNA at short, 20 micron (0.5×10^8 dalton), intervals. Recent experiments indicate that portions of these chromosomes replicate at discrete and programmed times during the DNA synthesis (S) period in the cell cycle.

[1]Congress presentation E–9–3.

REPAIR PROCESSES IN EUKARYOTES[1]

Chairmen: M. M. Elkind and S. Okada

[1]Symposium A–3

Repair Studies at the Molecular, Chromosomal, and Cellular Levels: a Review of Current Work in Japan[1]

S. Okada

Department of Radiation Biophysics
Faculty of Medicine
University of Tokyo
Hongo, Bunkyo-ku, Tokyo 113, Japan

Since the early days of radiobiology, the dose rate dependency and/or the dose fractionation effects have often been attributed to 'recovery' process(es) of radiobiological damage (*1*). In mammalian cells, the dose rate dependency and the dose fractionation effects have been observed in cell killing, chromosome aberrations, mutation, mitotic delay and depression of DNA synthesis (Table 1). Of these phenomena, cell killing and chromosome aberrations are two of

TABLE 1
Dose Rate and Dose Fraction Effects on Radiobiological Responses of Mammalian Cells

Radiobiological Response	Dose Rate Dependency	Dose Fractionation Effect	Type of Recovery
Reproductive death	Yes[1]	Yes[2]	Elkind type recovery and/or recovery from potential lethal damage
Interphase death	—	No[3]	—
Chromosome aberration	Yes[4]	Yes[4,5]	Restitution
Mutation	Yes[6]	Yes[7]	Repair?
Mitotic Delay	Yes[1,8]	Yes[9]	Re-entry to Mitosis and Division
Depression of DNA synthesis	Yes[10]	Yes[11]	Re-initiation of DNA synthesis
Formation of giant cell	—	—	—

—Not tested

1. E. C. Hall (A review) *Brit. J. Radiol.*, **45,** 81 (1972).
2. M. M. Elkind and W. K. Sinclair, *Current Topics in Radiat. Res.*, **1,** 167 (1965).
3. S. Sawada, unpublished data.
4. W. C. Dewey and R. M. Humphrey, *Exp. Cell Res.*, **35,** 262 (1964). J. G. Brewen, *Proc. Natl. Acad. Sci., U.S.*, **50,** 322 (1963).
5. H. J. Evans, in *"Radiation Research"*. (G. Silini, Ed.) p. 482. North-Holland Publishing Co., Amsterdam, 1967.
6. N. Suzuki and S. Okada, unpublished data.
7. C. F. Arlett and J. Potter, *Mutat. Res.*, **13,** 59 (1971).
8. J. S. Bedford and J. B. Mitchell, *Radiat. Res.*, **54,** 316 (1973).
9. S. Okada, *Radiation Biochemistry,* vol. 1. *Cells* (K. L. Altman, G. B. Gerber and S. Okada, Eds.) Academic Press, New York, 1969.
10. W. B. Looney and L. O. Chang, *Radiology,* **86,** 457 (1966).
11. W. B. Looney, *Nature,* **205,** 1334 (1965).

[1]Congress presentation A-3-1.

the effects in which 'recovery' and/or restitution processes have been clearly established. It is noted that other effects are those worthy of future exploration.

Recovery studies in Japan have been carried out by Chikako Sato of Aichi Cancer Center and her associates who investigated the dose response curves of three cell lines, and found that a Burkitt lymphoma (P3HR-1) cell line showed no shoulder, while a mouse melanoma (B16-C2W) line showed a relatively large shoulder (n=20) (Fig. 1) (*2*). The dose fractionation studies demonstrated little Elkind recovery in the lymphoma (*3*) and a prominent Elkind recovery in the mouse melanoma (Fig. 1). In other words, the Burkitt lymphoma of Sato's may be 'Elkind recovery deficient' while the melanoma is 'Elkind recovery proficient'.

The investigators focused their attention on the electrophoretic mobilities of the cells after irradiation. The overall negative charge on the cell surface decreased with time postirradiation (*3, 4*). The melanoma cells with Elkind recovery started to regain their negative charges after 4 hours post-irradiation, while the negative surface charges of the lymphoma cells with little Elkind recovery remained at a low level (*2, 4*) (Fig. 1). The addition of 0.005 μg/ml of actinomycin D, a known inhibitor of Elkind recovery (*5*), to the irradiated melanoma cells abolished the regaining of negative charges after irradiation. These results suggest that the alteration of the surface charge of the irradiated cells may be part of the dying syndrome of the irradiated cells, and could reflect the Elkind recovery process. The studies of molecular repair processes of the Burkitt lymphoma showed (i) a slow rejoining of single-strand breaks of DNA (T. Morita, personal communication), (ii) a rejoining of double-strand DNA cuts (S. Sawada, personal communication), and (iii) excision of UV dimers (T. Morita, personal communication). So far, no clear-cut relationship between regaining of surface charges and molecular repair processes have been demonstrated.

It is interesting to note that the dose response curves of mammalian cells to drugs are often composed of a fast sensitive slope followed by a slow resistant component. In a sense, this is quite the opposite to the sigmoidal survival curves of mammalian cells with X rays. Terasima et al. (*6*) and Sakamoto (personal communication) found such a reverse survival curve for Bleomycin, an anti-cancer antibiotic developed in Japan, in HeLa and Chinese hamster V79 cells. The fractionation studies revealed a pattern which one might call a 'reverse Elkind recovery' (Fig. 2).

Now, turning our attention to the restitution of chromosome aberrations, Sasaki and his associate of the Tokyo Medical and Dental University have been studying induction of chromosome aberrations in lymphocytes of patients with some cancer-prone diseases by X rays and other agents.

X rays induced dicentrics and rings in lymphocytes from individuals with Down's syndrome at a significantly higher yield than those from normal persons (*7*), and the restitution of chromosome aberrations in Down's lymphocytes was much faster than of those in normal lymphocytes with or without phytohemagglutinin stimulation (Fig. 3). We speculated that fibroblasts derived from Down's syndrome might be more radiosensitive than those from normals . A fibroblast culture developed from a Down's patient has now been

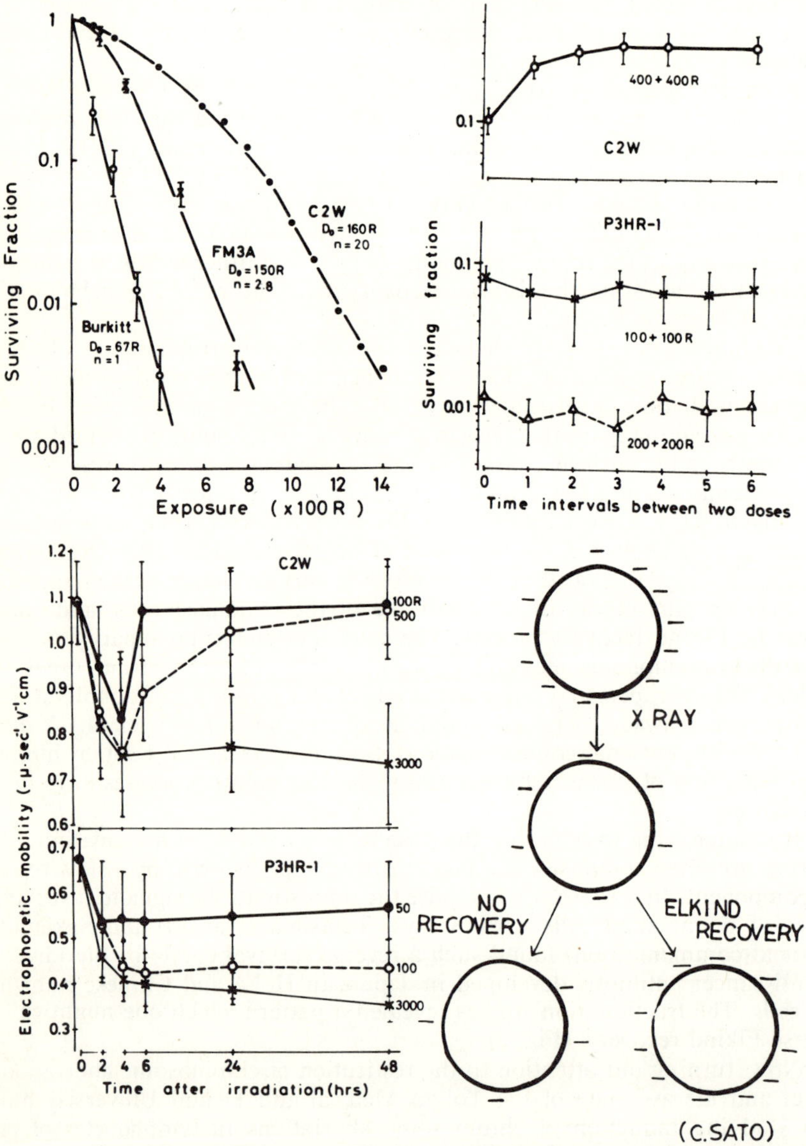

FIG. 1. Elkind recovery and electrophoretic mobilities of Burkitt lymphoma (P3HR-1), mouse melanoma (B16-C2W), and mouse mammary carcinoma (FM3A) cells.

Upper left figure: the survival curves of three cell lines.

Upper right figure: the survival of melanoma (B16/C2W) and lymphoma (P3HR-1) as a function of the time interval between radiation doses.

Lower right figure: the post-irradiation alteration of electrophoretic mobilities of melanoma (C2W) and lymphoma (P3HR-1) cells.

Lower left figure: a schematic model of post-irradiation changes in electrophoretic mobility of cells with and without Elkind recovery.

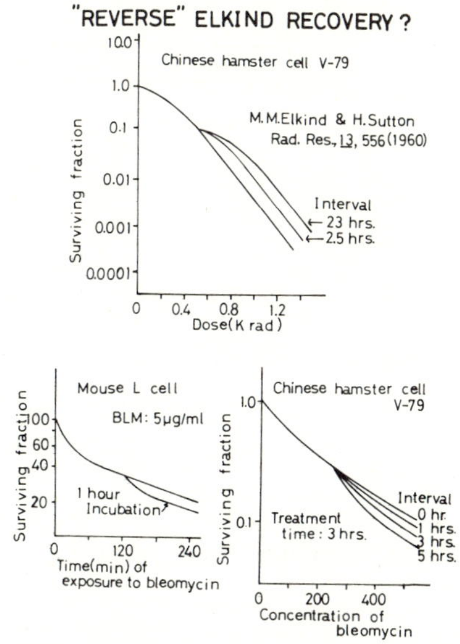

FIG. 2. A 'reverse' Elkind recovery? The survival curve of the left figure: mouse fibroblasts L were treated for various times with Bleomycin, 5 μg/ml. The curve marked "1 hour incubation" is the survival curve of L cells treated with Bleomycin for 120 minutes, left untreated for 1 hour and then treated for various times with Bleomycin at the same 5 μg/ml (*6*).

Right figure: the survival curve for Chinese hamster V79 cells treated with various concentrations of Bleomycin for three hours. In the dose fraction experiments, the cells were first treated with 250 μg/ml of Bleomycin for three hours, replenished with fresh medium and kept for the interval specified in the figure, and then treated with various concentrations (x μg/ml) of Bleomycin for three hours. The survival was plotted against a concentration of (250 μg/ml plus x μg/ml). (Sakamoto, unpublished data.)

examined. The X-ray-induced dicentrics and rings of Down's fibroblasts are greater than those of normal, as was the case for lymphocytes. However, the survival curve of Down's fibroblasts is shown to be the same as that of HeLa S3 cells (Fig. 3) or the Down's fibroblasts may be slightly more radioresistant than fibroblasts from normal persons (D_o of Down's and normals are 150 rads and 125 rads, respectively; Sasaki, personal communication).

Lymphocytes from Fanconi's anemia patients are the other type of the cells which Sasaki and his associates suspected to possess an abnormal repair capability (*8*). He has estimated the drug-induced chromatid aberrations in relation to the first cell cycle of phytohemagglutinin-stimulated lymphocytes (Fig. 4). He found that MMS (methylmethane sulfonate) showed a low yield of chromatid aberrations in G_1 stage cells of both Fanconi's lymphocytes and normal lymphocytes. MMC (Mitomycin C, a cross-linking agent) yielded aberrations in Fanconi's lymphocytes much higher than those of normal lymphocytes (Fig. 4). They suggested that lymphocytes from Fanconi's patient might be deficient in 'repair mechanism for cross-linkings'. Now, Sasaki developed fibroblast cultures from a Fanconi's anemia patient and obtained the dose response curve to various drugs. The dose response of Fanconi's fibroblasts to MMS was similar to that of normal, while the response of

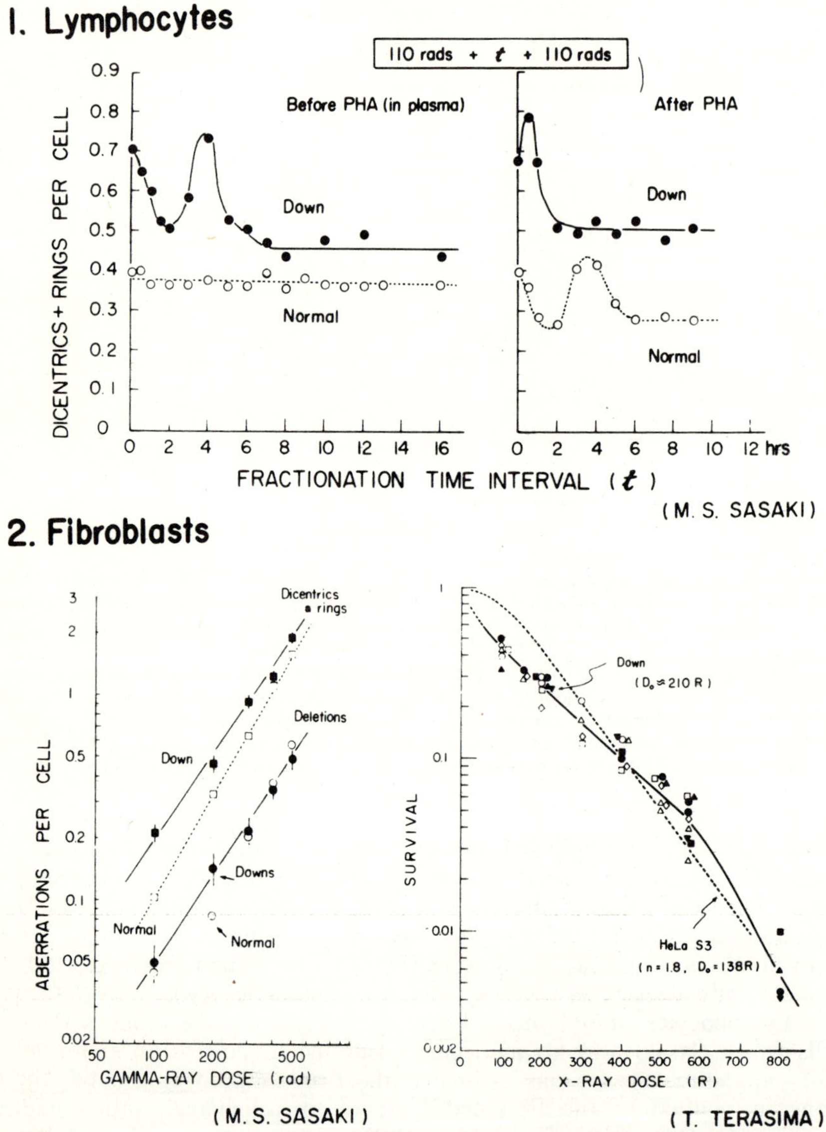

FIG. 3. X-ray sensitivities of cells derived from patient with Down's syndrome.

Upper left figure: lymphocytes were kept in plasma and two doses of X-rays were administered separated by a time interval (t). Lymphocytes were then stimulated by phytohemagglutinin (PHA) to score chromosome aberrations.

Upper right figure: lymphocytes were stimulated with PHA, and subjected to the dose fractionation scheme described above.

Lower left figure: fibroblasts derived from normal persons and patients with Down's syndrome were synchronized by removal of arginine, and the cells accumulated in G_1 and early S were irradiated with X-rays; chromosome aberrations were scored when the cells entered mitosis.

Lower right figure: the survival curves of HeLa S3 cells and fibroblast of Down's syndrome patient.

Fanconi's to Mitomycin C was far greater than that of normal (Fig. 4). A study to examine Sasaki's hypothesis, i.e. deficiency of a repair mechanism of cross links, is now under way.

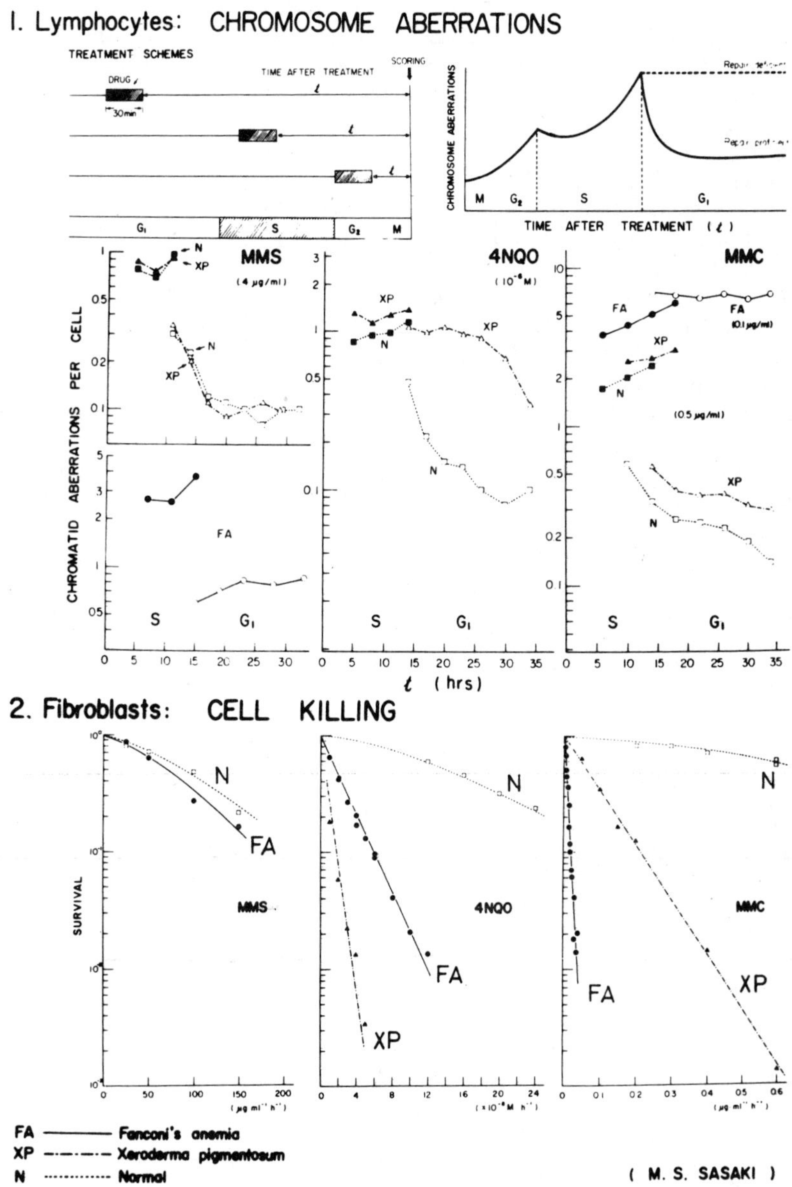

Figure 4.

The cell cycle dependencies of cultured mammalian cells to X rays and various drugs have been well established (e.g. *9*). However, as to the underlying causes, little is known. Horikawa and Watanabe of Kanazawa University found that HeLa S3 cells were the most UV-sensitive at the M and the middle-S stages, and the least sensitive at the G_1 and the late-S stages. The formation of thymine dimers in cellular DNA was found to be the highest at the M stage and the middle-S stage, and the lowest at the G_1 and the late-S stage. The removal of thymine dimers was found to be relatively independent of the cell cycle (*10*). Cell cycle dependency was also demonstrated by 4-NQO (4-nitroquinoline 1-oxide), a potent carcinogen developed in Japan. The sensitive stages were M–G_1–early-S stages and the resistant period was the late-S–G_2 stages. The binding of 4-NQO to DNA was the highest during the sensitive M–early-S period and the lowest at the resistant late-S stage-G_2 stage. The removal of DNA-bound 4-NQO was little dependent on the cell cycle (Watanabe and Horikawa, personal communication). These results indicate that the cell cycle dependency of HeLa S3 cells to UV and to 4-NQO is mostly due to the extents of the initial DNA damages and not due to their repair.

Now, I would like to describe some new methods for the repair studies developed in Japan. Ikuo Watanabe of the National Institute of Radiological Sciences has been applying the radioautographic methods to investigate a possible rejoining of double-strand cuts in DNA (personal communication). The other three methods are in vitro methods to detect repair and recovery of tissues of animals and man in vivo. Matsudaira and Furuno of the National Institute succeeded to reconstitute the rejoining system of single-strand

---

FIG. 4. Drug sensitivities of Fanconi's and normal cells.

Upper figures: the treatment scheme and its expected results of induced chromosome aberrations in lymphocytes.

Left. Lymphocytes stimulated with PHA are treated with a certain concentration of a drug for 30 minutes at various times of the cell cycle and chromosome aberrations are scored in their first entry into mitosis.

Right. A schematic relationship among frequencies of chromosome aberrations, their scoring time after a drug treatment, and cell stages at the time of the treatment in normal, repair-proficient cells and abnormal, repair-deficient cells. Xeroderma pigmentosum cells, known to be deficient in excision repair, were used to test the experimental scheme and to compare it with the cells of Fanconi's patients (see middle and bottom figures). It is noted that aberration frequencies of repair deficient cells (dotted line) are generally higher than those of repair-proficient cells (full line) when the cells are treated at the G_1 stage. Little or no difference in aberrations frequencies (full line) is found between two types of cells when they are treated at the S or the G_2 stage.

Middle figures:

Left. The MMS-induced chromosome aberrations of lymphocytes from normal individuals and xeroderma pigmentosum(XP) and Fanconi's patients.

Middle. The 4-NQO induced aberrations of lymphocytes from normal and xeroderma pigmentosum individuals.

Right. MMC induced aberrations of lymphocytes from normal individuals and xeroderma pigmentosum and Fanconi's patients.

Bottom figures:

Left. Survival curves of fibroblasts from normal and Fanconi's individuals after MMS treatments.

Middle. Survival curves of fibroblasts from normal individuals and xeroderma pigmentosum and Fanconi's patients after 4-NQO treatments.

Right. Survival curves of fibroblasts from normal individuals and xeroderma pigmentosum, and Fanconi's patients after MMC treatment.

TABLE 2
Recovery and Repair Capabilities of Tissues in Man and Animals

Tissues, Animals	Rejoining Of Single Strand Breaks	Unscheduled DNA Synthesis	Cellular Recovery: Elkind Type: Dose Fractionation	Cellular Recovery: Elkind Type: Electrophoretic Mobility	Cellular Recovery: Potential Lethal Damage
Neurons, mouse, dog	+[1]	−[12]			
Thymus, mouse, rat	+[2,3,4]			−[16]	
Spleen, mouse, rat	+[4,6]				
Bone Marrow, mouse, rat	+[4]		+[17]		
Liver, mouse, rat	+[3] ?[4]	+[12]			
Kidney, mouse, rat	?[4]	+[12]			
Intestine, mouse	+[7]		+[19]		
Skin, mouse, rat, pig, man		+[11,12]	+[20]		
Muscle, rat	?[4]				
Testis, mouse, man	+[6]	+[13,14]			
Spermatozoa, mouse, man	?[6]	−[13,14]			
Nucleated red blood cells, chicken	−[4]				
Circulating lymphocytes, man	+[5]	+[9]			
Fibrosarcoma NCTC-2472, mouse					+[15]
Ascites tumors, mouse	+[8]		+[18]	+[16]	
Spontaneous tumors, man		+[10]			

+ the presence of repair or recovery capability.
− the absence of repair or recovery capability.
? a repair or recovery capability uncertain.

1. K. T. Wheeler and J. T. Lett, *Radiat. Res.*, **52,** 59 (1972).
2. J. F. Scaife, *Int. J. Radiat. Biol.*, **21,** 197 (1972).
3. T. Ono and S. Okada, *Int. J. Radiat. Biol.* **25,** 291 (1974).
4. P. Karren and M. G. Ormerod, *Biochim. Biophys. Acta*, **299,** 54 (1973).
5. Y. Hashimoto, T. Ono and S. Okada, *Blood* (in press).
6. T. Ono and S. Okada, unpublished data.
7. T. Morita, unpublished data.
8. H. Matsudaira and I. Furuno, *Biochim. Biophys. Acta*, **272,** 202 (1972).
9. A. P. Spiegler and A. Norman, *Radiat. Res.*, **39,** 400 (1969).
10. A. Norman, R. E. Ottoman, P. Chan and I. Klisak, *Mutat. Res.*, **15,** 338 (1972).
11. J. H. Epstein, K. Fukuyama, W. B. Read and W. L. Esstein, *Science,* **168,** 1477 (1970).
12. N. W. Liebermann and P. D. Forbes, *Nature New Biology*, **241,** 199 (1973).
13. A. C. Chandley and S. Kofman-Alfaro, *Exp. Cell Res.*, **69,** 45 (1971).
14. S. Kofman-Alfaro and A. C. Chandley, *Exp. Cell Res.*, **69,** 33 (1971).
15. J. B. Little, G. M. Hahn, E. Frindel and M. Tubiana, *Radiology,* **106,** 689 (1973).
16. C. Sato, unpublished data.
17. J. E. Till and E. A. McCulloch, *Radiat. Res.*, **18,** 96 (1963).
18. S. Hornsey and G. Silini, *Radiat. Res.*, **16,** 712 (1962).
19. R. F. Hagemann, C. P. Sigdestad and S. Lesher, *Radiat. Res.* **47,** 149 (1971).
20. H. R. Withers and M. M. Elkind, *Radiat. Res.*, **38,** 598 (1969).

breaks from isolated nuclei, cell sap, NADH and ATP (*11*). Ono and I have set up the alkaline sucrose gradient centrifugation with the use of fluorometric DNA analysis of Kisane and Robbins (*12*). The same approaches have been developed independently by Wheeler and Lett (*13*) in U.S.A. and by Karran and Ormerod in U.K. (*14*). Sato of the Aichi Cancer Center is now examining the recovery of electrophoretic mobility of cells isolated from irradiated tissues. Table 2 summarizes some of these results of in vivo repair and recovery of various tissues of man and animals. It is interesting to note that a majority of dividing cells, as well as non-dividing cells possess the capability to repair some of the DNA damages. So far, a clear-cut relationship has not been seen between the radiosensitivities of tissues and their repair capabilities.

ACKNOWLEDGEMENT

The author expresses his appreciation to Drs. T. Terasima, I. Watanabe, F. Furuno and H. Matsudaira of the National Institute of Radiological Sciences, Dr. M. Sasaki of Tokyo Medical and Dental University, Drs. M. Watanabe and M. Horikawa of Kanazawa University, Dr. S. Sawada of Hiroshima University and Drs. K. Sakamoto, N. Suzuki and Mr. T. Ono of University of Tokyo for making available the unpublished data for the present talk. The author wishes to apologize to Drs. Y. Fujiwara, Y. Akita and Y. Doida for not describing their work because of a time limitation. Finally, the author acknowledges the Ministry of Education for providing funds for promoting the repair studies of mammalian cells.

REFERENCES

1. D. E. Lea, *Actions of Radiations on Living Cells,* 2nd Ed., the University Press, Cambridge, 1955.
2. C. Sato and K. Kojima, *Exp. Cell Res.*, **69,** 435 (1971).
3. C. Sato, K. Kojima, T. Matsuzawa, T. Sairenji and Y. Hinuma, *J. Radiat. Res.*, **15,** 25 (1974).
4. C. Sato, K. Kojima, M. Onozawa and T. Matsuzawa, *Int. J. Radiat. Biol.*, **22,** 479 (1972).
5. M. M. Elkind, G. F. Whitmore and T. Alescio, *Science,* **143,** 1454 (1964).
6. T. Terasima, Y. Takebe, T. Katsumata, M. Watanabe and H. Umezawa, *J. Nat. Cancer Inst., (U.S.)* **49,** 1093 (1972).
7. M. S. Sasaki and A. Tonomura, *Jap. J. Hum. Genet.* **14,** 81 (1969).
8. M. S. Sasaki and A. Tonomura. *Cancer Res.*, **33,** 1829 (1973).
9. M. Watanabe and M. Horikawa, *J. Radiat. Res.*, **14,** 258 (1973).
10. M. Watanabe and M. Horikawa, *Biochem. Biophys. Res. Comm.*, **58,** 185 (1974).
11. H. Matsudaira and I. Furuno, *Biochem. Biophys. Acta,* **272,** 202 (1972).
12. T. Ono and S. Okada, *J. Radiat. Res.*, **14,** 204 (1973).
13. K. T. Wheeler and J. T. Lett, *Radiat.* Res., **52,** 59 (1972).
14. P. Karran and M. G. Ormerod, *Biochem. Biophys. Acta,* **299,** 54 (1973).

DNA Damage and its Repair in Hyperthermic Mammalian Cells: Relation to Enhanced Cell Killing[1]

E. Ben-Hur and M. M. Elkind[2]
Department of Cellular Biochemistry
The Hebrew University
Hadassah Medical School
Jerusalem, Israel
and
Division of Biological and Medical Research
Argonne National Laboratory
Argonne, Illinois, 60439, U.S.A.

INTRODUCTION

In view of the size of the DNA molecule and its central role in information storage and transfer, it is not surprising that it has been implicated as a principal target of many radiation effects. Since the first reports that cells possess mechanisms by which they repair damage to their DNA (*1,2*), this field has been investigated intensively. Most of our information derives from studies with bacteria and much less is known relative to mammalian cells. A detailed understanding of the mechanisms by which cells preserve the integrity of their DNA has practical as well as theoretical importance. This is because in the absence of repair of DNA damage the probability of carcinogenesis is greatly enhanced (*3–6*). Also, treatments that kill cells by damaging their DNA—like radiation and alkylating agents—are used in cancer therapy and therefore repair of DNA, its absence, or misrepair is expected to play a major role in determining the efficacy of such treatments.

Recently, it was reported that cultured mammalian cells are killed more readily under hyperthermia compared to normal conditions (*7,8*). In this presentation we summarize observations relative to the sensitizing effects of hyperthermia and several agents that are known to induce DNA damage. Correlations between cell killing and DNA damage and repair are noted, and the possible use of this approach in cancer therapy is discussed.

ENHANCED KILLING OF HYPERTHERMIC CELLS

X-irradiation

X-irradiated cultured mammalian cells accumulate sublethal damage before they become lethally affected. This is evident from the presence of a shoulder on the survival curve (*9*). During postirradiation incubation, or during irradiation at a low dose rate at optimal temperatures for growth, cells

[1]Congress presentation A–3–2.

[2]Supported in part by a U.S. National Institute of Health Fellowship No. 7 FO3 CA52437-02 from the U.S. National Cancer Institute.

repair sublethal damage (for a review see Ref. *9*). Fig. 1 shows the results when Chinese hamster cells (a subclone of line *V79*) are irradiated at a low dose rate (3.3 rad/min) at different temperatures. The typical survival curve with a wide shoulder is observed at 0° C. At temperatures up to 37° C, the radiation is less effective in killing cells. The presence of a narrow shoulder under the latter conditions indicates that much less sublethal damage is accumulated since most of it is repaired during irradiation. When the temperature during irradiation is raised above 37° C, the response of the cells is progressively enhanced. At 41° C the survival curve is already close to that obtained at 0° C, and at 42° C cells are killed to an even greater extent than at 0° C. This suggests that more may be involved in the hyperthermic effect than just interference with the repair of sublethal damage.

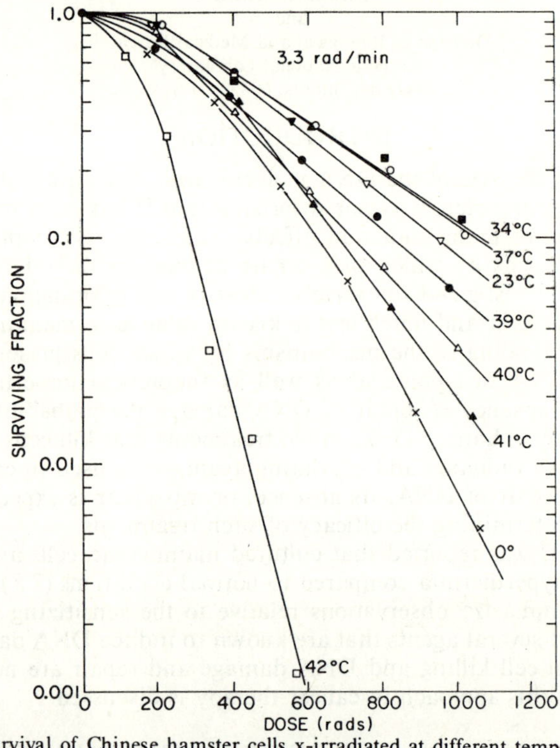

FIG. 1. Survival of Chinese hamster cells x-irradiated at different temperatures while suspended in buffer containing 20% growth medium (DUM-20) (250 kVp, 3.3 rad/min). 0° C, × ; 23° C, ▽ ; 34° C, ■ ; 37° C, o ; 39° C, ● ; 40° C, △ ; 41° C, ▲ ; 42° C, □. After irradiation, colony formation was assessed by plating cells in growth medium. (From ref. 7.)

Fig. 2 shows the effect of postirradiation hyperthermia. Under the conditions used, a slight reduction of the shoulder on the survival curve obtained at 0° C (dashed line) can be seen already for 37° C. This is because in the suspending medium used (see legends Figs. 1 and 2) cells do not repair as much sublethal damage as they do in complete medium. Raising the temperature to 40.5° C further reduces the shoulder without affecting the slope. At 41.5° C cell killing is further enhanced.

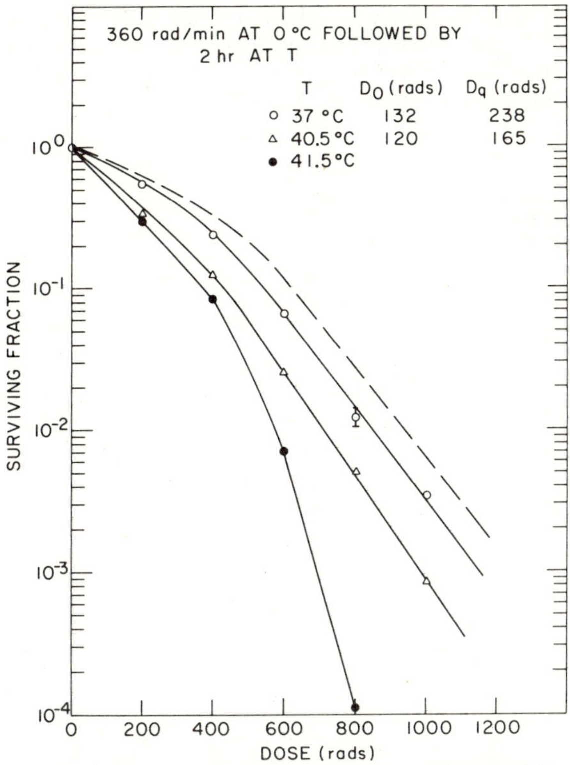

FIG. 2. Survival of Chinese hamster cells X-irradiated (250 kVp, 360 rad/min) while suspended in DUM-20 at 0°C and incubated for 2 hr at 37°C, o ; 40.5°C, △ ; 41.5°C, ● . After incubation the cells were plated in growth medium for colony formation. The dashed line is the survival curve obtained without postirradiation incubation in DUM-20. Standard errors are indicated where large enough to be shown. (From ref. *8*.)

The reduction of the shoulder at 40.5° C in Fig. 2 suggests that hyperthermia interferes with the repair of sublethal damage. A more direct indication of this is obtained from the fractionation survival curves in Fig. 3. In fractionated irradiation, frequently cells are incubated at 37° C after a conditioning dose and the repair of sublethal damage is then indicated by the reappearance of a shoulder after irradiation with graded second doses (*9*). From Fig. 3 it can be seen that in 2.5 hours at 37° C cells repaired an appreciable amount of sublethal damage even though they were suspended in a buffer containing only 20% medium. No repair is evident at 41° C. At 41.5° C the displacement of the survival curve downward indicates enhancement of lethal damage expression during the fractionation interval.

Alkylation

Methyl methanesulfonate (MMS) is a monofunctional alkylating agent which kills cells most probably by methylating DNA bases, mainly the N-7 position of guanine (*10*). Fig. 4 shows the survival curves when cells are treated with MMS for 6 hr at different temperatures (*11*). The effect of hyperthermia is clearly first to reduce the shoulder and at 41° C and above also to increase the final slope of the survival curve. This is consistent with the effects observed for X-irradiation; interference with the repair of sublethal damage up to 41° C and enhancement of lethal damage expression at higher temperatures.

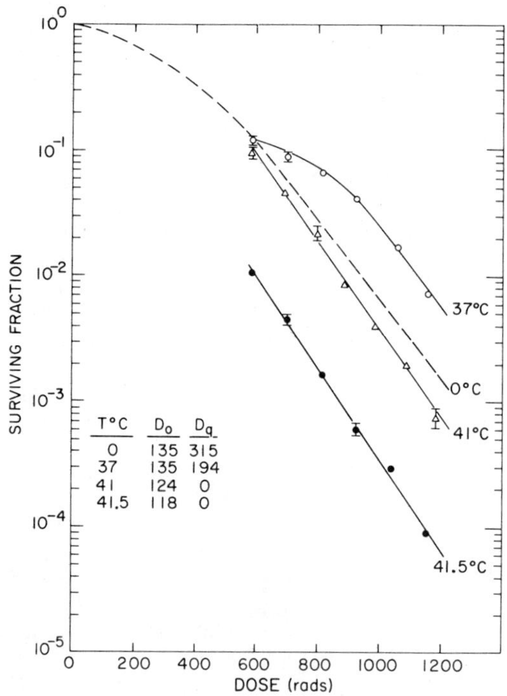

FIG. 3. Fractionation survival curves of Chinese hamster cells. Cells were exposed to a conditioning dose of 593 rads at 0° C and then incubated for 2.5 hr at: 37° C, o ; 41° C, △ ; 41.5° C, ● . They were then exposed to graded second doses at 0° C. The dashed line is the single dose survival curve; irradiation conditions as for Fig. 2.

Fox et al. (*12*) could not find evidence for repair of MMS-induced sublethal damage up to 4 hr after treatment. Fig. 5 shows that at 37° C Chinese hamster cells are able to repair sublethal damage after MMS treatment, but that this takes much longer time than after X-irradiation (12 hr for MMS treatment as opposed to 1–2 hr for X-irradiation for a comparable amount of repair). Incubation at 41° C between exposures to MMS reduces the extent of repair (D_q is reduced from 0.35 m*M* to 0.14 m*M* at 37° C and 41° C respectively, see legend for Fig. 4). At 41° C there is also enhancement of lethal damage expression as suggested by the displacement of the curves downward.

Suicide by incorporation of radionuclides

Although DNA is thought to be the target molecule when cells are killed by ionizing radiation, other cell structures are undoubtedly also affected. This is because of the indiscriminate nature of the bond breakage resulting from the random energy deposition involved, and the relatively long tracks of the secondary electrons. When β-emitters are incorporated specifically into DNA using labeled precursors like ^{3}H-thymidine, the radiation damage is more likely to be confined to the nucleus and its DNA. We used this approach to show that the effect of hyperthermia is due primarily to nuclear damage and probably to structures containing DNA. Fig. 6 shows the effect on cells of incubation at 42° C following the incorporation of ^{3}H-thymidine during the previous 10 hr at 37° C. Compared to 37° C after ^{3}H-thymidine incorporation, 5 hr at 42° C essentially eliminated the shoulder and decreased D_o by a factor of 1.8.

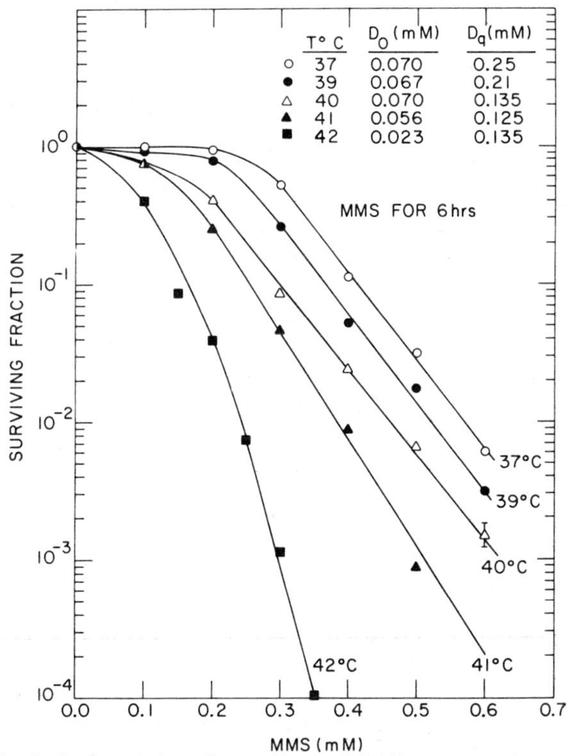

FIG. 4. Survival of methyl methanesulfonate (MMS)-treated Chinese hamster cells. Cells were incubated with growth medium containing MMS for 6 hr at: 37°C, o ; 39°C, ● ; 40°C, △ ; 41°C, ▲ ; 42°C, ■ . Survival was corrected for single cells (see chapters 2 and 3, ref. *9* for the method). D_o is the MMS concentration required to reduce survival by $1/e$ in the exponential region and D_q is the shoulder width, the MMS concentration on the abscissa intersected by a back extrapolate of the exponential region. (From ref. *11*.)

Results qualitatively similar to the above are obtained with cells containing ^{125}I-labeled iododeoxyuridine (^{125}IUdR) in their DNA (data not presented). The main difference is that ^{125}IUdR-labeled cells follow an exponential survival curve even at 37° C. Incubation at 42° C results in a steeper slope but only by a factor of 1.2. It is of interest that in addition to the lack of a shoulder, about 9 times fewer ^{125}I than ^{3}H disintegrations are required to kill a cell (D_o at 37° C 0.031 dmp/cell and 0.27 dmp/cell, respectively).

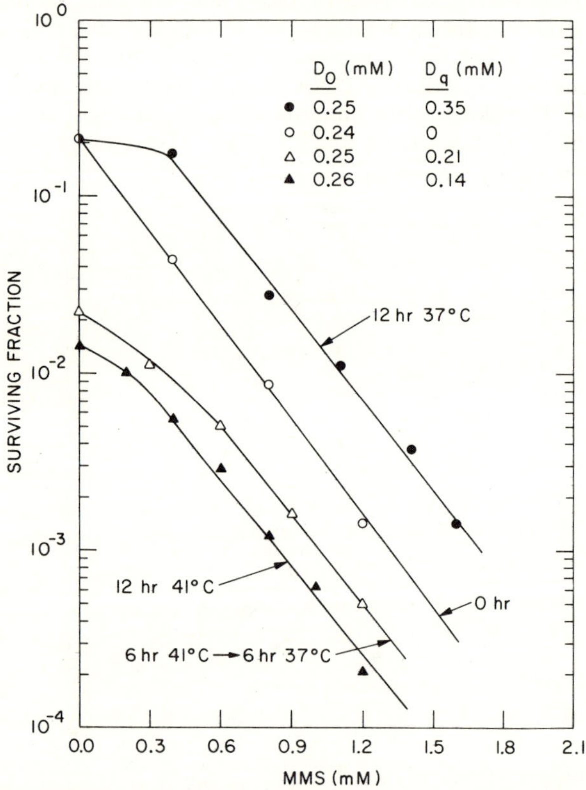

FIG. 5. Fractionation survival curves of Chinese hamster cells after treatment with MMS. The first dose was 1.5 m*M* MMS in medium for 1 hr at 37° C. Cells were exposed to graded second doses of MMS in medium (1 hr at 37°C) immediately after the first dose, o ; after 12 hr at 37°C, ● ; after 6 hr at 41°C followed by 6 hr at 37°C, △ ; and after 12 hr at 41°C, ▲ . Survival was corrected for single cells (see legend for Fig. 4). (From ref. *11*.)

UV-irradiation

UV-irradiation of cells produces cyclobutane-type pyrimidine dimers in their DNA which are thought to be the principal lesion leading to cell death (*13*). The survival curve of UV-irradiated Chinese hamster cells is of the threshold type (Fig. 7). Postirradiation incubation at 42° C for 5 hr reduces the shoulder and increases the final slope of the curve. Thus we have another example of enhanced killing response by hyperthermia following induction of lethal lesions in the DNA.

DNA DAMAGE AND REPAIR IN HYPERTHERMIC CELLS

X-irradiation

Using the alkaline sucrose gradient technique, X-ray-induced lesions resulting in scissions in single-strands of DNA and their subsequent rejoining in bacterial cells was first described by McGrath and Williams (*14*); subsequently, this technique was used by many others for studies with mammalian cells. Fig. 8 shows that at 42° C the rejoining of single-strand breaks following 5.8 krads X-rays is initially faster than at 37° C. However, after an hour cells have repaired about the same fraction of the initial breaks, about 90%.

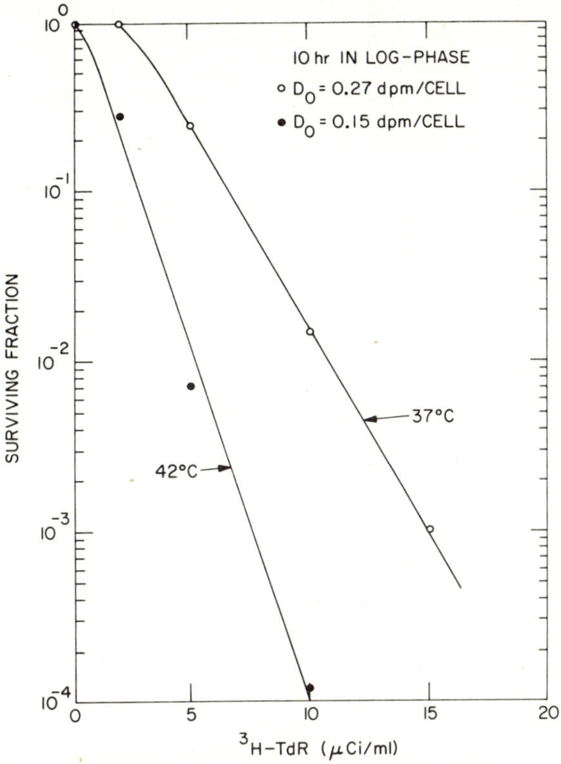

FIG. 6. Survival of Chinese hamster cells grown at 37° C for 10 hr in log phase with ^{3}H-thymidine and then incubated for 5 hr in DUM-20 (see legend for Fig. 1) at 37° C, o ; or at 42° C, ● .

When cells are incubated at 42° C for more than about 90 min, the rejoining of DNA strand-breaks is followed by DNA degradation (Fig. 9). The degradation is primarily endonucleolytic since there is very little acid-soluble material that appears concomitantly (*15*). This presumed endonuclease attack may affect both DNA strands since DNA fragments of comparable size were observed after sedimentation of hyperthermic, X-irradiated cells in neutral sucrose gradients (data not shown).

Under gentle lysis conditions of Chinese hamster cells, a DNA structure is observed with alkaline sucrose gradients which is termed the "DNA complex" since it cosediments with lipid and protein (*16–18*). The complex is much more sensitive to radiation than the single-stranded DNA, which is observed after longer lysis times, probably because its size is much larger. Consequently, the complex facilitates the study of the effects of small x-ray doses. In Fig. 10, panel *a* shows that a dose of 1700 rads resolves the complex (open-circle peak sedimenting near the bottom) into single-stranded DNA of number average molecular weight $\sim 2 \times 10^8$ daltons (*17*), which sediments near the top under the conditions used (mode at fraction 5). This dose produces very few single-strand breaks in DNA of this size (*16*). Incubation at 37° C after irradiation causes the DNA complex to reappear. After incubation at 42° C most of the single-stranded DNA disappears but a peak representing the complex does not reappear. Furthermore, the DNA material is spread almost uniformly through the gradient (Fig. 10*b*). Thus, while single-strand breaks are rejoined during hyperthermia, a normal repair of the complex is not observed.

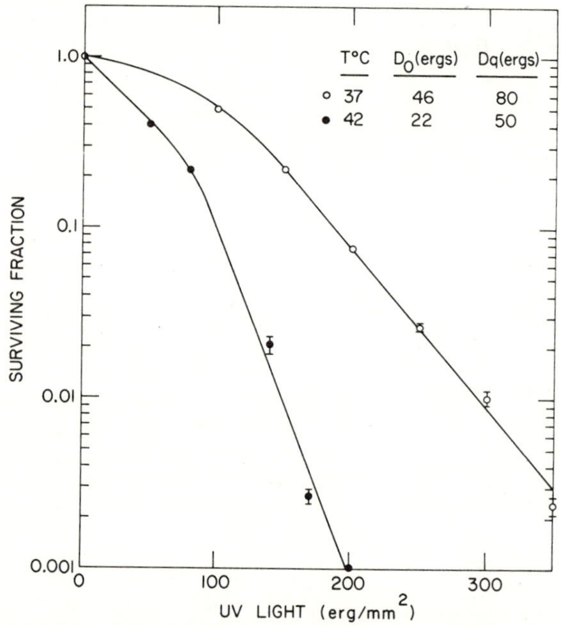

FIG. 7. Survival of Chinese hamster cells after irradiation with UV light. Cells were UV-irradiated (3.3 erg/mm^2-sec) while attached and incubated for 5 hr in growth medium at 37° C, o ; or at 42° C, •. Survival was corrected for single cells (see legend for Fig. 4).

Alkylation Treatment

MMS is a radiomimetic agent not only with respect to cell killing; it also produces lesions in DNA which result in single-strand breaks under alkaline conditions, and cells can repair such lesions (*19*). Fig. 11 shows that the yield of single-strand breaks produced by MMS treatment at 42° C is increased compared to 37° C and that this enhancement is much larger for a 6 hr than for a 1 hr treatment. Fig. 12 shows that at 42° C MMS-induced single-strand breaks are rejoined more slowly than at 37° C, which might be why apparently more breaks remain unrepaired at 42° C compared to 37° C. Also, the rate of repair of break-inducing lesions after MMS is much slower than after x-irradiation. This suggests that MMS lesions are different from x-rays-induced breaks, and consistent with the observation that hyperthermia does not slow the repair of the latter type of damage. In both cases, however, prolonged hyperthermia leads to DNA degradation.

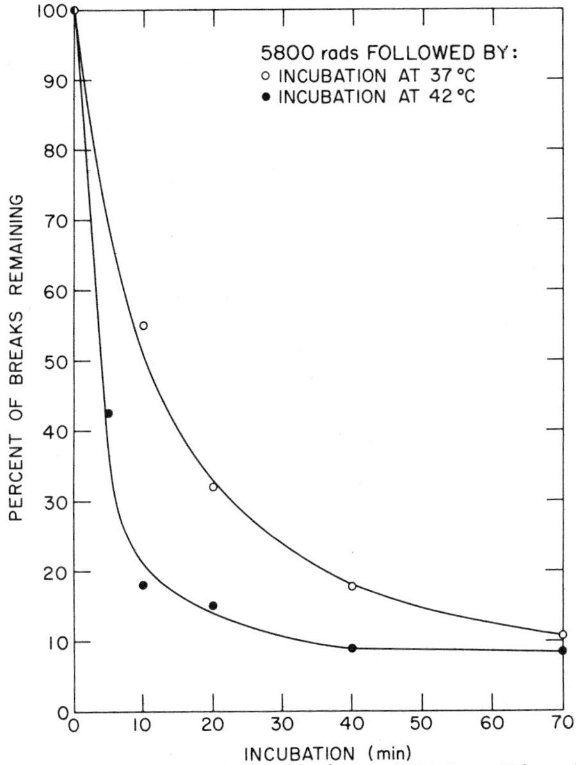

FIG. 8. The rate of repair of single-strand breaks in DNA from Chinese hamster cells. Cells were X-irradiated with 5.8 krads at 0° C and then incubated at 37° C, o ; or at 42° C, • . The number of breaks was determined after sedimentation in alkaline sucrose gradients (see legend to Fig. 9). (From ref. *15*.)

As was observed in the case of x-rays, hyperthermia also interferes with the repair of MMS-induced damage to the DNA complex. In Fig. 13, panel *a* shows that a sublethal dose of MMS (0.5 m*M* for 1 hr at 37° C) resolves the DNA complex into single-stranded DNA. At 37° C the repair of the complex takes about 6 hr (after 1700 rads x-ray, a dose that produces comparable damage, only about 45 min are required, Fig. 10*a*). At 42° C some repair of the complex is evident but to a lesser extent than at 37° C, and in addition, after 42° C more DNA is spread uniformly over the first 20 fractions than after 37° C. In panel *b* it can be seen that after a 0.9 m*M* MMS (1 hr, 37° C) a much longer time is required for the repair of the complex as is also found for repair of single-strand breaks after higher doses (Fig. 12). The effect of hyperthermia is more pronounced in this case since not only is less complex repaired but also some DNA is evidently degraded (material sedimenting at fractions 1-3).

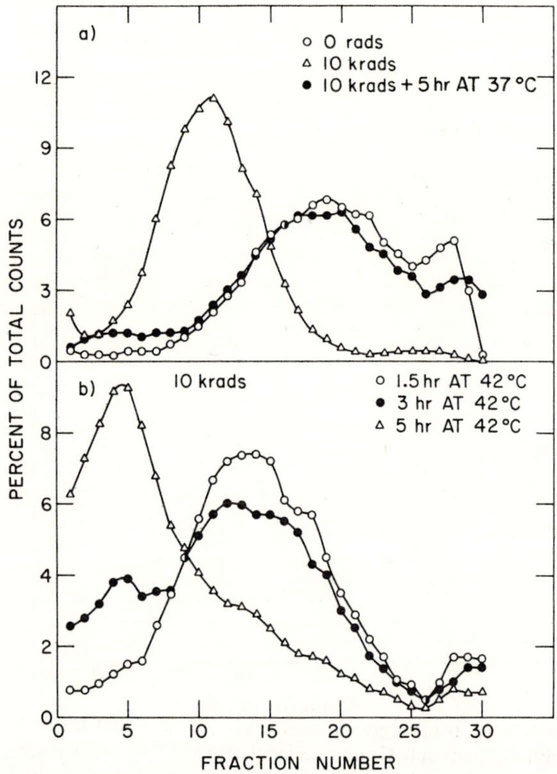

FIG. 9. Sedimentation in alkaline sucrose gradients of DNA from Chinese hamster cells. Cells (~5000), uniformly labeled with ^{3}H-thymidine, were lysed for 60 min at 42° C in a 0.25 ml top layer (0.45 *N* NaOH, 0.5 *M* NaCl, 0.01 *M* Na_2EDTA, pH 12.2) on top of 5-20% alkaline sucrose gradients (0.3 *N* NaOH, 0.7 *M* NaCl and 0.003 *M* Na_2EDTA). Sedimentation was at 11,000 rpm for 17.1 hr at 3° C in a Beckman SW-50.1 rotor.

a) Unirradiated cells, o ; cells exposed to 10 krads at 0° C, △ ; cells incubated for 5 hr at 37° C after exposure to 10 krads at 0° C, ● .

b) Cells exposed to 10 krads at 0° C and then incubated at 42° C for 1.5 hr, o ; 3 hr, ● ; or 5 hr, △ . Sedimentation from left to right (From ref. *15*.)

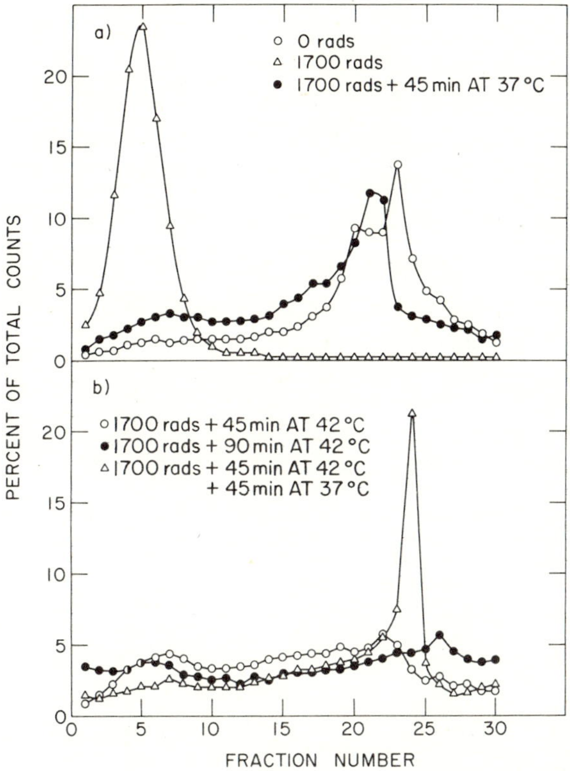

FIG. 10. Sedimentation in alkaline sucrose gradients of DNA labeled uniformly with ^{3}H-thymidine from Chinese hamster cells. Gradients and top layer characteristics were as described for Fig. 9. Lysis was for 10 min at 25° C, and sedimentation was at 5000 rpm for 16.25 hr at 0° C. a) Unirradiated cells, o ; cells exposed to 1700 rads at 0° C, △ ; cells incubated at 37°C for 45 min after exposure to 1700 rads at 0° C, ● .
b) Cells x-irradiated with 1700 rads at 0° C and then incubated at 42° C for: 45 min, o ; 90 min, ● ; or 45 min at 42° C followed by 45 min at 37° C, △. Sedimentation from left to right. (From ref. *15*.)

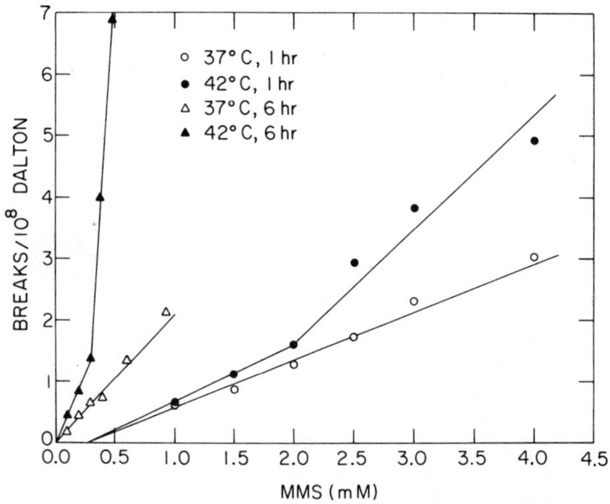

FIG. 11. Number of MMS-induced breaks in ^{3}H-thymidine labeled Chinese hamster cells per single-stranded DNA of size 1×10^8 daltons. Cells were treated with MMS for 1 hr at 37° C, o ; or at 42°C, o . Cells were also treated with MMS for 6 hr at 37°C, △ ; or at 42°C, △ . The number of breaks was determined after sedimentation in alkaline sucrose gradients (see legend for Fig. 9). (From ref. *11*.)

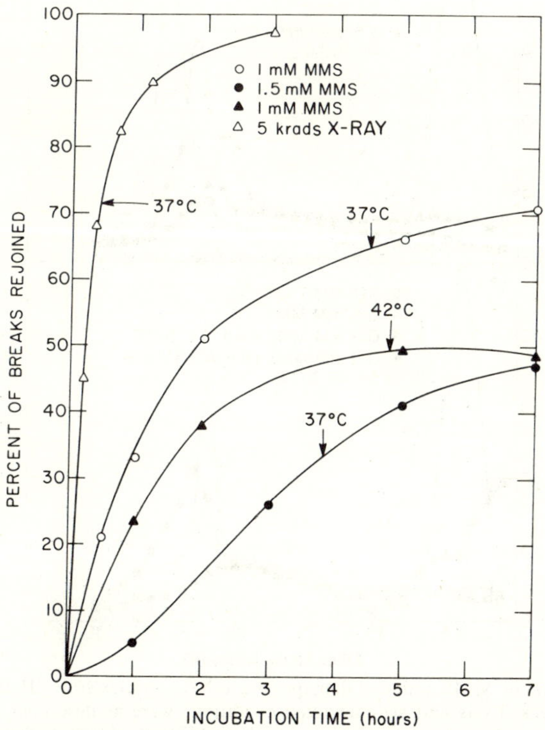

FIG. 12. The rate of repair of MMS-induced single-strand breaks in ^{3}H-thymidine labeled Chinese hamster cells. Cells were treated with 1 m*M* MMS, o ; or 1.5 m*M* MMS, o ; for 1 hr at 37°C and then incubated in medium without MMS at 37°C for varoius times. Cells treated with 1 m*M* MMS for 1 hr at 37°C and then incubated at 42°C without MMS, △ . Following incubation to allow repair the cells were lysed, sedimented in alkaline sucrose gradients, and the number of breaks determined. For comparison, the rate of repair at 37°C or X-ray-induced single-strand breaks following 5 krads is also included, △ . (From ref, *11*.)

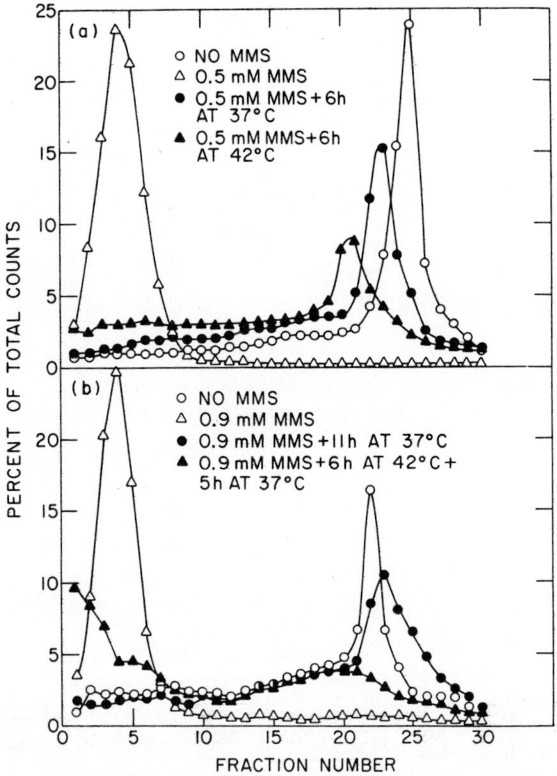

FIG. 13. Repair of MMS-induced damage to a DNA complex from ^{3}H-thymidine-labeled Chinese hamster cells. Lysis and sedimentation conditions were as described for Fig. 10. Cells were treated in medium for 1 hr at 37° C with or without MMS as shown. Sedimentation from left to right. (From ref. *11*.)

DISCUSSION AND CONCLUSIONS

Our results demonstrate that hyperthermia, up to temperatures which only marginally affect the viability of cells, strikingly enhances the cytotoxic effect of agents whose target probably involves DNA. The two effects observed at the cellular level—inhibition of sublethal damage repair and enhancement of lethal damage expression—are probably related to the effects at the molecular level; that is, the interference with the repair of the DNA complex and enhanced DNA degradation. The temperature at which these effects become evident is higher for enhancement of lethal damage expression and DNA degradation than for interference with repair. It seems reasonable, therefore, that inhibition of repair of sublethal damage could reflect interference with repair of the DNA complex while enhancement of lethal damage could result mainly from enhanced DNA degradation. These observations probably apply generally to mammalian cells since qualitatively similar results were obtained with x-irradiated HeLa cells in regard to cell killing and to DNA damage and its repair[3] (*20*), and also in regard to MMS toxicity and DNA damage in normal human fibroblasts (*21*).

Cancer cells appear to be more sensitive to hyperthermia than normal cells and some investigators have tried to exploit this property in cancer therapy by using heat treatment alone (*22* and references cited therein). We would expect that because of a higher thermal sensitivity, combined radiation and heat treatment would be more effective in killing cancer compared to normal cells. Of added importance in this regard are our results (not reported here, see *8*) which show that the return of a capacity for sublethal damage after radiation plus hyperthermia proceeds according to a much slower time scale than at 37° C. Since this lengthened time scale could result in incomplete repair in tumor cells between fractions in conventional radiotherapy, hyperthermic treatment with radiation fractionation could be particularly effective. Similar advantages may be possible for the chemotherapy of tumors.

ACKNOWLEDGEMENT

Part of this work was done under the auspices of the U.S. Atomic Energy Commission.

[3]C. Saladino and E. Ben-Hur, in preparation.

REFERENCES

1. R. B. Setlow and W. L. Carrier, *Proc. Nat. Acad. Sci. U.S.A.* **51,** 226 (1964).
2. R. P. Boyce and P. Howard-Flanders, *Proc. Nat. Acad. Sci. U.S.A.* **51,** 293 (1964).
3. W. B. Reed, G. I. Sugarman, B. H. Landing, J. E. Cleaver, and J. Melnyk, *J. Amer. Med. Ass.* **207,** 2073 (1969).
4. J. E. Cleaver, *Proc. Nat. Acad. Sci. U.S.A.* **63,** 428 (1969).
5. R. B. Setlow, J. D. Regan, J. German, and W. L. Carrier, *Proc. Nat. Acad. Sci. U.S.A.* **64,** 1035 (1969).
6. R. W. Hart and R. B. Setlow, *Abstracts. Amer. Soc. Photobiol. 1st Annu. Meet.,* Sarasota, Florida, p. 120 (1973).
7. E. Ben-Hur, B. V. Bronk, and M. M. Elkind, *Nature (London) New Biology* **238,** 209 (1972).
8. E. Ben-Hur, M. M. Elkind, and B. V. Bronk, *Radiat. Res.* **58,** 38 (1974).
9. M. M. Elkind and G. F. Whitmore, *The Radiobiology of Cultured Mammalian Cells,* Gordon and Breach, New York, 1967.
10. P. D. Lawley, *Prog. Nucl. Acid Res. and Mol. Biol.* **5,** 89 (1966).
11. E. Ben-Hur and M. M. Elkind, *Cancer Biochem. Biophys.* (in press).
12. M. Fox, C. W. Gilbert, L. G. Lajtha, and H. A. Nias, *Chem. Biol. Interact.* **1,** 241 (1969).
13. R. B. Setlow, *Science* **153,** 379 (1966).
14. R. A. McGrath and R. W. Williams, *Nature (London)* **212,** 534 (1966).
15. E. Ben-Hur and M. M. Elkind, *Radiat. Res.* **59,** 484 (1974).
16. M. M. Elkind and C. Kamper, *Biophys. J.* **10,** 237 (1970).
17. M. M. Elkind, *Biophys. J.* **11,** 502 (1971).
18. M. M. Elkind and C.-M. Chang-Liu, *Int. J. Radiat. Biol.* **22,** 75 (1972).
19. B. Strauss, M. Coyle, and M. Robbins, *Cold Spring Harbor Symp. Quant. Biol.* **33,** 277 (1968).
20. B. V. Bronk, R. J. Wilkins, and J. D. Regan, *Biochem. Biophys. Res. Commun.* **52,** 1064 (1973).
21. K. Overgaard and J. Overgaard, *Eur. J. Cancer* **8,** 65 (1972).

Use of a Purified Lesion-Recognizing Enzyme To Assay DNA Repair in Cultured Animal Cells[1]

M. C. Paterson
Biology and Health Physics Division
Atomic Energy of Canada Limited
Chalk River Nuclear Laboratories
Chalk River, Ontario, Canada

INTRODUCTION

The ability of living organisms to overcome detrimental effects caused by such environmental agents as ultraviolet (UV) radiation can be largely attributed to the presence of enzymatic processes capable of repairing damage to the deoxyribonucleic acid (DNA) (*1, 2*). Thus, despite exposure to these harmful agents, individual cells are still able to maintain the integrity of their DNA and thereby preserve the genetic information vital to survival and reproduction.

UV radiation induces the formation of cyclobutane dimers between adjacent pyrimidine bases within the same polynucleotide strand (*3*). Pyrimidine dimers are particularly well-suited to act as the reparable substrate in studies on DNA repair mechanisms (*4*) and consequently have been extensively used in this role. Two enzymatic processes mediating the repair of dimers are now understood in considerable detail in prokaryotic cells (*1, 5*) and, to a lesser extent, in eukaryotic cells (*2, 6*). In one mechanism, termed photoenzymatic repair (PER), the cyclobutane ring joining the two pyrimidine residues is cleaved chemically, thereby converting the dimer back to two normal monomers *in situ*. This process is catalyzed by the photoreactivating enzyme in the presence of 320 to 460 nm light. In a second mechanism, excision repair, dimers are selectively released from the UV-damaged DNA and then replaced with a normal nucleotide sequence complementary to that found in the opposite, undamaged strand. The most likely sequence of reactions in this light-independent process include: (i) a single strand incision near the dimer; (ii) a second single-strand nick resulting in the excision of the UV photo product, probably within an oligonucleotide, followed perhaps by the release of additional nucleotides; (iii) "repair synthesis" to fill in the gap; and (iv) ligation of the newly synthesized and pre-existing material to restore completely this stretch of DNA to its normal configuration. While pyrimidine dimers constitute the only known lesions subject to PER (*7*), excision repair mechanisms can apparently also operate on other UV photoproducts (*8*) as well as damage induced by ionizing radiation and various chemical agents (*5*).

[1]Congress presentation A–3–3.

METHODS FOR MONITORING DIMER REPAIR

Existing Assays

The ability of animal fibroblasts (cells cultured *in vitro* from tissue) to repair pyrimidine dimers has been routinely assessed by several criteria. The most direct approach entails the use of a radiochromatographic method to follow the selective disappearance of dimers from the DNA during subsequent incubation of UV-irradiated cells (*9*). This assay is specific for dimer repair and may be used to study both PER and excision repair. Unfortunately, however, this method yields imprecise measurements due to the small percentage of the total radioactively labelled thymine present in dimers at sublethal UV fluences (e.g., ~ 0.04% is found in dimers after the administration of 10 joules / m^2 to human cells; unpublished observation). Of course, as repair proceeds this already low value quickly approaches the background level (~ 0.01%), with the result that the kinetics of repair vary greatly from one experiment to the next.

A second approach involves measuring repair replication. This phenomenon reflects the sum total of small patches of new nucleotides inserted into the gaps left after excision of lesions (*10*). The extent of repair replication occurring in cells during post-UV incubation may be estimated from: (i) incorporation of ^{3}H-thymidine (dThd) into damaged DNA, determined by autoradiography (*10*); (ii) uptake of ^{3}H-bromodeoxyuridine into irradiated DNA, measured by isopycnic centrifugation (*11*); and (iii) decrease in the size of the irradiated DNA molecules as a consequence of selectively re-opening repaired lesions containing bromouracil by exposure to 313 nm light (*12*). Although all three techniques exhibit excellent sensitivity, none provide any information on the specific lesion(s) giving rise to the repair synthesis. Of course, these techniques cannot be used in PER studies.

Enzymatic Assays

We have recently described an in vitro enzymatic assay for DNA repair which affords *both* high sensitivity and excellent substrate specificity and is therefore superior to existing assays for most studies on PER and excision repair (*13, 14*). The assay utilizes the highly selective ability of a purified UV endonuclease to introduce single-strand breaks at sites containing pyrimidine dimers in DNA isolated from UV-irradiated cells. (This dimer-recognizing enzyme, purified more than 2,000-fold from *Micrococcus luteus,* presumably performs the initial incision step in excision repair in this bacterium.) The enzymatic assay is depicted schematically in Fig. 1. Animal fibroblasts, whose DNA has been previously radioactively labelled, are exposed to UV radiation and are then incubated for various times to permit dimer repair. After cell lysis, the DNA is purified and incubated with and (as control) without the UV endonuclease. The number of endonuclease-induced single-strand breaks, and hence the number of nuclease-susceptible sites (i.e., dimer-containing sites; ref. *13*) remaining in the extracted DNA, are subsequently measured by velocity sedimentation in alkaline sucrose gradients. The resulting time course of disappearance of nuclease-susceptible sites is considered to mimic the kinetics of the entire dimer-repair process occurring in vivo, since the incision step appears to be rate-limiting (*2, 5*).

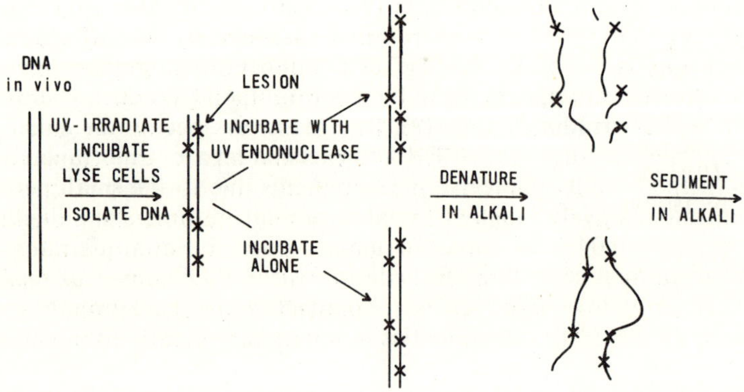

FIG. 1. General protocol illustrating the use of a UV endonuclease purified from *M. luteus* to measure the number of pyrimidine dimers in ^{3}H-labelled DNA of UV-irradiated animal cells. Although not shown here, ^{14}C-labelled DNA from undamaged cells is coextracted with UV-damaged DNA and accordingly serves as an internal control to monitor the number of non-specific strand breaks resulting from the nuclease treatment. (From Paterson et al., *Mutat. Res.* **19,** 245 (1973) with the permission of Elsevier Scientific Publishing Co.

Recently a second assay, similar in principle to ours, has been developed by Wilkins (*15*). (Another assay almost identical to that of Wilkins has also been described by Ganesan (*16*), but it has been employed solely in bacterial studies and consequently shall not be discussed here.) In this assay, after repair incubation, rather than isolating the UV-damaged DNA to allow the purified UV endonuclease to react with its purified substrate, the cells are kept intact but are made permeable to extracellular proteins, including the dimer-recognizing endonuclease whose entry is required for assay. After incubation to permit the foreign UV endonuclease to attack dimer-containing sites, the cells are then lysed directly on alkaline sucrose gradients. This in vivo assay procedure has the desired effect of increasing greatly the size of the sedimented DNA (e.g., control DNA is ~ 10-fold greater than that obtained when using our in vitro assay; refs. *13, 15*) and accordingly the sensitivity of the measurement. On the other hand, while the in vitro enzymatic assay permits the detection of all dimer-containing sites (*13*), it is estimated that approximately half of the sites go undetected when analyzed by the in vivo assay (*15*). Use of the in vivo enzymatic assay then presupposes that the dimer-containing sites detected are a random sample of the entire population in the UV-damaged DNA.

DIMER REPAIR AS MEASURED BY THE IN VITRO ENZYMATIC ASSAY

The sedimentation profiles in Fig. 2 typify those routinely obtained when DNA coisolated from UV-irradiated and (as an internal control) non-irradiated human cells is analyzed by the in vitro enzymatic assay. The results are as predicted by the scheme of Fig. 1. Incubation of the purified DNA in the presence of the UV endonuclease results in a dramatic decrease in the weight-average molecular weight (Mw) of the UV-damaged relative to the un-damaged DNA. (Mw is a measure of the average size of a given population of molecules.) This difference corresponds to 4.3 single-strand breaks per 10^7 daltons number-average molecular weight (Mn) in the irradiated compared with the non-irradiated DNA[2]. (The precision of this method is within ±0.1 sites per 10^7 daltons.) When the DNA extract is incubated without the *Micrococcus* enzyme, the difference in the Mw's of the two DNA's is negligible; that is, equivalent to 0.1 breaks per 10^7 daltons (Mn) in the irradiated DNA. Hence, we estimate that human DNA insulted in vivo with an incident UV fluence of 20 joules/m^2 contains, on the average, 4.2 sites per 10^7 daltons which are susceptible to attack by the dimer-recognizing endonuclease. This number of nuclease-susceptible sites agrees within experimental error with the number of pyrimidine dimers (~ 4.5 per 10^7 daltons) present in the same human DNA as measured by radiochromatography. (See ref. *17* for details concerning the calculation of dimer content.) Thus, as pointed out earlier, the enzymatic assay is apparently able to detect all dimer-containing sites.

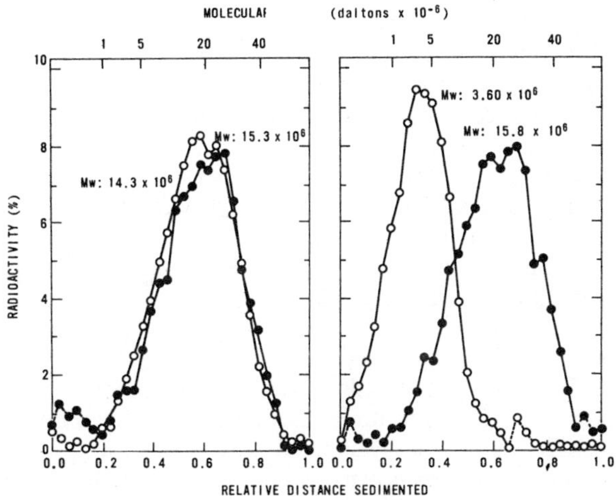

FIG. 2. Sedimentation profiles in alkaline sucrose gradients of radioactive human DNA co-extracted from ^{3}H-labelled, UV-irradiated (20 joules/m^2), (O) and ^{14}C-labelled, non-irradiated (●) AH fibroblasts after incubation in the absence (Left panel) or presence (right panel) of the *M. luteus* UV endonuclease. The portions of each profile drawn with a broken line denote those gradient fractions considered to be divorced from the main body of sedimented DNA and thus omitted in the computation of Mw. (For experimental details see ref. *13*).

We have used the in vitro enzymatic assay to assess the ability of human and embryonic chick fibroblasts to perform both excision repair and PER. To distinguish between the two repair processes, cultures of irradiated cells were incubated in duplicate, one in the dark (a condition permitting only excision repair) and the other under blacklight (a condition permitting PER in addition to excision repair). As may be seen from Fig. 3, human fibroblasts (AH) but not chick embryonic fibroblasts (CEF) are able to eliminate nuclease-susceptible sites when incubated in the dark. Nuclease-susceptible sites disappeared from the DNA of irradiated AH cells in two exponential phases; about 70% of the sites was removed in an initial phase extending up to 15 hr, and no more than an additional 10% was acted upon during the next 15-hr period. To obtain a more accurate measurement of PER in human cells, we used XP4 cells which are known to be defective in the removal of nuclease-susceptible sites by excision repair (*13*). These cells also failed to eliminate sites during post-UV incubation under blacklight, thereby confirming earlier reports demonstrating the inability of human fibroblasts to execute PER (*18*). In contrast, all nuclease-susceptible sites disappeared exponentially from the DNA of UV-irradiated CEF cells during ~ 1.5 hr of blacklight illumination. This rate of site removal is ~ 12 times faster than that of the fast component of the excision-repair process operating in AH cells. It would seem, therefore, that human and embryonic chick cells depend heavily, if not exclusively, on different enzymatic processes for the repair of pyrimidine dimers. While dimer-containing sites are eliminated from DNA of human cells by a light-independent process, undoubtedly excision repair (*13, 19*), embryonic chick cells rely on a light-dependent process, presumably PER (*8*), for removal of these UV photoproducts.

By obtaining for each cell strain a family of site-removal curves resulting from different UV fluences and then expressing for each fluence the absolute rate of site removal as a function of the initial number of sites, it is possible to ascertain the UV fluence needed to saturate the repair process under study. Such an analysis indicates that saturation of PER in chick cells requires ~ 50 joules/m^2 (*8*), a fluence ~ 5-fold greater than that (6.5–12.5 joules/m^2) needed to maximize the rate of excision repair in human cells (*13*).

The in vitro enzymatic assay has been employed to measure the proficiency with which enzymes associated with PER and excision repair in animal cells can operate in vivo on damaged DNA in foreign, as compared to native, chromatin. This has been done by following the disappearance of nuclease-susceptible sites in artificially fused cultures derived from both human and chick cells (*17*). The majority of cells in these cultures are giant heterokaryons containing as many as 15 intact nuclei as well as cytoplasm derived from both fusion members. The data strongly suggest that within these human/chick heterokaryons the chick photoreactivating enzyme and at least one human enzyme involved in excision repair (presumably the dimer-recognizing endo-

[2]Number of single-strand breaks per 10^7 daltons of irradiated DNA = $10^7[(1/Mn_i) - (1/Mn_o)]$ where Mn_i and Mn_o are the respective Mn values of the UV-irradiated and non-irradiated DNA's. Although strand breakage measurements are based on Mn determinations, rather than calculate Mn directly it is more accurate to compute Mw and then, assuming a random distribution, simply halve this value to obtain Mn (*12*).

nuclease) can remove nuclease-susceptible sites from foreign DNA in other nuclei with an efficiency comparable to that directed toward native DNA in their own nuclei. This would seem to indicate that enzymes mediating the chick and human repair processes can freely diffuse into other nuclei and, once therein, operate on UV-damaged chromatin independent of its origin.

The results of a recent investigation on human cells by Wilkins and Hart (*20*) are consistent with our finding that the kinetics of site removal performed by excision repair display a biphasic exponential decline. Their data also suggest that the fraction of nuclease-susceptible sites acted upon by the slow component of repair is located in stretches of DNA normally enclosed with nucleoproteins and, thus, less accessible to repair enzymes. Since, however, the chick photoreactivating enzyme can rapidly remove such sites from human DNA in the human/chick heterokaryons, this notion seems unlikely unless, of course, the photoreactivating enzyme can penetrate the suggested proteinaceous sheath.

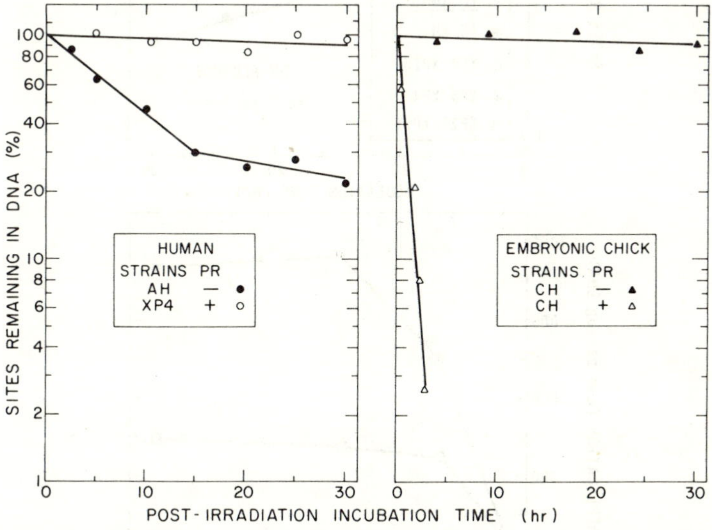

FIG. 3. Disappearance of nuclease-susceptible sites from the DNA of UV-irradiated (10 joules/m^2) human and embryonic chick cells during subsequent incubation in the dark (−PR) or under blacklight (+PR). (From Paterson et al., *Nature (London)* **248**, 50 (1974) with the permission of MacMillan Journals Ltd.)

USE OF EXISTING ASSAYS TO COMPLEMENT ENZYMATIC ASSAY

While the in vitro enzymatic assay has by itself yielded new and important information, it is often prudent to complement this assay with one of the conventional assays which monitor repair replication. For example, although apparently unable to excise dimer-containing sites (Fig. 3), CEF cells were found to perform a considerable amount of repair replication (*8*). Furthermore, the level observed was independent of blacklight illumination. Taken together, these data provide convincing evidence that chick cells possess an excision-type repair process which is operative on UV photoproducts other than dimers. This is noteworthy since it is generally accepted that the UV photoproducts giving rise to repair replication are almost exclusively pyrimidine dimers (*2, 21*).

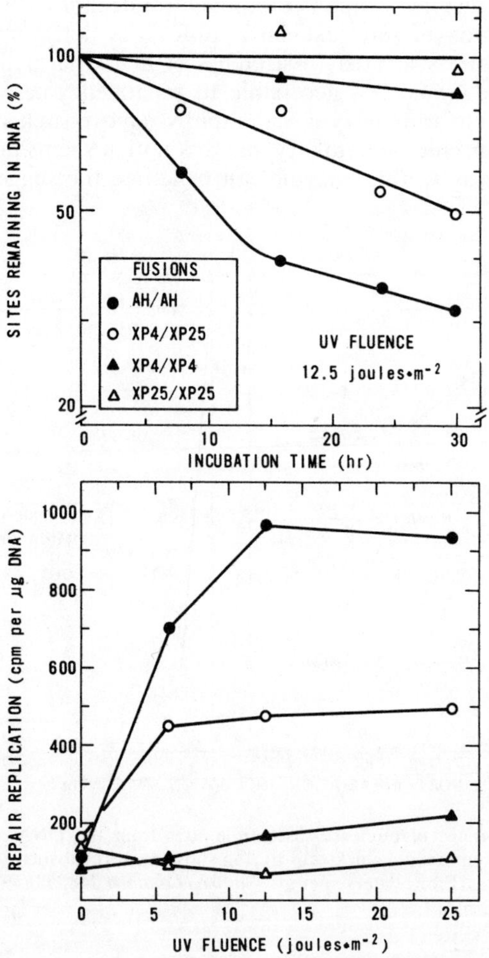

FIG. 4. Disappearance of nuclease-susceptible sites (upper panel) and occurrence of DNA repair replication (lower panel) in artificially fused human fibroblasts after exposure to indicated fluences of UV radiation. (From Paterson et al., *Nature* (*London*) **248,** 50 (1974) with the permission of MacMillan Journals Ltd.)

Xeroderma pigmentosum (XP) is a rare genetic disease of man characterized by hypersensitivity of the skin to the ultraviolet component of sunlight (reviewed in ref. *22*). Two clinical forms of the disease are known: (i) the classical form manifested by a high incidence of multiple cutaneous neoplasms; and (ii) the De Sanctis-Cacchione syndrome in which severe neurological abnormalities accompany the skin cancers. We have monitored both the disappearance of nuclease-susceptible sites and the occurrence of repair replication in fused cultures of XP4 and XP25 cells which were derived from patients suffering from both clinical forms (Fig. 4). Both the extent of site removal and the amount of repair replication are significantly larger in the XP4/XP25 hybrids than in self-fused XP4/XP4 and XP25/XP25 cultures, indicating that each cell type can complement the genetic deficiency of the other. However, for both repair parameters, the proficiency displayed by the XP4/XP25 hybrids is only ~50% of that observed in fused preparations of normal AH cells. This intermediate level of site removal and repair replication seen in XP4/XP25, as compared to AH-AH, cultures is an effect ascribed to the fact that only a fraction (~55%) of the total nuclei in the fused cultures are present in multinucleate cells (*19*). Hence, XP4/XP25 heterokaryons per se appear to exhibit normal kinetics for both site removal and repair replication, parameters believed to reflect initial and late steps in excision repair. Thus, in agreement with earlier studies reported by de Weerd-Kastelein and coworkers (*23*), we conclude that the two clinical forms of XP reflect mutations in different genetic complementation groups coding for the synthesis of a dimer-recognizing endonuclease responsible for initiating excision repair (*19*).

CONCLUSION AND PROSPECTUS

It is clear from the above data that the in vitro enzymatic assay, which utilizes a purified UV endonuclease as a dimer probe, not only complements existing assays for DNA repair, but also lends itself to a number of novel studies. In fact, preliminary data (unpublished) indicate that this assay may be gainfully employed to elucidate repair processes operating on a variety of physicochemical alterations in DNA including those induced by UV and ionizing radiations, as well as various chemical agents displaying mutagenic, carcinogenic, or teratogenic activity.

ACKNOWLEDGEMENTS

Much of the work reported here was performed when the author was an MRC of Canada Post-doctorate Fellow at the Laboratory for Molecular Genetics, Leiden State University, Leiden, Holland. I am deeply indebted to my Dutch collaborators, M. L. Sluyter, A. Westerveld, E. A. de Weerd-Kastelein, and especially P. H. M. Lohman. Thanks are extended to N. E. Gentner and B. P. Smith for their helpful commentary on the manuscript.

REFERENCES

1. I. P. Howard-Flanders, *Brit. Med. Bull.* **29,** 226 (1973).
2. J. E. Cleaver, in *Advances in Radiation Biology* (J. T. Lett, H. Adler and M. Zelle, Eds.) **Vol. 4,** pp. 1–75, Academic Press, New York, 1974.
3. R. B. Setlow, *Science,* **153,** 379 (1966).
4. R. B. Setlow, Progr. Nucl. Acid Res. Mol. Biol., 257 (1968).
5. L. Grossman, in *Advances in Radiation Biology* (J. T. Lett, H. Adler, and M. Zelle, Eds.) **Vol. 4,** pp. 77–129, Academic Press, New York (1974).
6. J. S. Cook, in *Photophysiology* (A. C. Giese, Ed.) **Vol. 5,** pp. 191–233, Academic Press, New York, 1970.
7. J. K. Setlow, M. E. Boling, and F. J. Bollum, *Proc. Nat. Acad. Sci. USA* **53,** 1430 (1965).
8. M. C. Paterson, P. H. M. Lohman, E. A. de Weerd-Kastelein, and A. Westerveld, *Biophys. J.* **14,** 454 (1974).
9. W. L. Carrier and R. B. Setlow, in *Methods in Enzymology* (L. Grossman and K. Moldave, Eds.) **Vol. 21** Part D, pp. 230–237, Academic Press, New York, 1971.
10. D. Bootsma, M. P. Mulder, F. Pot, and J. A. Cohan, *Mutat. Res.* **9,** 507 (1970).
1. P. C. Hanawalt and P. K. Cooper, in *Methods in Enzymology* (L. Grossman and K. Moldave, Eds.) **Vol. 21** Part D, p. 221, Academic Press, New York, 1971.
12. J. D. Regan, R. B. Setlow, and R. D. Ley, *Proc. Nat. Acad. Sci. USA* **68,** 708 (1971).
13. M. C. Paterson, P. H. M. Lohman, and M. L. Sluyter, *Mutat. Res.* **19,** 245 (1973).
14. M. C. Paterson and P. H. M. Lohman, in *Proc. ICN-UCLA Conference on Molecular Mechanisms for the Repair of DNA.* (P. C. Hanawalt and R. B. Setlow, Eds.) Plenum Press, New York, in press.
15. R. J. Wilkins, *Int. J. Radiat. Biol.* **24,** 609 (1973).
16. A. K. Ganesan, *Proc. Nat. Acad. Sci. USA* **70,** 2753 (1973).
17. M. C. Paterson, P. H. M. Lohman, A. Westerveld, and M. L. Sluyter, *Biophys. J.* **14,** 835 (1974).
18. J. E. Cleaver, *Biochem. Biophys. Res. Commun.* **24,** 569 (1966).
19. M. C. Paterson, P. H. M. Lohman, A. Westerveld, and M. L. Sluyter, *Nature (London)* **248,** 50 (1974).
20. R. J. Wilkins and R. W. Hart, *Nature (London)* **247,** 35 (1974).
21. R. B. Setlow and J. K. Setlow, *Annu. Rev. Biophys. Bioeng.* **1,** 293 (1972).
22. J. H. Robbin, K. H. Kraemer, M. A. Lutzner, B. W. Festosf, H. G. Coon, *Ann. Intern. Med.* **80,** 221 (1974).
23. E. A. de Weerd-Kastelein, W. Keijzer, and D. Bootsma, *Nature (London) New Biol.* **238,** 80 (1972).

Carcinogens And DNA Repair[1]

H. F. Stich and B. Laishes
Cancer Research Centre and Department of Zoology,
University of British Columbia, Vancouver, Canada

The capacity of microbial, plant and animal cells to correct DNA lesions has recently attracted considerable attention from many branches of biology and medicine. The time when studies of recovery mechanisms were a prerogative of radiobiologists and photobiologists has long passed (*1*). The original ideas, concerning the role of DNA repair in the process of recovery of bacterial cells, have diversified and expanded in great proportions. It therefore appears of general interest to present a short review of some of the current advances in the field.

Microbial and mammalian cells correct a multitude of DNA changes that are induced by different physical agents and chemical compounds of a diversity of molecular structures (Table 1). At present, the view prevails that compounds capable of eliciting a DNA repair synthesis in mammalian cells are potential carcinogens and mutagens (*2,3*). Most if not all of the compounds examined either generate electrophilic reactants spontaneously at physiologic pH, or can be changed into electrophilic metabolites within organs of mammals (*4,5*).

The thesis that chemical carcinogens have mutagenic properties is by no means a new one. However, convincing evidence in support of this notion has been gathered only recently. The long delay in proving or disproving this idea can be attributed mainly to a lack of understanding of the molecular changes that alter inactive compounds into electrophilic reactants. This difficulty is partly overcome by incubating chemicals with an activation mixture consisting usually of a 9,000xg liver preparation (*S–9*), fortified with an energy generating or supplying system (*6–8*). The in vitro activated compounds can induce gene mutations (*6,7*), or chromosome aberrations (*8*), elicit a DNA repair synthesis (*9*), or cause a shift in the molecular weight of DNA (*9*). In spite of these successes in activating several precarcinogens with the 9,000xg fraction of rat or mouse livers, the basic mechanism of this process remains only poorly understood, and likely varies from chemical to chemical. Similarly, a relationship between an organ-specific effect of a carcinogen and a possible organ-specific activation has not been established. A few examples may show some of the unsolved problems:

(1) S–9 preparations of mouse livers are potent activators of aflatoxin B1 although full-grown mice are resistant to this carcinogen (*10*) (Table 2).

[1]Congress presentation A-3-4.

TABLE 1
A DNA Repair Synthesis Occurs in Cultured Human Fibroblast After the Application of Following Chemical Carcinogen –

4 - NITROQUINOLINE 1 - OXIDE	BENZ (∝) ANTHRACENE - 5, 6 - EPOXIDE
4 - HYDROXYAMINOQUINOLINE 1 - OXIDE	20 - METHYLCHOLANTHRENE - 6, 7 - EPOXIDE
2 - METHYL - 4 - NITROQUINOLINE 1 - OXIDE	
3 - METHYL - 4 - NITROQUINOLINE 1 - OXIDE	N - HYDROXY - 2 - ACETYLAMINOFLUORENE
5 - METHYL - 4 - NITROQUINOLINE 1 - OXIDE	N - ACETOXY - 2 - ACETYLAMINOFLUORENE
6 - METHYL - 4 - NITROQUINOLINE 1 - OXIDE	
7 - METHYL - 4 - NITROQUINOLINE 1 - OXIDE	N - HYDROXY - 4 - ACETYLAMINOSTILBENE
8 - METHYL - 4 - NITROQUINOLINE 1 - OXIDE	N - ACETOXY - 4 - ACETYLAMINOSTILBENE
6 - n - BUTYL - 4 - NITROQUINOLINE 1 - OXIDE	
6 - t - BUTYL - 4 - NITROQUINOLINE 1 - OXIDE	N - HYDROXY - 2 - ACETYLAMINOPHENANTHRENE
6 - n - HEXYL - 4 - NITROQUINOLINE 1 - OXIDE	N - ACETOXY - 2 - ACETYLAMINOPHENANTHRENE
3 - FLUORO - 4 - NITROQUINOLINE 1 - OXIDE	
6 - CHLORO - 4 - NITROQUINOLINE 1 - OXIDE	N - HYDROXY - 4 - ACETYLAMINOBIPHENYL
6 - CARBOXY - 4 - NITROQUINOLINE 1 - OXIDE	N - ACETOXY - 4 - ACETYLAMINOBIPHENYL
4 - NITROQUINOLINE	
	N - ACETOXY - 3 - ACETYLAMINOFLUORENE
	N - MYRISTOYLOXY - 2 - ACETYLAMINOFLUORENE
3 - METHYL - 4 - NITROPYRIDINE 1 - OXIDE	N - ACETOXY - 2 - MYRISTOYLAMINOFLUORENE
3 - METHYL - 4 - HYDROAMINOPYRIDINE 1 - OXIDE	N - MYRISTOYLOXY - 2 - MYRISTOYLAMINOFLUORENE
2, 3 - DIMETHYL - 4 - NITROPYRIDINE 1 - OXIDE	
2, 3 - DIMETHYL 4 - HYDROXYAMINOPYRIDINE 1 - OXIDE	2 - NITROSOFLUORENE
2 - METHYL - 4 - NITROPYRIDINE 1 - OXIDE	4 - NITROSOBIPHENYL
3 - ETHYL - 4 - NITROPYRIDINE 1 - OXIDE	2 - NITROSOPHENANTHRENE
2, 3 - DIETHYL - 4 - NITROPYRIDINE 1 - OXIDE	2 - NITROSONAPHTHALENE
	1 - NITROSONAPHTHALENE
N - METHYL - N′ - NITRO - N - NITROSOGUANIDINE	
METHYLMETHANE SULFONATE	DIMETHYLNITROSAMINE (ACTIVATED)
ETHYLMETHANE SULFONATE	AFLATOXIN B1 (ACTIVATED)
NITROSOMETHYLUREA	STERIGMATOCYSTIN (ACTIVATED)
NITROGEN MUSTARD	
ICR - 191	
3′ - ACETOXY - SAFROLE	
1, 1 - DIPHENYL - 2 - PROPYNYL - n- CYCLOHEXYLCARBINOL	
1 - PHENYL - (3, 4 - XYLYL) - 2 - PROPYNYL - n- CYCLOHEXYLCARBAMATE	

TABLE 2
The Activation of Aflatoxin B1 and Dimethylnitrosamine by 9,000xg Supernatants of Liver, Kidney and Lung of Control and Phenobarbitol-Treated Syrian Hamsters
(Laishes and Stich, unpublished)

	Grains per nucleus (autoradiography)						
	Hamster			Hamster Pretreated			
	Liver	Kidney	Lung	Liver	Kidney	Lung	No Activation
Aflatoxin	25.3	21.3	20.6	21.9	23.3	21.8	0.8
DMN	8.7 9.3	3.1	2.2	24.0	2.8	0.6	1.7

(2) Sterigmatocystin induces, in rats, mainly hepatomas but no kidney or lung tumors (*11*). However, S–9 preparations from liver, kidney, and lung of rats activate sterigmatocystin at comparable levels.

(3) Pretreatment of rats or hamsters with phenobarbitol increases the activity of the S–9 fraction for some but not all precarcinogens.

The link between the carcinogenic properties of physical and chemical agents and their DNA-damaging action seem to implicate the DNA repair mechanism in one of the early steps of neoplastic transformation. The high frequency of actinic skin tumors of Xeroderma pigmentosum (XP) patients who have a reduced DNA repair capacity supports this view. An apparent contradiction to this assumption are XP patients with a normal level of DNA repair synthesis (*12,13*). However, cells of these patients have a reduced capacity to re-activate UV-irradiated viruses (*14*). Thus one cannot avoid the conclusion that this particular group of XP patients may have a deficient repair mechanism in spite of a normal capacity of DNA repair synthesis, as revealed by estimating the unscheduled uptake of ^{3}HTdR. However, the most convincing evidence that links a particular DNA alteration (pyrimidine dimers) and its repair, or absence of its repair, with neoplastic transformation was recently obtained by ingenious experiments with fish cells (*15,16*). The carcinogenic effect of UV-irradiation can be abolished when the fish cells are permitted to photo-reactivate the UV-induced pyrimidine dimers.

Considering the relationship between the action of physical/chemical carcinogens and DNA repair, the question should be raised as to a link between oncogenic viruses and repair processes. There is a paucity of information on this topic but the few published results show that this area merits greater attention than it has received in the past. A possible involvement of DNA repair in viral carcinogenesis becomes evident when virus and chemical carcinogens are jointly applied. For example, a short exposure of embryonal hamster cells to the potent chemical carcinogen 4NQO (4–nitroquinoline–1–oxide), prior to the addition of the oncogenic SA7 (Simian adenovirus) enhanced the transformation frequency up to 23-fold (*17*). The virus must be added during a 12-hour period following a single 90-minute application of 4NQO in order to obtain this enhancement effect (*8*). In this connection it is of interest to note that the "susceptible" time span coincides with the period of active DNA repair synthesis following the 4NQO pulse (*18*). Whether the single-stranded DNA sites, that may be present during the first 12 hours, facilitate the integration of viral DNA which in turn leads to a higher probability of transformation, is difficult to assess at present. In support of such an assumption, one can mention the elevated integration of SV40 genomes into hamster cells following 4NQO pretreatment (*19*). Furthermore, 4NQO derivatives that induce a low level of DNA repair or lack the ability to elicit DNA changes show either a lower or no enhancement effect respectively (Fig. 1).

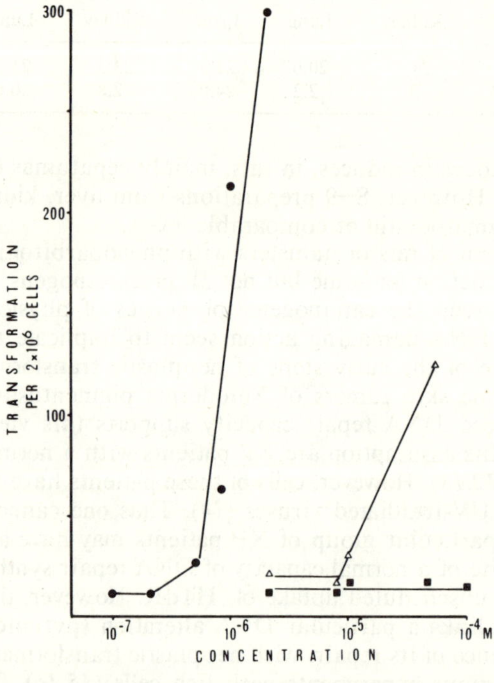

FIG. 1. Frequency of SA7-transformed colonies in Syrian hamster cell cultures that were exposed to the strong carcinogen 4NQO (●), the weak carcinogen 3-methyl-4NQO (△), and the non-oncogenic 6NQO (■) for 90 min prior to the addition of the virus (Stich and Casto, unpublished).

Another involvement of DNA repair in the expression of viral functions was uncovered by examining the reactivation of UV-irradiated viruses that contain DNA and replicate within the nucleus of the host cell. The replicative function of UV-irradiated SV40 (*20*), herpes simplex (*21*), AD2 (*14*), or AD12 (*22*) is restored by cells of control persons but shows a reduced recovery in cells of XP patients. The degree of reactivation of viral functions depends on the level of DNA repair capacity (*14*) (Fig. 2). Comparable results were obtained when the carcinogen 4NQO was applied to control cells and XP fibroblasts following viral (AD12) absorption (Fig. 3). The simplest interpretation of these results is to assume that UV- or 4NQO-induced DNA alterations in the viral genome are corrected by enzymes of the host cell and that a deficiency of the repair system is reflected in a lack of viral reactivation. The

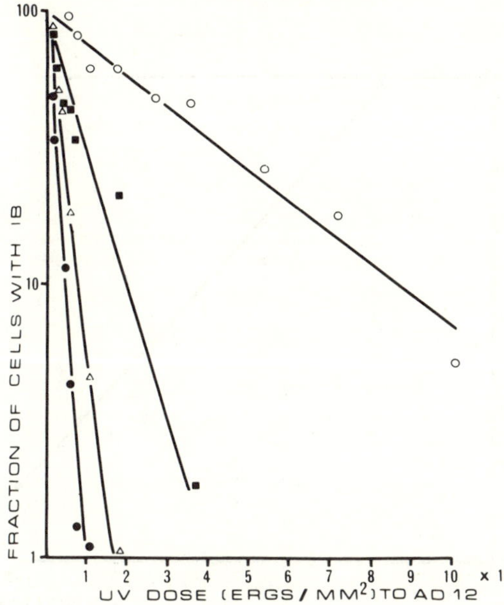

FIG. 2. The effect of UV-irradiation on the capacity of AD12 to form intranuclear inclusion bodies in cultured control cells and Xeroderma pigmentosum cells of various DNA repair levels. O controls, ● XP-E, △ XP-C and ■ XP-K. Samples were taken 72 hrs post-infection. The unscheduled uptake of ^{3}HTdR was about 21% for XP-E, 32% for XP-C and 56% for XP-K as compared to that found in three control subjects following UV-irradiation.

behavior of damaged viral genomes in repair-deficient host cells has a profound effect on the virus–cell interaction. For example UV-irradiated AD12 induces an extremely high frequency of chromosome aberrations in XP cells without affecting their proliferation, whereas the same virus preparation will inhibit cell division and will ultimately cause lysis of normal cells. Thus the outcome of an infection with damaged DNA viruses may significantly differ between cells with normal and impaired DNA repair system.

An understanding of DNA repair and its biological role is of academic interest as well as of practical importance. The capacity of chemical carcinogens to bind and alter DNA of mammalian cells can be utilized in the detection of compounds with carcinogenic properties (*2,3,23,24,25*). Since measurements of DNA repair synthesis are relatively fast, can be semimechanized and can be adapted to human lymphocytes (*26,27*) or cultured fibroblasts, this method may be applicable in large-scale screening programs of compounds for their carcinogenic or mutagenic properties. There are two great advantages to this approach that should not be overlooked. Firstly, DNA repair measurements can be performed on a battery of cells from "normal" persons and patients afflicted by various diseases. Thus the risk of normal and sensitive human cells to the toxic effect of chemicals can be assessed, a feature which

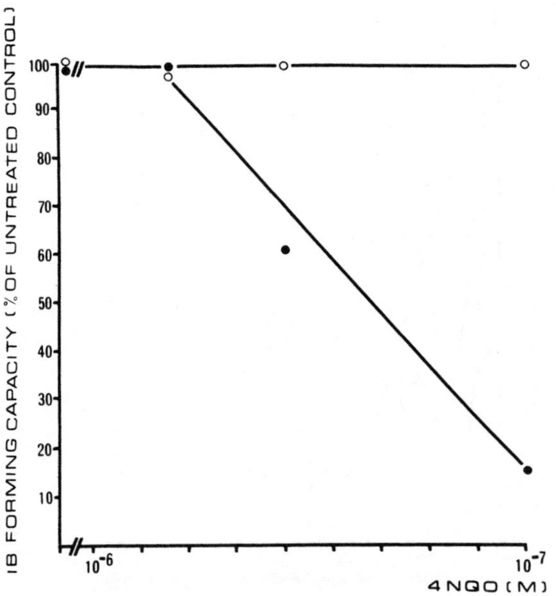

FIG. 3. The effect of 4NQO (90 min application following a 3 hrs virus absorption period) on the capacity of AD12 to form inclusion bodies in cultured control cells (O) and XP-E cells (●). Sample was taken 69 hr post-infection.

gives this bioassay an obviously greater relevance than can be obtained from microbial test systems. Secondly, DNA repair seems to be a general response to many different kinds of DNA alterations. Thus by estimating DNA repair synthesis, a wide spectrum of chemical and physical carcinogens can be detected without actually knowing, in each case, the precise type of interaction between DNA and oncogenic agent.

Evidence has recently been presented that DNA repair synthesis may also be used in a rapid identification of organs and tissues that will give rise to neoplasms. Organotropic carcinogens seem to induce DNA changes as measured by a shift in molecular weight (Table 3) and DNA repair synthesis only in target organs (*3,28*). Although these experiments were done by using only 4 different carcinogens (4NQO, dimethyl- and diethyl-nitrosamine, and MNNG) that were applied by two different routes (subcutaneous injection and force-feeding), the results are highly promising and show the possibility of identifying high-risk organs.

Finally, the use of DNA repair synthesis in the detection of DNA repair-deficient human groups needs a careful re-examination. The application of unscheduled DNA synthesis, and shifts in sedimentation profiles following centrifugation in alkaline sucrose gradients, revealed the repair deficiency of Xeroderma pigmentosum patients (e.g. *29*). In spite of extensive searching no other human repair deficiency has been uncovered. This lack of various repair-deficient mutants among the human population could possibly result

TABLE 3
Sedimentation Profiles (Alkaline Sucrose Gradients) of ^{3}H-DNA Released from Various Mouse Organs Following a Single Subcutaneous Injection of Dimethylnitrosamine (10μg/g)

		Control Region	Fragmentation Moderate	Extensive
Untreated	Lung	+		
	Kidney	+		
	Liver	+		
DMN 4 hrs	Lung		+	
	Kidney		+	
	Liver			+
DMN 16 hrs	Lung		+	
	Kidney		+	
	Liver			+
DMN 48 hrs	Lung	+		
	Kidney	+		
	Liver		+	

from a present inability to recognize defects at various steps of the entire DNA repair process. On the other hand, the possibility cannot be discarded that mammalian cells with pol-, lig-, rec-lesions do not survive. The application of UV-inactivated human viruses and their extent of reactivation should yield insight into this unresolved problem. As previously mentioned, this highly sensitive assay detects different levels of DNA repair capacities of XP patients. Furthermore, the exceptional XP cells with normal level of unscheduled DNA synthesis show a reduced capacity to reactivate UV-irradiated AD12 (*14*). This latter case may exemplify the great potentiality of this new method to estimate deficiencies in the DNA repair process. Since this essay depends on viral replication and on the restoration of the entire biological function, any defect along the repair pathway should be detectable.

REFERENCES

1. K. C. Smith and P. C. Hanawalt, *Molecular Photobiology*, Academic Press, New York, 1969.
2. H. F. Stich, in *Topics in Chemical Carcinogenesis.* (W. Nakahara, S. Takayama, S. Sugimura and S. Odashima, Eds.) pp. 17–28. University Park Press, Baltimore, 1973.
3. H. F. Stich, D. Kieser, B. A. Laishes, and R. H. C. San, in *Proc. 10th Can. Cancer Conf.*, Honey Harbor, 1973, (P. Scholefield, Ed.), pp. 83–101. University of Toronto Press, 1973.
4. J. A. Miller, and E. C. Miller, *Progr. Exp. Tumor Res.* **11** , 273 (1969).
5. J. A. Miller, *Cancer Res.* **30** , 559 (1970).
6. H. V. Malling, *Mutat. Res.*, **13** , 425 (1971).
7. R. C. Garner, E. C. Moller and J. A. Miller, *Cancer Res.* **32** , 2058 (1972).
8. H. F. Stich, in *Molecular Mechanisms for Repair of DNA* (R. B. Setlow and P. Hanawalt, Eds.), Plenum Press, New York (in press).
9. B.A. Laishes and H. F. Stich, *Biochem. Biophys. Res. Commun.* **52** , 827 (1973).
10. S.D. Vesselinovitch, N. Mihailovich, G. N. Wogan, L. S. Lombard and K. V. N. Rao, *Cancer Res.* **32** , 2289 (1972).
11. I. F. H. Purchase and J. J. Van der Watt, *Fed. Cosmet. Toxicol.* **8** , 289 (1970).
12. P. G. Burk, M. A. Lutzner, D. D. Clarke, and J. H. Pobbins, *J. Lab. Clin. Med.* **77** , 759 (1971).
13. J. E. Cleaver, *J. Invest. Dermatol.* **58** , 124 (1972).
14. R. S. Day, *Cancer Res.* (in press).

15. R. W. Hart and R. B. Setlow, Abstr. *First Annu. Meet. Amer. Soc. Photobiol.* p. 120 (1973).
16. R. B. Setlow, in *Proc. Symp. New Trends in Photobiol.*, Rio de Janeiro, 1973. (in press).
17. H. F. Stich, O. Hammerberg and B. Casto, *Can. J. Genet. Cytol.* **14** , 911 (1972).
18. H. F. Stich and R. H. C. San, *Mutat. Res.* **10** , 389, (1970).
19. K. Hirai, V. Defendi and L. Diamond, *Science* (in press)
20. S.A. Aaronson and C. D. Lytle, Nature **228** , 359 (1970).
21. C. P. Lytle, S. A. Aaronson and E. Harvey, *Int. J. Rad. Biol.* **22** , 159 (1972).
22. H.F. Stich, W. Stich and P. I. Lam, *Nature* (London) (in press)
23. H. F. Stich, R. H. C. San and Y. Kawazoe, *Nature* (London) **229** , 416 (1971).
24. H. F. Stich, and R. H. C. San, *Proc. Soc. Exp. Biol. Med.* **142** , 155 (1973).
25. H. F. Stich, R. H. C. San, J. A. Miller and E. S. Miller, *Nature* (London) *New Biology*, **238** , 9 (1972).
26. R. G. Evans and A. Norman, *Radiat. Res.* **36** , 287 (1968).
27. M. W. Lieberman, R. N. Baney, R. E. Lee, S. Sell and E. Farber, *Cancer Res.* **31** , 1297 (1972).
28. H. F. Stich and D. Kieser, *Proc. Soc. Exp. Biol. Med.* **145** , 1339 (1974).
29. J. E. Cleaver, in *Molecular and Cellular Repair Processes*, (R. F. Beers, Jr., R. M. Herriott, and R. C. Tilghman, Eds), pp. 195–211. Johns Hopkins, Baltimore, 1972.

Chemical Changes Induced in DNA by Ionizing Radiation and the Relationship of Their Repair to Survival of Mammalian Cells[1]

R. B. Painter
Laboratory of Radiobiology
University of California
San Francisco, California 94143 USA

INTRODUCTION

After exposure to ionizing radiation, chromosome aberrations seem to be responsible for most of the reproductive death that ensues in eukaryotic cells (*1–3*). The important question, then, is: In what component(s) of the chromosomes does the damage that leads to aberrations occur? It is very difficult to believe that it is in the proteins, because they are rapidly and easily replenishable from the relatively radioresistant cytoplasm. Ribonucleic acids are also in a constant flux and their transient existence within the nucleus eliminates them as serious candidates for the radiosensitive materials. Despite the recent interest in lipids and lipid-nucleic acid complexes, there are no confirmed reports of lipids in chromosomes. Therefore, there remains only one real candidate — deoxyribonucleic acid (DNA).

Mammalian cells are much more susceptible to damage induced by ionizing radiation than to that caused by ultraviolet light (UV); the factor for DNA damage is 500–1000 (*4*). This makes the search for the killing lesion(s) induced by ionizing radiation much more difficult than for those caused by UV because the chemical yield of products of irradiated DNA is extremely small at biologically effective doses. In general, the tack has been to irradiate with high doses and make the logical assumption that the products measured are not qualitatively different from those formed at low doses.

BASE DAMAGE

Almost no information was available for base damage from DNA irradiated in the cell until the recent work of Cerutti and his collaborators (*5*). Their experiments with *Micrococcus radiodurans* showed that products of thymine damage are removed from DNA in viable cells by a process that is consistent with operation of an enzymatic repair system (*5*). Later, Mattern et al. (*6*) examined the release of thymine products, and also the release of thymine, thymidine, and TMP, from the DNA of Chinese hamster cells irradiated with ^{137}Cs gamma rays. They found that the resulting number of damaged thymines formed was about the same as the number of single-strand breaks (SSB's)

[1]Congress presentation A-3-5

caused by radiation, and the kinetics of their release were similar to the kinetics of repair of SSB's. RotiRoti and Cerutti (*7*) and RotiRoti, Stern, and Cerutti (*8*) showed that the yield of base damage in mammalian cells was greatly suppressed by the addition of OH radical scavengers to the medium. The suppression of thymine damage was on the order of 90%, which is even greater than the 60–70% suppression for SSB's reported by Roots and Okada (*9*) for L5178Y cells. The kinetics of release of damaged thymine products suggest an active process of repair, but the time needed for apparent completion of repair—1 hr—is much shorter than the time needed for repair of base damage (dimers) induced by UV—at least 24 hr (*10*, *11*). This relatively short time for base damage repair after x-irradiation suggests that much of the activity that releases damaged bases (as well as undamaged ones) accompanies DNA strand break repair.

The role of base damage in mammalian cell killing by ionizing radiation is completely unknown. The notion that base damage may be of importance in cell killing comes from work with bacteriophage, where there are strong reasons to believe that base damage causes more inactivation of the phage than either single- or double-strand breaks after exposure to both low LET (*12*) and high LET radiation (*13*). Therefore, in assessing the kinds of DNA damage that may play a role in causing mammalian cell death, one must seriously consider base damage.

SINGLE-STRAND BREAKS

Repair

The best studied lesions in mammalian DNA are SSB's. After the introduction of the alkaline sucrose gradient technique by McGrath and Williams (*14*), the method was quickly adapted for use with mammalian cells (*15*) and has since been used by many workers. There is still some controversy over the interpretation of many profiles resulting from these gradients, but it is probably accurate to say that, given lysis times long enough (generally 3 hr or more) to form species with S values of 120–165 S in unirradiated controls, the effects of radiation on single-stranded DNA can be studied with high accuracy and reproducibility. The most important result of studies on the action of low LET radiation on mammalian DNA with this technique is that a very high fraction of SSB's is repaired. Indeed, Elkind and Kamper (*16*) claimed that the molecular weight of DNA from cells irradiated with doses below 5 krads returned all the way to control values. Since these data were collected using 36,000 rpm, where speed effects would be very marked (*17*), it was impossible for these authors to distinguish between profiles from controls and those from DNA with several breaks per cell. In our experience, after doses of 10 krads to WI-38 cells (*18*), the molecular weights of DNA do not return to control values and do not increase after 30 min of incubation (for up to 120 min). Similarly, C. Koch in our laboratory (personal communication) found that the molecular weight of DNA from Chinese hamster cells does not return to control values within 4 hr after irradiation with 5 krads. Dugle and Gillespie (*19*) concluded that in Chinese hamster cells irradiated with relatively low doses (55–550 rads), unrepaired strand breaks are a cause of reproductive death.

Thus, it seems very possible that unrepaired SSB's remain in irradiated cells for long times after irradiation and thus cause cell death.

Perhaps the best evidence that unrepaired SSB's are important in mammalian cell killing comes from recent work in our laboratory with ^{125}I incorporated into DNA (Young, Burki, and Painter, paper E–13–5 at this Congress). About one half of the SSB's caused by disintegrations of this isotope in Chinese hamster cell DNA are not repaired; in contrast, nearly all SSB's induced by x-rays or ^{3}H are repaired (*20*). The work of Burki et al. (*21*) showed that ^{125}I disintegrations in DNA are extremely lethal events. Even if the unrepairable lesions caused by ^{125}I disintegrations are unique kinds of double-strand breaks, this does not detract from the fact that this is the first example of a correlation of high numbers of unrepaired breaks with high lethality.

Chemistry

Unrepaired SSB's may be chemically different from those that are repaired. It will be a long time before this possibility can be tested, because the chemical characteristics of SSB's are only beginning to be investigated. Using very high doses and the enzyme polynucleotide kinase, Coquerelle et al. (*22*) showed that about 30% of the end groups formed in DNA of irradiated rat thymus cells were 5′OH and 5′ phosphoryl, with the latter in slight excess. [In contrast, when DNA is irradiated in solution, 90% of all groups formed as a result of SSB induction are 5′ phosphoryls (*23*).] The remaining 70% of the end groups probably are of more than one kind. Dr. Coquerelle will discuss her and her collaborators' more recent results with terminal deoxyribonucleotidyl transferase at this meeting (paper B–7–6). None of the data thus far obtained gives any information on the correlation of base damage and loss with the formation of the observed end groups.

Mechanisms of Induction

Several investigators have recently determined the efficiency of single-strand breakage to be the range of 30–40 eV/break (*19,24,25,* Roots and Smith, personal communication). Earlier reports of 55–60 eV/break were probably the result of the use of high centrifuge speeds, which cause overestimates of this measurement. Because the average energy loss per event is about 60 eV for direct effects in solids (*26*), the value of 30–40 eV/break is consistent with a significant contribution of indirect effects to SSB induction by low LET in mammalian cells. This is already borne out by the work of Roots and Okada (*9*), who demonstrated that OH scavengers in the medium depress the induction of SSB's in L5178Y cells. But this contribution of indirect effects to SSB formation is peculiar. One would expect that if DNA were fully accessible to OH radicals formed in a surrounding aqueous milieu, breaks would occur at the same rate as they do in solution. This clearly is not the case. Moreover, if one part of the DNA were freely accessible to water and another part were not, the expected result for SSB formation would be a two-component curve—a rapid formation at lower doses and a much slower formation at high doses, where direct action would predominate [for example, as observed for acetylcholinesterase in homogenates from the electric eel (*27*)].

Again, this is not what is observed; SSB induction is essentially linear with dose from the lowest testable doses to very high ones. This indicates that the availability of radicals is very limited and, therefore, either that there is only a very small amount of water in the immediate vicinity of the DNA or that the state of this water is such that radicals are not able to migrate readily to the DNA.

The oxygen enhancement ratio (OER) for SSB's in mammalian cells seems to be very close to 4 (*25*, *28*, Roots and Smith, paper B–7–1 at this Congress). Since the OER for killing of mammalian cells almost never exceeds 3, this discrepancy needs some scrutiny. If SSB's have no role at all in cell killing, then one need not be concerned about it. I do not believe this is true, however, and would explain the discrepancy by either of two similar hypotheses. In one, the unrepairable component of SSB's is a slightly higher fraction of total SSB's when cells are irradiated under hypoxia than when cells are irradiated in oxygen. Thus, although equal doses cause many fewer SSB's after irradiation under hypoxia than in air, at equal numbers of total SSB's, there would be more unrepairable ones induced under hypoxia. In the other hypothesis, unrepaired SSB's have equal effects under both conditions, but under hypoxia another kind of DNA damage, which is at least partially unrepairable, occurs in higher yields.

DOUBLE-STRAND BREAKS

It now seems that most double-strand breaks (DSB's) induced in mammalian cells by low LET radiation are repaired (*29*, *30*). If so, this will destroy the long-held belief that DSB's are *the* unrepairable lethal lesions formed by ionizing radiation. At first blush it is difficult to see how DSB's can be repaired if this lesion consists of breaks directly across from each other. However, if the two lesions are not in direct opposition, but a few nucleotides apart, it is possible that the two strands do not become separated in the cell and that two distinct, sequential events repair the two discontinuities. Since nothing is known about the chemistry or the actual physical structure of DSB's, we will not know the mechanisms of their repair until we have this information.

CONCLUSION

We still are ignorant of most of the details of DNA damage and repair that are necessary for determining the exact steps that lead to recovery after irradiation as well as the precise mechanisms by which radiation kills cells. From what is known, however, I am tempted to speculate that among the many kinds of lesions formed in mammalian DNA by ionizing radiation, certain ones have little or even no probability of being repaired. These may be of a kind that have both base damage and discontinuities in either or both DNA strands. It is possible that this kind of lesion—which, for example, might be a gap in the DNA with "sloppy" ends on one or both sides—is one with which the mammalian cell has rarely had to contend because, from its much more frequent encounters with longer wave length radiation, it has evolved mechanisms to repair damage of the kind typified by pyrimidine dimers. The cell repair systems, perhaps both excision and postreplicative, probably attempt to mend the damage, but the repair is either of poor fidelity or fails altogether. The end result, then, is a chromosome aberration that eventually leads to gene imbalance and cell death.

ACKNOWLEDGMENT

Work performed under the auspices of the U.S. Atomic Energy Commission.

REFERENCES

1. W. C. Dewey, S. C. Furman, and H. H. Miller, *Radiat. Res.* **43,** 561 (1970).
2. S. J. Grote, *Radiation-induced genetic damage and lethality in mammalian cells.* Ph.D. Dissertation, University of London, 1971.
3. A. V. Carrano, *Mutat. Res.* **17,** 355 (1973).
4. R. B. Painter, in *Molecular and Cellular Repair Processes* (R. F. Beers, Jr., R. M. Herriott, and R. C. Tilghman, Eds.), pp. 140–46. The Johns Hopkins University Press, Baltimore, 1972.
5. P. V. Hariharan and P. A. Cerutti, *J. Mol. Biol.* **66,** 65 (1972).
6. M. R. Mattern, P. V. Hariharan, B. E. Dunlap, and P. A. Cerutti, *Nature New Biol.* **245,** 230 (1973).
7. J. L. RotiRoti and P. A. Cerutti, *Int. J. Radiat. Biol.* in press.
8. J. L. RotiRoti, G. S. Stern, and P. A. Cerutti, *Biochemistry*, in press.
9. R. Roots and S. Okada, *Int. J. Radiat. Biol.* **21,** 329 (1972).
10. R. B. Painter, *Curr. Top. Radiat. Res. Quart.* **7,** 45 (1970).
11. H. J. Edenberg and P. C. Hanawalt, *Biochim. Biophys. Acta* ***324,*** *206 (1973).*
12. G. P. van der Schans, J. F. Bleichrodt, and J. Blok, *Int. J. Radiat. Biol.* **23,** 133 (1973).
13. R. C. Christensen, C. A. Tobias, and W. D. Taylor, *Int. J. Radiat. Biol.* **22,** 457 (1972).
14. R. A. McGrath and R. W. Williams, *Nature* (*London*) **212,** 534 (1966).
15. J. T. Lett, I. Caldwell, C. J. Dean, and P. Alexander, *Nature* (*London*) **214,** 790 (1967).
16. M. M. Elkind and C. Kamper, *Biophys. J.* **10,** 237 (1970).
17. B. H. Zimm, *Biophys. Chem.* **1,** 279 (1974).
18. J. M. Clarkson and R. B. Painter, *Mutat. Res.* **23,** 107 (1974).
19. D. L. Dugle and C. J. Gillespie, in *Proceedings of the Squaw Valley Conference on "Molecular Mechanisms for the Repair of DNA."* Plenum Press, New York, 1974.
20. J. E. Cleaver, G. H. Thomas, and H. J. Burki, *Science* **177,** 996 (1972).
21. H. J. Burki, R. Roots, L. E. Feinendegen, and V. P. Bond, *Int. J. Radiat. Biol.* **24,** 363 (1973).
22. T. Coquerelle, A. Bopp, B. Kessler, and U. Hagen, *Int. J. Radiat. Biol.* **24,** 397 (1973).
23. M. Ullrich and U. Hagen, *Int. J. Radiat. Biol.* **19,** 507 (1971).
24. B. Palcic and L. D. Skarsgard, *Int. J. Radiat. Biol.* **21,** 417 (1972).
25. C. J. Koch and R. B. Painter, *Radiat. Res.* (Submitted).
26. A. M. Rauth and J. A. Simpson, *Radiat. Res.* **22,** 643 (1964).
27. I. Serlin and G. C. Cotzias, *Radiat. Res.* **6,** 55 (1957).
28. M. Lennartz, T. Coquerelle, and U. Hagen, *Int. J. Radiat. Biol.* **24,** 621 (1973).
29. S. Sawada and S. Okada, *Int. J. Radiat. Biol.* **21,** 599 (1972).
30. P. M. Corry and A. Cole, *Nature New Biol.* **245,** 100 (1973).

CHEMICAL RADIO-SENSITIZATION OF MAMMALIAN CELLS[1]

Chairman: G. E. Adams

[1]Symposium D–2

Mammalian Cell Sensitization Repair and the Cell Cycle[1]

Warren K. Sinclair
Division of Biological and Medical Research
Argonne National Laboratory
Argonne, Illinois, USA

About three years ago I described a possible two-component model for the variation of the lethal radiation response in mammalian cells during the cell cycle (*1*). In this model it was postulated that two factors control the lethal response. The first factor was DNA synthesis and the second was a factor, identified as Q, which might be intracellular sulfhydryl, or some fraction of it, which varied during the cell cycle in the absence of DNA synthesis. These features are shown in Fig. 1. and I will review some of the main points of this model very briefly.

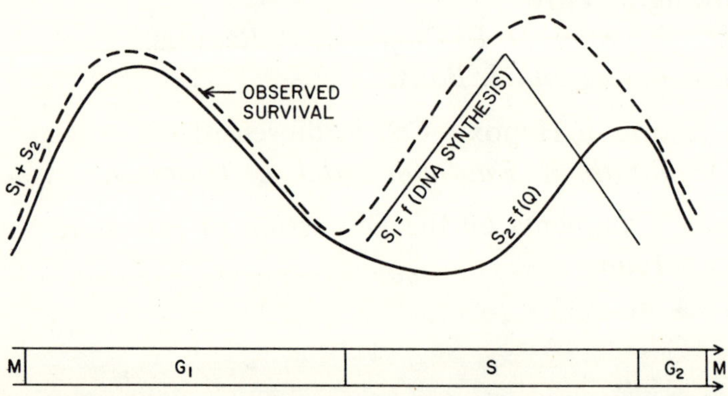

FIG. 1. Diagram of a two-component model representing the variation of two factors controlling lethal radiation response during the mammalian cell cycle.

DNA Synthesis

The intimate association between cell survival following exposure to ionizing radiation and DNA synthesis has been well established, initially by the work of Terasima and Tolmach (*2*) in HeLa cells, by myself (*3,4*) in Chinese hamster cells and by Terasima and Fujiwara (*5*) in L cells. Cell survival increases as soon as DNA synthesis begins and falls sharply as soon as DNA synthesis is completed. If the survival depends on the rate of DNA synthesis, as it may well do, and if this were the only factor controlling survival, maximum survival should occur in cells like Chinese hamster, about the middle of S.

[1] Congress presentation D–2–1.

Dependence on Q

In many cells however, survival is a maximum near the end of S. Furthermore, it has been clearly shown in Chinese hamster cells and L cells (*3,5*) that survival fluctuates even in the absence of DNA synthesis when an inhibitor such as hydroxyurea is applied to G_1 cells. A further example of this in HeLa cells obtained recently by A. Han (*6*), is shown in Fig. 2. Again the pattern of initial fall in survival as G_1 cells progress through the cycle, but cannot make DNA, is followed by a rise in survival late in the "S" period, presumably due to variations in Q.

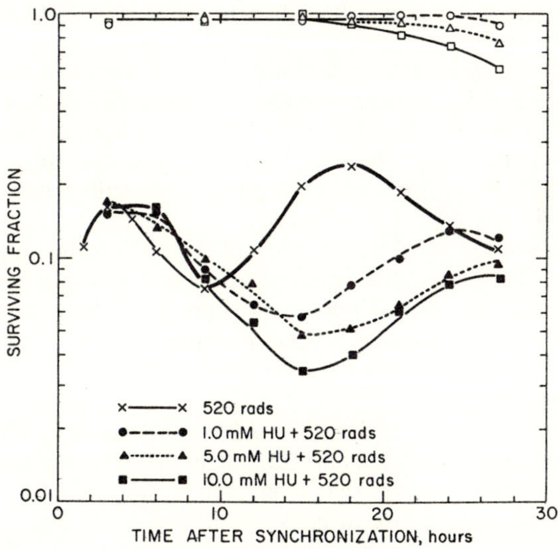

FIG. 2. The age response of HeLa cells after 550 R (520 rads) of 250 kVp X radiation. X——X X rays only. ■——■ 10 mM Hydroxyurea added in G_1.

The identification of Q still presents an intriguing problem. I have discussed (*1,4*) many possible factors that I have considered and rejected. I have also pointed out (*1*) that the discovery of the differential effects of cysteamine (*7*) implicates intracellular sulfhydryl in some manner and indeed, this has led to numerous investigations of the sulfhydryl content in mammalian cells at different stages of the cell cycle.

In the model depicted in Fig. 1. an important feature, which must be tested, is whether the identity of Q is the same throughout the cell cycle, that is, is there only one factor Q, and is the entity controlling response after DNA synthesis has been arrested in "S" the same as that controlling the well-known variation in survival during G_1 in long-G_1 cell lines? The only evidence in support of this, and it is suggestive rather than definitive, is that fluctuations in non-protein sulfhydryl during the cell cycle of HeLa cells were found, by Ohara and Terasima (*8*), to coincide almost exactly with fluctuations in X-ray sensitivity.

I shall present later some new evidence on this point from the work of Han, Sinclair and Kimler (*6*). My approach to the problem of identification of Q has been to follow the clue apparently provided by cysteamine that intracellular sulfhydryl was involved. I reasoned that a suitable SH binding agent should sensitize the cells most at a time when Q was most important, notably late in S and early in G_1: furthermore, in HU-inhibited cells, the rise in survival normally seen late in the "cell cycle" should be erased in the presence of the binding agent.

Sensitization by N-ethylmaleimide

N-ethylmaleimide (NEM) was selected for this purpose and I have already reported on many of my experiments on the effects of NEM before, during and after irradiation in Chinese hamster cells (*1,9*). Thus, I will only summarize the main points for these cells and then present some new data which has been obtained recently in HeLa cells (*6*).

First let me emphasize that in both cell systems, very low concentrations of NEM are used (0.75 μM), the agent is added to rinsed cells in the presence of PBS only, and the cells themselves are fully oxygenated and at room temperature during 30-minute exposures to NEM. In this respect the mode of action of NEM is quite different from its rapid effects upon anoxic cells when present during irradiation at relatively high concentration.

The sensitization of NEM during the cell cycle of Chinese hamster V79 cells is shown in Fig. 3. Sensitization is greatest at the end of S, negligible at the G_1-S border and slight in G_1. For longer exposures it is clear that there is some sensitization in G_1 as well (modifying factor about 1.2 cf. 1.6 in late S(9)). While the effect is mainly on the slope of the survival curve, there is some effect on the shoulder also, especially in G_1 (*9*).

The second important point is that NEM added to the inhibited cells not synthesizing DNA, greatly reduces the rise late in the cell cycle as shown in Fig. 4. These features strongly suggest that whatever NEM binds to, is a controlling factor in the radiation response of cells late in S.

Results in HeLa Cells

It is obvious that the short G_1 of the V79 Chinese hamster cells does not lend itself well to the examination of controlling factors in G_1. For this reason Antun Han, Bruce Kimler and I have undertaken experiments with HeLa cells (*6*). The results show, first of all, a remarkable similarity in toxicity and sensitization of NEM at 0.75 μM between HeLa cells and Chinese hamster cells. Second, the age response of synchronous HeLa cells in the presence of NEM show clearly (Fig. 5) that NEM sensitizes in G_1 so that the response is almost flat from mitosis to the G_1-S border (but note that specific experiments on mitotic HeLa cells have not yet been performed), there is no effect on late G_1-early S cells, and late S cells are sensitized very much like V79 cells.

Survival curves with and without NEM corresponding to the most interesting parts of this response, viz at 3 hours (in early G_1), 9 hours (G_1-S border) and 18 hours (very late S) are shown in Fig. 6. Both the shoulder and slope are affected.

Experiments are in progress on hydroxyurea-inhibited synchronous HeLa cells and the effects of NEM upon them.

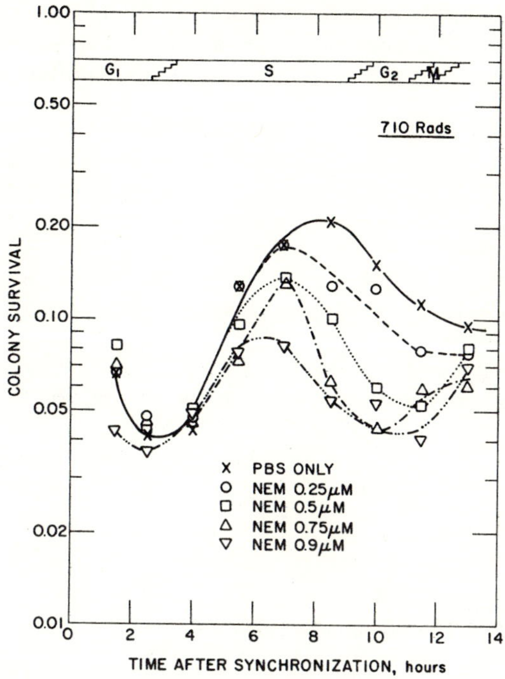

FIG. 3. Cyclic response of synchronous V79 Chinese hamster cells to 710 rads of 250 kVp X rays in the presence of NEM at different concentrations for 15 minutes. (The approximate phases of the cell cycle as determined by ^{3}HTdR pulse labeling are shown at the top).

Inhibition of Repair

Experiments with NEM in Chinese hamster V79 cells, have shown (Fig. 7) that NEM is equally effective when given just before, during or immediately after exposure (*3*). This, together with the facts reported earlier (*1*) that equimolar concentrations of cysteamine when added afterward could not reverse the action of NEM, suggests that NEM inhibits a repair mechanism in the cell, rather than modifying primary damage. Thus, cell survival in the presence of NEM is cell survival with all the primary damage expressed and no repair. It may be significant that in Chinese hamster cells, the level of survival is quite similar to that of mitotic cells after the same absorbed dose and that NEM has *no* effect upon mitotic cells.

Furthermore, the kinetics of repair are different in G_1 and S cells, as shown by the addition of NEM to Chinese hamster cells (see Fig. 7 of reference *3*) and similar results have recently been obtained by Han et al. for HeLa cells (Fig. 8). In both cases, recovery occurs much more rapidly in S cells than in G_1 cells, the irradiated S cells being insensitive to NEM after about 2 hours while irradiated G_1 cells require 6 to 8 hours. NEM may provide us with a useful means to examine repair kinetics in more detail. For example, the temperature-dependence of recovery seems also to differ between G_1 and S cells, recovery in S cells being more sensitive to temperature changes (*10*). Also, recent results in my laboratory indicate that S cells recover faster in PBS than in medium, a finding which may be related to the observation reported by Belli and Shelton (*11*) that cells can repair potentially lethal damage more readily in buffer than in medium.

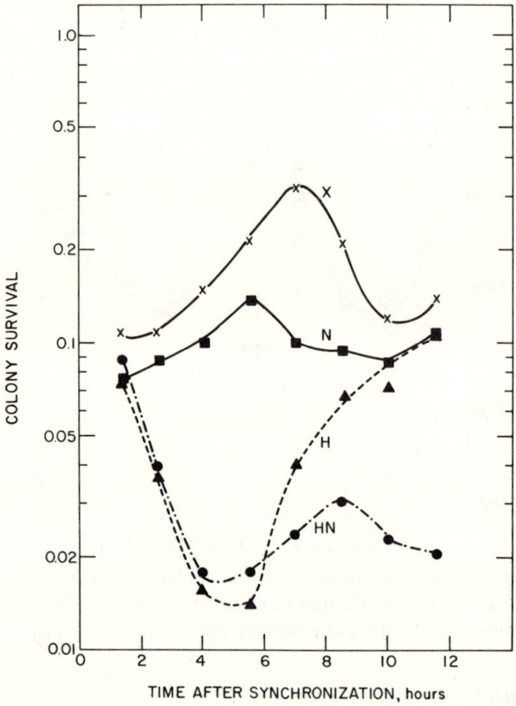

FIG. 4. Survival of synchronous V79 Chinese hamster cells at different stages of the cell cycle after 710 rads of 250 kVp X rays. X: 710 rads only; N: 710 rads in the presence of 7.5 μM NEM added to medium. H: 710 rads delivered to cells inhibited by 1 mM hydroxyurea added at 1 hour. HN: Cells inhibited by 1mM hydroxyurea added at 1 hour, irradiated with 710 rads in presence of 7.5 μM NEM.

Identification of Q

I have discussed earlier (*9,10*) evidence for Q being identified with a small fraction of the intracellular sulfhydryl which presumably is critically involved in repair processes in cells.

Recent measurements by Dr. Joy Archer (see ref. *10* for summary) on gross NPSH and total SH during the cell cycle, indicate that total SH yields a much better correlation with radiation survival than does NPSH. Also, NEM binding is greatest in late S. These findings tend to implicate a small fraction of the protein sulfhydryl rather than NPSH. This idea also avoids a conflict with the results of Harris and Power (*12*) who find that diamide, an agent which acts specifically on NPSH, does *not* sensitize oxygenated cells to radiation. It is also in agreement with recent results of Morita (*13*) who finds that the radioresistance of selected HeLa cells strains correlates well with their total SH but not with the NPSH content.

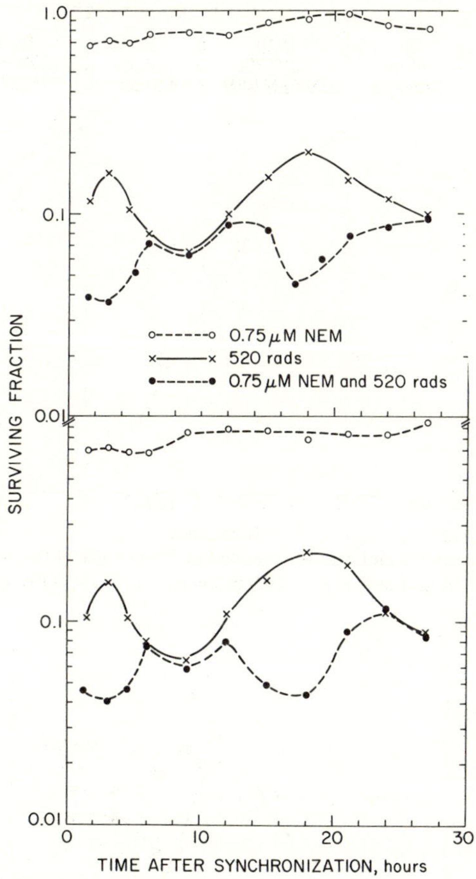

FIG. 5. Age response of HeLa cells after 520 rads of 250 kVp X radiation. X——X X rays only. ●—● in the presence of 0.75 μM NEM in PBS for 30 minutes.

It should also be mentioned that NEM inhibits DNA synthesis (Sinclair, Archer and Blakely, in preparation) and recent studies by J. Archer have shown that NEM binds to DNA. Whether these actions of the agent have any bearing on its control of radiation response, especially in a period such as G_1 when no DNA synthesis is in progress, is not known.

I have examined the effects upon the radiation response of Chinese hamster cells of some other molecules closely related to NEM. Neither succinimide nor N-ethyl succinimide bind to SH groups nor do they sensitize Chinese hamster cells (*10*). Maleimide hydrolyzes too rapidly to bind to SH groups and its sensitization capability has not been tested.

Another agent that, like NEM, binds rapidly to SH groups but, unlike NEM, its binding is easily reversible, is *p*-chloromercuribenzene sulphonic acid (PCMBS), PCMBS is more difficult to use in mammalian cell experiments than NEM, and cells must be very carefully rinsed prior to its application. Nevertheless, I have been able to show that, at a concentration of 3.5 μM

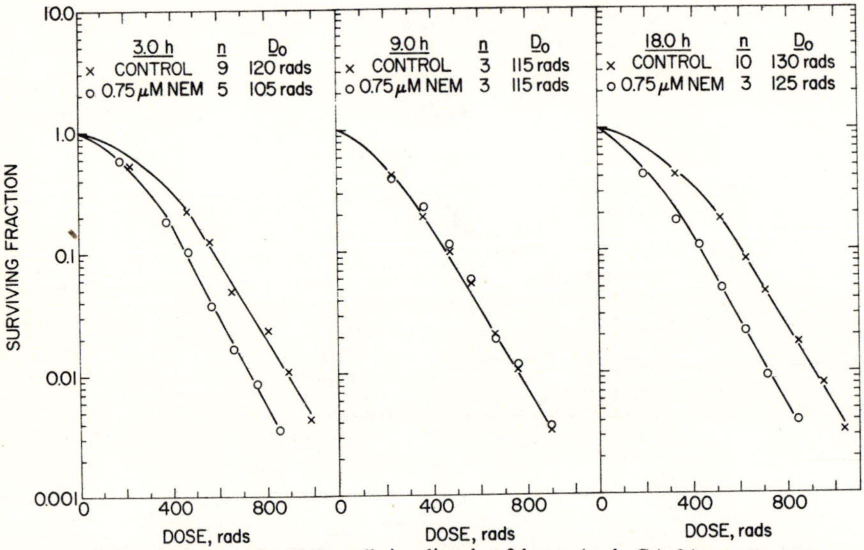

FIG. 6. Survival curves for HeLa cells irradiated at 3 hours (early G_1), 9 hours (G_1-S) border) and 18 hours (late S) with X rays only (x) or in the presence of 0.75μM NEM in PBS for 30 minutes. (O)

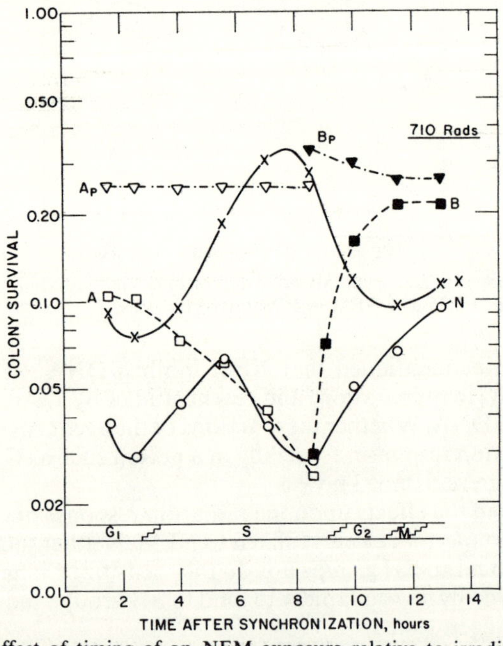

FIG. 7. The effect of timing of an NEM exposure relative to irradiation. The survival of synchronous Chinese hamster cells after 710 rads of 250 kVp X rays (alone, X) or during a 30-minute exposure to 0.75 μM NEM in PBS (curve N, O). In the curves labeled A (□) and B (■) cells were exposed to 0.75 μM NEM in PBS for 30 minutes at the times indicated, fresh medium was put on and *all* cells were irradiated at 8.5 hours. Curves Ap (▽) and Bp (▼) depict cells treated identically to A and B but PBS alone was used without NEM.

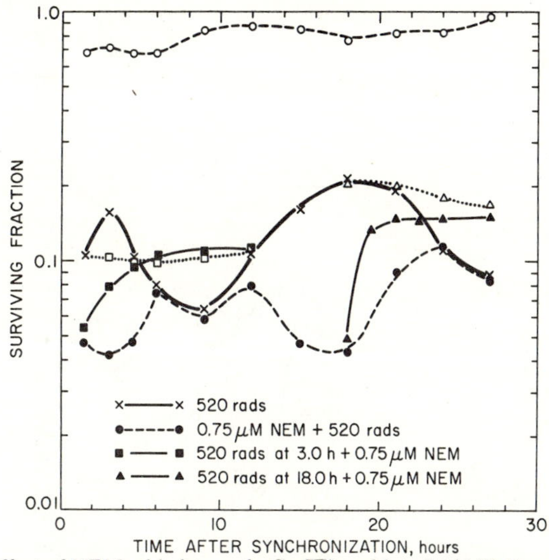

FIG. 8. The effect of NEM added to early G_1 (■) and late S (▲) HeLa cells at different times after irradiation, showing the rate of recovery from radiation damage.

PCMBS in PBS, the agent sensitizes Chinese hamster cells in late S *if present during irrradiation* (Fig. 9). If it is also present prior to irradiation as well as during, the effect is increased still more. If it is applied before or after irradiation and is not present during irradiation, it has no effect. While the action of PCMBS is clearly not quite the same as that of NEM, the fact that it sensitizes most in late S (as does NEM) and slightly in G_1 (see curve B of Fig. 9), as does NEM, again points to the significance of SH binding in controlling the radiation response during the cell cycle.

For all these reasons, I still favor the notion that NEM acts via SH binding, and that the relevant moiety is a small fraction of the PSH which is critically involved in the repair process following radiation damage. However, until this fraction is identified unequivocally the issue must remain unresolved, but if and when it is, we will have made a major step forward in the identification of repair processes in cells following ionization radiation damage.

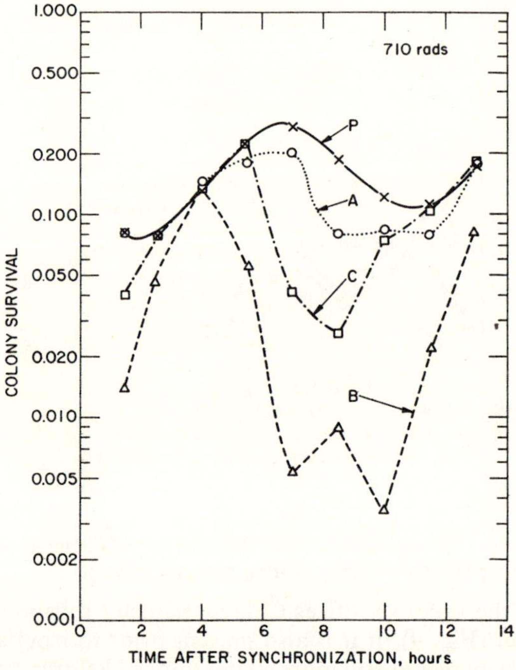

FIG. 9. Synchronous Chinese hamster cells irradiated with 710 rads of 250 kVp X radiation at different points in the cell cycle in the presence of PCMBS, 0.0035 mM.

Curve P—No PCMBS, 710 rads only.

Curve A—PCMBS present for 10 minutes during X-irradiation.

Curve B—PCMBS present for 10 minutes during + 20 minutes afterward.

Curve C—PCMBS present during (10 minutes) and before (20 minutes).

CONCLUSIONS

Experiments with hydroxyurea and with NEM, first in fully-oxygenated V79 Chinese hamster cells and more recently in HeLa cells, support the proposed two-component model for control of cell survival after x-irradiation during the cell cycle. One component is DNA synthesis; the other is determined by the action of NEM throughout the cell cycle and is probably a small fraction of the protein sulfhydryl located in critical molecules.

These results are generally consistent with the notion that during mitosis the cell is most sensitive because it lacks the ability to repair radiation damage. It may be noted that neither hydroxyurea nor NEM affects mitotic cells. At other stages of the cell cycle, hydroxyurea inhibits a repair mechanism that operates during DNA synthesis and thus appears to sensitize the cell, and NEM inhibits another repair mechanism that is usually most effective both early in G_1 and late in S. When both agents are used together, the response of cells is broadly (this is an oversimplification) flat through the cell cycle and roughly the same as that of mitotic cells.

REFERENCES

1. W. K. Sinclair, *Curr. Top. Radiat. Res.* **7**, 264 (1972).
2. T. Terasima and L. J. Tolmach, *Science,* **140**, 490 (1963).
3. W. K. Sinclair, *Proc. Nat. Acad. Sci., USA*, **58**, 115 (1967).
4. W. K. Sinclair, in *Time and Dose Relationships in Radiation Biology as applied to Radiotherapy,* 97–116, (NCI-AEC Conference, Carmel, California) Brookhaven National Laboratory, BNL 50203 (C-57), 1970. (Available from Clearinghouse for Federal Scientific and Technical Information, Springfield, Virginia 22151).
5. T. Terasima and Y. Fujiwara, in *Proceedings of International Conference on Radiation Biology and Cancer,* Kyoto 1966, pp. 3–8. Radiation Society of Japan, Tokyo 1967.
6. A. Han, W. K. Sinclair and B. F. Kimler, *Radiat. Res.* (submitted).
7. W. K. Sinclair, *Science,* **159**, 442 (1968).
8. H. Ohara and T. Terasima, *Exp. Cell. Res.* **58**, 182 (1969).
9. W. K. Sinclair, *Radiat. Res.* **55**, 41 (1973).
10. W. K. Sinclair, in *Control of Proliferation of Animal Cells,* (B. Clarkson and R. Baserga, Eds.) 985–994. Cold Spring Harbor Laboratory, N. Y. 1974.
11. J. A. Belli and M. Shelton, *Science,* **165**, 490 (1969).
12. J. W. Harris and J. A. Power, *Radiat. Res.* **56**, 97 (1973).
13. K. Morita, *Radiat. Res.* **56**, 405 (1973).

Chemical Radiosensitization Studies With Mammalian Cells Growing In Vitro[1]

J. D. Chapman, D. L. Dugle, A. P. Reuvers,
C. J. Gillespie, and J. Borsa
Medical Biophysics Branch
Whiteshell Nuclear Research Establishment
Atomic Energy of Canada Limited
Pinawa, Manitoba ROE 1LO
Canada

The study of chemicals which can sensitize cells to the lethal action of ionizing radiation has proved useful in the elucidation of cellular radiation mechanisms. Those which selectively radiosensitize hypoxic cells could have a role, in combination with radiotherapy, in an improved treatment of some solid tumors. The scope of this paper is restricted to those compounds which appear to mimic the radiosensitizing action of molecular oxygen in mammalian cells and have come to be known as electron-affinic (EA) sensitizers.

In hindsight, the radiosensitizing studies with the drug Synkavit, by J. S. Mitchell (*1*) and colleagues were the first experiments using EA compounds with mammalian cells. Of course, the bulk of the research effort in the 1950's and 60's, as regards the exploitation of the radiosensitizing effect of oxygen in radiotherapy, was directed towards hyperbaric oxygen usage and high LET radiations (*2*). The study of alternative chemicals which might radiosensitize hypoxic mammalian cells has only come of age in the 1970's. These studies are indebted 1) to the suggestion by Adams and Dewey (*3*) that the chemical property, electron-affinity, relates radiosensitizing effectiveness to chemical structure, and 2) to the improved understanding of cellular radiobiology resulting from recent studies of its chemistry (*4,5*). The only reports of chemical radiosensitization of hypoxic mammalian cells to appear by 1970 were studies with glyoxal derivatives (*6*) and with the stable free radical, triacetoneamine-N-oxyl (TAN) (*7*). The discovery that *p*-nitroacetophenone (PNAP) could radiosensitize hypoxic mammalian cells as well as bacterial cells (*8*) provided the impetus for numerous additional studies.

Criteria considered important in the selection of hypoxic radiosensitizers for application in radiotherapy included

a) selective radiosensitizing activity against hypoxic cells,
b) nontoxicity at radiosensitizing concentrations,
c) effectiveness against both proliferating and non-proliferating cell populations under various physiologic conditions,
d) resistance to rapid metabolization by tissue, and
e) diffusibility through tissue and into the hypoxic regions of a tumor within a few minutes.

[1] Congress presentation D-2-2.

Mammalian cells growing in vitro have been used to study the first three criteria in detail and other in vitro cellular systems may provide information about the last two criteria.

PNAP was shown to selectively radiosensitize hypoxic Chinese hamster cells in all phases of the cell cycle at concentrations producing no acute toxicity to cellular proliferative capacity (*9*). Its sensitizing activity reached a maximum enhancement ratio ($ER = k_{N_2+\text{drug}}/k_{N_2}$) of ~ 1.5 at concentrations of 500 μM and greater. A study with various acetophenone analogues of PNAP revealed that the ring-NO_2 group was essential for radiosensitizing activity (*10*). This study suggested that a threshold EA of at least that of nitrobenzene was required to produce radiosensitization in Chinese hamster cells made acutely hypoxic. Analogues of nitrobenzene, whose EAs were varied by ring substituents, showed radiosensitizing activity at 500 μM in complete growth medium which was proportional to EA (*11*). This study provided vivid confirmation of the EA principle stated 10 years earlier (*3*). Numerous radiosensitizers based on the nitrobenzene structure have been characterized in mammalian cells in vitro (*11,12*) and one of these structures shows positive sensitizing activity in animals whose cells have been made acutely hypoxic (*13*).

The demonstration that some antibacterial nitrofurans in current clinical use had excellent radiosensitizing activity in hypoxic mammalian cells (*14*) prompted the screening of several classes of nitroheterocyclic compounds for such activity. As well, this study generated some optimism as regards early clinical investigations with hypoxic radiosensitizers. Several nitrofuran derivatives were shown to selectively radiosensitize hypoxic mammalian cells in vitro in all phases of the cell cycle at nontoxic concentrations (*15*). The extent of sensitizing activity reached ERs of 2.0 and greater for several derivatives, with nifuraldezone and its water solubilized derivative (NF–167)—at 250 μM concentration in complete medium—giving the same sensitizing activity as oxygen under air-saturated conditions (*16*). This study showed that some nitrothiazole and nitroimidazole structures had selective hypoxic radiosensitizing activity as well as the nitrofurans. Further studies with nitroimidazole and nitropyrazole derivatives (*17,18*) have shown 2-nitroimidazole to be the most promising base structure as regards application in animal systems.

Most of the sensitizer screening procedures with mammalian cells in vitro have involved a limited number of cells ($\leqslant 10^4$) attached to glass petri dishes and irradiated in the presence of ~ 3/4 ml of medium containing sensitizer. Metabolism of the drugs, investigated to date, by this small number of cells would not be expected to significantly reduce the concentration of drug in the vicinity of the cellular targets. The study of drug metabolism and diffusion into tissue of the EA sensitizers may exploit two different in vitro cell systems. Agnew and Skarsgard (*19*) and Thomson and Rauth (*20*) have studied the effect of high cell density on the radiosensitizing activity of numerous compounds. In general, those compounds which are known to be highly EA and extremely chemically reactive are found to rapidly lose their sensitizing activity in the presence of cell densities greater than 10^7/ml. The mechanisms of sensitizer inactivation at high cell density are not completely understood, but compounds which are resistant to "dense cell" inactivation also demonstrate radiosensitizing effectiveness in animal tumor systems (*21*). A second in vitro cell technique which has proven useful in the study of chemical radiosensitizers is the cell spheroid system developed by Sutherland et al. (*22*). Various EA chemicals have been shown to radiosensitize the resistant component of the multi-component survival curve, characteristic of cell spheroids (*23,24*). As yet, it is not known to what extent the cell spheroid system might be exploited for the development of clinical sensitizers.

As well as being useful in the selection of radiosensitizing structures, mammalian cells have been used in the study of cellular radiosensitizing mechanisms. The groundwork for many of these studies had been established in the 1960's by radiation researchers whose concern it was to relate radiation chemical mechanisms to cellular radiobiology. Much of this work has been reviewed at this meeting by Professor Boag (*25*).

X- and γ-rays interact with the molecules of living cells, ejecting therefrom energetic electrons. These electrons distribute their energy through the ionization and excitation of molecules along their tracks. Ionizations and excitations which occur in vital cellular molecules and produce altered chemical structure and function have been classified as direct effects. Approximately 70% of a mammalian cell's weight consists of water, and energy deposited within the cellular water can also lead to lethal events. Various reactive free radical species generated in the radiolysis of water have been characterized along with their reactivities with numerous biomolecules (*26*). Alterations of vital cellular molecules resulting from reactions with these water radiolysis products have been classified as indirect effects.

Given the complexity of a mammalian cell and the multiplicity of possible chemical reactions induced by radiation, can one hope to experimentally test for chemical radiosensitization mechanisms with this system? Our approach to this challenge has been to restrict our mechanistic studies to the investigation of one extremely sensitive cell function, cellular proliferative capacity, and to one cellular structure, the DNA. Numerous studies which indicate that the integrity of the DNA is essential for normal cell proliferation have been reviewed (*27*).

The nitrofurans were shown to bind to various biomolecules and to whole cells upon irradiation in the absence of oxygen (*15*). This result suggested that the nitrofurans may compete with O_2 for reactive radical sites in various molecules. As well, several EA sensitizers, including O_2, were found to enhance the production of single-strand breaks in the DNA of Chinese hamster cells irradiated in hypoxia (*28–30*). The radiation chemical reactions by which ^{14}C-nitrofurazone became bound to DNA in irradiated solutions were investigated. The majority of the sensitizer-DNA adducts were generated by the indirect action of OH (*31*). By varying the relative concentrations of DNA and sensitizer in solution it was possible to show that the most important reaction resulting in addition products was that of unaltered nitrofurazone with DNA radicals.

Our attention was therefore directed to the role of OH in mammalian cell inactivation. Studies by Roots[2] and Roots and Okada (*32*) had shown the OH to be the most important water radical contributing to the indirect effect in L5178Y cells. This was true for both the inactivation of cellular proliferative capacity and the induction of DNA strand breaks. In Chinese hamster cells the OH was found

[2]R. Roots, Evaluation of radiation damage in mammalian cells due to the hydroxyl radical. Ph. D. Thesis, Dept. of Radiation Biology and Biophysics, Univ. of Rochester, 1970.

to be responsible for ~ 62% of the radiation inactivation rate (from the multitarget equation) for cells irradiated under air-saturated conditions (*33*). In hypoxia the OH was responsible for ~ 30% of the measured inactivation rate. These values were based on the maximum radioprotection afforded hamster cells by high concentrations of dimethylsulfoxide. Figure 1a shows the radioprotective effect of four different OH scavengers on the inactivation rate of hamster cells irradiated in air. Each compound has been tested up to the maximum concentration producing no toxicity in hamster cells (cells treated for 20 minutes at 21° C). Unfortunately, only with DMSO was a maximum in protective effect observed. If we assume this maximum protection represents the OH component to cell inactivation, we can obtain from these data the concentration of each protector effective in producing one half of the maximum effect. A competition process between the cellular targets and the protectors for OH will be governed by kinetics where the rates of each reaction will be equal at half-maximum effect. Figure 1b shows that the reciprocal of concentrations producing 50% of the maximum protection correlate well with the absolute rate constants of these protectors with OH (*34*). Studies with different EA sensitizers in hypoxia (Fig. 2) indicate that the concentration range over which DMSO protects against cell inactivation is independent of the sensitizer tested. These results suggest that in cells these compounds effect radiosensitization, in part, through reaction with target radicals generated by OH. From the data of Fig. 1 the rate of reaction of cellular target with OH can be computed to be $\sim 9 \times 10^{8}$ sec^{-1} and the diffusion distance of this radical is 15–17 Å. The data of Figs. 1 and 2 also indicate that after complete OH scavenging (DMSO at 1M and greater) there is an additional component of radiation damage which is sensitized by the EA compounds. We have attributed this component to target radicals generated by direct action (*33*) according to a scheme proposed by Willson and Emmerson (*35*) for the radiosensitizing mechanism of TAN. The radiosensitizing chemical reaction in this scheme for both indirect and direct action is the oxidation of a transient target radical by the EA chemical resulting in an altered target. This process in the cell competes with radical reduction processes which can restore the target to its original form. Our studies indicate that sulfhydryl compounds radioprotect mammalian cells at concentrations which are inadequate to compete effectively with the cellular targets for the radicals produced in water radiolysis (*33*). As well, the extent of protection by -SH compounds corresponded closely to the sum of damage resulting from the OH and the direct-action component. That various hydrogen-donating species can compete for DNA radicals with EA sensitizers has been conclusively demonstrated with our in vitro DNA system (*33,36*).

It was mentioned that Roots and Okada had demonstrated that OH were the most important water radicals involved in DNA strand breakage (*32*). We have studied the rate of single-strand break induction in Chinese hamster cells irradiated under numerous conditions of chemical radiosensitization and radioprotection. The radiosensitizing effect of O_2 occurs over the same concentration range and to the same extent for both cell inactivation and DNA strand breakage (*30*). Figure 3 shows the effect of various concentrations of DMSO on the rate of strand-breakage in air-saturated hamster cells. The kinetics of this radioprotective effect for this molecular lesion are indistinguishable from those measured for cell inactivation. Additional evidence (D. L. Dugle, to be published) indicates that with both -SH protectors and various OH scavengers the kinetics for protection of DNA strand breakage corresponds to the kinetics for protection against cell inactivation. These studies suggest that the radiation chemical model for cell inactivation, summarized in Fig. 4, is adequate to explain the bulk (~ 82% of the air-saturated inactivation rate) of the inactivating events induced in Chinese hamster cells by ionizing radiation. The fact that the same model can quantitatively account for the production by radiation of a specific lesion in the cellular DNA may not be coincidental.

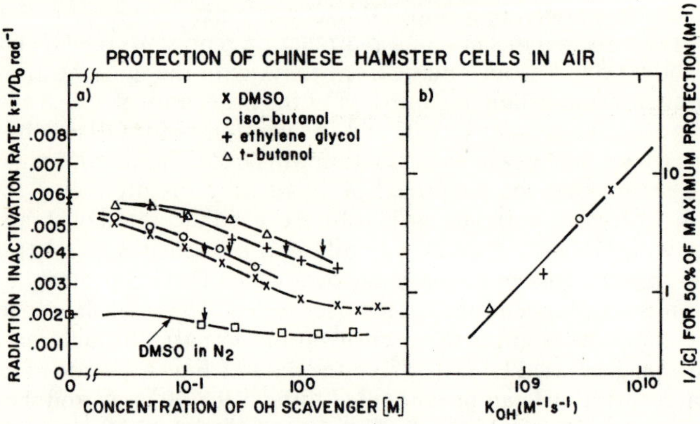

FIG. 1a. The chemical radioprotection by various OH scavengers of the multi-target inactivation rate of air-saturated Chinese hamster cells.

FIG. 1b. The correlation between the reciprocal of protector concentration effecting 50% of the maximum radioprotection and the absolute rate constants of OH with the specific chemical protector.

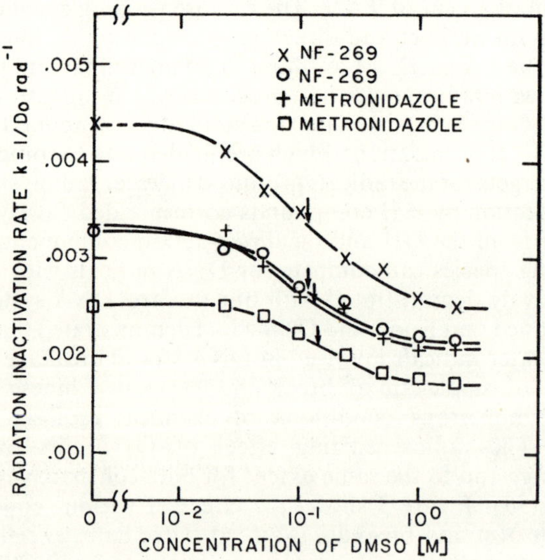

FIG. 2. The chemical radioprotection by DMSO of cells sensitized by various concentrations of a nitrofuran and a nitroimidazole hypoxic radiosensitizer.

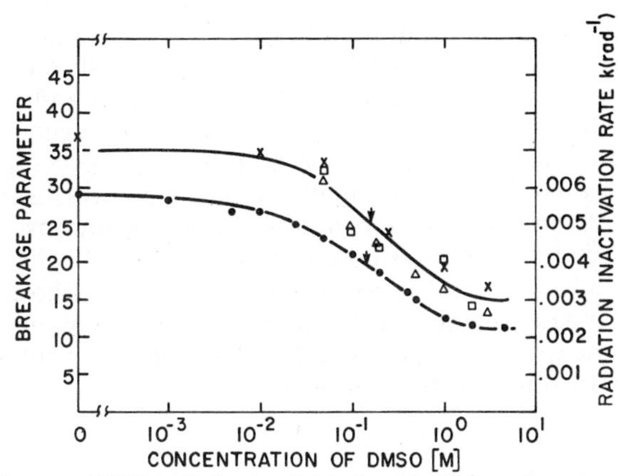

FIG. 3. The rate of DNA single-strand breakage in Chinese hamster cells as a function of DMSO concentration. The solid circles represent data replotted from Figure 1a for the radioprotection of cellular proliferative capacity. All other symbols correspond to independent determinations for the rate of single-strand breakage in the same cells.

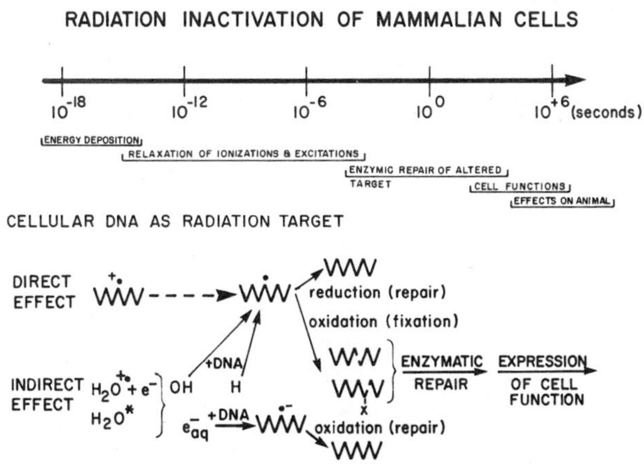

FIG. 4. A proposed working model for the inactivation of mammalian cells through the fixation of target radicals by radical-oxidizing chemicals.

While we were considering the significance of this correspondence between DNA strand breakage and cell inactivation, a paper by Chadwick and Leenhouts (*37*) appeared which proposed a model for cell inactivation. Their model defines the lethal molecular events as double-strand breaks in the DNA which can result from a single event as well as from the coincidence of two independent events on opposite DNA strands at or near the same base pair (double event). Experiments were performed to determine if the mathematical expression of this model was superior to the multitarget equation in fitting hamster cell inactivation data. For all cases tested, using both asynchronous and synchronous cell populations, the Chadwick-Leenhouts (C-L) equation gave a much superior fit to the experimental data (*38*). This model has been extended to determine the rate constants for single (α) and double (β) events for hamster cells irradiated under conditions of maximum radioprotection and radiosensitization (*39*) Typical survival data are shown in Fig. 5 along with the best-fitted curves using the C-L equation. Table 1 gives values for α and β determined for various conditions of radiosensitization by EA chemicals. It is apparent that these compounds effect cellular radiosensitization by predominately enhancing the double events or the expression of single-strand breaks. We find a much smaller, though significant, radiosensitization of single events by EA sensitizers and it has been demonstrated that OH contributes to the single events produced in air-saturated cells (*39*). This model for cell inactivation based on DNA strand breakage has been shown to account for cell inactivation data generated with radiations of different quality (*37*). We are presently assessing the implications this understanding of cell inactivation can have for radiotherapy and for microdosimetry.

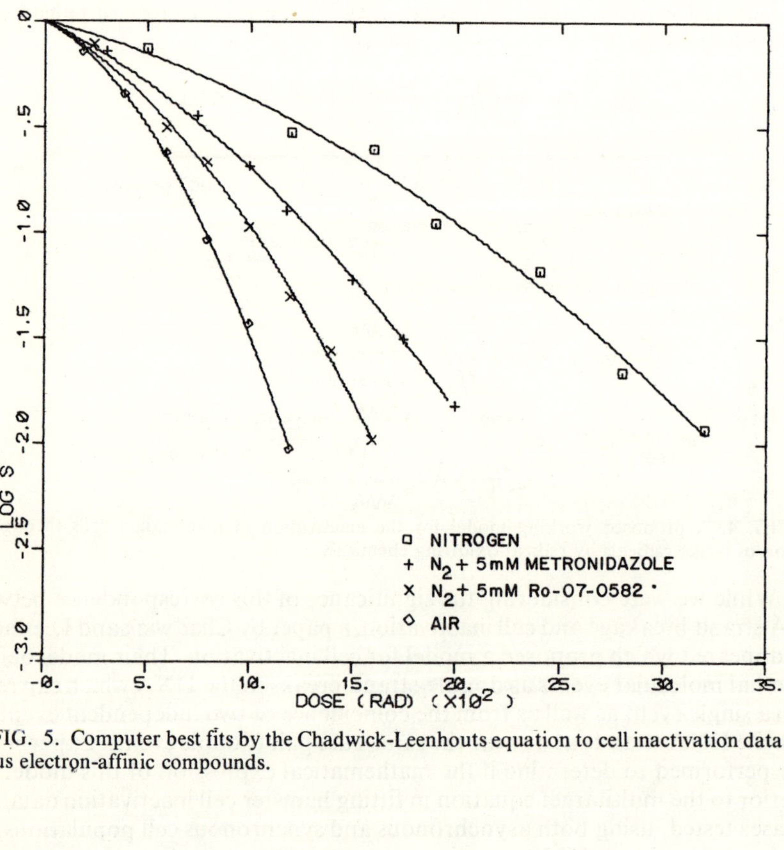

FIG. 5. Computer best fits by the Chadwick-Leenhouts equation to cell inactivation data with various electron-affinic compounds.

TABLE 1
Values for the α and β parameters from the Chadwick-Leenhouts equation for hypoxic Chinese hamster cells and for various hypoxic cell radiosensitizers

	$\alpha \times 10^3$ rad^{-1}	$\beta \times 10^6$ rad^{-2}
NITROGEN	.89 ± .24	.20 ± .09
N_2 + 5mM Metronidazole	1.09 ± .17	.51 ± .11
N_2 + 500μM NF–269	1.09 ± .25	1.26 ± .22
N_2 + 5mM Ro–07–0582	1.19 ± .28	1.03 ± .22
AIR	1.18 ± .39	2.52 ± .43

In conclusion, the salient characteristics of EA-sensitizer action upon which our model for cell inactivation and its sensitization has been constructed are:

1) the drugs' presence is required at the time of irradiation, implicating a rapid process,
2) to a first-order approximation their radiosensitizing effect is independent of temperature and cell cycle suggesting metabolic processes play a minimum role,
3) the compounds do not sensitize in the presence of O_2 and consequently may compete with O_2 for the same target radicals,
4) they produce lethal products in various biomolecules by target-radical oxidative processes,
5) they compete in the cell with radical-reducing species whose action can chemically restore the target radicals to their unaltered states,
6) they enhance the production of single-strand breaks in the cellular DNA of hypoxic cells,
7) they operate predominantly within the indirect action of OH, and
8) cell inactivation kinetics with the EA sensitizers are best fitted by the Chadwick-Leenhouts model.

These features, when synthesized into a complete picture, would indicate that the integrity of cellular DNA required for proliferation is interupted by the production of DNA double-strand breaks. At low radiation doses ($\leqslant$ 1000 rads in chemically sensitized cells) these breaks arise by single and double events. The double events (or single-strand breaks) arise predominantly by the abstraction of H from the surface of the DNA molecule (likely from the sugar) by OH. The resultant DNA radicals, if oxidized by EA compounds, can result in a phosphate ester cleavage. If two such breaks occur at or near the same base pair before enzymic repair occurs, a lethal event may result. The action of EA sensitizers and O_2 will consequently increase in importance with increasing dose. A smaller component of radiosensitization by EA compounds and O_2 is observed for single events (α) and could also result from reactions between the sensitizer and DNA radicals.

ACKNOWLEDGEMENTS

We thank our colleagues of the Medical Biophysics Branch for helpful support and discussions during the course of this work.

REFERENCES

1. J. S. Mitchell, in *Progress in Biochemical Pharmacology*, **Vol. I**, pp. 335–351, Karger, Basel/New York, 1965.
2. W. Duncan, *Brit. Med. Bull.* **29**, 33 (1973).
3. G. E. Adams and D. L. Dewey, *Biochem. Biophys. Res. Commun.* **12**, 473 (1963).
4. G. E. Adams, *Curr. Top. Radiat. Res.* **3**, 35 (1967).
5. G. E. Adams in *Radiation Protection and Sensitization* (H. L. Moroson and M. Quintiliani, Eds.) pp. 3-14. Taylor and Francis, London, 1970.
6. M. J. Ashwood-Smith, D. M. Robinson, J. H. Barnes, and B. A. Bridges, *Nature* (*London*) **216**, 137 (1967).
7. L. Parker, L. D. Skarsgard, and P. T. Emmerson, *Radiat. Res.* **38**, 493 (1968).
8. G. E. Adams, J. C. Asquith, D. L. Dewey, J. L. Foster, and R. L. Willson, *Int. J. Radiat. Biol.* **19**, 575 (1971).
9. J. D. Chapman, R. G. Webb, and J. Borsa, *Int. J. Radiat. Biol.* **19**, 561 (1971).
10. J. D. Chapman, J. A. Raleigh, J. Borsa, R. G. Webb, and R. Whitehouse, *Int. J. Radiat. Biol.* **21**, 475 (1972).
11. J. A. Raleigh, J. D. Chapman, J. Borsa, W. Kremers, and A. P. Reuvers, *Int. J. Radiat. Biol.* **23**, 377 (1973).
12. G. E. Adams, J. C. Asquith, M. E. Watts and C. E. Smithen, *Nature (London)* **239**, 23 (1972).
13. J. Denekamp and B. D. Michael, *Nature* (*London*) **239**, 21 (1972).
14. A. P. Reuvers, J. D. Chapman, J. Borsa, *Nature* (*London*) **237**, 402 (1972).
15. J. D. Chapman, A. P. Reuvers, J. Borsa, A. Petkau, and D. R. McCalla, *Cancer Res.* **32**, 2616 (1972).
16. J. D. Chapman, A. P. Reuvers, J. Borsa, *Brit. J. Radiol.* **46**, 623 (1973).
17. J. C. Asquith, J. L. Foster, R. L. Willson, R. Ings, and J. A. MacFadzean, *Brit. J. Radiol.* **47**, 474 (1974).
18. J. C. Asquith, M. E. Watts, K. Patel, C. E. Smithen, and G. E. Adams. *Radiat. Res.* **60**, 108 (1974).
19. D. A. Agnew and L. D. Skarsgard, *Radiat. Res.* **57**, 246 (1974).
20. J. E. Thomson and A. M. Rauth, *Radiat. Res.* **60**, 489 (1974).
21. A. M. Rauth and K. Kaufman, *Brit. J. Radiol.* **48**, (in press, 1975).
22. R. M. Sutherland, J. A. McCredie, and W. R. Inch, *J. Nat. Cancer Inst.* **46**, 113 (1971).
23. R. M. Sutherland and R. E. Durand, *Int. J. Radiat. Biol.* **22**, 613 (1972).
24. F. W. Hetzel, J. Kruuv, H. E. Frey, and C. J. Koch, *Radiat. Res.* **56**, 460 (1973).
25. J. W. Boag, These Proceedings.
26. G. E. Adams, *Adv. Radiat. Chem.* **3**, 125 (1972).
27. S. Okada, in *Radiation Biochemistry* (K. I. Altman, G. B. Gerber and S. Okada, Eds.) **Vol. I.** Academic Press, New York, 1970.
28. D. L. Dugle, J. D. Chapman, C. J. Gillespie, J. Borsa, R. G. Webb, B. E. Meeker, and A. P. Reuvers, *Int. J. Radiat. Biol.* **22**, 545 (1972).
29. B. Palcic and L. D. Skarsgard, *Int. J. Radiat. Biol.* **21**, 417 (1972).
30. J. D. Chapman, D. L. Dugle, A. P. Reuvers, B. E. Meeker, and J. Borsa. *Int. J. Radiat. Biol.* **26**, 383 (1974).
31. J. D. Chapman, C. L. Greenstock, A. P. Reuvers, E. McDonald, and I. Dunlop, *Radiat. Res.* **53**, 190 (1973).
32. R. Roots and S. Okada, *Int. J. Radiat. Biol.* **21**, 329 (1972).
33. J. D. Chapman, A. P. Reuvers, J. Borsa, and C. L. Greenstock, *Radiat. Res.* **56**, 291 (1973).
34. A. P. Reuvers, C. L. Greenstock, J. Borsa, and J. D. Chapman, *Int. J. Radiat. Res.* **24**, 533 (1973).
35. R. L. Willson and P. T. Emmerson, in *Radiation Protection and Sensitization* (H. L. Moroson, and M. Quintiliani, Eds.) pp. 73-79. Taylor & Francis, London, 1970.
36. C. L. Greenstock, J. D. Chapman, J. A. Raleigh, E. Shierman, and A. P. Reuvers, *Radiat. Res.* **59**, 556 (1974).
37. K. H. Chadwick and H. P. Leenhouts, *Phys. Med. Biol.* **18**, 78-87 (1973).
38. C. J. Gillespie, J. D. Chapman, A. P. Reuvers, and D. L. Dugle, *Radiat. Res.* (submitted).
39. J. D. Chapman, C. J. Gillespie, A. P. Reuvers, D. L. Dugle, and J. Borsa, *Radiat. Res.* (submitted).

In Vivo Testing of Hypoxic Cell Radiosensitizers[1]

A. M. Rauth, K. Kaufman and J. E. Thomson
Physics Division, Ontario Cancer Institute
and Department of Medical Biophysics,
University of Toronto,
500 Sherbourne St., Toronto 5,
Ontario, Canada

INTRODUCTION

Under the assumption that hypoxic cells play an important role in the radiocurability of human tumors, a number of different approaches have been made to deal specifically with these radiation resistant cells. One of these approaches involves the use of radiation sensitizers for hypoxic cells as described by Adams (*1*). The rationale behind their use is as follows. Oxygen diffusing from the capillaries is utilized by nearby cells and thus fails to reach some cells further away. Chemical sensitizers in the blood stream, if they are not metabolized, might diffuse uniformly to all parts of the tumor. When exposed to radiation, these hypoxic cells could be selectively sensitized.

Previous results as discussed and summarized by Chapman (*2*) have indicated that a number of nitro-group-containing compounds exist which specifically sensitize hypoxic mammalian cells to ionizing radiation in vitro. The question then arises as to which of these compounds, if any, will be effective in vivo. Some potential problems in going from the in vitro to the in vivo situation can be anticipated from the previous work of Hewitt and Blake (*3*), Hornsey (*4*) and Olive, Inch, and Sutherland (*5*) on in vivo testing of the nitroxyl group sensitizer triacetoneamine-N-oxyl (TAN). Some of these problems are as follows. First, there is the question of effective concentrations. Compounds can be used at their solubility limit in in vitro studies by making them up in the medium in which the cells are suspended, but for animal studies this is not possible. Assuming the drug is uniformly distributed in the animal, the maximum concentration attainable is likely to be about one-tenth or less of the concentration injected systemically or administered orally. Secondly, the mode of drug application may be important. Again in the in vitro situation the cells are suspended in the drug-containing medium. In the case of animals it may be administered by injecting it intravenously (i.v.), intraperitoneally (i.p.), directly into the tumor (i.t.), or given orally (p.o.) and the kinetics of drug distribution may be a function of the mode of administration. For three of the four modes of administration, the drug must pass via the blood to the tumor and diffuse to the site of hypoxic cells. On the way, it may be subject to differential absorption, metabolism, or binding. Once the drug

[1]Congress presentation D–2–3.

reaches the hypoxic cells it must be able to act at high cell density as discussed by Chapman (*2*). Thirdly, the toxicity of these drugs, which in vitro is only defined in terms of one cell type, becomes more difficult to predict in vivo considering the many cell types and organ functions that may limit the level of drug concentrations that can be used. Finally, the endpoint of drug effectiveness in vitro is generally that of selective sensitization of initially aerobic cells rendered suddenly hypoxic (acute hypoxia) while in solid tumors the target is cells which are hypoxic due to the gradual outgrowth of the tumor from its vascular supply (chronic hypoxia). There may well be biochemical and radiobiological differences between acutely and chronically hypoxic cells (*6*). Recent in vitro techniques such as the V79 hamster cell spheriods of Sutherland, McCredie and Inch (*7*) offer the possibility of chronically hypoxic cells in vitro, and acute hypoxia can be induced in vivo by clamping off the blood supply. Thus, possible differences in the radiation response of acutely hypoxic cells vs chronically hypoxic cells may be approachable by both in vitro and in vivo means.

A number of in vivo systems are currently in use for testing hypoxic cell sensitizers in mice such as: the skin clone system of Withers (*8*) as used for studies of the nitrobenzene N-dimethylparanitro-2-propriophenone (NDPP) by Denekamp and Michael (*9*), the tumor cure systems of Fowler and co-workers (*10*) and Inch and co-workers (*5*), the ^{125}IUdR release system and tumor regrowth assays of Begg, Sheldon, and Foster (*11*), the tumor regrowth assays of Denekamp, Michael, and Harris (*12,13*), and the leukemic cell system of Hewitt and Blake (*3*).

The system we have chosen to use is the KHT fibrosarcoma (*14*) growing in C3H mice in which cell survival can be assayed by the in vivo lung colony assay of Hill and Bush (*15*). Briefly, this assay involves removing tumors after treatment in vivo, making a single cell suspension and injecting aliquots i.v. into unirradiated C3H recipient mice. Eighteen to twenty-one days later the lungs of these mice are removed and the number of nodules or colonies on the surface counted. In this way survival curves can be determined.

The potential usefulness of this system in measuring the effectiveness of sensitizers in vivo is shown in Fig. 1. The dotted lines in this figure represent previous survival curves of Hill, Bush, and Yeung (*16*) for the KHT tumor cells under three irradiation conditions. When the tumor cells are irradiated as solid tumors in C3H mice breathing air, or in animals killed by breathing nitrogen, the curves labelled aerobic and anoxic, respectively, are obtained. When the cells are irradiated in vitro as a single cell suspension in the presence of air, the curve labelled fully oxygenated is obtained. Analysis of these curves has indicated that 10–30 per cent of the viable tumor cells in 0.5–1.0 cm^3 tumors are hypoxic and the final slope of the aerobic survival curve is characterized by the hypoxic cells. One can speculate that if sensitizer is given to animals breathing air at the time of irradiation and if it reaches all hypoxic cells so the tumor is fully sensitized, case 1 will be obtained. In the case shown, the enhancement ratio (E.R.) chosen is intermediate between that expected for nitrogen and oxygen. Alternatively, the sensitizer might only reach a fraction of the hypoxic cells resulting in sensitization of only part of the hypoxic population, case 2.

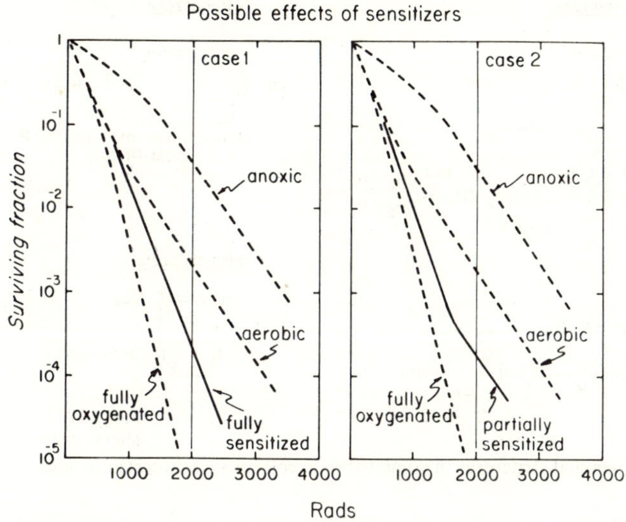

FIG. 1. Possible effects of sensitizers on KHT tumor cell survival in vivo in mice breathing air at the time of irradiation (solid lines). Dotted lines are from (*16*) and show the survival curves for KHT cells irradiated in vivo after the mice were killed by breathing nitrogen (anoxic), for mice breathing air (aerobic) or for cells irradiated in vitro under aerobic conditions (fully oxygenated). In case 1 the sensitizer reaches all hypoxic cells and in case 2 only a fraction of the hypoxic cells. [Reproduced from A. M. Rauth and K. Kaufman, *Brit. J. Radiol.* **48** (in press, 1975) with the permission of the British Institute of Radiology.]

RESULTS AND DISCUSSION

The advantage of the KHT system is its relative rapidity and degree of quantitation. With this system we have investigated a number of sensitizers that have proven effective in vitro. Some representative compounds are shown in Fig. 2. The nitrofurans were studied in collaboration with Dr. Chapman and his associates. As pointed out by them, this group of compounds is of special interest, since some of them are used clinically as antibiotics (*17,18*). In investigating a series of these compounds, we became aware of the desirability of the drugs being effective sensitizers well below their solubility limits so that we did not have dilution problems in in vivo studies. We failed to demonstrate significant sensitization in vivo with the nitrofurans tested, and this failure brought out the fact that high blood levels of a compound may not be a guarantee of sensitization (*19*). For example, as shown in Fig. 3, colorimetric estimation of plasma levels of NF-269 indicated that levels greater than 400 μM could be realized for appreciable periods of time. In vitro such levels yield an enhancement ratio of 2.1 (*18*) when tested at low cell density. However, we have shown, with techniques used previously for NF-167 (*20*), that the sensitizing ability of NF-269 decreases at high cell density in vitro and Chapman (personal communication) has shown that the tumor penetration of this drug is poor in vivo. Both these problems may contribute to our negative results.

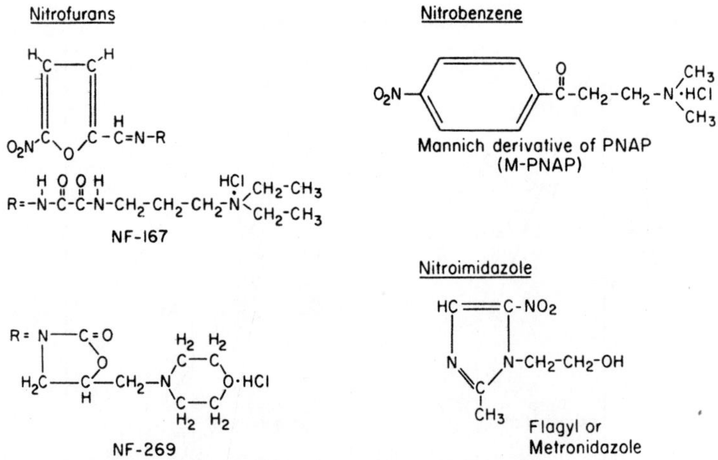

FIG. 2. Chemical structures of some hypoxic cell sensitizers tested in vivo.

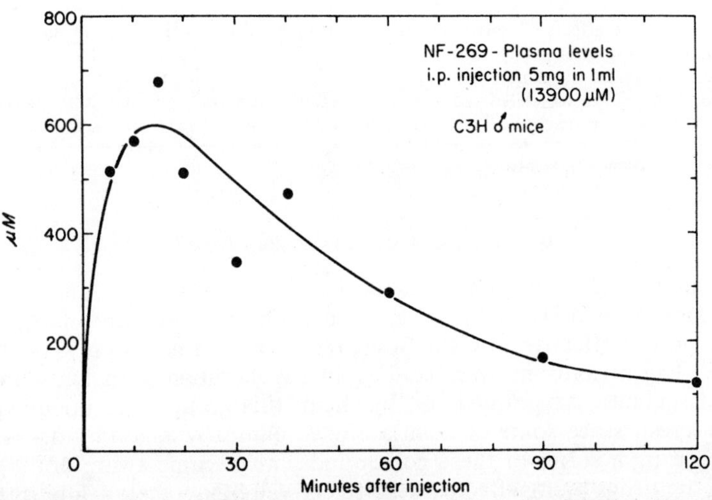

FIG. 3. Plasma levels of NF - 269 as a function of time after injection. Drug levels were measured as described previously (*19*).

Our first success in vivo was with NDPP which Denekamp and Michael showed gave sensitization in vivo using the skin clone assay under acutely hypoxic conditions (*9*). We were not able to obtain sensitization with this drug by i.p. injection. In an attempt to obtain a higher concentration of drug in the tumor, we injected NDPP directly into the tumor and irradiated as rapidly as possible after injection. Under these rather special conditions appreciable sensitization (E.R. 1.5) of chronically hypoxic cells in the KHT tumor was seen (*19*).

About this same time, we became aware of another compound, metronidazole (Flagyl), which was being studied by Foster and Willson (*21*) at the Gray Laboratory. Though this compound was shown to be a sensitizer in vitro, it was not as striking as some of the others in that maximum sensitization required drug concentrations in the 1–10 mM range and the maximum E.R. was only 1.7 as shown by Asquith et al. (*22*). However it was recognized by Willson that this compound has a number of properties desirable for in vivo studies such as high water solubility, low toxicity, long half-life in blood, and that it is already used clinically as an antiprotozoal agent. Our results with metronidazole in the KHT system, when it is given i.p. to tumor-bearing mice breathing air at the time of irradiation, are shown in Fig. 4. These results with mice given 1.5 mg/g metronidazole one hour before irradiation are compared with the survival curves for cells in the absence of sensitizer taken from Fig. 1. Significant sensitization is seen with an E.R. of 1.5, close to the in vitro maximum.

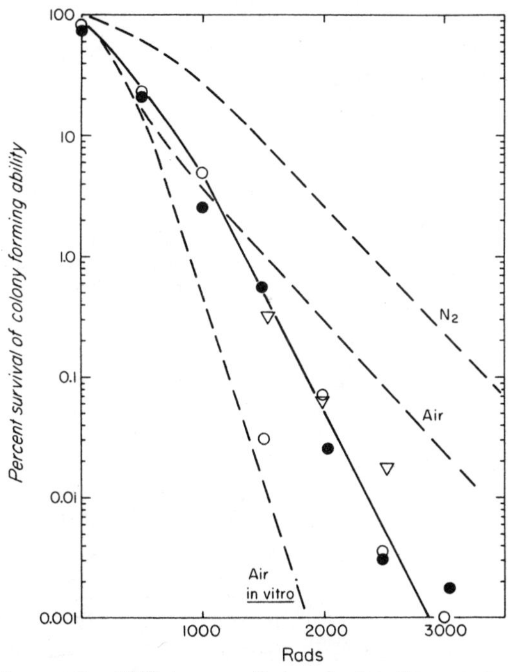

FIG. 4. Survival curve for KHT tumor cells in mice breathing air irradiated with ^{137}Cs γ rays (96 rads/min) one hour after i.p. injection of 1.5 mg/g of metronidazole (*19*). Different symbols indicate separate experiments. Dotted lines are the same as in Fig. 1. [Reproduced from A. M. Rauth and K. Kaufman, *Brit. J. Radiol.* **48** (in press, 1975) with the permission of the British Institute of Radiology.]

With the recognition that metronidazole was effective in vivo, other nitroimidazoles were investigated by ourselves and others. Tinidazole, in which the side chain at the 1-position of metronidazole is replaced by $-(CH_2)_2SOCH_2CH_3$, is an equally good sensitizer in vivo in the KHT system (*19*). The Gray Laboratory identified Ro 07–0582 as a sensitizer which was superior to metronidazole in that it gave a larger E.R. at lower concentrations in vitro (*23*). Although it is not used clinically at present, some pharmacology has been done with it in man (*24*). The structure of this 2-nitroimidazole is shown in Fig. 5. Using the highest tolerated dose of drug, 1.5 mg/g, we determined the survival curve for KHT tumor cells in vivo in mice breathing air. The results of this experiment are shown in Fig. 6, where an E.R. of ~2.0 was obtained. Once again these results are compared to the survival curves for KHT cells in the absence of sensitizer taken from Fig. 1 as well as to the results for metronidazole, Fig. 4. Ro 07–0582 at present gives the largest E.R. that we have observed in vivo with the KHT tumor system.

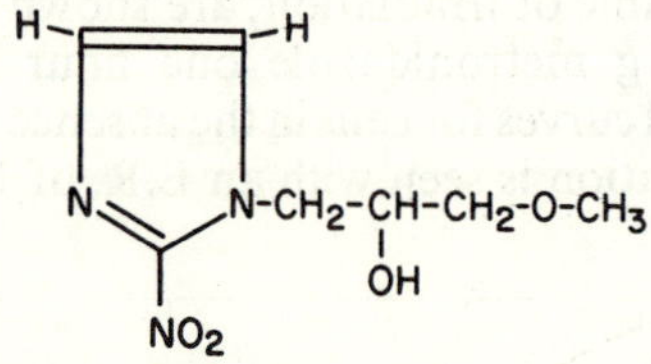

FIG. 5. Chemical structure of Ro 07–0582.

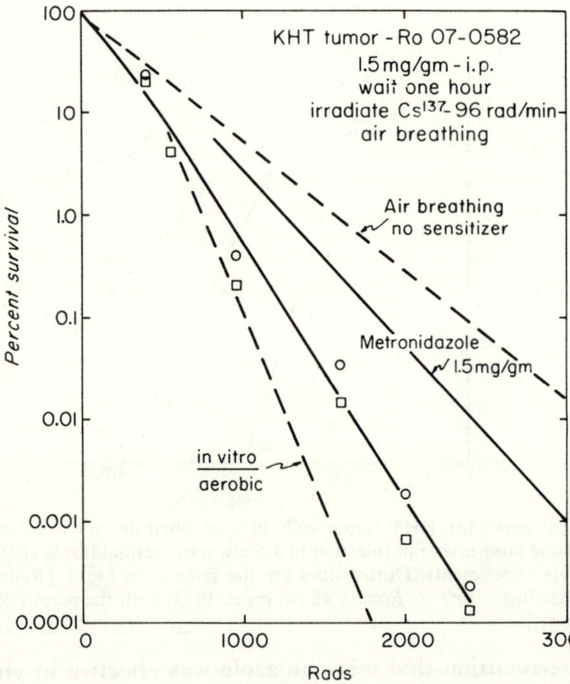

FIG. 6. Survival curve for FHT tumor cells in mice breathing air irradiated with ^{137}Cs γ rays (96 rads/min) one hour after i.p. injection of 1.5 mg/g of Ro 07–0582. Different symbols indicate separate experiments. Data are compared to previous data (Fig. 4) for metronidazole. Dotted lines are for the survival curves from Fig. 1 for KHT cells irradiated in vivo in mice breathing air, no sensitizer and in vitro under aerobic conditions.

These in vivo results with metronidazole and Ro 07–0582 can be compared with work from the Gray Laboratory in a number of in vivo systems as summarized in Tables 1 and 2. Here for a variety of mouse tumors irradiated under chronically or acutely hypoxic conditions as well as acutely hypoxic normal skin cells, in vivo sensitization was obtained with which our KHT results are in good agreement.

TABLE 1
Summary of In Vivo Sensitizer Testing
Gray Laboratory—London, England
Enhancement Ratios For Flagyl and Ro 07—0582
3–5 mM Blood Levels.

	System	Flagyl or Metronidazole	Ro 07-0582	Ref.
acute hypoxia	skin clones WH mice	1.5	1.9 - 2.1	Denekamp Michael Harris
acute hypoxia	^{125}IUDR loss and tumour regrowth CBA Sa "F"	1.3	2.1	Begg
acute hypoxia	leukemic cells in liver	-	2	Hewitt Peters
acute hypoxia	regrowth CBA Ca "NT"	-	2.1	Denekamp Harris
air breathing	^{125}IUDR loss and tumour regrowth CBA Sa "F"	1.3	1.6	Begg
air breathing	tumour care C3H mamm	1.4	1.9	Sheldon Foster Fowler
air breathing	regrowth CBA Ca "NT"	1.6	1.7 - 2.1	Denekamp Harris

TABLE 2
Summary of In Vivo Sensitizer Testing
Gray Laboratory—London, England
Enhancement Ratios for Ro 07—0582

System	0.2 mg / gm	1 mg / gm	Ref.
tumour cure WH Squa. Ca.	1.0	1.9	Peters
fast sarcoma CBA irradiate in vivo assay in vitro	1.3	2.2	McNally
squamous carcinoma a) mouse, single dose i.v.	1.0	-	Hewitt
b) repeated dose i.v. X 2	1.1	-	Hewitt

The success, to date, of these studies in terms of obtaining sensitizers which are effecive in vivo has been most encouraging. One always hopes new and better sensitizers are in the offing. We are currently carrying out limited testing of nitro-compounds using the KHT tumor as an in vivo screen. Some preliminary results are shown in Table 3. Ro 07–0207 is a 5-nitroimidazole which showed some promise in vitro (*23*), but with which we have not been able to show positive results in vivo. The reason for this is not clear at present. It may be a problem of tumor penetration and/or cell density effects since we observed plasma levels of greater than 2 mM in mice after irradiation was completed, and it has been shown (*23*) that in vitro at low cell density this concentration gives an E.R. of approximately 1.5.

TABLE 3
Preliminary data on the in vivo sensitizing ability of several nitroimidazoles and a nitropyrole
The amount of drug injected i.p. was in general either determined by the toxic or solubility limit of the particular drug tested

Exp.	Compound	Structure	Enhancement ratio*	Injected i.p.	Time before Irradiation
131	Ro 07-0207	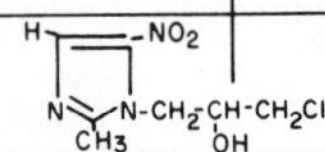	1.0	1 mg/gm	1 hr
130	R.P. 8532	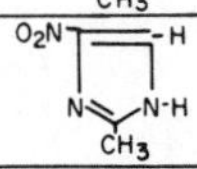	1.4	0.6 mg/gm	1 hr
128	R.P. 8979	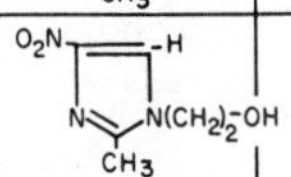	1.25	1.5 mg/gm	1 hr
136			1.0	1.5 mg/gm	15 min
129 and 135	HCNP	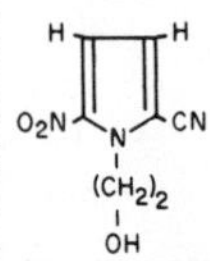	1.4	0.75 mg/gm	15 min

*Calculated from single survival point assuming drugs are dose modifying

Two compounds from Poulenc-Montreal were investigated since they were thought to be minor contaminants of metronidazole. These 4-nitroimidazoles are slightly less effective than metronidazole in vivo. The final compound, HCNP, is a nitropyrole under study in collaboration with Dr. J. Raleigh. It appears similar to the 4–nitroimidazoles in its in vivo sensitizing ability. This compound is a synthesis intermediate of a nitropyrole which has been tested as an antiprotozoal agent (*25*). Dr. Raleigh is currently systematically varying the substituents on the pyrole ring to correlate the role of electron affinity with its sensitizing ability in chemical systems. In vivo screening results should in time give an insight into the pharmacological properties of the drugs which are important for obtaining radiosensitization of chronically hypoxic cells in tumors.

The results with metronidazole have encouraged Urtasun, Sturmwind, Rabin, Band, and Chapman (*26*) and Deutsch and Foster (personal communication) to carry out preliminary pharmacological studies in humans with this compound. The fact that it is already widely used in humans as an antibiotic, albeit at much lower concentrations, has been an aid in anticipating problems and side effects. We have also carried out preliminary tests of plasma levels of the drug in humans in collaboration with R. S. Bush. The fact metronidazole can be identified by simple colorimetric (*26*) or polarographic tests (*27*) of serum make such studies routinely possible. As can be seen in Fig. 7, a patient given 1 g of metronidazole orally per day has an initial rapid rise in the plasma level of the drug and a slower decline, so that after 24 hours appreciable amounts of the drug remain, resulting in a gradual buildup of the maximum drug level. However, plasma levels of 100 μM are low compared to a predicted need of 1000 μM to give an E.R. of 1.1–1.2 after a single dose of radiation based on results on mice in vivo. Urtasun et al. (*26*) and Deutsch and Foster (op. cit.) have gone as high as 10 g single doses of metronidazole and under these conditions get serum levels in the range of 1000 μM. Some problems with nausea have been reported by these workers at high dose levels. This area of pharmacology is one of increasing importance as more consideration is given to possible clinical applications of sensitizers.

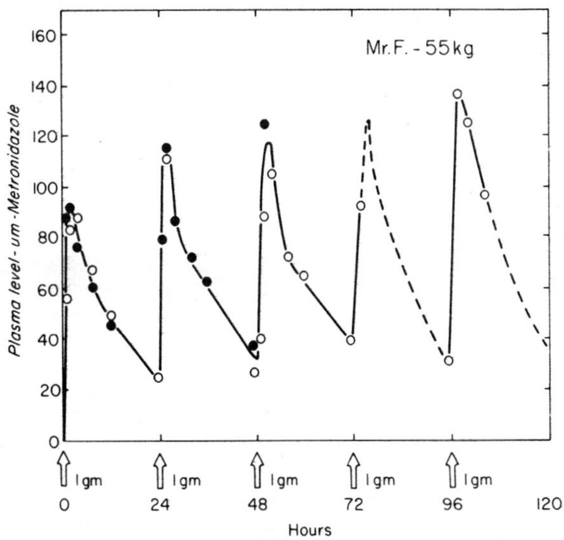

Fig. 7. Metronidazole plasma levels as measured by polarography in a patient given 1 gm daily orally for five days. Different symbols indicate plasma levels measured on two successive weeks in the same patient.

Besides the necessity of getting adequate drug levels to human tumors, the sensitizers must be effective with small doses of fractionated radiation. We have carried out a preliminary study of the effects of fractionated radiation on the KHT tumor in the presence and absence of sensitizer. In these experiments, mice bearing KHT tumors were injected daily for five days with 1 mg/g of Ro 07–0582 one hour before being given 500 rads whole body. The results of these experiments are shown in Fig. 8 in comparison with results for tumor-bearing mice not given sensitizer. The results with fractionation are also compared to the single dose survival data for this tumor taken from Fig. 1. The points for mice given sensitizer during fractionated radiation tend to fall reproducibly below the averaged points for radiation in the absence of sensitizer, but the effect of the drug is much less striking than in the single dose case in Fig. 6. In hindsight this might be expected, since the KHT tumor is thought to be a rapidly reoxygenating tumor (*6,28,29*) and when small doses of radiation are used one may not be giving sufficient radiation to reduce cell survival low enough so hypoxic cells are completely controlling tumor survival. Further tests with this and other tumors will be needed to verify this explanation.

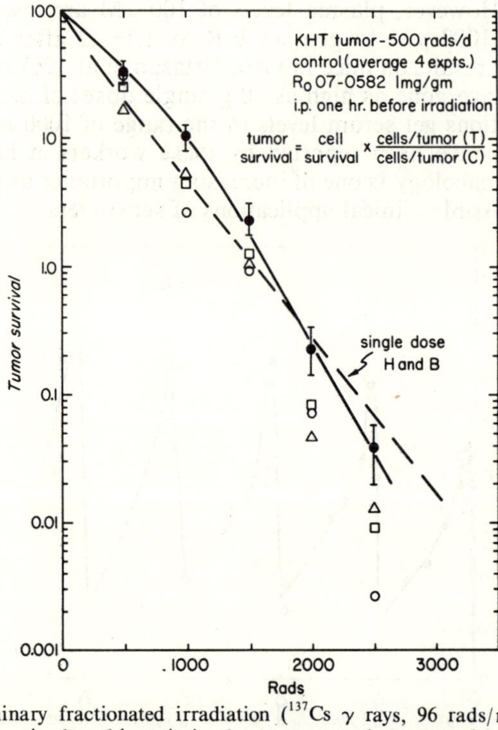

FIG. 8. Preliminary fractionated irradiation (^{137}Cs γ rays, 96 rads/min, whole body) for KHT tumor cells in mice breathing air in the presence and absence of Ro 07–0582. Groups of mice were either injected i.p. with buffer (solid circles) or 1 mg/g Ro 07–0582 daily 1 hour before irradiation with 500 rads (open symbols represent three separate experiments). Cell survivals were determined after each 500 rad dose and converted to tumor cell survivals as indicated in the figure (*31*). Cells per tumor (C) and cells per tumor (T) refer to the average number of cells recovered per tumor on day 0 for the control and the day of assay for treated tumors respectively. Rads refers to the total accumulated dose at the time of assay. Error bars are the standard error of the mean for the 'no sensitizer' groups. The dotted line labelled H and B is the single-dose survival curve for KHT cells irradiated in vivo in mice breathing air from Fig. 1.

These results presently available suggest the following areas for future research. First, screening for and development of new sensitizers with an aim towards making the ratio of the toxic concentration to the sensitizing concentration as large as possible. Second, testing of the present sensitizers with small doses of fractionated radiation to see if sensitizers are competitive with reoxygenation. In this context, perhaps sensitizers which modify the shoulder of the survival curve will turn out to be more useful than dose modifying agents. Third, in a more general context, the development of better animal models of human tumors, e.g. growth of human tumors in immunologically deficient mice (*30*). The relatively rapidly growing and reoxygenating mouse tumors may not be representative of the majority of the human tumor material.

This area of the in vivo application of hypoxic cell sensitizers has been an exciting one recently. Two to three years ago, only a few nitro-compounds were known that specifically sensitized hypoxic cells in vitro. In the intervening time, a large number of nitrobenzene, nitrofuran, nitroimidazole and nitropyrole compounds were shown to be effective in vitro (*2*). Some of these were effective in vivo in a wide variety of test systems and the fact that some of these were already used clinically as antibiotics has allowed them to be seriously considered as adjuncts to radiotherapy. We look forward to a continuation of this rapid progress over the next few years.

ACKNOWLEDGEMENTS

We would like to thank the following investigators for sharing their prepublication data and ideas. Drs. G. Adams, D. Agnew, R. S. Bush, J. D. Chapman, J. Denekamp, R. Hill, B. Michael, J. A. Raleigh, L. Skarsgard, G. F. Whitmore, and R. L. Willson, as well as Mr. J. Asquith, R. L. Foster, and S. Gulyas. Thanks are also extended to Mr. F. Ebetino of the Norwich Pharmaceutical Co., Norwich, N.Y., Mr. R. Sarrazin, Poulenc Ltd., Montreal, Quebec, Dr. H. Jackson of Ortho Pharmaceutical, Don Mills, Ontario, and Dr. C. Smithen, Roche Pharmaceutical, England, for their generous supplying of drugs and data. This work was supported in part by grants from the National Cancer Institute of Canada and the Medical Research Council of Canada, as well as the Ontario Cancer Treatment and Research Foundation.

REFERENCES

1. G. E. Adams, *Brit. Med. Bull.* **29**, 48 (1973).
2. J. D. Chapman, this symposium.
3. H. B. Hewitt and E. R. Blake, *Brit. J. Radiol.* **43**, 91 (1970).
4. S. Hornsey, *Int. J. Radiat. Biol.* **22**, 91 (1972).
5. P. L. Olive, W. R. Inch, and R. M. Sutherland, *Radiat. Res.* **52**, 618 (1972).
6. R. P. Hill and R. S. Bush, in *Advances in Radiation Research*, (T. F. Duplan and A. Chapiro, Eds.) Proceedings of the IVth International Congress of Radiation Research, Evian, France 1970 (in press).
7. R. M. Sutherland, J. A. McCredie and W. R. Inch, *J. Nat. Cancer Inst.* **46**, 113 (1971).
8. H. R. Withers, *Brit. J. Radiol.* **40**, 335 (1967).
9. J. Denekamp and B. D. Michael, *Nature (London)* **239**, 21 (1972).
10. J. F. Fowler, J. Denekamp, A. L. Page, A. C. Begg, S. R. Harris and K. Butler, *Brit. J. Radiol.* **45**, 237 (1972).
11. A. C. Begg, P. W. Sheldon, and J. L. Foster, *Brit. J. Radiol.*, **47**, 399 (1974).
12. J. Denekamp, B. D. Michael, and S. R. Harris, *Radiat. Res.* **60**, 119 (1974).
13. J. Denkamp and S. R. Harris, *Radiat. Res.* **61**, 191 (1974).
14. R. F. Kallman, G. Silini, and L. M. Van Putten, *J. Nat. Cancer Inst.* **39**, 539 (1967).
15. R. P. Hill and R. S. Bush, *Int. J. Radiat. Biol.* **15**, 435 (1969).
16. R. P. Hill, R. S. Bush, P. Yeung, *Brit. J. Radiol.* **44**, 299 (1971).
17. J. D. Chapman, A. P. Reuvers, J. Borsa, A. Petkau, and D. R. McCalla, *Cancer Res.* **32**, 2616 (1972).
18. J. D. Chapman, A. P. Reuvers and J. Borsa, *Brit. J. Radiol.* **46**, 623 (1973).
19. A. M. Rauth and K. Kaufman, *Brit. J. Radiol.*, **48**, in press (1975).
20. J. E. Thomson and A. M. Rauth, *Radiat. Res.*, **60**, 489 (1974).
21. J. L. Foster and R. L. Willson, *Brit. J. Radiol.* **46**, 234 (1973).
22. J. C. Asquith, J. L. Foster, R. L. Willson, R. Ings, and J. A. McFadzean, *Brit. J. Radiol.*, **47**, 474 (1974).
23. J. C. Asquith, M. E. Watts, K. Patel, C. E. Smithen and G. E. Adams. *Radiat. Res.*, **60**, 108 (1974).
24. J.A.F. de Silva, N. Munro, N. Strojny, *J. Pharm. Sci.* **59**, 201–210 (1970).

25. F. Bénazet, C. Cosar, P. Ganter, L. Julou, P. Populaire, and L. Guillame, *C. R. Acad. Sci. Ser. D.* **263**, 609 (1966).
26. R. C. Urtasun, J. Sturmwind, H. Rabin, P. R. Band, and J. D. Chapman, *Brit. J. Radiol.* **47**, 297 (1974).
27. P. O. Kane, *J. Polarogr. Soc.* **7**, 58 (1961).
28. R. F. Kallman, *Radiology* **105**, 135 (1972).
29. R. P. Hill and R. S. Bush, *Brit. J. Radiol.* **46**, 167 (1973).
30. L. Ozzello, B. Sordat, C. Merenda, S. Cauel, J. Huilimann, and J. P. Mach, *J. Nat. Cancer Inst.* **52**, 1669 (1974).

Chairman's Comments on the Discussion of Effects in Vivo[1]

G. E. Adams
Gray Laboratory, Cancer Research Campaign
Mount Vernon Hospital
Northwood, Middlesex HA6 2RN, England

Although there is much current interest in the development and possible application in radiotherapy of sensitizers which are specific for hypoxic cells, it has long been recognized that sensitizing agents of different types are useful for studying more general aspects of cellular radiobiology.

In setting up this symposium, it was decided to include two papers on sensitization in vitro, one dealing with the use of a sensitizer for the study of cell-cycle dependent phenomena and, by contrast, the other concentrating specifically on the differential sensitization of hypoxic cells by the oxygen-like sensitizers, the "electron-affinic" agents. The third paper dealt with electron-affinic sensitization in vivo.

In all three papers, the speakers discussed new experimental results and the present position in the respective areas of research. Each paper spoke for itself and any further comment would be superfluous. However, at the end of the program there was a general discussion which tended to concentrate on the potential applications of sensitizers in radiotherapy. Several questions were raised which in the Chairman's view, were sufficiently important to warrant the inclusion in the proceedings of a summary of part of this discussion.

SELF-CONSISTENCY IN IN-VIVO DATA

The Chairman noted that Dr. Rauth had included in his paper some of the recent in vivo data from workers in the Gray Laboratory. He asked Dr. Fowler if he would like to comment on the progress of these studies.

Fowler said he would first elaborate on the final results of the tumor control experiment referred to by Rauth, and then make a comment on the encouraging degree of consistency evident in the data from different laboratories and different experimental systems.

The data from local control experiments on first-generation transplants of mammary tumors in C_3H mice treated with single doses of X-rays while breathing oxygen at atmospheric pressure, are shown in Fig. 1. The dose required to control half of the tumors which contained 10% hypoxic cells, (TCD_{50}), was reduced by a factor of 1.8 when 1 mg/g of the 2-nitroimidazole, Ro-07-0582, was given i.p. to the mice 30 minutes before X-irradiation of the tumors. It is also apparent that the dose-response curve is steeper than the control curve by about the same factor. This would be expected if all the hypoxic cells in the tumor had been reached by the sensitizer and had been

[1]Symposium D-2: Chemical Radiosensitization of Mammalian Cells

fully sensitized. Fig. 2 compares the data on enhancement ratios for hypoxic cell sensitization in solid tumor systems in mice breathing either air or oxygen. Each capital letter represents an experimenter and a tumor system as listed in Table 1. The letters in the 'boxes' and the lower solid curve through the letters 'R' represent results obtained with metronidazole or "Flagyl"; all the remaining data are for Ro-07-0582.

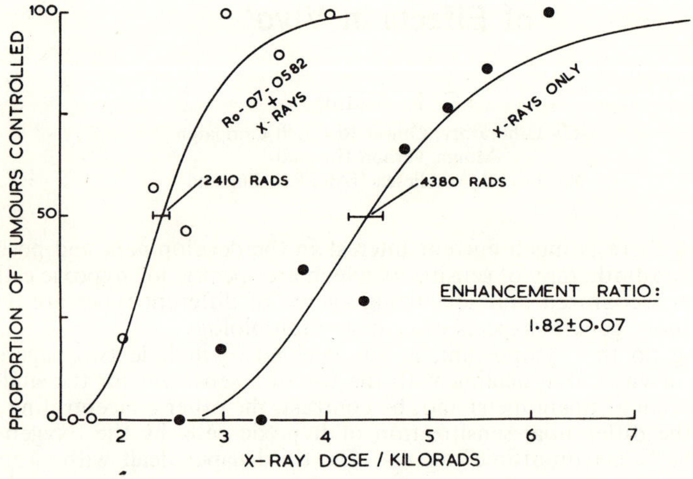

Fig. 1. 150-day local control of first-generation transplants of C_3H mouse mammary tumors irradiated with single doses of X-rays after injection of the drug, Ro-07-0582.

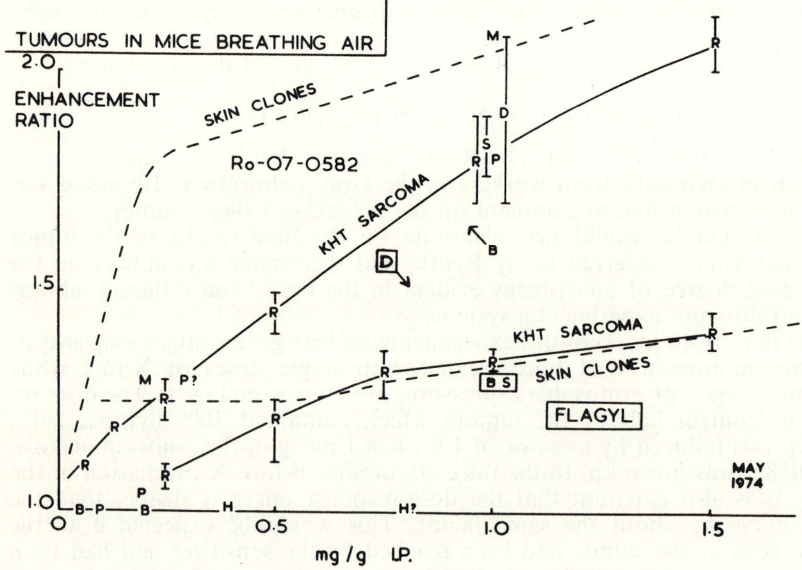

Fig. 2. Sensitization by Ro-07-0582: Comparison of in vivo data from different experimental systems.

The agreement between the data for the various tumor systems is, in general, remarkably good. The high enhancement ratios of about 1.8 for high doses of 0582 compare favorably with the hypoxic gain factors of about 1.7 found for neutrons or the figure of about 1.6 expected for negative pi mesons. At the lower concentrations more applicable to clinical use, i.e. up to about 0.2 mg/g in Fig. 1, the enhancement ratios of 1.0 to 1.4 are similar to those obtained with multiple small doses of neutrons in animal tumor experiments. Based on clinical experience with hyperbaric oxygen, both neutrons and hypoxic-cell sensitizers ought to be better. However, both might yet prove to give no advantage in many types of tumor.

TABLE I

ENHANCEMENT RATIOS FOR SOLID TUMORS IN VIVO
(mice breathing air or oxygen)

EXPERIMENTER		TUMOR	ASSAY	ENHANCEMENT WITH Ro-07-0582	
				0.2-0.3	1 mg/g
A. C. BEGG	B	CBA Fast Sarc F. 1 day* >10% hyp**	Regrowth, Loss 125-IUdR	1.0	1.6
J. DENEKAMP and S. R. HARRIS	D	CBA Carcinoma NT 3 day ~6%	Regrowth delay	–	1.7 to 2.1
H. B. HEWITT	H	WHT Squamous Carcinoma D 1 day 18%	Cell diln. in vivo	– 1.0	–
N. J. McNALLY	M	CBA Fast Sarc F. 1 day >10%	Cell diln. in vitro	1.3	2.2
L. J. PETERS	P	WHT Intradermal Squamous Ca G 1 day 0.3%	Local Control at 100 days	1.0-1.3	1.8
A. M. RAUTH (Toronto)	R	C_3H Sarc KHT 2 days 6%	Cell diln. Lung colonies	1.2-1.3	1.8
P. W. SHELDON J. F. FOWLER J. L. FOSTER	S	C_3H 1st gen. trans. of spont. mamm. Ca. 6 DAYS 10%	Local Control at 150 days	–	1.8

* Volume doubling time

** Estimated proportion of hypoxic cells

The authors named are at the Gray Laboratory except for Dr. A. M. Rauth who is at the Ontario Cancer Institute, Toronto, Canada.

TOXICOLOGY AND PHARMACOLOGY

Much discussion centered on the main toxicological problems of the two nitro-imidazoles. Generally, the animal experiments had been carried out with the drugs administered either i.p. or i.v. However, Dr. Wideroe raised the question whether direct injection of the sensitizer into the tumor might prove to be of advantage. Dr. Rauth replied that one sensitizer, NDPP, which was inactive in his system when given i.p., gave positive results when injected directly into the tumor immediately before irradiation.

The implications of the large drug doses used in the various animal experiments with either metronidazole or 0582 were recognized by several discussants. On the basis of the animal data, projected clinical doses sufficient to give comparable gain factors would probably be unrealistically high—perhaps several tens of grams. However, it was clear that both drugs are relatively nontoxic in mice, the Mount Vernon data indicating that the LD_{50} in mice is about 2.5 g/kg for metronidazole and about a factor of two less for the other drug.

With massive doses of metronidazole, an increase by about a factor of two in the incidence of lymphosarcomas and lung tumors has been reported for mice given daily doses of up to 830 mg/kg throughout life. No effect was seen in rats. This is probably not a major hazard in patients receiving the drug during radiotherapy only.

A more serious question arises from reports of cerebellar damage in dogs given repeated daily doses (50, 100 or 150 mg/kg) of metronidazole or the substituted 2-nitroimidazole, Ro-07-1051, not unlike in structure to 0582. Both drugs caused bouts of ataxia, continuous muscle spasm, cerebellar cell degeneration and finally death. If the drugs were withdrawn after the early symptoms and sedatives given, no further harmful effects were seen. No neurotoxicological symptoms were observed with similar doses of Ro-07-1051 in mice, rats, guinea pigs or rabbits, and it is suggested that dogs are atypically sensitive to this type of cerebellar damage. However, it is clear that the toxicology of 0582 itself requires much further investigation.

The main toxicological problem in the clinical use of metronidazole appears to be acute nausea, as reported by Dr. Urtasun in Edmonton and by Dr. Deutsch at Mount Vernon. Repeated doses of 4 to 10 grams (40 to 100 mg/kg) were used and the symptoms persisted up to two days after cessation of the drug. One would not expect, therefore, that significantly higher doses than these could be used in the clinic, particularly if the drug were given with each fraction during the course of radiotherapy.

The Chairman pointed out that some years ago, 0582 was investigated by Roche as a potential anti-flagellant and, in volunteer studies, daily doses of about 1 g had been used with no ill effect. However, these doses are considerably lower than the doses which would be necessary for its use as a radiosensitizer.

Dr. Fowler said that within the last month, some male volunteers from the staff of the Gray Laboratory and from Roche Products had taken oral doses of 1 g (4 subjects), 2 g (2 subjects) and 4 g (2 subjects) to obtain some information on maximum blood concentrations and retention times of the drug. Analysis of the results were not yet complete but they did suggest that somewhat higher levels are attained in man than are attained in mice for corresponding doses, provided the stomach was empty at the time of administration. Maximum concentrations occurred from 1- 2 hours later. Half-lives for disappearance of the drug ranged from 10-13 hours after 1 g, to 12-18 hours for 2 to 4 g. The two volunteers (one a clinician) who took 4 g doses complained of sleeplessness during the same night as if the drug had an amphetamine-like effect.

It seems that doses of about 5 to 7 g of 0582 will be necessary to achieve serum concentrations sufficient to give rise to enhancement ratios up to 1.3 or 1.4. Such enhancement ratios should be clinically detectable in situations where hypoxic cells control the overall radiation response of the tumor.

HOW IMPORTANT CLINICALLY ARE HYPOXIC CELLS?

As is usual in discussions of radiosensitizers, several contributors drew attention to the fact that whereas the presence of hypoxic cells has repeatedly been shown to be the limiting factor in overall radiosensitivity of some animal tumors, it has yet to be demonstrated convincingly that such cells constitute a barrier in the radiotherapy of human tumors. Dr. Wideröe commented that Eichhorn in East Berlin carried out clinical trials which were assessed by histological identification of malignant cells at autopsy. It was claimed that reoxygenation of 15-20% of the hypoxic cells present in the tumor at each fraction could be demonstrated. Dr. Fowler felt that the clinical trial of head and neck tumors carried out at Cardiff, provided more convincing evidence that hypoxic cells are present and do influence the clinical results. Patients treated in hyperbaric oxygen showed only 12 recurrences in 129 cases compared with 30/137 for patients treated in air. Fractionation schedules were identical in both series and normal tissue injury was similar. This can only mean that hypoxic cells do affect the response of these tumors. In addition, there are indications from the current neutron trial at Hammersmith, of statistically significant improvements over conventional radiotherapy.

Accepting that hypoxic cells are present in some human tumors, some discussants questioned whether *under the conditions of fractionated radiotherapy*, hypoxic cell sensitizers could offer much benefit. Dr. M. Brown said that there were three factors which he found disturbing. These were, firstly, in vivo enhancement ratios appear to be less than they are in vitro, secondly, it is clear that the doses which could be envisaged to be clinically realistic are lower than those required to give enhancement ratios of about 2 in animal tumors and, thirdly, there seems to be evidence that sensitization by both oxygen and the electron affinic sensitizers is smaller at low doses than it is at larger doses.

In answer to the first two points, Dr. Denekamp said that it should be remembered that chemical sensitization of hypoxic cells in vivo has been really established only within the last year or two and that few such compounds are, as yet, available. Progress has been very rapid and one could reasonably anticipate better results yet. More experience is needed, particularly with different kinds of tumor. Present results were as good as neutrons and, bearing in mind the huge cost of support for the high LET program, sensitizers promise an inexpensive and easily applicable alternative for the hypoxia problem.

With regard to the third point, the Chairman said that while he had heard evidence that enhancement ratios could be smaller at low doses, nevertheless, the various in vitro studies showed that they were still large. He asked Dr. Revesz whether he would comment on his own findings with 0582. Dr. Revesz said that with severely hypoxic cells in vitro he had observed no sensitization with 0582 up to doses of about 250 rads, although for higher doses he agreed with results of other workers showing enhancement ratios comparable to the OER.

In closing the symposium, the Chairman said that it was worth noting that while studies with micro-organisms have, over the last decade, produced many active sensitizers for these systems, it was only around the time of the last Congress that differential sensitization of hypoxic mammalian cells was convincingly demonstrated—and then, only in vitro. Clearly, many problems remain. However the present rate of progress is very encouraging and one can but speculate what the position will be by the time of the 6th Congress.

RESPONSE OF STEM CELLS TO SINGLE REPEATED AND CONTINUOUS IRRADIATION[1]

Chairman: T. M. Fliedner

[1]Symposium D–3

Concept of Human Stem Cell Kinetics[1,2]

Eugene P. Cronkite
Medical Research Center
Brookhaven National Laboratory
Upton, New York 11973 USA

INTRODUCTION

The concept of a stem cell for hemopoiesis goes back to at least the turn of the century. Maximow and Bloom (*1*) considered hemopoiesis to be divided into homoplastic hemopoiesis—the production of mature cells by young elements of the same type. Under physiologic conditions, the needs of the adult organism are supplied, usually by homoplastic hemopoiesis. When the requirements for hemopoiesis are increased after hemorrhage, during infection and regeneration from injury, homoplastic hemopoiesis does not suffice and new erythroblasts and myelocytes develop from a common stem cell and this is called heteroplastic hemopoiesis. To modern techniques for study of the stem cell and its properties, add new parlance and the stem cells are now divided into the pluripotent and committed stem cells (*2*). Almost all of the information on the stem cells derives from studies on mice.

The pluripotent hemopoietic stem cell (HSC) is the reserve precursor of at least the erythrocytic, granulocytic, and megakaryocytic committed stem cells. In the mouse the HSC is characterized by studying the colony formation in the spleen of mouse. From the elegant studies initiated by the Canadian group of Siminovitch, Till, and McCulloch (*3–5*) and confirmed by numerous investigators, the colony forming cell spleen (CFC_s) in the bone marrow has the following characteristics:

1) About 10–15 colonies are formed in the spleen per 10^5 bone marrow cells injected. With a seeding factor of about 0.17 (*6*) and rounding off, the abundance of CFC_s in murine bone marrow is ca. 1 per 1000.

2) "Thymidine suicide" studies show that the fraction of CFC_s varies among strains of mice from nearly 0 to 20%. Therefore, nearly 0–20% are in DNA synthesis and 80–100% are in G_0 (*7*).

3) According to early studies of Bond et al. (*8*) and Brecher and Cronkite (*9*) the pluripotent stem cell migrates through the blood and in fact has been isolated from the blood.

4) The morphologic identity is unclear. Its morphology may depend on its functional state (G_0 or in cycle). It has been described as a "small lymphocyte" (*10*), a monocytoid (*11*) transitional cell (*12*).

[1]Congress presentation D-3-2

[2]This research supported by the U.S. Atomic Energy Commission.

It is generally considered that the erythroid committed stem cell is the erythropoietin-responsive cell. Furthermore, the in vitro blood and bone marrow culture systems are believed to measure the abundance of the erythroid and granuloid committed stem cells (CSC) since both erythrocytic and granulocytic colonies are grown and enumerated in culture. The cell producing colonies in culture is called the colony forming cell (CFC_c).

The CFC_c has the following characteristics:

a) In the mouse—

1) Abundance in blood of about 1 per 10^4, and in bone marrow of 2 per 1000 bone marrow cells (*13*)

2) Thymidine suicide of about 0.35 in normal mice and 0.80 in mice with regenerating marrow (*13-15*)

b) In man—

1) Abundance of about 0.1 to 1 per 10^4 blood cells and 0.1 to 1 per 10^3 bone marrow cells (*16*)

2) Thymidine suicide of 0.35 (*17*)

The preceding numbers are rounded off from published papers.

Study of the human cell is very difficult. Human cells do not produce spleen colonies. It has been suggested by several investigators that the colonies formed in culture of human peripheral blood or bone marrow by CFC_c are CSC since they produce differentiated cells (erythrocytic, granulocytic, and macrophage colonies).

The organizing chairman of this symposium specifically asked for some information on the radiosensitivity of the stem cell and whether there are any new techniques for its study. My colleagues have produced data that bear on these subjects.

D_o OF HUMAN STEM CELLS

The D_o of murine CFU_s is circa 95 rad (*16*). However, the D_o varies from 100 rad for bone marrow and 140 rad for blood cells (*17*). The D_o of human CFC's in culture is ca. 137 rad and mouse CFC's ca. 160 rad (*18*). Another approach to the measurement of human stem cell D_o is the irradiation of human bone marrow and its growth in diffusion chambers implanted in the murine peritoneal cavity (*19*).

Details of this technique will not be discussed since all are published. Briefly, one makes dilutions of bone marrow containing fewer and fewer cells. Graded number of cells are placed in diffusion chambers and implanted into the peritoneal cavity of the mouse. After a stated number of days the diffusion chambers are removed, the cells harvested from the chamber, counted, smeared on slides and characterized morphologically. By irradiating the host mice containing human bone marrow in diffusion chambers and measuring cell recovery, one can establish a radiation dose survival curve for human granulocytic stem cells. Carsten et al. (*19*) have investigated the D_o of human stem cells by this technique. Figure 1 shows three studies of irradiation of human granulocytic stem cells. The D_o is 72.6 rad.

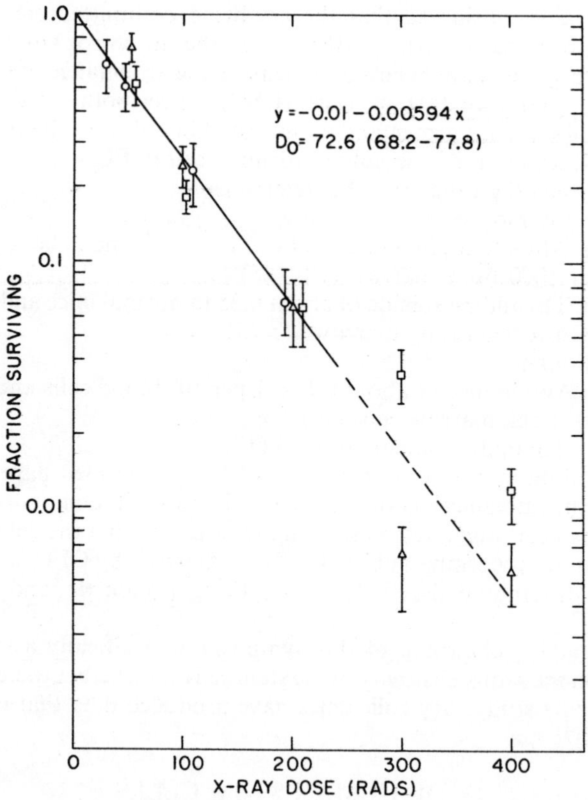

FIG. 1. The survival curve of human "stem cells" in irradiated bone marrow, measured by growth of granulocytic cells in diffusion chamber. The growth probably represents proliferation of pluripotent hemopoietic, committed stem cells and their amplification into recognizable granulocytic series.

A NEW APPROACH TO STUDY OF STEM CELLS

The existing techniques for study of stem cells have yielded a large mass of data and considerable insight into the PHSC and CSC in the mouse. An innovative approach for stem cell study has been introduced by Reincke et al. (*20*). This approach avails itself of the low energy and short range of Auger electrons from ^{55}Fe. By administering enough ^{55}Fe one can deliver a lethal intracellular dose of radiation to iron-incorporating erythropoietic precursors. The selective intramedullary death is demonstrated by absolute dose-dependent diminution of erythropoietic cells in the bone marrow and dose-dependent decrease in the subsequent incorporation of ^{59}Fe into circulating red blood cells. This dose-dependent hypoplasia of differentiated erythropoiesis makes possible the study of potential feedback loops from the differentiated marrow compartment to the erythroid CSC and to the PHSC. The ^{55}Fe erythropoietic cytocide continues because of reutilization of ^{55}Fe. This cytocide is followed by a prompt decrease in the number of CFU_s in the bone marrow of the ^{55}Fe-treated mice and a significant decrease in the capability of this marrow to protect against lethal irradiation injury of other mice, a measure of the number of pluripotent stem cells in the bone marrow.

It can be shown that the dose to other cells from ^{55}Fe in erythropoietic precursors is miniscule compared to the dose to the differentiated red cell precursors. At this moment one cannot eliminate the possibility that sufficient ^{55}Fe is incorporated into the PHSC to cause the CFU_s reduction. It is unlikely. It is tentatively concluded that the pluripotent stem cell pool was depleted because of an increased demand for differentiation along the erythropoietic pathway. This approach has considerable promise in testing the capability of the PHSC to undergo sustained mitotic activity by continual drainage down the erythropoietic pathway and in searching for intramedullary feedback loops regulating erythropoiesis and the relationship of erythropoiesis to granulopoiesis. One will also in principle be able to test whether Hayflick's hypothesis applies to PHSC since the latter will be forced to divide at a maximum rate for prolonged periods. This is a clinically important subject since repetitive courses of chemotherapy for leukemia destroys PHSC and leukemic cells in parallel, and hemopoietic regeneration requires extensive proliferation from a few surviving PHSC. If Hayflick's hypothesis applies to the PHSC one might expect marrow hypoplasia to take place at a later date after cessation of chemotherapy.

CHARACTERIZATION OF HUMAN STEM CELL POOLS FROM KNOWN KINETICS OF ERYTHRO- AND GRANULOPOIESIS

Another approach to study the human stem cell pool is to start from the peripheral blood where the turnover rates of red cells and granulocytes are well known. From the sum of these turnover rates, the structure of human bone marrow (amplification from stem cell to non-dividing cell), absolute cellularity, DNA synthesis time and fraction of cells in DNA synthesis, one can estimate the minimum flux of CSC into the erythrocytic and granulocytic differentiated pools in the marrow. The data from which the calculations will be made are reviewed (*21*) and summarized here:

1) Red cell average life span = 120 days
2) Red cell turnover rate (RTR) = 12×10^7/kg/hr (288×10^7/day) (*21*)
3) Granulocyte life span (random loss) half time of 6.8 hr (*22*)
4) Granulocyte turnover rate (GTR) = 6.8×10^7/kg/hr (163.2×10^7/day) (*22*)
5) RTR + GTR equals 18.8×10^7/kg/hr
6) Erythroid marrow cellularity (N^E) equals 536×10^7/kg (*23*)
7) Granulocytic marrow (N^G) = 1140×10^7/kg (*23*)
8) $N^E + N^G = 1676 \times 10^7$/kg (*23*)
9) DNA synthesis time in human bone marrow is about 12 hrs for erythrocytic and granulocytic proliferating pool (*24*)
10) Amplification from the committed stem cell to the non-dividing erythrocytic and granulocytic cell averages 16 (*21*)

The application of the above data to estimate the number of CSC in DNA synthesis, the CSC pool size and its potential relationship to a PHSC pool is shown in Fig. 2.

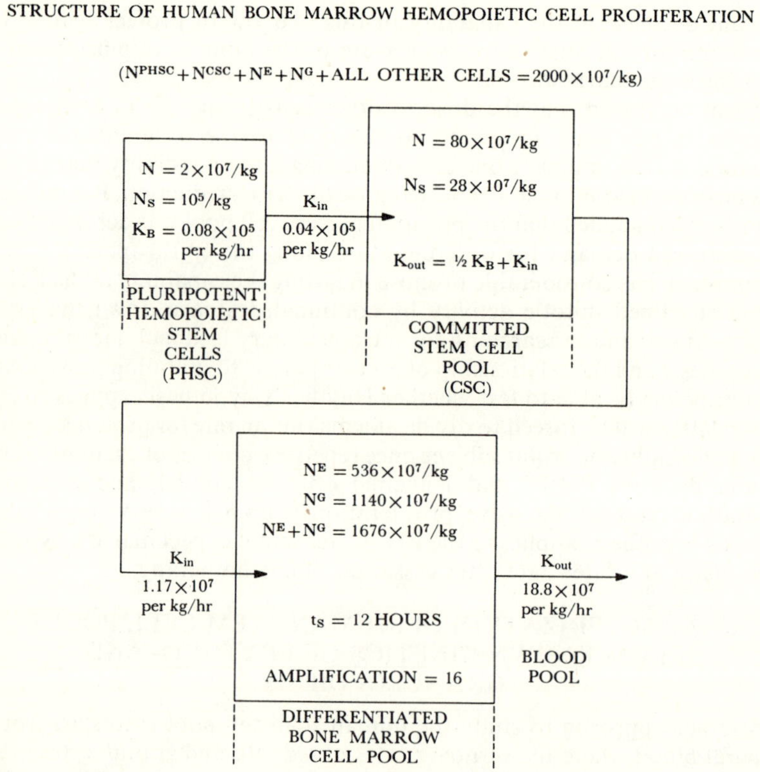

FIG. 2. A concept of human bone marrow cell proliferation constructed from the known erythrocytic and granulocytic turnover rates in human blood, absolute marrow cellularity, proportional cell distribution, and flow rates through the differentiated erythrocytic and granulocytic compartments.

If one assumes that detection of the most immature erythrocytic and granulocytic precursor establishes a cytologic boundary between the CSC pool and the differentiated pool, one can then estimate the flux of CSC into the differentiated cell pool (red and white) by dividing the RTR + GTR (K_{out} of marrow) by the average amplification of 16.

$$K_{out} = 18.8 \times 10^7/\text{kg}/\text{hr}$$

hence $\quad K_{in}$ from CSC pool $= 1.17 \times 10^7/\text{kg}/\text{hr}$.

In a stem cell pool the birth date (K_B) is twice the K_{out} (1.17×10^7/kg/hr or 2.35×10^7/kg/hr. From K_B and DNA synthesis time (t_s) one can calculate the number of cells in DNA synthesis (N_s^{CSC}). Since most human cells have a DNA synthesis time close to 12 hours, we assume this applies to CSC.

$$N_s^{CSC} = t_s \times K_B = 12 \times 2.34 \times 10^7 = 28 \times 10^7/\text{kg}$$

The total population of CSC (N^{CSC}) is obviously larger than the N_s^{CSC}.

Assume that thymidine suicide of 0.35 for colony forming cells pertains to the total CSC pool. Then

$$N^{CSC} = (28 \times 10^7)/0.35 = 80 \times 10^7/\text{kg}$$

Accepting this as the absolute value for CSC the ratio of

$$N^{CSC} \div (N^E + N^G) = 80 \div 1676 = 1/21$$

or there is 1 CSC to every 21 erythroid-granuloid cells.

Total bone marrow cellularity in man has been experimentally determined to be 1800×10^7/kg (*25*). Another approach to estimating the total bone marrow cellularity is from the mass of the red bone marrow of 1.3 kg in standard man and assuming an average volume for bone marrow cells of 1000×10^{-12} ml giving 2000 $\times 10^7$/kg. However, for this calculation Donohue's determined number is used. $N^E + N^G$ is 1676×10^7, hence other cells constitute 124×10^7/kg or 7% of the marrow. If 80×10^7/kg is the true value for CSC, then 4.4% of total marrow is CSC and over half of the none-erythroid non-granuloid cells are CSC. If true, we have all seen but didn't recognize what we saw.

Are these estimates valid? Does a thymidine suicide of 0.35 or 0.35 of the CSC in DNA synthesis apply? If the fraction in DNA synthesis is smaller, then the fraction of the total marrow occupied by CSC becomes larger. If the time for DNA synthesis is overestimated the N_s^{CSC} is overestimated. If the DNA synthesis time is really less than the measured time for differentiated marrow cells, the estimate of N_s^{CSC} is too high. Even reducing t_s to a probably unrealistically low value of 1 hour for man reduces N^{CSC} to 6.6×10^7/kg or 1 in 300 marrow cells are CSC. The above line of logic and assumption leads to a ratio of CSC to total marrow cells totally incompatible with the in vitro bone marrow culture of 5 per 10^4 colony forming cells in human bone marrow.

If the abundance of CFC_c in bone marrow represents the fraction of granulocytic CSC in bone marrow, one can then calculate the influx of CSC into the granulopoietic pool from the following data:

1) Abundance of CFC_c = 5 per 10^4 bone marrow cells (A)
2) Thymidine suicide for CFC_c = 0.35 (I_L)
3) DNA synthesis time of 12 hours (t_s)
4) GTR = 6.8×10^7/kg/hr
5) Absolute marrow cellularity of 1800×10^7/kg

The number of granulocytic CSC (N^{CSC}) is

$$N^{CSC} = N \times A = 1800 \times 10^7 \times 0.0005 = 0.9 \times 10^7/\text{kg}.$$

The number of granulocytic CSC in DNA synthesis (N_s^{CSC}) is

$$N_s^{CSC} = N^{CSC} \times I_L = 0.9 \times 10^7 \times 0.35 = 0.315 \times 10^7/\text{kg}.$$

The birth rate (K_B) of granulocytic CSC is

$$K_B = (N_s^{CSC})/t_s = (0.315 \times 10^7)/12 = 0.026 \times 10^7/\text{kg/hr}.$$

To maintain steady-state conditions the flux out of the CSC pool is the sum of influx and half of the K_B, or a minimum of 0.013×10^7/kg/hr assuming influx is zero.

The amplification is the ratio of the efflux from the granulocytic CSC pool to the GTR or

$$6.8 \times 10^7 \div 0.013 \times 10^7 = 520.$$

Such an amplification would require about nine serial mitoses in the proliferating granulocytic pool. It is reasonably well established that the time from the myeloblast to the metamyelocyte is 130 hours (*21*). If there is equal time for each successive multiplicative cell cycle, there would be a generation time of 14 hours. If mitotic time is 0.75 hours the mitotic index would be about 5% or 5 times that observed by Killman et al. (*26*). With a DNA synthesis time of 12 hours (*24*) the fraction of granulocytic proliferating cells in DNA synthesis would be 12/14 or 0.85 compared to the observations in man of 0.15–0.30 (*21*). The latter fraction is compatible with the number of myelocytes having a diploid DNA content. From these lines of reasoning and calculations based on experimentally determined data, the notion of nine serial mitoses in the proliferating granulocytic pool must be rejected along with the notion that the in vitro culture of bone marrow determines the fraction of CSC in human bone marrow. Although murine data for RTR and GTR are not as good as human data, similar calculations lead to similar conclusions on the mouse.

For the sake of argument, let us assume that human marrow contains one pluripotent stem cell (PHSC) per 1000 bone marrow cells. Then there are 2×10^7/kg. Feinendegen (*21*) has shown that there is a slowly turning over compartment of mouse bone marrow, presumably PHSC with turnover time of about 33 days. Cell cycle time in man is 2–3 times longer than in the mouse so an upper limit of PHSC turnover time in man may be 100 days and 1% per day go through the cycle. DNA synthesis time is assumed to be 0.5 day (12 hrs);

hence $\quad N_s = (N^{PHSC} \times t_s)/100 = 10^5/\text{kg}$

$$K_B = N_s/t_s = 10^5/12 = 0.08 \times 10^5/\text{kg}/\text{hr}$$

and the flux out is half the K_B to maintain steady-state size of the PHSC pool. The steady-state flux from PHSC to CSC is 0.04×10^5/kg/hr or about 2900 cells leave the CSC pool for every cell that enters it from PHSC pool under steady-state conditions.

CONCLUSION

Until one can refute the logic of this approach or find a method of directly assaying the PHSC and CSC in man, one must doubt the in vitro bone marrow culture results as an estimate of the abundance of CSC in human marrow.

REFERENCES

1. A. A. Maximow and W. Bloom, *Text-Book of Histology*. W. B. Saunders, Philadelphia, 1934.
2. T. J. Greenwalt and G. A. Jamieson, Eds., *Formation and Destruction of Blood Cells*. J. B. Lippincott Company, Philadelphia, 1970.
3. L. Siminovitch, J. E. Till and E. A. McCulloch, *Radiat. Res.* **24**, 482 (1965).
4. E. A. McCulloch and J. E. Till, *J. Cell. Comp. Physiol.* **61**, 301 (1963).
5. L. Siminovitch, J. E. Till and E. A. McCulloch, *J. Cell. Comp. Physiol.* **64**, 23 (1964).
6. J. E. Till and E. A. McCulloch, *Ser. Haematol.* **5**, 15 (1972).
7. F. Vassort, M. Winterholder, E. Frindel and M. Tubiana, *Blood* **41**, 789 (1973).
8. V. P. Bond, M. N. Swift, S. T. Taketa, G. P. Welch and C. A. Tobias, *Am. J. Physiol.* **174**, 259 (1953).

9. G. Brecher and E. P. Cronkite, *Proc. Soc. Exp. Biol. Med.* **77,** 292 (1951).
10. D. W. van Bekkum, M. J. van Noord, B. Maat and K. A. Dicke, *Blood* **38,** 547 (1971).
11. R. W. Caffrey Tyler and N. B. Everett, *Blood* **28,** 873 (1966).
12. D. J. Moffatt, C. Rosse and J. M. Yoffey, *Lancet* **9,** 547 (1967).
13. D. Metcalf, *Proc. Soc. Exp. Biol. Med.* **139,** 511 (1972).
14. N. N. Iscove, J. E. Till and E. A. McCulloch, *Proc. Soc. Exp. Biol. Med.* **134,** 33 (1970).
15. A. J. Becker, E. A. McCulloch, L. Siminovitch and J. E. Till, *Blood* **26,** 296 (1965).
16. J. S. Senn and E. A. McCulloch, *Blood* **35,** 56 (1970).
17. J. Gidali, this symposium.
18. M.A.S. Moore and N. Williams, in *Hemopoiesis in Culture,* 2nd Int. Workshop (William A. Robinson, Ed.) pp. 17–27. DHEW Publication No. (NIH) 74-205, May 23–26, 1973.
19. A. L. Carsten, A. Boyum and E. P. Cronkite, *Radiat. Res.* **59,** 48 (1974).
20. U. Reincke, H. Burlington, E. P. Cronkite and J. Laissue, *Blood* (in press).
21. E. P. Cronkite and P. C. Vicent, *Ser. Haematol.* **2,** 3 (1969).
22. G. E. Cartwright, J. W. Athens and M. M. Wintrobe, *Blood* **24,** 780 (1964).
23. D. M. Donohue, R. H. Reiff, M. L. Hanson, Y. Betson and C. A. Finch, *J. Clin. Invest.* **37,** 1571 (1958).
24. P. Stryckmans, E. P. Cronkite, T. M. Fliedner and J. Ramos, *Nature (London)* **211,** 717 (1966).
25. D. M. Donohue, B. W. Gabrio and C. A. Finch, *J. Clin. Invest.* **37,** 1564 (1958).
26. S. A. Killman, E. P. Cronkite, T. M. Fliedner and V. P. Bond, *Blood* **19,** 743 (1962).

Response of Stem Cell System to Whole Body and Partial Body Irradiation[1]

J. Gidáli

"Frédéric Joliot-Curie" National Research Institute for Radiobiology and Radiohygiene
Pentz K.u.5. Budapest 22., Hungary

The pluripotent hemopoietic stem cells (spleen colony-forming units: CFU_s (*1*)) though being distributed throughout the bones in the body (and in rodents in the spleen, too) form a homogeneous system as regards their responses. This homogeneity of the stem cell system involves two aspects of their reactions: on the one hand, stem cells react like parts of the same organ to various kinds of damage and, on the other hand, they also react to changes in the size of the committed stem pools (i.e. to the decrease in compartment size due either to damage or differentiation).

The co-ordination of the stem cell system is maintained partly by short range (local) and long range (humoral) regulations, and partly by the fact that, even under physiological conditions, both pluripotent and committed stem cells migrate from one hemopoietic site to the other. All these phenomena play a role in the response of the stem cell system to irradiation.

At any given time the circulating stem cells form an insignificant portion of the total stem cell compartment (0.25%)(*2*). However, the rapid clearance of circulating CFU_s from the blood suggests that 20–30% of the total CFU content of the organism may migrate through the circulation every day (*3*).

Our knowledge about both the physiological role of the circulating stem cells and the regulation of their level in the blood is rather limited. However, these factors undoubtedly play an important role in the regeneration of the hemopoietic system under certain pathological conditions, especially in regeneration after partial-body irradiation.

It is a well known radiobiological phenomenon that partial shielding of hemopoietic tissues during exposure protects the organism from the lethal effects of ionizing radiation (*4*). Although this protective effect is similar to the protective effect of bone marrow transplantation, only recent development in methodology during the last 10 years has enabled investigators to prove that stem cells migrate out of the shielded hemopoietic tissues and repopulate the bone marrow (*5–7,36*). Recently, Maloney and Patt furnished direct evidence showing that stem cell migration is mainly responsible for the regeneration of heavily irradiated bone marrow (*8*).

Stem cell regeneration in a partially irradiated animal seems to be a relevant model for studying the response of the stem cell system to irradiation involving

[1]Congress presentation D–3–3.

both the reaction of the irradiated part (including the complexity of cell destruction, regeneration, cell differentiation and homing of immigrated stem cells) and the reaction of the shielded part (as a function of cell demand in the irradiated sites). Most of these data are available on rodents only; therefore, unless otherwise stated, data will refer to murine or rat systems.

When studied early after irradiation, the CFU_s content of the irradiated bone marrow in the partially shielded (PS) mice exposed to sublethal doses did not differ significantly from that in the total-body irradiated (TBR) mice (*9*). The number of CFU_s in S phase of the cell cycle also behaves similarly in the irradiated marrow of TBR or PS mice. As early as 15 min after irradiation with 150 rad, a significant portion of the CFU population normally in G_0 (and not proliferating)(*10,11*) is in the S phase (*12,13*). This reaction seems to be relatively independent of the size of the irradiated areas. Surprisingly, the number of CFU_s in S phase soon returns to the initial level and starts to increase again only at 24 hr after irradiation. Then it begins to rise and stays at a high level indicating the start of real regeneration (*12,14*). Although one cannot exclude categorically the possibility of CFU immigration from the shielded part, the fact that no difference between PS and TBR mice was found suggests that these early phenomena are of local origin.

The reaction of the stem cells in the shielded area may serve as an indicator of the response of the intact stem cells forming a part of the stem cell system.

In the middle of the 1960's, several studies proved a quantitative correlation between the size of the shielded area and the survival and stem cell regeneration of the animal (*15*). Hanks even estimated the rate of CFU migration (*5*). Although the existence of stem migration from the shielded bone marrow or spleen has never been questioned, the kinetics of the migration are not clear. According to Hank's hypothesis, the migration is continuous and linear with time (*5*). Robinson and coworkers (*7*) as well as Hellman and Grate (*2*) proved that after an initial rapid release of stem cells, migration is continued at a constant but lower rate.

The observations that radiation damage of a certain area may induce the migration of stem cells from the unirradiated bone marrow support the concept of some form of humoral stimulus. However, relatively little is known about the mechanism itself. At least two kinds of mechanisms appear to be involved. The continuous migration may be well explained by the increased peripheral demand (feedback). However, such a feedback mechanism cannot be responsible for the early migration events since within the first postirradiation hours cell loss is not so marked that it could "per se" produce such a phenomenon.

The initial rapid release of CFU_s from the bone marrow may indicate an easily mobilizable CFU_s pool, perhaps specifically sensitive to the humoral factors initiating the migration.

One of the possible explanations for the induction of migration may be the postirradiation heparinemia due to disintegration of mast cells, since both heparin and heparinoids can stimulate CFU_s migration in normal mice as well (unpublished data).

This rapid outflow does not itself seem to be a specific reaction of stem cells to radiation, since a similar phenomenon may be induced also in normal mice, if 60 per cent of the total blood volume is rapidly exchanged with blood containing no viable CFU_s (i.e. blood from 1000 rad-irradiated isologous donors). Ninety sec after completing the procedure, the number of circulating CFU_s exceeded even the control value (unpublished data). A rather rapid mobilization of granuloid committed stem cells (CFU_c) was reported by Kovács and coworkers in dogs in which partial depletion of the blood CFU_c pool by repeated leukophereses seemed to be overcompensated in two days (*16*). Both observations seem to support the idea of a rapid mobilization of either pluripotent or committed stem cells provided the stimulus is adequate.

CFU migration at later dates after irradiation (in the constant phase) is possibly influenced by the mature cell demand of the body.

A rather good correlation between CFU_s depletion in the irradiated area and CFU_s loss in the shielded tibia could be detected in the following experiment. If one tibia of the mouse was locally irradiated with 300 rad, total CFU_s loss proved to be less than 5 per cent of the total CFU_s pool. This CFU_s loss did not result in any measurable reduction of CFU_s content in the contralateral, unirradiated tibia. If one tibia of 300 rad TBR mice was shielded, approximately 6 percent of the total CFU_s pool survived. In this case there was a more prominent and prolonged decrease in CFU_s count in the shielded tibia, which became even more pronounced if one irradiated the mice with 600 rad TBR with one tibia shielded (Fig. 1).

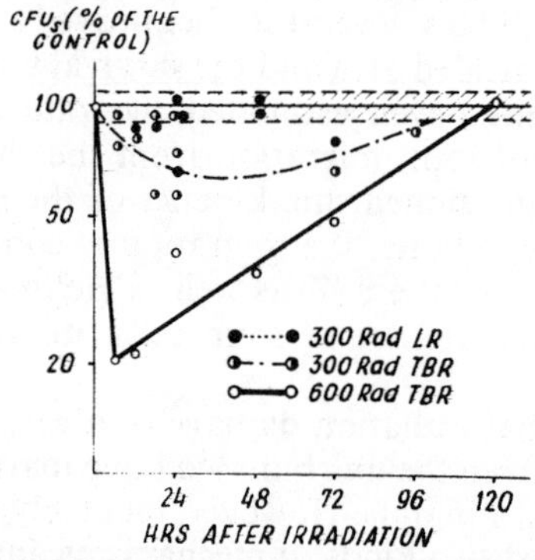

FIG. 1. Correlation between CFU level of the shielded tibia and CFU depletion of the irradiated parts.
300 rad LR: local irradiation of one tibia
300 or 600 rad TBR: total-body irradiation with one tibia shielded.
Each point represents the mean ± S.D. of 8–10 donors.

Although data on stem cell migration refer mainly to pluripotent stem cells, committed stem cells also migrate out of the shielded part (*17*). Even the idea of preferential migration of erythroid committed stem cells arose was suggested, since erythropoiesis in the irradiated area exceeded that of the shielded limb (*17,18*). This migration of either pluripotent or committed stem cells ought to be accompanied by a rapid repletion of the compartment size since one finds a normal CFU_s level, elevated CFU_c level and a rather short-term decrease in the number of erythroid precursors in the shielded area a few days after irradiation (*17,19*).

All the above-mentioned data definitely prove that circulating pluripotent stem cells play some role in stem cell regeneration in partially-irradiated animals. In spite of that, relatively few data are available on the properties of this easily mobilizable stem cell population. Several data suggest that stem cells derived from various tissues (bone marrow or spleen) differ in their characteristics, e.g. in repopulating ability, radiosensitivity, etc. (*20–22*).

Circulating CFU_s which migrate mainly out of the bone marrow, proved to be identical with bone marrow CFU_s in their basic parameters (pluripotency, differentiation rate)(*23*). However, if further parameters, which are well known characteristics for the bone marrow CFU_s population, are studied, some interesting differences are found. Higher number of CFU_s in S phase, lower radiosensitivity (D_o:140 R) and higher seeding efficiency were found for blood-derived CFU_s (Table 1). This suggested that this easily mobilizable CFU fraction may be a subpopulation of the pluripotential stem cell population (*24*). The higher seeding efficiency of circulating CFU_s may be of special interest since it may indicate their preference for repopulation.

TABLE 1
Some Characteristic Parameters of CFU Derived from Bone Marrow or Blood

Source of CFU	Bone Marrow	Blood
Radiosensitivity (D_0)	100 R	140 R
Seeding efficiency ("f" number)	12.9%	20.9%
Percentage of CFU in S phase (H^3-TdR sensitivity, in vitro)	2.3%	31.4%

Some correlation between circulating CFU_s level and CFU_s repopulation seemed obvious. Therefore, the repopulation rate of CFU_s in 650 R locally-irradiated tibia was studied as a function of blood CFU_s level in mice with various blood CFU_s levels.

In those groups in which the circulating CFU_s level was lower than in the control (rest of the body irradiated), 5–17 per cent CFU_s regeneration was found in the locally irradiated tibia on the 12th postirradiation day. If the circulating CFU_s count increased to 2–3 fold that of the normal (exdotoxin treatment), the CFU level of the bone marrow was 53 per cent of the normal on the 12th postirradiation day as opposed to the 38 per cent spontaneous CFU regeneration measured in the locally-irradiated tibia (Fig. 2).

These data seem to support the role of circulating stem cells in the regeneration of locally-irradiated bone marrow even if the subtotal damage of the bone marrow might also have allowed the local regeneration (*25*).

In spite of this stimulated regeneration, in repeatedly endotoxin-treated mice, only 2–3-fold increase in the circulating CFU level could be detected, though a similar endotoxin treatment induced an approximately tenfold increase in blood CFU level in normal mice (Fig. 3.). Since in these mice only a small

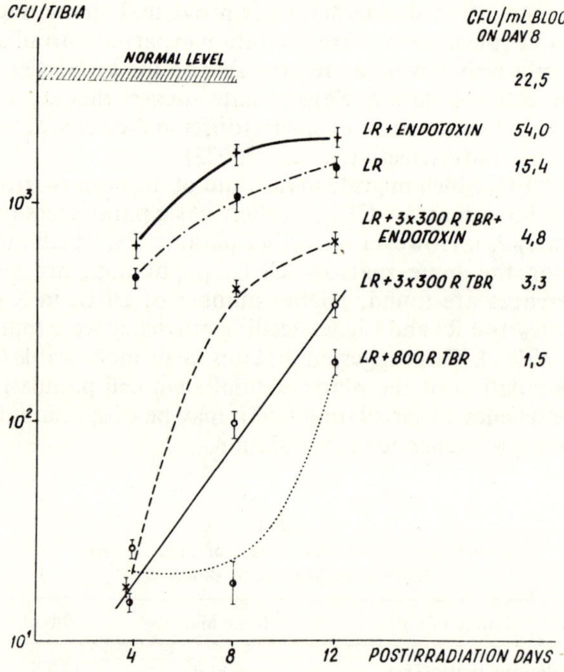

FIG. 2. CFU regeneration in 650 R locally-irradiated tibia in mice with various blood CFU levels.

Each point represents the mean ± mean S.D. of 6–8 donors.

LR: 650 R one tibia local irradiation on day 0.

800 R TBR: 800 R total-body irradiation with LR tibia shielded immediately after LR.

3 × 300 R TBR: total-body irradiation with LR tibia shielded every 3rd day. (The mice were not irradiated on the day of killing.)

Endotoxin treatment: 50 μg/ mouse *S. typhi* endotoxin s.c. every fourth day. (The mice were not treated on the day of killing.)

portion of the stem cells was irradiated, radiation damage cannot be held responsible for this poor reaction. The most probable explanation for this phenomenon seems to be the possibility of increased capture of circulating CFU by damaged bone marrow.

This experiment calls attention to the numerous pitfalls of this type of experiment.

In general, one has two possibilities (in addition to the indirect method of measuring the regeneration rate in the irradiated part) to determine CFU_s migration:

A. to detect CFU_s level in the shielded bone marrow or spleen.

B. to detect blood CFU count.

A. CFU level in the shielded bone marrow, in addition to the actual migration, may also be influenced by

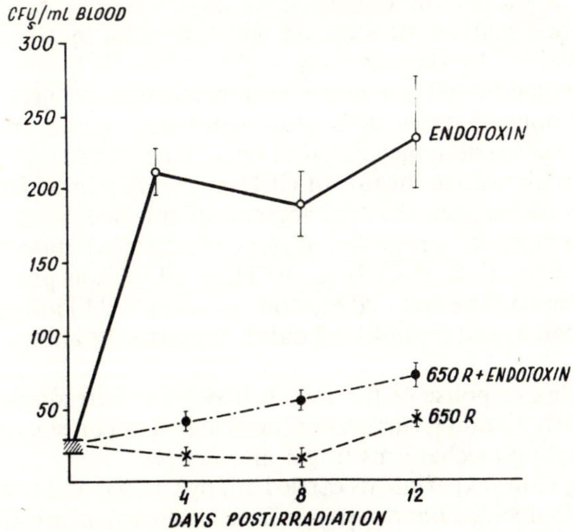

FIG. 3. Number of circulating CFU$_s$ in locally-irradiated, endotoxin-treated mice. Each point represents the mean ± S.D. of 6–8 donors.
Endotoxin treatment: 50μg/mouse S. *typhi* endotoxin s.c. every fourth day.

a. CFU loss due to scattered irradiation (this depends on the technique of shielding and may reach 0.2–8.0% of the dose absorbed by the irradiated part (*26*)).

b. CFU loss due to abscopal effects of irradiation.

c. Differentiation that can hardly be followed in this system. Recent methodology does not permit to measure committed stem cells locally, except the granuloid committed stem cells.

d. Finally, CFU proliferation compensating for CFU migration may also affect the number of CFU$_s$ in the shielded bone marrow.

B. Blood CFU level seems to be a resultant of CFU migration and seeding. Therefore, as mentioned above, a relatively low number of circulating CFU may be measured even at increased migration rate if the seeding is significant.

Irradiation of a hemopoietic area may induce not only CFU migration but also CFU proliferation in the shielded area (*27*). A reliable indicator of this phenomenon is the increase of the number of CFU$_s$ in S phase (*12,14*). This CFU proliferation wave may only be partially due to the compensatory proliferations induced by CFU migration (local control) since the number of CFU$_s$ in S phase returns to the normal when local CFU content is close to its normal level. At this time, however, the CFU content of the irradiated bone marrow may still be well below 10 per cent of the normal (*14*). In the shielded limb, the number of CFU$_c$ proved to have increased significantly, and was inversely related to the blood neutrophil level. This fact supports the idea that some humoral regulator of granulopoiesis takes part in this effect (*19*).

However, numerous data support the hypothesis that CFU proliferation of the shielded bone marrow may be the consequence of some nonspecific reactions.

For instance, a short-term cycling wave of CFU_s can be detected in the unirradiated bone marrow of a mouse with one tibia locally irradiated. This cannot be explained by the relatively slight (<5%) CFU loss. This triggering effect, however, can be well reproduced in normal mice, by injecting irradiated or heat-damaged bone marrow cells (*14*). Irradiated spleen fragments or the supernatant of spleen homogenate given i.p. (*28*) as well as foreign proteins (*29*) may result in an increase in the rate of CFU regeneration after irradiation. These data also support the idea that phenomena in shielded bone marrow are not representing some specific response of stem cells to irradiation, but rather some kind of general reaction of CFU to the effect of certain proteins (foreign or denatured proteins). The fact that injection of sheep red blood cells also triggers CFU into the cell cycle (unpublished data), supports the antigenic nature of the triggering agent.

The complex response of the stem cell system to irradiation is even more complicated if irradiation is carried out fractionated or continuously which allow some compensating mechanisms to get into action.

According to the experiments carried out by Croizat and coworkers (*12*), the CFU level in the shielded bone marrow decreased permanently in mice subjected to multiple TBR at intervals of one or several days. By the end of irradiation (11 × 150 rad), the CFU count in the shielded marrow decreased to 12 per cent of the control. CFU migration was so significant that by the end of the sessions, the CFU content in the irradiated leg was 1.6 per cent of the control. Comparing this to the total CFU content of the irradiated parts of the body, one can infer that in total the daily-irradiated bone marrow may contain more CFU than the shielded part. The authors explain this new state of equilibrium by the equality of CFU migration to and from the shielded site. Although this hypothesis seems plausible, it cannot explain the fact that, in spite of the low CFU level in the shielded leg, the number of CFU in S phase oscillates around the control level. One could offer as another possible explanation that in the supposed steady state, at this CFU level, CFU proliferation can only keep up with CFU loss owing to differentiation. (It should be noted that a constant significant differentiation is suggested indirectly by the fact that only a slight decrease in nucleated cells of the shielded bone marrow can be detected at the time of significant CFU decrease.) After completing the irradiations, this steady state proved to be unstable and 4 months after the last irradiation, the CFU count in the shielded bone marrow reached 300 per cent of the contralateral (irradiated) marrow. This also suggests an alteration in the function of some control mechanisms during repeated irradiations, as a consequence of which a decreased CFU level can be accepted as optimal and steady state can be maintained at this low level. The above-mentioned overshoot, when the steady state is no longer maintained, may also support this hypothesis.

Applying repeated sublethal TBR (3 times 440 rad separated by 7 days) Vos found out that the CFU duplication curve of the bone marrow reached a plateau at a subnormal level. When the interval between the 2nd and 3rd irradiation was extended to 35 days, the CFU duplication curve in the bone marrow agreed with that obtained after a single dose of 440 rad (*30*). This finding should also mean some change in the control of CFU population size. As a consequence of the increased cell demand, progressively more stem cells enter the committed compartments and this increased rate of differentiation may contribute to the relative depletion of surviving cells. Consequently, the self-renewal capacity of the rather limited number of surviving CFU decreases. If bone marrow is allowed to maintain its steady state by postponing the last irradiation, the normal self-renewal capacity of the CFU returns. Hendry and Lajtha (*31*) found a reduction in stem cell pool size after four repeated doses of 450 rad separated by 24 days. Since a similar growth curve was found for suboptimally growing femoral CFU as for normal CFU, and the number of CFU_s in S phase proved also to be similar in both populations, the authors assumed that not the CFU itself but some undefined milieu control may keep CFU proliferation at a suboptimal level. However, the migration of CFU from the bone marrow may also play a role in these experimental findings.

One of the most convincing examples of the new steady state of CFU_s after irradiation was measured in rats irradiated continuously with 45 rad/day (*32,33*) and in mice irradiated continuously with 70 rad/day (*34*). This new steady state may be characterized by the findings that, though 70 rad/day proved to stop population growth of a fully cycling CFU population, the activity of the erythropoietic system remained close to the normal and the number of leukocytes and thrombocytes is only slightly lower than the normal value. At the same time, the rate of production of recognizable erythroid precursors increases (*35*). At 1750 rad accumulated dose, the erythroid committed stem cell compartment (ERC) still remained at 10–15% of the control but the CFU level plateaued at 2 per cent of the control. This plateau for CFU and ERC populations indicates the establishment of a new steady state, in which CFU proliferation may to some extent compensate for both radiation-induced cell sterilization and differentiation. The feedback control which triggers CFU to enter the cell cycle seems to respond to a reduction of population size to 5 per cent of control.

CONCLUSIONS

1. The pluripotent stem cell system, though being distributed in the body, reacts homogeneously to irradiation. This homogeneity is controlled by short-range (local) and long-range (humoral) regulations acting primarily on pluripotent and committed stem cells. Migration of stem cells from unirradiated to irradiated areas may play a role in the regeneration processes even if local regeneration may also occur.

2. Migration induction as well as proliferation induction in the shielded area do not seem to be specific radiation-induced reactions. Both may be influenced either by some physiological regulators released after irradiation in a higher quantity or by some non-specific triggering agents.

3. Both repeated and continuous irradiation induce the establishment of a new steady state. In the steady state after repeated sublethal irradiations, the CFU count stays at a suboptimal level either as a consequence of an increased differentiation or of some undefined damage in milieu control. In the new steady state during continuous irradiation, the number of mature elements in blood is close to the normal while CFU population is reduced to less than 2 per cent of its original level.

REFERENCES

1. J. E. Till and E. A. McCulloch, *Radiat. Res.* **14,** 213 (1961).
2. S. Hellman and H. E. Grate, in *Effects of Radiation on Cellular Proliferation and Differentiation.* pp. 187–196. I.A.E.A. Vienna, 1968.
3. D. Metcalf and M.A.S. Moore, *Hemopoietic Cells.* North-Holland Publishing Company, Amsterdam. London, 1971.
4. L. O. Jacobson, E. K. Marks, M. J. Robson, E. O. Gaston, and R. E. Zirkle, *J. Lab. Clin. Med.* **34,** 1358 (1949).
5. G. E. Hanks, *Nature (London),* **203,** 1393 (1964).
6. S. Hellman, *Nature (London),* **205,** 100 (1965).
7. C. V. Robinson, S. L. Commerford, and J. L. Bateman, *Proc. Soc. Exp. Biol. Med.* **119,** 222 (1965).
8. M. A. Maloney and H. M. Patt, *Blood,* **39,** 804 (1972).
9. J. Gidáli, I. Fehér, A. Gallyas, and V. Várterész, *Radiat. Res.* **39,** 305 (1969).
10. L. G. Lajtha, *J. Cell. Comp. Physiol.* **62,** 143 (1963).
11. A. J. Becker, E. A. McCulloch, L. Siminovitch, and J. E. Till, *Blood* **26,** 296 (1965).
12. H. Croizat, E. Frindel, and M. Tubiana, *Int. J. Radiat. Biol.* **18,** 347 (1970).
13. S. K. Lahiri and L. M. van Putten, *Cell Tissue Kinet.* **5,** 365 (1972).
14 J. Gidáli and L. G. Lajtha, *Cell Tissue Kinet.* **5,** 147 (1972).
15. F.A.J. de Vries and O. Vos, *Int. J. Radiat. Biol.* **11,** 235 (1966).
16. P. Kovács, C. Bruch, E. Herbst, and T. M. Fliedner, *Blood,* (in press).
17. A. Morley, D. Howard, B. Bennett, and F. Stohlman, Jr., *Brit. J. Haematol.* **19,** 523 (1970).
18. S. Fujioka, K. Hirashima, T. Kumatori, F. Takaku, and K. Nakao, *Radiat. Res.* **31,** 826 (1967).
19. K. A. Rickard, A. Morley, D. Howard, and F. Stohlman, Jr., *Blood* **37,** 6 (1971).
20. A. L. Kretchmar and W. R. Conover, *Blood* **36,** 772 (1970).
21. J. F. Duplan, E. Legrand, and B. Guillemain, *Nouv. Rev. Fr. Hematol.* **11,** 821 (1971).
22. E. Guzman and L. G. Lajtha, *Cell Tissue Kinet.* **3,** 91 (1970).
23. J. P. Lewis, M. Passovoy, M. Freeman, and F. E. Trobaugh, Jr., *J. Cell. Physiol.* **71,** 121 (1968).
24. J. Gidáli, I. Fehér, and S. Antal, *Blood* **43,** 573 (1974).
25. I. Fehér, S. Antal, and J. Gidáli, *Radiat. Res.* **58,** 516 (1974).
26. I. Bojtor, K. Hiesche, and P. O. Schnell, *Acta Radiol.* **13,** 32 (1974).
27. W. Fried, C. W. Gurney, and M. Swatek, *Radiat. Res.* **29,** 50 (1966).
28. W. H. Knospe, W. Fried, S. A. Gregory, R. J. Sassetti, and F. E. Trobaugh, Jr., *J. Lab. Clin. Med.* **76,** 584 (1970).
29. D. R. Boggs, J. C. Marsh, P. A. Chervenick, C. R. Bishop, G. E. Cartwright, and M. M. Wintrobe, *J. Exp. Med.* **126,** 851 (1967).
30. O. Vos, *Int. J. Radiat. Biol.* **22,** 41 (1972).
31. J. H. Hendry and L. G. Lajtha, *Radiat. Res.* **52,** 309 (1972).
32. L. F. Lamerton, *Radiat. Res.* **27,** 119 (1966).
33. N. M. Blackett, *Brit. J. Haemat.* **13,** 915 (1967).
34. L. G. Lajtha, L. V. Pozzi, R. Schofield, M. Fox, *Cell Tissue Kinet.* **2,** 39 (1969).
35. B. I. Lord, in *Effects of Radiation on Cellular Proliferation and Differentiation* pp. 247–257. I.A.E.A., Vienna, 1968.
36. A. L. Carsten and E. P. Cronkite, *Proc. Soc. Exp. Biol. Med.* **137,** 948 (1971).

Characteristics of the Stem Cell Population Surviving a Sublethal Exposure to Ionizing Radiation[1]

J. F. Duplan
INSERM, Unité 117 et Fondation Bergonié
Bordeaux, France

INTRODUCTION

The purpose of this report is to analyze the properties of the hemopoietic stem cells that survive a sublethal exposure to ionizing radiation. Some aspects of this problem are closely related to the characteristics of stem cell kinetics in general, but others have a more specific radiobiological significance.

During the last ten years, there has been accumulating evidence that the fraction of the mouse stem cell pool known as CFU_s is heterogeneous with respect to its potentialities for multiplication and differentiation. It has been shown that the CFU_s belonging to different hemopoietic organs such as fetal liver, spleen or bone marrow (BM) bear different properties. Furthermore, the CFU_s which survive doses of 200–400 R of X-radiation display characteristics of growth and differentiation that differ from those of normal cells of similar origin.

The use of the spleen colony assay associated with other experimental techniques made it possible to measure with satisfactory accuracy some of their characteristics: radiosensitivity, spleen seeding factor, restorative or therapeutic efficiency, replication rate and doubling-time, preferential differentiation including responsiveness to the stimuli initiated by endotoxin and erythropoietin. It seems important to distinguish between the quantitative changes which are evaluated by the spleen colony assay and the qualitative modifications that are studied by cytologic and histologic methods or radioactive labeling. In most experiments the dose of radiation varies between 200 and 400 R; thus, depending on the strain and on the origin of the hemopoietic cells the survival of CFU_s will vary from 20 to 1%.

RADIOSENSITIVITY

McCulloch and Till (*1*) were the first to show that several days after irradiation, at a time when the spontaneous cell repair is completed, the surviving CFU_s are less radiosensitive than the unirradiated controls. This change is transient and disappears within 10 to 12 days. Similar results were obtained by others (*2,3*). Table 1 summarizes an experiment in which the radiosensitivity (D_o) of BM CFU_s has been determined from day 5 to 24 after a dose of 200 R. There is

[1]Congress presentation D–3–4.
[2]Work performed under contract n° 74.34.109.00.480.75.01 DRME

TABLE 1
Radiosensitivity of Normal and Irradiated (200 R) CFU_S as a Function of Time After Transplantation (femoral marrow)

Time in days	D_0 in R Normal	D_0 in R Irradiated
5	61	81*
7	64	72*
10	62	67
15	65	60
24	63	60

*Significantly different from normal value.

a significant increase of D_o at day 5 and 7, but from day 10 onwards the radiosensitivity returns to a normal value.

It must be noted that when normal BM CFU_s are irradiated with 200 to 400 R the slope of the dose-survival curve remains unchanged whereas it should decrease if a resistant cell population had been selected.

One could raise the objection that it is not justified to compare the radiosensitivity of normal resting CFU_s (steady state) with those of cycling elements belonging to a rapidly expanding population.

But the correlation between the radiation-sensitivity of the CFU_s and their mitotic activity is not satisfactory. According to in vivo and in vitro experiments with ^{3}HTdR or vinblastine the fraction of cycling CFU reaches 20% in the spleen whereas it stays below 10% in the BM, in spite of the fact that radiation sensitivity of spleen CFU_s is definitely higher than that of BM CFU_s. The correlation is apparently better for the fetal liver, but in this case the removal of the cycling cells by ^{3}HTdR suicide decreases the radiosensitivity (*4*), showing that factors other than mitotic activity influence the survival of irradiated CFU_s. This latter result is at variance with Silini's report (*5*) that radiosensitivity values of vinblastin-treated BM cells are similar to those of normal BM.

THERAPEUTIC EFFICIENCY

The pluripotential stem cells belonging to the CFU population have the capacity to restore lethally-irradiated syngeneic mice and it is possible to correlate the number of injected CFU_s to the survival of mice previously irradiated with a LD_{100} (*6*). The number of CFU_s which correspond to a survival of 50% (CFU/50) may be interpolated from the curve obtained by plotting the probit of lethality vs the log of the CFU_s. The therapeutic efficiency (TE) of a given cell suspension is defined as:

$$100/(CFU/50) = TE$$

Figure 1 gives the results of a series of experiments in which the TE of normal and irradiated (200 R) BM and spleen cells have been determined. Each curve is drawn from 4 points corresponding to the percentage of mortality at 30 days in groups of 64 mice (XVII strain) irradiated with 850 R and restored with

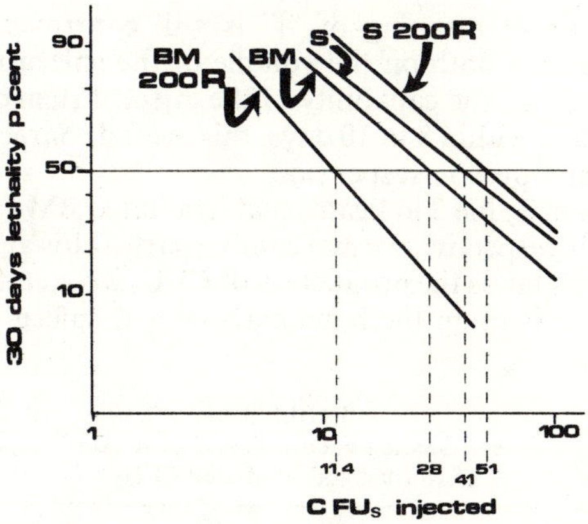

FIG. 1. Therapeutic efficiency of normal and irradiated (200 R) BM and spleen CFU_s. The CFU/50 values are indicated on the abcissa.

various numbers of normal or irradiated CFU_s. The values of the CFU/50 are 11.4 for irradiated BM, 38 for normal BM, 41 and 51 for normal and irradiated spleen CFU_s. The corresponding TE are given in Table 2. The TE of BM CFU_s surviving a dose of 200 R is significantly greater than that of their normal counterpart whereas irradiation does not appear to modify the TE of spleen CFU.

But the TE results from the interaction of several factors that need to be analyzed independently: spleen seeding factor (f), rate of replication, preferential orientation of differentiation, maturation and growth rate of the differentiated elements.

TABLE 2
Therapeutic Efficiency (T.E)
of Normal and Irradiated CFU_S

Origin	Dose in R	T.E
Bone marrow	0	3.6
Bone marrow	200	8.6
Spleen	0	2.4
Spleen	200	1.9

CFU_S/50: number of CFU_S to be injected to obtain 50% lethality in mice irradiated with LD_{100}.
T.E: 100/(CFU_S/50).

SPLEEN SEEDING FACTOR "f"

Retransplantation experiments have shown that only 16 to 20% of the BM cells capable of giving colonies (CFC) actually settle in the spleen. This "f" factor is only 12 to 14% when spleen CFU_s are injected and even less for fetal liver CFU_s

(*7,8*). Though the exact meaning of "f" is still controversial it is generally accepted that it depends both on the volume of the spleen independent of the origin of the cells and on the capability of the injected stem cell to self-replicate and to give sizable foci within 8 to 10 days; this second characteristic is related to the origin of the hemopoietic suspension.

The 3rd column in Table 3 indicates that irradiated BM and spleen cells have the same "f" as their respective normal counterparts. This at first sight supports the hypothesis that as far as the production of CFU_s is concerned, an exposure to 200 R has the same effect on the bone marrow and spleen stem cell pools.

TABLE 3
Spleen Seeding Factor (f) and 24 h "Dip"
of Normal and Irradiated CFU_S

Origin	Dose in R	"f" 3 h.	"Dip" 24 h.
Bone marrow	0	19*	7*
Bone marrow	200	17	2
Spleen	0	14	6
Spleen	200	13	2

*Per cent of injected CFU_S

To obtain the preceding values of "f," the spleens have been retransplanted 3 hours after the injection of the hemopoietic suspensions. When this delay increases over 6 h the number of recovered CFU_s diminishes and this apparent loss, which is known as the "dip" (*8,9*), is represented on the curves in Fig. 2. The values for normal BM and spleen CFU_s given in the last column of Table 3 are 7 and 6%, respectively, while they are only 2% for cells of both origins surviving a dose of 200 R. These results are in good agreement with those previously reported (*10*).

It must be stressed that the meaning of the "dip" is controversial. It has been attributed to the shrinkage of the irradiated spleen, to a specific sensitivity of some CFU_s to the transplantation procedure and/or to a high probability of differentiation of some seeded cells which is equivalent to a high probability of "death" of the CFU_s. The first hypothesis is unlikely to hold good in the present situation and the second is considered controversial. The last possibility is interesting because it leads to the problem of differentiation in cycling and resting cells. Several experiments carried out by Worton et al. (*11*) and by Moore and Metcalf (*12*) suggest that there is a fraction of the CFU_s compartment which is in G_o and has a high capacity for self-replication. Any selective killing of this subpopulation by X-radiation would lead to a new population of CFU, which, on the average, would express a relatively higher probability of differentiation than the normal CFU compartment and consequently a higher probability of "death" leading to an increased dip.

But if such an elimination has taken place, it becomes difficult to explain why, between day 2 and 3, the doubling time of the irradiated CFU is much shorter than that of the controls (Fig. 2). A transiently increased rate of proliferation of the surviving CFU_s could account for this effect.

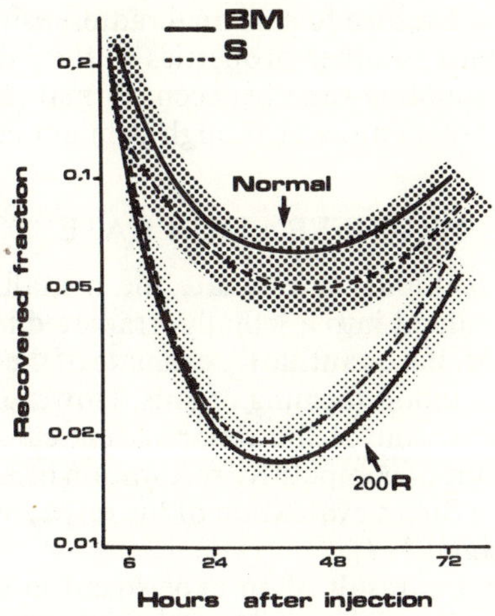

FIG. 2. Fraction of the injected CFU_s which is recovered in the recipient spleen at various times after transplantation (f and dip). Dotted areas correspond to the limits of significance.

DOUBLING TIME

The measure of the CFU_s doubling-time is based on the evaluation of the number of CFU_s produced per CFU_s injected within a fixed interval of time. The main factors involved in this estimation are the rate of multiplication and the self-replicative capacity of the CFU_s. The mitotic activity is apparently unstable during the early steps of regeneration and the exponential growth starts on day 3–4 to cease by day 8–10 after transplantation. Thus, comparable values of the doubling-time are unlikely to be obtained at different periods of the regeneration process. Such a discrepancy may be noted in Table 4 which gives estimate of the doubling-time corresponding to normal and irradiated BM and spleen CFU.

TABLE 4
Doubling-Time of Normal and Irradiated CFU_S

Origin	Dose in R	Doubling time in h.	
		day 5–9*	day 7–11**
Bone marrow	0	22.4	25
Bone marrow	200	19.3	–
Bone marrow	300	–	25
Spleen	0	26.1	–
Spleen	200	24.5	–

*From Duplan et al. (Unpublished observation).
**From Gidali et al. (*13*), (BALB/c × CBA)F_1.

With respect to irradiated BM CFU_s, there is significant decrease of the doubling-time (19.3 h versus 22.4 h) when it is calculated between days 5–9 whereas there is no difference (25 h in both cases) between days 7–11 (*13*). This might also indicate that the return to a normal radiosensitivity which takes place between days 7–10 extends to other properties of the irradiated CFU_s population. The difference in doubling-time between normal (26.1 h) and irradiated (24.5 h) spleen CFU is not significant though it might correspond to a similar situation.

PROLIFERATION RATE

There are several methods to estimate the overall proliferation rate of hemopoietic cells transplanted into a lethally-irradiated host. These include the spleen and thymus weight, the quantitative estimate of the circulating blood cells and the cellularity of the blood forming organs. However, these techniques do not allow a direct comparison of the repopulating capacity of various CFU populations. The technique of competitive restoration using cytogenetic markers has made possible such a direct evaluation of the respective proliferative ability of CFU_s of different origins (*14*).

Figure 3 represents the result of an experiment in which 125 CFU_s from normal AKR/T1ALD (36 acrocentrics and 2 metacentrics) marrow were injected into (AKR/T1 × AKR)F_1 (38 acrocentrics and 1 metacentric) 850 R-irradiated recipients together with 68 CFU_s from 200 R-irradiated AKR (40 acrocentrics) BM. The chimeras were sacrified 4, 9, 14 and 20 days after irradiation and the relative number of each caryotype was determined in BM, thymus, spleen and lymph nodes. Between days 4 and 9 the percentage of cells bearing the AKR caryotype (irradiated BM) increases from 36 to 68 at the expense of the T1 cells, and thereafter remains at a plateau near this value. This can be interpreted as proof that the production of differentiated elements is increased between days 4 and 9 after transplantation of irradiated BM CFU_s and

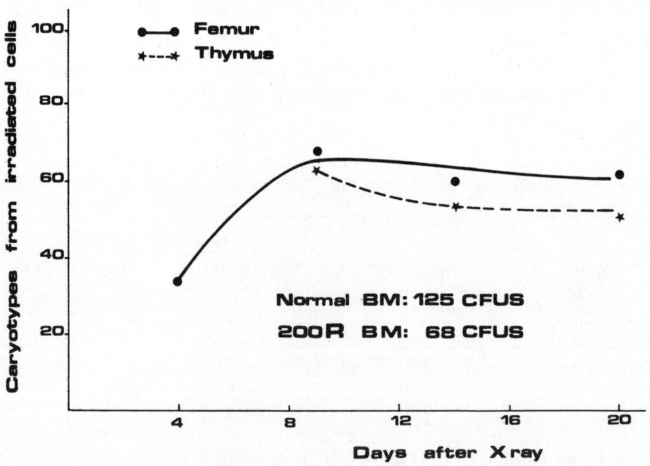

FIG. 3. Competition between irradiated AKR and normal T1ALD BM cells for the repopulation of femoral marrow and thymus. The curves correspond to the percentage of cells bearing the AKR caryotype (200 irradiated cells). Cells of the host type (Fi) have been excluded. They are very scarce (less than 10%) in the femora whereas they represent respectively 36 and 22% of the total thymic population at day 9 and 14.

that finally it is the overall proliferative capacity of the surviving irradiated stem cells which overcomes that of the normal corresponding elements. The thymus follows a pattern of repopulation that parallels that of BM, indicating that any shift of the differentiation towards the myeloid series does not take place at the expense of the lymphoid one.

DIFFERENTIATION PATTERN

From the previously reported observations one might question whether the various changes in the properties of the irradiated CFU_s are coincident with some specific modification of the differentiation pattern. This problem can be viewed from two angles: the cytological characteristics of the clones produced by irradiated CFU_s and the response of these CFU_s to various stimuli of differentiation.

Cytologic Analysis of the Spleen Nodules

The majority of the CFU_s belong to the compartment of the pluripotential stem cells and a small number might correspond to committed stem cells. This means that most colonies are theoretically mixed colonies though they actually show a preferential orientation of their differentiation specially if they are analyzed by days 8–9.

The predominant cytology of the spleen colonies obtained from normal and 300 R-irradiated BM CFU_s is given in Table 5 which summarizes a recent report of Gidali et al. (*13*). Analysis made 8 and 10 days after transplantation clearly indicates a fall of the erythrocytic colonies, the percentage of which is 42.9 instead of 56.7 (control) at day 8, and 36.8 instead of 47 (control) at day 10. On the 8th day, the granulocytic colonies are increased whereas on the 10th day the rise concerns the megacaryocytic clones so that the E/G ratio does not really account for the change.

TABLE 5
Differentiation Pattern of Normal (NBM) or 300 R Irradiated Bone Marrow (IBM) CFU_S at 8 and 10 Days After Transplantation
[from Gidali et al., (*13*)]

Group	Day after transplantation	Cytological type, %*				
		E	G	Me	U	E/G
NBM	8	56.7	21.5	8.1	0	2.64
IBM	8	42.9	32.2	10.7	7.1	1.33
NBM	10	47.0	20.0	19.6	1.0	2.35
CBM	10	36.8	18.3	30.6	4.1	2.01

*Mixed colonies have not been included; E: erythrocytic; G: granulocytic; Me: megacaryocytic; U: undifferentiated.

Similar results have been reported by Wolf and Trentin (*15*) and interpreted as an indication that pluripotential stem cells and a higher radiosensitivity than granulocytic committed stem cells, a fact which was confirmed by Hellman et al. (*16*) using a completely different method. Furthermore, Wolf and Trentin (*15*) have shown that the decreased E/G ratio ceased by day 9 and thus correlated well with the other transient changes in the properties of the CFU_s surviving a sublethal irradiation. In a pilot experiment Gidali et al. (*13*) provided important complementary information when they reported that more agar colony formers committed to the granulocytic series were present in spleen of animals treated with irradiated BM cells than in controls. Thus, irradiation would lead to a relative increase in the number of granulocytic committed stem cells. As these elements have a high probability of differentiation, the overproduction of granulocytic cells come to an end within 7–10 days after transplantation. This could also account for the higher TE provided platelet production is involved in the process as suggested by several reports. But this explanation does not hold true with respect to the rise of D_0 or the increased rate of self-replication, both changes being based on the production of pluripotential CFU_s and not of committed stem cells.

At this point it is important to stress that erythrocytic colonies which are located immediately beneath the spleen capsule are easily recognized and numbered. Owing to the deeper localization of the granulocytic colonies their number is less accurately determined and most of the megacaryocytic foci are lost. Thus, when the E/G ratio is high the spleen colony assay gives a satisfactory quantitative estimate of the actual number of CFU_s. This is not the case when the relative number of granulocytic and megacaryocytic colonies increases and many colonies are overlooked which may contain CFU_s that will later show up on retransplantation. Consequently, this leads to overestimate of the stem cells or the differentiated functional cells produced by macroscopically recognizable colonies. Most of the previous findings which were based on the spleen colony assay can be explained by this loss in colony count and they might be considered as artefacts related to the preferential differentiation of the irradiated CFU_s towards granulocytic and megacaryocytic series.

Looking at this problem from another angle, it could be envisaged that both recruitment and commitment of irradiated CFU_s are preferentially directed toward the granulocytic and megacaryocytic series. This would account for the production of granulocytic clones containing CFU_s as well as for this of committed cells. There are no experimental bases to decide whether such a specific orientation is related to the selection of a subpopulation, to a change in the environmental control of differentiation or to the release of a stimulating factor.

Effect of Endotoxin

Directly or indirectly endotoxins act on different compartments of the hemopoietic system. As a rule, they trigger into cycle the resting stem cells stimulate the CFU, favor their localization in the spleen, induce the release of granulocytes into the circulating blood and of CSF into the mouse serum. Their effect on the CFU content of normal spleen and BM is controversial, but they have a great influence on the hemopoietic regeneration of irradiated animals. The improved survival of endotoxin-treated mice is probably related to an earlier onset of CFU proliferation (*17,18*).

In the experiments to be reported, 1, 2, 4 and 8 μg of endotoxin from *Salmonella typhimurium,* which will be referred to as endotoxin, were injected i.p. into lethally-irradiated (850 R) XVII mice restored with normal or irradiated BM or spleen cells. The delay between hemopoietic cell injection and endotoxin treatment was 1–2 hr, and all animals were sacrified 8 days later.

The curves in Fig. 4 represent the ratio of the number of spleen colonies obtained in endotoxin-treated mice to that found in control animals. This ratio is significantly increased in mice injected with irradiated BM or spleen cells whereas normal cell suspensions exhibit a weak reaction within this endotoxin dose range.

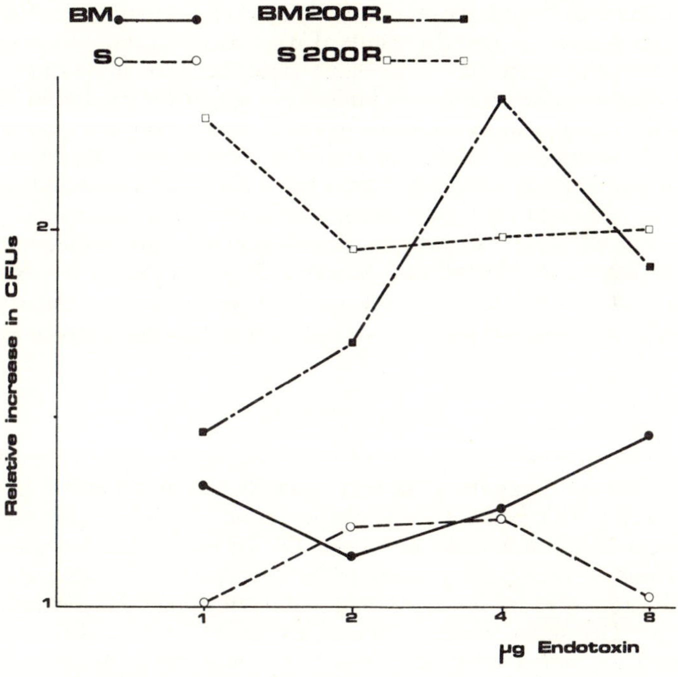

FIG. 4. Relative increase of spleen colonies in endotoxin-treated mice injected with equal number of normal or irradiated (200 R) BM or spleen CFU_s.

Under these experimental conditions, endotoxin acts mainly on the stem cell population and the precursor cells which have been transplanted, even though its effects might also extend to the earlier stages of differentiation and maturation. Thus, endotoxin could possibly affect either the spleen seeding efficiency, the survival or the growth and differentiation of the injected stem cells. If a change in "f" related to an increased spleen volume is responsible for this effect, it should apply to both irradiated and normal cells, which is not the case. As posttreatment with endotoxin is most unlikely to modify the actual survival of the CFU, it seems probable that endotoxin promotes the growth of descendants and/or the probability for self-replication of a stem cell subpopulation that would otherwise not have shown up in the spleen colony assay. Whatever the mechanism, direct or indirect, this result indicates that part of the population selected by irradiation has a high capacity to respond to the endotoxin-mediated stimulus and is only a small fraction of the overall stem cell population. For this reason the response of the unirradiated stem cell pool is much more limited.

Effect of Erythropoietin

It is now well documented that erythropoietin-sensitive cells (ESC) belong to a committed stem cell compartment and not to the pluripotential CFU population. Nevertheless, ESC are recruited into the CFU pool and this recruitment increases when the ESC population is depleted.

A good many authors have reported a delayed recovery of the erythropoietic cells and in the present context the results of Wolf et al. (*19*) and OKunewick et al. (*20*) are especially relevant. Both have described an impairment of the erythrocytic regeneration in spleen colonies developed from irradiated BM CFU. This delayed recovery can be corrected by the injection of erythropoietin. At the same time Porteous et al. (*21*) have provided experimental evidence that the response to low doses of erythropoietin is more radiosensitive than that to high doses. Thus, different cell populations respond to a hormonal stimulus depending on the dose of erythropoietin injected. This problem has been discussed at length by Metcalf and Moore (*22*). From their analysis it can be concluded that recruitment exhibits a lower sensitivity to radiation than commitment, meaning that the CFU_s are apparently less radiosensitive than the ESC.

DISCUSSION

The main and most undisputable effect of radiation on the stem cell pool is the shift of the differentiation pattern towards the granulocytic and megacaryocytic series. This is a transient phenomenon which lasts for about 7–10 days and is concomitant with an increase in D_o, TE and replication rate. For this reason it would be reasonable to assume that quantitative characteristics reflect the specific properties of this surviving stem cell population. Whether these changes depend on actual kinetics or on an artifact of the assay is a controversial matter. At this point it must be stressed that whereas stem cells which are localized in the spleen seem to escape the microenvironmental control, those that seed the BM do not exhibit significant changes.

In any case this preferential promotion of granulopoiesis and megacaryocytopoiesis may be related either to a selection based on differential radiosensitivity or to a new balance in the factors that control differentiation.

On the one hand, most of the new properties of the surviving CFU_s have been ascribed to the selection of a relatively radioresistant subpopulation; however, experimental evidence supporting this view is not established on firm radiobiological grounds. On the other hand, in the exocolony assay the CFU_s, whatever their origin and the treatment to which they have been subjected previous to their transplantation, are placed under the same control provided by the irradiated host. Therefore, the existing difference observed between normal and irradiated CFU_s should have had its origin in the donors, before the transplantation. This supports the hypothesis of a direct selective effect of irradiation on the stem cell pool. Nevertheless, as the hemopoietic cells are usually not harvested from the donor immediately after their irradiation, they might have been stimulated before transplantation and thus become less sensitive to the microenvironmental control of the spleen. Furthermore, owing to the small initial size of the stem cell compartment there may be a competition for recruitment and commitment in such a way that any preferential differentiation would take place at the expense of other pathways and must be compensated for at the level of the humoral factors.

REFERENCES

1. E. A. McCulloch and J. E. Till, *Radiat. Res.* **22,** 383 (1964).
2. E. Frindel, F. Charruyer, M. Tubiana, H. S. Kaplan and E. L. Alpen, *Int. J. Radiat. Biol.* **11,** 435 (1966).
3. G. Silini, R. Elli, G. Siracusa and R. F. Kallman, *Radiat. Res.* **31,** 603 (1967). (Abstract.)
4. J. F. Duplan, in *Hemopoietic Cellular Proliferation* (F. Stohlman, Jr., Ed.) pp. 189–196. Grune and Stratton, New York, 1970.
5. G. Silini, R. Elli, G. Siracusa and L. V. Pozzi, *Cell Tissue Kinet.* **1,** 111 (1968).
6. J. F. Duplan, E. Legrand and B. Guillemain, *Nouv. Rev. Fr. Hematol.* **11,** 821 (1971).
7. G. Silini, L. V. Pozzi and S. Pons, *J. Embryol. Exp. Morphol.* **17,** 303 (1967).
8. S. K. Lahiri and L. M. VanPutten, *Cell Tissue Kinet.* **2,** 21 (1969).
9. A. L. Kretchmar and W. R. Conover, *Proc. Soc. Exp. Biol. Med.* **129,** 218 (1968).
10. S. S. Fred and W. W. Smith, *Proc. Soc. Biol. Med.* **128,** 364 (1968).
11. R. G. Worton, E. A. McCulloch and J. E. Till, *J. Exp. Med.* **130,** 91 (1969).
12. M.A.S. Moore, T. A. McNeill and J. S. Haskill, *J. Cell. Physiol.* **75,** 181 (1970).
13. J. Gidali, I. Feher and V. Varteresz, *Radiat. Res.* **52,** 499 (1972).
14. H. S. Micklem, C. E. Ford, E. P. Evans, D. A. Ogden and D. S. Papworth, *J. Cell Physiol.* **79,** 293 (1972).
15. N. S. Wolf and J. J. Trentin, *J. Cell Physiol.* **75,** 225 (1970).
16. S. Hellman, H. E. Grate, J. T. Chaffey and R. Carmel, in *Hemopoietic Cellular Proliferation.* (F. Stohlman, Jr., Ed.) pp. 36–48. Grune and Stratton, New York, 1970.
17. G. E. Hanks and E. J. Ainsworth, *Radiat. Res.* **32,** 367 (1967).
18. F. Stohlman, Jr., P. Quesenberry, E. Niskanen, A. Morley, W. Tyler, K. Richard, M. Symann, F. Monette and D. Howard, in *Haemopoietic Stem Cells,* Ciba Foundation Symposium 13, pp. 205–219. Associated Scientific Publishers, Amsterdam, 1973.
19. N. S. Wolf, J. J. Trentin and V. Cheng, *Exp. Hematol.* **15,** 88 (1968).
20. J. P. OKunewick, K. M. Hartley and J. Darden, *Radiat. Res.* **38,** 530 (1969).
21. D. D. Porteous, S. C. Tso, K. Hirashima and L. G. Lajtha, *Nature (London)* **206,** 204 (1965).
22. D. Metcalf and M.A.S. Moore, *Haemopoietic Cells,* Frontiers of Biology, **Vol. 24,** North-Holland Publishing Company, Amsterdam, 1971.

CELL PROLIFERATION CHANGES IN TUMOR AND NORMAL TISSUE AS A RESULT OF IRRADIATION[1]

Chairman: M. Tubiana

[1]Symposium C–3

Changes in the Rate of Proliferation in Normal Tissues After Irradiation[1]

J. Denekamp
Cancer Research Campaign Gray Laboratory
Mount Vernon Hospital
Northwood, Middlesex HA6 2RN, England

In 1966, Lamerton (*1*) reviewed the data on changes in proliferation kinetics in the intestine and hemopoietic system under low dose-rate continuous irradiation. It was apparent that these tissues had a remarkable capacity to withstand constant irradiation and were able to compensate for the radiation-induced cell killing by increased rates of proliferation. Similar studies have since been performed on other normal tissues (e.g. *2–4*). It has been generally believed that a similar compensatory change in proliferation could occur in normal tissues during fractionated treatment with acute doses of radiation, but not in tumors, and that this could be one of the reasons why fractionation enables the radiotherapist to achieve more damage to a tumor relative to the normal tissue injury (e.g. *5*).

There is, however, relatively little detailed information about the changes in proliferation kinetics after acute irradiation in most normal tissues and, in particular, about the time at which such compensatory changes occur. Furthermore, recently tumor cell populations have also been shown to be capable of more rapid proliferation after irradiation (*6,7*). Most of the experimental work on normal tissues has been done on the hemopoietic system (*8*) and on the intestine and skin. Apart from these three tissues, all of which have fairly rapid turnover times in the unirradiated state, the information is very scanty. This is partly because this area of research falls outside the realms of most radiobiologists, cell kineticists and those interested in control mechanisms. The radiobiologist has provided some information using split-dose experiments and extending the interval between fractions to sufficiently long periods for repopulation to be assumed to take place. These experiments can sometimes be interpreted in terms of cell doubling times if various assumptions are made. The cell kineticist may well be interested in the changes that can occur in a population which has been damaged, but most of the cell kinetic techniques available depend for their interpretation on the assumption of a steady proliferation pattern during the period of observation. In addition, autoradiographic studies using thymidine can be greatly influenced by reutilization of the tracer as it is released from dying cells and by the problems of identifying the true radiation survivors as different from the cells which are doomed to die. For people interested in control mechanisms that operate to maintain the integrity of a tissue after damage, the

[1]Congress presentation C–3–1.

radiation lesion is so diffuse, both spatially and in its time of appearance, that more often they turn to chemicals, heat or surgery, where immediate cell death can be produced and the time course of the response to a shortage of cells can then be accurately studied.

The work that has been done using autoradiographic or functional tests to study changes in repopulation rates has been almost entirely restricted to rapidly renewing normal tissues, although the capacity for increasing proliferation rates should be much greater in tissues in which the turnover is normally very slow, e.g. lung, liver and kidney. Where studies have been performed on slowly renewing organs, they have often involved an additional stimulus to cause proliferation so that the damage is expressed and can be more readily observed, e.g. partial hepatectomy or unilateral nephrectomy in studies on liver and kidney.

Mitotic Delay

One of the first observable effects of radiation on cells both in vitro and in vivo is the prevention of cells from entering mitosis, often referred to as the G_2 block or mitotic delay (e.g. *9–15*). Cells which are in mitosis are not prevented from completing that division, but no new prophases are formed for a time that seems to be dependent upon the dose administered and the cell cycle time of the unirradiated cells. In Fig. 1 four sets of published data (*10,11,13*) are shown for mitotic delay as a function of dose of radiation for systems with widely varying cell cycle times. For cells in vitro it has often been quoted that 100 rad delays the cells by 1 hr or 10% of a cell cycle, although higher doses may be relatively less efficient at producing this effect (*9–11*). Fig. 1 shows that 1 cell cycle time of mitotic delay is achieved with 1000–1500 rad over a very wide range of cycle times, from 9.5 hr for the L5178Y cells of Doida and Okada (*11*), 16 hr for the L cells of Whitmore et al. (*10*), to 47 hr and 100 hr for plucked and unplucked skin of mice (*13*). It therefore appears that cells are delayed from entering mitosis for a time that is related to their normal cell cycle time rather than for an absolute number of hours per rad.

Time of Radiation-Induced Death

Another tenet of radiobiology, observed both in vitro and in vivo, is that most cells (other than lymphocytes) die of their radiation-induced damage at a subsequent mitosis (9,12,13). Furthermore, the number of divisions through which the cells may go before expressing their radiation-induced damage also appears inversely related to dose (*9*). Hence, most cells in vitro, after 1000 rad, will be delayed approximately 1 cycle time, will then progress through the G_2 block into mitosis and will die then, or at a subsequent mitosis if they have been lethally damaged. It seems likely that a similar process occurs in vivo and that, because the cell cycle time is much longer than in in vitro systems, the time of death is correspondingly delayed as judged from the time of appearance of pyknotic nuclei (12,13,16–18).

Time of Homeostatic Response

A feedback mechanism is known to operate in normal tissues after injury, but it is not clear whether this will be triggered by death of proliferative cells or by a shortage of cells in the differentiated compartment. It seems likely that the earliest time of triggering faster repopulation will be after the cells have died; i.e. it will be related to the cell cycle time of the dividing cells and to the size of the radiation dose. The latest time of triggering will depend on the transit time of cells through the differentiated compartment. Likewise, the time of expression of the tissue damage will relate to this transit time of cells through the differentiated compartment, and that transit time seems to be unaltered by radiation sterilization of the proliferating cells (*18,20*). Thus death from depletion of the crypt population of the small intestine occurs at 4–5 days and death from hemopoietic injury at about 30 days. Another illustration can be made from the time of appearance of maximum skin reactions in normal or plucked mouse skin, where it occurs at 20 days or at 12 days respectively (*13*).

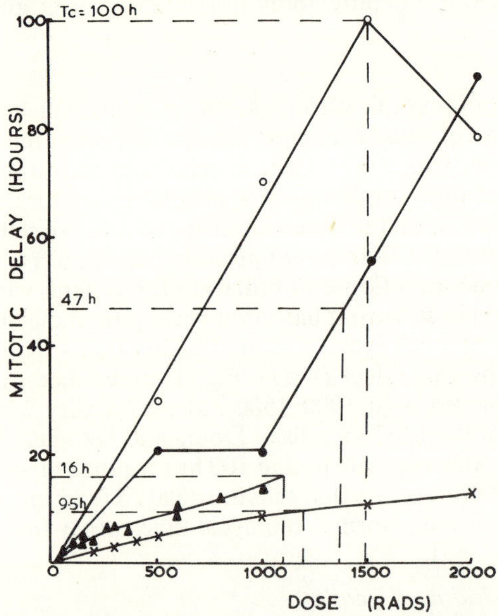

FIG. 1. Mitotic delay as a function of dose for 2 cell lines in vitro and 2 cell lines in vivo. A delay of 1 cell cycle time is achieved with 1100–1500 rad for L5178Y cells × (*11*), for L cells △ (*10*), for plucked skin ● (*13*), and unplucked skin O (*18*).

Autoradiographic Techniques

Several techniques have been developed for studying proliferation characteristics, involving the injection of tritiated thymidine, excision of the tissue at some time later, and autoradiography of the histological preparations. The most detailed technique involves following the progress of cells labelled in the S phase through several subsequent mitoses. By sequential sampling, a percent labelled mitoses curve can be constructed and the average intermitotic time or cell cycle time of the dividing cells can be measured from it, (T_c). Alternatively, a thymidine labelling index can be measured soon after injection and from this and a knowledge of the duration of the S phase, a potential doubling time, or apparent cell cycle time T_{ca} can be estimated. This is the time required for a number of cells to be born into the population, equal to the number present at the outset. It will only equal the cell cycle time if all the cells in the population are in the growth fraction. Therefore, from changes in this value, T_{ca} in a population which normally has a growth fraction that is less than unity, we cannot separate the effects of shortened cell cycle time and increased growth fraction. Most of the postirradiation proliferation data have been measured in this way and are yet further complicated at short times after irradiation by the presence of radiation sterilized, but not visibly dead cells, which will lead to underestimates of T_{ca}. A few studies have also used the technique of continuous ^{3}HTdR labelling to determine T_{ca}.

Skin—Autoradiographic Studies

The proliferation response of skin to single doses of radiation has been investigated by a number of workers, Table 1. In 1962, Devik (*19*) studied the response of hairless mice to 2700 rad using the technique of 'percent labelled mitoses' curves. He was unable to obtain a second wave of labelled mitoses either for unirradiated control skin or in the region that had been irradiated with 2700 rad 5 days previously, indicating a long cell cycle time. He did, however, observe a second wave in the epidermis bordering on to the irradiated zone and deduced a reduction in the cell cycle time to 8–10 hr. A similar rapid proliferation at the border of an irradiated area of guinea pig skin was observed by Yamaguchi and Tabachnick (*20*), but it did not occur until 14 days after irradiation with 3000 rad, when the apparent cell cycle time was estimated from labelling indices to be about 19 hr. The shortest cycle time measured was 15.5 hr at day 20 (from a PLM curve). By day 30 T_{ca} was reverting to a more normal value. This delay in the onset of rapid compensatory proliferation contrasts with surgical wounding where an increased labelling index can be observed within 1–4 hr extending to 1 mm from the edge. By day 50 and 150 there appeared to be slower proliferation in the irradiated zone, although delayed differentiation led to prolonged hyperplasia; a reduced skin thickness was not observed until 1 year (*20*). An increased proliferation at the irradiated border was also observed by Hegazy and Fowler (*13*). In plucked skin they found a minimum value for T_{ca} of about 22 hr, at 6–20 days after 1500 rad. An even lower value of 16 hr was estimated 20 days after 2000 rad (Table 1). The labelling indices in the irradiated area were difficult to interpret, particularly in the first 12 days when there were many degenerate cells present. At no time did the labelling index in the irradiated area reach as high a value as at the border (*13*).

Leith et al. (*21*) have studied the changes in the proliferation pattern in skin 1 year after irradiation with a range of doses of accelerated helium ions. After low doses, no difference could be detected in the labelling indices of irradiated and control skin. After doses in excess of 1200 rad, a slightly lower labelling index was observed, which they interpreted as an elongated apparent cell cycle time (Table 1), in accord with the data of Tabachnick and his colleagues (*17,20*).

Another epithelial tissue which has been studied by autoradiographic techniques is the buccal mucosa of the Hamster cheek pouch (*12*). Brown observed a slight shortening of the apparent cell cycle time when estimated using several ingenious approaches. A 20% shortening was observed 4–9 days after 500 rad and a 20–35% shortening between 4–17 days after 1000 rad (Table 1).

TABLE 1.
Changes in Proliferation Kinetics after Irradiation of Skin

	animal	T_c normal (hr)	irradiated T_c or T_{ca} (hr)	dose (rads)	days after irradiation
Devik (*19*)	hrhr mouse		8–10*	2700	5–7
Yamaguchi and Tabachnick (*20*)	guinea pig	91	18.8	3000	14
			16.2		17
			15.5*		20
			8		30
			118		50
			134		150
Hegazy and Fowler (*13*)	albino mouse plucked	47*	22	1500	6
			31		12
			25		16
			21		20
			69	2000	6
			22		12
			23		16
			16		20
Leith et al. (*21*) (^{4}He irradn.)	mouse	95*	95	75	1 year
			96	150	1 year
			92	300	1 year
			96	600	1 year
			102	1200	1 year
			106	1400	1 year
			110	2000	1 year
Brown (*12*)	Hamster cheek pouch	130–152	110	500	4–9
			87–117	1000	4–10
			118	1000	12–17

* from PLM curve

Skin-Fractionation Studies

The response of skin to single and fractionated doses of radiation has been studied for many years by scoring gross skin reactions and, more recently, by scoring the survival of epidermal clones (*22–24*). When radiation is given as 2 doses instead of as a single dose, additional dose is necessary to achieve the same level of damage. Over short intervals (24 hr) this additional dose is believed to be necessary to counteract the effect of repair of sublethal injury (Fig. 2). Beyond 24 hr the dose increment may be interpreted as a measure of the repopulation rate, since the D_o for epithelial cells has been measured as 135 rad in two strains of mice (*22,23*).

Various estimates of the doubling times of clonogenic cells from experiments involving two or more fractions and varying the overall time are shown in Table 2. Several authors have reported approximately 30 additional rad per day being necessary to counteract repopulation for skin reactions in mice, pigs and humans (*24–27*). A dose of 30 rad per day yields an estimated doubling time of about 3 days, in experiments covering a wide range of doses per fraction and overall times ranging from 1–5 weeks. Higher dose increments yielding shorter doubling times have been estimated by Comas (*27*) and by Fowler et al., (*28*). The Withers *in situ* cloning technique has also been used in this type of experiment, but for this the skin is stimulated by plucking 1 day before irradiation. Cell doubling times of less than 1 day and 2–3.3 days were estimated using this technique (*22,23*). Using similar clones which appeared in desquamated human skin, Dutreix (*21*) estimated a cell doubling time of 1–1.5 days from the rate of extension of the edge of the clones regrowing in irradiated areas.

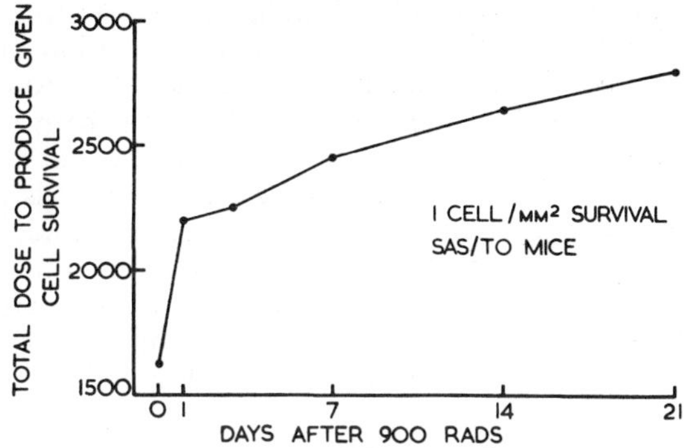

FIG. 2. Total dose to produce 1 surviving cell per mm^2 in the epidermis of albino mice as a function of the time interval between 900 rad and the range of test doses. In the first 24 hr a dose increment is necessary to counteract repair of sublethal injury. Beyond 24 hr, additional dose is necessary to counteract repopulation. [Reproduced from E. W. Emery, J. Denekamp, M. M. Ball and S. B. Field, *Radiat. Res.* **41**, 450 (1970)]

TABLE 2.
Changes in Proliferation after Irradiation of Skin—Fractionation Data

	animal	fraction number	overall time	increment rad/day	doubling time (days)
Fowler et al. (*25*)	pig	5	5–28	25	4
Denekamp et al. (*24*)	mouse	2	7–14	30	3.2
			14–21	29	3.2
Cohen (*26*)	human	up to 25	up to 35	28–34	3
Dutreix & Shaich (*29*)	mouse	15	17–35	32	3
Comas (*27*)	mouse	2	1–7	70	3.2
Fowler et al. (*28*)	mouse	5	4–9	90	1
Plucked skin					
Withers (*22*)	mouse	2	1–6	100	1
Emery et al. (*23*)	mouse	2	1–7	48	2
			7–14	42	2.3
			14–21	29	3.3
Clone extension					
Dutreix (*29*)	human		many doses to tolerance level		1–1.5

Recently, this type of experiment has been extended to multifraction irradiation with multiple small daily doses which simulate more closely what occurs in the radiotherapy clinic (*30,31*). These experiments were designed to determine both the extent to which repopulation could be stimulated and the *time* at which this compensatory mechanism operated. The doubling time of epithelial cells was estimated after 4, 9 and 14 daily fractions of 300 rad, using both the functional test of skin reaction and the continuous labelling with ^{3}HTdR and scoring the labelling index. The results are shown in Table 3 and Figs. 3 and 4. After 4 fractions there seemed to be little proliferation as judged from the skin reaction experiments. The labelling studies suggested an extended mitotic delay occurring for 5 days after the last fraction. After 9 fractions, the doubling time seemed to be reduced but it was not until 14 fractions had been administered that very rapid proliferation occurred with a doubling time of less than 1 day. These results support the hypothesis that the expression of radiation damage is slow, particularly during fractionated treatment, and that the response of faster proliferation is not triggered, at least until cell death has been registered. The more rapid proliferation after both 9 and 14 fractions is not apparent in the second week after the end of the fractionated treatment. This compensatory rapid proliferation ceases when the population has been restored to its original level. In more extended fractionated experiments on mouse skin, healing of the radiation reaction has been observed within the period of irradiation (6 weeks), presumably as a result of this compensatory proliferation (*32*). Similarly, healing of the mucosal reaction in patients during a course of radiotherapy has also been observed.

TABLE 3

Changes in the Rate of Proliferation after 4, 9 and 14 × 300 Rad

Albino mice (SAS/TO or WH)–control T_{ca}* Top 135 hr; Bottom 75 hr

	time	dose increments	doubling time (hr)	autoradiographic *T_{ca}(hr) top	autoradiographic bottom
4 × 300 rad	1–8 days	−10 rad/day	∞		
	4–7			152	76
	8–15	5 rad/day	450		
9 × 300 rad	1–8 days	58 rad/day	38		
	4–7			115	66
	8–15	22 rad/day	102		
14 × 300 rad	1–8 days	130 rad/day	17		
	4–7			26	33
	8–15	18 rad/day	125		

*T_{ca} derived from continuous labelling indices after repeated 6-hourly injections of ^{3}HTdR. The interfollicular skin of the upper and lower surface of the foot were scored separately

Similar values for dose increments have been observed after irradiation with 4, 9 and 14 × 100 rad/day of fast neutrons. From Denekamp (*30*) and Denekamp and Stewart (*31*)

Testes

Sixty years ago Regaud observed an increased sensitivity of the testes to fractionated doses relative to single doses of X-rays which, based on the premise that tumors would respond like the testes, led to the use of fractionation in radiotherapy. Recently, the proliferation aspects of the response of the testes to radiation have been studied by several workers. Fabrikant (*3*) found that the seminiferous epithelium, with its normal cycle time of 32–34 hr, could only maintain a normal turnover under continuous irradiation if the dose rate was less than 3 rad/day. This is a much lower dose rate than is tolerated by other normal tissues, e.g. cartilage (*2*), intestine and bone marrow (*1*). Split-dose studies with a variable time interval between two doses have also been used by two groups of workers. Withers et al. (*33*) found repair of sublethal injury after 600 rad of X-rays to be maximal at 4–8 hr; but no increased dose was necessary if the fractionation interval was increased to 14 days. Kramer *et al.* (*34*), using 1 MeV fast neutrons to eliminate the effects of repair of sublethal injury, found the testes more sensitive to two doses given at intervals ranging from 1 to 31 days, than to a single dose; again, they were unable to demonstrate any proliferation over this period. These studies involved the scoring of colonies of regenerating seminiferous epithelium 5 and 11 weeks after irradiation. It seems anomalous that no reproduction could be detected between the split doses, because the presence of clones means that regeneration must have occurred at least after the second dose. This may be an artefact of the scoring system because of the geometrical distribution of the tubules, or an hormonal response, or, as they suggest, a shift of the seminiferous cells into a more sensitive phase as a result of the first dose administered. This response contrasts with the sparing effects of fractionation and of increasing the overall time in other normal tissues.

Gastro-Intestinal Tract—Autoradiographic Studies

The other tissue which has been studied in some detail is the lining of the gastro-intestinal tract. Lesher and Bauman (*35*) have shown a shortening of the cell cycle time, measured in a very detailed study using 'percent labelled mitoses' curves after both continuous and acute irradiation (Table 4, Fig. 5). One day after an acute dose to the whole body of 300 rad, they noted a reduction of T_c to 10.4 hr (relative to the control value of 13.1 hr). This had returned to normal by 4–8 days. After 1000 rad they observed a remarkable shortening of T_c to less than 7 hr., a value which had been previously inconceivable in a mammalian cell system. They have studied, in some detail, the changes in the labelled nuclei per crypt, the radioactivity per crypt, and the mitotic activity per crypt after a wide range of doses from 75 to 1000 rad. However, these data are extremely difficult to interpret in terms of changed proliferation rates. Hageman, Sigdestad and Lesher (*36,37*) have, more recently, been studying the effects of repopulation on the functional integrity of the jejunum. They found that, if rapid repopulation commences within 3 days, it can be effective in preventing the animal's death by 5 days, whereas a later overshoot in repopulation is too late to be effective. They also found that the extent to which repopulation could be stimulated by radiation was dose dependent up to approximately 1 kilorad, at which stage it failed, presumably due to excessive cell killing.

An alternative approach has been used by Devik (*18*) who estimated the apparent cell cycle time from mitotic accumulation after vinblastine. He found that animals could recover from single doses of 1000 rad, after which they showed a prolonged T_{ca} at day 1, with progressive shortening at day 2 and 3. All his values are higher than those of Lesher, because they are apparent cell cycle times, not actual cell cycle times, but the conclusions are similar. Like Yamaguchi and Tabachnick for skin (*19*), Devik concluded that the rate of maturation of cells was unaffected by the radiation-induced killing of the crypt population.

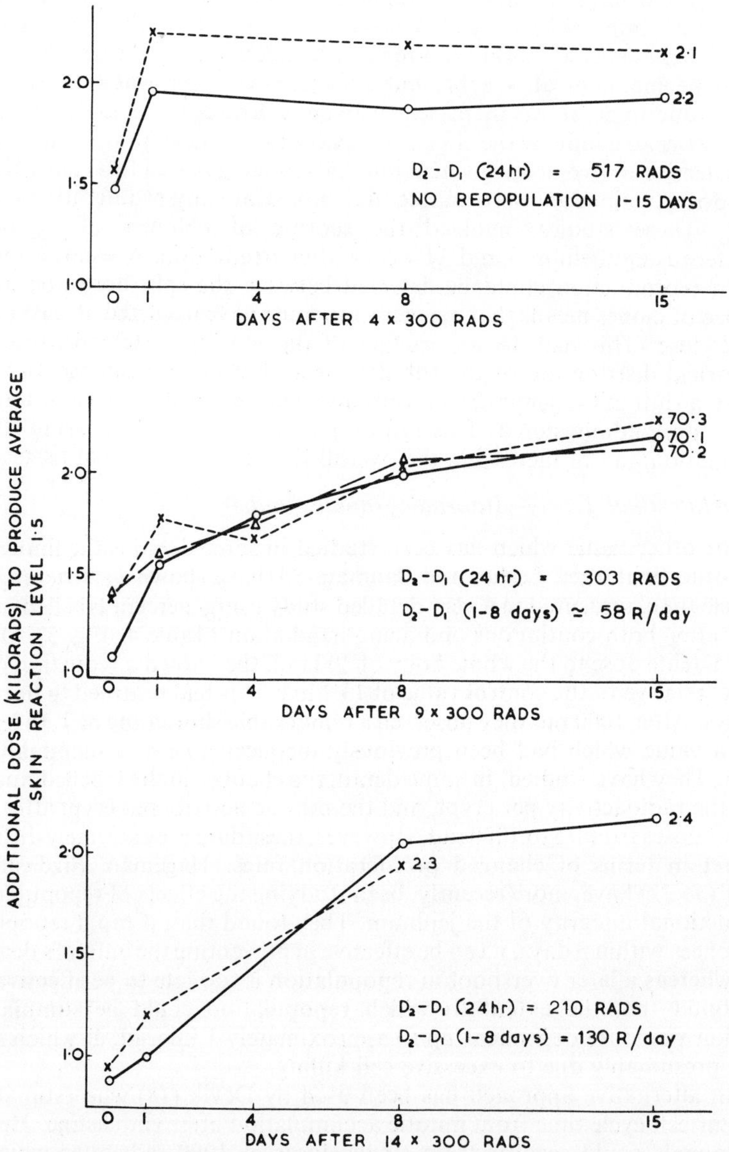

FIG. 3. The additional dose required to produce the same average skin reaction, when given at various times after 4, 9 or 14 daily fractions of 300 rad, plotted against the time interval. Each line represents a separate experiment. The average dose increments required for repair in the first 24 hr and for repopulation in the subsequent week are shown on each graph. [Reproduced from J. Denekamp, *Brit. J. Radiol.* **46**, 381 (1973) with the permission of the publisher.]

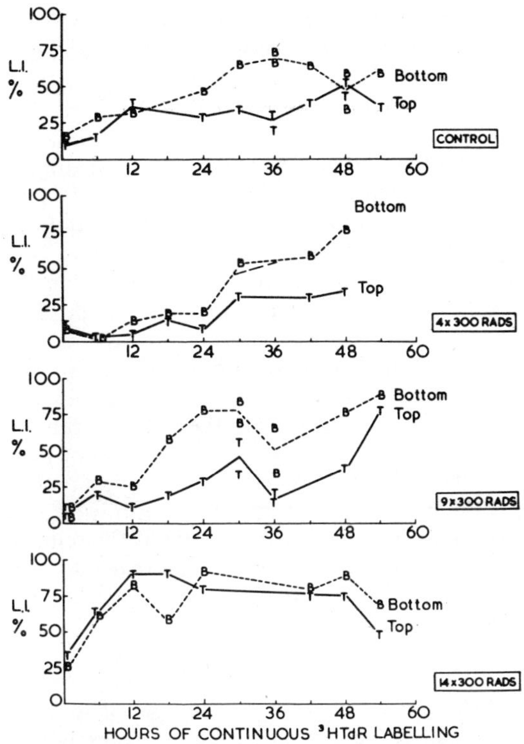

FIG. 4. Continuous labelling curves obtained from the interfollicular epidermis of mouse feet after 6 hourly repeat injections of ^{3}HTdR. 4 sets of data are shown — for controls and for animals injected 4 days after 4, 9 and 14 fractions of 300 rads. The labelling indices have been scored separately for the top and bottom (sole) surface of the foot, (Denekamp and Stewart, unpublished, 31).

TABLE 4
Changes in Repopulation of Small Intestine after Irradiation—Histological Studies of Jejunum

	T_c Control (hr)	T_c or T_{c_a} (hr)	dose (rad)	time (days)	method
Lesher and Bauman (*35*)	13.1	10.4	300	1	PLM
		11.0		2	curves
		13.5		4–8	
		7	1000	3	
Devik (*18*)	48	125	1000	1	vinblast.
		39	1000	2	+ M.1
		14.2	1000	3	

Gut—Fractionation Experiments

A beautiful series of split-dose experiments have been performed by Withers and his co-workers, who have scored clone formation after irradiation of exteriorized portions of the intestine at different levels (Fig. 6). In the jejunum, with its normal cycle time of 13–14 hr (*35*), they observed a lag of 2½ days after a dose of 600 rad before proliferation become really significant and then the cells doubled with a doubling time estimated at 4.2–8.4 hr (*38*). When the crypts had regained their normal size, this rapid proliferation ceased. In the stomach, in which the squamous epithelium has a normal T_c of 28 hr (*39*), repopulation appeared to be more gradual, beginning at day 1 and yielding a generation time of 43 hr (*40*). In the colon, with a normal T_c of 19 hr (*41*) there was a delay of almost 4 days before proliferation occurred, and then a doubling time of 12 hr was estimated (*16*). (For a summary see Table 5.) Thus, it appears that the response of similar cell populations at different sites within the intestinal tract can behave quite differently, possibly as a result of the differences in cell cycle time and transit time through the differentiated compartments, although it is not clear why such a high doubling time was observed for the stomach.

From the response of these rapidly turning over tissues, several factors can be deduced. Mitotic delay is proportional to the cell cycle time, as is also the time of death of injured cells. The time at which recognition of this injury occurs is proportional to the cell cycle time and/or the maturation time of differentiating cells. The transit time of cells through the maturation zone is unaltered immediately after irradiation, at least in skin and gut. The time of appearance of radiation injury as a functional impairment of the tissues also appears to depend upon the transit time through the maturation zone.

Slowly Dividing Tissues

In more slowly dividing tissues, therefore, the time of expression of damage should be later and this certainly seems to be true of lungs, heart, liver and kidney. It is often postulated that, in these tissues, where labelling studies have shown only 0.1% to 1.0% of the cells in DNA synthesis, the damage that is expressed 3–6 months after irradiation is caused by the damage and death of vascular endothelial cells and fibroblasts (*42,43*). Labelling studies have been performed on the endothelium of muscle, skin and bone after irradiation by Tannock and Hayashi (*44*). They observed no increase in LI for up to 3 weeks after irradiation, although a rapid increase in the labelling index could be induced by fracture of the femur. Fajardo and Stewart (*43*) did a more extended labelling study on the endothelial cells in preparations of irradiated rabbit hearts. They extended their periods of observation and finally observed increased LI at 40–70 days postirradiation, which correlated well with the time at which pericarditis became apparent and which they attributed to endothelial cell death.

It seems unlikely, in a system such as this, that repopulation will occur during a course of daily fractionated radiotherapy. Thus, an increased dose should not be necessary if the same number of fractions are spread over a longer period of time, if a non-proliferating organ is thought to be the limiting normal tissue. There is some radiological evidence, however, that this is not so. Table 6 shows the dose increments necessary if the overall treatment time is extended in fractionation studies on slowly dividing tissues. Field, Kutsutani and Hornsey (*45*) using the LD_{50} (180 days) (which has been shown by Phillips (*42*) to be associated with capillary endothelial cell death in the lungs), found that a greater dose was necessary if they increased the overall treatment time from 5 to 55 days for 2 fractions (Fig. 7). Their 2-fraction data for X-rays and neutrons and the data of Wara et al., (*46*) fit on the same isoeffect curve. There is a greater effect over the first few days than at longer time intervals and since proliferation seems unlikely to be occurring in this system, they attribute the effect to a slow repair process. If an average value for the dose increment is measured from their curve (Fig. 7) it is approximately 7 rad/day for Field et al. and 11 rad/day for Phillips, (Table 6). Hayashi and Suit (*48*), looking at callus formation after fracture of an irradiated limb, also found a greater total dose was necessary with increasing time, yielding a value of 10 rad/day. In functional studies on the kidney, Glatstein et al. (*47*) found a large time effect when 9 fractions were given in 18 days instead of 4 days. An extra 1000 rad was needed, corresponding to about 70 rad/day, although no proliferation could be detected in the 14-day period by labelling studies with ^{3}HTdR. It is possible that this slow repair process in slowly-dividing tissues may relate to the repair of potentially lethal damage observed by Van den Brenk (*49*) and Reinhold (*50*). These authors showed a fading of radiation damage to capillaries of the subcutis if time was allowed between single dose irradiation and the stimulation of the vascular system to form new vessels, with a half-time of several days. It seems possible that these studies are demonstrating an extremely slow repair process, taking days to weeks for completion, which does not occur in the normal tissues usually studied, either because of some inherent difference or, more likely, because there is not time for its expression before the next mitosis. It is also possible that the endothelial cells are not, after all, the critical cells in these organs that express late damage. However, if the expression of damage is related to the time of death of the critical cells, whatever they may be, it seems unlikely that rapid repopulation will occur within days in an organ that does not appear to express its damage for months, unless there is a long maturation process which does not require any cell divisions. Alternatively, the late damage may not be an expression of cell death directly, but of other pathological processes such as the increased permeability, increased fibrin deposition and increased collagen formation as postulated by Law and Thomlinson (personal communication).

Thus, in tissues where reproductive cell death is known to cause the functional tissue damage (e.g. intestine and skin), repopulation becomes important only after the death of the radiation-damaged cells. Since these tissues have a fairly rapid turnover, this can occur within a short period of time and can assist in the healing of tissues *during* fractionated therapy. However, in tissues which express their damage late, such as the lung, it is very unlikely that repopulation will be stimulated before cell death is manifested and this does not occur during the period over which fractionated radiotherapy is administered. Although repopulation may be of no importance in these tissues, e.g. lungs and kidneys, there appears to be some other 'repair' process which requires additional radiation dose to be administered to achieve the same endpoint if the overall time is increased.

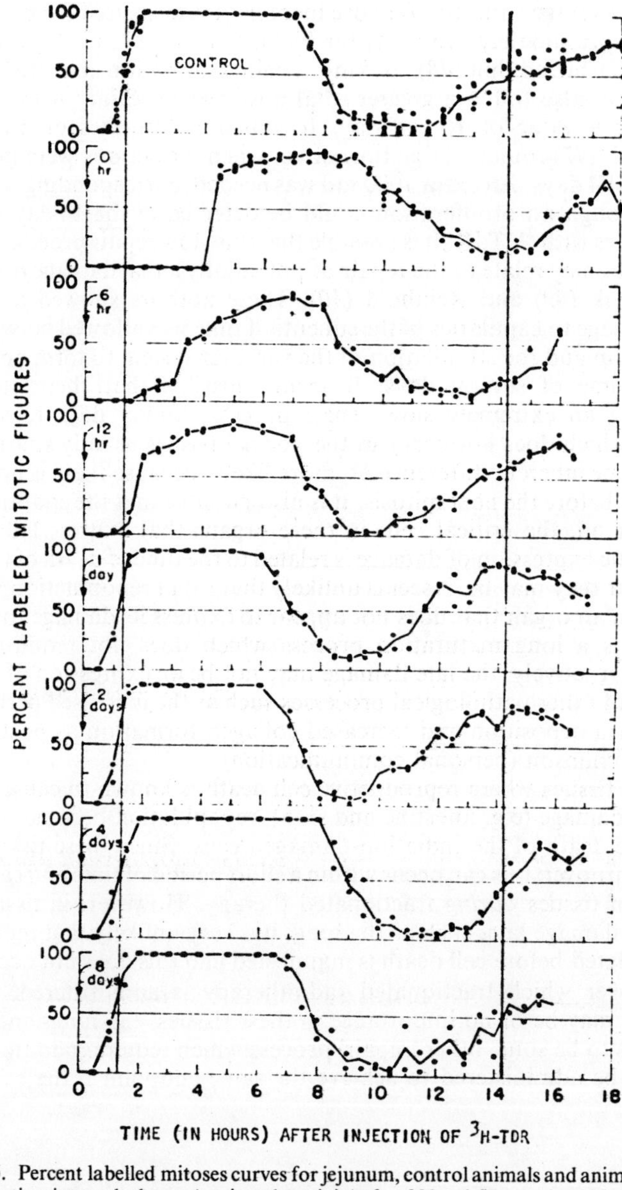

FIG. 5. Percent labelled mitoses curves for jejunum, control animals and animals injected with ^{3}HTdR at the time intervals shown (against the origin) after 300 rad. [Reproduced from S. Lesher and J. Bauman, *Nat. Cancer Inst. Monogr.*, **30**, 185 (1969).

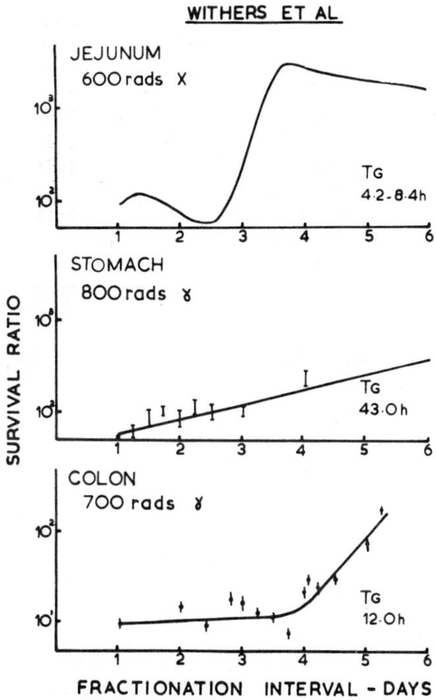

FIG. 6. Split dose experiments to determine the dose increments necessary to counteract repopulation if the interval between 2 doses is varied from 1–6 days, for jejunum (*38*), stomach (*40*) and colon (*16*). More rapid proliferation is observed in irradiated colon and jejunum (see text), but the generation time for the stomach is longer than the cell cycle time in controls (*39*).

TABLE 5
Changes in Repopulation of Intestine after Irradiation—Fractionation Studies Using 2 Doses

	T_c control (hr)	doubling time (hr)	dose (rad)	time after irradiation (days)
Jejunum	13.1 (*35*)	4.2–8.4 (*38*)	660	2.5–4
Stomach	28 (*39*)	43.0 (*40*)	800	1.8
Colon	19 (*41*)	12.0 (*16*)	700	3.5–5.2

TABLE 6
Dose Increments Necessary with Extended Time Intervals Between Fractionated X-Ray Doses to Slowly Dividing Tissues (Slow Repair Process)

	system	number of fractions	overall time (days)	average rad per day
Field et al. (*45*)	lung LD_{50}	2 or 5	2–55	7
Phillips (*42*)	lung LD_{50}	10	11–26	11
Hayashi & Suit (*48*)	callus formation	2	1–28	10
Glatstein et al. (*47*)	86 R_b extraction by kidneys	9	6–18	70

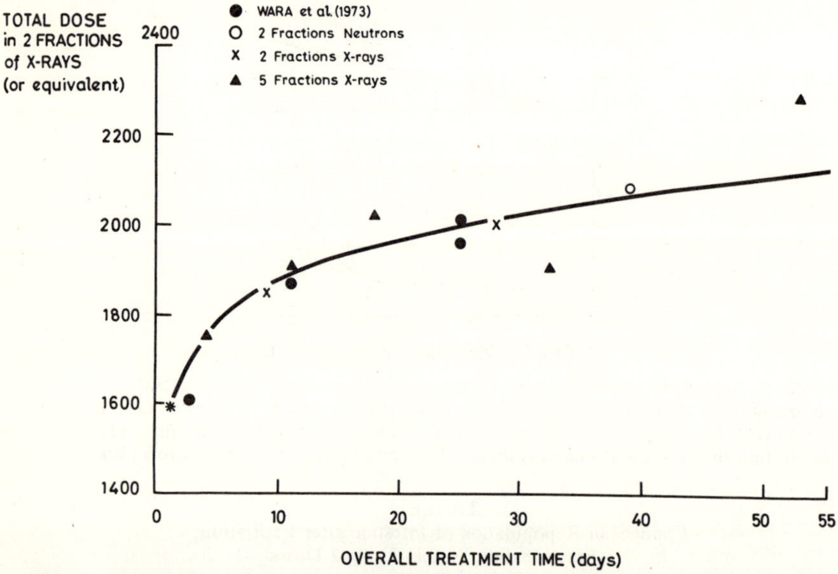

FIG. 7. Total dose in two fractions (or equivalent) to obtain 50% lethality at 180 days when the interval is varied from 1–53 days; from Field, Kutsutani and Hornsey (*45*). The solid line is drawn for a time exponent of T 0.07. This is believed to represent a slow repair process.

REFERENCES

1. L. F. Lamerton, *Radiat. Res.* **27**, 119 (1966).
2. N. F. Kember, in *Some Aspects of Internal Irradiation*, pp. 309–316. Pergamon Press, 1962.
3. J. I. Fabrikant, *Amer. J. Roentgenol. Radium Ther. Nucl. Med.*, **114**, 792 (1972).
4. P. Coultas, *Study of Cellular Proliferation in Various Growing and Continuously Irradiated Populations*. Ph.D. Thesis, Univ. of London (1973).
5. L. G. Lajtha and R. Oliver, *Brit. J. Radiol.*, **35**, 131 (1962).
6. A. F. Hermens and G. W. Barendsen, this Symposium.
7. E. P. Malaise, this Symposium.
8. B. Lord, this Symposium.
9. M. Elkind, A. Han and K. W. Volz, *J. Nat. Cancer Inst.*, **30**, 705 (1963).

10. G. F. Whitmore, J. E. Till and S. Gulyas, *Radiat. Res.*, **30**, 155 (1967).
11. Y. Doida and S. Okada, *Radiat. Res.*, **38**, 513 (1969).
12. J. M. Brown, *Radiat. Res.*, **43**, 240 (1970).
13. M.A.H. Hegazy and J. F. Fowler, *Cell Tissue Kinet.*, **6**, 587 (1973).
14. H.A.S. van den Brenk, *Amer. J. Roentgenol.*, **97**, 1023 (1966).
15. J. L. Ladinsky and H. W. Gruchow, *Cell Tissue Kinet.*, **3**, 175 (1970).
16. H. R. Withers, *Cancer*, **28**, 75 (1971).
17. H. Etoh, Y. H. Taguchi and J. Tabachnick, *J. Invest. Dermatol.* in press (1975).
18. F. Devik, *Acta Radiol. Ther. Phys. Biol.*, **101**, 181 (1971).
19. F. Devik, *Int. J. Radiat. Biol.*, **5**, 59 (1962).
20. T. Yamaguchi and J. Tabachnick, *Radiat. Res.*, **50**, 158 (1972).
21. J. T. Leith, W. A. Schilling, G. P. Welch and C. A. Tobias. In *Proc. of 13th Hanford Biology Symposium on The Cell Cycle in Malignancy and Immunity*, (J. C. Hampton, Ed.), 1975. (Available from Pub. Office of Public Affairs USERDA).
22. H. R. Withers, *Radiat. Res.*, **32**, 227 (1967).
23. E. W. Emery, J. Denekamp, M. M. Ball and S. B. Field, *Radiat. Res.*, **41**, 450 (1970).
24. J. Denekamp, M. M. Ball and J. F. Fowler, *Radiat. Res.*, **37**, 361 (1969).
25. J. F. Fowler, R. F. Morgan, J. A. Silvester, D. K. Bewley and B. A. Turner, *Brit. J. Radiol.*, **36**, 188 (1963).
26. L. Cohen, *Brit. J. Radiol.*, **41**, 522 (1968).
27. F. V. Comas, in *Dose Rate in Mammalian Radiation Biology*, (D. G. Brown, R. G. Cragle and T. R. Noonan, Eds.), pp. 18.1–18.10. Oak Ridge, 1968. A.E.C. Conf. 680410 (1968). Available from Clearinghouse for Technical and Scientific Information, Springfield, Va. 22151.
28. J. F. Fowler, J. Denekamp, C. Delapeyre, S. R. Harris and P. W. Sheldon, *Int. J. Radiat. Biol.*, **25**, 213 (1974).
29. J. Dutreix, M. Tubiana, A. Wambersie and E. Malaise, *Eur. J. Cancer*, **7**, 205 (1971).
30. J. Denekamp, *Brit. J. Radiol.*, **46**, 381 (1973).
31. J. Denekamp and F. A. Stewart, Changes in the proliferation rate in mouse skin after irradiation by continuous labelling studies. Submitted to *Cell Tissue Kinet.* (1975).
32. S. B. Field, C. Morris, J. Denekamp and J. F. Fowler, *Eur. J. Cancer*, in press, 1975.
33. M. F. Kramer, J.A.G. Davids, Th.P.A. van der Ven, *Int. J. Radiat. Biol.*, **25**, 253 (1974).
34. H. R. Withers, N. Hunter, H. T. Barkley and B. O. Rerd, *Radiat. Res.*, **57**, 88 (1974).
35. S. Lesher and J. Bauman. *Nat. Cancer Inst. Monogr.*, **30**, 185 (1969).
36. R. F. Hageman, C. P. Sigdestad and S. Lesher, *Radiat. Res.* **46**, 533 (1971).
37. R. F. Hageman, C. P. Sigdestad and S. Lesher, *Radiat. Res.*, **50**, 583 (1972).
38. H. R. Withers and M. M. Elkind, *Radiat. Res.*, **38**, 598 (1969).
39. O. S. Frankfurt, *Exp. Cell. Res.*, **46**, 603 (1967).
40. Kuang Y. Chen and H. R. Withers, *Int. J. Radiat. Biol.*, **21**, 521 (1972).
41. J. D. Thrasher, *Anat. Rec.*, **157**, 621 (1966).
42. T. L. Phillips, in *Conference on Time and Dose Relationships in Radiation Biology as Applied to Radiotherapy*, Carmel, California 1969 pp. 194–1999. BNL (C–57), Brookhaven National Laboratory, Upton, N.Y. 1970. (Available from Clearinghouse for Federal Scientific and Technical Information, Springfield, Va. 22151.)
43. L. F. Fajardo and J. R. Stewart, Lab. Invest. **29**, 244 (1973).
44. I. F. Tannock and S. Hayashi, *Cancer Res.*, **32**, 77 (1972).
45. S. B. Field and S. Hornsey, these Proceedings.
46. W. M. Wara, T. L. Phillips, L. W. Margolis and V. Suit, *Cancer*, **32**, 547 (1973).
47. E. Glatstein, R. C. Brown, G. D. Zanelli and J. F. Fowler, *Radiat. Res.* **61**, 417 (1975).
48. S. Hayashi and H. D. Suit, *Radiology*, **101**, 181 (1971).
49. H.A.S. van den Brenk, C. Sharpington, C. Orton, and M. Stone, *Int. J. Radiat. Biol.*, **25**, 277 (1974).
50. H. B. Reinhold, G. H. Buisman, *Brit. J. Radiol.* in press, (1975).

Cell Proliferation Changes in Hemopoietic Tissue As a Result of Irradiation or Drug Administration: The Control of Cell Proliferation in Hemopoietic Tissue[1]

B. I. Lord
Paterson Laboratories
Christie Hospital and Holt Radium Institute,
Manchester M20 9BX, Great Britain

CELL PROLIFERATION IN THE BONE MARROW

The hemopoietic system is largely divisible into two major classifications of cells; first the stem cell compartments which comprise the pluripotent and the committed precursor cells and second, the morphologically recognizable maturing cells which lead directly to the functional end cells. Normally these cells are all held in balance, and some degree of cell proliferation is maintained at each level. Controls over cell proliferation might therefore be expected to operate at any point throughout the sequence of development.

It is well recognized that irradiation and many drug treatments will deplete one or more of these compartments. Recovery to meet the animals' requirements, therefore, might be effected by increased proliferation of the maturing cells without an effect on the stem cells or by increased input from the stem cell compartment without affecting the normal rates of maturation or, more probably, by changing the proliferation characteristics of both the stem and maturing cells.

THE STEM CELL COMPARTMENT

The pluripotential stem cell is normally defined by the spleen colony assay test (*1*) as a colony forming unit (CFUs) while the stem cell committed to granulocyte production is defined by an in vitro colony test (*2, 3*) as a CFUc, and that committed to erythrocyte production is the erythropoietin responsive cell (ERC) defined by its ability to produce erythrocytes in the polycythemic mouse (*4*).

CFUs are found primarily in the bone marrow but they are not randomly distributed in the marrow spaces with respect to the total population. By fractionating femoral bone marrow longitudinally, we were able to demonstrate a square law distribution of CFUs across the marrow cavity (*5, 6*), many more CFUs lying close to the bone than along the femoral axis. CFUc, too, showed a well defined distribution which indicated their descendency from CFUs (*6*).

[1]Congress presentation C–3–2.

These distributions have some relevance to the response of the bone marrow to irradiation in that X-rays and neutrons produce a non-uniform dose absorption across the marrow space and will therefore change the CFUs distribution while γ-rays, which do give a uniform dose across the marrow, will not. Subsequent proliferation of the cells lies along the bone surface (*7, 8*) so that a second dose of radiation, depending upon its quality, may produce entirely different effects on the CFUs from the first one (*5*).

CFUs are largely non-proliferating (i.e. most cells are in G_o) (*9, 10*) or at least are very slowly turning over (i.e. long cycle time -t_c) (*11*). Suicide with tritiated thymidine (^{3}HTdR) (*12, 13*) indicates that less than 10% of CFUs are in DNA synthesis (S) under normal conditions. Since one cannot recognize CFUs morphologically, however, it is difficult to measure their cycle time and probably the only legitimate attempt can be ascribed to Vassort et al. (*14*) who used hydroxyurea (HU) or ^{3}HTdR to induce a partial sychronisation of CFUs and then measured their progression through S by the ^{3}HTdR suicide method. They determined a cycle of ~8 hrs with DNA synthesis lasting 4 hours. It must be remembered, however, that the system was initially disturbed in order to make the measurement so its validity is not certain. At the same time, since the number of cells in S is small, the disturbance is also small and 8 hours may therefore be a reasonable approximation. Lajtha et al. (*15*) calculated t_c = 6 hours but in a rapidly regenerating condition following transplantation so this may represent the shortest potential cycle time. Blackett on the other hand (*16, 17*) estimated t_c for cells with repopulating ability, cells which have commonly been considered analogous to CFUs, to be ~29 hours. It is probable that this technique includes G_o cells and at least some of the committed stem cells, and gives a turnover time rather than a cycle time for all the repopulating cells.

It seems likely, therefore, that most of the CFUs exist in some sort of G_o state while a small proportion of them are in cell cycle at any one time, and our recent work on the distribution of CFUs in the femoral marrow cavity has indicated that the majority of these cycling cells resides close to the bone surface (*6*). That the G_o phase does not occur before entering cycle at G_1, however, was demonstrated both in vivo and in vitro by radiation treatment (*18, 19*) and by hormone drug treatment (*20*). Within 10 minutes to 1 hour after treatment, cells not previously in DNA synthesis can be detected by ^{3}HTdR suicide measurements. Reports by Smith et al. (*21, 22, 23*) using vinblastine, vincristine, colchicine and endotoxin in conjunction with irradiation also demonstrated this readiness of the cells to enter a proliferative phase, and it appears therefore that the CFUs are probably held in a G_o located at the end of the G_1 phase and are ready to be triggered into S very quickly.

RECOVERY FROM DEPOPULATION OF THE CFUs COMPARTMENT

Irradiation and most standard drug treatments can initially only deplete CFUs numbers. Any depletion must be followed by recovery if the damage is not so great as to be lethal, i.e. there must be increased net proliferation. One sees this occurring in many situations, three of which are illustrated in Fig. 1. Following 300 rads X-rays, CFUs survival is initially 3-4% of control followed by a postirradiation dip to 2% at 1 day and then recovery to ~30% at 6 days

(*24*). one hundred rads X-rays initially leaves 30% survival of CFUs followed by a dip to 10% and recovery to ~30% at 3 days (Schofield, personal communication). On the other hand, isopropyl methane sulphonate (IMS), an alkylating agent, also produces an initial depopulation to ~30% of control but this is followed by a prolonged dip to <1% at 5 days before recovering to 30% at 12 days (*25*). So depletion is followed by recovery but the time scale of recovery does not depend solely upon the initial depopulation.

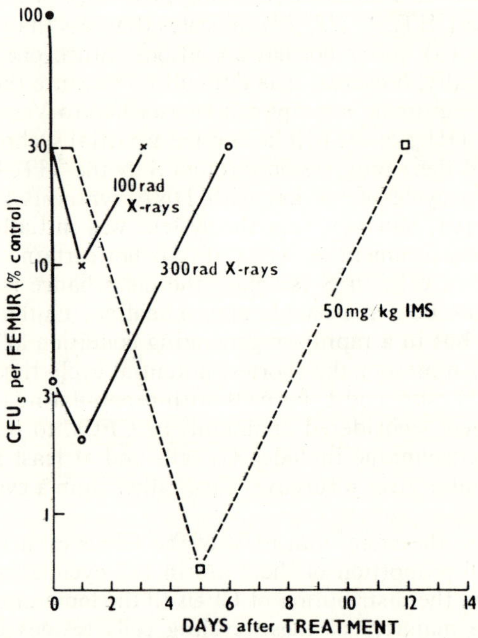

Fig. 1. Diagrammatic representation of the changes in femoral CFUs numbers following various treatments: (i) 100 rad X-rays from R. Schofield (personal communication), (ii) 300 rad X-rays from Guzman and Lajtha (*24*), (iii) 50 mg/kg isopropyl methane sulphonate from Schofield and Lajtha (*25*).

Guzman and Lajtha (*24*) showed that during the recovery phase, the net proliferation of CFUs is increased. Following 300 or 450 rad X-rays, the CFUs population grows with a doubling time of 27 hours in the femur, ~50% of them being in S. Subsequently, the proportion of cells in S decreases with the slowing growth rate as recovery becomes more complete. Frindel et al. (*26*) measured the D_o of CFUs after a primary dose of 250 rad and, on the grounds that D_o increases as a result of the loss of S phase cells, and vice versa, demonstrated the same fact.

One interesting factor associated with this recovery is that a complete recovery of CFUs to normal takes a long time. It is a common finding that when CFUs in the bone marrow have reached ~30% of normal, subsequent recovery is extremely slow (e.g. *24, 27, 28*). Schofield (personal communication) has shown that at doses as low as 50 or 100 rads, CFUs stay at about 30-40% control for at least 2 to 3 weeks. The rate of recovery, however, depends not only on the proliferation rate of CFUs but also on the rate of differentiation into the committed cell compartments. It has been shown, for example, that normal levels of erythropoiesis can be maintained from the very low levels of stem cells which survive during continuous irradiation (*29*). Lajtha et al. (*15*) showed also that when the CFUs population is proliferating at its maximum rate, continuous irradiation at a dose rate of 70 rads/day does not kill off any more CFUs. Proliferation and differentiation from the CFUs to the committed precursor cells are then exactly balanced. The low level of CFUs is capable of maintaining its own numbers while supplying the full requirements for differentiation. Chervenick and Boggs (*30*), using a split-dose irradiation technique, found that 90% reduction in the CFUs compartment represented something of a critical value in determining the nature of the response of CFUs. Below this level, CFUs do not differentiate. Self-replication is the only response operating. Once beyond the 10% threshold then, according to the requirements of the animal, a balance of CFUs differentiation and self replication is maintained. If further damage to the CFUs compartment is produced, such as under conditions of continuous irradiation, differentiation may account for all the production over the current level. If no further damage is imparted, differentiation may not account for all the CFUs production and hence the CFUs population can recover.

Some change can take place, therefore, in the control of CFUs proliferation which governs whether the population shall quickly regenerate or whether the return to normal is of secondary importance to the animal. It is clear that CFUs can be triggered directly into DNA synthesis but, on the other hand, Vassort et al. (*31*) and Pouillart and Huong (*32*) demonstrated a difference in action between phase-dependent drugs and irradiation, hormonal compounds and BCG. They found that irradiation, metonolone and BCG all trigger CFUs directly into S while with phase-dependent drugs, such as vincristine, HU and cytosine arabinoside, the time at which CFUs enter S is dependent on the type of drug and its dosage and that it is a phenomenon secondary to a fall in the number of committed stem cells.

One can obtain important information on the nature of the processes controlling these different effects on the CFUs from the work of Croizat et al. (*33*) and Gidali and Lajtha (*34*). Both groups used partial-body irradiation and studied depopulation, recovery and proliferation in both the shielded and irradiated parts of the body. The latter group was able to show that both local and humoral factors were acting to influence CFUs turnover during the subsequent recovery periods, and it appeared that depopulation has to be greater than a certain, perhaps critical, level before local regulation factors become operative. They concluded that it is differentiation of CFUs from the shielded areas which increases the demands on the CFUs, reducing their numbers and bringing the local controls into operation. Guzman and Lajtha (*24*) demonstrated this effect directly by stimulating differentiation along the erythroid pathway with erythropoietin, thus increasing the CFUs turnover. Mobilization of CFUs from the marrow such as described by Hanks (*35*) would have the same effect though this is probably not as important a factor as was previously thought (*34, 19*). Control over the initial differentiation processes on the other hand is more probably effected by long-range acting humoral factors such as erythropoietin. Thus, depletion of the maturing compartments, directly as a result of the treatment or indirectly as a result of a requirement for the production of end cells, causes the long-range humoral feedback control factors to increase differentiatiation and, eventually, proliferation of CFUs (see *14*).

The form taken by local control factors is unknown, but there is some evidence to suggest that they may take the form of a cell-to-cell interaction process. As such, Lord and Schofield (*36*) showed that live thymus cells can augment spleen colony formation in a specific manner under conditions in which the bone marrow populations have been damaged by sublethal irradiation. Frindel and Croizat (*37*) have also produced evidence of a similar cell-cell interaction phenomenon involving CFUs and thymus cells.

Having established a recovery process in the CFUs compartment, it has already been shown that recovery to normal becomes extremely protracted after ~30% levels are reached. Following depopulation, red cell production and the ERC population recover faster from radiation damage than the CFUs or repopulating cells (*29, 38*). Thus, when ERC are back to normal, differentiation pressure on the CFUs is released and CFUs turnover declines. This seems to occur at a time well before the CFUs population has fully recovered. In addition, there is now some evidence for a parallel situation in the CFUc compartment (Testa et al., personal communication). CFUc recovery starts later than CFUs recovery but is faster, returning to normal at a time when CFUs numbers are still very low.

Continuous irradiation at low dose rates can give further insight into the sensitivity of proliferation control to depopulation. Blackett and his colleagues (*29, 39*) showed that the repopulating ability of rat bone marrow is depressed by 45 rads/day continuous irradiation to 10% of normal, but that normal erythropoiesis is maintained. Lajtha et al. (*15*) then showed that under these conditions, sterilization of CFUs takes the form of loss of reproductive integrity and consequently, because they are only slowly proliferating, show little numerical change during the first few days, even up to ~90% sterilization levels. Only after about 5 days does the animal notice this depopulation and CFUs then start to proliferate. Even then, the population is not turning over to its full capacity. Blackett (*16*) reported that when the new steady state is set up, the repopulating cells have a cycle time of ~12 hours which is approaching the estimates for CFUs.

An analogous argument may explain the long slow fall of CFUs following the initial depopulation caused by IMS (*25*). In this case the secondary drop was from 30% to 1% over a period of 5 days (Fig. 1). A similar effect is probably seen for other alkylating agents, e.g. myleran (Josvasen, *40*). During this depopulation period, Schofield was unable to detect any significant increase in CFUs proliferation. Again, one has a situation where probably the drug has destroyed the reproductive integrity of the CFUs while not otherwise affecting it. Then, because of the normal slow turnover of CFUs, it is some days before the animal recognizes any significant run-down of the CFUs population.

In these considerations of the way in which CFUs can respond to depopulation, no consideration has so far been given to the role of the host environment. It is well known that in transplantation situations, microenvironmental factors may govern the direction in which stem cells may differentiate (*41, 42*), and two recent pieces of work by Hendry and his colleagues (*28, 43*) have shown how environmental changes resulting from irradiation may affect the proliferation of CFUs. They showed that CFUs regenerate with a doubling time of 24 hours following a single dose of 450 rads X-rays. After 450 rads X-rays repeated 4 times at 24 day intervals, they grow with a doubling time of 42 hours. That this was an environmental effect was clearly demonstrated by the fact that the cells grew at their normal fast rate if transplanted into a singly-irradiated mouse. At the same time, however, this repeated irradiation also produced some damage to the cells themselves. In the femur they were capable of growing up to about only 15% of normal, even on transplantation to a singly-irradiated host, before reaching a plateau. Furthermore, although they appeared incapable of increasing their numbers further, they were still in a state of rapid proliferation.

It is worth, at this point, introducing just one more of the complexities of the bone marrow cell populations. It is known that the CFUs in splenic colonies are capable of recolonizing further animals (*44*). Using this principle, Schofield assayed the number of CFUs/colony and accordingly defined two categories of CFUs: those with self-maintaining ability, CFUs-1 and those without, CFUs-2 (*26*). The CFUs-1 were those initially sterilized by IMS. The CFUs-2 were largely unaffected but since their supply line (i.e. the CFUs-1) was removed, the CFUs-2 simply run down until the depopulation is sufficient to stimulate the surviving CFUs-1 into proliferation. Acute irradiation has a general effect on all these CFUs, but drug treatments and, possibly, chronic irradiation regimes may be more specific in their effects on the cells and thus account for the different patterns of regeneration seen.

THE MATURATION COMPARTMENTS

The maturation compartments of the hemopoietic cell populations, comprising, in the main, the morphologically recognizable cells are the more completely understood parts of hemopoiesis, much of the information on the capacity of the cells to modify their proliferation characteristics coming initially from continuous irradiation studies and later from a chronic phenyl-hydrazine (PH)-induced anaemia. It was demonstrated that during a continuous daily X-ray dose regime of 50 or 84 rads/day, the erythroid and granulocytic cells in rat bone marrow compensate the depleted stem cell levels by increasing their net proliferation rates. This is achieved by increasing the number of cell cycles during maturation and by shortening the average cell cycle times, thus leading to a new enhanced level of steady state cell proliferation (*45, 46, 47*). As a result, more functional cells are produced thus maintaining the level of, for example, red cells in the circulating blood (*29, 39*). These observations have since been firmly established and extended with more refined measurements as described by Lamerton (*48*) who also cited Hanna et al. (*49*) to show a considerable reduction of the cycle times of these cells in response to a severe PH-induced hemolytic anemia.

The exact nature of the proliferative response depends, however, on the nature of the disturbance to the system. If the disturbancc represents a removal of end cells, such as induced by PH (*49*) or by bleeding (*50*), then one obtains increased proliferation of the maturing cells, but there is also the dependent stimulation of the earlier precursors, as discussed above, to introduce more differentiated cells into the system. A chronic stress, such as that given by low dose rate continuous irradiation, primarily depletes the stem cell compartments while the more rapidly proliferating recognizable cells are less affected. These cells then make up the deficiency of the stem cell supply by a greater output from this compartment.

The nature of the control processes operative on these cells is not completely understood. Erythropoietin has long been known as a direct stimulator of erythropoiesis at all levels. A similar compound has long been sought (unsuccessfully) to stimulate granulopoiesis. Currently the role of specific proliferation inhibitors of erythropoiesis (*51*) and granulopoiesis (*52*) are now attaining more prominence. In this respect, Patt and Maloney (*53*) demonstrated an inverse relationship of cell concentration in the rabbit femur and the uptake of tritiated thymidine by the cells, and we have now established that extracts of mature blood cells do have specific effects on developing hemopoietic cells which are compatible with proliferation inhibition (*54*) and which are completely reversible (*55*). Our current studies are showing that, used in vivo, these extracts are in fact capable of lowering the proliferation rates of the maturing hemopoietic cells (Lord- unpublished results). It is clear, therefore, that the maturing cell populations proliferate under a complex set of control processes.

REFERENCES

1. J. E. Till and E. A. McCulloch, *Radiat. Res.* **14,** 213 (1961).
2. T. R. Bradley and D. Metcalf, *Austr. J. Exp. Biol. Med. Sci.* **44,** 287 (1966).
3. D. H. Pluznik and L. Sachs, *J. Cell Comp. Physiol.* **66,** 319 (1965).
4. C. W. Gurney, L. G. Lajtha and R. Oliver, *Brit. J. Haemat.* **8,** 461 (1962).
5. B. I. Lord and J. H. Hendry, *Brit. J. Radiol.* **45,** 110 (1972).
6. B. I. Lord, N. G. Testa and J. H. Hendry, Submitted for publication, *Blood,* (1974).
7. M. Tavassoli and W. H. Crosby, *Science* **161,** 54 (1968).
8. M. A. Maloney and H. M. Patt, *Cell Tissue Kinet.* **2,** 29 (1969).
9. L. G. Lajtha, R. Oliver and C. W. Gurney, *Brit. J. Haemat.* **8,** 442 ((1962).
10. L. G. Lajtha, *J. Cell Comp. Physiol.* **62 (Suppl. 1)** , 143 (1963).
11. J. W. Byron, *Nature (London)* **228,** 1204 (1970).
12. A. J. Becker, E. A. McCulloch, L. Siminovitch and J. E. Till, *Blood,* **26,** 296 (1965).
13. B. I. Lord, L. G. Lajtha and J. Gidali, *Cell Tissue Kinet.* In press (1974).
14. F. Vassort, M. Winterholer, E. Frindel and M. Tubiana, *Blood* **41,** 789 (1973).
15. L. G. Lajtha, C. W. Gilbert and E. Guzman, *Brit. J. Haemat.* **20,** 343 (1971).
16. N. M. Blackett, in *Effects of Radiation on Cellular Proliferation and Differentiation,* p. 235. IAEA Vienna, 1968.
17. N. M. Blackett, *J. Nat. Cancer Inst.* **41,** 909 (1968).
18. S. K. Lahiri and L. M. van Putten, *Cell Tissue Kinet.* **5,** 365 (1972).
19. H. Croizat, E. Frindel and M. Tubiana, Submitted for publication, *Radiat. Res.* (1974).
20. J. W. Byron, *Nature (London)* **234,** 39 (1972).
21. W. W. Smith, G. Brecher, S. Fred and R. A. Budd, *Radiat. Res.* **27,** 710 (1966).
22. G. Brecher, W. W. Smith, S. Wilson and S. Fred, *Radiat. Res.* **30,** 600 (1967).
23. W. W. Smith, S. M. Wilson and S. S. Fred, *J. Nat. Cancer Inst.* **40,** 847 (1968).

24. E. Guzman and L. G. Lajtha, *Cell Tissue Kinet.* **3,** 91 (1970).
25. R. Schofield and L. G. Lajtha, *Brit. J. Haemat.* **25,** 195 (1973).
26. E. Frindel, F. Charruyer, M. Tubiana, H. S. Kaplan and E. L. Alpen, *Int. J. Radiat. Biol.* **11,** 435 (1966).
27. F. A. Valeriote, D. C. Collins and W. R. Bruce, *Radiat. Res.* **33,** 501 (1968).
28. J. H. Hendry and L. G. Lajtha, *Radiat. Res.* **52,** 309 (1972).
29. N. M. Blackett, P. J. Roylance and K. Adams, *Brit. J. Haemat.* **10,** 453 (1964).
30. P. A. Chervenick and D. R. Boggs, *Blood* **37,** 568 (1971).
31. F. Vassort, E. Frindel and M. Tubiana, *Bull. Cancer* **60,** 163 (1973).
32. P. Pouillart, T. Houng Thi Huong, *Exp. Haemat.* **1,** 263 (1973).
33. H. Croizat, E. Frindel and M. Tubiana, *Int. J. Radiat. Biol.* **18,** 347 (1970).
34. J. Gidali and L. G. Lajtha, *Cell Tissue Kinet.* **5,** 147 (1972).
35. G. E. Hanks, *Nature (London)* **203,** 1393 (1964).
36. B. I. Lord and R. Schofield, *Blood,* **42,** 395 (1973).
37. E. Frindel and H. Croizat, *Biomed. Express* **19,** 392 (1973).
38. D. D. Porteous and L. G. Lajtha, *Brit. J. Haemat.* **12,** 177 (1966).
39. N. M. Blackett, *Brit. J. Haemat.* **13,** 915 (1967).
40. N. Josvasen and A. Boyum, *Scand. J. Haemat.* **11,** 78 (1973).
41. N. S. Wolf and J. J. Trentin, *J. Exp. Med.* **127,** 205 (1968).
42. N. S. Wolf, *Cell Tissue Kinet.* **7,** 89 (1974).
43. J. H. Hendry, N. G. Testa and L. G. Lajtha, *Radiat. Res.* **59,** 645 (1974).
44. R. Schofield, *Cell Tissue Kinet.* **3,** 119 (1970).
45. B. I. Lord, *Brit. J. Haemat.* **10,** 496 (1964).
46. B. I. Lord, *Brit. J. Haemat.* **11,** 130 (1965).
47. B. I. Lord, in *Effects of Radiation on Cellular Proliferation and Differentiation,* p. 247. IAEA, Vienna, 1968.
48. L. F. Lamerton, in *XIII Int. Cong. Hemat., Munich,* p. 132 (1970).
49. I. R. A. Hanna, R. G. Tarbutt and L. F. Lamerton, *Brit. J. Haemat.* **16,** 381 (1969).
50. B. I. Lord, *Brit. J. Haemat.* **13,** 160 (1967).
51. E. Kivilaasko and T. Rytomaa, *Cell Tissue Kinet.* **4,** 1 (1971).
52. T. Rytomaa and K. Kiviniemi, *Cell Tissue Kinet.* **1,** 329 (1968).
53. H. M. Patt and M. A. Maloney, *Cell Tissue Kinet.* **5,** 303 (1972).
54. B. I. Lord, L. Cercek, B. Cercek, G. P. Shah, T. M. Dexter and L. G. Lajtha, *Brit. J. Cancer* **29,** 168 (1974).
55. B. I. Lord, L. Cercek, B. Cercek, G. P. Shah and L. G. Lajtha, *Brit. J. Cancer* **29,** 407 (1974).

The Importance of Proliferation Kinetics and Clonogenicity of Tumor Cells for Volume Responses of Experimental Tumors after Irradiation[1]

A. F. Hermens and G. W. Barendsen
Radiobiological Institute TNO, 151 Lange Kleiweg, Rijswijk (ZH), The Netherlands

INTRODUCTION

Curative treatments in cancer therapy aim at the ultimate destruction of the reproduction capacity[2] of all of the malignant cells without inducing permanent damage to the integrity and function of indispensable normal tissue. Consequently, in order to develop better therapeutic methods for improving the cure rates of cancer cases with poor 5-year survival rates, the therapeutic effects on both malignant and normal tissue must be evaluated. However, techniques for the direct measurement of cellular responses such as the time-course variations in the fraction of clonogenic cells in clinical tumors after therapy, among others, are not yet available. Up to now, this kind of information has been obtained in a sufficiently large quantity only for specially designed experimental tumor systems (*1–3*), a number of irradiated normal tissues in rodents (*4–7*), and a single approximate value has been obtained for irradiated human skin (*8*).

The radiation responses in patients are commonly evaluated from data on time-course variations in volume and in growth rate after therapy (*9–12*). Such data may be used to estimate the fractional surviving[3] clonogenic cells only for the theoretical situation such as is illustrated in Fig. 1, panel A. This is the situation represented by curve x_1 where all of the cells which have lost their reproductive capacity are eliminated at once, immediately after irradiation. The surviving cells then start to regrow the tumor without mitotic delay and without alterations in the cell proliferation characteristics of tumor growth. The fraction of surviving cells can then be derived directly from the residual volume observed after irradiation.

More realistic possibilities for interpreting radiation-induced changes in tumor growth are represented by the curves x_2, x_3, and x_4 which are shown in panel B. Volume responses which correspond to these curves are more difficult to

[1] Congress presentation C–3–3.

[2] In the further discussion, cells which have the capacity for unlimited reproduction will be designated as clonogenic cells.

[3] In the further discussion, the terms "surviving cells" and "nonsurviving cells" will refer to cells which maintain their clonogenic capacity and to cells which lose this capacity, respectively, after irradiation.

interpret and require information on different aspects of cellular responses including: 1. the fraction of surviving cells; 2. the duration of mitotic delay; 3. alterations in the rates of cell production and cell loss; and 4. alterations in the number of cells per unit of tissue volume.

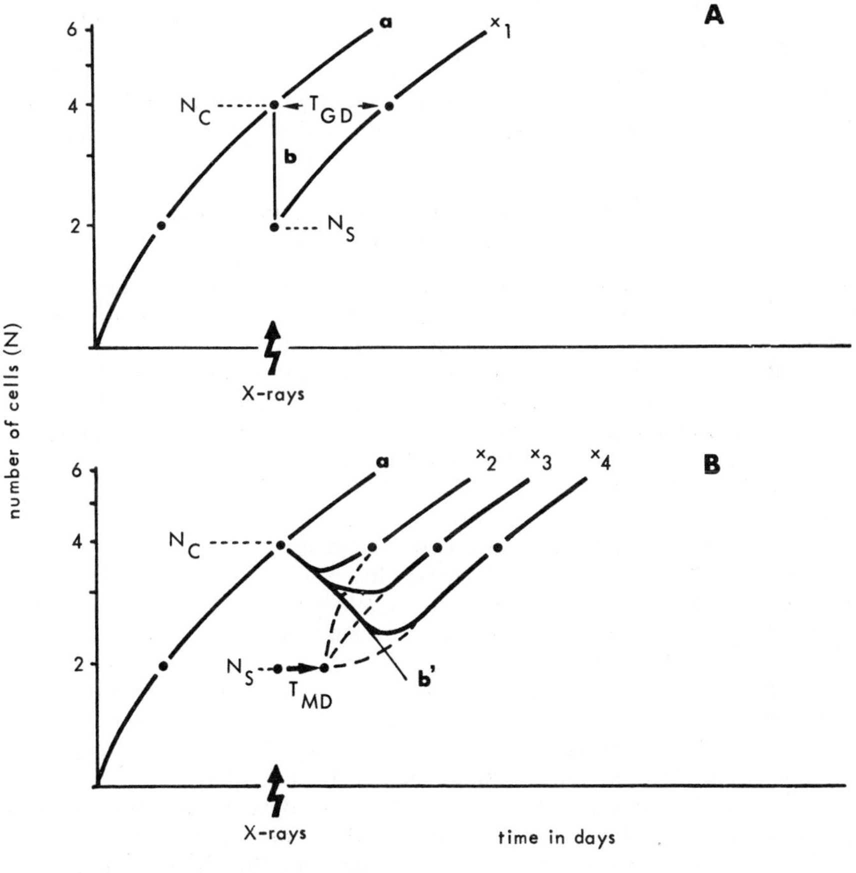

FIG. 1. Hypothetical changes in tumor volume after irradiation, assuming various rates for elimination of nonsurviving cells and for proliferation of surviving cells.

N_C: number of cells present in unirradiated controls;

N_S: number of surviving cells;

T_{GD}: duration of growth delay;

T_{MD}: duration of mitotic delay;

Arrow: X-irradiation;

Curve a: growth curve of unirradiated tumors;

Curves b and b′: represent acute (see: panel A) and gradual (see: panel B) loss of nonsurviving cells, respectively, after irradiation.

Curves $x_{1,2,3,4}$: represent different modes of regrowth of irradiated tumors. Curves x_1 and x_3 represent regrowth when resumption of cell proliferation by surviving cells takes place at a rate which is identical to that observed in unirradiated tumors containing N_S cells. Curves x_2 and x_4 represent regrowth when the cell production is resumed at larger and smaller rates, respectively, as is observed in unirradiated tumors of volume N_S.

It is clear that the number of surviving cells, N_s, cannot be derived either directly from the residual volume or indirectly by extrapolation from any of the curves shown in panel B.

The aim of this paper is to discuss several aspects of tumor cell kinetics in relation to cellular and "clinical" radiation responses. First, some parameters and theoretical models involved in the analysis of tumor growth will be briefly outlined. Subsequently, a review of data on radiation results obtained with a number of experimental tumors described in the literature and with the transplantable R-1 rat rhabdomyosarcoma will be given. These data will then be used to correlate volume changes with various cell proliferation parameters.

CELL KINETIC PARAMETERS AND MODELS FOR TUMOR GROWTH

It is important to note that models for solid tumor systems are considered in the discussion which follows. Changes in the volume of these tumors are supposed to result mainly from changes in cell number.

Data on cell kinetic parameters can be obtained experimentally by the method of pulse labelling of DNA-synthesizing tumor cells in vivo with ^{3}H-TdR (*13,14*). Autoradiographs are prepared from the labelled tumors for microscopic examination (*15*). By these combined techniques, it is possible to determine the labelling index, LI, and the per cent labelled mitoses, PLM, immediately after labelling and at subsequent time intervals after labelling, respectively.

Data on per cent labelled mitoses are plotted as a function of time and from these curves, designated as PLM curves, the median cell cycle time, T_C, the durations of the DNA synthesis period, T_S, and the relative position in the cell cycle of the latter period can be derived. This is done either directly from the PLM curve (*16,17*) or by computer (*18,19*).

The LI value determined for a given cell population shortly after pulse labelling with ^{3}H-TdR depends on the following cell kinetic parameters: 1. the fraction, α, of proliferating (P) cells produced per dividing cell; 2. the duration of T_C; and 3. the duration and timing of the DNA synthesis period (*20*).

These cell kinetic parameters are closely associated with mathematical models which are used for studying tumor growth. In principle, the mathematical model of choice is one which can properly describe the observed increase in tumor cell number on the basis of these parameters and which is as simple as possible. For this purpose, a number of models of increasing complexity should be considered, namely:

a. Model A: $N_t = N_0 \cdot e^{(t/T_C) \cdot \ln 2}$

where N_t and N_0 represent the number of tumor cells present at times t and t = 0, respectively. This model can describe the increase in cell number when: 1. all of the tumor cells are engaged in proliferation with an equally long cell cycle time of constant duration; 2. all cells start dividing in a random fashion; and 3. no cell loss is taking place. A satisfactory method of determining whether the model can indeed predict the actual tumor growth is by showing that the value for the tumor volume doubling time, T_d, is equal to that for the potential doubling time, $T_{d(Pot)}$

of the model[4] (*20*). From the mathematical expression of model A, it can be shown that $T_{d(Pot)} = T_C$. Consequently, model A is applicable when $T_d = T_C$; this has been observed mainly for cultured cells growing at low cell density (*22–24*).

b. For many experimental tumors, the duration of T_d is considerably longer than that of T_C, indicating that not all of the cells are engaged in cell proliferation in such tumors (*25,26*) and/or that cell loss is taking place (*27,28*). For these tumors, the increase in the number of proliferating cells may be described, to a first approximation, by model B (*20*):

$$\text{Model B: } N_t = N_0 \cdot e^{(t/T_C)\cdot \ln\alpha}$$

In this model, the dividing cells produce proliferative (P) cells as well as nonproliferative (Q) cells; the average numbers of P and Q cells produced per division are equal to α and $(2-\alpha)$, respectively, while $1 < \alpha \leqslant 2$. The potential doubling time (see footnote 4) corresponding to model B can be specified by equation 1:

$$T_{d(Pot)} = (\ln 2/\ln\alpha)\cdot T_C \qquad (1)$$

c. It frequently appears that, for a given stage of tumor growth, $T_d > T_{d(Pot)}$, implying that the calculated rate of increase in cell number still exceeds the actually measured rate of increase. However, equation 1 specifies the potential doubling time for a cell population for which it is assumed that every cycling cell should also succeed in producing two daughter cells at mitosis. Recent studies have indicated that cycling cells may be halted in the post-DNA-synthesis phase (*29*) or may disintegrate at mitosis (*30*). The increase in the number of proliferating cells for a population of P and Q cells in which a fraction, β, of cycling cells disintegrate at mitosis can be described by:

$$\text{Model C: } N_t = (1 - \beta/\alpha)\cdot N_0 \cdot e^{(t/T_C)\cdot \ln\alpha}$$

The formulation of this model is also based on the knowledge that in the R-1 rhabdomyosarcoma the cycle time and phase duration of the P cells are almost identical to those of cycling cells which die at mitosis; i.e., cells designated as terminal (T_e) cells which are completing their "terminal" cycle; e.g., Fig. 2.

According to model C, the the potential doubling time is specified by equation 2:

$$T_{d(Pot)} = (\alpha/\alpha - \beta)\cdot(\ln 2/\ln\alpha)\cdot T_C \qquad (2)$$

Equation 2 allows for a better approximation of the actual growth rate of a tumor containing P, Q, and Te cells than does equation 1. This can be illustrated by comparing the values for T_d and $T_{d(Pot)}$ which were derived for the natural growth of the R-1 rhabdomyosarcoma. From the data summarized in Table 1, it can be seen that, for tumors measuring 300 mm^3, the value for T_d is practically

[4] T_d defines the time that a tumor of a given volume, V_t, needs to double in size if the volume had increased exponentially in size at time t. This implies that, for tumors which do not grow exponentially (*21*), T_d is represented by the tangent, at volume V_t, to the growth curve obtained by plotting ln V_t as a function of time t.

$T_{d(Pot)}$ defines for a given mathematical model the time during which the number of cells increases from N_t to $2 \times N_t$ when there is no cell loss.

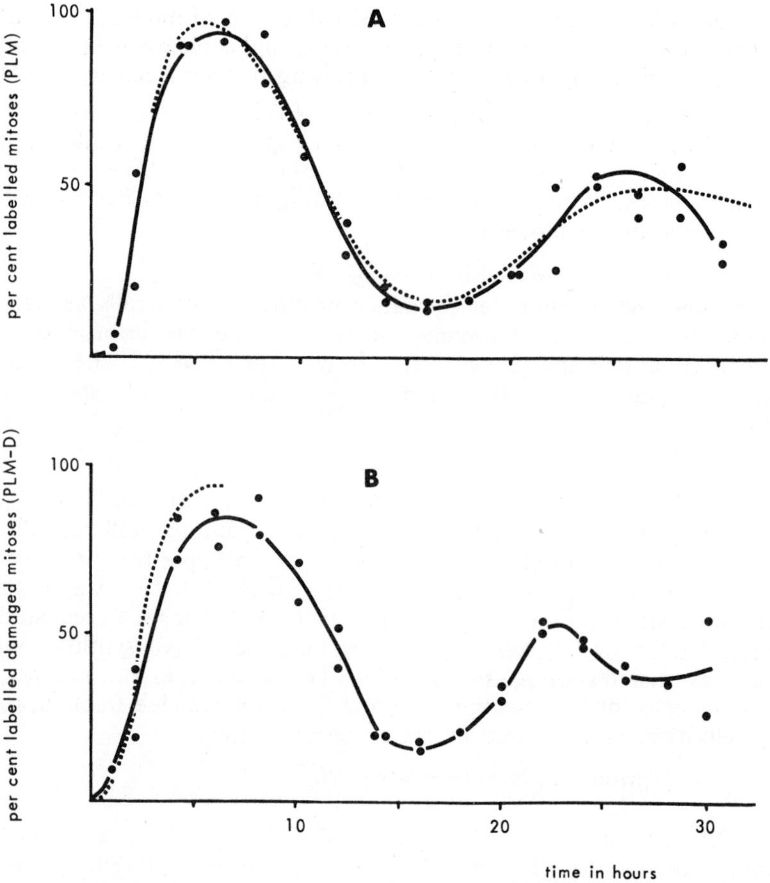

FIG. 2. Examples of variations with time in the per cent labelled intact mitoses (PLM curve shown in panel A) and in the per cent labelled aberrant mitoses (PLM-D curve shown in panel B) after injection of ^{3}H-TdR. The data were obtained for R-1 tumors measuring 1 to 5 mm^{3}. Each point on the solid curves represents the data calculated from a total of 100 labelled and nonlabelled mitoses of either type.

In panel A, the broken line represents a curve calculated by computer (*18*). The calculations were kindly performed by Dr. G. G. Steel.

In panel B, the broken line represents the ascending part of the first wave shown in panel A.

equal to that calculated for $T_{d(Pot)}$ by use of equation 2. However, it appears that $T_d > T_{d(Pot)}$ when tumors grow larger than 300 mm^3. This difference may be attributed to cell loss from the Q cell population and/or from the P cell population caused by factors other than death at mitosis.

The percentage cell loss relative to the production of cells can be calculated according to equation 3 (*20*):

$$\phi = [1 - (T_{d(Pot)}/T_d)] \cdot 100 \qquad (3)$$

TABLE 1
Values Calculated for the Potential Doubling Time, $T_{d(Pot)}$, Associated with Natural Growth of R-1 Tumors

Tumor volume (mm^3)	T_C (h)	T_d (h)	(1) α	(1) β	(1) $T_{d(Pot)}$ (h)	(2) α	(2) β	(2) $T_{d(Pot)}$ (h)
1 to 5	19.6	>41.4* <60.0	1.510	0.000	33.0	1.485	0.253	41.4
300	20.5	50.4	1.450	0.000	38.2	1.410	0.261	50.7
1500** to 3000	20.5	87.6	1.300	0.000	54.2	1.322	0.195	59.7

T_C : median cell cycle time of proliferating, P, cells.
T_d : measured volume doubling time.
α : number of newly produced P cells per cell division.
β : fraction of cells in cycle that cease to divide at mitosis.
$T_{d(Pot)}$: potential volume doubling time calculated according to
equation 1: $T_{d(Pot)} = [\ln 2/\ln\alpha]\, T_C$ see under (1), and to
equation 2: $T_{d(Pot)} = [\alpha/(\alpha-\beta)] \cdot [\ln 2/\ln\alpha] \cdot T_C$ see under (2).
* : this large range in volume doubling time is due to the inaccuracy of measurements of T_d for very small tumors.
** : data calculated for the complete tumor
The small variations of values for α in each row under (1) and (2) result from averaging values of α calculated for P cells and Te cells separately.

Finally, it should be mentioned that the growth rate of most experimental tumors is not constant as is assumed in models A, B, and C, but that it decelerates (*21*). Consequently, the mathematical models required for simulating tumor growth by use of cell kinetic parameters may be much more complicated than is outlined above (*30,31*).

Results of various studies on cell proliferation in a number of rodent tumors have been published in the literature. These tumors were investigated by the methods outlined above at different stages during their natural growth (*22,30,32–34*). The results of these studies which are also important for reasons of comparison with cell proliferation in irradiated tumors have revealed the following general trends:

1. the median cell cycle time may show regional variation in a tumor but has a rather constant value for a given tumor throughout its growth;
2. the value of α decreases with increasing tumor size; and,
3. the per cent cell loss rises from values of about 0–30 per cent to values of about 90 per cent or more when the tumor volume increases from 100 mm^3 to 10,000 mm^3.

CELL PROLIFERATION IN IRRADIATED TUMORS

In this review, results of analyses of cell proliferation as reported in the literature for different types of tumors which were irradiated with single doses of

X-rays will be presented. The results include data obtained with the following tumors: 1. the NCTC-2472 fibrosarcoma transplantable in mice (*35,36*); 2. a chemically induced squamous carcinoma of the cheek pouch of the Syrian hamster (*37,38*); 3. a mammary adenocarcinoma transplantable in mice (*11*); 4. three fibrosarcomas transplantable in the rat, namely the RIB5, the SSO, and SSB1, plus the C3H mammary carcinoma transplantable in mice (*39*); 5. the R-1 rhabdomyosarcoma (*2,30,40*); and 6. the murine adenocarcinoma 282 transplantable in C3H mice (*41*).

The conclusions which can be drawn from the results for each individual type of tumor may be summarized as follows:

1. After a dose of 2000 rad of 250 kV X-rays, the NCTC-2472 fibrosarcoma showed a volume reduction of 90% which is compatible with an estimated fractional surviving cells of 4.6% and a 100% removal of the dead cells within 10 days after irradiation. It was observed that, 16 days after irradiation, the average doubling time during the recurrence of the tumor was shorter than the doubling time of unirradiated controls. This reduction in volume doubling time was attributed to an increase in the growth fraction, since no change was observed in cell cycle time after irradiation.
2. For squamous epithelium carcinomas of the cheek pouch of Syrian hamsters, it was observed that one day after irradiation with a single dose of 500 rad and three days after a single dose of 1000 rad of 250 kV X-rays, the median cell cycle time was increased by 0.7 to 2.1 hr and by 1.8 to 3.1 hr, respectively, as compared with the cycle time of 10.7 hr for cells in unirradiated controls. Data on volume reduction and growth delay were not reported.
3. Studies with the mammary carcinoma were performed by using female F_1 (C57BL × CBA) mice in which lung metastases were induced. The metastases were produced by intravenous injection of cells isolated from a poorly differentiated adenocarcinoma which was detected in estron-treated male C57BL mice. After irradiation of lung metastases with a dose of 210 rad of 250 kV X-rays, it was observed that: a. the volume was reduced by 80% within 48 hr and that regrowth after this period took place with a shorter volume doubling time than was observed in unirradiated controls; b. this shorter volume doubling time resulted in a growth delay of 90.0 hr; c. 2 days after irradiation, the median cell cycle time was reduced by 3.4 hr to a value of 11.1 hr; d. the value for α had increased from 1.87 to 2.00 as measured in unirradiated controls and 2 days after irradiation, respectively.
4. Cell kinetic parameters of the three sarcomas and the one carcinoma were studied before and after irradiation with doses of either 1000 or 1500 rad of 250 kV X-rays. The results revealed that: a. reduction in tumor volume occurred only in the C3H carcinoma; b. the growth was delayed in three tumors; namely in the RIB5 and the SSB1 sarcomas and the C3H carcinoma; the SSO sarcoma continued to grow at a reduced rate and without apparent volume reduction; d. the growth rate after the period of delay was also reduced in the three other tumor types; e. significant changes in cell cycle times were not observed; f. changes in values for α could not be demonstrated except in the SSB1 sarcoma where the value for α might have decreased.

It should be noted that, in the studies described so far, cell kinetic characteristics for irradiated tumors were analyzed at only one stage during recurrent growth. This was usually at the stage when the tumors had regrown to the volumes which were measured at the time of irradiation.

5. Cell proliferation analysis in the R-1 rhabdomyosarcoma were performed at diferent time-intervals after irradiation with a dose of 2000 rad of 300 kV X-rays. The observed radiation responses of main interest are the following: a. the volume decreases from day t = 2 after irradiation until day t = 8 when it reaches a minimum value equal to 75% of that measured at the time of irradiation; b. the number of cells per unit tumor volume, i.e., the cell density, decreases by a factor of 2 approximately. This maximum decrease is observed at day t = 8; c. the median cell cycle time in the irradiated R-1 tumors is 7 to 10 hr shorter mainly because of the disappearance of the G_1 phase; d. the average value of α decreases from 1.322 to 1.200 as measured in unirradiated controls and in irradiated tumors at day t = 14, respectively; e. the fraction of cycling cells which disintegrate during mitosis, represented by β, increases to a maximum at day t = 8. At this time, it is larger by a factor of 4, as compared to that observed in unirradiated controls; f. regrowth and increase in cell density begins after day t = 8 and results in a growth delay[5] of about 12 to 14 days; g. the time for doubling in tumor cell number at day t = 14 is equal to that of unirradiated controls: h. calculations of potential doubling times indicate that the regrowth of irradiated R-1 tumors after day t = 10 and for a subsequent period of at least 4 days can be described most adequately by equation 2.

This latter observation is clearly supported by data summarized in Table 2 in which measured values for T_d derived from the time course variation in tumor

TABLE 2
Values Calculated for the Potential Doubling Time, $T_{d(Pot)}$, Associated With Variations in the Growth Rate of R-1 Tumors Irradiated with a Single Dose of 2000 Rad of 300 kV X Rays

t (days)	change in tumor volume	T_C (h)	T_d (h)	α	β	$T_{d(Pot)}$ (h)
0	natural growth (controls)	20.5	87.6	1.322	0.194	59.7
4	volume reduction	11.4	−81.6	1.260	0.775	90.0
8	minimal tumor volume reached	12.2	0.0	1.283	0.779	87.7
10	regrowth	12.5	121.6	1.217	0.776	121.6
14	regrowth	12.9	89.6	1.200	0.543	90.0

T_C: median cell cycle time of proliferating, P, cells.
T_d: measured volume doubling time.
α: number of newly produced P cells per cell division.
β: fraction of cells in cycle that cease to divide at mitosis.
$T_{d(Pot)}$: potential volume doubling time calculated according to equation 2:

$$T_{d(Pot)}=[\alpha/(\alpha-\beta)]\cdot[\ln 2/\ln\alpha]\cdot T_C.$$

[5] The growth delay has been defined here as the time elapsing between irradiation and when the recurrent tumor has regained a volume equal to that measured at the time of irradiation.

volume after irradiation are compared with values calculated for $T_{d(Pot)}$ by use of equation 2. The data in Table 2 also indicate that the values for $T_{d(Pot)}$ in unirradiated controls are smaller than those for $T_{d(Pot)}$ measured in irradiated tumors at the time intervals indicated. From this it can be deduced that the production of cells in R-1 tumors irradiated with a dose of 2000 rad fluctuates within a range of 50 to 68 per cent of the production in unirradiated controls between days t = 4 and t = 14. The fact that the volume still decreases between days t = 4 and t = 8 is the result of the high rate of irradiation-induced cell loss which temporarily exceeds the rate of cell production.

6. The adenocarcinoma 282 was used to study cell kinetic parameters at different time intervals after irradiation with both 600 rad and 1200 rad of 300 kV X-rays. The results indicate that the duration of the cell cycle increases after irradiation. In unirradiated controls, T_C = 12.0 to 13.6 h, while at day t = 1 after 600 rad, and at day t = 8 after 1200 rad, maximum values are measured equal to T_C = 17.6 to 19.3 h and T_C = 19.8 h, respectively. On the other hand, the fraction of proliferating cells increases by a factor of about 2 at three-to-four days after irradiation in tumors of each experimental group.

This review clearly illustrates that volume responses to radiation are apparently different for different types of tumors. These differences presumably depend on type-specific differences in intrinsic radiosensitivity and/or differences in proliferation characteristics of various types of tumor cells.

CORRELATION BETWEEN CELL KINETICS AND TUMOR VOLUME RESPONSES

When no cure can be achieved in the clinical therapy of radioresistant tumors, it is obviously of interest to induce the largest possible delay in tumor growth. Thus, even for the administration of palliative therapy, it is of fundamental importance to investigate which of the various proliferation parameters are involved in obtaining this clinical radiation effect. In order to use the results obtained with the tumors reviewed above for this purpose, the observed growth delay may be expressed in terms of "specific growth delay per rad of radiation dose." The "specific growth delay" will be defined as the ratio of the observed growth delay and the doubling time of unirradiated controls at time t = 0 of irradiation (*30*). Table 3 summarizes the data on the fraction of proliferating cells, F_P, and on percentage cell loss, ϕ, as measured before and sfter irradiation for the tumors under discussion; these data are correlated with data on specific growth delay per 100 rad of dose. The results are illustrated in Fig. 3 and indicate that: 1. there is a significant positive correlation between the fraction of proliferating cells present in the unirradiated tumors and the specific growth delay; and, 2. there is a less significant positive correlation between radiation-induced cell loss and specific growth delay.

Although the amount of data is small, these correlations indicate that the radiation-induced delay per rad of dose will be longer the higher the fraction of proliferating cells present at the time of irradiation, and, to some extent, the delay will be longer the higher the per cent radiation-induced cell loss.

TABLE 3
Values for Growth Parameters of Different Experimental Tumors
Before and After Irradiation

Tumor histological type & host			NCTC-2472 fibrosarcoma mouse		R-1 rhabdomyosarcoma rat		poorly differentiated adeno-carcinoma mouse		RIB-5 fibrosarcoma rat		SSO fibrosarcoma rat		SSB 1 fibrosarcoma rat		C3H mammary carcinoma mouse	
Source			1		2*		3		4		5		6		7	
	d	(rad)	2000		2000		210		1500		1000		1000		1500	
	t	(days)	0	19	0	14	0	2	0	7	0	10	0	28	0	14
Unirradiated controls	F_P	(%)	30.0†		30.0		87.0		45.0		48.0		60.0		37.0	
	T_C	(h)	17.5		20.5		14.5		13.2		20.5		39.0		15.0	
	T_d	(h)	84.0		87.6		24.0		24.0		34.0		101.0		108.0	
	$T_{d(Pot)}$	(h)	46.0†		54.2		16.1		24.6		36.3		57.3		33.0	
	ϕ	(%)	45.0†		38.0		33.0		0.0		0.0		43.0		70.0	
Irradiated tumors	F_P	(%)		–		20.1		100.0		57.0		55.0		43.0		40.0
	T_C	(h)		–		12.9		11.1		12.8		20.5		39.0		15.0
	T_d	(h)		96.0		89.6–154.0		10.1		55.0		90.0		24.0		132.0
	$T_{d(Pot)}$	(h)		–		50.0		11.1		19.7		32.4		75.7		30.9
	ϕ	(%)		–		56.0–67.6		0.0		64.0		64.0		68.0		76.0
	D	(h)	444.0		288.0		90.0		146.5**		(50.5)		454.5		262.6	

d : radiation dose in rad.
t : time of observation relative to time of irradiation.
F_P : fraction of proliferating cells calculated from the relationship $F_P = (\alpha - 1) \times 100$ (*20*)
T_C : median cell cycle time.
T_d : measured volume doubling time.
$T_{d(Pot)}$: potential volume doubling time.
ϕ : per cent cell loss.
D : growth delay (see also footnote 4).
† : data derived by interpolation from published data (see sources).

*: for reasons of comparison, values for proliferation parameters were calculated according to equation 1. The values for T_d are related to variations in cell number.
**: growth delay corresponding to the time interval between the moment at which the unirradiated and the irradiated tumors reached volumes of $V_x = 3.V_0$.

Sources: 1) Malaise and Tubiana (*35*); Tubiana et al. (*36*).
2) Hermens and Barendsen (*22,40*); Hermens (*30*).
3) Van Peperzeel (*11,48*).
4–7) Denekamp and Thomlinson (*39*).

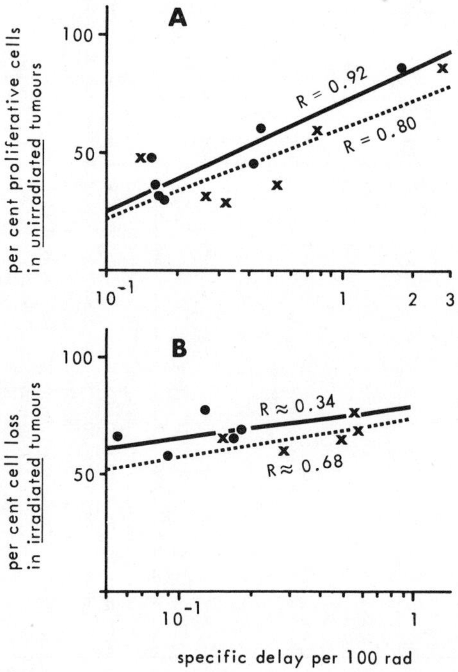

FIG. 3. Correlations between specific growth delay, DS, induced by radiation and growth parameters of different types of tumors. The data are given in Table 3.

Panel A: correlation between the fraction of proliferating cells, F_P, in unirradiated tumors and DS/100 rad of radiation dose. •: F_P versus $(D/T_d)\cdot(\text{dose in rad}/100)^{-1}$; x: F_P versus $(D/T_{d(Pot)})\cdot(\text{dose in rad}/100)^{-1}$.

Panel B: correlation between cell loss, ϕ, in irradiated tumors and DS/100 rad of radiation dose. •: ϕ versus $(D/T_d)\cdot(\text{dose in rad}/100)^{-1}$; x: ϕ versus $(D/T_{d(Pot)})\cdot(\text{dose in rad}/100)^{-1}$.

The solid lines and broken lines correspond with the closed circles and crosses, respectively. Values for the correlation coefficient, R, are indicated in the figures.

TIME COURSE VARIATION IN THE FRACTION OF SURVIVING TUMOR CELLS AFTER IRRADIATION

Tumor cure in radiotherapy may be achieved by the application of fractionated doses rather than by giving a single high dose of ionizing radiation. It is known that severe damage may be caused to the normal tissue by use of the latter method. Results in experimental tumor therapy obtained for instance with the R-1 rhabdomyosarcoma (*1,42*) have revealed that the probability of eradicating a tumor by fractionated therapy depends on a large number of factors (*43,44*). For a given radiation modality, the most important factors are: 1. the fraction of surviving cells after each session; 2. factors modifying cellular radiation response between two subsequent sessions, e.g., repair of accumulated damage, reoxygenation and changes in cell age distribution; 3. the rate of compensation for radiation-induced cell loss by repopulation of surviving clonogenic cells.

A reproducible method for the quantitative measurement of time course variations in fractional surviving cells after irradiation has been described earlier (*1*). The method involves the isolation of tumor cells by a suspension technique developed by Reinhold (*45*) and subsequent in vitro plating of known numbers of isolated cells. This technique was employed, among others, to assay fractions of clonogenic cells present at various time intervals after irradiation of R-1 tumors with a single dose of 2000 rad of X-rays. It was observed that fast repopulation by surviving clonogenic cells took place between days t = 4 and t = 12 after irradiation. From these results, it was concluded that the surviving clonogenic cells are mainly responsible for the regrowth of the tumor after irradiation with a high dose of X-rays (*46*). Further support for this conclusion is provided by results of calculations on the contribution to the cell production by clonogenic and nonclonogenic cells in irradiated R-1 tumors. These calculations are based on cell kinetic data of proliferating and terminal cells and pertain to time course variations in fractional clonogenic cells at the periphery and at the center of R-1 tumors after irradiation with a single dose of 2000 rad of X-rays (*30*). The results are shown in Fig. 4.

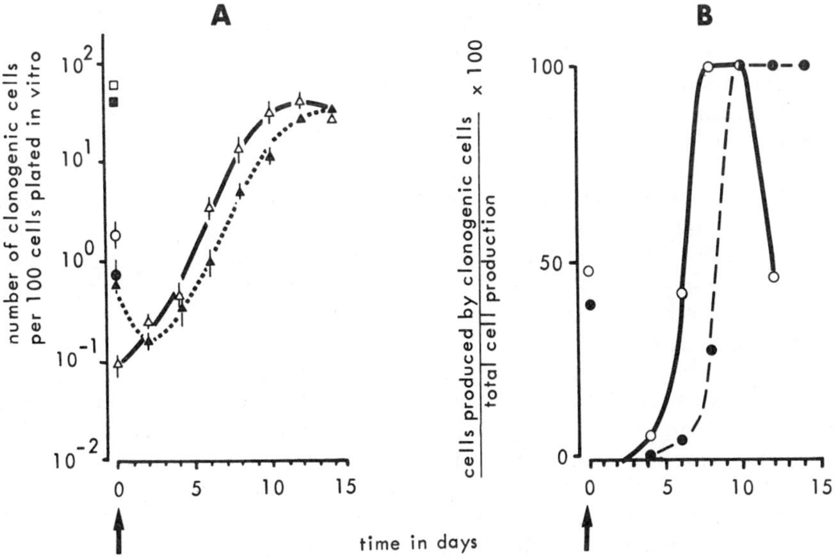

FIG. 4. Repopulation of R-1 tumors by clonogenic cells after irradiation with a single dose of 2000 rad of X-rays.

Panel A: Variations in fractions of clonogenic cells as a function of time after irradiation. △ and ▲ represent fractions of clonogenic cells per 100 cells present in cell suspensions prepared at different time intervals after irradiation from R-1 tumors irradiated in the living animal at time t = 0; ○ and ● represent fractions of clonogenic cells per 100 cells present in cell suspensions prepared from R-1 tumors within 15 minutes after irradiation in the dead animal; □ and ■ represent fractions of clonogenic cells per 100 cells present in cell suspensions prepared from unirradiated R-1 tumors at day t = 0. △, ○, □: data obtained for the periphery; ▲, ●, ■: data obtained for the center of the tumors.

Panel B: Variations with time after irradiation in the per cent cell production by clonogenic cells relative to the production of cells by clonogenic plus nonclonogenic cells calculated for different time intervals after irradiation. The data shown represent values for the per cent cell production calculated for the periphery (open circles) and the center (closed circles) of tumors.

Panel A shows a trend in the time course variation in fractional clonogenic cells similar to that reported by Barendsen and Broerse (*1*); the additional information obtained can be summarized as follows: 1. the fraction of surviving clonogenic cells at the periphery and in the center of tumors irradiated in the living animal are equal to 0.19% and 1.54%, respectively; 2. the fraction of hypoxic cells shows large differences with different locations of cells in the tumor and may range from 5% at the periphery to 80% in the tumor center; 3. the increase in the fraction of clonogenic cells after irradiation begins sooner at the periphery than in the center, notwithstanding the fact that the fraction of surviving clonogenic cells in the center is up to 8 times larger than that at the periphery. At the periphery, the fraction of clonogenic cells increases by a factor of about 300 within 12 days after irradiation, while there is an initial decrease in the center by a factor of about 3 during the first two days after irradiation. This is followed by a subsequent increase by a factor of 200 between days t = 2 and t = 14. This observation suggests that fast repopulation occurs in all instances from severely hypoxic cells and that the relative contribution by surviving cells from the tumor periphery to the repopulation of the tumor becomes more important with increasing time after irradiation.

Panel B shows that the contribution by clonogenic cells relative to the total production of tumor cells, i.e., by clonogenic and nonclonogenic cells, at the periphery increases sharply from 6% to 100% between days t = 4 and t = 8. A similar increase can be calculated for the center where the relative contribution of clonogenic cells to the total cell production increases from 5% to 100% between days t = 6 and t = 10. These results demonstrate that, for a short time during the period of tumor regrowth (namely, between days t = 8 and t = 10), the production of cells is entirely dependent upon dividing clonogenic cells. Interestingly, this rapid repopulation of irradiated R-1 tumors by clonogenic cells can be explained only by assuming that the values for α for clonogenic cells must be temporarily larger than can be calculated from data on per cent labelled clonogenic plus nonclonogenic cells.

A further implication of these results may be that, during the time interval between days t=8 and t=10, all of the P cells of the tumor are clonogenic cells while all of the noncycling cells produced by these P cells should be considered as being nonclonogenic Q cells. This raises the question as to whether only the P cells in the nonirradiated tumor have clonogenic properties and whether the Q cells have lost this property. In order to answer this question, experiments were performed with the R-1 rhabdomyosarcoma in which all of the proliferating P cells were labelled in vivo by injections of ^{3}H-TdR at 4-hour intervals during a total period of 16 hr. Duplicate tumors were excised after each injection and used to prepare single cell suspensions. In addition, a slice of tissue from each tumor was fixed in Bouin's solution and processed by histologic methods in order to prepare autoradiographs. The isolated cells were plated in vitro and cultured for 12 hr, 24 hr, 48 hr, etc., up to 144 hr. After culturing, the cells were fixed and autoradiographs were prepared from cells attached to the bottoms of the culture dishes. Grains of cells in "colonies" consisting of a minimum number of 1 to a maximum number of about 45 cells were counted. From these data, fractions of colonies containing labelled cells were calculated on the basis of Poisson probability for cells being labelled or nonlabelled. A summary of the results which have been detailed previously (*47*) is shown in Fig. 5. It can be seen that, for colonies of different sizes, the fractions of colonies originating from labelled and nonlabelled cells is about constant at a value of 50% within the limits of the statistical variations. This indicates that colonies of different sizes can develop in vitro from labelled P cells as well as from nonlabelled Q cells with approximately equal probability.

Because it is obvious from these experiments that Q cells may be capable of producing offspring when placed in a suitable environment, it may be speculated whether by a process of recruitment, the Q cells could be the main source of cells which repopulate a tumor after irradiation. This could be possible if the production of Q cells was due to local conditions associated with severe hypoxia in the unirradiated tumor. After irradiation of such a tumor with a single dose of 2000 rad of X-rays, practically all the surviving cells should be of hypoxic Q cell origin. The importance of hypoxia with respect to the P and Q cell status in R-1 tumors is still being investigated.

It may be clear from this discussion that studies on the cell kinetics of tumors before and after treatment are of fundamental importance to the understanding of the clinical effect of radiotherapy. Knowledge of changes in cell kinetics induced by therapy may also have a great bearing on the development of combined radiotherapy and chemotherapy. However, the amount of cell kinetic data for both the normal and tumor tissue should be further enlarged in order to sort out successful combinations of therapy for different types of tumors.

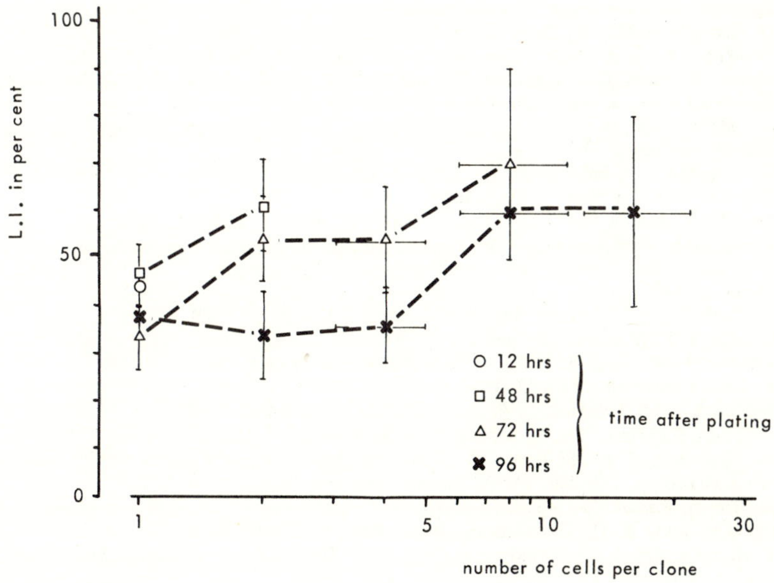

FIG. 5. Fractions of labelled colonies as a function of the number of cells per colony for cultures of different ages.

LI represents the per cent colonies containing labelled cells. Each value of LI is calculated by use of significantly labelled cells counted in colonies with different cell numbers as indicated by the horizontal bars. Vertical bars represent standard deviations of LI values.

REFERENCES

1. G. W. Barendsen and J. J. Broerse, *Eur. J. Cancer* **5**, 373 (1969).
2. S. Rockwell and R. F. Kallman, *Radiat. Res.* **53**, 281 (1973).
3. N. J. McNally, *Brit. J. Radiol.* **46**, 450 (1973).
4. E. A. McCulloch and J. E. Till, *Radiat. Res.* **16**, 822 (1962).
5. N. F. Kember, *Nature, (London)* **207**, 501 (1965).
6. H. R. Withers, *Brit. J. Radiol.* **40**, 187 (1967).
7. H. R. Withers and M. M. Elkind, *Int. J. Radiat. Biol.* **17**, 261 (1970).
8. K. Breur, in *Achttiende Jaarboek van Kankeronderzoek en Kankerbestrijding*, pp. 61–72. J. H. de Bussy, Amsterdam, 1968.
9. K. Breur, *Eur. J. Cancer* **2**, 157 (1966).
10. K. Breur, *Eur. J. Cancer* **2**, 173 (1966).
11. H. A. van Peperzeel, *Eur. J. Cancer* **8**, 665 (1972).
12. E. P. Malaise, A. Charbit, N. Chavaudra, P. F. Combes, J. Douchez and M. Tubiana, *Brit. J. Cancer* **26**, 43 (1972).
13. A. Howard and S. R. Pelc, *Heredity* **Suppl. 6**, 261 (1953).
14. J. E. Cleaver, *Thymidine metabolism and cell kinetics. North-Holland Research Monographs, Frontiers of Biology,* **vol. 6** (A. Neuberger and E. L. Tatum, Eds.). North Holland Publ. Co., Amsterdam, 1967.
15. A. W. Rogers, *Techniques of Autoradiography*. Elsevier Publ. Co., New York, 2nd Edition, 1973.
16. R. B. Painter and R. M. Drew, *Lab. Invest.* **8**, 278 (1959).
17. H. Quastler and F. G. Sherman, *Exp. Cell Res.* **17**, 420 (1959).
18. G. G. Steel and S. Hanes, *Cell Tissue Kinet.* **4**, 93 (1971).
19. M. L. Mendelsohn and M. Takahashi, in *The Cell Cycle and Cancer* (R. Baserga, Ed.), pp. 58–95. Marcel Dekker Inc., New York, 1971.
20. G. G. Steel, *Cell Tissue Kinet.* **1**, 193 (1968).
21. A. K. Laird, *Brit. J. Cancer* **18**, 490 (1964).
22. A. F. Hermens and G. W. Barendsen, *Eur. J. Cancer* **5**, 173 (1969).
23. G. M. Hahn and J. B. Little, *Curr. Top. Radiat. Res. Q.* **8**, 39 (1972).
24. F. Mauro, B. Falpo, G. Briganti, R. Elli and G. Zupi, *J. Nat. Cancer Inst.* **52**, 705 (1974).
25. M. L. Mendelsohn, *J. Nat. Cancer Inst.* **28**, 1015 (1962).
26. M. L. Mendelsohn, in *Cellular Radiation Biology*, pp. 498–513. The Williams and Wilkins Company, Baltimore, 1965.
27. S. B. Refsum and P. Berdal, *Eur. J. Cancer* **3**, 235 (1967).
28. G. G. Steel, *Eur. J. Cancer* **3**, 381 (1967).
29. J. Post, R. J. Sklarew and J. Hoffman, *J. Nat. Cancer Inst.* **50**, 403 (1973).
30. A. F. Hermens, *Variations in the Cell Kinetics and the Growth Rate in an Experimental Tumor during Natural Growth and after Irradiation*. Radiobiological Institute TNO, Rijswijk (ZH), The Netherlands, 1973.
31. S. B. Curtis, G. W. Barendsen and A. F. Hermens, *Eur. J. Cancer* **9**, 81 (1973).
32. E. Frindel, E. P. Malaise, E. Alpen and M. Tubiana, *Cancer Res.* **27**, 1122 (1967).
33. L. Simpson-Herren and H. H. Lloyd, *Cancer Chemother. Rep.* **54**, 143 (1970).
34. P. K. Lala, *Cancer Res.* **32**, 628 (1972).
35. E. Malaise and M. Tubiana, *C. R. Acad. Sci. Ser. D*, **263**, 292 (1966).
36. M. Tubiana, E. Frindel and E. Malaise, *Am. J. Roentgenol.* **102**, 822 (1968).
37. J. M. Brown and R. J. Berry, in *Effects of Radiation on Cellular Proliferation and Differentiation*, pp. 235–246. IAEA, Vienna, 1968.
38. J. M. Brown, *Radiat. Res.* **43**, 627 (1970).
39. J. Denekamp and R. H. Thomlinson, *Cancer Res.* **31**, 1279 (1971).
40. A. F. Hermens and G. W. Barendsen, in *Proc. Conf. Time and Dose Relationships in Radiation Biology as Applied to Radiotherapy*, Carmel, Calif., 1969, pp. 147–149. Brookhaven National Laboratory, Upton, BNL 50203 (C-57), 1970. (Available from Clearinghouse for Federal Scientific and Technical Information, Springfield, Va. 22151.)
41. L. Szcxepanski and K. R. Trott, *Brit. J. Radiol.* **48**, 200 (1975).
42. G. W. Barendsen and J. J. Broerse, *Eur. J. Cancer* **6**, 89 (1970).
43. G. W. Barendsen, *Eur. J. Cancer* **7**, 181 (1971).
44. G. W. Barendsen, *Eur. J. Cancer* **10**, 269 (1974).

45. H. S. Reinhold, *Eur. J. Cancer* **1**, 67 (1965).
46. G. W. Barendsen, in *Current Topics in Radiation Research* (M. Ebert and A. Howard, Eds.), **vol. IV**, pp. 293–356. North-Holland Publ. Co., Amsterdam, 1968.
47. G. W. Barendsen, H. Roelse, A. F. Hermens, H. T. Madhuizen, H. A. van Peperzeel and D. H. Rutgers, *J. Nat. Cancer Inst.* **51**, 1521 (1973).
48. H. A. van Peperzeel, *Patterns of Tumor Growth after Irradiation. A Comparative Study in Men, Dogs and Mice*. Thesis, Amsterdam, 1970.

Cell Proliferation Kinetics and Growth Rate of the Irradiated Human Tumors[1]

E. P. Malaise, N. Chavaudra, F. Pêne, J-M. Richard and M. Tubiana
Institut Gustave-Roussy, 94800 Villejuif, France

The study of the kinetics of human solid tumors should overcome both ethical and technical problems: one rarely finds tumors large enough to allow many biopsies after injection of tritiated thymidine or tumors whose growth can be measured for several weeks without treatment. These studies are even more difficult during or after treatment. This explains why the data are so scarce and relatively crude. Nevertheless, they deserve consideration since experimental transplanted tumors differ in many ways from human spontaneous tumors and since extrapolation from murine data to man is always a risky procedure.

The aim of this paper is to discuss 2 problems:

—firstly the relationship between the growth rate of a tumor and its radiosensitivity;

—second, the influence of irradiation on the cell kinetics of human tumors.

Breur in 1966 (*1*) measured the doubling time (DT) of lung metastases before and after irradiation. He found that the DT of a tumor was comparable before and after irradiation (doses delivered ranged between 1225 and 6000 rads). This conclusion remains valid if 2 courses of irradiation sessions separated by an interval of a few months are delivered to a tumor. Extrapolating the growth curve of the recurrent tumor back to the day of the first irradiation, Breur evaluated the radiosensitivity by using the ratio:

extrapolated volume/volume of the tumor before irradiation.

He found a statistically significant correlation between the radiosensitivity of lung metastases and the DT: tumors with a small DT are more radiosensitive whatever their histological types.

Comparable investigations were carried out more recently by Van Peperzeel (*2*) and ourselves (*3*). They confirmed the correlation between growth rate and radiosensitivity initially calculated by Breur (Fig. 1). When the results are plotted together, the slopes of the regression curves are comparable for the 3 sets of data (Table 1).

These data had been obtained on lung metastases. In order to obtain data on primary tumors, we studied 12 squamous cell carcinomas of the upper respiratory and digestive tract which were biopsied at various time intervals after

[1]Congress presentation C–3–4.

irradiation. The labeling index (LI) was measured in vitro before irradiation. It is well-known that there is a correlation between the LI and the growth rate (*4*). The radiosensitivity was measured by following the tumor cell depletion, that is by comparing the number of cells per unit of surface before irradiation and 15 days after 2 doses of 850 rads. This model does not take into account the changes in the volume of the tumors which are always difficult to evaluate for a human primary tumor. The cell depopulation may therefore be somewhat underestimated, but data for experimental tumors suggest that cell depopulation per unit of volume and tumor regression are correlated. Therefore, the trend of the results is probably not altered. In this investigation again, a statistically significant correlation between LI and depopulation was observed (Fig. 2).

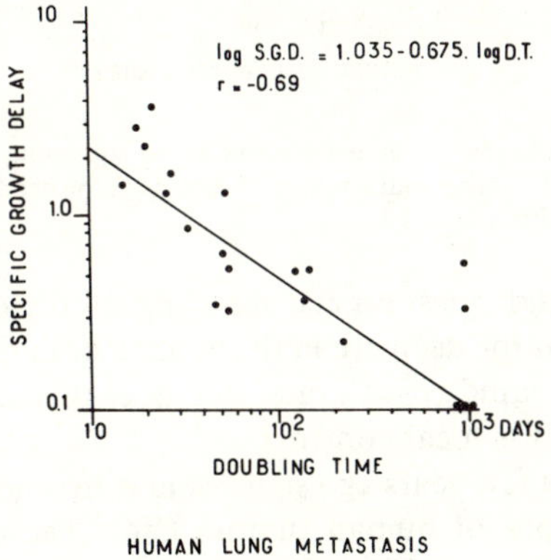

FIG. 1. Relationship between the growth specific delay and the doubling time. The growth specific delay is larger when the DT is smaller.

TABLE 1
Relationship Between Cell Kinetics and Radiosensitivity of Human Pulmonary Metastases

Authors	Number of Patients	Criterion	Relationship
Breur	16	Extrapolated volume	Log D37 = 1.780 + 0.652 log DT.
Van Peperzeel	11	Minimum volume	Log D37 = 1.347 ± 0.825 log DT.
Malaise et al.	13	Specific growth delay	Log D37 = 1.035 + 0.675 log DT.

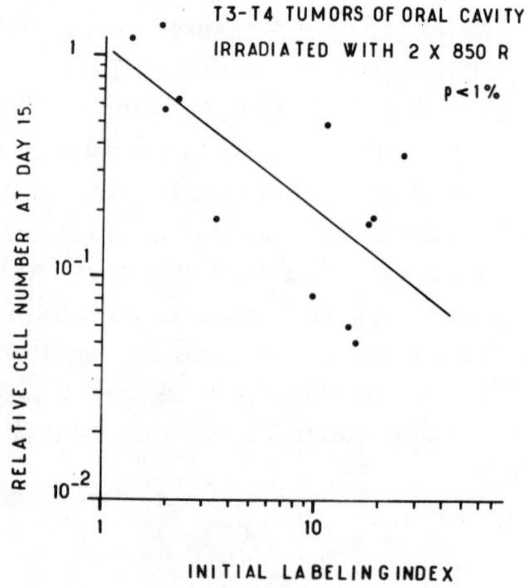

FIG. 2. Relationship between the labeling index of the squamous cell carcinomas before the treatment and the relative cell number in the tumor 15 days after the irradiation. The depopulation is relatively larger in tumors with a high LI.

However, one could question the meaning of these observations: these studies demonstrate that the decrease in the volume or in the number of cells are higher in tumors with a rapid growth rate. But, does this mean that these tumors require less irradiation for local control?

We demonstrated a few years ago significant differences between the DT of various histological types of human tumors (*5,6*). The pathological types of tumors with a short DT have a higher LI (*4*); they are also the tumors with the smallest tumor control dose (Table 2). This confirms that the most sensitive tumors are those with the highest mean proliferative fraction. We were able to calculate also, for the various histological types of tumors, the mean rate of cell loss and the mean daily cell turnover rate. The turnover rate varies widely for the various types of tumors. Of all the kinetic parameters considered, it seems to be the one which is the best correlated with the radiosensitivity or the chemosensitivity (Table 2).

A few years ago, J. Denekamp (*7*) found for experimental tumors that the shrinkage rate after irradiation was the highest for tumors with a high cell loss rate. Our data corroborate these findings for human tumors and suggest that shrinkage and radiosensitivity might be correlated. This may appear to be at variance with some clinical observations. However, recent data have demonstrated a highly significant correlation between the size of the residual tumor 3 weeks after the end of the course of radiotherapy and the local control of the tumor (*8*).

As previously discussed by Hermens (*9*), some experimental tumor studies have shown a repopulation after irradiation. Is this true also for human tumors?

By taking sequential chest X rays at short-term intervals after irradiation, Van Peperzeel (*2*) and we (*3*) were able to demonstrate an accelerated growth rate of human lung metastases during the first few weeks after irradiation. After this acceleration, the growth rate returns to its original value. The accelerated growth phase of the lung metastases has the following features:

1.—It is frequently observed. Van Peperzeel has found it in 10 patients out of a group of 11 patients irradiated. We have observed it in 26 out of 31 metastases which were irradiated in 13 patients.

2.—This acceleration can be characterized by the ratio:

DT before irradiation/DT during the accelerated phase.

The value of this ratio ranges from 2.5 to 5.0.

3.—The growth rate is accelerated more for those tumors which have a long DT before irradiation. A similar accelerated growth rate immediately after irradiation had formerly been observed for skin metastases (*10*).

4.—Repopulation tends to reduce the specific growth delay (SGD). This explains why the SGD is relatively short; for instance, it is only half a DT after a dose of 1000 rad (*3*).

TABLE 2
Mean Kinetic Parameters of Various Histological Types of Human Tumors
The numbers in brackets indicate the numbers of patients.

Histological Type	DT (days)	LI (%)	Growth Fract.	Cell Loss	Daily Turnover Rate	Average Clinical Tumor Control X-ray Dose
Embryonal tumors	27 (76)	30 (30)	90	93	49	2,500–3,000
Malignant lymphomas	29 (41)	29 (15)	90	93	47	3,500–4,500
Mesenchymal sarcomas	41 (87)	4 (32)	11	68	5.5	> 8,500
Squamous cell carcinomas	58 (51)	8 (68)	25	89	10	6,000–7,000
Adenocarcinomas	83 (134)	2 (121)	6	71	2	6,000–8,000

It would be essential to know whether accelerated repopulation also occurs in the primary tumors during a fractionated course of irradiation. In order to study this problem, cell kinetic studies are required.

Only a few workers have compared the cell cycle duration before and after irradiation. Several years ago, Tubiana et al. (*11*) measured the cycle time in the irradiated and the non-irradiated zones of 2 squamous cell carcinomas by the percentage of labeled mitoses (PLM) technique. In the 2 tumors, the mean T_c was slightly shortened in the zone of the tumor that had received 250 rad 48 hr before the study. More recently, Casale et al.[2] studied the cycle duration with the PLM method in a squamous cell carcinoma. After this investigation, the tumor was irradiated with 7,000 rad. Subsequently, the cycle was measured on the recurrence of this tumor. The mean cell cycle duration (T_c) decreased from 67 hr to 51 hr. Moreover, the spread of the individual values of T_c was reduced in the recurrent tumor. However, the ratio T_s/T_c was not changed; the changes of the LI therefore remained proportional to the proliferative fraction and increased from 31 to 45%.

In conclusion, the study of these 3 human tumors demonstrates only a small shortening of the cycle duration and suggests the possibility of an increase of the GF in a recurrent tumor. On the other hand, Hermens (*9*) has shown that in animal tumors, the cell kinetics of the viable cells and doomed cells are similar. If this is also true for human tumors, LI can give valuable information.

Hendrickson et al. (*12*) have followed in 7 tumors the LI after 300 rad delivered in one session (Fig. 3). In all the patients, the authors observed an early and large decrease of the LI which falls under 50% of the initial value. After the third day, the LI begins to increase but there are wide differences from one tumor to another. In the case of a highly radiosensitive tumor (Hodgkin disease), the LI remained very low until the last day of the investigation (Fig. 3).

We have ourselves studied with the same protocol 19 tumors.[3] The results are similar: 24 hr after an irradiation of 330 rad, the mean LI is decreased twofold; subsequently, the mean LI increases slowly, but by the 4th day, it is generally still lower than before irradiation. However, the results vary widely from one tumor to another. Nevertheless, in most of the tumors, an important decrease of the LI

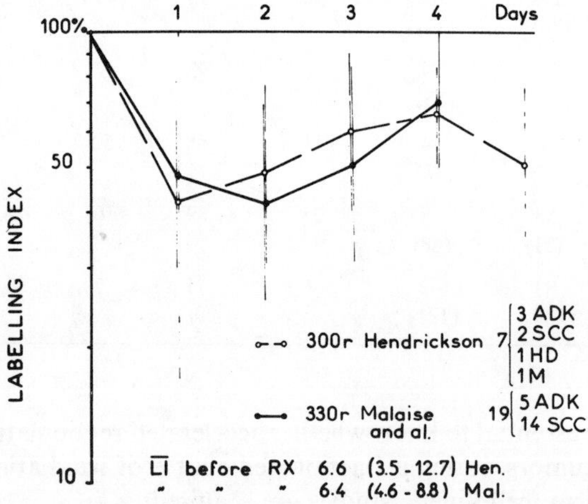

FIG. 3. Influence of an irradiation of 300 rads or 330 rads in one session on the LI of two groups of human tumors.

[2] C. Casale, E. Ziparo, R. Paoluzi, M. Benassi, C. Nervi and F. Bresciani, unpublished results.
[3] E. P. Malaise; N. Chavaudra, F. Pêne, J-M. Richard and M. Tubiana; unpublished results.

is observed. These data suggest that after 300 or 330 rad, the recruitment is probably low and, on the other hand, the cycling cells are probably blocked, the mitotic delay appearing to be longer than would be predicted from the data obtained on cultured cells. In vitro, 300 rad induce generally a mitotic delay of about 2 to 3 hours.

A few other studies have been performed. Breitenecker et al. (*13*) delivered 1500 rad with external radiotherapy to 10 patients with cancer of the cervix. Seventeen other patients were treated with interstitial therapy: they received 2100 rad. These authors observed a large and rapid decrease of the LI. However, in 4 patients the decrease was smaller or did not occur. The authors observed that in these 4 patients, the response of the tumor to radiotherapy was very poor.

We have delivered 2 doses of 850 rad with an interval of 2 days to 12 patients with T3-T4 tumors of the oral cavity[3] (Fig. 4). The LI was measured every 5 days for 20 days. Five days after irradiation, the LI is approximately half of the original value, that is to say that the magnitude of the decrease of the LI is comparable to that observed after 330 rad. In contrast, with 2 sessions 850 rad, the duration of the decrease is longer and is still observed 15 days after the irradiation. The decrease of the LI is generally larger for tumors with a high LI before irradiation (Fig. 5) but this is not always verified and in some tumors with a high initial LI, the decrease of the LI may be small. Later, the shrinkage and the inflammation make biopsies difficult. As previously discussed, the decrease in cancer cell number is larger when the original LI was high but it is also correlated with the fall of the LI. In some tumors, the decrease of the LI is small and, in these tumors, there is only a small decrease in cell number. This suggests that a large fall in the LI after irradiation correlates with radiosensitivity.

These data were obtained after irradiation regimens which were different from those used in conventional radiotherapy. They encouraged us to follow the LI during a course of conventional radiotherapy.

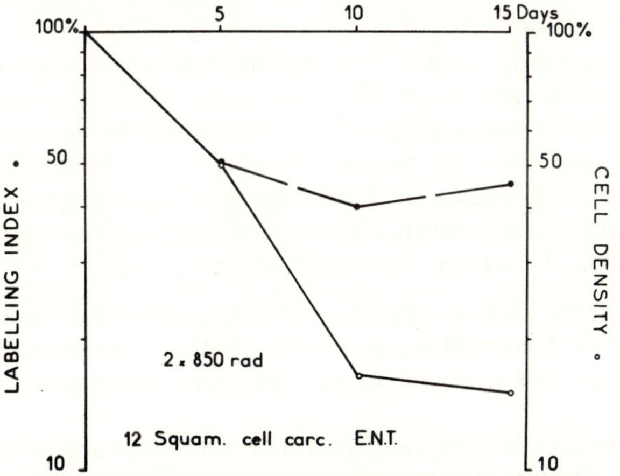

FIG. 4. Influence of an irradiation of 1,700 rads in two sessions of 850 rads on the LI and the cell number in 12 squamous cell carcinomas of the oral cavity.

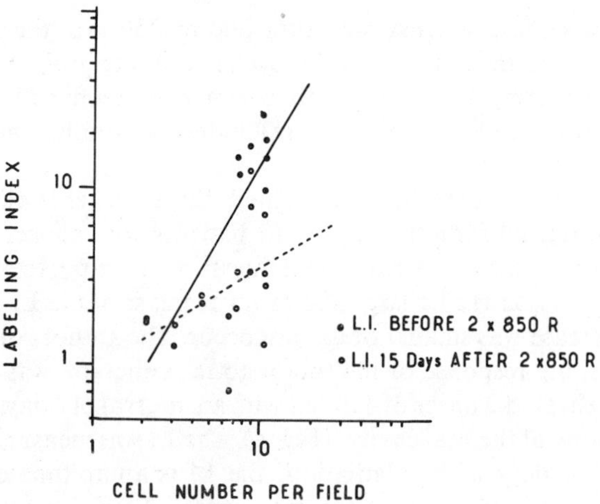

FIG. 5. Labeling index before and 15 days after 2×850 rads in 12 squamous cell carcinomas of the oral cavity.

We measured weekly the LI in 7 squamous cell carcinomas of the oral cavity and in 2 adenocarcinomas of the rectum, all these tumors being irradiated with conventional fractionation and receiving 1000 rad per week. During this treatment, the LI fell in all of the squamous cell carcinomas and the mean LI decreased progressively during the treatment (Fig. 6). The results observed in the 2 adenocarcinomas of the rectum were different: the LI increased during the treatment; it is well known that these tumors are much more radioresistant (Fig. 6). Here again, there seems to be a correlation between the time course of the changes of the LI after irradiation and the clinical response of the tumor. However, the data do not allow us to choose between 2 hypotheses: either the lack of effect of irradiation on the LI is due to a low cell radiosensitivity, or repopulation might be a factor of radioresistance during a fractionated course of irradition. Furthermore, it should not be forgotten that reassortment of cells among the various phases of the cell cycle could also influence the radiosensitivity of the cell population. Further work is needed but the present results already underline the prognostic value of the course of LI during irradiation and the importance of such studies. Close cooperation between experimental and clinical investigations is required for further progress.

In conclusion, 2 main ideas emerge from this review:

—A short DT, a high GF, a high rate of cell loss and in particular a high cell turnover rate are factors of radiosensitivity, both for metastases and primary human tumors. This is similar to what has been observed in experimental tumors (*9*).

—After a few sessions of X-rays and until the end of the course of radiotherapy, one generally observes in squamous cell carcinomas a large decrease in the cell proliferation rate, probably due to a mitotic delay. However, there are wide differences in the response of tumors, even within tumors of the same histology and location. In some tumors, the decrease of LI is small and the LI may even increase. This seems to be associated with clinical radioresistance. This conclusion seems to be valid for comparing both different histological types of tumors and various tumors within a same histological group.

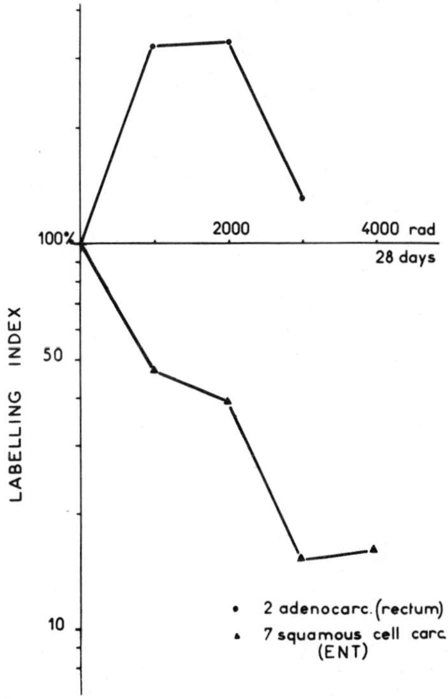

FIG. 6. Influence of a conventional radiotherapy on the labeling index in 7 squamous cell carcinomas of the oral cavity and 2 adenocarcinomas of the rectum.

ACKNOWLEDGEMENTS

We are indebted to Drs. J. P. Wolff, R. Le Fur, F. Eschwege and G. Schwaab for their cooperation in irradiating the patients and obtaining the tumor specimens. We would like to express our thanks especially to the Ear, Nose, and Throat Department. We appreciate the help of the pathologists and radiologists. We thank Mrs. F. de Langen for her valuable technical assistance.

REFERENCES

1. K. Breur, *Eur. J. Cancer,* **2**, 173 (1966).
2. H. A. Van Peperzeel, *Eur. J. Cancer,* **8**, 665 (1972).
3. E. P. Malaise, A. Charbit, N. Chavaudra, P. F. Combes, J. Douchez and M. Tubiana, *Brit. J. Cancer,* **26**, 43 (1972).
4. E. P. Malaise, N. Chavaudra and M. Tubiana, *Eur. J. Cancer,* **9**, 305 (1973).
5. A. Charbit, E. P. Malaise and M. Tubiana, *Eur. J. Cancer,* **7**, 307 (1971).
6. E. P. Malaise, N. Chavaudra, F. Battaglini, J. M. Richard and M. Tubiana, *Bull. Cancer,* **60**, 119, (1973).
7. J. Denekamp, *Eur. J. Cancer,* **8**, 335 (1972).
8. S. Kramer, Communication to symposium on laryngeal cancer. Toronto, 1974.

9. A. F. Hermens and G. W. Barendsen, *Eur. J. Cancer,* **5**, 173 (1969).
10. P. Rambert, E. P. Malaise, A. Laugier, M. Schlienger and M. Tubiana, *Bull. Cancer,* **55**, 323 (1969).
11. M. Tubiana, E. Frindel, E. P. Malaise, *Am. J. Roentgenol.* **102**, 822 (1968).
12. F. R. Hendrickson and G. G. Hibbs, *Radiology,* **83**, 131 (1964).
13. G. Breitenecker and G. Tatra, *Strahlentherapie,* **146**, 664 (1973).

CARCINOGENESIS: RADIATION AND OTHER AGENTS[1]

Chairman: R. J. M. Fry

[1]Symposium B-2

Carcinogenesis by Ionizing Radiation and Lessons For Other Pollutants[1]

R. H. Mole
Medical Research Council, Radiobiology Unit,
Harwell, Didcot, Oxon OX11 ORD, Great Britain.

INTRODUCTION

During the last 30 years an immense effort has been put into experimental work on carcinogenesis by ionizing radiation and into epidemiological surveys of exposed human populations. The facts are exhaustively and comprehensively surveyed at frequent intervals by authoritative individuals and committees, for example UNSCEAR (*1*) and BEIR (*2*) and many special symposium meetings are devoted to the subject, for example IAEA (*3*). Nevertheless, there may be things to say which are generally useful or novel. What follows is a severely pruned discussion of a small number of basic issues which the author finds important or interesting.

THE NATURE OF CANCER

A review for a general audience should begin by defining what it is about, but I know of no unambiguous and accurate definition of cancer. It is certainly not, in my view, simply a disorder of growth regulation or of cell division and, in particular, rapidity of growth is not a characteristic feature. These mistaken ideas seem to have developed in ignorance of the long latent periods of human tumors, commonly 25 years or more (*4, 5*) and because so much experimental use has been made of repeatedly transplanted tumor cells—repeated subculture, whether of bacteria, tumor cells in vitro or tumor cells in vivo, as in metastases, must inevitably select for the most rapidly growing and most frequently dividing cells. Moreover, it seems a logical error to imagine that study of the properties of established cancer cells will necessarily tell us anything at all about the nature of the process of induction, the transformation of normal into malignant cells. But cancer cells breed more or less true and have different properties from normal cells: "malignant transformation" must be mutation in some sense of that word of many meanings.

As is so often the case in biology, cancer is something which can be recognized by a trained observer with a high degree of accuracy although at the same time individual examples may be difficult or impossible to classify. The classification of cancer is a pragmatic exercise to allow cancers to be talked about and the frequencies to be measured and compared: there is little

[1]Congress presentation B-2-1.

fundamental or immutable about it. This elementary point needs stressing in relation to both experimentation and epidemiology, the two methods available for learning about carcinogenesis. Too many experimental papers leave the critical reader in doubt as to what kinds of tumor they are about, and even whether they are about cancer at all. Epidemiological surveys depend on statements by people other than epidemiologists and made for a variety of nonscientific purposes. It should not be surprising that errors of omission or of commission may be substantial and often not measurable.

IONIZING RADIATION AS A CARCINOGEN

There can be no question that ionizing radiation is a carcinogen in animals and in man. A great variety of tumors has been induced and the generalization seems justified that every tissue affected by spontaneous tumors is susceptible to radiation carcinogenesis. (Some tissues or cells seem never to suffer a carcinogenic transformation, such as the lens of the eye and the neurones of adult nervous tissue). There is no evidence that radiation acts by producing a diffusible chemical carcinogen, and direct mechanisms of carcinogenesis are often envisaged in terms of an interaction between radiation and a target or targets within a cell which is thereby "transformed." But direct scientific evidence exists only for indirect mechanisms whereby sufficient damage to one tissue leads to abscopal tumor development in another tissue, not itself subjected to irradiation at all. Two striking examples were provided some twenty years ago concerning pituitary tumors and thymic leukemia in mice (*6, 7*). The characteristic features of this kind of indirect mechanism are that severe damage is required in the irradiated target tissue if any tumors at all are to be seen in the reacting organ, that the dose-response is therefore highly curvi-linear, and consequently that experimentation on a relatively small scale may be sufficient to demonstrate its existence.

The nature of an indirect mechanism must be established in each case by detailed investigation and clearly involves growth-regulating influences as well as perhaps viruses. In 1974, however, the most pertinent comment might be that new examples of such indirect mechanisms for radiation carcinogenesis have not been uncovered in the last 20 years or so. Are they then just academic curiosities? If *severe* tissue damage is a necessary factor, they could well be relevant to carcinogenesis after radiotherapy but would not be particularly significant for questions about the possible carcinogenic effects of low levels of exposure.

CARCINOGENESIS BY LOW LEVELS OF EXPOSURE

Antenatal diagnostic radiography involves a whole-body dose to the human fetus of only a few rads, and the various surveys of cancer in childhood all agree, within the statistical limits of the observed cancer frequencies, that this kind of radiation exposure is indeed associated with an increased incidence of leukemia and other kinds of malignant disease (*8*). There has always been a genuine scientific reason for questioning whether the radiography caused the increase since only some fetuses are radiographed and selection for radiography could well be correlated with a higher than average expectation of cancer in childhood. However, this argument seems now to have lost any cogency it might once have had. In the United Kingdom twins and singletons differ 5–6 fold in the chance of being radiographed in utero but show little difference in the excess cancer rate in the radiographed as compared with those not radiographed (Table 1) (*9*).

There seems no alternative now but to accept that low doses of ionizing radiation, on the order of a few rads, are truly carcinogenic. It is difficult to imagine that this could have been demonstrated experimentally because, as Table 1 shows, the induction rate for all malignant disease by antenatal radiography was about 2.10^{-4} over a 10-year period (from birth to the 10th birthday) and the scale of a corresponding experiment would have had to be very large indeed. In contrast, observations on very large human populations can sometimes be manageable and relatively cheap. The Oxford Childhood Cancer Survey, from which the twin data already cited come, covers a population numbering so far 15 million.

TABLE 1
Deaths from malignant disease at 0–9 years of age after antenatal radiography

		Singleton births	Dizygotic twins	Monozygotic twins
Live born 1945–1964		14.8 million	254,000	99,000
Proportion X-rayed (per cent)		10	55	55
		Deaths from malignant disease per 100,000		
Not X-rayed	Leukemia	23	16	nil
	Solid cancers	27	21	11
Excess in X-rayed	Leukemia	12	9	29
	Solid cancers	12	10	18

Based on published data from Oxford Childhood Cancer Survey[9].

No conclusions whatever could have been drawn about antenatal radiography and carcinogenesis if it were not (i) that radiography is memorable and/or is recorded and (ii) that only some members of a population were exposed. It seems unlikely that exposure to environmental "pollutants" will ever be like this and, if so, ionizing radiation will remain a unique case where carcinogenic effects of low-level exposure have actually been demonstrated, not merely predicted by a working hypothesis derived from observations at much higher exposure levels. Academically, in relation to questions about mechanisms of carcinogenesis, and practically this is of fundamental significance: the opportunity to validate, or confound, a working hypothesis about the effects of low levels of exposure by direct appeal to observation, will exist for ionizing radiation but perhaps not for other potential or actual carcinogens, including environmental "pollutants."

LATENT PERIOD

Although cancer is a cellular disorder, recognition that a cancer is there depends on examining tissue: there must be a sufficient mass of abnormal cells. Under favorable conditions a tumor of volume = 1 mm^3 may be recognized in some tissues of a mouse, but such a volume already consists of about 10^6 cells (each with volume = 1000 μm^3). This number corresponds to about 20 successive cell divisions in which each cell divides into two. More than this minimum number of cell divisions will be required depending on the degree of cell loss during growth which in tumors is often substantial. Moreover tumor cells, like normal cells, require nourishment via the blood and a spherical mass of 1000 cells with a volume of 10^6 μm^3 is already too large to rely on diffusion of oxygen, sugar, etc. and must develop its own system of blood vessels to keep itself growing. These elementary considerations show that a minimum time-interval must exist between application of a carcinogenic "insult" of any kind and recognition that a tumor has developed. Is this "latent period" wholly accounted for by these growth processes, with rates subject to variation from a variety of causes? In the mouse irradiation of one ovary leads to tumor development in it only if the hormone balance of the body is sufficiently disturbed, for instance by operative removal of the other, non-irradiated ovary (*6*). Or, alternatively, is some of the latent period to be attributed to a supposed complexity of "transformation" and the necessity for a sequence of intra-cellular changes before the characters of malignancy can be fully expressed? When local application of a carcinogen is made to the skin of the mouse, the natural history of radiation-induced carcinogenesis, in contrast with that of chemical carcinogenesis, may suggest that transformation could be a single-step process (Table 2), the apparent complexity of carcinogenesis by chemical compounds being the result of specific features which are by-passed by ionizing radiation (see Concluding comment).

TABLE 2
Characteristic contrasts between carcinogenesis in mouse skin by chemical agents and by ionizing radiation

Chemical agent painted on skin	
Single application	No tumors occur.
Several successive applications	Warts may appear in a few weeks and then regress spontaneously after painting ceases.
Many successive applications	One of the warts begins to grow progressively and invasively and continues to do so even if painting ceases and the other warts regress.
Local irradiation	
Single application	If too small no tumors occur.
	If tumorigenic a tumor may first appear 2–3 years later, usually growing progressively and invasively from its beginning. Not preceded by warts which characteristically are not seen during the weeks after irradiation.

It is commonly said that latent period is inversely related to radiation dose in some way although there seems no very obvious reason why a radiation exposure should affect growth processes in vivo months or years later. The apparent inverse relation may indeed be an artefact of the experimental situation where relatively large radiation doses are used in order to get measurable frequencies of induced tumors in numerically small groups of animals. The only limit to the number of separately induced tumors originating in a given irradiated tissue is the number of transformable cells (or cell foci) in that tissue, yet only some of the induced tumors will be recognized as separately induced entities, the actual number depending on the way the experiment is carried out. Only a single tumor may be recognized because the animal dies or is killed before other tumors develop to the point of recognition, or because separately induced cancers may be unrecognizable in principle (as for leukemias where the leukemic cells migrate around the body, intermingling with each other, and do not form separate tumor masses at the original sites of transformation), or because the results are presented in terms of tumor-bearing animals instead of numbers of tumor per animal. In such circumstances, even if the mean latent period for all tumors is in fact constant, the data will appear to show that it is inversely dose dependent (Fig. 1). The same reasoning (and diagram) shows that dose-response relationships may be derived from experi-

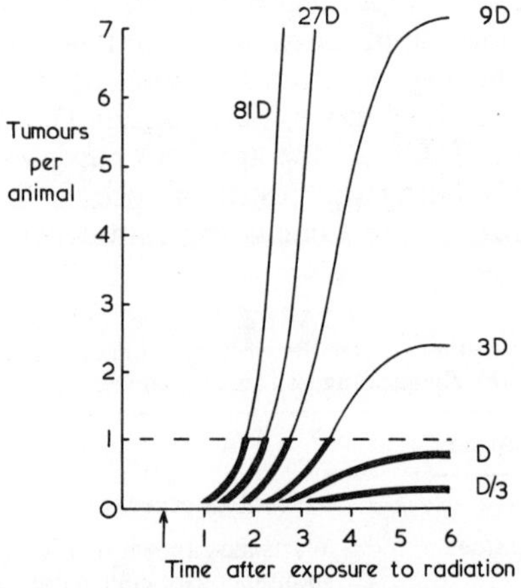

FIG. 1. Diagram to illustrate how an inverse relationship of latent period and radiation dose may be an observational artefact. Time t in arbitrary units after a brief exposure to radiation at $t = 0$. Radiation dose in multiples or fractions of an arbitrary value D.

Cumulative tumor incidence is assumed to be sigmoid with time, the shape of the curve independent of, and the final value linearly dependent on, the magnitude of the dose.

Observation of new tumors is assumed to cease when the mean number of tumors per animal = 1.0 (ignoring the likely Poisson distribution of tumors among animals). Thus, the only observable parts of the expected tumor distributions will be the segments below the horizontal dashed line as emphasized in the diagram. If natural mortality begins to become appreciable at $t = 3$-4, the uncorrected final cumulative tumor incidence actually observed for doses which are multiples of D will be unaffected but will be appreciably less for dose D and especially for D-3 and smaller doses.

mental groups, each with about 100% tumor incidence but differing in observed latent period according to radiation dose, only if some prior assumption is made about the distributions of latent periods of tumors which are not in fact observed. In practice, especially with large animals, the numbers in a given dose group in an experiment on carcinogenesis must be small, perhaps only 10-20, and, if dose varies by successive factors of 2 or 3, there will be only 2-4 dose groups with measured tumor incidences above zero and less than 100% (*10*). In human cases the greatest observed cumulative tumor incidence is likely to correspond to D/3 in Fig. 1 and most of the information will be about cancer frequencies below those normally accessible in experiments. Occasionally frequencies as high as 30-50% may be found in irradiated people and at these levels the latent period may appear to be inversely dependent on dose as in experimental work (Table 3).

TABLE 3
Body burden (μCi) and mean latent period (years) for tumors induced in man by Radium-226 (and Radium-228)

	Body burden (μCi Ra-226 or Pure Radium Equivalent)					
	0.1 –	0.5 –	1 –	2 –	5 –	10 –
Bone sarcoma	35 (3)	37 (2)	28 (6)	31 (5)	25 (4)	8 (2)
Head carcinoma	—	33 (4)	37 (3)	35 (4)	38 (2)	—

Number of cases given in parenthesis.

Latent period measured from start of exposure to data of diagnosis of tumor.

Values for latent period confined to subjects with recorded exposures to Radium lasting 3 years or less so as to minimize problems of comparability.

Data derived from case indices in Tables 4 and 5 of Rowland, Failla, Keane and Stehney (1970) in Argonne National Laboratory Report ANL-7760 part II, pp. 1–17, and from the detailed case histories given in Argonne National Laboratory Report ANL-7531 and ACRH-106 (1969) and in the Annual Progress Reports of the Radioactivity Centre, Massachusetts Institute of Technology, M.I.T.-952 nos. 1–7 (1963–1970).

THE RARITY OF THE CARCINOGENIC TRANSFORMATION

What kind of mechanism could account for carcinogenesis by only a few rads? There is almost no information about other radiobiological effects at such dose levels, and certainly none about familiar radiobiological phenomena such as cell sterilization or the production of chromosome aberrations. To avoid problems of extrapolation consider a dose of 200 rads. In Japanese bomb survivors this caused a leukemia incidence of about 0.6-0.7%: that is, one case of leukemia resulted from irradiation of 150 persons, 225 kg of active bone marrow (1.5 kg per adult) or 10^{14} cells (on the assumption that each marrow cell has a volume of 2250 μm^3 and density 1.00). If leukemia arises as the result of transformation of a single cell in the marrow, the probability per cell of this transformation is as low as 10^{-14} for a dose which is orders of magnitude more effective per cell for other kinds of radiobiological effect (Table 4). A direct comparison of probabilities per cell examined and per cell in tissue is unfair because in no tissue are all the cells likely to be susceptible to malignant transformation. However, even if only one bone marrow cell in a million were potentially transformable in this way, the probability per transformable cell would still be only 10^{-8}. A variety of conclusions may be suggested (*11*) but the point to stress here is as follows: if all we know about

TABLE 4
Comparative sensitivities of different radiobiological responses (*11*)

	Probability of effect per cell after 200 rads of low LET radiation at dose rates of 1–1000 rads/min.
Clonogenic sterilization	0.1–0.9
Visible chromosomal aberrations	0.2–0.4
Genetic effects: Dominant lethals in spermatogonia and adult dictyate oocytes	0.01–0.1
Specific mutations per locus	10^{-4}–10^{-6}
Leukemia: myeloid in mice	10^{-9}
per cell in hematopoietic tissue	
in man	10^{-14}

cellular radiobiology is about phenomena with a relatively high probability per cell, such as cell sterilization and cell mutation, can we expect our general radiobiological knowledge and understanding to tell us anything at all about the carcinogenic transformation, a phenomenon many orders of magnitude less probable per cell?

DOSE-RESPONSE RELATIONSHIPS AND CELL STERILIZATION

Since the carcinogenic transformation cannot be known to have occurred unless the transformed cell retains the capacity for indefinite cell division, and since transformation by a given exposure to ionizing radiation has a much smaller probability than cell sterilization, it is clear that all relationships between observed tumor frequencies and radiation dose must be the product of the two probabilities, for malignant transformation (induction of cancer) and for *not* being sterilized. Therefore the somewhat paradoxical conclusion is reached that observation of a linear dose-response at levels of dose which sterilize an appreciable fraction of cells must imply that the induction process cannot be linear but instead is a power function of dose. Furthermore, if induction was truly linear with dose, this could not be observed in animal experimentation or in Japanese bomb survivors because the doses involved are in the sterilizing range. The data on cancer induction by ionizing radiation in man seem all to be compatible with this conclusion (*11*) and may be regarded as suggesting that the induction process is commonly according to dose squared.

CARCINOGENESIS IN VIVO: AGE AND SENSITIVITY TO RADIATION

A common belief, reinforced by the results of antenatal radiography, is that the fetus in utero is more sensitive to cancer induction than the child, and the child more sensitive than the adult. However, the only human population in which individuals at all three stages of development received the same radiation exposure is the population of Japanese bomb survivors, and they show something different (Table 5). Cancers other than leukemia were 5-6 times less commonly induced in those aged 0-19 years at the time of exposure ATB than in adults, and overall induction of malignant disease was least in those exposed in utero. The ratio observed/expected was greater for exposure in childhood than in adult life (*12*), but in terms of numbers of additional cases adults have proved most sensitive and the fetus in utero least sensitive. There is in fact some evidence suggesting that sensitivity to carcinogens may in general increase with age at the time of exposure (*13*).

The apparent contradiction between the results of antenatal exposure to diagnostic radiography and to the atom bomb can be easily resolved if account is taken of cell sterilization by the larger radiation doses in Japan (*9*).

TABLE 5
Differences in mortality from neoplasia in Japanese bomb survivors according to age at time of exposure (*11*)

	Excess mortality per 10^6 person-years 1950–1970 for those receiving known doses 10+ T65 rads over expectation derived from those receiving 0–9 rads			
Age at time of irradiation by the bomb	In utero	0–19 years	20+ years	All postnatal
Leukemia		160	130	150
Other malignancies		50	280	180
Benign neoplasms and unspecified		4	14	10
All neoplastic diseases	−20[a]	190	420[b]	340

Calculated by author from basic data in Jablon and Kato (1970, 1971, 1972) and Kato (1971) see Mole 1974 (*9*)
[a]0–9 rads/group; 2 deaths in 795 subjects = 126 per 10^6 person–years.
10+ rads/group; 1 death in 461 subjects = 108 per 10^6 person–years.
[b]For those exposed at 20+ years the mean follow-up = 17.2 years; mean dose not available but likely to be close to the mean dose for all ages = 92 rads.

CONCLUDING COMMENT

This brief, telegraphic and perhaps too dogmatic outline may serve to illustrate that radiation carcinogenesis is becoming a quantitative subject with considerable potential for further development especially on the experimental side. The increasing volume of human data reveals striking differences between different human tissues in susceptibility to radiation-induced carcinogenesis and also in the RBE for carcinogenesis by fission neutrons (*11, 14*) for which there must be reasons. Practical questions about the quantitative effects of protraction and of physical dose rate have hardly ever been thoroughly investigated. Why have so many experiments failed to produce in animals the kind of tumor called lung cancer in man? Can experimental work on malignant transformation by irradiation in vitro be properly validated (*15, 16*)? But we must keep our feet firmly on biological ground and remember always that human information is irreplaceable, if it is allowed to disappear unexamined, and decisive, when it is available. The "hot particle" problem in the lung, for example, is certainly a genuine scientific problem but not the problem it may seem to be if we forget that Japanese bomb survivors at Hiroshima received

~1000 rems of whole lung irradiation but developed only a small frequency of induced lung cancer, and that children and young adults with certain forms of cancer who receive 1000 rads and more irradiation of the whole lung in the course of therapy, may survive for 15 years and more without, as yet, any report of induced lung cancer.

Lessons for other pollutants are clear but the language in which they are given may never be as lucid as for ionizing radiation. Ionizing radiation acts intracellularly and almost instantaneously, and species differences in response are minimal. In contrast, a chemical agent must penetrate a succession of permeability barriers, each with its own chemical specificity, in order to reach the intracellular targets; the intracellular concentration of the agent waxes and wanes with time and is never exactly known, and often the active principle is the product of a chemical transformation which occurs in some other part of the body: species differences must be expected for each of the processes mentioned. All these are highly relevant to the interpretation of observations on chemical carcinogenesis. Most fundamental of all is the lack, apart from cigarette smoking, of documented human exposure to a given chemical agent defining the time of occurrence and magnitude of the exposure. Take away from radiation carcinogenesis what has been learned about man and the significance of epidemiological studies for the applicability of experimentation is clear.

I conclude that for a number of reasons ionizing radiation is of quite basic importance in the search for a fuller understanding of the practically important and academically fascinating process of carcinogenesis.

REFERENCES

1. UNSCEAR, *Ionizing radiation: Levels and Effects* Vol. II *Effects.* A report of the United Nations Scientific Committee on the Effects of Atomic Radiation to the General Assembly. United Nations, New York, 1972.
2. BEIR, *The Effects on Populations of Exposure to Low Levels of Ionizing Radiation.* Report of the Advisory Committee on the Biological Effects of Ionizing Radiations, National Academy of Sciences – National Research Council, Washington, D.C. U.S.A., 1972.
3. *Radiation-induced Cancer. Proc. Symp. Athens,* 1969. International Atomic Energy Agency, Vienna, 1969.
4. A. W. G. Goolden and R. L. Morgan, *Acta Radiol.* **3,** 353 (1965).
5. L. H. Hempelmann, J. W. Pifer, G. J. Burke, R. Terry and W. R. Ames, *J. Nat. Cancer Inst.* **38,** 317 (1967).
6. A. Kirschbaum, *Cancer Res.* **17,** 432 (1957).
7. H. S. Kaplan, *Cancer Res.* **14,** 535 (1954).
8. B. MacMahon and G. B. Hutchinson, *Acta Unio. Int. Contra Cancrum* **20,** 1172 (1964).
9. R. H. Mole, *Brit. J. Cancer,* in the press (1974).
10. M. Goldman, L. S. Rosenblatt, N. W. Hetherington and M. Finkel, in *Radionuclide Carcinogenesis* (C. L. Saunders, R. H. Busch, J. E. Ballou, and D. D. Mahlum, Eds.) pp. 347–357. U.S. Atomic Energy Commission, CONF-720505, (1973). (Available from National Technical Information Service, U.S. Depart. of Commerce, Springfield, Va. 22151).
11. R. H. Mole, *Brit. J. Radiol.* in the press (1974).
12. S. Jablon, J. L. Belsky, K. Tachikawa and A. Steer, *Lancet,* i, 927 (1971).
13. R. Doll, in *Tenth International Congress Oncology 1970,* pp. 1–28. Year Book Medical Publishers Inc., Chicago, 1970.
14. *Radiosensitivity and Spatial Distribution of Dose.* ICRP Publication 14, Pergamon Press, Oxford, 1969.
15. C. Borek and E. J. Hall, *Nature (London)* **243,** 450 (1973).
16. J. C. Klein, *J. Nat. Cancer. Inst.,* **52,** 1111 (1974).

Mechanisms of Carcinogenesis[1]

George Klein
Department of Tumor Biology
Karolinska Institutet
S 104 01 Stockholm 60
Sweden

Mechanisms of carcinogenesis can be discussed at three main levels: the carcinogenic agent, the target cell, and the interaction of the transformed cell with the host. At the level of the agent, the main question revolves around the active group or part of the agent, chemical or viral, that is responsible for the transforming action. At the level of the host cell, it is most important to gain some insight into the mechanisms of "competence", i.e. the liability of a given cell to be transformed by a certain agent and the critical factors that separate transformability from non-transformability. At the level of the host, the main problem is to define the conditions under which transformed cells will be held in abeyance or will grow into frank malignancies.

In the fields of non-viral carcinogenesis, there has not been much progress during the last few years in defining the site of the carcinogenic agent responsible for transformation or the critical target in the cell. In contrast, there has been remarkable progress in defining the oncogenic portion of certain viral genomes, particularly the small DNA viruses polyoma and SV40 and some of the oncogenic adenoviruses as well.

This is essentially due to two important developments: restriction enzyme mapping of the small DNA viruses (*1, 2*), and the possibility to introduce DNA into eukaryotic cells with a high efficiency (*3*). The infectivity/transformation ratio of adenovirus 5 is about 4-5 log units, whereas there was only a 10-fold difference for the corresponding purified DNA (*3*). This is understandable, in view of the fact that only part of the viral DNA is required for transformation. It is important to define the part that is required. Using a restriction enzyme that cuts the circular polyoma DNA molecule into 8 specific fragments, Fried et al. have recently shown (*1*) that only one of the fragments has transforming activity. Similar developments are under way with regard to SV40 and some of the oncogenic adenoviruses. We can expect a fairly precise definition of the oncogenic information in the small and moderate-sized DNA viruses within the immediate future. When this happens, some of the most important questions concerning the gene products responsible for transformation will open up for direct experimental attack, both at the transcription and the translation level.

Important developments have also taken place in the RNA tumor virus field. Parallel genetic and biochemical analysis suggests that the C-type viral genome consists of three similar subunits and the true size of the haploid

[1]Congress presentation B–2–2.

genome is therefore not 10×10^6 as previously believed, but approximately 3×10^6. In other words, this is a polyploid, or more appropriately, polytenic virus. This brings the effective genome size down to the same order of magnitude as polyoma and SV40. As a consequence, the number of viral functions that can be considered as crucial for transformation becomes very small. The fact that transformed cells carry the viral genome as covalently integrated DNA copies further narrows the gap between the two virus groups. It has been also shown (*4*) that recombination is extremely frequent among RNA tumor viruses and probably occurs at the DNA level. This is probably an important fact for the understanding of the multitude and versatility of the C-type viruses. The main question is whether the oncogenic information is present in all forms or only in some very special forms, perhaps introduced by recombination. Avian sarcoma viruses contain "extra" information, not present in viruses with only helper (or leukemogenic) activity that cannot transform fibroblast monolayers (*5*). This transformation-specific information is lost when non-transforming variants arise from the transforming viruses. It was first identified by biochemical and later by genetic methods. Recent electron microscopic evidence (*6*) suggests that the transformation-associated extra information represents approximately 16% of the total RNA genome.

At least in its popularized form, the "oncogene" theory (*7*) has postulated the existence of ubiquitous, vertically transmitted C-type viruses, containing specific oncogenic information ("oncogenes"), as part of their genetic endowment. In terms of the theory most, if not all, oncogenic agents would act by activating these genomes. There are several reasons to doubt the generalizations of this theory, however. It was shown that endogenous C-type viral information of natural, outbred species, e.g. cat, fowl, or baboon, is present in every cell of every tissue of every individual (*8*). Since all cells carry the information, it is understandable that cells of the home species are resistant to superinfection with endogenous virus, in contrast to cells of other species that may be susceptible; this explains the experimental fact of the xenotropic viruses (*8—10*). Even more important is the fact, however, that none of the endogenous viruses found in natural species (e.g. RD-114 in the cat or the E subgroup viruses in the fowl) induced *any* tumors in any of the home or foreign species so far tested. This, together with the ubiquitous occurrence of the endogenous genomes in normal cells, suggests that they may not carry any oncogenic information at all. On the contrary, as suggested (*8, 11*) they may play some important role in normal cell physiology and development. If so, the evolution of oncogenic variants may be due to recombination between endogenous viruses and normal or abnormal cell genomes, or between endogenous and horizontally transmitted viruses. At least in broad outline, this would largely correspond to the theory of Temin (*11*). The evidence that a naturally oncogenic C-type virus is horizontally transmitted in the cat whereas the endogenous, vertically transmitted virus is not oncogenic has recently been summarized by Essex (*12*).

It may be objected that the inbred mouse strains provide quite a different picture. This is quite true, but the relevance of inbred, homozygous strains for the understanding of wild, natural populations is questionable in itself. The AKR mouse is often quoted as the most important example of "natural" vertical transmission of a leukemogenic virus. It must be remembered, however, that AKR and similar high leukemia strains have not only been inbred but also continuously selected for high leukemia incidence. As so often in the previous history of cancer research (cp. e.g. *13* for the genetic analysis of mammary tumor incidence in mice), inbreeding has resulted in the fixation of *several* genetic factors that favor leukemia development by quite different mechanisms (*14*). One locus, *AKv–1,* appears to represent the DNA copy of the leukemogenic Gross virus, now part of the germ line coded information of the AKR strain, not present at the same locus in other strains (*15*). In addition, the gene that influences the amplification and spreading of most C-type viruses in the mouse (Fv-1) has been fixed in the AKR strain in the allelic form (Fv-1^n) that is favorable for the multiplication of the N-tropic Gross virus. A third gene, Rgv-1, can normally convey a relative resistance against Gross virus-induced leukemogenesis, if present in its most common allelic form, Rgv-1^r. In AKR, Rgv-1 has been fixed in its relatively rare *s* form that conveys susceptibility, however, Rgv-1 is localized within the K-end of the H-2 complex, in the same area as known immune responsiveness (Ir) genes. It is therefore likely that it acts via an immune mechanism, although there is no direct evidence for this as yet. If it is an Ir-gene, the fixation of the susceptibility allele may be akin to genetically determined immunological unresponsiveness to certain defined antigens (*16*). It is thus clear that a whole constellation of genetic factors, all favoring high leukemia incidence, has been fixed as a result of inbreeding and selection and the list is probably far from complete (*14*). Nothing could be less representative for natural populations, particularly since natural selection can be expected to work in the opposite direction.

A most important question is whether the leukemogenic viral genome, e.g. the integrated Gross virus of the AKR mouse, carries a distinct piece of genetic information, required for leukemia induction, in analogy with the sarcomogenic "extra" piece or if it is essentially identical with any of the common endogenous viral genomes of the species. Recent nucleic acid hybridization experiments (*17*) suggest that while all mouse strains, including AKR, share 7–8 copies of endogenous C-type virus sequences per cell. AKR and a number of other high leukemia strains have 3–4 copies of an "extra" AKR specific information that is completely absent from a number of leukemia-free strains. Low leukemia strains like BALB-C have 1–2 copies. Although much remains to be done, this is at least consistent with the possibility that leukemogenic potency, like sarcomogenic transforming capacity, will be eventually associated with an extra piece of genetic information. One of the main reasons why this has not yet been achieved may lie in the fact that there are no readily available in vitro systems for the transformation of avian, murine or feline leukocytes into leukemia or lymphoma lines in vitro, in contrast to the readily available focus-forming assays for sarcomatous transformation.

Not much is known about the products of the oncogenic information in the transformed cells. Virus mutants, temperature sensitive for the transforming function, were useful in establishing the fact that a viral product, probably protein in nature, is responsible for the maintenance of the transforming phenotype (*18–22*), because the latter can be reversibly switched off by transfer to the non-permissive temperature. One interesting difference between the small oncogenic DNA viruses and the sarcoma-producing C-type viruses lies in the fact that the transforming function is required for the productive viral cycle in the former, but not in the latter system. This may be related to the fact that the "extra" piece of information required for the transforming effect of the C-type sarcoma viruses may have been picked up by recombination and would be non-essential for the replicative function. In contrast, there is no evidence for a similar heterogeneity within populations of, for instance, polyoma or SV40 particles. In the latter case, transformation is dependent on the non-permissive association of the virus and the target cell, because the productive cycle is invariably lytic. Transformation therefore appears more as a biological accident, resulting from the interplay of an integrated, but non-productively replicating viral genome and the appropriate host cell, rather than the presence of specific "transforming" information, acquired during the history of the particular virus strain.

All this suggests that the oncogenic part of transforming viruses will be rather precisely defined in the near future. This will be followed by the identification of its RNA and protein products in the transformed cell. This does not necessarily mean that transformation will have been understood, however. Different lines of experimental evidence concur in showing that the host cell can exert a decisive influence on the transformation event. For example, revertants isolated from SV40-transformed cells still carry the viral genome, express virally determined antigens and yield wild type virus in fusion-rescue type experiments, thereby proving that the reversion-inducing change was in the cellular, not the viral genome (*23*).

Direct evidence of cellular control over virally determined transformation is provided by the temperature-sensitive variants of SV40-transformed cells, isolated by Basilico (*24*). Although these cells were temperature-sensitive with regard to their transformed phenotype, they carry the wild type viral genome, showing that the transformation-controlling *ts* function is cellular, rather than viral. This contrasts with the previously mentioned viral *ts* mutants for transformation.

It thus appears that transformation depends on the interaction of viral and cellular functions that can influence its probability in a positive or negative direction. A similar conclusion was derived from the hybridization of normal and malignant cells (*25–27*). They behaved as non-malignant or low-malignant cells, provided they contained a nearly complete sum of the two parental karyotypes. After they lost chromosomes derived from the normal parent, the low malignant hybrids threw off high malignant segregants again. This suggests that normal parents introduced chromosomal determinants capable of suppressing the malignant behavior of the partner. Presumably, this is achieved by compensating some deficiency at the genetic or epigenetic level, associated with the malignant transformation. It is interesting to note that normal cells suppressed the malignant behavior of very diverse tumors, including methylcholanthrene-induced sarcomas, polyoma virus-induced sarcomas, Moloney virus-induced lymphomas and spontaneous mammary carcinomas. Different cells were active in a suppressive capacity on the normal side, including fibroblasts, lymphocytes and macrophages (*28*).

These findings suggest that the malignant cell has lost some function that is essential for normal cellular responsiveness to the growth-controlling mechanisms. How can we visualize such a loss of function in, for instance, polyoma-transformed cells where the integrated viral genome is assumed to control the transformed phenotype? It could be imagined that the polyoma information interferes with the structure and, as a consequence, the function of membrane-associated growth control receptors. The expression of polyoma-determined specific surface (TSTA) antigens illustrates this as a concrete possibility. Introduction of normal receptors by the karyotype of a normally differentiated cell may reestablish normal responsiveness, even though the antigen is maintained (*26*).

The influence of genetic factors on the proneness of different cells to undergo neoplastic transformation was also noted through another, now classical experimental design (*29*). In general, the same *normal* tissue was transplanted, from a mouse strain characterized by a high incidence of malignant tumors in that particular tissue and from a low incidence strain, to symmetrical sites in the same F_1 hybrid host. Since both parents were derived from highly inbred, homozygous strains, the hybrid hosts were compatible with the tissues of both. If the proneness to undergo neoplastic transformation is determined at the level of the target tissue, a larger number of tumors is expected to develop in the normal tissues of the high-tumor strain than in the tissues of the low-tumor strain. This was actually found to be the case for tumors of very diverse target tissues, including the mammary gland, the thymus (leukemia), the ovaries, the adrenals and the lung (adenomas). It is therefore clear that the probability of the neoplastic change can be influenced in a positive or negative direction by genes that act at the target tissue level.

These examples may have served to illustrate that carcinogenesis must be considered from the viewpoint of the causative agent (chemical reactivity for chemical carcinogens, specific information for oncogenic viruses) and also the target cell (specific "competence" for transformation). A third important level is the host organism.

The concept of immune surveillance postulates that all neoplastic cells are promptly recognized by the immune system, and tumor growth is therefore a consequence of a breakdown of the rejection mechanism or some malfunctioning of the effectors involved (e.g. "enhancement" or "blocking" by serum factors). Recent evidence suggests, however, that the surveillance concept has only limited validity (for review see *30*). While the evidence is strong for a large number of virus-induced tumors, it is much less impressive for chemically induced or "spontaneous" tumors.

Neonatal thymectomy, treatment with antilymphocytic serum (ALS) and other immunosuppressive agents increase the incidence of most virally induced experimental tumors (*31*) with two important exceptions: (i) Systems where the thymus-derived lymphocytes are the targets of the immunosuppressive action and of the neoplastic change as well. This can be exemplified by the thymus-derived mouse lymphomas induced by the Gross or the Moloney virus. (ii) The other exception is represented by tumors induced by vertically transmitted viruses that replicate productively, with viremia and antigen excess as the result. Gross virus-induced leukemia of AKR mice or MTV-induced mammary carcinomas of the C3H mice may serve to exemplify this.

The best evidence for surveillance concerns potentially neoplastic cells induced by ubiquitous viruses in their natural host species. Polyoma virus in mice is a good example. Under laboratory conditions, it can induce a large variety of tumors, provided this is given to baby mice during the first 24 hours of life and only if the mothers have been protected from viral infection. Although widespread in wild mice, the virus does not appear to cause any tumors in them (*32*), because infected mothers protect the young by passive antibody transfer during the critical, sensitive period. When infected later, the immunologically competent mice recognize the polyoma-induced tumor cells as foreign (*33*) and promptly reject them. This recognition is partly due to the fact that all polyoma-induced tumor cells share a common membrane change (usually referred to as TSTA) and partly because mice are apparently capable of recognizing this change as antigenically foreign, perhaps due to the selective fixation of appropriate immune responsiveness (Ir) genes (*16*).

While it is likely that most, and perhaps all, virus-induced tumor cells have analogous membrane changes, it does not follow that recognition is self-evident or that it is equally efficient and strong in different systems. In the mouse polyoma case and in other, similar examples (e.g. SV40 inoculated hamsters) the virus-inoculated newborn animals can be protected from most tumors, provided that a second virus inoculum is given towards the end of the oncogenic latency period. Recognition is more variable and less efficient in relation to another naturally contagious, oncogenic DNA virus, MDV, i.e. the herpesvirus that causes Marek's disease in chickens (*34*). In genetically resistant birds, the majority respond to virus exposure with a self-limiting lymphoproliferative disease whereas a minority gets malignant lymphomas. In genetically susceptible stocks, the majority develops the malignant lesion. Most malignant lesions can be prevented by an immunologically cross-reactive, but apathogenic turkey herpesvirus (HVT). Thus, it is not sufficient to give a second dose of the same virus in this case because recognition is less efficient and more variable. An attentuated vaccine virus is therefore necessary.

In the Marek system, genetic resistance is often due to a single, dominant gene. Since neonatal thymectomy can render genetically resistant birds susceptible (*35, 36*), it is likely that genetic resistance acts through a thymus-dependent, i.e. immunological mechanism.

If polyoma virus illustrates maximally efficient recognition and MDV represents a whole range from inefficient through partially efficient, to efficient rejection, depending on host genetics, Herpesvirus saimiri (HVS) illustrates complete recognition and rejection, and also its opposite, maximum susceptibility, depending on the host species (*37*). The natural host, the squirrel monkey, contains both virus and antibodies, but develops no tumors. This is thus similar to polyoma virus in wild mice. In two other species, the marmoset and the owl monkey, the virus induces 100% lethal lymphoproliferative malignancy. These animals are very difficult to vaccinate and it is therefore likely that their recognition of the virus-associated cellular antigens is very poor. This may well be due to their lack of exposure to the virus during their natural history and hence the absence of any preselection for Ir-genes that would condition their response to this particular group of antigens.

Another situation where a virally induced tumor provokes little if any host immune reaction is the murine sarcoma virus (MSV)-induced sarcoma, provided it is free of helper virus (MLV) (*38, 39*). If it is infected with MLV, it induces a very strong host response, leading to tumor rejection that is entirely directed against MLV-determined antigens. It must be realized, however, that MSV never occurs without MLV in nature, since it is a helper-dependent, defective virus. It would be understandable, therefore, if the host response would have evolved to recognize MLV, rather than MSV-induced surface changes since the former would be regularly encountered and common to MLV-induced leukemia and MSV (MLV)-induced sarcoma cells.

Taken together, this reasoning thus leads to the conclusion that immune surveillance is a reality, but mainly in relation to tumors that the host species has been selected to recognize during its evolution. Virus-induced tumors are obviously the main category here, since the same antigenic change is being induced repeatedly in a large number of target cells, in all or most individuals of the species, during reproductive age. In contrast, chemically induced tumors are often characterized by a remarkable antigenic diversity, even within groups of tumors induced by the same carcinogen, in the same cell type and in the same species. The immune system is apparently not prepared to recognize all antigenic varieties. It is therefore understandable that tumors induced by the same chemical carcinogen vary greatly with regard to immunogenicity and some are not detectably immunogenic at all. There are, moreover, entire groups of chemically induced neoplasms (e.g. the many different tumors induced by AAF, acetylaminofluorene) that do not seem to induce a rejection reaction in syngeneic hosts (*40*). The same applies to a variety of spontaneous tumors. It is still reasonable to assume that most if not all neoplastic changes are accompanied by changes in the structure and function of the cell membrane, but whether they are recognized as antigenic or not obviously depends on the question whether a corresponding recognition system exists within the "antigenic dictionary" of the host. It is important to distinguish, both conceptually and experimentally, between the breakdown of an existing anti-tumor response, prior to or in the course of tumor development, and a primary lack or deficiency of recognition. In the former case one has to look for the causes of the breakdown (like enhancing antibody, blocking antigen or complexes, quantitative inefficiency of an effector component, etc.). In the latter case, means will have to be found whereby the lack of recognition can be circumvented. There is ample evidence in the experimental literature concerning immune responsiveness (Ir) genes, showing that genetically unresponsive hosts can be rendered responsive by coupling a compound that is non-immunogenic on a certain genetic background, to a well recognized antigen (*16*). Similar approaches will have to be explored for tumor-associated antigens as well.

Are human tumors recognized as antigenic by the patient? Very little direct evidence exists on this point. The relevance of in vitro lymphocytotoxicity tests for in vivo rejection is very much in doubt (cp. *41*). It is obvious that direct extrapolations are not possible. The documented rejection of various subcutaneously implanted autologous tumors by the carcinoma patient (*42*) nevertheless suggests that at least a considerable number of human tumors are antigenic and grow in the presence of a certain "concomitant immunity" that can reject small numbers of tumor cells, but are unable to cope with large, established tumors. The same mechanism can account for the frequent clinical or histopathological observation that circulating tumor cells only rarely give rise to metastases.

The Epstein Barr virus (EBV) is an example of an ubiquitous, potentially oncogenic human virus that is probably kept under close immune surveillance under normal conditions (*43*). Between 80 and 90% of all adult human populations are seropositive for EBV antibodies. Under low socioeconomic conditions, the primary infection occurs most frequently during early childhood. In high socioeconomic groups, only a minor part of the population undergoes early infection and a larger part becomes infected later, usually during teenage or adolescence. At this age, infection causes mononucleosis in about half of the cases, whereas the other part undergoes "silent" seroconversion. Infectious mononucleosis in itself is often considered as a self-limiting neoplastic disease, terminated by a powerful immune reaction (*44*).

In vitro, EBV can transform normal B-lymphocytes that have a limited lifespan and do not grow spontaneously as established lines, into "immortal" lymphoblastoid cell lines that possess a malignant potential when inoculated into immunologically crippled animals, like nude mice or cortisonized hamsters (*45*). Similar, potentially neoplastic cell lines can be isolated from the peripheral circulation of EBV-positive individuals, but not from EBV negatives (*46*). This is in line with the assumption that EBV-transformed B-lymphocytes are present in normal virus-positive individuals, but are normally kept under immune surveillance.

In spite of increased serum titers against EBV in a number of diseases (*47*), there are only two neoplastic diseases in which the tumor cells are known to carry the viral genome, African Burkitt's lymphoma and anaplastic carcinoma of the nasopharynx. In both diseases, multiple EBV-genome equivalents are present per cell, together with the EBV-determined nuclear antigen (EBNA) (for review see *48*). African Burkitt's lymphoma represents the proliferation of an EBV-genome-carrying clone (*49*). Other lymphomas, including non-African Burkitt lymphomas (that do not show the same time-space clustering as the African disease), lymphosarcoma, Hodgkin's disease, chronic lymphatic leukemia etc., do *not* represent the proliferation of an EBV-genome-carrying neoplastic cell in any of the cases so far investigated (*48*). This argues against the possibility that the EBV-genome travels along in the African Burkitt lymphoma merely in the capacity of a passenger, because many of the genome-negative lymphomas actually come from highly seropositive (i.e. virus carrying) patients. From some of them, EBV-genome-negative cell lines have been established that can be superinfected with EBV in vitro (*50*). This can result in permanent conversion to a genome-positive line. The neoplastic cell is thus sensitive to superinfection, but has nevertheless failed to pick up the virus in vivo, presumably due to the presence of neutralizing antibodies.

If this reasoning is correct, the regular presence of the genome in the African Burkitt lymphoma must mean that it has been present from the beginning or, in other words, that the lymphoma has arisen in an EBV-genome-carrying cell. This leaves us essentially with two main alternatives. The lymphoma has either developed because the surveillance of the African patient has broken down, perhaps only temporarily, and allowed one clone to grow to irreversible size. Alternatively, some other agent was involved in promoting the full 'malignization' of what may have been only a premalignant, or conditioned neoplastic cell. It is not possible to distinguish between these alternatives at present, but the former appears to me simpler and therefore more attractive. Chronic holoendemic malaria has been suggested (*51*) as a possible immunosuppressive agent that would favor the neoplastic growth of the EBV-carrying cell. The distribution of chronic holoendemic malaria (not just malaria) is, in fact, quite similar to high endemic Burkitt lymphoma.

The significance of the genome in nasopharyngeal carcinoma is not yet clear. It has been shown recently (*52, 53*) however, that the genome is present in the carcinoma cells themselves, not in the contaminating lymphocytes as originally thought. This, together with the geographically uniform association of the virus with the anaplastic carcinoma, and its absence from differentiated, squamous cell carcinomas, localized in the nasopharynx or in adjacent regions, makes this association, together with African BL, a prime candidate for an oncogenic interaction.

ACKNOWLEDGEMENTS

The work of the author and his associates was supported by Public Health Service Research Grant No. 1RO1 CA14054-01 from the National Cancer Institute, the Swedish Cancer Society and the Damon Runyon Memorial Fund (DRG-1064-B).

REFERENCES

1. M. Fried, B. Griffin and E. Lund, *Cold Spring Harbor Symp. Quant. Biol.* **39** (1974), in press.
2. C.-J. Lai and D. Nathans, *Cold Spring Harbor Symp. Quant. Biol.* **39** (1974) in press.
3. F. L. Graham, P. J. Abrahams, S. O. Warnaar, C. Mulder, F. A. J. de Vries, W. Fiers and A. J. van der Eb, *Cold Spring Harbor Symp. Quant. Biol.* **39** (1974), in press.
4. P. H. Duesberg, P. K. Vogt, M. Lai and K. Beemon, *Cold Spring Harbor Symp. Quant. Biol.* **39** (1974), in press.
5. P. H. Duesberg and P. K. Vogt, *Virology* **54,** 207 (1973).
6. H.-J. Kung. J. M. Bailey, N. Davidson, P. Vogt. M. O. Nicolson and R. M. McAllister, *Cold Spring Harbor Symp. Quant. Biol.* **39** (1974), in press.
7. R. J. Huebner and G. J. Todaro, *Proc. Nat. Acad. Sci. (USA)* **64,** 1087 (1969).
8. G. J. Todaro, M. M. Lieber, C. J. Sherr and R. E. Benveniste, *Cold Spring Harbor Symp. Quant. Biol.* **39** (1974), in press.
9. J. A. Levy, *Science* **182,** 1151 (1973).
10. S. A. Aaronson, *Cold Spring Harbor Symp. Quant. Biol.* **39** (1974), in press.
11. H. M. Temin, *Proc. Nat. Acad. Sci. (USA)* **69,** 1016 (1972).
12. M. Essex, *Advan. Cancer Res.*, in press.
13. W. E. Heston, *Cancer Res.* **25,** 1320 (1965).
14. F. Lilly and T. Pincus, *Advan. Cancer Res.* **17,** 231 (1973).
15. W. P. Rowe, *Cancer Res.* **33,** 3061 (1973).
16. B. Benacerraf and D. H. Katz, *Advan. Cancer Res.*, in press.
17. S. K. Chattopadhyay, D. R. Lowy, N. M. Teich, A. S. Levine and W. P. Rowe, *Cold Spring Harbor Symp. Quant. Biol.* **39** (1974), in press.

18. W. Eckhart, *Cold Spring Harbor Symp. Quant. Biol.* **39** (1974), in press.
19. P. Tegtmeyer, *Cold Spring Harbor Symp. Quant. Biol.* **39** (1974), in press.
20. R. G. Martin, J. Y. Chou, R. Saral and J. Avila, *Cold Spring Harbor Symp. Quant. Biol.* **39** (1974), in press.
21. H. S. Ginsberg, M. J. Ensinger, R. S. Kauffman, A. J. Mayer and U. I. Lundholm, *Cold Spring Harbor Symp. Quant. Biol.* **39** (1974), in press.
22. J. A. Wyke and J. G. Bell, *Cold Spring Harbor Symp. Quant. Biol.* **39** (1974), in press.
23. I. MacPherson, *Advan. Cancer Res.* **13,** 169 (1970).
24. H. C. Renger and C. Basilico, *Proc. Nat. Acad. Sci. (USA)* **69,** 109 (1972).
25. G. Klein, U. Bregula, F. Wiener and H. Harris, *J. Cell Sci.* **8,** 659 (1971).
26. G. Klein and H. Harris, *Nature (London) New Biol.* **237,** 163 (1972).
27. G. Meyer, M. Berebbi and G. Klein, *Nature (London)* **249,** 47 (1974).
28. F. Wiener, G. Klein and H. Harris, *J. Cell Sci.* **15,** 177 (1974).
29. W. E. Heston, in *Genetics and Cancer*. 13th Annual Symposium on Fundamental Cancer Research, (R. W. Cumley, Ed.), pp. 226–240. University of Texas Press, Austin, 1959.
30. G. Klein, Immunological surveillance against neoplasia, in *Harvey Lectures,* Academic Press, New York, in press.
31. A. Allison, in *Scientific Foundations of Oncology,* (T. Symington and R. L. Carter, Eds.), Heinemann Med. Books, London, in press.
32. R. J. Huebner, *Ann. N.Y. Acad. Sci.* **108,** 1129 (1963).
33. H. O. Sjögren, *Progr. Exp. Tumor Res.* **6,** 289 (1965).
34. K. Nazerian, *Advan. Cancer Res.* **17,** 279 (1973).
35. L. N. Payne, in *Symposium on Oncogenesis and Herpes-type Viruses* pp. 21–37, Cambridge University Press, New York, 1972.
36. L. N. Payne, in *Oncogenesis and Herpesviruses*, Proc. Symp. Christ's College, Cambridge (P. M. Biggs, G. de Thé and L. N. Payne, Eds.) pp. 21–37. WHO, Geneva, 1972.
37. F. W. Deinhardt, L. A. Falk and L. G. Wolfe, *Advan. Cancer Res.* **19,** 167 (1974).
38. V. Strouk, G. Grundner, E. M. Fenyö, E. Lamon, H. Skurzak and G. Klein, *J. Exp. Med.* **136,** 344 (1972).
39. J. R. Stephenson and S. A. Aaronson, *J. Exp. Med.* **135,** 503 (1972).
40. R. W. Baldwin, *Advan. Cancer Res.* **18,** 1 (1973).
41. G. Klein, in *Scientific Foundations of Oncology,* (T. Symington and R. L. Carter, Eds.). William Heinemann Med. Books, London, in press.
42. A. Brunschwig, C. M. Southam and A. Levin, *Ann. Surg.* **162,** 416 (1965).
43. G. Klein, in *The Herpesviruses,* (A. S. Kaplan, Ed.), pp. 521–555. Academic Press, New York, 1973.
44. R. L. Carter, in *Oncogenesis and Herpesviruses,* (R. M. Biggs, G. de Thé and L. N. Payne, Eds.), pp. 230–238. Int. Agency Res. Cancer, Lyon, 1972.
45. G. Miller, *Yale J. Biol. Med.* **43,** 358 (1971).
46. V. Diehl, G. Henle, W. Henle and G. Kohn, *J. Virol.* **2,** 663 (1968).
47. G. Henle and W. Henle, *J. Infect. Dis.* **121,** 303 (1970).
48. G. Klein, *Cold Spring Harbor Symp. Quant. Biol.* **39** (1974), in press.
49. P. J. Fialkow, G. Klein, S. M. Gartler and P. Clifford, *Lancet* **7643,** 384 (1970).
50. G. Klein, B. Sugden, W. Leibold and J. Menezes, *Intervirology,* in press.
51. D. Burkitt, *J. Nat. Cancer Inst.* **42,** 19 (1969).
52. G. Klein, B. C. Giovanella, T. Lindahl, P. Fialkow, S. Singh and J. S. Stehlin, *Proc. Nat. Acad. Sci. (USA),* in press.
53. H. Wolf, H. zur Hausen and V. Becker, *Nature (London) New Biol.* **244,** 245 (1973).

Direct Evidence that Damaged DNA Results in Neoplastic Transformation—A Fish Story[1,2]

R. B. Setlow[3] and R. W. Hart[4]

Carcinogenesis Program, Biology Division, Oak Ridge National Laboratory
Oak Ridge, Tennessee 37830, USA

INTRODUCTION

The international concern about environmental pollutants and their mutagenic and carcinogenic activities is reflected by the many experiments on the interaction between such agents and the macromolecular components of cells. Our interest in ultraviolet (UV) radiation as a carcinogenic agent stems from the large amount of basic photochemical and photobiological information concerning UV damage and its repair (*1,2*) and from the fact that human skin cancer—which comprises over half of all the cancers of the white population of the United States (*3*)—is associated with UV radiation. Animal experiments indicate that it is the shorter wavelength component (<320 nm) of sunlight that is the carcinogenic agent; since this component increases with decreasing latitude, as does skin cancer, it is reasonable to suppose that human skin cancer arises from such wavelengths in the sun's spectrum (*4,5*). It is noteworthy that one environmental impact of a large fleet of supersonic transports could be a decrease in the ozone level in the stratosphere with an attendant increase in UV radiation and skin cancer (*6*). We are interested in knowing whether DNA is the target macromolecule, and, if so, what initial molecular changes in DNA result in neoplastic transformation. None of the experiments to be described elucidate the mechanism by which such changes result in transformation.

Individuals with the disease xeroderma pigmentosum (XP) are a very high risk population for UV-induced skin cancer (*7*). The observation that most of these individuals are not able to repair DNA damage (*2,8*)—they do not excise pyrimidine dimers from their DNA (*9*)—is evidence that DNA is the target and implies that damage to DNA is the causative agent. Moreover, a number of chemical carcinogens are also mutagens (*10*). The damage some of them do to DNA is repaired by a sequence of steps that mimics the repair of UV damage in normal human cells (*11*). XP cells are unable to repair DNA damage from these carcinogens (*11*). This finding is further evidence that DNA is the target for neoplastic transformation.

[1]Congress presentation B–2–3.

[2]This research was jointly sponsored by the Carcinogenesis Program of the National Cancer Institute and the U.S. Atomic Energy Commission under contract with Union Carbide Corporation.

[3]Present address: Biology Department, Brookhaven National Laboratory, Upton, New York 11973.

[4]Present address: Department of Radiology, Ohio State University, Columbus, Ohio 43210.

Since UV makes many changes in DNA (*1*), we do not know whether the biologically important photoproducts are cyclobutyl-pyrimidine dimers or one of the many other alterations induced in DNA. The absence of dimer excision in XP cells is not conclusive evidence that dimers are the important causative agent, because other uninvestigated changes may not be repaired either. Just as for alkylation damage, the most common product may not be the most important biologically (*12,13*). One particular repair system—enzymic photoreactivation (PR)—is specific for pyrimidine dimers (*14,15*). Of the many products made by UV irradiation of DNA, only dimers are affected by PR. The process monomerizes the dimers, yielding the original unaltered polynucleotide, but leaves the other products untouched. Therefore, PR can be used as a diagnostic tool. If formation of tumors by UV were prevented by PR, this finding would be evidence that cyclobutyl-pyrimidine dimers are a lesion in DNA that results in neoplastic transformation.

A number of years ago Rogers (*16*) showed that if embryonic mouse lung tissue was irradiated in vitro with 254-nm radiation, it would give rise to adenomas upon transplantation to homologous recipients. Such experiments are attractive since the treatment of cells in vitro permits one to make good physical or chemical dose measurements. However, since mice are deficient in PR activity (*15*), PR experiments cannot be used to assess the role, if any, of dimers in transformation. Marsupials do contain useful amounts of PR enzyme (*15*); therefore, in 1969 J. S. Cook, J. D. Regan, and I designed experiments using marsupials. The experiments were not begun because of the associated housekeeping problems. Meanwhile, a letter in *Science* (*17*) led J. D. Regan to suggest the use of clones of fish as an experimental system. Fish cells contain large amounts of PR enzyme (*15*). Furthermore, the fish we used, *Poecilia formosa,* may be grown in clones, permitting successful transplantation of cells from one animal to another (*18*).

EXPERIMENTAL METHODS

Initial Experiments

We used several clones of the naturally occurring gynogenetic (nonsexually reproducing) fish *Poecilia formosa* (*18,19*). The offspring of the fish are white and identical, and since the females are activated by the males of another species (black), the transmission of any male information is easily detected by the mottled skin color (*20*). The general experimental design is shown in Fig. 1. In the first series of experiments cells from larger fish (1 year old, 5 cm long) were obtained from liver. The tissue was homogenized to yield cell clumps (3–8 cells per clump), the clumps were treated as indicated in Fig. 1, and 20 μl containing 3–5 $\times$ 10^5 cells was injected into the dorsal muscle of isogenic recipients (~3 to 4 months old). The recipients were scored for tumors ~6 months later by B. Koesner of the Veterinary Pathobiology Department, Ohio State University.

Later Experiments

In later experiments designed to improve the histological analysis, cells were obtained from younger fish (3–4 months, 2 cm long), treated as in Fig. 1, and injected into the abdominal cavity of recipients. In these experiments several tissues were homogenized together. Hence, injected cells consisted of a mixture of liver, heart, and thyroid cells. The latter tissue represented ~40% of the injected cells. Fish were killed 6 to 9 months after the injection of cells and fixed in Bouin's solution. Tumors were scored by gross pathology and by the appearance of stained histological sections. We are deeply indebted to John Harshbarger of the Registry of Tumors in Lower Animals, Smithsonian Institution, Washington, D.C., for the histological analyses. It was he who first noticed that the fish injected with irradiated cells had thyroid carcinomas. His histological conclusions were confirmed by C. J. Dawe of the National Cancer Institute.

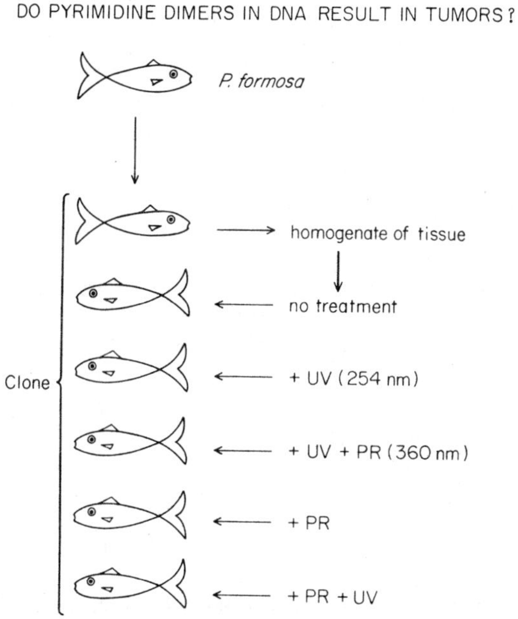

FIG. 1. Outline of an experiment to determine whether pyrimidine dimers in the DNA of irradiated cells result in neoplastic transformation. If they do, then there should be very few tumors in those fish injected with cells treated with UV plus PR compared with UV alone or compared with PR plus UV. The no-treatment and PR samples are controls.

UV Irradiation and Photoreactivation

The UV radiation, primarily 254 nm, was obtained from a germicidal lamp. The incident fluence rate on cell suspensions containing ~2×10^7 cells/ml was 1.0 $J/m^2/sec$. The irradiated samples were ~ 1 mm thick, and the average incident fluence rate through the sample was ~0.1 $J/m^2/sec$. PR illumination was supplied by black light lamps in an incubator at 37° C. Typical PR fluence rates were 700 $J/m^2/min$, in the wavelength range 320–400 nm.

Carcinogen Treatment

Approximately 2.5×10^7 cells/ml in medium containing 10% fetal calf serum (to approximate the experimental conditions used to compare UV and carcinogen damage to human cells in ref. *11*) were treated with the chemical carcinogen *N*-acetoxy-AAF (ref. *10*) at concentrations of 10^{-5} and 10^{-4} *M* for 1 hr. At the end of this time the cells were collected by centrifugation and resuspended in Ringer's solution; then 20 μl was injected into isogenic animals.

RESULTS

Initial Results

The results obtained by injecting treated liver cells into the back muscle of fish have been described briefly (*21*). After an incident UV influence of 20 J/m^2, ~10% of the fish had large tumors at the site of injection. Exposure of the cells to PR illumination after UV resulted in a large decrease in the numbers of tumors—a dose reduction factor of ~0.8. PR illumination by itself had little effect. Because such highly localized tumors were not suitable for histological examination, we turned to experiments in which treated cells were injected into the abdominal cavity.

Later Results

The initial attempts with cells injected into the abdominal cavity were not useful because of the presence of tuberculosis organisms in many of the fish. Many of the experimental animals contracted the disease and many had granulomas that might have been bacterially induced. After assuring that the colony was clean, we resumed the experiments. For an authoritative opinion on infections and on tumors in the fish, we enlisted the assistance of John Harshbarger (see Experimental Methods). He observed that most of the fish injected with UV-irradiated cells had large numbers of noninfectious granulomas and, in addition, had thyroid carcinomas which were invasive into both soft and hard tissue. Thyroid carcinomas were present in 100% of the fish injected with 5×10^5 cells exposed to an average incident fluence of 10–20 J/m^2. Smaller or larger fluences gave fewer tumors. Table 1 shows the effect of PR illumination before and after the UV exposure. Note that PR after UV irradiation results in a very large decrease in the number of thyroid carcinomas, whereas illumination before UV does not. A similar finding is observed for the granulomas. Thus, these data implicate pyrimidine dimers in DNA as the initial change that results in neoplastic transformation.

TABLE 1
Effects of PR Illumination[a] on UV-irradiated[b] Cells[c] of *P. formosa*

Treatment	Fraction of fish with granulomas	Granulomas per fish	Fraction of fish with thyroid carcinoma by	
			gross pathology	histology
UV	51/63	1.8	34/34	29/29
2.5 min PR + UV	49/57	2.0	26/26	22/22
5.0 min PR + UV	37/42	1.9	48/50	22/23
UV + 5.0 min PR	15/43	0.4	1/43	0/6

[a] 700 J/m^2/min (320–420 nm).
[b] 12 J/m^2 average incident fluence (254 nm).
[c] ~ 3×10^5 cells injected per fish.

Carcinogen-Treated Cells

Treatment of cells with *N*-acetoxy-AAF before injection resulted in the appearance of thyroid carcinomas in a large percentage of the fish, as shown in Table 2.

TABLE 2
Effect of *N*-Acetoxy-AAF[a] on Cells[b] of *P. formosa*

Concentration	Fraction of fish with thyroid carcinomas by gross pathology	histology
0	0/50	–
10^{-5} *M*	14/50	–
10^{-4} *M*	45/51	10/10

[a] Treatment for 1 hr at the indicated concentration.
[b] $\sim 5 \times 10^5$ cells injected per fish.

CONCLUSIONS

The initial experiments described here show that single exposures of fish cells to UV radiation result in neoplastic transformation, and that if the UV irradiation is followed by PR illumination, many fewer transformants arise. Our background knowledge of the photochemistry and photobiology (*1,14,15*) leads us to conclude that the initial change in macromolecules resulting from the UV irradiation has been in DNA and, moreover, that the particular change in DNA has been the formation of pyrimidine dimers. The data, of course, say nothing about the mechanism(s) by which such changes are converted into neoplastic transformations.

It is instructive to estimate the probability of a transformation per dimer in cellular DNA. We estimate (from data in this paper and also unpublished results) that an average incident fluence of 5 J/m^2 will result in approximately one transformed cell per 10^5 thyroid cells. This fluence converts ~0.02% of the thymines to thymine-containing dimers, which corresponds to ~1 dimer/10^7 daltons of DNA (*9*). Since there are about 10^{12} daltons of DNA per cell, there are 10^5 dimers/cell at 5 J/m^2. Hence, the probability of a random dimer resulting in a neoplastic transformation is $\sim 10^{-10}$.

Studies on the magnitude of DNA repair following treatment of normal human cells with UV and with *N*-acetoxy-AAF indicated that 10^{-5} *M* AAF was approximately equivalent to 5 J/m^2 (ref. *11*). In the experiments described here, 10^{-4} *M* AAF was equivalent to ~10 J/m^2. This comparison between measurements of repair in one species (human) and transformation in another (fish) may be farfetched, but the agreement isn't bad. We take these data as reinforcing the notion that damage to DNA is important to chemical carcinogenesis.

ACKNOWLEDGMENTS

This research was jointly sponsored by the National Cancer Institute and the U.S. Atomic Energy Commission under contract with the Union Carbide Corporation. We are indebted to C. P. Haskins who generously supplied us with the fish from which our clones were raised. We acknowledge the help of H. L. Livesey and B. J. Hendrix in the care and feeding of animals and the valuable assistance of F. Ahmed, G. E. Cosgrove, and T. L. Hoskins in the histological preparation and examination of specimens.

REFERENCES

1. R. B. Setlow and J. K. Setlow, *Annu. Rev. Biophys. Bioeng.* **1**, 293 (1972).
2. J. E. Cleaver, *Advan. Radiat. Biol.* **4**, 1 (1974).
3. J. Scotto, A. W. Kopf, and F. Urbach, *Cancer,* **34**, 1333 (1974).
4. H. F. Blum, *Carcinogenesis by Ultraviolet Light*. Princeton University Press, Princeton, 1959.
5. R. B. Setlow, *Proc. Nat. Acad. Sci. (USA)* **71**, 3363 (1974).
6. *Biological Impacts of Increased Intensities of Solar Ultraviolet Radiation*. National Academy of Sciences-National Academy of Engineering, USA, Washington, 1973.
7. J. H. Epstein, in *Photophysiology,* **Vol. 5** (A. C. Giese, Ed.), pp. 235–273. Academic Press, New York, 1970.
8. J. E. Cleaver, *Nature (London)* **218**, 652 (1968).
9. R. B. Setlow, J. D. Regan, J. German, and W. L. Carrier, *Proc. Nat. Acad. Sci. (USA)* **64**, 1035 (1969).
10. E. C. Miller and J. A. Miller, in *Chemical Mutagens,* **Vol. 1** (A. Hollaender, Ed.), pp. 83–119. Plenum Press, New York, 1971.
11. J. D. Regan and R. B. Setlow, in *Chemical Mutagens,* **Vol. 3** (A. Hollaender, Ed.), pp. 151–170. Plenum Press, New York, 1973.
12. R. F. Kimball, J. K. Setlow, and M. Liu, *Mutat. Res.* **12**, 21 (1971).
13. K. V. Shooter, R. Howse, S. A. Shah, and P. D. Lawley, *Biochem. J.* **137**, 303 (1974).
14. J. K. Setlow, *Res. Progr. Org. Biol. Med. Chem.* **3**, 335 (1972).
15. J. S. Cook, in *Photophysiology,* **Vol. 5** (A. C. Giese, Ed.), pp. 191–233. Academic Press, New York, 1970.
16. S. Rogers, *J. Nat. Cancer Inst.* **15**, 1001 (1955).
17. B. W. Agranoff, R. E. Davis, and R. E. Gossington, *Science* **171**, 238 (1971).
18. K. D. Kallman, *J. Genet.* **58**, 7 (1962).
19. C. L. Hubbs and L. C. Hubbs, *Science* **76**, 628 (1932).
20. R. W. Hart, H. R. Livesey, and R. B. Setlow, Reproduction and potential research used of the gynogenetic teleost, *Poecilia formosa. The Drum and Croaker,* in press.
21. R. W. Hart and R. B. Setlow, *Amer. Soc. Photobiol.* Abstr., p. 120, (1973).

IMMUNOLOGY OF CARCINOGENESIS AND ITS RELATION TO RADIATION[1]

Chairman: H. Cottier

[1]Symposium A-4

Mechanisms of Radiation Carcinogenesis[1]

Dirk W. van Bekkum
Radiobiological Institute TNO
Rijswijk, (ZH), the Netherlands

INTRODUCTION

This speculative review on radiation carcinogenesis deals mainly with its immunological aspects. It need not be emphasized that the role of immunology in carcinogenesis is not yet well understood. Immunological aspects of radiation carcinogenesis comprise a large number of different parameters on the part of the host as well as on the part of the tumor itself. Only two aspects, both related to radiation, will be discussed here. One is the way in which the carcinogenic exposure to ionizing radiation may affect the immune reactivity of the irradiated organism, thereby perhaps changing its responses against the malignant cells. The second aspect is the immunological properties of cells transformed by ionizing irradiation, which may provide the targets for a host-anti-tumor reaction.

IMMUNE SURVEILLANCE

The development of a malignant tumor, from its very initiation until the stage when the malignant cell population overwhelms the host defences by its infiltration into the surrounding tissues and by the growth of its distant metastases, is depicted schematically in Fig. 1. For most of the present discussion, only the first part of the natural history of the tumor is important, namely from the single transformed cell stage until the tumor becomes clinically manifest.

According to the theory of immune surveillance proposed originally by Thomas (*1*) and extended by Burnet (*2*), the organism controls aberrant cells —which may arise regularly under the influence of oncogenic viruses and chemical or physical carcinogens, or spontaneously— by the action of lymphocytes which police all tissues of the body. For this defence mechanism to be effective, eradication of those aberrant cells should, perhaps, take place before the stage of vascularization, i.e. when the tumor cell population comprises only a limited number of cells.

One crucial requirement for such a mechanism to become operative is the presence of new antigens on the surface of the aberrant cells, by which the lymphocytes can distinguish those cells from the normal tissues in such a way that the tumor cells can be rejected in a similar fashion as a foreign tissue transplant. It should be realized that the presence or absence of such tumor specific transplantation antigens (TSTA) has mainly been studied in established tumors, i.e. far beyond the stage of development that we are discussing now. There is, now, solid evidence that many established tumors possess antigens to which the host can react (*3, 4*). This certainly holds for the virus-induced animal tumors and for tumors induced by chemical carcinogens.

[1]Congress presentation A-4-1.

NATURAL HISTORY OF TUMORS

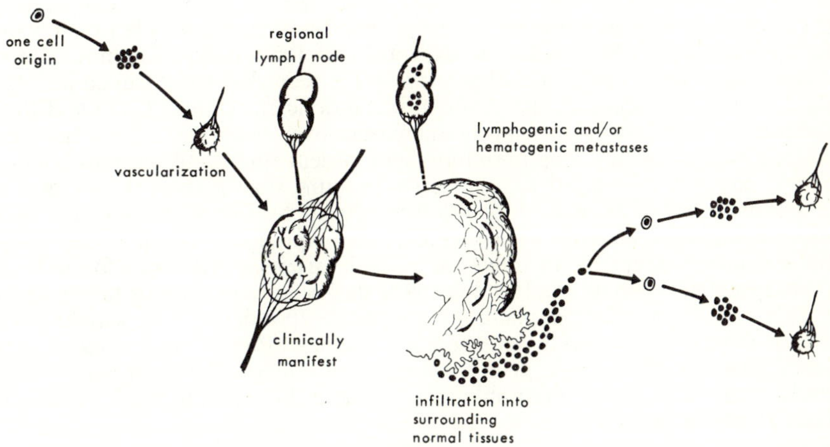

Figure 1.

We shall examine this point in more detail for radiation-induced tumors at a later stage; for the present it is assumed that radiation-induced tumors are not essentially different in this property from other tumors. But, even so, for the immune surveillance mechanism to be operative in preventing the establishment of tumors, it is necessary that the original few tumor cells already possess these TSTA. Specific antigenicity was demonstrated by Lappé (*5*) for papillomas which precede hydrocarbon-induced skin carcinomas and for the hyperplastic nodules preceding carcinogen-induced mammary carcinoma in the mouse (*6*). In this respect, the clinical studies of Byfield (*7*) are quite revealing, demonstrating the presence of both humoral and cellular immune responses in early stages of carcinoma of the cervix and even in the preneoplastic stage called dysplasia. However, antigenicity is not significant by itself. It should elicit an immune reaction of such a quality that it can destroy all transformed cells. We know that many induced tumors do possess tumor-specific antigens, which can evoke antibodies in the hosts, but which fail to induce a degree of transplantation immunity required for effective immunological surveillance.

If it is assumed for the sake of discussion that transformed cells can indeed elicit an effective immune response, it is obvious that the relation between tumor incidence and prolonged defective states of immunity is extremely relevant to revealing the role of immune surveillance mechanisms in tumor development. The highly increased tumor incidence reported in immune-deficient children (*8*) and in transplant recipients undergoing prolonged immune-suppressive treatment (*9*), seemed to provide crucial support for this hypothesis. But H. Cottier has pointed out (*10*) that there can be alternative interpretations of these observations. He drew attention to the preponderance of lymphoreticular tumors in immune-deficient children and suggested that these malignancies might be the result of a disturbance of the lymphatic system as a consequence of the deficient immunity, rather than due to a break-down of immune surveillance mechanisms.

A similar dilemma is encountered in radiation leukemogenesis, where exposure of a large part of the body might lead to both a temporary disturbance of proliferation homeostasis as well as to a temporary immune-deficient state.

There can be no doubt that the overall incidence of malignancies is enormously increased in certain immune-deficiency diseases in young children. In the various forms of immune-deficiency, it has been recorded as between 5 and 10% or even more. But, with the exception of Bruton type agammaglobulinemia, where all 5 cases of malignancy so far recorded were leukemias, epithelial tumors do occur in the other immune-deficient states (*9, 11*) and their incidence is certainly significantly increased over that in the normal age-matched control group. The preponderance of leukemias and lymphoreticular tumors may still be related to a defective immune surveillance, if we consider the possibility that such tumors may be caused by oncogenic viruses, as is common in many animal species. A defective immunity would render these children more susceptible to infection by such viruses, as has been found to be the case in experimentally induced immune deficiencies in rodents (as reviewed by Balner, *12*). Another explanation for the preponderance of leukemias in immune-deficient children is that these malignancies form a high proportion of all malignancies in this age group, so that the immune-deficient state would merely provide a more favorable environment for carcinogenic factors to which all children are exposed.

HUMORAL VERSUS CELLULAR SURVEILLANCE

The immune system operates by way of two effector mechanisms—a humoral one (the production of circulating antibodies) and a cellular one (the production of sensitised lymphocytes capable of killing target cells). Because the bone marrow is mainly involved in generating cells responsible for the humoral immunity and the thymus for cells involved in the cellular reactions, these systems are now generally designated as the B and T cell systems, respectively. Because the mechanism of immune surveillance has been associated with the mechanism of transplant rejection, the T cells have generally been implicated as being the predominant effectors of immune surveillance against tumors. Therefore, the absence of a high tumor incidence at an early age in the nude mouse (congenital thymic aplasia), which is essentially lacking T cells, has been used as an argument against the immune surveillance hypothesis. However, these arguments ignore the possibility that T cell-independent humoral immunity, rather than cellular immunity may constitute the main defence of mammals against the persistence of small numbers of malignant cells. It should be recalled that humoral antibodies are effective in the inactivation of leukemic cells (*13*), but not of established solid tumors. We like to speculate (Fig. 2) that in the initial phases of tumor growth (as well as in the final stage of metastatic spread), transformed cells might be similarly susceptible toward cytotoxic antibodies as leukemic cells. When the tumor mass becomes organized and vascularized, the tumor antigens, which are shielded by the malignant cells, reach a high concentration in the periphery of the tumor so that circulating antibodies are bound before they reach the cells located in the more central parts of the tumor. The absence of an increased incidence of malignancies in leprosy patients (*14*) who have a severely decreased T cell activity, also seems to argue against the crucial role of T lymphocytes in the surveillance against the first few transformed cells. The fact that imuran—a notable suppressor of antibody synthesis—is the major common denominator in transplant patients undergoing prolonged immune suppression, could therefore account for the increased incidence of both carcinomas and lymphoreticular tumors in this category of patients.

NATURAL HISTORY OF TUMORS

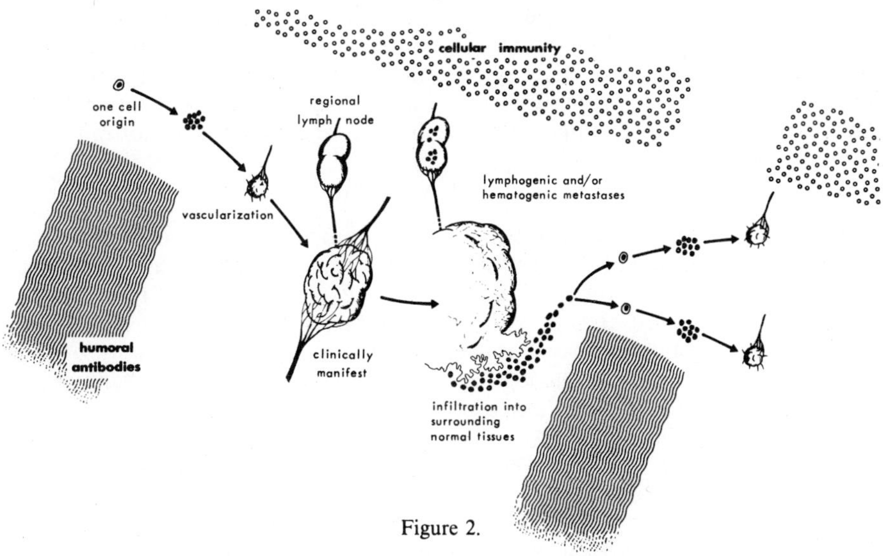

Figure 2.

IMMUNE SUPPRESSIVE EFFECTS OF RADIATION

We can now ask ourselves, which role a suppression of immunity might play in radiation carcinogenesis. In all instances where a large dose whole-body or partial-body irradiation is the carcinogenic stimulus, severe immune suppression of limited duration may be expected to occur (*15, 16*). Such seems certainly true for instance with the Kaplan (*17*) schedule of irradiation (4 × 180 rad/week) which results in a high incidence of thymic lymphomas in certain mouse strains. There is very little exact information about the degree of depression of T and B cell immunity, respectively, by such regimens, simply because the systems of measurement and comparison of the many immune parameters are still in the stage of development. The establishment of such a comprehensive evaluation of immune responses has been the subject of a working group called the Development Immune Monitors Subgroup (DIMS) which operates under the auspices of EULEP in Europe, and we feel that this kind of approach has to be applied to various carcinogenic radiation regimens before the role of immune depression in the process of tumor development can be properly assessed.

When radiation is administered by radioactive isotopes which are localized diffusely among large parts of the lymphohemopoietic system, e.g. in the case of the induction of osteosarcoma with ^{90}Sr in mice, a more prolonged immune-suppressed state may be expected. This has indeed been reported by Stjernswärd (*18*), who found a diminished number of plaque forming cells in the spleen for a period of 7 months following the administration of the isotope. As mentioned before, immune suppression promotes the appearance of tumors following infection of animals with oncogenic viruses.

In two instances where diffuse radiation throughout the body is operative as the carcinogenic stimulus, namely radiation leukemogenesis (*19–22*) and induction of osteogenic sarcomas by isotopes, one or more oncogenic viruses have been implicated (*23*). A more recent third example (*24*) is the provocation by whole-body irradiation at an early age of mammary tumor development in the O_{20} mouse strain, which is associated with the activation of an endogenous mammary tumor virus by a breakdown in prevailing immunity. Although these examples suggest that an inhibitory effect of radiation on the immune system can be an intricate and essential part of the process leading to tumor development, this does not imply that immune suppression is a prerequisite for all forms of radiation carcinogenesis.

LOCAL IRRADIATION

There are now some very well-studied examples of irradiation of target tissue, where exposure of the host organism was precluded by transplanting the irradiated tissue to a non-irradiated host. Barendsen et al. (*25*) transplanted pieces of X-irradiated skin of WAG rats to F_1 (WAG × BN) recipients and observed the development of basal cell carcinomas of WAG genotype in a proportion of the cases. An analogous procedure was employed by Shellabarger (*26*) who exposed mammary tissue of rats in vitro to 800 R of X-irradiation followed by regrafting of the pieces into the autologous donor rat. In 16 out of 110 autografts a tumor developed, compared to only 1 tumor in 110 sham irradiated autografts. Finally, the fact that in vitro irradiation of cultured cells has resulted in malignant transformation (*27*) and proven transplantable tumors (*28, 29, 30*), argues against immune suppression as an absolute requirement for radiation carcinogenesis. The question whether immune suppression can consistently and substantially contribute to the incidence of radiation-induced tumors cannot yet be answered, except perhaps for those tumors that are clearly induced via an oncogenic virus.

TUMOR ANTIGENS

The second immunological issue that we have to examine is the antigenic properties of radiation-induced tumors, and it is at this point that we should consider the mechanisms involved in the carcinogenic processes more closely. Neoplastic cells frequently—and perhaps always—show antigenic deviation from normal cells. TSTA in oncogenic virus-induced tumors cross-react (*3, 4*) between the tumors induced by the same virus. They are separable from structural virus antigens, produced by the same cells. The chemically induced tumors usually possess strong TSTA with individual specificity, although some chemical carcinogens have been shown to produce tumors of weak antigenicity.

The two main mechanisms of radiation carcinogenesis that have dominated the literature of the past decades are the somatic mutation hypothesis and the theory of virus activation. Malignant transformation of cells induced by way of somatic mutations may be expected to result in antigenic changes analogous to those observed in chemically induced tumors, since in both cases the localization of the damage to the DNA of the cell is random. One might expect that such a mechanism would lead to a variety of antigenic differences between the resulting tumors. On the other hand, in the case that ionizing radiation acts by activating a dormant oncogenic virus, each category of tumors would be expected to show cross-reactivity specific for the etiologic virus. The latter situation seems indeed to prevail in the case of radiation-induced leukemias (*31–33*) and probably also for certain radiation-induced mammary tumors in mice. The available information on the antigenicity of other types of radiation-induced tumors is unfortunately very limited. Nilsson and Révész (*34*) investigated the antigenicity of osteogenic sarcomas induced in mice by ^{90}Sr. In general, they found these tumors to be weakly antigenic, but the antigens were individually different. Moore and Williams (*35*) could evoke immunity against ^{90}Sr-induced osteogenic sarcomas in only 6 out of 15 tumors, and the resistance induced was weak. These authors did not investigate the individual specificity of the tumor antigens.

The lack of information on TSTA in radiation-induced tumors does not allow us to decide for either one of the two mechanisms of carcinogenesis. It is true that oncogenic viruses have been demonstrated to fulfill the role of an intermediary agent, but in other tumors no evidence has been obtained for such a mechanism and the patterns of antigenicity described for the osteosarcomas at first sight do not seem to fit the viral theory. However, there is recent confirmation that virus-induced tumors can also display unique antigens (*36*), which renders the immunological differences between the various categories of tumors less clear.

A PROPOSED MECHANISM FOR RADIATION CARCINOGENESIS

Although it is clear that much additional data are to be collected before the antigenic properties of tumors can be seriously employed as an argument in selecting for different mechanisms of carcinogenesis, two sets of observations recently made in our Institute may throw some light on the carcinogenic mechanism. The first one is an observation by Barendsen (*25*) in irradiated transplanted pieces of skin, referred to above. He recorded the appearance of carcinomas in the grafts following exposure to doses as high as 3000 rad, which according to cell survival calculations would have left only very few if any of the original epithelial cells to survive. On the basis of such calculations, Barendsen proposed that cells which have lost the capacity for unlimited proliferation as a result of radiation damage may be rescued to regain that capacity by the uptake of genetic material from other damaged cells. By this same process, some of these cells might be transformed into tumor cells.

The other observation was made by J. C. Klein (*30*) while studying malignant transfomation in cultured mouse fibroblast lines under the influence of X-irradiation. She observed the induction of malignant transformation in two sublines by a single exposure to 300 rad. This exposure caused an extensive amount of cell killing. When she induced a similar degree of damage by mechanical means, or by repeated freezing and thawing of cells, the cultures were rendered equally capable of producing tumors upon injection into syngeneic mice. Of course, the first explanation that comes to mind is that an oncogenic virus or provirus is involved. The mechanism of provirus activation by radiation is presently envisaged as shown in Fig. 3 (upper part). In the experiments of Klein, an extensive search was made for the appearance of virus, but without any success. Therefore, another hypothesis is needed to explain the observation that a large amount of damaged cells may provide the conditions under which other cells may become transformed. It is postulated that this transformation occurs following the uptake of small pieces of DNA which function as genes involved in proliferation of normal cells (Fig. 3). If such genes were incorporated at the wrong sites in the DNA, sufficiently distant from the appropriate repressor genes, they could be imagined to release the cell from its normal inhibitions for uncontrolled proliferation. The incorporation of the transformed genes might perhaps occur during normal DNA synthesis, but the chances for such mishappenings to occur during repair of DNA following radiation damage (misrepair) would obviously be much increased. The antigenic properties of such transformed cells would be expected to be different from one tumor to another, since the process of DNA damage and repair by incoming DNA fragments could be expected to be variable, if not random. This mechanism of radiation carcinogenesis seems to accomodate nicely the clinical observation that tumors arise frequently in tissues severely damaged by frequent or high dose irradiation. It also explains the recent observations of Zajdela and Latarjet (*37*), who found caffeine—a known inhibitor of DNA repair—to decrease the tumor incidence following u.v. irradiation of mouse ear skin, as well as the findings of Borek and Hall (this conference) that fractionated irradiation causes a higher incidence of transformed cells in vitro than

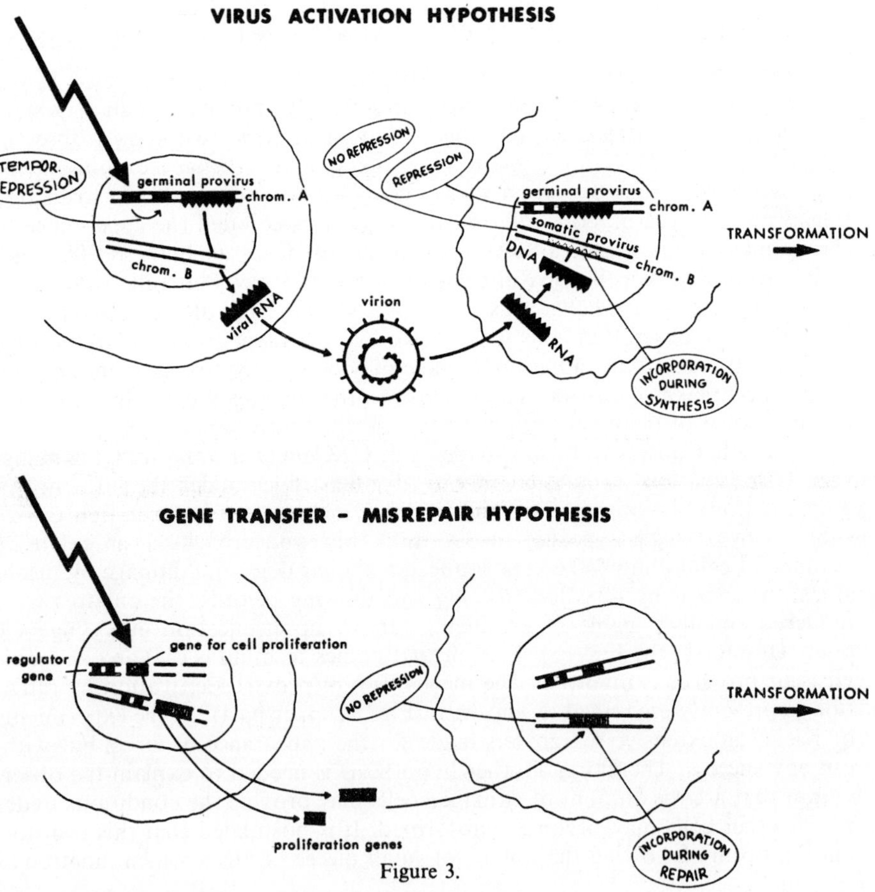

Figure 3.

single irradiation. One consequence of this mechanism seems to be that there would be a threshold for tumor induction, namely at the doses which produce death of a sizable proportion of the irradiated cells. One attraction of this gene transfer-misrepair hypothesis is that this mechanism shares a number of pathways with the oncogenic virus theory. In addition, the proposed mechanism seems at present more accessible to experimental verification than the somatic mutation hypothesis.

ACKNOWLEDGEMENT

The author wishes to thank Dr. P. Bentvelzen for his contribution to the development of the gene transfer-misrepair hypothesis and for his critical reading of the manuscript.

REFERENCES

1. L. Thomas, in H. S. Lawrence: *Cellular and Humoral Aspects of the Hypersensitive State*. p. 529. Hoeber, Medical Books, New York, 1959.
2. F. M. Burnet, *Progr. Exp. Tumor Res.*, **13,** 1 (1970).
3. R. W. Baldwin, D. Glaves, and M. V. Pimm, in *Progress in Immunology*, (B. Amos, Ed.), pp. 907–920. Academic Press, New York and London, 1971.
4. H. O. Sjögren and S. C. Bansal, in *Progress in Immunology*, (B. Amos, Ed.) pp. 921–937. Academic Press, New York and London, 1971.
5. M. A. Lappé, *J. Nat. Cancer Inst.* **40,** 823 (1968).
6. R. T. Prehn and G. L. Slemmer, in *Endogenous Factors Influencing Host-Tumor Balance*, (R. W. Wissler, T. L. Dao and S. Wood Jr., Eds.), pp. 185–190. The University of Chicago Press, Chicago and London, 1967.
7. J. Byfield, in *Interaction of Radiation and Host Immune Defense Mechanisms in Malignancy* (Conference, March 1974, Greenbrier, W. Va.), Brookhaven National Laboratory Report, BNL 50418, 1974.
8. R. A. Good, *Proc. Nat. Acad. Sci. (USA)* **69,** 1026 (1972).
9. I. Penn, *Recent Results in Cancer Research: Malignant Tumors in Organ Transplant Recipients*. Springer-Verlag, Berlin, Heidelberg, New York, 1970.
10. H. Cottier, in *Interaction of Radiation and Host Immune Defense Mechanisms in Malignancy* (Conference, March 1974, Greenbrier, W. Va.), Brookhaven National Laboratory Report, BNL 50418, 1974.
11. B. K. Walder, M. R. Robertson and D. Jeremy, *Lancet* **7737,** 1282 (1971).
12. H. Balner, *Rev. Europ. Etudes Clin. Biol.* **15,** 599 (1970).
13. P. Gorer and D. B. Amos, *Cancer Res.*, **16,** 338 (1956).
14. A. Oleinick, *J. Nat. Cancer Inst.* **43,** 775, (1969).
15. W. Rosenau and H. D. Moon, *Cancer Res.*, **27,** 1973, (1967).
16. E. R. King, L. S. DeGiorgi, R. L. Royster, J. S. Wolf and S. H. Levitt, *Radiology*, **93,** 297, (1969).
17. H. S. Kaplan, in *BEGG Proc. 2nd Canadian Cancer Research Conference, 1956* **vol. 2,** p. 127, Academic Press, New York, 1957.
18. J. Stjernswärd, *Antibiot. Chemotherapia*, **15,** 213 (1969).
19. R. Latarjet and J. F. Duplan, *Int. J. Radiat. Biol.*, **5,** 339 (1962).
20. A. Peled and N. Haran-Ghera, *Immunology*, **26,** 323 (1974).
21. N. Haran-Ghera, *Int. J. Cancer*, **1,** 81 (1966).
22. M. Lieberman and H. S. Kaplan, *Science* **130,** 387 (1959).
23. M. P. Finkel and B. O. Biskis, *Progr. Exp. Tumor Res.*, **10,** 72 (1968).
24. A. Timmermans, P. Bentvelzen, P. C. Hagemen and J. Calafat, *J. Gen. Virol.*, **4,** 619 (1969).
25. G. W. Barendsen, in *Radiation-Induced Cancer*, pp. 413-424. I.A.E.A., Vienna, 1969. (STI/PUB-228).
26. C. J. Shellabarger, *Proc. Soc. Exp. Biol. Med.*, **136,** 1103 (1971).
27. C. Borek and L. Sachs, *Nature (London)* **210,** 276 (1966). Quoted by Borek and Hall in: *Transformation of Mammalian Cells in vitro by Low Doses of X-Rays.*

28. J. C. Klein, in *Radiation-Induced Cancer*, pp. 57–66. I.A.E.A., Vienna, 1969. (STI/PUB-228).
29. C. Borek and E. J. Hall, *Nature (London)* **243**, 450 (1973).
30. J. C. Klein, *J. Nat. Cancer Inst.*, **52**, 1111 (1974).
31. N. Haran-Ghera, *Nature (London)* **222**, 992 (1969).
32. C. Mazurek and E. Estienne, *Bull. Cancer*, **55**, 503 (1968).
33. N. K. Clapp and J. M. Yuhas, *J. Nat. Cancer Inst.*, **51**, 1211 (1973).
34. Nilsson and Révész (unpublished). Quoted by H. O. Sjögren in *Progr. Exp. Tumor Res.*, **6**, 289 (1965).
35. M. Moore and D. E. Williams, *Brit. J. Cancer*, **26**, 90 (1972).
36. J. Vaage, T. Kalinovsky and R. Olson, *Cancer Res.*, **29**, 1452 (1969).
37. MM. Zajdela and R. Latarjet, *C. R. Acad. Sci. (Paris)*, **277**, 1073 (1973).

The Interplay of Viruses and Radiation in Carcinogenesis[1]

Arthur C. Upton
Health Services Center
State University of New York at Stony Brook, N.Y. 11790, USA

INTRODUCTION

The discovery by Gross (*1, 2*) and by Lieberman and Kaplan (*3*) that leukemias induced by irradiation in mice may be transmissible by cell-free filtrates pointed to the existence of an interplay between the effects of radiation and those of an otherwise latent oncogenic virus, which has since been the subject of intensive study. In the years intervening since these pioneer observations, efforts have been made to elucidate the nature and mechanisms of the interplay and to determine whether other neoplasms also may be elicited by an interaction between virus and radiation. The results of these efforts, although still highly preliminary, are surveyed below in the light of recent advances in viral oncology.

COMPARATIVE IMPORTANCE OF VIRUSES IN THE ETIOLOGY OF NEOPLASMS

Since the discovery of the myeloblastosis virus in chickens, by Ellerman and Bang in 1908 (*4*), many other oncogenic viruses have been identified. The lengthening list of such agents now encompasses viruses and viral tumors of virtually every type, with examples in nearly all of the commonly studied species of animals (Table 1). Viewed historically, the evolution of viral oncology can be divided into four major phases (*5*): 1) the first period, from 1910 to about 1935 was dominated by studies on avian tumor viruses and efforts to establish etiologic relationships in terms of Koch's postulates, which were often frustrated by expectations that are now known not to be strictly applicable to oncogenic viruses; namely, that the virus causes rapid onset of the disease and remains at the height of the disease. At that time, however, the failure to fulfill Koch's postulates consistently, .especially in any mammalian tumors, prevented the viral hypothesis from gaining wide credence. 2) The second period, from about 1935 to 1960, saw the discovery of a growing number of mammalian tumor viruses and the important realization that such viruses might not necessarily act alone but might require the interaction of specific genetic, physiologic, or environmental cofactors. It also came to be recognized during this period that the induced neoplasms might develop only after a long latency, appear in but few of the infected animals, and be noninfectious themselves. Because of these restrictions, viruses continued to be viewed as having

[1]Congress presentation A-4-2.

TABLE 1
Viruses that induce neoplasms in animals (from Ref. *8*)

Common name of virus	No. of major isolates	Host of origin	Produces neoplasia in:	Tumor type in animals
Mouse leukemia (MLV)*	16	Mouse	Mouse, rats, hamsters	Leukemia, lymphoma
Mouse sarcoma (MSV)	6	Mouse	Mouse, rats, hamsters, cats, tissue culture	Sarcomas
Polyoma (Py)	2	Mouse	Mouse, hamsters, tissue cultures	All types except leukemia
Mammary tumor (MTV)	2	Mouse	Mouse	Carcinoma
Chicken leukemia (ALV)	4	Chicken	Chicken, tissue culture	Leukemia
Twiehaus	1	Chicken	Chicken, quail, hamster	Reticuloendotheliosis
Chicken sarcoma (Rous et al.) (RSV)	9	Chicken	Chicken, quail, turkey, duck, hamster, monkey, snake, tissue culture	Sarcoma
Marek's (MHV)	1	Chicken	Chicken	Lymphoma
CELO	1	Chicken	Hamster	Sarcoma
Cat leukemia (FLV)	4	Cat	Cat	Leukemia, lymphoma
Cat sarcoma (FSV)	3	Cat	Cat, rat, dog, monkey, tissue culture	Sarcoma
G. pig leukemia	1	G. pig	G. pig	Leukemia
G. pig herpes	1	G. pig	G. pig	Sarcoma
Deer fibroma	1	Deer	Deer	Fibroma
Squirrel fibroma	1	Squirrel	Squirrel	Fibroma
Shope fibroma	1	Rabbit	Rabbit	Fibroma
Shope papilloma	1	Rabbit	Rabbit	Papilloma
Dog sarcoma	1	Dog	Dog, tissue culture	Sarcoma
Dog mast cell	1	Dog	Dog	Carcinoma
Lucké	1	Frog	Frog	Carcinoma
Human adeno	31[a]	Man	Hamster, mouse, tissue culture	Sarcoma-lymphoma
Wart	1	Man	Man	Papilloma
Hybrids[b]	7	Monkey Man Cat Mouse	Hamster, cat, tissue culture	Sarcoma, lymphoma
Yaba	1	Monkey	Monkey, man	Histiocytoma
H. saimiri	1	Monkey	Monkey	Lymphoma
Simian adeno	6	Monkey	Hamster, tissue culture	Sarcoma-lymphoma
SV40[c]	1	Monkey	Hamster, mouse, tissue culture	Lymphosarcoma
Graffi hamster	2	Hamster	Hamster	Lymphoma, papilloma
Bovine papilloma	1	Cow	Cow, horse, mouse, hamster	Papilloma, fibroma sarcoma
Bullhead papilloma	1	Fish	Fish (bullhead catfish)	Papilloma
	110			

*Strains of mouse leukemia virus (MLV) are usually designated by names of original investigator GLV = Gross leukemia virus. FLV = Friend leukemia virus. RLV = Rauscher leukemia virus.
[a]As of May 1970 approximately 12 of 31 human adenoviruses induce malignancies in hamsters. These 12 and the remaining 19 induce discrete foci of transformed (apparently cancerous) cells in tissue cultures.
[b]Hybrid = Genotypic recombinates of 2 different viruses, e.g., SV40 + adeno; cat leukemia + mouse sarcoma.
[c]Simian virus 40.

little general importance in the etiology of cancer. 3) During the third era, in the early 1960's, it became apparent that tumorigenesis by DNA viruses was characteristically accompanied by integration of viral genes into the genome of the host cell, after which the viral genes were transmitted vertically and could be expressed without necessarily any production of infectious virus. These observations served to reconcile the hitherto puzzling lack of correlation between patterns of cancer incidence and epidemiological evidence for an infectious causative agent. The observations also suggested the possibility that any DNA virus might conceivably exert carcinogenic effects under appropriate circumstances, thus stimulating search for oncogenic activity among common viruses, which has since led to demonstration of tumorigenesis by adenoviruses and implication of herpes simplex virus in the pathogenesis of certain types of cancer. 4) The fourth phase of viral oncology has seen the emergence of the revolutionary concept that RNA tumor viruses ("oncornaviruses"), like their DNA counterparts, contribute to the genome of the host cell genetic information which becomes part of the inheritance of the affected cell. With this latest development in viral oncology, tumor viruses have come to be considered more properly as endogenous, rather than exogenous, to the host, requiring for their elucidation adaptation of the approaches used in research on molecular genetics and cell regulation (*5*).

Although, with the exception of the wart virus, viral agents have yet to be implicated conclusively in the pathogenesis of neoplastic lesions in humans, the susceptibility of human cells to virus-induced "neoplastic" transformation in vitro is amply documented (Table 2). Indirect evidence implicating viruses in human neoplasia is also mounting: e.g. 1) the frequent occurrence of characteristic virus particles in the cells of certain malignancies (*6, 7*); 2) the association of group-specific, possibly viral or virus-mediated, antigens with the cells of some neoplasms (*8, 9*); 3) the presence of reverse transcriptase of oncornavirus-type in certain cancer cells (*10*); and 4) the presence in certain cancer cell nuclei of DNA base sequences complementary to the base sequences of known or suspect tumor viruses (*11*). These findings, which are analogous to those associated with virus-induced neoplasms in animals, strongly suggest the involvement of viruses in the etiology of certain human cancers, although such an interpretation remains to be corroborated.

BIOLOGY OF TUMOR VIRUSES

In order to envision how radiation might interact with a virus to cause cancer, it is necessary to consider what is known of the biology of oncogenic viruses. These viruses, as a family, differ widely in size, shape, structure, and chemical composition (Table 3). It is evident, moreover, that dissimilar types of viruses may give rise to highly similar neoplasms and that, conversely, a single virus (polyoma) can give rise to neoplasms that are dramatically different, involving virtually every type of cell and tissue in the body. Although this seems paradoxical at first glance, it becomes explicable when the virus is recognized to consist for the most part of machinery concerned with viral packaging, transport, attachment, penetration, and replication, with only a small part of its genome being responsible for oncogenic activity (*11*).

TABLE 2
"Tumor" viruses that replicate in and/or transform human cells (from Ref. *8*)

Common name of virus	No. of major isolates	Type	Host of origin	Replication/ transformation*
A. "Known Tumor Viruses"				
Chicken sarcoma	3	RNA	Chicken	T
CELO	1	DNA	Chicken	T
Mouse sarcoma	2	RNA	Mouse	R, T
Mouse leukemia	1	RNA	Mouse	R
H. saimiri	1	DNA	Monkey	R
Yaba	1	DNA	Monkey	T
SV40	1	DNA	Monkey	R, T
"Hybrids"[a]	7	DNA	Monkey–man	T
Cat sarcoma	2	RNA	Cat	R, T
Cat leukemia	1	RNA	Cat	R
Human adeno (type 12)	1	DNA	Man	R, T
EBV	2	DNA	Man	R, T
B. "Suspect Tumor Viruses"[b]				
Influenza	2	RNA	Man	R
Sarcoma particle	2	("C")	Man	R
Herpes - 2	1	DNA	Man	R
Herpes - 1	1	DNA	Man	R
Human breast particle	3	"RNA"	Man	R
C. Other				
Shope papilloma[c]	1	DNA	Rabbit	R
Monkey breast particle	1	"RNA"	Monkey	R

*R = Replication of intact detectable virion; T = transformation as represented by nonlytic morphologic alteration and prolonged life.

[a]Genotypic recombinates of 2 different viruses, e.g., SV40 + Adenovirus 12; cat leukemia + mouse sarcoma viruses.

[b]Viruses or virus-like particles reported in "consistent" association with cancers or which enhance (or are enhanced by) physical or chemical agents. See text.

[c]This virus may replicate in man as judged by significant antibody levels and low serum arginine levels in some exposed laboratory personnel.

Prevailing evidence favors the view that the virus exerts its oncogenic effects through integration of genetic information encoded in its nucleic acid into the genome of the infected host cell. In the case of the DNA viruses, the integration and subsequent transcription of viral nucleic acid may be envisioned to be analogous to processes that have been best characterized in lysogenic bacteriophages (*11*). In the case of the RNA viruses, on the other hand, the process of integration is considered to involve a DNA intermediate, synthesized from viral RNA through the action of a virus-specified RNA-directed DNA polymerase, or "reverse transcriptase" (*11, 12*).

The viral information that is integrated into the genome of the host cell is viewed in each of the two currently prevailing hypotheses—the "oncogene" hypothesis (*13, 14*) and the "protovirus" hypothesis (*12*)—as constituting part of the normal inheritance of the cell, with the virus-derived genes being sub-

TABLE 3
Types of Oncogenic Viruses (from Ref. *8*)

Classification of virion	Poplar group designation	Site of maturation	Prototype virus	Type of lesion	Proliferative lesion transmitted under natural conditions
DNA					
Large Membrane-bound	Pox	Cytoplasmic inclusion	Shope rabbit fibroma	Near-neoplasia	Yes
Medium Membrane-bound	"Herpes-type"	Nucleus	Lucké frog kidney carcinoma*	Solid malignant neoplasia	Yes
Naked	Adeno	Nucleus	Adenovirus 12	Solid malignant neoplasia	No
Small Naked	Papova	Nucleus	1. Human wart	Near-neoplasia	Yes
			2. Shope rabbit papilloma	Solid benign neoplasia	Yes
			3. S-E polyoma	Solid malignant neoplasia	No
RNA					
Medium Membrane-bound	"Myxo-like"[a]	Cytoplasm	1. Rous sarcoma	Solid malignant neoplasia	—[b]
			2. Bittner mouse mammary cancer[c]	Solid malignant neoplasia	Yes
			3. Avian leukosis	Leukemia and solid lymphoma	Yes
			4. Gross mouse leukemia	Aleukemic leukemia	Yes

*Frog kidney carcinoma virus of Lucké. Tentative classification on a basis of: (a) reproduction and maturation in the nucleus; and (b) morphologic similarity to "herpes-like" viruses, known to be of DNA type.

[a]So-called because of budding from cell membranes, as is characteristic for the myxoviruses. Hemagglutination, which is characteristic of the myxoviruses, does not occur with the RNA tumor viruses.

[b]Laboratory strain isolated from a particular spontaneous tumor. Other similar tumors occur spontaneously in chickens with low frequency, but the relationship among etiologic agents is unknown.

[c]Mammary gland carcinoma virus of Bittner. Tentative classification on a basis of budding from cytoplasic membranes, as is characteristic for the RNA myxoviruses and "myxo-like" tumor viruses.

ject to regulation by mechanisms of repression and derepression similar to those controlling normal genes. Viewed in this context, the viral theory of carcinogenesis and the genetic theory of carcinogenesis become merged into a single unifying framework.

The "oncogene" hypothesis postulates that the genomes of C-type RNA viruses consist of 1) "virogenes", which code for replication of the virus, and 2) "oncogenes", which code for neoplastic transformation of the host cell. Such viral genomes are held to be widespread, if not universal, in vertebrates, being transmitted vertically in the germ line and playing a functional role in normal cellular growth and differentiation by coding for alloantigens, or differentiation antigens, on the cell surface (*11, 14, 15*). According to this hypothesis, cancer is envisioned to result from derepression of viral oncogenes, either through the action of external carcinogens or through spontaneously occurring stochastic somatic mutational events. Derepression of virogenes, leading to virus production, is not considered necessary for the induction of neoplasia (*11*).

The "protovirus" hypothesis postulates that genetic information can be transmitted within or among somatic cells, from the DNA in "protovirus" regions of the genome into RNA intermediates and thence, via reverse transcriptase, back into DNA sequences which are reinserted into the genome. Through this mechanism, existing genes are conceived to be amplified and new DNA sequences evolved in differentiation, without affecting the stability of the germ line. "Misevolution" of protoviruses, either by mutation of their base sequences or their faulty integration into the wrong sites in the genome, is postulated to result in neoplastic transformation of the affected cell (*11*).

Knowledge of the genetic composition of the oncogenic viruses and of the mechanisms regulating their expression is still meagre; however, it is clear that viral mutants exist which differ to varying degrees in their ability to transform host cells, host cell range, ability to replicate, specification of virus-associated cell surface antigens, etc. Depending on the nature of the mutation, the affected form of the virus may be said to be "defective" for the property in question. The occurrence of genetic interactions between defective and nondefective viruses, and between defective mutants and host cell genomes, are indicated by "marker rescue" phenomena. As yet, however, genetic mapping of the viruses and characterization of their genetic interactions are at a rudimentary stage (*11*).

The genetic complexity and heterogeneity of tumor viruses is paralleled by heterogeneity of host cells in susceptibility to transformation by a given virus, further reflecting the complexity of interactions between viral and host cell genomes. In avian cells, as well as mouse cells (Table 4), several genetic loci have been found to influence the control of spontaneous and induced virus activation, expression of viral antigen, susceptibility to spreading of viral infection, and the host's ability to respond immunologically to virus-induced antigens in the transformed cells (*5, 11, 16*). These observations demonstrate unequivocally that the expression and host range of endogenous viruses can be controlled in different ways and at more than one level in the viral life cycle; however, the extent to which the mechanisms are comparable in different tumor viruses and different host species, including man, remains to be determined.

TABLE 4
Genes Affecting Endogenous Murine Leukemia Virus Expression (from Ref. *11*)

Linkage Group	Location	Designation	Allele	Expression	Phenotype	Mouse strain
I	viral?	v_1	v_1	dominant	N-tropic MLV release	AKR
			v_1	recessive	no MLV release	
unknown not I, VIII, IX	viral?	v_2	v_2	dominant	N-tropic MLV release	AKR
			v_2	recessive	no MLV release	
unknown not VIII	viral?	Ind	Ind+	dominant	N-tropic MLV release	Balb/c
			Ind−	recessive	no MLV release	NIH-Swiss
unknown	?	none	+	dominant	gs antigen	AKR
			−	recessive	no gs antigen	C57L
unknown	?	none	+	dominant	MLV release if gs+ gene is present	AKR
			−	recessive	no MLV release	
unknown	?	Mlv-1	$Mlv\text{-}1^a$	recessive	gs antigen	B10, D2
			$Mlv\text{-}1^b$	dominant	no gs antigen	
VIII	host	Fv-1	$Fv\text{-}1^n$	recessive	susceptibility to N-tropic MLV	AKR, C58
			$Fv\text{-}1^b$	recessive	susceptibility to B-tropic MLV	C57BL/6 Balb/c
IX	host	H-2	$H\text{-}2^k$	semidominant	early leukemia	AKR
			$H\text{-}2^b$	semidominant	late leukemia	AKR/$H\text{-}2^b$
IX	viral?	Tla	Tla+	dominant	epistatic for TL antigen expression	A, C58
			Tla−	recessive	no TL antigen	AKR, C57Bl/6
IX	viral?	G_{ix}	G^+_{ix}	semidominant	G_{ix} antigen	129
			G^-_{ix}	semidominant	no G_{ix} antigen	
unknown not IX	viral?	none	+	dominant	G_{ix} antigen	129
			−	recessive	no G_{ix} antigen	

*It is not known whether the H-2, Tla and G_{ix} loci affect MLV replication directly or whether they affect leukemogenesis induced by MLV. Both G_{ix} loci are unlikely to represent structural genes for G_{ix} antigens; either or both loci may have epistatic effects on the structural gene.

The development of rapid in vitro assays has greatly facilitated systematic classification and screening of cultured cells for their ability to support the growth of inoculated viruses and to release latent endogenous viruses upon appropriate stimulation. Based on early experiments with embryo cells from NIH strain mice, as compared with BALB/c mice, certain viruses were designated as N-tropic or B-tropic, respectively, depending on the strain in which they showed preferential growth. Embryo cells from mice of other strains were subsequently shown to be broadly classifiable as being either of "N-type" or "B-type", depending on whether they resembeled the NIH strain or the BALB/c strain in preferential growth of N- or B-tropic viruses, respectively. Susceptibility to N- and B-tropic viruses has since been correlated with the *Fv*-1 locus (Table 4), heterozygotes at this locus not permitting replication of either type of virus. Although naturally occurring murine C-type viruses have heretofore generally been considered to be either N-tropic or B-tropic, based largely on the properties of the viruses isolated from leukemic mice, viruses recovered recently from the spleen cells of normal weanling mice of low-leukemia strains, by induction with 5-bromodeoxyuridine in vitro, are infective preponderantly for cells of nonmurine origin, and have thus been designated as xenotropic or "S-tropic" viruses (*17*). The widespread occurrence of S-tropic viruses in weanling mice has been viewed as an indication that such viruses may have a normal physiologic function; e.g., it has been suggested that viruses activated during cellular neoplastic transformation provide a protective antigenic signal for the host's immune defenses, a role in which viruses that are unable to reinfect other cells of the host, such as the S-tropic viruses, would presumably function best (*17*). Mice of high-leukemia strains have been observed to yield viruses capable of growing on more than one indicator cell line, suggesting that the cells of such mice may commonly harbor multiple endogenous C-type viruses, which differ in genetic composition and in regulation (*17*).

VIRUSES IMPLICATED IN RADIATION CARCINOGENESIS

In the pathogenesis of radiation-induced leukemias, C-type viruses have been implicated repeatedly in mice of several strains (*1–3*, *18–21*) and also tentatively in swine (*22*). In mice of the AKR strain, the majority of which die from naturally occurring viral leukemia, whole-body irradiation early in life has been observed to hasten the onset of the disease (*23-25*).

In the pathogenesis of radiation-induced bone tumors, viruses have likewise been implicated in Xgf and CF1 mice (*26*) and tentatively in C3H/H mice (*27*). Morphological and immunological evidence implies, moreover, that the viruses involved in the skeletal neoplasms of the mouse belong to a leukemia-sarcoma virus complex (*28, 29*) comparable in many respects to the avian C-type leukosis–sarcoma virus complex (*8*, *15*), the sarcoma viruses tending to require interaction with leukemia ("helper") viruses for replication.

An interaction between radiation and virus has also been envisioned to play a role in the pathogenesis of radiation-induced breast tumors in mice, along with hormones and other cofactors (*30–32*). As yet, it is not clear whether viruses enter into the induction of breast tumors by irradiation in other species, but C-type particles have been identified in mammary tumors arising in nonirradiated rat (*33*), monkey (*8*), and human (*8*) breast tissue. Mammary fibroadenomas have, furthermore, been induced in rats by neonatal injection of adenovirus (*34*). In light of evidence implicating cell–cell interactions in the induction of mammary neoplasms by fast neutrons in Sprague-Dawley female rats (*35*), it is tempting to postulate that the intercellular interactions reflect the transmission of induced virus from one cell to another.

In general, the viruses isolated from radiation-induced neoplasms have been relatively less "virulent" than those isolated from serially transplanted tumors, such as the Friend, Moloney, and Rauscher viruses (*36*). During early passages, the radiation tumor viruses characteristically have shown a narrowly restricted host range, a tendency to produce neoplasms in only a small percentage of inoculated recipients, and a long induction period for neoplasia (*36, 37*). Such viruses, moreover, have been inconsistently recoverable from radiation-induced neoplasms, as judged by the variable oncogenic activity of cell-free filtrates of neoplastic tissue assayed in vivo (*26, 36*). In fact, it has been observed that viruses are recoverable more readily from the tissues of nonleukemic mice shortly after irradiation than from the leukemias that develop later in such animals (*21, 38*).

INTERACTION OF RADIATION AND VIRUS AT THE CELLULAR LEVEL

Several investigators have reported that X-irradiation increases the susceptibility of cells to viral "neoplastic" transformation in vitro (*39–41*). It has also been reported that X-irradiation can induce the production of latent leukosis virus in cultured chick embryo cells (*42*).

Whether the "neoplastic" transformation observed in hamster embryo cells irradiated in vitro (*43, 44*) and the tumorigenic effects observed in rat mammary gland explants irradiated in vitro (*45*) involve the activation of a latent oncogenic virus is yet to be determined.

Activation and/or release of latent leukemia virus by irradiation in vivo has been suggested by the appearance of leukemogenic activity in cell-free extracts of tissues obtained from nonleukemic mice 5–10 days after whole-body X-irradiation (*21, 38*). Likewise, the activation of latent mammary tumor virus (MTV) by X rays in vivo is suggested by the appearance of B-type virus particles in breast tumors induced in 0_{20} strain female mice by procedures involving whole-body X-irradiation (*30, 31*). Release of virulent MTV by radiation in vivo has not been detected in mice of other low-cancer strains, with the exception of BALB/c mice; however, whole-body irradiation has led to the appearence of MTV antigens (detected by immunofluorescence) and MTV-related RNA (detected by use of the DNA-product of MTV as a probe) in such animals (*32*). The release and/or partial expression of leukemia virus and of mammary tumor virus in response to radiation in vivo appear to represent accentuation of activation phenomena that are uncommonly observed spontaneously except in mice of advancing age (*32, 38*).

The induction of virus release by radiation, chemical carcinogens, and chemical mutagens (*5, 11, 42*) is yet to be fully explained. In this process, iododeoxyuridine (IUdR) and bromodeoxyuridine (BUdR) have been found to be vastly more effective than X rays, the former inducing the release of murine leukemia virus (MuLV) in about 10^{-2} cells, as compared with 10^{-4} to 10^{-5} cells in the case of X rays (*46*). The release of endogenous MuLV is also more readily induced in the cells of established lines and lines that have become transformed than in newly cultured or nontransformed cells. The release has been inferred to involve the incorporation of BUdR into host cell DNA, and the persistence of induced virus to require genetic susceptibility to horizontal spreading of the infection or continued presence of BUdR in the cells at risk. These observations have been interpreted to suggest that induction of C-type RNA viruses involves excision of the integrated viral DNA, in this respect resembling induction of lambda phage and other integrated viral genomes; however, it is not clear why merely transcription of the integrated DNA should be insufficient for virus production in viruses that possess their own reverse transcriptase (*11*).

The ability of radiation to cause strand breakage and other lesions in DNA, which then undergo modification through the action of host cell repair processes (*47*), provides ample basis in theory for myriad genetic interactions involving radiation, viruses, and host cells. The enhanced susceptibility to viral transformation of cells from patients with chromosomal aberrations (*48*) suggests that the chromosome-breaking potency of radiation could be an important factor in radiation-virus carcinogenesis.

ROLE OF IMMUNOLOGIC FACTORS

The widespread occurrence of tumor-specific transplantation antigens (TSTA) in diverse types of neoplastic cells (Table 5) has led to the hypothesis

TABLE 5
Tumor-specific antigens in tumors of experimental animals (from Ref. *9*)

Tumor	Predominant antigen
Chemically induced tumors (polycyclic hydrocarbons)	Transplantation antigens specific for each tumor
Tumors induced by DNA viruses (polyoma, SV40, adenovirus)	Transplantation antigens and neoantigens common to tumors induced by the same virus; tumors usually lack virion antigens and do not release infective virus.
Tumors and leukemias induced by RNA viruses (avian and murine leukemiasarcoma viruses, mammary tumor virus)	Transplantation antigens and viral antigens common to cells transformed by the same virus; cells release infective virus (except in circumstances in which the virus is defective.

that such antigens may exist in all neoplasms (*9*). In fact, it has even been proposed that such antigens may favor the growth of tumor cells by eliciting stimulatory effects of immunity early in the process of neoplasia, or as long as the immune reaction remains mild, although such antigens would be expected to elicit inhibitory effects ("immunological surveillance") under other circumstances (*49*). Stimulatory effects of immunity have yet to be documented in radiation-induced neoplasia, but there is evidence of inhibitory effects in several such systems, examples of which are cited below.

Leukemogenesis by whole-body irradiation in C57Bl mice appears to involve radiation-induced depression of immunity to an endogenous leukemia virus (*50*) that is immunogenic to nonirradiated animals, in which it emerges rarely except during senescence (*51, 52*). Consistent with this interpretation are observations that mice of the same strain can be protected against radiation leukemogenesis by interferon (*53*), immune sera (*54*), and a methanol extraction residue fraction of BCG (*55*). Radiation-induced immunosuppression has, likewise, been implicated as a factor in radiation leukemogenesis in RF mice, based on the reciprocal correlation between the incidence of lymphoid neoplasms and the severity of chronic glomerular lesions containing antiviral antibody in such animals (*56*). The occurrence of a similar chronic immunological response against endogenous C-type viruses has been demonstrated in nonleukemic AKR (*57*) RF (*58*), and B6C3F1 (*59*) mice, indicating that tolerance to endogenous viral antigens is incomplete in these animals. Based on such evidence, and the observation that viral antigens are detectable in mouse tissues most consistently either prior to immunological maturity or during immunological senescence (*58, 60*), it has been suggested that susceptibility to leukemogenesis is inversely correlated with antiviral immunity in mice of low-leukemia strains (*58, 61*).

As in the case of leukemia induction, cited above, induction of bone tumors by strontium-90 in mice has been observed to be inhibited by injection of BCG (*62*).

In addition to immunodepressive effects of irradiation per se, which are widely recognized (*63*), some further impairment of immunity may also occur through the action of the murine leukemia viruses (*58, 64, 65*) and even the mammary tumor virus (*66*). In C57Bl mice, a radiation leukemia virus that causes tumors of thymic lymphoid cells has been found to depress thymus-dependent lymphocytes ("T" cells) preferentially (*67*), in this respect resembling the Moloney leukemia virus (*65*). Another leukemia virus derived from an irradiated C57Bl mouse, which does not characteristically induce thymic tumors, appears to alter immunoglobulin synthesis through preferential effects on bone-marrow-derived lymphocytes ("B" cells) (*68*). Both situations have been observed in leukemias induced by virus in SJL/J mice, which include B-cell as well as T-cell types (*69*), indicating that different forms of immunosuppression may be induced by leukemia viruses, depending on their target cells. Whether the immunosuppressive effects of the viruses play any significant role in their oncogenic effects remains to be determined; however, no obvious correlation between virus-induced changes in the immune response and the onset of viral leukemias have been detected in C57Bl mice (*52, 67*).

The appearence of leukemia virus in BALB/c mice immunosuppressed by antilymphocyte serum, but only in the concomitant presence of allogeneic A/J skin grafts, points to the possibility that immunologic reactions to histocompatibility antigens on the cell surface may activate latent virus in some as yet unknown way (*70*). This interpretation, if corroborated, would clearly have implications of special significance in radiation chimeras.

OVERVIEW

Evidence which has accumulated in recent years, implying that oncogenic viruses are endogenous to many, if not all, species of animals, with the viral genes being integrated into the genome of the host, has served to unify the viral and genetic theories of carcinogenesis. Viewed in this perspective, the interplay between virus and radiation becomes a special aspect of somatic cell radiation genetics. Study of radiation carcinogenesis in vivo, however, is complicated by the complexity of regulatory dynamics that operate at the supracellular level; e.g. the action of such conditioning factors as the immunological response, hormonal stimulation, inhibition by chalones, cell–cell interactions, etc. Hence, as has been stressed previously, the interplay between radiation and virus may constitute merely one of many possible effects through which radiation can influence the probability of neoplasia (Fig. 1), the predominant effect in any given instance depending on the constitution of the irradiated individual and the particular conditions of irradiation (*71*).

Because of the complexity of the neoplastic process in vivo and the failure of many virus-induced neoplasms to yield virus detectable by traditional bioassays, future research on the role of viruses in radiation carcinogenesis must rely heavily on newer, more sensitive techniques, such as have revolutionized modern viral oncology. Until such techniques have been developed to maximal effectiveness and applied systematically to all types of radiation-induced neoplasms, any assessment of the role of viruses in radiation carcinogenesis must remain tentative.

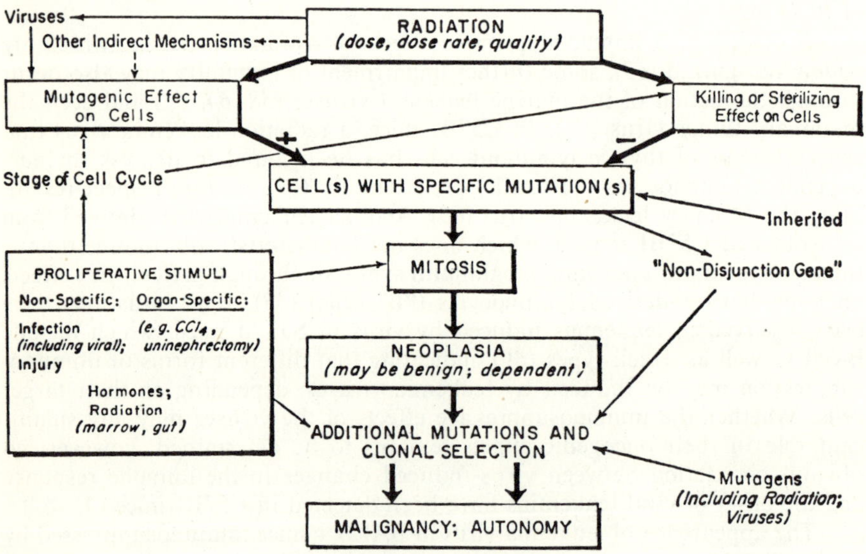

FIG. 1. Different types of radiation effects postulated to affect the initiation, promotion, and progression of neoplasia [From L. J. Cole and P. C. Nowell, *Science* **150**, 1782 (1965).]

REFERENCES

1. L. Gross, *Acata Haematol.* **19**, 353 (1958).
2. L. Gross, *Proc. Soc. Exp. Biol. and Med.*, **100**, 102 (1959).
3. M. Lieberman, and H. S. Kaplan, *Science*, **130**, 387 (1959).

4. V. Ellerman, and O. Bang, *Centralbl. f. Bakt., Abt. I, (Orig.)* **46,** 595 (1908).
5. W. P. Rowe, *Cancer Res.* **33,** 3061 (1973).
6. A. J. Dalton, *J. Nat. Cancer Inst.* **48,** 1095 (1972).
7. A. J. Dalton, *J. Nat. Cancer Inst.*, **49,** 323 (1972).
8. F. J. Rauscher, Jr., and T. E. O'Connor, in *Cancer Medicine,* (J. F. Holland and E. Frei, III., Eds.) pp. 15–44. Lea and Febiger, Philadelphia, 1973.
9. H. F. Oetgen, and K. E. Hellstrom, *ibid*, pp. 951–990.
10. R. C. Gallo, *Amer. J. Clin. Path.* **60,** 80 (1973).
11. J. Tooze, (Ed.), *The Molecular Biology of Tumor Viruses,* Cold Spring Harbor Laboratory, 1973.
12. H. M. Temin, and D. Baltimore, *Advan. Virus Res.* **17,** 129 (1972).
13. R. J. Huebner and G. J. Todaro, *Proc. Nat. Acad. Sci. (USA)* **64,** 1087 (1969).
14. R. J. Huebner, G. J. Kelloff, P. S. Sarma, W. T. Lane, H. C. Turner, R. V. Gilden, S. Orszlan, H. Meier, D. D. Myers, and R. L. Peters, *Proc. Nat. Acad. Sci. (USA)* **67,** 366 (1970).
15. T. Aoki, *J. Nat. Cancer Inst.* **52,** 1029 (1974).
16. F. Lilly, and T. Pincus, *Advan. Cancer Res.* **17,** 231 (1973).
17. M. M. Lieber, A. J. Sherr, and G. J. Todaro, *Int. J. Cancer* **13,** 587 (1974).
18. A. C. Upton and G. E. Cosgrove, Jr. in *Experimental Leukemia,* (M. A. Rich, Ed.), pp. 131–158. Appleton-Century-Crofts, New York, 1968.
19. J. F. Duplan and R. Latarjet, *Cancer Res.* **26,** 395 (1966).
20. R. Ribbacchi, and G. Giraldo, *Nat. Cancer Inst. Monogr.* **No. 22,** 701 (1966).
21. N. Haran-Ghera and A. Peled, *Brit. J. Cancer* **21,** 730 (1967).
22. M. E. Frazier, R. N Ushijima, J. R. Pratt, T. K. Andrews, and R. Rosario, in *Radionuclide Carcinogenesis,* (C. L. Sanders, R. H. Busch, J. E. Ballou, and D. D. Mahlum, Eds.) pp. 377–390. U. S. Atomic Energy Commission, CONF-720505, 1973. (Available from National Technical Information Service, U.S. Department of Commerce, Springfield, Va. 22151.)
23. J. Reverdy, G. Rudali, J. F. Duplan, R. Latarjet, *Sang* **29,** 796 (1958).
24. J. F. Duplan, *C. R. Soc. Biol. F.* **159,** 1277 (1965).
25. A. Leonard, G. Deknudt, and G. Linden, *C. R. Soc. Biol. F.* **167,** 1055 (1973).
26. M. Finkel and C. A. Reilly, Jr. in *Radionuclide Carcinogenesis,* (C. L. Sanders, R. H. Busch, J. E. Ballou, and D. D. Mahlum, Eds.), pp. 278–288. U.S. Atomic Energy Commission, CONF-720505, (1973).
27. E. L. Lloyd, J. E. Loutit, and F. Mackevicis, in *Radiological and Environmental Research Division Annual Report.* ANL-8060 Part II. pp. 51–80. Center for Human Radiobiology, Argonne National Laboratory, Argonne, Illinois, 1973.
28. J. A. Levy, J. W. Hartley, W. P. Rowe, and R. J. Huebner, *J. Nat. Cancer Inst.* **51,** 525 (1973).
29. D. B. Jones and M. Moore, *Brit. J. Cancer* **29,** 158 (1974).
30. A. Timmermans, P. Bentvelzen, C. Ph. Hageman, and J. Calafat, *J. Gen. Virol.* **4,** 619 (1969).
31. L. M. Boot, P. Bentvelzen, J. Calafat, G. Röpcke, and A. Timmermans, in *Oncology,* Vol. 1-A, *Cellular and Molecular Mechanisms of Carcinogenesis,* (R. L. Clark, Ed.), pp. 434–440. Year Book Medical Publishers, Chicago 1971.
32. P. Bentvelzen, in *Abstracts of the Breast Cancer Task Force,* p. 45. Nat. Cancer Inst., Washington, D.C., 1974.
33. K. H. Hollmann, and J. M. Verley, *Pathol. Biol.* **20,** 83 (1972).
34. J. Ankerst, N. Jonsson, L. Kjellén, E. Norrby, and H. O. Sjögren, *Int. J. Cancer* **13,** 286 (1974). (1974).
35. H. H. Rossi, and A. M. Kellerer, *Science* **175,** 200 (1972).
36. H. S. Kaplan, *Cancer Res.* **27,** 1325 (1967).
37. L. Gross, *Oncogenic Viruses,* Oxford, England, Pergamon Press, 1961.
38. N. Haran-Ghera, in *Immunity and Tolerance in Oncogenesis*, (L. Severi, Ed.) pp. 585–596. Division of Cancer Research, Perugia, 1970.
39. M. Stoker, *Nature (London)* **200,** 756 (1963).
40. E. J. Pollock, and G. J. Todaro, *Nature (London)* **219,** 520 (1968).
41. J. H. Coggin, Jr., *J. Virol.* **3,** 458 (1969).
42. R. A. Weiss, R. R. Friis, E. Katz, and P. Voght, *Virol.* **46,** 920 (1971).
43. C. Borek, and L. Sachs, *Nature (London)* **210,** 276 (1966).
44. C. Borek, and E. J. Hall, *Nature (London)* **243,** 450 (1973).

45. C. J. Shellabarger and R. W. Schmidt, *Nature (London)* **218,** 192 (1968).
46. D. R. Lowy, W. P. Rowe, N. Teich, and J. W. Hartley, *Science* **174,** 155 (1971).
47. B. W. Fox, and L. G. Lajtha, *Brit. Med. Bull.* **29,** 16 (1973).
48. R. W. Miller, and G. J. Todaro, *Lancet* **1,** 81 (1969).
49. R. T. Prehn, *Science* **176,** 170 (1972).
50. N. Haran-Ghera, A. Peled, *Israel J. Med. Sci.* **4,** 1181 (1968).
51. M. Haran-Ghera, in *Unifying Concepts of Leukemia,* Bibl. Haematol. No. 39 (R. M. Dutcher, and L. Chieco-Bianchi, Eds.) pp. 671–676. Karger, Basel, 1973.
52. N. Haran-Ghera, in *Dynamic Aspects of Host-Parasite Relationship,* Vol. II (A. Zuckerman, Ed.). Israel Universities Press—Wiley Imprint, 1975. (in press).
53. M. Lieberman, T. C. Merigan, and H. S. Kaplan, *Proc. Soc. Exp. Biol. Med.* **138,** 575 (1971).
54. J. F. Ferer, M. Lieberman, and H. S. Kaplan, *Cancer Res.* **33,** 1339 (1973).
55. N. Haran-Ghera, and D. W. Weiss, *J. Nat. Cancer Inst.* **50,** 229 (1973).
56. N. K. Clapp, and J. M. Yuhas, *J. Nat. Cancer Inst.* **51,** 1211 (1973).
57. M. B. Oldstone, T. Aoki, and F. J. Dixon, *Proc. Nat. Acad. Sci. (USA)* **69,** 134 (1972).
58. M. G. Hanna, Jr., M. J. Snodgrass, R. W. Tennant, J. M. Yuhas and B. L. Batzing, in *Virus Tumorigenesis and Immunogenesis,* (W. S. Ceglowski and H. Friedman, Eds.) pp. 59–89. Academic Press, New York, 1973.
59. B. L. Batzing, M. Yurconic, and M. G. Hanna, *J. Nat. Cancer Inst.* **52,** 117 (1974).
60. R. L. Peters, J. W. Hartley, G. Spahn, L. S. Rabstein, C. E. Whitmire, H. C. Turner and R. J. Huebner, *Int. J. Cancer* **10,** 283 (1972).
61. J. M. Yuhas, R. W. Tennant, M. G. Hanna, Jr., and N. K. Clapp, in *Radionuclide Carcinogenesis,* (C. L. Sanders, R. H. Busch, J. E. Ballou, and D. D. Mahlum, Eds.), pp. 312–321. U.S. Atomic Energy Commission, CONF-720505, (1973).
62. A. Nilsson, L. Révész, and J. Stjernswärd, *Radiat. Res.* **26,** 378 (1965)
63. *Ionizing Radiation; Levels and Effects,* Vol. II: *Effects.* A Report of The United Nations Scientific Committee on the Effects of Atomic Radiation to the General Assembly, Official Records of the General Assembly, Twenty-Seventh Session, Supplement No. 25 (A/8725), pp. 303–378. United Nations, New York, 1972.
64. A. Peled, N. Haran-Ghera, *Int. J. Cancer* **8,** 97 (1971).
65. L. Chieco-Bianchi, F. Sendo, T. Aoki, O. L. Barrera, *J. Nat. Cancer Inst.* **52,** 1345 (1974).
66. D. E. Griswold, G. H. Heppner, P. Calabresi, *J. Nat. Cancer Inst.* **50,** 1035 (1973).
67. A. Peled, N. Haran-Ghera, *Immunology* **26,** 323 (1974).
68. A. Sassen, F. Vander Plaetse, M. Janowski, J. R. Maisin, *J. Nat. Cancer Inst.* **52,** 539 (1974).
69. G. M. Shearer, E. Mozes, N. Haran-Ghera, Z. Bentwich, *J. Immunol.* **110,** 736 (1973).
70. M. S. Hirsch, D. A. Ellis, P. H. Black, A. P. Monaco, and M. L. Wood, *Science* **180,** 500 (1973).
71. A. C. Upton, in *Frontiers of Radiation Therapy and Oncology,* Vol. 6, (J. M. Vaeth, Ed.) pp. 470–478. S. Karger, Basel 1972.
72. L. J. Cole, P. C. Nowell, *Science* **150,** 1782 (1965).

Roles of Cellular and Humoral Immunity in Malignancy[1,2]

T. Juhani Linna and Cheng-po Hu
Department of Microbiology and Immunology
Temple University School of Medicine
Philadelphia, Pennsylvania 19140 USA

The relationship between malignancy and the immunological system seems to be very complex. On the one hand, oncogenic agents such as tumor viruses, and chemical carcinogens, and even tumors themselves, influence immune responses (*1–4*) and, on the other, the immunological system influences tumor development (*5–7*). The subject of this presentation will be mainly this latter aspect — the influence of immunity on tumor development.

Tumor immunology was a popular discipline from the turn of the century until the thirties, with many studies done on the influence of immunity on tumor rejection, non-specific boosting of immunological responsiveness to accomplish tumor cure, etc. However, the development by Strong, Little, Snell and others of inbred strains of mice demonstrated that most of the early studies were impossible to interpret, since there had been histocompatibility differences between donors and recipients and one could therefore interpret tumor rejection as allograft rejection (*8*, *9*). Tumor immunology came into disrepute. Although a couple of studies in the forties (*10*) and early fifties (*11*) strongly suggested that tumor-specific antigens did exist, no conclusive evidence of this was available. More than ten years ago, Prehn and Main (*12*), and Klein et al. (*13*) demonstrated clearly that tumors possess specific, transplantation-like antigens (TATA). These findings brought on a renaissance of tumor immunology. If tumors possess unique antigens capable of eliciting a host-protective specific response, it seems that these antigens could provide a "handle" that could be used to influence tumor development, possibly to obtain tumor cure.

It seems that tumors caused by chemical carcinogens all possess unique TATA (*14*), while virus-induced tumor transplantation antigens in different animals, even in different species, cross-react, if they are caused by the same agent (*15*). However, these findings should be somewhat modified. If one looks very carefully at chemical carcinogen-induced tumors, one can find some cross-reactivity (*14*), and careful studies also reveal some "private" antigens on virus-induced tumors (*15*). The virus-induced transplantation-like antigens are expressed on the cell surface, and are different from the T-antigens of DNA viruses and from the RNA virus structural antigens (*16*).

[1]Congress presentation A-4-3

[2]Work supported by NIH grant CA 13347 from the National Cancer Institute.

Most, if not all, tumors possess transplantation-like antigens. Most spontaneous tumors have little antigenicity, while rapidly arising tumors caused by most laboratory tumor viruses and many chemical carcinogens are strongly antigenic (*7*).

How does the immunological system use the tumor-associated transplantation antigens to influence tumor development? Obviously, if tumors are immunogenic, the immunological system should be able to recognize the antigen and mount an immune response. It was suggested by Thomas many years ago that maybe the main function of cell-mediated immunity was a surveillance function — the detection and destruction of malignantly transformed cells (*17*). This suggestion won much acceptance among immunologists, and is still a very attractive idea, although the complex tumor-host relationship may not lend itself to the concept of surveillance in its simplest form.

One can make a strong case for an immunosurveillance function for cell-mediated immunity. It is not necessary to state the evidence for it in detail, since the subject has been extensively reviewed and discussed in recent years (*4–6*, *18*). The main experimental support for surveillance comes from work with tumor development in immunodeficient and immunosuppressed laboratory animals. Immunosuppression can be accomplished by irradiation, immunosuppressive drugs, antilymphocyte sera, or a combination of these procedures, with or without the addition of neonatal thymectomy. In most studies, such immunosuppressed and neonatally thymectomized mice, deficient mainly in cell-mediated immunological functions, develop tumors more frequently than do their controls (*19–23*). An especially interesting animal model is the "nude" mouse. This pleomorphic defect consists in the homozygous animal, besides defective development of hair follicles, of an absence of the thymus, an almost complete absence of the thymus-dependent cell population, and a profound deficiency in thymus-dependent, cell-mediated immunity (*24–26*). Such mice can readily accept and retain allogeneic, and even xenogeneic, tumors, including tumors of human origin (*27*). Although spontaneous tumor development has been described in these animals, it seems to be rare (*28*), possibly at least partly due to the short life-span of these highly immunodeficient animals in the conventional environment. The "nude" mice have been reported to have approximately the same susceptibility to a chemical carcinogen (3-methylcholanthrene) as the heterozygous controls (*29*). On the other hand, "nude" mice develop certain virus-induced tumors much more frequently than do the heterozygous controls (*A. C. Allison, personal communication, 1974*). Long-term studies under germ-free conditions seem essential for obtaining information on the importance of thymus-dependent, cell-mediated immunity with regard to spontaneous and induced tumors in this animal model.

In man, the association with decreased cell-mediated immunological responsiveness is well documented in certain malignancies, such as Hodgkin's disease. Data pertaining to immunodeficiency and oncogenesis are also available in two other groups of patients—the iatrogenically immunodepressed ones, such as kidney transplant recipients, and the primary immunodeficiencies, "experiments of nature". While it is quite clear that immunodeficient patients have a much higher frequency of malignancies than the general population of the same age, it is also clear that most of their malignancies are of the lymphoreticular tissues, i.e. of the already perturbed immune system (*4, 18*). However, this is not always the case. Solid epithelial tumors also are much overrepresented in certain immunodeficiencies (*30*, and S. Barandun, personal communication, 1973). It is clear, however, that most immunodeficient patients developing malignancies are not only deficient in cell-mediated immunity, but in humoral immunity as well.

We have been interested in looking at the role of cell-mediated immunity in the control of tumor development in an animal model, namely the chicken. The reason for choosing to work with the chicken is the rather clear-cut delineation of its immunological responsiveness. Cell-mediated immunity in this species is under thymic control, while humoral immunity — antibody-producing and immunoglobulin-forming capacity — in the chicken is under the control of the hind-gut lymphoid organ unique for these birds, the bursa of Fabricius (*31*, *32*).

The tumor which we have mainly worked with is avian reticuloendotheliosis, strain T. This is an RNA tumor virus, which was isolated by Twiehaus and Robinson (*33*, *34*), and which causes tumors of the reticuloendothelial system, leading rapidly to tumor death in a vast majority of the birds, if enough tumor virus is used. When newly hatched chickens are thymectomized and they and controls are infected a few days later intraperitoneally with RE tumor virus, a significantly increased tumor mortality in thymectomized birds is seen (*35-37*). When development of local RE tumors is induced by administration of RE virus into the wing web, in most normal birds the local tumors regress completely and the birds survive the tumor challenge, while significantly more chickens which were thymectomized in the newly hatched period developed progressively growing local tumors and died from RE tumors (*37*). Thus, our thymectomy data in this tumor system in the chicken strongly support the concept of surveillance, as do most similar studies in mammalian tumors induced by many oncogenic laboratory viruses and chemical carcinogens (*19–23*), using strongly antigenic tumors. Thus, the case for surveillance by cell-mediated immunity is a strong one, but it should be pointed out here that it is by no means conclusive.

The role of humoral immunity in oncogenesis is even more controversial than that of cell-mediated immunity. One influence, which has been much in the focus of discussion in the past few years, is enhancement by serum factors. The concept of enhancement, and the experimental laboratory evidence for enhancement by antibodies induced by certain immunization schedules, has been described a long time ago (*38*, *39*). In the past few years, a tumor-protective role for antibodies and/or antigen-antibody complexes in tumor development has been very forcefully advocated (*15*, *40*). Although the chemical nature of such "blocking" factors has not been well defined, there is clearly something in the serum of many patients and laboratory animals with progressively growing tumors that can interfere with in vitro killing by immune lymphocytes of tumor target cells. These "blocking" factors also have tumor specificity. Most of the work on "blocking" factors has been done in vitro. Very scanty evidence for the importance of "blocking" by antibodies or antigen-antibody complexes in oncogenesis in vivo is available.

We have studied the effect of the antibody-forming system on RE tumor development using the chicken model. Successful bursectomy in the newly hatched period deprives the animals of antibody-producing capacity and, consequently, of the blocking antibodies or antigen-antibody complexes. Both development of systemic tumors and localized wing web tumors were studied.

Treatment with high doses of cyclophosphamide in the newly hatched period is known to permanently and irreversibly destroy the antibody-producing capacity in most treated chickens (*32*, *41*) with concomitant destruction of the bursa of Fabricius (*32*), without affecting cell-mediated immunity (*32*). This treatment was therefore used to abolish antibody-producing capacity. We found that such chemically bursectomized birds had increased tumor mortality after both local and systemic RE virus infection (*35–37*). Most local tumors grew progressively and were ultimately fatal, while most local tumors in non-bursectomized animals regressed completely (*37*). Thus, these data indicate that, at least in this malignancy, the main function of the antibody-forming system is in protection of the host, not of the tumor.

To exclude the possibility that we were dealing with an effect of the cyclophosphamide which was unrelated to its effect on the bursa and on the antibody-forming system, we also studied animals which were subjected to surgical bursectomy in the newly hatched period. This treatment alone does not regularly produce agammaglobulinemia, although most animals are deficient in antibody-producing capacity. The fact that this treatment resulted in similar results as the cyclophosphamide-bursectomy studies — significantly increased mortality, and most animals locally infected with tumor virus having progressively growing tumors, rather than regressing tumors, as was the case in normal birds — certainly indicates that the effect is bursa-dependent.

Thus, all these studies indicated that the bursa of Fabricius, and the antibody-forming system, are important in protecting the host and not the tumor. We have never had data compatible with enhancement of tumor growth by the antibody-forming system. We feel, therefore, that not only thymus-dependent, cell-mediated immunity, but also the antibody-forming system, is important for host defense against tumors, at least in this tumor system. The data do not exclude the existence of antibodies with tumor-enhancing function, but it seems that if they do exist, their function is overshadowed by the host-protective effects of the antibody-forming system as a whole.

The above studies elucidate the effects of deprivation of part of the immunological system. An important complement to these studies are experiments providing a normal, or an immunodeficient, animal with serum from animals immunized against the tumor. Immune serum was obtained from animals which has survived sublethal challenges with the tumor virus. This serum was used to intensively treat with wing web tumors normal and bursectomized, immunodeficient animals. At the time of initiation of treatment, 60% of the birds had wing web tumors, and later virtually all animals had wing web tumors. Control groups of animals were given the same dose of RE virus into the wing web and were treated with normal serum, according to the same schedule, or were left untreated. In immune serum-treated birds significantly more local tumors regressed, and significantly more birds survived and were definitely cured from the tumor. This was the case both in normal tumor-bearing animals (Table 1) and in cyclophosphamide-bursectomized tumor-bearing animals, where the host's humoral immune system would contribute little to the immune defense (Table 2).

TABLE 1
Effect of Normal and Immune Sera on the Development of RE Tumors in Normal Chickens

Treatment	Frequency of regression of local tumors	p	Frequency of survival	p
None	38% (6/16)		25% (4/16)	
Normal serum	19% (3/16)		19% (3/16)	
Immune serum	94% (15/16)	<0.01[a]<0.002[b]		<0.002[a,b]

[a] Comparison with no treatment.
[b] Comparison with normal serum treatment.

TABLE 2
Effect of Normal and Immune Sera on the Development of RE Tumors in Cyclophosphamide-bursectomized Chickens

Treatment	Frequency of regression of local tumors	p	Frequency of survival	p
None	50% (8/16)		12% (2/16)	
Normal serum	53% (10/19)		37% (7/19)	
Immune serum	90% (18/20)	<0.005[a]		<0.01[a]

[a] Comparison with no treatment or normal serum treatment.

There is the possibility that the bursectomies result in the absence or significant decrease of antiviral antibodies. This would then allow the tumor viruses to infect more target cells and, consequently, increase mortality in the bursectomized groups. The evidence from the locally induced RE tumors, which regress in most normal birds and grow progressively in most bursectomized birds, argues against this possibility, since it is difficult to imagine that antiviral antibodies would be decisively instrumental in local tumor regression.

To further elucidate the role of antiviral antibodies in this system, we absorbed the immune serum 5 times with RE virus, until no precipitin line between the immune serum and a virus preparation could be detected by counter-immunoelectrophoresis with the naked eye. When such serum was evaluated for its immunotherapeutic capacity, it was found to work at least as well as non-absorbed serum (Table 3). We could therefore definitely conclude that antiviral antibodies do not play an important part in the tumor cure in this system.

We do not have definite evidence that the curative effect is conveyed by antibodies directed against the tumor-associated transplantation antigens. However, we know that the therapeutic effect of immune serum has some specificity; it has no effect on the development of a laboratory strain of Marek's disease.

TABLE 3
Effect of Non-absorbed and Absorbed Immune Sera on the Development of RE Tumors

Treatment	Frequency of regression of local tumors	p	Frequency of survival	p
None	23% (3/13)		15% (2/13)	
Non-absorbed immune serum	65% (11/17)	N.S.[a]	59% (10/17)	≤0.05[a]
Absorbed immune serum	80% (12/15)	<0.02[a]	73% (11/15)	<0.01[a]

[a] Comparison with no treatment.

The data presented here pertain to only one tumor system, and it is much too early to conclude that they describe a phenomenon of general relevance in tumor immunity. However, we obtained results compatible with a host-protective function for the antibody-forming system using a transplantable chemical carcinogen-induced tumor in bursectomized chickens as well. If one looks at data in other species, one finds much support for the concept of a host-protective effect for the antibody-forming system in oncogenesis, especially in lymphoreticular malignancies. The increased frequency of malignancies in Bruton's type of agammaglobulinemic children, where the defect is specific for humoral immunity, is well known (*4*, *18*, *30*). However, most of these malignancies are leukemias, that is, of the already diseased system. More convincing is the fact that in the common variable form of immunodeficiency, including that in patients who have an intact cell-mediated immunity, one has a vast overrepresentation of malignancies, including malignancies of epithelial origin (*18*, *30*).

Also, Gorer's classical experimental studies on mouse leukemia show that the influence of immune serum on tumor development is complicated, but clearly includes host protection (*42*, *43*). Recent data leading to similar conclusions are now available from different laboratories for other experimental tumor models (*42–46*).

It is also interesting to note that most studies demonstrating an increased tumor frequency in immunocrippled animals or patients are not on selective defects of cell-mediated immunity, but on impaired humoral immunity in addition. Thus, the concept that not only cell-mediated immunity, but humoral immunity as well, is important for host protection in oncogenesis, seems well tenable.

The evidence for surveillance is clearest in laboratory tumors which are strongly antigenic. However, most spontaneously arising tumors are little antigenic (*7*, *47*, *48*). Recognizing this, Prehn has built up a substantial case, from evidence in the literature and from his own studies, for an immunostimulation hypothesis in oncogenesis (*7*, *47*, *49*). The immunological system seems oftentimes to be stimulating tumor growth rather than inhibiting it, when the antigenic differences between tumor and host is a slight one, that is, in the initial phase of tumor development. This concept should be considered not as a replacement of surveillance, but as a complement to it. Immunity may well have a dual influence on tumor development. In the initial stages of spontaneous tumor development, when the tumor mass is small and its immunogenic load only slight, the immune system may stimulate rather than inhibit the growth of the neoplasm. However, when the tumor mass is larger and the antigenic stimulation greater, a surveillance function may be operating. Whether the immunosurveillance function is sufficient at this stage to effectively help the host is still an open question.

New findings on the escape mechanisms of tumors from destruction by host defense mechanisms have recently come to the fore. Thomson et al. (*50*, *51*) and Baldwin et al. (*52*) have studied closer the factor in tumor-bearer serum, which is protecting the tumor cells in vitro from attack by immune lymphocytes. It seems that the protection — "blocking," if you will — is afforded by tumor antigen, rather than by antibodies, and when it is by antigen-antibody complexes, they are in antigen excess. Nowotny and Grohsman have also presented data on tumor enhancement by tumor cell membrane fragments in mouse TA-3 tumors (*53*, *54*). Thus, the phenomenon of blocking by serum factors should be regarded in a new light. The main biological importance in the escape from immunosurveillance may well be attached to tumor antigen, surrounding tumors and present in serum as a consequence of the turnover of the glycocalix of these cells. Tumor antigen can then specifically bind to receptor sites on sensitized lymphoid cells or antibodies, and in this way interfere with their cytotoxic capacity.

REFERENCES

1. J. Stjernswärd, *Antibiot. Chemotherap.* (*Basel*) **15,** 213 (1969).
2. H. Friedman and W. Ceglowski, *Progress in Immunology I,* pp. 815–829. Academic Press, New York, 1971.
3. C. M. Southam, *Cancer Res.* **28,** 1433 (1968).
4. R. A. Good, *Proc. Nat. Acad. Sci. (USA)* **69,** 1026 (1972).
5. R. T. Smith, and M. Landy, *Immune Surveillance,* pp. 1–536. Academic Press, New York, 1970.
6. F. M. Burnet, *Immunological Surveillance.* Pergamon Press, Oxford, 1970.
7. R. T. Prehn, *Progr. Exp. Tumor Res.* **14,** (1971).
8. C. C. Little, *The Biology of the Laboratory Mouse, pp. 279–309.* The Blakiston Co., Philadelphia, 1941.
9. G. D. Snell, E. Russell, E. Fekete, and P. Smith, *J. Nat. Cancer Inst.* **14,** 485 (1953).
10. L. Gross, *Cancer Res.* **3,** 326 (1943).
11. E. J. Foley, *Cancer Res.* **13,** 835 (1953).
12. R. T. Prehn, and J. M. Main, *J. Nat. Cancer Inst.* **18,** 769 (1957).
13. G. Klein, H. O. Sjörgren, E. Klein, and K. E. Hellström, *Cancer Res.* **20,** 1561 (1960).
14. R. W. Baldwin, *Advan. Cancer Res.* **18,** 1 (1974).
15. K. E. Hellström, and I. Hellström, *Advan. Cancer Res.* **12,** 167 (1969).
16. H. O. Sjögren, and S. C. Bansal, *Progress in Immunology* I, pp. 921–938. Academic Press New York-London, 1971.
17. L. Thomas, *Cellular and Humoral Aspects of Hypersensitive States*, pp. 529–532. Hoeber-Harper, New York, 1961.
18. H. Cottier, M. W. Hess, H. U. Keller, P. Luccieti, and B. Sordat, in *Symposium on Interaction of Radiation and Host Immune Defense Mechanisms in Malignancy*, White Sulphur Springs, W. Virginia, in press, 1974.
19. G. A. Grant, and J. F. A. P. Miller, *Nature* (*London*) **205,** 1124 (1965).
20. R. L. Kirchstein, A. S. Rabson, and E. A. Peters, *Proc. Soc. Exp. Biol. N.Y.* **117,** 198 (1964).
21. L. W. Law, *Cancer Res.* **26,** 551 (1966).
22. J. F. A. P. Miller, A. Grant, and F. J. C. Roe, *Nature* (*London*) **199,** 920 (1963).
23. A. C. Allison, and L. W. Law, *Proc. Soc. Exp. Biol. N.Y.* **127,** 207 (1968).
24. E. M. Pantelouris, *Nature* (*London*) **217,** 370 (1968).

25. E. M. Pantelouris, *Immunology* **20,** 247 (1971).
26. P. R. Pennyculk, *Transplantation* **11,** 417 (1971).
27. J. P. Mach, S. Carrel, C. Merenda, B. Sordat, and J. C. Cerottini, *Nature* (*London*) **248,** 704 (1974).
28. R. P. Custer, H. C. Outzen, G. J. Eaton, and R. T. Prehn, *J. Nat. Cancer Inst.* **51,** 707 (1973).
29. O. Stutman, *Science* **183,** 534 (1974).
30. J. H. Kersey, B. D. Spector, and R. A. Good, *Int. J. Cancer* **12,** 333 (1973).
31. M. D. Cooper, R. D. A. Peterson, M. A. South, and R. A. Good, *J. Exp. Med.* **123,** 75 (1966).
32. T. J. Linna, D. Frommel, and R. A. Good, *Int. Arch. Allergy* **42,** 20 (1972).
33. M. Sevoian, R. N. Larose, and D. M. Chamberlain, *Avian Dis.* **8,** 336 (1964).
34. G. H. Theilen, R. F. Zeigel, and M. J. Twiehaus, *J. Nat. Cancer Inst.* **37,** 731 (1966).
35. T. J. Linna, C. Hu, and K. Thompson, *Virus Tumorigenesis and Immunogenesis,* p. 141, Academic Press, New York-London, 1973.
36. K. D. Thompson, and T. J. Linna, *Nature New Biol.* **245,** 10 (1973).
37. T. J. Linna, C. Hu, and K. D. Thompson, *J. Nat. Cancer Inst.* **53,** in press, 1974.
38. N. Kaliss, *Cancer Res.* **18,** 992 (1958).
39. N. Kaliss, *Ann. N.Y. Acad. Sci.* **101,** 64 (1962).
40. K. E. Hellström, and I. Hellström, *Ann. Rev. Microbiol.* **24,** 373 (1970).
41. S. P. Lerman, and W. P. Weidanz, *J. Immunol.* **105,** 614 (1970).
42. P. A. Gorer, *Ann. N.Y. Acad. Sci.* **73,** 707 (1958).
43. P. A. Gorer, *Adv. Immunol.* **1,** 345 (1961).
44. B. W. Calnek, *Infect. Immunity* **6,** 193 (1972).
45. P. Hersey, *Brit. J. Cancer* **28, Suppl. I,** 11 (1973).
46. M. Woodruff, N. Dunbar, and A. Ghaffar, *Proc. R. Soc. London,* **B 184,** 97 (1973).
47. R. T. Prehn, Perspectives in oncogenesis: Does immunity stimulate or inhibit neoplasia? *J. Recticuloendoth. Soc.* **10,** 1 (1971).
48. A. Zimmermann, M. Bertschmann, B. Roos, M. W. Hess, and H. Cottier, in *Symposium on Interaction of Radiation and Host Immune Defense Mechanisms in Malignancy.* White Sulphur Springs, W. Virginia, in press, 1974.
49. R. T. Prehn, *Science* **176,** 170 (1972).
50. D. M. P. Thomson, K. Steele, and P. Alexander, *Brit. J. Cancer* **27,** 27 (1973).
51. D. M. P. Thomson, S. Eccles, and P. Alexander, *Brit. J. Cancer* **28,** 6 (1973).
52. R. W. Baldwin, J. G. Bowen, and M. R. Price, *Brit. J. Cancer* **28,** 16 (1973).
53. J. Grohsman, and A. Nowotny, *J. Immunol.* ***109,*** *1090 (1972).*
54. A. Nowotny, J. Grohsman, A. Abdelnoor, N. Rote, C. Yang, and R. Waltersdorff, *Eur. J. Immunol.* **4,** 72 (1974).

Macrophage Functions in Immune Responses to Tumors[1,2]

H. Cottier, H. Bürki, M. W. Hess, H. U. Keller and B. Roos
Institute of Pathology
University of Bern, Bern, Switzerland

INTRODUCTION

In considering host resistance to tumor growth, one is confronted with a number of different possible mechanisms involved. Immunity, in its broader sense, encompasses a) non-specific, and b) specific, antigen-directed phenomena. Macrophage functions within the defensive apparatus of the macroorganism relate to both these categories of so-called cancer surveillance. As information regarding experimental and clinical tumor immunology continues to accumulate, it becomes apparent that the macrophage plays an important role as an effector cell against neoplastic tissue (for reviews, see Refs. *1–3*). The role of adherent cells is not, however, restricted to killer functions, but may be equally important in the initiation of immune response to tumor-associated antigens.

THE MACROPHAGE CELL LINE(S)

It is generally agreed that most—if not all—macrophages originate in the bone marrow. Lymphoid, non-phagocytic, non-adherent "blast" cells seem to develop into promonocytes and monocytes, both of which exhibit certain macrophage functions and are peroxidase-positive (*4*). Circulating monocytes usually are not in a state of proliferation. Their potential, however, to reenter DNA synthesis and mitosis as they leave the blood stream and become histiocytes, does not appear to be permanently lost. The term "macrophage" is often used as a synonym for histiocyte, although the former may only be recognized as such in tissue sections when macrophagocytosis has set in. It is of interest that histiocytes, in contrast to promonocytes and monocytes, do no longer display noticeable peroxidase activity. In the course of macrophage differentiation these cells develop various types of surface receptors, e.g. for immunoglobulins (Ig) (in the human: IgG_1 and IgG_3), certain complement components (C3), migration inhibitory factor (MIF), and migration stimulating factor (MSF) (*5*). As macrophages become activated by various stimuli, energy production switches more and more from anaerobic glycolysis to oxidative metabolism as reflected in the numerical increase of mitochondria, and the cells are able to perform many functions with greater efficiency (for review, see *6*). Thus, one is not surprised to find in a population of adherent cells from rabbit peritoneal exudate several

[1]Congress presentation A–4–4.
[2]Supported by the Swiss National Foundation for Scientific Research.

subpopulations of macrophages showing distinct functional and structural characteristics (*7*). Certain cell types are difficult to classify, e.g. the so-called "K" ("killer") cell. We largely ignore whether, to what extent, and under what conditions immature (blast) macrophage precursors can leave the bone marrow and reach the tissues without immediate differentiation. Most macrophages are rather short-lived (days), but some seem to survive for many months (for review, see *6*). It is not yet decided if histiocytes can transform into collagen-producing fibroblasts. The so-called "dendritic" cells do not seem to belong to the macrophage system (*8*, *9*); however, they also seem to play some role in immune responses.

RESPONSES OF THE MACROPHAGE CELL LINE(S) TO STIMULI

Macrophage responses to various types of stimuli consist not only of so-called cell activation, but also involve the rate of production of precursor cells and migration characteristics of these elements. Using the soft agar system for macrophage cloning, it was shown that agents such as incomplete (IFA) or complete (CFA) Freund's adjuvant induce a dramatic rise in the number of macrophage precursor cells (*10*). In fact, almost every inflammatory reaction is accompanied by increased monocyte production. In mice, colony-forming macrophage precursors reach maximum numbers within approximately one week after administration of CFA (*11*). Recent observations suggest that fluctuations in the concentration of serum colony-stimulating factor (CSF) or—inversely—CSF-inhibitor precede and/or accompany increased or reduced precursor proliferation, respectively (for review, see *12*). Directional migration of macrophages towards sites of action is most probably influenced by chemotactic factors (cytotaxins). Macrophage-specific cytotaxins can be found in blood plasma incubated with immune complexes (*13*) and may be released by sensitized T lymphocytes. This lymphocyte-derived cytotaxin is distinct from the migration inhibition factor (MIF) (*14*). In several model systems, tissue macrophages carry out their functions without reentering a state of proliferation; however, there are numerous exceptions to this rule, particularly in so-called granulomatous reactions (*15*; for review, see *16*). In contrast to tissues, the peritoneal fluid of mice always contains DNA-synthesizing macrophages (*17*). Nonetheless, at least some macrophages seem to reach a stage in which they lose the capacity for further proliferation.

NON-SPECIFIC ACTIVATION OF MACROPHAGES

It is well established that agents evoking macrophage activation need not have detectable antigenic properties (for review, see *18*). The mechanisms mediating macrophage activation are not yet fully understood; however, endogenous metabolites such as cyclic AMP and cyclic GMP seem to act as regulators for various functions, e.g. random migration (*19*) and others. The use of chemicals interacting with macrophages has greatly helped us to better understand the process involved in activation of these cells. Some examples may suffice to emphasize this approach. Phagocytosis can be completely inhibited by 10^{-2}M sodium fluoride (*20*). Cell movements are disturbed by the microfilamentaffecting cytochalasin B, and the microtubular-disruptive agents, colchicine and vinblastine sulfate. Certain doses of synthetic polyanions, such as dextran sulfate, heparin and lower molecular weight Pyran copolymer, can enhance random migration of macrophages and, in part, stimulate phagocytosis. Macrophage motility also depends upon the presence of serum and may be increased by mild treatment with trypsin or pronase, or the reduction of calcium and magnesium ion concentration in the medium (for review, see *19*). Controversial results have been obtained regarding the effect of hydrocortisone, depending upon the system and dose used. Intracellular killing of microorganisms can be amplified by the administration of so-called lysosomotropic drugs, e.g. complexes of high-molecular DNA and antibiotics.

THE ROLE OF MACROPHAGES IN THE INITIATION OF A SPECIFIC IMMUNE RESPONSE

From in vitro studies on mouse spleen cells exposed to sheep erythrocytes, a three-cell interaction hypothesis has emerged with regard to humoral antibody responses to this antigen (*21*). The exact role of macrophages in such systems is still largely unknown. Different possibilities have been proposed: antigen-macrophage interaction resulting in the production of "informational" RNA; formation of antigen-RNA complexes or some other types of "processed" antigen with subsequent release of immunogenic material; grouping of lymphoid cells into clusters around macrophages which facilitates cell-to-cell interaction; prevention of antigen excess from reaching lymphocytes directly (reviewed in *16*). However, evidence is still lacking that macrophages are essential for the initiation of *all* types of specific immune responses. The function of these cells may be quite different from one system to another with respect to removal from extracellular fluids, concentration, presentation and/or modification of antigen (*22*). More recent observations have emphasized at least two distinct functions of macrophages in early phases of antibody responses, i.e. promotion of lymphocyte viability and antigen presentation (*23*). The latter appears to imply a two-step, T lymphocyte-dependent process leading first to binding of antigen to the macrophage surface which is followed by an active sequestration of bound antigen from the latter (*24*). It was also learned that—at least in certain systems—the macrophage ingests considerable amounts of antigen of which it conserves a small pool in a form or in a compartment providing for slow catabolism, and that from this pool minute amounts of antigen or fragments thereof are released over a prolonged period of time, most probably via exocytosis (*25*). Possibly, phagocytosis and digestion of antigenic material, and induction of antibody formation by the B lymphocyte line on the one hand, and/or stimulation of T lymphocytes on the other hand, are functions exerted by two different types of adherent cells (*26*). Furthermore, macrophages may be instrumental in the trapping of lymphocytes after antigenic stimulation (*27*).

The role of macrophages in the initiation of *cell-mediated immune responses* is also not fully understood. Several findings support the hypothesis that antigen presentation by "adherent" cells augment immunogenicity not only for B cells but also for T lymphocytes. Recent experiments have documented that infection with BCG, which is known to result in marked macrophage production and activation, causes an increase in the label of T cell activity. For some as yet poorly understood reason, antibodies formed in BCG-infected animals do not exercise the normal inhibitory (regulatory) effect on the production and/or the efficiency of activated T cells (*28–30*).

As a result of interaction between antigen, T lymphocytes, B cells and/or macrophages, a number of cellular products may be released. Some of these factors appear to function as *mediators* for cellular immunity during initiation and/or expression of the immune response. Most such substances were detected under conditions in vitro, i.e. after addition of antigen to specifically sensitized lymphocytes—with or without other cell types—and are commonly referred to as lymphokines (e.g. macrophage migration inhibitory factor (MIF), macrophage migration stimulatory factor (MSF), chemotactic factors, mitogenic (blastogenic) factor, transfer factor, cytotoxic factor and others). Interferon production in response to viral infection should also be mentioned in this context. Some of these mediators have been tested in vivo (e.g. transfer factor, permeability generating factor, skin reactive factor) and it is generally agreed that they are products of sensitized, activated T lymphocytes. Once such factors are generated and secreted, they do not require the continued presence of antigen for deployment of their function nor are most of them antigen-directed. They are, therefore, called non-specific T dependent factors. Their number and chemical composition are not yet defined, and the relevance of their action in vivo still largely escapes our knowledge.

Non-specific factors (e.g. T lymphocyte activating factor (*31*) and others) may be produced by macrophages, and the latter seem to favor, or to be a prerequisite for, the production of certain non-specific T-dependent factors. Macrophages appear to be able to influence T cell functions in the sense that they may stimulate the "helper" activity of the latter, e.g. by presenting adequate amounts and types of immunogenic material fixed to their surface, possibly by cytophilic antibody. Conversely, the absence or numerical/functional insufficiency of macrophages could result in the presence of supraoptimal concentration of antigen in situ and enhanced T cell suppressor activity. However, these considerations are still largely speculative and we are far from understanding the details of cellular interactions (for review, see *32*).

THE EFFECTOR ROLE OF MACROPHAGES AGAINST TUMOR CELLS

Macrophages can participate in tumor cytolysis in different ways (for review, see *33*): 1. Through activation by specifically sensitized or non-specifically triggered T lymphocytes; 2. through activation by nonspecific agents such as peptone (*18*); 3. with the help of an antigen-specific "cytophilic arming factor" (*34*); 4. as effector cells in antibody-mediated target cell lysis (reviewed in *35*); and 5. by increasing the dissemination ("firing") of toxic factors (*36*).

The sequence of events charcterizing host reaction to injected syngeneic leukemia cells in guinea pigs involves a) accumulation of neutrophils; b) vascular proliferation and degeneration; c) focal histiocyte infiltration at the beginning of tumor growth; and d) a marked increase in the number of macrophages and large lymphoid cells at the time when regression sets in (*37*). These findings are consistent with the assumption of a cooperative cytotoxic effect of lymphoid cells and macrophages.

Experiments in vitro using peritoneal exudate cells from hyperimmune guinea pigs showed that maximum activation of macrophages after addition of sensitized lymphocytes and antigen is reached within approximately one week (*38*). For certain immune reactions, such as skin allograft rejection, two types of T lymphocytes and B cells seem to be required for full deployment of the effector apparatus (*39*). Furthermore, there appears to be a cooperative effect of activated T lymphocytes and macrophages which can both be cytotoxic (*40*). The mechanisms underlying target cell lysis by macrophages are not yet fully understood; however, recent evidence points to strong extracellular cytotoxic activity of these cells. The latter was shown to depend upon a) proximity between macrophages and target cells; b) intact energy metabolism and membrane functions of macrophages; and c) previous activation of macrophages. In contrast, extracellular target cell lysis seems to be possible in the absence of DNA, RNA and protein synthesis (*41*). It is of interest in this context that macrophages synthesize and excrete substantial amounts of lysozyme (*42*) and other enzymes. We ignore, as yet, to what extent lysosomal enzymes are important for the extracellular cytotoxic activity of macrophages, since the pH in the intercellular space may not be optimal for their action.

THE USE OF NONSPECIFIC MACROPHAGE STIMULANTS AS ANTI-CANCER AGENTS

From the foregoing it becomes apparent that—specific or nonspecific—activation of macrophages and their proximity to the target tumor cells seem to represent the essential prerequisites for successful anti-cancer activity of these elements. In theory, one could think of many different means by which macrophages may be activated and thus rendered cytotoxic for tumor cells. Peptone-elicited peritoneal exudate contain activated macrophages which inhibit tumor cell growth in vitro (*18*) and in vivo (*43*). The most widely used agents, however, are certain Mycobacteria, in particular BCG, and Corynebacteria (*parvum, granulosum*). The mode of action of BCG has been analyzed by Mackaness and his coworkers (*28–30*), using sheep erythrocytes instead of tumor cells as antigen. Their findings may be summarized as follows: it was first demonstrated that an immune response to BCG itself was an initial and necessary step in a sequence of events resulting in tumor suppression. Anti-target cell antibody together with target cell antigen in circulation has an inhibitory effect on T cell activity, and BCG, although it may enhance antibody production, causes a large increase in numbers and activity of T cells. Thus, this agent allows for a higher and more sustained level of cell-mediated immunity to develop, by stimulating the production of specifically sensitized T cells and by inhibiting antibody-mediated suppression of the T cell system. Experiments using BCG or Corynebacteria to stimulate the defense against tumor cells have clearly demonstrated the importance of bringing tumor cells and microorganisms—and, therefore, activated macrophages—into close contact. It was shown that intratumor or regional injection of these agents was more efficient in suppressing neoplastic growth than systemic application, and that intravenous injection of the microorganisms was particularly successful in hindering the development of lung metastases. Satisfactory results were only obtained, however, if a) the tumor burden was small; b) adequate numbers of viable microorganisms were used; and c) the animal gave an immune response to these bacteria as well as to tumor-associated antigens (for review, see *44*). The common morphological feature of reactions to BCG and Corynebacteria is the development of epitheloid cell granulomas which represent accumulations of intensely activated macrophages. Models for the use of these microorganisms in cancer therapy have taken into consideration that it lasts at least two weeks for epitheloid cell granulomas to develop fully, and that repeated stimulation produces a more marked reaction than a single injection.

RADIATION-INDUCED MODIFICATION OF MACROPHAGE FUNCTIONS AGAINST TUMOR CELLS

It is well documented that ionizing radiation damages the monocyte-macrophage system primarily by killing radiosensitive precursors in the bone marrow (for review, see *45*). Differentiated macrophages are quite radioresistant. In fact, the enhancing effect of mouse peritoneal macrophages on immunocompetent cells was not abolished by as much as 5000 R delivered in vitro (*46*). It should be emphasized, however, that radiosensitivity in vitro does not necessarily reflect in vivo-conditions. Whole-body X-irradiation of mice with 550 R abolishes the ability of peritoneal macrophages to establish a normal, close contact with nonirradiated lymphocytes. This phenomenon may contribute to the failure of irradiated animals to mount immune responses (*47*). Whole-body irradiation also modifies cell traffic (248). Carbon clearance may be a misleading parameter for testing macrophage functions in vivo, since removal of these inert particles is usually reduced by radiation while that of injected *E. coli* is not (*49*). The combined application of radiation therapy and "immunotherapy" of cancer will have to take into account the kinetics of macrophage stimulation, radiosensitivity of cells involved, and the time course of postirradiative regeneration of the monocyte-histiocyte line, if optimal schemes for timing and dosage shall be arrived at.

REFERENCES

1. G. Biozzi, C. Stiffel, B. N. Halpern, et D. Mouton, *Compt. Rend. Soc. Biol.* (Paris) **153,** 987 (1959).
2. L. J. Old, D. A. Clarke, and B. Benacerraf, *Nature* **184,** 291 (1959).
3. D. L. Morton, F. R. Eilber, and W. L. Joseph, *Ann. Surg.* **172,** 740 (1970).
4. J. Cline, and M. A. Sumner, *Blood* **40,** 62 (1972).
5. R. A. Fox, D. S. Gregory, and J. D. Feldman, *J. Immunol.* **112,** 1867 (1974).
6. B. Roos, in *Handbuch der Allgemeinen Pathologie* (H. W. Altmann, F. Büchner, H. Cottier, E. Grundmann, G. Holle, E. Letterer, W. Masshoff, H. Meessen, F. Roulet, G. Seifert, G. Siebert, A. Studer, Eds.), pp. 1–128. Springer-Verlag, Berlin, Heidelberg, New York, 1970.
7. W. S. Walker. *Immunology* **26,** 1025 (1974).
8. R. M. Steinman, and Z. A. Cohn, *J. Exp. Med.* **137,** 1142 (1973).
9. R. M. Steinman, and Z. A. Cohn, *J. Exp. Med.* **139,** 380 (1974).
10. D. H. Pluznik, V. Rotter, and R. Scheinman, *J. Reticuloendothel. Soc.* **11,** 154 (1972).
11. T. A. McNeill, W. A. Fleming, S. F. McClure and M. Killen, *Immunology* **25,** 91 (1973).
12. S. H. Chan and D. Metcalf, *Cell Tissue Kinet.* **6,** 185 (1973).

13. J. F. Borel, and E. Sorkin, *Experientia* **25,** 1333 (1969).
14. L. C. Altman, R. Snyderman, J. J. Oppenheim and S. E. Mergehagen, *J. Immunol.* **110,** 801 (1973).
15. U. Kissling, B. Roos, F. Joos, H. Bürki, K. Bürki, und J. Laissue, *Beitr. path. Anat.* **139,** 115 (1969).
16. S. Düby-Nunes, B. Roos, H. Cottier, R. Schindler, M. W. Hess, and R. D. Stoner, *Immunology* **26,** 23 (1974).
17. F. Joos, B. Roos, H. Bürki, K. Bürki, und J. Laissue, *Z. Zellforsch.* **95,** 68 (1969).
18. Keller, R., *J. Exp. Med.* **138,** 625 (1973).
19. E. Pick and S. Manheimer, *Cell. Immunol.* **11,** 30 (1974).
20. K. Yokomuro and T. Nozima, *J. Reticuloendothel. Soc.* **11,** 579 (1972).
21. D. E. Mosier, and L. W. Coppleson, *Proc. Nat. Acad. Sci.* (USA) **61,** 542 (1968).
22. E. R. Unanue, *Adv. Immunology* **15,** 95 (1972).
23. C. W. Pierce, J. A. Kapp, D. D. Wood, and B. Benacerraf, *J. Immunol.* **112,** 1181 (1974).
24. J. A. Waldron, Jr., R. G. Horn, and A. S. Rosenthal, *J. Immunol.* **112,** 746 (1974).
25. J. Calderon, and E. R. Unanue, *J. Immunol.* **112,** 1804 (1974).
26. J. Watson, M. Thoman, P. Ralph, and E. Trenkner, *J. Immunol.* **112,** 1873 (1974).
27. P. Frost, and E. M. Lance, *Immunology* **26,** 175 (1974).
28. G. B. Mackaness, D. J. Auclair, and P. H. Lagrange, *J. Nat. Cancer Inst.* **51,** 1655 (1973).
29. T. E. Miller, G. B. Mackaness, and P. H. Lagrange, *J. Nat. Cancer Inst.* **51,** 1669 (1973).
30. E. Hawrylko, and G. B. Mackaness, *J. Nat. Cancer Inst.* **51,** 1683 (1973).
31. I. Gery, and R. E. Handschumacher, *Cell. Immunol.* **11,** 162 (1974).
32. M. Feldmann, in *Series Hematologica—Relationship between Lymphocytopoiesis and the Immune Response* (E. Cronkite, and H. Cottier, Eds.), in press.
33. H. Cottier, M. W. Hess, H. U. Keller, P. Luscieti, and B. Sordat, in Symposium on *Interaction of Radiation and Host Immune Defense Mechanism in Malignancy* (V. Bond, S. Hellman, S. Order, H. Suit, R. Withers, Eds.), The Greenbrier, White Sulphur Springs, West Virginia, March 23–27, 1974. Brookhaven National Laboratory, Upton, N.Y., in press.
34. R. Evans, and P. Alexander, *Nature* **236,** 168 (1972).
35. K. Resch, E. W. Gelfand, and M. Prester, *J. Immunol.* **112,** 792 (1974).
36. S. Ariyan, and R. K. Gershon, *J. Nat. Cancer Inst.* **51,** 1145 (1973).
37. D. G. Feldman, T. Ehrenreich, and L. Gross, *Cancer Res.* **34,** 901 (1974).
38. D. O. Adams, J. L. Biesecker, and L. G. Koss, *J. Reticuloendothel. Soc.* **14,** 550 (1973).
39. M. C. Gelfand, and A. D. Steinberg, *Cell. Immunol.* **11,** 221 (1974).
40. J. Zighelboim, B. Bonavida, and J. L. Fahey, *Cell. Immunol.* **12,** 280 (1974).
41. H. Melsom, *J. Exp. Med.* **139,** 1049 (1974).
42. S. Gordon, J. Todd, and Z. A. Cohn, *J. Exp. Med.* **139,** 1228 (1974).
43. R. Keller, and M. W. Hess, *Br. J. Exp. Path.* **53,** 570 (1972).
44. V. Bond, S. Hellman, S. Order, H. Suit, and R. Withers (Eds.), Symposium on *Interaction of Radiation and Host Immune Defense Mechanisms in Malignancy*, The Greenbrier, White Sulphur Springs, West Virginia, March 23–27, 1974. Brookhaven National Laboratory, Upton, N.Y., in press.
45. T. M. Fliedner, und W. Calvo, in *Handbuch der Allgemeinen Pathologie* (H. W. Altmann, F. Büchner, H. Cottier, E. Grundmann, G. Holle, E. Letterer, W. Masshoff, H. Meessen, F. Roulet, G. Seifert, G. Siebert, A. Studer, Eds.), pp. 375–495. Springer-Verlag, Berlin, Heidelberg, New York, 1969.
46. F. W. Falkenberg, A. Sulica, G. M. Shearer, E. Mozes, and M. Sela, *Cell. Immunol.* **12,** 271 (1974).
47. R. Gallily, and Z. Ben-Ishay, *Cell. Immunol.* **11,** 314 (1974).
48. R. W. Gillette, and E. M. Lance, *J. Reticuloendothel. Soc.* **14,** 18 (1973).
49. S. Order, in Symposium on *Interaction of Radiation and Host Immune Defense Mechanisms in Malignancy* (V. Bond, S. Hellman, S. Order, H. Suit, R. Withers, Eds.), The Greenbrier, White Sulphur Springs, West Virginia, March 23–27, 1974. Brookhaven National Laboratory, Upton, N.Y., in press.

NEW WAYS OF ESTIMATING GENETIC RISKS IN MAN[1]

Chairman: S. Abrahamson

[1]Symposium B–3

The Use Of Chromosome Aberrations for Predicting Genetic Hazards to Man[1]

J. G. Brewen and R. J. Preston
Biology Division, Oak Ridge National Laboratory[2],
Oak Ridge, Tennessee 37830, USA

INTRODUCTION

Almost twenty-seven years ago, in November 1947, to be precise, W. L. Russell began work in Oak Ridge to establish the mouse as the mammalian model for estimating the genetic hazards from ionizing radiations. The ensuing twenty-seven years have resulted in a prodigious amount of research, not only by Russell and colleagues but by many other workers (*1*) in mammalian radiation genetics and an equally prodigious amount of skepticism, from some workers, about the work; a skepticism that has had the recurring theme, "A mouse is not a man, etc., etc. . .". This skepticism was, and in part still is, reinforced by the findings that closely related, in addition to diverse, mammalian species exhibited quantitatively different genetic responses to ionizing radiations. Thus it is to the fundamental problem of establishing a "bridge," as it is called, between mouse and man that we address ourselves in this paper.

The genetic consequences of exposure to ionizing radiation are diverse and, simultaneously, manifest. Those effects that are demonstrably genetic in nature include gene mutations, reciprocal translocations, chromosomal inversions, dominant lethality expressed as either pre- or post-implantation death of the zygote, and chromosome loss or addition. Of all of these genetic effects, the only one that has not been clearly demonstrated to consist entirely, or predominately, of structural chromosome damage is the effect generally alluded to as specific locus mutations, and data are rapidly being collected that show that a significant portion of the radiation-induced specific locus mutations are actually minute chromosome deletions. It is obvious, therefore, that structural damage to the chromosomes constitutes the major portion of the genetic damage produced by ionizing radiations.

The advent of the short-term tissue culture of human peripheral leukocytes by Moorhead et al. (*2*) enabled cytogeneticists to begin measuring production of chromosomal aberrations, by ionizing radiations, in a homogeneous population of human cells. The original 'qualitation' and 'quantitation' of aberration production was done by Bender and Gooch (*3,4*), and in the subsequent twelve years a copious literature has been amassed on this subject. Two

[1]Congress presentation B-3-1

[2]Research sponsored by the U.S. Atomic Energy Commission under contract with Union Carbide Corporation.

excellent reviews (*5,6*) are available on human chromosome aberration production by ionizing radiation, and it suffices to mention that the accumulated data show that the general results found in the classical cytological material, e.g. *Vicia faba* and *Tradescantia,* are also found in the human studies. These include (1) the presence of a dose-square component in the kinetics of interchange production by low LET radiations, (2) fractionation effects with low LET radiations, (3) dose rate effects with low LET radiations, (4) linear dose response kinetics with high LET radiations, and (5) a lack of a fractionation, or dose rate, effect with high LET radiations.

Although the data on chromosome aberration production in human cells were very extensive, it was extremely difficult to use them for predicting heritable genetic effects. The reasons for this were (1) the cells used were of somatic origin and the possibility existed of quantitative differences between somatic and germ cells, (*7*), and (*2*) the aberration types analyzed were generally recognized as being cell (*8,9*) and organism lethal (*10*) and thus of no genetic consequences. It was obvious to radiation biologists that if cytogenetic studies could be performed on mouse leukocytes, and the data so derived correlated to genetically recoverable events in the mouse, a reasonable extrapolation from human somatic cell chromosome aberration data to true genetic effects might be made. Although many laboratories endeavored to develop a method for culturing mouse leukocytes, very little progress was made until Buckton and Nettesheim (*11*) reported success by growing the leukocytes in cell-impermeable diffusion chambers placed in the peritoneal cavities of other species.

In the meantime, however, aberration data on other mammalian species were underscoring the fact that diverse species had diverse quantitative, if not qualitative, responses in terms of aberration production by ionizing radiation (*12–15*). These findings made it even more imperative that data be obtained on the mouse if the human data were to be used, with any confidence, in predicting genetic hazards.

The one seemingly consistent pattern that emerged from the studies on different species was that the species with the higher chromosome numbers were the more sensitive to chromosome interchange production. Qualitatively speaking, this observation had been made earlier in plant materials where it was observed that increases in ploidy resulted in increases in sensitivity to interchange production (*16–18*). In these instances, however, the DNA content also varied, and hence it was not clear whether the increased sensitivity was due to increased chromosome number or DNA content per nucleus. The observation by Griffin et al. (*19*) that the interchange frequency in amphibians remained constant as DNA content varied but the chromosome number remained the same, suggested that sensitivity to interchange production was influenced by chromosome number.

Following these studies, Brewen et al. (*20*) reported a comprehensive inter-specific comparison of dicentric and deletion production by acute X-rays in six mammalian species. This study was then followed by a comparison of the yields of dicentrics produced in peripheral leukocytes and reciprocal translocations produced in spermatogonial stem cells (*21*). Based on this work, and that of several other investigators, we will attempt to create a working model for estimating various genetic hazards to man from ionizing radiations. Each of the various aspects of the model and the pertinent data will be discussed in separate sections.

INTERSPECIFIC COMPARISONS OF DICENTRIC AND DELETION PRODUCTION IN PERIPHERAL LEUKOCYTES

The dose response characteristics of dicentric chromosome and deletion production in six mammalian species were studied in a series of experiments (*12–14,20*). The data from these experiments are summarized in Tables 1 and 2. It was observed that both aberration types increased exponentially with dose in all species (Fig. 1). The data also show that the species with the higher number of chromosome arms available for participation in interchange formation are the most sensitive to dicentric production. This correlation between chromosome arm number and sensitivity to exchange induction appears to be linear and, consequently, human cells with an "effective arm number" of 81 are twice as sensitive as mouse cells with an "effective arm number" of 40 (Fig. 2). Recent data of Bajerska and Liniecki (personal communication) show that the rabbit and man are equally sensitive to dicentric induction. The rabbit has an "effective arm number" approximately equal to man. There is no obvious karyological effect of this type on the sensitivity to deletion induction and, in fact, the human and mouse cells appear equal in sensitivity.

TABLE 1
Frequencies of dicentric chromosomes per 100 cells observed in the peripheral leukocytes of different species following irradiation with various doses of X rays

Dose (R)	*W. bicolor* 19[a]	*C. griseus* 38	*M. musculus* 40	*S. scrofa* 64	*H. sapiens* 81	*S. fuscicollis* 83
0	0	0	0.3 ± 0.3	0	0	0
50	0.3 ± 0.3	1.6±0.6	2.3 ± 0.9	1.3 ± 0.7	4.3 ± 0.8	6.5 ± 1.8
100	1.0 ± 0.6	6.0±1.5	6.7 ± 1.5	10.0 ± 1.8	13.2 ± 1.5	12.7 ± 2.0
150	5.4 ± 0.9	—	—	19.7 ± 2.6	20.8 ± 1.9	34.3 ± 3.4
200	9.5 ± 1.5	12.6±1.6	14.3 ± 2.2	27.0 ± 3.0	33.7 ± 2.4	50.3 ± 4.1
300	23.0 ± 2.8	30.3±2.8	31.3 ± 3.2	61.0 ± 4.5	67.5 ± 3.4	80.0 ± 5.2
400	51.3 ± 4.1	51.4±3.8	55.0 ± 3.7	92.7 ± 5.6	120.5 ± 4.5	—

[a]The numbers below the species' names indicate the "effective arm number."

TABLE 2
Frequencies of deletions per 100 cells observed in the peripheral leukocytes of different species following irradiation with various doses of X rays

Dose (R)	*W. bicolor* 19[a]	*C. griseus* 38	*M. musculus* 40	*S. scrofa* 64	*H. sapiens* 81	*S. fuscicollis* 83
0	2.6 ± 0.6	0.5±0.4	0	0	0.3 ± 0.2	0
50	5.3 ± 1.1	2.4±0.7	2.0 ± 0.8	4.7±1.2	4.2 ± 0.8	5.0 ± 1.6
100	6.7 ± 1.5	8.3±1.4	3.0 ± 1.0	7.7±1.6	6.0 ± 1.0	11.0 ± 1.9
150	15.1 ± 1.5	–	–	22.0±2.7	8.8 ± 1.2	24.3 ± 2.8
200	23.8 ± 2.4	15.8±1.8	16.7 ± 2.4	32.3±3.3	15.8 ± 1.6	32.0 ± 3.3
300	50.0 ± 4.1	38.8±3.1	29.3 ± 3.1	69.3±4.8	33.8 ± 2.4	53.7 ± 3.3
400	114.7 ± 6.2	61.4±4.2	52.8 ± 3.5	105.0±5.9	61.8 ± 3.2	–

[a]The numbers below the species' names indicate the "effective arm number."

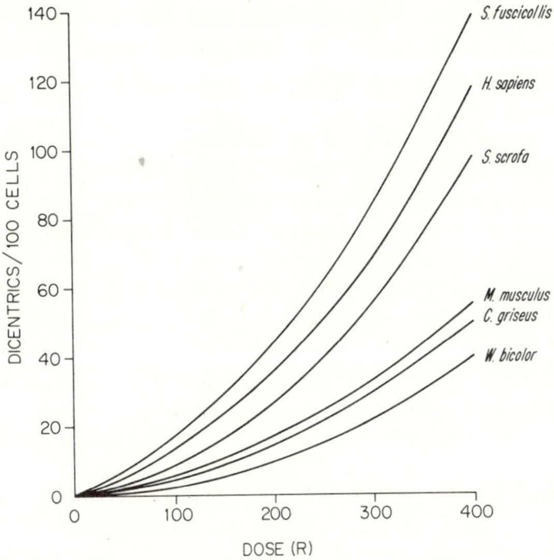

FIG. 1. Best-estimate regression curves for dicentric population in six mammalian species. (Reproduced from J. G. Brewen and N. Gengozian, *Mutat. Res.* **17**, 245 (1973), with permission of Elsevier Publishing Co.)

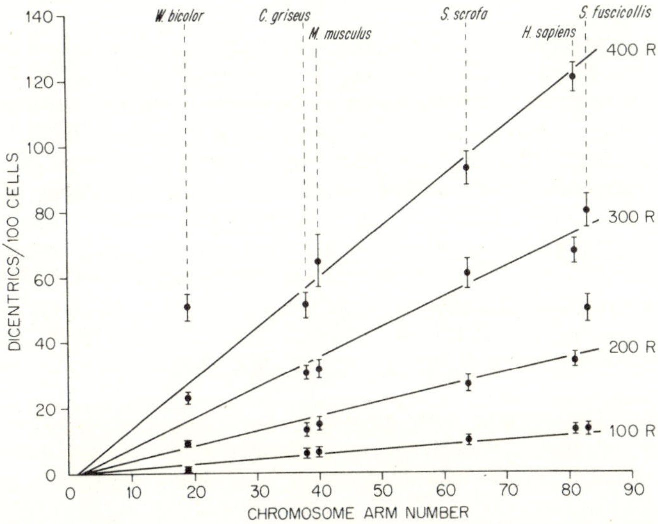

FIG. 2. Best-estimate regression curves for dicentric yields vs. chromosome arm number in six mammalian species. Observed yields are plotted with standard errors at each of the four doses considered. (Reproduced from J. G. Brewen and N. Gengozian, *Mutat. Res.* **17**, 245 (1973), with permission of Elsevier Publishing Co.)

Since a dicentric is one of the two possible interchange configurations that are formed following irradiation and a reciprocal translocation is the other (*22*), it seems reasonable to conclude that man is twice as sensitive as the mouse to the induction of reciprocal translocations. It is not clear that dicentrics and reciprocal translocations are induced in equal frequencies per unit dose. Heddle (*23*) has presented evidence, from studies in *Vicia faba,* that the two events occur at equal frequencies and Caspersson et al.'s (*24*) data on human leukocytes, though not as clear as Heddle's, suggest the two events occur with approximately equal frequencies. Recently, however, several authors (*25,26*) utilizing the new band-staining techniques, have presented data showing that dicentrics occur more frequently than reciprocal translocations. Not all reciprocal translocations are expected to be seen, even with the new staining techniques (*27*), and consequently an excess of dicentrics should not be considered proof that the two events do not occur at equal frequencies. However, even if the two exchange types do not occur with equal frequency, the dicentric data can be used to (1) demonstrate inequalities in sensitivity to interchange induction between species, and (2) as a common index of mouse and man for comparing events in somatic cells to events in germ cells.

THE COMPARISON OF DICENTRIC YIELDS IN LEUKOCYTES TO RECIPROCAL TRANSLOCATION YIELDS IN SPERMATOGONIA

The production of reciprocal translocations in spermatogonial stem cells by X-rays was studied in young adult male mice and Chinese hamsters by irradiating the animals with various doses of 250kV X-rays, waiting an appropriate time, and analyzing diplotenemetaphase I primary spermatocytes for multivalent configurations (*21,28*). The frequencies of reciprocal translocations were then compared to the frequencies of dicentrics observed in leukocytes over the same range of doses (*21,29*). Fig. 3 summarizes the data. The mouse and hamster have been demonstrated to be equally sensitive to dicentric induction and Fig. 3 shows that they are also equally sensitive to reciprocal translocation induction. It is also seen in Fig. 3 that the ratio of dicentrics to reciprocal translocations is approximately 3–4:1, and the slopes of the two curves are the same, i.e. about 1.5, indicating the presence of a two-track process in exchange formation. This latter observation is in contrast to other published data in that other workers have consistently reported linear dose kinetics for reciprocal translocation induction (*30–33*). The dose rate effects and fractionation effects that have been demonstrated argue for the presence of a two-track component (*31,34–37*), however, and the present result is therefore not surprising.

In order to determine if the interspecific variations in radiosensitivity to interchange induction demonstrated in the peripheral leukocyte system extended to the germ cells, several male marmosets[3] and guinea pigs were testicularly irradiated with various doses of X-rays. The results to date are presented in Table 3 and Fig. 4. The data clearly show that both the marmoset and guinea pig are more sensitive than the mouse and Chinese hamster to the production of reciprocal translocations at the lower (25–200 R) doses. There is an indication that both species demonstrate the unusual "humped" dose response curve and the peak yield occurs at a lower dose than in the mouse.

We have some data on the induction of reciprocal translocations in testicularly irradiated men. The data come from three dose points, namely 78, 200,

[3]In collaboration with N. Gengozian, Oak Ridge Associated Universities, Oak Ridge, Tenn.

and 600 R. The yields of translocations observed are 4.0± 1.0%, 7.0 ± 1.3%, and 6.1 ± 1.8%, respectively. The number of cells analyzed, however, is small (371, 300, and 180) and the data carry a large standard error. The data indicate nevertheless, that, as in the case of the marmoset, man is more sensitive than the mouse to translocation induction at lower doses.

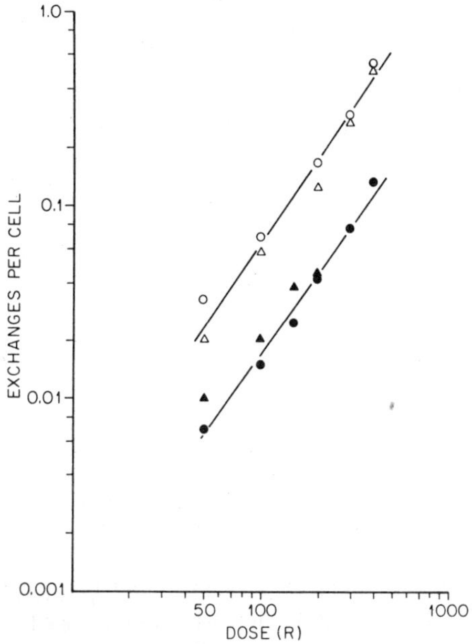

FIG. 3. Log-log plot of chromosome exchange yield and X-ray doses for the pooled Chinese hamster and mouse leukocyte data and the mouse spermatogonial stem cell data. O———O, Mouse leukocytes; Δ———Δ, hamster leukocytes; ●———●, mouse spermatogonia; ▲———▲, hamster spermatogonia. (Reproduced from J. G. Brewen and R. J. Preston, *Nature New Biology* **244**, 111, (1973) with permission of MacMillan Journals Ltd.).

TABLE 3
Frequencies of Reciprocal Translocations in Primary Spermatocytes After Spermatogonia Irradiation of Several Mammals

Dose (R)	Mouse	Ch. hamster	Guinea pig	Guinea pig[a]	Rabbit[a]	Marmoset
0	0	0	0	0	0.15 ± 0.2	0
25	–	–	–	–	–	2.8 ± 0.7
50	0.7 ± 0.3	1.0 ± 0.6	1.4 ± 0.6	–	–	3.4 ± 0.8
100	1.5 ± 0.5	2.0 ± 0.8	2.7 ± 0.9	0.85 ± 0.6	–	7.5 ± 1.1
200	4.3 ± 0.8	4.3 ± 1.2	6.1 ± 1.1	4.62 ± 0.9	–	7.3 ± 2.2
250	–	–	–	–	1.17 ± 0.4	
300	7.8 ± 1.1	–	8.0 ± 2.0	–	5.79 ± 0.9	6.7 ± 1.5
400	13.5 ± 1.3	–	–	0.31 ± 0.2	–	–
500	17.0 ± 1.5	–	–	–	0.23 ± 0.2	–
600	19.5 ± 1.6	–	–	–	0	–

[a]Data of Lyon and Smith (*38*).

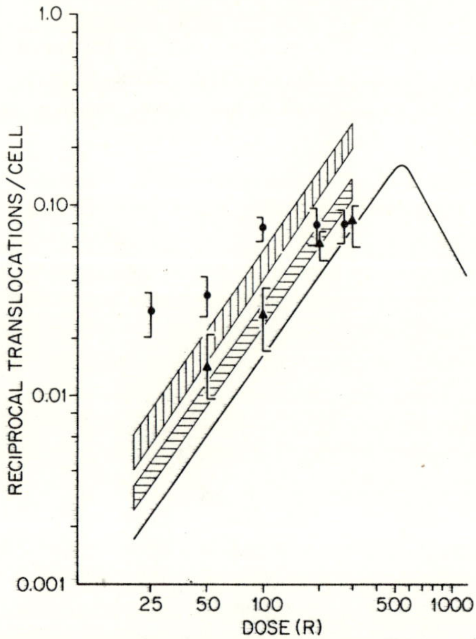

FIG. 4. Log-log plot of reciprocal translocation, induced in spermatogonia and observed in primary spermatocytes, vs. X-ray dose for the mouse (solid line), guinea pig (▲) and marmoset (●). The shaded areas represent the range of expected yields for guinea pig (▤) and marmoset or man (▥).

RECOVERY OF SPERMATOGONIA INDUCED TRANSLOCATIONS IN THE SONS OF IRRADIATED MALES

Ford et al. (*39*) have reported that, following a dose of 1,200 R separated into two equal fractions of 600 R administered eight weeks apart to male mice, the yields of reciprocal translocation bearing primary spermatocytes were 41.2% and 32.6%. Subsequent testing of sons of identically irradiated males for semi-sterility and translocations revealed five translocation heterozygotes in 150 F_1 males. Using the distribution of translocations scored cytologically in the primary spermatocytes, Ford et al. (*39*) calculated an expected recovery of translocations twice that observed. Ford et al. (*39*) concluded that a selection process may be in operation that affects the haploid spermatids or sperm. This observation in itself would permit geneticists to conclude that the risk of recovering a radiation-induced translocation in the F_1 is only one-half of the expectation of 1/4, or 1/8 of the observed yield in primary spermatocytes. Using the previously presented data on the ratio of reciprocal translocations in primary spermatocytes to dicentrics in peripheral leukocytes of the mouse, i.e. 1/4, we conclude that the yield of dicentrics in human leukocytes after radiation exposure can be corrected by approximately 1/32 (i.e. 1/4 x 1/8) to arrive at an estimate of recovered translocations in the offspring of irradiated men.

There are two lines of evidence, however, that must be considered. First, in surveys of families known to carry a balanced translocation the ratio of normal plus balanced heterozygote births to abortions exceeds the expected 1:1

ratio in that there are far too few abortions (*40,41*). It can be argued that most of the abortions occur before implantation and consequently the mother is not aware of her pregnancy. This question must be investigated more thoroughly. Second, and perhaps more significant, are the findings of Generoso (*42*) and Brewen et al. (*43*). Generoso studied the incidence of semisterility and sterility in F_1 males of spermatogonial stem cell irradiated fathers. The doses and frequencies of semi-sterile and semi-sterile plus sterile sons are presented in Table 4. All of the semi-sterile sons were confirmed cytologically, by Brewen and Preston, to be translocation heterozygotes. The cytological constitution of the steriles has been studied by Cacheiro and Russell (personal communication) and four of the males were shown to be translocation heterozygotes. The data can be used to estimate the frequency of translocation recovery. If these frequencies are compared to the frequencies of cytologically scored translocations in primary spermatocytes at the same doses, it is seen that the probability of recovery varies with the dose (Table 4). In fact, at the lowest dose (150 R) approximately 25%, or the 'expected' number, are recovered. These observations are preliminary in nature and are in the process of being expanded. If the variation in recovery rate as a function of dose is confirmed, we will have to reassess our estimates of hazard from translocations.

TABLE 4
Frequencies of cytologically scored and genetically recovered reciprocal translocations following irradiation of spermatogonial stem cells

Dose (R)	% Translocations scored[a]	No. F_1 males tested	% Translocations Heterozygotes Minimum[b]	Ratio recovered/scored
0	0.0	2,633	0.0	—
150	2.5 (2.5)	1,122	0.62	.25
300	7.8 (5.7)	1.053	0.84	.11
600	17.2 (16.2)	1,062	1.90	.11
1,200	4.5 (3.8)	1,038	0.58	.13

[a]Numbers in brackets are percentages of cell with a single translocation.
[b]Minimum frequency of translocation heterozygotes are the cytologically confirmed semi-steriles.

CYTOLOGICALLY AND GENETICALLY DETECTED DELETIONS

Data were presented (Table 2) that indicated that mouse and human leukocytes were equally sensitive to the induction of microscopically visible deletions by X-rays. The deletions scored in those experiments represent size variations from loss of entire chromosome arms to minute intercalary deletions, and there is no *a priori* reason to believe that this apparent equivalence in sensitivity to deletion induction cannot be extended to include deletions that are submicroscopic in size. If we accept this premise, we can postulate that human and mouse cells are equally sensitive to the production of chromosome deletions of the order of one map unit and less. Russell (*44*) has shown that 13.5% of the specific locus mutations produced at the *d-se* loci in spermatogonia are in fact minute deletions. The proportion of the mutations known to be associated with deletions ranges from 13.5% to 42.3%, the variations being determined by the male germ cell stage irradiated and the LET of the radiation. More recent studies by Russell (personal communication) indicate that a significant number of mutants at the *c* locus are deficiencies in that they are homozygous lethals. Erickson et al. (*45*) have also recently reported the existence of deletions for the structural gene *MOD*–2 in existing mutant stocks at the *c* locus. These mutants were induced by X-rays. Based on the cytological data derived from the leukocyte studies we conclude that mouse and man are equally sensitive to the production of those gene mutations that result from small deletions.

In addition, there are some data on X-ray-induced autosomal recessive lethals in mouse spermatogonia (*46*). It is generally assumed that these result from small deficiencies, and the mean overall rate of production is about 2×10^{-4}/gamete/rad. Again, if the results from the leukocytes are indicative of an equality in the sensitivities of mouse and human cells to deletion induction, these data can be used to make an estimate of expected autosomal recessive frequencies in man. Finally, it has been assumed that the preponderance of dominant lethality produced in postmeiotic germ cell stages results from large deficiencies, (*47,48*). It is not known what proportion is contributed by simple deletions and what proportion is the result of unbalanced segregants of interchanges, but the existing data allow us to estimate that man might be anywhere from equally to twice as sensitive as the mouse.

None of these assumptions take into consideration the possibility of varying degrees of selection operating against the aberration-bearing cells in different species. There is evidence that such selection occurs in the mouse (*49*) and the possibility of lesser, or greater, selection in man introduces an uncertainty in the extrapolations.

CURRENT GAPS IN OUR KNOWLEDGE AND PROBLEMS OF FUTURE INTEREST

The great majority of data on the cytological effects of ionizing radiation on germ cells has been accumulated from studies on males and very little information is available on the effects on female germ cells. New techniques (*50–52*) for analyzing oocyte meiosis will enable researchers to collect data on translocation and deletion induction in both mature and embryonic females.

Although there is a plethora of data on the induction of translocations by chronic exposures in the male mouse, no comparable data exist for other mammalian species. Although the mouse-data show that the translocation frequency is lower at reduced dose rates at the lower ($<$ 400 R) doses (*53,54*), the possibility exists that the well-documented loss of translocation-bearing cells following acute exposures might be greater in other species. This could conceivably result in spuriously low yields with acute exposures and a higher recovery rate with chronic exposures due to the complexities introduced by the interaction of cell killing and translocation induction.

Just recently two articles appeared that report a failure to induce translocations in embryonic male mice, (*55,56*). We have data that is contrary to these reports in that we observed no significant differences in the translocation yields observed in irradiated 13 1/2–day fetuses, newborns, and young adults. More information is needed here.

Comprehensive cytogenetic studies on the effects of low dose and chronic dose rate exposures are possible because of the relative ease of analyzing thousands of cells. Such studies on the mouse, coupled with more interspecific comparisons, should enable us to make highly meaningful predictions of man's hazards. It appears that the first "bridge" between mouse and man has been made and the next few years should result in the filling of many of the gaps in our knowledge.

REFERENCES

1. *United Nations Scientific Committee on the Effects of Atomic Radiation*, General Assembly Document, 17th Session, **Suppl. No. 16** (A-5216). United Nations, N. Y., 1962.
2. P. S. Moorehead, P. C. Nowell, W. J. Mellman, D. M. Battips and D. A. Hungerford, *Exp. Cell Res.* **20,** 613, 1960.
3. M. A Bender, and P. C. Gooch, *Proc. Nat. Acad. Sci. U.S.* **48,** 522, 1962.
4. P. C. Gooch, M. A Bender, and M. L. Randolph, in *Biological Effects of Neutron and Proton Irradiation*. Vol. I, pp 325–342. I.A.E.A., Vienna, 1964.
5. *United Nations Scientific Committee on the Effects of Atomic Radiation*, General Assembly Document, 24th Session, **Suppl. No. 13** (A/7613). United Nations, N. Y., 1969.
6. M. A Bender, in *Advances in Radiation Biology*, L. G. Augenstein, R. Mason, and M. Zelle, Eds. Vol. 3, pp. 215–275, Academic Press, New York, 1969.
7. A. L. Brooks, and F. W. Lengemann, *Radiat. Res.* **32**, 587, 1967.
8. A. D. Conger, *Radiat. Bot.* **5,** 81, 1965.
9. S. J. Grote, and S. H. Revell, in *IVth Intern. Congr. Radiation Res.*, Abstract 332. Evian, France, 1970.
10. H. J. Muller, in *Radiation Biology,* (A. Hollaender Ed.), Vol. I, pp. 351–626, McGraw-Hill, New York, 1954.
11. K. E. Buckton, and P. Nettesheim, *Proc. Soc. Exp. Biol. Med.* **128,** 1106, 1968.
12. J. G. Brewen, and R. D. Brock, *Mutat. Res.* **6,** 245, 1968.
13. J. G. Brewen, and N. Gengozian, *Mutat. Res.* **13**, 383, 1971.
14. R. J. Preston, J. G. Brewen and K. P. Jones, *Int. J. Radiat. Biol.* **21,** 397, 1972.
15. J. F. P. Clemenger, and D. Scott, *Int. J. Radiat. Biol.* **24,** 487, 1973.
16. A. D. Conger, and A. H. Johnston, *Nature* **179,** 271, 1956.
17. M. S. Swaminathan, and A. T. Natarajan, *Nature* **179,** 479, 1957.
18. H. J. Evans, *Genetics* **46,** 257, 1961.
19. C. S. Griffin, D. Scott, and D. G. Papworth, *Chromosoma* **30,** 228, 1970.
20. J. G. Brewen, R. J. Preston, K. P. Jones, and D. G. Gosslee, *Mutat. Res.* **17,** 245, 1973.
21. J. G. Brewen, and R. J. Preston, *Nature New Biol.* **244,** 111, 1973.
22. K. Sax, *Genetics* **25,** 41, 1940.
23. J. A. Heddle, *Genetics* **52,** 1329, 1965.
24. T. Caspersson, U. Haglund, B. Lindell, and L. Zech, *Exp. Cell Res.* **75,** 541, 1972.
25. M. Seabright, *Chromosoma* **40,** 333, 1973.
26. C. San Roman, and M. Bobrow, *Mutat. Res.* **18,** 325, 1973.
27. A. C. Chandley, S. Christie, J. Fletcher, A. Frackiewicz, and P. A. Jacobs, *Cytogenetics* **11,** 516, 1972.
28. R. J. Preston, and J. G. Brewen, *Mutat. Res.* **19,** 215, 1973.
29. J. G. Brewen, and R. J. Preston, *Mutat. Res.*, in press.
30. A. Léonard and Gh. Deknudt, *Radiat. Res.* **40,** 276, 1969.
31. M. F. Lyon, and T. Morris, *Mutat. Res.* **8,** 191, 1969.
32. E. P. Evans, C. E. Ford, A. G. Searle and B. J. West, *Mutat. Res.* **9,** 501, 1970.
33. S. Muramatsu, W. Nakamura, and H. Eto, *Japan. J. Genet.* **46,** 281, 1971.
34. A. Léonard, and Gh. Deknudt, *Radiat. Res.* **45,** 72, 1971.
35. M. F. Lyon, T. Morris, P. Glenister, and S. E. O'Grady, *Mutat. Res.* **9,** 219, 1970.
36. T. Morris, and S. E. O'Grady, *Mutat. Res.* **9,** 411, 1970.
37. A. G. Searle, C. V. Beechey, E. P. Evans, and C. E. Ford, *Mutat. Res.* **15,** 169, 1972.
38. M. F. Lyon, and B. D. Smith, *Mutat. Res.* **11,** 45, 1971.
39. C. E. Ford, A. G. Searle, E. P. Evans and B. J. West, *Cytogenetics* **8,** 447, 1969.
40. C. E. Ford, and H. M. Clegg, *Brit. Med. Bull.* **25,** 110, 1969.

41. J. Lejeune, B. Dutrillaux, and J. de Crouchy, in P. A. Jacobs, W. H. Price and P. Law (Eds.) *Human Population Cytogenetics*, pp. 82-87. Williams and Wilkins, Baltimore, 1970.
42. W. M. Generoso, *Biology Division Annual Progress Report*, Oak Ridge National Laboratory, in preparation.
43. J. G. Brewen, R. J. Preston and W. M. Generoso, *ibid*.
44. L. B. Russell, *Mutat. Res.* **11,** 107, 1971.
45. R. P. Erickson, G. M. Eicher and S. Gluecksohn-Waelsch, *Nature* **248,** 416, 1974.
46. K. G. Lüning, and A. G. Searle, *Mutat. Res.* **12,** 291, 1971.
47. E. Lifschytz, and R. Falk, *Mutat. Res.* **6,** 235, 1968.
48. *United Nations Scientific Committee on the Effects of Atomic Radiation*, General Assembly Document, 21st Session, **Suppl. No. 14,** (A-6314), United Nations, N. Y., 1966.
49. G. B. Gerber and A. Léonard, *Mutat. Res.* **12,** 175, 1971.
50. R. P. Donahue, in *4th International Congress of Human Genetics*, Paris, 1971, *Excerpta Medica* 233, 1971.
51. I. Hansmann, *Mutat. Res.* **20,** 353, 1973.
52. P. C. Hoppe and S. Pitts, *Biol. Reprod.* **8,** 420, 1973.
53. A. G. Searle, E. P. Evans, C. E. Ford, B. J. West and D. G. Papworth, *Mutat. Res.* **6,** 427, 1968.
54. A. G. Searle, C. V. Beechy, M. J. Corp and D. G. Papworth, *Mutat. Res.* **15,** 89, 1972.
55. B. Ivanov, A. Léonard, and Gh. Deknudt, *Mutat. Res.* **18,** 89, 1973.
56. W. S. Tsuchida, and I. A. Uchida, *Mutat. Res.* **22,** 277, 1974.

Mutation and the Amount of Human Ill Health[1]

Howard B. Newcombe
Biology and Health Physics Division
Atomic Energy of Canada Limited
Chalk River, Ontario, Canada

INTRODUCTION

Geneticists in the past have tended to pay special attention to any social development that might conceivably affect the quality of the human germ plasm. This is still true today.

Major changes in our way of life have been predicted as a result of depletion of the reserves of fossil fuels. And we are currently witnessing the beginnings of such changes as relating in particular to petroleum. As a result it is now generally recognized that without alternative means for the production of abundant power it will be difficult indeed to ensure a stable future for human society. Most informed persons have thus come to accept as inevitable a continuing proliferation of nuclear reactors in the more densely populated areas of the world.

The challenges and problems arising from these developments are reflected in the programs for almost all of the plenary sessions of this International Congress. In the present symposium, we are considering one of these problems, which has to do with the protection of man from the hereditary changes that can be induced in his germ plasm by ionizing radiation. Much is known about the nature of these genetic changes, but surprisingly little about their importance in terms of the amount of human ill health that they are likely to cause. And yet, without such knowledge it is difficult to determine the most appropriate standards for radiation protection.

GENERAL PRINCIPLES

Geneticists from whom advice has been sought have tended, because of the lack of relevant data, to quote general principles. Few of them have voluntarily attempted to make quantitative predictions of risks.

With commendable foresight, however, they warned at an early stage that the human germ plasm is our most precious possession. And they insisted on our moral obligation to future generations to protect it from all but the most essential and unavoidable exposures to mutagenic influences such as ionizing radiation.

Because of the social importance of this matter, atomic energy programs have, since their inception, been among the most generous patrons of basic genetic research. The belief was held by both parties that from the develop-

[1]Congress presentation B-3-3.

ment of the science of heredity as a whole would come answers to the important practical questions. Currently, decisions are being made regarding the most desirable levels of protection for the human germ plasm. As a basis for such judgments geneticists are being asked, "Just how important *are* the risks arising from an increase in the mutation rate due to a given radiation exposure?"

It is not too difficult to estimate the increase in mutation rate that would be caused by various levels of radiation. There are, for example, good data from experiments with mice, and reasons to believe that the increases in man would not be so very different. But knowing the increase in mutation rate does not tell us the increase in human disease. This brings us to a major area of fundamental genetic ignorance. Mutations are assumed to be predominantly harmful, but the total impact of the harm they cause has not been quantified. An increase in the mutation rate would undoubtedly cause a proportional increase in the amount of mutation-maintained disease normally present in the population, but we can only guess what this is. That geneticists are reluctant to make quantitative statements about genetic risks is thus not surprising.

This reluctance, however, has sometimes placed them in a paradoxical and undignified position. I recall that in the mid 1950's a group of geneticists was being questioned by the Secretary of the British Medical Research Council. He asked, "Suppose the mutation rate in man were doubled, would the effect on human health be important or would it be trivial?" The hesitation that followed indicated their dilemma. In principle, any increase in the mutation rate must be viewed as harming the germ plasm; but in practice the effect of even a doubling of the mutation rate might perhaps be minute in terms of human ill health. No one was really sure.

The question is scarcely academic. Decisions have to be made now between alternative levels of radiation protection. The cost of additional safety measures becomes important where any substantial part of society's limited resources is at stake. The degree of future prosperity could depend on our answer, and positive harm could result if needed sources of power were denied development and construction because of feared risks that had been overestimated. This brings us back to the scientific problem. Increased support for basic genetic research as a whole is not the answer. We have here a specific area of ignorance, relating to the total impact of mutation on health. Other more fashionable kinds of research will not solve this particular problem.

METHODS OF ESTIMATING RISKS

The estimates of radiation risks that geneticists have been persuaded to prepare have necessarily been based on a mixture of biological data and plausible assumptions (*1, 2*).

Where an answer in terms of human disease is sought, it is usual to start with the natural frequency of hereditary disorders in human populations, and the proportions of these that are inherited in various ways. Surprisingly, there is only one available body of data suitable for the purpose, and this was obtained largely from a survey of Northern Ireland carried out over 15 years ago (*3, 4, 5*). From these data it would seem that about 6 people out of every 100 born alive will have more or less serious trouble at some time in their lives due to conditions that are largely or in part hereditary in origin (Table 1). One out of the 6 per 100 is regarded as having some dominant disease, and 4 out of the 6 as having irregularly inherited conditions.

TABLE 1
Conventional estimates of the frequencies of different kinds of hereditary disease

Kind of Disease	Per 100 Liveborn[a]
single gene (e.g. retinoblastoma)	1[b]
chromosomal (e.g. Down's syndrome)	1
congenital malformation (e.g. cleft palate)	2.5
constitutional (e.g. diabetes)	1.5
Total	6.0

[a]From UNSCEAR 1958 and 1966, (refs. *4,5*).
[b]Believed to be due mainly to dominant genes.

It is generally agreed that, if the mutation rate were permanently doubled, all of the simple dominant diseases, at least, would double in frequency. These are diseases caused by the presence of a single gene, and they are transmitted on the average to half of the offspring from affected persons. Geneticists are less sure about the effect of a change in the mutation rate on the collectively more numerous irregularly inherited conditions (i.e. congenital malformations and constitutional disorders). But to be on the safe side it has in the past been commonly assumed that as many as half of these might also double in frequency. If this were true, the combined frequency for all serious genetic disease would perhaps rise from 6 per 100 liveborn to as many as 9 per 100 (Table 2). (Recessive and chromosomal diseases are omitted from discussion in this connection, as unlikely to affect the argument).

TABLE 2
Conventional estimate of the effect of a radiation-induced doubling of the mutation rate[a]

Kind of Disease	Cases per 100 Liveborn		
	Natural	Additional	New Level
single gene	1	1	2
chromosomal	1	very small	1
congenital malf. constitutional	4	up to 2	up to 6
New Total			up to 9

[a]See UNSCEAR 1972 and BEIR 1972 (*1,2*).

This calculation probably substantially overestimates the risk, however, and I will present reasons to believe that it does so by more than an order of magnitude. The belief that hereditary disease is maintained in the population largely by repeated mutations, which reintroduce those harmful mutant genes lost from the population by deaths and failures to reproduce, is probably correct only as applied to the simple dominant conditions. The prevalence of these seems now to have been overestimated, by more than 10-fold.

Recurrent mutation is certainly not the only possible cause of the continued prevalence of genetic disease. The seemingly harmful genes for sickle cell anemia and Mediterranean anemia, for example, remain common in northern Africa and southern Europe because the carriers who are not themselves sick tend to be resistant to malaria. Hereditary traits due to single genes that are known to be maintained in such a fashion by natural selection are common, and are called balanced polymorphisms. Many, and perhaps most, recessive conditions fall in this category.

Similar selective mechanisms are responsible also for the prevalence of irregularly inherited conditions which are determined by a multiplicity of genes that have beneficial effects in appropriate settings. A frequently fatal disease of infants, known as pyloric stenosis, is a good example. It is associated with overdevelopment of the pyloric sphincter muscle at the bottom of the stomach, and death results from starvation when food is prevented from passing into the intestine. The same genes, however, have a markedly beneficial effect on the whole body musculature, and infants who survive (due to surgery) tend to become excellent athletes. Their siblings will of course share these benefits since they have many of their genes in common. Natural selection apparently maintains these genes in some sort of favorable equilibrium in the population. Such polygenic polymorphisms are common, and they probably account for the bulk of the irregularly inherited diseases. But the subject is only gradually attracting attention among geneticists.

An increase in the mutation rate would seem unlikely to disturb appreciably the frequencies of irregularly inherited diseases if most of them are maintained in this fashion. It is not surprising therefore that studies with laboratory mammals mostly fail to demonstrate such an effect. Geneticists have argued, however, largely on the basis of certain theories, that a non-specific deterioration of health and well-being, as distinct from actual disease, must result from any sustained increase in mutation rate. But the relevant tests with mice, rats, pigs etc., do almost nothing to support this prediction (for reviews see references *6, 7*). For example, one of the studies with mice involved 15 generations of exposure to just over 300 rad per generation (*8*), and another study was based on 35 generations of exposure to 200 rad per generation (*9*). The design of the last mentioned experiment has been criticized (*10*), but the results of most such studies are in general agreement; the ability to survive and develop normally does not appear to deteriorate. Among experiments with mammals, I can think of only two exceptions that did show harmful effects, but in both cases the irradiations were of mature spermatozoa, and this special circumstance would favor production of gross chromosomal damage that is largely absent following irradiations at all other stages of the germ cells (*11, 12*). These exceptional results are therefore of doubtful relevance to protection from chronic exposures.

The bulk of the most directly pertinent experimental studies thus fail to demonstrate any important effect of irradiation on the irregularly inherited diseases, or on general health and wellbeing. It is always difficult to *disprove* a possible effect; but, where neither current theory nor 25 years of research and empirical testing provide any strong reason to believe that such an effect must occur, it seems unreasonable to go on indefinitely assuming it to be real and important.

DOMINANT DISEASES

If one accepts that the collectively numerous, irregularly inherited diseases of man are unlikely to be substantially increased in frequency by exposure of his germ plasm to radiation, it becomes especially important to estimate accurately the combined natural frequency of the simple dominant diseases, since these at least are generally believed to depend on recurrent mutation for their continued prevalence in the population.

Over the past 15 years, the data from Northern Ireland on the "load" of hereditary disease have played a unique role in the estimation of genetic risks from radiation. The lists of diseases and their frequencies have been criticized, but until recently no geneticist has seriously attempted to improve upon them for this particular purpose. According to these data, one individual out of every hundred born alive is severely affected by diseases which at that time were presumed to be caused by single genes, and the bulk of the one per cent was regarded as made up of dominant traits.

In the period since the survey, however, more has become known about the inheritance of the conditions listed as dominant, and other sources of data on the frequencies of hereditary diseases have become available. This new evidence indicates that the collective frequency of dominant disease is much lower than was previously supposed, perhaps by more than an order of magnitude.

Two conditions, for example, account for a third of the cases in the list of presumed dominants, namely hydrocephalus and cataracts of the elderly. However, these would today not be regarded as simply inherited. The usual risk of the same condition appearing in brothers and sisters of persons with dominant traits is in the vicinity of 50 per cent, but for hydrocephalus the risk among siblings of cases is known to be only 3 per cent in the Canadian province of British Columbia, thus strongly indicating irregular inheritance (*13*). In fact, simple dominant inheritance now seems probable for only a small fraction of the total cases that were included in this category fifteen years ago.

To settle the matter, data on incidence and inheritance are needed for all cases of presumed dominant disease in an entire population. Fortunately such information can be obtained for the two million people in British Columbia, where an exceptionally well organized register of handicapped children and adults has been integrated with the vital registration system. Recently, Trimble and Doughty (*14*) have carried out tabulations from this register, indicating only 0.06 cases of disease known to be dominant per 100 liveborn. To this might perhaps be added a few cases from a list of hereditary conditions of uncertain etiology where there exists a possibility of dominant inheritance. However, the numbers from this source are thought unlikely to increase the figure by more than 0.005 per 100 liveborn.

For the purpose of deriving risk estimates I shall take 0.1 per hundred liveborn as the frequency of dominant disease, although this value may have to be revised downward again at a later time. Since only these conditions are likely to increase in direct proportion to the mutation rate, a sustained doubling of that rate would be expected to raise the total amount of genetic disease by just 1/10th of a case per hundred liveborn (Table 3). This is less by about 30-fold than the increase as commonly estimated and as given in the previous Table.

TABLE 3
Revised estimates of the effect of a radiation-induced doubling of the mutation rate

Kind of Disease	Cases per 100 Liveborn		
	Natural	Additional	New Level
single gene[a]	0.1	0.1	0.2
chromosomal } congenital malf } constitutional }	1	very small	1
	4.9	0[b]	4.9
	——	——	——
Totals	6	0.1	6.1

[a]Based on new data on the prevalence of dominant disease (*14*), and omitting the relatively rare recessive diseases for the sake of simplicity.
[b]The multifactorial traits are regarded as largely selection-maintained, and unlikely to be greatly affected by an increase in the mutation rate (see text).

THE COST OF PREVENTING A GENETIC DEFECT

For genetic risk estimates to be of practical value they must be considered in relation to actual problems of radiation protection. Currently a choice is being made between two quite different standards of radiation protection.

In the United States, while it is legally permissible now for people in the vicinity of a nuclear reactor to receive an average exposure of 167 millirem in a year, a much more restrictive safety standard has been proposed by the USAEC as an engineering design objective. This is expected to limit the average exposure in sizable population groups to 1 millirem in a year, i.e. to 1 per cent above the natural background level of 100 to 125 millirem per year (Table 4) (*15*). While these two safety standards differ by a factor of more than 100, the change in actual exposure that will result from adoption of the more restrictive 1 mrem per individual is much less than this because present day reactors are operating at one or two orders of magnitude below the legal limit.

The proposed design objectives are estimated by the AEC to restrict the integrated population dose from a reactor of 1000 megawatts electrical generating capacity to less than 400 man-rem per year. Actual population exposures from existing nuclear power reactors, both in the United States and in Canada, normally fall well below this figure (Wilson, personal communication). Nevertheless, we may assume, as has Terrill (*16*), that a population dose reduction of 100 man-rem per year, or 5000 man-rem over the working lifetime of such a reactor, might be purchased for something like 1 to 10 million dollars spent on additional containment of wastes, the higher cost figure being the more probable. This is not out of line with a Canadian estimate that for each million dollars spent to reduce releases of radioactive noble gases there would be a reduction to 80 per cent of the previous level (Siddall, personal communication). (With liquid wastes a higher percentage reduction of the releases per unit cost may be possible, but this is less easy to relate to the population exposures.)

TABLE 4
Standards of radiation protection; present legal limit versus new design objective

	Annual Exposures	
	Present Limit	New Design Objective
Average to groups	167 rem	1 mrem[a]
Population exposure from a 1000 MWe reactor	—[c]	400 man-rem[b]
Population exposure avoided by new design objectives	—	100 man-rem[d]

[a]1 per cent of natural background.
[b]As estimated by USAEC, Federal Register 1971 (*15*)
[c]Actual population exposures vary widely for different reactors, but are far below the legal limit—see Terrill (1972) (*16*).
[d]As inferred by Terrill (*16*).

The real genetic benefit of the increased safety is, of course, better expressed in terms of the numbers of hereditary disorders avoided. To derive this we may use the revised estimate of the effect of a doubling of the mutation rate in man, as given in Table 3, and the knowledge that 100 rem of chronic exposure, in a generation, is required to double the mutation rate in laboratory mice, the response being directly proportional to the dose. The calculation is shown in Table 5. For each 100 man-rem "input" into a population there will be an estimated "output" of 0.1/100th of a case of a genetic disease. Thus, the benefit from a reduction of 5000 man-rem in the population dose would be 1/20th of a case of a genetic disease avoided. If one wishes arbitrarily to introduce a safety factor of 10, the estimate becomes 1/2 of a case per 5000 man-rem; it would then be more nearly in line with current official estimates.

The expected benefit, in these terms, per million dollars spent on the additional safety, is given in Table 6. This estimate has a spread of 100-fold, because a ten-fold factor for uncertainty is assumed with respect to both the benefit and the cost. It ranges from 1/200th part of a case of hereditary disease to 1/2 of a case, per million dollars. The lower estimate is regarded as the more realistic. (We have not considered here the somatic risk of cancer in the exposed individuals, but the picture would perhaps remain much the same if the two kinds of risk were treated together, provided that all safety factors are brought into the open and are not concealed in conservative assumptions – see Mays et al. (*17*).

In view of the small size of the expected benefit one is inclined to wonder if there might be better ways of spending a million dollars on safety. Geneticists have commonly regarded protection of the human germ plasm from environmentally induced mutations as of special importance in comparison with other more routine kinds of safety activity, but the available evidence does not seem to lend quantitative support to such a view.

TABLE 5
Revised estimate of the number of genetic defects produced per man-rem of exposure

Genetic defects/man-rem =

A) the number of genetic defects per person, in each generation, resulting from a sustained doubling of the mutation rate.

divided by

B) the number of man-rem per person, in each generation, required to cause a sustained doubling of the mutation rate.

Where A = 0.1/100 (from Table 3 of this paper),
and B = 100 (from mouse data for chronic exposure)*,

A/B will equal

0.1×10^{-4} genetic defects per man-rem

*See UNSCEAR 1958 and 1966 (*4,5*)

TABLE 6
Genetic defects prevented by an expenditure of one million dollars on special containment of radioactivity

	[a]Reduction in exposure per $\$10^6$	
[b]Genetic risks per man-rem	500 man rem	5000 man rem
(no safety factor) $0.1/10^4$	$\frac{0.1 \times 500}{10^4} = 1/200$	$\frac{0.1 \times 5000}{10^4} = 1/20$
[c](safety factor of 10) $1/10^4$	$\frac{1 \times 500}{10^4} = 1/20$	$\frac{1 \times 5000}{10^4} = 1/2$

[a]From Terrill (16).
[b]From Table 5 of this paper.
[c]Comparable with conventional estimates from BEIR and UNSCEAR reports.

CONCLUSIONS

As a geneticist I have a special interest in applications of the science of heredity to social problems. I am also concerned, however, that these applications be of a credible and responsible sort. The particular social problem which we have been discussing has arisen because of the advent and necessary expansion of nuclear power production. But analagous problems, associated with environmental mutagens of a chemical nature, are also being discovered and emphasized by large groups of geneticists who have banded together specifically for this purpose.

A recently published statement by one such group (*18*) reads as follows:

> "Genetic material is the most important substance in living organisms. The importance of protecting it against undesirable changes induced by the environment cannot be overemphasized."

The statement is not quite the self-evident truth it appears to be. But it is a proposition of some scientific and social significance, and well worth the effort of supporting it with quantitative data from strictly relevant studies. In the absence of such support, however, the statement may eventually prove to be more a product of enthusiasm than of the scientific method.

The actual research interests of the authors of the statement relate more to the mechanisms of mutation, as indicated by nearly 100 abstracts of papers published in the same journal issue (*19*). What is disturbing is that none of these reports attempted to assess the collective importance, or triviality, of the effects of the mutations against which it was urged that we must be protected.

Fashion in the science of genetics, and the need for a socially important umbrella under which to do fashionable studies of the mutation process, could account for the anomaly. But the public image and credibility of geneticists in the future may depend on the extent to which their own research reflects a more directly relevant approach to the social problems about which they voice special concern.

If the arguments advanced in this paper are correct, the statement quoted earlier is essentially untrue. We could already have over-emphasized the importance of protecting the human genetic material, and possibly at the expense of other activity in the field of human safety.

The matter deserves attention because, if the human germ plasm really does need greater protection than the present argument suggests, this ought to be demonstrated without delay by strictly pertinent genetic studies. But until such evidence is produced, society must consider also the risks of over-protection, and of unnecessary delay in the replacement of energy sources that are no longer readily available.

REFERENCES

1. UNSCEAR, *Ionizing Radiation: Levels and Effects*. A Report of the United Nations Scientific Committee on the Effects of Atomic Radiation to the General Assembly. Volumes I and II. United Nations Publication Sales Numbers E. 72.IX. 17 and E. 72.IX. 18, United Nations, New York, 1972.
2. BEIR, *The Effects on Population of Exposure to Low Levels of Ionizing Radiation*, Report of the Advisory Committee on the Biological Effects of Ionizing Radiation. National Academy of Sciences-National Research Council, Washington, D.C. (U.S. Government Printing Office: 1972, 0-489-797).
3. A. C. Stevenson, *Radiat. Res.* **Suppl. 1,** 306 (1959).
4. UNSCEAR, *Reports of the United Nations Scientific Committee on the Effects of Atomic Radiation*. General Assembly Office Records: Thirteenth Session, Supplement No.17 (A-3838), pp. 197–200. New York, 1958.

5. UNSCEAR, *Report of the United Nations Scientific Committee on the Effects of Atomic Radiation*. General Assembly Official Records: Twenty-First Session, Supplement No. 14 (A/6314), pp. 121–127. New York, 1966.
6. E. L. Green, Genetic effect of radiation on mammalian populations. *Ann. Rev. Genet.* **2,** 87 (1968).
7. H. B. Newcombe, *Genet.* **16,** 239 (1971).
8. E. L. Green, *Mutation Res.* **6,** 437 (1968).
9. J. F. Spalding, M. R. Books and G. L. Tietjen, *Genetics* **63,** 897 (1969).
10. K. G. Luning and A. G. Searle, *Mutation Res.* **12**, 291 (1971).
11. H. B. Newcombe and J. F. McGregor, *Genetics* **40,** 1065 (1964).
12. V. H. Ehling, *Genetics* **54,** 1381 (1966).
13. H. B. Newcombe, in *Chemical Mutagens* (A. Hollaender, Ed.) Volume 3, pp. 5577, Plenum Press, New York, 1973.
14. B. K. Trimble and J. H. Doughty, The amount of hereditary disease in human populations. *J. Prevent. Social Med.* (in press).
15. Federal Register, June 9, 1971. Quoted in J. G. Terrill, Ref. (*16*). (See below.)
16. J. G. Terrill, *Amer. J. Public Health* **62,** 1008 (1972).
17. C. W. Mays, R. D. Lloyd and J. H. Marshall, in *Proceedings of the Third International Radiation Protection Association*, Washington, D. C., 1937 (in press).
18. EEMS, *Mutation Res.* **21**, 245 (1973).
19. Abstracts of papers presented at the Third Annual Meeting of the European Environmental Mutagen Society, and the First International Conference on Environmental Mutagens, *Mutation Res.* **21,** 183–205 and 207–244, respectively (1973).

Use of the Mouse to Fill Gaps In Our Risk Assignments[1]

A. G. Searle
Medical Research Council, Radiobiology Unit,
Harwell, Oxon. OX11 ORD, U.K.

INTRODUCTION

In the last few years, the natural scientific caution of radiation geneticists has come under considerable strain. Most large-scale studies in this field have, of course, been bound up with the problems of assessing genetic hazards to man. It has always been clear that, sooner or later, some detailed estimates of the risks of gene and chromosomal damage would have to be made on the basis of these studies and the growth in our knowledge of human population genetics, however great the remaining uncertainties. That time has now arrived and the latest BEIR (*1*) and UNSCEAR (*2*) reports show the results. All who read and compare them will realize how much we still don't know.

If we are to improve our estimates (and the process of reassessment is a continuous one) I think we must begin by a careful reappraisal of the main types of genetic risk in the light of these two documents and subsequent work, to see whether any of the wide discrepancies which were revealed can now be resolved, to identify the most important remaining gaps in our knowledge, and to propose ways in which these could be filled. This is really a job for a team, in which human genetics should be well represented, but perhaps I can make a few suggestions from the point of view of a mouse-worker.

DOMINANT MUTATIONS

Let us begin by considering dominant gene mutations, which are of obvious importance because their effects are often severe and are felt in the first generation after exposure. The UN committee estimated that 2 extra dominants per million progeny per rad would be induced by chronic or low dose irradiation of spermatogonia, while the BEIR committee thought the right figure was 50–500 per million in the first generation after 5 rem chronic exposure of both parents, or 10–100 per million after 1 rem, i.e. much higher than UNSCEAR even though the yield from female exposures was thought to be virtually nil.

Both committees agreed that the probable number of human loci capable of mutating to give dominant Mendelian traits was about 1000 (*3*). UNSCEAR used an estimate (*4*) of the actual rate of induction of dominant visible mutations in the mouse spermatogonia, based on four separate experi-

[1]Congress presentation B-3-4.

ments at Harwell, in which each mouse was carefully and systematically classified at weaning by two people. This estimate was 5×10^{-7} per R per genome. Since about 75 loci are known in the mouse at which visible dominant mutations, similar to those detected in specific locus experiments, have occurred, an upper estimate of the per locus rate of induction of these dominant visibles is 6.7×10^{-9} per R. This is at acute exposures; UNSCEAR estimated a rate of 2.2 $\times 10^{-9}$ for chronic exposures, which is decidedly higher than the estimate based on actual figures from chronic gamma exposures (*5*).

These per locus rates are only 4% of rates per specific locus under comparable conditions of exposure. Yet, surprisingly, the BEIR Committee used the mouse specific locus rate in one of their risk estimates for dominant traits, even though they state in a previous paragraph: "... in mice the rate of induction of dominant visible mutations is lower than that for recessives by at least an order of magnitude." They state elsewhere: "We prefer, however, to err on the safe side ..." but that is no reason for using an irrelevant estimate. The other estimate of the BEIR Committee with respect to dominant traits is based on (i) the assumption of a 20–200 rad doubling dose for chronic or low dose exposure, (ii) a population incidence of 10,000 autosomal dominant traits per million live-born, and (iii) the assumption that the average dominant mutant persists for five generations, so that the mutation frequency is about 2000 per million. This is probably a considerable over-estimate, since Trimble (*6*) has reported that in British Columbia "the incidence of simple dominant conditions is probably no more than 1 in every 1000 individuals born alive," i.e. one tenth of the previous estimate (*7*). For both specific locus and dominant visible mutations the doubling dose for chronic exposures is about 100 rad. If we substitute these revised figures in the calculations we arrive at an estimate of two extra newborn with dominant traits per million per rad paternal (spermatogonial) irradiation. This is exactly the same value as that given by UNSCEAR, although arrived at by an entirely different route. So I think we can be fairly confident that it is near the truth.

Our ability to predict the genetic consequences of female exposures is far less satisfactory. No information whatever has been published on the yield of dominant visible mutations after acute irradiation of female mice. Two experiments (*8*, *9*) at low dose-rates have given low frequencies of dominants but in that of Batchelor et al. (*8*) nearly all the exposure was to oocytes in immature follicles. The considerable differences between these dictyate oocytes of the mouse and human diplotene oocytes have often been stressed. Even if it is assumed that dominant visible mutations will have exactly the same mutagenic properties as specific locus recessive ones the uncertainties do not end, because of the heterogeneity in response of even the maturing oocyte. Thus Russell (*10*) found that mutation frequencies in second litters from older females irradiated at 0.8 R/min were significantly higher than those from first litters and in fact were comparable with those obtained from acute irradiation at 90 R/min. More work needs to be done on this very interesting observation and we are attempting to obtain more specific locus and other information from female mice irradiated when older. Recently, we have found marked differences in frequencies of chromosome aberrations in oocytes and dominant lethals in F_1 between the 1st and 3rd weeks after X-ray treatment of female mice (*11*).

The need for more experimental work on the radiation genetics of female mammals possessing oocytes more closely resembling human ones is obvious. In view of all the uncertainties, the decision of the BEIR Committee to assume that virtually no dominant or recessive mutations would be induced by irradiation of human oocytes at low doses or dose-rates is difficult to reconcile with their stated wish "to err on the safe side."

RECESSIVES AND THEIR HETEROZYGOUS EFFECTS

The treatment of induced recessive mutations has changed greatly from the days when they were regarded as "genetic deaths" and viewed with considerable alarm. As Newcombe (*12*) and others have pointed out, there is such uncertainty over the proportion of genetic deaths which would lead to social harm that the concept is no longer useful. A high proportion of induced recessives in the mouse appear to be lethals acting around the time of implantation (*13*) and with no detectable heterozygous disadvantage (*14*). These are much less of a hazard than the types of mutation responsible for recessive diseases in man. The BEIR Committee considered that the frequencies of these diseases were not generally maintained by recurrent mutations but instead by such factors as heterozygous advantage. The high frequencies of some of these conditions lend support to this view. It should be remembered, however, that radiation may induce recessive mutations (including deficiencies) with extreme lethal effects at the loci responsible for recessive diseases. If, in the gene-pool of a human population, these induced mutant "alleles" combined with recessive alleles for the disease, situated at the same locus, then the compound would probably show the phenotype of the disease and not an extreme lethal phenotype. The probability of this event is admittedly very low but would be increased if (i) the disease allele had a high frequency because of heterozygous advantage, and (ii) heterozygous disadvantage of the induced allele or small deficiency was nil or very slight, so that it was not rapidly eliminated.

UNSCEAR laid some stress on possible deleterious heterozygous effects of recessive lethals and detrimentals with the assumption that these were associated with a 2–5% reduction in fitness on the average. This was taken from work on *Drosophila* with, as yet, little or no confirmatory evidence in mice. If such mutations are induced in man, they would presumably contribute mainly to the category of "diseases of more complex etiology" of the BEIR report or to a general, rather unspecified, increase in ill-health. These very important categories are particularly difficult to study in mice because it is by no means obvious what are the most appropriate end-points to use. In many population experiments which have been carried out (*15*) the characters mainly studied have been indicators of fitness in a Darwinian sense, such as lifetime productivity, embryonic death or post-natal survival. Irradiation over several generations has given little evidence for the accumulation of deleterious mutations affecting these characters. Perhaps, instead we should consider more direct effects of genetic (or partially genetic) ill-health. For instance, it may lead to a change in activity pattern; either a reluctance or slowness to move or (more rarely) to hyperactivity. Mutations at very many loci might be expected to affect activity, yet it has been used in very few studies on the genetic effects of radiation on mammalian populations. Thiessen et al. (*16*) have shown that mutants at a number of coat color loci affected a wide range of behavioral activities involving a series of simple tests. Learning ability was also tested by comparing two successive water escape performances. It should be pointed out here that a significant reduction in learning ability of rats in a maze after paternal spermatozoal irradiation was reported by Newcombe and McGregor (*17*). This positive result has never been followed up, probably because the time taken to test each offspring must be substantial. Perhaps a simpler quick test of the type used by Thiessen et al. might be more feasible and could help to plug a very serious gap.

THRESHOLD VARIANTS

All the evidence obtained so far suggests that frequencies of the "threshold" type of phenodeviant are not increased by irradiation of mice. Ehling (*18*) found no marked change in the frequencies of minor skeletal variants of this type after paternal irradiation even when there was a significant increase in frequency of unique multiple abnormalities. Searle (*19*) found that the rate of subline divergence with respect to 23 threshold and 4 metrical variants tended to decrease rather than increase after low level ancestral irradiation over many generations, as compared with controls. Some data extracted from a specific locus experiment (*20*) on the comparative effectiveness of chronic neutron and gamma exposures to mouse spermatogonia may also be relevant. In each series any offspring deviating from the expected wild type was kept for inheritance tests. Table 1 shows that while the frequency of dominants and of specific locus mutations was much higher in the neutron series (despite the lower dose) the yields of non-heritable phenodeviants were very similar in large homogeneous samples from the two series. Many of these non-heritable variants were presumably of the threshold type.

TABLE 1

Comparative Frequencies of Specific Locus, Dominant Visible and Non-Heritable Variants After Chronic Gamma and Neutron Exposures to Mouse Spermatogonia. Partly from Batchelor et al. (*20*).

	606 rad γ-rays			214 rad fast neutrons		
Variant	No. affected	Total	Frequency/gamete	No. affected	Total	Frequency/gamete
Specific locus	16	58795	2.7×10^{-4}	67	41875	16.0×10^{-4}
Dominant	6	58795	1.0×10^{-4}	24	41875	5.7×10^{-4}
Non-heritable	33	22073	15.0×10^{-4}	21	16897	12.4×10^{-4}

USE OF INVERSIONS

As already mentioned, the most important types of mutation from the human hazard point of view are the detrimentals rather than the early acting lethals. Both types of mutations have long been studied in *Drosophila* by crossover suppression stocks incorporating an inversion. Roderick (*21*) has now successfully induced inversions in mice and has used a particular paracentric inversion of chromosome 1 with associated markers, splotch and leaden to look for recessive lethals after paternal irradiation (*22*), through absence or reduction of the non-splotch class. He has found 1 presumptive recessive lethal in 49 mice tested after doses of 800–900 R X-rays to spermatozoa. The inversion eliminates recombination in about 3.5% of the genome. The efficiency of the test would be increased if a larger inversion, or a combination of several inversions, could be used. From this point of view some of the Robertsonian translocations, in which two normal acrocentric mouse chromosomes are combined to form a metacentric, are likely to prove particularly useful. For instance, the Robertsonian translocation Rb1Bnr combines chromosomes 1 and 3 which comprise over 13% of the haploid chromosome complement. Even if only half of the length could be incorporated into one or two inversions with suitable markers, a substantial gain in efficiency would be achieved. Two paracentric inversions would probably be better than a single long pericentric one because (i) double crossovers would occur within the long pericentric inversion loop, leading to unwanted recombination between induced lethal or detrimental and marker, and (ii) it seems probable that heterozygotes for pericentric inversions (unlike paracentrics) would show semisterility like translocation heterozygotes. This is because they too produce gametes with duplications and deficiencies, while heterozygotes for paracentric inversions produce germ-cells with dicentric and acentric chromosomes, which are apparently eliminated before they can be transmitted (*21*).

SOMATIC MUTATIONS

As De Mars (*23*) has shown in this symposium, recent advances in biochemical genetics and in culture techniques have greatly advanced our knowledge of mutation and mutagenesis in human somatic cells. Before we can know what these mean in terms of genetic hazards, we must bridge the gap between somatic cells and germ-cells, by comparing mutation induction in both kinds of cell, at identical loci. This can be done at certain specific loci, where recessive coat color alleles have an autonomous action on the pigment cells carrying them. This action may lead to a mutant patch of coat color if mutation has occurred in a heterozygote and if the clone of mutant pigment cells is significantly large. L. B. Russell and Major (*24*) scored such mosaic patches after embryonic irradiation of heterozygotes for brown (*b*), pink-eyed dilution (*p*), chinchilla (c^{ch}), dilute (*d*) and short-eared (*se*). The first three of these are all known to act autonomously on melanocytes. The authors calculated that the rate of induction of somatic mutations in these embryonic pigment cells was roughly three times the rates for the same loci in spermatogonia.

These pioneer studies should be followed up, as they present great possibilities for development on the basis of our increasing knowledge. First, thousands of individual pigment cells (melanocytes) are clearly visible in the skin and developing hair follicles 1–2 days after birth. Second, they can be rendered more clearly visible as individual cells by having a dilute (*dd*) or leaden (*lnln*) genotype, since these genes shorten the melanocyte's dendritic processes and clump the pigment inside them. Third, the retinal pigment epithelium can be examined too, since the retinal melanocytes (derived from the optic cup) form a pavement monolayer between choroid and retina, in which pigmentary mosaicism can be made clearly visible (*25*). For this purpose it is important to avoid choroidal pigmentation either by using the right genetic background, e.g. piebald spotting which removes choroidal pigment, or by examining the retinal epithelium before choroidal pigment has arrived, e.g. at birth.

Thus there are good possibilities that an in vivo cellular assay for somatic point mutations can be developed. At present, we are investigating the possibility of using the pink-eyed unstable allele, p^{un}. As Wolfe (*26*) has shown, this has a high spontaneous rate of back-mutation from *p* to +, a change which greatly increases the amount of pigmentation in each cell and is thus easily visible. Use of this interesting mutant may increase the sensitivity of the test system.

Another possible way of greatly increasing the sensitivity of this test system for somatic mutations is to go from pigment cell to pigment cell product, namely the pigment granule. Each melanocyte injects thousands of these into developing hairs, epidermal cells etc. Some coat color mutations like *b* and *p* act by altering the size and shape of pigment granules (*27*). This effect is particularly marked in *pp* mice, in which the normal ovoid wild-type granule, about 2μm in length, becomes a much smaller and more irregular particle. This is the result of a change in the protein framework of the granule on which melanin is deposited. A sensitive method for detecting a very small proportion of mutant granules in very large numbers of wild type ones could be of great value in mutagenesis studies.

LOSSES AND GAINS OF CHROMOSOMES

It is now realized that a substantial proportion of man's genetic load stems from the structural and numerical changes which arise in human chromosomes. We are still unsure of the proportion of these which are truly spontaneous and the proportion which are precipitated by some preventable event like exposure to man-made radiation or other environmental mutagen. Some of the biggest gaps in our knowledge relate to gains and losses of chromosomes and in particular to the phenomenon of non-disjunction. In man, there is evidence (*28*, *29*) of an increased risk of Down's syndrome (trisomy-21) in the offspring of women exposed to diagnostic irradiation before pregnancy. The risk seemed greater with increasing maternal age. Although the data of Stevenson et al. (*30*) showed a fairly normal incidence of Down's syndrome after maternal diagnostic irradiation pre-conception, the mean radiation dose of 0.9 rad was lower than in the Uchida series (*28*). Taken as a whole, the data suggest the urgent need for more work on this important problem, both epidemiological and experimental.

So far, there is little relevant evidence from experimental work on the mouse, although it is known from *Drosophila* that radiation can increase non-disjunction in oocytes (*31*, *32*). Recently, Yamamoto et al. (*33*) have reported a significant increase in the frequency of aneuploid fetuses after exposing old female mice to only 5 R. There has been some criticism of their conclusions (*34*), partly because a number of the aneuploids appeared to be mosaic, so confirmatory data are clearly needed. There is good evidence from several laboratories for increasing spontaneous non-disjunction in the mouse with maternal aging (*35*, *36*, *37*).

Let us consider what sorts of evidence can help to throw light on this problem (Table 2). The probable source of the non-disjunction is the failure of some pairs of chromosomes to form bivalents at 1st meiotic division, with resultant independent segregation of the univalents, leading to losses and gains of chromosomes. Such a tendency would be detectable by a decrease in overall chiasma frequency and a concomitant increase in the frequency of univalents. We have looked at the frequency of univalents in metaphase-1 oocytes of young female mice in the first and third week after doses of 100 and 400 rad and found no significant differences from control values (*11*) although the mean chiasma count was significantly reduced in the first week after 400 rad. We now intend to make similar observations on older female mice.

TABLE 2
Some Methods for Detection of Induced Non-Disjunction

Type of method	Stage examined	Information required
Cytological	Metaphase I	Frequency of univalent chromosomes. Overall chiasma frequency.
	Metaphase II	Frequency of aneuploid chromosome counts.
	Pre-implantation embryos	Proportion of aneuploidy and mosaicism.
	Post-implantation embryos	As above.
Dominant lethal	Late foetuses	Frequency of lethals in relation to maternal age etc.
Genetic	F_1 weanling	Frequency of proximal marker phenotype after intercrosses of heterozygotes for reciprocal translocations (adjacent-2 disjunction). Frequency of marker phenotype after intercrosses of heterozygotes for Robertsonian translocations (numerical non-disjunction). Frequency of presumptive XO and XXY mice after crosses with sex-linked markers.

Chromosome counts can also be made at 2nd meiotic metaphase and (with careful timing) at the pronuclear stage, when it is sometimes possible to tell the female from the male set of chromosomes. Cytological methods are also available (*38*, *39*) for counting the chromosomes of both pre-implantation and post-implantation embryos of a wide range of ages, so as to detect aneuploidy. Although both monosomy and trisomy are found in pre-implantation mouse embryos, the monosomics usually die soon after implantation and most of the trisomics around mid-term[1]. Therefore, an increased frequency of non-

[1]M. D. Burtenshaw, H. Clegg, E. P. Evans and C. E. Ford, Personal Communication. *Mouse News Letter* **48,** 40 (1973).

disjunction will result in more dominant lethals and more late-acting ones. Dominant lethals are induced by irradiation of young female mice (*11, 40, 41*); if the yield increased with maternal age then the induction of non-disjunction would be suspected.

Some of the most promising methods for detecting increased frequencies of non-disjunction are genetic rather than chromosomal. A sex-linked marker, such as tabby *Ta*, can be used to detect XO and XXY conditions (*42*). The XXY condition is very rare in mice and no definite effect of irradiation has yet been established, although no work has yet been done with older female mice or with the immature resting oocyte. This is so sensitive to killing that only low doses can be used. We are now giving doses of 25 or 50 rad to female mice, mating them to tabby males and allowing them to breed as long as possible to see whether the earlier oocytes, from older mothers, show greater radiosensitivity with respect to chromosome losses. If so, this would contrast with the resistance of exposed immature oocytes to specific locus mutation induction (*43, 44*).

Genetic methods for studying autosomal non-disjunction are based on the presence of marker genes and on the phenomenon of complementation. It has been shown (*45*) that if heterozygotes for reciprocal translocations are intercrossed, and if one was homozygous for a recessive marker on a translocated chromosome while the other was homozygous dominant wild type, then some offspring may have the recessive phenotype instead of being the expected wild type. This is because a duplication of the marker-carrying part of the chromosome in one gamete has met a deficiency of the same area, leading to a balanced complementation zygote. In this way the frequency of adjacent-2 disjunction (involving non-disjunction of homologous centromeres) can be calculated. If the same procedure is used with Robertsonian (whole-arm) translocations then the frequency of numerical non-disjunction for a marked chromosome involved in the translocation can be determined. For certain of the Robertsonian translocations (13 of which are now known in the mouse) cytological methods show that this frequency is quite high (*46*), making them very suitable subjects for studying the effects of irradiation of various germ-cell stages on the frequency of non-disjunction. By use of the complementation method, Lyon and co-workers[2] have calculated a spontaneous frequency of about 15% non-disjunction in each chromosome arm of Rb2Bnr and Rb6Bnr.

One further advance might be the use of isochromosomes, i.e. metacentric chromosomes in which the two arms are homologous. In *Drosophila*, these have proved very valuable for non-disjunctional studies (*32*). A mouse in which a homologous pair of chromosomes was present as an isochromosome would normally produce gametes with two copies of the chromosome concerned or none at all. Therefore, balanced zygotes could only be produced when these gametes met complementary nullosomics or disomics, i.e. the products of non-disjunction or of chromosome loss by some other means. By use of the invaluable Robertsonian translocations it may be possible eventually to con-

[2]M. F. Lyon, G. M. Simpson and H. C. Ward, Personal Communication. *Mouse News Letter* **51**, 20 (1974).

struct such isochromosomes in the mouse, but a number of snags must be overcome first.

These are just a few of the ways in which the many gaps in our knowledge of genetic hazards might be narrowed by use of the mouse. I feel we are still a long way from the stage at which our assessments are so firmly based that any small additions to our knowledge are not worth the effort involved. Nevertheless, it is now more than ever important to consider carefully which particular experiments are likely to be especially relevant.

ACKNOWLEDGEMENTS

I am grateful to Drs. C. E. Ford, F.R.S. and Mary F. Lyon, F.R.S. for allowing me to quote unpublished information. I have benefitted greatly from stimulating discussion and correspondence with Dr. Howard B. Newcombe.

REFERENCES

1. Advisory Committee on the Biological Effects of Ionizing Radiations, *The Effects on Populations of Low Levels of Ionizing Radiation* (BEIR Report). National Academy of Sciences, National Research Council, Washington, 1972.
2. United Nations Scientific Committee on the Effects of Atomic Radiation, *Ionizing Radiation: Levels and Effects. Vol. II: Effects (UNSCEAR Report).* United Nations, New York, 1972.
3. V. A. McKusick, *Mendelian Inheritance in Man.* The Johns Hopkins Press, Baltimore, 1971.
4. K. G. Luning and A. G. Searle, *Mutation Res.* **12,** 291 (1971).
5. A. G. Searle, *Adv. Radiat. Biol.* **4,** 131 (1974).
6. B. K. Trimble, (in preparation).
7. A. C. Stevenson, *Radiat. Res.* Suppl. **1,** 306 (1959).
8. A. L. Batchelor, R. J. S. Phillips and A. G. Searle, *Brit. J. Radiol.* **42,** 448 (1969).
9. T. C. Carter, *Brit. J. Radiol.* **3,** 407 (1958).
10. W. R. Russell, in *Repair from Genetic Radiation Damage* (F. H. Sobels, Ed.) pp. 205217. Pergamon, Oxford, 1963.
11. A. G. Searle and C. V. Beechey, *Mutation Res.* **24,** 171 (1974).
12. H. B. Newcombe, *Advan. Genet.* **16,** 240 (1971).
13. M. F. Lyon, R. J. S. Phillips and A. G. Searle, *Genet. Res.* (*Camb.*) **5,** 448 (1964).
14. K. G. Lüning and W. Sheridan, *Mutation Res.* **3,** 340 (1966).
15. E. L. Green, *Ann. Rev. Genet.* **2,** 87 (1968).
16. D. D. Thiessen, K. Owen and M. Whitsett, in *Contributions to Behavioral Genetic Analysis. The Mouse as a Prototype* (G. Lindzey and D. D. Thiessen, Eds.) pp. 161204. Appleton Century Crofts, New York, 1970.
17. H. B. Newcombe and J. F. McGregor, *Genetics* **50,** 1065 (1964).
18. U. H. Ehling, *Genetics* **54,** 1381 (1966).
19. A. G. Searle, *Genetics* **50,** 1159 (1964).
20. A. L. Batchelor, R. J. S. Phillips and A. G. Searle, *Mutation Res.* **3,** 218 (1966).
21. T. H. Roderick, *Mutation Res.* **11,** 59 (1971).
22. T. H. Roderick, *Mutation Res.* **21,** 349 (1973).
23. R. DeMars, Congress presentation B-3-2 (not available).
24. L. B. Russell and M. H. Major, *Genetics* **42,** 161 (1957).
25. M. S. Deol and T. K. Whitten, *Nature New Biol.* **238,** 159 (1972).
26. H. G. Wolfe, in *Genetics Today* Vol. I. (S. J. Geerts, Ed.) p. 251. Pergamon, Oxford, 1963.
27. A. G. Searle, *Comparative Genetics of Coat Colour in Mammals.* Logos/Academic, London and New York, 1968.
28. I. A. Uchida, R. Holunga and C. Lawler, *Lancet* **ii,** 845 (1968).
29. E. Alberman, P. E. Polani, J. A. Fraser Roberts, C. C. Spicer, M. Elliott and E. Armstrong, *ann. Hum. Genet.* **36,** 195 (1972).
30. A. C. Stevenson, R. Mason and K. D. Edwards, *Lancet* **ii,** 1335 (1970).
31. H. Traut, *Mutation Res.* **10,** 125 (1970).

32. A. M. Clark and F. H. Sobels, *Mutation Res.* **18,** 47 (1973).
33. M. Yamamoto, T. Shimada, A. Endo and G. Watanabe, *Nature New Biol.* **244,** 206 (1973).
34. R. G. Gosden and D. E. Walters, *Nature* **248,** 54 (1974).
35. S. A. Henderson and R. G. Edwards, *Nature* **218,** 22 (1968).
36. R. G. Gosden, *J. Reprod. Fertil.* **35,** 351 (1973).
37. M. Yamamoto, A. Endo and G. Watanabe, *Nature* **241,** 141 (1973).
38. A. K. Tarkowski, *Cytogenetics* **5,** 394 (1966).
39. E. P. Evans, M. D. Burtenshaw and C. E. Ford, *Stain Technol.* **47,** 229 (1972).
40. L. B. Russell and W. L. Russell, in *Progress in Radiobiology* (J. S. Mitchell, B. E. Holmes and C. C. Smith, Eds.) pp. 187192. Oliver and Boyd, Edinburgh, 1956.
41. R. G. Edwards and A. G. Searle, *Genet. Res.* (*Camb.*) **4,** 389 (1963).
42. L. B. Russell and C. L. Saylors, in *Repair from Genetic Radiation Damage* (F. H. Sobels, Ed.) pp. 313342. Pergamon, Oxford, 1963.
43. W. L. Russell, *Proc. Nat. Acad. Sci.,* (*USA*) **54,** 1552 (1965).
44. W. L. Russell, *Proc. IV U.N. Conf. Peaceful Uses Atomic Energy* **13,** 487 (1971).
45. A. G. Searle, C. E. Ford and C. V. Beechey, *Genet. Res.* (*Camb.*) **18,** 215 (1971).
46. B. M. Cattanach and H. Moseley, *Cytogenet. Cell Genet.* **12,** 264 (1973).

COMPARISON OF THE MUTAGENIC EFFECTS OF CHEMICALS AND IONIZING RADIATION[1]

Chairman: F. H. Sobels

[1]Symposium C–2

A Comparison of the Mutagenic Effects of Chemicals and Ionizing Radiation (Chairman's Remarks)[1]

F. H. Sobels
Department of Radiation Genetics and Chemical Mutagenesis
University of Leiden, Netherlands

INTRODUCTION

The main purpose of this symposium is to bring into focus and compare the mutagenic effects of chemicals and ionizing radiations in various biological systems ranging from microbes to mammals, and perhaps gain an insight into the problems that lie ahead of us in mutagenesis work. The fact that radiation mutagenesis and chemical mutagenesis are conceptually interrelated and the major problems are common to both fields hardly needs to be stressed.

We have the background and experience of over four decades of radiation mutagenesis work; a solid body of knowledge has been built up dealing with the mutagenic effects of ionizing radiations, their dose-response relationships, repair processes that intervene between the induction of primary genetic damage and its realization as scorable mutational events, effects of radiations of different LET's, the nature, magnitude and dynamics of radiation-induced genetic loads in populations and so on. Studies of these aspects in higher organisms, particularly experimental mammals, have been especially helpful to estimate, by extrapolation, the possible genetic hazards to man from exposure to ionizing radiation. A recent analysis of published mutation data in several organisms by Abrahamson and others (*1*) has even raised the possibility that we can extract more information from these data, and draw more meaningful conclusions about mutational response and genetic hazards from radiation or, as the authors put it, "extrapolate with confidence from mutation rates obtained in experimental organisms to man." Similarly, the work of Brewen and colleagues (*2*) has led to the suggestion that a technique is now at hand which will permit us to evaluate radiation hazards in man for at least one kind of chromosome aberrations, namely reciprocal translocations. Impressive as these developments are, one should honestly admit that there are still several gaps in our knowledge of the genetic effects of radiation—an agent whose tissue dose can be precisely quantitated—and we are still many steps removed from the ability to predict precisely and biologically meaningfully the genetic consequences of radiation exposure at the human level.

Chemical mutagenesis is *not* a new field of scientific inquiry; its origins date back to the early 40's when Auerbach and Robson (*3*, *4*, *5*) conducted their first successful experiments on the mutagenic effects of mustard gas in *Drosophila*.

[1] Congress presentation C-2-1.

The choice of this substance for testing was dictated by the consideration that both X-rays and mustard gas produce burns that heal slowly and then tend to break open again. These first tests produced spectacular results with sex-linked lethal frequencies of up to 24%. Since then, the field has been expanding, and hundreds of chemicals have been tested in several biological systems and, at least for a number of different kinds of chemical mutagens, a great deal of information has been gathered about their mode of action. It is not until recently, however, that chemical mutagenesis studies have acquired a new dimension and a perspective, stemming from the growing awareness that a formidable array of chemical substances are already present and are constantly introduced into our environment, and that such substances pose a potential threat to man's heredity. We are thus faced with the compelling need to evaluate the effects of chemical mutagens on man and their concomitant genetic hazards.

DIFFERENCES BETWEEN THE EFFECTS OF CHEMICALS AND IRRADIATION

While, as I said earlier, the problems in chemical mutagenesis and radiation mutagenesis are conceptually interrelated, there are special difficulties that confront us in chemical mutagenesis. There is no doubt that ionizing radiations produce all types of genetic damage in all types of cells and what we have to decide is, how much damage of a given kind is expected in a given cell type. The problem here, therefore, is basically a quantitative one. With chemicals, however, the problems are both qualitative and quantitative.

The early work of Auerbach in *Drosophila* revealed two major differences between the effects of mustard gas and those of ionizing radiation. One was the relative shortage of translocations compared with X-ray doses that yielded comparable frequencies of sex-linked lethals, and the other was the high incidence of "delayed mutations" whose appearance was separated from the time of treatment by one or more cell cycles. The former peculiarity may well be a consequence of the latter, since the formation of a rearrangement requires the simultaneous presence and availability of two open breaks in the same cell: if two potential breaks open up in different cell cycles, a potential rearrangement would be lost, as a dominant lethal. This postulate was confirmed by the finding (i) that dominant lethals occurred at relatively higher frequencies after mustard gas treatment, than with mutagenically equivalent doses of X-rays (*6*) and (ii) by storage experiments with alkylating agents such as TEM or DEB, which showed that the translocation frequencies drastically increased when spermatozoa were stored in the female for a week or more (*7–11*); by storing of the sperm more breaks became available simultaneously, and the probability of the formation of a translocation is greatly increased.

Detailed cytological comparisons by Slizynska (*10*) of rearrangement patterns produced by chemicals and irradiation revealed great differences before storage. However, after storage of the chemically treated sperm for 7–9 days, the spectrum of chromosomal aberrations resulting from TEM became remarkably similar to that produced by X-rays.

Chemical mutagens, apart from producing comparatively more mosaic mutations and fewer translocations and inversions than comparable doses of X-rays, in addition induce more duplications and deficiencies (*12*). The observation that chemical mutagens produce such high frequencies of duplications is of interest for various reasons: (i) from an evolutionary standpoint, since duplications are assumed to provide a mechanism for the evolution of genes with a new function, and (ii) because duplications have been shown to underlie various cases of spontaneous instabilities (*13*)—premutational changes that can replicate in the unstable state. Replicating instabilities thus constitute a new kind of genetic hazard that is specific for chemicals and has not so far been encountered with ionizing irradiation (*14*).

THE RELATIONSHIP BETWEEN MUTATIONS AND CHROMOSOME ABERRATIONS

Let me now turn to a slightly different theme, namely, the relationship between mutations and chromosome aberrations. In the storage experiments in *Drosophila* that I mentioned earlier, one of the other findings was that storage of sperm after chemical treatment did not increase the frequency of recessive lethal mutations. In my opinion, this provides a simple demonstration that the quantitative relationship between mutations and chromosome aberrations is subject to much greater variation with chemical treatment than after irradiation, as the simple storing of *Drosophila* sperm may vary the ratio of recessive lethal mutations to translocations by more than one order of magnitude.

It is exactly this question of correlation between different kinds of genetic effects that constitutes one of the important problems in the testing of environmental mutagens. In other words, will one particular kind of test produce results that are equally representative for other kinds of genetic damage. Relevant genetic damage consists of gene mutations, small deficiencies and duplications, eucentric translocations which throw off unbalanced chromosome combinations resulting in abnormality or prenatal death, non-disjunction and genetic recombination. Ionizing radiation can produce all these effects in all cells of the body, though the relative frequencies may differ depending on organism, sex, cell type, and experimental conditions. With chemicals, the situation is different. Some chemical substances produce chromosome breakage in plants or mammalian cells, but few or no rearrangements (for references see *14*); ethylene glycol, which makes the chromosome sticky, induces dominant lethals but no recessive lethals in *Drosophila* (*15*); again in *Drosophila* diethylnitrosamine at LD50 concentrations produces 20% recessive sex-linked lethals in spermatozoa (equivalent to some 6000 R of X-rays) but no dominant lethality or translocations (*16*); similarly the antischistosomal agent hycanthone produces up to 9% recessive lethals in late spermatids but no dominant lethals or translocations (*17*). These few examples illustrate the main point that correlation between different kinds of genetic damage with chemicals is complex, that systems designed to detect either chromosome breakage per se, or only one kind of effect such as dominant lethals or translocations alone, may generate erroneous conclusions.

SPECIFICITY

Turning now to the problem of specificity of response, it is known that chemicals may have widely different genetic effects on different species, strains, sexes, even individuals, and, within the same individual, on different cells and different stages of the cell cycles. While this is to some extent true of radiation as well, the differences here are mainly quantitative, whereas with chemicals the question is both quantitative and qualitative. A chemical may act as a strong mutagen on one organism or type of cell and be completely harmless for another organism or another type of cell. Urethane, for example, produces mutations in *Drosophila* and translocations in plants, but has no effect in *Neurospora* (*14*, *15*, *18*, *19*). Formaldehyde, which could present a potential mutagenic hazard (because it is mixed with protein used to feed cows so that beef and milk with reduced content of unsaturated fatty acids can be obtained) is an example of specificity. In *Drosophila,* formaldehyde can be applied in two different ways: as injection into adults and as admixture to the food of the larvae. Both types of treatment are effective in males but ineffective in females; a typical example of sex-specificity. In adults, formaldehyde acts predominantly on mature spermatozoa and peroxides presumably play a part as intermediates; their breakdown by catalase is in part responsible for the differential sex-effect (*20*). Formaldehyde given in the medium remains without effect in adults, but when fed to the larvae, it has a potent effect only in the early spermatocytes, in fact the mutation frequencies are comparable to those obtained for *Drosophila* sperm after irradiation with 3000-4000R (*14,15,18,19*). Moreover, formaldehyde acts only in the presence of adenosine (*21,22*) but crossing-over can be induced in earlier stages of spermatogenesis (*23,24*) and does not depend on adenosine (*25*).

This raises the interesting question about the possible role that nutritional regimens or local variations in the ionic composition of the drinking water may play in the response of germ cells to mutagens. Striking examples of ionic interactions have been reported for EMS, the chromosome-breaking activity of which can be enhanced by Cu or Zn ions in the plant *Nigella*, whereas the yield of chlorophyll mutations is only slightly affected (*26*). Vogel (*16*) observed that NaF added to *Drosophila* medium resulted in the complete elimination of chromosome breakage effects of various potent chemical mutagens such as Trenimon, whereas the yield of point mutations is unaffected. In contrast, NaF injected prior to x-irradiation of adult *Drosophila* males led to a pronounced potentiation of radiation induced sex-linked lethals in sperm (*27*). I cite these examples to illustrate some of the unexpected complexities that interactions between different agents present.

Differences in the response to ionizing radiation of the different germ cell stages of the mouse at different dose rates and fractionation schedules have been of crucial importance in the evaluation of genetic radiation hazards. For chemicals, the first data of this kind are beginning to appear (*28,29*). Thus Ehling (*28*) uses dominant lethal induction as a first approximation of mutagenic effectiveness and then complements it with the much more laborious specific locus method. With this procedure, interesting correlations have been obtained; for instance the alkylating agent MMS induces both dominant lethals and specific locus mutations in only sperm and spermatids, whereas the mutagenic activity of mitomycin-C, an inhibitor of DNA synthesis, is mainly restricted to spermatogonia.

In addition to germ cell stage specificity, specificity for certain chromosomal regions has been described for several chemical mutagens—such as maleic hydrazide, formaldehyde, phenol, and calf-thymus DNA (for references see *15, 18,19*). Needless to say, that here the findings are entirely different from those with radiation. Besides, there is the problem of mutagenic specificity at the genic level as illustrated by the work of Auerbach and her coworkers (for review see *13*).

All these problems make the testing for mutagenicity of chemicals more complex than is the case with radiation. For a critical evaluation of the mutagenic potential of any given comparison a proper understanding of the various steps involved in the process of mutation and how these can possibly be modified is obviously of critical importance.

DOSE EFFECT RELATIONSHIPS

Specific locus tests with mice have provided the scientific foundation on which the estimation of genetic risks to man for exposure to irradiation is based. These large scale breeding tests are both time consuming and laborious, and the excessive cost involved in the determination of a dose-effect curve for even one chemical mutagen would make this method of limited practical value.

The determination of dose-effect curves, however, is absolutely essential for a proper evaluation of mutagenic activity of environmental mutagens. The problem here is the lower limit of the dose at which genetic effects can be scored. Not only will labor and cost increase exponentially with decreasing doses, but also it becomes virtually impossible to detect increments over the spontaneous level. Auerbach (*14*) has argued that when a dose-effect curve is linear throughout its testable dose range, it is cautious to assume that there will be no lower threshold. However, when multi-hit curves are obtained, particular caution is required in extrapolation to lower dose levels. In asynchronous, metabolizing cell populations that are capable of repair, multi-hit curves almost certainly do not reflect the primary effect of the mutagen on the DNA, but may include various other components. Thus one can imagine dose dependence at levels of penetration through the cell wall or nuclear membrane, diffusion within the cell, combination with other cell constituents than DNA, or interference with a process responsible for repair of premutational damage. If the efficiency of repair is subject to damage by the mutagen, more premutational lesions will result in mutation at high than at low doses, and this may result in bi- or multi-phasic curves. The work of Auerbach and Ramsay (*30*) on adenine reversions in *Neurospora* provides an example how a linear dose effect curve changes in shape at high doses, presumably by inhibition or saturation of a repair enzyme. Consequently, multi-hit curves can be misleading and extrapolation for setting risks against benefit should be based on extrapolation from the low-dose component of such curves.

A question that may arise at this point is to what extent perhaps the ABCW (Abrahamson et al., *1*) method could be applied to extrapolate from findings in lower forms of life, where large numbers can be easily scored to man. Briefly, the method rests on the idea that the per locus mutation frequency as induced by irradiation remains nearly constant if related to the amount of DNA per nucleus. The remarkable constancy which is then observed in the range from bacteria to the mouse has already been used to point at the possible future redundancy of the costly, large-scale mouse experiments. Conger (*31*), one of the authors of the above ABCW paper, as I call it for convenience, has taken the analysis one step further: he finds that the total mass of chromatin per unit mass of DNA is not very different over the span from prokaryotes to higher eucaryotes, and speculates cautiously that "the target could be DNA alone but equally as well, it could be total chromatin mass on the basis of present information". In case of chemical mutagenesis, we are not yet in the position to make a similar projection; caution should, I believe, be exercised in view of the large areas of ignorance and the extreme cases of specificity, modification and interaction which have come to light.

On the problem of dose and dose-dependence our experience with radiations tells us that although the sources of ionizing radiations are diverse, qualitatively different radiation doses can be summed in fairly reliable ways to produce estimates of total mutagenic effects; summation is likely to be more difficult, however, in the case of chemical mutagens, for there are both experimental and theoretical grounds for anticipating that significant interactions will occur among such compounds. This is, doubtless, one of the important problems that we have to tackle in the years to come.

THE MUTATION SPECTRUM

Throughout many years of radiation genetic studies, a considerable body of knowledge has been accumulated on what is commonly called the mutation spectrum. The ground-work was laid by Muller many years ago, who showed that after exposure to 150 R, less than one dominant, some 5 recessive visibles, 25 recessive lethals, but 100 recessive detrimentals in a total of 1000 progeny can be expected. Both recessive lethals and detrimentals are characterized by a limited degree of some 2–5% dominance, which in fact determines their persistence throughout future generations in the population. Later, Crow and coworkers extended and confirmed these data, and these figures on dominance of recessive detrimental mutations figure still, for example, as one of the pillars for assessing the expression of mutational damage among first generation offspring of exposed generations.

From what we know from even a few chemical mutagens, the quantitative relation between induced mutations and chromosome aberrations may exhibit extreme variation. Even less is known, I believe, about the quantitative relationship concerning the various kinds of mutations, their relative degree of harmfulness, and the measure in which they may become expressed in the heterozygous condition. In this context some recent observations (Langley, Mukai, personal communication) may be cited, which show that mutations obtained after the application of low doses of EMS exhibit a high degree of dominance, while those induced by high doses of EMS are characterized by a lower degree of dominance.

CONCLUDING REMARKS

Let me come back to one of the main theses of this presentation: the resurgence of interest in the problem of genetic hazards from chemical mutagens has brought to the forefront the need for consolidated effort to assess the nature, magnitude and consequences of exposure to chemical mutagens; I have given a few examples to illustrate the complexity of the problem, although none of these are new or have not been stated succinctly by others. We have behind us the experience of over four decades of radiation research, which has provided us with the possibilities and tools to attempt to solve the problem before us. Several test systems are currently available or are being developed for environmental chemicals; we know what the relevant kinds of genetic damage are and the relative contribution each test system can make to the overall objective of hazard evaluation, and why we have to consider these as complementary rather than competing. The tier-system of mutagen-testing protocols as proposed by Bridges (*32*) and Flamm (*33*) certainly has intrinsic merit in that it puts an order into the array of needs and possibilities that are to be handled or faced. With the growth of knowledge in this field, we can meaningfully employ the approaches used in radiation to set limits to population exposure, bearing in mind the complex needs of our modern technological society. Sooner or later we have to think seriously about units of measurement for risk evaluation for chemicals, and several suggestions have already been put forward in the literature, such as the RED (Roentgen equivalent dose) of Crow (*34*) or RADEQ (of Bridges) or the ReC (*35*), all of which are approaches to define the effect of chemical mutagens in terms of the doses of ionizing radiation which would produce the same effect. In the larger view, atomic energy and its attendant hazards, environmental chemicals and their hazards are but aspects of the sophisticated and complex environmental challenges of modern life. Let us hope that the experience that has been gained in coping with the hazards arising from exposure to ionizing radiation, may guide us in facing some of the problems in the field of environmental mutagenesis.

ACKNOWLEDGEMENT

I wish to express my sincere thanks to my colleague Dr. K. Sankaranarayanan for his constructive criticism on the manuscript and his generous contribution to get it into final shape.

REFERENCES

1. S. Abrahamson, M. A Bender, A. D. Conger and S. Wolff, *Nature* **245**, 460 (1973).
2. J. G. Brewen, R. J. Preston, K. P. Jones and D. G. Gosslee, *Mutation Res.* **17**, 245 (1973).
3. C. Auerbach and J. M. Robson, *Nature*, **154**, 81 (1944).
4. C. Auerbach and J. M. Robson, *Nature* **157**, 302 (1946).
5. C. Auerbach, J. M. Robson and J. G. Carr, *Science* **108**, 243 (1947).
6. G. E. Nasrat, W. D. Kaplan and C. Auerbach, *Z.f. indukt. Abstamm.-u. Vererbungslehre* **86**, 249 (1954).
7. L. A. Snyder, *Z. Vererbungslehre* **94**, 182 (1963).
8. W. A. F. Watson, *Z. Vererbungslehre* **95**, 374 (1964).
9. W. A. F. Watson, *Mutat. Res.* **3**, 455 (1966).
10. H. Slizynska, *Mutat. Res.* **8**, 165 (1969).
11. R. J. Sram, *Mutat. Res.* **9**, 243 (1970).
12. H. Slizynska, *Genet. Res.* (*Cambridge*) **4**, 154 (1963).
13. C. Auerbach and B. J. Kilbey, *Ann. Rev. Genet.* **5**, 163 (1971).
14. C. Auerbach, *Newsletter of the Environmental Mutagen Society*, pp. 16–20 (Nov. 1971).
15. C. Auerbach, in *Chemical Mutagens Principles and methods for their detection* (A. Hollaender, Ed.) Vol. 3, pp. 1–19 Plenum Press, New York, 1974.

16. E. Vogel, *Mutat. Res.* (in press) (1974).
17. A. G. A. C. Knaap and P. G. N. Kramers, *Mutat. Res.* **22,** 55 (1974).
18. C. Auerbach, *Science* **158,** 1141 (1967).
19. C. Auerbach, *Transact. Kansas Acad. Science* **72,** 273 (1959).
20. F. H. Sobels, *Nature* **177,** 979 (1956).
21. T. Alderson, *Nature* **187,** 485 (1960).
22. T. Alderson, *Nature* **191,** 251 (1961).
23. F. H. Sobels, D. Bootsma and A. D. Tates, *Drosophila Information Service* **33,** 161 (1969).
24. W. E. Ratnayake, *Mutation Res.* **9,** 71 (1970).
25. T. Alderson, *Nature* **215,** 1281 (1967).
26. J. Gilot, M. Moutschen-Dahmen and J. Moutschen-Dahmen, *Experientia* **23,** 673 (1967).
27. R. N. Mukherjee and F. H. Sobels, *Mutat. Res.* **6,** 217 (1968).
28. U. H. Ehling, *Mutation Res.* (in press) (1974).
29. R. B. Cumming, *Mutat. Res.* (in press) (1974).
30. C. Auerbach and D. Ramsay, *Molec. Gen. Genetics* **103,** 72 (1968).
31. A. D. Conger, *Mutat. Res.* (in press) (1974).
32. B. A. Bridges, *Environ. Health Perspect.* **Nr. 6,** 221 (1973).
33. W. G. Flamm, *Mutat. Res.* (in press) (1974).
34. J. F. Crow, *Environ. Health Perspec.* **Nr. 6,** 1 (1973).
35. Environmental mutagenic hazards: recommendations concerning detection and prevention. A report of the American Environmental Mutagen Society (in preparation, to appear in *Science*).

Determinants of the Mutagenic Specificity of Chemical and Physical Agents in Microorganisms[1]

Brian J. Kilbey
Department of Genetics
University of Edinburgh
Edinburgh, EH9 3 JN, Scotland

THE PROBLEM

The term 'mutagen specificity' can be applied to a variety of different situations. A mutagen may, for example, be confined in its action to certain species, to certain cell types or to certain stages during the growth cycle of the cell. In this presentation, however, I shall restrict myself to a consideration of specificity which usually interprets differential activity as indicating a propensity on the part of the mutagen to attack certain regions of the genome rather than others. The regions concerned may be clusters of genes, individual genes or particular alleles at a locus. Of course, we never observe events occurring at the DNA level directly, all our conclusions are inferences drawn from observations at the phenotypic level; we count colonies or observe morphological variants. Obvious though this is, it is well worth emphasizing since it leads inevitably to the appreciation that physiological and metabolic events may play an important role in the determination of the observed specificity, augmenting, reducing or, conceivably, even reversing the primary specificity at the DNA level. Once a mutagen has reacted with the DNA, the reaction product must be converted into an altered base sequence in the face of repair processes which would eradicate it. After fixation the mutated gene must gain phenotypic expression and, after expression the mutant cell must compete favorably with the many non-mutant cells around it for nutrients and space to form a mutant clone which can be observed. It would be very surprising if mutants of all types were able to surmount these hurdles with equal facility but it is only if they can that the basic specificity will be preserved.

It is the object of this presentation to examine some of the evidence which is accumulating concerning the determination of the mutagen specificity of some chemical and physical mutagens. My main reason for doing this is my belief that by attempting to explain specificity and its modification we are forced inevitably to examine some of the very cellular factors which must be understood if we are to account adequately for the other types of specificity mentioned above, and to make meaningful quantitative comparisons between chemical and physical mutagens in the same, and in different organisms.

For convenience, I would like to discuss separately those determinants which appear to act (a) at the DNA level itself, (b) at the pre-mutation fixation

[1]Congress presentation C-2-2.

stages and (c) at the post-mutation fixation, or expression stages. Our understanding of the mechanisms involved is often severely limited and consequently it is difficult to assign any one factor with certainty to a particular level of operation. However, I hope that, by attempting this, attention will be drawn to those aspects of the problem which must require further investigation.

SPECIFICITY AT THE DNA LEVEL

We are really asking here what are the factors which determine whether a particular gene or allele will be susceptible to a mutagenic agent. Among the many we could consider, three immediately spring to mind: (a) The state of the gene; that is, whether it is being actively transcribed or replicated at the time of treatment, (b) the local base sequences in the neighborhood of the allele in question, and (c) the types of alteration required for the reversal of the mutant in relation to the changes which the agent can produce. Enough has been said about (c) elsewhere and I shall concentrate on the first two possibilities.

Cerda-Olmedo et al. (*1*) were able to show that nitrosoguanidine can mutate genes in sequence around the chromosome of *E. coli* when synchronous cultures are exposed to short bursts of the mutagen. Although a direct relationship with DNA synthesis has not been established, the simplest interpretation of the results is that nitrosoguanidine acts at the replication fork. A similar study has been made with yeast (*2*). These authors made use of the zonal centrifuge to separate cells of different ages from each other on the basis of size. Nuclear mutations to erythromycin- and oligomycin-resistance were induced by nitrosoguanidine at restricted and slightly separated times in the cell cycle. Mitochondrial mutants with these phenotypes occured at a much later stage in the cycle than the nuclear mutants. It may thus be possible, with some mutagens, to achieve a selective mutagenesis of certain parts of the genome by synchronizing the population treated and restricting the treatment to a small part of the cell cycle.

The effect of the activity of a gene on its susceptibility to mutagenesis has also been studied in several laboratories including our own (*3*). The general conclusion reached from these studies is that, depending on the mutagen, the active gene is more mutable than the inactive one. Brock (*4*) has shown that mutants at the β-galactosidase locus of *E. coli* are induced at higher frequencies by the alkylating agents diethylsulfate, ethyl methane sulfonate and nitrosoguanidine when the gene is active than when it is inactive. In contrast, the two states, active and inactive, are equally responsive to the mutagenic action of γ rays and base analogues. Similar results have been reported, also in *E. coli*, for ICR-171 which is predominantly a frame-shift mutagen (*5*).

As early as 1958, Benzer and Freese had noted that highly mutable sites or 'hotspots' exist at the rII locus of T4 bacteriophage (*6*) and had suggested that the probability of a mutational change in a particular base pair might be strongly influenced by its neighbors. In 1971 direct evidence for such a prediction was obtained by Koch (*7*) again using bacteriophage T4 and studying 2-aminopurine-induced transitions in an ochre nonsense codon. Ochre can be converted into amber by a single transitional change in the third base, UAA→UAG, and the influence of this change on the frequency of transitions at the first and second positions can be determined. As a result of substituting A with G in the third position of the codon, the transition rate at the second base increased 23-fold while at the first base it increased 3-to-4-fold. Sherman and his collaborators (*8*) have also studied this problem in yeast. Five chain-terminating codons in the gene coding for iso-1-cytochromes-C have been identified mainly on the basis of the amino acids inserted in revertants of these mutants and by their response to suppressor mutations. There is no indication that any of the possible amino acid insertions resulting from single base changes in these codons is discriminated against—an important requirement if an unbiased picture of the primary specificity of the mutagens subsequently to be tested is to be obtained. Nitrous acid (NA), which for a number of reasons is believed to selectively induce AT↔GC transitions (*9,10*), was used to induce intragenic reversion of these amber and ochre mutants and the amino acid substitutions in the revertants determined. On the basis of these substitutions the induced base changes were identified. Approximately half of the reversions of the amber and ochre mutants (excluding ochre *cyc 1-72*) arose by means of AT→GC transitions. The remainder resulted from various transversions mainly of the middle AT base pair. In the mutant *cyc 1-72*, however, only lysine insertions were found. In other words, these reversions were exclusively AT→TA transversion of the first base pair. NA did not induce lysine insertions in any of the other amber or ochre mutants nor did it produce leucine or tyrosine replacements in *cyc 1-72*. These would have arisen by transversion in the 2nd and 3rd positions, respectively. In other words, the action of NA is not simply determined by the presence of an AT base pair; that base pair must be in the right position within the codon and the codon must occupy a particular position within the gene.

This complexity in the determination of DNA specificity also extends to UV irradiation. *cyc 1-9*, another ochre mutant, reverts strongly with UV compared with other ochre mutants (e.g. *cyc 1-2*) and amber mutants (e.g. *cyc 1-179*). Furthermore, the high mutability is linked to an almost exclusive insertion of glutamine. In contrast, glutamine is absent in revertants induced by UV in *cyc 1-2* and *cyc 1-76*, another amber mutant. Clearly, a unique neighboring nucleotide sequence is a likely candidate for explaining the high and specific response of *cyc 1-9* to UV although there is evidence that repair may also play an important role (see later).

FACTORS OPERATING BEFORE FIXATION

The examples I would like to draw on here all refer to repair. The question is whether repair itself can act specifically at either the locus or the allele level. Two observations have been made which may indicate that repair activity could be locus-specific under certain circumstances. Kölsch and Starlinger (*11*) have shown that the galactokinase-forming ability of *E. coli* K-12 can be inactivated by UV and, in the absence of the inducer, can be partially photoreactivated. If the inducer is added after irradiation but before attempts at photoreactivation, no photoreactivation is observed. This example is clearly not concerned with mutation, but my second example is. Witkin (*12*) has shown, along with others, that if trace amounts of a mixture of amino acids or of broth are added to the post-irradiation plating medium, the frequency of suppressor mutations of certain amino acid auxotrophs is selectively enhanced in *E. coli.* If the supplementation is omitted, or simply the required amino acid for the particular auxotroph is added, the mutations are rapidly lost. The phenomenon is referred to as 'mutation frequency decline' or MFD. It can be inhibited by both caffeine and acriflavin and is absent in excision-deficient strains. Witkin has interpreted MFD as the excision of specific premutational lesions at suppressor loci which become susceptible to repair by virtue of their repressed state when the cells are plated either without supplementation or simply with the amino acid required by the auxotroph. The prevention of MFD by supplementation with traces of an amino acid mixture or by broth is interpreted as resistance to repair in suppressor genes which become active when protein and RNA synthesis resumes on plating. An attempt has been made to test this interpretation directly: Bockrath and Cheung (*13*) have shown that protein synthesis is the same in cells grown on minimal medium enriched with broth or with the amino acid required by the auxotroph-conditions which promote different MFD responses (see also Haak and Bockrath this Congress[2]). If, however, Witkin's interpretation does stand up to further tests, it may be an example of a case in which similar findings are to be expected in other organisms also. So far our attempts to demonstrate it for super-suppressors in yeast have failed, although this may simply be attributed to their low induction rates with UV in this organism.

At the allele level, photoreactivation has been shown to be specific. Resnick (*14*) studied the reversion of a number of mutants in yeast following UV-treatment. The strain used was *rad*$^+$ and, for most mutants, photoreactivation of mutational and lethal damage occurred to the same extent. However, in one case, reversion of the allele *arg 4-27*, mutational damage was repaired far less effectively than lethal damage. Since only one allele at this locus was tested, it is impossible to know whether this is a characteristic of the *arg 4* locus or simply of this allele. In the case of *Neurospora* an allele-specific effect could be demonstrated (*15*). Earlier experiments with this organism had shown that photoreactivation could repair all classes of point mutation to a similar extent (*16*). The same conclusion had also been reached for bacteriophage T4 by Drake (*17*). However, during the course of some UV experiments using the diauxotroph *ad-3A 38701; inos 37401,* it was noticed that photoreactivation of the lesions causing the inositol allele to revert was far less marked than photoreactivation both of lesions giving adenine reversions and lethal lesions. Other inositol alleles show normal photoreactivation (*15*). Whether this difference between the alleles reflects some intrinsic differences between the photo-products producing the revertants or whether some local condition, e.g. a neighboring base sequence or a masking protein, prevents the repair enzyme from gaining access to its substrate,

[2]Congress presentation B–5–4. *Radiat. Res.* **59**, 77 (1974).

cannot at present be decided. Whatever the reason the net result is a repair-mediated modification in specificity.

Photorepair has also been tested in *Neurospora* for its ability to discriminate between UV-induced lesions which generate point mutations on the one hand and recessive lethal mutations, probably deletions, on the other (*18*). In contrast to its general lack of specificity in removing different classes of point mutation, photoreactivation shows a definite tendency to repair damage giving deletions more readily than damage-producing point mutations.

One well-studied case of mutagen specificity in *Neurospora* can be attributed, at least in part, to the activity of a repair process. The bifunctional alkylating agent diepoxybutane shows a marked preference in its mutagenic action in the diauxotrophic strain just referred to: Reversions of the adenine allele *3A-38701* occur very much more frequently than reversions of the inositol allele *37401*, and the specificity becomes more marked as the dose increases. Kølmark and Kilbey (*19,20*) made a detailed study of the kinetics of induction of adenine reversions and were able to show that dose-rate is an important determining factor. At high dose-rates upward-bending mutation induction curves were obtained while at low dose rates a linear relationship was found. The explanation put forward by them on the basis of earlier ideas of Haynes (*21*) is that DEB inactivates a repair system which normally eradicates some of the mutational damage it, DEB, produces. At low dose rates the time of treatment is sufficiently extended to allow replacement of the damaged repair components and a linear dose-effect relationship is observed. Since the initial observation, several other pieces of evidence have been found which support this interpretation. They have been summarized and discussed elsewhere (*22*) but briefly they are as follows: DEB-treated cells become sensitive to further mutagenic treatments (*23*); this sensitization rapidly disappears when the treated cells are kept in water (*24*). Loss of the sensitive state during storage—presumably the consequence of replacement of the damaged repair system—can be prevented if protein synthesis is inhibited (*25*) and, finally, if replacement is prevented during low dose-rate exposure to DEB the dose-effect curve is converted from a linear to an upward bending one (*26*). It is interesting, and I think very important, to note here that a similar system is apparently operating in yeast (Olzewska and Kilbey, in preparation). Auerbach and Ramsay (*27*) adopted this explanation to account in part for the high specificity for adenine reversion shown by DEB since, as far as can be ascertained, inositol reversion is not susceptible to this repair system—the induction of mutation being linearly related to dose or even less so. An alternative possibility, to which I will return in the next section, is that repair is not specific for adenine reversions but that other, as yet unknown factors, limit the expression of inositol reversions after DEB treatment.

So far as I am aware this is the first case in which an example of a dose-rate effect for point mutations has been so intensively studied in a micro-organism, and it is interesting to note that it has been achieved with a chemical and not with ionizing radiation.

Repair processes are of course also implicated in the fixation of mutations produced by UV (*12*) and by some chemicals (*28,29*); so far, however, the studies are mainly of a quantitative type: By this I mean that the influence of repair differences on the spectrum of mutational types produced has not been widely studied. One case, however, deserves mention as possibly falling into this category: I have already referred to the high mutability with UV of the ochre mutant *cyc 1-9* of yeast and the almost exclusive insertion of glutamine in the iso-1-cytochromes C of the revertants. In the presence of *rad-6*, a UV-sensitive mutation which reduces UV mutability, this preference for glutamine insertion is relaxed (*8*). It will be interesting to study other UV-sensitive mutants and other mutagenic chemicals in this regard. One difficulty in yeast is that the nature of the repair deficiency in the several mutants is largely unknown at present.

FACTORS OPERATING AFTER FIXATION

I have no firm conviction that the data I shall discuss here are correctly inserted into the framework I have tried to develop. However, there is no direct evidence that repair is involved and it is unlikely that the changes in specificity to be described reflect changes in the susceptibility of the DNA itself to the mutagens. The examples refer to the same diauxotrophic *Neurospora* strain already described.

It is now firmly established that the allele *inos 37401* of *Neurospora* responds apparently only marginally towards DEB. Other alleles at this locus show the same low responsiveness. Although possible, it is unlikely that all respond poorly because there is a failure in the first step in the process, the interaction between DNA and DEB. It seems more likely that, after DEB treatment, the cells are less able to support the expression of potential inositol reversions. Various other pieces of evidence also suggest this. Other alkylating agents, e.g. EMS, revert inositol mutants (*30*). When the diauxotrophic strain is grown in the presence of low concentrations of DEB, there are indications that the *inos 37401* allele shows a slightly better response (*31*). DEB can depress the mutagenic activity of other mutagens towards the inositol allele, e.g. UV (*23*). However, final vindication for the view that the low response of *inos 37401* to DEB arises at a stage of expression and not at the primary DNA-level comes from experiments in which the same alleles were tested in a cytoplasmic environment modified by the *poky* mutation (*32,33*). This is a maternally-inherited trait which interferes with mitochrondrial and cytochrome function in *Neurospora*. The type of result obtained is shown in Fig. 1. In the presence of *poky*, the inositol allele *37401* now responds better than the adenine allele to DEB and although in absolute terms the response of the adenine allele is reduced, the increase in the response of *inos 37401* appears real. The removal of *poky* causes the response to return to normal (*33*). The reason for this change in behavior is not yet determined but the system has the advantage of offering clear background biochemical information from which to proceed in the analysis. One of our current studies involves testing the response of other adenine and inositol alleles to DEB and other adenine-specific mutagens when in the *poky* background.

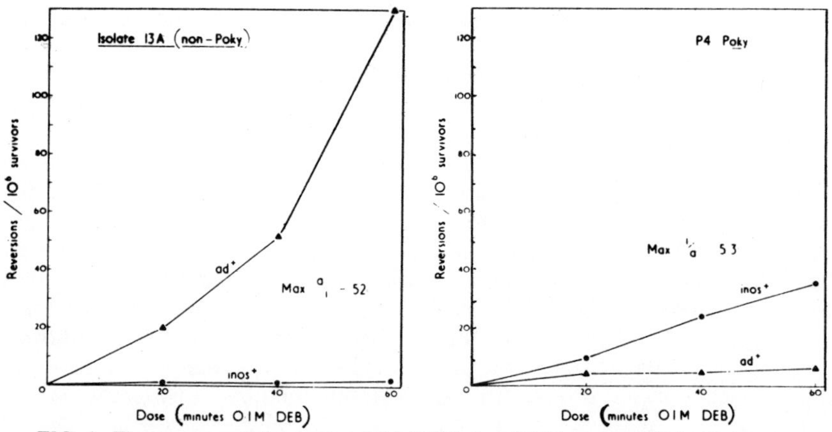

FIG. 1. The response of the strain *ad-3A 38701; inos 37401; cot* to DEB in the non-*'poky'* and the *'poky'* backgrounds.

My second example of the modification of specificity by a factor possibly operating at later stages in expression concerns the action of UV in *Neurospora*. For some time we have studied the action of this mutagen on the same pair of auxotrophic markers with the object of discovering ancillary conditions which modify its specificity. UV produces characteristically 2 to 3 times as many inositol as adenine reversions. If the temperature during irradiation is lowered to 0 to 2°C the specificity is reduced because the adenine reversion frequency is specifically enhanced (*34*). If the cells are irradiated in phosphate buffer instead of water, the specificity is enhanced because inositol reversion is specifically enhanced (*35*). As we saw earlier, photoreactivation also enhances specificity by specifically reducing the frequency of adenine reversions. However, the most dramatic factor so far discovered is presumably attributable to a physiological difference between two substrains of an original culture of the diauxotroph, one maintained for nearly 10 years by Kølmark in Uppsala by serial culture and the other maintained in Edinburgh by us on silica gel crystals. When the UV responses of these two strains are compared, the results shown in Fig. 2 are obtained. The Edinburgh sub-strain shows the characteristic response but in the Uppsala isolate the specificity is completely reversed. The alleles can be shown to be the same in the two strains since it is possible to recover them by outcrossing the Uppsala substrain and to show that the reversion behavior returns to normal. The picture is complicated still further by our recent demonstration that culture age is of paramount importance in the determination of specificity in the Uppsala substrain (Figs. 3a and 3b). Cultures which are barely a week old show a specificity pattern like that of the Edinburgh substrain; however, this rapidly changes with culture age and, within 3 to 4 weeks, the specificity reverses and the Uppsala pattern is obtained. The Edinburgh substrain is far less susceptible to culture age. It seems very unlikely that specificity changes of such a dramatic type can occur at the DNA level and it is more likely that the effect is the consequence of physiological differences between the strains which develop with age. What these are and why they have arisen in the first place is not at present clear, but preliminary data suggest that they result from a single gene difference which now exists between the two substrains (Kilbey and Purdom, in preparation).

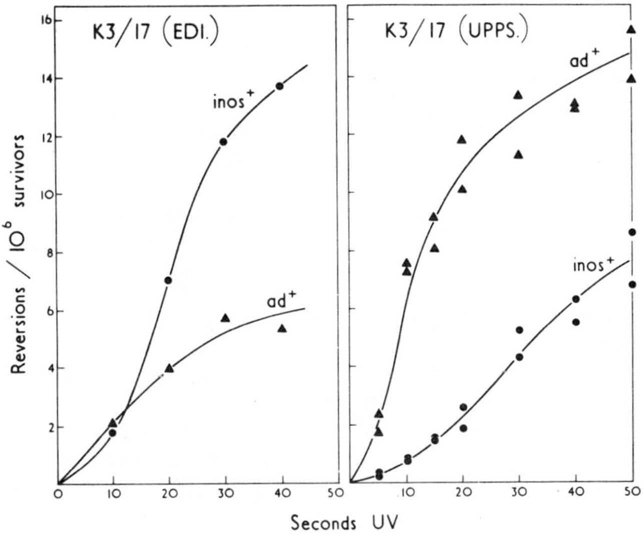

FIG. 2. The response of the *ad-3A 38701; inos 37401; cot* to UV, (a) in the K3/17 EDI substrain and (b) in the K3/17 UPPS substrain.

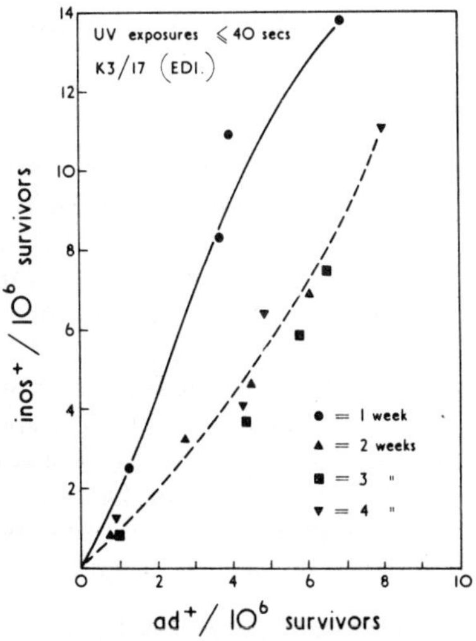

FIG. 3a. Changes in mutagen specificity of UV with age of culture in the K3/17 EDI strain. Lack of specificity in this graph would be indicated by a line drawn at 45° to the abscissa and intercepting it at zero.

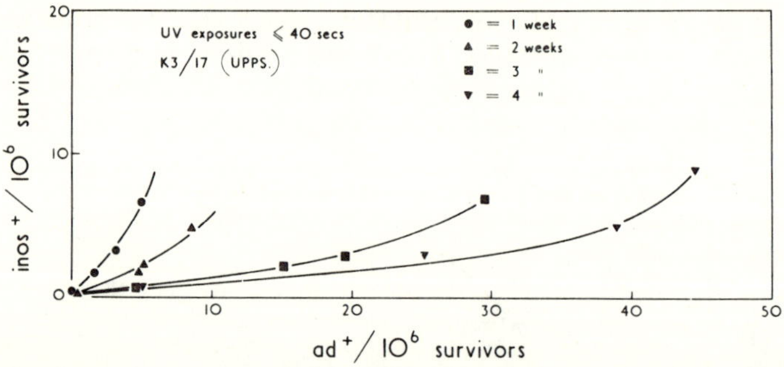

FIG. 3b. Changes in mutagen specificity of UV with age of culture in the K3/17 UPPS strain. Lack of specificity in this graph would be indicated by a line drawn at 45° to the abscissa and intercepting it at zero.

It is unfortunate that very little of what I have said has had much to do with ionizing radiation. This reflects the paucity of data concerning the mutagenic specificity of ionizing radiation and its modification in micro-organisms. Alderson and Scott (*36*) have shown that, in *Aspergillus nidulans,* the proportion of *2-TxR* and *XDH* mutations induced by γ rays can be modified by the provision of anoxic versus oxic conditions during exposure. However, in both bacteria and in *Neurospora* the difference between oxic and anoxic conditions has produced only small differences in mutation yield (*37,38*) and modification of mutational specificities has not been studied. Auerbach and Ramsay (*27*) have shown that the initial specificity of X rays for adenine reversions in the diauxotrophic system they use tends to disappear with increasing X-ray dose, and Zimmerman (quoted in *27*) has shown that the degree to which this occurs depends on dose-rate. Both observations suggest that secondary factors are thus able to influence the observed specificity but their nature and mode of action have not been examined. This area may well repay further more extensive investigation.

CONCLUSIONS

These examples emphasize beyond question the importance of considering mutation induction and consequently mutagen specificity as a process integrated into the total network of cellular events and not as an isolated series of DNA-limited reactions. However, there seem to be two important steps which must be taken before this type of information can help us with the sort of practical problems referred to at the commencement of the talk. First, we need a far better understanding of the determinants in biochemical and biophysical terms—it is even possible that the study of these aspects of mutagen specificity might point to biochemical and physiological connections within the cell which were hitherto unsuspected. Second, we must try to move from an *ad hoc* treatment of individual cases to a perception of generalities and principles. In a few of the cases described here these may already be starting to emerge.

ACKNOWLEDGEMENT

The work from the author's own laboratory has been made possible by research grants from the Medical Research Council of the United Kingdom. I am indebted to Hugh Patterson for permission to use Figure 1.

REFERENCES

1. E. Cerda-Olmedo, P. C. Hanawalt and N. Guerola, *J. Mol. Biol.* **33**, 705 (1968).
2. I. W. Dawes and B.L.A. Carter, (in press).
3. A.W.B. Johnston, *Mutagen Specificity and Potency in Microorganisms,* Ph.D. Thesis, University of Edinburgh, 1974.
4. R. D. Brock, *Mutat. Res.* **11**, 181 (1971).
5. B. K. Herman and N. B. Dworkin, *J. Bacteriol.* **106**, 543 (1971).
6. S. Benzer and E. Freese, *Proc. Nat. Acad. Sci. U.S.A.* **44**, 112 (1958).
7. R. E. Koch, *Proc. Nat. Acad. Sci. U.S.A.* **68**, 773 (1971).
8. F. Sherman and J. W. Stewart, *Proc. XIII Int. Congr. Genet.* 1973 (in press).
9. E. Bautz-Freese and E. Freese, *Virology* **13**, 19 (1961).
10. E. Freese, Brookhaven Symp. Biol. **12**, 63 (1959). (Available from Ofc. of Tech. Serv., Dept. of Commerce, Wash., D.C.).
11. E. Kölsch and P. Starlinger, *Z. Vererbungsl.* **96**, 304 (1965).
12. E. M. Witkin, *Annu. Rev. Genet.* **3**, 525 (1969).
13. R. Brockrath and M. K. Cheung, *Mutat. Res.* **19**, 23 (1973).
14. M. A. Resnick, *Mutat. Res.* **7**, 315 (1969).
15. B. J. Kilbey, *Molec. Gen. Genet.* **100**, 159 (1967).
16. B. J. Kilbey and F. J. De Serres, *Mutat. Res.* **4**, 21 (1967).
17. J. W. Drake, *J. Bacteriol.* **92**, 144 (1966).
18. F. J. De Serres and B. J. Kilbey, *Mutat. Res.* **12**, 221 (1971).
19. H.C. Kølmark, B. J. Kilbey and S. Kondo, in *Genetics Today,* **Vol. I.** pp. 61–62, Pergamon Press, London, 1963.
20. H. G. Kølmark and B. J. Kilbey, *Molec. Gen. Genet.* **101**, 89 (1968).
21. R. H. Haynes, *Radiat. Res., Suppl.* **6**, 1 (1966).
22. B. J. Kilbey, *Mutat. Res.* **26**, 249 (1974).
23. C. Auerbach, in CIBA Symposium on *Mutation as a Cellular Process*, pp. 66–86. Churchill, London, 1969.
24. C. Auerbach and D. P. Ramsay, *Molec. Gen. Genet.* **109**, 285 (1970).
25. B. J. Kilbey, *Molec. Gen. Genet.* **123**, 62 (1973).
26. B. J. Kilbey, *Molec. Gen. Genet.* **123**, 73 (1973).
27. C. Auerbach and D. P. Ramsay, *Molec. Gen. Genet.* **103**, 72 (1968).
28. S. Kondo, H. Ichikawa, K. Iwo and T. Kata, *Genetics* **66**, 187 (1970).
29. I. Pietrzykowska, *Mutat. Res.* **19**, 1 (1973).
30. B. J. Kilbey, *Microbiol. Genet. Bull.* **18**, 13 (1962).
31. C. Auerbach and D. P. Ramsay, *Jpn. J. Genet.* **43**, 1 (1968).
32. C. Auerbach and D. P. Ramsay, *Microbiol. Genet. Bull.* **35**, 4 (1973).
33. H. P. Patterson, *Mutat. Res.* (in press).
34. B. J. Kilbey, *Z. Vererbungsl.* **94**, 385 (1963).
35. B. J. Kilbey, in CIBA Symposium on *Mutation as a Cellular Process,* pp. 50–62. Churchill, London, 1969.
36. T. Alderson and B. R. Scott, *Nature (London) New Biol.* **230**, 45 (1971).
37. B. A. Bridges, *Annu. Rev. Nucl. Sci.* **19**, 139 (1969).
38. H. G. Kølmark, *Mutat. Res.* **2**, 222 (1965).

Comparison of the Mutagenic Effects of Chemicals and Ionizing Radiation Using *Drosophila melanogaster* Test Systems[1]

William R. Lee
Department of Zoology and Physiology
Louisiana State University
Baton Rouge, Louisiana 70803, USA

INTRODUCTION

In the comparison of the mutagenic effects of ionizing radiation with chemical agents we are immediately faced with the dissimilarity in the development of a study of these two types of mutagens, as, for example, the quantitative approach in the study of ionizing radiation *vs.* the qualitative approach to chemical mutagenesis. Evidence that X-rays induced mutations was presented by Dr. H. J. Muller in two 1927 papers (*1, 2*) in which he introduced a technique for detecting sex-linked recessive lethals in *Drosophila*. The sex-linked recessive lethal test devised by Muller was particularly suitable for mutation testing because not only was it sensitive in that it detected any lethal mutations on 1/5 of the genome, but it also automatically screened out pre-existing lethal mutations. Scoring was delayed until the F_2 thus assuring the heritability of observed events. These publications were quickly followed by the work of numerous investigators who demonstrated in a variety of test systems that various forms of ionizing radiation were mutagenic. By 1934 a review article written by Dr. C. P. Oliver (*3*) showed that various forms of ionizing radiation were mutagenic in a large variety of test systems.

During the following decade quantitative experiments were begun on the induction of mutations by ionizing radiation, and in Muller's 1940 paper (*4*) the dose-response relationship of translocations involving the 3/2 power rule was demonstrated and explained in terms of the increased probability of dominant lethals being formed as higher doses were reached. We have had since this time an extension of these quantitative methods to the numerous forms of ionizing radiation that the nuclear age has made readily accessible to the experimentalist.

During the time that *Drosophila* was being extensively used to study mutations induced by ionizing radiations, the genetic tests using this organism were greatly extended. Tests were developed which made possible the identification of exchanges between non-homologous chromosomes, between homologous chromosomes in the male (where cross-over does not occur), loss and exchange of dominant markers that were translocated to such heterochromatic

[1]Congress presentation C-2-3

regions as the Y-chromosome, as well as other numerous tests for visible mutations which could be detected either as completes or mosaics. Methods for detecting non-disjunction were also perfected. *Drosophila* workers have labored to develop many stocks, referred to as multipurpose stocks, to be used for these various tests, as reviewed by Drs. H. J. Muller and I. I. Oster (*5*). This accomplishment essentially makes *Drosophila* analogous to a spectrometer; instead of detecting one or a few points in the mutational spectrum it is possible by studying progeny from the same treated males to detect a variety of genetic changes ranging from gross chromosomal aberrations to changes that are present in only some of the daughter cells which produce mosaics. As a result of these sophisticated test schemes a properly organized *Drosophila* experiment today can give a number of points across the mutational spectrum, all within the same experiment.

Chemical mutagenesis had its birth in 1942 as a result of some unusual occurrences involving wartime work with mustard gas (*6*). The initial finding was classified and not published until after the war. By then Dr. C. Auerbach (*7*), using the numerous tests that had been devised for *Drosophila*, had shown that mustard gas was capable of inducing the wide spectrum of genetic changes produced by ionizing radiation. In this initial work, however, it was apparent that the proportions of genetic change were not the same for this chemical mutagen as for ionizing radiation. There was a higher proportion of visibles and sex-linked recessive lethals to translocations than in the case of ionizing radiation, and mosaics were quite frequent, in contrast to Muller's earlier experiment with ionizing radiation where mosaics had not even been significant.

There followed in the later 1940's and has been continuing until the present the testing of a large variety of compounds. It was soon evident that mustard gas was only one of a class called alkylating agents which are capable of adding chemical groups to DNA. From the study of alkylating agents alone a large number of mutagens have been found and one book by Dr. A. Loveless entitled *"Genetic and Allied Effects of Alkylating Agents"* (*8*) deals entirely with the mutational studies of this class of compounds. Other classes of chemical mutagens were found to intercalate into the DNA thus causing a frame shift in the reading of the genetic code during replication. Unfortunately, the precise mode of action of many chemicals shown to be mutagenic in *Drosophila* is not known.

CELL STAGE SPECIFICITY

Different germ cell stages vary in sensitivity to chemical mutagens as they do to ionizing radiation, but the magnitude of differences is much larger for chemical mutagens. Some chemical mutagens like mustard gas, for example, can be shown to affect germ cells in all stages of their development, but other mutagens are very stage-specific. For instance, the antischistosomal drug hycanthone methanesulfonate was found to be mutagenic only in late spermatids of male *Drosophila melanogaster* (*9*).

Another interesting example of stage specificity of a chemical mutagen occurs with the alkylating agent ethyl methanesulfonate. This is one of our most effective mutagens in late spermatid and mature spermatozoa of *Dro-*

sophila melanogaster, but the mutation frequency drops to near the spontaneous level when gonial cells are treated (*10*). An interesting contrast is the related compound, chloroethyl methanesulfonate, which induces a higher mutation frequency in spermatogonia than in postmeiotic stages (*11*). Because of such cases of cell stage specificity the problem of screening for chemical mutagens, as well as the problem of extrapolating the results of such screening to man, becomes far more difficult for chemical agents than it is for ionizing radiation.

The greater variation in cell stage sensitivity to chemical mutagens illustrates one of the hazards of applying the established "rule of thumb" for ionizing radiation to chemical mutagens. It has been well established that the spermatid stage of germ cell development is an order of magnitude more sensitive to ionizing radiation than gonial cells (*12, 13*), but the rate of induction of mutations in gonial cells is what is important if one extrapolates to species like man with a long generation interval. Consequently, some investigators have concluded that only the rate of induction of mutations in gonial stages should be considered in evaluating the hazard of chronic exposure. This conclusion is valid for ionizing radiation because there is only slightly over one order of magnitude difference in the sensitivity between gonial cells and the most sensitive cell stage, but does not necessarily hold true for chemical mutagens that may produce a greater difference in mutation frequency between gonial cells and the most sensitive cell stage.

In my laboratory, where we have been using the method of Lewis and Bacher (*14*) to treat *Drosophila* with ethyl methanesulfonate, we have found the sex-linked recessive lethal frequency to vary by at least two orders of magnitude between late spermatids (with a frequency in excess of 50%) and spermatogonia (with a frequency of less than 0.5%). If one extrapolates this difference in response among cell stages to the long life cycle of man, the short sensitive stage then becomes important. For example, if the sensitive cell stage is only 1/3 of the spermatogenesis period of 74 days in man (*15*), then the sensitive period is 2×10^{-3} of the 30 year generation interval. If this sensitive stage is two orders of magnitude more sensitive than the remainder of the 30 year generation interval, then at least 20% of the mutations would be induced during the 25 day sensitive stage. If there are three orders of magnitude difference between stages, then twice as many mutations would occur in the short sensitive stage. Consequently, in testing for the mutational hazard of chemical agents we must be on the alert for sensitivity in any stage in the development of the germ line.

MOSAICS

One of the first differences between ionizing radiation and chemical mutagens noted by Dr. C. Auerbach was the high frequency of mosaics induced by mustard gas (*7*, *16*, *17*). It was later found that monofunctional agents like ethyl methanesulfonate and chlorethyl methanesulfonate induce even higher frequencies of mosaics (*11*, *18*). For example, we have studied mosaics induced at the yellow locus of *Drosophila* by feeding ethyl methanesulfonate (*19*). These mosaics were the consequence of both mutant and nonmutant nuclei being produced during the cleavage process. Mutations at the yellow locus of *Drosophila* that become fixed during cleavage, can readily be studied because following cleavage a large sample of the nuclei migrate to the periphery of the blastodermic epithelium (*20*). Regions spaced systematically around the periphery of the epithelium differentiate into the imaginal discs which, during the pupal stage, develop into the adult structures that can be scored (*19*, *21*, *22*). Since, in the case of mutations of the yellow locus, the pigment affected by the yellow-1 allele of this locus is visible in most of the regions of the adult insect, it is possible to identify from the color pattern of the adult insect the location of the affected imaginal disc. From the location of these imaginal discs the pattern of mosaicism and the frequency of mutant nuclei in late cleavage can be determined. These studies in my laboratory have shown that approximately 1/4 of the nuclei are mutant in late cleavage after treatment with EMS (*19*). By using a different analytical approach involving the white locus or the dumpy locus and supplementing data from other laboratories, we found a consistent pattern of approximately 1/4 mutant nuclei (*19*).

The probability of detecting a mosaic mutation depends upon the developmental history of the structure scored for mutation or, in the case of transmissible mutations (such as the sex-linked recessive lethal test), it would depend on developmental history of the germ line. For example, relative to the yellow locus which can be scored in 48 body regions, only 21% of the mutations at the white locus and only 75% of the mutations at the dumpy locus would be observed as mosaics (*22*). Therefore, the ratio of mutants at different loci and the ratio of visible mutations detected to lethals will be expected to change as the proportion of mosaics induced changes (*19, 22, 23*). It should be emphasized that the detection of mosaics that become fixed during cleavage is dependent on an embryology such as that found in insects where a stratified sample of cleavage nuclei occurs. One would not expect to often observe this type of mosaicism with the mammalian type of embryological development.

In addition to the alteration in the mutational spectrum that can be explained by different proportions of mosaic induction, there is also a considerable difference in the mutational spectrum with regard to chromosome aberrations versus point mutations and among the types of chromosomal aberrations that are observed. For example, ethyl methanesulfonate induces a high frequency of sex-linked recessive lethal mutations but with the same treatment a low frequency of translocations is observed relative to what would be expected with X-rays (*24*, *25*). We have found, however, that the frequency of EMS-induced chromosome loss, as measured by loss of dominant markers, is influenced by adjacent heterochromatin (*26*). This increase in loss of dominant markers due to their close proximity to heterochromatin could be due to either an increase in breakage in the heterochromatin or to the more viable nature of individuals with aberrations in a largely heterochromatic chromosome. In either case, altering the position in the genome of the dominant marker studied alters the spectrum observed. Consequently, differences in the mutational spectrum can be attributed to differences in the structure of the genome as well as to the other causes I have discussed.

During the last three decades of work on chemical mutagenesis an extensive amount of data has been collected on mutagenic activity of numerous chemical agents and the response of a large variety of test systems to these chemicals. No attempt is being made here to give an encyclopediclike treatment of this field because reviews of this material have been written by Loveless (*8*), Fishbein, Flamm and Falk (*27*) and Hollaender (*28*), and an environmental mutagen information center has been established (*29*) to catalog this information.

DOSIMETRY

In spite of the large amount of data collected in the past thirty years on chemical mutagenesis, the status of the information seems remarkably similar to Dr. C. P. Oliver's 1934 review (*3*) of ionizing radiation where numerous agents had been tested for mutagenicity in a variety of test systems and the data cataloged in encyclopedic fashion. The striking contrast between chemical mutagenesis and radiation mutagenesis is that within a decade of the discovery of the mutational effect of X-ray we had excellent quantitative data including dosage response curves for various mutational test systems. In contrast, 30 years of chemical mutagenesis has not produced a body of acceptable quantitative results.

Attempts have been made to establish dosage response curves for chemical mutagens, but unfortunately little significance can be attached to them because there is no distinction made between dose and exposure. Actually, what is referred to as *dose* in the chemical mutation literature is merely *exposure* of the organism. For example, the animal is exposed by being treated to a gas of certain composition for a certain period of time or, alternatively, is fed a defined composition for a period of time, or perhaps a fly is injected with a certain volume of a certain concentration of a chemical agent. These so-called measures of dose only describe the exposure of the whole animal to the mutagen and in no way describe the amount of chemical which actually reaches the germ line.

In chemical mutagenesis I will define dose as the amount of the chemical that reacts with a selected target molecule, such as DNA in the germ line.

When we consider all of the possible physiological effects involved in absorption, circulation, transport across numerous cell membranes in multicellular organisms, the metabolic activity of the intact animal and enzymatic degradation of the compound, it can be readily seen that there is probably very little relationship between exposure and dose. In fact, there is no justification in assuming linearity between exposure and dose when the multitude of physiological conditions and fates of chemical mutagens are considered. Furthermore, when exposure is by feeding, the physiology of the microflora of the fly's ventriculus must also be considered. Without knowing the dose, the relation among mutational classes can be determined but the quantitative relation of mutation frequency to chemical events in the germ line remains unknown. The disparity between measurements of exposure and the unknown dose received by the germ line in the fly and other multicellular test systems has caused chemical mutagenesis to remain a qualitative, descriptive science as opposed to the quantitative work done in the study of ionizing radiation.

The ability to determine whether a substance is mutagenic depends upon the relationship of mutagenicity to the level of toxicity of a particular chemical. For example, the usual experimental protocol of testing a new compound is to determine its toxic level and then carry out appropriate genetic tests with slightly lower levels of the compound. Therefore, whether a substance is determined to be mutagenic depends upon the difference between toxicity and mutagenicity at near toxic level of exposure.

The most difficult problem in chemical mutagenesis is the reliability of extrapolating from existing data in experimental test systems to man, because of the differences in physiology that exist between test mammals and man, and the even greater differences in non-mammalian test systems. Fortunately, this difficulty can be largely avoided if one measures as dose the reaction of the chemical agent or its metabolic product with the target molecule DNA instead of the amount fed or injected—which is defined here as exposure. Therefore, the proper distinction between exposure and dose to the DNA in the germ line could in effect separate the problems of comparative physiology from the problems of comparative mutagenesis.

It is unfortunate that the concept of dose to the area of interest, the germ line, as opposed to exposure has not been widely applied in chemical mutagenesis. Perhaps this is because of the technical difficulties involved, and we do not have at this time an adequate data based on dose-response curves that can be used. Nevertheless, the situation is not hopeless, for we have been able to show in recent experiments in my laboratory that at least for one class of mutagens it is possible to measure dose in terms of direct reactions with DNA in the germ cell. For alkylating agents, the obvious measure of dose is the amount of alkylation of DNA in the germ line (*30, 31*). This measurement of dose can be extended to intercalating agents by measuring the amount of intercalation into DNA and, in fact, to anything that forms an adduct onto the DNA. This measurement of dose has not yet been extended to other classes of mutagens; however, in considering alkylating agents as well as intercalating agents, we cover a large proportion of the known mutagens.

In our current work we are measuring the alkylations of *Drosophila* spermatogonial DNA by feeding male larvae on media containing ^{14}C-methyl thymidine and then, as adults, feeding these males ^{3}H-2-ethyl methanesulfonate (0.025M EMS in a 1% sucrose solution for 24 hours at 25°C) (*14*). Female *Drosophila,* frozen after mating with treated males, were used for DNA extraction and for sperm cell counts of the seminal receptacles.

Approximately 60 frozen female flies were dissected for visual counts of sperm cells in seminal receptacles. After counting approximately 4,000 sperm cells, the receptacles were washed onto a millipore filter, digested, and a cocktail made for scintillation counting. A random sample of about 100 of the inseminated female flies from the group used for sperm counts was used for a modified phenol extraction of the DNA (*32*). Only the tips of the abdomens containing receptacles and spermatheca were used in this procedure. After purification the DNA was dissolved in SSC and the purity and concentration was checked spectrophotometrically. Finally, its specific activity was determined by scintillation counting.

Alkylations per sperm cell have been calculated as follows: The number of sperm cell equivalents in a purified sample of DNA can be determined from the ^{14}C DPM in the DNA sample and the ^{14}C DPM per cell was calculated from the receptacle sperm cell count. The number of ^{3}H DPM per sperm cell equivalent can then be calculated from the ^{3}H DPM in the purified DNA sample and the sperm cell equivalents of this sample. Alkylations per sperm cell are then calculated from the ^{3}H DPM per sperm cell and the specific activity of the ^{3}H-EMS.

These calculations give the alkylations of DNA that are retained until after insemination of the female. ^{3}H-DPM per sperm cell is also determined at the time ^{14}C-DPM per sperm cell is determined. ^{3}H-DPM/^{14}C-DPM from the purified DNA divided by ^{3}H-DPM/^{14}C-DPM in the whole cell digest gives the proportion of total alkylations in DNA. Therefore, we also have a measurement of total alkylations in cell constituents other than DNA.

A total of ten separate experiments gave a mean of $4.3 \pm 0.5 \times 10^6$ alkylations of DNA per sperm cell. The confidence limit reported is one standard deviation determined from the experimental variation among repeated experiments. Recent control experiments with ^{14}C-arginine show that the earlier procedure (*31*) did not remove all the labeled arginine, whereas the current procedure left no residual ^{14}C labeled amino acids.

A chromatograph of the purified DNA showed 91 to 100 percent of the tritium co-chromatographing with 7-ethyl guanine. Taking an average value of 3×10^8 nucleotides per sperm cell (*31*), 21% of which is guanine in *Drosophila* DNA (*33*), we find 7% ± 1% of the guanines ethylated at the 7 position. This figure applies only to the mature spermatozoa and late spermatids, for all our experiments using early spermatids and spermatogonia had too low a level of alkylation to permit detection with the size samples and specific activity used. This high level of ethylation following the conventional treatment (*14*) with EMS led us (*31*) to hypothesize that ethylation of a rare site or combination of sites in the sperm DNA is responsible for the mutagenic action of EMS. At the present time we have dosage determined for only one level of exposure; therefore, we will not speculate on dose-response relationship.

Because of difficulties in dosage determination, the status of chemical mutagenesis in *Drosophila* remains comparable to the status of mutagenesis by ionizing radiation at the time of Oliver's (*3*) review in 1934. However, enough work has been done with alkylation agents to show that determination of dose to *Drosophila* germ cells is possible.

ACKNOWLEDGEMENT

This investigation was supported in part by PHS Research Grant No. ES 00320–07 from the National Institute of Environmental Health Sciences and the Atomic Energy Commission Contract No. AT-(40–1)-3728.

REFERENCES

1. H. J. Muller, *Science* **66,** 84 (1927).
2. H. J. Muller, The problem of genic modification. Read before 5th Int. Genetic Cong. Berlin, 1927. *Publ. der Zeitschrift fur induktive Abstammungsund Verrerbungslehre* **Suppl. 1,** 234–260 (1928) (Reprinted, *Studies in Genetics—The Selected Papers of H. J. Muller*. Indiana Univ. Press, 1962).
3. C. P. Oliver, *Quart. Rev. Biol.* **9,** 381 (1934).

4. H. J. Muller, *J. Genet.* **40,** 1 (1940).
5. H. J. Muller and I. I. Oster, in *Methodology in Basic Genetics* (Walter J. Burdette, Ed.) pp. 249–274. Holden-Day, Inc. 1963.
6. C. Auerbach, in "Chemical Mutagens: Principles and methods for their detection." (A. Hollaender, Ed.) Vol. 3, pp. 1–19 (1974).
7. C. Auerbach, *Proc. Roy. Soc. Edinburgh Sect. B.* **62,** 211 (1945).
8. A. Loveless, *Genetic and Allied Effects of Alkylating Agents.* Penn. State Univ. Press, University Park, Penn. 1966.
9. A. G. A. C. Knapp and P. G. N. Kramers, *Mutat. Res.* **22,** 55 (1974).
10. J. B. Jenkins, *Genet.* **57,** 783 (1967).
11. C. Mathew, *Mutat. Res.* **1,** 163 (1964).
12. I. I. Oster, in "Radiation Biology" (J. H. Martin, Ed.) *Proc. Second Australian Conf. Rad. Biology,* pp. 253–267, Butterworths Scientific Publ. London 1959.
13. F. H. Sobels, in *Proc. 11th Int. Cong. Genetics, The Hague* **2,** pp. 235–255. Pergamon Press, Oxford, 1965.
14. E. B. Lewis and F. Bacher, *Drosophila Information Service* **43,** 193 (1968).
15. C. G. Heller and Y. Clermont, *Recent Progr. Hormone Res.* **20,** 545 (1964).
16. C. Auerbach, *Cold Spr. Harbor Symp. Quant. Biol.* **16,** 199 (1951).
17. E. Altenburg and L. S. Browning, *Genet.* **46,** 203 (1961).
18. T. Alderson, *Nature (London)* **207,** 164 (1965).
19. W. R. Lee, G. A. Sega and J. B. Bishop, *Mutat. Res.* **9,** 323 (1970).
20. B. P. Sonnenblick, in *Biology of Drosophila* (M. Demerec, Ed.) pp. 62–167, Hafner Co. N. Y., 1965.
21. J. T. Patterson and W. Stone, *Gynandromorphs in* Drosophila melanogaster. *Texas Univ. Publ.* No. 3825 (1938).
22. W. R. Lee, C. J. Kirby and C. W. Debney, *Genet.* **55,** 619 (1967).
23. L. S. Browning and E. Altenburg, *Genet.* **46,** 1317 (1961).
24. S. K. Hotchkiss and J. K. Lim, *Drosophila Information Service* **43,** 116 (1968).
25. S. Abrahamson, W. C. Kiriazis and E. M. Sabol, A storage effect of ethyl methanesulfonate (EMS) on the induction of translocations in Drosophila sperm. *Drosophila Information Service* **44,** 110 (1969).
26. J. B. Bishop and W. R. Lee, (EMS). *Mutat. Res.* **21,** 327 (1973).
27. L. W. Fishbein, G. Flamm and H. L. Falk, *Chemical Mutagens.* Academic Press, New York, (1970).
28. A. Hollaender, *Chemical Mutagens: Principles and Methods for their Detection.* Vol. I and II. Plenum Press, New York, (1971).
29. H. V. Malling, *Newsletter of the Environmental Mutagen Society* **4,** 11 (1971).
30. W. R. Lee and G. A. Sega, *Newsletter Environ. Mut. Soc.* **3,** 37 (1970).
31. G. A. Sega, P. A. Gee and W. R. Lee, *Mutat. Res.* **16,** 203 (1972).
32. G. A. Sega, R. B. Cumming and M. F. Walton, *Mutat. Res.* **24,** 317 (1974).
33. J. R. B. Hastings and K. S. Kirby, *Biochem J.* **100,** 532 (1966).

Comparison of the Mutagenic Effect of Chemicals and Ionizing Radiation in the Spermatogenic Cells of the Mouse[1]

B. M. Cattanach
Medical Research Council Radiobiology Unit
Harwell, Didcot, Oxfordshire, OX11 ORD, Great Britain

The problems involved in chemical mutation work are both complex and varied. With ionizing radiation it is known that genetic effects are produced in all organisms, in all cells, and the assessment of genetic risk is thus one of ascertaining how much damage is being induced: It is a quantitative problem. By contrast, chemical mutagens may be highly mutagenic in one organism, or one cell type, but totally ineffective in another. In cases where mutagenic effects are produced, all those classes of mutation recovered following radiation may occur, or only a few. Where no mutagenic effects can be detected, this may be due to any one of a variety of different causes.

These complexities stem primarily from two properties that may be considered characteristics of chemical mutagens and which distinguish them from ionizing radiation. One is the great specificity of action, the other is the ability to produce delayed mutation. Specificity of action is responsible for the variable response obtained with different organisms, and the classic example of this is the clear mutagenic effectiveness of urethane in *Drosophila* (*1, 2*) and plant material (*3*), contrasting with its ineffectiveness in *Neurospora* (*4*). Specificity also accounts for the cell-stage sensitivity differences observed among *Drosophila* spermatogenic cells. Here, the most extreme example is perhaps that of formaldehyde which exclusively produces genetic effects upon the early spermatocyte of the larval testis (*5*). Specificity in response to chemical mutagen treatment is also found at the sub-cellular level: Non-randomness in the distribution of lesions along the chromosomes has been demonstrated in several organisms and, especially in plant material, breakage events have been found to occur preferentially in heterochromatic regions (*6-11*). Locus specificity has been indicated in *Drosophila* and plant species (*12-18*) but has been most convincingly demonstrated in such organisms as *Neurospora* (*19-20*) or *Aspergillus* (*21*). There is even evidence of allele specificity in bacteriophage (*22*). Chemical mutagen specificity has been discussed at length by Auerbach (*23-26*), and she has argued that the specificity operates not at the level of primary mutagenic action but upon those other cellular processes that take place before an initial pre-mutational event can yield the final mutation.

This leads to the second property of chemical mutagens, namely, that of causing delayed mutation. Pre-mutational lesions can be carried through many

[1]Congress presentation C–2–4.

cell generations before yielding the final mutation. In an organism such as the fission yeast (*Schizosaccharomyces pombe*) this instability leads to the appearance of mosaic sectors in colonies derived from single cells (*27*). Following the replication of cells from non-mutant sectors, these replicating instabilities may be further transmitted and eventually yield secondary mosaics. In *Drosophila*, mosaicism of both soma and gonad may be found not only in F_1 flies but even in F_2 or F_3 generations (*5, 28–31*).

One characteristic of chemical mutagens that has been observed in *Drosophila* and attributed to delayed mutation is the low ratio of translocations and deletions to point mutations (*32, 33*). Some mutagens, e.g. ethylmethanesulfonate (EMS), produce almost exclusively point mutations (*34*). This is not due to any lack of ability by those compounds to break chromosomes, for dominant lethals are produced, but from a failure of broken chromosomes to rejoin (*33, 35, 36*). It is thought that the breaks may not be open at the same time to allow rejoining to occur.

The purpose of this paper is to discuss the results of some of the chemical mutation studies that have been carried out in mammals, and I shall be dealing almost exclusively with the mouse. Most of this work has been carried out on the spermatogenic cells of the male and I shall concentrate on this material. The approach will be to try and assess the data with the aim of ascertaining whether those properties of chemical mutagens, observed in other organisms, are also to be found in mammals.

First, we can look for evidence of specificity: At the most gross level, we find we have evidence of strain and sex differences in response to mutagen treatment. The extensive studies of Generoso (*37, 38*) have demonstrated quite large differences in dominant lethal induction by EMS in the oocytes of female mice. A puzzling aspect of the data is that the greater response with the most sensitive strain was achieved primarily by higher levels of pre-implantation loss. This is a criterion for dominant lethality which can be confused with such non-genetic factors as failure of fertilization or of implantation. However, Generoso did not find any indication that this might have been due to toxic effects upon the treated females and his histological studies provided evidence of genetic damage to the early embryos. The basis for these strain differences is not yet known; metabolic differences, differences in transport to the germ cells, differences in the susceptibility of the germ cells to chromosome damage, differences in rejoining ability or repair have been discussed. Parallel studies were also carried out with males and these did not show such extensive strain differences. This contrasting response thus provides evidence of a sex difference in response to the mutagen. To my knowledge strain differences in the genetic response to radiation have not been found. Leonard and Deknudt (*39*) did not find any significant differences between strains in a study of translocation-induction by X-rays in male mice.

A second level of chemical mutagen specificity in the mouse is found, as in *Drosophila*, among the array of spermatogenic cells. Many chemicals have been tested in this respect by the dominant lethal test (*40*), but data from three which have been subjected to a variety of tests will be selected to demonstrate the differences in response that may be obtained and how they all differ from X-rays. Summarizing the X-ray results first (Table 1), it is known that the spermatids are more sensitive than the spermatozoa to chromosome damage whether assessed by dominant lethals (*41*) which result primarily from chromosome breakage, translocations (*42, 43*) which also require the reunion of broken ends, or by sex chromosome loss (*44*). The greater sensitivity of the spermatids is also found in *Drosophila* and has been interpreted in terms of differences between the repair processes in the two cell stages, with oxygen tension being a major factor in each (*45*). Dominant lethals and translocations are also induced in the earlier spermatocyte stages (*43*) but, like most of the various classes of spermatogonial cells (*46*), these are highly sensitive to killing. The consequence of this is that a period of sterility due to aspermia occurs several weeks after higher radiation doses. The most resistant of the spermatogonia is the stem cell (*47*). A proportion of these survive most radiation treatments to repopulate the testis and eventually restore fertility. Translocations (*47*) and point mutations (*48*) are induced among surviving stem cells and may be propagated and transmitted throughout the life-time of a treated animal to the progeny. Point mutations may also be recovered from post-gonial stages and, here, the yield is about twice as high per dose as that from spermatogonia (*49*). The main conclusion that I wish to draw from the X-ray work is that both genic and chromosomal changes can be recovered from all germ cell stages.

Turning to the examples of chemical mutagens, the first that has been fairly extensively investigated is triethylenemelamine (TEM). It can be seen from Table 1 that this compound, like X-rays, produces its maximal effect in the spermatids. The increase in sensitivity relative to that of the spermatozoa is much more marked than for X-rays, however, and this has been demonstrated with dominant lethals (*50, 51*), translocations (*52*), and sex chromosome losses (*53*). Point mutations have also been recovered from TEM-treated spermatids (*53*). In contrast to X-rays, only the late spermatids show this elevated sensitivity; early spermatids are much more resistant and spermatocytes are also more resistant to both killing and dominant lethal induction. Evidence for the breakage and reunion of spermatocyte chromosomes is provided by the histological studies of Moutschen (*54*). The spermatogonial stem cell is even less sensitive to genetic damage by TEM for, although both translocations (*55, 56*) (less than 1%) and point mutations (*57*) have been recovered, extremely high doses appear to be necessary. On the basis of length of sterile period, some of the treatments cause as much stem cell killing as about 600 rads of X-rays (*57, 58*). For TEM, therefore, we can conclude that all classes of genetic damage may be recoverable from all germ cell stages, but the relative sensitivities of the individual stages vary much more widely than for X-rays.

A more extreme response is obtained with the second chemical mutagen listed in Table 1. Ethylmethanesulfonate (EMS) produces its maximum dominant lethality in early spermatozoa and the latest spermatids (*59, 60*), i.e. in later stages than either TEM or X-rays. A parallel response is obtained with translocations (*59, 61*), and point mutations have also been recovered (*62*). By contrast, no genetic effects, chromosomal (*59, 60, 63*) or genic (*64*), have clearly been detected in spermatocytes or spermatogonia and, although histological evidence of spermatogonial killing has been found (*65*), even with the highest doses that can be tolerated there is no indication of a sterile period that would indicate appreciable stem cell killing. It may be concluded that mutation tests with EMS on different germ cell stages can give what appear to be qualitatively different results. However, from the TEM studies one might anticipate that these might still be quantitative. If high enough doses could be administered the amount of stem cell killing might be increased sufficiently to bring about a sterile period and genetic damage might be detectable among the surviving cells. This view might be consistent with the finding that while EMS does not produce translocations in *Drosophila* when administered by injection, it does so when fed (*34*), a procedure which may allow a higher dose.

TABLE 1
Differential germ cell stage sensitivities to genetic effects of X-rays and chemical mutagens

	Spermatozoa		Spermatids			
Treatment	Mature	Immature	Late	Early	Spermatocytes	Spermatogonia
X-rays	+	+	++	+++	+	+
TEM	+	+	+++	+	(+)	(+)
EMS	+	+++	++	+	−	−
MC	−	−	−	+	+	+

The third chemical mutagen listed in Table 1 produces its maximal effects in quite different stages from TEM, EMS or X-rays. Mitomycin C (MC) produces dominant lethals only in early spermatids and spermatocytes, and spermatocytes and spermatogonia are sufficiently sensitive to killing to give clearly recognizable sterile periods with most doses (*66*). Point mutations are also recovered from MC treatments and these have almost exclusively been derived from spermatogonia (*67*). Evidence of translocation induction has only been sought in spermatogonia and here the situation is confused: The standard technique of screening spermatocytes for translocations coming through from spermatogonial stem cells has so far yielded clear negative results (*68*), but studies on mitotically-dividing spermatogonia within a few hours of treatment have revealed evidence of breakage and reunion (*69, 70*). According to Adler (*69*) the breakage and reunion tends to occur in the centric regions and, hence, exchanges between the acrocentric chromosomes of the mouse would not be expected to bring about multivalent associations in spermatocytes, the criteria normally taken as evidence for the presence of translocations. Such concept would be consistent with the observation that treatment of spermatogonia brings about a slight reduction of fertility after the sterile period has passed, for this could be attributed to dominant lethality generated from translocations present in spermatogonial populations (*66, 67*). However, Adler's (*69*) interpretation of the data is only valid for acrocentric chromosomes; with metacentric chromosomes exchanges in the centric regions would still lead to multivalent associations during meiosis. We have recently undertaken a test for translocation-induction by MC in a stock of mice carrying three pairs of metacentrics and have not observed any multivalents. The translocation yield with an X-ray control was similar to that obtained with ordinary mice. Thus, it would seem that either MC rearrangements are selectively eliminated before meiosis or the mitotic cells in which aberrations were detected (*69, 70*) were primarily spermatogonial types other than the stem cell proper.

The results with these three chemicals, I think, adequately demonstrates the variety of response that can be obtained with chemical mutagens in spermatogenic cells. The main difference from X-rays is that the sensitivity differences between the cell stages is far larger and with some chemicals these appear to be qualitative rather than quantitative, at least with the dose levels it is possible to use. The basis for the germ cell stage sensitivity differences with chemical mutagens is not understood, but it is possible that the relative ineffectiveness of chemicals upon the spermatogonial stem cell may be related to the fact that their action upon mitotically-dividing cells is limited to those cells undergoing DNA synthesis. Since the stem cell has a very long cell cycle time—about 8 1/2 days for the mouse (*71*)—most of which will be G_1, the effective target for chemical mutagens may be much smaller than that for X-rays (*72*).

Let us now look for evidence of mutagen specificity operating at the subcellular level. Such evidence clearly exists for chromosome aberrations induced in cultured somatic cells (*73*). Breakage is non-random both within and between chromosomes and tends to be localized in regions of heterochromatin. Natarajan and co-workers (*74, 75*) have shown that MC preferentially causes breaks in the constitutive heterochromatin of both the Chinese hamster (*Cricelius griseus*) and field vole (*Microtus agrestis*). In these species the heterochromatin is located both as large blocks in the sex chromosomes and in smaller amounts in the vicinity of the centromeres. Similar findings have been made in *Rattus natalensis* (*76*). More recent studies (*77*) with the tobacco mouse (*Mus poschiavinus*) have shown breakage to occur preferentially in the centric regions which, as in its close relative the laboratory mouse (*Mus musclus*), are the principal sites of constitutive heterochromatin. Adler's (*69*) finding that MC preferentially causes damage in the centric regions of spermatogonial chromosomes is thus fully consistent with these other studies, and further demonstrates that the specificity of MC action is not limited to somatic cells, i.e. it occurs in mammalian germ cells also.

Further data that might be taken as evidence of a localization of chemically-induced chromosome breaks comes from studies upon translocations recovered in F_1 animals. Male heterozygotes for most X-ray-induced translocations are fertile and meiotic studies show the most common multivalent associations to be rings-of-four (*47*). Male heterozygotes for some translocations are completely sterile, however, and meiotic and mitotic studies have demonstrated that in these cases the exchanges have been asymmetrical (*78, 79*). Extra large and/or ultra-small rearranged chromosomes may be detected and, at meiosis, these most often form chains-of-four or other forms of incomplete pairing with the normal homologues. Sterility results from a breakdown in meiosis. Following both TEM (*52*) and EMS (*69*) treatments, F_1 male sterility occurs much more commonly than following X-irradiation and the presence of a translocation has been shown to be the most probable cause of the sterility in many cases. This suggests that these chemicals are causing more breakage near the ends of chromosomes than X-rays. Perhaps, like MC, breakage tends to occur at or near the centromeric regions and the failure of such translocations to survive meiosis provides one reason for their near-absence following the chemical-treatment of spermatogonia.

There is also some evidence of locus specificity in the mouse. Point mutations are most commonly detected by a method which screens F_1 animals for changes at seven selected loci. Data for spermatogonial mutation is available only for TEM (*57*) and MC (*67*) and, although limited, both strongly suggest that the distribution of mutations among the seven loci differs from that following X- or γ-rays (*80*) (Table 2). Together, TEM and MC have yielded three *a* locus mutations, which are extremely rare following radiation treatments, and none of the very frequently occurring *s* locus mutations. It is tempting to relate this difference to the near-absence of chromosome changes from chemically treated spermatogonia. The implication would be that most of the *s* locus mutations produced by radiation represent chromosomal rather than genic changes.

TABLE 2
Distribution of point mutations induced by chemicals and ionizing radiations at seven tested loci

	Locus							
Treatment	a	b	c	p	d	d and se	se	s
TEM[a]	2	0	0	2	3	0	0	0
MC[b]	1	0	1	0	2	0	0	0
X- or γ-rays[c]	5	45	27	38	42	1	3	108

[a]Data of Cattanach (1966); [b]Data of Neuhauser and Ehling (1973); [c]Data of Russell (1964).

Finally, we come to the question of whether or not delayed mutation occurs in mammalian species. Definite evidence one way or another is lacking. In *Drosophila,* as described earlier, evidence of delayed mutation comes from the appearance of mosaic files among the F_1 or subsequent generations (*28–31*). In the mouse, mosaicism should be most readily recognized in point mutation tests upon post-meiotic germ cell stages. Since very few complete mutations have been recovered from this type of test (*62*), the data are too limited to be meaningful. However, no case of mosaicism has yet been reported.

The low ratio of translocations to point mutations obtained with chemical mutagens in *Drosophila* post-meiotic cells (*32, 33*) is also thought to be a consequence of delayed mutation. Aside from the shortage of point mutation data in the mouse, comparisons with the radiation data are made extremely difficult by such problems as germ cell stage sensitivity differences and the lack of a common level of any one response that could be used as a standard. However, as the data stand, they provide little reason to suspect that the mouse responds in a similar way to *Drosophila*. It is evident that some chemicals cause high frequencies of translocations, e.g. about 20% for given doses of

TEM and EMS (*52, 59*) but the point mutation frequencies observed—5/6500 for both TEM and EMS (*62*)—are probably no higher than would be expected with X-ray doses (about 400 rads) giving the same translocation yields (*79*). It is from spermatogonial cells that we find a shortage or absence of translocations with chemical mutagens (*55,56,63,68*) and we do not know at this time to what extent this might result from selection or from a failure of chemicals to produce exchanges in this cell type.

A further criterion for delayed mutation that could be considered is the ratio of dominant lethals to translocations. If as in *Drosophila* (*33, 35, 36*) chemically-induced breaks tend not to rejoin, then dominant lethality will be high relative to the translocation frequency. Comparisons of the chemical mutagen and radiation data are again difficult for the reasons described above, but from the rough estimates shown in Table 3 it can be seen that there is no reason to suspect that chromosome breaks produced by chemicals in the mouse are less liable to rejoin than those produced by X-rays.

TABLE 3
The ratio of dominant lethals to translocations following X-ray and chemical mutagen treatments

Treatment	Matings post-treatment (days)	Dominant lethals % (DL)	Translocations % (T)	Ratio DL:T
X-rays[a]	1–7	23	6.8	3.4:1
	8–14	23	10.4	2.2:1
	15–21	42	28.3	1.5:1
TEM[b]	1–3	27	8.1	3.3:1
	10–14	72	29.7	2.4:1
EMS[c]	1–3*	4	10.0	0.4:1
	4–5*	75	13.6	5.5:1
	6–7	74	50.0	1.5:1
	8–11*	78	27.6	2.8:1

[a] Data of Leonard and Deknudt (1968); [b] Data of Cattanach (1959); [c] Data of Cattanach et al., (1968);
* Dominant lethals and translocations scored in separate experiments.
Dominant lethality assessed as 1 − (live embryos in treated series/live embryos in control) × 100.
Translocations frequency based on assumption that sterile F_1 males carry a single translocation unless demonstrated otherwise.

To conclude, we can say that there is good evidence of chemical mutagen specificity operating in the mouse at the level of strain, sex, cell stage, chromosomal region and possibly even at the gene locus. So far, we do not have definitive evidence concerning delayed mutation. The implications of the specificity of chemical mutagens must be taken into consideration when assessing the genetic risks of environmental mutagens.

REFERENCES

1. I. I. Oster, Z. Vereblehre **89,** 1 (1968).
2. M. Vogt, *Publ. Staz. zool. Napoli.* **Suppl. 22,** 114 (1950).
3. F. Oehlkers, *Z. Vereblehre* **81,** 313 (1943).
4. K. A. Jensen, I. Kirk, G. Kolmark and M. Westergaard, *Cold Spring Harbor Symp. Quant. Biol.* **16,** 245 (1951).
5. C. Auerbach and H. Moser, *Z. indukt. Abstramm.- u. Vererblehre* **85,** 479 (1953).
6. H. J. Belitz, *Z. indukt. Abstramm.- u. Vererblehre* **88,** 434 (1957).
7. O. G. Fahmy and M. J. Fahmy, *J. Genet.* **54,** 146 (1956).
8. J. McLeish, in *Symposium on Chromosome breakage. Heredity* **Suppl. 6,** 125 (1953).
9. A. J. Natarajan and M. D. Upadhya, *Chromosoma (Berlin)* **15,** 156 (1964).
10. S. H. Revell, Chromosome breakage by X-rays and radiomimetic substances in *Vicia.* in *Symposium on Chromosome breakage. Heredity.* **Suppl. 6,** 107 (1953).
11. W. C. J. Ross, *Nature* **165,** 808 (1950).
12. H. K. Jain and P. T. Shukla, *Mutat. Res.* **14,** 440 (1972).
13. L. Ehrenberg, A. Gustafsson and U. Lundquist, *Hereditas,* **47,** 243 (1961).
14. A. Gustafsson, *Hereditas* **50,** 211 (1963).
15. M. Jacobs, *Radiat. Bot.* **9,** 251 (1969).
16. U. Lundqvist, P. Von Wettstein-Knowles and D. Von Wettstein, *Hereditas* **59,** 473 (1968).
17. J. Mackey, *Acta Agr. Scand.* **4,** 419 (1954).
18. L. M. Mouti, *Mutat. Res.* **5,** 187 (1968).
19. C. Auerbach and D. Ramsay, *Micro. Genet. Bull.* **26,** 3 (1967).
20. G. Kølmark, *Hereditas* **39,** 270 (1953).
21. T. Alderson and C. Scazzocchio, *Mutat. Res.* **4,** 567 (1967).
22. C. Auerbach and B. J. Kilbey, *Ann. Res. Genet.* **5,** 163 (1971).
23. C. Auerbach, *Proc. Roy. Phys. Soc.* **XXIX,** 1 (1966).
24. C. Auerbach, in *CIBA Foundation Symp. on Mutation as Cellular Process* (Ed. G. E. W. Wolstenholme and M. O'Conner, Eds.), pp. 66–78. J. and A. Churchill Ltd., 194, Gloucester Place, London W.1., 1969.
25. C. Auerbach, *Trans. Kansas Acad. Sci.* **72,** 273 (1970).
26. C. Auerbach, *Environmental Mutagen Society Newsletter* **4,** 16 (1971).
27. A. Nasim, *Mutat. Res.* **4,** 753 (1967).
28. C. Auerbach, *Proc. Roy. Soc. (Edin.)* **B62,** 211 (1946).
29. L. S. Browning and E. Altenburg, *Genetics* **47,** 945 (1962).
30. E. A. Carlson and J. L. Southin, *Genetics* **48,** 663 (1963).
31. C. Mathew, *Mutat. Res.* **1,** 163 (1964).
32. H. Slizynska, *Mutat. Res.* **8,** 165 (1969).
33. G. E. Nasrat, W. D. Kaplan and C. Auerbach, *Z. indukt. Abstramm.-u. Vererblehre* **86,** 249 (1954).
34. J. K. Lim and L. A. Snyder, *Mutat. Res.* **6,** 129 (1969).
35. O. G. Fahmy and M. J. Fahmy, *J. Genet.* **52,** 603 (1954).
36. R. J. Sram, *Mol. Gen. Genet.* **106,** 286 (1970).
37. W. M. Generoso, *Genetics* **61,** 461 (1969).
38. W. M. Generoso and W. L. Russell, *Mutat. Res.* **8,** 589 (1969).
39. A. Leonard and G. Deknudt, *Canad. J. Genet. Cytol.* **8,** 520 (1966).
40. A. J. Bateman and S. S. Epstein, in *Chemical Mutagens. Principles and Methods for their Detection.* (A. Hollaender, Ed.) Vol. II. Ch. 21, pp. 541-567. Plenum Press, New York, 1971.
41. A. J. Bateman, *Heredity* **12,** 213 (1958).
42. C. Auerbach and B. M. Slizynski, *Nature* **177,** 376 (1956).
43. A. Leonard and G. Deknudt, *Can. J. Genet. Cytol.* **10,** 495 (1968).
44. L. B. Russell and C. L. Saylors, *Genetics* **47,** 7 (1962).
45. F. H. Sobels, *Proc. 12th Int. Congr. Genet.* **3,** 205 (1969).
46. E. F. Oakberg, *Mutat. Res.* **11,** 1 (1971).
47. A. Léonard, *Mutat. Res.* **11,** 71 (1971).
48. A. G. Searle, *Adv. Rad. Biol.* **4,** 131 (1974).
49. W. L. Russell, J. W. Bangham and J. S. Gower, *Proc. Tenth Int. Conf. Genetics* **2,** 245 (1958).

50. A. J. Bateman, *Genet. Res.* **1,** 381 (1960).
51. B. M. Cattanach and R. G. Edwards, *Proc. Roy. Soc. Edinburgh Sect. B.* **67,** 54 (1958).
52. B. M. Cattanach, *Z. Vererblehre* **90,** 1 (1959).
53. B. M. Cattanach, *Mutat. Res.* **4,** 73 (1967).
54. J. Moutschen, *Genetics* **46,** 291 (1961).
55. B. M. Cattanach and C. E. Pollard, *Mutat. Res.* **13,** 371 (1971).
56. I. D. Adler, *Mutat. Res.* in press (1974).
57. B. M. Cattanach, *Mutat. Res.* **3,** 346 (1966).
58. B. M. Cattanach, *Mutat. Res.* in press (1974).
59. B. M. Cattanach, C. E. Pollard and J. H. Isaacson, *Mutat. Res.* **6,** 297 (1968).
60. U. H. Ehling, R. B. Cumming and H. V. Malling, *Mutat. Res.* **5,** 1 417 (1968).
61. W. M. Generoso, W. L. Russell, S. W. Hoff and S. K. Stout, *Mutat. Res.* **21,** 32 (1973).
62. B. M. Cattanach, Specific locus mutations in mice. in *Chemical Mutagens. Principle and Methods for their Detection.* (A. Hollaender, Ed.) Vol. II. Ch. 20, pp. 535–539. Plenum Press, New York, 1971.
63. A. Léonard, G. Deknudt and G. Linden, *Mutat. Res.* **13,** 89 (1971).
64. U. H. Ehling and W. L. Russell, *Genetics* **61, Suppl. 2, part 2,** 14 (1969).
65. R. B. Cumming and M. F. Walton, *Mutat. Res.* **10,** 365 (1970).
66. U. H. Ehling, *Mut. Res.* **13,** 433 (1971).
67. A. Neuhauser and U. H. Ehling, *Mutat. Res.* **21,** 198 (1973).
68. N. Gilliavod and A. Leonard, *Mutat. Res.* **13,** 274 (1971).
69. I. D. Adler, *Mutat. Res.* **21,** 184 (1973).
70. E. Schleiermacher, *Mutat. Res.* **21,** 47 (1973).
71. E. F. Oakberg, *Mutat. Res.* **11,** 1 (1971).
72. B. M. Cattanach, in A. G. Searle, *Mutat. Res.* **11,** 133 (1971).
73. F. Funes-Cravioto, K. N. Yakovienko, N. P. Kuleshov and V. S. Zhurkov, *Mutat. Res.* **23,** 87 (1974).
74. A. T. Natarajan and W. Schmid, *Chromosoma (Berlin)* **33,** 48 (1971).
75. A. T. Natarajan, G. Ahnström and R. P. Sharma, *Mutat. Res.* **22,** 73 (1974).
76. C. C. Huang, *Chromosoma (Berl.)* **23,** 162 (1967).
77. A. T. Natarajan and T. Raposa, *Mutat. Res.* in press (1974).
78. M. F. Lyon and R. Meredith, *Cytogenetics* **5,** 335 (1966).
79. A. G. Searle, C. E. Ford, E. P. Evans, C. V. Beechey, M. D. Burtenshaw and H. M. Clegg, *Mutat. Res.* **22,** 157 (1974).
80. W. L. Russell, in *Genetics Today,* Proc. XI Int. Cong. Genetics, The Hague, Netherlands, 1963; pp. 257–264 (1964).

THEORETICAL APPROACHES TO RADIATION BIOLOGY[1]

Chairman: A. M. Kellerer

[1]Symposium B–4

Biophysical Implications of Radiation Quality[1,2]

Harald H. Rossi
Radiological Research Laboratories
Department of Radiology
College of Physicians & Surgeons of Columbia University
630 West 168th Street, New York, N.Y. 10032, USA

Almost as soon as the biological effects of ionizing radiation were discovered, attempts were made to explain some of their basic characteristics in terms of the physical process of energy absorption in the irradiated tissues. The classical target theory and a variety of other approaches, although occasionally useful in simple situations, have had little success when effects on higher organisms were under consideration. It seems apparent that such failures were due to the fact that many of the radiobiological experiments were designed from a phenomenological point of view; that the radiations employed were usually X or gamma rays which have complex inactivating mechanisms; and that the physical concepts employed in the analysis (such as absorbed dose and LET) were inadequate. The results of systematic experiments in which the effects of neutrons and electromagnetic radiations were compared coupled with the development of microdosimetry have in recent years resulted in advances which are the subject of this presentation.

Perhaps the most important biophysical attribute of neutrons (particularly those having energies between 0.1 and 1 MeV) is that their secondaries can singly injure and kill individual cells. On the basis of microdosimetric determinations it can readily be concluded that, if cells of higher organisms are irradiated by neutrons in this energy range, their nuclei will on the average be traversed by one neutron secondary (usually a proton) when the absorbed dose is typically 25 rads. It follows that, at least at doses that are substantially less than 25 rads, the number of cells traversed must be proportional to dose and that consequently the frequency of any cellular radiation effects must also be proportional to dose provided that there is no interaction between irradiated cells. Conversely, if one finds that the incidence of some biological effect is not proportional to dose at such low traversal frequencies, it follows that the effect depends in some way or other on an interaction of several cells.

This situation is illustrated in Fig. 1 which deals with the induction of mammary neoplasms in the rat as observed by Shellabarger et al. *(1)*. Because of the large range of doses covered by this plot, the axes are logarithmic. In such a representation a linear (proportional) relation between the two variables of incidence and absorbed dose appears as a line of slope +1 (i.e. one that has an

[1] Congress presentation B–4–1.

[2] This investigation was supported by contract AT–(11–1)–3243 for the U.S. Atomic Energy Commission and Research Grant Ca 12536 from the National Cancer Institute.

angle of 45° with respect to either axis). It is evident that for neutrons the slope is considerably less than 1 at doses that range as low as 100 millirads. It may, therefore, immediately be concluded that the induction of these tumors must be a complex process in that it is not due to the transformation of independent individual cells, but must arise by a mechanism in which several cells are involved.

At the lowest tumor incidences covered in Fig. 1 only about one in 250 cell nuclei is traversed by a neutron secondary. On the other hand, since the

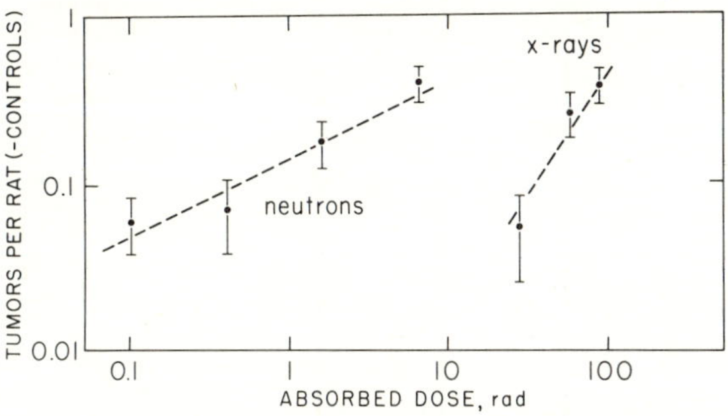

FIG. 1. Incidence of mammary neoplasms in the Sprague-Dawley rat as a function of dose of 430 keV neutrons and 250 kV X rays as observed by Shellabarger et al. *(1)*.

RBE is about 100 and since the energy deposited in an electron traversal is about 100 times less than in a proton traversal, the frequency of electron traversals is 10,000 times larger than that of proton traversals at low effect levels. Thus, each nucleus is traversed by some 40 electrons, a number that is so large that the specific energy *(2)*, z, is not likely to differ appreciably from the absorbed dose, D. This is plainly not so in the case of neutrons where at an absorbed dose of 100 mrads nuclei receive a specific energy that is usually zero, but, if not zero, on the order of 25 rads.

With either radiation the affected cells should be essentially randomly distributed throughout the irradiated tissues. It would seem plausible that once either X rays or neutrons have affected equal numbers of cells, the ensuing processes depend no longer on radiation quality. Support for this belief can be derived from the fact that the temporal pattern of tumor incidence is the same for both radiations and that there is no evidence that the two pathologically distinct types of tumors observed (fibromas and adenocarcinomas) are produced in different proportions by the two radiations. Thus, a comparison of the relative effectiveness with which X rays and neutrons induced mammary neoplasms is likely to be a comparison of the effectiveness with which they induce cellular damage regardless of the complexity of the process of tumor induction.

The relative biological effectiveness (RBE) is defined as the ratio of the absorbed doses of reference radiation and radiation of interest that are required for equal biological effect. From the foregoing remarks one may conclude that if one considers the RBE of neutrons relative to X rays, one is in fact considering the biological effectiveness of various doses of X rays because it is apparent that the number of cells affected must be strictly proportional to the neutron dose (in the dose range under discussion) and that the RBE must thus be an indication of the absorbed dose of X rays required per cell affected. Fig. 2 is a logarithmic representation of the RBE versus neutron dose and it shows

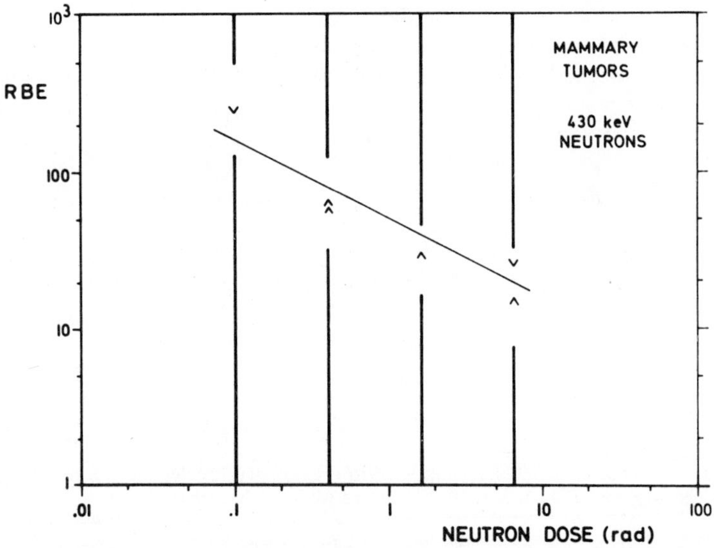

FIG. 2. The dependence of RBE on neutron dose for the induction of mammary neoplasms in the rat. The vertical bars represent values that are excluded with 95% confidence; the arrow heads represent limits that are excluded with lesser confidence. (Statistical tratment by Kellerer and Brenot *(12)*).

that the X ray dose per inactivated cell decreases with increasing numbers of affected cells. It also indicated that this relation is a very simple one since, over a one hundred fold range of neutron dose, the function has a slope that must be quite close to − 0.5. It may readily be shown *(3)* that this is equivalent to the statement that the number of cells affected must be proportional to the square of the absorbed dose of X rays. This surprisingly simple result applies not only to the induction of mammary neoplasms in the rat but also to the induction of lenticular opacities in the mouse *(4)*, to somatic mutations in Tradescantia *(5)* and Zea mays *(6)*, the frequency of chromosome aberrations in human lymphocytes *(7)*, and quite probably to the induction of leukemia in humans *(8)*.

As Kellerer and I have pointed out in our formulation of the *theory of dual radiation action (9)*, this relation is consistent with the assumption that these and many other biological effects of radiation are due to second order reactions in the sense that pairs of what we term *sublesions* in *loci* combine to produce the *lesions* responsible for cellular radiation effects. In an analysis which we have presented elsewhere *(9)* we have shown that if neutron secondaries and electrons are equally effective in producing these sublesions, the RBE of neutrons relative to X rays should approach 1 at large doses where multiple track inactivation operates not only for electrons but also for neutrons and that it should reach a certain ratio at extremely low doses where single trace inactivation must operate not only for neutron secondaries but for X and γ ray secondaries as well. This ratio should be the ratio of dose mean specific energies, ζ, for individual traversals by the secondaries of the radiations. ζ depends on the diameter of the site volume, i.e. the region in which the postulated second order reaction takes place. It can be shown (*9*) that the site volume although smaller than the nucleus is of the same order of magnitude. For such dimensions the *ratio* of ζ values for neutrons and X rays is largely independent of size. In the case of neutrons under consideration (which have an energy of 430 keV) and electrons secondary to 250 kVp X rays the ratio should be about 50.

The fact that the RBE apparently exceeds 100 in the case under discussion (and in at least one more) implies that energy concentration on an entirely different scale also influences biological effectiveness. The most obvious significant region is the locus which we believe to have dimensions on the order of nanometers. At this level substantial differences in energy deposition can occur at high LET when the delta rays produced in primary ionizations are separated by distances comparable to their range. The domains under discussion here have dimensions comparable to those of the DNA molecule, and energy transfer over such minute distances probably is affected by the chemical environment. Thus, the action of sensitizing and protective substances may well operate at this level. These considerations indicate the need for physical and biological research designed to investigate the nature and the effects of energy concentration at the nanometer scale.

Some of the effects that appear to follow the postulates of the theory of dual radiation action are chromosome aberrations or are known to be due to chromosome changes. In this connection it is of especial interest that the kinetics of many *(10)* if not all *(11)* chromosome aberrations are virtually identical to those postulated by our theory. Investigations exploring close correlations between radiation effects on chromosomes and damage recognized in other forms must therefore be another important objective of future research.

REFERENCES

1. C. J. Shellabarger, R. D. Brown, A. R. Rao, J. P. Shanley, V. P. Bond, A. M. Kellerer, H. H. Rossi, L. J. Goodman, and R. E. Mills, in *Biological Effects of Neutron Irradiation* (IAEA-SM-179/26), pp. 391-401. International Atomic Energy Agency, Vienna, 1974.
2. International Commission on Radiation Units and Measurements, *Radiation Quantities and Units.* ICRU Report **19** (1971).
3. H. H. Rossi, *Phys. Med. Biol.* **15**, 255 (1970).
4. J. L. Bateman, H. H. Rossi, A. M. Kellerer, C. V. Robinson, and V. P. Bond, *Radiat. Res.* **51**, 381 (1972).
5. A. H. Sparrow, A. G. Underbrink, and H. H. Rossi, *Science* **176**, 916 (1972).
6. H. H. Smith, N. C. Combatti, and H. H. Rossi, in *Neutron Irradiation of Seeds II*, pp. 3-8. International Atomic Energy Agency, Vienna (1968).
7. M. T. Biola, R. LeGo, G. Ducatez, and M. Bourguignon, in *Advances in Physical and Biological Radiation Detectors*, pp. 633–645. International Atomic Energy Agency, Vienna, 1971.
8. H.H. Rossi and A. M. Kellerer, *Radiat. Res.* **58**, 131 (1974).
9. A. M. Kellerer and H. H. Rossi, *Curr. Top. Radiat. Res. Quart.* **8**, 85 (1972).
10. K. Sax, *Genetics* **23**, 494 (1938).
11. S. H. Revell, in *Radiobiology Symposium 1954* (Z. M. Bacq, and P. Alexander, Ed.), pp. 243–253. Butterworths Scientific Publications, London, 1955.
12. A. M. Kellerer and J. Brenot, *Radiat. Environ. Biophys.* **11**, 1 (1974).

Analysis of Radiation-Induced Chromosome Aberrations[1]

J. M. Brenot and N. Parmentier
Service de Protection Sanitaire, Départment de Protection
Commissariat a l'Energie Atomique, Fontenay aux Roses, France

INTRODUCTION

Ionizing radiation produces lethal and mutagenic effects on the cell. Among these effects, chromosome damage has been studied in particular detail; indeed, the chromosome anomalies represent changes in the cell's genetic information and it is believed that the cell's ability to survive and function is affected by them. However, the consequences of chromosome anomalies and their relationship to manifest radiation effects are not as yet well understood. Recently, some experiments on potentially lethal radiation damage in stationary phase cell cultures suggest that extrachromosomal mechanisms may be involved in the loss of proliferative ability. Little (*1*), for example, assumes that a mechanism related to the cell membrane could play a role in repair processes.

The study of radiation-induced chromosome anomalies has always been an important topic in experimental and theoretical radiobiology. The reason for this has been that one deals with a well-defined effect which is closely related to the primary sensitive structures in the cell. Moreover, such studies are simplified by the wide use of the peripheral blood lymphocyte culture method and by the fact that anomalies in the chromosome number and/or chromosomal structural changes are readily visible under an ordinary microscope if the cells are fixed in metaphase. Because of the theoretical as well as practical implications of the studies, an attempt will be made in the following to survey the experimental techniques, the biophysical implications, and the statistical analyses which are relevant in this context.

Structural changes of chromosomes are of different types according to the stage of the mitotic cycle during which the cell is irradiated. In particular, a distinction is made between chromosome aberrations (cells irradiated in G1 or early S-phase) and chromatid aberrations (cells irradiated in late S or G2). This paper is concerned with chromosome aberrations, a classification of which was given by Lea (*2*). Some types are easily recognizable under the microscope, for example, fragments which include terminal and interstitial deletions, dicentrics, centric and acentric rings. Other types, such as translocations and inversions classified as symmetrical exchanges, are not in general observable, and the setting up of the karyotype by classical method—culture and staining—is not precise enough for a good study of these structural rearrange-

[1]Congress presentation B-4-2.

ments. It is therefore of interest to consider some new experimental developments which may remove some of these experimental limitations.

PROGRESS IN OBSERVATIONAL METHODS

Autoradiography was a first improvement; it consists of introducing into the cell culture tritiated thymidine which is incorporated into the DNA during the S-phase. One observes characteristic replication patterns on a photographic emulsion. This technique was employed by Norman and Sasaki (*3*) when they studied, with human lymphocytes, how a given chromosome was involved in the formation of dicentrics. The technique is not easy to use and induces a certain, albeit rather low, rate of aberrations as artefact.

The real progress came in 1968 with the findings of Caspersson et al. (*4–7*). They realized that DNA-binding fluorescent agents can be used to indicate differentiations in the chemical structure along metaphase chromosomes of plants and animals. Under the fluorescence microscope each chromosome presents a characteristic and reproducible pattern of fluorescent bands. In this Q-banding[2], strongly fluorescent bands coincide with heterochromatic regions. Numerous other techniques have been proposed since 1970 after the observation of Pardue and Gall (*8*) of the differential staining properties of Giemsa in mouse chromosomes when submitted to a process of denaturation and renaturation of DNA. The methods differ in details of the denaturing and annealing treatments and in the denaturing agents. Gall and Pardue (*9*) as well as Arrighi and Hsu (*10*) propose a banding which gives precise information in the centromeric area[3]. Sumner et al. (*11*), Drets and Shaw (*12*), Patil et al. (*13*), Schnedl (*14*), Lomholt and Mohr (*15*), Seabright (*16*), Shiraishi and Yoside (*17*) obtain a G-banding[4] pattern which is remarkably similar to Q-banding: the more intensely stained G-bands correspond to the more fluorescent Q-bands. A reverse banding has been obtained by Dutrillaux and Lejeune (*18*). With all of these techniques, the resolution depends to some extent on the state of contraction of the chromosomes. The various techniques do not give exactly the same information, but each one allows a full identification of the chromosome.

Automatic karyotyping has been developed mainly by Caspersson et al. (*19–21*). Fluorescent intensities in different chromosome parts are measured with a high resolution fluorometer and the fluorescence curves can be analyzed by computer. Such automatic analysis is not as yet highly developed but TV-based procedures for quick identification of chromosomes and chromosome regions are being utilized.

The banding has been promptly employed for the analysis of radiation-induced aberrations. A limited study by Caspersson et al. (*22*) with human lymphocytes exposed to X-rays indicates the possibilities offered by the banding technique, namely the possibility to recognize the various types of symmetrical and asymmetrical exchanges and therefore to determine their relative frequencies, and the analysis of the position at which breaks and exchanges occur. Breaks lie in interband regions which are considered to be euchromatic. Seabright (*23*) in accordance with previous findings by Norman and Sasaki (*3*)

[2]The fluorescent agent is Quinacrine
[3]This banding is called C-banding
[4]G for Giemsa stain

has confirmed this result. The distribution of breaks among the autosomes is related to their relative amount of interband material or to their relative length; the unexpected deficiency of aberrations in/or involving both the X and Y chromosomes indicates the possibility of nonrandom involvment of some chromosomes (Seabright (*24*), San Roman and Bobrow(*25*)).

Comparing the many banding techniques, and each month brings a new one, one can observe that the resulting patterns are very similar despite wide differences in procedure. There is, by now, general agreement that banding patterns are not chemical artefacts but that they reflect chromosomal structure. They allow recognition of particular chromosome characteristics and chromosome identification. This explains why banding has brought considerable progress to radiation cytogenetics.

MODELS FOR ABERRATION PRODUCTION

The "breakage first hypothesis" was proposed by Stadler (*26*) in 1931 to explain the formation of aberrations. The hypothesis states that the passage of an ionizing particle—an event—through a chromatin thread or its vicinity produces a physical break—called primary break—in the interphase chromosome. These breaks can remain open and thereby result in the terminal deletions. Or they can rejoin in one of two ways, either restituting or reuniting with other breaks and thereby resulting in exchanges (two-break aberrations). This hypothesis has been challenged by Revell (*27–31*) who proposed the "exchange hypothesis." It suggests that an event in a chromosome or in its vicinity causes a localized activation, which is not an actual break; all chromosome aberrations result from the initiation of an exchange between two activated points. According to the "exchange hypothesis," terminal deletions are products of incomplete exchanges. This hypothesis has been discussed by Evans (*32*), Scott and Evans (*33*), Brewen and Brock (*34*), and Conger (*35*).

In the following we will refer to the "breakage first" hypothesis which at present is the most widely accepted. The main arguments for breakage followed by reunion, come from the experiments of Sax (*36–39*) on *Tradescantia* microspores irradiated by X-rays. The one break class—i.e. terminal deletions—is dependent only on breakage and its yield increases linearly with dose and is dose rate independent. The exchanges increase approximately with the square of the absorbed dose, as one has to expect if they are produced by two independent breaks—i.e. two breaks induced by two independent ionizing tracks. Therefore they are dose rate dependent. However, the "two-track" formation of exchanges is not the only possible mechanism: experiments at low dose rate have shown that one deals also with a linear dependence on dose. This led Catcheside et al. (*40*) to propose that some two-break aberrations are produced by a single ionizing track; then the relation between yield, $\epsilon(D)$ and dose D is linear-quadratic:

$$\epsilon(D) = \alpha D + \beta D^2 \qquad (1)$$

The spontaneous rate is neglected, αD and βD^2 respectively are the "one-track" and the "two-track" contributions to the yield. The relation holds for low LET radiation and at intermediate and high doses the "two-track" component is the dominating term. For high LET radiation the relation is still valid, but the "one-track" term is dominant. This is easily understood because it must be assumed that densely ionizing particle tracks can cause a large number of breaks in close vicinity (Giles (*41*, *42*)). Accordingly, the "one-track" term αD is proportional to LET and is independent of dose rate and dose fractionation. In both modes of formation of exchange aberrations one deals with a distance factor. This factor exhibits itself in the resulting non-randomness in the rejoining of the broken ends of chromosomes. This was first realized by Sax when he compared the number of dicentrics to the number of rings. The relative excess of rings is explained by the fact that breaks close to each other have an increased probability to rejoin.

Classical Models

Target theory considerations have been used by Lea (*2*) to explain that the linear-quadratic relation between dose and yield of exchanges results from two basic mechanisms: an "intratrack" action (one-track or single-event or one-hit) and an "intertrack" action (two-track or two-event or two-hit). One deals with the following situation: the primary breaks occur randomly in the nucleus and some of the breaks are able to join other breaks—i.e. they are called joinable breaks. The probability for two joinable breaks to exchange depends on their distance. The concept of rejoining distance *d*, which determines the size of the target, is often based on the simplifying assumption that this probability is constant up to the distance *d* and zero for larger distances. This is the so-called "distance model." The equation $\epsilon(D) = \alpha D + \beta D^2$ appears to be an approximation of the equation obtained on the previous basis when the mean number of joinable breaks inside a target for a dose D is low. With this formulation, numerous attempts have been made to determine the distribution, the size and the number of targets within the cell in order to correlate such postulated targets with defined structures in the cell.

In a series of papers, Wolff et al. (*43*), Wolff (*44, 45*), Atwood (*46*), found in *Tradescantia* microspores and *Vicia faba* seeds that the relative frequencies of cells with multiple exchanges were different from the frequencies expected under the hypothesis of a random distribution of breaks. For this reason, they proposed the "site-model." It postulates that the two-break aberrations do not occur randomly throughout the nucleus but that there is only a limited number of sites where two chromosomes are close enough to produce an exchange when broken. Then, the yield of exchanges induced by intratrack and intertrack mechanisms involves the "site number," N and is given by the following equation:

$$\epsilon (D) = N E \cdot \left\{ 1 - e^{k_2 D} \cdot \left[1 - \left(1 - e^{-k_1 D} \right)^2 \right] \right\} \quad (2)$$

where E is the probability to score the exchange in the site, k_1 is the mean number of breaks in a site per rad and k_2 is a constant. When $k_1 D$—the mean number of breaks in a site at dose D—is low, Eq. (2) reduces to Eq. (1) with: $\alpha = k_2$ and $\beta = k_1^2$. The site model some difficult questions about he site number and currently two positions have been chosen. The first one, chosen by Read (*47*), and by Norman and Sasaki (*3*) rejects the assumption of a limited number of sites and states that the number of dicentrics and centric rings per cell is only limited by the number of centromeres. The second position taken by Neary et al. (*48*), Savage (*49*), Savage and Papworth (*50, 51*) is that the number of sites per cell is variable within a population of cells and N is considered as a parameter of its distribution.

A compromise between the distance model and the site model is proposed in an important paper of Neary (*52*). He assumes that a primary damage (or sublesion) to a micromolecular target is produced by a single energy loss along the track of a charged particle; a site for exchange initiation contains two lengths of either the same or different chromosomes which are the two damaged targets. The exchange then occurs at the site with a given probability. One notes that the sublesion is not specified. The yield of exchanges is again of the form of Eq. (2) but the constants k_1 and k_2 are linked explicitly to the geometry of the targets at the site and to the linear energy transfer (LET) of the radiation. In this treatment too, Eq. (2) reduces to Eq. (1) if the mean number of sublesions produced in the target by one particle is low, that is, $\ll 1$.

Microdosimetric Treatment of the Dual Radiation Action

The classical models are subject to essential limitations. One of the main limitations is inherent in the use of LET or of one of its average values. The quantity LET is according to its definition merely a mean value (1970 ICRU, *53*). It disregards the statistical fluctuations of energy deposition along the particle tracks and the finite length of the tracks, and, as discussed by Rossi and Rosenzweig (*54*) and by Rossi (*55–57*), the concept is therefore of limited applicability. A general limitation is that, in these models, the primary damage—breaks or sublesions—which must depend on the magnitude and frequency of local energy depositions in microscopic regions, is related to quantities such as absorbed dose or LET which are merely average values and which do not fully account for such microscopic fluctuations.

For a more rigorous treatment, one must study the dependence of the yield of dual lesions—chromosome exchanges—on the actual local concentration of absorbed energy. For this purpose the microdosimetric quantity, specific energy z, has been introduced. This quantity, absorbed energy divided by the mass of the reference volume, is the statistical variable whose mean value is the absorbed dose (1971 ICRU, *58*). In the present context it is neither possible nor necessary to consider the various quantities relevant to microdosimetry (*59, 60*); however an attempt will be made to indicate the main arguments and the main results which are relevant to the analysis of chromosome aberrations. The experimental curve, mean yield of exchanges per cell vs dose, is a graphic representation of a relation between two expectation values, namely that of the yield of exchanges per cell and that of the specific energy z. Ideally, it would be desirable to obtain the relation between the two statistical variables directly but there is no experimental way to measure the specific energy in the cell's nucleus itself. Therefore, one can only reach indirect conclusions concerning the dependence of the yield of chromosome aberrations on the energy actually absorbed in the nucleus of the cell or in certain of its sensitive substructures. Such indirect conclusions must be deduced from the analysis of the dose-effect curves.

The fluctuations of the specific energy at a given absorbed dose D are described by the probability distributions $f(z;D)$. These distributions depend on the shape and the size of the reference volume and on the radiation quality. They can be experimentally obtained and have, at least in some cases, also been derived theoretically. For the smallest volumes and for the most densely ionizing radiations the fluctuations are greatest. For large volumes and for large doses of sparsely ionizing radiations the values of specific energy are very close to the values of the absorbed dose. It is known that the fluctuations of specific energy are about twice as large for low and medium energy X-rays than for γ-rays or fast electrons. However, at intermediate doses of a few hundred rad it has been found that most cellular effects of γ-rays and X-rays differ only slightly, and one may therefore conclude that for these radiations and for the dose range of hundred to several hundred rad the fluctuations of absorbed energy are irrelevant to the effect. This means that the site where the exchange occurs is large enough so that the fluctuations of the specific energy z may be neglected in these cases. Then the dependence of the yield of chromosome aberrations on specific energy can be identified with the observed dependence on absorbed dose. Since in the dose range of a few hundred rad one obtains quadratic dose-reponse functions for the production of dicentrics by sparsely ionizing radiations, one may conclude that one deals with a quadratic dependence of the effect on specific energy:

$$\epsilon(z) = k\, z^2 \tag{3}$$

The constant k is supposed to have the same value for all radiation qualities. In fact, it is possible that the submicroscopic distribution of energy within the site influences the coefficient k (Bateman et al., *61*). But actually no definite statement about this dependence on radiation quality can be made.

While the dependence of the effect on absorbed dose is very close to the quadratic dependence for sparsely ionizing radiations at sufficiently high doses, one must expect to find deviations for densely ionizing radiations and also for sparsely ionizing radiations if one deals with very small doses. Such deviations must be expected because the mean yield observed at a certain absorbed dose represents the integral response to a spectrum of values of specific energy z which actually occur at this dose. It has been shown that by averaging Eq. (3) over the probability distribution $f(z;D)$ one obtains the linear-quadratic dependence on absorbed dose:

$$\epsilon(D) = k\,(\zeta D + D^2) \tag{4}$$

Zeta is the "energy mean" of the increments of specific energy produced in the site by an event, i.e. the average increment of z produced by a single charged particle crossing the site. This microdosimetric quantity has the dimension of a dose, and according to Eq. (4) it is equal to the dose at which the intratrack and the intertrack terms of the dose-effect relation are equal. The microdosimetric quantity ζ depends on the radiation quality and on the size of the site; some numerical values are represented in Fig. 1. A value of ζ inferred from an observed dose dependence of dicentrics as ratio of the linear and the square component, can be used to estimate the size of the site which is involved in the formation of exchanges by reference to a plot of ζ versus diameter d (Fig. 1). The values of d which have been deduced on this basis for X-rays, fast electrons and neutrons are of the order of one to several micrometer. It is of interest to note that these values are of the same order as the estimates obtained in the earlier works of Lea. They are somewhat larger than values obtained on the basis of the LET concept in later works.

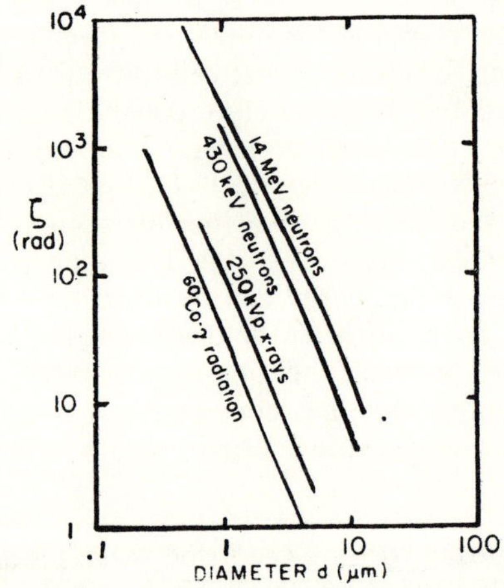

FIG. 1. Energy mean, ζ, of specific energy produced in an event as a function of the diameter, d, of the tissue sphere. [*Reproduced from A. M. Kellerer and H. H.* Rossi, *Curr. Top. Radiat. Res.* **8**, 85 (1972) with the permission of North-Holland Publ. Co., Amsterdam.]

The LET concept is at least roughly applicable for densely ionizing particles, because in this case the statistical fluctuations of energy loss are of less significance. In this case the treatment in terms of LET is nearly equivalent to the more rigorous microdosimetric analysis. The "distance model" and the "site model" are both subject to the same treatment: the intratrack and intertrack interactions are evaluated from the LET distribution and the interaction probability in the former model, and from the LET and chord-length distributions in the latter. In these two cases ζ is approximated by 22.9 $\bar{L}_D/d^2$ where L_D is the dose average LET (in keV/μm), and d (in μm) can either be understood as the effective diameter of the sphere of potential interaction around a sublesion produced by an ionization or as the diameter of the site (Kellerer and Rossi, *62*).

Chromosome exchanges are a specific illustration of one type of radiation effect for which Eq. (3) and Eq. (4) hold. This type is defined as the induction of a biological effect by a dual action, i.e. an action involving first and second order kinetics. It has been found, in this context, that, on the basis of the analysis of the relative biological effect (RBE) as a function of high-LET dose, this same quadratic dependence of the effect on specific energy holds for various other cellular effects on higher organisms (*62*).

STATISTICAL ANALYSIS OF DOSE-RESPONSE FUNCTIONS

The linear quadratic relation $\epsilon(D) = \alpha D + \beta D^2$ arises from a general characteristic of the underlying mechanism, namely a second order interaction of primary radiation products. The coefficients α and β characterize a biological effect obtained under particular irradiation conditions and have well-defined biophysical meaning. For this reason, there is a considerable interest in a rigorous statistical analysis. Such analysis is facilitated by the fact that the equation is linear in α and β; this property allows a more detailed statistical treatment than the usual curve fitting. In curve fitting, one computes only the parameters α and β. In a more detailed statistical treatment, one obtains estimates of these parameters whose true values are unknown. In the following, the statistical procedure will be considered and related to the study of chromosome exchanges dicentrics— although this procedure is suitable for other respons functions (Kellerer and Brenot, *63*). Usually, to estimate the coefficients the least squares method is used; the estimates minimize the sum of squares of errors where the errors are the differences between the observed and the theoretical effect levels. A general presentation of this well-known method can be found in classical text books such as the one by Draper and Smith (*64*) and only some basic considerations will be given here.

It is not uncommon that in a least-squares evaluation data are uncritically used with equal weight. This is unjustified because the standard deviations associated with the observations can be substantially different. In practical cases, such as the analysis of dicentrics, one always deals with different variances at different dose levels. Accordingly, one has to weigh each observation and this weighing has to be based on variances estimated from experimental data. The criterion is then to minimize the sum of squares of the errors divided by the corresponding standard deviations.

The fitted equation permits to check whether the theoretical model is to be rejected or not. TIn this connection, it must be pointed out that the statistical procedure is concerned only with the linear-quadratic dependence. The procedure can only decide whether this dependence must be rejected or whether, on the basis of the observed data, it cannot be rejected; the word "accepted" should not be used since one cannot by statistical methods alone establish positive proof for the absolute validity of a hypothesis. For dicentrics, the same set of data—dependence of the mean yield on dose—can often be statistically consistent with several different models. Some of these have clear biophysical meaning; others, such as proportionality of the yield to a power of dose intermediate between 1 and 2, lack such meaning.

Confidence statements about the estimates of parameters are possible only if the theoretical model is "accepted." Accordingly, it must be assumed for such analysis that the errors are entirely due to statistical fluctuations and not to systematic deviations of the experimental results. Moreover, one must assume that the errors are independent and normally distributed. Commonly, the standard errors or confidence ranges are given individually for the parameters. This can, however, be misleading insofar as the estimates are not independent. In the linear-quadratic relation a larger estimate of one of the parameters would imply a decrease of the optimum estimate of the other. In reality, one deals with a joint confidence region; and this region can be shown to be an ellipse in the case of two parameters, and an ellipsoid for more parameters. Examples can be found in the literature (*65–67*). Such regions are particularly meaningful when one has to compare different equations. In this context, a similar information is often given, namely the confidence bounds of the fitted dose response curve. The confidence intervals defined by these bounds for each dose are not as wide as those associated to the experimental points; this difference originates from the fact that the former confidence intervals take into account the information contained in the whole data set while the latter intervals use only the information in one point.

The least squares estimations of k and ζ in Eq. (4) are, respectively, the estimates of β and the ratio of the two estimated values of α and β.

In general, it is desirable that the statistical treatment is clearly described. Too often, substantial details are omitted and this makes it difficult or impossible to judge the validity of the statistical analysis.

CONCLUSION

The topic of this survey has been advances on three levels: experimentation, model building and statistical assessment. In the experimental technique, banding is one of the most promising new developments. In the context of model building, considerable progress has been made due to our better understanding of the physics of energy deposition on the microscopic scale. Finally, improved statistical techniques have been considered which are applicable to the assessment of the experimental data and to the attempt of relating them to biophysical models.

ACKNOWLEDGEMENTS

The authors wish to thank Prof. A. M. Kellerer for helpful comments on a preliminary draft of this paper. They are grateful also to Dr. Haag and Dr. Berger for fruitful discussions on radiation cytogenetics.

REFERENCES

1. J. B. Little, *Radiat. Res.*, **56,** 320 (1973).
2. D. E. Lea, *The Actions of Radiation on Living Cells*, 2nd Ed., Cambridge Univ. Press, Cambridge, 1962.
3. A. Norman and M. S. Sasaki, *Int. J. Radiat. Biol.*, **11,** 321 (1966).
4. T. Caspersson, S. Farber, G. E. Foley, J. Kudynowski, E. J. Modest, E. Simonsson, U. Wagh and L. Zech, *Exp. Cell Res.*, **49,** 219 (1968).
5. T. Caspersson, L. Zech, E. J. Modest, G. E. Foley, U. Wagh and E. Simonsson, *Exp. Cell Res.*, **58,** 128 (1969).
6. T. Caspersson, L. Zech, C. Johansson and E. J. Modest, *Chromosoma* (Berl.), **30,** 215 (1970).
7. T. Caspersson, G. Lomakka and L. Zech, *Hereditas*, **67,** 89 (1971).
8. M. L. Pardue and J. G. Gall, *Science*, **168,** 1356 (1970).
9. J. G. Gall and M. L. Pardue, in *Methods in Enzymology* (L. Grossman and K. Moldave, Eds.), **vol. 31,** pp. 470–480. Academic Press, New York, 1971.
10. F. E. Arrighi and T. C. Hsu, *Cytogenet.* **10,** 81 (1971).
11. A. T. Sumner, H. J. Evans and R. A. Bukland, *Nature New Biology*, **232,** 31 (1971).
12. M. E. Drets and M. W. Shaw, *Proc. Nat. Acad. Sci.* USA, **68,** 2073 (1971).
13. S. R. Patil, S. Merrick and H. A. Lubs, *Science*, **171,** 821 (1971).
14. W. Schnedl, *Chromosoma*, **34,** 448 (1971).
15. B. Lomholt and J. Mohr, *Nature New Biology*, **234,** 109 (1971).
16. M. Seabright, *Chromosoma*, **36,** 204 (1972).

17. Y. Shiraishi and T. H. Yosida, *Chromosoma*, **37**, 75 (1972).
18. B. Dutrillaux and J. Lejeune, *C. R. Acad. Sci. (Paris)*, **272**, 2638 (1971).
19. T. Caspersson, J. Lindsten, G. Lomakka, H. Wallman and L. Zech, *Exp. Cell Res.*, **63**, 477 (1970).
20. T. Caspersson, K. Castleman, G. Lomakka, E. Modest, A. Moller, R. Nathan, R. Wall and L. Zech, *Exp. Cell Res.*, **67**, 233 (1971).
21. T. Caspersson, P. Issler and G. Lomakka, *Exp. Cell Res.*, **75**, 543 (1972).
22. T. Caspersson, U. Haglund, B. Lindell and L. Zech, *Exp. Cell Res.*, **75**, 541 (1972).
23. M. Seabright, *Chromosoma*, **40**, 333 (1973).
24. M. Seabright, *Cytogenet. Cell. Genet.*, **12**, 342 (1973).
25. C. San Roman and M. Bobrow, *Mutat. at. Res.*, **18**, 325 (1973).
26. L. J. Stadler, *Sci. Agric.*, **2**, 557 (1931).
27. S. H. Revell, *Heredity*, **6**, 107 (1953).
28. S. H. Revell, in *Radiobiology Symposium 1954*, Liege (Z. M. Bacq and P. Alexander, Eds.), pp. 243–253. Butterworth, London, 1955.
29. S. H. Revell, *Proc. Roy. Soc. (London), Ser.B*, **150**, 563 (1959).
30. S. H. Revell, in *Radiation-Induced Chromosome Aberrations, Ser. B.* (S. Wolff, Ed.) pp. 41–72. Columbia University Press, New York, 1963.
31. S. H. Revell, *Mutat. Res.*, **3**, 34 (1966).
32. H. J. Evans, in *General Aspects of Radiosensitivity: Mechanisms of Repair*, pp. 31–48. IAEA, Vienna, 1966.
33. D. Scott and H. J. Evans, *Mutat. Res.*, **4**, 579 (1967).
34. J. G. Brewen and R. D. Brock, *Mutat. Res.*, **6**, 245 (1968).
35. A. D. Conger, *Mutat. Res.*, **4**, 449 (1967).
36. K. Sax, *Genetics*, **23**, 494 (1938).
37. K. Sax, *Proc. Nat. Acad. Sci. (USA)*, **25**, 225 (1939).
38. K. Sax, *Genetics*, **25**, 41 (1940).
39. K. Sax, *Cold Spring Harbor Symp. Quant. Biol.*, **9**, 93 (1941).
40. D. G. Catcheside, D. E. Lea and J. M. Thoday, *J. Genet.*, **47**, 137 (1946).
41. N. H. Giles, *Proc. Nat. Acad. Sci. (USA)*, **26**, 567 (1940).
42. N. H. Giles, *Genetics*, **28**, 398 (1943).
43. S. Wolff, K. C. Atwood, M. L. Randolph and H. E. Luippold, *J. Biophys. Biochem. Cytol.*, **4**, 365 (1958).
44. S. Wolff, *Radiat. Res.* **Suppl. 1**, 453 (1969).
45. S. Wolff, *J. Theoret. Biol.*, **3**, 304 (1962).
46. K. C. Atwood, in *Radiation-Induced Chromosome Aberrations*, (S. Wolff, Ed.), pp.73–86. Columbia University Press, New York, 1963.
47. J. Read, *Int. J. Radiat. Biol.*, **9**, 53 (1965).
48. G. J. Neary, J. R. K. Savage and H. J. Evans, *Int. J. Radiat. Biol.*, **8**, 1 (1964).
49. J. R. K. Savage, *Int. J. Radiat. Biol.*, **9**, 81 (1965).
50. J. R. K. Savage and D. G. Papworth, *J. Theoret. Biol.*, **11**, 1 (1966).
51. J. R. K. Savage and D. G. Papworth, *J. Theoret. Biol.*, **22**, 493 (1969).
52. G. J. Neary, *Int. J. Radiat. Biol.*, **9**, 477 (1965).
53. ICRU, *Linear energy transfer, Report 16*. Int. Commission on Radiation Units and Measurements, Washington D.C., 1970.
54. H. H. Rossi and W. Rosenzweig, *Radiat. Res.*, **2**, 417 (1955).
55. H. H. Rossi, *Radiat. Res.*, **10**, 522 (1959).
56. H. H. Rossi, *Advan. Biol. Med. Phys.*, **11**, 27 (1967).
57. H. H. Rossi, in *Radiation Dosimetry*, **vol. 1** (F. H. Attix, H. Frank, and W. C. Roesch, Eds.), p. 74. Academic Press, New York, 1968.
58. ICRU, *Radiation Quantities and Units*, Report 19. Int. Commission on Radiation Units and Measurements, Washington D.C., 1971.
59. O. Hug and A. M. Kellerer, *Stochastik der Strahlenwirkung*, Springer, New York (1966).
60. A. M. Kellerer and H. H. Rossi, in *Proc. 2nd Symposium on Microdosimetry*, p. 843. Stresa, Italy, 1969.
61. J. L. Bateman, H. H. Rossi, A. M. Kellerer, C. V. Robinson and V. P. Bond, *Radiat. Res.*,
61. J. L. Bateman, H. H. Rossi, A. M. Kellerer, C. V. Robinson, and V. P. Bond, *Radiat. Res.*, **51**, 381 (1972).
63. A. M. Kellerer and J. M. Brenot, *Radiat. Environ. Biophys.*, **11**, 1 (1974).

64. N. R. Draper and H. Smith, in *Applied Regression Analysis*, Wiley, New York, 1966.
65. E. J. Hall, H. H. Rossi, A. M. Kellerer, L. Goodman, and S. Marion, *Radiat. Res.*, **54**, 431 (1973). 431 (1973).
66. A. H. Sparrow, A. G. Underbrink, H. H. Rossi, *Science*, **176**, 916 (1972).
67. J. Brenot, M. Chemtob, D. Chmelevsky, P. Fache, N. Parmentier, R. Soulie, M. T. Biola, J. Haag, R. Le Go, M. Bourguignon, D. Courant, J. Dacher, G. Ducatez, in *Proc. 4th Symposium on Microdosimetry*, Verbania, Pallanza, Italy 1973, (K. Booz, H. G. Ebert, R. Eickel and A. Waker, Eds.), p. 545. Euratom Publication EUR-5122-d-e-f, Brussels, 1974.

Cell Cycle Kinetics and Radiation Therapy [1,2]

Mortimer L. Mendelsohn
Biomedical Division
Lawrence Livermore Laboratory
University of California
Livermore, California 94550, USA

INTRODUCTION

Radiation therapy as currently practiced involves the subtle, largely empirical art of balancing the recurrence of cancer due to undertreatment against severe damage to local tissues due to overtreatment. Therapeutic results too often fall short of desired success rates; yet, the therapist is continually tantalized to the likelihood that a slight shift of therapeutic ratio favoring normal tissue over cancer would have a profoundly beneficial effect. The application of cell cycle kinetics to radiation therapy is one hope for improving the therapeutic ratio; but, as I will try to show, kinetic approaches are complex, poorly understood, and presently too elusive to elicit confidence or to be used clinically. Their promise lies in their diversity and in the magnitude of our ignorance about how they work and how they should be used.

Potentially useful kinetic approaches to therapy can be grouped into three classes. The first class takes advantage of intracyclic differential sensitivity, an effect involving the metabolism and biology of the cell cycle; its strategies are based on synchronization of cells over intervals of hours to days. The second class involves the distinction between cycling and noncycling cells; its strategies are based on the resistance of noncycling cells to cycle-linked radiation sensitizers and chemotherapeutic agents. The third class uses cell repopulation between fractions; its strategies are based on the relative growth rates of tumor and relevant normal tissue before and after perturbation.

INTRACYCLIC DIFFERENTIAL SENSITIVITY

Pattern, Amplitude and Phasing

The radiation sensitivity of cells varies with relative cell age (*1*). Such age responses are roughly sinusoidal and fall into two patterns depending on whether there are one or two resistant periods per cell cycle (Fig. 1). Conceivably the differences in pattern could be used therapeutically. For example, if one were dealing with tumor cells having a single resistant period per cell cycle and normal cells having a double resistant period, then the therapeutic ratio might be

[1] Congress presentation B–4–3.

[2] Work performed under the auspices of the U.S. Atomic Energy Commission with the support of USPHS Grant 14533.

improved by synchronizing the cells early in the cycle and treating at the time of the first resistant peak. But realistically, almost all cells that have been studied are of the two-period-per-cycle type; much of the seemingly dramatic difference in pattern disappears when considered in real time because the cells with the single period of resistance generally have very short cycle lengths with little or no G_1 period.

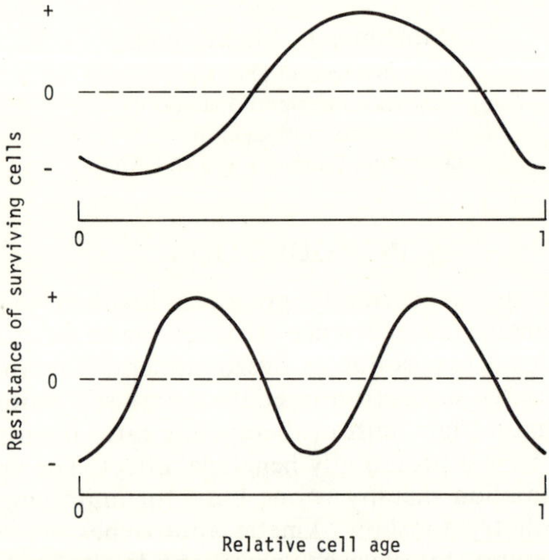

FIG. 1. Pattern of radiation age-response. A cell's relative age ranges from zero when the cell emerges from mitosis to one when it subsequently divides. Cells fall into two roughly sinusoidal patterns depending on whether there are one or two periods of resistance per cell cycle.

The *amplitude* of the sinusoidal age response describes the magnitude of selective killing effects. When two populations such as those in Fig. 2 differ in their amplitude of age response, a therapeutic advantage is achievable. In general, such strategies will focus on the large-amplitude population. For example, if the normal cell response has the large amplitude, then timing of therapy would be oriented toward the resistant peaks. If the tumor-cell response has the large amplitude, then one aims for the sensitive troughs. Large differences in amplitude have been reported among cell types, with peak-to-trough survival ratios ranging from threefold to seventyfold (*2*). Ten-to-twentyfold differences are present in the one-period-per-cycle response of V79 Chinese hamster cells studied by Sinclair (*1*) and shown in Fig. 3. Amplitude generally increases with dose; in the V79 cells it plateaus at 750 rad, a single dose larger than those usually given in clinical radiotherapy.

In Fig. 4 threefold differences are shown by Rockwell and Kallman in the two-periods-per-cycle response of EMT6 mouse mammary tumor cells grown in vitro (*3*); and here the full effect is manifest at 600 rad. So far as I know, systematic differences in amplitude have not been found for tumor vs normal cells. What is more, much of the reported differences in amplitude could be technical, reflecting methods and quality of synchronization rather than inherent biological differences in age response.

For exploitation of amplitude or any other age response the cells must be synchronized. In examples such as Fig. 3 or 4, a first radiation dose to asynchronous cells gives the age-response survival pattern and thus partially synchronizes the survivors into the resistant parts of the cycle. Alternatively, the cells can be synchronized by chemical blocking agents, but then to capitalize on differential sensitivity one must understand how the *phasing* of the age response relates to the site of drug action.

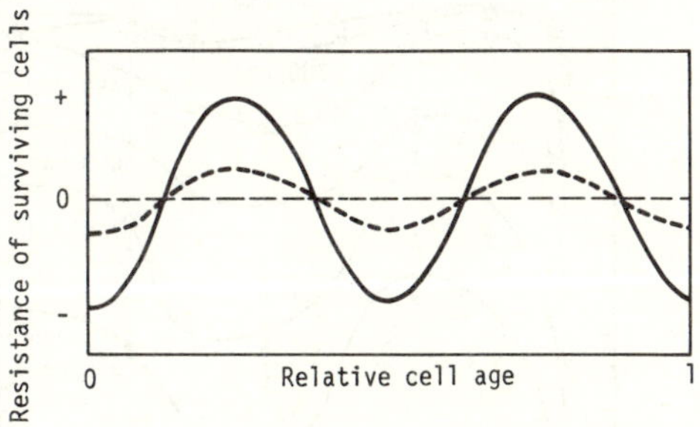

FIG. 2. Amplitude of radiation age-response. A therapeutic advantage can be gained when two populations, depicted as the solid and broken lines, differ in the amplitude of their age response.

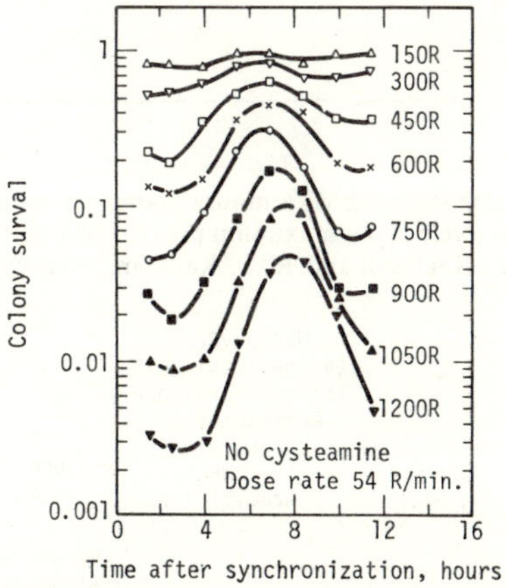

FIG. 3. Radiation age-response of V79 Chinese hamster cells in vitro. Note the consistent pattern of one resistant period per cycle. The amplitude increases with radiation dose and reaches a maximum peak-to-trough difference of twenty-fold in cell survival. The cells are in mitosis at zero hour. [Reproduced from W. K. Sinclair, *Curr. Top. Radiat. Res. Q.* 7, 264 (1972) with the permission of North-Holland Publishing Co.]

Figure 5 combines the radiation age-response, the phases of the cell cycle, and the site of action of some representative blocking agents. The radiation age-response is loosely coupled to the phases, with maximum sensitivity at mitosis and resistance rising during the S (DNA synthetic) phase. Drug effects are more directly tied to the phases of the cycle, with cell killing occurring predominantly in S and cell blocking predominantly at the G_1-S boundary and at mitosis (*4*). Use of blocking agents to synchronize cells might greatly sharpen the therapeutic ratio of a subsequent treatment with radiation; but there are many complications to this approach, including systemic effects of such drugs, uncertainty about the kinetics as the block is released, and the interaction of chemical and radiation effects on cell survival (*4*).

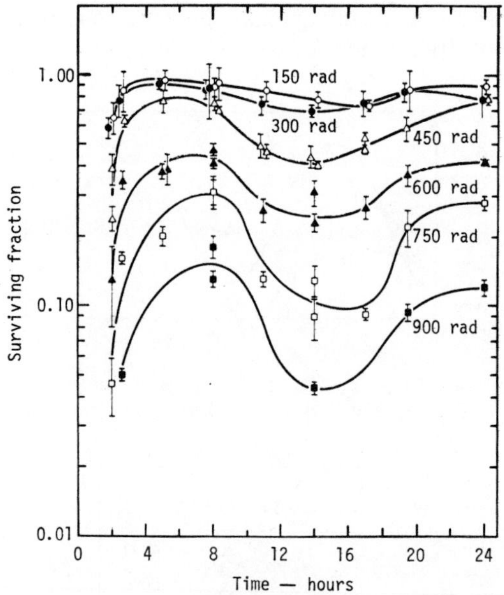

FIG. 4. Radiation age-response of EMT6 mouse mammary tumor cells in vitro. These cells have two resistant periods per cycle and a maximum peak-to-trough difference of threefold in cell survival. [Reproduced from S. Rockwell and R. F. Kallman, *Radiat. Res.* **57**, 132 (1974).]

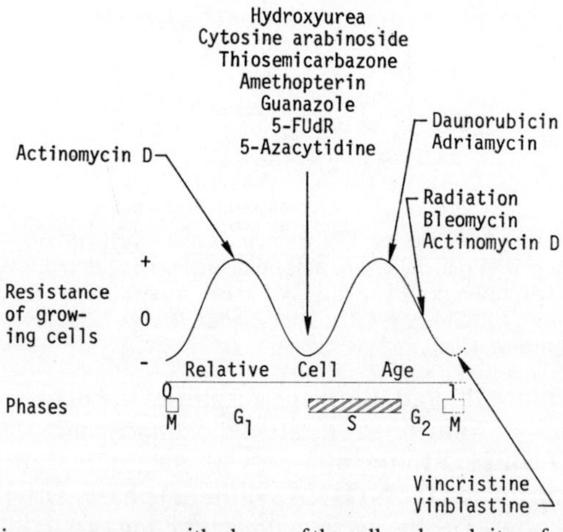

FIG. 5. Radiation age-response with phases of the cell cycle and sites of action of representative agents which block cell cycle progression.

Figure 6 shows an example of the complex pattern of sensitivity found in combination treatment of mouse leukemia cells with 5FU and X-rays (*5*). Such patterns are poorly understood at present, but they do differ in at least one nontumor system—normal bone-marrow colony-forming cells. Extension to other tissues, other tumors, and other combinations may someday make this a formidable approach to cancer therapy.

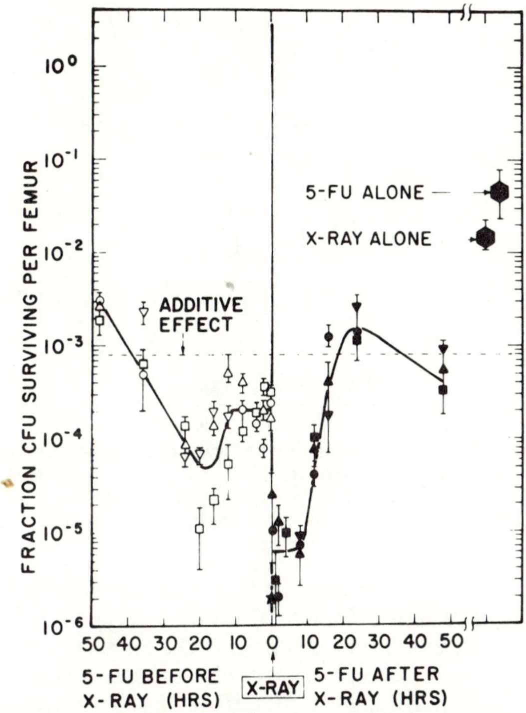

FIG. 6. Combined effect of X rays and 5-fluorouracil (5FU) on asynchronous leukemia cells in AKR mice. Depending on the timing of the two agents, there is a complex pattern ranging from simple additivity of effect to 100-fold synergism. [Reproduced from T. Vietti, F. Eggerding, and F. Valeriote, *J. Nat. Cancer Inst.* **47**, 865 (1971).]

Cell Cycle Time

To make the transition from relative age to real time, we now consider the duration of the cell cycle. Under idealized steady-state conditions, a stable difference in the mean cycle times of the critical normal and tumor populations leads directly to strategies of selective killing of the cancer cells. The upper panel of Fig. 7 shows two age responses with identical patterns and amplitudes but with cycle lengths in the ratio of 2:3. The resistance ratio of the two responses shown in the lower panel is a compound sinusoid with much structure and dramatic peak-to-trough ratios. Regardless of which response is that of the normal cell population, one could in principle optimize the therapeutic ratio from such a diagram. As a rule the interval between peaks and troughs of therapeutic ratio is a quarter of the shorter cell-cycle time—and hence may be only a few hours. The construction of such a diagram in an actual therapeutic situation would require estimates of each cycle time to within plus or minus 10 percent, a difficult feat with current methods.

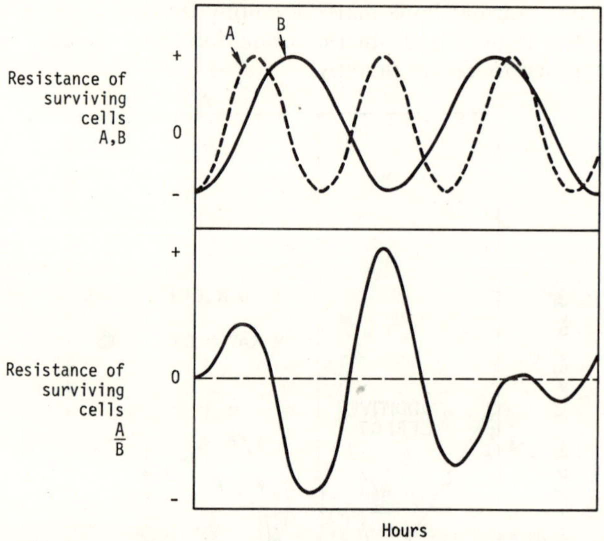

FIG. 7. Cell time and radiation age-response. In the upper panel the curves marked A and B are the age responses of two populations with cell times in the ratio of 2:3. The comparative resistance or survival ratio of the two populations is shown in the lower panel. Complex changes in amplitude are present in the compound response, but the intervals between major peaks and troughs are determined largely by the population with the shorter cycle time.

The cell cycle times of tumor and normal tissues overlap throughout a broad range. The shortest cycles are found in intestinal crypts, hair follicles, and exceptionally aggressive experimental tumors. The longest measured cycles are in corneal epithelium, skin, and epithelial human tumors. Since current methods of selecting tumors and analyzing cycle times are biased toward short cycles, we probably have a distorted view of the range and distribution of cell cycle times.

For several reasons it will be unrealistic to plan therapy on the basis of unperturbed steady-state estimates of the cell cycle time. Each radiation exposure is followed by a *G_2 block* which effectively prolongs the cycle. The duration of the block in V79 cells as a function of dose and cell age is shown in Fig. 8 as measured by Sinclair (*6*). The impact of this block on the timing of peaks and troughs of sensitivity is shown in Fig. 9 taken from Young and Fowler (*7*) and based on Sinclair's data. The location of peak and trough and the interval between them varies with dose. Obviously, if one were selectively killing V79 cells in the presence of yet another cell population, differences of a few hours would have major effects on the therapeutic ratio.

Typically, the G_2 block in a variety of cell systems lasts for about one hour per 100 rad (*8*). Differences amongst cell types may relate to cycle length (*9*), but consistent differences between tumor and normal tissues are yet to be described. Such differences are worth exploring because G_2 phases generally are longer in tumors than in normal tissues (*10,11*). This is the only consistent cell cycle difference between the two that has so far been uncovered, and to my knowledge it has not been exploited for therapy.

Another complication of cell cycle time is *variability*. Regardless of how cells have been synchronized, the population progressively disperses and typically returns to the asynchronous state within one or two cycles. Thus the sinusoidal responses already described are actually decaying in amplitude with time. Such damping works against the therapist by forcing him to treat quickly, but it can also work for him when the rates of damping differ greatly between tumor and critical normal tissue.

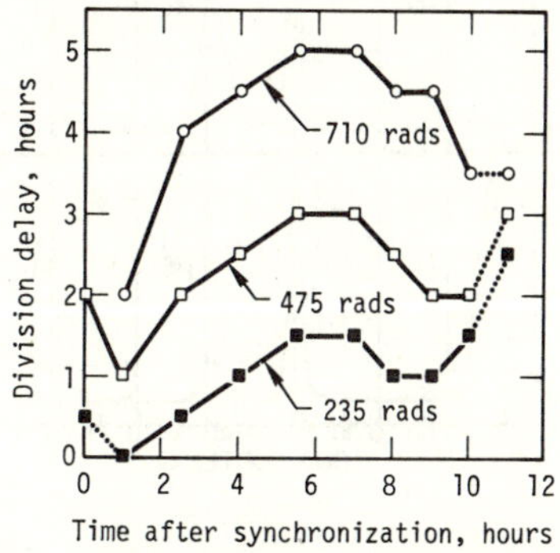

FIG. 8. Division delay in V79 Chinese hamster cells irradiated in vitro. Delay was measured in terms of the peak incidence of mitoses at the first division after exposure. [Reproduced from W. K. Sinclair, *Curr. Top. Radiat. Q.* 7, 323 (1972) with the permission of North Holland Publishing Co.]

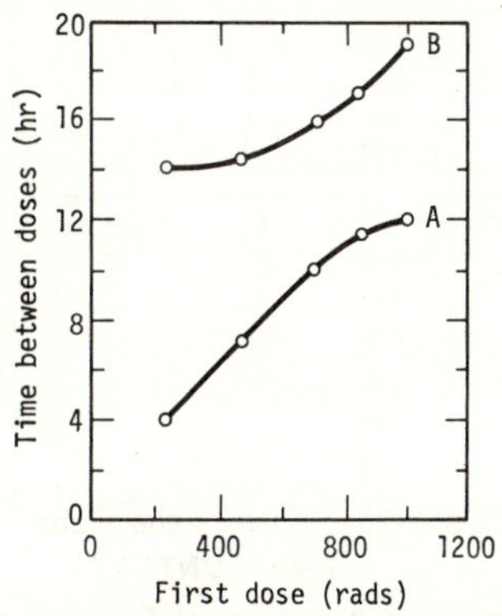

FIG. 9. The timing of peak and trough of resistance of V79 Chinese hamster cells after sychronization by irradiation in vitro. Curve A is the time after a priming dose when the surviving cells are maximally sensitive. Curve B gives the timing of the subsequent peak of resistance dose [Reproduced from J. M. Young and J. F. Fowler, *Cell. Tissue Kinet.* 2, 95 (1969) with the permission of Blackwell Scientific Publications, LTD.]

Flow microfluorimetry (or impulse cytophotometry) of DNA per cell is the most direct and powerful way to measure synchronization and desynchronization in perturbed cell populations. The upper block of panels in Fig. 10 shows a time sequence of experimentally determined distributions of DNA content of CHO cells after their release from block in early G_1 (*12*). At 2-1/2 hours after release the cells peak in early S, then later they are in mid-S, in G_2 and G_1, and on into the second cycle with much broadening of the peaks due to desynchronization. The lower panel of Fig. 10 is a computed simulation of the sequential DNA distributions of the CHO cells (*13*). Good estimates of the rates of dispersion as well as the transit times of the synchronized cells can be obtained from such simulations.

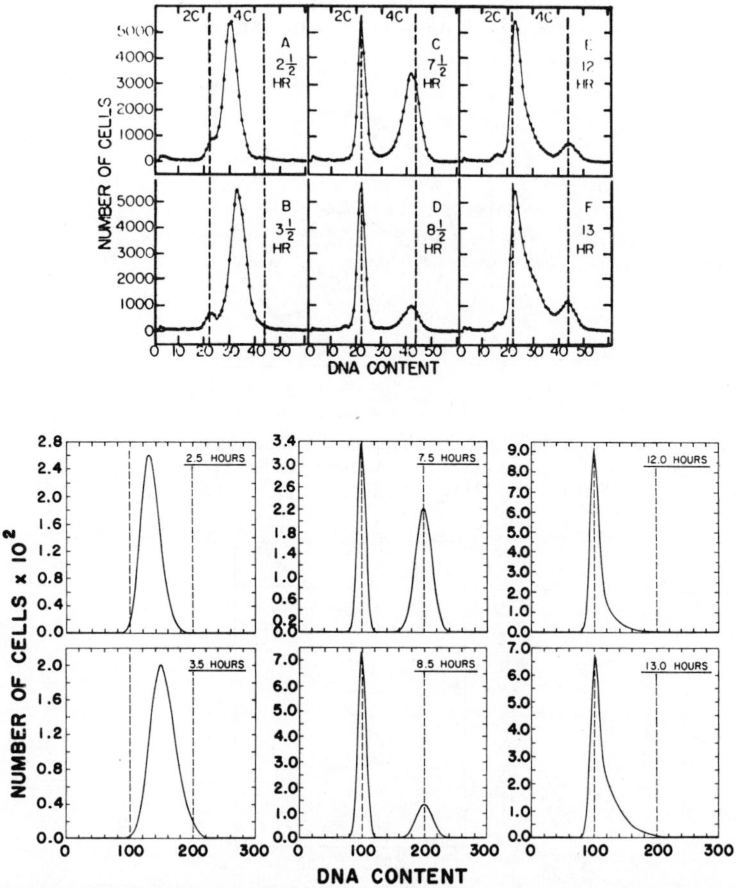

FIG. 10. Flow microfluorimetric measurement of dispersion of synchronized cells. The upper block of panels shows distributions of DNA content for CHO cells at various times after release from a G_1 block (*12*). Such distributions are readily obtained by flow microfluorimetry; they show the progression and dispersion of cells as they proceed through DNA synthesis, mitosis and again into DNA synthesis. In the lower block the same distributions have been simulated by a computer model which, when properly fit to the experimental data, provides estimates of the rates of progression and dispersion of the cells. [Reproduced from J. W. Gray, *J. Histochem. Cytochem.* **22**, 642, (1974) with the permission of The Williams & Wilkins Co.]

In unperturbed asynchronous populations, the transit times and standard deviations of transit can be taken from conventional curves of fraction of labeled mitoses (FLM). Very likely the dispersion of *truly* synchronized cells is the same phenomenon as the damping of FLM curves due to desynchronization of cells labeled in S, although it may be that truly synchronized cells are subject to dispersive pressures or have abnormal kinetics secondary to the synchronizing event. Direct comparisons of FLM curves and fluorimetry on the same cell system would readily settle this question.

If FLM curves are valid indicators of damping, then wide differences in desynchronization rates exist. Computer analyses of FLM data provide estimates of the standard deviation and coefficient of variation of cell transit through each phase and through the entire cycle. These cycle-based or phase-based statistics are straightforward estimates of desynchronization when comparing populations with similar cycle and phase lengths, but when transit times are dissimilar the statistics are difficult to interpret. For example, two populations with equal coefficients of variation of cycle time but unequal cycle lengths will have different rates of desynchronization. For some applications dispersion per cycle or per S phase is of interest; in other situations the most useful descriptor of desynchronization would be the dispersion that takes place over a stated time interval. By disregarding phase-specific differences in dispersion and by assuming statistical independence of the aging process throughout the cycle, a simplified estimate of time-based dispersion σ_t can be derived from dispersion per cycle σ_c:

$$\sigma_t/\sqrt{t} = \sigma_c/\sqrt{T_c}$$

where t is the interval over which the dispersion is being estimated and T_c is the mean cell cycle time. A useful special case is the hourly dispersion σ_d, when $t = 1$.

Figure 11 shows the behavior of hourly dispersion as a function of cell cycle time for 49 computer-analyzed well-behaved FLM curves. These curves are culled from the literature and have been fitted by Gordon Steel (*14*), by our own program (*15*), and by several other methods (*16,17*). The data base and its other kinetic properties have been partially described (*11*). The clear regression of hourly dispersion on cycle time suggests that the control of progression loosens as the cycle lengthens. Very likely the sensitive peaks and troughs also lengthen with cycle length and hence, from the therapist's viewpoint, the ideal parameter might be dispersion per resistant peak duration, an ordinate that perhaps would eliminate much of the regression in Fig. 11. But it is interesting to note that tumors and normal tissues have an overlapping distribution whereas ascites cell tumors lie slightly below. In general, ascites cells have unique kinetic attributes; in this case the data suggest that the fluid environment of ascites cells may minimize dispersive effects.

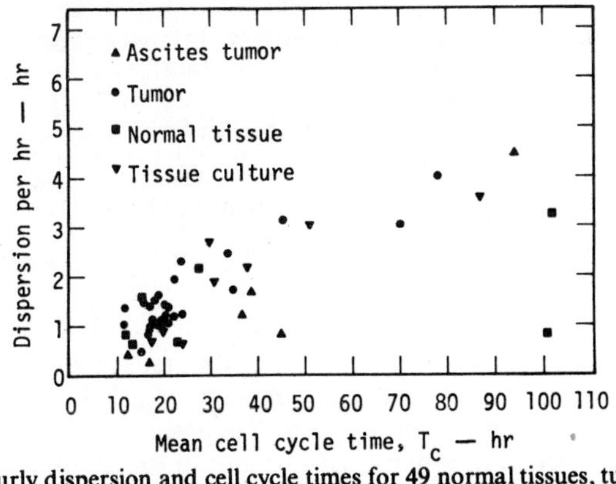

FIG. 11. Hourly dispersion and cell cycle times for 49 normal tissues, tumors, ascites tumors, and tissue culture populations based on computer-analyzed, well-behaved FLM curves.

The therapeutic impact of dispersion is summarized in Fig. 12 taken from Gray (*13*). The three panels show the effect of coefficients of variation of cycle time (and hourly dispersion) of 19 percent (0.78 hours), 31 percent (1.27 hours) and 62 percent (2.53 hours) on cells initially fully synchronized in G_1. The curves, which give the oscillations of the fraction of cells in each phase, are realistic indicators of what a therapist might have to work with if, for example, he was optimizing the killing of cells in the S phase.

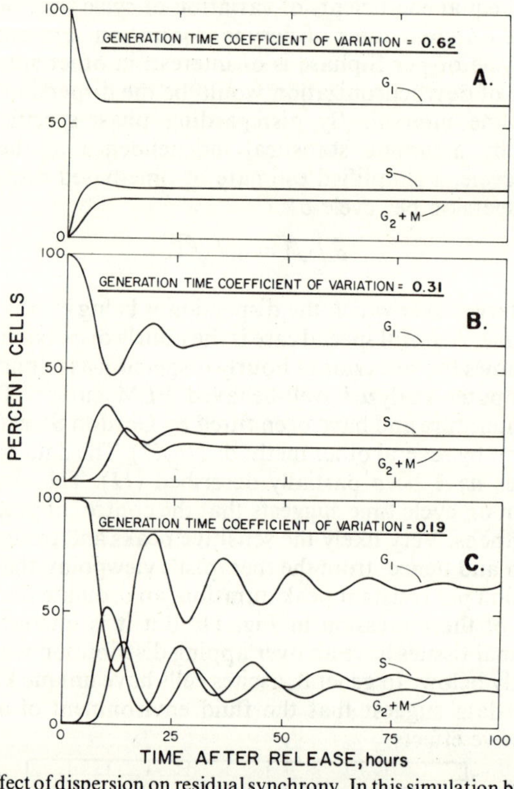

FIG. 12. The effect of dispersion on residual synchrony. In this simulation by Gray (*13*) cells are synchronized in G_1 and then allowed to cycle with transit times of 9.1 hours for G_1, 4.1 hours for S, 3.0 hours for G_2, and 0.5 hours for M. The three panels show the percent of cells in G_1, S and $G_2 + M$, based on different coefficients of variation of cell cycle time. In panel A, an hourly dispersion of 2.53 hours effectively damps the synchrony within 5 hours and offers little hope of using intracyclic differential sensitivity. In panel C the hourly dispersion is 0.78 hours and significant oscillations are available to the therapist for at least 75 hours. Panel B is intermediate with an hourly dispersion of 1.27 hours and about 25 hours of reasonable synchrony. [Reproduced from J. W. Gray, *J. Histochem. Cytochem.* **22**, 642, (1974) with the permission of The Williams & Wilkins Co.]

There are other complications of cycle time, such as diurnal variation (*18*), and the kinetic distinctions between oxygenated versus hypoxic or clonogenic versus nonclonogenic cells, but to me the most interesting complication is the *change in cycle time as therapy progresses*. Cell depletion secondary to treatment causes reactive hyperplasia involving recruitment of noncycling or G_o cells *and* shortening of the cell cycle times. In solid tumors, the cycle shortening is sometimes present and sometimes not; at best it is a twofold effect (*19*). In skin, greater than a fivefold shortening of the cycle has been measured after irradiation (*20*), and significant shortening occurs in gut and other rapidly turning over tissues (*21*). What little is known of these effects in normal tissue suggests that they involve tissue-specific feedback controls driven by the depletion of functioning cells.

The contrasting effects of cell depletion on the cycle time of tumor, rapidly turning-over tissue, and slowly turning-over tissue are shown in Fig. 13. Cells are reproductively killed by radiation in much the same way in all three cases, but the depletion of what might be called the functioning fraction depends on the turnover time. In rapidly turning-over tissue the functioning fraction lags slightly behind and then parallels the survival curve. The rapid depletion of such normal cells elicits a brisk corrective kinetic response in cycle time. The tumor-cell cycle may be changing as well, although not as dramatically and not on the same time scale as the normal tissue. When such rapidly-responding tissues are dose-limiting and intracyclic sensitivity is to be used throughout therapy, one must understand the full kinetics of the corrective responses. Slowly turning-over tissues are much more likely to be dose-limiting, and in these tissues cell depletion lags greatly behind cell survival. Although this simplifies the dynamics of the intracyclic response, it seriously compromises these tissues in other ways, as will be shown below. It should be emphasized that Fig. 13 is largely conjectural because we are presently limited to data on only a few tumors (*19*) and a few normal tissues (*9*) soon after single doses.

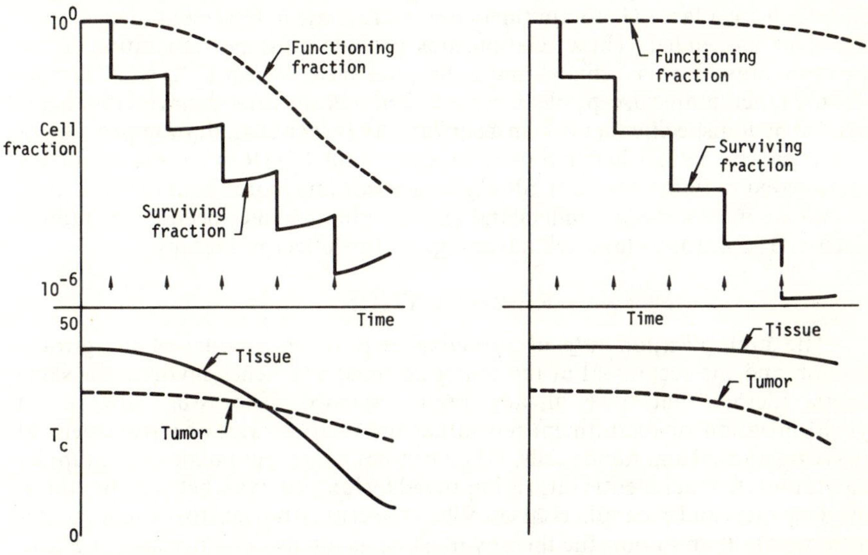

FIG. 13. Cell survival, cell turnover and cell cycle time (T_c) in rapidly and slowly turning-over tissue and tumor. In this hypothetical situation the arrows represent fractions of radiotherapy. On the left the critical normal tissue has a rapid turnover; thus the functioning fraction of cells closely tracks the surviving fraction and there is a prompt shortening of the normal cell cycle time. On the right a slowly turning-over normal tissue shows little change in functioning fraction or in cell cycle time.

NONCYCLING CELLS

The effects of intracyclic differential sensitivity do not apply to out-of-cycle or G_0 cells. Thus the proportion of cells in G_0 limits a population's susceptibility to cycle-linked drug effects and to strategies based on synchronization. This immunity to therapy can work for or against the therapist. The G_0 tumor cell resists chemical radiosensitization by BUdR (*22*) and is a major factor in the nonresponsiveness of solid tumors to all cycle-linked chemotherapy (*23*). The G_0 normal cell protects bone marrow, liver, kidney, and many other tissues from chemotherapy and radiosensitization. In radiotherapy the out-of-cycle normal cells can be a major problem because they are vulnerable to radiation but are not protected by repopulation.

The G_0 commitments range greatly in normal tissues, from essentially zero in skin and intestinal crypts to essentially 100 percent in liver and other facultatively or slowly turning-over tissues. In microscopic foci of tumor the G_0 commitment is small but increases rapidly with subsequent growth. By the time a solid tumor is macroscopic, the commitment stabilizes with some 25 to 80 percent of cells out of cycle.

The G_0 cells are dramatically responsive to depletion of the population and are probably controlled in much the same way as the feedback mechanisms that change the cell cycle. How G_0 commitment and recruitment might work during therapy is shown in Fig. 14. In rapidly turning-over tissues the few G_0 cells that are present are soon recruited into cycle. Tumors probably respond more slowly but before therapy ends they too are likely to have recruited all their surviving cells into cycle. Slowly turning-over tissues, the ones most likely to be dose-limiting, have a large G_0 commitment and recruit late in treatment or even after treatment has ended. These relationships provide an important rationale for chemotherapy, for as Schabel and colleagues have shown (*24*), solid tumors become much more susceptible to cycle-linked therapy after their cell number is reduced by nonspecific therapy. In a similar way Brown et al. show improvement of therapeutic ratio when radiosensitization with BUdR is continued during fractionated radiotherapy (*22*). Slowly we are learning more about the biology of G_0 and we may someday understand enough about tissue-specific and tumor-specific G_0 control to take full advantage of this effect in therapy.

REPOPULATION

The multiplication rate of surviving cells is the product of the growth fraction and the reciprocal of the cell cycle time; and hence involves the same kinetic factors that have already been discussed. The issue now is not synchronization or recruitment but rather the relative rates of replacement of surviving normal and tumor cells. When normal tissue repopulates more rapidly than tumor, the therapeutic ratio is improved by long intervals between fractions, low dose rates and even split courses. When the critical normal tissue repopulates more slowly than tumor, the therapy must be as intense as other factors allow. Most likely the relative rates of repopulation will change as therapy progresses; and thus the intensity of optimum therapy will also have to change. To obtain maximal repopulation rates for normal tissue, the therapist must force a functional impairment of the tissue. For slowly turning-over tissue, it is conceivable that optimal therapy would involve a deliberate insult a month or more before definitive therapy in order to activate repopulation in the critical dose-limiting cells. Such strategies oriented toward maximizing normal regrowth are incompatible with those oriented toward the protected status of normal G_0 cells.

The accelerated repopulation of clonogenic tumor cells is shown dramatically in the data of Hermens and Barendsen in Fig. 15 (*25*). After 2000 rad, the clonogenic rhabdomyosarcoma cells show a roughly tenfold shortening of their doubling time and at their peak response appear to be doubling twice a day. Figure 16 shows the comparable response of jejeunal and colonic epithelium as measured by Withers using the split-dose approach to colony assay in vivo (*26*). In jejeunum after a delay of almost three days, the data suggest two logs of repopulation with a doubling time of four to eight hours. The response in colon is later and less brisk, with doubling times roughly similar to those of the repopulating rhabdomyosarcoma cells. Withers' results using similar techniques give a doubling time of 22 hours for skin (*27*), 42 hours for stomach, and 70 days for testis (*28*). Such experiments need corroboration and extension to many tumors and to other normal tissues of relevance to the therapist, but it is obvious already that the patterns will not be simple.

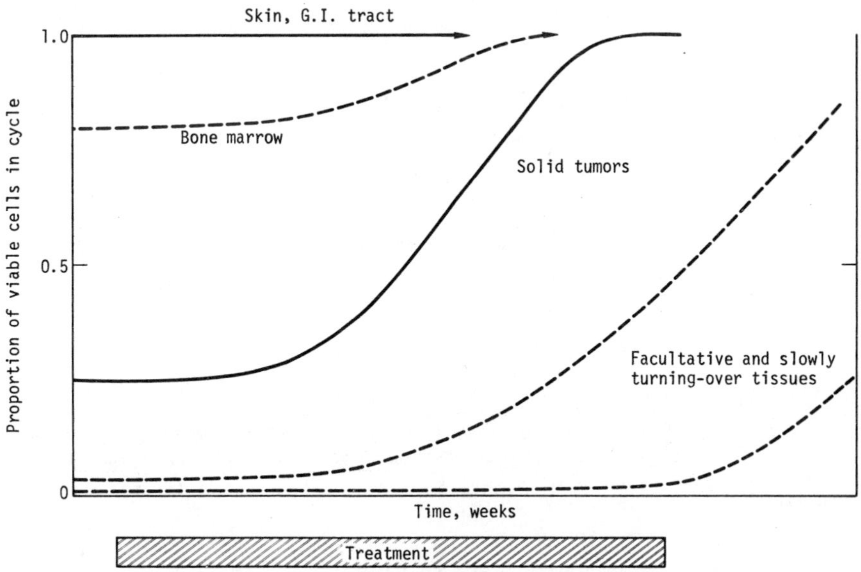

FIG. 14. The commitment of G_0 cells during a full course of radiotherapy. This oversimplified and hypothetical diagram shows the contrasting G_0 commitment of rapidly turning-over tissues, tumors, and facultative or slowly turning-over tissue.

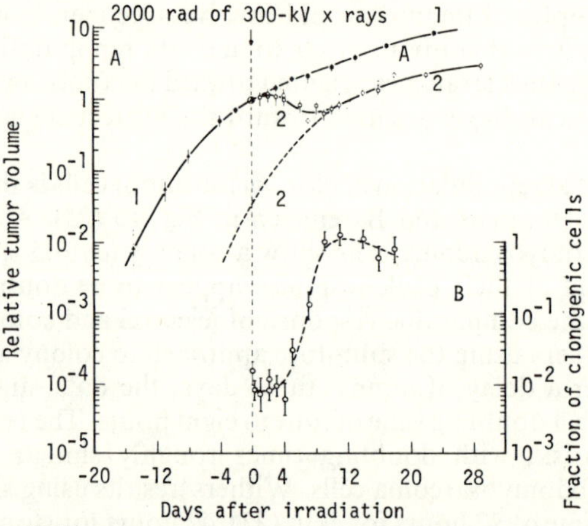

FIG. 15. Effect of 2000 rad on the volume and clonogenicity of rhabdomyosarcoma. Between days 4 and 8 post-therapy the clonogenic cells increase their number 100-fold, accelerate their growth rate tenfold, and at the peak response are doubling approximately twice a day. [Reproduced from A. F. Hermens and G. W. Barendsen, *Eur. J. Cancer* **5**, 173 (1969) with the permission of Pergamon Press, Ltd.]

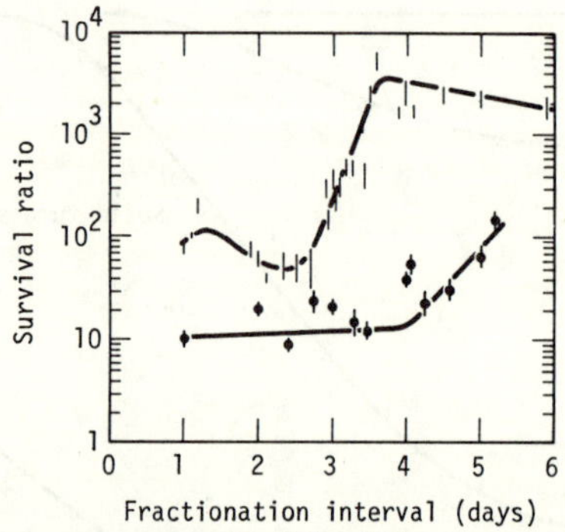

FIG. 16. Split-dose recovery curves for jejunal (upper curve) and colonic (lower curve) clonogenic cells after a primary exposure of 660 and 700 rad respectively. For jejunum the 100-fold rise between days 2-1/2 and 3-1/2 is interpreted to be repopulation of clonogenic cells and implies a doubling time of 4 to 8 hours. In colon the response begins 4 days after the priming exposure and has a doubling time of about half a day. [Reproduced from H. R. Withers, *Cancer* **28**, 75 (1971) with the permission of the American Cancer Society.]

CONCLUSION

By now the message should be clear: cell cycle effects impinge on therapy in many ways, so many that to optimize therapy in kinetic terms will be an enormous conceptual and biological task. Conceptually we are dealing with many cell types, eight or more dynamic properties of the cell cycle, multiple modalities of treatment, and several other important noncyclic aspects of therapy such as repair of sublethal damage, reoxygenation, and cell mobility. Problems of this complexity cannot be handled intuitively or with blackboard logic. They must be modeled, and one can readily predict that with this many variables the modeling will have to be done with great sophistication. There are now dozens of computer models of cancer therapy, with a wide and healthy diversity of biological assumptions and methematical techniques (*29*).

But much more importantly, before we can establish the credibility of such models we need a far better understanding of the key biological phenomena. We must know the cell kinetic properties of a wide variety of tissues and tumors—not just unperturbed and not just after a single perturbation, but throughout entire treatment courses. We must identify and understand the specific dose-limiting cells for each radiotherapeutic region of the body. We must extend such studies to man. And to do these things we must have new techniques, particularly more accurate and more practical methods of kinetic and clonogenic analysis.

The near-term prospects for kinetic optimization are rather poor, but I believe the long-term chances are too good to pass up. Unless cancer is cured by some other means, we may well be exploiting cell kinetics routinely in the years to come. What form might this take?

> — Perhaps an all-radiotherapeutic approach but with doses and dose intervals that change greatly over the course of history.
>
> — Perhaps combined therapy in which each fractionated episode includes a relatively nontoxic blocking agent to synchronize the cells followed by radiation at a carefully defined interval.
>
> — Perhaps with an armamentarium of chalones and antichalones such that the therapist is in full control of the kinetic variables for all the relevant tissues.

But first there is much science to be done.

REFERENCES

1. W. K. Sinclair, *Curr. Top. Radiat. Res. Q.* **7,** 264 (1972).
2. G. F. Whitmore, S. Gulyas and J. Botond, in *Cellular Radiation Biology,* 423. The Williams and Wilkins Company, Baltimore, 1965.
3. S. Rockwell and R. F. Kallman, *Radiat. Res.* **57,** 132 (1974).
4. F. A. Valeriote, in *The Cell Cycle in Malignancy and Immunity* (CONF-731005), National Technical Information Service, Springfield, Va. 22151 (in press).
5. T. Vietti, F. Eggerding and F. Valeriote, *J. Nat. Cancer Inst.* **47,** 865 (1971).
6. W. K. Sinclair, *Curr. Top. Radiat. Res. Q.* **7,** 323 (1972).
7. J. M. Young and J. F. Fowler, *Cell Tissue Kinet.* **2,** 95 (1969).
8. S. Okada, in *Radiation Biochemistry,* Volume **1,** (K. I. Altman, G. B. Gerber, and S. Okada, Eds.), pp. 189-246. Academic Press, New York, 1970.
9. J. Denekamp, Congress presentation C–3–1 (these *Proceedings*).
10. R. Baserga and F. Wiebel, *Int. Rev. Exp. Pathol.* **7,** 1 (1969).
11. M. L. Mendelsohn, in *The Cell Cycle in Malignancy and Immunity* (CONF-731005), NTIS, Springfield, Va. (in press).
12. P. M. Kraemer and R. A. Tobey, *J. Cell Biol.* **55,** 713 (1972).
13. J. W. Gray, *J. Histochem. Cytochem.* **22,** 642 (1974).
14. G. G. Steel, *Cell Tissue Kinet.* **5,** 87 (1972).
15. M. Takahashi, J. D. Hogg, Jr., and M. L. Mendelsohn, *Cell Tissue Kinet.* **4,** 505 (1971).
16. P. J. Brockwell, E. Trucco, and R.J.M. Fry, *Bull. Math. Biophys.* **34,** 1 (1972).

17. P. Dombernowsky and N. R. Hartman, *Cancer Res.* **32,** 2452 (1972).
18. J. N. Izquierdo and S. J. Gibbs, *Cell Tissue Kinet.* **7,** 99 (1974).
19. A. F. Hermens and G. W. Barendsen, Congress presentation C–3–3 (these *Proceedings*).
20. T. Yamaguchi and J. Tabachnick, *Radiat. Res.* **50,** 158 (1972).
21. J. Lesher and S. Lesher, *Radiat. Res.* **57,** 148 (1974).
22. J. M. Brown, D. R. Goffinet, and J. E. Cleaver, *J. Nat. Cancer Inst.* **47,** 75 (1971).
23. H. E. Skipper and S. Perry, *Cancer Res.* **30,** 1883 (1970).
24. F. M. Schabel, Jr., *Cancer Res.* **29,** 2384 (1969).
25. A. F. Hermens and G. W. Barendsen, *Eur. J. Cancer* **5,** 173 (1969).
26. H. R. Withers, *Cancer* **28,** 75 (1971).
27. H. R. Withers, *Radiat. Res.* **32,** 227 (1967).
28. H. R. Withers, N. Hunter, H. T. Barkley, Jr., and B. O. Reid, *Radiat. Res.* **57,** 88 (1974).
29. J. Aroesty, T. Lincoln, N. Shapiro and G. Boccia, *Math. Biosc.* **17,** 243 (1973).

Theoretical Aspects and Implications of the Oxygen Effect[1]

Jürgen Kiefer
Strahlenzentrum der Justus Liebig-Universität
Giessen, Germany

INTRODUCTION

It has been known for more than fifty years that many biological systems are more resistant against the damaging effects of X rays if they are irradiated under anoxic conditions (*1*). This finding is of considerable importance for radiotherapy since many tumors contain oxygen-depleted fractions which counteract the desired goal to kill the malignant cells by treatment with ionizing radiation. The so-called "oxygen-effect" attracted many research workers. This led to significant progress in our knowledge of the mechanisms involved, but we are still far from a complete understanding. We shall consider in this paper only the loss of reproductive integrity as biological endpoint which is normally assayed by plating of single cells in vitro.

Figure 1 shows as an example survival curves of diploid yeast irradiated with 12 MeV electrons in aerated and anoxic aqueous suspensions. The dose dependence of colony-forming ability has the usual shape with a shoulder (characterized by the extrapolation number n or the "threshold dose" D_q) and a terminal exponential part (characterized by the inverse of the slope, D_o) if plotted semi-logarithmically.

It appears that both curves in Fig. 1 extrapolate to the same n, which means that they can be superimposed if the dose scale is multiplied by a constant factor. The ratio "dose for a given effect in nitrogen/dose for the same effect in oxygen," which is constant under this condition, has been termed "oxygen enhancement ratio" (OER).

Alper and Howard-Flanders (*2*) have shown that the dependence of OER on oxygen concentration can be described by the following formula

$$\mathrm{OER}(O_2) = (\mathrm{m}\,[O_2] + \mathrm{k})/([O_2] + \mathrm{k}) \qquad (1)$$

where $[O_2]$ denotes the oxygen concentration, m and k are constants. m is the maximum OER for high oxygen tensions and k, which has the dimension of a concentration, is the O_2-concentration where the OER reaches half its maximum value, i.e. where

$$\mathrm{OER}\ (O_2) = (m + 1)/2$$

Equation (1) may be transformed in a simple way:

[1]Congress presentation B–4–4.

$$1/[\text{OER}(O_2) - 1] = (1 + K/[O_2])/(m - 1) \tag{2}$$

If one plots $1/\text{OER}-1$ versus $1/[O_2]$, a linear relationship is obtained which facilitates the test greatly. *K* and *m* may be read off the graphical representation as seen in Fig. 2 which was constructed from data of Howard-Flanders and Alper (*3*): The y-axis intercept is $1/m-1$, the x-axis intercept is $-1/K$.

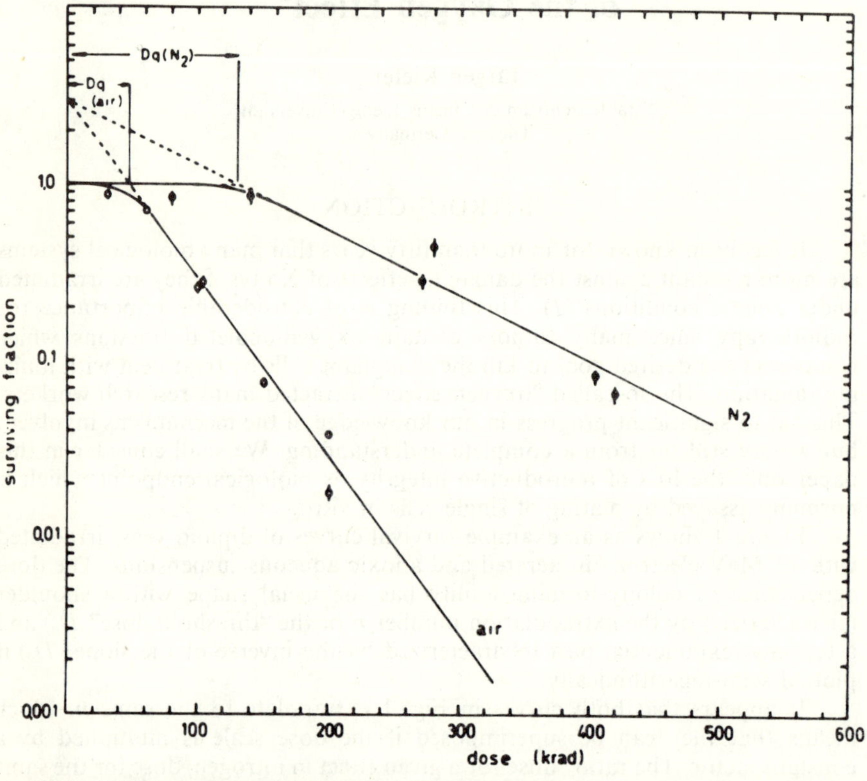

FIG. 1. Colony forming ability of diploid yeast irradiated with 12 MeV-electrons in air and nitrogen.

The validity of the formula (1) has been verified for many systems and with various biological endpoints. For the case of sparsely ionizing radiation and the loss of colony-forming ability, relevant data are summarized in Table 1. It may be seen that *m* and *k* do not differ to a great extent although the systems studied are taken from the whole range of the biological hierarchy and differ considerably in radiation sensitivity. The main features of the oxygen effect may be summarized as follows:

1. Oxygen must be present *during* irradiation; if added only a few milliseconds after exposure it does not show any action (*4*).
2. Oxygen appears to act in a dose-modifying way.
3. The oxygen effect is limited to ionizing radiation, it is not observed after exposure to ultraviolet light or alkylating chemicals.

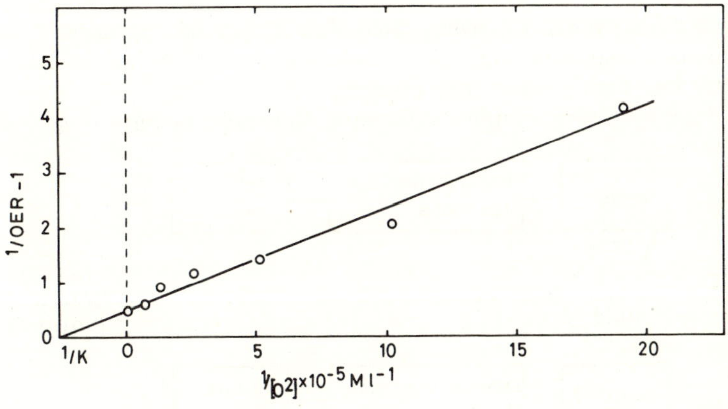

FIG. 2. The dependence of sensitization on oxygen concentration plotted according to equation (2). Data from reference (*14*) (Shigella flexnari).

TABLE 1
Parameters of Survival Curves and the Oxygen Effect with X-Rays for Different Biological Systems. The Endpoint Was Always the Loss of Reproduction Integrity. If the Parameters Were Not Explicitly Given in the References, They Were Estimated from the Figures.

Organism	m	k (Ml^{-1})	D_0 (krads) (air)	D_q (krads) (air)	n	Reference
Bacteria:						
Shigella	2.9	4.0	3.4	0	1	2
E.coli B/r	3.1	4.7	4.4	8.7	7	2
uvr^+ rec^+	6.0	8	not	given		(*19*)
uvrA rec^+	6.7	8	not	given		(*19*)
uvr^+ recA	1.85	8	not	given		(*19*)
uvrA recA	1.5	8	not	given		(*19*)
Yeast (haploid)	2.4	5.8	2.4	0	1	2
Algae (Oedogonium)	2.9	17.6	0.8	1.1	4	27
Mammalian (Chinese hamster cells)	3.14	8	0.19	0.44	10	29

4. The extent of the oxygen effect depends on radiation quality; it is reduced with increasing LET (*5,6, vide infra* for further discussion).
5. Protecting and sensitizing agents generally lower the maximum obtainable OER (*7*).
6. The extent of the oxygen effect depends on culture conditions after exposure and on the genetic background as far as radiosensitivity is concerned (*7*).

ALTERNATIVE MODELS

As a basis for the following discussion it may be appropriate to recall that radiation exposure triggers a whole sequence of events in cells which ultimately lose their reproductive integrity:
Figure 3 shows a very simplified diagram: Generally we have to assume that

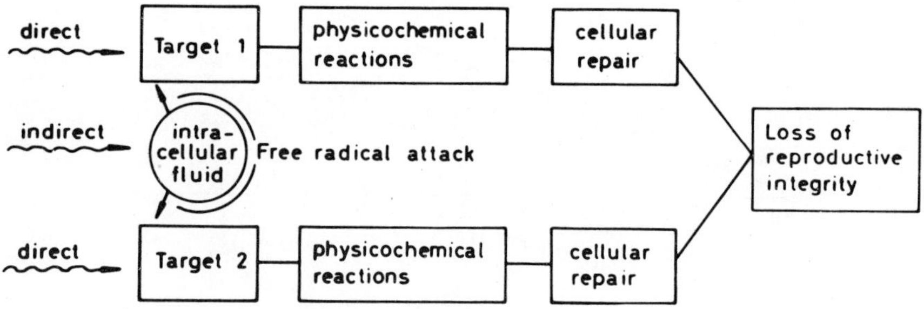

FIG. 3. Simplified diagram of damage development in cells.

there are one or more types of target molecules in the cell whose integrity is essential for reproductive integrity. Radiation may alter their structure either by direct attack or indirectly *via* free radicals formed by energy absorption in their close environment. The primary damage may be modified by subsequent physico-chemical reactions or cellular repair processes. "Repair" is used throughout this paper in a very general way, i.e. as removing or bypassing of damage, without restriction to any particular mechanism. Oxygen can be thought to interfere at any of these levels. Thus four possibilities for the explanation of the oxygen effect can be considered:

1. Oxygen changes the inherent sensitivity of different types of target molecules, i.e. there are target molecules which are e.g. very sensitive in oxygen and relatively resistant in its absence. This means that different target molecules are responsible for the ultimate biological effect if exposures are given in oxygen or in nitrogen; we may therefore distinguish between "type O" and "type N" damage (*8*).
2. Oxygen may react with primary radicals before they reach the target molecules, thus changing the extent of indirect action.
3. Oxygen may react with the target molecule immediately after it is altered, thus changing qualitatively and quantitatively the physicochemical reactions to follow (*9*).
4. Oxygen may interfere with cellular repair processes, e.g. *via* energy metabolism.

Before we discuss the alternative models it must be pointed out that a change at an early step in the pathway depicted in Fig. 3 will generally also influence the following ones. If for instance, altered physico-chemical reactions lead to different chemical products in the target molecule, the capability of repair processes may be changed. On the other hand, if initial damage is reversed at the physico-chemical level, cellular repair processes may not be very important. Let us now consider in more detail the four hypotheses put forward.

Alternative 1: Type O and Type N Damage:

If there are essential target molecules in the cell showing different sensitivity in the presence and the absence of oxygen, then it is hard to reconcile with this that oxygen acts as dose modifying. It would imply that the (mathematical) function which links the ultimate biological effect to the initial damage in the different targets is always the same; however, this appears quite improbable. In addition, there is direct evidence from experiments with fractionated exposure: If a first dose, D_1, is given in oxygen and a second one, D_2, in nitrogen and if the surviving fractions are Y_1 and Y_2, respectively, then the surviving fraction $Y(D_1, D_2)$ after both irradiations should be the product of Y_1 and Y_2 if independent target molecules are damaged with the two regimes. On the other hand, if the same target molecules are involved, only damaged to a different extent, then the surviving fraction should be as if one single dose

$$D = D_1 + D_2/\text{OER}$$

were given in oxygen. Experiments show that the second result is obtained (*10*). This seems to rule out the assumption of independent type O and type N damage.

Alternative 2: Oxygen Effect *via* Indirect Radiation Damage:

Although this hypothesis is still widely accepted, it appears not very probable on the basis of radiation chemical data: If one presumes that a primary radical R which is harmless under anoxic conditions is transformed by reaction with oxygen into a damaging species R′, one may write

$$dR/dt = -(K + K'[O_2])R \tag{3}$$

where K is the decay constant of R and K′ its reaction rate constant with oxygen. Also

$$dR'/dt = K'[O_2] \cdot R \tag{4}$$

If the primary yield of R is R_o, then we obtain for the final yield of R′:

$$R'(R = O) = R_0 \cdot K'[O_2]/(K + K'[O_2]) \tag{5}$$

and for the oxygen concentration where $R' = R_o/2$:

$$[O_2]_{1/2} = K/K' \tag{6}$$

which should correspond to the experimental K-values in the Alper formula (equation (1)). The hydrated electron e_{aq} is the most likely candidate for R because it is formed with sufficient yield upon irradiation and reacts readily with oxygen. The rate constant has been measured to be

$$K' = 1.9 \cdot 10^{10}\ M^{-1}\ 1\ sec^{-1}\ (11).$$

If we insert this value into equation (5) we may calculate K and hence the lifetime τ of e_{aq} in the cell. We obtain $\tau = 7.3 \cdot 10^{-6}$ sec. Cerček and Cerček (*12*), however, determined τ experimentally in egg white which is considered to be quite representative for the conditions in the cell and found a value of $\tau = 4.5 \cdot 10^{-8}$ sec, which is about two orders of magnitude shorter than anticipated; therefore it is fair to conclude that the cellular oxygen effect does not take place *via* reactions with primary radicals since the lifetime of other primary species is certainly not considerably longer and the reaction rate constants with oxygen are not higher than for e_{aq}.

Alternative 3: This is discussed in detail below.

Alternative 4:

Interference of oxygen with metabolic processes appears unlikely *a priori* because of the short time which is required for its action. There is, however, also direct experimental evidence that the metabolic state of the cell (e.g. respiring versus non-respiring) does not influence the oxygen effect to any extent (13). It has to be taken into account, however, if the oxygen concentration within a cell suspension or, even more, within a tissue is to be determined (14), since oxygen consumption will lead to a buildup of concentration gradients.

A Basic Model:

We are left with alternative 3 and shall first formulate it in a more qualitative manner. Therefore we start with the following general assumption (Fig. 4):

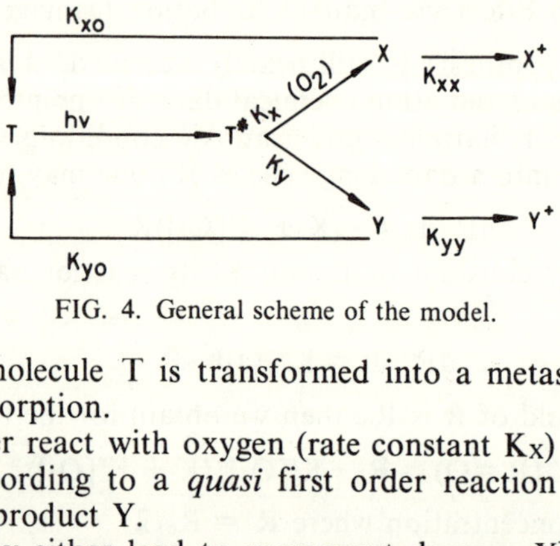

FIG. 4. General scheme of the model.

1. The target molecule T is transformed into a metastable form T* upon radiation absorption.
2. T* may either react with oxygen (rate constant K_X) to form a product X or decay according to a *quasi* first order reaction (decay constant K_Y) leading to a product Y.
3. X and Y may either lead to permanent damage X^+ or Y^+, respectively, (rate constants K_{XX} and K_{YY}) or be rendered harmless by reactions with constants K_{XO} and K_{YO}, respectively.

This model which is depicted diagrammatically in Fig. 4 may be formulated mathematically (*15*) and leads to the following equation for the oxygen enhancement ratio:

$$\text{OER}\,(O_2) = \frac{[(1 + K_{YO}/K_{YY})/(1 + K_{XO}/K_{XX})][O_2] + (K_Y/K_X)}{[O_2] + K_Y/K_X} \tag{7}$$

Details of calculations are not given here; they are quite elementary and may be found elsewhere (*15*).

By comparison with equation (1) it is obvious that equation (7) has the general form of the Alper formula but the parameters now bear special meanings:

$$K = K_Y/K_X \tag{8}$$

and

$$m = (1 + K_{YO}/K_{YY})/(1 + K_{XO}/K_{XX}) \tag{9}$$

K is the ratio of the "decay constant" of T* in the absence of oxygen (or an oxygen-like sensitizer) and the reaction rate constant of T* with oxygen. *m* is related to the "repair probabilities" of X and Y.

Reaction rate constants of oxygen with radiation-induced transients in biomolecules have been determined for free radicals (*16*) and triplet states (*17*). They approach those of diffusion-controlled reactions and lie on the order of $2 \cdot 10^9$ M sec $^{-1}$. Using equation (8) and the data from Table 1 we may calculate K_Y and hence the lifetime τ of the transient T*. With $K = 5 \cdot 10^{-6}$ M $\cdot$ 1^{-1} one obtains $\sphericalangle = 7{\cdot}10^{-5}$ sec which compares very satisfactorily with the experimental value of 90 μsec determined by Epp's group (*18*).

Values of m > 1 imply that anoxic damages are more readily repairedthan oxic ones. Since with certain compounds (e.g. tetramethyl-piperiminol N-oxyl, TMPN (*19*)) a higher sensitization is seen than with oxygen, one has to conclude that K_{XO}/K_{XX} is greater than unity. A lower limit may be estimated for *E.coli* if one assumes that TMPN at high concentrations suppresses repairability of transients completely in the *rec A*-strain 2463: As shown below in the presence of high concentrations of the sensitizer the maximum OER is smaller than unity, i.e. oxygen acts as a protector. One obtains an estimate of $K_{XO}/K_{XX} = 1.6$. The maximum OER without sensitizer, is normally on the order of 3.5 or lower; also, on the basis of this number an estimate of the repair probability for "anoxic" damage is possible:

$$(1 + K_{YO}/K_{YY})/2.6 = 3.5 \tag{10}$$

$$K_{YO}/K_{YY} \approx 8 \tag{11}$$

It must be pointed out that these figures constitute only lower estimates since *m* gives only the *ratio* for oxic and anoxic damage. The relation (8) predicts that *m* depends on repair capabilities of the cell; it is smaller if repair is inhibited.

This is the case even if K_{XO}/K_{XX} and K_{YO}/K_{YY} are reduced by the same factor. As an example we may take the above given figures and assume that repair is only one tenth for both types of damage. We then have

$$m = (1 + 0.8)/(1 + 0.26) = 1.46$$

If repair of anoxic damage is selectively reduced there will be an even more drastic effect on *m*.

This is nicely illustrated by an experiment with differently sensitive mutants of yeast (*20*): It has been shown by Town et al. (*21*) that single-strand breaks are rapidly repaired—or are not even formed—in the absence of oxygen in the cells. Normally they are not held to be lethal for the cell since they are almost completely rejoined, but strand break rejoining may be important as part of other repair mechanisms. Wild type cells remove genetic damage by the well-known excision pathway. Some mutants cannot perform the first incision step and are therefore very UV sensitive but not X-ray sensitive since radiation induces single-strand breaks—which are not formed upon UV action and which replace the enzymatic incision. If they are repaired, however, before the enzymatic progress begins, as in the case of exposures in nitrogen, the damage remains in the genome and cannot be excised. In terms of our model K_{YO} is reduced and hence the maximum OER lowered. This is exactly found in Averbecks S 2094 mutant carrying a mutation in the rad 2-gene which is responsible for the incision function: m is 2.7 in the wild type and 1.5 in the mutant with the sensitivity in oxygen being the same.

SENSITIZATION

Since the oxygen effect plays an important role in practical problems of radiotherapy there has been a long-lasting search for "sensitizers to replace oxygen." Principally they should show no sensitization under fully oxygenated conditions. This is generally not found; with many sensitizers the OER is lowered and the action in oxygen less pronounced but still significant. A list of examples may be found in T. Alper's paper (7). One exception is $KMnO_4$ which will be dealt with below. Our model of the oxygen effect would predict two types of sensitization: one by reaction with the metastable target molecule transient T* in a manner similar to oxygen, and another one by interfering with the restitution of the intermediate products X or Y. The two cases will be treated separately:

Sensitization by Competition with Oxygen:

The basic scheme of this kind of sensitization according to our model is shown in Fig. 5: T* may decay with the constant K_Y, react with oxygen (K_X) or with the sensitizer (K_Z) to form products X,Y,Z, respectively.

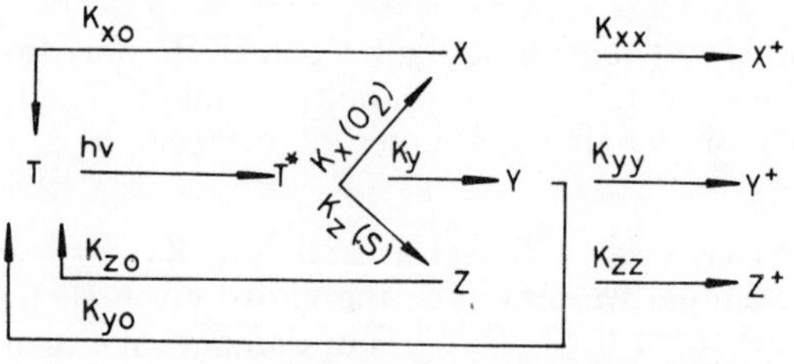

FIG. 5. Sensitization by competition with oxygen.

If no oxygen is present, we may define a "sensitizer enhancement ratio" (SER) whose dependence on concentration is given by an analog of the Alper formula (equation 1) in which $[O_2]$ is replaced by sensitizer concentration [S] and where the constants are appropriately changed. The maximum obtainable SER which we call m_s is then

$$m_s = (1 + K_{YO}/K_{YY})/(1 + K_{ZO}/K_{ZZ}) \tag{12}$$

For a substance which truly replaces oxygen, $K_{XO}/K_{XX} = K_{ZO}/K_{ZZ}$, i.e. the product Z should be essentially the same as X. This does not generally appear to be the case but has been found for $KMnO_4$ (*22*) in bacterial spores. We have studied the action of this compound in vegetative yeast cells, and were able to confirm the results obtained in bacterial spores. Figure 6, plotted according to equation (2), shows the dependence of sensitization on concentration. It may be seen that the general relationship holds, the K-value measured is $2 \cdot 10^{-4}$ M l^{-1} which is about 40 times larger than with oxygen, implying that the reaction with T* is slower by the same factor.

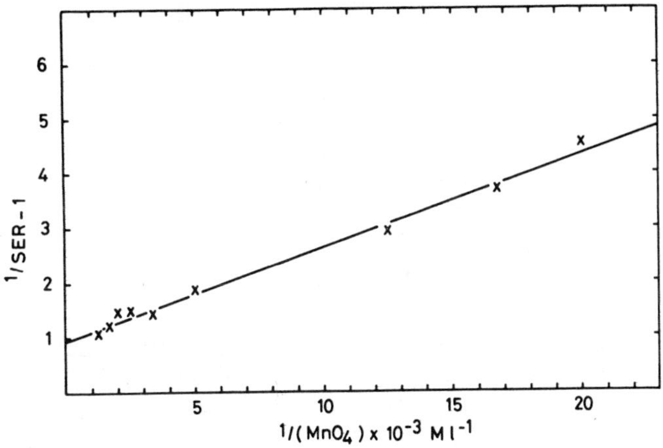

FIG. 6. Sensitization of anoxic yeast to $KMnO_4$.

If sensitizer and oxygen are present during irradiation, the OER may be described by the following formula:

$$OER(O_2) = \frac{m'[O_2] + (K_Y + K_Z[S])/K_X}{[O_2] + (K_Y + K_Z[S])/K_X} \tag{13}$$

if we still assume $K_{XO}/K_{XX} = K_{ZO}/K_{ZZ}$.

It is seen that we again have the general relationship of equation (1) but the maximum OER and the K-value are changed. It is

$$m' = m(1 + K_Y/K_Z[S])/(m + K_Y/K_Z[S]) \tag{14}$$

with m given by equation (9).

Generally, it may be stated that the presence of a sensitizer competing with oxygen causes an increase of the K-value and a reduction of maximum OER. If the reaction of the sensitizer with T* leads to products Z whose repair probabilities differ from those of X (see Fig. 5), the mathematical formulation for OER (O_2) will be more complicated (ref. *15*, eq. (8)). If irradiaton occurs, however, in high concentrations of S, "pure" anoxic reactions play only a negligible role, and we may write

$$OER[O_2] = \{(1 + K_X[O_2])/K_Z[S]\}\{(1 + m''K_X[O_2]/K_Z[S])\} \qquad (15)$$

where

$$m'' = (1 + K_{ZO}/K_{ZZ})/(1 + K_{XO}/K_{XX}) \qquad (16)$$

TMPN may be an example of this kind of sensitizer. It was studied extensively by Johanson (19). Using his data one calculates m″ to be 0.39, i.e. lower than unity, which means that oxygen in the presence of high concentrations of TMPN acts as a *protecting* agent.

Sensitization by Interference with "Repair"

We have already seen that OER may be changed if the relative repair probabilities, K_{XO}/K_{XX} or K_{YO}/K_{YY}, are altered. In this case the K-value in equation (1) remains the same. This statement gives an experimental tool to decide on what level sensitization occurs, either by competition with oxygen or by interference with repair processes. An example is again given by Johanson (*19*). He measured the maximum OER in four strains of *E.coli* with different radiosensitivities. The value m, as defined by equation (9), varies between 6.3 in the most resistant strain and 1.45 in the most sensitive one, but *k* was always measured to be $8{+}10^{-6}$ M $l^{-1}O_2$.

Changes in repair probabilities may also be brought about by the addition of chemicals. It is implied by the model that fast physico-chemical reactions occur with the transient products X and Y, either to render them harmless or to transform them into permanent damage. Sensitizers may directly react with X or Y, thus reducing or increasing the probability of either pathway.

It is known that a sensitization action of oxygen is normally only seen in cells but not in isolated chemical systems (*23*). It appears likely that cell constituents or their radiation-induced intermediates are able to reconstitute the damaged molecule X or Y, e.g. by donating an electron. If this process is hindered by electron acceptors interrupting the electron donation, the probability for physico-chemical repair would be lowered. This appears to be an attractive hypothesis originally put forward by Adams (*24*); it is based on the finding that the extent of sensitizing and electron affinity are related (*24*). A hypothetical scheme of this idea is depicted in Fig. 7. It differs from that

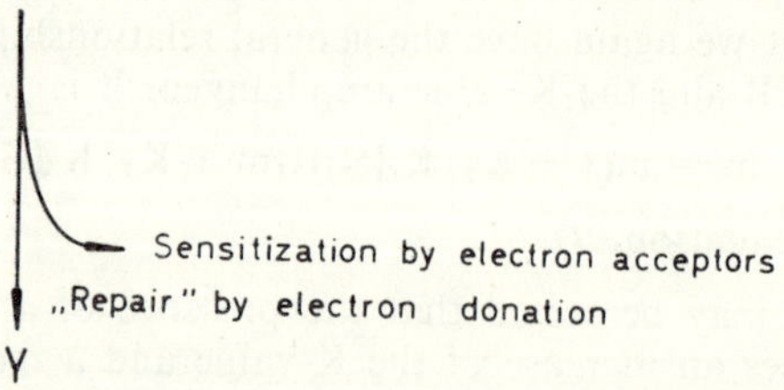

FIG. 7. "Repair" inhibition by electron acceptors.

suggested by Adams (*24*) insofar as it does not assume that the sensitizer reacts with the target molecule but rather that it interrupts electron donation to it. The reason for this difference is that many sensitizers react very fast with electron-adducts of DNA — like oxygen — but the K-value for sensitization is smaller by about two orders of magnitudes.

LET-EFFECTS

No exhaustive discussion of the problem is attempted in this section, but some facets will be presented. In most cells increasing LET leads to a decreased survival probability at equal radiation dose. The effect is more pronounced when irradiation takes place under anoxic than under oxic conditions. This generally accepted result may be explained by several hypotheses which are not all mutually exclusive:

1. "Oxygen in the track" —hypothesis (*25, 26*):
   It has been suggested that within the track of a densely ionizing particle oxygen or oxygen-like substances (which may be short-lived radicals) are formed.

   This is the same as if a certain amount of sensitizer is present and equations (13) and (14) apply. It is seen that m is lowered and k increased. To test the hypothesis K-values have to be measured for high LET-irradiations as accurately as possible. Unfortunately, data are scarce and contradictory (Table 2).

TABLE 2
OER and K-Values for Neutrons in Comparison to That of X-Rays

System	m(x)	m(u)	k(X) [μMI^{-1}]	k(n) [μMI^{-1}]	Reference
Bacteria	3.0	2.0	2.6	3.4	(*28*)
Algae	2.9	1.7	17.6	8.7	(*27*)
mammalian (bone marrow)	2.3	1.5	k(X)/k(n) = 1.5		*

* J.Hendry, personal communication

2. Oxygen depletion in the track:
   If cells are irradiated with ionizing radiation, the products formed react with oxygen and remove it from the site if there is not enough time for rediffusion. This may well happen in and close to the track and would result in a *lower* K-value, as was actually found in algae (*27*) or mammalian cells (F. Hendry, personal communication) with neutrons.

3. Interference with repair:
   With the two hypotheses given above LET would be expected to act strictly as a dose-modifying variable since the hypotheses are based on the assumed formation or removal of oxygen which is itself a dose modifying agent.
   But with increasing LET not only the terminal slope but the entire shape of survival curves is altered. This suggests that repair is impaired by increased LET which would — as we tried to show in the preceding sections

— also explain the lower OER generally found. The determination of *k* is again crucial. An involvement of repair may be indicated by the finding that with various LET the values of OER and D_q are linearly related (5).

There is probably more than one mechanism responsible for the LET effects. It is important to note that with densely ionizing radiation the local concentrations of reactants are increased; accordingly, the reactions that take place will be changed quantitatively and qualitatively. To estimate the actual concentrations involved one must know how much energy is deposited in volumes which are on the order of the diffusion distances of the important species. This problem links the exploration of the oxygen effect with the problems of microdosimetry.

CONCLUSIONS

It has been stressed several times that the knowledge of k-values is very important for the test of various hypotheses concerning the oxygen effect under a given set of conditions. Radiation-chemical as well as biological data have to be linked to estimate the time-course of fast reactions in irradiated cells. But there are also some practical implications for radiotherapy: Tumor tissues are seldom completely anoxic; they usually contain traces of oxygen. Therefore, the maximum OER may not be the only essential parameter for radiotherapy but the dependence on oxygen tension is also of importance.

ACKNOWLEDGEMENTS

The experiments with $KMnO_4$ in yeast were carried out by Miss Wilma Weyrather in the author's laboratory. I should like to thank her, and Drs. Alma Howard and J. Hendry (Paterson Laboratories, Manchester) for the permission to quote their unpublished results. Part of this work was financially supported by the Science Ministry of the Federal Republic of Germany.

REFERENCES

1. H. Holthusen, *Pflügers Arch. Gesamte Physiol.* **187,** 1 (1921).
2. T. Alper and P. Howard-Flanders, *Nature (London)* **178**, 978 (1956).
3. P. Howard-Flanders and T. Alper, *Radiat. Res.* **7**, 518 (1957).
4. P. Howard-Flanders and D. Moore, *Radiat. Res.* **9**, 422 (1958).
5. G. W. Barendsen, *2nd Symp. Microdosimetry* (H. G. Ebert, Ed.); pp. 82–106. EURATOM EUR 4452 d-f-e, 1970.
6. P. E. Bryant, *Int. J. Radiat. Biol.* **23**, 217 (1973).
7. T. Alper, *2nd Symp. Microdosimetry*, (H. G. Ebert, Ed.), pp. 5–50. EURATOM EUR 4452 d-f-e, 1970.
8. T. Alper, *Nature (London)* **217,** 862 (1968).
9. P. Howard-Flanders, *Adv. Biol. Med. Phys.* **6**, 554 (1958).
10. J. Kiefer and D. Averbeck, *Int. J. Radiat. Biol.* **24**, 325 (1973).
11. E. J. Hart and M. Auber, *The Hydrated Electron*, Wiley-Interscience, New York-London-Sydney-Toronto, 1970.
12. B. Cerček and L. Cerček, *Int. J. Radiat. RBiol.* **24,** 137 (1973).
13. Z. M. Bacq and P. Alexander, *Fundamentals of Radiobiology*, Pergamon Press, New York-Oxford-London-Paris, 1961.
14. J. W. Boag, *Curr. Top. Radiat. Res.* **5**, 141 (1969).
15. J. Kiefer, *3rd Symp. Microdosimetry* (J. Boaz, H. G. Ebert, R. Eickel, A. Walker, Eds.) pp. 441–461. EURATOM EUR 5122 d-e-f, (1974).
16. G. E. Adams and R. L. Willson, *Trans. Faraday Soc.* **65**, 2981 (1969).
17. R. Bensasson, C. Chachety, E. J. Land, and E. Salet, *Photochem. Photobiol.* **16**, 27 (1972).

18. E. R. Epp, H. Weiss, N. A. Kessaris, H. Santomasso, J. Heslin, and C. C. Ling, *Radiat. Res.* **54**, 171 (1973).
19. I. Johnansen, in: *Modification of Radiosensitivity in Biological Systems*, IAEA, Geneva (in press).
20. D. Averbeck and M. Ebert, in *Mutat. Res.* **19**, 305 (1973).
21. C. D. Town, K. C. Smith, and H. S. Kaplan, *Curr. Top. Radiat. Res. Q.* **8**, 305 (1973).
22. A. Tallentire and A. B. Jones, *Int. J. Radiat. Biol.* **24**, 345 (1973).
23. J. J. van Hemmen, W.J.A. Menling, G. P. van der Schans, and J. F. Bleichrodt, *Int. J. Radiat. Biol.* **25**, 399 (1974).
24. G. E. Adams, in *Radiation Protection and Sensitization* (H. L. Moroson and M. Quintiliani, Eds.), pp. 3–14. Taylor and Francis, London, 1970.
25. Y. L. Shekhtman, *Role of peroxides and oxygen in the first stages of the radiobiological effect*, Akad. Neuk., Moscow, 1960.
26. G. J. Neary, *Int. J. Radiat. Biol.* **9**, 477 (1965).
27. A. Howard, C. W. Gilbert, and D. Green, *Radiat. Bot.*, 1974 (in the press).
28. T. Alper, J. L. Moore, and P. Smith, *Radiat. Res.* **32**, 780 (1967).
29. M. M. Elkind, R. W. Swain, T. Alescio, H. Sutton, and W. B. Moses, in *Cellular Radiobiology*, pp. 442–461. Williams and Wilkins Co., Baltimore, 1965.

THE SIGNIFICANCE OF LET IN RADIOBIOLOGY[1]

Chairman: G. W. Barendsen

[1]Symposium A–2

Fundamental Aspects of LET in Radiobiology[1]

J. F. Fowler

Gray Laboratory of the Cancer Research Campaign, Mount Vernon Hospital, Northwood, Middlesex, HA6 2RN, Great Britain

INTRODUCTION

Studies with radiations of different Linear Energy Transfer (LET) are important in radiobiology because they illuminate the field from a new direction, bringing both answers and more questions than can come from studies with low LET radiation only. As LET increases from gamma rays of 0.2 keV per micron of electron track to heavy charged nuclei of over 1000 keV/μm, the relative biological effectiveness (RBE) for many effects in plant and mammalian cells rises to a peak at about 100 keV/μm and then falls (Fig. 1). At low LET, oxygen and other modifiers can alter the response to irradiation, but at values of LET exceeding the peak of RBE little or no such modification occurs (*1*). Dr. Barendsen has asked me to emphasize "areas of doubt and further work" and Fig. 1 illustrates also the unsolved question "is the biological target only DNA?" The RBE for single-strand breaks in DNA falls steadily as LET increases, as for enzymes and other small target molecules where a single "event" (i.e. passage of a particle track) is sufficient to inactivate the molecule. Fig. 1 shows that the RBE for double-strand breaks in DNA does not rise significantly with LET until 100 keV/μm is exceeded (*2*) both for dry DNA and for DNA in phages. Therefore the peak in plant and mammalian RBE at about 100 keV/μm *cannot be due simply to DNA breakage,* but to interaction with other components of the cell. We hope more light will be thrown on this question during the present Congress.

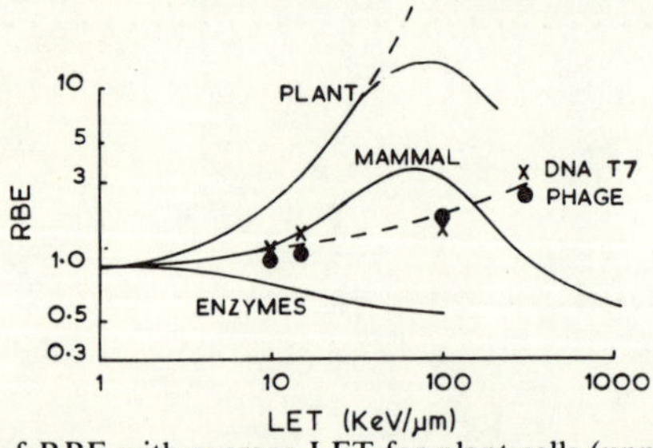

FIG. 1. Variation of RBE with average LET for plant cells (upper curve), mammalian cells (middle full curve), and enzymes (lowest curve). The dotted curve is for double-strand breaks in DNA: crosses, extracted dry; dots, in phage. The RBE for single-strand breaks in DNA follows the enzyme curve.

[1] Congress presentation A–2–1.

The Twenty Questions of Radiobiology might be put as listed in Tables 1, 2 and 3 from the three areas of Radiation Hazards, Radiotherapy, and Fundamental Mechanisms, respectively.

TABLE 1

Radiation Protection
1. Can predictions replace some experiments?
2. Does RBE level off at low doses?
3. Does RBE level off at low dose rates?
4. How large can RBE values become?
5. What values apply to different human tissues?
6. Accidental neutron exposures.
7. Other ecological systems.

TABLE 2

Radiation Therapy
1. Can predictions save injury?
2. Can predictions guide machine design?
3. Does RBE level off at low doses per fraction?
4. How different can RBE be, for various critical normal tissues?
5. How different can RBE be, for different types of tumor?
(3, 4 & 5 require knowledge of relative repair, X rays vs. high LET)
6. How does OER vary in different types of beam?
7. How much do OER and RBE vary with depth in tissue?
8. Is low dose rate of X rays equivalent to high LET?

TABLE 3

Basic Questions

Comparisons over a wide range of LET are essential.

1. TARGET VOLUMES
   Size, shape and energy required.
2. REPAIR
   What are the conditions for "loss of repair" in various cells?
   Is it very variable in different types of mammalian cell?
3. DOSE-RATE
   Is there then no dose-rate effect?
4. MODIFICATION
   Why does OER go down with increasing LET? Do both components of OER go down? What about other radiosensitizers? Radioprotectors?
5. BIOLOGICAL EXPERIMENTS
   More data are needed before theories can be tested properly.
6. RADIATION CHEMISTRY
   Pulse radiolysis, rapid mix—much to do.

Concerning Q. 3 in Table 3, one wonders whether the absence of a dose rate effect at high-LET also applies to the "single-hit" component of low-LET radiation, i.e. to the low-dose region of X ray dose-response curves. Concerning Q. 4, recent studies using time-resolving techniques have demonstrated an *after*-effect of oxygen (*3,4*), so that at least two different processes contribute to sensitization by oxygen. The details of why OER decreases with increasing LET are not even yet understood.

TRACK SEGMENT EXPERIMENTS

Let us start with an example of the simplest kind of LET experiment, attempting to answer a basic question: what is the size, shape and energy requirement of target volumes which are critical for mammalian cell sterilization? Fig. 2 shows the series of cell survival curves obtained when kidney cells, as a monolayer cuture, were exposed to beams of X rays, deuterons, or helium nuclei (*5*). Each cell intercepts only a short segment of the particle track and the LET is as well-defined as is ever possible in a biological experiment: but statistical fluctuations of LET along the track and the presence of delta rays spoils even this simplicity. As the LET is increased, the shoulder is first removed from the survival curves (i.e. "repair" is abolished) and then the high-dose part of the survival curve also steepens. This is interpreted as an increase in the proportion of cell-killing by "one event," which deposits so much energy in the molecular lesion at one track passage that there is no chance of repair. Even for low-LET X-rays, a small proportion of this type of damage apears to be present because the initial slope of the cell survival curve is finite, not zero. We shall return to this point later.

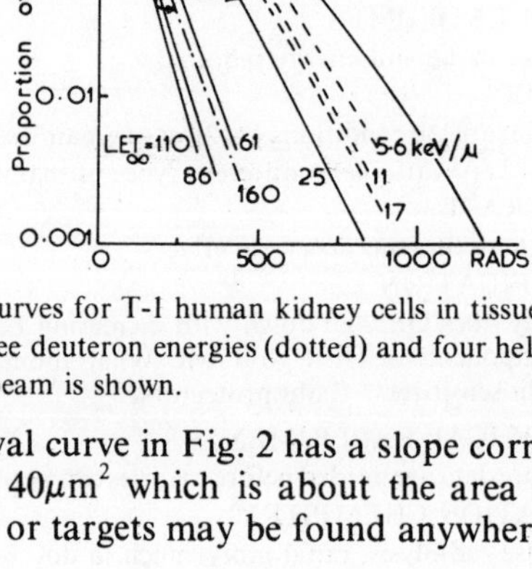

FIG. 2. Cell survival curves for T-1 human kidney cells in tissue culture exposed to: (top to bottom) 250 kV X rays, three deuteron energies (dotted) and four helium nuclei energies (*5*). The total-particle LET of each beam is shown.

The steepest survival curve in Fig. 2 has a slope corresponding to 1 helium particle track through $40 \mu m^2$ which is about the area of the cell nucleus. It suggests that the target or targets may be found anywhere within the volume of the cell nucleus.

The change of biological effect with LET can be analyzed by plotting RBE against mean LET (here the total particle LET, i.e. LET_∞[2]) as in Fig. 3. Several interesting conclusions can be drawn from these results.

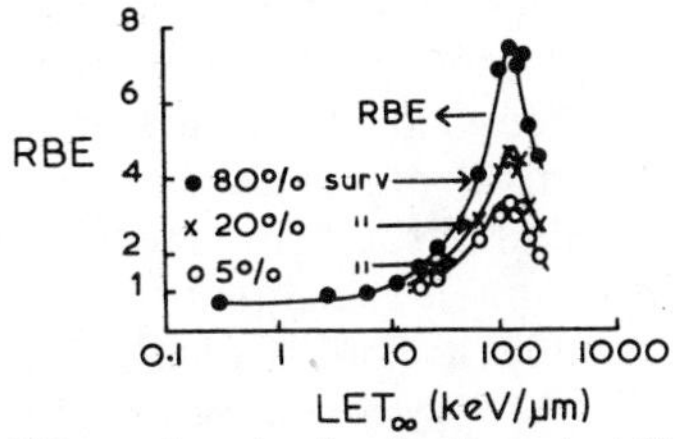

FIG. 3. R.B.E. vs log LET_∞ at three levels of cell survival (from Fig. 2).

1. *Rate of rise of RBE with LET* is linear, so 2 events are required (Neary, *6–8*).
2. *The amount of rise of RBE* is by a factor of about 7, so 10 to 20 events must have been deposited within the sensitive site (*5*). If it had been by only a factor of 2, then about only 2 events would be required instead.
3. *The LET at the peak of RBE* is 100 keV/μm; and for the 10–20 events just determined, this means a site thickness of 30 to 100 Å thick, in which 200–600 eV of energy has to be deposited, to kill an average of one cell (*5*).
4. But since only 2 of the 10–20 events appear to actually cause the injury, the whole site must contain smaller target regions some 2–5 Å thick. Neary (*6*) proposed the model of two strands of critical molecules (both DNA? or one DNA and one membrane?) each of which is damaged by one ionization; but chromosome damage only follows if the second is damaged within a limited distance of the first, so that chromosome interchange can occur. These distances (site diameters) may be as large as 0.2 μm in plant systems. Ionizations occurring elsewhere in the site are largely irrelevant, although other theories give them a proportion of damage to contribute by free radicals which diffuse to either of the ionized critical molecules to add to, or fix, the damage caused by the first ionization.

The picture we get is of very small strands, occupying any position in the cell nucleus, two of which have to be damaged almost simultaneously by ionizations occurring within 0.2 microns of each other; or less, perhaps a hundred Ångströms or so, in mammalian systems. The time scale of "almost simultaneous" has not been well established and we can look to pulse radiolysis and rapid-mix methods to elucidate this further. Now, why do we talk about 0.2 microns as the site diameter or interaction distance? (*8*)

MICRODOSIMETRY

Another way of analyzing the biological effectiveness of radiations of different LET or "quality" is to consider how much energy is deposited in small volumes of tissue by the different types of radiation (*9*). The fluctuations of

[2] LET is defined as energy lost per unit length of track; included are all secondary (delta) particles up to the energy cut-off of Δ eV if the LET is expressed as LET_Δ. Here LET_∞ is the same as stopping power. The *track average LET* is useful if RBE decreases inversely as LET increases. The *dose average LET* is useful if RBE increases proportionally with LET. It is clear that no single average of LET can represent the effects of a composite spectrum of LET, as is usually present.

energy density (i.e. dose) from one small volume to another are due to the total energy deposited by several charged particle tracks in some volumes, by one track in some, and by none in others.

Now, when the absorbed dose is extremely small, the probability of more than one track passing through any small volume is negligible. One track will pass through some and none through most volumes. In this "low dose" condition, each increment of energy in a small volume gives the same distribution of "LET" (if divided by the mean track length through the small volume), or of "microdose" (if divided by the mass of the volume). These quantities, which vary stochastically from one small volume to another, are called *lineal* energy (y) and *specific energy imparted* (z) (ICRU, ref. *10*). At very low doses, each increment of energy is a single event only, so the distribution of lineal energy y, analogous to LET, is very simply related to the distribution of specific energy z, i.e. the microscopic distribution of dose. The main difference between y and LET is that y has a geometrical cut-off—the size of the spherical volume—whereas LET has an energy cut-off. The y spectrum can be measured experimentally for volumes of about 0.1 to 1 micron (*9,11*). The result leads to the spectrum of specific energy z for low doses, which in turn must be related to the biological effect.

For example, y spectra for ^{60}Co and for fast neutrons are shown diagrammatically in Fig. 4. ^{60}Co gives more low-y events whereas fast neutrons generate more high-y events. Now, from a knowledge of these spectra, and of the RBE at low doses, deductions can be made about the size and critical energy requirement of the biological targets.

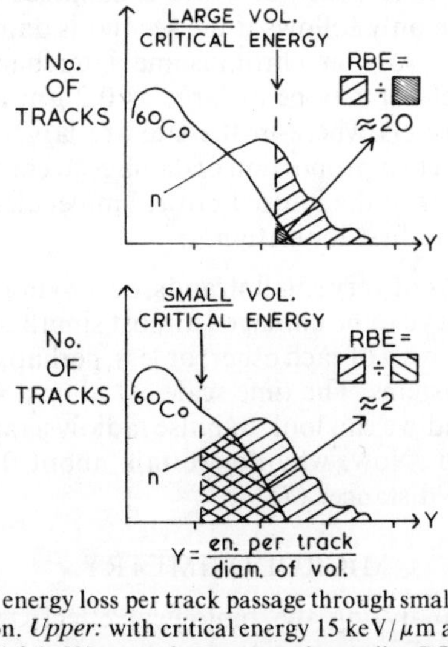

FIG. 4. Distribution of energy loss per track passage through small volumes in terms of "micro-LET" or Y in keV per micron. *Upper:* with critical energy 15 keV/μm and consequently large RBE. *Lower:* with critical energy 1.2 keV/μm and consequently smaller RBE. (*9*).

If a certain critical energy must be exceeded before the biological effect is seen, then, as indicated by the dotted vertical lines in Fig. 4, the proportion of small volumes receiving more than this energy is represented by the area under each curve to the right of the dotted line. If the biological effect is simply proportional to the absorbed dose—as it should be for the very low doses we are discussing—then RBE will be proportional to the ratio of the two shaded areas in Fig. 4. The two graphs of Fig. 4 illustrate how this model can be used to indicate large or small volumes. The upper graph represents a large critical energy, and thus probably a large volume. The dotted line is towards the right and the expected RBE would be about 20. The lower graph represents a small critical energy, with the dotted line towards the left and a smaller RBE of 2. From experimental values of RBE at low doses and from measured y distributions, a crude estimate of the threshold energy and size of sensitive site in a given biological system can be obtained. In this way, using the high RBE's found for various effects in plant systems, Rossi (*12,13*) found that sensitive sites may be as large as 0.2 μm, the figure which Neary (*6–8*) also obtained by quite different geometrical considerations about LET and track geometry, from experimental results on chromosome abberations in plants.

The RBE's of mammalian cells for cell sterilization seldom rise above 8 and are consistent with sensitive sites much smaller than in plants, by possibly a linear factor of ten, i.e. 100 to 200 Å. Nevertheless, certain mammalian systems do show very high RBE values: up to about 20 for cancer induction by fission neutrons in rats (*14*) and 60 for opacification in the lens of the mouse eye (*15*), for neutron doses of 3 and 0.5 rads respectively (*16*).

The detailed application of microdosimetry distributions of y and z to explain biological results is still an area requiring further work and the question of small (100 Å or less) or large sites (2000 Å, i.e. 0.2 μm) cannot be regarded as settled.

VARIATION OF RBE WITH ABSORBED DOSE

Dose-response curves for high-LET radiation tend to be straighter than those for low-LET; less repair of sublethal injury is evident. The consequence is that RBE rises as dose (and level of effect) falls. Fig. 5 shows typical dose-response curves for X rays and cyclotron neutrons derived from our skin reaction experiments (*17*).

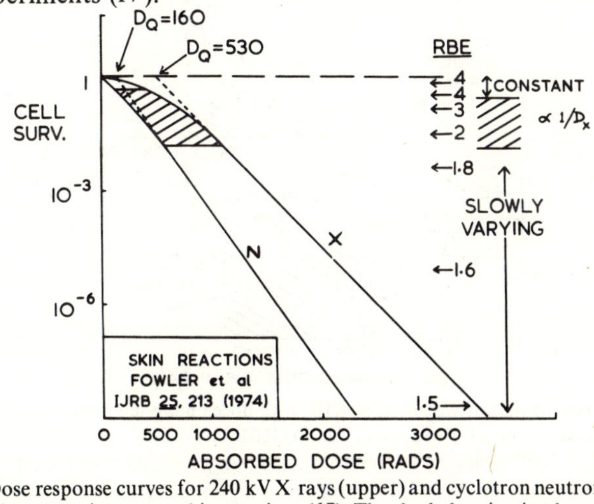

FIG. 5. Dose response curves for 240 kV X rays (upper) and cyclotron neutrons (lower) derived from multiple dose experiments on skin reactions (*17*). The shaded region is where the X ray curve is bending most rapidly.

It is clear from Fig. 5 that the higher the dose, the nearer the RBE comes to reaching a constant value (at doses so large that D_Q[3] is negligible). For very low doses, on the other hand, RBE only becomes constant if both X ray and high-LET survival curves have a straight exponential region of finite slope at very low doses. This has now been demonstrated for X rays in a number of mammalian systems (*17–20*).

The levels of damage over which the X ray curve continues to bend (shown shaded in Fig. 5) cover a dose region where the RBE is changing rapidly. Rossi (*16*) has pointed out that if the X ray response can be represented by a term proportional to dose plus one proportional to dose squared—a classical idea in radiobiology—then:

$$S_L = 1 - aD_L - bD_L^2$$

and

$$S_H = 1 - cD_H$$

where S_L, S_H are cell surviving fractions after low and high LET radiation respectively; D_L and D_H are the absorbed doses of low and high LET radiation; and *a, b* and *c* are coefficients. When $S_H = S_L$, then the RBE is, by definition, D_L/D_H. At very small doses, RBE would be constant and equal to c/a. At large doses, RBE would also tend to become constant, as explained above. In the intermediate range, RBE should be proportional to $1/D_L$ and to $1/\sqrt{D_H}$. Fig. 6 illustrates this idealized relationship between RBE and dose. In the lower part the log-log slope of $-^1/_2$ corresponding to the square root law is shown. The region of constant RBE at the low dose end defines what is mean by a "very low dose." Here only single-track events contribute to the biological effect, both for low and high LET radiation. In addition to low dose, the use of low dose rates of radiation helps this condition because the coincidence of a second track through the same critical volume within a time short enough to add to the injury is very unlikely.

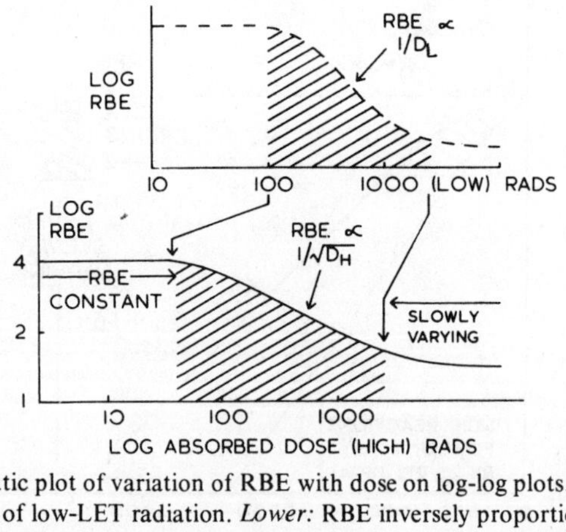

FIG. 6. Schematic plot of variation of RBE with dose on log-log plots. *Upper:* RBE inversely proportional to dose of low-LET radiation. *Lower:* RBE inversely proportional to square roots of high-LET radiation. RBE tends to become constant at very low and very high doses (*16,19*).

[3] D_Q is the intercept of the exponential (high-dose) part of the survival curve on the 100% survival abcissa.

Thus the physical measurements of y and z are relevant to low dose and low dose rate RBE values, regardless of any time factor.

How do experimental results compare with this theory? As Rossi (*16*) has shown, there is an excellent square root relationship for Smith's results in corn seeds down to a few rads (*13*), for lens opacities in mice down to less than a rad (*15*), and for induction of mammary carcinoma in rats (*14*). In the latter results, the curve of log RBE vs. log dose is becoming flatter at neutron doses above 20 rads, suggesting that the transition from "one-hit" to "multiple event" damage is beginning to occur for the low LET radiation at the corresponding dose (about 240 rads of X rays).

If, however, the RBE for skin reactions in mice, rats, pigs or human patients areplotted against neutron dose as in Fig. 7, the slope is −0.03 instead of −0.5(*21*). This is because the neutron dose-response curve is not straight, but has some curvature at the same level of damage as with X rays (the shaded region in Fig. 5).

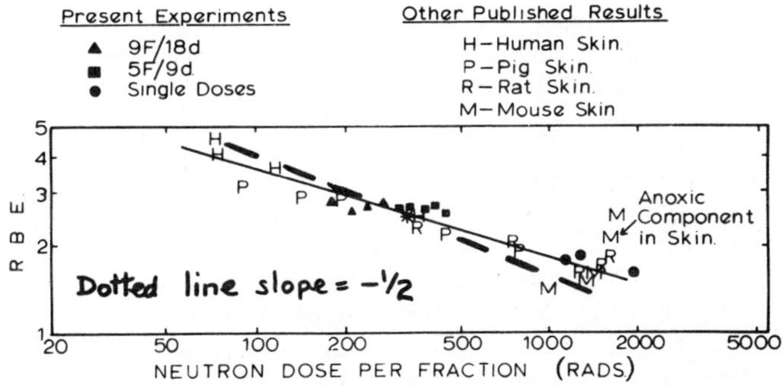

FIG. 7. Experimental results of log RBE vs log neutron dose for skin reaction in mice (M), rats (R), pigs (P) and human patients (H) (*21*). The dotted line of slope −1/2 is significantly steeper because the neutron dose-response curve in Fig. 5 is not straight (see text).

CONSEQUENCES OF REDUCED REPAIR AT HIGH LET

The reduced repair capability of cells after high-LET radiation has important implications in radiation therapy. Let us first consider the case of X rays. A straight exponential region of the dose-response curve with a finite slope, at very low doses, could be explained either by a proportion of cells peculiarly sensitive to single event killing (e.g. cells in mitosis), or by a proportion of damage due to single events of sufficient local energy density to overcome repair (i.e. the "high-LET" component of low-LET radiation, as mentioned above). Irrespective of whichever of these explanations is true, then as the dose per fraction is decreased in multifraction treatments, the total dose necessary to produce a given level of damage increases at first, until each dose per fraction is small enough to be on the straight initial region, if any. Thereafter the total dose rises no further, even when a greater number of still smaller fractions is used. This is illustrated by Dutreix's curve (*22*) in Fig. 8, which plateaus at about 16 fractions. If, by contrast, further repair could occur even after the smallest X ray doses (i.e. the dose-response curve keeps bending as it approaches zero dose), then the total dose would keep rising (e.g. $N^{0.24}$ in Fig. 8). Our own experiments (Douglas and Fowler, *18*) indicate an approach to a plateau which is higher than that of

Dutreix et al. (*22*) but which does indicate a proportion of 15 to 20% of single-hit injury. This is significantly less than Dutreix's 40% for skin reactions (*22*).

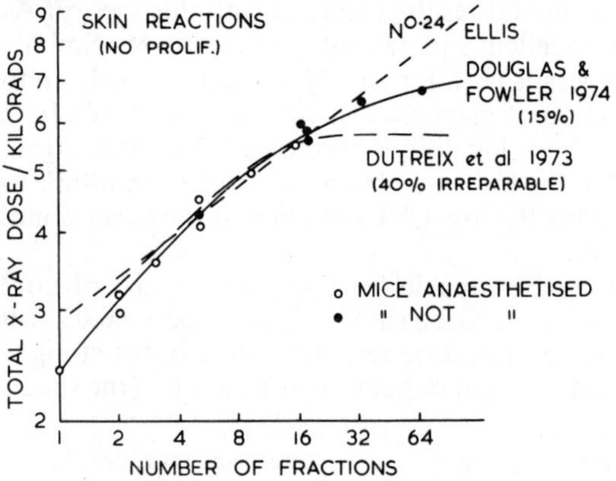

FIG. 8. Total dose of X rays required to cause a given skin reaction versus number of dose fractions. Results of our experiments shown as points (*18*).

The important point is that for X rays the total dose rises by a factor of at least 2.5 (ref. *22*) or more probably 4 (ref. *18*), as the number of fractions is increased. With cyclotron neutrons, however, representing a broad spectrum of moderately high-LET radiation, the increase is less and a plateau of total dose is reached after only 4 or 5 fractions and with an increase of only 1.5 times the single dose (Fig. 9, derived from Fig. 5). Here the proportion of single-hit injury is about 60%. The actual increase will be somewhat greater for higher energy neutrons (*23*), but there will be little or no change in total dose as fraction number is changed in the likely range of use. The practical importance is that different fractionation schedules will be easier to prescribe than with X rays. (As with X rays there will be a separate time-dependent factor, of course.)

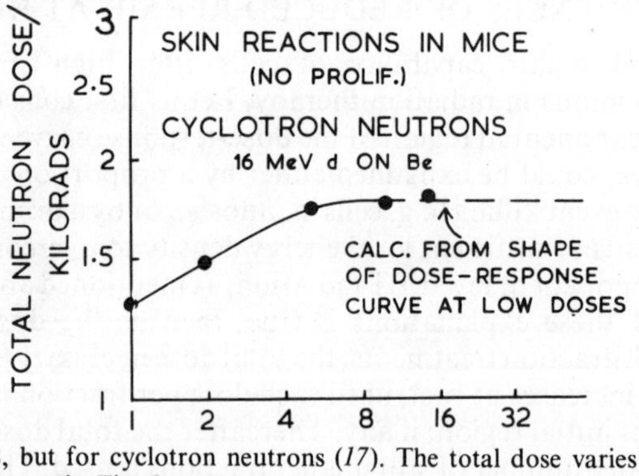

FIG. 9. As Fig. 8, but for cyclotron neutrons (*17*). The total dose varies less with fraction number than for the X rays in Fig. 8.

A recent result published by Howlett, Thomlinson and Alper (*24*) indicates that large differences may occur for different tumors in the relative fall in repair on changing from X rays to neutrons. The experiments were carried out with doses for which the effects of hypoxic cells were demonstrably absent so that the differences in RBE observed must be due to the relative sizes of the shoulders (i.e. repair capabilities) for X ray vs. neutron irradiation. This finding suggests that some types of tumor may show an advantage when treated with neutrons, independently of the reduced oxygen effect which is the main rationale for neutrons, but that other types may even show a disadvantage as pointed out earlier (*1*). Until we know more about selecting tumors with such differences in mind, clincial trials of neutrons may be diluted by unsuitable clinical material. This is an area where more fundamental cell biology is required, with respect to radiation injury and repair.

THE USE OF HIGH LET RADIATION TO DISTINGUISH BETWEEN REPAIR AND REPOPULATION

An interesting application of low-energy neutrons (mean energy 1 MeV from a reactor fission plate) has recently been reported by Kramer, Davids and Van der Ven (*25*). The increase in radiosensitivity of spermatogonial epithelium when fractionated radiation was given has been known since the days of Regaud. The reason for this has been investigated using 1 MeV neutrons, and the surviving stem cells appeared to enter a very radiosensitive phase of their cell cycle. The absence of repair after the neutron irradiation was essential for distinguishing between repair and proliferation. More applications along these lines are likely in future radiobiology.

THE OXYGEN ENHANCEMENT RATIO

Some questions concerning OER were posed at the beginning of this review. It has been known for some time that the OER for a broad range of neutron beams was constant at about 1.6; from Hammersmith's 16 MeV to TAMVEC's 50 MeV deutrons, on beryllium targets. This constancy of OER is in contrast to the small but definite decrease of RBE as neutron energy is increased (*23*). It must mean that OER falls more steeply than might have been expected some years ago, as the "average LET" of a beam is increased. Recent experiments by Raju and Jett (*26*) and by Railton et al. (*20*) have demonstrated this change clearly. Figs. 5, 10 and 11 show curves of RBE and OER plotted against the proportion of high-LET radiation in a virtually simultaneous irradiation with X rays and high-LET radiation. The OER falls steeply at first, reaching a (usefully) low value for only a modest proportion of high LET radiation, whereas RBE rises steadily throughout. Although these results are plotted for 10% or 5% cell survival, quite similar conclusions are demonstrated if the "single-hit" components at the 80% survival are used instead. It is not yet known exactly why oxygen sensitization is so vulnerable to a small proportion of high LET radiation in a mixed beam. However, the result is entirely in agreement with the observation from Fig. 2 that, as LET is increased, the "repairable shoulder" (i.e. the modifiable component, including by oxygen) is eliminated first, and then the main slope is steepened. This would all be expected intuitively from an increase in the proportion of "single-event" damage as LET increases.

Uncertainty still surrounds the question of lower OER's at depth in tissue than at the surface. This was first reported by Berry and collaborators (*27*) although it could not be confirmed using other biological systems at the

Hammersmith cyclotron (*28*). The discrepancy may be due to the use of different cell systems, that used by Berry (*3*) losing its oxygen effect at lower values of LET than that used by Barendsen (*1*), but further evidence is required.

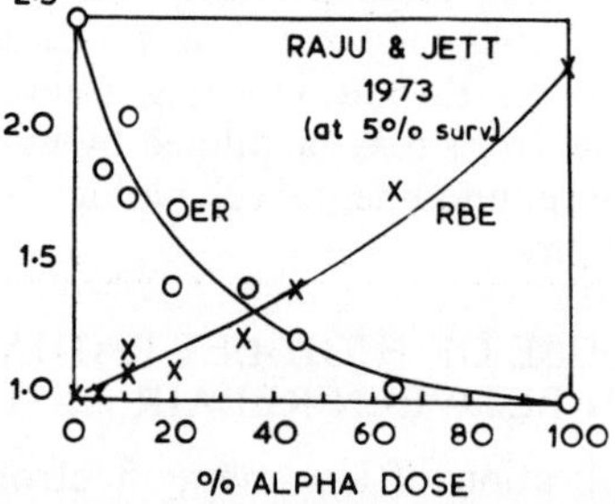

FIG. 10. Variation of OER (left scale) and RBE (right scale) with proportion of plutonium alpha particle dose in a mixed irradiation with X rays and α-particles. Measured at 5% cell survival (*26*).

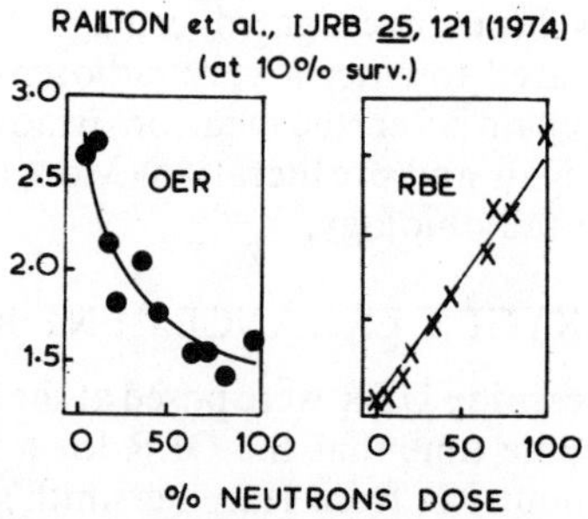

FIG. 11. As Fig. 10, but plotted against proportion of 14 MeV neutrons in a mixed irradiation with ^{60}Co and neutrons. Measured at 10% cell survival (*20*).

OER AND DOSE RATE

Dr. Eric Hall said that "Low LET at low dose rate may be the poor man's high LET radiation." This seems at first sight to be true; and there is a possible extension to it that seems important to investigate. As X ray dose rate is decreased, the dose-reponse curves become flatter, i.e. less senstive (*29*), and one can imagine that only the initial slope, the "single-hit" component, is relevant because all damage that could be repaired, has been repaired. It is an experimental finding that the OER is then also reduced (*29*) as shown by the shaded area in Fig. 12. There is less capability for further repair, and a smaller OER, in this relatively "non-reparable" single-hit component. It might indeed be expected, from the simplest biophysical consideration of the nature of "single-event" damage, that there would be no modification by oxygen at all, i.e. the OER would fall to unity. Experimentally, however, the finding is simply one of a significantly reduced OER.

If this experimental result is a true bill, it follows logically that the low-dose region of X ray survival curves should also show a reduced OER. (This would be expected if this "single-event" region were due to the "high LET" component present even in low LET radiation. It would not be expected if instead it were due to a proportion of cells in a sensitive phase of the cell cycle.) An objection to the low dose rate experiments in which OER was measured arises because the maintenance of hypoxia for many hours is likely to slow up the repair of sublethal injury, thus causing the hypoxic dose-response curve to be spuriously steep and the OER to be wrongly found low. In spite of this objection, it seems a real area of uncertainty, and preliminary experiments by McNally in our laboratory indicate that the OER does appear to be somewhat lower for the initial low-dose region of the X ray survival curve for Chinese hamster cells than for the high-dose region.

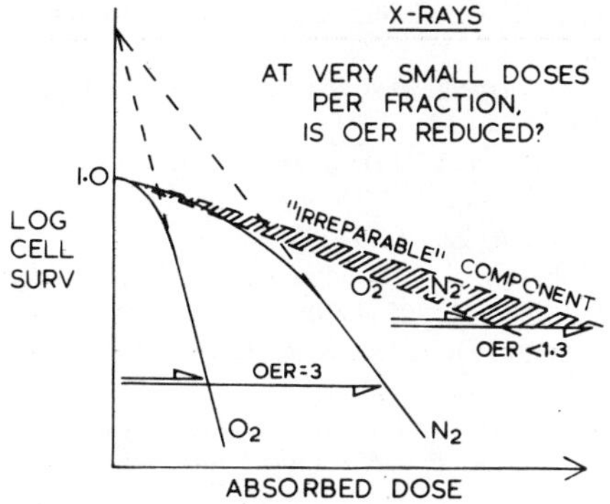

FIG. 12. Schematic diagram showing reduced OER for the low-dose, single-event, component of X ray response (*29*).

It is indeed possible that very low dose rate therapy, from interstitial radium sources, has always enjoyed this benefit of a reduced OER.

CONCLUSIONS

Some of the radiobiological features of radiation studies using radiations with different distributions of LET or *y* have been discussed, with emphasis on the areas of doubt. A considerable amount of data can be explained reasonably well, but some very basic questions remain unanswered, such as how much radiation injury is due to damage to DNA and why does OER go down as LET rises? Further studies using radiations of different LET are essential to help solve basic problems in radiobiology.

ACKNOWLEDGEMENTS

It is a pleasure to thank Drs. G. W. Barendsen, H. H. Rossi, S. B. Field, D. K. Bewley, G. E. Adams, E. J. Hall and J. Denekamp for stimulating conversations over several years. Any errors of interpretation are my responsibility and not theirs.

REFERENCES

1. G. W. Barendsen, in *Current Topics in Radiation Research,* Vol. IV. (M. Ebert and A. Howard, Eds.) pp. 293–365. North-Holland, Amserdam, 1968.
2. G. J. Neary, V.F.W. Simpson-Gildemeister and A. R. Peacocke, *Int. J. Radiat. Biol.* **18**, 25 (1970).
3. B. D. Michael, G. E. Adams, H. B. Hewitt, W. B. Jones and M. E. Watts, *Radiat. Res.* **54**, 239 (1973).

4. M. A. Shenoy, J. C. Asquith, G. E. Adams, B. D. Michael and M. E. Watts, *Radiat. Res.* **62**, (1975), in press.
5. G. W. Barendsen, H.M.D. Walter, J. F. Fowler and D. K. Bewley, *Radiat. Res.* **18**, 106 (1963).
6. G. J. Neary, *Int. J. Radiat. Biol.* **9**, 477 (1965).
7. G. J. Neary and J.R.K. Savage, *Int. J. Radiat. Biol.* **11**, 209 (1966).
8. G. J. Neary, R. J. Preston and J.R.K. Savage, *Int. J. Radiat. Biol.* **12**, 317 (1967).
9. H. H. Rossi, in *Radiation Dosimetry,* Vol. I (F. H. Attix, W. C. Roesch and E. Tochilin, Eds.), Academic Press, New York, 1967.
10. I.C.R.U. REPORT 19, *Radiation quantities and units.* International Commission on Radiation Units and Measurements, 7910 Woodmont Avenue, Suite 1016, Washington, D.C. 20014 (1971).
11. D. Srdoc, this symposium.
12. H. H. Rossi, in *Biophysical Aspects of Radiation Quality,* Technical Report Series No. 58, pp. 161–165. I.A.E.A, Vienna 1966.
13. H. H. Smith, N. C. Combatti and H. H. Rossi, in *Neutron irradiation of seeds II.* Technical Report Series No. 92, pp. 3–8. I.A.E.A., Vienna, 1968.
14. H. H. Vogel, *Nature (London)* **222**, 1279 (1969).
15. J. L. Bateman, H. A. Johnson, V. P. Bond and H. H. Rossi, *Radiat. Res.* **35**, 86 (1968).
16. H. H. Rossi, *Phys. Med. Biol.* **15**, 225 (1970).
17. J. F. Fowler, J. Denekamp, C. Delapeyre, S. R. Harris and P. W. Sheldon, *Int. J. Radiat. Biol.* **25**, 213 (1974).
18. B. G. Douglas and J. F. Fowler, in Proc. 6th L. H. Gray Conference on *"The Initial Shapes of Cell Survival Curves: Clinical and Theoretical Implications".* Pub. Institute of Physics, London (1974).
19. E. J. Hall, in *Proc. 37th Annu. Meet. Can. Assn. Radiologists*, Vancouver, B.C., 1974.
20. R. Railton, D. Porter, R. C. Lawson and W. J. Hannan, *Int. J. Radiat. Biol.* **25**, 121 (1974).
21. S. B. Field and S. Hornsey, *Eur. J. Cancer* **7**, 161 (1971).
22. J. Dutreix, A. Wambersie and C. Bounik, *Eur. J. Cancer* **9**, 159 (1973).
23. J. Broerse, this symposium.
24. J. Howlet, R. H. Thomlinson and T. Alper, *Brit. J. Radiol.* (in press).
25. M. G. Kramer, J.A.G. Davis and Th. P. A. van der Ven, *Int. J. Radiat. Biol.* **25**, 253 (1974).
26. M. R. Raju and J. H. Jett, *Radiat. Res.* (in press).
27. R. J. Berry, in *Radiobiological Applications of Neutron Irradiation.* Panel Proceedings Series—STI/PUB/325. I.A.E.A., Vienna, 1972.
28. N. J. McNally and D. K. Bewley, *Brit. J. Radiol.* **42**, 289 (1969).
29. E. J. Hall, *Brit. J. Radiol.* **45**, 81 (1972).

Energy Deposition in Small Volumes in Relation to Linear Energy Transfer (LET)[1]

D. Srdoč
Institute "Rudjer Bošković", 41001 Zagreb, POB 1016
Yugoslavia

INTRODUCTION

The final result of interaction of ionizing radiation with matter depends to a greater or lesser extent on the patterns of energy deposition within the matter. Many phenomena in physics (recombination of charges in gaseous and liquid media, light output of scintillators), in radiation chemistry (chemical yield), and in radiobiology (cell survival, gene mutation) are functions of the absorbed dose as well as the spatial distribution of the energy deposition. The term "energy deposition" involves ionization and all other forms of energy dissipation such as excitations, formation of radicals etc, which very often lead to subsequent processes. These processes are not necessarily confined within the so-called track structure, determined primarily by the ionization distribution, and may involve ionization, excitation, radical formation and dissociation at an appreciable distance from the original track. It is evident from this brief introduction that the problem of an adequate description of energy deposition patterns is very difficult and may involve a complex formalism. Several attempts have been made in the past to describe adequately the energy deposition structures. Lea calculated the primary ionization densities, the spectra of secondary ionization tracks (or delta rays), and stopping powers and published data for electrons, protons and alpha particles in water in his pioneering book on the action of radiation on living cells (*1*). Gray (*2*) proposed in 1947 the parameter *mean linear ion density* defined as $\overline{E}_o / R_{\overline{E}_o} \cdot W$, where

$\overline{E}_o$ = average initial kinetic energy of primary electrons

$R_{\overline{E}_o}$ = range of electrons of energy $\overline{E}_o$

W = average energy expended per ion pair formed in a gas.

Zirkle with coworkers (*3*) introduced in 1952 a relatively simple concept of *linear energy transfer, $L = dE/dt$,* which is generally accepted in an elaborated form defined by the ICRU (*4*). Rossi (*5*) however pointed out the deficiency of the LET concept when very small structures—on the order of 1 μm in diameter or less—are considered. LET is an *average* quantity which does not contain information on the energy deposition spectrum within the volume of interest. Factors influencing such a spectrum as well as the experimental methods to measure

[1] Congress presentation A-2-2

energy deposition distribution in small volumes will be described in subsequent sections.

ENERGY DEPOSITION IN SMALL VOLUMES

The fundamental difference between the LET concept, which gives the average value, and the energy deposition spectrum when small volumes are considered can be illustrated by an idealized experiment. A high pressure, parallel plate, pulsed ionization chamber could be used to measure the LET of a heavy particle, crossing the chamber perpendicularly to the plates. The measurement with a properly collimated and aligned beam would give a relatively narrow peak corresponding to the energy loss over a rather long path. Thus, the average dE/dl could be obtained experimentally. In order to obtain the data on energy loss over a much smaller distance ($\sim 1-10 \mu m$) one should reduce the distance between the plates, or (in practice) reduce the pressure. Providing that the system works down to a very small pressure, one would obtain a very broad energy distribution spectrum, in contrast to a rather narrow distribution at high pressure representing the average LET. Thus, the broad energy deposition spectrum reflects the real sequence of ionization events when a small volume is exposed to radiation. It should be emphasized that the actual measurement of energy deposition spectra is far more complicated due to the very small amount of electric charge formed in a small volume. The experimental technique of measurements of energy deposition spectra in small volumes is described in the next section.

It should be pointed out, however, that all microdosimetric data are based on measurements of electric charge (or number of electrons) in the experimental volume. Hence, the deposited energy calculation depends on W, i.e. energy required to form an ion pair. Although the W values are surprisingly constant over a large range of energy and vary within a few percent for common types of particles, some precautions have to be taken for extremely low energy particles and for heavy nuclei. Next, the W values are measured in gases, not in tissues or tissue-equivalent solid or liquid material. Finally, an important point should be stressed: On the average, W values are twice as large as the ionization potential of gases. It follows that (at least in gases) about 50 percent of the energy is released in forms other than ionization. In gases it is possible to measure spectroscopically and calculate other forms of energy dissipation, such as excitations, formation of metastables etc. Much less is known about mechanisms of energy dissipation in complex organic material, not to mention the repercussion of these processes on living structures. Also, when considering "locally imparted energy" it should be kept in mind that ionization and other forms of energy dissipation need not be confined to the same volume, even though originated along the path of the same particle.

A relatively complex formalism (when compared to the definition of the LET) has to be applied for specification of energy deposition distribution in small volumes of tissue. The basic definitions, as defined by Rossi and his coworkers (*6*, *7*) are simple and straightforward; however, the complete description of shapes of distributions for various types of radiation, their theoretical derivation and relationship between various definition and units is more complex. Rossi has defined the term "*event size*", Y, as the energy deposited in a single event (i.e. a crossing of a charged particle, or a pair formation, or any other ionizing interaction with matter) in a small sphere divided by its diameter. The dimensions of Y are the same as for LET, i.e. E/l. The "*event size*", Y, represents the actual energy deposited in the specified volume of interest, and the $P(Y)$ shows the probability distribution of all the events, expressed in Y units. Thus, the description of energy deposition fluctuation in a small volume is obtained and it does not require corrections for delta ray escape, track curvature, LET variation, simultaneous events etc. Furthermore, such distributions can be measured and checked by theoretical calculations. A number of various probability distributions can be derived from a measured Y spectrum. The fluctuation in the number of events at a certain dose is given by probability distribution of Z, $P(Z)$, where Z is defined as the *local energy density*, i.e. energy absorbed in a sphere divided by the mass of the sphere.

Theoretical considerations on energy deposition straggling are presented by Kellerer (*8*). According to the theory there are several factors which contribute to the spread of energy deposition spectra in small volumes. The most convenient method of handling and processing the complex distributions is to calculate the relative variance of contributing factors. The relative variance of measured spectra is the sum of relative variances of all contributing factors. The most important factors which bring about enormous spread in energy depositions are: (1) amount of energy delivered per primary collision, (2) number of collisions per particle crossing the specified volume, (3) track distribution in the specified volume, (4) number of electrons per deposited energy ΔE, and (5) LET distribution. In addition, there is an instrumental broadening of the measured spectrum due to the preamplifier noise and gas multiplication.

EXPERIMENTAL METHOD

Direct measurement of energy deposited within a small volume of tissue (e.g. a cell) when irradiated by ionizing radiation is not possible by any existing technique. Indirect methods are developed by simulating the tissue size and composition. The basic principle of experimental methods consists of measurement of electric charge developed within the simulated volume as a result of energy deposition. Sometimes an indirect method of detecting charged particles is adopted such as cloud chamber or track etching methods.

The first and most successful experimental technique for measurement of energy deposition distribution in small irradiated volumes was developed by Rossi and Rosenzweig (*9*). The device consists of a specially designed spherical proportional counter. The counter wall as well as the gas filling are chosen to match as close as possible the elemental composition of tissue. Other specific features of a typical Rossi counter are: (1) The anode wire is surrounded by a helix to keep the electric field along the anode as uniform as possible; (2) a gas flow system maintains the gas composition and pressure constant (the gas multiplication is extremely sensitive to gas contamination and pressure variation); (3) a calibration source is essential to convert the pulse height spectrum into the energy deposition spectrum. Details of the experimental technique and inherent limitations of the method has been presented by Srdoč (*10*). The experimental data for various types of radiation and different tissue size are scattered in the literature. No attempt has been made to compile and scrutinize the microdosimetric data. The following list of references is composed from the available literature collected by the author:

Alpha particles:	Ref. (*11*) to (*13*)
Beta particles:	Ref. (*14*) to (*17*)
X and gamma rays:	Ref. (*18*) to (*27*)
Protons:	Ref. (*28*) to (*30*)
Neutrons:	Ref. (*31*) to (*46*)
Heavy particles:	Ref. (*47*) to (*50*)

Most of the data were measured by using Rossi's original or modified method.

There are several other experimental methods for measurement of electric charge developed in a small volume from which microdosimetric data can be derived. Bengtsson (*50*) developed a method which consists of measurement of current fluctuations in an ionization chamber. Burlin (*51*) recently proposed measurement of secondary electrons emitted from a solid surface. Cloud chamber photographs also revealed patterns of ion pair distribution along particle tracks in a gas (*52*). Microdosimetric data can be obtained by studying spatial distribution of grains in a developed photographic emulsion (*53*) and by so-called track etching technique (*54*).

Another experimental technique—closely connected to microdosimetry—is the measurement of quality factor for radiation protection purposes. An average LET could be derived from microdosimetric data. Instrumentation for radiation protection has been developed by Baum et al. (*55*) and the author and his coworkers (*56*). Both devices use a Rossi counter as the radiation detector. Another principle is used in the recombination chamber quality factor meter developed by Zielczynski (*57*) and by Sullivan and Baarli (*58*). Here columnar recombination is a function of track density and the difference between the ionization current in the saturation region and below saturation gives information on the quality factor of the measured radiation.

CONCLUSION

It becomes evident that data on the energy deposition in tissue on a microscopic scale are extremely valuable and essential for interpretation of mechanisms of radiation damage in living organisms. Regardless of how successful are the present hypotheses or proposed mechanisms of radiation damage to cells, the microscopic distribution of energy deposition cannot be ignored. Another fact is that energy deposition in living structures is not the only factor which has to be taken into account when trying to interpret experimental data in radiobiology because of complexity of the systems and interplay of several factors. Radiation damage mechanisms based on microdosimetric concepts were recently reviewed by Booz (*59*).

The evaluation of radiation hazard in neutron or heavy particle fields is based on the LET-QF relationship as recommended by the ICRP (*60*). Very often radiation composition in a mixed field is not known and therefore it is impossible to attribute any reasonable LET value to such a radiation. The measurement of microdosimetric quantities, particularly the *Y* spectrum, may help in such circumstances, even though the calculation of the average LET or a LET distribution from a *Y* spectrum is based on several assumptions which may not be fulfilled in all cases (*61*). A more direct application of microdosimetric data to the subject of radiation protection has been suggested by the author (*62*).

Concerning high LET radiation radiotherapy, which is very promising, the application of microdosimetric methods will certainly help to establish patterns of energy distribution at different depths in the body. Such measurements are essential to evaluate the dose equivalent at different depths and particularly at the site of interest. Thus, microdosimetric technique could prove to be essential in depth dose measurement in phantoms and in patient treatment.

REFERENCES

1. D. E. Lea, *Actions of Radiations on Living Cells*, Second Edition, Cambridge University Press, London, 1956.
2. L. H. Gray, *Brit. J. Radiol.* **Suppl. No. 1**, 7, (1947).
3. R. E. Zirkle, D. F. Marchbank and K. D. Kuck, Exponential and sigmoid survival curves resulting from alpha and x-irradiation of Aspergillus spores, *J. Cell. Comp. Physiol.* **39, Suppl. 1**, 75, (1952).
4. International Commission on Radiation Units and Measurements (ICRU). Report 11. Radiation Quantities and Units. International Commission on Radiation Units and Measurements, Washington, (1968).
5. H. H. Rossi, *Radiat. Res.* **10**, 522 (1959).
6. A. M. Kellerer and H. H. Rossi, Summary of quantities and functions employed in microdosimetry, Second Symposium on Microdosimetry (Euratom), Stresa, Italy, (October 1969), See also (7).
7. H. H. Rossi, in *Radiation Dosimetry*, Second Edition, Vol. I. (Frank H. Attix and William C. Roesch, Eds.), p. 43. Academic Press, New York, 1968.
8. A.M. Kellerer, in *Biophysical Aspects of Radiation Quality*, pp. 89–103, IAEA, 1968.
9. H. H. Rossi and W. Rosenzweig, *Radiology* **64**, 404 (1955).
10. D. Srdoč, *Radiat. Res.* **43**, 302 (1970).
11. S. B. Field and K. Wilson, in *First Symposium on Microdosimetry (Euratom)*, pp. 147–160. Ispra, Italy, 1967.
12. W. Gross, in *Second Symposium on Microdosimetry (Euratom)*, p. 249. Stresa, Italy, 1969.
13. B. Hogeweg and G. W. Barendsen, in *Third Symposium on Microdosimetry (Euratom)*, p. 857. Stresa, Italy, 1971.
14. L. A. Braby and W. H. Ellett, *Ionization in Microscopic Volumes Irradiated by Energetic Photons*, Radiation Center, Oregon State University, 1971.
15. M. J. Berger, in *Second Symposium on Microdosimetry (Euratom)*, p. 541. Stresa, Italy, 1969.
16. J. P. Patau et al., in *Fourth Symposium on Microdosimetry*, p. 755. Verbania Pallanza, Italy, 1973.
17. L. Lindborg, in *Fourth Symposium on Microdosimetry (Euratom)*, p. 799. Verbania Pallanza, Italy, 1973.
18. M. H. Biavati and E. Boer, *Annual Report on Research Project, NYO-2740-3*, pp. 87–101. Columbia University, 1966.
19. J. F. Dicello and D. Srdoč, *Annual Report on Research Project, NYO-2740-6*, pp. 131–135. Columbia University, 1969.
20. B. Hogeweg and G. W. Barendsen, in *Second Symposium on Microdosimetry (Euratom)*, pp. 171–181. Stresa, Italy, 1969.

21. D. Srdoč, in *Second Symposium on Microdosimetry (Euratom)*, pp. 153–170. Stresa, Italy, 1969.
22. L. G. Bengtsson, in *Third Symposium on Microdosimetry (Euratom)*, pp. 483–496. Stresa, Italy, 1971.
23. W. H. Ellett and L. A. Braby, in *Third Symposium on Microdosimetry (Euratom)*, p. 471. Stresa, Italy, 1971.
24. L. Lindborg, in *Fourth Symposium on Microdosimetry (Euratom)*, pp. 799–821. Verbania Pallanza, Italy, 1973.
25. D.M. J. Benstock, T. E. Burlin and J. A. Simmons, in *Fourth Symposium on Microdosimetry (Euratom)*, pp. 855–868. Verbania Pallanza, Italy, 1973.
26. R. Dvorak and W. Gross, *Annual Report on Research Project, COO-3243-1*, pp. 60–68. Columbia University, 1972.
27. R. Dvorak and W. Gross, *Annual Report on Research Project, COO-3243-2*, pp. 52–60. Columbia University, 1973.
28. W. Gross and R. C. Rodgers, in *Third Symposium on Microdosimetry (Euratom)*, p. 873. Stresa, Italy, 1971.
29. N. A. Baily and J. E. Steigerwalt, *Radiat. Res.* **56,** 213 (1973).
30. J. E. Steigerwalt and N. A. Baily, *Radiat. Res.* **53, 1** (1973).
31. M. H. Biavati, H. H. Rossi and E. Boer, *Annual Report on Research Project, NYO-2740-2*, pp. 41–76. Columbia University, 1965.
32. R. S. Caswell, *Radiat. Res.* **27,** 92 (1966).
33. M. H. Biavati and E. Boer, *Annual Report on Research Project, NYO-2740-4*, pp. 74–80. Columbia University, 1967.
34. K. S. J. Wilson and S. B. Field, in *Second Symposium on Microdosimetry (Euratom)*, pp. 137–152, Stresa, Italy, 1969.
35. U. Oldenburg and J. Booz, in *Second Symposium on Microdosimetry (Euratom)*, pp. 269–281. Stresa, Italy, 1969.
36. S. B. Field and K. Wilson, in *First Symposium on Microdosimetry (Euratom)*, pp. 147–160. Ispra, Italy, 1967.
37. H. Bichsel, P. Wootton and P. Heintz, in *First Symposium on Neutron Dosimetry in Biology and Medicine*, pp. 93–109. Neuherberg München, 1972.
38. U. Oldenburg and J. Booz, in *Third Symposium on Microdosimetry (Euratom)*, p. 511. Stresa, Italy, 1971.
39. J. F. Dicello, U. Kraljevic and W. Gross, *Annual Report on Research Project, NYO-2740-7*, p. 91. Columbia University, 1970.
40. R. C. Rodgers, W. Gross and H. H. Rossi, *Annual Report on Research Project, NYO-2740-8*, pp. 38–46. Columbia University, 1971.
41. R. S. Caswell and J. J. Coyne, in *Fourth Symposium on Microdosimetry (Euratom)*, p 967. Verbania Pallanza, Italy, 1973.
42. J. Booz and M. Coppola, *Ibid.*, p. 983.
43. M. Coppola, D. Pirrwitz and J. Booz, *Ibid.*, p. 1001.
44. R. C. Rodgers and W. Gross, *Ibid.* p. 1027.
45. R. Rodgers, J. Dicello and W. Gross, *Annual Report on Research Project,COO-3243-2*, pp. 87–101. Columbia University, 1973.
46. R. Rodgers, J. Dicello and W. Gross, *Annual Report on Research Project, COO-3243-1*, pp. 52–59. Columbia University, 1972.
47. R. C. Rodgers, J. F. Dicello and W. Gross, *Radiat. Res.* **54,** 12 (1973).
48. J. F. Dicello, U. Kraljevic and W. Gross, *Annual Report on Research Project, COO-3243-2*, pp. 64–86. Columbia University, 1973.
49. J. E. Turner, R. N. Hamm and H. A. Wright, in *Fourth Symposium on Microdosimetry, (Euratom)*, pp. 75–91. Verbania Pallanza, Italy, 1973.
50. L. G. Bengtsson, in *Second Symposium on Microdosimetry (Euratom),* p. 375. Stresa, Italy, 1969.
51. T. E. Burlin, in *Fourth Symposium on Microdosimetry (Euratom)*, p. 35. Verbania Pallanza, Italy, 1973.
52. D. C. Hodges, H. J. Delafield and J. A. Dennis, in *Third Symposium on Microdosimetry (Euratom)*, p. 455. Stresa, Italy, 1971.
53. R. Katz, in *Second Symposium on Microdosimetry (Euratom)*, p. 615. Stresa, Italy, 1969.
54. H. G. Paretzke, in *Fourth Symposium on Microdosimetry (Euratom)*, p. 141. Verbania Pallanza, Italy, 1973.

55. J. W. Baum, A. V. Kuehner and R. L. Chase, *Health Phys.* **15**, 195 (1968).
56. D. Srdoč and B. Breyer, in *International Congress on Protection against Accelerator and Space Radiation*, p. 371. CERN, Geneva, 1971.
57. M. Zielczynski, (in Russian) in *Biophysical Aspects of Radiation Quality*, p. 125. IAEA, Vienna, 1968.
58. A. H. Sullivan and J. Baarli, Report 63-17, CERN Geneva, 1963.
59. J. Booz, in *Biological Effects of Neutron Irradiation*, p. 119. IAEA, Vienna, 1974. (SM-179/37.)
60. *Recommendations of the International Commission on Radiological Protection*, ICRP Publication 4. Pergamon Press, 1964.
61. ICRU Report 16, *Linear Energy Transfer*, 1970.
62. D. Srdoč, *Practical use and application of microdosimetric concepts in field and personnel monitoring*; Report of the Joint IAEA/WHO Consultants' Meeting on Practical Applications of Quantities, Units and Measurement Methods in Radiation Dosimetry for Radiological and Protection Purposes, CERN Geneva, 1971. (IAEA/MSW/6.)

The Dependence on LET of Various Types of Damage in Phage DNA in Relation to the Inactivation Efficiency[1]

R. C. Christensen
Departments of Microbiology and Radiology
College of Medicine
The Milton S. Hershey Medical Center of
The Pennsylvania State University
Hershey, Pennsylvania 17033, USA

TYPES OF DNA DAMAGE

Ionizing radiation produces several types of lesions in DNA. Some of these lesions are moderately well characterized in a physical sense, others can be detected at present only by their biological consequence. This paper will consider DNA lesions as falling into one of four categories: single-strand breaks, double-strand breaks, crosslinks of various types, and "other" nucleotide damage. The first two categories have been investigated in some detail; however, we still do not know as much as we should about the chemistry of break formation. Furthermore, the consequences to the organism of such breaks generally is not clearly understood, even for the simplest phages. Radiation-induced crosslinks have been measured by a few groups during the past five or six years, but their biological consequence is not clear. A very few groups, using simple, well-defined phage systems, have been able to directly or indirectly measure "other" nucleotide damage, which empirically refers to that component of lethality which is not accounted for by the other three categories.

Most measurements of DNA strand breakage have been made under alkaline denaturing conditions. With the discovery by Bopp and Hagen (*1*) four years ago that some lesions expressed as breaks under alkaline assay conditions are not so expressed at neutral pH, these alkaline-based data were found to yield single-strand breakage values which frequently were about 30% higher than breakage measurements at neutral pH, and (depending on the milieu) occasionally double the neutral values. Indeed, with BUdR-substituted DNA, UV-irradiated samples of supercoiled bacteriophage λ DNA were found to give five-fold higher values for single-strand breakage with an alkaline assay than with a neutral assay (*2*). Clearly, a single-strand break is not always the same entity. Just as clearly, it should not be unexpected to find considerable variation in strand breakage values in the literature, variation which is a function both of irradiation conditions and of assay conditions.

[1]Congress presentation A-2-3.

THE LET-DEPENDENCE OF DNA DAMAGE

Keeping these remarks in mind, let us now see what is known about the LET-dependence of these various lesions. Neary et al. (*3, 4*) and Christensen et al. (*5*) have systematically measured strand breakage over a broad range of LET. Figure 1, taken from reference *5*, displays a composite of pooled data from these investigators, plus other high-LET data which is available in the literature. Forgetting for the moment the absolute amplitude of these curves, note that for double-strand breaks the data appear reasonably consistent with the hypothesis that increasing LET causes an enhanced probability per rad for double-strand breakage. Only Dewey et al. (*7*) and Christensen et al. (*5*) show data taken at LET's above 3000 MeV-cm^2/g (or 300 KeV/micron), and this data is suggestive of a peak in biological effect somewhere between 3000 and 10,000 MeV-cm^2/g followed by a decrease in RBE at higher LET values. Quite possibly this represents a saturation effect, in which (crudely speaking) more energy is packaged per micron of particle track, than is needed for most efficient double-strand breakage. (It must be noted that the data of Christensen et al. (*5*) measured at pH 8 with supercoil ϕX-174 RF DNA which had been irradiated under anoxic conditions in a highly radioprotective broth containing AET- a sulfhydryl protective agent-, displayed cross -sections some five-fold less than those of Neary et al. (*3,4*) which were taken with dried T7 DNA and assayed under alkaline conditons.) The RBE maximum for double-strand breakage which Christensen et al. (*5*) saw, occurred at higher LET than the peaking observed in the RBE vs. LET curves for yeast (*8*) and human kidney cell viability, (*9,10*) which occurred in the region 1000–2000 MeV-cm^2/g.

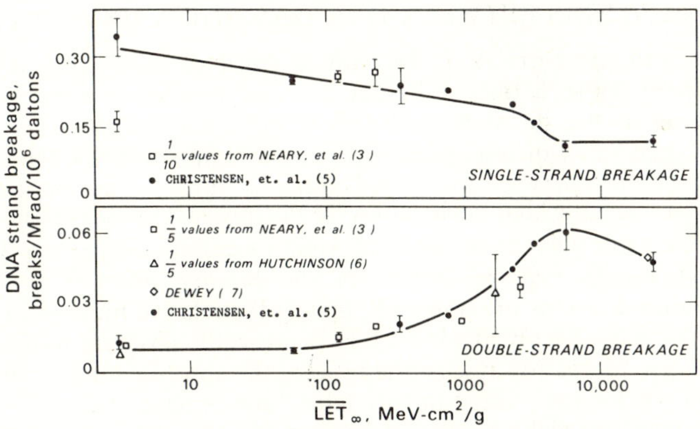

Fig. 1. DNA strand breakage radiosensitivity versus LET$_\infty$, various investigators. Note that values from Neary et al. (1970) and Hutchinson (1965) have been multiplied by scaling factors to allow display in this figure. Values from Dewey (1967) and this paper were measured in thick liquid samples mixed during irradiation. Error bars represent standard deviations of experimental mean values. [From Ref. 5]

With regard to single strand breaks, the data of Christensen et al. (*5*) show a gradual three-fold decrease in cross-section per rad when going to higher LET values. Neary et al. (*3,4*) found, on the contrary, that over their somewhat more restricted range in LET, single-strand break cross-sections appeared to increase a factor of about 2.5 with increasing LET. These differences remain to be resolved.

No studies have been performed on the LET dependence of crosslinking. Indeed, few low-LET data exist. Bohne et al. (*11*), irradiating Tl phage in salt-histidine solution, found for this dense packing situation that the probability for DNA-DNA crosslinks was about 0.005 crosslinks per Mrad per 10^6 daltons, roughly a factor of 10 less than the probability for double-strand breakage in this system. Other authors (e.g. *12*), using dry DNA systems or DNA gels, obtained values for DNA-DNA crosslinks some ten-fold higher. In accounting for the remainder of a rapidly-sedimenting DNA fraction, Bohne et al. (*11*) were able to assign a maximum value for DNA-protein crosslinks in T1 phage, of about 0.01 per Mrad per 10^6 daltons, twice that seen for DNA-DNA crosslinks.

Crosslinking and other nucleotide damage in DNA as a function of LET has not been investigated systematically. Some fragmentary data is available from the work of Christensen et al. (*5*) and van der Schans et al. (*13*), but only in terms of *lethal* nucleotide damage; specific lesions were unknown and only their lethal effect could be assayed. (Tools with which to probe for specific nucleotide damage are being developed; an example is the use of biological enzymes or other reactive chemicals which recognize only certain types of nucleotide damage and either translate them into physical damage we can measure directly, such as DNA strand breakage, or make the damage assayable by other means.)

CORRELATIONS BETWEEN DNA DAMAGE AND VIABILITY

In attempting to correlate DNA damage and biological effect, it is useful to discriminate between three different ways of presenting data. First, and most common, is the statistical correlation, in which one measures physical damage independently of biological effect and attempts to draw some conclusion by using a ratio between the sets of numbers. Table 1 is a typical example. It is important to note that such tables do *not* prove a causal relationship in any way. This table gives us several important pieces of information: 1) not all single-strand breaks can be lethal, 2) the *relative* amounts of various radiation-induced lesions are apparent, and 3) the relative number of double-strand breaks rises sharply at high LET. Because the absolute number of lesions of any given type (or at least the ratio of the absolute number of lesions) must be known for each entry in this table, any damage measured solely in terms of its effect on viability, cannot be entered. Thus, we see no entry under "other nucleotide damage."

A second way of tabulating such data is to provide an "event-specific" ratio which gives the true probability that a lethal event will arise as the result of a given physical event. This is sometimes called a "biological inactivation efficiency." Van der Schans et al. (*13*) and Christensen et al. (*5*), because of unique experimental design, are able to give some information of this type (Table 2). These investigators both used initially supercoiled free DNA's which, after irradiation, could be sorted into classes possessing a double-strand break or one or more single-strand breaks. These sorted classes were then individually tested for biological activity, and the direct effect on viability of either strand breakage or non-break nucleotide damage could be measured. In both of these experimental systems, a double-strand break was invariably lethal. Van der Schans et al. (*13*), under indirect action conditions, found that about one single-strand break in fifty was lethal. Christensen et al. (*5*), with direct action conditions, were unable to correlate single-strand breaks with viability at low LET, but by pooling all data taken above $\overline{\mathrm{LET}}_\infty > 2000$ MeV-cm^2/g, they determined that, in this LET region, possibly one single-strand break event out of five was lethal (0.20 ± 0.14 lethal events per single-strand break in the class of molecules possessing one or more single-strand breaks). Data of this type for crosslinks and other nucleotide damage is not available.

TABLE 1
Statistical relationships between phage DNA lesions and loss of biological activity, for several phages (number of physical lesions per biological inactivation)[a]

Type of DNA lesion	Bohne et al. (*11*)	van der Schans et al. (*13*)	Christensen et al. (*5*)
Double-strand breaks	0.17	0.045	0.12 (0.5)[b]
Single-strand breaks	3.5	4.3	2.1 (1.0)[b]
DNA-DNA crosslinks	0.03	—	—
DNA-protein crosslinks	0.06	—	—
Other nucleotide damage	—	—	—

[a]No causal relationship is implied by presentation of these data.
[b]Parentheses enclose high LET values ($\overline{\mathrm{LET}}_\infty > 2000$ Mev-cm^2/g).

TABLE 2
Biological inactivation efficiencies of various physical lesions in phage DNA (lethality per physical event)

Type of lesion	Christensen et al. (*5*)	van der Schans et al. (*13*)
Double-strand break	1.00 (1.00)[a]	1.00
Single-strand break	— (0.20 ± 0.14)[a]	0.02
Crosslink	—	—
"Other" nucleotide damage	—	—

[a]Parentheses enclose high LET values ($\overline{\mathrm{LET}}_\infty > 2000$ Mev-cm^2/g).

The preceding two tables have given us the inactivation efficiency per lesion, and the ratio of total number of lesions. Now, let us translate these data into a table showing the % contribution to lethality each class of lesions provides (Table 3). Double-strand breaks, though 100% efficient at causing biological inactivation in the systems of van der Schans et al. (*13*) and Christensen et al. (*5*), cause a very small fraction of the total lethal events at low LET, 5–8% in the systems viewed. At high LET, however, this proportion rises to 50% in the system of Christensen et al. (*5*). Single-strand breaks are also relatively unimportant at low LET, causing about as much net lethal damage as do double-strand breaks. At high LET, single-strand breakage is somewhat more important, causing perhaps 20% of the net lethal damage. Note that single- and double-strand breakage account for much less than 20% of the total lethal damage at low LET in the system of van der Schans, but this rises sharply with increasing LET in the system of Christensen et al. (*5*) to account for about 2/3 of the total lethal damage at LETs greater than 2000 MeV-cm^2/g.

TABLE 3
% Contribution of various categories of DNA lesions to lethality in phage DNA

Type of lesion	van der Schans et al. (13)	Christensen et al. (5)
Double-strand breaks	~5	~8 (~50)[a]
Single-strand breaks	~8	— (~20)[a]
Crosslinks	—	—
Other nucleotide damage	87	⩾90[b] (~30)[a]

[a]Parentheses enclose high LET values ($\overline{\text{LET}}_\infty > 2000$ Mev-cm^2/g)
[b]Includes lethal single-strand breaks

Although crosslink damage is frequently considered lethal, there is little data to support this. Neither van der Schans et al. (*13*) nor Christensen et al. (*5*) assayed for crosslinks.

Other nucleotide damage comprised the major component of lethality at low LET in the experiments of van der Schans et al. (*13*) and Christensen et al. (*5*) (Since the low LET data of Christensen et al. (*5*) were insufficient to separate out "lethal other nucleotide damage," the table reports the *sum* of lethal other damage and lethal single-strand break damage.) At high LET, interestingly, other nucleotide damage becomes relatively less important, comprising perhaps 30% of the total biological inactivation.

It is worth digressing to mention *how* lethal other nucleotide damage was measured by Christensen et al. (*5*) Those gradient-separated ϕX−174 RF DNA molecules which did not receive a single- or double-strand break during irradiation (the RFI peak), nonetheless displayed a loss in viability when measured on a per molecule basis (i.e. a decrease in the experimentally measured ratio of plaque-forming ability of all molecules in the RFI peak to total ^{3}H-labeled DNA in the RFI peak.) This loss of "specific viability" displayed one-hit kinetics when plotted as a function of dose. Since neither single-strand breaks nor double-strand breaks were involved in this loss of viability, it fell into the category of lethal other nucleotide damage.

CONCLUSION

We have seen what little is known about how strand breakage and nucleotide damage contribute to overall lethality in phage or phage DNA irradiated under direct action conditions (Christensen et al. (*5*)) and indirect action conditions (van der Schans et al. (*13*)) and assayed in repair-proficient bacteria. Repair-deficient organisms would naturally display abnormal sensitivity to those lesions affected by the defective repair system. Most bacteria, including those used in the studies cited here, have no known double-strand break repair. Organisms such as *Micrococcus radiodurans* and (at least some) mammalian cells which possess repair capability for double-strand breaks, can be expected to show less double-strand break sensitivity, providing the breaks are repaired with fidelity. (It is also interesting to speculate about the possibility for new discoveries of repair systems acting on the as yet uncharacterized, but obviously very important, "other" nucleotide damage). The nature and LET dependence of lethal non-break nucleotide events has scarcely been investigated, and the problem badly needs clever new ideas and better physical and chemical probes.

REFERENCES

1. A. Bopp and U. Hagen, *Biochim. Biophys. Acta* **20,** 320 (1970).
2. R. R. Hewitt and K. Marburger, *15th Ann. Biophys. Soc. Meet. Abstr.*, p. 27a (1971).
3. G. J. Neary, V. F. W. Simpson-Gildemeister, and A. R. Peacock, *Int. J. Radiat. Biol.* **18,** 25 (1970).
4. G. J. Neary, V. J. Horgan, D. A. Bance, and A. Stretch, *Int. J. Radiat. Biol.* **22,** 525 (1972).
5. R. C. Christensen, C. A. Tobias, and W. D. Taylor, *Int. J. Radiat. Biol.* **22,** 457 (1972).
6. F. Hutchinson, in *Cellular Radiation Biology,* p. 86. Williams and Wilkins, Baltimore, 1965.
7. D. L. Dewey, *Int. J. Radiat. Biol.* **12,** 497 (1967).
8. J. A. Sayeg, A. C. Birge, C. A. Beam, and C. A. Tobias, *Radiat. Res.* **10,** 449 (1959).
9. G. W. Barendsen, *Int. J. Radiat. Biol.* **8,** 453 (1964).
10. P. Todd, *Reversible and irreversible effects of ionizing radiations on the reproductive integrity of mammalian cells cultured* in vitro, Ph.D. thesis, University of California at Berkeley, 1964.
11. L. Bohne, Th. Coquerelle, and U. Hagen, *Int. J. Radiat. Biol.* **17,** 205 (1970).
12. J. T. Lett and P. Alexander, *Radiat. Res.* **15,** 159 (1961).
13. G. P. van der Schans, J. F. Bleichrodt, and J. Blok, *Int. J. Radiat. Biol.* **23,** 133 (1973).

The Dependence of RBE And OER on Neutron Energy For Damage to Mammalian Cells And Plant Systems.[1,2]

Eric J. Hall

Radiological Research Laboratory, College of Physicians and Surgeons of Columbia University, New York, N.Y. 10032 USA

INTRODUCTION

Neutrons are of considerable interest from at least two points of view. First, they are a useful tool in basic research because they are relatively densely ionizing, especially at lower energies, so that a cell traversed by a single secondary particle is inactivated with a high probability. Second, neutrons are being introduced into clinical radiotherapy as an alternative to X or gamma rays.

Neutrons are uncharged particles and as such are indirectly ionizing. Intermediate fast neutrons (100 keV to 20 MeV) give up their energy by elastic or inelastic scattering with nuclei of the absorber. In tissue, the dominant process is the production of fast recoil protons due to the interaction of the incident neutron with a hydrogen nucleus; in addition, other heavier recoil nuclei are produced. With increasing neutron energy the recoil protons have a longer range and (on the average) their energy is deposited at lower LET. This trend is reversed for neutrons below about 300 keV, because then many of the recoil protons have an energy less than 100 keV, which corresponds to the proton Bragg peak where LET is a maximum. An additional process becomes important at higher energies. In tissue, for example, a neutron may interact with a carbon nucleus to produce three α-particles, or with an oxygen nucleus to give rise to four α-particles. These are the so-called spallation products, which become important for energies in excess of about 8 MeV.

Compared with (for example) low energy alpha particles which are *densely ionizing*, or gamma rays which are *sparsely ionizing*, neutrons are said to be intermediate in ionization density. However, the deposition of energy from a neutron beam is essentially a *mixture* of high and low LET. For example, 14 MeV neutrons generated by the d $\rightarrow$ T process give rise to secondary particles that deposit energy over an enormous range of LET values. LET values from 1 to 100 keV/μ are due largely to recoil protons, but some energy is deposited by spallation products at LET's above 100 keV/μ. This represents a relatively small fraction of the dose, but it is biologically very important

[1]Congress presentation A-2-4.

[2]This investigation was supported by Contract AT-(*11-1*)-3243 from the U.S. Atomic Energy Commission and Public Health Service Research Grant CA-12536-01 from the National Cancer Institute.

because it is characterized by a high relative biological effectiveness (RBE) and a low oxygen effect (OER). Although responsible for only a modest fraction of the dose, these heavy recoils dominate the biological response. They are the saving grace of high energy cyclotron-produced neutrons which would otherwise have too high an OER for them to be of much interest in radiotherapy.

RECENT EXPERIMENTAL RESULTS

Within the last few years, we have made an extensive study of the effects of neutrons over a wider range of energies than was previously available to most investigators. The results will be reviewed briefly and then a comparison made with previous data in the literature. The biological systems used were mammalian cells cultured in vitro (for RBE determinations over a wide dose range) and the inhibition of root growth of *Vicia* seedlings (for measurement of OER and RBE over the low dose range).

Most of the measurements were made at the Radiological Research Accelerator Facility (RARAF) at Brookhaven National Laboratory. This consists of a large van de Graaff accelerator, capable of accelerating protons or deuterons up to an energy of 4.2 MeV. By bombarding suitable targets, and using neutrons emitted at an appropriate angle from the incident charged particle beam, essentially monoenergetic neutron beams can be generated ranging in energy from 15.4 MeV down to 220 keV. The energies chosen are listed in Table 1. In addition, two lower energy beams were used, each of which consists of a wide spectrum of energies, described as the "110 keV spectrum" and the "60 keV spectrum."

TABLE 1
Characteristics of Neutron Beams Studied

Facility	Neutron Energy	Maximum Energy Spread	Neutron Production Process	Average Dose Rate	
				For Vicia	For Cells
TAMVEC	50 MeV max.	0–50 MeV	d→Be	70 rad/min.	70 rad/min.
NRL	35 MeV max.	0–35 MeV	d→Be	50 rad/min.	65 rad/min.
RARAF	15 MeV	±4%	d→T	280 rad/h	1500 rad/h
	5.8 MeV	±8%	d→D	450 rad/h	475 rad/h
	2.0 MeV	±7%	p→T	140 rad/h	450 rad/h
	1.0 MeV	+15% −12%	p→T	70 rad/h	625 rad/h
	680 keV	±13%	p→T		550 rad/h
	430 keV	±15%	p→T	40 rad/h	300 rad/h
	340 keV	±15%	p→T	30 rad/h	250 rad/h
	220 keV	+28% −25%	p→T	16 rad/h	130 rad/h
	110 keV spectrum	0–110 keV	p→T	35 rad/h	85 rad/h
	60 keV spectrum	0–60 keV	p→T	15 rad/h	

To extend the energy range beyond that available at RARAF, experiments were performed at the Texas A&M Variable Energy Cyclotron (TAMVEC) and the Naval Research Laboratory (NRL) cyclotron. In these machines neutrons are generated by bombarding a beryllium target with accelerated deuterons; the resultant neutron spectra have a wide range of energies with maxima of 50 and 35 MeV, respectively, for the two facilities.

Experiments with Chinese Hamster Cells

Survival curves for Chinese hamster cells exposed to neutrons at RARAF are shown in Fig. 1. Together with each neutron energy an X-ray survival

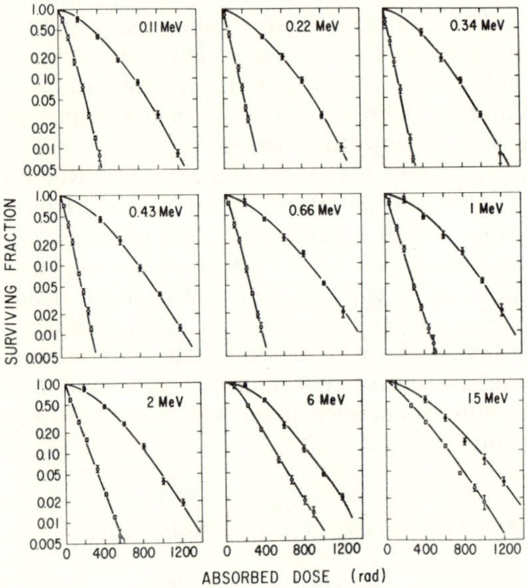

FIG. 1. Survival curves for Chinese hamster cells exposed to neutrons of various energy at the Radiological Research Accelerator Facility and compared with 250 KVp x-rays. In each panel, the open symbols refer to data with neutrons of the energy specified, while the closed symbols refer to x-ray data obtained in the same experiment.

curve was also established on the same day using cells from a common suspension. Data from the two high energy cyclotrons are shown in Fig. 2.

The survival data for neutrons of intermediate energies are closely fitted by exponential functions of dose. At higher energies, where part of the energy is deposited by protons which are relatively sparsely ionizing, or at low energies, where many protons have initial energies below 100 keV, the survival curves are characterized by an initial shoulder. This trend is very clearly evident from Figs. 1 and 2.

The RBE values, calculated for a cell surviving fraction of 37%, are plotted as a function of neutron energy in the *bottom* panel of Fig. 3.

Experiments with Vicia *Seedlings*

The range of energies over which experiments were performed with *Vicia* was similar to that with mammalian cells except that an additional low energy spectrum could be studied because the plant system is more radiosensitive and can be used at lower dose rates than mammalian cells. Dose-response relationships, under aerated and hypoxic conditions, for the wide range of neutron energies studied at RARAF is shown in Fig. 4. Corresponding data obtained on the two high energy cyclotrons are shown in Fig. 5. From these data, OER is plotted as a function of neutron energy in Fig. 6 (bottom panel) and RBE versus neutron energy in Fig. 3 (second panel from the top).

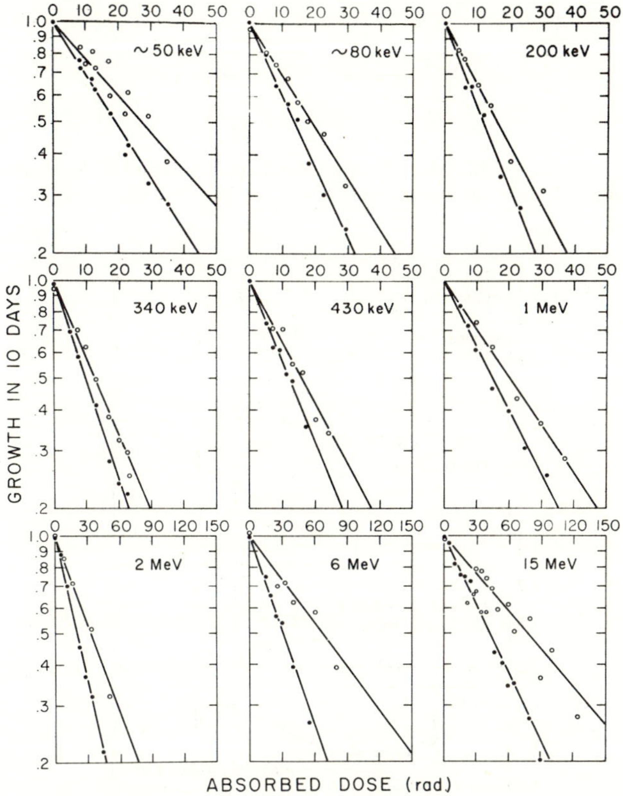

FIG. 4 Dose response relationships for *Vicia* seedlings exposed to neutrons of various energy at the Radiological Research Accelerator Facility. Open and closed symbols refer to hypoxic and aerated conditions respectively.

COMPARISON WITH PREVIOUS DATA IN THE LITERATURE

There are a number of reports in the literature where OER has been measured for several neutron energies, using a variety of biological systems. The data have been replotted in Fig. 6. Scoring somatic mutations in *Tradescantia*, Dennis and Boot (*1*) found a high OER at 440 keV which they attributed to oxygen resonance. This has not been confirmed by the more recent

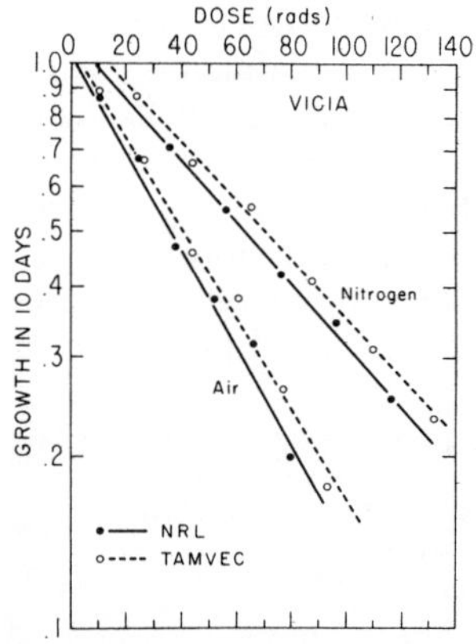

FIG. 5. Dose response relationships for *Vicia* seedlings exposed to 35 MeV neutrons at the NRL cyclotron or 50 MeV neutrons at TAMVEC.

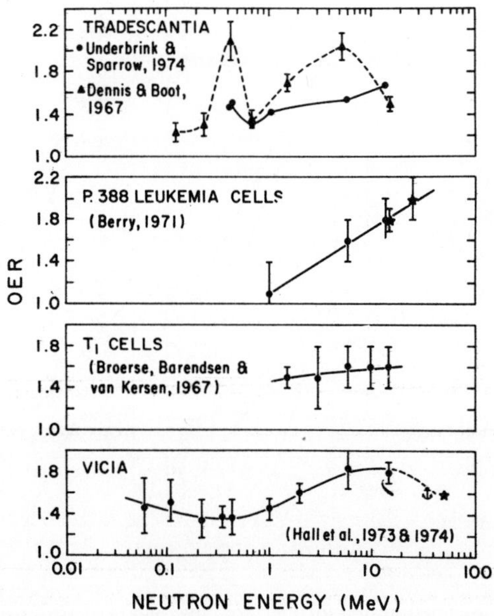

FIG. 6. Oxygen Enhancement Ratio (OER) as a function of neutron energy as reported in the literature for a number of biological test systems.

measurements of Underbrink and Sparrow using the same system (*2*). The lymphocytic leukemia cells used by Berry (*3*) appear to show a more rapid change of OER with neutron energy (and indeed with radiation quality in general) than any other system tested. The data for T_1 human kidney cells by Broerse et al. (*4*) show little change of OER over the limited energy range

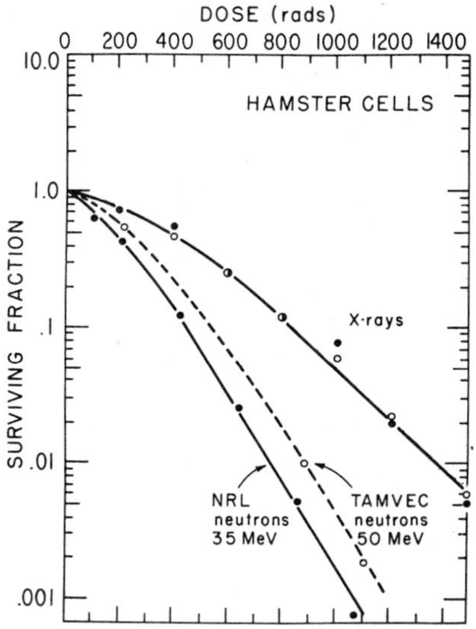

FIG. 2. Survival curves for Chinese hamster cells irradiated with 35 MeV neutrons at the NRL cyclotron or 50 MeV neutrons at TAMVEC. The x-ray survival data, obtained in the same experiments, is shown for comparison.

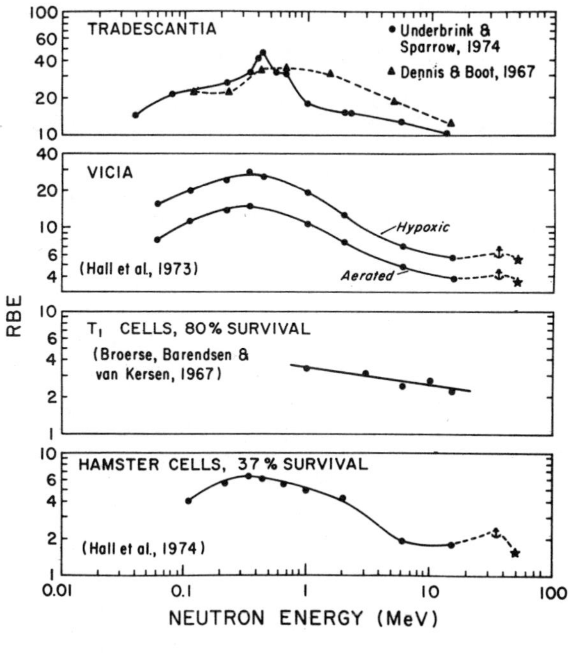

FIG. 3. Relative Biological Effectiveness (RBE) as a function of neutron energy as reported in the literature for a number of biological test systems.

tested, but the error bars are such that they do not disagree with the *Vicia* data (*5*).

The various RBE determinations as a function of neutron energy are compared in Fig. 3. The two most recent sets of data, with *Vicia* seedlings and Chinese hamster cells (*5, 6*), include a wider energy range than previous reports. There is good agreement on the *shape* of the RBE versus neutron energy relationship, but, of course, absolute RBE values differ according to the biological system used and also on the dose level at which the comparison is made.

The picture is complicated by the fact that some experiments are performed with monoenergetic neutrons, while others are performed with cyclotron-produced or fission neutrons which involves a wide energy range; it is then an arbitrary decision to use mean, maximum, modal energy etc. However, these difficulties aside, a clear picture emerges. The most biologically effective neutrons have an energy of about 340 keV. RBE falls off below this energy because many of the recoil protons have an energy less than 100 keV, necessary for the maximum proton LET at their Bragg peak. Above 340 keV, RBE falls off because the neutrons produce recoil protons of higher and higher energy which are relatively sparsely ionizing. This process is mitigated to some extent above about 8 MeV by the increasing importance of spallation products.

Thus, the variation of both OER and RBE with neutron energy are reasonably well documented and tolerably well understood in terms of the basic physics of energy deposition.

REFERENCES

1. J. A. Dennis and S. J. Boot, Nature **215,** 310, 1967.
2. A. G. Underbrink and A. H. Sparrow, in *Biological Effects of Neutron Irradiation*, IAEA, Vienna, 1974.
3. R. J. Berry, Europ. J. Cancer **7,** 145, 1971.
4. J. J. Broerse, G. W. Barendsen and G. R. van Kersen, Int. J. Radiat. Biol. **13,** 559, 1967.
5. E. J. Hall, H. H. Rossi, A. M. Kellerer, L. Goodman and S. Marino, Radiat. Res. **54,** 431, 1973.
6. E. J. Hall, J. K. Novak, A. M. Kellerer, H. H. Rossi, S. Marino and L. J. Goodman, Radiat. Res. In press.

RBE Values of Fast Neutrons for Damage to Organized Tissues in Experimental Animals[1]

J. J. Broerse
Radiobiological Institute TNO
151 Lange Kleiweg, Rijswijk (ZH).
The Netherlands

INTRODUCTION

The characteristic parameter of a type of ionizing radiation which determines its relative biological effectiveness for a specified effect is the spatial distribution of the energy deposition in tissue, which can, to a first approximation, be described by the distribution of dose as a function of the linear energy transfer (LET) of the ionizing particles. Directly ionizing particles of well-defined LET can be employed for RBE studies with single cells or thin layers of cells (*1*); however, in general the more densely ionizing radiation has insufficient penetration to be used for studies of RBE-LET relationships for relatively large multicellular systems including experimental animals. This difficulty can be overcome by employing beams of fast neutrons or negative pions, which have sufficient penetrating power and deposit a considerable part of their energy through secondaries of high LET.

A first objective of studies of biological effects of fast neutrons on normal tissues is the accumulation of information about basic mechanisms through which damage in mammalian tissues is induced and expressed. The comparison of dose-effect relations for fast neutrons and X- or gamma-rays provides the possibility to analyze the influence of factors which are different for these two type of radiation, e.g. repair of sublethal damage or the oxygenation status of cells. A further aim of studying neutron-induced normal tissue damage is to obtain data on the tolerance of normal tissues to single and fractionated doses, which is of importance for the application of fast neutrons in radiotherapy.

Neutrons entering a biological medium release energy by different types of interactions; viz. elastic scattering, inelastic scattering (interactions with the bombarded nucleus in which the neutron is promptly remitted and is generally accompanied by the emission of a nuclear de-excitation gamma-ray), non-elastic scattering (reactions with the bombarded nucleus which result in the emission of particles other than a single neutron), capture processes and spallation reactions. As a result of these different nuclear interactions, the energy deposition is characterized by a complex spectrum of secondary charged particles. The energy deposition characteristics of fast neutron beams are discussed in relation to the distribution of the fast neutron dose in LET.

Over the past fifteen years a number of techniques have been developed to allow quantitative determinations of the radiation damage to normal tissues. A

[1]Congress presentation A-2-5.

number of these experimental methods have also been applied to study effects induced by fast neutrons and to derive RBE values for these specific biological end-points. In the present contribution special attention will be given to the following normal tissues in rodents: bone marrow, intestinal mucosa, oesophagus, skin, vascular endothelium, spinal cord and lung. Since the absolute neutron dosimetry performed at different laboratories can show significant differences (*2*), the effects on normal tissues will generally be illustrated with the results of experiments carried out in the Radiobiological Institute TNO at Rijswijk, the Netherlands, using 15 MeV neutrons.

Up to the present, the majority of studies on normal tissue response has been carried out with fission spectrum fast neutrons, neutrons produced by the $^{9}Be(d,n)^{10}B$ reaction and monoenergetic neutrons produced by the $T(d,n)^{4}He$ reaction. The response of the same type of normal tissue has been studied with neutron beams of different energy and the RBE values for neutron beams of various qualities will be discussed.

ENERGY DEPOSITION CHARACTERISTICS OF FAST-NEUTRON BEAMS

The energy dissipation of fast neutrons in biological material occurs from recoil protons, from heavy recoil nuclei, such as carbon, oxygen, nitrogen and from products of nuclear reactions, e.g. alpha particles. The relative contributions to the kerma or dose in soft tissue from these interactions depend on the neutron energy. Calculations of mass energy transfer coefficients (*3*), show that the contribution from hydrogen decreases from 97% of the total kerma at a neutron energy of 10 keV to 69% at an energy of 18 MeV (see Fig. 1). Due to the large contribution from recoil protons to the neutron kerma, energy deposition in various biological materials is determined mainly by the hydrogen-content. Thus, in soft tissue containing about 10% by weight of hydrogen, a 1% deviation in the total hydrogen-content will change the kerma due to protons by approximately 10%.

As can be seen from Fig. 1, relatively small amounts of energy are dissipated through heavy charged particles. Nevertheless, in order to assess the contribution of the various fractions to the final biological effect, it is necessary to consider the LET contributions of the various types of secondary particles.

Based on stopping power data for the various recoil particles involved, Bewley (*4*) calculated LET spectra for the irradiation of thin layers of tissue with 14.6 MeV neutrons, cyclotron-produced neutrons and fission-neutrons. In the LET distributions for the various neutron beams, presented in Fig. 2, two important regions can be distinguished which correspond to the energy dissipated through protons and through interactions with the heavier nuclei, respectively. Protons set in motion by neutrons through elastic collisions dissipate energy with a distribution of the dose in LET which extends from low values of a few keV/μm to a maximum of about 96 keV/μm of the Bragg peak. Energy dissipated by fast neutrons through interactions with C, O and N nuclei is deposited mainly in the LET region between 100 and 1000 keV/μm.

In irradiations of mono-layers of cells, a first approximation can be made to distinguish experimentally between the biological damage produced by the protons and by the heavier nuclei. This can be achieved by providing or

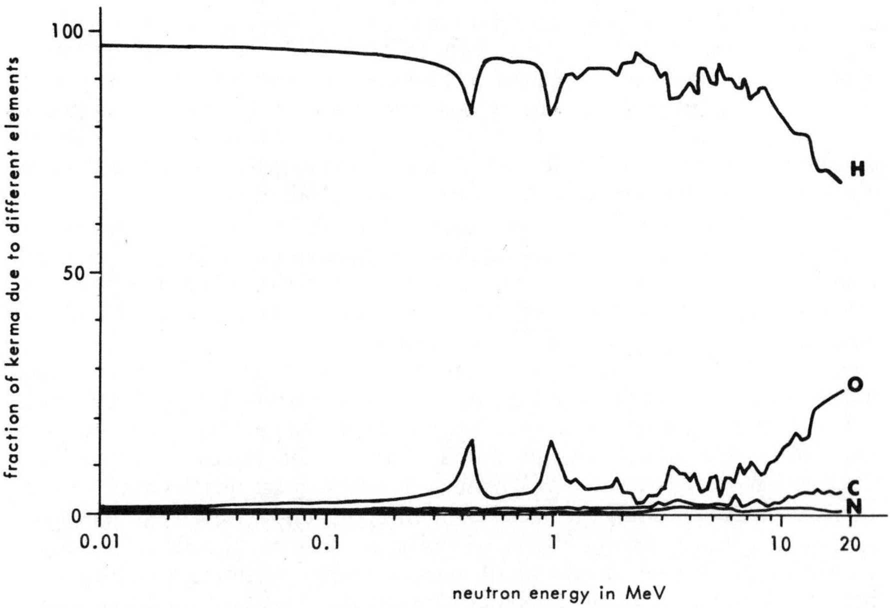

FIG. 1. Relative contribution from interaction processes with different elements to kerma in soft tissue based on the calculations of Bach and Caswell (*3*).

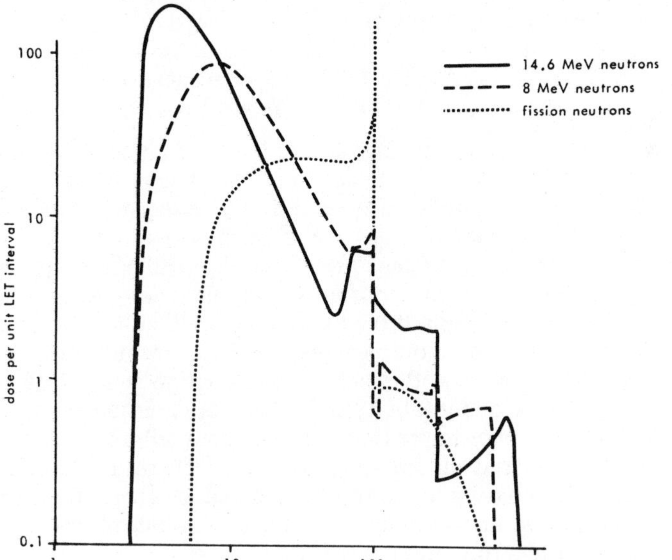

FIG. 2. LET∞ spectra as calculated by Bewley (*5*) for three different beams of fast neutrons. Reproduced from Broerse and Barendsen, *Curr. Top. Radiat. Res. Q.* 8, 305 (1973) with the permission of the North-Holland Publishing Co., Amsterdam.]

deliberately avoiding, the secondary particle equilibrium for the neutron irradiations of the cells. Studies with 15 MeV neutrons (*5*) and cyclotron neutrons with a mean energy of 8 MeV (*6*) have shown that the proton component has a lower RBE and a higher OER than the radiation component consisting of heavier nuclei. This trend is qualitatively in accordance with the dependence of RBE and OER on LET for monoenergetic charged particles as obtained from experiments with cultured mammalian cells (*7*).

Experimental data on the microscopic parameters of fast-neutron beams have been obtained from measurements with tissue-equivalent proportional counters (*8,9*). The distribution of dose in LET derived from the frequency distributions of the lineal energy of individual particles are generally in good agreement with the calculated LET distributions.

It is unfortunate that as yet a direct correlation cannot be made between the microscopic distribution of energy absorption and the final biological effect. To permit a direct comparison of the biological effects of fast-neutron beams with the effects of charged particles, different mean LET values are derived. As discussed in detail elsewhere (*10*), neither the dose average nor the track average LET can serve as a single parameter to interpret the biological effects of neutron radiation in terms of the effects of radiations having discrete LET values. Hopefully, further developments of microdosimetric techniques will lead to a better understanding of fundamental physical processes and will provide more suitable physical quantities or average values for an appropriate interpretation of the biological response after neutron irradiation. However, up to the present the best and most unambiguous quality specification of fast neutron beams is in terms of the neutron energy spectrum.

QUANTITATIVE DETERMINATION OF NORMAL TISSUE DAMAGE

Studies of the effects of ionizing radiation on rapidly proliferating tissues, such as bone marrow, intestinal mucosa and esophagus, have initially been concerned with the radiation syndromes leading to the death of the animal. Although death of an animal provides one of the most distinct criteria, it should be realized that complex phenomena are involved. Furthermore, the quality of the animals employed is of great importance since bacteriological complications may obstruct the interpretation of the biological results. The introduction of in vivo assay methods has therefore been essential for an understanding of the cellular responses associated with some of the radiation syndromes.

Only recently have quantitative studies been initiated on the responses of tissues with much longer turnover times of functional cells, such as the vascular endothelium, the lung and the spinal cord. The effects of irradiation on these systems in which cells have a low turnover rate and which are frequently dose-limiting in clinical radiotherapy, may only appear after long observation periods of up to one year after irradiation.

In this section the methods of assessing normal tissue damage will be summarized and some RBE values will be given for single-dose irradiations carried out with 15 MeV neutrons at the Radiobiological Institute.

Mouse bone marrow

Death due to the hemopoietic syndrome occurs mainly between 8 and 14 days after irradiation. The $LD_{50/30d}$ values for X- and neutron-radiation are usually taken to determine the RBE for the occurrence of the hemopoietic syndrome. The survival of the hemopoietic stem cells can be investigated by the spleen colony technique introduced by Till and McCulloch (*11*). The RBE value of 1.1 for the bone marrow syndrome obtained with 15 MeV neutrons (660 rad) is in agreement with the value of 1.03 obtained for the survival of hemopoietic stem cells (*12*). Recently, higher RBE values have been reported (*13,14*), which might be attributed to difference in the qualities and characteristics of the animals employed or to differences in the neutron dosimetry and in the spectrum of the fast neutron beams.

Mouse intestinal mucosa

Death due to the intestinal syndrome occurs within a 3 to 5 days interval after irradiation. RBE values are generally determined on the basis of the $LD_{50/5d}$ values. The survival of intestinal crypt cells can be investigated with the microcolony assay introduced by Withers and Elkind (*15*). The assessment of crypt survival is based on a quantitative scoring of clusters of regenerating crypt stem cells in the intestinal mucosa after total-body irradiation. The RBE of 1.4 observed for the occurrence of the intestinal syndrome after 15 MeV neutron irradiation (920 rad) is in good agreement with the values obtained for the survival of jejunal crypt cells (*12,16*).

Skin

Several methods of scoring damage to skin have been employed. Fowler et al. (*17*) introduced a method for estimating the reactions in skin of pigs, based on visual observations and a scoring according to an artibrary scale. Similar methods have been used for rat skin and mouse skin (*18*). A skin transplantation technique was developed by Barendsen (*19*) in which shrinkage of the area of an irradiated piece of white skin was measured after transplantation onto brown F_1 hybrid rats. For a dose of 1250 rad of 15 MeV neutrons an RBE of 1.4 was obtained. The survival of stem cells in mouse skin can be studied with a clone technique introduced by Withers (*20*). Employing this technique Denekamp et al. (*21*) found that the RBE for cell survival was very similar to that derived from gross skin reaction in the same strain of mice irradiated with cyclotron neutrons with a mean energy of 8 MeV.

Esophagus

Mortality within 8 to 28 days after irradiation can be taken as an end-point for esophageal damage in mice, when the exposure area is restricted to the thoracic region (*22*). The pathological changes observed in the irradiated esophagus at death are edema of the submucosa, massive inflammation in the esophageal wall and perforation of the wall leading to death of the animal. It is of importance that the determination of the $LD_{50/28d}$ will be supplemented with a quantitative scoring of the histological changes. For cyclotron neutrons with a mean energy of 8 MeV, Field and Hornsey (1973, private communication), obtained RBE values which were higher than those for damage to other normal tissues at the same neutron dose. Results obtained by Phillips et al. (*14*) for 15 MeV neutrons, also indicated a comparatively high RBE for esophageal lethality.

Vascular endothelium

Radiation induced late changes in normal tissue are often attributed to damage to the blood vessels. A number of morphological and physiological methods, including the xenon-clearance technique, have been applied to quantitatively study the effects on vascular endothelium (*23–25*). Reinhold and Buisman (*26*) developed a technique in which capillaries are induced to proliferate in a two-dimensional system in subcutaneous tissue of the rat. Quantification of the effect is performed with an angiographic technique, employing microfluorometry. Dose-effect curves have been obtained for single dose irradiations with collimated beams of X-rays and 15 MeV neutrons; for neutron doses between 250 and 500 rad, the RBE varies only slightly from 1.9 to 1.8 (*27*).

Spinal cord

Investigation of damage to the spinal cord in the rat after irradiation with X-rays and 15 MeV neutrons have been carried out by Van der Kogel and Barendsen (*28*), using muscle atrophy and paralysis of the hind legs as a measure of the induction of myelopathy. For motor damage observed after a minimum latent period of about 3 months, an RBE of 1.1 can be derived for a single dose irradiation with 1900 rad of 15 MeV neutrons.

Lung

According to the studies of Phillips and Margolis (*22*), mortality between 100 and 160 days can be taken as an end-point for pulmonary damage in mice after thoracic irradiation. The time period in which mortality can occur will be different for various mouse strains. In studies with a different mouse strain (Broerse and Hollander, 1973, unpublished), mortality connected with pulmonary damage occurred about 200 days post-irradiation. It will be necessary to correlate the pulmonary LD_{50} values with a quantitative scoring of the late radiation fibrosis which seems to be connected with capillary injury in the alveolar septa. The process is characterized by a distinct thickening of the septa and a reduction of the alveolar air spaces. Preliminary results indicate an RBE of about 1.2 for the $LD_{50/210d}$ for a single dose irradiation with 1200 rad of 15 MeV neutrons. This RBE is slightly lower than the RBE values obtained for epithelial tissues in this range of neutron doses.

In some studies of radiation-induced damage in normal tissues it has been possible to investigate the underlying cellular mechanisms. The survival curves of different types of clonogenic cells show a considerable variation of intrinsic radiosensitivities for X-rays, bone marrow stem cells being the most sensitive and intestinal crypt stem cells being very resistant. The differences involve shoulder widths as well as the slope of the curves for surviving fractions below 10%.

Furthermore, it can be concluded that similar but somewhat smaller differences are observed for irradiations with 15 MeV neutrons.

RBE values of 15 MeV neutrons for damage to organized tissues in experimental animals are presented in Fig. 3 as a function of the dose or dose per fraction. The data show a general tendency for an increase of the RBE with decreasing dose or dose per fraction, which is to be ascribed to the smaller influence of accumulation of damage after neutron irradiation. It can further be concluded that different RBE values are observed for effects on various normal tissues. These differences are even more pronounced for neutrons with lower energies (*29*).

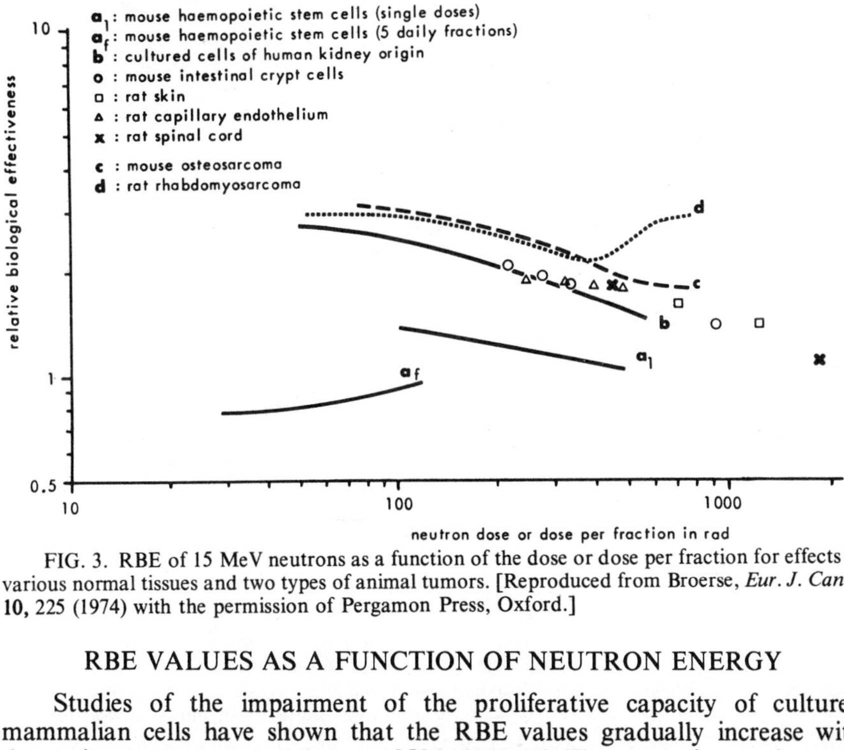

FIG. 3. RBE of 15 MeV neutrons as a function of the dose or dose per fraction for effects on various normal tissues and two types of animal tumors. [Reproduced from Broerse, *Eur. J. Cancer* **10,** 225 (1974) with the permission of Pergamon Press, Oxford.]

RBE VALUES AS A FUNCTION OF NEUTRON ENERGY

Studies of the impairment of the proliferative capacity of cultured mammalian cells have shown that the RBE values gradually increase with decreasing neutron energy, down to 350 keV (*5, 30*). The conclusion can be made that for decreasing neutron energy, the decrease of the contribution from heavy recoils with relatively high LET to the total energy dissipation, is more than compensated for by the increase in effective LET of the protons.

A similar increase of the RBE with decreasing neutron energy has been observed for the occurrence of the hemopoietic syndrome and the intestinal syndrome as shown in Fig. 4. The RBE values for the two radiation syndromes show distinct differences for fission spectrum fast neutrons with a mean energy of 1 MeV (*31*), cyclotron neutrons with a mean energy of 8 MeV (*32*), and 15 MeV monoenergetic neutrons (*33*).

Recently the survival of jejunal crypt cells has also been investigated with neutron beams with higher energies produced at TAMVEC by bombarding beryllium with 50 MeV deuterons (mean neutron energy 22 MeV) and has been compared with the response to neutrons produced by 16 MeV deuterons (mean neutron energy 7 MeV). For different neutron doses, it has been found that the RBE for the lower energy neutrons is greater than that for the higher energy by a constant ratio of 1.33 (Fletcher, Hussey and Withers, 1974, private communication). Studies with mammalian cells have shown a similar tendency, although the quantitative differences in RBE for the two neutron energies are smaller (Gragg, Meyn and Humphrey, 1974, private communication).

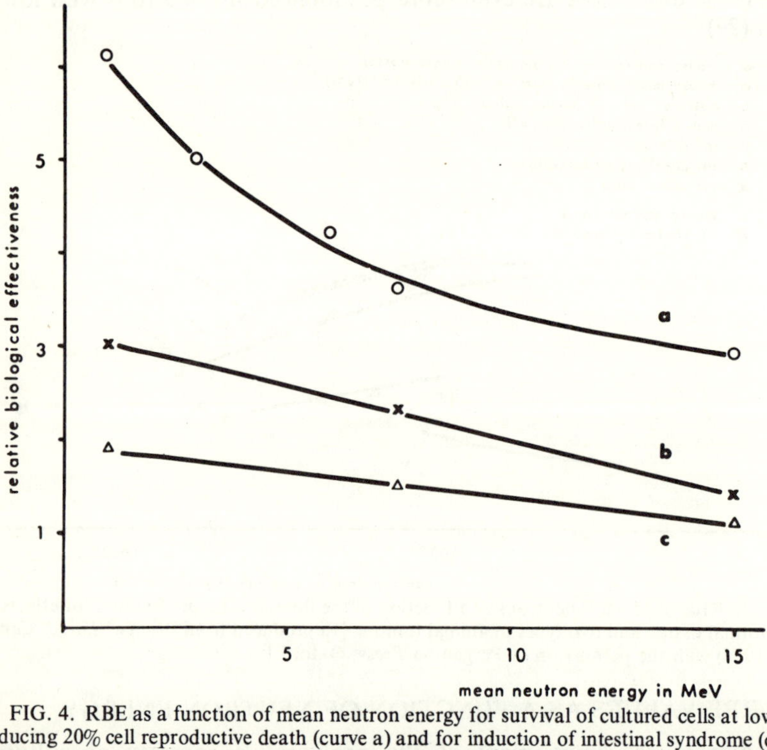

FIG. 4. RBE as a function of mean neutron energy for survival of cultured cells at low doses, producing 20% cell reproductive death (curve a) and for induction of intestinal syndrome (curve b) and hemopoietic syndrome (curve c) in mice. [Reproduced from Broerse, *Eur. J. Cancer* **10**, 225, (1974) with the permission of Pergamon Press, Oxford.]

CONCLUSION

For various normal tissue endpoints it has been shown that the RBE of fast neutrons depends on a number of factors, e.g. neutron energy spectrum, dose, fractionation scheme and dose rate. General relationships between RBE and neutron dose have been established. It can be concluded that, in general, extrapolations of RBE values from one tissue to another are not justified. The differences in RBE values are considered to be due to differences in intrinsic radiosensitivity of the clonogenic cells in the tissues concerned.

Additional information is required on effects of fractionated irradiations with fast neutrons and on the response of other types of normal tissues. In view of the clinical applications of fast neutrons, it will be important to obtain information on normal tissue response in monkeys with the aim of extrapolating these data to the human. Studies with Rhesus monkeys performed at TAMVEC indicate that late changes including mandibular osteoradionecrosis are greater than would be expected on the basis of acute oral mucosal radiation reactions for a large-dose twice-weekly fractionation schedule (Fletcher, Hussey and Withers, 1974, private communication). Similar studies should be initiated for more conventional fractionation regimen and for other neutron energies.

With regard to the differences in RBE values reported by various authors for effects on the same type of normal tissue for neutrons produced by the same type of reaction, it will be important to obtain information on the uncertainty of the RBE values quoted. The total uncertainty will result from errors in the biological techniques employed and in the assessment of the total absorbed dose. Since the n/γ ratio of various neutron beams can show considerable variations, it will be essential to provide quantitative information on both radiation components. Finally, it has to be stressed that variations in radiation quality can introduce significant differences in biological effectiveness. Although the quality is correlated with the distributions of microdosimetric parameters, it is recommended that the quality specification of fast neutron beams is given in terms of the neutron energy spectrum.

REFERENCES

1. G. W. Barendsen, H. M. D. Walter, J. F. Fowler and D. K. Bewley, *Radiat. Res.* **18,** 106 (1963).
2. R. S. Caswell, L. J. Goodman and R. D. Colvett, These Proceedings.
3. R. L. Bach and R. S. Caswell, *Radiat. Res.* **35,** 1 (1968).
4. D. K. Bewley, *Radiat. Res.* **34,** 437 (1968).
5. J. J. Broerse, G. W. Barendsen and G. R. Van Kersen, *Int. J. Rad. Biol.* **13,** 559 (1968).
6. D. K. Bewley, N. J. McNally and B. C. Page, *Radiat. Res.* **58,** 111 (1974).
7. G. W. Barendsen, C. J. Koot, G. R. Van Kersen, D. K. Bewley, S. B. Field and C. J. Parnell, *Int. J. Radiat. Biol.* **10,** 317 (1966).
8. K. S. J. Wilson and S. B. Field, *Phys. Med. Biol.* **15,** 657 (1970).
9. H. Krüger and E. H. Graul, *Proc. 2nd Euratom Symposium on Microdosimetry,* Stresa (Italy), 813 (1969).
10. J. J. Broerse and G. W. Barendsen, in *Curr. Top. Radiat. Res. Q.* **8,** 305 (1973).
11. J. E. Till and E. A. McCulloch, *Radiat. Res.* **14,** 213 (1961).
12. J. J. Broerse, G. W. Barendsen, G. Freriks and L. M. Van Putten, *Eur. J. Cancer* **7,** 171 (1971).
13. J. H. Hendry, *Brit. J. Radiol.* **45,** 923 (1972).
14. T. L. Phillips and L. Margolis, *Eur. J. Cancer* (1974). In press.
15. H. R. Withers and M. M. Elkind, *Int. J. Radiat. Biol.* **17,** 261 (1970).
16. H. R. Withers, J. T. Brennan and M. M. Elkind, *Brit. J. Radiol.* **43,** 796 (1970).
17. J. F. Fowler, R. L. Morgan, J. A. Silvester, D. K. Bewley and B. A. Turner, *Brit. J. Radiol.* **36,** 188 (1963).
18. S. B. Field, T. Jones and R. H. Thomlinson, *Brit. J. Radiol.* **41,** 597 (1968).
19. G. W. Barendsen, in *Current Topics Radiation Research* (M. Ebert and A. Howard, Eds.) Vol. **4,** pp. 293–356. North Holland, Amsterdam, 1968.
20. H. Withers, *Brit. J. Radiol.* **40,** 187 (1967).
21. J. Denekamp, J. E. Fowler, K. Kragt, C. J. Parnell and S. B. Field, *Radiat. Res.* **29,** 71 (1966).
22. T. L. Phillips and L. Margolis, *Front. Radiat. Ther. Oncol.* **6,** 254 (1972).
23. P. J. Lindop, in *Current Topics in Radiation Research* (M. Ebert and A. Howard, Eds.) Vol. **6,** pp. 293–324. North Holland, Amsterdam, 1970.
24. J. De Ruiter and L. M. Van Putten, in *Annual Report* REP-TNO, 1972, p. 11.

25. S. P. Stearner and E. J. B. Christian, in *Annual Report,* Argonne National Lab. ANL-7970, 1972, p. 42.
26. H. S. Reinhold and G. H. Buisman, *Brit. J. Radiol.* **46,** 54 (1973).
27. J. J. Broerse, H. S. Reinhold, G. H. Buisman and A. C. Engels, *Radiat. Res.* **56,** 180 (1973).
28. A. J. Van Der Kogel and G. W. Barendsen, to be published (*Brit. J. Radiol.*).
29. S. B. Field and S. Hornsey, *Eur. J. Cancer 7,* 161 (1971).
30. E. J. Hall, this symposium.
31. J. A. G. Davids, *Int. J. Radiat. Biol.* **17,** 173 (1970).
32. S. Hornsey, S. Vatistas, D. K. Bewley, and C. J. Parnell, *Brit. J. Radiol.* **38,** 878 (1965).
33. J. J. Broerse, *Int. J. Radiat. Biol.* **15,** 115 (1969).

THE USE OF HEAVY PARTICLES IN RADIOTHERAPY[1]

Chairman: J. F. Fowler

[1]Symposium E–4

Biological Basis of Heavy Particle and Fast Neutron Radiotherapy[1]

G. W. Barendsen
Radiobiological Institute TNO, 151 Lange Kleiweg, Rijswijk (ZH), The Netherlands.

INTRODUCTION

Treatments of cancer with fractionated doses of X rays, gamma rays or electrons can achieve local control with frequencies up to 90 or 95% for some types of tumors, depending on the size and location, while for other types of neoplasms much lower frequencies are observed (*1*). In cases where treatment is successful, it is implied that complete impairment of the proliferative capacity of all tumor cells has been attained at doses whereby the critical normal tissues can retain or restore their integrity and function. In cases where a local recurrence is observed after a total dose equal to the tolerance dose of the limiting normal tissue, the implication is that some of the tumor cells have retained the capacity for unlimited proliferation. Such a failure can be ascribed to a variety of factors. Firstly, the normal tissue tolerance may be significantly smaller for some locations as compared to others because of a high intrinsic sensitivity or, especially with large lesions, because the normal tissue may have been destroyed and replaced by tumor tissue to such an extent that even if the tumor is eradicated, adequate restoration of normal tissue function is impossible. The second cause of failure to attain local tumor control is an insufficiently large lethal effect on the tumor even with the relatively large doses which are tolerated by several normal tissues, e.g. with total doses of 6000 to 7000 rad of ^{60}Co gamma rays administered in a schedule of 200 rad per day, 5 fractions per week. Factors which are known to influence responses of cells in tumors, rendering treatments less effective, include the intrinsic cellular radiosensitivity, as related to the shapes of survival curves, repair of sublethal and potentially lethal damage, as well as the presence of severely hypoxic cells and their kinetics of reoxygenation, the presence of noncycling cells and distributions of cell ages in the proliferation cycle (*2*). In addition, responses of tumors to fractionated treatments will depend on cell proliferation kinetics during and after protracted treatments.

Fast neutrons and heavy particles could provide a therapeutic advantage over X rays or gamma rays on account of two characteristics. Firstly, it is possible that the relative biological effectiveness (RBE) of the high-LET radiation for damage to the proliferative capacity of the tumor cells is larger than the RBE for damage to those normal tissues which are dose-limiting in the

[1]Congress presentation E-4-1.

treatment considered. Thus for the same location in normal tissues, tumors for which the largest RBE values are observed, would offer the highest probability to obtain a radiotherapeutic advantage from the use of heavy particles and fast neutrons. Secondly, with heavy particles and especially negative pions, the high dose required for tumor eradication can be confined to the prescribed volume with sharp boundaries, which can be an important advantage for specific locations.

The purpose of the present review is to evaluate and illustrate the relative influence of the various radiobiological factors on the therapeutic margin, which can be defined as the region of doses where tumor eradication is attained without undue damage to normal tissues.

DIFFERENCES IN INTRINSIC CELLULAR RADIOSENSITIVITY TO X RAYS AND HIGH-LET RADIATIONS

In Fig. 1, a number of survival curves are presented, obtained with 300 kV X rays and 15 MeV neutrons for cells from different normal tissues and tumors, investigated in the Radiobiological Institute TNO at Rijswijk (*2*). These curves illustrate that especially with X rays, a considerable variation in values of D_0 and D_Q is observed. For X rays, D_0 values vary by a factor of about 2.3, while D_Q values show an even larger spread, with differences up to a factor of 5.2. With fast neutrons a reduction in shoulder width is generally obtained in comparison with X rays, i.e. D_Q/D_0 ratios are smaller by factors ranging from 1.5 to 2.4. The D_0 values are in general also smaller for neutrons than for X rays, but these differences, expressed as $(D_0)_X/(D_0)_n$, range for the data represented in Fig. 1 from 1.0 to 1.5, i.e. the extent of the reduction in slope at large doses, is smaller than for the shoulder width. This indicates that the differences between high-LET radiations and X rays may be more pronounced with respect to accumulation of damage than with respect to the sensitivity at large doses. Values of D_0 and D_Q do not define the exact shape of the shoulder, however, whereas this low dose region is the most important for responses to fractionated radiotherapy.

It is further illustrated in Fig. 1 that no specific differences are observed between shapes of survival curves of cells from tumors as compared with cells from normal tissues. This observation leads to an important conclusion with respect to the interpretation of differences in radiation responses observed among various types of tumors, because it implies that even if only the intrinsic cellular radiosensitivity determined their responses, some tumors are expected to be curable with a dose of X rays or gamma rays, which is only one half of that required for others. With heavy particles and fast neutrons these differences may be expected to be reduced.

The first survival curve for mammalian cells obtained with high-LET alpha radiation indicated already that the RBE of a given type of radiation is not constant, but depends on the level of damage or the dose considered (*3*). RBE values derived from the curves presented in Fig. 1, are tabulated in Table 1. These values show that in general at larger doses or lower fractions of surviving cells, RBE values are smaller than at small doses or high surviving fractions. In addition, a considerable spread in RBE values by a factor of about 2 is observed among different types of cells. The spread in RBE values illustrated in Fig. 1 for

15 MeV neutrons could be even larger for neutrons of lower energies or for heavy particles of very high LET.

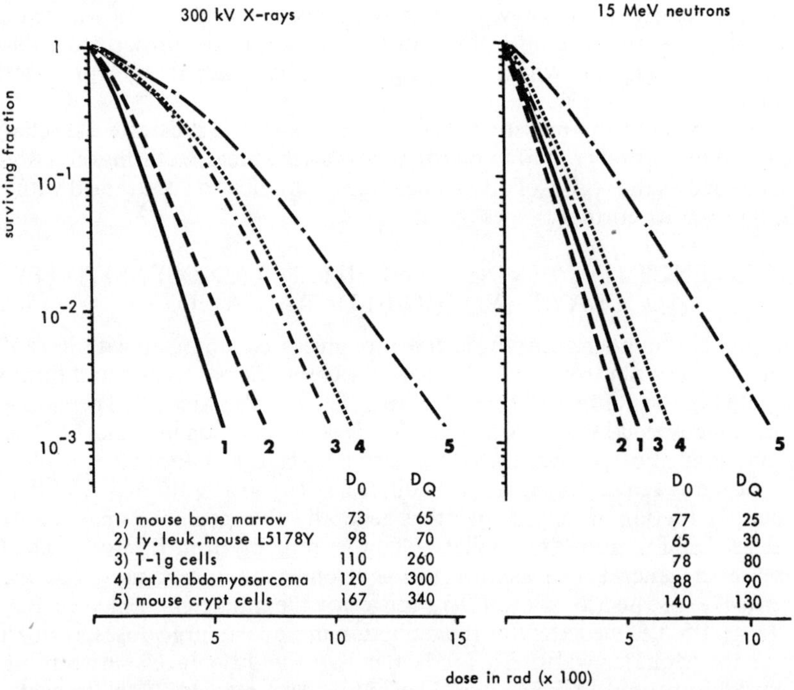

FIG. 1. Survival curves of different types of clonogenic cells, irradiated with 300 kV X rays or 15 MeV neutrons. [Reproduced from G. W. Barendsen, *Curr. Top. Radiat. Res. Q.* **9**, 101 (1973) with the permission of North-Holland Publishing Co.]

TABLE 1

RBE Values Derived from Survival Curves of Different Types of Cells

	at 50% survival	at 10% survival	at 100 rad of 15 MeV neutrons	at 200 rad of 15 Mev neutrons
mouse hemopoietic stem cells	1.5	1.1	1.3	1.1
mouse lymphocytic leukemia cells L5178Y	2.1	1.6	1.8	1.6
cultured cells of human kidney origin	2.7	2.0	2.5	2.2
rat rhabdomyosarcoma cells	2.6	2.0	2.5	2.2
mouse intestinal crypt stem cells	1.8	1.6	1.9	1.8

For radiotherapeutic applications the comparison of RBE values at a fixed dose level is of greater interest than RBE values at a specified level of survival. In radiotherapy the dose per fraction, the fractionation intervals and the total dose will be determined mainly by the tolerances of the normal tissues which for a given location show, in general, a smaller variation among patients than tumor responses. RBE values given in Table 1 for two dose levels show a similar spread as for fixed levels of survival.

With respect to the influence of repair of sublethal damage it is important to note that D_Q/D_0 ratios provide only a measure of the relative width of the shoulder of the survival curve, but the precise shapes of the curves at low doses, corresponding to the region between 100% and 10% survival, are frequently inaccurately known, especially with regard to the contribution of the linear component in the dose-effect relationship (*3*). Because in radiotherapeutic applications doses are commonly administered in daily fractions of 200 to 400 rad of X rays or gamma rays, RBE values of heavy particles or fast neutrons for this region are the most important and these values should be determined with better accuracy than the data presently available for most types of cultured cells, experimental tumors and normal tissues.

It can be concluded that if survival curves of cells in human tumors show a similar spread in RBE values as observed for the types of cells represented in Fig. 1, then large variations in RBE values by a factor of 2 must be expected, on account of differences in intrinsic radiosensitivities of these cells to X rays and high-LET radiations. This must imply that for some types of tumors the radiobiological properties of fast neutrons or heavy particles might not provide an advantage over X rays with respect to the therapeutic margin, while for other types of tumors the gain might be significant. It should be a main object of radiobiological and clinical studies to obtain information about the types of tumors for which a useful gain can be obtained. The data represented in Table 1 and data in the literature, as well as results from preliminary clinical studies, suggest that for cells from tumors originating from hemopoietic and lymphatic tissues, relatively low RBE values are derived. If this applies to human tumors, these types of tumors would be less likely to show an advantage from the use of heavy particles or neutrons over X rays or gamma rays than tumors originating from other tissues.

HYPOXIC CELLS AND REOXYGENATION IN TUMORS

The presence of hypoxic cells in tumors, and the observation that oxygen enhancement ratios (OER) of heavy particles and fast neutrons are smaller by a factor of 1.3–1.8 than the OER of X rays, has long been considered to constitute the main rationale for the use of high-LET radiations in radiotherapy (*4,5*). The ratio $(OER)_X/(OER)_n$ has been denoted "gain factor" for neutrons, but more careful examination of the influence of hypoxic cells in tumors has led to the conclusion that this ratio represents a maximum gain factor which would apply only if all cells in a tumor were severely hypoxic and remained hypoxic throughout a fractionated treatment course. However, the effective gain factor due to the lower OER for heavy particles and fast neutrons is in general smaller than the maximum gain factor, because not all cells in a tumor are severely hypoxic and because in fractionated treatments reoxygenation may occur (*6,7*).

Studies of the time-dependence of reoxygenation have shown that significant differences exist among various types of tumors (*7*). Consequently, in order to eliminate the influence of hypoxic cells as much as possible, tumors would require individually optimalized fractionation regimens. However, no ways are presently available to devise such schemes for human tumors. The use of fast neutrons, negative pions or heavy particles with their low OER values might eliminate this requirement partly and thus provide a more uniform response with standardized fractionation regimens (*8*).

It is important to note in addition that even in tumors in which reoxygenation is relatively rapid, in such a way that at an interval of 24 hours after a given dose the fraction of hypoxic cells has returned to the value present before the last dose, nevertheless the semi-continuous presence of a proportion of hypoxic cells may significantly increase the total dose required for tumor eradication. This phenomenon has been demonstrated for an experimental rat rhabdomyosarcoma as shown in Table 2 (*9*). In this tumor during a fractionated treatment with 300 rad of X rays per day, the fraction of hypoxic cells remains constant at about 0.15–0.30 although experimental accuracy is limited. Based on the survival curves determined for these cells after irradiation in culture or in the tumor, respectively, it can be calculated that the total dose necessary for tumor eradication with this fractionation regimen is about 15% larger as compared with the dose required if no hypoxic cells were present. With fast neutrons having an OER of 1.6, similar calculations show that only about 7% more total dose would be required. This implies that, due to the low OER alone, a gain factor of about 1.1 may be obtained for these neutrons, even for a rapidly reoxygenating tumor. This value is consistent with the observation that the overall gain factor actually derived for the R–1 rhabdomyosarcoma treated with daily doses of 70 to 100 rad of 15 MeV neutrons for several weeks, in comparison with daily doses of 200 to 300 rad of X rays, is about 1.2 to 1.3 (*8*). Many other types of tumors might reoxygenate more slowly, however, as compared with the R–1 rhabdomyosarcoma, and the gain factor of fast neutrons or heavy particles due to the OER alone, might be larger than the value of 1.1 derived for the R–1 rhabdomyosarcoma. It would be extremely important for further evaluations of the possible advantages of high-LET radiations if the mechanisms involved in reoxygenation of cells in tumors were elucidated, because this might allow the selection of types of tumors likely to reoxygenate slowly. It is important to note however, that even small differences by a factor of 1.1 may be significant, depending on the steepness of the relationship between tumor control rate and total dose.

DIFFERENCES IN RADIOSENSITIVITY AS A FUNCTION OF CELL AGE IN THE PROLIFERATION CYCLE

Variations in radiation responses of mammalian cells depending upon their age in the cycle have been shown to be smaller for high-LET than for low-LET radiations (*10*). The OER does not change with cell age, however. If tumor cells could be sychronized by a specific dose-fractionation regimen, and each fraction could be given to tumor cells in the most sensitive phase without similar sensitization being induced in cells from normal tissues, then a therapeutic advantage would be largest in the case of X rays or other low-LET radiations. However, in the view of the variations in cell cycle times of cells in tumors and the relatively small dose fractions commonly employed in radiotherapy, a significant advantage from synchronization for treatments of tumors in patients is unlikely to be obtained in a predictable way. A possible exception might be the accumulation of cells in the sensitive phase before mitosis during low dose rate irradiation as described for the R–1 rat rhabdomyosarcoma (*11*).

TABLE 2
Fraction of Anoxic Cells in R–1 Rhabdomyosarcoma During Treatment with Daily Doses of 300 Rad of 300 kV X rays

Days after start of treatment	accumulated dose (in rad)	mean fraction of clonogenic cells in tumor	fraction of anoxic cells
0	0	100	0.15
3	900	0.07	0.17
5	1500	0.04	0.08
9	2100	0.005	0.22
12	3000	0.002	0.40
16	3600	0.002	0.30
18	4200	0.002	0.30
19	4500	0.002	0.16

PROLIFERATION OF CELLS IN INTERVALS OF FRACTIONATED TREATMENTS AND THE PRESENCE OF NONPROLIFERATING CELLS

To a first approximation it can be assumed that the RBE of fast neutrons for effects on tumors will not depend on cell proliferation in intervals between dose fractions or during low dose rate treatments (*8*). It should be pointed out, however, that, if long overall treatment times have a favorable influence on the therapeutic margin because of a more rapid repopulation in normal tissues as compared to tumors, and if neutrons were most effective with a small number of fractions, then fast neutron treatments might have to be given with intervals in excess of one day, in order to obtain optimal results.

In the case of the experimental R–1 rhabdomyosarcoma it has been shown that with treatments of five fractions of X rays per week, surviving cells proliferate more rapidly than before treatment, starting after the second week of treatment (*9*). Similarly increased cell proliferation rates may occur in normal tissues as well, and much more detailed information is required to evaluate whether a therapeutic advantage can be obtained by selecting specific overall times of treatments for various types of tumors.

In addition to proliferating cells (P cells), tumors may contain nonproliferating cells (Q cells), which can be clonogenic and have to be killed if tumor eradication is to be achieved. Little is known, however, about the shapes of survival curves of Q cells in tumors and the RBE of neutrons for these cells. For some types of cells in culture, the X ray survival curve of nonproliferating cells is characterized by a smaller shoulder as compared with proliferating cells (*12*). We might then expect relatively low RBE values for fast neutrons for these nonproliferating cells. For other types of cultured cells such a reduction in shoulder width has not been observed, however (*12*).

RBE VALUES OF 15 MeV NEUTRONS FOR RESPONSES OF PULMONARY METASTASES OF TUMORS IN PATIENTS

From the discussion of variations of RBE values of heavy particles and neutrons, related to various factors, it can be concluded that similarly large differences may be encountered with tumors in patients. In order to obtain some insight into the spread of RBE values for clinical applications, volume changes of pulmonary metastases in patients were measured from serial chest radiographs after irradiation with single doses of 15 MeV neutrons or ^{60}Co gamma rays and 250 kVp X rays (*13*). The neutron generator at Rijswijk, The Netherlands, producing 15 MeV neutrons through the D–T reaction, was used to deliver single doses ranging from 150 to 220 rad of neutrons to lung metastases from different types of tumors in a number of patients. Comparison was made with doses ranging from 380 to 510 rad of ^{60}Co gamma rays or 250 kVp X rays given to the same or other metastases in the same patients. RBE values were derived for growth delay of metastases in 13 patients. The values related to a standard dose of 150 rad of 15 MeV neutrons ranged from 1.2 to 4.0. As shown in Fig. 2, in general the largest RBE values were obtained for slowly growing metastases. This may provide some indication about the types of tumors for which high-LET radiations might provide an advantage, but the radiobiological aspects of this observation remain to be elucidated. For comparison, the dotted line indicates the mean value of RBE derived for several normal tissues in experimental systems for this dose region.

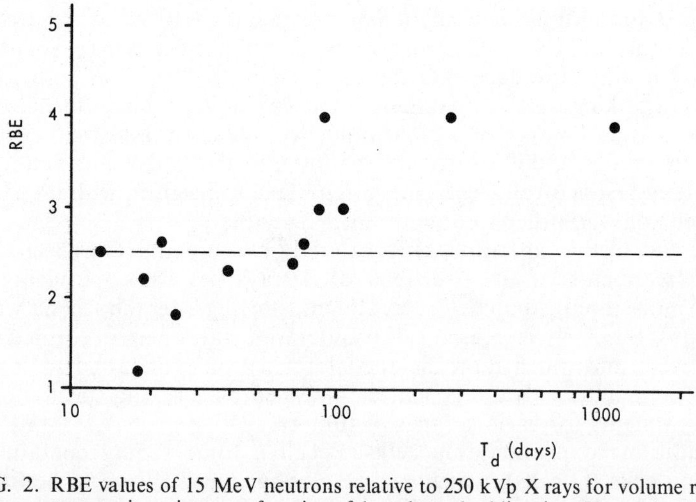

FIG. 2. RBE values of 15 MeV neutrons relative to 250 kVp X rays for volume reduction of pulmonary metastases in patients as a function of the volume doubling time. [Reproduced from H. A. van Peperzeel, K. Breur, J. J. Broerse and G. W. Barendsen, *Eur. J. Cancer,* **10**, 349 (1974) with the permission of Pergamon Press, Ltd.]

CONCLUDING REMARKS

Several factors have been demonstrated to be of importance with respect to the RBE of high-LET radiations for effects on tumors and normal tissues in animals. The relative influence of these factors in experimental systems as well as in clinical situations remains to be investigated in much more detail, in order to provide a basis for the selection of types of tumors for which the application of heavy particles and fast neutrons might provide an advantage over the commonly employed low-LET radiations.

REFERENCES

1. H. D. Suit, in *Proc. Conf. Time and Dose Relationships in Radiation Biology as Applied to Radiotherapy,* Carmel, Calif., 1969. BNL Report 50203 (C–57), pp. vii-x (1970). (Available from Clearinghouse for Federal Scientific and Technical Information, Springfield, Virginia 22151.)
2. G. W. Barendsen, *Curr. Top. Radiat. Res. Q.* **9**, 101 (1973).
3. G. W. Barendsen, *Nature, London,* **193**, 1153 (1962).
4. J. F. Fowler, in *Current Topics in Radiation Research* (M. Ebert and A. Howard, Eds.), **Vol. II**, pp. 303–364. North-Holland Publ. Co., Amsterdam, 1966.
5. D. K. Bewley, in *Current Topics in Radiation Research* (M. Ebert and A. Howard, Eds.), **Vol. VI**, pp. 249–292. North-Holland Publ. Co., Amsterdam, 1970.
6. L. M. van Putten and R. F. Kallman, *J. Nat. Cancer Inst.,* **40**, 441 (1968).
7. R. H. Thomlinson, in *Proc. Conf. Time and Dose Relationships in Radiation Biology as Applied to Radiotherapy,* Carmel, Calif., 1969. BNL Report 50203 (C–57), pp. 242–247 (1970). (See ref. *1* for availability)
8. G. W. Barendsen, *Eur. J. Cancer,* **7**, 181 (1971).
9. G. W. Barendsen and J. J. Broerse, *Eur. J. Cancer,* **6**, 89 (1970).
10. R. P. Bird, *Sensitivity of Synchronized Mammalian Cells to High-LET Radiation.* Ph. D. Thesis, Berkeley, University of California, LBL–968 (1972).
11. H. B. Kal and G. W. Barendsen, *Brit. J. Radiol.,* **45**, 279 (1972).
12. G. M. Hahn and J. B. Little, Plateau-phase cultures of mammalian cells. *Curr. Top. Radiat. Res.Q.,* **8**, 39 (1972).
13. H. A. van Peperzeel, K. Breur, J. J. Broerse and G. W. Barendsen, *Eur. J. Cancer,* **10**, 349 (1974).

The Middle Atlantic Neutron Therapy Trial[1]

Charles Rogers, M.D.
Division of Therapeutic Radiology and Oncology
Medical College of Virginia
Virginia Commonwealth University
Richmond, Virginia 23298, USA

INTRODUCTION

A consortium of therapeutic radiologists in the Middle Atlantic States and physicists at the Naval Research Laboratory has been established to investigate the use of fast neutron beams in the control of some tumors. Many radiobiology experiments have indicated that neutron beams may have an advantage in the control of local tumors over that of conventional forms of radiotherapy (*1*). In preparation for clinical radiotherapy trials, extensive measurements have quantified the various physical characteristics of the NRL cyclotron-produced neutron beam (*2*). Techniques have been developed for the absolute determination of delivered dose at depth in tissue for this beam, accounting for the relatively small component of dose delivered by gamma rays, as well as that by the neutrons. A collimator system has been designed to allow the precise field definition necessary for optimum therapy treatment planning. A dose control and monitor unit has been engineered and has demonstrated a reproducibility of 0.2 percent. The relative biological effectiveness of this neutron beam has been studied with several biological systems to aid in determining proper radiotherapeutic dose levels. The objective of these studies is a full-scale clinical radiotherapy trial to test neutron effectiveness as compared to that of conventional radiotherapy, after a pilot study in man.

MODIFICATION OF THE NRL CYCLOTRON FACILITY

At the beginning of this program in September 1971, a preliminary decision had to be made as to the type of neutron beam to be initially generated and studied for possible radiotherapeutic applications. A variety of neutron-generating interactions are possible with a flexible accelerator such as the NRL sector-focusing cyclotron, which can accelerate protons to 75 MeV, helium-3 particles to 110 MeV, etc. The choice was rapidly narrowed down to the use of a 35-MeV, 10-microampere deuteron beam colliding with a thick beryllium target to produce neutrons having an average energy of approximately 15 MeV. This reaction has the primary advantage of providing a fast-neutron dose rate in human tissue of about 60 neutron rad per minute at a convenient target-to-skin distance of 125 cm, with very little gamma-ray contamination.

The penetrating power of this beam is somewhat superior to that of cobalt-

[1]Congress presentation E–4–3.

60 gamma rays at 80 cm SSD with an adequate "skin-sparing" effect (i.e., the dose to the skin is less than that at a depth of several millimeters because of the building of secondary protons as the neutrons collide with hydrogen atoms in the tissue). Also, the target stability is excellent and its lifetime practically infinite. This neutron beam comfortably satisfies or exceeds all criteria specified by radiotherapists for a neutron beam to be suitable for clinical trials.

The first modifications of the NRL Cyclotron facility for the MANTA program involved installation of magnetic lens elements to provide optimum focusing of the deuteron beam. A target assembly was designed and constructed, including monitoring ionization chambers for controlling the dose delivered by the beam. Additionally, a Benelex (pressed wood) shield, collimator-receptacle, and an inventory of collimators were fabricated for defining a variety of field sizes. A 30 centimeter box of Lucite, when filled with tissue-equivalent fluid, simulates the body for studies of neutron beam dose in depth. Under the control of an EMR-6050 computer, automatic mapping of doses under a variety of conditions for studying beam penetration, collimator effectiveness, as well as perturbations due to wedge filters and collimator inserts has been accomplished.

The second group of facility modifications for this program relates to the patient-treatment plans. A new beam line for the radiotherapy room has been designed by using ion-optic computer codes. This design includes a beam-switching magnet and transport equipment such as beam stops, viewers, and collimators. Also, quadrupole elements and a vacuum-pumping station are being fabricated.

The shielding and radiation properties of various materials exposed to this specific neutron beam have been determined, and a suitable clinical neutron-beam shield and collimator have been designed.

A patient-handling annex has been constructed. The design strategy includes two parallel flow paths for patients, comprised of separate examination rooms and setup rooms in order to optimize the utilization of the neutron beam facility. Furthermore, the use of a shielding maze instead of a massive shielding door will minimize the time required for patient transport into the irradiation area. The design includes considerations for allaying patient anxiety and affording comfort.

PHYSICAL MEASUREMENTS AND DOSIMETRY OF THE NRL BEAM[2]

Neutron Beam Dose-Rate Calibration

The techniques currently available for fast-neutron dosimetry are not as straightforward as those routinely employed in radiotherapy with X and gamma radiation. This is partly because the interaction probabilities with the elements comprising human tissue (carbon, hydrogen, oxygen, and nitrogen) are not as well known for neutrons as they are for X and gamma rays. Also, neutrons are always accompanied by gamma rays. Since gamma rays have a smaller biological effect than an equal dose of fast neutrons, the two dose components must be measured separately at each point of interest in the irradiation field.

Fortunately, the gamma-ray "contamination" of the NRL neutron beam is

[2]This constitutes Phase 1 of the planned clinical trial. For the overall time schedule of the clinical trial, consult Chart 1.

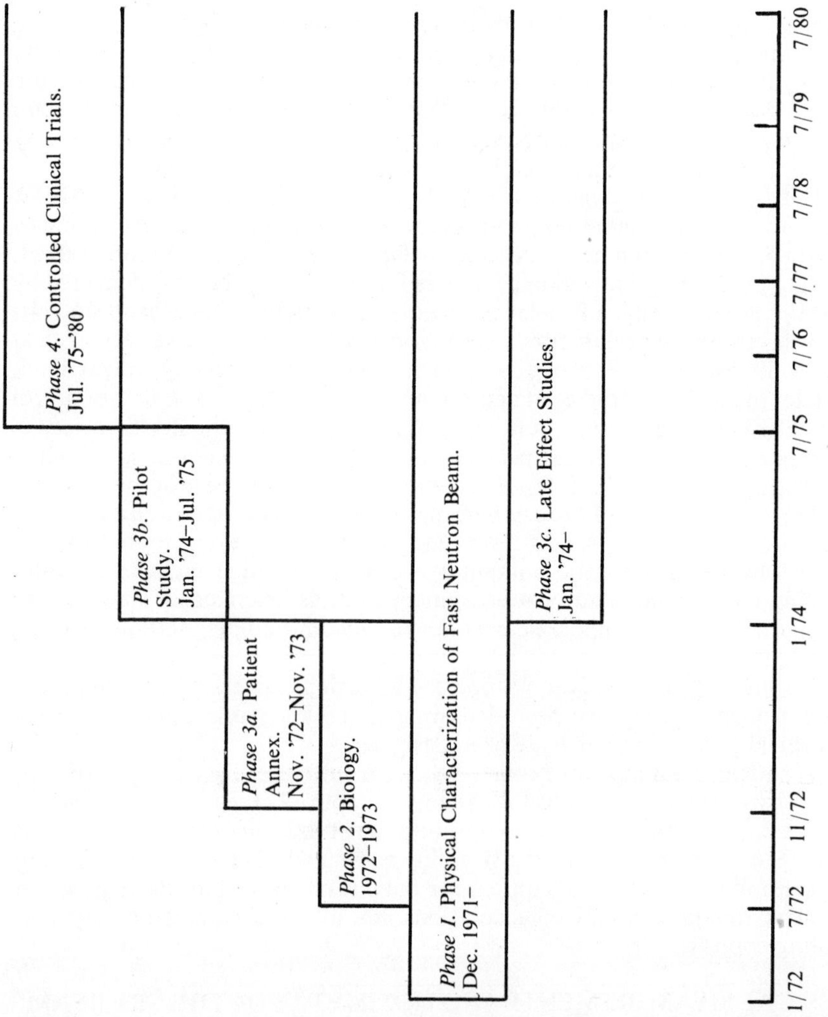

Chart 1—Time-line schedule for various phases of the planned clinical trial.

usually only a few percent of the total dose, which allows the use of a differential dosimetry technique. This approach involves two measurements: one (by a tissue-equivalent (TE) plastic ionization chamber) of the total neutron-plus-gamma dose, and a second (by a thermoluminescent (TL) dosimeter) of the gamma-ray dose, which can then be eliminated to obtain the neutron dose alone. However, this method is complicated by the facts that the available TL dosimeters (such as lithium fluoride or calcium fluoride) do have some sensitivity to fast neutrons, and that the most suitable TE plastic for making the ionization chambers is not precisely tissue-equivalent with respect to neutron interactions. Calculations estimate that TE plastic over-responds by about 5% in comparison with standard wet tissue (*2*).

Considerable effort has been spent at NRL in unraveling this problem, and excellent progress has been made. As a result, it is believed that the absolute neutron beam dose rate in wet tissue is probably correct within 5 percent and is capable of further improvement. Direct measurements of the neutron beams at the MRC cyclotron at Hammersmith Hospital and at the Texas A&M cyclotron have indicated tentative agreement within 3% with the results obtained from the dosimetry techniques employed at those centers (*2*).

Beam Monitoring and Control

The absolute calibration of dose in the neutron beam described above would be pointless without some means of guaranteeing delivery of the same dose from one exposure to the next. In other words, the neutron tissue-dose delivery capability of the beam must be continuously monitored and precisely controlled. The main features of the target array are as follows:

1. The steering ring aperture through which the deuteron beam is focused by nulling the current collected on it, and which guarantees that all the deuterons strike the Be target.
2. The electron suppressor ring which is biased 200V negatively to drive secondary electrons back to the target for accurate measurement of the deuteron current.
3. The Be target disc (0.864 gm/cm^{-2} thick) backed by watercooled aluminum plate.
4. A flat fission proportional counter.
5. Twin, flat ionization chambers, with three 3-mm graphite-coated polystyrene walls and two 6-mm air gaps.

The information obtained from this monitoring system is processed in a control console and is displayed in a time-integrated digital form. Each monitoring channel is calibrated nominally to indicate neutron dose (tissue rads) in free space on the 10×10 cm beam axis 125 cm from the target.

Thumbwheel switches allow the setting of a desired neutron dose to the nearest 0.1 rad, at the 125 cm reference position in air. Auxiliary measurements and/or calculations are used to relate this dose to those at other locations, in or out of a phantom, including or excluding gamma rays, or under other special conditions. The number-one ion chamber is used to control the cutoff of the beam, when the down-counter reading equals the preset dose. The number-two chamber and the beam-current channel are set to cut off the beam at a slightly larger dose if the first ion chamber fails to do so.

Neutron Dose at Depth in Tissue

For treatment planning, physicians require accurate depth-dose distributions of the beam for each collimator, together with the perturbations due to various wedge filters and field-modifying inserts. This information requires the accumulation and processing of an enormous number of data points. To do this, an automated dose-mapping system interfaced with an EMR-6050 computer has been developed and provides overall accuracy and position reproducibility of approximately 0.5 mm. In the computer, a software algorithm generates any desired sequence of x,y values as command position coordinates. The dose at the prescribed coordinates is measured automatically, together with the monitor dose values for normalization. The data are displayed on a cathode-ray tube and a line printer, and are written on magnetic tape for subsequent processing. A computer program has been developed for translating these data points into isodose contours, which are automatically drawn on graph paper. Additionally, cylindrical collimators having radii of 1, 3, 5, 9, 13, and 17 cm have been fabricated, and a data base has been obtained which will allow computation of depth-dose distributions for any collimator of arbitrary cross section by separating the dose into components due to primary and scattered radiation.

Collimation

Defining neutron field sizes by adjustable interleafed shielding shutters similar to those used in cobalt-60 gamma-ray teletherapy units is not feasible because of the large amount of shielding material required for fast neutrons. However, clinically acceptable collimation of the NRL beam has been achieved through the use of 75-cm thick Benelex (pressed wood) collimator inserts. Measurements at 125 cm from the target have revealed that the total neutron-plus-gamma dose at a position 4 cm out of the beam and into the collimator shadow is only 1.4% of the corresponding dose at the beam center. Moreover, measurements have shown that there is no significant radiation hazard associated with these collimators due to induced radioactivity.

Spectrum Determination

Neutron time-of-flight techniques have been used to determine the spectral distribution of the NRL beam in air and the modification of this spectrum as the beam penetrates TE material. This method has been reported (*2*). The energy spectrum changes very little with TE material thickness, the average energy of the primary and the forward-scattered neutron component being increased by only 8% in penetrating 24 cm of the TE fluid. This result implies that the relative biological effectiveness (RBE) of the beam should be essentially independent of tissue depth (assuming constant dose levels), which is desirable from the standpoint of treatment planning. We are attempting to verify this finding in radiobiological experiments.

BIOLOGICAL STUDIES UTILIZING THE NRL NEUTRON BEAM

Radiobiological investigations in support of the proposed fast neutron clinical trial have been divided into two phases. Studies of a variety of acute responses to neutron irradiation were undertaken first (Phase 2) in order to establish quantitative biological information about the NRL beam to be used in planning its practical application in the therapeutic situation. These experiments have been completed and the results reported elsewhere (*3–6*). Investigations of the long-term effects of neutron irradiation on normal tissues have recently been initiated (Phase 3c). These experiments are considered to be necessary in view of the adverse late effects in normal tissues encountered in the Stone-Larkin neutron therapy trial(*7*).

Biological Characterization of the NRL Neutron Beam

The acute effects studies were designed to provide the following biological information about the NRL neutron beam:

1) The RBE of the neutron beam compared to conventional therapy sources for a variety of endpoints in order to provide a basis for the choice of practical therapeutic doses. RBE values have already been established with the use of various biological systems at other neutron facilities and serve as useful guides for the NRL beam. However, since RBE depends in part on linear energy transfer (LET) and because the gamma-ray contamination and neutron energy spectrum (and thus the LET spectrum) may vary from one heterogeneous neutron beam to another, this biological parameter should be determined for the specific beam to be employed in the clinical trial. Since it has been shown that the RBE of neutrons for radiation effects based on cell lethality increases with decreasing neutron fraction size (*8*), it was of particular importance to establish this relationship for the NRL beam in order to arrive at the pertinent RBE values in the therapeutic dose range.

2) The oxygen dependence (OER) of the beam in order to estimate the therapeutic advantage that it might afford over conventional sources.

3) The relationship between RBE and OER and depth of beam penetration in tissue. The physical data indicating that the energy spectrum of the NRL beam changes very little with depth predicted that these biological parameters should be found to vary insignificantly with depth.

4) The effect of fractionated exposures. The reduced capacity for repair of sublethal damage observed for other high LET radiation sources should be substantiated for the NRL beam. Experiments to quantitate this component of recovery as well as the repopulation component were included to provide information for decisions regarding proper fractionation schedules for human treatment.

5) Biological measurements of beam uniformity and collimator effectiveness for comparison with physical measurements of dose. In selecting the specific experiments to provide the above information, the following points were considered:

a) Each experiment should provide a well-tested and quantitatively reproducible endpoint over a range of doses.

b) Some experiments should be included which have been performed on other fast neutron beams in order to provide a basis for comparison between beams.

c) A variety of biological systems having a range of radiosensitivities should be employed.

Small Animal, Single Dose Experiments

Initial biological studies performed with the NRL beam involved the examination of several endpoints in whole-body irradiated small animals. These experiments were carried out by Drs. J. L. Montour and J. D. Wilson of the Medical College of Virginia. RBE values were determined for the responses of hematopoietic tissue (measured by 30-day mortality and endogenous spleen colony production in $CD2F_1$ mice), gastrointestinal epithelium (measured by 5-day mortality in mice), lymphatic tissue (measured by weight loss in mouse, thymus and spleen), and vascular endothelium (measured by 30-hour mortality in the young chicken). The RBE values obtained in these experiments and the neutron dose ranges employed are shown in Table 1 (taken from reference *3*). Also included in Table 1 for comparative purposes are RBE values for endpoints in other biological systems obtained by MANTA collaborators. Figure 1 shows the relationship between RBE and neutron fraction size for tissue responses studied in the small animal experiments.

TABLE 1
RBE Values And Neutron Dose Range For NRL Beam Studies

End Point	Neutron Dose (rads)	Gamma Dose (rads)	RBE	Dose Range
Mouse Mortality				362–1047
$LD_{50/5}$	703	1179	1.68	
$LD_{50/5}$	639	1149	1.80	
$LD_{50/30}$	484	873	1.80	
Mouse Endogenous Spleen Colony				285–493
Colonies at 484 rads of neutrons	484	765	1.58	
D_0, dose-response curves	68	102	1.50	
Mouse Thymic Weight Depression				26–675
Maximum involution (ED_{50})	84	122	1.45	
Sensitive component (ED_{50})	50	68	1.36	
Insensitive component (ED_{50})	365	784	2.15	
Mouse Splenic Weight Depression				26–679
Maximum involution (ED_{50})	139	195	1.40	
Sensitive component (ED_{50})	53	70	1.32	
Insensitive component (ED_{50})	532	830	1.56	
Chick Early Mortality				341–638
$LD_{50/30}$ hour	519	939	1.81	
Murine Lymphocytic Leukemia*				100–600
0.1 survival level	520	1025	1.97	
Mouse Jejunal Crypt Cells**				100–233
D_0, dose-response curves			2.6	
Vicia faba Root Growth***				10–90
Inhibition at 70 rads of neutrons	70	266	3.8	
V79 Chinese Hamster Cells***				100–900
0.1 survival level	450	840	1.87	
T-1 Kidney Cells****				50–500
Survival at 200 rads of neutrons	200	474	2.37	

* Raymond U, Duke University Medical Center, personal communication BW 5147 cells irradiated in vivo and assayed in vitro.

** Withers(*5*)

*** Hall et al.(*6*) *Vicia* RBE increases to about 5.8 at approximately 10 rad. V79 RBE increases to 3.0 at 100 rad.

**** Todd et al.(*4*) RBE increases to 3.1 at 105 rad.

The eight-day post-irradiation response of mouse spleen and thymus indicated that the time course and rate of cellular repopulation in these organs are similar in neutron and gamma-irradiated animals. The finding suggests that the repopulation component of recovery in tissue of patients receiving a fractionated course of treatment with the NRL beam will not be different from that observed in patients receiving a similar course of conventional therapy.

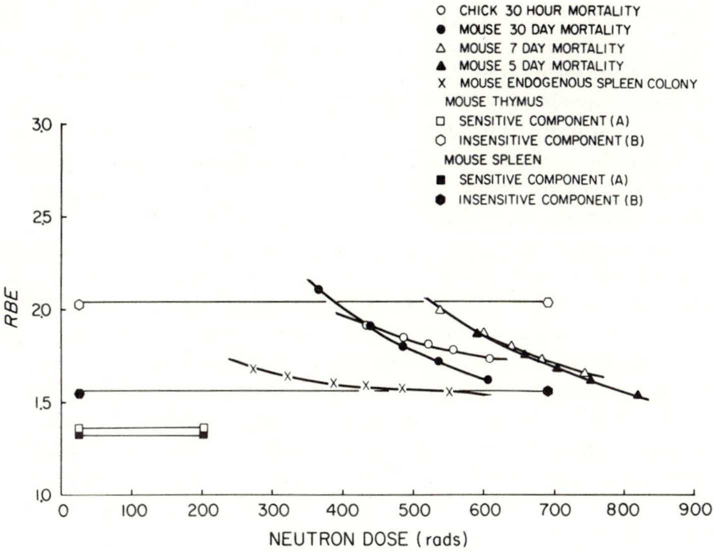

In Vitro Mammalian Cell Studies

The survival of T-1 human kidney cells irradiated at various positions in a tissue-equivalent fluid phantom was used to determine the spatial distribution of cell lethality in the NRL beam (*4*). These experiments were performed by Dr. Paul Todd of Pennsylvania State University. An RBE of 2.4 for lethal damage in this cell line was obtained at about 200 rad of neutrons. As expected from physical measurements of the neutron spectrum, there is no significant change in RBE with depth up to 12 cm. Similar experiments with hypoxic cells indicated that the OER of the NRL beam is about 1.6 and also appears to be independent of depth.

T-1 cell survival was also used to evaluate the transverse uniformity of the beam and collimator effectiveness. As an example of the type of information obtained in these experiments, it was found that a dose of 230 neutron rad measured on the axis of a beam defined by a 10×10 cm collimator resulted in 30% survival across the full width of the beam at a tissue depth of 5 cm, 80% survival 1 cm outside the beam, and more than 90% survival 2 cm outside the beam. Survival data resulting from experiments of this type were used by Dr. Todd to construct isosurvival curves which may be useful in treatment planning for human exposures.

The survival of Chinese hamster cells provided data for neutron RBE determinations at the NRL and TAMVEC facilities (*6*). These studies, conducted by Dr. Eric Hall of Columbia University, indicated that the 35 MeV NRL beam is biologically more effective than the 50 MeV TAMVEC beam by a factor of 1.1. For the NRL beam, the RBE for lethality in this cell line decreases from a value of

3 at 90 rad to a limiting value of about 1.8 at higher doses.

Experiments with Fractionated Neutron Doses

The survival of jejunal stem cells in the intestine of CDF mice following whole-body exposure was measured by Dr. H. Rodney Withers of M. D. Anderson Hospital using the NRL and TAMVEC neutron beams (*5*). The purpose of these experiments was to determine the RBE for intestinal stem cell lethality and to measure the degree of recovery occurring with fractionated exposures. From the data obtained at NRL, RBE values of approximately 2 and 2.6 were found at neutron doses of 500 and 180 rad, respectively. Experiments in which equal total doses were given in 2, 3, or 5 fractions over a time interval during which cellular proliferation in the crypts was minimal, indicated that the fractionated exposures are only slightly less effective than single doses. Similar results were obtained for the response of lymphocytic leukemia cells (BW 5147) exposed in vivo to fractionated doses of neutrons at NRL by Dr. Raymond U of Duke University (to be reported at these meetings).[3] Both of these studies suggest that little if any sparing can be achieved in the clinical situation by increasing the number of neutron fractions. However, over longer periods of time, sparing should occur as a result of repopulation (as discussed above, the rate of sparing due to repopulation may be similar for neutrons and conventional radiations) and thus it will be important to consider the overall treatment time in deciding on a fractionated schedule for the clinical trial.

Plant Cell Studies

Inhibition of root growth in the broad bean (*Vicia faba*) is a reliable indicator of radiation exposure. The advantages of this system are its simplicity and its relatively high radiation sensitivity making it responsive at doses in the therapeutic range. It has been used extensively in biological studies of pi-mesons and other charged particles and was used in an intercomparison of the NRL and TAMVEC neutron beams by Dr. Eric Hall(*6*). These experiments are in agreement with the in vitro mammalian cell data which indicate the NRL beam to be about 10% more effective than the TAMVEC beam. Measurements of OER and of RBE as a function of neutron dose size were also carried out with the *Vicia faba* system. An OER of 1.6 was obtained which agrees with the OER value determined in the T-1 cell studies of Todd mentioned above.

Summary of the Initial Phase of Radiobiological Investigations

The initial biological experiments performed with the NRL fast neutron beam indicated that it has sufficient uniformity of dose, effectiveness, and oxygen dependence to be used with confidence for therapy purposes. The depth-dose profile and collimation are at least as good as can be obtained with conventional cobalt-60 gamma ray sources. Information was developed about the relationship between RBE and fraction size for a variety of biological systems and about the capacity for recovery with fractionated exposures which aided in the selection of dose and treatment schedule to begin the human exposures.

[3]Raymond U et al., *Radiat. Res.* **59,** 125 (1974).

Late Effect Studies

A pilot study on the late effects of 35 MeV neutrons on the brain and spinal cord of rabbits has been in progress at NRL since May 1974. These experiments are being carried out by Drs. Eileen Bradley and John G. Maier of George Washington University Medical Center and Dr. Kenneth Mossman of Georgetown University Medical School. Animals were irradiated twice weekly for six weeks with doses ranging from 140 to 560 neutron rad per fraction under conditions maintained as close as possible to those used in the treatment of head and neck patients. A second group of animals was exposed to fractionated doses of X-rays using a 4 MeV linear accelerator in the Radiation Therapy Department of George Washington University Medical Center. Data from this late effect study are not yet available and will be reported at a later date.

HUMAN CANCER PILOT STUDY

Prior to the initiation of a controlled clinical trial of fast neutrons a practical and acceptable treatment schedule must be determined, as well as the development of treatment techniques. This would include the determination of dose, (both fraction-size and total dose), fraction number, overall time, and the development of technical apparatus and procedures, such as wedge filters, treatment planning, tissue-equivalent material for bolus, patient-positioning, and the logistics of transportation.

Choice of Treatment Techniques

The beam is fixed and horizontal. Patients are treated either lying on a couch, or in a sitting position, or kneeling position in a chair with armrests, back, chest, and head supports which can be adjusted to suit each patient. Both the chair and the couch can be rotated, lowered and raised. For all regions treated with more than one direct field, treatment plans are constructed in the usual way. Surface and depth doses are checked by placing aluminum pellets on the patient's skin, beneath compensating material and molds, and in cavities such as the mouth. Fast neutrons transmute ^{27}Al into ^{27}Mg; this latter isotope decays with a half-life of 9.54 minutes, emitting gamma-rays of 0.84 and 1.01 MeV, and can be counted with an error of less than 0.5%, on a one-gram sample. One pellet, placed at the center of the incident field, receives a known dose of radiation, and the doses received by the other pellets are related to this.

After extensive radiobiology work on the MANTA beam, and some comparisons with other beams, it was decided that an attempt would be made to simulate a megavoltage photon effect of 200 gamma-rad daily, five days a week, giving a total of 1000 gamma-rad per week. The first alteration required changing this for twice-weekly fractionation. It was estimated that (partly relying on the Ellis concept) 400 gamma-rad twice-weekly (800 rad per week) would be equivalent to 1000 gamma-rad in 5 fractions. The biological work for fraction sizes of this order shows an apparent RBE of the MANTA beam to be 2.4 to 2.6, compared with Cobalt-60 gamma-rays. In order to err on the low side, 2.6 was chosen for the RBE. This would indicate that the appropriate fraction size for twice-weekly treatment would be 153.9 neutron rad. However, the then current dose for the M. D. Anderson neutron project was 160 neutron rad, and comparative studies by Hall (*6*) had indicated that the MANTA beam was 10% more effective than that at TAMVEC (using 50 MeV deuterons on beryllium).

The initial MANTA patients were treated using 140 neutron rad, which was purposely chosen to be on the low side, for each of the above considerations.

The schedule assumed that 140 neutron rad twice per week (280 neutron rad per week) was equivalent to 900 to 1000 gamma-rad per week when given for five days at 180 to 200 rad. When treating large volumes, thorax, abdomen, or subclinical disease, the schedule called for 140 neutron rad twice weekly, for 5 weeks, giving a total of 1400 neutron rad. This was estimated to be equivalent to 5000 gamma-rad, given in 25 fractions, over 5 weeks.

The schedule for treating more modest volumes, and when dealing with the head and neck, called for 140 neutron rad twice weekly, for 7 weeks, reducing the field sizes after five and six weeks.

After normal tissue response to this schedule was found to be minimal, and after several "in-field" failures, the dose has been increased to 145 neutron rad daily, with the same schedules as above.

The doses of radiation received by the patient's eyes are measured by placing neutron film badges as near to the eyes as possible during the treatment. In patients whose teeth have been in the irradiated field, examinations of the teeth are performed during the course of treatment, and at regular intervals after the completion of treatment.

Assessment of the Response of Normal Tissue

Values for the RBE of fast neutrons for several different tissues should be determined, but skin and mucous membranes are practical choices. These can be subjected to visual macroscopic observations. At MANTA, the skin reactions are observed by the radiation therapist two times weekly, during the course of treatment, and at regular intervals after treatment. Color photographs are taken at weekly intervals during treatment.

TABLE 2
THE SCORING OF THE SKIN REACTIONS

THE DEVELOPING REACTION:

0 No visible reaction
1.0 Slight but definite erythema
2.0 Moderate erythema
3.0 Severe erythema (red or pink)
4.0 First sign of acute exudative response
5.0 Exudative response over less than half the field
6.0 Exudative response over more than half the field
7.0 Exudative response over entire field
If the area were showing signs of scaling 0.5 could be added to the erythema score.
+ or − could be added to a score and counted for a third

FADING REACTION:

4.0 Whole area covered by dry scab
3.0 More than half the area covered by dry scab
2.0 less than half the area covered by dry scab
1.0 Small area of dry scab
The addition of + or − could be added and counted for a third

A numerical scoring system, designed for the skin reactions of mice (*9*), has been modified and is used for normal tissue reactions (Table 2). Scoring is performed by a physician and a biologist, at various sites chosen for their biological significance by the clinician. This is done twice weekly. The system used for scoring reactions clearly has its limitations. For example, the extent of several different types of change, including edema of the tissues, increased vascularity and desquamation, have been reduced to one number. Also, in healing (fading) reactions, a scab may have covered several conditions. Also, the healing score is not biologically equivalent to the numerical equivalent of the ascending score. However, scoring is an approach to a quantitative method, and it is thought that the RBE obtained in this way might have greater precision than that achieved by clinical description.

Strict comparison is only possible where regions of skin are anatomically and physiologically alike. For example, there are differences between the extensor and the flexor aspects of the skin, and between surfaces which are exposed and those which are covered.

At MANTA, this scoring system has now been modified by the author and by Dr. Kenneth Mossman of Georgetown University. The notations for the modified system are shown in Table 3. It should be noted that, first of all, developing and fading scores are recorded with different notations. More importantly, erythema, pigmentation, moist desquamation, and dry desquamation are considered as separate phenomena and are scored and recorded separately. This scoring system will be used on the next set of patients and hopefully can be used to evaluate both neutron-treated patients and low LET controls. Obviously, the normal tissue reactions of the trial and control patients must be equivalent in order to demonstrate an increased therapeutic ratio.

TABLE 3

MOSSMAN AND ROGERS—July 1974

Reaction	Key
Erythema Reaction DE.0—no visible reaction DE.1—slight erythema DE.2—definite erythema DE.3—moderate erythema DE.4—severe erythema FE.0—no visible reaction FE.1—slight erythema FE.2—definite erythema FE.3—moderate erythema FE.4—severe erythema	DE—developing erythema FE—fading erythema
Pigmentation Reaction DP.0—no pigmentation DP.1—pigmentation <1/2 field DP.2—pigmentation >1/2 field DP.3—pigmentation over entire field FP.0—no pigmentation FP.1—pigmentation <1/2 field FP.2—pigmentation >1/2 field FP.3—pigmentation over entire field	DP—developing pigmentation FP—fading pigmentation
Moist Desquamation Reaction DM.0—no visible reaction DM.1—moist desquamation <1/2 field DM.2—moist desquamation >1/2 field DM.3—moist desquamation over entire field FM.0—no visible reaction FM.1—scabbing <1/2 field FM.2—scabbing >1/2 field FM.3—scabbing over entire field	DM—developing moist desquamation FM—fading moist desquamation
Dry Desquamation Reaction DD.0—no visible reaction DD.1—dry desquamation <1/2 field DD.2—dry desquamation >1/2 field DD.3—dry desquamation over entire field FD.0—no visible reaction FD.1—flaking <1/2 field FD.2—flaking >1/2 field FD.3—flaking over entire field	DD—developing dry desquamation FD—fading dry desquamation

SUMMARY

MANTA, a consortium of physicians, physicists, biologists, engineers and technologists, has developed a neutron therapy facility, characterized the neutron therapy beam and initiated a pilot study to develop a practical regimen for neutron therapy which can be tested in a controlled clinical trial. Presently, 145 neutron rad twice weekly is believed to cause equivalent normal tissue response to 200 gamma rad five times weekly. A typical curative dose for squamous carcinoma of the upper aerodigestive areas would be 2030 neutron rad in fourteen fractions in seven weeks. Treatment volumes are reduced during treatment. No judgment as to the value of neutron therapy relative to low LET therapy can yet be made from this effort.

REFERENCES

1. British Institute of Radiology Symposium, 1962: *Pretherapeutic Experiments with the Fast Neutron Beam from the Medical Research Council Cyclotron. Brit. J. Radiol.* **36**, 77–121 (1963).
2. R. B. Theus, R. O. Bondelid, F. H. Attix, L. S. August, P. Shapiro, R. E. Surratt and C. C. Rogers, *Cancer* **34**, 17 (1974).
3. J. L. Montour, J. D. Wilson, C. C. Rogers, R. B. Theus, and F. H. Attix, *Cancer* **34**, 54 (1974).
4. P. Todd, C. B. Schroy, F. H. Attix, and R. B. Theus, *Cancer* **34**, 33 (1974).
5. H. R. Withers, K. Mason, B. O. Reid, N. Bubradsky, H. T. Barkley, B. W. Brown, and J. B. Smathers, *Cancer* **34**, 39 (1974).
6. E. J. Hall, J. K. Novak, and S. A. Marino, *Brit. J. Radiol.* **47**, 882 (1974).
7. R. S. Stone, *Am. J. Roentgenol. Radium Ther.* **59**, 771 (1948).
8. S. B. Field, *Radiology* **93**, 915 (1969).
9. J. F. Fowler, K. Kragt, R. E. Ellis, P. Lindop and R. J. Berry, *Int. J. Radiat. Biol.* **9**, 241 (1965).

A Preliminary Report of the MDAH-TAMVEC Neutron Therapy Pilot Study[1,2,3]

David H. Hussey, Gilbert H. Fletcher, and Jesus B. Caderao

The University of Texas System Cancer Center
M. D. Anderson Hospital and Tumor Institute
Department of Radiotherapy
Houston, Texas, USA

INTRODUCTION

The pilot study of neutron therapy using the Texas A&M Variable Energy Cyclotron (TAMVEC) was started in October 1972. Treating only patients with advanced cancer in this initial phase, the objectives were: (1) to evaluate acute and late effects on normal tissues, (2) to develop satisfactory dosage schedules and treatment policies based on the histological type, location, and extent of tumor, and (3) to define a suitable patient population for future clinical trials. This paper is a preliminary report of the results of the pilot study.

Initially, neutrons produced by bombarding beryllium with 16-MeV deuterons (7.6-MeV mean neutron energy) were used since there was clinical experience at Hammersmith Hospital with this energy. However, since March 1973, a neutron beam generated by 50-MeV deuterons (21-MeV mean neutron energy) has been employed because it has a better skin sparing effect and greater depth dose.

DOSAGE SCHEDULES

Dosimetric Differences at TAMVEC and Hammersmith

As the pretherapeutic studies with TAMVEC progressed, it became evident that the relative biological effectiveness (RBE) was less than that reported by Hammersmith. With the 50-$MeV_{d\rightarrow Be}$ beam, a lower RBE was anticipated since this beam presents a broader linear energy transfer (LET) spectrum than does the 16-$MeV_{d\rightarrow Be}$ beam (lower LET secondary protons;

[1]Congress presentation E-4-4.

[2]MDAH – The University of Texas System Cancer Center, M. D. Anderson Hospital and Tumor Institute, Houston, Texas; TAMVEC – Texas A & M Variable Energy Cyclotron, Texas A & M University, College Station, Texas.

[3]This investigation was supported by Public Health Service Research Grant CA 12542 from the National Cancer Institute.

high LET secondary alpha particles and recoil ions). However, the RBE of the 16-MeV beam was also less than had been expected.

It appears now that the radiobiology results with TAMVEC were related to the following dosimetric differences:

(1) The Hammersmith doses do not include the gamma contribution (*1*). At TAMVEC the doses have been determined by measurements with a tissue equivalent (TE) ionization chamber immersed in TE liquid, and expressed in rads including both the neutron and gamma components ($rad_{\eta\gamma}$). With the 16-MeV beam, the gamma dose is approximately 6% of the total dose at D_{max} and increases to 11–13% at 10 cm depth.

(2) In addition, dosimetry intercomparisons between the two institutions (*2,3*) have indicated a 10% difference in the value of the neutron rad (rad_{η}) (i.e., a quoted dose of 100 rad_{η} at Hammersmith would be 110 rad_{η}, or 117 $rad_{\eta\gamma}$ at TAMVEC).

Dosage Schedules at Hammersmith

The fractionation schedule used at Hammersmith is 1440 rad_{η} in 12 fractions in 26 days (*4*). This is approximately 1680 $rad_{\eta\gamma}$/26 days as determined at TAMVEC (17% difference in dosimetry). To compare this dosage schedule with those at TAMVEC, additional corrections must be made to account for differences in fractionation and RBE. Using Field's (*5*) modification of the NSD formula [$D=NSD\ N^{.04}t^{.11}$] and a ratio of RBE's [$RBE_{16\ MeV}/RBE_{50\ MeV}$] of 1.1, the equivalent dose as stated at TAMVEC is 1995 $rad_{\eta\gamma}$ in 14 fractions in 47 days.

Evolution of Dosage Schedules at TAMVEC

Dosage schedules at TAMVEC have progressed through three phases: (1) the initial schedules based on RBE determinations of the TAMVEC radiobiology studies, (2) a subsequent 10% dosage increase based on a clinical evaluation of acute mucous membrane reactions, and (3) a subsequent 10% dosage reduction based on a clinical evaluation of late radiation sequelae. With head and neck and breast tumors, the aim has been to deliver the equivalent of 1000 rad per week to a total equivalent dose of 6000 rad in 6 weeks for moderately advanced tumors and 7000 rad in 7 weeks for massive cancers.

(1) Initial Dosage Schedules

The initial assumption, based on NSD calculations, was that 1000 rad per week with ^{60}Co delivered in 5 fractions (1000 rad/5 Fx/wk) is equivalent to 800 rad per week in 2 fractions. The pretherapeutic TAMVEC studies had indicated that the RBE's relative to fractions of 400 rad with ^{60}Co were 2.65 for the $16\text{-MeV}_{d\rightarrow Be}$ beam and 2.4 for the $50\text{-MeV}_{d\rightarrow Be}$ beam (Table 1). Initially, equivalences were derived as follows:

a) Standard treatment at MDAH:
   1000 rad/5 Fx/wk (^{60}Co)
b) Equivalent to:
   800 rad/2 Fx/wk (^{60}Co)
c) Equivalent to:

300 $rad_{\eta\gamma}$/2 Fx/wk (16 MeV; RBE 2.65)
or
330 $rad_{\eta\gamma}$/2 Fx/wk (50 MeV; RBE 2.4)

TABLE 1
RBE of TAMVEC Neutron Therapy Beams
Relative to 400 ^{60}Co Rad

	16-MeV d→Be Neutrons	50-MeV d→Be Neutrons
Chinese hamster cells (CHO) (*11*)	2.6	2.7
Human melanoma cells	2.6	2.5
Human kidney T-1 cells	2.0	1.8
Mouse jejunal crypts (*10*)	3.7	2.8
Canine oral mucosa (acute)	--	2.2
Monkey oral mucosa (necrosis)	--	2.5[b]
Pig skin (*12*)	< 4	1-2
Spontaneous animal tumor therapy	~ 2.4[a]	~ 2.2[a]
Human oral mucosa (acute)	2.6[b]	2.4[b]

[a]Clinical impression
[b]Relative to 425 ^{60}Co rad

(2) Dosage Increase Based on Acute Radiation Reactions

Mucosal reactions of the faucial arch were chosen to determine equivalences with ^{60}Co because a 10% change in dosage produces a clinically detectable difference in reaction intensity (*6*). The majority of patients treated with ^{60}Co with weekly doses of 1000 rad in 5 fractions develop a studded or confluent mucositis 12–17 days after the beginning of treatment.

Initially, three patients were treated with the 16-MeV beam; 300 $rad_{\eta\gamma}$ per week were given through parallel opposing fields covering the oropharynx and faucial arch (Fig. 1-A). None developed the studded-to-confluent mucositis seen when ^{60}Co irradiation is given at 1000 rad/5 Fx/wk. Two of these patients had continued tumor growth (the growth was histologically confirmed at three months in both). With an increase to 330 $rad_{\eta\gamma}$, the majority of patients developed a studded-to-confluent mucositis (Fig. 1-B). When treatment with the 50-MeV beam was initiated, a weekly dose of 330 $rad_{\eta\gamma}$ was used. The first eight patients developed only a mild erythema (Fig. 1-C), but when the dosage was increased to 360 $rad_{\eta\gamma}$, a studded-to-confluent mucositis resulted (Fig. 1-D).

Because the lower dosage schedules failed to produce the expected reaction pattern, we tested whether this difference was due to an error in RBE determination, or an incorrect assumption that 800 rad/2 Fx/wk with ^{60}Co is equivalent for acute mucosal reactions to 1000 rad/5 Fx/wk. The latter was found to be the cause of the discrepancy.

Seven patients were treated palliatively with ^{60}Co, given 800 rad/2 Fx/wk through parallel opposing portals covering the oropharynx and faucial arch. This produced only a mild erythema similar to that produced by 900 rad/5 Fx/wk with ^{60}Co. One patient was treated with 900 rad/2 Fx/wk. This patient

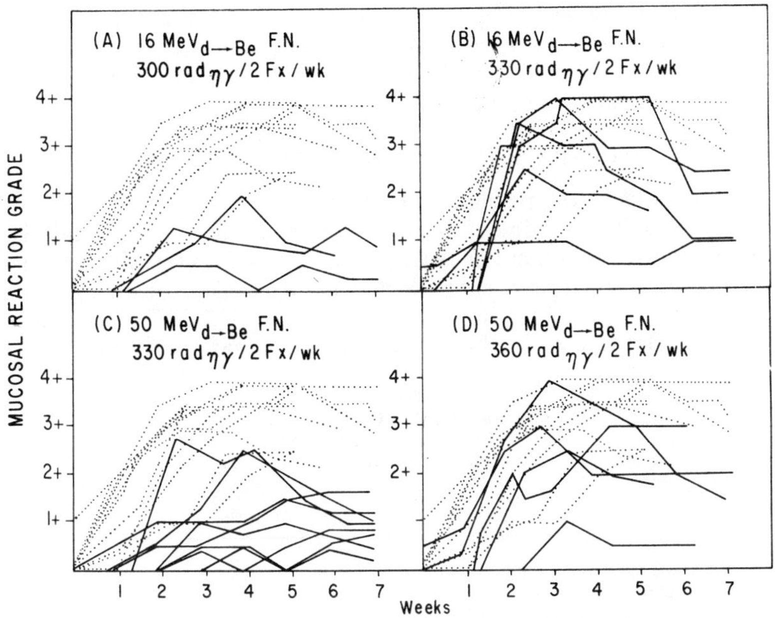

FIG. 1. A comparison of acute oromucosal reaction patterns produced by twice-weekly neutron irradiation (solid lines) and five times weekly ^{60}Co irradiation (1000 rad/5 Fx/wk) (dotted lines).
Mucosal reaction grades: 1+) mild erythema, 2+) brisk erythema, 3+) studded fibrinous mucositis, 4+) confluent fibrinous mucositis.

developed a greater mucosal reaction than would have been expected with 1000 rad/5 Fx/wk. Three patients were treated with 850 rad/2 Fx/wk. All three developed the studded-to-confluent mucositis pattern expected with 1000 rad/5 Fx/wk.

Based on these findings, the RBE's for acute mucosal reactions are similar to those determined by the pretherapeutic radiobiology experiments:

$$\text{RBE}_{16\text{ MeV}} = \frac{850\text{ rad }(^{60}\text{Co, 2 Fx/wk})}{330\text{ rad}_{\eta\gamma}\text{ (FN, 2 Fx/wk)}} = 2.6$$

$$\text{RBE}_{50\text{ MeV}} = \frac{850\text{ rad }(^{60}\text{Co, 2 Fx/wk})}{360\text{ rad}_{\eta\gamma}\text{ (FN, 2 Fx/wk)}} = 2.4$$

(3) Dosage Reduction Based on Late Radiation Sequelae

A review of the clinical data, in September 1973, showed that the late radiation changes were greater than would be expected on the basis of the acute reactions observed. The greater incidence of complications and the severity of fibrosis may have been caused by (1) inherent qualities of the neutron beam, and/or (2) the large-dose twice-weekly fractionation schedules used.

At present, a weekly dose of 320 $rad_{\eta\gamma}$ in 2 fractions is used in most clinical situations. The majority of patients receive tumor doses in the range of 1900 $rad_{\eta\gamma}$/6 wk to 2150 $rad_{\eta\gamma}$/7 wk depending on the histological type, location, and extent of the tumor.

PRELIMINARY EVALUATION OF LOCAL CONTROL

As of July 1974, 185 patients with advanced cancer had been entered into the pilot study. One hundred fifteen were alive at the time of analysis. No patient was refused treatment because the disease was so far advanced that he was thought to be incurable; six patients were rejected because they were too debilitated to tolerate the twice-weekly trips to TAMVEC (100 miles, one way). Table 2 shows the distribution of lesions by anatomical site.

TABLE 2
Clinical Material: TAMVEC Neutron Therapy Trial
(October 1972 - July 1974)

	Included for Analysis[a]	Total No. Treated
Head and neck carcinoma	47	68
Breast carcinoma	15	22
Superior sulcus bronchogenic carcinoma	11	14
Esophageal carcinoma	7	11
Gynecological carcinoma	14	27
Soft tissue sarcoma	8	9
Glioblastoma multiforme	5	6
Miscellaneous: (7-colon, 4-stomach, 2-pancreas, 3-bone sarcoma, 2-melanoma, 2-hypernephroma, 1-bladder, 2-prostate, 5-palliation only)	16	28
Total	123	185

[a]Exclusions: 55 patients treated less than 6 months previously, 6 in whom XRT was stopped in the first 2 weeks of treatment; 1 suicide without followup.

The preliminary results are outlined in Tables 3 and 4, where 123 patients have been included for analysis. Sixty-two patients were excluded from the analysis including 55 patients treated recently (<6 months follow-up), six patients in whom radiotherapy was interrupted in the first two weeks of treatment, and one patient who committed suicide without follow-up.

Disease-free intervals have been computed from the first day of treatment. Patients were scored as 'tumor locally controlled' (1) if they were locally free of cancer at the time of analysis (6–21 months), (2) if they died of distant metastasis or intercurrent disease later than six months with no clinical evidence of local disease, or (3) if they died earlier than six months with no local cancer found at autopsy. Patients dying within six months without autopsy are listed as having local failure even if there was no clinical evidence of local cancer.

Head and Neck and Breast Cancer

Treatment Planning

With head and neck and breast tumors, treatment planning has included elective irradiation of the regional lymphatics because of the likelihood of subclinical spread. For this elective irradiation, a dose of 4500 rad/15 Fx/5 wk (^{60}Co) was used since it was easily coordinated with twice-weekly neutron irradiations.

In treatment for gross masses, two approaches have been employed:

(1) When both the primary site and regional lymphatics contained bulky disease, the entire treatment has been given with neutrons. If feasible, the fields were reduced for the final 3–4 treatments.

(2) If coverage of the first echelon of lymph nodes required larger fields than those needed for gross masses, a midline tumor dose of 5000 rad in 5 weeks was delivered with ^{60}Co through fields encompassing the gross tumor and the clinically uninvolved regional lymphatics. Then the dose to palpable disease was boosted with neutrons with a dose equivalent to 2000 rad in 2 weeks.

Results in Head and Neck Cancer

Eighteen patients were treated for late T_3, early T_4, or early N_3 tumors who would have been treated for cure with photon irradiation prior to the neutron trial. However, the majority (29 of 47 patients) presented with late T_4 or late N_3 lesions who would have been palliatively irradiated or even not treated (Table 3). Subsequent histological confirmation has been obtained in 16 patients—eight with no evidence of residual cancer and eight with residual tumor.

All 18 patients with moderately advanced lesions have been scored as having local control of their tumors. Fifteen patients are NED[4] at 6–21 months, and three are dead of intercurrent disease or complications (9, 12, and 13 months) with no tumor found at autopsy.

Of 29 patients with massive tumors, 10 tumors have been listed as locally controlled. Seven patients are NED (6–17 months), and three are dead with no evidence of local disease clinically or at autopsy (6–9 months). Nineteen patients have been listed as having local failure (two are NED following surgical salvage, four are alive with cancer and 13 are dead from cancer).

Although the local control rate was higher in patients treated with conventional radiotherapy followed by a fast neutron "boost" (nine of 12 patients) than in patients treated with neutrons only to the bulky disease (19 of 35 patients), there was no difference when the data were analyzed by extent of disease (Table 3). The results are listed by tumor site in Table 4. With this small number of patients there was no correlation between tumor site and local control. However, patients with large tumors of the floor of mouth showed a higher incidence of complication (three of six patients).

[4]Alive with no evidence of disease clinically.

Results in Breast Cancer

All patients had Stage D disease by Haagensen's criteria (*7*), but they were divided into 'moderately advanced' and 'massive' categories based on the size of the primary tumor and of the axillary nodes. Six patients presented with primary tumors $\leqslant$ 7 1/2 cm and axillary nodes $\leqslant$ 5 cm in size, and nine presented with primary tumors > 7 1/2 cm and/or axillary nodes > 5 cm (Table 3).

TABLE 3
Preliminary Evaluation of Local Control
(6–21 months)
(October 1972–January 1974)

	Neutrons Only[a]	Conventional XRT & Neutrons[b]	TOTAL
Head and Neck Carcinoma			
Moderately large (late T_3, early T_4, or early N_3)	11/11	7/7	
Massive (late T_4, late N_3, or palliation only	8/24[c]	2/5	
	19/35	9/12	28/47 (59.6%)
Breast Carcinoma			
Moderately large (primary $\leqslant$7-1/2 cm; and ax nodes $\leqslant$5 cm)	2/2	2/4	
Massive (primary >7-1/2 cm; or ax nodes >5 cm)	6/8	1/1	
	9/10	3/5	11/15
Superior Sulcus Bronchogenic Carcinoma		8/11	8/11
Squamous Carcinoma of Esophagus	2/5	2/2	4/7
Gynecological Carcinoma	4/9	3/5	7/14
Soft Tissue Sarcoma	5/6	1/2	6/8
Glioblastoma Multiforme	1/4	0/1	1/5
Miscellaneous	6/15	1/1	7/16
TOTAL	45/84 (53.6%)	27/39 (69.2%)	72/123 (58.5%)

[a]Treatment modality to the bulky tumor (many of these patients received conventional radiotherapy to the regional lymphatics).
[b]In most cases neutron therapy was given as a "boost" after conventional radiotherapy.
[c]Two patients listed as failures were surgically salvaged.

Of the six patients with moderately advanced cancers, four have lesions locally controlled for 7–19 months. Two patients have developed locally recurrent cancer. Both were treated with a neutron "boost" to bulky axillary nodes following ^{60}Co irradiation.

Of nine patients with massive breast carcinomas, seven have lesions scored as being locally controlled. Three patients are NED, and one is alive with distant metastasis (12–16 months). Three are dead because of distant spread with local control clinically (one marginal recurrence). Two patients are listed as having local failure (one alive at 13 months with local recurrence and one dead at 5 months because of distant metastasis with the primary tumor regressing).

TABLE 4
Head and Neck Cancer
(October 1972–January 1974)

	Number of Patients		Local Control (6–22 mo)		Complications: Tumor Related[a]		Complications: Nontumor Related[b]	
Squamous carcinoma predominant at primary site:								
Floor of mouth	6		4		2		1	
Oral tongue	5		2		1		–	
Base of tongue	9		6		1		2	
Anterior tonsillar pillar	3		2		1		1	
Soft palate	3		2		1		–	
Buccal mucosa	1		1		–		–	
Nasopharynx	1		0		1		–	
Pharyngeal wall/ pyriform sinus	4		2		1		–	
		32		19		8		4
Salivary gland carcinoma (major and minor)		4		3		–		1
Squamous carcinoma predominant in neck		9		6		1		1
Adenocarcinoma predominant in neck		2		0		–		–
TOTAL		47		28		9		6

[a]6 mucosal ulcerations; 1 skin ulceration; 2 massive hemorrhages.
[b]5 mucosal necroses; 1 osteoradionecrosis.

Other Tumor Sites

Superior Sulcus Bronchogenic Carcinoma

All 11 patients were treated with ^{60}Co irradiation to the primary tumor, mediastinum, and ipsilateral supraclavicular area (4000–4500 rad/4–4 1/2 wks) followed by a neutron "boost" to the primary site (630–940 rad$_{\eta\gamma}$/2–3 1/2 wks). All patients were well palliated, and eight have been scored as having local control of their disease. Three patients are NED at 6, 19, and 21 months; two are alive and three dead (6–17 months) with distant metastases but local control clinically. Only one patient has developed locally recurrent cancer as determined clinically, but autopsies of two other patients showed residual tumor at the primary site.

Squamous Carcinoma of the Esophagus

The majority of patients with squamous carcinoma of the esophagus received doses in the range of 1800–1900 rad$_{\eta\gamma}$/5 1/2–6 wks with no correction for tissue inhomogeneity. All seven patients were initially well palliated with radiographic evidence of significant tumor regression. Four have been scored as having local control (two are NED at 8 and 10 months; and two are dead at 4 and 5 months after treatment with no local tumor found at autopsy). Three patients have been scored as having local failure—two with residual tumor clinically and one with residual tumor at autopsy.

Gynecological Cancer

Fourteen patients with a variety of gynecological neoplasms have been treated. The majority received doses in the range of 1900–2050 $rad_{\eta\gamma}/6$–7 1/2 wks. Five patients are NED (6–9 months) and two are dead with local control clinically. Seven patients have been scored as having local failure. One of these seven, treated with preoperative neutron therapy (only 1320 $rad_{\eta\gamma}$), was found to have local tumor and distant metastases at laparatomy.

Soft Tissue Sarcoma

The majority of the eight patients with soft tissue sarcomas received doses in the range of 1900–2250 $rad_{\eta\gamma}/6$–7 1/2 wks although the fractionation schedules varied. Six patients have been scored as having local control (6–15 months); three of the six have had subsequent resections which revealed no residual cancer.

Glioblastoma Multiforme

The results in five patients with glioblastoma multiforme have been uniformly disappointing. One patient treated with 2180 $rad_{\eta\gamma}/7$ 1/2 wks is scored as NED (6 months), whereas four patients treated with lower doses (2000 $rad_{\eta\gamma}$ /7 wks) have developed clinical signs of an early local recurrence (4–10 months). Two of these recurrences have been confirmed histologically (at a second craniotomy or at autopsy).

PRELIMINARY EVALUATION OF RADIATION SEQUELAE

Fibrosis

Fibrosis was a concern at the initiation of the clinical trial because (1) neutrons are selectively absorbed in fat resulting in approximately 18% higher dose to subcutaneous tissue (*8*), (2) photon treatment schedules employing fewer fractions produce more fibrosis than more fractionated schedules giving equivalent acute reactions (*9*), and (3) the TAMVEC pretherapeutic studies had shown a significant fibrosis in canines and Rhesus monkeys irradiated with 16 or 50-$MeV_{d\rightarrow Be}$ neutrons. In the pilot study, significant to severe fibrosis has already appeared in a number of patients (approximately 1/3 of patients in the head and neck and breast categories). In one breast patient, fibrosis has been severe enough to be classified as a major complication. This is ominous, since when fibrosis occurs within a few months following irradiation, it will increase to severe proportions in following years.

Gastrointestinal Symptoms

Nausea and diarrhea were concerns because large daily doses to the upper abdomen with ^{60}Co are often poorly tolerated, and high RBE's have been reported for intestinal mucosa (*10*). So far, acute gastrointestinal symptoms have not been a problem since only 4 of 47 patients treated with large abdominal fields have developed nausea or diarrhea. In one patient, treatment was interrupted for one week to allow the diarrhea to subside. In the other three, it was controlled with Lomotil, and was less severe than that seen with conventional radiotherapy.

Major Complications

Twenty-six patients have developed major complications (Table 5). Eleven of these occurred in patients also classified as having residual cancer locally. The complications have been divided into two categories:

TABLE 5
COMPLICATIONS

Tumor-Related	Not-Tumor-Related
15	11
6—mucosal ulcerations	5—mucosal necrosis
2—skin ulcerations	1—osteoradionecrosis
3—massive hemorrhages	3—skin necrosis
2—esophageal fistulae	1—severe fibrosis
2—vesicovaginal fistulae	1—intestinal obstruction

(1) Tumor-related Complications

Fifteen patients have developed complications arising where eradication of the neoplasms resulted in residual ulceration leading to hemorrhage and/or infection and necrosis. Nine occurred in patients with massive head and neck tumors (six mucosal ulcerations, one skin ulceration, two massive hemorrhages), three in patients with extensive esophageal cancers (two fistulas, one massive hemorrhage), two in patients with massive gynecological cancers (vesicovaginal fistulae), and one in a patient with a soft tissue sarcoma with extensive skin involvement (skin ulceration).

(2) Nontumor-related Complications

Eleven patients have developed complications which could not be attributed to involvement of the covering epithelium by cancer. These complications arose in the vicinity of the neoplasm after epithelial healing or in tissues not involved by tumor. These included five patients with mucosal necrosis and one with osteoradionecrosis (head and neck cancer patients), three with skin necrosis (one breast and two soft tissue sarcoma patients), one with severe fibrosis (breast carcinoma patient), and one with intestinal obstruction (preoperative irradiation and surgery in a patient with carcinoma of the cervix).

TIME-DOSE RELATIONSHIPS

One must use a relatively homogeneous population to evaluate time-dose relationships. For local control, the clinical material was restricted to patients with head and neck and breast tumors where the bulky disease was treated entirely with neutrons. In addition, 15 patients with very massive tumors were excluded since these were considered impossible to control even before treatment. The data are shown in Fig. 2 where patients scored as having local

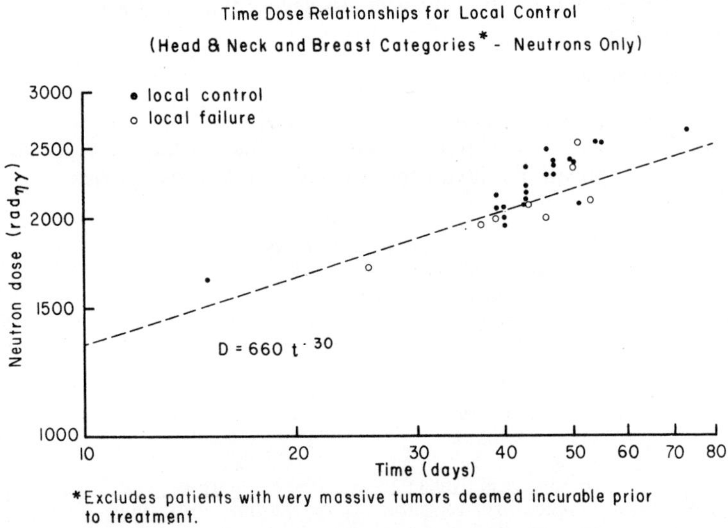

FIG. 2. Time-dose relationships for local control. Thirty patients with head and neck and breast tumors whose bulky disease was treated entirely with neutrons. The doses for patients treated with the 16-MeV$_{d\rightarrow Be}$ beam have been normalized for the equivalent 50-MeV$_{d\rightarrow Be}$ neutron dose by multiplying by 1.1 (RBE $_{16\text{-MeV}}$/RBE $_{50\text{-MeV}}$ = 1.1). (Exclusions: Fifteen patients with massive tumors who were deemed incurable prior to the start of treatment.)

control or local failure are plotted as a function of tumor dose in rad$_{\eta\gamma}$ (ordinate) and over-all treatment time in days (abscissa). A line defined by the equation D=660 t$^{.30}$ shows the best separation. Above this line there were only two local failures in 20 patients; below this line, there were local failures in six of ten patients.

The time-dose relationship for major complications is illustrated in Fig. 3. The clinical material included all patients who developed complications in areas treated entirely with neutrons. The best separation for nontumor-related complications is shown by the line defined by the equation D=715 t$^{.30}$. Six of seven nontumor-related complications occurred with doses above this line. This time-dose relationship was not demonstrated for tumor-related complications since five of these complications occurred above and seven below the line. Tumor-related complications are thought to be a result of massive epithelial involvement by cancer rather than the radiation dose administered.

With longer follow-up and a more uniform patient population, the time-dose relationships are expected to be more accurately determined. The slopes may be erroneously steep (exponent = 0.30) since the tumor population was not uniformly homogeneous and patients with more advanced cancers were selectively treated to higher doses.

The preliminary data indicate that the range of acceptable neutron doses is narrow. For a 43-day treatment schedule (6 1/2 weeks) the lines are separated by only 170 $rad_{\eta\gamma}$. In certain sites (for example, floor of mouth), the chance of locally controlling massive tumors without complications is limited. Such tumors are suitable for inclusion in pilot studies to determine dose schedules and normal tissue effects. However, they should be excluded from later clinical trials evaluating the effectiveness of fast neutron therapy in the local control of malignant disease.

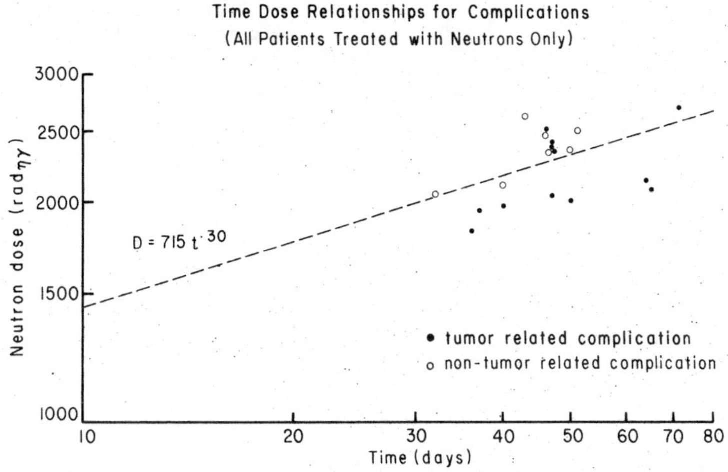

FIG. 3. Time-dose relationships for tumor-related and nontumor-related complications. Nineteen paatients developed complications in areas treated entirely with neutrons. The doses for patients treated with the 16-MeV d→Be beam have been normalized to the equivalent 50-MeV d→Be neutron dose by multiplying by 1.1 (RBE 16-MeV/RBE 50-MeV = 1.1).

REFERENCES

1. D. K. Bewley, *Eur. J. Cancer* **7,** 99 (1971).
2. F. H. Attix, personal communication, 1972.
3. A. R. Smith, J. Smathers, F. Attix, R. Theus, P. Wootton, H. Bichsel, D. Bewley and J. Parnell, *Med. Phys.* **1,** 101 (1974). (Abstract)
4. M. Catterall, *Eur. J. Cancer* **7,** 227 (1971).
5. S. B. Field, *Proc. Roy. Soc. Med.* **65,** 835 (1972).
6. G. H. Fletcher, W. S. MacComb, *Radiation Therapy in the Management of Cancer of the Oral Cavity and Oropharynx.* C. C. Thomas, Springfield, Illinois, 1962.
7. C. D. Haagensen and E. Colley, *Ann. Surg.* **157,** 157 (1963).
8. D. K. Bewley, *Brit. J. Radiol.* **36,** 81 (1963).
9. G. Arcangeli, M. Friedman and R. Paoluzi, *Brit. J. Radiol.* **47,** 44 (1974).
10. H. R. Withers, K. Mason, B. O. Reid, N. Dubravsky, and H. T. Barkley, Jr., and A. M. Chu, *Cancer*, **34,** 39 (1974).
11. R. L. Gragg, R. E. Meyn, and R. M. Humphrey, Presented at the meeting of the Radiologic Society of North America, November 1973, Chicago.
12. D. Hightower, personal communication, 1973.

Clinical Experience of Therapy With Cyclotron Neutrons[1, 2] At The University of Washington

R. G. Parker, A. J. Gerdes, H. C. Berry & M. D. Soronen
University of Washington, Seattle, Washington 98195, USA

INTRODUCTION

The ultimate objective of all of the clinical fast neutron programs is the vigorous appraisal of this treatment modality in the management of humans with cancer. Whether fast neutron teletherapy proves of value or not, it is important that any judgment be based on the best possible evidence. The history of the evaluation of new cancer therapies does not support the conclusion that a well-considered evaluation comes easily.

Concurrent with this focus on a clinical answer, there is a unique and possibly momentary opportunity to study neutron radiobiology in humans.

PHYSICAL PARAMETERS

The University of Washington fast neutron teletherapy beam is generated at the Department of Physics' fixed energy 60" cyclotron from a 2.0 x 0.6 cm focal spot by 40 μA of 21 MeV deuterons incident on a "thin" (70% of incident deuterons absorbed), directly water-cooled beryllium target. The mean of the beam energy spectrum is about 8 MeV. Dose rates measured in air can reach 40 rads (n+γ) per minute 150 cm from the target (1 rad per μA per minute). The 50% central axis depth dose at 150 cm TSD exceeds 10.0 cm in a tissue-equivalent phantom ($p = 1.10$ gm/cm^3) for a 15 x 15 cm field. The treatment beam cross-section is adjusted by custom-made borated water-extended polyester (B-WEP) collimators which are interchangeable, rotate about the long axis of the beam and average about 35 pounds. The largest field size at 150 cm TSD is 23 x 23 cm and the uniformity of dose measured in air is ± 5% over 80% of the beam width. (*1*).

PRECLINICAL RADIOBIOLOGICAL STUDIES

Prior to clinical application, the fast neutron teletheraphy beam was biologically characterized by: (1) determination of the Relative Biological Effectiveness (RBE) for early and late damage to normal tissues of importance; and (2) examination of the RBE for acute injury in normal tissues as a function of dose size and depth in a tissue-equivalent medium.

[1]Congress presentation E-4-5.

[2]Supported by National Cancer Institute Grant CA 12441

In the mouse, there was no difference between the RBE for acute intestinal death measured as $LD_{50/5\ day}$ after total body irradiation (1.8) and that for late death from irradiation-produced obstruction of the terminal ileum, measured as $LD_{50/90\ day}$ after irradiation of a surgically exteriorized 4 cm segment (1.9). (*2*).

Late damage to the mouse spinal cord was studied by correlating the time of onset of hindquarter paralysis to a single radiation dose to a surgically exposed lower cord segment (*3*). The RBE was approximately 1.0 for single doses ranging above 1700 rad and seemed to increase with decreasing neutron doses to approach 1.9 at a dose of 675 rad.

Other RBE's determined by colorometric measurement of DNA were: in vitro interphase thymocyte death—1.0 (*4*); cell loss from the thymus—1.2; cell loss from the spleen—1.2; and mouse testicular damage—3.0. Although the single dose size did not alter the RBE for cell loss from the thymus or spleen, fractionation of total dose altered the RBE for the thymus but not the spleen (J. P. Geraci, personal communication).

Comparison of the effects of fast neutrons and X-rays on animal tumors and selected normal tissues at the cellular level were related to a series of nonconventional fractionation patterns and the possible role of cell repopulation during interfraction intervals (*5*).

The RBE's and D_n-D_1 values for single and fractionated doses were not found disproportionate for the late effects on mouse skin and the production of foot deformity compared to early effects (J. S. R. Nelson, personal communication).

C3HBA mouse mammary tumors were irradiated with: (1) 2 increments of neutrons over 5 days; (2) 5 increments of 250 KVP X-rays over 5 days; and (3) 2 increments of neutrons and 3 increments of X-rays in various patterns over 5 days. The mixed neutron and X-ray fractionated irradiation was intermediate in effectiveness between 2 fractions of neutrons and 5 fractions of X-rays. However, the 2 neutron increments were more effective than predicted by the RBE of 2.25 used to calculate the individual neutron dose (*5*). Studies, either ongoing or planned for the immediate future, include: the neutron RBE for oral death in mice; cell repopulation rates of various normal tissues following neutrons and X-ray irradiation; the rate of short term recovery in vivo of injury produced by neutron or X-ray irradiation; changes in OER and RBE in the treatment volume with different geometries of clinical collimators; the RBE and potential for recovery from injury of nongrowing and exponentially growing cells in culture; the response of cultured cells in plateau and logarithmic phases to different patterns of irradiation; the RBE for neutron-produced obstruction of different segments of mouse gut; the RBE for damage to nongrowing cartilage; the RBE for mouse spinal cord injury with multiple neutron dose fractions as used clinically; the RBE for kidney damage; the RBE for leukemia induction in mice; neutron RBE for mitotic delay compared to neutron RBE for reproductive cell death in the X-180 mouse ascites tumor; neutron RBE for delay in mouse skin wound healing compared to neutron RBE for preventing regrowth of a surgically excised C3HBA mammary tumor, when the radiation is given preoperatively; tumor regrowth rate, cell cycle kinetics of regrowing tumors and cell survival in the C3HBA mammary tumor and the 6C3HED lymphosarcoma of the C3H mouse, following single doses of X-rays or neutrons; and cell survival tumor growth, cell cycle kinetics and tumor histology during and after fractionated neutrons, X-rays or mixed neutron-photon radiation in the EMT-6 mammary tumor of the BALB-c mouse.

CLINICAL STUDIES

The first patient was treated with the University of Washington fast neutron beam on 9-10-73. As of 6-20-74, 38 selected patients have entered the study. Of these, 5 have not completed treatment. Immediate objectives include: (1) observation of the frequency, rate, completeness and permanence of response of various tumors; (2) study of early and late changes in various normal tissues of clinical interest (i.e. skin, mucosa, subcutaneous tissue, soft tissues of the neck, bone and teeth, eye, brain, spinal cord, salivary glands, cartilage, bowel, bladder, lung); (3) study of various patterns of irradiation; and (4) development of the mechanics of treatment application. Ultimately, cooperative, randomized clinical trials should include all ongoing programs in the United States.

Patterns of Application

It is imperative that several patterns of fast neutron irradiation be investigated prior to the institution of large-scale clinical trials. A treatment pattern based on limited cyclotron availability or tolerance of personnel may not be the best choice biologically and might reflect unfavorably on fast neutron teletherapy.

The proximity of the University of Washington medical center to a cyclotron made available through the good offices of the faculty of the Department of Physics, has made it possible to randomize patients to treatment either 2 or 3 times weekly without interruption. To date, no difference has been noted either in tumor responses or normal tissue tolerances between patients treated 2 times per week with 150 rad ($n+\gamma$) increments or 3 times per week with 100 rad ($n+\gamma$) increments to the same total doses.

Based on Nelson's (*5*) studies in our laboratory suggesting a disproportionate contribution of 2 increments of fast neutrons mixed with 3 increments of conventional photon irradiation, this pattern is being tested.

Inasmuch as fast neutrons might be an effective and minimally morbid "boost" treatment for tumor cells frequently persisting at the conclusion of conventional radiation therapy, this use is being tested. This pattern would allow treatment of a large number of patients with limited facilities.

Inasmuch as many cancers are treated by a combination of surgery and conventional irradiation, reallocation to radiation therapy only, even to test a new modality, may be difficult. Preoperative irradiation with fast neutrons may increase the opportunity to study tumor responses and normal tissue tolerances without encroaching on the numbers and types of tumors available for treatment with neutrons only.

Tumor Responses

Tumors currently being studied are: epidermoid carcinomas arising in the oral cavity, nasopharynx, oropharynx, laryngopharynx, or paranasal sinuses; metastatic cervical adenopathy; glioblastoma multiforme; salivary gland tumors; melanomas; Pancoast tumors; epidermoid carcinomas of the thoracic esophagus. In the near future, our studies will be extended to carcinomas of the breast, bladder and uterus.

Thirteen of the 18 patients with *epidermoid carcinomas primary in the head and neck* are eligible for preliminary assessment. Primary sites include: oral tongue—2; pharyngeal tongue—2; entire tongue—2; faucial arch—1; oropharynx—1; nasopharynx—1; laryngopharynx—1; piriform sinus—1; maxillary sinus—1, and undetermined—1. All tumors have been very extensive locally. All tumors treated entirely with fast neutrons have responded with nearly all becoming at least 50% smaller in volume. However, only 3 of these have responded such that local tumor control ultimately might result: (1) a well-differentiated epidermoid carcinoma involving the tonsillar fossa, ipsilateral anterior and posterior pillars, posterior buccal mucosa and adjacent tongue with cervical adenopathy in a patient without evidence of tumor 5-1/2 months following start of treatment; (2) a well-differentiated epidermoid carcinoma involving the oral and pharyngeal tongue, and palatine arch with cervical adenopathy which was no longer palpable 4-1/2 months after the start of treatment when the patient died of other cause; and (3) a well-differentiated epidermoid carcinoma arising in the maxillary sinus and extending to the orbit, nasal fossa, oral cavity and soft tissues of the face, which was not detectable grossly at the conclusion of treatment. Doses have ranged from 1500 rads $(n+\gamma)$ in 13 increments over 34 days to 2200 rads $(n+\gamma)$ in 13-19 increments over 45 days.

Two patients have completed neutron "boost" treatment after conventional ^{60}Co teletherapy. One with an extensive cancer involving the junction of oral and pharyngeal tongue and glossopharyngeal sulcus and midjugular neck nodes, remains without clinically detectable tumor 7 months after the beginning of the "boost." The other patient has just had a laryngectomy for suspected persistent tumor which initially involved the base of the epiglottis, aryepiglottic fold, false cord and nodes in both sides of the neck.

Although 4 patients entered the study for preoperative irradiation with fast neutrons, none ultimately has had definitive surgery for various reasons. In 3 of the 4, neutron irradiation was interrupted and then resumed with a consequent form of "split course" treatment with higher total doses protracted over considerably longer times.

Ten patients are eligible for preliminary assessment of fast neutron irradiation of *metastatic cervical adenopathy*. All masses, including some up to 8.0 cm in greatest dimension, became smaller with 9 of 10 being reduced to less than half by volume and 4 of 10 becoming nonpalpable. Three of the 4 necks which became clinically tumor-free, initially had masses larger than 3.5 cm in greatest dimension associated with well-differentiated epidermoid carcinomas arising in the tongue or tonsillar region. Doses and patterns were the same as for the primary epidermoid carcinomas in the head and neck except when "boost" doses were added.

Twelve patients with *glioblastoma multiforme* have entered the study. The 8 patients completing treatment to date have tolerated doses ranging from

1550 rad (n+γ) in 10 fractions over 34 days to 1850 rad (n+γ) in 17 fractions over 43 days with little difficulty. In order to compare the results of patients treated with fast neutrons to other studies using ^{60}Co or high energy X-ray teletherapy, all patients have been accepted into the study regardless of general condition. One functional class IV patient (*6*) died during treatment 1 month after it started. One functional class II patient, who received only 1100 rads (n+γ) in 11 fractions over 24 days because treatment was discontinued following an episode interpreted as intracranial hemorrhage, remains alive with nearly complete neurological recovery 7 months after the start of treatment.

Two *melanomas* have been irradiated with fast neutrons. A massive recurrence following orbital exenteration was markedly reduced at the conclusion of treatment with 2050 rads (n+γ) in 19 increments over 49 days and continues to regress 4 months after the start of treatment. The other patient with trismus incident to extension into the pterygoid space just finished treatment without any relief.

Two patients with *Pancoast tumors* developed clinically detected metastases during conventional radiation therapy, prior to scheduled neutron boost. A third patient just finished treatment. A single patient with locally advanced and metastatic breast cancer withdrew from treatment.

Normal Tissue Tolerance

Even if better local tumor control can be obtained with fast neutron teletherapy, the gain could be dissipated by a concurrent increase in treatment-related sequelae. An evaluation of fast neutron teletherapy side effects must precede large-scale clinical trials.

All of our patients irradiated above the clavicle have had mild-to-moderately brisk skin and mucosal reactions. Treatment was interrupted in 2 patients because of our concern. All skin and mucosal reactions promptly healed. These reactions are difficult to assess for they could be related to the physical characteristics of the beam (comparable to ^{137}Cs), the nonconventional patterns of irradiation, the large tissue volumes treated, as well as to some unusual biological effect. These skin reactions, using the scalp as an example, usually have become visible by the 17th-20th day, have reached their peak by the 25th-35th day and have grossly completely healed by the 45th-70th day. The estimated dose to the scalp for a midplane intracranial dose of 1800 rads (n+γ) is about 1850 rads (n+γ), consisting of 1320 rad entrance field dose (at 0.2 cm) and 530 rad exit dose.

All patients treated for glioblastoma mutiforme have been epilated. Hair loss is first noted about the 18th-21st day and is complete about the 30th-40th day. It is too early to assess hair regrowth.

All patients who will be irradiated to the brain and/or spinal cord are examined neurologically before treatment and at intervals thereafter. No findings unattributable to the brain tumors themselves have been noted with doses up to 1850 rads (n+γ) in 17 increments over 43 days and observation periods as long as 7 months.

All patients with head and neck tumors have baseline ophthalmological exams and interval exams thereafter. No acute changes have been noted following doses up to 757 rads (n+γ). The follow-up period is too short for the very difficult assessment of radiation-related cataract development (*7*). This group is about to have findings recorded photographically.

When the oral cavity, including the salivary glands, has been irradiated, taste alteration and salivary suppression have slowly recovered as expected after conventional radiation therapy. Sophisticated tests to qualify and quantify these changes have been developed.

Those patients treated for cancers above the clavicles receive an estimated dose of 20 rads of mixed neutron and gamma radiations to the trunk and extremities. However, these neutrons may have a relatively high RBE as noted by Geraci et al. (*8*). Two of our patients had very slight decreases in peripheral WBC counts near the completion of treatment and another had a slightly decreased platelet count at the finish of treatment. Treatment was not discontinued because of these findings and "recovery," measured by peripheral blood counts, was spontaneous and prompt.

Although all patients have had locally advanced cancers, and several have had concurrent metastases and many have been receiving pharmacological treatment, there has been no evidence of systemic "radiation sickness."

Development of Treatment Techniques

Use of a fixed horizontal beam defined by interchangeable collimators is considerably removed from conveniences expected in modern supervoltage/megavoltage radiation therapy. Although a fixed horizontal beam may be used for conventional treatment of head and neck cancers with the patients sitting, tumors within the trunk and pelvis may have to be irradiated unconventionally with the patients upright rather than reclining. Radiation exposure and other factors, i.e. inconvenience, may dictate rotation of personnel assignments with consequent difficulty in exact repositioning of the patient. It is important to account for these variations from standard operating procedures when comparing fast neutron teletherapy to modern conventional photon teletherapy.

Cooperative Clinical Trials

Barring the unexpected discovery of dramatic improvement in local tumor control or disastrous treatment-produced sequelae in current Phase I (human radiobiology) studies, the evaluation of fast neutron teletherapy will require large-scale clinical trials. Such trials, which must be preceded by careful definition of the target tumors, normal tissue tolerances and optimal irradiation patterns and doses, may be possible only with dedicated fast neutron generators installed in large research-oriented medical centers. The number of patients required for study will be related directly to the minimal degree of improvement considered acceptable (*9*). If this hoped-for improvement is modest, the large number of patients necessary for study will require close cooperation between all viable research programs. Such cooperation already exists between currently operational programs and should be a prerequisite for public funding of any additional programs.

SUMMARY & CONCLUSIONS

1. Proper clinical evaluation of fast neutron teletherapy of human cancers will be difficult and time-consuming and will require extension over many years.
2. Fast neutron teletherapy may not have significant advantage over modern conventional supervoltage/megavoltage radiation therapy. A negative answer will be acceptable only if based on sound evidence.
3. If fast neutron teletherapy has an advantage over modern conventional radiation therapy, the advantage may be small and tumor- and site-specific. Detection of a small advantage will require study of a large number of patients and this will be possible only in carefully planned cooperative clinical trials.
4. The study of a large number of patients over a long period will require installation of dedicated neutron generators in research-oriented medical centers.
5. Cooperative clinical trials must be preceded by careful definition of target tumors, normal tissue tolerances, reasonable doses and advantageous patterns of application.
6. Any improvement in local tumor control attributed to fast neutron teletherapy might be dissipated by a concurrent increase in treatment-related sequelae.
7. Much related physics and radiobiology remains to be done and must be supported concurrent with the clinical programs.

ACKNOWLEDGEMENTS

The efforts of: P. Wootton, H. Bichsel, J. Eenmaa, K. Weaver, K. Alvar & D. Williams of the Division of Medical Radiation Physics; K. Jackson, G. Christensen & J. Geraci of the Division of Radiation Biology; and J. S. R. Nelson of the Division of Radiation Oncology (Experimental) are appreciated.

REFERENCES

1. P. Wootton, H. Bichsel, R. G. Parker, K. Weaver, D. L. Williams & W. G. Wyckoff, The University of Washington Fast Neutron Beam Facility. Presented at the Boston Workshop on Practical Clinical Criteria for a Fast Neutron Generator, May 28–29, 1973.
2. J. P. Geraci, K. L. Jackson, G. M. Christensen, R. G. Parker, M. S. Fox & P. S. Thrower, *Europ. J. Cancer* (in press)
3. J. P. Geraci, P. D. Thrower, K. L. Jackson, G. M. Christensen, R. G. Parker & M. S. Fox, *Radiat. Res.* **59**, 496 (1974).
4. J. P. Geraci, P. D. Thrower, K. L. Jackson, G. M. Christensen & M. S. Fox, The RBE of cyclotron fast neutrons for interphase death in rat thymocytes in vitro. *Int. J. Radiat. Biol.* **25**, 403 (1974).
5. J. S. R. Nelson & R. Carpenter, *Radiat. Res.* **59**, 92 (1974). Abstract B-12-5.
6. S. Order, S. Hellman, C. von Essen & M. Kligerman, *Radiology* **91**, 149 (1968).
7. R. G. Parker, L. Burnett, P. Wootton & D. McIntyre, *Radiology*, **82**, 794 (1964).
8. J. P. Geraci, K. L. Jackson, G. M. Christensen, R. G. Parker, M. S. Fox & P. D. Thrower, *Radiat. Res.* **59**, 247 (1974). Abstract E-5-5.
9. H. R. Withers, P. R. Almond & B. W. Brown, *Europ. J. Cancer* **7**, 211 (1971).

The RBE for Fast Neutrons: The Link Between Animal Experiments and Clinical Practice[1]

S. B. Field
MRC Cyclotron Unit
Hammersmith Hospital
Ducane Road
London W12 OHS, Great Britain

and

Shirley Hornsey
MRC Experimental Radiopathology Unit
Hammersmith Hospital
Ducane Road
London W12 OHS, Great Britain

INTRODUCTION

The radiobiological effectiveness (RBE) of a neutron beam is defined as the dose of X-rays to that of neutrons to produce the same level of damage. In general, the cell survival curve after X-irradiation has a "shoulder" region, becoming exponential at large doses, whereas after a high LET radiation, such as neutrons, the curve usually has a much smaller "shoulder" and is more nearly exponential throughout. Because of this difference in shape the RBE varies, being largest at low doses and decreasing as the dose and level of damage is increased. This variation in RBE was first pointed out in 1940 by Gray, Mottram, Read and Spear (*1*) who observed the inhibition of mitosis in tissue cultures of chick embryo fibroblasts and in the root of the broad bean after irradiation with X-rays or neutrons. In the early 1960's Barendsen and his colleagues (*2*) clearly demonstrated the relationships between the shapes of survival curves for cells irradiated in vitro with beams of different LET. Fowler and Morgan (*3*) showed that the RBE for damage to a tissue in vivo (pig skin) increased with increasing number of fractions, because each small fraction was within the shoulder region of the survival curves, whereas larger fractions were not. In the early trial of neutron radiotherapy, from 1938 to 1942 (*4*), the RBE was initially assessed for the single doses which produced an erythema, whereas the treatments were given in many fractions. Fowler and Morgan (*3*) suggested that overdoses of radiation would have resulted from an underestimate of RBE and would account for the necroses reported (*4*). It is now difficult to assess the exact part played by several different factors in the early trial, but it is clear that for a variety of reasons overdoses were given (*5*) and great care must be taken to ensure that the mistakes

[1] Congress presentation E–4–6.

made are not repeated.

Since survival curves obtained for tissues in vivo (*6–10*) appear to be different from those obtained in vitro (*11*) mainly by virtue of the larger extrapolation numbers (i.e. larger "shoulders") obtained in vivo, it is not reasonable to assume that RBE values obtained from measurements on cultured cells would apply to tissues. In order to rationalize RBE data obtained from irradiating tissues in vivo, Field (*12*) suggested plotting RBE as a function of dose per fraction of neutrons. The justification for this approach is (a) that the response of many tissues is qualitatively the same after both low and high LET radiations and (b) that any cellular proliferation between fractions occurs at the same rate, regardless of the radiation. There are numerous examples in the literature demonstrating the validity of this approach (for a review of such data see Field and Hornsey (*13*)), together with the evidence that plotted in this way RBE is no longer correlated with the number of fractions, but only with the dose per fraction. Initially, all available data from normal tissues was plotted together (*12*). The data for skin, gut and cartilage were initially not considered to be different (*12*), and the resulting composite curve relating RBE to dose per fraction was used in estimating the doses given in the early trial of neutron therapy (*4,5*). However, it was clear (*12*) that the RBE for damage to hemopoetic tissues was significantly less than that for the other tissues, the reason being that the survival curve for cells of the hemopoetic system has a small shoulder for X-rays (*11*) so there can be little difference between the shapes of the curve for low and high LET radiations in this case. When more data became available, however, particularly for damage to the mouse intestine, it was shown that the RBE also varied from tissue to tissue (*14*), the RBE for the intestine being significantly greater than that for skin. This observation led us to investigate the RBE for a variety of different tissues, the results of which will be discussed later.

An alternative approach to the comparison of RBE was suggested by Alper (*15*). By plotting RBE as a function of the reciprocal dose per fraction, it was shown that for large doses, beyond the shoulder region of the cell survival curves, the slope of the RBE curve, designated ΔD_Q was the D_Q for X-rays minus R times D_Q for neutrons (where R is that value of the RBE at infinite dose, found by extrapolation of the RBE curve, and D_Q is the dose at which the exponential region of a survival curve extrapolates to the dose axis(*16*)). The advantage of this approach is that it clearly shows that where ΔD_Q is larger for a tumor than for the limiting normal tissues, then neutrons would be therapeutically more effective than X-rays, based simply on the capacities of the different tissues to repair low and high LET radiation damage. Of course, where ΔD_Q is smaller for the tumor, then X-rays would be the treatment of choice. The disadvantage of this approach is that it is difficult to obtain sufficiently accurate data in the useful dose range (i.e. high doses) to estimate ΔD_Q with reasonable certainty. This is particularly true for tumors, in which the presence of hypoxic cells and their possible reoxygenation between fractions will modify the curve and may affect the value of ΔD_Q.

The RBE data for various normal tissues will now be considered. All the data has been obtained using neutrons from the Hammersmith cyclotron.

SKIN

After Stone (*4*) reported severe late skin damage in patients treated with neutrons, it was necessary to obtain data on animals before any further irradiation on man could be carried out. The initial experiments were performed on pig skin (*17*) because it is structurally similar to human skin. With this experience some measurements were made on man, and subsequently a great deal of data has been accumulated for mouse skin (*18*) and rat skin (*19*). These are shown in Fig. 1, in which RBE is plotted as a function of dose per fraction. It is seen that the same smooth curve is a good fit to all the skin data regardless of species. Of course, the observations on pigs and on man were necessary before a new clinical trial could begin but, in retrospect, the same results, quantitatively as well as qualitatively, could have been obtained far more quickly and at far less cost by experiments on rodents alone. These data provided the initial estimate for RBE used in the clinical trial at Hammersmith, and current clinical impressions (Catterall, personal communication, Eichhorn, personal communication) have confirmed their validity. This experience at Hammersmith leads us to recommend the measurement of RBE for the skin of rats or mice over a wide range of doses per fraction, as a preliminary to any patient treatments using a new beam of high LET radiation. We must point out, however, that it is only for skin that such data are available for more than one animal system. It may be that, for other tissues, the RBE is not independent of species, and this should be investigated where possible.

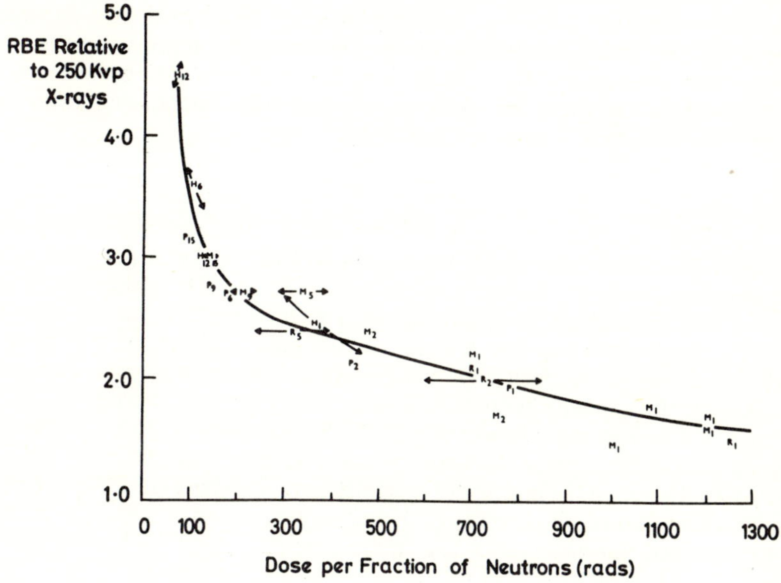

FIG. 1. The RBE for skin as a function of dose per fraction of neutrons. H is for human skin, P for pig skin, R for rat skin and M for mouse skin. The subscripts refer to the number of fractions used.

Late skin reactions have been the subject of considerable interest since the report of Stone (*4*). It was noted that the RBE for late fibrosis in pig skin was not different from that for early damage, for pigs kept up to 6 years after irradiation (*20*). For rat skin, Field (*21*) reported no difference in RBE for early skin reactions, late skin reactions and deformities at 6 months. It was shown, however, that a small overdose of radiation would lead to a small change in early skin reaction, but a massive increase in late damage and deformity. This observation has also been shown in mouse feet for up to 15 fractions of X-rays and fast neutrons (*22*). The cause of this late damage to rodents' feet is not clear, but evidence is mounting that it is causally related to the early reactions. The rate at which more rapid repopulation is initiated during a fractionated course of X-rays appears to be the same for both the early and late endpoints (*23*), and experiments with high dose rates, at which oxygen depletion occurs, indicate that the oxygen concentration is also similar for whichever cells contribute to both endpoints (*24*). We do not know whether or not the late damage observed in rodents' feet is similar to that seen in man.

Repopulation in skin has been assumed to be the same after equally effective doses of X-rays and fast neutrons, regardless of fractionation. This was demonstrated by the rate of rise of the skin reaction and its subsequent healing being similar in both cases (*12,18*). Recently Denekamp (*23*) investigated the kinetics of repopulation in mouse skin after single doses of X-rays and after fractionated doses of 300 rad each. Initially experiments with neutrons revealed no difference in repopulation after one, four or nine fractions of the radiations, but after 14 fractions it appeared that the skin repopulated more slowly after the high LET radiation (*25*). This anomaly was not confirmed in a subsequent experiment, in which the repopulation rate after 14 fractions was somewhat greater than after X-rays (Denekamp, personal communication). We conclude from these data that there is no difference between the repopulation rates in skin after irradiation with X-rays and neutrons.

INTESTINE

The RBE has been obtained with the use of three techniques to evaluate damage to the mouse intestinal mucosa, i.e. LD/50 at 5 days (*26*), survival of cells with the clonal assay (*7*), and survival of crypts (*27*). The RBE values obtained do not vary with the method used, but only with the dose per fraction (*14*). There is a complication in assessing the response of the gut, however, in that there is a marked dependence of sensitivity to low LET radiation on dose rate, due to these cells having the capacity to repair sublethal damage very rapidly indeed, i.e. during a 10- to 20-minute irradiation (*28*). This has been taken into account by keeping irradiation times constant where radiations of different LET are compared. The RBE for mouse gut is significantly greater than for skin (Fig. 2). This was first shown by Hornsey (*14*), and more detailed recent measurements have confirmed this observation. A similar dose rate effect has been looked for, but not found in rat skin (*24*).

The gut is also, perhaps, rather unusual in that the RBE for megavoltage X-rays relative to 250 kVp X-rays is 1.0, compared with about 0.85 for other tissues (*29*).

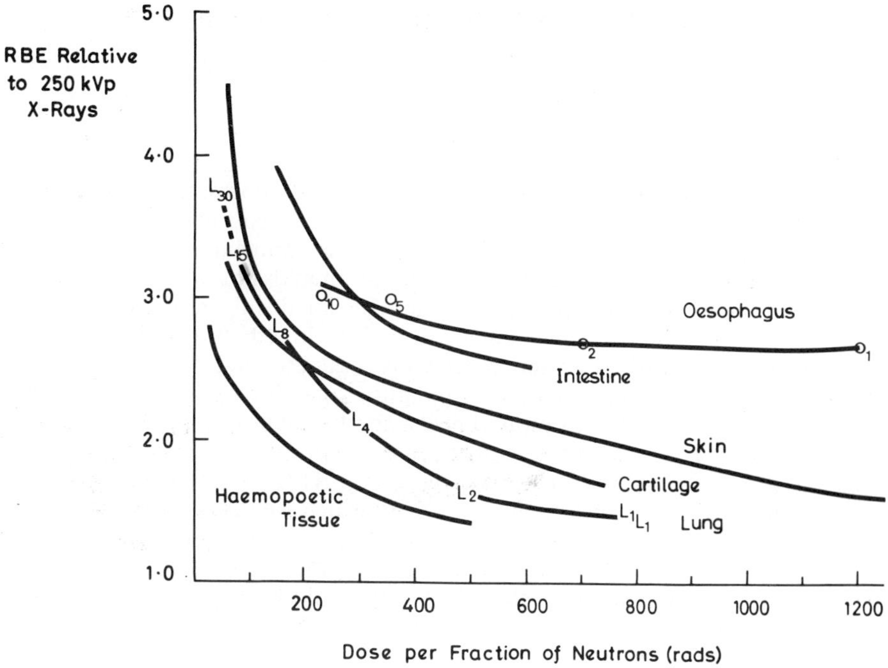

FIG. 2. The RBE for various tissues, as a function of dose per fraction. L is for lung and O for esophagus.

The high RBE values obtained from damage to the intestinal epithelium of the small bowel may indicate a limited use for neutrons in radiotherapy at those sites where the early acute reaction of the small bowel restricts the dose. At sites where little of the small intestine was included in the field, e.g. in treatment for carcinoma of the stomach, for secondary tumors arising at the site of the colostomy and in the inguinal glands after removal of a carcinoma of the colon, Dr. Catterall (personal communication) using the Hammersmith neutron beam obtained no undue problems and a good response to the fractionation schedule of 1440 rad of neutrons in 12 fractions in 26 days.

There is a clear impression that radiation sickness after large abdominal fields including the stomach is less after neutrons than after X-rays (Catterall, personal communication). Experiments on rats (*30*) indicated a particularly low RBE in comparison with other tissues for the slowing of the passage of barium through the stomach, an endpoint which has been associated with radiation sickness.

However, it is often the late reaction of the gut, particularly the large bowel, which limits many treatments in radiotherapy, so that RBE values are required for late changes and perhaps late death after abdominal irradiation. Such experiments are now in progress at Hammersmith.

CARTILAGE AND HEMOPOIETIC TISSUE

No measurements have been made beyond those reported previously (*12*). The available data are included in Fig. 2.

ESOPHAGUS

Phillips and Margolis (*31*) showed that, after irradiation of the thorax. mice died between days 10 and 30 from esophagitis, Kurohara and Casarett (*32*) suggested that the primary site of damage was the epithelium, but death was probably due to starvation and dehydration, the animals not eating or drinking because of the discomfort. Information on RBE for this form of damage was obtained by irradiation of mice under normal air-breathing conditions, with single doses, 2, 5 and 10 daily fractions of X-rays or neutrons (Hornsey and Field, in preparation). The RBE was found to be approximately constant with dose per fraction (Fig. 2). This might simply reflect a similar shape of cell survival curve after X-rays and fast neutrons, with a dose modification factor of about 3. However, there are reasons for suspecting that the RBE value observed with the large single doses is affected by hypoxia (*31*), which would make it artificially high. The same might also, but to a lesser extent, be true for the fractionated exposures, which would tend to give an apparent uniformity of RBE with dose per fraction. In either case, the value of RBE at low doses per fraction, for which hypoxia is least likely to affect the situation, is probably reliable.

LUNG

At lower doses to the mouse thorax than those which cause esophageal death, mice die after several months from lung damage (*31*). Phillips (*33*), using electron microscopy techniques, suggested that this was due to endothelial cell damage. Kurohara and Casarett (*32*) thought death was due to pulmonary infection superimposed on developing fibrosis, whereas Maisin (*34*) believes death is due to both endothelial and epithelial cell damage.

In our experiments, we used as an endpoint the LD/50 between 40 and 180 days to determine the RBE for various fractionation schedules (*35*). For single doses and two fractions, the RBE was about half that for the esophageal endpoint, although the physical arrangement was identical in the two cases. This is a similar situation to the RBE values found after whole-body irradiation, with a high RBE for gut damage (LD/50 at 5 days) and a low RBE for hemopoetic tissue damage (LD/50 at 30 days). RBE values for lung damage for a range of doses per fraction, obtained by giving up to 15 fractions of X-rays and neutrons, are shown in Fig. 2.

In man, gross physiological changes in irradiated lung are maximal at about 4 months after radiotherapy (Peckham, personal communication). This is similar in time scale to the time of death of mice from lung damage after irradiation of the thorax. Also, the relationship between total dose, number of fractions and overall treatment time in radiotherapy is similar for the incidence of radiation pneumonitis in man and LD/50 in mice (*36*). Thus, the RBE for lung damage in mice may be relevant to that in man.

These experimental results may be of wider interest than those for lung damage specifically. Late radiation damage limits the dose that can be given to many tissues in radiotherapy, and may result primarily from vascular injury. The same may also be true for lung damage (*33*), and if so these RBE values may be relevant to late damage in other tissues.

When two doses of X-rays are delivered to the mouse thorax separated by a few hours, there is evidence of a considerable degree of repair of sublethal damage (*31,35*). With neutrons, allowing for RBE, this repair is about half. If the time interval between fractions is increased beyond one day there is further repair, but the dose increment per day becomes progressively smaller with increasing time. From the difference between 10 fractions given in 11 or 25 days Wara, Phillips, Margolis and Smith (*36*) suggested that total dose was proportional to $T^{0.06}$. Our own data suggest $T^{0.07}$ and are plotted in Fig. 3 together with the data from Wara et al. (*36*). Where more than two fractions were used, the data was normalized to two fractions.[2] It is unlikely that this effect is due to cellular repopulation. Experiments designed to measure cell turnover in lung and endothelial cells indicate that it is very slow. Estimates of labelling index with tritiated thymidine range from 0.1 to 1% (*37,38*). Furthermore, no stimulation of capillary endothelial cell turnover has been observed until 40–70 days after irradiation (*39*). A very slow turnover is also consistent with death not occurring until many months after irradiation. We thus describe the increase in dose required as the time interval is increased (Fig. 3) as slow repair.[2] It is significant

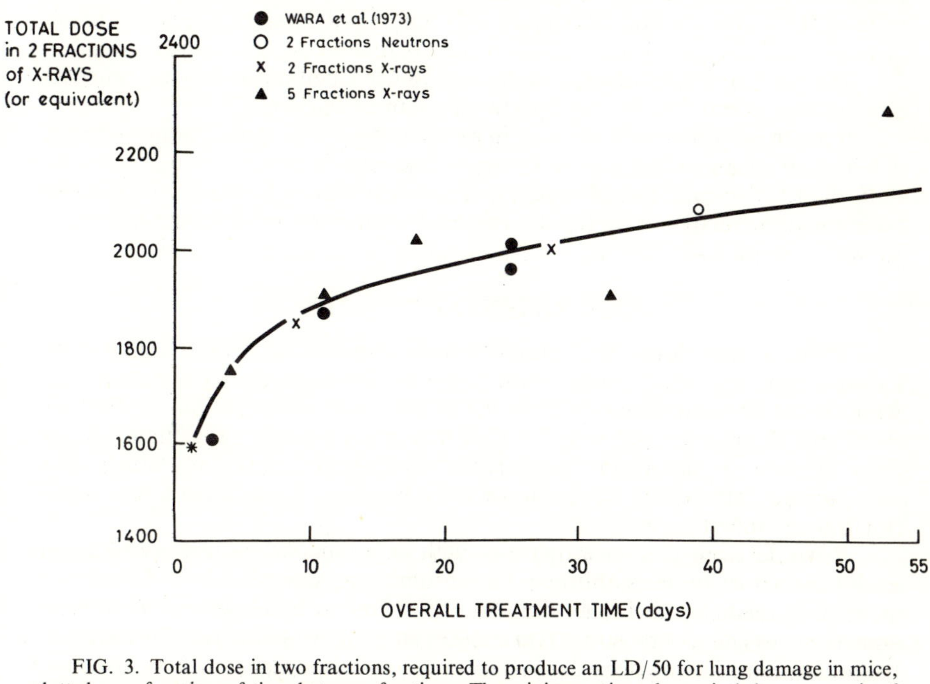

FIG. 3. Total dose in two fractions, required to produce an LD/50 for lung damage in mice, plotted as a function of time between fractions. The minimum time shown is 1 day, so repair of sublethal damage is expected to be complete. The full line is computed from $T^{0.07}$.

[2]S. B. Field, K. Kutsutani and S. Hornsey, Effects of overall treatment time and number of fractions on mouse lung (in preparation).

that after allowing for RBE, this slow repair is the same for neutrons and X-rays. This is contrasted with repair of sublethal damage, complete in a few hours, which is about half for neutron compared with X-rays, and implies a difference between the two processes. This slow repair process may also be characteristic of other slowly proliferating tissues, such as kidney (*40*).

RBE AS A FUNCTION OF NEUTRON ENERGY

Using human kidney cells, in vitro, Broerse and Barendsen (*41*) have demonstrated a marked dependence of RBE on "mean" neutron energy. The RBE falls with increasing energy from 6 at 2 MeV to about 3 at 15 MeV. With the increasing use of neutrons in radiotherapy, it is important to know how the RBEs for various tissues vary with neutron energy.

Field, Hornsey and Bewley (unpublished) have measured the RBE for skin reactions in mouse feet with neutrons produced on a cyclotron by 16 or 40 MeV deuterons onto a beryllium target. The higher energy beam gave an RBE value of about 0.85 relative to the 16 MeV deuteron-produced beam, and independent of neutron dose from 500 to 1500 rad.

Mouse intestine has also been the subject of comparative RBE measurements of this type. Withers (personal communication) reports values for cyclotron neutrons generated by 16 and 50 MeV deuterons in the ratio of 0.74 and Field, Hornsey and Bewley (unpublished) for 16 and 40 MeV deuterons in the ratio of 0.75. Broerse and Barendsen (*41*) report RBE values of 3.0 for fission neutrons and 1.4 for 14 MeV (D-T) neutrons relative to 250 kVp X-rays.

The ratio of RBE values from 50 and 16 MeV deuterons for oral mucosal damage was about 0.9 (Hussey, personal communication).

If neutrons prove to be of value in radiotherapy, the higher energy neutrons will be used because of their superior depth dose characteristics. In this case more RBE data for normal tissues is required for such beams. Perhaps a reasonable estimate of the relative RBE for 40 or 50 MeV deuterons on beryllium relative to 16 MeV is about 0.85. This value will be assumed in the section on clinical results.

EXPERIMENTAL TUMORS

There is little doubt that tumors whose response to X-rays is limited by hypoxic cells are better treated with neutrons. This was clearly demonstrated with the rat fibroscarcoma (RIB_5) (*19*,*42*). However, investigations by Fowler, Field and Denekamp (*43*) with the C_3H mammary carcinoma, a tumor which does contain a significant proportion of hypoxic cells but which also reoxygenates efficiently, have shown no advantage for neutrons for multifraction treatment regimes.

It would seem that investigations with experimental tumors are likely to enable principles to be established, but absolute measurements of RBE will not necessarily relate to the human situation. Experiments by Howlett, Thomlinson and Alper (*44*) have shown that two experimental tumors which are histologically quite similar, having been derived from the same spontaneous fibrosarcoma, exhibit very different values of RBE. They also showed, with 5 tumors, no correlation between tumor growth rate, proportion of hypoxic cells and RBE. Until the biological factors affecting RBE can be established, there is no way of predicting the relative response of a neoplasm to low and high LET radiation.

CLINICAL ASPECTS

In the neutron trial at Hammersmith, 1440 rad in 12 fractions in 26 days is normally given (*45*) and is currently found acceptable for a wide range of tumor treatments. This dose may be converted to an equivalent megavoltage X-ray treatment of 30 fractions in 39 days, the dose being $1440 \times R_N \times R_\gamma \times C$. R_N is the RBE for neutrons at 120 rad per fraction, which from Fig. 2 is 3.1 for skin. The value 3.1 was also estimated from observations of the skin of patients undergoing radiotherapy (Field and Catterall, unpublished). R_γ is the RBE for megavoltage X-rays, which we take to be 1.15. C is the conversion from 12 fractions in 26 days to 30 fractions in 39 days, which by means of the Ellis formula (*46*) is 1.303. Thus, the equivalent dose of megavoltage X-rays is 6,690 rad in 30 fractions in 39 days. If we were to use the RBE for lung of 2.9 at 120 rad per fraction, instead of that for skin, the equivalent megavoltage dose would be 6,260 rad.

At Hammersmith there has been no special indication of any adverse effects of neutron therapy at these doses. Controlled clinical trials are in progress, and the early results from the trial of advanced head and neck tumors will be reported in the near future.[3] Until such results are available, firm conclusions cannot be drawn regarding the therapeutic value of fast neutrons, but the clinical impression is that neutrons are being effective in many cases which are selected for neutron therapy because they are not expected to respond to conventional treatment (Catterall, personal communication).

In order to relate the dose given at Hammersmith to that at other centers, three factors must be taken into account: (1) The fractionation scheme used; (2) the system of dosimetry employed; and (3) the relative RBE for the two different beams. For treatment with the Houston, Texas beam of neutrons produced by 50 MeV deutrons on beryllium, the Hammersmith dose would need upgrading as follows: (1) to accommodate the Houston treatment given in 13 fractions over 43 days the Hammersmith dose would require an increase to 1533 rad, calculated by means of a modified Ellis formula for neutrons (*47*); (2) the dose would need a further increase to 1780 rad to allow for dosimetric differences at the two centers; (3) the difference in RBE between the two beams is about 0.85 (see above), so the equivalent dose at Houston would be 2,100 rad. This dose is within the range given in the trial at Houston (Hussey, personal communication). The MANTA project, using the National Radiation Laboratory cyclotron with 33 MeV deuterons, currently employs about 2,000 rad neutrons in 14 fractions in 46 days, usually with a shrinking field technique (Rogers, personal communication). The dosimetry is measured on the same basis as that in Houston. The RBE for the MANTA beam might be slightly higher than that at Houston, so that the biological effects of the doses given are probably similar to those at Houston and Hammersmith.

CONCLUSIONS

Radiotherapy with fast neutrons is no longer associated with severe late damage. This is a result of careful selection of the dose given to patients. We estimated the doses used to be biologically similar at Hammersmith, Houston and Washington. In extrapolating experience with X-rays to treatments with any

[3]M. Catterall (in preparation for the Brit. Med. J.).

new beam, a knowledge of RBE is required as a function of dose per fraction. With neutrons, such data for mouse skin can safely be used on man, but we do not know if this is true for other tissues. We suggest experiments be done to measure RBE in different species, for tissues other than skin. Careful experiments are also required to measure RBE for tissues as a function of beam energy.

We do not know whether the above conclusions are valid for other high LET radiations such as π mesons. A similar series of experiments to those performed at Hammersmith, on a range of normal tissues, are required before patients can be treated with confidence.

With normal tissues the radiation response can, at least in some cases, be extrapolated from experiments on animals directly to man. However, with tumors the RBE shows no clearly discernible pattern with any recognizable tumor characteristic and it is therefore not possible to make any direct extrapolation from animal to human tumors. In order to choose the most effective beam or fractionation schedule in a particular case, a knowledge is required of the factors which affect the response of a tumor to radiation. This can only be achieved by basic research, not by analogy as appears to be the case for normal tissue.

REFERENCES

1. L. H. Gray, J. C. Mottram, J. Read and F. G. Spear, *Brit. J. Radiol.* **12,** 371 (1940).
2. G. W. Barendsen, T. L. Bensker, A. J. Vergroesen and L. Budke, *Radiat. Res.* **13,** 841 (1960).
3. J. F. Fowler and R. L. Morgan, *Brit. J. Radiol.* **36,** 115 (1963).
4. R. S. Stone, *Am. J. Roentgenol. Radiat. Ther.* **59,** 771 (1948).
5. G. E. Sheline, T. L. Phillips, S. B. Field, J. T. Brennan and A. Raventos, *Am. J. Roentgenol. Radiat. Ther.* **111,** 31 (1971).
6. H. R. Withers, *Radiat. Res.* **32,** 227 (1967).
7. H. R. Withers and M. M. Elkind, *Radiat. Res.* **38,** 598 (1969).
8. E. W. Emery, J. Denekamp, M. M. Ball and S. B. Field, *Radiat. Res.* **41,** 450 (1970).
9. K. Y. Chen and H. R. Withers, *Int. J. Radiat. Biol.* **21,** 521 (1972).
10. S. Hornsey, *Radiat. Res.* **55,** 58 (1973).
11. G. F. Whitmore and J. E. Till, *Annu. Rev. Nucl. Sci.* **14,** 347 (1964).
12. S. B. Field, *Radiology* **93,** 915 (1969).
13. S. B. Field and S. Hornsey, *Eur. J. Cancer* **7,** 161 (1971).
14. S. Hornsey, *Radiology* **97,** 649 (1971).
15. T. Alper, *Brit. J. Radiol.* **45,** 39 (1972).
16. T. Alper, J. F. Fowler, R. L. Morgan, D. D. Vonberg, F. Ellis and R. Oliver, *Brit. J. Radiol.* **35,** 722 (1962).
17. D. K. Bewley, J. F. Fowler, R. L. Morgan, J. A. Silvester, B. A. Turner and R. H. Thomlinson, *Brit. J. Radiol.* **36,** 107 (1963).
18. J. Denekamp, J. F. Fowler, K. Kragt, C. J. Parnell and S. B. Field, *Radiat. Res.* **29,** 71 (1966).
19. S. B. Field, T. Jones and R. H. Thomlinson, *Brit. J. Radiol.* **40,** 834 (1967).
20. D. K. Bewley, S. B. Field, R. L. Morgan, B. C. Page and C. J. Parnell, *Brit. J. Radiol.* **40,** 765 (1967).
21. S. B. Field, *Radiology* **92,** 381 (1969).
22. S. Hornsey and S. B. Field, *Eur. J. Cancer* **10,** 231 (1974).
23. J. Denekamp, *Brit. J. Radiol.* **46,** 381 (1973).
24. S. B. Field and D. K. Bewley, *Int. J. Radiat. Biol.* **26,** 259 (1974).

25. J. Denekamp and S. B. Field, *Eur. J. Cancer* **10,** 241 (1974).
26. S. Hornsey, S. Vatistas, D. K. Bewley and C. J. Parnell, *Brit. J. Radiol.* **38,** 878 (1965).
27. H. R. Withers and M. M. Elkind, *Int. J. Radiat. Biol.* **17,** 261 (1970).
28. S. Hornsey and T. Alper, *Nature (London)* **210,** 212 (1966).
29. R. Patterson, *Brit. J. Radiol.* **30,** 354 (1957).
30. B. A. Turner and J. F. Fowler, *Brit. J. Radiol.* **36,** 101 (1963).
31. T. L. Phillips and L. Margolis, *Front. Radiat. Ther. Oncol.* **6,** 254 (1972).
32. S. S. Kurohara and G. W. Casarett, *Radiat. Res.* **52,** 263 (1972).
33. T. L. Phillips, *Radiology* **87,** 49 (1966).
34. J. R. Maisin, *Radiat. Res.* **44,** 545 (1970).
35. S. B. Field and S. Hornsey, *Eur. J. Cancer* **10,** 621 (1974).
36. W. M. Wara, T. L. Phillips, L. W. Margolis and V. Smith, *Cancer* **32,** 547 (1973).
37. I. F. Tannoch and S. Hayashi, *Cancer Res.* **32,** 77 (1972).
38. M. B. Shimkin, T. Sasaki, M. McDonough, R. Baserga, D. Thatcher and R. Wieder, *Cancer Res.* **29,** (1969).
39. L. F. Fajardo and J. R. Stewart, *Radiology* **101,** 429 (1971).
40. E. Glatstein, R. C. Brown, G. D. Zanelli and J. F. Fowler, *Radiat. Res.* **61,** 417 (1975).
41. J. J. Broerse and G. W. Barendsen, *Curr. Top. Radiat. Res. Q.* **8,** 305 (1973).
42. S. B. Field, T. Jones and R. H. Thomlinson, *Brit. J. Radiol.* **41,** 597 (1968).
43. J. F. Fowler, S. B. Field and J. Denekamp, *Eur. J. Cancer* (in press) 1975.
44. J. F. Howlett, R. H. Thomlinson and T. Alper, *Brit. J. Radiol.* **48,** 40 (1975).
45. M. Catterall, C. Rogers, R. H. Thomlinson and S. B. Field, *Brit. J. Radiol.* **44,** 603 (1971).
46. F. Ellis, *Clin. Radiol.* **20,** 1 (1969).
47. S. B. Field, *Brit. J. Radiol.* **45,** 315 (1972).

HZE PARTICLE EFFECTS IN MANNED SPACEFLIGHT[1]

Chairman: D. Grahn

[1]Symposium C–4

Studies on the Effects of Cosmic HZE-Particles on Different Biological Systems in the Biostack Experiments I and II Flown on Board of Apollo 16 and 17[1]

H. Bücker and G. Horneck

Arbeitsgruppe für biophysikalische Weltraumforschung
Universität Frankfurt
Frankfurt/Main, West Germany

INTRODUCTION

The investigative program of Biostack aims to ascertain the kind and extent of damage produced by HZE-particles of cosmic radiation in biological matter. HZE-particles are those primary cosmic ray particles that possess a charge of $Z > 2$ and a sufficient high energy loss over an extended track length to penetrate at least 1 mm of shielding (*1*). Biostack experiments (I and II) had been accommodated in the command modules Apollo 16 and 17 for spaceflights outside of the geomagnetosphere, one specimen has participated in a balloonflight, and an additional Biostack is assigned for the Apollo-Soyuz-Test-Mission.

In the Biostack experiments I and II, the total dose of cosmic ionizing radiation was measured by means of LiF thermoluminescence dosimeters (TLD). For the 226-hour Apollo 16 flight a total dose of 530 mrad to 600 mrad was determined, and for the 305-hour Apollo 17 mission the total dose ranged from 715 mrad to 750 mrad. The dose decreased from that part of the Biostack directed towards the cabin due to absorption by the Biostack material itself (Table 1).

TABLE 1
Total Dose of Cosmic Ionizing Radiation
to which the Biostack Experiments I and II Were Exposed
Data from LiF-TLD

Flight	Dose (mrad) Bottom[+)]	Dose (mrad) Middle	Dose (mrad) Top	Mean dose rate (mrad/h)
Apollo 16	600		530	2.34
Apollo 17	750	730	715	2.41

[+)]directed toward space

[1]Congress presentation C-4-1.

Only 0.5% of this total dose is ascribed to the HZE-particles of cosmic radiation. A dose of 3.3 mrad has been calculated from nuclear emulsion data of Biostack I and II for particles with an energy loss exceeding 80 keV/μm. However, this conventional dosimetric method is not considered to be adequate for appraising the biological effectiveness of HZE-particles. Considering the high local concentration of the energy deposition along the path of an HZE-particle, a nearly 10,000 times higher local dose is obtained, if the total dose is assumed to be deposited inside a cylinder 10 μm in diameter. Microdosimetric concepts should consider this strong local concentration as well as the strong temporal concentration of absorbed energy produced by an HZE-particle penetrating biological matter.

In the past, a considerable amount of effort has been expended to clarify the biological importance of HZE-particles. However, in most experiments no definite conclusion could be reached as to whether the observed biological damages were attributable to cosmic ray HZE-particles. To establish the direct involvement of HZE-particles it is absolutely necessary to isolate the precise biological region which has been penetrated by an HZE-particle. The Biostack experiments meet this requirement, just as the Biocore experiment flown on board of Apollo 17.

CONCEPTION OF THE BIOSTACK EXPERIMENTS

In the Biostack experiments, the demand for a precise determination of the hit biological area has been accomplished by using a special sandwich construction (Fig. 1). The biological objects, all of them being in resting state, were fixed in monolayers between physical track detectors (Fig. 2). Thus, it was achieved to trace the track of an HZE-particle through the biological layer.

FIG. 1. Biostack software composed of biological layers and sheets of radiation detectors.

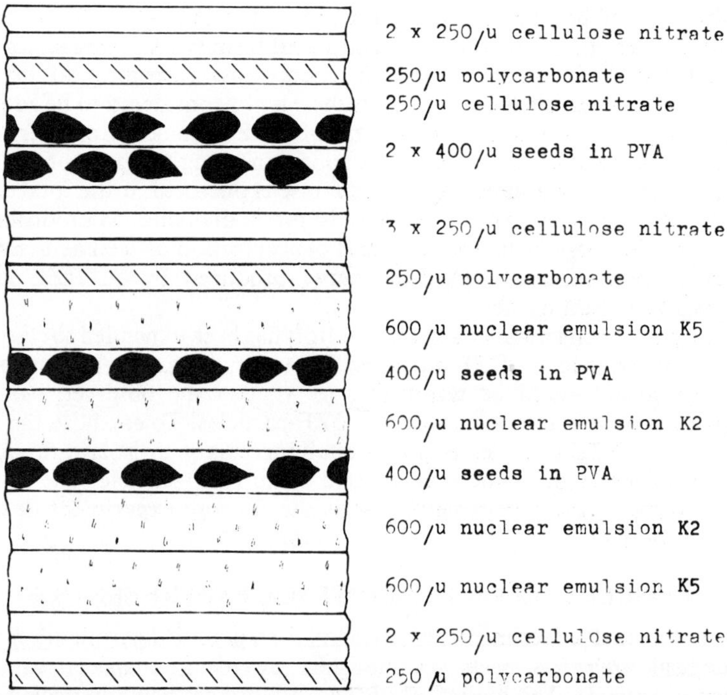

FIG. 2. Typical configuration of biological layers and detectors in the Biostack.

Different physical track detectors were used (Table 2). These detectors, the nuclear emulsions, plastic detectors and AgCl-crystals, complement one another in their recording characteristics of HZE-particles as well as in the localization of the biological region hit. Special methods were developed for optimal localization of the point of penetration in the biological layer. The accuracy in determining this penetrating point reached $\pm$ 1 μm. Therefore, in case of the animal eggs or of the seeds, which all exceeded 50 μm in diameter, even the hit region inside the biological organisms could be identified. For these objects, a sufficient high accuracy was obtained with all three kinds of detectors. In case of the bacterial spores, however, which were 1.5 μm in diameter, usually several spores covered the determined target area with different probabilities of being hit. For determination of the target area inside the spore layer, plastic detectors of cellulose nitrate (CN) were used. The CN sheet was in fixed contact with the biological layer (Fig. 3). This contact was maintained during flight. during postflight etching and track measurements, and even during the growth studies. During etching, the biological layer was protected against the toxic etching solution (NaOH), resulting in only one etch cone at the pure CN side of the sheet which was not covered with biological material. The trajectory of an HZE-particle in the biological layer was extrapolated from this etch cone. The AgCl-crystals have the advantage that the track is observed nearly up to the biological layer.

The nuclear emulsions, which were attached to the biological layers, received the pattern of the biological objects by weak optical illumination during postflight disassembly. Thus, the biological objects hit were identified directly from the developed emulsion, which showed the HZE-particle track together with the faint photography of the biological objects and, in addition, of a coordinate grid (Fig. 4).

TABLE 2
Radiation Detectors in the Biostack Experiments

BIOLOGICAL EXPERIMENTS IN THE BIOSTACK

BIOLOGICAL SYSTEM		INVESTIGATOR	ORGANIZATION
MONOCELLULAR	SPORES OF BACILLUS SUBTILIS	G. HORNECK	UNIVERSITY OF FRANKFURT, GERMANY
	CYSTS OF COLPODA CUCULLUS	H. PLANEL, J.P. SOLEILHAVOUP	UNIVERSITY OF TOULOUSE, FRANCE
PLANT	SEEDS OF ARABIDOPSIS THALIANA	E. REINHOLZ	MPI FÜR BIOPHYSIK FRANKFURT, GERMANY
	RADICULAE OF VICIA FABA	W. SCHEUERMANN	T. UNIVERSITY OF HANNOVER, GERMANY
ANIMAL	EGGS OF ARTEMIA SALINA	W. RÜTHER, E.H. GRAUL	UNIVERSITY OF MARBURG, GERMANY
		H. PLANEL, J.P. SOLEILHAVOUP	UNIVERSITY OF TOULOUSE, FRANCE
	EGGS OF TRIBOLIUM CASTANEUM EGGS OF CARAUSIUS MOROSUS	W. RÜTHER	UNIVERSITY OF MARBURG, GERMANY

Besides identification of the biological area hit, the evaluation of the different track detectors resulted in extensive information on the flux and angular incidence of the cosmic ray particles, on their absorption by the wall of the spacecraft and the Biostack material, and on the spectral distribution of their charge, energy and energy loss.

The influence of the HZE-particles was studied on several biological systems at the molecular, cellular and organic level. The criteria for their selection were to combine species of different organizational levels, of different radiation sensitivities (as known from X-ray irradiation), of adequate shelf-life, and of a high resistance to additional environmental factors of spece flight. They should be typical representatives of at least one kind of radiation damage of genetical or modifying nature. The radiation effects studied were mainly changes in cellular and organic growth, damages to cell nuclei or other subcellular systems, and induction of mutations leading to genetic changes of biological significance. The variety of biological specimens under investigation ranged from procaryotic unicellular systems, the *Bacillus subtilis* spores, via the protozoon *Colpoda cucullus* to plant embryos, like *Vicia faba* radiculae and *Arabidopsis thaliana* seeds, and to animal eggs of the shrimp *Artemia salina* and the insects *Tribolium castaneum* and *Carausius morosus* (Table 3). The Biostack I experiment aboard Apollo 16 consisted of seven different experiments, the Biostack II aboard Apollo 17 of ten. They were conducted by twenty-five European scientists in eleven different institutes.

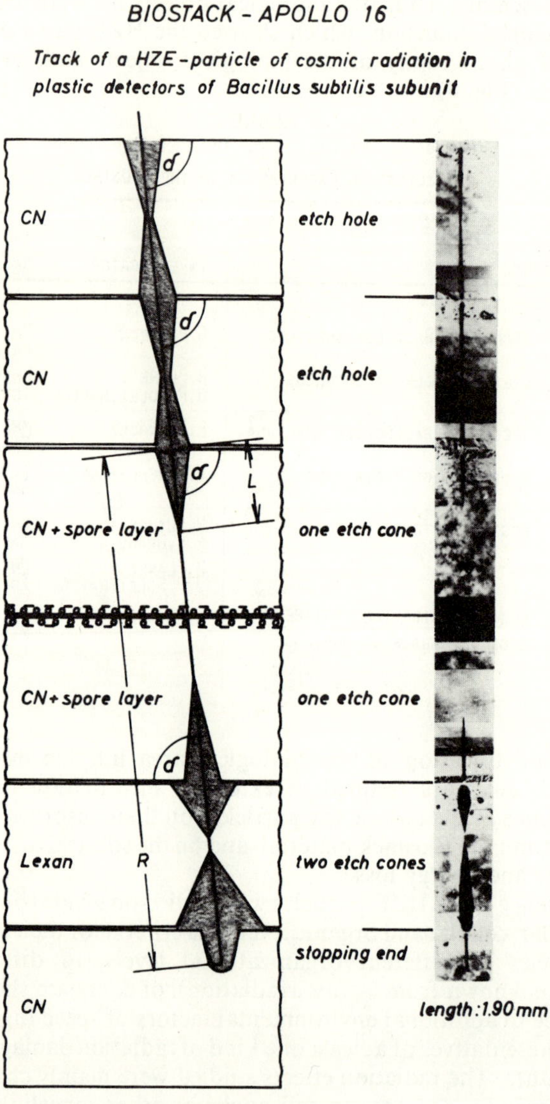

FIG. 3. Track of an HZE-particle stopping in the *Bacillus subtilis* unit of Biostack I, sketch (left) and micrograph (right).

RESULTS

Spectrum of Biological Effects Caused by HZE-Particles

Processes not remarkably influenced by a hit

In each Biostack experiment, several thousands of biological objects were found to be hit by an HZE-particle. Their response to 'enders' (stopping

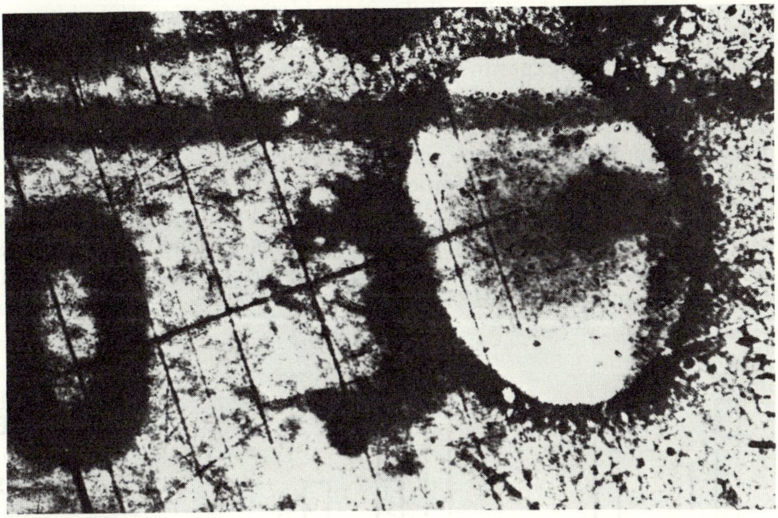

FIG. 4. Micrograph of an HZE-particle track in K2 nuclear emulsion of Biostack I together with a negative of an *Arabidopsis thaliana* grain and the coordinate grid. [Reproduced from H. Bücker et al.: *Biostack experiment*, in *Apollo 16, Preliminary Science Report*, NASA SP–315, p. 27–1 (1972).]

TABLE 3
Biological Objects in the Biostack Experiments

element	Li_3	Be_4	B_5	C_6	N_7	O_8	F_9
number	2	4	5	37	16	71	1
element	Ne_{10}	Na_{11}	Mg_{12}	Al_{13}	Si_{14}	P_{15}	S_{16}
number	15	7	24	2	5	-	4

particles) or 'through-shots' (non-stopping particles) of HZE-particles was studied. The result was a widespread spectrum of HZE-particle-induced effects in biological matter.

In bacterial spores and plant seeds, germination was found to be highly resistant to an HZE-particle hit. During germination, the initially 'phase bright' bacterial spores (*Bacillus subtilis*) became dark because of a change in the refractive index, which is probably caused by excretion of dry matter, slight swelling and redistribution of water within the spore. This process has been proved to be highly radiation resistant. X-ray doses of approximately 400 krad, which reduced the surviving fraction of colony formers to about 10^{-4}, did not influence the germination process. Much higher doses, approximately 2000 krad, are necessary to induce "pseudogermination", which is correlated with an increased permeability of the cell wall. The germinating fraction of the hit spores was more than 90% in the Biostack I experiment (Fig. 5) and reached nearly 100% in the Biostack II experiment. This indicates the high resistance of this process to HZE-particle bombardment. Pseudogermination was not observed in the spores hit. Likewise, the seeds of *Arabidopsis thaliana* that were hit germinated with the same frequency and rate as the controls. The growth of *Vicia faba* radiculae also did not differ significantly from that of the controls (Fig. 5). It is likely that the surrounding intact cells have replaced the destroyed cells, if any destruction did occur. Even cytological investigations, dealing with achromasie of the nuclear material, or repairability of damages in the nuclear DNA, did not reveal remarkable influence of HZE-particles.

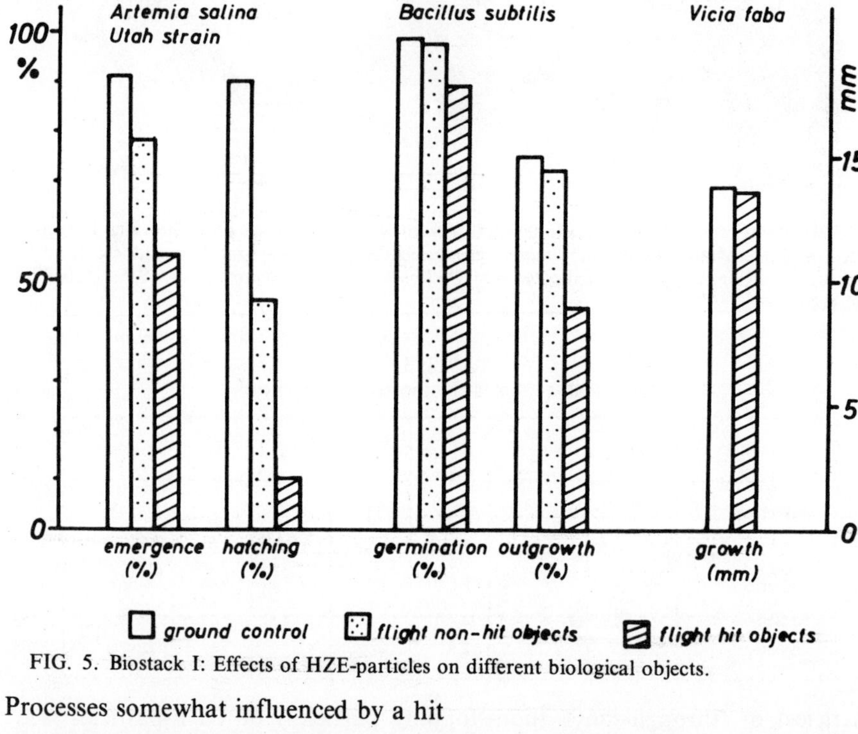

FIG. 5. Biostack I: Effects of HZE-particles on different biological objects.

Processes somewhat influenced by a hit

The developmental stages, following the germination process, turned out to be more sensitive to a hit by an HZE-particle. The outgrowth of the *Bacillus subtilis* spores of Biostack I that were hit was reduced (Fig. 5). During spore outgrowth the spore cases rupture and the embryo vegetative cell emerges, capable of developing into the fully grown vegetative cell. After irradiation with X-rays, the outgrowing fraction decreased with irradiation dose. A dose of 350 krad produced a surviving fraction of 37% (D_{37}) of outgrowing cells. Only 45% of the spores hit were able to grow out compared to a 72% outgrowth of the flight control. Growth was observed over a 4-hr period. The spores hit, which failed to grow out simultaneously with the flight control, did not resume their development during the whole time of incubation (Fig. 6).

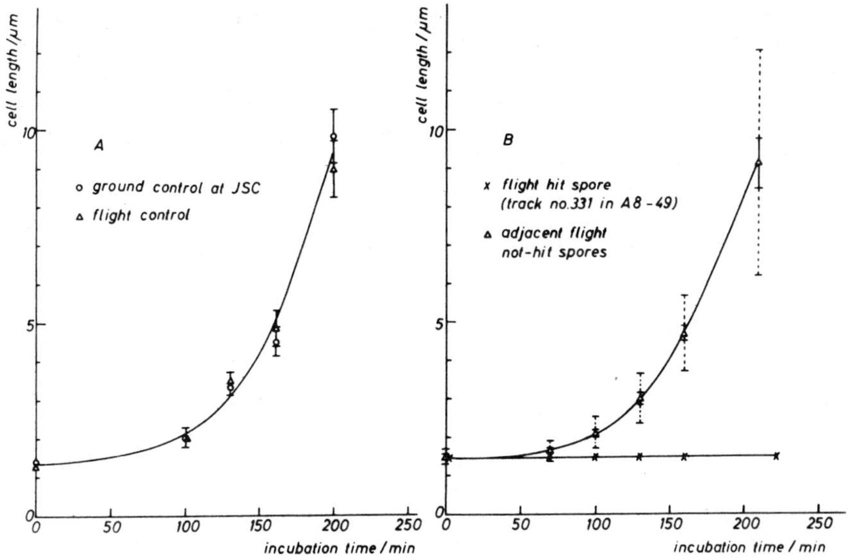

FIG. 6. Biostack I: Cellular elongation curves of the first vegetative cells of *Bacillus subtilis*.

During the development of *Arabidopsis thaliana* plants, grown from seeds hit, the frequency of multicaulous individuals was remarkably increased compared with the controls. These multicaulous forms are assumed to have developed from seeds in which cells of the vegetation cone have been destroyed by a penetrating HZE-particle.

Only 55% of the *Artemia salina* eggs hit were able to pass the first developmental stage, the emergence (Fig. 5). During this process the egg shell cracks open and the larva, the nauplius, emerges still enclosed in the egg membrane. In most of these injured eggs, no development at all was detected. By clearing up the egg shell with antiformin, the undeveloped gastrula became visible.

Processes highly distributed by a hit

The animal eggs were the most sensitive biological objects accommodated in the Biostack experiments in regard to HZE-particle hits. Whereas irradiation of *Artemia salina* eggs with gamma-rays, neutrons, electrons and even helium ions ($Z = 2$) resulted in a sigmoidal dose-effect curve in terms of development to a swimming larva, irradiation with oxygen ions of 160 MeV resulted in an exponential curve. It is assumed that the passage of one single particle of high energy loss may damage a cellular area large enough to disturb embryogenesis.

Only 10% of the *Artemia salina* eggs that were hit developed to a swimming larva, the nauplius, compared to 90% of the ground controls and 46% of the flight controls not hit. The larvae that were derived from eggs hit had a high mortality. Only few of them reached maturity; but none of them was completely normal in further growth and behavior. The surviving nauplii grew up retardedly and remained stunted in growth (Fig. 7); they did never reach the normal length of 12 mm. Maturity in terms of pair mating was delayed (Fig. 8). The time until deposition of the eggs took twice as long as in the case of ground controls. The number of broods varied from 0 to 2, and the number of descendants in the F1 generation was essentially reduced. The occurrence of malformations was increased by a factor of approximately 10. Shortened extremities or abnormal thorax or abdomen (Fig. 9) were the main abnormalities caused by HZE-particle hits. These results show that an HZE-particle-induced damage in cells of the encysted blastula may be manifest in the gastrula stage or in later steps of development of the nauplius, or even in the adult. This indicates the inability of the intact cells to replace the function of the destroyed cells.

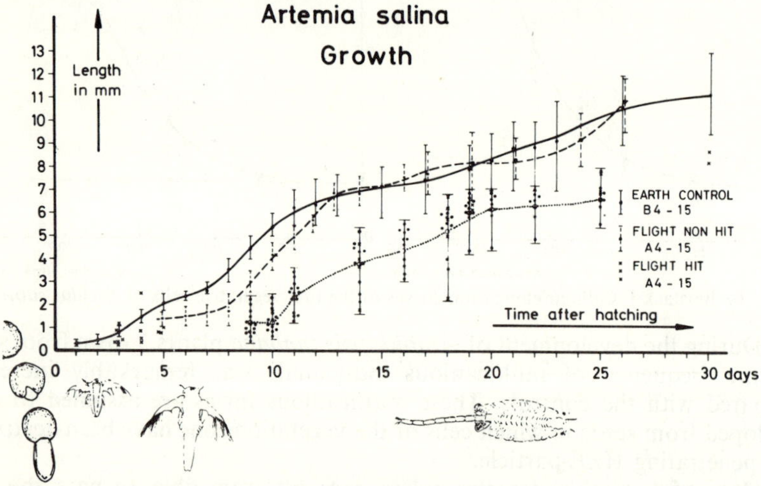

FIG. 7. *Artemia salina,* Biostack I: Growth curve. [Reproduced from W. Rüther, E. Craul, W. Heinrich, O. C. Allkofer, R. Kaiser, R. Pfohl and H. Bücker: "Radiobiological Results of the Biostack experiment on board of Apollo 16 and 17", Life Sci. Space Res. 13 (1975), in press.]

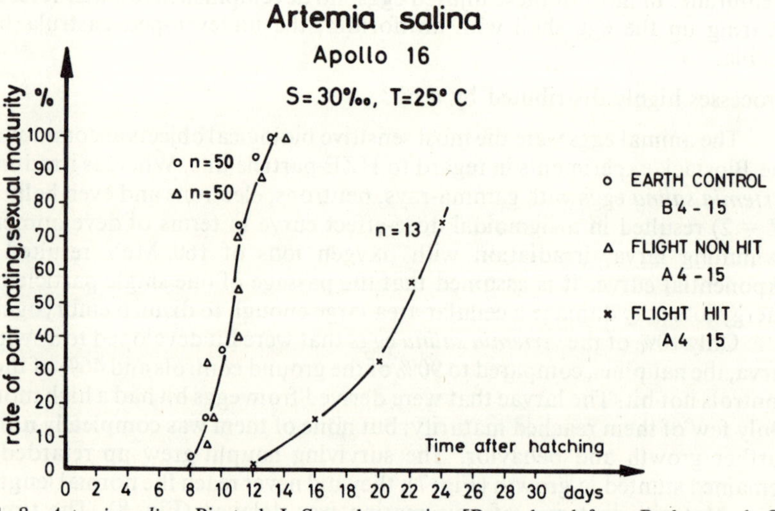

FIG. 8. *Artemia salina,* Biostack I: Sexual maturity. [Reproduced from Rüther et al. Cf. Fig. 7.]

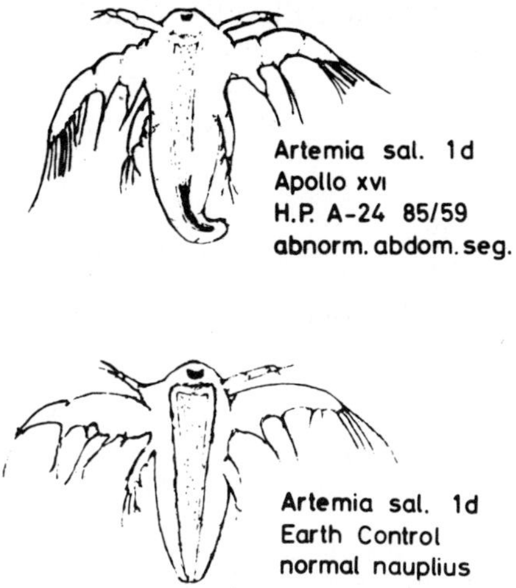

FIG. 9. *Artemia salina,* Biostack I: Malformation. [Reproduced from Rüther, et al. Cf. Fig. 7.]

Similar effects were found during the development from hit *Tribolium castaneum* eggs. The hatching frequency was significantly lowered, and during the first two days after hatching a high lethality was observed. The frequency of abnormalities was increased by a factor of 20. The most frequent malformations were curved abdomina, fused segments of the abdomen or the antenna, and split or shortened elytra.

Likewise, the hatching of *Carausius morosus* eggs hit was essentially reduced. Many of the larvae died during the first two weeks after hatching. Curved abdomina, fused segments or shortened legs were the main abnormalities in the descendants of the eggs hit.

These are the main radiobiological results obtained up to now in the Biostack experiments. They already show that the degree of injury produced by a penetrating HZE-particle depends on whether the cell that was hit is replaceable or not.

Physical Characteristics of the Hitting HZE-Particles

The energy loss of the HZE-particles is one physical quantity of high importance in regard to the biological efficiency. The integral distribution function of the REL (restricted energy loss) was obtained from plastic detector analysis of Biostack I (Fig. 10). This REL-spectrum agrees satisfactorily with that obtained in the MEED experiment (*2*), which was stowed within the command module for nearly the whole mission. The personal detectors of the astronauts, however, who had partially left the command module for lunar landing, recorded higher fluxes (*3*).

In the biological studies of Biostack, special emphasis was placed on the effects of HZE-particles of very high energy loss. Therefore, it was agreed to restrict the studies to hits with an REL above a lower threshold of 1.8 GeV cm^{2}g^{-1}. For each hitting particle, the REL in the biological layer was determined. In CN sheets, the length L (Fig. 3) of the single etch cone gives the REL of the particle at that point of its path. In spores hit, REL values up to 5.0 GeV cm^2 g^{-1} have been determined.

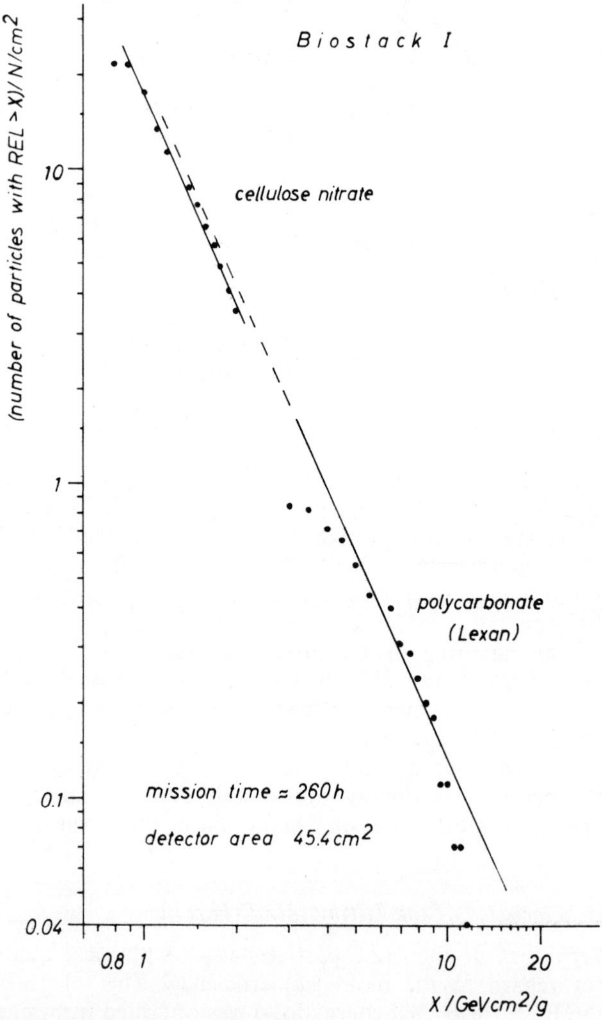

FIG. 10. CN-detectors, Biostack I: Integral REL-spectrum of heavy ions.

The charge of the HZE-particle is the other physical quantity of high interest. In the CN sheets, it is determined from the relation between the cone length L and the residual range R (Fig. 3). The charge distribution based on 200 individual track measurements is seen in Table 4. The charge of each hitting particle has been estimated. Special attention was given to very heavy particles. In the evaluation of *Bacillus subtilis* spores, the charge of the hitting particles ranged from $Z > 12$ to $Z \leq 26$.

All particles that reached the Biostack had penetrated the approximately 3 g cm^{-2} shielding of the spacecraft wall. A mean flux of approximately 0.1 particles

$cm^{-2}h^{-1}$ of $Z \geqslant 4$ has been found for Biostack I as well as for Biostack II. The flux was diminished remarkably from outside to inside of the Biostack software, due to the absorption of the Biostack material itself, which was approximately 16 g cm^{-2} (Fig. 11). An adequate decrease was observed in the cosmic background radiation. The level of radiation—measured at the inner side of the Biostack—which is still relatively high, clearly demonstrates the difficulty to protect by shielding the crew and payload inside Apollo spacecrafts against HZE-particles.

TABLE 4

CN-Detectors in Biostack I: Measured Abundances of the Individual Elements

RADIATION DETECTORS IN THE BIOSTACK EXPERIMENT

COSMIC RADIATION COMPONENT	DETECTOR	RANGE OF INFORMATION ON Z AND LET	THRES-HOLD	TISSUE EQUI-VALENCE	BACK-GROUND NOISE	INVESTIGATORS	ORGANIZATION
IONS	NUCLEAR EMULSION K2, K5	VERY BROAD	NO	NO	HIGH	R.KAISER, J.P.MASSUÉ, R.PFOHL	LPC - CRN STRASBOURG, FRANCE
	PLASTICS CELLULOSE NITRATE, POLY-CARBONATE	MEDIUM	YES	YES	LOW	W.ENGE, O.C.ALLKOFER, K.BARTHOLOMÄ, R.BEAUJEAN, W.HEINRICH, K.FUKUI	UNIVERSITY OF KIEL, GERMANY
						E.V.BENTON	UNIVERSITY OF SAN FRANCISCO
	AGCL-CRYSTALS ILLUMINATED	BROAD	YES	NO	MEDIUM TO LOW	E.SCHOPPER, G.HENIG, J.U.SCHOTT	UNIVERSITY OF FRANKFURT, GERMANY
γ-RAYS, X-RAYS, PROTONS	LIF THERMO-LUMINESCENCE DOSIMETER	INTEGRATING DOSIMETER				H.FRANCOIS, G.PORTAL	CEA, PARIS, FRANCE

Influence of the Attendant, Not Specifically Tested Spaceflight Factors

The conditions of spaceflight complicate the radiobiological research on HZE-particle effects. Launch vibration, weightlessness and cosmic background radiation are the main inevitable factors acting on the biological material. Therefore, in each biological study of Biostack, flown but not hit specimens were taken as flight controls, in addition to the ground controls.

The *Bacillus subtilis* spores were shown not to be influenced by the space flight environment. Germination and outgrowth of the flight controls agreed with those of the ground controls (Fig. 5) as did the rate of cellular elongation (Fig. 6). Likewise, there was no difference in the kinetics of germination of *Arabidopsis thaliana* flight control and ground control seeds. Slight damage was observed, however, in the *Artemia salina* flight control eggs. The percentage of emergence and hatching was reduced as compared to the ground controls (Fig. 5). Those flight control individuals that were able to hatch afterwards developed completely in accordance with ground controls. This slight damage to the flight controls has been assumed to be caused either by the vibrational stress or by the cosmic background radiation. Experiments to clarify this phenomenon are under way.

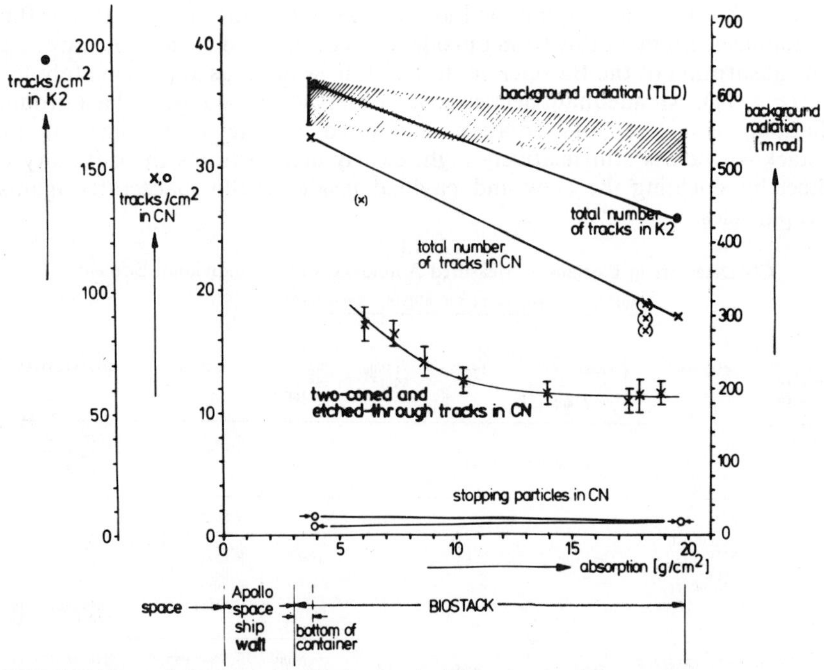

FIG. 11. Absorption of cosmic radiation by the Biostack material.

CONCLUSIONS FOR MANNED SPACEFLIGHT

Owing to the conception of the Biostack experiment, it was possible, for the first time, to prove significantly the relationship between cosmic ray HZE-particles and their biological effects. Special emphasis should be placed on the fact that the dose causing the biological damages during the Apollo spaceflights was less than 35 mrem. This dose is lower than the maximum permissible dose for man on earth according to the recommendations of the International Committee on Radiation Protection. Up to now, the question concerning the critical nature of HZE-particles to man in long-duration spaceflights cannot be answered satisfactorily, but it looks as if the risk can be sufficiently lowered if, in the construction of future space vehicles, the ensuring of maximum possible shielding will be taken into consideration. Further biophysical experiments are necessary to establish the upper limit of HZE-particle fluence to be tolerated inside spacecrafts in long-duration missions.

The data of the Biostack experiments I and II confirm the assumption that HZE-particle-induced damage might become manifest if non-replaceable cells are destroyed. In manned spaceflight, the prime target tissue in this connection is the central nervous system, which consists of highly differentiated non-replaceable cells. The question arises what number of cells might be destroyed by each hit compared to the number of cells that form a functional unit. It is likely that a large number of HZE-particle hits to the same area of the brain would be required to destroy that particular function.

REFERENCES

1. D. Grahn, (Ed.), *HZE-particle Effects in Manned Spaceflight*, Report of the National Academy of Sciences, Washington, 1973.
2. E. V. Benton and R. P. Henke, in *Proceedings of the Microbial Response to Space Environment Symposium,* (G. R. Taylor, Ed.), pp. 179–189. NASA TM X–58103, 1973.
3. E. V. Benton and R. P. Henke, *Heavy Cosmic Ray Exposure of Apollo 16 Astronauts,* University of San Francisco Report, USF-TR-72-20, 1972.

Effects of Cosmic Heavy Ions on Artemia Egg Development[1]

H. Planel, Y. Blanquet, J. P. Soleilhavoup, R. Kaiser and B. Pianezzi

Laboratoire de Biologie Médicale-Faculté de Médecine, Toulouse
and Laboratoire de Physique Corpusculaire—C.E.N., Strasbourg, France

INTRODUCTION

Primary cosmic radiations are made up of protons, alpha particles and heavy ions. In spite of their relatively low flux (1% of cosmic particles), it appears that heavy ions could be a source of hazard for living organisms during space flights, especially during the long duration ones. Indeed, cosmic heavy ions are very high LET particles and it can be assumed they may induce important damage when passing through living matter. However, this biological effect is insufficiently known although several biological responses already observed could have been ascribed to these cosmic particles, especially in experiments where genetic aberrations occurred in organisms not exposed to on-board gamma radiations (1). But during space flights, only some parts of living organisms are hit directly by cosmic ions: so, it appears that every study on the biological effect of cosmic heavy ions must be based upon an experimental method allowing a good correlation between the biological objects and the particle tracks.

Biostack experiments have been carried out using such a method with resting state biological objects embedded in plastic layers.

Artemia salina eggs appear to be a very suitable material for this research:

—*Artemia* eggs (called also *Artemia* cysts) are very resistant. They do not need any life support;

—*Artemia* cysts may be embedded in plastic and distributed in thin layers.

The *Artemia* cyst is a diapause embryonic organism, blocked in the gastrula stage. Segmentation takes place in adult females. They can survive for many years under desiccated laboratory conditions and may undergo emergence and hatching even after they have been exposed to very low temperatures. Each cyst (egg) is made up of about 4000 cells (*2*). Endoblastic cells have been differentiated but an obvious digestive lumen has not yet been noticed. The cyst is surrounded by a shell produced in female oviducts. It is made up of several layers as has been shown by electron microscopy (*3*).

After hydration, the eggs are allowed to develop. It is obvious that no cell division occurs at this time and the beginning of the development is merely related to the hydration phenomenon.

[1]Congress presentation C–4–2.

The breaking up of the shell begins about 12 hours after egg rehydration. Emergence will occur until about the 96th hour and gives rise to an embryo enclosed within a transparent hatching membrane. Later, this membrane is torn open by embryo appendiculae.

After hatching, the embryo gives rise to a free-swimming nauplius. this hatching period ends on the 4th day.

During the embryonic phase of development, cell divisions occur and the number of cells increases. The complete development lasts 3 or 4 weeks; the animals may be considered adults one month after hatching.

The maximum life span is of about 10 months. It must be noted that variation in developmental rate and in emergence percentages are observed depending on the strains and on the experimental conditions.

The present study was designed to investigate the developmental ability of Artemia eggs flown in Apollo XVI and XVII in order to determine the individual effects of cosmic heavy particles combined with the influence of other space environmental factors. Therefore, this study is a part of Biostack I and II experiments (Dr. Bücker, principal investigator).

MATERIALS AND METHODS

A) Dry *Artemia salina* eggs (Brine Shrimp Sales Co., Hayward, California) were used and tested before embedding. Eggs were mixed with a polyvinyl-alcohol 20–30% aqueous solution. The mixture was poured on a horizontal glass sheet and allowed to desiccate at room temperature. After drying, each layer included about five thousand eggs. Four biological layers have been sandwiched with Ilford nuclear K2 and K5 emulsions (Fig. 1).

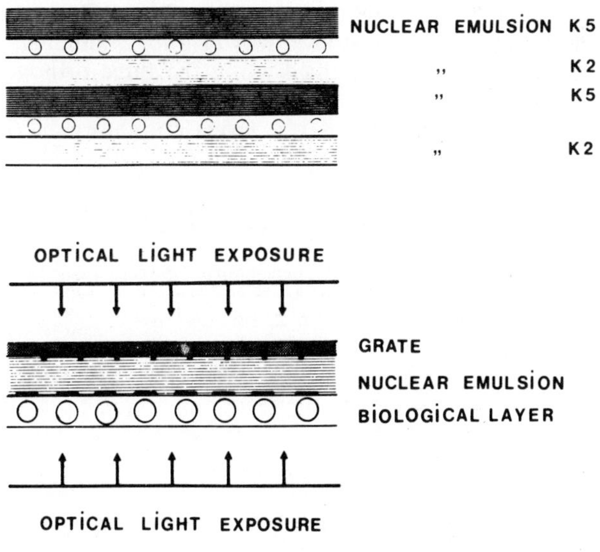

FIG. 1. General scheme of Biostack experiment.

After recovery, Biostack was disassembled and each *Artemia* layer and each nuclear emulsion was covered with a millimetric grid. An exposure to weak light allowed egg and grid shadows to be obtained.

After processing, heavy ion trajectories were geometrically reconstructed and the hit eggs were located.

Based upon nuclear emulsion scanning, we have picked up the hit and non-hit eggs with a scalpel after the polyvinyl-alcohol was cut around the eggs.

The eggs were placed individually in crystal dishes with 2 ml of culture medium. The medium was prepared in the following manner just before the experiment:

—Bidistilled water	250 ml
—"Saltemia blue"—(Tetrawerke-Melle-Germany)	tablet 1
—Aquarium filtered water	60 ml

The medium was oxygenated by means of air flow for at least 6 hours.

After the polyvinyl was dissolved, the eggs were washed in the same medium and new medium was poured into culture dishes.

The cultures were placed in an incubator at 27±0.5° C and were exposed to daylight. The cultures were checked every three hours from the 10th hour on. The numbers of emergences, of hatchings and live nauplii were determined 4–5 days after the cultures had been set up.

B) In order to determine the effects of flight dynamic factors, the eggs were submitted to vibration and acceleration tests.

The frequency and the amplitude were similar or three times higher than during the Apollo flight.

1.—Acceleration tests were performed along the three orthogonal axes in a centrifuge:

—the simulated Apollo flight test level was 7 g for 3 minutes

—the predetermined overtest level was 21 g for 3 minutes

2.—Vibration tests were performed along the three orthogonal axes on electrodynamic vibrators.

—the simulated Apollo flight test levels are shown in the following tables:

A–Sinusoidal vibration tests

Axis	Level	Frequency range	Sweep rate
-X, Y - Y	1.3 g	5–27 Hz	
Z - Z	0.036 g	.27–52 Hz	2 oct/min
	5 g	52–500 Hz	

B–Random vibration tests

Axis	P.S.D levels	Frequency range	Acceleration	Duration
X, Y-Y	0.067 G^2/Hz	80–350 Hz	8.4 g-rms	5 min
Z - Z	±3 db	20–2000 Hz		

—the predetermined overtest levels were the same as above, multiplied by a factor of 3.

IRRADIATION EXPERIMENTS

In order to evaluate the effects of cosmic rays, the eggs have been irradiated by gamma rays and heavy ions.

1)—Cosmic background simulation experiments have been carried out with embedded and non-embedded eggs, of emergence percentages greater than 90%, with various embedding conditions (10° C or 30° C) and 48 hours to 8 months after PVA embedding. Eggs were irradiated with 950 mrad gamma rays delivered by a ^{60}Co source, for 11 days.

2)—Embedded eggs have been irradiated with carbon and oxygen ions (Berkeley Laboratory) 1.05 and 2.1 GeV/nucleon. Biological layers were sandwiched with nuclear emulsions, and particle tracks were determined using the same methods as those employed for Biostack. Histological studies have been performed after a 24 hour fixation with AFA fixative and paraffin embedding.

3)—The determination of the total energy loss in biological objects requires the knowledge of track lengths in eggs and of the specific energy loss. For a heavy ion having a known charge in a definite medium (water), an energy E and an energy loss per length unit $dE/dx = f(R)$ are associated with each residual range R by Barkas equations (1964).

On the other hand, microphotometric measures of calibrated heavy ion tracks (Berkeley accelerator) show that opacity varies as the energy loss/length unit

$$Op = G\,(dE)/dx$$

independently, in a first approximation, of considered charge.

Thus, energy loss by ionization induced in a living organism by cosmic heavy ion with an unknown energy and charge can be calculated as the product of trajectory ion length by ionization. The ionization itself was determined, according to the experimental curve in Fig. 2, from the measurement of ion track opacity in the nuclear emulsions.

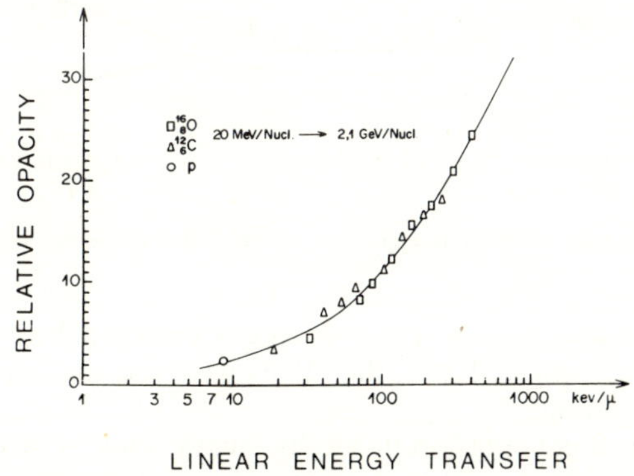

FIG. 2. Differential opacity as function of energy loss in K2 nuclear emulsion.

RESULTS

Biological Studies

The following results concerning the studies on egg developmental capacity after the Apollo XVI flight have been obtained (Fig. 3):

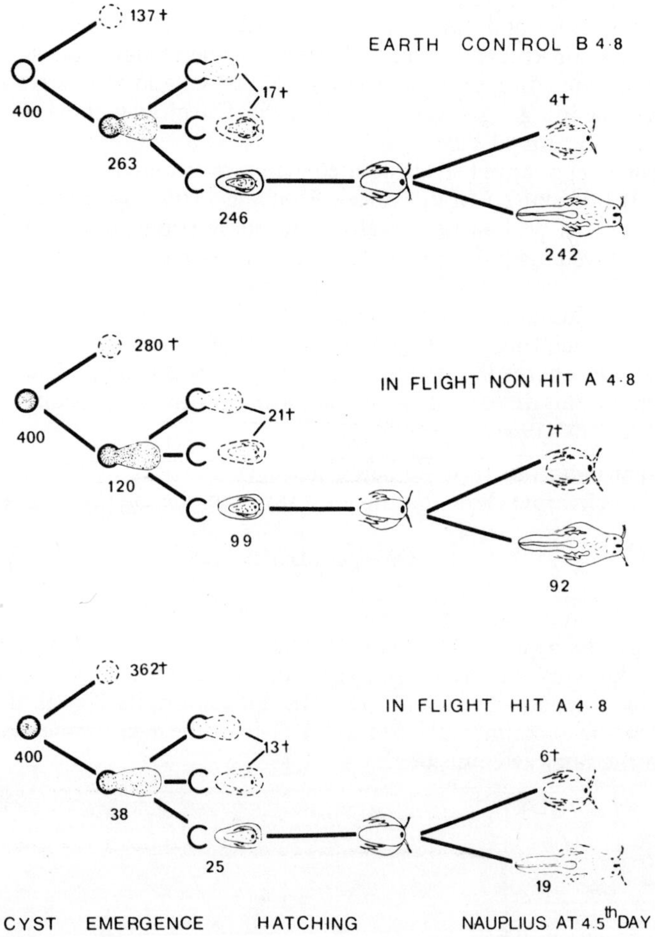

FIG. 3. Development of inflight eggs, hit or non-hit by cosmic heavy ions (Biostack I experiment).

—for 400 ground control eggs, 263 eggs emerged but 17 emerged eggs stopped their development. Later, the others gave rise to nauplii, 4 nauplii died 4–5 days after hatching;

—for 400 in-flight eggs, not hit by heavy ions, 120 emergences were observed. After emergences, 21 eggs died and 7 nauplii had disappeared 4–5 days thereafter;

—for 400 in-flight eggs, hit by heavy ions, 38 underwent a beginning of emergence. However, total emergence was obtained for 25 specimens only and 6 nauplii later died.

Different percentages of emerged eggs and of hatched eggs, and the number of alive nauplii on the 4th–5th day of experiment, have been plotted in Fig. 4.

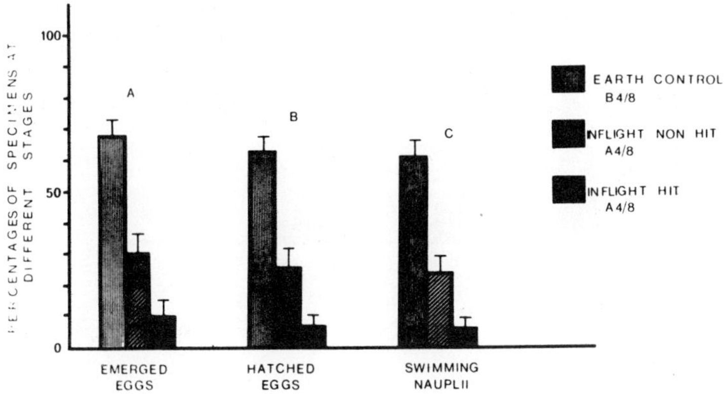

FIG. 4. Histograms comparing percentages of emergences, hatchings and alive nauplii in control and Biostack eggs.

Figure 5 shows the percentages of egg emergences from the 18th hour to the 5th day. On the 18th hour, 20% of control eggs had emerged while emergences were only starting for in-flight specimens.

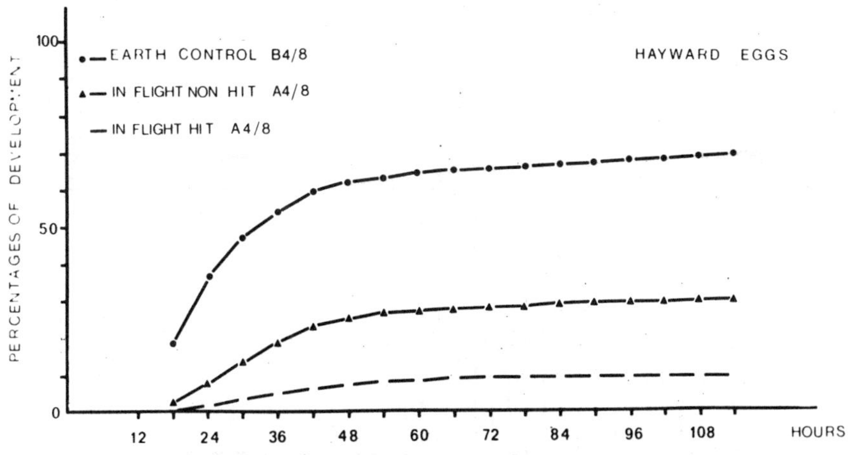

FIG. 5. Relation between emergence percentages and times in hours after rehydration.

Modifications in egg development percentages after the Apollo XVII flight (Biostack II) are summarized in Fig. 6.

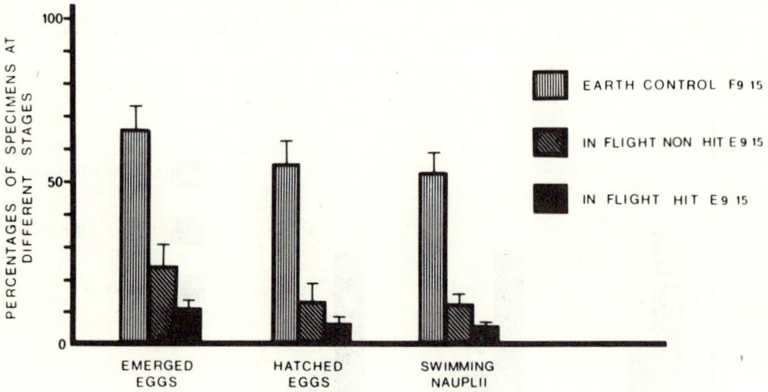

FIG. 6. Development ability of control and inflight eggs (Biostack II experiment).

Energy Loss Determination

Total energy loss by heavy ions in *Artemia* eggs was determined for 70 eggs by photometry, compared to well-known tracks and according to the track lengths. Most tracks cross a large portion of eggs; they are about 180 μm long (Fig. 7). The eggs, which can undergo either a beginning of emergence or a complete development, have received the lowest energy loss, less than about 50 MeV (Fig. 8).

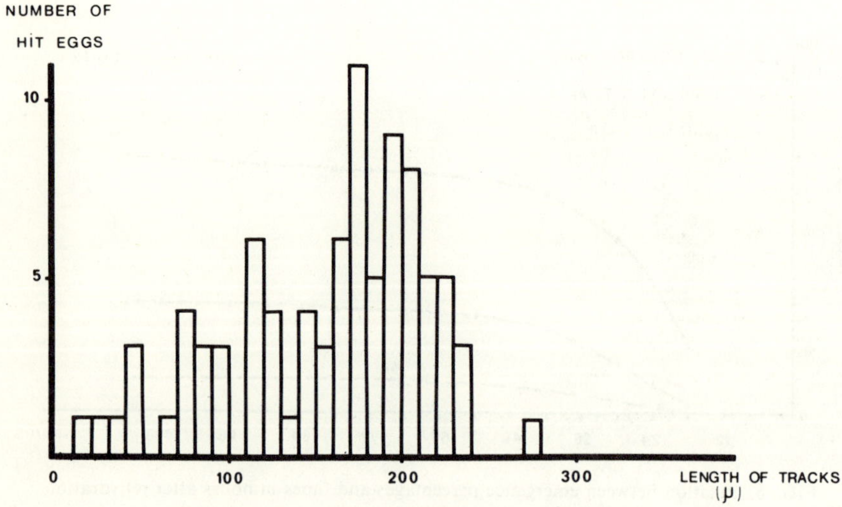

FIG. 7. Lengths of heavy ion tracks in *Artemia* eggs.

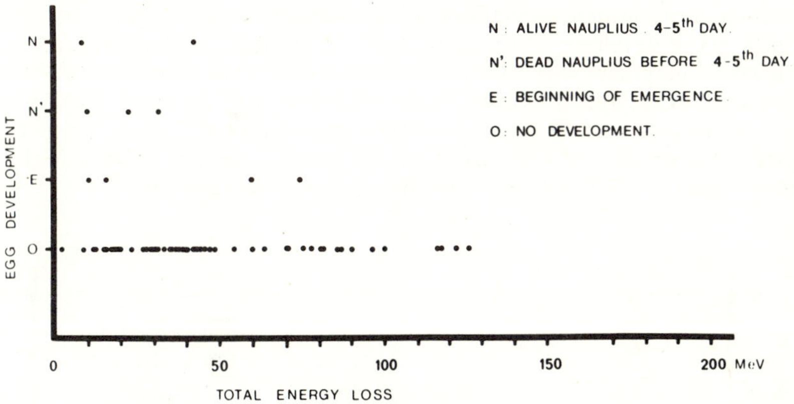

FIG. 8. Relation between total energy loss from cosmic heavy ions and the egg development capacity.

Simulation Experiments

1)—Embedded eggs were submitted to vibration and acceleration similar to, or three times more intense than, those that could occur in Apollo spacecraft. Acceleration does not result in egg development modifications. A very slight but statistically significant emergence decrease is observed after vibration tests or after combined vibration and acceleration experiments (Fig. 9).

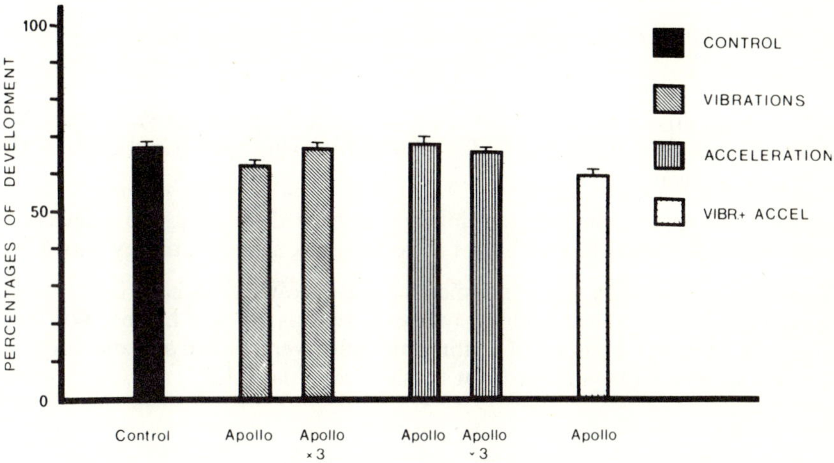

FIG. 9. Effects of acceleration and vibrations on *Artemia* egg emergence.

2)—Very low doses of ^{60}Co gamma radiations did not result in egg development modifications. Emergence percentages of different irradiated or non-irradiated lots ranged from 89.5 to 94%.

3)—In order to compare Biostack eggs to accelerator-irradiated ones, the relative percentages of development for eggs irradiated by heavy ions are plotted in Fig. 10. Each development study was carried out on eggs hit by a single track. Relative emergence percentages are, respectively, 31.4% for 1.05 GeV and 56.5% for 2.1 GeV carbon ions, and 49.3% for 1.05 GeV and 46.1% for 2.1 GeV oxygen ions. Relative percentages of live nauplii on the 4th/5th day are 8.7 and 33.3% for carbon ions and 23.9 and 18.5% for oxygen ions, respectively.

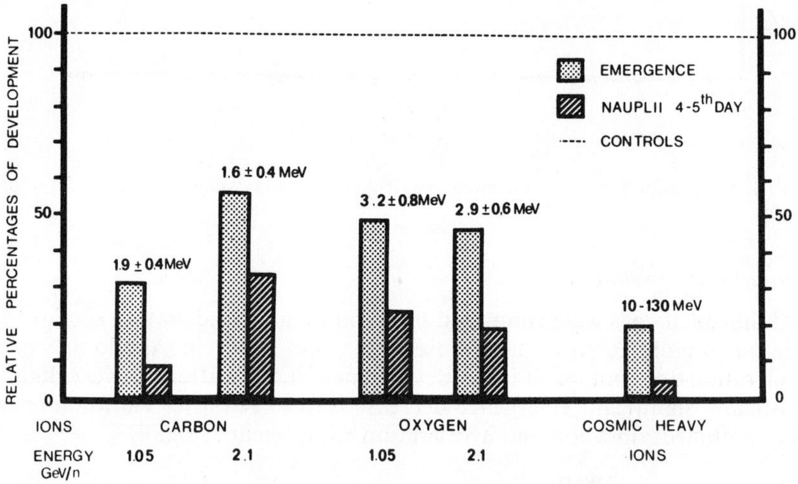

FIG. 10. Effects of *Artemia* egg irradiation by carbon, oxygen and cosmic heavy ions.

Total energy losses were 1.9±0.4 MeV (1.05 GeV carbon ions), 1.6±0.4 MeV (2.1 GeV carbon ions), 3.2±0.8 MeV (1.05 GeV oxygen ions) and 2.9±0.6 MeV (2.1 GeV oxygen ions).

Relative emergence percentage of in-flight eggs used for energy loss determination was 19.5%; relative number of alive nauplii on the 4–5th day was 4.7%.

A morphological study was performed on eggs irradiated with 2.1 GeV/nucleon oxygen ions. In eggs hit by one or two ions, serious damage has been observed (Fig. 11). A great number of embryonic cells were destroyed and large egg sections seemed to be stamped out by the heavy particles.

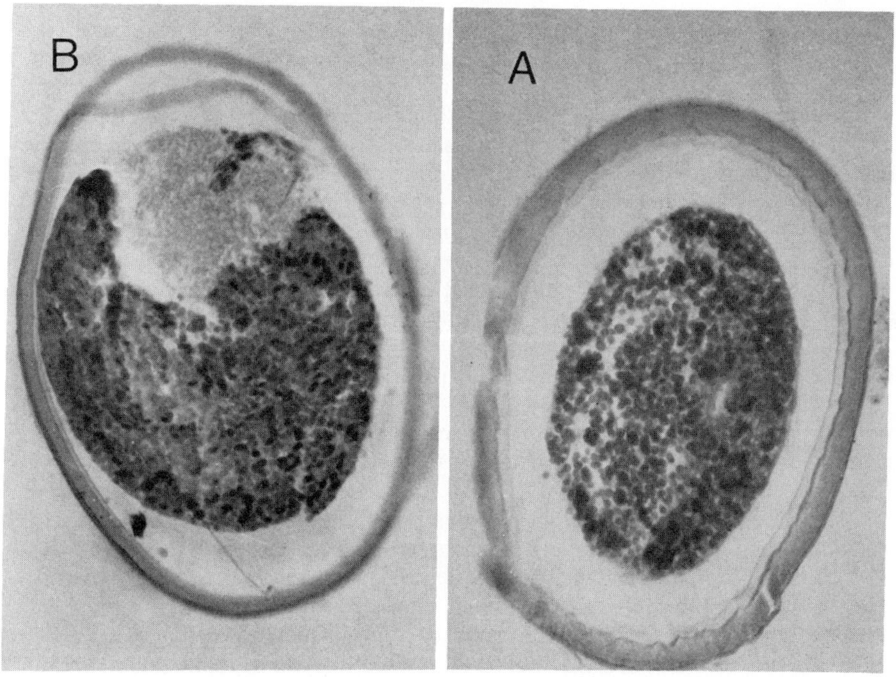

FIG. 11. *Artemia* egg A: normal egg. B: egg irradiated with 2.1 GeV oxygen ions.

DISCUSSION

Physiological or morphological changes observed in space biological experiments have often been ascribed to irradiation by cosmic rays. Indeed, it has been suggested that irradiation by heavy particles of cosmic rays may explain an increase in the mutation frequency in some species, as was observed with Drosophila in the Biosatellite II experiments (*5*).

When nuclear emulsions or photographic plates were used in balloon flight experiments (*6*) or in "Discoverer 32" (*7*), the lack of an accurate correlative method and the very low flux of heavy ions did not allow the evaluation of the biological effect of these particles. Yet, data here reported show cosmic heavy ion effects on living organisms and demonstrate their capacity to induce great developmental modifications in an embryonic organism. These results could explain the hazards related to cosmic irradiation in long duration space flights.

The extent of histological damage observed in simulation experiments with oxygen ions is correlated with the high LET of heavy particles. Their extent, compared to the track thickness, shows that the biological response is much more widespread than it could be assumed.

Indeed, if the biological effect of heavy particles may be related to energy deposition very close to the tracks (*8*), it is obvious that the production of secondary electrons (*9*) and free radical diffusion must be taken into account.

On the other hand, heavy ions are HZE particles and correlative methods have shown that the passage of a single ion may result in complete inhibition of egg development. A similar phenomenon has been described (*10*), but in these experiments several ions were necessary to inhibit an *Artemia* cyst development. Furthermore, when compared on the basis of equal energy absorption, heavy ions appear to be more effective than gamma or X-rays (*11*). From a technical point of view, the fact that the greatest inhibition occurs in eggs assumed—according to physical determinations—to be hit by cosmic heavy ions, demonstrate the value of the correlative technique we have used.

On the other hand, when developmental inhibition was also observed with carbon or oxygen ions (Berkeley accelerator), our experiments indicate that the inhibitory effect of heavy ions does not vary exactly with the energy loss: the relative emergence percentage of in-flight hit eggs, corresponding to a 10–130 MeV energy loss, is thus only about two times less than the emergence percentages of eggs irradiated with oxygen and carbon ions, i.e. for an energy loss ranging from 1.6 to 3.2 MeV. These results require new investigations using different kinds of heavy ions, with LET similar to those of cosmic ions. These investigations would be necessary to determine a possible synergetic effect of heavy ion irradiations and other environmental space factors.

Biostack experiments also show a decreased developmental ability in non-hit eggs. This result is not consistent with those of von Borstel et al. (*12*) who had reported a strong trend in the emergence of eggs which had flown in Biosatellite II. Differences in the hardware used and in the duration of flights may explain these variations in experimental results.

At the present time, the inhibition of non-hit eggs cannot be explained. Very slight variations in hatchability after vibration and acceleration tests suggest that this developmental inhibition would not be directly related to dynamic space flight factors. On the other hand, it seems unlikely that developmental inhibition might be induced by the cosmic background, i.e. by the cosmic component made up of protons.

Indeed, it is well known that dried *Artemia* cysts are highly radio-resistant (*13–14*) and that the total absorbed dose during the Apollo flights was less than 1 rad. Our experimental results obtained with ^{60}Co sources cannot demonstrate the part played by the cosmic background. Nevertheless, it has been established that space flight factors may exhibit a modifying influence on radiation effects (*15*). So, it could be suggested that developmental inhibition of non-hit eggs may be related to a synergetic effect of the cosmic background and of other space factors such as weightlessness. New experiments carried out on the ground and in space would be necessary to explain this phenomenon and to demonstrate a possible lethal effect of the cosmic background irradiation on some living organisms during a space flight.

REFERENCES

1. J. F. Saunders, in Observations from the United States Space Biology Mission. Proc. Colloquium Space Biol. related Post Apollo Progr.-ESRO-Ed., pp. 100–169 (1971).
2. Y. H. Nakanishi, T. Iwasaki, T. Okigari and H. Kato, *Annot. Zool. Jpn.* **35**, 223 (1962).
3. J. E. Morris, *J. Ultrastruct. Res.* **25**, 64 (1968).
4. W. H. Barkas, *Nuclear Research Emulsions*. Academic Press, N.Y. 1963.

5. L. S. Browning, in *The Experiments of Biosatellite II.* NASA SP 204, pp. 55–78 (1971).
6. J. Eugster, *J. Aviat. Med.* **24**, 222 (1953).
7. H. J. Curtis and H. H. Smith, *Science,* **141**, 158 (1963).
8. J. Fain, M. Monnin and M. Montret, *Radiat. Res.* **57**, 379 (1974).
9. R. Katz, B. Ackerson, M. Homayoonfar and S. C. Sharma, *Radiat. Res.* **47**, 402 (1971).
10. S. S. Easter and F. Hutchinson, *Radiat. Res.* **15,** 333 (1961).
11. R. A. Deering, F. Hutchinson and P. E. Schambra, *Aerosp. Med.* **32**, 915 (1961).
12. R. C. Von Borstel, R. H. Smith, A. R. Whiting and D. S. Grosch, *The Experiments of Biosatellite II.* NASA SP 204. pp. 17–40 (1971).
13. T. Iwasaki, *Int. J. Radiat. Biol.* **9**, 573 (1965).
14. P. Metalli, E. Ballardin, *Curr. Top. Radiat. Res. Q.* **7,** 181 (1972).
15. Y. G. Grigoriev, Y. Y. Kovalev. *Physical and Radiobiological Investigations on Artificial Earth Satellites.* NASA TT F. 724 (1971).

High-LET Particles in Manned Spaceflight—Preliminary Report[1]

E. V. Benton, R. P. Henke, D. D. Peterson
Physics Department, University of San Francisco
San Francisco, California 94117, USA

J.V. Bailey
NASA-Johnson Spacecraft Center
Houston, Texas 77058, USA

C. A. Tobias
Donner Laboratory, University of California
Berkeley, California 94720, USA

INTRODUCTION

During space missions, the radiation environment of the astronauts is quite complex. The primary galactic radiation consists primarily of protons and ^{4}He nuclei with less than 2% of the particles being high-Z, heavy nuclei. The heavy, high-Z nuclei consist primarily of nuclei of carbon (Z = 6) through iron (Z = 26) (*1*). However, recently the presence of small fluxes of superheavy nuclei such as lead (Z = 82) and uranium (Z = 92) has been experimentally established (*2*). The intensity and the energy spectra of these particles vary with time and position within the spacecraft. The long-term time variations come about due to the changes in the eleven-year solar cycle; the rapid short-term variations are due to geomagnetic effects and solar emissions, such as the solar flares. The radiation environment inside the spacecraft is also significantly influenced by the shielding offered by the various components of the spacecraft and, indeed, by the bodies of the astronauts themselves (*3*). For example, since the Apollo Command Module shielding varies from about 2 to 260 g/cm^2, depending on the direction, the incoming radiation is highly anisotropic.

Background low-LET (linear energy transfer) radiation is a part of everyday life; however, the high-LET radiation produced by the high-Z particles is unique to the space environment. While the radiobiological action of this radiation is poorly understood, particularly when it comes to the possible long-term, delayed effects, certain definite effects on biological systems have been observed. For example, it has been demonstrated that high-LET radiation is particularly effective in cell inactivation of certain non-regenerative cells [human kidney (T1) cells, for example] (*4,5*). A more dramatic effect of multi-charged particles has been the light-flash phenomenon observed by the Apollo and Skylab astronauts (*6-12*). It becomes important then to measure

[1] Congress presentation C-4-3.

the exposure of individuals to these radiations so that this information will be available as part of the medical history and subject to future interpretations.

The technique for the measurement of dose due to protons is well understood; however, the dosimetry of HZE particles is much more complicated. Indeed, in this case the whole concept of dose is open to question. The energy associated with heavy ions is concentrated within a limited volume of tissue, so that direct effects along the particle track are far more pronounced than for lighter particles even when the absorbed dose, measured macroscopically, is the same. A knowledge of how such particle tracks are distributed within a given biological volume now becomes a significant dosimetric parameter.

DOSIMETRIC PROCEDURES

On the Apollo and Skylab missions, the individual astronauts' HZE particle exposure was measured by means of plastic nuclear track detectors (*3, 13, 14*). On Apollo, the detectors were located in the passive dosimetry packs carried on the chest, thigh, and ankle of each astronaut. On Skylab, each astronaut wore a single passive dosimeter on either the wrist or the ankle. On Apollo missions 8-12, each plastic packet consisted of two 190-μm-thick layers of type 8070-112 Lexan. On Apollo 13 and subsequent missions, three layers were used. The addition of the third layer was found to improve significantly the charge resolution of the detector stack. Each detector has an area of about 8 cm^2. The layers are held stationary with respect to each other by means of a heat seal on each edge. In order to reduce the number of variables, all Lexan detectors flown to date have been taken from the same batch of Lexan. In the Skylab, in addition to Lexan several layers of cellulose nitrate (CN-USF 4b) were added. This allowed the measurement of lower-LET particles.

The track recording characteristics of CN and Lexan detectors are summarized in Table 1. Here E_{max} represents the maximum energy (in MeV/nucleon) of representative particles capable of registering a track in the given detector. The quantity R_{reg} represents the corresponding range of particles (*15*).

TABLE 1
E_{max} and R_{reg} for various ions in CN and Lexan

Particle Type	E_{max}[a] (MeV/nucl)		R_{reg} (cm)	
	CN	Lexan	CN	Lexan
^4He	1.5	0.2	0.004	0.0003
^{14}N	30	7	0.19	0.02
^{24}Mg	110	25	1.36	0.10
^{40}Ca	650	80	14.8	0.47
^{56}Fe	>1000	150	>20	1.18

[a] Corresponds to $LET_{350}^{(crit.)}$ in CN and Lexan of ~75 keV/μm·CN and ~230 keV/μm·Lex, respectively.

All detectors were processed in NaOH solution to produce conical particle tracks. Tracks are located by scanning with an optical microscope usually at 200× magnification. Each detector is scanned independently by at least two different observers. For LET determination, measurements are performed at 600× in each of the surfaces where a visible etch cone is found. Particle identification is accomplished either by measuring the particles' LET at a given residual range (for particles stopping in the detector), or by measuring the LET and its rate of change with the residual range. It is usually possible to determine the particles charge, energy, and its direction of travel.

For each particle which does not stop in the Lexan packet (the typical situation) the information available about this particle is the track etch rate as a function of position along the trajectory. Calibrating the detector packet consists of relating the etch rate to some function of particle atomic number, Z, atomic weight, A, and residual range, R. Here, rather than the energy, the residual range is used as an independent variable because it is more closely related to the measurement situation. Theoretical reasoning indicates that the function of interest is approximated by the particle linear energy transfer, LET_w.[2] It is found that LET_{350} represents the experimental situation well.

APOLLO MEASUREMENTS

From the HZE particle dosimetry point of view the most heavily instrumented Apollo mission was that of Apollo 17. In addition to the personnel passive dosimeters, five other biologically related experiments were instrumented to contain plastic nuclear track detectors. The five experiments included the HZE dosimeter, the Biostack, the Alfmed, and the Biocore. In addition to the Lexan, the dosimetry for these experiments also included several layers of cellulose nitrate. The detailed results from the Lexan detectors contained in the passive personnel dosimeters of Apollo 17 appear elsewhere (*16*); the results from the CN detectors are shown in Table 2. The approximate average shielding of each experiment is listed underneath the experiment name in column one. In columns two, three, and four appears information on the detector layer number and the measured experimental data. In columns five and six are shown the absolute planar particle track fluences. In the last column are listed the absolute, planar particle fluxes for particles with $LET_{350}^{CN} \geqslant 80$ keV/μm which corresponds to an $LET_{\infty}^{Tissue} \geqslant 105$ kev/μm. The particle flux is in units of tracks/cm^2-day where the time is taken as the effective time for the mission. This is defined as the trans-earth plus trans-lunar time plus one-half of the total time spent in lunar orbit.

The data show clearly the influence of shielding on the HZE exposure with the lightly shielded HZE dosimeter recording nearly four times the flux recorded with the Biocore detectors. At the same time it is clear that even for a very heavily shielded situation, such as the Biocore (30-40 gm/cm^2 is a conservative estimate), there are still a significant number of HZE particle hits. Consequently, complete shielding from the galactic cosmic rays does not appear practical due to spacecraft weight limitations.

[2]The response of plastic nuclear track detectors more closely follows the LET_{350} rather than LET_{∞} of particles. This is why the quantity LET_{350} is used throughout this work. Over a wide range of particle velocities LET_{350} can be converted to LET_{∞} by multiplying by a factor of ~1.6.

TABLE 2
HZE particle track fluences on Apollo 17[a]

Experiment	Detector layer number	Measured: 2 cones	Measured: Etched through holes	Absolute planar track fluence (tr/cm²): $LET_{350}^{CN} >$ 170 keV/μm[b]	Absolute planar track fluence (tr/cm²): $LET_{350}^{CN} >$ 80.0 keV/μm[c]	Flux[d] (tr/cm²-day): $LET_{350}^{CN} >$ 80 keV/μm
HZE dosimeter (2 g/cm²)	C5	108.5	30.8	26.1 ± 3.9	153 ± 23	16.4
Biostack (10 g/cm²)	E2–1	60.7	19.4	16.2 ± 2.3	95 ± 14	10.2
	E13–25	55.0	18.0	13.4 ± 1.8	78 ± 11	8.4
ALFMED (~20 g/cm²)	bottom sheet	26.6	14.7	11.7 ± 1.5	69 ± 9	7.4
Biocore[e] (30–40 g/cm²)	implanted mouse brain dosimeters	31.0	8.7	7.0 ± 1.6	41 ± 9	4.4

[a] All measurements are from cellulose nitrate of the type USF 4b processed for 10.0 hr at 40°C in 6.25N NaOH.
[b] Corresponds to an $LET_{350}^{Lexan} >$ 150 keV/μm in Lexan or $LET_{\infty}^{Tissue} >$ 225 keV/μm.
[c] Corresponds to an $LET_{\infty}^{Tissue} \geqslant$ 105 keV/μm.
[d] Flux is in effective days in interplanetary space; mission duration 301.5 hr or 9.33 effective days.
[e] The Biocore numbers are average for the four live mice.

The data obtained from the Lexan detectors contained in the Apollo passive personnel dosimeters are shown in Fig. 1 (*11*); all data represent the average for each mission. In Fig. 1 is shown the particle track fluence for particles with $LET_{350}^{Lexan} > 150$ keV/μm as a function of mission number. A significant increase in the fluence was observed starting with Apollo 14. Some of the increase is due simply to the increase in mission time. The absolute planar particle track flux as a function of mission date for the nine lunar missions is also shown in Fig. 1. Here the time is the effective mission time as defined previously. The flux as well as the fluence is observed to increase with the Apollo 17 flux being a factor of ~2.4 higher as compared with the Apollo 8. This increase correlates well with solar modulation (*17*). Unfortunately, Apollo 17 occurred well before the next solar minimum (during which the cosmic-ray flux becomes maximum) which will occur sometime in 1975-1976.

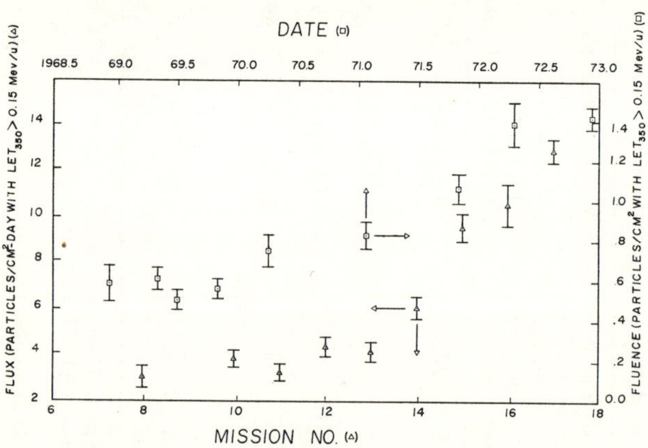

FIG. 1. The average measured particle fluence as a function of Apollo mission number (squares) and the average measured Apollo particle flux as a function of mission date (triangles).

SKYLAB D-008 HZE MEASUREMENTS

In addition to the personnel HZE dosimetry, which was performed on all three Skylab missions, a set of plastic nuclear track detectors was included in the D-008 experiment (*14*). This experiment consisted of five individual dosimeters distributed throughout the volume of the Command Module. The D-008 experiment was flown only on the SL2 mission with the dosimeters remaining in the Command Module throughout the 28-day mission.

The measured Skylab D-008 HZE particle planar fluence and flux are listed in Table 3. Here, the first column designates the dosimeter, the second the approximate shielding. The third and fourth columns contain the particle fluence data with the planar fluence being corrected for detector efficiency. The planar flux is shown in the last column and represents the fluence divided by the total mission time. The data clearly show the effect of shielding on the particle flux with the flux changing from a high of 3.33 to a low of 0.75 particles/cm^2-day. The average planar flux for the five dosimeters for particles with $LET_{350}^{CN} \geqslant 80$ keV/μm ($LET_{\infty}^{Tissue} \geqslant 105$ keV/μm) is 1.90 ± 0.30 particles cm^2-day.

TABLE 3
HZE particle planar fluence and planar flux in the Skylab (D-008-SL2) Command Module [a]

Dosimeter (layer 11)	Approximate shielding (g/cm^2 of Al)	Fluence (particles/cm^2)		Planar flux[b] (particles/cm^2-day)
		observed	planar[b]	
F1	~10	58.2 ± 7.6	93.8 ± 9.7	3.33 ± 0.38
F2	>20	35.1 ± 5.9	27.5 ± 5.2	1.01 ± 0.19
F3	>20	26.1 ± 5.1	21.1 ± 4.6	0.75 ± 0.19
F4	~10	53.4 ± 7.3	85.9 ± 9.3	3.08 ± 0.31
F5	~20	52.8 ± 7.3	37.0 ± 6.1	1.32 ± 0.19

[a] The data represents measurements performed with type USF 4b cellulose nitrate.
[b] Corresponds to particles with $LET_{350}^{CN} \geqslant 80$ keV/μm which is equivalent to $LET_{\infty}^{Tissue} \geqslant 105$ keV/μm.

COMPARISON OF SKYLAB AND APOLLO HZE DOSIMETRY

The HZE particle environment of the Skylab missions differed significantly from that of the Apollo. Being an earth orbital mission, the Skylab was shielded by the physical presence of the earth as well as the geomagnetic field. Both of these factors considerably attenuate the HZE particle flux.

In Table 4 is shown the comparison between the Apollo 17 and Skylab (D-008) measurements. Since the Skylab utilized a slightly modified version of an Apollo Command Module, the comparison between the two is meaningful. The Apollo flux of particles with $LET_{350}^{CN} \geqslant 80$ keV/μm is a factor of nearly 5 times higher than observed for the Skylab. A similar ratio for particles with $LET_{350}^{CN} \geqslant 170$ keV/μm is about 12. As stated previously, the differences arise from earth shielding and the geomagnetic effect; however, there may also be a solar modulation factor. The differences in the ratios for the different LET's also imply differences in the integral LET particle spectra for the two kinds of missions. The integral LET particle spectrum is definitely steeper in the case of the Skylab. This reflects the lower incidence of the highest LET particles which are filtered out, because of their generally lower energy, by the geomagnetic field.

TABLE 4
Comparison of the average HZE particle planar flux in Skylab (SL2) and Apollo 17

Mission	Average HZE particle planar flux density (cm^{-2}·day^{-1})	
	($LET_{350}^{CN} \geqslant 170$ keV/μm)[a]	($LET_{350}^{CN} \geqslant 80$ keV/μm)[b]
Skylab (SL2) (Command Module)	0.13 ± 0.04	1.90 ± 0.30
Apollo 17	1.58 ± 0.22	9.27 ± 1.33

[a] Corresponds to $LET_{350}^{Lex} \gtrsim 150$ keV/μm; $LET_{\infty}^{Tissue} \gtrsim 225$ keV/μm.
[b] Corresponds to $LET_{350}^{Lex} \gtrsim 70$ keV/μm; $LET_{\infty}^{Tissue} \gtrsim 105$ keV/μm.

SUMMARY

Several important results and observations can be realized from the HZE dosimetry. Because of the relatively close orbit of Skylab and the presence of the earth's geomagnetic field, the Apollo flux of high-LET, HZE particles is considerably higher than that of Skylab. However, a comparison of track fluences measured on later Apollo missions with those of earlier missions indicates an increase in the HZE cosmic-ray flux partially due to a decrease in the degree of solar modulation.

In addition to the variations with regard to time or the nature of the flight, considerable variation has been found to exist in recorded high-Z fluences as a function of detector location on the astronaut's body and in the spacecraft, caused by variations in the amount of shielding. The shielding also shifts the measured Z spectrum toward the lighter particles even though the plastic track detectors are heavily biased toward the higher-Z particles. Fewer $Z = 26$ (Fe) particles are observed than expected as compared with lighter particles. This suggests a breakup of Fe particles in passage through the spacecraft shielding.

From the radiobiological point of view, it is important to note that only the fraction of the $Z \geqslant 6$ particle flux that corresponds to essentially stopping particles was measured. These are the highest LET particles, $\mathrm{LET}_{\infty}^{\mathrm{Tissue}} \gtrsim 100$ keV/μm (and probably the most biologically significant) that either stop in the detector or the astronaut. From the Z spectrum measurements, stopping particle densities in the form of integral Z distributions can be obtained and integral LET spectra can be measured and used in estimating biologically meaningful quantities. Finally, the frequency of the light flash phenomenon reported by the astronauts is considerably higher than the frequency with which the $\mathrm{LET}_{\infty}^{\mathrm{Tissue}} \geqslant 100$ keV/μm particles are incident on tissue as recorded with plastic detectors. This implies that lower LET particles are at least partially responsible.

ACKNOWLEDGMENT

The authors would like to thank John McGinnis for help in preparation of the manuscript.

REFERENCES

1. D. Grahn, *National Academy of Sciences Report* (1973).
2. P. H. Fowler, *Proc. 13th International Cosmic Ray Conference, Denver*, **5**, 3627 (1973).
3. E. V. Benton and R. P. Henke, *Air Force Weapons Laboratory Report,* AFWL-TR-72-5 (1972).
4. P. Todd, *Radiat. Res. Suppl.* **7**, 196 (1967).
5. *University of California Radiation Laboratory Report* UCRL-17357; *Lawrence Berkeley Laboratory Reports*, LBL-529 (1971) and LBL-2016 (1973).
6. C. A. Tobias, *University of California Radiation Laboratory Report*, UCRL-8915 (1958).
7. G. G. Fazio, J. V. Jelley, and W. M. Charman, *Nature (London)* **228**, 260 (1970).
8. C. A. Tobias, T. F. Budinger, and J. T. Lyman, *Nature (London)* **230**, 596 (1971).
9. T. G. Budinger, C. A. Tobias, J. T. Lyman, P. K. Chapman, L. S. Pinsky, H. Bichsel, J. D. Denney, and W. B. Nelp, *Lawrence Berkeley Laboratory Report*, LBL-31 (1971).
10. T. F. Budinger, J. T. Lyman, and C. A. Tobias, *Nature (London)* **239**, 209 (1972).
11. P. J. McNulty, V. P. Pease, L. S. Pinsky, V. P. Bond, W. Schimmerling, and K. G. Vosburgh, *Science* **178**, 160 (1972).

12. L. S. Pinsky, W. Z. Osborne, J. V. Bailey, R. E. Benson, and L. F. Thompson, *Science* **183**, 957 (1974).
13. E. V. Benton and R. P. Henke, *University of San Francisco Reports*, USF-TR-71-15 (1971). USF-TR-72-20 (1972), USF-TR-73-26 (1973).
14. D. D. Peterson and E. V. Benton, *University of San Francisco Report*, USF-TR-73-29 (1973).
15. E. V. Benton, *U.S. Naval Research and Development Laboratory Report*, USNRDL-TR-68-14 (1968).
16. E. V. Benton and R. P. Henke, *Health Physics* (1974) (in press).
17. E. V. Benton and R. P. Henke, *University of San Francisco Report*, USF-TR-74-30 (1974).

Some Studies on Visual Perception and Pathologic Effects of Accelerated Heavy Ions[1]

Cornelius A. Tobias, Thomas F. Budinger, John T. Leith,
John T. Lyman, Abdel M. Mamoon, and T. C. H. Yang
Donner Laboratory/Lawrence Berkeley Laboratory
Berkeley, California 94720, USA

Nonrelativistic, individual accelerated particles, including helium, nitrogen, oxygen, and neon ions, as well as secondaries from fast neutrons, produce light sensations of stars and streaks in the dark-adapted human eye. The particles usually do not leave a visible track if their rate of energy loss is less than 10 keV per micron. Many similar events have been seen by astronauts on American lunar and orbital flights. X rays down to a dose of 90 microrads or less also produce light flashes, but these are diffuse light phenomena different in character from events seen by astronauts. The efficiency of perception of heavy-ion-initiated events is related to the physiological state of the individual, such as dark adaptation, and also to the particle energy deposited in and near the retina.

The pathologic effects of accelerated nitrogen and oxygen ions on the rabbit and rat retina appear earlier and are more severe than those produced by comparable X-ray doses. These include degeneration of the outer segments of visual rods and other nerve tissues, as revealed by scanning electron microscopy. Degeneration of the rods appears to be related to derangement of protein synthesis in the inner segment as studied by autoradiography. Accelerated heavy ions have an immediate effect on membranous structures of neurons in culture, which is evidenced by osmotic imbalance and swelling of the cells. Heavy-ion injury to differentiated neurons manifests itself in reduction of life-span and eventual lysis. We are searching for the sites of primary injury and the energy thresholds for irreversible central nervous system cell damage. (This work is supported by the U.S. Atomic Energy Commission and NASA.)

[1]Congress presentation C–4–4.

Biological Effects of Accelerated Boron, Carbon, and Neon Ions[1]

Yu. G. Grigoryev, N. I. Ryzhov, V. I. Popov,
V. N. Benevolensky, B. S. Fedorenko, S. V. Vorozhtsova,
R. D. Govorun, E. A. Krasavin, E. I. Kudryashov,
N. Ya. Savchenko, V. N. Gerasimenko, D. Ya. Oparina,
L. K. Vekshina, L. A. Koshcheyeva
Institute of Biomedical Problems, USSR
Ministry of Health, Moscow, USSR

INTRODUCTION

Investigations of the radiation environment in space during the last two decades have shown that manned space flights may be hazardous from the point of view of ionizing radiation. Calculations have demonstrated that during long-term space flights, the main radiation hazard is due to heavy nuclei of galactic space radiation (GSR). It is necessary, therefore, to obtain comprehensive information about possible injuries to both men and life support systems by GSR heavy ions during prolonged space missions.

At present the therapeutic use of heavy ions for oncological treatment is intensively studied. Of no less significance for general radiation biology are investigations of the biological effects of heavy multicharged particles of high energies.

It is, therefore, very important to discuss at length the results obtained to date, in view of the fact that the pertinent studies are few and the possibilities to carry out the experiments in space or with accelerators are limited.

MATERIALS AND METHODS

Mammalian cells and tissues with different levels of proliferative activity were used: corneal epithelium of mice, liver of young and adult rats, and human lymphocytes, as well as yeast cells, unicellular algae (*Chlorella*), and hydrogen bacteria. The damaging effect of heavy ions was determined on the basis of quantitiative comparisons of cell division inactivation, cytologic and cytogenetic effects. Particular attention was paid to studying the postradiation recovery and also the manifestation of biological effects of heavy ions under the influence of modifying agents.

Biological objects were exposed to heavy ions from the Accelerator of Multicharged Ions (AMI-300) in the Laboratory of Nuclear Reactions, the Joint Nuclear Research Institute, Dubna, USSR. The following ions have been used: ^{10}B, ^{11}B, ^{12}C, ^{20}Ne, ^{22}Ne and ^{40}Ar with energies of 7.5–3.5 MeV/nucleon. The dose rates for the irradiation of mammalian cells and tissues ranged from 1 to 10

[1]Congress presentation C–4–5 (Not delivered).

rads/sec; and that for the irradiation of yeast cells, *Chlorella*, and hydrogen bacteria ranged from 100 to 1000 rad/sec. Gamma rays of ^{60}Co, ^{137}Cs and 180–200 kVp X-rays were used as reference. The exposure conditions with the reference radiations were the same as with the heavy ions.

RESULTS AND DISCUSSION

Experiments on hydrogen bacteria

A culture of hydrogen bacteria *Hydrogenomonas eutropha* Z-1, grown under autotrophic or hetertrophic conditions, was used. Cells were irradiated on the surface of potato-glycerol agar or in the Schlegel liquid medium. Figure 1 shows the survival of bacteria–grown under autotrophic conditions before the irradiation–after exposure to gamma-rays or to accelerated boron and carbon ions. With all radiations used, the survival-dose dependence was exponential. This points to a one-hit mechanism of the inactivation process in the bacterial cells. As compared to ^{60}Co gamma-rays, the RBE value for heavy ions decreased, this being in accordance with the hit principle and the small target dimensions in bacterial cells (Table 1).

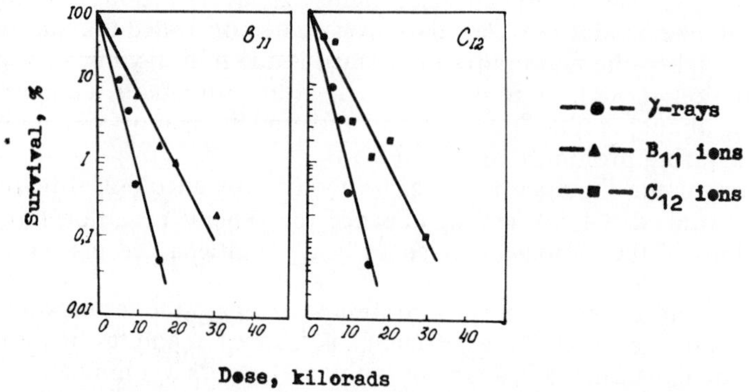

FIG. 1. Survival of hydrogen bacteria grown under autotrophic conditions after exposure to accelerated boron and carbon ions.

TABLE 1
RBE Values for Exposure of Hydrogen Bacteria to Multicharged Ions

Type of radiation	LET (MeV·cm^2/g)	Exposure conditions	Factor $1/D_{37}$ (10^{-4} rad^{-1})	P	RBE
γ-rays	3.0	On the PGA[a] surface	4.76 ± 0.16	—	1.0
B_{11}	1500	"	2.44 ± 0.11	0.001	0.5±0.1
C_{12}	3500	"	2.50 ± 0.11	0.001	0.5±0.1

[a]PGA – potato-glycerol agar

The modifying action of anoxia and radioprotectors—glycerol and inositol—on the survival of bacterial cells irradiated with gamma rays, 645 MeV protons or accelerated heavy ions was investigated. When exposed to gamma-rays or protons in a nitrogen or argon atmosphere, hydrogen bacteria showed significantly decreased radiosensitivity. However, with accelerated boron ions the oxygen effect diminished markedly (Table 2). Glycerol and inositol were used in concentrations of 1 *M* and 0.28 *M*, respectively. Glycerol and inositol exhibited a fairly good protection of hydrogen bacteria against the inactivating action of gamma-radiation. However, the protective effect decreased when the cells were exposed to densely ionizing radiations.

TABLE 2
Survival and Oxygen Enhancement Ratios for Various Radiation Types and Exposure Conditions

Type of radiation	Exposure conditions	Atmosphere	Factor $1/D_{37}$ (10^{-4} rad^{-1})	P	Oxygen enhancement ratio (OER)
γ-rays	PGA surface	air nitrogen	3.84 ± 0.13 1.45 ± 0.03	0.001	2.6 ± 0.2
γ-rays	Schlegel medium	air argon	5.16 ± 0.16 1.66 ± 0.05	0.001	3.1 ± 0.2
Protons	Schlegel medium	air argon	5.26 ± 0.24 1.05 ± 0.05	0.001	4.0 ± 0.4
^{11}B	PGA surface	air nitrogen	2.44 ± 0.11 1.82 ± 0.08	0.001	1.3 ± 0.1

Investigation of the biological effects of heavy ions on yeast cells

In the studies of the biological effects of heavy ions on yeast cells, attention was concentrated on the distribution of different forms of inhibition of cell division and the kinetics of the recovery that takes place during the first postirradiation mitotic delay. These studies were performed with diploid cells of *Saccharomyces ellipsoides*, strain Megri 139-B. The survival of cells and the relative number of inactivation forms were determined by the microcolony technique. The recovery curves were plotted with respect to a regular count of the number of surviving and inactivated cells in cultures incubated after exposure in a non-nutrient medium at 30°C.

The distributions of inactivation forms produced by exposure to both sparsely and densely ionizing radiations were studied with equi-effective doses of both types of radiation. Fig. 2 shows the results of the comparison performed during an exposure of cells to ^{60}Co γ-rays or ^{12}C ions. Fig. 2 presents the percentage of microcolonies with cell numbers of 1 to 4 (M_2-class) vs. the percentage of colonies with 16 cells (S-class). The number of S-class microcolonies is used as an indicator of cell survival. As can be seen, an exposure to heavy ions gives rise to a greater amount of M_2-class microcolonies than does an exposure to γ-irradiation. Hence, an exposure to densely ionizing heavy ions brings about a more severe injury in terms of inactivated cells; this is the difference between the response to densely and sparsely ionizing radiations.

The studies of intracellular recovery of yeast cells exposed to ^{12}C ions are summarized in Fig. 3. As suggested by the analysis of inactivation forms, the recovery process after heavy ion and gamma-irradiation follows a course of

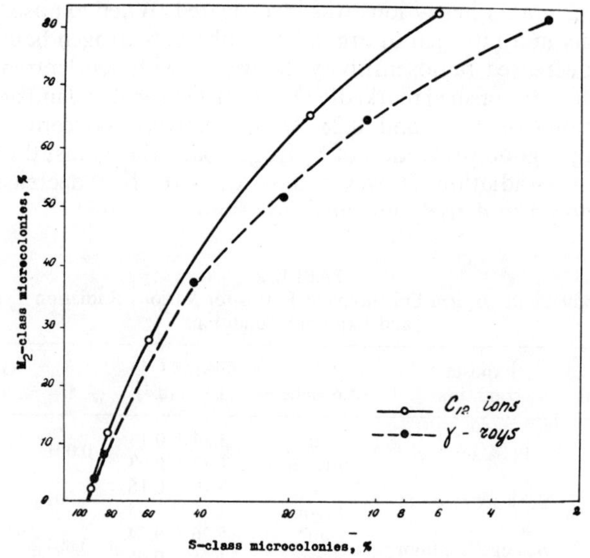

FIG. 2. Inhibition of cell division in yeast cells exposed to γ-rays of ^{60}Co or ^{12}C ions.

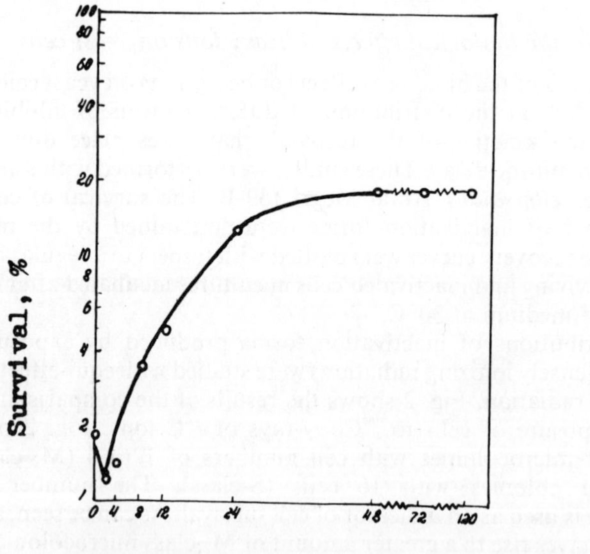

FIG. 3. Postirradiation recovery in yeast cells exposed to 60 kilorads of ^{12}C ions.

gradual elimination of primary elementary injury and is described by the well-known formula:

$$D_{ef} = D_o \cdot k + (1 - k)e^{-\beta t} .$$

A comparison of the k and β values for cells irradiated with gamma rays and ^{12}C ions has shown that the recovery rate is lower and the amount of irreversible damages is higher in the case of ^{13}C ions. These differences may be related to the different initial lesion of the reproductive apparatus of the cell or to the different degree of damage to the enzymatic systems responsible for intracellular recovery.

Experiments on Chlorella

In these experiments cells of *Chlorella vulgaris* Beijor, strain LARG-1,in synchronous cultures consisting of 95–97% of young autospores in the G_1 phase, were used. The survival of cells was evaluated by the microcolony technique after a three-day cultivation on an agarized mineral medium in a luminostat. The *Chlorella* cells that contributed to the final count had completed the second division and the subsequent sporulations were classified as survival. All other forms of microcolonies were considered as inactivated. Simultaneously the spectrum and level of visible morphological mutations were determined in the culture. Irradiation of the cultures was performed on the surface of an agarized medium.

The studies have shown that the survival curves, found to be sigmoidal after gamma-irradiation, become exponential after heavy ion irradiation. With respect to the lethality index, the RBE values of various ions varied significantly, increasing, as a rule, with decreasing dose. The maximal value of RBE was observed with an exposure to carbon ions (Fig. 4).

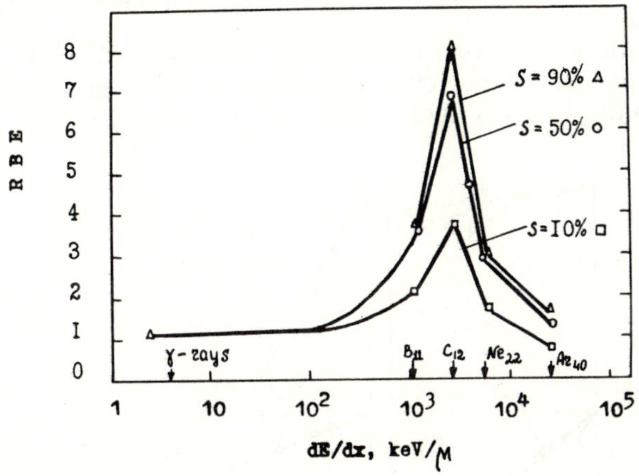

FIG. 4. RBE values vs. dE/dx of various radiation types for different levels of survival (S) in *Chlorella* cells.

It was also found that the inactivation cross-section (δ) for *Chlorella* cells increased linearly with an increase of dE/dx up to the value of about 1000 MeV·cm^2/g. At higher values of specific energy losses δ increased more rapidly and then attained a constant level, forming a plateau when the dE/dx value was greater than 3000 MeV·cm^2/g (Fig. 5).

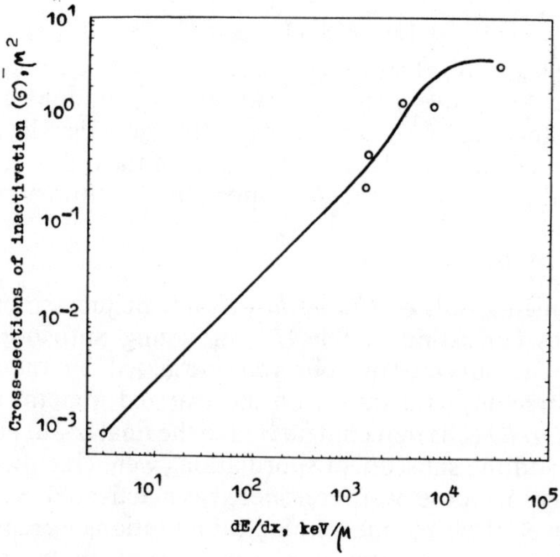

FIG. 5. Cross-sections of inactivation in *Chlorella* cells vs the dE/dx values of the radiations.

A change in the dose rate by two orders of magnitude had a slight influence on the RBE values. In *Chlorella* cells, as well as in yeasts, heavy ion irradiation led to an earlier inactivation of cell division. Due to this, the relative number of microcolonies composed of few cells, including those which underwent no division (the interphase form of death), was higher than that in gamma-irradiated cells (Fig. 6).

The genetic effect of heavy ions was the greatest with LET = 2850 MeV · cm^2/g; in this case the dose-effect curves had a maximum, whereas with gamma irradiation they were of linear shape.

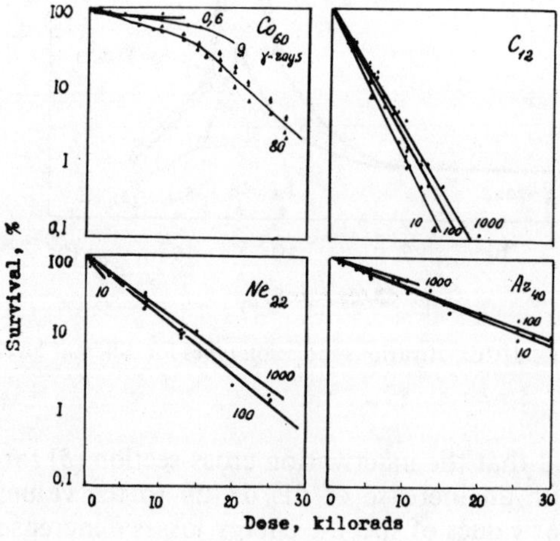

FIG. 6. Survival of *Chlorella* cells exposed to various radiation types, the doses being delivered at dose rates ranging from 0.6 to 1000 rad/sec.

Radiobiological effects in human lymphocytes

Disturbances of the chromosome apparatus in human cells exposed to heavy ions and gamma rays were examined and compared. The frequency and types of chromosome aberrations in human blood lymphocytes were studied. Leukocytes suspended in blood plasma, obtained from fresh donor blood, were irradiated with 50 rad of boron-10, carbon-12 or neon-22 ions.

The results demonstrated that exposure of leukocytes to heavy ions and gamma rays damaged substantially the chromosome apparatus of cells: this included a significant increase in the number of cells with chromosome aberrations and a change in the ratio of the various types of aberrations (Table 3). After exposure to heavy ions the content of aberrant cells was 3.6–5.3 times, and the number of aberrations 4.0–5.3 times as high as the spontaneous level. With

TABLE 3.
Number of Chromosome Aberrations in Human Lymphocytes After an Exposure to 50 Rad of HZE Particles or γ-rays

Type of radiation	Cell count	Aberrant cells, %	Chromosome aberrations, %	One-break aberrations, %		Two-break aberrations, %		Chromosome breakages, %
				total	chromosome fragments	total	dicentrics + rings	
Controls	915	2.8	3.3	3.3	2.2	0	0	3.3
γ-rays	200	9.0	10.0	7.5	5.0	2.5	2.5	12.5
Ions of:								
Boron	510	13.3	17.5	15.9	11.6	1.6	1.4	19.1
Carbon	500	10.0	12.8	11.2	6.2	1.6	0.8	14.4
Neon	500	14.8	17.2	16.8	7.0	1.1	0.4	17.6

gamma irradiation the ratio of aberrant cells and chromosome aberrations increased 3-fold. Irradiation of cells also resulted in a change in the ratio of the various types of chromosome aberrations. The yield of one-break aberrations with heavy ions was 3.5–5.0 times higher than the spontaneous level, whereas with gamma radiation it was only twice as large. It was also accompanied by a high yield of chromatid aberrations, which was 4, 5 and 9 times higher with boron, carbon and neon ions, respectively, than the control values.

In a non-irradiated culture and in a gamma or boron ion irradiated culture the one-break aberrations were mainly paired acentric fragments; after an exposure to carbon ions the numbers of chromatid and chromosome aberrations were equal and after an exposure to neon ions chromatid deletions prevailed. This was never observed with X- and γ-rays or with high energy protons. An irradiation with heavy ions was followed by a distinct decrease in the yield of every type of exchange chromosome aberrations. These manifest differences in the formation of various types of chromosome aberrations due to irradiation with gamma-rays and heavy ions brought about a sharp shift in the ratio of the one- and two-break aberrations. As compared to γ-rays, for which this ratio was 3, it was 3 times higher for boron and carbon ions and 10 times higher for neon ions.

A comparison of the yields of genetic abnormalities due to gamma and heavy ion irradiations indicated that—as suggested by the total number of chromosome aberrations, the number of breakages, and the number of aberrant cells per unit dose—the biological effectiveness of boron and neon ions was 1.4–1.5 times, and that of carbon ions 1.2 times as great as the effectiveness of gamma rays. (Table 4). This ratio varied slightly for individual aberration types. The effectiveness of boron, carbon, and especially neon ions was significantly lowered with respect to the yield of dicentrics and rings and the total yield of two-break aberrations, and was substantially increased in regards to the yield of chromatid fragments.

TABLE 4
Effectiveness of Accelerated Multicharged Ions as Compared to Gamma Radiation

Criteria	Radiation types		
	Boron ions	Carbon ions	Neon ions
Aberrant cells	1.4	1.1	1.4
Total number of chromosome aberrations	1.7	1.3	1.5
Chromosome breakages	1.5	1.2	1.2
One-break aberrations:			
Total number	2.1	1.5	2.0
Chromosome fragments	2.3	1.2	1.1
Chromatid fragments	1.8	2.0	4.0
Two-break aberrations:			
Total number	0.6	0.6	0.2
Dicentrics and rings	0.6	0.3	0.2

Effect on the liver tissue

The damaging effect of multicharged boron and carbon ions and of X-rays on the liver tissue of rats irradiated with doses of 50–3000 rad was studied. White non-inbred male rats weighing 180–200 g were used. The left lobe of the liver of etherized animals, withdrawn onto the anterior abdominal surface, was irradiated. Fifteen and 30 days later the animals were sacrificed for a pathological study of the liver cells. The results obtained showed that irradiation with heavy ions at doses of 50–200 rad did not reduce the number of liver cells but decreased significantly the diameter of the cell nuclei (Table 5). An exposure of the liver to boron-10 and carbon-12 ions with a dose of 300 rad led to the death of 40–50% of hepatocytes. Similar changes were induced by an exposure to 1000 rad of boron-10 ions, whereas X-irradiation with doses of 300 and 1000 rad caused no significant decrease in the number of liver cells. X-irradiation with a dose of 3000 rad decreased the number of hepatocytes. The decline in the number of hepatocytes in irradiated animals was followed by degenerative changes of the liver. In the early period of observations necrotic and necrobiotic areas, that were later replaced by the connective tissue, were detected. The dose dependence of these pathological changes was revealed. An exposure to heavy ions at a dose of 50–200 rad caused only protein dystrophy. Because the irradiation with heavy particles was less penetrating, more severe dystrophic changes developed in the surface layers of the liver, located in its capsule. After X-irradiation the dystrophic changes were distributed diffusely.

TABLE 5
Number of Hepatocytes and the Diameter of Their Nuclei
After Local Liver Irradiation With Heavy Ions or X-Rays

Radiation type	Dose, rads	Days after exposure 15: number of cells	15: diameter of nuclei	30: number of cells	30: diameter of nuclei
Unexposed animals	—	123 ± 5	8.35 ± 0.08	143 ± 9	8.12 ± 0.12
180-kVp X-rays	300	—	—	187 ± 12	8.29 ± 0.22
	1000	90 ± 6	7.84 ± 0.24	118 ± 14	6.97 ± 0.14
	3000	—	—	87 ± 11	8.41 ± 0.16
Boron-11 ions	50	—	—	180 ± 6	7.57 ± 0.21
	100	—	—	176 ± 12	7.55 ± 0.25
	200	—	—	157 ± 30	7.47 ± 0.14
Boron-10 ions	300	—	—	87 ± 6	7.48 ± 0.12
	1000	—	—	69 ± 3	7.67 ± 0.12
Carbon ions	300	94 ± 8	6.96 ± 0.09	67 ± 4	7.21 ± 0.09

The RBE values calculated for the survival of hepatocytes by the 30th postradiation day were 7 ± 0.23 and 9 ± 0.02 for boron-10 and boron-11 ions, respectively.

The effect of heavy ions on the cells of the regenerating liver also included disorders in the mitotic activity and the appearance of aberrant mitoses. The deleterious effect of carbon ions on the genetic apparatus of the cell was more pronounced than that of gamma irradiation, as was obvious from an analysis of the dose dependence of the development of aberrant mitoses. The study of cytogenetic effects revealed the usual spectrum of mutations: bridges of the chromosome and chromatid type, single and double fragments, bridges+fragments; however, the ratio of the main types of aberrations changed with an increase in the number of bridges.

Effects of postirradiation recovery in eye corneal epithelium cells exposed to heavy ions

To obtain information concerning the injury and repair processes after an exposure to heavy ions, eye corneal epithelium of mice was used. The rats and types of structural injuries to the chromosome apparatus of cells after single and fractionated exposures to boron-10, boron-11 and carbon-12 ions at doses of 25 to 250 rad were studied in early, late and delayed stages. Fractionated irradiation was applied with intervals of 24, 36 or 72 hours. The mice were sacrificed at 24 hours, or at 1, 2 or 6 months following the exposure: corneal preparations were made from the enucleated eyeballs.

The investigation of biological effects of a single irradiation showed that heavy ions cause the same disturbances in the chromosome apparatus as X-rays. However, the rates of the disturbances and the ratios of individual types of chromosome breaks were different (Table 6). An exposure to heavy ions led to the formation of more aberrant mitoses than observed with X-irradiation. The rate of induced mitoses was dependent on the dose, LET, and the post-irradiation

TABLE 6
Frequency of Aberrant Mitoses in Corneal Epithelium Cells of Mice Exposed to Heavy Ions or X-Rays

Radiation type	Dose, rads	Aberrant mitoses, %		
		24 hours	2 months after irradiation	6 months
X-ray	50	14.9 ± 0.5	1.3 ± 0.3	0.9 ± 0.2
	100	22.5 ± 0.5	2.1 ± 0.1	1.3 ± 0.4
	250	41.5 ± 2.1	4.4 ± 0.1	2.3 ± 0.3
	500	59.3 ± 2.9	8.1 ± 0.7	4.7 ± 0.6
	control	0.7 ± 0.01	1.0 ± 0.03	0.5 ± 0.02
Carbon ions	25	20.5 ± 0.6	22.3 ± 1.3	14.6 ± 1.1
	50	35.5 ± 0.8	—	—
	100	52.1 ± 1.0	34.3 ± 1.6	22.8 ± 1.9
	250	78.6 ± 0.9	—	—
	500	80.3 ± 1.1	—	—
	control	0.8 ± 0.01	0.8 ± 0.01	0.9 ± 0.02
Boron ions	25	23.3 ± 0.8	20.2 ± 0.7	13.9 ± 0.5
	50	39.2 ± 1.8	30.8 ± 0.5	18.3 ± 0.3
	100	56.1 ± 2.8	35.5 ± 0.6	22.4 ± 0.1
	250	68.7 ± 2.8	59.8 ± 1.0	27.8 ± 0.3
	control	0.9 ± 0.01	0.6 ± 0.03	0.9 ± 0.02

time. With doses of 5–100 rad this dependence was exponential. With higher doses it became a more complex power function.

For heavy ions, the RBE values derived from the number of aberrant mitoses 24 hours after the exposure were 1.5–6.0 and were highest with boron-11 ions.

The experiments revealed significant differences in the effects of fractionated exposures to heavy ions and to gamma radiation. As may be seen from Table 7, with a split-dose gamma irradiation the yield of cells with chromosome aberrations was lower than with a single exposure, whereas in similar experiments with boron ions no differences in these effects after single and fractionated irradiation were observed.

TABLE 7
Aberrant Mitoses in Eye Corneal Epithelium Cells of Mice Exposed to Fractionated Doses of Boron Ions or Gamma Rays

Irradiation conditions	Boron ions		^{60}Co γ-rays	
	Dose, rads	Aberrant mitoses, %	Dose,rads	Aberrant mitoses, %
single	25	24.9 ± 2.7	—	—
twice	25 + 25	42.0 ± 2.5	250 + 250	46.7 ± 1.3
single	50	42.7 ± 1.9	500	56.1 ± 1.3
twice	50 + 50	55.8 ± 3.7	300 + 300	50.3 ± 1.7
single	100	56.3 ± 2.5	600	64.0 ± 1.2
twice	100 + 100	75.0 ± 2.5	400 + 400	65.4 ± 1.4
single	250	80.3 ± 1.1	800	76.4 ± 2.6

The decrease in the number of aberrant mitoses noted after a fractionated irradiation or at a late post-irradiation stage is usually attributed to the elimination of deficient cells or to their repair. The lack of this decrease in the case of a fractionated exposure to heavy ions is evidence that none of these processes takes place. This was also suggested by an additional study, in which the kinetics of the basic cytologic changes was followed for 6 months after a single exposure to boron and carbon ions. As shown in Table 6, during the whole 6-month period a great number of cells with aberrant mitoses persisted. On the other hand, with high doses of X-rays the rate of aberrant mitoses did not differ from that seen in intact animals. The tabular data show that six months after an exposure to boron-10 and carbon-12 ions at a dose of 25 rad the rate of aberrant mitoses was 3 times as large as with an exposure to X-rays at a dose of 500 rads. A comparison of the doses which induced 4.5% of aberrant mitoses 6 months after the exposure gave RBE values of 80 to 100.

These observations seem to indicate serious disorders in the reproductive ability of cells and, consequently, disturbances in the natural movement of cells from the basal to the cornifying layer.

Biophysical aspects of the problem

The radiobiological effects in a living cell are stochastic processes determined by physical stochastics and biological variability. Many mathematical models have been built to describe the biological effects of heavy ions. One of the models, proposed by K. Günther and others, postulates that the final radiation effect is related to damage in the DNA molecules. To describe the formation of individual DNA lesions the target theory is used. It is assumed that for a given number of injuries in the nucleus, they are randomly distributed along the DNA molecule. This holds true for microorganisms and cells of higher organisms in the interphase of the cell cycle.

The mathematical formulation of this model is extremely complex; therefore, to date only the biological effects of exposure of DNA molecules and bacteria to radiations with various LET was evaluated: the $1/D_{37}$ values and oxygen enhancement ratios were calculated for bacteria *E. coli* B, which were in good agreement with the experimental data (Table 2). This agreement supports the validity of the assumptions made to derive the model.

The results of the radiobiological experiments with heavy ions form the starting point of our studies in this field. The experience gained promises successful development of investigations to assay the biological effects of heavy ions. It is also necessary to carry out experiments simulating the effect of heavy ions of GSR with respect to such parameters as the absorbed dose, its micro- and macrodistribution in the irradiated object, as well as its time distribution.

TRANSPORT AND FATE OF RADIONUCLIDES[1]

Chairman: S. I. Auerbach

[1]Symposium A–5

Radioecological Studies Applied to Electro-Nuclear Siting[1]

A. Grauby
C.E.A. Radioecology Laboratory
Saint Paul Lez Durance, France

For many years, radioecological studies have been oriented towards laboratory studies or studies of nuclear sites in operation. Very little of this work has been carried out in locations where future nuclear sites are envisaged; likewise, in reports concerned with the study of future sites, the application of radioecological research has been very limited.

However, the ecological diversity of the natural environment entails a considerable variability in the critical pathways of transfer of radioelements. This variability exists both in space and in time (due, in the latter case, especially to human activities). In the face of the inevitable increase of electro-nuclear power stations, the study of natural sites suitable for nuclear installations constitutes a new sphere for radioecology to master.

The processes of transfer between an electro-nuclear power station and the surrounding environment cannot be understood from a simple extrapolation of the research results obtained from experiments of a basic nature. For example, the concentration factor of radionuclides obtained in the laboratory for life-forms or agricultural crops, represent simply the orders of magnitude which must be verified "in situ" as a function of the local pathways of transfer and the physical chemistry of the water, the soil, etc..

Thus, besides the indispensable traditional research on transfer mechanisms we feel it is necessary, in the context of a preliminary study for the installation of an electro-nuclear site, to undertake a systematic series of studies and experimental interventions applied to local conditions and to use the same techniques and the same teams of radioecologists.

This "applied" contribution of Radioecology exists on different levels:

–*To determine,* on the basis of predictions about waste products which will be released in the air or the water, what will become of the pollutants through the various elements of the environment and the food cycle;

–*To quantify* the predicted concentrations, in the medium and long term, for each radionuclide released, as a function of human activities and the "quality" of the environment (the compound effect of the pollutants);

–*To evaluate* the toxic effects on the environment of the chemical and thermic load contained in the waste products;

–*To formulate,* as a function of the identified transfer pathways, suggestions for effective and economical control of the environment of the site, proposing

[1]Congress presentation A–5–1.

bio-indicators for radio-contamination;
–*To reassess,* every two or three years after the initiation of a new installation, the radioecological balance of the environment of the site.

The sphere of action embraces all the aspects of the natural conditions which can play a part in the protection of Man and the Environment. Thus, studies should be carried out for every site on atmospheric deposits on the vegetation and soil, transfer by fresh or sea water as it affects land or marine ecology, transfer in the soil and underground watertables, and the part played by farming methods in relation to livestock and crops, and the technology of food production.

The methodology developed for this type of study involves three successive stages: the Inventory, the Simulation and the Ecological Synthesis.

The Inventory concerns essentially the *biological constituents of the environment:* natural vegetation, agriculture, fishing and livestock; the *physical constituents:* hydrology, sedimentology, pedology (soils), the physical chemistry of the water and soil; the *utilization of the water* : agricultural, industrial, and domestic.

The Inventory comprises two parts: the description of the present situation and an attempt at predicting the future in terms of projected installation of equipment, particularly in the field of agriculture (for example, irrigation).

The Simulation involves creating, on the basis of a knowledge of the principal parameters of the site, a model ("in situ" or in the laboratory) of the conditions under which waste products are released into the air or the water, and observing the consequent effects on the environment (concentration or dispersion, toxicity).

These simulations are based on the following aggregates:

water–soil–water-table–plants
water–sediments
water–sediments–aquatic life
air–soil–plants

The Ecological Synthesis is carried out by assembling the results of the Inventory and the Simulation and applying the knowledge acquired by general radioecological research; this leads to a better understanding of the complex phenomena controlling the dispersion and concentration of the pollutants.

This synthetic operation results in the pinpointing of the dominating critical parameter which will either create a potential danger or set a limit to the installation's waste disposal and suggest measures for the control of the immediate and more distant environment of the site.

These studies can also be linked to the study of an inland or coastal basin in order to define the possible interaction between neighboring installations (industrial or urban waste).

Using two sites chosen as models, it is possible to show schematically the mechanisms of transfer and to specify the contribution of Radioecological Research of Sites to Radioprotection.

The first model represents an electro-nuclear site of the P.W.R. type, the second model a fuel-treatment plant.

On both sites we introduce a standard radioelement, for example: cobalt for

P.W.R. sites and iodine for fuel-plants. We apply these models to two types of European sites, both in two different waste-release situations: controlled release into the water for the electro-nuclear plant and accidental release into the air for the fuel-treatment plant.

The application of the ecological model to two electro-nuclear sites reveals that for the same type of installation, using the same radionuclide (cobalt), the consequences for the Environment and Man are very different (Figs 1, 2, and 3).

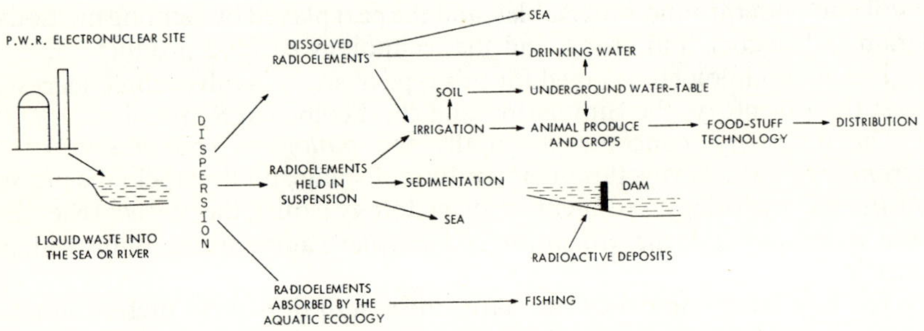

FIG. 1. Radioecological model of an electronuclear site.

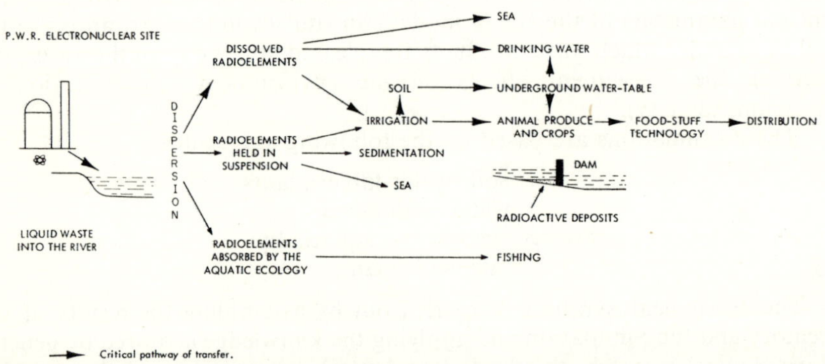

FIG. 2. Application of the model to a controlled discharge of cobalt on a site on the Rhine.

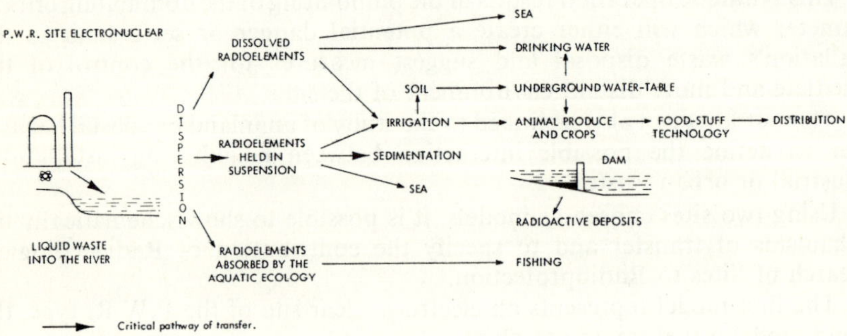

FIG. 3. Application of the model to a controlled discharge of cobalt on a site on the Rhone.

In the case of the site on a northern European river (for example, the Rhine, Fig. 2), the chemical quality of the water favors the solubility by complexing of the radionuclide and, in consequence, accelerates the transfer of cobalt through all the different elements of the Environment, finally reaching Man. This solubility originates in the industrial waste discharged into this river. In the site on a southern European river (for example, the Rhone), because of the high concentration of suspended matter in the river, the cobalt will be confined to the river bed and zones of active sedimentation, i.e. dams. In this case the Ecology—including Man—will be little if at all affected by the discharge of cobalt.

For identical liquid discharges of waste, this difference in the critical pathways is not linked to the physical criteria of diffusion, as opposed to the case of atmospheric discharges, but depends upon the chemical characteristics and the utilization of the water. In the two proposed examples, then, it is human activity (industrial waste, dam construction, and irrigation) that constitutes the critical parameter.

Analysis of the Ecological Model reveals, for an accident involving an emission of iodine-129, various critical pathways of transfer for coastal and inland sites (Figs. 4, 5, and 6).

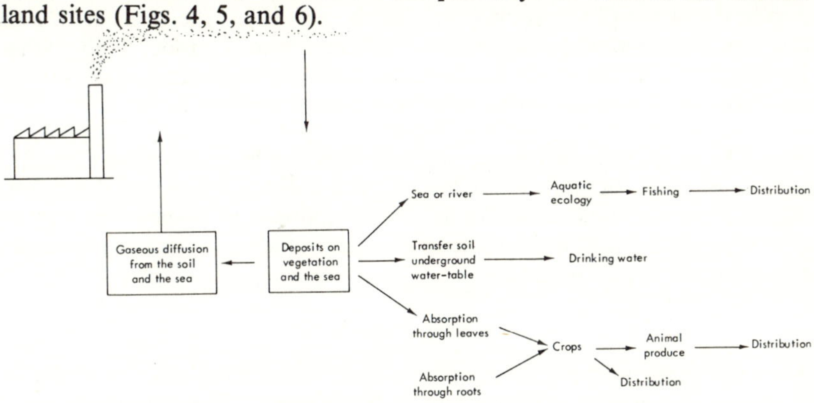

FIG. 4. Radioecological model of the site of a fuel-treatment plant.

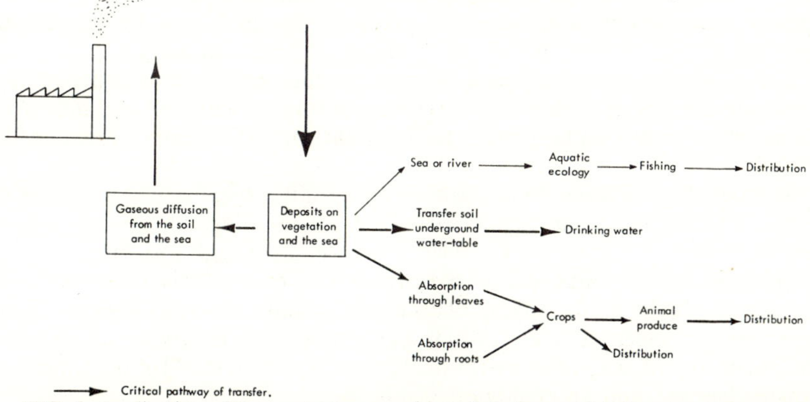

FIG. 5. Application of the model to an accidental discharge of iodine on a coastal site.

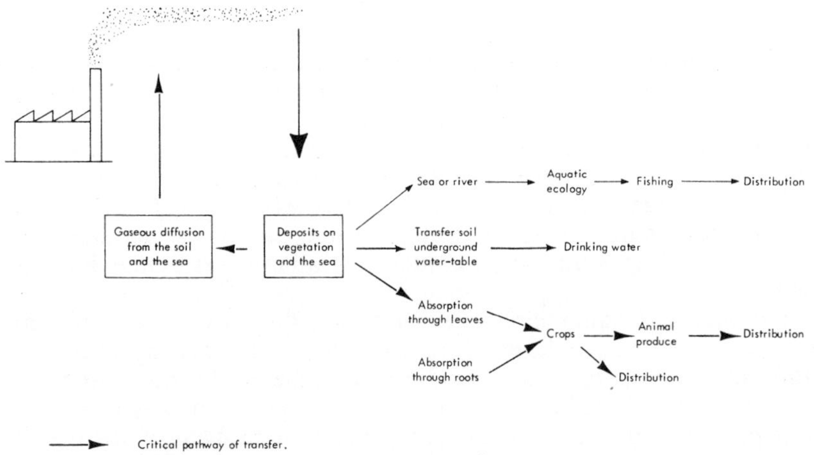

FIG. 6. Application of the model to an accidental discharge of iodine on an inland site.

In the case of the Coastal Site, the characteristics of the environment, the climatology and the physical chemistry of the soil favor the transfer of iodine towards underground water tables. For the Inland Site, the most likely pathway for the transfer of iodine will be through plant-life, depending upon the local climatology and the mineral and organic elements present in the soil.

It is to be noted that in the case of an accidental release of iodine, the action of the environment (in particular the physico-chemical form of the natural atmospheric components) modifies from the moment of release the physico-chemical form of the radiopollutant; so transfer in the biological constituents is effected along different pathways.

CONCLUSION

The ecological diversity of the natural environment entails a considerable variability in the critical pathways of radioelements. This variability exists equally in time in relation to the future development of sites. Thus, the two proposed examples demonstrate the necessity for each site to identify the parameter, or parameters, which are a source of potential danger.

The possibilities for the release of waste products and the consequences for a nuclear site are linked to human activities, notably agricultural and industrial; these activities may well take place far from the immediate environment of the site.

Radioecological research can equally have for its objective integration of the site into a basin or coastal region for which the radiological capacity is to be determined, starting from the knowledge acquired through our research.

The information that can be derived from these examples is of various kinds. First, as regards the methodology of an investigation concerned with the choice of a site, it seems essential to conform to a general scheme involving: an inventory of the environment, a simulation (in the laboratory or "in situ") of the conditions of waste-disposal, and an ecological synthesis.

This work rests on a solid bibliography and laboratory research, but the pinpointing of the *critical characteristic* of the site cannot be made without a previous study of the terrain. It is a question of applying a common methodology to particular cases.

Next, as regards the choice itself, one cannot emphasize too strongly the importance of integrating the radioecological team as soon as possible with the team of planners.

If it is rare, in practice, that radioecological considerations constitute the decisive parameter for the choice of the site, it is nevertheless certain that these considerations could facilitate the choice, if a compromise is necessary; the more calmly and logically it is reached, the more suitable the compromise will prove to be.

Finally, it must be borne in mind that the safety of a site cannot be ensured once and for all. The evolution of the characteristics of the installation, of the hydrographic conditions, of cultural practices and the customs of the population, etc., necessarily entails frequent revision (every three years, for example) of the report on that site.

To sum up, the radioecologist must, by laboratory experiments and, above all, by investigation "in situ," ascertain a critical pathway particular to the site, and make sure that it is permanent. This is to say there is no one universal method which allows us to choose a site without ensuring, on the spot, the correct application of the rules for radiological protection.

GENERAL REFERENCES

1. P. Bovard, A. Grauby, L. Foulquier and Ph. Picat, in *Environmental Behavior of Radionuclides Released in the Nuclear Industry,* pp. 507–523. IAEA Proc. Series STI/PUB/345, 1973.
2. J. Delmas, A. Grauby and R. Disdier, cf. ref. *1,* pp. 321–332.
3. A. Saas and A. Grauby, cf. ref *1,* pp. 255–269.
4. P. Picat, P. Bovard, A. Grauby, M. Ijuin, J. Reveillault, J. Faure and J. Andres, in *Radioecology applied to the protection of man and his environment,* pp. 139–174. Comm. of the Europ. Commun., EUR 4800 d-f-i-e, 1972.
5. P. Guegueniat and J. Ancelin, cf. ref. *4,* pp. 1221–1237.
6. J. Ancellin and P. Bovard, *Rev. Int. Oceanogr. Med.* **21,** 85 (1971).
7. P. Bovard and A. Grauby, in *Disposal of Radioactive Wastes into the Seas, Oceans and Surface Waters,* pp. 3–10. IAEA Proc. Series STI/PUB/126, 1966.
8. P. Picat, P. Bovard, and A. Grauby, *International Congress on Radioprotection of the Environment Owing to the Peaceful Development of Nuclear Energy,* March 14–16, 1967.
9. A. Saas, A. Grauby, and J. Andres, cf. ref. *4,* pp. 381–393.
10. M. Ijuin, P. Picat, A. Saas, A. Grauby, *Health Phys.* **24,** 665 (1973).
11. L. Foulquier, in technical collaboration with A. Lambrechts. French Atomic Energy Commission Report.
12. P. Bovard, A. Grauby, and J. Delmas, in *International Congress on "Chemical Engineering in the Service of Man."* School of Arts and Manufacturers, Paris, Sept. 3–9, 1972.

Effects of Limnological Variables on Bioaccumulation Factors[1]

H. A. Vanderploeg[2] and J. R. Kercher[3]

Environmental Sciences Division
Oak Ridge National Laboratory
Oak Ridge, Tennessee 37830, USA[4]

INTRODUCTION

The bioaccumulation factor of an organism or tissue, *i*, is the steady-state ratio of radionuclide concentration in *i* to that in water:

$$B(R)_i = [R]_i/[R]_w \tag{1}$$

where

$B(R)_i$ = bioaccumulation factor of radionuclide R in organism or tissue *i*,

$[R]_i$ = radionuclide concentration (μCi/g fresh weight) in *i*, and

$[R]_w$ = radionuclide concentration in water (μCi/g).

Bioaccumulation factors are used to predict radionuclide concentrations in whole organisms or their tissues from knowledge of $[R._w$ for chronic releases of radionuclides. Bioaccumulation factors also appear in expressions that are used to predict the transient dynamics of radionuclide concentration in organisms (*1*). Therefore, an understanding of variables that affect bioaccumulation factors is central to understanding steady-state and transient dynamics of radionuclide concentration in aquatic organisms.

Bioaccumulation factors are determined from concentrations of radionuclides or their stable element analogues in water and biota in laboratory and natural water bodies. Since conditions in the laboratory are often unnatural in terms of chemical and ecological relationships, laboratory-derived bioaccumulation factors are often seriously biased. Thus, we restricted our review to bioaccumulation factors determined for natural water bodies. Our review is further restricted to longer-lived isotopes of Cs, Co, Mn, and Sr, which are the most important dose contributors to aquatic biota and man from liquid releases of light-water-cooled nuclear power stations (*2*). Bioaccumulation fac-

[1] Congress presentation A-5-2.
[2] Present Address: National Oceanic and Atmospheric Administration, Great Lakes Environmental Research Laboratory, Ann Arbor, Michigan. 48104.
[3] Present Address: Bio-Medical Division, Lawrence Livermore Laboratory, Livermore, California. 94550.
[4] Operated by Union Carbide Corporation under contract with U.S. Atomic Energy Commission.

tors reported for these metallic nuclides exhibit wide variation among water bodies (*3*).

In this paper we attempt to relate these variations to certain limnological variables. Our review emphasizes fishes since more data have been collected for them than for other taxa. We review the simple models that relate the bioaccumulation factor to concentrations of the stable-element analogue and nonisotopic carrier element in water. We then review the role of physicochemical form of the nuclide and other factors which may affect or cause deviations from these simple relations.

IDEALIZED BIOACCUMULATION PATTERNS

Bioaccumulation factors are often portrayed as being related to ambient concentration of their stable isotopes or non-isotopic carrier element analogues according to three idealized patterns. The first pattern is that $B(R)_i$ is constant regardless of stable element or carrier element concentrations:

$$B(R)_i = \text{const.} \tag{1}$$

The second pattern is that the concentration of an element in an organism or tissue is homeostatically maintained at a constant concentration despite different concentrations of the element in water. This implies that $B(C)_i$, the bioaccumulation factor of the stable element, of a homeostatically controlled element is inversely proportional to its concentration in the water:

$$B(C)_i = \Sigma_i/[C]_W , \tag{2}$$

where Σ_i = concentration of stable element in organism or tissue *i*, a constant (μg/g fresh weight), and

$[C]_W$ = concentration of stable element in water (μg/g).

Assuming that the specific activity, that is, the ratio of the quantity of radionuclide to that of the corresponding total element (μCi/g stable element), of *i* is equal to that of the water, $B(R)_i$ is also inversely proportional to the concentration of the stable element:

$$B(R)_i = \Sigma_i/[C]_W \tag{3}$$

Taking the logarithm of each side of Eqs. (2) and (3) results in the linear relation:

$$\ln B(R)_i = \ln B(C)_i = \ln \Sigma_i - \ln [C]_W , \tag{4}$$

which implies that a plot of ln $B(R)_i$ vs. ln $[C]_W$ has a slope of -1.

The third pattern is that $B(R)_i$ is inversely proportional to the concentration of a non-isotopic carrier element in water. By a non-isotopic carrier element we mean an element that is nearly chemically similar to but occurs in higher concentrations than the stable element analogue. The derivation of this relationship follows.

The bioaccumulation factor of R is related to $B(C^*)_i$, the bioaccumulation factor of the carrier element C^*, by

$$B(R)_i = B(C^*)_i[B(R)_i/B(C^*)_i] , \tag{5}$$

which may be rewritten as

$$B(R)_i = [(R/C^*)_i/(R/C^*)_w] \cdot B(C^*)_i \ , \tag{6}$$

where $(R/C^*)_i$ and $(R/C^*)_w$ are ratios of radionuclide concentration to carrier element concentration found in i and in water, respectively. Assume that the non-isotopic carrier element is homeostatically controlled in the organism, that is:

$$B(C^*)_i = \Sigma_i^*/[C^*]_w \ , \tag{7}$$

where

Σ_i^* = concentration of non-isotopic carrier element in organism or tissue i, a constant (μg/g fresh weight), and

$[C^*]_w$ = concentration of non-isotopic carrier element in water.

Combining Eqs. (6) and (7) and letting $q_i = (R/C^*)_i/(R/C^*)_w$,

$$B(R)_i = q_i\Sigma_i^*/[C^*]_w \ . \tag{8}$$

Taking the logarithm of each side of Eq. (8),

$$\ln B(R)_i = \ln q_i\Sigma_i - \ln[C^*]_w \ . \tag{9}$$

Since Σ_i is a constant, a plot of $\ln B(R)_i$ vs. $\ln[C^*]_w$ will have a slope of -1 if q_i is constant.

If water were the immediate source of the radionuclide and carrier element to i, q_i would be called the "observed ratio" (*4,5*). Fleishman (*6*) has used the terminology "coefficient of deviation" for q_i, which we will use here. On the basis of empirical data on ^{137}Cs and its carrier K, Fleishman (*6*) suggested that q_i is not a constant but rather a function of $[C^*]_w$ (Appendix A).

Because the number of sites where steady-state bioaccumulation factors can be determined from release of radionuclides is limited, bioaccumulation factors of the radionuclides are often estimated from distributions of their stable element analogues. Therefbre, the requirement of equal specific activities in organisms and water that we made for Eq. (3) applies as well to Eqs. (1) and (8) or any relation in which stable element data are used to estimate bioaccumulation factors.

PHYSICOCHEMICAL FORMS OF RADIONUCLIDES AND THEIR AVAILABILITY TO BIOTA

Metals exist in a wide variety of physicochemical forms in natural waters (*7,8*) and their different physicochemical forms have different availabilities to aquatic biota. The major physicochemical forms of trace metals that Gibbs (*8*) reports for rivers illustrate this diversity:

1. Dissolved as ionic species and inorganic associations.
2. Complexed with organic molecules in solution.
3. Adsorbed to solids.

4. Precipitated and coprecipitated on solids (metallic coatings).
5. Incorporated in solid biological materials.
6. Incorporated in crystalline structures.

Both unicellular and multicellular algae, which generally form the base of the aquatic food web, accumulate elements from the soluble phase. Likewise, some radionuclides are directly accumulated by aquatic animals from the soluble phase. Aquatic vascular plants, which may be important as food to consumers in some systems, accumulate elements from the soluble phase of water and also from interstitial water of the sediment. It has been demonstrated that radionuclides and metals complexed to organic molecules may have profoundly lower availabilities to algae and animals than free ions (*9,10*).

Aquatic animals, especially filter feeders, may accumulate radionuclides from the suspended phase. The availability of different forms of radionuclides in the suspended phase has not been studied. It can be said that adsorbed (exchangeable) radionuclides and radionuclides in solid biological materials are available to some aquatic consumers (*8,11,12*). Metals in the lattice structure of clay minerals and other lithogenous detritus are almost completely unavailable (*8*).

Radionuclides enter food webs not only from the water but also from bottom sediments, and availability of nuclides to the food web varies with sediment type. Benthic invertebrates accumulate radionuclides from bottom sediments. Fishes may accumulate radionuclides from bottom sediment by ingestion of these invertebrates, and also directly by incidental ingestion of sediment with prey (*11,12*). Due to differences in sediment characteristics, concentrations of radionuclides in sediments varies greatly (*13,14*). Availability of these isotopes to bottom feeding biota varies with sediment characteristics (*11,12*). Thus, for two separate reasons, there can be a wide variation in uptake by biota.

BIOACCUMULATION FACTORS AND AVAILABILITY OF RADIONUCLIDES

For those elements that are homeostatically maintained at constant concentrations in the organism, the concentration of stable element in the organism is independent of concentrations of stable element or its availability from prey, sediment, and different physicochemical forms in water. If the radionuclide in i has the same specific activity as that in water, the bioaccumulation factor of the radionuclide will be given by Eq. (3). In contrast, differences in availability of radionuclides in different sediments and different physicochemical forms in water can lead to marked deviations from the idealized pattern of a constant bioaccumulation factor, Eq. (1).

For the bioaccumulation factor pattern of Eq. (8), we pointed out that q_i is a function $[C^*]_w$. We note too that if sediment is an important source of radionuclide to the animal, the concentration of carrier in sediment relative to food and water will have an influence on q_i (Appendix A). Moreover, differences between the proportions of radionuclie and carrier element appearing in various physicochemical forms can lead to differences in q_i since the radionuclide or carrier could be preferentially accumulated by the food web.

BIOACCUMULATION FACTOR RELATIONS FOR Cs, Sr, Co, AND Mn IN FISHES

Sr and Cs

Calcium is a non-isotopic carrier element for strontium, as is potassium for cesium; in the soluble phase these elements exist primarily as free ions. In fishes, both calcium and potassium are homeostatically maintained at constant concentrations. For both Ca (*5,15–17*) and K (*5,17,18*) in fishes, linear regressions of ln $B(C^*)_i$ vs. ln $[C^*]_w$ had high correlation coefficients (nearly 1), and the slopes of the regressions were very close to −1. Thus, both Ca and K are homeostatically maintained, and their bioaccumulation factors can be accurately predicted from knowledge of $[C^*]_w$. Strontium bioaccumulation factors for muscle and bone in freshwater fishes appear to vary inversely with ambient Ca concentrations since ln $B(\mathrm{Sr})_i$ vs. ln $[\mathrm{Ca}]_w$ regressions were highly correlated and had slopes near −1 (*5,15–17*). Thus, q_i appears to be constant for strontium.

The q_i of ^{137}Cs is not constant and may be highly variable. Relatively high correlation coefficients for regressions of ln $B(^{137}\mathrm{Cs})_i$ vs. ln $[\mathrm{K}]_w$ for trout flesh (*17*) and for some whole fishes (*5,17*) from the study of Kolehmainen, Häsänen, and Miettinen (*18*) were reported; however, slopes were closer to −2 than −1, which implies that q_i decreases with increasing $[\mathrm{K}]_w$. Table 1 gives bioaccumulation factors of cesium in fishes and $[\mathrm{K}]_w$'s from some later studies. Variance of $[\mathrm{K}]_w$ does not explain the variance of these bioaccumulation factors. Other examples of this lack of correlation can be found in the tables of Vanderploeg et al. (*19*).

Table 1
Bioaccumulation factors of cesium as related to potassium and suspended solids concentrations[a]

K concentration in water (ppm)	Suspended solids concentration (ppm)	Bioaccumulation factors of Cs: Bluegill (*Lepomis macrochirus*) ^{137}Cs	Bluegill Stable Cs	Largemouth bass (*Micropterus salmoides*) ^{137}Cs	Largemouth bass Stable Cs	Catfish (*Ictalurus*) ^{137}Cs	Catfish Stable Cs	Ref.
1.3	Very high (25-185)		140				160	(*20*) (*21*)
1.4	Low (1.3)	900		1200		1200		(*22*) (*23*)
1.8	High	270	310	350	620			(*24*)
9.8	Low							(225)

[a] Taken from flesh or whole fish bioaccumulation factor tabulated by Vanderploeg et al. (*19*).

The question that naturally arises is: Why is the q_i of ^{137}Cs so variable? This variability would seem especially curious since both the bioaccumulation factors of Ca and K are accurately predicted by Eq. (7). This question can be answered in part by examining the physicochemical forms of strontium, cesium, and their carriers in water. For the water bodies listed in Table 2, the proportion of radiostrontium in water appearing in the particulate phase is

Table 2
Mean percentages of ^{137}Cs and radiostrontium in water in the particulate phase.

Water body	% in particulate phase: Cesium	Strontium	Suspended solids concentration (ppm)	Reference
White Oak Lake, Tennessee	69	2		(*25*)
Clinch River, Tennessee[a]	82–92	2–9	25–185	(*25*)
Tennessee River, Tennessee[a]	19–30	9–10	9–22	(*25*)
Experimental pond[b]	58	0.9	21	(12)

[a] Range of means of three river stations.
[b] Based on five samples between 24 and 80 days after introduction of radionuclides.

low for all water bodies, varying between 0.9 and 10%, whereas the percentage of ^{137}Cs that is particulate varies between 19 and 92%. Also in Table 2, there appears to be a rough correlation between the percentage of ^{137}Cs in the particulate phase and the concentration of suspended solids. As can be seen, for example, in Table 3, cesium is strongly sorbed to suspended clay particles;

Table 3
Sorption of ^{134}Cs to varying concentrations of natural clay suspended in distilled water (*26*).

Concentration of clay (ppm)	% of ^{134}Cs sorbed	K_d[a]
16	20	15,600
32	31	13,600
64	43	11,800
128	58	10,800
256	65	7,300

[a] Concentration of element (cesium) sorbed per concentration in water.

and an increased clay particle concentration, which usually represents the bulk of the suspended solids, implies an increase in the proportion of cesium sorbed. The proportion of calcium in the particulate phase is about the same as that of strontium, since the distribution coefficients, K_d's, that is the concentrations of elements sorbed to solid per concentrations in water, of calcium and strontium on suspended sediments are about the same (*26,27*). In contrast, the K_d of cesium is very much larger than that of potassium (*27*). Thus, a greater proportion of cesium is in the suspended phase, and the proportions of cesium and of cesium relative to potassium in the suspended phase will vary with concentration of suspended solids. Since the food chain is probably the most important source of cesium to fishes, it follows that q_i and $B(Cs)_i$ in fishes will vary with suspended solids concentration. This mechanism may explain much of the lack of correlation between $B(Cs)_i$ and $[K]_w$ in Table 1, where highest $B(Cs)_i$'s are seen for waters of low suspended solids concentrations. Possibly, the slope of −2 mentioned above for regressions of ln $B(Cs)_i$ vs ln $[K]_w$ for some fishes results from ac increase in concentration of suspended solids coincident with increasing $[K]_w$. Also, increased intake of K from water and sediment with increased $[K]_w$ may have lowered q_i with increasing $[K]_w$. This requires that q_i available from prey be larger than the average q_i of water by definition equals 1.

Variations of sediment type may contribute to some of the Cs variation. Gallegos et al. (*11*) reported that concentrations of ^{137}Cs in trout in a montane lake were higher than could be explained by the trouts' accumulation from their prey. Laboratory experiments indicated that this difference could be explained by the trout ingesting small amounts of sediment. In fishes, absorption of ^{137}Cs from ingested clay varies with clay type (*12*).

Mn

The review of Stumm and Morgan (*7*) and Wangersky (*28*) indicate that the aquatic chemistry of Mn is complicated. Manganese (II) is soluble in water and complexes easily with organic material. In oxidizing environments, Mn (II) may be transformed to precipitates of MnO_2. The proportion of Mn in water that is soluble varies considerably among water bodies and among seasons but is often low (*23*). In the particulate phase, Mn appears primarily as a metallic coating or as a part of crystalline sediments (*8*). Thus, the availability of Mn to the biota varies considerably from one environment to another and from season to season.

Manganese is homeostatically maintained at constant concentrations in fishes (*19*). Table 4, taken from the data summarized by Vanderploeg et al.

Table 4
Means and coefficients of variation of Mn concentration (ppm) in water and freshwater fishes (fresh weight).

	Water	Flesh	Whole Fishes
Arithmetic Mean	0.028±0.034	0.34±.17	7.9±5.5
Coefficient of Variation	119%	50%	70%
N	7	16	14
Geometric Mean	0.014 4.43	0.32 1.45	6.7 1.84
Coefficient of Variation	343%	45%	84%
N	7	16	14

(*19*), gives means and coefficients of variation for concentrations of Mn in freshwater fishes and in the water from which they were taken. The coefficients of variation for fishes are very small relative to those of the water. Much of the variation in the concentration of Mn in the fishes results from interspecies differences (*19*). In Fig. 1, the bioaccumulation factors of stable Mn in fishes are plotted against concentration in water. The lines were fit by dividing

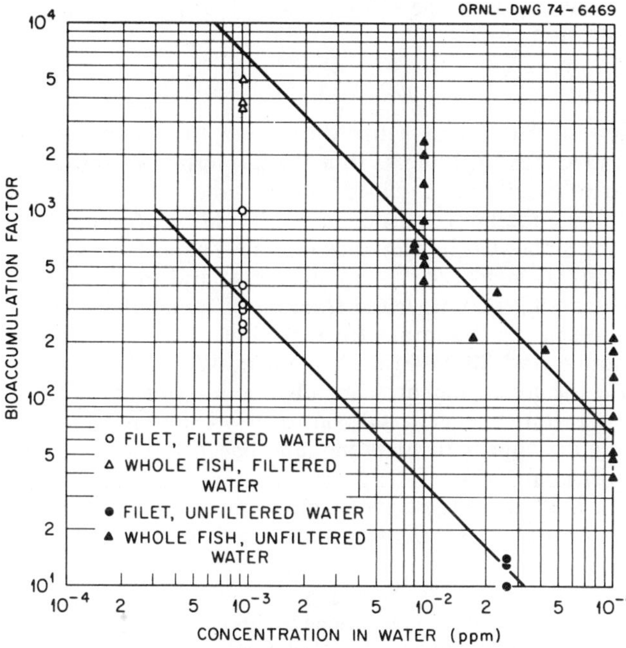

Fig. 1. Manganese bioaccumulation factors of freshwater fishes plotted against water concentration of Mn (ppm).

the geometric mean of Mn concentration in fishes by ambient concentrations to give

$$B(\mathrm{Mn})_i = 6.7/[\mathrm{Mn}]_w , \tag{10}$$

where i = whole fish and

$$B(\mathrm{Mn})_i = 0.32/[\mathrm{Mn}]_w , \tag{11}$$

where i = fish flesh.

To apply these formulas to radiomanganese assumes that the specific activity of radiomanganese in i is the same as that in water. This assumption will clearly be violated for waters in which a great proportion of the stable manganese is associated with the crystal structure of suspended sediments, where it is unavailable to biotw. Assuming that the radiomanganese enters as a soluble nuclide, the specific activity of radiomanganese in the crystal structure may be very low compared to the soluble phase since migration of ions into the structure is very slow; therefore, the specific activity of the biota will be greater than that of the water. Also, the specific activities of radiomanganese in precipitates could be somewhat lower than that of the soluble phase. To prevent underprediction of the bioaccumulation factor of radiomanganese in turbid environments, where a large fraction of stable Mn may be in the crystal structure, we recommend that $[Mn]_w$ appearing in Eqs. (10) and (11) be based on concentrations of Mn in the soluble phase, that is, Mn concentration in the water after filtration.

Co

The review of Fukai and Murray (*29*) and the studies of Marshall and LeRoy (*23*) and Gibbs (*8*) indicate that the proportion of stable cobalt that is particulate varies greatly among water bodies. In a reservoir of low turbidity, 22% of the stable cobalt was particulate (*23*). Gibbs (*8*) reported that greater than 98% of the stable Co in the Amazon and Yukon Rivers was particulate, the Co appearing as adsorbed, in organic solids and in crystalline sediments. Fukai and Murray (*29*) contrast the fraction of radiocobalt appearing in the particulate phase in the Columbia River to that in the Clinch River, Tennessee; the Columbia had 80-95% (at Vancouver, Washington) and the Clinch, 2-30%, despite the much higher suspended solids concentration in the latter river. Cobalt is readily chelated and controversy exists as to whether naturally occurring chelators appear in high enough concentration to chelate a significant fraction of cobalt (*29*).

In Table 5 are summarized mean bioaccumulation factors, derived from both stable element and ^{60}Co distributions, for fishes from eutrophic and mesotrophic water bodies (*19*). A possible explanation for the much lower bioaccumulation factor in the eutrophic environments is that dissolved Co may be associated with organic matter and is, therefore, less available to the food chain. Lack of sufficient data on stable Co concentrations in the eutrophic waters prevents us from ruling out homeostatic control as a possible explanation. In addition, the effect of bottom sediments on the bioaccumulation factor of Co needs to be examined.

Table 5
Cobalt bioaccumulation factors of fishes from eutrophic and mesotrophic waters.[a]

Eutrophy of water body	Bioaccumulation factors	
	Flesh	Whole fish
Mesotrophic waters (2)[b]	300 ± 43(9)[c]	376 ± 63(9)
Eutrophic waters (4)	26 ± 3.8(12)	51 ± 12(11)

[a] Based on water concentration after filtration to remove particulates.

[b] Number of water bodies.

[c] Number of species averaged.

CONCLUSIONS

Simple relations between bioaccumulation factors of radionuclides and concentrations of stable- and carrier-element analogues explain much of the difference of bioaccumulation factors among water bodies for Mn and Sr. For Co and Cs, differing sediment types and differing proportions of the element appearing in different physicochemical forms may lead to great variance from these patterns.

We were limited in developing quantitative relations between bioaccumulation factors of Co and Cs and important limnological variables because these variables were not often measured. Few studies on bioaccumulation factors have been coupled to studies that characterize the physicochemical forms of the element in water. Nor has enough attention been given to the differential availability of elements in these forms to the biota. Hopefully, aquatic ecologists and chemists will work closely together in the future to better define bioaccumulation factor relations.

APPENDIX A. EFFECT OF CARRIER CONCENTRATION ON THE COEFFICIENT OF DEVIATION.

The purpose of this appendix is to show that q_i may vary with $[C^*]_W$ if i accumulates some of the radionuclide from water. Also derived is a general expression for the relation between q_i of the organism and the q_i's of its sources of radionuclide.

Treating i as a single compartment, time rate of change in amount of radionuclide R in an organism or tissue i is

$$\dot{R}_i = I_T - R_i k \,, \tag{A1}$$

and at steady state the amount, C_i^*, of carrier element C^* in i is given by

$$C_i^* = I_T^* / k^* \tag{A2}$$

where

R_i = amount of radionuclide, R, in compartment i (μCi).
I_T = rate of input of radionuclide from all sources (μCi/day)
I_T^* = rate of input of carrier element from all sources (μg/day),
k = excretion rate coefficient for the radionuclide (day^{-1}),
k^* = excretion rate coefficient of C^* at steady state concentration (day^{-1}), and
$(\cdot)$ = time rate of change.

At steady state, Eqs. (A1) and (A2) imply that

$$(R/C^*)_i = (R_i/C_i^*) = I_T k^* / I_T^* k \tag{A3}$$

For the case where uptake is from a single prey, j, and from water:

$$I_T = I_W + I_j, \text{ and} \tag{A4}$$

$$I_T^* = I_W^* + I_j^*, \tag{A5}$$

where

I_W and I_j are uptake rates of radionuclide from water and j, respectively; and I_W^* and I_j^* are uptake rates of carrier elements from water and j, respec-

tively.
For uptake from water:

$$I_W = a[R]_W \text{ and,} \tag{A6}$$

$$I_W^* = a^*[C^*]_W \tag{A7}$$

where

a = uptake rate coefficient for radionuclide from water (g/day), and
a^* = uptake rate coefficient for carrier element from water (g/day).

For uptake from food:

$$I_j = b[R]_j Q \text{ and,} \tag{A8}$$

$$I_j^* = b^*[C^*]_j Q \tag{A9}$$

where

b = absorption efficiency of radionuclide uptake from food (unitless),
b^* = absorption efficiency of carrier element uptake from food (unitless), and
Q = feeding rate (g/day).

Thus,

$$(R/C^*)_i = (k^*/k)\left[\frac{a[R]_W + b[R]_j Q}{a^*[C^*]_W + b^*[C^*]_j Q}\right] \tag{A10}$$

$$= (k^*/k)\left[\frac{a[C^*]_W(R/C^*)_W + b[C^*]_j(R/C^*)_j Q}{a^*[C^*]_W + b^*[C^*]_j Q}\right], \tag{A11}$$

where $(R/C^*)_W \equiv [R]_W/[C^*]_W$
Dividing Eq. (A11) through by $(R/C^*)_W$,

$$q_i \equiv \frac{(R/C^*)_i}{(R/C^*)_W} = \frac{k^*}{k}\left[\frac{a[C^*]_W + b[C^*]_j q_j Q}{a^*[C^*]_W + b^*[C^*]_j Q}\right], \tag{A12}$$

where q_j is the coefficient of deviation in the prey. Eq. (A12) may be rewritten as

$$q_j = (k^*/k)\left[\frac{a^*(a/a^*)[C^*]_W + b^*(b/b^*)[C^*]_j q_j Q}{I_T^*}\right] \tag{A13}$$

$$= (k^*/k)\left[(I_W^*/I_T^*)(a/a^*) + (I_j^*/I_T^*)(b/b^*)q_j\right] \tag{A14}$$

Because of homeostatic control, $[C^*]_j$ is constant. Thus, as can be seen from Eqs. (A13) and (A14), q_i should vary with $[C^*]_W$.

Equation (A14) may be generalized for any set of sources:

$$q_j = (k^*/k) \left[\sum_{j \neq i} (I_{ij}^* / I_T^*)(a_{ij}/a_{ij}^*) q_j \right] \quad \text{(A15)}$$

where

I_{ij}^* = intake rate of carrier element by i

from $j(\mu g/\text{day})$, and

a_{ij} = absorbtion effiency of uptake of radionuclide from source j at surfaces of i (unitless), and

a_{ij}^* = absorption efficiency of uptake of carrier element from source at j at surfaces of i (unitless).

If water is a source, where $j = 1$ for water, a_{i1} is absorption of the radionuclide brought to the body surfaces and $q_1 \equiv 1$.

LITERATURE CITED

1. H. A. Vanderploeg, R. S. Booth, and F. H. Clark, in *Mineral Cycling in Southeastern Ecosystems*, Symposium held 1–3 May 1974, Augusta, Georgia.
2. S. V. Kaye, in *Symposium on Environmental Behavior of Radionuclides Released in the Nuclear Industry,* pp. 649–661. IAEA Symposium, Aix-en-Provence, France, 14–18 May 1973.
3. S. M. Jinks, and M. Eisenbud *Radiat. Data and Reports* **13,** 243 (1972).
4. C. L. Comar, R. H. Wasserman, and M. M. Nold, *Proc. Soc. Expl. Biol. Med.* **92,** 859 (1956).
5. H. T. Peterson, Jr., in *Proceedings of the Fifth Annual Health Physics Society Midyear Topical Symposium: Health Physics Aspects of Nuclear Siting,* **Vol. 2.** pp. 289–328. AEC CONF–701106, 1970.
6. D. J. Fleishman, in *Radioecology.* (V. M. Klechkovskii, G. G. Polikarpov, and R. M. Aleksakhin, Eds.) pp. 347–369. John Wiley and Sons, New York, 1973.
7. W. Stumm, and J. J. Morgan. *Aquatic chemistry.* Wiley-Interscience, New York, 1970. (583 pp.)
8. R. J. Gibbs, *Science* **180,** 71 (1973).
9. N. V. Timofeev-Resovskii, E. A. Timofeeva-Resovskaya, G. A. Milyutina, and A. B. Getsova, *Dokl. Nauk SSSR* **132,** 369 (1960).
10. M. Bernhard, and A. Zattera. in *Symposium on Radioecology,* (D. J. Nelson and F. C. Evans, Eds.), pp. 389–398. AEC-CONF–670503, 1969.
11. A. F. Gallegos, F. W. Wicker, and T. E. Hakonson, in *Proceedings of Fifth Annual Health Physics Society Midyear Topical Symposium: Health Physics Aspects of Nuclear Siting.* **Vol. 2.** pp. 477–497. AEC CONF–701106. 1970.
12. L. D. Eyman, in *Proceedings of the XIX Congress of the International Association of Limnology* (*SIL*). Winnipeg, Canada, August 22–29, 1974. (in press).
13. A. G. Friend, in *Transport of Radionuclides in Freshwater Systems.* (B. H. Kornegay et al., Eds.). pp. 43–60. TID–7664, 1963.
14. T. F. Lomenick, and D. A. Gardiner, *Health Phys.* **11,** 567 (1965).
15. P. O. Agnedal, in *Radioecological concentration Processes.* (B. Aberg and F. P. Hungate, Eds.), pp. 879–895. Pergamm Press, New York, 1967.
16. W. L. Templeton, and V. M. Brown, in *Int. J. Air. Water Pollut.* **8,** p. 49 (1964).
17. A. Preston, D. F. Jefferies, and J. W. R. Dutton, *Water Research* **1,** 475 (1967).
18. S. Kolehmainen, E. Hasanen, and J. K. Miettinen, in *Radioecological Concentration Processes,* (B. Abearg and F. P. Hungate, Eds.), pp. 913–919. Pergamon Press, New York, 1967.
19. H. A. Vanderploeg, W. H. Wilcox, J. R. Kercher, D. C. Parzyck, and S. V. Kaye, Bioaccumulation factors for radionuclides in freshwater biota. U.S. AEC Report ORNL–5002, Oak Ridge National Laboratory, Oak Ridge, Tennessee, 1975.

20. D. J. Nelson, in *Symposium on Radioecology,* (D. J. Nelson and F. C. Evans, Eds.), pp. 240–248. AEC CONF–670503. 1969.
21. E. G. Struxness, P. H. Carrigan, Jr., M. A. Churchill, K. E. Cowser, R. J. Morton, D. J. Nelson, and F. L. Parker. Comprehensive report of the Clinch River study. U.S. AEC Report ORNL–4035, Oak Ridge National Laboratory, Oak Ridge, Tennessee, 1967.
22. R. S. Harvey, *Health Phys.* **10**,243 (1964).
23. J. S. Marshall and J. H. LeRoy, Iron, manganese, cobalt, and zinc cycles in a South Carolina reservoir, in *Radionuclides in Ecosystems,* **Vol. 1**,pp. 465–473. AEC CONF–710501–P1. 1973.
24. S. E. Kolehmainen, *Health Phys.* **23**,301 (1972).
25. L. D. Eyman, Cesium–137 and stable cesium in a hypereutrophic lake. Ph.D. Thesis, Michigan State University. 1972. (98 pp.)
26. K. Garder, and O. Skulberg, *Int. J. Air Wat. Poll.* **8**, 229 (1963).
27. T. D. Reynolds, and E. F. Gloyna, in *Proc. Second International Conf. on Advances in Water Poll. Res.* **Vol. 1** (O. Jaag, Ed.). pp. 151–172. Pergamon Press, New York, 1964.
28. P. J. Wangersky, in *Radioecology* (V. Schultz and A. W. Klement, Jr., Eds.), pp. 499–508. Keinholt Pub. Corp., New York, 1963.
29. R. Fukai, and C. N. Murray, in *Symposium on Environmental Behavior of Radionuclides Released in the Nuclear Industry.* IAEA Symposium, Aix-en-Provence, France. 14–18 May 1973. (in press).

Effect of Some Important Physio-ecological Factors on the Accumulation of Radionuclides by Freshwater Fish[1,2]

A. Berg
G. D. Research, Science and Education
Commission of the European Communities
EURATOM Joint Research Centre, 21020 Ispra, Italy

INTRODUCTION

The increased production of nuclear energy to be expected in the near future requires that more reliable predictions be made concerning the levels of radionuclide accumulation in the environment. A considerable effort in radio-ecological research in the last two decades has provided the radioprotectionist with values of the parameters of radionuclide transfer at various steps in the food chains. But these parameters are often obtained under empirical laboratory conditions and suffer, therefore, from the difficulty of their extrapolation to field conditions.

A preliminary requisite for such an extrapolation is that it be made on the basis of a better knowledge of the effect of those intrinsic and extrinsic factors (pertaining to the aquatic organism and to its environment respectively) which are likely to intervene in determining the level of radioaccumulation in relation to time. In this regard, several physio-ecological factors play an important role that will be illustrated here as far as freshwater fish is concerned.

GENERAL CONCEPTS AND TERMINOLOGY

It seems useful to apply to fish the concepts widely used for mammals in mineral balance studies by means of isotopic tracers (*1*). In freshwater fish, the accumulation of radionuclides is the result of absorption processes of the corresponding stable element which take place at two different sites, the gills and the intestine, from two different sources, water and food. In Fig. 1, the rates of absorption of the stable element from these two sources are represented by the arrows $fe_{,d}$ and $fe_{,i}$ (fluxes of entry). By exchange processes which take place between blood and tissues, a part of the absorbed elements are retained by the tissues: we can define a bidirectional exchange rate (*ER*) which compensates for losses from preformed tissues, and an eventual growth rate (*GR*) which increases the pool by the growth of tissues. In such a scheme,

[1]Congress presentation A–5–3.
[2]Contribution No. 1024 of the Biology Programme, Directorate General XII of the Commission of the European Communities.

three successive steps are considered: absorption, exchange and pool formation. On the other hand, excretion occurs from the blood to the outside.

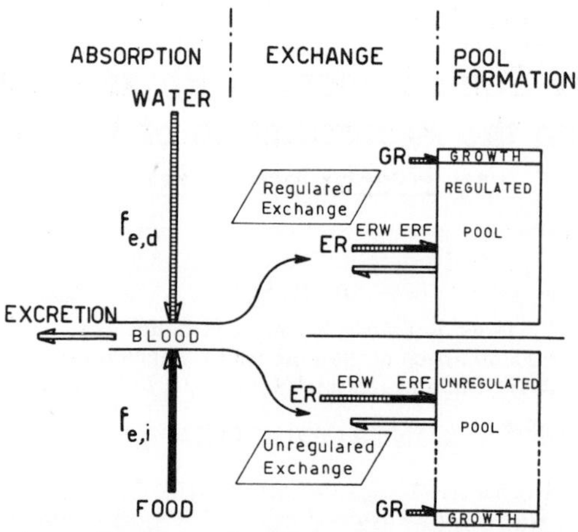

FIG. 1. General scheme of exchange of an element between fish and its medium (water, food).

The use of tracers allows the determination of the exchange rate *ER* but not of the absorption rates *fe*. In a compartment in rapid equilibrium like blood, it may be assumed that the proportions between the rates of direct and intestinal absorption are maintained in the exchange rate which can thus be theoretically split into two components *ERW* and *ERF* (exchange rates from water and from food). We can then define *CDA* (contribution of direct absorption) as the ratio:

$$ERW/\ (ERW + ERF) = f_{e,d}/\ (f_{e,d} + f_{e,i})$$

and *CIA* (contribution of intestinal absorption) the complementary parameter (1−*CDA*).

CAPACITY OF METABOLIC REGULATION

On Fig. 1 two kinds of pools are represented. On the upper side of the figure, the pool is regulated at a constant level as a consequence of the regulation of the exchange rate *ER*, independently from the absorption rates. On the lower side of the figure, the pool is variable (unregulated) since *ER* may vary with the importance of the absorption rates. Essential elements like Ca, K, P and Zn are metabolically regulated by fish, at least within a certain range of concentration in the medium. On the other hand, other elements which hardly seem to play any physiological role are exchanged and accumulated by fish in proportion to their availability in the environment (water and food). This seems to be the case of pollutants like Cr and Pb.

In a third category, one finds elements which are not regulated as such but are linked by their metabolism with an essential element which is itself regulated. This is the case of Sr and Cs, whose metabolism is linked with that of Ca and of K, respectively. According to the category to which they belong, the radionuclides may reach quite different levels of accumulation which can be described by adequate formulations.

Absence of Metabolic Regulation

This case can be illustrated by chromium for which the level of accumulation by fish is the result of the sum from the two sources of entry (water and food) proportional to the respective concentrations of the element. In the absence of availability from food, the concentration in fish at equilibrium may be represented by the simple relationship:

$$[X]_F = k[X]_w \tag{1}$$

where $[X]_F$ and $[X]_w$ are the concentrations of the element in fish and in water respectively, and k is a factor of proportionality which expresses the capacity of the gills to pump the element from water together with the capacity of the tissues to retain the element from blood.

If the element is also available from food, the resulting level of accumulation by fish may be considered as the sum of two partial concentrations which are interrelated through the parameter CDA:

$$CDA = \frac{[X]_{F,w}\ \text{partial, from water}}{[X]_{F,w}\ \text{partial, from water} + [X]_{F,f}\ \text{partial, from food}} \tag{2}$$

Therefore

$$[X]_{F,\,\text{total}} = [X]_{F,w}/CDA = k\,[X]_w/CDA \tag{3}$$

If a given radioactivity of the same nuclide X is present in water (R_w), it may be assumed that the specific radioactivity in food organisms (Rs_f) is, at equilibrium, identical to that prevailing in water, i.e. $Rs_w = R_w/[X]_w$, and therefore

$$R_F = [X]_{F,\,\text{total}} \times Rs_w = k[X]_w/CDA \times R_w/[X]_w = kR_w/CDA\ . \tag{4}$$

According to this formulation, the following statements may be advanced as far as the effect of physio-ecological factors is concerned:

- under growing conditions, the increase in food consumption by fish causes a strong decrease of CDA which may still be enforced by a parallel increase of the concentration of the element X in the food organisms. On the other hand, the higher temperature characteristic of the growing season probably brings an increase of the branchial capacity of pumping the element from water (factor k). Therefore, at a given radioactivity of the element in water (R_w), the fish radioactivity will sharply increase through a simultaneous decrease of CDA and an increase of k in formula (4);
- the change in fish radioactivity induced by changes in the radioactive contamination of water will not depend on the degree of water pollution by the element X. Actually, in the case of an increase in $[X]_w$ which would not be accompanied by a corresponding increase in radioactivity (case of independent sources of conventional and radioactive pollution), the consequent increase in $[X]_F$ would not cause any increase in R_F because of the compensatory decrease in the specific radioactivity in water.

Metabolic Regulation

The case of Zn can be taken as an example. In this case, the level of accumulation of the stable element by fish is independent from its concentration in water or in food, but is regulated at a constant value in the fish body ($[X]_F$).

The corresponding level of radioactivity in fish is determined by the value of the specific radioactivity of the medium (water and food). Under equilibrium conditions, one may assume that the specific radioactivity of fish tissues is identical to that of water (*2,3*) and of food organisms. Therefore, one may write:

$$R_F = [X]_F \times Rs_w = [X]_F \times R_w/[X]_w \ . \tag{5}$$

According to this formulation, the level of radioactivity accumulated by fish is (aside from the evident relation to R_w) directly proportional to the regulated concentration of the stable element in fish and inversely correlated to its concentration in water. In no way is the level of fish radioactivity dependent on the relative contribution of direct and intestinal absorption. The ingestion of food organisms which generally exhibit a great capacity for nuclide (both stable and radioactive) accumulation does not bring, as one would expect at first sight, an increase in the level of fish radioactivity. In other words, the regulation processes by themselves prevent the two sources of Zn to be cumulative at the site of retention by the tissues even if they are cumulative at the sites of absorption (gills and intestine).

The only exception would be in the case of a specific radioactivity which would be different in food organisms and in the surrounding water, a condition which could occur, among other things, in the absence of equilibrium between the physico-chemical forms of the stable element and of its radionuclide in water. Such a situation is described for $Zn/^{65}Zn$ (*4*) and, strangely enough, for a nuclide like $Sr/^{90}Sr$ which is expected to be present in water only in the ionic form (*5*). A concentration factor which would vary in inverse relation to the level of radioactivity (*6*) could also induce a similar situation.

If this case is to be considered, it can be easily shown that if the specific radioactivity of food organisms (Rs_f) is n times higher than that of water (Rs_w), the specific radioactivity of fish is given by the formula:

$$Rs_F = Rs_w[1 + (n - 1)(1 - CDA)] \tag{6}$$

and the fish radioactivity:

$$R_F = X_F \times Rs_F = X_F \times Rs_w[1 + (n - 1)(1 - CDA)] \tag{7}$$

According to this formulation, if the specific radioactivity of food organisms is 10 times higher than that of water and in case of a low value of *CDA* (for example 0.5 in growing conditions), the fish radioactivity would be 5.5 times higher than normally expected from formula (5).

Linkage With a Metabolically Regulated Element

This condition can be illustrated by the case of the couple Sr/Ca. The indirect regulation of Sr via Ca metabolism permits us to apply formula (5) for the prevision of radiostrontium accumulation by fish, considering Ca (and not Sr) as the corresponding stable element X and introducing a correction factor which takes into account the discrimination exhibited by fish between Sr and Ca (factor called "observed ratio"):

$$R_{F,\,Sr} = ([Ca]_F/[Ca]_w)\, R_{w,\,Sr} \times \text{O.R.} \tag{8}$$

Since discrimination is not identical in indirect and intestinal absorption, the observed ratio is a function of the following parameters:
- DF_w and DF_f, i.e. the discrimination factors in processes of direct and intestinal absorption;
- *CDA* and *CIA*, i.e. the contributions of direct and intestinal absorption;
- *d* or the observed ratio exhibited by the food organisms;

according to the formula:

$$\text{O.R.} = DF_w \times CDA + DF_f \times d \times CIA \quad (9)$$

Therefore:

$$R_{F,\,\mathrm{Sr}} = ([\mathrm{Ca}]_F/[\mathrm{Ca}]_w)R_{w,\,\mathrm{Sr}}\,(DF \times CDA + DF_f \times d \times CIA) \quad (10)$$

The following physio-ecological factors are thus likely to induce variations of radiostrontium accumulation by fish:
- concentration of Ca in water, which presents the well-known inverse relationship with radiostrontium accumulation.
- concentration of Ca in fish: this factor is to be considered as constant for a given species and is responsible for strong differences, between the species, of radiostrontium activity in the whole fish (on the order of 1 to 7 in Lake Maggiore fishes (*7*).
- variations in feeding regimen may cause a limited amount of variation of fish radioactivity (probably by a maximum factor of 2–3): a decrease of *CDA* (as the one occurring during growth processes for example) brings a decrease of the observed ratio and of the radioactivity in the growing tissues since the corresponding increase of *CIA* is compensated by a generally low observed ratio *d* exhibited by the food organisms.
- the presence of chelating agents in water may cause an overall disturbance of the processes of radiostrontium accumulation. In the case of EDTA, for example, the higher stability constant of the Ca-complex relative to Sr ($K_{\mathrm{CaZ}} = 10^{10.7}$; $K_{\mathrm{SrZ}} = 10^{8.6}$) causes a drop in Ca availability which is compensated by a preferential utilization of Sr (*8*).

In other words, the discrimination factor DF_w increases from a value generally lower than 1 to a value far greater than unity (in one experiment, DF_w increased from 0.4 to about 9, thus causing a 25-fold increase of fish radioactivity).

GROWTH PROCESSES

The impact of growth processes on the intensity of radioaccumulation has been illustrated in the preceding section in relation to metabolic regulation. It is worthwhile to consider now to what extent fish radiocontamination resulting from a discontinuous nuclear discharge (as in the case of a nuclear accident) is influenced by growing conditions.

Absence of Growth as Point of Reference

In this situation, radioaccumulation results only from the exchange of the element between fish and its medium. If we call ER_x the total exchange rate (from water and from food, expressed as μg X/g fish/day) of the stable element X and *k* the turnover coefficient (= ER_x: $[X]_F$) which expresses that fraction of the fish pool which is exchanged daily, we may write that the fraction of the pool exchanged at the time t (in days) is equal to:

$$b = 1 - e^{-kt} \tag{11}$$

Therefore, if we assume that equilibrium condition is reached when the specific radioactivity of fish is equal to that of water ($Rs_F = Rs_w$), and on the basis of a ratio n ($= Rs_2 : Rs_1$) between the specific radioactivity of water after and before a radioactive discharge, the evolution of the specific radioactivity of fish is given by the formula:

$$Rs_{F2} = Rs_{w1}\,(1 - b) + n\,Rs_{w1}\,b = Rs_{w1}\,[1 + (n - 1)\,b] \tag{12}$$

where the two factors of the sum represent the contributions of the unexchanged and of the exchanged fractions of the body pool, respectively. Using the value of b from formula (11), one finally gets:

$$Rs_{F2} = Rs_{w1}\,[n - (n - 1)\,e^{-kt}] \tag{13}$$

The relative factor of increase in the specific radioactivity of fish is thus given by the value:

$$F_{EX} = n - (n - 1)\,e^{-kt} \tag{14}$$

This relative increase is represented in Fig. 2 in relation to time, for a 50-fold increase in the specific radioactivity of water, using a value of $k = 0.003$/day for the turnover coefficient (for Zn, for example).

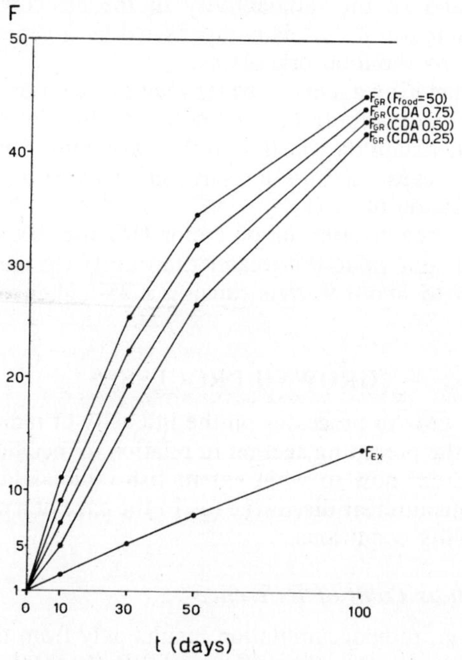

FIG. 2. Evolution of fish radioactivity with time, in case of a sudden increase of water radioactivity (from 1 to 50), under subsistence conditions (F_{EX}) and under growth conditions (F_{GR}).

Presence of Growth

In this situation, radioaccumulation will result from the exchange of the element by the preformed tissues and from the net accumulation by the growing tissues. If fish feed on organisms in the first trophic level which rapidly reach equilibrium with their medium, one may assume the same equilibrium conditions for the growing tissues of fish. Defining g as the ponderal growth coefficient of fish per day (i.e. relative to the initial weight P_o) during the period considered, the specific radioactivity of the preformed tissues (of weight P_o) will increase according to formula (12) while that of the growing tissues (of weight $P = P_o[e^{gt} - 1]$) will be identical to that of water ($Rs_{w2} = n\ Rs_{w1}$).

The resultant specific radioactivity of fish in growing conditions will be:

$$Rs_{F2} = \frac{P_o Rs_{w1}[1 + (n-1)b] + P_o(e^{gt} - 1)n\,Rs_{w1}}{P_o e^{gt}}$$

$$= Rs_{w1}\ \frac{1 + (n-1)b + n(e^{gt} - 1)}{e^{gt}} \qquad (15)$$

Using the value of b from formula (11), one finally gets:

$$Rs_{F2} = Rs_{w1}\,[n - (n-1)\,e^{-(g+k)t}] \qquad (16)$$

The relative factor of increase of the fish specific radioactivity is thus given by the formula:

$$F_{GR} = n - (n-1)\,e^{-(g+k)t}, \qquad (17)$$

which differs from F_{EX} only for the inclusion of the parameter of growth g.

This relative increase is represented in Fig. 2 in comparison to F_{EX}, using a value of g equal to 0.02/day for the ponderal growth coefficient. This example puts in evidence the acceleration of radionuclide transfer which takes place in an ecosystem when active growth is going on in the various steps of a food chain (*9*). This acceleration is particularly evident (from formulae (17) and (14)) in the case of a slow turnover (low value of k, as encountered for elements like Ca and Zn) under conditions of intensive growth (high value of g, as encountered in temperate regions at the beginning of the growth season (*10*). Such a situation is illustrated by the heavy contamination by ^{55}Fe which was described in the years of intensive fallout for rapidly growing marine species (*11*). On the other hand, a slow growth rate may be responsible for several years' delay in the attainment of equilibrium between fish and water (*12*).

The preceding formulae may be easily adapted to the case of fishes feeding on organisms whose specific radioactivity is not in direct equilibrium with that of water but follow the same pattern as F_{GR}. In this case, the relative increase of the specific radioactivity of fish will depend on the contribution of intestinal absorption (*CIA*). Such values have been reported in Fig. 2 for *CIA* 0.25 − 0.75. As one would have expected, one notices a gradual slowing up of the curves with an increase in the importance of intestinal absorption.

CONCLUSIONS

The foregoing discussion serves to illustrate the fact that a background knowledge is needed if more reliable predictions are expected on radionuclide accumulation in fish. Among the factors of a physiological and ecological nature which have been shown to play a significant role, the following have been considered: intensity of metabolic turnover, capacity of metabolic regulation, contribution of direct and intestinal absorption, position in the trophic chain, intensity of growth and length of the growing season, chemical composition of water and eventual contamination by the stable element corresponding to the radionuclide. The discharge of a vast spectrum of polluted effluents into the water system may contribute to render the whole problem still more complicated because of variations of radionuclide availability to the organism from water and from food.

In radioecological research, future effort should take into account those factors which may affect radioaccumulation in freshwater organisms in order to allow an adequate extrapolation from experimental results to the actual conditions encountered in fresh waters.

REFERENCES

1. P. Besançon et L. Gueguen, *Ann. Biol. Anim. Biochim. Biophys.* **9,** 537 (1969).
2. I. L. Ophel, in *Advances in Water Pollution Research Proc. 2nd Int. Conf. Tokyo, 1964,* **Vol. 3,** 275 (1965).
3. F. Ya. Rovinskii and A. L. Agre, *Prediction of the Accumulation of Strontium-90 in Fish.* Translation from Russian, National Lending Library for Science Technol., Boston Spa, England, 1965.
4. F. A. Cross, J. N. Willis, and J. P. Baptist, *J. Fish. Res. Bd. Can.* **28,** 1783 (1971).
5. Z. Kalnina, [Ratio $^{90}Sr/Sr$ in plankton of Latvian lakes of various types] (in Russian). *Radioecologiya vodnykh organizmov, Zinatne, Riga:* 81 (1973).
6. I. A. Shekhanova and V. L. Pechkurenkov, [Accumulation of ^{90}Sr and ^{90y} dissolved in water by loach spawners and its effect on their offspring] (in Russian). Voprosy Ikhtiologii **9,** 338 (1969).
7. A. Berg, *Mem. Ist. Ital. Idrobiol.* **29,** 145 (1972).
8. A. Berg, *Mem. Ist. Ital. Idrobiol.* **26,** 257 (1970).
9. M. Bernhard, in *Proc. Symp. on Nuclear Detonations and Marine Radioactivity,* pp. 137–150. Kjeller, Norway, 1963.
10. A. Berg and E. Grimaldi, *Mem. Ist. Ital. Idrobiol.* **18,** 25 (1965).
11. H. E. Palmer and T. M. Beasley, in *Proc. Int. Symp. on Radioecological Processes, Stockholm,* p. 8. Pergamon Press, New York, 1967.
12. I. L. Ophel, in *Radioecology* (Schultz and Klement, Eds.) pp. 213–216. Reinhold Publishing Corporation, New York and The American Institute of Biological Sciences, Washington, D. C., 1963.

TRANSURANIUM ELEMENT TOXICITY— DOSE-RESPONSE RELATIONSHIPS AT LOW EXPOSURE LEVELS[1]

Chairwoman: P. W. Durbin

[1]Symposium C–9

Metabolism and Biological Effects in Rodents of Plutonium and Other Actinide Elements[1,2]

Yu. I. Moskalev, T. I. Levdik, E. R. Lyubchanskii, A. P. Nifatov, R. A. Erokhin, L. A. Buldakov, V. K. Lemberg, N. A. Koshurnikova, L. G. Filippova, and I. A. Ternovskii

Biophysics Institute, U.S.S.R. Ministry of Public Health, Moscow, USSR.

TISSUE DISTRIBUTION OF INHALED ACTINIDES

The available information on metabolism and biological effects of Pu and other actinides in rodents has been reviewed (*1,2*). The actinides are poorly absorbed from the gastrointestinal tract, and when their soluble compounds are inhaled, significant amounts are translocated to skeleton and liver.

Deposition in Lung

In rats about 26% of inhaled Pu[3] or Am[3] is deposited in the lung, and the deposited fraction is independent of the actinide used, its solubility or the form of its chemical compound, and its concentration in the inhaled air over the range 0.02 to 2.0 μCi/1 (see Fig. 1). However, as shown in Fig. 2, clearance of these elements from rat lung depends on the physical-chemical properties of the inhaled compound. On the day after inhalation and thereafter, the lung content of polymeric Pu or Pu compounds that readily hydrolyze in vivo [Pu (IV) nitrate; Pu(III) chloride; Pu(IV) pentacarbonate dissolved in water, PPC_2] is two to three times greater than the lung content of stable Pu complexes [Pu(IV) citrate; Pu(VI) acetate; Pu (IV) pentacarbonate dissolved in 10% $(NH_4)_2CO_3$, PPC_1] or of the Am and Np compounds tested.

Clearance from the lung

Beginning on the first day after inhalation, clearances of actinide compounds from rat lung can be described by exponential functions (Np and Pu, two components; Am, three components). The fractions (α) cleared from the

[1]Congress presentation C–9–1 (not delivered).

[2]This paper was submitted in Russian. It was translated by Martha Gerrard, Technical Information Division, Oak Ridge National Laboratory. The figures were relabeled and prepared for publication by John Flambard, Technical Information Division, Lawrence Berkeley Laboratory. The translation was edited by Dr. Patricia W. Durbin, Donner Laboratory and Division of Biology and Medicine, Lawrence Berkeley Laboratory. Publication schedule did not permit the authors to review the edited translation.

[3]Ed. note. The radionuclides, ^{237}Np, ^{239}Pu, and ^{241}Am, were used in all of the authors' studies. In order to save space and simplify printing the chemical symbols, Np, Pu, and Am without the mass numbers, have been used to designate those particular isotopes throughout this paper. Mass numbers are shown when reference is made to the work of others, or if needed for clarity.

lung and their corresponding biological half-lives ($T_{1/2}$) differ by factors of two to three depending on the tendency of the compounds to be converted to insoluble forms in vivo (see Table 1). In general, the $T_{1/2}$'s of the lung clearance equations were similar for Np and soluble Pu compounds; clearance of the Am compounds was more rapid. As can be seen from Fig. 2 and Table 1, not only the rates of lung clearance but also the amounts of actinide cleared from the lung are determined by the rates at which their compounds can be transported to the blood and fixed in organs of secondary deposition.

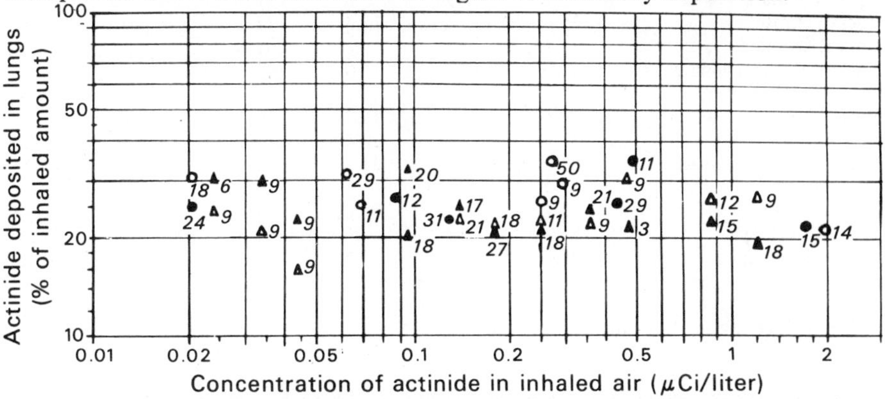

FIG. 1. Deposition in rat lung of soluble compounds of Pu and Am. Pu(IV) citrate, open circle; Pu (IV) pentacarbonate, filled circle; Am(III) citrate, open triangle; Am(III) nitrate, filled triangle. The number of rats analyzed is shown next to each data point.

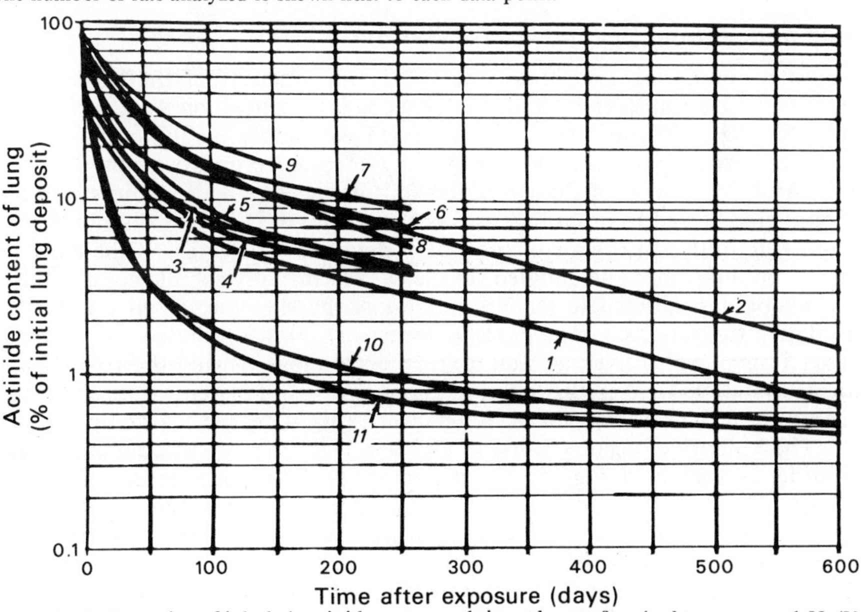

FIG. 2. Retention of inhaled actinide compounds in rat lungs after single exposures. 1-Np(V-VI) nitrate, 2-Np(IV) oxalate, 3-Pu(IV) citrate, 4-Pu(VI) acetate, 5-Pu(IV) pentacarbonate complex (PPC_1), 6-Pu(III) chloride, 7-Pu(IV) nitrate, 8-Pu(IV) pentacarbonate (PPC_2), 9-Pu(IV) polymeric, 10-Am(III) nitrate, 11-Am(III) citrate.

TABLE 1
Clearance of inhaled compounds Np, Pu, or Am from rat lung

Element and compound	Parameters of clearance equations[a]					
	α_1 (%ILD)[b]	T_1 (days)	α_2 (%ILD)[b]	T_2 (days)	α_3 (%ILD)[b]	T_3 (days)
Np(V-VI) nitrate	26.0	15.4	8.1	168	—	—
Np(IV) oxalate	43.8	22.3	22.6	151	—	—
Pu(IV) citrate	35.8	13.3	11.3	172	—	—
Pu(VI) acetate	34.3	14.5	8.8	209	—	—
Pu(IV) pentacarbonate complex (PPC_1)[c]	48.7	17.8	11.0	167	—	—
Pu(III) chloride	79.7	5.0	19.7	169	—	—
Pu(IV) nitrate	57.4	15.1	20.4	212	—	—
Pu(IV) pentacarbonate (PPC_2)[d]	64.0	18.1	23.3	122	—	—
Pu (IV) colloid	52.4	16.4	36.0	131	—	—
Am(III) nitrate	51.9	6.9	2.6	74.5	0.81	855
Am(III) citrate	49.4	8.0	2.6	64.2	0.61	1300

[a] Clearance equation of the form, lung content (%ILD) = $\sum_{i=1}^{i=n} \alpha_i e^{-0.693/T_i}$

[b] ILD; initial lung deposit.

[c] PPC_1; ammonium plutonium pentacarbonate complex dissolved in 10% ammonium carbonate.

[d] PPC_2; ammonium plutonium pentacarbonate dissolved in water.

Translocation from lung to liver and skeleton

After inhalation as nitrates the maximum amounts of the initial lung deposits (ILD) transferred to organs of secondary deposition were as follows: Np, 72%; Am, 52%; and Pu, 10%. If Np (IV) oxalate was inhaled, the maximum amount of Np transferred from lung to other organs was only 30% of ILD. After inhalation of Pu complexes (citrate, acetate, PPC_1), the maximum amount transferred to other organs was about 30% of ILD. The distribution of inhaled actinides in organs of secondary deposition depends on the element but not on the compound inhaled (in the case of initially soluble compounds). The major deposition site for Np and Pu is the skeleton; at all times after inhalation the ratio, skeleton (%ILD)/liver (%ILD) $\geqslant 3$. Inhaled Am is transferred from lung to liver as well as to the skeleton. During the first 8 days after inhalation of Am the ratio, skeleton/liver < 1; later, as the Am was cleared from the rat liver, the skeleton/liver ratio increased markedly.

The kinetics of accumulation in and loss from the rat skeleton of inhaled actinides is shown in Fig. 3. The maximum skeletal deposits and $T_{1/2}$'s are given in Table 2. Inhaled Np(V-VI) nitrate accumulated in the skeleton very rapidly (59% of ILD at 6 hr after exposure), whereas the maximum skeletal deposits of Np (IV) oxalate and of the Am compounds and soluble Pu complexes were 23 to 30% of ILD and were reached at 8 to 23 days after exposure, respectively. After inhalation of readily hydrolyzable Pu(IV) compounds (PPC_2 or nitrate), maximum skeletal deposits were less than 10% of ILD and were not reached until 64 to 128 days after exposure. The $T_{1/2}$'s of actinides in the skeleton were greater than the life-spans of the animals (see Table 2). A portion of inhaled Np, Pu, or Am compounds deposited in lung was translocated to the liver, and maximum accumulations in liver were 6, 8, and 25%, respectively. Most of the maximum liver accumulation was cleared fairly rapidly. The remainder, ranging from 0.6 to 3.8%, was lost at a slow rate which was independent of the element or compound inhaled.

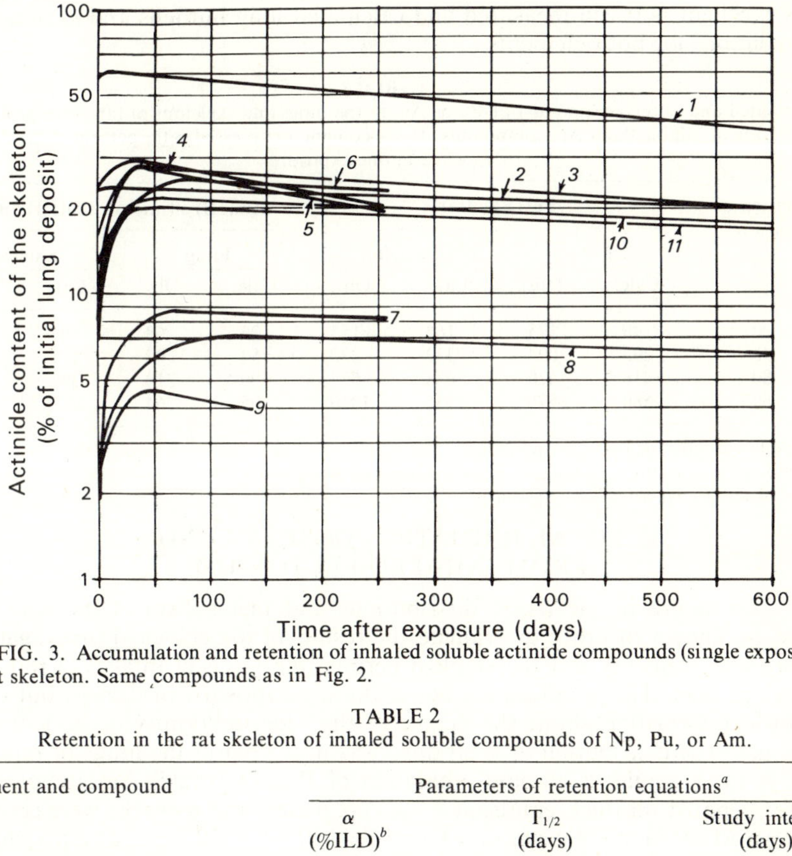

FIG. 3. Accumulation and retention of inhaled soluble actinide compounds (single exposure) in rat skeleton. Same compounds as in Fig. 2.

TABLE 2
Retention in the rat skeleton of inhaled soluble compounds of Np, Pu, or Am.

Element and compound	Parameters of retention equations[a]		
	α (%ILD)[b]	$T_{1/2}$ (days)	Study interval (days)
Np(V-VI) nitrate	61.0	845	512
Np(IV) oxalate	23.4	2310	650
Pu(IV) citrate[c]	29.1	1080	846
Pu(III) chloride	26.0	1320	256
Pu(IV) nitrate	8.8	2680	256
Pu(IV) PPC_2[c,d]	7.4	2360	834
Am(III) nitrate	21.8	1930	650
Am(III) citrate	20.3	2170	650

[a] Skeleton (%ILD) = $\alpha e^{-0.693/T_{1/2}}$
[b] ILD, initial lung deposit.
[c] Observed for lifespan
[d] PPC_2; ammonium plutonium pentacarbonate dissolved in water.

Tissue distributions after single and chronic inhalation

Studies were performed in which rats inhaled Pu(IV) citrate, Am(III) nitrate, or Am(III) citrate for 10 min daily, 6 days per week. The lung and skeleton contents of Pu or Am, measured after varying periods of chronic inhalation, are compared in Table 3 to the organ contents calculated with the use of metabolic parameters derived from single inhalations. The agreement was excellent between measured and calculated lung burdens and skeletal accumulations for both elements.

TABLE 3
Calculated and observed amounts of Pu or Am in the lungs and skeletons of rats after prolonged inhalation of soluble Pu or Am compounds. Tissue content is expressed as the percent of the amount deposited in one exposure.

Duration of exposure[b]	Pu(IV) citrate				Am(III) citrate or Am(III) nitrate			
	Lung		Skeleton		Lung		Skeleton	
	Calc.	Obs.	Calc.	Obs.	Calc.	Obs.	Calc.	Obs.
20	190	175	166	155	123	68	99	98
40	293	294	332	222	154	90	199	193
80	416	420	638	537	180	210	398	395
160	580	576	1236	1040	205	264	783	694[c]

[a] Combined data for both compounds.
[b] Inhalation exposures, 10 minutes each, 6 times per week.
[c] Measured 8 days after last exposure.

LOCAL RADIATION DOSE IN LUNG FROM INHALED PLUTONIUM

The nature of the tissue distributions and metabolism of the actinides must be known in order to evaluate the results of toxicological investigations with these elements and to establish permissible levels of intakes of radionuclides in man. The actinides are not uniformly deposited in tissues, but there is little information about the degree of that non-uniformity or its temporal changes. The distribution of local dose was measured in the lungs of rats 1 to 1.5 yr after single or chronic inhalation of Pu(IV) citrate. Autoradiographs were prepared on thick-emulsion Type A-2 plates, and α-tracks were counted using a MBM-6 microscope in a visual field of 3.8×10^{-5} cm^2 at a magnification of $\times$ 1575. Lungs were taken from three rats at each sacrifice interval, and 20 visual fields in each section of rat lung were studied. Visual fields were selected such that peripheral, central and radial areas were represented in each sample, taking account of the fact that some fields did not contain tracks. Because the area of the visual field was the same as that of a central cross-section of a point source of Pu, i.e., a "star", the energy absorbed in the selected volume was the same for a given number of α-tracks whether diffusely distributed or oriented as a "star".

Local dose rates were calculated according to ref. (*3*).

$$\text{DR (rad/day)} = \frac{0.021\text{NE}}{\text{TAh}\zeta}, \qquad \text{Eq. (1)}$$

where N is the number of α-tracks in a visual field; E is the ^{239}Pu α-particle energy, 8.22×10^{-6} erg; T is the time the autoradiographic plate was exposed in days; A is the area of the visual field, $3.8 \times 10^{-5} cm^2$; h is the thickness of the section, 7×10^{-4}cm; and ζ is the density of soft tissue, 1.1 g/cm^3.

The measurements show that local dose rates calculated from Eq. (1) do not differ very much from average radiation dose rates (ADR) calculated for a uniform concentration of the nuclide in the lung. The number of visual fields that did not contain α-tracks increased exponentially with time, which means that the amount of unirradiated lung (NIR) increases with time after a single inhalation. However, the Pu is unevenly distributed in the irradiated lung mass (IR). The fractions of lung tissue that were irradiated at local dose rates ranging from 0.25 to 5 times ADR were measured at various times after a single inhalation or the chronic inhalation of Pu(IV) citrate. These measurements have been converted to the nomograms in Fig. 4 (single inhalation) and Fig. 5 (chronic inhalation), from which it is possible to determine at any time up to 800 to 900 days after cessation of inhalation the amount of lung tissue irradiated and the dose rate distribution. For example, using Fig. 4 one can determine that 100 days after a single inhalation of Pu(IV) citrate 60% of rat lung is being irradiated (line a), less than 15% of the lung is being irradiated at a rate greater than ADR (line b), and only 0.5% of the lung is being irradiated at a rate greater than 5 × ADR (line c). By 600 days after inhalation only 6% of the lung is being irradiated (line d), and 1.3% of the lung receives a dose rate more than ADR (line e). From Figs. 4 and 5 and Table 4 it can be seen that both the amount of lung tissue irradiated and the fractions irradiated at the higher dose rates decline sharply with time after inhalation. Further, at the same times after the last inhalations of Pu(IV) citrate, the amounts of lung being irradiated and the distributions of the local dose rates are nearly the same for both single and chronic exposures.

Microscopic Distribution in Lung of Inhaled Plutonium

Autoradiographic studies demonstrate the high concentration of ^{239}Pu (leading to high dose rates) in the subpleural zones of the lung (*4,5*). Such "hot spots" of α-irradiation may lead to metaplasia of the epithelium, sclerosis, and finally to alveolar-bronchiolar carcinoma.

Inhaled $^{239}PuO_2$ particles deposited in the alveoli of rats are taken up by alveolar macrophages within a few hours after exposure. Two days after inhalation more than 95% of the $^{239}PuO_2$ particles were found in phagocytic cells. More than 85% of the ^{239}Pu was present in only 2% of collected macrophages, creating local doses that are significantly greater than ADR in small volumes of lung tissue. One must consider that at the maximum permissible average dose rate to lung of 15 rem/yr, 2% of lung tissue may be receiving a dose as high as 700 rem/yr (*4*).

Significant amounts of $^{239}PuO_2$ and other insoluble actinide compounds are always found in the tracheo-bronchial lymph nodes (*4,6*). However, in spite of the high concentrations of actinides, development of tumorous processes has not been observed in those lymph nodes.

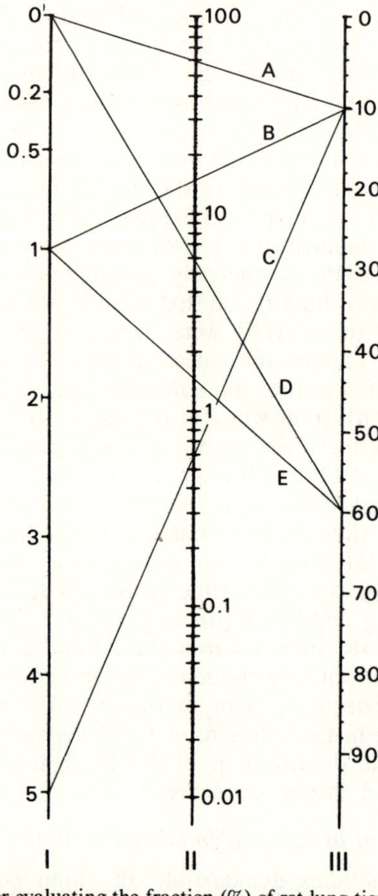

FIG. 4. Nomogram for evaluating the fraction (%) of rat lung tissue irradiated at a local dose rate (multiple of ARD) at different times after a single inhalation of a soluble Pu compound. Scale I. Local dose rate (multiple of ARD). Scale II. Fraction of lung irradiated (%). Scale III. Time after exposure (days).

The Problem of Relating Effects to Dose Distribution

Microscopic distribution of the actinides is also always non-uniform in other tissues such as kidney and skeleton. Thus, the local tissue doses in "hot spots" may be significantly greater than the average dose. The degree of non-uniformity of irradiation of tissue and organs containing actinide deposits has not yet been adequately evaluated, nor has non-uniform dose distribution been correlated with the development of damage. The available information on those points helps to focus attention on one of the fundamental questions of modern radiobiology: Which is the more dangerous—the equal irradiation of all tissues and organs or irradiation only of parts of them but at proportionately larger doses?

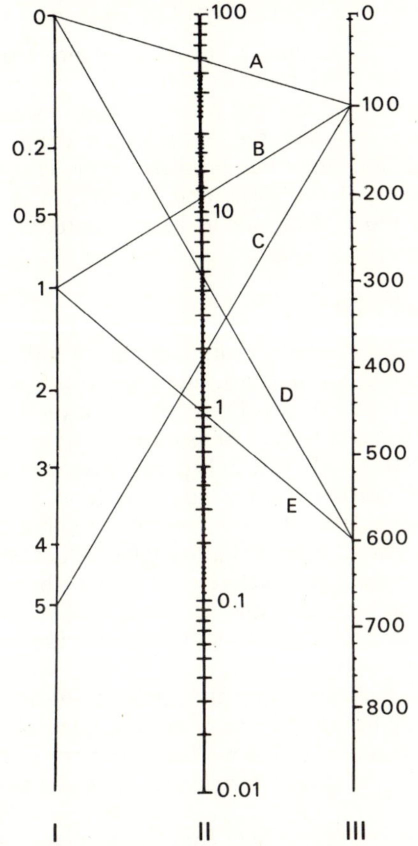

FIG. 5. Nomogram for evaluating the fraction (%) of rat lung tissue irradiated at a local dose rate (multiple of ARD) at different times after cessation of chronic inhalation of soluble Pu compounds.

TABLE 4

Temporal changes in the amount of lung tissue irradiated (IR) and unirradiated (NIR) and the distribution of local dose rates relative to the average radiation dose rate (ADR)[a] after cessation of single or chronic inhalation of Pu(IV) citrate by rats.

Time after last exposure (days)	Exposure pattern	Percent of lung irradiated at local dose rate						
			Local dose rate (multiple of ADR)[a]					
		NIR	<0.25	0.25–0.5	0.5–1	1–2	2–5	>5
100	Single	40	22	13	10	9	5.5	0.5
	Chronic	40	28	15	4	7	4.3	1.7
300	Single	77	8	4	5	3.7	2.1	0.2
	Chronic	80	10.5	3	2.3	2	1.6	0.6
600	Single	94	2.5	1	1	0.9	0.55	0.05
	Chronic	96	2	0.5	0.6	0.4	0.36	0.14

[a] ADR is the radiation dose a tissue would receive if all the radioactivity it contained were distributed uniformly.

LATE EFFECTS OF DEPOSITED ACTINIDES

Late effects in experimental animals of deposited ^{237}Np, ^{238}Pu, ^{239}Pu, ^{241}Am, ^{244}Cm, ^{252}Cf, or ^{253}Es include both non-tumorous processes and neoplasia. Both tumor development and non-tumorous processes, chiefly sclerotic, are observed in tissues in which the actinides are deposited or through which they are transported. Non-tumorous processes include such tissue reactions as inflammation, cirrhosis of the liver, and pneumo- and nephrosclerosis. The neoplasms observed most often are leukemia, osteosarcoma and cancers of lung, kidney and liver (*1,2,4,6–13*).

Induction of Osteosarcoma

The experiments conducted in our Laboratory indicate that the long-term sequelae in experimental animals are often the result of comparatively small accumulated average tissue doses. For example, after inhalation of Pu(IV) as the citrate or the PPC_1 complex, osteosarcomas were observed in 2/157 rats (average bone doses, 16.6 rad) and 2/192 rats (average skeletal doses, 3.6 and 6.4, rad, respectively). Moreover, the rats that developed osteosarcomas after accumulating average skeletal doses of less than 100 rad lived as long or sometimes longer than controls. After inhaling 0.08 μCi of Pu(IV) citrate or 0.045 μCi of the PPC_1 complex, the skeletons of rats with osteosarcomas had accumulated average doses of 16.6 and 25.2 rad, respectively, and had lived on the average 787 $\pm$ 69 and 735$\pm$ 35 days; the average lifespan of the controls rats was 570 $\pm$ 8 days.

After a single intravenous or intratracheal administration of Np(V-VI) nitrate or Np(IV) oxalate, the minimum accumulated skeletal doses were 4 to 5 and 5 to 13 rad, respectively. The average lifespans of tumor bearers ranged from 710 to 794 days—the same as or longer than the average control lifespan of 700 days.

The longest-lived irradiated rats [one each in the groups given 0.017 or 0.083 μCi/kg of Np (V-VI) nitrate] accumulated skeletal doses of 5 and 23 rad and died on Day 1030 and 1036, respectively. The last control rat died on Day 904. The average lifespans of the rats given intravenous injections of 0.017 μCi/kg or 0.083 μCi/kg of Np (V-VI) nitrate were 711 $\pm$ 20 and 692 $\pm$ 20 days, respectively—the same as that of the controls, 700 days (*7,14*).

Other workers have observed osteosarcomas in rodents that had accumulated small skeletal doses from administered actinides. ^{239}*Pu*: Two rats developed osteosarcomas with accumulated bone doses of 80 and 200 rad after intraperitoneal injection of $^{239}PuO_2$ particles (*15*). The incidence of osteosarcomas in female CF-1 mice was increased from 1.1% in controls to 3.9% after intravenous injection of 0.16 μCi/kg of ^{239}Pu(IV) citrate. The average bone dose was 40 rad, 140 days before death, and the average survival of the injected group was 460 days compared to 450 days for controls (*16,17*). ^{241}*Am*: Osteosarcomas were induced in rats at accumulated bone doses of 17 rad from injected ^{241}Am (*11*). ^{253}*Es*: After intratracheal administration of 10.7 μCi/kg of $^{253}EsCl_3$ to rats, osteosarcomas were seen in 8.3% (4/48 rats); 150 days after injection skeletal doses were 600 rad. As of 607 days after injection no ill effects had been seen at a dose of 0.21 μCi/kg of ^{253}Es and an accumulated bone dose of 15 rad (*12*).

Induction of Tumors of the Lung and Other Soft Tissues

The incidence of malignant lung tumors (squamous cell carcinoma, hemangioma, adenocarcinoma) in rats that inhaled small amounts of Pu(IV) citrate or the PPC_1 complex was 12 to 30 times greater than in controls. At accumulated lung doses of 44 and 210 rad, incidences of lung cancers were 4.7 and 10.8%, respectively; incidence was 0.4% in controls (*18*).

Plutonium-238 is also highly carcinogenic in lung. Rats inhaled a "soluble" ^{238}Pu aerosol (72% ultrafilterable, count median diameter 0.02μm; prepared by soaking crushed $^{238}PuO_2$ microspheres in saline for 3 months). Lung cancer incidence was 23.3% at an accumulated lung dose of 32 rad; control incidence was 1.1%. Other soft-tissue tumors were increased after deposition in lung of 5 or 18 nCi of ^{238}Pu. Osteosarcomas developed in three rats with accumulated bone doses of 51, 91, and 152 rad. Myelogenous leukemia was observed only in controls (*19*).

Pneumosclerosis Following Inhalation of Actinides

We have studied the incidence of pneumosclerosis in rats after intratracheal instillation of Np(V-VI) nitrate or Np(IV) oxalate, Pu(IV) nitrate, or Am(III) nitrate (*5,14*). When soluble actinide compounds are deposited in the lungs and their long-term sequelae are compared on the basis of radiation doses in the lung, the actinides are found to be almost equally effective in inducing pneumosclerosis.

Relationship of Late Effects to Dose Rate and Dose Distribution

As shown in Fig. 6, the frequency of pneumosclerosis induction (%/rad) decreases with increasing total dose. For the three actinides tested the frequencies of development of lung cancer (Fig. 7) or osteosarcoma (Fig. 8) were comparable, and the frequencies of those late effects also decreased with increasing total dose. A relatively high biological effectiveness per unit dose at small radiation doses has also been observed for the late effects of internally deposited β-emitters (*20*). The relatively lower biological effectiveness (for late effects) of high doses is apparently due to the fact that such doses are lethal to cells and markedly shorten the lifespan. Thus, late-appearing sclerogenic or oncogenic effects, which are the result of more subtle changes in surviving cells, cannot be revealed. The change with dose in the frequency of induction of late effects has practical importance. In comparing the radiation sensitivities of different species of animals and of people of different ages, it is necessary to compare the effects produced by the same total doses.

The question still remains as to what volume of lung tissue must be irradiated by α-particles or what amount of $^{239}PuO_2$ must be deposited in lung to produce a lung cancer. Transthoracic injection of $^{239}PuO_2$ particles into rat lungs has been found to cause lung cancers adjacent to the fibrous zones containing the Pu particles (*21*). Thus, cancers can be induced by α-irradiation of a limited volume of lung tissue.

If $^{239}PuO_2$ articles of submicroscopic size are introduced into the peritoneal cavities of rats, mesotheliomas and a variety of abdominal sarcomas develop (*15,22*). For example, after injection of 2.88 μCi of $^{239}PuO_2$, which delivers 33,500 rad to the omentum, median survival was 350 days, and 27% and 38% of the rats had developed at least one mesothelioma or sarcoma, respectively. Despite the fact that the concentration of Pu in the mediastinal lymph nodes was higher (delivering a dose of 7600 rad/day) than in the fibrous nodules of the omentum (dose rate, 380 rad/day), tumors did not arise in the lymph nodes.

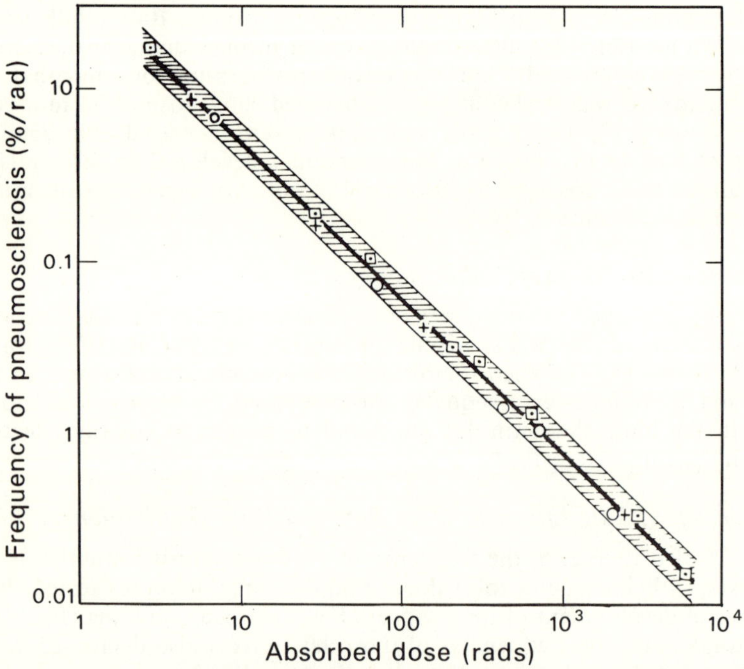

FIG. 6. Frequency of pneumosclerosis (%/rad) as a function of average radiation dose to rat lung (rad) after intratracheal instillation of actinide nitrates. Np(V-VI), cross; Pu(IV), open square; Am(III), open circle. Hatched area shows 95% confidence limits.

ASSESSMENT OF THE LATE EFFECTS ARISING FROM LOW DOSES OF ALPHA-EMITTERS

Interspecies Comparison

The actinides, Np, Pu, and Am, are selectively deposited on bone surfaces, where active bone cells (endosteal cells) are located. Damage in the endosteal cells is presumed to be the initiating event in formation of osteosarcoma. Table 5 is a summary of our data on the oncogenic action of those α-emitting nuclides deposited in the bones of rats. The short-lived α-emitter, ^{224}Ra, which also delivers most of its energy to bone surfaces, was administered therapeutically to human beings (*17,26*), and the bone tumor incidence in that group is included.

Comparison Over a Similar Range of Doses

The effects of internal α-irradiation are best compared for the same or nearly the same total radiation doses. In spite of the large differences in the lifespans of rats and men, similar average accumulated bone doses from Np, Pu, and Am in adult rats and from ^{224}Ra in juvenile and adult human beings appear to produce similar carcinogenic effects. The frequency of development of osteosarcoma in the rat per unit of accumulated bone dose (0.05%-rad, on the average for the three nuclides) is six and four times the tumor frequency in human adults and juveniles, respectively.

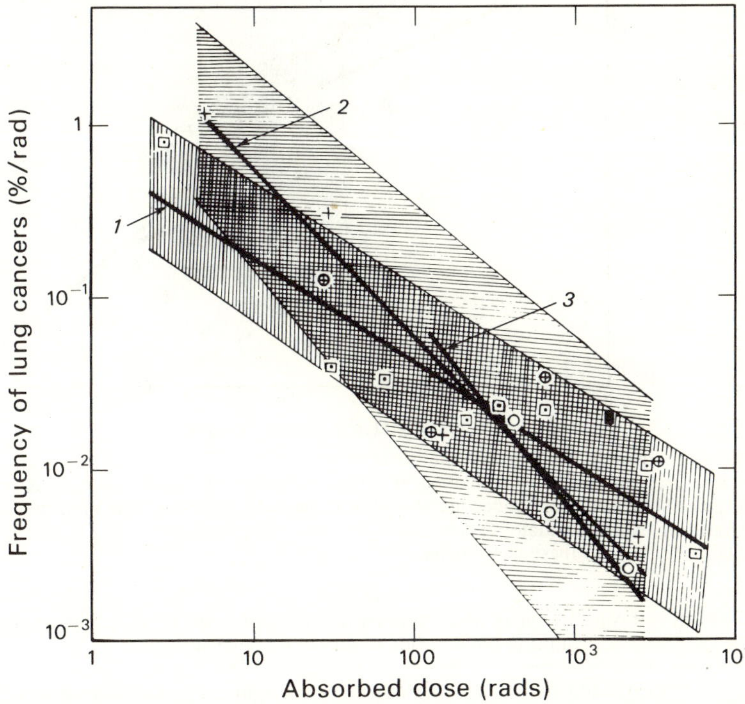

FIG. 7. Frequency of lung cancer (%/rad) as a function of average radiation dose to rat lung (rad) after intratracheal instillation of actinide compounds. Curve 1-Pu(IV) nitrate, open square, and Pu(VI) acetate, filled square. Curve 2-Am(III) nitrate, open circle. Curve 3-Np(V-VI) nitrate, cross, and Np(IV) oxalate, circle with cross. Hatched areas show 95% confidence limits (vertical, Pu(IV) compounds; horizontal Np(IV), Np(V-VI), and Am(III) compounds).

Comparisons Based on Linear Extrapolation

However, if oncogenic effect is related to absorbed dose using linear extrapolations over varying ranges of dose, different results are obtained. Data used are from ref. (*17*) in which osteosarcoma incidences were plotted on linear coordinates as functions of absorbed dose; the slopes of the straight lines obtained by least-squares fitting are the tumor frequencies which are shown in Table 6. Osteosarcomas appear to be 10 to 30 times more frequent in young adult dogs given single injections of ^{228}Ra, ^{228}Th, or ^{239}Pu than in either adult or juvenile human beings exposed to ^{224}Ra. Osteosarcoma incidences in the dogs given those nuclides were two to six times higher than in adult rats injected with ^{239}Pu.

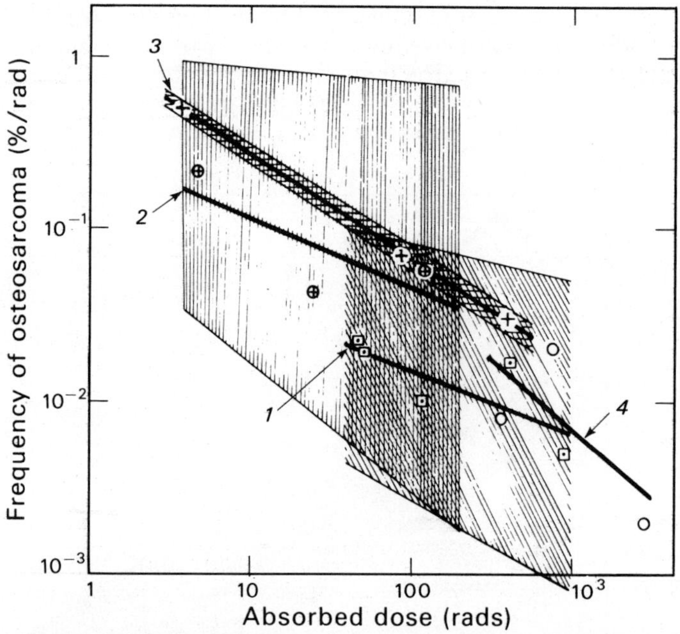

FIG. 8. Frequency of osteosarcoma (%/rad) as a function of average radiation dose (rad) to the rat skeleton after intratracheal instillation of actinide compounds. Same symbols as in Fig. 7. Hatched areas show 95% confidence limits. Curve 1, diagonal hatching, Pu(IV) nitrate. Curve 2, vertical hatching, Np(IV) oxalate. Curve 3, horizontal hatching, Np(V-VI) nitrate. Curve 4, Am(III) nitrate; limits lie beyond these coordinates.

TABLE 5
Frequency of osteosarcomas which develop in rats and human beings as the result of α-emitting radionuclide deposited on bone surfaces.

Species	Element and compound	Dose (μCi/kg)	Skeletal dose (rad)	Osteosarcomas (%)	Osteosarcomas (%/rad)	Ref.
Rat	Np(V-VI) nitrate	2.0	520	34.6	0.067	—[a]
	Pu(IV) citrate	1.3	503	19.2	0.038	23
	Am(III) citrate	1.16	422	17.3	0.048	24
Man						
Juvenile	^{224}Ra "Peteosthor"[b]	—	1100[c]	13.5	0.012	25
Juvenile	^{224}Ra "Peteosthor"	—	1100[d]	20	0.018	25
Adult	^{224}Ra "Peteosthor"	—	139–650	1.3–5.45	0.008	26
Juvenile	^{224}Ra "Peteosthor"	—	363–3329	5.7–36.3	0.013	26

[a] Yu. I. Moskalev et al., unpublished.
[b] "Peteosthor" is the name of a proprietary preparation of ^{224}Ra.
[c] Bone of children irradiated at 160 rad/mo. for 6.8 mo.; 14/104 developed osteosarcomas.
[d] Bone of children irradiated at 70 rad/mo. for 15.6 mo.; 21/104 children developed osteosarcoma.

Dose Distributions in Skeletons of Different Species

The greater tumorigenicity of ^{239}Pu in the bones of dogs (compared with rats) may be the result of a relatively greater dose to surface structures in dog bone for the same injected dose of ^{239}Pu. After injection of 1 μCi/kg of ^{239}Pu, the Pu concentrations on surfaces of lumbar vertebral trabeculae were 140, 258, and 444 pCi/cm^2 in rats, rabbits, and dogs, respectively (*27*). Tissue doses in the dog bones are, therefore, three times higher than in rat bones. The dose to bone surfaces may not be the entire answer, because the concentration of ^{239}Pu on human vertebral trabeculae after injection of 1 μCi/kg is estimated to be 800 pCi/cm^2—higher than in either dogs or rats. From these kinds of calculations we would expect a higher frequency of osteosarcomas in the human beings than in the experimental animals, but the reverse is actually the case.

TABLE 6
Frequencies of osteosarcoma induction after administration of bone-seeking α-emitters to various mammalian species (see ref. *17*).

Nuclide	Mammal	Osteosarcoma frequency (%/rad)
^{226}Ra + ^{228}Ra	man, adult	0.0046
^{224}Ra	man, adult	0.009
^{224}Ra	man, juvenile	0.015
^{226}Ra	dog, young adult	0.043
^{228}Ra	dog, young adult	0.13
^{228}Th	dog, young adult	0.28
^{239}Pu	dog, young adult	0.38
^{239}Pu	rat, adult, wistar	0.06
^{226}Ra	mouse, adult, CF-1[a,c]	0.043
^{239}Pu	mouse, adult, CF-1[a,c]	0.10
^{224}Ra	mouse, adult, NMRI[a,c]	0.12
^{224}Ra	mouse, adult, NMRI[b]	0.0073

[a] Female.
[b] Male.
[c] Spontaneous osteosarcoma incidences of CF-1 and NMRI female mice are 1.1 and 1.3%, respectively.

FACTORS THAT INFLUENCE DEVELOPMENT OF LATE EFFECTS

Normal Incidence of Radiation-Induced Pathology

There are data indicating that the carcinogenic action of α-irradiation in bone depends on the spontaneous incidence of osteosarcomas in a given species or strain. For example, in female NMRI mice the normal incidence of osteosarcomas is 1.3%, and in ^{224}Ra-treated mice it is 0.12%/rad. On the other hand, in male NMRI mice spontaneous osteosarcomas have not been observed, and the frequency after ^{224}Ra treatment is 0.0073%/rad, much lower than in the females. The spontaneous incidence of osteosarcomas in dogs is 10^{-4}, an order of magnitude higher than in man, 10^{-5}, which is in accord with the higher frequency of α-radiation-induced osteosarcomas in the dog.

Protraction of Exposure

There is still no ready explanation for the observation of an enhanced oncogenic activity in the rat skeleton of actinides that are deposited over an extended interval. Osteosarcomas also developed more frequently in the juvenile human beings that were exposed to ^{224}Ra for a longer time at a lower dose rate (see Table 5).

Our studies with Np show that the same total skeletal dose induced 1.5 to 4 times more osteosarcomas when the Np was administered chronically than when it was given as a single dose. (see Fig. 9). The different osteosarcoma frequencies appear to be related to differences in the microscopic distributions of Np in the bones after the two patterns of administration. The unevenness of Np distribution in bone structures is most pronounced after a single injection—in the femur the site of greatest Np concentration is the trabecular surfaces immediately beneath the epiphyseal cartilage, and little Np is present in the diaphysis. After daily administration there is a nearly uniform distribution of Np in the metaphyseal trabeculae, and Np is present in the diaphysis as a result of bone growth. In the chronic case more bone is irradiated, and damaged structures continue to be irradiated, which may contribute to the higher frequencies of osteosarcomas and leukemia [8.9%, chronic administration; 4%, single administration (*14*)].

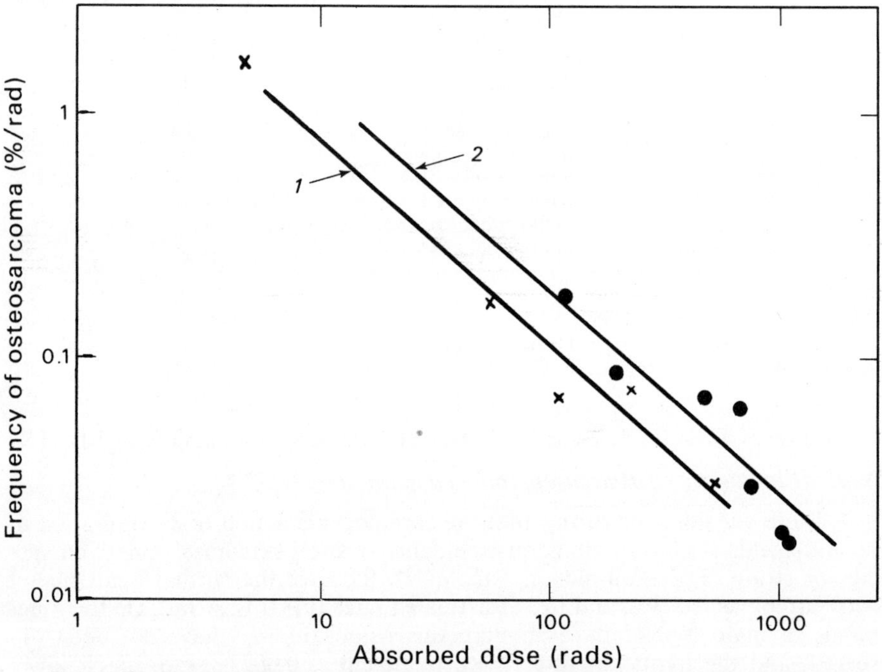

FIG. 9. Frequency of osteosarcoma (%/ rad) as a function of average radiation dose (rad) to the rat skeleton after intraperitoneal administration of Np(V-VI) salts. Curve 1-short-term administration, symbol "x". Curve 2-chronic administration, filled circle. Curves were fit by least squares using the "NAIRI-K" computer.

CONSIDERATIONS OF TIME OF OBSERVATION AND LIFESPAN

In addition to such considerations as the temporal dose delivery pattern and the concentration of nuclide on various bone structures, we must also remember that osteosarcomas arise throughout the lives of animals given Np,

Pu, or Am, and that the experimental animals were observed over their entire lifespans. Therefore, it seems likely that the actual frequency of osteosarcomas induced in people will be higher than is shown in Tables 5 and 6 and closer to the osteosarcoma frequencies of the rats given the actinide nuclides.

Radiosensitivity of Critical Cells

The data summarized above demonstrate the high carcinogenic action of the α-radiation of internally deposited actinides and apparently confirm the hypothesis that mammalian endosteal cells have a common radiosensitivity for osteosarcoma induction (*28*). However, the radiation sensitivity of bone cells with respect to tumor induction also appears to depend on radiation quality, on the physiological status (age and sex), and on the general predisposition of a population of animals to the induction of tumors.

The Problem of Repair

It should be noted that β-radiation is less efficient in inducing osteosarcomas than α-radiation. The frequencies of osteosarcoma induction in ^{90}Sr-treated animals is 0.0023, 0.0081, 0.0043, and 0.0014%/rad in rats, CF-1 mice, dogs, and miniature swine, respectively (*29*). Furthermore, the oncogenic action of α-radiation compared with the β-radiation of ^{90}Sr or ^{114}Ce has been observed over a wider dose range, after accumulation of smaller doses, and after shorter regimens of chronic administration. Induction of osteosarcomas by low LET radiation occurs over a narrow range of doses and decreases markedly if the dose is delivered in small amounts over a long period of time. These observations support the view that there is some repair of the cell damage caused by low LET radiation delivered at low dose rates, but little, if any, repair of the cellular damage caused by high LET radiation (*8,17,20*).

Physiological Factors

The effect of physiological factors on the action of the actinides has been studied very little. After injection of soluble (monomeric) ^{239}Pu, more osteosarcomas and chloroleukemias developed in male than in female rats, and at 16 months after injection there were more tumors in rats injected as adults than in those exposed in utero, at birth, or at weaning. On the other hand, liver tumors were seen only in rats exposed in utero or at birth (*30*).

EVALUATION OF SAFE LEVELS OF DEPOSITED ACTINIDES

The effort to evaluate safe levels of actinides in animals and man is beset by many difficulties. We have called attention to the fact that after incorporation of small amounts of those nuclides, some animals may live a full normal lifespan. In order to determine the tolerance level of incorporated radionuclides, one must know the deposited quantities of them that do not change the lifespan or the frequency and rate of development of tumors. In most cases the lifespans of rats with a particular type of tumor depend on the amount of radionuclide administered. A straight line is usually obtained from a plot of the lifespan (age at death) of animals with a particular tumor or pathology as a function of the logarithm of the administered activity (μCi/kg) or the accumulated dose in the critical organ (for examples, see Figs. 10, 11). The points at which the experimental dose-response line crosses the average and maximum lifespans of normal controls gives the indifference or non-effective doses—i. e., the amounts of radioactivity or accumulated radiation dose that will not shorten the average or maximum lifespans of the animals. Those doses may be considered to be measures of the tolerance of the organism to a particular radiation effect. In the case of a specific pathology, obviously the tolerance dose will be that dose which does not cause the incidence of that pathology in the experimental animals to exceed the control incidence. For example, the quantity of injected ^{252}Cf that does not reduce the lifespan or change the incidence and rate of appearance of mammary gland tumors in male rats is 0.178 μCi/kg. Over the lifespan of the animals, about 500 days, that amount of ^{252}Cf would deliver radiation doses of 90 and 10 rad to the bone and liver, respectively (*13*). Using the criteria of induction of soft-tissue tumors and maximum lifespan, the indifference dose of Np in rats is 0.06 μCi/kg; similarly, for induction of leukemia, liver tumors, or endocrine gland tumors, the indifference doses of Np are 0.025, 0.018, and 0.4 μCi/kg, respectively.

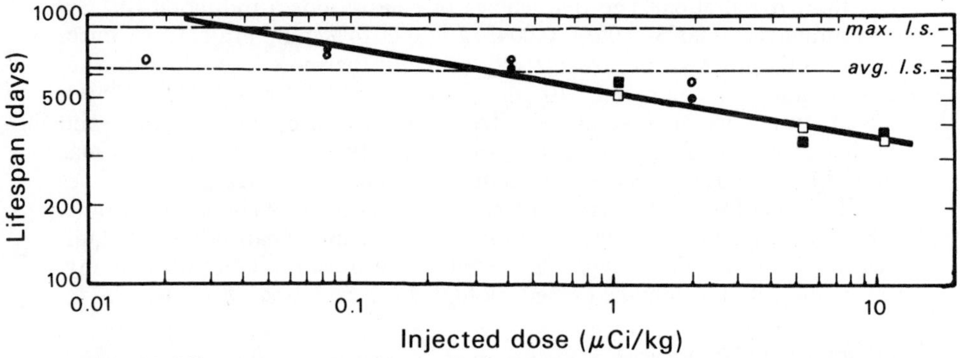

FIG. 10. Lifespan (age at death, days) of the first rat in each dose group that developed an osteosarcoma as a function of injected dose (μCi/kg) of Np compounds. Np(V-VI) nitrate, single dose, open circle; chronic administration, filled circle; Np(IV) oxalate, single dose, open square; chronic administration, filled square. Maximum (max. l.s.) and average (avg. l.s.) lifespans of normal controls are shown as broken lines in this and Figs. 11–14.

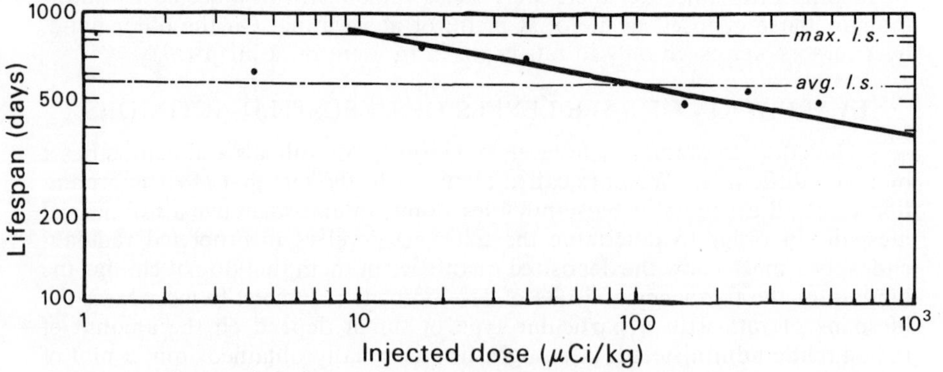

FIG. 11. Lifespan of the first rat in each dose group that developed an osteosarcoma as a function of the average radiation dose (rad) to the skeleton delivered by Pu(IV) compounds.

The doses of several actinides (obtained by the extrapolation technique just described) which should not increase the incidence of neoplasia or shorten the average or maximum lifespans of rats are shown in Table 7. Taken all together, those values indicate that a minimum effective dose based on the criterion of maximum lifespan is more rigid than that based on average lifespan. The values of the minimum effective doses in rats based on induction of osteosarcoma and maximum lifespan of controls for internally deposited ^{237}Np, ^{239}Pu, ^{241}Am, and ^{252}Cf are 6, 10, 12, and 10 rad to bone, respectively. If the criteria are induction of leukemia or liver tumors and maximum lifespan, the minimum effective doses to bone and liver from deposited ^{241}Am are 10 and 15 rad, respectively. It would be expected that if those average doses were accumulated over the animals' lifetimes, their maximum lifespans would not be shortened.

TABLE 7
Minimum effective absorbed doses from α-emitting actinide elements in rats based on the combined criteria (a) tumorigenesis and (b) shortening of the average or maximum lifespan.

Element and compound	Tumor type	Method of analysis of tumor data	Minimum effective dose (rad) based on	
			average lifespan	maximum lifespan
Np(V-VI) nitrate or Np(IV) oxalate	Osteosarcoma	Time of death of first tumor-bearer	63	6
^{252}Cf(III) nitrate	Osteosarcoma	Time of death of first tumor-bearer	100	~10
Pu(IV) citrate or pentacarbonate (PPC$_1$)	Osteosarcoma	Time of death of tumor-bearers	100	~10
Am(III) nitrate	Osteosarcoma	Avg. lifespan of tumor-bearers	320	12
Np(V-VI) nitrate or Np(IV) oxalate	Osteosarcoma	Avg. lifespan of tumor-bearers	120	12
Am(III) nitrate	Leukemia	Avg. lifespan of tumor-bearers	70	10
Am(III) nitrate	Kidney tumor	Avg. lifespan of tumor-bearers	60	5

On the basis of the above discussion, we conclude that determination of the tolerance levels of deposited radionuclides is complex and requires evaluation not only of the induction of a particular type of neoplasm but also of the effect on the length of life. Thus, the maximum permissible amount of a deposited radionuclide can be taken as that amount which does not cause harm (i.e., does not induce tumors of bone, lung or other critical tissues) to an animal in its maximum lifetime.

REFERENCES

1. L. A. Buldakov, E. R. Lyubchanskii, Yu. I. Moskalev, and A. P. Nifatov, *Problemi toksikologii plutoniya*. Atomizdat, Moscow, 1969. English translation LF-tr-41.
2. J. Vaughan, B. Bleaney, and D. M. Taylor in *Handbook of Experimental Pharmacology, Vol. 36, Uranium, Plutonium, Transplutonic Elements* (Hodge, Stannard, and Hursh, Eds.), pp.

349–502. Springer-Verlag, Berlin, 1973. W. J. Bair, J. Ballou, J. F. Park, and C. L. Sanders, ibid., pp. 503–568. P. W. Durbin, ibid. pp. 739–898.
3. F. E. Hoecker and P. G. Roofe, *Radiology* **56**, 89 (1951).
4. C. L. Sanders, *Health Phys.* **22**, 607 (1972).
5. R. A. Erokhin, N. A. Koshurnikova, V. K. Lemberg, A. P. Nifatov, and A. A. Puzyrev, Some remote aftereffects of intratracheal administration of chemically soluble plutonium-239 compounds. In *Otalennye posledstviya luchevykh porazhenii* (Yu. I. Moskalev, Ed.), pp. 315–333, Atomizdat, Moscow (1971). English translation AEC-tr-7387.
6. J. F. Park, W. J. Bair, and R. H. Busch, *Health Phys.* **22**, 803 (1972).
7. T. I. Levdik, V. K. Lemberg, L. A. Buldakov, E. R. Lyubchanskii, and V. M. Pesternikov, *Health Phys.* **22**, 643 (1972).
8. Yu. I. Moskalev, *Health Phys.* **22**, 723, (1972).
9. N. A. Koshurnikova, V. P. Aristov, V. K. Lemberg, G. S. Mushkacheva, M. G. Poplyko, and I. A. Tseveleva, *Health Phys.* **22**, 753 (1972). Abstract.
10. A P. Nifatov, L. A. Buldakov, and V. I. Matveev, *Health Phys.* **22**, 875 (1972). Abstract.
11. Yu. I. Moskalev. E. I. Rudnitskaya, L. A. Buldakov, and A. P. Nifitov, in *Collected abstracts,* Third Intl. Congress Intl. Radiation Protection Ass'n, Sept. 9–14, 1973. Washington, D.C., p. 11.
12. J. E. Ballou, and W. G. Morrow, Long-term biological effects of intratracheally instilled $^{253}EsCl_3$ in rats. *Pacific Northwest Laboratory Annual Report*, Vol. 1, Pt. 1, BNWL-1750, pp. 32–33 (1973).
13. Yu. I. Moskalev, G. A. Zalikin, V. N. Strel'tsova, I. K. Petrovich, A. I. Semenov, and E. S. Zhorova, *Radiobiologiya* **14**, 270 (1974). English translation AEC-tr-7571.
14. T. I. Levdik, V. K. Lemberg, L. A. Buldakov, E. R. Lyubchanskii, and V. M. Pesternikov, *Radiobiologiya* **11**, 747 (1971). English translation AEC-tr-7306.
15. C. L. Sanders and T. A. Jackson, *Health Phys.* **22**, 755 (1972).
16. M. P. Finkel and B. O. Biskis, *Health Phys.* **8**, 565 (1962).
17. C. W. Mays and R. D. Lloyd, in *Radiobiology of Plutonium* (B. J. Stover and W.S.S. Jee, Eds.), pp. 409–430. The J. W. Press, Salt Lake City (1972).
18. N. A. Koshurnikova, V. K. Lemberg, and E. R. Lyubchanskii, Late effects of inhalation of soluble plutonium-239 compounds. In *Otalennye posledstviya luchevykh porazhenii* (Yu. I. Moskalev, Ed.), pp. 305–314. Atomizdat, Moscow (1971). English translation AEC-tr-7387.
19. C. L. Sanders, Carcinogenicity of inhaled plutonium-238 microspheres. *Pacific Northwest Laboratory Annual Report*, Vol. 1, Pt. 1, BNWL-1750, pp. 28–31 (1973).
20. Yu. I. Moskalev, V. N. Strel'tsova, and L.A. Buldakov, Late effects of radionuclide damage. In *Delayed Effects of Bone-Seeking Radionuclides* (C. W. Mays et al., Eds.), pp. 489–509, University of Utah Press, Salt Lake City (1969).
21. C. L. Sanders and J. F. Park, in *Inhaled Particles III* (W. H. Walton, Ed.), pp. 489–497. Unwin Brothers, Ltd., Surrey, England, 1971.
22. C. L. Sanders, in *Radionuclide Carcinogenesis* (C. L. Sanders, R. H. Busch, T. E. Ballou, and D. D. Mahlum, Eds.), pp. 138–159. CONF-720505, U.S. Atomic Energy Commission Office of Information Services, 1973.
23. Yu. I. Moskalev, L. A. Buldakov, N. A. Koshurnikova, A. P. Nifatov, and G. N. Reshetov, *Radiobiologiya* **5**, 836 (1965). English translation AEC-tr-6603.
24. A. P. Nifatov, L. A. Buldakov, and L. G. Filippova, Blastomogenic action of ^{241}Am. In *Biologicheskoe deistvie vneshnikh i vnutrennikh istochnikov radiatsii* (Yu. I. Moskalev and V. S. Kalistratova, Eds.), pp. 340–345. "Meditsina", Moscow (1972). English translation AEC-tr-7457.
25. C. W. Mays, *Radiat. Res.* **55**, 593 (1973). Abstract.
26. H. Spiess and C. W. Mays, *Health Phys.* **19**, 713 (1970).
27. A. C. James and N. F. Kember, in *Radiobiology of Plutonium* (B. J. Stover and W.S.S. Jee, Eds.), pp. 281–291. The J. W. Press, Salt Lake City (1972).
28. R. H. Mole, in *Delayed Effects of Bone-Seeking Radionuclides* (C. W. Mays et al., Eds.), pp. 249–261, Univ. Utah Press, Salt Lake City (1969).
29. C. W. Mays and R. D. Lloyd, in *Biomedical Implications of Radiostrontium Exposure* (M. Goldman and L. K. Bustad, Eds.), pp. 362–370. AEC Symposium Ser. No. 25 (1972). (Available at CONF-710201, National Technical Information Service, Springfield, Va. 22151).
30. M. R. Sikov and D. D. Mahlum, *Radiat. Res.* **55**, 514 (1973). Abstract.

Late Effects of Inhaled Plutonium in Dogs[1]

James F. Park
Biology Department, Battelle, Pacific Northwest Laboratory
Richland, Washington 99352, USA

INTRODUCTION

The effects of inhaled $^{239}PuO_2$ and of inhaled $^{238}PuO_2$ in beagle dogs were summarized in 1972 (*1*). More recently the effects of inhaled plutonium in dogs compared with effects in other species (*2–6*) and with effects of other radionuclides have been reviewed (*7*). This report is an overview summarizing effects of inhaled $^{239}PuO_2$ and of inhaled $^{238}PuO_2$ in dogs with emphasis on dose-response relationships at low exposure levels.

EFFECTS OF INHALED $^{239}PuO_2$ IN BEAGLES

The following summary of the effects of inhaled $^{239}PuO_2$ was reported previously (*1*). Beagle dogs, 12–43 months old, were given 10–30 min inhalation exposures to $^{239}PuO_2$ aerosols and held for life-time observations. The $^{239}PuO_2$ was obtained by calcining the plutonium oxalate in air at 300–350° C. The CMD (count median diameter) of the aerosol ranged from 0.1 to 0.5 μm.

The dogs died or were sacrificed when death was imminent between 55 and 4068 days after exposure. The relationship between the quantity of plutonium initially deposited in the lower respiratory tract or alveoli, expressed as nCi/g of bloodless lung, and the survival time of the dogs is shown in Fig. 1. The curve, fitted to all the data by least squares analyses, was extrapolated to 15 yr postexposure (about 16 yr of age, the maximum expected life-span of these dogs). This suggests that deposition of more than 5 nCi per g of lung might be expected to cause premature death due to pulmonary neoplasia and pulmonary fibrosis-induced respiratory insufficiency. Between 55 and 1600 days postexposure, 36 dogs died due to plutonium-induced pulmonary edema, fibrosis, and bronchiolar and alveolar epithelial hyperplasia and metaplasia, resulting in severe respiratory insufficiency characterized by progressive hypercapnia and hypoxemia (*8*). Twenty-four of the dogs that survived more than 1000 days after exposure had pulmonary neoplasia in addition to the fibrotic and metaplastic lesions. Twenty of the 21 dogs that survived more than 1600 days postexposure had pulmonary neoplasia. None of the control dogs, some of which are still alive, had pulmonary neoplasia. The incidence of primary pulmonary neoplasia in dogs has been reported to be 0.2 to 0.6% (*9*).

Since mortality was due to two causes, pulmonary fibrosis and neoplasia, a curve was also fitted by least squares analysis to the data for only those dogs that showed pulmonary neoplasia (Fig. 1). The slope of the curve for all the dogs

[1]Congress presentation C–9–2.

differed from that for the tumor dogs. The intercepts at 15 yr postexposure were about 5 and 3 nCi/g, respectively. Considering the spread of the data points and the uncertainty in extrapolating the curves to 15 yr, a conservative estimate is that deposition of more than 1 nCi/g may cause premature death due to pulmonary neoplasia, or from neoplasia and pulmonary fibrosis-induced respiratory insufficiency.

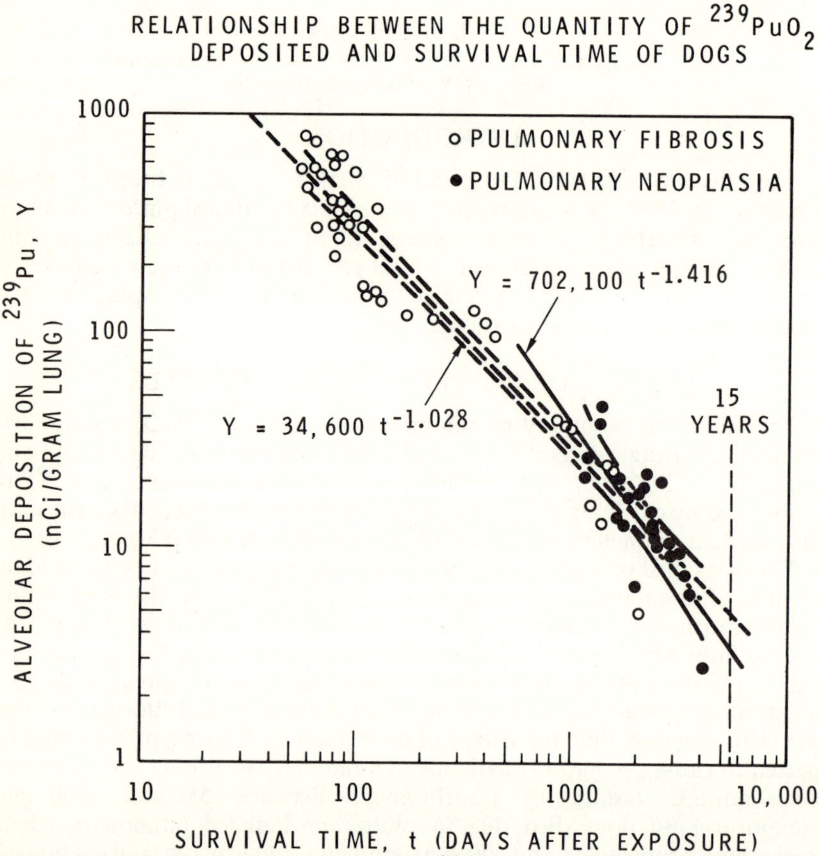

FIG. 1. Relationship between the quantity of $^{239}PuO_2$ deposited and survival time of dogs with pulmonary neoplasia and/or pulmonary fibrosis (---) compared with dogs that showed pulmonary neoplasia (——). (From ref. *1*.)

The estimated initial alveolar deposition in dogs with plutonium-induced pulmonary tumors was 0.2 to 3.3 μCi or 3 to 45 nCi/g bloodless lung. This amount of plutonium is 100–1500 times the estimated maximum permissible lung burden (the quantity at equilibrium resulting in a mean dose of 0.3 rem/week) for man, which is 0.016 μCi or about 0.03 nCi/g lung, assuming that the bloodless lung of man weighs 500 g. The conservative estimate of 1 nCi/g initial deposition ($\simeq$70 nCi in the total lung), causing premature death in the dog, is more than 30 times the concentration equivalent to the maximum permissible lung burden for man. This assumes that the induction of pulmonary tumors at low doses is related to plutonium concentration and not to total plutonium. If tumor induction is related to the total amount of plutonium, total number of particles, or total cells at risk, then the size of the human lung compared to the dog lung may not be important and 70 nCi in the dog lung, causing premature death due to pulmonary cancer, may be only about 5 times the amount equivalent to the maximum permissible lung burden for man.

The morphology of these lung tumors was described previously (*10,11*). Most of the tumors were bronchiolo-alveolar carcinomas of peripheral pulmonary origin. However, squamous cell carcinomas and epidermoid carcinomas were found in the lung, as well as bronchiolo-alveolar carcinomas. The majority of the tumors were peripheral in primary site of origin, usually appeared to be multicentric, and were rather slow to invade the lymphatics or vascular system. Metastases were common to the tracheobronchial and mediastinal lymph nodes and lymphatics, and were less commonly seen in kidney, diaphragm, bone marrow, liver, adrenal and mesenteric lymph nodes.

The terminal bronchiolar and alveolar epithelium were the most frequent target cells for damage and neoplastic transformation, due to the residual localization of ^{239}Pu particles in the peripheral, especially subpleural, alveolar walls and associated scar tissue. The induced lesions consist primarily of pulmonary fibrosis, bronchiolo-alveolar epithelial hyperplasia and metaplasia, alveolar histiocyte proliferation, pleural fibrosis, and alveolar-cell tumor formation. Tumors were observed radiographically as early as 2 yrs. prior to death (*8,12*).

The retention and translocation of alveolar-deposited plutonium is shown in Fig. 2. Approximately 10% of the alveolar-deposited plutonium was retained in the lungs at 11 yrs after exposure, having delivered an accumulated average radiation dose to the lungs of 2000–12,000 rad in the tumor-bearing dogs. Forty to fifty percent of plutonium was translocated from the lung to the tracheobronchial and mediastinal lymph nodes (thoracic lymph nodes): 10–15% to the liver, 5% to the skeleton and 5% to the abdominal lymph nodes (hepatic and splenic lymph nodes). About 85% of the plutonium initially deposited in the lower respiratory tract or alveoli was retained in the dogs 9–10 yrs after exposure. A dynamic simulation model for the biological disposition of inhaled $^{239}PuO_2$, developed with the aid of hybrid computer techniques and the data from these dogs, suggests that the clearance rate from the pulmonary compartment changes slowly and is best described mathematically by a power function (*13*). However, after the first 100 days, clearance of plutonium from the lung could be approximated by a half-time of 1000 days.

The distribution of plutonium in the tissues of dogs 2500–3500 days after exposure is shown in Table 1. The highest concentration occurred in the tracheobronchial and mediastinal lymph nodes, followed in descending order by abdominal lymph nodes, lungs, liver, other lymph nodes, spleen and skeleton. At 7–10 yrs after exposure, as much as 66% of the plutonium in the dogs was associated with the lymphatic system. Pathology in the plutonium-containing lymph nodes consisted of severe necrosis of lymphoid tissue with histiocytic proliferation and scarring.

The primary clinical signs associated with the fibrotic, metaplastic and neoplastic changes in the lungs and with the fibrosis of the lymph nodes were

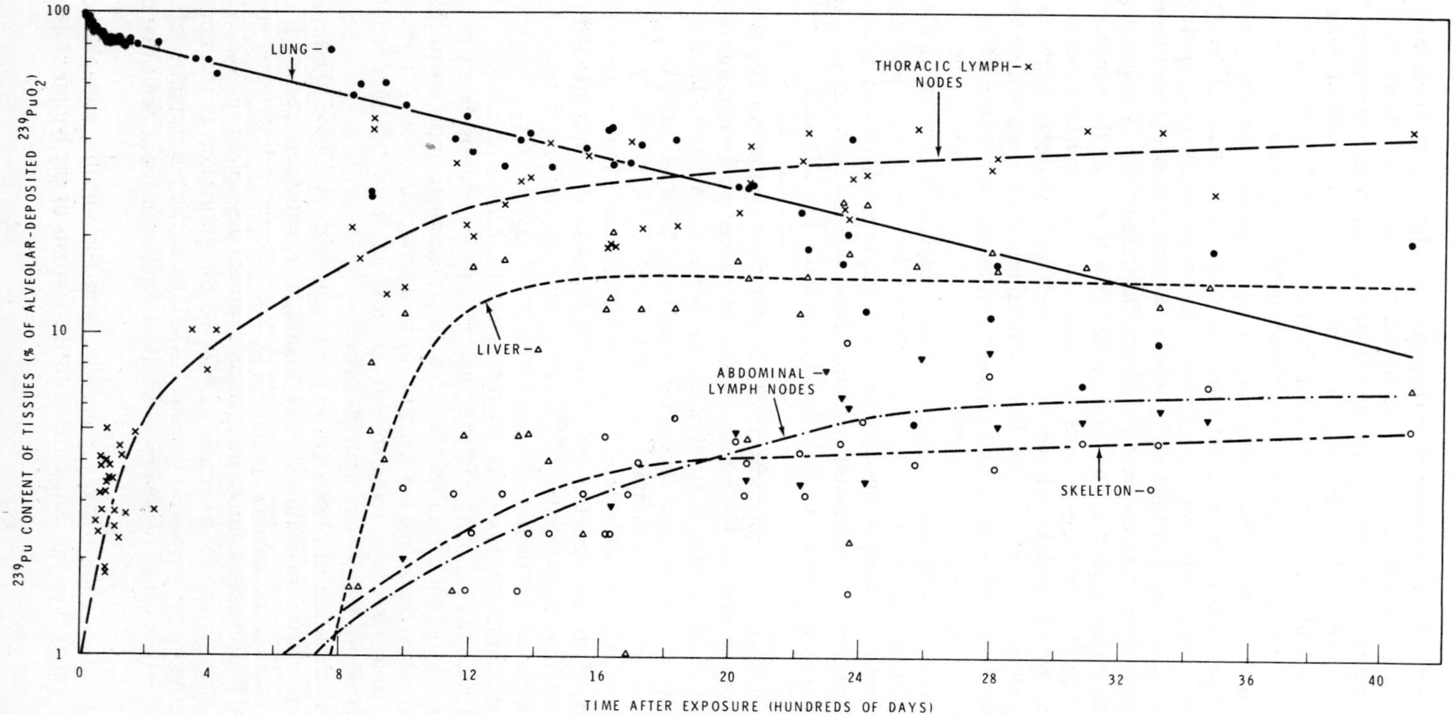

FIG. 2. Retention and translocation of alveolar deposited $^{239}PuO_2$ in dogs. (From ref. *1*.)

respiratory insufficiency and lymphocytopenia. Fig. 3 shows the mean leukocyte values of 14 dogs with body burdens of 0.2–1.0 μCi. The mean lymphocyte count of the plutonium-exposed dogs was less than that of the control dogs 6 months after exposure, and continued to be 30–50% of the control-dog values for the subsequent period. The mean total leukocyte count of the plutonium-exposed dogs was frequently less than that of the control dogs, primarily due to the decrease in lymphocytes. Leukocyte counts were frequently elevated prior to death when the lungs were severely damaged due to pulmonary neoplasia. No other significant changes were observed in the hemogram of the plutonium-exposed dogs. The critical plutonium containing tissues related to lymphocytopenia is unknown. Yuile et al. (*14*) suggested that the irradiation of circulating cells by plutonium in the lung and lymph nodes was responsible for lymphocytopenia, rather than a general depression of hematopoiesis and lymphopoiesis. There is a potential relationship between plutonium-induced lymphocytopenia, lymph node pathology, decreased immunological capability, and the pathogenesis of plutonium-induced pulmonary neoplasia.

TABLE 1[a]

Plutonium in Tissues of Dogs After Inhalation of $^{239}PuO_2$

Survival (days after exposure)	Terminal Body Burden (μCi)	Pu Concentration (nCi per gram wet tissue)[b]					
		Lungs[c]	Liver	Skeleton	Spleen	Thoracic Lymph Nodes[d]	Abdominal Lymph nodes[e]
2565	1.5	1.1	1.1	0.11	0.3	3700	160
		(70)	(21)	(5)	(0.6)	(56)	(9)
2792	0.8	1.2	0.5	0.11	0.4	1200	210
		(14)	(23)	(10)	(1.4)	(41)	(10)
2809	0.6	1.5	0.4	0.03	0.2	2600	92
		(21)	(21)	(5)	(0.7)	(45)	(6)
3313	0.4	0.7	0.3	0.04	0.1	420	90
		(13)	(16)	(6)	(0.6)	(56)	(7)
3441	0.5	1.2	0.3	0.03	0.03	180	71
		(24)	(22)	(9)	(0.7)	(36)	(6)

[a]From Ref. *1*.
[b]Percentage of the terminal body burden of plutonium in each tissue given in parentheses
[c]Estimated normal weight of the lung was used to calculate concentration since actual weight of the lung was 2–3 times the normal weight due to lesions
[d]Highest concentration in any tracheobronchial or mediastinal lymph node
[e]Highest concentration in any hepatic or splenic lymph node

The livers of some of the dogs that died showed passive congestion with central lobular degeneration. This may be related to circulatory changes caused by severe lung pathology. No significant pathological changes were observed in skeletons that could be related to plutonium in the skeleton. No primary neoplasia of the skeleton or liver, as was described for intravenously administered ^{230}Pu citrate in beagle dogs (*15*), was observed following inhalation of $^{239}PuO_2$.

In the experiments described above, the incidence of pulmonary neoplasia in beagle dogs after inhalation of $^{239}PuO_2$ was nearly 100% at average plutonium doses 100–1500 times the maximum permissible level for man. Table 2 shows the design of experiments initiated in 1970 to study the dose-effect relationship of inhaled $^{239}PuO_2$ in beagle dogs. Eighteen-month old beagles were given 5 to 20 min exposures to aerosols of $^{239}PuO_2$ prepared by calcining the oxalate at 750° C for 2 hours. The mean activity median aerodynamic diameter (AMAD) was 2.3 μm with a mean geometric standard deviation (GSD) of 1.9. The six dose-level groups ranged from 3 to 5800 nCi mean initial alveolar burden. The lowest deposition level corresponds to an average lung dose of about 0.3 rem/week, the maximum permissible level for man. The dogs will be held for life-time observation to determine the incidence of pulmonary neoplasia and evaluate other effects in beagle dogs at these low exposure levels.

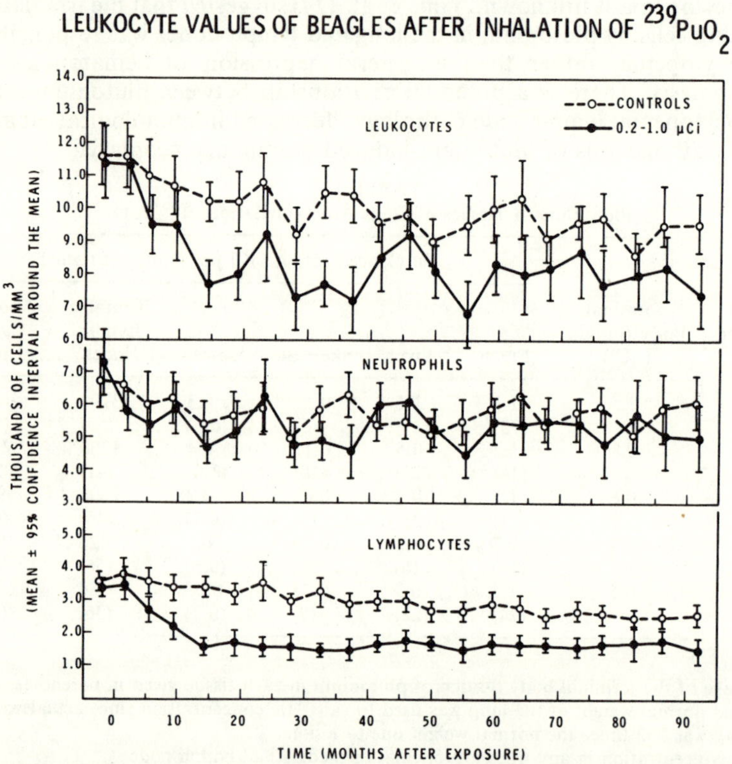

FIG. 3. Leukocyte values of dogs showing lymphocytopenia after inhalation of $^{239}PuO_2$. (From ref. *1*.)

During the first two postexposure years, dogs in the highest dose-level group were euthanized when death was imminent due to respiratory insufficiency. These dogs showed pulmonary fibrosis and fibrosis of the tracheobronchial and mediastinal lymph nodes. Clinical changes included increased respiratory rate, hypoxemia, anorexia, body weight loss and lymphocytopenia. The effects and plutonium distribution in these dogs were similar to that shown in Figs. 1 and 2 for dogs 1 to 2 years after exposure. Twelve to thirty-four percent of the plutonium was in the thoracic lymph nodes, 64 to 88% in the lungs and less than 3% translocated to other tissues.

TABLE 2
Dose-Effect Studies with Inhaled $^{239}PuO_2$ in Beagles

Dose Level Group	Number of Dogs		Initial Alveolar Deposition[a]	
	Male	Female	nCi[b]	nCi/g Lung[b]
0	10	10	0	0
1	10	10	3.5 ± 1.3	0.04 ± 0.02
2	10	10	22 ± 4	0.3 ± 0.05
3	10	10	79 ± 14	1.1 ± 0.2
4	10	10	303 ± 62	3.9 ± 0.7
5	10	10	1083 ± 167	14.1 ± 2.0
6	3	5	5827 ± 3282	85.6 ± 43.9
	63	65		

[a]Estimated from external thorax counts at 14 and 30 days postexposure and estimated lung weights.
[b]Means ± 95% confidence intervals around the means.

A dose-related lymphocytopenia was observed in the $^{239}PuO_2$ dogs (Fig. 4). The lymphocytopenia appeared to be related to initial alveolar burden, with a greater depression in lymphocyte counts in the high dose-level groups; and to time after exposure, with the lower dose-level groups showing lymphocytopenia later than the high-dose-level groups. The groups with mean initial alveolar burdens of 80 nCi and higher showed lymphocytopenia during the first 2 years after exposure, while the 22 nCi and lower dose-groups were similar to controls. The higher dose-level groups also showed a leukopenia, primarily due to the decrease in lymphocytes. No significant abnormalities were seen in other hematological and clinical chemistry measurements. The lymphocytopenia in the 80 nCi group at 2-years postexposure is the lowest dose-level at which an effect of inhaled plutonium in dogs has been observed, to date. In previous experiments (described above), dogs at the lowest level studied died due to pulmonary neoplasia 8 to 11 years after initial alveolar deposition of 200 to 800 nCi $^{239}PuO_2$. These dogs also showed lymphocytopenia. The 80 nCi dose-level group is 5 times the 16 nCi lung deposition which delivers an average radiation dose of 0.3 rem/week to the human lung and is equivalent to a dose of about 12 rem/week to the lung of a dog, 40 times the maximum permissible dose for humans.

EFFECT OF INHALED $^{238}PuO_2$ IN BEAGLES

The acute toxicity and biological effects of inhaled $^{238}PuO_2$ (calcined at 700° C, CMD 0.05 μm) in beagle dogs, up to 185 days post-exposure, was similar to that of $^{239}PuO_2$ at comparable radiation doses to the lungs. The dogs died due to pulmonary fibrosis-induced respiratory insufficiency. However, due to faster translocation of ^{238}Pu from the lung, the initial plutonium deposition required to produce the effect may differ (*16*). The dogs exposed to $^{238}PuO_2$ showed a higher rate of plutonium translocation to skeleton, up to 13% of the body burden after 6 months (Table 3), compared with less than 1% for $^{239}PuO_2$-exposed dogs. Liver and other tissues also reflected similar difference in translocation between the two plutonium isotopes.

Beagle dogs were exposed to aerosols of $^{238}PuO_2$ calcined at 350° C (CMD 0.1 μm) or $^{238}PuO_2$ crushed microspheres (CMD 0.1 μm) to determine the long-term disposition and biological effects. Eight of the 10 dogs exposed to 350° C calcined $^{238}PuO_2$ and eight of the 12 dogs exposed to $^{238}PuO_2$ crushed microspheres were euthanized when death was imminent during the 6-year postexposure period. Table 4 shows the causes of death and the ^{238}Pu distribution in the tissues.

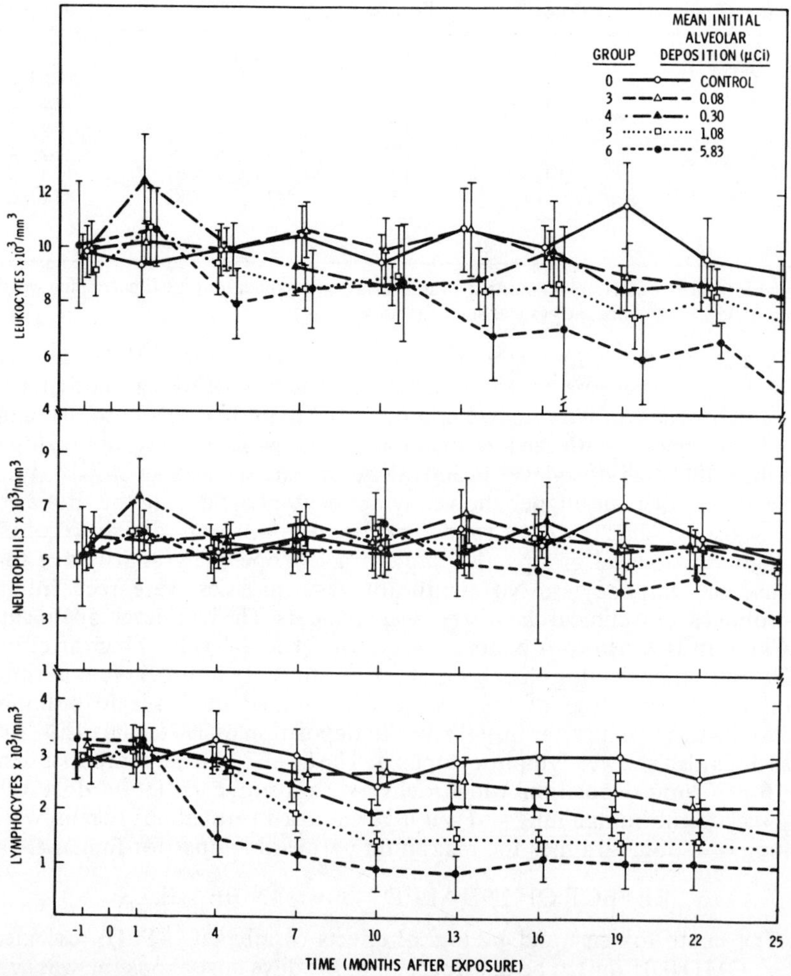

FIG. 4. Leukocyte values of dogs showing lymphocytopenia dose-response relationships after inhalation of $^{239}PuO_2$. (Means ±95% confidence intervals around the means)

Two of the dogs were euthanized because of respiratory insufficiency related to plutonium-induced pulmonary fibrosis. Eight dogs were euthanized because of plutonium-induced bone tumors, all of the bone tumors were osteosarcomas. Dog 481 F had a bone tumor, observed at necropsy in addition to myelogenous leukemia, and dog 405 M had a lung tumor which was evident 15 months prior to euthanasia because of a vertebral bone tumor. Five dogs died of causes thought to be unrelated to plutonium exposures at the present time. In addition to the lesions causing death, the dogs had fibrotic tracheobronchial lymph nodes, focal pulmonary fibrosis and nodular hyperplasia in the liver. None of the 17 control dogs in the study have died.

TABLE 3

Dog Mortality and Tissue Distribution After Inhalation of $^{238}PuO_2$

Dog Number	Survival (days after exposure)	Terminal Body Burden (μCi)	Plutonium Distribution (Percent of Terminal Body Burden)			
			Lungs	Lymph Nodes[a]	Skeleton	Liver
390 M	37	261	92	4	2	1
403 F	30	167	94	2	2	2
408 F	35	168	93	1	3	2
395 M	56	112	92	3	4	1
450 F	56	74	91	4	4	1
393 M	61	140	94	2	3	1
460 M	70	84	90	3	4	2
442 F	76	88	90	4	3	2
449 F	94	44	91	3	4	1
413 F	125	17	80	9	7	3
400 M	180	25	77	4	13	5

[a]Tracheobronchial, mediastinal and sternal lymph nodes

The dogs showed a persistent lymphocytopenia (Figs. 5 and 6), which became apparent in both groups by 200 days after exposure. In the crushed microsphere-exposed dogs, the mean lymphocyte count of the six surviving, exposed dogs remained below that of controls at 6-years postexposure. The total leukocyte count of the exposed dogs was frequently lower than that of controls, primarily due to the low lymphocyte counts and a small but consistent reduction in neutrophils (Fig. 6). Lymphocytopenia in the 350° C calcined plutonium-exposed dogs was evident until 5 years postexposure when only four of the dogs with the lowest body burdens remained alive (Fig. 5). The dogs in this group also showed a reduction in neutrophils which was apparent 1 month to 2 years after exposure, when the higher dose-level dogs were still alive. The dogs also showed a leukopenia due to the reduction in neutrophils and lymphocytes. No significant changes were observed in the values for other leukocytes or hemoglobin. The lower initial mean lymphocyte values in the dogs exposed to crushed microspheres, compared to the 350° C calcined plutonium-exposed group, may be due to an age difference. The dogs exposed to 350° C calcined $^{238}PuO_2$ were 3 years old while the dogs exposed to $^{238}PuO_2$ crushed microspheres were 8 months old at the time of exposure. Older dogs generally have lower lymphocyte counts.

No differences were observed in serum blood urea nitrogen, creatinine, glucose, alkaline phosphatase, or glutamic oxalacetic transaminase. Mean serum glutamic pyruvic transaminase (SGPT) levels were not significantly elevated compared to controls; however, the percentage of elevated SGPT values (>55 units) in dogs receiving ^{238}Pu was higher than in the controls. These changes are suspected to be related to plutonium-induced lesions in the liver.

Five years after exposure, the largest fraction of plutonium retained in the body was in the skeleton followed in descending order by liver, thoracic lymph nodes and lung (Table 4). The highest plutonium concentration occurred in the lymph nodes followed in descending order by liver or lung and skeleton. The biological effects observed during the 4 to 6 years following exposure to $^{238}PuO_2$ were primarily related to skeletal deposition. Nine dogs had osteosarcomas. Assuming a linear translocation of plutonium from the lung to the skeleton over the postexposure period, the estimated accumulated radiation dose to the skeleton in bone tumor-bearing dogs was 100 to 300 rads.

TABLE 4
Mortality and Tissue Distribution in Dogs After Inhalation of $^{238}PuO_2$

Dog Number	Survival (months after exposure)	Terminal Body Burden (μCi)	Plutonium Distribution (Per cent of Terminal Body Burden)				Cause of Death
			Lungs	Lymph Nodes[a]	Liver	Skeleton	
		(After Inhalation of $^{238}PuO_2$ Calcined at 350° C)					
492 F	23	3.0	4	4	23	64	Bone Fracture
404 M	36	8.1	32	10	23	32	Respiratory Insufficiency
467 M	38	7.0	15	11	13	57	Respiratory Insufficiency
469 M	54	2.6	34	5	17	41	Bone Tumor
445 M	58	2.5	6	10	23	55	Bone Tumor
438 M	60	2.3	7	11	33	43	Bone Tumor
453 M	62	2.2	17	9	22	47	Bone Tumor
405 M	70	2.6	23	10	28	34	Bone Tumor (Lung Tumor)[b]
		(After Inhalation of $^{238}PuO_2$ Crushed Microspheres)					
485 M	22	3.1	72	7	7	12	Encephalitis
500 M	34	1.1	39	21	12	24	Hemorrhage
497 F	52	0.2	16	3	31	46	Intestinal Obstruction
481 F	60	0.5	13	23	23	37	Myelogenous Leukemia (Bone Tumor)[b]
489 F	62	2.5	7	26	27	32	Bone Tumor
482 F	70						Bone Tumor
494 F	75						Fibrosarcoma
488 F	76	1.4	9	15	29	41	Bone Tumor

[a]Tracheobronchial, mediastinal and sternal lymph nodes
[b]Other lesions

In studies with inhaled $^{239}PuO_2$, plutonium was retained primarily in the lungs and thoracic lymph nodes, and dogs died 3 to 11 years after inhalation exposure due to lung tumors. No bone tumors were observed. Differences in plutonium translocation from the lung to other tissues and the influence of the radiochemical and/or physical state of plutonium on this translocation are evident in these results. Studies on the in vivo and in vitro behavior of $^{238}PuO_2$ compared to $^{239}PuO_2$ suggest that $^{238}PuO_2$ behaves as a more "soluble" compound than $^{239}PuO_2$ (*17*).

Experiments were initiated in 1972 to study the dose-effect relationships of inhaled $^{238}PuO_2$ beagles. Table 5 shows the experimental design. Eighteen-month-old beagles were exposed to aerosols of $^{238}Pu^{16}O_2$ (mean AMAD 1.8. μm, mean GSD 1.9) prepared by calcining the oxalate at 700°C and subjecting the product to $H_2\,^{16}O$ steam in Ar exchange at 800° C for 96 hours. The mean initial alveolar burden in the six dose-level groups ranged from 2 to 5200 nCi. The lowest deposition level corresponds to an average lung dose of about 0.3 rem/week, the maximum permissible exposure for man. This experiment and the previously described inhaled $^{239}PuO_2$ dose-effect relationship experiment will provide information on possible differences in the disposition and biological effects of inhaled $^{239}PuO_2$ and $^{238}PuO_2$, two similar alpha-emitting isotopes that differ by a factor of about 280 in specific activity.

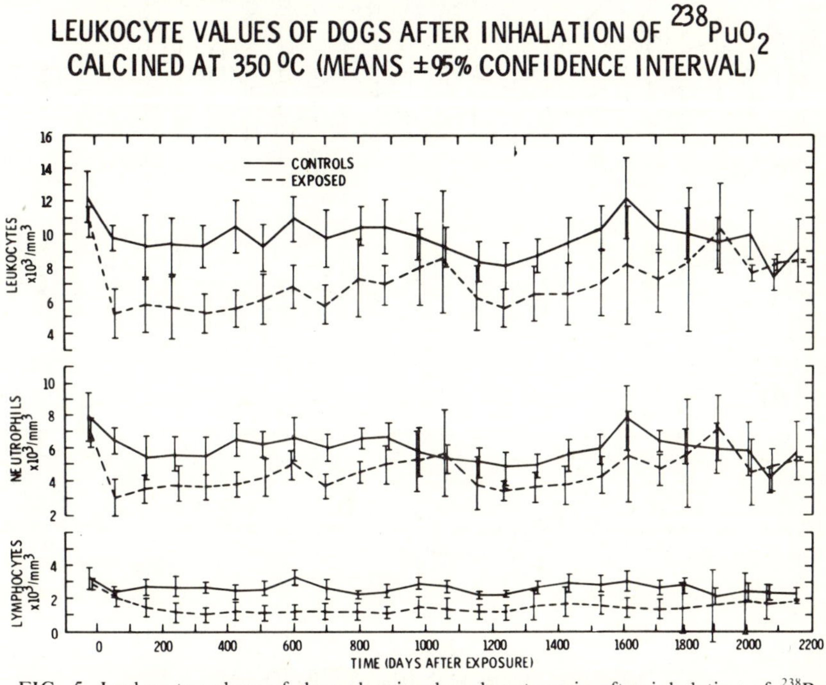

FIG. 5. Leukocyte values of dogs showing lymphocytopenia after inhalation of $^{238}PuO_2$ calcined at 350°. (Means ±95% confidence intervals around the means)

During the first postexposure year, none of the dogs exposed to $^{238}Pu^{16}O_2$ have died. The $^{238}Pu^{16}O_2$-exposed dogs also showed a dose-related lymphocytopenia and leukopenia during the first year postexposure. The data in Fig. 7, obtained only from those dogs that were at least 10 months postexposure at the time of this writing, indicated that groups with mean initial alveolar deposition of 140 nCi and higher showed lymphocytopenia, while the lower-dose groups were similar to controls.

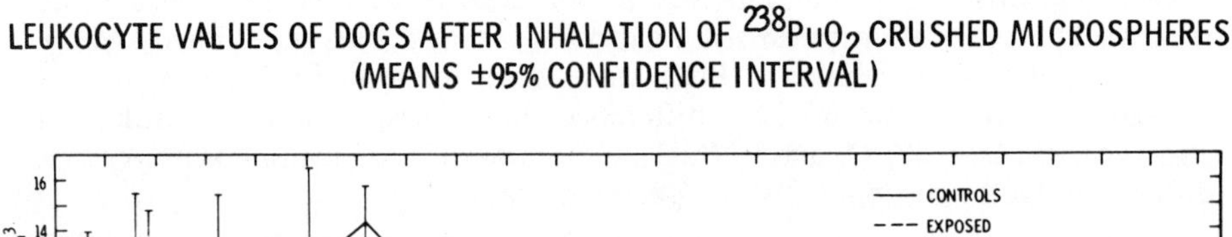

FIG. 6. Leukocyte values of dogs showing lymphocytopenia after inhalation of $^{238}PuO_2$ crushed microspheres. (Means ±95% confidence intervals around the means)

TABLE 5
Dose-Effect Studies With Inhaled $^{238}Pu^{16}O_2$ in Dogs

	Number of Dogs		Initial Alveolar Deposition[a]	
Dose Level Group	Male	Female	nCi[b]	nCi/g lung[b]
0	10	10	0	0
1	10	10	2.3± 0.8	0.02 ± 0.01
2	10	10	18 ± 3	0.2 ± 0.04
3	10	10	77 ± 11	1.1 ± 0.4
4	10	10	354 ± 81	4.4 ± 0.9
5	10	10	1308 ± 270	16.9 ± 2.9
6	7	6	5192 ± 1372	69.2 ±17.8
	67	66		

[a]Estimated from external thorax counts at 14 and 30 days postexposure and estimated lung weights
[b]Means ± 95% confidence intervals around the means

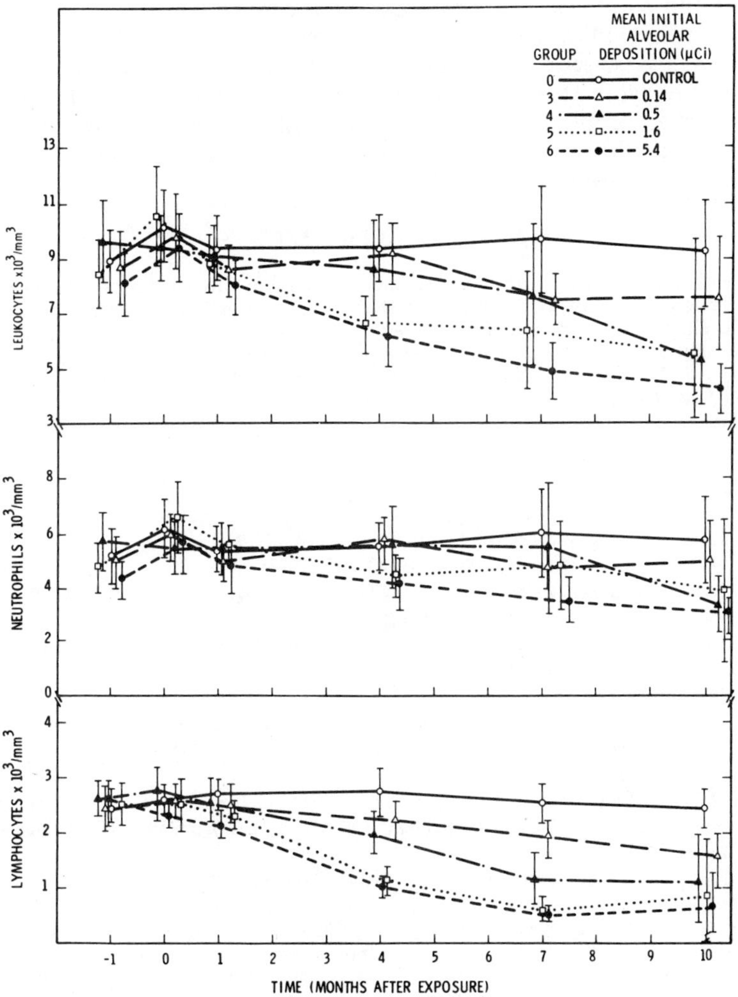

FIG. 7. Leukocyte values of dogs showing lymphocytopenia dose-response relationships after inhalation of $^{238}PuO_2$. (Means ±95% confidence intervals around the means)

CONCLUSIONS

With the dose levels studied to date, pulmonary neoplasia was the primary cause of death in beagles 5 to 10 years after inhalation of $^{239}PuO_2$. None of the dogs exposed to $^{239}PuO_2$ developed bone tumors. Bone neoplasia was the primary cause of death 5 to 6 years after inhalation of $^{238}PuO_2$; however, one of the dogs had pulmonary tumors in addition to bone tumors.

The earliest indication of a biological effect after the inhalation of $^{239}PuO_2$ or $^{238}PuO_2$ was dose-related lymphocytopenia, occurring during the first two postexposure years at dose-levels lower than those which have thus far been shown to cause neoplasia. The experiments now in progress will provide further information on the dose-response relationships of inhaled $^{239}PuO_2$ and $^{238}PuO_2$ at low exposure levels.

ACKNOWLEDGEMENTS

The author is indebted to many scientists and technicians involved in these studies, in particular to E. F. Blanton, D. L. Catt, D. K. Craig, J. R. Decker, G. L. Webb, and E. L. Wierman for exposure of dogs to plutonium aerosols; to J. S. Barnett, A. C. Case, J. P. Herring, and G. J. Powers for plutonium analyses; to H. A. Ragan for hematology; to P. L. Hackett for clinical chemistry; to G. E. Dagle and J. E. Lund for histopathology; to M. G. Brown and S. E. Rowe for animal care; to R. J. Olson and J. M. Thomas for statistical assistance; to W. J. Bair for evaluation of the results and to P. M. Bresina, D. L. Felton, E. H. Groff, and J. A. Harrison for assistance in preparation of the manuscript.

REFERENCES

1. J. F. Park, W. J. Bair, and R. H. Busch, *Health Phys.* **22**, 803 (1972).
2. W. J. Bair and R. C. Thompson, *Science* **183**, 715 (1974).
3. W. J. Bair, in *Advances in Radiation Biology* **Vol. 4** (J. T. Lett, H. Adler, and M. Zelle, Eds.), pp. 255–315. Academic Press, Inc., New York, 1974.
4. R. C. Thompson, *Effects of Plutonium in Animals.* United States Atomic Energy Commission Advisory Committee on Reactor Safeguards Special Report. (in press).
5. R. C. Thompson, *Health Phys.* (in press).
6. W. J. Bair, in *Seminar on Radiological Safety Evaluation of Population Doses and Application of Radiological Safety Standards to Man and the Environment* (L. A. Self, Ed.). Portoroz, Yugoslavia. (in press).
7. W. J. Bair, J. E. Ballou, J. F. Park, and C. L. Sanders, in *Uranium Plutonium Transplutonic Elements* (H. C. Hodge, J. N. Stannard, and J. B. Hursh, Eds.), pp. 503–568. Springer-Verlag, New York, 1973.
8. J. F. Park, W. J. Clarke, and W. J. Bair, *Health Phys.* **10**, 1211 (1964).
9. S. W. Nielsen, in *Morphology of Experimental Respiratory Carcinogenesis* (P. Nettesheim, M. G. Hanna, Jr., and J. W. Deatherage, Jr., Eds.) p. 123–145. USAEC Document CONF 700501, (Available from National Technical Information Service, Springfield, Va. 22151.)
10. W. J. Clarke, J. F. Park, J. L. Palotay, and W. J. Bair, *Health Phys.* **12**, 609 (1966).
11. E. B. Howard, in Morphology of Experimental Respiratory Carcinogenesis (P. Nettesheim, M. G. Hanna, Jr., and J. W. Deatherage, Jr., Eds.), p. 147–160. USAEC Document CONF 700501, (Available from National Technical Information Service, Springfield, Va. 22151.)
12. J. F. Park, E. B. Howard, B. O. Stuart, A. P. Wehner, and J. V. Dilley, in *Morphology of Experimental Respiratory Carcinogenesis* (P. Nettesheim, M. G. Hanna, Jr., and J. W. Deatherage, Jr., Eds.), pp. 417–436. USAEC Document CONF 700501, (Available from National Technical Information Service, Springfield, Va. 22151.)
13. B. O. Stuart, P. J. Dionne, and W. J. Bair, in *Proc. 11th AEC Air Cleaning Conference* (M. W. First and J. M. Morgan, Eds.), pp. 721–737. USAEC Document CONF 700816, (Available from

National Technical Information Service, Springfield, Va. 22151.)
14. C. L. Yuile, F. R. Gibb, and P. E. Morrow, *Radiat. Res.* **44**, 821 (1970).
15. C. W. Mays and T. F. Dougherty, *Health Phys.* **22**, 793 (1972).
16. J. F. Park, E. B. Howard, and W. J. Bair, in *Pacific Northwest Laboratory Annual Report for 1968*, pp. 3.6–3.13. BNWL-1050, Vol. 1, Pt. 1. Battelle-Northwest, Richland, WA, 1970.
17. J. F. Park, D. L. Catt, D. K. Craig, R. J. Olson, and V. H. Smith, in *Proc. of Third International Congress of the International Radiation Protection Association*. Washington, D. C. (in press).

Current Status of Information Obtained from Plutonium Contaminated People[1]

C. R. Richmond[2]
Los Alamos Scientific Laboratory
University of California
Los Alamos, NM 87544, USA

Earlier in this Congress Dr. Alvin Weinberg asked the following question, "Can man learn to live with fission?" Perhaps another way of phrasing this important question would be to ask if man can learn to live with plutonium and other actinide elements. There has been much interest and concern expressed about the potential toxicity and effects of plutonium and the actinide elements on man. Although there exists a considerable data base on the biological effects of plutonium on experimental animals, there is a relative paucity of data existing for exposed human subjects. Therefore, we are placed in a situation in which we must extrapolate or project in some manner the effects of plutonium on experimental animals to man. Nevertheless, we do have some information concerning human subjects who were exposed to plutonium through occupational exposures. I shall attempt to review this information for you today. We should keep in mind that this kind of information, even though very limited, is human experience of the most relevant kind for establishing value judgments where inadequate data exist for formulating risk evaluations.

There have been no recorded substantiated incidents of cancer production in man from the internal deposition of any plutonium isotope in the nuclear energy industry, even though internal depositions have occurred. This excellent record has resulted from effective control measures. Although some exposures occurred 30 years ago, we must keep in mind the relatively short time since exposure for many of these cases. There has been a reasonable number of individuals exposed to quantities of plutonium at or near the maximum permissible lung or body burden. Although the exact levels of internal contamination are uncertain, it is clear that this body of information represents a valuable resource for current and future study. Some work has been done on specific groups and other studies are in the process of gathering pertinent data for future use as these populations age.

Another source of information on the effects (or lack thereof) is a group of persons who received plutonium during the mid-1940's. These are particularly interesting as data accrued for these persons provided the basis for the human excretion equations derived by Langham at Los Alamos. In addition, the injected levels were reasonably high and were retained in one case for 21 years. Still another source of information exists in the form of virtually the entire

[1] Congress presentation C-9-3.

[2] Present address: Oak Ridge National Laboratory, P. O. Box X, Oak Ridge, TN 37830, USA

world's population, which has accumulated plutonium, although rather inefficiently, from previous and present atmospheric detonations of nuclear weapons. Past and current autopsy programs involving both occupationally and non-occupationally exposed persons will also furnish useful information on the uptake and concentration of plutonium in various body tissues. These data provide us with some insight on the efficiency of the human organism for incorporating plutonium and on its distribution among body tissues. In the case of occupational exposures, this data base will also allow us to make comparisons of the body burden with estimates based on bioassay data obtained during the life of the subjects in question.

The message I would like to leave with you is that we have not observed any serious biological consequences in man from occupational exposure to plutonium. The next logical question is, "How do you know?" or "Have you looked?" We have indeed looked and the record to date is quite reassuring. However, I would like to point out that the past is not always prologue, cheerful as that thought might be. Although the final answers are not in, I would hasten to add that we now have information on people which has spanned almost three decades so that the argument of latency for certain late effects (how long does it take to produce a tumor) is losing impact in at least some of the cases I will discuss.

MANHATTAN PROJECT EXPOSURES

Since the discovery of plutonium over three decades ago, (*1*), some personnel exposures have occurred. Some have been studied and reported on in varying degree, both during life and after death. (*2–9*). One of the most interesting groups, because of both time since exposure and level of exposure, is that of the Manhattan Project plutonium workers (*10*).

Twenty-five male subjects who worked with plutonium during World War II under very crude working conditions, as judged by today's standards, have been followed medically for the intervening years. Within the past several years, 21 of these men have been examined at the Los Alamos Scientific Laboratory. In addition to physical examinations and laboratory studies (complete blood count, urine chemistry profiles and urine analysis), roentgenograms were taken of the chest, pelvis, knee and teeth. Chromosomes of lymphocytes cultured from peripheral blood and cells shed from the pulmonary tract were also studied.

TABLE 1

Biomedical Work Related to Manhattan Project Plutonium Workers

Medical History and Examination
Radiology
Karyology
Pulmonary Cytology
Urine Radiochemistry
Chest Counting (Uranium L X-Rays)
Blood Chemistry Profiles
Hematology

Urine specimens assayed for plutonium yielded calculated body burdens which ranged from 0.005 to 0.42 μCi. These estimates of body burden, on the average, are higher than earlier estimates based on radioassay of urine samples collected in the past, perhaps reflecting uncertainties in the models used to estimate body burden from excretion data. Table 1 indicates the kinds of information obtained from the Manhattan Project plutonium workers. Most, but not all, of these tests have been conducted every four to five years since the group has been studied. This appears to be a very healthy group of men in their early to mid-fifties. There are no adverse biological effects that have been detected which can be attributed to the radiation exposure. Except for the ailments that one would expect in a group of men of this age, all subjects examined were in remarkably good health. One man had a coronary occlusion but had recovered and was well compensated, and another of the original group died in 1959 of a coronary at age 38. Another had a hamartoma of the lung surgically removed without complication in 1971. A third had a melanoma of the chest wall (regional lymph nodes were negative). A fourth had a partial gastrectomy free-bleeding ulcer. Several had mild hypertension and moderate obesity and one had gout. All men were actively working, mostly as successful executives. More detailed information on this particular group of workers has been published (*11–12*). It is important to realize that these men worked under very crude working conditions as judged by today's standards. Figure 1 shows the building in which these men worked in 1944 and 1945 as part of the program to make the first nuclear weapons. I can assure you

FIG. 1. Original wooden D Building which housed the chemists and metallurgists who worked with plutonium at Los Alamos in 1944–45.

that there is probably a factor of 115,000 between safety constraints in this building and those modern facilities now used for working with plutonium. Things have changed quite dramatically with the goal of providing more protection for the worker and more complete containment of the plutonium within the facility. The health protection record of plutonium workers has been quite good and I think it is clear that if we stick to our controls and regulations as regard plutonium, we are not destined to repeat the kinds of things that happened with the use of radium earlier in this century.

Table 2 shows the current status of several LASL plutonium study groups.

TABLE 2
Los Alamos Scientific Laboratory
Plutonium Study Groups

Group 1[a]	
	25 men exposed during 1944–1945
	Periodic biomedical follow-up for three decades
	Reports available in literature[b]
Group 2	
	42 men also exposed during Manhattan Project
	25 located and responded to questionnaires
Group 3	
	190 early and current plutonium workers
	175 identified by Social Security Number
	2 identified by military service number

[a]Manhattan Project Plutonium Workers
[b]References *10–12*

Group 2, which is now in the early stages of study, will expand the size of the original cohort by perhaps 40 people. Twenty-five of the 40 men who have been identified have been located and have responded to questionnaires. Groups 1 and 2 collectively represent the 'creme de la creme' of the early plutonium exposures. Once again, these are extremely important subjects to study intensively as approximately three decades have elapsed since their exposures. Group 3 will include a larger spectrum of exposures including more recent accidents and exposures to plutonium 238. All are estimated to have systemic plutonium burdens of 4 or more nanocuries. Table 3 shows the current systemic body

TABLE 3
Plutonium Systematic Body Burden Estimates
for Selected Manhattan Project Plutonium Workers
at Three Different Times[a]

Case Code	$^{239-240}$Pu (nCi) 1953	1962	1972
1	30–60	10	210
3	80	130	420
4	80	140	260
5	80	140	180
6	60	70	140
7	60	80	150
17	40	90	130

[a]persons with more than 120 nCi $^{239-240}$Pu systemic burden in 1972.

burden estimates based on urine excretion for some of the men in the Manhattan Project plutonium worker group. Only those with more than three maximum permissible body burdens (or 120 nCi) in 1972 are shown. Again, the original exposures were in the early 1940's so that Case 3, who now has approximately 10 times the allowable occupational bone burden, has carried this estimated 410 nCi of plutonium for approximately three decades. The selected cases shown in Table 3 represent systemic plutonium burdens ranging from 0.13 to 0.42 μCi, which correspond to annual bone doses of approximately 2 to 6 rad. I won't dwell on the earlier estimates for 1953 and 1962 other than to point out that these are somewhat uncertain. The model that is used to estimate body burden from the urine analysis data does contain some uncertainty (*13*) and cannot be applied accurately to each individual. It is also true that some of the body burden estimates are based on relatively few data points. Table 4 contains information which is detailed in an earlier publication (*10*). The data are for the plutonium 239–240 content for some tissues that were removed from Case 2 who developed a non-malignant growth (hamartoma) in the lung. This particular medical condition, which was found as part of the medical follow-up study, afforded an opportunity to obtain some hamartoma tissue, lymph node tissue, rib tissue and normal lung tissue for radiochemical analyses.

The concentration of $^{239-240}$plutonium was approximately the same in both the tumor and normal lung tissue. The lowest plutonium concentration was found in a rib sample and the highest in the lymph node tissue consistent with experimental findings in dogs exposed to plutonium dioxide by inhalation. If one assumes a total lung weight of 1000 grams, a respiratory lymph node weight of 20 grams, and a homogeneous distribution of plutonium throughout these tissues, the total plutonium burden is approximately 8 nCi, roughly equally divided between the lung and lymph node. Estimates of the chest burden of plutonium in Case 2, based on chest counting procedures, are in reasonable agreement (a factor of about 2) with the estimate based on extrapolation from analysis of lung and lymph node tissue.

TABLE 4
$^{239-240}$Plutonium Content of Tissues Removed From Case No. 2 in May 1971

Tissue	Wet Weight (g)	$^{239-240}$Plutonium dpm/g	pCi/g
Lung	70.85	8.48	3.85[a]
Lymph node	1.25	451.00	205.00
Hamartoma	0.77	7.47	3.40
Rib	20.00	3.55	1.61

[a]Contemporary level from weapons detonations is about 0.0004 pCi/g.

Figure 2 is a photomicrograph of an autoradiograph of a plutonium particle in a lymph node section removed from Case 2 in May 1971. Observations of other plutonium particles in the lymph node tissue examined suggested a very non-uniform radiation dose distribution from the plutonium particles. Attempts have been made to estimate the number of particles inhaled by the Manhattan Project plutonium workers. By making certain assumptions for the mass median diameter, geometric standard deviation of the distribution and the particle density, one can show that the mass fraction for plutonium dioxide particles larger than 0.6 μm diameter in the distribution is approximately 15%. Further calculations indicate that approximately 10^7 particles larger than 0.6 μm diameter could have been retained by the 25 subjects during their exposures in 1944 and 1945 (*14*). If the lung cancer risk for such "hot particles" were 5×10^{-4} per particle, as suggested by the National Resources Defense Council Report (*15*), then the 10^7 particles should yield about 5000 lung cancers for this group of men. The observed lung cancer incidence after almost 30 years since exposures is zero.

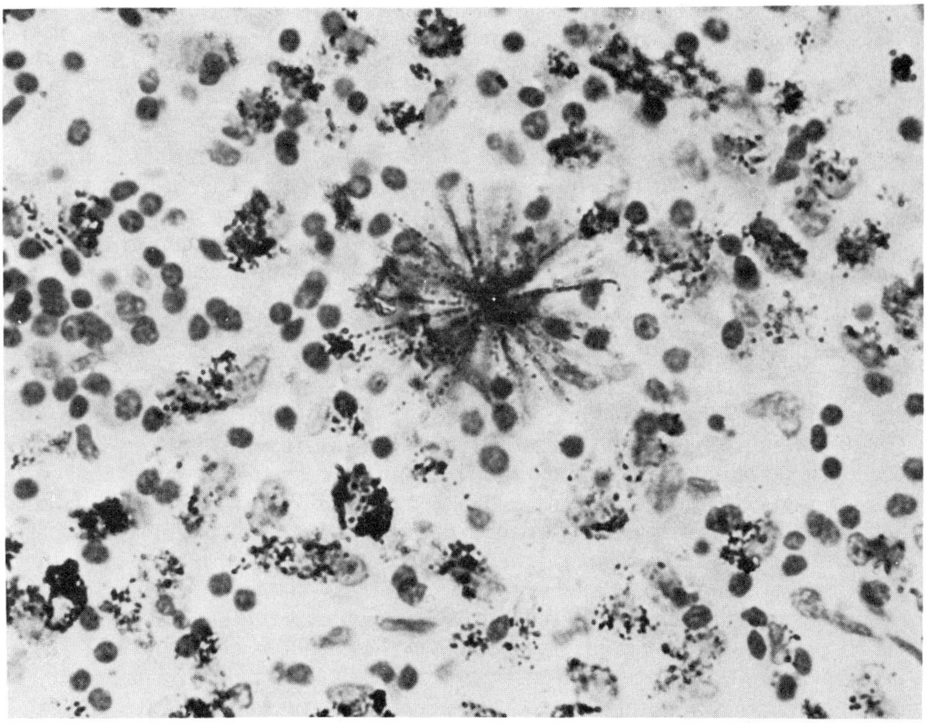

FIG. 2. Photomicrograph of an autoradiograph of a lymph node section removed from early Manhattan Project plutonium worker (Case No. 2).

PLUTONIUM ADMINISTRATION STUDIES IN HUMAN SUBJECTS

In an attempt to determine relationships between urinary excretion, total excretion, and body content of plutonium, 18 persons received plutonium parentally in 1945 and 1946 (*16–17*). Fifteen of the 18 were older than 45 years, and all but two of the 18 were given ^{239}Pu only (one received ^{238}Pu and ^{239}Pu and another received only ^{238}Pu). The amounts of plutonium administered ranged from about 0.3 to about 6 μCi (Table 5). For comparison, the current occupational maximum permissible body burden for ^{239}Pu is 0.04 μCi.

TABLE 5
Plutonium Recipients[a]

- Eighteen persons received plutonium in 1945 and 1946.
- Fifteen of the 18 were older than age 45.
- All but 2 given ^{239}Pu; one received ^{238}Pu only and one received ^{238}Pu and ^{239}Pu
- Amounts ranged from ~0.3 to ~6 μCi (current maximum permissible occupational body burden is 0.04 μCi)
- Data from group provided basis for plutonium excretion functions
- One male received 5.2 μCi ^{238}Pu and 0.12 μCi ^{239}Pu and $PuO_2(NO_3)_2$. Later had total gastrectomy and splenectomy. Died 21 years later of cardiovascular disease.
- Some of original group still alive three decades later.

[a]Reference *17*

Although these subjects were thought to be hopelessly ill, four of the group were alive in November 1973, almost three decades after receiving the plutonium. Excretion data from some of the survivors was presented earlier at this symposium (*18*). This particular situation represents the unique opportunity to check on the validity of the excretion data equations that are used currently by radiation protectionists to estimate body burdens. It is of considerable interest that much of the data used to establish the excretion equations came originally from this group of 18 subjects during the relatively short time (several months in most cases) they were studied. To establish the unique excretion curves, Langham used additional data obtained over a five-month period from three of the plutonium recipients and additional excretion data from several Los Alamos occupational exposure cases for about 300 days and one for about 1700 days. These retention equations have proved extremely useful for over several decades and appear to be somewhat conservative in most instances when estimated body burdens can be checked with data obtained at postmortem. When initially reported in 1950 (*16*), Langham and coworkers stated, "Admittedly the observations made during this study provide no evidence of what may happen in 10–30 years. It may be said, however, that these studies and those of other investigators indicate that the intravenous injection of a single dose of 5 to 100 microgram of plutonium was without acute subjective or objective clinical effects."

One of the original 18 plutonium recipients is of particular interest. He was a white 58 year old male who was suffering from gastric carcinoma and gastrointestinal hemorrhage when he received 5.19 μCi of ^{238}Pu and 0.12 μCi of ^{239}Pu as $PuO_2(NO_3)_2$ by intravenous injection on 14 May 1945. He later received a total gastrectomy and a splenectomy but did not die until 9 January 1966, some 21 years later, from cardiovascular disease. We can obtain a very rough estimate of the bone dose by assuming 40% deposition in the skeletal tissues with no subsequent loss. Under these circumstances, the skeletal dose over the 21-year period would be approximately 900 rad. The annual dose rate to the skeletal tissues would be approximately 40 rad, a factor of approximately 70 higher than the annual skeletal dose rate of 0.6 rad delivered by the maximum occupational bone burden of 40 nCi of ^{239}Pu.

U.S. TRANSURANIUM REGISTRY (USTR)

During the summer of 1968, the United States Atomic Energy Commission authorized the establishment of the National Plutonium Registry which was later renamed as the United States Transuranium Registry. The Registry is part of the organizational structure of the Hanford Environmental Health Foundation in Hanford, Washington, and functions as a repository for all pertinent available information that may be used to identify any radiation hazard to the worker that may exist from exposure to transuranium elements (*19–21*). Major AEC contractors and some licensees using plutonium and other transuranium elements have agreed to endorse the program and to recommend that their affected employees support this program. Cooperation with the USTR is completely voluntary on an individual basis and includes release of medical and health physics data. Permission is also obtained on a voluntary basis for postmortem analyses of certain organs.

The employees remain under the health physics and medical programs of the AEC contractor until termination of employment, at which time the USTR assumes responsibility if the employee has authorized postmortem examination.

The major criterion used by the USTR to determine inclusion of an individual into the Registry is as follows: persons for whom the employer has provided a routine surveillance program because of a reasonable probability that a deposition could occur. This rather broad criterion allows for the different methods used to estimate the extent of contamination (e.g., urine analysis, chest counting, or air concentration data) and for uncertainties in estimating burdens under certain conditions (e.g., chronic inhalation of insoluble plutonium). It also removes the emphasis on the larger contamination cases.

At autopsy, comparisons can be made between estimates of the body burden based upon tissue analyses and estimates made previously on the basis of health physics and operational data. In addition to a medical history of a disease or abnormal conditions, employees can often provide information on their work history, smoking habits, exposure to toxic materials, and other pertinent data.

In addition to AEC contractors, the USTR receives health physics information from the AEC's Health and Mortality Study and from the Assistant for Workman's Compensation and Radiation Records of the AEC.

Preliminary findings for the first fourteen autopsy cases reported by the USTR appear in the proceedings of the 12th Hanford Biology Symposium (*22*). To date, information obtained by the USTR indicates that estimates of the plutonium systemic burden based on urine analysis have been on the conservative side, that is, higher than those estimates based on analysis of tissue obtained at autopsy (*22*). Workers in the United Kingdom have also found this to be true (*23*). In addition, the USTR reports that, "to our knowledge, no employee has ever sustained serious harmful effects due to the internal deposition of any of the transuranium elements" (*24*).

Table 6 indicates the status of the USTR as of June 1974. To date, most of the USTR activities have been confined to Hanford, Los Alamos, and Rocky Flats. The interested reader is directed to a recent USTR publication for details of the level of cooperation between the USTR and the other AEC contractors shown in Table 6 (*25*).

TABLE 6
Current Status of United States Transuranium Registry[a]

AEC Contractor	Workers Identified	Record Releases[b]	Autopsy Agreements	Autopsies Performed	Autopsy Reports Completed
Hanford	2199	2132	525	12	8
Los Alamos	259	259	127	1	0
Rocky Flats	1504	1489	167	30	21
Savannah River	1559	0	0	0	0
Mound Laboratory	322	0	0	1	1
Oak Ridge	0	0	0	1	1

[a]Reference *25*
[b]Health Physics and Medical Records

ACCIDENT CASES

A considerable amount of information has been obtained from accidental occupational exposures to plutonium. However, the total number of accident cases has been relatively small. Information obtained from the AEC's Division of Operational Safety as shown in Table 7 indicates that during the period 1957 to 1970 there have been on the order of 200 contractor personnel exposed to 25% or more of the maximum permissible body burden (MPBB) for plutonium. These data also indicate that inhalation is the major portal of entry and that more than half the exposure cases represent plutonium burdens that are less than 50% of the

TABLE 7
Internal Plutonium Depositions Exceeding 25% of the Occupational Maximum Permissible Body Burden Among AEC Contractor Personnel During the Period 1957–1970

By Year		By Activity		By % of Permissible Body Burden	
1957	12	Research	25	25 to 50%	118
1958	6	Production	109	50 to 75%	35
1959	10	Maintenance	17	75 to 100%	13
1960	10	Development	4	100 to 200%	15
1961	16	Health Physics	18	200 to 500%	15
1962	20	Construction	2	500 to 1000%	7
1963	9	Analytical	1		
1964	29	Recovery	5		
1965	22	Unknown	22		
1966	27				
1967	10				
1968	8	By Route of Entry		By Therapy Employed	
1969	6	Inhalation	131	Chelate	37
1970	3	Wound	48	Excision	21
??	15	Both	8	Both	10
Total	203	Unknown	16	None	135

MPBB. Table 7 also shows that of the 203 accidental depositions, approximately 18% were treated by chelation therapy and 54% were related to production activities. Eighteen percent of the 203 exposures shown in Table 7 were associated with plutonium activities greater than one occupational permissible body burden. It may be instructive to look at a specific instance of an industrial accident which was reported in the open literature. A fire in a plutonium fabrication plant resulted in a large-scale spread of plutonium oxide on October 15, 1965 (*3*). Following the accident, the Rocky Flats body counter was used to measure the plutonium in the lungs of all employees working in the area at the time of the accident. Of the approximately 400 employees counted, 25 were found to have enough plutonium in their lungs to deliver a dose of 15 rem/year or greater. These data are shown in Table 8. Estimates of the plutonium were obtained by using two scintillation detectors placed in contact with the subject's chest. A 60 keV photon peak from ^{241}Am was used. The ^{241}Am content of the material released during the fire was determined and the amount of plutonium in

TABLE 8
Rocky Flats ^{239}Pu Inhalation Exposures from a Single Accident[a]

Number of workers	Multiples of occupational maximum permissible lung burdens
1	17
1	10
1	7
3	3–5
19	1–2

[a]Reference *3*
[b]Derived MPLB is 0.016 μCi

The plutonium consisted of "high fired" PuO_2 and particle size measurements made on air samples collected after the fire indicated a mass median diameter of 0.32 μm with a geometric deviation of 1.83. On the average, about 30% of the material initially deposited in the lungs was cleared in two to three months. Subsequent lung measurements indicate a much slower clearance. All of the 25 individuals shown in Table 8 have one or more maximum permissible lung burdens for occupational workers. The quantities of plutonium the subjects was then determined from calibrations done on a chest phantom containing material with a similar ^{241}Am to ^{239}Pu ratio.
for this group of 25 individuals range from 16 to 272 nCi. It is not clear how many individuals received measurable body burdens less than 16 nCi ^{239}Pu. Hopefully, these cases will ultimately be studied by the U.S. Transuranium Registry. One would also think that this particular cohort of individuals should be carefully studied and followed throughout their lives in a manner similar to the Manhattan Project workers mentioned earlier.

Another example of accidental exposure of workers is of interest because the isotope involved was ^{238}Pu (*2*). This particular evaluation deals with three personnel who were contaminated as a result of an explosion inside a glove box at Mound Laboratory which resulted in significant inhalation exposures. Chest counting measurements made at another laboratory indicated that the employees had lung burdens of 134, 141, and 91 nCi. Radiochemical analysis of urine samples obtained from these three individuals during the first half-year following exposure suggested elimination patterns which differed from those observed following previous inhalation exposures to ^{238}Pu. The long-term systemic body burdens were estimated to be 290, 210, and 30 nCi. Because of the isotope involved and the relatively high levels of contamination, these individuals might also represent useful additions to long-term biomedical follow-up investigations.

One case of contamination resulting from a puncture wound is extremely interesting as it has been interpreted by some as being an example of cancer produced in man by plutonium. The lesion was first described in the open literature more than ten years ago (*26*) and was also included with other information on plutonium wounds at a later time (*4*). The radiation dose around the plutonium implanted in the palmar skin was estimated to be 75 million rads. However, this kind of dose estimate is probably meaningless as we do not know which cells were exposed or for what time periods. The entire lesion was very small, being on the order of 2.8×10^{-5} cm^3. The author stated, "Although no malignancies of the skin of man have ever been shown autoradiographically to be associated with such alpha-emitting foreign bodies, the changes here would seem to indicate that the development of such a lesion is possible." The lesion is shown in Fig. 3. The 5 nCi particle of plutonium was surgically excised from the individual's palm approximately four years after the accident. The pathologist involved in the study carefully described the cell changes as having "a similarity to known precancerous epidermal cytologic changes" (*26*). More recently, Tamplin and Cochran correctly acknowledged this wording when they first refer to it in their report. However, as pointed out by Dr. Dolphin[3], in the second reference the wording is altered slightly to "pre-cancerous changes in human tissue," and in the third reference another word change is made and it becomes "particle-induced cancer" (*15*). This particular lesion appears to be about the most severe insult that we have seen to date in terms of potential harm to man from plutonium.

TISSUE ANALYSIS PROGRAMS

For many years the Los Alamos Scientific Laboratory (*27*) and several other AEC contractor laboratories (*28–31*) have conducted tissue analysis programs to determine plutonium levels in various tissues of both exposed occupational personnel and individuals of the general population who are not engaged in work with plutonium. These programs were started in the 1940's in the Hanford complex and at the Los Alamos Scientific Laboratory. One report contains information on approximately 350 autopsies (*28*). Additional reports from these and other groups are available (*32–34*). Table 9 shows plutonium concentrations as determined for lung, liver, lymph nodes, kidney, and bone for the period 1959–1971 for non-occupationally exposed persons from several parts of the United States and for occupationally exposed persons (*35*). Data for the plutonium concentration of gonadal tissue which appears in the original publication is not included in Table 9, as errors associated with a change in the

[3]G. W. Dolphin, *Hot Particles*, National Radiological Protection Board, Radiological Protection Bulletin No. 8, pp. 8–10, 1974.

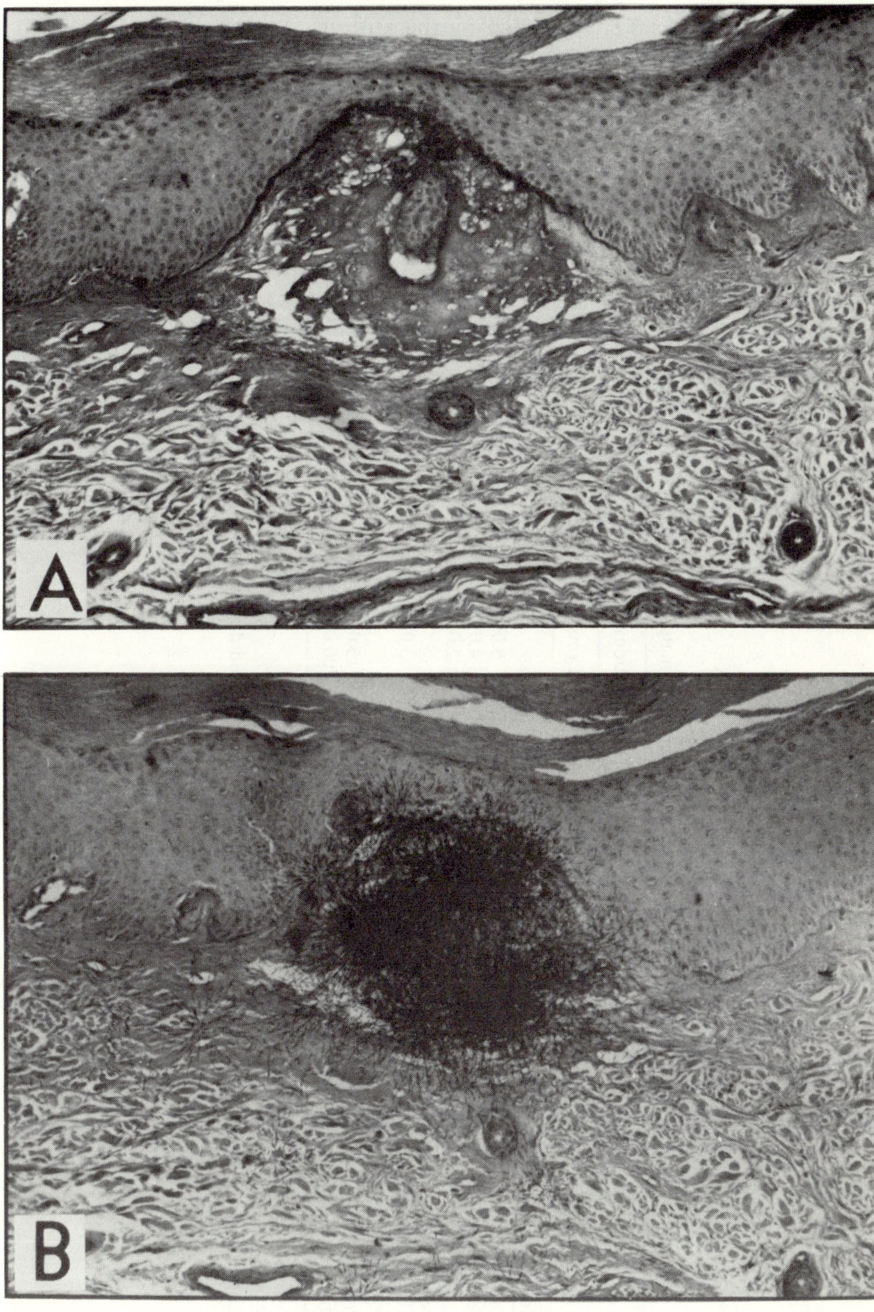

FIG. 3. *a* Photomicrograph of skin section removed from palmar surface of plutonium worker's hand about four years after accident. *b* Photomicrograph of autoradiograph prepared from tissue close to that shown in *a*.

analytical procedures were detected by the authors subsequent to its original publication. Similar data shown in Table 10 have been obtained for non-occupationally exposed persons and represent anaylses made during the period 1972–1973 (*35*). The average lung concentration for the data shown in Table 10 is about 0.3 pCi for a 1000 gram lung, and the lumph node concentration is about 11pCi/kg. No special concentration of plutonium in gonadal tissue has been observed in this particular study. Recent analysis of the gonadal data suggest

TABLE 9.
50th Percentile Distribution of Plutonium in Human Tissue (1959–1971)

	Plutonium Disintegrations per Minute per Kilogram				
	Lung	Liver	Lymph Node	Kidney	Bone
Nonoccupationally Exposed:					
Los Alamos	1.3 (57)[2]	1.1 (58)	5.0 (52)	0.1 (54)	0.4 (35)
New Mexico and U.S.	1.0 (76)	0.9 (73)	4.0 (66)	0.2 (66)	0.5 (41)
Colorado	0.5 (66)	1.7 (60)	2.0 (46)	1.4 (45)	0.9 (65)
New York	0.4 (26)	1.7 (26)	b	b	2.0 (25)
All Populations	0.8 (217)	1.4 (217)	3.0 (164)	0.6 (163)	0.6(166)
Occupationally Exposed:[c]					
Potential	4.0 (44)	1.0 (41)	15.0 (42)	0.1 (42)	0.3 (25)
High Potential	100.0 (15)	100.0 (15)	700.0 (14)	10.0 (13)	50.0 (11)

[a]Number of samples (in parentheses)
[b]Samples not requested
[c]Data cannot be compared as a group because of differences in type and duration of exposure

plutonium concentrations of about 0.18 pCi/kg for non-occupationally exposed persons.

The higher plutonium concentrations in lymph node tissue in Table 10 as compared with those in Table 9 for non-occupationally exposed individuals are not thought to represent real increases, but rather an improvement in the dissection techniques used to separate lymph node tissue from lung.

The AEC's Health and Safety Laboratory (HASL) in New York recently has used information obtained from the International Commission on Radiological Protection to model the intake and body burden from fallout plutonium and to estimate the radiation dose to man from this source (*36*). The cumulative lung and bone dose for the period 1954–2000 is 16 and 34 mrem, respectively. The

TABLE 10

50th Percentile Distribution of Plutonium in Human Tissue (1972–1973)

	Plutonium Disintegrations per Minute per Kilogram				
	Lung	Liver	Lymph Node	Kidney	Vertebrae
Nonoccupationally Exposed:					
Los Alamos	0.8 (8)[a]	1.6 (5)	35 (4)	0.2 (5)	1.6 (5)
New Mexico and U.S.	0.4 (17)	0.7 (10)	20 (15)	1.2 (10)	0.4 (16)
Colorado	0.7 (29)	1.8 (25)	15 (22)	3.0 (25)	1.1 (25)
New York[b]	0.3 (34)	1.4 (31)	c	c	0.7 (32)
Savannah River	0.4 (20)	1.2 (14)	40 (6)	2.2 (11)	0.7 (12)
All Populations[d]	0.6 (74)	1.5 (54)	25 (47)	1.5 (51)	0.7 (58)

[a] (n) number of samples (in parentheses)
[b] New York data for 1972 include all data analyzed
[c] Samples not requested
[d] All population data for 1972 do not include the New York data analyzed during 1972

HASL group has also compared the body burden based on its own model with that actually obtained from the tissue sampling programs. The agreement between the Colorado-New Mexico tissue data and the model predictions as shown in Table 11 for 1970–1971 is quite good. The comparison based on the New York tissue sampling data is not as good and probably represents the fact that only about 24 autopsies were included in the analysis.

TABLE 11
Plutonium-239 in Man[a]

	Colorado-New Mexico (1970–1971) Concentration (dpm/kg)	Burden (pCi)	Computed Burden (pCi)
Lung (1.0 kg)	0.6 (96)[b]	0.30	0.3
Lymph (0.015 kg)	5.0 (73)	0.03	0.6
Liver (1.7 kg)	1.8 (88)	1.40	0.8
Kidney (0.3 kg)	1.1 (73)	0.10	—
Bone (5.0 kg)	0.6 (96)	1.40	0.9
		3.20	2.6
	New York (1968)		
Lung (1.0 kg)	0.4	0.20	0.6
Liver (1.7 kg)	1.7	1.30	0.7
Bone (5.0 kg)	2.0	4.50	0.8
Lymph (0.015 kg)	—	—	0.7
		6.00	2.8

[a]Reference *36*
[b]Number of samples (in parentheses)

Plutonium is present in extremely small quantities in various organs of contemporary humans. Although most of the plutonium was produced from atmospheric testing of nuclear weapons by several countries prior to the 1963 limited test ban treaty, some material from contemporary atmospheric weapons testing by France and the People's Republic of China adds to the total human burden. The current lung burden estimated for persons in the United States is about 0.3 pCi, ^{239,240}Pu, and a very rough estimate of the total amount in the body is perhaps 5 pCi as shown in Table 12. Estimates of the total amount of plutonium produced via nuclear weapons testings vary, but a value of about 0.5 megacurie probably is quite reasonable. If, of this amount, 0.4 megacurie has returned to the biosphere, then very little (about 10^{-8}) has found its way into the earth's population (3×10^9 people). Another way of looking at this is to simply divide the estimated average human burden (5×10^{-12} curies) by the estimated amount in the biosphere (0.4×10^6 curie); in this instance the average accumulation per person is about 10^{-17}.

Results of the tissue sampling programs for occupationally exposed plutonium workers has also given us the opportunity to compare the body burden found at autopsy with that estimated during life on the basis of bioassay data. Almost without exception, workers in the USA and United Kingdom have found less plutonium by a factor of approximately 10 at autopsy as compared with that amount predicted by bioassay data. Thus it would appear that estimates

TABLE 12
Plutonium in Man from Atmospheric Nuclear Weapons Tests

A. Plutonium produced from atmospheric weapons tests ~0.5 MCi
B. Plutonium now on earth's surface ~0.4 MCi
C. Plutonium in contemporary man

	$dpm \cdot kg^{-1}$	$pCi \cdot kg^{-1}$	kg	pCi·organ
Lung	0.8	0.4	1.0	0.4
Liver	1.4	0.6	1.7	1.0
Bone	0.6	0.3	10.0	3.0
Other	—	—	—	0.6
			Total	~5.0 pCi

D. Each person on earth accumulated about 10^{-17} of B

$$\frac{5 \times 10^{-12}\ \text{Ci} \cdot \text{person}^{-1}}{0.4 \times 10^{6}\ \text{Ci}}$$

E. About 10^{-8} of the earth's inventory found its way into the 3×10^9 persons on earth

$$\frac{(3 \times 10^9\ \text{persons})(5 \times 10^{-12}\ \text{Ci} \cdot \text{person}^{-1})}{0.4 \times 10^{6}\ \text{Ci}}$$

of the body burden made during life are conservative in that they predict more plutonium than is actually present in the body. Because a considerable amount of data on this subject is now available, it might be profitable for responsible persons in the radiation protection areas to evaluate this finding in terms of current radiation protection practices and guides followed in the nuclear energy industry.

Recently several researchers have examined the United Kingdom experience in terms of the medical aspects of radiological protection of workers handling plutonium. They conclude that the information to date cannot conclusively show that the presently accepted working levels for plutonium are adequate, but they do allow for a certain amount of cautious optimism. They also state, "it is true to say that after 30 years' experience in the USA and 22 years' in this country, no disease attributable to plutonium toxicity has been diagnosed in any worker concerned in the production or manipulation of plutonium" (*23*).

It is important to realize that none of the information presented in this paper means that we will never see an adverse biological effect of plutonium in man. Nevertheless, the record of human experience with plutonium to date is reassuring in the sense that the radiation protection standards which have been followed for many years are very probably quite adequate and do not misrepresent the potential risk by many orders of magnitude, as has been suggested.

REFERENCES

1. G. T. Seaborg, in *Radiobiology of Plutonium*, (B. J. Stover, and W. S. S. Jee, Eds.), pp. 1–21. The J. W. Press, University of Utah, Salt Lake CIty, 1972.
2. H. F. Anderson, W. E. Sheehan, J. R. Mann and R. W. Bistline, *Health Phys.* **18**, 631 (1970).
3. J. R. Mann and R. A. Kirchner, Health Phys. **13**, 877 (1967).
4. C. C. Lushbaugh, R. J. Cloutier, G. Humason, J. Langham and S. Guzak, *Ann. N. Y. Acad. Sci.* **145**, 791 (1967).

5. W. D. Norwood, J. A. Norcross, C. E. Newton, Jr., D. B. Hylton and C. Lagerquist, in *Radionuclide Carcinogenesis* (C. L. Sanders, R. H. Busch, J. E. Ballow, and D. D. Mahlum, Eds.), pp. 465–474. AEC Symposium Series No. 29, USAEC Office of Information Services, June 1973.
6. C. R. Lagerquist, S. E. Hammond and D. B. Hylton, *Health Phys.* **22**, 921 (1972).
7. W. H. Langham, J. N. P. Lawrence, Jean McClelland and L. H. Hempelmann, *Health Phys.* **8**, 753–760 (1962).
8. C. E. Newton, Jr., M. V. Larson, K. R. Heid, I. C. Nelson, P. A. Fuqua, W. D. Norwood, S. Marks and T. D. Mahoney, in *Diagnosis and Treatment of Deposited Radionuclides*, (H. A. Kornberg and W. D. Norwood, Eds.), pp. 460–468. Excerpta Medica Foundation, New York, 1968.
9. H. Foreman, W. Moss and W. Langham, *Health Phys.* **2**, 326 (1960).
10. L. H. Hempelmann, W. H. Langham, C. R. Richmond, and G. L. Voelz, *Health Phys.* **25**, 461 (1973).
11. L. H. Hempelmann, C. R. Richmond, and G. L. Voelz, *A Twenty-seven Year Study of Selected Los Alamos Plutonium Workers*, Los Alamos Scientific Laboratory Report LA-5148-MS, 31 pp. (1973b).
12. L. H. Hempelmann, W. H. Langham, G. L. Voelz, and C. R. Richmond, in *Proc. Third Congress of the Int. Radiation Protection Assoc.*, Washington, D.C. 1974 pp. (Sept. 9–14, 1973).
13. J. N. P. Lawrence, *Health Phys.* **8**, 61 (1962). See also, Los Alamos Scientific Laboratory Report LA-2329 (April 1960).
14. W. J. Bair, C. R. Richmond and B. W. Wachholz, *A Radiological Assessment of the Spatial Distribution of Radiation Dose*, USAEC WASH-1320, 47 pp., (Sept. 1974).
15. A. R. Tamplin and T. B. Cochran, *Radiation standards for hot particles: A report on the inadequacy of existing radiation protection standards related to internal exposure of man to insoluble particles of plutonium and other alpha-emitting hot particles.* Natural Resources Defense Council Report, Washington, D.C., 1974.
16. W. H. Langham, S. H. Bassett, P. S. Harris, and R. E. Carter, *Distribution and Excretion of Plutonium Administered Intravenously to Man.* Los Alamos Scientific Laboratory Report LA-1151, p. 16 (Sept. 1950).
17. P. W. Durbin, in *Radiobiology of Plutonium*, (B. J. Stover and W. S.S. Jee, Eds.) pp. 469–530. The J. W. Press, University of Itah, Salt Lake City, 1972.
18. J. Rundo, P. M– Starzyk and J. Sedlet, *Radiat. Res.* **59**, 86, (1974).
19. H. D. Bruner, in *Diagnosis and Treatment of Deposited Radionuclides*, (H. A. Kornberg and W. D. Norwood, Eds.), pp. 661–665. Excerpta Medica Foundation, New York, 1968.
20. W. D. Norwood, in *Radiobiology of Plutonium*, (B. J. Stover and W.S.S. Jee, Eds.), pp. 531–537. The J. W. Press, University of Utah, Salt Lake City, 1972.
21. J. A. Norcross and C. E. Newton, Jr., *Health Phys.* **22**, 887 (1972).
22. W. D. Norwood, J. A. Norcross, C. E. Newton, Jr., D. B. Hylton and C. Lagerquist, Preliminary autopsy findings in U.S. Transuranium cases, in *Radionuclide Carcinogenesis*, (C. L. Sanders, et al., Eds.), AEC Symposium Series **29**, pp. 465–474. U.S. Atomic Energy Commission, Washington, D.C. (Available as CONF-720505 from National Technical Information Service, Springfield, Va. 22151.)
23. G. B. Schofield and G. W. Dolphin, *Brit. J. Occup. Med.* (in press).
24. W. D. Norwood, in *XVI International Congress on Occupational Health*, Tokyo, Sept. 22–27, 1969, *Abstracts*, pp. 691–693.
25. W. D. Norwood and C. E. Newton, Jr., *United States Transuranium Registry Summary to June 30, 1974 to USAEC Division of Biomedical and Environmental Research.* Hanford Environmental Health Foundation Report HEHF No. 22, 28 pp., (1974).
26. C. C. Lushbaugh and J. Langham, *Arch. Dermatol.* **86**, 461–464 (1962).
27. E. E. Campbell, M. F. Milligan, W. D. Moss and H. F. Schulte, *History of the Plutonium Bioassay Program at the Los Alamos Scientific Laboratory, 1944*–1972. USAEC Document LA-5008, 7 pp., (1972).
28. I. C. Nelson, K. R. Heid, P. A. Fuqua, and T. D. Mahony, *Health Phys.* **22**, 925 (1972).
29. C. R. Lagerquist, S. E. Hammond, D. L. Bokowski and D. B. Hylton, *Health Phys.* **25**, 581 (1973).
30. C. R. Lagerquist, D. L. Bokowski, S. E. Hammond and D. B. Hylton, *Am. Ind. Hyg. Assoc. J.* **30**, 417 (1969).
31. E. E. Campbell, B. C. Eutsler, J. McClelland, and H. M. Ide, *Health Phys.* **22**, 931 (1972).

32. C. E. Newton, Jr., I. C. Nelson, K. R. Heid and H. V. Larson, *Transuranium Elements and Thorium in Man, Assessment, Applicability, of Biologic Models and Needed Research.* Battelle Northwest Laboratory Report BNWL-SA-279, (August 1965).
33. C. E. Newton, Jr., K. R. Heid, H. V. Larson and I. C. Nelson, *Tissue Sampling for Plutonium through an Autopsy Program.* Battelle Northwest Laboratory Report BNWL-SA-918 (Sept. 1966).
34. E. E. Campbell, M. F. Milligan, W. D. Moss, H. F. Schulte, and J. F. McInroy, *Plutonium in Autopsy Tissue.* Los Alamos Scientific Laboratory Report LA-4875, 47 pp., (January 1973).
35. *Annual Report of the Biomedical and Environmental Research Program of the LASL Health Division for 1973 to the USAEC, Division of Biomedical and Environmental Research.* Los Alamos Scientific Laboratory Report LA-5633-PR, (C. R. Richmond and E. M. Sullivan, Eds.), (1974).
36. B. G. Bennett, in *Fallout Program, Quarterly Summary Report.* Health and Safety Laboratory Report HASL-278 (January 1, 1974).

Dose-response Relationships for Beagles Injected with ^{239}Pu(IV) or ^{241}Am(III)[1,2]

Betsy J. Stover
Department of Pharmacology
University of North Carolina
Chapel Hill, N. C. 27514 USA
and
Division of Radiobiology
Department of Anatomy
University of Utah
Salt Lake City, Utah 84132 USA

INTRODUCTION

That ionizing radiation could adversely affect living systems was recognized within a few years after the discovery of radioactivity by Becquerel in 1896. In 1974, the two deleterious long-term effects of intermediate and low dose level α-irradiation from ^{239}Pu(IV) in mammalian systems emerge as life-shortening and induction of osteosarcomas. Theory has been derived from basic principles to interpret life-shortening in general (*1—4*). However, there is not yet a theory of the radiation dose-carcinogenic response relationship that has been unequivocally verified by experimental or epidemiological data.

The results from two long-term studies of the toxicology of ^{239}Pu in beagles and of a companion study using ^{241}Am will be presented from this point of view. All of the beagles in the first ^{239}Pu study have succumbed. Hence, data on life-shortening and occurrence of osteosarcomas are complete. The second ^{239}Pu and the ^{241}Am studies are in progress.

MATERIALS AND METHODS

The isotopic composition[3] of the plutonium is approximately 95% ^{239}Pu and 5% ^{240}Pu.. Neither ^{241}Pu nor its decay product ^{241}Am were detected. Plutonium has four valence states which can co-exist in significant amounts in aqueous solution. Further, Pu(IV) forms stable oxygen-bridged polymers as the pH is increased. Thus, ^{239}Pu(IV) was prepared and stabilized by complexing it first with nitrate and then with citrate. (*5-7*). The americium was ^{241}Am(III); it was also complexed with citrate. Approximately 9 ml of the radionuclide in 0.08 M total citrate at pH 3.5 was given in a single intravenous injection when the beagles were young adults (15 to 18 months). The

[1]Congress presentation C-9-4.

[2]Supported by the U.S. Atomic Energy Commission.

[3]This "old-fashioned" plutonium was obtained through the helpful efforts of the late Dr. Wright H. Langham.

injected doses and numbers of dogs are given in Tables 1, 2 and 3. Details of the specific analyses, measurements, and veterinary care have been published elsewhere (see, for example, (*8, 9*).

RESULTS

First ^{239}Pu Study

Following injection of ^{239}Pu(IV) as a citrate complex the concentration in blood plasma decreases slowly as a consequence of the transfer of ^{239}Pu(IV) from citrate to transferrin, the iron transport protein (*5, 10, 11*). There is no significant association of ^{239}Pu(IV) with the blood cells. Concentration of ^{239}Pu in plasma decreases continuously throughout the life-time of the beagle and has been measured through 12.5 years after injection. It can be described by a sum of four exponentials with half-periods of 0.66, 1.5, 8.7, and 780 days (*12*). The initial distribution and excretion of ^{239}Pu(IV) are independent of dose level over the range of dose levels studied. At 22 days after injection, (48.9±3.4)% of the injected ^{239}Pu(IV) is in the skeleton, (32.5±5.2)% is in the liver, 0.3% is in the spleen, 0.3% is in the kidneys, roughly 2% is ubiquitously distributed in other organs and tissues, and (12.1±2.1)% has been excreted (*13-15*).

The rate of decrease of hepatic retention is dependent on dose level from 2.9 to 0.095 μCi/kg (*6, 12, 15, 16*). The approximate half-periods for decrease in the amount of ^{239}Pu(IV) in the liver are 600, 1300, and 3100 days at 2.9, 0.91, and 0.30 μCi/kg, and 3800 days at 0.095 μCi/kg and lower. Cochran et al. and Taylor et al. have determined the microdistribution of ^{239}Pu(IV) in the liver by autoradiography (*17, 18*). Initially there is a uniform diffuse distribution in the parenchymal cells. At the subcellular level a significant fraction of the early deposit of ^{239}Pu is bound by ferritin, the iron storage protein (*7, 19, 20*). Thus, ^{239}Pu(IV) follows the transferrin-ferritin part of the metabolic pathway of iron.

At 2 to 3 months after injection of 2.9 μCi/kg translocation of ^{239}Pu from hepatic cells to reticuloendothelial (RE) cells was observed autoradiographically (*17,18*). This transfer to RE cells also occurred at the lower dose levels at progressively longer times after injection. Concurrently there was an increase in serum glutamic pyruvic transaminase (SGPT) which reflected death of hepatic cells (*21*). Later a second transfer of ^{239}Pu from RE cells to free macrophages occurred. The macrophages were usually observed in the portal space and occasionally intravascularly. Cellular necrosis was observed histologically, and in some beagles at the 2.9 μCi/kg dose level it resulted in a marked atrophy of the liver. Thus, hepatic cellular necrosis leads to transfer of ^{239}Pu from one cell type to another, its eventual exit from the liver, and is the basis for the observed effect of dose level on hepatic retention of ^{239}Pu(IV). The average radiation doses to liver are given in Table 1.

A second kind of displacement of ^{239}Pu in the liver occurred as the result of focal expansion of regenerative hepatic epithelium (*18*). There is little or no ^{239}Pu in the regenerating cells, and the older cells with their initial deposit of ^{239}Pu are compressed toward the periphery of the acinus. The regenerative tissue was observed at all six dose levels, 2.9 to 0.016 μCi/kg.

TABLE 1
Toxicology of ^{239}Pu, First Study

Injected Dose (μCi/kg)	Number of Beagles	Dose to Liver (rad)	Dose to Skeleton (rad)	Life Shortening (%)	Occurrence of Osteosarcoma (%)
2.9	9	5400	6400	63	78
0.91	12	2450	1950	61	100
0.30	12	970	720	55	100
0.095	11	570	260	33	91[a]
0.048	13	340	160	19	69
0.016	12	120	60	0.5	33

(2.9/0.016) = 180

[a] one squamous cell carcinoma, frontal sinus

Hepatic cell tumors were not observed, but ten bile duct tumors were (*18*). Two of these were malignant and eight were incidental findings at autopsy. The occurrence of bile duct tumors is 4.2% in control beagles and 10.4% in the beagles given 0.016 to 2.9 μCi/kg. The increased occurrence with respect to ^{239}Pu appears equivocal since the bile duct does not have a high deposition of ^{239}Pu(IV). Taylor et al. have discussed the possibility that these tumors arise from the hepatic epithelium.

Since ^{239}Pu(IV) deposits on the surfaces of bone, the cells of the marrow are at risk as well as those of bone. The hematological effects of ^{239}Pu(IV) in the beagle have recently been summarized by J. H. Dougherty (*22*). There was an early depression of leukocytes at dose levels 2.9 to 0.095 μCi/kg, and the degree of depression was dependent on dose level. Polymorphonuclear leukocytes, monocytes, and eosinophils reached lowest levels at 14 to 21 days and returned toward normal values during the first year after injection. The lymphocytes were depressed at 2.9 and 0.91 μCi/kg and did not recover during the first year. Platelets were depressed at 14 to 21 days at 2.9 and 0.91 μCi/kg, but returned to normal at two to three months. Three beagles at 2.9 μCi/kg showed a terminal thrombocytopenia. There was anemia at two to six months in beagles at 2.9 μCi/kg and a corresponding increase in reticulocytes, and terminal anemias were observed in beagles at 2.9 to 0.30 μCi/kg. No significant changes in leukocytes were observed after about four years following injection. This finding correlates in time with observations of the microdistribution of ^{239}Pu(IV) in bone, which are presented later in this paper.

Taylor et al. have summarized skeletal changes that were observed clinically. Skeletal fractures that were a result of radiation damage occurred at 2.9 and 0.91 μCi/kg. Accelerated loss of teeth was observed at dose levels 2.9 to 0.048 μCi/kg as well as involution of the maxilloturbinate bone. Skeletal changes that were observed radiologically at high dose levels include abnormal thickening and thinning of cortical bone and coarsening of trabecular bone. No changes of this kind that could be distinguished from changes that occur with aging were observed at the 0.016 μCi/kg dose level.[4] Histopathologic changes observed in bone of beagles at the 2.9 μCi/kg dose level include

[4] Wm. R. Christensen, private communication.

necrosis of endosteal cells and osteocytes within the range of the α-particles, abnormal remodeling of trabecular bone, and peritrabecular fibrosis at 1 week to 3 months after injection.

Skeletal retention of ^{239}Pu(IV) in the beagle varies with dose level in a manner opposite to that of hepatic retention but for the same basic reason, cellular necrosis (*12*). Removal of ^{239}Pu(IV) from the skeleton is dependent upon bone remodeling processes which cease at high dose levels as a result of death of osteoclasts and osteoblasts. The equations for skeletal retention are as follows:

$$B\ (2.9\ \mu\text{Ci/kg}) = 6.9e^{-0.0011t} + 42.8 \quad (1)$$

$$B\ (0.91\ \mu\text{Ci/kg}) = 11.0e^{-0.0011t} + 38.6 \quad (2)$$

$$B\ (0.31\ \mu\text{Ci/kg}) = 15.2e^{-0.0011t} + 34.4 \quad (3)$$

$$B\ (0.095\ \mu\text{Ci and lower}) = 28.6e^{-0.0011t} + 21.0 \quad (4)$$

where B is per cent of injected ^{239}Pu(IV) and t is days after injection. Using Eqs. (1–4), the measured injected doses, the finding that skeletal weight equals 7.5% body weight at age 15 to 18 months, and appropriate conversion factors, the average radiation doses to the skeleton have been calculated. The results appear in Table 1. The hepatic dose relative to the skeletal dose changes appreciably with decreasing dose level.

The skeletal retention equations Eqs. (1–4) are empirical best fits that utilize all the available data (*12,13*). The autoradiographic and histological findings of Jee and Arnold at the highest dose level and those of Jee at the lower levels provide a biological basis for the empirical retention equations (*24,25*). Initially ^{239}Pu is deposited nonuniformly on endosteal, periosteal and haversian system surfaces. The major deposition is on trabecular surfaces which correlates with the macro-observation that initial deposition is highest in bones with a high proportion of trabecular bone. The continuous remodeling of bone removes ^{239}Pu from the surface by new bone formation and by resorption of the surface. The new bone which covers the initial high surface concentration of ^{239}Pu has a low concentration of ^{239}Pu and the deposition is diffuse. When ^{239}Pu on the surface is resorbed it is temporarily contained in the osteoclasts and then transferred to macrophages in the marrow. The macrophages are subsequently cleared from the marrow (*25*). Since some ^{239}Pu is found diffusely in new bone, it is also possible that some ^{239}Pu from the disintegrating dead osteoclasts is bound by transferrin and then redeposited on bone surfaces. Thus, the exponential of the skeletal retention equation is related to the removal of surface deposits of ^{239}Pu(IV), and the long term constant term is related to burial of surface deposits and redeposition of ^{239}Pu in new bone.

Jee's findings include quantitative data on the microdistribution of ^{239}Pu on bone surfaces, in bone surfaces that have been covered with new bone, in new bone, and in macrophages at times that range from 5 to 2257 days after injection (*25*). Of particular interest are the data for beagles at the 0.016 μCi/kg dose level since there is no measurable effect of dose level over the range 0.0055 to 0.095 μCi/kg on total skeletal retention, and it can be assumed that bone remodeling processes are proceeding at normal rates. These data and measurements of total skeletal retention have been used to evaluate the rate constants in the kinetic model of skeletal retention of ^{239}Pu(IV) (*26*). In this model the initial skelctal ^{239}Pu(IV) is $A_0 = A + B + C + D$ which are given as follows:

On bone surfaces

$$A = A_0 e^{-(k_1 + k_2)t} \tag{5}$$

Buried and in new bone

$$B = \frac{k_1 A_0}{k_1 + k_2} \left\{ 1 - e^{-(k_1 + k_2)t} \right\} \tag{6}$$

In macrophages

$$C = \frac{k_2 A_0}{k_1 + k_2 - k_3} \left\{ e^{-k_3 t} - e^{-(k_1 + k_2)t} \right\} \tag{7}$$

Eliminated from bone and bone marrow

$$D = \frac{k_2 k_3 A_0}{k_1 + k_2 - k_3} \left\{ \frac{1 - e^{-k_3 t}}{k_3} - \frac{1 - e^{-(k_1 + k_2)t}}{k_1 + k_2} \right\} \tag{8}$$

Skeletal retention at any time t is then given by $A + B + C$. From the skeletal retention and autoradiographic data at the dose level 0.095 μCi/kg and lower the following values for the constants are estimated: the initial skeletal deposition, A_0 = 50% of the injected ^{239}Pu(IV); the combined rate constant for burial of ^{239}Pu(IV) and for its deposition in new bone, $k_1 = 0.0025\text{d}^{-1}$; the rate constant for resorption of ^{239}Pu(IV) from bone surfaces, $k_2 = 0.0035\text{d}^{-1}$; and the rate constant for the egress of macrophages, $k_3 = 0.0013\text{d}^{-1}$. Substituting these values into Eqs. (5–8) gives

$$A = 50e^{-0.0060t} \tag{9}$$

$$B = 21\,(1 - e^{-0.0060t}) \tag{10}$$

$$C = 37\,(e^{-0.0013t} - e^{-0.0060t}) \tag{11}$$

$$D = 37\,(1 - e^{-0.0013t}) - 8(1 - e^{-0.0060t}) \tag{12}$$

$$A + B + C = 37e^{-0.0013t} - 8e^{-0.0060t} + 21 \tag{13}$$

Eqs. (9–11) in combination with bone morphometric data should give more realistic estimates of radiation dose to the skeleton.

The Steady State Theory of Mutation Rates has been used to interpret survival data for the beagles at the six dose levels of ^{239}Pu(IV) (*1–4,27,28*). In this theory statistical survival of a population is given by

$$S = (1 + e^{-k'(\tau - t)})^{-1} \tag{14}$$

the death rate by

$$-dS/dt = k'S(1 - S) \tag{15}$$

and the mortality rate by

$$-d\ln S/dt = k'(1 - S) \tag{16}$$

In Eq. (14) τ is the time at which 1/2 the population is surviving, $k'\tau$ is identified as the cellular reserves against the ravages of time and specific environmental factors (in this case ^{239}Pu(IV)), and k' is the difference between the sum of the rate constants for deleterious processes and the sum of the rate constants for repair or removal of the effects of the deleterious processes. This theory is derived from basic principles and provides, among other things, a formalism for the calculation of life-shortening. This now supersedes methods based on the widely used but nevertheless empirical Gompertz equation. Per cent life-shortening is calculated from this equation

$$\text{life shortening } (\%) = 100\ (\tau_c - \tau)/\tau_c \tag{17}$$

where τ_c = 4778 days is the age at which 1/2 of a population of 32 contemporary beagles is surviving and τ is that age for the beagles at a given dose level of ^{239}Pu(IV). The results are given in Table 1. Life-shortening is essentially the same at dose levels of 2.9 and 0.91 μCi/kg. Since the age at injection is about 12% of τ_c, this suggests that about 20% of τ_c is required for the induction and growth of an osteosarcoma. Life-shortening at dose levels 0.30, 0.095, and 0.048 μCi/kg, as a first approximation, appears to be proportional to the logarithm of the average radiation dose to the skeleton. The most important finding, however, is that there is no life-shortening at the 0.016 μCi/kg dose level.

Osteosarcomas were induced at all six dose levels. At the highest dose level liver degeneration, skeletal fractures, and gingivitis were important factors in the animals' demise. Dose levels of 0.91, 0.30, and 0.095 μCi/kg are optimal for carcinogenesis. One beagle at 0.095 μCi/kg developed a squamous cell carcinoma of the frontal sinus. In view of the proximity of the tissue to the skeletal surface and the experience with humans exposed to ^{228}Ra and/or ^{226}Ra (see, for example, Evans et al. (*29*)) it is considered that this tumor was induced by ^{239}Pu(IV). Thus, only at two dose levels are data on occurrence of osteosarcoma available for calculation of classical dose-response relations or to test theoretical predictions (*3,4*). Thus, the results of twenty-four years of research are inadequate to determine a radiation dose-carcinogenic response relationship!

Second ^{239}Pu Study

A second study of the toxicology of ^{239}Pu (IV) in the beagle is in progress. The 0.016 μCi/kg dose level is being repeated with an additional 12 beagles and four lower dose levels are included as shown in Table 2. Thus, the two studies cover ten dose levels which vary from highest to lowest by a factor of 4500. The lowest dose level, 0.0064 μCi/kg, is essentially equal the maximum permissible level in humans which is 0.04 μCi/70kg = 0.0057 μCi/kg.

TABLE 2
Toxicology of ^{239}Pu, Second Study as of March 31, 1974

Injected Dose (μCi/kg)	Number of Beagles	Expected Dose to Liver (rad)	Expected Dose to Skeleton (rad)	Longest Exposure (days)	
0.016	12	120	60	1713	
0.010	36	80	40	1713	No
0.0055	25	40	20	3610	osteosarcomas
0.0019	36	15	7	3610	observed
0.00064	22	5	2	3679	

(0.016/0.00064) = 25 (2.9/0.00064) = 4500 (0.04 μCi/70 kg) = 0.00057 μCi/kg

Since no effect of dose level on either hepatic or skeletal retention was observed at dose levels 0.095, 0.048, and 0.016 μCi/kg, the retention and average radiation dose rate equations should apply at the lower dose levels, and this has already been confirmed at the 0.0055 μCi/kg dose level (*12*). Since no life-shortening was observed at 0.016 μCi/kg in the first study, none is anticipated at the dose levels of the second study. Therefore, the time over which the dose rate equations should be integrated is known, and the average hepatic and skeletal doses can be estimated in advance. This has been done and the results are given in Table 2.

The only effects at the 0.016 μCi/kg dose level of the first study that were detected were the increase in occurrence of osteosarcoma above that of controls, focal volumes of regenerative tissue in the liver, and, perhaps, carcinoma of the bile duct. In addition, two beagles succumbed with carcinoma of the thyroid. (The concentration of ^{239}Pu (IV) in the thyroid is low but measurable.) Carcinoma of the thyroid has been observed in only four other beagles of this colony: a control, one given ^{228}Th, and two given ^{90}Sr.

The stimulation of liver regeneration by ^{239}Pu (IV) is of importance in hepatic retention of ^{239}Pu (IV), and both retention and regeneration will be determined at these low dose levels.

The principal data anticipated in this second study are the occurrence of osteosarcoma at these five dose levels. None has been observed as of March 31, 1974. The longest exposure time for each dose level is shown in Table 2. However, many of the beagles have been injected quite recently and it is anticipated that ten to fifteen more years will be required to obtain definitive results.

The companion ^{241}Am study

The retention, distribution, and biological effects of ^{241}Am in the beagle are being studied at eight dose levels which range from 2.8 to 0.0018 μCi/kg and vary between highest and lowest by a factor of 1600 (Table 3). The ^{241}Am (III) ion has a lower charge and a greater ionic radius than does ^{239}Pu (IV). The chemical effect of this is that ^{241}Am (III) in general forms less stable complexes and other chemical associations and undergoes a lesser degree of hydrolysis than ^{239}Pu (IV). ^{241}Am (III) does not form stable oxygen-bridged polymers as ^{239}Pu (IV) does. The retention and distribution of ^{241}Am (III) is qualitatively similar to that of ^{239}Pu (IV), but there are important quantitative differences.

TABLE 3
Toxicology of ^{241}Am as of March 31, 1974

Injected Dose (μCi/kg)	Number of Beagles	Number Dead	Number Dead with Osteosarcoma and Radiation Dose (rad)
2.8	2	2	0
0.91	12	10	6, 1900
0.30	13	7	7, 720
0.096	12	1	1, 250
0.048	13	0	0
0.016	13	1	0
0.0053	14	0	0
0.0018	14	1	0

The concentration of ^{241}Am (III) in blood plasma decreases much more rapidly than that of ^{239}Pu (IV), and the rate is comparable with those of divalent alkaline earth ions (*7,30*). During the first 22 days after injection 11.2% is expected. The major sites of deposition are the liver and skeleton (*31,32*). Initially, 48.7% of the injected ^{241}Am (III) is found in the liver and 28.8% is in the skeleton. ^{241}Am (III) also deposits on skeletal surfaces, but the deposition pattern as shown by autoradiography is more uniform than that of ^{239}Pu (IV) (*32*). In the liver ^{241}Am (III) is bound by ferritin as is ^{239}Pu (IV) (*33*). Both radionuclides are also associated with all subcellular structures and with the connective tissue of the liver. Early deposition in the kidneys is 0.65% of the injected ^{241}Am (III). Although the total amount in the thyroid is only 0.05%, the concentration of ^{241}Am (III) in the thyroid is exceeded only by that in the liver.

The two beagles at the 2.8 μCi/kg dose level suffered extreme liver degeneration. The average radiation doses to the liver were estimated to be 5500 and 7600 rads, and the cellular necrosis was so great that a shift of ^{241}Am from liver to the skeleton was measured (*31,32*). Six of the ten dogs at the 0.91 μCi/kg dose level that are dead developed osteosarcomas. Degeneration of the liver, kidneys, and thyroid was also observed (Table 3). Osteosarcomas have occurred at the next two lower levels, and the average radiation doses to the skeleton are comparable with those at dose levels of ^{239}Pu (IV) at which the occurrence of osteosarcomas is high. Definitive results at lower dose levels will require about ten more years.

UNANSWERED QUESTIONS AND FUTURE RESEARCH

1. *What is the radiation dose–carcinogenic response relationship for ^{239}Pu(IV) in the beagle?*

Twenty-four years of research have not yielded sufficient data to answer this question. In late 1969 this author proposed that additional dose levels between 0.095 and 0.016 μCi/kg be interpolated in the experiment. In 1974 it is recommended that 12 beagles be injected at dose levels of 0.032, 0.064, and 0.080

μCi/kg which are 2, 4, and 5 times 0.016 μCi/kg. (0.048 and 0.095 are 3 and 6 times this value.) On the basis of current results an increase in the occurrence of osteosarcomas above that for control beagles is anticipated at all three dose levels. Since life-shortening is anticipated at 0.064 and 0.080 μCi/kg and is highly likely at 0.032 μCi/kg, definitive results from this proposed third study of ^{239}Pu(IV) in the beagle could be available at about the same time as the definitive results from the current second study in which the dose levels are such that no life shortening is anticipated. The results from these three studies would provide much needed information on statistical variation, but of chief importance is the fact that they would provide a superb set of experimental data to test basic theories such as the Steady State Theory of Mutation Rates (*1–4*). If an extensive, accurate set of data is available to prove or disprove a theory, and if there is found a one-to-one correspondence between the experimental data and a given theory, then that theory can be used wih confidence in the interpretation of less extensive sets of data. The effect of this would be to remove the dose-response relationship question from the political arena and restore it to its proper place in the scientific community.

2. *What is the effective radiation dose parameter?*

It is generally agreed among those deeply involved with internal irradiation carcinogenesis that the effective radiation dose parameter depends on the source of the irradiation, and that, in general, the parameters have not been identified and measured. In the case of ^{239}Pu(IV) in the beagle, the continuing autoradiographic and morphometric studies by Jee of bone from control beagles and/or beagles at low dose levels and throughout the life time of the beagle most likely will provide the experimental basis to answer the dose parameter question[5], (*25*). These results in conjunction with retention studies and kinetic models will provide sets of dose parameters to test with experimental data (*26*).

3. *Are dose and time separable?*

At the 0.016 μCi/kg dose level no life-shortening was observed but the occurrence of osteosarcomas, which are rare tumors in beagles, is 33%. This suggests that radiation dose and time are not separable in the analysis of the carcinogenic response to internal irradiation from ^{239}Pu(IV) when an appreciable portion of, or all of, the beagle's expected remaining life span is required for the carcinogenic response to make itself manifest. The statistical survival equation given by the Steady State Theory of Mutation Rates serves as the first step in answering this question. Since this survival equation appears to be generally applicable, it raises the interesting general question—to which the question about ^{239}Pu(IV) may be a corollary—of whether dose and time are separable in any long-term carcinogenic response to an environmental agent.

4. *What is the entire role of the macrophage in the toxicology of ^{239}Pu(IV)?*

The role of the macrophage in the egress of ^{239}Pu(IV) from both the skeleton and the liver has been shown to be one of the critical steps in interpreting

[5]W. S. S. Jee, D. B. Kimmel, E. G. Hashimoto, Rebecca B. Dell, and L. H. Woodbury, Quantitative studies of beagle lumbar vertebral bodies, Research in Radiobiology, C00–119–248, pp. 255–273, 1973.

the skeletal and hepatic retention of ^{239}Pu(IV) (*18,25*). What is the fate of these macrophages, does the ^{239}Pu that they contain become completely available for excretion or is some ^{239}Pu deposited in tissues that might be critical target tissues in a beagle of a different age or in an animal of a different species? What is the immunological role of the macrophage in the skeleton with respect to osteosarcomas, might the time when the macrophages are engaged in transporting ^{239}Pu(IV) out of the bone marrow (given by Eq. (11)) be the critical time in which a viral agent succeeds in causing a change in one or more cells that is the prologue to the osteosarcoma? Perhaps the same question might be asked with respect to hepatic macrophages and carcinoma of the bile duct.

5. *What is the effect of age?*

The deposition, distribution, and retention of ^{239}Pu(IV) as functions of time are determined by basic biological processes such as binding to proteins, bone remodeling, phagocytosis, division of cells, cellular necrosis. The rates of these processes vary with age. Studies to date are limited mainly to an injection age of young adulthood. Less extensive companion studies at low and intermediate dose levels in which developing and maturing beagles are given both single and multiple injections of ^{239}Pu(IV) are greatly needed.

6. *How can the findings in beagles be interpreted in terms applicable to humans?*

Determination of the time scales or rates of all the basic biological processes involved in the toxicology of ^{239}Pu in humans relative to those in beagles would provide the ideal basis for translating the findings in beagles into expected findings for humans. These expected findings could then be tested against observations in humans (which, however, appear to be so limited that they may not provide an adequate test).

7. *What are rational scientific bases for therapy that is intended to protect an exposed person?*

Only the first step has been taken in developing therapy of this kind. Chelation therapy succeeds in removing ^{239}Pu from the blood if the association of ^{239}Pu with the chelating agent is significantly greater than with other ions or molecules under the specific conditions and if the resulting ^{239}Pu chelate is excreted by the kidney. Since the amount of ^{239}Pu is very small on a molar basis, only low doses of the chelating agent are required, and high doses should be avoided to minimize undesired side effects.

An agent designed to remove ^{239}Pu that is bound intracellularly or to bone surfaces must also fulfill the above two requirements. Further, it must have a lipophyllic part so that it can penetrate the appropriate membranes, and the compound that results from its having bound ^{239}Pu must possess characteristics such that existing mechanisms will transfer it from the cell or from the volume close to the bone surface to the circulation. A different kind of agent is one that would stimulate the biological process that leads to the removal of ^{239}Pu from bone or liver. Perhaps the most realistic approach to therapy would be to determine the site of the carcinogenic action of the α-irradiation of ^{239}Pu and then seek an agent to minimize the residence time of ^{239}Pu in this site.

In conclusion, much work remains to be done to determine the radiation dose–carcinogenic response relationship for ^{239}Pu(IV) in the beagle and to translate the result into terms applicable to humans. However, basic mechanisms have been identified and are being quantitated so that kinetic models and theories can be formulated. Future research in this field need not be empirical, for sufficient progress has been made that it can be planned and guided by prior results, models, and theories.

ACKNOWLEDGEMENTS

The author thanks the many persons who over many years have contributed to these results. She especially wishes to acknowledge the role of the late Prof. Thomas H. Dougherty, who served as Director from 1955 to 1974, and that of the late Dr. Wright H. Langham, who was an initiator of the studies and served as a Consultant.

REFERENCES

1. B. J. Stover and H. Eyring, *Proc. Natl. Acad. Sci. (USA)* **66**: 132 (1970).
2. H. Eyring and B. J. Stover, *Proc. Natl. Acad. Sci. (USA)* **66**: 441 (1970).
3. B. J. Stover and H. Eyring, *Proc. Natl. Acad. Sci. (USA)* **66**: 672 (1970).
4. F. H. Johnson, H. Eyring and B. J. Stover, *The Theory of Rate Processes in Biology and Medicine*. John Wiley and Sons, Inc., New York, 1974, pp. 549–598.
5. B. J. Stover, D. R. Atherton and N. Keller, *Radiat. Res.* **10**: 130 (1959).
6. B. J. Stover, D. R. Atherton, F. W. Bruenger and D. W. Buster, *Health Phys.* **8**: 589–597 (1962).
7. B. J. Stover, W. Stevens, and F. W. Bruenger, in *Radiobiology of Plutonium* (J. Stover and W. S. S. Jee, Eds.), pp. 129–140. J. W. Press, Salt Lake City, 1972.
8. T. F. Dougherty, B. J. Stover, J. H. Dougherty, W. S. S. Jee, C. W. Mays, C. E. Rehfeld, W. R. Christensen and H. C. Goldthorpe, *Radiat. Res.*, **17**: 625 (1962).
9. B. J. Stover and W. S. S. Jee, (Eds.), *Radiobiology of Plutonium*. J. W. Press, Salt Lake City, 1972.
10. B. J. Stover, F. W. Bruenger and W. Stevens, *Radiat. Res.* **33**: 381 (1968).
11. W. Stevens, F. W. Bruenger and B. J. Stover, *Radiat. Res.* **33**: 490 (1968).
12. B. J. Stover, D. R. Atherton, F. W. Bruenger, and W. Stevens, Submitted to *Radiat. Res.*
13. B. J. Stover, D. R. Atherton, F. W. Bruenger and D. S. Buster, in *Delayed Effects of Bone-Seeking Radionuclides;* (C. W. Mays, W. S. S. Jee, R. D. Lloyd, B. J. Stover, J. H. Dougherty and G. N. Taylor, Eds.), pp. 109–124. University of Utah Press, Salt Lake City (1969).
14. B. J. Stover, D. R. Atherton and D. S. Buster, in *Radiobiology of Plutonium* (B. J. Stover and W. S. S. Jee, Eds.), pp. 149–169. The J. W. Press, Salt Lake City, 1972.
15. B. J. Stover, D. R. Atherton, F. W. Breunger and D. S. Buster, *Health Phys.* **14**: 193 (1968).
16. B. J. Stover, D. R. Atherton and D. S. Buster, *Health Phys.* **20**: 369 (1971).
17. T. H. Cochran, W. S. S. Jee, B. J. Stover and G. N. Taylor, *Health Phys.* **8**: 699 (1962).
18. G. N. Taylor, W. S. S. Jee, J. L. Williams, and L. Shabestari, in *Radiobiology of Plutonium* (Betsy J. Stover and Webster S. S. Jee, Eds.) pp. 105–127. J. W. Press, Salt Lake City, 1972.
19. F. W. Bruenger, B. J. Stover, W. Stevens and D. R. Atherton, *Health Phys.* **16**: 339 (1969).
20. F. W. Bruenger, B. J. Stover and W. Stevens, *Health Phys.* **21**: 679 (1971).
21. C. J. Nabors, Jr., W. Stevens, and R. E. Maxwell, in *Radiobiology of Plutonium,* (Betsy J. Stover and Webster S. S. Jee, Eds.) pp. 87–103. J. W. Press, Salt Lake City, 1972.
22. J. H. Dougherty, in *Radiobiology of Plutonium* (B. J. Stover and W. S. S. Jee, Eds.) pp. 75–86. J. W. Press, Salt Lake City, 1972.
23. G. N. Taylor, W. R. Christensen, L. Shabestari, and W. S. S. Jee, in *Radiobiology of Plutonium* (B. J. Stover and W. S. S. Jee, Eds.) pp. 59–74. J. W. Press, Salt Lake Cty, 1972.
24. W. S. S. Jee and J. S. Arnold, *Lab. Invest.* **10**, 797 (1961).
25. W. S. S. Jee, in *Radiobiology of Plutonium* (B. J. Stover and W. S. S. Jee, Eds.) pp. 195–202. J. W. Press, Salt Lake City, 1972.
26. B. J. Stover and D. R. Atherton, *Radiat. Res.*, in press.
27. B. J. Stover, H. Eyring and D. R. Atherton, in *Radiobiology of Plutonium* (B. J. Stover and W. S. S. Jee, Eds.), pp. 385–407. J. W. Press, Salt Lake City, 1972.
28. B. J. Stover, *Health Phys.* **22**: 823 (1972).

29. R. D. Evans, A. T. Keane, and M. M. Shanahan, in *Radiobiology of Plutonium* (B. J. Stover and W. S. S. Jee, Eds.) pp. 431–468. J. W. Press, Salt Lake City, 1972.
30. F. W. Bruenger, W. Stevens and B. J. Stover, *Radiat. Res.* **37**: 349 (1969).
31. R. D. Lloyd, C. W. Mays, G. N. Taylor, and D. R. Atherton, *Health Phys.* **18,** 149 (1970).
32. R. D. Lloyd, W. S. S. Jee, D. R. Atherton, G. N. Taylor, and C. W. Mays, in *Radiobiology of Plutonium* (B. J. Stover and W. S. S. Jee, Eds.) pp. 141–148. J. W. Press, Salt Lake Cty, 1972.
33. B. J. Stover, F. W. Bruenger and W. Stevens, *Radiat. Res.* **43**: 173 (1970).
34. W. Stevens, B. J. Stover, F. W. Bruenger and G. N. Taylor, *Radiat. Res.*, **39**: 201 (1969).

Transuranium Element Toxicity—Dose-Response Relationships at Low Exposure Levels[1]

Summary and Speculative Interpretation Relative to Exposure Limits

Roy C. Thompson
Battelle Pacific Northwest Laboratory
Richland, Washington, USA

The setting of exposure limits for internally deposited radionuclides has been traditionally—and properly, I believe—entrusted to such bodies as the International Commission on Radiological Protection, and to corresponding national organizations, such as the National Council on Radiation Protection and Measurements in this country. This presentation should not be construed as a specific proposal for changing these limits. Any such changes should result from careful deliberation by many experts of diverse training and experience.

It is certainly appropriate, however, to summarize our information on transuranium element toxicity in terms of how well this information jibes with currently established exposure limits. In fact, I believe that in no other way can one assess the significance or adequacy of this information. The establishment of limits is, after all, our primary reason for gathering the information. To prevent the occurrence of effects from transuranium elements in people, we must limit their exposure to some "safe" level; it being understood that some degree of exposure is practically unpreventable and that some incidence of effect, perhaps undetectable, may be inevitable.

For most radionuclides, occupational exposure limits are set in such a manner as to limit the radiation dose to the critical organ to some "acceptable level." This acceptable level is derived from our total exposure with human irradiation, which is mostly experience with uniform irradiation from external sources. These limits, such as 5 rem per year to total body, and 15 rem per year to organs like liver and lung, bear little experimental relationship to irradiation by internally deposited radionuclides.

Plutonium and the other transuranium elements are particularly difficult to fit into this scheme of limit setting. As we have heard in this Symposium, there is frequently not a clear choice of critical organ. Under the same conditions of exposure, either lung or bone tumors may develop; under other conditions, either bone or liver tumors; and the radiation dose to lymph nodes may frequently exceed greatly the dose to any other tissue. It is also clear that the radiation dose from plutonium is always very non-uniformly distributed within an organ. This is due to the natural predilection of plutonium for certain

[1]Congress presentation C–9–5.

deposition sites, as for example on bone surfaces or in the reticuloendothelial cells of liver; it is also due to the particulate nature of most forms of plutonium that gain entry to the body; and to the aggregation of plutonium by macrophage engulfment, and other processes, subsequent to initial deposition.

For all of the above reasons, and a few more, it is difficult to know how to calculate a biologically relevant radiation dose from deposited plutonium. One may average the dose over the total body, over an organ, over some anatomically defined portion of an organ, or one may calculate the dose within some chosen volume of tissue immediately adjacent to the deposited plutonium. However one chooses to average, measure, or calculate dose, the numbers obtained are only numbers, and they have meaning for limit setting purposes only insofar as they can be related to a biological effect.

Another way of saying what I have just said is that the evaluation of radiation dose is a non-essential intermediate step between exposure and biological effect. It is exposure that we must control in order to prevent biological effect, and if we know the relationship between exposure and effect, we need not be concerned about radiation dose. But, when there are deficiencies in our understanding of exposure-effect relationships, then it may be useful and helpful to generalize the problem and to consider this intermediate stage of radiation dose. It *may* be useful. It may also be confusing, especially when the distribution of radiation dose is as complex as it is with plutonium. I stress this point because I believe that much of the current confusion over plutonium exposure limits, particularly that with regard to "hot particles," is due to a misplaced emphasis on radiation dose.

To make my point more specifically let me tell a very hypothetical story—a fable, you might call it. Once upon a time, an experiment was conducted, using precisely people-equivalent animals which showed that a certain level of biological effect resulted from a certain level of inhaled plutonium. This measured, very low level of biological effect was considered an acceptable risk and the permissible limit for exposure was therefore set to correspond to the exposure employed in this ideal experiment. However, those who wished to appear more sophisticated calculated radiation doses and they found that the average dose to the total lung was 1 rad, the dose to lymph nodes was 1000 rads, and the dose within alpha range of the deposited plutonium was 115,000 rads. And they decided that it would look more scientific if they expressed their basic limit in terms of radiation dose—but which dose? They were all equivalent, since they resulted from the same exposure. So they decided to keep it simple and set the limit at 1 rad to the total lung. Well, things went along fine for many years until one day someone said, "Look, the radiation dose limit for plutonium in the lung is one rad, right?" And everyone agreed, "that's right." "But around these hot particles the dose is 115,000 rads, right?" And everyone agreed, "that's right". "So, we're overexposing the lung and should reduce our limit by 115,000, shouldn't we?" And a lot of people were confused.

Now, that was a fable. Fables are designed to make a point by oversimplifying, and I'm sure the oversimplification was evident; but I hope the point was also evident. Radiation doses have no relevance to the setting of limits apart from the effects that result from these doses. And now I will attempt to summarize our information on plutonium toxicity with that "moral" in mind.

We are probably on the soundest ground with regard to plutonium effects in bone, because of the extensive data on radium effects in man and our experimental animal data comparing effects of radium and plutonium. The studies of plutonium-induced bone tumors in beagle dogs at the University of Utah, summarized by Dr. Stover in this Symposium, were paralleled by similar studies with ^{226}Ra. The primary objective of the whole program was to determine an experimental relationship between plutonium and radium toxicity (*1*).

The 0.1 μCi limit for deposition of ^{226}Ra in man is based directly on the observation of effects in humans. It is a quantity of radium somewhat less than the smallest quantity that has caused detrimental effects in the hundreds of people that have been studied following radium exposure. The 0.04 μCi limit for deposition of ^{239}Pu in man is a quantity of plutonium that is assumed to produce the same biological effect in bone as 0.1 μCi of radium, as judged from comparative studies of biological effects in animals. No consideration of radiation dose is necessary in order to arrive at this limit. If, however, one chooses to calculate average radiation doses to bone, corresponding to these limits, one finds that on this basis plutonium is assumed to be more toxic than radium by a factor of five. This assumption was made more than 20 years ago on the basis of limited studies in rodents and a few dogs (*2*).

The Utah results indicate that plutonium in beagles may be as much as 10 times more toxic than radium on the basis of an average radiation dose to bone (*3*). It has been suggested that this toxicity ratio may be even higher in man than in the beagle dog, because of differences in bone surface-to-volume ratios and differences in the rate of burial of surface deposited plutonium (*4*). This suggestion is supported by autoradiographic studies of plutonium in human bone, as reported at this Congress (*5*).

A substantial body of rodent data also points to the probability that our present limit for plutonium is less conservative than our limit for radium. Figure 1 portrays all of the plutonium-induced bone cancer data (*6*). Each point represents a group of animals exposed in a particular way, in a particular experiment. The experiments employed mice, rats, rabbits, and dogs, exposed by all manner of injection, ingestion, and inhalation procedures. This is a very gross way of looking at a complex collection of data, but I hope it makes the point that there is a substantial amount of data—more than 60 points, from 15 studies—and that these data are not too incoherent. The 0.04 μCi limit for deposition of plutonium in man corresponds to an average radiation dose to bone, over 50 years, of about 30 rads. If we compare this permissible dose with the similarly calculated doses in these experiments I think we must conclude that our limit is uncomfortably close to dose levels that cause an increased incidence of bone tumors in experimental animals.

I would say that there is considerable argument for decreasing the permissible level of plutonium deposition in bone. However, a decrease of no more than 2- to 5-fold would seem adequate to bring this limit in line with the limit for radium.

PLUTONIUM-INDUCED OSTEOSARCOMA IN EXPERIMENTAL ANIMALS

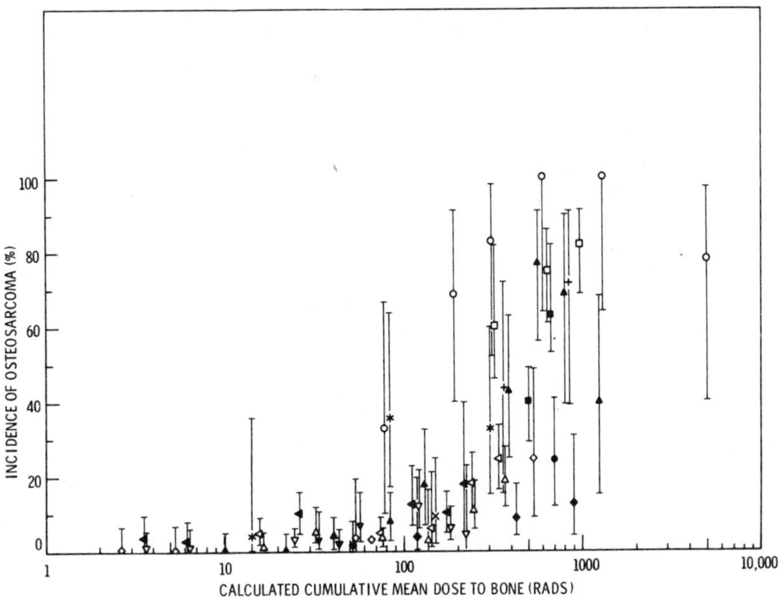

FIG. 1. Plutonium-induced osteosarcoma in experimental animals. Mean incidence and radiation dose values are those reported in the referenced literature. Binomial 95% confidence limits were estimated (*6*).

- ○ inhaled ^{239}Pu citrate, dogs (*7*)
- △ inhaled ^{239}Pu citrate, rats (*8*)
- ▽ inhaled ^{239}Pu pentacarbonate, rats (*8*)
- ◊ sub- and intracutaneous ^{239}Pu nitrate, rats (*9*)
- ▼ oral (daily) ^{239}Pu citrate, rats (*10*)
- • intratracheal ^{239}Pu triacetate, rats (*11*)
- ▲ intravenous ^{239}Pu citrate, mice (*12*)
- □ intravenous ^{239}Pu citrate (monomeric), mice (*13*)
- ■ intravenous ^{239}Pu citrate (polymeric), mice (*13*)
- × inhaled $^{238}PuO_2$, rats (*14*)
- ♦ intratracheal ^{239}Pu nitrate, rats (*11*)
- + intratracheal ^{239}Pu nitrate, rabbits (*15*)
- * inhaled ^{239}Pu pentacarbonate, rabbits (*15*)

Figure 2 shows the relevant data on plutonium-induced lung cancer (*6*). One should be cautioned as to the arbitrary nature of the plotted "dose," since, to an even greater degree than in bone, the distribution of plutonium in the lung is grossly non-uniform, and greatly influenced by the chemical and physical nature of the plutonium compound inhaled. Perhaps the most remarkable thing about these data is the reasonably consistent pattern of dose-effect relationship, despite the probably great variation in dose distribution, since these data are from studies with soluble plutonium citrate, insoluble plutonium oxide, plutonium-238 and plutonium-239—a wide variety of chemical and physical forms.

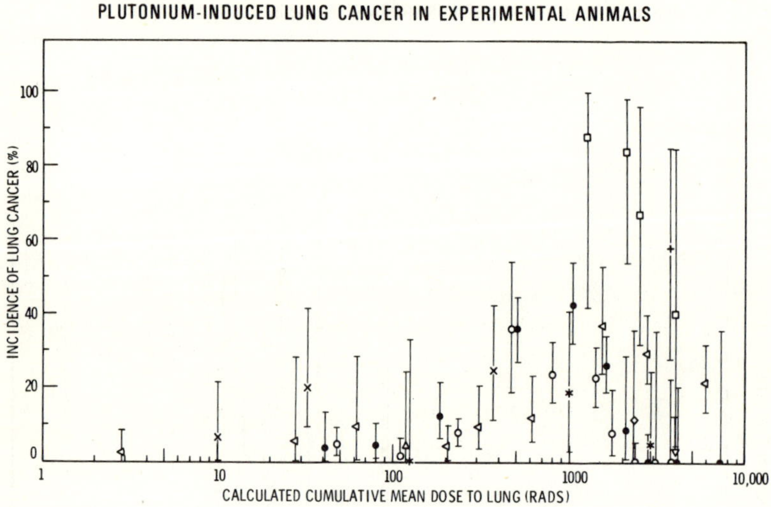

FIG. 2. Plutonium-induced lung cancer in experimental animals. Mean incidence and radiation dose values are those reported in the referenced literature. Binomial 95% confidence limits were estimated (*6*).

□ $^{239}PuO_2$, dogs (*16*)
▽ $^{239}PuO_2$, mice (*17*)
△ $^{239}PuO_2$, mice (*17*)
◇ $^{239}PuO_2$, mice (*18*)
○ ^{239}Pu citrate, rats (*8*)
● ^{239}Pu pentacarbonate, rats (*8*)
× ^{238}Pu (soluble), rats (*14*)
◁ ^{239}Pu nitrate, rats (*11*)
+ ^{239}Pu nitrate, rabbits (*15*)
* ^{239}Pu pentacarbonate, rabbits (*15*)

The highest points in Fig. 2 are the dog data described earlier in this Symposium by Dr. Park. It is probably significant that the experiment showing the highest lung cancer incidence in the lower-dose ranges was the only experiment employing plutonium-238, and probably the experiment in which the plutonium was most nearly uniformly distributed (14).

In arriving at a permissible limit for deposition of plutonium in the lung, we have no such relevant human data for comparison as that of radium in bone. We have only the general radiation dose limit of 15 rems per year, which, for an alpha emitter, translates to 1.5 rads per year, or 75 rads in 50 years. If we look at 75 rads in Fig. 2, we see that it is a higher dose than has caused lung cancers in several experiments, and is within a factor of ten of many experimental groups showing a substantially increased tumor incidence. It would seem to me, comparing this lung cancer data with the bone cancer data of Fig. 1, that the limit for plutonium in lung is less conservative than the limit for plutonium in bone, perhaps by a factor of ten. It is clear that we need more data, from more animals, at lower dose levels; data such as should be eventually forthcoming from the experiments described by Dr. Park and Dr. Sanders earlier in this Symposium.

More could be said about other organs that might be of concern because of their plutonium content. Liver, lymph nodes, gonads, bone marrow, do not seem to be "as critical" as bone or lung, but they bear watching. I also have said nothing specifically of transuranium elements other than plutonium. I will pass these subjects by, in order to spend the remaining time on the human data, and on the subject of exposure limits for large populations.

There is little to be said of the data on effects in humans, except that there are no data—which is, of course, the most encouraging data one could have. As described earlier in this Symposium by Dr. Richmond, we have information on a substantial number of workers with plutonium depositions exceeding the current permissible limit. The absence of demonstrable effects in these people, many of whom received their exposures more than 25 years ago, is certainly presumptive evidence that our exposure limits are not grossly inadequate. The accumulating information on distribution of plutonium in man, from autopsy data, also forms a very important link between animal experimentation and human prediction.

Turning now to the question of exposure limits for world populations, I think the most important point to be made is that neither the data from exposed humans, nor any conceivable study in animals, can possibly provide us with direct experimental data on biological effect at the exposure levels of concern. One *must* extrapolate from the exposure levels, and from the levels of effect, that are experimentally accessible; to the very much lower exposure levels and effect levels, that may be unmeasurable, but are nevertheless of concern when one must consider the exposure of several billion people. This is a problem that is not unique to the transuranium elements, nor to other radioactive pollutants, although they may offer some uniquely challenging complications in this extrapolation process. Our present capabilities for extrapolating dose-effect relationships seem to be limited to the setting of upper limits, based on assumptions of linearity and considerations related to natural background. It is important that we not lose sight of the fact that these are upper limits, and that we are, in fact, ignorant of the actual dose-effect relationship at the levels of concern. We can gain knowledge in that area only through studies of the mechanism of radionuclide-induced carcinogenesis—and that is another symposium.

REFERENCES

1. T. F. Dougherty, B. J. Stover, J. H. Dougherty, W.S.S. Jee, C. W. Mays, C. E. Rehfeld, W. R. Christensen, and H. C. Goldthorpe, *Radiat. Res.* **17**, 625 (1962).
2. W. H. Langham, *Health Phys.* **22**, 943 (1972).
3. C. W. Mays and T. F. Dougherty, *Health Phys.* **22**, 793 (1972).
4. E. Lloyd and J. H. Marshall, in *Radiobiology of Plutonium* (B. J. Stover and W.S.S. Jee, Eds.), pp. 377–383. J. W. Press, Salt Lake City, 1972.
5. R. A. Schlenker and B. G. Oltman, *Radiat. Res.* **59**, 55 (1974).
6. W. J. Bair, *Advances in Radiation Biology*, **Vol. 4**, (J. T. Lett, H. I. Adler and M. R. Zelle, Eds.) pp. 255–315. Academic Press, New York, 1974.
7. W.S.S. Jee, *Health Phys.* **22**, 583 (1972).
8. L. A. Buldakov and E. R. Lyubchanskii, *Experimental basis for maximum allowable load (MAL) of ^{239}Pu in human organism, and maximum allowable concentration (MAC) of ^{239}Pu in air at work locations.* Translated in Argonne National Laboratory Document, ANL-tr-864 (1970).

9. L. A. Buldakov, A. P. Nifatov, R. A. Erokhin, and L. G. Filippova, in *Remote Aftereffects of Radiation Damage* (Y. I. Moskalev, Ed.), pp. 350–355. Atomizdat, Moscow, 1971. Translated in AEC-tr-7387, pp. 381–387.
10. L. A. Buldakov, E. R. Lyubchanskii, Y. I. Moskalev, and A. P. Nifatov, *Problems of Plutonium Toxicology,* Atomizdat, Moscow, 1969. Translated in Lovelace Foundation Report, LF-tr-41.
11. R. A. Erokhim, N. A. Koshurnikova, V. K. Lemberg, A. P. Nifatov, and A. A. Puzyrev, in *Remote Aftereffects of Radiation Damage* (Y. I. Moskalev, Ed.), pp. 315–333. Atomizdat, Moscow, 1971. Translated in AEC-tr-7387, pp. 344–363.
12. M. P. Finkel and B. O. Biskis, *Health Phys.* **8**, 565 (1972).
13. M. W. Rosenthal and A. Lindenbaum, *Radiat. Res.* **31**, 506 (1967).
14. C. L. Sanders, *Radiat. Res.* **56**, 540 (1973).
15. N. A. Koshurnikova, V. K. Lemberg, and E. R. Lyubchanskii, in *Remote Aftereffects of Radiation Damage* (Y. I. Moskalev, Ed.), pp. 305–314. Atomizdat, Moscow, 1971. Translated in AEC-tr-7387, pp. 334–343.
16. J. F. Park, this Symposium.
17. L. A. Temple, D. H. Willard, S. Marks, and W. J. Bair, *Nature (London)* **183**, 498 (1959).
18. R. W. Wager, N. L. Dockum, L. A. Temple, and D. H. Willard, in *Biology Research Annual Report* 1955, pp. 61–72, Hanford Report HW-41500, 1956.

RADIATION AND PHOTOCHEMICAL PROCESSES IN THE ATMOSPHERE AND THEIR ECOLOGICAL EFFECTS[1]

Chairman: E. J. Hart

[1]Symposium B–9

Some Aspects of Airborne Particles and Radiation in the Atmosphere[1]

G. M. Hidy[2]
Rockwell International Science Center
Thousand Oaks, California, 91360, USA

INTRODUCTION

Over the past decade, considerable effort has been devoted to the significance of changes in the Earth's atmospheric solar heat balance resulting from changes in trace constituents in the air. As a result of absorption of energy in the infrared by increasing amounts of carbon dioxide from anthropogenic sources, for example, meteorologists have predicted systematic surface temperature increase (e.g., Manabe and Wetherald, ref. *1*). Similarly, the scattering of visible and infrared radiation from atmospheric aerosols may result in cooling at the Earth's surface, as noted by many investigators and reviewed by Schneider and Kellogg (*2*).

It is well known that both natural processes and man's activities contribute to the aerosol burden of the atmosphere. On a global basis, workers have estimated that natural emissions of airborne particles or aerosol producing gases far exceed those from anthropogenic sources (e.g., Hidy and Brock, ref. *3*). However, locally in urban air the reverse is the case. In many places, aerosol formation resulting from atmospheric chemical reactions of sulfur dioxide, nitrogen oxides, and hydrocarbon vapors dominate the airborne particle mass concentration. Thus, control of haze becomes a complicated problem linked with gaseous air pollutants as well as direct particulate emissions.

The role of aerosols in influencing sunlight in the atmosphere is important not only from the aesthetics of visibility, but also from its potential for creating climatological change. In this paper, the physics of radiative scattering and absorption by aerosols is summarized, and aspects of its impact on climatology are noted.

SIMPLE MODEL OF AEROSOL-RADIATION INTERACTION

The attenuation of solar radiation by atmospheric constituents can be described in terms of the Beer's Law. For a change dI_λ in intensity I_λ of a parallel monochromatic beam,

$$dI_\lambda = I_\lambda \sigma_\lambda \rho ds,$$

where ρ is the density of the attenuating medium, σ_λ is the extinction coefficient

[1]Congress presentation B–9–1.

[2]Present address, Environmental Research and Technology, Westlake Village, California 91360.

per unit mass, and ds is the path length. Integration of this expression gives

$$I_\lambda = I_{o\lambda} \exp\left(-\int_o^S \sigma_\lambda \rho \, ds\right), \qquad (1)$$

where $I_{o\lambda}$ is the intensity of radiation at the point s=0. The exponential term in Eq. (1) is the monochromatic transmission function or transmittance, and the integral in the exponential term is the monochromatic optical thickness of the medium along the path 0 to s. In practice, the relations for monochromatic radiation intensity can be extended easily to the multi-wavelength case. (See also Middleton, ref. *4*).

The extinction of a direct beam of sunlight is the result of the combined effort of absorption and scattering. Absorption in a gas element results if all of the scattered and transmitted energy summed over all angles is less than the incident beam energy. In the absence of absorption, the incident energy is redistributed in many different directions.

The radiative absorption and scattering properties depend on the wavelength of the radiation as well as the optical properties of the medium. For a layer of aerosol particles, the optical properties involve the complex index of refraction of the particles, their size distribution and the shapes of particles. The particle index of refraction depends on chemical composition, and particle size as well as shape. These vary significantly from source to source. Any change or variability in these properties, whether of natural or anthropogenic origin, may significantly influence the Earth's radiation field.

ORIGINS AND PROPERTIES OF ATMOSPHERIC AEROSOLS

There has been a considerable increase in knowledge of the sources and properties of atmospheric particles over the past ten years. On a global basis, workers have estimated that the principal contributors are natural sources, as indicated in Table 1. The production of aerosols from secondary sources involving atmospheric chemical transformation exceeds the material from primary sources on a tonnage emission basis.

In the cities, the aerosol population generally is dominated by material from anthropogenic sources, as indicated in Fig. 1. In the case of Los Angeles, the mass coming from secondary sources related to the smog chemistry represents a significant portion of the aerosol.

Generally, the mass concentration of airborne particles in American cities is 100 $\mu g/m^3$ or less for a 24-hour averaged sample. This is three to five times that observed in non-urban areas.

Measurements of particle size distribution of aerosols vary widely from place to place. Often, aerosol distributions are pictured in terms of the number distribution, where

$$\text{the No. of particles per unit volume } N = \int_o^\infty n(D_p)\, dD_p, \qquad (2)$$

D_p being the particle diameter, and $n(D_p)\,dD_p$ the number unit per volume in the range D_p and $D_p + dD_p$.

The moments of the number of distribution can be defined as

$$\text{surface area per unit volume } S = \int_o^\infty \pi D_p^2 n(D_p) dD_p, \qquad (3)$$

$$\text{and Volume Fraction } V = \int_o^\infty \frac{\pi}{6} D_p^3\, n(D_p) dD_p. \qquad (4)$$

TABLE 1

Estimates of the Weight of Particles Smaller Than 20μm Radius Emitted into or Formed in the Atmosphere (10^6 metric ions yr^{-1})

Natural	
Soil and rock debris*	100–500
Forest fires and slash-burning debris*	3–150
Sea salt	(300)
Volcanic debris	25–150
Particles formed from gaseous emissions:	
Sulfate from H_2S	130–200
Ammonium salts from NH_3	80–270
Nitrate from NO_x	60–430
Hydrocarbons from plant exudations	75–200
Subtotal	773–2200
Man-made	
Particles (direct emissions)	10–90
Particles formed from gaseous emissions:	
Sulfate from SO_2	130–200
Nitrate from NO_x	30–35
Hydrocarbons	15–90
Subtotal	185–415
Total	958–2615

*Includes unknown amounts of indirect man-made contributions. [Reproduced from S. H. Schneider and W.W. Kellogg, in *Chemistry of the Lower Atmosphere* (S. I. Rasool, Ed.) pp. 203–249. Plenum Press, New York, 1973, with the publisher's permission.]

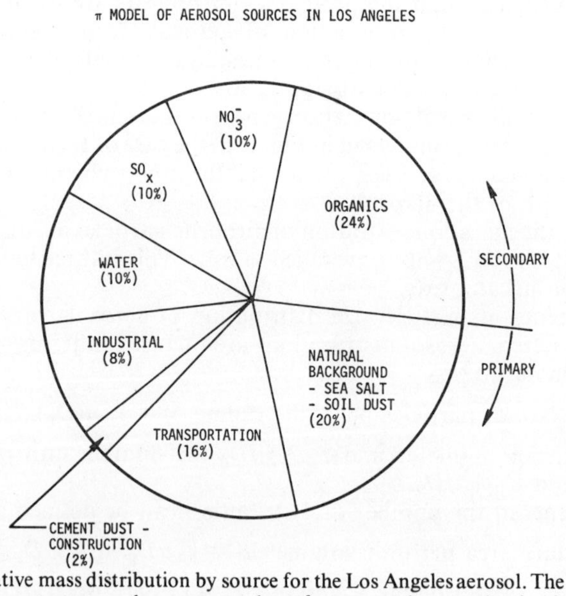

FIG. 1. Relative mass distribution by source for the Los Angeles aerosol. The lower portion are particles from primary sources; the upper are those from secondary production in the atmosphere. (Basis: Data taken in 1969 and 1972 at Pasadena, CA.)

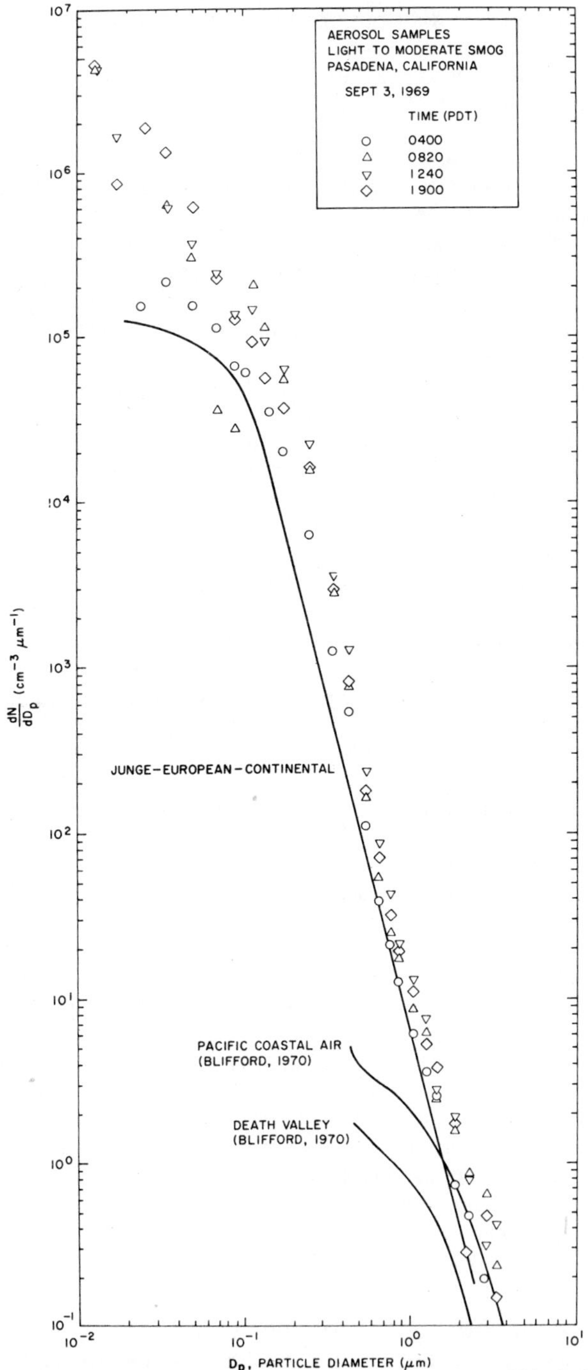

FIG. 2. Size-number distributions for troposheric aerosols. [From Hidy & Friedlander, (*5*).]

Typical number-size distributions for aerosol samples taken at the ground are shown in Fig. 2. The similar behavior in the 0.1 μm to 1 μm range has led many workers to use a power law description for the aerosol distribution. This distribution is attributed to C. Junge (*6*), where

$$n(D_p) \sim D_p^{-a}, 3 \lesssim a \lesssim 4 . \tag{5}$$

Observational evidence recently has accumulated that the simple power law description is inadequate. Moments of the number-distribution exhibit a multimodal structure, as suggested in the data in Fig. 3 and 4. Fig. 3 shows volume distributions measured near different types of sources. These results suggest that natural or quasi-natural sources of blowing dust, as well as sea spray, contribute mainly material in the supermicron fraction, while combustion sources dominate the submicron portion.

Observations of aerosol behavior in the Los Angeles area show a strong tendency for particles to grow by chemical reactions in the 0.1 μm–1 μm range. This is illustrated by two volume distributions shown in Fig. 4. Note the marked increase in volume concentration in the submicron range during the buildup of smog from early morning to midday. This behavior also has been observed and documented by other investigators (e.g., Whitby et al., (*8*)).

The distribution of chemical constituents of the Los Angeles aerosol follows the particulate sources reasonably well. The distribution of constituents is shown schematically in Fig. 5. Here, the soil and sea salt material is in the supermicron fraction, while the submicron fraction contains lead and carbon from combustion sources. The secondary constituents, sulfate, nitrate and (part of the) carbon are smeared over a broad range below a few μm diameter.

Measurements of aerosol concentrations indicate that most of particulate material is found near the ground. Such observations are illustrated in Fig. 6. Here the total number concentration of aerosols, as depicted by Aitken nuclei, decreases strongly with height. However, the number concentration of particles larger than 0.3 μm diameter show less variation with height. Although not well shown in this diagram, there is a persistent layer of particles in the submicron size range between 0.1 and 1.0 μm present in the stratosphere at about 20 km altitude. This is the well-known "Junge layer".

The stratospheric size distribution of aerosols is less known than for the troposphere. There is some speculation that the number distribution is of bimodal form with a large number of very tiny particles and a few larger than 0.1 μm (e.g., Fig. 7 after Friend, ref. *9*). Other investigators have suggested that the distribution is more uni-modal, following a power law form similar to the one proposed for the tropospheric case.

The composition of the stratospheric aerosol is poorly documented. A synthesis of a few observations suggests the stratospheric particles have a distribution of constituents as listed in Table 2. By far, the dominant portion of mass is sulfate. Whether or not the particles are primarily sulfuric acid or a neutralized species has not been determined.

AEROSOLS AND THE EXTINCTION COEFFICIENT

Comparatively little is known about the role of aerosol chemical composition and the variation of light scattering in the atmosphere. However,

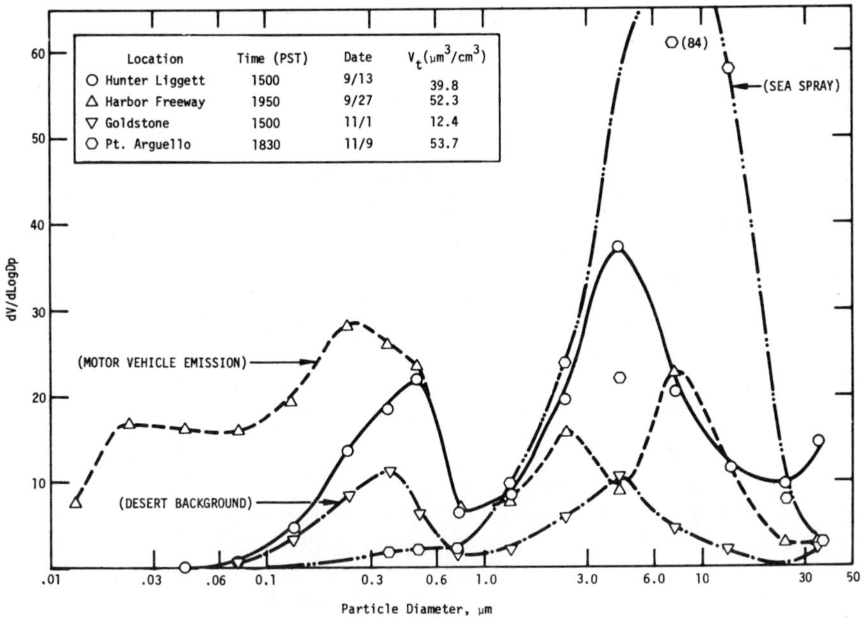

FIG. 3. Volume-size distributions of source dominated aerosols [from Hidy et al. (7).]

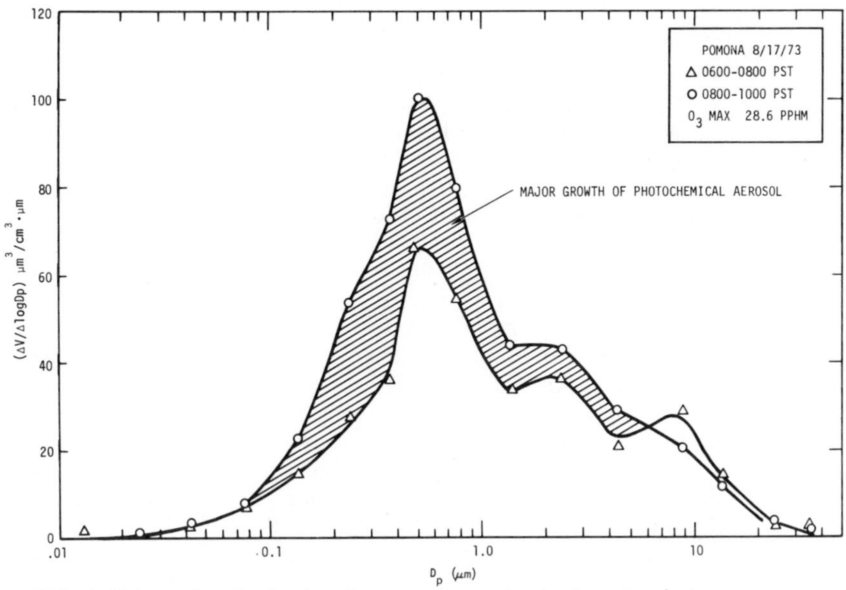

FIG. 4. Volume-size distributions for aerosols growing in photochemical smog.

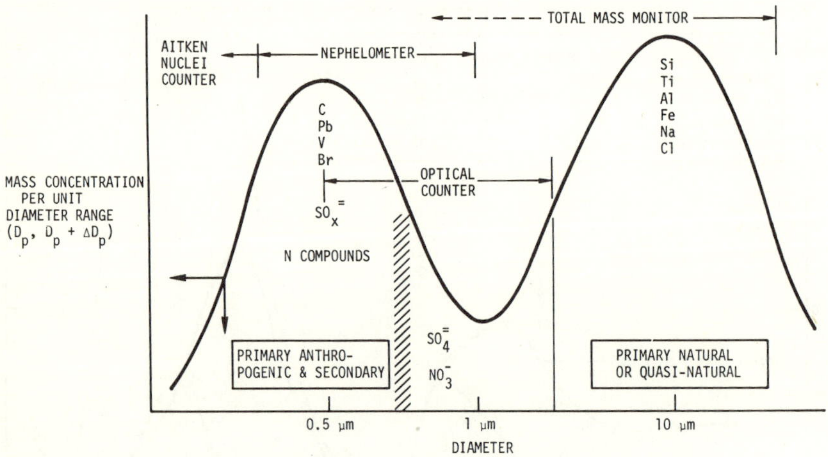

FIG. 5. Schematic diagram of chemical constituents distributed by particle size.

recent measurements of the extinction coefficient for scattering of visible light (b_{scat})[3] have suggested a correlation of this parameter with mass concentration (TMC) of suspended particles (e.g., Charlson et al., ref. *11*). The uniqueness of any relationship between b_{scat} and TMC depends on the uniformity of the size distribution and composition of atmospheric aerosols. Studies such as those of Samuels et al. (*12*) have suggested that such a correlation will be useful for measurement of TMC only in certain geographical locations.

The extinction coefficient for (visible) light scattering empirically has been shown to be dominated by aerosol particles in the 0.1–1.0 μm diameter range. An example of such a correspondence is shown in Fig. 8. These results, taken from the ACHEX experiment, illustrate well that b_{scat} is an important measure of the particle *volume* concentration in the 0.1–1.0 μm diameter range. The calculation for particle total volume, or cumulative volume $<$1.0 μm showed a correlation coefficient significantly less than the 0.948 listed in Fig. 8. Thus, it is anticipated that the principal influence on light scattering in the atmosphere, and consequently on visibility, will be linked to airborne particles in the submicron size range.

The relationship between the changes in visibility and aerosol growth or formation by chemical reaction can be fully appreciated now, with the use of the measure b_{scat}. In the case of photochemical smog evolution in the Los Angeles area, for example, it is well known that visibility is reduced substantially as the smog reactions take place during the day. Using ozone as a measure of the "completion" of the smog reactions, one finds a strong relation between b_{scat} and maximum midday ozone concentration. This is shown for ground-based data taken in 1973 (Fig. 9). The variation in b_{scat} with ozone was confirmed by a blimp flight in a north-south path across the Los Angeles Basin during the midday hours when ozone is at maximal concentration over the city. From such data, it is clear that the aerosol behavior is an integral part of the photochemical smog evolution and must be considered as part of this complicated system of

[3] b_{scat} is effectively equal to the sum of σ_λ over all wavelengths of visible light.

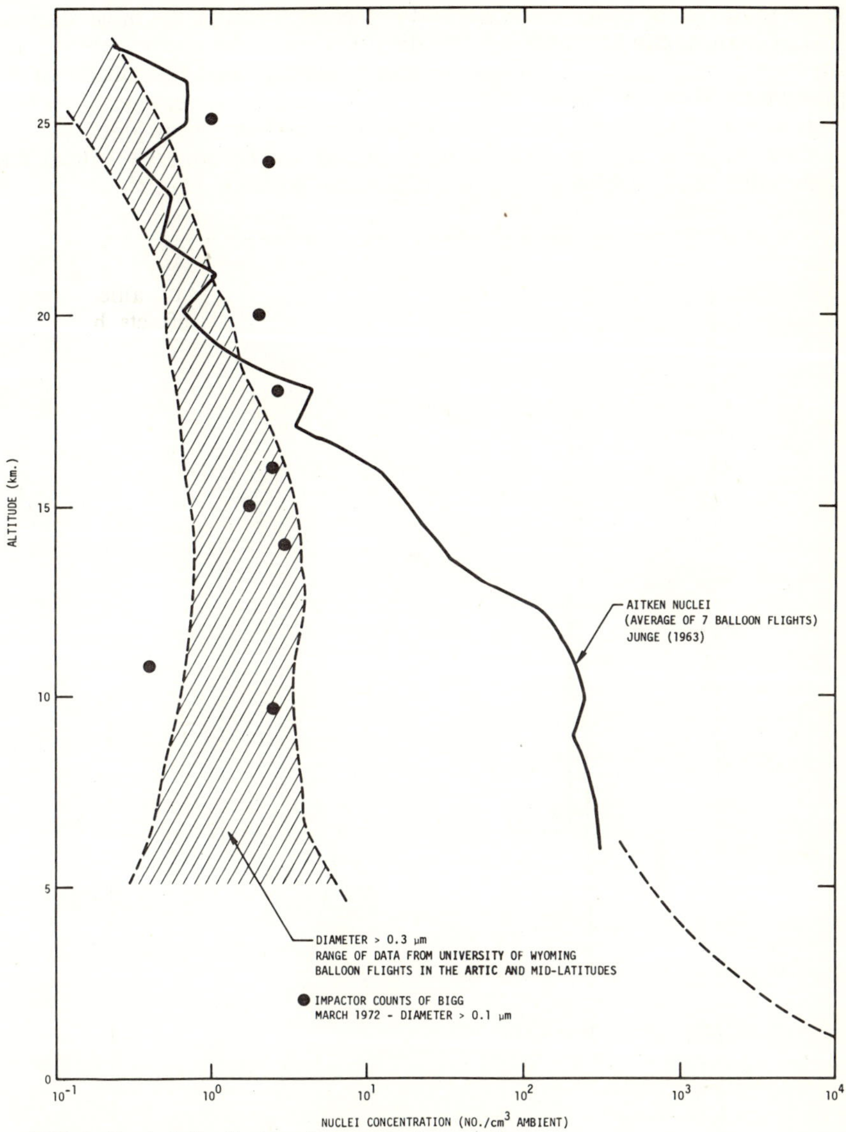

FIG. 6. Vertical distribution of atmospheric aerosols.

chemically reactive air modulated by solar radiation and meteorological variations.

CLIMATOLOGICAL AND METEOROLOGICAL INTERACTIONS

At first thought, the trace constituents in the atmosphere should have a negligible influence on weather and climate. However, theoretical studies have suggested that a variety of changes can result from emitting more material into the air than can be removed by natural processes. Perhaps the most straightforward changes can be identified with disturbances in the Earth's temperature distribution as a result of changes in solar radiation and the albedo of the atmosphere. However, many other potentially significant effects have been identified that range from microscale changes in cloud behavior to global shifts in snow and ice cover, or to widespread dynamic coupling of radiative changes with air motion (e.g., Pruppacher (*13*); Schneider and Kellogg, (*2*)).

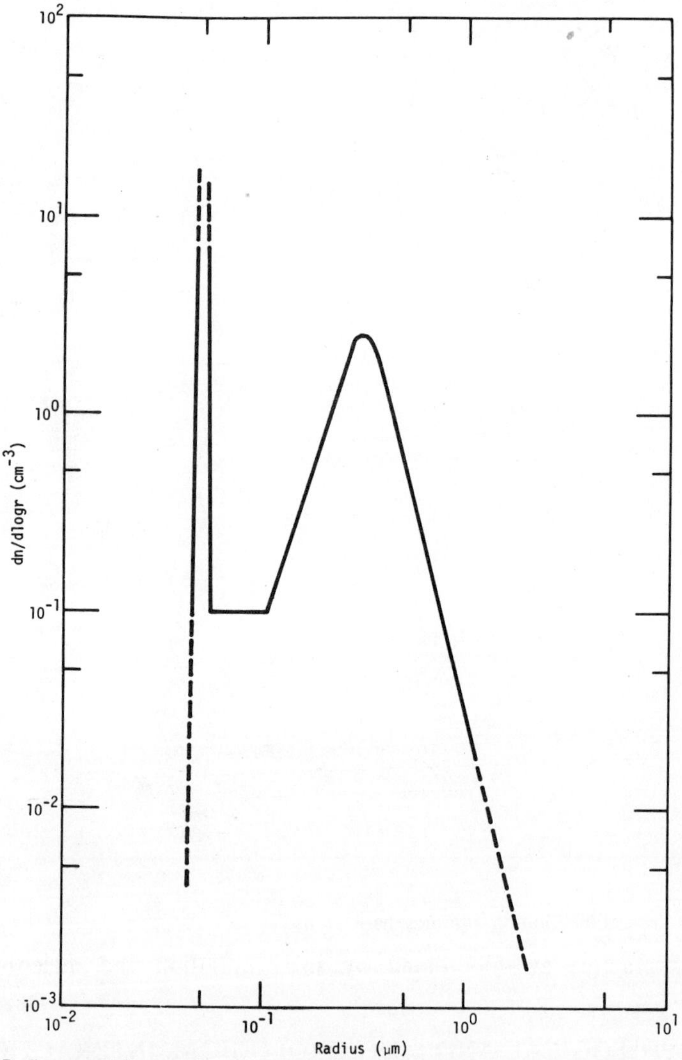

FIG. 7. Proposed size-number distribution for stratospheric aerosols [after Friend, (*9*).]

TABLE 2
Model chemical composition of stratospheric particles at 20 km altitude based on data taken during the period 1969–1973.

Substance	Concentration in air $\mu g/m^3$ (ambient)	Observed concentration range $\mu g/m^3$ (ambient)
Sulfate	0.6[c]	0.01–4 (Agung peak)[c]
Basalt[a] (calculated from Si)	0.05	0–0.7
NH_4^+	0.005	0–0.01
NO_3^-[b]	0	0
NO_2^-	0	0
Na	0.01	0.001–0.05
Cl	0.04	0.002–0.09
Br	0.002	0–0.003
Total	0.71	0.01–1

[a]This includes all other components of the basalt, e.g., Al, Ca, Mg.
[b]Particulate NO_3^- as contrasted with HNO_3 vapor.
[c]For eight year averaged data at 18 km [(from Castleman et al. (*10*).]

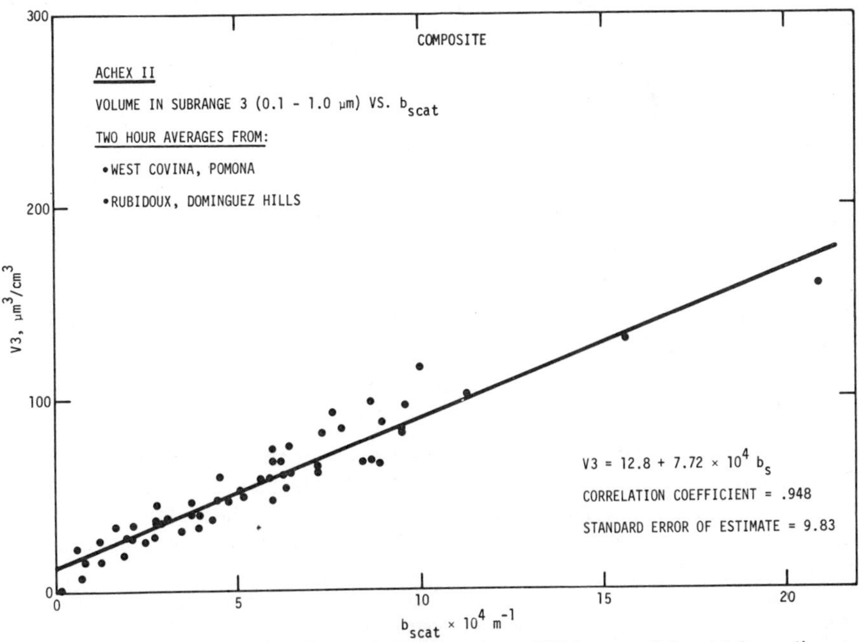

FIG. 8. Correlation between b_{scat} and particle volume (V3) between 0.1 and 1.0 μm diameter based on ACHEX data (analysis by W. E. Clark).

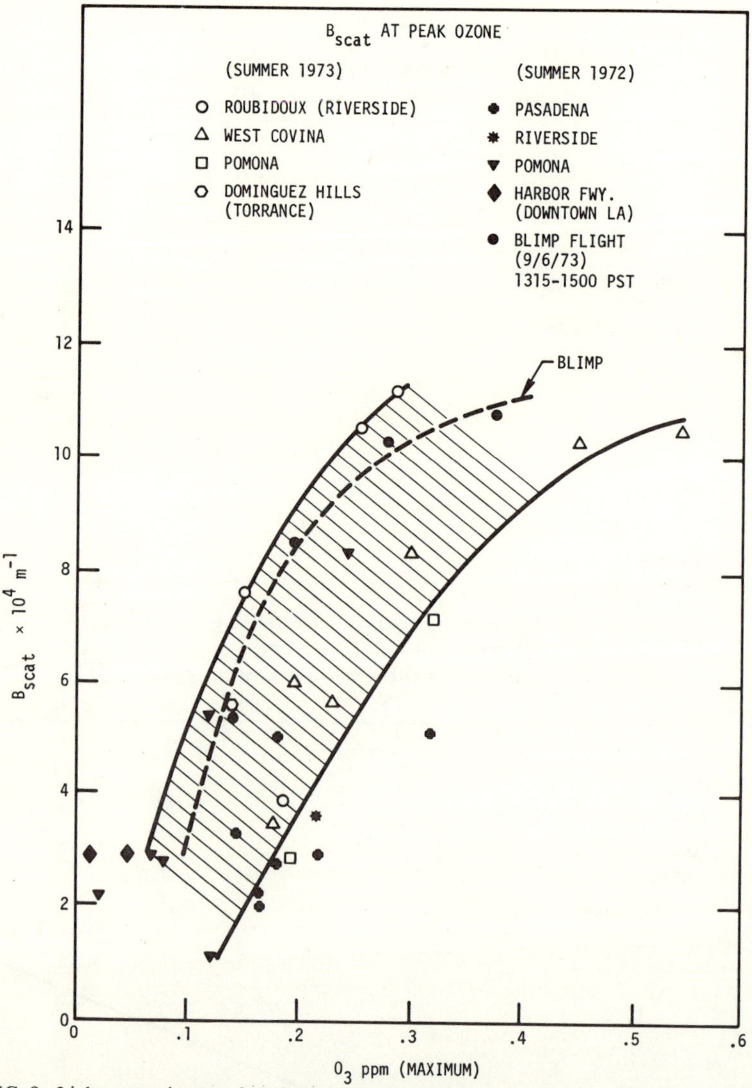

FIG. 9. Light scattering as a function of ozone concentration in photchemical smog. B_{scat} value corresponds to value observed at maximum levels of ozone.

For illustration, let us examine the simple case of changing the albedo of the Earth by addition of aerosols to the atmosphere. The influence of an aerosol layer in the lower atmosphere is shown schematically in Fig. 10. The surface reflectivity is α_s, whereas α_{Bs} is the fraction of sunlight backscattered to space from the aerosol layer. The transmission of radiation through the layer is $t = 1 - \alpha_{Bs} - \alpha$, where α is the fraction of the flux absorbed by the particles. There the effective albedo of the Earth's aerosol layer system is

$$\alpha_E = \alpha_{BS} + [t^2 \alpha S / (1 - \alpha S \alpha_{BS})] \quad (8)$$

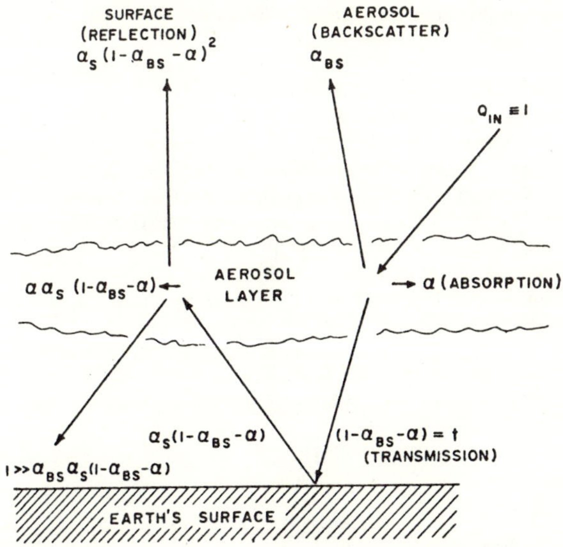

FIG. 10. Schematic diagram showing how the presence of an absorbing and scattering aerosol layer in the lower atmosphere can change the effective albedo of the Earth's surface. [Reproduced from S. H. Schneider and W. W. Kellogg, in *Chemistry of the Lower Atmosphere* (S. I. Rasool, Ed.) pp. 203–249. Plenum Press, New York, 1973, with the publisher's permission.]

Calculation of the influence of changes in absorption and backscatter fractions suggests that the influence of the aerosol generally is to *increase* the albedo of the Earth's surface. The albedo of the cloudless fraction of the Earth is about 0.14 for a global average, so the global effect of an equally absorbing and scattering aerosol layer concentrated below cloud tops would increase the albedo of the cloudless part of the Earth's surface. Roughly, a change of 1% in the albedo corresponds to an average surface temperature decrease of 2° C. This change will tend to be offset by a predicted increase in surface temperature resulting from increases in carbon dioxide in the atmosphere.

As a result of relatively simple considerations of the kind discussed here (see also SMIC, ref. *14*), there are a number of projects now under way to better estimate potential changes in climate associated with pollution in the troposhere and the stratosphere. The combination of theory and measurement soon will provide a much better basis for control strategy for air resources management for our future.

ACKNOWLEDGEMENT

This work is sponsored, in part, by the California Air Resources Board under Contract ARB No. 358, and partly by the Department of Transportation, Climatic Impact Assessment Program via an interagency grant to the Environmental Protection Agency (Contract 68-02-1081).

REFERENCES

1. S. Manabe, R. T. Wetherald, *J. Atmos. Sci.* **24** (1967).
2. S. H. Schneider, and W. W. Kellogg, in *Chemistry of the Lower Atmosphere* (S. I. Rasool, Ed.) pp. 203–249. Plenum Press, New York, 1973.

3. G. M. Hidy and J. R. Brock, in *Proc. 2nd IUAPPA Clean Air Congr.*, pp. 1088–1097. Academic Press, New York, 1971.
4. W.E.K. Middleton, *Vision through the Atmosphere,* Univ. of Toronto Press, Toronto, 1952.
5. G. M. Hidy and S. K. Friedlander, in *Proc. 2nd IUAPPA Clean Air Congr.*, pp. 391–404. Academic Press, New York, 1971.
6. C. E. Junge, *Air Chemistry and Radioactivity,* Academic Press, New York, 1963.
7. G. M. Hidy (Editor), *"Characterization of Aerosols in California* (ACHEX)." Interim Report, Phase 1. Science Center SC524.191R, Rockwell Science Center, Thousand Oaks, Calif., 1973.
8. K. T. Whitby et al., in *Aerosols and Atmospheric Chemistry* (G. M. Hidy, Ed.), p. 237. Academic Press, New York, 1972.
9. J. P. Fried, *Tellus.* **18**, 465 (1966).
10. A. W. Castleman, R. H. Munkewitz, and B. Manowitz, *Tellus.* (1974), in press.
11. R. J. Charlson, N. C. Ahlquist, H. Horvath, *Atmos. Environ.* **2,** 455 (1968).
12. H. J. Samuels, S. Twiss, E. Wong, *Visibility, Light Scattering and Mass Concentration of Particulate Matter.* State of Calif. Air Resources Board, Sacramento, July, 1973.
13. H. R. Pruppacher, in *Chemistry of the Lower Atmosphere* (S. I. Rasool, Ed.), pp. 1–67. Plenum Press, New York, 1973.
14. Study of Man's Impact on Climate (SMIC), *Inadvertent Climate Modification.* MIT Press, Cambridge, Mass., 1971.

The Oxides of Nitrogen with Respect to Urban Smog, Supersonic Transports, and Global Methane[1]

Harold S. Johnston and Edward Quitevis
Department of Chemistry and
Inorganic Materials Research Division,
Lawrence Berkeley Laboratory
University of California
Berkeley, California 94720, USA

INTRODUCTION

By now it is well established that the oxides of nitrogen from the exhaust gases of large fleets of supersonic aircraft would seriously reduce stratospheric ozone (*1–12*) and admit extra, biologically damaging, ultraviolet radiation to the earth's surface (*13*). For over 20 years it has been established that the oxides of nitrogen and hydrocarbons in urban areas from ozone through the photochemical smog reaction (*14*), and this ozone is toxic, damages materials, and kills plants (*15*). When these two statements are put side by side, they seem to say that an increase of nitrogen oxides in the stratosphere would reduce ozone there but an increase of nitrogen oxides in cities would increase ozone—and both changes have deleterious effects. But then one asks, How can this be? How can nitrogen oxides destroy ozone in one part of the atmosphere and form ozone in another part? This article reviews the properties of urban photochemical smog, the role of nitrogen oxides in the stratospheric ozone balance, and the role of methane as a source of smog in the global troposphere. It is shown that a common set of chemical reactions with known rate constants, which gives reaction rates at all elevations, does predict a crossover between formation and destruction of ozone by nitrogen oxides at about 13 km.

OZONE FORMATION IN URBAN SMOG

The photochemical nature of Los Angeles smog was established by Haagen-Smit and co-workers (*14*) in the early 1950's. By means of a simple specific test for ozone, Haagen-Smit showed that air containing trace amounts of hydrocarbons and nitrogen oxides react in sunlight to produce ozone. For the same initial concentrations of reactants, the rate of reaction depended strongly on the nature of the hydrocarbon; methane was found to be the least reactive of all hydrocarbons. It was shown that the role of nitrogen oxides was complex: with zero nitrogen oxides there was no smog; with a moderate amount there was strong, rapid production of smog; but with somewhat larger amounts of nitrogen oxides the production of smog was inhibited.

[1]Congress presentation B–9–2.

A number of aspects of photochemical smog can be shown from recent observations in the atmosphere at a number of sites in California. In five widely separated cities in the central valley of California (*16*), oxidant (mostly ozone) builds up simultaneously during the morning, it reaches a broad maximum in the early afternoon, and it falls to low values at night (Fig. 1). There is a strong seasonal effect with maximum smog during the summer and early fall and much less during the winter and spring months (Fig. 1). Over almost a decade, 1963–1971, photochemical oxidant showed a pronounced decrease in downtown Los Angeles, but there was an equally pronounced increase 50 miles inland in the

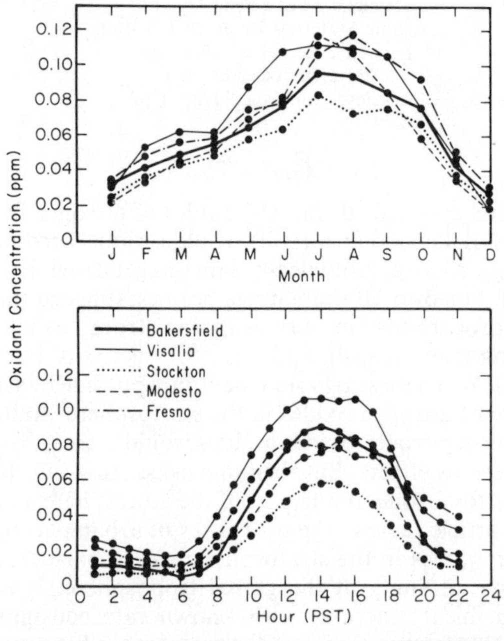

FIG. 1. A. Photochemical oxidant (primarily ozone) as a function of hour of day in 5 cities in the California Central Valley up to 200 miles apart. The ozone forms and decays at approximately the same time at each site, indicating *in situ* photochemical formation. Average hourly oxidant concentration during the months July, August, and September. B. Maximum hourly oxidant concentration as a function of month of the year. Note the high concentrations in summer and early fall.

same basin at Riverside (*17*) (Fig. 2). In the middle period covered by these trends at Los Angeles and Riverside (starting in 1966), new automobiles were required to reduce hydrocarbon and carbon monoxide emissions, but nitrogen oxides increased by 50 per cent as a result of the higher combustion temperatures used to reduce hydrocarbons. The decrease of ozone in Los Angeles, the increase of ozone in Riverside, and the simultaneous increase in nitrogen oxides is one of many examples of the complex response of atmospheric ozone to nitrogen oxides and other factors.

The essential features of photochemical smog can be produced in the laboratory (*18–20*). Light sources simulate sunlight by providing radiation above

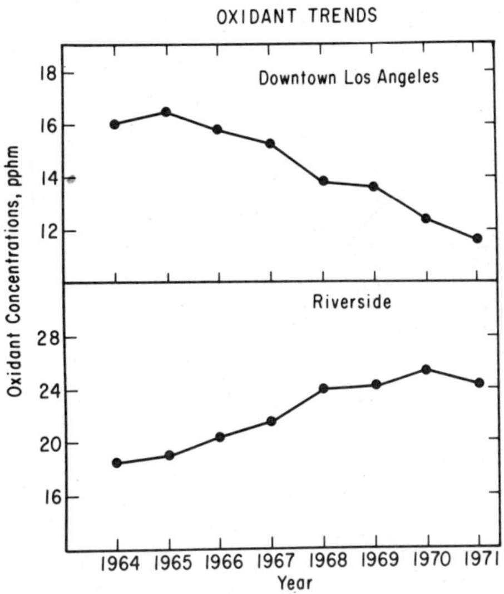

FIG. 2. Long-term trends in photochemical smog (oxidant) at Downtown Los Angeles (close to the Pacific Ocean) and Riverside (50 miles downwind and inland). Three-year moving average of daily one-hour maximum concentration for July-August, and September. Over the same period of time, oxidant decreased in Downtown Los Angeles and increased in Riverside.

300nm. Small amounts of hydrocarbons and nitrogen oxides are added to scrupulously purified air in a large chamber, typically made of glass. An example of such a "smog chamber" experiment is given by Fig. 3, where initial reactants were 3 parts per million (ppm) of propylene, 1.3 ppm of nitric oxide (NO), and about 0.2 ppm of nitrogen dioxide (NO_2). When the light was turned on, propylene decreased, NO was converted to NO_2, and later ozone built up while NO_2 decreased. Many other products, such as formaldehyde, other aldehydes, peroxyacetyl nitrate (PAN), and nitric acid, were also formed.

The multi-faceted role of nitrogen oxides (*21*) is illustrated by Fig. 4. Three smog-chamber experiments are presented, in which the initial propylene was 3 ppm in each case but the initial NO_x($NO + NO_2$, mostly NO in these cases) was 3, 1.5, and 0.5 ppm in the 3 cases. During the first half of the reaction, the rate of destruction of propylene was fastest for the least nitric oxide and slowest for the most nitric oxide. Nitrogen oxides above 0.5 ppm, in this case, acted to slow down or inhibit the initial rate of the smog reactions. However, in the second half of the destruction of propylene, the roles reversed. With 0.5 ppm of NO_x the rate became much slower and after five hours much propylene remained unconsumed. With 1.5 ppm of NO_x, the rate accelerated until the rate became much faster than for the other cases, and by five hours the propylene was virtually all consumed.

Similar data, but in this case focusing on the oxidant produced, are given in Fig. 5. A series of experiments was carried out, each starting with 2 ppm propylene and with various initial concentrations of nitrogen oxides. One set of experiments progressed for 6 hours, and one set of experiments was carried out

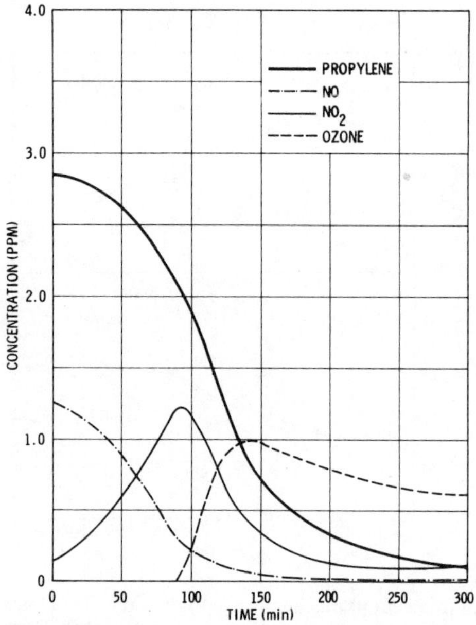

FIG. 3. A typical smog chamber experiment starting with 3 ppm propylene, 1.3 ppm NO, and 0.2 ppm NO_2 in air. The time scale is from turning on artificial sunlamps. Propylene decreases, NO is converted to NO_2, then ozone builds up and slowly decays. Other products such as aldehydes and nitrates are not shown.

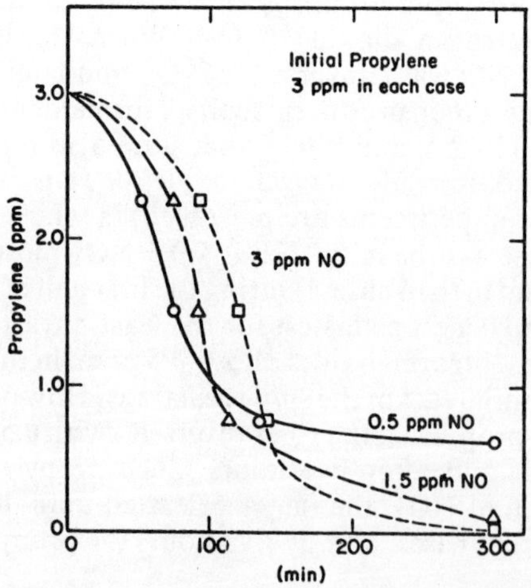

FIG. 4. Three smog-chamber runs with the same initial concentration of propylene (3 ppm) and with three different initial concentrations of nitrogen oxides. Note the rapid initial and slow long-term reaction with low initial (0.5 ppm) NO_x, and the slow initial and fast long-term reaction with high initial (1.5 ppm) NO_x.

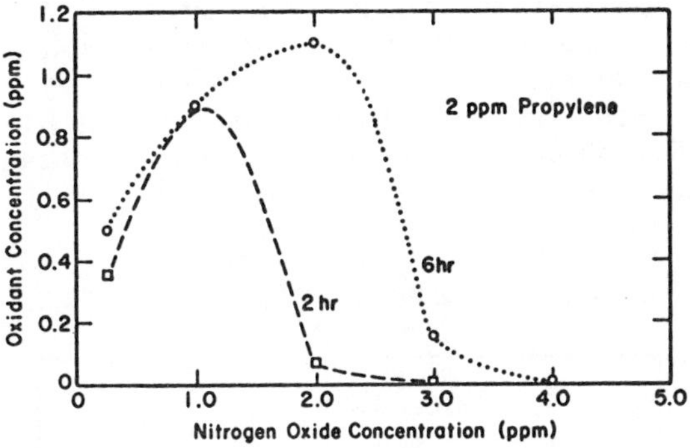

FIG. 5. Maximum oxidant produced with the same initial propylene (2 ppm) and various initial concentrations of nitrogen oxides. One series went for two hours and the other series for six hours. The "inhibition" of the smog reaction by NO_x occurred in each case but with quite different ratios and total effect between the two and six-hour experiments.

for 2 hours. Maximum oxidant is plotted against initial nitrogen oxide concentration. The two-hour experiments show the classic "inhibition" of smog by NO_x; low NO_x gave low oxidant, 1 ppm of NO_x gave maximum oxidant, 2 ppm or higher gave very little oxidant. The experiments that went for 6 hours showed a similar qualitative pattern, but the quantities are quite different. The maximum oxidant occurred with 2 ppm initial NO_x for the 6-hour experiment, and the maximum amount of oxidant is greater than that for the 2-hour case. The "inhibition," for 6-hour runs, occurs at and above 3 ppm of NO_x.

The laboratory data shown in Fig. 5 provide a possible explanation for the long-term trends in Los Angeles and Riverside shown in Fig. 3. Seabreezes take about 2 hours to reach downtown Los Angeles and 6 or 8 hours to reach Riverside, which is 50 miles inland. Picture both Los Angeles and Riverside as being situated in 1964 somewhat like the point at 1.0 ppm NO_x in Fig. 5. In the next few years there was a large increase of NO_x at each station, both as population increased and as the automobile emission index for NO_x increased by 50%. Moving from 1.0 to 1.5 ppm NO_x in Fig. 5 results in a large decrease in smog in the 2-hour experiment (Downtown Los Angeles) but in a large increase in smog in the 6-hour experiment (Riverside).

This example may reflect a general effect: increasing NO_x may cause little or no increase of smog in the traffic-heavy urban centers, but it may cause large increases of smog in suburban and rural areas up to hundreds of miles downwind. There have been many recent examples of newly-discovered, high concentrations of ozone in non-urban areas (*22–25*).

The broad aspects of the chemistry of photochemical smog are understood, and the major features can be reproduced by means of large-scale computer programs.

OZONE DESTRUCTION IN THE STRATOSPHERE (*1–12*)

The stratosphere is high, dry, cold, and penetrated by ultraviolet radiation with wave lengths down to 190nm. Radiation below 242nm dissociates oxygen to produce ozone. Ozone, in turn, strongly absorbs solar radiation below 300nm. The only effective shield of the surface of the earth against biologically damaging radiation between 300 and 250nm is ozone. An average world-wide vertical profile of ozone is given by Fig. 6. The troposphere is typically 0 to 15 km, and the stratosphere is 15 to 50 km. The height of maximum ozone concentration is about 20 to 25 km.

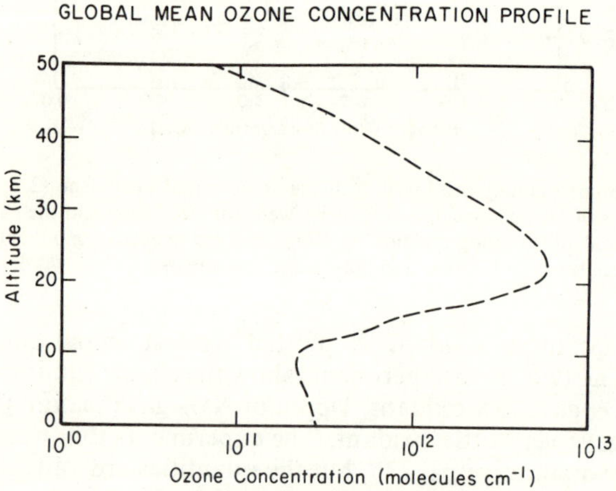

FIG. 6. An average world-wide vertical profile of ozone concentration as a function of height above sea level. The stratosphere is between about 15 and 50 km.

In the stratosphere ozone is formed almost exclusively by the photolysis of oxygen; curve A in Fig. 7 shows a vertical profile of the rate of formation of ozone (*9*). In the natural stratosphere, ozone is destroyed by a long list of reactions, the major ones of which are given in Fig. 7: B. NOx catalytic cycle; C, Ozone destruction by ozone ($O + O_3 \rightarrow O_2 + O_2$); D, E, F, G, H. Reactions of free radicals (H, HO, HOO) based on water. As can be seen from Fig. 7, the NOx catalytic cycle is far more important than all other mechanisms for ozone destruction. On a world-wide basis the balance sheet for ozone formation and destruction is given by Table 1.

By now the sources and sinks of natural NOx are recognized and evaluated. The artificial source of NOx from 500 Boeing SST as projected in 1971 (fuel flow 60 metric tons of fuel per hour per SST, emission index 15 grams NO per kilogram of fuel, 7 hours per day at cruise height of about 20 km) would be about twice the natural source of NOx; and 500 Concordes or Tupolevs (one third the rate of fuel consumption as the Boeing SST) would add about two-thirds as much NOx to the stratosphere as the natural source. It is now widely recognized that these large perturbations of stratospheric nitrogen oxides would cause a large reduction of stratospheric ozone.

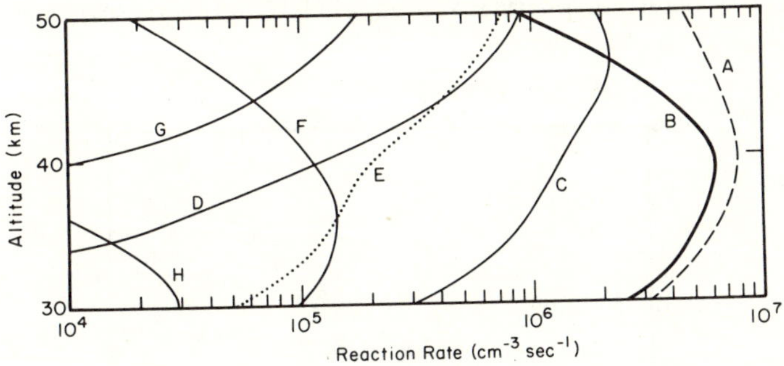

FIG. 7. Relative rates of photochemical formation of ozone (Curve A) and various mechanisms of destruction of ozone in the natural stratosphere:

B. The NO_2 catalytic cycle

$$NO + O_3 \rightarrow NO_2 + O_2$$
$$NO_2 + O \rightarrow NO + O_2$$

$$\text{net:} \quad O_3 + O \rightarrow O_2 + O_2$$

C. The elementary reaction $O + O_3 \rightarrow O_2 + O_2$.
D. $HO + O \rightarrow H + O_2$.
E. $HOO + O \rightarrow HO + O_2$.
F. $HO + O_3 \rightarrow HOO + O_2$.
G. $H + O_3 \rightarrow HO + O_2$.
H. $HOO + O_3 \rightarrow HO + O_2 + O_2$.

Curves D through H are the "water reactions."

TABLE 1
Balance Between Ozone Formation and Destruction in the Natural Stratosphere (Global Averages).

Mechanism	Relative ozone rate
$O_2 + h\nu$ (below 242nm)	+100%
$O + O_3 \quad O_2 + O_2$	− 17%
HO_x reactions	− 12%
Transport to earths surface	− 1%
$NO + O_3 \rightarrow NO_2 + O_2$ $NO_2 + O \rightarrow NO + O_2$ net: $O_3 + O \quad O_2 + O_2$	− 70%

The magnitude of the expected reduction of ozone by supersonic transports has been estimated by a series of model calculations that have included various degrees of atmospheric motions. The percentage reduction of ozone as a function of the percentage increase of stratospheric nitrogen oxides, as calculated by several different groups, is given by Fig. 8.The various modelers tend to agree that doubling the stratospheric column of NO_x would decrease the ozone column by about 20%. There is some disagreement as to how much a given fleet of SST would increase the stratospheric inventory of NO_x and decrease ozone, as shown in Table 2.

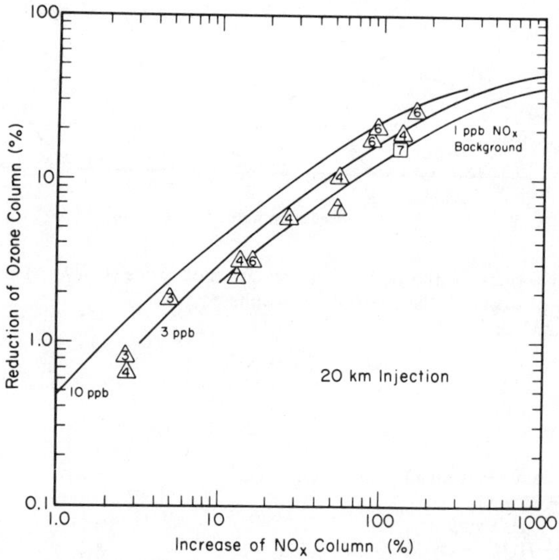

FIG. 8. Model calculations of percentage ozone reduction as a function of percentage NO_x increase according to various investigators. The numbers enclosed in the symbols refer to work by different authors:

3—reference *5* of the present paper
4—refs. *6* and *7* of the present paper
6—reference *9* of the present paper
7—(in the square): ref. *10* of the present paper.

Numerals 7 in triangles refer to earlier publications by the same authors. The fully drawn lines are from reference *1*.

TABLE 2

Results of Model Calculations for Worldwide Average Reduction of Ozone by 500 Large SST (See Text for Detailed Properties), Including Maximum "Corridor Effect" in Some Cases.

Calculated Ozone Reduction, per cents			
Global	Local maximum	Author	Ref.
23	50	Johnston 1971	(*1–2*)
8		Crutzen 1974	(*3–4*)
7		Chang 1973	(*6–7*)
11		McElroy et al. 1974	(*9*)
21		Shimazaki et al. 1974	(*11*)
12	25	Cunnold et al. 1974	(*12*)

The calculation by Cunnold et al. (1974) involved a three dimensional model (*12*) of atmospheric motions and ozone photochemistry; the other calculations were primarily one-dimensional (vertical, global average). The model involving three-dimensional motions gives a result somewhat in the middle of the models including only one dimension. The reduction of ozone as a function of latitude as predicted by Cunnold et al. is given by Fig. 9. The world-wide average ozone reduction was 12%, that in the northern hemisphere (where all traffic was assumed to be) was 16%, and the maximum ozone reduction near the flight corridor was 25%.

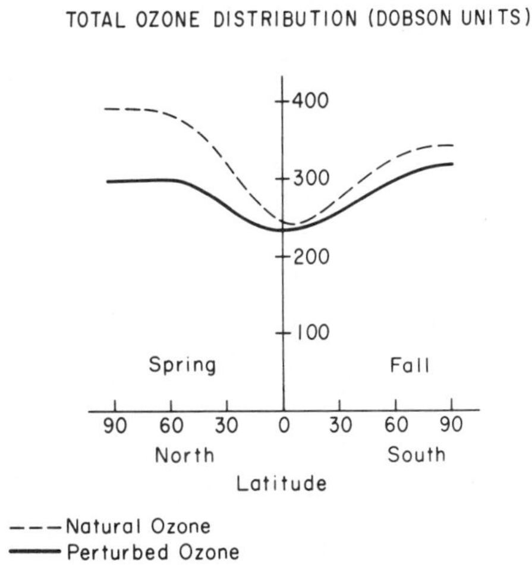

FIG. 9. Latitude dependence of ozone reduction when all SST flights occurred between 40 and 50° North, according to MIT three dimensional model, reference *12*. The rate of injection corresponds to 500 Boeing SST as projected in 1971 (but with reduced NO_x emission index, 15 g NO per kg fuel). The world-wide average ozone reduction is 12%; the local maximum near the flight corridor is 25%. One Dobson unit is 2.68×10^{16} molecules cm^{-2}.

The results of modelers of stratospheric motions and photochemistry, the considerations of the magnitude of natural vs. artificial sources of NO_x, and the recognized importance of nitrogen oxides in the natural ozone balance strongly confirm the proposition that supersonic transports would seriously deplete ozone in the stratosphere. Quantitative considerations (*2*) of the effects on ozone of nitric oxide produced by cosmic rays and by nuclear bombs further support this thesis.

CONTINUITY OF OZONE PHOTOCHEMISTRY FROM THE GROUND TO THE TOP OF THE STRATOSPHERE

The two sections above reviewed the photochemistry of urban smog and stratospheric perturbation by supersonic transports. By examples and references these sections stated each horn of the dilemma posed in the Introduction: NO_x does form ozone in urban smog and destroy ozone in the stratosphere. This section considers where in the atmosphere the crossover occurs between these two opposing trends.

A series of calculations (*26*) was carried out with set of 23 species and 36 reactions from the ground to the top of the stratosphere. Standard temperature and concentration of air at mid-latitudes was used. Standard observed profiles of ozone, water, and methane were used (Fig. 10). A profile of total NO_x ($NO + NO_2 + HNO_3$) was deduced from calculations and observations, and the

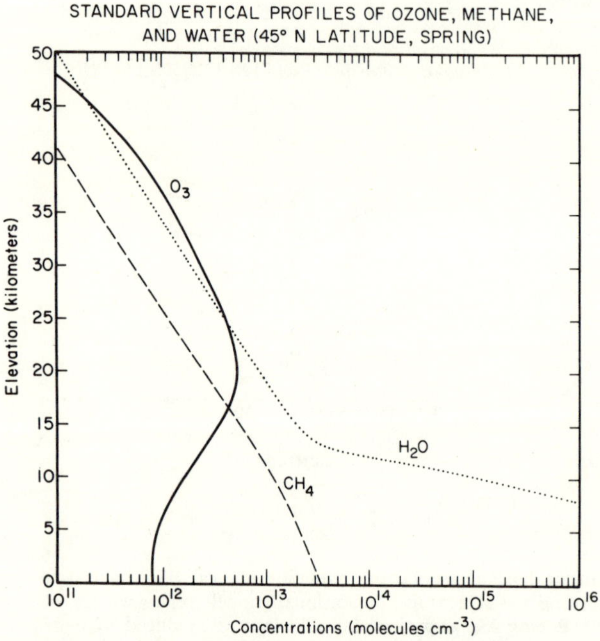

FIG. 10. Vertical profiles of ozone, water, and methane.

partitioning between NO, NO_2, and HNO_3 at midday is shown in Fig. 11. The calculated concentration of oxygen atoms is also included in Fig. 11.

In urban situations methane is considered not to form smog in the few available hours per day (Fig. 1). However, in the global troposphere (and stratosphere) some ozone is formed from air, methane, NO_x, and sunlight (*26*,27). The reaction is initiated by an attack by hydroxyl radicals, HO, or singlet oxygen atoms, $O(^1D)$, on methane to form methyl radicals:

$$\begin{pmatrix} HO \\ O(^1D) \end{pmatrix} + CH_4 \rightarrow \begin{pmatrix} H_2O + CH_3 \\ HO + CH_3 \end{pmatrix}$$

The methyl radicals add molecular oxygen to form a peroxyl radical that converts NO to NO_2:

$$CH_3OO + NO \rightarrow CH_3O + NO_2$$

Photolysis of nitrogen dioxide leads to ozone formation

$$NO_2 + h\nu(\text{below 400nm}) \rightarrow NO + O$$
$$O + O_2 + M \rightarrow O_3 + M$$

Subsequent reactions of the radical CH_3O lead to one to three perhydroxyl radicals, which also can convert NO to NO_2:

$$HOO + NO \rightarrow HO + NO_2$$

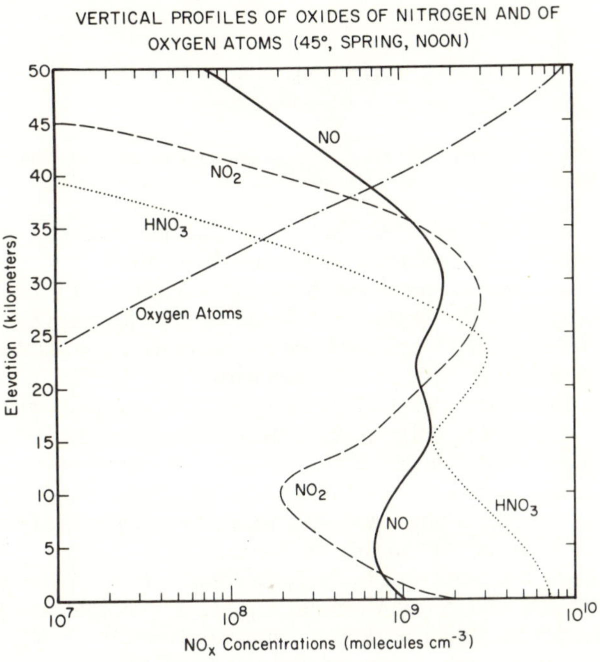

FIG. 11. Vertical profile of natural NO, NO_2, HNO_3 and oxygen atoms.

The net effect is that the slow, photochemical combustion of methane forms from one to four molecules of ozone. Starting with zero ozone, the rate of formation of ozone by the methane smog reaction is given by Fig. 12 for two cases in terms of molecules cm^{-3}: (1) 10^9NO, $10^{11}NO_2$; (2) 10^{11}NO, 10^9NO_2. The initial rapid formation of ozone is simply the photolysis of NO_2 to form NO and O_3. The subsequent rise in ozone with a doubling time of about one month is the methane-smog reaction. The final amount of ozone produced in each case is 2 or 3×10^{11} molecules cm^{-3}, comparable to observed, global, tropospheric ozone.

With the distributions of species given in Figs. 10 and 11, a model calcu-

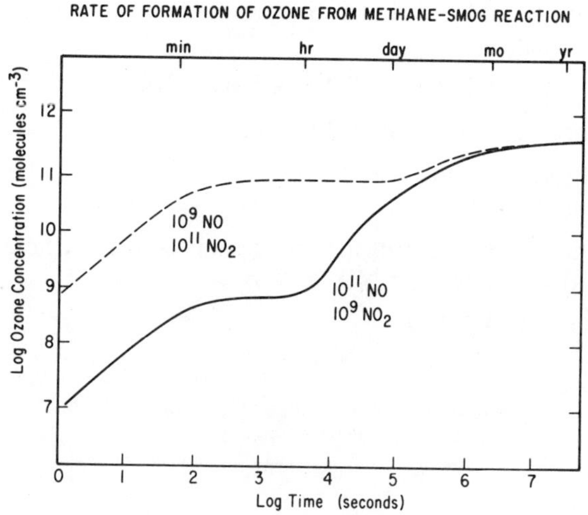

FIG. 12. The rate of formation of ozone by the methane–NO_x–smog reaction for two initial cases: (1) 10^9 NO, 10^{11} NO_2 molecules cm^{-3}; (2) 10^{11} NO, 10^9 NO_2 molecules cm^{-3}.

lation was made for a full set of reactions by a modified Gear routine (*6*). The calculation was run for 10^4 seconds, long enough to obtain a steady concentration of atomic and free radical species but short enough to avoid change of the background concentration of species. The relative rate of gross formation of ozone (HOO + NO and CH_3OO + NO) from the methane-smog reaction is given by Fig. 13. The rate of this reaction increases with O_3, H_2O, CH_4, and NO_x. As can be seen from Figs. 10 and 11, the last 3 are relatively constant with height in the stratosphere, and thus the rate of ozone formation from the methane-smog reaction roughly parallels the ozone profile there. In the troposphere, water increases very rapidly as one moves down from the tropopause to the ground, and the rate of the methane-smog reaction increases by a factor of 100 between 15 and 0 km, largely because of tropospheric water.

The rate of ozone destruction by NO_x increases simply as the product of concentration of NO_2 and oxygen atoms. Fig. 12 shows that oxygen atoms rapidly increase with height. The rate of ozone destruction by NO_x, Fig. 13, similarly shows a strong decrease as one moves from the troposphere to the ground.

In the stratosphere, the rate of ozone destruction by the NO_x catalytic cycle is more than 100 times faster than the rate of formation of ozone from the methane-smog reaction (Fig. 13). At about 13 km these two rates become equal. Below 13 km, the methane-smog reaction is faster than the catalytic destruction of ozone by NO_x. (For hydrocarbons in urban areas that are much more active than methane in forming ozone and smog, the difference would be even greater than that shown for methane). Thus a given set of reactions directly predicts that NO_x has a net effect of destroying ozone in the stratosphere and of forming ozone in the lower troposphere. The distribution of species (Figs. 11 and 12) and temperature is such that the relative effect of NO_x on ozone changes strongly with elevation.

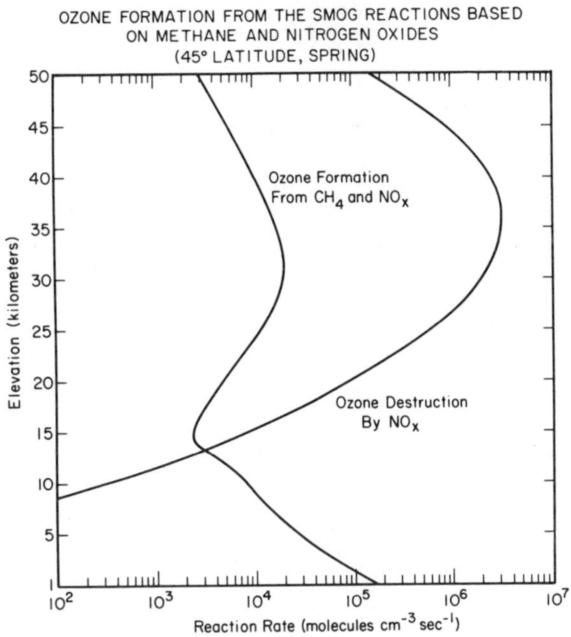

FIG. 13. Vertical profiles of the rate of formation of ozone from the methane–NO_x–smog reaction and of rate of ozone destruction from the NO_2 catalytic cycle. The former reaction increases with background O_3, CH_4, H_2O (Fig. 10) and the latter reaction is proportional to the concentration of oxygen atoms (Fig. 11) and NO_2 (Fig. 11). For a given set of 36 reactions and the profiles shown in Figs. 10 and 11, there is a crossover at about 13 km between ozone destruction and ozone formation so far as NO_x is concerned.

Figure 13 shows a cross-over between ozone formation and destruction by NO_x at about 13 km with the standard NO_x profile given by Fig. 12. Model calculations were carried out with twice as much and one-half as much as the standard NO_x profile. The cross-over points are shown in Fig. 14 in an enlarged scale. It can be seen that the cross-over point is not sensitive to the NO_x concentration; an increase in NO_x increases both ozone formation and destruction and the cross-over is seen to be about 13.3±0.2 km for all three cases.

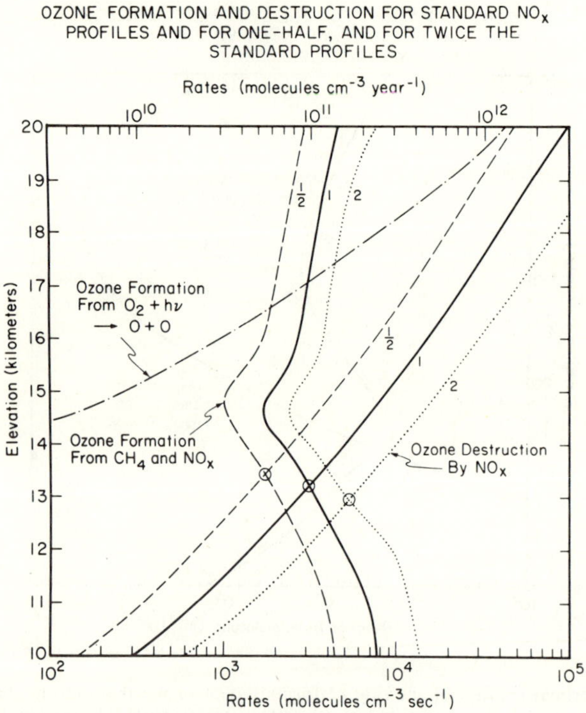

FIG. 14. Similar to Fig. 13 except for an enlarged scale between 10 and 20 km. Curve 1: NO_x as in Fig. 11. Curve 1/2: NO_x as in Fig. 11 reduced by a factor of 2. Curve 2: NO_x in Fig. 11 multiplied by a factor of 2. Note that the elevation of the crossover point is almost the same for these three widely different NO_x profiles.

ACKNOWLEDGMENT

This work was supported in part by the Climatic Impact Assessment Program by means of an interagency agreement between the Department of Transportation and the U.S. Atomic Energy Commission through the Inorganic Materials Research Division, Lawrence Berkeley Laboratory. Appreciation is expressed to Dr. Gary Whitten for assistance in carrying out the calculations on methane photochemistry.

REFERENCES

1. H. S. Johnston, *Science* **173**, 517 (1971).
2. H. S. Johnston, *Acta Astronaut.* **1**, 135 (1974).
3. P. J. Crutzen, *Ambio* **1**, 41 (1972).
4. P. J. Crutzen, *Can. J. Chem.* **52**, 1569 (1974).
5. E. Hesstvedt, *Can. J. Chem.* **52**, 1592 (1974).
6. J. S. Chang, A. C. Hindmarsh, and N. K. Madsen, *Simulation of Chemical Kinetics Transport in the Stratosphere*. Report UCRL-74823. Lawrence Livermore Laboratory, Livermore, California, 1973.
7. J. S. Chang and H. S. Johnston, *The Effect of NO_X Effluents on Ozone*. Report UCRL75801. Lawrence Livermore Laboratory, Livermore, California, 1974.

8. R. W. Stewart, *Response of Stratospheric Ozone to Simulated Injection of Nitric Oxide.* Paper presented at the Fall American Geophysical Union Meeting, San Francisco, California, 1973.
9. M. McElroy, S. Wofsy, J. Penner, J. McConnell, *J. Atmospheric Sci.* **31**, 287 (1974).
10. R. C. Whitten and R. P. Turco, *The Effect of SST Emission on the Earth's Ozone Layer.* Paper presented at the IAMAP-IAPSO First Special Assemblies, Melbourne, Australia, 1974.
11. T. Shimazaki and T. Ogawa, *Theoretical Models of Minor Constituents' Distributions in the Stratosphere and the Impact of the SST Exhaust Gases.* Paper presented at the IAMAP/IAPSO First Special Assemblies, Melbourne, Australia, 1974.
12. D. M. Cunnold, F. N. Alyea, N. A. Phillips, and R. G. Prinn, *First Results of a General Circulation Model Applied to the SST-NO_X Problem.* Paper presented at meeting of the American Meteorological Society, San Diego, California, 1974.
13. *Biological Impacts of Increased Intensities of Solar Ultraviolet Radiation,* National Academy of Sciences, National Academy of Engineering, Washington, D.C. 1973.
14. A. J. Haagen-Smit, C. E. Bradley, and M. M. Fox, *Ind. Eng. Chem.* **45**, 2086 (1953).
15. *Project Clean Air*, **Volumes 1 and 2**. University of California, Berkeley, California, 1970.
16. J. R. Kinosian, J. Poskind, and R. Selfridge, *Air Quality in the San Joaquin Valley Air Basin.* California Air Resources Board, Division of Technical Services, Sacramento, California, 1973.
17. J. R. Kinosian and S. Duckworth, *Oxidant Trends in the South Coast Air Basin,* 1963–1972. California Air Resources Board, Division of Technical Services, Sacramento, California, 1973.
18. A. P. Altshuller and J. J. Bufalini, *Photochem. Photobiol.* **4**, 97 (1965).
19. A. P. Altshuller, S. L. Kopczynski, W. A. Lonneman, T. L. Becker, and R. Slater, *Environ. Sci. Technol.* **11**, 899 (1967).
20. B. Dimitriades, *On the Function of Hydrocarbons and Nitrogen Oxides in PhotochemicalSmog Formation.* Report 7433, TN23.U7. Bureau of Mines, United States Department of the Interior, 37 pp. (1970).
21. National Academy of Sciences, *A Critique of the 1975–76 Federal Automobile Emission Standards for Hydrocarbons and Oxides of Nitrogen*, pp. 28–39, May 1973.
22. L. A. Ripperton, J.J.B. Worth, C. E. Decker, and D. R. Johnston, *High Ozone Concentrations in Non-urban Air.* Research Triangle Institute, Report for EPA 68-02-1077, March 1974.
23. W. N. Stasiuk and P. F. Coffey, *J. Air Pollut. Control Assoc.*, 1974 (in press).
24. P. R. Miller, M. H. Cutchan, and H. D. Milligan, *Atmos. Environ.* **6**, 623 (1972).
25. F. D. Hall, Jr., C. E. Duchon, L. G. Lee, R. R. Hagan, *Mon. Weather Rev.* **101,** 404 (1973).
26. Paul Crutzen, *Pure Appl. Geophys.* **106–108**, 1385 (1973).
27. William Chameides and J.C.G. Walker, *J. Geophys. Res.* **78**, 8751 (1973).

Gas-to-Particle Conversion in the Atmospheric Environment by Radiation-Induced and Photochemical Reactions[1]

K. G. Vohra
Division of Radiological Protection,
Bhabha Atomic Research Center,
Trombay, Bombay-400085, India

INTRODUCTION

Radiation chemistry and photochemistry of the earth's atmosphere deal with complex reactions involving a fascinating variety of reactive species and energy sources. We shall be concerned here with the reactions which lead to the formation of particles from the gas phase. Small particles, ranging in size from a few angstroms to about a micrometer, control a vast majority of atmospheric processes. In spite of this, today we know very little about the source, chemical composition and formation mechanisms of the particles which take part in some of the most vital phenomena in the atmosphere. For example, we do not know much about the nature and source of a variety of cloud nuclei and ice nuclei formed in the atmosphere. Gas-to-particle conversion by radiation and photochemical processes is the likely atmospheric source of these nuclei. Small particles also affect the radiation balance and ionization balance in the atmosphere. The radiation balance which controls the global climate depends on the scattering and absorption of solar radiation by numerous small particles and reflection by clouds formed with these particles acting as nuclei.

Measurements carried out during the last decade have shown that the concentration of small particles from man-made sources has been continuously increasing in the atmosphere (*1*). These particles, formed mainly by gas phase reactions, can act as nuclei for the formation of cloud droplets (cloud nuclei) or for the formation of ice crystals (ice nuclei). Higher concentrations of nuclei of either type can lead to the formation of stable fog, which inhibits precipitation and also reduces incoming solar radiation. This can lead to the inadvertent modification of the local and global climate. We shall not go into much details of the mechanism of involvement of small particles in the whole gamut of atmospheric processes, directly or indirectly; the possible formation mechanisms of these particles by atmospheric photochemistry and radiation chemistry will be mainly considered.

Although much has been written in the air pollution literature about atmospheric photochemistry in particle formation, the role of ionizing radia-

[1]Congress presentation B–9–3.

tions in the processes has not received much attention. It may be intriguing to know that as early as 1899, C.T.R. Wilson published a paper entitled, "On the condensation nuclei produced in gases by the action of Roentgen rays, uranium rays, UV light and other agents" and suggested this as a source of atmospheric nuclei (*2*), but no serious study of the effect was taken up until quite recently. The main sources of ionizing energy input are cosmic rays and natural atmospheric and terrestrial radioactivity. The magnitude of this energy is extremely small compared to the solar energy input. In spite of this, the effect of this source in particle formations could be comparable to the photochemical effect when the two are considered together. We shall review the field briefly and consider in detail the possible reactions and results of some of the laboratory studies on the role of radiation chemistry and photochemistry in gas-to-particle conversion. In further discussion in this paper, gas-to-particle conversion will be frequently abbreviated as 'GPC' and the term 'aerosols' will be used for small particles of different type.

CHEMISTRY, PHYSICS AND TERMINOLOGY OF GAS-TO-PARTICLE CONVERSION

Atmospheric trace gases responsible for particle formation are generally present in the concentration range of a few parts per billion to a few parts per million of air by volume. At these low concentrations, no significant reactions can be expected via the conventional chemistry. Chemistry of GPC is unique, with a complex sequence of chemical and physical changes aiding each other to form the embryo and the fully developed aerosol. The initial chemical change of the trace gas takes place under the action of UV, visible light or ionizing radiations, generally through a step of free radical formation followed by steps of oxidation and hydration. We shall consider these mechanisms in detail along with the role of ion-molecule reactions. Some of the commonly used terms in GPC are as follows. If a system (gas mixture) is initially completely free of all colloidal particles, and through some physical or chemical change one species attains supersaturation and starts condensing (without any foreign particles), the process is called homogeneous nucleation, generally referred to as nucleation. If condensation takes place only as a result of a chemical change, it is called chemical nucleation. If the system initially contains some particles and a new condensible species is formed chemically, it will condense on the existing particles which will grow in size. Normally this is termed as a growth process. If, however, the growth process is rapid enough to lead to sudden formation of large particles such as cloud, fog or smog from superheated vapors, the process is called heterogeneous nucleation. The nucleation term called heteromolecular nucleation, represents formation of particles consisting of two or more types of molecules, for example, H_2SO_4 and H_2O. Thus we can have heteromolecular homogeneous, or heteromolecular heterogeneous nucleation, depending on whether foreign particles are present or not.

ION-INDUCED GAS-TO-PARTICLE CONVERSION—BRIEF HISTORY

Small particles formation by the effect of ionizing radiations on air and other gas mixture is a field not yet explored, even though the laboratory studies carried out by Cameron and Ramsay as early as 1908 provided the indication of the effect (*3*). Subsequent studies by Chamberlain et al. showed that irradiation of air by doses in the range of 1–15 rad can produce particle concentrations on the order of 10^4–10^5 cm^{-3} (*4*). Bricard et al. (*5*) studied the increase in particle formation when air was mixed with thoron, the estimated

TABLE 1
BRIEF HISTORY OF NUCLEUS FORMATION BY IONIZING RADIATIONS

Year	Authors	Gas Mixture	Dose	Remarks
1908	Cameron and Ramsay (*3*)	Carbon Monoxide	100 mCi radon	Solid carbon and C_3O_2 formed Exposure – 1 day M/N = 1.7
1957	Chamberlain et al. (*4*)	Filtered air + thoron	1–15 rads	10^4 to 10^5 particles cm^{-3} formed
1969	Bricard et al. (*5*)	Air + thoron	100 millirads	Particle concentration increased from 7×10^4 to 1.5×10^5 cm^{-3}
1969	Vohra et al. (*6*)	Filtered air	20 microrads per hour	10^{-5} particles formed per ion pair
1971	Mercer and Stowe (*25*)	Room air + radon	1 μCi/cm^3 of radon	Increase from $2 \times 10_3$ to 10^4 particles cm^{-3}
1972	Vohra et al. (*7*)	$N_2 + O_2^* + SO_2$ (5 ppm)	7 millirads per hour	10^4 to 10^5 particles cm^{-3} formed
1973	Hart et al. (*26*)	Tank air + SO_2 (1 ppm)	1 mCi of Sr-90	4×10^5 particles cm^{-3}

dose for nearly doubling the particle concentration in a sample of air was 100 millirads. Vohra et al. showed that dose rates as low as 20 microrads per hour, with a total dose of 1 millirad, were effective in particle formation (*6*). They estimated ion-induced particle formation rate of 10^{-5} per ion pair in the room air without exposure to UV or visible light. These studies showed that the rate of formation of particles by the action of ionizing radiations alone is too small to account for the production of numerous condensation nuclei in the atmosphere. Recent studies have shown that synergistic effects of ionizing and solar radiations could be more important under atmospheric conditions (*7*) where the ionizing radiation doses are small, in the range of about 10 μrad/hr (near ground level) to around 20 mrad/day in the stratosphere at higher latitudes. The role of sunlight has been suggested to be in the formation of certain species which aid the ionic effect. Table 1 gives a chronological summary of some of the important studies on ion-induced particle formation.

The Photochemical Effects in Gas-to-particle Conversion

The solar radiation reaching the lower atmosphere is limited to wavelengths above 2900 Å, the shorter wavelengths being completely absorbed by ozone above the troposphere. In the stratosphere and mesosphere, more energetic solar radiations are present, down to nearly 2000 Å. The energy of solar radiations in these atmospheric regions determines the nature of the photochemical effects produced. In the lower atmosphere only trace gases, for example, halogens, NO_2, SO_2, ozone and aldehydes, absorb the radiation above 2900 Å; nitrogen does not absorb in this wavelength region and molecular oxygen absorbs only very weakly in the red end of the visible spectrum. In the stratosphere, there is significant absorption of the UV component by oxygen and water vapor. Absorption of solar radiations may lead to the formation of excited molecules of the absorbing species, or to the primary photochemical effect yielding new reactive species, mainly consisting of free atoms and radicals in the ground or excited state. Further participation of these species in atmospheric chemical reactions is termed secondary photochemical effects. Table 2 gives the primary photochemical reaction of NO_2, SO_2, O_3, H_2O, O_2, HCHO, I_2 and Cl_2, the main absorbing species of atmospheric interest. The table also gives important reactive primary products and the secondary reactions leading to the formation of aerosols. It is seen that the primary species react with abundant molecular oxygen and water vapor in the atmosphere, forming a new set of species, mainly acids and peroxides. This is the state at which particle formation is initiated. For example, photochemically produced H_2SO_4 molecules form particles by heteromolecular nucleation involving water molecules. In the $H_2SO_4 + H_2O$ system, the concentrations of acid vapor needed to nucleate droplets are in the range of 4.10^{-5}–$1.3.10^{-2}$ ppm only (*8*). In the case of HNO_3 the concentrations needed for nucleation are much higher. The acid particles can further grow by chemical reactions of trace gases such as ammonia on the particle surface. Liquid and solid aerosols consisting of pure or mixed organic substances may also be formed photochemically. Formation of such aerosols is initiated by an oxidation process followed by free radical reactions, including polymerization and chain reactions of dif-

TABLE 2
Important Photochemical Processes of Gas-to-Particle Conversion

Number	Primary photochemical processes	Important primary products	Suggested secondary reactions forming aerosols
1	$NO_2 + h\nu \xrightarrow{<4000\ A} NO + O$	O	$O + O_2 + M \rightarrow O_3 + M$
	$O_2 + h\nu \xrightarrow{<2424\ A} O + O$	NO	$O_3 + NO_2 \rightarrow NO_3 + O_2$ $NO_3 + NO_2 + N \rightarrow N_2O_5 + M$ $N_2O_5 + H_2O \rightarrow 2HNO_3$
2	$SO_2 + h\nu \xrightarrow{3400\text{–}4000\ A} SO_2^*$ $SO_2^* + O_2 \rightarrow SO_4$	SO_4	$SO_4 + O_2 \rightarrow SO_3 + O_3$ $SO_3 + H_2O \rightarrow H_2SO_4$
3	i) $O_3 + h\nu \rightarrow O_2^* + O$		$SO_2 + O^* \rightarrow SO_4$
	ii) $NO_2 + h\nu \xrightarrow{>4000\ A} NO_2^*$ $NO_2^* + O_2 \rightarrow O_2^* + NO_2$	O_2^*	$SO_4 + H_2O \rightarrow HSO_4 + OH$ $HSO_4 + H_2O \rightarrow H_2SO_4 + OH$
4	i) $H_2O + h\nu \xrightarrow{<2200\ A} OH + H$	H	$H + O_2 + M \rightarrow HO_2 + M$
	ii) $HCHO + h\nu \xrightarrow{<3400\ A} CH\dot{O} + H$	OH	$SO_2 + HO_2 \rightarrow SO_3 + OH$
5	$I_2 + h\nu \xrightarrow{<8000\ A} I + I$	I	$H_2O + I \rightarrow HI + OH$
	$Cl_2 + h\nu \xrightarrow{<5000\ A} Cl + Cl$	Cl	$H_2O + Cl \rightarrow HCl + OH$

Important Secondary Products—O_3, NO_3, SO_3, SO_4, HSO_4, OH, HO_2

ferent type. The mechanisms involved in photochemical formation of such aerosols are so complex that in spite of extensive research very little is known about them.

MECHANISMS BASED ON IONIC REACTIONS

The primary photochemical reactions considered in the last section involve the breaking of chemical bonds by photons of a few electron volts energy. The primary processes in radiation chemistry involve the breaking of bonds as well as 'knocking off' of electrons from the molecules, producing free radicals and atoms, excited species, and ions of a large variety. These provide many fascinating possibilities of nucleus-forming reactions. However, the reaction mechanisms so far proposed in the literature are speculative, and there is an urgent need for establishing definite reaction schemes. Some of the primary reactions give products similar to those from photochemical reactions.

For example, O_3 and NO formed by the radiolysis of moist air can react further as follows,

$$NO + O_3 \rightarrow NO_2 + O_2$$

$$NO_2 + O_3 + M \rightarrow NO_3 + O_2 + M$$

and lead to the formation of HNO_3. Similarly, radiation chemistry of SO_2 in moist air yields H_2SO_4 nuclei by the following net reaction:

$$2\ SO_2 + O_2 + 2\ H_2O \rightarrow 2\ H_2SO_4$$

However, the above purely radiolytic reactions are inefficient in nucleus formation and are of little significance in the atmosphere where the available ionizing radiation dose rates are very small. We shall consider a few of the specific ionic mechanisms that have been proposed.

Ionic Recombination Mechanism

Vohra et al. have proposed that the continuous formation and recombination of hydrated ions can be a continuous atmospheric source of condensation nuclei (*9*). The following general reaction has been proposed:

$$H_3O^+(H_2O)_n + X^-(H_2O)_N \rightarrow H_3O^+(H_2O)_{n+N}X^-$$

The product of such a reaction can be a hydrated molecule of H_2SO_4, HNO_3 or HCl, depending on the nature of the negative ion X. This mechanism has been established by experiments using SO_2, NO_2, and Cl_2 as trace gases. It is suggested that these trace gases form electronegative species HSO_4, NO_3 and Cl by the photochemical reactions shown in Table 2, which can form hydrated negative ions taking part in the above recombination reaction. This is a truly synergistic mechanism involving radiation and photochemical processes.

Explosive Growth Reactions

Coffey and Mohnen have proposed an explosive growth reaction scheme in which $NH_4^+(H_2O)_n$ ion clusters catalyze the growth of hydrated ammonium chloride and ammonium nitrate (*10*). The following reactions have been suggested.

$$H^+(H_2O)_n + NH_3 \rightarrow NH_4^+(H_2O)_n$$

$$NH_4^+(H_2O)_n + HCl \rightarrow NH_4\ Cl(H_2O)_m + H_3O^+(H_2O)_{n-m}$$

$H_3O^+(H_2O)_n$ rapidly attaches an ammonia molecule to form $NH_4^+(H_2O)_{n+1}$ which can repeat the above sequence. In this way, each positive ion forms many ammonium chloride embryos. The embryo size is suggested to be around 5 Å, and they grow to particles of 100 Å size.

Aerosols Formed by Chemical Reactions on Ion Centers

This mechanism is essentially based on ion-molecule chemical reactions. In the formation of aerosols, certain charge-specific complex ions are created by chemical bonding, e.g. hydrogen bonding. These complex ions may constitute embryos which can grow by further chemical reactions on the surface involving polar gases such as SO_2, NO_2 etc. (McKnight, 1972)[2]. Castleman has proposed the following clustering reactions (*11*):

$$H_3O^+ + nSO_2 \rightarrow H_3O^+(SO_2)_n$$

$$NO^+ + nSO_2 \rightarrow NO^+(SO_2)_n$$

The measured bond energies of these ion clusters were found to be larger than theoretical values based on electrostatic bonding.

Aerosols Formed by Clustering Around Negative Ions

There have been a number of independent studies which suggest that negative ions at thermal energies attach certain polar molecules which lower their mobility. Tyndall (see Loeb, (*12*)) has reported the formation of clusters of H_2O, CH_3OH and C_2H_5OH selectivity on negative ions. Moruzzi and Phelps have suggested the formation of hydrated ions in the earth's atmosphere wherever there is significant concentration of water vapor (*13*). Vohra et al. have also studied the formation of nuclei up to a size of 50–60 Å using thermal negative ions and vapors of H_2O, C_6H_6 and C_2H_5OH for clustering (*14*). However, so far there is no satisfactory mechanism proposed for the growth of embryos to nuclei. Loeb has suggested that such clustering is possible only by the formation of chemical bonds, and has strongly opposed any suggestion of formation of aerosols by physical aggregation of molecules around the ions (*15*). This field is open for further research.

Nucleus Formation by Hydrogen Peroxide From Natural Ionization

In cloud chamber studies of homogeneous nucleation, Allen et al. have reported the existence of a large array of intermediate clathrate type agglomerates in equilibrium with the ambient vapor (*16*). Some of these appeared to be associated with the heterogeneous nuclei that cause condensation to occur at saturations slightly less than that for cloudy condensation. This has been explained on the basis of the existence of embryos formed by background radiations. The mechanism suggested is radiolytic generation of H_2O_2 by background radiations and its role in the formation of clathrate structures with up to 80 molecules of water. No attempt has been made to identify the nucleating substance.

Mechanism Based on Hydrated Electrons

In the studies on the condensation of water vapor by corona discharge, Alberto et al. have reported an enhancement of cloudy condensation by N_2O (*17*). N_2O is known to be an efficient scavenger of hydrated electrons, yielding hydrated negative ions of oxygen atoms, through the following reaction (*18*):

$$N_2O + e^-_{aq} \rightarrow N_2 + O^-_{aq}$$

The enhancement can be due to O^-_{aq} through a heterogeneous nucleation mechanism. In the earth's atmosphere significant concentrations of N_2O exist, and it may be worthwhile to investigate the likely importance of the reaction in par-

[2]McKnight, L. G., Bell Telephone Laboratories, Whippany, New Jersey 07981, *Clustering of Ions: Atmospheric Chemistry and the Aerosol Particulate Background* (Private Communication, 1972).

ticle formation. In the presence of oxygen, the following very fast reaction (rate constant 2×10^{10} M^{-1} sec^{-1}) can also lead to the formation of hydrated O_2^- molecules:

$$e^-_{aq} + O_2 \rightarrow O_2^-$$

In view of the above fast reaction it is not possible to detect hydrated electrons in the air. Nevertheless, the mechanism is likely to have an important role in the atmosphere.

EXPERIMENTAL STUDIES

The experiments carried out in the author's laboratory for the study of the mechanism of nucleation induced jointly by ionizing radiations and photochemical reactions will be reviewed briefly. These studies included spontaneous nucleation in the filtered air, nucleation in the presence of added trace gases, and synergistic effects of UV or visible light and ionizing radiations. A 12 liter pyrex reaction vessel has been used for the study of particle formation by gas phase reactions, described earlier (*7,9*). All these studies have been carried out at atmospheric pressure in moist air or gas mixture. For the study of ionic reactions at low dose rates, a ^{60}Co source was used for gamma irradiation and a tritium source was used for beta irradiation. For the study of combined effects of ionizing and non-ionizing radiations, the air or the gas mixture was irradiated by a UV lamp, through quartz or pyrex, depending on the photon energy range of interest. The results of a few representative experiments will be presented here.

In the reaction vessel used, particle formation starts when air or an appropriate gas mixture is introduced, rises to a peak level and declines, as found in almost all the experiments. This behavior has been explained in detail by Walter (*19*). A typical study, with a sample of ground level air containing 0.02 ppm of SO_2 introduced into the reaction vessel during the afternoon on a bright sunny day, is shown in Figure 1. The lower curve shows particle formation under the normal background dose of ionizing radiations and the normal effect of solar radiation. The upper curve gives particle concentrations obtained with a gamma exposure rate of 5 mR/hr, showing almost an order of magnitude increase. For the study of the effect of different exposure rates, the reaction vessel was filled with air partially irradiated by UV light and filtered, to simulate the effect of sunlight. SO_2 was introduced into the reaction vessel to give a concentration of 0.1 ppm. By this procedure, the formation of particles in the reaction vessel was found to be extremely sensitive to the dose of ionization. Figure 2 gives the particle concentrations in the reaction vessel for different dose rates. Increase in concentration is seen even at an exposure rate as low as 0.35 mR/hr. Without UV-irradiation of the air, no significant increase could be observed even up to an exposure rate of 65 mR/hr.

In another typical experiment, the reaction vessel was filled with tank nitrogen with 0.1 ppm of chlorine. A tritium source giving an ionization rate of 10^6 ion pairs cm^{-3} was suspended in the center of the reaction vessel, and simultaneous irradiation was carried out with light above 2950 Å. The results are shown in Fig. 3. Peak particle concentration of 10^6 is reached at the end of 10 minutes.

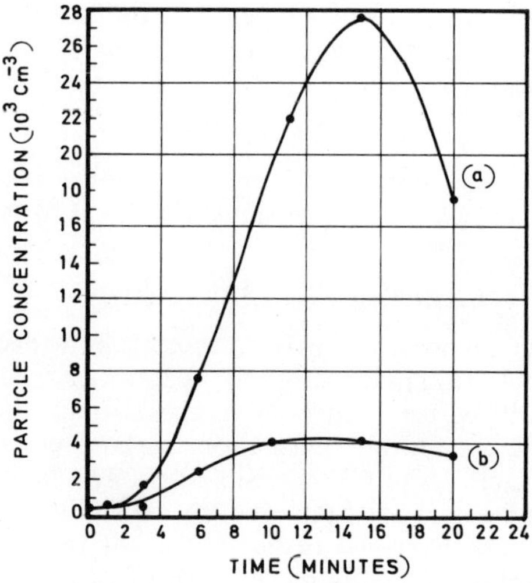

FIG. 1. Particle formation in the filtered air during daytime, (a) with an exposure rate of 5 mR/hr and (b) under normal conditions.

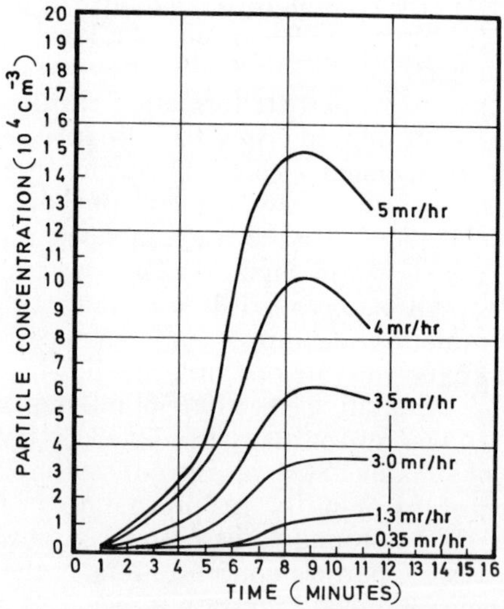

FIG. 2. Particle formation in the filtered air with 0.1 ppm SO_2, partial UV-irradiation and different dose rates.

For the study of correlation between the formation of nuclei in the filtered air and natural radon content of the air, measurements were carried out during the afternoon hours in the months of May and June, 1972. The results are shown in Fig. 4. There is a close correlation between radon and particle concentration.

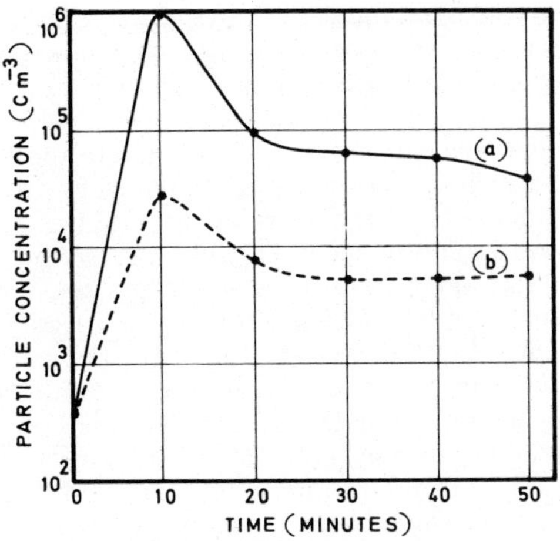

FIG. 3. Particle formation in nitrogen with 0.1 ppm of chlorine and irradiation with light above 2950 Å. (a) with tritium beta irradiation; (b) without beta irradiation.

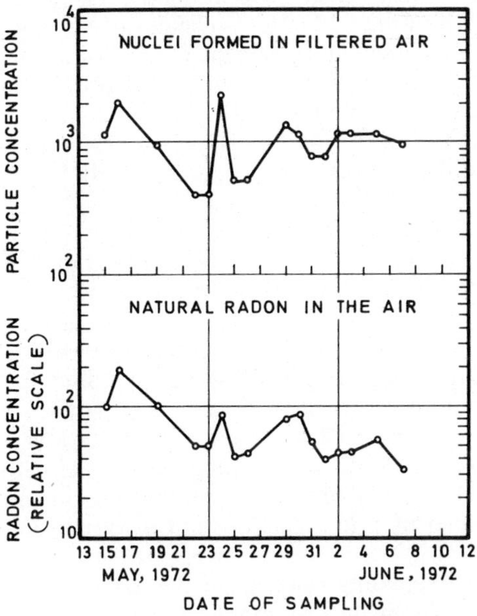

FIG. 4. Correlation between particle formation in the filtered air and radon content of the air during daytime.

DISCUSSION AND CONCLUSIONS

The experiments provide a clear evidence of the importance of synergistic effects of radiation chemistry and photochemistry in gas-to-particle conversion. The effects produced at very low dose rates are particularly important in the atmosphere. The photochemical effects summarized in Table 2 have been fairly well known. The striking increase in particle concentration when ionizing radiation exposure is given after or along with the exposure to UV and visible light needs to be explained on the basis of ionic mechanisms presented earlier.

Nearly all the ionic mechanisms proposed relate to the formation of prenucleation embryos which can grow by continual physical aggregation and chemical reactions. Photochemistry provides the reactive molecules and free radicals (Table 2), and radiation chemistry aids aggregation by providing the charge. Thus, the initial aggregation of polar molecules, such as H_2O, SO_2, NO_2 etc., on ion centers by electrostatic forces could lead to chemical reactions. The new products formed, e.g. H_2SO_4 from SO_2, can grow further by hydration. The process can continue until recombination stabilizes the particle. Further growth in size can take place by coagulation. Thus, the ionic recombination scheme proposed by Vohra et al. and the growth on ion centers proposed by McKnight, both can be important in ion-induced nucleation. The hydrated electron reactions can also provide clustered ions which could participate in the scheme of nucleation.

The peak particle concentration in Fig. 3 at the end of 10 minutes shows that nearly 1 particle is formed per 600 ion pairs. The estimated particle size corresponds to something like 10^4–10^5 molecules of the nucleating substance, presumably HCl in this case. Thus, several molecules of HCl are formed per ion pair which could be due to a chain reaction involving photolytic and ionic processes, giving very high chain yield. No definite mechanism can be proposed yet. It seems, however, that the ion-induced nucleation mechanisms could be different for different trace gases, particularly for polar and nonpolar molecules.

We shall now consider briefly the possible formation of nuclei and aerosols in the atmosphere. In a recent paper Meszaros and Vissy have suggested that aerosol particles in the global background mainly consist of ammonium sulphate and sulphuric acid, formed from the trace gases (*20*). The Junge layer of sulphate particles centered at 20 kilometers is also known to have the same composition. In addition to particles larger than about 0.1 micrometer, the stratosphere also contains high concentrations of Aitken nuclei (*21*). Hoffmann et al. have reported the formation of distinct layers of aerosols just above the tropopause (*22*); this is also the region of peak ionization intensity due to cosmic rays. It seems likely, therefore, that photochemical oxidation of SO_2 to sulphate is aided by ionic reactions in the formation of aerosols in this region.

It would also be interesting to consider the formation of cloud nuclei and ice nuclei by these mechanisms. Ammonium sulphate particles formed from the gas phase are known to be very efficient cloud nuclei, and may be the atmospheric source of such nuclei. Ice nuclei for the formation of cirrus clouds in the upper troposphere are also likely to be formed by gas reactions involving certain organic substances, but detailed information on the nature of such nuclei is lacking. A model has also been proposed for the formation of nuclei for noctilucent clouds by ionic and photochemical reactions (*23*).

An understanding of the mechanisms of formation of these nuclei would also enable an assessment of the impact of pollutant gases reaching the upper atmosphere on the global climate. In the polluted urban atmosphere there are several efficient chemical and photochemical processes of particle formation. In spite of this, it is likely that nucleation is considerably aided by natural radon and thoron emanating from the soil. This conclusion is supported by the observation reported in Fig. 4, where a very good correlation has been obtained between the nucleating ability of the filtered air and the radon content of the air during the afternoon hours at Trombay, an industrial suburb of Bombay. This observation is also significant in the assessment of combined effects of radioactive and nonradioactive releases to the environment in the nuclear era (*24*).

REFERENCES

1. V. J. Schaefer, in *Global Effects of Environmental Pollution* (S. F. Singer, Ed.), pp. 158–174. D. Reidel Publishing Company, Holland, 1970.
2. C. T. R. Wilson, *Philos. Trans. R. Soc. London, Ser. A* **192,** 403 (1899).
3. A. T. Cameron and W. Ramsay, quoted in *The Chemical Effects of Alpha Particles and Electrons* (S. C. Lind, Ed.), pp 152–154. The Chemical Catalog Co., New York, 1928.
4. A. C. Chamberlain, W. J. Megaw and R. D. Wiffen, *Geofis. Pura Appl.* **36**, 233 (1957).
5. J. Bricard, F. Billard and G. Madelaine, in *Planetary Electrodynamics*, (S. C. Coroniti and J. Hughes, Ed.), pp. 231–241. Gordon and Breach, New York, 1969.
6. K. G. Vohra, M. C. Subba Ramu and K. N. Vasudevan, in *Planetary Electrodynamics*, (S. C. Coroniti and J. Hughes, Ed.), pp. 127–135. Gordon and Breach, New York, 1969.
7. K. G. Vohra, P. V. N. Nair and T. S. Muraleedharan, *Aerosol Sci.* **3**, 225 (1972).
8. P. Mirabel and J. L. Katz, *Aerosol Sci.* **5**, 115 (1974).
9. K. G. Vohra, K. N. Vasudevan and P. V. N. Nair, *J. Geophys. Res.*, **75**, 2951 (1970).
10. P. Coffey and V. A. Mohnen, *24th Annual Meeting of the Gaseous Electronics Conference*, Gainesville, Fla., 1971. (Available from the Secretary of the Gaseous Electronics Conference, ARC Symposium, Department of Physics, University of Florida, Gainesville, Florida 32601.)
11. A. W. Castleman, *Aerosol Sci.*, **5**, 116 (1974).
12. L. B. Loeb, *Basic Processes of Gaseous Electronics*. University of California Press, Berkeley, 1955.
13. J. L. Moruzzi and A. V. Phelps, *J. Chem. Phys.* **45**, 4617 (1960).
14. K. G. Vohra, M. C. Subba Ramu and K. N. Vasudevan, *Atmos. Environ.*, **3**, 99 (1969).
15. L. B. Loeb, *Aerosol Sci.* **2**, 133 (1971).
16. L. B. Allen and J. L. Kassner, Jr., *J. Colloid. Interface Sci.* **30**, 89 (1969).
17. A. C. Montefinale and H. M. Papee, *J. Colloid. Interface Sci.* **22**, 296 (1966).
18. F. S. Dainton and D. C. Walker, *Proc. R. Soc. London, Ser. A.*, **285**, 339, (1965).
19. H. Walter, *Aerosol Sci.* **4**, 1 (1973).
20. A. Meszaros and K. Vissy, *Aerosol Sci.* **5**, 101 (1974).
21. J. M. Rosen, *Proc. Int. Conf. Aersosp. Aeronaut.*, Boston, Mass., May, 1972, 205 (1972).
22. D. J. Hoffmann, J. M. Rosen and T. J. Pepin, in *5th Conference on Laser Radar Studies*, Williamsburg, Virginia, June 1973.
23. K. G. Vohra, in *IAMAP Symposium on Atmospheric Electricity*, XV IUGG, Moscow, August 1971. (Available from the Secretary of IAMAP, Dr. W. L. Godson, Department of the Environment, 4905 Deffrin, Downsview, Ontario, Canada.)
24. K. G. Vohra, in *IAEA Symposium on the Physical Behaviour of Radioactive Contaminants in the Atmosphere*, SM-181/1, Vienna, November 1973.
25. T. T. Mercer and W. A. Stowe, in *Inhaled Particles* **III** (W. H. Walton, Ed.), pp. 839–851. Unwin Bros., Ltd., England, 1971.
26. Edwin J. Hart, Klaus H. Schmidt and Keelathur N. Vasudevan, *Science*, **180**, 1064 (1973).

Effects of Air Pollution On Ecological Processes[1]

Norman R. Glass
and
David T. Tingey
Environmental Protection Agency
National Ecological Research Laboratory
Corvallis, Oregon 97330 USA

INTRODUCTION

A variety of ecological processes are affected and altered by air pollution, both gaseous and airborne particulate. Other pollutants such as ionizing radiation, noise, and waste heat also exert an effect on ecological processes. Such processes as community succession and retrogression (*1-9*), nutrient biogeochemical cycling (*10*), photosynthetic activity, primary and secondary productivity (*1*), species diversity and community stability (*12*, *13*), community respiration (*11*), dominance or abundance of key species within a community (*14*), and a variety of other attributes and processes may all be affected by air pollutants. These ecological processes may be mediated or caused by alterations in a number of physiological processes or mechanisms or by population changes which may occur in the dominant or otherwise important species in a particular ecosystem. The biochemical and physiological basis for change resulting from pollution in the rate and amount of nitrogen fixation in plant communities, as well as other processes which are important to ecosystem change, will be discussed below.

Before embarking on a more thorough discussion of some of these features of ecosystems which are affected by air pollutants, two related points should be mentioned. First, Table I shows an estimate of nationwide emissions of air pollutants by source category for 1970 in millions of tons/year (*15*). From this table, it is easy to see that the larger sources of the critical air pollutants such as SO_x, NO_x, and hydrocarbon are in large part stationary sources and transportation oriented. Furthermore, given the turn of the country toward a coal-based economy as an energy source and away from our more historical dependence upon petroleum fuels such as oil and gas, SO_x and other emissions from stationary sources will increase substantially over the coming years. In the near term (the decade 1975-1985), SO_x and other airborne emissions from power plant sources which are burning coal can be expected to more than double the present nationwide loading of these pollutants.

Second, the public welfare (non health-related) cost of air pollutants has been variously estimated between 50 million dollars/year and several hundred

[1]Congress presentation B-9-4.

million dollars/year. These estimates in general have included only those relatively easily measured considerations such as the known losses to traditional

TABLE I
Estimates of Nationwide Emissions, 1970
(Million tons/year)

Source Category	CO	Part	SO_x	HC	NO_x	Total
Transportation	11.0	0.7	1.0	19.5	11.7	143.9
Fuel combustion in stationary sources	0.8	6.8	26.5	0.6	10.0	44.7
Industrial process losses	11.4	13.3	6.0	5.5	0.2	36.4
Solid waste disposal	7.2	1.4	0.1	2.0	0.4	11.1
Agricultural burning	13.8	2.4	Neg.	2.8	0.3	19.3
Miscellaneous	4.5	1.5	0.3	4.5	0.2	11.0
Total	148.7	26.1	33.9	34.9	22.8	266.4

agriculture from acute air pollution episodes or the cost of having to repaint surfaces more frequently because of air pollution exposure (*16*). No substantial estimates of losses to forest productivity, natural ecosystem productivity which is tapped by domestic grazing animals and wildlife, and other significant dollar losses have been made nationwide. This is partly due to the fact that such measurements are very difficult to make. However, the dollar losses which have been put forward as the consequence of air pollution damage are inordinately low. My estimate is that we could easily conservatively double, or possibly triple, present estimates of dollar loss and still be well within the probable range of costs resulting from air pollution damage to the public's welfare. In particular, dollar losses incurred from long-term, chronic, low level exposure to air pollutants remain virtually unmeasured.

PHYSIOLOGICAL MECHANISMS UNDERLYING ECOLOGICAL EFFECTS

When an air mass containing pollutants such as ozone or sulfur dioxide is transported into a plant community, the gases can be absorbed by the plant (*17*). The plant community can provide a significant sink for gaseous pollutants. If the pollutant occurs in a high concentration for short duration, visual injury will usually be apparent within 24 to 48 hours (acute injury). If the pollutant exposure occurs at a lower concentration and for an extended duration, the effect does not become apparent for varying periods of time depending on the concentration and the length of exposure (chronic injury). Injury from sulfur dioxide can appear as bifacial necrosis that is interveinal and can range from ivory white to red-brown or black lesions(*18*). Ozone injury occurs first in the palisade layer of the mesophyll cells of the upper leaf surface and can appear as discrete pigmented or chlorotic lesions in the interveinal area (*19*). As injury becomes severe, the lesions coalesce and become bifacial. Both species and cultivar differences in foliar sensitivity to pollutant injury exist and the different level of response to a common pollutant dose can alter other plant responses (*18, 19, 20, 21*).

When ozone or sulfur dioxide causes foliar injury on a plant, an alteration in plant growth or other critical growth process such as the ability to reproduce seeds, etc. can occur. In general, the effect of foliar injury will be the greatest if the injury occurs during bloom, fruit set, seed maturation or during the early stages of the growth of storage organs such as roots (*22, 23*). Chronic levels of either ozone or sulfur dioxide that cause little or no visual injury can also cause significant reductions in growth (*24, 25, 26, 27*). In some restricted cases, when plants are deficient in sulfur, low levels of sulfur dioxide that do not cause visual injury can correct the sulfur deficiency and, in this case, aid plant growth (*28, 29*).

The reduction in plant growth induced by ozone or sulfur dioxide is associated with alterations in fundamental plant processes. When either gas diffuses into the leaf, it can cause a depression in photosynthesis rate (*30, 31, 32*). If the pollutant concentration is not high enough to cause visual injury, the photosynthetic rate will gradually return to the control level when the pollutant is withdrawn. Respiration rates are also increased by these pollutants (*32, 33*). The pollutant-induced reduction in photosynthesis and the stimulation of respiration would reduce the amount of carbon available to the plant for use in growth processes and, by implication, the availability of carbon to the next higher trophic level. In those instances where necrotic areas of leaves comprise a significant fraction of the total plant leaf area, or when the pollutant causes premature leaf abscission, the direct loss of photosynthetic surface will decrease primary productivity.

Ozone, even at low levels, is known to induce stomatal closure, reduce transpiration and decrease the water use of crops as well as native vegetation (*30, 34*). In contrast, low levels of sulfur dioxide increase stomatal opening, lower the leaf resistance to water loss, and increase transpiration of plants (*35, 36*). This should also increase the amount of water required to produce crops and the amount of water required to maintain a given plant community in nature. At high sulfur dioxide doses, the stomata are closed, but if injury develops, transpiration is increased until the leaf dies (*37*).

In addition to the alterations in photosynthesis and respiration caused by ozone, it will also alter carbohydrate pools within the plant. Soluble sugars and starches tend to accumulate in the foliage of plants exposed to ozone at levels that do and do not cause visual injury (*38*). Concurrent with the accumulation of carbohydrates in the leaves, there is a reduction in the levels of soluble sugars and starches in the roots. These alterations in carbohydrate pools may also effect the growth and stress tolerance of plants.

Ozone and sulfur dioxide can cause an increase in the amino acid levels in leaves which in some cases is reflected in the seed (*39, 40, 41, 42*). In addition, in plants exposed to sulfur dioxide, there is an increase in the levels of cysteine and glutathione (*37*). Ozone treatment is also known to increase the levels of phenols, anthocyanins, and glycosides in exposed tissue. These compounds, in addition to their association with pigmented lesions, can also effect plant quality.

The pollution-induced changes in metabolite pools are reflected in several additional changes in plant health. Leachates from fescue leaves exposed to ozone were inhibitory to the nodulation of clover (*46*). Rabbits fed on clover exposed to sulfur dioxide did not grow as large as rabbits fed on unexposed clover (*37*). Root exudates from fescue plants exposed to ozone reduced the top and root dry weight of clover plants and decreased the number of nodules (*46*).

Chronic levels of ozone can reduce the number of flowers formed on the plants, reduce the germination of pollen and, therefore, reduce the number of seeds (*47, 48, 49*). Sulfur dioxide can also reduce pollination and pollen germination (*37*). In addition, the stigma can also be directly injured by sulfur dioxide (*37*).

Both acute and chronic ozone exposures have been reported to reduce or inhibit the nodulation of legumes (*50, 51*). The nodules that form on the exposed plants are similar in size and leghemoglobin content, suggesting that the nitrogen fixation efficiency of each individual nodule was not impaired, but with fewer nodules/plant there would be less nitrogen fixed per plant. In pinto beans, concurrent with the decrease in nodule number there is a quantitative increase in the number of fungi on the root and hypocotyl, but there were no qualitative changes in the fungal association.

Following exposure to ozone, competition among pasture species such as clover and a variety of grasses, results in altered competitive relationships and community composition. This leads to changes in the final plant community which results from plant competition depending upon exposure and the levels of exposure of that community to oxidant air pollution.

From this brief description of several possible physiological mechanisms whereby ecological changes may occur, it is possible to infer a number of causal mechanisms which may be operating to alter ecological relationships.

ECOLOGICAL EFFECTS

A number of ecological effects, some manifestations of the above physiological processes, are known to occur following exposure to a variety of pollutant types. One of the most well-known studies of ecosystem changes which result from pollution is that conducted in the Brookhaven Experimental Forest (*5–9*). In this study over the past few years, documentation has been developed for the retrogression of an eastern forest community from a climax broadleafed forest form to a progressively simpler and lower growth habit community in the immediate vicinity of a ^{137}Cs source. Similar cases of retrogression of plant and soil communities near concentrated pollution sources such as power plants, smelters, and other heavy industrial sources are also known.

This simplification of plant, animal, and soil communities is approximately the reverse of the successional stages which occurred at that site in earlier times. There is a simplification of the biotic community in the sense of a reduction in species diversity, standing crop, productivity, and other ecosystem parameters.

Succession in terrestrial communities tends to have a number of characteristics, many of which are reversed or in other ways altered or simplified because of air pollution effects. Some of these characteristics are (*2, 10, 12, 52*):

1. A progressive increase in the height of large growth forms such as the dominant trees in a forest;

2. An increase in the types of species present (an increase in species diversity);

3. An increase in the structural complexity of the community, usually in the form of more clearly defined community strata such as overstory, low shrubs, and ground cover;

4. A progressive increase in soil depth and organic matter content;

5. An increase in community productivity and respiration;

6. Increasing biomass;

7. Increasing stability of species populations;

8. An increase in the amount of nutrients and other material tied up in the biomass (*53*).

A second example of an ecological process which is adversely affected by air pollution impact is that of nitrogen fixation. Nitrogen, of course, is fixed by bacterial processes in legume root nodules, by free living bacteria, by nitrogen-fixing algae in the algal portion of epiphytic and other lichens, and is fixed by nonleguminous plants such as red alder. While there are several aspects of the nitrogen cycle in ecosystems which are important (for example, the conversion of atmospheric nitrogen to a form which is usable by biota, the transport, distribution and uptake of nitrogen in an appropriate form, etc.) we will deal here only with the nitrogen fixation process itself, a subelement of the nitrogen cycle. As mentioned above under the Physiological Effects section, nitrogen fixation in terms of total quantity as well as rate is affected by oxidant air pollution by means of a reduction in legume root nodule number. Since the number of root nodules is reduced, the total biomass of nitrogen-fixing bacteria is also reduced and the total quantity of nitrogen fixed per unit time from the atmosphere is probably thereby reduced as well. Another aspect of nitrogen fixation which has been given little attention in the literature conerns the fixation of atmospheric nitrogen by several varieties of lichen.

For example, *Lobaria pulmonaria,* a nitrogen fixing lichen, is affected even by such a low level of sulfur dioxide exposure as one-fourth of the annual average sulfur dioxide concentration considered to be safe and still protect the public welfare. This amount of SO_2 has been sufficient to discolor and alter the functioning of this epiphytic lichen. On the western slope of the Cascades in old growth Douglas-fir forest, this lichen represents essentially the sole source of nitrogen fixation. A more detailed accounting of this process and of its implications for forest production in the Northwest will be made in the very near future (Dr. William Denison, personal communication). However, the International Biological Program, based at Oregon State University, has determined that the balance of nitrogen in the Andrews State Experimental Forest near Eugene is in steady state with respect to nitrogen input and output except for the input of atmospheric nitrogen fixed by *Lobaria pulmonaria*. We believe that it is safe to conclude that if this lichen were to disappear from those forest areas where it is the sole nitrogen fixer, that growth in terms of above ground tree growth would virtually stop. In other words, the effect of an atmospheric air pollutant such as sulfur dioxide (and probably oxidant air pollution as well) could alter forest production by means of an indirect effect on the nitrogen fixation process.

In a recent study, Whittaker, et al. (*1*) estimate that, for the Hubbard Brook forest in New Hampshire during the period 1956-1960, net ecosystem production was 350 $g/m^2/yr$ above ground and 85 g/m^2yr below, whereas in the period 1961-1965 above ground production was 238 $g/m^2/yr$ with 52 $g/m^2/yr$ below. The 32 percent above ground and 39 percent below ground reductions in production were attributed to possible air pollution (increasing acidity of rainfall) effects and drought.

SUMMARY

It is clear that oxidant pollution can affect ecosystems by adversely influencing the nitrogen fixation process and in other ways as well (*54*). SO_x, as shown above, also influences nitrogen fixation but, as with oxidant pollution, affects a number of other ecosystem processes in addition.

Oxidant air pollution and SO_x emissions are the major air pollutants from transportation-oriented and stationary sources, respectively. In the Western United States, the preponderant air pollution problem is associated with oxidants. In the Eastern United States, the dominating forms of air pollution are SO_x, NO_x, and oxidant. Ecosystems in all parts of the country are being affected in one way or another by these two types of air pollution emissions (*1, 54*). The mechanisms whereby air pollutants exert an impact on natural and man-made or managed ecosystems are numerous. Two possible mechanisms which have been examined in the field in the laboratory were presented above.

In order for an ecosystem to remain extant, a measure of system stability must be achieved and retained. Stability is related to the number and diversity of living forms in conjunction with their abiotic environment, and any perturbing influence such as occurs with pollution exposure tends to disrupt or distort local ecosystem stability. While widely differing types of pollution (radiation, dissolved chemicals, gaseous, etc.) generally tend to produce the same type of ecosystemic change in structure and function, air pollution provides a specific example of the means whereby a local ecosystem may undergo retrogression, simplification, instability, lowered productivity, and approaches the successional state (perhaps not in detail) which existed at that place at an earlier time.

Since animals, including man, depend upon system stability and are themselves a part of that stability, it is clear that if communities and whole ecosystems are to remain in viable, relatively constant form, the input of atmospheric and other pollutants must be minimized. If stability in ecosystems is desirable then the level of loading of pollutants to those systems must be reduced.

REFERENCES

1. R. H. Whittaker, F. H. Bormann, G. E. Likens, and T. G. Siccama. *Ecol. Monog.* **44**:233 (1974).
2. R. H. Whittaker, *Ecol. Monog.* **21** :41 (1953).
3. R. H. Whittaker, Brookhaven Symp. Biol. **22** :178 (1969).
4. R. H. Whittaker, *Communities and ecosystems*. MacMillan, New York, 1970.
5. G. M. Woodwell, Science **138** :572 (1962).
6. G. M. Woodwell, Science **156** :461 (1967).
7. G. M. Woodwell, Science **168** :429 (1970).
8. G. M. Woodwell, and R. H. Whittaker. Quart. Rev. Biol. **43** :42 (1968).

9. G. M. Woodwell, in *Ecological and Biological Effects of Air Pollution*, pp. 10–17. MSS Information Corp. 1973.
10. E.P. Odum, Science **164** :262 (1969)
11. G. E. Likens, F. H. Bormann, N. M. Johnson, D. W. Fisher and R. S. Pierce. *Ecol. Monog.* **40** :23 (1970.
12. R. Margalef, Amer. Nat. **97** :357 (1963).
13. R. Margalef, *Brookhaven Symp. Biol.* **22** :25 (1969).
14. R. H. Whittaker, *Science* **147** :250 (1965).
15. J. H. Cavender, D. S. Kircher, and A. J. Hoffman. Nationwide air pollutant emission trends 1940-1970. Publ. No. AP-115, U. S. Environmental Protection Agency, 1973.
16. T. E. Waddell, The economic damages of air pollution. U. S. Environmental Protection Agency, 1974.
17. A. C. Hill, 1971. *APCA Journal* **21** :341 (1971).
18. T. W. Barrett, and H. M. Benedict, in *Recognition of Air Pollution Injury to Vegetation: A Pictorial Atlas* (J. S. Jacobson and A. C. Hill, Eds.). Air Pollution Control Association, Pittsburgh, Pennsylvania, 1970.
19. A. C. Hill, H. E. Heggestad and S. N. Linzon, in *Recognition of Air Pollution Injury to Vegetation*: A Pictorial Atlas, (J. S. Jacobson, and A. G. Hill, Eds.). Air Pollution Control Association, Pittsburgh, Pennsylvania, 1970.
20. R. A. Reinert, D. T. Tingey, and H. B. Carter. *Amer. Soc. Horticult. Sci.* **97** :711 (1972).
21. V. L. Miller, R. K. Howell, and B. C. Caldwell. *Environ. Qual.* **3** :35 (1974).
22. A. H. Heagle, 1974. Personal communication.
23. D. T. Tingey, J. A. Dunning, and G. M. Jividen, in *Proceedings of the Third International Clean Air Congress* A154-A156, 1973.
24. D. T. Tingey, W. W. Heck, and R. A. Reinert. *J. Amer. Soc. Hortic. Sci.* **96** :369 (1971).
25. R. Guderian and H. Stratmann, *Forschungberichte des Landes Nordrhein-Westfalen*, No. 1118, pp. 7–102. Westdeutscher Verlag, Köln u. Opladen, 1968.
26. J. N. B. Bell, and W. S. Clough. *Nature* **241** :47 (1973).
27. D. T. Tingey, and R. A. Reinert. *Environ. Poll.* (in press).
28. M. Faller, *Landwint. Forsch. Sonderh.* **25** :48 (1970).
29. C. Setterstrom, P. W. Zimmerman, W. Crocker. *Boyce Thompson Inst. Contrib.* **9** :179 (1938).
30. A. C. Hill, and N. Littlefield. *Environ. Sci. Technol.* **3** :52 (1969).
31. J. H. Bennett, and A. C. Hill. *Environ.* Qual. **2** :526 (1973).
32. R. L. Barnes, *Environ. Poll. 3 :133 (1972).*
33. M. Katz, *Effect of Dilute Sulfur Dioxide on Alfalfa. Trail Smelter Question.* Appendix DDI, 265 pp. National Research Council of Canada, 1935.
34. C. R. Thompson, O. C. Taylor, M. D. Thomas and J. O. Ivie. *Environ. Sci. Tech.* **1** :644 (1967).
35. T. V. Biscoe, M. H. Unseworth, H. R. Pincney. *New Phytologist* **72** :1299 (1973).
36. T. A. Mansfield. *Commentaries in Plant Science* **2** :11–20; in *Current Advances in Plant Science* Vol. 2. Robt. Maxwell, Oxford, 1973.
37. K. Garber, 1967. *Luftverunreinigung und ihre Wirkungen.* Gebruder Born Traeger. Berlin-Nikolassee. 279 p.
38. D. T. Tingey, in *Air Pollution Related to Plant Growth* (W. M. Dugger, Ed.), American Chemical Society Symposium Series (in press).
39. D. T. Tingey, R. C. Fites, C. Wickliff. *Physiol. Plant. 29 :33 (1973).*
40. I. P. Ting, and S. K. Mukerji. *Am. J. Bot.* **58** :497 (1971).
41. U. Arndt, *Staub-Reinhalt. Luft* **30** :28 (1970).
42. J. E. Frey, *The Effects of Ozone on the Synthesis and Storage on Nutrients in Soybean Seeds.* Ph.D. Thesis. University of Northern Colorado, Greeley, Col., 1972.
43. R. K. Howell in *Air Pollution Related to Plant Growth* (W. M. Dugger, Ed.), American Chemical Society Symposium Series (in press).
44. J. Koukol, W. M. Dugger, Jr. *Plant Physiol.* **42** :1023 (1967).
45. H. A. Menser and J. F. Chaplin. *Science* **14** :169 (1967).
46. M.Kochhar, *Plant-Plant Interactions and the Growth of Ladino Clover and Tall Fescue as Modified by Ozone and/or Rhizoctonia solani.* Ph.D. Thesis. North Carolina State University, Raleigh, N. C., 1973.
47. W. A. Feder, *Science* **160** :1122 (1968).

48. W. A. Feder and F. J. Campbell. *Phytophathol.* **58** :1038 (1968).
49. W. A. Feder and F. Sullivan. *Science* **165** :1373 (1969).
50. D. T. Tingey, and U. Blum. *J. Environ. Qual.* **2** :341 (1973).
51. W. J. Manning, W. A. Feder, P. M. Papia, and I. Perkins. *Environ. Pollut.* **1** :305 (1971).
52. R. H. Whittaker and G. M. Woodwell, in *Ecosystem Structure and Function,* (John A. Wiens, Ed.), p. 137, Oregon State University Press, 1972.
53. G. E. Likens and F. H. Borman, in *Ecosystem Structure and Function,* (John A. Wiens, Ed.), p. 25, Oregon State University Press, 1972.
54. O. C. Taylor and P. R. Miller. *California Air Environ.* **4** :1 (1973).

SUMMARY PLENARY SESSION[1]

Chairman: G. F. Whitmore

[1]Summary Plenary Session F–1

Physics[1]

H. E. Johns
Ontario Cancer Institute
Toronto, Canada

As I understand it, my task is to interpret for you the implications of the physics papers which have been presented at this conference. In many cases, an arbitrary decision has been made as to whether a paper falls into physics, chemistry or radiobiology. To illustrate this, I remember at one session on solvated electrons, I found Baxendale, who accused me of piracy! I decided to leave the solvated electron to him.

I cannot possibly summarize every Physics paper, so I have chosen to review a few areas of Physics and try to give you an inkling of the status of these fields and where developments are likely to occur in the next few years.

Nuclear Energy

There are three papers at this conference that dealt with nuclear power and its implications. Dr. A. M. Weinberg (*1*) gave a summary of the nuclear power program of the United States. All of us are intensely aware of the problem of energy shortage, and I think no one at this meeting would disagree that we must continue a vigorous program of development of nuclear energy through fission. In fact, I think we would also agree we have no other practical way of meeting our energy needs for the next half century. Dr. Weinberg reviewed other forms of energy production, but concluded, to quote him "we cannot yet count on these eventualities and it seems at least highly plausible that man will have no choice but to learn to live comfortably with fission breeders."

Dr. Weinberg then discussed a number of possible ways of dealing with the environmental and health problems arising from breeder reactors, and suggested that if the actinides were separated from the lanthinides the problem of long-lived activities in the environment would certainly be eased. He also put forward the idea that reactors should be bunched together in nuclear energy parks. Finally, he mentioned a rather far out idea that man's biological system might be taught to repair ionizing radiation damage the way certain bacteria repair ultraviolet induced damage. As a Canadian, I would have been happy to have heard Dr. Weinberg talk about heavy water reactors, since our Canadian press has led us to believe that these are the most advanced in the world.

We were privileged to hear a discussion on the future of fusion by Dr. R. L. Hirsch (*2*) and Dr. S. Yoshikawa (*3*). These two papers gave us a picture of the status of fusion relative to controlled power production. We were told that fusion will indeed be scientifically feasible sometime within the next 25 years.

[1]Congress presentation F-1-1 (Summary).

To achieve fusion it is necessary to satisfy three conditions simultaneously. One needs a high temperature plasma, a large enough volume of plasma at this temperature, and these conditions for a sufficient length of time. Pilot machines have achieved pairs of these conditions but no machine has yet been created in which all three conditions are satisfied simultaneously. In one class of machines call the Tokamak, the plasma is contained in a toroid, which is excited by very large currents in a superconducting winding. The heat energy from the device is transferred to a turbine with hot helium gas. In the Japanese machine, for example, the inner diameter of the toroid is 2 meters, the outer diameter 6 meters. The current in the superconducting coils is 3×10^6A and the plasma has a temperature of 50×10^6 °C. The Japanese are hoping to have their prototype machine operating by 1980.

Both speakers emphasized the advantages of fusion power. Some of these are: an essentially infinite fuel supply at low cost; safety—no runaway; no chemical combustion products; low radioactivity. In particular, Dr. Yoshikawa pointed out that Japan is a small country with a large population where the energy problems are very severe—it needs fusion power. At the meeting there was some hostile heckling suggesting that enough money was not being spent on fusion. Dr. Hirsch suggested that more money might be a hindrance to rapid development, since a crash program is not feasible until the basic problems of fusion are understood.

Platzman Physics

At this conference there were groups of physics papers which I can best refer to as 'Platzman Physics'. Bob Platzman was a member of *Radiation Research Society* for many years and was well known to most of you. His untimely death on July 2, 1973, is mourned by us all. His program is continuing under the able direction of his students and friends, and Dr. Inokuti, one of his students, was asked to organize a symposium dedicated to his memory. Now, what is 'Platzman Physics' all about? For many years, he studied *in detail* the ionization of gases. All of us have used ionization in gases as a tool for measuring exposure, but Bob was interested in the problem at a much more detailed level. He wanted to know all of the types of ions produced, when either an electron or a photon bombards an atom or molecule. He wanted to know how the electron bounced from the collision, where it went, how much energy it lost, how far it went to the next collision, what types of excited states were left behind, etc. In much of radiation biology we have focused on the amount of energy absorbed—this, of course, is the basis of the rad—Platzman was interested in *how* the energy was absorbed. In all of these calculations, the nature of the electronic orbitals is important and the wave functions for the atom must be used—this is complicated enough when one deals with simple atoms, but when it is extended to molecules and molecular orbitals the problem becomes very difficult. Although most of Platzman's work dealt with the gaseous state he had started to extend his ideas to the liquid state which is, of course, of more relevance to radiobiology. None of the questions outlined above is easy to answer, but they were wrestled with in the Symposium on "The Basic Physics of the Interaction of Radiation with Matter" by M. Inokuti, J. Berkowitz, M. J. van der Wiel, Y. K. Kim, A. Niehaus, F. T. Smith, C. Ottinger and P. Kebarle (these Proceedings). These people dealt with specialized aspects of the problem which are very difficult to summarize, but one had the feeling that workers in this field will soon be able to catalogue the numbers and types of fragments set in motion when ionization is produced in biological molecules in condensed systems.

Auger Explosions

In several of the sessions [e.g. Berkowitz (*4*)], frequent references were made to "Auger explosions," so rather than discuss these papers individually, I will talk about the problem in general. In the Auger effect schematically represented in Fig. 1, a photoelectron is first ejected from the K or L shell by a photon, leaving a K or L "hole". In high Z material the hole is *usually* filled by an outer electron with the *emission* of fluorescent radiation. In low Z material, however, fluorescent radiation is not usually emitted but rather an Auger electron. One can think of the fluorescent photon, on trying to emerge from the atom, as ejecting an electron from one of the outer shells and this electron is called an Auger electron. Of course, its ejection in turn leaves a second "hole." Now, each of these two holes may again be filled, giving rise to other Auger electrons. *Thus, one K absorption may give rise to many Auger electrons and this is called an Auger shower.* Dr. Y. Feige (*5*) presented a slide for xenon which showed that up to 23 Auger electrons could be ejected from xenon with the most probable value being 8–10. The ejection of 10 Auger electrons occurs in about 20% of the xenon atoms after ejection of a K electron.

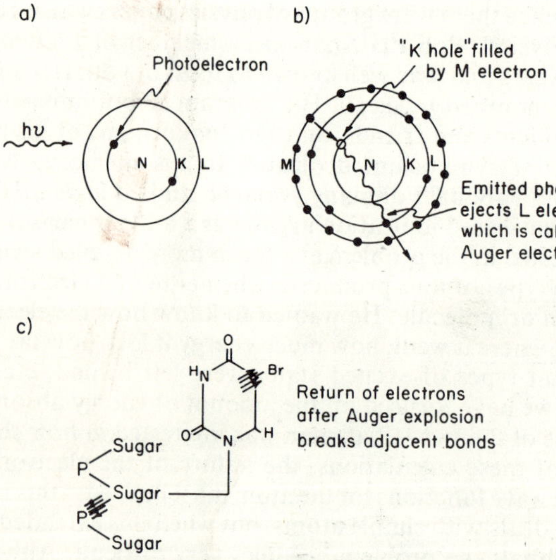

Figure 1.

The possible consequences of Auger explosions are just now being appreciated—some of the papers at this Congress which dealt with this problem are J. Berkowitz (*4*), Y. Feige and A. Gavron (*5*); H. J. Burki (*6*), B. R. Young, H. J. Burki and R. B. Painter (*7*); B. Diehn, A. Halpern and G. Stöcklin (*8*); A.

Halpern and G. Stöcklin (*9*); L. E. Feinendegen, G. Tisljar-Lentulis and V. P. Bond (*10*). There are two major effects which could arise from the Auger explosion.

a) Large number of low energy electrons. The Auger explosion gives rise to many low energy electrons which have a high LET—these could produce extra damage and should be taken into account in radiobiology calculations. For example, equal absorption of energy from incorporated ^{131}I and ^{125}I may give rise to very different biological effects since ^{125}I decays by internal conversion giving rise to Auger electrons which have a high LET, while ^{131}I decays by β^- emission.

b) Disruption of bonds. A more important effect may be a chemical alteration resulting from the Auger cascade. This is illustrated in Fig. 1c. The cascade gives rise to the ejection of many electrons—when these return to the parent atom the neighboring electron environment will be altered and bonds broken as suggested in the figure. In DNA, for example, one can preferentially excite the phosphorus atom by ejection of a K electron and so bring about an Auger shower which can break the sugar phosphate bond. Similarly, if the DNA were altered by the substitution of 5-bromouracil for thymine, specific damage could be produced by excitation of the K electron of bromine giving rise to an Auger cascade. We can look forward to exciting developments in this field in the next few years.

An interesting symposium arranged by the ICRU took place. The first paper by R. S. Caswell, L. J. Goodman and R. D. Colvett (*11*) dealt with neutron dosimetry. Fourteen organizations came to Brookhaven with a variety of measuring equipment and measured the neutron flux at 5 neutron energies. Their dosimetry agreed to within about ±10%. It was my impression that the tissue equivalent chamber was the most favored instrument. In addition to the paper on the intercomparison, there were a number of other papers dealing with certain aspects of neutron dosimetry.[2] One paper by N. A. Atari, K. V. Ettinger and J. R. Mallard of the University of Aberdeen (*12*) and given by Dr. Dennis was rather intriguing. It described lyoluminescence—a new term to me. Certain materials on being exposed to radiation in the crystal state store up charge in the color centers. When the crystals are later *dissolved* in water or other solvents light is emitted—and this emitted light is proportional to the radiation dose. Lyoluminescence is less sensitive than thermoluminescence by a factor of about 10^3, but has the advantage of being present in many "biological" molecules such as sugars and so offers real tissue equivalence.

Dr. R. Oliver (*13*) reviewed the progress made by a working committee of the ICRU in writing a revised handbook on Radiological Dosimetry. He made a plea for comments and suggestions concerning their handbook before its final publication.

Finally, we heard an interesting paper on particle dosimetry using track etching by Dr. R. L. Fleischer (*14*). The passage of ionizing particles through many insulating materials causes local damage along the track which can be made visible by chemical treatment. The length of treatment to make a track visible depends upon the type of particle which caused the damage. Thus, the method can be used to determine the *number* and *type* of particles which bombarded the foil. Dr. Fleischer gave a number of interesting applications of

[2]For instance, Session D-30: Dosimetry—Neutrons. *Radiat. Res.* **59**, 226–228 (1974).

the method. He showed how track counting could be used to detect the escape of radon from the soil and thus locate uranium deposits. The method has been used to study radiation hazards in space. The inside of the astronaut's plastic helmet can be treated chemically to determine the number and type of particles which passed through the helmet and the astronaut's head during flight.

A symposium on Particle Penetration Phenomena was presented.[3] This is a complex problem of rather specialized nature, but involved a number of exciting talks. Dr. P. Sigmund (*15*) laid the foundation by describing how a particle of energy E on passing into a foil emerges with a distribution of energies about $E-\overline{\Delta E}$ where $\overline{\Delta E}$ is the average energy loss. The chairman, Dr. W. Brandt, and Dr. R. H. Ritchie (*16*) discussed how the energy losses took place in the track. In a region of radius 20Å about an electron track some 60% of the energy is absorbed—this region is called the infrastructure. Outside this core and extending to about 1000 Å is the ultrastructure where some 40% of the energy is absorbed. This outer portion is the region dealt with by the Platzman method. It is nice that the two groups have divided the available "energy absorption pie" into more or less equal halves. When an ionizing particle passes through water, the region of the infrastructure (the core) oscillates, giving rise to an energy loss of about 21 eV—this energy loss is called the plasma peak. This is a complex problem, but one had the feeling that the group was getting a real understanding of the passage of ionizing particles through matter.

Dr. S. Datz (*17*) gave an interesting talk on channeling. If a charged particle is directed along or nearly along an axis of a crystal, it passes through the crystal between the crystal planes. The transmission of a beam through a crystal thus depends on the orientation of the crystal. The transmission can be measured and used to accurately align the crystal axes. By directing a charged particle beam at the crystal, impurities can be imbedded at specific points. The method can be used to study atomic potential inside a crystal and to detect impurities by studying backscatter.

Dr. J. Remillieux (*18*) showed how charged species such as H_2^+ ions with a few MeV energy can be passed through gold foils a few hundred angstroms thick. The yield is low (10^{-6}) but the surprising thing is that any molecular particles emerge from the foil. The charged particles pass through the crystal along channels and the optimum channel is not necessarily the widest one. Experiments suggest that at the first surface the molecular ion breaks into two protons which pass down the channel independently and then recombine at the rear surface of the foil. It is an intriguing subject.

There was one paper on X-ray spectra by N. K. Sherman, K. H. Lokan, R. M. Hutcheon and W. Funk (*19*) that deserves mention. These workers have developed an accurate method to obtain the spectrum of X-rays emitted from linacs. They bombard the target material with high energy electrons and remove the transmitted electrons from the beam by a magnetic dump. The X-ray photon of energy $h\nu$ then strikes a liquid deuterium target where a proton, p, and neutron, n, are emitted with equal energy in opposite directions. The neutron energy is measured by a time of flight spectrometer set up at right angles to the X-ray beam. Under these conditions

[3]See these Proceedings.

$$h\nu = 2.2 \text{ MeV} + (E_n + E_p) = 2.2 \text{ MeV} + 2\ E_n$$

With this spectrometer and the known photodisintegration cross sections as a function of energy, they were able to determine the spectra of the X-ray beam produced with different targets with different filters, and so determine the optimum conditions for X-ray production in linacs.

This problem is of considerable practical interest since in Toronto we have shown that a 25 MeV linac with a thick tungsten target and tungsten filter gives the same depth dose as a 16 MeV betatron using a thin tungsten target plus aluminium filter. If a 25 MeV linac is made with a *thick aluminium target* and *aluminium filter* its dose rate is essentially unchanged but, from a depth dose point of view, it is equivalent to a 25 MeV betatron. [For details see recent issues of the *Am. J. Roentgenol.*]

A review of the physics papers at this conference would not be complete without some mention of negative π mesons. As most of you know, there are four planned facilities in the world with sufficiently intense beams to be useful in radiotherapy. These are at Los Alamos, Zürich, Vancouver and Stanford. Numerous papers were presented on this and related subjects,[4] but here I report on a conversation with Dr. Dicello of the Los Alamos Group who showed me some of the unreported results of their biomedical group. Their π^- beam gives 7 rad/min at 60 cm with some 15% electron and 15% muon contamination. The surface dose is about 40% of the peak dose at 18 cm depth, and the RBE in the peak is about 2. These results were in agreement with the expected values so it is probable that the other facilities will also obtain beams according to their plans. We must await the next conference to hear about the usefulness of π^- beams in radiotherapy.

In spite of the many interesting papers on physics which were presented, I would like to draw attention to a number of fields of radiation research and physics which were not discussed at all at this conference. No papers were presented on new methods of using X-rays as a diagnostic tool—to produce images of internal organs. Exciting developments have taken place in this field in electrostatic imaging. Another field which received no attention was the new method of scanning, using X-ray pencils as in the EMI scanner. Perhaps such topics will be covered at the next Congress.

REFERENCES

1. A. M. Weinberg, these Proceedings. [Congress presentation A-1-1.]
2. R. L. Hirsch, these Proceedings. [B-1-1.]
3. S. Yoshikawa, these Proceedings. [B-1-2.]
4. J. Berkowitz, these Proceedings. [B-10-2.]
5. Y. Feige and A. Gavron, these Proceedings. [D-7-3.]
6. H. J. Burki, *Radiat. Res.* **59,** 262 (1974). [E-13-4.]
7. B. R. Young, H. J. Burki and R. B. Painter, *Radiat. Res.* **59,** 262 (1974). [E-13-5.]
8. B. Diehn, A. Halpern and G. Stöcklin, *Radiat. Res.* **59,** 273 (1974). [E-18-4.]
9. A. Halpern and G. Stöcklin, *Radiat. Res.* **59,** 279 (1974). [E-22-1.]
10. L. E. Feinendegen, G. Tisljar-Lentulis and V. P. Bond, *Radiat. Res.* **59,** 298 (1974). [E-30-8.]
11. R. S. Caswell, L. J. Goodman and R. D. Colvett, these Proceedings. [D-7-1.]

[4]See Symposium on "The Use of Heavy Particles in Radiotherapy" (these Proceedings) and Session E-26: "Dosimetry—Heavy Charged Particles," *Radiat. Res.* *59,* 288–291 (1974).

12. N. A. Atari, K. V. Ettinger and J. R. Mallard, *Radiat. Res.* **59,** 224 (1974). [D-29-5.]
13. R. Oliver, these Proceedings. [D-7-2.]
14. R. L. Fleischer, these Proceedings. [D-7-4.]
15. P. Sigmund, these Proceedings. [E-2-1.]
16. R. H. Ritchie and W. Brandt, these Proceedings. [E-2-4.]
17. S. Datz, these Proceedings. [E-2-2.]
18. J. Remillieux, these Proceedings. [E-2-3.]
19. N. K. Sherman, K. H. Lokan, R. M. Hutcheon and W. Funk, *Radiat. Res.* **59,** 224 (1974). [D-29-4]

Chemistry[1]

J. H. Baxendale
Chemistry Department, University
Manchester, Great Britain

BIOLOGICAL SYSTEMS

Examination of the program of the Congress leaves little doubt that among the wide variety of systems to which a radiation chemistry approach has been applied, most activity is being shown in those areas involving biological materials. In fact, about 40% of the contributions in radiation chemistry have been of this nature. This, I believe, reflects the appreciation of the fact that now the basic species and processes of aqueous radiation chemistry are more or less understood, a powerful tool is available for application to biological systems.

There have been papers on the whole range of biological materials, from sugars to enzymes, which have exploited the reactions of the species HO·, H· and e^-_{aq} both in steady state and pulse radiolysis conditions. For example, there is interest in the redox behavior of various iron-heme-protein systems (*1–3*), and activity continues in the investigation of the active sites of enzymes by pulse radiolysis studies of their reactions with electrons and free radicals (*4–9*).

A promising development is the attempt to simulate the behavior of the chemically complex biological membranes by using micellar systems formed from synthetic surfactants such as sodium lauryl sulfate or cetyltrimethyl ammonium bromide. These micelles, which are aggregates of, say, 100 molecules, present a non-polar interior of hydrocarbon surrounded by a polar surface made up of positive or negative ions. Reactive solutes are added which when non-polar are incorporated inside the micelle and when polar may concentrate in the ionic surface.

Two types of experiment have been done. In one the effect of the micelles on the radiation chemistry of the solutes has been studied (*10*), and it is reported, for example, that the yields of radiation products from uracil and thymine are significantly affected by the micellar environment. The second type involves the use of flash photolysis or pulse radiolysis to follow the reactions of materials (*11–12*), either inside or on the micellar surface, with say e^-_{aq} or with additives (e.g. quenchers of excited states). Effects of surface charge can be observed and the rates of diffusion through the outer layer measured. Preliminary extensions to the study of more complex systems, e.g. micelles incorporating globular proteins and cell membranes themselves, have already been made, and further development in this field seems likely.

[1]Congress presentation F-1-2 (Summary).

FREE RADICALS

A major part of the information we have about transient free radicals in the liquid phase has been obtained using radiation chemistry methods and in particular pulse radiolysis. Free radical properties such as reactivities, spectra, acid-base behavior, etc. form the basic information which is used to understand the radiolysis behavior of many systems. Another technique introduced recently and reported on here promises to provide knowledge of the very important oxidation-reduction properties of free radicals, i.e. their tendency to give and take electrons. This is the application of the well-established polarographic analysis to short-lived species produced by pulse radiolysis.

Conventional polarography follows the effect of electrical potential on the ability of a stable solute to give and take electrons from a mercury surface. This has now been extended to cover transient solutes which exist only for microseconds (*13*). For a free radical R the following reactions may occur at the mercury surface:

$$R^+ \underset{}{\overset{-e^-}{\rightleftarrows}} R \overset{+e^-}{\rightleftarrows} R^-$$

and the processes are followed by measuring the current which flows to the surface as a function of the applied potential, so that a current-voltage curve (or polarographic wave) is generated.

The form of these curves is determined by several factors, but in general the more negative the potential—ideally the half-wave potential—which is required to prevent electrons being donated by R, the higher is the reducing potency of R. Similarly, the more positive the potential required for R to accept an electron, the higher its oxidizing potency. Unfortunately, these potentials are not thermodynamic quantities since the processes in general are not reversible, but half-wave potentials of stable molecules have in the past proved extremely useful and promise to be equally so for free radicals. Figure 1 shows examples of the current-voltage curves for the free radicals generated by the HO· oxidation of cyclopentane and caproic acid.

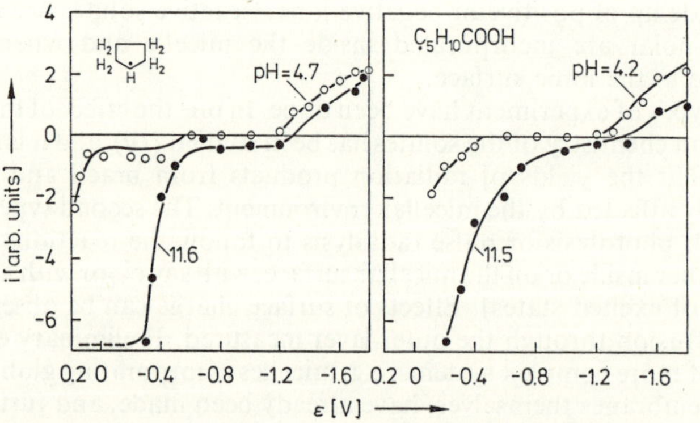

FIG. 1. Polarograms at various pH of the cyclopentyl radical (left) and the mixture of radicals produced from caproic acid. [Reproduced from K. M. Bansal, A. Henglein and R. M. Sellars, *Ber. Bunsenges. Physik. Chem.* **78**, 569 (1974), with the authors' permission.]

ELECTRONS IN GLASSES AT LOW TEMPERATURES

Electrons in a variety of systems continue to attract a lot of attention, and in addition to symposia specifically devoted to electrons in liquids, there have been many other contributions concerned with electron properties and behavior.

New work on the reactions of electrons (*14–16*) trapped in low temperature glasses—aqueous, alcohols and hydrocarbons at 77° K or below—studied by light absorption, emphasizes the importance now attributed to the ability of electrons to move from one place to another by a quantum mechanical tunnelling process as opposed to normal diffusion. The evidence for this is that in these conditions electron reactions extend over a very long time scale—several orders of magnitude—and have very little temperature dependence. Figure 2 shows the kinetics of electron transfer from the anion of biphenyl, Ph_2^-, to triphenyl ethylene in an ethanol glass at 77° K, a reaction which is also considered to proceed by electron tunnelling (*17*).

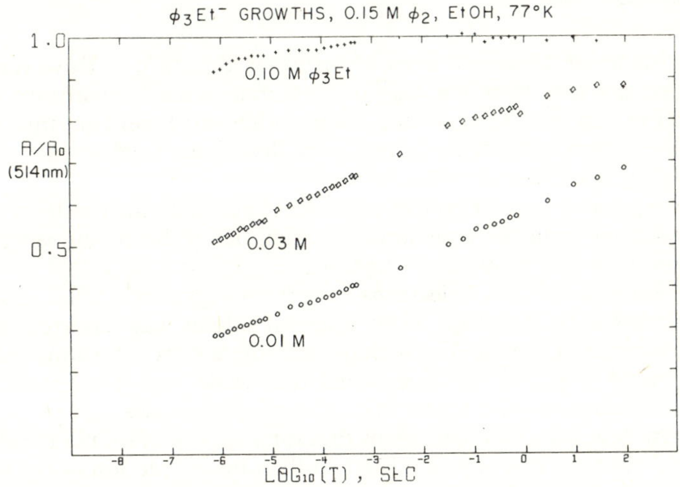

FIG. 2. Kinetics of transfer of e^- from biphenyl anion to triphenyl ethylene in an ethanol glass at 77° K. [Courtesy J. R. Miller, cf. ref. *17*.]

ELECTRONS IN WATER

In liquids at room temperature, observations on electrons which extend into the picosecond time range have produced interesting observations on both yields and reactivities (*18–20*).

Figure 3 summarizes the data on hydrated electron yields in water at different times following a pulse of radiation. After microseconds, $G(e_{aq}^-) = 2.7$ is now generally agreed as the homogeneously distributed (or "free") ion yield. At shorter times the yield increases since e_{aq}^- in the spur is now being measured, but this appears constant from some tens of picoseconds up to 1 ns. The value $G(e_{aq}^-) = 4.0 \pm 0.2$, as in the figure, was agreed on originally in two laboratories but it was reported that this may now be 4.5.

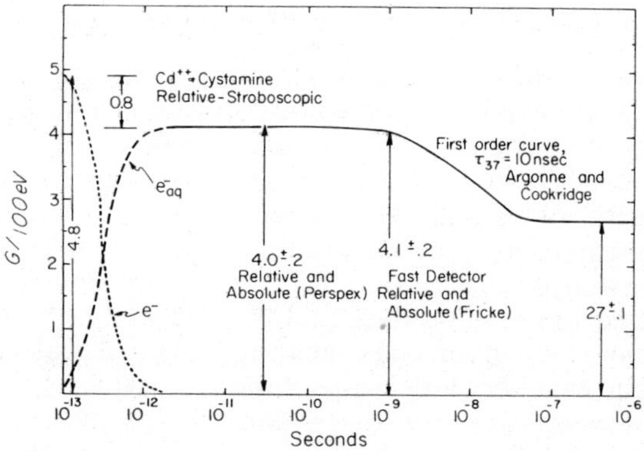

FIG. 3. Time scale of electron yields in water at various times. [Reproduced from R. K. Wolff, J. E. Aldrich, T. L. Penner and J. W. Hunt, *J. Phys. Chem.* (in press) with the permission of the American Chemical Society.]

These data are obtained from absorptions due to e^-_{aq}. However, it was reported that if a scavenger, e.g. Cd^{++} or cystamine, is used and absorptions due to the electron adducts are measured, then at high scavenger concentration (0.1 M and higher) these adducts are formed with 20% higher yields than e^-_{aq}, i.e. 4.8 if $G(e^-_{aq}) = 4.0$ or 5.4 if $G(e^-_{aq}) = 4.5$.

This extra yield is considered to arise from electrons which, in the absence of solute, recombine with the concomitant cation before becoming solvated and hence before they can show in absorption.

Such reactions of "dry" electrons have been suggested earlier, and at this meeting further extension to a wider range of solutes was reported—some of biological interest. The evidence is that high solute concentrations reduce the immediate yield of e^-_{aq} on the picosecond time scale, and it is concluded that solvation and reaction with solute are competing processes.

The new data presented here show that for a variety of solutes the efficiency of the "dry" electron reaction is related to the rate of reaction of the hydrated electron provided the latter are measured at the high concentrations, e.g. 0.1–1.0 M.

Dry electron reactions would be very significant in the reactions of electrons in cells where solute concentrations are of this order of magnitude. However, there was not complete acceptance of the "dry" electron interpretation of the observations. An alternative explanation in terms of the special kinetics which apply in conditions of high solute concentrations was presented (*21*) and the resolution of these different views is a matter of some importance.

ELECTRONS IN NON-POLAR LIQUIDS

The meeting revealed an increasing interest and activity in the behavior of electrons, i.e. excess electrons, in non-polar liquids such as liquefied noble gases and hydrocarbons (*22-26*).

Kinetic investigations such as measurements of homogeneous rates of recombination with cations, show that in these liquids the electrons conform to the theory based on randomly diffusing species as they do in polar liquids. Moreover, they also absorb light as in polar liquids but the absorption spectra are further in the infra-red.

However, quantitatively there are considerable differences in electron reaction rates both from the polar liquid system and among the non-polar liquids themselves. Thus, for normal diffusion-controlled reactions with solutes, rate constants of ca. 10^{10} $M^{-1}s^{-1}$ might be expected and in fact are observed in water and alcohols. In hexane, however, ca. 10^{12} $M^{-1}s^{-1}$ is found, in neopentane ca. 3×10^{13} $M^{-1}s^{-1}$, and in liquid methane 5×10^{14} $M^{-1}s^{-1}$ has been reported.

These numbers imply very high diffusion rates for electrons in these systems and direct measurements of mobilities, i.e. velocities under the influence of an applied field, confirm that this is in fact the case. Mobilities—or diffusion coefficients (they are related by a simple equation)—range over ca. five orders of magnitude in going from water through the hydrocarbons to liquid methane or liquid argon.

The factors which determine these values are being actively investigated, and we have had reports on a large variety of liquid systems representing changes in various molecular parameters and also on parallel work in the gas phase. At the moment it seems that the only factor which appears to correlate qualitatively is molecular shape, in that the more spherical the molecule—e.g. neopentane, tetramethyl silane, methane, argon—the higher is the electron diffusion rate.

Other factors, such as slight changes in molecular structure or change of state, sometimes produce strange anomalies in mobilities, which need to be explained on any theory, and it would seem that there are likely to be interesting developments in this whole area.

MODELS FOR ELECTRONS IN POLAR LIQUIDS

Progress has been reported in the theoretical study of electrons in polar media (*27*).

Absorption spectra are now available for electrons in a wide range of liquids (*28*) and any theoretical model should account for the experimental observations such as the variation in the wavelength of the absorptions for different liquids, the effect of temperature and pressure on the spectra, and also the band widths of the spectra. In addition, heats of solution of electrons from the gas phase can be estimated for some liquids and must fit the theoretical picture.

A common starting point for most theoretical models of solvated electrons is the electron surrounded by a number (4, 6 or 8) solvent molecules with the positive end of their dipoles oriented towards the electron. This structure is then considered immersed in a cavity in the liquid which is then treated in the simplest case as a homogeneous dielectric. Recent electron spin resonance work reported at this meeting (*29*) allows the atomic separation distances in such a structure to be estimated in the case of water.

Treatments of varying degrees of sophistication have been applied to this type of model, taking into account electron-molecule interactions, and a number of methods have been used to calculate the minimum total energy of the system and the cavity size. From these the energies of optical transitions, i.e. the positions of absorption bands, can be derived. The results are encouraging in that the latter agree reasonably well with experiment for a range of liquids. In the particular cases of water and ammonia, which have been treated in most detail, the calculated heats of solution are also of the right order. However, the broad nature of the absorption bands found in every liquid, which is a striking feature of the absorption spectra, is still unaccounted for. This, and the absorption spectra in non-polar liquids and in mixed solvents, are some of the outstanding theoretical problems.

The very bulk of the material makes it impossible to give a complete picture of the radiation chemistry presented at the Congress, and the above has necessarily been an individual view of what constitutes the major items of interest. No doubt, a different picture would be seen through other eyes, but there is little doubt that all would agree on the generally high quality of the contributions and the many excellent discussions which have taken place throughout the meeting.

REFERENCES

1. M. J. Person and G. A. Salmon, *Radiat. Res.* **59**, 104 (1974). [Congress presentation B-17-4.]
2. N. N. Lichtin, A. Shafferman and G. Stein, *Radiat. Res.* **59**, 107 (1974). [B-18-6.]
3. J. Butler, G. G. Jayson and A. J. Swallow, *Radiat. Res.* **59**, 193 (1974). [D-15-5.]
4. J. R. Clement, D. A. Armstrong, N. V. Klassen and H. A. Gillis, *Radiat. Res.* **59**, 29 (1974). [A-14-3.]
5. M. Z. Hoffman and E. Hayon, *Radiat. Res.* **59**, 106 (1974). [B-18-3.]
6. J. L. Redpath, L. I. Grossweiner and J. Ovadia, *Radiat. Res.* **59**, 271 (1974). [A-14-3.]
7. G. M. Gaucher, W. S. Lin and D. A. Armstrong, *Radiat. Res.* **59**, 272 (1974). [E-18-1.]
8. M. Saito and H. Maki, *Radiat. Res.* **59**, 273 (1974). [E-18-2.]
9. W. D. Felix and O. R. Kalkwarf, *Radiat. Res.* **59**, 274 (1974). [E-18-7.]
10. J. H. Fendler, *Radiat. Res.* **59**, 29 (1974). [A-14-4.]
11. M. Grätzel and J. K. Thomas, *Radiat. Res.* **59**, 29 (1974). [A-14-5.]
12. M. Grätzel, M. Cooper and J. K. Thomas, these Proceedings. [C-10-3.]
13. K. M. Bansal, *Radiat. Res.* **59**, 233 (1974). [D-33-5.]
14. Symposium on *Radiation Effects in Frozen Media*, these Proceedings. [A-6-1 to 7.]
15. B. G. Ershov and F. Kieffer, *Radiat. Res.* **59**, 35 (1974). [A-17-6.]
16. A. I. Mikhailov and V. I. Goldanskii, *Radiat. Res.* **59**, 36 (1974). [A-17-7.]
17. J. R. Miller, *Radiat. Res.* **59**, 35 (1974). [A-17-5.]
18. K. Y. Lam and J. W. Hunt, these Proceedings. [C-10-4.]
19. C. D. Jonah, E. J. Hart, J. R. Miller and M. S. Matheson, *Radiat. Res.* **59**, 110 (1974). [B-20-3.]
20. M. S. Matheson, E. J. Hart, J. R. Miller and C. D. Jonah, *Radiat. Res.* **59**, 110 (1974). [B-20-4.]
21. G. Czapski, these Proceedings. [E-10-4.]
22. Symposium on *Electrons in Liquids (Part B)*, these Proceedings [E-28-1 to 4.]
23. O. A. Allen, R. A. Holroyd and T. E. Gangwer, *Radiat. Res.* **59**, 198 (1974). [D-18-1.]
24. A. J. Robinson and M. A. J. Rodgers, *Radiat. Res.* **59**, 198 (1974). [D-18-3.]
25. T. Kimura and G. R. Freeman, *Radiat. Res.* **59**, 276 (1974). [E-20-4.]
26. E. C. Gregg and R. D. McCreary, *Radiat. Res.* **59**, 277 (1974). [E-20-6.]
27. N. R. Kestner, these Proceedings. [E-10-2.]
28. L. M. Dorfman and J. F. Gavlas, these Proceedings. [E-10-1.]
29. D. F. Feng and L. Kevan, *Radiat. Res.* **59**, 236 (1974). [D-35-2.]

Cellular and Subcellular Biology[1]

M. M. Elkind[2]
Division of Biological and Medical Research
Argonne National Laboratory
Argonne, Illinois 60439 USA

INTRODUCTION

A broad scope of material was presented at the 5th International Congress of Radiation Research concerned with cellular and sub-cellular effects. In thinking about how to summarize the various reports, I considered simply identifying highlights. For example, I thought of describing the experiments of the Berkley group—Tobias, Budinger, Leith, Lyman, Mamoon, and Yang (*1*)—who have reduced to laboratory investigation the visual light flashes experienced by astronauts in space flight and have made progress in the explanation of the radiobiology and physiology involved. Or, as another example, I considered reporting on the remarkable chromosome analysis and sorting device developed by the Livermore group—Gray, Steinmetz, Van Dilla, and Mendelsohn (abstract not available)—a device which not only can identify individual metaphase chromosomes in suspension and sort them by size, but also can do this at very high speed. However, it seemed to me that while recounting highlights would result in a broad mosaic, it would not help identify central themes. Since central themes, or "growing points," do exist in cellular radiobiology, I elected to structure my review around such major issues. What I will describe, therefore, are four areas of ferment which I view as among the more important topics discussed at the Congress. Foregoing a broad sweep of all the material presented has as its compensation the possibility of appreciating where we are going in a few of the more important areas.

REPAIR STUDIES

As in many meetings, symposia, and conferences in the past decade, repair studies figured importantly in the Congress. In the main, these studies relate on the one hand to repair of biological function, and on the other, to repair of macromolecules, principally DNA. All told, one symposium plus 7 sessions of proffered papers dealt specifically with repair and, of course, observations involving repair came up in many other sessions. In view of this sustained and burgeoning interest, it is difficult to present an overview without being

[1]Congress presentation F-1-3 (Summary).

[2]Work supported in part by U.S. Atomic Energy Commission and in part by a Special Fellowship from the National Institutes of Health No. FO 3 CA54327-02 awarded by the National Cancer Institute.

superficial. To help avoid this, at the outset I focus on mammalian cells and ultraviolet radiation.

In contrast to their earlier work, Regan and Carrier reported (*2*) that with improved techniques—involving the extraction and purification of the DNA—up to 100% of the dimers induced in human cells in culture by small UV doses is excised. For doses above 30–50 ergs/mm^2, there is no further increase in the absolute amount of dimers excised. Since complete excision was obtained after doses in the high survival region, a principal role for so-called "dark repair" in survival after UV is supported.

The work of Moustacchi (*3*) offers an example of DNA repair not associated with dimer excision. UV-induced dimers in mitochondrial DNA in yeast are not selectively removed. Still, the yield of cytoplasmic mutants is not fixed after UV and, in fact, Moustacchi finds that the variation in the yield of such mutants through the cell cycle reflects variation in repair ability. These results are, therefore, an example of repair in eukaryotic DNA not dependent upon excision repair, possibly involving recombinational repair.

Another example of repair capabilities after UV not specifically related to excision repair comes from the work of Takebe (*4*) who used different lines of *Xeroderma pigmentosum* cells. Individuals suffering from this autosomal, recessive disease show an enhanced susceptibility to skin cancers, on surfaces normally exposed to the sun, which are frequently contracted in the second decade of life and usually lead to death by the third decade. Takebe examined the UV survival of *herpes simplex virus* assayed on normal or one or another line of XP cells. While all the XP cells led to steeper survival curves for the virus compared to its survival curve when normal human cells were used for the assay—indicating deficient host-cell reactivation—there was no correlation of the virus sensitivity with the individual abilities of these various XP cells to perform non-scheduled DNA synthesis. Non-scheduled DNA synthesis is considered to be a reflection of excision repair and, hence, Takebe's results indicate that repair processes, in addition to excision repair, play a role in repairing viral DNA.

An example of the progress that has been made in the understanding of UV repair in mammalian cells comes from the work of M. C. Paterson (*5*). Building on experiments started with R. B. Setlow, Paterson has used an endonuclease extracted from *Micrococcus luteus* to examine the complementation of UV repair processes in hybrid mammalian cells. The concept behind his approach is sketched in Fig. 1, page 720.

Starting at the left:

1) Cells are UV irradiated and their DNA extracted.

2) This is followed by incubation with the endonuclease which has the effect of inserting single-strand breaks near *intra*-strand pyrimidine dimers. The subsequent denaturation of the extracted DNA to obtain single-strands leads to degraded DNA—that is, a reduction in molecular weight.

3) If, however, cells are incubated without the enzyme, denaturation does not lead to a drop in molecular weight since dimers by themselves do not produce single-strand breaks.

An example of how Paterson's system works is shown in the DNA sedimentation patterns in Fig. 2, page 721. The closed circles represent DNA extracted from unirradiated normal human fibroblasts; the extraction and purification procedure is responsible for the reduction of the size of the DNA to a weight average of 15×10^6 daltons. In the left panel, incubation in the absence of the endonuclease shows very little reduction in molecular weight while the right panel shows that 3.3 breaks were introduced by incubation with the enzyme.

To use this procedure to determine the competance of a cell line to effect the removal of dimers, it is simply necessary to allow for such removal after irradiation and before the DNA is extracted. Examples of this in respect to dark repair and photoreactivation are shown in Fig. 3, page 723. If dimers are removed, the number of enzyme-sensitive sites remaining in the DNA will decrease. In the left panel, incubation of normal fibroblasts, i.e. AH cells, shows this, but repair-deficient *X. pigmentosum* cells do not. The left panel shows also that dimers persisted in the XP cells even though they were illuminated with photoreactivating light after the UV. This illustrates the well-known fact that cells from placental mammals lack, or have very little, photoreactivating enzyme, an enzyme which mediates the photochemical opening of dimers. On the other hand, the right panel shows that embryonic chick cells do contain the photoreacting enzyme since, when they are illuminated, an appreciable loss of endonuclease sensitive sites results. However, at the same time, they lack the ability that AH cells have to excise dimers because incubation in the dark does not result in a loss of such sites.

Paterson used observations such as the foregoing to examine the complementation of repair capabilities in artifically fused, hybrid cells. He found that fusions of chick and normal human cells can effect dimer removal, showing that chick cells do not inhibit dark repair; fusions of chick and XP cells do not remove dimers, showing that the two cell types do not complement each other; and fusions of XP and normal cells do remove dimers, showing that the lack of dimer excision in the XP cells used was not because of the possible production of gene products which are inhibitory.

In mammalian cells, excision repair is not restricted to UV irradiation. As shown by the early work of Painter and his collaborators—and since confirmed by others—while excision repair occurs after X- or gamma-irradiation, the amount of repair per lesion is much less than after UV. At this meeting Remsen, Mattern and Cerutti (*6*) added an important additional feature to this picture. Using the technique for detecting pyrimidine radiation products of the 5,6-dihydroxy-dihydrothymine type developed in Cerutti's laboratory, they showed that the nuclei of HeLa and Chinese hamster cells have the ability to remove damaged thymines from irradiated phage DNA or from the synthetic polynucleotide, poly d(A-T), after treatment with OsO_4. The latter reagent is specific for the production of the hydration product of thymine.

Convening this Congress on the West Coast of North America made it somewhat easier for scientists to attend from countries like Japan. In addition to Takebe's work already mentioned, Okada undertook the difficult job of reviewing a number of repair studies by Japanese scientists (*7*). An example of the information he presented—in this case the work of C. Sato—is shown in Fig. 1, page 696. The upper left figure shows X-ray survival curves for cells derived from a Burkitt lymphoma, a mouse mammary carcinoma (FM3A), and a mouse melanoma (C2W). To begin with, these data illustrate survival parameters over a range of values even though all the cells are malignant in origin. The Burkitt lymphoma cells, however, have the property characteristic of the normal cells from which they derive; namely, a very modest D_0 and a very small shoulder width. Consistent with the lack of a shoulder on the survival curve, little if any sublethal damage repair is evident when dose fractionation is used (upper right panel). The lack of repair was found by Sato to be associated with an inability to redevelop a normal amount of surface charge as measured by electrophoretic mobility. Whereas melanoma cells seemed to do this rather well, between 6 and 24 hours after irradiation, Burkitt lymphoma cells appeared deficient. One can speculate that the change in plasma membrane surface charge may be secondary to the biochemical activity needed to support repair, but in addition these changes are worth keeping in mind in respect to the possibility that they may influence cell communication as in the cell spheroids in culture used by Sutherland and Richardson (*8*) and Durand (*9*), or in tissues in vivo.

Single-Strand Breaks

In the past 8 years or so, a very considerable interest developed in the measurement of strand scissions in the DNA of cells. This resulted from the introduction of a technique for examining the sedimentation of long strands of DNA. McGrath and Williams (*10*) showed that cells lysed by strong alkali on top of an alkaline sucrose gradient would release their DNA in lengths 20–100× longer than yielded by the most gentle methods of chemical extraction. We know now that the latter methods introduced strand breaks in numbers comparable to those produced by 50 krads or more. The method of McGrath and Williams became generally applied to studies with microbes and mammalian cells, and in connection with various kinds of treatments including radiation, even though the extraction at high pH usually led to denaturation of the DNA. Similar techniques for the examination of damage to duplex DNA have evolved more slowly although data along these lines were presented at this Congress, as I note presently.

Naturally, the question was soon asked; Do single-strand breaks and/or their repair have a direct connection with cell killing? In certain bacterial repair-deficient mutants this now appears to be the case. I regret to report, however, that the situation is far from clear in respect to mammalian cells. Here are some examples of the arguments presented *for* and *against* a connection between DNA single-strand damage and cell killing.

Chapman, Dugle, Reuvers, Gillespie, and Borsa (*11*) have measured the dependence of the killing of Chinese hamster cells on oxygen concentration and a number of other sensitivity modifiers. Close correlations were observed between the sensitizing action of oxygen, other electron affinic agents, and radioprotectors when cell survival was compared with single-strand break induction (e.g., Fig. 3, page 757). Therefore, these investigators concluded that the production of single-strand breaks is closely associated with, if not responsible for, cell killing.

Somewhat in contrast to the preceding, Painter (*12*) argues that unrepaired breaks, possibly due to base damage, are the cause of cell death. In addition to the observations by Cerutti and his coworkers relative to base damage already noted, the production of alkali-labile lesions—not due to frank strand scissions and therefore presumably a reflection of base damage—are now well documented as illustrated by the report of Coquerelle, Lennartz, and Hagen (*13*). Their data show that perhaps as many as 1/4 to 1/2 of the breaks observed under alkaline conditions are due to base damage. As I will show, many of these breaks must be repaired although we know very little about the fidelity of this repair at the nucleotide level.

Still, there are examples of other ways of modifying cell killing which do not support a close connection between strand breakage and cell killing (e.g. *14, 15*). One of these was reported by Ben-Hur and Elkind (*16*) who studied the enhanced killing that results when Chinese hamster cells are X-rayed under hyperthermic conditions. For present purposes, the data which are particularly apropos refer to the enhanced killing when cells are made hyperthermic between dose fractions. Fig. 3, page 706 shows this point. The dashed curve is obtained when single doses are delivered at ice temperature. The results for 37° C between first and second doses are typical of fractionation survival data observed with V79 and other cells. However, if the temperature is raised between dose fractions, the survival is reduced. At 41° C, there is a small reduction in survival after the first dose alone, but the shoulder on the fractionation curve—indicative of a capacity for sublethal damage—does not reappear. At 41.5° C, the degree of lethal damage due to the first dose alone is enhanced but otherwise the result is the same; that is, only a shoulder suppression with essentially no effect on the final slope. Since Ben-Hur and Elkind also found that hyperthermia before a single dose did not affect survival, these data indicate that hyperthermia between dose fractions inhibits the repair of sublethal damage.

Ben-Hur and Elkind then looked into the effect of hyperthermia on single-strand break repair using cells lysed in alkali; that is, conditions which would be expected to cause various forms of base damage to lead to strand breaks. Fig. 8, page 711 shows that after 5800 rads–a modest strand-breaking dose, but still a supralethal dose—breaks are repaired more rapidly at 42° C than at 37° C. At both temperatures, about 10% of the breaks remain unrepaired after 70 min which indicates that at least a considerable amount of base damage must have been repaired. Clearly, these data do not support an interference with single-strand repair due to hyperthermia although enhanced *misrepair* cannot be ruled out. I return to misrepair or error-prone repair later.

Relative to cell killing, little additional support in favor of a direct involvement of single-strand breaks—or for that matter double-strand breaks—comes from the report of Shonka and Cole (*17*). They used gamma-rays and compared them to American alpha-particles having a near-maximally effective energy for cell killing; that is, 3–4 MeV. In addition to an RBE for double-strand breaks too low to correlate with that for cell killing, extensive if not complete repair of single-strand breaks was observed after gamma- or alpha-irradiation; complete repair of double-strand breaks after gamma-irradiation; and almost complete double-strand repair after alpha-irradiation.

In spite of the conflicting conclusions that these various data suggest, one very important point emerges. An irradiated cell that survives very likely is a cell that repaired a considerable amount of damage. A mammalian cell has an average of about 5×10^{12} daltons of DNA (i.e., $\sim 9 \times 10^{-12}$g). If we consider the killing effect of, say, 500 rad, from the radioenergitics of strand breakage, we would have to conclude that while only 4 out of 5 cells are killed, immediately after irradiation all the cells had an average 5,000–10,000 single-strand breaks, and 100–500 double-strand breaks. Clearly, these expectation values are too large to be consistent with 1 cell in 5, the survivor, escaping without damage. From this it must follow that either large numbers of single- and/or double-strand breaks are compatible with survival, or large numbers of lesions must be repaired. The likelihood and the magnitude of this repair points to the importance that modulations of processes such as these could have on the expression of potentially lethal damage.

All this leads to the question: Are there targets in a cell other than just duplex DNA *as a whole* which are reasonable candidates for those considered to be primary relative to cell killing by ionizing radiation? Judging from work with bacteria as well as with mammalian cells, at this juncture the answer should be *yes*. It is now well established that in bacteria the growing point of DNA replication resides in a DNA-membrane complex which includes some, if not all, of the required enzymes for DNA replication. Building on the ideas of T. Alper (e.g., *18*) in respect to her proposal of membrane-associated, oxygen-sensitive sites in bacteria, Cramp and his collaborators (e.g., *19*) have developed an impressive set of observations in support of the DNA-membrane complex as the site of "O"-type, or oxygen sensitive-type, damage. While precise analogies with bacteria are not justified, the report of A. Cole and his collaborators (*20*) supports the nuclear periphery as the anatomical domain of at least a class of sensitive sites in mammalian cells. They used low energy electrons and compared the effects of energy depositions in or near the nuclear envelope to those dispersed throughout the cell. Consistent with what we know from the bacterial work of Cramp, the results of Cole and collaborators, as well as those of others, favor the confluence of DNA and lipid at the nuclear envelope as a principal target; a confluence not necessarily involving DNA replication in mammalian cells. Single- or double strand DNA breakage in this confluence could be the target as opposed to such breakage in the nuclear DNA as a whole.

CARCINOGENESIS

Ultraviolet Induction

The mechanism of carcinogenesis is complex as the incisive discussions of Robin Mole and George Klein so well illustrated. Nevertheless, as major advances in science have often shown, posing the right question, in the right context can lead to meaningful insights. This seems to be the case in respect to the so-called "fish story" being developed by Setlow and Hart (*21*).

As I have already noted, the disease *Xeroderma pigmentosum* is characterized at the molecular level by an inability to repair UV lesions, and at the tissue level by the development of skin cancer that usually results in early death. Setlow and Hart inquired if the repair of such lesions—in particular repair of pyrimidine dimers—would affect the incidence of tumors in an appropriate experimental test system. Specificity for dimers could be insured by using the UV photoreactivating system—i.e., the light-dependent enzymatic reversal of dimer formation—which restricted their choice of material for reasons already noted.

UV produces a number of different types of lesions in DNA in addition to $\widehat{TT}$, $\widehat{CT}$, and $\widehat{CC}$ dimers. Demethylations, hydration products, and DNA-protein crosslinks are also produced. UV followed by photoreactivating light—so-called black light—removes the dimers if the photoreactivation enzyme is present and if treatment goes to completion. The question posed by these investigators was, therefore, whether or not tumor induction could be shown to depend upon a photoreactivating treatment following UV.

For a number of practical reasons, Setlow and Hart chose to work with the mollie, *Poecelia formosa*. One advantage of this organism is its clonal breeding capabilities; that is, all the offspring are female diploids having the same genetic and therefore immunological construction.

Figure 1, page 881 shows the layout of the experiment. A tissue homogenate was prepared from which cells were injected into isogenic hosts after the treatments shown. About 10^5 cells were injected either intramuscularly, or into the abdominal cavity. A number of months later, the number of tumors was scored. In addition to the observation of gross tumor formation, the more recent results of Setlow and Hart have been confirmed by histological examination.

The induction of thyroid tumors and granulomas with UV light is shown in Table 1, page 882. The effect of photoreactivating light is also shown. It is clear that black light illumination after UV markedly decreased the proportion of thyroid tumors and, to a lesser extent, the proportion of granulomas.

Since it is well-known that one effect of photoreactivation is to render once again viable cells that would have been sterilized by the UV, the influence of the survival rescue of cells should be considered. This was raised by van Bekkum, a point to which I return presently.

Ionizing Radiation

A valid mammalian cell culture system to help unravel the complexities of carcinogenesis in mammals would be quite useful, particularly in respect to ionizing radiation. Two studies along these lines were presented which, unfortunately, leave some major issues unsettled.

The first study is in some respects the more comprehensive. Using hamster embryos, Borek and Hall (*22*) measured the induction of cell transformation in primary cultures. *Cell transformation* is a term that connotes a change in a number of in vitro properties. In the work of Borek and Hall, it refers to a change in the colonial morphology produced by viable cells. Normal or untransformed cells usually produce sparse colonies of "contact inhibited," not piled-up, cells; when 1×10^7 of these are inoculated into appropriate hosts malignant tumors do not form. In contrast, transformed cells produce denser colonies the cells of which frequently overlap; colonies of transformed cells are identified, therefore, on morphological grounds. When $1-10 \times 10^6$ of these cells are implanted in vivo, tumors form in most instances.

Borek and Hall have amassed an extensive amount of data with this system. Their results with single doses of 210 kV X rays have been published (*23*) and show that the frequency of transformation rises from about 0.01% at 1 rad to about 1% at 150–300 rad. At the Congress they reported new data with fast neutrons and with X-ray dose fractionation. Neutrons of 430 keV have an RBE of about 5–10 for doses up to about 75 rad. X-ray dose fractionation of 50 and 75 rad total doses (5 hr intervals) increased the frequency of transformants by a factor of about 1.7 in each case. The enhanced transformation resulting from fractionation is unexpected since, in animals, dose protraction or reduced dose rate irradiation with the low LET radiation usually results in lower tumor yields. Still, insofar as the fractionation results of Borek and Hall are concerned, the enhanced transformation—which is associated with a small increase in survival—may be indicative of the role of *misrepair* in cell transformation.

J. C. Klein has also studied in vitro neoplastic transformation (*24*). She used mouse spleen cells, cultivated for long periods of time. Thus, in two respects the cells Klein used differed very considerably from those employed by Borek and Hall, and these differences may be critical as noted by Henry Kaplan. Klein found that neoplastic growth in vivo after 300 rad was only occasionally achieved. Further analysis convinced her that not only was radiation required but also the presence of dying and therefore degenerating cells. In fact, she also found that mechanically killed cells—e.g., by repeated quick freezing and thawing—admixed in the same proportion with viable cells as would have been the case after 300 rad, produced comparable numbers of tumors when the equivalent of $1-10 \times 10^6$ cells was inoculated (*25*). Thus, Klein concluded that cell debris is an essential concomitant for neoplastic change, possibly because a transforming agent—e.g., a virus or a provirus—must be released from the damaged, degenerating cells and incorporated into the normal cells.

D. W. van Bekkum supports this view (*26*) and pictorializes the process as shown in the lower portion of Fig. 3, page 892. He proposes that if genes responsible for proliferation are released from one cell and incorporated into another at sites sufficiently removed from the recipient's appropriate repressor genes, then uncontrolled growth and presumably transformation to malignancy could be effected. In this scheme, the reason for the requirement for irradiation of the recipient cell is that it must be undergoing repair—that is, a process of opening and of closing gaps in the DNA—in order that the foreign, transforming DNA can be integrated. This same point was also raised by van Bekkum in reference to the data of Setlow and Hart (*21*) since they used inocula of at least 10^5 cells. If much cell killing resulted from the UV doses employed—killing then reversed by black light exposure—qualitatively, the conditions for the mechanism proposed by Klein and van Bekkum would be fulfilled.

Thus, in a gross sense, the notions of Klein and van Bekkum are in the pattern of error-prone repair. Error-prone repair is also favored by Zajdela and Latarjet (*27*) to explain the greater UV induction of skin cancer in mice under conditions of no repair inhibition compared to conditions involving repair inhibition. Consistent with this, Stich and Laishes (*28*) report that DNA repair synthesis occurs apparently only in those tissues in which neoplasms arise when animals are treated with chemical carcinogens. But the critical point which remains unresolved is whether or not radiation-induced neoplastic transformation requires a direct effect in the target cell and is independent of interactive, collateral effects in nearby cells.

CHEMICAL SENSITIZERS

Chemicals that modify radiation response have a two-fold attraction. To begin with, as with other response modifiers, they make possible insights into how radiation acts if the chemical action of the modifier is known. But in addition, modifiers can be potentially useful in practical terms either for reasons of radiation safety or in connection with radiation therapy. Information presented at the Congress illustrates both applications.

NEM Studies

Guided by the cell-cycle dependence of the protection afforded by cysteamine, W. K. Sinclair)*29*) undertook to examine the effect of the sulfhydryl poison N-ethylmaleimide (NEM). Fig. 3, page 745 shows that in V79 Chinese hamster cells, NEM sensitizes in G_1 but mainly in late S; that is, in the DNA presynthetic phase and in the latter part of the synthetic phase, respectively. Figure 7, page 748 shows that this comes about very likely because repair processes are inhibited. The curve marked "B" was obtained by irradiating cells at 8.5 hours and then adding 0.75 μM NEM for 30 min at the times indicated. "B" shows clearly that the sensitization of NEM fades rapidly if its administration is delayed and supports the conclusion that the resistance observed in late S results from a repair process that can be inhibited by NEM.

To test if NEM acts as just described in cells in addition to V79 cells, Sinclair extended his study to HeLa cells in collaboration with A. Han and B. F. Kimler (*30*). Since HeLa cells have a longer G_1 period, as well as a prominent maximum survival in G_1, it was also possible to examine more easily the post-irradiation influence of NEM in G_1. Figure 8, page 749 shows that expectations were satisfied. NEM, 0.75 μM, clearly reduces survival only in G_1 and S (dashed curve). The closed triangles trace the post-irradiation loss of sensitization of S cells due to NEM, as in the case of V79 cells, and the closed squares show the same result for G_1 cells.

From these results, Sinclair reasons that a major part of the structure in the age-response pattern of mammalian cells reflects a repair process mediated by a sulfhydryl agent. Since Archer and Sinclair (see *31*) found that, in V79 cells, total SH content correlates with X-ray response better than the non-protein SH moiety, Sinclair proposed that protein-bound SH is the active principle. In sum, these data suggest that the repair of damage in at least two regions of the age-cycle involves protein-bound sulfhydryl, the influence of which can be reduced by sulfhydryl poisons.

Sensitizers of Hypoxic Cells

By way of illustrating some of the practical uses of chemical radiation sensitizers, I will review some of the information presented relative to the sensitization of hypoxic cells. J. W. Boag, in his Failla Memorial Lecture (*32*), reviewed the background of this work. He noted the important concepts that came from the studies of G. Adams and D. Dewey and discussed a number of the recent results relative to the mechanism of the oxygen effect in mammalian cells. These have come mainly from the rapid mix experiments of Adams and his associates at the Mt. Vernon Hospital near London. There is little that needs to be added to Boag's review.

A few words are in order, however, in respect to a recent sensitizer and some of the early results with it in vivo. The structure of a new material synthesized for G. Adams, designated Ro 07-0582, is shown in Fig. 5 of the presentation by Rauth et al. (page 766). Adams and associates (*33–35*) tested this material in vitro with V79 Chinese hamster cells and showed that it had no effect on aerobic cells but sensitized hypoxic cells almost as completely as does air.

Although Ro 07-0582 has a structure similar to metronidazole, it sensitizes to a greater degree but is less toxic. These properties prompted Rauth, Kaufman and Thomson (*36*) to test Ro 07-0582 in vivo using the KHT tumor system of Hill and Busch (*37,38*). This system allows a solid fibrosarcoma to be treated in vivo but assayed for cell survival in a second host not exposed to any treatment. Rauth and Kaufman found that Ro 07-0582 is appreciably more effective than metronidazole and almost as effective as full air saturation would be (see Fig. 6, page 766).

This is only one example of what now is an active and frequently collaborative effort of a number of groups. Because of the enhanced survival conferred by hypoxia, and because a number of lines of evidence indicate that tumors have regions of hypoxic, viable cells, considerable interest has developed around the in vivo testing in animal systems of candidate hypoxic radiation sensitizers. A number of data along these lines were presented at this Congress. Some of the principals involved in the cell and tumor studies are: J. Fowler, J. Denekamp, H. Hewitt, L. Peters, and N. McNally at the Mt. Vernon Hospital near London; G. F. Whitmore and A. M. Rauth at the Ontario Cancer Institute in Toronto; J. D. Chapman, C. L. Greenstock and their collaborators at Atomic Energy of Canada in Pinawa; and L. D. Skarsgard and B. Palcic at the British Columbia Cancer Institute in Vancouver.

THE SMALL DOSE REGION

As the last example of a growing point in cellular radiobiology, I draw upon material which brings together questions of the epidemiological aspects of radiation, damage registration and repair, and the use of radiation in radiotherapy. In a way, this topic covers the gamut of our current interests. Also, this example will illustrate how cell studies become the crossroads, so to speak, of fields as diverse as the physics of microdosimetry and the physiology of tumors and normal tissues.

In 1956, when Puck and Marcus (*39*) published the first low LET mammalian cell survival curve—the survival curve of cells derived from Helen Lane's cervical carcinoma, so-called HeLa cells—first impressions quite naturally dwelt upon the major features of the curve. These were a rather small D_0, 100–150 rad, and a shoulder giving rise to an extrapolation number of about 2. In the past few years, we have started to focus once again on the features of the curve, in particular on its initial slope. To illustrate what I mean consider Fig. 1. On the left, curve A is intended to be typical of the damage accumulation type. Survival, as you know, measures "unsuccessful events." The right panel shows what a curve like A would look like if "successful events" were plotted, or 1-*Survival.* The ordinate here is simply labeled "frequency" because it could stand for mutation induction, cancer induction, and so on.

From the large body of data relative to the repair of sublethal damage, it may be inferred that a zero initial slope means no killing at small doses, and no killing at very low dose rates. A corresponding inference might be expected to hold in regard to the induction of endpoints of epidemiological importance. On the other hand, a non-zero initial slope in both cases would mean some effect

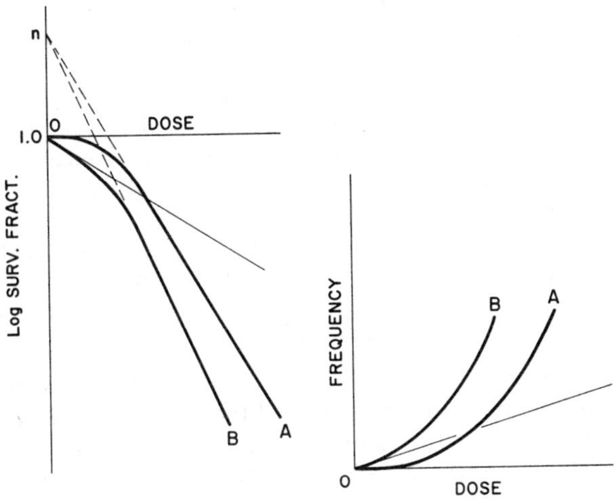

FIG. 1. Survival and incidence curves having zero initial slope, curves A, and non-zero initial slopes, curves B.

even at small doses and even at low dose rates.

Our general lack of concern with the shape of the initial region of the survival curve (or the induction curve) was disturbed in the last few years by H. H. Rossi and A. M. Kellerer (see *40*). Reasoning from microdosimetric measurements and a body of RBE data carried to smaller and smaller doses, Rossi and Kellerer proposed that a mammalian cell is killed when two targets, separated by distances on the order of the nuclear dimensions—i.e., by microns—are both hit. They reasoned that the *effect* would depend on the square of the *energy concentration* from which it follows that in addition to a dose square dependence, *effect* would also depend on a term linear in dose. At low doses, the contribution from the D^2 term becomes small compared to the linear term and, hence, they predicted that the initial part of a survival curve, for example, would have a negative slope reflecting a dependence on the first power of the dose, for doses up to about 40 rad for X rays and 1000 rad for fast neutrons.

Quite at the other end of the spectrum of radiation studies, J. Dutreix and A. Wambersie (*41*)—two radiotherapists—were led to a similar qualitative conclusion about the shape of the low LET survival curve but, as would be expected, for quite different reasons. Dose fractionation data are of critical importance to a practicing radiotherapist, both in respect to normal and neoplastic tissue. From a series of measurements involving different dose fractionation regimes—some of which were reported at the Congress—Dutreix and Wambersie were forced to conclude that some cell killing results even at very small doses and, hence, that the survival curve has a non-zero initial slope like curve B in Fig. 1.

For the radiotherapist, these considerations lead to the inference that for small repeated doses, the total dose to sterilize a tumor approaches a maximum at a plateau value. And for the epidemiologist, they imply that very low dose rates may still give rise to non-zero effects.

A number of questions follow:

1) Do all survival curves—or all dose incidence curves—have non-zero initial slopes?
2) Are the initial slopes the same and dictated only by biophysical considerations like the physical separation of targets in the nucleus?
3) For a given population of cells, are initial slopes fixed and not dependent upon biochemical or hormonal signals from the host?

Clearly, the shape of the initial portion of the dose-effect curve bears on a number of questions of fundamental importance, not only in the context of radiation effects but also in respect to environmental and chemical agents in general.

CONCLUSION

The field of *radiation research*, from outward appearances, must seem to be a conglomeration of unrelated topics. Still, if there is one area I would identify as being a focus or crossroads of most of the others, it would be *cellular* and *subcellular biology*. Here the physics and chemistry of elementary absorption processes converge on studies of radiation effects in molecules, viruses, microbes, and even mammalian cells. And, it is from here, with cellular effects in mind, that the physiologist and physician seek to understand tissue and organ responses. Thus, our knowledge of the biochemistry and biology of cell effects *on the one hand* draws upon the physics of absorption processes and the chemistry following therefrom, and *on the other hand* is essential for the understanding of ecological and environmental effects of radiation, and for the improved use of radiation in medicine. The attraction of the gamut of science to the cell, in my opinion, has been a mainstream in the 5th Congress of Radiation Research.

REFERENCES

1. C. A. Tobias, T. F. Budinger, J. T. Leith, J. T. Lyman, A. M. Mamoon, and T.C.H. Yang, *Radiat. Res.* **59**, 150 (1974). [See also these Proceedings.]
2. J. D. Regan and W. L. Carrier, *Radiat. Res.* **59**, 176 (1974).
3. E. Moustacchi, *Radiat. Res.* **59**, 176 (1974). [See also these Proceedings.]
4. H. Takebe, *Radiat. Res.* **59**, 177 (1974).
5. M. C. Paterson, *Radiat. Res.* **59**, 7 (1974). [See also these Proceedings.]
6. J. Remsen, M. Mattern and P. A. Cerutti, *Radiat. Res.* **59**, 177 (1974).
7. S. Okada, *Radiat. Res.* **59**, 6 (1974). [See also these Proceedings.]
8. R. M. Sutherland and J. Richardson, *Radiat. Res.* **59**, 159 (1974).
9. R. E. Durand, *Radiat. Res.* **59**, 93 (1974).
10. R. A. McGrath and R. W. Williams, *Nature (London)* **212**, 534 (1966).
11. J. D. Chapman, D. L. Dugle, A. P. Reuvers, C. J. Gillespie, and J. Borsa, *Radiat. Res.* **59**, 165 (1974). [See also these Proceedings.]
12. R. B. Painter, *Radiat. Res.* **59**, 7 (1974). [See also these Proceedings.]
13. T. Coquerelle, M. Lennartz, and U. Hagen, *Radiat. Res.* **59**, 83 (1974).
14. W. U. Shipley and M. M. Elkind, *Radiat. Res.* **48**, 86 (1971).
15. M. M. Elkind and C.-M. Chang-Liu, *Int. J. Radiat. Biol.* **22**, 313 (1972).
16. E. Ben-Hur and M. M. Elkind, *Radiat. Res.* **59**, 6 (1974). [See also these Proceedings.]
17. F. J. Shonka and A. Cole, *Radiat. Res.* **59**, 60 (1974).
18. T. Alper, in *Biophysical Aspects of Radiation Quality*, Proceedings of the IAEA Symposium, Lucas Heights, Australia, pp. 171–184. IAEA-SM-145/1, Vienna, 1971.
19. W. A. Cramp and P. E. Bryant, *Radiat. Res.* **59**, 215 (1974).
20. A. Cole, W. T. Tobleman, F. J. Shonka, and W. G. Cooper, *Radiat. Res* .**59**, 180 (1974).
21. R. B. Setlow and R. W. Hart, *Radiat. Res.* **59**, 73–74 (1974). [See also these Proceedings.]

22. C. Borek and E. J. Hall, *Radiat. Res.* **59**, 91 (1974).
23. C. Borek and E. J. Hall, *Nature (London)* **243**, 450 (1973).
24. J. C. Klein, *Radiat. Res.* **59**, 91 (1974).
25. J. C. Klein, *J. Natl. Cancer Inst.*, **52**, 1111 (1974).
26. D. W. van Bekkum, *Radiat. Res.* **59**, 8 (1974). [See also these Proceedings.]
27. F. Zajdela and R. Latarjet, *C. R. Acad. Sci.* **277**, 1073 (1973).
28. H. F. Stich, *Radiat. Res.* **59**, 7 (1974). [See also these Proceedings.]
29. W. K. Sinclair, *Radiat. Res.* **59**, 164 (1974). [See also these Proceedings.]
30. A. Han, W. K. Sinclair, and B. F. Kimler, *Radiat. Res.* **59**, 79 (1974).
31. W. K. Sinclair, in *Control of Proliferation of Animal Cells* (B. Clarkson and R. Baserga, Eds.), pp. 985–994. Cold Spring Harbor Laboratory, Cold Spring Harbor, 1974.
32. J. W. Boag, these Proceedings.
33. G. E. Adams and B. D. Michael, *Radiat. Res.* **59**, 162 (1974). [See also these Proceedings.]
34. J. C. Asquith, M. E. Watts, and G. E. Adams, *Radiat. Res.* **59**, 158 (1974).
35. B. D. Michael, G. E. Adams, and M. E. Watts, *Radiat. Res.* **59**, 156 (1974).
36. A. M. Rauth, *Radiat. Res.* **59**, 165 (1974). [See also these Proceedings.]
37. R. P. Hill and R. S. Bush, *Int. J. Radiat. Biol.* **15**, 435 (1969).
38. R. P. Hill and R. S. Bush, *Radiat. Res.* **59**, 92 (1974).
39. T. T. Puck and P. I. Marcus, *J. Exp. Med.* **103**, 653 (1965).
40. H. H. Rossi, *Radiat. Res.* **59**, 75 (1974). [See also these Proceedings.]
41. J. Dutreix and A. Wambersie, *Radiat. Res.* **59**, 251 (1974).

Tissue and Animal Biology[1]

Giovanni Silini
Laboratorio di Radiobiologia Animale
CNEN, C.S.N.-Casaccia
Roma, Italy

To discuss in half an hour the 420 papers presented at the Congress on the subject of "Tissue and whole animal radiation biology" is clearly a hopeless task and in preparing this review I often asked myself what prompted me to accept it. Since this paper is bound to be scientifically unsatisfactory both to myself and to the audience, and I cannot expect therefore anything but bad reputation, I was forced to conclude that my acceptance is due purely to a perverse desire for self-punishment. No advance apologies therefore, but only my gratitude to the organizers for helping this insane attitude of mine.

Profound changes of interest in the field that I will review have taken place in the course of years, and I will attempt to point them out as I go along. I thought at first about very fancy ways to group the various reports, but then I decided that a more conventional subdivision might be better followed and might offer the advantage of wider perspectives and conclusions, since it is obvious that, under the circumstances, I must point to generalities rather than to the resolution of details.

In the field of the *early somatic effects*, cell population changes in normal tissues were in great evidence and were discussed for a variety of in vivo systems. It is quite clear that at present this line of research has added to its previous classical aim, namely, the understanding of effects and syndromes in mammals for radiation protection purposes, another practical objective related to experimental radiotherapy with the ultimate scope of improving our dosage schedules for organized tissues especially with fractionated treatments and non-conventional radiations.

The vast majority of the contributions dealt with self-renewing tissues and in particular with haemopoietic stem cells, their response to whole- or partial-body exposure and their reactions to various drugs that are applied therapeutically in conjunction with radiation or may be used to manipulate their division and differentiation properties. Technical refinements in this area of study have been outstanding in recent years: methods of biological assay of the stem cells in vivo and in vitro; cell fractionation techniques to isolate or to enrich the pluripotent cell fractions; the use of treatments to experimentally recognize different classes of pluripotential or committed stem cells; drug treatments or biological procedures to influence their differentiation pathways or their renewal characteristics; all these have accounted for the major advances in the

[1]Congress presentation F-1-4 (Summary).

understanding of the basic biological properties of stem cells not only in experimental animals but even, as we have heard, in the human. The killing, recovery and repopulation phases of their reaction to irradiation may now be described rather precisely. Control mechanisms remain to be investigated and it is interesting to see how many reports were actually directed to their elucidation under physiological conditions or after radiation or drug treatments. It seems quite obvious that the implications of this research are projected beyond the immediate radiobiological and hematological interest towards the basic understanding of the homeostatic mechanisms of cell division and of functional tissue regulation.

The hemopoietic system presents of course great technical advantages and is probably the best example where radiation-induced cell proliferation changes may rather easily be followed in vivo. But other tissues are being increasingly investigated; the intestinal epithelium, the skin, the lens of the eye, the testis, the lymphatic tissues, the salivary glands were also considered for their response to radiation through cloning or isotope-labelling techniques for the quantitative analysis of cellular and population kinetics. Methods are also being developed to follow radiation effects on two radiotherapeutically important tissue components: the connective and the vascular tissues. Up to a few years ago we knew much less about normal tissues than about neoplastic ones and it must be recognized that radiobiologists have made considerable efforts in this direction.

But it is even more interesting to note that all this new information has been obtained without any appreciable loss of interest for tumor tissues. The analysis of irradiated primary and transplanted tumor cell populations is in fact being very vigorously pursued at present, so that it has recently been possible to collect data directly on human tumors. The picture of the tumor functional structure emerging from these studies is very complex indeed: profilerative and non-proliferative, clonogenic and differentiated cell compartments are gradually being recognized in suitable tumor model systems by the use of mixed in vitro - in vivo methods of analysis; cell kinetic parameters and rates of recruitment between these compartments are also being accurately measured. The state of the various cell classes with respect to oxygen tension, and the rate of death and removal of the radiation-inactivated cells are being worked out. The changes in each of these parameters following single and fractionated radiation courses, cytotoxic drugs or various combined treatments may now be followed in a variety of tumor systems with different growth characteristics and histotypes. The practical aim is that, through a systematic collection of these data, regularities and relationships generally applicable to all tumor types might at some stage be identified which might improve the results of tumor therapy.

Many radiotherapists remain rather skeptical as to the ultimate impact of this knowledge in everyday practice, and probably it is true that so far the most outstanding results in this field have rather been of a fundamental nature in uncovering the structure and organization of tumors than of a practical nature in suggesting new schemes for radiotherapy. It is also possible that many of the subtle changes that radiobiologists are now describing might balance each other in the course of the overall tumor reaction to radiation, and that the space for further dramatic improvements in radiotherapy dictated by a better knowledge of tumor proliferation kinetics is at present rather limited. But I still maintain that there is merit and intellectual value in rationally justifying empirical results and in making sure that we are not proceeding in the right direction for the wrong reasons.

It would of course be wrong to leave the audience under the impression that cell population changes were the only immediate effects of radiation deserving attention but, under the circumstances, I cannot do much more than just mention a few of the other titles. From the effects on the central nervous system where morphological, biochemical and functional lesions were described to the modifications of tissue and whole-body radiation response by chemical protectors or sensitizers; from the metabolic effects concerning various changes of enzymes, hormones, nucleic acid precursors and other biochemical compounds to the effects on differentiating tissues or the pathology of pre-natal irradiation. Papers on immunology and transplantation, which traditionally were extensively represented at our meetings, appeared this time considerably fewer, but it would be difficult to say whether this reflects a decreasing interest in this particular field or simply a change of direction of immunologists towards problems more relevant to other scientific societies.

Among the *non-neoplastic late effects,* life-shortening in relation to various causes of death was considered after acute, chronic, fractionated and protracted external irradiation and after contamination with α- and β-emitting nuclides. In the dog, typical dose-rate-dependent syndromes showing at different exposure levels following life-time external irradiation have been identified and described, especially in the hemopoietic tissue, but also in the liver, the eye lens, the ovary and testis. In the rodent, it is becoming increasingly evident by the use of refined techniques of survival analysis that the age-specific rate of appearance of various neoplastic diseases may account for the bulk, if not the entire, life-span shortening of irradiated animal populations. Long-term effects on the vascular tissue were also analyzed by quantitative morphological and functional techniques together with other changes of the kidney vascular structure.

But in the field of *late radiation action, carcinogenesis* stands out as a most important effect and it was dealt with both in the symposia and in the open sessions from many different view-points.

Concerning the basic mechanisms of tumor induction, we heard with great interest a symposium on the role of the immune system in cancerogenesis, hoping that we might gather a better insight into an area which appears at present rather promising. The possible role of the immune system has been realized only recently with the notion that cellular immunity may act as a general surveillance system to control and inactivate potentially malignant clones and with the subsequent discovery of tumor-associated antigens. Even though these antigens seem less strong than those of chemically-induced tumors, and even though the data on the antigenicity of transplanted tumors are by far more extensive than those on autochtonous ones, it appears nevertheless likely that these antigens may stimulate immunological responses. The significance of these in the overall process of radiation carcinogenesis is still rather hazy, except for the notion that they act only at the level of promotion. Cellular as well as humoral reactions may be involved. On the one hand, radiation may unspecifically depress the lymphoid system, thus causing a decreased reactivity of the host at a critical time of a tumor life-history when the transformed cells carrying specific antigens might otherwise be recognized and inactivated. On the other hand, circulating soluble tumor antigens or antigen-antibody complexes might interfere with the recognition and the inactivation of the transformed clone by killer cells, thus favoring its further development.

In some experimental tumor systems (leukemia, osteogenic sarcoma) RNA-viruses have definitely been identified as etiological agents, but their role in the process of carcinogenesis with respect to the virus' own antigens or to virus-induced cellular antigens is still far from understood. The interaction of radiation and viruses in carcinogenesis suggests a number of possible mechanisms, from very specific ones at the molecular level analogous to phage induction in lysogenic bacteria to very unspecific ones affecting the establishment and subsequent growth of the malignant cells as it may depend not only on immunological but also on hormonal or other influences.

In the open session on immunology and oncogenesis were described a number of possible models involving transplantation of tumor cells and oncogenic murine viruses, and these experiences should be pursued in order to build from a wide spectrum of observations some general scheme that may perhaps be extrapolated to man where this type of experimentation appears impossible. For the time being, however, the significance of the immune reactions described in the experimental animals to human diseases, and their relevance at the low doses and dose-rates (which are the conditions of practical concern), remain to my mind as the most difficult questions of principle to be solved.

The basic mechanisms of radiation cancerogenesis represent, however, only one of the possible experimental approaches to these late effects. Even in the absence of refined information about the etiology and the pathogenesis of these diseases, it is still possible (it is actually required by policy decisions) to deal with the problem on different levels. In one of the general sessions the cellular mechanisms of carcinogenesis were considered, and it was pointed out that the major difficulties in the interpretation of the carcinogenic effect stem from the fact that the probability of cell transformation by radiation is lower than the killing probability and both these probabilities have different dose-dependencies. Accurate evaluations of these probabilities "per cell" have of course important implications for interpreting the experimental data on whole animals, for example in terms of linearity. We are now at a point where it may be possible to obtain in vitro and in vivo estimates of the cellular dose-induction rate, as we have done for the killing effect over the last fifteen years.

To this end, a profound change in the strategy of our experimentation will probably be required with the abandonment of large-scale epidemiological studies on the whole animal and the development of suitable model systems in vivo and in vitro, allowing a better control of some parameters of the cells at risk such as their number, distribution, radiosensitivity, as well as the kinetic properties of their proliferation and differentiation. These biological parameters, which cannot be resolved in the complex interactions taking place in the whole-body irradiated animal, are no less important than the physical parameters of carcinogenesis. The analysis of transformation probability, as it depends on all these variables, should be strongly pursued in the future. It may thus be possible within a few years to gain a better insight into the cellular mechanisms of carcinogenesis and to improve our risk estimates for cancer induction.

A large variety of models of radiation carcinogenesis were illustrated; in view of the impossibility to describe them all and to point out advantages and limitations, I will only mention the tissues in which tumor induction may now be to some extent quantitatively analyzed: the skin and its annexes, the ovary, the thymus, the reticulo-endothelial and the hemopoietic systems, the gastrointestinal tract, the mammary and salivary glands, the liver, the thyroid, the brain and nervous tissues, and the lung. Among the classical biophysical variables of carcinogenesis, the dose, dose-rate, fractionation, LET, external vs internal irradiation were considered. Reports were also presented concerning the interactions of various physical and chemical agents with radiations for the induction of particular forms of malignancy. Among the biological variables, the age at irradiation and the influence of cell proliferation on the final yield of tumors in various animal species, including man, were also treated extensively.

At present, however, follow-up studies in selected human populations remain as the only source of information to guide public policy decisions and should be given the highest priority in the future for very obvious reasons. The numerical estimates of cancer induction available in man after external exposure are, according to some of the reports, sufficiently close to justify some confidence in their precision. But in other areas, like cancerogenesis after exposure in utero, data are rather ambiguous and there is disagreement among estimates. In other cases also, as after internal irradiation by α-emitters, uncertainties remain. As we know, the hypothesis of linearity of risk with dose is commonly used to extrapolate data in man from the fairly well known high-dose range down to the very low doses and intensities. It is important that linearity be verified by the study of populations where doses range over several orders of magnitude, but there are great limitations to these studies. It is possible that below a certain dose limit we will just have to accept linearity, but it is imperative that this limit should be determined with great accuracy.

At last, neoplastic diseases were also discussed in the context of the relative risks posed by radiation and by other environmental agents. Similarities and differences in the mechanisms of action of radiation and chemical carcinogenesis were pointed out in many papers.

With regard to similarities, it has been reported that in chemical carcinogenesis, a variety of molecules are activated in vivo through intermediate reaction steps and give rise eventually to the active forms of the carcinogens which are characterized by their strong electrophilic properties. These compounds may interact with electron-rich centers of cellular informational macromolecules and the ensuing modification of their functional properties results in the cell's malignant transformation. Such a general mechanism may apply equally well to ionizing radiations, which are known to produce electrophilic species in the course of their chemical action on water or on other biomolecules. Mutagenesis and cancerogenesis are also common effects to radiation and to chemical compounds, although their causal relationship, which has been many times implied through direct or indirect mechanisms, is by no means settled.

Differences emerge, however, about the quantitative aspects of the carcinogenic effect by the two classes of agents. Ionizing radiations have in the past provided a unique and relatively simple situation, but in chemical carcinogenesis the analysis of the dose distributions as a function of time and at the level of the various target tissues, cells and molecules is rendered far more complicated by external and by metabolic internal factors. Moreover, the presence of human populations exposed to known radiation doses under special circumstances has made it possible to validate in man a number of qualitative or semi-quantitative experimental hypotheses. It seems very doubtful, on the other hand, that these circumstances will be encountered in chemical carcinogenesis, since human populations subjected to known exposures of a specified agent may be difficult to identify and to quantitatively analyze.

We now come to the *genetic effects of radiation*. The comparative aspects of radiation and chemical mutagenesis and the methodological aspects quantitation of genetic effects in man formed the subjects of two symposia, while the open sessions dealt with various problems of cytogenetics and formal genetics in germ cells and in the whole animal.

With regard to chemical mutagenesis, many contributors pointed out the infinitely greater complexity in the analysis of mutational events by environmental chemicals than by ionizing radiation. Some of these difficulties are common to the field of somatic effects, such as that of assessing the mutagenic action of the chemical compound itself or of some metabolically-activated form, that of studying the concentration of the active substance at the target tissue as a function of dose and time after single exposure, and that of calculating the overall active level of the compound upon chronic administration. But for the genetic effects specifically, qualitative and quantitative differences were illustrated. Metabolic differences among different test species make it sometimes very difficult to assess whether a substance which is mutagenic for a particular species may be mutagenic at all for another. The ratio between gene mutations and chromosomal rearrangements, which is in general higher in chemical than in radiation mutagenesis, proves to be extremely variable from one substance to another. Cell cycle stage specificity is also more pronounced and variable than in radiation mutagenesis. And when, in addition, one considers the further complication of possible synergistic effects, one is forced to conclude that the knowledge of the mutation spectrum and of the dose-effect relationships which are essential to the evaluation of the genetic risk will hardly be possible for each potentially mutagenic chemical. There is an increasing gap between the pressing demand for empirical philosophies to adopt on regulatory action and the absence of systematic information about the hazards of chemical mutagens in man; no one has, to my knowledge, firm proposals for a comprehensive solution to which scientific as well as political responsibilities are presumably called upon to contribute. The crux of this solution remains a scientific one, however: that is, whether the total generation dose of 5 rem should be made to include the possible mutagenic action of environmental chemicals, in addition to radiation. The solution requires either the establishment of commensurable mutagenicity indexes for physical and chemical agents or the development of a completely new, but as yet unknown, approach.

Data were presented in the course of this symposium, mainly in *Drosophila* and in the mouse, and they centered about the problems of stage sensitivity, the variation of the mutational spectrum and the site of chromosome breaks for chemical treatments, in comparison to ionizing radiation.

The second symposium in radiation genetics dealt with new techniques for the estimation of genetic hazards in man. Technical difficulties, doubts about the applicability to man, and the lack of human population genetic and cytogenetic information were identified as the major areas of uncertainty which prevent the refinement of the present estimates of genetic risk.

With regard to the first point, some of the papers offered good examples of how technical progress may lead to more advanced knowledge. The mouse specific locus method, which still remains the corner stone of our estimates in man, is now being integrated with other techniques of complementation analysis. They are providing, in the mouse, a direct estimate of the loci in the entire genome, a very important parameter for the calculation of genetic risks. These methods are also used for a further characterization of radiation-induced and spontaneous mutations and for the study of their effects in heterozygous conditions. Chromosomal structural changes constitute a major portion of the genetic damage produced by ionizing radiation, and on this basis interspecific comparative studies on chromosome interchanges and deletions in somatic and germinal cells were carried out in order to correlate cytogenetic events on somatic cells with genetically-recoverable events. It is hoped that this approach will allow some projection from aberration data in human somatic cell chromosomes to true genetic effects. Another system for the modeling of the mutational response of human cultured somatic cells was also described. Because they enable the investigator to screen a large number of mutants, to control the biochemical events influencing radiosensitivity and repair, and analyze the lethal and mutagenic response concomitantly, these systems in vitro may offer considerable advantages over the traditional in vivo experimentation.

As to the applicability to man of data obtained in other species and the inherent extrapolation difficulties in the absence of basic information on human genetics, recent data were referred to concerning the relative weight of the simple dominant versus the irregularly-inherited recessive conditions in man. A re-evaluation of the role of radiation in increasing the overall mutation rate based on such information has been reported, suggesting that present estimates of radiation-induced genetic defects might have been overestimated by a factor of 30.

Finally, arguments were given by some of the speakers in favor of developing techniques and programs for the study of detrimental mutations which were so far rather underrated by comparison with early-acting dominant and recessive lethals, but are probably more meaningful in terms of human ill-health and suffering. Although it may be true that in the future the public image of radiation geneticists might depend on their willingness to take up these new problems of social relevance, it does seem to me that there is still disagreement concerning some of the data and their interpretation which should be further clarified.

In view of the projected increase in the production of transuranium elements (by the end of this decade Pu^{239} in the USA is estimated to be on the order of 2×10^4 kg) and in consideration of the large uncertainties existing about their toxicity, the Congress organizers quite wisely set up a Symposium on *Transuranium element toxicity*. Since it was centered mainly on the effects of low doses, it was inevitable that carcinogenesis had a prominent place here too.

As we know, tumor induction by internal emitters is traditionally treated separately from carcinogenesis by external irradiation, not because of any recognizable intrinsic differences in the basic induction mechanisms, but because of the complexities introduced by the large variations in the spatial and temporal distribution of internally-absorbed doses. The chemical form of the nuclide and the various possible routes of entry, metabolism, deposition and excretion render in this case the establishment of precise dose-effect relationships very uncertain. A number of valuable contributions emphasized the variations of these parameters and the many difficulties in correct dose calculations owing to the inhomogeneities of distribution, dose-rate and quality factors among the various organs, tissues and cells. Any evaluation of the carcinogenic hazards of α- and β-emitting radionuclides rests upon an accurate elucidation of all these variables in appropriately selected animal species. Precious new information was presented on the effects of Pu, Np and Am inhaled by or injected into rodents and large mammals, with rather accurate estimates of the deposition of these elements in various target tissues, description of the pathology and of rate of tumor appearances. Bone tumors are unequivocally increased in dogs down to levels of 10^{-2} μCi/kg of Pu^{239} and these experiments are now extended down to levels of 10^{-3} or 10^{-4} μCi/kg, which doses are approaching (and perhaps disturbingly so) the maximum permissible burden for radiation workers. Information in man from internally deposited Pu is very scanty, but the human cases exposed in the course of industrial accidents to doses on the order of the MPBB or higher are being followed, have yielded some rough data for the evaluation of the excretion rate and in due course, as they age, will provide information on dose-related risk of cancer induction.

Although it would seem likely that man is the most sensitive link in the biosphere and that protection standards for human populations should therefore be most conservative even in comparison to other organisms, the protection of the natural environment where and through which man lives is also the object of much interest. All possible aspects of the *environmental radioactive pollution* have by now been considered with such a wealth of general statements and detailed analyses that they have become not only a matter of concern within the highest political circles, but almost a commonplace of our modern cultural background. I would not presume therefore to say anything new and original on this subject, but I think it is appropriate here to underline one fact. The fact is that people working in the nuclear energy field have since the very beginning set a model for these hazard evaluations by reasonable empirical philosophies. Examples of how these ever-evolving mechanisms of scientific evaluation, followed by recommendations of practical rules, are carried out through the analysis of experimental data were given here during the symposium on transport and fate of radionuclides in the environment and during the many sessions on radioecology. Techniques and programs of environmental monitoring, studies of radiation effects on ecological systems, transfer pathways for a wide range of soluble and insoluble radioactive contaminants. evaluations of doses to the food chains and ultimately to man, were debated in great detail. The symposium itself was concerned with various physio-ecological aspects of radioactive environmental contamination. The need was pointed out that the data on the ecological impact of plants in operation should be applied to the correct siting of future installations so that, from the present era of passive monitoring, we may soon enter a new era of more active research and advice for the location of plants respectful of the natural characteristics which represent in themselves cultural values.

In this process, laboratory investigations will have to be complemented by field studies, since the large diversity of the environments gives rise to a great variability of the critical steps and pathways for various radioactive contaminants. Model systems of environmental situations will also be used extensively, but their capacity to predict the actual transfer of the pollutants in nature will depend essentially on the accuracy with which the relevant experimental variables will be determined. Examples were discussed of how aquatic ecosystems, which are of primary concern for effluent discharge, may be studied: various causes of variability of the bioaccumulation factors were identified in connection with the physicochemical form of the contaminants, their concentration and equilibrium with stable forms, the limnological characteristics and the growth of the contaminated tissues.

As a non-ecologist, I dare to point out the danger that these studies may grow too scattered: efforts for a better identification of the really critical pathways for radiation protection and for the integration of information into coherent working models will ensure the best utilization of the data in practice. And let us also remind ourselves that the problem of the collective doses and dose commitments to the whole population is not exhausted by the study of the critical steps or critical groups, because it also should take into account the very small dose contributions given to large numbers of individuals throughout all the ecological chains.

Having almost reached the end of my presentation, I find, as expected, that the papers I did not refer to are greatly in excess of those I attempted to review. But before finishing I would like to mention the symposium on the *technological applications of radiations* where at least three of the contributions touched upon subjects of biological interest. Two of them discussed the use of radiation for the preservation of food and feed products and for the sterilization of medical supplies and other materials. In the case of food it seems quite clear that an extensive application of radiation technologies is often discouraged by health considerations connected with the need for the exhaustive and expensive testing of the wholesomeness of irradiated food required by international authorities. Such considerations do not apply to the radiosterilization of medical supplies where radiation has proved to be a competitive way of industrial processing. Advantages and limitations were also discussed of the use of radiation in plant breeding: here the realistic evaluation and the critical acceptance of radiation as a tool for the induction of progressive mutations and chromosomal rearrangements is gradually prevailing over the skeptical or negative attitude. On the other hand, radiation is no longer expected to provide the success that only the talent and the experience of the plant breeders may achieve in the selection and isolation of favorable genetic traits. A good number of contributions were also submitted dealing with various hereditary or radiobiological *effects on plant materials*. With regard to genetic effects, studies of mutation rates and their comparison among different species as a function of the amount of nuclear material, already initiated in the past, have been pursued and are providing useful models for the extrapolation of mutational indexes to various species and for the analysis of comparative radiosensitivity of plant material. I wish to draw special attention to the fact that in plants, contrary to mammalian systems, it is now possible to obtain haploid somatic cells: these techniques have potential interest not only in a variety of problems of basic and applied plant genetics, but also as possible models for the screening of induced mutational events, until haploid mammalian lines will become available.

After this process of analysis, affected of necessity by many personal and objective limitations, I should now proceed to the synthesis and attempt to draw some conclusions and to indicate future perspectives for this area of work. But in view of the great diversity of the interests covered, and of the methodological and technical problems specific to each area, I do not feel I can offer any firm conclusion but only some considerations or, rather, personal feelings about the meaning of this Congress. And I hope you will accept them for what they are worth.

It seems to me that we are at present experiencing in many fields of science, and in our field too, a time of transition during which the previously established guide-lines are being reconsidered in an effort to find new ways to proceed. This is quite natural in science where only through a continuous scrutiny of previous knowledge real advancement is brought about. The peculiarity of the situation is not in the transition itself, therefore, but rather in the fact that this process of reconsideration is taking place at a time when many values of our life in present society are also being reconsidered. I will cite as an example the emphasis paid by the Congress to the effects on man and on the environment by radioactive and chemical pollutants, which is a reflection of the larger dispute between those who demand production of abundant energy to satisfy the needs of our society and those who—by emphasizing the potential hazards that man, in so doing, is creating for himself—strongly oppose such course of action and hypothesize a zero derivative for our growth curve. My feeling is that the solution to these tremendous problems is unlikely to come from any dramatic structural change of our society but only from responsible evaluations based on hard experimental facts: it is our responsibility as scientists to provide the necessary evidence and to help with its interpretation. In this connection our Congress has provided a good opportunity to discuss the scientific data which are essential to any reasonable decisions.

But before being scientists we are also (as our President, Dr. Bond, put it the other night) remarkably similar to human beings and, as such, often rather reluctant at the idea of adapting ourselves to new situations in a rapidly evolving world. Well, we must convince ourselves that changes are not necessarily for the worse, provided we undertake them rationally and we see them in a long perspective. Here, history and the humanities may probably help us better than our everyday science and I am very tempted, just to give you the right sort of perspective, to recall the words of a great poet who lived in Rome about two thousand years ago. In my broken English translation his words remind us that man's virtue "stands in the middle and there are certain boundaries outside which truth cannot be found." No extreme solutions, therefore, nor the acceptance of a dignified mediocrity, but only the difficult research of these limits and the constant effort to progress within them.

Needless to say, Mr. Chairman, that to a "lazy Italian" this outlook of hard work is the worst possible one.

AUTHOR INDEX

SUBJECT INDEX

Only the initial page of each presentation is given.

H

I

J

L

M

N

O

P

R

S

T

U

A 5
B 6
C 7
D 8
E 9
F 0
G 1
H 2
I 3
J 4